D1476923

UNDERGROUND
Mining Methods

*Engineering Fundamentals
and International Case Studies*

*Edited by William A. Hustrulid
and Richard L. Bullock*

Published by the
Society for Mining, Metallurgy, and Exploration, Inc.

Society for Mining, Metallurgy, and Exploration, Inc. (SME)
8307 Shaffer Parkway
Littleton, Colorado, USA 80127
(303) 973-9550 / (800) 763-3132
www.smenet.org

SME advances the worldwide minerals community through information exchange and professional development. SME is the world's largest professional association of minerals professionals.

Disclaimer
The papers contained in this volume are published as supplied by individual authors. Any statement or views presented here are those of individual authors and are not necessarily those of the Society for Mining, Metallurgy, and Exploration, Inc. The mention of trade names for commercial products does not imply the approval or endorsement of SME.

Cover images contributed by The Itasca Consulting Group, the Climax Molybdenum Company, and LKAB.

ISBN 0-87335-193-2

Library of Congress Cataloging-in-Publication Data

Underground mining methods : engineering fundamentals and international case studies /
edited by William A. Hustrulid and Richard L. Bullock.
    p. cm.
  Includes bibliographical references and index.
  ISBN 0-87335-193-2
  1. Mining engineering. I. Hustrulid, W.A. II., Bullock, Richard L.

TN145.U53 2001
622.2--dc21   2001027301

Contents

This book is dedicated to the memory of Marianne Snedeker,
who was the guiding light and the driving force
behind SME book publications for many, many years.

Preface

In 1556, Georgius Agricola wrote *De Re Metallica*, one of the first and probably the most famous of mining reference works. In Book 1, he reflects rather elegantly on mining and miners.

Many persons hold the opinion that the metal industries are fortuitous and that the occupation is one of sordid toil, and altogether a kind of business requiring not so much skill as labour. But as for myself, when I reflect carefully upon its special points one by one, it appears to be far otherwise. For a miner must have the greatest skill in his work, that he may know first of all what mountain or hill, what valley or plain, can be prospected most profitably, or what he should leave alone; moreover, he must understand the veins, stringers, and seams in the rocks. Then he must be thoroughly familiar with the many and varied species of earths, juices, gems, stones, marbles, rocks, metals, and compounds. He must also have a complete knowledge of the method of making all underground works. Lastly, there are the various systems of assaying substances and of preparing them for smelting; and here again there are many altogether diverse methods.

Furthermore, there are many arts and sciences of which a miner should not be ignorant. First there is Philosophy, that he may discern the origin, cause, and nature of subterranean things; for then he will be able to dig out the veins easily and advantageously, and to obtain more abundant results from his mining. Secondly, there is Medicine, that he may be able to look after his diggers and other workmen, that they do not meet with those diseases to which they are more liable than workmen in other occupations, or if they do meet with them, that he himself may be able to heal them or may see that the doctors do so. Thirdly follows Astronomy, that he may know the divisions of the heavens and from them judge the direction of the veins. Fourthly, there is the science of Surveying that he may be able to estimate how deep a shaft should be sunk to reach the tunnel which is being driven to it, and to determine the limits and boundaries in these workings, especially in depth. Fifthly, his knowledge of Arithmetical Science should be such that he may calculate the cost to be incurred in the machinery and the working of the mine. Sixthly, his learning must comprise Architecture, that he himself may construct the various machines and timber work required underground, or that he may be able to explain the method of the construction to others. Next, he must have knowledge of Drawing, that he can draw plans of his machinery. Lastly, there is the Law, especially that dealing with metals, that he may claim his own rights, that he may undertake the duty of giving others his opinion on legal matters, that he may not take another man's property and so make trouble for himself, and that he may fulfill his obligations to others according to the law.

It is therefore necessary that those who take an interest in the methods and precepts of mining and metallurgy should read these and others of our books studiously and diligently; or on every point they should consult expert mining people, though they will discover few who are skilled in the whole art. As a rule one man understands only the methods of mining, another processes the knowledge of washing, another is experienced in the art of smelting, another has a knowledge of measuring the hidden parts of the earth, another is skilful in the art of making machines, and finally, another is learned in mining law. But as for us, though we may not have perfected the whole art of the discovery and preparation of metals, at least we can be of great assistance to persons studious in its acquisition.

In continuing along the path of Agricola, the Society for Mining, Metallurgy and Exploration, Inc. (SME), has provided the focus, stimulus, and support for a number of books covering the different interests and needs of its membership, the mining community as a whole, and the public it serves. Although a rather recent effort had produced the *SME Mining Engineering Handbook*, in 1974, some people felt that a book devoted strictly to underground mining was needed, one which could be used for the training of new mining engineers as well as serving as a reference book for those already in the field. The SME Book Publishing Committee agreed, and the *Underground Mining Methods Handbook* emerged as a title and objective. An editorial advisory board was formed, an outline was developed, and section editors were selected.

The book that emerged in 1982 consists of eight sections: General Mine Design Considerations, Stopes Requiring Minimum Support, Stopes Requiring Some Additional Support Other Than Pillars, Caving Methods, Underground Equipment, Financial Considerations, Foundations for Design, and Mine Ventilation.

It has been nearly 20 years since the *Handbook* (which over the years has become affectionately known as the "Blue Book" because of the color of its cover) was published and much has taken place in the interim. Some of us felt that it was time for the next volume in this series of books stretching back to Agricola. A proposal submitted to the SME Book Publishing Committee to do just that was approved by the SME Board of Directors at its August 1999 meeting, and the project to produce a companion volume to the Blue Book was underway.

From conception to birth, the Blue Book required nearly 8 years. In the present case, the book was put on a fast track, and publication was scheduled for 2001. This new volume includes revised sections on general mine design, mining methods, and foundations for design. A new section, "Underground Mining Looks to the Future," is included so that readers might spend at least a little time contemplating future courses rather than focusing strictly on today's task of getting "rock-in-the-box."

Beginning with the Phoenician traders more than 3,000 years ago, mining today is an international business. This book has tried to reflect this international character by presenting a snapshot of the world of mining through a series of case studies.

Unfortunately, recent years have been a time of depressed prices for many commodities and a downsizing of staff. The efforts made by mining companies, consulting companies, educational institutions, and individuals to provide their contributions to this book are greatly appreciated.

Some argue that the long-standing role of books in the development of our society is changing to the point where they will go the way of the dinosaur. This may be true, but we believe that books still offer a very special way of capturing and presenting information in a way unequaled by other forms of media.

The decision taken in 1982 to publish the Blue Book in a loose-leaf format was done in recognition that a user might choose to rearrange the contents, add material to the various sections, and/or update as appropriate. The book was intended to be a dynamic one that would continue to grow and thrive with the mining industry. The original concept for the continuous renewal and rejuvenation of the Blue Book through publication of new articles in magazines such as *Mining Engineering* was good, even though it turned out to be an idea before its time and was not implemented. Now, however, via the wonders of computers and the Internet, there are far-reaching opportunities to make this new book a living document. We look forward with anticipation to this exciting development.

For all who have and will contribute to this volume, we are most grateful. Special appreciation is extended to Jane Olivier at SME, and to Priscilla Wopat, Spokane Research Laboratory of the National Institute for Occupational Safety and Health, who worked so diligently and carefully on the editing and publication of this book. Cheryl Bradley and Diane Christopherson of the Department of Mining Engineering at the University of Utah have helped keep track of the many manuscripts and have made corrections as needed. The significant support provided by The Itasca Consulting Group in the initial stages of this project is gratefully acknowledged.

William A. Hustrulid and Richard L. Bullock

General Mine Design Considerations

CHAPTER 1

Underground Mining Methods and Applications

Hans Hamrin[*]

1.1 INTRODUCTION

Ore is an economic concept. It is defined as a concentration of minerals that can be exploited and turned into a saleable product to generate a financially acceptable profit under existing economic conditions. The definition of ore calls for afterthoughts. Ore does not properly exist until it has been labeled as such. To name a mineral prospect an ore body requires more information than needed to establish metal grades. Sufficient knowledge of the mineral deposit, mining technology, processing methods, and costs is needed for undertaking a feasibility study and proving the prospect worthy of being developed into a mine.

The expression "existing economic conditions" deserves an explanation. "Run-of-mine" ore is a mix of valuable minerals and worthless rock in which each ingredient is priced separately. Run-of-mine ore is treated in the dressing plant and processed into different concentrates. Where the ore contains more than one metal of value, separate concentrates of, for example, copper, zinc, and lead are produced. The value of in situ ore can be calculated by applying market prices to metal content and deducting costs for treatment and transportation of concentrates and smelter fees. The balance must cover direct mining costs and leave a margin for the mine operator.

Metal prices are set on international metal market exchanges in London and New York and fluctuate from day to day, depending on the supply-and-demand situation. An over-supply builds stocks of surplus metal, which is reflected in a drop in the market price. The profit margin for a mine decreases as the values of its products drop. As costs for processing, transport, smelting, and refining remain constant, the mine must adjust to a reduced income. The mine operating on a narrow margin must be prepared to survive periods of depressed metal prices.

One tactic to deal with such a situation is to adjust the boundaries of the area being mined and draw these boundaries at a higher cut-off grade. This will increase the value of the run-of-mine product, and the mine will maintain its profit. Another way is to increase the efficiency of mine production. Modifying the mining method and introducing new, more powerful machines are actions that should raise the efficiency of work procedures. The mine must remain a profit generator, which is not a simple task in an environment of increasing labor costs and demands for better living.

This chapter describes and explains methods for the underground mining of mineral deposits. The descriptions are generalized and focus on typical applications. Examples chosen illustrate types of mining practices as of 1999. However, every mineral deposit, with its geology, grade, shape, and volume, is unique. As methods are described here, please bear in mind that rock is variable, miners have ideas, and the world of mines will always display special features.

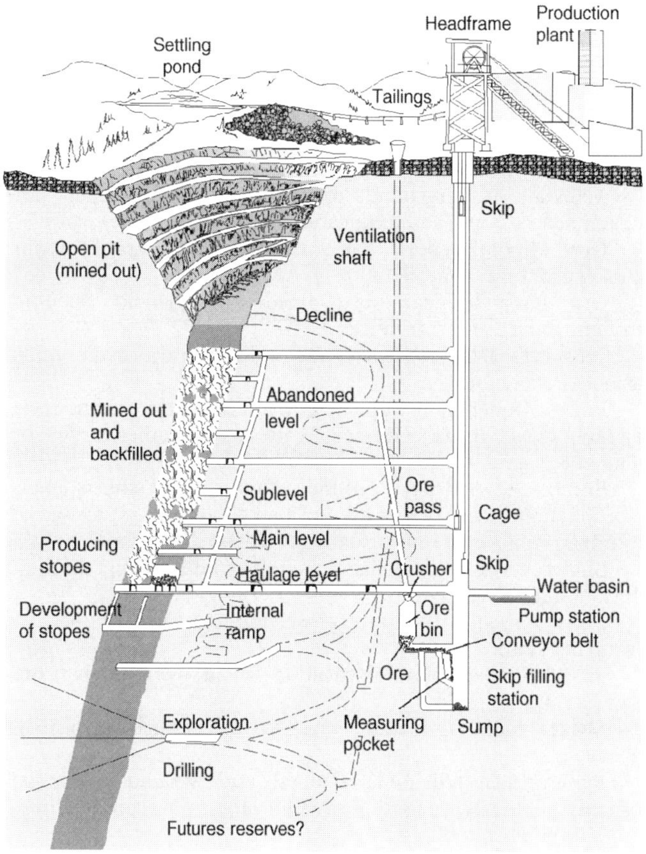

FIGURE 1.1 The underground mine—basic infrastructure

1.2 DEFINITION OF TERMS

To better understand the material presented herein, some of the more common mining terms are defined in the following paragraphs. Figure 1.1 further clarifies some of the terms.

Adit: Horizontal or nearly horizontal entrance to a mine.

Back: Roof or overhead surface of an underground excavation.

Chute: Loading arrangement that utilizes gravity to move material from a higher level to a lower level.

Cone: Funnel-shaped excavation located at the top of a raise used to collect rock from the area above.

---

[*] Retired, Atlas Copco, Stockholm, Sweden.

Crosscut: Horizontal or nearly horizontal underground opening driven to intersect an ore body.

Dip: Angle at which an ore deposit is inclined from the horizontal.

Drawpoint: Place where ore can be loaded and removed. A drawpoint is located beneath the stoping area, and gravity flow transfers the ore to the loading place.

Drift: Horizontal or nearly horizontal underground opening.

Finger Raise: Typically, a system of several raises that branch together to the same delivery point. Used for transferring ore.

Footwall: Wall or rock under the ore deposit.

Grizzly: Arrangement that prevents oversized rock from entering an ore transfer system. A grizzly usually consists of a steel grating for coarse screening or scalping.

Hanging Wall: Wall or rock above an ore deposit.

Level: System of horizontal underground workings connected to the shaft. A level forms the basis for excavation of the ore above or below.

Manway: Underground opening that is intended for personnel access and communication.

Ore: Mineral deposit that can be worked at a profit under existing economic conditions.

Ore Pass: Vertical or inclined underground opening through which ore is transferred.

Prospect: Mineral deposit for which the economic value has not yet been proven.

Raise: Underground opening driven upward from one level to a higher level or to the surface; a raise may be either vertical or inclined (*compare winze*).

Ramp: Inclined underground opening that connects levels or production areas; ramps are inclined to allow the passage of motorized vehicles. Ramps usually are driven downward.

Shaft: Vertical or inclined underground opening through which a mine is worked.

Slot: Vertical or inclined ore section excavated to open up for further stoping.

Stope: Underground excavation made by removing ore from surrounding rock.

Strike: Main horizontal course or direction of a mineral deposit.

Sublevel: System of horizontal underground workings; normally, sublevels are used only within stoping areas where they are required for ore production.

Wall Rock: Wall in which an ore deposit is enclosed.

Waste: Barren rock or rock of too low a grade to be mined economically.

Winze: Vertical or inclined underground opening driven downward from one level to another level or from the surface to a level (*compare raise*).

1.3 MINING METHODS

1.3.1 Introduction

Once an ore body has been probed and outlined and sufficient information has been collected to warrant further analysis, the important process of selecting the most appropriate method or methods of mining can begin. At this stage, the selection is preliminary, serving only as the basis for a project layout and feasibility study. Later it may be found necessary to revise details, but the basic principles for ore extraction should remain a part of the final layout.

With respect to the basic principles employed, relatively few mining methods are used today. Because of the uniqueness of each ore deposit, variations on each of these methods are nearly limitless. It is impossible to include even the major variations in

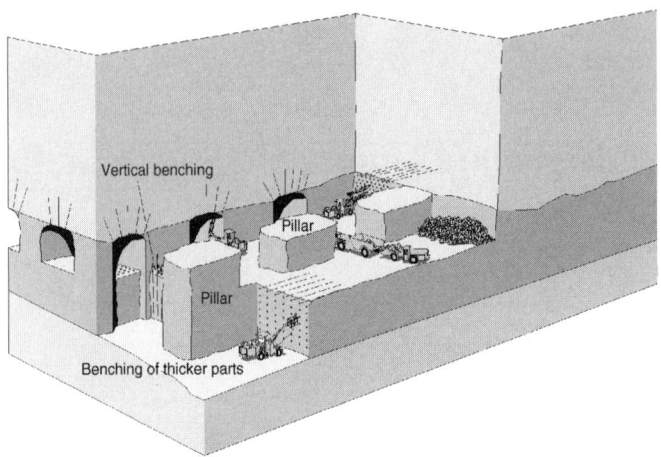

FIGURE 1.2 Classic room-and-pillar mining

this chapter; the goal of this chapter is to summarize briefly the characteristics of the major mining methods.

1.3.2 Room-and-Pillar Mining

Room-and-pillar mining is designed for flat-bedded deposits of limited thickness, such as copper shale, coal, salt and potash, limestone, and dolomite. This method is used to recover resources in open stopes. The method leaves pillars to support the hanging wall; to recover the maximum amount of ore, miners aim to leave the smallest possible pillars. The roof must remain intact, and rock bolts are often installed to reinforce rock strata. Rooms and pillars are normally arranged in regular patterns. Pillars can be designed with circular or square cross sections or shaped as elongated walls separating the rooms. Minerals contained in pillars are nonrecoverable and therefore are not included in the ore reserves of the mine. Differing geological conditions give rise to variations in room-and-pillar mining. Three typical variations are described in the following text.

Classic room-and-pillar mining (Figure 1.2) applies to flat deposits having moderate-to-thick beds and to inclined deposits with thicker beds. Mining the ore body creates large open stopes where trackless machines can travel on the flat floor. Ore bodies with large vertical heights are mined in horizontal slices starting at the top and benching down in steps.

Post room-and-pillar mining (Figure 1.3) applies to inclined ore bodies with dip angles from 20° to 55°. These mines have large vertical heights where the mined-out space is backfilled. The fill keeps the rock mass stable and serves as a work platform while the next ore slice is mined.

Step room-and-pillar mining (Figure 1.4) is an adaptation of trackless mining to ore bodies where dip is too steep for rubber-tired vehicles. A special "angle" orientation of haulage drifts and stopes related to dip creates work areas with level floors. This allows trackless equipment to be used in drilling and mucking. Mining advances downward along the step room angle.

Classic Room-and-Pillar Mining. In classic room-and-pillar mining, only a minimum of development work is required to prepare a flat-bedded deposit for mining. Roadways for ore transport and communication are established inside production stopes. Excavation of roadways can be combined with ore production, and mined-out stopes can serve as transport routes.

Ore production involves the same drill-blast techniques as in normal drifting where drift dimensions equal the width and height of the stope. Where geological conditions are favorable, stopes can be large, and big drill jumbos can be used for mechanized drilling.

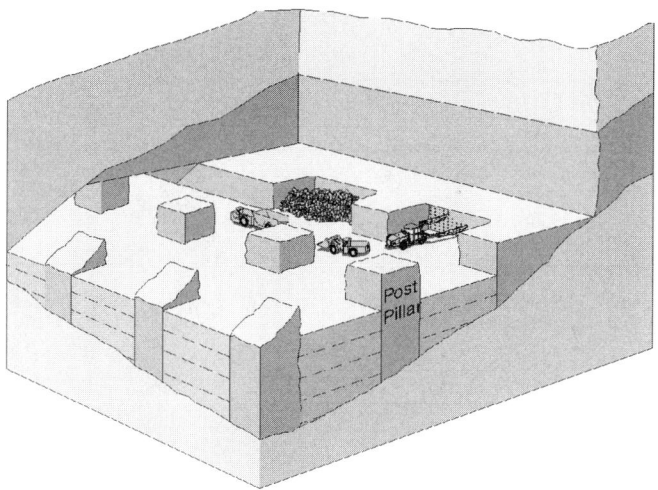

FIGURE 1.3 Post room-and-pillar mining

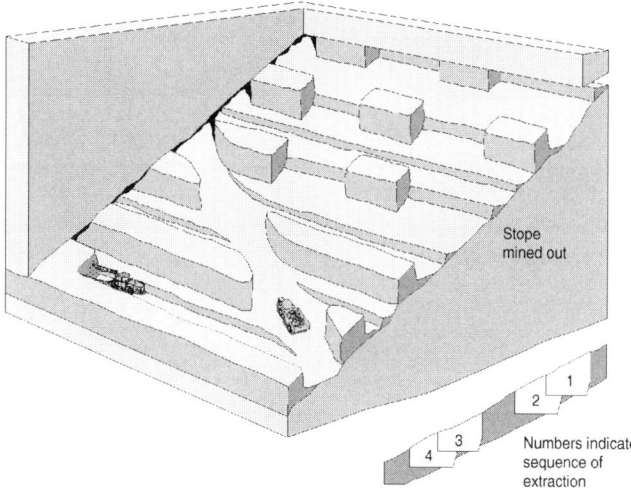

FIGURE 1.4 Step-room mining of inclined orebody

Deposits with large vertical heights are mined in slices. Mining starts at the top below the hanging wall. At this stage, rock bolts are installed for roof control with the back at a convenient height. Sections below are recovered in one or more steps by benching. Standard crawler rigs are used for drilling vertical holes and for conventional bench blasting. Horizontal drilling and "flat" benching are more practical alternatives because the same drill jumbo can be used for both topheading and drilling flat bench holds.

The blasted ore is loaded at the muckpile with diesel-driven front-end loaders. Different transport systems are used, depending on stope height and transport distance. Where the opening is high enough, the common dump truck provides economical transport from stopes to collection points. In thin ore bodies, specially built low mine trucks are available from manufacturers. Stopes with very little headroom can be cleaned by load-haul-dump (LHD) machines, and muck can be transferred onto trucks parked in special loading bays for transport over longer distances.

Mobile mechanized equipment is ideal in flat or slightly inclined ore bodies. In the room-and-pillar layout, several production areas can be established. Communications are straightforward and simple. These factors set the stage for the high utilization of both men and machines in an efficient ore recovery system.

Post Room-and-Pillar Mining. Post room-and-pillar mining (or "post-pillar" mining) is a combination of room-and-pillar and cut-and-fill stoping. With this method, ore is recovered in horizontal slices starting from the bottom and advancing upward. Pillars are left inside the stope to support the roof. Mined-out stopes are hydraulically backfilled with tailings, and the next slice is mined by machines working from the fill surface. Pillars continue through several layers of fill. Sandfill provides the possibility of modifying the stope layout and adapting the post-pillar method to variations in rock conditions and ore boundaries. Both backfill and sandfill increase the support capability of the pillar, permitting a higher rate of ore recovery than does classic room-and-pillar mining.

Post-pillar mining combines the advantages of cut-and-fill mining—that is, allowing work on flat, smooth floors—with the spacious stopes offered by room-and-pillar mining. Easy access to multiple production points favors the use of efficient mechanized equipment.

Step Room-and-Pillar Mining. Step room-and-pillar mining is a variation in which the footwall of an inclined ore body is adapted for efficient use of trackless equipment. Although applications cannot be fully generalized, step room-and-pillar mining applies to tabular deposits with thicknesses from 2 to 5 m and dips ranging from 15° to 30°.

The method features a layout in which stopes and haulageways cross the dip of the ore body in a polar coordinate system. By orienting stopes at certain angles across dip, stope floors assume an angle that is comfortably traveled by trackless vehicles. Transport routes cross in the opposite direction to establish roadway access to stopes and to transport blasted ore to the shaft.

The main development of step room-and-pillar mining includes a network of parallel transport drifts traversing the ore body in predetermined directions. Drift floors are maintained with grades that allow the use of selected trucks.

Stopes are excavated from transport drifts branching out at a predetermined step-room angle. The stope is advanced forward in a mode similar to drifting until breakthrough into the next parallel transport drive. The next step is to excavate a similar drift or side slash one step downdip and adjacent to the first drive. This procedure is repeated until the roof span becomes almost too wide to remain stable. Then an elongated strip parallel to the stopes is left as a pillar. The next stope is excavated the same way, and mining continues downward step by step. The numbers in Figure 1.4 indicate the sequence of extraction.

1.3.3 Vein Mining

In vein mines (Figures 1.5, 1.6, and 1.7), the dimensions of mineral deposits are highly variable. An ore body can be anything from a large, massive formation several square kilometers in surface area to a 0.5-m-wide quartz vein containing some 20 g/tonne of gold. Miners aim to recover the mineral's value, but prefer to leave waste rock in the hanging wall and the footwall intact. In the thicker deposits, a machine operates within the ore body walls without problems. When the mineralized zone narrows to a few meters, machines may be too wide to fit inside the ore boundaries. To excavate rock only to permit the machine to fit produces waste, which dilutes the ore. The alternative is to use manual labor to recover high-grade ore. However, labor is costly, and manual mining techniques are inefficient. Also, it is difficult to find people who accept working with hand-held rock drills and using muscle power.

Today, a selection of standard slim-sized machines is available, allowing mechanized mining in 2-m-wide drifts. These slim-sized machines include the face jumbo for narrow drifts matched with a longhole rig of the same size. The small drifter jumbo and longhole rig complemented with an LHD with a 2-m^3 bucket provides everything needed for the mechanized mining of a 2-m-wide vein.

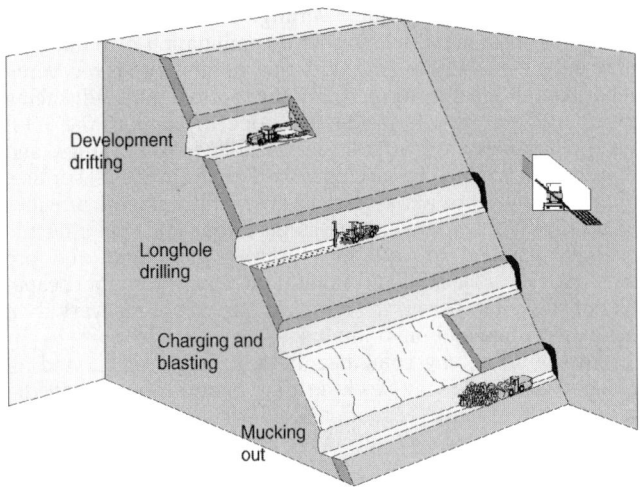

FIGURE 1.5   Mining a narrow vein with steep dip

FIGURE 1.6   Small size drill jumbo, Dome mine, Canada

FIGURE 1.7   Mini-rig for longhole drilling, Stillwater platinum mine, Montana, USA

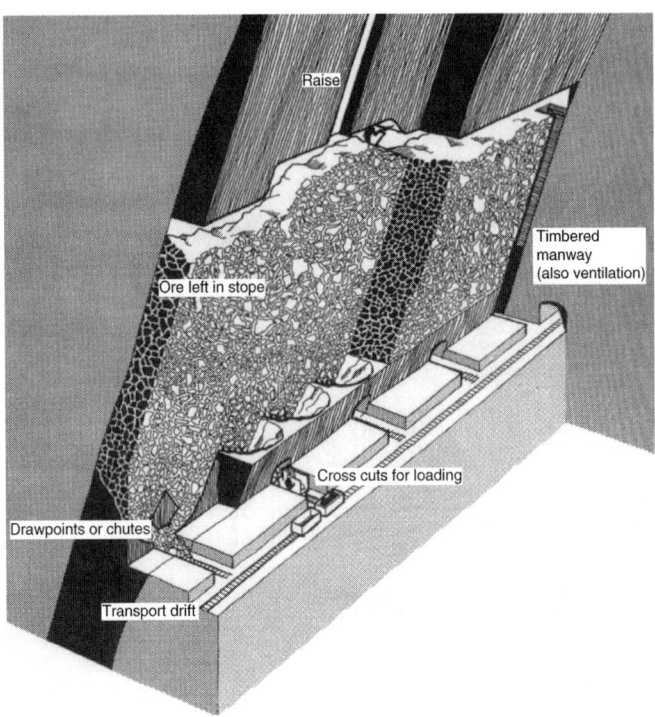

FIGURE 1.8   Shrinkage stoping with cross-cut loading

1.3.4   Shrinkage Stoping

In shrinkage stoping (Figure 1.8), ore is excavated in horizontal slices, starting from the bottom of the stope and advancing upward. Part of the broken ore is left in the mined-out stope, where it serves as a working platform for mining the ore above and to support the stope walls.

Through blasting, rock increases its occupied volume by about 50%. Therefore, 40% of the blasted ore must be drawn off continuously during mining to maintain suitable headroom between the back and the top of the blasted ore. When the stope has advanced to the upper border of the planned stope, it is discontinued, and the remaining 60% of the ore can be recovered.

Smaller ore bodies can be mined with a single stope, whereas larger ore bodies are divided into separate stopes with intermediate pillars to stabilize the hanging wall. The pillars can generally be recovered upon completion of regular mining. Shrinkage stoping can be used in ore bodies with

- Steep dips (the dip must exceed the angle of repose),
- Firm ore,
- Comparatively stable hanging wall and footwall,
- Regular ore boundaries,
- Ore that is not affected by storage in the stope (certain sulfide ores tend to oxidize and decompose when exposed to the atmosphere).

The development for shrinkage stoping consists of

- A haulage drift along the bottom of the stope,
- Crosscuts into the ore underneath the stope,
- Finger raises and cones from the crosscuts to the undercut,
- An undercut or complete bottom slice of the stope 5 to 10 m above the haulage drift, and

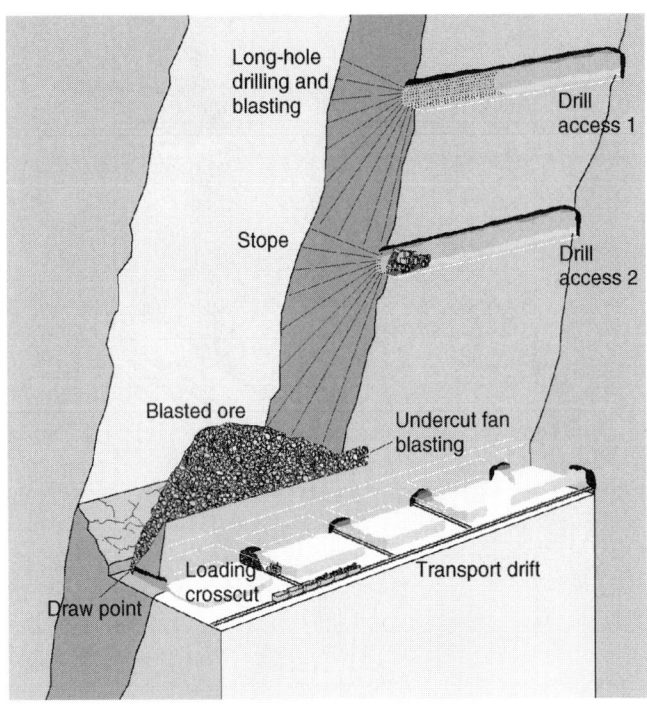

FIGURE 1.9   Sublevel open stoping

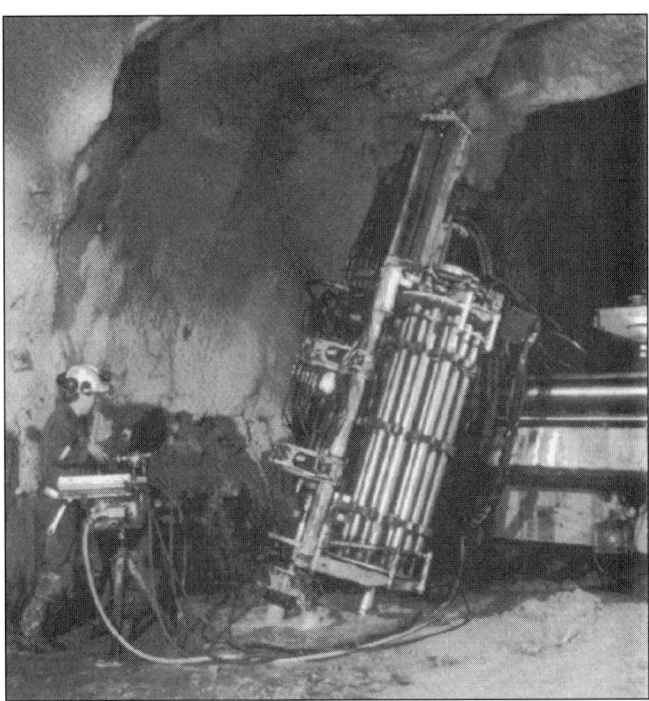

FIGURE 1.10   Longhole rig with slide positioning, remote control, and tube carousel, Williams gold, Canada

- A raise from the haulage level passing through the undercut to the main level above to provide access and ventilation to the stope.

The development of the bottom section of the stope can be simplified in the same way as for sublevel stoping—the finger raises are deleted, and the cross-cuts are developed for drawpoint loading.

Drilling and blasting are carried out as overhead stoping. The rough pile of ore in the stope prevents the use of mechanized equipment. Standard practice is to use air-leg rock drills and stoper drillers.

The traditional ore handling system in shrinkage stoping entails direct dumping of ore into rail cars from chutes below the finger raises. Shovel loaders are more effective in conjunction with a drawpoint loading system.

Shrinkage stoping was a common and important method in the days when few machines were employed in underground mining. Its advantage is the fact that the ore could be dumped directly into cars through the chutes, eliminating hand-loading. This is of little importance today, and the drawbacks—that is, the method is labor intensive, working conditions are difficult and dangerous, productivity is limited, and the bulk of the ore remains stored in the stope for a long period of time—have resulted in the replacement of shrinkage stoping by other methods. Sublevel stoping, vertical retreat stoping, sublevel caving, and cut-and-fill mining are methods that usually can be applied under similar conditions.

Shrinkage stoping remains, however, as one of the methods that can be practiced with a minimum of investment in machinery and yet is still not entirely dependent on manual capacity.

1.3.5   Sublevel Open Stoping

In sublevel open stoping (Figures 1.9 and 1.10), ore is recovered in open stopes normally backfilled after being mined. Stopes are often large, particularly in the vertical direction. The ore body is divided into separate stopes. Between stopes, ore sections are set aside for pillars to support the hanging wall. Pillars are normally shaped as vertical beams across the ore body. Horizontal sections

of ore, known as crown pillars, are also left to support mine workings above the producing stopes.

Enlarging stope dimensions influences mining efficiency. Miners therefore aim for the largest possible stopes. The stability of the rock mass is a limiting factor to be considered when selecting the sizes of stopes and pillars. Sublevel stoping is used for mining mineral deposits with following characteristics:

- Steep dip—the inclination of the footwall must exceed the angle of repose,
- Stable rock in both the hanging wall and the footwall,
- Competent ore and host rock,
- Regular ore boundaries.

Sublevel drifts for longhole drilling are prepared inside the ore body between main levels. These are strategically located since these are the points from which the longhole rig drills the blast pattern. The drill pattern specifies where blastholes are to be collared and the depth and angle of each hole, all of which must be set with great precision to achieve a successful blast.

Drawpoints are excavated below the stope bottom for safe mucking with LHDs, which may be combined with trucks or rail cars for longer transport. Different layouts for undercut drawpoints are used. The trough-shaped stope bottom is typically accessed through loading drifts at regular spacings.

Developing the set of drifts and drawpoints underneath the stope is an extensive and costly procedure. A simpler layout is gaining in popularity as an alternative to the conventional drawpoint-and-muck-out system. Here, the loading level is integrated with the undercut. Mucking out is done directly on the stope bottom inside the open stope. The LHD works inside the open stope and, for safety reasons, is operated by radio control by an operator based inside the access drift.

Sublevel stoping requires a regular shape of stopes and ore boundaries. Inside the drill pattern, everything qualifies as ore. In larger ore bodies, the area between the hanging wall and the footwall is divided into modules along strike and mined as primary and secondary stopes.

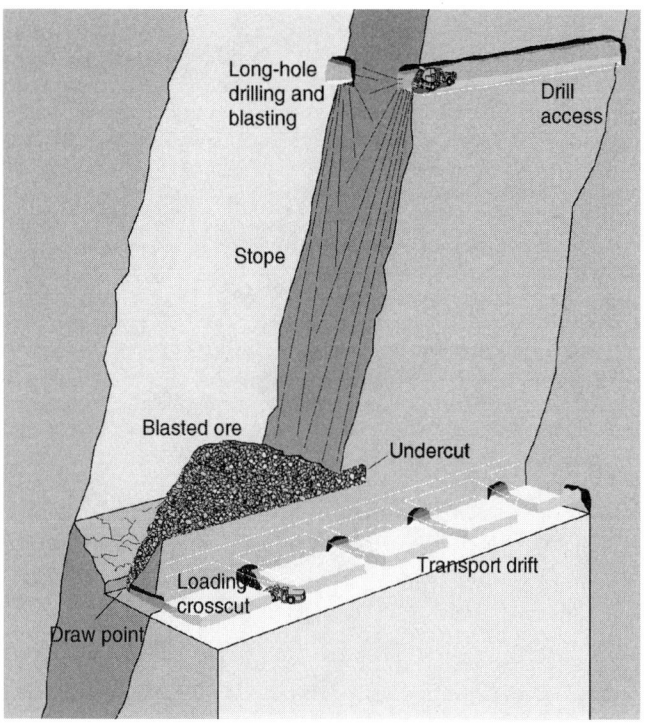

FIGURE 1.11 Bighole open stoping

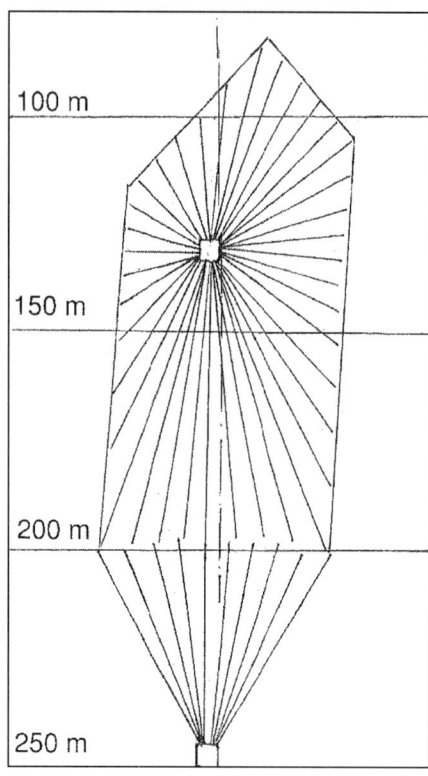

FIGURE 1.12 Bighole sample pattern, Mount Charlotte, Australia

1.3.6 Bighole Stoping

Bighole stoping (Figures 1.11–1.13) is a scaled-up variant of sublevel open stoping in which longer blastholes with larger diameters (140 to 165 mm) are used. The holes are normally drilled using the in-the-hole (ITH) technique. Hole depths may reach 100 m, which is double the length that can be drilled with tophammer rigs. Blast patterns are similar to those used in sublevel open stoping. The 140-mm-diameter blasthole breaks a rock slice 4 m thick with a 6-m toe spacing.

The advantage of bighole stoping as compared to sublevel stoping is the scale factor. The ITH-drilled holes are straight, and drilling accuracy can be exploited. For instance, vertical spacings between sublevels can be extended from 40 m with sublevel open stoping to 60 m with bighole stoping. Risks of damage to rock structures is a factor to be considered when bighole stoping is used.

1.3.7 Vertical Crater Retreat

Vertical crater retreat (VCR) mining (Figures 1.14, 1.15, and 1.16) is a method originally developed by the Canadian mining company INCO. Today, VCR is an established mining method used by mines all over the world that have competent, steeply dipping ore and host rock. VCR is based on the crater blasting technique in which powerful explosive charges are placed in large-diameter holes and fired. Part of the blasted ore remains in the stope over the production cycle, serving as temporary support for the stope walls.

The sequence of development of VCR stopes is

- A haulage drift is excavated along the ore body at the drawpoint level,
- A drawpoint loading arrangement is created underneath the stope,
- The stope is undercut,
- An overcut accesss is excavated for drilling and charging.

FIGURE 1.13 Bighole drill rig with automatic controls and tube carousel for 50 m long holes, Mount Charlotte, Australia

The ore in a stope block is drilled with ITH drill rigs positioned in the overcut. Holes are drilled downward until they break through into the undercut. Vertical holes are preferred wherever possible. Hole diameters vary from 140 to 165 mm, although holes 205 mm in diameter have been tried in a few mines. For 165-mm-diameter holes, a hole pattern of 4 by 4 m is typical.

Holes are charged from the overcut using powerful charges contained in a short section of blast hole. These crater charges

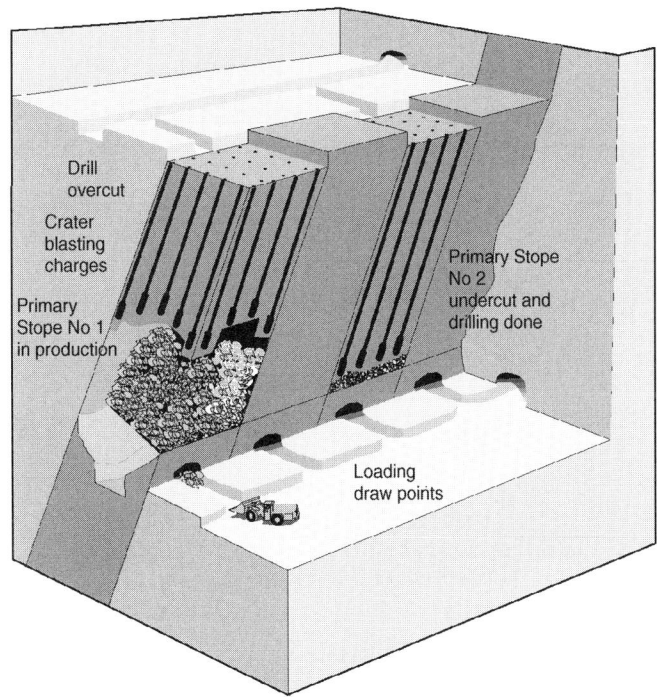

FIGURE 1.14   VCR mining, primary stopes

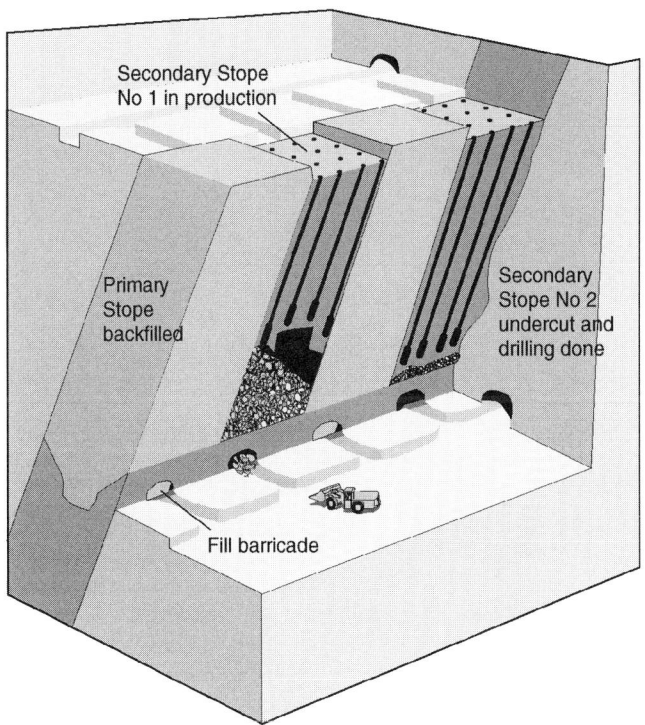

FIGURE 1.16   VCR mining, recovery of secondary stopes

FIGURE 1.15   ITH drilling, 165 mm blast holes, INCO, Canada

are placed a specified distance above the free surface. Holes are grouped so that charges will be at the same elevation and depth. First, the hole depth is measured. Then the hole is blocked at the proper height. Explosive charges are lowered, and the hole is stemmed with sand and water placed on top of the charge. Adjacent explosive charges aid in breaking the rock, normally loosening a 3-m slice of ore that falls into the void below. Charging

requires a trained crew for successful blast results, and records are necessary to keep track of the blasting progress in each hole. The ore is mucked from stopes through the undercut using remote-controlled LHDs or recovered by a drawpoint system underneath the stope as in sublevel stoping. The stopes may or may not be backfilled.

VCR mining is applicable in conditions similar to those in which sublevel open stoping is used. VCR is technically simpler with ITH drilling compared to tophammer drilling. ITH holes are straight, and hole deviations are minimal. The charging of the blastholes is complex, and techniques must be mastered by the charging team. The powerful VCR charges involve higher risks for damaging the surrounding rock than sublevel open stoping.

1.3.8   Cut-and-Fill Stoping

Cut-and-fill mining (Figures 1.17 and 1.18) removes ore in horizontal slices, starting from the bottom undercut and advancing upward. Ore is drilled and blasted, and muck is loaded and removed from the stope. When the stope has been mined out, voids are backfilled with hydraulic sand tailings or waste rock. The fill serves both to support the stope walls and provide a working platform for equipment when the next slice is mined.

Cut-and-fill mining is used in steeply dipping ore bodies in strata having good-to-moderate stability and comparatively high-grade ore. It provides better selectivity than the alternative sublevel stoping and VCR mining techniques. Hence, cut-and-fill is preferred for ore bodies having an irregular shape and scattered mineralization. Cut-and-fill allows selective mining, separate recovery of high-grade sections, and the leaving of low-grade rock behind in stopes.

The development for cut-and-fill mining includes

- A haulage drive along the footwall of the ore body at the main level,

- Undercutting the stope area with drains for water,

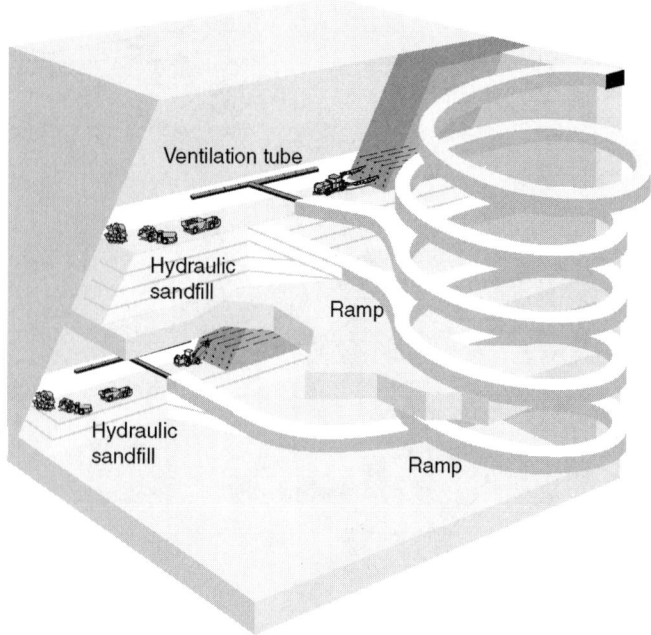

FIGURE 1.17   Mining with cut and fill

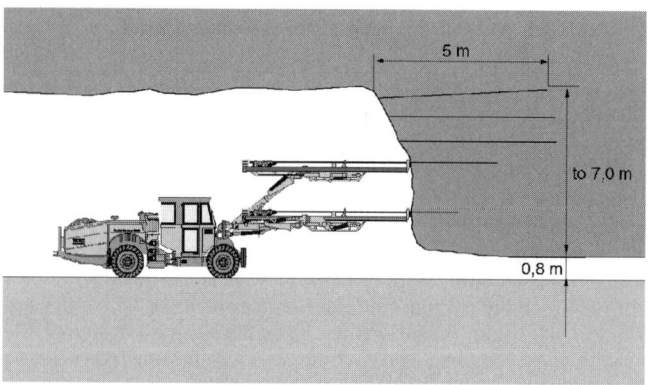

FIGURE 1.18   Face drilling in cut-and-fill stope

- A spiral ramp in the footwall with access drive to the undercut,
- A raise connecting to levels above for ventilation and filling material.

Modern cut-and-fill mines drill the stope face with a jumbo. The face appears to be a wall across the stope with an open slot at the bottom above the fill. The face is drilled with breasting holes, charged, and blasted. The slot underneath provides space into which the blasted rock can expand.

The drill pattern can be modified before each round to follow variations in ore boundaries. Sections with low-grade ore are left in place or separated while mucking out. Mining can be diverted from the planned stope boundaries to recover pockets of minerals in the host rock.

A smooth fill surface and controlled fragmentation are ideal for LHDs. Tramming distances inside the stope to the ore pass are convenient for LHD cycles. Ore passes made from steel segments welded to form a large tube can be positioned inside the stope, by-passing layers of sandfill. The ore pass can also consist of a raise excavated in rock close to the stope-ramp access.

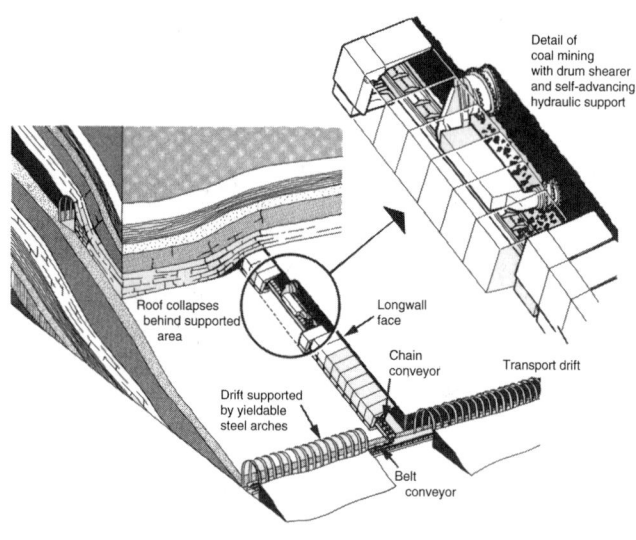

FIGURE 1.19   Longwall mining in soft rock (coal seam)

Hydraulic sandfill is often used with cut-and-fill mining. The fill—deslimed sand tailings from the mine's dressing plant—is mixed with water to 60% to 70% solids and distributed to stopes via a network of pipes. Before filling, stopes are prepared by barricading entries, and drainage tubes are laid out on the floor. The sand fills the stope to almost its full height. As a harder fill is required on the surface, cement is added in the last pour. When the water has drained, the fill surface is smooth and compact. It forms a good base for mobile machines while mining the next slice of ore.

Cut-and-fill mining is a versatile method and preferred by mines that require the capability of mining selected ore pockets and adaptability to variations in the rock mass.

1.3.9 Longwall Mining

Longwall mining applies to thin-bedded deposits of uniform thickness and large horizontal extent. Typical deposits are represented by coal seams, potash layers, or conglomerate reefs mined by the South African gold mining companies. Longwall mining applies to both hard and soft rock as the working area along the mining face can be artificially supported where the hanging wall tends to collapse.

The longwall mining method extracts ore along a straight front having a large longitudinal extension. The stoping area close to the face is kept open to provide space for personnel and mining equipment. The hanging wall may be allowed to subside at some distance behind the working face.

Development of longwall mines involves the excavation of a network of haulage drifts for access to production areas and transport of ore to shaft stations. As the mineralized zone extends over a large area, haulage drifts are accompanied with parallel excavations to ventilate mine workings. Haulage drifts are usually arranged in regular patterns and excavated in the deposit itself. The distance between two adjacent haulage drifts determines the length of the longwall face.

Longwall mining (Figure 1.19) is a common method for extracting coal, trona, and potash from seams of various thickness. It can be mechanized almost to perfection. The soft material does not require drilling and blasting, but can be cut loose mechanically. Special machines shaped as cutting plows or rotating drums with cutters run back and forth along the faces, each time cutting a fresh slice of the seam. The coal or mineral falls onto a chain conveyor that carries the mineral to the haulage

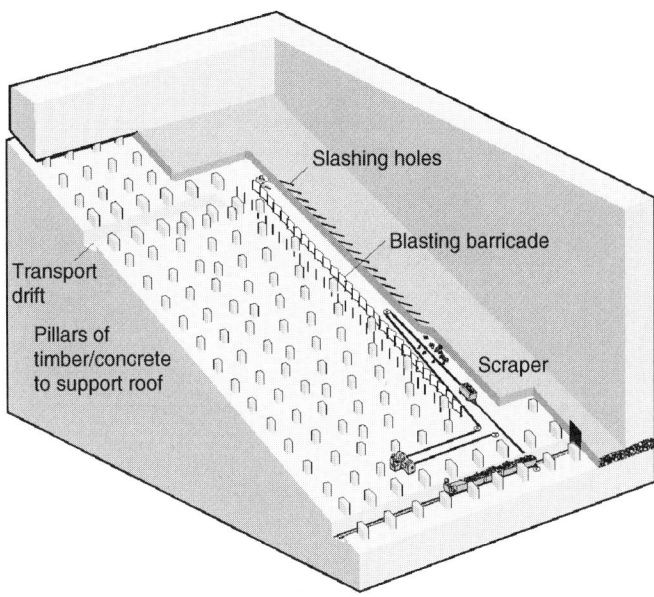

FIGURE 1.20   Longwall mining in gold reef

FIGURE 1.21   Drilling the reef with hand-held rock drill, East Rand Properties, South Africa

drift, from where it is transported for hoisting out of the mine. Conveyor belts are frequently used to transport material, as belts are adaptable to the almost continuous flow of material from the production areas. The roof along the longwall face is supported and the working area completely protected by a system of hydraulically operated props. The supports move forward as mining advances, and the roof behind can be allowed to collapse.

Longwall mining is also used for mining thin, reef-type deposits. The gold reef conglomerates are very hard and difficult to mine. South African gold mines have developed their own techniques based on manpower and the use of hand-held pneumatic rock drills. Figures 1.20 and 1.21 show mining of a reef approximately 1 m thick. The width of the mineralized section might be even less, but there must be space for miners crawling on their knees. Pillars of timber and concrete are installed to support the roof in very deep mines.

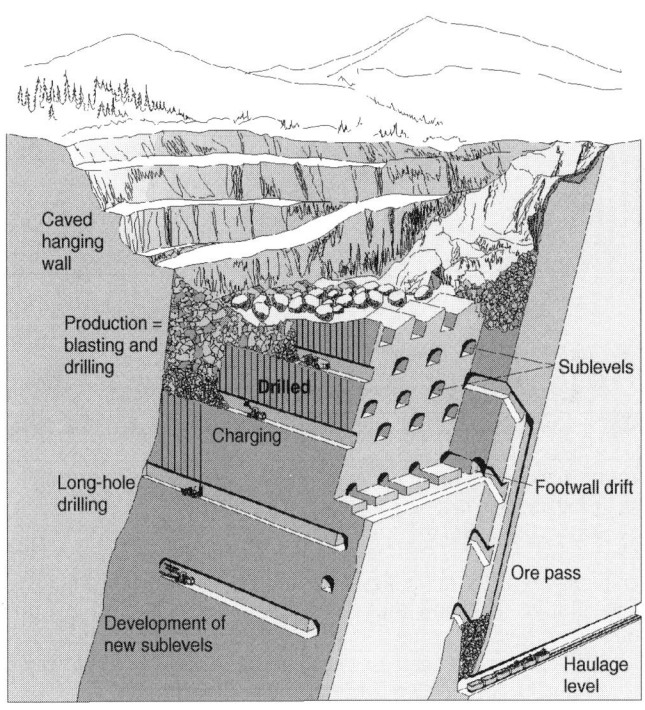

FIGURE 1.22   Mining by sublevel caving

1.3.10   Sublevel Caving

In sublevel caving, the ore is extracted via sublevels developed in the ore body at regular intervals. Each sublevel features a systematic layout with parallel drifts along or across the ore body. In a wide ore body, the sublevel drifts start from the footwall drift and are driven until they reach the hanging wall. This is referred to as transverse sublevel caving (Figure 1.22). In ore bodies of lesser width, longitudinal sublevel caving is used. In this variant, drifts branch off in both directions from a center crosscut.

Sublevel caving is used in large, steeply dipping ore bodies. The rock mass must be stable enough to allow the sublevel drifts to remain open with just occasional rock bolting. The hanging wall should fracture and collapse to follow the cave, and the ground on top of the ore body must be permitted to subside.

Caving requires a rock mass where both the ore body and the host rock fracture under controlled conditions. As mining removes rock and the mined-out area is not backfilled, the hanging wall keeps caving into the voids. Continued mining results in subsidence of the surface, and sinkholes may appear. Continuous caving is important to avoid creation of cavities inside the rock where a sudden collapse could be harmful to mine installations.

The amount of development needed to institute sublevel caving is extensive as compared to other mining methods. However, development primarily involves drifting to prepare sublevels. Drifting is a simple and routine job for a mechanized mine. Development of sublevels is done efficiently in an environment where there are multiple faces on one sublevel available to drill rigs and loaders.

A ramp is needed to connect different sublevels and communicate with the main transport routes. Ore passes are also required at strategic locations along the sublevels to allow LHDs to dump ore to be collected and transported to the haulage level below.

A drawing showing sublevel drifts is almost identical for every second sublevel, which means that drifts on the first sublevel are positioned right on top of drifts on the third sublevel, while drifts on the second sublevel are located underneath pillars

FIGURE 1.23 Twin boom rig for fan drilling, with operator's cabin, tube handling, and drill automatics, Kiruna, Sweden

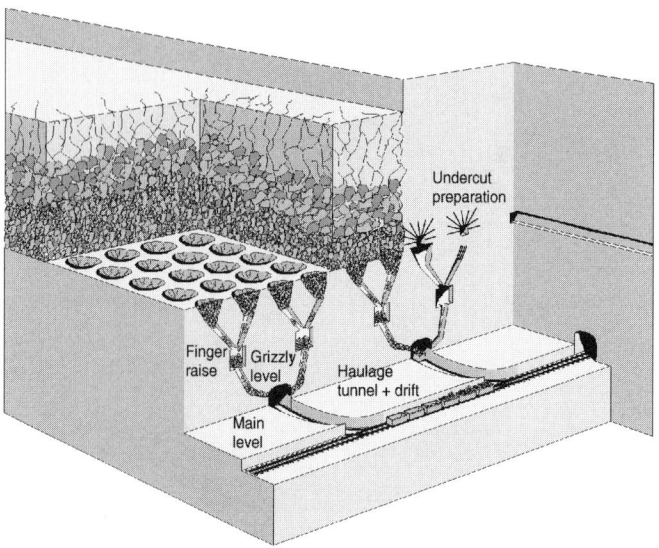

FIGURE 1.24 Block caving with finger raises, grizzly treatment and chute loading

between the drifts on sublevels 1 and 2. A section through the sublevel area will show that the drifts are spread across the ore body in a regular pattern both vertically and horizontally. A diamond-shaped area can be traced above one drift and indicates the volume of ore to be recovered from each drift.

The ore section above the drift is drilled in a fan-shaped pattern with longhole drills (Figure 1.23). Drilling can be done independently of other procedures, often well ahead of charging. Thus, drilling, charging, and blasting longholes can be timed to suit the mine's production schedules. Blasting on each sublevel starts at the hanging wall and mining retreats toward the footwall. The cave line should follow an approximately straight front, and hence adjacent drifts should be mined at a similar pace. A section through the cave shows the upper sublevels one step ahead of the sublevels underneath.

Blasting the longhole fan breaks the ore in the slice. Most of the blasted ore remains in place while some falls down into the drift opening. Mucking out with LHDs creates a cave pattern of ore and waste from above. Loading continues until the operator decides there is too much dilution, stops mucking, and moves to another heading. With the heading vacated, the charging team moves into the heading and charges and fires the next ring of longholes.

Ore handling involves mucking out the blasted material at the front, transporting it on the sublevels, and dumping the ore into ore passes. These are ideal conditions for LHDs as they can be kept in continuous operation. When one face is mucked clean, the LHD is moved to a nearby drift heading and mucking continues. Sublevels are designed with tramming distances matched to particular sizes of LHDs.

Dilution and ore losses are drawbacks for sublevel caving. Extensive scientific investigations have been made to determine the flow of ore in a cave and to identify means of reducing ore losses and minimizing dilution. Dilution varies between 15% and 40%, and ore losses can be from 15% to 25%, depending on local conditions. Dilution is of less influence for ore bodies with diffuse boundaries where the host rock contains low-grade minerals or for magnetite ores, which are upgraded by simple magnetic separators. Sulfides, in contrast, must be refined by costly flotation processes.

Sublevel caving is repetitive both in layout and working procedures. Development drifting, production drilling, charging, blasting, and mucking are all carried out separately. Work takes place at different levels, allowing each procedure to be carried out continuously without disturbing the others. There is always a place for the machine to work.

1.3.11 Block Caving

Block caving is a technique in which gravity is used in conjunction with internal rock stresses to fracture and break the rock mass into pieces that can be handled by miners. "Block" refers to the mining layout in which the ore body is divided into large sections of several thousand square meters. Caving of the rock mass is induced by undercutting a block. The rock slice directly beneath the block is fractured by blasting, which destroys its ability to support the overlaying rock. Gravity forces on the order of millions of tons act on the block, causing the fractures to spread until the whole block is affected. Continued pressure breaks the rock into smaller pieces that pass through drawpoints where the ore is handled by LHDs.

Block caving is a large-scale production technique applicable to low-grade, massive ore bodies with the following characteristics:

- Large vertical and horizontal dimensions,
- A rock mass that will break into pieces of manageable size, and
- A surface that is allowed to subside.

These rather unique conditions limit block caving to particular types of mineral deposits. Looking at worldwide practices, one finds block caving used for extracting iron ore, low-grade copper, molybdenum deposits, and diamond-bearing kimberlite pipes. The large tonnage produced by each individual mine makes block-caving mines the real heavyweights when compared to most other mines.

The development of block caving when conventional gravity flow is applied (Figure 1.24) involves

- An undercut where the rock mass underneath the block is fractured by longhole blasting,
- Drawbells beneath the undercut that gather the rock into finger raises,
- Finger raises that collect rock from drawbells to the grizzlies,
- A grizzly level where oversized blocks are caught and broken up,

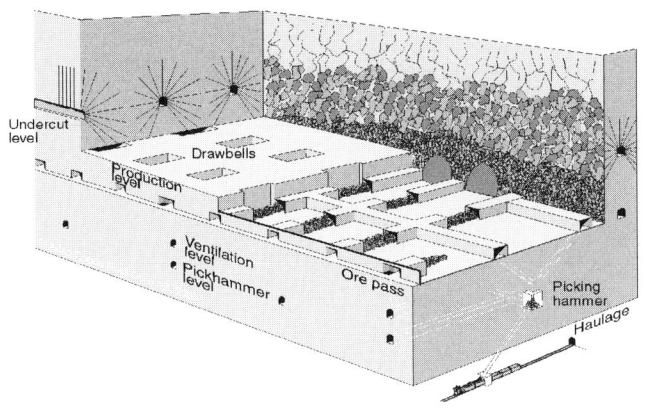

FIGURE 1.25   Block caving LHD loaders and pick-hammer control, El Teniente, Chile

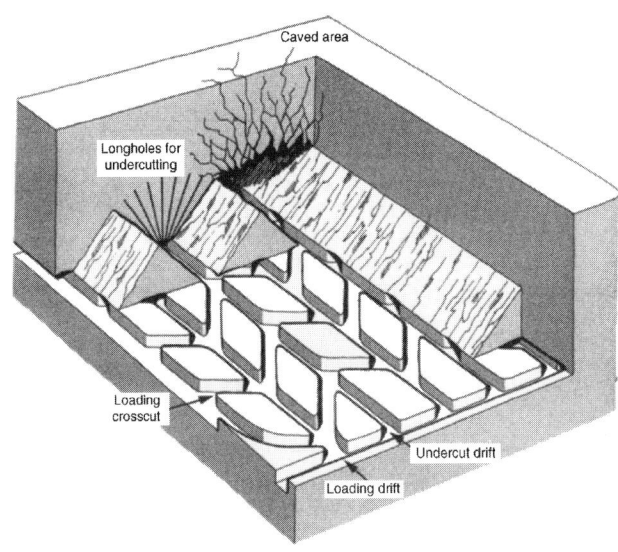

FIGURE 1.26   Block caving with one-level development, longhole undercutting and drawpoint loading

■ A lower set of finger raises that channel ore from grizzlies to chutes for train loading. The finger raises are arranged like branches of a tree, gathering ore from a large area at the undercut level and further channeling material to chutes at the haulage level.

■ A lowermost level where ore is prepared for train haulage and chute loading.

Openings underneath the block are subject to high internal stresses. Drifts and other openings in a block-caving mine are excavated with minimal cross sections. Heavy concrete liners and many rock bolts are necessary to secure the integrity of mine drifts and drawpoint openings.

After completion of the undercut, the rock mass above begins to fracture. The blocks are gathered by drawbells and crates and funneled down through finger raises. The intention is to maintain a steady flow from each block. Miners keep records of the volume extracted from individual drawpoints. Theoretically, no drilling and blasting are required for ore production. In practice, it is often necessary to assist rock mass fracturing by longhole drilling and blasting in widely spaced patterns. Boulders that must be broken by drilling and blasting frequently interrupt the flow. Large blocks cause hang-ups in the cave that are difficult and dangerous to tackle.

Originally, block-caving techniques relied 100% on gravity flow to deliver ore from the cave into rail cars. The ore was funneled through a system of finger raises and ore passes, ending at trough chutes at the main haulage level. As chute loading requires controlled fragmentation, the rock had to pass through a grizzly before it entered the ore pass. The grizzly-man with a sledgehammer used to be a bottleneck in old-style block-caving mines. Now it is common to use hydraulic hammers for breaking the boulders.

Today, block-caving mines have adapted trackless mining in which LHDs are used to handle the cave in the drawpoints (Figures 1.25 and 1.26). As a consequence, ventilation must be added to development preparations to clear the production level of diesel exhaust. The LHDs are able to handle large rocks while oversized boulders are blasted in the drawpoints.

Block caving is an economical and efficient mass-mining method when rock conditions are favorable. The amount of drilling and blasting required for ore production is minimal while the development volume is immense. The behavior of the rock mass and conditions for caving are difficult to predict when a block-caving mine is planned. The extensive development required and time lag before production starts are also factors to consider when block caving is being compared to other methods.

1.4   MECHANIZATION AND EFFICIENCY

1.4.1   Preparing for the Future

The mining industry exists in a competitive environment. The only way to survive in the long term is to ensure that each ton of ore mined leaves a profit after all cost factors are deducted. All of us face a climate of escalating labor costs and tougher environmental rules that increase the burden on production costs. Development of new equipment and improved efficiency help us compensate for these increasing cost factors.

1.4.2   Mechanization—Automation—Robotics

Labor represents a major share of the production costs in underground mines. Replacing labor with powerful machines is a natural way to counteract escalating costs. Mechanization has proven itself by a steady increase in production in underground mines over the last few decades. More duties are being taken over by machines, more powerful machines increase output, and more sensitive controls are able to handle dangerous and complex procedures to produce a higher quality product.

Elaborate machines in the hands of skilled operators turn mine production into an efficient industrial process. Any mine will have the potential to introduce new equipment and improve existing standards by exploiting the potential of modern technology.

1.4.3   Quality and Grade Control

Economic mining is a matter not only of production efficiency, but also the quality or grade of the run-of-mine product. Thus, the degree of selectivity achieved by equipment and methods becomes a prime consideration. Mechanized cut-and-fill mining is highly selective. The cut-and-fill method is adaptable to variations in ore body boundaries and rock conditions, which makes it interesting for mine prospects and rehabilitation projects. Cement-consolidated backfill has improved recovery during open stoping so that this method is comparable with cut-and-fill mining.

1.4.4 Efficiency Ratings

Efficiencies of mining operations are rated in tons per man-shift or kilotons per man-year (with reference to underground workers only). Efficiency varies from mine to mine and should not be given much weight except when considered as a general characteristic. Each mining method provides certain conditions for efficiency, from 1 ton per man shift for a complex method to 100 tons per man-shift for an efficient room-and-pillar mine. Efficiency relates to costs per ton. Where the ratio of tons per man-shift is low, ore grades must be high.

1.4.5 Utilization and Output

Work Time and Schedules. Work time in mines is often based on schedules of 8 hours per shift, three shifts a day, 7 days a week all year around. Other schedules may include a 6-day week or 11-hour shifts with two shifts per day. Evidently ambitions are to keep production going by utilizing the available time in the most efficient manner, considering public and religious holidays.

Even if the work schedule is continuous, few machines operate more than 70% of the time. Machines that can be kept in operation much of the time are those directly involved with production—longhole drill rigs, LHDs, and mine trucks. For other demands, such as development drifting and rock bolting, 30% use would be considered normal. The machine itself is a tool for doing things in a practical way that minimizes manual efforts.

Multipoint Attack. The mining method influences conditions for use of mechanized equipment. Drill-blast rock excavation involves a cycle featuring frequent changes of techniques and machinery. Where several attack points lie within a reasonable distance of one another, machines can be kept busy shuttling from one workplace to the next. A drill rig, charging truck loader, and other machines should be scheduled to minimize delays while they are changing places.

Work Specialization. Carrying out the continuity and volume of a specific procedure without interruption is important. Sublevel stoping, VCR mining, and sublevel caving are methods in which drilling, charging, blasting, and mucking out are independent procedures. The longhole drill rig keeps hammering in the same drift for long periods of time, promoting output from an elaborate and capital-intensive machine. As longhole drilling is independent of mucking out, blastholes can be drilled and maintained until the production schedule calls for more tonnage.

1.4.6 The Machine—A Versatile Tool

Selecting Machines. Selecting a machine for a specific purpose is a complex procedure. The application is rarely straightforward, but includes special requirements and restrictions. The machine itself is a complex device with capabilities that are difficult to explain in specification sheets. A dialogue between a mine's technical staff and a manufacturer's representative prevents misunderstandings and is the best way to the correct choice.

The development of new products and improvements to existing equipment is enhanced through regular contacts between the end-user and the manufacturer. A continuing dialogue ascertains that introduced products meet requirements from both management and operators.

Output and Size. Typical mining machines are offered in a range of sizes distinguished by output or ability to perform specified procedures. The potential buyer should be able to make a first choice of a machine for his application without too much of a problem. It is advisable to select a larger rather than a smaller model. Extra capacity is always a safety margin, adding to flexibility in production planning, and a heavy-duty design increases resistance to wear and tear in a tough underground environment.

Capability. Qualifications of a machine are documented in the technical specification pages. Here, the important data, such as engine power, weight, length, width, etc., should be found. The specification sheet should be studied carefully before a decision is made on a new machine. The machine must meet expectations, in terms of both performance and capability, and it must fit inside the mine's drift openings.

Options and Extras. Specifications come with a list of options and special features that can be added to a basic machine at additional cost. Each option should be checked and analyzed with regard to selecting the basic machine. Some options will be necessary for the proper function of the basic unit, while others may seem to be fancy gadgets without real justification. As the basic machine represents the major share of the capital to be invested, the options may add valuable qualities to the basic unit at marginal extra cost. The value of option-services integrated into the basic unit should be assessed by the end-user.

1.5 SUMMARY

In the preceding sections, the author has tried to present conventional underground mining methods as clearly as possible. Naturally, there are additional considerations that cannot be included within the scope of this text.

Some readers may miss the inclusion of more detailed figures in the text. However, the variations in ore deposits are so great and the state of mining technology so dynamic that being too specific could mislead the reader. Every ore body is unique. The successful application of a mining method requires more than textbook knowledge; it also requires practical reasoning with a creative mind that is open to new impressions. The application of a mining method is a distinct challenge to any mining engineer.

The author wishes to express his gratitude to Atlas Copco MCT AB for permission to use the material included herein and for assistance in preparing the illustrations.

CHAPTER 2

General Planning of the Noncoal Underground Mine

Richard L. Bullock[*]

2.1 INTRODUCTION TO UNDERGROUND MINE PLANNING

Many details must go into the planning of an underground mine, and information must come from several sources. Geological, structural, and mineralogical information must first be collected and combined with data on resources and reserves. This information leads to the preliminary selection of a potential mining method and sizing mine production. Development planning is done, equipment is selected, and mine manning is completed, all leading to an economic analysis of the foregoing scenario of mine planning.

However, one cannot assume that the planning sequence just described will guarantee the best possible mine operation unless it is the best possible mine planning done correctly. Any sacrifice in best-possible mine planning introduces the risk that the end result may not achieve the optimum mine operation desired. Planning is an iterative process that requires looking at many options and determining which yields the optimum solution in the long run.

This chapter addresses many of the factors to be considered in the initial phase of all mine planning. These factors have the determining influence on mining method, size of operation, size of mine openings, mine productivity, mine cost, and, eventually, economic parameters used to determine whether the mineral reserve should even be developed.

2.2 PHYSICAL AND GEOTECHNICAL INFORMATION NEEDED

2.2.1 Technical Information Needed for Preliminary Mine Planning

Assuming that the resource to be mined has been delineated with diamond-drill holes, the items listed in the following paragraphs need to be established with respect to mine planning for the mineralized material. If this is an exploration project that has been drilled out by the company exploration team, the following information should have been gathered during the exploration phase and turned over to the mine evaluation team or the mine development group. Such information includes—

- Property location and access.
- Description of surface features.
- Description of regional, local, and mineral deposit geology.
- Review of exploration activities.
- Tabulation of potential ore reserves and resources.
- Explanation of ore-reserve calculation method.
- Description of company's land position.
- Description of the company's water position.

- Ownership and royalty conditions.
- History of the property.
- Any special studies by exploration teams (metallurgical, geotechnical, etc.).
- Any social issues or environmental issues that have surfaced while exploration was being completed.

This isn't to say that more information on each of these subjects will not have to be gathered, but if it can be started during the exploration phase of the project, much time will be saved during the feasibility-evaluation and development phases.

2.2.2 Geologic and Mineralogic Information

General knowledge of similar rock types or structures in existing mining districts is always helpful. The first mine in a new district is far more likely to run the risk of making costly errors during development than are mines that may be developed later.

The geologic and mineralogic information needed includes the following:

- The size (length, width, and thickness) of the areas to be mined within the overall area to be considered, including multiple areas, zones, or seams.

- The dip or plunge of each mineralized zone, area, or seam and the maximum depth to which the mineralization is known.

- Continuities or discontinuities within each mineralized zone.

- Any swelling or narrowing of each mineralized zone.

- Sharpness between grades of mineralized zones within the material considered economically minable.

- Sharpness between the ore and waste cut-off, specifically—

  — Whether this cut-off can be determined by observation or must be determined by assay or some special tool,

  — Whether this cut-off also serves as a natural parting, resulting in little or no dilution, or whether the break between ore and waste must be induced entirely by the mining method, and

  — Whether the mineralized zone beyond (above or below) the existing cut-off represents a current submarginal economic value that may become economic at a later time.

- Distribution of various valuable minerals making up each of the potentially minable areas.

- Distribution of the various deleterious minerals that may be harmful in processing the valuable mineral.

* University of Missouri, Rolla, MO.

- Interlocking of identified valuable minerals with other fine-grained minerals or waste material.

- Presence of alteration zones in both the mineralized and the waste zones.

- Tendency for the ore to oxidize once broken.

- Quantity and quality of ore reserves and resources with detailed cross sections showing mineral distribution, fault zones, or any other geologic structure related to the mineralization.

2.2.3 Structural Information (Physical and Chemical)

The needed structural information includes—

- Depth of cover.

- Detailed description of cover, including type, structural features in relation to mineralized zone, structural features in relation to proposed mine development; and information about any water, gas, or oil that might have been found.

- Quality and structure of the host rock (back, floor, hanging wall, footwall), including—

  - Type of rock.
  - Approximate strength or range of strengths.
  - Any noted weakening structures.
  - Any noted zones of inherent high stress.
  - Noted zones of alteration.
  - Porosity and permeability.
  - Presence of any swelling clay or shale interbedding.
  - Rock quality designation (RQD) throughout the various zones in and around all the mineralized area to be mined out.
  - Rock mass classification of the host rock.
  - Temperature of the zones proposed for mining.
  - Acid-generating nature of the host rock.

The structure of the mineralized material, including all of the factors in the above plus—

  - Tendency of the mineral to change character after being broken (e.g., oxidizing, degenerating to all fines, recompacting into a solid mass, becoming fluid, etc.).
  - Silica content of the ore.
  - Fiber content of the ore.
  - Acid-generating nature of the ore.

2.2.4 Property Information

The needed property information includes—

- Details on land ownership and/or lease holdings, including royalties to be paid or collected according to mineral zones or areas.

- Availability of water and its ownership on or near the property.

- Details of surface ownership and surface structures that might be affected by surface subsidence.

- Location of the mining area in relation to any existing roads, railroads, navigable rivers, power, community infrastructure, and available commercial supplies.

- Local, regional, and national political situations observed with regard to the deposit.

2.2.5 Need for a Test Mine

From this long list of information, all of which is badly needed to do a proper job of mine planning, it is evident that data acquired during the exploration phase of the operation will not be enough. Nor is it likely that all the needed data can all be obtained accurately from the surface. If this is the first mine in a mining area or district, then what is probably needed during the middle phase of the mine feasibility study is development of a test mine. While this may be an expense that the ownership was hoping that it would not have to endure up-front, the reasons for a test mine are quite compelling.

From a mining point of view, a test mine will—
1. Verify expected ore continuity, thus eliminating disastrous surprises.
2. Allow rock strength to be assessed accurately, which will allow prudent planning and sizing of the commercial mine opening.
3. Allow mining efficiency and productivity to be verified as it relates to drilling, blasting, and materials handling.
4. Permit more reliable studies of the nature of mine water inflows, which will allow for adequate water-handling equipment to be installed before problems are encountered.
5. Better quantify mine ventilation friction factors and requirements.
6. Confirm the character of the waste product and how waste will be handled in the commercial operation.

From a metallurgical point of view—
1. Verify and optimize the metallurgical flow sheet with a pilot plant process that is continuous lock cycle testing.
2. Determine size and type of equipment optimal for recovering the resource.
3. Determine type and amount of reagents that will lead to the best recoveries and concentrate grades.
4. Determine the required amount of water and how to achieve a water balance.
5. Predict concentrate grade, moisture content, and impurities more accurately.
6. Using bulk samples, assess the work index to a much better degree than with small samples.

From an environmental perspective—
1. Demonstrate the ability to control the operation in such a manner that it will not do harm to the environment.
2. Allow the project to study the waste characterization completely and determine any future problems.
3. If water discharge is involved, study the difficulty of settling the discharged water and determine what might be necessary to mitigate any future problems and determine if zero discharge is possible.

From the engineering design perspective—
1. Improve the ability to make more accurate cost estimates on the basis of better knowledge of the abrasivity of the rock and control of stopes and pit walls. This could actually lower the cost estimate since less contingency funds may have to be used.
2. Improve labor estimates so the productivity of each unit operation will be better understood.
3. Predict a more accurate schedule on the basis of unit productivities.
4. Lower the overall risk of the project in every aspect.

From the perspective of expediting later mine development—

1. Explore early access to develop the commercial mine, thus shortening the overall schedule from the end of the feasibility study to the end of construction.
2. Consider how openings can be completely utilized as part of the commercial mine operation.
3. Because access to the underground opening will already exist, evaluate whether some shafts may be able to be raise-bored and then expanded by mechanical excavation rather than by the more expensive conventional shaft sinking methods.
4. Consider a test mine as an ideal training facility prior to start-up of the commercial mine.

2.3 OTHER FACTORS OF EARLY MINE PLANNING

2.3.1 Sizing Production of a Mine

There is a considerable amount of literature available on the selection of a production rate to yield the greatest value to the owners (Carlisle 1955; Lama 1964; Tessaro 1960; Christie 1997). Basic to all modern mine evaluations and design concepts is the desire to optimize the net present value or to operate the property in such a way that the maximum internal rate of return is generated from the discounted cash flows. Anyone involved in the planning of a new operation must be thoroughly familiar with these concepts. Equally important is the fact that, *solely from the financial aspects of optimization*, any entrepreneur planning a mining operation and who is not familiar with the problems of maintaining high levels of concentrated production at low operating costs per tonne over a prolonged period is likely to experience unexpected disappointments in some years when returns are low (or there are none).

Other aspects of optimizing mine production relate to the potential effect of net present value. Viewed from the purely financial side, producing the product from the mineral deposit at the maximum rate yields the greatest return. This is because of the fixed costs involved in mining, as well as present-value concepts of any investment. Still, there are "practical limitations to the maximum intensity of production, arising out of many other considerations to which weight must be given" (Hoover 1909). There can be many factors limiting mine size, some of which are listed here.

- Market conditions.
- Current price of the product(s) versus trend price.
- Grade of the mineral and corresponding reserve tonnage.
- Effect of the time required before the property can start producing.
- Attitudes and policies of local and national governments, their taxes and laws that affect mining, and their degree of stability. Even with the potential instability of a government in the future, a company might consider developing a smaller, high-grade mine in the beginning until the objective return has been received, then use the income from the existing property to expand to mine out lower-grade ores having much less return.
- Availability of a source of energy and its cost.
- Availability of usable water and its cost.
- Costs and methods of bringing in supplies and shipping production.
- Physical properties of the rock and minerals to be developed and mined.
- Amount of development required to achieve the desired production related to the shape of the mineral reserve.
- Size and availability of the work force that must be obtained, trained, and maintained.

While all of the above factors must be taken into consideration, another approach to sizing the mine is to use the Taylor formula (Taylor 1977). Taylor studied over 200 mining properties, after which he developed a formula for sizing a mine using regression analysis. Actually, this formula is a very good place to start, but it must be tested against all the other variables.

$$\text{Life of mine} = 0.20\chi^4\sqrt{(\text{Expected ore tonnes})} = \text{Life of mine (years)} \pm 1.2 \text{ years}$$

Not only does the resource tonnage affect mine size, but the distribution of grade can certainly affect mine planning. However, unless mining a totally homogeneous mass, it may make a considerable economic difference as to which portion is mined first and which is mined last. Furthermore, no ore reserve has an absolute fixed grade-to-tonnage relationship; tradeoffs always must be considered. In most mineral deposits, lowering the mining grade cut-off means there will be more tonnes available to mine. But cut-off must be balanced by every type of cost supported by the operation. Even in bedded deposits such as potash or trona, the ability or willingness to mine a lower seam height may mean that more tonnes can eventually be produced from the reserve. In such cases, the cost per unit of value of the product generally increases.

In considering the economic model of a new property, all of the variables of grade and tonnage, with related mining costs, must be tried at various levels of mine production that, in the engineer's judgment, are reasonable for that particular mineral resource. At this point in the analysis, the external restraints of production are ignored to develop an array of data that illustrate the return from various rates of production at various grades corresponding to particular tonnages of the resource. At a later time, probability factors can be applied as the model is expanded to include other items.

2.3.2 Effects of Timing on Mine Production

For any given ore body, the development required before production start-up generally is related to size of the production and dependent on mining method. For example, a very large production may require a larger shaft or multiple hoisting shafts, more and larger development drifts, the simultaneous opening of more minable reserves, and a greater lead time for planning and engineering all aspects of the mine and plant. The amount of development on multiple levels for a sublevel caving operation or a block caving operation will be extensive compared to the simple development of a room-and-pillar operation. In combination, all of these factors could amount to a considerable difference in the development time of a property. In the past, this has varied from 2 to 8 yr. In turn, this could have two indirect economic effects.

1. Capital would be invested over a longer period of time before a positive cash flow is achieved, and
2. The inflation rate-to-time relationship in some countries has been known to inflate costs by as much as 10% to 20% a year, thereby eliminating the benefits of economies of scale of large projects.

To aid the engineer in making rough approximations of the time parameters related to the size and depth of mine shafts, Table 2.1 lists the average times for excavating, lining, and equipping vertical concrete circular shafts and slopes with a 0.26-radian (15°) slope for mines varying in size from 4535 to 18,140 tonne/d (5000 to 20,000 ton/d) (Dravo 1974).

For mines that must have extensive development at depth compared to those that are primarily developed on one or two levels and have extensive lateral development, the intensity of development can be much different. Lateral development on each level of a room-and-pillar mine opens up new working places, and the mine development rate can be accelerated each time a turn-off is passed, provided there is enough equipment and

TABLE 2.1   Construction schedule comparison (in weeks) between production shafts and slopes

Depth, m (ft)	Shaft, 6.1 m (20 ft) diam			Slope, −0.26 rad (−15°) (240 ft²)		
	Construction	Equipment	Total	Construction	Equipment	Total
4536 tonne/d (5000 ton/d)						
305 (1000)	43.7	20.3	64.0	46.2	5.8	52.0
610 (2000)	65.0	25.0	90.0	85.0	9.0	94.0
914 (3000)	85.0	28.0	113.0	124.1	16.9	141.0
9072 tonne/d (10,000 ton/d)						
305 (1000)	43.7	26.3	70.0	46.2	6.8	53.0
610 (2000)	65.0	31.0	96.0	85.0	11.0	96.0
914 (3000)	85.0	34.0	119.0	124.1	19.9	144.0
13,608 tonne/d (15,000 ton/d)						
305 (1000)	43.7	30.3	74.0	46.2	7.8	54.0
610 (2000)	65.0	35.0	100.0	85.0	13.0	98.0
914 (3000)	85.0	39.0	124.1	124.1	23.9	148.0
18,144 tonne/d (20,000 ton/d)						
305 (1000)	43.7	34.3	78.0	46.2	8.8	55.0
610 (2000)	65.0	39.0	104.0	85.0	15.0	100.0
914 (3000)	85.0	46.0	131.0	124.1	26.9	151.0

Source: Dravo 1974

hoisting capacuty available. This is in contrast to mines that have a very limited number of development faces per level but more levels. As a rule of thumb, it is normal that after the initial mine development for these mines, about 30 m (100 ft) of vertical mine can be developed per year.

Timing of a cost often is more important than the amount of the cost. Timing is an item that must be studied in a sensitivity analysis of the financial model for the mine being planned. In this respect, any development that can be postponed until after a positive cash flow is achieved without increasing other mine costs certainly should be postponed.

2.3.3   Government Attitudes, Policies, and Taxes

Government attitudes, policies, and taxes generally affect all mineral extraction systems and should be considered as they relate to mining method and mine size. Assume that a mine is being developed in a foreign country and that the political scene is currently stable, but that stability is impossible to predict beyond 5 to 8 yr. In such a case, it would be desirable to keep the maximum amount of development within the mineral zones, avoiding development in waste rock as much as possible. That would maximize return in the short period of political stability. Such a practice is usually possible in a room-and -pillar mine, but it may result in a less-efficient mine in later years. Also, it might be desirable to use a method that extracts the better ore at an accelerated rate to get an early payback on the investment; if the investment remains secure at a later date, the lower-grade margins of the reserve may be exploited.

Some mining methods, such as room-and-pillar, allow the flexibility of delaying development that does not jeopardize the recovery of the mineral remaining in the mine, whereas other, such as such as block caving or longwall mining, might be jeopardized by such delays.

Similar situations might arise as a result of a country's tax or royalty policies. Policies are sometimes established so as to favor mine development and provide good benefits during the early years of production, but in later years, such policies change. This would have the same effect as the preceding case; again, flexibility of the mining rate and system must be considered.

2.4   SPECIFIC PLANNING RELATED TO PHYSICAL PROPERTIES

The physical nature of the extracted mass and the mass left behind are very important in planning many of the characteristics of the operating mine. Four aspects of any mining system are particularly sensitive to rock properties.

1. The competency of the rock mass in relation to the in situ stress existing in the rock determines open dimensions of unsupported roof unless specified by government regulations. It also determines whether additional support is needed.

2. When small openings are required, they have a great effect on productivity, especially in harder materials for which drill and blast cycles must be used.

3. The hardness, toughness, and abrasiveness of the material determines the type and class of equipment that can extract the material efficiently.

4. If the mineral contains or has entrapped toxic or explosive gases, the mining operation will be controlled by special provisions in government regulations.

2.4.1   Preplanning from Geologic Data

The reader is referred to Section 10, "Foundations for Design," and the chapters on rock mass strength by Evert Hoek and the MRMR index by Dennis H. Laubscher. This author will not overlap the design factors reported there; however, it is important to discuss in a general fashion some of the common-sense rules-of-thumb that can usually be applied to help the designer to check or design a basic layout rapidly without the need for lengthy investigation when the conditions do not warrant such a study.

The following is taken from the works of Spearing (1995), which is taken from and represents years of operating experience in South African deep mines. While it may be true that all of this information applies to rock under considerable stress, it would apply equally to weaker ground under low-to-moderate stress.

Spacing of Excavations.   The following rules are based on the theory of stress concentrations around underground

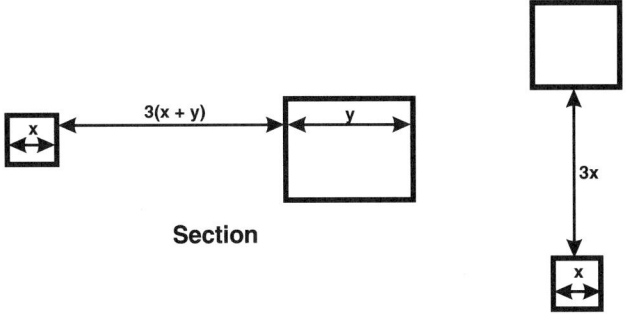

FIGURE 2.1   Square cross section

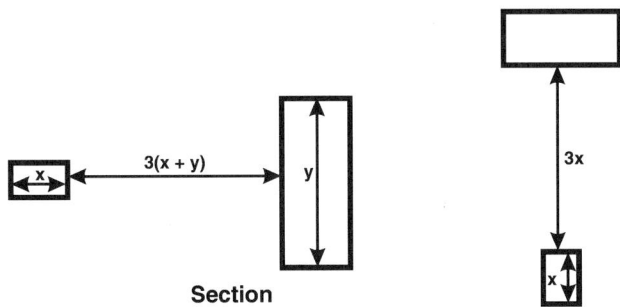

FIGURE 2.2   Rectangular cross section

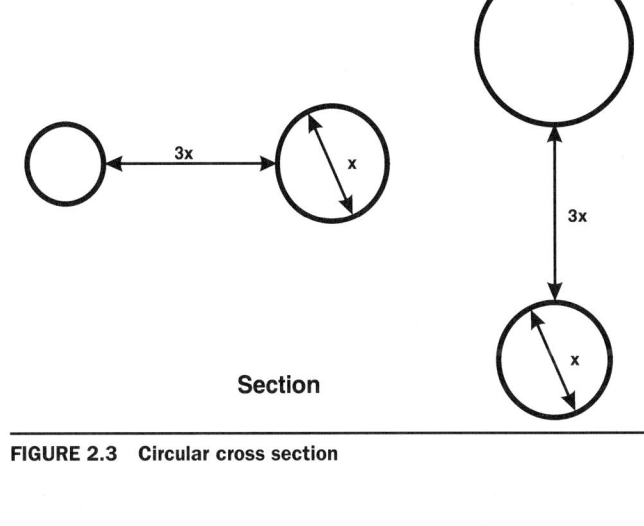

FIGURE 2.3   Circular cross section

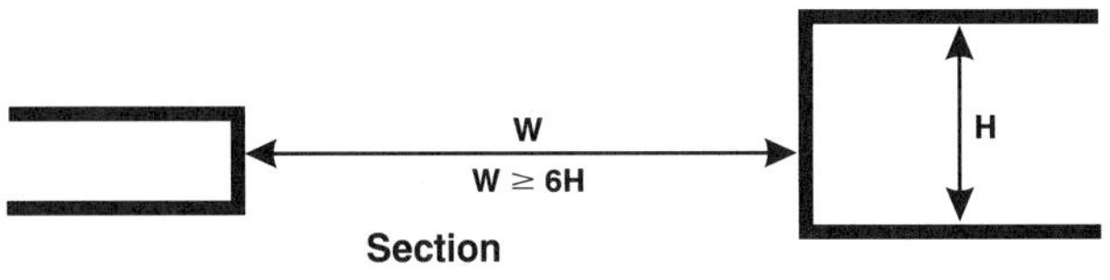

FIGURE 2.4   Pillar height-to-width relationship

openings and the interaction of those stress concentrations. The usefulness of these guidelines has been borne out by experience obtained underground. It should be noted that the accompanying sketches are not necessarily to scale. Stress interaction between excavations can obviously be controlled by an increase in the installed support, but costs will also increase significantly. If there is adequate available space, it is generally more cost effective to limit stress interaction between excavations.

Flat Development

1. Square cross section (Figure 2.1)

   A. Spaced horizontally at three times the combined width of the excavations.

   B. Spaced vertically at three times the width of the smaller excavation, provided that the area of the larger excavation is less than four times the area of the smaller.

2. Rectangular cross section (Figure 2.2)

   A. Spaced horizontally at three times the combined maximum dimensions of the excavations.

   B. Spaced vertically at three times the maximum dimension of the smaller excavation provided that the height-to-width ratio of either excavation does not exceed 2:1 or 1:2.

3. Circular cross section (Figure 2.3)

   A. Spaced horizontally at three times the diameter of the larger excavation.

   B. Spaced vertically at three times the diameter of the smaller excavation provided that the area of the excavation is less than four times the area of the smaller.

Vertical Development (e.g., shafts)

1. Square cross section at three times the combined widths of the excavation.

2. Rectangular cross section at three times the combined diagonal dimensions of the excavations.

3. Circular cross section at three times the diameter of the larger excavation.

Pillar Sizes. The pillar between irregularly shaped excavations should maintain a height-to-width ratio of a least 1:6, i.e., pillar width must exceed six times the maximum pillar height (Figure 2.4). For a pillar design under conditions of high stress (e.g., at depth), the height of the excavation should include an approximation of the fractured rock in the immediate vicinity of the excavation. (It is the author's opinion that the "pillar sizes" criterion applies, as Spearing states, to areas of high stress and thus does not apply to the many room-and-pillar operations in rather shallow environments, such as the many limestone properties and lead-zinc mines in the mid-continental United States).

Shape of Excavation. *Cross-sectional shape.* To achieve a given cross-sectional area for an excavation, it is often better to utilize a square shape. High, narrow excavations lead to excessive sidewall slabbing, which requires more intensive support, and low, wide excavations lead to large unsupported hanging wall spans that generally require very long supports. Therefore, as a guideline, the width and height of excavations should be kept to the absolute minimum, and width and height should be as equal as possible. However, under conditions of very high stress, a low, wide excavation is generally the best (i.e., the shape of a horizontal ellipse).

Uniform shape. Sharp comers in excavations lead to unnecessarily high concentrations of stress, with a likelihood of excessive fracturing and premature failure. Therefore, the shape of openings should be kept as regular as possible, and any changes in shape should be "contoured" (Figure 2.5).

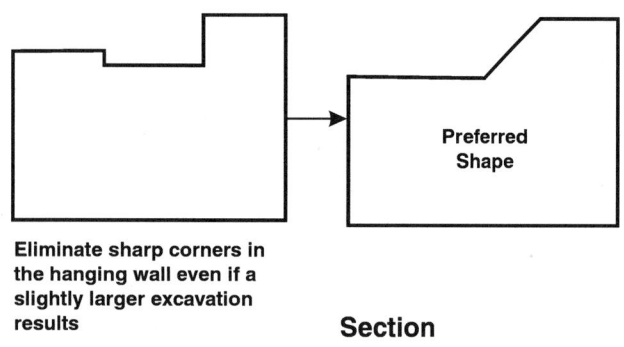

Eliminate sharp corners in the hanging wall even if a slightly larger excavation results

Section

FIGURE 2.5 Preferred opening shapes

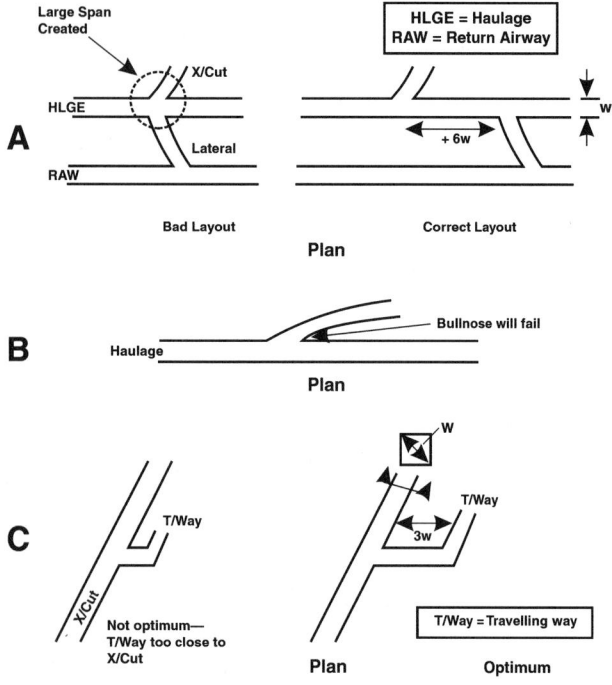

FIGURE 2.6   Stable and unstable breakaways. A, Offsetting breakaways; B, angle of breakaways; C, distance between breakaway and larger crosscut

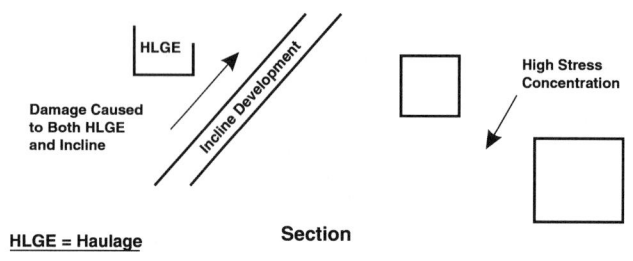

FIGURE 2.7   Preferred orientation of development openings

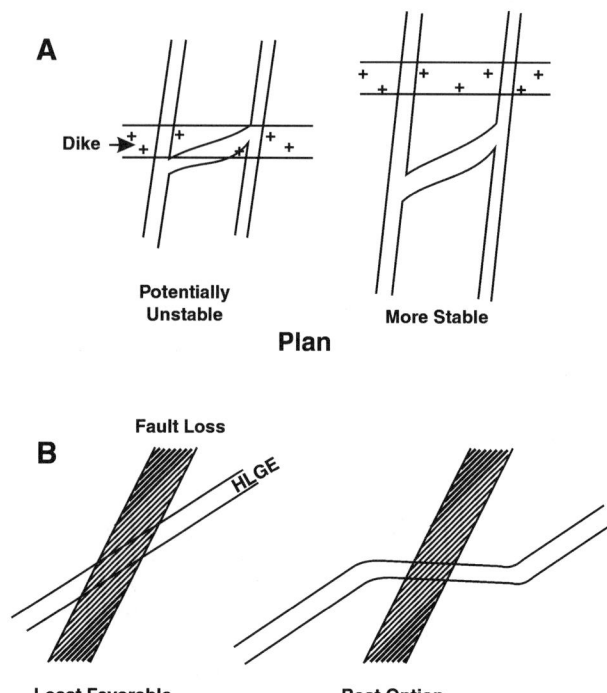

FIGURE 2.8   Preferred approach to and intersections with (A) dikes and (B) faults.

Breakaways. Multiple breakaways should be avoided to reduce dangerously large hanging wall spans. The rule is that breakaways should be spaced at six times the width of the excavation between successive tangent points (Figure 2.6A).

Acute breakaways (i.e., less than 45°) should be avoided since these result in "pointed" bullnoses, which are unstable (Figure 2.6B). Fracture and failure of a bullnose results in large unsupported hanging wall spans.

The breakaways for inclines are often brought too close to the connecting crosscut. The length of the connection between an incline and the flat should be three times the diagonal dimension of the flat end (Figure 2.6C).

Orientation of Adjacent Excavations. The most highly stressed part of an excavation is the corner. Therefore, positioning of development in unfavorable orientations, as shown in Figure 2.7, should be avoided. In plan also, similar precautions should be taken to avoid this type of unfavorable orientation, especially where existing development is slipped out for excavations such as substations and battery bays.

Geology. In all cases, the geology of the area should be taken into consideration.

1. Known weak geological horizons, for example, the Upper Shale Marker and the Khaki beds associated with the Orange State gold-bearing reef structures, should be avoided even at the expense of longer crosscuts.

2. In permanent excavations (sumps, settlers, hoist chambers, etc.), the position of faults and the orientation of joints sets are critical to the stability of the development. Layouts must therefore cater to such geological features.

3. All excavations should be kept away from dikes where possible (Figure 2.8A). When a dike is traversed, this should be done by the most direct route. Breakaways should not be sited in dikes, even at the expense of extra development.

4. Haulages should not be positioned in fault losses, and development should not occur alongside a fault. A fault should always be intersected at an angle as near to normal as possible (Figure 2.8B).

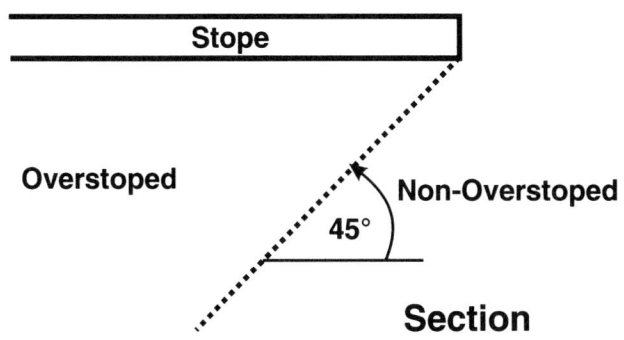

FIGURE 2.9 Diagram of 45° overstopping angle for destressing haulageway

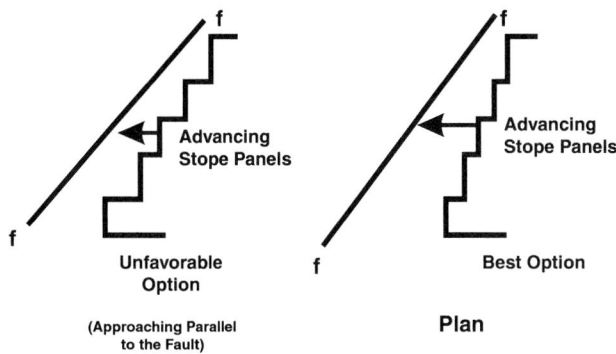

FIGURE 2.10 Preferred approach to stoping through geologic feature

Overstoping. When haulages are positioned beneath mined-out areas, consideration should be given to the 45° destressing guideline (Figure 2.9). An overstoping angle of 45° is generally required to destress a haulage. At an angle greater than 45°, stress concentrations are higher. This rule applies to reefs of 0° to 20°. Dips greater than this often call for computer modeling to show the extent of overstoping needed. Haulages should not be laid out too close to remnant pillars, which are very highly stressed.

Stoping. Stoping as applicable to advancing longwall panels and room-and-pillar stoping should not be carried out toward adverse geological features with the overall face shape parallel, or near-parallel, to the feature. Mining should take place toward (and through) the geologic feature at as large an acute angle as possible (Figure 2.10).

Using geologic and rock property information obtained during preliminary investigations of sedimentary deposits, mine planners should construct isopach maps to show the horizons to be mined and those that are to be left as roofs and floors. Such maps show variances in the seam or vein thickness and identify geologic structures, such as channels, washouts (wants), and deltas. Where differential compaction is indicated, associated fractures in areas of transition should be examined. Areas where structural changes occur might be the most favored mineral traps, but they are usually areas of potentially weakened structures. Where possible, locating major haulage drifts or main entries in such areas should be avoided; if intersections are planned in these areas, they should be reinforced as soon as they are opened to an extent greater than is ordinarily necessary elsewhere in the opening.

Again referring to flat-lying-type deposits, extra reinforcement (or decreasing the extraction ratio) may also be necessary where the ore becomes much higher in grade than is

typical, and the rock mass, therefore, has much less strength. Where the pillars already have been formed prior to discovering structural weaknesses, it will probably be necessary to reinforce the pillars with fully anchored reinforcing rock bolts or cables. It is advisable to map all joint and fracture information obtained from diamond-drill holes and from mine development to correlate structural features with any roof falls that might occur.

2.4.2 Hardness, Toughness, and Abrasiveness of Extracted Material

The hardness, toughness, and abrasiveness of the material determine whether the material is extracted using some form of mechanical cutting action, by drilling and blasting, or by a combination of both methods.

Technological advances in hard metal cutting surfaces, steel strengths, and available thrust forces allow increasingly harder and tougher materials to be extracted by continuous mining machines. The economics of continuous cutting or fracturing as compared to drilling and blasting gradually are being changed for some of the materials that are not so tough or abrasive. However, for continuous mining (other than tunnel-boring machines) to be competitive with modern high-speed drills and relatively inexpensive explosives, it appears that rock strengths must be less than 103,400 to 124,020 kPa (15,000 to 18,000 psi) and have low abrasiveness.

However, rock full of fractures is a great aid to mechanical excavation. In one case, this author knows where a roadheader is being used in a welded volcanic tuff containing lots of fractures even though the rock strength is well over 137,800 kPa (20,000 psi). This entire subject is covered in an article on the gradual trend toward mechanical excavation (Bullock 1994).

At times, reasons other than simple economics favor the use of one mining system over another. Using a mechanical excavation machine is nearly always advantageous in protecting the remaining rock where blasting might be prohibited. This was seen in the development openings driven by Magma Copper in developing the Kalamazoo ore body (Chadwick 1994; Snyder 1994) and Stillwater Mining Company in developing its original ore body, as well as for the East Boulder ore body (Tilley 1989; Alexander 1999). A continuous boring tool may also be desirable for totally extracting an ore body without personnel having to enter the stoping area. Certainly continuous mining machines can be much easier to automate than cyclic drilling and blasting equipment. The automation of the 130 mobile miner at the Broken Hill Mine in Australia is a case in point (Willoughby and Dahmen 1995).

2.5 TYPES OF MINE DEVELOPMENTS

It is extremely difficult to describe each and every type of mine development because of confusion in terminology and because openings can be driven in any direction, in all sizes, and by many different methods. Table 2.2 is provided in an attempt to show that there are three basic reasons for which all developments are driven—to obtain information, access production areas, and provide services.

For each development, decisions must be made pertaining to each of the five physical variables—length, direction, inclination, size, and method of ground control. In the final analysis, each development must serve the purpose for which it was intended and usually named. A good general discussion of development planning can be found in the *Mining Engineering Handbook* (Unrug 1992).

2.5.1 Information for Mine Planning

During the very early stages of evaluation and planning, a great deal of information is needed to make the correct decisions concerning the property. Therefore, the initial opening might be driven to test a particular aspect of the mineral reserve that is

TABLE 2.2 Various types of mine developments classified according to purpose

Information	Production accessway	Service accessway
	EXAMPLES OF FUNCTIONS	
Quality of mineral	Expose areas of minable material	Normal movement of workers
Quantity of mineral	Open accessways for production	Emergency movement of workers
Continuity of mineral	Movement:	Normal movement of supplies
Physical and structural properties of mineral or rock surrounding it	Horizontally on strike	Movement of air intake
Stress condition of mineral or rock surrounding it	Horizontally against strike	Movement of air discharge
Mineralogical-metallurgical nature of mineral	Vertically down	Water drainage or discharge
	Vertically up	Water storage and pumps
		Cleanout for other openings
		Rooms for supplies, repairs, and offices
	PHYSICAL DEVELOPMENT VARIABLES	
Inclination	Inclination	Inclination
Size	Size	Size
Depth or distance	Depth or distance	Depth or distance
Arrangement when completed	Arrangement when completed	Arrangement when completed
Method of development	Method of development	Method of development
	EXAMPLES OF DEVELOPMENT NAMES	
Exploration adit	Main haulage drift	Miner and material decline
Prospect decline	Ore shaft	Ventilation shaft
Exploration shaft	Belt slope	Main sump
Prospect drift entry	Ore pass	Shaft cleanout raise
Experimental entry intersection	Production roadway	Ventilation "submain"
Exploration winze	"Main" belt-line entry	Main pump station
	Skip-loading station	Supply house and shop

considered critical for successful exploitation. More often, this is termed a "test mine." This could be from an adit, ramp, or shaft. In any case, permitting is usually required that may very well delay the project.

Most of the technical information listed in the preceding paragraphs can be obtained from diamond-drill holes, particularly where the mineralized zone is fairly shallow, is continuous, and is flat. Unfortunately, it is common that the true economic potential of many mineral resources cannot be determined until after an exploration development or test mine is driven. From the development opening, underground prospecting can further delineate the materials present, a bulk sample can be taken for metallurgical testing, and/or true mining conditions can be revealed. This procedure involves considerable financial exposure and, unfortunately, the delineated resource may turn out not to be economically viable. Nevertheless, there may be little choice about driving a test development opening if the potential of the resource is to be proven and the risk to larger downstream capital expenditures is to be minimized.

Regardless of the reason for driving a preliminary development, three basic principles should be observed.

First, the development should be planned to obtain as much of the critical information as possible without driving the opening in a location that will interfere with later production. For most mining methods, it usually is possible to locate at least a portion of the exploration development where it can be utilized in the developed mine that may come later, even if it becomes necessary to enlarge openings to accommodate production equipment at a later date.

Second, the preliminary development should be held to the minimum needed to obtain the necessary information, and third, the development should be completed as quickly as possible.

Most exploration drifts for metal mines are only large enough to accommodate small loading and hauling units (from 2.4 by 3 m to 3 by 4.3 m [8 by 10 ft to 10 by 14 ft]). If geologic information is needed, it often is possible to locate the developments in proximity to several of the questionable structures without actually drifting to each one. With modern high-speed diamond drills having a range of 213 to 305 m (700 to 1000 ft), these structures usually can be cored, assayed, and mapped at minimum cost.

Once the preliminary development has been done to satisfy a specific need, the development should be used to maximize all other physical and structural information to the maximum extent that will not delay the project.

There are no rules covering the size of the shaft or the entry for exploration development. The size depends upon many factors, including the probability of the mine being later developed to the production stage. If the probability favors follow-up with full mine development immediately after exploration development, and if the shaft depth is not excessive (e.g., 244 to 366 m [800 to 1200 ft]), it would be logical to sink a shaft sized for later use in production development. This is particularly true of operations normally using small shafts for production.

Using a small shaft of 3.7 to 4.0 m (12 to 13 ft) in diameter is not uncommon in hard-rock mining in the United States. For example, most of Doe Run Company's shafts in the Viburnum area are this size, as well as many of the shafts in the Tennessee zinc district. The Elmwood Mine originally opened by New Jersey Zinc is a good example of this practice. Initially a 3.7-m (12-ft) diameter conventional shaft was sunk for exploration. After the continuity and quality of the ore were proven by drifting and underground prospecting, the shaft was used for further development of the mine and was eventually used as the production shaft. The Savage Zinc Company's mine at Carthage, Missouri, in the United States is another example of a 3.8-m shaft originally sunk as a test mine and then later opened as an operating shaft.

In contrast, some companies choose smaller drilled shafts to speed shaft sinking and minimize early expenses. Such shafts usually are completed considerably faster than conventional shafts, but not as fast as a raise-bored shaft (where an opening already exists, the hole is reamed upward, and the cuttings fall back into the mine). Because this section is directed toward initial mine openings, removal of cuttings must be upward through the hole and out the collar. Various methods involve the use of air or mud with either direct or reverse circulation. There are a large number of blind-bored shafts in the industry. Although bored shafts may stand very well without support, most of these shafts contain either a steel or a concrete lining.

Shotcreting bored shafts has become acceptable and probably results in substantial cost savings. Remote-controlled jumbos have been developed to help place the shotcrete in this difficult application (for example, Shelob shaft shotcreting units, Caledonian Mining Co., Ltd.).

2.5.2 Planning for Production Accessways and Systems

The basic considerations in planning and developing production accessways and systems are related directly to the design parameters of the development openings. These are functions in production flow, inclination of the opening, size of the opening, and the arrangement and support of the opening.

Production Development. Each opening must be considered in the production flow process, and each must be planned to be compatible with the elements of flow on both sides of it. In a room-and-pillar mine, a typical sequence of development openings for material flow from the production face might be a preliminary drift for load-haul-dump (LHD) units, an ore pass, the main haulage drift, the main ore pocket, an ore pass to the crusher and/or skip loading station, and a shaft for hoisting material to the surface. In a caving operation, the typical sequence might be the undercut or slot raise, the LHD drawpoint, the first ore pass, the main ore pass, the main haulage system, the ore pocket, the crusher station, and the hoist. Although other configurations and production flow sequences are used, this sequence is typical. The sequence of development production openings is discussed in Chapter 4.

In the following paragraphs, production development from the surface is considered first, followed by descriptions of internal mine development.

Inclination from the Surface. As described earlier, inclination determines the name of the development. Inclination is determined (1) by a combination of the relative elevations of the surface and the mineralized zone and (2) by the economics of driving the development and operating the production system.

Drift entry (adit). An adit, or drift entry, is the most economical approach when the minable material is above the floor elevation of a nearby valley. Most of the early mines in the western United States used this type of development, as did most of the limestone and dolomite mines throughout the Missouri-Mississippi River basin. The exposed rock could be opened through horizontal adits into the sides of mountain bluffs. When the minable material goes to a depth lower than the surrounding terrain, either a slope (decline) or a shaft must be developed.

Slopes and ramps (declines) versus shafts. There is no general rule governing whether a shaft or a slope is the proper type of entry to be developed; each has advantages and disadvantages. Slopes usually are limited to relatively shallow mines because, for a given vertical depth, they require approximately four times the linear distance required for a shaft. Thus, to reach a depth of only

460 m (1500 ft), a normal entry slope would be approximately 1840 m (6000 ft) long. Although the cost per unit distance for a decline is usually lower for a slope than for a vertical shaft, the overall cost usually is higher and the development time usually is longer as a result of the greater length required. The break-even depth is somewhere in the 300- to 450-m (1000- to 1500-ft) range, depending on conditions.

Most new shafts being sunk today are lined with concrete, which offers excellent ground stability. Most slopes have a rectangular shape, with the widest span for the floor and roof. In incompetent ground, slopes can be difficult to develop and difficult to maintain over the life of the property. In ground that must be frozen prior to excavation, a much larger volume of material must be frozen for a slope than for a shaft, and a slope would undoubtedly would cost far more than a shaft.

However, slopes do offer many advantages.

1. Where a conveyor brings material to a central point, it is logical to continue material transfer to the surface by means of a conveyor up a slope. Where slopes and conveyor systems are used to move the material to the surface, the mine generally is geared for a consistent high rate of production.

2. Slopes provide easier access for mobile equipment entering or leaving the mine. In many cases, equipment can be transported or driven into the mine intact. With a conventional shaft, large pieces of trackless equipment will usually have to be disassembled completely and then reassembled underground, which can be particularly difficult if the unit is a large truck and the bed must be split and rewelded. A similar advantage is found when using rail haulage in a slope for transportation of supplies—the material can remain in its original container for transport from the surface.

3. Where minor ground movements are a problem, a production slope offers considerably less difficulty in maintenance than does a vertical production shaft.

The economics of using declines becomes obvious only where conveyor belts are used throughout the operation or where the mine is very shallow and rubber-tired vehicles can be driven from the working area to the surface.

Despite the advantages of slopes, the disadvantages of cost and development time appear to have the greater influence on noncoal mines deeper than 30 to 60 m (100 to 200 ft). Although most noncoal mines using room-and-pillar mining are reasonably shallow, vertical shafts continue to be used during development. The usual exception to this is when the mine is fairly shallow and plans to use a crusher-conveyor system to transport ore to the surface. Several notable examples fit into the situation mentioned above; for example, the White Pine copper mine (when it was operating), formerly of the Copper Range Co., the Maysville Mine, and the Black River Mine in Kentucky, and the Lost Creek Mine and the Gordonsville Mine of the Tennessee zinc district. Of these, the first three used conveyor belts to transport the material from the operating areas to the surface.

The size of the opening should be tailored to accommodate the specified purpose. For production shafts, the shaft cross section is influenced (but not dictated) by the skip capacity, and skip capacity is determined by the variables of mineral density, hoist capacity, hoisting speed, skip loading time, skip dumping time, and allowable hoisting time. Most conventional shafts are 3.7 to 8.5 m (12 to 28 ft) in diameter, while most drilled shafts are 1.8 to 5.5 m (6 to 18 ft) in diameter.

Arrangement of Facilities in Production Openings. The production shaft or slope must accommodate the conveyances for production material and may, at times, also contain some of the services, such as an emergency mancage and/or a ladderway. Whether such services are located in the production opening is dependent upon the number of other openings from the mine to the surface. Usually, the service facilities are contained inside pipes that may be embedded in the lining of a concrete shaft (Lama 1964). Drawings of general shaft arrangements are contained in other literature (Dravo Corp. 1974; Gerity 1975).

Design procedures for compartment size and compartment arrangement can be found in the *Mining Engineering Handbook* (Edwards 1992). Most new conventional shafts are lined with 0.3 to 0.6 m (1 to 2 ft) of concrete. The use of timber for shaft support generally is limited to deepening older rectangular shafts or to small mine openings what are not using mechanized shaft sinking. Hoisting in a shaft is accomplished either with two containers (skips) that balance each other or with one large skip that is balanced by a counterweight.

Most slopes (declines) utilize roof bolting and/or shotcrete with a concrete portal. The three arrangements for production slopes are (1) one open compartment, (2) two side-by-side compartments, or (3) two over-and-under (stacked) compartments. The latter two designs are used primarily when the one opening is serving in the development phase of the operation as both ventilation intake and discharge.

The decision to use a conventional hoisting system or a Koepe hoisting system is usually based on considerations of depth and how many hoisting levels are required. Where the mine is hoisting from only one level and is fairly shallow, conventional hoisting still prevails in the United States. However, part of the reason for this is that used hoist equipment is available that is still in good condition. So the answer, as is often the case, is one of economics.

The other major arteries needed for production (i.e., drifts, crosscuts, raises, and ramps) are specific to mining method and will be discussed under such methods.

Ore Storage Pocket. Some surge capacity must be provided between the normal mine production gathered from various parts of the mine and the conveyance that transports the material to the surface. The need arises because the material usually has many production paths to reach the shaft, slope, or adit, but usually only one path out of the mine. Therefore, a single-path conveyance is vulnerable to being down for maintenance and repair, but such downtime cannot be allowed to disturb the rest of the mining cycle. Similarly, multipath production setups can operate intermittently (one or two shifts per day), while material can flow from the mine continuously.

The correct size of the surge pocket depends on what it is intended to accomplish. For example, if management decided to try to hoist 20 shifts per week, but ran the stoping operations with only two shifts per day in a mine producing 4535 tonne/d (5000 ton/d) and stopped mine production between midnight Friday and 7:00 a.m. Monday, the minimum pocket size would be found from the equation—

$$C = [Sm/Sh] \times Smd \times Tms,$$

where   C = ore pocket capacity needed in tonnes,

     Sm = number of mine shifts per week,

     Sh = number of shifts per week the hoist operates,

    Smd = number of shifts the mine is down,

and   Tms = number of tonnes produced per mine shift.

Continuing with the above example—

     Sm = 10 hr,

     Sh = 20 hr,

    Smd = 6 shifts,

and   Tms = 1267 tonnes per shift (2500 tons per shift).

In this example, capacity would be calculated as—

$$C = [10/20] \times 6 \times 1267 = 3801 \text{ tonnes}$$
$$(C = [10/20] \times 6 \times 2500 = 7500 \text{ tons}).$$

In this example, storage was calculated to be 150% greater than the daily mine capacity. Although this is typical for some room-and-pillar operations, it is much larger than the requirement for many others. A survey several years ago summarized hoisting practices of room-and-pillar mines, listing 11 hoisting rates from 227 to 1089 tonne/hr (250 to 1200 ton/hr) and capacities of storage pockets from 227 to 13,608 tonnes (250 to 15,000 tons). Assuming that hoisting is done 21 hr/d, storage capacity of these 11 operations varied from 3% to 191% of the hoisting capacity.

Obviously, there is no general agreement on the optimum size of an ore pocket; both the nature of the operations and management priorities differ. However, the two major considerations are—

1. The size of the storage and whether interruptions before the next step are critical, and
2. Whether there is storage area at the discharge end of the hoisting operation before the ore proceeds to the next step in the flow.

Another reason for using ore storage pockets before hoisting is to keep various products or waste materials segregated for separate hoisting. The products may be different minerals, or they may be the same mineral but owned by different parties. Owners might insist that the material be kept separate until after crushing, weighing, and sampling so that royalties on the value would be paid correctly. Such practices are quite common in mines where government ownership of the mineral is involved.

2.5.3 Planning for Services

Any mine opening is a service development if it is not developed specifically to gain information about the mineral reserve and mining conditions, to obtain a bulk sample, or to handle the production system. Service developments are subject to the same basic considerations as are the production openings, i.e., their functions, inclinations, sizes, and physical arrangements. The name of the development is a combination of function and inclination, size, and arrangements, thus, ventilation raises, personnel and materials service shafts, sump clean-out drifts, emergency escape shafts, etc. A service development may serve a combination of functions or change its function over the life of the property.

Basically, the design rationale for service developments follows that of production developments, modified by special considerations associated with the service function.

Ventilation Openings. The current practice in ventilation raise and shaft development is for the opening to be bored or reamed upward, rather than drilled or blasted. Such holes can be completed by an outside contractor or by a mine crew who may own or lease a boring machine.

The size of the opening is determined by the ventilation needs. However, size is a trade-off between the cost of the larger raise opening versus the cost of the power that will be required. The trend should be to enlarge the opening size to reduce ventilation power consumption and to move the needed amount of air at minimum cost. This is a problem of optimization that has to be solved for each individual user. In horizontal developments or slopes, ventilation tubes or ducts are used in most metal mines, though various types of stopping, such as brattice cloth or canvas, are also used.

Shop and Storehouse. The efficiency of a modern mining system depends heavily upon the productivity and availability of the equipment used to extract the material. Much of the energy used to extract the mineral product and move it to the processing facility is consumed by mechanical equipment.

Because the mined material may be very heavy and/or abrasive, and because the environment of an underground mine imposes adverse operating conditions, underground mining equipment requires a great deal of maintenance and repair. One of the most serious and most prevalent errors made in designing a mine is the failure to provide adequate space and equipment for necessary maintenance and repair work. Thus, most underground mines today must have well-developed underground shop areas. The total area usually used as a shop can range from 465 to 1394 m^2 (5000 to 15,000 ft^2).

The amount of service, if any, provided underground depends upon several factors. The degree of difficulty in moving equipment into or out of the mine is a major consideration. If, for example, the property is a limestone mine with adit entrances and a good shop on the surface, it probably would not be advantageous to duplicate facilities and personnel underground, at least immediately. However, when the mine is developed several thousands of meters from the initial entry, it may make good sense to move all maintenance facilities for mobile equipment closer to the operating faces.

Underground shops should provide a safe and good working environment. In mines that are gassy or carry a gassy classification, the construction of underground shop facilities may be much more difficult or may not be practical. However, the advantages are the same even for gassy and nongassy mines; therefore, some room-and-pillar operations, such as trona mines, have very good, large, and well-equipped underground shops even when classified as gassy.

If the active working area will be abandoned totally after a short life and operations will be moved to another such area nearby, it may not pay to invest in an extensive underground shop area.

The important features of a shop design are—

- One or two large bridge cranes over the motor pits,
- Easy access from several different directions for access to and around the cranes when they are in service for extended periods of time,
- A motor pit and service area separate from the break-down motor pits for scheduled lubrication performed as part of a preventive maintenance program,
- A separate area for welding operations,
- A separate area and equipment for tire mounting and repair,
- A separate enclosed area for recharging batteries,
- Various work areas equipped with steel worktables or benches,
- A separate area for washing and steam-cleaning equipment,
- An office for the shop foreman, computers, records, manuals, catalogs, drawings, and possibly a drafting table. The main pit areas should be visible from the office windows.

The shop area should also be in proximity to the main underground supply room. The size of this supply room also depends upon several factors. Some mines have no supply room, whereas others have very large storage rooms with as much as 110 m^3 (12,000 ft^2). Factors influencing inventory policy include the frequency and ease with which the mine receives supplies, whether a supply system is installed at the collar or portal of the mine, and the dependability of suppliers in supplying parts for the equipment being used.

Sump Area and Pump Station. In mines below surface drainage, areas must be provided in which to store water before it is pumped to the surface and discharged. There is no general rule of thumb to determine the capacity for water storage. The water flow that enters into mines in the United States has historically varied from 0 to 2208 L/s (0 to 35,000 gal/min).

Adequate pumping capacity is the only permanent solution to removing water from a mine. However, fluctuations in water inflow and/or the time periods when some or all of the pumps are inoperable must be handled with an adequate sump capacity. Sumps also are needed as settling areas in wet room-and-pillar mines. This occurs because the rubber-tired vehicles traveling on the roadways create fine material that eventually collects in the water ditches if the roadways have drips or continuous streams of water running through them. Although this situation should be corrected to keep water off the roadways, it invariably occurs or reoccurs. A place must be provided for the fines to settle out of the water so that they do not damage clear-water, high-head pump impellers. Settling sometimes can be done effectively in a small catch basin that is cleaned out frequently. However, if water flows are substantial, the main sump receives the bulk of the fine material. As a result, provisions must be made to—

- Clean the sump with conventional or special equipment,
- Divert the water into another sump while one sump is being cleaned,
- Provide a place to put the running slimes ("soup") removed from the sump, and
- Devise a method of transporting the slimes away from the sump.

One of the best methods of cleaning a sump is to use conventional front-end loaders to remove the material. However, a ramp must be provided down to the sump floor level. If deep-well impeller sections are suspended from a pump station down into the sump, considerable care must be taken to avoid bumping the impellers with the loader. Other systems using slurry pumps, spiral- or cavity- type pumps, scrapers, or systems to divert the sludge material into skips have been tried, but there still is no good way to clean a sump.

Diverting water flow into one sump while cleaning out another can also be a problem unless at least twice the pumping capacity normally required is provided to the two sump-pumping installations. Half the pumps have to handle all the inflowing water as one sump is being pumped down and cleaned. Mines that generate large amounts of water also usually produce large amounts of fine material that fill the sumps much faster than anticipated. The sump of a new mine in the development stage cannot be cleaned too soon; even if the mine is not producing much water at a particular stage of development, the sump still should be kept clean. It is always possible that the next shift may bring on more water than can be handled with half the pumps, and once that point is reached, there will be no way to get to the muck in the bottom of the sump for cleaning.

Two methods often are used to divide the sumps for cleaning. One is to position a concrete wall between the two areas and provide a means of closing off one side and making the water flow to the other side. The wall must be anchored properly on the top and bottom with reinforcing steel into the rock. This provides a safety factor to prevent the wall from collapsing into the side being cleaned. The other system is to develop two physically separate sumps. Both sumps must be provided with pumps, a method of stopping water inflow and diverting it to the other sump, and a means of access for cleaning.

Some mines locate the pumping station below the sump chambers, using horizontal centrifugal pumps instead of deep-well turbine pumps. These mines avoid the need to install a vacuum priming system because the overflow feeds the main pump (Schwandt 1970). Although this system provides easier access to the sumps and makes cleaning easier, the risk of losing the pump station by flooding is increased.

TABLE 2.3   Productivity of mining methods, tonnes per manshift

Stoping method	Normal	High
Room-and-pillar	30–50	50–70
Sublevel caving	20–40	40–50
Block caving	15–40	40–50
Sublevel caving	15–30	30–40
Cut-and-fill	10–20	30–40
Shrinkage stoping	5–10	10–15
Square set	1–3	

These numbers are for hard-rock mines. Limestone aggregate producers go as high as 1000 tonnes per manshift, and evaporite producers go as high as 170 tonnes per manshift (Bullock 1982).

If power for the pumps originates from a source not maintained by the mine (i.e., public power with or without tie lines), the mine may be without power for extended periods of time. Therefore, every protective measure should be provided. Using deep-well pumps on one of the upper or middle levels helps protect the facility for the maximum length of time. All major pumping stations should be well equipped to replace the pumps and motors by having small overhead cranes on a monorail above the pumps. If deep-well pumps are used, sufficient headroom is needed to remove the multistage impellers.

Other Service Developments. A variety of other rooms or drifts also must be developed. For example, a lunch area or room is nearly essential around an underground shop. Office space should be provided for the privacy of the underground supervisor and the security of records and office equipment. Strategically located mine rescue chambers are needed throughout the mine. Sometime these are developed and placed within stub-drifts, or sometime in mined-out stope areas.

Ramps driven strictly for access and not production are important. Normally, these can be driven steeper than haulage ramps because the equipment using them will not be loaded and time to negotiate the ramp should not be as critical. However, the ramp must not be so steep as to prevent unloaded equipment from having good traction. Equipment manufacturers can advise the mine planner on this point.

A small area at the bottom of the shaft will need to be provided for the equipment for cleaning the shaft bottom. Sufficient maneuvering room is needed for equipment such as an overshot loader and a bucket that lifts spill rock from the bottom of the shaft up the skip loading level. Ideally, a drift intersects the bottom of the shaft and a ramp leads back up to the bottom level of the mine. In such cases, a steel deflection door can be pushed into place for shaft cleaning while hoisting is in progress. Shaft cleaning may also be accomplished by installing a remote-controlled shaft mucking unit, such as a Cryderman or an Alimak mucker, on the shaft wall or on the shaft steel below the skip loading facility but above the shaft bottom.

2.6   PLANNING THE ORGANIZATION AND REQUIRED EQUIPMENT

The amount of equipment or personnel for the needs of all mines cannot be specified in general terms. However, personnel and equipment productivity ranges can serve as guidelines for planning. Table 2.3 was published over 20 years ago (Hamrin 1982) and has been referenced in many books since the original estimates were made, but nothing more recent has been published.

2.6.1   Work Force and Production Design

It is necessary to consider several factors concerning planning the work force to operate the mine. Many questions needing investigation will be difficult to answer, but they have profound effects on the financial success of any mining project and eventually must be faced.

- Is the supply of labor adequate to sustain the production level dictated by other economic factors? If not, can the needed labor be brought in, and at what cost?

- What is the past history of labor relations in the area? Are the workers accustomed to a 5-day work schedule and, if so, how will they react to a staggered 6- or 7-day schedule?

- Are local people trained in similar production operations or must everyone be trained before production can achieve full capacity?

- Can people with maintenance skills be attracted to the property or will the maintenance crew have to be built up through an apprenticeship program?

Apprenticeship programs are very slow processes. Accordingly, some state laws restrict the number of people who can be trained each year in such programs. That one item could cause a mine designed and equipped for a very large daily production to fall far short of its goals.

It is important that the "people problems" be investigated at the same time that the property is being evaluated and designed. This will allow adequate time for specialized training, minimize unexpected costs, and prevent basing economic projections on policies that, if implemented, could destroy employee morale or community confidence. The productivity and profitability difference between an operation with good morale and good labor relations and an operation with poor morale and poor labor relations (with many work stoppages) can make the difference between profit or loss of the mining operation. Of all the items involved in mine design, this one is the most neglected and can be the most disastrous.

2.6.2   Field-Tested Equipment

The selected equipment should be produced by manufacturers that field-test their equipment for long periods of time before marketing it to the mining industry. Too many manufacturers build a prototype machine and install it in a customer's mine on the contingency that they will stand behind it and make it work properly. Eventually, after both user and manufacturer redesign, rebuild, reinforce, and retrofit, a workable machine finally is obtained. However, the cost in lost production is imposed on the mine operator, not on the manufacturer. The manufacturer then can proceed to sell the "field-tested" retrofitted model to the entire industry, including competitive mines. In the case of an LHD unit introduced several years ago, there were 53 design changes between the prototype and the final production model, all made by the mining company and all adapted by the manufacturer. In another case, a prototype drill jumbo was modified so extensively over a 2-yr period that almost every auxiliary component was retrofitted to the user's operation. However, even after 2 yr, the drill jumbo still was not performing up to the specifications at the time of purchase. Even though manufacturers do need help and cooperation from the mining industry to develop equipment, the manufacturer must pursue longer periods of testing in the industrial environment, rather than selling the units to the industry and then pursing continuous research on the prototype models.

2.6.3   Equipment Versatility

The equipment selected for the mining operation should be as versatile as possible. For example, in one mining operation (Bullock 1973), the same high-performance rotary percussion drilling machines were used for drilling the bluff or brow headings and were mounted on standard drill jumbos for drilling holes for burn cut and slabbing rounds in the breast headings.

Because the drills on these jumbos penetrate extremely fast, they also were used to drill the holes for roof bolts and, in some cases, to drill holes for reinforcing pillars. The same front-end loader was used to load trucks in one stope and to perform as an LHD in another stope. By switching working platforms, the same forklift tractors served as explosives-charging vehicles and as utility service units for handling air, water, and power lines. They also served as standard forklifts for handling mine supplies. This philosophy results in the following advantages:

- There is less equipment to purchase and maintain.

- Less training is required for operators and maintenance personnel. At the same time, all personnel have a better chance of becoming more efficient at their jobs.

- Having fewer types of machinery means having less inventory to obtain and maintain.

Some of the possible disadvantages of this philosophy include—

- A more efficient machine may be available to do the job being done by the versatile machine. Therefore, if a great amount of a specific type of work is to be done, it may be advisable to use a specialized machine.

- The mine may become too dependent upon a single source to supply its equipment.

As an example of the first disadvantage, the mine previously cited was drilling a great number of high bluffs, some brows, and breakthrough pillars. As a result, the mine eventually switched over to air-track drills because those machines were more efficient for the particular applications. Likewise, a high-speed roof-bolting jumbo was eventually acquired because it was much more efficient for installing roof bolts.

2.6.4 Equipment Acceptance

The equipment selected should have a very broad acceptance, and it is desirable if the equipment is in common use throughout both the mining and construction industries. Since mines impose a headroom restriction not encountered on the surface, this is not always possible. However, where headroom is not a problem, selecting a standard piece of equipment means that the components will have endured rigorous use in the construction industry. Furthermore, parts for such equipment normally are off-the-shelf items in distributors' warehouses across the country.

2.6.5 Application Flexibility

The selected equipment should be flexible in application. That is, the equipment should be able to accelerate and move rapidly, have good balance and control at high speeds, be very maneuverable, and have plenty of reserve power for severe conditions. Both trucks and loaders should have ample power to climb all grades in the mine and to accelerate quickly to top speed on long straight hauls.

2.7 PLANNING PERSONNEL MOVEMENT TO AND FROM THE WORK FACE

During the early stages of mine planning, consideration should be given to the methods and problems involved in transporting personnel and materials into the mine and to the working face.

Obviously mines with multiple levels and sublevels can have a much more difficult time getting workers to the face in a timely manner than mines that are spread out on one or two levels and have a roadway going to every heading. In both cases, how you transport workers and their materials quickly deserves a lot of study when the mine is being planned.

From a study made several years ago (Theodore Barry and Associates 1975), it was demonstrated that even in the simple situation of a room-and-pillar mine, time lost getting workers to the face took from 8% to 14 % of the 7.5-hr shift once the miners were underground and before they collectively moved to the surface. In deep, multiple-level mines, this time can easily go up to 25%.

The reasons for excessive personnel and material transportation time are believed to be—

- Typically, equipment transport speeds are not up to the level of performance currently achievable through existing technology.

- Nonmechanized means of travel, such as walking and climbing, are significantly slower than mechanized transport. Underground environments typically make walking or climbing slow, tedious, and hazardous.

- General mine conditions, such as poor track and rough or crooked roadways, adversely affect the operational efficiency of personnel transport systems.

- In some cases, equipment design impedes efficient loading and unloading of personnel, thereby increasing the transportation time.

To improve personnel movement, the following recommendations are made.

- To optimize personnel transport, each mine should identify specific groups of personnel according to job functions and list ultimate destinations, intermediate transport requirements, etc.

- The system component dependence should be handled by personnel scheduling techniques.

- The horizontal system should accept the personnel unit on a predetermined schedule.

- An engineering study of all personnel movement elements should be conducted to establish present and future requirements for a balanced transport system.

- Crews should be transported to a point as close as possible to the actual work, and they should be picked up on a predetermined schedule at the same point for outbound transport.

- Supervisory staff should use mechanized transport wherever possible.

- An incentive or supervision system must be incorporated to assure that the time gained at the face is utilized productively. It cannot be assumed that putting a miner at a working place a given number of minutes early will automatically increase an individual's daily output. A miner must be shown some personal benefit in actually using those extra minutes.

2.8 REFERENCES

Alexander, C. 1999. Tunnel Boring at Stillwater's East Boulder Project. *Mining Engineering*, Vol 51, No. 9, pp. 15–24.

Bullock, R.L. 1973. Mine Plant Design Philosophy Evolves from St. Joe's New Lead Belt Operations. *Mining Congress Journal*, Vol. 59, No. 5, May, pp. 20–29.

Bullock, R.L. 1994. Underground Hard Rock Mechanical Mining. *Mining Engineering*, Vol. 46, No. 11, November, pp. 1254–1258

Caledonian Mining Co., Ltd., Shelob Shaft Shotcreting Units. Carlton Works, Carlton-on-Trent, Newark, Notts, England.

Carlisle, D. 1955. Economic Aspects of the Definition of Ore. *Transactions, Institution of Mining and Metallurgy*, pp. 64–95.

Chadwick, J. 1994. Boring into the Lower Kalamazoo. *World Tunnelling*, North American Tunnelling Supplement, May, pp. N17–N23.

Christie, M.A. 1997. Ore Body Definition and Optimisation. International Conference on Mine Project Development, AusIMM, Preston, South Victoria, pp. 47–56.

Dravo Corp. 1974. Analysis of Large Scale Noncoal Underground Mines. Contract report SO 122-059, Bureau of Mines, Jan., 605 pp.

Edwards, F.A. 1992. Hoisting Systems. *Mining Engineering Handbook,* Hartman, H.L., ed. SME, Littleton, CO, pp. 1646–1678.

Gerity, C.E., ed. 1975. Hoisting Conference, AIME, New York.

Hamrin, H. 1982. Choosing an Underground Mining Method. Section 1, *Underground Mining Methods Handbook,* Hustrulid, W., ed. SME, Littleton, CO, pp. 88–109.

Hoover, H.C. 1909. *Principles of Mining.* McGraw-Hill, New York, pp. 199.

Lama, R.D. 1964. Some Aspects of Planning of Deep Mines. *Colliery Engineering,* June, pp. 242–245.

Spearing, A.J.S. 1995. *Handbook on Hard-Rock Strata Control.* South African Institute of Mining and Metallurgy, Special Publication Series, Johannesburg, pp. 89–93.

Schwandt, A.J.W. 1970. The Magmont Operation. AIME World Symposium on the Mining and Metallurgy of Lead and Zinc. D.O. Rausch and B.C. Mariacher, eds. AIME, New York.

Snyder, M.T. 1994. Boring for the Lower K. *Engineering and Mining Journal,* April, pp. 20ww–24ww.

Taylor, H.K. 1977. Mine Valuation and Feasibility Studies, Mineral Industry Costs. Northwest Mining Association, Spokane, WA, pp. 1–17.

Tessaro, D.J. 1960. Factors Affecting the Choice of a Rate of Production in Mining. *Canadian Mining and Metallurgical Bulletin,* Nov., pp. 848–856.

Theodore Barry and Associates. 1975. Analysis of Men, Material, and Supplies Handling Systems in Underground Metal and Nonmetal Mines. Vol. 1, Phase II, NTIS PB 249131.

Tilley, C.M. 1989. Tunnel Boring at the Stillwater Mine. Proceedings, RETC, R. Pond and P. Kenny, eds. SME, Littleton, CO, pp. 449–460.

Unrug, K.F. 1992. Construction of Development Openings. *Mining Engineering Handbook,* H.L. Hartman, ed. SME, Littleton, CO, pp. 1580–1643.

Willoughby, R., and N. Dahmen. 1995. Automated Mining With the Mobile Miner. Mechanical Mining Technology for Hard Rock. Short Course, Colorado School of Mines, Golden, CO.

CHAPTER 3

Planning the Underground Mine on the Basis of Mining Method

Richard Bullock* and William Hustrulid†

3.1 INTRODUCTION

The previous chapter concentrated on defining those areas that must be addressed for any and all types of noncoal underground mines. This chapter will address those areas of planning and design applicable to specific mining methods.

3.2 ROOM-AND-PILLAR STOPING

When one considers the number of underground mines in limestone, dolomite, salt, trona, potash, and gypsum, as well as all the Mississippi Valley-type lead and zinc mines there are, then it should not be surprising to realize that approximately 60% to 70% of all noncoal underground mining in the United States is done with some form of room-and-pillar mining. According to a 1998 survey from the National Stone Association, for the aggregate industry alone there were approximately 92 underground mines, all of which are room-and-pillar. At the present time, there are between 30 and 40 more room-and-pillar underground aggregate mines being planned.

This portion of the chapter will only touch on some of the more significant aspects of the design of a room-and-pillar mine and will refer to those case studies later in this book to be read for the particular applications in the various commodities.

3.2.1 Access to Room-and-Pillar Mines

The discussion in Chapter 2 (2.4.2.2) adequately covers the subject of where and how to access a new mine. But to repeat some of the basic concepts for a room-and-pillar mine—

- If it is possible to develop the resource from a hillside adit, this will obviously be the least expensive method of entry.
- If a shaft is sunk, then—
  - The production shaft should be sunk somewhere close to the center of gravity of the ore body
  - The shaft depth should be sunk to where most of the ore will be hauled down grade to reach the shaft dump pockets
  - The shaft depth should be sunk deep enough to accommodate adequate dump pockets, skip loading, and the crusher station
  - For aesthetic reasons, the shaft should be in a position where the headframe will be out of sight of the general public.
- If a decline is driven that will be used for—
  - Trackless haulage, then the maximum grade recommended is 8%

  - Conveyor belt haulage, but the decline will be negotiated by rubber-tired trackless equipment on a regular basis, then 15% is the maximum grade recommended
  - Conveyor belt haulage only, then the maximum grade could theoretically be 0.26 to 0.44 radians (15° to 25°), depending on the type of material conveyed.

However, one must remember that equipment will be driven alongside the belt to clean up spill rock. It is this activity that may limit the decline grade unless manual shoveling is planned for cleanup.

3.2.2 Orientation of Rooms and Pillars

Pillar Orientation with Consideration of In Situ Stress. As in all mining methods, the planner of a room-and-pillar operation must be aware of probable in situ stress within the rock prior to mining. If indeed there is a significant amount of maximum horizontal stress in a particular direction, then the mine planner should take this into account by orienting room advance and the direction of rectangular pillars to give the most support in that direction. In the very early phase of development, research should be done to determine the magnitude and direction of inherent stress levels. When one does not know the direction of horizontal stress, at least in the middle and eastern areas of the United States, pillars should be aligned at right angles with rows at N 79° E to best cope with natural horizontal stresses in the earth's crust. While this is necessary and is considered good operating practice, many very shallow room-and-pillar operations may have very little horizontal stress and need not be concerned with the direction and orientation of rectangular or barrier pillars. This is particularly true if the mine has been opened from a hillside with adits, where nature could have relieved horizontal stresses eons ago.

However, in sharp contrast to this condition, some deep room-and-pillar mines have tremendous problems, not only with high horizontal stress levels, but with rock that will absorb a large amount of energy before it fails violently. In such operations, not only is pillar orientation important, but the sequence of extraction and how it takes place is important. Such an operation is well documented as a case study by Korzeniowski and Stankiewicz, "Modifications of Room-and-Pillar Mining Method for Polish Copper Ore Deposits," (Section 2, Chapter 9).

Room-and-Pillar Orientation with Consideration of Dip. While most room-and-pillar mining is done in fairly flat strata, the method does not necessarily have to be limited to flat horizons. With dips up to about 5% to 8%, there is little difference in the layout of a room-and-pillar stope, except that the rooms should be laid out in such a manner that loads are being

* University of Missouri, Rolla, MO.
† University of Utah, Salt Lake City, UT.

hauled downgrade instead of up. As to orientation of rectangular pillars, it would probably make sense to orient the long side of the pillar in the dip direction.

The other alternative is to mine a series of parallel slices in steps following a level contour of the dipping ore body. Thus, as each round is blasted, much of the rock will cascade down to the next level, where it can then be loaded (hence the term "cascade mining"). The method was first published by Mufulira (Anon. 1966). However, in this case, the pillars were also removed almost immediately in a second cycle of mining, which then allowed the hanging wall to cave as mining retreated along strike.

When dip becomes very steep, 35% to 45%, for example, and it is too difficult to operate trackless equipment, a few manufacturers have proposed back-mounted, cog-wheel-driven jumbos that could drill stope rounds under these conditions. However, ore would then have to be removed by a scraper. At this angle, ore would not flow by gravity, and it would be too steep for any other type of loading equipment.

3.2.3 Room-and-Pillar Mine Haulage Development

Normally, the production shaft is developed somewhere near the centroid of the ore body. The objective of production development is to minimize the cost of hauling ore to the shaft. If this is a trackless haulage operation, other considerations will help minimize haulage costs.

- Grades should be as low as possible, and long hauls under 8%.

- The road should be as straight as possible. This means keeping pillar locations from causing the road to deviate around newly formed pillars. Therefore, all main haulage roads should be laid out prior to mining, and pillars should be laid out from this plan.

- Haul roads should be maintained in excellent condition with adequate amounts of crushed stone, graded, and kept dry. Water not only causes potholes, but lubricates rock to cut the tires.

For laying out the mine development plan, keep the rules of thumb given in Chapter 2 (2.3.1) by Spearing in mind. These include keeping intersections off the main drift six widths apart, avoiding acute-angle turnouts that create sharp bullnose pillars, and keeping nearly parallel ramps and declines apart from the main drift by at least three times the diagonal widths of the main drift.

For planning the trackless, rubber-tired haulage system, one must consider the various methods of moving rock from the working face to the crushing and/or hoisting facility. There are many ways to do this. In fact, with today's modern hydraulic excavators and powerful rubber-tired machines, equipment is so versatile and flexible that it sometimes creates the dilemma of how to be sure that the optimum method is always used.

When a room-and-pillar mine first starts out, either from the bottom of a decline or a shaft, the haul distances are short, and unless material is being hauled up a decline to the surface, load-haul-dump (LHD) machines with a rubber-tired loader will probably be the best way to go.

As the mine is gradually worked out from the dump location, at some point a front-end loader should start loading ore into a truck. At an intermediate point, when more than one truck is required for any given tonnage produced, the loader can load the truck, load itself, and follow the truck to the dump point. Eventually, as the distance increases even more, the loader should stay in the heading and a sufficient number of trucks should be used that will keep the loader busy.

When more than one level is involved within a mining zone and the main haulage is on the bottom level, then various

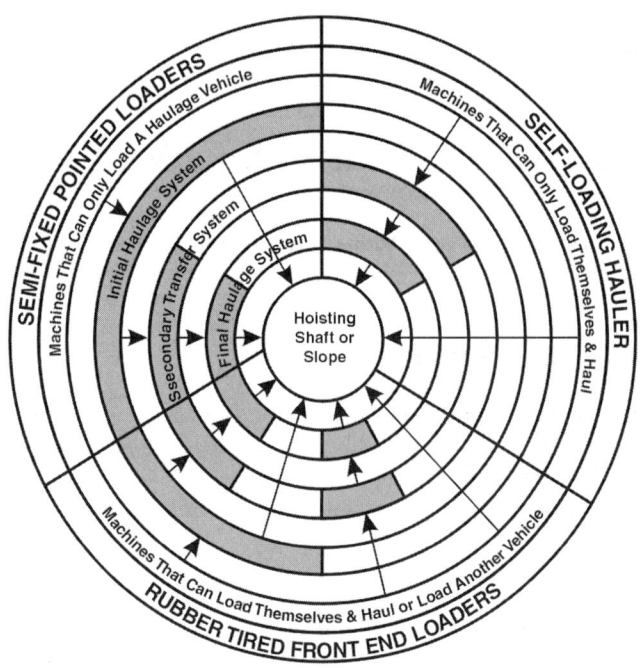

FIGURE 3.1 Various methods of moving ore from the room-and-pillar stope to hoisting facility

combinations of LHDs, front-end loaders, and/or trucks can be used to haul ore to the ore pass. At the bottom of the ore pass, automatic truck-loading feeders can transfer ore to trucks or rail-mounted trains. If the upper ore body is rather small, an automatic ore chute cannot be justified, and the ore can simply be deposited on the ground, from where an LHD or a front-end loader can load the ore into another truck. Figure 3.1 illustrates all the possibilities for moving material from a room-and-pillar face to the final ore pocket at the shaft.

The mine planner needs to be aware that in an ever-expanding room-and-pillar mine, the optimum method or combination of moving ore is constantly changing and that for every condition and distance, there is only one optimum, least-cost method. All of the above can and have been demonstrated using computer simulations of the underground mine environment (Bullock 1975, 1982; Gignac 1978).

One other concept used for some room-and-pillar mines is well worth noting. In some of the older Tennessee mines, haulageways were driven somewhat small. Thus, as the haul distance became longer, it would have been desirable to use larger haulage trucks. However, it was not possible to go to larger trucks because of the small haulageways. To address this problem, Savage Zinc, in its Gordonville Mine, began hooking side-dump truck-trailers together to make a minitruck train. Their use increased the payload per trip for the long hauls, thus optimizing production for this specific loading and hauling condition.

If a conveyor system is used for the main distance haulage, then the above advice for a mine layout still applies, except the conveyor will start from the decline of the mine entrance. In these cases, the mine usually has a semimoveable crusher or breaker underground at a point before the ore is loaded to the feeder to the conveyor system. The system of haulage, either LHD or front-end loader and truck, will haul broken rock from the faces to the central receiving point at the crusher. The crusher then feeds the conveyor. The crusher must be moved periodically, say, every six to 12 months, depending on how fast mine production moves away from the central haul point.

There was a time when rail haulage was the principle method of transporting ore from the faces to the production shafts. In the Old Lead Belt of Missouri, over 556 km (300 miles) of interconnected railroad once brought all ore from about 15 mines into two main shafts. Today there are very few room-and-pillar mines using rail haulage.

3.2.4 Room-and-Pillar Extraction Methods

One of the advantages of modern room-and-pillar mining systems is that every task can be mechanized to some degree. Mechanization minimizes the operating labor force and makes staffing the operation easier. The high-capacity equipment for modern room-and-pillar operations is reasonably simple to learn and operate. Anyone who has operated any heavy machinery in construction work, the military, or even on a farm has little trouble adapting to loading and hauling equipment in a room-and-pillar operation.

While most room-and-pillar mining is done by drilling and blasting, particularly in the aggregate and metal-mining businesses, a very large portion of room-and-pillar mining is also done by mechanical excavation. Many trona, potash, and some salt mines excavate all openings by mechanical excavation. The reader is directed to the case studies in this publication, which illustrate those types of room-and-pillar mines using drilling and blasting techniques and those using mechanical excavation techniques. With the power of today's mechanical excavating machines and with the improvements being made in tools such as disk and pick cutters, the possibility of mechanical excavation should be at least studied and considered for any rock under 100 MPa (15,000 psi) (or even up to 136 MPa [20,000 psi]) if it is highly fractured and is low in silica content. The advantages of various types of mechanical excavation, where applicable, are well documented (Bullock 1994). Furthermore, for long developments, full-face tunnel-boring machines are proving their worth under certain conditions (Snyder 1994; Alexander 1999).

The advantages to mechanical excavation (Ozdemer 1990) are—

- Improved personnel safety
- Minimal ground disturbance
- Less ground support needed
- Continuous, noncyclic operations
- Few ground vibrations and no air blast
- Uniform muck size
- Less crushing and grinding required at the mill
- Reduced ventilation requirements
- Automation capability

Where mechanical excavation is truly feasible, it adds up to higher production rates and reduces mine operating costs.

Many room-and-pillar mines must rely on drilling and blasting the face. The initial pass of mining by drilling and blasting is usually either done by drilling "V-cut" patterns or "burn cut" patterns. These types are well documented in the literature (Bullock 1961, 1982; Langefors and Kihlstrom 1963; Hopler 1998). However, one aspect often overlooked is that only about 40% of the rock should be broken with drilled rounds breaking to only one free face, but about 60% of the rock should be broken by slabbing to multiple free faces as they are exposed (Figure 3.2). This minimizes the cost per ton of rock broken and maximizes productivity.

3.2.5 Single-Pass or Multiple-Pass Extraction

There are two approaches to mining the existing thickness of the ore body or valuable rock: taking the entire thickness in one pass (mining slice) or removing the ore body using multiple passes or

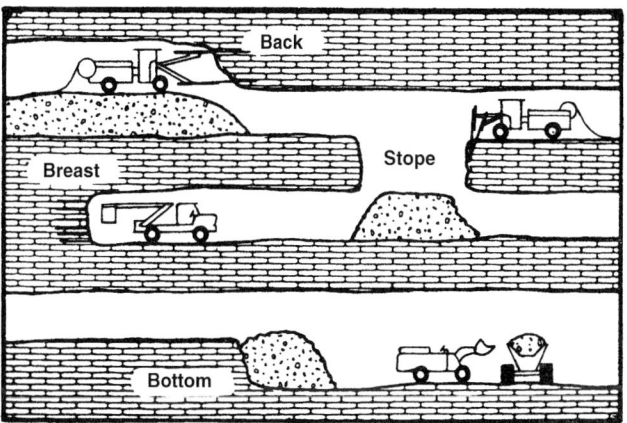

FIGURE 3.2 Plan view of room-and-pillar stope showing advance by drilling a combination of drill pattern rounds with one free face and slabbing to an open second free face

slices. Which option chosen is, of course, related to the overall thickness to be extracted.

Typically in mining bedded aggregate deposits, the total thickness of the desired horizon is known and the decision about how thick a slice to take can be made in advance. However, in metal ore deposits, particularly Mississippi Valley- or collapsed breccia-type deposits, the mine planner generally does not know how thick the total mining horizon is going to be except in those places where each diamond-drill hole has passed through the formation. A few feet away from that hole, conditions may very well be different.

For both an aggregate producer and a metal ore producer, the best approach is to mine the top slice through the ore body first. The thickness of the slice to be taken depends on what equipment is available or what is going to be purchased. It also depends on what height of ground can be mined and maintained safely and efficiently. Mining the top slice first allows whatever back and rib-pillar scaling and reinforcement are required to be reached easily and safely. In the authors' opinion, this thickness should not exceed 8.5 to 9.8 m (28 to 32 ft). However, there are aggregate producers that will mine 12-m- (40-ft-) thick and thicker slices in one pass using a high-mass jumbo and extendable-boom roof-scaling equipment.

After the first pass is completed in a metal mine, for a given stoping area, the back and floor should be "jackhammer-" prospected to identify what ore remains in the back and floor that will need to be removed by other mining slices. If ore is found in the back and floor, then the ore in the back should be mined first. After this ore is removed and additional back-prospecting reveals no more ore and the back is again made secure and safe, then ore in the floor can be taken.

Methods of mining the ore in the back vary somewhat depending on the thickness of the ore yet to be mined and the original stope height. If the original stope height was no more than 7.6 m (25 ft) and the thickness of the back slice is to be no more than 2 m (7 ft), then most extendable-boom face jumbos can be used to drill the brow with breast (horizontal) drilling. By drilling horizontal holes, a miner has a better chance of leaving a smooth back that requires less maintenance than if the upper holes had been drilled to break to the free face of the brow. This is especially true if careful blasting is practiced. The authors do not recommend either tilting the jumbo feed up to drill nearly vertical holes or drilling uppers to break to a brow, particularly if a bedded deposit is being drilled, which is the case in many room-and-pillar mines.

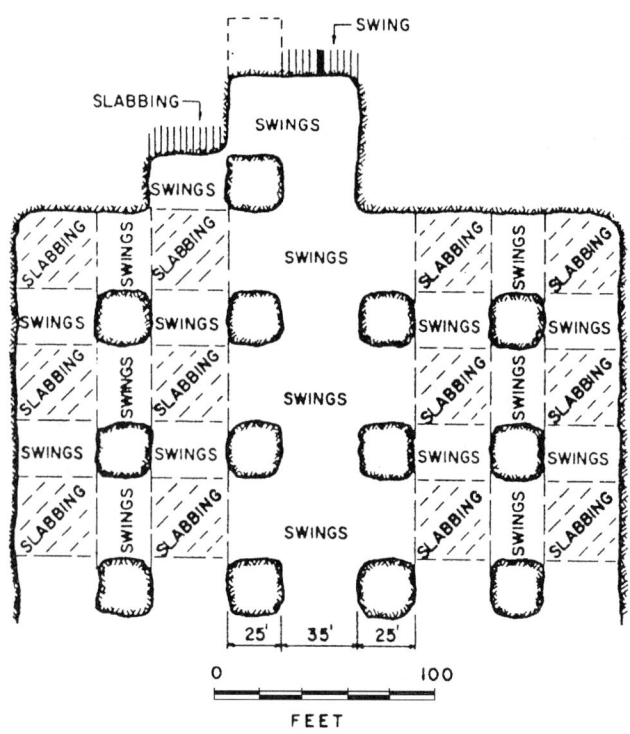

FIGURE 3.3 Multipass stoping methods of thick ore bodies in room-and-pillar mining

If the ore thickness in the back is greater than 2 m (7 ft), or the room height is already at the maximum height that can be safely maintained, then the approach to mining the back slice is to first cut a slot in the back at the edge of the entrance to the stope. This slot should be between the pillars and should reach the height of the next slice or the top of the ore body. Care should be taken not to damage what will be the rock that forms the top of the new pillars. Smoothwall blasting could be used to advantage in this area. A mine dozer (usual a small one such as a D-4 or D-6) should begin pushing the rock up and making a roadway for the jumbo on the rock pile. A front-end loader can also do a reasonable job of building the roadway if no dozer is available. From here, the jumbo can drill breast (horizontal) holes in the brow and follow the ore zone throughout the stoped area. Even if it goes into solid rock beyond the original stoped area, a problem will not be created other than the loader will now have to load out the ore as it is broken. About 75% of the ore will have to remain in the stope rock pile until the back-mining job is completed. This can be a disadvantage if the mine needs the production immediately, or it can be a big advantage if the ore can be moved when many miners leave on summer vacations.

When the top of the ore is finally reached, the back can be made safe since the miners are still up very close to the back. This type of mining has been practiced in both the Tennessee zinc and the Viburnum lead-zinc districts. It isn't uncommon for several passes of the ore to be mined from broken rock piles. In these cases, it will be necessary to load the ore from the bottom edges of the rock piles to make room for new broken rock or a loader will have to go upon the rock pile and load out the excess rock (Figure 3.3).

If a very thick, continuous ore zone (say, 15 to 20 m [50 to 65 ft] or more) is found above this first slice after the first slice is taken through the stope, then an entirely different approach may be taken. In such a case, it be better to drive a development ramp to the top of the known ore zone and mine out the top slice from

this new ramp. A slot raise can then be put through from the bottom level to the top level that can be slabbed out to make room for longhole drilling and blasting the ore to the level below. Again, the pillars should be presplit or smoothwall blasting should be done to protect them.

Once all the back ore is removed from the stope and the final back is made completely secure, the bottom ore can be removed from the floor. This is best done by first cutting a ditch or short decline in the floor at the entrance to the stope to the depth desired to carry the bluff. Bluffs can be carried very thick, limited only by the height considered safe for loading equipment. It is common practice to carry bluffs up to 9 m (30 ft) high in the lead mines of the Viburnum area using 7- to 11-tonne (8- to 12-ton) loaders. However, beyond this height, loader operator safety may become an issue.

Drilling and blasting the bluffs is usually done with downholes using small surface quarry drills if the bluffs are at least 4 m (13 ft) thick. If the bluffs are thinner, breasting or horizontal holes may be drilled out with face jumbos.

The above procedure can be repeated over and over until the bottom to the ore is reached (Figure 3.3). However, in areas where this type of mining is expected to take place, the pillar width *must* be large enough to accommodate considerable height for the initial pass if it should be needed. If additional ore is not found, then the pillars can be slabbed down to a smaller size.

However, one precaution is that pillars must be protected with smoothwall blasting or presplitting around the pillars both when removing the back ore and taking up the bottom. Pillar design will be briefly discussed below, but the point here is that for any given pillar width assumed by the mine planner, the designed safe width-to-height ratio will be exceeded by multiple passes unless there is a very large safety factor in the pillar design or it is planned to take down the back and take up the bottom. It can not be overemphasized that everything should be done to protect the integrity of the pillars during the first pass of what may become multiple passes.

3.2.6 Room Widths

For productivity reasons, room widths should be as wide as practical and safe. The wider the rooms, the more efficient the drilling and blasting and the more efficient the loading and hauling equipment. However, room width for any given mine environment will be limited by the strength of the ore body, the back, and the floor compared to the stress levels induced into the rock. It is inappropriate to design room-and-pillar widths simply from elastic theory without taking into account rock mass strength. On the other hand, because the rock and pillars can be reinforced and thus affective rock mass strength can be increased, it may becomes a matter of economics as to how wide the room can be.

There have been many papers written on how to design roof spans. For a complete discussion of the rock mechanics aspects of this problem, turn to Section 10, "Foundations for Design." What is important to consider at this point is what information will be needed for the design and how much of the needed information is already at hand. In Chapter 2 (2.1.1 and 2.1.2), there is a general summary of the geologic and structural information to be obtained during exploration of an ore body. Unfortunately, most exploration groups spend little time or money in trying to determine the information needed to construct a rock mass classification of the mineralized area and the rock surrounding the mineralization. It may be that a "best guess" rock mass analysis will have to be done with nothing more than the exploration information. In any case, it is hoped that there would be enough information about underground structures from core logs, surface mapping, mapping of surface outcrops of the same formation underground, and geophysical information that a crude rock mass classification could be constructed.

What is really needed is an underground test section. In Chapter 2 (2.1.4), 10 good reasons for developing a test section during the middle phase of a mine feasibility study were provided. One of the most critical reasons for doing this is to obtain better geophysical information on which to base mine planning: in this case, room widths and pillar widths and heights, which will greatly affect mine operating costs. Structures can be seen and mapped, and joints and fractures accurately measured. In situ stress measurements can be taken, and larger core samples can be collected for laboratory testing.

3.2.7 Pillar Width

One cannot discuss pillar width without relating it to pillar height. The overall strength of a pillar is related to its height, i.e., it is a matter of the ratio of pillar width W to pillar H. The amount of load that the pillar can safely carry is proportional to the W:H ratio. Thus it is that a pillar with a ratio of 4:1 has a much larger safety factor than a pillar of 4:4 or 1:1.

The actual load that pillars can really carry can only be measured. The theoretical load as calculated by the overburden load distributed to the pillars may or may not be the load actually being carried. There is a good chance that the load may be arching over some of the interior pillars of the stope and transferred to barrier pillars or waste areas. In such cases, it may be that the interior pillars can be made smaller as yielding pillars. If the stopes are very wide, then a row of large rectangular barrier pillars should be left at regular intervals. In areas of very large lateral extent, this will prevent cascading pillar failure of the entire area (Zipf and Mark 1997).

The reader is referred to the case studies on room-and-pillar stoping to examine how different mines approach this problem, to the design theory expressed in Section 10, and particularly to the paper by Zipf in Chapter 59, Section 10, on catastrophic failure of large room-and-pillar areas where proper precautions have not been taken.

3.2.8 Ventilation

It is not the intent to go into the design of a noncoal room-and-pillar mine ventilation system in this chapter. There are books that the mine planner should refer to for specific information on mine ventilation planning (Hartman et al. 1997; Ramani 1997; Tien 1999). However, most of the literature on ventilation design for room-and-pillar mining has been written for coal mines. Thus it may be advisable to mention a few points unique to many metal and aggregate room-and-pillar mines.

- Everything is larger. It is not uncommon to have entry drifts 9 by 9 m (30 by 30 ft) and rooms stoped out to 9 by 21 m (30 by 70 ft). A lot more air is needed to meet minimum velocities across the working face in these conditions.

- Stoppings are difficult to build, and the total force against a stopping can be enormous. Thus many room-and-pillar mines rely on auxiliary fans to pick up air from the main ventilation drifts and carry it through vent tubing in sufficient quantities to serve the needs of the active faces.

- Ventilation doors are more like airplane hangar doors than a small ventilation door. Again, the total force of air against these doors is enormous and must be considered, particularly when the doors are controlled automatically.

- Air stratification in large stopes can be a problem.

- Diesel equipment is used extensively in these mines, and diesel exhaust must be diluted.

- Ventilation fans can be placed underground if it is beneficial to do so.

3.2.9 Pillar Robbing

Pillar removal should be a planned part of overall mining in areas where the economic value of the ore remaining in the pillars justifies and warrants their extraction. As an example, it is not uncommon for some very high-grade pillars in the lead-zinc-copper mines of the Viburnum Trend to have a value of over a million dollars a pillar. For more on pillar removal, the reader is referred to a case study in Chapter 8, by W.L. Lane et al.

Thus if future pillar extraction is planned, whether it be partial slabbing of pillars, removal of only a few high-grade pillars, or complete removal incorporating some system of backfilling, then what is left out of the initial mine design will highly influence what may be able to be done in the future.

From the experience of the authors in planning and supervising slabbing and removing hundreds of pillars, the first and most important thing to do is to install a complete network of convergence stations throughout the area in question. This procedure is applicable to room-and-pillar stoping where continuous pillars are left without interim barrier pillars. Some mines have back-floor convergence of as much as an inch per month without the back breaking up and failing (Parker 1973). At other mines, such as some mines of the Viburnum trend, back-floor convergence on the order of a few thousands of an inch per month becomes very significant, and a convergence of 0.003 inch per month indicates a serious problem, although controllable with immediate action. Therefore, convergence must be considered in relation to what is typical at each individual mine, and each mine much decide how much convergence indicates the need for corrective steps.

There are several methods of pillar removal in room-and-pillar stoping. In broad terms, these are—

- Slabbing some ore off each pillar containing the high-grade portion of the ore as the area is retreated.

- Completely removing a few of the most valuable pillars, leaving enough pillars untouched to support the back;

- In narrow stopes, completely removing all the pillars in a controlled retreat.

These methods were used in final mining of the Old Lead Belt of Missouri over a period of approximately 25 yr, but much more intensely over the last 10 yr. This was a room-and-pillar mining district that was mined for 110 yr before finally being shut down.

Another approach is to place cemented backfill around pillars all the way to the back to support the stope properly between the solid rock and the backfill to create a "pressure-arch" over the pillars in between. These pillars can then be removed. Economics permitting, this area could be backfilled if necessary, and the pillars encapsulated or trapped in the original fill could then be mined from a sublevel beneath the pillar and blasted into the sublevel area (Lane et al. 1999). The total backfill would prevent any future subsidence.

All of the above methods have been used and proven to be profitable. Each method must be analyzed from an economic feasibility point of view, as well as from a technical ground control-rock mechanics point of view, before planning and executing pillar removal practices.

Sometimes when partial pillar removal takes place without the proper planning, or when the pillars left in the first pass of mining are too small, then they begin breaking up, and serious convergence begins to accelerate that can not be controlled. One of two things will have to take place to save the area: either massive pillar reinforcement will have to take place (if there is time prior to collapse) or massive amounts of backfill will have to be placed in the entire area. The author has been involved with both methods of stopping convergence and eventual catastrophic collapse of large areas.

The first solution involved placing fully grouted rebar bolts in over 300 pillars in the room-and-pillar mines of the Viburnum Trend. This major project is well documented by Weakly (1982) and covers the method employed, reinforcing pattern, cement grout mixture, convergence instrument use, and results. In the Old Lead Belt areas, the pillars needing reinforcing were wrapped with used hoist cable, with a load of 5.4 tonnes (6 tons) placed on each wrap (Wycoff 1950). However, this method of reinforcement was not as fast, economical, or effective as installing fully grouted rebar bolts in the pillars.

The second solution was used in the Leadwood mines of the Old Lead Belt and involved placing huge amounts of backfill to mitigate the potential for a massive catastrophic failure. These mines were only about 132 m (425 ft) deep. However, the back was thin-bedded dolomite interbedded with shale and glauconite and was badly fractured and leached. "Roof bolts" were originally developed in these mines in the early 1930s as a means of tying the layers together to act as a beam (Weigel 1943; Casteel 1973). The roof bolts were used with channel irons to form a crude truss. Even though the rock in the pillars was equally as bad for support as the back because it also contained bands of high-grade galena, as a means of economic survival pillar removal and slabbing took place over a period of 25 yr. Occasionally, local cave-ins would occur after an area had been "pillared." Since these cave-ins were beneath uninhabited St. Joe-owned land, they were of no real concern. However, when slabbed pillars between two of these smaller cave-ins began to fail and a third and fourth cave-in occurred in more critical areas, there was a considerable amount of concern. The initial extraction of some of these areas involved multiple-pass mining and room heights that were mostly 6.1 to 12.2 m (20 to 40 ft). Final mining of the area had resulted in ore extraction of approximately 95%. Around 1962, to stop the caving in the third and fourth areas, over a million tons of uncemented, cycloned, sand tailings were put into the mines, filling the rooms nearly to the back. The results were very successful in controlling the converging, failing ground.

In retrospect, the end result might be compared with the overhand cut-and-fill practice of deliberately mining the pillars to very small dimensions and immediately filling in around them in what is known as the "post-pillar" mining system. Such a system is used by Falconbridge Nickel Mines (Cleland and Singh 1973) and the Elliot Lake uranium mines (Hedley and Grant 1972) and sometimes by San Martin and Niaca in Mexico and San Vincente in Peru. The end result is that the two mining systems, with small pillars encapsulated in sand tailings, look similar.

3.2.10 Room-and-Pillar Mining as It Relates to Changing Market Conditions

An added advantage of room-and-pillar stoping in metal mining (or with any commodity that is gradational in value) is related to changing market conditions. The flexible mining conditions of room-and-pillar stoping can usually be adjusted to the elastic nature of markets if mine planners are always aware of current market trends. The extreme flexibility of the room-and-pillar mining method allows the mine operation to react to market needs faster than other mining systems.

Another of the advantages is that new faces are continuously being opened if the ore body is continuous. Even if the stope is only four pillars wide (Figure 3.3), at any one time that stope may have as many as 12 to 15 faces open for drilling and blasting. One can imagine if the stope were 10 pillars wide how many faces would be exposed to drilling and blasting. For metal mines, this offers a lot of flexibility to mine the grade of ore most desirable for any current price of the metal being mined. For short periods of time in each stope, it is usually possible to work only the higher- or lower-grade faces, depending upon the market. This usually has a drastic effect on the grade within a few days. For example, after the high-grade Fletcher Mine had been operating for about 3 or 4 yr, there were approximately 50 to 70 faces open for mining on any given day, but only 10 to 12 would actually be worked. It becomes a matter of face selectivity to maintain a grade of ore that can best be handled by the concentrator and still optimize the financial objectives of the mine.

Similarly, spare equipment can be put into reserve stopes to increase production if the remaining materials flow can take the added capacity. However, if these practices are carried on too long or too often, mine development must also be accelerated. If maintained, old stopes can be reactivated quickly to mine lower-grade minerals that become minable because of economic cycles.

Another aspect of the room-and-pillar mine is that often lower-grade resources are left in the floor or the back of the stopes. When price rises, new reserves are readily available for quick mining. This technique is often overlooked by individuals not accustomed to planning room-and-pillar metal mines where the mineral values are gradational. There is the option of mining through the better areas of the mineral reserve and maintaining a grade of ore that satisfies the economic objectives at that time. At a later time, when mining economics may have changed, the lower-grade areas left as remnant ore reserves can be mined

However, in spite of the above, even in room-and-pillar mines, drastic changes in the rate of mining (momentum) cannot be assumed to be free. It often takes several months with an increased labor force to regain a production level that seemed easy to maintain before a mine production cutback. If spare equipment is used to increase production, maintenance probably will convert to a "breakdown" overtime schedule compared to the previous preventive maintenance schedule on shift, at least until permanent additional equipment can be obtained. Nevertheless, the necessary changes can be made.

As discussed under the section on "Pillar Robbing," the other technique is that of slabbing or removing high-grade pillars. Thus, even in the later years of the mining operation, some of the "sweetener" is left to blend with the lower-grade ore. Although not unique to room-and-pillar mining, this technique is certainly easier to accomplish in a room-and-pillar operation than in other, more complex mining systems.

3.3 SUBLEVEL STOPING

3.3.1 Introduction

As discussed in the previous section, the room-and-pillar mining system is applied to subhorizontal ore bodies of relatively uniform thickness. A portion of the ore body is removed in the form of rooms, and pillars are left to support the overlying strata. Mining may be done in a checkerboard pattern of rooms and pillars or long rooms may be created with rib pillars left between. The strata making up the roof and floor are competent, as is the ore. The deposits mined range from thin to moderately thick. As the dip of the strata increases and/or the thickness of the ore body increases, other methods must be employed.

Consider a moderately thick deposit that would be mined by the rib pillar system if flat dipping. Now increase the dip to 90°. In this case, loading on the pillars would come from the horizontal direction, and the blasted ore would fall downward to be collected at bottom. Although the general geometry is the same as for the room-and-pillar method, the generic name given the system is sublevel stoping. Blasthole stoping, vertical crater retreat, and vein mining also fall under this general heading. The shrinkage stoping method is a special form of sublevel stoping. In general, the method is applied to ore bodies having dips greater than the angle of repose of broken material (greater than about 50°), so that material transport to collection points occurs by gravity. For massive deposits, stopes with vertical walls are created, and the overall dip of the deposit is immaterial.

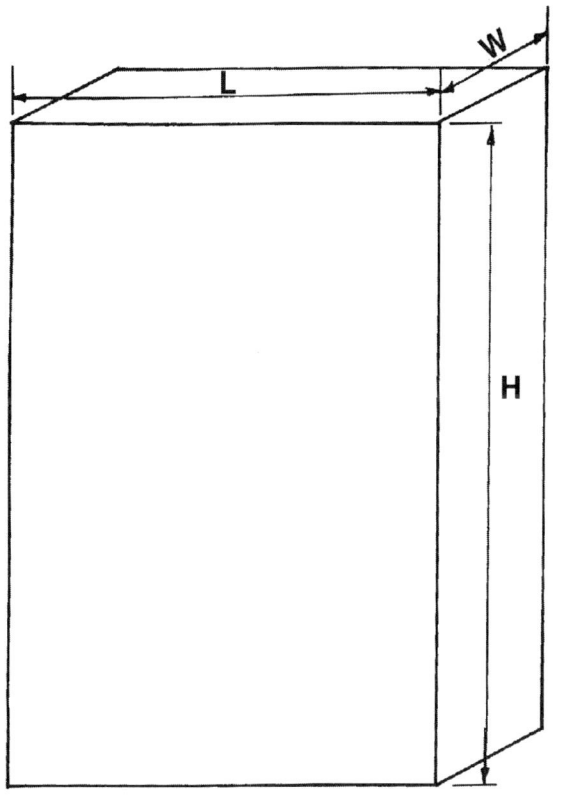

FIGURE 3.4 Diagrammatic representation of extraction block

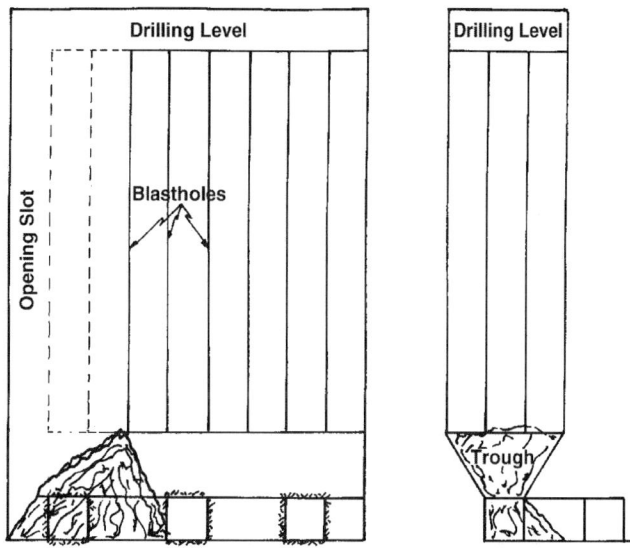

FIGURE 3.5 Diagrammatic representation of blasthole stoping

The criterion for applying the method is that the openings created remain open during extraction. These openings may be later filled or left open. The pillars left between stopes may be extracted at a later time or left in place. In this section, some typical layouts used for extracting ore will be briefly presented. It is assumed that mobile equipment is used with ramp access.

3.3.2 Extraction Principles

Consider an ore block of width W, length L, and height H as shown diagrammatically in Figure 3.4. For simplicity, it has been drawn vertical. This block will be mined using a number of sublevel stoping techniques. The blasted ore will fall to the bottom of the block and be removed using LHDs. There are various designs for the extraction level. Here it will be assumed that a trough is created using fans blasted toward an opening slot. The LHDs travel in a footwall haulage drift running parallel to the trough. Access to the trough is from the side. The location and number of access points (drawpoints) are such as to provide full extraction coverage.

Blasthole Stoping. Blasthole stoping will be the first method considered for mining the block. From the drilling level at the top of the block (Figure 3.5), rows of parallel blast holes are drilled down to the top of the extraction trough. A raise is driven at one end of the block and slashed to full stoping width to form a slot. The rows of blastholes are now blasted as one row or as several rows at a time toward the open slot. The blasting design and layout are very similar to that used in bench blasting.

Hole diameters vary widely, but typically lie in the range of 76 to 165 mm (3 to 6.5 in). For wide blocks, 165-mm (6.5-in) in diameter holes are often used. Hole straightness is an important design consideration that affects fragmentation, ore loss, and dilution. In general, one would select the largest hole diameter possible for the stope geometry since straight hole length is

strongly dependent on hole diameter. The specific development (amount of development required to exploit a certain volume of ore) is inversely proportional to block height. Since the cost of development is significantly higher than costs for stoping, one wants to have the highest possible number of extraction blocks associated with a given extraction and a given drilling level.

Sublevel Stoping. If geomechanics studies indicate that very high blocks (heights exceeding the straight drilling length from one drill location) can be extracted using the same extraction level, then several drilling levels at various heights within the block must be created. Because of the multiple drilling levels, or sublevels, this method is called sublevel stoping. The layout is very similar to blasthole stoping with an extraction level and an opening slot, but now there are multiple drilling levels.

Mining can take place overhand, in which the lower drilling blocks are extracted before the upper, or underhand, in which the extraction of the upper drilling blocks precedes those beneath. Here it is assumed that overhand stoping is employed.

The simplest approach is to repeat the drilling layout for one-level blasthole stoping. This is shown in Figure 3.6. The ore body thickness is assumed to be such that the full width is undercut and becomes available for drilling access. Parallel holes can be drilled in this case. An alternative is to drill fans of holes (Figure 3.7) rather than parallel holes from the sublevels. Furthermore, there may be one or multiple drill drifts on each sublevel, and the rings may be drilled downward, upward, or in full rings. Selection is based upon a number of factors, a full discussion of which is beyond the scope of this chapter.

As indicated, the application of the sublevel stoping technique assumes good stability of the openings created. Stability surprises can mean the partial or even full collapse of partially extracted stopes. Production may be stopped completely because of the presence of large blocks in the drawpoints. In the best case, there is ore loss and dilution. Reinforcement of the footwall, hanging wall, and roof can be done prior to or during mining. These extraction blocks can be oriented along the strike of the ore body (longitudinally) or transversely.

Vertical Crater Retreat. In the cases just discussed, rings of holes were blasted toward a vertical slot. In vertical crater retreat or vertical retreat mining systems, the need for a slot to connect the drilling and extraction levels has been eliminated, thus simplifying development. The vertical slot is replaced by a horizontal slot (undercut) created at the bottom of the block on

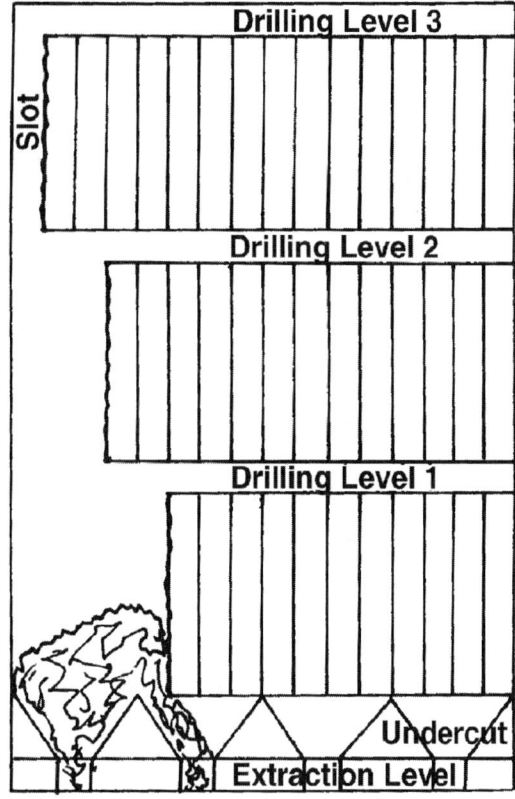

FIGURE 3.6   Multilevel blasthole stoping

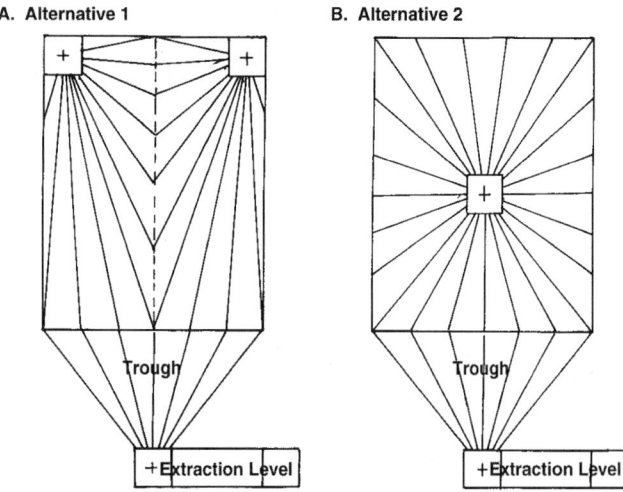

FIGURE 3.7   Typical fan patterns for sublevel stoping

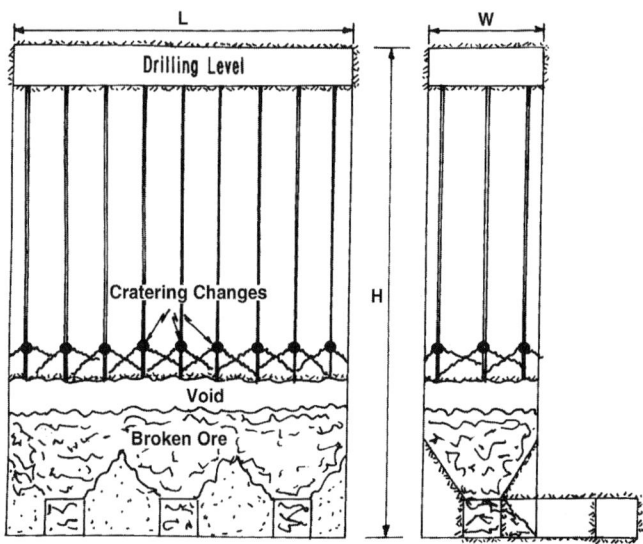

FIGURE 3.8   Diagrammatic representation of vertical retreat mining

the extraction level. Although a real trough may be created, it is not necessary.

From the drilling level, large-diameter (on the order of 165 mm [6.5 in]) parallel holes are drilled downward to the undercut level (Figure 3.8). Short explosive charges (length = six hole diameters) are lowered to positions slightly above the top of the undercut. These "spherical" charges are detonated, dislodging a crater or cone-shaped volume of rock into the underlying void. As each layer of charges is placed and detonated, the stope retreats vertically upward, hence the name "vertical crater retreat" mining.

The design of the blasting pattern is based upon full coverage of the block cross section by the adjacent craters. Normally, the blasting pattern is tighter (holes spaced closer) than would be the case in large-hole blasthole stoping, and hence the powder factor is larger. When blasting under these confined conditions, fragmentation is generally finer than with blasthole stoping.

Prior to charge placement, care must be taken in determining the location of the free surface. Special tests are performed to determine crater dimensions. In this system, the level of broken rock remaining in the stope can be controlled to provide varying levels of support to the stope walls. If the stope is kept full except for a small slot to provide a free surface and room for swell volume for the blasted rock in the slice, it is classified as a shrinkage method. In this case, the remaining ore would be drawn out at the completion of mining.

Vein Mining.   Another approach to extracting the ore block is called vein mining. At the highest level of the block to be extracted, a connection is made to the ore body. It will be assumed that the access is located on the footwall side and the connection is made in the middle of the extraction block. On the extraction level, an undercut or an extraction trough is prepared. A raise is driven between the extraction level and the upper access point using the Alimak technique (Figure 3.9). Here, the raise will be assumed to also be located in the footwall a short distance from the ore-footwall contact.

The next step in the process is the drilling of subhorizontal fans of blastholes from the Alimak platform in such a way that the plan area of the extraction block is fully covered. Hole diameter is determined by the capacity of the drilling machine, but should be as large as possible since the toe spacing and the burden (distance between fans) is determined by hole diameter and the explosive used. Once the drilling of the entire extraction block has been completed, the fans are charged and blasted one or more at a time working off the Alimak platform. Access to the block is now only from the upper level, and the Alimak guides are removed as the stope is retreated upward.

Ore in the stope can be removed after each blast or it can be left in place, removing only enough to provide for swell volume for the next slice(s). If required, rock reinforcement can be installed in the hanging wall from the Alimak platform during drilling the production holes.

This method allows the extraction of very high ore blocks with a minimum of development (upper access point, the

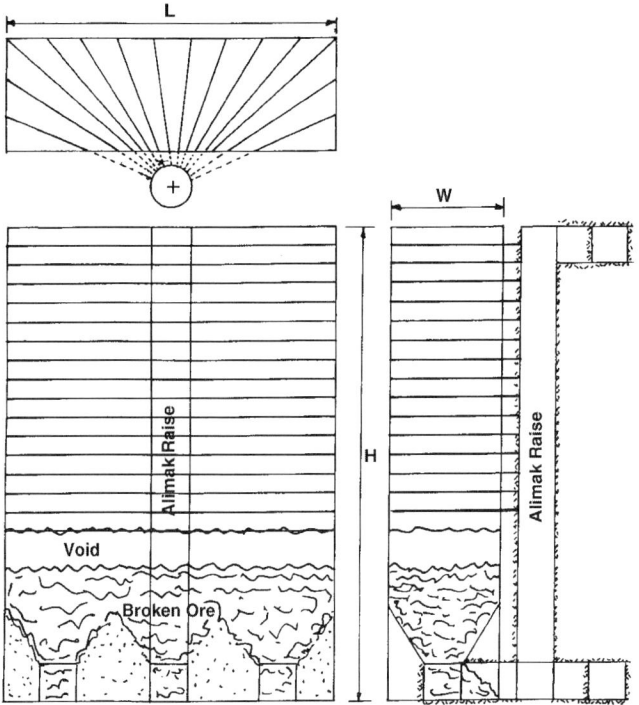

FIGURE 3.9 Diagrammatic representation of vein mining

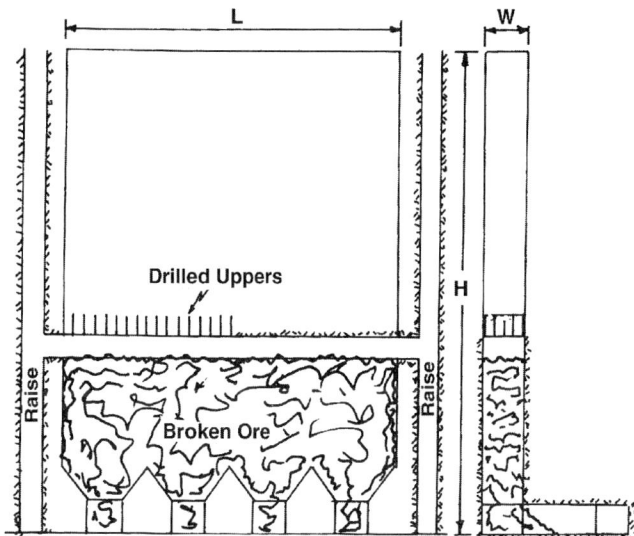

FIGURE 3.10 Diagrammatic representation of shrinkage stoping

extraction level, and the connecting raise). The overall length of the extraction block is determined by the straight-hole drilling length of the available drilling equipment. The disadvantage of the method is that drilling and charging must be done from a raise environment, which traditionally has not been pleasant. Major advances have, however, been made in the mechanization and automation of the rigs used for drilling.

Shrinkage Stoping. The final method to be considered under this category is shrinkage stoping. Although normally considered as a separate method, it is logical to include it here. The method is generally applied to very narrow extraction blocks that have traditionally not lent themselves to a high degree of mechanization. Here, a very simplified layout (Figure 3.10) is presented to illustrate the steps.

The extraction block is laid out longitudinally because of the very narrow nature of the ore body to be recovered. An extraction drift is established in the footwall with loading crosscuts positioned at regular intervals. Raises are driven at each end of the extraction block connecting to the above-lying level. An initial horizontal extraction slice is driven across the block from raise to raise.

Extraction troughs are created by drilling and blasting the rock between the extraction level and the underlying extraction points. When the extraction system has been created, short vertical holes are drilled into the roof of the first extraction slice using the raise access. The miners stand on the broken ore, which forms the working floor. Jackleg or stoper drills are used for drilling small-diameter holes. The holes are charged, and then ore is extracted from the stope to provide room for the blasted material. The blast is initiated, and the miners reenter the newly created void to drill out the next slice. The process continues working upward one slice at a time. Upon reaching the upper end of the extraction block, the ore is drawn out. Until that time, the stope is filled with broken ore.

3.3.3 Summary

In summary, depending upon the geometry of the ore body, several varieties of sublevel stoping can be employed. The ore

bodies must have strong wall rocks and competent ore either naturally or helped by the emplacement of reinforcement, since large openings are created in the process of ore removal.

The extraction block used to illustrate the layouts for the different mining systems can now be duplicated and translated laterally and vertically in the ore body, leaving pillars to separate adjacent blocks. The size and shape of the extraction block can be adjusted to fit ore body geometry and mine infrastructure. The openings created during primary mining may be filled with various materials or left unfilled. Filling materials may be cemented or uncemented, depending on the next stage of recovery envisioned. Various methods are used to recover the remaining reserves tied up in the pillars. These secondary recovery methods should be studied at the same time the primary system is designed. Although for simplicity the basic extraction block was considered to be vertical, the process could obviously be repeated for ore bodies having various dip conditions.

3.4 CUT-AND-FILL MINING

3.4.1 Introduction

In the previous section, it was assumed that the rock mass properties were such that large openings could be created. Because of the way ores are emplaced, there are many instances where the ore and/or the wall rocks are weak, and hence both opening size and the allowable time between ore removal and filling the excavation are strictly limited.

There are a number of different extraction designs that can be applied in weak rock, all of which fit under the general category of cut-and-fill mining. It is a very versatile method and can be adapted to the extraction of any ore body shape. With some exceptions, all ore is removed via drifting, and the drifts created are then filled. As a result, mining costs are high compared to the other methods. On the other hand, recovery is high also, and dilution is generally low. Thus, it is an appropriate approach to the extraction of high-grade ore bodies.

3.4.2 Extraction Principles

For simplicity, an extraction block of the same type as described in the previous section will be assumed. Access will be via a ramp driven into the footwall, and mobile equipment will be used. Typically, the drifts used in mechanized cut-and fill mining are on the order of 5 m (16.4 ft) high.

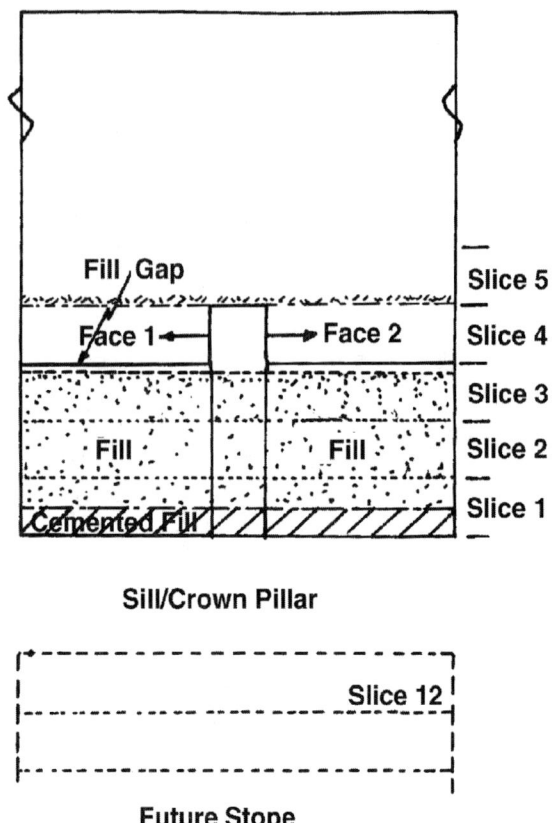

FIGURE 3.11   Diagrammatic representation of cut-and-fill mining

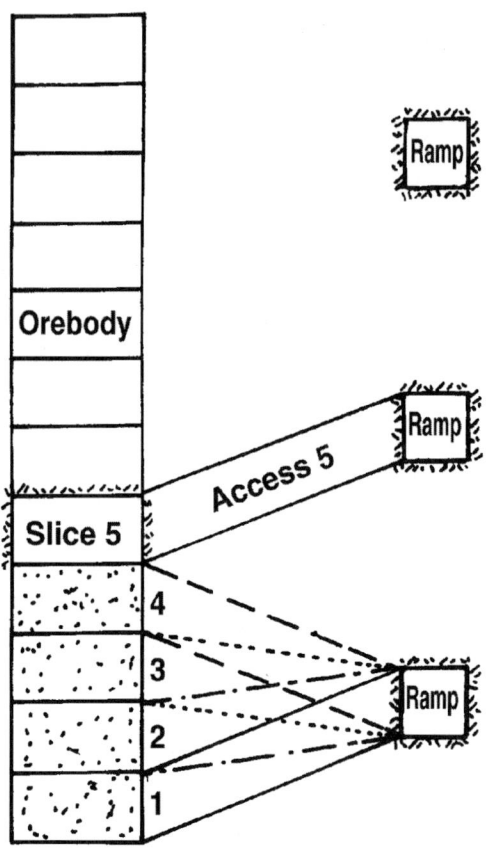

FIGURE 3.13   Ramp access for cut-and-fill mining

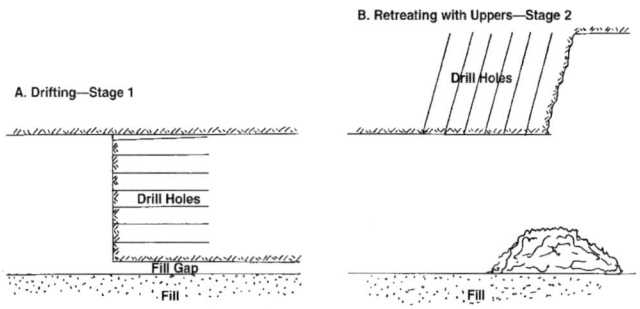

FIGURE 3.12   Drifting on advance and retreating with uppers

To begin the discussion, it will be assumed that the ore block to be extracted is vertical and has a width that can be removed during normal drifting. When ore body strength is fairly good, overhand cut-and-fill mining is normally applied (Figure 3.11). This means starting at the bottom of the block and working upward. Ideally, access to each level will be via crosscuts originating at the mid-length position of the block In this way, two headings can be operated at any one time.

Typical drift rounds consisting of drilling, blasting, loading, scaling, and installing rock reinforcement are used. This progression of operations can lead to delays unless carefully planned. Drilling of the second heading is carried out while the other operations are being done at heading 1. When the slice has been completed, filling is done. The fill is placed, leaving a small gap to the overlying ore (Figure 3.12A). On the next slice, this

gap forms the free surface for blasting. The process continues upward slice by slice to the top of the block. Several such extraction blocks can be in operation at any one time to meet production requirements. The horizontal pillar created between two such stacked extraction blocks is called the crown pillar with respect to the underlying stope and the sill pillar for the stope above. Normally the first cut of the extraction block above the sill pillar is filled with a cemented fill to facilitate later extraction of the pillar.

In some cases, the wall rock is strong enough to allow a double slice to be open at any given time. Here the first slice is mined by drifting and then, rather than filling directly, uppers (upward-oriented drill holes) are drilled the length of the slice (Figure 3.12B). Once drilling has been completed, the several rows of holes are charged and blasted, beginning at the ends of the extraction block and retreating toward the access. The ore is extracted by LHDs after each blast and transported to the ore pass. In this way, efficiency can be improved by changing the typical cycle to one in which all drilling is done first, followed by charging and loading. Then one lift can be filled, followed by drilling uppers, and so on, or both lifts can be filled, followed by drifting, followed by drilling of uppers.

Access to this one-drift-wide cut-and-fill stope is from an access ramp located in the footwall. Often four slices are accessed from a given point on the ramp. This is shown in Figure 3.13. In overhand cut-and-fill, crosscut 1 is made first. When the slice is completed, the roof of the crosscut is slashed down to form crosscut 2. This continues for the four slices, at which time a higher point on the ramp is selected as the origin of the crosscuts. Generally the maximum crosscut inclination is about 20°. This sets the position of the ramp with respect to the ore body.

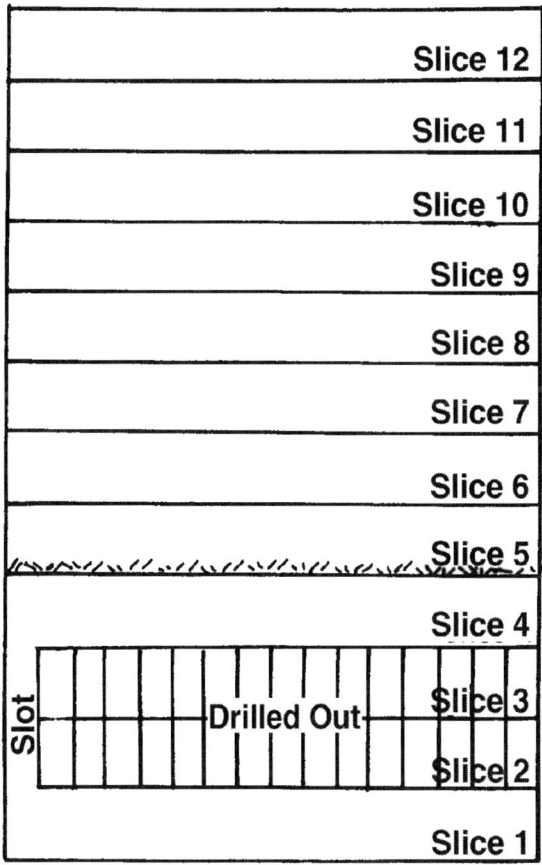

FIGURE 3.14   Initial geometry for rill or Avoca mining

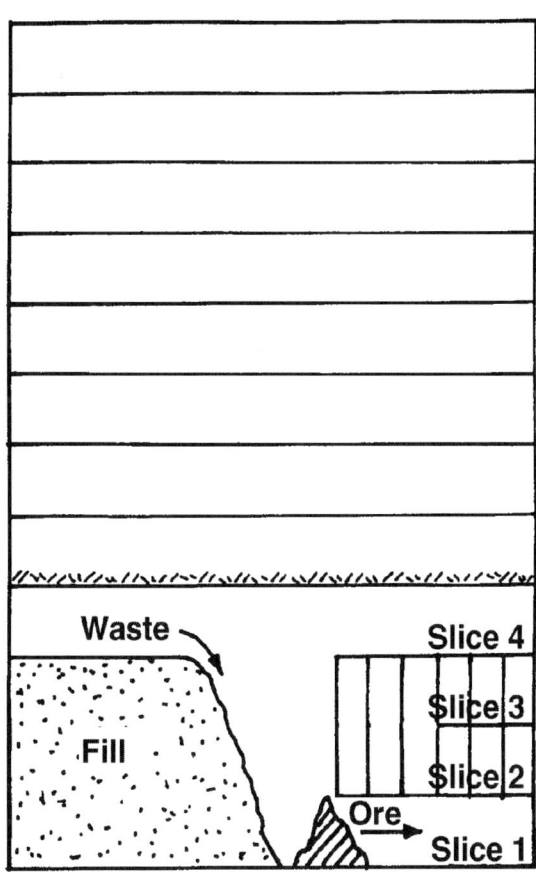

FIGURE 3.15   Extraction and filling operations

If the strengths of the wall rock and the ore are quite good, then spans of more then two lifts can be created. For example, Figure 3.14 shows the case in which slices 1 and 4 are extracted by drifting. Rows of vertical blastholes are then drilled from the floor of slice 4 to the roof of slice 1. A vertical slot is created, and the rows of holes are blasted one or more at a time toward the slot. The ore is extracted by an LHD operating in slice 1. At the same time that retreat extraction is underway, fill is being placed from the opposite end of the stope (Figure 3.15). A gap is maintained between the extraction and filling fronts to minimize dilution. When completed, slice 7 is removed by drifting (Figure 3.16). Slices 5 and 6 are now removed using slice 7 as the drilling level and slice 4 as the extraction level. The method is called rill mining or the Avoca method.

If the extraction block is quite wide, then the cut-and-fill method can still be used, but now several drifts are driven side by side (Figure 3.17). This is similar to room-and-pillar (rib pillar) mining where the rooms are filled first, and then the pillars are extracted.

One technique is to shape the primary extraction drifts so that they have downward slanting sides (Figure 3.17A, B). The drifts are then filled with uncemented fill or fills containing only a small amount of cement. With care, the secondary drifts shown in Figure 3.17B can be extracted with little dilution. These variations are called drift-and-fill mining or inclined wall mining. A more typical layout is shown in Figure 3.17C and D. Here straight walls are used, and every other drift is removed in a primary mining phase. Cemented fill is used to avoid dilution during removal of the interlying drifts. One option is to make the primary drifts narrow and the secondary drifts wide to minimize the need for cemented fill.

If the strength of the ore is very poor, then the underhand cut-and-fill method may be used (Figure 3.18). The first slice is taken, and then various techniques are used to prepare a layer that will become the roof when extracting the slice below. In the past, a timber floor pinned into the walls was the main technique. Today, it is more common to pour a layer of cemented fill with or without reinforcement. The remainder (upper portion) of the drift may be left open or filled with uncemented fill. The next slice is then extracted under the constructed roof. From the same access level, some mines use overhand cut-and-fill working upward from this level while employing underhand cut-and-fill working downward. This doubles the number of working faces in operation at any one time from a given level.

Wide ore bodies can also be mined using underhand cut-and-fill (Figure 3.19). The process is the same as described earlier, but now cold joints are allowed between the individual drift floor pours. Generally, miners try to avoid positioning drifts of the underlying layer directly under those above by shifting the drifts sideways or driving at an angle to the drifts above. In the latter case, a basket weave pattern results.

In the thick, inclined ore bodies appropriate for overhand cut-and-fill mining, vertical pillars are sometimes left to provide additional support between the hanging wall and the footwall (Figure 3.20). If possible, these pillars are located in the internal waste or low-grade areas. On the lowest slice, a room-and-pillar mine is created. The rooms are then filled. A second slice is then taken, continuing the vertical upward extension of the pillars. This level is then filled, and the process is repeated. The mining system is called post-pillar mining because the pillars appear as vertical posts surrounded by fill (see 3.1.9). Some authors include the method under room-and-pillar mining and others

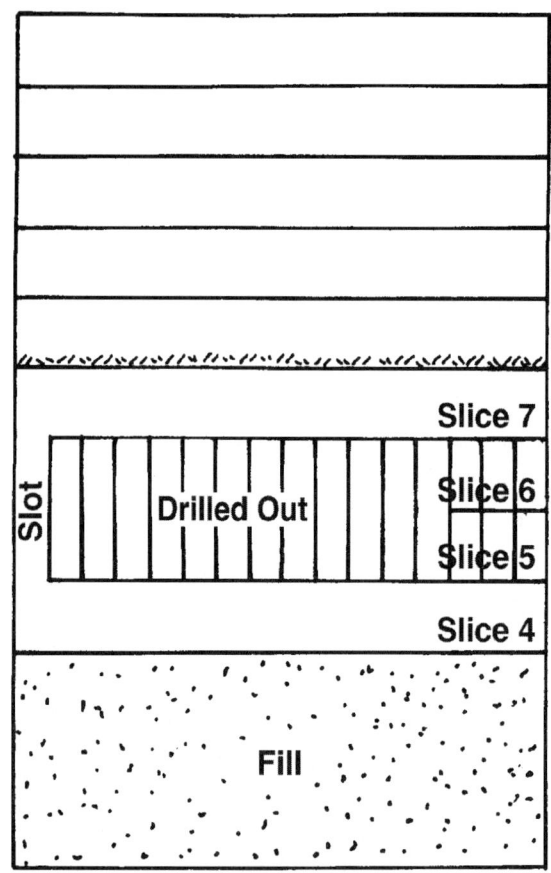

FIGURE 3.16   Initial geometry for next extraction phase

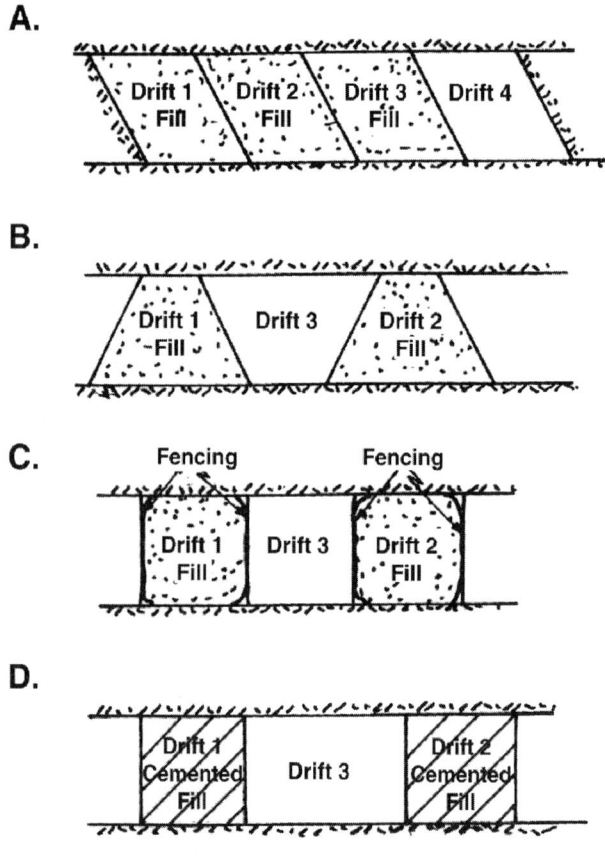

FIGURE 3.17   Different geometries for drift-and-fill mining

under cut-and-fill. Because of the presence of the surrounding fill, even very tall and slender pillars can be quite strong.

3.5   SUBLEVEL CAVING

3.5.1   Introduction

The origin of sublevel caving has been attributed to a scaling-up of the now largely extinct top-slicing method. In this latter method, a horizontal slice of ore was extracted from the top of an ore body of relatively large extent. A timber mat was constructed on the floor of the opening, and the timbers supporting the roof were either blasted or otherwise removed, allowing the overlying waste to fall down onto the timber mat. The miners would then mine beneath this timber mat and extract a second slice. The supporting timbers would be removed, the overlying waste would cave, and the process would be repeated.

The first modification of this system was to skip one drifting slice. Instead of extracting each slice by drifting, a layer of ore was left in the form of a roof between the horizontal slices. In retreating from the periphery of the ore body on the drifting slice, the ore in the slice above would be allowed to cave. It was then extracted from the drifts. In this way, miners would extract two slices from one development level. This retreating process would continue back to the main access, and a new development slice would be taken, leaving the slice to be caved in-between.

Since sublevels and ore caving were involved, it was logical to name the technique sublevel caving. A natural extension of this technique was to increase the number of slices removed by caving for every slice removed by drifting. Since a series of individual mining "blocks" were used to extract these large ore bodies, the method received the name block caving.

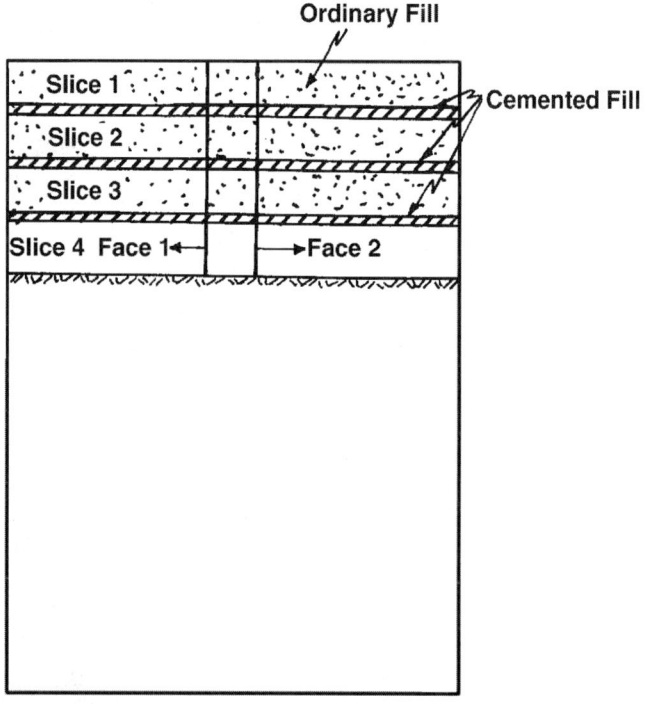

FIGURE 3.18   Underhand cut-and-fill mining

A. Staggered Cuts (Section View)

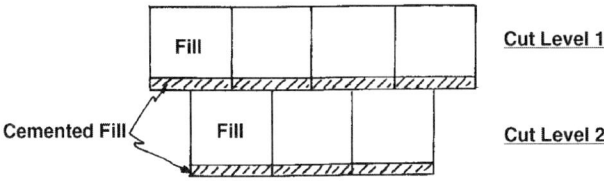

B. Aligned Cuts (Section View)

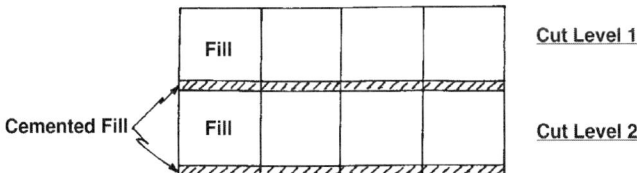

C. Basket Weave (Plan View)

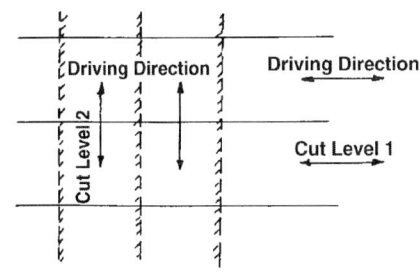

FIGURE 3.19   Alternate cut arrangements

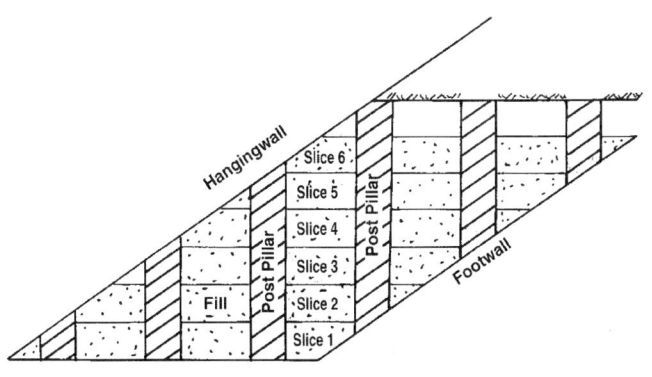

FIGURE 3.20   Post-pillar mining

Sublevel caving was initially applied in extracting the soft iron ores found in the iron ranges of Minnesota and Michigan. The sublevel caving as practiced today is significantly different from this early version and should probably be given another name, such as sublevel retreat stoping, continuous underhand sublevel stoping, or something similar that would better reflect the process. Sublevels are created at intervals of between 20 and 30 m (66 and 98 ft), beginning at the top of the ore body and working downward. On each sublevel, a series of parallel drifts are driven on a center-to-center spacing that is of the same order as the level spacing. From each sublevel drift, vertical or near-vertical fans of blast holes are drilled upward to the immediately overlying sublevels. The distance between fans (the burden) is on the order

of 2 to 3 m (6.5 to 10 ft). Beginning typically at the hanging wall, the fans are blasted one by one against the front-lying material, which consists of a mixture of ore from overlying slices and the waste making up the hanging wall and/or footwall. Extraction of the ore from the blasted slice continues until total dilution reaches a prescribed level. The next slice is then blasted, and the process continued. Depending on ore body geometry, the technique may be applied using transverse or longitudinal retreat.

Today, the sublevel caving technique is applied in hard, strong ore materials in which the hanging wall rocks cave readily. The key layout and design considerations are to achieve high recovery with an acceptable amount of dilution. The uncertainties of fragmentation and ore cavability present in panel caving (discussed in the following section) are removed because each ton of ore is drilled and blasted from the sublevels. The method has been used most for mining magnetic iron ores that can be easily and inexpensively separated from the waste. However, it has been and can be applied to a wide variety of other ore types.

3.5.2   Sublevel Caving Layout

As indicated, ore is recovered both through drifting and through stoping. Because the cost per ton for drifting is several times that for stoping, it is desirable to maximize stoping and minimize drifting. This has meant that through the years, the height of the sublevels has steadily increased until today they are up to 30 m (98 ft). Whereas approximately 25% of the total volume was removed by drifting in the early designs, today that value has dropped to about 6% in the largest-scale sublevel caving designs. The sublevel intervals have changed from 9 m up to nearly 30 m (30 to 98 ft). The key to this development has been the ability to drill longer, straighter, and larger-diameter holes.

Sublevel caving is an underhand method with all of the blastholes drilled upward. The ore moves down to the extraction and drilling drift under the action of gravity.

There are a number of factors that determine the design. The sublevel drifts typically have dimensions (W/H) of 5 by 4 m, 6 by 5 m, or 7 by 5 m (16 by 13, 20 by 16, or 23 by 16 ft) to accommodate LHDs. In the example used to illustrate the layout principles, it is assumed that the drift size is 7 by 5 m (23 by 15 ft). The largest possible blasthole diameter (from the viewpoint of drilling capacity and explosive charging) is normally chosen; today, this is 115 mm (4.5 in) based largely on the ability to charge and retain explosive in the hole. These large holes may be drilled using either in-the-hole (ITH) or tophammer machines. The large diameters and large drift sizes permit the use of tubular drill steel of relatively long lengths (thereby minimizing the number of joints and maximizing joint stiffness), so that the required long, straight holes can be produced. The largest ring designs incorporate holes with lengths up to 50 m (164 ft).

The distance between slices (burden B) depends both on hole diameter (D) and the explosive used. For initial design when using ANFO as the explosive, the relationship is B = 20D. For more energetic explosives (bulk strength basis), the relationship is B = 25D. Assuming that D equals 115 mm (4.5 in) and an emulsion explosive is used, B would equal about 3 m (10 m). Typically, the toe spacing- (S) to-burden ratio is about 1.3. Hence the maximum S would equal 4 m (13 ft) in this case. To achieve a relatively uniform distribution of explosive energy in the ring, the holes making up the ring would have different uncharged lengths. Both toe and collar priming initiation techniques are used.

The sublevel drift interval is decided largely on the ability to drill straight holes. In this example it will be assumed that the sublevel interval based upon drilling accuracy is 25 m (82 ft) (Figure 3.21). Once the sublevel interval has been decided, it is necessary to position the sublevel drifts. In this example, the drifts are placed so that the angle drawn from the upper corner of the extraction drift to the bottom center of drifts on the overlying

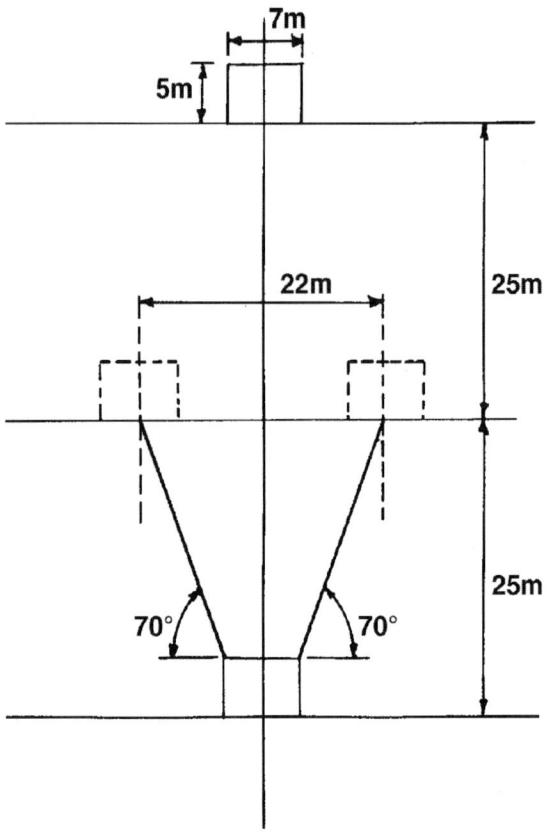

FIGURE 3.21   Initial design step for sublevel caving

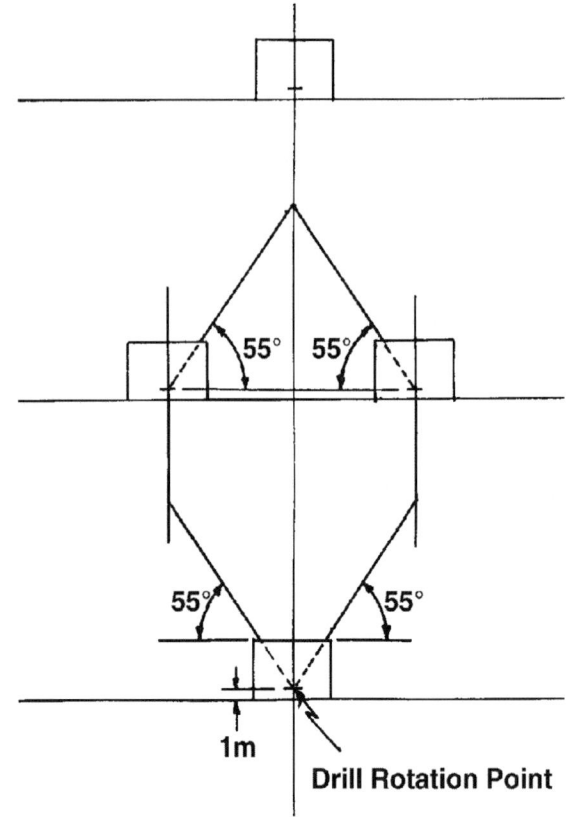

FIGURE 3.22   Addition of side hole locations to sublevel caving design

sublevel is 70°. This is approximately the minimum angle at which the material in the ring would move to the drawpoint. The resulting center-to-center spacing is 22 m (72 ft). A one-boom drill is assumed to drill all the holes in the ring. The rotation point is shown in Figure 3.22. The inclination of the side holes has been chosen as 55°, although holes somewhat flatter than this can be drilled and charged. The function of holes drilled flatter than 70° is largely (1) to crack the ore, which is then removed from the sublevel below and (2) to reduce the maximum drill hole length. Holes flatter than 45° are difficult to charge because of the angle of repose of the ore at the extraction front. In Figure 3.23, the locations of the individual drillholes are shown. A buffer zone 1 m (3.3 ft) wide has been left between the ends of the blastholes and the boundary to the overlying drifts and outer fan holes.

In Figure 3.24, an extraction ellipse has been superimposed. The layout is very similar to that obtained using the theory of bulk flow as described by Kvapil (1982, 1992). The fans may be drilled vertically or inclined from the horizontal at an angle, typically 70° to 80°. Inclining the fans improves brow stability and access for charging the holes. In the example, the inclination of the fans is 80° and the burden is 3 m (10 ft).

To initiate mining a new sublevel, an opening slot must be made toward which the fans can be blasted. Several techniques are used: blind hole boring and slashing, fan drilling using an increasing inclination angle until the production fan inclination is reached, creation of an opening slot longitudinally along the hanging wall, to name a few. In transverse sublevel caving, upon reaching the footwall, the inclination of the fans is sometimes steepened to permit recovery of additional ore and to minimize waste extraction.

In Figure 3.23 the importance of drilling precision is easily seen. If the forward or backward angular position caused by

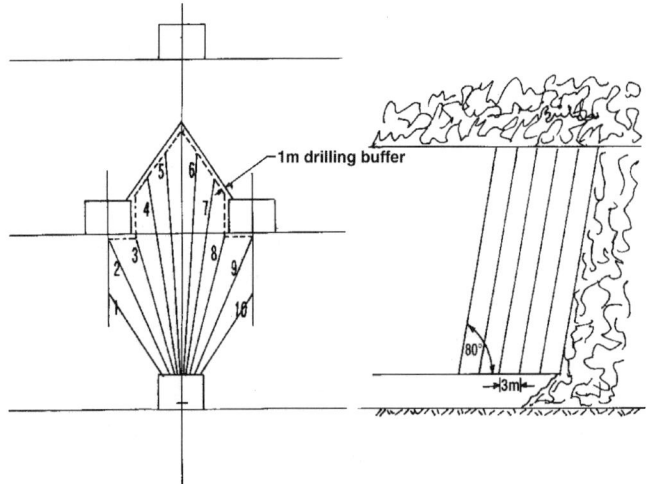

FIGURE 3.23   Final sublevel caving ring design

incorrect initial alignment or in-hole deviation exceeds 2°, the ends of the longest holes find themselves in the wrong ring. Side-to-side angular deviations can mean that the fragmentation is poor because of too little explosive concentration, dead-pressing of explosive, etc. Thus, careful drilling is of utmost importance for successful sublevel caving.

3.5.3   Recovery and Dilution

Sublevel caving lends itself to a very high degree of mechanization and automation. Each of the different unit

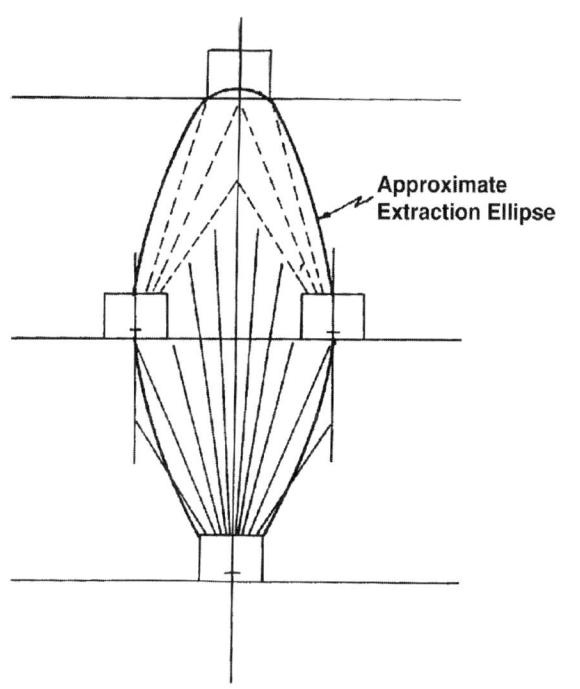

FIGURE 3.24 Approximate extraction ellipse superimposed

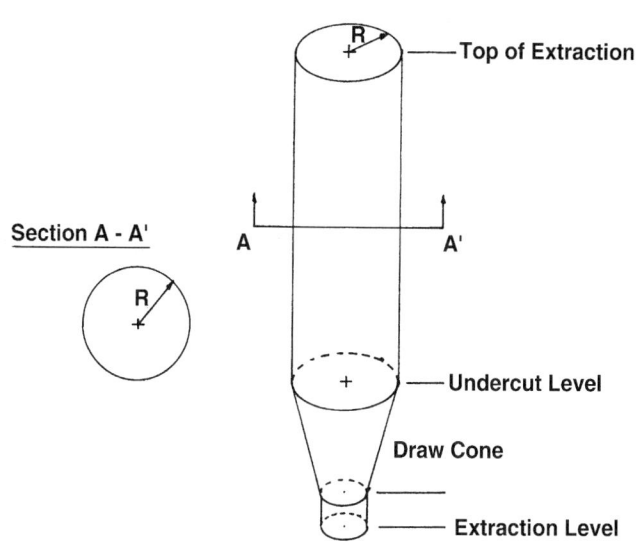

FIGURE 3.25 Diagram of flow

operations of drifting, production drilling, blasting, and extraction can be done largely without affecting another operation. Specialized equipment and techniques can and have been developed leading to a near-factorylike mining environment. As indicated earlier, because every tonne of ore is drilled and blasted, there are not the same uncertainties regarding cavability and fragmentation present with block caving. However, a very narrow slice of blasted ore surrounded by a mixture of waste and ore must be extracted with high recovery and a minimum of dilution. As can be easily visualized, the ore at the top part of the ring in the example is more than 40 m (130 ft) away from the extraction point, whereas the waste-ore mixture lies only the distance of the burden in front of the ring (on the order of 3 m [10 ft]). With care, recoveries on the order of 80% with dilution held below 25% can be achieved.

3.6 PANEL CAVING

3.6.1 Introduction

In this section, the term "panel caving" will be used to represent both "block caving," suggesting the mining of individual blocks, and "panel caving," indicating a laterally expanding extraction. There are a great number of variants of this system, and it is impossible to do them all justice in a very short discussion such as this. The intention is to provide the reader with an introduction to some of the more important layout considerations. The emphasis will be on development and extraction.

In panel caving, the three most important elements of the extraction system are the undercut level, which removes the support from the overlying rock column, the funnel through which the rock is transported downward to the extraction level, and the extraction level itself.

The basis for system design and performance is the degree of fragmentation present as the rock blocks enter the top of the funnel. The impact of fragmentation will be discussed in more detail as the section proceeds. An overview of the panel-block caving system has been presented in Chapter 1 of this book. In the early days of block caving, the materials were soft and caved

readily. Today the trend is to use cave mining on ever harder and tougher ores. The result is that an engineer must thoroughly evaluate the ore body and tailor the design so that a successful extraction will result. This is the least expensive of the mining systems as measured on an extracted-tonnage basis.

3.6.2 Extraction Level Layout

Assuming the use of LHDs, the major development on the extraction level consists of extraction drifts, drawpoints, and extraction troughs and bells. To simplify the discussion, it is assumed that all drifts have the same cross section.

Design is an iterative process, and it is always a question as to where design begins. In this case, it is with knowing or assuming the size of the material to be handled. The physical size of the loading equipment is related to the required scoop capacity, which, in turn, is related to the size of the material to be handled. If fragmentation is expected to be coarse, then a larger bucket size and a larger machine are required than if fragmentation is fine. Knowing the size of the machine, one arrives at a drift size. In sizing ore passes, it is expected that ore pass diameter should be three to five times the largest block size to avoid hang-ups. If this same rule is applied to the size of extraction openings, then the size of the extraction opening should be of the order of 5 to 7 m (16 to 23 ft) for block sizes with a maximum dimension of 1.5 m (5 ft). Depending upon density and shape, such a block would weigh 5 to 10 tonnes (5.5 to 11 tons). A large piece of equipment is required to be able to handle such blocks. It is typical for extraction drifts to be sized (W:H expressed in meters) according to the ratios 4:3, 5:4, and 6:5. For the machine in the example used here, drift size would be on the order of 5 by 4 m (16 by 13 ft) or larger.

To begin the design of the extraction level, a grid of extraction drifts that will accommodate the LHDs and the lines of associated drawpoints is created. The actual caving and draw behavior is quite complicated. The simplified geometry shown in Figure 3.25 is assumed to be representative. In practice, a series of circles of radius R corresponding to the draw radius of influence on the undercut level is drawn first. Figure 3.26 shows one such pattern for staggered coverage with the locations of the extraction and drawpoint drifts superimposed. It has been found that the value of R depends upon the degree of fragmentation. If fragmentation is coarse, the radius will be larger than if fragmentation is fine. This presents a design problem since in the

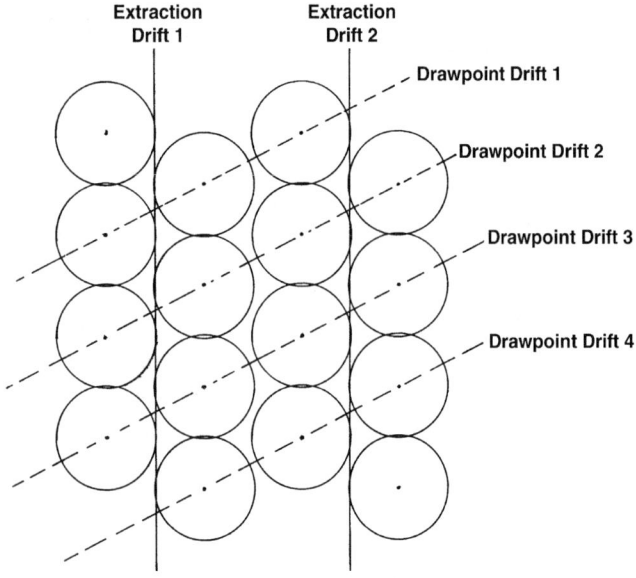

FIGURE 3.26   Initial panel caving design step with a staggered pattern

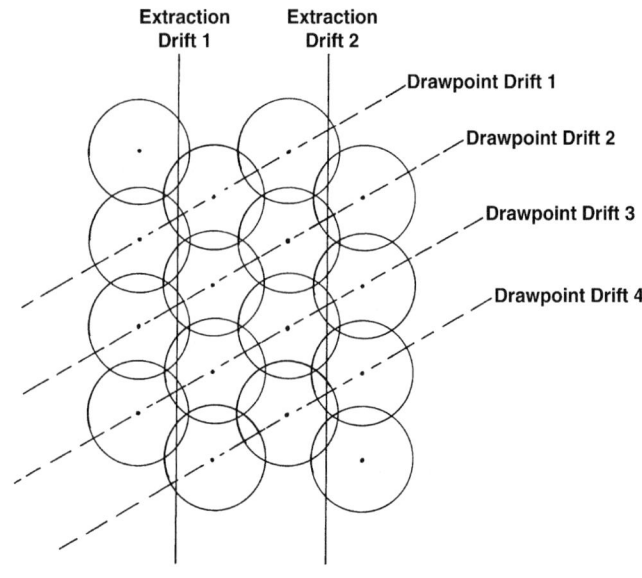

FIGURE 3.28   Full-coverage design

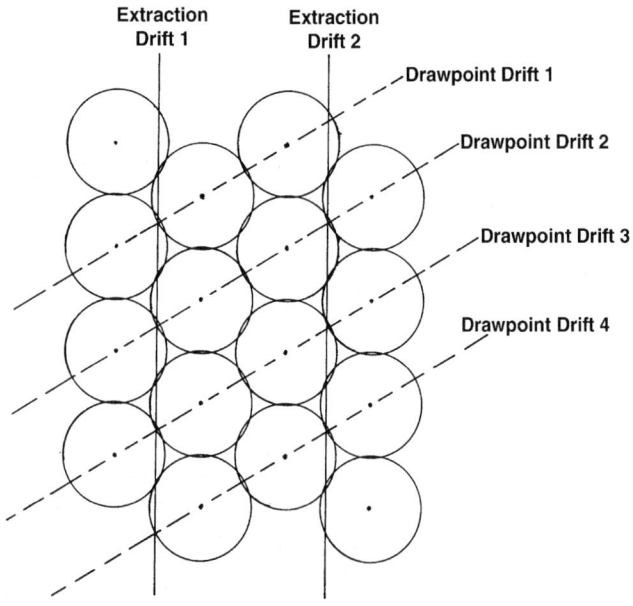

FIGURE 3.27   "Just-touching" design

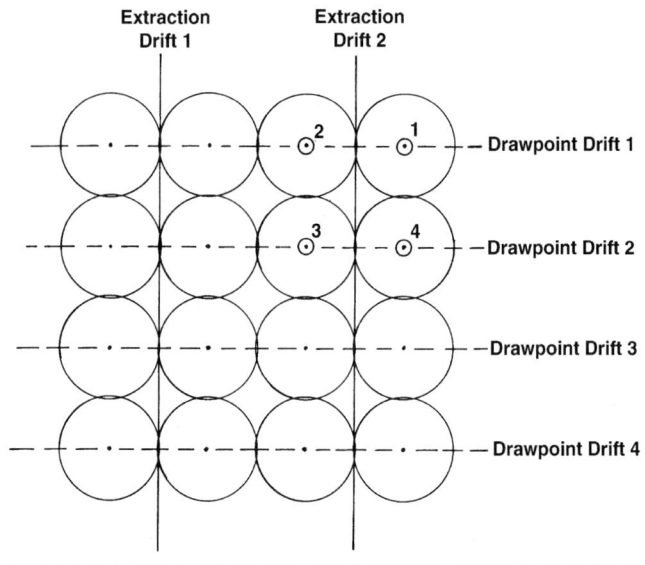

FIGURE 3.29   Square layout of extraction and drawpoint drifts

initial stages of draw, fragmentation will generally be larger than at later stages.

The degree of desired coverage is one of the design factors. The "just-touching" case is shown in Figure 3.27 and the "total coverage" case in Figure 3.28. In the example, it is assumed that R = 7.5 m (25 ft, and a square "just-touching" drawpoint pattern is used. Shown in Figure 3.29 are the locations of the extraction drifts and the drawpoint drifts on the extraction level. For extraction drift 2, drawpoints 1 and 2 are associated with drawpoint drift 1, whereas drawpoints 3 and 4 are associated with drawpoint drift 2.

The orientation of the drawpoint drift with respect to the extraction drift must then be decided. Figures 3.30 and 3.31 show two possibilities involving the use of a 45° angle. A careful

examination of these figures reveals that the choice affects both loading direction and the ease with which the openings can be driven.

A drawpoint entrance made at 60° to the axis of the extraction drift is very convenient angle from the loader operator's point of view. Some designs involve the use of 90° angles (square pattern). In this case, loading can be done from either direction. The 90° pillars provide good corner stability, but the loading operation is more difficult. When considering the different drawpoint design possibilities, LHD construction must be taken into account. It is important for the two parts of the LHD to be aligned when loading to avoid high maintenance costs and low machine availability.

As indicated, the design of the extraction level is made in response to the type of fragmentation expected. For coarse fragmentation, the openings have to be larger to permit

Extraction
Drift 1   Extraction
Drift 2

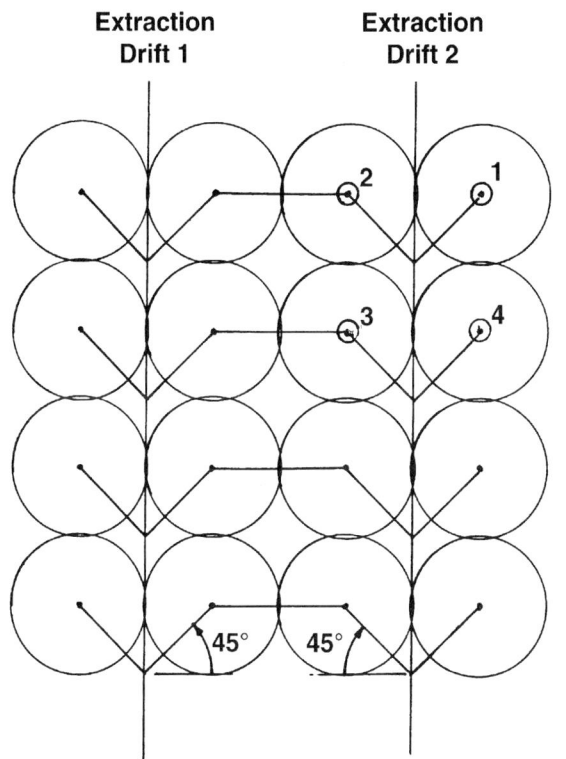

FIGURE 3.30   Herringbone pattern of drawpoint drifts

Extraction
Drift 1   Extraction
Drift 2

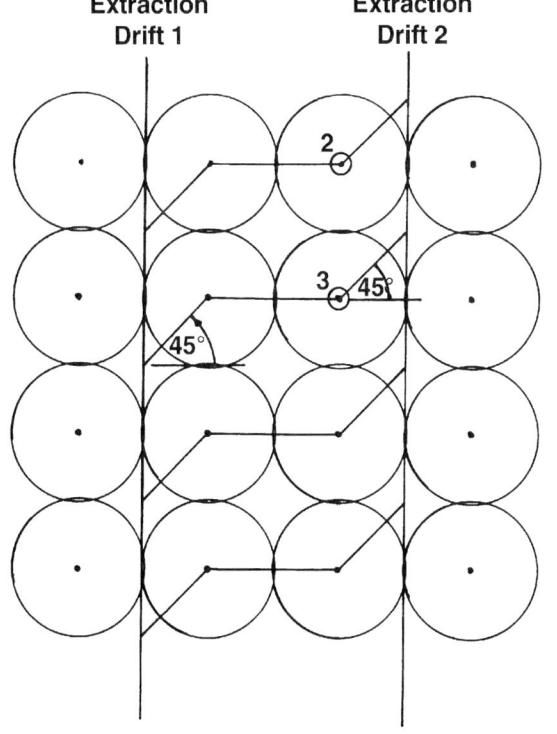

FIGURE 3.31   Flow-through pattern of drawpoint drifts

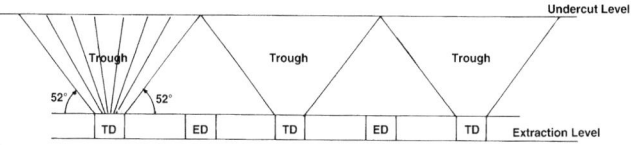

FIGURE 3.32   Single-level extraction and undercutting

extraction of the blocks. However, larger openings present possibilities for stability problems, and since these openings must last for the time required to extract the overlying column of ore, the design, creation, and reinforcement of the openings must be carefully made. Fortunately the type of rock in which one expects coarser fragmentation is also stronger, providing a better construction material. In softer rocks yielding a finer fragmentation, openings can be smaller. The need to protect the integrity of the openings is of highest importance. This will be discussed in more detail under the undercutting heading.

There are, as indicated, a great number of different design possibilities for the extraction level. All involve the basic components of fragmentation, radius of influence, draw coverage, machine size, and drift size examined roughly in that order

3.6.3   Undercutting and Formation of the Extraction Trough

In the undercutting process, a slice of ore forming the lower portion of the extraction column is mined. As the drilled and blasted material is removed, a horizontal cavity is formed beneath the overlying intact rock. Because of the presence of this free surface, subhorizontal side stresses, and the action of gravity, the intact rock undergoes a complex process involving loosening, crushing, and caving. The ease with which the intact rock transforms into a mass of fragments is reflected in its characteristic "cavability."

One approach to addressing a material's cavability is to describe the size and the shape of the area that must be undercut to promote caving. The other, and more important, part of cavability is the description of fragment size distribution. This is much more difficult to predict, but ultimately of more importance from a design viewpoint.

The simplest design is to combine undercutting and the trough-formation process into a single step. As described in the previous section, a series of parallel extraction drifts is driven. The center-to-center spacing of these drifts is determined by the size of the influence circles. In this example, the plan layout of Figure 3.29 is used. The center-to-center spacing of the extraction drifts is 30 m (100 ft) (4R). A series of parallel trough drifts is driven between the extraction drifts. Starting at the far end of the extraction block, fans of holes are drilled and then blasted toward opening slots. In the case shown in Figure 3.32, the side angles of the fans have been chosen as 52°, and the resulting vertical distance between the extraction level and the top of the undercut is 20 m (66 ft).

Note that the trough drifts and the troughs can be created either before or after driving the extraction drifts. The latter case would be termed "advance" or preundercutting. An advantage with this design is that all the development is done from one level. An example of the use of this design has been presented by Weiss (1981).

Most mining companies using panel caving have separate undercut and extraction levels. Figure 3.33 shows the same cross section as shown in Figure 3.32, but now a separate undercut is constructed. As seen in Figure 3.34, the undercut level has been designed as a rib pillar mine. The rooms are 5 by 4 m (16 by 13 ft), and the room center-to-center spacing is 15 m (50 ft). In step 2 of this design, the interlying pillars are drilled and blasted. In step 3,

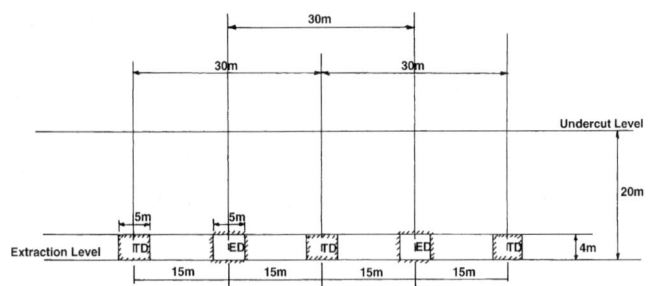

FIGURE 3.33   Separate level undercutting and extraction

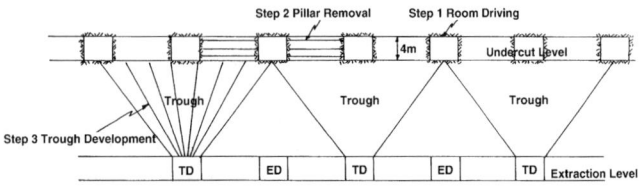

FIGURE 3.34   Undercutting and trough development

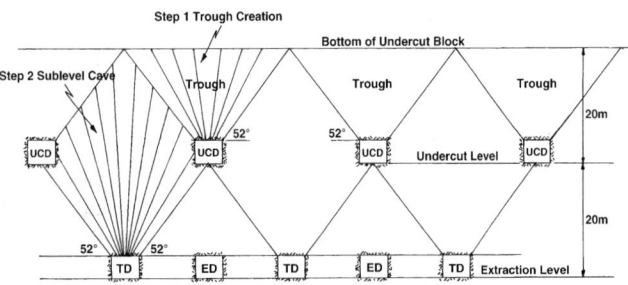

FIGURE 3.35   Two-level undercutting

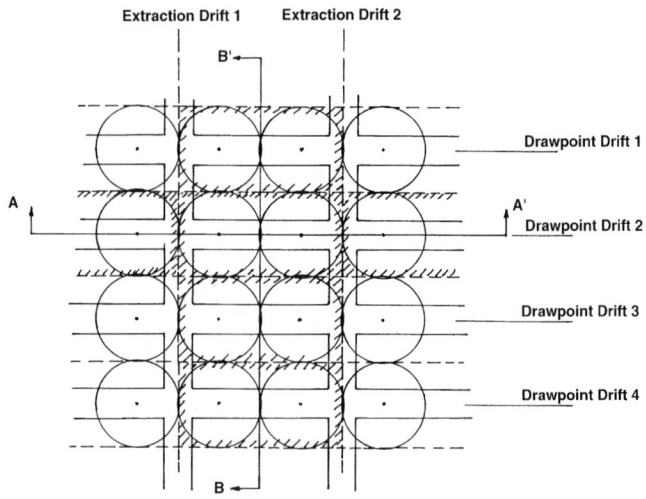

FIGURE 3.36   Plan view of traditional bell development

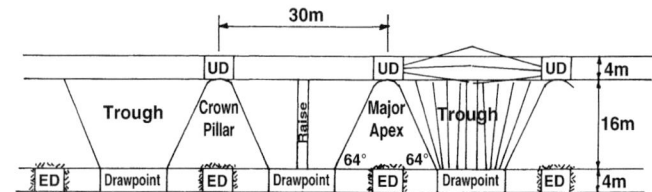

FIGURE 3.37   Section view of bell development

the extraction troughs are created to complete the undercut-trough development. It is possible and often desirable to develop the undercut level first and then do the development on the extraction level.

Figure 3.35 is an alternative design for the same basic extraction level layout. A separate undercut level is used with the undercut drifts spaced on 30-m (100-ft) centers. From these drifts, fans of holes are drilled to form a trough. The angle of the side holes is 52°. As can be seen, the undercut drifts are positioned directly above the underlying extraction drifts. Once the undercut has been created, a sublevel caving type of fan pattern is drilled from the trough drifts on the extraction level. This completes the development. The total height of the undercut in this case is 40 m (132 ft), which has some advantages in the caving of harder rock types.

Figures 3.36 and 3.37 are the plan and section views of a more traditional undercutting and bell layout for panel caving. In the previous examples, an extraction trough is used primarily to demonstrate the principles involved. A trough has the advantage of simplicity of construction, but the disadvantage that additional rock is extracted during the development process. This rock, if left in place, could provide extra stability to both the extraction drifts and the drawpoints. Drawbells are created rather than troughs.

The first step in drawbell construction is driving a drawpoint drift connecting adjacent extraction drifts. A raise is driven from this drift up to the undercut level. Fans of drillholes are then drilled from the drawpoint drift around the opening raise to form the bottom of the drawbell. Fans of holes are also drilled from the undercut drifts to complete the bell formation. A disadvantage with this design is that the amount of development and the level of workmanship required is higher than if the trough design is used. As a result, it is more difficult to automate.

In all designs, it is important that a complete undercut be accomplished. If this is not done, then very high stresses can be transmitted from the extraction block to the extraction level, causing major damage. Traditionally, the extraction level has been prepared first, followed by creation of the undercut and completion of the drawbells. This procedure does have a number of advantages.

Unfortunately, very high near-vertical stresses are created just ahead of the leading edge of the undercut. These stresses are transmitted through the pillars to the extraction level and can induce heavy damage to the newly completed level. The result is that repairs must be made before production can begin. The concrete used for making the repairs is generally many times weaker than the rock that has been broken, and structural strength can never be completely restored.

An alternative to this procedure is to create the undercut first (advance undercutting), thereby cutting off vertical stress. The extraction level is then created under this stress umbrella. Where this has been done, conditions on the extraction level have been markedly improved over those in which undercutting has been done afterward. There are pros and cons with both techniques, but advance undercutting will be the way of the future for most mines.

3.6.4 Size of Block

The size of the block refers both to the height of the extracted column and to the plan area. In the early days of block caving, the height of the blocks was on the order of 30 to 50 m (98 to 164 ft). Over time, this height has progressed to the point where extraction heights of several hundred meters are being used or planned. Obviously, as the specific development is inversely proportional to the height of the block, there are pressures to make the extraction units as high as possible.

Naturally, there are limits imposed by ore body geometry, mineral types, etc. There are also limits imposed by the life of the extraction points. If the reasonable life of the extraction point is, for example, 100,000 tonnes (110,000 tons), there is no point in selecting a block height yielding 200,000 tonnes (220,000 tons) per drawpoint. Drawpoints can and are rebuilt, but it is best if they can last the life of the draw.

As indicated in the introduction, most caving today is done in the form of panel caving rather than the caving of individual blocks. Once the initial cave is started, lateral dimensions are expanded. Cavability is an issue affecting the size of the undercut that must be created to get a sustainable cave. Relationships have been developed relating rock mass characteristics, hydraulic radius (area/perimeter), and ease of caving.

It is possible, unfortunately, to have initial caving followed by the formation of a stable arch. The undercut area must then be expanded and/or other techniques, such as boundary weakening, must be used to get the cave started once again. With a large enough undercut area, caving can be induced in any rock mass. Although necessary, it is not sufficient for successful block caving.

The other factor is the degree of fragmentation that results. As the method is being considered for application to ever stronger rock types, both of these factors, cavability and fragmentation distribution, must be satisfactorily addressed prior to selection of any method. Unfortunately, the database upon which such decisions are made is very limited.

3.6.5 Cave Management

Cave management refers to keeping control over how much is extracted from each drawpoint each day. It involves a number of different factors. The rate of draw is an important parameter in planning the required area under exploitation. As loosening of the fragments appears to be a time-dependent process, this must be recognized in planning the draw. The rate must not be so rapid that a large gap is formed between the top of the cave and the bottom of the block. A sudden collapse of the rock above can result in disastrous air blasts. In high-stress fields, it has been observed that too rapid a draw can result in the creation of rock-bursting conditions.

In section, there is a zone in which the height of the column under draw increases from near zero (where extraction is just beginning) to the full column height. This is followed by a zone in which the height of the ore column decreases to near zero where extraction is complete. It is important to maintain the proper height of draw versus distance slopes in these two sections to avoid the early introduction of waste from above. Poor cave management can also mean the build-up of high loads in various areas and subsequent stability problems. Typical rates of draw as taken from the available literature are on the order of 0.3 to 0.6 m/d (1 to 2 ft/d).

The proper sequencing of undercut and extraction is a very important aspect of cave management. Unfortunately, design guidelines are difficult to obtain from the literature in this regard.

An important design consideration for the extraction level is the means by which oversized material will be handled. There are a number of different problems to be addressed. The first concern is management of true hang-ups at the extraction points. Sometimes these can simply be blasted down by the careful placement of explosives. At other times, boulders must be drilled first. This is not a simple procedure and involves dangers to men and machines. The second concern is where and how to handle the "movable" oversize. These blocks can be (1) handled at the extraction points, (2) moved to a special gallery for blasting, (3) moved to an ore pass equipped with a grizzly and handled there, or (4) directly dumped into an ore pass for later handling. All variations are used, and each company has its own philosophy in this regard.

Initially, the sizes of the blocks arriving at the drawpoints are the result of natural jointing, bedding, and other weakness planes. As the blocks separate from the parent rock mass, they displace and rotate within the loose volume occupying a larger volume than the intact rock. The swell volume is extracted from the extraction points, thereby providing expansion room for the overlying intact rock. Loosening eventually encompasses the entire column.

As the column is withdrawn, the individual blocks abrade and split, resulting in finer fragmentation than that in the early part of the draw. The initial fragmentation, corresponding to that resulting from initial fractures in the rock, has been termed primary fragmentation. As the column moves downward and new breakage occurs, the resulting fragmentation has been termed secondary fragmentation. Data concerning this transition from primary to secondary fragmentation are very difficult to obtain.

3.7 SUMMARY

This chapter has presented some of the design and layout aspects of the major mining systems used in underground mining. With this background, it is hoped that the reader will better understand the mining systems described in some detail in the case studies portion of this book.

3.8 REFERENCES

Alexander, C. 1999. Tunnel Boring at Stillwater's East Boulder Project. *Mining Engineering,* Vol. 51, No. 9, pp. 15–24.

Brackebusch, F.W. 1992. Cut-and-Fill Stoping. Chapter 19.1, *Mining Engineering Handbook*, H.L. Hartman, ed. SME-AIME, New York, pp. 1743–1748.

Bullock, R.L. 1961. Fundamental Research on Burncut Drift Rounds. *The Explosive Engineer,* Vol. 1 and Vol. 2, Jan. and March.

Bullock, R.L. 1975. *Optimizing Underground, Trackless Loading and Hauling Systems.* Doctor of Engineering Dissertation, University of Missouri-Rolla, Rolla, Missouri, 132 pp.

Bullock, R.L. 1982. General Mine Planning. Section 1, Chapter 1, Subsection 7, *Underground Mining Methods Handbook,* W.A. Hustrulid, ed. SME, Littleton, CO, pp. 113–154.

Bullock , R.L. 1994. Underground Hard Rock Mechanical Mining. *Mining Engineering,* Vol. 46, No. 11, November, pp. 1254–1258.

Casteel, L.W. 1973. Open Stopes Horizontal Deposits. *SME Mining Engineering Handbook*, A.B. Cummins and I.A. Given, eds. Vol. 1, Sec. 12, AIME, New York, pp. 12–123 to 12–135.

Cleland, R.S. and K.H. Singh. 1973. Development of Post-Pillar Mining at Falconbridge Nickel Mines, Ltd. *CIM Bulletin,* Vol. 66, No. 732, April.

Cokayne, E.W. 1982. Sublevel Caving. Chapter 1, *Underground Mining Methods Handbook,* W.A. Hustrulid, ed. SME, Littleton, CO, pp. 872–879.

Doepken, W.G. 1982. The Henderson Mine. Chapter 4, *Underground Mining Methods Handbook,* W.A. Hustrulid, ed. SME, Littleton, CO, pp. 990–997.

Gignac, L.P. 1978. *Hybrid Simulation of Underground Trackless Equipment.* Doctor of Engineering Thesis, University of Missouri-Rolla, Rolla, Missouri, 119 pp.

Gould, J.C. 1982. Climax Panel Caving and Extraction System. Chapter 2, *Underground Mining Methods Handbook*, W.A. Hustrulid, ed. SME, Littleton, CO, pp. 973–981.

Haptonstall, J. 1992. Shrinkage Stoping. Chapter 18.3, *Mining Engineering Handbook*, H.L. Hartman, ed. SME-AIME, New York, pp. 1712–1715.

Hartman, H.L., et al. 1997. *Mine Ventilation and Air Conditioning,* 3rd ed. John Wiley and Sons, New York, 730 pp.

Haycocks, C. 1973. Sublevel Caving. Section 12.15, *Mining Engineering Handbook*, A.B. Cummins and I.A. Given, eds. SME-AIME, New York, pp. 12–222 to 12–228.

Haycocks, C. 1973. Sublevel Stoping. Section 12.12, *Mining Engineering Handbook*, A.B. Cummins and I.A. Given, eds. SME-AIME, New York, pp. 12–140 to 12–147.

Haycocks, C. 1973 . Minor Stoping Systems—Top Slicing, Breast Stoping, Underhand and Overhand Stoping. Section 12.13, *Mining Engineering Handbook*, A.B. Cummins and I.A. Given, eds. SME-AIME, New York, pp. 12–150 to 12–161.

Haycocks, C., and R.C. Aelick. 1992. Sublevel Stoping. Chapter 18.4, *Mining Engineering Handbook*, H.L. Hartman, ed. SME-AIME, New York, pp. 1717–1724.

Hedley, D.G.F., and F. Grant. 1972. Stope and Pillar Design for the Elliot Lake Uranium Mines. *CIM Bulletin,* Vol. 65, No. 723, July.

Hoover, A. 1973 . Shrinkage Stoping. Section 12.11, *Mining Engineering Handbook*, A.B. Cummins and I.A. Given, eds. SME-AIME, New York, pp. 12–135 to 12–138.

Hopler, R.B., ed. 1998. *Blasters Handbook.* International Society of Explosive Engineers, pp. 455–460.

Julin, D.E. 1992. Block Caving. Chapter 20.3, *Mining Engineering Handbook*, H.L. Hartman, ed. SME-AIME, New York, pp. 1815–1826.

Julin, D.E., and R.L. Tobie. 1973. Block Caving. Section 12.14, *Mining Engineering Handbook*, A.B. Cummins and I.A. Given, eds. SME-AIME, New York, pp. 12–162 to 12–167.

Kvapil, R. 1982. The Mechanics and Design of Sublevel Caving Systems. *Underground Mining Methods Handbook*, W.A. Hustrulid, ed. SME, Littleton, CO, pp. 880–897.

Kvapil, R. 1992. Sublevel Caving. Chapter 20.2, *Mining Engineering Handbook*, H.L. Hartman, ed. SME-AIME, New York, pp. 1789–1814.

Lacasse, M., and P. Legast. 1981. Change From Grizzly to LHD Extraction System. Chapter 10, *Design and Operation of Caving and Sublevel Stoping Mines*, D.R. Stewart, ed. SME-AIME, New York, pp.107–118.

Lane, W.L., et al. 1999. Pillar Extraction and Rock Mechanics at the Doe Run Company in Missouri 1991 to 1999. *Rock Mechanics for Industry, Proceedings of the 37th Rock Mechanics Symposium,* Vail, CO, June, pp. 285–292.

Langefors, U. and B. Kihlstrom. 1963. *The Modern Technique of Rock Blasting.* Almquist & Wiksell, Stockholm, pp. 230–257.

Lawrence, B.W. 1982. Considerations for Sublevel Open Stoping. *Underground Mining Methods Handbook*, W.A. Hustrulid, ed. SME, Littleton, CO, pp. 364–374.

MacMillan, P.W., and B.A. Ferguson. 1982. Principles of Stope Planning and Layout for Ground Control. *Underground Mining Methods Handbook*, W.A. Hustrulid, ed. SME, Littleton, CO, pp. 526–530.

Marklund, I. 1982. Vein Mining at LKAB, Malmberget, Sweden. *Underground Mining Methods Handbook*, W.A. Hustrulid, ed. SME, Littleton, CO, pp. 441–442.

Murray, J.W. 1982. Undercut-and Fill Mining. *Underground Mining Methods Handbook*, W.A. Hustrulid, ed. SME, Littleton, CO, pp. 631–638.

Osborne, K., and V. Baker. 1992. Vertical Crater Retreat Mining. Chapter 18.5, *Mining Engineering Handbook*, H.L. Hartman, ed. SME-AIME, New York, pp. 1732–1740.

Owen, K.C. 1981. Block Caving at Premier Mine. Chapter 15, *Design and Operation of Caving and Sublevel Stoping Mines*, D.R. Stewart, ed. SME-AIME, New York, pp. 177–187.

Ozdemir, L., 1990. Principles of Mechanical Rock Cutting and Boring. *Mechanical Excavation Short Course,* Colorado School of Mines, Golden, CO, p. 2.

Panek, L.A. 1981. Comparative Cavability Studies at Three Mines. Chapter 9, *Design and Operation of Caving and Sublevel Stoping Mines*, D.R. Stewart, ed. SME-AIME, New York, pp. 99–106.

Parker, J. 1973. How Convergence Measurements Can Save Money. Part 3 of Practical Rock Mechanics of the Miner. *Engineering & Mining Journal,* August, 1973, pp. 92–97.

Paroni, W.A. 1992. Excavation Techniques. Chapter 19.2, *Mining Engineering Handbook*, H.L. Hartman, ed. SME-AIME, New York. pp. 1749–1755.

Ramani, R.V. ed. 1997. *Proceedings of the 6th International Mine Ventilation Congress*, SME, Littleton, CO, 551 pp.

Richardson, M.P. 1981. Area of Draw Influence and Drawpoint Spacing for Block Caving Mines. Chapter 13, *Design and Operation of Caving and Sublevel Stoping Mines*, D.R. Stewart, ed. SME-AIME, New York, pp. 149–156.

Snyder, M.T. 1994. Boring for the Lower K. *Engineering and Mining Journal,* April, pp. 20ww–24ww.

Tien, J.C. ed. 1999. *Proceedings of the 8th US Mine Ventilation Symposium,* University of Missouri-Rolla, Rolla, Missouri, June, 1999, 732 pp.

Tobie, R.L., and D.E. Julin. 1982. Block Caving, Chapter 1. General Description. *Underground Mining Methods Handbook*, W.A. Hustrulid, ed. SME, Littleton, CO, pp. 967–972.

Tobie, R.L., Thomas, L.A., and H.H. Richards. 1982. Chapter 3, San Manuel Mine. *Underground Mining Methods Handbook*, W.A. Hustrulid, ed. SME, Littleton, CO, pp. 982–989.

Vera, S.G. 1981. Caving at Climax. Chapter 14, *Design and Operation of Caving and Sublevel Stoping Mines*, D.R. Stewart, ed. SME-AIME, New York, pp. 157–176.

Ward, M.H. 1981. Technical and Economical Considerations of the Block Caving Mine. Chapter 11, *Design and Operation of Caving and Sublevel Stoping Mines*, D.R. Stewart, ed. SME-AIME, New York, pp. 119–142.

Waterland, J.K. 1982. Introduction to Open Cut-and-Fill Stoping. *Underground Mining Methods Handbook*, W.A. Hustrulid, ed. SME, Littleton, CO, pp. 523–525.

Weakly, L.A. 1982. Room and Pillar Ground Control Utilizing the Grouted Reinforcing Bar System, *Underground Mining Methods Handbook,* W.A. Hustrulid, ed. SME, Littleton, CO, Section 7, Chapter 2, pp. 1556–1560.

Weigel, W. 1943. Channel Irons for Roof Control, *Engineering & Mining Journal,* May, Vol. 144, No. 5, pp. 70–72.

Weiss, P.F. 1981. Development System for Block Caving Under Severe Conditions. Chapter 12, *Design and Operation of Caving and Sublevel Stoping Mines*, D.R. Stewart, ed. SME-AIME, New York, pp. 143–146.

White, T.G. 1992. Hard-Rock Mining: Method Advantages and Disadvantages. Chapter 21.2, *Mining Engineering Handbook*, H.L. Hartman, ed. SME-AIME, New York, pp. 1843–1849.

Wycoff, B.T. 1950. Wrapping Pillars With Old Hoist Rope. *Trans. AIME,* New York, Vol. 187, pp. 898–902.

Zipf, R.K, and C. Mark. 1997. Design Methods to Control Violent Pillar Failures in Room-and-Pillar Mines. *Trans. Instn. Min. Metall. Sect. A: Mining Industry*, Vol. 106, Sept.–Dec., 1997, pp. A124–A132.

CHAPTER 4

Cost Estimating for Underground Mines

Scott A. Stebbins* and **Otto L. Schumacher†**

4.1 PRINCIPLES OF ESTIMATION

4.1.1 Introduction

Mine cost estimating is often referred to as an art. Unfortunately, this reference turns many would-be evaluators away from what is really a logical, straightforward process. Yet the description as art is an understandable misconception. Cost estimating, as with any predictive process, requires the evaluator to envision and quantify future events, or in other words, to be creative. This is a far better description; mine cost estimating is a *creative endeavor*. Fortunately, in mining, most of the information that must be predicted stems either from measurable entities, such as deposit configurations, or from well-understood, practiced engineering relationships. In actuality, mine cost estimating is a process of matching values obtained through simple engineering calculations with cost data, a process made even easier in recent years with the availability of printed and electronic information databases.

Perhaps, also, mine cost estimating is referred to as an art because no widely accepted, rigorous approach to the process exists. This is because, unlike other cost estimating tasks such as those used in construction, the process varies noticeably from one evaluation to the next, not only in approach but also in scope.

A complete estimate cannot be fully detailed in the few pages available. The following is presented primarily to minimize the intimidation felt by many geologists and engineers when they undertake a cost estimate. The basic premise here is that anything can be estimated. And the approach detailed is one in which complete listings of labor, supply, and equipment requirements are derived using information about the deposit and the proposed mine. These listings are then used in conjunction with documented salaries, wages, supply costs, and equipment prices to arrive at estimates of mine capital and operating costs. This method, most often referred to as an abbreviated itemized approach, is much easier than it might initially appear. Although there are several other methods available, including parametric equations, factoring, cost models, and scaling, itemizing has the advantage of providing thorough documentation of all assumptions and calculations used in the estimate. Consequently, the results are much easier to evaluate and adjust, and so they are more useful. Evaluators are often surprised to find that itemizing, since it relies on much of the same information required to do a proper job using the other methods, can be accomplished with some expedience.

Early in any mine cost estimate—long before the evaluator begins to worry about the cost of a load-haul-dump (LHD)—the scope of the evaluation must be determined. To accomplish this, the purpose of the estimate must first be defined. If used to determine which of several deposits should be retained for future exploration expenditures, then the estimate will be less thorough than one used to determine the final economic feasibility of a proposed mine or to obtain funding for development. Coincidently, the level of information available regarding deposit specifics also determines the level of the estimate. As the level of information increases, so does the scope of the estimate and the reliability of the results.

A note here regarding reliability and accuracy: Remember that accuracy is a measure of predicted (or measured) versus actual values. It can't really be quantified until well after the project is underway and the estimated costs can be compared with actual expenditures. Instead, cost estimators work more in terms of reliability, which is a measure of the confidence that can be placed in estimated costs. Reliability is determined by the level of effort involved in the evaluation and the extent of information available regarding the deposit. Simply put: The more information available (items such as the deposit boundary locations and the rock characteristics), the greater the reliability of the estimated costs. If an evaluator has a firm grasp on deposit specifics and works to estimate all costs associated with development and production, a highly reliable estimate should result.

Determining the potential economic success of developing a mineral deposit can be broken into four steps.

1. Design the underground workings to the extent necessary for cost estimating.

2. Calculate equipment, labor, and supply cost parameters associated with both preproduction development and daily operations.

3. Apply equipment costs, wages, salaries, and supply prices to the cost parameters to estimate associated mine capital and operating costs.

4. Examine the estimated costs with respect to anticipated revenues and economic conditions pertinent to the project (using discounted cash-flow techniques) to determine project viability.

4.1.2 Preliminary Mine Design

The purpose of the mine design, as it relates to estimating costs, is to determine equipment, labor, and supply requirements both for preproduction development and daily operations. The extent to which the mine is designed is important—the process is one of marginal propensity. The first 10% of the engineering involved in designing the mine probably provides data sufficient to estimate 90% of the costs. The more detailed final engineering aspects of mine design (such as those needed to provide adequate structural protection for the workers and properly ventilate the underground workings) very seldom have more than a minor impact on the overall mine costs.

---
* Aventurine Mine Cost Engineering, Elk, WA.
† Western Mine Engineering, Inc., Spokane, WA.

At this stage of the process, the key element is distance. In the preliminary design, the critical values are those associated with accessing the deposit, either by shaft, adit, or ramp. Most of the costs associated with preproduction development are directly tied to the excavations required to access the deposit. The length (or depth) of these excavations and their placement provide several cost parameters directly—the lengths of pipe, wire, rail, and ventilation tubing. These lengths also provide a more indirect path to estimating preproduction consumption values for several other items, including explosives, drill bits, rock bolts, shotcrete, and timber. And finally, they impact many subsequent calculations used to determine the sizes of pumps, ore haulers, hoists, and ventilation fans.

The method used to recover ore from the stopes is typically dictated by the configuration of the deposit and the structural nature of the ore, footwall, and hanging wall. Method selection is discussed in great detail in other sections of this book, so not much will be said here.

Development associated with the stopes is as important as the stoping method itself. Some basic calculations provide the lengths of drifts, crosscuts, ramps, and raises needed for each stope. After the amount of ore available in the stope is estimated, these lengths can be used to determine the daily advance rates necessary to achieve the desired daily production of ore.

In *Sherpa Cost Estimating Software for Underground Mines* (Stebbins 2000), stope development requirements are based upon simple stope models used in conjunction with the deposit dimensions. For example, the diagram in Figure 4.1 begins the process of stope design for the traditional cut-and-fill method. Next, the following relations (Example 4.1) are used to establish the design parameters for the stope.

Example 4.1   Stope Design Parameters

Stope alignment: Stopes are aligned along the strike of vein deposits.
Stope length: Maximum suggested stope length = 90 meters.
(1) If the deposit length is greater than 90 meters, then the suggested stope length is provided by—
 Deposit length ÷ rounded integer of (deposit length ÷ (maximum stope length × 0.75)).
(2) If deposit length is less than 90 meters, then the suggested stope length is equal to deposit length.
Stope width: Maximum suggested stope width is—
(Ore strength + footwall strength + hanging wall strength) ÷ **4,619,665 + 1.132**
 Where ore strength = ore compressive strength (psi) × ore rock quality designation (%), footwall strength = footwall compressive strength (psi) × footwall rock quality designation (%),
 and hanging wall strength = hanging wall compressive strength (psi) × hanging wall rock quality designation (%).
(1) If the adjusted deposit width is greater than the maximum stope width, then the suggested stope width is provided by—
 Adjusted deposit width ÷ rounded integer of (adjusted deposit width ÷ (maximum stope width ÷ 0.75)).
Where adjusted deposit width = deposit width ÷ sine (deposit dip).
(2) If the adjusted deposit width is less than the maximum stope width, then the suggested stope width is provided by—
 Deposit width ÷ sine (deposit dip).

Stope height: Maximum suggested stope height = 60 meters.
(1) If the adjusted deposit height is greater than 60 meters, then the suggested stope height is provided by—
 Adjusted deposit height ÷ rounded integer of (adjusted deposit height ÷ (maximum stope height × 0.75)),
Where adjusted deposit height = deposit height × sine (deposit dip).
(2) If the adjusted deposit height is less than 60 meters, then the suggested stope height is provided by—
 Deposit height × sine (deposit dip).
Face width: Maximum suggested face width = 5 meters.
(1) If the stope width is greater than 5 meters, then the suggested face width is provided by—
 Stope width ÷ rounded integer of (stope width ÷ (maximum face width × 0.75)).
(2) If the stope width is less than 5 meters, then the face width is—
 Equal to stope width.
Face height: Initially, face height is estimated by—
 (0.364 × face width) + 1.273.
(1) If this value is less than 2 meters, then the estimated face height is set to 2 meters.
(2) If this value is greater than 3 meters, then the estimated face height is set to 3 meters.
Actual face height is then provided by—
 Stope height ÷ rounded integer of (stope height ÷ estimated face height).
Advance per round:
(1) If face width is greater than face height, then the suggested advance per round is provided by—
 0.85 × face height.
(2) If face width is less than face height, then the suggested advance per round is provided by—
 0.85 × face width.
Pillar length (sill pillar): Pillar length = stope length.
Pillar width (sill pillar): Pillar width = stope width.
Pillar height (sill pillar): 2 × face height.
Development requirements(advance per stope):*
(1) Haulage drifts:
 Length = stope length
 Advance per day = length × (daily production from stope ÷ total metric tons in stope)
 Location = footwall
(2) Haulage crosscuts:
 Length = 30 meters
 Advance per day = length × (daily production from stope ÷ total metric tons in stope)
 Location = footwall
(3) Stope access raises:
 Length = (stope height ÷ sine(deposit dip)) − haulage crosscut height
 Advance per day = length × (daily production from stope ÷ total metric tons in stope)
 Location = ore
(4) Initial cut:
 Length = stope length
 Advance per day = length × (daily production from stope ÷ total metric tons in stope)
 Location = ore

*The cut-and-fill model applies to vein deposits dipping more than 60°.

Using the above relations, the stope design can be applied to a multitude of deposit configurations. Similar relations can be developed for any stoping method, thus enabling the evaluator to determine the development requirements of the stope and, as a consequence, the pertinent cost parameters.

Sketching out (usually with a three-view drawing) the deposit access headings, stopes, and underground excavations (shops, pump stations, lunch stations, hoist rooms, etc.) provides most of the preliminary mine design information needed for a cost estimate. Remember that only the lengths of the excavations are needed at this point. Values determined while calculating the subsequent cost parameters provide the information needed to define the cross sections of these openings.

4.1.3 Cost Parameters

The process of defining the parameters necessary for a cost estimate is a wonderful (perhaps only for an engineer) progression of simple mathematical calculations, with one value usually leading to and interconnecting with the next. These calculations branch in ways that create many logical paths to a complete compilation of the parameters, but all paths do eventually lead to such a compilation. In the next few pages, one generalized path is illustrated. Please be aware, however, that most of the procedures are interchangeable and many paths exist. Also, it is not the intent here to work through a complete estimate, step by step, since such an example would apply only to a finite number of deposit types and would not fit in the allowed space. The intent instead is to provide insight into the estimation process and remove any mystery that might create a hesitation to proceed. Successful estimators do not necessarily possess skills or knowledge in excess of others; but they do show a willingness to use their imagination to suggest values for as-yet unknown parameters.

Most (if not all) parameters required for a cost estimate fall into one of three categories: labor, supplies, and equipment. These categories represent the items that cost money, that is, the items for which the checks must be written. Consequently, all the work in this phase of the estimate is geared toward specifying the workers, supplies, and equipment necessary to mine the deposit. And one of the keys to specifying these items lies in determining how much time (how many hours) it takes to perform the individual tasks of mining.

Most operations in an underground mine are either cyclic or continuous in nature, and most are designed to displace material—ore, waste, air, water, workers, supplies, etc. The operations that are not cyclical or continuous (such as equipment repair or opening maintenance) are typically in place solely to service ones that are.

The rate at which ore is produced provides a good place to start when the cost parameters are defined. It's based upon the desired life of the mine and the size of the resource. Resource size is a known (or has at least been approximated, hence the evaluation) and the following relationship—known as *Taylor's Rule*—is often used to pinpoint a possible mine life.

Mine life (years) = $0.2 \times \sqrt[4]{}$ resource size (metric tons)

With values for the mine life and the resource known, the daily production rate is determined by—

Production rate (metric tons/day) = resource (metric tons) ÷ (mine life (years) × operating schedule (days/year))

Of course, many factors influence the rate of ore production (such as market conditions, deposit configuration, and profit maximization), so this derived rate may need adjustment. Once the production rate is determined, it can be used in conjunction with haul distances (gleaned from mine design) to estimate the required equipment capacities—those of the machines used to collect ore in the stopes, transport it through crosscuts and drifts,

and then finally haul it (through an adit, ramp, or shaft) to the surface. The dimensions of these machines (rear-dump haulers, LHD machines, rail cars, conveyors, or hoists) provide the basis for determining the cross-sectional area of all the openings through which the machines must travel. The height and width of the openings, when applied to the density of the rock, provide the amount (in metric tons) of rock that must be removed during excavation and, subsequently, the amount of explosives needed to liberate the rock. Significantly, when one design parameter is determined, its value usually provides the information needed to determine many more.

With hauler size in hand, company literature (often available at their websites) can be referenced to determine machine speeds in relation to various haul conditions. With those speeds, the operating time required to meet production can be calculated; and with that, the necessary number of machines and workers can be estimated.

The basis for determining most of these values is provided through cycle-time calculations (Example 4.2), one of the more important concepts of any cost estimate. Cycle-time calculations are used whenever the number of machines required to perform a cyclic operation must be determined.

Example 4.2—Cycle Time Calculations

Consider a case where a 20-metric-ton capacity, articulated rear-dump truck hauls ore to the surface. Ore is placed in the truck by a 6.1-cubic-meter capacity, remote LHD near the entrance of the stope. The truck hauls the ore 550 meters along a nearly level drift, then 1,450 meters up a 10% gradient to the surface. Once on the surface, the truck travels another 200 meters to the mill, where the ore is dumped into a crusher feed bin.

First, the speeds of the machine must be estimated over the various segments of the haul route. Using information from technical manuals supplied by equipment manufacturers (in this case the Wagner Mining Equipment manual) approximate speeds over the following gradients are—

- Up a 10% gradient, loaded @ 6.4 kilometer/hour
- Over a level gradient, loaded @ 16.1 kilometer/hour
- Over a level gradient, empty @ 20.3 kilometer/hour
- Down a 10% gradient, empty @ 15.8 kilometer/hour

These values account for a rolling resistance equivalent to a 3% gradient. The estimated time for this haul profile is then—

HAUL

550 meters ÷ (16.1 kilometers/hour × (1,000 meters/ kilometer) × 60 minutes/hour = 2.05 minutes

1,450 meters ÷ (6.4 kilometers/hour × 1,000 meters/ kilometer) × 60 minutes/hour = 13.59 minutes

200 meters ÷ (16.1 kilometers/hour × 1,000 meters/ kilometer) × 60 minutes/hour = 1.04 minutes

RETURN

550 meters ÷ (20.3 kilometers/hour × 1,000 meters/ kilometer) × 60 minutes/hour = 1.63 minutes

1,450 meters ÷ (15.8 kilometers/hour × 1,000 meters/ kilometer) × 60 minutes/hour = 5.51 minutes

200 meters ÷ (20.3 kilometers/hour × 1,000 meters/ kilometer) × 60 minutes/hour = 0.59 minutes

TOTAL TRAVEL TIME = 24.41 minutes

Evaluators may wish to further tune the above estimate by accounting for items such as altitude duration, acceleration, and deceleration. However, the effort spent should be proportionate to the purpose of the estimate and the reliability of the available information. Specifically, if acceleration and deceleration increase the overall cycle time by half a minute (or 2%), but the mill has not been firmly sited and the overall haul distance may yet change by as much as 10%, then the effort spent fine-tuning the estimate would be futile because it would not increase the reliability of the results.

In addition to traveling, the truck expends time as it's loaded, as it dumps its load, and as it maneuvers into position for each of these tasks. In this example, the cycle time for the LHD must also be estimated to figure the time needed to load the hauler. Assume that the volume capacity of the LHD is 2.7 cubic meters and the weight capacity is 5.44 metric tons. On any given load, the bucket is typically 85% full, and the material in its blasted condition weighs 2.85 metric tons per cubic meter. If the round trip from the dump point to the stope and back takes 2.5 minutes and the LHD takes 0.4 minutes to dump the load, then the following series of calculations provides the time necessary to load the truck.

First, the volume capacity of the LHD is examined.

2.7 cubic meters/load × 2.85 metric tons/cubic meter × 0.85 = 6.54 metric tons.

Since the weight capacity of the LHD is 5.44 metric tons, the load is limited by weight. Consequently, the time required to load the truck is—

20 metric tons ÷ 5.44 metric tons/load = 3.68 loads.
3.68 loads × 2.9 minutes/load = 10.67 minutes to load truck.

Assuming that the truck dump mechanism takes 0.4 minutes to cycle and 2.25 minutes are spent turning and maneuvering during each cycle, the total cycle time estimated for the truck is—

LOAD = 10.67 minutes, TRAVEL = 24.41 minutes,
DUMP = 0.40 minutes, MANEUVER = 2.25 minutes
TOTAL CYCLE TIME = 37.73 minutes

If the mine operates 2 shifts per day, 10 hours per shift, then one truck can deliver—

2 shifts × 10 hours/shift × 60 minutes/hour ÷
37.73 minutes/cycle × 20 metric tons/cycle =
636 metric tons/day.

If the mine production rate is to be 4,000 metric tons per day, then the required number of trucks would be—

4,000 metric tons/day ÷ 636 metric tons/truck = 6.29 or 7 trucks,

And if the production rate per truck is—

20 metric tons/cycle ÷ 37.73 minutes/cycle × 60 minutes/ hour = 31.80 metric tons/hour.

Then total daily truck use is—

4,000 metric tons/day ÷ 31.80 metric tons/hour = 125.79 hours.

Now, for the number of operators. Typically, operators work noticeably less than the total number of hours for which they are paid. When time spent at lunch and breaks (in addition to time lost in traveling to and from the work face) is factored into the estimate, an average of 83% of the operator's time is actually spent working. Consequently, the total amount of labor time required to meet production is—

125.79 hours ÷ 0.83 = 151.55 hours.

If each shift is 10 hours long, then the following number of truck drivers are required:

151.55 hours ÷ 10 hours/shift = 16 workers

Looking back to the number of trucks initially selected, it can be seen that (after accounting for worker efficiency) more trucks will be required.

16 workers ÷ 2 shifts/day = 8 trucks

So, in examining the truck as it operates over the designed haul profile, several cost parameters unfold. These include the number of trucks, the number of operators, daily truck use requirements (hours per day), and the number of hours that the drivers must work. Note that the two latter values differ. Each value is used to determine a different cost, so each must be estimated separately.

Properly determining the hourly workforce goes a long way toward ensuring an acceptable level of reliability. Wages often account for over half the total operating cost, so if the workforce estimate is solid, the cost estimate is probably more than halfway home. Conversely, the cost of operating underground equipment typically represents only a minor portion of the total underground cost. However, since the size and configuration of the workforce is closely related to the equipment requirements (and since equipment *purchase* costs can be significant), the importance of determining those requirements should not be minimized. The information directly impacts the reliability of estimated costs.

Other cyclic operations (drilling, mucking, loading, hauling, hoisting, etc.) are modeled in a manner similar to the truck example above, with the same cost parameters developed for each. As is evident, the actual cycle time calculations are quite simple. The most difficult task is usually finding the rate (speed) at which the machine operates (drill penetration rates, mucker transport speeds, hoist velocities, etc.), but even this information is often readily available. The most common sources include literature from the manufacturer, references such as this book, information databases such as *Mining Cost Service* (Western Mine Engineering, Inc., 2000), or statistical compilations such as the *Canadian Mining Journal's Mining Sourcebook* (Southam 2000). Exclusive of these, speed is very often easy to estimate through observation. With the machine speed and a bit of imagination, the evaluator can provide a more than reasonable estimate of the cost parameters associated with almost any cyclic operation.

Most noncyclic underground operations are based on the continuous movement of materials (ore, waste, air, water, workers, etc.), and the cost parameters can be estimated accordingly. Conveyors, generators, pumps, and ventilation fans all fall under this category. In the following example, the parameters associated with draining the mine and pumping the water to the surface are determined to illustrate the process as it applies to continuous-flow operations.

Example 4.3—Continuous Flow Calculations

A mine makes 400 liters of water per minute. In examining engineering references, it is apparent that a flow rate of roughly 1.0 meter per second represents a reasonable value for the velocity of liquid pumped through a conduit. With this information, the following series of calculations provides an estimate of several of the required cost estimation parameters.

Pipe diameter:

(1) (400 liters/minute ÷ 60 seconds/minute) ÷ 1 meter/ second = 6.66 liters/meter; therefore,

(2) 6.66 liters/meter × 0.001 cubic meter/liter = 0.0066 cubic meter/meter.

(3) Using the relation that area of a circle = π × diameter^2 ÷ 4, then (0.0066 cubic meter × 4 ÷ π)$^{0.5}$ = 0.092 meters or 9.2 centimeters.

So the shaft must be fitted with a 9.2-centimeter-diameter (inside) pipe to remove water from the mine. Since pumps (and most machines) used in underground operations are rated in terms of horsepower and since horsepower costs money (fuel or electricity), it is useful to remember that (for cost-estimating purposes), 1.0 horsepower = 550 foot pounds/second.

In other words, if the weight of the material, the distance that it must be moved, and the speed at which it travels are all known, the size of the motor required for the task can be estimated as can the subsequent costs. So, for our drainage problem (using English units because horsepower is most often reported as such),

    (1) the volume of water in the pipe at any one point of time is 250 meters × 3.281 feet/meter × π × (9.2 centimeters × 0.03281 feet/centimeter)^2 ÷ 4 = 58.70 cubic feet,

    (2) the velocity of the water is 1.0 meter/second × 3.281 feet/meter = 3.281 feet/second,

    (3) and the weight of the water in the pipe is 58.70 cubic feet × 62.4 pounds/cubic foot = 3,663 pounds,

    (4) so the horsepower required to move the water up the shaft is (3.281 feet/second × 3,663 pounds) ÷ 550 foot pounds/second = 21.8 horsepower.

This value can be used (for cost-estimating purposes) to size the pump and estimate the power required to move the water. It is important here, however, to differentiate between actually selecting the equipment—as an engineer would do in the advanced stages of designing the mine—and estimating parameters for a cost estimate. Certainly, there is more to selecting pumps and determining their horsepower than is illustrated here. But the evaluator must keep in mind the purpose of the estimate and the reliability of the available information. Well-established equipment selection procedures for items such as pumps are available elsewhere in this manual, and their use is encouraged whenever appropriate. But the principles presented above, or variations of those principals, can be used to estimate the cost parameters associated with almost any continuous-flow operation found underground, whether it's pumping water, conveying ore, blowing ventilation or compressed air, or transporting backfill. And remember that the premise of this approach is that anything can be estimated. Since, in the early stages of a deposit evaluation, items such as drainage requirements are sketchy at best, the evaluator must remember that additional complication does not necessarily lead to additional reliability.

For comparison purposes, the horsepower of the pump (or pumps) required to transport water up a shaft is typically determined using a relationship similar to the following:

HORSEPOWER = (GALLONS PER MINUTE × Σ HEAD LOSSES × 8.33) ÷ (33,000 × PUMP EFFICIENCY)

For our example,

    (1) the flow rate is 400 liters/minute ÷ 3.785 liters/gallon = 106 gallons/minute,

    (2) discharge head loss is 250 meters × 3.281 feet/meter = 820 feet,

    (3) friction head loss is 0.24 × 820 feet × 3.281 feet/second^2 ÷ (0.302 feet × 64.4 feet/second^2) = 33.2 feet,

    (4) and suction head loss is 10 feet,

    (5) so if the pump efficiency is 68%, the pump horsepower is 106 gallons/minute × (820 feet + 33 feet + 10 feet) × 8.33 pounds/gallon ÷ (33,000 × 0.68) = 34.0 horsepower.

By examining the two relationships and the associated results, it can be seen that the primary difference lies in the pump efficiency value—which, of course, should be the case unless the head loss due to friction is excessive.

So the above series of calculations provides the size of the drain pump and the diameter of the associated pipe, as well as a basis for determining the cost parameters associated with any

continuous-flow operation. Daily use specifications (hours per day) for these operations are apparent, usually 24 hours per day.

Ventilation requirements represent one of the more difficult aspects of the cost estimate. Typically, the nature (length, perimeter, and roughness) of the openings used to access the deposit are taken in combination with the flow of air through the stopes to determine the energy required to move the fresh air. Roughly, the volume of air required for underground operations is based upon the number of workers, the amount of air required to dilute diesel fumes (which is based upon the horsepower ratings of the underground diesel equipment) and any volume losses through rock structure or old workings. The energy and volume values, when considered along with the natural ventilation properties of the designed workings, determine the size and horsepower requirements for the fans (which determines the costs). Ventilation requirements are usually beyond the scope of most early-stage studies. However, they really must be estimated since they represent the one item that can profoundly change the size (or number) of the deposit access openings. If ventilation horsepower requirements are excessive (for instance, if large volumes of air must be pushed through small openings), then operating costs increase dramatically, and the openings should be enlarged or other openings added.

Once most of the equipment and associated labor parameters have been determined, the evaluator should begin tabulating supply needs. Equipment operating parameters provide the basis for estimating the consumption of several supply items, including fuel, electricity, repair parts, lubricants, and tires. However, the more popular cost services such as *Mine and Mill Equipment Operating Costs, an Estimator's Guide* (Western Mine Engineering, Inc. 2000) and the *Cost Reference Guide* (Primedia 2000) include these as equipment operating costs, and they are, therefore, excluded from supply cost estimates.

When defining supply cost parameters, explosives consumption represents a reasonable point from which to begin. For estimation purposes, explosives consumption is usually determined using a powder factor. At operating mines, powder factors are defined through testing and trial and are dynamic. Since this is impossible during the stage at which most cost estimates occur, a historic factor from a mine using a similar stoping method provides a reasonable compromise. Factors at operating mines may be found in other sections of this handbook or in statistical compilations, such as the previously referenced *Canadian Mining Journal's Mining SourceBook* (Southam 2000). Remember that explosives are also used in each development opening, and the associated powder factors will differ from one to the next, depending upon face area, opening configuration, and rock characteristics. Powder factors for development blasting in confined headings are usually higher than those for production blasting in stopes.

After explosive consumption (pounds per metric ton or pounds per day) has been established, the value can be used in conjunction with the density of the powder (plus a hole loading factor) to provide a rough estimate of the length of blast hole drilled each day. Daily blasthole requirements, when divided by a drill penetration rate (again available from sources such as this handbook) provide drill use (hours per day) and, subsequently, the required number of drills and drillers. Blasthole requirements also provide the information necessary to estimate blasting cap and fuse needs. And, when used in conjunction with drill bit and steel wear rates, daily drilling requirements provide the basis for estimating the consumption of these.

Requirements for many supplies are tied specifically to the advance rates of the development openings in which they are placed. Daily consumption for pipes used to supply compressed air and fresh water, in addition to pipes used to remove drain water from the mine and items such as electrical cable and ventilation tubing, obviously mimic the advance rates of the development openings (both prior to, and during, production). Rock bolt, shotcrete, and timber requirements are also closely linked to development opening advance.

As you may have surmised, estimating consumption rates for supplies and daily use requirements for equipment is not nearly as difficult as knowing and understanding all the implications of the differing specifications of these supplies and machines (tapered drill bits versus rope threaded, ANFO versus high explosives, friction-set rock bolts versus resin, hydraulic drifters versus pneumatic, etc.). The level of reliability that can be placed in estimated costs is proportional to the level of effort placed in researching the implications of these differing specifications.

4.1.4 Cost Estimates

Once mine cost parameters have been established, the estimating process is one of simple calculations and tabulations. Since most costs—both capital and operating—are tied to either equipment use, supply consumption, wages, or salaries, the process from here on is finding (or recognizing) the best source for the cost information, applying the costs to the derived parameters, and tabulating the results. This process is best demonstrated by examining the cost models in Tables 4.6 through 4.13.

Equipment operating costs are calculated by multiplying use (hours) by the hourly operating costs of the machine, which can be found in one of several sources, including *Mine and Mill Equipment Operating Costs, An Estimator's Guide* (Western Mine Engineering, Inc. 2000) and the *Cost Reference Guide*, (Primedia 2000). These costs are typically itemized in terms of repair parts and labor, fuel or electricity, lubricants, tires, and ground engaging components.

Equipment operating costs associated with excavating underground openings prior to development (preproduction development costs) are calculated similarly, except the total use (hours) is multiplied by the hourly costs of the machine.

Labor costs are determined in a similar manner. To arrive at daily labor costs, multiply the number of workers assigned to any one discipline by the number of hours worked per shift, and then multiply the result by the hourly wage (factored for burden). Sources for wages include *Mining Cost Service* (Western Mine Engineering, Inc. 2000) and *U.S. Metal and Industrial Mineral Mine Salaries, Wages, and Benefits* (Western Mine Engineering, Inc. 2000).

Wages must be factored for the additional cost incurred by the employer for having the worker as a full-time employee. Employers typically contribute to Social Security taxes, worker's compensation and unemployment insurance, retirement plans, and medical benefit packages. In addition, either the wages or the workforce must be factored to account for the expense of vacation and sick leave, shift differential pay, and overtime. Publications available from Western Mine Engineering, Inc., extensively detail the cost of these benefits (most often referred to as burden) at more than 300 operations.

The costs for salaried workers are calculated in a manner similar to those for the hourly workers. *U.S. Metal and Industrial Mineral Mine Salaries, Wages, and Benefits* (Western Mine Engineering, Inc. 2000) details salaried staffs at many operations.

And finally, supply costs are determined by simply multiplying daily consumption by the price of the commodity. Commodity (supply) prices are available either from individual vendors or from *Mining Cost Service* (Western Mine Engineering, Inc. 2000). As with equipment operation and labor costs, supplies costs contribute both to development and operating costs. Items such as pipe, rail, ventilation tubing, and rock support installed prior to production should be tallied with the preproduction development costs associated with each development opening, along with the costs of the drill bits, explosives, and caps expended during excavation.

Note that up to this point, most cost parameters have been tied to daily values. The convenience of this approach comes into play at this stage of the estimate. Operating costs are typically reported in dollars per metric ton of ore, whereas preproduction development costs are reported as total sums (in dollars). To arrive at operating costs in the appropriate units, the daily values are simply divided by the metric tons of ore produced each day. For development costs, the project duration (for instance, the days required to complete an access adit) is multiplied by the daily costs to arrive at the total. Alternately, evaluators often compute the total number of hours required for each machine and worker to complete an underground excavation, then multiply the results by the hourly rates. Supply costs are then added by computing the total consumptions for the opening, then multiplying the results by the respective unit prices.

Operating costs often include a miscellaneous charge to account for items too numerous to list separately or for unscheduled and unanticipated (yet predictable) tasks. Although evaluators often include allowances for such costs when developing the mine parameters, it is generally more acceptable to include this as one value at the end of the compilation so that those relying on the estimate can see the exact impact of the allowance.

Capital costs should include a contingency. As opposed to the miscellaneous allowance for the operating costs, this is an actual cost that represents a fund set aside for any additional, unforeseen costs associated with unanticipated geologic circumstances or site conditions. It is not intended to account for inadequacies in the cost estimate or discrepancies in the design. Companies assume that the money in the contingency fund will be spent.

Other items that must be included in the capital costs tabulation include those associated with feasibility, engineering, planning, construction supervision and management, administration, accounting, and legal services. These are most often factored from the total capital cost figure, and values for the factors can be found in a number of references. Some often-used factors include—

- Feasibility, engineering, and planning: 4% to 8%

- Construction supervision and project management: 8% to 10%

- Administrative, accounting, permitting, and legal services: 8% to 14%

Alternately, the costs of many of these items can be based upon salaried staff requirements prior to production. However, preproduction functions such as these are often contracted, and estimated costs should be adjusted accordingly.

4.1.5 Economic Evaluation

To conduct a discounted cash-flow economic evaluation, the estimated costs of the operation must be examined with respect to anticipated revenues and the economic conditions associated with the project (taxes, royalties, and financing). Particularly for tax purposes (but also to a lesser extent, royalties and financing purposes), estimated costs are divided into the capital and operating categories alluded to above. Operating costs are those that can be directly expensed against revenues when figuring income for tax purposes and include expenditures pertaining to wages, salaries, operating the equipment, and purchasing supplies. For underground mines, capital costs include those associated with purchasing equipment, excavating underground workings prior to production, and constructing of surface facilities. Most costs that are not directly related to purchasing equipment fall under the category of preproduction development and typically include the expenses of—

- Excavating the shafts, adits, drifts, crosscuts, and raises used to access the deposit.

- Excavating the drifts and raises needed to prepare the initial stopes for production.

- Constructing and installing services (compressors, hoists, ventilation, water supply, offices, shops, and worker change houses).
- Engineering and feasibility studies.
- Supervising and managing construction.
- Administrating, accounting, and legal services.
- Establishing a contingency fund.

Operating costs can be categorized in several ways. Production-oriented evaluators are often most comfortable breaking expenses down into categories of development or production costs per metric ton of ore recovered. Since the checks written can usually be traced to either supply, equipment operation, or labor categories, many evaluators like to see costs broken down as such. The decision as to approach is based as much upon intended use as evaluator preference. Since most predevelopment economic evaluations are concerned only with total operating costs, the breakdown is not critical, only the results.

As with the cost estimate, discounted cash-flow analysis techniques are too extensive to describe in detail here. Suffice it to say, however, that several economic aspects other than estimated costs are critical to determining project viability properly. For instance, project revenues are not simply the product of mineral grade and commodity price. First, the grade must be factored for recovery losses and dilution at the mine and for concentrating inefficiencies in the mill. Smelting and refining charges must be included, as must penalties for deleterious minerals and deductions at the smelter. Federal and regional taxes related to income, as well as sales, property, and severance taxes, detract from anticipated profits. And if the project is financed externally, or if royalties must be paid to partners, property owners, or other entities, project economics are further influenced.

One closing note. Don't take your own estimates too seriously, especially in the early stages of an evaluation. Experienced estimators know that the easiest part of cost estimating is finding errors in other people's work. Cost estimates evolve, and you can be assured that your estimates will be challenged. Ultimately, every cost an evaluator estimates is proven wrong. Your goal is not to be wrong by much.

4.2 MINE COST MODELS

4.2.1 Introduction

A series of mine cost models is presented in Tables 4.6 through 4.13 to aid in making preliminary order-of-magnitude estimates based on limited deposit information. The models are theoretical engineering estimates and do not represent any specific mining operation.

The reader is cautioned against relying too heavily on these or any other models for making significant economic decisions. A cost model, no matter how carefully prepared, is only a model and should not be expected to represent projected costs for a specific mine with a degree of reliability beyond order-of-magnitude. However, cost models can be very useful for comparison purposes or for acquisition and exploration decisions. They are commonly used to establish cut-off grades for preliminary reserve estimates. As further information is developed about the deposit, which allows the design of realistic mine plans, the cost model can be used as the starting point for more detailed estimates.

4.2.2 How the Models Were Developed

The models were developed using *Sherpa Cost Estimating Software for Underground Mines* (Stebbins 2000). This copyrighted software is commonly used to provide prefeasibility cost estimates. It uses standard cost engineering techniques (similar to those described in the preceding paragraphs) to estimate capital

TABLE 4.1   Cost items not included in the cost models

Exploration
Permitting
Off-site access roads, powerlines, pipelines or railroads
Home office overhead
Taxes (except sales taxes)
Insurance
Depreciation
Contingencies
Townsite construction or operation
Off-site transportation of products
Incentive bonus premiums
Overtime labor costs
Sales expenses
Smelting, refining and milling costs
Interest expense
Start-up costs (except working capital)

and operating costs for a proposed underground mine based on specific mine design parameters. Selected unit costs used in the models are listed in Tables 4.3 through 4.5. These unit costs are the most recent values from *Mining Cost Service* (Western Mine Engineering, Inc. 2000).

4.2.3   What is Included in the Models

The models include all labor, material, supply, and equipment operating costs incurred at the mine site, including those for supervision, administration, and on-site management. The costs of purchasing and installing all equipment are included, as well as those associated with preproduction development and the costs of construction of surface facilities.

Equally important to this process is an understanding of the items that are not included in the models, because these are the items that must be added when appropriate in order to produce a complete cost estimate. The most significant of these are listed in Table 4.1.

4.2.4   Model Criteria

All of the models include at least two routes of access. Mines producing less than 4,000 tonnes of ore per day are accessed by one primary shaft and a secondary raise that serves to complete the ventilation circuit and provide an emergency egress route. Larger mines are accessed by two primary shafts and at least one secondary raise.

Preproduction development costs are sufficient to allow access to enough ore so that initial production rates match the design production rate. Sufficient development costs required to access enough ore in the future to sustain the design production rate are included in operating costs. All shops, offices, change-houses, warehouses, and mine plants are located on the surface. Working capital included in the estimates accounts for two months of operation. Sales tax at the rate of 6.75% is added to all equipment and nonfuel supply prices. All units are metric, and all costs are in first-quarter, year-2000 dollars. An idealized sketch of the stope layout for each of the models is shown in Figures 4.1 through 4.8.

4.2.5   Rock Characteristics

Ore and waste rock densities are 0.367 cubic meters per tonne and 0.401 cubic meters per tonne respectively for all the models. Swell factors are 55% for ore and 45% for waste. Rock quality designations and compressive strengths vary from model to model.

TABLE 4.2 Rock characteristics

	Rock quality designation (percent)	Compressive strength (kilopascals)
Cut-and-Fill Mining		
Ore	50	68,950
Waste	35	51,700
Shrinkage Mining		
Ore	65	103,425
Waste	80	172,375
End Slice Mining		
Ore	75	137,900
Waste	80	172,375
VCR Mining		
Ore	75	137,900
Waste	80	172,375
Sublevel Longhole Mining		
Ore	55	82,700
Waste	75	137,900
Room-and-Pillar Mining		
Ore	75	155,135
Footwall waste	65	137,900
Hanging wall waste	55	120,660

TABLE 4.3 Salaries for selected managerial, supervisory, technical and administrative personnel used in the cost models

Job title	Annual salary* >100 hourly employees	<100 hourly employees
Manager	$ 129,400	$ 105,800
Superintendent	101,700	80,120
Foreman	70,550	63,850
Engineer	74,300	68,220
Geologist	67,620	63,730
Shift Boss	63,980	56,250
Technician	49,820	53,290
Accountant	67,030	62,860
Purchasing	72,410	55,320
Personnel	107,600	60,660
Secretary	39,780	35,160
Clerk	38,930	33,480

*Including a 36% burden factor.
Source: Leinart, J.B. 2000

4.2.6 Unit Costs

Tables 4.3 through 4.5 list unit costs used in constructing the cost models. The wages and salaries are U.S. national averages as reported in *U.S. Metal and Industrial Mineral Mines Salaries and Wages and Benefits, 2000 Survey Results* (Leinart 2000). In keeping with the results of the survey, wages and salaries for the smaller mines are less than those for the larger ones. For purposes of this study, small mines are defined as those with 100 or fewer employees, and larger mines are those with more than 100 employees. The equipment and supply prices are taken from the latest *Mining Cost Service* (2000) survey. All the unit costs are resident in the *Sherpa* program which was used to generate the models. The tables are presented here to provide details of the model estimates and to serve as references for the reader's own cost estimates.

4.2.7 Equipment Costs

Unit prices for the equipment used in each of the models are embedded in Tables 4.6 through 4.11 for reference purposes. The prices are extracted from *Mine and Mill Equipment Costs, An Esti-*

TABLE 4.4 Wages for selected hourly workers used in the cost models

Job title	Hourly wages >100 hourly employees*	<100 hourly employees†
Stope miner	$ 23.70	$ 21.79
Development miner	20.28	21.79
Mobile equipment operator	23.22	21.07
Hoist operator	22.14	22.07
Motorman	16.67	17.79
Support miner	23.70	21.79
Exploration driller	22.14	19.26
Crusher operator	22.21	14.49
Backfill plant operator	19.71	21.24
Electrician	24.29	23.16
Mechanic	24.62	21.86
Maintenance worker	20.64	17.76
Helper	19.12	15.99
Underground laborer	18.87	16.41
Surface laborer	18.23	15.88

*Including a 41% burden factor
†Including a 38% burden factor
Source: Leinart, J.B. 2000

TABLE 4.5 Unit prices for selected supply items used in the cost models.

Item	Cost
Diesel fuel	$ 0.144/liter
Electricity	0.068/kilowatt
Explosives	1.01/kilogram
Caps	1.20 each
Boosters	2.75 each
Fuse	0.49/meter
Timber	296.65/cubic meter
Lagging	250.03/cubic meter
Steel liner material	1.17/kilogram
Cement (for backfill)	90.94/tonne

Source: Mining Cost Service 2000

mator's Guide (Western Mine Engineering, Inc. 2000). Because of space restrictions, only minimal specification information is provided in the tables. More complete specifications can be found in the Guide. The prices are undiscounted budget prices for standard-equipped units. Actual selling prices for the equipment may vary considerably from those listed.

4.3 REFERENCES

Canadian Mining Journal's Mining Sourcebook, published annually, Southam Inc., Don Mills, Ontario.
Cost Reference Guide, 2000 and periodically updated, Primedia Information, Inc., Hightstown, New Jersey.
Leinart, J.B. 2000. *U.S. Metal and Industrial Mineral Mine Salaries, Wages, and Benefits, 2000 Survey Results*. Western Mine Engineering, Inc., Spokane, Washington, www.westernmine.com.
Mine and Mill Equipment Costs, an Estimator's Guide, 2000. updated annually, Western Mine Engineering, Inc., Spokane, Washington, www.westernmine.com.
Mining Cost Service, 2000. O.L. Schumacher, ed., periodically updated, Western Mine Engineering, Inc., Spokane, Washington, www.westernmine.com.
Stebbins, S.A. 2000. *Sherpa Cost Estimating Software of Underground Mines*. Aventurine Mine Cost Engineering, Elk, Washington, www.aventurine.net.

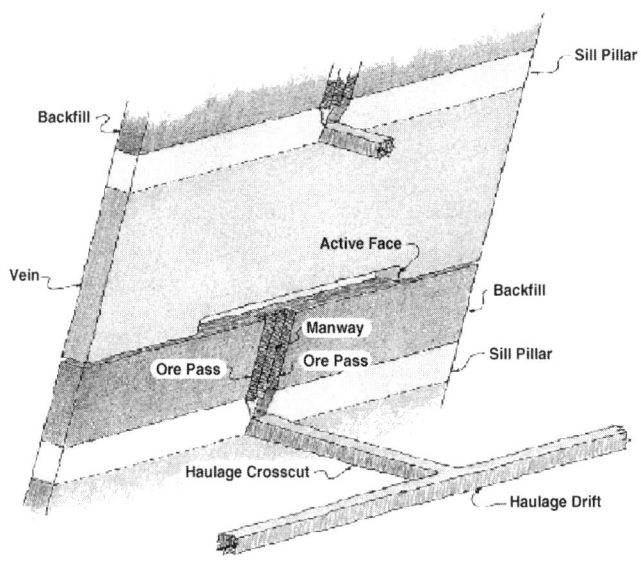

FIGURE 4.1   Stope development: cut and fill mining

TABLE 4.6   Cost models: cut and fill mining

These models represent mines on steeply dipping veins, 2.5, 3.0, or 3.5 meters wide respectively, and 500, 1,400, or 1,900 meters along strike. Access is via shaft, 524, 719, or 863 meters deep. Haulage to the shaft is via rail. Stoping includes drilling and blasting with jackleg drills, slushing to ore chutes, and sand filling. A secondary access/vent raise extends to the surface. See Figure 4.1.

Daily ore production (tonnes)	Units	200	1,000	2,000
Production				
Hours per Shift		8	8	10
Shifts per Day		1	2	2
Days per Year		320	320	320
Deposit				
Total Minable Resource	tonnes	704,000	4,231,700	9,874,000
Dip	degrees	75	75	75
Average Strike Length	meters	500	1,400	1,900
Average Vein Width	meters	2.5	3.0	3.5
Average Vertical	meters	200	350	500
Stopes				
Stope Length	meters	71	70	68
Stope Width	meters	2.6	3.1	3.6
Stope Height	meters	48	48	44
Face Width	meters	2.6	3.1	3.6
Face Height	meters	2.2	2.4	2.6
Advance per Round	meters	1.9	2.1	2.2
Sill Pillar Length	meters	71	70	68
Sill Pillar Width	meters	2.6	3.1	3.6
Sill Pillar Height	meters	4.4	4.8	5.2
Development Openings				
Shafts				
Face Area	square meters	11.9	18.6	22.6
Preproduction Advance	meters	524	719	863
Cost	dollars/meter	$6,300	$7,986	$10,264
Drifts				
Face Area	square meters	6.8	7.6	7.8
Daily Advance	meters	1.3	5.4	10.5
Preproduction Advance	meters	103	435	836
Cost	dollars/meter	$562	$602	$626

(continued)

TABLE 4.6   Cost models: cut and fill mining (continued)

Daily ore production (tonnes)	Units	200	1,000	2,000
Crosscuts				
Face Area	square meters	6.8	7.6	7.8
Daily Advance	meters	0.3	1.2	2.3
Preproduction Advance	meters	22	95	188
Cost	dollars/meter	$471	$497	$522
Access Raises				
Face Area	square meters	2.4	2.9	3.4
Daily Advance	meters	0.4	1.8	3.2
Preproduction Advance	meters	33	142	253
Cost	dollars/meter	$726	$764	$788
Ore Passes				
Face Area	square meters	0.9	1.6	2.5
Daily Advance	meters	0.2	0.7	1.2
Preproduction Advance	meters	193	338	483
Cost	dollars/meter	$303	$373	$433
Ventilation Raises				
Face Area	square meters	2.0	3.5	5.3
Daily Advance	meters	0.1	0.5	0.7
Preproduction Advance	meters	493	688	833
Cost	dollars/meter	$414	$486	$567
Hourly Labor Requirements				
Stope Miners		10	32	36
Development Miners		6	12	16
Equipment Operators		1	1	1
Hoist Operators		2	4	4
Motormen		—	—	—
Support Miners		4	4	4
Diamond Drillers		2	6	6
Backfill Plant Operators		1	2	4
Electricians		5	6	6
Mechanics		9	18	19
Maintenance Workers		2	4	6
Helpers		3	8	9
Underground Laborers		2	5	8
Surface Laborers		2	4	6
Total Hourly Personnel		49	106	125
Salaried Personnel Requirements				
Managers		1	1	1
Superintendents		1	1	2
Foremen		1	2	4
Engineers		1	2	3
Geologists		1	2	3
Shift Bosses		3	10	10
Technicians		2	4	6
Accountants		1	2	2
Purchasing		1	2	4
Personnel		2	5	6
Secretaries		2	4	6
Clerks		2	5	8
Total Salaried Personnel		18	40	55
Supply Requirements (daily)				
Explosives	kilograms	155	711	1,344
Caps	each	86	370	676
Boosters	each	71	315	591
Fuse	meters	450	1,729	2,973

(continued)

TABLE 4.6 Cost models: cut and fill mining (continued)

Daily ore production (tonnes)	Units	200	1,000	2,000
Supply Requirements (daily) (continued)				
Drill Bits	each	1.02	4.17	7.76
Drill Steel	each	0.07	0.3	0.56
Backfill Pipe	meters	1.6	6.6	12.8
Fresh Water Pipe	meters	2.0	8.4	16.0
Drainage Pipe	meters	—	—	—
Compressed Air Pipe	meters	2.0	8.4	16.0
Electric Cable	meters	2.0	8.4	16.0
Ventilation Tubing	meters	2.0	8.4	16.0
Rail	meters	3.2	13.2	25.6
Steel Liner Plate	kilograms	30	175	365
Steel Beam	kilograms	—	—	—
Rock Bolts	each	40	171	300
Cement	tonnes	11.2	56.0	111.8
Shotcrete	cubic meters	—	—	—
Timber	cubic centimeters	559,358	2,638,659	5,178,191
Buildings				
Office	square meters	460	1,022	1,405
Change House	square meters	569	1,231	1,452
Warehouse	square meters	181	207	261
Shop	square meters	338	397	517
Mine Plant	square meters	181	222	222
Equipment Requirements (number–size)				
Stope Jacklegs	centimeter	7 – 3.175	10 – 3.175	11 – 3.175
Stope Slushers	centimeter	6 – 122	9 – 152	10 – 183
Stope Locomotives w/Cars	tonne	2 – 9.1	3 – 9.1	4 – 18.1
Vertical Development Stopers	centimeter	2 – 2.870	2 – 2.870	2 – 2.870
Horizontal Development Jacklegs	centimeter	2 – 3.175	3 – 3.175	4 – 3.175
Development Muckers	cubic meter	2 – 0.14	2 – 0.14	3 – 0.14
Development Locomotives w/Cars	tonne	2 – 9.1	2 – 9.1	3 – 18.1
Production Hoists	centimeter	1 – 152	1 – 152	1 – 152
Rock Bolt Jacklegs	centimeter	3 – 3.81	3 – 3.81	3 – 3.81
Drain Pumps	horsepower	6 – 11	8 – 46	9 – 95
Fresh Water Pumps	horsepower	2 – 0.5	2 – 0.5	2 – 0.5
Backfill Mixers	horsepower	1 – 3	1 – 15	1 – 15
Backfill Pumps	horsepower	2 – 5	2 – 8	2 – 8
Service Vehicles	horsepower	7 – 75	11 – 82	13 – 130
Compressors	cubic meters per minute	1 – 85	1 – 142	1 – 142
Ventilation Fans	centimeter	1 – 122	1 – 122	1 – 122
Exploration Drills	centimeter	1 – 4.45	2 – 4.45	2 – 4.45
Equipment Costs (dollars/unit)				
Stope Jacklegs		$5,670	$5,670	$5,670
Stope Slushers		36,800	41,900	49,400
Stope Locomotives w/Cars		248,800	248,800	444,000
Vertical Development Stopers		5,880	5,880	5,880
Horizontal Development Jacklegs		5,670	5,670	5,670
Development Muckers		55,600	55,600	55,600
Development Locomotives w/Cars		248,800	248,800	444,000

(continued)

TABLE 4.6 Cost models: cut and fill mining (continued)

Daily ore production (tonnes)	Units	200	1,000	2,000
Production Hoists		544,624	672,542	794,446
Rock Bolt Jacklegs		5,670	5,670	5,670
Shotcreters		44,650	50,850	50,850
Drain Pumps		9,230	10,820	12,150
Fresh Water Pumps		5,810	5,810	5,810
Backfill Mixers		14,280	29,250	29,250
Backfill Pumps		7,980	8,190	8,190
Service Vehicles		71,820	79,990	91,010
Compressors		125,500	200,800	200,800
Ventilation Fans		107,120	107,120	107,120
Exploration Drills		37,670	37,670	37,670

COST SUMMARY

Operating Costs (dollars/tonne ore)

		200	1,000	2,000
Equipment Operation		$3.69	$2.17	$1.89
Supplies		12.56	11.73	11.53
Hourly Labor		41.74	19.16	14.90
Administration		16.31	8.33	5.99
Sundries		7.43	4.14	3.43
Total Operating Costs		**$81.73**	**$45.53**	**$37.74**

Unit Operating Cost Distribution (dollars/tonne ore)

		200	1,000	2,000
Stopes		$18.04	$12.02	$9.90
Drifts		7.21	3.23	3.08
Crosscuts		0.54	0.42	0.46
Ore Passes		0.65	0.30	0.25
Access Raises		2.05	0.77	0.60
Vent Raises		$0.55	$0.24	$0.19
Main Haulage		4.78	2.37	1.85
Backfill		5.98	5.43	5.50
Services		6.64	4.68	3.89
Ventilation		0.75	0.15	0.07
Exploration		1.66	1.14	0.72
Maintenance		12.77	4.75	3.46
Administration		12.68	5.89	4.35
Miscellaneous		7.43	4.14	3.42
Total Operating Costs		**$81.73**	**$45.53**	**$37.74**

Capital Costs

	200	1,000	2,000
Equipment Purchase	$3,729,000	$5,028,000	$7,884,000
Preproduction Underground Excavation			
Shafts	3,299,000	5,739,000	8,862,000
Drifts	58,060	261,600	523,000
Crosscuts	10,370	47,030	98,010
Access Raises	24,060	108,100	199,500
Ore Passes	58,570	126,200	208,900
Ventilation Raises	204,200	334,700	472,300
Surface Facilities	1,716,000	2,769,000	3,481,000
Working Capital	871,900	2,428,000	4,027,000
Engineering & Management	1,183,000	1,874,000	2,825,000
Contingency	910,000	1,441,000	2,173,000
Total Capital Costs	**$12,064,160**	**$20,156,630**	**$30,753,710**
Total Capital Cost per Daily Tonne Ore	**$60,321**	**$20,157**	**$15,377**

Source: Sherpa Cost Estimating Software for Underground Mines (Stebbins 2000)

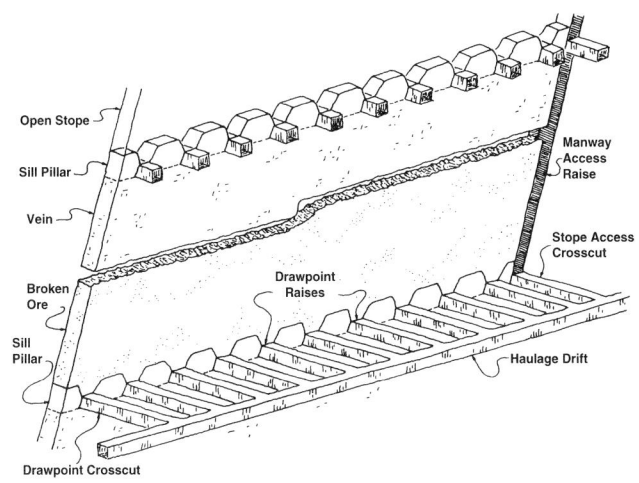

FIGURE 4.2   Stope development: shrinkage mining

TABLE 4.7   Cost models: shrinkage mining

These models represent mines on steeply dipping veins, 2.5, 3.0, or 3.5 meters wide respectively, and 500, 1,400, or 1,900 meters along strike. Access is via shaft, 524, 717, or 863 meters deep. Haulage to the shaft is by rail. Shrink stoping includes drilling and blasting a breast face with jackleg drills and drawing ore to the level below, with no sand filling. A secondary access/vent raise extends to the surface. See Figure 4.2.

Daily ore production (tonnes)	Units	200	1,000	2,000
Production				
Hours per Shift		8	8	10
Shifts per Day		1	2	2
Days per Year		320	320	320
Deposit				
Total Minable Resource	tonnes	704,000	4,231,700	9,874,000
Dip	degrees	75	75	75
Average Strike Length	meters	500	1,400	1,900
Average Vein Width	meters	2.5	3.0	3.5
Average Vertical	meters	200	350	500
Stopes				
Stope Length	meters	71	70	68
Stope Width	meters	2.6	3.1	3.6
Stope Height	meters	48	48	44
Face Width	meters	2.6	3.1	3.6
Face Height	meters	2.4	2.4	2.4
Advance per Round	meters	2.1	2.1	2.1
Sill Pillar Length	meters	71	70	68
Sill Pillar Width	meters	2.6	3.1	3.6
Sill Pillar Height	meters	9.7	9.7	9.8
Development Openings				
Shafts				
Face Area	square meters	11.9	18.6	22.6
Preproduction Advance	meters	524	719	863
Cost	dollars/meter	$6,302	$7,554	$10,413
Drifts				
Face Area	square meters	6.8	7.6	7.8
Daily Advance	meters	1.5	6.1	11.9
Preproduction Advance	meters	117	489	948
Cost	dollars/meter	$611	$628	$687

(continued)

TABLE 4.7   Cost models: shrinkage mining (continued)

Daily ore production (tonnes)	Units	200	1,000	2,000
Crosscuts				
Face Area	square meters	6.8	7.6	7.8
Daily Advance	meters	1.7	6.9	13.5
Preproduction Advance	meters	132	553	1,076
Cost	dollars/meter	$531	$536	$590
Access Raises				
Face Area	square meters	2.4	2.9	3.4
Daily Advance	meters	0.5	2.1	3.7
Preproduction Advance	meters	39	165	298
Cost	dollars/meter	$659	$678	$716
Draw Points				
Average Face Area	square meters	10.5	12.9	15.3
Daily Advance	meters	0.6	2.4	4.7
Preproduction Advance	meters	48	195	377
Cost	dollars/meter	$361	$382	$427
Ore Passes				
Face Area	square meters	0.9	1.6	2.5
Daily Advance	meters	0.2	0.7	1.2
Preproduction Advance	meters	193	338	483
Cost	dollars/meter	$344	$368	$422
Ventilation Raises				
Face Area	square meters	2.0	3.5	5.3
Daily Advance	meters	0.1	0.5	0.7
Preproduction Advance	meters	493	688	833
Cost	dollars/meter	$410	$392	$409
Hourly Labor Requirements				
Stope Miners		6	20	28
Development Miners		8	22	32
Equipment Operators		1	1	1
Hoist Operators		2	4	4
Support Miners		4	4	4
Diamond Drillers		2	6	6
Electricians		4	6	6
Mechanics		6	13	16
Maintenance Workers		2	4	6
Helpers		3	8	11
Underground Laborers		2	5	8
Surface Laborers		2	4	6
Total Hourly Personnel		42	97	128
Salaried Personnel Requirements				
Managers		1	1	1
Superintendents		1	1	2
Foremen		1	2	4
Engineers		1	2	3
Geologists		1	2	3
Shift Bosses		3	6	10
Technicians		2	4	6
Accountants		1	2	2
Purchasing		1	2	4
Personnel		2	5	6
Secretaries		2	4	6
Clerks		2	5	8
Total Salaried Personnel		18	36	55

(continued)

TABLE 4.7   Cost models: shrinkage mining (continued)

Daily ore production (tonnes)	Units	200	1,000	2,000
Supply Requirements (daily)				
Explosives	kilograms	223	959	1,755
Caps	each	130	576	1,103
Boosters	each	114	512	993
Fuse	meters	525	2,180	4,036
Drill Bits	each	2.63	10.03	19.10
Drill Steel	each	0.19	0.72	1.38
Fresh Water Pipe	meters	3.7	15.1	29.1
Compressed Air Pipe	meters	3.7	15.1	29.1
Electric Cable	meters	3.7	15.1	29.1
Ventilation Tubing	meters	3.7	15.1	29.1
Rail	meters	6.4	26.0	50.8
Rock Bolts	each	15	67	125
Shotcrete	cubic meters	—	1.0	2.0
Timber	cubic centimeters	722,208	3,306,584	6,447,957
Buildings				
Office	square meters	460	920	1,405
Change House	square meters	488	1,126	1,486
Warehouse	square meters	184	213	258
Shop	square meters	345	411	510
Mine Plant	square meters	181	222	222
Equipment Requirements (number–size)				
Stope Jacklegs	centimeter	4 – 3.175	7 – 3.175	9 – 3.175
Stope Mucker	cubic meter	4 – 0.13	6 – 0.13	8 – 0.13
Stope Locomotives w/Cars	tonne	2 – 9.1	2 – 9.1	3 – 18.1
Vertical Development Stopers	centimeter	2 – 3.175	3 – 3.175	4 – 3.175
Horizontal Development Jacklegs	centimeter	3 – 3.175	5 – 3.175	6 – 3.175
Development Muckers	cubic meter	2 – 0.14	4 – 0.14	5 – 0.14
Development Locomotives w/Cars	tonne	2 – 9.1	4 – 9.1	5 – 18.1
Production Hoists	centimeter	1 – 152	1 – 152	1 – 203
Rock Bolt Jacklegs	centimeter	3 – 3.81	3 – 3.81	3 – 3.81
Drain Pumps	horsepower	6 – 11	8 – 46	9 – 95
Fresh Water Pumps	horsepower	2 – 0.5	2 – 0.5	2 – 0.5
Service Vehicles	horsepower	6 – 75	11 – 82	15 – 130
Compressors	cubic meters per minute	1 – 85	1 – 142	1 – 142
Ventilation Fans	centimeter	1 – 122	1 – 122	1 – 152
Exploration Drills	centimeter	1 – 4.45	2 – 4.45	2 – 4.45
Equipment Costs (dollars/unit)				
Stope Jacklegs		$5,670	$5,670	$5,670
Stope Muckers		85,900	85,900	85,900
Stope Locomotives w/Cars		248,800	248,800	444,000
Vertical Development Stopers		5,880	5,880	5,880
Horizontal Development Jacklegs		5,670	5,670	5,670
Development Muckers		55,600	55,600	55,600
Development Locomotives w/Cars		248,800	248,800	444,000
Hoists		609,384	721,549	938,550

(continued)

TABLE 4.7   Cost models: shrinkage mining (continued)

Daily ore production (tonnes)	Units	200	1,000	2,000
Rock Bolt Jacklegs		5,670	5,670	5,670
Shotcreters		44,650	50,850	50,850
Drain Pumps		9,230	10,820	12,150
Fresh Water Pumps		5,810	5,810	5,810
Service Vehicles		71,820	79,990	91,010
Compressors		125,500	200,800	200,000
Ventilation Fans		107,120	107,120	117,820
Exploration Drills		37,670	37,670	37,670
COST SUMMARY				
Operating Costs (dollars/tonne ore)				
Equipment Operation		$3.20	$2.24	$2.21
Supplies		8.42	7.24	7.35
Hourly Labor		34.71	16.12	14.61
Administration		16.37	6.25	5.62
Sundries		6.27	3.18	2.98
Total Operating Costs		**$68.97**	**$35.03**	**$32.77**
Unit Operating Cost Distribution (dollars/tonne ore)				
Stopes		$11.30	$7.34	$7.00
Drifts		7.31	3.96	3.95
Crosscuts		6.74	3.97	4.04
Draw Points		1.92	1.00	1.00
Access Raises		2.06	1.00	0.85
Ore Passes		0.63	0.31	0.23
Vent Raises		0.47	0.20	0.13
Main Haulage		4.32	2.42	2.13
Services		3.91	2.30	2.76
Ventilation		0.75	0.15	0.08
Unit Operating Cost Distribution (dollars/tonne ore)				
Exploration		1.63	1.00	0.72
Maintenance		8.95	3.42	2.72
Administration		12.72	4.80	4.18
Miscellaneous		6.27	3.19	2.98
Total Operating Costs		**$68.98**	**$35.06**	**$32.77**
Capital Costs				
Equipment Purchase		$3,913,000	$5,629,000	$8,976,000
Preproduction Underground Excavation				
Shafts		3,300,000	5,428,000	8,991,000
Drifts		71,730	307,700	652,100
Crosscuts		70,240	296,500	635,100
Draw Points		17,460	74,410	160,800
Access Raises		25,720	112,000	213,300
Ore Passes		$66,490	$124,300	$203,700
Ventilation Raises		202,200	269,600	340,700
Surface Facilities		1,659,000	2,630,000	3,351,000
Working Capital		735,800	1,868,000	3,495,000
Engineering & Management		1,212,000	1,933,000	3,058,000
Contingency		933,000	1,487,000	2,352,000
Total Capital Costs		**$12,206,640**	**$20,159,510**	**$32,428,700**
Total Capital Cost per Daily Tonne Ore		**$61,033**	**$20,160**	**$16,214**

Source: Sherpa Cost Estimating Software for Underground Mines (Stebbins 2000)

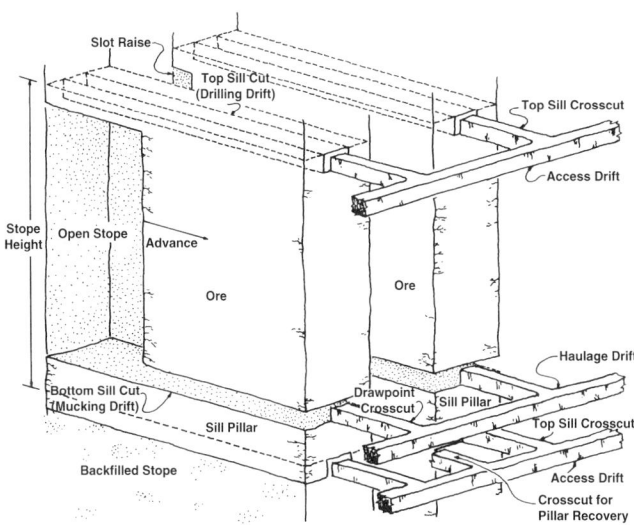

FIGURE 4.3   Stope development: end slice mining

TABLE 4.8   Cost models: end slice mining

These models represent mines on steeply dipping veins, 40, 60, or 80 meters wide respectively, and 200, 360, or 400 meters along strike. Access is via one or two shafts, 477, 685, or 872 meters deep, and a secondary access/vent raise. Haulage from the stopes is by load-haul-dump unit. Stoping includes driving top sill and haulage level crosscuts across the vein, excavating a bottom sill cut and a slot raise, and end slice drilling and blasting using down-the-hole blasthole drills, followed by sand filling. See Figure 4.3.

Daily ore production (tonnes)	Units	800	2,000	4,000
Production				
Hours per Shift		8	10	10
Shifts per Day		2	2	2
Days per Year		320	320	320
Deposit				
Total Minable Resource	tonnes	2,822,400	9,873,900	22,574,000
Dip	degrees	75	75	75
Average Strike Length	meters	200	360	400
Average Vein Width	meters	40	60	80
Average Vertical	meters	100	160	250
Stopes				
Stope Length	meters	41	62	41
Stope Width	meters	11.1	10.9	10.8
Stope Height	meters	32	31	35
Face Width	meters	11.1	10.9	10.8
Face Height	meters	24.7	23.2	26.7
Advance per Round	meters	4.2	4.1	4.1
Sill Pillar Length	meters	41	62	41
Sill Pillar Width	meters	11.1	10.9	10.8
Sill Pillar Height	meters	4.3	4.3	4.3
Development Openings				
Shafts				2 ea.
Face Area	square meters	17.5	22.6	27.5
Preproduction Advance	meters	477	685	872
Cost	dollars/meter	$7,456	$9,736	$10,713
Drifts				
Face Area	square meters	11.5	13.9	15.4
Daily Advance	meters	2.4	5.9	11.3
Preproduction Advance	meters	192	475	907
Cost	dollars/meter	$609	$659	$703

(continued)

TABLE 4.8   Cost models: end slice mining (continued)

Daily ore production (tonnes)	Units	800	2,000	4,000
Crosscuts				
Face Area	square meters	11.5	13.9	15.4
Daily Advance	meters	0.7	1.2	3.3
Preproduction Advance	meters	56	99	265
Cost	dollars/meter	$539	$596	$640
Ore Passes				
Face Area	square meters	1.4	2.5	4.3
Daily Advance	meters	0.03	0.1	0.1
Preproduction Advance	meters	97	155	241
Cost	dollars/meter	$370	$362	$557
Ventilation Raises				
Face Area	square meters	3.1	5.3	9.0
Daily Advance	meters	0.1	0.1	0.2
Preproduction Advance	meters	447	655	841
Cost	dollars/meter	$570	$1,047	$1,598
Hourly Labor Requirements				
Stope Miners		4	8	12
Development Miners		8	8	12
Equipment Operators		1	1	2
Hoist Operators		4	4	8
Support Miners		2	2	2
Diamond Drillers		2	2	2
Backfill Plant Workers		2	4	4
Electricians		4	5	5
Mechanics		5	7	10
Maintenance Workers		4	6	10
Helpers		3	3	5
Underground Laborers		5	8	12
Surface Laborers		4	6	10
Total Hourly Personnel		48	64	94
Salaried Personnel Requirements				
Managers		1	1	1
Superintendents		1	2	3
Foremen		2	4	6
Engineers		2	3	4
Geologists		2	3	4
Shift Bosses		2	4	4
Technicians		4	6	8
Accountants		1	2	4
Purchasing		2	4	5
Personnel		2	3	5
Secretaries		4	6	10
Clerks		5	8	12
Total Salaried Personnel		28	46	66
Supply Requirements (daily)				
Explosives	kilograms	494	1,231	2,513
Caps	each	57	134	278
Boosters	each	54	127	265
Fuse	meters	195	331	594
Drill Bits	each	1.5	2.2	3.9
Drill Steel	each	0.11	0.15	0.27
Backfill Pipe	meters	3.1	7.1	14.6
Fresh Water Pipe	meters	3.1	7.1	14.6
Compressed Air Pipe	meters	3.1	7.1	14.6
Electric Cable	meters	3.1	7.1	14.6

(continued)

TABLE 4.8   Cost models: end slice mining (continued)

Daily ore production (tonnes)	Units	800	2,000	4,000
Supply Requirements (daily) (continued)				
Ventilation Tubing	meters	3.1	7.1	14.6
Rock Bolts	each	4	10	21
Cement	tonnes	26	63	126
Buildings				
Office	square meters	715	1,175	1,737
Change House	square meters	557	743	1,092
Warehouse	square meters	255	419	551
Shop	square meters	505	873	1,170
Mine Plant	square meters	105	142	181
Equipment Requirements (number–size)				
Stope DTH Drills	centimeter	2 – 14.0	3 – 14.0	4 – 14.0
Stope Load-Haul-Dumps	cubic meter	2 – 3.1	3 – 6.5	4 – 9.9
Horizontal Development Drills	centimeter	2 – 3.5	2 – 3.5	3 – 3.81
Development Load-Haul-Dumps	cubic meter	2 – 3.1	2 – 6.5	2 – 9.9
Raise Borers	meter	1 – 2.4	1 – 3.0	1 – 3.7
Production Hoists	centimeter	1 – 152	1 – 152	2 – 152
Rock Bolt Jacklegs	centimeter	1 – 3.81	1 – 3.81	1 – 3.81
Drain Pumps	horsepower	6 – 35	8 – 86	18 – 94
Fresh Water Pumps	horsepower	2 – 0.5	2 – 0.5	4 – 0.5
Backfill Mixers	horsepower	1 – 15	1 – 15	1 – 25
Backfill Pumps	horsepower	2 – 8	2 – 8	2 – 12
Service Vehicles	horsepower	3 – 82	4 – 130	6 – 130
Compressors	cubic meters per minute	1 – 22.7	1 – 42.5	1 – 85.0
Ventilation Fans	centimeter	1 – 122	1 – 122	1 – 213
Exploration Drills	centimeter	1 – 4.45	1 – 4.45	1 – 4.45
Equipment Costs (dollars/unit)				
Stope DTH Drills		$215,300	$215,300	$215,300
Stope Load-Haul-Dumps		232,200	469,500	570,200
Horizontal Development Drills		397,400	397,400	397,400
Development Load-Haul-Dumps		232,200	469,500	570,200
Raise Borers		2,429,700	3,855,800	5,192,700
Hoists		609,384	758,130	794,446
Rock Bolters		657,150	657,150	657,150
Drain Pumps		10,320	12,150	12,150
Fresh Water Pumps		5,810	5,810	5,810
Backfill Mixers		29,250	29,250	33,890
Backfill Pumps		8,190	8,190	9,230
Service Vehicles		79,990	91,010	91,010
Compressors		70,280	90,360	125,500
Ventilation Fans		107,120	107,120	180,290
Exploration Drills		37,670	37,670	37,670

(continued)

TABLE 4.8   Cost models: end slice mining (continued)

Daily ore production (tonnes)	Units	800	2,000	4,000
COST SUMMARY				
Operating Costs (dollars/tonne ore)				
Equipment Operation		$1.99	$1.79	$2.07
Supplies		4.59	4.43	4.49
Hourly Labor		9.72	6.44	4.69
Administration		5.98	3.92	2.86
Sundries		2.23	1.66	1.41
Total Operating Costs		**$24.51**	**$18.24**	**$15.52**
Unit Operating Cost Distribution (dollars/tonne ore)				
Stopes		$2.45	$2.47	$2.00
Drifts		4.33	2.46	2.04
Crosscuts		1.22	0.49	0.57
Ore Passes		0.01	0.01	0.01
Vent Raises		0.06	0.05	0.05
Main Haulage		2.31	1.63	1.84
Backfill		3.37	3.32	3.10
Services		0.87	1.34	1.14
Ventilation		0.19	0.08	0.12
Exploration		0.39	0.20	0.10
Maintenance		2.04	1.36	0.84
Administration		5.04	3.17	2.30
Miscellaneous		2.23	1.66	1.41
Total Operating Costs		**$24.51**	**$18.24**	**$15.52**
Capital Costs				
Equipment Purchase		$7,352,000	$10,950,000	$16,440,000
Preproduction Underground Excavation				
Shaft #1		3,557,000	6,669,000	9,341,000
Shaft #2		—	—	9,319,000
Drifts		116,900	313,200	637,700
Crosscuts		30,050	59,140	169,300
Ore Passes		35,700	55,930	134,600
Preproduction Underground Excavation				
Ventilation Raises		$254,400	$685,300	$1,344,000
Surface Facilities		2,019,000	2,920,000	3,892,000
Working Capital		1,046,000	1,946,000	3,310,000
Engineering & Management		2,005,000	3,248,000	6,192,000
Contingency		1,337,000	2,165,000	4,128,000
Total Capital Costs		**$17,753,050**	**$29,011,570**	**$54,907,600**
Total Capital Cost per Daily Tonne Ore		**$22,191**	**$14,506**	**$13,727**

Source: Sherpa Cost Estimating Software for Underground Mines (Stebbins 2000)

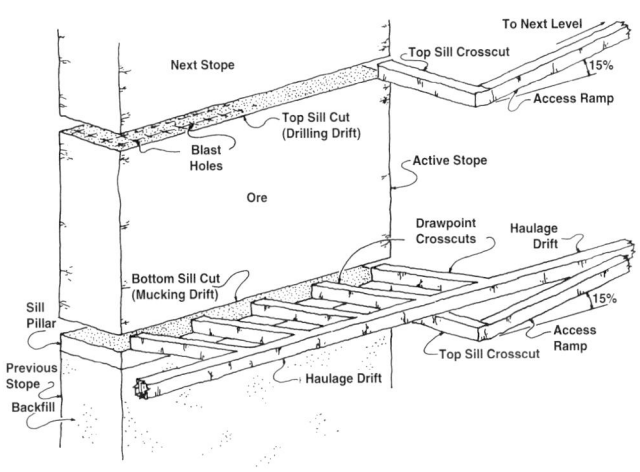

FIGURE 4.4   Stope development: vertical crater retreat (VCR) mining

TABLE 4.9   Cost models: VCR mining

These models represent mines on steeply dipping veins, 8.5, 10, or 11.5 meters wide respectively, and 450, 675 or 900 meters along strike. Access is by one or two shafts, 627, 823, or 1,219 meters deep and a secondary access/vent raise. Haulage from the stopes is by load-haul-dump units. Stoping includes excavating top and bottom sill cuts across the stopes, and VCR drilling and blasting using down-the-hole blasthole drills, followed by sand filling. See Figure 4.4.

Daily ore production (tonnes)	Units	800	2,000	4,000
Production				
Hours per Shift		8	10	10
Shifts per Day		2	2	2
Days per Year		320	320	320
Deposit				
Total Minable Resource	tonnes	2,822,400	9,873,900	22,574,000
Dip	degrees	80	80	80
Average Strike Length	meters	450	675	900
Average Vein Width	meters	8.5	10	11.5
Average Vertical	meters	250	500	750
Stopes				
Stope Length	meters	38	38	36
Stope Width	meters	8.6	10.2	11.7
Stope Height	meters	27	27	28
Face Width	meters	8.6	10.2	11.7
Face Height	meters	3.0	3.0	3.5
Advance per Round	meters	38	38	36
Sill Pillar Length	meters	38	38	36
Sill Pillar Width	meters	8.6	10.2	11.7
Sill Pillar Height	meters	3.0	3.0	3.4
Development Openings				
Shafts				2 ea.
Face Area	square meters	17.5	22.6	27.5
Preproduction Advance	meters	627	823	1,219
Costs	dollars/meter	$7,417	$10,476	$11,514
Drifts				
Face Area	square meters	11.5	13.9	15.4
Daily Advance	meters	5.7	12.3	22.5
Preproduction Advance	meters	459	984	1,801
Costs	dollars/meter	$609	$678	$728

(continued)

TABLE 4.9   Cost models: VCR mining (continued)

Daily ore production (tonnes)	Units	800	2,000	4,000
Crosscuts				
Face Area	square meters	11.5	13.9	15.4
Daily Advance	meters	2.8	5.9	10.5
Preproduction Advance	meters	222	472	843
Costs	dollars/meter	$532	$608	$653
Ore Passes				
Face Area	square meters	1.4	2.5	4.3
Daily Advance	meters	0.14	0.4	0.65
Preproduction Advance	meters	246	492	739
Costs	dollars/meter	$370	$366	$564
Ventilation Raises				
Face Area	square meters	3.1	5.3	9.0
Daily Advance	meters	0.17	0.32	0.4
Preproduction Advance	meters	596	792	1,189
Costs	dollars/meter	$571	$1,056	$1,610
Hourly Labor Requirements				
Stope Miners		16	24	36
Development Miners		10	14	20
Equipment Operators		2	2	3
Hoist Operators		4	4	8
Support Miners		2	2	2
Diamond Drillers		2	2	4
Backfill Plant Workers		2	4	4
Electricians		5	6	6
Mechanics		12	15	19
Maintenance Workers		4	6	10
Helpers		5	7	10
Underground Laborers		5	8	12
Surface Laborers		4	6	10
Total Hourly Personnel		73	100	144
Salaried Personnel Requirements				
Managers		1	1	1
Superintendents		1	2	3
Foremen		2	4	6
Engineers		2	3	4
Geologists		2	3	4
Shift Bosses		6	8	10
Technicians		4	6	8
Accountants		1	2	4
Purchasing		2	4	5
Personnel		4	5	7
Secretaries		4	6	10
Clerks		5	8	12
Total Salaried Personnel		34	52	74
Supply Requirements (daily)				
Explosives	kilograms	651	1,564	3,054
Caps	each	162	354	645
Boosters	each	154	338	618
Fuse	meters	654	1,427	2,158
Drill Bits	each	3.97	8.06	12.91
ODrill Steel	each	0.20	0.38	0.65
Backfill Pipe	meters	8.5	18.2	33.0
Fresh Water Pipe	meters	8.5	18.2	33.0
Compressed Air Pipe	meters	8.5	18.2	33.0
Electric Cable	meters	8.5	18.2	33.0

(continued)

TABLE 4.9   Cost models: VCR mining (continued)

Daily ore production (tonnes)	Units	800	2,000	4,000
Supply Requirements (daily) (continued)				
Ventilation Tubing	meters	8.5	18.2	33.0
Rock Bolts	each	9	21	41
Cement	tonnes	24	60	120
Buildings				
Office	square meters	869	1,328	1,891
Change House	square meters	848	1,161	1,672
Warehouse	square meters	349	708	1,042
Shop	square meters	716	1,523	2,276
Mine Plant	square meters	181	222	265
Equipment Requirements (number–size)				
Stope DTH Drills	centimeter	5 – 14.0	8 – 14.0	11 – 15.2
Stope Load-Haul-Dumps	cubic meter	5 – 3.1	7 – 6.5	10 – 9.9
Horizontal Development Drills	centimeter	3 – 3.5	4 – 3.5	5 – 3.8
Development Load-Haul-Dumps	cubic meter	2 – 3.1	3 – 6.5	3 – 9.9
Raise Borers	meter	1 – 2.4	1 – 3.0	1 – 3.7
Production Hoists	centimeter	1 – 152	1 – 152	2 – 152
Rock Bolt Jacklegs	centimeter	1 – 3.81	1 – 3.81	1 – 3.81
Drain Pumps	horsepower	7 – 38	9 – 91	26 – 88
Fresh Water Pumps	horsepower	2 – 0.5	2 – 0.5	4 – 0.5
Backfill Mixers	horsepower	1 – 15	1 – 15	1 – 25
Backfill Pumps	horsepower	2 – 5	2 – 8	2 – 12
Service Vehicles	horsepower	7 – 82	10 – 130	14 – 130
Compressors	cubic meters per minute	1 – .85	1 – 142	1 – 227
Ventilation Fans	centimeter	1 – 122	1 – 152	1 – 244
Exploration Drills	centimeter	1 – 4.45	1 – 4.45	1 – 4.45
Equipment Costs (dollars/unit)				
Stope DTH Drills		$215,300	$215,300	$215,300
Stope Load-Haul-Dumps		232,200	469,500	570,200
Horizontal Development Drills		397,400	397,400	397,400
Development Load-Haul-Dumps		232,000	469,500	570,200
Raise Borers		2,429,700	3,855,800	5,192,700
Hoists		672,542	794,446	875,414
Rock Bolters		480,200	480,200	480,200
Drain Pumps		10,320	12,150	12,150
Fresh Water Pumps		5,810	5,810	5,810
Backfill Mixers		29,250	29,250	33,890
Backfill Pumps		7,980	8,190	9,230
Service Vehicles		79,990	91,010	91,010
Compressors		125,280	200,800	316,300
Ventilation Fans		107,120	136,340	234,510
Exploration Drills		37,670	37,670	37,670

(continued)

TABLE 4.9   Cost models: VCR mining (continued)

Daily ore production (tonnes)	Units	800	2,000	4,000
COST SUMMARY				
Operating Costs (dollars/tonne ore)				
Equipment Operation		$2.86	$2.76	$3.44
Supplies		7.05	6.71	6.36
Hourly Labor		15.11	11.06	7.92
Administration		7.33	5.25	3.68
Sundries		3.24	2.58	2.14
Total Operating Costs		**$35.59**	**$28.36**	**$23.54**
Unit Operating Cost Distribution (dollars/tonne ore)				
Stopes		$7.27	$6.30	$4.93
Drifts		4.36	3.49	3.02
Crosscuts		2.00	1.59	1.35
Ore Passes		0.04	0.05	0.06
Vent Raises		0.08	0.13	0.14
Main Haulage		2.76	2.03	2.44
Backfill		3.19	3.14	2.94
Services		2.71	2.08	1.66
Ventilation		0.19	0.12	0.22
Exploration		0.41	0.24	0.24
Maintenance		3.82	2.60	1.55
Administration		5.52	4.01	2.85
Miscellaneous		3.24	2.58	2.14
Total Operating Costs		**$35.59**	**$28.36**	**$23.54**
Capital Costs				
Equipment Purchase		$9,426,000	$15,760,000	$23,950,000
Preproduction Underground Excavation				
Shaft #1		4,648,000	8,620,000	14,040,000
Shaft #2		—	—	14,000,000
Drifts		279,400	667,300	1,312,000
Crosscuts		118,350	287,140	550,400
Preproduction Underground Excavation				
Ore Passes		$91,090	$180,200	$416,400
Ventilation Raises		340,400	836,400	1,914,000
Surface Facilities		2,658,000	4,000,000	5,403,000
Working Capital		1,519,000	3,025,000	5,024,000
Engineering & Management		2,634,000	4,553,000	9,239,000
Contingency		1,756,000	3,035,000	6,159,000
Total Capital Costs		**$23,470,240**	**$40,964,040**	**$82,007,800**
Total Capital Cost per Daily Tonne Ore		**$29,338**	**$20,482**	**$20,502**

Source: Sherpa Cost Estimating Software for Underground Mines (Stebbins 2000)

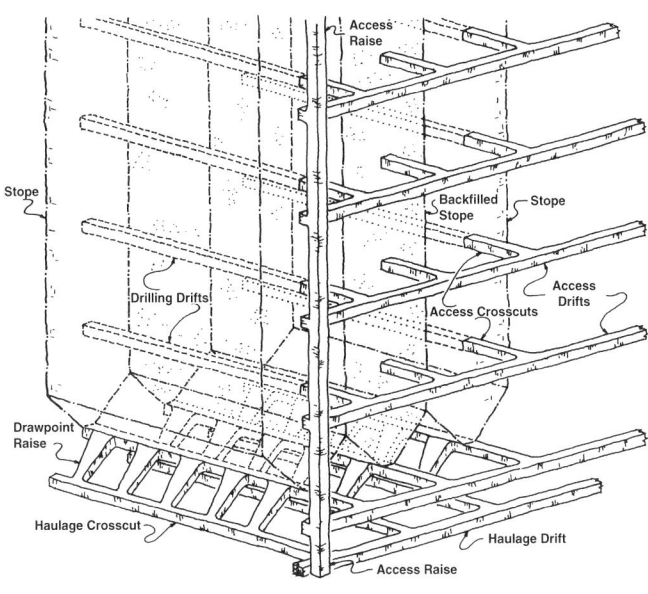

FIGURE 4.5   Stope development: sublevel longhole mining

TABLE 4.10   Cost models: sublevel long hole mining

These models represent mines on steeply dipping ore zones, 100x100, 250x250, or 320x320 meters in plan view respectively. Access is by one or two shafts, 380, 550, or 980 meters deep, and a secondary access/vent raise. Haulage from the stopes is by load-haul-dump units. Stoping includes excavating haulage cross cuts and drawpoints at the base of the stope and drill access crosscuts into the stope, followed by ring drilling using longhole drill jumbos, blasting, and sand filling. See Figure 4.5.

Daily ore production (tonnes)	Units	800	4,000	8,000
Production				
Hours per Shift		8	10	10
Shifts per Day		2	2	2
Days per Year		320	320	350
Deposit				
Total Minable Resource	tonnes	2,822,400	22,574,000	55,553,000
Dip	degrees	90	90	90
Average Strike Length	meters	100	250	320
Average Vein Width	meters	100	250	320
Average Vertical	meters	100	125	200
Stopes				
Stope Length	meters	100	125	160
Stope Width	meters	12.5	12.5	12.8
Stope Height	meters	100	62.5	66.7
Face Width	meters	12.5	12.5	12.8
Face Height	meters	25.0	20.8	22.2
Advance per Round	meters	4.5	4.5	4.5
Sill Pillar Length	meters	100	125	160
Sill Pillar Width	meters	12.5	12.5	12.8
Sill Pillar Height	meters	13.7	13.7	13.7
Development Openings				
Shafts			2 ea.	2 ea.
Face Area	square meters	17.5	27.5	33.4
Preproduction Advance	meters	380	555	980
Cost	dollars/meter	$7,403	$10,649	$12,371

(continued)

TABLE 4.10   Cost models: sublevel long hole mining (continued)

Daily ore production (tonnes)	Units	800	4,000	8,000
Drifts				
Face Area	square meters	11.5	15.4	17.8
Daily Advance	meters	0.6	5.5	9.7
Preproduction Advance	meters	52	443	848
Cost	dollars/meter	$502	$570	$598
Crosscuts				
Face Area	square meters	11.5	15.4	17.8
Daily Advance	meters	0.1	0.9	1.2
Preproduction Advance	meters	10	70	108
Cost	dollars/meter	$487	$541	$569
Access Raises				
Face Area	square meters	2.7	4.6	6.8
Daily Advance	meters	0.3	1.2	1.8
Preproduction Advance	meters	22	96	158
Cost	dollars/meter	$752	$857	$1,035
Draw Points				
Average Face Area	square meters	33.2	48.5	57.2
Daily Advance	meters	0.3	2.7	4.8
Preproduction Advance	meters	24	213	419
Cost	dollars/meter	$639	$844	$1,120
Ore Passes				
Face Area	square meters	—	4.3	7.9
Daily Advance	meters	—	0.02	0.03
Preproduction Advance	meters	—	250	400
Cost	dollars/meter	—	$487	$1,390
Ventilation Raises				
Face Area	square meters	3.1	9.0	16.3
Daily Advance	meters	0.10	0.09	0.14
Preproduction Advance	meters	350	525	950
Cost	dollars/meter	$497	$1,390	$1,395
Hourly Labor Requirements				
Stope Miners		4	8	12
Development Miners		6	12	12
Equipment Operators		1	2	3
Hoist Operators		4	8	8
Support Miners		4	4	4
Diamond Drillers		2	2	2
Backfill Plant Workers		2	4	6
Electricians		4	5	5
Mechanics		5	7	10
Maintenance Workers		4	10	14
Helpers		2	4	5
Underground Laborers		5	12	18
Surface Laborers		4	10	14
Total Hourly Personnel		47	88	113
Salaried Personnel Requirements				
Managers		1	1	1
Superintendents		1	3	4
Foremen		2	6	10
Engineers		2	4	5
Geologists		2	4	6

(continued)

TABLE 4.10   Cost models: sublevel long hole mining (continued)

Daily ore production (tonnes)	Units	800	4,000	8,000
Salaried Personnel Requirements (continued)				
Shift Bosses		2	4	6
Technicians		4	8	10
Accountants		1	4	5
Purchasing		2	5	8
Personnel		2	4	6
Secretaries		4	10	14
Clerks		5	12	18
Total Salaried Personnel		28	65	93
Supply Requirements (daily)				
Explosives	kilograms	341	1,647	3,191
Caps	each	24	198	360
Boosters	each	23	188	343
Fuse	meters	179	793	1,422
Drill Bits	each	0.68	2.81	4.87
Drill Steel	each	0.05	0.20	0.35
Backfill Pipe	meters	0.7	6.4	10.9
Fresh Water Pipe	meters	1.0	7.6	12.7
Compressed Air Pipe	meters	1.0	7.6	12.7
Electric Cable	meters	1.0	7.6	12.7
Ventilation Tubing	meters	1.0	7.6	12.7
Rock Bolts	each	1	8	14
Cement	tonnes	25	114	227
Shotcrete	cubic meters	—	3.0	6.0
Timber	cubic centimeters	328,062	1,803,162	3,325,465
Buildings				
Office	square meters	715	1,661	2,376
Change House	square meters	546	999	1,312
Warehouse	square meters	255	481	551
Shop	square meters	505	1,012	1,170
Equipment Requirements (number–size)				
Stope Longhole Drills	centimeter	2 – 6.35	3 – 6.35	4 – 7.00
Stope Load-Haul-Dumps	cubic meter	2 – 3.1	3 – 10	4 – 10
Vertical Development Stopers	centimeter	2 – 4.14	2 – 5.08	2 – 5.72
Horizontal Development Jumbos	centimeter	2 – 3.51	2 – 3.81	3 – 3.81
Development Load-Haul-Dumps	cubic meter	2 – 3.1	2 – 10	2 – 10
Raise Borers	meter	1 – 2.4	1 – 3.7	1 – 3.7
Production Hoists	centimeter	1 – 152	2 – 152	2 – 203
Shotcreter	cubic meter/hr	1 – 32	1 – 32	2 – 53
Drain Pumps	horsepower	3 – 68	12 – 96	18 – 206
Fresh Water Pumps	horsepower	2 – 0.5	4 – 0.5	4 – 0.5
Backfill Mixers	horsepower	1 – 15	1 – 25	1 – 40
Backfill Pumps	horsepower	2 – 8	2 – 12	2 – 12
ANFO Loader	kilogram per minute	1 – 272	1 – 272	2 – 272
Service Vehicles	horsepower	3 – 82	4 – 130	6 – 210
Ventilation Fans	centimeter	1 – 122	1 – 183	1 – 244
Exploration Drills	centimeter	1 – 4.45	1 – 4.45	1 – 4.45
Equipment Costs (dollars/unit)				
Stope Longhole Drills		$405,700	$405,700	$405,700
Stope Load-Haul-Dumps		232,200	570,200	570,200
Vertical Development Stopers		5,880	5,880	5,880
Horizontal Development Jacklegs		397,400	397,400	703,700

(continued)

TABLE 4.10   Cost models: sublevel long hole mining (continued)

Daily ore production (tonnes)	Units	800	4,000	8,000
Development Load-Haul-Dumps		232,200	570,200	570,200
Raise Borers		2,429,700	5,192,700	5,192,700
Hoists		606,069	709,499	1,028,709
Shotcreters		50,850	50,850	65,600
Drain Pumps		10,820	12,150	49,390
Fresh Water Pumps		5,810	5,810	5,810
Backfill Mixers		29,250	33,890	47,200
Backfill Pumps		8,190	9,230	9,230
ANFO Loader		29,750	29,750	29,750
Service Vehicles		79,990	91,010	214,100
Ventilation Fans		107,120	135,420	219,890
Exploration Drills		37,670	37,670	37,670

COST SUMMARY

Operating Costs (dollars/tonne ore)

Equipment Operation		$1.62	$1.45	$1.99
Supplies		3.89	3.79	3.69
Hourly Labor		9.55	4.26	3.00
Administration		5.98	2.73	2.04
Sundries		2.10	1.22	1.07
Total Operating Costs		**$23.14**	**$13.45**	**$11.79**

Unit Operating Cost Distribution (dollars/tonne ore)

Stopes		$2.66	$1.64	$1.49
Drifts		1.83	1.09	0.79
Crosscuts		0.34	0.17	0.10
Draw Points		0.74	0.43	0.32
Access Raises		0.90	0.25	0.15
Vent Raises		0.05	0.03	0.02
Main Haulage		2.06	1.47	1.54
Backfill		3.33	2.82	2.74
Services		$1.46	$1.26	$1.33
Ventilation		0.19	0.04	0.07
Exploration		0.39	0.10	0.06
Maintenance		2.04	0.68	0.46
Administration		5.04	2.25	1.65
Miscellaneous		2.11	1.22	1.07
Total Operating Costs		**$23.14**	**$13.45**	**$11.79**

Capital Costs

Equipment Purchase		$7,062,000	$14,500,000	$20,080,000
Preproduction Underground Excavation				
Shaft #1		2,817,000	5,915,000	12,130,000
Shaft #2		—	5,896,000	12,100,000
Drifts		25,970	252,100	507,400
Crosscuts		4,842	38,100	61,540
Draw Points		15,360	179,500	469,100
Access Raises		16,360	82,430	163,300
Ore Passes		—	121,600	556,000
Ventilation Raises		174,000	729,700	1,325,000
Surface Facilities		1,874,000	3,413,000	4,265,000
Working Capital		987,700	2,870,000	5,505,000
Engineering & Management		1,559,000	4,046,000	6,715,000
Contingency		1,199,000	3,113,000	5,165,000
Total Capital Costs		**$15,735,232**	**$41,156,430**	**$69,042,340**
Total Capital Cost per Daily Tonne Ore		**$19,669**	**$10,289**	**$8,630**

Source: Sherpa Cost Estimating Software for Underground Mines (Stebbins 2000)

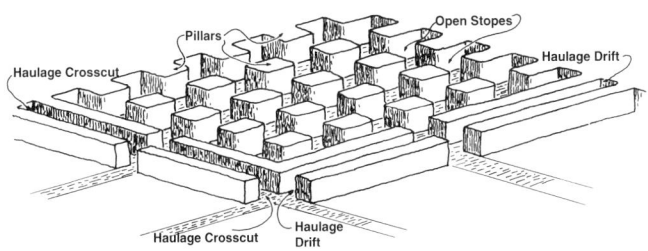

FIGURE 4.6   Stope development: room and pillar mining

TABLE 4.11   Cost models: room and pillar mining

These models represent mines on flatlying bedded deposits, 2.5, 5.0, or 10 meters thick, respectively, with extensive areal dimensions. Access is by two shafts, 281, 581, or 781 meters deep, and a secondary access/vent raise. Ore is collected at the face using front-end-loaders and loaded into articulated rear-dump trucks for transport to a shaft. Stoping follows a conventional room and pillar pattern, with drilling accomplished using horizontal drill jumbos. See Figure 4.6.

Daily ore production (tonnes)	Units	1,200	8,000	14,000
Production				
Hours per Shift		8	10	8
Shifts per Day		2	2	3
Days per Year		320	350	350
Deposit				
Total Minable Resource	tonnes	5,080,300	43,208,000	86,419,000
Dip	degrees	5	5	5
Average Strike Length	meters	1,000	2,000	2,400
Average Vein Width	meters	700	1,500	1,500
Average Vertical	meters	2.5	5.0	10
Stopes				
Stope Length	meters	59	59	60
Stope Width	meters	43.5	44.4	45.3
Stope Height	meters	2.5	5.0	10.0
Face Width	meters	4.3	4.4	4.5
Face Height	meters	2.5	5.0	10.0
Advance per Round	meters	2.3	3.0	3.9
Pillar Length	meters	6.9	6.9	7.0
Pillar Width	meters	5.1	5.2	5.3
Pillar Height	meters	2.5	5.0	10.0
Development Openings				
Shafts		2 ea.	2 ea.	2 ea.
Face Area	square meters	19.6	33.4	39.1
Preproduction Advance	meters	281	581	781
Cost	dollars/meter	$7,834	$12,411	$11,440
Drifts				
Face Area	square meters	12.5	17.8	19.9
Daily Advance	meters	6.1	20.0	17.2
Preproduction Advance	meters	490	1,748	1,501
Cost	dollars/meter	$666	$746	$819
Crosscuts				
Face Area	square meters	12.5	17.8	19.9
Daily Advance	meters	4.5	15.0	12.9
Preproduction Advance	meters	360	1,311	1,125
Cost	dollars/meter	$637	$715	$780

(continued)

TABLE 4.11   Cost models: room and pillar mining (continued)

Daily ore production (tonnes)	Units	1,200	8,000	14,000
Ventilation Raises				
Face Area	square meters	3.9	16.3	27.2
Daily Advance	meters	0.3	1.53	2.19
Preproduction Advance	meters	250	550	750
Cost	dollars/meter	$464	$1,299	$1,328
Hourly Labor Requirements				
Stope Miners		16	56	96
Development Miners		12	24	24
Equipment Operators		2	6	14
Hoist Operators		8	8	12
Support Miners		2	2	2
Diamond Drillers		2	4	6
Electricians			5	7
Mechanics			12	25
Maintenance Workers		5	14	19
Helpers			5	14
Underground Laborers		6	18	25
Surface Laborers		5	14	19
Total Hourly Personnel		80	192	282
Salaried Personnel Requirements				
Managers		1	1	1
Superintendents		2	4	4
Foremen		4	10	21
Engineers		2	5	7
Geologists		2	6	8
Shift Bosses		6	16	27
Technicians		4	10	14
Accountants		2	5	7
Purchasing		3	8	11
Personnel		4	10	14
Secretaries		5	14	19
Clerks		6	18	25
Total Salaried Personnel		41	107	158
Supply Requirements (daily)				
Explosives	kilograms	959	5,975	10,208
Caps	each	389	1,591	1,582
Boosters	each	357	1,497	1,510
Fuse	meters	1,529	6,643	7,773
Drill Bits	each	8.65	43.59	58.02
Drills Steel	each	0.62	3.15	4.19
Fresh Water Pipe	meters	10.6	35.0	30.1
Compressed Air Pipe	meters	10.6	35.0	30.1
Electric Cable	meters	10.6	35.0	30.1
Ventilation Tubing	meters	10.6	35.0	30.1
Rock Bolts	each	61	309	455
Buildings				
Office	square meters	1,047	2,734	4,037
Change House	square meters	929	2,230	3,275
Warehouse	square meters	269	657	748
Shop	square meters	536	1,409	1,614

(continued)

TABLE 4.11 Cost models: room and pillar mining (continued)

Daily ore production (tonnes)	Units	1,200	8,000	14,000
Equipment Requirements (number–size)				
Stope Jumbos	centimeter	5 – 3.49	17 – 4.13	20 – 5.72
Stope Front-End Loader	cubic meter	5 – 1.1	16 – 1.1	18 – 1.5
Stope Rear-Dump Truck	tonne	1 – 15	4 – 35	6 – 35
Horizontal Development Jumbos	centimeter	3 – 3.49	6 – 3.81	5 – 4.14
Development Front-End Loader	cubic meter	2 – 1.1	4 – 1.1	3 – 1.5
Development Rear-Dump Truck	tonne	2 – 15	4 – 35	3 – 35
Raise Borers	meter	1 – 2.4	1 – 3.7	1 – 3 .7
Production Hoists	centimeter	2 – 152	2 – 203	2 – 305
Rock Bolters	centimeter	1 – 3.81	1 – 3.81	1 – 3.81
Drain Pumps	horsepower	8 – 25	14 – 164	18 – 288
Fresh Water Pumps	horsepower	4 – 0.5	4 – 0.5	4 – 0.5
ANFO Loader	kilogram per minute	2 – 272	5 – 272	6 – 272
Service Vehicles/Scalers	horsepower	7 – 82	20 – .210	28 – 210
Ventilation Fans	centimeter	1 – 122	1 – 244	1 – 274
Exploration Drills	centimeter	1 – 4.45	1 – 4.45	1 – 4.45
Equipment Costs (dollars/unit)				
Stope Jumbos		$703,700	$703,700	$726,500
Stope Front-End Loader		101,700	101,700	118,100
Stope Rear-Dump Truck		210,700	433,700	433,700
Equipment Costs (dollars/unit) – continued				
Horizontal Development Jumbos		$397,400	$703,700	$703,700
Development Front-End Loader		101,700	101,700	118,100
Development Rear-Dump Truck		210,700	433,700	433,700
Raise Borers		2,429,700	5,192,700	5,192,700
Hoists		532,054	963,214	1,286,841
Roof Bolters		480,200	480,200	657,150
Drain Pumps		9,600	33,640	49,390
Fresh Water Pumps		5,810	5,810	5,810
ANFO Loader		29,750	29,750	29,750
Service Vehicles/Scalers		79,990	214,100	181,605
Ventilation Fans		107,120	219,890	250,890
Exploration Drills		37,670	37,670	37,670

(continued)

TABLE 4.11 Cost models: room and pillar mining (continued)

Daily ore production (tonnes)	Units	1,200	8,000	14,000
COST SUMMARY				
Operating Costs (dollars/tonne ore)				
Equipment Operation		$1.86	$1.80	$2.04
Supplies		3.98	2.79	2.05
Hourly Labor		11.01	5.30	3.56
Administration		5.93	2.42	2.04
Sundries		2.28	1.23	0.97
Total Operating Costs		**$25.06**	**$13.54**	**$10.66**
Unit Operating Cost Distribution (dollars/tonne ore)				
Stopes		$5.62	$4.47	$3.74
Drifts		3.05	1.26	0.59
Crosscuts		2.09	0.92	0.43
Vent Raises		0.08	0.20	0.17
Main Haulage		2.57	1.22	1.37
Services		2.04	1.25	1.26
Ventilation		0.12	0.07	0.04
Exploration		0.28	0.12	0.08
Maintenance		2.54	0.99	0.62
Administration		4.39	1.81	1.39
Miscellaneous		2.28	1.23	0.97
Total Operating Costs		**$25.06**	**$13.54**	**$10.66**
Capital Costs				
Equipment Purchase		$12,700,000	$37,290,000	$43,000,000
Preproduction Underground Excavation				
Shaft #1		2,199,000	7,210,000	8,939,000
Shaft #2		2,201,000	7,211,000	8,937,000
Drifts		326,500	1,305,000	1,230,000
Crosscuts		229,100	936,700	877,600
Ventilation Raises		115,900	714,400	996,000
Surface Facilities		2,459,000	5,329,000	6,980,000
Working Capital		1,604,000	6,324,000	8,702,000
Engineering & Management		2,631,000	7,800,000	9,224,000
Contingency		2,024,000	6,000,000	7,096,000
Total Capital Costs		**$26,489,500**	**$80,120,100**	**$95,981,600**
Total Capital Cost per Daily Tonne Ore		**$22,075**	**$10,015**	**$6,856**

Source: Sherpa Cost Estimating Software for Underground Mines (Stebbins 2000)

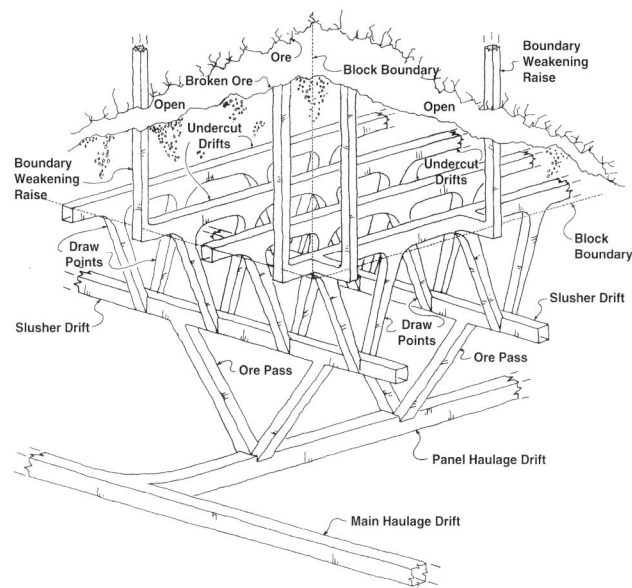

FIGURE 4.7   Stope development: block cave mining

TABLE 4.12   Cost models: block cave mining

These three models represent mines on large, bulk deposits, roughly 450, 525, and 600 meters square. Access is through 3 to 5 shafts, 430, 530, or 630 meters deep, and by secondary access/ventilation raises. Ore is collected using slushers, and haulage from the stopes is by diesel locomotive. Stope development includes driving drifts (haulage, slusher, and undercut) and raises (stope draw, ore pass, and boundary weakening). Caving is initiated by blasting on the undercut level. See Figure 4.7.

Daily ore production (tonnes)	Units	20,000	30,000	45,000
Production				
Hours per Shift		8	8	8
Shifts per Day		3	3	3
Days per Year		365	365	365
Deposit				
Total Minable Resource	tonnes	84,000,000	147,000,000	252,000,000
Average Maximum Horizontal	meters	450	525	600
Average Minimum Horizontal	meters	450	525	600
Average Vertical	meters	150	200	250
Stopes				
Block Length	meters	150	175	200
Block Width	meters	150	175	200
Block Height	meters	150	200	250
Development Openings				
Shafts		3 ea.	4 ea.	5 ea.
Face Area	square meters	43.2	48.4	54.2
Preproduction Advance	meters	1,290	2,120	3,150
Cost	dollars/meter	$11,694	$12,290	$13,226
Drifts				
Face Area	square meters	9.6	9.6	9.86
Daily Advance	meters	2.6	3.1	3.4
Preproduction Advance	meters	1,350	1,925	2,400
Cost	dollars/meter	$785	$788	$819
Crosscuts				
Face Area	square meters	9.6	9.6	9.86
Daily Advance	meters	3.4	4.3	4.9

(continued)

TABLE 4.12   Cost models: block cave mining (continued)

Daily ore production (tonnes)	Units	20,000	30,000	45,000
Preproduction Advance	meters	1,800	2,625	3,400
Cost	dollars/meter	$693	$695	$719
Draw Points				
Face Area	square meters	100	100	100
Daily Advance	meters	2.3	3.0	3.1
Preproduction Advance	meters	1,215	1,815	2,160
Cost	dollars/meter	$3,211	$3,416	$3,837
Ore Passes				
Face Area	square meters	18.8	27.9	41.5
Daily Advance	meters	0.91	1.22	1.41
Preproduction Advance	meters	480	750	990
Cost	dollars/meter	$1,254	$1,797	$2,699
Boundary Raises				
Face Area	square meters	13.6	19.3	27.8
Daily Advance	meters	11.4	15.7	18.6
Preproduction Advance	meters	6,000	9,600	13,000
Cost	dollars/meter	$526	$657	$899
Ventilation Raises				
Face Area	square meters	38.1	56.3	83.7
Daily Advance	meters	0.19	0.20	0.21
Preproduction Advance	meters	400	500	600
Cost	dollars/meter	$1,862	$2,743	$4,195
Hourly Labor Requirements (workers/day)				
Undercut Miners		66	96	144
Development Miners		30	42	50
Motormen		5	6	8
Hoist Operators		18	24	30
Support Miners		4	4	4
Diamond Drillers		6	10	18
Electricians		7	8	9
Mechanics		28	36	46
Maintenance Workers		24	30	38
Helpers		17	24	33
Underground Laborers		30	38	48
Surface Laborers		24	30	38
Total Hourly Personnel		259	348	466
Salaried Personnel Requirements (workers)				
Managers		1	1	1
Superintendents		4	4	4
Foremen		30	42	63
Engineers		8	10	12
Geologists		9	11	14
Shift Bosses		21	27	39
Technicians		16	20	24
Accountants		9	11	14
Purchasing		14	17	22
Personnel		13	17	23
Secretaries		24	30	38
Clerks		30	38	48
Total Salaried Personnel		179	228	302
Supply Requirements (daily)				
Explosives	kilograms	545	726	754
Caps	number	429	595	740

(continued)

TABLE 4.12 Cost models: block cave mining (continued)

Daily ore production (tonnes)	Units	20,000	30,000	45,000
Supply Requirements (daily) (continued)				
Boosters	number	412	575	717
Fuse	meters	3,407	4,923	6,913
Drill Bits	number	18.70	26.78	35.87
Drills Steel	number	1.042	1.481	1.915
Fresh Water	meters	17	23	27
Compressed Air Pipe	meters	17	23	27
Electric Cable	meters	17	23	27
Ventilation Tubing	meters	17	23	27
Rockbolts	number	30	37	48
Shotcrete	cubic meters	1	1	1
Concrete	cubic meters	5	10	17
Rail	meters	12	15	17
Buildings				
Office	square meters	4,573	5,825	7,515
Change House	square meters	2,961	3,983	5,342
Warehouse	square meters	369	462	597
Shop	square meters	761	971	1,274
Mine Plant	square meters	222	265	265
Equipment Requirements (number–size)				
Undercut Drills	centimeter	14 – 4.76	20 – 5.08	29 – 5.72
Undercut Slushers	centimeter	13 – 213.4	18 – 213.4	27 – 213.4
Development Jumbos	centimeter	2 – 3.17	3 – 3.17	3 – 3.49
Raise Drills	centimeter	5 – 4.76	7 – 5.08	8 – 5.72
Overshot Muckers	cubic meter	3 – 0.3	5 – 0.3	5 – 0.3
Diesel Locomotives w/ cars	tonne	3 – 31.8	5 – 31.8	5 – 31.8
Hoists	horsepower	3 – 3,176	4 – 3,979	5 – 5,518
Rockbolt Drills	centimeter	3 – 3.81	3 – 3.81	3 – 3.81
Shotcreters	cubic meters per hour	1 – 53	2 – 0.53	2 – 53
Fresh Water Pumps	horsepower	6 – 0.5	8 – 0.5	10 – 0.5
Drain Pumps	horsepower	9 – 593	16 – 550	20 – 781
Service Vehicles	horsepower	16 – 210	24 – 210	33 – 210
Compressors	cubics meters per minute	1 – 142	1 – 227	1 – 227
Ventilation Fans	centimeter	1 – 152	1 – 183	1 – 213
Exploration Drills	centimeter	1 – 4.4	2 – 4.4	3 – 4.4
Equipment Costs (dollars/unit)				
Undercut Jackleg Drills		$5,670	$5,670	$5,670
Undercut Slushers		61,680	61,680	61,680
Development Jumbos		397,400	397,400	397,400
Raise Drills		5,880	5,880	5,880
Development Muckers		66,700	66,700	66,700
Diesel Locomotives w/ cars		948,080	971,940	995,800
Hoists		1,130,064	1,170,385	1,321,056
Rockbolt Drills		5,670	5,670	5,670
Shotcreters		65,600	65,600	65,600

(continued)

TABLE 4.12 Cost models: block cave mining (continued)

Daily ore production (tonnes)	Units	20,000	30,000	45,000
Fresh Water Pumps		5,810	5,810	5,810
Drain Pumps		83,110	83,110	102,570
Service Vehicles		214,100	214,100	214,100
Compressors		200,800	316,300	316,300
Ventilation Fans		117,820	135,420	150,320
Exploration Drills		37,670	37,670	37,670
COST SUMMARY				
Operating Costs (dollars/tonne ore)				
Equipment Operation		$0.92	$1.05	$1.17
Supplies		0.67	0.65	0.61
Hourly Labor		2.20	1.98	1.78
Administration		1.53	1.29	1.15
Sundries		0.53	0.50	0.47
Total Operating Costs		**$5.85**	**$5.47**	**$5.18**
Unit Operating Cost Distribution (dollars/tonne ore)				
Stopes		$1.42	$1.35	$1.31
Drifts		0.10	0.07	0.05
Crosscuts		0.12	0.10	0.07
Draw Points		0.13	0.11	0.07
Ore Passes		0.03	0.04	0.03
Boundary Raises		0.26	0.25	0.21
Vent Raises		0.01	0.01	0.01
Main Haulage		0.86	0.92	0.99
Services		0.83	0.80	0.82
Ventilation		0.11	0.10	0.09
Exploration		0.06	0.06	0.09
Maintenance		0.34	0.29	0.24
Administration		1.05	0.87	0.73
Miscellaneous		0.53	0.50	0.47
Total Operating Costs		**$5.85**	**$5.47**	**$5.18**
Capital Costs (total dollars spent)				
Equipment Purchase		$16,940,000	$26,380,000	$33,180,000
Preproduction Underground Excavation				
Shafts		15,092,000	26,062,000	41,659,000
Drifts		1,059,000	1,516,000	1,965,000
Crosscuts		1,248,000	1,823,000	2,445,000
Draw Points		3,901,000	6,200,000	8,288,000
Ore Passes		602,000	1,348,000	2,672,000
Boundary Raises		3,158,000	6,309,000	11,690,000
Ventilation Raises		744,700	1,371,000	2,517,000
Surface Facilities		6,601,000	8,198,000	10,320,000
Working Capital		7,127,000	9,989,000	14,170,000
Engineering & Management		6,415,000	10,300,000	14,920,000
Contingency		4,935,000	7,921,000	11,470,000
Total Capital Costs		**$67,822,700**	**$107,417,000**	**$155,296,000**
Total Capital Cost per Daily Tonne Ore		**$3,391**	**$3,581**	**$3,451**

Source: Sherpa Cost Estimating Software for Underground Mines (Stebbins 2000)

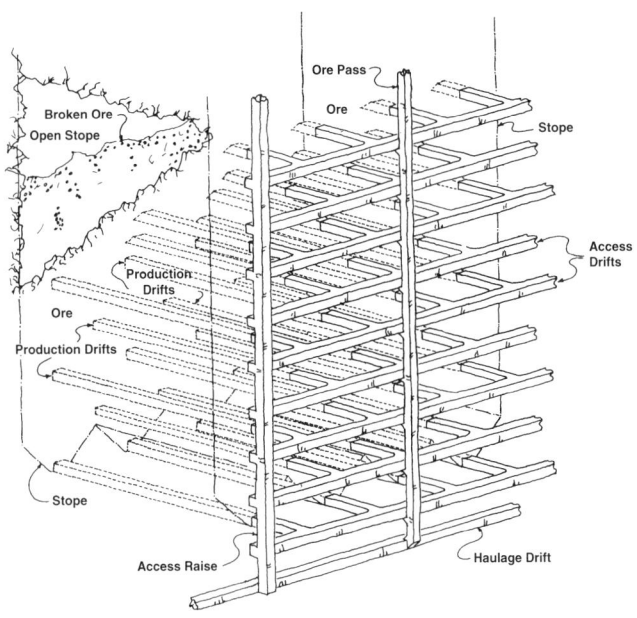

FIGURE 4.8 Stope development: sublevel cave mining

TABLE 4.13 Cost models: sublevel caving

These three models represent mines on steeply dipping veins, 40, 60, and 80 meters wide and 450, 600, and 800 meters along strike. Access is through 2 to 4 shafts, 775, 874, or 974 meters deep, and by secondary access/ventilation raises. Haulage from the stopes is by load-haul-dump. Stoping includes driving production drifts and access cross-cuts. Development includes ore passes, access raises, haulage drifts, and ventilation raises. The entire ore zone is blasted to induce caving. See Figure 4.8.

Daily ore production (tonnes)	Units	4,000	8,000	14,000
Production				
Hours per Shift		8	8	8
Shifts per Day		3	3	3
Days per Year		350	350	350
Deposit				
Total Minable Resource	tonnes	16,800,000	39,200,000	78,400,000
Dip	degrees	80	80	80
Average Maximum Horizontal	meters	450	600	800
Average Minimum Horizontal	meters	40	60	80
Average Vertical	meters	350	400	450
Stopes				
Stope Length	meters	40.6	60.9	81.2
Stope Width	meters	450	600	800
Stope Height	meters	350	400	450
Face Width	meters	12.5	12.5	12.5
Face Height	meters	23.0	21.9	21.9
Advance per Round	meters	12.5	12.5	12.5
Development Openings				
Shafts		2 ea.	3 ea.	4 ea.
Face Area	square meters	27.5	33.4	39.1
Preproduction Advance	meters	1,550	2,622	3,896
Cost	dollars/meter	$9,478	$10,343	$11,164
Drifts				
Face Area	square meters	15.4	17.8	19.9

(continued)

TABLE 4.13 Cost models: sublevel caving (continued)

Daily ore production (tonnes)	Units	4,000	8,000	14,000
Daily Advance	meters	20.4	40.4	65.6
Preproduction Advance	meters	1,787	3,539	5,743
Cost	dollars/meter	$545	$566	$605
Crosscuts				
Face Area	square meters	15.4	17.8	19.9
Daily Advance	meters	9.5	14.3	18.8
Preproduction Advance	meters	830	1,256	1,648
Cost	dollars/meter	$466	$480	$503
Access Raises				
Face Area	square meters	4.60	8.5	12.8
Daily Advance	meters	7.2	10.3	13.5
Preproduction Advance	meters	626	902	1,184
Cost	dollars/meter	$822	$963	$1,113
Ore Passes				
Face Area	square meters	4.3	7.9	13.4
Daily Advance	meters	0.16	0.24	0.40
Preproduction Advance	meters	345	394	443
Cost	dollars/meter	$605	$1,379	$1,625
Ventilation Raises				
Face Area	square meters	9.0	16.3	27.2
Daily Advance	meters	0.18	0.17	0.34
Preproduction Advance	meters	745	844	943
Cost	dollars/meter	$1,428	$1,754	$2,228
Hourly Labor Requirements (workers/day)				
Stope Miners		24	28	48
Development Miners		28	42	60
Equipment Operators		4	6	11
Hoist Operators		12	18	24
Support Miners		4	4	4
Diamond Drillers		2	2	4
Electricians		6	5	7
Mechanics		15	17	23
Maintenance Workers		10	14	19
Helpers		9	11	19
Underground Laborers		12	18	25
Surface Laborers		10	14	19
Total Hourly Personnel		136	179	263
Salaried Personnel Requirements (workers)				
Managers		1	1	1
Superintendents		3	4	4
Foremen		9	15	21
Engineers		4	5	7
Geologists		4	6	8
Shift Bosses		9	9	15
Technicians		8	10	14
Accountants		4	5	7
Purchasing		5	8	11
Personnel		7	8	13
Secretaries		10	14	19
Clerks		12	18	25
Total Salaried Personnel		76	103	145

(continued)

TABLE 4.13 Cost models: sublevel caving (continued)

Daily ore production (tonnes)	Units	4,000	8,000	14,000
Supply Requirements (daily)				
Explosives	kilograms	2,285	4,353	7,249
Caps	number	556	971	1,494
Boosters	number	524	919	1,421
Fuse	meters	1,088	1,646	2,358
Drill Bits	number	4.06	5.83	8.00
Drill Steel	number	0.277	0.391	0.525
Fresh Water Pipe	meters	37	65	98
Compressed Air Pipe	meters	37	65	98
Electric Cable	meters	37	65	98
Ventilation Tubing	meters	37	65	98
Shotcrete	cubic meters	17	33	53
Timber	cubic centimeters	12,000,000	21,000,000	33,000,000
Steel Liner Plate	kilograms	109	206	591
Buildings				
Office	square meters	1,942	2,759	3,704
Change House	square meters	1,579	2,241	3,054
Warehouse	square meters	490	689	892
Shop	square meters	1,032	1,481	1,938
Equipment Requirements (number–size)				
Stope Drills	centimeter	5 – 13.97	4 – 13.97	10 – 13.97
Stope Load-Haul-Dumps	cubic meter	5 – 3.1	4 – 4.6	9 – 6.1
Raise Drills	centimeter	3 – 3.17	3 – 3.17	3 – 3.49
Development Jumbo Drills	centimeter	4 – 3.81	6 – 3.81	9 – 4.13
Development Load-Haul-Dumps	cubic meter	3 – 3.1	4 – 4.6	6 – 6.1
Raise Borers	meter	1 – 4.0	1 – 4.0	1 – 4.0
Hoists	horsepower	2 – 1,915	3 – 2,655	4 – 3,692
Shotcreters	cubic meters per hour	2 – 32	3 – 53	5 – 53
Fresh Water Pumps	horsepower	4 – 0.5	6 – 0.5	8 – 0.5
Drain Pumps	horsepower	17 – 88	27 – 124	40 – 162
Service Vehicles	horsepower	9 – 130	10 – 210	18 – 210
ANFO Loaders	kilogram per minute	2 – 272	2 – 272	3 – 272
Ventilation Fans	centimeter	1 – 152	1 – 213	1 – 244
Exploration Drills	centimeter	1 – 4.4	1 – 4.4	1 – 4.4
Equipment Costs (dollars/unit)				
Stope Drills		$405,700	$405,700	$405,700
Stope Load-Haul-Dumps		313,100	403,500	403,500
Raise Drills		5,880	5,880	5,880
Development Jumbo Drill		397,400	703,700	703,700
Development Load-Haul-Dumps		313,100	403,500	403,500
Raise Borers		5,192,700	5,192,700	5,192,700
Hoists		758,130	938,550	1,008,064
Shotcreters		50,850	65,600	65,600

(continued)

TABLE 4.13 Cost models: sublevel caving (continued)

Daily ore production (tonnes)	Units	4,000	8,000	14,000
Fresh Water Pumps		5,810	5,810	5,810
Drain Pumps		12,150	31,240	33,640
Service Vehicles		91,010	214,100	214,100
ANFO Loader		29,750	29,750	29,750
Ventilation Fans		117,820	150,750	178,550
Exploration Drills		37,670	37,670	37,670

COST SUMMARY

Operating Costs (dollars/tonne ore)

		4,000	8,000	14,000
Equipment Operation		$2.02	$2.17	$2.21
Supplies		3.65	3.28	2.98
Hourly Labor		5.87	4.15	3.22
Administration		3.47	2.46	1.86
Sundries		1.50	1.21	1.03
Total Operating Costs		**$16.51**	**$13.27**	**$11.30**

Unit Operating Cost Distribution (dollars/tonne ore)

		4,000	8,000	14,000
Stopes		$2.36	$1.87	$1.56
Drifts		2.41	2.29	2.14
Crosscuts		1.18	0.88	0.67
Access Raises		0.74	0.49	0.37
Ore Passes		0.02	0.04	0.04
Vent Raises		0.06	0.03	0.05
Main Haulage		2.06	1.95	1.85
Services		2.40	2.00	1.73
Ventilation		0.05	0.04	0.03
Exploration		0.09	0.05	0.05
Maintenance		1.03	0.61	0.42
Administration		2.61	1.81	1.36
Miscellaneous		1.50	1.21	1.03
Total Operating Costs		**$16.51**	**$13.27**	**$11.30**

Capital Costs (total dollars spent)

	4,000	8,000	14,000
Equipment Purchase	$16,480,000	$28,230,000	$37,480,000
Preproduction Underground Excavation			
Shafts	14,468,000	27,124,000	43,510,000
Drifts	974,700	2,003,000	3,472,000
Crosscuts	386,600	602,800	829,500
Access Raises	514,700	868,500	1,318,000
Ore Passes	208,400	543,200	720,200
Ventilation Raises	1,063,000	1,480,000	2,101,000
Surface Facilities	4,038,000	5,414,000	6,905,000
Working Capital	3,852,000	6,189,000	9,228,000
Engineering & Management	4,958,000	8,614,000	12,520,000
Contingency	3,814,000	6,626,000	9,633,000
Total Capital Costs	**$50,757,400**	**$87,694,500**	**$127,716,700**
Total Capital Cost per Daily Tonne Ore	**$12,689**	**$10,962**	**$9,123**

Source: Sherpa Cost Estimating Software for Underground Mines (Stebbins 2000)

CHAPTER 5

Mineral and Metal Prices: Mechanisms, Instability, and Trends

Peter A. Howie* and Roderick G. Eggert*

5.1 INTRODUCTION

Mineral and metal markets differ significantly from one another in terms of the degree of competition that prevails between and among buyers and sellers, the extent of market fragmentation, and whether there are close substitutes for a particular mineral or metal, as well as many other factors. Thus, it is difficult to develop a single model that explains all the important features of mineral and metal pricing. Nevertheless, there are features and characteristics common to most mineral and metal markets. These similarities are the focus of this paper, which surveys pricing mechanisms, instabilities, and trends.†

The discussion is organized in three parts. The first outlines the principal pricing arrangements and mechanisms, including transfer pricing, producer- and user-dictated prices, bilateral contracts, and commodity-exchange prices. The second part focuses on short- and medium-term instability in mineral and metal prices and the underlying market characteristics that lead to instability. The third part examines long-term trends in prices and the economic and technical determinants of these trends.

5.2 PRICING ARRANGEMENTS AND MECHANISMS

Mineral and metal prices are formed through a variety of arrangements and mechanisms. The discussion here progresses from those that are the most private and opaque to those that are the most public and transparent.

5.2.1 Transfer Pricing

Transfer pricing occurs when the producer and the consumer of a particular mineral or metal are part of the same vertically integrated firm. Consequently, prices are internal to the firm and can be arbitrarily set for the benefit of the company. A profit-maximizing firm has an interest in setting transfer prices so as to minimize tax payments. These prices are based on production costs and perceptions of prevailing market price levels, as well as a comparison of the tax codes of the various jurisdictions in which a firm operates. Transfer prices typically appear only in the accounts of a company and ordinarily are not published. Bauxite, the principal source of aluminum metal, offers an excellent example of transfer pricing. Much of the bauxite "sold" in international markets represents internal corporate deals (Radetzki 1990).

5.2.2 Producer- and User-Dictated Prices

Producer-dictated prices can occur in mineral markets where the number of producers is relatively small and where each sells to relatively many customers. As a result, producer pricing may imply a certain degree of market power (i.e., an individual producer can affect the market price by varying its output). The mineral is commonly sold on a take-it-or-leave-it basis, and the level of bargaining with each consumer is negligible. In the situation with the roles reversed, i.e., few consumers and many producers, user-dictated price agreements can exist.

During much of the twentieth century prior to 1970, nonferrous metal sales in North America were conducted under producer-pricing agreements. The dominant firms in the industry (e.g., INCO in nickel) set producer prices. Since the early 1970s, many North American producers that continue to use producer pricing have linked their list prices to futures trading on the Commodity Exchange of New York (now part of the New York Mercantile Exchange [NYMEX]) or the London Metal Exchange [LME] (Nappi 1985).

Producer-dictated pricing sometimes is associated with producer collusion (i.e., the presence of cartels, which are groups of producers that agree to control prices, production, or marketing of a commodity). The international tin agreements that were managed by the International Tin Council from 1956 to 1985 supported the price of tin by assuring a sufficient flow of tin to consumers while guaranteeing a decent profit for the producers. In the 1980s and 1990s, the world output of cobalt was dominated by the state-owned mining firms in the Democratic Republic of Congo, formerly Zaire (Gècamines), and in Zambia (Zambian Consolidated Copper Mines).

User-dictated prices are less common. An example is rough-cut gem diamonds. The Central Selling Organisation (CSO), the marketing arm of De Beers, is dominant, buying two-thirds or more of the rough-cut gem diamonds sold each year. The CSO's market power permits it to set the terms of purchase.

5.2.3 Bilateral Contracts

Bilateral contracts are the predominant agreement in the international commodity trade (Radetzki 1990). They involve a pair of agents who agree to the terms of the agreement that will apply to the trade that occurs between them. The key characteristics common to most bilateral contracts are specification of the commodity, quantity, time and place of delivery, and price. Other terms are contract specific; for example, some contracts relate to a single transaction, while others concern repeated deliveries stretching over several years.

---

* Colorado School of Mines, Golden, CO.

† Most of the information on trends, production, and other aspects of world metal markets not individually identified comes from he following sources:
   American Metal Market, various years, Metal Statistics. Fairchild, New York.
   London Metal Exchange. 2000. Historical Data: Daily LME Official Offer Prices. Data online.
   Schurr, S.H. 1960. Historical Statistics of Minerals in the United States. Resources for the Future, Washington, D.C.
   U.S. Bureau of Mines, various years. Minerals Yearbook. Vol. I: Metals and Minerals. Washington, D.C.
   U.S. Geological Survey, various years. Minerals Yearbook. Vol. I: Metals and Minerals. Washington, D.C.

In recent years, prices in bilateral contracts have tended to mirror commodity exchange prices for those products traded on exchanges, such as copper and lead. For other products not traded on exchanges, such as iron ore, pricing is less public and observable to parties not involved in a trade. In some such markets, conventions have emerged to standardize pricing procedures across an industry in the absence of an organized exchange. For example, in the manganese market, where most of the trade is transacted through annual bilateral contracts, the practice is for a major supplier to enter into negotiation with a major customer; once agreement is reached, the rest of the industry adopts the agreed-upon price as a guideline for their own negotiations (Radetzki 1990).

In some cases, the price transparency in bilateral contract markets is quite limited. Even though trade associations and specialized journals publish prices or price ranges, commonly these prices are for spot sales and do not properly reflect prices applied to longer-term contracts.

5.2.4 Commodity Exchanges

In the 1970s and 1980s, radical changes occurred in the pricing of many metals. During this period, producer-price contracts not linked to commodity exchange prices of copper virtually vanished; publication of producer prices for aluminum, nickel, and molybdenum ceased; and aluminum, tin, and nickel contracts were introduced on the LME.

Commodity exchanges are institutions permitting buyers and sellers (or their agents) to meet and enter into agreements to buy or sell commodities traded on that particular exchange. The most important international metals exchanges are in London (i.e., LME, London Bullion Market) and New York (i.e., NYMEX). Trading on the LME began in 1876. Today LME quotations set global benchmark prices for aluminum, copper, lead, nickel, silver, and tin. Trading in the precious metals gold, silver, platinum, and palladium (as well as copper) is active on the NYMEX.

Commodity exchanges are principally hedge and investment (i.e., speculative) markets; however, spot or cash transactions that result in the physical turnover of a mineral product are substantial. Hedging is aimed at reducing risk—practices taken by a producer to lock in a future sales price (guarding against a possible future price decline) or by a consumer to lock in a future purchase price (guarding against a future price increase). Investment (or speculation), on the other hand, is aimed at making profits by taking advantage of price changes over time. In spot or cash markets, commodities are traded for immediate delivery.

The LME was established as a marketplace for international metals where cargoes could be hedged by trading them on an arrival-of-ship basis; thus it actively trades both spot and forward contracts (Edwards and Ma 1992). A forward contract is a contract initiated at one time requiring the exchange of one asset for another at a later time. The price and quantity of the transaction are set at the time of the initial contracting. Actual payment and delivery of the good occurs later. In contrast, trading on the NYMEX is predominantly futures contracts. A futures contract is essentially a standardized forward contract with very specific contract terms. The principal differences between forward and futures contracts are that futures are (1) always traded on an organized exchange, (2) significantly more standardized than forward contracts, (3) guaranteed by the futures exchange, (4) backed by good faith deposits that traders are required to post to indicate an ability to fulfill all financial obligations that may arise, and (5) regulated by an identifiable government agency (Edwards and Ma 1992).

Futures prices reveal information today about expected cash-market prices in the future. Thus futures contracts can be thought of as a price-discovery mechanism. Futures prices on commodity exchanges also are publicly available and contribute to market transparency. Commodity exchanges permit continuous buying and selling. As a result of all of these characteristics of commodity exchanges, commodity exchange prices are widely accepted as representative of overall market conditions. These prices also serve as the basis for many bilateral agreements.

5.2.5 Price Formation

Significant changes have occurred in recent years in the pricing practices of many producers of mineral commodities. These changes are mirrored by the evolution of dominant agreement types, from largely opaque to quite transparent. In earlier years, the dominant agreement types were transfer pricing, user- and producer-dictated prices, and prices in bilateral agreements not linked to exchange prices. However, since the middle 1970s and the expansion of commodity exchange trading, increasing numbers of producers in an increasing number of mineral and metal sectors have linked their prices to exchange prices. This evolution in pricing and market transparency coincides with the declining market power of producers, especially in aluminum, copper, lead, nickel, and zinc. It is important to note, however, that markets in many nonmetallic industrial minerals continue to be characterized by a small number of producers and consumers and by highly differentiated products (i.e., significant differences in product specifications from one source to another); under such conditions, commodity exchanges are less likely to develop and thus producer- and user-dictated prices and bilateral agreements continue to dominate.

5.3 SHORT- AND MEDIUM-TERM PRICE INSTABILITY

Mineral and metal prices are notoriously unstable. Figure 5.1 presents daily-price series for four common nonferrous metals for the period of January 1995 to February 2000. The figure illustrates two types of instability. The first occurs over periods of days, weeks, or months, and can be called short-term volatility, for example, the ups and downs of nickel prices in 1995. The second type of fluctuation is year to year in character, referred to here as medium-term instability. Examples are generally falling copper prices over the period 1995-1998, followed by rising prices in 1999, and falling nickel prices between 1995 and 1998, followed by rising prices in 1999 and early 2000.

5.3.1 Short-Term Volatility

Short-term volatility, Brunetti and Gilbert (1995) suggest, is due to a combination of three factors. The first is new information. That is, as new information arrives in a market, agents in the market (buyers, sellers, speculators, and so on) react quickly. For example, news of a possible strike or potential production problems at an important mine may lead market agents to anticipate higher future prices, which in turn leads to actions that drive prices up now. A second source of volatility is hedging or speculative pressure. Activity by one group (say, hedgers) leads to the necessary price change that results in the other group (in this case, speculators) accommodating the activity. The third source of short-term volatility is the physical availability of a mineral or metal. Inventories play a critical role here. When an unanticipated increase in demand occurs at a time when inventories are low, prices will rise to balance the market. On the other hand, a similar unanticipated increase in demand when inventories are large will result in a drawdown of inventories and a much smaller increase in price (or perhaps no increase in price at all).

5.3.2 Medium-Term Instability

Medium-term (year-to-year) instability in mineral and metal prices is due primarily to the strong link between (1) mineral and metal demand and (2) the state of the macroeconomy. The level of macroeconomic activity tends to fluctuate from year to year, in large part due to swings in investment spending, that is, spending

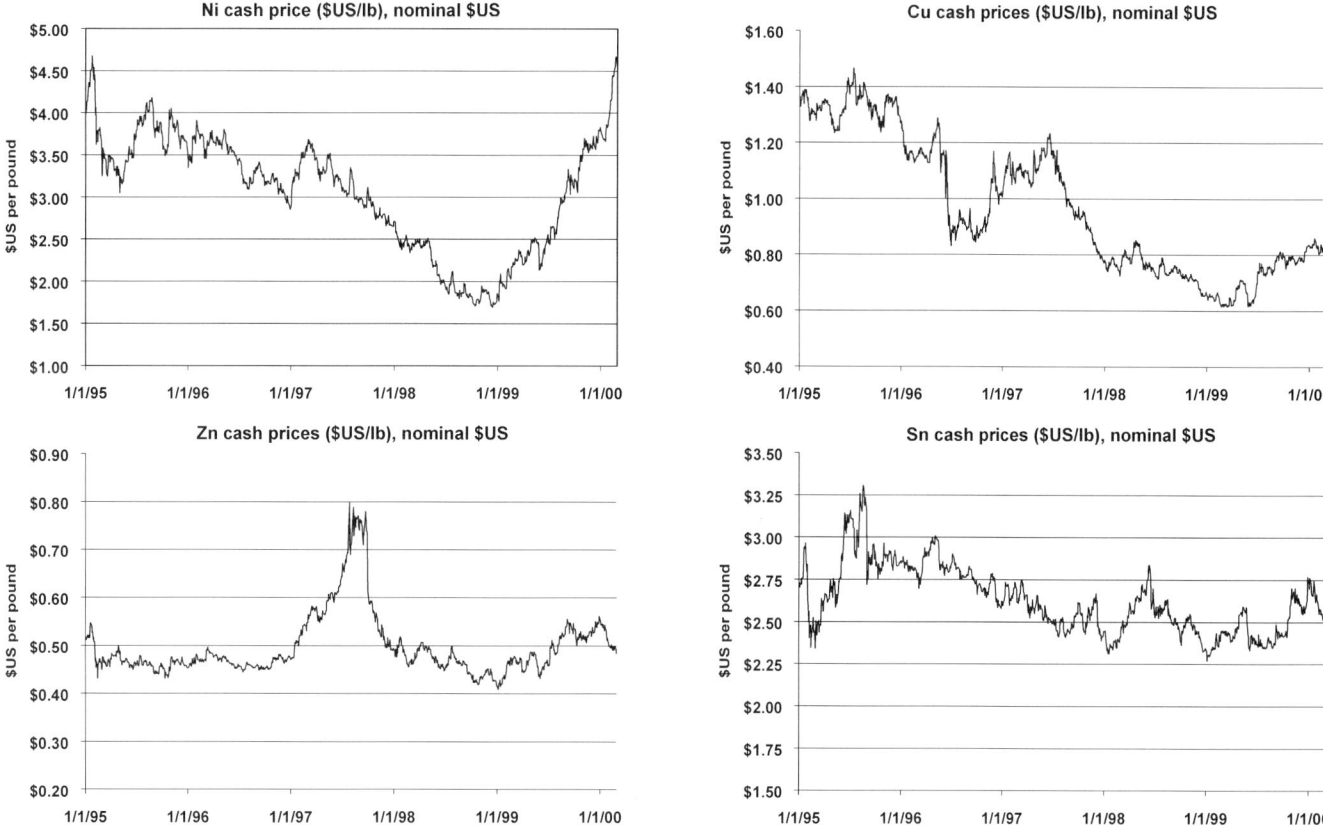

FIGURE 5.1   Daily metal prices, 01/1995–02/2000
Source: London Metal Exchange

on buildings and structures, transportation equipment, heavy machinery, and consumer durables. It is in exactly these sectors of the economy that demand for most minerals and metals is concentrated. Thus, fluctuating mineral and metal demand over the course of the macroeconomic business cycle leads to associated fluctuations in prices.

Two other factors contribute to this form of price instability. First, once production approaches capacity as demand increases, it becomes increasingly difficult and costly to increase output further. Second, consumers of most minerals and metals have little ability to respond to price changes in the short run because of the importance of minerals or metals in the production process (e.g., in the short run, demand for steel is relatively unaffected by changes in the price of steel because existing plant and equipment—and automobile designs—require a certain amount of steel per auto produced, and it takes a year or more to modify automobile designs and production processes to substitute another material for steel—or any other material, for that matter). Both factors tend to exacerbate medium-term fluctuations in prices that are due primarily to the macroeconomic business cycle.

5.3.3   Economic Theory

Both short-term volatility and medium-term fluctuations in prices can be understood better with the help of some simple economic theory. The starting points are the concepts of demand and supply. Demand focuses on the behavior of consumers and the driving forces behind their decisions. Analogously, supply focuses on producers and the determinants of their decisions. Demand (consumers) and supply (producers) interact to determine price. Changes in demand and supply lead to changes in price and, over short periods of time, inventories. Supply-demand interactions

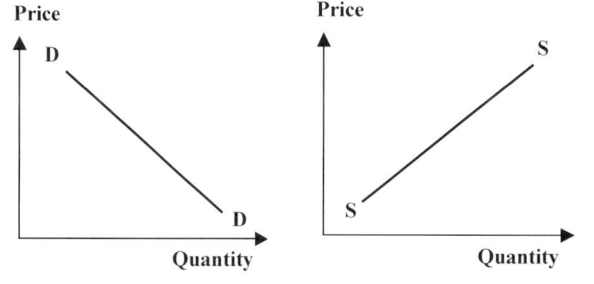

FIGURE 5.2   Demand and supply curves

incorporate the behavior of all market agents, including not just mining companies and metal buyers, but also speculators (or investors) who hold inventories in the expectation of making profits.

Typically consumers demand more of a mineral the lower its price, while producers supply more of a mineral the higher the price, as depicted in Figure 5.2 (market demand and supply curves).

These curves show what quantities would be demanded and supplied at each price. In a balanced market, the actual market equilibrium price Pe will occur where the two curves intersect (Figure 5.3). The quantity produced and sold would be Qe.

With this simplified diagram, one can visually examine the effect on price of changing demand and supply. An increase in demand due to macroeconomic expansion (a rightward shift in

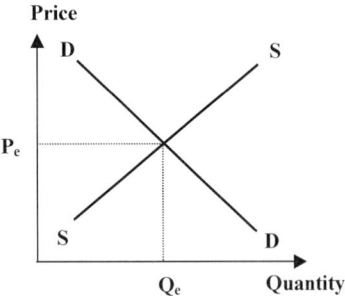

FIGURE 5.3 Equilibrium condition

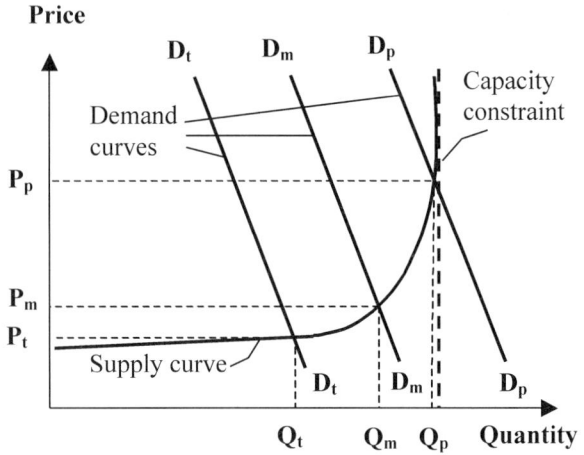

FIGURE 5.4 Short-run supply and demand curves
Source: **Tilton 1992**

the demand curve) results in a higher price. Likewise, a decrease in supply due to strikes at a number of mines (a leftward shift of the supply curve) also leads to a higher price.

The slopes of the demand and supply curves indicate how sensitive consumers and producers are to changes in price over short periods of time. The steeper the slope, the smaller the effects of a price change on rates of production and consumption. Demand curves for most minerals are steep if we consider the immediate and near-term effects of a price change on quantity demanded; as noted above, most users of minerals and metals, such as equipment manufacturers and automobile companies, are constrained by their existing plant and equipment. Their ability to substitute another mineral or metal usually is quite limited unless they invest in new equipment or production facilities. Over the longer term, however, consumers can invest in new equipment and can alter production processes to use less of a material that has become more expensive. Similarly, supply curves tend to be steep, at least as output approaches capacity in the immediate- and near-term (adjustment periods so short that producers are unable to expand existing facilities or bring new operations into production). But over longer periods, capacity and inventory constraints are not as critical because producers add to capacity by developing existing or newly discovered reserves.

Economists also are interested in how responsive demand is to a change in income. For example, how do changes in national income influence steel demand? As noted earlier, national income importantly influences mineral and metal demand. Fama and French (1988) showed that there were large increases in the price of all nonferrous metals before the business cycle peak of January 1980 and large declines afterward. Copper and lead doubled their January 1978 values and then fell about 40% off their peak values by July 1982.

A variety of events cause demand and supply curves to shift. Demand for nearly all minerals and metals tends to shift to the right (to larger quantities) when macroeconomic expansion occurs and investment spending increases; it tends to shift to the left (to lower quantities) during periods of recession. Supply shifts to the right when investment in new production capacity more than offsets the retirement of existing capacity. Supply shifts to the left (to lower quantities) when plant retirement more than offsets the amount of new capacity and, notably, when short-term supply disruptions occur because of, for instance, technical production problems, strikes, and civil disturbances. In 1996, an unofficial strike at Rustenburg, Republic of South Africa, caused a decline in output of an estimated 100,000 ounces of platinum (approximately 2% of the 1996 world production). Political disturbances and infighting have cut the supply of many types of mineral commodities. In 2000, the inability of the Russian Gokhran (which manages the issuance of export licenses) to sign sales quotas led to an increase in the price of palladium from about $450 per ounce at the start of January to

more than $800 per ounce during the last week of February. (In 1998 and 1999, Russia accounted for 65%-70% of the worldwide palladium supply [Johnson Matthey 1999].)

With this basic theory of supply and demand, we can better explain the causes of short-term volatility and medium-term fluctuations in mineral and metal prices noted at the beginning of this section. Consider first short-term instability due to new information, hedging or speculative activities, or physical availability. Each of these sources of instability can be represented as a shift in either supply or demand. New information suggesting that economy-wide investment spending will increase next year could lead to a rightward shift in demand, reflecting increased demand today in anticipation of higher future demand and prices; the result is higher prices today. Similarly, speculative buying can be thought of as an increase in demand leading to increased prices.

As for medium-term price fluctuations, consider Figure 5.4 for a mineral sold in a competitive market, that is, a market in which the number of producers and consumers is large enough so that no single producer or consumer acting alone has an appreciable influence over price.

The supply curve has a shape typical of supply over short periods of time when capacity is fixed—relatively horizontal when an industry is producing at much less than full capacity, increasingly steep as output approaches capacity. The three demand curves reflect shifting demand for minerals during different phases of the business cycle. Curve DDt reflects demand at the trough of the cycle curve DDm demand at the midpoint of the cycle, and curve DDp demand at the peak of the cycle. As the figure shows, the consequence of a mineral operating in a competitive market (e.g., copper, gold) is the severe fluctuations in price. Another consequence of operating in a competitive market is that market instability causes coincidental changes in the quantity sold and the price.

5.3.4 Additional Determinants of Demand and Supply

Three additional factors importantly influence mineral and metal prices: joint production (by-products and co-products), secondary production (recovery of material from scrap), and exchange rates.

More complex price-determination processes are involved when a mineral is produced as a single or main product at some mines and as a by-product or co-product at other mines. A main product is a mineral so important to the viability of the mine that its price alone determines the mine's output of ore. A by-product is so unimportant that its price has no influence on the output of ore at a mine. When two or more minerals affect the output of the mine, then they are co-products.

In understanding by-product supply, two points are critical. First, the supply of a by-product is constrained by the output level of the main product. Consider gold produced as a by-product at a copper mine. The amount of ore produced, and in turn the amount of available gold, is a function of conditions in the copper market. Thus gold supply is partly a function of copper supply. Second, by-products tend to be a lower-cost source of supply than main-product (or individual-product) output of the same mineral. Again, consider gold, which is produced as a by-product of copper at some mines and as an individual product at other mines. Costs of by-product gold production typically will be lower because the by-product supply benefits from sharing the mining costs of copper. Thus, the price of gold needs only to be high enough to cover the additional costs of extracting gold from the copper ore to induce the supply of by-product gold. In contrast, at mines producing only gold, the price of gold needs to be high enough to cover all costs of gold production.

Co-products represent a case intermediate between main products and by-products. Co-product situations occur when production of two or more minerals is necessary to justify mining. In other words, the prices of both or all co-products influence the output of ore. The combinations lead-zinc and copper-molybdenum often are produced as co-products. Since a co-product mineral has to only bear a proportion of the joint costs, it may be available at a lower cost than a main-product supply of the same mineral.

Now consider secondary production (recovery from scrap), which for some minerals and metals constitutes an important source of supply. For example, secondary lead represented 76% of all refined lead produced in the United States in 1997. For copper, the figure was 37% (USGS 1997). Secondary production comprises the recycling of both new and old scrap. New scrap is generated during the process of manufacturing a new good. In contrast, old scrap consists of consumer and producer goods that have come obsolete, worn out, or for some other reasons are no longer of use. Secondary supply has determinants of supply and a cost structure that are significantly different from those that apply to primary supply (Radetzki 1990; Tilton 1999). The availability of secondary supply has a tendency to increase a mineral's responsiveness to price changes and may reduce the market power of a primary producer. Secondary producers tend to operate in a highly competitive market.

Finally, consider exchange rates. Globally, prices for many minerals and metals are quoted in U.S. dollars; however, mining costs are paid in many currencies. Consequently, U.S. dollar appreciation has the effect of increasing revenues in other currencies (or of reducing production costs relative to revenues). In this case, profit margins on mineral production increase, often leading to increases in output from these countries. Australian base-metal producers are forecasting a boom in 2000, especially in the aluminum and nickel industries, because producers are benefiting from the falling Australian dollar. A one-U.S.-cent change in the value of the Australian dollar changes profits significantly for Comalco, the Australian-listed aluminum subsidiary of Rio Tinto, by A$21 million (Wyatt 2000).

5.4 LONG-TERM PRICE TRENDS

For producers examining the feasibility of expanding existing operations or opening new mines, it is important to recognize the influence of secular trends in mineral prices on profitability. Moreover, it is also important to realize that the determinants of these trends may be different than the driving forces behind short- and medium-term instabilities. Consequently, this section takes a longer-term view and analyzes prices over the last 130 years.

When more than 100 years of data are examined, systematic patterns emerge, and useful generalizations about price trends can be made. Usually long-run price series are expressed in constant-dollar terms to eliminate the effects of inflation (overall changes in prices throughout the economy) and to focus attention on those factors unique to minerals and metals. Constant-dollar values are hypothetical and represent constant purchasing-power dollars for a given base year obtained by adjusting current dollar values at the inflation rate (Stermole and Stermole 1996). Current dollar prices, which refer to actual dollars of revenue or cost, tend to be highly correlated to the economy-wide rate of inflation and, for this reason, can lead to misleading results.

Figure 5.5 presents constant-dollar price paths for the same four nonferrous metals (whose short-run price paths are shown in the last section). Superimposed on the paths are fitted quadratic price trends, which reveal three types of basic long-term trends: declining, stable, and rising. Most minerals have experienced all three trends, with the period of minimum prices being mineral specific. The driving forces behind these trends are changes in demand and supply. More specifically, growth in mineral demand associated with overall economic growth puts upward pressure on prices. On the supply side, depletion of high-quality deposits also puts upward pressure on prices; on the other hand, technological advances in exploration, mining, and metallurgy and the discovery of new deposits work in the opposite direction and put downward pressure on prices. A final supply-side feature, changes in market structure, works in both directions. If a market becomes less competitive over time because of, for example, collusion among producers, then prices tend to rise. If a market becomes more competitive over time, prices tend to fall.

Consider periods of falling prices. These often occur early in a mineral's economic life. When a market is new, mines and smelters tend to be too small to be fully efficient (Slade 1988). Mining and processing operations commonly have the ability to substantially reduce average costs of output by expanding the level of output. Consequently, as mineral demand increases from new uses and increased needs from traditional users, firms can increase their scale of operation, with the associated reduction in average costs.

Technological change can occur in both production and consumption. The integration of new technologies in production is commonly termed process innovation; there is a reduction in the quantity of inputs used by a firm to produce a given level of output. The use of new technologies on the consumption side can either be process innovation or product innovation. Product innovation is a change made by a firm in its saleable product that arises from new technology. When new technologies on the consumption side increase demand, mining firms may be able to increase capacity and lower average costs, which exerts a downward pressure on prices.

Finally, the discovery of new ore bodies can increase mineral supply, shifting the long-run supply curve and causing prices to fall.

A period of stable prices implies that any changes in demand (e.g., increased demand due to long-run economic growth) are balanced by enhancements on the supply side (e.g., technological innovation, discovery).

A period of rising prices may be caused by demand growth coinciding with the depletion of high-quality deposits. Newly discovered ore bodies may have lower metal content, be more deeply buried, or have ore that is more complex metallurgically. Also, if prices do start to rise, cost-lowering technologies may be insufficient to offset the rising costs.

Figure 5.5 shows that the price path for each of the four metals is unique. The real (inflation-adjusted) price of nickel fell between 1870 and 1950, but has remained fairly stable since then. Technological advances along with increasing capacity have

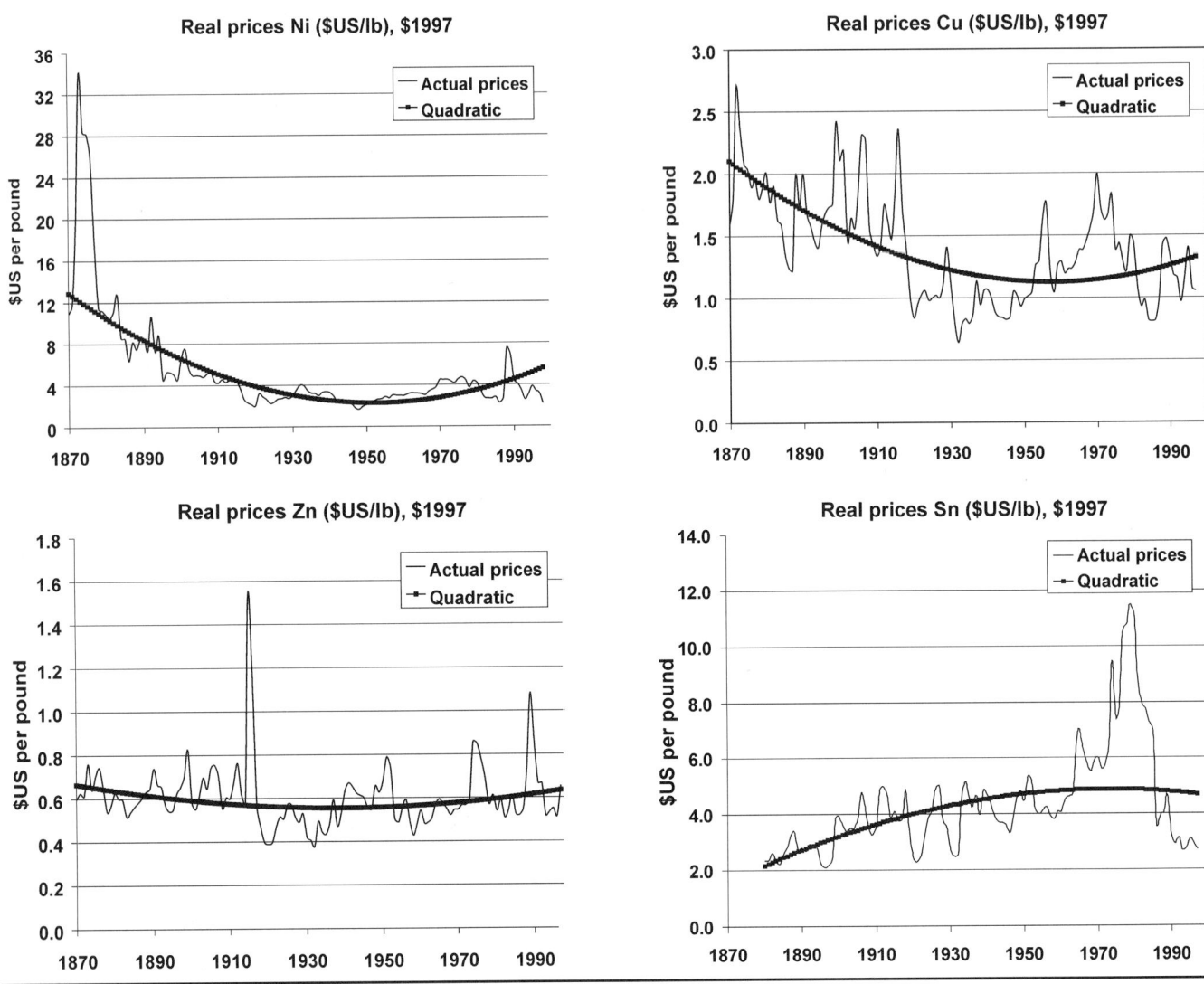

FIGURE 5.5 Average annual metal prices, 1870–1997
Source: **American Metal Market, Manthy, United States Bureau of Mines, United States Geological Survey**

made this possible. In contrast to nickel, copper displays a well developed U-shaped price path, initially falling, followed by a period of stable prices, but eventually rising. Zinc prices have been virtually trendless over 130 years. Technological breakthroughs and grade declines essentially have offset each other in determining production costs. Finally, tin prices consistently increased until 1985, when prices fell sharply. The price path of tin between the mid-1950s and mid-1980s was caused by long-term agreements, which collapsed in 1985. During the period 1956–1985, a continuous series of agreements managed by the International Tin Council supported tin prices via restricting production levels (USGS 1999).

Petersen and Maxwell (1979) analyzed the effect of technology innovation and changes in ore grade on the cost of producing silver, tin, lead, zinc, and iron. They found that these two factors had been much more important in lowering costs, and in turn prices, than the discovery of new deposits.

It worth digressing for a moment to examine the role that changes in market structure have on price trends. MacAvoy (1988) concluded that only about two-thirds of the variation in metal prices could be explained by growth in demand, changes in mining and metallurgical technology, and discovery of new deposits. As a result, a significant part of long-term price behavior

was left unexplained, and market structure has to be considered a likely candidate in accounting for the unexplained one-third.

In a market with a small number of producers (a highly concentrated industry), extraction rates tend to be lower than in markets with many producers, and thus prices tend to be higher. The same is true when producers explicitly or implicitly cooperate to restrict output or divide up a market. However, both high industry concentration and cooperation among producers tend to be unstable over time. The presence of "high" prices attracts the entry of new producers. Any time individual producers restrict output for the overall benefit of a group of producers, there is an incentive to cheat (i.e., produce more than the terms of cooperation permit). Furthermore, consumers have the incentive, when prices are high for a specific mineral or metal, to substitute away from this mineral or metal. The result is a loss of market power among producers.

Another factor influencing long-term trends in mineral prices is transportation cost. These costs can constitute a significant proportion of the delivered cost of minerals and metals. Nonferrous metal ores normally contain 5% or less metal, so transporting the unprocessed ore involves volumes at least 20 times greater than the metal content. Thus virtually all such ores are concentrated at the mine site to increase the metal content to 25% to

30%. Dramatic advances in maritime technology lowered freight rates by approximately two-thirds over the latter half of the twentieth century (Lundgren 1996). As a result the geographic extent of markets has broadened considerably. What once were regional or national markets now are global markets; what once were markets with some degree of market power in the hands of producers have become much more competitive. The overall effect is to put downward pressure on prices.

5.5 CONCLUSIONS

For much of the twentieth century, metals were sold using two dominant pricing systems: prices set by the major firms in the industry (producer-dictated agreements) and prices determined on metal exchanges such as the LME. Prior to the 1970s, producer-dictated agreements dominated in North America, whereas exchange-based prices prevailed in the rest of the world. Since the 1970s, a combination of factors has weakened the North American producer-pricing system, leading to increased reliance on metal exchanges for price formation. To be sure, other pricing mechanisms (such as producer pricing) prevail today in particular markets, especially for the nonmetallic industrial minerals.

Prices for many minerals and metals, particularly those traded on commodity exchanges, exhibit short-term volatility. This daily, weekly, and monthly volatility is due primarily to new market information, hedging and speculative behavior, and physical availability of product from inventories. Medium-term (yearly) instability in prices, in contrast, is due largely to the strong link between mineral and metal demand and the macroeconomic business cycle, and the inability of both producers and consumers to make significant changes to producing and consuming behavior over relatively short periods of time (they are constrained by the existing production capabilities and by their existing mineral- and metal-using equipment and processes, respectively). Finally, over the longer term, whether prices rise or fall depends largely on the balance between two countervailing forces—the price-increasing effects of demand growth and deple-tion of high-quality deposits, and the price-reducing effects of technological innovation, discovery of new deposits, and enhanced competition.

5.6 REFERENCES

Brunetti, Celso, and Christopher L. Gilbert. 1995. Metals Price Volatility, 1972-95. *Resources Policy*, vol. 21, no. 4, pp. 237–254.

Edwards, F., and C. Ma. 1992. *Futures and Options*. McGraw-Hill, New York

Fama, E.F., and K.R. French. 1988. Business Cycles and the Behavior of Metals Prices. *Journal of Finance*, vol. 43, no. 5, pp. 1075–1093.

Johnson Matthey. 1999. Platinum 1999 Interim Review, M. Steel, ed. London, UK.

Lundgren, Nils-Gustav. 1996. Bulk Trade and Maritime Transport Costs: the Evolution of Global Markets. *Resources Policy*, vol. 22, nos. 1/2, pp. 5–32.

MacAvoy, P.W. 1988. *Explaining Metals Prices*. Kluwer Academic Publ., Boston.

Manthy, R. 1978. *Natural Resource Commodities—A Century of Statistics*. Johns Hopkins University Press, Baltimore.

Nappi, C. 1985. Pricing Behavior and Market Power in North-American Nonferrous-Metal Industries. *Resources Policy*, vol. 11, pp. 213–224.

Petersen, U., and R.S. Maxwell. 1979. Historical Mineral-Production and Price Trends. *Mining Engineering*, pp. 25–34.

Radetzki, M. 1990. A Guide to Primary Commodities in the World Economy. Basil Blackwell, Cambridge, Massachusetts.

Slade, M. 1988. Pricing of Metals. Centre for Resource Studies, Queen's University, Kingston, Ontario.

Stermole F.J., and J.M. Stermole. 1996. Economic Evaluation and Investment Decision Methods. Investment Evaluations Corp., Golden, Colorado.

Tilton, J.E. 1992. Economics of the Mineral Industries. In *SME Mining Engineering Handbook,* ed. by H. Hartman. Society for Mining, Metallurgy, and Exploration, Littleton, Colorado, vol. 1, pp. 43–62.

Tilton, J.E. 1999. The Future of Recycling. *Resources Policy*, vol. 25, no. 3, pp. 197–204.

U.S. Geological Survey. 1997. Metal Prices in the United States Through 1998. Washington, D.C.

Wyatt, S. 2000. Australian Base Metal Producers Tap a Rich Exchange. *Financial Times*, March 23.

Room-and-Pillar Mining of Hard Rock

. . . .

CHAPTER 6

Mining Methodology and Description: The Immel Mine

William R. Begg* and Nikolai A. Pohrivchak*

6.1 INTRODUCTION

The Immel Mine is an underground stope-and-pillar zinc mine located in east Tennessee about 15 miles east of Knoxville (Figure 6.1). Immel is owned and operated by Asarco, Inc., a fully owned subsidiary of Grupo Mexico. Development of the site was started in 1966, and production commenced in 1968 under the American Zinc Company. The mine was purchased by Asarco, Inc., in 1971.

Current production consists of 2,000 t/d with an average anticipated grade of 3.72% zinc for the year 2000. Annual production is anticipated as approximately 680,000 tons. Development is maintained at approximately 3,800 ft/yr. Immel operates three shifts a day 7 days a week, with one nonproduction shift utilized as a maintenance shift.

The Immel Mine maintains all safety standards and was honored with the Sentinels of Safety award for being the safest underground metal/nonmetal mine in the United States in 1992. Safety is a primary concern for all Asarco employees and is regarded as the highest priority at all operations.

6.2 GEOLOGY

The Immel Mine is situated in the western portion of a 30-mile-long zone of Mississippi Valley-type sphalerite mineralization within the Mascot-Jefferson City zinc district of east Tennessee. Zinc ore is hosted by interbedded, brecciated dolostone and dolomite of the Upper Kinsport and Lower Mascot formations, which are Early Ordovician in age. Ore bodies are typically low- to high-domed structures in which the dissolution of limestone has resulted in mineralized breccias from 15 to 120 ft thick and 50 to 600 ft wide. Bedding is controlled by a northeast-plunging anticlinal structure; most of the zinc mineralization lies within the northern limb. Because of the plunge of the structure, dips as steep as 45°E have been noted within the mining area.

As a result of dip, ground control is a major issue, and all headings must be bolted. Another issue resulting from the steep dips is that ore dilution must be controlled at the flanks of the ore bodies as they reach the upper or lower extent of the beds.

Sphalerite is virtually the only zinc mineral present and is generally associated with secondary fracture filling in the dolomite. Pyrite is found in varying amounts throughout the deposit. Silica is also present in the form of bedded or nodular chert or jasperoid. Galena has been noted in some areas, but quantities are too limited to be of economic importance.

6.3 GEOTECHNICAL INFORMATION

Because the Immel Mine is situated within the Valley-and-Ridge Province, Alleghenian activity has superimposed a general northeast fracture trend paralleling the trend of tectonic events. At this bearing, a compressive stress of 3,000 psi has been observed

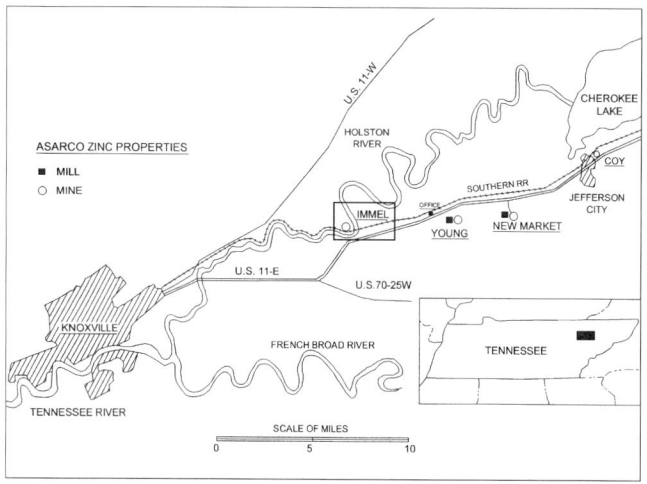

FIGURE 6.1   Location of the Immel Mine and other ASARCO, Inc. properties of east Tennessee

(Bickel and Dolinar 1975). Studies were carried out in 1976 to determine compressive strengths of roof rock and ore rock (Bickel 1976). Those results indicate the strength of roof rock at 18,400 psi at a specific gravity of 2.75, and the strength of ore rock at 28,800 psi at a specific gravity of 2.90.

The trend of the fracture pattern in the pillars of older stopes has led to some ground control problems. Typically, however, these fractures appear after the larger stopes have been mined and are only of concern in those areas where roadways may be built.

6.4 MINE LAYOUT

Active and completed mining at the Immel Mine extends laterally approximately 9,600 ft with a maximum width of 3,200 ft (Figure 6.2). Mining depths range from 100 ft above to 900 ft below sea level (1,000 to 1,800 ft below the surface). Ore is crushed and hoisted at the main shaft, which also serves as a service hoist and man cage. Two air shafts ventilate the mine, one in the center of the mine and one at the northeast end of the mine. The central air shaft also serves as an emergency escape shaft, but has no ore hoisting capabilities.

The mine consists of three levels designated by track haulage. The −484 level is the most active and contributes 95% of all production from the mine. The −242 level is currently inactive because of the lack of reserves on this level. The −7 level yields about 5% of all production and contains a limited amount of reserves for future mining.

---

* Immel Mine, Asarco, Inc., TN.

83

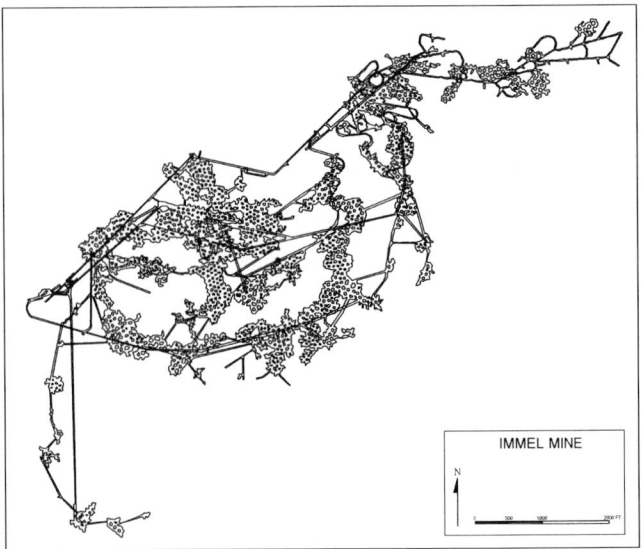

FIGURE 6.2   Geberakuzed layout of the Immel Mine, ASARCO, Inc.

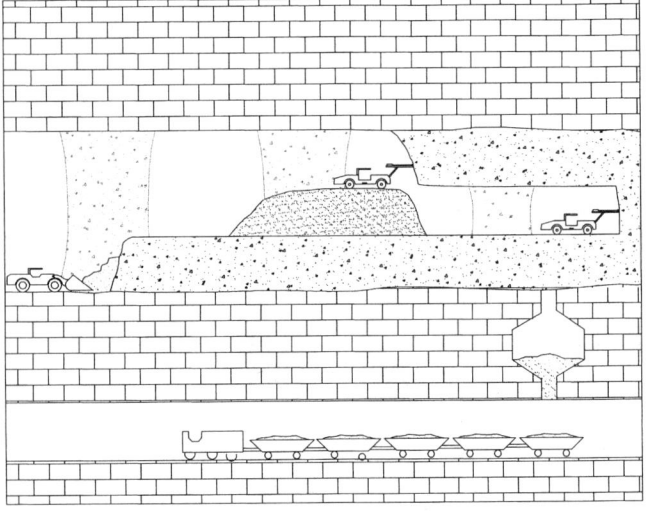

FIGURE 6.3   Typical mining methods employed at the Immel Mine

Development is ongoing on the −484 level because most of the known reserves are on this level. Limited development is planned on −7 level as additional reserves are exploited. Because of the dip of the ore bodies in the area, development must be maintained not only to access new ore bodies, but also to gain access to additional mining passes in existing stopes.

6.5   MINING METHODOLOGY

Ore at the Immel Mine is extracted using the random room-and-pillar (stope-and-pillar) method (Figure 6.3). In this method, large rooms (stopes) are mined out, and pillars are left to support the back. The method is modified to accommodate dips up to 45°E. Pillars are generally 35 to 50 ft in diameter and are offset to maintain an acceptable distribution of stresses from overburden. Generally, pillars are spaced 35 ft apart with a pillar-to-ore ratio of about 1:5.

Depending on the thickness of the ore body, mining is carried out in a series of passes. The typical sequence for multiple-pass, random room-and-pillar mining at the Immel Mine consists of an initial cutting-in pass, followed by overhand

stoping of ore in the back and horizontal benching of ore in the floor. The initial pass (cut) is usually initiated toward the top of the ore body. This first phase helps define the lateral extent of the ore body.

Once the initial phase is complete, preparations can be made to extract ore from the back using overhand stoping. This pass is initiated by drilling a steeply inclined starter round into the back. Once shot, the broken ore from the back will serve as the footing for subsequent mining. Additional mining is carried out by horizontally drilling and blasting the brow. This process continues until the lateral boundaries of the ore body are reached for this particular pass. The overhand stoping passes of the ore body will continue upward until the top of the ore body is reached.

Once the overhand stoping phase is completed, all broken ore is mucked out in preparation for the final phase of multiple-pass, room-and-pillar mining, that is, benching. Horizontal benching is carried out in much the same fashion as overhand stoping, only downward. If benches are not already accessed by planned development or subsequent mining, then starting rounds are drilled at steeply declining angles until the desired bench height is reached. Once the bench is of the desired height, mining continues laterally to the boundaries of the ore body.

This process continues for as many passes as are needed to fully extract the ore in the floor. Additional cutting-in may be done off both overhand stopes and bench levels if suitable ore grades warrant mining, although the lateral extent of the ore body is generally defined during the first phase of mining. At the Immel Mine, approximately 40% of production tonnage comes from the initial cutting-in phase of stoping, whereas overhand stoping and benching yield 30% and 25% of the production, respectively. The remaining 5% of the ore is typically extracted from development headings as they are advanced.

6.5.1   Drilling and Blasting

Production drilling at the Immel Mine is done 7 days a week, totaling up to 35 manshifts to maintain the production level of 2,000 t/d. Under current local union labor agreements, a driller is classified as a machineman, which involves both drilling and blasting. Depending on requirements, the machineman may either drill or blast or both drill and blast during a scheduled shift.

The current drill fleet consists of three Eimco-Secoma Pluton 17, diesel-hydraulic, single-boom jumbos and one Tamrock Monomatic, diesel-hydraulic, single-boom jumbo. The Eimco-Secoma drill rigs are equipped with Hydrastar 300 hammer drills, while the Tamrock is equipped with an HL500 hammer drill. All three rigs are capable of drilling a 12-ft-deep hole. Production holes are generally drilled 11 ft deep with a 14-ft-long carbonized hexagonal drill rod and a $1^3/_4$-in button bit.

The typical sequence for drilling involves an initial inspection of the equipment. This usually involves a walk-around inspection, as well as necessary fluid and safety checks. Once the equipment is deemed safe and operable, the driller will inspect the workplace to evaluate ground and any other potentially unsafe working condition. Unsafe ground is reported to the shift supervisor and corrective action is taken.

Once the operator feels that the work area is safe, drilling is started. If mining the stope is in the first phase (the cutting-in phase), the driller will start with a burn, which is typically a 16-hole pattern with two central 3-in in diameter holes, which will allow an opening for additional holes to break to when shot. Once the burn is drilled, additional holes are added to complete the round (Figure 6.4).

In a typical heading, a cut varies from 15 to 25 ft wide and 16 ft high. Headings may be taken higher in some places to aid in loading trucks. A typical heading contains an average of 49 to 51 holes, depending on the dimensions of the cut. Once adequate

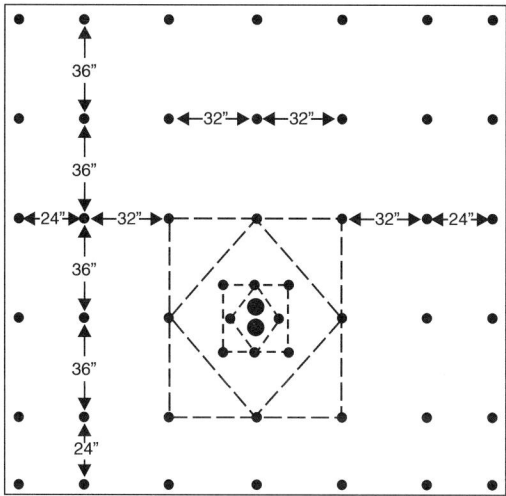

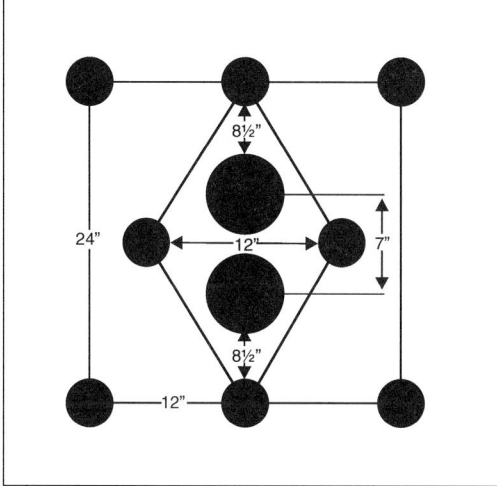

FIGURE 6.4 Standardized drilling pattern showing a detail of burn hole placement, Immel Mine

breakage has been achieved from the cut, the operator drills a series of rows to the left and right of the original area to turn off and begin setting pillars. These rows are typically the same height as the cut and are slightly fanned to gain the maximum breakage from the minimum number of holes. Locally called slabbing, this method is extremely efficient in breaking the maximum number of tons per hole. Therefore, cuts are kept to a minimum and only taken when the break set-up is not sufficient for slabbing.

The method for drilling brows in overhand stopes differs from drilling cuts in that there is no burn drilled, since a break is already set up. Typically, a brow is drilled out in rows of holes spaced about 4 ft apart. Particular attention is paid to hole location and direction adjacent to pillars. Benches are drilled in much the same manner, only in an inverted sequence.

Blasting at the Immel Mine is carried out with nonelectric delay blasting caps initiating either a 1½- by 16-in emulsion primer or primed directly with a 20-g booster. The remainder of the explosion column is loaded pneumatically with ammonium nitrate and fuel oil (ANFO), except wet holes, which are loaded completely with stick powder. The shot is initiated with a 4-min safety fuse and cap. The mine operates one of two Getman A-64 loading rigs, maintained to be available for either production or development.

ANFO is currently received on-site in 400-bag lots, although future packaging will include 2,000-lb bags, which are easier to

transport and lower in cost. The ANFO is conveyed underground from the surface storage facility in either a specially equipped Ford tractor or a protected wagon to the underground magazine. All stick powder (1½ by 16 in) is received in 55-lb boxes and is stored in the surface magazine and hauled underground as needed.

6.5.2 Loading, Hauling, and Hoisting

Loading is performed by diesel-powered loaders equipped with 9-yd^3 buckets with a 10-t payload capacity. Broken rock from the stope areas is loaded into 30-ton haul trucks or trammed to designated ore passes connected to one of three rail haulage levels. Loading equipment includes four Wagner ST8-A Scooptrams, one EJC 300, and one Toro 1250 D. Haulage trucks include two EJC 430s and two Tamrock 430 Ramp Rabbits. The latter trucks were purchased to increase cycle times on haulage grades that range from 15% to 20%. Trucking is essential to mine operations in that maximum haulage distances are as much as 3,500 ft to the central ore raise.

Ore is conveyed to the production shaft via rail. Currently, one haulage level is used (the −484) with a haulage distance of approximately 5,800 ft. A maximum of six 12-t-capacity cars are pulled with one of three locomotives. These include two Brookville 12-t diesel locomotives and one Plymouth 12-t diesel locomotive.

Ore is loaded directly into ore cars by a pneumatically operated chute system from the ore pass. The broken ore is then transported to the production shaft, where it is dumped into one of two ore passes leading to the main crusher. The mine was originally designed with two ore passes on each level to accommodate the separation of ore and waste, but since all waste is now dumped in mined-out stopes, either raise can be used to dump ore.

Once ore is received at the crusher station, it is fed to a 36-by 48-in Traylor jaw crusher by a 48- by 12-ft, $^3/_4$-in NICO apron feeder, where it is sized for hoisting to the surface. Once crushed, rock is fed to a hydraulically driven NICO Hydrastroke feeder to the skip.

The skip system is fully automated and consists of an Ingersoll-Rand, tower-mounted friction hoist with a 104-in diameter drum. Ore is conveyed to the surface via a 15-t counterbalanced skip at a rate of approximately 1,400 ft/min. Typically, ore is hoisted on the second and third shift, and the day shift is used for hauling supplies and general maintenance and shaft inspection.

Ore is dumped into a bin, from which it is conveyed by a belt to the ore storage pile. Two feeders under the storage pile convey rock to a truck loading point, where it is then transported to the Young Mill for processing.

With the various steps involved in getting the ore from the stope to the mill, all phases must be maintained so there will be no break in the flow of production. The shaft is inspected weekly and railways are inspected daily for any problems that could cause a disruption in conveyance. Because of long hauls and the age of the equipment, other haulage and conveyance systems are being considered.

6.5.3 Grade Control and Dilution

Because of the low grade of the east Tennessee zinc deposits, grade control is of great importance to the viability of the operation. Estimates are derived through the use of a chart that references inches of sphalerite exposed in the face versus height of the face. Calculations are based on the ratio of the specific gravity of sphalerite (as well as zinc content) and the specific gravity of the surrounding rock (limestone and dolostone). With this information and other geologic information obtained from sources such as prospect holes, decisions are made as to take a heading or not. The Immel Mine maintains a cutoff grade of 2.50% zinc, but in all mining areas the grade must be sufficient to warrant taking headings of lower grade. Grades are discussed by the mine geologist

TABLE 6.1   Historical and predicted direct mining costs per ton of crude ore, US dollars

Activity	1995	1996	1997	1998	1999	2000	Average
Development	0.48	1.09	1.03	0.76	0.66	0.60	0.77
Ore breaking	2.37	2.17	2.48	2.28	2.46	2.64	2.40
Underground haulage	1.11	1.32	1.25	1.50	1.42	1.23	1.31
Crushing	0.18	0.24	0.15	0.18	0.17	0.23	0.19
Hoisting	0.28	0.43	0.59	0.63	0.52	0.37	0.47
Surface haulage	1.18	1.15	1.07	1.08	1.05	1.06	1.10
Maintenance	0.02	0.01	0.01	0.01	0.00	0.00	0.01
Ground support	0.51	0.80	0.91	0.88	1.04	0.93	0.85
Shaft and station repairs	0.06	0.11	0.03	0.05	0.06	0.03	0.06
Waste disposal	0.01	0.01	0.01	0.07	0.08	0.01	0.03
Underground pumping	0.31	0.51	0.19	0.20	0.17	0.07	0.24
Main ventilation	0.01	0.01	0.02	0.01	0.00	0.01	0.01
General	2.45	2.84	2.93	2.92	3.51	3.60	3.04
Total	8.97	10.69	10.67	10.57	11.14	10.78	10.47

and production foremen daily and are reported by stope or development to derive a weighted average grade. Immel must maintain an average grade of 3.72% zinc to maintain cost efficiency.

The disseminated nature of the ore results in some dilution in grade from the grade originally derived for a given mining area. Some of this dilution is accounted for in reserve calculations by adding a 20% overbreak factor to the total tonnage of an ore block. Other factors leading to dilution include lower grades at the fringes of the ore bodies and lower grades of ore at the up-dip or down-dip boundaries during the particular phase of mining being completed in a given stope.

One method to control dilution is the use of selective mining and pillar placement. Once all available geologic information is gathered for a given stope or ore body, pillars are placed in areas of lower grade as often as possible to maximize the affected grade for a given work area. As stated earlier, waste material is dumped into mined-out stopes so not to affect production grade.

6.5.4   Stope Development and Preparation

As with any mining operation, proper stope development must first start with knowledge of the ore body being mined. When an area is drifted, information from surface core holes or underground core holes must be augmented. This task is achieved with percussion longholing at the face to make an initial determination of the lateral extent of the ore. The Immel Mine operates one Joy pneumatic longhole rig equipped with a Cannon VCR460 drill. Drilling is done one shift a day with an anticipated advance of 100 ft/d. The longhole rig is used again after the initial phase of mining to further delineate the vertical extent of mineralization. Once all information is gathered, then and only then can an adequate stope design be considered.

Stope preparation involves steps necessary to ensure both the safety of workers and the effectiveness of stope availability. Ground control is a major issue at the Immel Mine because of the mine's depth and the steep dip of the bedding. Before any production work can be started, adequate ground control and ventilation must be achieved.

If ground cannot be scaled manually with a ground bar, it must be scaled mechanically. This task is done with either a Dux DS-30 RB diesel-hydraulic scaler or a Getman S-300-24 diesel-hydraulic scaler. For high backs, Immel also operates one of two boom lifts for manual scaling. These include a Getman snorkel with a 95-ft reach and a Cline C-4520 snorkel with an 11-ft reach.

Once all ground is thought to be sound, the back is bolted with either 5-ft-long Swellex or 7-ft-long Split-Set rock bolts. The Swellex bolts are installed with one of two Eimco-Secoma automated hydraulic bolters, while the Split-Set bolts are installed

with a Cannon automated hydraulic bolter. Ventilation is advanced as needed with 40- and 75-hp auxiliary fans supplying 30- and 36-in vent tubing to the working areas. Water and compressed air lines are also advanced within a reasonable distance to active headings. Although air lines have limited use, they are needed by air pumps and longholing rigs.

6.6   PRODUCTION STATISTICS

The Immel Mine currently employs a unionized labor force of 100 and a salaried team of 12. With this staffing level, the mine is expected to produce 2,000 t of crude ore a day having an average grade of 3.72% zinc. In 1999, production totaled 487,800 t having an average grade of 4.10%. With the recent expansion to a seven-day work schedule, tonnages are expected to approach 680,000 t in the year 2000.

Production expectations depend greatly on equipment availability, and the Immel Mine has instituted various preventive maintenance programs to maintain adequate and acceptable equipment availability. For most of the active fleets, machines should be available for, on the average, 75% of the time, while use is expected for 80% of the time. These figures are reported daily along with major repairs. A maintenance schedule is prepared for machinery and operations. All equipment hours are tracked as well, and major changes in components are planned before equipment fails.

6.7   SAFETY AND HEALTH

At the Immel Mine, safety is first and foremost among all aspects of the operation. Management and supervisory staff are committed to providing a safe and healthful workplace. Success depends on maintaining an active program of safety education for all employees at the mine. All new employees receive a 40-hour training course conducted by the safety department. An extensive review of health, safety, and environmental responsibility is covered to ensure employee awareness. An annual 8-hour safety and first aid refresher course is mandated for all mining personnel. Foremen hold two 1-hour safety meetings per month with their crews to discuss current and pertinent safety concerns and issues. Preshift inspections of workplaces and equipment are carried out every shift before any work is to commence. This is a requirement by the Mining Safety and Health Administration (MSHA) and Asarco company policy. Periodically, safety personnel conduct inspections both in the underground areas of the mine and at surface facilities. At the Immel Mine, under no circumstances will unsafe equipment be operated or will employees be allowed to work in unsafe places.

6.8 OPERATING COSTS

Since ore grade is low at the Immel Mine compared to other similar operations, operating costs must be kept to a minimum. With direct mining costs expected to be $10.78 per ton of crude ore in the year 2000, and the average grade expected to be 3.72%, every effort must be made to maintain acceptable limits on spending. Table 6.1 shows a breakdown of actual past and predicted direct mining costs per ton of crude ore. Through improvement programs and close attention paid to spending details, the Immel Mine has proven itself to be a viable and economical property in all aspects of the operation.

6.9 FUTURE PLANS

With the Immel Mine's limited reserve base of 5 to 6 years (5.8 million tons), additional surface drilling will be instituted to reach an acceptable and confident level of resources to justify expenditures in the near future. For example, the surface holing program will test areas between the Immel Mine and the Young Mine (another Asarco property) and known resources northeast at the Mitchell Bend property. Long-term developments may include a tie-in to the Young Mine and exploitation of those reserves defined by the Mitchell Bend property.

In addition to surface drilling, underground core drilling will commence in the second quarter of 2000 for a 6-month period to delineate those intercepts of mineralization indicated by surface drilling. Additional underground drilling is planned for the following 4 years at 6-month intervals. One possible improvement is the conversion of the Gann air shaft to a full-production shaft with hoisting capabilities. This upgrade would substantially decrease haulage distances. The possibility of a shaft central to both the Immel Mine and Young Mine is another option to be considered.

Because the future depends on expanding and defining additional growth areas, the Immel Mine is committed to a program that subscribes to both looking forward and learning from the past.

6.10 REFERENCES

Bickel, D.L., and D.R. Dolinar. 1975. Report on the Determination of In Situ Stress in the Immel Mine, Knoxville, Tennessee. Report to Asarco, Inc., by U. S. Bureau of Mines.

Bickel, D.L. April 22, 1976. Internal memorandum. U. S. Bureau of Mines.

CHAPTER 7

The Viburnum Trend Underground—An Overview

Jon Carmack,* Bob Dunn,* Michael Flach,* and Greg Sutton*

7.1 INTRODUCTION

The Viburnum Trend is a world-class lead, zinc, and copper ore body located in southeast Missouri, USA (Figure 7.1). The ore trend is approximately 64 km (40 miles) long by an average of 152 m (500 ft) wide. Ore thickness runs from less than 3 m (10 ft) to over 30.5 m (100 ft), with an average around 12 m (40 ft). Average depth of the ore body is 0.3 km (1000 ft). Originally, in-place mineral resources were in the neighborhood of 450 million tonnes (500 million tons). During the past 40 years, five different companies have operated 11 different mines within the trend. Today, this number has dropped to eight mines and one company: SEMO Mining and Milling Division of the Doe Run Resources Corporation. Presently (as of April 1, 2000), SEMO employs approximately 890 people at eight different under-ground mines, six mills, and various support offices. This chapter discusses the current practices and some unique aspects of the Viburnum Trend underground mines.

7.2 SAFETY FIRST

One feature of the Viburnum Trend is its outstanding safety record. The underground mines of the Viburnum Trend are known as some of the safest mines in operation throughout the world. The Sentinels of Safety Award is presented to the operation that works the most man-hours in a year without a single lost-time accident in its respective mining classification. The Sentinels Award for underground metal mining has been presented to various Viburnum Trend underground mines 20 times in the last 28 years (Table 7.1).

SEMO's safety philosophy is a belief that employee behavior is just as vital to safety as working conditions. SEMO encourages its employees to take a proactive approach to their own safety through preshift safety meetings, equipment preshift and work area inspections, monthly safety discussions, and safety programs such as the SEMO roof and ground control policy.

Mechanical scalers eliminate the exposure of people during most scaling operations. In addition to mechanical scaling, hand scaling is also performed. Hand scaling from aerial platforms is done where mechanical scalers can no longer reach the back, or roof. Also, hand scaling, sounding of the back with a steel bar, and visual inspections are performed behind the mechanical scaler. These steps are important to check for, and correct, any change in ground conditions subsequent to mechanical scaling of the stope.

These programs, along with rewards for outstanding safety milestones, contribute to the safe working environment in SEMO operations. During 1999, SEMO experienced a total of 28 MSHA-reportable accidents, seven of which were lost-time accidents, leaving the division with a lost-time incident rate (lost time per 200,000 man hours) of 0.80 and an average 5-year lost-time inci-

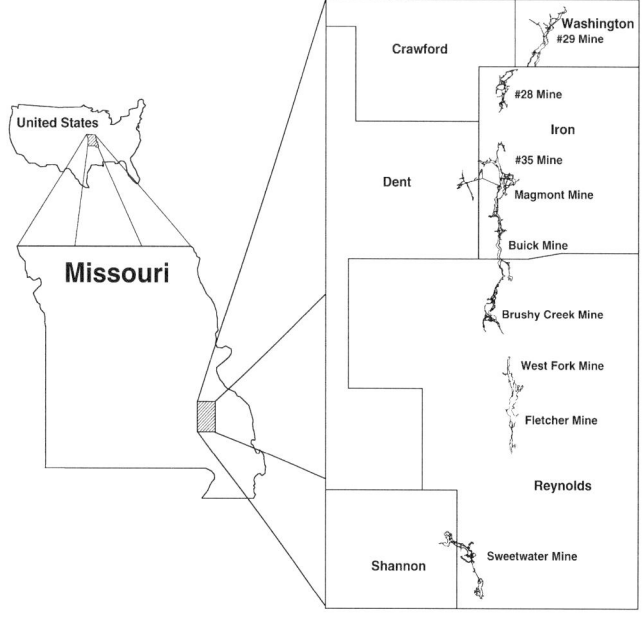

FIGURE 7.1 Viburnum trend mines

dent rate of 0.60 (Table 7.1). The employees of the SEMO Mining and Milling Division pride themselves on these accomplishments, helping SEMO remain an industry leader in safety.

7.3 GEOLOGY

The Viburnum Trend is part of the greater southeast Missouri Lead district that includes Indian Creek, Lead Belt, Mine La Motte and Fredericktown. These deposits, hosted in Upper Cambrian carbonate, plus minor amounts of sandstone and shale, wrap around the St. Francois uplift where a core of igneous basement rocks (granites and rhyolite porphyry) of Precambrian age (1.3 billion years) is exposed (Figure 7.2). The Viburnum Trend portion extends from approximately 64 km (40 miles) north of the town of Viburnum to just south of the town of Reynolds. It is a narrow mineralized zone striking predominantly north–south with a maximum mineralized width of about 600 m (2,000 ft). The ore consists of galena (PbS), sphalerite (ZnS), and chalcopy-rite ($CuFeS_2$). There are minor amounts of bornite and other copper sulfides, as well as siegenite and other cobalt/nickel sulfides. Noneconomic sulfides include pyrite and marcasite.

The minerals listed above make up ores containing lead, zinc, copper, silver, cobalt, and nickel. The silver is primarily

---

* The Doe Run Company, Viburnum, MO.

TABLE 7.1 Viburnum Trend Sentinels of Safety Winners

Year	Mine	Company
1971	Ozark Lead Mine (Sweetwater Mine)	Ozark Lead Company
1972	Ozark Lead Mine (Sweetwater Mine)	Ozark Lead Company
1973	Indian Creek No. 23 Mine	St. Joe Minerals Corporation
1975	Magmont Mine	Cominco American Incorporated
1978	Magmont Mine	Cominco American Incorporated
1979	Brushy Creek Mine	St. Joe Minerals Corporation
1980	Fletcher Mine	St. Joe Minerals Corporation
1981	Brushy Creek Mine	St. Joe Minerals Corporation
1982	Magmont Mine	Cominco American Incorporated
1983	Buick Mine	Amax Lead Company of Missouri
1984	Fletcher Mine	St. Joe Minerals Corporation
1986	Magmont Mine	Cominco American Incorporated
1988	Sweetwater Unit	Asarco Incorporated
1989	Buick Mine	The Doe Run Company
1990	Fletcher Mine and Mill	The Doe Run Company
1993	Sweetwater Unit	Asarco Incorporated
1994	Buick Mine and Mill	The Doe Run Company
1995	Fletcher Mine and Mill	The Doe Run Company
1996	Casteel-Buick Mine and Mill	The Doe Run Company
1998	Fletcher Mine and Mill	The Doe Run Company

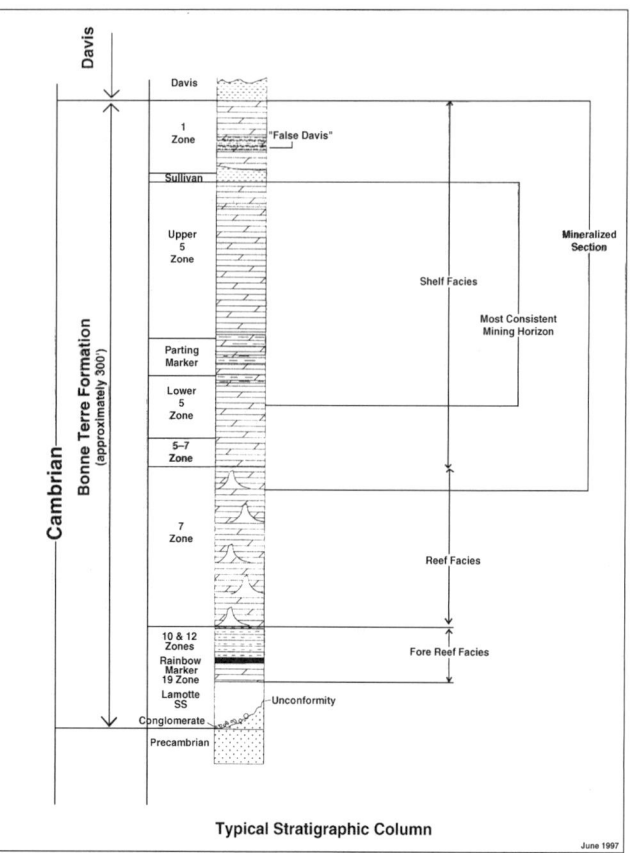

Typical Stratigraphic Column

June 1997

FIGURE 7.3   Stratigraphic column

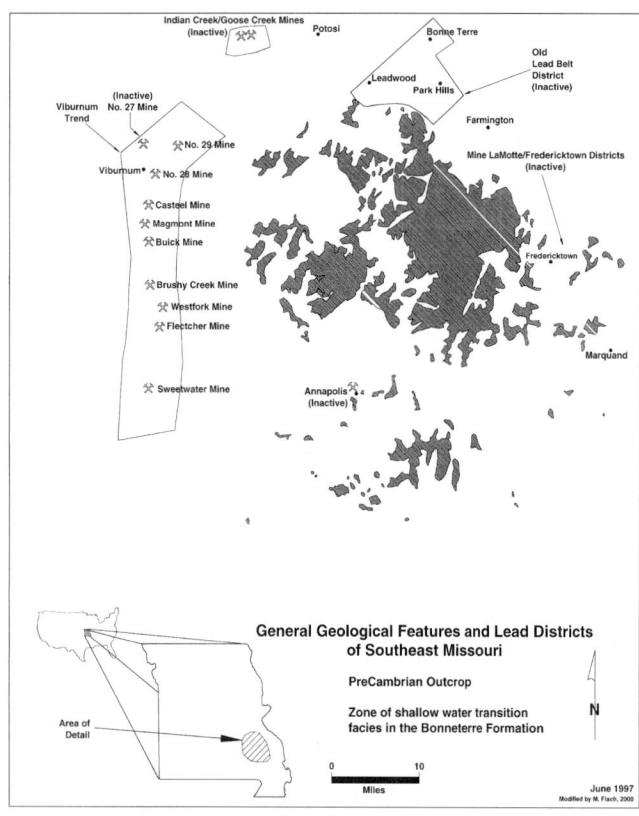

FIGURE 7.2   The Viburnum Trend

contained in zinc concentrates and is in solid solution in with the sphalerite. Gangue minerals are calcite, dolomite, quartz, and dickite. Ore bodies can be stacked and/or brecciated, blanket, and algal reef.

The mineralization occurs as horizontal bands and disseminations, ore matrix breccia, disseminations in rock matrix breccia, veins and fillings in stockwork breccia, and massive sulfides.

In the Viburnum Trend, ore is found in the Bonne Terre Formation. The stratigraphic section in Figure 7.3 shows typical rock zones and that portion of the formation that hosts most of the ore-grade mineralization. The total thickness of the Bonne Terre in the Viburnum Trend is about 91 m (300 ft). Upper Cambrian sediments drape over an irregular unconformity formed on the Precambrian igneous rocks. In places, the Lamotte Sandstone and other Upper Cambrian rocks pinch out against positive irregularities on this unconformity. These pinchouts may control the location of ore in the rocks at or above the pinchout. Ore occurs only in dolomitized Bonne Terre. Limestone facies are found west of the Viburnum Trend and occasionally in the Trend. The "white rock" facies, a white-to-light bluish-gray or light gray crystalline dolostone, lies east of or wedges out in the Trend. Relationships of limestone and white rock to the Viburnum Trend are discussed by Gerdemann and Myers (1972).

Essentially all ore that has been mined or will be mined is in the upper 46 m (150 ft) of the Bonne Terre Formation. Ore thickness in the Bonne Terre usually ranges from 0.6 m (2 ft) to 24 m (80 ft). Some ore-grade mineralization occurs in the base of the overlying Davis Shale and the underlying Lamotte Sandstone. This mineralization makes up a small percentage of the ore and to date is largely uneconomic.

Currently mining is restricted to minimum mining heights of 3.5 to 4 m (12 to 14 ft) but much of the remaining ore could be optimized with low-profile equipment capable of mining to 2.4 m (8 ft). Mining conditions are usually excellent with the rock dipping at less than 5°. However, some areas are vuggy or fractured, making drilling and charging difficult. Roof conditions are generally good, but layered beds of dolomite can separate and be a problem. Some areas have breccia and/or shale in the roof and present a challenge. Algal reef dolomite in the roof is massive, lumpy, or knotted. Where the reef contacts other reef units or layered units, the contacts can be weak. These units are often separated at the contacts by weak shaley or argillaceous material. In some places massive or thick-bedded dolomite can feather out to thin, layered beds, presenting a problem.

Faults with vertical displacements of more than a few inches are rare in the Viburnum Trend, while joints are common and occur in several sets. Most of the faulting in the Viburnum Trend and the region is thought to be wrench faulting and may have occurred during several periods of movement. Observations indicate that most of the last faulting episode was postore deposition.

Hydrology in the Trend is related to two regional fresh water aquifers, the Ozark aquifer and the St. Francois aquifer. The Ozark aquifer extends from the surface to the Davis Shale, an aquatard that overlies the Bonne Terre Formation. The St. Francois aquifer occupies strata, primarily the Lamotte Formation, beneath the Davis Shale but above Precambrian strata. All mining is in the upper part of the St. Francois aquifer, which is dewatered in areas of previous mining. Large inflows of water occurred in the early stages of mining, but decreased as a cone of depression was established in the St. Francois aquifer. Fractures are oriented N 55° E and N 80° E and are often water bearing.

Geologic conditions in the Viburnum Trend contribute in many ways to the reduction of environmental impacts. The strong, flat-lying strata are composed of carbonate rock that helps prevent the formation of acid. The high-pH water associated with the carbonate rocks generally does not carry significant amounts of heavy metals in solution. Water quality is good. Overall, the amount of iron sulfides is low. There are no dangerous gases and little faulting. The relatively low topographic relief and an extensive network of access roads reduce environmental impacts and keep the costs of surface prospect drilling low.

7.4 MINE DIMENSIONS

The standard rock quality designation used is between 90 and 100. Room widths are held to a maximum of 11 m (36 ft), but are typically 9.8 m (32 ft); final heights are preferred to be less than 15 m (50 ft). Roof bolting typically has been done in every intersection and elsewhere on an as-needed basis. Long-term accesses and likely areas for pillar extraction are also bolted to ease extraction and reduce dilution. Pillar sizes have historically been 8.5 by 8.5 m (28 by 28 ft) on 20-m (60-ft) centers. Panel pillar sizes of 11.6 by 23 m (38 by 76 ft) have been adopted for wider areas where pillar extraction is planned. Relative pillars strengths are visually monitored using a rating system of 1 to 6, where 1 signifies no visual stress and 6 signifies pillar failure (Figure 7.4). These ratings are used in the NFOLD numerical modeling programs currently employed for predicting pillar stability.

7.5 MINING METHOD

Because of the relatively flat-lying nature of the ore, pattern-type room-and-pillar mining methods are employed. When several companies were operating in the Trend, several different pillar patterns and extraction methods were employed, but currently only three extraction phases are used: primary, secondary, and tertiary.

Pillar rating	Pillar Condition	Appearance
1	No indication of stress-induced fracturing. Intact pillars.	
2	Spalling on pillar corners, minor spalling of pillar walls. Fractures oriented sub-parallel in walls and are short relative to pillar height.	PLAN
3	Increased corner spalling. Fractures on pillar walls more numerous and continuous. Fractures oriented sub-parallel to pillar walls and lengths are less than pillar height.	
4	Continuous, sub-parallel, open fractures along pillars walls. Early development of diagonal fractures (start of hourglassing). Fracture lengths are greater than half of pillar height.	
5	Continuous, sub-parallel, open fractures along pillar walls. Well developed diagonal fractures (classic hourglassing.) Fracture lengths are greater than half the pillar height.	
6	Failed pillar, may have minimal residual load carrying capacity and be providing local support to the stope back. Extreme hourglassed shape or major blocks fallen out.	OR

FIGURE 7.4 Pillar rating guide

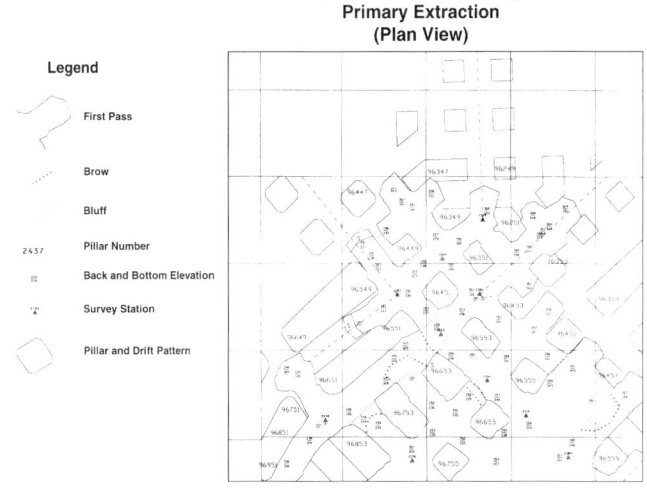

Stope and Pillar Mining
Primary Extraction
(Plan View)

Legend

First Pass

Brow

Bluff

2437   Pillar Number

Back and Bottom Elevation

Survey Station

Pillar and Drift Pattern

FIGURE 7.5 Stope and pillar mining (plan view)

7.5.1 Primary Extraction Phase

Primary extraction is the development and first-pass phase. It is generally done in the horizon with the highest ore grade in the ore body as determined on the basis of results of diamond drill holes from the surface. Typically, the plan view of a primary extraction stope looks like a V, with development beginning in the middle of the V and directed either north or south. Pillars are created behind the V on either side (Figure 7.5).

Drill rounds are generally 5.5 m (18 ft) high by 9.8 m (32 ft) wide by 4.3 m (14 ft) deep and break approximately 560 tonnes (620 tons) of rock. Sixty to sixty-five holes 44 mm (1¾-in) in diameter are drilled per round with a standard burn cut (Figure 7.6) used for relief.

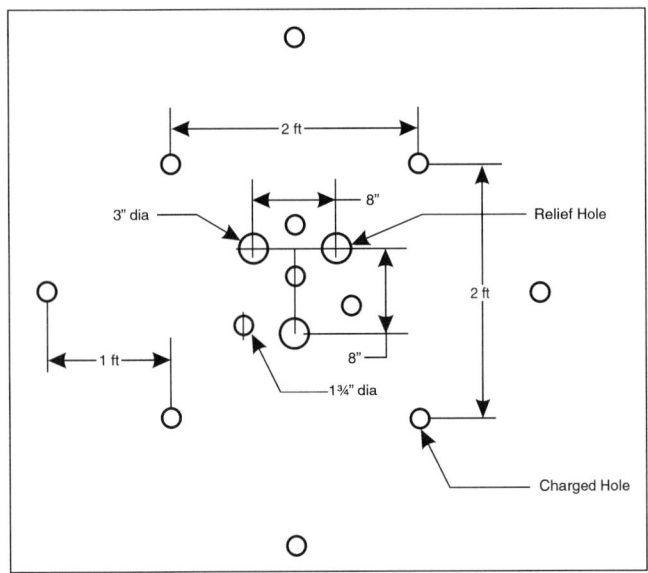

FIGURE 7.6   Standard burn cut

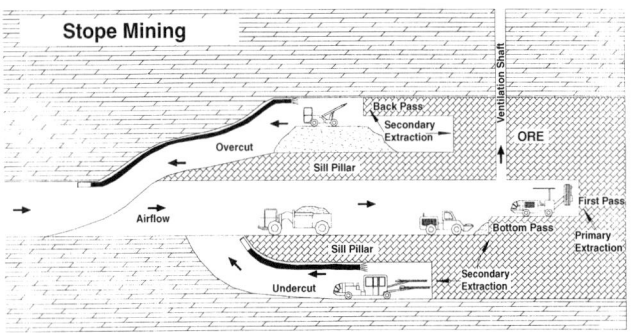

FIGURE 7.7   Stope and pillar mining (side view)

Blasting is primarily done with ANFO, stick dynamite, and nonelectric long-period-delay blasting caps. Powder factors are around 0.65 kg/tonnes (1.3 lb/ton), and average pull is around 85%. Because of the wet nature of the rock, only about 50% ANFO is used. Bulk emulsion is used in some extremely wet areas (Figure 7.6).

7.5.2   Secondary Extraction Phase

Following completion of the primary extraction phase, the mine roof and floor are drilled to delineate all ore to be extracted in the secondary extraction phase. These holes are 15.2 to 30.5 m (50 to 100 ft) long and completed on 18.3-m (60-ft) centers. Some longhole, low-angle drilling is performed in front of mining on occasion, but it is of limited value because of the flat-lying nature of the deposit. Small-diameter, diamond-drill coring is preferred, but rotary percussion drilling is also used.

The secondary extraction phase involves mining the back or bottom or overcut or undercut portion of the ore body (Figure 7.7). An overcut is a pass cut above the primary extraction horizon, and an undercut is a pass cut below the primary extraction horizon. These elevation changes are gained via ramp access normally driven at ±10%. Ventilation raises are cut between levels approximately every 213 linear meters (700 ft). The decision about which practice to follow is based on total ore thickness or the amount of separation of ore by nonore-bearing strata. Ore with thicknesses of

FIGURE 7.8   Pillar extraction

less than 15.2 m (50 ft) is mined as a back or bottom from the first pass. If the ore is over 15.2 m (50 ft) thick or is separated by thick beds of nonore-bearing rock, an overcut or undercut is taken above or below the main pass, and a horizontal sill pillar is left between the passes. To maintain stability, a sill pillar is typically no less than 4.6 m (15 ft) thick. In areas where the sill pillar is of sufficient grade, it is generally mined out during the tertiary extraction phase along with the vertical pillars (Figure 7.7).

Drill rounds are matched to the width of the existing drift, usually 9.8 m (32 ft) wide and 1 to 6 m (3 to 20 ft) high and 4.3 m (14 ft) deep. Holes are 44 mm (1¾ in) in diameter. Drill patterns average 1 by 1.5 m (3 by 5 ft), and no burn cut is needed for relief.

The back and bottom faces are typically drilled by either breasting up or down with flat holes. Breasting has its disadvantages: 60% of the rock broken during a back pass must be held in inventory until the faces have advanced sufficiently to move the ramp ahead, and bottom passes must be hand scaled because of the low roof. However, even with these disadvantages, this form of secondary extraction is the most efficient and cheapest way to mine ore in the Viburnum Trend. Blasting is essentially the same as for primary extraction except that one big factor—water—must be considered. Primary extraction tends to draw down the water table immediately above and below the main pass. ANFO, which is approximately one-sixth the cost of dynamite per pound, is almost exclusively used in these back and bottom rounds. Powder factors typically average 0.45 kg/tonne (0.90 lb/ton), and average pull is 95%. ANFO can be used at a higher percentages, and 75% or higher ANFO usage is common, especially during overcut mining.

Normally, more than one secondary extraction method is employed in the same stope area because of the varied nature of ore geometry. For instance, after the primary extraction phase is complete in a given stope, a back pass may be taken on the primary extraction horizon while an undercut pass is being taken or planned. Another possible combination is a bottom pass being taken on an undercut level while an overcut level is being planned or taken. Several different combinations are possible for any given ore geometry, and no set combination works everywhere.

7.5.3   Tertiary Extraction Phase

Tertiary extraction is the pillar recovery phase (Figure 7.8). The amount of pillar recovery possible for a given stope is highly dependent upon ore grade and the as-mined configuration of the stope. In areas of the highest ore grade, it is possible to extract nearly 100% of the pillars with the aid of backfill. While it is not economic to backfill areas of average grade, 20% to 75% extraction of the pillars is possible, depending upon geometry (Figure 7.8).

Pillar recovery drill patterns are 1.2 by 1.8 m (4 by 6 ft). Holes are 64 mm (2½ in) in diameter and vary from 7.3 to 12.2 m (24 to 40 ft) long. Fan patterns are drilled using a combination of flat and angled holes; that is, flat holes are drilled as high as the jumbo can reach, and all holes above are drilled to form a fan to cover the remaining top portion of the pillar.

Blasting is done primarily with ANFO (80%) and nonelectric millisecond-delay blasting caps. Powder factors are around 0.38 kg/tonne (0.75 lb/ton). Average break is around 90%.

Tertiary extraction without backfill is one of the cheapest mining methods employed in the Viburnum Trend and were it not for the extra expense associated with the maintenance of remote-controlled load-haul-dump (LHD) machines, costs would be as low as those for removing back or bottom ore.

7.6 MINING EQUIPMENT

High degrees of underground mining mechanization have long been the trademark of Viburnum Trend mines. The drilling equipment used for both primary and secondary extraction is essentially the same throughout the Viburnum Trend. Two-boom electric hydraulic jumbos with 4.3- or 4-m (14- or 16- ft) feeds are the standard and, with only a couple exceptions, are 1980s-vintage Joy machines with JH-2 drifters. All jumbos have been rebuilt several times over the years, but there are no plans to replace them over the remaining life of the mines. Penetration in ore averages about 3 m/min (10 ft/min) for a 44-mm- (1$\frac{3}{4}$-in-) diameter bit. Penetration rates in waste are slightly slower.

Drilling equipment for pillar recovery varies from mine to mine. Generally, all jumbos have rod changers and run either 3- or 1.8-m (10- or 6-ft) drill rod. Fully self-contained jumbos that do not require electricity or water are preferred for pillar drilling because pillar recovery usually covers extensive areas of stope quickly. Many of these stopes may also be lacking such utilities, having been mined for 30 or 40 years.

Charging equipment is essentially the same at all mines. Center-articulated boom trucks with man-lifts and 454-kg (1000-lb) ANFO pots are standard. Booms normally reach 6 m (20 ft) high, which is sufficient to reach most drill holes. Normet, Tamrock, and Pettibone are some of the manufacturers of this equipment, but the Getman A64 is the most common machine.

Mucking during primary and secondary extraction is done mainly with construction-type front-end loaders. Loader sizes vary from 4.2 to 4.6 m^3 (5.5 to 6 yd^3) with the preferred model being the Caterpillar 980 series. These loaders provide fast cycling times and good operator comfort. Typically, these loaders are replaced or rebuilt after 15,000 hours when their maintenance requirements start to increase unacceptably.

Mucking for tertiary extraction is done with remote-controlled LHDs. Sizes range from 5.4 to 8.4 m^3 (7 to 11 yd^3) and most have ejector-style buckets to aid in truck loading. The most commonly used model is the Wagner ST8B. Maintenance on these units is significantly higher in this environment than for an LHD operating in a nonremote situation. The practice in some Viburnum Trend mines is to load directly into haulage trucks, while other mines stockpile rock out in a safe area for conventional loaders to load into haulage trucks. Which practice is chosen depends on mine site management preferences and equipment specifics.

Haulage trucks in the Viburnum Trend vary from mine to mine. Sizes are generally in the 36- to 45-tonne (40- to 50-ton) class. Haulage distances in excess of 3.2 km (2 miles) are common at most of the mines now, so operator comfort and tire cost are major factors in truck selection. Historically, Caterpillar 621 and 631 scraper tractors with modified rock beds were used as the primary movers for ore haulage. During the 1980s, several of these units were purchased throughout the Trend because of their durability and dependability, and many are still in service today. Because of current economic constraints, most new haulage trucks are now leased. Consequently, the most common new trucks are six-wheel-drive, articulated, surface construction trucks. These trucks provide good service for about 8,000 to 10,000 hours, at which point they must be replaced. Such a point is typically at the end of their lease.

With mechanical scalers, it is not uncommon for one person to scale three freshly blasted headings in one shift. The most common scaler used at the mines is the Getman version with a 9-m- (30-ft-) long boom and a claw.

Roof bolting is normally done with 1.8-m (6-ft) resin-grouted rebar bolts or galvanized Split-Set bolts. Roof bolters at most mines are single boom with automated bolt handlers. An average of around 50 bolts can be placed by a single operator per shift.

7.7 PRODUCTION DETAILS

Mine production as of April 1, 2000

Mine	Name plate capacity, tonnes/day	Current production, tonnes/day	Current number of employees
No. 29	3,600	1,200	28
No. 28	3,600	1,300	32
No. 35	3,600	1,900	42
Viburnum Mill			31
Buick Mine/Mill	6,500	4,000	132
Brushy Creek Mine/Mill	4,500	4,700	91
Westfork Mine/Mill	3,600	3,400	85
Fletcher Mine/Mill	4,500	4,000	83
Sweetwater Mine/Mill	6,500	3,500	119
Total	36,400	24,000	643

7.8 WORKFORCE

Most of the workforce in the Trend is local. Several are second-generation miners. The mines are the single biggest source of employment and provide some of the highest-paying jobs in the area. These factors combine to make a very talented and dedicated group of employees. In recent years, low metal prices and the remaining reserves of the Trend mines have challenged the economic viability of the mines. Without a workforce of the quality found in the Trend, most of these mines might not have survived the latest low points of the lead market.

7.9 CONCLUSIONS

The Viburnum Trend mines have a long history of using diverse and unique underground mining methods. Over the years, the geologic conditions and different ownership philosophies have led to many variations in underground stope and pillar mining. Today, pillar recovery is still effecting changes to current mining methods and giving birth to many new and cutting-edge technologies. The technologies and methods developed throughout the years, along with the excellent safety record and unique geological features, truly make the Viburnum Trend a world-class ore body.

7.10 REFERENCES

Gerdemann, P.E., and H.E. Myers. 1972. Relationships of Carbonate Facies Patterns to Ore Genesis in the Southeast Missouri Lead District. *Economic Geology*, v. 67, pp. 426–433.

Lane, W.L., T.R. Yanske, and D.P. Roberts. 1999. Pillar Extraction and Rock Mechanics at the Doe Run Company in Missouri 1991 to 1999. Doe Run Co.

Lloyd, W.G. May 2000. Phone Conversation. The Doe Run Company SEMO Central Services Office, Viburnum, MO.

Mount, W.H. May 2000. Phone Conversation. General Manager, The Doe Run Company SEMO Operations. SEMO Central Office, Viburnum, MO.

Paarlberg, N. 1979. *Guidebook, The 26th Annual Field Trip of the Association of Missouri Geologists*.

Snyder, F.G., and P.E. Gerdemann. 1968. *Geology of the Southeast Missouri Lead District, in Ore Deposits of the United States, 1933–1967*, v. 1. New York: AIME, pp. 326–358.

CHAPTER 8

Pillar Extraction and Rock Mechanics at the Doe Run Company in Missouri 1991 to 2000

W.L. Lane,* **T.R. Yanske,*** **L.M. Clark,*** **and D.P. Roberts†**

8.1 INTRODUCTION

8.1.1 Background

The Doe Run Company's lead mines are located in southeast Missouri, USA. The company, owned by the privately held RENCO Group, is the largest integrated lead producer in North America, providing 85% of the total primary lead production in the United States. In Missouri the company owns and operates eight underground lead-zinc-copper mines, six mills, two primary smelters, and a secondary smelter. In addition, Doe Run owns the La Oroya smelter and Cobriza Mine in Peru, lead fabrication businesses in the United States, and several greenfield properties in North America and South Africa. The eight Missouri mines are the Fletcher, Brushy Creek, Casteel, No. 28, and No. 29, which were previously owned by St. Joe Minerals Corp.; the Buick Mine, which is a combination of the old Homestake-Amax Mine and the Cominco-Dresser Magmont Mine; and the Sweetwater and West Fork mines, which were purchased from Asarco in 1998.

Mining operations in Missouri take place in a flat-lying, tabular ore body 150 to 350 m below the surface. The ore body is from 1 to 40 m thick and 10 to 600 m wide and has a trend length of 65 km. The mine workings are located below an aquifer, from which it is separated by 30 m of impermeable shale. The ore trend traverses both the Mark Twain National Forest and privately owned lands.

8.1.2 Primary Mining

The primary mining method is a highly mechanized room-and-pillar method incorporating two-boom drill jumbos, 7- to 9-tonne-capacity loaders, and 27- to 45-tonne-capacity haul trucks. Thicker ore zones are mined first using an initial pillar pass followed by a varied combination of back, bottom, undercut, and overcut passes. Ultimately, pillars range in height from 4 to 37 m.

Until 1993, pillars were cut 9 by 9 m with 10-m drift openings between pillars. Since 1993, larger panel pillars have been utilized where stopes are wider than 50 m. These measure 11.5 by 23 m with 10-m drift openings between pillars. Current pillar reserves are estimated to be 16,000,000 tonnes, much of which is high grade.

8.1.3 Secondary Mining

Secondary mining in the form of pillar extraction is an extremely important component of the total production at Doe Run. In fiscal year 2000, pillars will be 16% of the tonnage and 25% of the metal. The ability to successfully extract pillars allows Doe Run to nearly double its remaining mine life.

Pillar recovery planning at Doe Run is coordinated by the Technical Service Department working with mine personnel. All recovery plans are modeled using Golder Associates' stress analysis computer program NFOLD. Golder's expertise has proven invaluable to the success of pillar extraction at Doe Run.

Pillar recovery is being maximized by the use of backfill, which varies from 2 tonnes of backfill per tonne of pillar ore mined to no backfill in narrow, restricted areas. Of primary importance in designing a backfill plan is to ensure that any potential pillar failures are contained within the backfilled area so that there is no propagation into adjacent pillar areas. In addition, backfill limits the extent of back failures, thus avoiding air blasts, surface subsidence, and breaching of aquifers above the mine workings.

In the older areas of the mines, extracting the smaller 9- by 9-m pillars has proven to be difficult, since most of these areas are over 100 m wide with 20-m-high pillars (an example is shown in Figure 8.1). When it is economical, backfill is placed to allow for the extraction of additional pillars. The backfill eliminates the potential for catastrophic domino-type pillar failures and extensive collapses of the back. With the introduction of the larger (11.5 by 23 m), stronger panel pillars in the wider areas of the ore trend, more efficient fill schemes and higher extraction ratios have been obtained. Unfortunately, most of the pillar reserves are in the older 9- by 9-m pillar areas.

8.2 BACKFILL OPERATIONS

8.2.1 Buick Mine Shaft Area Backfill Plant

The first backfill plant at the Buick Mine went into operation in 1993. The plant is composed of an 8-m^3 Denver mixer fed by two conveyor belts, one of which feeds cement and fly ash from two 63-tonne bins and the other which feeds cycloned tailings from a storage hopper (Figure 8.2). Cement and fly ash are delivered to the underground bins from surface via a 20-cm-in-diameter, 305-m-deep cased hole. Cycloned tailings (Table 8.1) are dropped from the surface down another 20-cm-in-diameter hole and deposited in a sump area, where the water is decanted. Once the cycloned tailings are allowed to dry, they are picked up by a Caterpillar 966 loader and fed into the tailings hopper. The mixer is capable of cycling a batch of slurry in 7 min. For normal backfilling, the slurry batch mix consists of the following materials.

1. 450 kg cement
2. 540 kg fly ash
3. 1140 L water
4. 4550 kg cycloned mill tailings

* The Doe Run Company, Viburnum, MO.
† Golder Associates, Ltd., Burnaby, British Columbia, Canada.

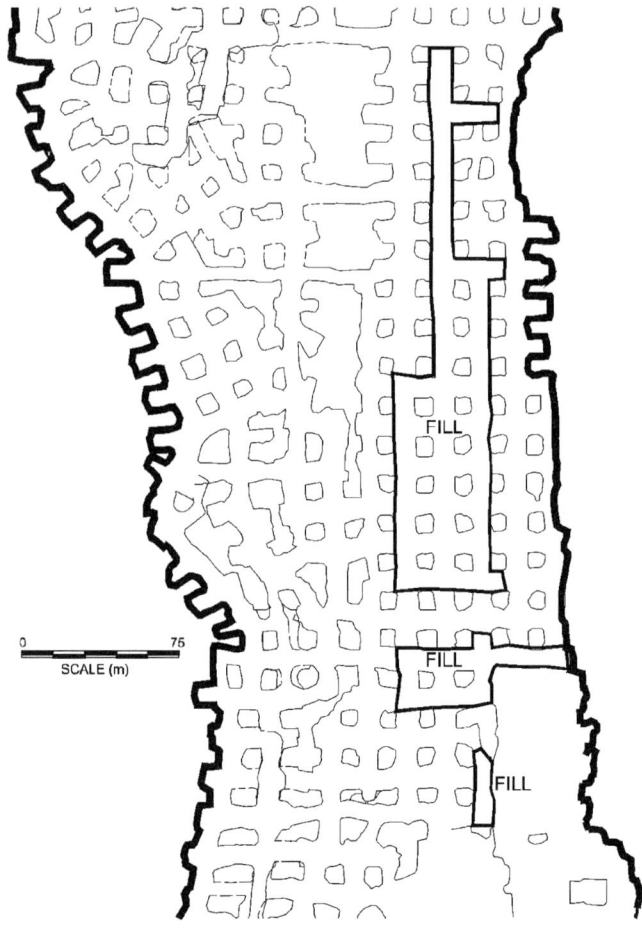

FIGURE 8.1   Cemented backfill layout at the Buick Mine

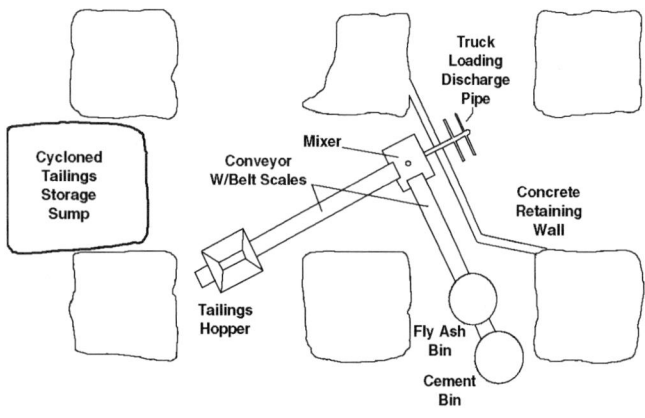

FIGURE 8.2   Underground batch plant at the Buick Mine

For final topping off the backfill, the mix is modified as follows:

1. 540 kg cement
2. 630 kg fly ash
3. 1420 L water
4. 910 kg cycloned mill tailings

Once the normal backfill slurry is mixed, it is dumped into a 45-tonne truck that has previously been loaded with 36 tonnes of –1-m run-of-mine rock that has come either from development

TABLE 8.1   Size of cycloned tailings from the Buick Mill

Size range (mesh)	Percentage range
>20	0.0
20–40	0.2
40–70	19.0
70–100	21.4
100–140	21.2
140–200	14.1
<200	24.1

waste rock or from the underground quarry. The loaded truck then is trammed to a sump area near the backfill site where the load is deposited. Next, a Wagner ST-8B, Caterpillar 980C, or MTI 700M loader mixes the rock and slurry in the sump. Once mixing is complete, the loader then transports the material into the area to be backfilled and places it in a 1-m lift. This sequence of lifts continues to 3.5 m from the back. At this time, a JCI 400M loader mounted with a 5-m-long boom with a 30- by 90-cm pusher blade at the end pushes additional –1-m rock to within 0.5 to 1 m of the back, creating a bench that is 1 to 2.5 m deep.

The backfill operation then switches from regular backfill to topping-off fill. Topping-off fill is composed of –5-cm rock mixed with a modified slurry. The slurry is hauled in a truck from the batch plant to the sump at the backfill area, where it is dumped on top of –5-cm rock. A loader then mixes the material in the sump. The loader transports this mix into the backfill work area, where it is loaded into a 9-tonne Paus slinger truck that fills in (tops off) the last 0.5 to 1 m of open drift. This is accomplished by the slinger mechanism at the base of the rear of the truck, which throws the slurry-rock mixture at a velocity of 30 m/s. The slinger mechanism can be raised and lowered and swung from left to right, enabling it to place the topping-off fill so that it makes almost 100% contact with the back.

In an effort to lower backfill cost, uncemented, cycloned mill tailings are sometimes placed down the centers of the drifts as the backfill lifts are placed. The sides of the drifts next to the pillars are backfilled with cemented backfill for a distance of 3 m from the pillars. This leaves 4 m of uncemented tailings in the middle. Both materials are placed by loaders. This process continues until the fill reaches 3.5 m from the back, after which regular topping-off takes place. The backfill sequence is modified when trapped pillars (pillars that have backfill on all four sides) are involved. This modification involves placing uncemented cycloned tailings on one side of the trapped pillars starting from the floor and progressing to the back as subsequent lifts are placed. The uncemented cycloned tailings are later removed through an undercut access drift from which a raise is driven up into the sand. This creates a void into which the trapped pillars can be blasted.

8.2.2   Buick Mine North Backfill Plant

A second backfill plant was constructed at the Buick Mine in 1998 to service backfill placement at the far north end of the mine. This fill plant is similar in construction and operation to the Buick shaft plant. The only difference is that uncemented classified tailings had been previously placed in this area of the plant by Cominco to stabilize pillars. These same tailings are now being hauled a short distance to the plant where they are used in the slurry mix.

8.2.3   Fletcher Mine Backfill Plant

The Fletcher backfill plant is located on the surface (Figures 8.3 and 8.4). The plant consists of a Denver high-intensity mixer, which is fed cement and fly ash stored in 64- and 45-tonne bins, respectively. Cement and fly ash are transferred by screw conveyors to a bucket elevator, which then elevates the cement

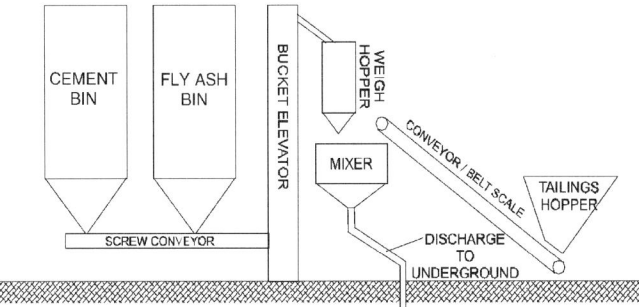

FIGURE 8.3   Surface batch plant at the Fletcher Mine

FIGURE 8.4   Backfill batch plant at the Fletcher Mine

and fly ash to a weigh hopper that feeds the material into the mixer by gravity. Dried, cycloned mill tailings are also fed into the mixer via a conveyor belt from a 45-tonne tailings storage bin. A batching cycle takes 12 min to complete, and batch composition is similar to that produced at the Buick Mine.

Once the batch is complete, it is dumped down a 305-m-long cased hole through a pipe of 15-cm inside diameter into a storage bin. The cemented slurry is then loaded into a 7-m^3 Redi-Mix truck, which transports the slurry to the fill area and dumps it into a sump on top of 23 m^3 of run-of-mine waste rock (–1 m). This cycle takes 20 min. Placement of backfill from this point forward is similar to that previously described for the Buick backfill operation.

8.2.4   Fill Fences

To minimize the amount of backfill to be placed, fill fences 10 m wide by 6 to 18 m high are placed between pillars (Figure 8.5). The fences are constructed on top of the backfill once 3 m of fill material has been placed. Fences are built in 6-m lifts, which is the maximum height that the equipment can reach. The steps used in fence construction are as follows.

1. Anchor bolt holes, 3.8 cm in diameter and 90 cm deep, are drilled on 90-cm vertical spacings in the corners of adjacent pillars.
2. A 90-cm-long expansion-shell-type rock bolt is installed into each hole with an angle bracket at the collar of the hole.
3. Cables 1.3 cm in diameter are strung horizontally across the drift and anchored to the angle brackets with two 1.3-cm Crosby clamps at each bracket.

FIGURE 8.5   Fill fence at the Buick Mine

4. Nongalvanised, 9-gauge, chain-link fence 1.8 m wide is then suspended from the top cable on the fill side. The fence is attached at the top and allowed to hang down to the fill level, where it is cut to length, allowing at least 1 m of the chain link to lie on the fill.
5. Three-millimeter in diameter aircraft cable is then used to lace the adjoining lengths of chain link fence together. The cable is also used to attach the chain link to the angle brackets at the sides.
6. Fill material is then placed at the bottom of the fence onto the chain link lying on the fill.
7. Fill material is then dumped, but not pushed, next to the fence as each subsequent 1-m-high lift is placed. The fill placed at the fence is always kept higher than the rest of the fill in order to provide a berm to prevent equipment from driving off the edge of the fill.
8. Once the fill has been placed to 1 m of the top of the fence, the fill fence construction sequence is begun again.

This method is a typical one for building fences. Alternative methods have included the use of "landslide" steel netting or wire baskets filled with backfill (gabions). We continue to look for cheaper, more efficient ways to construct backfill fences.

8.3   PILLAR MINING

8.3.1   Economics

Technical Services performs an economic evaluation of every pillar extraction area in order to provide a pillar plan that both profitable and technically feasible. The evaluation is done in the following manner.

1. The mine first provides Technical Services with tonnage and grade information for all pillars in a designated pillar removal area. To ensure that the grades are as accurate as possible, Doe Run has a database on pillars constructed with information about previous face grades, jackhammer grades, and diamond-drill-hole grades. Where data are lacking, a pillar-washing program provides additional information. In pillar washing, a high-pressure water spray containing sand and mounted on a boom lift truck is used to wash a strip down each side of the pillar. A visual estimate of grade is then made.
2. Technical Services inputs the tonnage and grade data into a spreadsheet to determine profits and losses based on current metal prices and production costs. The pillars are then sorted and ranked by value.

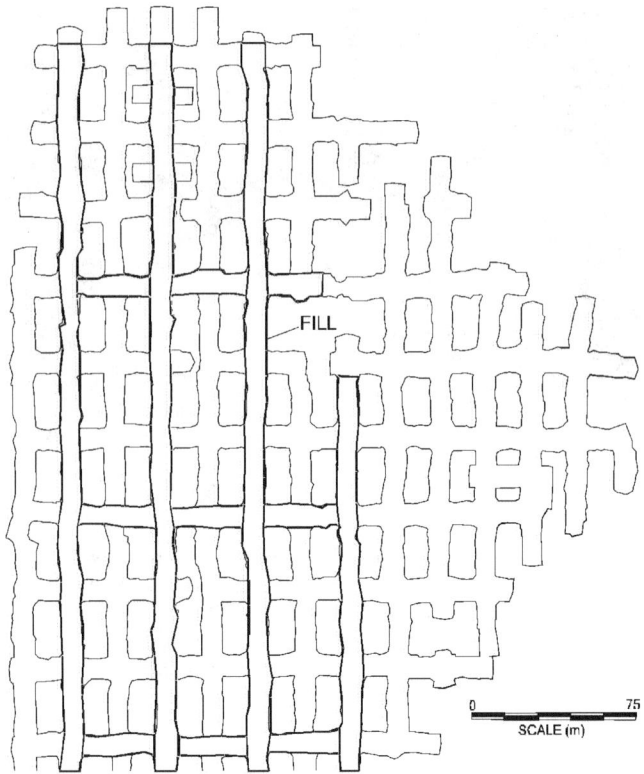

FIGURE 8.6   Cemented backfill layout at the Fletcher Mine

3. A preliminary pillar mining plan is constructed, with the pillars of highest value selected first. Back spans resulting from pillar extraction are limited to 46 m in the short dimension. Backfill is often used to allow larger spans between pillars.

4. A pillar stability evaluation is done that includes computer modeling. To aid with this evaluation, the condition of underground pillars is rated so that the model can be calibrated to actual conditions. Narrow pillar areas are generally not modeled. In high-grade areas, an additional evaluation is done to determine the benefit of placing fill to allow for additional pillar extraction.

5. A final, detailed plan is provided that describes what pillars are to be taken and what fill is to be placed.

8.3.2   Pillar Extraction Method

Freestanding pillars are drilled with Tamrock and Atlas Copco pillar drills from the floor level. Some shorter pillars are drilled with conventional two-boom Joy drill jumbos. Drill holes are 64 mm in diameter on 120- (horizontal) by 180-cm (vertical) spacings. Normally, one or two pillars are blasted at a time using ANFO as the explosive. Single-level pillars up to 18 m high and multilevel pillars up to 37 m high have been shot to date. Freestanding pillars with cemented fill on one or two sides are drilled from the top of the fill as the fill is placed.

Trapped pillars (that is, pillars with fill on all four sides) are drilled and blasted from an undercut drift below the pillars. The pillars are then shot into a void created by the removal of the previously placed uncemented mill tailings. Blasted pillar rock is loaded by remote-controlled loaders. Wagner ST8-A & Bs, Elphinstone, and MTI 700M loaders are usually equipped with Cattron, Hetronix, and Nautilus remote-control units.

At the Fletcher Mine, larger panel pillars are being extracted (Figure 8.6). Cemented fill is placed in alternating drifts that run

the length of the pillar area. This leaves an opening every other drift into which the panel pillars can be blasted. The pillars are shot two at a time using the same drilling pattern as previously mentioned. Once six pillars have been blasted and loaded out, the open stope is sealed at the end, and uncemented cycloned mill tailings are piped in to fill the void, thus eliminating any potential air blast. The pillar extraction sequence then begins again.

8.4   SAFETY

Safety is of prime importance to the Doe Run Company. As evidence of this, the company has been the proud recipient of the nation's highest safety award, the Sentinels of Safety Award, a total of 13 times. The goal with regard to pillar extraction is that there be no lost-time accidents. This has been achieved since beginning the pillar extraction process. If a safe way can't be found to do the work, it won't be done.

To ensure that safety is not compromised during pillar extraction, Doe Run has incorporated several procedures.

- Back spans are limited to 46 m in the short dimension in pillar extraction areas (this may be increased through the use of fill). This limits the potential for large back failures that can cause air blasts, which are dangerous to employees working in the area.

- Visual classifications of pillar stability are made regularly to monitor changing conditions. In those areas found to have stability problems, the extraction process is modified to ensure that the extraction can be done safely.

- A set of rules has been defined that lists the "do's and don'ts" of working in a pillar area.

- A safe means for retrieving disabled loaders in nonaccess open stope areas has been developed. A second remote loader is available at each operation to assist with the rescue of the other. A small remote-controlled tractor with a camera has is also available for making cable hook-ups so that nonremote-controlled trucks and loaders can be used for loader retrieval. Creative means of making cable hook-ups using scaler rigs and other equipment have been used at times. To date, no loaders have been lost.

- Intermine pillar audits are conducted on a rotating schedule at each mine in which pillars are extracted. When an audit takes place, each mine sends a representative (salaried or hourly) to attend the audit. Normally, a meeting is first held on the surface to review current pillar production from each mine. Discussion of problems and solutions are also held at this time. Following this meeting, the audit attendees are taken on an underground tour of pillar areas at the mine being audited. While underground, the attendees look for safety issues that need attention and review the pillar extraction procedures being used at the audited mine.

- Rock mechanics instruments have been installed to monitor ground conditions for potential unstable and unsafe conditions. These instruments are piano-wire extensometers and multipoint in-hole vibrating wire transducers. Pressure cells, embedment strain gauges, and biaxial stressmeters have also been used.

8.5   ROCK MECHANICS

8.5.1   Modeling

Numerical modeling at the Doe Run Company's mines using the NFOLD computer program has been ongoing for a number of years and much experience has been gained. This has given a reasonable amount of confidence in the model's use as a method of predicting pillar stability as mining progresses.

NFOLD is based on the displacement-discontinuity stress analysis method whereby the ore body is modeled as a planar feature with the thickness of the ore body located within an infinite elastic material (the host rock). The method has its greatest application in determination of stresses and displacements associated with tabular ore bodies, such as the Missouri lead belt.

The elements characterizing the mining horizon may be either elastic (in which failure of the pillar is suppressed), nonlinear with postfailure properties, or representative of backfill material. The strengths assigned to the elements representing the pillars depend primarily on the strength of the rock and the geometry of the pillars. In the Missouri lead belt models, if failure is initiated in a freestanding pillar, the pillar is assumed to follow a falling stress-strain relationship to complete failure, with little or no residual strength. This is consistent with observations of pillars that have actually failed. The effect of trapping a pillar with backfill on the postfailure response of the pillar is one of the points of discussion of the following sections.

Pillar strengths for NFOLD modeling were initially based loosely on Hedley's formula, in which

$$\text{Strength} = K \times UCS \times (W^{1/2}/H^{3/4}),$$

where   Strength = ultimate pillar strength

             K = constant

        UCS = uniaxial compressive strength of rock

          W = pillar width

          H = pillar height

This strength equation was found to represent satisfactorily the observed response of Missouri lead belt pillars between 10 and 15 m high. However, when the equation was used for pillars outside this range, it was thought to be conservative, particularly for pillars greater than 15 m high. This was resolved by further calibration.

8.5.2 Calibration

Numerous NFOLD models, including models of previous pillar failures, have provided the opportunity to calibrate estimated strength with actual underground observations. This calibration involves two independent steps.

The first step is to rate the condition of the pillars through underground observations. For this purpose, a pillar rating system has been developed whereby each pillar (or in some cases portions of pillars) is assigned a number between 1 and 6. A rating of 1 represents a pillar for which no signs of stress are evident while a 6 is given to pillars that have completely failed

The second step is to model the area in question using the NFOLD program. This allows a pillar load to be estimated for the pillars, or portions of pillars, that have been assigned a pillar condition rating.

Once pillar condition ratings and pillar loads have been established, pillar load as a ratio of the UCS value of the rock is plotted against pillar height for standard pillar widths. Figure 8.7 shows the curves that have been developed by applying this method to hundreds of pillars in various areas at four of Doe Run's mines.

This calibration is used to predict the progress of pillar deterioration and ultimately complete failure as adjacent pillars are mined.

8.5.3 Instrumentation

To facilitate pillar extraction, in the early 1990s The Doe Run Company and Golder Associates began to investigate the implications of using cemented backfill to surround, or trap, pillars. It was thought that although these pillars would likely fail regardless of the backfill, their residual strength might provide support to both stabilize the back and reduce the loads on surrounding pillars.

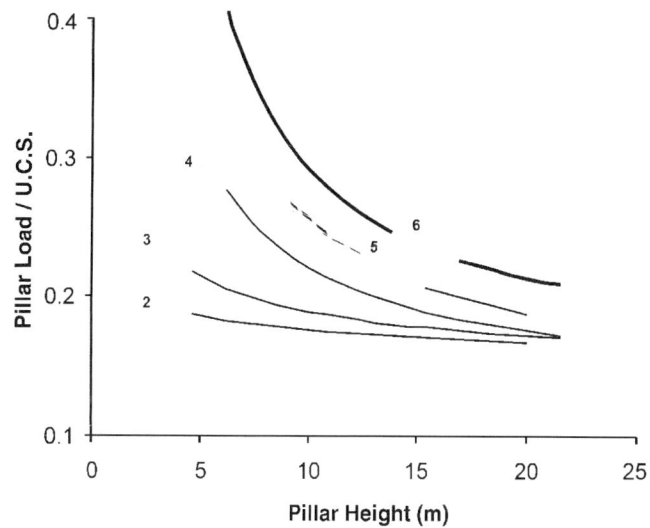

FIGURE 8.7   Pillar height versus load and uniaxial compressive strength

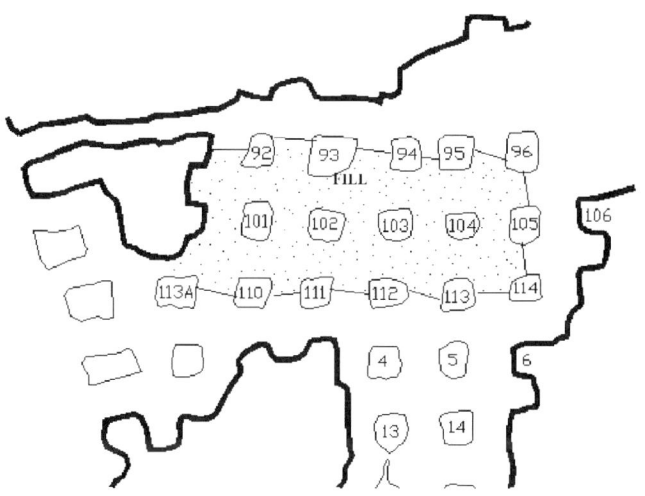

FIGURE 8.8   Test area fill plan

To test this theory, an isolated area in the Buick Mine was chosen where four pillars (101, 102, 103, and 104) were trapped, and the surrounding pillars were extracted (Figure 8.8). With the assistance of personnel from the U.S. Bureau of Mines, 38 instruments were placed in the back, pillars, and backfill and monitored as extraction proceeded. The instruments included earth pressure cells, embedment strain gauges, stressmeters, and extensometers. In addition, NFOLD modeling of the test area was carried out by Golder Associates to simulate the extraction sequence.

The purpose of the modeling was to assess the pillar strength relationship in as much detail as possible and to develop other pillar response properties consistent with the results of the monitoring program. Of particular importance was the response of the trapped pillars to increased loads as extraction proceeded. From January 1992 to December 1993, 24 pillars were removed. This included the four trapped pillars, which were mined last. These were accessed for drilling, blasting, and extraction using an undercut development.

The results of the instrumentation program were mixed. In general, the extensometers produced the best data, giving results

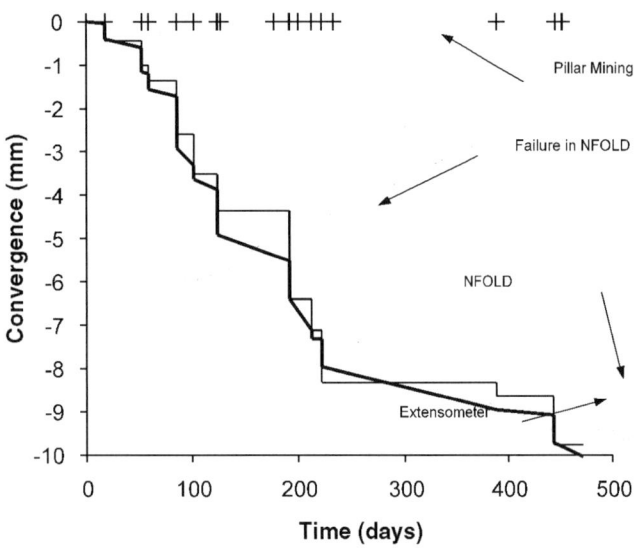

FIGURE 8.9 Convergence, NFOLD model versus extensometer readings

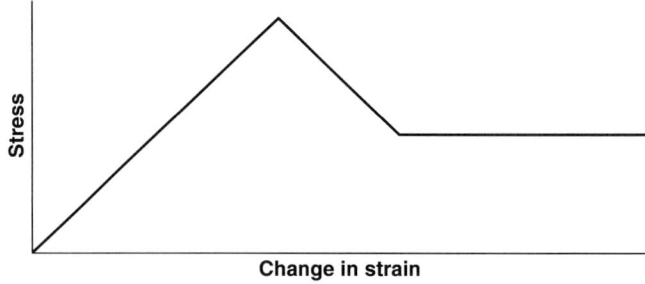

FIGURE 8.10 NFOLD model response for trapped pillars

that were both consistent with each other and with elastic theory (where applicable).

The amount and quality of the data obtained from the extensometers allowed the vertical convergence in the NFOLD model to be calibrated to an unexpectedly high degree. This applied not only to the results obtained from the truly vertical extensometers installed in the trapped pillars from the undercut development, but also to the results obtained from reduction of data from horizontal and inclined extensometers installed in the free-standing pillars.

In the test area, the computer model simulated the measured movements of the instrumented pillars very closely. Figure 8.9 shows the convergence of trapped pillar 104. This pillar's failure was precipitated in the model both by the removal of surrounding pillars and by the failure of an adjacent trapped pillar. Extensometer measurements from pillar 104 and other pillars confirmed both the failure of the pillar and the preceding failure of the adjacent trapped pillar.

Conclusions drawn from the instruments in the test area are—

- The prefailure modulus of elasticity used for pillars required no adjustment.
- Trapped pillars were found to have the same prefailure modulus of elasticity as untrapped pillars.
- Peak strength values previously used for pillars above a height of approximately 15 m were found to be conservative.
- Trapped pillars were not found to have higher peak strengths than untrapped pillars.
- The postfailure modulus of elasticity used for untrapped pillars required no adjustment.
- Trapped pillars were found to have a significantly different postfailure modulus of elasticity than untrapped pillars.

Figure 8.10 shows how the shape of the NFOLD load deformation response was changed for the trapped pillars. The significance of the reduced postfailure modulus of elasticity for trapped pillars cannot be overstated. Where cemented fill is used to trap pillars, far greater extraction rates are achieved than would

otherwise be possible due almost entirely to the improved postfailure characteristics of the trapped pillars.

A good example of the effect of this change is the test area itself. If the extraction sequence is modeled using the original steeper postfailure elastic modulus, all the failed pillars go directly from the prefailure part of the load deformation response to their residual strength (conservatively assumed to be none). However, when the flatter postfailure elastic modulus is used for the trapped pillars, none of the failed trapped pillars reached residual strength, assumed, but not proven, to be 50% for trapped pillars. In the test area, the result was that even once all the surrounding pillars were extracted, those trapped by fill were still carrying 65% to 80% of their maximum load immediately prior to being extracted. This had a significant impact on the number of pillars that could safely be extracted.

8.6 CURRENT AND FUTURE PROJECTS

The Doe Run Company continues to be very proactive in its efforts to improve pillar extraction and backfilling techniques. Current projects include—

- Further modeling using a variety of two- and three-dimensional programs to better understand multilevel mining situations. The NFOLD program has proven to be a reliable tool for predicting pillar conditions in single-level situations. However, where there are multiple mining levels separated by thin sills, the model cannot be calibrated as well.
- The continued development of a geographical information system to assist with pillar extraction planning, particularly for performing economic evaluations and giving graphical presentations of the large amounts of data associated with these areas.
- The evaluation of a microseismic monitoring system. A system was purchased from ISS in 1998 and installed at the Buick Mine in an active pillar extraction area. The system has had equipment problems, but continues to be evaluated as to its usefulness versus conventional instruments.
- Working with the University of Missouri-Rolla on refining trapped pillar strengths.
- Implementation of a new type of slinger for topping off. This new equipment will allow for more efficient placement of topping-off material.
- Implementation of a remote drilling and charging rig for shooting large boulders in open stope areas.
- Implementation of a remote charging arm for charging sills at the edge of caved stopes.

Future research projects include paste fill, undercut-and-fill, and different methods of topping-off the fill.

8.7   CONCLUSIONS

Pillar extraction has proven to be very successful for the Doe Run Company and will play a prominent role in future production. The company will be increasingly reliant on the revenue produced from pillars and is committed to the continued development of pillar extraction procedures and technologies. In support of this, rock mechanics will remain a critical aspect of the pillar extraction process. The Doe Run Company is continually striving for ways to improve the accuracy of the models. These improvements will allow for the development of more cost-effective fill processes, enhanced revenues, increased mine life, and continuation of the excellent safety record achieved by the employees of the Doe Run Company.

8.8   ACKNOWLEDGMENTS

The authors gratefully acknowledge the assistance of Mr. Doug Tesarik and Mr. Brad Seymour of NIOSH (formerly the U.S. Bureau of Mines), Mr. Darren Kennard and Dr. Ross Hammett of Golder Associates, Mr. Greg Sutton of Doe Run's Buick Mine, and Mr. Dave Olkkonen of Doe Run's Fletcher Mine.

8.9   REFERENCES

Hedley, D.G.F., and F. Grant. 1972. Stope and Pillar Design for the Elliot Lake Uranium Mines, *CIM Bulletin*, July.

Tesarik, D.R., J.B. Seymour, T.R. Yanske, and R.W. McKibbin. 1995. Stability Analyses of a Backfilled Room-and-Pillar Mine. USBM RI 9565.

CHAPTER 9

Modifications of the Room-and-Pillar Mining Method for Polish Copper Ore Deposits

Waldemar Korzeniowski* and **Andrzej Stankiewicz†**

9.1 INTRODUCTION

Copper mining in Sudeten Mountains began in the thirteenth century, however intensive exploratory geological works began only about the middle of twentieth century, when a copper ore-bearing seam, 1,000 m below surface with over 0.5% content of copper (Cu) was confirmed, and led to beginning of construction the first underground mines: Lena and Nowy Kościól near Zlotoryja, and Lubichów and Konrad near Boleslawiec, (Rydzewski 1996). These mines, now closed, are considered as the Old Basin. Today the New Basin, i.e., Legnicko-Glogowski Okreg Miedziowy (LGOM), situated on south-west region of Poland, Figure 9.1, is based on three mines with various dates of starting their construction: ZG Lubin (since 1960), ZG Rudna (since 1970), ZG Polkowice–Sieroszowice (since 1996). The last mine resulted from the joining of two formerly single mines: ZG Polkowice (since 1962) and ZG Sieroszowice (since 1974). All the mines belong to a joint stock company KGHM Polska Miedź SA located in Lubin. This a company formed in 1991 after commercialization, from formerly state-owned enterprise KGHM (Kombinat Górniczo-Hutniczy Miedzi) after its deep restructurization, upon background of the general political-economical transformation in Poland started in the 1990s. Now, the Company is the largest in Europe and fifth in the world in the production of copper from its own reserves. In silver production, it is also a world leader. KGHM Polska Miedź SA exploits, produces and trades with nonferrous metals, mainly copper and silver. The basic products are refined copper (cathode), copper rod, round runners and refined silver.

9.2 GEOLOGICAL CONDITIONS

The LGOM is situated in The Presudeten Monocline. Foundation of the monocline is composed of Proterozoic crystalline rocks and sedimentary rocks belonging to Carboniferous system. Permian and Triassic sedimentary rocks occur above the formation, and next the Tertiary and Quaternary. The Permian rocks are represented by sandstones with clayey-limestone binding or locally with gypsum-anhydrite binding, conglomerates and shales of total thickness up to a dozen and so meters, (Klapacin´ski and Peryt 1996). The sandstones are mineralized with copper sulfides. The Triassic is an assemblage of fine- and medium-crystalline sandstones with grains of diameter 0.1 to 0.5 mm, sometimes 1 mm; marls; clay-shales and dolomites. The Tertiary is created with sands and lime sandstones, quartz and quartzite gravels, quartz sands, clays, and brown coal. The Quaternary consists of sands, quartz gravels and clays that are shown in Figure 9.2.

FIGURE 9.1 Location of Polish copper mines

9.2.1 Tectonics and Water

The Presudeten Monocline is separated from the Presudeten Block with a fault of the medium Odra River, Figure 9.1. Many deformations and tectonic dislocations are confirmed directly in the underground excavations. The dominant strike of the fault is NW–SE that occurs at the Perm-Presudeten Block contact. Groups of summing (en echelon) faults are observed very often with a total throws up to 100 m and expansion to a few hundreds kilometers. About 60% of the faults are characterized with the throws less than 1 m, 35% of faults have throw between 1 to 10 m (Salski 1996). The greatest throws reach 50–60 m. Dip of the fault planes ranges from 30° and 90° dominant inclinations 71° to 75°, but significant variations even within the same fault are observed. There is a characteristic that in dolomite-limestone veins of gypsum, with a thickness of 4 to 6 mm and more, which runs along the slip planes. In Figure 9.3 the representative directions of faults are drawn and Figure 9.4 shows the cross section of groups of the faults. In Tertiary and Quaternary formations one can distinguish two water-bearing complexes (Kleczkowski and Kalisz 1996). The first is a groundwater reservoir with regional significance that occurs down

* University of Mining and Metallurgy, Cracow, Poland.
† KGHM Polska Miedz, S.A., Polkowice, Poland.

TABLE 9.1   Average range of rock properties

Type of rock	Density per unit volume $\rho[kg/dm^3]$	Compression strength $R_c[MPa]$	Tensile strength $R_r[MPa]$	Bending strength $R_g[MPa]$	Shear strength $R_t[MPa]$	Deformation modulus $E_d[GPa]$	Young's modulus $E_e[GPa]$	Poisson's ratio $v[-]$	Energetic Rockburst Index $W_{et}[-]$
Dolomites	2.46–2.68	55.8–182.5	3.7–9.1	7.5–17.3	12.3–25.9	21.5–83.8	25.3–86.0	0.22–0.25	2.46–7.56
Dolomitic limestones	2.47–2.67	65.9–144.4	3.9–7.7	8.5–15.9	14.7–21.8	27.6–68.4	32.2–72.3	0.22–0.24	2.41–6.16
Shales	2.48–2.64	35.1–115.2	3.2–7.8	5.3–12.9	8.5–19.2	13.3–35.1	18.5–42.1	0.21–0.23	1.81–3.94
Sandstones	1.98–2.73	15.1–103.2	0.7–5.5	1.8–12.1	3.0–21.2	3.8–42.9	6.6–48.6	0.12–0.22	0.71–2.79

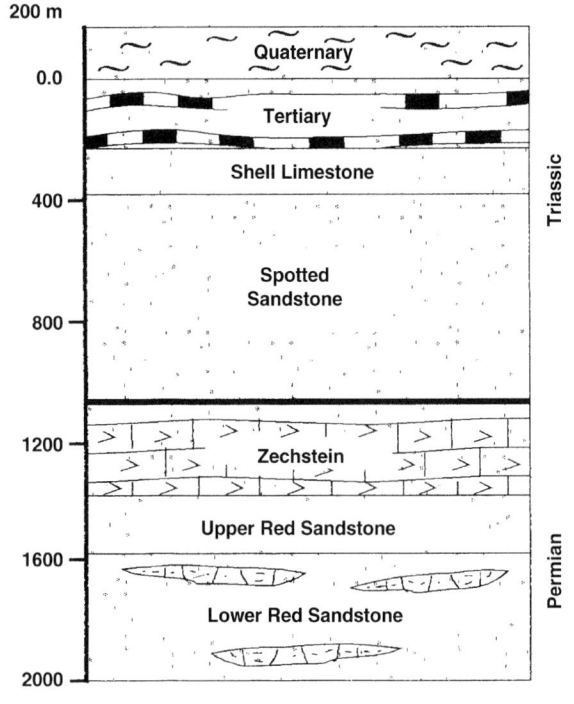

FIGURE 9.2   Sample of geological profile

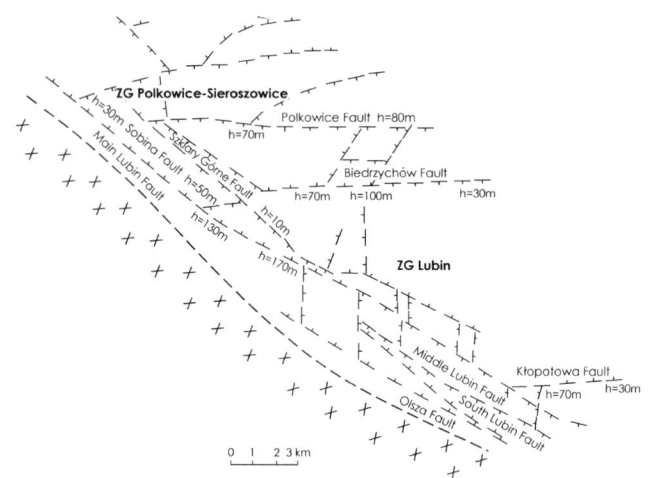

FIGURE 9.3   Directions of faults in the mining area

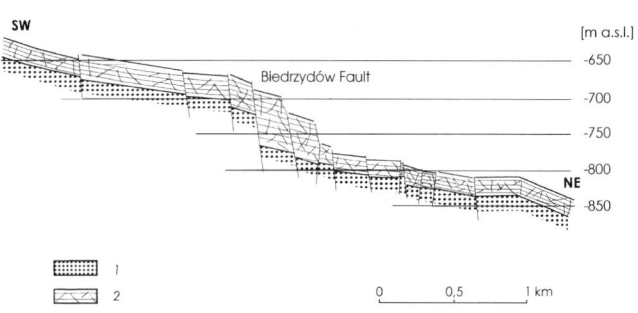

FIGURE 9.4   Faults in the cross section of the mining area

to 230–340 m from the surface; the second, a Triassic-Permian complex with water in carbonate rocks, especially within the dislocation zones.

9.2.2   Copper Content and Main Minerals

The copper ore deposit consists of sedimentary rocks with sulfide accumulation. The basic minerals that occur in LGOM are chalcocite, digenite, djurleite, anilite, covellite, bornite, and chalcopiryte (Piestrzyński 1996). The deposit is of irregular shape, with slight dip up to about 6°. The maximum thickness of the ore-bearing series is 26 m (Nieć and Piestrzyński 1996) assuming 0.7% cut off grade of copper. Distribution of copper contents in vertical profile is variable. Higher copper contents are characteristic for thin seams, usually in mineralized shales. In the Lubin mine the average contents of copper is less than 2.2% (in sandstone). In the other mines, the values are higher, but rarely above 3.5%. In Polkowice-Sieroszowice mine, the mean contents is even over 6%. The richest ore are shales with average contents over 10%. In limestones the contents range from 1 to 3%.

9.2.3   Basic Geotechnical Data

Sedimentary rocks, mainly limestones, dolomites, sandstones, shales, and sometimes anhydrites, occur in profiles of excavations.

Their basic parameters (Kunysz et al. 1996) are shown in Table 9.1. Also shown are their properties such as lamination, fracture sets, and thickness of the single layers, which are very different from place to place. An especially dangerous feature of the rock is its ability for accumulation of high amounts of energy, which is the most important factor that can cause rock burst. Even within a strong roof, in some places, one can recognize thin but very weak layer of shales, that essentially decrease bearing capacity of the roof. This is the main reason the bolting design should be extremely precise and based as much as possible on the database. In the mines, there are several kinds of rock mass classifications for different purposes. Philosophy of the classifications is very different. The original coefficient W_{et} (Szecówka et al. 1973), defined as a ratio of elastic strain energy accumulated in the specimen due to its loading, to the loss of energy for permanent strain of the specimen, is especially useful for rock burst hazard

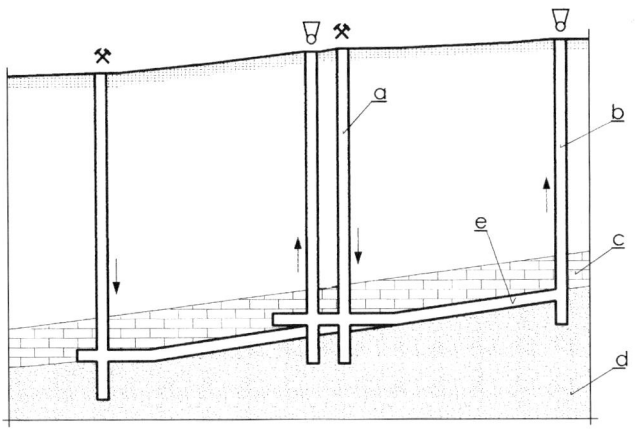

FIGURE 9.5   Model of access of the copper deposit in LGOM

estimation. Other categorizations include well-known RQD rate, fracture systems, etc. Actually, a new classification is under preparation, particularly regarding a more effective bolting designing procedure.

9.3   ACCESS WORKINGS

The deposit is developed with vertical shafts and horizontal drifts, Figure 9.5, (Butra et al. 1996). Depths of the shafts are varied from 632 m in ZG Lubin to 1120 m in ZG Rudna and are of 6 to 7.5 m in diameter. Peripheral shafts are 3 to 5 km away from the main shafts. In the near future there are plans to sink the deepest shaft, for ventilation purpose, down to about 1250 m. The total number of shafts in the three mines is 26, including ventilation, production and for other purpose. It is interesting that in the LGOM the overburden freezing method for shaft sinking was applied.

Access to the deposit from the shafts and preparatory workings are realized with drift networks located directly under the strong dolomite roof and upon the sandstone, along the dip direction or strike line. All the preparatory works that divide the deposits into levels are developed inside the mineralized zone. Transportation, ventilation and technological require that the gate roads to be driven in a multi-line system: generally double and triple, but also four and exceptionally five gate roads run in parallel. The excavations in the multiline system are connected with breakthroughs made in each 40 to 100 m.

9.4   MECHANIZATION IN THE MINING TECHNOLOGY

9.4.1   Drilling and Bolting

Different drilling techniques are used in the copper mining technology for the different production stages of

- Production blasting

- Release blasting

- Bolting

Diesel-powered, tired jumbos with hydraulic drills are the basic equipment for the functions above. Diameter of boreholes ranges from 28 mm, occasionally 25 mm, (Korzeniowski 1998) to 130 mm in case of cut boreholes for blasting purpose. Length of the boreholes depend on bolt length and blasting needs, and are varied from 1.8 m for standard bolting to 5–7 m for cable bolts. Set of a basic technical data for some models (Pawlik et al. 1996) of the machines is presented in Tables 9.2, 9.3, and 9.4.

9.4.2   Blasting Techniques

In the past the mines tried to use dynamites, the water-resistant explosives of high density and energy concentration. However,

TABLE 9.2   Technical data of drill jumbos

Parameters/Type	SWW-2H	SWWN-4H	SWWN-5H	SWW-1/HT	SWW-1Hk
Manufacturer	Boart-Lena Wilków	Boart-Lena Wilków	Boart-Lena Wilków	Zanam Polkowice	ZD Lubin
Mass, [kg]	16,500	16,900	14,000	17,200	16,300
Length, [mm]	12,500	12,000	12,000	15,050	11,600
Width, [mm]	← 2,200 →			2,500	2,450
Height, [mm	2,200	1,800	1,800	2,200	2,200
Diameter of borehole, [mm]	← 38–76 →			45–64	38–76
Length of borehole, [mm]	← 3,200 →			3,200	3,200
Max. working area-width, [mm]	← 6,500 →			5,200	6,500
Max. working area-height, [mm]	5,500	4,500	4,500	5,900	4,500
Acceptable dip, [deg]	← 15 →				
Boom	BDS B 40Mk2 or WT-1	BDS B 40Mk2 or WT-1	WTH-1000	ZRU 700	WT-1 Tamrock
Boring machine	HD 65 Boart	HD 150 or HL 538 or COP 1238	HD 150 or HL 538 or COP 1238	HL 500 Tamrock	TMK-1 or COP1238 or HD 150

TABLE 9.3   Technical data of drilling-bolting jumbos

Parameters/Type	SWWK-4	SWWK-1Hk	SWWK-2L	SWWK-2Hz/w
Manufacturer	Boart-Lena Wilków	ZD Lubin	ZD Lubin	ZD Lubin
Mass, [kg]	18,500	16,300	20,000	16,300
Length,[mm]	12,400	11,600	10,050	11,600
Width, [mm]	2,200	2,540	2,500	2,540
Height, [mm]	2,200	2,200	2,400	2,200
Diameter of borehole, [mm]	38–45	28–45	45–64	28–45
Length of borehole, [mm]	2,200	2,600 or 2,800	15,000	1,600–2,600
Length of bolts, [mm]	180	1,600 or 1,800	to 7,000	160–2,600
Min. height of excavation, [mm]	2,500	2,800	3,800	2,800
Min. width of excavation, [mm]	3,500	3,700	3,700	3,700
Acceptable dip, [deg]	← 15 →			
Boom	BOS B40 MK2 Boart	WT-1	WT-1	WT-1
Type of bolt	expansive	bond-cable	cable, Swellex	expansive, bond

due to their great sensitivity for detonation and lack of possibility for mechanical loading of the explosives into blast holes, they are no longer in common use. A better solution was the ammonium nitrate explosives. They are weaker in effect but less sensitive to detonation and are now used in all but few percent of the total amount. Loose poured and emulsion of ammonium nitrate and fuel oil mixtures, known as ANFO, are commonly used. The most important advantages of these bulk explosives is the possibility of pneumatic loading and more than triple increase in power as compared with cartridges. Initiation of the blasting occurs with electric delay detonators and in the blast holes longer than 6 m, with detonating cord (pentryt, 11 g/m). The delay interval of the

TABLE 9.4 Technical data of bolting jumbos

Parameters/Type	SWK-3B KOT	SWK-1/1 LIN	SWKL-1/1	SWK-2Hz
Manufacturer	Boart-Lena Wilków	Zanam Polkowice	Zanam Polkowice	ZD Lubin
Mass, [kg]	18,000	19,000	20,000	16,300
Length, [mm]	12,500	11,050	10,050	11,200
Width, [mm]	2,260	2,500	2,500	2,540
Height, [mm]	2,200	2,400	2,400	2,200
Length of bolts, [mm]	1,800	1,800	to 15,000	1,600–2,600
Diameter of borehole [mm]	38	28–44	45–64	28–38
Length of borehole, [mm]	1,860	1,890	15,000	1,690–2,690
Min. height of excavation, [mm]	2,500	3,800	3,600	2,600
Min. width of excavation, [mm]	3,500	3,500	3,500	3,700
Acceptable dip, [deg]	15	15	15	15
Type of bolt	Expansive	expansive, bond, cement	cable	expansive, bond, frictional
Boom	WT-1	ZRU 707 Tamrock	FRC 1007	WT-1
Boring machine	HD 65 Boart	HL 300 S Tamrock	HL 500 S Tamrock	TMK-1, HD 65, HL 300S

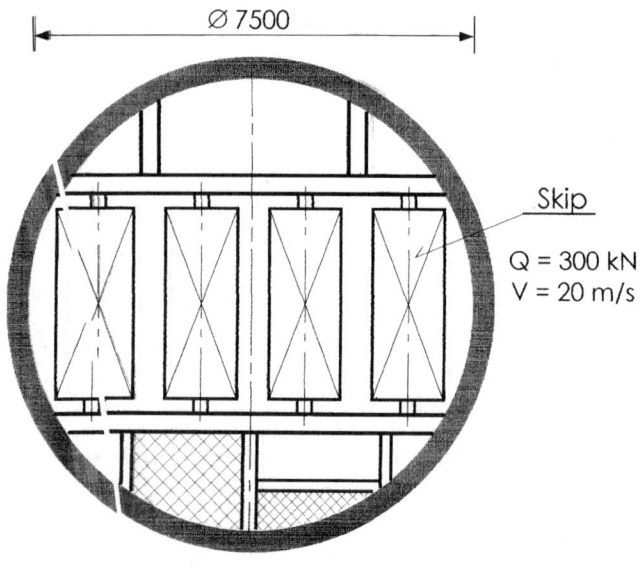

FIGURE 9.6 Cross section of a shaft shield

0.5 s detonators ranges from 0 to 10 and sometimes 15 s for bigger cross sections, (Janowski et al. 1996). The electric detonators are successively being replaced by nonelectric method, NONEL. The blasting patterns are based on the following diameter of the blast holes: 41, 45, 64 mm and wedge or parallel hole cuts of two types:

- Central, large diameter (90–130 mm) empty hole with parallel charged holes
- Parallel, holes close to each other, alternately empty and charged holes of the same diameter as the rest of the drilling.

The heading rounds achieved with such cuts are usually 4 to 5 m, and in single cases 6 to 7 m deep that the blasted drilling broke. Since 1997, the blasting vehicle jumbo SWS-5 from Nitro-Nobel, for pneumatic jet or pressure charging was introduced.

9.4.3 Haulage, Transportation, and Hoisting

Haulage and transportation equipment is typical for many mines and is based on LHD machines, and belt conveyors. An example set of technical data of the loaders used in the mines is shown in Table 9.5. Equipment used in the shafts is varied and depends on the purpose of the shaft. The most modern shaft in Rudna Mine is Koepe hoist equipped (Kawecki and Hildebrand 1996) with in two, two-skip hoisting installations. Each of the skips is of 300 kN hoisting capacity and their transportation velocity: 20 m/s. The depth for the loading level is 1,022 m. The original solution applied in the construction is a double-side, simultaneous loading of the skip from a belt conveyor and bunker, while the skip is braking, but still lower. Each of the equipment is powered with four-line hoisting machine and the transmission wheel of 5.5 m diameter and 3,600 kW engines. In Figure 9.6 an example shaft shield (shaft arrangement in a cross-section) is presented. The other shafts designed for transport of men and accessories are equipped with cages and skips of different hoisting capacity.

TABLE 9.5 Technical data of bucket loaders and LHDs

Parameters/Type	LK2NC LK2NCD	LK2c LK2CD	LK1ACD	LKP-0401	LKP-803	TORO 501D	ST8B
Manufacturer	FMB-Fadroma, Wroclaw →			← Zanam-Polkowice →		Tamrock	Atlas Copco Wagner
Mass, [kg]	18,600	20,500	13,000	11,950	20,500	35,650	38,913
Length, [mm]	9,825	9,985	8,900	8,972	10,350	10,680	10,287
Width, [mm]	2,870	2,820	2,340	2,380	2,830	3,040	2,921
Height, [mm]	2,050	2,625	2,050	2,450 or 2,200	2,635 or 2,300	2,710	2,791
Loading capacity, [kg]	7,000	8,000	4,000	4,000	8,000	14,000	13,600
Bucket capacity, [m³]	3,5	3,5	2,0	2,0	3,5	7,0	7,6
Run velocity without loading, [km/h]	14,0	12,9	13,3	38,0	12,9	12,9	13,3
Acceptable dip, [deg]	11	14	15,1	15	15	26,7	32,0
Engine	SW680/73 or Deutz F8L413FW	SW680/73 or Deutz F8L413FW	Deutz F8L413FW	SW400/L2 or 6CT107-2/L2	Deutz BF6M1013	Detroit Diesel 6067 WK60	Detroit Diesel seria 6011
Gear box	Clark C5402	Clark C5402	Clark 13, 7HR28319	ZM130N/L2	Clark	Clark 21D 3960	Clark 508

9.5 MINING METHODS

9.5.1 Evolution of Mining Methods

Exploitation of Deposits Up to 5 m Thick. After experience with first shafts sinking and a recognized water threat, the only mining method being consideration at that time was backfilling technology. Then, long-wall methods with walking mechanized hydraulic support, and armoured and belt conveyors were applied. Very soon, after negative results and unsatisfactory efficiency, they decided to use room and pillar method with bolting techniques and LHD heavy equipment that could assure mass production and better output concentration. With time and production experience, room and pillar mining methods with roof caving have become safer and more and more effective, since it enabled full mechanization to be introduced. The caving methods were more competitive due to low production costs as compared with backfilling techniques. Initially, for the exploitation with roof caving, two stages of pillar cut were typically used. In the first stage the exploitation area was divided into bigger pillars 25×35 m. Only in the second stage each of the pillars, beginning from the abandoned line, was cut into many smaller pillars. From the potential rock burst threat point of view, the two-stage method is worse because the pillars in the first stage show a dangerous tendency for accumulation of energy.

In recent years, the larger area of the exploited space and greater depth of the deposit increases rock mass pressures with negative effect. It has been shown to be an urgent need to improve mining work safety and rock burst prevention.

Until 1983, the room and pillars methods were based on an assumption that all the pillars should be of the same dimensions regardless of changes the geomechanical parameters of the rock mass. After 1983 in ZG Rudna Mine, the engineers decided to fit dimension of the pillars to local conditions, so their bearing capacity were differentiated in relation to geomechanical properties of the rock mass. Further modifications of the room and pillar method were connected with

- Changes of angle between direction of rooms and lanes
- Alternating directions of driving stopes
- Back-driving
- Extension of the roof opening up to 150 m (distance from abandoned area line to the face line).

Exploitation of Deposits from 5 to 7 m Thick. Until recently the deposit over 5 m thick used to be mined only with backfilling technology. The newest technology (to 7 m) is based on the hypothesis of advance fracturing and post failure bearing capacity of pillars. The roof opening reaches 150 m. The longer edges of pillars are situated perpendicular to the exploitation front line.

What happens within caved areas, the upper layers of roof are not fully supported with broken rock. Such a situation creates real threat of rock burst, roof fall or local stress relief of strata. As a result, secondary scaling and bolting are very often necessary. Also, other effects are dilution in the broken ore influencing increase of production costs and of course decrease work safety level. The above reasons caused the different philosophy for abandoned workings to be undertaken in this mine. They completely stopped the practice of blasting residual large-size barrier pillars. The completion and abandonment of the old workings occurs, instead of caving, as an effect of gentle roof sag upon the residual pillars being under post failure stage of work, and finally self-caving phenomenon. This technology assures protection of the weak immediate roof and advantageous stress distribution, due to increased support of the upper strata, reinforced with the progressing self-caved area. Since 1989, this mining method (R-UO) has been applied in the mine, especially under difficult conditions of tectonics and high stress concentration.

Exploitation of Deposit from 5 to 15 m Thick with Backfilling. The most important factors that should be considered for such a situation are as follows:

- Roof control
- Protection of the high wall sides and roofs
- Proper ventilation and air conditioning
- Special roof falls prevention.

The first mining method used for the thick deposit (5 to 7 m) was two-layer method with hydraulic backfilling and 10 m filling modulus for 120 m length of the exploitation front line (UZG method in ZG Lubin Mine).

Some experience from the UZG method has been transferred to the Rudna Mine, in the seventies, where mineralized zone up to 10 m thick were to be mined. In Rudna-1 method the filling zone has been elongated from 10 to 32 m, which resulted in triple decrease of filling dams (interesting solution of cable dams were implemented) and timber use with increase of production concentration. Post filling water was drained off into setting channels through perforated pipelines. Such a solution completely eliminated construction of special inclines for that purpose used in the past.

In 1984 a new mining method—Rudna-5—for deposit up to 15 m has been introduced. In the two-layer method, direction of exploitation was from the upper layer towards the lower one. The maximum allowable filling length was 17.5 m. Roof and high wall sides were protected with 2.6 m length resin bolts. An alternative method for thickness, 13–18 m, Rudna-4, was used for exploitation of the protection pillar for Polkowice town. The mine achieved production of 5,000 mg/day under very rigorous limitation connected with requirements of surface protection. Recent methods are based on the solutions worked out from the above experience

9.5.2 Rock Burst Threat

Since 1972, when the first rock burst event in Polkowice Mine was reported, this threat was considered as the most significant one in Polish copper mines. The basic reasons for increasing rate of the hazard are as follows (Kleczek et al. 1996):

- Great depth of mining
- Large area of exploitation
- Difficult tectonics
- High strength parameters of rocks and their high ability for accumulation and releasing of energy
- Large extension of works and huge number of pillars.

The most characteristic dimension of pillars that usually are demolished, when rock bursting occurs, is 20 to 25 m width. Smaller and greater dimension are less able for rock bursts (Kunysz and Mrozek 1992). One can distinguish two types of rock burst:

1. *Seam type*, when center of the event and failure location overlap each other
2. *Roof or tectonic type* when the event and location do not overlap each other.

The very high energy of rock bursts can reach value of 10^9 J. Thus it is very important to apply an effective prevention method for safe mining. In the copper mine, when the dominant mining method is room and pillar with its many modifications, pillar-yielding technique is used. Immediately, after creation of the first line of pillars in direct vicinity of the rock mass, the pillars are loaded with pressure induced by exploitation and because of their appropriate small dimensions, transit from precritical to postfailure stage of work. As a result, when the mining front line is moving ahead, the partially destroyed pillars in the exploitation field must still support the roof and assure stability of the rooms. Their ability to accumulating energy is significantly decreased. One can obtain good effects by changing the orientation of the rectangle cross-section pillars within the front line. If the longer sides of pillars are perpendicular to the front line, the number of rooms is greater as compared with situation when the shorter sides are perpendicular (assuming the same length of the exploitation front line). In such situation, we also achieve extra yielding effect. To relax wall sides of drifts simple niches or recesses are cut into the rock mass as they are driven.

An active prevention method is based on advance inducing rock mass stress release with explosives blasted in long boreholes of diameter about 70 mm, at many working faces at the same time, usually no more than 15 stopes. The concentrated explosions, and eight hours waiting time following blasting allowed the most number of tremors to take place during this period.

9.5.3 Mining of Semi-Seam Deposits Up to 3 m Thick

Especially in the Polkowice-Sieroszowice Mine, most of the area occupies thin, semi-seam deposits with thickness less than 3 m. For exploitation of this kind of deposit the special mining method has been developed. The ore is extracted in a selective way. The idea of the method is shown in Figure 9.7.

The mining field is typically encountered with double or triple entries preparatory workings. Rooms, entries, and pillars are basically of 7 m width. Work in the faces consists of two phases concerning thickness of the layers: barren rock and mineralized ore. (In United States, it is known as a "reusing" method of stoping.) First, the upper (adjacent to the roof) ore-bearing layer is worked out and hauled out to special chutes and main transportation system. In the second phase of the working, the barren rock adjacent to the floor is worked out and placed in other rooms (not hauled and transported out) as entries of a dry back fill. Each of the entries covers at least two rows of pillars plus one room. The back fill width is 14 m and maximum length of the mining front is about 49 m. No more than three rows of pillars at the same time period, not covered with the back fill are allowable in the mining area. During extraction in the last row of pillars, working occurs only in the ore-bearing layer until the cross section of pillar achieves approximately 21 m² area. The completion of the pillar mining process work is based on roof sag resting upon the dry back fill entries prior to abandoning the area.

9.5.4 Room-and-Pillar Mining Method with Roof Sag and Controlled Yielding of Pillars for Semi-Seam Deposits Up to 7 m Thick

Idea of the Method. The method is especially useful in barrier pillars of drifts, heavy faulted zones, and in direct vicinity of abandoned areas. Maximum allowable dip of the deposit is up to 8 deg, and thickness is 3.5 to 7 m; however, local increase of thickness is acceptable.

The exploitation area is encountered with double gate roads, as presented in Figure 9.8. For thickness over 4.5 m, the gate roads are drifted in the roof of the ore-bearing layer. The boundary excavations must be connected with other active excavations in the mine. An optimal length of the mining front line ranges from 50 to 600 m. The ore is extracted with 7-m-wide and 7-m-high rooms. Roof is supported by pillars of 7 to 10 m × 8 to 38 m. While advancing the mining front line, the pillars from the last row are cut until their final dimension, 7 to 10 m × 2.5 to 4.5 m. Afterwards, the smaller pillars (cubes) are successively decreased in diameter up to the contact surface area 12 to 20 m², so-called residual pillars. The roof that has been opened must be immediately bolted. The next stage is the floor working, down to the boundary of the mineralized zone. The extracted part is closed for people and equipment with timber posts or chocks. Length of the blast-holes in the upper layer is 2.5 to 3.0 m, but in the lower

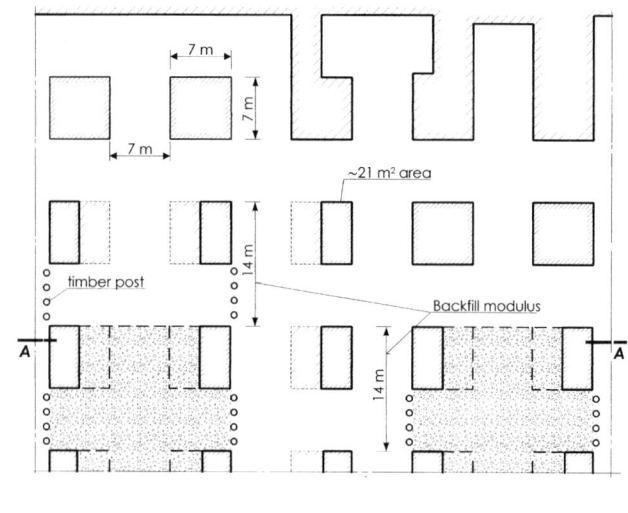

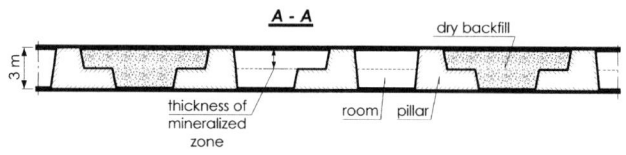

FIGURE 9.7 Room and pillar mining method for 3 m thickness

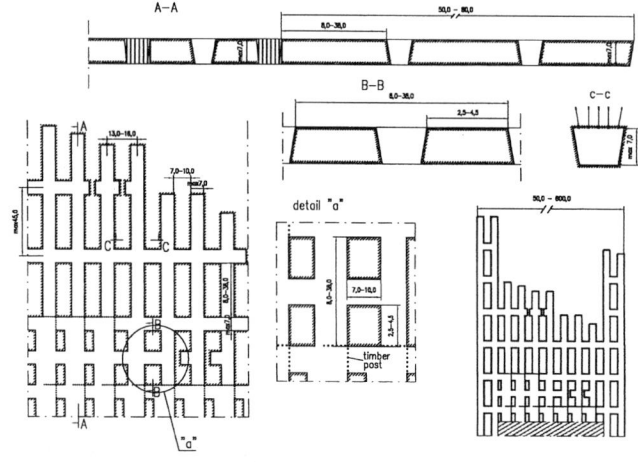

FIGURE 9.8 Room-and-pillar mining method for 7 m thickness, with controlled roof sag and yielding pillars

layer is equal to the distance to the ore boundary. Broken ore is loaded with bucket loaders onto hauling machines or with LHD to chutes and belt conveyors. Work in the mining front line is fully mechanized and organized in serial-parallel cycles within three segments of the front:

1. In one segment of the entry, a roof part of the ore is extracted

2. In the second segment of the entry, a floor part of the ore is extracted

3. In the third part, the technological (the larger) barrier pillars are slabbed.

9.5.5 Room-and-Pillar Mining Method with Hydraulic Back Fill for Semi-Seam Deposits Up to 15 m Thick

General Idea of the Method. Range of application of the method is limited with 15 m thickness and inclination up to

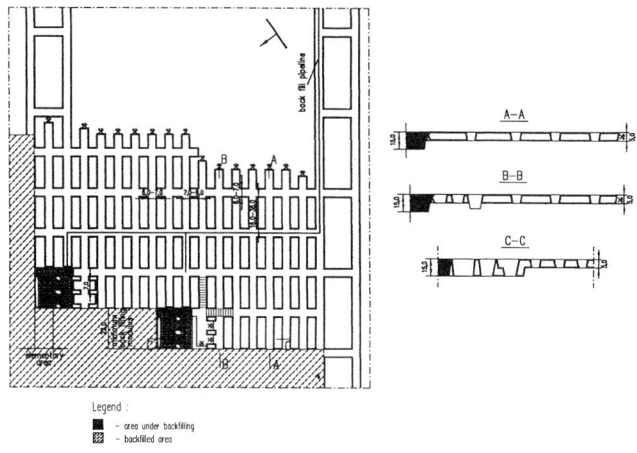

Legend :
■ - area under backfilling
▨ - backfilled area

FIGURE 9.9   Room-and-pillar mining method for 15 m thickness with backfill

8 deg. The field to be exploited must be encountered with double or triple system gate roads driven immediately upon the top of the mineralized zone. The boundary excavations must have connection with other active excavations in the mine. Exploitation is conducted in one or two layers with faces along the same line or with separation larger size barrier blocks, where the single operations are performed alternately.

Order of the operations that occur in two layers is shown below and in Figure 9.9:

1. Division of the exploitation area in the upper layer with 6–7-m-width and 5-m-high rooms and entry (in perpendicular direction each other) into larger barrier pillars. Dimensions of the pillars vary from 7 to 9 m by 16 to 38 m, depending on the local geological conditions. The longer side of the rectangle is perpendicular to the front line (exploitation line) assuring large number of rooms and great concentration of faces winning simultaneously along the front line.

2. Width of the excavation in the upper layer can reach 10 m due to dynamic activity of the rock mass and need of the wall sides that should have pre-assumed inclination after scaling.

3. Mining in the lower layer proceeds for its full thickness with many faces by means of floor extraction. Haulage of the broken ore takes place in each second room in inclines with 5 to 8 deg dip.

4. At least two rows of the barrier pillars should be maintained within the front line and the distance between the front line and the back line where the stoping is abandoned, have to be greater than 54 m.

5. With the progress of the mining, the last row of the pillars (in the back line) is completed with 7 m width cuts. The residual pillars are slabbed successively until the contact with the roof achieves 12 to 20 m², and occasionally even less under suitable conditions.

6. Finally, the abandoned area is filled up to the roof with hydraulic fill.

7. The mining of the next barrier pillar can be extracted after the elementary backfill field is completely finished.

Blasting works are made with 4.5 m long boreholes drilled vertically downwards into the lower layer from the floor of the upper layer or from the face of the lower layer. Loading and haulage are realized with bucket loaders and diesel powered,

rubber-tired, container type vehicles or LHD to belt conveyors or special bunkers. Roof bolting is typical for the mine.

Hydraulic backfilling runs for the full thickness of the each separated part of the area and post filling waters are drained off through pipeline collectors into setting bunkers. The minimum distance (length) for the fill is 22 m.

The working organization at the faces are cyclic in series. Conducting of faces (rooms and lanes) and back filling proceed at each stage of mining phase. Bulk and emulsion explosives are loaded mechanically from blasting purpose vehicles (SWS) while explosives in charges are loaded manually.

9.6   COSTS AND SOME PRODUCTION STATISTICS

- Total number of employees, underground in the three mines, is below 10,000 men.

- According to *Raport kwartalny za IV kw. 1999. KPWiG, system Emitent,* cost of copper production in 1999 was 1469 US$/t (67USc/lb) and the cash cost (without amortization) 1287 US$/t (59 USc/lb).

- Copper production in Poland in 1999 was 470,494 t of electrolytic copper and 1,092,639 kg of metal silver.

- Prognosis for 2000 year forecast increase of copper production up to 478,000 t and silver 1,070,000 t. Unit production cost for copper will be 1,445 USD/t.

9.7   FUTURE PLANS OF MODERNIZATION OF COPPER MINING

The recent technology used for extraction the copper ore is very effective for the specific geological conditions in Polish mines and rock burst threats, however it is under continuous development and modifications. We would like to emphasize that we only explained the mining methods according to thickness criterion, i.e. for thin, medium and thick deposits but many very interesting, individual solutions applied in some of the mines are not described here due to editorial limitations and general goal of the paper. Our depth of experience in mining within protection pillars under urbanized areas, in old residual parts of deposits and extraction of old barrier pillars to release rock mass stress and other techniques are not included here. In addition, the bolting practice is very highly developed in the mines and we could devote a separate chapter for that problem. We had assumed that the basic ideas of room and pillar methods, typical mechanization, and roof support systems are well known to the readers and there was no need to include it. In a contemporary modern mine the rock mass and every element of the technological chain should be under permanent control to assure maximum safety for miners and economical effectiveness to owner. To improve the two fundamental items the following measures are undertaken in the mines:

- Further developing the rock mass monitoring system

- Changes in work organization

- Developing a new construction of rock burst proof bolts

- Decrease of diameter of boreholes for roof bolts

- Implementation of special construction of jumbos for thin (less than 1.5 and 2 m) deposits

- Modernization of mining methods by further minimizing ore wastes and dilution

- Projects for access to deeper strata, below 1,200 m

9.8   REFERENCES

Annual Report '98. KGHM Polska Miedź. 1999.
Butra, J., Bugajski, W., Piechota, S., Gajoch, K. 1996. Poziome wyrobiska udostępniające i przygotowawcze. In *Monografia KGHM Polska Miedź S.A.*, ed. CBPM Cuprum Sp. z o.o. , Wrocław.

Janowski, A., Cieszkowski, H., Dąbski, J., Katulski, A., Piechota, S., Zarski, R. 1996. Systemy eksploatacji, sposoby urabiania oraz rodzaje obudowy wyrobisk. In *Monografia KGHM Polska Miedź S.A.*, ed. CBPM Cuprum Sp. z o.o. , Wroclaw.

Kawecki, Z., Hildebrand, J. 1996. Transport szybowy. In *Monografia KGHM Polska Miedź S.A.*, ed. CBPM Cuprum Sp. z o.o. , Wroclaw.

Klapciński, J., Peryt, T.M. 1996. Budowa geologiczna monokliny przedsudeckiej. In *Monografia KGHM Polska Miedź S.A.*, ed. CBPM Cuprum Sp. z o.o. , Wroclaw.

Kleczek, Z., Mrozek, K., Siewierski, S. 1996. Prowadzenie robót górniczych w warunkach zagrozenia tąpaniami. In *Monografia KGHM Polska Miedź S.A.*, ed. CBPM Cuprum Sp. z o.o. , Wroclaw.

Kleczkowski, A.S., Kalisz, M. 1996. Hydrogeologia. In *Monografia KGHM Polska Miedź S.A.*, ed. CBPM Cuprum Sp. z o.o. , Wroclaw.

Korzeniowski, W. 1998. Poszukiwanie nowych rozwiązań obudowy wyrobisk w kopalniach węgla na tle doświadczeń kotwienia w kopalniach rud. In *Przegląd Górniczy nr 11,* Katowice.

Kunysz, N., Kijewski, P., Korzeniowski, W., Lis, J. 1996. Geomechaniczne wlasności skal serii zlozowej. In *Monografia KGHM Polska Miedź S.A.*, ed. CBPM Cuprum Sp. z o.o. , Wroclaw.

Kunysz, N.M., Mrozek, K. 1992. Problematyka utrzymania stateczności stropu wyrobisk górniczych w górnictwie rud miedzi. In *Prace Naukowe Instytutu Geotechniki i Hydrotechniki Politechniki Wroclawskiej*, nr 63. Konferencje nr 32.

Nieć, M., Piestrzyński A. 1996. Forma i budowa zloza. In *Monografia KGHM Polska Miedź S.A.*, ed. CBPM Cuprum Sp. z o.o. , Wroclaw.

Pawlik, K., Ogrodniczek, R., Miluch, J., Krajewski, J. 1996. In *Monografia KGHM Polska Miedź S.A.*, ed. CBPM Cuprum Sp. z o.o. , Wroclaw.

Piestrzyński, A. 1996. Okruszcowanie. In *Monografia KGHM Polska Miedź S.A.*, ed. CBPM Cuprum Sp. z o.o. , Wroclaw.

Raport kwartalny za IV kw. 1999. KPWiG, system Emitent-raporty spólek publicznych. *Biznes.Onet.pl,(internet)*.

Rydzewski, A. 1996. Historia odkrycia Nowego Zaglębia Miedziowego. In *Monografia KGHM Polska Miedź S.A.*, ed. CBPM Cuprum Sp. z o.o. , Wroclaw.

Salski, W. 1996.: Tektonika zloza. In *Monografia KGHM Polska Miedź S.A.*, ed. CBPM Cuprum Sp. z o.o. , Wroclaw.

Szecówka, Z., Domzal, J., Ozana, P. 1973. Wskaźnik energetyczny sklonności naturalnej węgla do tąpań. In *Prace GIG. Komunikat nr 594*, Katowice.

CHAPTER 10

Underhand Room-and-Pillar Mining as Applied at the Aurora Mine, Charcas Unit, Grupo Mexico

Marco A. Perez G.* and Abel González V.***

10.1 INTRODUCTION

10.1.1 Location and Access

The Charcas operating unit of Grupo Mexico is located 120 km from the city of San Luis Potosí in the region known as the Altiplano (high plains) (Figure 10.1). The nearest town to the unit is Charcas, located 4 km to the northwest. There are two main roads to the property—an 89-km-long secondary paved road from Matehuala that connects the unit to federal highway 57 and a 100-km-long secondary paved road that connects the unit to federal highway 49. The Los Charcos railway station is 14 km away and connects the unit to the México-Laredo railway. There is a landing strip for small planes 5 km from the mine site.

10.1.2 History

The origin of Charcas dates back to 1574 when Don Juan de Oñate founded the town. It was given the name Real de la Natividad. The very first workings were begun in 1583 when silver ore was mined from the oxidized zones of the Santa Isabel and Leones veins. These veins, previously discovered by natives, constituted the first mine in the region, the Mina Tiro General. There were a number of different owners through the years, but in 1859 the properties were unified by the Frenchman Mereo Sescosse who worked these veins.

The silver ore, with grades over 1500 gr/mt, came exclusively from the oxidized zone. It was treated using the Patio process. Around 1870, the oxides were exhausted as the workings reached the sulfide zone. Operations were suspended because of the lack of an economic method of concentrating sulfides. With the advent of selective flotation, the deposits could be worked again.

For many years, these mines were controlled by Compañia Minera Tiro General, S.A. In 1911, ASARCO and the Compañía Metalúrgica Nacional acquired the exploitation rights. ASARCO acquired all the properties and mining rights in 1924 and began the construction of a beneficiation plant to process 725 mt/d. In 1965, the Mexicanization of mining prompted ASARCO to become ASARCO Mexicana, S.A., with 51% of its shares being held by Mexican investors. The name of the company was changed to Industrial Minera México, S.A., in 1974.

Production capacity has evolved in the steps shown in Table 10.1.

10.1.3 Charcas Unit Today

Today, the Charcas Unit has among its goals the exploration, mining, and processing of silver, lead, copper, and zinc ores to produce lead, copper, and zinc concentrates. The ores are mined

FIGURE 10.1 Location of Charcas Unit

TABLE 10.1   Evolution of production capacity of the Charcas Unit

Period	Enterprise	Capacity, mt/d
1925-1969	ASARCO	725
1970–1984	ASARCO Mexicana, S.A.	950
1975	Industrial Minera México, S.A.	1,030
1976–1982	Industrial Minera México, S.A. de C.V.	1,250
1983–1985	Industrial Minera México, S.A. de C.V.	2,450
1986–1988	Industrial Minera México, S.A. de C.V.	2,600
1989–1991	Industrial Minera México, S.A. de C.V.	3,300
1992–2000	Industrial Minera México, S.A. de C.V.	4,500

* Grupo Mexico, Charcas, S.L.P., Mexico.

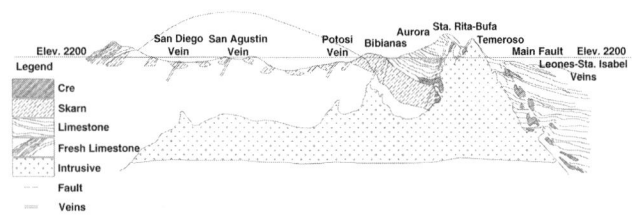

FIGURE 10.2   Geologic (section N 70° W) across Charcas district

from the Aurora, Rey, Reina, and San Bartola ore bodies, with daily production rates of 652, 3,422, and 815 mt, respectively. Mining methods include underhand room-and-pillar at the Aurora ore body and post-pillar cut-and-fill at the Rey, Reina, and San Bartolo ore bodies. Installed milling capacity is 1,200 mt/d at plant 1 and 3,300 mt/d at plant 2, for a total of 4,500 mt/d. Ore is hoisted through the San Bartolo and Leones shafts, with capacities of 1,400 and 3,800 mt/d, respectively.

10.2   GEOLOGY

10.2.1   Regional Geology

The present structural features show that there were three tectonic stages. The first stage involved regional metamorphism of Triassic rocks, while the second involved the Laramide Orogeny. In the third stage, tension stresses during the Pliocene gave rise to a system of northwest-southeast fractures and north-south-striking normal faults that affected the earlier structures. Figure 10.2 shows the mineralized fringe of the Charcas district.

Regionally, the rocks range in age from Lower Cretaceous to Recent. The oldest represent platform and basin sediments and shallow basin sediments. Spills and structures in the rhyolite overlying the older formations are Tertiary. The younger rocks are continental conglomerates discordant on the Cretaceous rocks. Finally, clastic sediments of Quaternary age are distributed widely within the valley. Figure 10.3 shows the stratigraphic column.

10.2.2   Structural Geology

The main structure observed in the area is the doubly dipping San Rafael anticlinorium whose axis has an approximate north-south orientation. In addition, there are several localized anticlines and synclines, either in Triassic rocks or on the flanks of the anticlinorium.

The intrusive body of greatest importance in the area is called El Temeroso. This granodioritic stock was intruded during the middle Eocene and is classified as a quartz monzonite. In the sedimentary formations contiguous to this intrusive, a metamorphic aureole was formed.

The system of fractures and faults was the result of, on one hand, orogenic movements during the Laramide event, and on the other, tensional stresses caused by the intrusion of the granodioritic stock. In both cases, mineralizing solutions filled the structures.

Three systems of mineralized structures have been defined.

1. The vein system formed by the Leones and Santa Isabel veins (Santa Rosa, La Viejita, Santa Ines, Veta Nueva, San Rafael, Progreso) and the filled fractures in Las Bibianas and El Potosí. This set displays a preferential northwest-southeast strike with some east-west variations as occurs at Las Margaritas, El Potosí, and branches of the San Rafael vein.

2. A set of veins and faults oriented about northeast-southwest, as, for example, the San Salvador and San Sebastián veins.

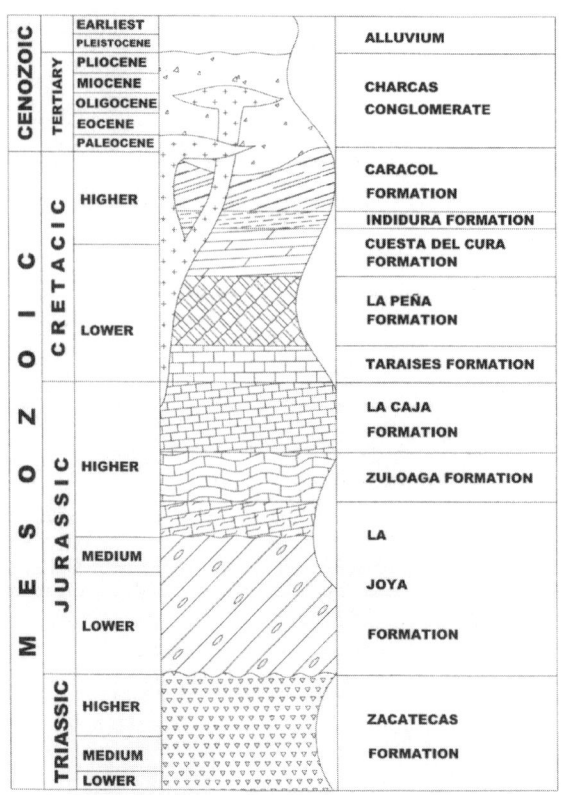

FIGURE 10.3   Stratigraphic column

3. One set of concentric faults and mineralized fractures, which are at the margins of the El Temeroso stock. It is in this system that the main mineralized replacement ore bodies are found.

10.2.3   Local Geology

The rocks in the area correspond to all three major types of rock: sedimentary, igneous, and metamorphic. The Cuesta del Cura Formation is the most recent one in the environs of the unit. Cropping out in the hills immediately to the south and north of the unit, it consists of clear gray limestone in thin, folded layers. It is characterized by the stratification of lenses and bands of black chert. With an approximate thickness of 300 m, it dates between the last periods of the Lower Cretaceous and the beginning of the Upper Cretaceous. Sediments were deposited in a slightly tilting basin environment, as indicated by the presence of terrigenous interbeds in the limestone. The Aurora ore bodies lie in this formation.

10.2.4   Zoning

1. Locally, both vertical and horizontal zoning can be observed.

   Vertical zoning. In the mineralized structures, both sphalerite and galena decrease with depth, whereas chalcopyrite increases.

   Horizontal zoning. Both sphalerite and galena decrease from the northeast to the southwest, while chalcopyrite increases considerably.

2. Regionally, horizontal zoning is seen. High-temperature minerals such as chalcopyrite and bornite are found toward the nucleus, whereas only low-temperature minerals such as stibnite and antimony oxide are observed at the edges of the San Rafael anticlinorium.

TABLE 10.2   Results of geomechanical tests

Rock type	Estimated compressive strength, point loading test, MPa	Results of uniaxial compression tests	
		Density, mt/m^3	Compressive strength, MPa
Limestone (hanging wall)	234	2.98	235
Ore	148	3.41	87
Sterile skarn (footwall)	181	2.54	146

10.2.5   Mineralization Controls

In the mineralized areas, the solutions were controlled by lithological and physical-chemical factors. Replacement ore bodies were controlled by the purity of the limestone, clay content, and degree of recrystallization. If the amount of recrystallization is high, then the possibilities for replacement are high. On the other hand, the contact between the intrusive and the sedimentary rocks served as structural and lithological traps for the deposition of mineral fluids. Some rhyolitic dikes that intruded the limestone served as ground preparators and simultaneously acted like receivers of mineralizing solutions.

The main mineralized structures strike N 70° W (Figure 10.2) and dip toward the northeast and the southwest. These structures are intercepted by another system striking N 20° E that controlled the mineralization. A noticeable stratigraphic control does not exist, because both the sedimentary sequence and the mineral bodies cross the Cuesta del Cura and Zuloaga formations.

10.2.6   Mineralogy

In the economic deposits of the district, the minerals correspond to the sulfide group. Thus, the minerals that constitute the ores are—

- Zinc ore–sphalerite (zinc sulfide)

- Lead ore–galena (lead sulfide)

- Copper ore–chalcopyrite (copper-iron sulfide)

- Silver ore–mainly argentite (silver sulfide) and, in smaller proportions, diaphorite (considered as silver sulfosalt)

- Iron ore–pyrite (iron sulfide) and chalcopyrite

In addition, other minerals exist that are not considered of economic importance, but because of their interest from a mineralogical point of view, it is good to know them. The most outstanding are bornite, covellite, malachite, chrysocolla, smithsonite, tetrahedrite, cerussite, arsenopyrite, diopside, wollastonite, epidote, grossularite, calcite, quartz, and silicoborates represented by danburite and datolite.

10.3   GEOMECHANICS

Several geomechanical studies have been conducted in the mine area. Samples of the three main rock types found in the Aurora ore body have been subjected to point-load and uniaxial compression tests. The results are given in Table 10.2.

Maps of rock mass classification values indicate that the ground has a good rock mass quality with a rock mass rating (RMR) of > 65. From the mapping of discontinuities mapping in a small area of the mine, three joint sets were identified. Joint surfaces are rough to very rough. The discontinuities in limestone are open, but those in skarn contain fill material. Fracture frequencies in the skarn are higher than in limestone. These characteristics agree with the excellent ground conditions for both the roof and the pillars, as observed in the open stope areas.

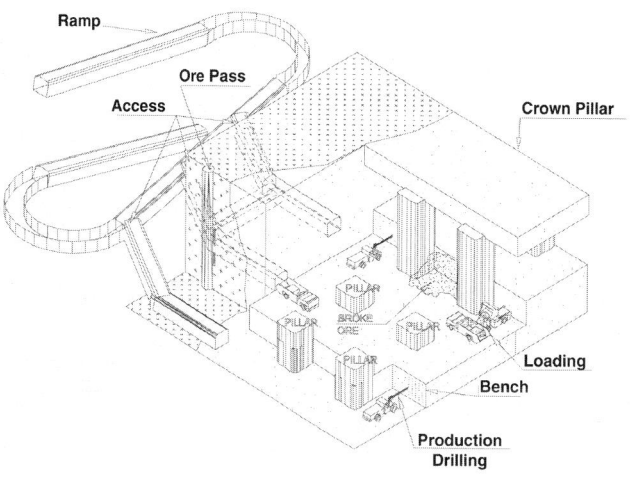

FIGURE 10.4   Three-dimensional view of mine layout for underhand room-and-pillar method

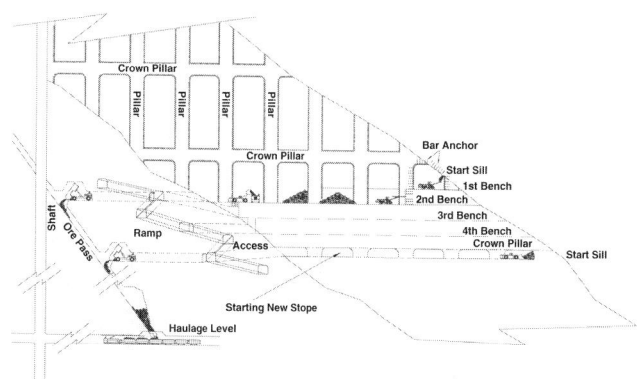

FIGURE 10.5   Cross section showing general layout for underhand room-and-pillar method

The stopes require, therefore, a minimum of support, and the amount of time and effort spent in ground control are small.

10.4   UNDERHAND ROOM-AND-PILLAR MINING

The underhand room-and-pillar mining method is applied in the Aurora ore body. Rock mechanics analyses have been used to define the operating parameters needed to achieve the goals of productivity and safety.

Ore body thickness varies from 15 to 60 m, height varies from 15 to 80 m, and dip varies from 20° to 35°. Both the wall rock and ore body itself are of good quality. The pillars are square in cross section (6 by 6 m) with a maximum height of 30 m. It is a rule to reinforce the roof, pillars, and stope walls with rock bolts from the beginning of the stope. Once a pillar reaches its maximum height, a crown pillar 8 m thick is left to protect the mining operations in the next stage. Figure 10.4 shows an isometric view of mining layout, and Figure 10.5 is a section view.

The method has both advantages and disadvantages.

Advantages

- Low cost

- Immediate availability of blasted ore

- High productivity

- Possible pillar recovery

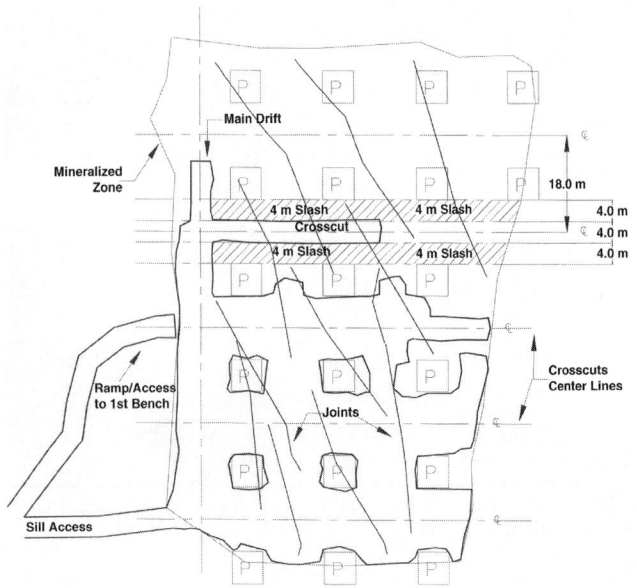

FIGURE 10.6   Plan view of typical stope showing crosscuts driven perpendicular to joint orientation

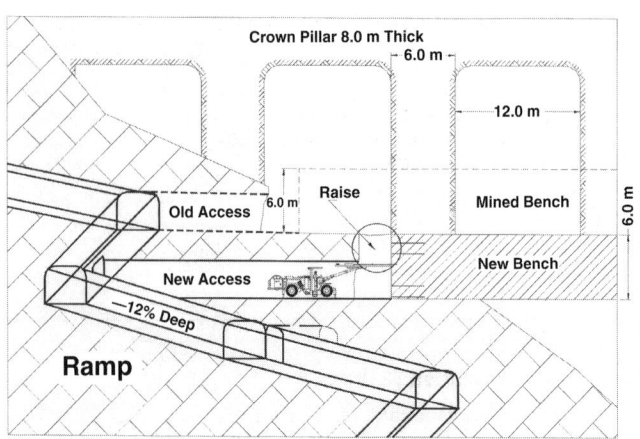

FIGURE 10.7   Starting a new bench

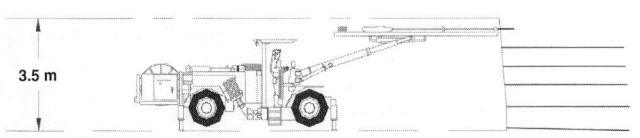

FIGURE 10.8   Jumbo drilling the face on the sill level

Disadvantages

- Dilution up to 10%
- Ore recovery limited to 80%

To define the mining sequence of a stope, several factors, such as ore grade, number of benches in the stope (or attack fronts), and ground stability, need to be considered.

10.4.1 Development

Once the geological boundaries and potential for the ore body have been defined, a new stope begins with the excavation of the sill. This consists of a main drift driven in the footwall of the ore body. Its direction is parallel to the strike of the ore body, but its length is variable depending on the longitudinal dimensions of the ore body. Crosscuts are driven perpendicular to this longitudinal drift and extend to the hanging wall contact. The crosscuts have a section of 4 by 3.5 m and a center-to-center separation of 18 m.

The orientation of the primary systems of joints and fractures is a very important consideration when preparing pillar pattern design and pillar definition. These structures should be perpendicular to the crosscut orientation (Figure 10.6). Next, the walls of the crosscuts are slashed 4 m to give the specified room dimension. Upon reaching the mid-point of the crosscuts, a communications drift is driven between and perpendicular to the crosscuts. Slashing is done as necessary to complete the shape of pillars. Rock bolts are installed immediately after opening the roof, walls, and pillars to prevent weakening of the ground after repeated blasts. It is very important to have this reinforcement 100% completed because with bench advance, the roof height will be out of the reach of both people and machines.

Simultaneously with driving of these openings, a ramp is driven with a grade of −12% (Figure 10.7). From this ramp, an access is driven at the benching level. Located 6 m below the sill (as measured from floor to floor), this yields the optimum operational bench height. Once the access cuts the mineralized contact and is under the sill floor, a 3-m in diameter raise or slot is driven to the sill. This is later slashed until the full bench dimensions are obtained; benching then begins (Figure 10.7).

The ramp will continue downward, and accesses will be driven every 6 m until the maximum pillar height has been reached. Generally, there are one sill 3.5 m high and four 6-m-high benches each (see Figure 10.5).

10.4.2 Drilling

Three different drilling operations are performed. The first is development drilling (ramps, crosscuts, and drifts) as shown in Figure 10.8; the second is production drilling (benching); and the third is for rock reinforcement. All three operations are performed using electro-hydraulic jumbos. The rig can execute both development and production drilling, depending on the needs of the area. The diameter of the bit used is 44 mm, and the effective round length is 4.27 m for both drifting and production. Bolt hole diameter is 38 mm, and hole length is 2.4 m.

10.4.3 Development or Drift Blasting

These blasts utilize safety fuses as the starting line (with No. 6 detonator) and detonating cord as the initiation line (Nonel LP as the detonator). The bottom charge is hydrogel, and the column charge is ANFO. No stemming is used. Figures 10.9 and 10.10 show a typical 4- by 3.5-m face drill pattern and cut hole arrangement.

10.4.4 Production or Bench Blasting

Bit diameter is 44 mm with an effective hole length of 4.27 m. Bench dimensions and the drilling pattern are shown in Figure 10.11. As can be seen, the holes are horizontal and parallel. Just as in development blasting, the starting line is a safety fuse with a No. 6 detonator, the initiation line is detonating cord, and the detonators are Nonel MS. Hydrogel is used as the bottom charge and ANFO as the column charge. Bench blasts are loaded with a Getman mechanized loader, which offers great versatility and operating speed.

As the Figure 10.11 shows, benches are 12 m wide and 6 m high. The mean advance is 3 m. Normally, the hole burden is 1 m, and hole spacing is 1.5 m. On the sides, spacing is reduced to

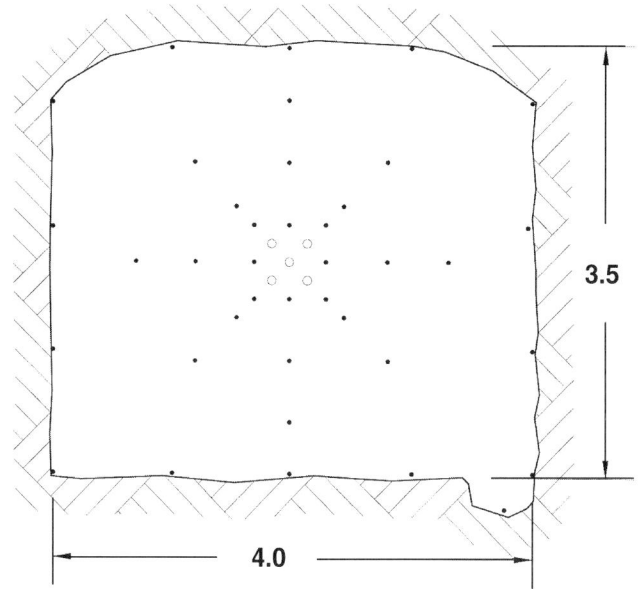

FIGURE 10.9   Drill pattern for 4- by 3.5-m drift face

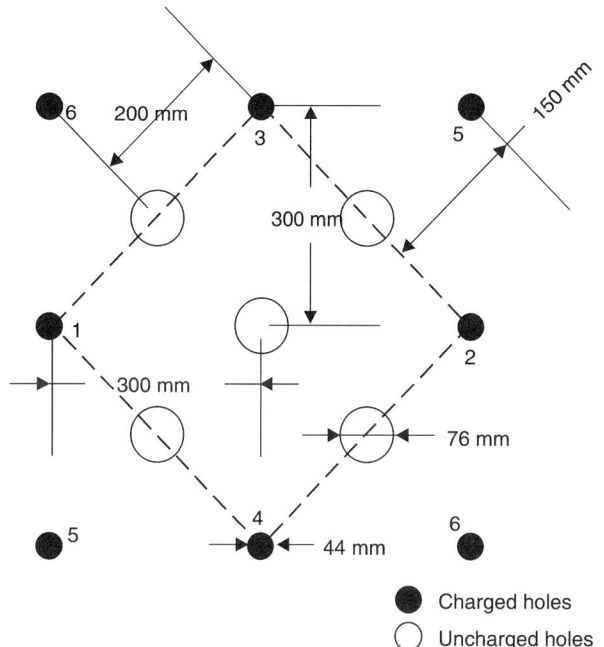

FIGURE 10.10   Cut arrangement for drifting round

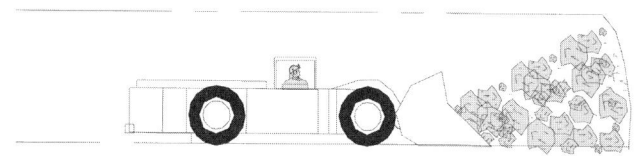

FIGURE 10.12   Mucking face

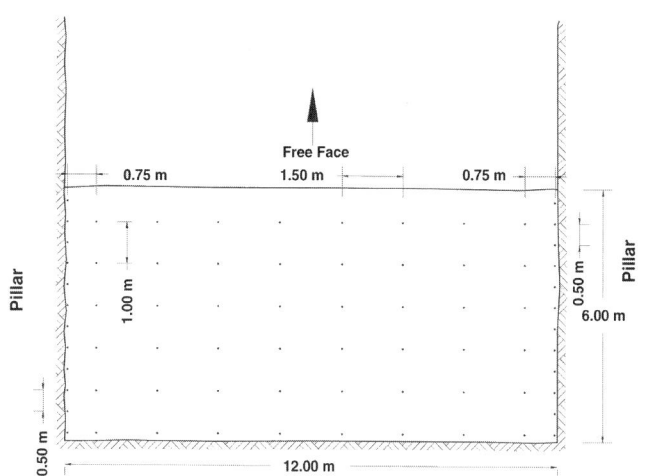

FIGURE 10.11   Bench dimension and drilling pattern

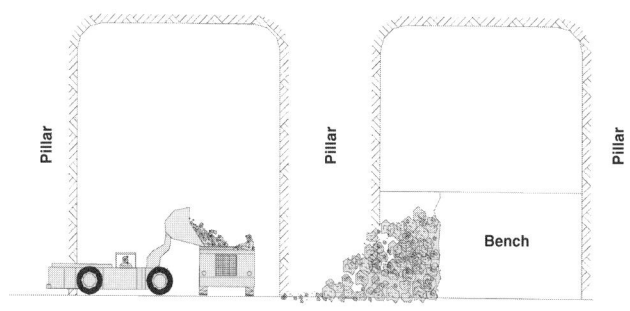

FIGURE 10.13   Loading trucks

0.75 m. Along both sides of the bench, a presplit line is created with holes on 0.5-m spacings. The purpose of the presplitting is to shape the pillars and reduce blasting damage.

Preparations for a blast consist of drilling, flushing, charging, and connecting. All are done during the shift. Blasting takes place at the end of the shift to allow for the dissipation of the fumes produced by the explosion. In the following shift, the roof, walls, pillars, and bench faces are scaled, followed by mucking, loading, transporting, and hoisting.

10.4.5   Dilution

Dilution control is closely related to the amount of available geological information. The geological data from the upper bench are very helpful for planning the mining of the next bench. Other primary factors for dilution control are close supervision of blasting and handling the blasted rock. Nevertheless, there are occasionally periods when it is necessary to mine low-grade zones or even excavate barren rock when developing toward economically attractive zones. The waste is separated and used as fill in other areas. For most of the blasts, dilution can be controlled by relating the extracted volume to the information available from diamond-drill holes, sill development, and previous benches. Dilution has been found to range from 6% to 10% of total extraction.

10.4.6   Mucking, Hauling, and Transportation

After blasting, the open areas are checked, and any loose rock is scaled to avoid risks to people and machines. Mucking is done with 0.17-m^3 (6-yd^3) scoop trams (Figure 10.12).

For hauling, there are four low-profile trucks, each with a capacity of 26 mt. They are loaded by the scoop trams (Figure 10.13) and haul the ore an average distance of 500 m to an ore pass (Figure 10.14). The ore falls down to a bin on the

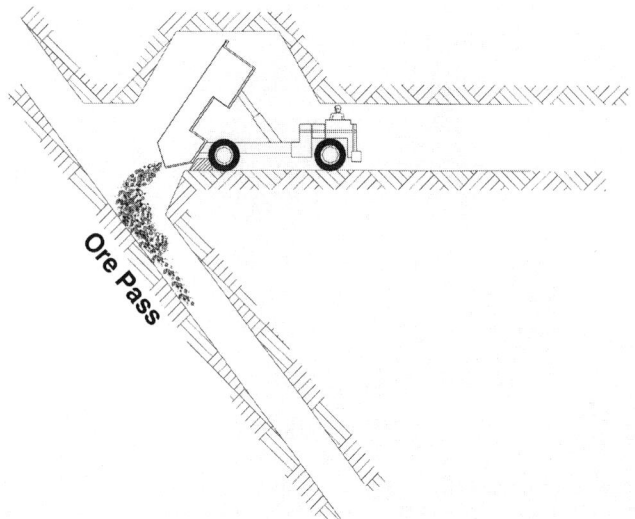

FIGURE 10.14   Trucks dumping ore into ore pass

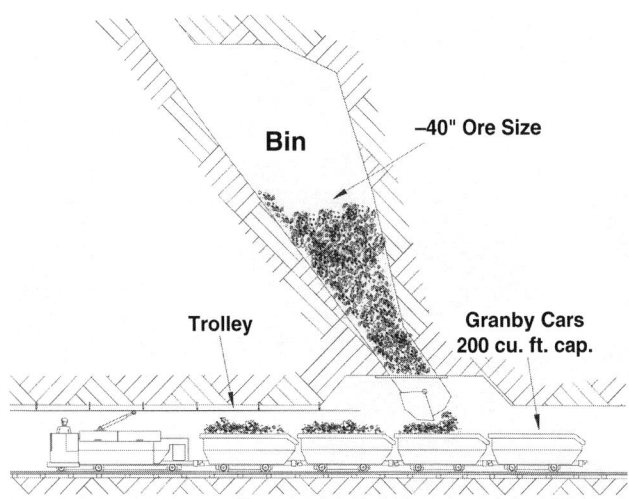

FIGURE 10.15   Bin loading of trains on level 10

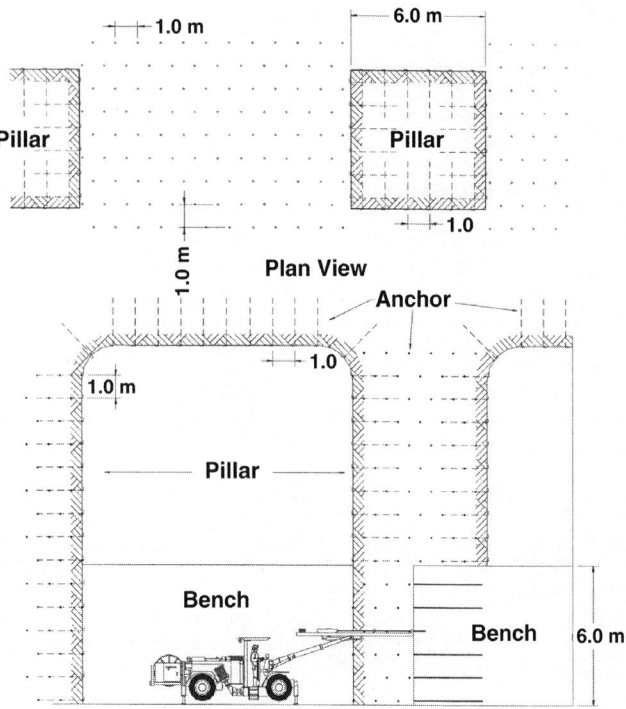

FIGURE 10.16   Plan and longitudinal sections showing reinforcement pattern for roof and pillars

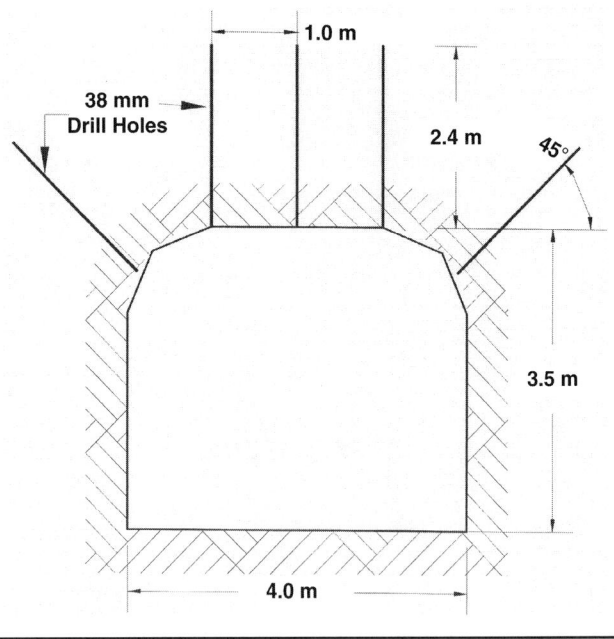

FIGURE 10.17   Ground reinforcement for horizontal and ramp development using threaded rebar.

general haulage level from which 6.6-m³-capacity Granby cars are loaded (Figure 10.15). A trolley locomotive transports the ore to the general bins of the Leones shaft, from where it will be hoisted to the surface.

10.5   GROUND SUPPORT

The ground support parameters are defined through rock mechanics studies. All workings have to be bolted immediately after opening to avoid ground deformation and to prevent unexpected rock falls.

Reinforcement of the stope roof is of paramount importance since the stopes reach considerable heights and a defective or misplaced bolt becomes dangerous to both personnel and equipment circulating under the roof up to the last day of the stope's life. On starting a stope, the sill roof, walls, and pillars are bolted once their limits are known. Later, as the benches advance and the pillars are delimited vertically, bolting continues on pillars and walls (Figure 10.16). Figure 10.17 shows a typical drift bolting pattern.

Hole diameter is 38 mm and hole length is 2.4 m. Reinforcement steel 16 mm) in diameter and 2.4 m long is used.

This includes a steel plate (150 by 450 mm) and a fixture nut. It is installed with cement cartridges 27 mm in diameter and 450 mm long with steel mesh contained in the cartridge (Figure 10.18). The total material cost is about $4.87 US. Bolting work is highly mechanized and bolting equipment is automated (Figure 10.19) so that personnel are not exposed to unsupported conditions during drilling and bolting.

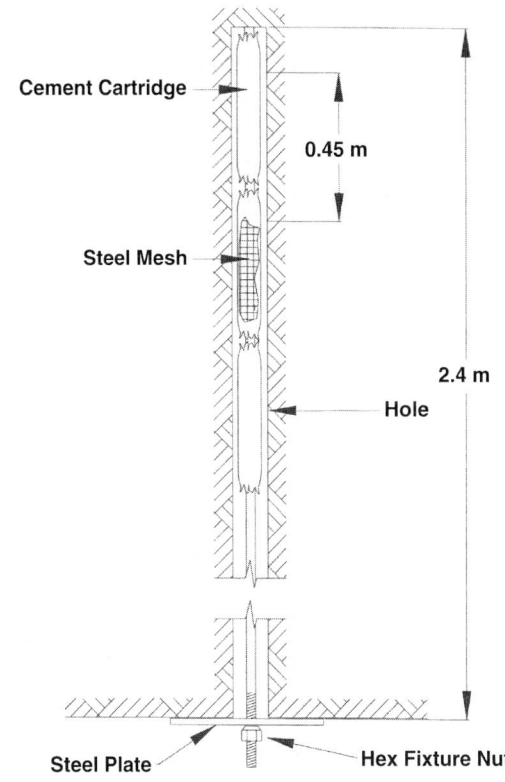

FIGURE 10.18   Mechanized ground reinforcement

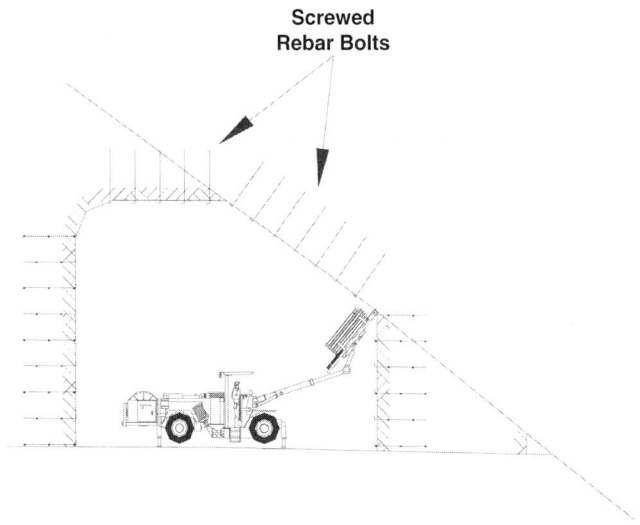

FIGURE 10.19   Sample rebar bolt installation

10.6   CRUSHING AND HOISTING

At the shaft station, the ore is dumped onto a grizzly with an opening size of 61 cm mounted over a bin. A hydraulic hammer breaks the oversized rock. This bin feeds a 0.9- by 1.2-m (36- by 48-in) jaw crusher that reduces rock to −15-cm (−6-in) size. The crushed ore is deposited into a bin that discharges onto a belt conveyor that in turn discharges into the skip filling station. The

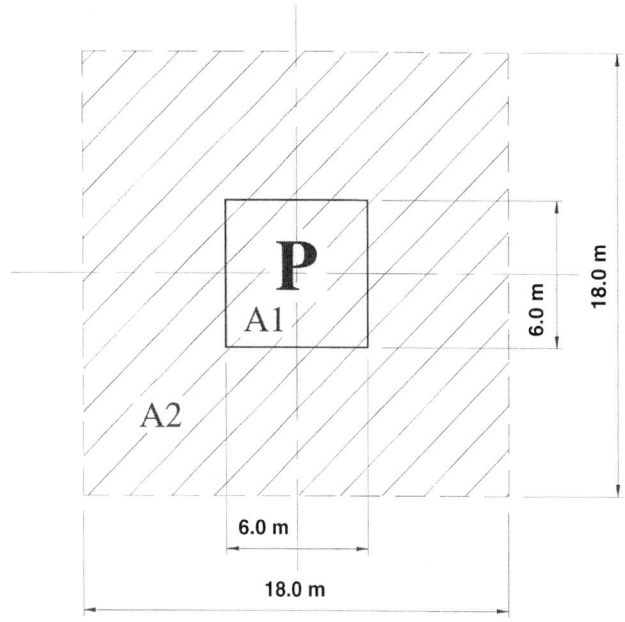

FIGURE 10.20   Tributary area method used to calculate ore recovery

skips are Jetto type with a nominal capacity of 7.3 mt. The Hepburn friction hoist is 2.3 m in diameter, is equipped with a 746-kW (1,000-hp) motor, and has a total suspended weight capacity of 50 mt.

10.7   VENTILATION

There is good natural convective ventilation in the mine. The ramps and shafts form an air intake circuit that continues through the stopes; air finally exits through raises strategically located with the aim of achieving a good environment in the mine. Nevertheless, with increasing depth, this is not enough, and it is necessary to have exhaust fans on the surface (3680 m^3/min capacity) to satisfy the ventilation requirements as specified by Mexico's regulatory agencies.

10.8   ORE RECOVERY

Ore recovery ranges between 80% to 88%, since a portion of the ore body must be left as pillars to support the roof. Ore recovery also depends on the ground conditions, which restrict pillar dimensions as well as room width. The effective control of blasting and bolting helps to achieve high recoveries. The recent rock mechanics study made of the Aurora ore body indicates the possibility of some pillar recovery. This will increase the overall ore recovery to 95%, but this is recommended only after primary mining of the Aurora ore body has been completed. Using the tributary area method to calculate ore recovery (Figure 10.20), one finds that with current dimensions—
    A1 = Pillar area = 6 × 6 = 36 m^2
    A2 = Tributary area = 18 × 18 = 324 m^2
    Mined area = 324 − 36 = 288 m^2
    Extraction ratio = 288 ÷ 324 = 0.8888
    Hence the percentage of recovery = 0.8888 × 100 = 88.88%.

10.9   PRODUCTION STATISTICS

The ore grades and products for the entire Charcas Unit are given in Table 10.3. The includes the production from the Rey, Reina, and San Bartola ore bodies as well as the Aurora. Operating costs for the Aurora Mine are summarized in Table 10.4, and productivity is summarized in Table 10.5.

TABLE 10.3 Ore grades and products for the Charcas Unit in 1999

Production	Tonnage	Silver, m/mt	Lead, %	Copper, %	Zinc, %
Ore (run-of-mine)	1,338,444	50.9	0.21	0.32	4.0
Lead concentrate	2,271	4,275.4	42.56	10.41	6.47
Copper concentrate	9,859	2,120.9	8.47	22.58	9.72
Zinc concentrate	80,818	132.6	0.33	0.81	57.52

TABLE 10.4 Operating costs for the Aurora Mine only for 1999, U.S. dollars per ton[1]

Item	Mine	Mill	Indirect costs	Total
Labor	1.2	0.3	2.17	3.67
Materials	3.69	1.64	0.51	5.84
Indirect costs	0.67	1.58	1.21	3.46
Total	5.56	3.52	3.89	12.97

[1] Rate: $1 = 9.5605 pesos.

10.10 REFERENCES

Blake, W., S. McKinnon, W.A. Hustrulid, and J. Méndez. 1999. Geotechnical and Mining Study for Charcas Unit. Phase I: Mar. 1999; Phase II: Dec. 1999.

TABLE 10.5 Productivity for the Aurora Mine only for 1999

Item	Average
Tons of ore blasted per man-shift	80
Meters of linear advance per man-shift	0.9
Meters of linear advance per blast	3.5
Tons per blasted bench	250
Diesel equipment availability, percent	75
Diesel equipment utilization, percent	90
Tons moved by scoop tram per hour	150
Tons moved by truck per hour	120
No. of 2.4-m-long holes charged per hour, Getman autoloader	50
No. of bolts installed per hour, mechanized bolter	25

Hamrin, H. 1986. Construction and Mining Methods. Underground Mining Guide, Underground Mining, Methods, and Applications. Stockholm, Sweden.

Hustrulid, W.A. 1982. Room-and-Pillar Open-Stope Methods. In *Underground Mining Methods Handbook*, Section 2. SME-AIME.

Nava, R. 1997. Rock Mechanics for Stopes and Drifts at Charcas Unit: Technical Report.

Operation. Topographic and Geological Registries of Charcas Unit.

Puhakka, T. 1997. *Underground Drilling and Loading Handbook*. Tamrock, Finland.

. . . .
CHAPTER 11

Underground Mining of Frozen Placers

Michael G. Nelson[*]

11.1 INTRODUCTION

A placer is a deposit of sand, gravel, or similar detrital material in which a valuable mineral has been concentrated. Such a concentration almost always occurs because the valuable mineral has a high specific gravity and is therefore selectively deposited by the action of wind, flowing water, or moving glacial ice. Minerals recovered from placer deposits include native gold, silver, and platinum; oxides of tin, titanium, zirconium, hafnium, and tungsten; and gemstones such as diamonds, rubies, garnets, and others.

Placer deposits have been found in virtually every earthly setting, from the vast beaches of mineral sands in India and Australia to offshore deposits of cassiterite in Indonesia to frozen gold-bearing gravels in Siberia, Alaska, and the Yukon. In most cases, the detrital nature of these deposits makes them suitable for artisanal mining by simple hand-excavation methods or bulk mining by either direct excavation or dredging. The frozen gravels in northern latitudes are exceptional for two reasons. First, because the material is frozen, it must either be thawed or blasted before it can be loaded and processed. Second, many deposits exist at depths too great for economic mining by surface methods.

11.2 HISTORICAL BACKGROUND

Mining of frozen gravel in modern times began with the exploitation of gold-bearing gravel bars and stream banks in the rivers of Alaska and the Yukon Territory. As early as 1880, typical practice in the Forty-Mile district of Alaska was to thaw gravel in the winter and wash it in the summer (Gates 1994). Miners would build a large wood fire on the surface deposit and remove layers of thawed gravel after the fire burned out. Portions of each thaw level were typically panned, and gravel was stockpiled as pay or waste. The pay gravel was washed in the summer, usually in an in-stream sluice.

During the Klondike Gold Rush in 1897 and 1898, and the later rush to the Fairbanks area in 1902, the deposits that could be exploited by this method were rapidly exhausted. Following the lead of miners in California (Peele and Church 1948), Alaskan miners turned to underground mining methods (Purington 1905). However, in contrast with practice on the Mother Lode, the buried placers in the North were accessed and mined by thawing the frozen gravel with steam points. Descriptions of these mines are given by Gibson (1914), Fleming (1917), and Wimmler (1927). Wimmler also discusses the use of underground hydraulic mining but does not refer to any specific location. However, as also noted by Wimmler, drift mining was almost obsolete by 1927 as a result of the application of large-scale dredging, which was more economic.

Some buried placers were too deep to be dredged, and there were attempts to develop them by other methods. In the 1930s,

the Fairbanks Exploration Co., operator of numerous dredges in Alaska and elsewhere, sank a shaft into the buried placer on Little Eldorado Creek near Fairbanks. A large hydraulic monitor set up in the stope was used to wash gravel to a sump, from whence it was pumped to the surface for washing. Apparently, several thousand yards of gravel were processed in this way, but the handling of large amounts of water and slurry underground was difficult at best. The temperature of the frozen gravel was only 1° or 2° below the freezing point, and as it took on heat from the water, it began to thaw and became unstable. Thus the effort was abandoned (Cope 1994).

When Josef Stalin took control of the Soviet Union, he decided to develop the gold deposits in the Soviet Far East and Yakutia. The placer deposits were exploited on a huge scale, often by the slave labor of political prisoners. Several Alaskans worked in the Soviet Far East during this time (Littlepage and Bess 1938), and in the late 1940s, dredging technology was transferred from Alaska to the Soviet operations (Bogdanov 1992). With the collapse of the Soviet system in the late 1980s, there were numerous visits and technical exchanges between Alaska and the former Soviet Far East. Besides the extensive surface operations (Barker 1990), it was found that there were many underground operations in Russian and Siberian territories (Skudrzyk and Barker 1986).

Placer mining operations in Alaska and the Yukon increased in both size and number after the U.S. government allowed the domestic gold price to follow that of world markets. With these increases, deeply buried frozen deposits were re-examined. For example, Albanese (1981) suggested underground operations for a large deposit on Wilbur Creek in the Livengood district, and in 1986, room-and-pillar operations were begun at Wilbur Creek (Skudrzyk et al. 1987). Because placer miners are usually independent operators, it is difficult to quantify the extent of underground mining in Alaska and the Yukon since 1986. However, it is known that by 1990, at least five successful operations in Alaska and the Yukon had been developed, all of them using variations on the room-and-pillar method.

This article describes underground mining practices for frozen gravel in northern North America and summarizes the technology as practiced in the former Soviet Far East and Yakutia. It also describes cooperative research carried out by investigators in Alaska and the Commonwealth of Independent States.

11.3 MINING METHODS

The methods used in the underground mining of frozen gravel are closely related to those used for underground mining of similar bedded deposits. Both regular and irregular room-and-pillar methods have been employed, as have extended-face longwall methods similar to those used in the reefs of South

---

[*] University of Utah, Salt Lake City, UT.

119

FIGURE 11.1 Thawing an underground placer with steam points (from Gibson 1914)

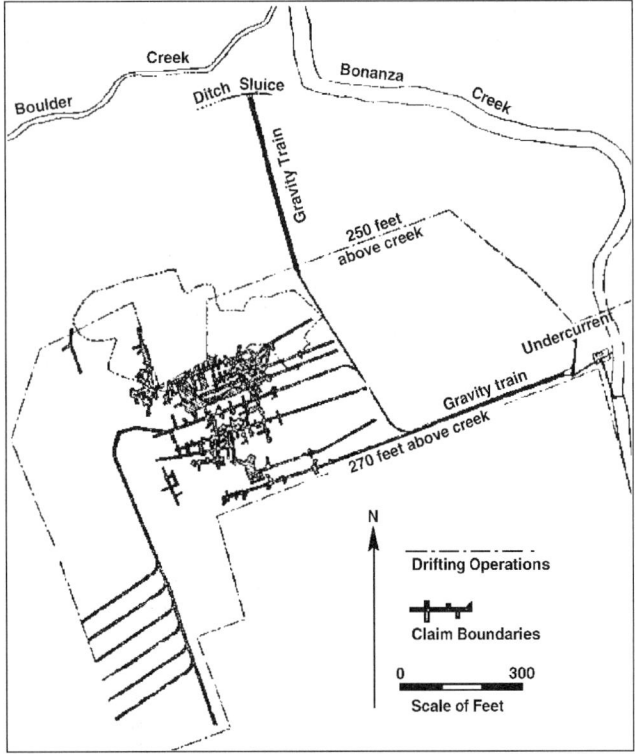

FIGURE 11.2 Underground placer mine in the Klondike (from Purington 1905)

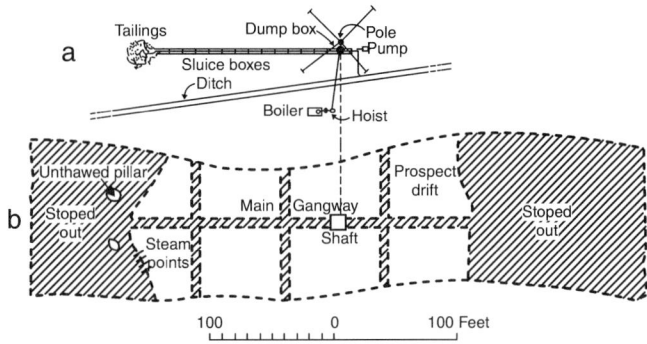

FIGURE 11.3 Room-and-pillar underground placer mine (from Prindle and Katz 1914)

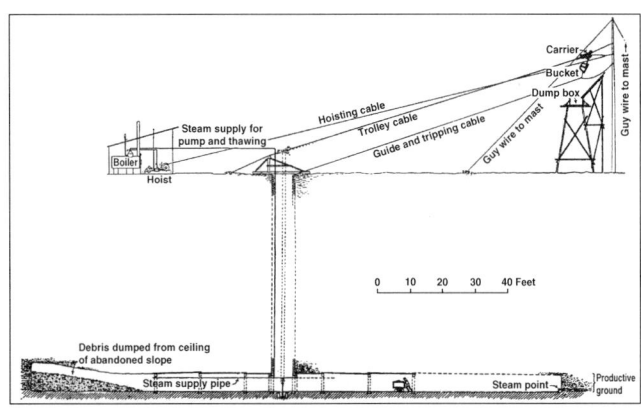

FIGURE 11.4 Shaft arrangement for an underground placer mine (from Prindle and Katz 1914)

Africa. Access to the workings is by shaft or adit, depending on the requirements of the deposit.

11.3.1 Early Practice in North America

In early mining of frozen placers, the ground was often thawed using steam points. Figure 11.1 shows miners in the Nome, Alaska, district using steam thaw points. Figure 11.2 shows a plan view of drifting operations conducted by the Anglo-Klondike Mining Company on Solomon Hill in the Klondike. This mine is described in detail by Purington (1905). The irregular room-and-pillar development results mainly from the fact that some unsystematic work had been done by independent operators before Anglo-Klondike acquired the mining rights to the property. After the ground was thawed, it was supported with 10-cm (4-in) timbers set on 1.68-m (5.5-ft) centers.

Figure 11.3 shows the general arrangement of a room-and-pillar mine in the Fairbanks district, with access by a shaft. Shafts were developed by thawing or picking, then excavating the frozen material. The shafts usually extended 2 to 4 m (6.5 to 13 ft) into bedrock and were timbered as required. Figure 11.4 shows the arrangement on the surface. Depending on the extent of the deposit, drifts were driven from the shaft bottom in either two or four directions for distances of 60 to 90 m (200 to 295 ft). Crosscuts were driven at 30-m (100-ft) intervals. Extraction near 100% was common, with rock-filled cribs being left to provide support as mining retreated toward the shaft. In production, gravel and bedrock were usually picked down and shoveled into wheelbarrows and transported to the shaft bottom to be hoisted out of the mine. The operation of such a mine is described in detail by Hefflinger (1980).

11.3.2 Practice in the Soviet Far East and Yakutia

Under the Soviet system, mining was much more systematic, as would be expected in a planned economy with numerous technical experts. Mining method selection and operating details were selected based on ground control conditions, as described below. Room-and-pillar methods were the most commonly used. Figures 11.5 and 11.6 show room-and-pillar layouts for narrow and wide deposits, respectively.

Declines were favored over shafts. Figure 11.7 shows a typical timbered decline. In Yakutia, placers deeper than 20 m are mined by underground methods. Deposits at depths of 20 to 40 m (65 to 130 ft) are mined seasonally, while those deeper than 40 m (130 ft) are operated year-round. In untimbered operations,

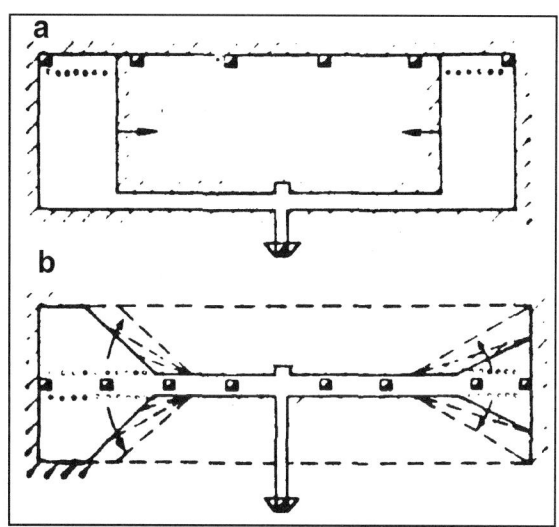

FIGURE 11.5   Room-and-pillar layouts for narrow paystreaks as used in Soviet mines. a, Haulageway on the side of paystreak; b, haulageway on axis of paystreak (from Emelanov et al. 1982).

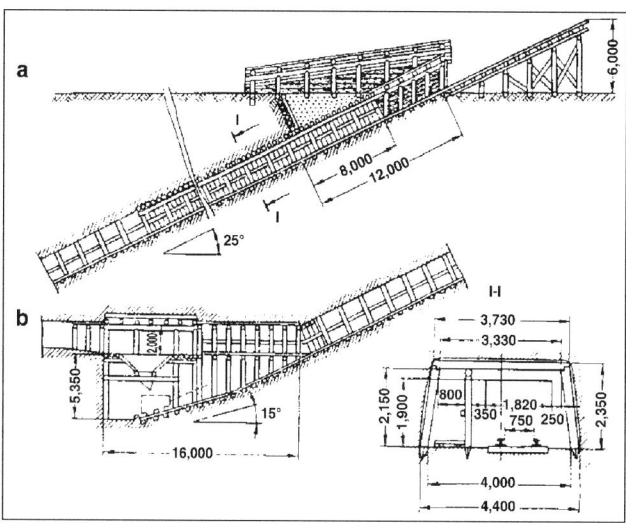

FIGURE 11.7   Timbered decline used in Soviet placer mines. a, Portal with elevated skip dumping point; b, loading point, I-I cross section through decline. All dimensions are in millimeters (from Emelanov et al. 1982).

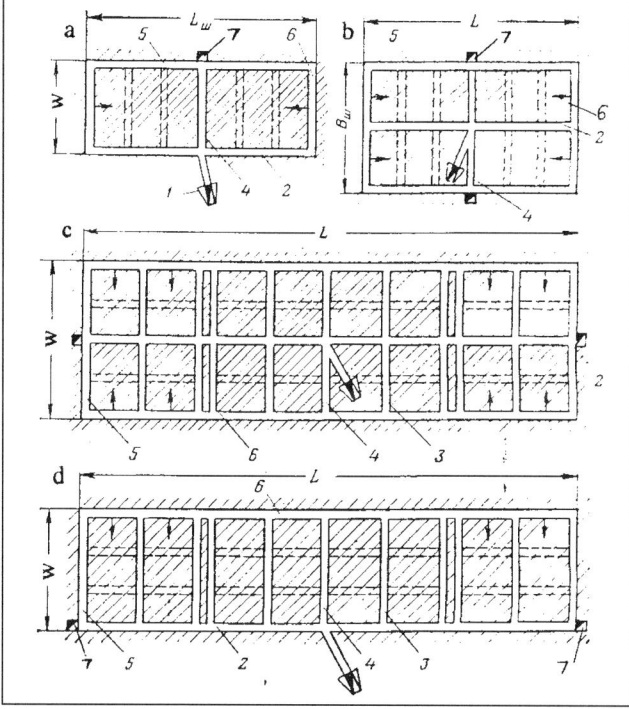

FIGURE 11.6   Room-and-pillar layouts for wide deposits as used in Soviet mines. a. Single, one-sided; b, double, two-sided; c, multiple, two-sided; 1, decline; 2, main access drift; 3, panel crosscut, 4, central crosscut; 5, ventilation drift; 6, extraction opening; 7, ventilation shaft (from Emelanov et al. 1982)

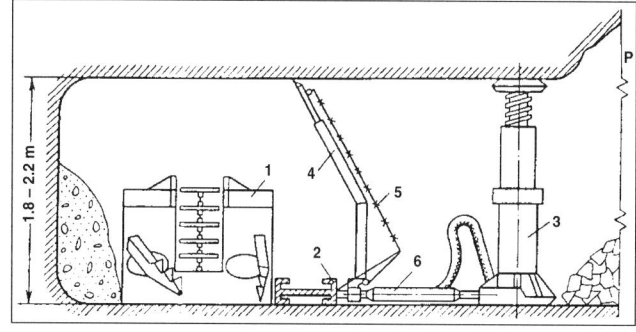

FIGURE 11.8   Typical longwall configuration for mining frozen placers in Soviet mines. 1. Gathering arm loader; 2, armored face conveyor; 3, Sputnic-type roof support; 4, barrier frame; 5, screen; 6, advance ram (from Emelanov et al. 1982)

Longwall methods began to be used about 1980. The most successful methods combined coal loading and handling equipment (an armored face conveyor, a gathering-arm loader, and individual prop-type roof supports) with drilling and blasting techniques typical of those use*d in the deep gold mines of South Africa. Figure 11.8 (Emelanov et al. 1982) shows a cross section through a face using such equipment. Packaged ammonia dynamite and water-resistant aluminized dynamite were used to blast down the face. Barrier frames with wire or plastic screens were used to contain the rock during the blast. The equipment transferred from coal mining suffered very high wear. For example, armored face conveyors that lasted 5 years in coal lasted only 0.5 to 3 years in the underground placer mine (Skudrzyk et al. 1991).

Recently, there have been attempts to use longwall shearing machines similar to those used in coal mines. Modifications made included reducing cutting tool velocities, increasing power, and increasing resistance to abrasion. It was also determined that rotating cutter heads could be severely damaged when they encountered large boulders. Picks broke off, pick blocks were cracked or broken, and gear teeth were damaged. A shearer was

10% to 12% of the gold is left behind in pillars. In timbered mines, 6% to 8% of the gold is unrecovered, but 16 to 18 m (53 to 60 ft) of cribbing are used for every 1,000 m³ of gravel removed. Production for seasonal mines is from 20,000 to 40,000 m³/yr (15,000 to 30,580 ft³/yr); for year-round mines, it is 50,000 to 70,000 m³/yr (38,000 to 53,500 ft³/yr) (Milyuta et al. 1992).

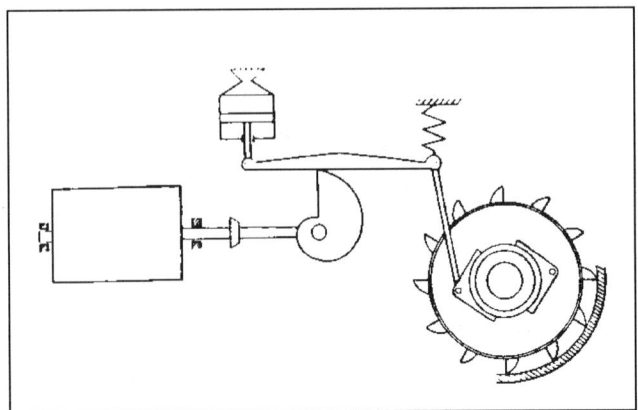

FIGURE 11.9 Soviet design for a shearing machine to cut frozen gravel (from Skudrzyk et al. 1991)

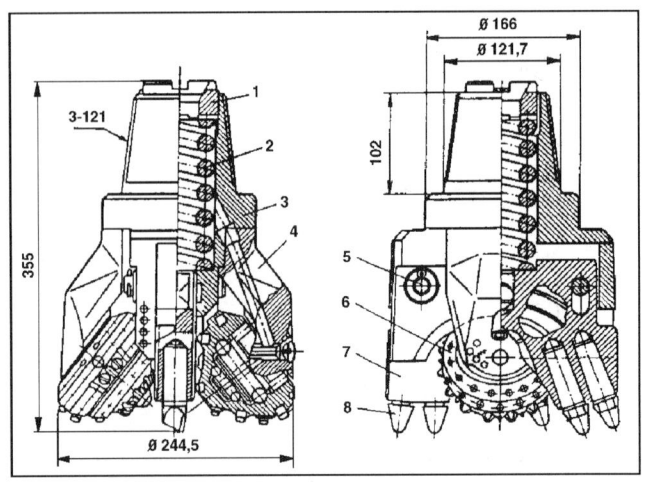

FIGURE 11.10 Soviet design of a modified tricone bit for drilling in frozen ground. 1, Nut; 2, spring; 3, bit body; 4, cone arm; 5, pin; 6, cone; 7, dragfinger bit block; 8, dimensions of dragfinger bit in millimeters (from Skudrzyk et al. 1991).

modified in 1992, as shown in Figure 11.9 (Skudrzyk et al. 1991), to overcome these problems (Kudlai 1992). Rotation speed was reduced from the typical range (43 to 58 rpm) to only 16 rpm. The main drive transmission was fitted with a mechanical clutch. When one or more of the picks encountered a rock too hard to break on first contact, the torque of the mechanical clutch was exceeded, and the drum's rotational motion became percussive, with energy derived from a spring accumulator in the clutch. This design was expected to double shearer production from 100 to 200 m^3 (76 to 153 ft^3) per shift.

11.3.3 Recent Practice in North America

Recent practice in Alaska and the Yukon has included only room-and-pillar methods. One mine accessed the paystreak through a shaft and developed adits by using steam points (Groppel and Madonna 1989). All other mines used access declines. Material is broken by drilling with jumbos or jacklegs and blasting using ANFO slurry. Loading and hauling to the surface are by underground load-haul-dump (LHD) machines or surface front-end loaders modified for underground operation. Haulage efficiency is sometimes improved by using underground trucks. One operation is described in detail below.

Coal mining equipment has also been tested in North America, but with little success. A continuous mining machine was tested on Jackson Hill in the Klondike Mine in 1981, but the costs of replacing worn cutting bits was too high (Prins 1990). A roadheader was tested in Dome Creek Drift Mine in the Fairbanks district in 1991. It also showed high bit wear and the cutter head was damaged when it hit boulders larger than about (10 in) (Voytilla 1992).

11.4 UNIT OPERATIONS

11.4.1 Drilling

Drilling technology for underground mining of frozen gravel has been adapted from two sources: exploration drilling in frozen material from the surface and conventional underground drilling. A wide variety of drill types has been used for surface exploration in frozen ground, including churn drills, augers, sonic drills, and dual-tube reverse circulation drills. The latter have been used successfully with both rotary and hammer action on the bits (McDonald 1990).

Basic drilling research in frozen muck and gravel was conducted by the Bureau of Mines (Chester and Frank 1969) and the Cold Regions Research and Engineering Laboratory (CRREL) of the U.S. Army Corps of Engineers (Mellor and Sellman 1970). It was found that a standard jackleg drill operating with a 3.81-cm-

(1.5-in-) diameter bit could attain penetrations rates of 0.5 to 1.5 m/min (1.6 to 5 ft/min). A larger mechanical drill, model T-650 manufactured by Chicago Pneumatic, attained a penetration rate of 0.75 m/min (2.5 ft/min) using a 20.32-cm (8-in) in diameter bit. In all cases, drilling rates decreased with increasing hole depth.

Wet drilling, dry drilling, and dry drilling with a dust collector were all tested. Dry drilling alone is not recommended because it produces high ambient dust levels. With wet drilling, there was difficulty in extracting the tool and steel when drilling was completed. This problem was worse with smaller drill steels and was thought to occur for one of two reasons: Either the material thawed around the tool and refroze behind it, or the water used in drilling continually thawed the hole, allowing bits of material to fall in behind the tool. Dry drilling with a dust collector was judged successful.

Soviet experience is primarily with hand-operated jacklegs. A penetration rate of 0.74 m/min (2.4 ft/min) was reported with a 4-cm-diameter bit (Emelanov et al. 1982). Again, drilling rates decreased with increasing depth. A promising new bit design was reported by Skudrzyk et al. (1991). This design is similar in concept to the modified longwall shearer described above. It attempts to address the widely varying materials encountered in frozen gravel, particularly those related to the occasional encounter with a large hard boulder. As shown in Figure 11.10, a typical rotary tricone bit for drilling hard rock is combined with a chisel-type drag bit, which performs best in soft formations. When drilling in soft ground, the spring is not fully compressed, and the drag bits are in primary contact with the material. When hard material is encountered, the increased thrust compresses the spring and the cones engage the rock.

Operating mines in North America have reported results for underground drilling with jumbos, jacklegs, and air-track drills. In the Wilbur Creek Mine, near Livengood, Alaska, Ziegler (1987) measured penetration rates for a Secoma ATH-12, single-boom jumbo fitted with an RPH-200 hydraulic drifter. Cross-bits with carbide inserts were used, and rates were calculated for penetration to 1.2 m (4 ft) and to a full depth of 2.6 m (8.5 ft). The results are shown in Table 11.1.

The decreased penetration rate at depth in bedrock was attributed to the poor condition of the drill. In practice, a 2.5-cm (1-in) bit was used to facilitate the flow of cuttings out of the holes.

TABLE 11.1   Drilling rates in frozen alluvium (Ziegler 1987)

Bit diameter	Depth		Penetration rate, all holes		Penetration rate, gravel holes		Penetration rate, bedrock holes	
	m	ft	m/min	ft/min	m/min	ft/min	m/min	ft/min
4.5 cm (1¾ in):	1.2	4.0	0.89	2.93	0.96	3.14	0.87	2.85
	2.6	8.5	0.81	2.66	0.90	2.95	0.71	2.34
5 cm (2 in):	1.2	4.0	0.67	2.20	0.56	1.83	0.78	2.57
	2.6	2.6	0.55	1.79	0.49	1.60	0.71	1.98
6.4 cm (2.5 in):	1.2	4.0	0.89	2.93	0.96	3.14	0.87	2.85
	2.6	8.5	0.81	2.66	0.90	2.95	0.71	2.34

TABLE 11.2   Comparison of blasting parameters in frozen alluvium (Chester and Frank 1969)

	V cut	Burn cut	Result
Delay	Slow delays resulted in better breakage	Millisecond	
Explosive	40% special gelatin	60% high-density ammonia dynamite produced better breakage	More uniform muck pile with fewer large blocks, so that mucking time was decreased

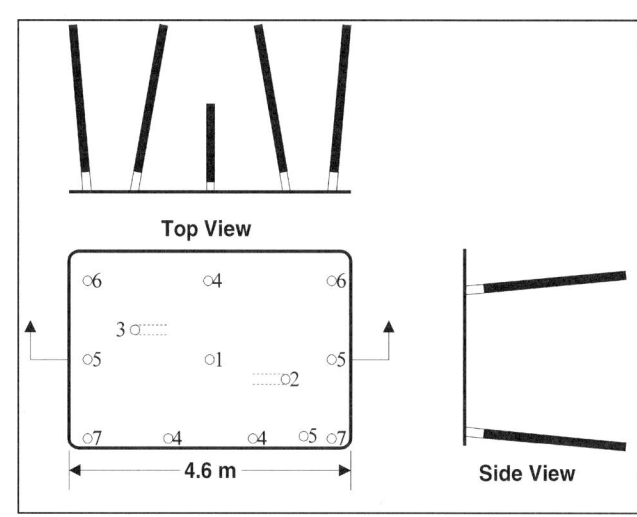

FIGURE 11.12   Modified V cut (from Ziegler 1987)

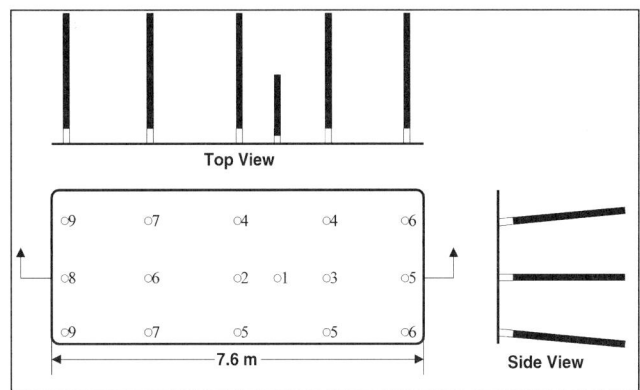

FIGURE 11.11   Modified burn cut (from Ziegler 1987)

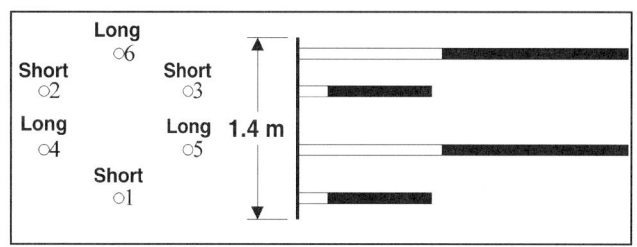

FIGURE 11.13   Prismatic cut (from Ziegler 1987)

11.4.2   Blasting

Frozen silt or gravel is more plastic than rock, because of the ice present. Thus, it is generally desirable to use an explosive with a slower detonation velocity when blasting these materials. This will allow a longer time for the energy of the explosion to be transferred into the ground, producing a larger fracture zone. Chester and Frank (1969) conducted a 2^3 factorial-design experiment on blasting in frozen ground, as described in Table 11.2.

At the Wilbur Creek Mine, Ziegler (1987) confirmed that slow delays and lower-energy explosives (in this case ANFO slurry) produced better breakage. However, he found that variation in fragmentation and throw between V and burn cuts did not affect mucking time. It was noted that overall face advance was faster with 1.2-m (4-ft) than with 2.4-m (8-ft) rounds.

Ziegler also tested 12 different round designs whose configurations were suggested in the Soviet literature. A burn cut with a single unloaded hole gave very poor results because the high swell factor of the frozen material tended to freeze the round. Three were found to work best. The modified burn cut, shown in Figure 11.11, worked well, but required many holes. The modified V cut, shown in Figure 11.12, had a "buster" hole in the center and was preferred because it required less drilling. In

this round, the buster hole is fired first, cratering to the surface and providing a free face for subsequent blasting. The prismatic cut, shown in Figure 11.13, worked well with either six or four holes (alternating long and short around the perimeter). In all the round designs, a minimum burden of 0.6 m (2 ft) was used to avoid dead-pressing the ANFO. This conforms with Soviet practice, where the minimum recommended burden is 0.75 m (2.5 ft) (Emelanov et al. 1982)

11.4.3   Cutting

Mechanical cutting of frozen gravel and silt has not been successful. The unsorted nature of almost all deposits means that, sooner or later, the cutting device will encounter a large, hard rock. As mentioned above, this has defeated attempts to use roadheaders, continuous miners, and longwall shearers. The results obtained in the Soviet Far East with the modified shearer are not known.

TABLE 11.3 Soviet classification of frozen intermediate roof materials up to 15 m thick

Stability class	Composition	Temperature, °C	Ice content, wet wt %	Texture
I. Highly stable	1. Alluvial deposits with pebbles, cobbles, and occasional boulders with matrix of sand, silt, and clay. Complete saturation of pores by ice. Stratified with single homogeneous stratum at least 10m thick.	< –6	≤ 25	Massive
II. Stable	1. Alluvial and lacustrine deposits similar in composition and thickness to the above.	–6 to –3	< 25	Massive
	2. Sandy and loamy despots with up to 30% cobbles, pebbles, and pea-gravel. Ice-saturated pores.	< –3	< 25	Massive
	3. Homogeneous silty and clayey deposits.	< –4	25 to 50	Massive, stratified
	4. Alluvial, lacustrine glacial, and shallow water coastal marine sediments of interbedded layers of large- and fine-grained materials with poor stratification, homogeneous layers 5 to 10 m thick.	< –3	< 25 for large-grained material, 25 to 50 for fine-grained material	Massive, stratified
III. Medium stable	1. Alluvial and lacustrine coarse-grained deposits, composition as in Class I.	–3 to –2	< 25	Massive
	2. Sandy deposits with up to 30% cobbles, pebbles, and pea-gravel.	–3 to –2	25 to 50	Stratified
	3. Homogeneous silty and clayey sediments.	–4 to –3	25 to 50	Stratified lattice
	4. Interbedded layers of large-grained to clay-size material with horizontal stratification, homogeneous layers up to 2 m thick.	–2 to –1	< 25 for large-grained material, 25 to 50 for fine-grained material	Massive layered lattice
	5. Ground ice and clear ice.	< –6	> 60	Ataxic basal
IV. Poorly stable	1. Alluvial and lacustrine coarse-grained deposits, composition as in Class I.	–2 to –1	< 25	Massive
	2. Sandy deposits with up to 30% cobbles, pebbles, and pea-gravel.	< 6	25 to 50	Stratified
	3. Homogeneous silty and clayey sediments.	–2 to –1	25 to 50	Stratified lattice
	4. Interbedded layers of large-grained to clay-size material with horizontal stratification, homogeneous layers up to 2 m thick.	–2 to –1	25 to 50	Massive, porous, stratified lattice
	5. Eluvial-solifluction silty formations; loess-like clays of lake-swamp and marine lagoon lakes.	–3 to –15	25 to 50	
	6. High ice content, salt content > 0.25%, clay formations of lake and swamp, offshore marine, and lagoon genesis.	–6 to –3	> 50	Blocky ataxic
	7. Ground ice and clear ice.	–6 to –3	> 60	Ataxic basal
V. Stable	1. Plastic frozen alluvial materials of any grain size distribution with silty and clayey matrix.	> –1.5	< 50	Any
	2. As above, with sandy matrix.	> –1	< 50	Any
	3. Unconsolidated and poorly cemented by ice.	Any	< 3	Massive, porous
	4. Ground ice and clear ice.	> –3	> 60	Ataxic basal

11.4.4 Haulage

All known operations in North America have used mine trucks, LHDs, or modified front-end loaders for haulage. The Soviet literature shows drift layouts for rail and belt haulage, but gives no details on these methods.

11.4.5 Ground Control

The properties of frozen gravel and silt have been studied extensively. Those properties depend on many variables, including in situ temperature, ice content, particle size and composition, and stratification. Tabulations of material properties for frozen alluvium are found in Chester and Frank (1969), Sellman (1972), Haynes et al. (1975), and Skudrzyk (1983).

Soviet researchers have defined stability classes for frozen materials, as shown in Table 11.3 (Emelanov et al. 1982). They also defined stable spans and recommended pillar sizes for each of those stability classes, as shown in Tables 11.4 and 11.5.

Table 11.5 shows recommended pillar sizes for given widths of rooms, based again on the stability criteria in Table 11.3.

The most important characteristic of these materials is their tendency to creep, and they may exhibit considerable deformation before failure. Extensive measurements in the Dome Creek Drift Mine, near Fairbanks, Alaska, showed that roof-to-floor closure depended on the width of the entry, proximity to active mining, and elapsed time. The roof usually moved as a

unit, creeping slowly into the entry until slabs developed along silt layers or other planes of weakness. Steady-state closure in inactive areas approached a constant rate of 0.42 mm/d, but closure rates as high as 75 cm/d were recorded in active areas. At mining spans of 45 to 60 m (150 to 200 ft), entries 2.5 to 3 m (8 to 10 ft) high experienced complete closure within 4 to 7 days (Seymour et al. 1996). Interestingly, this closure was not considered a problem by the miners because it occurred slowly and, for the most part, predictably.

Bandopadhyay et al. (1996b) completed a finite-element analysis of roof-to-floor convergence for entries in frozen material that accounted for both material characteristics and heat transfer. It was shown that convergence was substantially dependent on ambient air temperature. The first rapid deformation was followed by a period of secondary creep in which closure rate was approximately constant. The relationship between convergence and span was found to be linear, even though the model accounted for nonlinear creep in the material. Figures 11.14 and 11.15 show typical results from the analysis. The simulations shown were made with an air temperature of –2.8°C and a ground temperature of –4°C.

Roof bolts can be used to slow the separation of frozen material along bedding planes. This practice is common in Yakutia and was also used in the Dome Creek Drift Mine. Bandopadhyay et al. (1995; 1996a) completed a finite-element analysis of the temperature regimes around various roof-bolting

TABLE 11.4   Spans of stable extraction openings in frozen ground

| | Thickness of affected strata, m | | Maximum span of extraction openings, ft | | | | | |
| | | | Room-and-pillar (non-caving roof) | | Longwall (caving roof) | | | |
Stability class	Monolithic roof	Stratified roof	Monolithic roof	Stratified roof	Monolithic roof	Stratified roof
I	14–20	13–18	35–45	30–40	26–37	26–37
II	13–16	10–12	25–35	23–27	22–30	22–30
III	10–13	7–9	20–25	15–20	19–24	19–24
IV	7–10	4–6	10–15	8–12	8–12	8–12
V	4–7	2–4	6–10	5–8	5–7	5–7

TABLE 11.5   Recommended pillar sizes depending on thickness of extracted pay gravel in mines in frozen ground, meters

| Room width | 36.6 | | | | 30.5 | | | | 20.1 | | | | 10.1 | | | | 5.03 | | | |
| | | | | | | | | Thickness of extracted pay gravel, m | | | | | | | | | | | | |
Stability class	1.4	1.8	2.4	3.0	1.4	1.8	2.4	3.0	1.4	1.8	2.4	3.0	1.4	1.8	2.4	3.0	1.4	1.8	2.4	3.0
I	3.0	3.6	4.4	5.2	2.3	2.7	3.2	3.9	2.0	2.4	2.9	3.4								
II					2.5	2.9	3.6	4.1	2.2	2.6	3.1	3.5								
III									2.3	2.6	3.3	3.8								
IV									2.5	2.9	3.6	4.1	2.0	2.4	3.0	3.5				
V													2.3	2.6	3.3	3.8	1.8	2.2	2.9	3.5

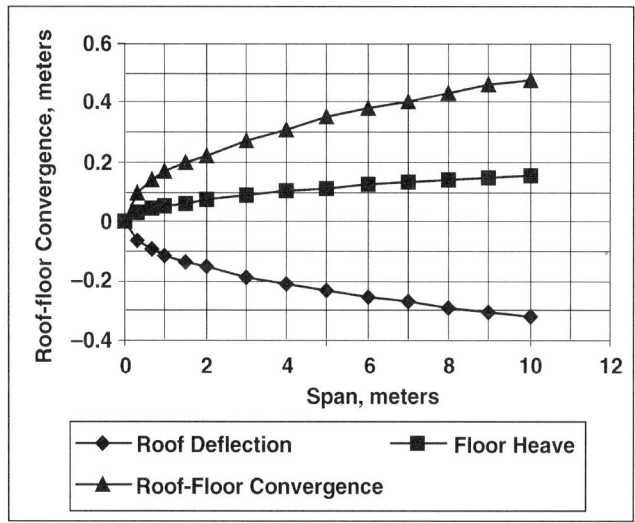

FIGURE 11.14   Roof-floor convergence versus time (from Bandopadhyay et al. 1996b)

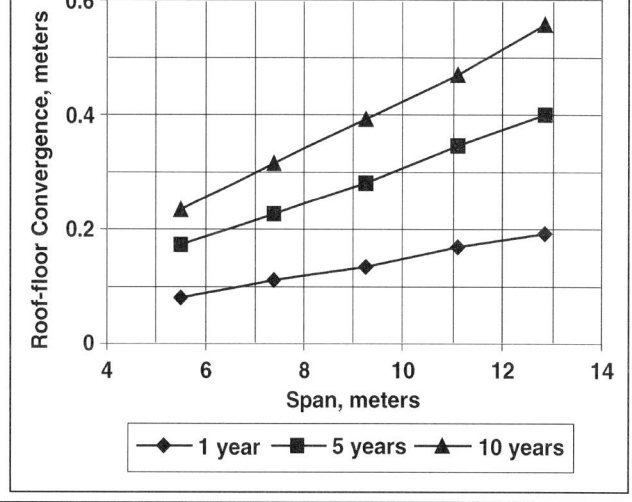

FIGURE 11.15   Roof-floor convergence versus span (from Bandopadhyay et al. 1996b)

systems. It was shown that heat conducted through the bolts into the roof contributed to roof failure, but that the temperature of the ambient air was a much more important factor.

11.4.6   Ventilation and Climate Control

Temperature control is especially important in mining frozen placers. Obviously, if the temperature of material surrounding the mine openings rises above freezing, it becomes unstable, with ribs and roof failing rapidly. Mines in North America typically rely on the circulation of cold winter air to keep the entry perimeters frozen. These mines operate in the winter and are sealed in the summer. Mines in the former Soviet Union also use this method, but some also use chilled air during summer operation.

Bandopadhyay et al. (1997; 1999) performed studies of ventilation and climate control in underground placer mines using a finite-difference algorithm. These studies showed that

during the winter months, underground temperature could be controlled adequately by circulating cold, outside air. In some cases, preheating of outside air could be required to maintain temperatures comfortable for workers.

In another important study (Bandopadhyay et al. 1996c), it was shown that thermosyphons can be used to remove ventilation-induced heat in underground placer mines. A thermosyphon (sometimes called a heat pipe) is a passive device with very high thermal conductance resulting from the repeated cycle of evaporation and condensation in its working fluid (Heuer et al. 1985). Thermosyphons can also be used to freeze thawed ground rapidly, preventing water inflow and entry instability.

11.5   EXAMPLE B THE WILBUR CREEK MINE

The Wilbur Creek placer property near Livengood, Alaska, is a buried stream channel. Auriferous gravels are covered with 20 to

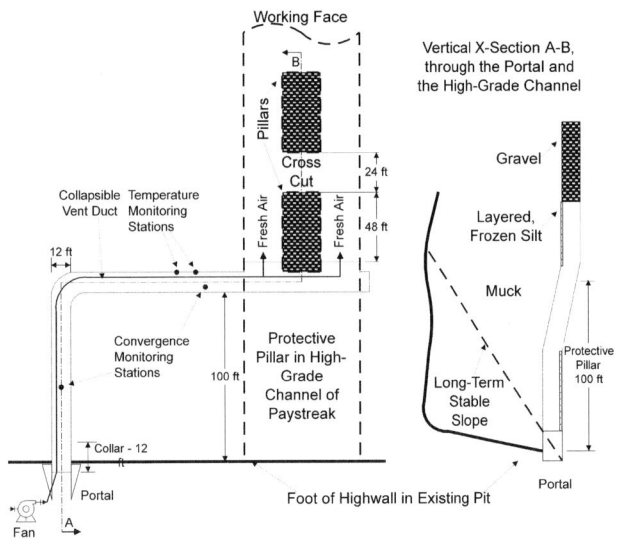

FIGURE 11.16   General layout of Wilbur Creek Mine

TABLE 11.6   Typical production cycle at the Wilbur Creek Mine

Operation	Time (hr:min)
Drilling	2:25
Cleaning and loading holes	1:30
Post-blast ventilation	0:30
Mucking and haulage	3:20
TOTAL	8:00

11.5.2   Ventilation

The mine was ventilated under pressure with a 7.5-kW (10-hp) fan providing air through 0.6-m- (24-in-) diameter flexible tubing. This system was adequate for the equipment and production cycle described above.

11.5.3   Economics

The economics of the operation were calculated on the basis of mining and processing 8,180 bank m^3 (10,700 yd^3) of gravel per year. Costs are summarized in Table 11.7.

11.6   REFERENCES

Albanese, T. 1981. A Hypothetical Underground Placer Operation at Livengood. MIRL Report No. 52. Mineral Industries Research Laboratory, University of Alaska, Fairbanks, pp. 61–84.

Bandopadhyay, S., K. Biswas, and M.G. Nelson. 1995. Evaluation of a Roof Bolt Support System in Arctic Placer Mines. *Transactions of SME-AIME*, vol. 298, pp. 1801–1808.

Bandopadhyay, S., X. Wang, and M.G. Nelson. 1996a. Effect of Roof Bolting Parameters on the Stability of an Opening in a Frozen Ground Placer Mine. *Transactions of SME-AIME*, vol. 300, pp. 140–145.

Bandopadhyay, S., X. Wang, and M.G. Nelson. 1996b. Roof Span Stability in an Underground Placer Mine in the Arctic. In Mining in the Arctic. Proceedings of the 4th International Symposium, ed. by A. Myrvang (Svalbard, Norway, July 1996).

Bandopadhyay, S., H. Wu, and M.G. Nelson. 1997. Ventilation Design Alternatives for Underground Placer Mines in the Arctic. In Proceedings of the 6th International Mine Ventilation Congress, ed. by R.V. Ramani. SME, Littleton, Colorado.

Bandopadhyay, S., H. Wu, M.G. Nelson, and V. Izaxson. 1999. A Finite Element Model of Heat Transfer in a Shallow Placer Mine in the Arctic, ed. by J.E. Udd and A. J. Keen. In Mining in the Arctic. Proceedings of the 5th International Symposium (Yellowknife, Northwest Territories, June 14–17, 1998). Balkema, Rottterdam, Netherlands.

Bandopadhyay, S., H. Wu, M.G. Nelson, and V. Izaxson. 1996c. Thermosyphons for Removal of Ventilation-Induced Heat in an Underground Placer Mine in the Arctic. *Transactions of SME-AIME*, vol. 300, pp. 1915–1921.

Barker, J.C. 1990. Surface Placer Mining in the Soviet Far East. Presentation at 12th annual Alaskan Conference on Placer Mining, Fox, Alaska, March 22–24. Fairbanks Placer Mining Conference Committee, Fairbanks, Alaska, pp. 4–6 (abridged compilation).

Bogdanov, E. 1992. Personal interview, Fairbanks, Alaska, July 23.

Chester, J.W., and J.N. Frank. 1969. Fairbanks Placers Fragmentation Research, Final Report, Heavy Metals Program. Authorization No. 9-1115-33, Twin Cities Mining Research Center, Bureau of Mines, Minneapolis, Minnesota, 52 pp.

Cope, D.W. 1994. Personal interview, Fairbanks, Alaska, February 26.

Emelanov, V.I., U.A. Maneau, and E.D. Kudlai. 1982. Underground Mining of Frozen Placers. Ch. 4 in *Excavation and Support of Mine Development Openings*. Nedra, Moscow, U.S.S.R., pp. 1–18 (translated copy in possession of author).

Fleming, E.E. 1917. Block Stoping and Timbering in Deep Placer Mining. *Mining & Scientific Press*, September 15, p. 378.

Gates, M. 1994. Gold at Fortymile Creek: Early Days in the Yukon., UBC Press, Vancouver, BC.

Gibson, A. 1914. Thawing Frozen Ground for Placer Mining. *Mining & Scientific Press*, January 17, pp. 143–145.

11.5   (continued)

40 m (65 to 135 ft) of frozen overburden composed of silt and gravel. The pay section includes 1.5 m (5 ft) of gravel and 0.9 m (3 ft) of fragmented bedrock. The pay section averages 60 m (200 ft) wide, but a 25-m- (82-ft-) wide, high-grade zone contains about 80% of the values.

The claims had been mined by hydraulicking from 1961 to 1985, when enforcement of the 1972 Clean Water Act made hydraulic mining impracticable. At that time, financial grants from the state of Alaska made the design and implementation of underground mining possible by recirculating 100% of the water to meet the standards of the Clean Water Act.

11.5.1   Development and Mining

The mine was planned as a room-and-pillar operation. Because of the narrow width of the pay section, it was designed with two rooms along either side of a single row of pillars. A general layout is shown in Figure 11.16.

A portal was established in the existing highwall, and an access drift was driven into the pay section. The portal was above the elevation of the pay section and did not decline into the pay section until it was 30 m (100 ft) back from the highwall. This provided a protective pillar for the access drift and also isolated the new workings from the highwall.

An ATH12 Secoma jumbo fitted with a single RPH 200 hydraulic drifter was used to drill blast holes 3.6 m (8 2 ft) long by 4.4 cm (1¾ in) in diameter. The RPH 200 drifter is designed to drill holes 3.19 to 4.4 cm (1¼ to 1¾ in) in diameter for production holes and 8.9-cm (3½ in) in diameter for cut holes with water flushing. The hydraulic drifter generated no mist. The noise level was 105 dB at 0.9 m (3 ft). Drilling rates were generally good, at 1.5 to 3 m/min (5 to 10 ft/min).

Several blasting patterns were tried based on local experience and reference to the Soviet literature. As shown in Figure 11.12, a modified V-cut in which a short center hole is fired first was the preferred pattern because it required less drilling. The biggest problems in drilling and blasting were dead pressing of ANFO slurries. Fragmentation was always good, and there was no time lost in excessive floor cleanup. Mucking and haulage were done with a Wagner ST-2D LHD fitted with a 1.5 m^3 (2 yd^3) bucket. The LHD delivered ore to an outside stockpile for processing during the summer.

A typical production cycle is summarized in Table 11.6.

TABLE 11.7 Capital and operating costs for the Wilbur Creek placer mine in 1987 dollars (Skudrzyk et al. 1987)

Cost category	Item	Cost, $	Dollars per unit m³	yd³
PROCESSING				
Capital equipment and facilities (cost of ownership):*				
	Sluice box	15,000	0.071	0.054
	E-Z panner	3,750	0.018	0.014
	Diesel pump	25,000	0.238	0.182
	D-8 bulldozer	40,000	0.477	0.365
	Settling pond construction	4,000	0.020	0.015
	Miscellaneous	20,000	0.094	0.072
	Total		0.92	0.70
Operating costs:				
	Equipment maintenance	25% of owning cost	0.235	0.180
	Fuel	$1/gal	0.209	0.160
	Settling pond clean-up	$1.25/yd³ of silt handled	0.759	0.580
	Labor	$20/hr	0.981	0.750
	Total		2.18	1.67
TOTAL PROCESSING			3.10	2.37
MINING				
Capital equipment and facilities (cost of ownership):				
	Shop and living quarters	10,000	0.127	0.097
	LHD	36,000	0.268	0.205
	Jumbo	28,000	0.089	0.068
	Generator	13,880	0.409	0.313
	Fan and ductwork	2,500	0.038	0.029
	Dump truck	15,395	0.115	0.088
	Portal	25,000	0.122	0.093
	Miscellaneous	25,000	0.122	0.093
	Total		1.29	0.99
Operating:				
	Equipment maintenance	50% of owning cost	0.521	0.398
	Fuel	$1/gal	1.556	1.190
	Explosives	†	2.420	1.850
	Labor	$20/hr	9.116	6.970
	Total		13.61	10.41
TOTAL MINING			14.90	11.40
TOTAL PRODUCTION COSTS			18.00	13.77

*Ownership costs calculated by straight-line depreciation over equipment life, with no salvage value.
†Explosives use calculated for 19 holes per round, each hole using 10 lb of ANFO at $0.33/lb, 1 stick of TOVEX 220 at $1.37 per stick, and 1 cap at $1.75.

Groppel, C., and J.A. Madonna. 1989. Drift Mining at Tenderfoot Creek. In *Mining in the Arctic, Proceedings of the 1st International Symposium on Mining in the Arctic*, ed. by S. Bandopadhyay and F.J. Skudrzyk (Fairbanks, Alaska, July 17–19). Balkema, Rotterdam, pp. 211–215.

Haynes, F.D., J.A. Karalius, and J. Kalafut. 1975. Strain Rate Effect on the Strength of Frozen Soil. CRREL Report RR-350. Cold Regions Research and Engineering Laboratory, U.S. Army Corps of Engineers, Hanover, New Hampshire.

Hefflinger. 1980. Drift Mining. Presentation at 2nd annual Alaskan Conference on Placer Mining, Fox, Alaska, April 7–8. MIRL Report 46. Mineral Industries Research Laboratory, University of Alaska, Fairbanks, Alaska, pp. 163–168 (abridged compilation).

Heuer, C.E., E.L. Long, and J.P. Zarling. 1985. Passive Techniques for Ground Temperature Control. In *Thermal Design Considerations in Frozen Ground Engineering*, ed. by T.G. Krsewinski and R.G. Tart. American Society of Civil Engineers, pp. 72–154.

Kudlai, 1992. Prospects for Underground Mining of Frozen Placer Deposits in the North-East of Russia. Presentation at 13th annual Alaskan Conference on Placer Mining, Fairbanks, Alaska, March 4–7. Fairbanks Placer Mining Conference Committee, Fairbanks, Alaska, pp. 7–8 (abridged compilation).

Littlepage, J.D., and D. Bess. 1938. *In Search of Soviet Gold*. Harcourt, Brace, New York, 310 pp.

McDonald, R. 1990. Drilling in Frozen Ground. Presentation at 12th annual Alaskan Conference on Placer Mining, Fox, Alaska, March 22–24. Fairbanks Placer Mining Conference Committee, Fairbanks, Alaska, pp. 4–6 (abridged compilation).

Mellor, M., and P.V. Sellman. 1970. Experimental Blasting in Frozen Ground. CRREL Report SR-153. Cold Regions Research and Engineering Laboratory, U.S. Army Corps of Engineers, Hanover, New Hampshire, pp. 1–33.

Milyuta, B.I., V. Kuznetsov, and V.A. Sherstov. 1992. Placer Deposits of Yakutia: Technical Improvements and Mechanization. Presentation at 13th Annual Alaskan Conference on Placer Mining, Fairbanks, Alaska, March 4–7. Fairbanks Placer Mining Conference Committee, Fairbanks, Alaska, pp. 9–11 (abridged compilation).

Peele, R., and J.A. Church. 1948. *Mining Engineers' Handbook*, 3rd ed. John Wiley & Sons, New York, pp. 10-606–10-614.

Prindle, L.M., and F.J. Katz. 1914. A Geologic Reconnaissance of the Fairbanks Quadrangle, Alaska. U.S. Geological Survey Bulletin No. 328.

Prins, M. 1990. Underground Placer Mining in the Yukon Territory and Alaska. Presentation at 12th annual Alaskan Conference on Placer Mining, Fox, Alaska, March 22–24. Fairbanks Placer Mining Conference Committee, Fairbanks, Alaska, pp. 84–86 (abridged compilation).

Purington, C. 1905. Gravel and Placer Mining in Alaska. U. S. Geological Survey Bulletin 263, pp. 82–95.

Sellman, P.V. 1972. Geology and Properties of Materials Exposed in the USA CRREL Permafrost Tunnel. CRREL Report SR-177. Cold Regions Research and Engineering Laboratory, U.S. Army Corps of Engineers, Hanover, New Hampshire.

Seymour, J.B., D.R. Tesarik, and R.W. McKibbin. 1996. Stability of Permafrost Gravels in an Alaskan Underground Placer Mine. In Mining in the Arctic. Proceedings of the 4th International Symposium, ed. by A. Myrvang. Svalbard, Norway, July 1996.

Skudrzyk, F. 1983. Impact of Permafrost on Placer Mining. In Proceedings of the 5th Annual Alaska Conference on Placer Mining, ed. by B.W. Campbell, J.A. Madonna, and M.S. Husted. MIRL Report No. 68,

Mineral Industries Research Laboratory, University of Alaska, Fairbanks, Alaska, pp. 71–80.

Skudrzyk, F.J., S. Bandopadhyay, and S.C. Rybachek. 1987. Experimental Underground Placer Mining on Wilbur Creek. Presentation at 9th annual Alaska Conference on Placer Mining, Fairbanks, Alaska, March 18–25, pp. 89–120.

Skudrzyk, F., and J.C. Barker. 1986. Underground Placer Mining in Siberia. Presentation at 8th Annual Alaska Conference on Placer Mining, Fairbanks, Alaska, April 2–5, pp. 81–90.

Skudrzyk, F.J., J.C. Barker, D.E. Walsh, and R. McDonald. 1991. Applicability of Siberian Placer Mining Technology to Alaska. Final Report for the Alaska Science and Technology Foundation (Draft). ASTF Project 89-1-147, MIRL Report No. 89-10. Mineral Industries Research Laboratory, University of Alaska, Fairbanks, Alaska, February, pp. 35–40, 45–50.

Voytilla, E. 1992. Video presentation , 13th biennial Conference on Placer Mining, Fairbanks, Alaska, March 4–7 (copy in possession of author).

Wimmler, N. L. 1927. Placer Mining Methods and Costs in Alaska. Bureau of Mines Bulletin 529, pp. 113–133.

Ziegler, D., 1987. Drilling and Blasting in Permafrost at the Wilbur Creek Mine. In Proceedings of the Ninth Annual Alaska Conference on Placer Mining, ed. by M. Albanese and B.W. Campbell, pp. 229–234.

Room-and-Pillar Mining of Soft Rock

CHAPTER 12

Mining at IMC Potash Carlsbad

P. Livingstone,* J. Purcell,* D. Morehouse,* and D. Waugh*

12.1 INTRODUCTION

The IMC Kalium Carlsbad Potash Company (IMC Potash), a wholly owned subsidiary of IMC Global, Inc., owns and operates a potash mine 26 km (16 miles) east of Carlsbad, a small city (population 25,000) in the southeastern corner of New Mexico, USA. The mine utilizes seven shafts and holds potassium leases for over 28,800 ha (71,200 acres). Two types of ore, one containing sylvite (KCl) and the other containing langbeinite ($K_2SO_4 \cdot 2MgSO_4$), are mined at IMC Potash. These ores are processed through two separate refineries. In addition, the operation includes a langbeinite granulation plant, a muriate (sylvite) compaction plant, a plant that produces potassium sulfate (K_2SO_4), warehouses, and rail and truck load-out facilities.

During the mid-1970s, there were seven companies operating mines and processing plants within an area about 29 km (18 miles) east-west by 38 km (24 miles) north-south. Today, after consolidations and closures in the area, there are just two companies remaining, IMC Potash and Mississippi Potash, Inc.

The potash produced at Carlsbad is predominantly used as a source of potassium nutrients in agricultural fertilizers. Potash was first discovered in the Carlsbad area in 1925 during examination of cuttings taken while drilling for oil beneath salt beds. The first potash mine, belonging to the U.S. Potash Company, began operations in 1931. The Potash Company of America started another mine in 1933. The Union Potash and Chemical Company started sinking shafts for a third mine in the basin in 1936. Before the shafts and developmental work were completed, the International Minerals & Chemical Corporation (now IMC Potash) purchased the property. The first product from IMC Potash, two railroad cars of refined langbeinite, was shipped in October 1940.

Since opening, over 225 million tonnes (248 million tons) of material has been hoisted and processed through IMC Potash's refinery. This mine has produced material from four different ore zones (mining levels) over an area 14.5 km (9 miles) east-west and 19 km (12 miles) north-south. Approximately 4,800 km (3,000 miles) of drift have been developed at this operation.

Today, the IMC Potash operation has a capacity of over 1.36 million tonnes (1.5 million tons) of product from 8.15 million tonnes (9 million tons) hoisted each year. At 23,500 tonnes/day (25,000 t/d), IMC Potash is one of the highest capacity underground mines in the nation. Until last year, plant capacity was limited to an average production of 20,000 tonnes/day (22,000 t/d). In August 1999, IMC Potash commissioned a new 12,700-tonnes/day (14,000-t/d) langbeinite-processing plant and installed a second production hoist on the No. 2 shaft to feed the new plant.

Safety in an underground mine is very important. Significant time and resources are dedicated to developing a safety awareness culture throughout the operation. IMC Potash has implemented a behavior-based safety program directed and staffed by representatives from the hourly workforce. In 1999, IMC Potash achieved a total of 1.26 million work hours with only one lost-time accident. There was a 40% reduction in the number of accidents over the last 3 years, and a 35% reduction in workers' compensation costs over the same period. The Mine Safety and Health Administration's (MSHA) incident rate for reportable accidents was 2.31 for the first 9 months of 1999 compared with a rate of 7.46 for the mining industry nationwide, 4.63 for the Dallas district of MSHA, and 4.11 for the Carlsbad area.

12.2 GEOLOGY AND HYDROLOGY

The Carlsbad mining area is part of an evaporite sequence deposited in the Delaware basin of southeast New Mexico and southwestern Texas. The Upper Permian deposit in the Delaware basin is characterized by a thick accumulation of evaporites that have been subdivided bottom to top into the Castile, Salado, and Rustler formations (Figure 12.1). These evaporites were buried by the Dewy Lake red beds that marked the close of the Permian era.

The Salado Formation is essentially a halite unit interlayered with anhydrite, polyhalite, potassium salts, clay, and minor amounts of sandstone and siltstone. Its thickness varies in the Carlsbad area from zero along the dissolution boundary east of the city of Carlsbad to more than 600 m (2,000 ft) 24 km (15 miles) to the east. The Salado is divided into three units: the Lower (oldest), McNutt, and Upper (youngest). Economic potassium minerals are confined to 11 units in the McNutt Member. The most common potassium minerals are polyhalite, sylvite, langbeinite, carnallite, kainite, and leonite. The identified ore reserves of the McNutt Member are confined to an area east of Carlsbad, as delimited by the federal government's "Known Potash Leasing Area." This area covers more than 1,100 km^2 (425 mi^2), although potassium mineralization is estimated to cover an area ten times greater.

The anhydrite and polyhalite beds of the Salado Formation are persistent throughout the area. Forty-three beds were identified as formal marker beds by the U.S. Geological Survey (USGS) with beds 117 through 126 occurring within the McNutt Member. Several informal polyhalite intervals are present, such as the 2.5-cm (1-in) unit separating the third zone from the fourth. These markers are critical to stratigraphic correlation of the potash zones because of their persistence.

Potash zones were subdivided by the various mining companies into stratigraphic intervals separating mineralized beds from halite and clay-halite. The potash zones typically have a muddy interval at the bottom with varying upper boundaries between the halite, polyhalite, and/or clay. The mineralized intervals vary from centimeters to 1 m (3.3 ft) thick, and several intervals make up a potash zone. Stratigraphically, the potash

* IMC Potash, Carlsbad, NM.

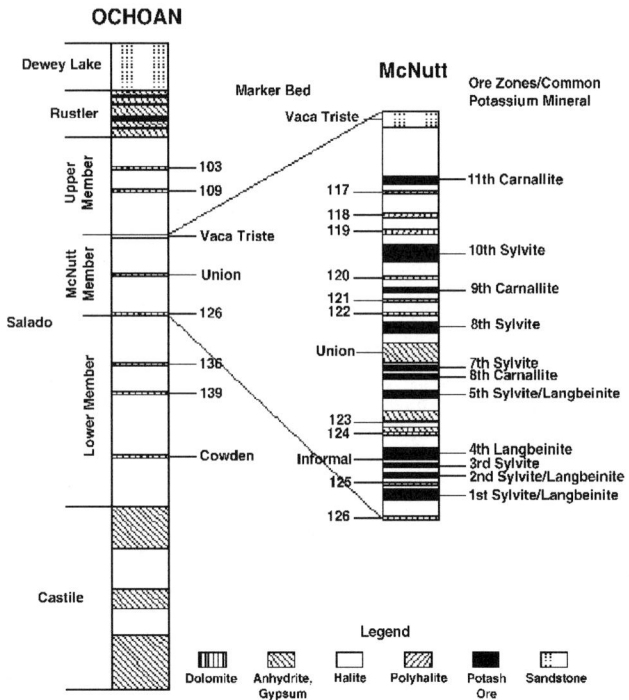

FIGURE 12.1   Stratigraphic column Castile/Salido/Rustler with detailed McNutt member of the Salado

zones vary from 1 to more than 6 m (3.3 to 20 ft) thick. Mining horizons are selected from these zones with regard for ground control, mining equipment constraints, and economic potash mineralization.

The principal economic minerals are sylvite and langbeinite. Economic quantities are confined to intervals 1 through 5, 7, and 10, of which 4, 5, and 10 are mined at IMC Potash (interval 1 was mined from 1940 until 1987). The fourth interval is the main langbeinite zone, although langbeinite mineralization occurs to some extent throughout most of the zones.

The 11 ore zones are consistent in their stratigraphic relationship to the marker beds; however, their thickness, mineralogy, and potassium mineralization are extremely variable. Regional variations in thickness and mineralogy occur gradually over thousands of meters, while local variations occur rapidly over intervals between a few meters to several hundred meters. Local features such as potassium-barren zones (locally called "salt horses") are common throughout all potash horizons. These barren zones may occur within individual mineralized beds or across entire potash zones, and their lateral extent may range from a few to thousands of meters. Stratigraphically, the clays, polyhalite stringers, and halite beds show little to no disturbance, with only the potash mineralization missing. Frequently, increased potassium mineralization and higher concentrations of less common minerals, such as kainite, loeweite, leonite, or kiesersite, are found along the margins of these barren zones. Recrystallized halite and blue halite are also common along these contacts.

The limits and distribution of potassium mineralization is interpreted to be the result of a series of primary and secondary processes. The present potassium distribution is mainly the result of secondary deposition from migrating brines. There is evidence that the brines migrated through the unconsolidated evaporite sequence controlled by aquitards such as the clay beds. Brines would flow horizontally until a channel was intersected that

allowed the brine to continue through the overlying strata. This erratic origin accounts for the irregular distribution of the potassium. This same process was responsible for the formation of the barren zones.

The strata overlying the mining intervals vary from zone to zone, but generally persist laterally over an individual zone for several kilometers. The overlying stratum is essentially halite with varying distributions of stringers and beds of polyhalite, clay, and clay-halite. When these stringers and beds are found near the back of an excavation, they act as planes of weakness and thus of possible separations.

Within the Salado Formation, isolated pockets of brine and air are routinely encountered. These pockets are most frequently found along polyhalite-clay units or markers, although pockets are intersected within or near the potash horizons mined. Although brine flows are small (less than 4 L/min [1 gal/min]), they may drip for years. Air, which is frequently associated with the brine pockets, is composed mainly of nitrogen. These pressurized air pockets are a mining concern as they may exert load on the overlying strata, contributing to poor ground conditions or rock failure.

The Salado Formation is overlain by the Rustler Formation in the Carlsbad Mining District. The Rustler Formation is composed mainly of dolomite, siltstone, anhydrite, and halite. The dolomite and siltstone units are the main water-bearing rocks.

Three aquifers exist in the Rustler Formation: the Salado-Rustler contact, the Culebra Dolomite, and the Magenta Dolomite (Hill 1996). These three zones are hydrologically separate with evaporite aquitards between them. The Salado-Rustler aquifer is estimated to produce only a few liters per day with transmissivities of only 1×10^{-2}/d (Mercer and Orr 1979; Mercer and Gonzales 1981) with total dissolved solids of >300,000 ppm. The Culebra and Magenta units have low transmissivities with maximums ranging from 1×10^2 m²/sec in the Culebra to 5×10^{-2} m²/d in the Magenta. The Magenta aquifer is the most likely to provide high-quality groundwater. The three aquifers offer little concern to mining in the Carlsbad basin.

12.3   GEOTECHNICAL INFORMATION

12.3.1   Rock and Ore Strength

Samples of langbeinite ore have yielded unconfined compressive strength values of 38 to 44 MPa (5,500 to 6,400 psi). Sylvite ores typically give values that are lower and are greatly reduced by the amount of insols, carnallite, and moisture in the sample. Approximate pillar stresses, in order of magnitude, can be calculated by the following formula:

Average pillar stress = (unit weight of the rock mass × overburden depth) ÷ (1 – percent extraction).

In areas where long-term use is desired, maximum pillar load is limited to 21 MPa (3,000 psi) in sylvite and 27 to 31 MPa (4,000 to 4,500 psi) in langbeinite. In-mine observations have confirmed that pillars will begin to deteriorate rapidly at greater than 40 MPa (6,000 psi).

12.3.2   Ground Support

Secondary ground support is accomplished primarily by the installation of mechanical roof bolts. Typical bolt diameters are 1.6 and 1.9 cm (⅝ and ¾ in), and lengths are 60 to 180 cm (2 to 6 ft). The bolts are grade 75 and use expansion shell anchors with shell lengths of 7 to 8 cm (2⅞ to 3¼ in). Roof bolt plates are generally 15 by 15 by 0.6 cm (6 by 6 by ¼ in).

Each production area has roof bolt and air relief hole drilling plans that are determined by mining dimensions and overlying strata. Plans are reviewed and updated as needed by mine production, engineering, and safety personnel. Bolting patterns are based on yield strength of the rock bolt and involve the beam or slab concept for bedded rock to determine the dead weight of the rock. A safety factor of at least 2 is used when formulating roof-bolting plans.

12.3.3 Air Relief Holes

Pockets of confined air under pressure are found within the rock strata in the mine. This "air," which is mostly nitrogen, occurs primarily in mud seams and at the base of anhydrite layers. Mud seams act as conduits for air to travel. The back in the underground excavations serves as a free face for the release of the air pressure. A confined air pocket can exert enough force on the formations above or below the workings to induce ground failure. To relieve the air pressure in the strata above the back, air relief holes extending up to 9 m (30 ft) are drilled. The 3.5-cm (1$\frac{3}{8}$-in) diameter holes are drilled by a roof bolting machine using extension steels. Typically, holes are drilled in a checkerboard pattern with one in every other intersection across a production panel.

12.3.4 Extraction Rates and Pillar Determinations

Extraction rates are primarily designed on the basis of the depth of the ore. Primary extraction is empirically designed to stay at or below pillar loads of 21 MPa (3,000 psi) in sylvite areas and 27 MPa (4,000 psi) in langbeinite areas.

12.4 MINING METHODS

IMC Potash has adopted several variations on the basic systematic room-and-pillar method common to coal and many nonmetal mines. The two basic classifications, which are best defined by the type of equipment they employ, are conventional and continuous mining. Continuous mining is further categorized into car haulage and continuous conveyor haulage. All three variations are contingent on a relatively horizontal ore body. Physical dimensions and the degree of consistency of a particular ore body are important factors in dictating which variation is used. Prior to 1994, conventional mining had been used almost exclusively for production, with the exception of some early-generation boring and ripper machines, circa 1960. Up to seven conventional sections were in production at one time. Currently sylvite ore production consists of five continuous miner sections, while langbeinite production consists of four continuous miners and one remaining conventional section.

12.4.1 Conventional Mining Method

As in coal mining, conventional mining at IMC Potash is defined as drill-and-blast mining. It is accomplished through the following five basic operations: drilling, undercutting, blasting, loading and hauling, and roof support. With the advent of improved continuous mining technology, only one conventional mining panel remains active.

The layout of a typical conventional mining section consists of either a single conveyor belt feeder with four rooms each to the left and right (a total of nine advancing rooms), or a two-belt variation of this layout (18 advancing rooms). Various room-and-pillar dimensions have been used as conditions have dictated. Rooms are typically mined from 8.5 to 9.8 m (28 to 32 ft) wide and 2 to 4 m (6.5 to 13 ft) high. Pillars have typically been square, ranging from about 9 by 9 m (30 by 30 ft) to 16 by 16 m (52 by 52 ft), although rectangular configurations have also been used.

12.4.2 Continuous Mining Method

To understand the application of continuous mining at IMC Potash, a brief history is in order.

As technological advances were made in continuous miner design, it became clear that continuous miner production had an economic advantage over conventional mining. Drum-type

miners have been cutting sylvite ore in the Carlsbad area since the mid-1960s. These proved to be effective in the softer sylvite beds with relatively little sulfate, but less productive in higher-sulfate ore bodies. The sylvite ore bodies at IMC Potash have high sulfate contents; thus the transition to continuous miners has been slower than in many other potash mines.

The use of conventional equipment requires a minimum mining height of about 2 m (6.6 ft), which limited the economic recovery of sylvite reserves. In 1994, two Joy 14PM 10-drum miners were placed into production in a low-height, low-sulfate area of the USGS's 10th ore zone. Within 3 years, two newer, more robust models of drum miners, the Joy 14PM-15 and 12HM-27, were acquired for mining sylvite ore containing more sulfate minerals as well as langbeinite ore.

12.4.3 Continuous Mining with Continuous Haulage

As in the case of conventional mining, much of the equipment and technology used in continuous mining at IMC Potash has been adapted from the coal industry. Since the first two continuous miners were justified by their ability to mine at lower heights than conventional equipment, face haulage equipment was also selected to meet this constraint. Chain-conveyor-type continuous haulage was chosen because of its advantage for lower mining heights. As larger, more robust continuous miners were selected to mine langbeinite ore, the concept of continuous haulage was carried over to the low-seam portion of the langbeinite reserves. Currently, there are five active continuous mining-continuous haulage sections.

Each panel consists of a drum-type continuous miner and a multiunit, mobile chain conveyor system. The haulage system consists of three 7.6-m- (25-ft-) long, crawler-mounted mobile conveyor units with three 12.2-m- (40-ft-) long bridge conveyor units piggybacked onto the mobile units. Each of the six components has separate conveyor chains, but they are pinned together at transfer points. The unit closest to the miner is the "hopper" mobile unit, which is physically detached from the conveyor discharge boom of the continuous miner. The final bridge conveyor is pinned to a dolly that rolls on the panel belt structure frame. This frame is typically about 67 m (220 ft) long and includes the panel belt tailpiece. The tailpiece gives the haulage system the capability of backing alongside the panel belt. When fully extended, the haulage system reaches about 61 m (200 ft) from the panel belt.

There have been two panel designs used: production layout and development.

A typical production layout consists of a sequence that advances to the limits of the ore body or property boundary and is then retreated. The advancing sequence (Figure 12.2) has an extraction ratio of about 40%. After these rooms are completed, the belt tailpiece and rigid frame structure is advanced 49 m (160 ft), and the sequence is repeated. Belt storage units are set in-line near the head of the panel belt to expedite moves. The retreat sequence (Figure 12.3) is characterized by a chevron pattern. The first belt move is a 24-m (80-ft) retreat. After the first retreat pattern is mined, subsequent moves of the retreating belt are on 49-m (160-ft) centers. The extraction ratio of the retreat pattern shown is also about 40%, resulting in an overall extraction of 80%. The individual panel layouts are normally arranged side-by-side so that rooms driven perpendicular to the belt line will connect for purposes of ventilation.

Development mining patterns with continuous haulage have been used to open up main conveyorways and roadways. Development is typically within the ore body using an advance-only sequence. Development extraction is lower than in the production sequence because these rooms must remain accessible for long periods of time (Figure 12.4). Two separate development panels are driven parallel to each other, with rooms

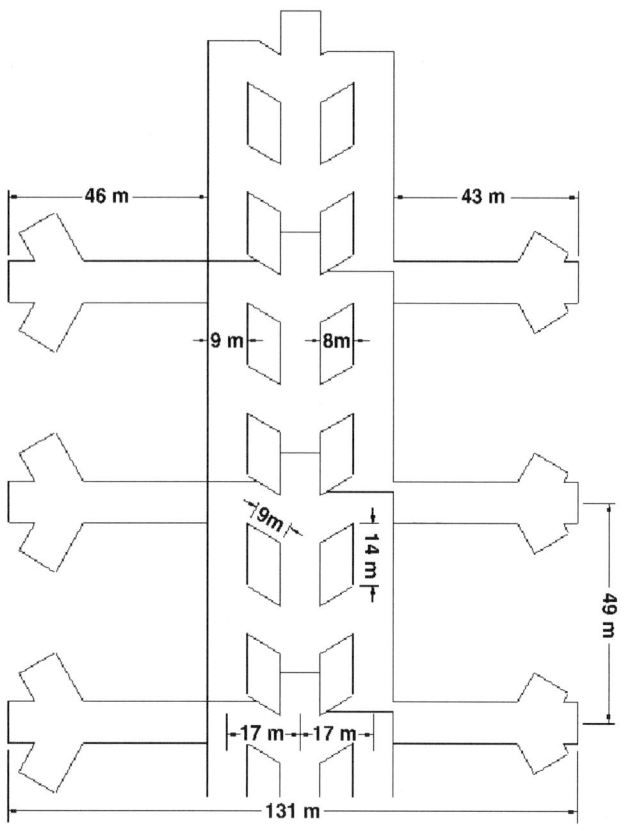

FIGURE 12.2 Continuous haulage advance sequence—three entry, 17 m × 49 m centers, 9 m rooms, 60 degree chevron showing three advance sequences

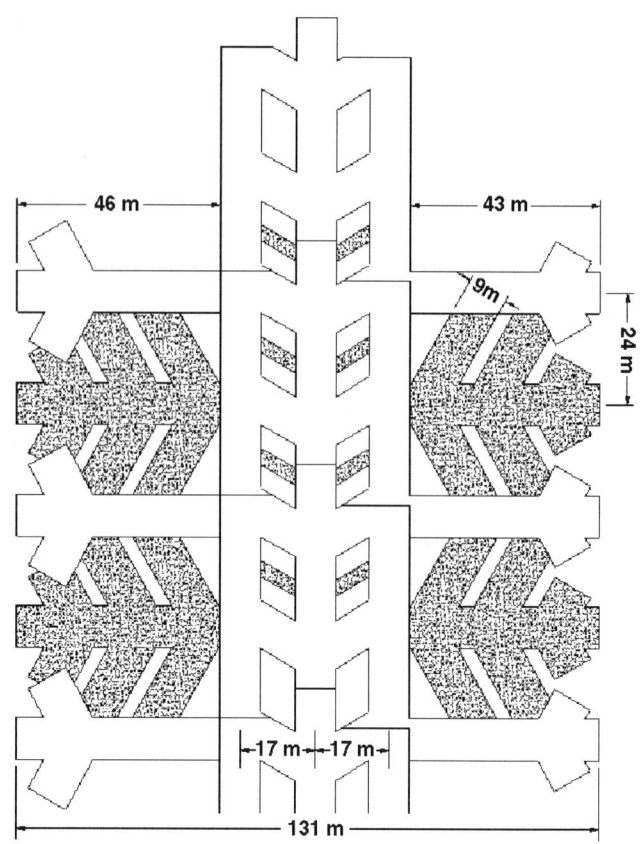

FIGURE 12.3 Continuous haulage retreat sequence—three entry, 17 m × 49 m centers, 9 m rooms, 60 degree chevron showing two retreat sequences

driven at right angles to the beltline designed to line up with each other. This overcomes an inherent problem with continuous haulage systems—the distance mined to the sides of the panel belt is not enough to set up the haulage system and belt structure to turn off a panel. The panel belt of the first development pass becomes a main belt, while the second pass opens up ground to give sufficient distance between the main belt and the faces.

12.4.4 Continuous Mining with Car Haulage

Continuous mining with a drum miner and diesel cars involves a layout nearly identical to that for a conventional section. A single panel belt with a belt feeder is positioned in the middle of the panel, typically nine rooms wide. Room and pillar dimensions are essentially the same as those in conventional panels. Mining progresses by cycling through the faces from right to left across a section. Each face is advanced a predetermined distance that varies between panels and is determined by bolting and ventilation restrictions. Car haulage has the advantage of increased flexibility in comparison to continuous haulage. This flexibility allows mining to adjust quickly to erratic variations in mineralization and the presence of salt horses. Mining can be more selective because mobility is greater. In addition, waste can be cut and gobbed to maximize the ore feed grade.

12.4.5 Underground Materials Handling and Hoisting

The ore is transported from the working panels over an extensive underground belt conveyor system consisting of over 56 km (35 miles) of belt. With two types of ore hoisted from the mine at two different shafts, two completely separate belt systems are used, and two types of storage devices are employed underground (raises and horizontal ore storage [stackers]). Raises are large

vertical openings in the salt with capacities of 80 to 1,300 tonnes (90 to 1,400 t) of ore. A feeder at the bottom of each raise discharges ore onto a conveyor. Problems with ore bridging in these raises have repeatedly required explosives to break material free.

To increase storage capacity and reduce the ongoing problems with raises, IMC Potash designed and installed a succession of three stackers. Stackers are ore storage systems that windrow ore on the floor parallel to a conveyor and then reclaim that ore upon demand. A stacker weighs up to 45 tonnes (50 t) and trams itself horizontally on rail in a gallery up to 600 m (2,000 ft) long. It is composed of a tripper (which removes the ore from the belt), a stack belt (which deposits ore into the center of a room), a feeder (which picks the ore back up), and a reclaim belt (which returns the ore to the conveyor). Stackers can reclaim ore at a controllable rate independent of the rate of incoming ore. The largest stacker underground holds about 4,500 tonnes (5,000 t) of ore, can stack ore at the rate of about 45 tonnes/min (50 t/min), and recovers ore at a precise rate to meet hoisting requirements.

The sylvite ore is transported to the No. 1 shaft by way of a conveyor system that uses 107-cm (42-in) wide belts, two stackers, and two raises. Ore from the langbeinite areas is fed onto 107-cm (42-in) wide feeder belts to a 137-cm (54-in) wide main conveyor. The main conveyor belt carries the ore to the No. 2 shaft 10.5 km (6.5 miles) away.

The new 1.37-m (54-in) belt system has single and dual 225-kW (300-hp) FMC drives. These drives utilize alignment-free gear boxes. The entire belt line is suspended from the back to facilitate cleanup. The new production hoist at the No. 2 shaft is actually a 1950 model, 3.35-m (11-ft) diameter, double-drum

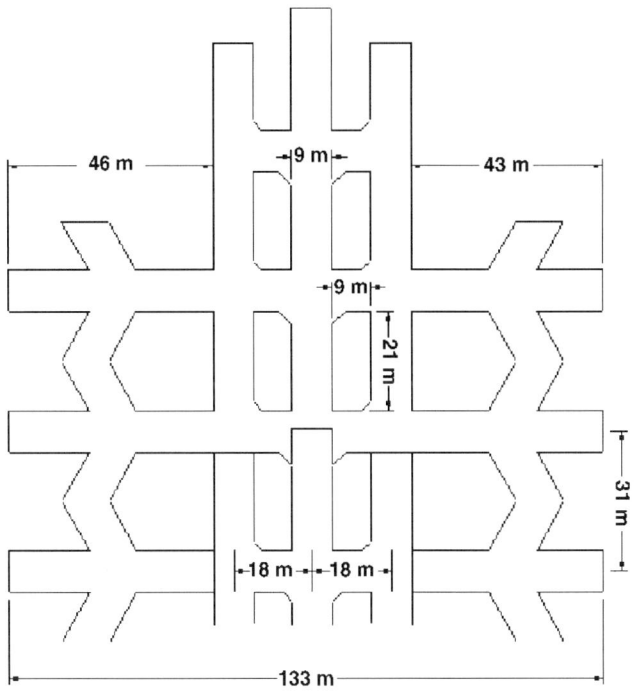

FIGURE 12.4 Continuous haulage development sequence–three entry, 18 m × 31 m centers, 9 m rooms, 60 degree chevron showing three advance sequences

Nordberg hoist. Two 745-kW (1,000-hp) dc motors, power supply, and controls were purchased new from ABB. The hoist is automatic, and the hoistman is stationed about 1.5 km (1 mile) away at the No. 1 shaft. The hoist lifts the ore from the new skip loader approximately 290 m (950 ft) to a point where it is discharged into the new 900-tonne (1,000-t) mass flow ore bin. This hoist uses 41-mm (1⅝-in) ropes, 13-tonne (14-t) capacity skips, an Ultimate Hoist Monitor from ABB, and Nordberg hydraulic braking controls.

12.4.6 Grade Control and Dilution

Grade control in each mining section is based on visual inspection in combination with the collection of belt samples during each shift. Sylvite ore zones typically consist of both red and white sylvite. Some zones are a relatively pure sylvite and halite mixture, but most have at least some langbeinite, kainite, loeweite, leonite, or kieserite present. Likewise, langbeinite ore zones typically contain sylvite and at least trace amounts of the other minerals. Visual estimates of potassium content present as sylvinite or langbeinite is, therefore, often challenging. Gamma-radiation measuring devices, which are being used in some potash mines as "potassium meters," have not been adopted at IMC Potash because of the complex mix of potassium minerals within the ore.

Consistent mud seams or mineral marker beds, such as polyhalite, either within the ore body or in the immediate roof or floor, are usually present within each ore zone and aid in controlling mining intervals. In conventional sections, such a marker bed usually provides a good discontinuity for the blast fracture zone and allows for a smooth back. In continuous miner panels, such a marker bed in the immediate roof must be cut out or it may create the potential for separations and roof falls.

The advent of continuous miners has been a benefit in controlling dilution. In addition to reducing mining height, the back and floor can be held to closer tolerances than with conventional mining.

12.4.7 Productivity

The underground operation employs 45% of the total hourly and salaried workforce at IMC Potash. Maintenance and electrical workers account for 30% of the underground employees.

The mine is organized by mining level. The sylvite and langbeinite operations have separate production and maintenance personnel, but a common main shop and electrical and belt groups. The long distances and travel times limit substantial sharing of labor resources between levels. The two extreme production areas are 32 km (20 miles) apart, which requires over 1 hour of travel time.

Since the deployment of continuous mining systems, employment levels have been reduced by 7% while production has increased by over 25%. The annual production capacity is 8.1million tonnes (8.9 million tons) of combined sylvite and langbeinite ore.

The operation is organized around two 10-hour production shifts per day with the 4 hours between shifts used for a preventive maintenance program. The shifts are staggered to maximize the maintenance "window" available. This window is also used for belt work, moves of electrical equipment, and any fill-in bolting. Each production area has one scheduled downshift per week for major maintenance repairs and routine preventive maintenance. The continuous miners are rebuilt offsite when they have produced approximately 3 million tonnes (3.3 million tons) of ore. A completely rebuilt continuous miner is placed into service prior to hoisting a worn miner to the surface for shipment.

12.4.8 Mine Ventilation

The underground ventilation system at IMC Potash is extensive and complex. Open mine workings are spread over 14,245 ha (55 mi²) of land on four working levels. There are two separate ventilation systems. The north system, providing approximately 2,800 m³ (100,000 ft³) of air per minute, uses the No. 1 shaft as the intake shaft and the No. 2 shaft as the exhaust shaft.

The south system uses the No. 5 and No. 7 shafts as intakes, providing approximately 21,000 m³ (750,000 ft³) of air per minute. The air is exhausted from the No. 3, No. 4, and No. 6 shafts. The entire mine, including old workings, is ventilated. Stoppings consist of wooden or muck bulkheads and, closer to production panels, curtain lines. Stoppings normally separate the intake air entries and main travelways from the exhaust rooms and conveyor belt lines. Ventilation doors are located at most belt drives to provide access to the belt lines.

Ventilation to the working areas is accomplished with bulkheads, curtains, and waste rock walls foamed at the top to separate intake from exhaust air. Production panels utilizing diesel-powered haulage equipment require auxiliary fans in the last open breakthrough. There are usually three 120-cm (48-in), 11-kW (15-hp) fans at even intervals across the panels to help direct air. In addition, each heading or room is equipped with a 60-cm (24-in), 2.2-kW (3-hp) fan to direct air to each working face.

In the continuous miner panels utilizing continuous haulage, a 90-cm (36-in), 22-kW (30-hp) fan is positioned in the exhaust side, and a curtain line is carried up close to the face. A 60-cm (24-in), 2.2-kW (3-hp) portable fan is used on the intake side.

12.5 SAFETY- AND HEALTH-RELATED ISSUES

IMC Potash is subject to the Federal Mine Safety and Health Act (MSHA) of 1977. The mine must abide by regulation 30CFR, Part 57, which sets forth mandatory safety and health standards for underground metal and nonmetal mines. MSHA conducts quarterly inspections of each working area in the mine.

The mine has been has been designated as a Category IV "gassy" mine by MSHA. Category IV means that, based on the history of the mine, the ore is noncombustable and that the concentration of methane is neither explosive nor capable of forming explosive mixtures with air. Testing for methane is required at least once a shift prior to starting work at each face and upon initial release of gas into the mine atmosphere from boreholes. However, the use of permissible equipment (30CFR, Part 18) is not required.

12.5.1 Oil and Gas Drilling Regulations

Below the Salado Formation are commercial oil and gas deposits. The Bureau of Land Management (BLM) and the State of New Mexico, while promoting multiple use and the conservation of resources, have established regulations protecting potash deposits from oil and gas drilling and from the associated risk of explosive quantities of methane entering the mine workings. Current regulations protect both the open mine workings and "life of mine reserves" (LMR) by restricting any oil and gas drilling in those reserves and adjacent buffer zones. Current mine maps and LMR reserve maps are updated annually and filed with the BLM and New Mexico. All data on the reserves are held in strict confidence.

12.5.2 Training

In addition to the 40-hour new miner training as required by MSHA, all new underground employees, including those with experience, receive a minimum of four additional weeks of training when hired. The training is "hands on" and is taught in a production area designated as a training panel that does not have to meet daily production requirements. The time spent with equipment is used to become familiar with the safe operation of machinery and to learn standard operating practices and procedures. All training is the under the direction of a training supervisor and is taught by qualified instructors.

Supervisor development is accomplished through on-going training programs. Quality management teams, made up of both hourly and salaried personnel, have been formed to look at various issues and bring about improvements throughout the mine.

12.5.3 Safety Programs

The safety of all employees is a primary concern to IMC Potash. An employee-driven, behavior-based system incorporating observation and positive feedback is one innovative process used at IMC Potash to increase safety awareness. Daily tool-box meetings, monthly safety meetings, quarterly area inspections, supervisor safety tours, a personal safety plan program, pre-operation equipment check lists, and quarterly incentive award programs are also part of the overall safety program. An 8-hour, off-site annual refresher course is given once a year to all employees. The result of these activities is a 1999 MSHA reportable-incident rate of 2.31, compared to the nationwide rate of 7.46. Working over one million man-hours without a lost-time accident has been achieved several times in the recent history of the mine.

12.6 COSTS

12.6.1 Operating Costs

Total mining costs represent 35% of total plant costs. Mine maintenance accounts for less than 30% of the operating mine cost. With the use of continuous miners, mining costs from production materials, maintenance materials, labor, and dilution of ore grade were reduced by approximately 25% from conventional mining. The use of continuous miners also resulted in improved ground conditions, which in turn meant a reduction in time spent barring down ribs and backs and less spot-bolting off-pattern.

Everyone from foremen up takes part in mine budgeting and cost control. The total workforce is held accountable for costs within their areas, along with safety, tonnage and grade control.

12.6.2 Recent Capital Expenditures

Recent capital projects, totaling about $150 million, have improved the capacity, recovery, and operating costs of producing the langbeinite product. These projects included the acquisition of the contiguous Western Ag-Minerals mine for $45 million, construction of a granulation plant for over $25 million, and construction of a new langbeinite processing plant for $55 million. Approximately $16 million was spent on the underground portion of this project, which included the 1.37-m (4.5-ft) mainline conveyor, 4,500-tonne (5,000-t) stacker, skip loader, refurbishment of the No. 2 shaft (including conversion to steel guides), moving and refurbishing the hoist, new headframe, and a 900-tonne (1,000-t) surface bin. The new langbeinite plant was commissioned in August 1999.

12.7 CONCLUSION

IMC Potash approached the new century with requirements to optimize the reserve base, increase productivity, and reduce cost per ton of ore. This was successfully achieved by utilizing continuous miners and adapting mine layouts and support equipment to match the complexity of the ore body. Maintenance and production schedules were organized around production windows using 10-hour shifts. The management organization was flattened and direct supervision reduced by placing more responsibility in the hands of the crew leaders. The behavior-based safety program—combined with in-depth, root-cause investigations of all incidents—and supervisor safety tours has improved safety and successfully fostered other employee-driven programs.

IMC Potash's mining operations will continue to find ways to improve safety, increase productivity, and reduce costs. Immediate goals include increasing cutting time per continuous-miner shift, extending the rebuild life of the continuous miners, identifying and testing ways to reduce manpower on continuous haulage units, and continued testing to reduce bit costs.

12.8 REFERENCES

Hill, Carol A. 1996. Geology of the Delaware Basin Guadalupe, Apache, and Glass Mountains New Mexico and West Texas. Permian Basin Section, SEPM. Publication No. 96–39.

Mercer, J.W., and D.D. Gonzalez. 1981. Geohydrology of the Proposed Waste Isolation Pilot Plant Site in Southeastern New Mexico. In Environmental Geology and Hydrology in New Mexico, S. G. Wells and W. Lambert, eds. New Mexico Geological Society, Spec. Publication 10, pp. 123–131.

Mercer, J.W., and B.R. Orr. 1979. Interim Data Report on the Geohydrology of the Proposed Waste Isolation Pilot Plant Site, Southeastern New Mexico. U.S. Geological Survey, Water Resources Investigation Report 79–98.

CHAPTER 13

Mississippi Potash, Inc.'s, Underground Operations

Victoria Herne* and Tom McGuire*

13.1 INTRODUCTION

Mississippi Potash, Inc., located in the Carlsbad potash district, is one of the largest producers of potash in the United States. Potash is the common industrial term that refers to various types of potassium salts, which are vital nutrients essential for plant growth. The company currently produces potash from two underground mine operations, the East Mine and the West Mine. Mississippi Potash, Inc., produces approximately 1.1 million tons of ore annually, including both red and white potash. Approximately 60% of Mississippi Potash, Inc.'s, final product is used for agricultural purposes, and 40% of its final product is used for industrial purposes. The agricultural products are primarily fertilizers, and the industrial products include animal feed, potassium nitrate, and sodium hydroxide.

Prior to the First World War, potash was primarily an imported item. After the war started, it became essential that the United States develop a domestic supply to ensure the availability of fertilizer. In 1925, the Snowden-McSweeney Oil Company found crystals of the potash mineral sylvite in oil-well bailings near Carlsbad, New Mexico. This discovery led to the establishment of the present potash industry near Carlsbad.

The first company to begin work in the Carlsbad basin was the United States Potash Company. The first shaft was started in 1929 and completed in 1931 to a total depth of 986 ft. The first carload was shipped in 1931 and consisted of manure salts containing 25% K_2O. In 1932, construction of the refinery was started near the Pecos River, about 13 miles southwest of the mine. This location was selected for two reasons: it had a dependable water supply and it was close to the Santa Fe Railroad. A narrow-gage railroad was constructed to connect the mine to the refinery. This landmark mine is still operating today and is now Mississippi Potash, Inc.'s, West Mine.

Throughout the years, the potash industry has experienced market highs and lows, which resulted in shutdowns and cutbacks in several mines within the basin. Today the small independent operators have become extinct, leaving only two operations, Mississippi Potash, Inc., and International Mining Corporation-Kalium.

The acquisition of Mississippi Potash, Inc.'s, operations started in 1974, when its parent company, Mississippi Chemical Corporation, purchased the Mississippi Chemical Mine from Teledyne Corporation. In the fall of 1996, Mississippi Chemical Corporation purchased two additional mining operations from Trans Resources, Inc.: the New Mexico Potash Mine and the Eddy Potash Mine.

In 1997, the Eddy Potash Mine was shut down. At this time also, the New Mexico Potash Mine was renamed the Mississippi Potash East Mine, and the Mississippi Chemical Mine was renamed the Mississippi Potash West Mine. The new names were derived

from the mine's locations. Both mines are east of Carlsbad on Highway 62-180. The West Mine is only 24 miles east of Carlsbad, while the East Mine is 35 miles east of Carlsbad. Today, Mississippi Potash, Inc.'s, underground operations employ 515 people.

13.2 SAFETY

Safety is a key aspect of mining that has become essential for productive operations. Mississippi Potash, Inc., has always emphasized the importance of safety and strives to ensure that all of its operations meet high standards. The company's commitment to safety has been acknowledged by both the New Mexico State Mine Inspector's Office and the Mine Safety and Health Administration. In recent history, Mississippi Potash has been awarded for its outstanding safety record. Citations include three Sentinels of Safety awards in the underground nonmetal division and six awards for Operator of the Year for Large Underground Mines.

The Carlsbad potash operations are relatively free of the hazards of underground mining due primarily to the nongassy salt deposits in which the ore body is located. Convergence and floor heave are controlled by engineering mine layouts, opening widths, and pillar sizes to ensure safe and efficient extraction of the potash. When clay seams occur in the salt formation, they form zones of weakness that are controlled during mining operations by roof bolts or cribs.

It is also common practice to ensure that the back, or roof of the mine, consists of a salt layer rather than mud layers. This is often accomplished by either mining the clay, which results in a diluted grade of potash, or dumping (gobbing) the clay in mined-out areas. A salt layer in the back is preferable to clay because, unlike mud, the salt layers do not easily separate from the upper strata. Air pockets associated with clay seams occur and are routinely relieved by drilling small-diameter holes into the upper zones.

13.3 GEOLOGIC DESCRIPTION

Southeast New Mexico is characterized by a desolate landscape of mesquite bushes and native grasses. The general rolling topography is marked by scattered, irregularly shaped sinkholes. The West Mine plant lies on the edge of sandy plains, with lowlands to the south and west, in the northwest arm of the Y-shaped Nash Draw. This is a topographic depression approximately 19 miles long and 4 miles wide bounded by the Quahada Ridge, the Livingston Ridge, and the Maroon Cliffs. The East Mine plant lies east of Nash Draw on the northwest shelf of the Delaware basin (Figure 13.1). The drainage pattern is generally to the southwest. The red sandstone marking the landscape is of Triassic age. In Nash Draw, the Dewey Lake redbeds crop out in the draws and gullies (Figure 13.2).

---

* Mississippi Potash Inc., Carlsbad, NM.

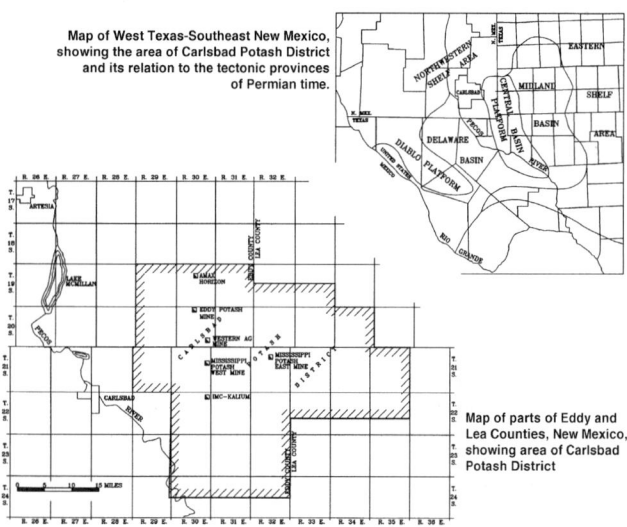

FIGURE 13.1   Map of the East Mine

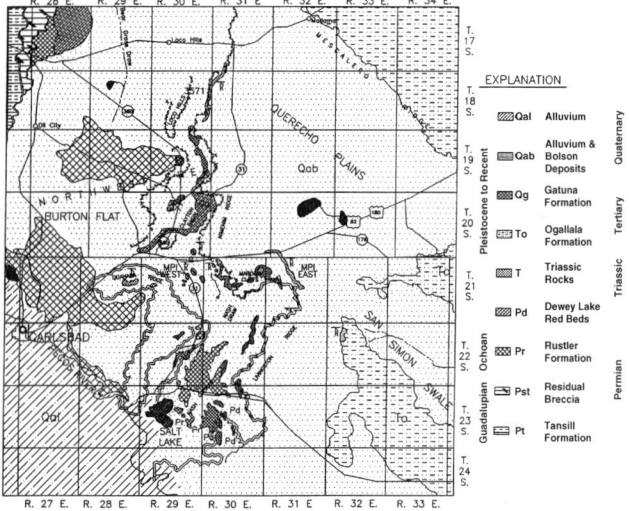

FIGURE 13.2   Geologic map of southeastern New Mexico

Few people who pass through this area are aware that approximately 1,500 ft beneath the Chihuahuan desert lies a vast infrastructure of room-and-pillar mines in the potassium-bearing salt beds. This region is called the Carlsbad Potash Enclave. The U.S. Geological Survey has identified 12 potassium-bearing ore zones within the Carlsbad Potash Enclave. The two mines are currently operating in three of these ore zones. The West Mine is mining in the fifth and seventh ore zones, while the East Mine is mining in the tenth ore zone. The Salado and Rustler formations primarily characterize these ore zones. Variations in the mineralogy of individual formations are common.

The ore is an intimate intergrowth of crystalline halite (sodium chloride-NaCl) and sylvite (potassium chloride-KCl) in various proportions. Most of the ore contains 50% more halite then sylvite. The sylvite ore is milky or faintly bluish gray, but many grains are stained red by iron oxide. The halite is clear, grayish, or orange-yellow with occasional red staining. Occasionally, blue halite is found associated with the sylvite. The easiest way to distinguish between the two ores is to taste them.

The potash salts will leave a distinct burning sensation, and the halite will taste salty.

The Rustler Formation is composed chiefly of anhydrite, commonly altered to gypsum. Other beds within the formation are dolomite, sand and gravel, limestone, and shale. The thickness and characteristics of the Rustler Formation vary. At the East Mine, it has a thickness ranging between 35 and 360 ft, with an average thickness of 270 ft. At the West Mine, the same bed is 512 ft thick at the main shaft. Although it is almost entirely covered by sand, gypsite, caliche, and soil, the Rustler Formation forms the bedrock in the vicinity of the West Mine.

The Salado Formation is composed of hundreds of distinct beds, which may be assigned to five principal rock groups. These groups are argillaceous halite (10% clay), clean halite, calcium sulfates, clay, and potash beds. The potash beds are a combination of sylvite, langbeinite, leonite, kainite, and carnallite. Clean and argillaceous halites represent at least 80% of the formation. Each rock type has a considerable range of thickness. Individual clean and argillaceous halite beds may be a few inches to 20 ft thick. Sulfate beds range from thin, discontinuous seams to 30 ft with an average thickness of about 1 ft. Clay beds are generally only a few inches thick. The thickness of the potash beds is between 6 in and 14 ft.

The potash deposits at the West Mine consist of mixed sylvite and halite ores in two distinct zones within one of the flat-lying halite beds. This bed is located near the middle of the Salado Formation. Thin zones of enriched potash-bearing minerals are located within the 150-ft-thick deposit. Mining activities currently take place in the fifth and seventh ore zones. Thickness, grade, and tonnage limit exploitation of the other potash-bearing beds. In most parts of the deposit, the vertical change from ore to barren salt is abrupt, while the lateral transition at the edges of the ore body is gradual. Barren masses of halite, known as "salt horses," are scattered irregularly throughout the ore body.

The East Mine is located primarily in the tenth ore zone. This zone is subdivided into three distinct layers: top, bottom, and middle. The layers are 2, 3, and 4.5 ft thick. The bottom member is the almost always mineralized, the middle member is almost always barren, and the top member may or may not be mineralized. The common minerals found in the tenth ore zone are halite, sylvite, clay (montmorillonite), sulfate minerals, and carnallite. The eastern sections of the mine have large deposits of carnallite and kieserite. The tenth ore zone is also characterized by isolated pods of barren clays ranging from a few square feet to several hundred-thousand square feet in area. These pods appear at random, and no way has been discovered to predict their occurrence.

13.4   MINING METHODS

Although mining techniques vary between the East and West mines, some aspects of the underground operations are similar. The roof control and underground haulage operations are very similar. In addition, both operations have areas developed for maintenance facilities, including a main shop, a diesel shop, an electrical shop, a warehouse, and production offices.

Roof control is an essential part of the underground mining operation. As mining advances into new areas, the back and ribs are scaled to remove any loose materials. After the scaling is completed, a series of sounding tests are conducted to determine the competency of the overlying strata. Vertical holes are drilled into the roof in each crosscut or mine intersection to relieve pressure in the overlying mud seams. All holes are drilled and all bolts are installed by either a diesel or electrical roof bolting machine. The West Mine often uses a set pattern of bolting on 5-ft centers in alternating rows of four and five bolts. Pattern bolting is utilized primarily in the fifth ore zone. The other areas in both

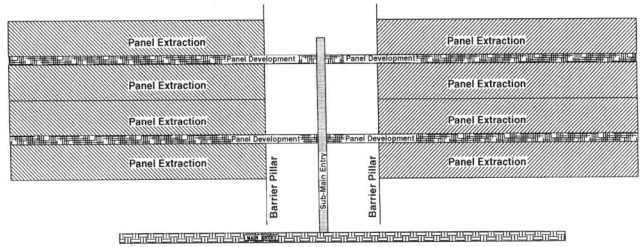

FIGURE 13.3   Mining sequence

mines bolt only when weak areas are found. The rock bolts are 2 to 6 ft long and $5/8$-in diameter. Rock anchorage for the bolt is achieved by an expansion shell gripping the competent strata. After inserting the bolts into the holes, the bolts are subjected to a torque of 180–200 ft-lb.

All areas of both mines are equipped with roof-mounted, cable-supported belt conveyors. The East Mine belts terminate within 500 ft of the active mining faces. These belts are 42 and 48 in wide and travel 450 to 600 ft/min. Troughing idlers are spaced every 5 ft along the cable, and return idlers are spaced on 10-ft centers. Near the shaft, the main line belts discharge into one of three 750-ton-capacity storage pockets. The feed into the storage pockets is automatically switched by means of a tripper system that fills each pocket in succession. The potash is then moved from the ore pockets by means of vibrating feeders and a sloped conveyor belt to the automatic skip loading station at the production shaft. At this point, ore is fed into skips and hoisted to the surface.

The West Mine belts terminate within 250 ft of the active mining faces. The belts are 36 and 42 in wide and travel 430 to 730 ft/min. Troughing idlers are spaced every 5 ft along the cable, and return idlers are spaced on 15-ft centers. Near the shaft, the main line belts discharge into a 125-ft storage area. From this point, the potash is transferred to one of two 120-ton ore pockets. These pockets discharge ore into skips, which are hoisted to the surface.

13.4.1   West Mine Mining Methods

Mississippi Potash's West Mine is laid out in a chevron pattern (Figure 13.3). The submains and panel development consist of two entries, Room 1 and Room 2. These entries are connected by break-throughs. The breakthroughs in the submains and panels are cut on 125- and 46-ft centers, respectively. Room 1 is used as the primary exhaust airway and a secondary escapeway. Room 2 is used for the primary airway (30,000 ft³/min), travelway, and escapeway. It is also used for primary ore haulage. The sequence for developing and mining a new area consists of four basic steps: developing the submains, developing the panels, extracting the panels, and extracting the barrier pillars (Figure 13.4).

The submains are located in the center of the ore zone and extend for the length of the deposit. Ideally, the submains are completely developed before panel work starts. The average mining heights for both the submains and panel development is 6.25 ft. The preferred location of the first panel is in the center of the submains. It is essential for safety reasons that a panel is not mined between two mined-out panels. By locating the first panel in the center of the submains, mining can proceed in either direction without violating this safety rule. The panel length is estimated using the ore reserve models; however, grade and ore thickness are continually monitored to determine the final length of the panel.

After the panel is developed, miners start retreating the extraction panels. Unlike in development areas, the heights of the extraction panels are variable, depending on the overall thickness of the ore layer and what is above it. Typically, either a salt layer

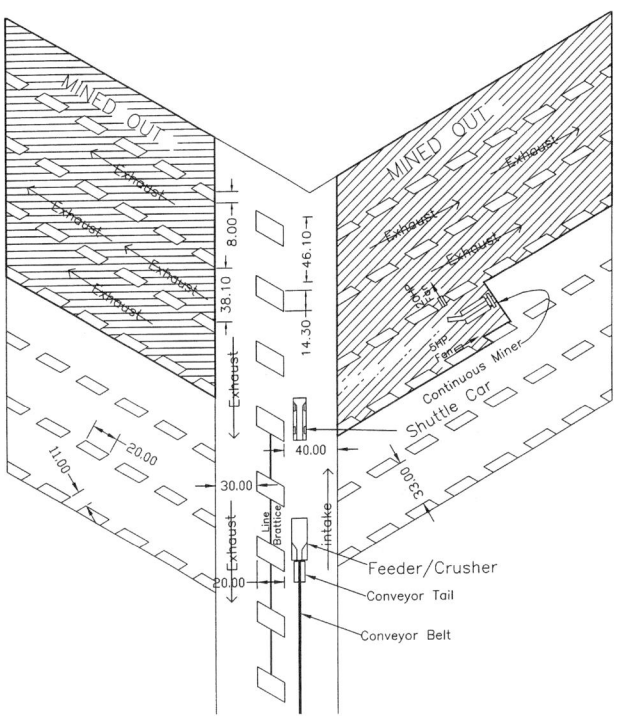

FIGURE 13.4   Typical configuration

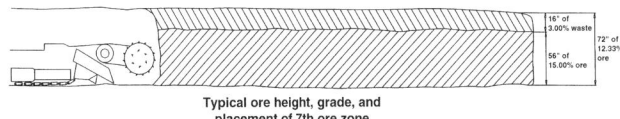

Typical ore height, grade, and placement of 7th ore zone

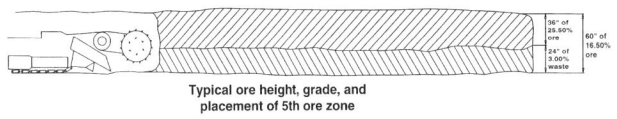

Typical ore height, grade, and placement of 5th ore zone

FIGURE 13.5   Mining heights or ore zones

or a mud layer will be found above the ore. Salt layers provide a sound back and do not need to be cut out. On the other hand, the mud layers do not provide a stable back and must be mined.

The seventh ore zone tends to have thick mud seams above the ore. In this area, the mud is not dumped onto the belts. Instead it is gobbed or placed in mined-out areas. The height of extraction ranges between 5 and 9 ft because of the varying thicknesses of the mud seams. In the fifth ore zone, the mud seam is absent or minimal, and is usually mined with the ore. Here, the mining height is much lower and ranges between 4.7 and 5 ft (Figure 13.5).

The final step of the mining sequence is extraction or robbing the barrier pillars. The barrier pillars are robbed after all the panels have been extracted and mining is going to retreat permanently from the area.

Generally, Mississippi Potash, Inc., does not have a cut-off grade or minimum ore thickness. Each area is monitored as it is developed, and the decision to start extracting is based upon the overall mine conditions at that time. As the panel is developed, ore samples from the belt lines are tested continually to determine the grade.

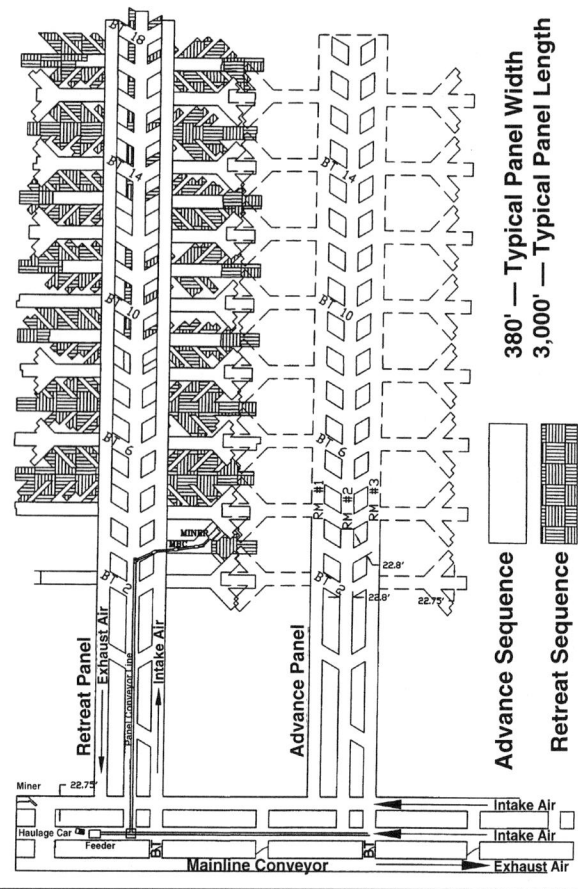

FIGURE 13.6   East Mine's three-entry system

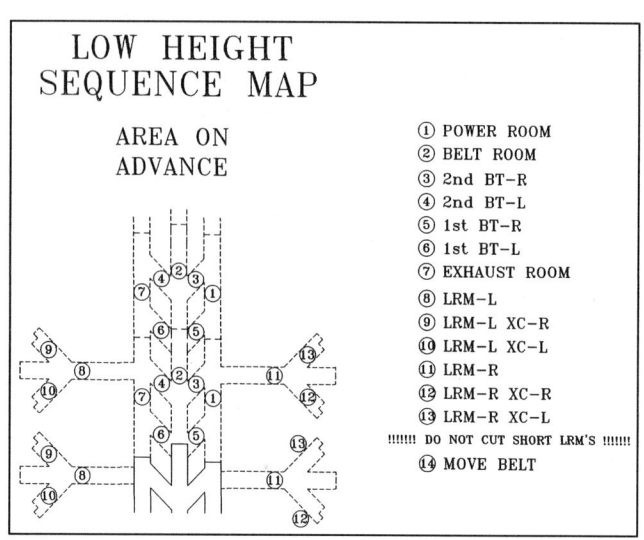

FIGURE 13.7   A basic mine advance sequence

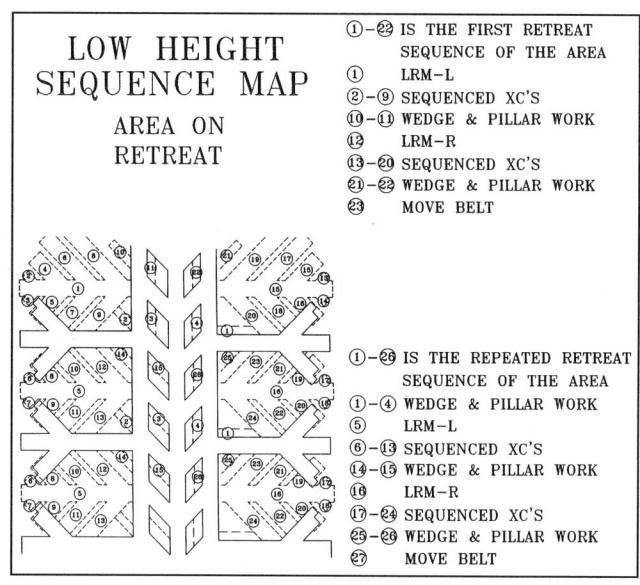

FIGURE 13.8   A basic mine retreat sequence

The ore body has been modeled with Techbase software; however, due to the corehole spacing, the model is not always able to predict small salt horses in the ore body. Therefore, if the sample has dropped in grade, and if the face does not show reasonable ore thickness, development will be stopped. Once extraction is started, the ore grade will be higher because of the dilution experienced during development. Development mining height is higher than the ore height, thus diluting the grade.

Currently, the west mine has six operating panels–five primary panels and one backup panel. Typical panel equipment consists of a miner, shuttle cars, fans, a roof bolter, a stamler, a belt line, and a miner transformer (miner pot). The equipment varies in make and model from panel to panel. All of the six operating miners are Joy miners. A variety of cars are used, including Joy 10 SC shuttle cars, Long Airdox Un-A-Haulers, and Eimco battery cars. Each panel has two of the same type of shuttle car. The roof bolters and fans are also a mesh of different models. The overall availability of a panel is around 72%, and the overall availability of the miners is 95%.

Ore hoisting is accomplished in a timber-lined production shaft. The shaft is used for hoisting raw ore, supplying intake air (127,000 ft^3/min), and as an emergency escapeway. The manually operated hoist has a capacity of 338 t/hr. Currently, the mine hoists 24 hr a day at an average of 1,477 skips per day, or 8,123 t/d.

13.4.2   East Mine Mining Methods

Mississippi Potash's East Mine also utilizes the chevron mining pattern, known in the Potash basin as a modified longwall method. The mine uses a three-entry system. Room 1 is used for exhaust and travel. Room 2 is used for ore haulage and intake. Room 3 is used for additional intake and travel (Figure 13.6).

Mine conditions are highly variable, requiring adaptability of mine sequencing. The primary variations result from height of the ore body and limitations in the initial haulage systems. The East Mine utilizes both continuous haulage systems and cyclical haulage systems. The continuous haulage systems can be operated in extremely low areas, but they are not easily maneuvered. A basic mine sequence consists of developing submains, developing panels, and extracting panels (Figures 13.7 and 13.8).

The East Mine also has six active mine panels. Panel equipment varies because of the two separate haulage systems. The cyclical haulage panels consist of an Eimco miner, Long-Airdox-Una-haulers, roof bolter, and fans. The continuous haulage panels consist of an Eimco miner, a Long-Airdox mobile bridge conveyor system, roof bolter, and fans. Overall panel availability is 82%, and overall miner availability is 95%.

Ore hoisting is accomplished in a 15-ft-diameter production shaft. This shaft is concrete lined from top to bottom and is approximately 1,650 ft deep. The shaft is used for hoisting raw ore, supplying intake air (210,000 ft^3/min), and as an emergency escapeway. The automated hoist has a capacity of 460 t/hr. Currently, the mine hoists 24 h/d for an average of 716 skip loads per day or 9,317 t/d.

13.5 CONCLUSIONS

Mississippi Potash, Inc., has established itself as a productive and safe operation within the mining community. In order to maintain its competitive edge, Mississippi Potash, Inc., will continue to improve its Carlsbad operations. Increasing current production levels, developing the indicated reserves, expanding the reserve base, and improving milling processes will accomplish this.

13.6 REFERENCES

Jones, C.L., C.G. Bowles, and A.E. Disbrow. 1952. Generalized Columnar Section and Radioactivity Log, Carlsbad Potash District. U.S. Geological Survey Open-File Report, 25 pp.

Eddy Potash, Inc. 1996. Internal archives, geologic map of southeastern New Mexico.

Mississippi Potash, Inc. 2000. Internal archives, figures 3–8.

CHAPTER 14

Advanced Mine-Wide Automation in Potash

Stephen J. Fortney[*]

14.1 INTRODUCTION

The Potash Corporation of Saskatchewan (PCS), Inc., is a publicly traded company with five potash mines in Saskatchewan and two in New Brunswick, in addition to extensive phosphate and nitrogen facilities throughout the world. The corporate head office is located in Saskatoon, Saskatchewan. These mines have an annual potash production capacity of 12 million tonnes and produce muriate of potash exclusively.

Since 1983, PCS has actively pursued automation through several programs, many of which were tested at the Rocanville Division. One mining machine at Rocanville was successfully automated by 1989. With years of experience in automation, Rocanville implemented a mine-wide automation and control system that was completed in 1994. This paper focuses on the program at Rocanville and reviews the effects of automation on production personnel, maintenance programs, and operating costs.

14.2 HISTORY OF ROCANVILLE

The Rocanville operation is located in the southeast corner of Saskatchewan near the Manitoba border. The plant was put into production in 1971 utilizing a room-and-pillar system of underground mining. The ore body lies at a depth of 1,000 m in the Esterhazy Member of the Prairie Evaporite Formation. The active underground operation covers 145 km^2 and has 950 km of tunnels, of which the farthest from the shaft is 11 km.

The operation is highly mechanized and uses five Marietta 780-AW4 continuous borers (Figure 14.1) to cut a room-and-pillar pattern. The miner initially cuts a 1,825-m-long first pass (Figure 14.2) that is 8 m wide by 2.4 m high. The miner is turned around, the tub removed, a cross-conveyor attached, and an additional 6-m-wide pass (the "second pass") is cut for the full 1,825 m. The miner then crosses the belt and cuts an additional 6 m off the opposite wall (the "third pass"). This completes the room, which is now 20 m wide by 2.4 m high by 1,825 m long. The miner and extensible belt are moved 45 m down the panel conveyor and the process is repeated, leaving a 25-m pillar. The cycle typically takes 1 month and produces 180,000 tonnes of ore.

After the ore leaves the extensible belt, it passes along 10 mainline conveyors to be deposited via a tripper into seven bins having a total capacity of 6,000 tonnes. The ore from the storage bins is then conveyed to a 300-tonne-capacity surge bin and into two loading pockets that feed two 27-tonne-capacity skips. A friction hoist lifts the ore 1,000 m to the surface, where it is transported to a storage building. A hoist cycle is completed every 103 s.

Prior to automation, two supervisors drove from miner to miner visually inspecting the progress of the machines. If a

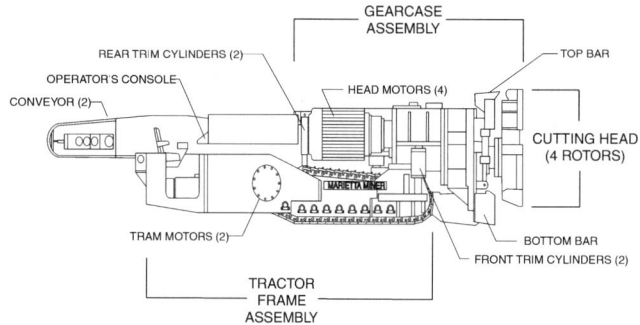

FIGURE 14.1 Marietta continuous borer

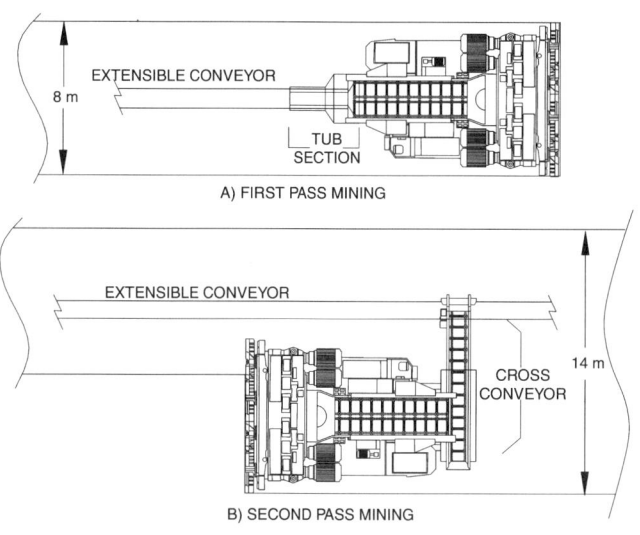

FIGURE 14.2 First- and second-pass mining cuts

problem occurred, the operators would contact the supervisors, who would arrange for the required maintenance. This was difficult as the supervisor was often traveling and could not be easily reached.

Some of the mining machines are as far as 11 km from the shaft. Travel time required to and from the shafts often left the miners unproductive for periods as long as 3 hr between shifts. A combination of travel time, lunch, and coffee breaks resulted in

---
[*] Potash Corp. of Saskatchewan, Inc., Rocanville, Saskatchewan, Canada.

the miners being idle for approximately 8 hr/d. This time did not include time required to retool the cutting heads and perform routine maintenance. The primary goal of the automating the miners was to optimize miner production.

PCS had successfully automated all five miners between 1989 and 1992 so many of the benefits of miner automation had been proven. Up to 1992, automation had been targeted at the mining face with little consideration of how to integrate the miners into a total control system. In 1992, a multiphase program was initiated to automate the mine fully and realize the benefits of using both miners and mine-wide automation.

14.3 CONSIDERATIONS AND GOALS

With the apparent deficiencies of a conventional control system, a list of goals for the new control system was compiled.

1. Increase utilization of the miners by running through shift changes and lunch breaks.

2. Reduce the number of production and maintenance personnel needed to operate the mine by—

    a. Coordinating maintenance with production requirements.

    b. Scheduling and prioritizing miners.

3. Reduce hoist downtime.

4. Reduce wear and tear on equipment through improved alarming and computer control.

5. Maximize underground bin usage.

6. Monitor key locations within the mine visually.

7. Maintain system with little or no increase in labor. The system must be technically manageable.

8. Assure flexibility of the entire control system, as mine plan changes were often unforeseen.

9. Control and monitor all systems from a central location, preferably one on the surface.

10. Assure that no single point of failure would disable the entire mine's production.

14.4 COMMUNICATIONS OVERVIEW

The ideal system would use one communication medium from the main control room through to the miners. This was not possible, so a combination of technologies was used. The backbone of the communication system consisted of a broadband signal transmission system installed along the mainline and panel belts and up the shaft. This required an installation totaling 33 km. Because of the mobility of the miners, it is not practical to install the system to the mining face. This portion of the communication link is handled by a digital radio link installed on each miner.

The data handled by this system (Figure 14.3) consist of video transmissions from the five miners and five key belt transfers to the control room. A standard computer link (RS232) and telephony are also required from the miners. The mine has five extensible belts and 15 mainline and panel belts, each of which requires an RS232 link. The main computer in the shaft area acts as a data concentrator and controls the loading pocket, four shaft belts, eight pan feeders, and 20 vibrators. The main control station is on the surface along with the main programming facilities. A secondary control station is located underground and is used only in the event of a failure in the shaft communication cable. The head end for the broadband system is located underground in the shaft area so the system would not be dependent on the shaft cable.

14.4.1 Broadband Transmission System

The broadband transmission system is essentially a cable television network (CATV) with the capability of passing television signals in either direction. The underlying technology

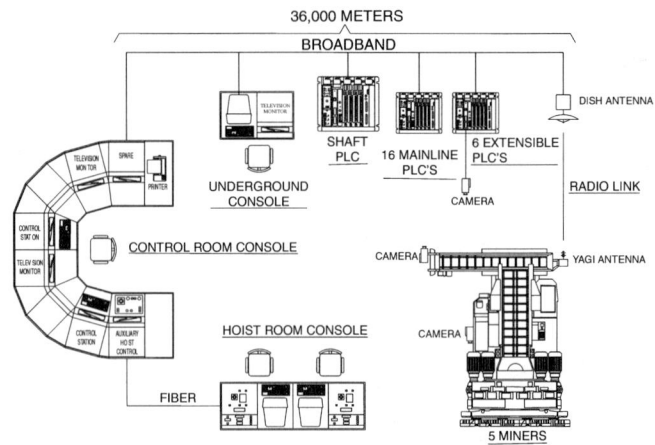

FIGURE 14.3   Overview of underground automated mining

is based on standard CATV. Therefore, all components are inexpensive and can be purchased readily from several established suppliers. Broadband systems are highly immune to electrical noise, which is an important feature as the cable often parallels power feeders in the mine. Bandwidths are typically 30 television channels in both directions. One TV channel can be converted into 60 RS232 links. System configuration can be easily changed by adding branches at almost any point to handle any size of mine.

A high-pass system was installed at Rocanville, where outbound frequencies range from 5 to 186 MHz, and inbound frequencies range from 210 to 450 MHz. The current configuration has 33 km of 19-mm coaxial cable strung along all the conveyors. Every 800 m, an in-line amplifier reestablishes the signal levels. The power for the amplifiers is passed through the coaxial cable. Every 8 km, a new power supply must be installed. At any point along the first 550 m past an amplifier, a drop tap can be installed. A drop tap is a passive device, and once installed, the full capabilities of the broadband system can be used at that point.

14.4.2 Radio Link to Miner

One of the most crucial links in the mine is between the miner and the extensible belt drive. The mining machine can advance as much as 500 m in a 24-hr period, making broadband transmission impractical. A radio link provides communication between the broadband system along the mainline conveyor and the miner. This radio link allows control of the miner from the surface and handles the interlocks between the miner and the extensible belt drive. The distance covered by this link varies from 60 to 1,825 m. In the first pass, the cross section of the drift is only 2.4 m high by 8 m wide. Previous tests indicated that only frequencies in the 2- to 4-GHz range are able to cover the required distance. Frequencies in this range couple to the drift and use the tunnel as a wave guide. Although the drift is straight in plan view, undulations in the ore body result in a nonline-of-sight propagation path.

The radio link has a mobile unit and a fixed unit. The mobile unit is located at the miner and the fixed unit is located at the extensible drive. Each unit consists of one box into which three cameras, telephone, RS232 link, remote mounted antenna, and power supplies, including a battery backup, are connected. All the items except for the main radio link are readily available off-the-shelf components. The cameras comply with cable television industry standards, as does the video output from the extensible drive end. The central control room operator sends commands to the programmable logic controller (PLC) on the miner, and the

PLC controls which camera is being viewed. This allows the control room operator to disable certain views if they are not required or modify the viewing sequence as dictated by the operating mode of the miner. At the stationary (or drive) end, a 0.75-m dish antenna is used, while a smaller but more rugged Yagi antenna is used at the miner end. Using dish antennas at both ends would provide slightly better results, but a dish antenna is susceptible to damage from vibration and operator mishaps.

To control the miner properly from the surface, at least two video-camera views from the miner to the main control room are desirable. In addition, bidirectional RS232-link data and telephony are required. Several options were reviewed, and a decision was made to convert the video and telephony to digital signals and use a radio link to pass all three signals over a single digital carrier.

Digital video transmission has several inherent advantages over frequency-modulated video transmission. It requires a significantly narrower bandwidth, which means that for the same radio output, the signal will travel farther. Because of the proximity of the operators, keeping the radio signal strength as low as possible was an important factor. Digital video transmission can be demodulated at a significantly lower signal strength, which allows even further penetration down the drift. Digital video transmission is a relatively new technology and carries a higher cost than frequency-modulated video transmission.

Several technical issues must be overcome when telephony, the RS232 link, and video are integrated into a single digital stream. All three signals must be multiplexed and then demultiplexed at the opposite end. The video and RS232 can easily be split up and then demodulated at the opposite end, but in order for the telephony to operate properly, the data must be transmitted almost continuously. This places great demands on the communications computer, since telephony is broken down into several smaller packets and reassembled at the receiving end. If the data packets are too large, telephone conversations will be sent intermittently. If the data packets are too small, the overhead will overload the communications computer, and the entire radio link will not work properly.

14.5 GENERAL

Control of the mine from a central location required the total rewiring of all underground controls. All miners and belts were completely dismantled and rewired using PLCs for control and an alpha-numeric display for status and alarms. This involved 29 PLCs throughout the mine. The central control room monitors operation of the mine and issues commands to the PLCs, which determine if the command is allowed before executing the function. The information presented to field personnel at a piece of equipment is identical to the information available in the central control room. All interlocks between the PLCs are hard-wired to ensure that a communication failure does not affect operation of the mine. In the event of a total communications failure, the mine can be operated in the same fashion as before it was automated, except that alarms and status are clearly displayed at each control point.

An underground production supervisor was relocated to the central control room on the surface. All production and applicable maintenance are monitored and approved from this location. Having a central point of reference has improved coordination within the mine, which has generated several additional benefits. For example, in the past, maintenance crews would go to a miner and wait until production requirements were met sufficiently to allow the miner to be stopped for routine maintenance. Direction from the central control room operator has allowed maintenance crews to work in the shop and travel to the miners only when the machine is about to become available.

The central control room operator is also responsible for control of peak power. All production crews are informed when to operate and when to take breaks. This has resulted in a major reduction in the demand portion of the electrical bill. If there is sufficient ore underground, the control room operator will shut down part of the mine to reduce electrical costs and send the operating crews to perform auxiliary duties. The control room operator will also shut down all belts that are not required.

In addition to tracking the maintenance and production personnel, the central control room operator has cameras that monitor key locations throughout the mine. The most important points are the belt transfers immediately behind the miners. The cameras allow the central control room operator to detect belt spillage and alert operators before such spillage becomes a major problem and damages the belt.

The control room operator has the capacity to monitor a miner's operation during the second and third pass and make minor adjustments to cutting if required. It is possible to start and stop the miners, but an operator is still required to change the cutting head bits every 4 hr.

The central control room operator has complete control over the bin area and hoists. This information is used to control the flow of ore into the bin area, so bins do not overfill and cause the mainline belts to shut down fully loaded. All information on bin levels are fed into a PLC, which then draws the bins down evenly. If desired, the central control room operator can manually control one or all of the bins.

The production and cage hoists are fully automated, and no operator intervention is required. Prior to initiating the program, each hoist had a dedicated control station in the main hoist room. During the day shift, the shaft crew assigned a hoist operator to the main hoist room. During the night shift, the security guard filled in as hoist operator in the event the hoist required manual intervention. This often caused unnecessary delays when the security guard was performing other duties and was unable to respond immediately.

Part of the new program was the conversion of both hoists to computer control. Individual controls for the production and cage hoists are still located in the main hoist room. In the central control room, a dual-function hoist console can be switched to control either hoist. However, at no time can both locations contend for control over the same hoist. During the day shift, a hoist operator is still assigned to the main hoisting room, but during the night shift, the central control room operator can operate either hoist.

14.5.1 Miners

All underground production is generated by five 780-AW4 Marrieta four-rotor miners. These miners weigh 250 tonnes each and have a gearcase powered by four 300-kW motors. The miners are propelled forward by individually powered calked track pads. The track pads, conveyors, and cylinder functions are all hydraulically driven. The hydraulic power is generated by a 190-kW electric motor mounted on the side of the miner. All of the miner functions are initiated from the operator's console. When the operator initiates a function, the PLC checks to see if all the permissives are set, then executes the command or displays an alarm indicating why the function cannot be carried out. This is true under all modes of cutting.

14.5.2 Manual Cutting

Under manual cutting, the operator utilizes two basic actions to control the miner, the track pads, and the side jacks. The forward advance rate and direction are both controlled by the track pads. The operator has two independent levers, one for each track. By adjusting these levers, the miner can be made to cut in the proper direction and with the optimum advance rate, which is determined by torque on the cutting heads. A reliable method of determining torque is to measure the amperage required by the four head motors. If the amperage is too low, the operator moves both tram levers forward, which in turn increases the pressure applied to both tracks; if the amperage is too high, the pressure is reduced.

As the miner is cutting, the face shears away in sections and causes the miner to surge forward and then cut heavily as a full face is again encountered, which requires adjustments every few seconds. This function leads to much of the operator fatigue. After a few hours, the operator's natural tendency is to allow for larger surges or cut at a reduced rate so the peak of the surges is within an acceptable range, thereby requiring fewer adjustments. If large surges occur, more maintenance on the miner is required. If cutting is done at a reduced rate, productivity of the miner is reduced.

The other function of the tram levers is to determine the heading of the miner. In the first pass, the heading is determined by a laser mounted on the back and pointed toward the miner. To control the heading, the operator maintains a laser dot on a predetermined location on the miner. To adjust the heading, the operator changes the ratio of the left tram to the right tram while maintaining the overall average so as to maintain the penetration rate. In the second and third passes, the heading is determined by following the wall of the first pass. The object in the second and third pass is to cut as wide as possible while producing an even back.

The second major function controlled by the operator is the front trim cylinders. These independently controlled cylinders raise and lower either side of the gearcase, which is connected to the cutting heads, which then determines the pitch and roll of the miner. In the first pass, the pitch of the miner is determined by the lay of the ore body. An on-board ore analyzer provides the operator with a graphical representation of the cross section of the ore body. The operator then guides the miner through the ore body and adjusts the front trim cylinders to adjust the pitch of the gearcase to follow the ore body and maximize the ore grade mined.

In the second and third passes, the pitch of the miner is set so the backs of the first and second or third pass are at the same elevation. In all three passes, the miner cuts a level back. An angle sensor on the gearcase indicates the roll of the gearcase. To maintain the level back, the operator moves one trim cylinder in one direction and the opposite one in the other direction until the angle sensor indicates zero roll, at which time the trim cylinders are equalized.

The operators are also responsible for controlling the backup equipment. In the first pass the tub (Figure 14.2) follows the miner. The extensible belt is drawn from the take-up by the tub, which positions the belt immediately behind the miner. The tub serves to align the belt behind the miner and drills holes for posts in the floor for the extensible conveyor. These functions are handled by the backup operators, who install the extensible conveyor structure and ventilation brattice. In the second and third passes, the tub is removed, and a cross conveyor is attached behind the miner. The cross conveyor takes the ore from the rear of the miner and conveys it onto the extensible belt installed during the first pass.

14.6 COMPUTERIZED CONTROLS

All functions on the mining machine are controlled by a PLC. The operator's control levers are all wired into the PLC, which in turn makes certain all interlocks are met before the corresponding cylinder or tram function is executed.

During first-pass cutting, the operator used to guide the miner through the ore body to obtain the best grade of ore. The ore body has slight undulations, and the operator followed a gray clay band 1 m above the floor. This clay seam was only a rough reference, and often the miner would cut too high or too low. As shown in Figure 14.4, the ore is most highly concentrated in the

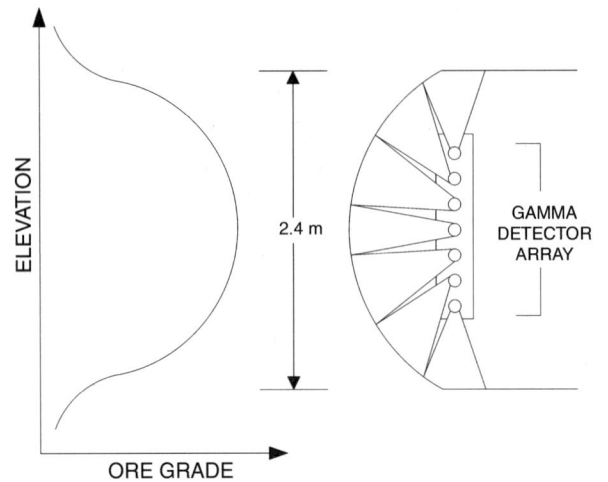

FIGURE 14.4   Schematic of ore concentration

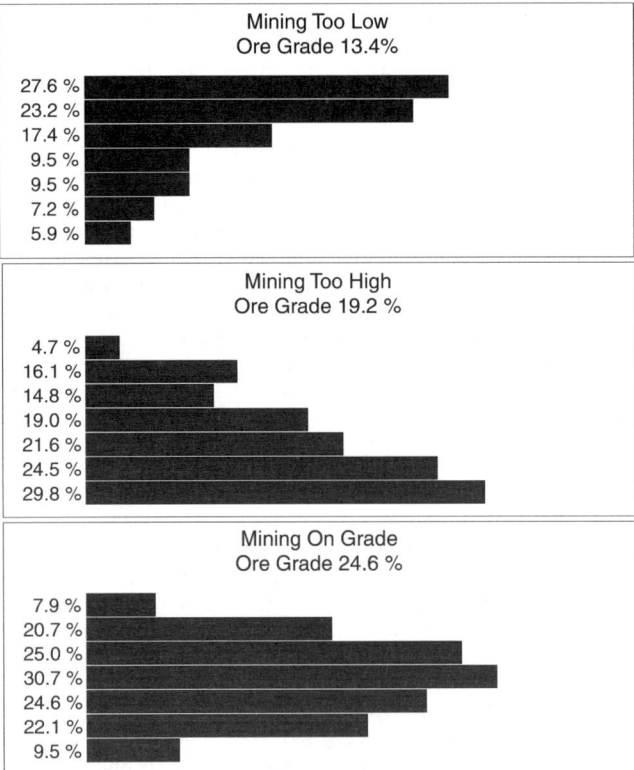

FIGURE 14.5   Ore grade profile as read by gamma detection array

center of the ore body and tapers off in both directions. To maximize the recovered ore, PCS developed an on-line ore analyzer.

Potash contains a K_{40} isotope that emits gamma radiation. To give the operator a proper representation of the ore body, a seven-probe gamma detection array is used. The seven probes are mounted in an 1100-kg lead enclosure that shields the probes from unwanted radiation. The ore analyzer is connected to the PLC, which converts the raw data and displays an ore grade profile on the operator's display station. The operator then uses these data to determine if the miner is cutting too high or too low, as shown in Figure 14.5.

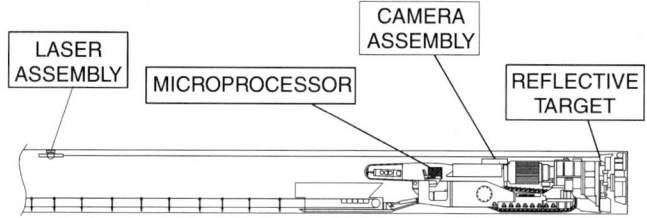

FIGURE 14.6 Components of automated miner system

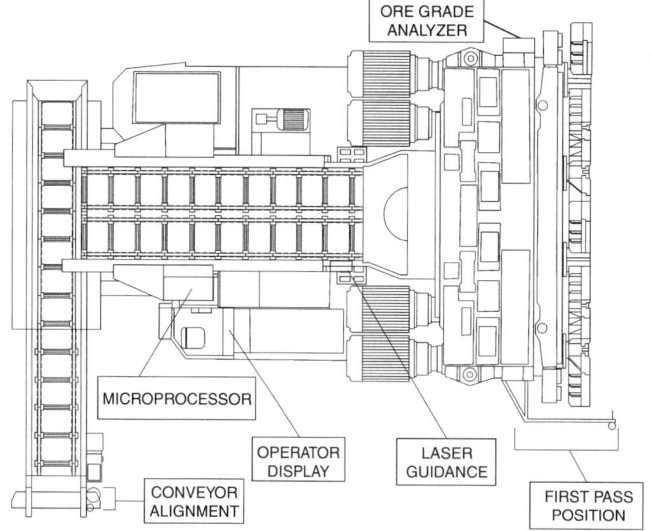

FIGURE 14.7 Layout for automated mining

There are essentially four modes of operation for the miner. These modes offer various amounts of operating freedom, from nearly total manual control to fully automatic control. The modes are manual, semiautomatic, first automatic, and second and third automatic. All modes are selected by using a four-position selector switch in the operator's control console.

14.6.1 Manual Mode

When cutting under manual mode, the operator places the selector switch in the manual position. Under this mode, the miner performs as it does without any computerized controls, except that the PLC will automatically scale back the inputs from the tram joysticks if the miner head amperage exceeds the allowable limit.

14.6.2 Semiautomatic Mode

To cut under semiautomatic mode, the operator places the selector switch in the semiautomatic position so the miner cuts with the desired head amperage. At this time, the miner is essentially cutting under manual control. When the desired head amperage is reached, the operator presses a button on the control panel, and the PLC scales the inputs from the tram joysticks to maintain the desired cutting amperage. The miner is still steered by the operator while the PLC maintains the ratio of the tram lever positions and scales each lever by an equal amount. The operator can adjust the amperage while cutting by using a combination of existing controls. The miner can be returned to fully manual control by turning the selector switch to another mode or pulling either tram lever into the neutral position.

14.6.3 First Automatic Mode

To cut using the first automatic mode, the operator places the selector switch in the first automatic position. Then the operator sets the miner to cutting with the desired head amperage, the laser in roughly the correct position, and the ore analyzer reading near-optimum ore recovery (Figure 14.6). Again the miner is essentially cutting under manual control. When these three conditions are satisfied, the operator presses a button on the control panel, and the PLC assumes control over all functions of the miner. That is, the PLC (1) adjusts the ratio of the tram lever inputs to steer the miner and maintain the laser dot at a predetermined position on the miner, (2) scales the tram lever inputs to maintain the desired cutting amperage, which can be adjusted by the operator, and (3) interprets the readings from the ore analyzer and adjusts the miner side jacks to recover the optimum ore grade and cut a horizontally level drift.

As the miner moves vertically to follow the ore body, the PLC automatically repositions the laser to follow the miner. All of the tub functions are handled manually by the operators. As when operating in semiautomatic, the miner can be returned to full manual control by turning the selector switch to another mode or pulling either tram lever into the neutral position.

14.6.4 Second and Third Automatic Mode

To cut under the second or third automatic mode, the operator places the selector switch in the second automatic position. The miner is started with the desired head amperage and with the second pass back even with the first pass back and the width of the second pass as wide as possible without leaving a center cusp. The operator also positions the cross conveyor so it is centered over the extensible conveyor and the ore is deposited correctly on the extensible belt (Figure 14.7). The miner continues to cut under manual control. When all the conditions are satisfied, the operator presses a button on the control panel. The PLC locks on to the readings presented by the ultrasonic system from the second pass arm and cross conveyor and assumes control over all functions of the miner and cross conveyor. The PLC adjusts the ratio of the tram lever inputs to steer the miner and maintain the same width of cut as originally set when the miner was placed in automatic. As when in the other modes, the PLC scales the tram lever inputs to maintain the desired cutting amperage.

All three functions of the cross conveyor are controlled by the PLC in order to dump the ore properly on the belt. Unlike during the first pass, the operator is free to make adjustments to all settings while the miner is cutting. Once it has been set up and under ideal conditions, the miner is able to cut unattended for several hours. The maximum amount of continuous cutting has been 7 hours, at which time the miner was stopped to change bits. As in other modes, the miner can be returned to full manual control by turning the selector switch to another mode or pulling either tram lever into the neutral position.

14.7 SAVINGS AND IMPROVEMENTS IN EFFICIENCY

Mine automation has numerous paybacks, some of which are obvious and can be easily measured. Other benefits are less obvious.

With regard to safety, there has been only one rock fall in a mining panel since completion of miner automation. The improvement is attributed to the even back resulting from automated cutting; that is, an even back does not leave a stress point and reduces the risk of a rock fall.

The frequency of miner overhauls has decreased. Previously, a miner was overhauled on the average of every 2.5 million tonnes; now the average is every 6 million tonnes. Miner automation has assisted in extending the period of overhaul frequency by reducing peak cutting loads and by generating more even cutting forces. During this period also, there have been numerous improvements to the mining machine outside of the automation program.

Automated cutting allows straighter alignment of extensible belting on level ground. This increases belt life and reduces spillage in the panel. It appears that a 15% increase in life expectancy will be achieved. In addition, all belts and bin levels are controlled by the control room operator who ensures that the mainline belts are not stopped while loaded or run empty for extended periods of time. Early indications are that this also will increase belt life.

Using cameras positioned at major trouble spots throughout the mine, the control room operator has identified potential problems and prevented damage to conveyors on several occasions.

By scheduling all mining and maintaining bin levels, the mine can be operated in a fashion that reduces peaks in electrical consumption. This has resulted in substantial savings in electrical costs. For example, the control room operator shuts down the mainline conveyors when they are not required. During a typical month, the conveyors are shut down for 204 hours when otherwise they would have been left operating.

Automating hoist operation has increased hoisting capacity by 5% to 7% while allowing for more maintenance time on the hoists. Because the hoist is the bottleneck in the entire plant, this has led to a 3% increase in plant production.

Automated miner operations reduce the level of manpower required at each machine. Automated cutting is less demanding for the operators, so the miner is stopped only to add power cable and change bits. Operations such as developing new panels and reclaiming hardware are still labor intensive. Nevertheless, through attrition, production crews have been reduced from an 11-person crew to a 10-person crew.

The conversion of the miner to automatic control and, more specifically, the addition of the on-line ore analyzers have increased the ore grade fed to the mill by 0.5%, resulting in lower unit costs of production.

14.8   CONCLUSION

The automation program at PCS Rocanville encompasses every aspect of the mine. The central control room utilizes 29 PLCs connected to 3,100 points throughout the mine and at the hoists. Over 600 control functions can be executed, and 3,350 alarms can be generated. The high degree of control improves safety and increases efficiency.

First-pass cutting is seldom performed under full automation. Most commonly, the operator uses the on-board ore analyzer to determine the optimum pass through the ore. The operator then sets the laser on the desired line, and the miner tracks the laser in both the horizontal and vertical planes. Full

gearcase control relies heavily on the laser. The mining machine will routinely cut at distances 160 m from the laser. Dust in the air blocks the signal to a point at which full control is not possible. In the near future, laser ring gyroscopes will become accurate enough to replace a laser. Once this technology replaces the laser, full automated control will be possible.

Video equipment that transmits images from the miner to the control room has been installed on four of the mining machines. The original purpose was to allow the control room operator to monitor the condition of the mining machine during the second and third pass and make adjustments as required. After years of operation, automation procedures have been refined to the point where adjustments from the control room are seldom required. The digital video signal is being removed from the radio links, and the cameras are being removed from the miners.

Radio links are used on all mining machines. The reliability of the radio equipment has steadily improved, and removing the video links will reduce the complexity of the system and dramatically increase the radio power-to-data ratio. This increase will increase the performance of the radio links.

Broadband transmission has proven extremely reliable. Most of the failures result from physical damage to equipment in active mining areas. The original configuration in 1995 used 33 km of broadband cable, while the current configuration has 21 km of cable. As new areas are developed, the cable will reach 30 km.

Automatic second- and third-pass cutting has proven to be reliable and practical. Ninety percent of the cutting is performed with the computer operating the miner. The original plan was to have one operator move between two miners to change cutting bits. In practical terms, the mine is too large for one operator to service two machines. One operator is assigned to each second- and third-pass machine. This operator performs routine tasks in the area of the miner and monitors the machine's advance. To maintain the ore supply underground during the second or third pass, a miner is operated through shift changes about 60% of the time. The miner routinely operates unattended for 2 to 3 hours and has operated without intervention for 7 hr. Under normal conditions, the miner must be shut down every 4 hr to change bits. Maintaining the ore supply results in a controlled operating environment.

Future improvements to the system will include automatic reporting of miner and crew performance and improved monitoring of ventilation.

This project was realized through the combined efforts of all departments, including operators, tradesmen, supervisors, and management at PCS Rocanville. The benefits provided enhance the competitiveness of PCS within the potash industry.

CHAPTER 15

Closed-Circuit Mechanized Cut-and-Fill Mining at PCS New Brunswick Division

D.R. Roberts*

15.1 INTRODUCTION

15.1.1 Historical Background

Potash Company of America's (PCA) long history in the potash business began in 1932 when, in Carlsbad, New Mexico, the company became the second potash producer in North America. Among other innovative accomplishments, PCA designed and built the first commercially successful noncoal continuous miner, and the Carlsbad operation became the first to mine potash successfully with continuous miners. In 1934, PCA became the first company to utilize the flotation process for the recovery of potassium chloride on a commercial basis. In 1955, a continuous conveyor system was developed to operate with the continuous miners, a system that was the forerunner to conveyor systems at a number of PCS divisions.

In 1973, PCA was one of the companies invited by the New Brunswick Department of Natural Resources to bid on exploration, and subsequent development rights, for a selected potash reservation area in the Sussex area (Hamilton 1971). This invitation was of immediate interest to PCA management for two reasons: the Carlsbad ore reserves were marginal and the New Brunswick reservation area was less than 80 km (50 miles) from ice-free tidewater shipping facilities and had excellent access to potash markets along the Atlantic.

The potash reservation area was one of three in New Brunswick discovered in the early 1970s. Two holes drilled into an elongated salt structure about 5 km (3 miles) east of Sussex met a significant intersection of potash with an aggregate thickness of 20 m (66 ft) of sylvinite averaging 23.7% K_2O at a depth between 227 and 304 m (745 and 997 ft) below the surface (Webb and Barnett 1986).

Following a difficult exploration program and signing of a mining lease in 1978, management approved the development of a potash facility near Sussex, New Brunswick, in 1980. Various problems during startup of the potash facility resulted in initial levels of production considerably below the rated capacity, and by the end of 1984, production had reached only 40% of design capacity. In 1985 deficient areas were identified, and an upgrade program was subsequently completed.

In January 1986, Rio Algom, Ltd., agreed to purchase PCA. Rio Algom was committed to overcome the production problems in the New Brunswick Division by investing additional capital in an upgrade program. This investment allowed the New Brunswick Division to reach design capacity.

Potash Corporation of Saskatchewan (PCS), Inc., purchased the potash assets of Rio Algom in 1993. These assets included PCA's New Brunswick Division, now called the PCS New Brunswick Division.

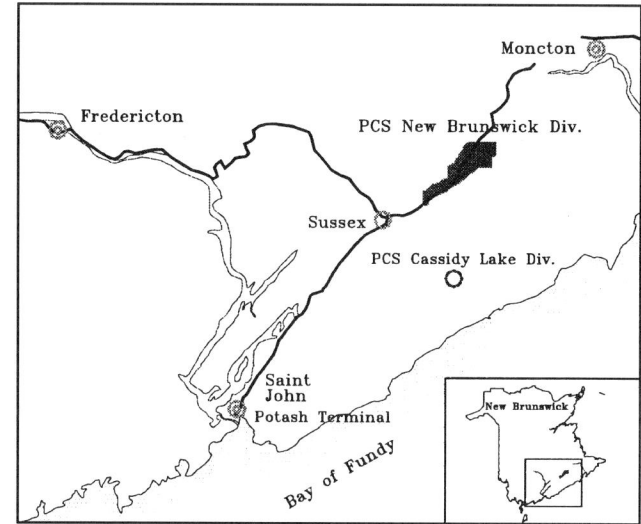

FIGURE 15.1   Location of PCE Potash New Brunswick Division lease area

15.1.2 Location

The PCS New Brunswick Division facility consists of a mine and concentrator in the Sussex-Petitcodiac area of southeastern New Brunswick (Figure 15.1) and a potash terminal 80 km (50 miles) away on the Bay of Fundy in Saint John. The potash terminal is owned by PCS and operated by a stevedore contractor; it is used to load potash and road salt.

The mine site is located in a lush green area surrounded by dairy farms and bordered by a salmon river to the north and the old TransCanada Highway to the south. By the most direct surface route, the Bay of Fundy is 50 km (31 miles) from the site.

The lushness of the environment results from an annual rainfall of 107 cm (42 in) and an evaporation rate only 53 cm/yr (21 in/yr). By comparison, other potash-producing regions in North America are either arid or semiarid. From the beginning, it was clear that the methods of handling waste tailings and waste brine in New Mexico and Saskatoon were not appropriate for New Brunswick because of the geological structure and the climatic conditions.

15.1.3 Geology

The PCS ore body is contained within a northeasterly trending anticline, or salt pillow. This anticline is at least 27 km (17 miles) in

---
* Potash Corp. of Saskatchewan, Inc., Saskatoon, Saskatchewan, Canada.

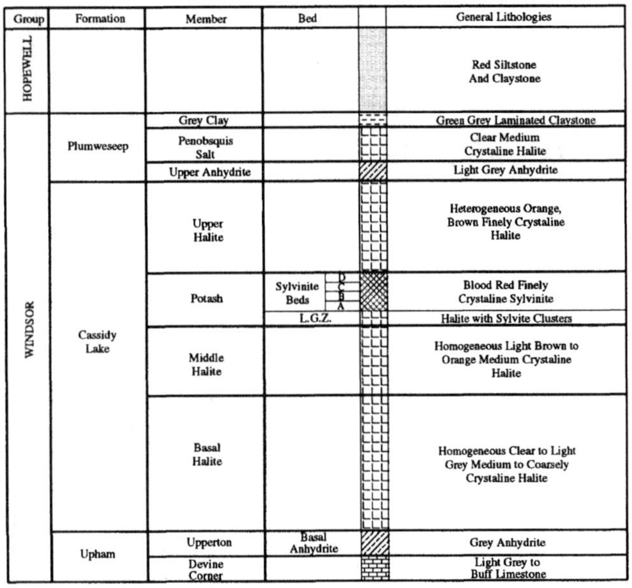

Group	Formation	Member	Bed		General Lithologies
HOPEWELL					Red Siltstone And Claystone
		Grey Clay			Green Grey Laminated Claystone
	Plumweseep	Penobsquis Salt			Clear Medium Crystaline Halite
		Upper Anhydrite			Light Grey Anhydrite
WINDSOR		Upper Halite			Heterogeneous Orange, Brown Finely Crystaline Halite
		Potash	Sylvinite Beds D C B		Blood Red Finely Crystaline Sylvinite
	Cassidy Lake		L.G.Z. A		Halite with Sylvite Clusters
		Middle Halite			Homogeneous Light Brown to Orange Medium Crystaline Halite
		Basal Halite			Homogeneous Clear to Light Grey Medium to Coarsely Crystaline Halite
	Upham	Upperton	Basal Anhydrite		Grey Anhydrite
		Devine Corner			Light Grey to Buff Limestone

FIGURE 15.2 Stratigraphic column of evaporite sequence at PCS lease area

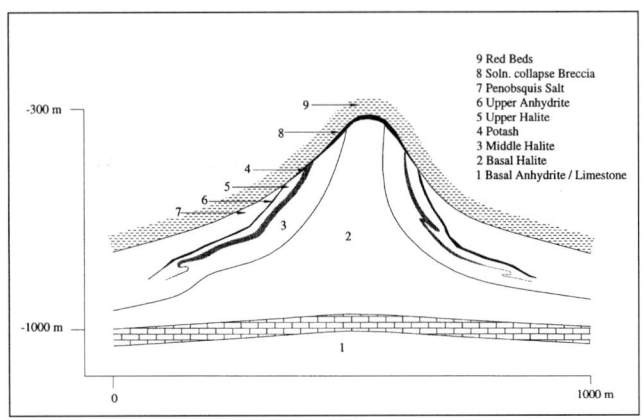

FIGURE 15.3 Cross section of salt dome

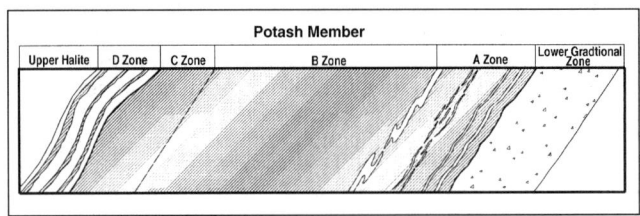

FIGURE 15.4 Typical ore zone section at potash stope face

length, up to 1,000 m (3,280 ft) thick, and comes to within 100 m (328 ft) of the surface, where it displays early stages of diapirism (Roulston and Waugh 1983).

The stratigraphy and structure of the deposit were determined through a surface drilling program initiated by the New Brunswick Department of Natural Resources after the deposit's initial discovery in 1971 (Kingston and Dickie 1979). To evaluate the accuracy of the geological interpretations and to establish a proven ore reserve, an exploration shaft was completed in April 1980, and subsequently 1,600 m (5,250 ft) of development and extensive underground exploration drilling were completed.

15.1.4 Stratigraphy

The evaporites (Windsor Group) were formed as a result of two marine incursions into what is otherwise a clastic red-bed sequence. The first marine cycle deposited a thick sequence of carbonates and sulphates (Upham Formation) followed by higher order evaporates (Cassidy Lake Formation), including the potash ore bed. The second cycle (Plumweseep Formation), which is a similar but less well-developed sequence, does not include potash mineralization and is capped by a grey claystone (Figure 15.2).

The lowermost unit of the Cassidy Lake Formation—the Basal Halite Member—is composed of clean, clear halite with minor zones of anhydrite laminae. This is overlain by the argillaceous Middle Halite Member, which contains weakly bedded, scattered inclusions of grey-green to red-brown clay making up approximately 5% of the total lithology. The contact with the overlying Potash Member is gradational. This member is made up of a lower gradational zone overlain by sylvinite beds that can be subdivided into the A, B, and C zones and a D zone of interbedded sylvinite and halite. The hanging wall salt, or Upper Halite, is largely fine- to medium-grained, clear halite with zones of weakly bedded argillaceous halite. Higher in the section, toward the top of this evaporite cycle, the lithology becomes more heterogeneous, with many thin argillaceous beds present.

The second evaporite cycle is preserved on the lower flanks of the salt structure. On the crest of the anticline, the salt has been partially eroded and the Upper Anhydrite is intermixed with the overlying grey claystones. Erosion has also led to the

development of a residual bed representing the preserved insoluble fraction of the various members of the Cassidy Lake Formation.

Friable, jointed, water-bearing claystones of the Hopewell Group are the predominant lithology overlying the evaporates.

15.1.5 Structure

As indicated earlier, the evaporates form a northeasterly trending anticlinal feature that plunges at 5° to the northeast. While the flanks of the anticline have a nominal 45° dip, large-scale, open, recumbent folding is also present (Figure 15.3). Within the evaporite sequence, halokenesis has led to folding on various scales. The Basal Halite is highly foliated and extensively recrystallized; tight isoclinal folding is found near the axial plane of the main anticline, while on the limbs, open drag folds with a wavelength of a few tens of meters are common. This style of folding extends into the Middle Halite and Potash members. However, in the ore zone, evidence of differential flowage is ubiquitous, particularly near the upper and lower contacts where dislocation and boudinage of the halite beds are common.

Typically, the A zone of the Potash Member is quite regular and contains small-scale drag folds with an amplitude of 25 cm (10 in) and a wavelength of less than 1 m (39 in). The upper part of the C zone and the D zone are characteristically more chaotic with irregular drag folding on a much larger scale. The degree of deformation decreases in the Upper Halite, although large-scale open drag folds are typically present. A typical ore zone section as seen at a mining face is shown in Figure 15.4.

In the Potash Member, it is common to find extensive areas where the potash mineralization has been completely leached out. These barren zones of discoloured halite (known as salt horses) generally occur in the A zone and the upper C and D zones, and although aerially extensive, they are usually thin. However, they sometimes affect the complete ore zone, in which case more extensive deformation is typical, together with an associated irregular halo of very coarse-grained, recrystallized, euhedral, clean sylvite and halite (Stirling et al. 1988).

Occasional slickensides have been identified in the Potash Member with indications of movement parallel to dip. The relationship of these features to the other structures identified in this member is not known at this time.

15.1.6 Comparison with Other North American Deposits

The New Brunswick Division potash deposit is quite different from those in either Saskatchewan or New Mexico, occurring in thicknesses up to 60 m (197 ft) at depths seldom exceeding 1,000 m (3,280 ft). At first glance, the most obvious difference is their contrasting sizes and extent. The potash resources in Saskatchewan are vast; published estimates of in situ potash resources vary from up to 500 billion tonnes (550 billion tons) K_2O, including up to 34 billion tonnes (37 billion tons) K_2O mineable through conventional and solution-mining techniques (Mayrhofer 1983). By comparison, the New Brunswick resource is much smaller. For comparative purposes, an estimate of less than 1 billion tonnes (1.1 billion tons) would not be unreasonable.

Apart from the relative size of the resource, the two most significant differences are in structural complexity and mineralization of the ore. Unlike the flat-lying New Mexico and Saskatchewan deposits, the Sussex ore body occurs with large-scale, open, recumbent folds within the flanks of the anticline (Figure 15.3). In fact, early geological interpretations led to the perception by many in the industry that the complexity of folding, the structure, and the environmental conditions posed a deterrent to cost-effective mining.

Although the ore grade is comparable to Saskatoon deposits, the extremely fine sylvite crystal size and the effects of folding have resulted in significantly greater milling difficulties. The size distribution of the sylvite varies from 40 to 4,000 micrometers (Stirling 1989). For 90% liberation, the optimum grind was predicted to be 270 micrometers. Part of the 1986 upgrade program involved the installation of a regrind ball mill to liberate the fine sylvite crystal from the middling halite fraction.

15.2 CLOSED-CIRCUIT MECHANIZED CUT-AND-FILL MINING

The previously noted environmental, geological, and mineralogical conditions existing in New Brunswick presented challenges that differed significantly from those experienced in mining other potash deposits in North America. These conditions led to the development of a unique mining method. This method is not limited to the mine, but is encompassed by the entire facility. Simply stated, the mine and mill form a closed circuit in which only muriate of potash, rock salt, and cooled condensate leave the facility without the necessity of significant surface storage facilities for mill wastes. Because of a small brine inflow in 1998, surplus brine is trucked to the PCS Cassidy Lake Division for disposal through its systems.

15.2.1 Mine Layout

Access to the mine is through two, 5 m (16 ft) in diameter shafts sunk on the north flank of the anticline. These shafts are concrete lined to the contact with the top of the salt at approximately 470 m (1,542 ft). Below this elevation to the main development level (1,900 level) at 580 m (1,903 ft), the shaft is screened. Overlapping pillars with a radius of 300 m (984 ft) are left around each shaft, within which no production takes place. However, the underground mine shop complex, together with salt and potash storage bins, the salt crusher, and the screening area, are developed within the shaft area. Most of this development is on the order of 3 m (10 ft) high by 8 m (26 ft) wide.

Potash and salt stopes extend eastward from the shafts. The high-extraction mining method does not permit gaining access to the mine from within the ore zone. Rather access is from two

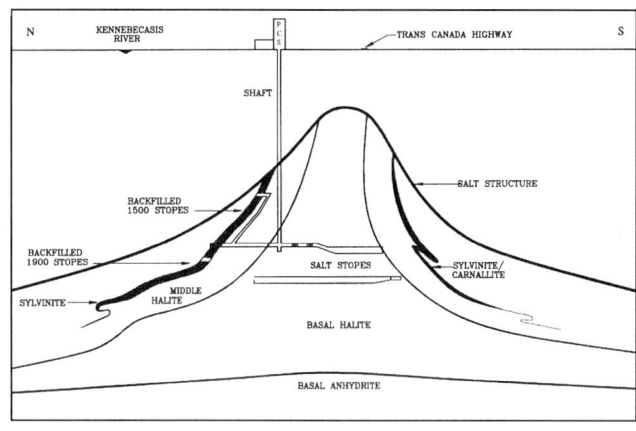

FIGURE 15.5   Section through salt dome

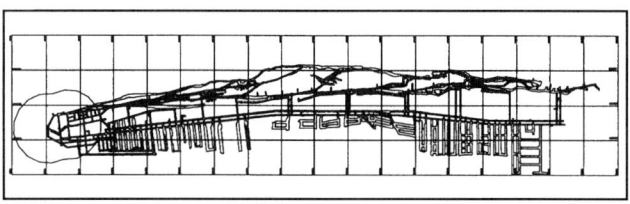

FIGURE 15.6   Composite mine plan

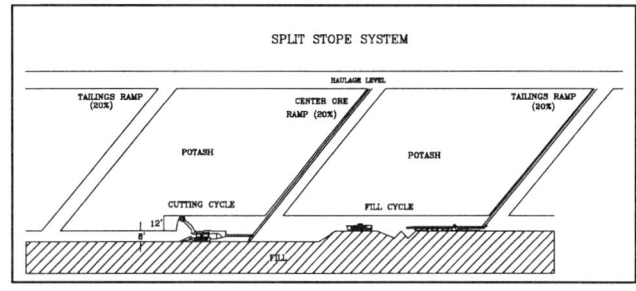

FIGURE 15.7   Typical ramp and stope layout for lower-level potash stopes

parallel drifts developed in the Basal Halite running close to the contact with the Middle Halite (Figure 15.5).

The two parallel drifts form the permanent infrastructure of the mine and contain three separate conveyor systems. The main ore conveyor and tailings conveyors are hung along each side of the north drift. The main salt conveyor is hung on the side of the south drift. Conveyor drives, as well as all other structures, are hung from the roof, resulting in efficient conveyor transfer.

Exploration is carried out from a drift in the ore zone at approximately the same elevation as the access drifts. A composite mine plan indicating the major mining elements is provided in Figure 15.6.

15.2.2 Potash Stopes

Cut-and-fill potash stopes are developed both above and below the main level and connected by crosscuts through the Middle Halite to the access drifts. A typical stope is on the order of 900 m (2,953 ft) long with a vertical height of 120 m (393 ft). Such a block of ore is defined by 20% tailings ramps at either end, connected by a basal sill cut, and split by a central 20% ramp used to convey ore from the stope (Figure 15.7).

Two salt tailings placement systems are employed, the first a dry system and the second a hydraulic fill system. With the first system, salt tailings from the milling process are conveyed into the end of the stope that has just been mined and spread out and compacted by scoop trams to within 2.4 m (8 ft) of the back. The hydraulic fill system, introduced in 1999, takes dry mill tailings and reslurries the tailings for hydraulic placement. While this fill is being placed in one half of the stope, the continuous mining machine is taking another cut in the other half of the stope. The AM 100 roadheader-type continuous mining machine trams on recently placed fill and cuts a 4.3-m- (14-ft-) high brow from hanging wall to footwall. The width mined varies according to the width of the ore and can range from less than 10 m (33 ft) to as much as 40 m (130 ft). The profile of the cut is normally dictated by the dip of the ore zone contacts. Structural irregularities in the hanging wall, together with the need to mine the D zone for ground control purposes in areas of low dip, lead to deviations from the plan in certain areas. Broken ore is picked up by the mining machine as it advances and conveyed by a continuous conveyor system out of the stope, up the ore ramp, and along the crosscut ore conveyor in the mainline access drift.

Ore is mined from stopes above the main level in a manner similar to that used in the lower stopes. Using the exploration drift on the main level as the sill cut, the stope is mined progressively higher while ramps are advanced in the footwall salt.

This mechanized cut-and-fill method of potash mining allows for the continuous placement of mill waste while allowing for extraction of a high percentage of the ore.

Only a single 12-m (39-ft) crown pillar is left between the upper and lower series of stopes. Although existing individual stopes within the series were developed without intervening pillars, future stopes will have an intervening pillar. Pillars 30 m (98 ft) in radius are also left around surface exploration drill holes.

15.2.3　Salt Stopes

South of the two main access drifts, salt stopes are mined in the Basal Halite. As they were initially mined for tailings and slimes storage during early development of the potash stopes, and later as storage for mill slime and mill brine in excess of plant evaporator capacity, they are developed below the main access level. A completed stope is on the order of 300 m (984 ft) long, 23 m (75 ft) wide, and 18 m (59 ft) high, with pillar widths of some 20 to 60 m (66 to 197 ft). These stopes are mined using roadheader machines as employed in the potash stopes and take progressively lower cuts in the floor to the final dimensions noted above. As in the potash stopes, no artificial ground control is required.

Future development will progress toward the east as an extension of the present mine layout. However, this will be modified as required by the geology of the ore body. Further east, the ore extends over a greater dip length than in the area presently being mined. The ore also becomes wider, so that additional modifications of the high-extraction mining techniques may be required.

The development of salt stopes to the east has kept pace with potash stope development. However, given the increasing distance from the shaft for the pumping brine and slimes to the salt stopes and for the return of clean brine, a series of deeper salt stopes has been developed. This decreases pumping distances and allows mining of some of the high-quality salt that exists beneath the present 1900 level. The requirement to mine salt stopes must be kept in balance with the storage requirements of mill slime and surplus mill brine.

15.2.4　Closed Circuit

The potash ore and rock salt mined form the first part of a complex closed-circuit materials balance. Mill tailings, mill slime, fine rock salt, waste evaporator salt, excess brine, and fine potash

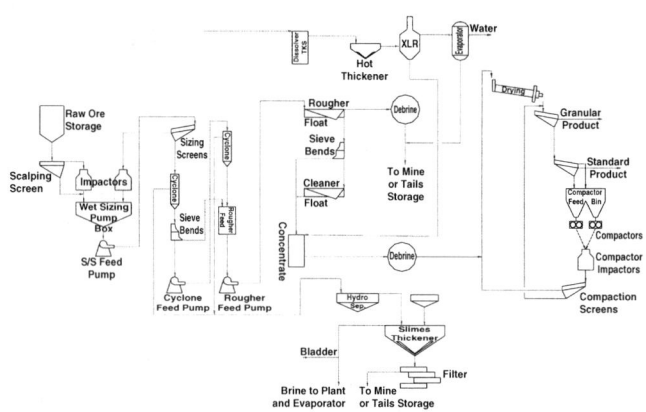

FIGURE 15.8　Simplified diagram of closed circuit at Sussex facility

dust are all important components. Figure 15.8 shows a simplified closed-circuit flow diagram for the Sussex facility.

Water Balance.　Water is used in potash concentrators to cool equipment and to leach the flotation concentrate to enhance the grade of the product. Any water that comes in contact with salt becomes brine and must be disposed of within the closed-circuit facility.

In 1998, a small (1514 L/min [400 gal/min]) inflow of brine resulted in additional pressure on the brine balance. One consequence was that the brine was trucked to a sister division at Cassidy Lake for disposal. A three-dimensional seismic exploration program was carried out over the mine area. The objective of the survey was to identify grouting and injection well targets.

With the relatively fine size of the crystal structure in New Brunswick ore, a greater reduction in size is necessary to liberate the sylvite crystal from the halite crystal. Still, the concentrate from recleaner flotation is normally only 59.0% K_2O. With a product specification at a minimum of 60.0% K_2O, additional upgrades with leaching water are required to maintain product grade. The greater the amount of excess brine produced, the greater the loss in mill recovery.

Brine Disposal.　The feasibility of pumping the excess brine from the site to the Bay of Fundy, a distance of 50 km (31 miles), across five significant tributaries and a series of valleys and ridges with elevation changes up to 300 m (1,000 ft) was studied (Davies et al. 1979), but was rejected by the original owner. However, the second potash mine in New Brunswick, now owned by PCS, is located less than 30 km (19 miles) from the Bay of Fundy in favourable terrain and has constructed a pipeline to the Bay. Surplus brine is trucked to this facility for disposal.

An evaporator with two 2,000-hp mechanical vapour recompressors is used to evaporate excess salt brine. In the process, fine evaporator salt, principally sodium chloride, but with some potassium and magnesium impurities, is produced. This evaporator brine is fully saturated in potassium chloride and becomes part of the feed stock to the crystallizer circuit. Figure 15.9 shows the simplified materials balance.

In the crystallizer, the hot brine is cooled and high-purity potassium chloride is recovered from the brine. Thus, three products are removed from mill excess brine: muriate of potash product, an acceptable salt tailings for the potash stope, and a hot condensate. Eventually, after cooling, the condensate and the product leave the closed circuit.

Brine in excess of the evaporator capacity must remain in the closed circuit and is presently pumped underground to the salt stope along with excess mill slime. Any loss of evaporator running time increases the amount of surplus brine in the closed

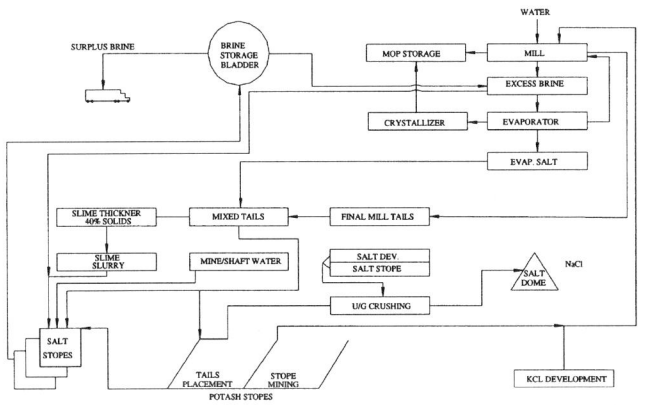

FIGURE 15.9 Simplified materials balance

circuit. Increases in mill feed increase the amount of brine being created. Changes in crystal size result in changes in leaching water and in the amount of brine being created. This surplus brine must be disposed of through a reduction in water consumption, improved evaporator performance, trucking to Cassidy Lake, pumping into injection wells, or storage in salt stopes.

Potash Development Mining. In earlier years, a significant percentage of the potash ore feed to the mill came from development mining in potash ore. For every 2.7 tonnes (3 tons) of ore mined, 1.8 tonnes (2 tons) of tailings needed to be placed either in the potash stope or in a salt stope. Since ore development entries are not filled, the mill waste from the development ore represents an additional burden to the materials balance. Even today, care must be taken to ensure that a minimum of 80% of the ore mined comes from fillable potash stopes.

Rock Salt Mining. Rock salt is mined for two purposes. The first is to develop the infrastructure of the mine. The second is to produce a product that can leave the facility, thereby creating additional room within the closed circuit for the storage of excess brine and slime.

In the process of producing road salt, fine rock salt waste is produced that must be placed in the potash stope as part of the fill. This leads to the consideration of how much infrastructure development can be completed without interfering with the volume of the potash stope available for fill.

Mill Tailings. Normally the debrined flotation waste salt (or mill tailings), along with evaporator salt and a portion of the extremely fine debrined slime waste, are immediately conveyed to the service shaft, where it is dropped through one of two drop pipes to an underground bunker and conveyed to the potash stope.

A covered 3,600 tonne (4,000 ton) emergency tailings storage building provides for up to 24 hours of storage when delays to the mine tailings system occurs.

The mill tailings, to be useable by the mine as potash stope fill, must meet quality requirements established by the mine. If the tailings contain more than 7% moisture, the brine will not drain from the surface of the placed fill. The rolling action of the scoop brings the brine to the surface and can literally shut down the filling operation in a quagmire.

Mill Slime. The extremely fine waste salts along with the 0.5% of insoluble clays present a separate problem for disposal.

As only a portion of the slime can be sufficiently debrined to be mixed with the mill tailings to form a satisfactory mine fill material, the remainder must be pumped to salt stopes.

Since the slime waste also contains fine potassium chloride or sylvite crystals and potash dust, further crystallizer recovery will effectively reduce the amount of fine slime returning to the mine while improving mill recovery and producing more muriate of potash.

Crystallizer Circuit. Since startup, the crystallizer circuit was upgraded to increase crystal production from 2.7 to 10.9 tonnes/hr (3 to 12 tons/hr). Further upgrades of the facility were completed with the installation of a third-stage crystallizer vessel along with greater dissolver capacity for greater recovery of fine potash dust by the crystallizer's scavenger circuit. This improvement upgraded the scavenger crystallizer circuit from 11 to 16 tonnes/hr (12 to18 tons/hr), thereby both resulting in more product and removing more slime from the closed circuit.

15.3 TRANSPORTATION

The additional components of work required at the New Brunswick Division to produce muriate of potash results in production costs higher than those of our North American competitors. However, the economics of the potash industry, like other commodities, is affected by a combination of both production and transportation costs. Clearly without the transportation advantage provided by being 80 km (50 miles) from tidewater, New Brunswick Division potash could not compete in today's market.

Potash is shipped by rail to the potash terminal in Saint John, where it is stored in one of two large sheds for vessel loading. The potash terminal is capable of loading ships at a rate of 3,000 tonnes/hr (3,300 tons/hr) onto vessels capable of lifting up to 50,000 tonnes. From Saint John, potash is shipped to Atlantic Ocean markets.

15.4 REFERENCES

Davies, G.S., G. Pisarzowski and J.H. Allen. 1979. Route Selection and Environmental Assessment on A Waste Brine Pipeline, Sussex to Bay of Fundy. MacLaren Atlantic, Ltd.

Hamilton, J.B. 1971. Request for Submissions for Exploration of a Salt-Potash Prospect, Sussex Area, Kings County, New Brunswick. New Brunswick Department of Natural Resources, Open File Report.

Kingston, P.W., and D.E. Dickie. 1979. Geology of New Brunswick Potash Deposits. *Canadian Mining and Metallurgical Bulletin,* vol. 72, no. 208, pp. 134–141.

Mayrhofer, H. 1983. World Reserves of Mineable Potash Salts Based on Structural Analysis. In Sixth International Symposium on Salt, ed. by B.C. Schreiber and H.L. Harmer (Toronto, Ontario). Vol. 1, pp. 141–159.

Roulston, B.V., and D.C.E. Waugh. 1983. Stratigraphic Comparison of the Mississippian Potash Deposits in New Brunswick, Canada. In Sixth International Symposium on Salt, ed. by B.C. Schreiber and H.L. Harmer (Toronto, Ontario). Vol. 1, pp. 115–129.

Stirling, J.A. 1989. The Petrology, Geochemistry and Mineralogy of Potash Ore Mill Products from the Potash Company of America Mine and the Denison-Potacan Potash Mine. DSS Contract Serial No. OSQB4-00443 for CANMET.

Stirling, J.A., B.V. Roulston, and D.C.E. Waugh. 1988. Preliminary Results of Bromine Distribution and Partitioning in Salt Deposits of Sussex, New Brunswick. Atlantic Geoscience Society Colloquium, Antigonish, Nova Scotia.

Webb, T C., and D.E. Barnett. 1986. New Brunswick Potash: Discovery to Production. Canadian Institute of Mining and Metallurgy, Fifty District 1 Meeting, Halifax, Nova Scotia.

Longwall Mining of Hard Rock

Underground Mining at Impala Platinum, Ltd., in the Northwest Province of South Africa

G.R. Ackerman* and **A.W. Jameson†**

16.1   INTRODUCTION

The Impala operations are situated some 20 km to the north of the town of Rustenburg in the Northwest Province of South Africa. The overall mining lease covers an area of 247 km^2 on the western lobe of the Bushveld Complex (Figure 16.1).

The Rustenburg area is generally one of low relief, lying between the Magaliesberg range to the south and west, the Pilanesberg massif to the northwest, and the hills of the Bushveld Main Zone Gabbro to the north and east. Isolated conical hills of hortonlite, dunite, and ultramafic pegmatoid rise above the open flats. The average surface elevation is approximately 1,120 m above sea level. Rock exposures are generally poor because of the almost ubiquitous cover of black turf, a lithomorphic vertisol derived from the underlying basic rocks. The vegetation is mainly grass cover and dwarf acacia, interspersed with isolated trees and patches of acacia scrub.

Following the initial discovery of platinum north of Warmbaths in 1923, a discovery on the eastern lobe of the Bushveld Complex was brought to the attention of Dr. Hans Merensky in 1924. During the latter half of 1925, prospectors working under the direction of Dr. Merensky discovered the platinum "reef" in the Rustenburg area. Once the reef's position relative to the well-known chromitite layers was appreciated, the suboutcrop beneath the black turf was easily traced in the whole area.

Consumption of platinum group metals increased in the 1960s, brought on by the introduction of catalytic converters to the automobile industry. Following 2 years of exploration and evaluation of the area, Impala Platinum was created in September 1967 to produce no less than 100,000 oz of platinum per year. In July 1969, platinum group metals and associated copper and nickel were first produced. Expansion started to bring production up to the current level of 1 million oz of platinum per year.

To maintain the 1.2 million tonnes per month of ore for the next 20 years, 13 operational vertical shaft systems are in place with five declines from the existing shafts being developed or in place. Although gradual deepening of the workings will occur, the average depth in 20 years time will only be 1,100 m, which is relatively shallow by South African standards. Current reserves and resources can sustain production for 30 to 35 years.

Ore is transported to two treatment plants on the property by diesel or electric locomotives operating on 85 km of national gauge track. Ore is milled and a flotation concentrate is smelted on the mine site to create a matte that is treated at the company's refinery at Springs, east of Johannesburg. At Springs, all platinum group metals are produced for sale along with nickel, copper, gold, silver, and cobalt.

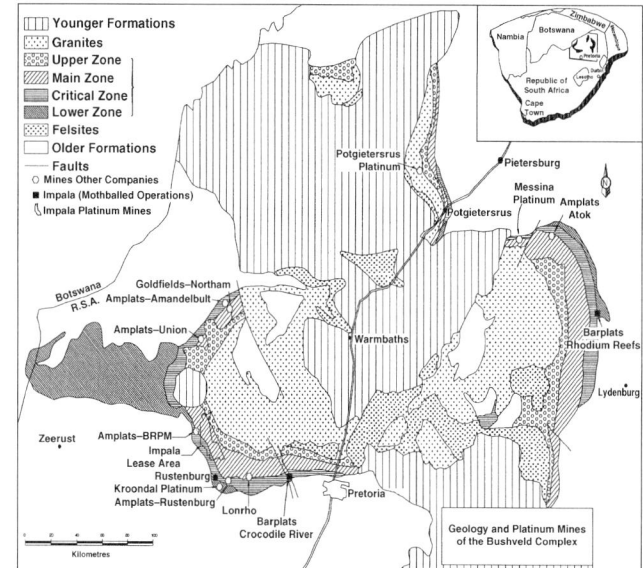

FIGURE 16.1   Bushveld Complex

A labour force of 28,700 is employed by the company on the 13 shafts, two processing plants, the smelter, and refineries. This total has been reduced by 30% over the last 5 years to make Impala an extremely cost-competitive primary platinum producer. It is worth noting that the deepest shaft (the 12 shaft) is also the lowest-cost producer, putting pay to the belief that going deeper must also add costs.

16.2   GEOLOGY

The Bushveld Complex is an approximately 2000-million-year-old, layered igneous intrusion into the Transvaal Supergroup. The intrusion has a large surface area with outcrop extremities of about 450 km east-west and 300 km north-south (Figure 16.1). It is quadrilobate in shape, with the lobes having the form of inverted cones. The complex comprises an array of diverse igneous rocks from plutonic to volcanic with a range in composition from mafic to felsic. The well-layered ultramafic-to-mafic succession comprises the Rustenburg Layered Suite, in which both the Merensky Reef and the UG2 chromitite layer are found. These two horizons are located in the Upper Critical Zone of the Rustenburg Layered Suite and are exploited for their platinum-group-element content and associated metals.

---

* Impala Platinum Ltd., Marshalltown, Republic of South Africa.
† Consulting Mining Engineer, Chichester, West Sussex, United Kingdom.

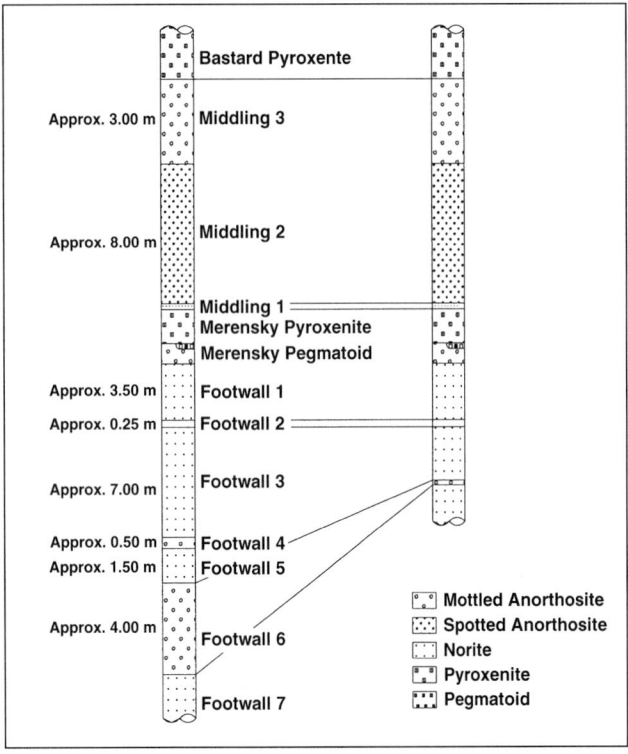

FIGURE 16.2   Geologic section, Merensky Reef

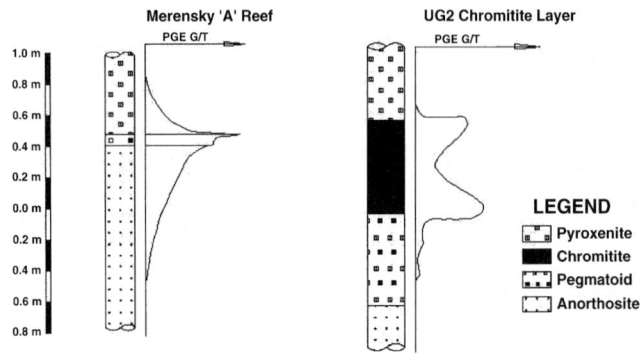

FIGURE 16.3   Generalised histogram showing typical platinum group element values in the Merensky Reef and UG2 chromitite layer

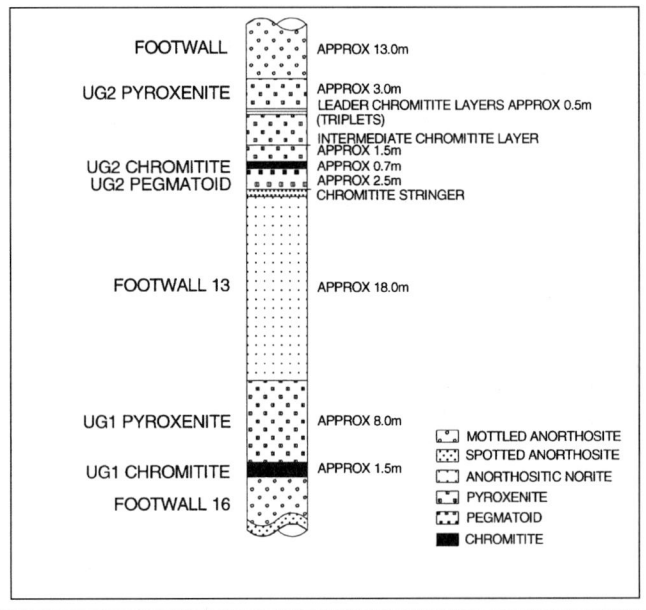

FIGURE 16.4   Geologic section, UG2 chromitite

16.2.1   Merensky Reef

Two Merensky Reef types are recognised, a pyroxenite and a pegmatoid reef. The pyroxenite reef has a basal chromitite layer up to 30 mm thick resting directly on the footwall layers (Figure 16.2). The pegmatoid reef lies below the basal chromitite layer and may also have another chromitite layer at the contact with the footwall. The pegmatoid is a very coarse crystalline pyroxenite. The Merensky Reef defines the level of a major erosional surface. The reef thus regularly transgresses the footwall layers, a phenomenon frequently ascribed to potholing.

The economic mineralisation of the Merensky Reef is concordantly correlated with nickel and copper sulphides, which are commonly concentrated at the chromitite markers and the pegmatoid. The mineralisation extends into the overlying Merensky pyroxenite as well as the underlying anorthositic footwall (Figure 16.3).

16.2.2   UG2 Chromitite Layer

A coarse plagioclase-orthopyroxene pegmatoid layer ranging in thickness from zero to 150 cm represents the base of this unit. The absence of this unit usually indicates potholing of the UG2 chromitite layer. Overlying the pegmatoid, with an irregular bottom contact, is the UG2 chromitite layer, which varies between 60 and 75 cm thick (Figure 16.4). The chromitite layer has a sharp contact with the overlying 8-m-thick porphyritic pyroxenite in which three leader chromitite layers are always present. Known locally as the "leader triplets," they occur within a thickness of 45 cm and may lie from a few centimetres to a few metres above the UG2 chromitite layer.

An intermediate leader chromitite layer a few centimetres thick may occur between the leaders and the UG2 chromitite layer. The proximity of the leader layers affect the stability of the hanging wall in mining operations.

The higher-grade platinum group elements occur at the upper and basal contacts of the UG2 chromitite layer (Figure 16.3)

16.2.3   Structural Aspects

North and northwest strike-slip faults are common at Impala; however, dip faults and isolated areas with complex faulting and folding also occur. The scale of faulting varies from 1 to 150 m of vertical displacement. Ground conditions are usually adversely affected by the faults.

Northwest- and west-trending dolerite dykes ranging in excess of 25 m thick are present. In addition, thin (±1 m) lamprophyre dykes have intruded into zones of weakness, e.g., faults, joints, or potholes, and are widespread throughout the area. Pegmatite veins or dykes are also common.

Potholes are slump structures in which the reef horizon or the overlying pyroxenite layer lies at a lower horizon than normal. They are roughly circular in shape and may vary from tens to hundreds of metres in diameter; depth varies, but can be in excess of 100 m. Composite potholes are sometimes found, resulting in an irregular shape because of the coalescence of two or more individual potholes.

Potholes are found on the Merensky Reef and the UG2 chromitite layer and have been noted at other stratigraphic levels. There is no apparent correlation among potholes on the various stratigraphic levels; however, these have a significant effect on ore extraction and are often not mineable.

Small, magnetite-dunite pegmatoid pipes or plugs are known in the area. Irregularly shaped and scattered bodies of magnetite-rich ultramafic pegmatoid are also widespread, lying above, below, and at the reef horizons. These bodies are not intrusive and where intersected in underground workings, show evidence of replacement of the host rock. In general, the replacement process seems more common in anorthositic rocks and could adversely affect mining conditions.

16.3 GEOLOGICAL PARAMETERS

16.3.1 Area Classification

1A. Merensky Reef, 30 to 1,000 m

- Potential for plug failure to surface.

- Potential for backbreak failures up to the Bastard pyroxenite contact.

1B. UG2 Seam, 30 to 1,000 m

- Potential for plug failure to surface.

- Potential parting failure between Merensky and UG2 horizons.

- Potential for beam failures within the pyroxenite layer up to the first chromitite stringer.

2A. Merensky Reef, 1,000 to 1,500 m

- Potential for backbreak failures up to the Bastard pyroxenite contact.

2B. UG2 Reef, 1,000 to 1,500 m

- Potential parting failure between Merensky and UG2 horizons.

- Potential for beam failures within the pyroxenite layer up to the first chromitite stringer.

In the context of geotechnical areas, the virgin stress state is of major importance as it plays an important roll in hanging wall stability. Stress measurements indicate that the k-ratio can be as high as 2 to depths of 500 m and at least 1 up to 1,000 m. K ratios assist with hanging wall stability, but in tunnels can sometimes lead to hanging wall slabbing where the host rock mass strength is lower.

16.3.2 Rock Mass Classification

Impala uses a rock mass classification system called the Q system. The idea behind the Q system is that enough information obtained from underground observations is included to provide a realistic assessment of the rock mass strength and hence the stability of excavations developed in that rock mass. The following parameters are taken into account.

- Rock quality designation (RQD)–joints per linear metre.

- Joint number (Jn)–number of joint sets.

- Joint roughness (Jr)–quality of the surface of the joint.

- Joint alteration (Ja)–thickness of the joint fill.

- Joint water (w)–presence of water.

- Stress reduction factor (SRF), specific conditions for Merensky and UG2 horizons are included.

The equation used to determine the Q rating for the rock mass is—

$$Q = RQD/Jn \times Jr/Ja \times Jw/SRF.$$

TABLE 16.1 Q rating system and ground conditions

Q rating	Ground conditions
0.005–0.9	Very poor
1–5	Poor
6–10	Good
10+	Very good

Section From Reef Outcrop to 11 Shaft

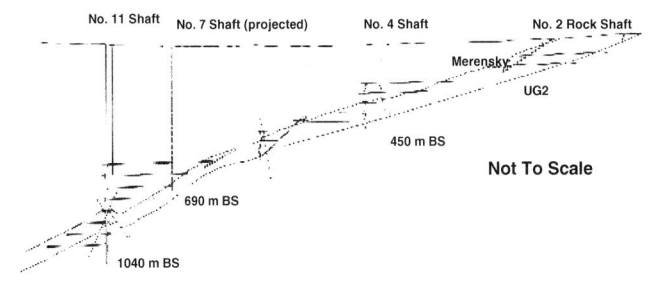

FIGURE 16.5 Locations of shafts

Table 16.1 shows the provisional scale that can be used to provide an estimate of rock mass conditions associated with the Q rating system. Rock mass ratings are measured on a daily basis and are used to identify critical areas.

16.3.3 Rock Fracturing

Presently stress fracturing is not a major problem at Impala. However, there are areas in the hanging wall that consist of a spotted anorthosite, and, because of the brittle nature of this rock, fracturing does occur, resulting in the following failure modes.

Development. Fracturing of the hanging wall as result of the high K-ratio, i.e., higher horizontal stresses.

Stoping. Tensile fractures in the immediate hanging wall, resulting in flat fractures and/or large flat falls of ground between the face and the permanent support. These fractures can extend up to 2.5 to 3 m into the hanging wall.

To provide for the safest working environment, the support pattern and types of support are selected based on placing active support as close to the working face as practicable. This is in contrast to the passive support installed in the deep-level gold mines of the Witwatersrand.

16.4 DEVELOPMENT

16.4.1 Shaft Systems

Initial mining was carried out from tracked inclines developed on the reef plane; however, with the early expansion of operations, steeper declines (25°) that provided faster hoisting speeds and higher capacity production units were installed. Later, short (400-m) shafts were sunk some 1,000 m apart from which declines were developed to extend their production lives.

Second-generation shafts were sunk to a depth of 750 m and spaced 3,000 to 3,500 m on strike. Third-generation shafts with similar strike spacings and sunk to 1,000 m provide the base from which conveyor declines are being developed into the deeper sections of the mining area. In virtually all cases, all shafts are able to exploit both reefs (Figure 16.5).

The horizontal level spacing is mainly 55 m that, along with the local dip of the reefs, provides raise lengths on the order of 300 m. Mining blocks are from 120 to 180 m on strike. The reefs generally dip at about 10° in this area, but can flatten out or steepen to 15° on occasion.

LAYBYE, MATERIAL BAY, TRAVELLINGWAY, STEP-OVER,
RAISE/WINZE, OREPASS LAYOUT

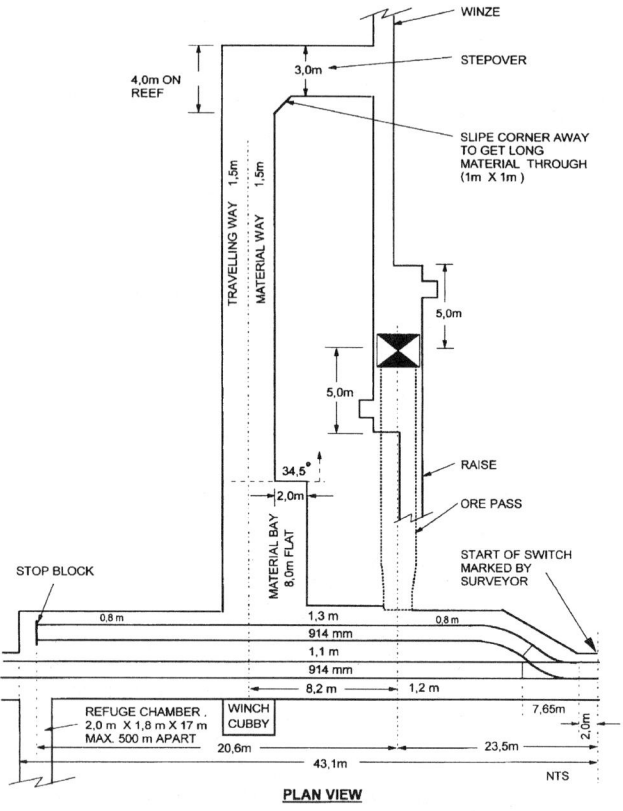

FIGURE 16.6   Typical laybye layout at Impala

TRAVELLING WAY/STEPOVER
INTERSECTION

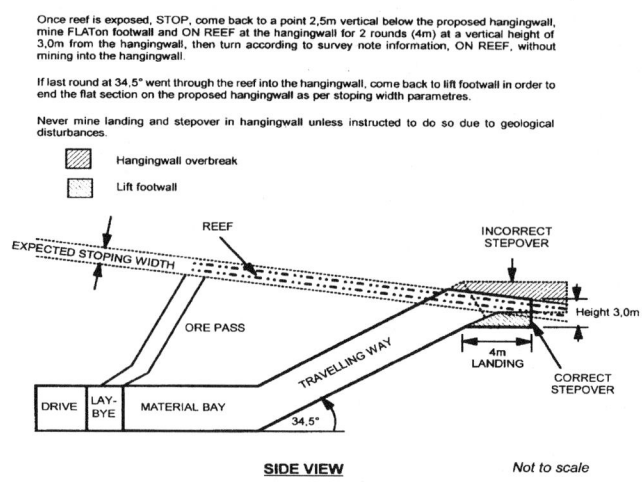

FIGURE 16.7   Travelling way and ore pass layout at Impala

16.4.2   Off-Reef Development

Following station development, crosscuts to the reef are developed
on each level. Footwall drives are developed about 18 m below the
reef on a specific geologic horizon, where possible, to permit easy
identification. The position of footwall drives is often determined
by the prevailing stratigraphy, which can affect ground conditions
in the hanging wall. Footwall drives are frequently developed in a
ubiquitous unit; diamond drilling is used to guide development.
Supervisory personnel can, in certain instances, easily identify
whether the development end is traversing normal or abnormal
geologic conditions. As soon as abnormal geology is intersected, a
geologist can be called in for assistance and advice.

Use is made of a laybye, with an ore pass and travellingway
developed up to the reef horizon (Figure 16.6). In this instance, the
footwall drive is widened (on the downdip side) to allow for the
installation of a rail turnout and duplicate track system. The laybye
will be about 40 m long. The ore pass is developed to intersect the
reef, after which the travellingway is laid out and developed.

16.4.3   On-Reef Development

When the travellingway reaches the reef position, a short "step-
over" is blasted to intersect the reef raise position about 20 m
below the ore pass position (Figure 16.7). This 20-m raise is
blasted to hole into the ore pass and extended for 6 m beyond the
ore pass. The broken ore is used to fill up the ore pass to facilitate
the safe construction of the 30- by 75-cm grizzly.

Boxfront construction and chute installation are completed
prior to holing the raise and ore pass. Seventy-five-kilowatt
winches are then installed to facilitate the development of both the
raise and winze on reef between the different levels (±300 m).

Reef raises are developed on reef where possible to
accommodate both sampling and future stoping from level to
level. These raises (1.2 m wide by 3 m high) are then ledged and
equipped for stoping to commence.

Development is carried out by drilling 2-m rounds in flat
ends 3 m wide by 3.2 m high using hand-held jackhammers with
airlegs from drill platforms. A nine-hole burn cut is standard with
ammonium nitrate-based explosives used in all but bottom and
wet holes. In decline development and capital projects, trackless
drill rigs and load-haul-dump (LHD) machines load material onto
an advancing conveyor.

16.5   STOPING

16.5.1   Layout

After holing the raise or winze, ledging is carried out to prepare
the new line for stoping operations. On completion of ledging,
permanent stope services (power, compressed air, water, etc.) are
installed together with the scraper winches required to move the
broken ore from the stope faces to the centre gully. The original
development winch installed in the raiseline remains in place as
the centre gully scraping winch for stoping operations.

Panel lengths vary according to local condition: 36 m
maximum on the Merensky Reef and 30 m maximum on the UG2
horizon. Strike gullies are developed at an inclination of 25° or
more above strike to assist with negotiating rolling reefs and
potholes and to eliminate waterlogging.

Crush pillars 6 m long on strike and 3 m on dip with 2-m-
wide holings between them are left in place as part of the overall
support of the stoping horizon (Figure 16.8).

Advanced strike gullies (ASGs) are carried between 4 and 6 m
ahead of the stoping face. The ASG is kept as narrow as possible
(±1.2 m) to lessen the span across the gully, which is typically
supported with grouted rebar.

16.5.2   Drilling

The height of the gully is normally 2.5 m for ore capacity whilst
scraping during cleaning operations. ASGs are advanced at the
same rate as the stoping face and are drilled with some 34 holes.

Stope faces are carried at 90° to the ASG and are mined
from as low as 80 cm up to 150 cm, depending on the type of reef
and local valuations. Typically, the UG2 reef is mined at the
minimum possible practical width, while the Merensky is mined
at an average of about 100 cm.

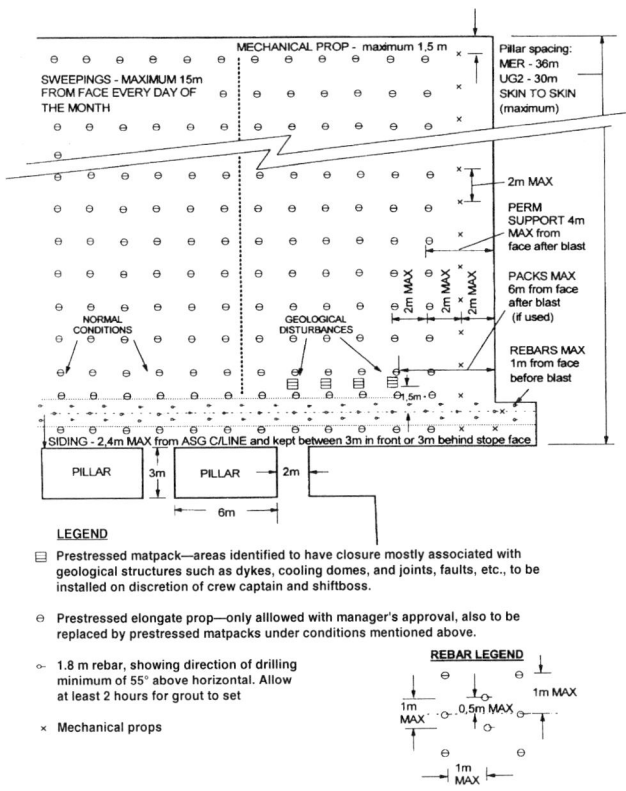

FIGURE 16.8   Panel layout at Impala

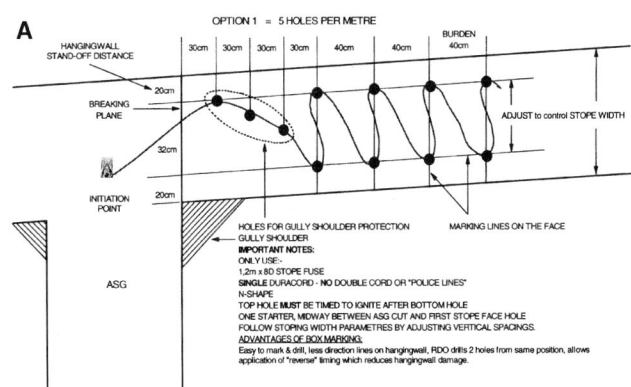

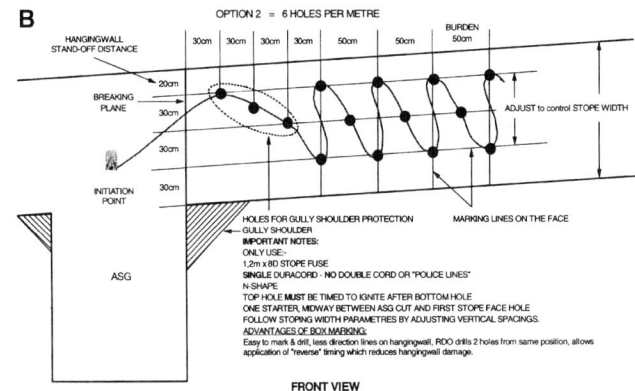

FIGURE 16.9   Timing of stope blasting: A, Option 1; B, option 2

Faces are drilled with 1.2-m-long steel using hand-held jackhammers with short airlegs. On the UG2 reef, the holes are spaced 40 cm apart, and two rows are drilled. On the Merensky, three rows of holes are drilled; the top and bottom rows are identical and spaced 50 cm apart. A third row is inserted to form a diamond pattern between the upper and lower rows. The low gauge loss in drilling the pyroxenites and norites of the Bushveld Complex means that drill-steel life is long (>500 m). Knock-off bits of 38 mm are used.

16.5.3   Blasting and Cleaning

Charging up is done with low-density ammonium nitrate explosives. Capped fuses with igniter cord complete blasting preparations. Timing of blasting the face holes is achieved by the burning front of the igniter cord (Figure 16.9). Faces are connected to a central blasting system on each shaft, and blasting is done at the end of the morning shift when the shift is clear. A 4-hr period is observed in areas affected by blasting fumes before the night shift is allowed to enter to clean.

Cleaning is carried out with 56-kW face-scraping winches pulling a 1-tonne scoop into the ASG, then strike scraping to the centre gully. The 75-kW centre gully winch pulls the ore to the stope ore pass.

16.5.4   Support

Stability pillars and in-stope crush pillars carried on strike next to each panel are the primary supports. In-stope support is by means of elongate supports prestressed to about 20 tonnes and placed at 2 m apart on dip and strike. Mechanical props are used as temporary support at the stope faces whilst drilling and charging up. The mechanical props are removed at the conclusion of the shift and stored behind the blasting barricade. Blasting barricades are typically made up of low-density

polyethylene (LDPE) pipe attached to the elongate supports by means of 7-mm chain. Cleaning of the back areas (sweepings) is maintained up to behind this barricade.

16.5.5   Redevelopment

Because of geological disturbances (mainly potholes), redevelopment may be required. Once the geometry of the disturbance has been determined, an ASG will be continued to a predetermined position before reraising is carried out. Stoping will commence when the raise has been holed.

16.6   GRADE CONTROL

From the grade distribution profiles in Figure 16.3, it can be appreciated that changes in platinum group metal proportions within the ore and metal prices do change the width of the Merensky Reef that is economic to mine. With all platinum values in the UG2 chromitite layer lying within this narrow layer, every effort is made to keep mining widths to a minimum.

Sampling of on-reef development is carried out every 20 m by using a compressed-air-powered diamond saw to cut 8-cm triangular sections through the reef. The samples in the vertical direction are organised in 10-cm-thick layers, and this is known as a sample set. As the stope face advances, a sample is cut in the ASG for each 1,600 m^2 of mining.

Block values are determined by ordinary kriging of values within each sample layer, and economic widths are determined for each block by applying marginal cut-off grades to the sample layers, taking a minimum mining width of 80 cm. Mining parameters are then generated by using the chromitite layer at the base of the Merensky pyroxenite as the marker from which channel parameters are measured. Mining of the UG2 involves taking the whole chromitite layer, with allowable overbreak being taken in the footwall pegmatoids.

As with all narrow-reef mining operations, grade control revolves around the measurement and control of the sources of dilution and loss generated within the stope. Techniques include off-reef mining through potholing and/or rolling reef, ASG development, travellingways within the stoping area, and overbreaking in the hanging wall or footwall along with underbreaking into the planned mining take-out. Surveys of face positions are carried out monthly, and stope width measurements are taken at least twice during the month for each mining panel.

Go belt hammer sampling on the conveyors feeding the plant provides a means of accounting for production from each shaft as each surface train of railcars is weighed.

16.7 VENTILATION

In the Bushveld igneous complex, rock temperatures 1,000 m below the surface are approximately 46°C. The geothermal gradient is typically 2.2°C per 100 m, which is virtually double the gradient at the Witwatersrand gold mines. For the third-generation shafts, the air entering the mine requires cooling. Surface ammonia refrigeration plants of 9.5 MW with bulk air cooling provide downcast air at 12° to 15°C at the bottom working level; chilled service water is not required. The downcast velocity is typically between 10 and 12 m/s. Doors in series on the upper levels control air flow. The air is then distributed to the various stope lines from the footwall haulages and along the reef horizon to the surface fans.

Air flow in the raise is directed to the stope face by means of conveyor belt strips hung in the raise and polypropylene curtains hung on the updip side of ASGs. Downdip of the ASG, all except the last pillar holing is closed off to ensure that the air reaches the working face. Raise-bored holes to the surface provide upcast facilities with velocities of between 18 and 22 m/s.

The quantity of air required for mining is approximately 3 m³/s per kiloton per month (reef and waste) at mean rock-breaking depth. To ensure that air distribution on the shaft is effective, old worked-out areas are closed off with brick seals. The shafts are designed for a reject wet bulb temperature of 27.5°C.

In areas where trackless equipment is being used, increased air volumes are provided to ensure sufficient dilution of exhaust gases. Local regulations require that individual self-rescuers are provided to all personnel working in such an area or in the downcast air from such area. In all other areas, refuge bays are provided.

Methane does occur sporadically in the workings, and all development ends not holed into the main ventilation circuit are regarded as possible sources.

16.8 ORE AND MATERIAL TRANSPORT

Ore is drawn from stope ore passes into 4-tonne rail cars. Ore trains of 6- by 4-tonne cars at Merensky and 5- by 4-tonne cars at UG2 are pulled by 6-tonne, battery-powered locomotives to the station tips. Separate ore passes for each ore type and waste create three separate rock streams. Balanced 11-tonne skips deliver the ore and waste up the shaft into separate shaft headgear bins. Waste dumps are built at each shaft, and 50-tonne railcars deliver ore to the processing plant.

Men and material are hoisted to and from the working levels using multideck cages that hold up to 90 people per trip (Figure 16.10). Material is placed on rail cars for transportation into the working section. Mono rope winches haul material up the travellingway into the stope. Where material is to be used down a decline, pallets of material are moved by overhead monorail to the required level. Chair lifts transport personnel to the workings below the bottom shaft level.

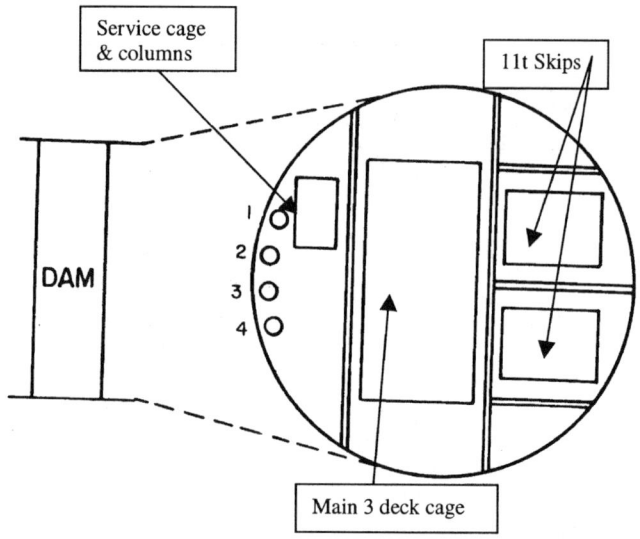

FIGURE 16.10   Cross section through typical third-generation shaft station

TABLE 16.2   Five-year production record

	1999	1998	1997	1996	1995
Tonnes milled (000's)	14638	14509	13775	13475	13703
Percentage of UG2 mined	48.1	45.9	45.6	51.8	48.4
Headgrade for platinum group elements plus gold, gm/tonne	5.31	5.17	5.22	5.33	5.45
Square metres per stope employee	41	40	36	34	32
Tonnes per employee	51	48	44	42	41
Number of employees (000's)	28.7	29.5	31.0	31.1	33.1
Fatality accident rate per million man hours	0.09	0.20	0.14	0.18	0.10
Lost-time injury rate per million man hours	10.9	21.2	20.5	29.4	27.4

16.9 PRODUCTION STATISTICS

Overall production statistics for the past 5 years are shown in Table 16.2.

Over the past decade, Impala has undergone a revolution in its operating philosophy and practice that has resulted in the group transforming itself from the highest-cost platinum operation to the lowest-cost primary producer in the industry. It has the best productivity statistics of all South African hard-rock mines, gold and platinum combined.

Achieving this change required a reduction in the labour force along with a complete revision of how Impala ran its business from mining to processing. Success started with the introduction of the participative "one team, one vision" approach through an operational process known as Fixco. Through this process, employees were given a more meaningful say in how the business was run.

A policy of transferring best practice is being implemented whereby top-producing teams train underperforming teams. Currently, the top-producing team performed at 110 m² per stope employee compared to a mine rate of 41 m² per stope employee in 1999.

TABLE 16.3   Operating cost breakdown (R/$ exchange rate = 6.08)

Mining:		
Labour	13.87	
Stores	6.13	
Utilities	1.13	
Total		21.12
Processing		4.59
Refining		2.96
Selling and administration		0.90
Total costs		29.58

16.10   OPERATING COSTS

A breakdown of the overall cost structure for the last financial year is given in Table 16.3.

16.11   HEALTH AND SAFETY

Impala Platinum, Ltd., has undergone a series of planned health and safety-related processes, beginning in 1993-1994.

- An audit to establish the validity of Impala's injury statistics commenced at the end of 1993. The outcome was that injury reports were standardised at all Impala mines in terms that were accurate and legally valid.

- In 1995, the draft legislation for the Mine Health and Safety Act No. 29 of 1996 was used to establish a health and safety agreement between Impala management and the workforce. This was a historic move to change from a prescriptive to participative style of health and safety management. The benefits thereof are being experienced today.

- The implementation of the Mine Health and Safety Act in 1996 had a positive impact on safety performances on Impala. The fact that there was a need to comply with additional legislative requirements resulted in a refocused approach to managing health and safety issues.

- In October 1996, Impala introduced the concept of best practices within its shaft operations. This had a major impact not only in terms of production efficiencies, but also on safe production practices.

- An Impala risk management process was entrenched by late 1999. The purpose of the process is not only to manage the risks associated with the mining, processing, and refining operations, but also to provide a foundation for the proper management of health and safety issues.

- In 1999, an environmental management programme document was accepted by the relevant legislative authorities. Impala has subsequently elected to go for an ISO 14001 accreditation. The process is well underway, with the current year as the target year for accreditation.

- In the occupational health area, Impala Medical Services has a licensed occupational health bureau that targets the certification of employees to work in certain underground and surface environments, as well as to monitor the health of employees. Hearing conservation and education in AIDS are currently major occupational health issues that are receiving increased attention.

Impala's mineral processes and refineries are currently both 5-star safety organisations that function on the National Occupational Safety Association (NOSA) MBO system. Impala mining operations are in the process of adapting to the NOSA system.

16.12   FUTURE PROGRESS

Impala is constantly examining ways to maintain its status as the lowest-cost primary producer. The following highlight a number of developments being pursued to determine if they are applicable to the platinum mining industry.

- Selective blasting within the stope

- Nonexplosive techniques for rock breaking (cone fracturing, disc cutting, etc.)

- Reengineering layouts for increased production per level

- Increased mechanisation to improved development efficiency

16.13   ACKNOWLEDGMENTS

The authors wish to thank the Impala projects, services, and best practices departments for their contributions and assistance.

16.14   REFERENCES

Jagger, L. 1999. Shallow Platinum Mining in South Africa. South African Institute of Mining and Metallury School on Narrow Tabular Mining in South Africa, 21–22 October.

Baker, M.E. 1993. The Geovaluation System at Impala Platinum Limited. Geostatistical Association of South Africa Symposium, August.

CHAPTER 17

Mining Practice for Intermediate-Depth, Narrow Tabular Ore Bodies with Particular Reference to the Free State and Klerksdorp Goldfields

P.R. Carter*

17.1 INTRODUCTION

17.1.1 Description of Mines

The Tshepong Mine on the northwest margin of the Free State Goldfields is a large modern mine still in the ore build-up phase. The reef is overlain by a fairly thick shale bedding plane in the hanging wall that must be undercut to below the Leader Quartzite middling for the mine to be economically viable. The ore is of average grade, so mining methods are focused on low stoping widths, strategies to maintain the undercut hanging wall, low working costs, and high-efficiency work teams.

Of interest is that the Jeanette Mine to the northwest has never been mined because of problems associated with the thick shale bedding. Conventional undercutting techniques are unsuitable because of blast damage to the thin Basal Quartzite beam. Here, some form of ultralow stoping width, nonexplosive stoping, or reuse open-stope mining methods would have to be employed.

Matjhabeng is an old mine and consists of the old high-grade Free State Geduld, Western Holdings, and the low-grade Welkom Mines. The ore body encompasses the Basal Reef suboutcrop area, which trends northwest from the St. Helena Mine in the south to the old Lorraine Mine to the north. The ground against the outcrop is extensively thrust faulted, and the gains of reef are known as the "overlap area." Mining techniques in this area include vertical and steep stoping. Gold-bearing pyrite stringers occur widely in the Basal Quartzite hanging wall and have been mined extensively at high stoping widths. Extensive mining of stope pillars and extraction of high-grade shaft pillars have taken place. The mine continues east to the Stuurmanspan fault and consists of a series of upthrow faults that contain the reef to intermediate depths.

The Harmony Mine area beyond the De Bron-Homestead faults is characterized by a reef package with very thin shale bedding or the absence of shale. A syncline trends north-south through the lease area. The limbs of the syncline incline east to the Homestead fault and up to the west. Generally, the reefs are flat or gently dipping, steepening toward the edges of the basin. The reef has average- to low-grade values so that mining is focused on highly productive, low-cost, open-stope mining methods.

The Great Noligwa Mine, formally known as Vaal Reefs No. 8 shaft, is fairly typical of the intermediate mining horizon and gives a good representation of mining practice of the Vaal River Reef in the area. The reef is highly faulted, which severely affects efficient mine design layouts. The mine is a high-grade operation typical of a high-gold-recovery shaft.

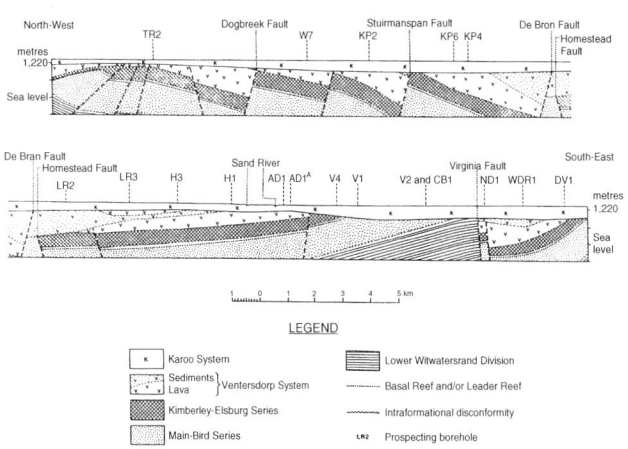

FIGURE 17.1   Generalized section across the Free State Goldfields (Source: *The Mineral Resources of South Africa*)

17.1.2 Overview of Geology at Free State Mines

The Matjhabeng, Tshepong, and Harmony mines are situated in the Free State Goldfields. The main economic reef mined is the Basal Reef of the Main Bird series (Figures 17.1, 17.2, and 17.3). The Matjhabeng Mine is situated on the northwestern margin of the Free State Witwatersrand basin. It extends from the suboutcrop of the Main Bird series of the upper Witwatersrand Division against the Karoo System and extends to the southeast to the Stuurmanspan fault. The reef dips to the southeast. The central area of the mine is characterized by a series of major upthrow normal faults, the biggest of which is the Dagbreek fault.

The reef beyond the Stuurmanspan fault was mined by the old President Brand-President Steyn Mine. The No. 4 shaft of the President Steyn Mine was sunk to exploit the deeper Basal Reef to the De Bron fault some 3,200 m below the surface. The original lease area of the Harmony Mine is situated to the southeast beyond the De Bron and the Homestead faults, which form a graben of the Ventersdorp and the lower portions of the Main Bird series. The Basal Reef is faulted out in this zone. Beyond the Homestead fault, the Basal Reef inclines gently to the southeast to crop out beyond the Sand River against the Karoo System. The Virginia Complex mines extend from the Virginia fault, which throws the reef down to 2,500 m below the surface. The reef rises sharply to the northeast to crop out again unconformably against the Karoo System.

---

* Witwatersrand Technikon, Johannesburg, South Africa.

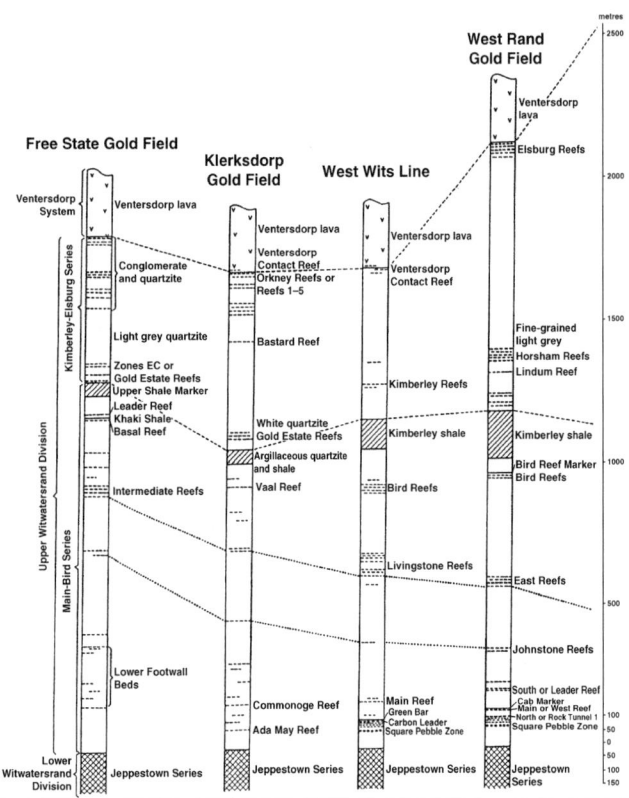

FIGURE 17.2 Comparative stratigraphic columns (Source: *The Mineral Resources of South Africa*)

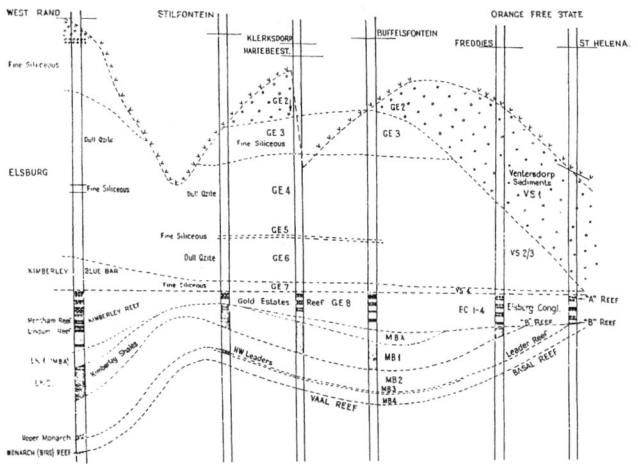

FIGURE 17.3 Geological section of West Rand, Klerksdorp, and Orange Free State (Source: J.N. Morgan, *Mining of the Vaal Reef in the Klerksdorp Area*)

In the major fault zones, i.e., the overlap area against the western margin, the Dagbreek, and the De Bron, the reef has been dragged up or down in the direction of movement of the fault. Some steep-to-vertical portions of the reefs are common in the shear zones.

17.2 MINING PRACTICE AT FREE STATE MINES

17.2.1 Tshepong Mine

The Tshepong Mine lies just northeast of the town of Odendaalsrus. The shaft was sunk in the early 1980s and was the

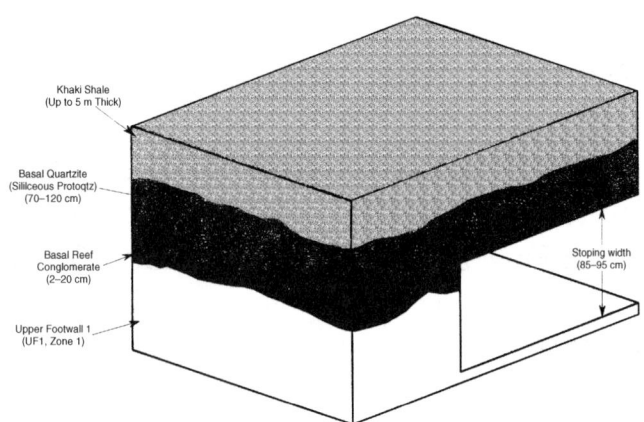

FIGURE 17.4 Three-dimensional schematic section showing optimum stoping cut

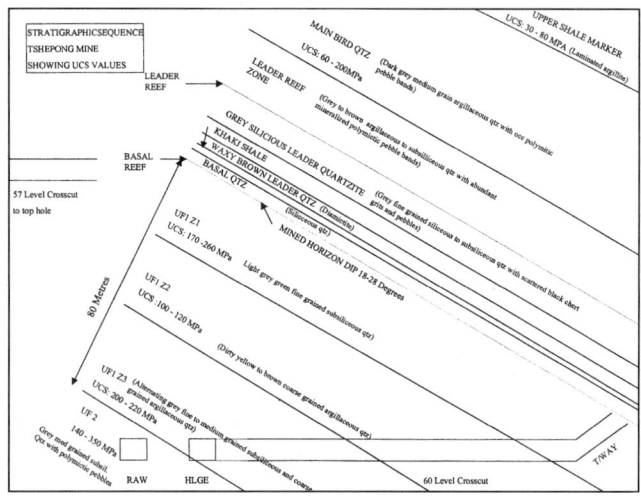

FIGURE 17.5 Stratigraphic sequence at Tshepong Mine showing values for uniaxial compressive strength (UCS)

last big project for Anglo-American in the Free State, Welkom area. The shaft was sunk to exploit the ground on the northwestern margin of the Free State basin.

The shaft does not have its own mine hostel for its migrant workers. In the village of Khutsong, a joint-venture housing development project sponsored by the Free State government, the municipality of Odendaalsrus, and the mine has been established. As it is close to the shaft, the workers of the mine and their families reside here. Today, the village of Khutsong is a large town and is a model of what can be achieved in providing housing for local inhabitants.

Geology. The reef at the Tshepong Mine dips 18° to 20° to the southeast. The immediate hanging wall above the reef consists of the Basal Quartzite, which is 70 to 120 mm thick, overlain by the Khaki Shale, which varies in thickness up to 5 m. Because of the width of the Khaki Shale band, the stope must undercut the shale to keep the face payable (Figure 17.4).

Footwall development is carried 80 m below the reef in the competent zone of the UF1 zone 3 (UF1-Z3) quartzites just above UF2 quartzites. The uniaxial compressive strength of the UF1-Z3 quartzite is between 200 to 220 MPa, while the uniaxial compressive strength of the UF2 varies from 140 to 350 MPa (Figure 17.5). Stope-induced mining stresses at these depths do not normally exceed the primitive stress 80 m below the stoping

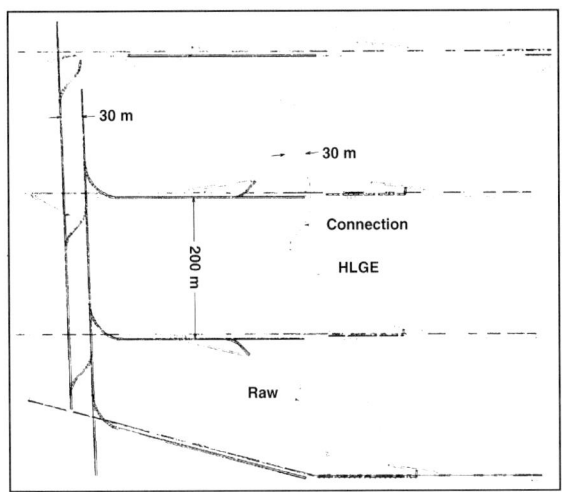

FIGURE 17.6 Typical Basal Reef mine design layout, plan view

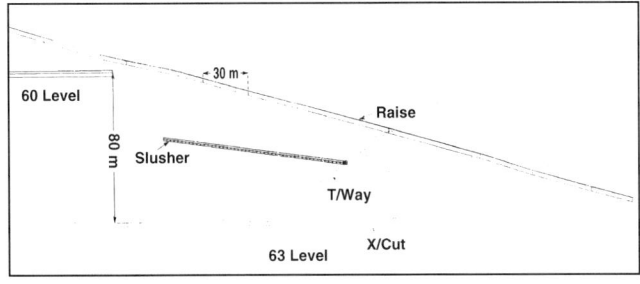

FIGURE 17.7 Typical Basal Reef mine design layout, section view

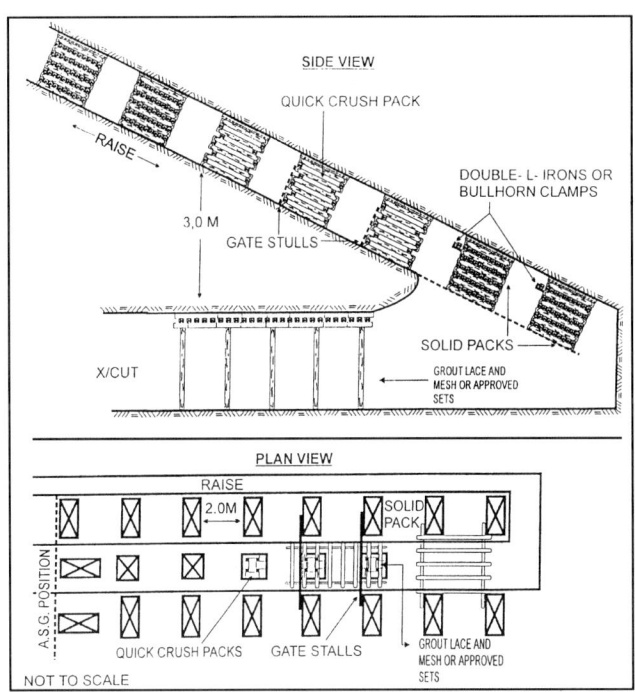

FIGURE 17.8 Overstoping crosscut

excavations, although good mining practice is to overstope footwall tunnels first.

Mining Practice. *Mine Layout and Stoping Practice.* The mine was originally laid out as a trackless mine. As a result, the levels were spaced 80 m apart vertically, and the crosscuts were spaced 400 m apart. Subsequently, the decision to mine using trackless methods was reversed because of the high capital purchase and maintenance costs of trackless vehicles. Distances between the crosscuts has been reduced to 200 m so that scraper winches are able to pull the advance strike gullies.

To mine the 330-m-long raises, a travellingway has been laid out to access the middle of the stope from the bottom crosscut so that 10 panels lie below the intersection and 12 panels lie above. This travellingway improves the transport of miners and material into the stope. From this travellingway, a slusher is developed 30 to 40 m updip below the reef. This slusher serves as an access tunnel into which ore passes developed from the footwall crosscut below hole into the ore passes developed to the reef above. Boxholes developed longer than 60 m are not feasible for hand-held conventional equipment as advance is very slow in the last 20 m because of the difficulty of traveling and installing ventilation and services. The slusher reduces the vertical distance and allows the boxholes above and below the slusher to be developed simultaneously at a faster rate. Boxholes are laid out so that each two panels have their own tip. Figures 17.6 and 17.7 show the grid layout for haulage, crosscuts, boxholes, and raises in plan and section views, respectively.

Development. Development is done with conventional hand-held equipment and air-powered loaders. Development was previously done with trackless equipment.

Ledging. The emphasis in the ledging operation at the Tshepong Mine is to maintain the undercut below the Basal Quartzite beam. It is therefore important to maintain good drilling, blasting, and support practice. Once a holing is made to the Khaki shale horizon, the shale cannot be supported, and attempts must be made to contain the shale in a localized area with stulls and packs. Mining takes place around the fall. If the fall is in the face, the panel needs to be reestablished by means of a wide raise.

Ledging is carried out by means of breast or downdip ledging. If the previous two methods are unsuitable for the area, the panel is cubby ledged.

The process of ledging commences with preledging. During this process, planning is completed, geology studies are finalized, and an up-to-date 1:200-scale mining layout plan is prepared by the mine overseer for approval. Prior to the commencement of ledging, the top and bottom crosscuts with all services must be constructed, as well as all boxfronts and grizzlies. Care must be taken to install good water-handling arrangements in the bottom crosscut as the muddy water caused by the mining process and shale exposures soon result in muddy footwall conditions, which severely hamper tramming.

Mining commences with overstoping of the crosscut. This activity involves ledging the raise from the intersection of the crosscut with the reef to the No. 1 ore pass situated at the first panel. The overstoped portion of the raise is known as the crosscut pillar and is normally mined from the level below. The ledge is mined on either side of the raise. Once it has been secured on both sides with the first line of support (packs), the crosscut side of the raise is ledged first so as to destress the crosscut below. Ledging is advanced over the crosscut until the 45° stress cutoff point. Once the overstoping process is completed, the raise is cleaned out, services extended, and ledging of the raiseline by means of breast, down, or cubby ledging commences, depending on ground conditions or the width of the Basal Quartzite beam (Figures 17.8–17.12).

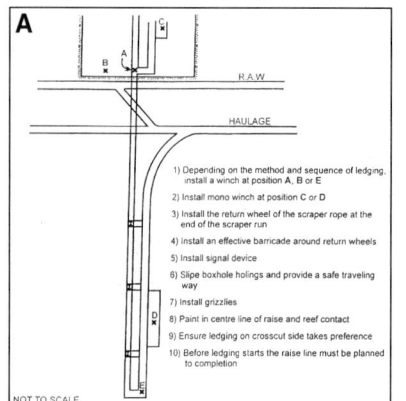
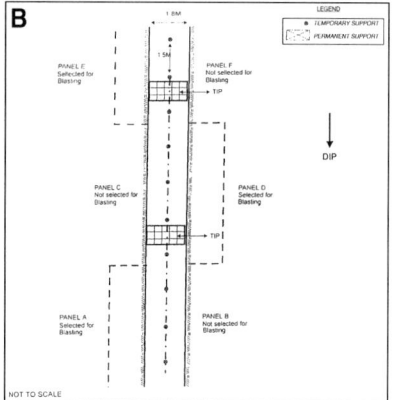
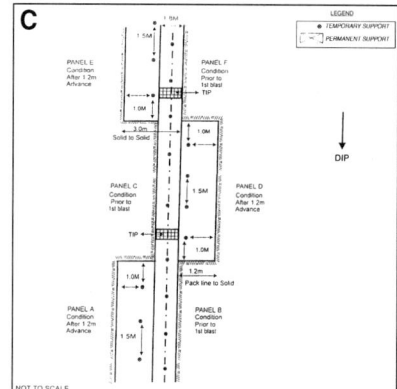
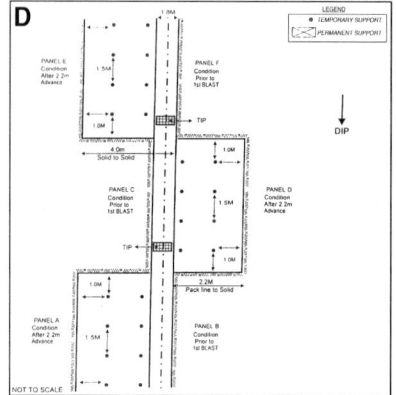
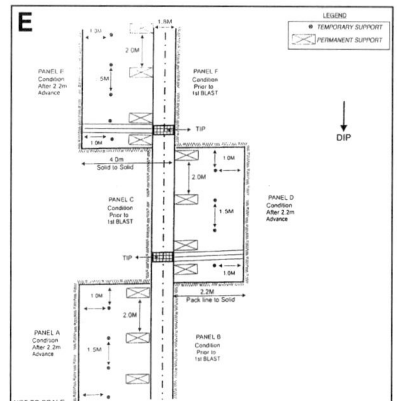
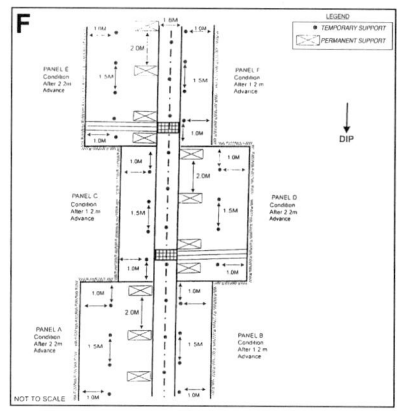
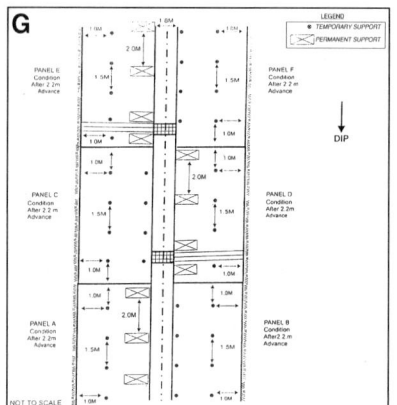
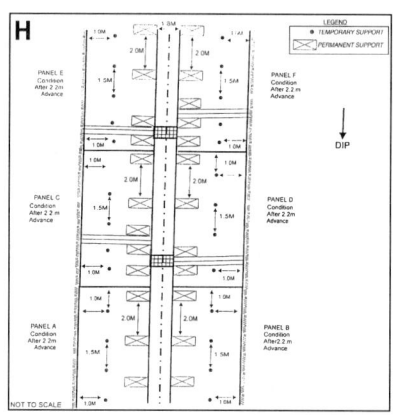
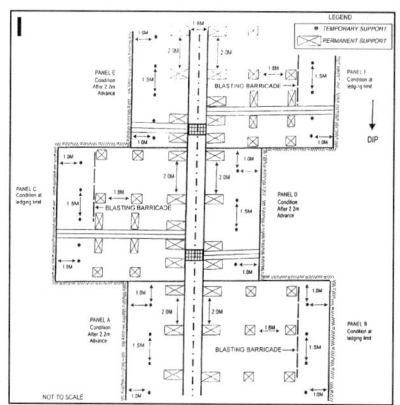
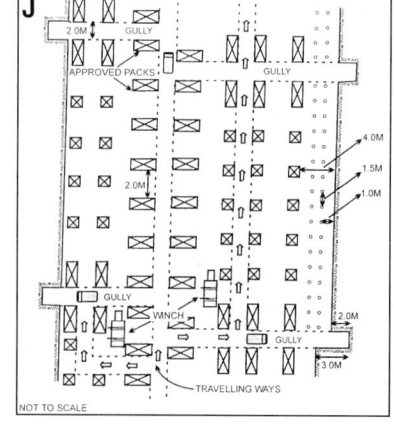

FIGURE 17.9   Breast ledging process. A, Preparation for breast ledging; B, Stage 1; C, Stage 2; D, Stage 3; E, Stage 4; F, Stage 5; G, Stage 6; H, Stage 7; I, Stage 8; J, Fully ledged conventional stope.

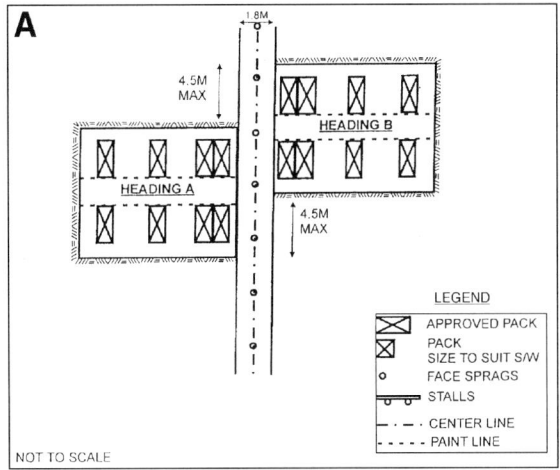

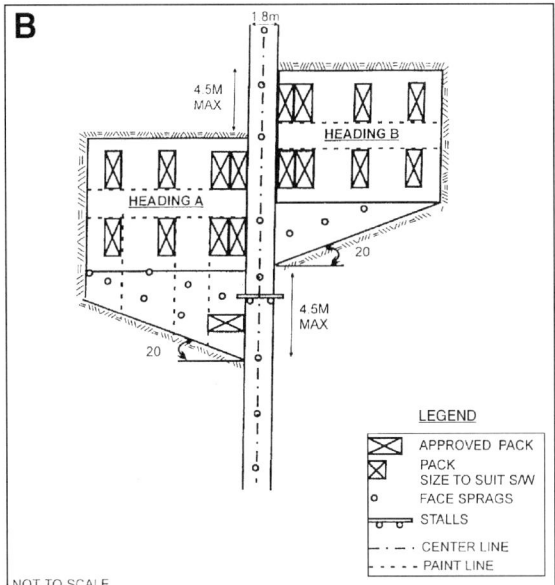

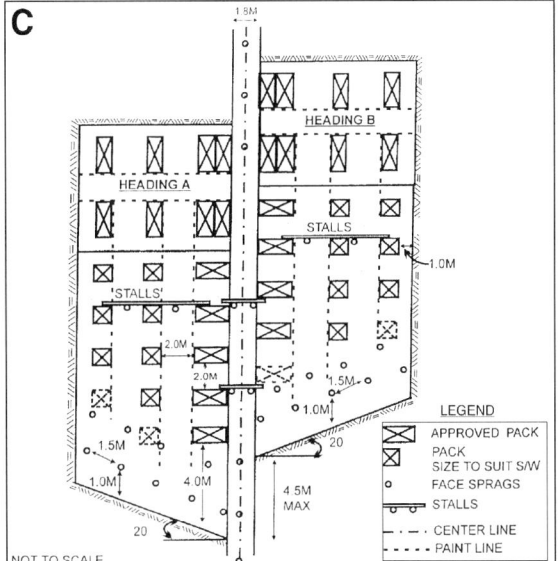

FIGURE 17.10 Stages of downdip ledging process. A, Stage 1;
B, Stage 2; C, Stage 3.

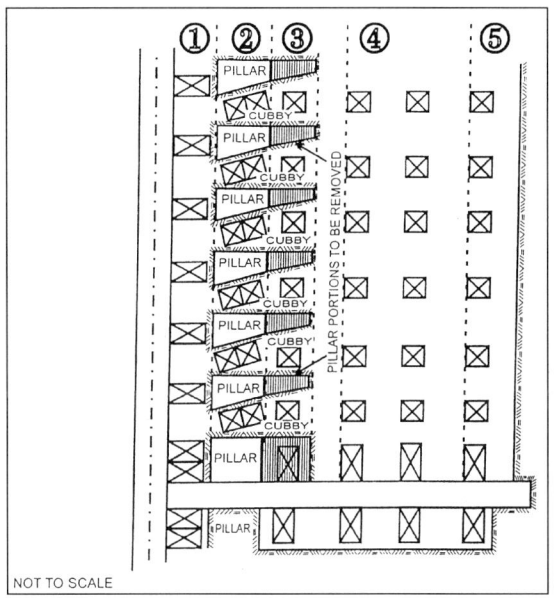

FIGURE 17.11 Stages of cubby ledging

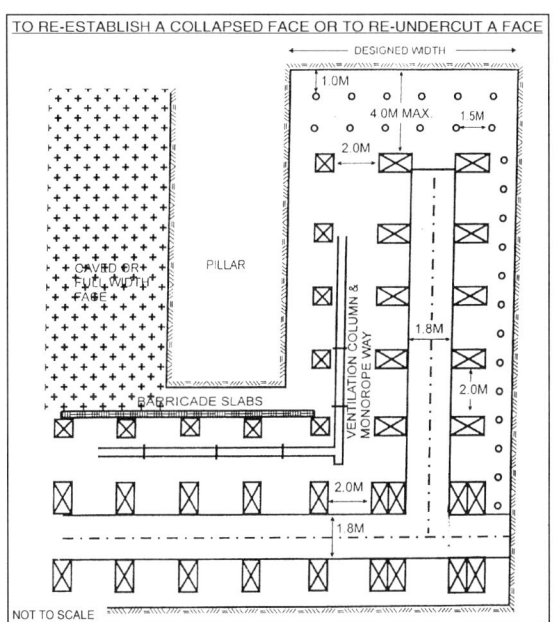

FIGURE 17.12 Wide raising

General. A feature of the mine its "change culture" program, which aims to instill a "zero tolerance" attitude in its employees. Zero tolerance is an attitude whereby management and employees collectively embrace a mindset of no tolerance for deviations from safety and health standards and procedures. The aim is to improve safety and production by working to standards.

Self-directed work teams have been established at each panel managed by a panel leader. The intention is to appoint a blasting certificate holder per panel. The efficiencies of the specially trained teams are considerably better than for teams that have not been through the training process. A specially designed worker incentive bonus scheme based on worker productivity is part of the program The self-directed work team receives 3 weeks of training on the

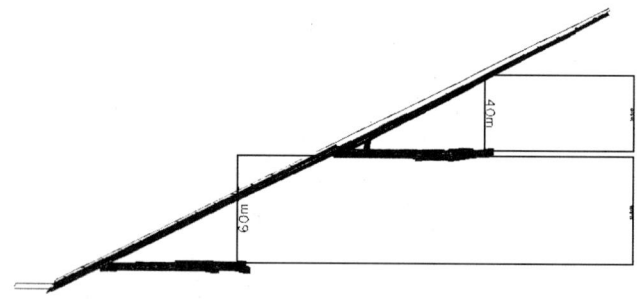

FIGURE 17.13 Cross section of raise layout

TABLE 17.1 Pipe sizes for various services, millimetres

	Air	Water	Drinking water	Vent column, minimum	Water blast
Haulage	200 minimum	150	25 or 50	570	25
Crosscuts	150	100	25	570	25
RAWs and connections	100	50	—	570	25
Maximum distance to the face, m	15	15	60	12	12

surface during which team-building programs, problem-solving, and other planning and mining training, as well as multitask training, are done. Although multitask traning is unpopular with the mining trade union because of job preservation issues, increased worker productivity and hence earnings of individuals have tempered this resistance. The training is designed to empower the teams to identify their own production problems and solve them if they can. If the problem can not be solved at the team level, team members are encouraged to get it fixed quickly with the help of their line supervisors.

Costs and Efficiencies.

Cash cost, dollars per ounce	227
Total employees costed	5,200
Square metres mined per month	29,900
Number of stope employees	1,607
Metres developed per month	1,890
Square metres per stope employee	18.6
Square metres per total employee costed	5.96
Metres of development per employee	3.6
Stoping cost, R/m^2	353
Development cost, R/m	1,523
Working cost, R/m^2	1,308
Cash cost, R/tonne milled	377
Ratio of m^2:m	15.98
Face advance, m	8.18
Recovery grade, gm/tonne	8.0
Tonnes milled per month	104,000
Stoping width, cm	92

17.2.2 Matjhabeng Mine

Mining Practice. *Layout.* The mining layout is scattered. The crosscut grid is spaced 150 m apart, and levels are spaced 60 m vertically. Haulages are situated 40 m below the reef in the UF1 quartzite (Figures 17.13).

Development. All development is done conventionally with hand-held equipment, such as rock drills and airlegs, using a constructed drilling platform. All haulages are developed 3 m wide and 3.3 m high. Grade lines are carried 1.8 m above the footwall elevation, and rail elevation is 1.5 m below grade. Tramming spurs are blasted at suitable intervals, but not closer than 50 m on the drain side of the tunnel. Crosscuts and return airways are established every 150 m. Every crosscut has a timber bay established prior to overstoping operations. In haulages and crosscuts, columns are supported on 0.5-m spreader bars and suspended 2.4 m above the rail with 10-mm short-link chain attached to grouting hooks. Pigtail eyebolts are installed every 30 m for extra strength.

Standards exist for both drag and burn-cut rounds, but most shafts use only the 46 hole burn round in preference to a 42 drag round, as it is easier to mark off and drill. Experience has shown that better advances are achieved with the drag round. Smooth-wall blasting is done only in exceptional circumstances.

All direction lines, grade lines, and holes to be drilled are marked before each blast. All suspension holes for pipes and cables are drilled 30 cm deep and are spaced 6.1 m apart in the hanging wall. Plastic pipes are suspended every 2.5 m to ensure a level pipe.

Table 17.1 provides a list of services and pipe sizes showing maximum distances to the face. Air and water columns become permanent once development is completed.

Twelve sleepers support each 9-m length of 30-kg rail at a gauge of 762 mm with a double sleeper rail at joints. Permanent track construction is maintained 27 m from the face. Concreting of tracks is discretionary, but mandatory below boxes. Most footwall crosscuts to the reef are concreted so as to cope with mud from the stoping operations. Vertical spindle pumps are installed at each crosscut to handle water, particularly if a water jet is used in the stope cleaning operation. The use of a water jet is discouraged for cleaning the high-carbon Basal Reef due to loss of fine gold into crevices or pack support.

Before any work commences, each development end is supported with two mechanical props. Primary support is grouted 12-mm- by 2.3-m-long shepherd's crook rods. For haulages or excavations wider than 3.5 m, grouted shepherd's crooks 16 mm by 3.4 m are used in a 32-mm hole.

For primary support, medium- to fast-setting grout capsules are specified. Permanent support uses pumpable grout cement as it is cheaper. All development areas are grouted using a 2- by 2-m pattern. Special precautions of installing three extra grouted rods on either side of a discontinuity are specified. Grout lacing and meshing is done with 75-mm diamond mesh and 8- to 12-mm-thick, destranded hoist rope clamped with U clips or Crosby clamps.

The end is loaded by means of air-operated loaders into 6-tonne hoppers equipped with either 5- or 8-tonne, battery-operated locomotives.

Basal and Leader Reef Connections. A distinction must be made between open mining (shale is removed to the Leader Quartzite), undercut mining (mining below the shale and a Basal Quartzite beam), and Leader Reef mining (Figures 17.14, 17.15, and 17.16).

The Basal Reef is mined by undercutting. The horizontal stresses, which are 0.3 to 0.5 of the vertical stress, clamp the hanging wall blocks in place. However, undercutting is not always possible because of the proximity of the reef to the shale or because of the fractured, friable nature of the hanging wall beam, which has poor cohesion to the shale or the Leader Quartzite.

The aim of the undercut stoping is to prevent exposing the shale bedding in the hanging wall. Once the shale is exposed, it cannot be supported, and the fallout must be isolated by packs because the shale runs out between the Basal and Leader quartzites. The panel must be reestablished by wide-raising methods or mined with the open method. The open method involves removing the shale up to the Leader Quartzite beam. If the shale is thick, it can make mining uneconomic because of the dilution of the reef by the shale and waste. Once a panel is on

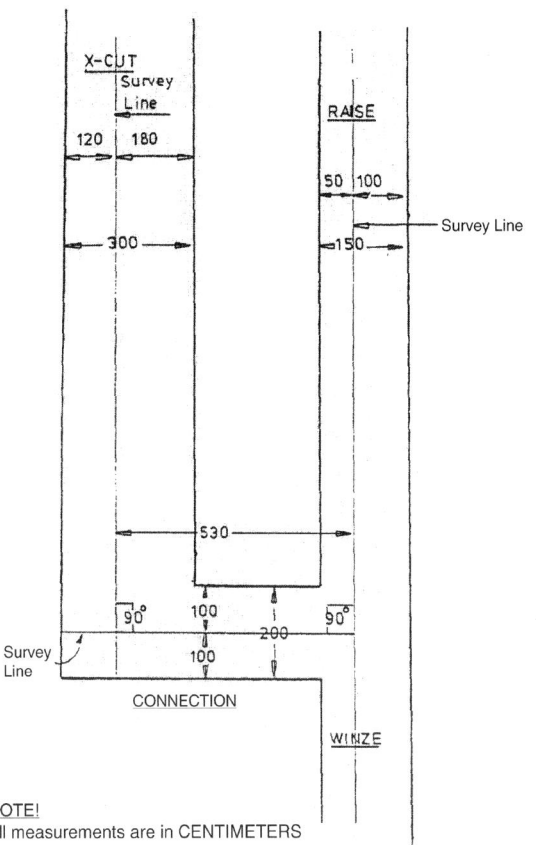

FIGURE 17.14   Raise or winze breakaway

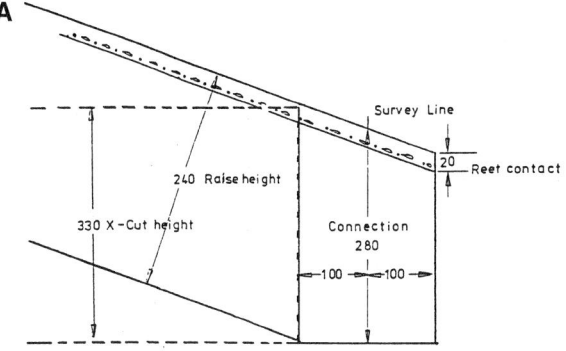

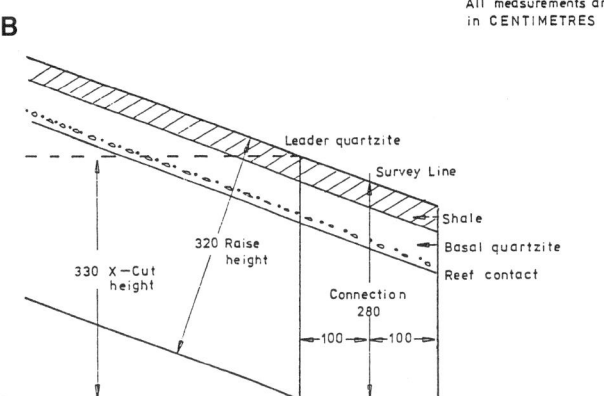

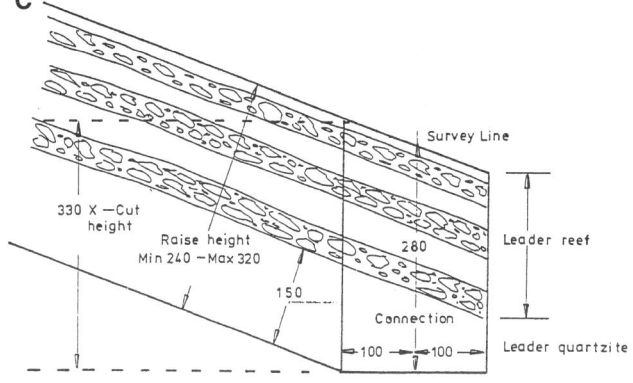

FIGURE 17.15   Raise connections. A, Basal Reef, undercut mining; B, Basal Reef, open mining; C, Leader Reef.

open mining and conditions change to make undercut mining feasible, then the face is undercut again using cubby methods or a wide raise.

Undercut mining is a technique that miners must master to maintain the undercut. An incorrectly marked panel or a carelessly drilled hole can result in loss of the undercut. A miner going on leave sometimes needs to reundercut his panels on his return because the panel has fallen to open mining during his absence.

Open-mining techniques are characterized by high stoping widths and muddy shale conditions that make handling the ore in ore passes, hoppers, and tips a problem. It is advisable to manage the mix of ore from undercut and open mining in a section to minimize the clogging effect of the shale. However, because of the high grades in the Matjhabeng area, open mining methods are still payable.

The Leader Reef lies approximately 17 m above the Basal Reef. The Leader Reef is entirely different from the Basal Reef and, as a result, must be mined in a different way. It can consist of one or more pebble bands (conglomerates) in which the gold is randomly distributed. These conglomerates can vary in thickness over very short distances. They can pinch out and then reappear without much warning. The footwall of this reef is either a waxy brown or a grey siliceous quartzite (Leader Quartzite). To avoid "off reef" mining, the shift boss is required to bring a piece of the footwall to the surface for inspection by a geologist to ensure that an upper band is not being mined. Mapping by the geologist is required every 20 m.

As the values of the Leader Reef are scattered and variable, careful evaluation of payability is required. The reef is prospected during the mining of the Basal Reef by means of drilling from the

crosscuts. Where values are promising, crosscut extensions are mined to the Leader Reef to obtain samples. Raises are developed where grades warrant and then carefully sampled and mapped to expose the pay trends. Regional sampling has highlighted payable shoots that have been mined successfully over wide areas.

The Leader Reef hanging wall is more competent than the Basal Quartzite hanging wall, but can prove treacherous over larger spans because the Leader reef is mined in an environment of destressed tensile strengths. Because of the lack of cohesion between beds, the hanging wall can fall over a large area if not supported adequately. It is imperative that good temporary support be installed before any work is carried out.

Basal raises are 1.5 m wide and a maximum of 2.4 m high. In open mining, raises are 1.5 m wide and 3.2 m high. A Leader Reef raise is 2.4 m wide and 3.2 m high. Connections from the

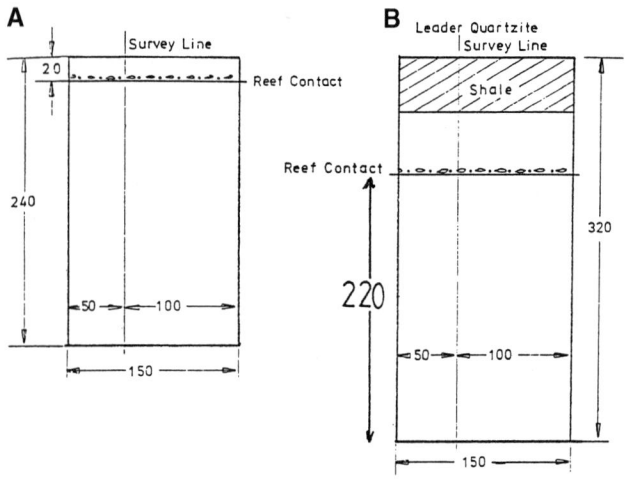

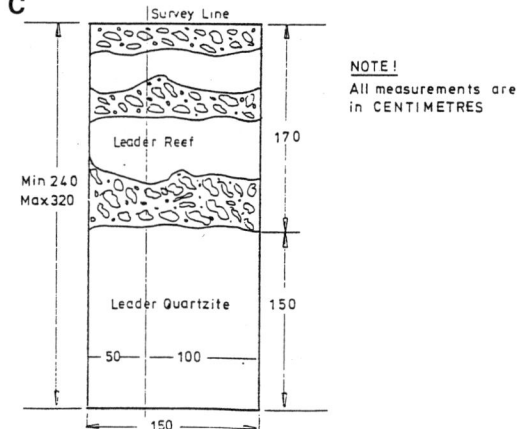

FIGURE 17.16 Raise dimensions. A, Undercut mining; B, Open mining; C, Leader Reef.

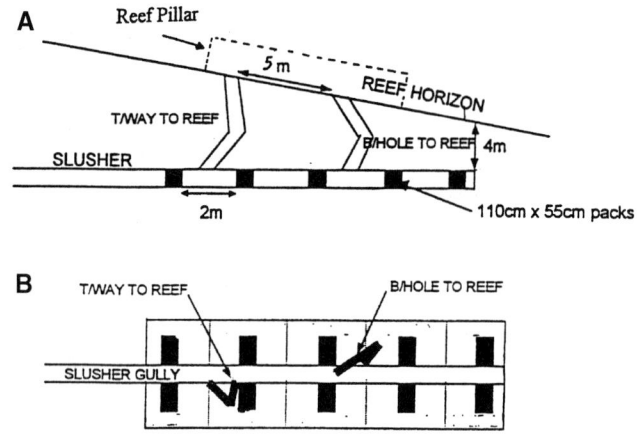

FIGURE 17.17 Mushrooming procedure. A, Section view; B, Plan view.

crosscuts are 2 m wide by 2.8 m high. All raises are equipped with grizzlies prior to ledging.

Ledging. In the ledging sequence, the crosscut reef connections are overstoped first. Ledging is then commenced using breast ledging. The undercut can be lost during the ledging process or where the middling between the Basal Quartzite and the shale is small. The reef is sometimes ledged on open mining and then reundercut using the wide-raising or cubby-ledging method once mining recommences after the equipping phase.

Pillar Ledging. Mining of pillars makes up a large proportion of the Matjhabeng Mine's production. Access to the old pillars is not always easy because of closure of the surrounding stopes or collapse of the access or travellingways. Where this situation arises, the mushroom method of ledging the pillar is used (Figure 17.17). This involves developing a wide end slusher from the nearest access point to where the middling between the slusher and the reef is not less than 4 m. A boxhole and a travellingway from the slusher are developed into the lower portion of the pillar. The holing distance between them on the reef horizon is 5 m. A wide raise is then developed between the two holing points. A grizzly is installed, and the wide raise is continued on-reef into the pillar. On the updip side of the pillar, a travellingway from the slusher is developed for ventilation purposes and serves as a second outlet.

Stoping. Stoping is carried out on breast panels. Conventional methods are used. All packs, regardless of size, are spaced at 2-m intervals on strike and dip. All packs are prestressed

to improve stability and reduce creep. Pack sizes depend on stoping width. For stoping widths less than 120 cm, 55- by 55-cm packs are specified; for stopes less than 170 cm, packs are 75 by 75 cm; and for stope widths above 170 cm, 110-cm packs are used. Temporary support on panels are face sprags 8 to 10 cm in diameter or mechanical props spaced 1 m from the face on strike and 1.5 m on dip.

Use is made of the pack in a pipe system. This support system can best be described as round steel rings with wire mesh sides. On installation, the cage is lifted and secured by timber poles. A bag placed in the middle is then filled with grout. These packs are installed at the same spacings as normal packs (Figure 17.18).

Costs and Efficiencies.

Cash cost, dollars per ounce	275
Total employees costed	8,821
Square metres mined per month	37,800
Metres developed per month	860
Square metres per stope employee	15
Square metres per total employee costed	4.0
Metres of development per employee	3.5
Development cost, R/m	2,164
Working cost, R/m²	1,757
Working cost, R/tonne milled	371
Ratio of m²:m	42.6
Face advance, m	7.15
Recovery grade, gm/tonne	7.51
Tonnes milled per month	173,793
Stoping width, cm	129

17.2.3 Harmony Gold Mine

Geological Overview. Structure is generally quite simple, with few major faults apart from the bounding De Bron and Lava fault system in the west. The strata generally strike slightly east of north, dipping west at 10° to 20°, with a swing in strike in the south of the lease area to east-west, dipping north at 10° to 15°.

The ore body is bounded on the west by the De Bron fault system. In the east of the lease area, a progressive onlap results in the younger beds cutting off older beds. The Basal Reef abuts the Leader Reef, and the Leader Reef is truncated by Karoo sediments in the far east of the lease.

Table 17.2 describes the reefs.

Mining Practice. The Harmony gold mine has extended its mining area from the original Merriespruit and Virginia shafts with the purchase of the Unisel Mine; Saaiplaas 3, 4, and 5 shafts; and the President Brand 5 shaft.

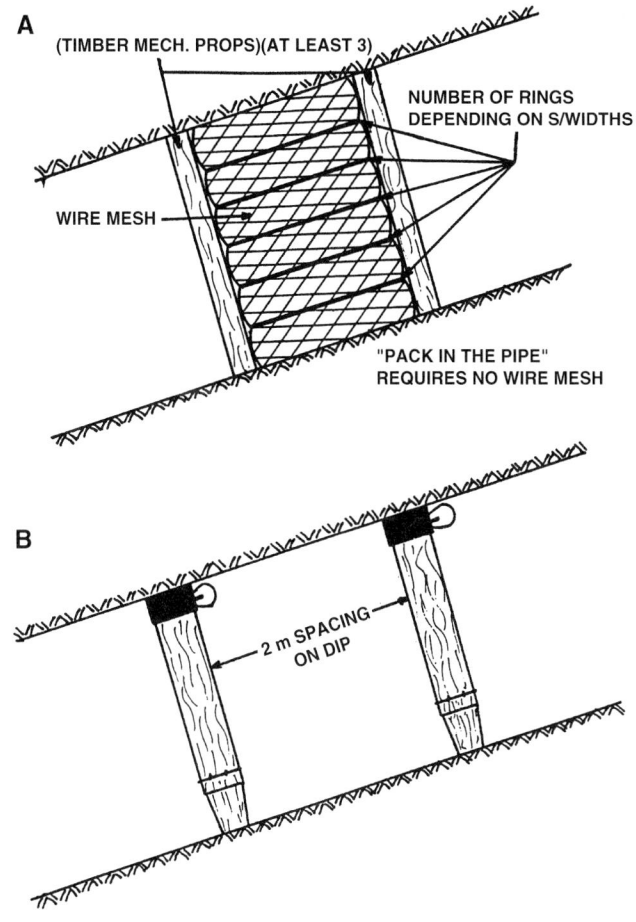

FIGURE 17.18   Pack-in-a-pipe support. A, Temporary support; B, Elongate support.

The working area, which is typical of the area under consideration, is the original Harmony lease and the Saaiplaas/Erfdeel (renamed Masimong) areas. From scrutiny of mine standards, the Harmony 2, 3, and 4 shafts, the Virginia Complex, and the Masimong 1 and 2 shafts have been incorporated into a common mining standard, while the Unisel and Brand 5 shaft have retained their own set of standards. Only standards from the original lease area and the Masimong Mine are included in this paper.

Development.   Development is with conventional hand-held equipment. Air-powered loaders are used for cleaning. Raises, winzes, and travellingways are cleaned by scraper winch. Standard layouts call for development ends to be blasted with elliptical hanging walls.

Standard sizes of haulages and crosscuts are 3 m wide by 3.7 m high. The grade line is carried 1.5 m above the rail, while the footwall is blasted 0.5 m below the rail.

Temporary support consists of two mechanical props or mine holes spaced 1.5 m apart at a maximum distance of 1 m from the face. The standards make distinctions between development ends in "stress" or "destressed" conditions. "In stress" tunnels require support, which consists of 9- by 1.8-m by 16-mm grouted rebar to be installed in a ring emanating from the centre of the tunnel 0.6 m above grade line. The rings are 1.5 m apart and are installed at a maximum of 5 m from the face after the blast. Destressed tunnels require staggered rings of three holes using 1.8-m-long, 16-mm in diameter grouted rebar drilled

TABLE 17.2   Description of reefs in vicinity of Harmony Mine

Name and location	Description
A Reef	
Hanging wall	Medium-grained argillaceous quartzite with pyrite stringers and scattered conglomerates.
Reef	Uitsig. Oligomictic, small-to-medium-pebble conglomerate 20 to 150 cm thick. Mineralisation is pyritic.
	Internal quartzite. Ranges from 0 to 3 m thick. Argillaceous, medium-grained quartzite; not very competent.
	Witpan and reworked big pebble conglomerate. Ranges from 10 to 200 cm thick. Oligomictic medium-pebble-to-cobble conglomerate. Mineralisation is pyritic. Both conglomerates have sub-siliceous to siliceous quartzite matrix.
Footwall	Argillaceous, medium-grained quartzite. Can have a footwall conglomerate that is polymictic in nature and causes confusion in the minds of the mining people as to where the footwall contact actually is.
Leader Reef	
Hanging wall	Upper Leader. Polymictic conglomerates, small-to-medium-pebble conglomerates. Subsiliceous matrix.
Reef	Tends to be polymictic conglomerates and quartzites that vary from small-to-large-pebble conglomerates with siliceous quartzite matrix. Channels from 15 to 150 cm. Varies from thin pebble bands on top of or under quartzites to massive gravels. Highly variable sedimentation. Internal quartzites are also variable, from quite argillaceous to siliceous in the conglomerate matrix. Grades vary from being concentrated on the footwall contact in a carbon seam to being dispersed throughout the massive conglomerates.
Footwall	Incompetent, coarse-grained, argillaceous quartzites. Commonly called "waxy brown quartzites."
Basal Reef	
Hanging wall	Khaki Shale or "waxy brown quartzite." Shale 10-15 cm thick on average where it exists. In south and east of lease area, generally no shale present. In the northwest part of lease area shaft, shale can reach 100 cm thick.
Reef	Reef generally consists of the Steyn facies. Polymictic to oligomictic, small-to-large-pebble conglomerate. Varying amounts of sand and gravels in different proportions and sequences. Can vary from a 10-cm, thin conglomerate overlain by 100 cm of siliceous quartzite below the shale to massive gravels 150 to 200 cm thick. Three subfacies: B1, B2, and B3 (the youngest unit). They are all conglomerates with subtle differences that are macroscopically apparent. The oldest B1 unit is the main gold carrier, consisting of one or more quartz and chert pebble bands ranging from a few centimetres to 100 cm thick. The B2 is ±80 cm thick and is not as good a gold carrier as the B1 unit. B2 also tends to be polymictic in nature. The B3 unit is generally nonpay across the Harmony lease area.
	Gold mineralisation is associated with "buckshot" pyrite and carbon in "flyspeck" and seam form. The best grades are associated with seam carbon at base of B1 unit.

from the centre of the tunnel 0.6 m above grade to the same standards as the stress situation. Raises are blasted 1.5 m wide and 2.7 m high. A single mine pole or mechanical prop serves as temporary support in the middle of the raise. Permanent support consists of 1- by 2-m staggered patterns of rock bolts.

Additional support densities are specified in situations where stress levels exceed 75, 91, or 120 MPa, respectively. Such support consists of grout lacing and mesh according to a pattern specified by the rock mechanics department. Where high stress has resulted in hanging wall and sidewall sag in tunnels, timber sets or 30-kg arches are required.

Ledging.   Staggered breast ledging is practiced. Panels are 30 m long and are ledged 4.5 m from the centerline in six stages.

Stoping. From the mining standards, five types of breast stoping were identified: pack support system with a wide heading, pack support system with a narrow heading, pack support system −5° underhand, pillar system type P1, and pillar system type P2.

1. Pack support system with a wide heading. A feature of this panel is a heading 5.2 m wide over the advance strike gullies that is carried a maximum of 5 m ahead of the panel. Pack support is 2 by 2 m skin-to-skin; the size of pack depends on stoping width. A travellingway-monoropeway is carried on strike between the packs in the centre of the panel.

2. Pack support system with narrow heading. As above, but with a narrow, conventionally carried heading for a maximum of 2 m ahead of the face.

3. Pack support system −5° underhand. This method allows the face to be carried 5° underhand. The panel incorporates a wide gully and a mono-travellingway as in method 1. A feature of this method is a split row of packs above and below the mono-travellingway, provided that the packs below the monoway are no more than 4.8 m from the face after blasting.

4–5. Pillar system types P1 and P2. A feature of these systems is that the gullies are supported on the downdip side by a pillar left 2 m skin-to-skin for a minimum of 3 m wide and 8 m long. An integrated, elongate pack or only an elongated stope support is called for in this case.

All gullies are grouted with a staggered 1- by 2-m pattern with 1.8-m-long grouted rods and rings a maximum of 1 m apart. Where dip exceeds 25°, gully sidings are not required, provided that 1.8-m grouted rebar bolts are installed 0.5 m apart on the hanging wall and sidewall.

Temporary support in panels consists of two lines of mine poles or mechanical props installed a maximum of 1 m on strike and 2 m apart on dip not more than 1 m from the face.

Good results are achieved using 0.9-m-long rock studs in the hanging wall in place of conventional temporary support. They are installed at a maximum of 1 m apart on strike and 2 m on dip for a maximum of 1.2 m from the face. After blasting, the rock studs are normally installed on strike at an angle of 70° on face advance. Rock studs are particularly effective in preventing dilation of low-angle stress fractures behind the face. The clamping effect improves hanging wall integrity considerably.

At Masimong, the gully support standard has been revised to prevent gully sidewall failure in the very weak upper shale marker of the B Reef footwall. The gully packs have been moved 0.5 m from the ledge, i.e., 1.5 m from the centreline. The hanging wall is supported by a staggered pattern of two-by-three rows of rock bolts anchored with quick-setting grout. The ring of three bolts is positioned between the packs. The rings are 1.8 m apart, which coincides with gully pack spacing (Figure 17.19).

A set of standards for steep stopes exist for reef dips greater than 35°. The panel is carried 5° underhand. The standard specifies the use of climbing chain on the face with gate stulls on temporary supports to prevent stope workers from slipping or falling and also to protect them from rolling rocks. Designated travellingways are required in the stope.

Costs and Efficiencies.

Cash cost, dollars per ounce	259
Total employees costed	11,493
Square metres mined	94,335
No. of stope employees	5,807
Square metres per stope employee	18.25
Square metres per total employee costed	8.2
Metres of development per employee	3.5
Stoping cost, R/m²	430

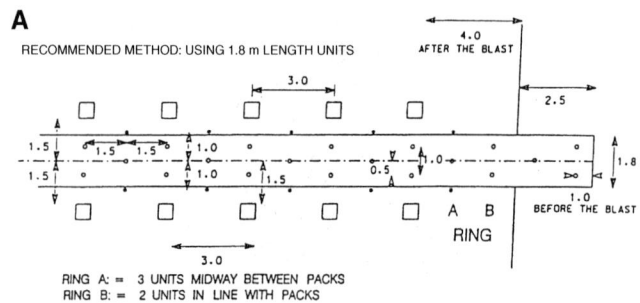

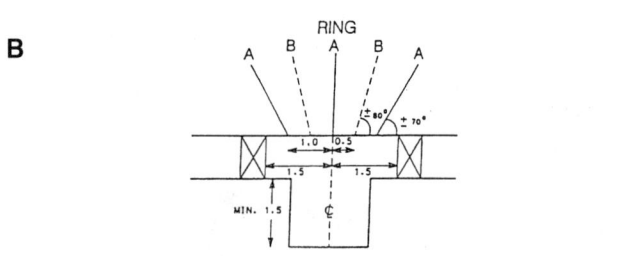

FIGURE 17.19 Grouted rod or rock stud support of gullies. A, Plan view; B, Section view.

Development cost, R/m	2,164
Working cost, R/m²	1,900
Working cost, R/tonne milled	248
Ratio of m²:/m	26.5
Face advance, m	7.13
Recovery grade, gm/tonne	4.69
Tonnes milled per month	44,581

17.3 MINING PRACTICE AT GREAT NOLIGWA MINE

Great Noligwa forms a major part of Anglo-Gold's Vaal River operations (formerly known as Vaal Reefs Exploration and Mining Company). Sinking of the shaft commenced in 1967, and the shaft came into production in early 1972. The mine consists of a twin shaft system that extends from the surface down to the 81 level at a depth of 2,400 m below datum. The first gold extracted was poured in late 1972. To date, 354,291 m² of ground has been mined, producing 23,360 kg of gold per year. Current production is around 37,320 m²/month, yielding approximately 2,800 kg of gold per year.

The mine is the largest gold producer within Vaal River Operations, contributing 59% of total gold production. It is estimated to contribute about 5% of total South African gold production and 1% of world production.

17.3.1 Geological Structure and Mineralisation at Great Noligwa Mine

The Great Noligwa Mine is situated within the Klerksdorp Goldfields approximately 20 km to the south of Klerksdorp.

The Vaal Reef is a planar body dipping approximately 25° to the southeast. The Great Noligwa lease area is traversed by northeast-trending normal faults dipping to the northwest. These have displaced the ore body to mineable depths (2,800 m below datum). They are present in the study area as the north-dipping Zuiping faults and in the south as the southerly dipping Jersey fault, which throws the reef down by approximately 1,000 m to the south into the Moab lease area. Other structural features are northeast-trending intrusives of Klipriviersberg, Platberg, and Karoo age.

The Ventersdorp intrusives have been displaced in places by the faults described above, while the younger Karoo age dykes have a cross-cutting relationship with respect to the surrounding stratigraphy. The study area is bounded in the north by the Zuiping fault and in the south by the Jersey fault.

Vaal Reef. The economic horizons exploited in the Great Noligwa area fall within the Strathmore Formation. The formation is composed of the Krugersdorp and the Booysens conglomerates. The Krugersdorp is subdivided into the MB (Main Bird) 4, MB 3, and MB 2 units. The Vaal Reef forms the base of the MB 4 unit.

The Vaal Reef placer is the main gold- and uranium-bearing ore body in the Klerksdorp Goldfields. The Vaal Reef covers an area of approximately 260 km². Histograms of channel width for the Vaal Reef in the Great Noligwa area indicate that the placer varies between 70 and 150 cm.

Generally, the Vaal Reef consists of a single oligomicitic conglomerate that rests directly on MB 5 quartzites. Occasionally, it is seen with a basal carbon seam. In the southern part of the lease area, it rests on Mizpah quartzites and conglomerates. The remainder of the Vaal Reef zone above the reef conglomerate consists of subsiliceous to argillaceous, light-grey orthoquartzites that grade upward into the polymicitc Zandpan Conglomerate marker (1 to 2 m thick). In the south of the mining area, the reef package varies from a single band of conglomerate to a multiple conglomerate composed of an upper oligomictic zone, a lower oligomictic zone, and a polymictic zone, named the G V Bosch, Zaaiplaats, and Mizpah conglomerates, respectively.

The conglomerate bands tend to be bottom loaded, but despite this phenomenon, lesser amounts of metal are spread throughout the reef conglomerates. This holds true for both the G V Bosch and the Zaaiplaats facies. The Mizpah Conglomerate is subeconomic.

Carbon is present within the Vaal Reef package and is associated with gold and uranium. It occurs as a thin, impersistent seam usually less than 5 mm thick. It is found predominantly as columnar carbon, with occasional flyspeck carbon present. Most often, these seams are situated at the base of the G V Bosch Conglomerate; however, seams have been seen within the conglomerate as well. Occasionally, carbon seams can also be present at the basal contact of the Zaaiplaats Conglomerate.

Mining Practice. The mining layout is a grid pattern with raises 180 m apart and levels 90 m apart. Mining is scattered stoping largely because of extensive faulting. Major faults trend northeast-southwest along the strike of the reef with the effect that the length of the raiselines are affected. The main levels are the 61, 64, 68, 70, and 71. Mining is concentrated mainly between the 64 and the 68 levels north of the shaft and between the 70 and the 73 levels south of the shaft. Because of upthrow faulting, multiple reef intersections are present on these levels. Figure 17.20 shows the numerous faults.

Subsequent to planning the original layout, extensive faulting was discovered to the south of the lease area, necessitating cutting of the 51 level to access the reef. Downdip of this area is the new Moab Khotzong Mine, also known for its numerous faults.

Ledging. During the breast ledging process the installation of support over the raiselines results in large spans that can result in falls of ground if the temporary support is inadequate. Multiple fatalities have occurred during this critical stage. For this reason, the method of breast ledging was largely abandoned in 1987. Currently, approximately 90% of all ledging is done downdip. This method has been found to be largely successful provided that the angle between the downdip face and the raiseline is greater than 70%. This area is known locally as the Devil's Triangle. At angles of less than 70%, it is difficult to install temporary or permanent support in the acute angle next to the raiseline.

The hanging wall of the Vaal Reef is extremely friable and weak and is characterized by well-defined partings that lack

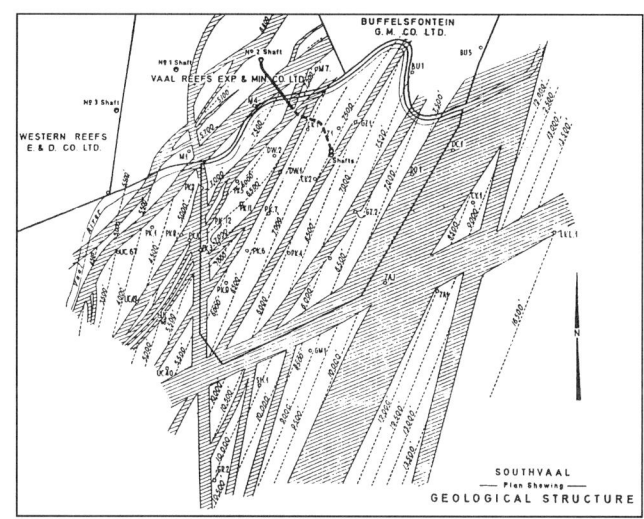

FIGURE 17.20   Geologic structure in the Great Noligwa Mine area

cohesion over relatively short spans. The footwall quartzites are also well bedded, and the immediate footwall is brittle and glassy. High levels of convergence in relatively short spans of 30 m are also common. These sets of conditions have influenced mining practice.

The problem of ledging the Vaal Reef was well described by J.N. Morgan in a paper entitled, "Mining the Vaal Reef in the Klerksdorp Area," in which he highlighted the fact that supporting the hanging wall was a problem. An extract from the paper illustrates the problem of the friable footwall and hanging wall conditions.

> *Ledging.* It was found at an early stage in stoping operations that unless the unsupported span in the main travelling way was kept to a minimum, the hanging-wall deteriorated rapidly and serious delays resulted owing to damage to air and water columns and electric cables, in addition to falls of ground in the central scraper gully. These travelling ways soon deteriorated into the most dangerous areas in the working place. Various methods were tried to overcome the difficulty, such as the building of concrete piers on which to install timber support. This was an expensive and cumbersome method. Pillar support on each side of the original raise was also unsatisfactory owing to the rapid rate of subsidence, which caused the hanging-wall to fracture along the pillars, which, being small, eventually crushed and caused more dangerous conditions.
>
> Normal stoping operations always resulted in the footwall breaking away on each side of the central gully, with the result that mat-packs could not be built at close enough intervals on each side. As previously mentioned, the width of reef raises has now been reduced to 5 ft. and a method of ledging introduced that does not break the footwall on each side of the central gully.
>
> Only one face is ledged at a time, as it has been found that by ledging opposite faces simultaneously, the hanging-wall deteriorates rapidly and collapses before it can be adequately supported. The method consists fundamentally of developing narrow development ends 30 in. high and 6 ft. wide to a depth of 10 ft. Two-foot mat-packs are then installed in each bay, and stoping holes are drilled 12 in. apart along the faces between bays. These holes are drilled parallel and on strike and do not tear away the footwall.
>
> Although this method is somewhat expensive in respect of drilling and breaking costs, it has enabled us to overcome the support problem in stope travelling ways. An additional important advantage is that it has enabled us to establish a satisfactory stoping width from the commencement of stoping operations.

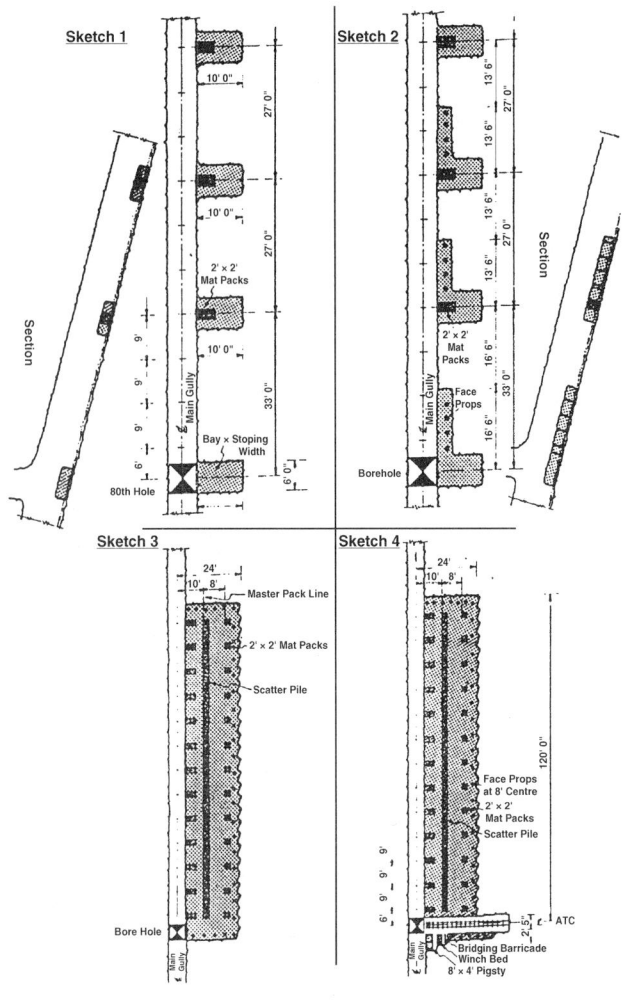

FIGURE 17.21 Cubby ledging

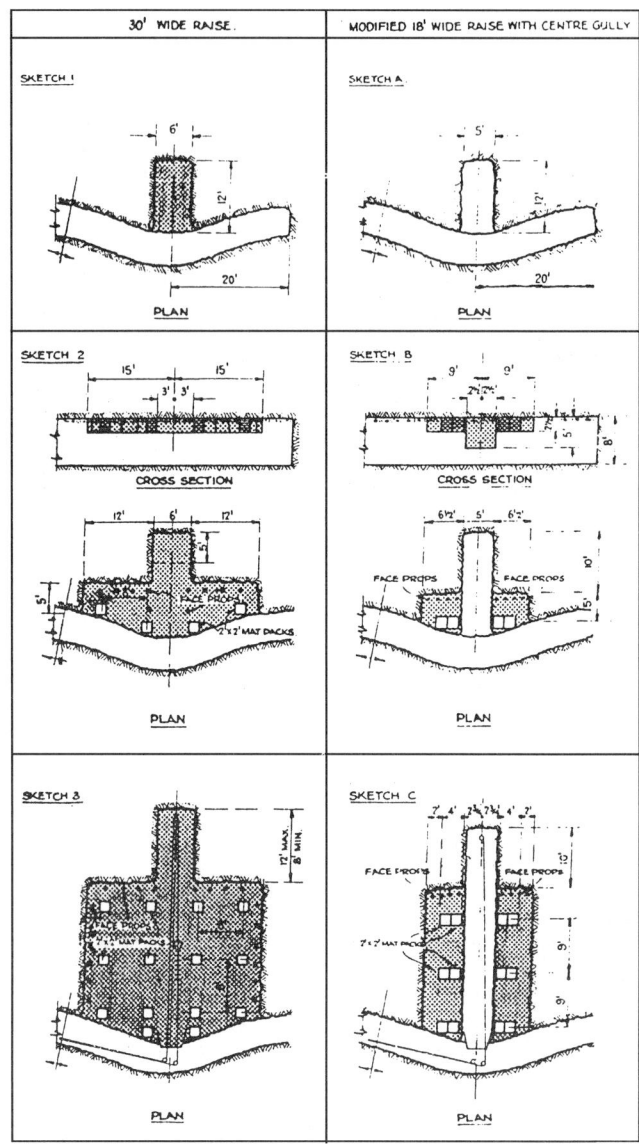

FIGURE 17.22 Wide raise

Of interest is that their experience showed that when the ledging span was reduced from 9 to 5.5 m, hanging wall conditions improved because of the increased rates at which raises were advanced. Care was always taken to protect the gully from blast damage. This was achieved by leaving the gully full to prevent the sidewalls from slabbing when the gully shoulder holes were blasted. The first shoulder holes were also drilled close to the hanging wall to minimize blast damage to the side walls (Figures 17.21 and 17.22).

Current practice at Noligwa with downdip ledging appears to have resolved the difficulties with breast ledging and confirms the mining practice of narrow-span, wide raising pioneered by others in mining the Vaal River Reef.

Stoping. As noted, because of the extensive faulting, the stoping method employs scattered mining. The raises are laid out on 180-m centres with 30-m-long panels on both sides of the raiseline, each with its own scraper winch. Each panel has its own face winch to clean the face. The emphasis is on maintenance of a stope on the order of 90 cm wide (Figure 17.23).

A critical point in the stoping process is the maintenance of a stable strike gully. This is potentially the most hazardous working area as gullies are subject to falls of ground because of the increased spans and blast damage (fracturing) to the hanging wall during gully development. The fundamentals of this problem were described by Mason et al. in their paper, "Stability of Strike Gullies

in the Mines of the Klerksdorp District." The main problem highlighted was the presence of nonstandard sidings, which lagged behind the mining face with inadequate support. This practice allows the buildup of stress and increases the intensity of fracturing on the unsupported gully sidewall. Their recommendations were that a gully should not be advanced ahead of the stope fracture zone and that blast damage around gullies could be minimized by footwall lifting gullies behind the blasted panel. Here the concept of the wide heading has application.

An interesting feature of the mine support standards at Noligwa is that they categorise the various ledging and stoping panels in terms of risk ratings, from a low-risk category 1 to a high-risk category 5. The assumption is that each panel operation is subject to the same risk in terms of the support installed. For example, a panel with a risk rating of 5 would have a high level of stress and be subject to rock bursts; as a result, the standard requires that a high density of support capable of withstanding rock burst conditions be installed. In turn, a panel with a risk rating of 1 would need a lower support density because of its lower inherent risk.

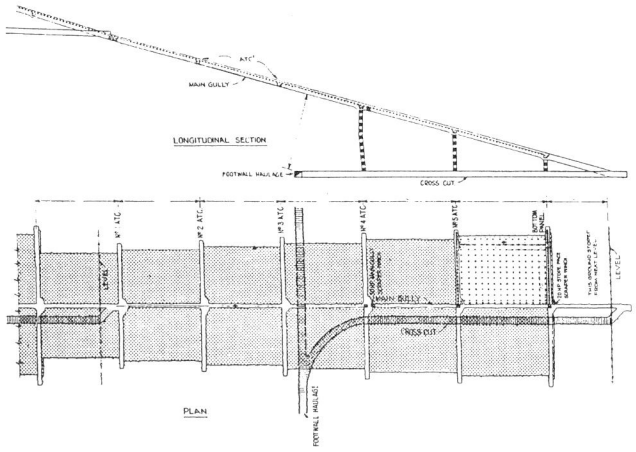

FIGURE 17.23   Cross-sectional layout and plan of stope

Equipping.   Equipping a stope for mining at the completion of ledging is a most time-consuming process. The standard of stope equipping and installation of equipment determines the ultimate productivity of the stope. In the normal equipping process, the rehabilitation of the breast-ledged raiseline must be done prior to installation of winches, pipes, and the mono-winch. This involves replacement of damaged supports and resupport of the hanging wall in centre gullies and tip areas, and removal of broken rock from centre gullies and ledges. Adoption of the downdip ledging method obviates these problems as the ledging raiseline can be cleared and equipped from the top down during the ledging phase. The ledged raise is virtually equipped ready for mining after ledging reaches the overstoped bottom travellingway.

The use of hydraulic props for breast ledging has declined because of the unpopularity of the system with stope workers, who find them heavy to handle and therefore become fatigued during continuous removal and reinstallation. There is also a tendency for many props to be left in the panels or back areas where they become frozen and, if not blasted out, result in props being abandoned. If not installed with headboards, hydraulic props punch into the weak Vaal River hanging wall and footwall beams. The standards, however, still require their use in panels with risk a rating of 5 to ensure the support resistance of 200 kN required for rock burst conditions. Certain mines that are required to use props because of seismicity have found that the task of prop installation and control can be effectively performed by outside contractors, who tend to have a more focused approach.

The practice of ledging with hydraulic props was described by J. M. Langer and H. M. D. Hobday in their paper, "A Technical Note on Ledging Operations with the Aid of Hydraulic Props at Vaal Reefs Exploration and Mining Company Ltd." In this paper, the authors stressed the importance of strata control in the ledging process.

Development.   Development of haulages, raises, and box-holes is done conventionally with hand-held machines and airlegs and cleaned by air-powered loaders. It is generally accepted that boxhole and raise development is one of the more hazardous operations in local mines. Therefore, Noligwa is experimenting with the Swedish Alimak raise climber for boxhole development. The raise climber is used internationally, and more than 2,500 units are operational. The raise climber runs on a guide rail using a rack-and-pinion system anchored to the hanging wall. It serves the dual purpose of transporting workers

into the raise and serving as a platform from which workers can drill holes. It is anticipated that 1.8 m of advance will be achieved per shift. This machine eliminates the need for boxhole equipment and carries its own services, such as compressed air, water, and the ventilation column, up in the supporting guide rail. The forced-air ventilation is provided by a high-pressure fan blowing through a 100-mm pipe.

Costs and Efficiencies.

Cash cost, dollars per ounce	242
Total employees costed	10,500
Square metres mined per month	34,333
Metres per month	2,038
Square metres per stope employee	12.66
Square metres per total employee costed	3.26
Metres of development per employee	2.73
Full cost, R/m^2	2,575
Working cost, R/m^2	2,279
Cash cost, R/tonne milled	387
Ratio of m^2:m	17.21
Face advance, m	12
Recovery grade, gm/tonne	13.57
Tonnes milled per month	200,000
Stoping width, cm	95

17.4   OBSERVATIONS

From the standards and practices prevalent at the mines, one can make the following observations.

1. Dips are generally on the order of 20° to 25°, with the steeper reefs present next to major faults.

2. Reefs are generally highly faulted, with upthrow normal faulting being the main mode of uplift.

3. The stoping method of choice is scattered mining laid out in a grid pattern of crosscuts generally 180 m apart. Where faulting is prevalent, spacings can be reduced to 150 m. Spacing is largely determined by the distance of pull of a 37-kW winch. Longer pulls result in buildup of ore in the gullies and slower face advance. Travelling and material transport are also difficult. Scattered mining also allows for greater flexibility to negotiate the numerous faults of the Free State and Vaal River areas.

4. The mining methods and practices at the mines are similar for similar activities largely because of the mobility of personnel and management between areas and recognition of best practices.

5. The trend of support systems is away from solid timber packs to elongate rock props or lightweight, design cement foam packs, such as Grinaker's Durapack. Mines are starting to consider "pack-in-a-pipe" support to minimise transport costs and timber handling. Durapacks, although a more costly type of support, are showing good results in highly stressed, weak hanging walls because of their uniform deformation characteristics. The use of hydraulic props has declined in favour of prestressed blast-on elongates coupled with backfill in high-stress areas.

6. In the areas where the reef is overlain by shale, the critical activities are raising and ledging. Mine standards are very detailed regarding ledging methods. Procedures to reestablish undercutting are also well documented. This activity takes know-how from miners and takes time. There are no shortcuts.

7. Of interest are the efforts to secure the gullies and gully sidewalls. At Masimong, the practice of spacing packs 0.5 m from the gully sidewall and grouting the hanging wall with a staggered pattern of roof bolts, coupled with

the installation of rock studs in the stope panel hanging wall, has shown an improvement in hanging wall conditions. Installation of rock studs is not effective in stopes with widths below 1 m or where shale is present in the hanging wall because water from drilling weakens the shale beam.

17.5 ACKNOWLEDGMENTS

The author wishes to thank the general managers of Tshepong, Matjhabeng, Harmony, and Great Noligwa mines for their assistance in providing information for the preparation of this paper.

17.6 REFERENCES

J.N. Morgan. Mining the Vaal Reef in the Klerksdorp Area. *Papers and Discussions, 1945–1955*. A.M.M.

CHAPTER 18

Extraction of a Wide Ore Body at Depth in the SV2/3 Area at Placer Dome Western Areas Joint Venture

N. Singh[*] and A.J. MacDonald[*]

18.1 INTRODUCTION

The Placer Dome Western Areas Joint Venture (PDWAJV), formed by the merger of Placer Dome, Inc., and Western Areas, Ltd., is situated approximately 50 km southwest of Johannesburg near the town of Westonaria (Figure 18.1).

The existing ore body, because of its geometry and disposition, provides a challenge that must be met productively and safely. The rock engineering considerations for deep-level mining coupled with irregular gold distribution require flexibility in both mine planning and mining method. The ability to manage a highly stressed mining environment through stress modification is of paramount importance. Consideration of in-stope stability is critical to the successful extraction of the ore body.

18.2 GEOLOGIC SETTING

The PDWAJV site lies on the southern limit of the West Rand Goldfields with the economic horizons lying within the Central Rand Group. The gold-bearing conglomerates exploited are the Ventersdorp Contact Reef (VCR) and both the Elsburg Massives and the Elsburg Individuals of the Upper Elsburg conglomerates. It is the Upper Elsburg that constitutes the wide reef ore body.

The uppermost reef being mined at PDWAJV is the VCR, which is found over the entire property. Dividing the property in a north-south direction is a subcrop known locally as the "shoreline" that defines the distinct single-reef zone to the west and the wide multiple-reef zone to the east (Figures 18.2 and 18.3). These wide reefs form a diverging wedge east of the shoreline. Moving east from the shoreline, reef horizons diverge, and the percentage of quartzite middlings increases. As the reefs become more distal, they increase in thickness and decrease in payability.

The VCR displays a variable nature in both deposition and value, ranging from poorly developed pebbles with no grade to multilayered and channeled conglomerates containing high gold values.

Individual reef widths vary from about 0.8 m up to 10 m. Multi-reef, payable mining widths of up to 18 m have been achieved in the past and are being targeted in the South Deep massive mining area for the near future.

The most dominant geological feature is the Gemsbokfontein No. 2 dyke, which divides the mining area. The thickness of the dyke varies from 25 to 30 m and together with the 12-m-wide bracket pillars on either side, forms a 50-m-wide regional pillar.

18.3 CURRENT STOPING OPERATIONS

Current mining areas are depicted in Figure 18.4. Current stoping being carried out involves either a primary conventional cut in

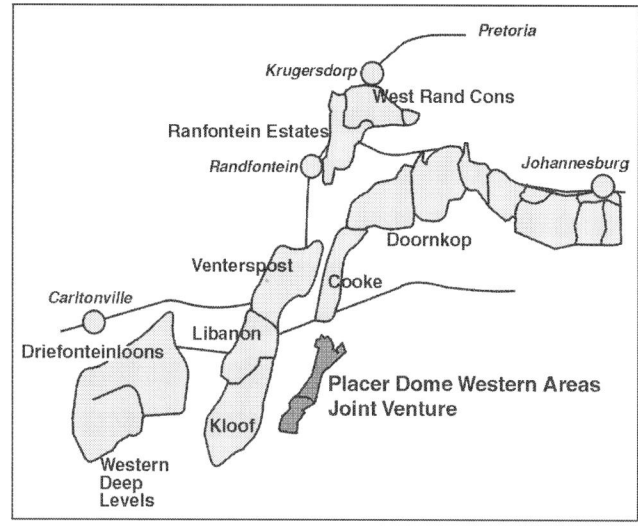

FIGURE 18.1 Map of PDWAJV

the form of mini-longwalls or scattered single-reef cuts in some of the VCR areas. This is followed by secondary conventional and tertiary cuts (depending on payability). This constitutes the wide multiple-reef mining that predominates at PDWAJV.

Secondary extraction involves methods similar to those mentioned above or conventional updip staggered drift-and-fill (CUSDF) mining. In the future, plans are to mine the SV2/3 area with trackless mechanized mining methods (TM³). The South Deep section will be extracted using a modified cut-and-fill method and trackless mining machinery (MacDonald 1998).

18.3.1 Area 1: East of the Gemsbokfontein No. 2 Dyke

To the east of the Gemsbokfontein dyke, the Upper Elsburg massives are being mined in three selected cuts. The uppermost reef horizon (MB reef) is being mined as a primary cut. This destresses the various underlying reef packages (refer to Figure 18.3). The MI reef is approximately 3 m directly below the MB reef and is being extracted as a secondary cut. The EC wide reef (approximately 16 m below the MB horizon) will be mined using TM³ in the near future.

18.3.2 Area 2: Gemsbokfontein No. 2 Dyke to the Shoreline

To the west of the dyke, the bottommost EC reef is being mined as a primary cut. This destresses the overlying reef packages

* Placer Dome Western Areas Joint Venture, Gauteng, Republic of South Africa.

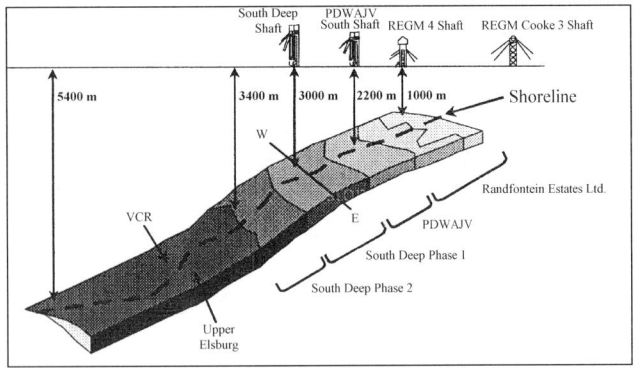

FIGURE 18.2   Ore body at PDWAJV

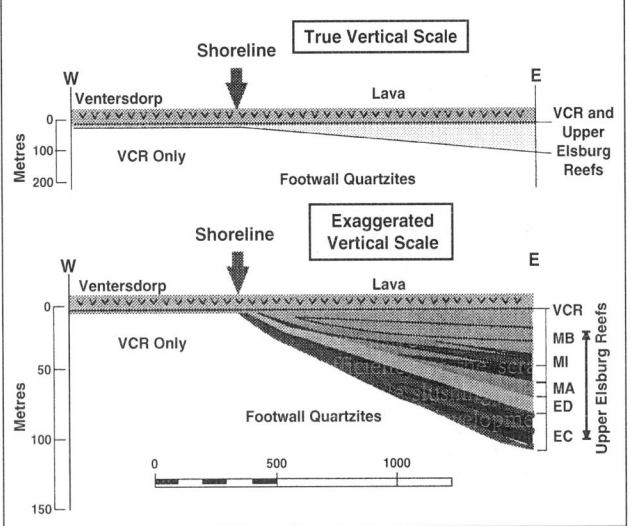

FIGURE 18.3   Upper Elsburg shoreline and widening of the ore body

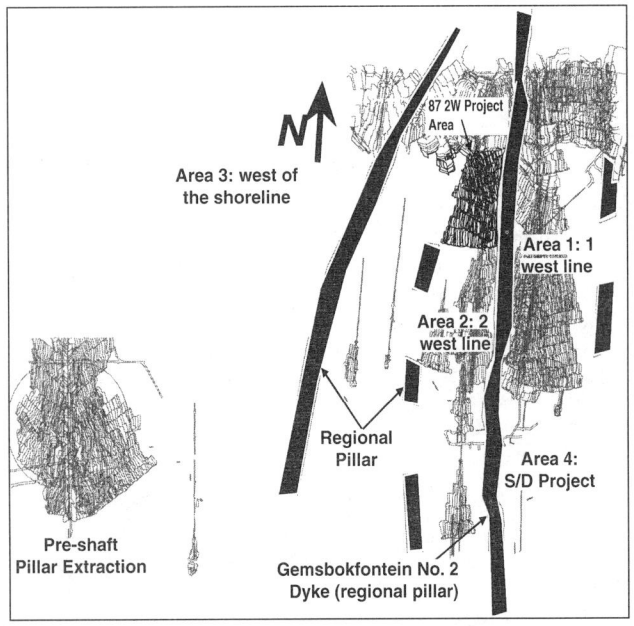

FIGURE 18.4   Current stoping operations at PDWAJV

(Figure 18.3). The wide ore body lies above the EC across this area, but all the packages narrow and converge to the shoreline. Wide ore body mining is currently underway in the 87 2 West project area (Figure 18.4).

18.3.3   Area 3: West of the Shoreline

To the west of the shoreline, only the VCR reef is developed. Mining in this area will consist primarily of a backfilled 1.5- to 2-m cut on the reef horizon.

18.3.4   Area 4: The South Deep Project

Contiguous to the SV2/3 area is the southern extension of the ore body, which forms part of the South Deep section. This project is based on mining a wide ore body at stope widths of up to 30 m. The area will be preconditioned by an underlying destress cut.

The mining method suggested for the extraction of the South Deep section is a modified form of the cut-and-fill method. This is a cyclical mining method that, as applied in the South Deep section, requires only a small extent of face exposure at any time and only for short periods. Extraction of the ore body is done in horizontal slices and progresses upward from the bottom of the defined stope block. Each horizontal slice is extracted using a staggered drift-and-fill operation on either side of the access box cut driven from the ore body ramp system. The utilization of ore body geometry and stress modification results in a concentrated, highly flexible, and practical mining method capable of providing safe mechanized extraction (MacDonald 1998).

Further to the west, the VCR reef horizon is being stoped for shaft pillar pre-extraction (Raffield 1993).

18.4   REGIONAL STABILITY

The concept of regional support in deep mines aims to reduce rock burst hazards directly by using support of such strength and stiffness so as to restrict back area volumetric closure and thereby reduce levels of face stress, energy release rates, excess shear stress, and the accompanying seismicity (COMRO 1988). Areas in which possible fault loss zones and nonpay features occur (e.g., the Gemsbokfontein No. 2 dyke) are used to site these pillars.

To ensure regional stability in the SV2/3 area, 50-m-wide dip-stabilizing pillars, fully constrained by backfill, have been sited at spans 280 m apart. The rock engineering design characteristics used to determine the siting and sizes of these pillars are—

Energy Release Rate (ERR). This concept was introduced in the 1960s and is a convenient and easily calculated measure of stress concentration and closure, and their effects in terms of severity of mining conditions and the expected incidence of seismicity. The upper bound limit used in PDWAJV is 30 MJ/m^2 based on elastic modelling using a Young's modulus of 50 GPa and underground instruments.

The industry standard of 70 GPa for Young's modulus is based on back-analysis using the Witwatersrand quartzitic rocks. Lower Young's modulus increases the strain induced by a level of stress and thus increases the ERR, which is a product of closure and stress. The standard goes further to state that in fact, this makes the 30 MJ/m^2 energy criterion about 30% conservative and thus acts as a safety margin (James 1998).

Average Pillar Stress (APS). The APS on regional pillars must be less than 2½ times the uniaxial compressive strength (UCS) of the rock mass i.e., APS < 2.5. This number equates to an upper bound limit of 500 MPa; however, to remain on the conservative side, a factor of safety was built in, reducing the upper bound limit to 400 MPa.

Numerical modelling techniques are used to determine if the two criteria are satisfied. An example of the results obtained from such an exercise is shown in Figure 18.5.

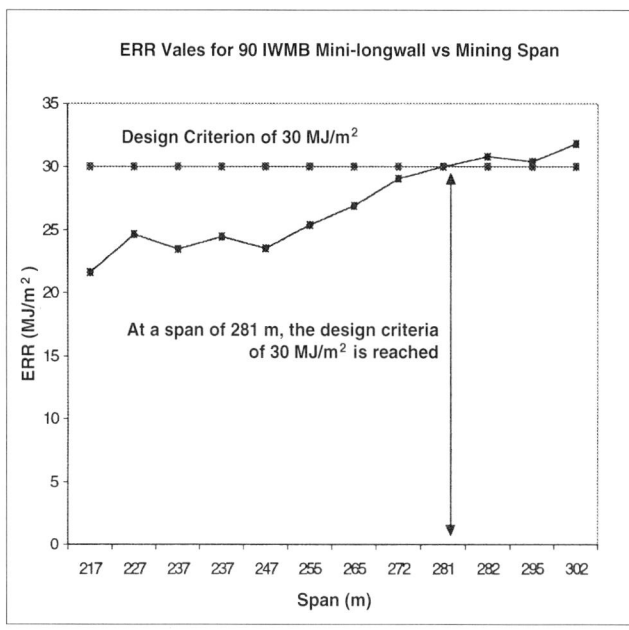

FIGURE 18.5 Results from numerical modelling used to determine strike span of regional pillars

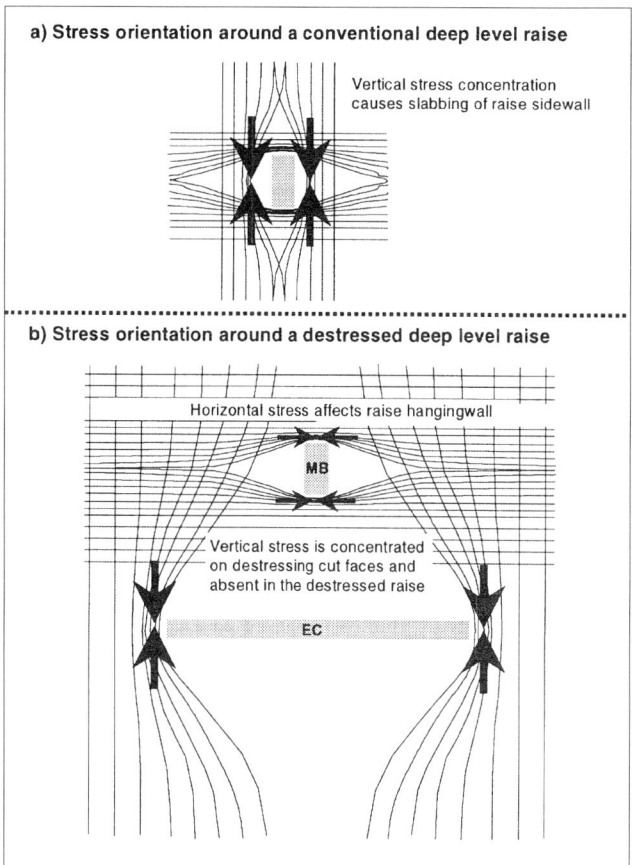

FIGURE 18.6 Principles of destressing

18.5 DESTRESSING PHILOSOPHY

The current wide ore body mining operations in the 87 2 West project area are taking place at depths of 2450 to 2550 m. The vertical virgin stress is within the range of 70 to 75 MPa in the SV2/3 and South Deep areas (Smallbone 1992). In high stope widths at these depths, confinement of the rock ahead of the face is small; hence, the strength of the fractured zone ahead of the face is low. This increases the probability of face buckling and pillar instability, which is further aggravated by the increased probability of geological weaknesses to be found in a larger volume of rock. It is therefore relatively easy for large blocks of ground to be forced out of the face. This significantly reduces the stability of the excavation.

This problem can be overcome by manipulating the stress field to create a "pseudo-shallow" mining environment (i.e., reduce the vertical field stress from 70–75 MPa to 15–25 MPa). Such a reduction is achieved by either understoping or overstoping the relevant area and mining within the destressed area.

In each of these options, the result of the destressing is that the vertical stress component is significantly reduced whilst the horizontal component remains largely unaffected by distressing (Figure 18.6). As a result of the change in the maximum principal stress orientation, fracturing is most likely to occur in the hanging wall, thus requiring the installation of rigorous support.

Backfill as support reduces closure, face stress levels, and hence ERR. Apart from these functions, it is also essential for the following reasons:

- In understope cuts, placing backfill tightly against the hanging wall minimizes the potential risk of relaxation of the hanging wall by allowing a regenerated stress state to be established. This limits propagation of bedding plane separations into the overlying reef horizons.

- In overstope cuts, by limiting closure in the top cut, backfill inhibits total closure, thus preventing the regeneration of stress to values greater than or equal to virgin stress levels. This ensures that large excavations remain in a destressed environment.

18.6 DESIGN CONSIDERATIONS FOR WIDE ORE BODY MINING AT DEPTH

18.6.1 Introduction

The inherent problems that exist with mining at depth necessitate the improvement of the designs of stress-tolerant layouts, support systems, and the use of efficient mining methods. There are numerous and interdependent design parameters that affect the design of deep, wide-ore-body extraction methodologies.

Previous experience at PDWAJV and Randfontein Estates, Ltd. (REL), has shown that wide ore body mining is successful at depths between 800 and 1200 m where the vertical stress component is between 20 and 30 MPa. In the Cooke 2 and 3 sections at REL in particular, high extraction percentages were achieved with the use of mechanized mining and extensive backfill placement. TM[3] was successfully carried out at the REL 4 shaft at intermediate depths and at depth on PDWAJV in destressed areas. An overview of the design parameters for the 87 2 West project area follows.

18.6.2 EC Destress Cut

Destress mining of the wide reefs will target the lowest possible reef horizon. Typically this will be a portion of the EC reef. The underlying EC reef horizon is being mined conventionally in an overhand minilongwall configuration. The stope width varies from 1.5 m to a maximum of 2 m. Because of the requirements of the destressing cuts, mining is continuous (no pillars are left). This is so as to precondition the whole of the overlying wide reef package, which necessitates mining through nonpay or fault-loss zones.

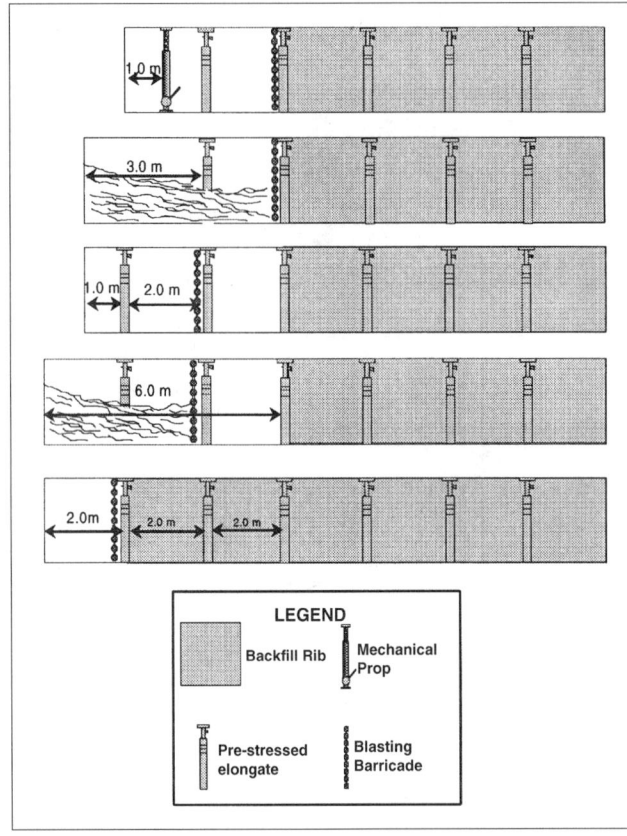

FIGURE 18.7   Cross section of support for EC destress cut

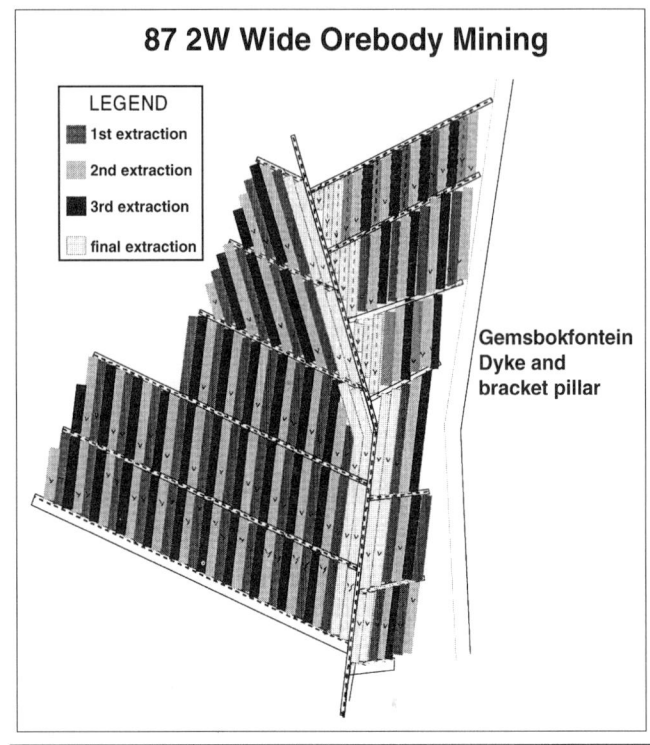

FIGURE 18.8   Plan view of wide ore body mining for 87 2 West project area

Support Requirements. The stope width for the destress cut is restricted to 2 m maximum because of the increased probability that the face will buckle at the high face widths (as discussed in section 18.4).

The support system implemented on the destress cut consists of prestressed elongates and cemented comminuted classified tailings (cemented CCT) backfill as face support. The gullies are footwall lifted behind the face and are protected using prestressed timber packs. Backfill is placed in concurrent ribs with the ultimate aim of achieving 60% backfill placement by area. Figure 18.7 shows a cross section of the support system implemented on the EC horizon.

The placement of backfill is necessary to ensure that the propagation of bed separation to the uppermost reef horizons will remain at a minimum. In addition, the placement of backfill also offers confinement on regional pillars and reduces possible shear failure and related seismic activity on the Gemsbokfontein dyke.

18.6.3   Wide Ore Body Mining Method for 87 2 West Project Area

The CUSDF mining method is being utilized for the extraction of the wide ore body. A raise is developed conventionally in the centre of the ore body from which reef drives (strike gullies) are spaced 30 m apart. From each reef drive, 4-m-wide drifts are mined updip along the top reef contact to hole into a reef drive above. Once this is complete, the drift is then footwall lifted to expose the reef package fully. Pillars whose widths are multiples of the drift width, as shown in Figures 18.8 and 18.9, separate the drifts.

Once the initial drifts are mined out, backfill is then placed until the drifts are filled tightly. The second scheduled drifts are then mined adjacent to the backfilled drift. It is important to note that these drifts are mined using hand-held jackhammers and cleaned by winch-operated scrapers.

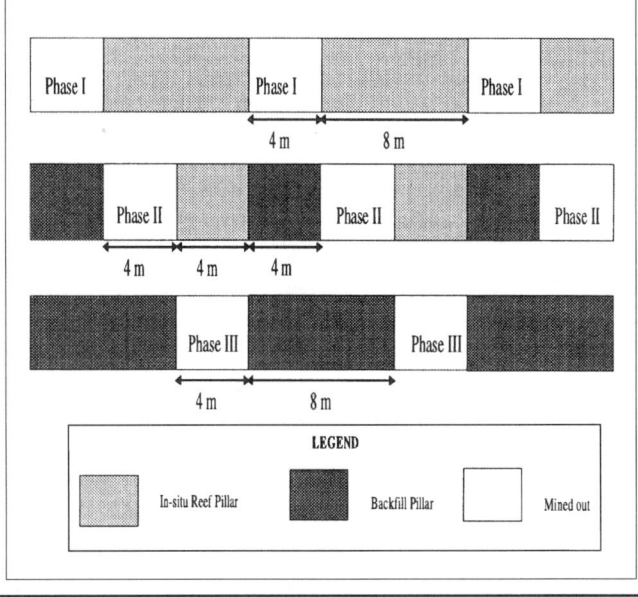

FIGURE 18.9   Cross section of wide ore body mining for 87 2 West project area

Support Requirements. The support system used in wide ore body mining consists primarily of dowels and tendons. The hanging wall is supported with 2.4-m by 20-mm, full-column, grouted dowels, while sidewall support consists of 1.2-m by 16-mm, end-anchored rock studs. Due to the nature of the CUSDF mining method, pillars will be created after Phase I extractions (Figure 18.9). Once mined out and vamped, these drifts are then backfilled.

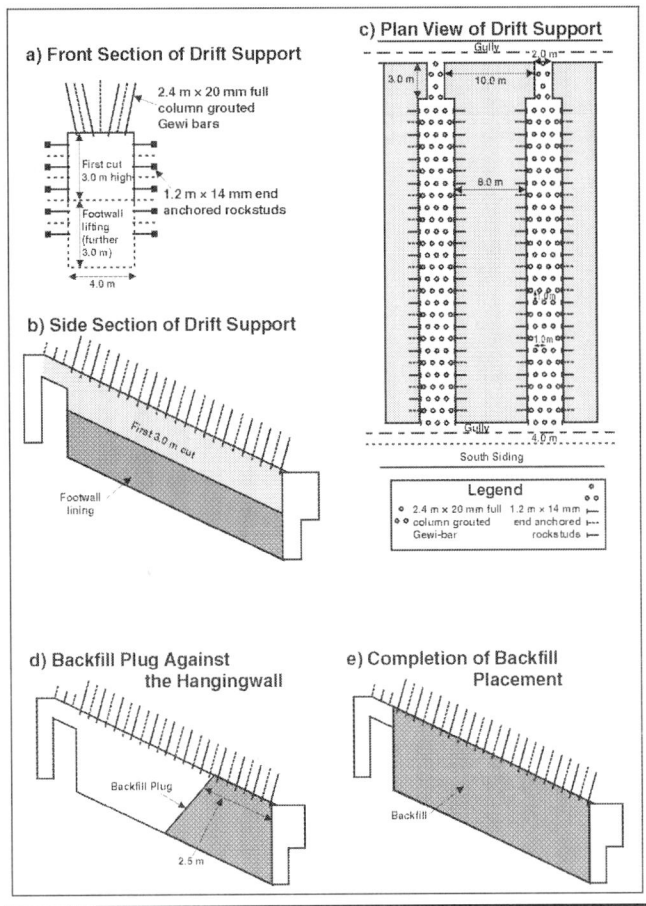

FIGURE 18.10 Support system for wide ore body

The backfilling process requires careful design of bulkhead barricades to allow for hydrostatic pressure when the backfill is placed. This is done using pigtail eye bolts and destranded hoist rope. Backfill is placed in 3-m lifts until a 2.5-m-long plug is formed against the hanging wall. The drift is then completely backfilled (Figures 18.10d and 18.10e).

The backfill stands for 28 days to allow curing. During this time, no mining takes place adjacent to any backfilled drifts. The use of backfill is essential to increase the percentage of extraction in the area and also to stabilize the pillars by providing lateral confinement. Shotcrete is occasionally used in areas where friable ground conditions are present. Cemented CCT backfill is placed in 66% of the drifts along each drive.

Sequence of Extraction. As a result of the backfilling requirements, each drift will be mined adjacent to an 8-m-wide pillar. Pillar material may vary in nature and may consist of 8-m-wide reef pillars (during Phase I), 4-m-wide backfill and 4-m-wide reef pillars (during Phase II), or 8-m-wide backfill pillars (after Phase II) (Figure 18.9). This sequence is critical to ensure the overall stability of the project area.

18.7 PILLAR STABILITY

18.7.1 Design Criteria

The success of wide ore body mining is highly dependent on the stability of the pillars created after the Phase I and II extraction. The design criterion used to assess the stability of these pillars is the factor of safety.

Factor-of-safety is the ratio of pillar strength to average pillar stress. An acceptable factor-of-safety as used in the South African

mining industry is between 1.5 and 1.6. In effect, this means that the pillar must be 66% stronger than the applied stress.

Pillar strength (σ_s) is determined empirically (Equation 1) using a modified version of the power formula proposed by Salamon and Munro (1967).

$$\sigma_s = K(w^\alpha \times h^\beta)\text{MPa} \qquad \textbf{(18.1)}$$

where K = laboratory-determined cubic rock mass strength, w = pillar width, h = pillar height, and α and β = constants. The commonly used values for the exponents in the formula are $\alpha = 0.5$ and $\beta = -0.75$, based on the work done by Hedley and Grant (1972) from back-analysis studies. This equation takes into account the width-to-height ratio of the pillar as it affects pillar strength.

Spearing (1985) found K to be between 110 and 140 MPa from back-analysis work done at REL. Because of the similarity of the reef mined at PDWAJV and REL, this was found to be an acceptable range for K. In the assessment of pillar stability, a value of K = 110 MPa was used to simulate a worst-case scenario.

To calculate average pillar stress, numerical modelling was carried out using the Besol/MS suite of programs.

18.7.2 Factor-of-Safety Calculations

The analysis of pillar stability was carried out on the 8- and 4-m-wide pillars developed after Phases I and II, respectively. Phase I operations will be carried out in solid ground when there are no pillars so that strata control issues are the only concern. The calculations were done for extraction during Phases II and III.

8-M-Wide Pillar Stability. Only three pillars (6%) failed to satisfy the design criterion after Phase I. This was, however, because of inadequate understoping, which, in these drifts, were in the abutment from the EC stopes.

4-M-Wide Pillar Stability. After the extraction during Phase II, the percentage of pillars that failed to satisfy the design criterion increased significantly, to 40%. To determine an optimum height at which these pillars could be extracted, factors of safety were calculated for 9-, 6-, 5-, and 3-m-wide stopes. At a stope width of 6 m, most of the pillars satisfied the design criterion, with the exception of the three pillars in the abutment from the underlying EC stope.

18.7.3 Conclusions

From the numerical modelling and factor-of-safety calculations, it was concluded that Phase I could be mined at a height of 9 m, while Phases II and III could be mined at a maximum height of 6 m.

The success of the massive mining operation is totally dependent on the understoping on the EC horizon. Because of the size of the excavations and the depth below the surface, all massive mining must be excavated in destressed ground. Thus, the understoping cut was extended farther to ensure that the entire project area fell within the destressing limit.

This design may be considered conservative because the design criterion specified in wide ore body mining is used in long-term stability designs and excludes the effects of backfill confinement.

Various experts in the field of pillar design were consulted about the design criteria used for the project area. The consensus was that the methodology was sound; however, the value for the cubic strength K may be conservative. It was further emphasized that laboratory tests must be carried out to determine the in situ value of K.

18.8 DEFINING INPUT PARAMETERS AND REVIEW OF DESIGN CRITERIA

18.8.1 Estimating Strength of Pillars

Because of insufficient data and literature on in-stope pillar stability in deep-level gold mines, extensive laboratory tests of

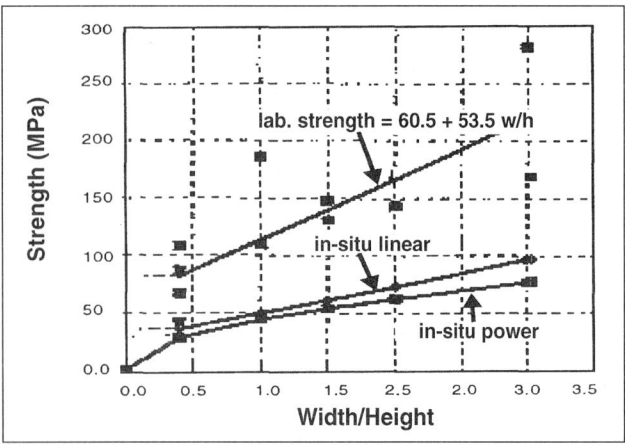

FIGURE 18.11   Laboratory data and estimated in situ pillar strength (Ryder and others 1997)

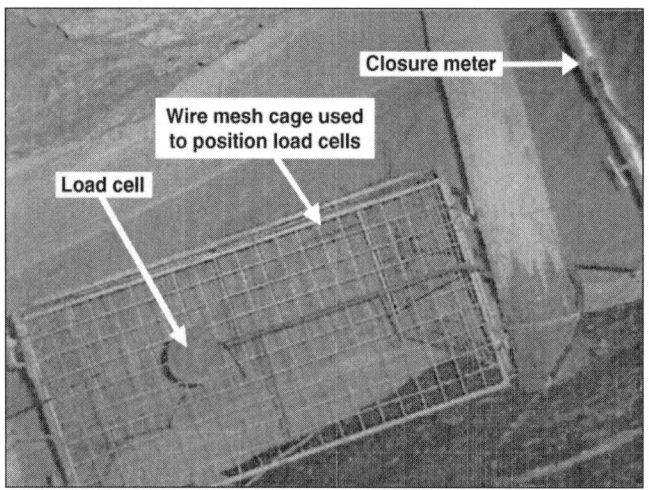

FIGURE 18.12   Backfill stress-measuring instruments

the Elsburg reef are of paramount importance. The laboratory tests will provide an indication of in situ cubic strength, uniaxial compressive strength, and the effects of the width-to-height ratio.

Industry practice has been to relate K to the unconfined compressive strength (UCS) of the reef rock as determined from laboratory tests. Estimates in the literature range from K= 0.25 × UCS to K = UCS.

Ryder and others (1997) proposed that a direct estimation of K be achieved by means of back-analysis using numerical modelling. Their underlying assumption was that the sidewall of an existing pillar is loaded under uniaxial compression and can be used to estimate in situ unconfined compressive strength (UCS_i) of the rock mass. UCS_i can be determined from numerical modelling and can be used as a lower bound for the vertical stress acting on the sidewalls of these pillars. The assumption goes further to say that the value of K can be obtained from UCS_i by applying a geometrical correction factor obtained from laboratory testing.

To review the design criterion in determining the pillar strength, it is suggested that this procedure, together with the results from the laboratory tests, be applied to the formula proposed by Obert and Duval (1967), that is—

$$\sigma_s = K_i\left(A + B\frac{w}{h}\right)MPa \qquad (18.2)$$

where K_i = in situ cubic rock mass strength and A and B = constants that describe the effect of the width:height ratio on pillar strength.

To verify this hypothesis, Ryder and others (1997) obtained results from laboratory tests of core from a mine in the Rustenburg area (Figure 18.11). The values obtained for the geometrical correction factor constants A and B were substituted in Equation 2 to obtain a value for pillar strength. These were then compared to the values obtained using the power formula in Equation 1. They concluded that the two formulae were comparable with regard to pillar strength obtained. The value for pillar strength using the linear formula was preferred because it was based on laboratory data from the actual rock involved. On this basis, the use of the linear formula was advocated, since it is probably the more realistic of the two.

18.8.2   Laboratory Testing

The need to determine the cubic strength of the rock mass from laboratory tests was deemed to be essential to verify the design of pillars in the wide ore body mining method. Because of the conglomerate nature of the reef (particularly the large pebbles)

large-diameter core samples had to be obtained to avoid bias. Initially a core 150 mm in diameter was drilled; however, this proved to be inadequate as it did not satisfy the criterion that the diameter must be at least 10 times larger than the largest grain size. Cores with diameters of 250 mm were then required. This research, in conjunction with CSIR-Miningtek, is ongoing. Results of the tests were not completed at the time this paper was published.

18.8.3   Instrumentation

To facilitate the process of back-analysis and to aid in future design of such mining methods, an extensive instrumentation program has been scheduled for both the EC reef horizon and the wide ore body. To date, a backfill station consisting of load cells for stress measurements and closure meters has been installed in the distress cut on the EC horizon (Figure 18.12). The results show that closure is occurring at 0.5 cm/m of advance. No significant readings were recorded on the load cells because the cable had been damaged.

Petroscopes and extensometers are planned to be installed in the wide ore body. The petroscopes will be used to determine the performance of the reef pillars during the various extraction phases. The extensometers will be installed in the hanging wall to determine the extent of bedding plane separation caused by closure in the underlying EC stope.

18.8.4   Numerical Modelling

Most of the numerical modelling being carried out at PDWAJV uses the Besol/MS suite of three-dimensional boundary-element programs. This system, although limited to elastic rock mass responses, offers quantitative results that have been continually tested by back-analysis.

Extensive work has been done in the South Deep area using the 2D FLAC program. This program allows for a much wider choice of constitutive models and postfailure behaviour. James (1998) states that the results obtained from 2D FLAC modelling remains conceptual rather than exact in its results.

It is important to note that numerical modelling is not a stand-alone tool for use by rock engineers. It must be combined with continuous in situ monitoring (visually and with instruments) and back-analysis to verify designs and ground conditions (James 1998).

18.9   CONCLUSION

Mining in the 87 2 West project area has been ongoing since July 1997. Current extraction is primarily focussed on the Phase I extractions with some heights exceeding 9 m in places. The pillars show no signs of spalling or slabbing. The destress cut is functioning as designed, since there are no visible stress fractures in the sidewalls of the existing pillars.

The design criteria are considered to be conservative for the following reasons.

- First, the pillar strength formula used in this exercise is mainly used for long-term pillars, e.g., board-and-pillar layouts. The destressed pillars in this layout are temporary, as they will be extracted later in the life of the operation.
- Second, the effects of the pillars being confined by backfill are not taken into account when determining pillar strength.

To refine the design criteria used, the value of K and the geometrical constants A and B are needed, which must then be applied to the linear pillar strength formula. To supplement this, instrument results and direct observations need to be carried out on an ongoing basis.

Because of the nature of the ore body at PDWAJV, it has been established that the wide ore body mining method (conventional or with the aid of trackless equipment) is the most economical method to adopt. Current mining will provide invaluable data for future wide ore body mining in the SV2/3 area, as well as in the South Deep project.

18.10 ACKNOWLEDGMENTS

Permission to publish this material by senior management of Placer Dome Western Areas Joint Venture is gratefully acknowledged.

18.11 REFERENCES

COMRO. 1988. *An Industry Guide to Methods of Ameliorating the Hazards of rockfalls and rockbursts. October 1988. User Guide No. 12* Chamber of Mines. Republic of South Africa.

Hedley, D.G.F. & Grant, F. 1972. Stope Pillar Design for the Elliot Lake Uranium Mines. *Bull. Can. Inst. Min. Metal, 65.* pp 37–44.

James, J.V. 1998. Geotechnical Influences upon the Design and Operation of a Deep Level Wide Ore body Gold Mine. PhD Thesis, Cardiff School of Engineering, University of Wales, Cardiff.

James, J.V. & Leih, D.L. 1990. Rock Mechanics Aspects of Mechanized and Bulk Mining Methods at Western Areas Gold Mine. *Static and Dynamic Considerations in Rock Engineering.* Balkema, Rotterdam. ISBN 90 6191 153 2.

James, J.V. & Raffield, M.P. 1996. Rock Engineering Design Considerations for Massive Mining in the South Deep Section, Western Areas Gold Mine. *S. Afr. Inst. Min. Metall. Colloquium: Massive Mining Methods,* July 1996.

James, J.V. & Raffield, M.P. & Pethö, S.Z. 1998. Monitoring of the performance of a crushed waste/classified tailing backfill for shaft pillar pre-extraction in the South Deep Section, Western Areas Gold Mine, Mine fill 1998.

MacDonald, A.J. 1998. Design Considerations for the Extraction of a Wide Ore Body in a South African Deep Level Gold Mine. PhD Thesis, Cardiff School of Engineering, University of Wales, Cardiff.

Personal communications. York, G. & Ryder, J.A. & Ozbay, M.U.

Placer Dome Western Areas Joint Venture. 1997. Code of Practice to Combat Rockfalls and Rockburst.

Raffield, M.P. 1993. Stability Investigation Associated with Shaft Reef in a Deep Level Gold Mine. PhD Thesis, Cardiff School of Engineering, University of Wales, Cardiff.

Ryder, J.A. & de Maar, W. & Ozbay, M.U. 1997. A methodology for estimating the strength of hard rock support pillars. *SARES 97 1st South Afr. Rock Eng. Symposium Proceedings.* pp 435–438.

Salamon, M.D.G. and Munro, A.H. 1967. A Study of the strength of Coal Pillars. *J.S. Afr. Inst. Min. Metall.* Vol. 68, pp 55–67.

Singh, N. 1997 Backfill Barricade Design for 87L Massives. *Unpublished internal report. Placer Dome Western Areas Joint Venture.*

Singh, N. 1997. 87 2W MB and MI Support Requirements. *Unpublished internal report. Placer Dome Western Areas Joint Venture.*

Singh, N. 1998. Assessment of Pillar Stability in the Wide Ore Body Mining on 87 2W. *Unpublished internal report. Placer Dome Western Areas Joint Venture.*

Smallbone, P.R. 1992. In-situ monitoring crusher excavation. JCI South Deep. Quarterly Report No. 3, University of Wales, Cardiff.

Spearing, A.J.S. Handbook on Hard Rock Strata Control. 1995. Published by SAIMM.

Shrinkage Stoping

CHAPTER 19

Shrinkage Stoping at the Mouska Mine

Robert Marchand,* Patrick Godin,† and Chantale Doucet*

19.1 INTRODUCTION

The Mouska gold mine is located 80 km west of Val-d'Or and 20 km east of Rouyn-Noranda near the provincial border of Quebec and Ontario in northwestern Quebec, Canada. This area is approximately 420 km northeast of Montreal. The mine lies 5 km north of the Cadillac Fault in the Bousquet Mining District, one of Canada's most prolific gold-mining camps.

Gold was first discovered on the Mouska property in 1936, and over a dozen companies have explored and produced gold from the property. Cambior, Inc., acquired 100% of the Mouska property in 1986 in connection with the partial privatization of the assets of Soquem, a mining and exploration company owned by the Quebec government, and in 1987 an extensive exploration and development program led to the delineation of reserves. Commercial production began in July 1990.

Development work included infrastructure construction, shaft sinking, underground development, and underground exploration drilling. As a result of continued mineralization at depth, the shaft was deepened at the beginning of 1995. Further exploration has led to the development of an internal shaft to facilitate underground access.

The current mine plan is based on the production of 100,000 tonnes of ore per year, with the mine operating on a 5-day-a-week schedule. Seventy-two percent of the ore produced comes from shrinkage stopes, 20% from longhole mining, and the remaining 8% from development work. Daily production is 400 tonnes of ore and 200 tonnes of waste.

The shrinkage stope method is labour intensive and thus is likely to generate more accidents than more mechanized mining methods. Furthermore, skilled and experienced miners are becoming harder to find. In 1999, the mine's overall compensable accident frequency was 3.1 accidents per 200,000 hours worked compared to a frequency of 3.5 for the entire mining sector in the province. Even though these statistics show a better-than-average performance, the number of temporary work assignments is important. The main injuries to be reduced in shrinkage stoping are related to physical effort, mostly back pain and head injuries (hit by object).

19.2 GEOLOGY

19.2.1 Regional Geology

The Mouska deposit lies in the southern part of the Abitibi subprovince in the Bousquet Mining District. It is hosted by volcanic and plutonic rocks of the Blake River Group (BRG). The BRG is bordered to the north and south by sedimentary rocks of the Kewagama and Cadillac groups, respectively. The contacts between the BRG and the adjacent sedimentary groups are characterized by extensive zones of deformation; that is, the Lac

Parfouru fault zone to the north and the Dumagami fault zone to the south.

The Bousquet Mining District is one of the largest gold-producing districts of the Abitibi Greenstone Belt, Quebec's most productive metal-mining region. The belt hosts several large deposits, including La Ronde, Bousquet 2, Bousquet 1, Doyon, and Mouska. These deposits lie in an intensely altered and deformed corridor of the BRG known as the Doyon-Dumagami deformation zone. Combined historical production and current reserves and resources from these five mines exceed 21 million ounces of gold.

19.2.2 Local Geology

The Mouska deposit is described as a lode-type deposit, in which economic mineralization is confined to narrow (average width of 30 cm) quartz veins having good lateral and vertical continuity. Veins strike N 110° E with a subvertical dip and contain 5%–15% pyrite-pyrrhotite and 5%–10% chalcopyrite. Economic lenses are found both in andesites and in the Mooshla diorite. Reserves and resources at Mouska are calculated using cross sections every 10 m. Minimum widths of shrinkage and longhole stopes are 1.6 and 1.8 m, respectively.

Reserves are located primarily within the Mooshla intrusive in three main zones: 40, 50, and 60. Zone 50 contains most of the reserves. The cut-off grade for reserves is 9.8 g/t of gold. Five percent dilution is added at zero grade in shrinkage stopes and 14% dilution at zero grade in longhole stopes. Dilution is based on an ore-to-waste ratio and is defined as the material extracted over the minimum mining width included in the reserves.

Year-end 1999 reserves were 303,000 tonnes with a grade of 15.2 g/t of gold, totaling 148,000 oz in situ. Gold price was US$325/oz. Year-end 1999 resources were evaluated at 329,000 tonnes with a grade of 9.7 g/t of gold, totaling 102,500 oz in situ. On the average, 40,495 oz of gold were renewed per year since 1992, which corresponds to approximately 1 year of production. Renewal cost for the reserves is US$10/oz.

19.3 EXPLORATION

Exploration is done both from the surface and underground. Underground exploration drilling is done with BQ-size (3.65-cm) diamond-drill holes on a 40-by 40-m pattern using electric-powered drills. Special exploration drifts are required to position the drills better. Definition drilling is done with ATW-size (3.05-cm) diamond-drill holes on a 10-by 15-m diamond pattern using air-powered drills. A 15-m-long cross-cut driven perpendicular to the haulage drift will enable good coverage of the stope. Shorter holes could be drilled directly from the haulage drift. At Mouska, 95% of the core is recovered.

Table 19.1 shows typical drilling data required every year to maintain the reserves level.

---

* Cambior, Technical Services, Val-d'Or, Quebec, Canada.
† Cambior, Mouska Mine, Rouyn-Noranda, Quebec, Canada.

TABLE 19.1 Diamond drilling costs

	Average m/yr	US$/m	Total US$/yr
Exploration, surface	7,500	47.75	358,000
Exploration, underground	10,500	23.75	249,400
Definition drilling	6,650	21.60	143,600
Total	24,650		751,000

TABLE 19.2 Employee distribution per department

Department	Number
Management	2
Purchasing	1
Human resources	3
Engineering	5
Geology	5
Maintenance	12
Underground	63
Total	91

TABLE 19.3 Rock properties of intact diorite

Parameter	Value
Uniaxial compressive strength	175 MPa
Young's modulus	60,000 MPa
Poisson's ratio	0.15
Rock quality designation (RQD)	80%
Ore density	$2.8\ t/m^3$

19.4 GENERAL MINE DESCRIPTION

A four-compartment, 485-m-deep shaft provides access to the Mouska Mine. The mine has been developed with levels on 60-m centres from the surface. The levels are equipped for rail haulage with 18-kg gauge rails having 76-cm inside spacings. Haulage drifts are developed in waste parallel to the veins at an average distance of 10 m from the stope. Ore is extracted from drawpoints excavated perpendicular to the ore.

The average width of shrinkage stopes is 1.6 m. Drilling in the stopes is done with hand-held drills (jacklegs and stopers), and ore is loaded into the rail cars with rubber-tired CAVO 320 or LM56 mucking machines. On the active haulage levels, $5\frac{1}{2}$-tonne battery-powered locomotives, pulling seven 5-tonne cars, move the broken ore from the stopes to the ore pass near the shaft. The ore is directed to the hammer and grizzly installation near the bottom of the shaft, where it is loaded into 6-tonne skips to be hoisted to the surface.

To access deeper levels, an internal shaft is being excavated 800 m east of the main shaft. This new 318-m shaft has three compartments and a hoisting capacity of 600 t/d.

As of January 1, 2000, 91 employees were on Mouska's payroll. Their distribution within the different departments is shown in Table 19.2.

19.5 GEOTECHNICAL INFORMATION

The host rock mass is diorite. Its intact rock properties are listed in Table 19.3 and were obtained from tests performed on diamond-drilled core samples.

Structural mapping of the underground openings identified three major joint sets that are relatively constant within the diorite: (1) N 110° E strike, subvertical dip (parallel to veins), (2) N 45° E strike, 60° to 90° dip, and (3) subhorizontal. The joints are smooth, moderately undulating, and lightly chloritized.

Bieniawski's Rock Mass Rating system (RMR) and NGI's Tunneling Quality Index (Q) were both used to classify the rock mass. Typical RMR and Q values for Mouska are 79 and 11, respectively, and both numbers qualify the rock mass as good.

Mineralized veins have good continuity. The sinuosity of the veins usually results in a dip between ±5° from vertical and ±10° from the orientation direction (N 110°E) within the same stope. These variations are mostly contained within the designed stope width of 1.6 m, but can sometimes cause changes in the stope profile. In some cases, fault zones intersect the veins and can offset the veins to distances wider than actual stope width. The kinks and bends formed when the stopes are offset are often the source of instability and dilution. Most of the ground control problems can therefore be attributed to—

- Diorite behaving as a brittle material under pressure,
- Unstable blocks formed by the intersection of major joints,
- Major changes in dip or strike of the veins.

19.5.1 Ground Support

The length of ground support within the stopes is controlled by the width of each stope. Each cut typically measures 2.4 m high by 1.6 m wide. At a minimum, hanging walls, footwalls, and backs are bolted on a 1.2- by 1.2-m diamond pattern with 1.2-m-long rock bolts on each cut. Chain link screen is added to the back of the last cut.

Drawpoints are 2.7 by 2.9 m in cross section and are supported with 1.5-m-long, resin-grouted bolts (rebar) on a 1.2- by 1.2-m pattern on the backs and walls. No. 9-gauge, welded mesh screen (4 by 4 in) is also bolted on the back of the drawpoints. The footwall drift measures 2.7 by 2.7 m in cross section. It is supported with 1.5-m-long rock bolts on a 1.2- by 1.2-m diamond pattern and No. 9-gauge, welded, 4- by 4-in mesh screen on the back. The walls are supported with 1.5-m-long, resin-grouted bolts on a 1.2- by 1.2-m-square pattern. Raises are supported with 1.2-m-long friction bolts (Split-Set rock bolts) on a 1.2- by 1.2-m-square pattern and chain link screen.

A 5-m-high sill pillar is left at the top of every stope to separate the stope being mined from the one above. Mucking can therefore proceed in the top stope while mining advances in the bottom stope. This procedure also provides additional ground support for the drawpoints, pillars, and footwall drift when the stope is empty. Pillars are planned in every stope, but their location may vary according to ground conditions and drawpoint locations. Typical pillar designs used at Mouska are illustrated in Figure 19.1.

19.5.2 Stress Distribution

No in situ stress measurements were conducted at Mouska, but measurements were collected at Doyon, the nearest active mine. These measurements indicate that the major principal stress (σ_1) should be oriented N 45°E, i.e., nearly perpendicular to the veins. This correlates well with observations underground. The stress gradients and orientations used for modeling are listed in Table 19.4.

19.6 STOPE PREPARATION

Once the diamond-drilling programs are completed and the characteristics of the ore (geometric shape, continuity, tonnage, grade, etc.) indicate that an economic stope is feasible, the next step is to develop and prepare the stope for mining. The development of the ore drift will confirm the exact position and continuity of the vein and enable planners to finalize stope design. This step will include excavation and installations (access, ventilation, compressed air, water, equipment) required to extract the ore in a safe and economic manner.

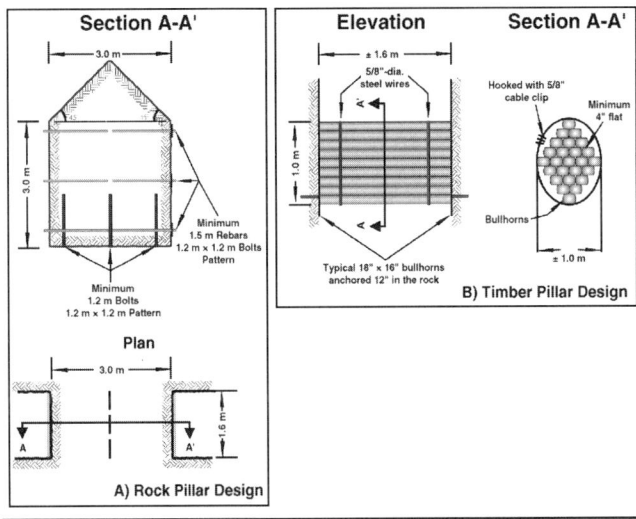

FIGURE 19.1 Typical pillar designs used at the Mouska Mine

TABLE 19.4 In situ stress gradients and orientations

Stress	Gradient, m	Orientation
Maximum horizontal stress (σ_1)	$0.059 \times$ depth	N45°E (perpendicular to vein)
Minimum horizontal stress (σ_2)	$0.053 \times$ depth	Parallel to vein
Vertical stress (σ_3)	$0.032 \times$ depth	

While different development configurations and construction arrangements are possible for shrinkage stoping, at Mouska drawpoints are developed perpendicular to the haulageway at a distance of 10 m from the ore. The spacing of these drawpoints is 10 m centre to centre, which has a direct influence on the effectiveness of the mucking operation. With this type of arrangement, less construction work is required, the mining of the stope can start at the same elevation of the haulage way, and most of the ore is recovered. This arrangement, while increasing the amount of waste, provides access to a number of other stopes on the same level without construction delays. It also provides the opportunity to use a diamond drill to complete, if required, definition of the boundary of the reserves.

19.6.1 Ore Drift (Sill)

It is well known that before development starts, the ore will not be found in a straight line on a plan view. Although the diamond-drill holes give some information on the location and shape of the vein, it is common practice to develop the ore drift (sill) ahead of the footwall drift (or haulage drift). This approach allows for a better consistency in the length of drawpoints, defines the ideal location of the footwall drift with respect to ore, and preserves the integrity of the pillars between drawpoints.

The sill is usually driven at a maximum of 30 m ahead of the footwall drift (Figure 19.2). Since these excavations, including the drawpoints, are done with the same crews, they provide more working faces and increase productivity. This practice also allows more time for the geologist to properly align the development crew in the ore, especially if the vein is displaced by a fault, shear, or other feature.

It is important to develop the ore drift with great care. When the first four lifts of the stope are mined, the most important dilution factor is related to the width of the sill drift. If the first opening in the ore is too wide, it will be very difficult to reduce the width afterward when the successive cuts are mined. At Mouska, ore drifts are excavated 1.5-m wide with the help of

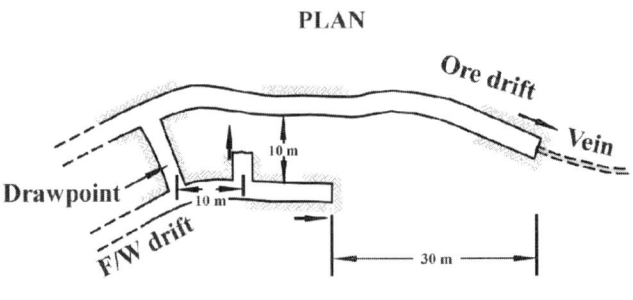

FIGURE 19.2 Development of ore and footwall drifts

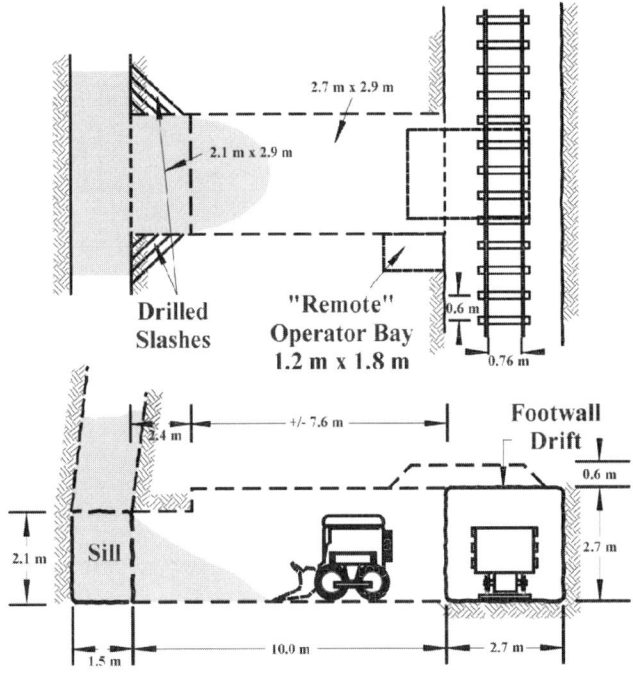

FIGURE 19.3 Typical drawpoint configuration

jacklegs and slushers equipped with 60- to 80-cm-wide blades. Ore is then slushed to the first drawpoint intersection, where it will be loaded with Cavo or muck machines into wagons.

19.6.2 Drawpoints

Figure 19.3 shows a typical drawpoint configuration. The following factors are used in designing drawpoints.

- Drawpoints are 10 m long and driven perpendicular to the haulageway to facilitate ore loading into cars. As mentioned before, they are spaced 10 m apart centre to centre.

- The cross section is 2.7 by 2.9 m.

- At the intersection of each drawpoint with the haulage drift, the roof is excavated a little higher than the normal height of the drift to allow proper loading into cars.

- Since there is no boxhole driven to direct the flow of ore toward drawpoints, remote-controlled muck machines are used to recover ore between drawpoints. At the stope preparation stage, a small bay on the side of each drawpoint is excavated for the operator of remote-controlled equipment.

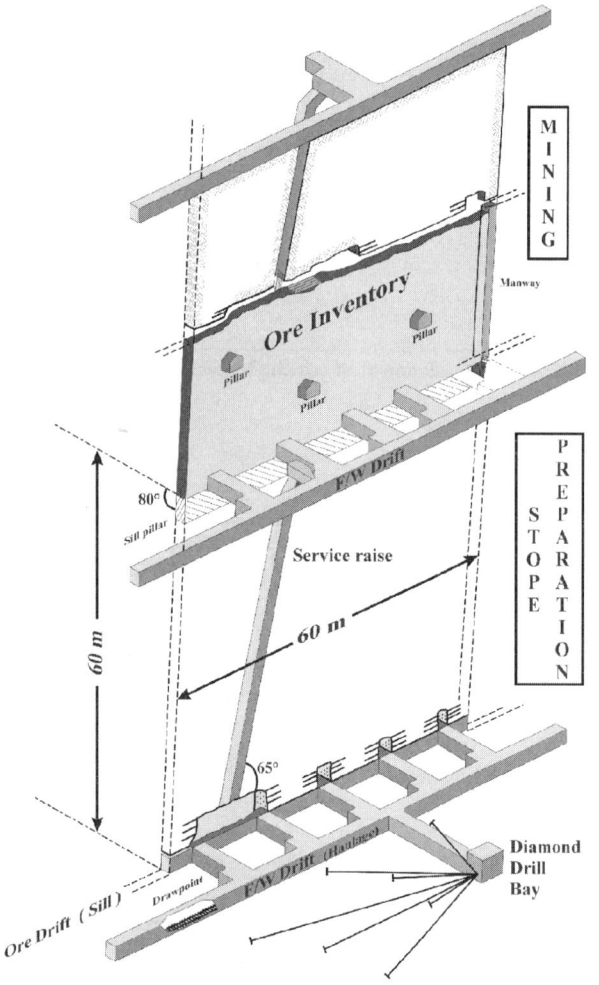

FIGURE 19.4   Schematic of different mining steps in shrinkage stoping at the Mouska Mine

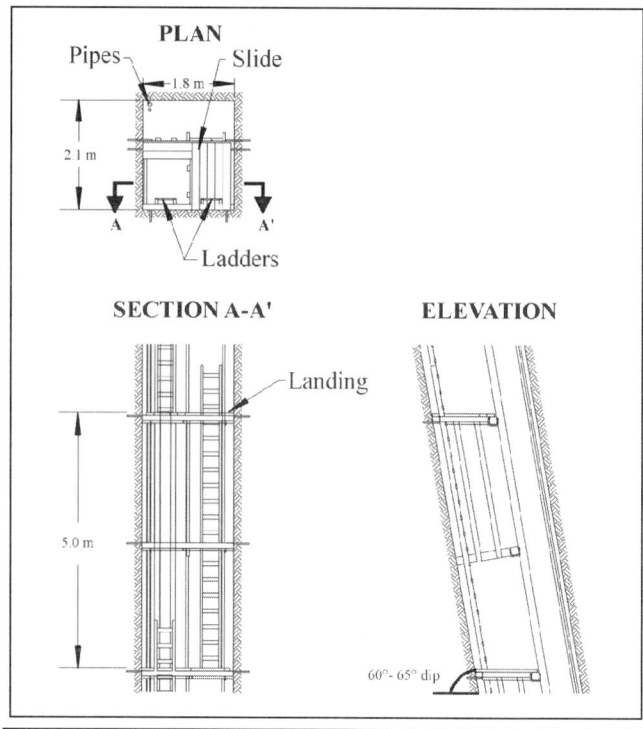

FIGURE 19.5   Typical arrangement for service raise

- To ease movement of remote-controlled equipment, slash holes are drilled on each side of the drawpoint at the stope entrance. These holes are blasted only if remote mucking of the stope is necessary.

- Finally, the last drawpoint round at the entrance of the stope requires a lower back to establish a brow that will prevent ore from spreading too far into the drawpoint.

19.6.3   Service Raise

Once the ore drift (sill), drawpoints, and footwall drift are driven along the future stope, a vertical connection is needed to reach the upper level. The purpose of this raise is to bring all services (compressed air, water, ventilation, access) to the working faces during the mining phase. Services are therefore connected to the level above. This raise is driven conventionally in ore on a straight line and gently offset to clear the 5-m sill pillar to exit in a drawpoint on the next level 60 m above (Figure 19.4). The raise dips at 60° to 65° and has a cross section 2.1 m high by 1.8 m wide.

Once excavation of the raise is completed, the timbering phase can begin. Construction of the service raise involves installing a ladderway with resting pads at a maximum distance of 7 m. Each ladder segment needs to be offset from each other for safety reasons (Figure 19.5). A straight raise also allows for the installation of a wooden slide that gives miners the opportunity to use a winch and bucket for transporting all

supplies required during the mining phase (rock bolts, screens, explosives, steels, tools, etc.).

19.7   MINING

The overall mining sequence is from top to bottom, but in any given stope, it is obviously from the bottom up. The choice of shrinkage mining at Mouska is based on the following factors.

- It is a selective method that allows daily follow-up on the vein being mined.

- It is a flexible method that permits better recovery of the ore in the extremities of the stopes.

- Ore inventory left in the stopes during the mining phase provides additional wall support.

- Pillars can occasionally be left inside the stope, improving ground stability and mining grades.

The minimum mining width included in reserve calculations is based on the following items:

- Working space required.

- Effectiveness in following changes in vein direction.

- Minimum space required to install ground support in both walls.

To consistently produce 400 t/d, 10 active stopes are required, two in the preparation phase, six in the mining phase, and two in the final pull phase. Planning of further work must take into account these proportions to avoid too much ore feed fluctuation.

19.7.1   Mining Description

Mining shrinkage stopes at the Mouska Mine is done by taking successive 2.4-m-high horizontal lifts starting at the elevation of the haulage drift. A brow is required for drilling horizontal holes and blasting toward the floor. This technique, known as breasting, was chosen for the following reasons.

- Miners are not exposed to the ground they actually drill, as they would be if they were drilling vertical holes.

- The vein can be followed more easily.

- Bootlegs or missholes can be taken care of after each blast.

- A smoother blasting practice can be followed, especially to the back of stope.

- Wall ground support is installed at the best time, i.e., immediately after each blast.

Each blast of approximately 25 tonnes requires about 12 holes each 2.4 m long. In shrinkage stoping, the time required to bring material and equipment to the face, scale, and set up is relatively long compared to the time it actually takes to drill a breast, which is quite labour intensive for the tonnage produced. This shows the importance of being well organized in a stope to be as efficient as possible and focus on quality.

19.7.2   Startup of the Stope

Stope startup gives miners the opportunity to multiply the number of faces. At the intersection of each drawpoint with the ore sill, a raise round 1.8 m high by 2.1 by 1.5 m is taken. The ground is then supported, and the brow of the drawpoint is protected with bolted steel straps. This vertical opening provides two available faces for breasting at each drawpoint. Miners can access these faces directly from the drawpoints and eventually establish a level opening all along the stope. A second and last lift can be taken using the same approach. Some mucking of ore is required to enter the stope.

Once this second lift is taken, the service raise excavation will start (see the section on "Service Raise").

The remainer of the stope will be mined using three working faces, two from each side of the service raise and one from the manway side. This type of arrangement, with a two-man crew, leads to an overall performance of 18.5 t/m-s. Each 60-m cut contains 645 tonnes and will take approximately 35 manshifts (8 hours) to complete.

19.7.3   Manway

A shrinkage stope needs two accessways available at all times. The service raise already in place provides access from the top of the stope. This raise will shorten as mining progresses up.

An access from the bottom of the stope also needs to be established. This manway will often be positioned at one end of the stope and accessed by a side opening on the last drawpoint. This location will allow fresh air to circulate through the stope. It is built out of wood on one side and follows the rock profile on the other three sides. It is important to establish a perfectly vertical wood wall. It may sometimes be necessary to use wood on the other sides, depending on the shape of the stope extremity. This manway will be raised one lift at a time following the mining cycle. To establish a working face from the manway, it is necessary to excavate the first round using uppers (raise) aside from the manway so it can be protected from the blast. Synchronization of activities within the stope is required since access during this phase is from the service raise. Figure 19.6 illustrates a typical manway arrangement used at Mouska.

19.7.4   Ore Inventory

Because a given amount of broken rock takes up more space than solid rock, some material must be removed as the stope advances. To keep a working floor from which to mine the next lift, only 35% of the broken ore is removed during the mining stage (swell). This ore stays in the stope until successive cuts bring the elevation of the stope to 5 m below the next level above. This inventory is gradually trapped in the stope for a period that can extend to 8 months and can reach 10,000 tonnes before the final

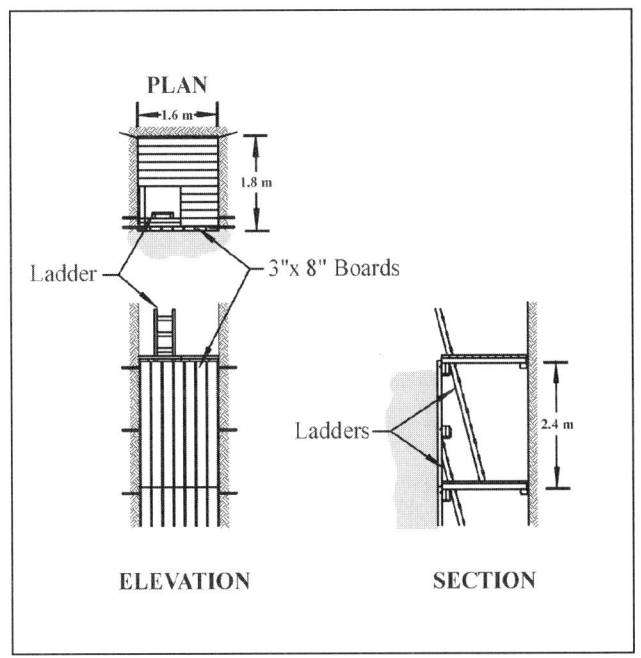

FIGURE 19.6   Typical manway arrangement

draw of the stope. At 15 g/t, the value of this inventory can reach US$1.4 million per stope. This amount of "trapped inventory" is the main drawback of the method when compared with the more-productive longhole method.

19.7.5   Working Floor

During mining, the broken ore leaves uneven ground from which to work and, in combination with the ore pulled from drawpoints, makes it necessary to establish a more level work floor. A slusher or a blow pipe is used to spread the ore more evenly or a local wooden platform is built. Monitoring the distance between drawpoints and the tonnage being pulled from each drawpoint are important factors governing floor quality in the stope. Very strict safety procedures need to be followed when pulling ore from a shrinkage stope.

19.7.6   Hangups

Because of the narrowness of the stopes, it is common that ore will hang up in the stope during the mucking phase, even though the ore is finely broken, which prevents further mucking from taking place. An intervention is required to make ore flow again. This situation can occur during either pulling during the mining cycle or at the final pull of the stope. During the mining cycle, this situation is closely monitored by miners, and if after mucking 10 cars (50 tonnes) of ore and the stope floor has not moved, mucking is halted and the ore is washed or blasted.

It used to be that the blasting or spraying water were done from from a catwalk on the last upper lift of the stope. This catwalk provides access for miners to enter the open stope and unwedge ore below. However, instead of this system, miners at Mouska install a wooden rail along the stope back during the last lift. A small, remote-controlled carrier moves along this rail and brings a high-pressure water hose to the desired location. This system enables efficient unblocking of ore without the need to send miners into an open stope. One-hundred percent of the broken ore is recovered, and the need to blast hang-ups, which can damage the walls, is eliminated. Once the stope is empty, it is not backfilled. Mining recovery, taking into account the pillars, is 95%.

TABLE 19.5   Operating costs at Mouska Mine (average, 1995–1999)

	Cost per tonne milled	Cost per ounce
Definition drilling	1.85	4.63
Stope preparation[1]	12.72	31.80
Mining[2]	22.00	55.00
Services[3]	28.70	71.75
Administration[4]	10.11	5.28
Transportation	1.30	3.25
Milling	0.00	5.00
Total	US$86.68/t	US$216.70/oz

[1] Stope preparation includes sill drift, drawpoints, service raise.
[2] Mining includes drilling, blasting, ground support, and mucking
[3] Services includes supervision, hoist, level maintenance, pumps, ventilation, equipment maintenance, material handling, power, surface costs, etc.
[4] Administration includes geology, engineering, management, human resources, purchasing, etc.

19.8   COSTS

At 400 t/d, Mouska Mine is a low-tonnage operation, so the proportion of total operating costs relative to fixed costs (mostly services and administration) is relatively high. Any variation in tonnage has a sizeable effect on total costs, especially when analyzed on a unit cost basis. Shrinkage stoping is a method focusing on quality, rather than quantity. Therefore, any comparison with other types of operations on a cost-per-tonne basis is misleading. It is much preferable to analyze costs using the end-product, such as cost per ounce. Table 19.4 shows different cost areas in both costs per tonne and cost per ounce (U.S. dollars at an exchange rate of $1U.S. = $1.45 Canadian).

These figures represent an average cost. Improvements made in the operation over the years show decreasing costs per ounce. The forecast for year 2000 is US$160/oz.

Table 19.5 shows the breakdown of unit costs per metre of development and stoping and mucking costs per tonne. Unit costs include base wage, fringe benefits, and incentive system, which is based on quality and productivity and respect for all operating standards. It also includes distribution of equipment maintenance costs.

19.9   CAPITAL EXPENDITURES

Capital expenditures include long-term development (footwall drifts, ore and waste passes, ventilation raises), construction, exploration diamond drilling, and equipment purchasing (replacement and addition). On the average, over the past 5 years investments at Mouska totaled US$1.7 million per year. This amount does not include special nonrecurrent projects, such as the internal shaft. Table 19.6 lists the distribution of these expenditures.

TABLE 19.6   Unit costs in U.S. dollars at an exchange rate of $1.00 U.S. to $1.45 Canadian

	Costs per metre			Costs per tonne	
	Track drift	Ore drift	Raise	Stoping	Mucking
Labour	334	338	414	15.23	3.52
Supplies:					
Drilling	68	43	30	2.48	—
Blasting	90	63	46	2.07	—
Ground support	62	46	38	2.28	—
Track	52	—	—	—	—
Timber	10	12	65	0.39	—
Miscellaneous	40	69	17	0.18	0.07
Assays	—	2	7	0.14	0.28
Equipment maintenance	100	76	52	2.07	2.07
Total	US$757/m	US$648/m	US$669/m	US$24.83/t	US$5.93/t
Performance	1 m/m-s	1 m/m-s	1 m/m-s	18.5 t/m-s	60 t/m-s

TABLE 19.7   Distribution of expenditures in U.S. million dollars per year

Development (deferred)	1.0
Diamond drilling (exploration), surface and underground	0.6
Equipment	0.1
Total	1.7

19.10   CONCLUSIONS

Mining operations at the Mouska Mine demonstrate that shrinkage stoping is an old method that still has its place in this type of environment. It is obvious that with the shortage of skilled and experienced miners, improvements to the method (mostly related to equipment) are required. Efforts are being made to change and improve the efficiency and safety of the method using the know-how and experience of the people on-site. With 91 employees, this small "family" can quickly make changes to improve the items it controls, which is everything but the price of gold.

Surprisingly, shrinkage stoping is a method that still uses basically the same equipment as was used 40 years ago. Modernization of the method must involve a combination of improvements in method and equipment. The end result may very well be a combination of partly shrinkage, partly longhole, with a little less labour-intensive work, which is the case at Mouska.

CHAPTER 20

Shrinkage Stoping Practices at the Schwartzwalder Mine

Bruce Norquist[*]

20.1 INTRODUCTION

The Schwartzwalder Mine is located in Jefferson County, Colorado, approximately 24 miles west of Denver and 7 miles north of Golden. Comprising 558 acres, the property is privately owned by the Cotter Corporation. Mine workings range in elevation from 4,250 ft at the lowest point to 7,200 ft in the uppermost stopes. The mine portal and facilities are located at an elevation of 6,590 feet in the floor of a steep, narrow, mountain valley. A well-maintained dirt road to the site allows continuous mining operations year round.

The Schwartzwalder Mine is the most significant uranium deposit to be mined in the Front Range mineral belt of Colorado. It has produced nearly 18 million pounds of U_3O_8 in the span of 50 years.

Prospectors scouring the hillsides in search of gold originally found copper at the site in the late 1800s. In the late 1940s, Fred Schwartzwalder discovered uranium in several of the old prospects and surrounding hillsides. The first recorded production at the Schwartzwalder Mine dates back to 1953. By 1958, approximately 36,500 tons of uranium ore had been produced. Large-scale mining commenced in the early 1960s and continued until uranium prices fell in the mid-1980s. The mine went through a period of shut-down from 1987 to 1995. In 1995, projected uranium price increases renewed interest in the Schwartzwalder Mine. Development operations commenced in mid-1995, and ore was again being produced by mid-1996. By the end of 1999, nearly 2 million pounds of U_3O_8 had been produced during this most recent period.

20.2 GEOLOGY OF THE SCHWARTZWALDER MINE

The Schwartzwalder ore bodies occur in Precambrian metamorphic rock terrain formerly known as the Idaho Springs Formation. A generalized geologic map (Figure 20.1) shows the major rock units surrounding the Schwartzwalder Mine. The major lithologic units in the mine area are a lime-silicate hornblende gneiss to the north and a mica schist to the south. These units were deposited as subsea volcanics and shales and then deeply buried, folded, and steeply tilted. Later in the Precambrian era, the region underwent brittle deformation and major faulting. Numerous pegmatites were intruded into these rocks. The two principal faults in the mine vicinity, which influenced local structure, are known as the East and West Rogers faults. They are about 3,500 ft apart, trend northwest, and dip steeply to the northeast. The Illinois fault, a more northerly fault system, runs diagonally between these two faults and dips steeply westward.

Between the Illinois fault and the West Rogers fault runs a thin band of siliceous and iron-rich metamorphic rocks. This zone consists of quartzite and garnet-biotite gneiss. The quartzite grades into a quartz-biotite schist, and the garnet-biotite gneiss

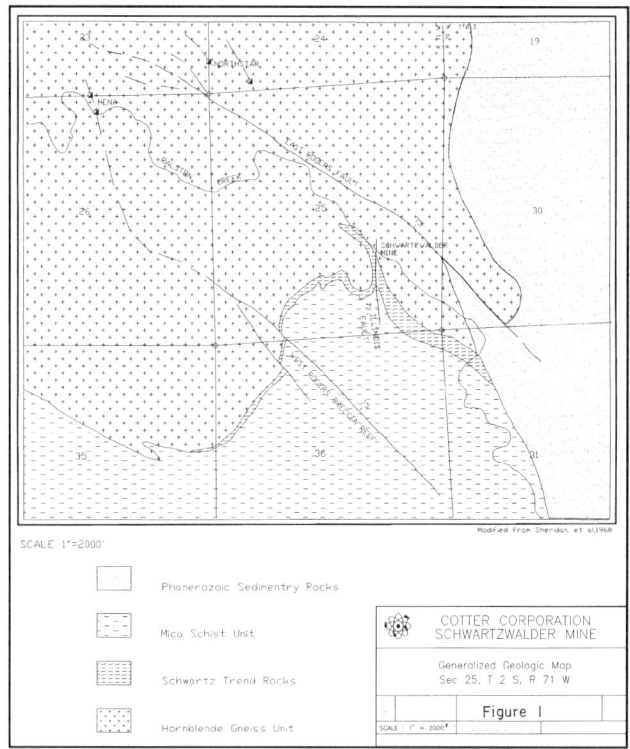

FIGURE 20.1 Generalized geologic map

interfingers with pyritized mica schist. These four rock types are collectively referred to as the "Schwartz Trend" and are the principal host rocks of the uranium ore bodies in the Schwartzwalder Mine. The south-dipping Schwartz Trend band is oriented nearly vertically and is about 100 ft thick. The Schwartz Trend is found in and between the Illinois and West Rogers faults and has been mapped on every level of the mine.

Uranium mineralization in the Schwartzwalder Mine occurred during the Laramide orogeny (~70 mya). Compressional forces during this time reactivated the Illinois and West Rogers faults. These forces created tension and shear fractures in and between these two faults. Uranium-bearing hydrothermal fluids ascended the faults, permeated the fracture systems, and deposited the uranium minerals as a breccia vein filling. Pitchblende is the dominant uranium mineral.

Uranium deposition tended to occur almost exclusively in the open fractures of the siliceous and brittle "Schwartz Trend"

---

* Cotter Corporation, Schwartzwalder Mine, Golden, CO.

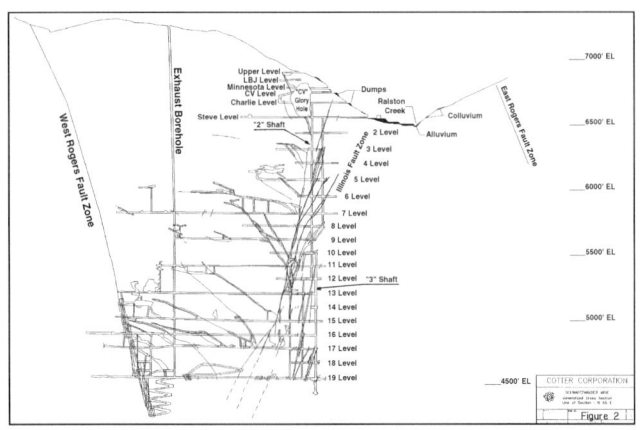

FIGURE 20.2   Generalized cross section of Schwartzwalder Mine

host rock. The bounding mica schist and hornblende gneiss rock units deformed in a plastic manner that did not allow for open fractures. Fractures in the brittle band of Trend rock tend to be flat to intermediate in dip and extend for great distances, on the order of 2,000 ft. As these fractures emanate from the Illinois fault and terminate in or near the West Rogers fault, they are referred to as horsetail structures.

Because of the thickness of the Trend rock, ore bodies tend to be 100 to 150 ft in extent. Vein thicknesses are on the order of 1 to 4 ft and extend discontinuously updip and downdip. Length of most mineable high-grade veins is about 200 ft. The higher-grade ore bodies tend to be in the thickened root zones associated with the intersections between the horsetails and the Illinois and West Rogers faults. Changes in local structure and rock type significantly influence ore continuity. Vein thicknesses tend to vary greatly, particularly where Trend rock interfingers with the mica schist or hornblende gneiss rock units.

20.3   GENERAL MINE LAYOUT

Above the Steve level are the four abandoned "original" levels, now used only for exhaust ventilation. The active portion of the mine, from the Steve level down, is composed of 20 levels, with average level spacings of 110 ft (Figure 20.2). Two internal shafts provide access to these levels, the No. 2 shaft serving the upper 11 levels and the No. 3 shaft serving the lower nine levels. A decline from the end of the 1900 level provides access to the 2200 level, the deepest developed level in the mine.

Stoping is performed by miners using conventional, narrow-vein stoping techniques. Both rubber-tired and rail-mounted haulage equipment is used to move ore from the stopes to the shaft pockets. Rail haulage is utilized in the older No. 2 shaft levels. In rail haulage, muck is trammed from the stope to the shaft pocket in 1.5-ton side dump cars pulled by 3-ton, battery-powered locomotives. Rubber-tired haulage is used in the No. 3 shaft levels. In rubber-tired haulage, muck is trammed from the stopes to a system of ore passes by 1-yd³ load-haul-dump (LHD) units or 5-ton mine trucks. From these ore passes, muck is trammed to the 1900 shaft pocket in 3.5-ton Granby- type cars pulled by a 7-ton, battery-powered locomotive.

Ore is hoisted to the surface in two stages. Ore from the 1900 pocket is hoisted up the No. 3 shaft to the 1100-level shaft pocket. From here, the ore is then hoisted to the Steve-level pocket up the No. 2 shaft. All ore from the upper levels is also hoisted up the No. 2 shaft to the Steve-level pockets. The last stage in the muck circuit involves tramming ore from the Steve-level pocket to the surface ore bins using 1.5-ton side-dump cars pulled by a 5-ton, battery-powered locomotive.

Both the No. 2 and No. 3 shaft hoists are 200-hp, double-drum units. Hoisting distance in each shaft is around 1,070 ft. At a hoist speed of 80 ft/min and using two 2.4-ton-capacity skips per hoist, hoisting capacity is 81 tons/hr. On average, 150 tons per shift is hoisted. Because the hoists are the only means of access to the different levels, much of the daily hoist operation is dedicated to hoisting men and materials, a time-consuming process.

Ventilation at the Schwartzwalder is unique. Because of radon, a push-pull ventilation system is utilized to provide good control of airflow. Present fan capacity is 250,000 ft³/min of air. Push-pull ventilation is accomplished by using two intake and two exhaust fans on the surface. To ensure that radon is forced away from the active areas, the two intake fans are arranged so that under positive pressure, they deliver fresh air into the shafts. The two exhaust fans create negative pressure on the back side of the workings to pull radon away from the active areas. Air flows across the levels and through both active and inactive workings in its course to the exhaust borehole, an 8-ft-diameter, 2,400-ft-high bored raise that exhausts to the surface. The intake fan arrangement enables air to be routed independently through the upper and lower levels.

20.4   MINING PRACTICES

20.4.1   Ground Control

The ground is extremely competent at the Schwartzwalder Mine. No technical ground control is practiced. Unstable ground is dealt with as it is encountered. In both stopes and development, the greatest concern has been the support of slabby ground. In nearly all cases, slabby ground is supported with 4-ft-long Split-Set-type rock bolts. Bolts are installed either in a random pattern or on regular 5-ft spacings. The Split-Set-type rock bolt system tends to be an easy and familiar bolting system for all miners at the Schwartzwalder Mine.

Rock bolts are inadequate for the support of large, open stopes. In these situations, stope pillars are relied upon to provide large-scale support. The practice is to use a 20-ft pillar spacing. Planned pillar size is 8 by 8 ft. As a rule of thumb, a 1-to-1 pillar height-to-width ratio is practiced. This fits well with usual stope mining heights of 6 to 9 ft. These pillar dimensions (spacing, size, and height), over time, have been proven to promote stable stope conditions. Pillars are left in waste as much as possible, which ultimately influences spacing. Decisions on pillar location are generally left to the lead miner or shift boss, as production is the primary goal.

Three primary stoping methods are used at the Schwartzwalder Mine: shrinkage stoping, random room-and-pillar stoping, and a variation of longwall stoping. Random room-and-pillar stoping is practiced in the flat veins, which vary in dip from 0° to 30°. These stopes are associated with veins at the ends of horsetail structures, where the vein rolls flat before pinching out.

A longwall stoping variant is practiced in intermediate stopes where dip is between 30° and 70°. In these stopes, the dip is sufficiently flat that ore does not run easily, yet too steep to mine without having some amount of stope fill or staging. Intermediate stopes are associated with the portions of horsetail veins rolling out of the steep veins. Mucking relies heavily on slushing techniques rather than free-pull through drawpoints, as the vein dip is too flat for muck to flow easily by gravity.

Shrinkage stoping is practiced in veins that have dips of 70° or more. Shrinkage stopes are associated with the two major vertical fault systems of the Schwartzwalder Mine, the West Rogers Breccia Reef, and the Illinois fault. Shrinkage stopes associated with the fault systems tend to be composed of numerous narrow stringers that congregate to create wide stopes, often up to 15 ft thick.

The ideal conditions for shrinkage stoping include competent ore and host rock, stable footwalls and hanging walls, regular stope boundaries, propensity for overhand methods, and qualities that can tolerate long-term storage in the stope without significant degradation (Hustrulid 1982). Except for irregular stope boundaries, all of these conditions are found at the Schwartzwalder Mine.

20.4.2 Development

Typical development at the Schwartzwalder Mine includes haulage drifts, crosscut drifts, chutes and drawpoints, raises, and in-stope development such as slushing and ventilation scrams. The size of the development is dependent on the equipment and mining method to be used.

The practice of drifting, scramming, and drawpoint or chute construction is standard from level to level at Schwartzwalder. Because of widely differing stope geometries, several different raise methods have been adopted in shrinkage stope development. Several factors must be considered when selecting a particular raise method for a particular stope geometry; stope height, development speed, excavated waste volume, cost of raise materials, and crew skill. Typically, raise height and costs of raise material tend to be the key factors in choosing a raise method at the Schwartzwalder Mine.

Minimum development requirements for a typical ore body include driving a drift from the main haulage to the ore body, raising into the ore body, driving the stope sill scram, and finally, installing drawpoints or chutes. Whether development or production mining is being performed, three basic components compose the excavation cycle: drilling, blasting, and mucking.

All drilling is performed with Gardner-Denver GD-83 jackleg rock drills using 6.5-ft drill steels and cross-shaped, $1\frac{3}{8}$-in-diameter bits with tungsten-carbide inserts. Because of the highly siliceous nature of the Trend rock, average drilling rates are 2 ft/min. A standard drift round takes about 2.5 hours to drill out.

Blasting methods utilize NONEL initiation. Holes are pneumatically loaded with an ammonium nitrate-fuel oil (ANFO) mixture. Hole priming is achieved by using half a stick of emulsion-based powder and a NONEL cap. Once the loaded round has been blasted, the crew must wait 30 min before reentering the area to allow smoke to clear and to avoid injury from possible misfires. Blasting takes, on average, 3 hours to perform. After blasting, the face is scaled for loose rock, and the muck pile is watered down to reduce radon and dust. The scaling or barring down stage takes about 2 hours.

Once blasting is complete, the mucking cycle begins. In the headings where rubber-tired equipment is used, mucking of the face is performed by either 1-yd^3 Wagner LHD's or 1-yd^3 Eimco 911 LHDs. For tramming distances exceeding 150 ft, either 5-ton Elmac D5-4A haul trucks or 3-ton Young buggies are used to transport muck to dump points. Productivity in the drifts is 2.5 ft per manshift.

In drift headings where rails are used, an overshot rail mucker is used to load train cars. At the Schwartzwalder Mine, Eimco 12-Bs and 21-Bs are used. Track gauge is commonly 24 in, but some 18-in gauge is still used. When required, curves with a 50-ft radius are employed in the rail drifts. The rail mucking cycle takes about 2 hours. Overall, productivity in a rail drift is about 2 ft per man-shift. With either rail or rubber-tired haulage, waste is preferentially gobbed into conveniently located abandoned stopes. When no other options exist, the waste must be trammed to the shaft to be hoisted to the surface.

Because both rail-mounted and rubber-tired equipment are used at the Schwartzwalder Mine, two common drift sizes are employed—a 5- by 7-ft drift for rail-mounted equipment and an 8- by 9-ft drift for rubber-tired equipment. Drifts are developed parallel to the long dimension of the stopes. A typical 8- by 9-ft

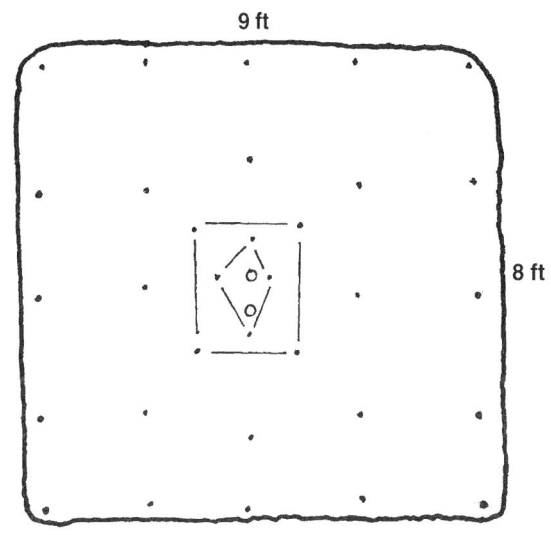

FIGURE 20.3   LHD drift (8 by 9 ft)

drift for rubber-tired equipment is shown in Figure 20.3. In this drift size, 34 holes are drilled. Powder factor for a 6-ft-deep round is about 4 lb/ton. A typical 5- by 7-ft rail development drift is shown in Figure 20.4. In this drift size, 32 holes are drilled, and the powder factor for a 6-ft-deep round is about 5.5 lb/ton.

Drawpoints and chutes are considered part of drift development. In the construction of either, a short raise is driven into the ore body and connected to the stope sill scram to create a functional drawpoint. Once the basic drawpoint is excavated, a chute can be constructed if desired. On average, construction of a single chute will require two experienced men about four shifts to complete. Because broken muck has an angle of repose of 40°, chute beds are inclined at least 43°. To facilitate stope draw-down, chutes or drawpoints are spaced about 25 ft along the drift.

Scram rounds, while not shown in any figures, are variations of drift design. A drilling pattern for a common 4- by 7-ft scram round is similar to that of the 5- by 7-ft drift round (Figure 20.4), but with one column of holes removed. For a 4- by 7-ft scram, 26 holes are drilled. The powder factor for a 6-ft-deep round is about 5 lb/ton. In general, most headings are designed around a powder factor of 4. Because of the confined geometry of the smaller headings and the need to provide adequate hole spacing, powder factors tend to be high.

20.4.3 Raise Methods

Raises come in several common sizes: a 4- by 4-ft raise for ore passes and muckways, a 5- by 8-ft size for double-compartment raises, and a 6- by 6-ft raise for single-compartment raises (man and material raises). In Figure 20.5, a typical 4- by 4-ft raise round is shown. In this round, 17 holes are drilled. Powder factor for a 4.5-ft-deep round is about 7 lb/ton.

Raise methods used at the Schwartzwalder Mine are bald-headed, Roybal, "tin can," and two-compartment timbered raises. Ideally, a raise should be inclined at 70°, which allows for comfortable climbing by men. MSHA regulations require landings to be installed on 30-ft vertical intervals in all steeply inclined raises. Landings prevent personnel from falling more than 30 ft and reduce the potential for injury by falling objects, such as rock or tools.

Bald-Headed Raises. A typical bald-headed raise is shown in Figure 20.6. Bald-headed raises are the simplest, fastest, and least costly raise system to drive. This method is ideally suited for

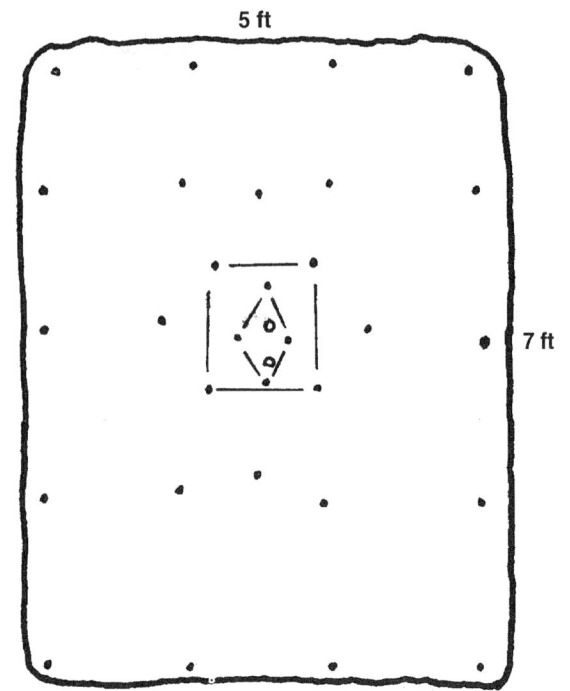

FIGURE 20.4   Rail drift (5 by 7 ft)

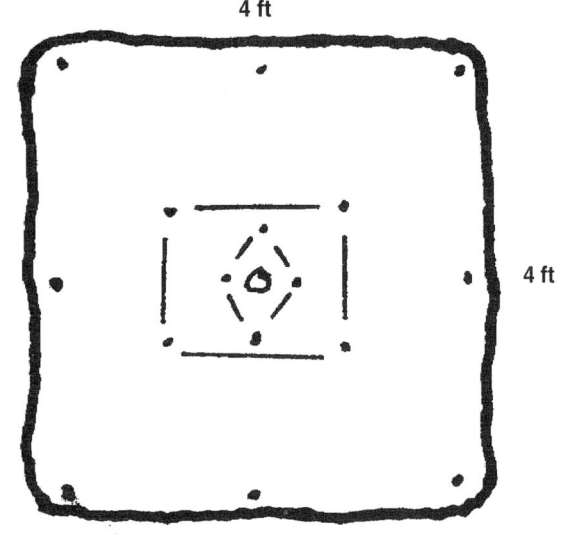

FIGURE 20.5   Raise round (4 by 4 ft)

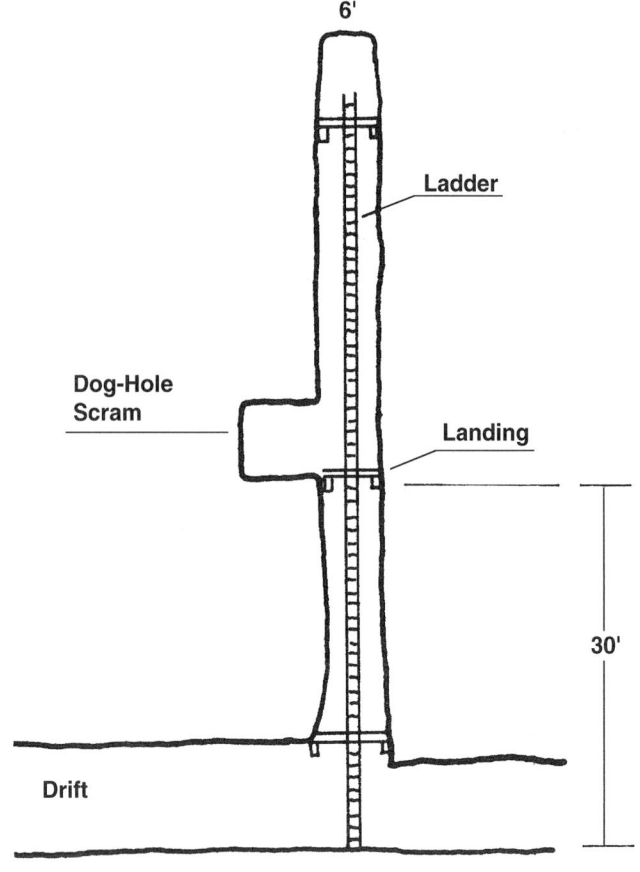

FIGURE 20.6   Bald-headed raise

raises less than 50 ft high. Above this height, construction and removal of work decks and temporary staging significantly reduce productivity. A disadvantage of this system is that personnel must enter the raise under potentially unsafe conditions created by high, unscaled back.

In the bald-headed raise method, a single-compartment raise is developed. Raise equipping is minimal, as only a temporary work deck and ladder access are needed. The work deck, a frame of 4- by 6-in timbers, is suspended 7 ft below the back by eyebolts and chains. A platform of 3- by 12- in lagging is nailed to this deck. Temporary landings must be installed every 30 vertical feet. The raise is advanced by drilling and blasting each round. Blasted muck

falls to the level below for subsequent mucking. Before drilling commences, the deck and landings must be installed. Prior to blasting, the deck and landings must be removed.

Average advance is roughly one round a day for distances up to about 30 ft. Beyond this, deck and landing construction and removal begin to slow productivity. The average advance rate for raises between 30 and 100 ft is about one round every 2 to 3 days. In practice, the first 24 ft can be excavated by working from broken muck with no need for staging. By driving a 6- by 6- by 6-ft-deep dog-hole scram at convenient intervals, equipment can be kept close to the face. For each 10 ft of raise height, staging installation requires 1 hour of work.

Roybal Raises. The Roybal raise system utilizes a variation of the two-compartment raise. A good description of this method can be found in Wright et al. in the January 1984 issue of *Mining Engineering*. In this method, two bald-headed raises are developed separated by a 6-ft rock pillar (Figure 20.7). The circled numbers indicate development sequencing. In this system, a 6- by 6-ft manway and a 5- by 5-ft muckway are developed in a leap-frog manner. The raises are connected on 24-ft intervals by 6- by 6-ft dog-hole scrams. The rock pillar serves as a "muck wall" similar to that of the timbered raise muck wall. Because rock is used to provide the raise lining, a minimal amount of timber is used. Raise development time is much faster in the Roybal raise than in a timbered raise for this reason.

In the first stage, both raises are simultaneously advanced 30 ft above the drift (Figure 20.7A). A dog-hole scram is excavated at the top to connect the two raises. Development footage is quick at this stage, as work is performed from the broken muck filling the raise. Upon completion of the first 30 ft of raise development, muck is removed from the left-hand raise,

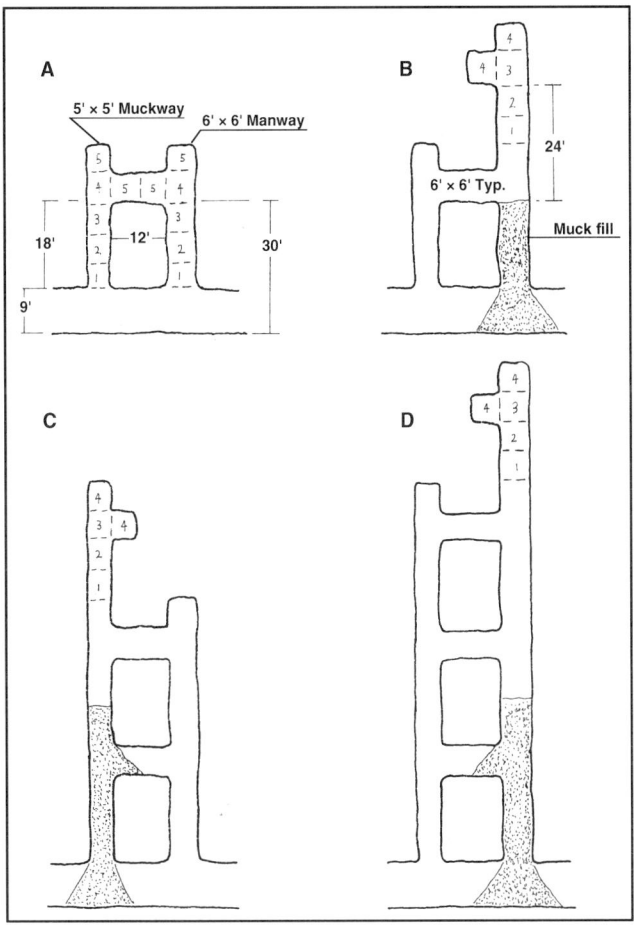

FIGURE 20.7 Roybal raise method. A, First stage, initial advance; B, second stage, advancing right-hand raise; C, third stage, advancing left-hand raise; D, full raise height

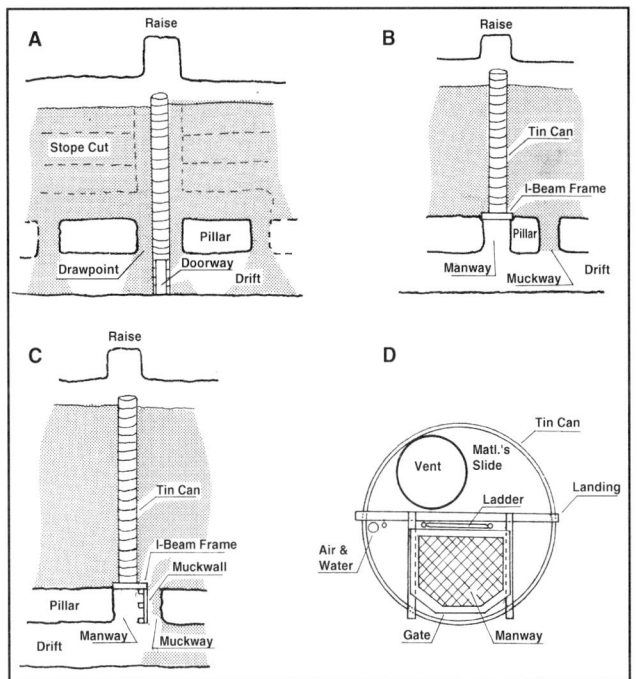

FIGURE 20.8 Tin can raise mounted on A, level and B, stope sill. C, Showing muckwall. D, Detail of tin can raise

and ladders are installed to the elevation of the dog-hole scram for access.

In the second stage, the right-hand raise is advanced 30 ft higher (Figure 20.7B). Because not enough muck is available to work from in the raise, temporary decks must be erected to reach the advancing face. While the construction of this stage slows productivity somewhat, it does not slow it nearly as much as it would in a bald-headed raise driven to a similar height. At the top of the right-hand raise, now 60 ft up, another dog-hole scram is excavated. Upon completion, the right-hand raise is mucked out and ladders installed for access.

The third stage involves advancing the left-hand raise 30 ft higher, topping out at the same elevation as the right-hand raise (Figure 20.7C). Development of the left-hand raise proceeds in a manner similar to the development of the right-hand raise. The right-hand raise serves as access. Upon completion of this raise, the dog-hole scrams are connected. Ladders are then advanced in the left-hand raise.

This alternating cycle of raising is continued from side to side in 30-ft development heights until the full intended raise height is reached (Figure 20.7D). The Roybal raise method is ideally suited to raise heights exceeding 60 ft because of improved crew safety and a reduced need for the installation and removal of temporary staging between rounds. Development crews are never more than 30 ft from unscaled back. With each blast, only one temporary landing need be installed or removed as access is provided by the adjacent raise. Additionally, the dog-hole scrams provide a

convenient place to store equipment (within 30 ft of the advancing face). These factors enable productivity to be higher for the Roybal raise than the bald-headed raise when raise heights are excessive. A drawback to the Roybal raise method is the need for additional development required to construct the dog-hole scrams. Typical productivity for Roybal raises at the Schwartzwalder Mine is about 1.3 ft per manshift.

Tin Can Raises. Liner ring raising is the second most commonly used method at the Schwartzwalder Mine (Figure 20.8) because of ease of construction and simplicity in design. Safety is greatly improved over all other raise methods, as the rings form an enclosed, protective structure. At the Schwartzwalder Mine, they are affectionately referred to as "tin cans." Liner rings are curved, stamped metal plates that bolt together to form 4-ft in diameter circular rings. Each liner ring is composed of four 16-in-high, 3/8-in-thick pieces. There is no limit to the height of a tin can raise; however, above 100 ft, climbing the tight manway becomes tiring. The tin can method is ideal for raises exceeding 60 ft in height.

Development of the tin can raise begins by advancing a 5- by 8-ft raise into the stope. As the raise is advanced, ring sections are bolted one on top another to form a continuous closed structure. Inside the tin can structure, ladders and landings are installed to form a manway. The muckway is created from the space between the outside of the tin cans and the raise walls. As the tin can structure occupies about 5 ft of the 8-ft raise dimension, a 3-ft open area remains for the muckway. In practice, tin can rings are kept about 2 ft above the stope muck. Raise development leads tin can installation by two rounds to provide access to the stope. The tin can rings provide a convenient mount for in-stope slushers. Normally, the tin can raise must be anchored to the raise wall to prevent it from leaning as muck is drawn around it.

The simple ladder and landing installation is shown in Figure 20.8D. Two compartments are created by the ladder arrangement, one for the manway and the other for material access and ventilation. Ladders and landings are prefabricated on the surface using standardized patterns. Four ring sections can be

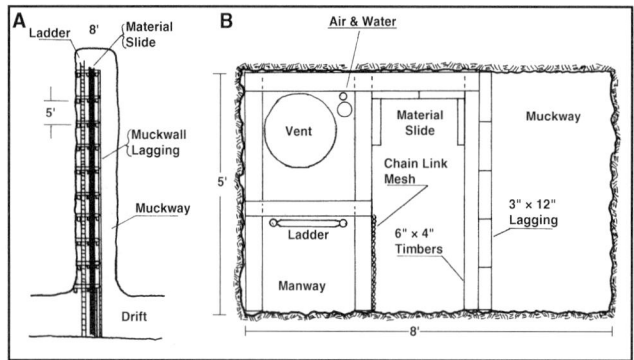

FIGURE 20.9   Timbered raise. A, Two-compartment raise; B, detail of two-compartment raise

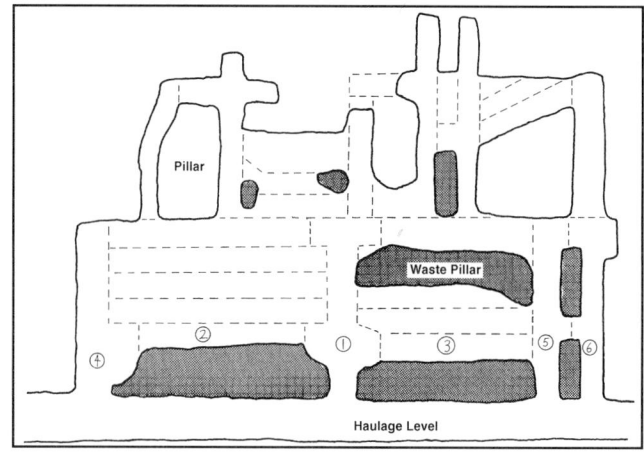

FIGURE 20.10   Shrinkage stope development sequence

assembled in an hour. With ladders and landings, about 1 linear foot of liner installation an hour can be expected. Raise advance, including drilling and blasting, is 3 ft per manshift up to about 50 ft. Beyond this, raise advance averages 2 ft per manshift. This method of raising works well at the Schwartzwalder Mine as minimal crew skills are needed. Typical material costs for new tin can liner rings are $225 per linear foot. While the cost per foot of tin can rings is higher than that of timbered raises, the tin can rings can be salvaged and reused. This practice of reusing the ring sections ultimately reduces the cost of tin can raising below the costs of timbered raises. Tin can raises can be advanced much faster than timbered raises because of their ease of construction.

Timbered Raises. In terms of overall productivity, timbered raises tend to be the most costly and their construction the slowest at the Schwartzwalder Mine. To be constructed efficiently, a high level of crew experience and framing knowledge is required. Nipping of timber materials requires additional manpower and time, particularly with regard to the hoisting complexities. Timbered manways are commonly used in narrower stopes, where high-grade end pillars are recovered. There is no limit to raise height, but moving materials into the taller raises becomes a bottleneck in the construction process.

Typically, a 5- by 8-ft raise is advanced into the stope. Timber sets are constructed every 5 ft (see Figure 20.9) to form the two-compartment raise. A timber set consists of three horizontal 8- by 8-in timber beams hitched into the hanging wall and the footwall as shown. Two 4- by 6-in timber cross members are nailed to the top of the timber beams to provide support for the manway ladder and timber slide. The muck wall divider forming the muckway compartment is constructed of 5-ft-long 3- by 12-in boards. These boards are nailed edge-to-edge to the outside of the muckway compartment's timber beams and span the 5 ft between two successive timber sets. Landing gates are installed on 30-ft spacings. Chain-link mesh is nailed to the manway side to prevent men from falling into the material slide.

As the raise advances, timbered sections are brought up, usually within two rounds of the face. As the stope advances, timbering is kept slightly above the muck. This reduces the amount of rock falling into the manway and allows for slushing from the last timbered cross member. Construction of a typical two-compartment raise requires about 230 board feet per 5 ft of raise, which equates to $30/ft of raise (1999 dollars). Typical timber raise advance rates are on the order of 2 ft per manshift, particularly as raise distances go beyond 30 or 40 ft.

20.5   GENERAL SHRINK STOPING SEQUENCE

In shrinkage stoping at the Schwartzwalder Mine, overhand mining techniques are employed. These techniques involve excavating the ore in horizontal slices. As these slices are mined,

stoping progresses updip. Figure 20.10 shows the typical sequence in the development and mining of a shrinkage stope. The stope shown is the 75°-dipping 1333 stope. Stope width is 4 ft, a minimum practical width for men to work.

The stope development sequence begins with the excavation of the central raise (1). The purpose of this raise is to define the size, continuity, and orientation of the ore body. This raise will become the central stope muck-raise chute when stoping commences. From this raise, the stope sill scrams (2) and (3) are next developed to identify the lateral extent of the ore and to create the stope sill pillar. The bald-headed end raises (4) and (5–6) are driven upward to intersect with the stope sill scram to provide for primary man and muck raises and to provide flowthrough ventilation. These raises are initially developed to only one round above the stope sill scram, as they will advance with the stope as it progresses. This shortens development time and allows production to begin sooner. Because of the short height of the 1333 stope, bald-headed raises allow rapid development with minimal material cost.

Stoping commences with drilling of a slab round for the first stope cut. In the stope sill scram back, two rows of 6-ft-deep holes are drilled updip, one along the hanging wall and one along the footwall-vein contacts, for the length of the stope. Hole spacing is 1.5 ft. By maintaining the end raises one round ahead of the active stope face, a free face becomes available for the slab round to initially break to. The stope sill scram provides additional expansion volume for the broken muck.

Once the slab round is blasted, broken muck is drawn out through the central and end muck raises. Because the broken ore swells to nearly 50% of its in-place volume, one-third of the broken muck must be drawn out to provide access for excavation of the next stope cut. The remaining two-thirds of the ore remains in the stope as a working floor. Slushers are used in the stopes to control the level of the working floor.

Each subsequent stope cut is mined in a similar manner. As the stope cuts advance, the end raises are advanced, always leading the stope cut by one round.

Once the stope is mined out and production mining ceases, a "free pull" phase is begun. In the free pull phase, productivity is high, as the only activity being performed is mucking of ore. At this time, the remaining two-thirds of the muck is drawn from the stope in the quickest manner possible. During this stage, men do not usually reenter the stope, as ore drawdown prevents safe access. Should hang-ups or other conditions require men to reenter the stope, then overhead protection must be installed, reducing productivity.

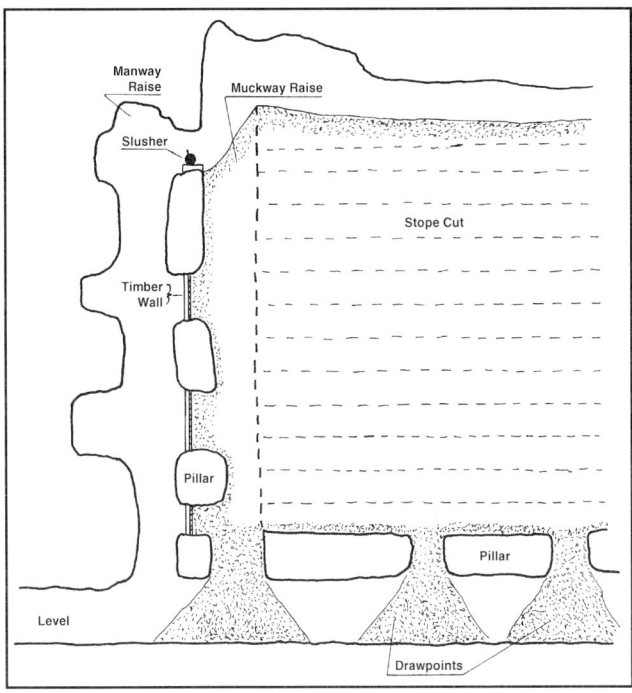

FIGURE 20.11 Stope with Roybal raise

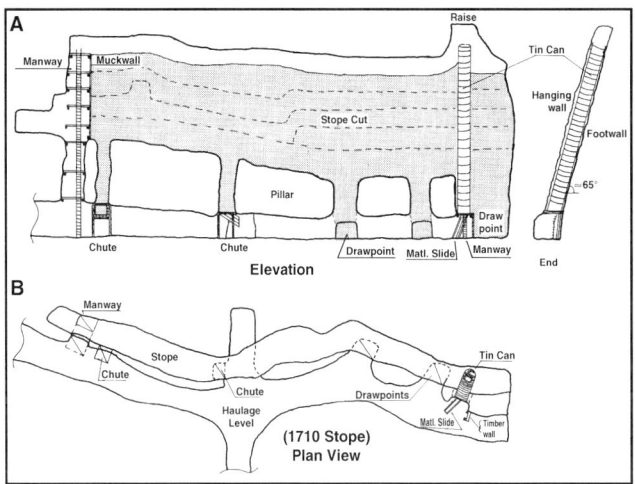

FIGURE 20.12 Shrinkage stope with tin can raise. A, Section view; B, plan view.

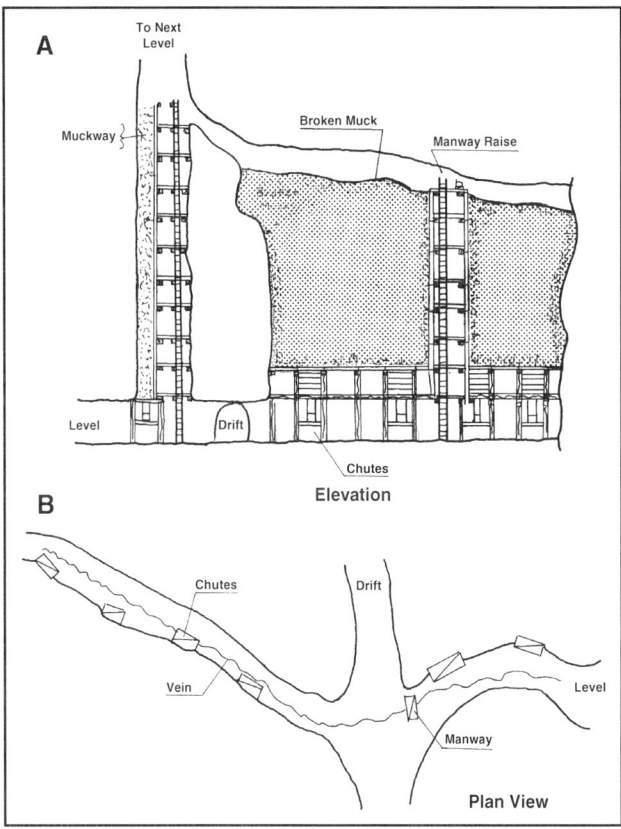

FIGURE 20.13 Rail-type shrinkage stope with timbered raise. A, Section view; B, plan view.

In the 1333 stope, muck tends to hang up because of the narrowness of the stope. Because of this, men must reenter the stope during free pull activities to wash down and bomb the muck. To provide a safe work area and to reduce the hazards of falling rock in the unsupported stope, overhead protection in the form of chain-link mesh is installed. Before free pull begins, mesh is bolted horizontally between the hanging wall and the footwall over the length of the stope near the top. As free pull continues, additional chain-link mesh is installed horizontally across the stope on 30-ft intervals vertically. This mesh becomes cost effective when considering the improvement in ore recovery in narrow veins.

Figure 20.11 shows the Radon 1 shrinkage stope. This stope was nearly vertical and averaged 7 ft in thickness for a height of over 80 ft. The ore block itself was short in lateral dimension, being about 60 ft. Mucking was performed with rubber-tired equipment. Drawpoints were spaced about 25 ft apart to facilitate draw control from the stope. A Roybal raise method was chosen for developing this stope, as the stope height was too high for efficient bald-headed raise methods. A timbered raise was not selected because the raise material requirements and costs were excessively high when compared to the short stope length and small nature of the ore body.

Figure 20.12 shows the 70°-dipping 1710 shrinkage stope. Stope thickness averaged 6 ft and height was 90 ft. Rubber-tired equipment was used to muck the stope from drawpoints and chutes developed on 40-ft centers. Because of the stope height, a tin can raise was utilized. Using this method simplified drawpoint construction and allowed recovery of stope sill pillars without compromising manway access. A tin can raise was preferable to a timber raise system because of the availability of salvaged tin can liner rings and the lack of crew skills.

Figure 20.13 shows an Illinois-type shrinkage stope above the Steve-level rail haulage. As can be seen, muck raises and chutes were laid out parallel with the vein on 20-ft centers to facilitate stope drawdown. A timbered, two-compartment manway was driven in the middle of the stope. As the vein intersected the haulageway, muck raises and manway raises were driven up from the level on ore. Because of the high grade of ore,

no stope sill pillar was left between chutes. A timber stope floor was constructed instead to contain the muck in the stope and to funnel ore to the chutes. This practice allowed for nearly 100% recovery of ore from the stope. A timber raise was selected because of stope height and ore grade.

Figure 20.14 shows the 1606 shrinkage stope. This was a timbered stope averaging 65° to 70° in dip. Stope height was 110 ft. Mucking was performed with rubber-tired equipment loading ore from a chute as well as a drawpoint. So the mine

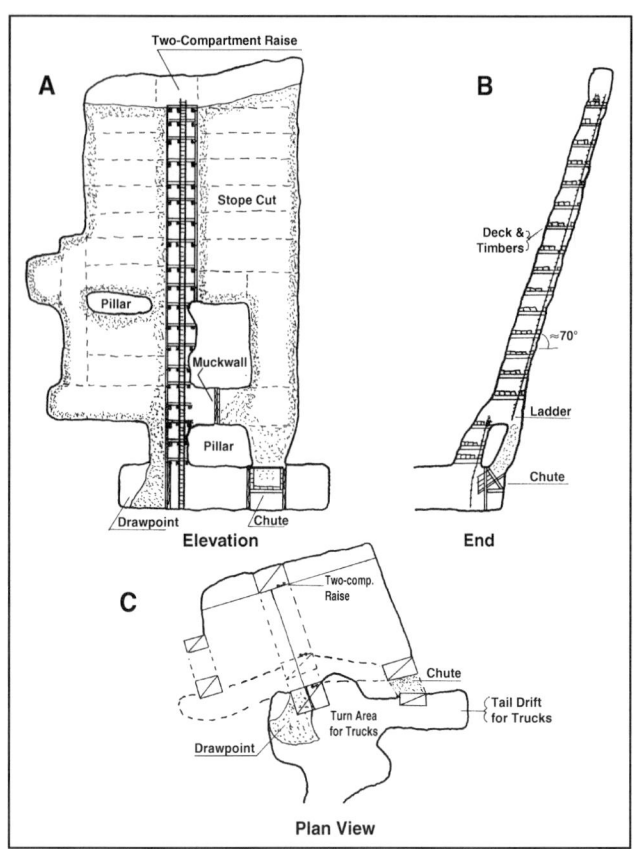

FIGURE 20.14   Stope with timbered raise for rubber-tired equipment. A, Section view, broadside; B, section view, end; C, plan view

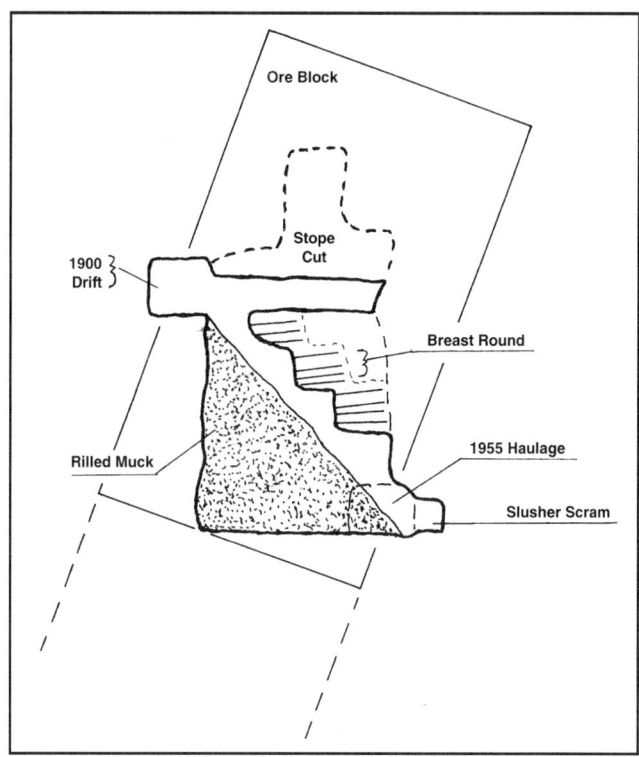

FIGURE 20.15   Rill-type shrinkage stope

haulage trucks would have sufficient clearance to back in under the chute, a T intersection was developed at the end of the access drift and to provide enough turning radius for the equipment, the intersection was excavated 30 ft long and 15 ft wide.

In the 1606 stope, the short lateral length of the ore block required only a central manway and one muck raise. A two-compartment timbered raise was developed for this purpose. Timber was chosen because no tin can liner was available at the time development started. Stope height again precluded the use of bald-headed raises. In practice, muck is pulled from both sides of the manway. As with tin can raises, muckways can be used on both sides of the timbered raise. This is because both of these methods utilize raise structures that are enclosed.

Figure 20.15 shows a rill-type shrinkage stope characteristic of underhand mining techniques. This method of stoping is not common at the Schwartzwalder Mine, but good stoping efficiency was achieved. The shrinkage stope shown is the 1919. This was a small, 5-ft-thick ore body dipping at 75°. Stope height was on the order of 60 ft. In the rill method, a sill scram was driven perpendicular to the 1955 haulage. A slusher station was set up across the haulage, and a slusher mucked ore into the haulage. An LHD loaded the ore from the haulage floor into 5-ton Elmac mine trucks. In this method, the stope sill scram is first cut. All subsequent stope cuts are mined at a 45° angle towards the 1955 haulage level. As muck fills the stope, men work off of the muck, drilling 8-ft-deep breast rounds. As mining progresses, the muck pile is slushed off to provide access. The face is mined in the stair-step manner shown. This method produces a high degree of productivity. Because of the low stope height, raising was not necessary. Raise development costs could not be justified because of the small size of this ore body.

20.6   PRODUCTION STATISTICS AND MINING COSTS

During the most recent mining period at the Schwartzwalder Mine (1995 through 1999), peak production occurred in 1997. At this peak, 50 men were employed underground on two shifts. Average stoping productivities were around 6.5 tons per manshift. Mine-wide productivity averaged 2.9 tons per manshift. Powder factors were in the neighborhood of 2.5 to 3.5 lb/ton. Monthly out-the-portal production averaged 4,500 tons, or an average of 225 ton/d. Costs averaged around $150/ton mining and $220/ft development (all types). These costs include all direct and indirect costs on site. In 1996, uranium prices were sufficiently high that a cut-off grade of 0.5% U_3O_8 was used. By 1999, prices had fallen far enough that the cut-off grade was raised to 0.8% U_3O_8.

Table 20.1 shows stope productivity for selected stopes producing since 1995. The stopes are grouped by stoping method and are sorted in order of increasing productivity. It can be seen that the shrinkage stope mining methods had the lowest productivity. Productivity in the shrinkage stopes, as practiced at the Schwartzwalder Mine, average 8.5 tons per manshift. This tends to be in the middle of industry averages, ranging between 5 and 10 tons per manshift (Hustrulid 1982).

Table 20.2 shows individual development productivity in feet per manshift by development type and size. Productivities are lowest for raise development, with the exception of the 2111 level, a 4- by 4-ft, 50-ft-high bald-headed raise. No information was available on drift development.

Tables 20.3 and 20.4 show 1999 direct unit costs for mining and development in selected stopes. Costs have been derived from time studies, hourly costs, and general departmental costs. As shown in these tables, shrinkage stoping tends to be a higher-cost activity, primarily because of lower productivities and greater material needs. Overall average productivities were 9.8 tons per manshift in the selected stopes. Average shrinkage stope productivities were 7.3 tons per man-shift, while flat stopes averaged 10 tons per manshift. Rubber-tired haulage costs were

TABLE 20.1 Typical production productivity

Stope	Haulage	Delivery	Access	Stoping method	Tons per manshift
535	Rail	Chute	Tin can raise	Flat	6.5
1446	LHD	Chute	Tin can raise	Flat	10.3
779	Rail	Drawpoint	Drift	Intermediate	10.7
1240	Rail	Chute	Timbered raise	Intermediate	11.5
1919	LHD	Drawpoint	Drift	Shrinkage	4.7
1710	LHD	Both	Tin can raise	Shrinkage	4.9
1333	Rail	Chute	Timbered raise	Shrinkage	5.0
1606	LHD	Both	Timbered raise	Shrinkage	6.7
2021-stoping	LHD	Both	Tin can raise	Shrinkage	6.8
2021-free pull	LHD	Both	Tin can raise	Shrinkage	34.3

TABLE 20.1 Typical raise development productivity

Stope	Dimensions, ft	Raise lining	Raise height, ft	Distance per manshift, ft
1240	5 by 5	Bald-headed	120	0.6
1792	5 by 10	Tin can	100	1.1
1837	5 by 5	Bald-headed	80	1.5
2111	4 by 4	Bald-headed	40	2.7

TABLE 20.1 Typical stoping costs

Stope	Stoping method	Haulage	Access	Tons per manshift	Direct costs, 1999 dollars per ton							
					Tramming*	Maintenance†	Drilling	Powder	Stope labor	Operating supplies	Tramming‡	Total
535	Flat	Rail	Tin can raise	10.9	0.70	2.23	1.55	4.42	16.05	6.58	6.30	37.83
1727	Flat	LHD	Drawpoint	9.7	0.70	4.29	1.55	4.42	18.04	6.58	6.30	41.88
779	Intermediate	Rail	Drift	13.2	0.70	2.23	1.55	4.42	23.33	6.58	6.30	45.11
1937	Intermediate	Rail	Roybal raise	12.3	1.15	2.23	1.55	4.42	14.17	6.58	6.30	36.40
1333	Steep	Rail	Timbered raise	4.0	1.20	2.23	1.55	4.42	43.75	6.58	6.30	66.03
1606	Steep	LHD	Timbered raise	7.6	0.70	4.29	1.55	4.42	23.03	6.58	6.30	46.87
2021	Steep	LHD	Tin can raise	10.3	1.32	4.29	1.55	4.42	16.40	6.58	6.30	40.86

*Tramming = Stope to main haulage.
†Maintenance = Underground labor and parts.
‡Tramming = In-mine and hoist to surface.

TABLE 20.1 Typical development costs

Stope	Development type	Haulage	Raise type	Tons per manshift	Direct costs, 1999 dollars per ton							
					Tramming*	Maintenance†	Drilling	Powder	Stope labor	Operating supplies	Tramming‡	Total
2021	Drift	LHD	NA	8.5	0.32	4.29	1.55	4.42	20.59	6.58		37.75
535	Scram	Rail	NA	3.6	0.70	2.23	1.55	4.42	48.61	6.58		64.09
1937	Scram	Rail	NA	10.6	0.45	2.23	1.55	4.42	16.51	6.58		31.74
761	Raise	Rail	Alimak	6.1	0.93	2.23	1.55	4.42	56.47	6.58	6.30	78.87
1240	Raise	Rail	Bald-headed	3.4	0.32	2.23	1.55	4.42	51.47	6.58	6.30	72.87
1333	Raise	Rail	Timbered	2.4	0.50	2.23	1.55	4.42	72.92	6.58		88.20

*Tramming = Stope to main haulage.
†Maintenance = Underground labor and parts.
‡Tramming = In-mine and hoist to surface.

higher compared to rail haulage costs because of equipment operating and maintenance costs.

In Table 20.3 it can be seen that direct stoping costs are all on the same order, with the exception of the 1333 stope. Average direct cost for the selected stopes, excluding 1333, was $41/ton. With regard to flat stopes, direct mining costs averaged $40/ton, with rail haulage being slightly less costly. In the intermediate stope category, direct mining costs averaged $41/ton. The high cost associated with the 779 stope was due to updip slushing rather than the more conventional downdip slushing usually practiced. Efficiency of updip slushing is only about half because of the inability of the slusher bucket to maintain a full load.

In the shrinkage stope category, average direct cost was $44/ton. The cost associated with mining in the 1606 stope was relatively high because of the lower productivity associated with timbered raises. The 1333 stope had the highest cost of the selected shrinkage stopes. This stope became extremely labor intensive because of the significant amount of raising required to mine the discontinuous, irregular ore body. When studying the costs in the table, one will note that the most significant direct mining cost was stope labor. This cost was heavily influenced by productivity. This is often the case in narrow-vein conventional mining where mechanization cannot be easily adapted.

In Table 20.4, development productivity was calculated on the basis of tons per manshift rather than the more usual feet per manshift. Because of the large differences in development excavation size, this approach was taken to ensure cost comparisons were uniformly applied. Overall, raise development costs were highest because of the low productivity and large quantities of material needed. Scrams tend to be the least expensive development method, but this depends on the efficiency of the scram operation. Scram efficiency falls when multiple slushing is employed. This is indicated by the high cost in the 535 development scram.

Raise development costs averaged $80/ton, with the most costly raise method being the timbered raise. The bald-headed raise method was the least expensive because of the simplicity of the operation and minimal material requirements.

20.7 CONCLUSIONS AND SUMMARY

The Schwartzwalder Mine has approached shrinkage stope mining in a unique way. Because of the variable nature of the veins, many development and stoping methods have been practiced over the years. Because of the numerous factors involved in the selection of practical stoping methods, no single method can be claimed as the best. The prevailing attitudes have been to use methods that have historically produced good results. When problems arise, these "tried and true" methods are modified to better suit the conditions at hand. In practice, the most efficient stoping methods are the ones that are simple in approach and layout. Most important, good planning and good stope layout tend to minimize costs and maintain high productivity.

Several factors influence the ability to keep operations efficient: stope geometry and size, ore reserve definition, geologic parameters, and crew experience, to name a few. All of the above factors affect cost. The greatest influence on cost at a nonmechanized, small, narrow-vein mine like the Schwartzwalder is labor.

At the time of this writing, mining operations at the Schwartzwalder Mine have ceased. Plans are now in place for reclamation and mine closure. Significant low-grade reserves still exist, but mining is not economic at today's low uranium prices. Near-future increases in uranium price are not expected. Further deep drilling could identify higher-grade reserves. Other opportunities have been investigated, including the possibility of operating an underground quarry. While this project is only in the feasibility stage, there still appears to be potential for future mining at the Schwartzwalder site.

20.8 REFERENCES

Hustrulid, W.A., ed. 1982. *Underground Mining Methods Handbook.* SME, 485 pp.

Judy, Susan. 1999. Internal geology reports and personal conversations, Schwartzwalder Mine, Cotter Corp.

Soule, J.H. 1960. Mining Methods and Costs, Schwartzwalder Uranium Mine, Jefferson County, Colorado. Bureau of Mines Information Circular IC 7963.

Wright, J., F. Roybal, and D. Suttie. 1984. Roybal Raise: An Alternative Two-Compartment Raise. *Mining Engineering*, January, pp 44–46.

Sublevel Stoping

CHAPTER 21

Ground-Stability-Based Mine Design Guidelines at the Brunswick Mine

Patrick Andrieux and Brad Simser

21.1 INTRODUCTION

Noranda's Brunswick No. 12 Mine is a 9,500-tonne/d underground zinc-lead-copper-silver mine that has been in continuous operation since 1964. The property is located near Bathurst in northeastern New Brunswick in Atlantic Canada. The ore body is in the form of a complex arrangement of up to 10 subparallel, massive, ore-bearing sulphide lenses striking north-south and dipping about 75° to the west. Strike length is close to 1,200 m, width is up to 200 m , and depth from the surface is about 1,200 m. The ore body is a vent-proximal massive sulphide deposit that formed above a feeder pipe by venting of metal-laden hydrothermal fluids at the bottom of the sea (Luff et al. 1992). Five phases of structural deformation have shaped the deposit to its current form. An intrusive porphyry dyke is also present throughout the mine. Over 15 distinct geological horizons are commonly encountered throughout the mine (Luff 1977).

The deposit is hosted by metamorphosed volcanic clastic sediments and tuffs that overlie a sequence of felsic volcanic rocks. These units show a wide range of mechanical properties. The chloritic and sericitic footwall metasediments that constitute the older geological units tend to be more competent than the younger hanging wall sequence of chloritic sedimentary units intercalated with felsic tuffs and rhyolites. This increased competency is partially due to sulphide stringer mineralization and associated silicification and chloritization.

Sharp mechanical contrasts exist between the massive sulphides and the metasediments. The sulphides are heavy (specific gravity [ρ] of 4.3), stiff (Young's modulus [E] of 70 GPa), and strong (unconfined compressive strength [UCS] of up to 210 MPa), while the metasediments are much lighter (ρ of 2.9 to as little as 2.6 for the hanging wall units), softer (E down to 10 GPa), and weaker (UCS of 60 MPa on the footwall side to as little as 30 MPa locally on the hanging wall side). The metasediments, particularly those on the hanging wall side, are very schistose with phyllitic partings; slickensided surfaces are very common. The rock mass properties of the dyke also vary widely, from strong in massive horizons to weak in narrower regions. Structural characteristics throughout the mine are complex as a result of successive folding episodes that created four distinct sets of structures, particularly within the stiff sulphides (Godin 1987). These structures are both subhorizontal, which tends to affect the stability of the stope backs in some areas, and subvertical, and some play major roles in seismic activity. Small, thin bands of metasediments are also locally sandwiched within the hard sulphides, further contributing to difficult ground condition settings.

The far-field principal stress (σ_1) is subhorizontal and oriented east-west, perpendicular to the strike of the ore body. Its virgin gradient is 0.052 MPa per metre of depth. The far-field intermediate stress (σ_2) is also subhorizontal, but oriented north-south, with a virgin gradient of 0.044 MPa per metre of depth. The far-field minor stress (σ_3) is subvertical, with a virgin gradient of 0.028 MPa per metre of depth. The deviatoric stress ratio (σ_1 over σ_3) thus stands at about 1.9.

21.2 STRESS-RELATED MINING DIFFICULTIES

The extraction methods used over the life of the mine have been mechanised cut-and-fill, primary-secondary open stoping with delayed backfilling, end slicing (modified avoca), and, more recently, pyramidal pillarless open-stope mining with rapid paste backfilling.

As mining evolved, the operation became more and more susceptible to stress-related problems. By the early 1990s, high-stress mining conditions had become common, principally because of an extraction ratio in excess of 70%, deeper mining (to 1,150 m below the surface), and high virgin ground stresses. This caused increasingly difficult mining conditions, which culminated in late 1995 with the loss of entire active mining zones because of ground instability and bouts of intense seismic activity. In 1998 alone, 1.2 million tonnes of scheduled ore were affected to some degree, from delayed to written off, because of ground-condition-related difficulties. This represented close to 35% of the yearly production. Such difficulties create havoc on the operation as planning must be quickly and significantly readjusted, and alternate horizons must be quickly brought forward to achieve production targets.

The stress-related problems encountered at the mine typically take on one of three forms.

- High stress levels leading to seismic activity.

- High stress levels leading to aseismic failures.

- Decreasing stress levels leading to loose conditions.

21.2.1 High Stress Leading to Seismic Activity

Seismicity is the result of the sudden release of energy due to the brittle failure of part of the rock mass or sudden movement along a structural feature. Depending on the magnitude and location of the seismic event, the consequences can vary from no damage to the complete collapse of excavations, as shown in Figure 21.1.

Shaking damage is the predominant mode of seismically induced failure at Brunswick Mine; rock bursting, or damage occurring in the form of violent expulsion of large volumes of rock, does occur (Simser 2000), but it is rare. Seismicity at Brunswick Mine is almost always located within the hard and brittle sulphide horizons. Many different mechanisms are at play, the principal one being high stress concentration in pillars, high stiffness contrasts between adjacent rock types, and slippage along discrete structural features (Simser and Andrieux 1999).

* Noranda, Inc., Brunswick Mine, Bathurst, New Brunswick, Canada.

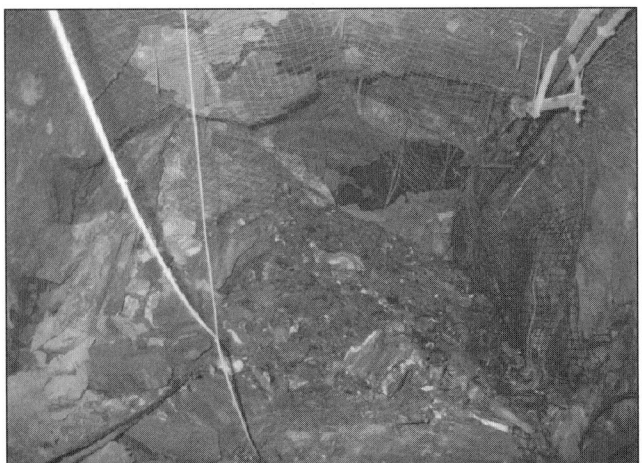

FIGURE 21.1   Complete collapse of a hanging wall access drift because of seismic activity on the 1,000-m level

FIGURE 21.2   Severe squeezing of weak metasediments along hanging wall of stope on the 1,000-m level

Seismicity became a serious concern in the mid-1980s, and the mine's first seismic monitoring system was put in place in 1986. Successive system upgrades have led to the 1995 installation of a state-of-the-art, 90-channel ISS microseismic monitoring network sensitive to local Richter magnitude 2.8 events (Hudyma 1995). To put things in perspective, a total of 24,596 seismic events have been recorded in 1999 alone, of which 283 were above a local Richter magnitude 0.2. Seismic events as powerful as local Richter magnitude 2.7 have been recorded underground. Every recorded seismic event is manually processed to fine-tune location and provide reliable quantitative source parameter data (Simser 1996). These records, in turn, are used to maintain safe mining conditions for the underground workers, understand the reaction of the rock mass to mining, and aid in the design of future mining (Simser et al. 1998).

21.2.2   High Stress Leading to Aseismic Failures

High-stress conditions in weak and soft material generally do not lead to seismicity, but rather to squeezing conditions by which the rock progressively deforms and fails. Such failure can be serious and can result in severe infrastructure damage, as shown in Figure 21.2. The various ground support systems tried to date

FIGURE 21.3   Gravity-driven ground fall because of loss of clamping between strata in an access drift on 575-m level

at the mine under these circumstances have not been able to control this type of failure satisfactorily.

21.2.3   Decreasing Stress Leading to Loose Conditions

Significant reduction in ground stresses can also lead to unstable conditions, particularly in highly laminated or broken ground, as it takes away the clamping forces keeping the strata together. This situation is often encountered at the Brunswick Mine during the later stages of mining, when remnant areas sitting behind mined-out horizons are extracted. Figure 21.3 shows such a ground fall in a drift sitting behind a completely mined-out lens.

Regions subjected to a stress cycle (i.e., to a stress increase while they sit in the high-stress abutment ahead of mining, followed by a stress decrease once mining has progressed past them and put them in the stress shadow of the excavation) are prone to this type of failure, as the high-stress component tends to break up the rock mass prior to its being declamped by the subsequent stress decrease. As the amplitude of the stress cycle increases, the likelihood of ground instability increases. This type of instability, however, can usually be controlled with adequate ground support.

21.3   REMEDIAL APPROACHES

By mid-1996, the Brunswick Mine was facing increasing difficulties in maintaining its production because of ground instability problems. Accesses to crucial production areas were being damaged beyond repair, scheduled stopes were caving in, and high seismicity was preventing access to active production horizons.

The only possible short-term remedy that would allow us to respect our commitments toward our concentrate customers was to focus mining on difficult horizons that had been vacated in the past because of tough ground conditions. These included old secondary stopes, abutments, and remnant pillars, all of which were associated with either high-stress conditions or low-stress regimes following high-amplitude loading cycles. Furthermore, ground support elements in most of these older areas were aging and corroding and of questionable efficiency.

Even though the extraction of these difficult areas was successful, due in considerable part to the introduction on a large scale of shotcrete for ground support (Lessard and Andrieux 1998), the long-term viability of the Brunswick Mine involved drastic changes in mining approach. Three major transformations took place: the introduction of paste backfill, a radical change in the mining method, and the implementation of new ground support systems.

Paste backfill, which was introduced primarily because of severe rockfill delivery problems, has allowed tight filling of

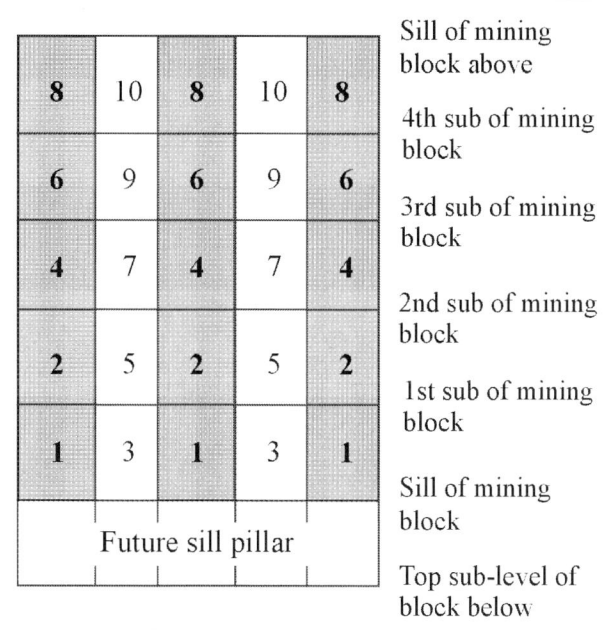

FIGURE 21.4 Conceptual longitudinal sketch of primary-secondary mining, looking west. Numbers shown in blocks represent the mining steps. Shaded blocks represent primary panels. A backfilling cycle is needed between each step.

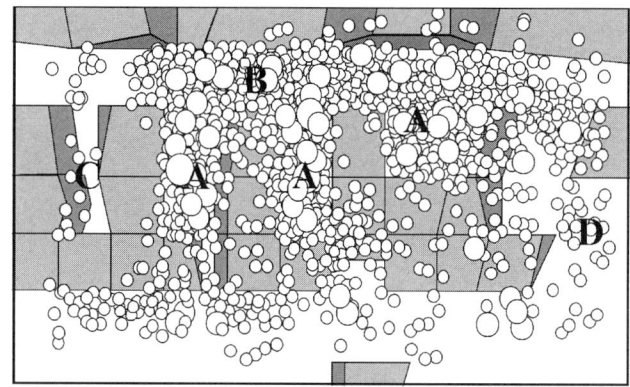

FIGURE 21.5 Seismic activity associated with primary-secondary mining in the 1,000-m level north region. Longitudinal section looking west. Each sphere represents a seismic event with diameter of sphere scaled to Richter magnitude.

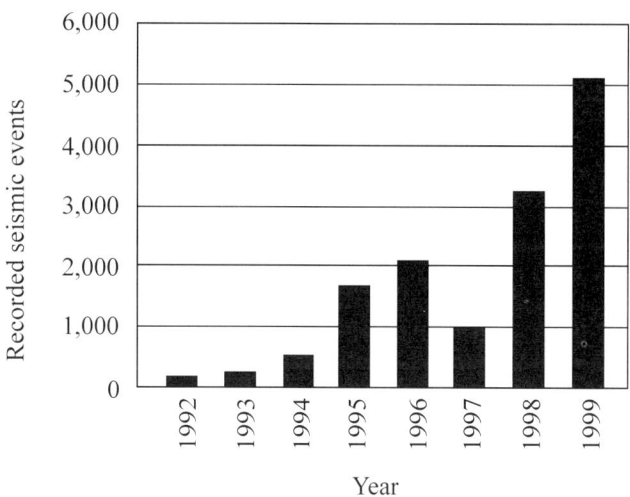

FIGURE 21.6 Number of seismic events recorded yearly from 1992 to 1999 in the 1,000-m level north region.

difficult-to-access caved areas. This, in turn, has stabilized those areas on a regional scale, allowing mining to resume there. The chronic difficulties experienced with the underground delivery of rockfill from the surface quarry had caused numerous backfilling delays that led to large voids remaining open underground for extended periods of time. This further contributed to the ground instability problems. Details of the conversion to paste backfill at the Brunswick Mine can be found in Moerman et al. (1999) and Ouellette et al. (1999).

21.4 MINING METHODS

The change of mining method consisted of a switch from primary-secondary mining to pyramidal pillarless mining. Figure 21.4 illustrates the concept of primary-secondary mining schematically.

Maintaining primary-secondary mining as the principal extraction method was given a lot of thought. It yields both high production and high productivity through its typically big stopes—75,000 tonnes and over—coupled with high flexibility early in the cycle and minimum preparatory work. Furthermore, the mining areas already developed were designed to accommodate this type of mining.

However, the disadvantages of the method eventually prevailed over its advantages. Easy primary mining was usually followed by more difficult and, even more worrisome planning-wise, more unpredictable secondary mining. Larger secondary stopes designed with a squat shape typically exhibited high-stress conditions (generally associated with seismic activity) over long periods of time. Those that eventually failed under the load rarely did so in a timely fashion, i.e., prior to being extracted. On the other hand, smaller secondary stopes designed with a thin shape typically failed early, developing hour-glass shapes that significantly reduced the tonnage left in them for later recovery. These failures usually occurred in the form of large blocks detaching from the abutments, which impaired mucking operations in the primary blocks.

Another problem with primary-secondary mining was the amount of ground open at the same time, which could more easily lead to massive failures. Primary mining itself became increasingly difficult in the second and third lifts of each block as mining progressed upward toward the bottom of the stoping horizon, effectively pinching more and more stress in an ever-shrinking sill pillar. Figure 21.5 illustrates these points by showing seismic activity recorded in 1999 in the north end of the 1,000-m level. Interestingly, and as seen in Figure 21.6, 1999 was by far the most seismically active year for this region, even though very little mining took place during this period.

Most of the 1999 seismicity in this area was associated with secondary and sill pillars that had been cut for at least 3 years. Parts of this region were so near equilibrium that any mining in the lower north end of the mine (on the 850-m level above and the 1,125-m level below) caused seismicity to flare up. The various regions highlighted on Figure 21.5 warrant further discussion.

Region A. Region A corresponds to secondary stopes of widely differing ages (as shown by the lifts they sit in, lower being older) that still show extensive seismicity. These pillars clearly have not failed and still possess strong cores carrying

FIGURE 21.7 Back failure in the 351-8 primary stope above the 1,000-3 subelevation looking west. Note pronounced stress-induced fractures in the back.

8	10	15	18	20
6	9	13	16	19
4	7	11	14	17
2	5	8	12	15
1	3	6	9	13

Future sill pillar

Sill of mining block above

4th sub of mining block

3rd sub of mining block

2nd sub of mining block

1st sub of mining block

Bottom of mining block

Top sub-level of block below

FIGURE 21.8 Conceptual longitudinal sketch of pyramidal pillarless mining, looking west. Shaded blocks show mining front at end of step 7.

substantial loads. Mining in these areas is quite unpredictable; some stopes fail very rapidly (sometimes with the first slot blasts), while others remain very seismically active throughout extraction, making it very difficult to plan for consistent production out of them.

Region B. Region B represents a very seismically active, 30-m-thick sill pillar that spans four mined-out primary stopes. Seismicity has resulted in the backs above each of these primaries failing severely, caving to a height of up to 12 m in some areas. Figure 21.7 shows such a failure.

These four back failures, lined up above 1000-3 sub, the last lift before the bottom of the mining horizon on the 850-m level, had made regional stability of the mining floor on the 850-m level above a concern. Indeed, any mining in the secondaries between had the potential to bridge these back failures, significantly increasing the span of the undercuts and in turn increasing the chance that the caving front would progress higher. The result was that mining of this part of the 1,000-m level north region was suspended for nearly 3 years, until paste backfill could be used to fill the voids. Backfilling enabled the mine to attempt the recovery of the secondaries between. In light of this problem, a mining method such as pyramidal that does not create numerous open back spans to line up between pillars at the same time is preferable. One can imagine how difficult the recovery of this sill pillar would be.

Region C. Region C represents a pair of secondary stopes that failed completely under stress. Swift and severe stress-induced hour-glassing has not left enough intact rock material in place to take any significant load. Consequently, this area never underwent any significant seismicity over long periods of time. Such broken areas typically require extensive ground support and yield few tonnes of ore.

Region D. Region D is an example of a large squat pillar that has not yet been subjected to very high stress concentrations and that is moderately seismically active. The core of this pillar is still strong and easily capable of carrying the ground stresses currently put on it.

Pyramidal pillarless mining was introduced at the Brunswick Mine in 1997. Figure 21.8 illustrates the concept schematically.

Other mining fronts can be pushed south (toward the left on Figure 21.8) and either east or west or both at the same time (if the ore lens is thick enough), increasing the number of stopes active at any given time. A backfilling cycle is required between each step.

Pyramidal pillarless mining unarguably has intrinsic problems. The sequence is not only more complex, it is also much more rigid and requires discipline. Stoping is pushed upward prior to being mined sideways to limit back spans to one block at any given time, which allows a maximum of two blocks to be mined simultaneously. This significantly reduces flexibility and considerably increases the length of time required to complete mining. As seen when comparing Figures 21.4 and 21.8, ten steps are required to complete extraction with the primary-secondary method, while as many as 20 are needed with the pyramidal approach. Pyramidal mining can be carried out in the third dimension as well, if the ore lens is sufficiently wide, which further complicates the extraction sequence. However, this has the advantage of providing additional mining faces. Figure 21.9 illustrates this concept of three-dimensional pyramids.

The issue of the additional mining steps has been made worse at the Brunswick Mine by the need to reduce the average stope size from about 75,000 to around 40,000 tonnes to improve the stability of both hanging walls and backs. These smaller stopes are indisputably less productive because all the nonproducing preparatory steps have to be repeated much more often, yielding less tonnes of ore each time.

Delays in the sequence can also be easily encountered because of the highly serial nature of the process and again, its lack of flexibility. Slower mining, combined with reduced flexibility, susceptibility to delays, and lower productivity of the active fronts, results in the need to maintain substantially more working fronts simultaneously for the same output than does primary-secondary mining. This, in turn, requires highly complex planning and a tight follow-up of all processes underground.

In spite of all these drawbacks, the pyramidal pillarless method was retained at the Brunswick Mine as the only viable approach because of the way the method pushes ground stresses toward "virgin" abutments. Indeed, the method eliminates all pillars in which stress can concentrate and make subsequent mining arduous and unpredictable. Although stress levels do increase as mining progresses upward and creates a shrinking sill pillar, less wide-ranging stress conditions prevail throughout the extraction of a given mining block than with the primary-secondary approach. The smaller stopes are also emptied faster

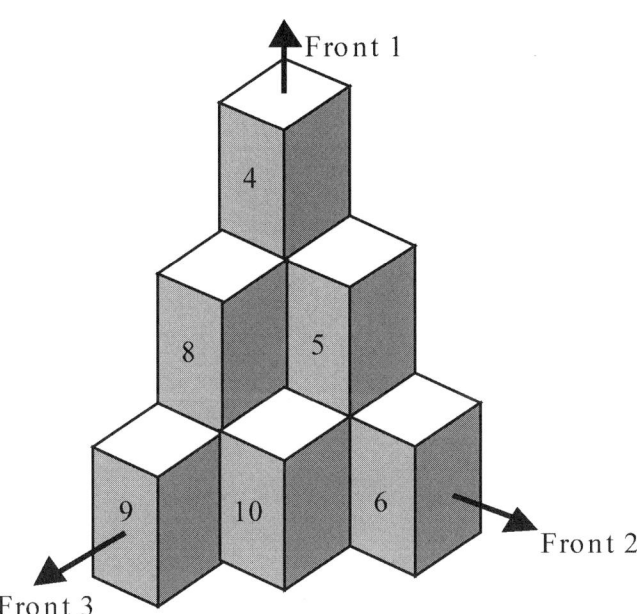

FIGURE 21.9 Conceptual isometric sketch of three-dimensional pyramidal pillarless mining, looking southwest. Numbers on blocks indicate their position in the sequence. As with two-dimensional pyramids, a backfilling cycle is required between each step.

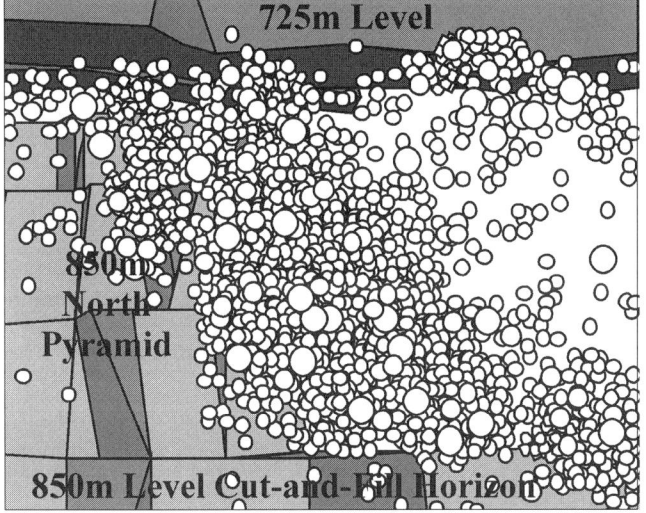

FIGURE 21.10 Seismic activity recorded in 1999 associated with three-dimensional pyramidal pillarless mining in the 850-m level north region; longitudinal section looking west

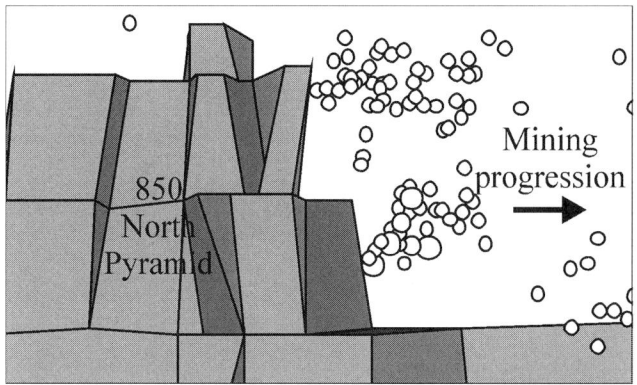

FIGURE 21.11 Seismic activity recorded in January 2000 on 850-m level north pyramidal pillarless mining horizon. Longitudinal section looking west (same view as in Figure 21.10). Note distance of seismicity ahead of active mining.

and are thus backfilled quicker and more efficiently, reducing the length of time voids remain open. This, combined with the fact that such smaller stopes are inherently more stable, has resulted in better control of the excavations.

Figure 21.10 shows the seismic activity recorded in 1999 in the 850-m level north mining horizon, the first area to be converted to pyramidal pillarless mining. The bottom of the mined-out 725-m-level horizon above is visible at the top of the figure, while the mined-out cut-and-fill horizon on which the base of the pyramid sits is visible at the bottom. The angled dashed line shows the location of the north leading edge of the pyramid at the end of 1999; a second leading edge progresses east. The seismic cluster

clearly shows how the seismicity generated by mining in this high-stress abutment is pushed ahead of the lead stopes toward virgin ground. As mining advances, seismic activity advances as well, affecting virgin ground further ahead.

Figure 21.11 is similar to Figure 21.10, but shows only the seismicity recorded in January 2000 (the 725-m level blocks have been removed for added clarity.) Figure 21.11 shows how the bulk of the seismicity is not located in the immediate abutment of the active stopes, but well ahead of them. This is a key point on how pyramidal mining has behaved so far at the Brunswick Mine, whereby the seismicity the method generates tends to occur one to two stope spans away inside intact rock as a result of fracturing in the immediate abutment of the lead stopes. Such behaviour has correlated well with a series of nonelastic numerical stress modelling studies (Pierce et al. 1999).

Careful consideration is given to the directions along which a given pyramidal mining horizon will be advanced. Whenever possible, the sequence is started right against a mined-out region in order to leave no pillars and is pushed outward toward virgin ground. In wide areas where three-dimensional pyramids are implemented, the emphasis is on designing a sequence that will stress-shadow the bulk of the reserves, as schematically shown in Figure 21.12.

The concept consists of pushing lead stopes (blocks 1, 2, 3, 4, and 5 in Figure 21.12) to deflect stress away from the centre of the pyramid, where most of the reserves sit. Because the outer abutment of these lead stopes lies in high-stress areas, they are made smaller for added stability. As illustrated in Figure 21.12, centre stopes (such as block 6) can be larger because they will be essentially destressed; their design becomes more an issue of static stability with regard to their size, geometry, and orientation. This approach provides less output initially, but yields a higher productivity once centre stopes come on-line.

The orientation of the ore lens with respect to the principal stresses and its location with regard to the existing voids and the infrastructures already in place dictate the directions along which the lead stopes should be pushed.

Figure 21.13 shows the result of the implementation of these guidelines in the 850-m level north pyramid. This plan view is cut near the base of the three-dimensional pyramid. The hatched areas represent the mined stopes, while the spheres show the location and magnitude of the seismic events recorded in the region during the month of July 1998. The recorded seismicity in the abutments of the lead stopes at this elevation correlates well with the areas of high stress concentration at the tip of these stopes, as represented by the dashed stress flow lines. Zones of

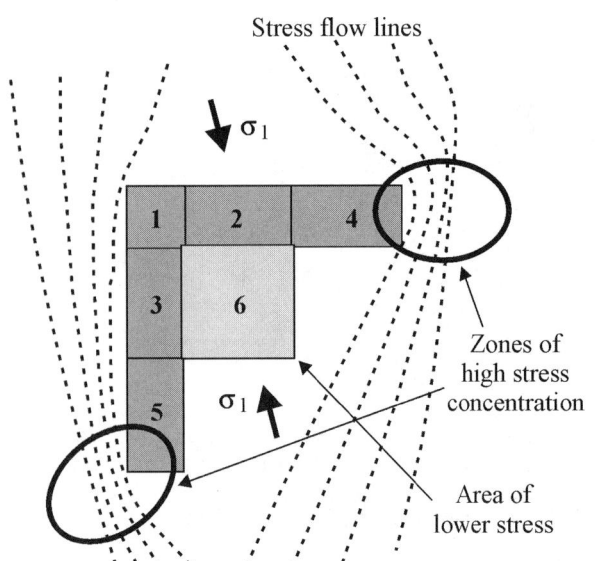

FIGURE 21.12 Schematic plan view cut through three-dimensional pyramid showing how lead stopes can be used to stress shadow bulk of tonnage in the middle of block

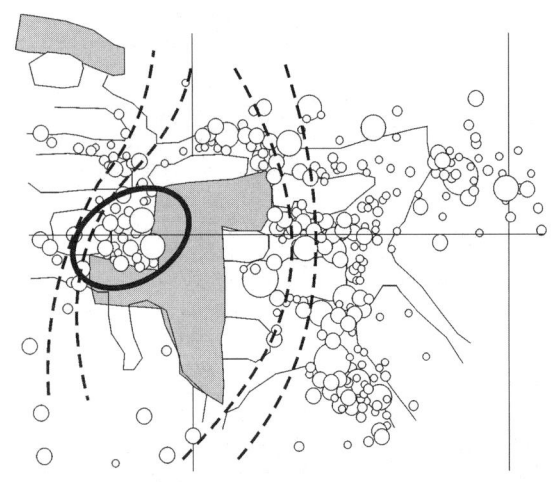

FIGURE 21.13 Plan view showing seismicity recorded in July 1998 on the 850-m level north pyramid region. North is to the right. Dashed lines represent major principal stress flow. This horizontal section is cut near base of pyramid just above 850-2 sub. (The mined-out cut-and-fill horizon on which the pyramid sits prevented stresses from being redistributed immediately below its base.)

low stress (indicated by their low seismicity) are visible in the shadow of the lead stopes. These low-stress stopes have been relatively easy to mine with adequate ground support. The cluster highlighted with an ellipse (left in the figure) occurred above the south lead stope, while the clusters far in advance of stoping (right) were associated with known structural features.

To compensate for the intrinsically slower extraction rate associated with pyramidal mining, several such sequences must be exploited simultaneously. However, pyramidal fronts progressing toward each other do create shrinking pillars in which increasing levels of stress are concentrated, as schematically illustrated in Figure 21.14. Multiple converging pyramids allow mining to progress faster and maintain higher

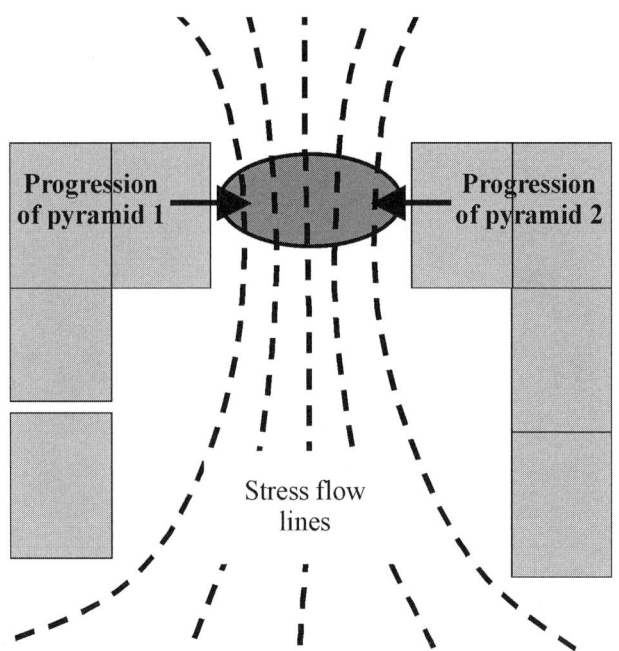

FIGURE 21.14 Schematic plan view showing how two pyramidal fronts marching toward each other create a shrinking pillar where stress levels increase as both sequences progress. Grey ellipse shows area of highest stress concentration.

production rates by providing more working faces at the same time, but at the cost of sterilizing pillars. The size of these sterilized pillars can be assessed using numerical stress modelling to gauge at which point load concentrations will become too high and prevent safe and economic extraction of the pillar.

The value of the ore reserves that are written off, or at least put in jeopardy, in such pillars must be weighed against the financial benefits brought forth by the quicker extraction and higher production achieved while the converging pyramids are producing and put in perspective with the life-of-mine plan. A net present value analysis needs to be performed to determine the optimum approach.

Other steps can be implemented with regard to stress management.

- In areas where high stress is anticipated, long-term development should be located as far away from the stopes as practical. This will help minimise the amplitude of the stress cycles as mining progresses.

- Such long-term development should also be driven as parallel to the major principal stress as practical, which will help reduce stress levels in the back of the drifts.

- Panel destressing can be successfully used to reduce stress levels in certain areas. This approach was tried successfully on a large scale at the Brunswick Mine in October 1999 (Brummer et al. 2000).

21.4.1 Support Systems

Ground support is another possible alternative when confronted with ground instability problems. As described in the previous sections, the emphasis at the Brunswick Mine has been on managing ground stresses in order to prevent loads from becoming too high and to minimise the amplitude of stress cycles. This choice resulted from the belief that no ground support system could conceivably withstand the levels of seismicity sustained or the amount of squeezing experienced at the mine indefinitely. Nonetheless, ground support has played, and continues to play, a critical role in the Brunswick Mine's ground control program.

Shotcrete was introduced on a large scale in 1996. A total of 60,570 tonnes had been placed for ground support by dedicated crews by the end of 1999. Shotcrete is used both "straight" (i.e., applied over welded wire mesh) and reinforced with steel fibres; the ratio of use of shotcrete-and-mesh to fibre-reinforced shotcrete is currently 60:40. Shotcrete is used over rock surfaces and in structures such as pillars (Hadjigeorgiou et al. 1999) and arches (Lessard and Andrieux 1998). Even though extremely efficient under many circumstances, such as in broken ground under moderate stress and in regions subjected to moderate levels of seismicity, shotcrete has not been successful under extreme conditions of high seismicity or heavy squeezing (Lessard et al. 2000). Cable bolts are also used extensively at the mine.

21.4.2 Other Remedial Measures

A number of other remedial measures have been implemented that have contributed to control of ground instability problems. Some of the most noteworthy are as follows.

- Stopes are developed drift-width as much as possible. Even though not ideal for drilling and blasting, this has proven much better in terms of stability during the extraction cycle.

- Mining is carried out from outside the ore lens when very high stress conditions are prevalent or excessively wide back spans are present.

- Drift intersections are offset as much as possible to avoid the creation of wide back spans. Whenever possible, intersections between tunnels are in the form of Ts rather than Xs.

- Development is designed to avoid running subparallel to contacts between rock types of sharply contrasting mechanical properties. Such contacts are cut at 90° whenever possible.

- Development is driven to accommodate the natural jointing of the rock as closely as possible. Depending on local conditions, drifts are driven either with an arched back (in massive, moderately jointed, or blocky ground) or a shanty back (in heavily foliated formations).

21.5 CONCLUSIONS

Since the early 1990s, the Brunswick Mine has had to rethink its mining approach as it encountered increasingly difficult ground conditions. Serious problems were caused by some major issues, among which were—

- Stress issues associated with secondary, sill, and remnant pillars,

- Instabilities resulting from wide spans supported by aging ground-support elements, and

- Severe difficulties supplying rock fill from the surface quarry, which resulted in numerous large voids remaining open for long periods of time.

The result was increasingly frequent ground falls, which prompted numerous emergency changes in the short-term mining plans to replace lost scheduled ore.

A series of remedial measures that have had a significant impact has been implemented at the mine site since 1996. A drastic change in the mining method, from primary-secondary to pyramidal pillarless mining, and a series of stress-management measures such as stress shadowing, panel destressing, and

TABLE 21.1 Number of falls of ground recorded at the Brunswick Mine from 1996 to 1999

Year	Falls of ground, total	Seismically induced falls of ground	
		Total number	No. per 1000 seismic events
1996	45	31	1.85
1997	21	7	0.66
1998	13	1	0.07
1999	11	4	0.15

adjustments of stand-off distance have had very beneficial effects on ground control. Efforts to minimise the size of the openings have also paid off.

The introduction of shotcrete on a large scale and the increased use of cable bolts have also been instrumental in lessening the ground control problems at the mine. Paste backfill, even though not as strong as rockfill, has also contributed. Its more reliable delivery, its capacity to fill areas difficult to reach, and its capability of providing immediate confinement to exposed surfaces have allowed the operation to catch up on the backfilling backlog and to resume mining in particularly difficult areas.

Monitoring has also played a critical role. State-of-the-art, real-time quantitative seismic monitoring has allowed a much better understanding of the reaction of the rock mass to mining and consequently to adjust techniques. Mining decisions based on microseismic activity have now become common practice at the Brunswick Mine (Alcott et al. 1998; Simser et al. 1998). Real-time nonseismic monitoring whereby ground movement monitors and multipoint extensometers are read automatically through the seismic system communication infrastructure has added further insight. A ground control on-call system that provides coverage 24 hours a day, 365 days a year, coupled with the systematic preshift analysis of all seismic and nonseismic data recorded over the past 12 hours, has contributed to safer mining conditions.

Table 21.1 summarises how the number of ground falls has changed over the past years. These falls of ground are unexpected ground failures that occurred in areas accessible to personnel, i.e., that had the potential to cause injury. Ground falls in areas barricaded off or inside open stopes are not considered. Seismically induced ground falls are those resulting from shake damage and/or strain-bursting failures, but does not include those driven by gravity or blast-induced vibrations in wide spans or in areas inadequately supported. Column 3 is expressed as the ratio of seismically induced ground falls over the number of seismic events recorded, or, more conveniently, as the number of seismically induced ground falls for every 1,000 seismic events recorded.

In 1999, the total energy released by seismicity was close to 2.5 times the energy released in 1998 (102 MJ in 1999 versus 41 MJ in 1998). Comparing the number of seismically induced ground falls and the number of recorded seismic events (of all magnitudes) on a weekly basis shows that, even though seismicity increased significantly in 1999, the resulting number of ground falls did not. Figure 21.15 illustrates this point and shows that, even though seismic levels were much higher in 1999 than in 1996, the number of ground falls in 1999 was much smaller.

Importantly, pyramidal pillarless mining is not aimed primarily at reducing the level of seismicity. Because the method was established in highly stressed areas, it has so far been associated with relatively high seismicity. What pyramidal pillarless mining has done so far is essentially concentrate this seismic activity *ahead* of mining in areas not affecting production.

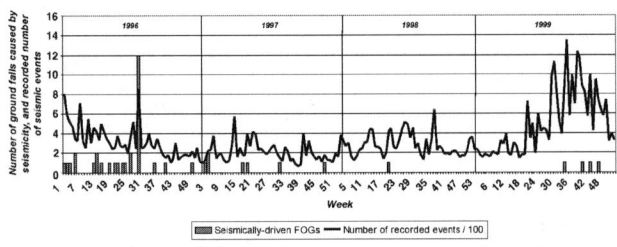

FIGURE 21.15   Number of seismic events and seismically induced falls of ground recorded weekly between January 1, 1996, and December 31, 1999

21.6   ACKNOWLEDGMENTS

The authors would like to thank the Society of Mining Engineers for the opportunity to publish this paper and the management of the Brunswick Mine for permission to present these data. The significant contributions to this paper by Jean-Sébastien Lessard, Alexandre Proulx, Terry MacDonald, and Brent Fisher of the Brunswick Mine Ground Control Group are gratefully acknowledged. Many thanks are also due to Glen Crowther, Arie Moerman, and Tim Babin of the Brunswick Mine and to Richard Brummer of Itasca Consulting Canada for their careful proofreading of the manuscript and their constructive comments.

21.7   REFERENCES

Alcott, J., P. Kaiser, B. Simser, and D. Peterson. 1998. Microseismic Risk Assessment: Integration of Microseismic Data into Ground Control Decision Making Process. Presentation at 100th CIM AGM. Montreal, Quebec, Canada.

Brummer, R., M. Board, A. Mortazavi, P. Andrieux, and B. Simser. 2000. Report on the Destress Blasting Trial in the 29-9 Pillar at Brunswick Mine. Report to CAMIRO. Sudbury, Ontario, Canada.

Godin, R. 1987. Structural Model of Brunswick #12 Mines as an Aid to Mining and Exploration. Presentation at annual meeting of the CIM's New Brunswick Branch. Bathurst, NB, Canada.

Hadjigeorgiou, J., F. Habiyaremye, J.-S. Lessard, and P. Andrieux. 1999. An Investigation into the Behaviour of Shotcrete Pillars. International Symposium on Rock Support and Reinforcement Practice in Mining. Kalgoorlie, Western Australia, Australia. Balkema.

Hudyma, M. 1995. Seismicity at Brunswick Mining. 10th Ground Control Colloque of Quebec Mining Association. Val d'Or, Quebec, Canada.

Lessard, J.-S., and, P. Andrieux. 1998. Shotcrete Pillars at Brunswick Mine (in French). 13th Ground Control Colloque of the Quebec Mining Association. Val d'Or, Quebec, Canada.

Lessard, J.-S., A. Proulx, and, P. Andrieux. 2000. Recent Developments in the Usage of Shotcrete at Brunswick Mine (in French). 15th Ground Control Colloque of the Quebec Mining Association. Val d'Or, Quebec, Canada.

Luff, W. 1977. Geology of Brunswick #12 Mine. *CIM Bulletin*, Vol. 70, No. 782, pp. 109–119.

Luff, W., W. Goodfellow, and S. Juras. 1992. Evidence of a Feeder Pipe and Associated Alteration at the Brunswick No. 12 Massive Sulphides Deposit. *Exploration and Mining Geology*, Vol. 1, No. 2, pp. 167–185.

Moerman, A., K. Rogers, and M. Cooper. 1999. Paste Backfill at Brunswick—Part I: Technical Issues in Implementation. 14th CIM Underground Operators' Conference. Bathurst, NB, Canada.

Ouellette, G., A. Moerman, and K. Rogers. 1999. Paste Backfill at Brunswick—Part II: Underground Construction and Implementation. 14th CIM Underground Operators' Conference. Bathurst, NB, Canada.

Pierce, M., M. Board, and R. Brummer. 1999. 3DEC Modelling of Alternative Sequences for the 1000/1125 Block at Brunswick Mine. Technical Report to Brunswick Mine. Itasca Consulting Group, Minneapolis, MN, USA.

Simser, B. 1996. Seismic Monitoring at Brunswick Mine. ISS International Seminar on Advances in Seismic Monitoring. Western Deep Levels Mine Village, Carletonville, Republic of South Africa.

Simser, B. 2000. Numerical Modelling and Seismic Analysis of Events Leading Up to a Violent Wall Burst. ISS International Seminar on Modelling with Data. Stellenbosch, Republic of South Africa.

Simser, B., P. Andrieux, D. Peterson, T. MacDonald, and J. Alcott. 1998. Advanced Monitoring and Analysis of Microseismic Activity as an Aid to Mining at Brunswick Mines. 3rd North American Rock Mechanics Symposium (NARMS 3). Cancun, Quintana Roo, Mexico.

Simser, B., and P. Andrieux. 1999. Seismic Source Mechanisms at the Brunswick Mine. 14th CIM Underground Operators' Conference. Bathurst, NB, Canada.

CHAPTER 22

Stoping at the Pyhäsalmi Mine

Pekka Pera,* Seppo Tuovinen,* Jyrki Korteniemi,* Marko Matinlassi,* and Sami Niiranen*

22.1 INTRODUCTION

The Pyhäsalmi Mine is located in central Finland, 4 km southeast of the village of Pyhäsalmi (Figure 22.1). This underground mine produces 3100 mt/d of copper-zinc-sulphur ore with a head grade of 1% copper, 1.8% zinc and 38% sulphur. The ore body, which typically dips at 70°, extends from the surface to a depth of 1,400 m. Maximum length of the ore body is 650 m, and width can reach 100 m. The current ore reserves are 20 million tonnes, and total production to date has been over 30 million tonnes.

Current mining methods are sublevel stoping and bench stoping. Extensive rock support is required to permit mining under very high horizontal rock stresses. Measurements indicate that they exceed 70 MPa.

The ore is currently (year 2000) hoisted to the surface in three stages. Trucks transport the ore to an underground crusher on the +660 level from where it is hoisted to the surface in two stages via skip counterweight friction hoists. The Pyhäsalmi New Mine Project (to be finished in 2001) includes a new shaft directly connecting the +1400 level to the surface.

The Pyhäsalmi mill produces copper, zinc, and sulphur concentrates using a flotation process.

The high-horizontal-stress environment characterizes underground mining in Pyhäsalmi and therefore maintaining safe working conditions has been a high priority since the beginning of the operation.

The Pyhäsalmi Mine employs 270 people.

22.2 HISTORY

The ore body was discovered in 1958 by a local farmer who was digging a well in his back yard. Outokumpu Oy immediately commenced geological and geophysical exploration. Diamond drilling proved the well to be located almost in the center of the deposit. After a 1-year period of intensive prospecting, Outokumpu Oy made the decision to go ahead with mine development.

A 2½-year construction period followed. The mine commenced operations on the 1st of March 1962 as an open pit. The final depth of 125 m was reached in 1967. This same year saw phasing in of underground operations. Total production from the open pit was 6.8 million tonnes of ore and 5.6 million tonnes of waste.

In 1967, after the expansion and automation of the processing plant, the production rate was increased from 600,000 tonne/yr to 800,000 tonne/yr. During this phase, the quality of the concentrates improved and recovery increased. The current production is 1.1 million tonnes a year.

22.3 GEOLOGY

The Pyhäsalmi ore deposit is a low-grade, volcanic-hosted massive sulphide deposit surrounded by mainly felsic and mafic volcanic rocks. The ore body has a complex shape resulting from multiphase deformation and, in the upper levels, is surrounded by a large alteration zone. At the deeper levels, this alteration zone is absent. The alteration zone has caused major mining problems as it contains weaker rock types such as sericite and talc-schists.

The supracrustal rocks in Pyhäjärvi are a part of the northwest-trending Paleo-Proterozoic Svecofennian Schist belt. The volcanism started with felsic volcanism that was followed by mafic volcanism in a rifted marine environment. Large-scale hydrothermal alteration was associated with this stage resulting in ore formation near the centers of the mafic volcanism.

At the surface, the ore body is 650 m long and up to 80 m wide. The ore continues down to 1,400 m below the surface. The contacts of the ore body are usually sharp and conformable.

The Pyhäsalmi ore is massive and coarse grained. The total amount of sulphide minerals is 80% to 100% in copper-rich ore types and massive pyrite ore. In zinc ores, the sulphide content is usually lower, lying in the range of 60% to 70%. Mineable copper and zinc ores usually change gradually to nonmineable massive pyrite ore. Waste rock inclusions in the ore are common.

The pyrite content of the ores is usually 60% to 90%, while the sphalerite and chalcopyrite contents are 5% and 2.5%, respectively. Average grade is 0.8% copper, 2.5% zinc, and 37% sulphur. The specific gravity of the ore varies between 4.0 and 4.5.

Dilution has decreased from 30% to 5% because of more selective mining methods and a better understanding of the deposit's geology.

22.4 ROCK MECHANICS

Rock mechanics factors have a strong influence on the mining methods applied at the Pyhäsalmi Mine. The most important factors are (1) the difference in strength between the country rock and the ore and (2) the high horizontal stress field.

According to the stress measurements, the major principal stress increases with depth. At the 400-m level, it is around 45 MPa, and on the 1,125-m level, it is around 66 MPa. It is estimated that the major principal stress will exceed 80 MPa in the deepest parts of the mine. The orientation of the major principal stress on the 1125 level is at about 308°, and the stopes are designed to parallel this direction to avoid large stress failures.

The ore is usually very competent pyrite. Uniaxial compressive strength varies from 92 to 202 MPa. The hanging wall and footwall rocks are mainly felsic volcanics with narrow mafic interlayers and mafic volcanics. The uniaxial compressive strength of the volcanics is about 180 MPa. Talc-schists occur near the footwall contact and partly inside the ore in the upper parts of the ore body, whilst in the lower parts, the talc-schists are

---

* Pyhäsalmi Mine, Pyhäsalmi, Finland.

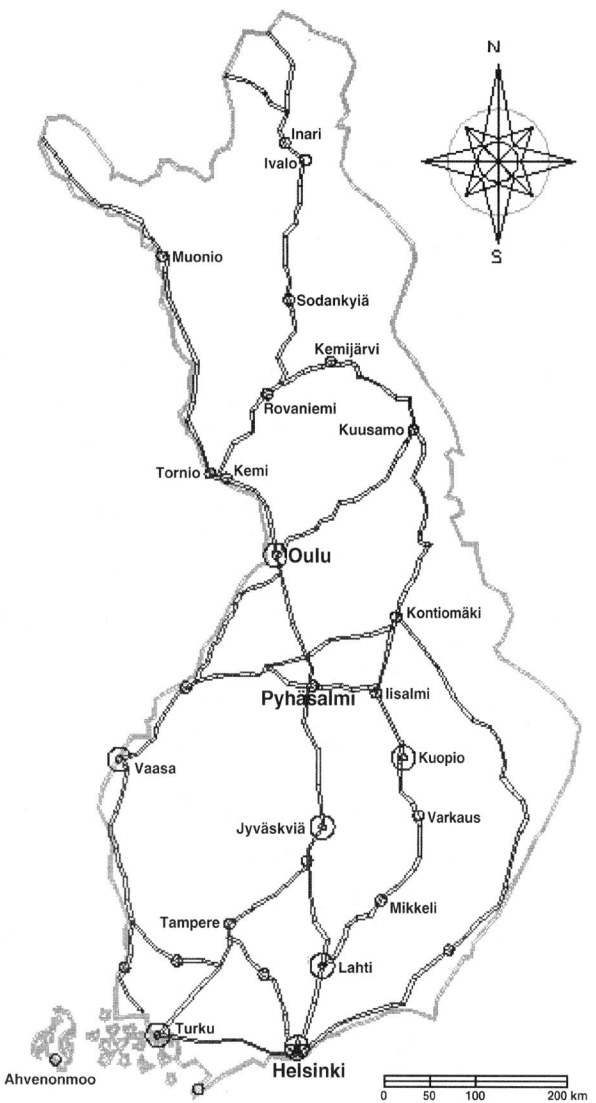

FIGURE 22.1   Location of Pyhåsalmi Mine

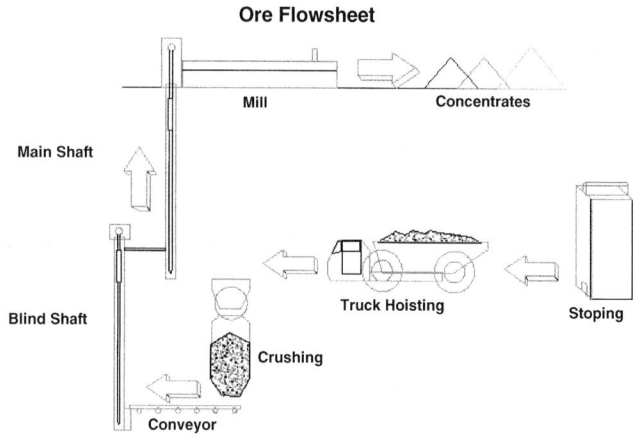

FIGURE 22.2   Simplified flow sheet

found only inside the ore. The uniaxial compressive strength of the talc-schist is only 62 MPa.

The Q′-values of the host rock vary from good to very good. The rock quality designation (RQD) is typically around 90. Jointing usually relates to schistosity, which is stronger and more vertical close to the ore but becomes weaker and more horizontal further away.

Ground conditions in the ore zone are quite variable. Massive pyrite has very good to extremely good Q′-values, but in the vicinity of the talc-schists, the values range from fair to poor.

Stability monitoring is carried out mainly with rod extensometers. HI-Cells are used to measure changes in the stress field.

22.5   DESCRIPTION OF MINING

22.5.1   Introduction

The mine is accessed via a decline with a 1:7 grade. The decline reached the 1,400-m level early in 2000. The main shaft connects the surface down to the 470-m level, and a blind shaft connects the 470-m level to the 702-m level.

The underground workshop is located on the 990-m level, and two lunchrooms are located on the 990- and 1,250-m levels.

More than 80% of the ore is mined below the 1,050-m level with an average trucking distance of 3.5 km (in 1999). A simplified flow sheet is shown in Figure 22.2.

22.5.2   Mine Development

The amount of lateral development required annually is 3.5 to 4 km for the production rate of 1.1 Mt. Mine development is carried out by two Tamrock drilling rigs, the Para 306 and the Paramatic 305. Charging in waste rock is carried out with a Normet ANFO loader and in the ore with a Normet emulsion loader.

22.5.3   Ground Control

Almost all drifts have to be supported with a layer of shotcrete and end-anchored, grouted rock bolts. Some 10,000 rock bolts are installed annually, and the amount of wet-mix shotcrete applied exceeds 6000 m³.

Shotcreting is carried out with one spraying unit. The shotcrete is transported underground by truck from the batching plant located in a nearby village. Rock bolting is carried out with two Tamrock units that drill the holes and then install and grout the bolts with cement. A slick-line system for transporting the wet mix from the surface to the 1,230-m level is under construction.

Cable bolts are employed in the ore drifts to support the stopes. The cable bolts are installed in a 1.5- to 2-m grid with effective lengths of 5 to 10 m, depending on rock conditions and stope dimensions. Annually, some 30,000 m of cable bolt holes are being drilled with the two Tamrock cable bolting rigs.

22.5.4   Mining Methods

The current mining methods in use are bench stoping and sublevel stoping with backfilling. Stopes are developed along strike (longitudinal) when the ore thickness is less than 20 m. In wider parts, transverse stopes are developed.

Above the 1,050-m level, the primary method used is longitudinal bench stoping with bench heights of 40 m. Below this level, transverse stoping is mainly carried out because of the wider ore body. In the primary stage, the method is sublevel stoping with stope heights of 50 m. In the secondary stage, the stopes are generally 25-m bench stopes.

Consolidated backfill is used in the primary stopes to facilitate mining of the pillars between the stopes, e.g., the secondary stopes. Therefore, the idea is to minimize the primary stope width in order to save on backfill costs. The mining sequence is shown in Figure 22.3. Today, the stope size averages 40,000 tonnes, but this will be increased to above 50,000 tonnes in the future.

Mining Sequence

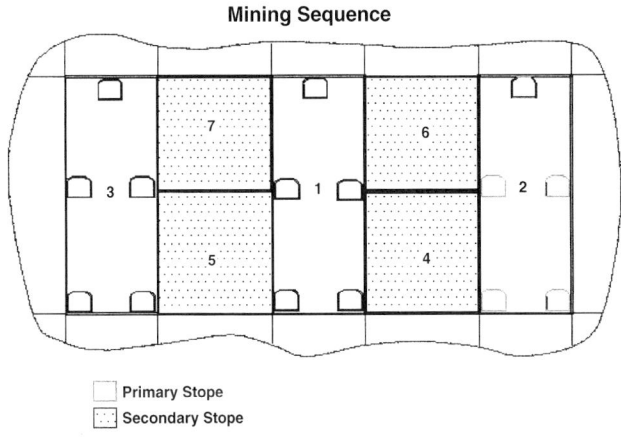

☐ Primary Stope
▨ Secondary Stope

FIGURE 22.3   Mining sequence

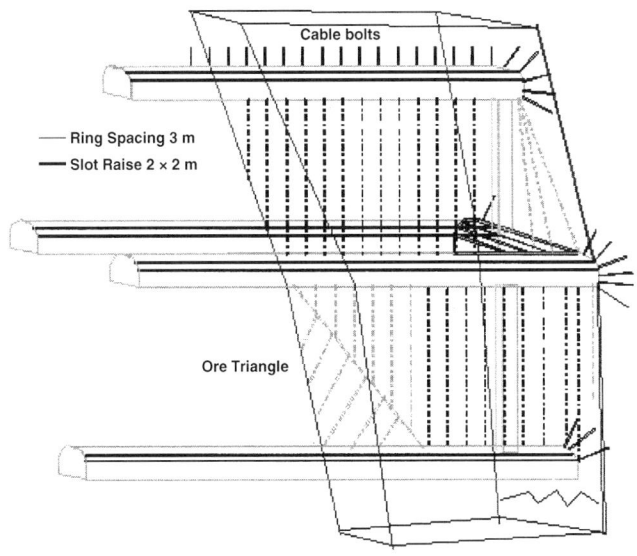

FIGURE 22.4   Schematic sublevel stope layout

In sublevel stoping, a single 4.5- by 4.5-m drift is driven on the top level. Two parallel drifts are driven on the sublevel and on the loading level. In some areas the top level drift is wider (11 to 12 m) to create an oval stope profile to improve stability and increase mining recovery. A schematic sublevel stope layout is presented in Figure 22.4.

22.5.5   Production Drilling and Blasting

Longhole raises are used to open the stopes. In the 40-m bench stopes, 25 m is drilled and charged with downholes and the remaining 10 m with upholes from the loading level. In the sublevel stopes and in the 25-m bench stopes, the slots are 25-m downhole raises.

Blasthole diameter is 76 mm, the distance between production rings is 3 m, and the maximum hole toe spacing is limited to 2.5 m. The explosive in use is a nitrate-based emulsion with an average explosive consumption of 150 gm/tonne.

Two Tamrock Solo 689 rigs are used for production drilling. The mine could satisfy the requirements with only one unit, but the high horizontal stresses tend to break and block the holes.

22.5.6   Backfilling

Primary stopes are backfilled with consolidated material, whereas the secondary stopes are filled with waste rock only. The consolidated backfill consists of classified mill tailings, furnace slag, and lime. The slurry is pumped and drained into the stopes via diamond-drill holes. The method has proven to be reliable and relatively inexpensive.

All development waste rock is dumped into the stopes. Additional waste rock is excavated on the surface and sent underground via a fill pass to be used as backfill material. Waste filling is carried out by load-haul-dump (LHD) machines.

22.5.7   Loading and Hauling

Loading is carried out with four Toro 501 LHDs, three of which are equipped with remote controls and one with video control. Annually, 900,000 tonnes of ore are recovered from the stopes. Of this amount, 15% is loaded with a remote-controlled loader. Video-controlled loading is utilized under conditions of limited visibility. The mine has been able to reduce remote-controlled loading by 25% by leaving an ore triangle above the loading level (see Figure 22.6). The triangle is mined after the "normal" ore has been completely recovered.

Currently, the biggest mining cost is hauling, because the ore has to be hauled from the production areas to the crusher on the 660-m level by dump trucks. On the average, the haulage distance is 3.5 km. In the new mine, this cost item will disappear. The haulage fleet consists of seven Tamrock 40D dump trucks.

22.5.8   Hoisting

After crushing on the 660-m level, the ore is conveyed to the measuring pocket and hoisted to the +470 level with a skip counterweight friction hoist. There, the ore is hoisted up to the surface via the main shaft in 10.8-tonne-capacity skips. The underground crushing and conveying is controlled with TV cameras from the control room on the +470-m level.

22.6   PROCESSING

Processing operations are divided into five stages: screening, grinding, flotation, dewatering, and tailings disposal. The fineness of the grind is about 60% minus-0.074 mm. At this particle size, individual grains contain only a single mineral. The ore is slurried with water during grinding. In the flotation cells, the three concentrates are separated from the pulp with chemicals and air bubbles. The products, in order of separation, are copper concentrate, zinc concentrate, and pyrite concentrate. The concentrates are thickened, filtered, and dried to a moisture content of 4% to 7%. All concentrates are transported from Pyhäsalmi by train. Copper and zinc concentrates are supplied to the company's own smelters. The pyrite concentrate is sold domestically and abroad.

The coarser fraction of the tailings is separated for mine backfill. The fines are pumped into a tailings ponds having an area of 150 ha.

The Pyhäsalmi plant was one of the first in the world to apply computers for process control, and it is extensively automated.

The annual output varies as follows.

Copper concentrate	25,000–30,000 tonnes
Zinc concentrate	35,000–40,000 tonnes
Pyrite concentrate	700,000–800,000 tonnes

22.7   MAINTENANCE

The maintenance cost is about 35% of the total Pyhäsalmi Mine annual cost. The maintenance function is divided into two operational units–mine and mill maintenance–which are assigned to the respective operational departments. Mine maintenance is divided into two parts: mobile equipment maintenance and fixed plant maintenance. The number of maintenance personnel has decreased since the early 1990s and, as shown in Figure 22.5, totaled 71 at the end of 1999.

During the 1990s, maintenance was systematically developed toward prevention, which had a remarkable influence on both costs and equipment parameters. For example, the availability of the mill remained below 90% in early 1990s, but because of changes geared to preventive maintenance, availability has gone up to 93% during the last 5 years.

Considerable improvements were carried out with mobile equipment in the late 1990's and included a cooperative agreement with the manufacturer and the implementation of a new bonus salary system. Figure 22.6 presents the progress in availability and degree of utilization with the Toro 501 loaders since 1990. Availability has varied from 80% up to 87.5% during the decade.

Availability and degree of utilization values for the Toro 40D trucks are shown in Figure 22.7. During the 1990s, availability has mainly varied between 80% and 90%. The average degree of utilization was 54% between 1993 and 1997 and then increased to 82% by the end of the millennium.

22.8 NEW MINE PROJECT

The Pyhäsalmi New Mine is being built below the existing mine between depths of 1,050 and 1,450 m. Total investment cost of the project, in U.S. dollars, is $50 million over the period of 1998-2001. Major investments include a new 1.4-km-long vertical shaft and a new ore-handling system, e.g., ore passes, crushing plant, hoisting equipment, and conveyors and mobile equipment.

Hoisting via the new shaft will commence in July 2001 with a production rate of 1.3 million tonnes a year. The investment will extend the mine's life another 13+ years. The target of the New Mine project is to utilize the latest technology to improve overall productivity with a special emphasis on safe working conditions.

Labor		Staff	
1981	169	1981	33
1982	177	1982	34
1983	166	1983	35
1984	154	1984	34
1985	139	1985	32
1986	130	1986	33
1987	135	1987	31
1988	121	1988	26
1989	107	1989	26
1990	101	1990	26
1991	93	1991	24
1992	89	1992	21
1993	81	1993	19
1994	74	1994	19
1995	68	1995	17
1996	60	1996	17
1997	59	1997	16
1998	62	1998	17
1999	57	1999	14

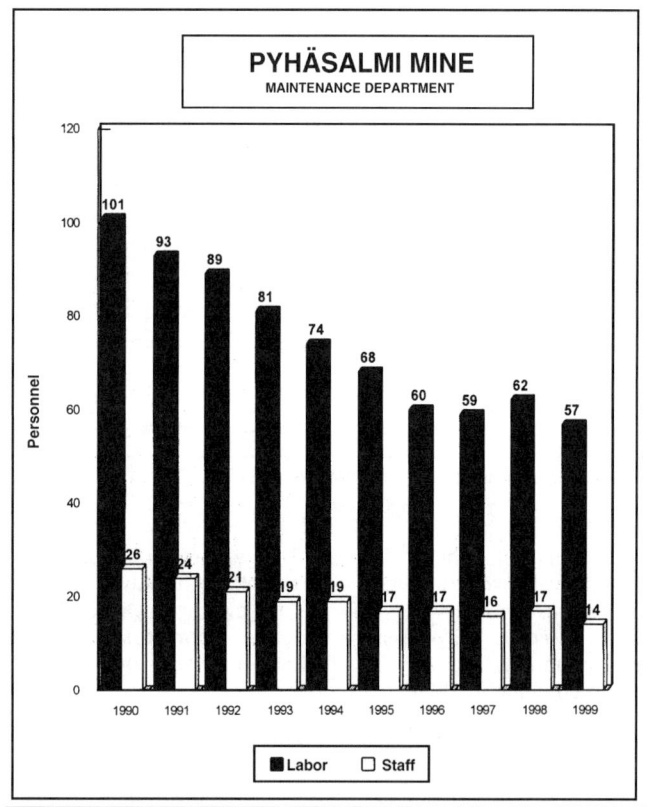

FIGURE 22.5   Number of maintenance personnel

Availability

1990	1991	1992	1993	1994	1995	1996	1997	1998	1999
79.0	78.9	77.7	76.3	80.9	83.1	76.7	76.8	83.2	87.5

Degree of Utilization

1990	1991	1992	1993	1994	1995	1996	1997	1998	1999
44.3	46.4	48.0	47.6	56.4	47.9	41.3	35.3	37.4	58.2

Availability

1993	1994	1995	1996	1997	1998	1999
92.0	88.9	89.4	82.6	75.9	85.9	89.1

Degree of Utilization

1993	1994	1995	1996	1997	1998	1999
50.9	60.7	50.1	54.8	52.4	61.6	82.1

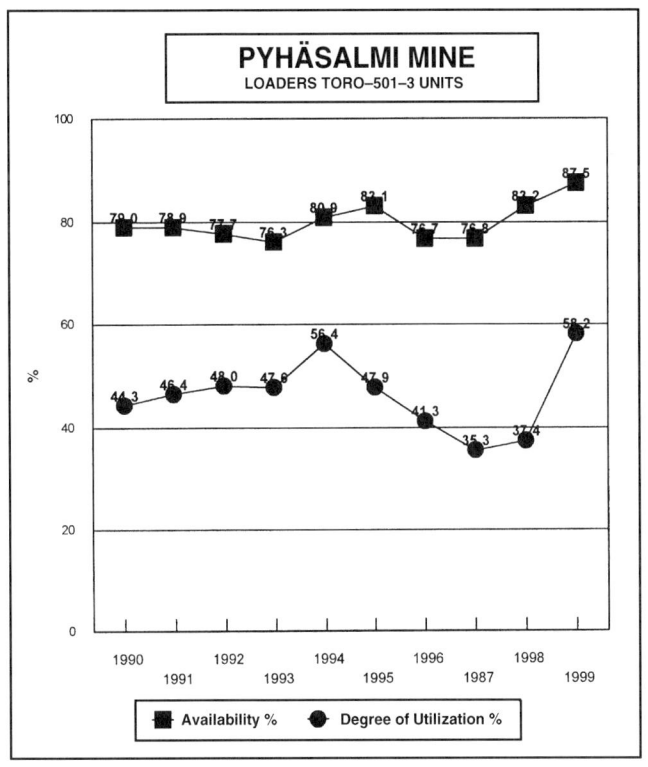

FIGURE 22.6   Availability and degree of utilization of Toro 501 loaders

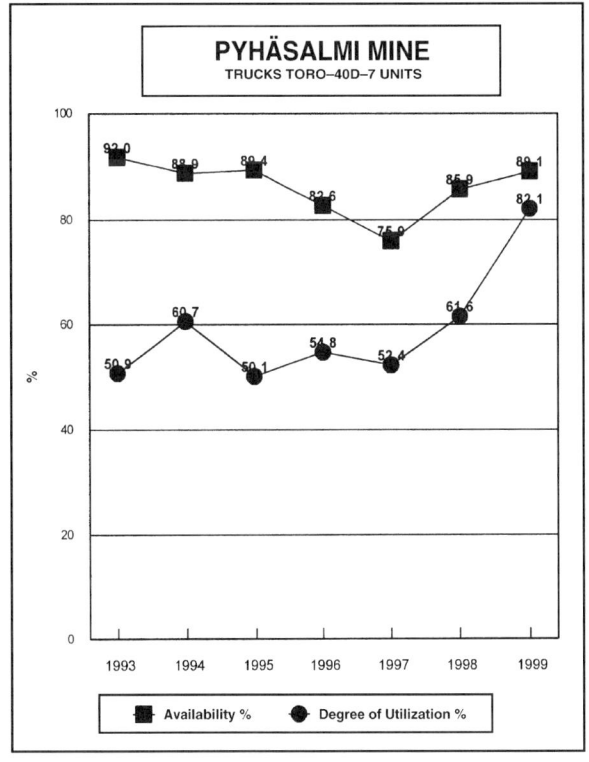

FIGURE 22.7   Availability and degree of utilization of Toro 40D trucks

CHAPTER 23

Narrow Ore Mining in Zinkgruvan, Sweden

M. Finkel,* M. Olsson,† H. Thorshag,‡ J. Wernström,§ and G. Johansson**

23.1 INTRODUCTION

For many years, the mining of narrow ore bodies has attracted worldwide interest. Among the reasons for this are—

- High metal grades, especially precious metals in vein deposits

- Requirements from governmental authorities to extract the natural resources fully

The development of underground mining techniques over the last decades has primarily been directed to large-scale methods and the associated mechanized equipment. Production has been focused on large productive units with a high tonnage capacity. The application of open-pit techniques to underground mines has contributed to their survival in a highly competitive environment. The mines applying large-scale methods have been successful.

For narrow ore bodies, the situation is different as space limitations hamper the economic application of large-scale, low-cost techniques.

Even in operating mines, relatively high-grade, but narrow, ore bodies are often left behind. The main reason for this is that the dimensions of the available equipment are so large that the stope would have to be made much wider than the mineralization in a high-grade vein.

There is no clear-cut definition as to what a "narrow ore body" means. The concept will be interpreted differently by different mining companies and from country to country. Often the concepts of small scale and narrow vein mining are interchanged. Discussions with mining engineers in Sweden and other countries suggest that one should distinguish between vein-type mineralization and compact, coherent ore bodies. For the latter, a "narrow" ore body is defined as mineralization less than 4.0 m wide.

23.2 PROJECT DESCRIPTION

23.2.1 Introduction

An investigation made in the 1980s (Olsson and Thorshag 1987) showed that the then-current practices for mining narrow ore bodies were labor intensive with low productivities. Furthermore, excessive dilution was a problem that greatly reduced the value of the ore product. Introducing a method that lent itself to mechanization had the potential to increase productivity. Likewise, better control of stope boundaries would improve grades in the recovered ore.

Mining companies both in Sweden and other countries have on several occasions expressed a strong interest in more mechanized and productive mining techniques applicable to narrow ore bodies. This initiated the research project, "Mining of Steeply Dipping Narrow Ore Bodies," undertaken as full-scale trials at the Zinkgruvan Mine. This mine is owned by Vieille Montagne, Sweden (from 1998, Zinkgruvan Mining AB).

23.2.2 Objectives

The goal of the research project was to demonstrate the technical and economic feasibility of applying mechanized sublevel stoping for mining steeply dipping, narrow ore bodies through a field test conducted at the Zinkgruvan Mine.

23.2.3 Participants

The participants in the project were the companies Vieille Montagne (Zinkgruvan Mine), Atlas Copco MCT AB, Sandvik Rock Tools, Boliden Mineral, Nitro Nobel, and SveDeFo.

23.3 TEST AREA

The Zinkgruvan Mine is located in the southern part of the Bergslagen region in central Sweden (Figure 23.1). Mining in Bergslagen has traditions since the Middle Ages; the ore deposit at Zinkgruvan was already known in the 16th century. Mining concessions were acquired in 1857 by the Belgian company Vieille Montagne, which continued exploitation at the Zinkgruvan Mine until 1998. At this point in time, the owner is North, Ltd. (Australia).

The ores at Zinkgruvan are of the so-called Åmmeberg type and consist of sphalerite and galena. They occur as stratified, calciferous, leptite impregnations. Grades vary from 6% to 10% zinc, 1.5% to 5.5% lead, and about 45 gm/tonne of silver.

The tabular zinc-lead ore bodies occur in a 5- to 25-m-thick stratified zone in the upper part of a metavolcanic sedimentary group. The ore body is about 5 km long and known to a depth of about 1,300 m. Strike is east-west, and the ore body dips to the north. A subvertical north-northeast fault system has divided the mining area into large blocks. Knalla Mine lies to the west and Nygruvan to the east. Ore thickness varies between 0.5 to 10 m, and dip varies between 70° and 75°.

The criteria for the choice of the test site were—

- Suitable size

- Possibility of testing at least two mining methods in a similar way

- Well-defined ore body with no extensive additional evaluation necessary

---

* Retired, Stockholm, Sweden.
† SveBeFo, Stockholm, Sweden.
‡ Boliden Contech, Stockholm, Sweden.
§ Atlas Copco TM AB, Sweden.
** Vieille Montagne, Sweden.

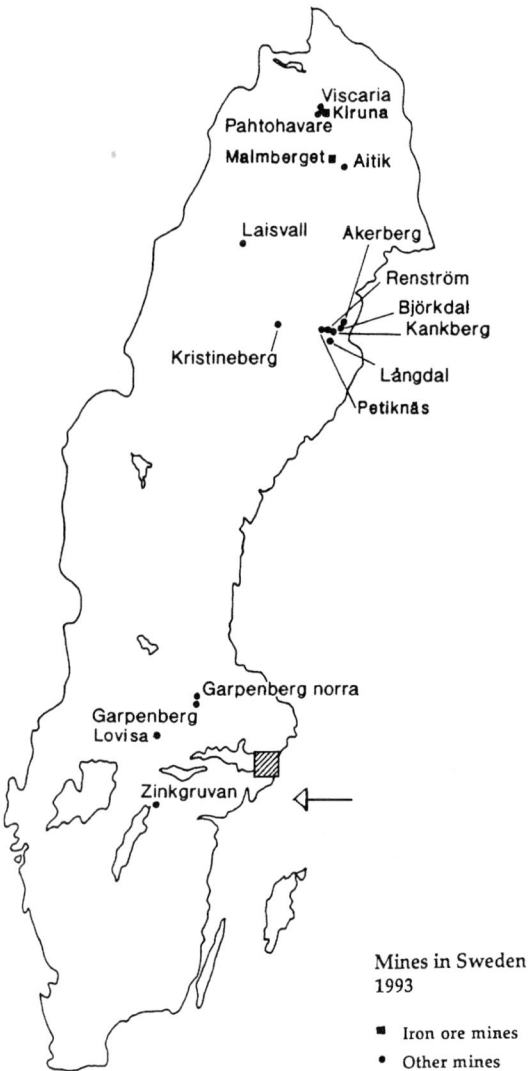

FIGURE 23.1   Location of Zinkgruvan Mine

- Small amount of new development work required

- No interference with the mine's normal planning

- Test mining could start quickly.

An ore block between the 446- and 495-m levels in the eastern part of Nygruvan was determined to satisfy the above-named criteria and was chosen. The area is characterized by distinct layering in both the strike and dip directions and distinct ore boundaries. Mineralization is continuous and regular, and consists, in part, of a compact ore made up of sphalerete with galena and silver impregnations, together with a leptite with stratified zinc impregnations. The wall rock closest to the ore on the footwall side consists of calcareous skarn-stratified leptite with chlorite-filled openings. The hanging wall consists of a homogeneous gneiss-leptite of good strength. The so-called parallel ore lies about 4 to 10 m farther out into the hanging wall.

The ore block above the test area is mined out, as is part of the ore on one side. The stress state is therefore assumed to be representative of that for a typical mining block. Maximum principal stress is horizontal, oriented perpendicular to the strike, and of relatively high magnitude. From a rock mechanics viewpoint, rock quality is considered good.

TABLE 23.1   Comparison of cut-and-fill mining and sublevel stoping

	Cut-and-fill	Sublevel stoping
Recovery, ktonne	29	3
Productivity, tonne/manshift	12	26
Round size, tonne	11	39
Mining cost, SEK/tonne	169	95
Total dilution, %	10	21

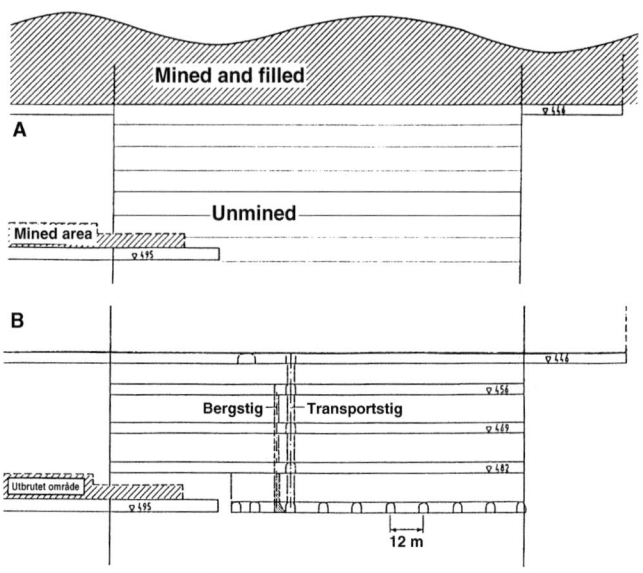

FIGURE 23.2   Cross section through mining block. A, Projected mining block; B, Longitudinal view.

The test area was 150 m long and 50 m high with a thickness varying between 0.5 and 3.5 m, with the average being 1.3 m. The average in situ grade for the ore block was calculated to be 15.2% zinc, 1.45% lead, and 50 gm/tonne silver.

23.4   FIELD TEST ARRANGEMENTS

23.4.1   Selection of Method

In a SveDeFo report DS 1985:4 entitled, "Techniques for Mining of Narrow Ore Bodies" (Olsson and Thorshag), the technical and economical potential of four different mining methods—sublevel stoping, cut-and-fill mining, mechanized shrinkage stoping, and raise mining—were compared. Of these, sublevel stoping and cut-and-fill mining were considered to have the best potential for application to narrow ore bodies. Both methods were well proven and selected for closer study.

Assuming a 10% ore loss and a 10% primary method-related dilution, an evaluation of cut-and-fill and small-scale sublevel stoping gave the results shown in Table 23.1.

The small-scale sublevel stoping variant with small-diameter blast holes was identified as being the most favorable for mining the narrow ore body, both economically and practically. Conventional cut-and-fill mining with roof blasting was being used in another area of the mine (room 24/500) to provide a technical and economic comparison.

23.4.2   Mining Layout

The sublevel stoping layout was planned with sublevels at 456, 569, and 482 m and a bottom level at 495 m. At the bottom level, drawpoints were prepared with spacings of 12 m. The plan and section views for the stope test area are given in Figures 23.2 through 23.4.

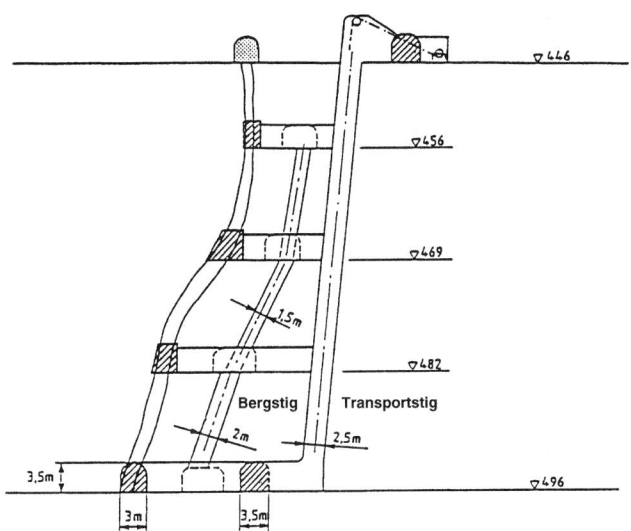

FIGURE 23.3   Cross section through mining block, end view

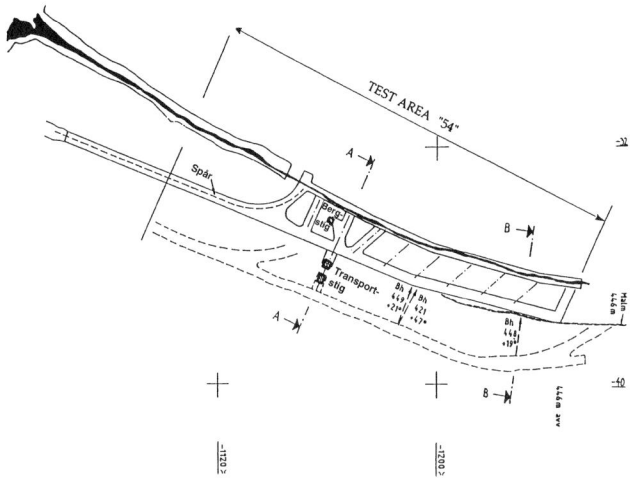

FIGURE 23.4   Plan view showing development on 496- and 446-m levels

The layout was designed with the thought of being able to mine out a portion of the horizontal ore pillar above level 456. In addition, the layout included an ore pass extending between the levels for development rock, together with a transport raise. The transport raise would be equipped with a hoist located on level 446. The basic data regarding the ore block are summarized in Table 23.2.

23.4.3   Choice of Explosive

Introduction.   To mine narrow ore bodies, it is important that dilution be kept low. Dilution can be reduced by, among other means, preventing blast damage to the walls. Therefore, during project planning, it was decided to select a low charge concentration. The smallest useable borehole diameter in which ammonium nitrate-fuel oil (ANFO) mixtures could be used and still get a relatively limited damage zone was found to be 38 mm. Theoretically, when using ANFO in this diameter, one would get a damage zone of about 1 m.

In both drift driving and production blasting, the contour holes should be charged with a weaker explosive. This means, however, that the damage zone from the more strongly charged holes cannot exceed the damage zones from the contour holes.

TABLE 23.2   Basic data for ore block

Mining factor		Values
Dip angle, average		74°
Height, m		50
Length, m		150
Volume, m³		8.950
Ore content, tonne		26,850
Method-related primary dilution, estimated tonnes	16%	(5,100)
Secondary dilution (not planned), estimated tonnes	10%	(2,985)
Total dilution, estimated tonnes	25%	(8,085)
Ore losses, targeted tonnes	10%	(2,685)
Recoverable ore, targeted tonnes		32,250

TABLE 23.3   Damage zone and dilution for different hole diameters

Hole diameter, mm	Ore width, m	Outside of contour row, m	Dilution, %
38	1.3	0.35	35
	1.5	0.25	25
	2.0	0.25	20
51	1.3	0.85	57
	1.5	0.75	50
	2.0	0.50	33

The size of hole diameter in the damage zone and thereby dilution for different ore widths is shown in Table 23.3. The damage zone outside of the ore is considered to be dilution. Here it is assumed that there is an ANFO-charged hole in the middle of the ore and that the damage zone from the contour hole is 0.25 m.

As can be seen from Table 23.3, at least theoretically, the 51-mm in diameter hole yields a significantly larger damage zone and hence also higher dilution than does the 38-mm hole. The damage zone outside of the contour is about 0.3 m for the 38-mm hole and about 0.8 m for a 51-mm hole charged with ANFO. The choice, from the blasting point of view, was 38-mm in diameter holes for both drift driving and production blasting.

The above result assumes that the contour holes are charged with tube charges containing a weak blasting agent of the Gurit type. Pipe charges such as Gurit are, however, not possible from a handling point of view in long upward holes and have therefore not been used in slope blasting.

Explosives for Drift Driving.   For the drift driving, it was planned to use smooth wall blasting. A weak explosive would be used in the contour holes and ANFO (Prillit) would be used in the other holes. At Zinkgruvan, Kimit 22 is normally used for charging the contour holes.

Since the mine had had good experience with the use of "Spoon-ANFO" earlier, it was possible to test even this technique here. At the same time Nitro Nobel had developed a new explosive, Emulet, which was also possible to test in the contour holes.

Spoon-ANFO is a charging method developed in Grängesberg to reduce explosive strength in horizontal holes (Niklasson and Olsson 1989). The charging equipment is the same as for ANFO except that a special device, the so-called "ladle," is screwed to the outer end of the charging tube. The spoon (see Figure 23.5) consists of a slit, strengthened charging tube that distributes the ANFO out in a string along the bottom of the hole during charging.

Emulet is a blowable explosive from Nitro Nobel that consists of a mixture of ANFO, Styrofoam prills, and a smaller amount of emulsion. It is possible to charge Emulet in all directions.

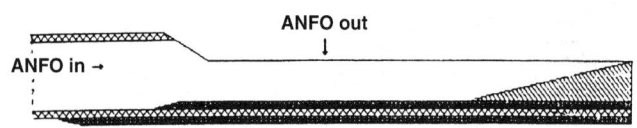

FIGURE 23.5   Diagrammatic representation of ANFO spoon

TABLE 23.4   Factors in explosive charges

Explosive type	Charge concentrations, kg/m	Weight strength relative to ANFO	Velocity of detonation, m/s
Prillit	0.96	1	3,500
Spoon-ANFO	0.45	1	3,500
Emulet 50	0.57	0.89	2,650

In a 38-mm in diameter hole, the data in Table 23.4 apply for these explosives.

Explosives for Stoping.   For stope blasting, it was planned to use ANFO in the middle hole and a weaker explosive at the contour. Early charging and initiation with 3 gm/m detonating cord running along the hole was recommended. Using early charging (in advance of requirements), the charging work can be carried out from a sheltered place, i.e., one does not need to stand under the stope breast.

During stope blasting where the holes are closely spaced, one should use short interval times to avoid, for example, destruction and dead-pressing of the explosive. Theoretically, one should have very short times here when the hole spacing is ≤1 m. The shortest time available with the present blasting caps is 17 ms (Unidet). In a mining situation, it is more common to use Nonel GT/MS. This cap has a delay time of 25 ms between intervals. Initiation with this cap was therefore recommended.

With the new electronic blasting caps, one can get times of 1 ms between intervals. This naturally opens quite different possibilities.

23.4.4   Choice of Equipment

Mining of narrow ore bodies exerts special requirements for the equipment to be used, namely—

■ Compact design with high capacity and precision

■ Easy disassembly, which simplifies transport in the shaft

■ Versatility, such as the capability for both horizontal and vertical drilling

In conjunction with Zinkgruvan, the project's machine requirements were determined and compared with existing machines in the park. It was determined that all the equipment, with the exception of that required to drill the 38-mm in diameter holes for both drifting and stoping, was available at the mine.

As no standard drill rig for 38-mm holes was available, Atlas Copco developed the Trakker 526, a drill rig for small-hole drilling that was adapted to narrow vein mining (Figure 23.6). The Trakker 526 is a very compact drill rig, only 1,100 mm wide and, with the protection roof lowered, 1,850 mm high at maximum. Low height becomes an advantage in a narrow stope and dipping ore body, as the drill can be set up close to the hanging wall and the blast holes can be collared inside the ore body. After development drifting was completed, the Trakker 526 was rebuilt and fitted with a feed for drilling long upholes (Figure 23.7).

A new type of extension drill steel for the longhole drilling of 38-mm in diameter holes was developed by Sandvik Rock Tools in cooperation with Atlas Copco. A special instrument for precision alignment of the feed beam was developed by Director AS,

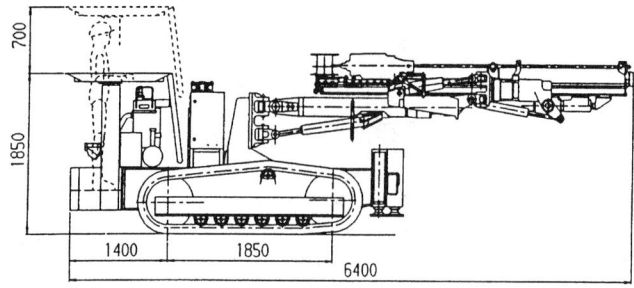

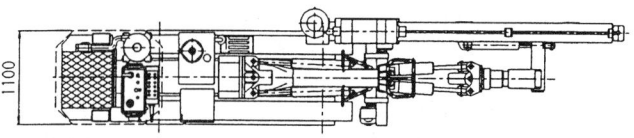

FIGURE 23.6   Trakker 526 drill rig used for development drifting

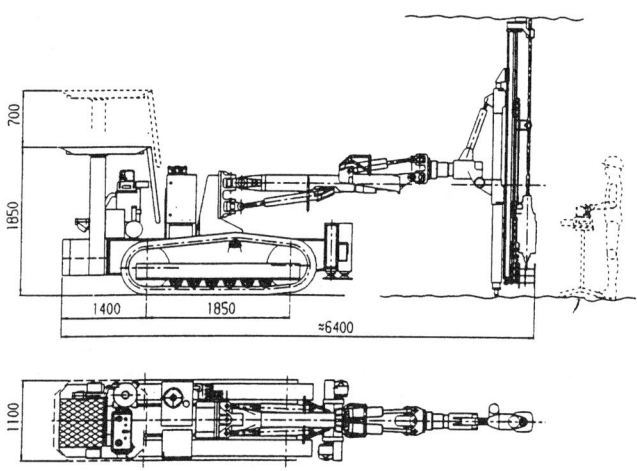

FIGURE 23.7   Trakker 526 drill rig converted to longhole drilling

Norway, together with Atlas Copco. A computerized system was built into the Trakker's electro-hydraulic controls (Figure 23.8).

A Toro 150 E load-haul-dump (LHD) machine was used for loading and transporting rock and ore. The loader was 0.3 m wider than the drill rig. If a loader with the same width as the drill rig had been available, the drifts could have been smaller and dilution reduced.

23.5   TEST MINING AND RESULTS

23.5.1   Development

Initially, the sublevels were developed using the Trakker 526 drill rig (Figure 23.9). This was later replaced by the Boomer H104 drill rig (Figure 23.10). The reason for this change was a request from Atlas Copco to test a new, narrow-type drill rig equipped with a hydraulic rock drill and mounted on a rubber-tired chassis. Using the Boomer H104 for development drifting made the Trakker available for conversion to longhole drilling earlier than originally planned. This shortened the time required to complete the project.

Various explosives for cautious blasting were tested in connection with the development of the 469-m level. The best results, both in terms of advance and minimization of blast damage to the host rock, were achieved with ANFO.

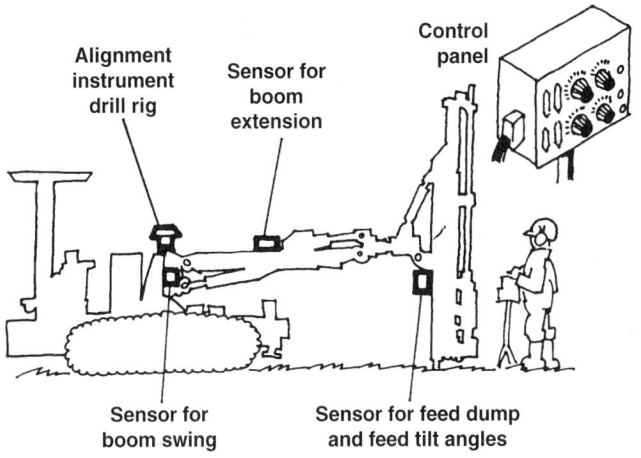

FIGURE 23.8   Instruments for setting drill hole angle

FIGURE 23.9   Trakker 526 being used for drift driving

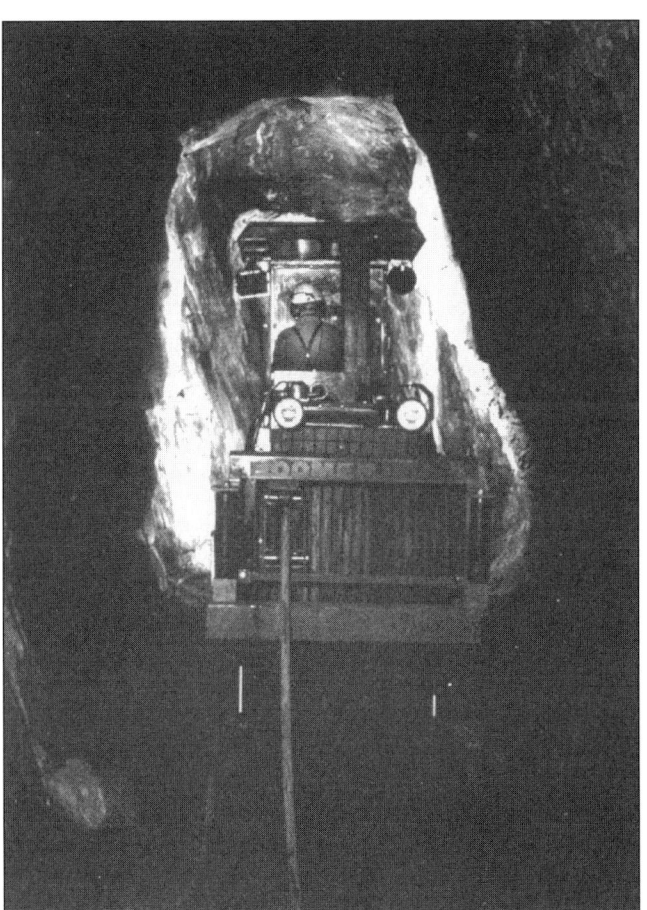

FIGURE 23.10   Boomer 104 being used for drift driving

A summary of data from the development drifting on the 456-, 469-, 482-, and 495-m levels is shown in Table 23.5.

23.5.2   Mining

Knowledge of the ore configuration between each level was important for planning longhole drilling. Ore contours were continuously mapped along the sublevel drifts. A geologist mapped the roof while the drillers themselves recorded the face after each blast. Drill patterns were designed based upon the mapped ore configuration. The location of each row of holes was recorded.

Each row of holes consisted of three blast holes with spacings of 0.6 to 1 m. Overburden was 0.8 m. The contour holes were placed about 0.1 m inside the ore boundary to avoid damage to the waste rock in the hanging wall and footwall. Normally, drilling was stopped just before breakthrough to the level above. A few holes were drilled through to check accuracy. These parallel holes were drilled at angles of 65° to 83° from the horizontal. The initial intention was to tilt each row of holes 10° forward. Because of collaring difficulties, most holes were drilled without a forward tilt.

The longholes were drilled with the Trakker 526 rig. The average hole depth was 9.6 m. There were 367 rows with a total

TABLE 23.5   Summary of development data

Item	456-m level	469-m level	482-m level	495-m level	Total
Drift length, m	131.0	129.5	132.0	90.0	482.5
Drift width, m	2.53	2.67	2.89	3.50	
Ore width, m	1.59	1.57	1.35	1.72	1.54 (avg)
Waste dilution, %	37.1	41.2	53.3	50.8	45.2
Specific drilling, m/m³	5.7	5.6	4.9		
Drill capacity, m/shift	87	116	198		
Specific charge, kg/m³	4.0	3.9	3.4		
Charging capacity, kg/shift	132	143	124		
No. of bolts	408	287	221		
Bolt capacity, bolt/shift	17	17	35		
Ore production, tonnes	3,323	4,285	4,104	3,700	15,385
Loading capacity, tonne/shift	65	116	171		
Time used, manhours	2,005	2,785	1,169		
Production, manhours	1,902	1,749	1,108		
Development, manhours	1,491	1,220	766		
Development/productivity, %	78	70	69		
Productivity, tonne/manshift	13.1	15.2	28.0		
Advance, m/manshift	0.50	0.44	0.83		

TABLE 23.6 Distribution of rows and cap interval numbers

Row number	Cap interval numbers		
1	4	3	4
2	8	7	8
3	15	14	15
4	20	18	20

of 11,556 drill meters. The excavated ore volume of 4,001 m^3 gives a drilling density of 2.9 m/m^3.

Hole deviations inside the rock were minimal. Only two of 50 holes drilled through were not straight enough for a flashlight to be seen. These curved holes were redrilled later.

Occasionally errors of substantial magnitude in hole alignment were recorded, distorting the target picture on the level above. Off-target deviations of 1.5 m were recorded in several instances. This means that holes in different rows may cross each other, sometimes ending in the side walls. Several holes were drilled with the incorrect inclination angle. This can possibly be explained by poor agreement between the drill pattern and the actual ore configuration, or malfunctions in the angle-setting instrument.

ANFO was the main explosive used for longhole blasting. Several systems—Nonel GT/T, Unidet, and Nonel GT/MS—were tested for initiation of the blast. The best results were achieved with the Nonel GT/MS system. The safest initiation was achieved with a 3 gm/m detonating cord inserted in the longhole. The detonating cord functioned both to trigger the detonation and to reduce the detonation velocity of the ANFO. The latter lowers the explosive pressure inside the borehole and enhances the cautious blasting result.

Three to four rows of holes were usually blasted at one time. With Nonel GT/MS, the cap interval numbers were distributed as shown in Table 23.6.

At times, mining was delayed because of blasting mishaps and failures that caused irregularities in rock breakage.

The average specific charge for the sublevel blasting was 4 kg/m^3. This is a very high value compared to normal longhole blasting. Some of the high explosives consumption can be explained by practice in the bottom slice, which was to drill and blast with 51-mm holes while retaining the same drill pattern as for 38-mm holes.

The longhole blasting came out well, despite stuck drill steel and blasting problems. The walls in the mined-out stopes were smooth (Figure 23.11), the fragmentation was excellent, and no problems with oversized rock at the drawpoints were recorded.

Loading and transport of rock on the sublevels was done with the Toro 150 E LHD loader. The same machine was used for loading longhole-blasted ore through drawpoints at the bottom level. The LHD drawpoint loading worked without problems. A certain part of the ore was loaded out lengthwise inside the stope. Here, the loader entered the open stope guided by radio remote control, which worked well in this application. In total, 14,521 tonnes of ore were loaded.

Rock bolting, scaling, and charging were done manually. Mechanization of these operations is desirable for future projects.

Communication and transport between levels was done through a raise equipped with an Alimak hoist. The vertical raise transport became a troublesome bottleneck. Ramp transport is recommended for future projects.

23.5.3 Performance Data

Performance data are summarized in Table 23.7.

23.6 RESULTS

The total productivity for the test area was calculated to be 19 tonne/manshift. However, an optimization of productivity

FIGURE 23.11 Walls of mined stope

TABLE 23.7 Performance data for drifting and longhole mining

Level, m	Advance, m/round	Specific drilling, m/m^3	Specific charging, kg/m^3	Drilling, m/shift	Loading, tonne	Productivity, m/manshift
456	2.8	5.7	4.0	87	3,323	0.5
469	2.7	5.6	3.9	116	4,258	0.4
482	2.7	4.9	3.4	198	4,104	0.8
Mining	—	2.9	4.0	100	14,521	27

in which available work time for production is considered together with mechanical availability of machines results in an estimated productivity of 25.4 tonne/manshift.

23.6.1 Ore Recovery and Waste Dilution

Total ore recovery was 29,209 tonnes, of which 15,209 tonnes was produced by development excavations and the remaining 14,521 tonnes by sublevel stoping. Ore losses for the longhole stopes were calculated to be 1683 tonnes or 10.4% of the estimated ore reserves. Dilution for development drifting was calculated to be 45%, while for longhole stoping it was 5%. Combining unavoidable dilution by waste rock from the oversized development drifts and rock coming loose from the stope walls, the average dilution over the test area amounted to 25%.

TABLE 23.8 Comparison of expected results and actual outcome, sublevel stoping

Item	Expected result	Actual outcome
Ore recovery, tonnes	28,600	29,730
Advance, m/manshift	1.4	0.7
Productivity, tonne/manshift	22.2	25.4
Mining cost, SEK/tonne	208	214
Total dilution, %	23	25
Ore losses, %	25	10

TABLE 23.9 Comparison of expected results and actual outcome, cut-and-fill mining

Item	Expected result	Actual outcome
Productivity, tonne/manshift	11.5	10.5
Mining cost, SEK/tonne	329	316
Total dilution, %	10	31
Ore losses, %	10	4

23.7 EXPECTATIONS AND RESULTS

The test results were compared with model expectations presented in 1985 by SveDeFo in the report "Development of Techniques for Mining of Narrow Ore Bodies" (Olsson and Thorshag).

23.7.1 Sublevel Stoping

An evaluation related to sublevel stope mining of the narrow ore body is shown in Table 23.8. An examination of the data indicates a close relationship between expected results and the actual outcome of the full-scale test.

23.7.2 Cut-and-Fill Mining

An evaluation related to cut-and-fill mining of a narrow ore body is shown in Table 23.9. The low productivity is characteristic for cut-and-fill mining. Waste dilution is high despite the use of narrow equipment.

23.8 CONCLUSIONS AND RECOMMENDATIONS

Important conclusions from the study of narrow vein mining at Zinkgruvan are summarized as follows:

1. The width of the sublevel drifts is very important for selective mining; in the test, 50% of the ore was recovered from drifting.
2. Longhole drilling and blasting provide high selectivity—only 5% dilution at test.
3. Minimum stope width was identified as 1.2 m.
4. Small hole drilling (38 mm) is required for the smooth blasting of stope walls.
5. Precision drilling of 38-mm in diameter holes provides 10-m-long straight holes.
6. Precision in charging and blasting is required to avoid remnant rock sections in the stopes.
7. Careful preinvestigations and planning are required.
8. Machinery with compact dimensions is required to work in narrow drifts.

For additional information, see Olsson et al. (1992) and Finkel et al. (1992).

23.9 ACKNOWLEDGMENTS

The authors wish to thank Hans Hamrin at Atlas Copco and William Hustrulid at the University of Utah for having translated this report into English, thus making it possible to spread the results outside Sweden.

23.10 REFERENCES

Olsson, M, and H. Thorshag. 1985. Teknikutveckling för smalmalms-brytning (in Swedish). SveDeFo Report DS 1985:4; Technique for Mining of Narrow Ore Bodies (summary in English). SveDeFo Report DS 1987:1.

Olsson, M., H. Thorshag, M. Finkel, and H. Mariena. 1992. New Techniques for Mining of Steep Narrow Orebodies. World Mining Conference, Madrid.

Niklasson, B., and M. Olsson. 1989. Tests with a New Technique for Charging of Bore Holes—Spoon-ANFO. SveDeFo Report DS 1989:3G (in Swedish).

Finkel, M., M. Olsson, H. Thorshag, J. Wernström, and G. Johansson. 1992. Narrow Ore Mining: Full Scale Tests in the Zinkgruvan Mine (in Swedish). BeFo 250:1/92, Stockholm.

CHAPTER 24

Mining Operations at Pea Ridge Iron Ore Company—A Case Study

Jason Ovanic*

24.1 INTRODUCTION

The Pea Ridge Mine is located 105 km (65 miles) southwest of St. Louis, in Washington County, Missouri, and is currently the only underground iron mine operating in the United States. The deposit was discovered by diamond drilling in 1953 after initial aeromagnetic surveys in 1949 and 1951 indicated a large anomaly in the area. In 1957, the St. Joe Lead Company entered into a joint venture agreement with Bethlehem Steel Corporation to form the Meramec Mining Company specifically to mine the Pea Ridge deposit and produce iron ore pellets for blast furnace feed. Shaft sinking began late in 1957, with the first production stopes being mined on the east end of the 2275 level in 1963. The first trainload of pellets was shipped on March 16, 1964. The initial capital cost of the project was $52 million.

Meramec Mining operated the property continuously from 1957 to 1977. During that time, hourly and some salaried personnel were represented by the United Steel Workers. During those years, the total work force peaked at 1,064 employees, and the company produced 21,087,500 tonnes (23,245,000 gross tons) of pellets, requiring hoisting of 31,723,000 tonnes (34,969,000 tons) of broken ore. The largest production year was 1972 when 2,771,500 tonnes (3,055,000 tons) of ore was hoisted.

At the end of 1977, with falling steel prices, Bethlehem determined that it no longer needed pellets from Meramec and negotiated a termination of its agreement with St. Joe. Pea Ridge was placed on standby during 1978, with only essential personnel retained for idling maintenance and reduction of stockpiled pellet inventory. In 1979, St. Joe made the decision to resume operations as an independent iron ore producer, and on January 16, 1979, the Pea Ridge Iron Ore Company was formed as a subsidiary of St. Joe Minerals with a nonunion workforce that peaked at 337 employees in 1984. The company began selling pellets on the spot market, as well as expanding production of high-purity iron oxide, phosphorus, iron sulfide, and trap rock byproducts.

In developing the policies under which Pea Ridge Iron Ore Company would operate, industrial hygiene and safety were given significant consideration. Management believed that safety performance had to be better than Meramec's previous record, and therefore took the opportunity to develop a formalized safety program. Today, safety involves every employee, including members of the management team. Safety meetings are held on a monthly basis, and a review of safety goals and concerns is presented in an open forum format. Safety incentives are a part of the overall program, but success has been achieved through employee involvement and management's commitment to each employee's effort to make the workplace safe.

In 1981, St. Joe was purchased by the Fluor Corporation, which in turn sold Pea Ridge to a group of private investors in 1990. These investors have since formed Woodridge Resources Corporation, which currently holds Pea Ridge as a wholly owned subsidiary.

During the 1979–1990 period, Pea Ridge Iron Ore produced 9,706,900 tonnes (10,700,000 tons) of magnetic iron concentrate from 16,423,700 tonnes (18,104,000 tons) of hoisted ore.

Today, Pea Ridge produces specialty iron oxide products for the coal processing, ferrite magnet, coatings, heavy aggregate, water treatment, and pigment industries. The company has 103 employees and produces between 453,600 and 544,000 tonnes (500,000 and 600,000 tons) of ore annually. The current focus is on the recent installation of new calcining technology coupled with attrition grinding for the production of ultrafine-grained iron oxide powders (1 µ+). Also at this time, Pea Ridge is conducting a long-term program of refurbishing the mining infrastructure and upgrading its drilling and haulage fleets to ensure its mining capabilities well into the future.

24.2 GEOLOGY

24.2.1 Regional Geology

The Pea Ridge deposit is hosted in felsic volcanic rocks of the Middle Proterozoic St. Francois terrane, a remnant core complex of volcanic structures. The St. Francois terrane is mostly silicic igneous rocks, mainly rhyolitic ash flow tuffs and lava flows, and associated structures such as granitic plutons and volcanic ring complexes. These features are fully described by Kisvarsanyi (1981). The Pea Ridge deposit is located within one of the ring complexes and is therefore thought to be underlain by a granite pluton. The steep dip of the hosting volcanics indicates that they may be part of a collapse block created as a result of a caldera collapsing into one of these structures, referred to as the Pea Ridge pluton (Marikos et al. 1989; Kisvarsanyi 1981).

24.2.2 Ore Body Geology

The steeply dipping, discordant, iron oxide body is hosted by a suite of anorogenic, Middle Proterozoic, rhyolite porphyries. The volcanics are unconformably overlain by 335 to 427 m (1,100 to 1,400 ft) of Cambrian and Ordovician sediments. The host porphyries are a series of rhyolitic flows and ash flow tuffs that strike N80°W and dip 75° to 90° northeast. Emery (1968) identifies five individual porphyry units that have been intercepted by mine workings. For detailed descriptions of these units, see Emery (1968, p. 361).

The magnetite ore body is roughly tabular, striking N55°–60°E and dipping 75° to 90° southeast. Dips tend to flatten below the 2475 level; however, the lower extent of the ore body is yet to be determined. The dimensions of the known extent of the ore body are approximately 760 m (2,500 ft) in strike length

* Pea Ridge Iron Ore Company, Washington County, MO.

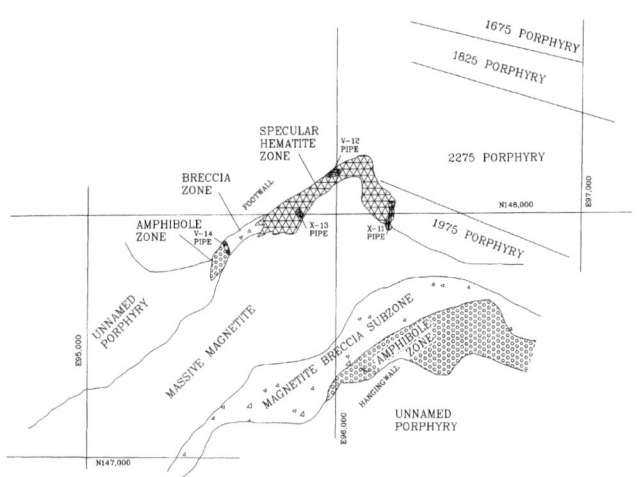

FIGURE 24.1   Geology of the 2275 level (*Source:* Nuelle and Seeger 1990)

and 183 m (600 ft) at its thickest point in cross section. It extends from the 1375 level to below the 2735 (in excess of 400 m [1,300 ft]). Typically, magnetite grades run above 45% magnetic iron, with some grades above 65%, and grades appear to increase with depth. Pyrite also occurs in economically recoverable quantities, from dispersed grains to large veins within the magnetite ore zone. The magnetite zone is the economically important unit within the ore body; however, other zones of geologic and geotechnical interest include a specular hematite zone, a silicified zone, a porphyry breccia, and a quartz-amphibole zone, as well as mafic and aplite dikes. All of these zones can be seen in a plan view of the geology on the 2275 haulage level (Figure 24.1).

In addition to these zones, Marikos et al. (1989) and Nuelle et al. (1991) have documented four major rare-earth breccia pipes. These pipes occur along the eastern footwall between the ore and the other lithologic zones. The pipes are generally ovoid in plan view and are up to 61 m (200 ft) wide and 15 m (50 ft) thick. The vertical lengths of these pipes have yet to be determined, but some have been traced up to 122 m (400 ft). These pipes dip steeply with the ore body, with dips generally greater than 60°. The main rare-earth minerals in the pipes are monazite, xenotime, and rarely bastnaesite. Gold also occurs in these pipes, generally in concentrations of less than 1 ppm. However, nugget-effect gold has been found in concentrations as high as 371 ppm (337g/tonne [11.9 oz/ton]) (Hussman 1989). There was some interest in developing these pipes in the early 1990s (Tucker and Roberts 1991). However, in concentrating the monazite, which contains thorium as a constituent, a low-level radioactive material would be produced. Difficulties in processing the monazite ore, as well as potential waste disposal costs, made the project unattractive.

24.3   HYDROLOGY, DRAINAGE, AND PUMPING

The Lamotte and the upper portion of the Bonne Terre formations are the two major aquifers that occur in the sedimentary units lying above the ore body. The Lamotte Formation is a relatively clean quartz sandstone in excess of 61 m (200 ft) thick, and the Bonne Terre is a dolomitic limestone, sandy at the bottom and silty at the top. Potable water and drill water are obtained from 50-mm- (2-in-) diameter holes drilled from the 1375 level up into the Lamotte Formation. The 12 operating packered drill holes deliver up to 2,450 L/min (650 gal/min) at pressures in excess of 689 kPa (100 psi). Prior to pillar recovery, total mine flow was approximately 2,000 L/min (530 gal/min); today, after caving through both the Lamotte and Bonne Terre formations (but not

yet to the surface), total mine flow is 3,700 L/min (977 gal/min). Although the ore body and host rhyolites have very low permeability, they still convey considerable amounts of water through fractures and shear zones. However, workings that intercept the caved zone created by pillar recovery provide the greatest conduits for fluid transport. Water is collected in roadway ditches and directed to drainholes, which divert water to the nearest sump through collection pipes. Prior to water storage, solids are removed from the water via a laundering system in an effort to protect sumps and pumping equipment.

The main pumping process occurs in two stages, from the 2275 to the 1375 level and from the 1375 level to the tailings thickener on the surface. Originally, the pumps were a combination of 186- and 372-kW (250- and 500-hp) vertical turbine pumps in parallel on the 2275 and 1375 levels. Today, the vertical turbine pumps are used for backup purposes only. Currently, three 186-kW (250-hp) centrifugal pumps are used in series on both the 2275 and 1375 levels. The third pump in the series on the 1375 level has a variable speed drive to increase pumping efficiency. Secondary pumping for lower level development and shaft dewatering consists of 75-kW (100-hp) twin volute centrifugal pumps and 3.75-kW (5-hp) submersible pumps. Work was completed in 1998 on a new 304-mm (12-in), schedule 40XH steel pipe, rising main in the No. 1 service shaft. The cost for this contracted construction, including tie-ins on the surface and at the 1375 and 2275 pumping stations, was approximately $275,000.

24.4   GEOTECHNICAL INFORMATION

24.4.1   Rock Properties

Table 24.1 shows the geotechnical properties of various rock types.

The high densities of the magnetite and hematite affect the haulage of these materials. The buckets of the main haulage units have been reduced from 6 to 4.6 m³ (8 to 6 yd³) to reduce the load to an average 9 tonnes (10 tons). The relatively high strain energies of the porphyries and the quartz-amphiboles contribute to occasional minor rock bursts. Another property of interest is the relatively low strength of the mafic dikes. These dikes cause localized ground control problems when intercepted by mine workings.

24.4.2   In Situ Stress Field

In 1977, the U.S. Bureau of Mines (USBM) performed overcoring tests to determine the state of in situ stresses at Pea Ridge in both the ore zone (2505 level) and in the footwall porphyry (2475 level). The stress isometry shown in Figure 24.2 can be developed from the USBM data (Meramec Mining Co. 1977).

24.4.3   Ground Control

Ground control is achieved through mining sequencing, controlled blasting techniques, and artificial support and rock fixtures. Because the ore is relatively strong and is structurally massive in nature, the ground conditions within the ore body are generally very good. Although there have been limited sill and pillar failures between levels, which prompted the backfilling of some stopes, the original stope and pillar design is structurally sound. While many of the stopes have been standing unsupported for 35 years, the mining sequence is most important to ground control in the recovery of sills and pillars. As a result of experience in sublevel caving gained in mining the western end of the ore body (see Tucker 1981), the pillar recovery sequence is from the hanging wall to the footwall, from west to east, and from the top down. This allows the in situ stresses to move to the east or beneath the stoping complex into unmined ground. Stress relief achieved by mining across the ends of pillars at the hanging wall greatly reduces the ground control problems experienced as mining retreats to the footwall.

TABLE 24.1 Mean values of rock properties

Rock type				Rock properties			
	Specific gravity	Modulus of rupture, MPa (psi)	Young's modulus, GPa (psi × 10^6)	Poisson's ratio	Tensile strength, MPa (psi)	Compressive strength, MPa (psi)	Strain energy, N-m/cm^3 (in-lb/in^3)
Magnetite	4.89	22.0 (3,191)	94.5 (13.70)	0.19	7.5 (1,086)	212.5 (30,823)	30.1 (30.3)
Hematite	4.35	25.5 (3,695)	149.0 (21.61)	0.28	14.3 (2,068)	162.0 (23,490)	41.2 (41.5)
Rhyolite porphyry	2.61	33.5 (4,860)	86.3 (12.52)	0.24	11.6 (1,686)	311.9 (45,231)	78.2 (78.8)
Quartz amphibole	2.70	33.9 (4,922)	120.5 (17.48)	0.16	14.1 (2,046)	309.9 (44,944)	42.1 (42.4)
Mafic dike	2.78	30.9 (4,477)	74.7 (10.84)	0.25	14.7 (2,127)	133.6 (19,370)	11.6 (11.7)

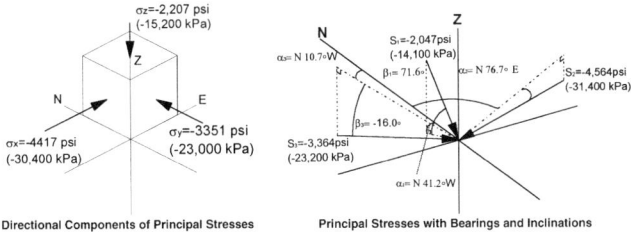

FIGURE 24.2 Isometry of in-situ stresses

Ground control in development drifting is also achieved by employing "smooth wall" blasting techniques. The use of 25-gr detonating cord in the blasting agent column in all back and rib holes (as needed) works well under existing conditions. The detonating cord runs the entire length of these perimeter holes and is inserted with the blasting cap and 8-gm booster before charging with ANFO. The detonating cord, having a faster velocity of detonation, densifies the ANFO just prior to its detonation. This densification reduces the explosive energy at the perimeter of the drift, thus reducing damage to the back and ribs.

As standard practice, rock bolts are installed in all drifts, generally with a combination of 2.1-m- (7-ft-) long, 15.9-mm (5/8-in) diameter, point-anchor bolts with 35 mm (1⅜ in) of mechanical expansion and 35-mm- (1⅜-in-) diameter friction bolts. When conditions dictate, 10-gauge welded wire mesh or metal strapping may be used. Due to acid water conditions within the workings, the mechanical bolts have a slightly longer life expectancy than the friction bolts. Therefore, judgment must be exercised regarding ground and water conditions as well as the life expectancy of the drift being bolted. When an extended life is needed, a 19-mm- (3/4-in-) diameter resin-rebar bolt is used. Although cost varies with the type of bolt used, current average bolt installation costs are $25 per bolt, including material, labor, and equipment maintenance costs.

In areas of weak or overstressed sills, 15.2-mm (0.6-in), 1 × 7 wire-strand cable bolts up to 10 m (30 ft) long have been grouted between sublevels. A 64-mm (2½-in) hole is drilled downward, breaking out in the back of a drift in the sublevel below. The cable is inserted, and a bearing plate and reaction block are attached from below. The cable is pulled up and tensioned to 2 tons. A mixture of flyash and cement is then gravity-fed into the annulus around the cable. The same type of cable is used in conjunction with grouted friction anchors to wrap and secure areas of weak or deteriorating pillars. The combination of these appliances is referred to as a "cable sling."

24.5 MINING METHODS

24.5.1 Mine Development

In general, the Pea Ridge ore body has been developed from the top down using various forms of sublevel stoping and sublevel

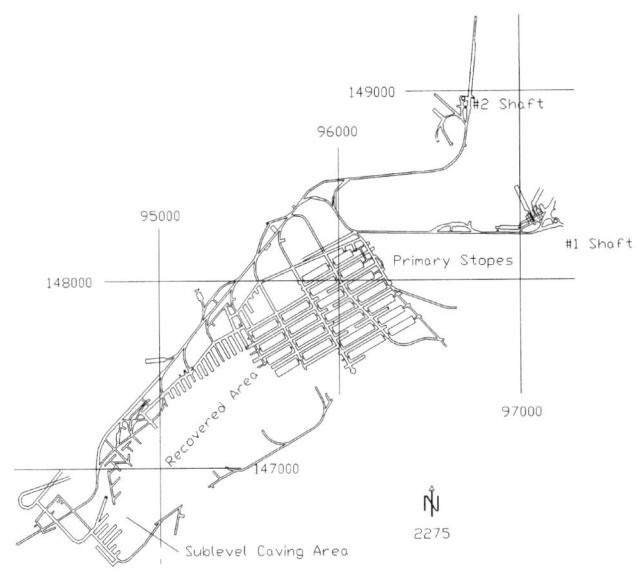

FIGURE 24.3 Mine map of the 2275 level

caving production methods. The mine is entered via two 5.8-m- (19-ft-) diameter, concrete-lined shafts. Both shafts are approximately 762 ft (2,500 ft) deep. The No. 1 shaft is fitted with two 746-kW (1,000-hp), cylindrical, double-drum hoists. The primary hoist moves a 45-person man cage, and the secondary hoist moves a 27-person, triple-deck, auxiliary man cage counterbalanced by a 7-tonne- (8-ton-) capacity rock skip. The No. 2 shaft carries two 16-tonne- (17.5-ton-) capacity bottom dump ore skips moved by two 1,680-kW (2,250-hp) Koepe friction hoists. A second means of exiting is provided at the No. 2 shaft with a 15-person auxiliary cage.

The original ventilation system used shaft fans located near the No. 1 shaft landings on each major level. These fans forced air across each level to the main ventilation raises. The air was then exhausted to the 1375 level and upcast via the No. 2 shaft. Because of the complexity of the mine, this system was difficult to control and maintain. In the early 1980s, the vent system was upgraded by moving the fans from near the No. 1 shaft to ventilation raises added in the footwall. Today, seven large fans having a total of 870 kW (1165 hp) are positioned on various levels throughout the mine. These fans draw approximately 5,660 m³/min (200,000 ft³/min) of ventilation air down the No. 1 shaft and exhaust it through the No. 2 shaft via the 1675 and 1375 levels. Both shafts are located in the footwall on the eastern end of the deposit. Figure 24.3 shows the location of the No. 1 and No. 2 shafts in relation to the production zones.

Seven stoping-haulage levels have been developed between the 1375 and 2475 levels, with partial development on 2675. Intervals are spaced 46 m (150 ft) between the 1375 and

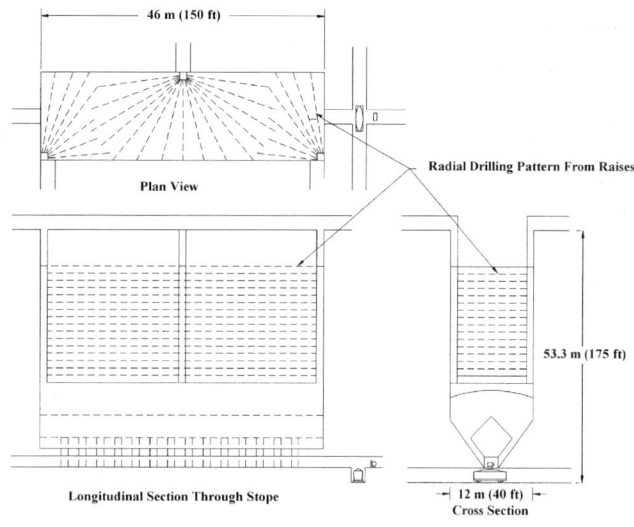

FIGURE 24.4   Modified shrinkage stope

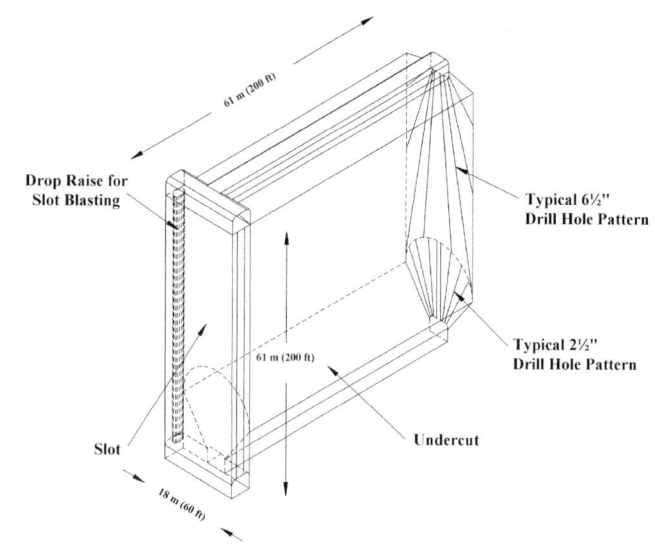

FIGURE 24.5   Mechanized sublevel stope

2275 levels and 61 m (200 ft) between the 2275 and 2675 levels. Mainline haulage, mine services, and facilities are all located in the footwall on the north side of the deposit. In the older sections of the mine, crosscuts are located at stoping-haulage levels roughly 61 m (200 ft) apart.

Most of the upper levels were developed by pneumatic feedleg drilling or pneumatic rail jumbos. Original track haulage drifts were cut 2.7 by 3.3 m (9 by 11 ft). All development is currently performed with pneumatic two-boom jumbos on diesel-trammed carriers, and typical development headings are 4.8 by 4 m (16 by 13 ft). Although drifting costs vary with rock type and ground conditions, the current average is $985 per meter ($300 per foot).

24.5.2   Production Methods

Modified Shrinkage Stoping.   Initial mine production came from modified shrinkage stopes. These stopes measured 46 to 90 m (150 to 300 ft) long by 12 to 18 m (40 to 60 ft) wide by 46 m (150 ft) high, with pillars of similar dimensions left between them. These stopes had various orientations, running either parallel or perpendicular to the strike of the ore body. After mainline and crosscut train drifts were constructed, a 1.8- by 2.4-m (6- by 8-ft) scram (a.k.a. scraper or slusher) drift was cut on-center below the base of the stope just above the train drift. A 2.4- by 2.4-m (8- by 8-ft) undercutting drift was also driven 6 m (20 ft) directly above the scram drift.

The undercutting drift was used for initial longhole drilling access in the base of the stope to create lateral undercuts from which stope ore would be drawn. Draw fingers were then excavated from the scram drift by typically raising and undercutting with small pneumatic feedleg drills. Production drilling was performed from three raises developed in two corners and at the midpoint of the opposing stope wall. Horizontal longholes were drilled in a radial pattern from these raises. Horizontal slices were then blasted into the undercut. Figure 24.4 shows stope geometry.

Only the bulked volume was drawn from the stope, leaving most of the ore in the stope to prevent air blasts from the successive blasting of horizontal slices. Following the last sequence of longhole blasting, a stope would be emptied by wire-rope slushers pulling the ore from the draw fingers into train cars positioned below the scram station.

There were several disadvantages to this type of stoping. Development and production costs were high, ventilation was difficult, most of the ore was left in the stope until final stope blasting, and resource recovery was only 33%. The draw finger structures also presented fragmentation problems during pillar recovery efforts.

Mechanized Sublevel Stoping.   The disadvantages of the modified shrinkage method led to the adoption of a more mechanized sublevel stoping method with one or more sublevels, a slot at one end of the stope, and a typical mechanized undercut. Sublevel drilling drifts provide access for "in-the-hole" production drilling. The slot at the end of the stope provides a vertical free face for end slicing the stope through blasting a radial pattern of 165-mm (6½-in) diameter holes drilled from the sublevels, while the undercut provides a horizontal free face and void for shot movement and expansion.

Haulage level development proceeds as sublevel drifting is being conducted. On the haulage level, stope undercutting is performed by twin-boom fan drills. Undercut patterns are laid out in five to seven 64-mm- (2½-in-) diameter holes with each boom (left and right). Radial angles increase from 0° (vertical) to 30° left and right of center. The ring of drill holes is usually pitched forward 20° from vertical. Undercut blasting commences at the slot and retreats axially beneath the stope. To access the broken ore, drawpoints are driven into the base of the undercut from access drifts driven into the base of the pillars left between stopes (Figure 24.3).

Stope widths and intervals have been maintained at 12 to 18 m (40 to 60 ft) widths (Figure 24.5), but the vertical interval between levels was increased to 61 m (200 ft) because of the increased drilling capability of the downhole hammer drills now being used for production drilling. Productivity has markedly increased because this generation of stopes is amenable to a higher level of mechanization; however, to date, drilling has been done entirely with pneumatic percussive-type drills.

There are several advantages to sublevel stoping at Pea Ridge. Stope development is in ore, which aids in recovering development costs quickly. Grade dilution can be controlled better with open stoping, also resulting in lower costs. The method is amenable to a high level of mechanization and automation. Ore fragmentation can be better controlled, resulting in the need for less secondary blasting.

The disadvantages are that the method results in only partial primary recovery, a complex development system with specialized procedures and equipment is required, extensive development must be completed before stope production can begin, and pillar recovery requires the maintenance of pillar access for extended periods of time.

Sublevel Caving. In hopes of finding a unified total recovery method to replace open stoping, Pea Ridge began experimenting with sublevel caving in 1967. The method was initially tried in the upper portion of the west end of the ore body. In order to use existing sublevels that had been built for adjacent stoping operations, caving was conducted on-strike and retreated east toward existing open stopes. Increasing stress as the caving area approached the open stopes caused ground failure in lower sublevels to the extent that blastholes had to be redrilled, and finally some ore had to be abandoned because of unstable ground.

In the next evolution of sublevel caving, the direction of mining remained on-strike, but was reversed by retreating west away from the stope system. This method essentially called for retreating into the hanging wall and underneath areas where ground failure had already taken place. This approach also proved unsuccessful. A third approach, that of mining perpendicular to strike, proved to be technically feasible. However, the economic result was not equally successful. Additional development of the hanging wall slot, a reduced interval for sublevels, and difficulties in ventilation made the approach unattractive as a production method. By the late 1970s, sublevel caving in unmined ground had been phased out. (For further details on Pea Ridge sublevel caving see Mason 1973 and Tucker 1981).

Pillar Recovery. The experience gained from sublevel caving indicated that the method might be modified to recover the system of massive pillars left after completion of primary open stoping on the upper levels. Pillars and sills were drilled from primary stope access drifts running the length of the base of each pillar. Finger and crown structures left in the base of each stope and the lower portion of the pillar above the access drift were drilled overhead with fan drills and single-boom, tophammer, longhole jumbos using 64- and 76-mm- (2½- and 3-in-) diameter bits. In-the-hole drills were used to drill 165-mm- (6½-in-) diameter holes into the adjacent sills and the upper portion of the pillar below the access drift.

Several different types of explosives and explosives handling systems have been tried over the years. Hand-loading of packaged emulsions, prilled ANFO, and ANFO slurries have been tried in downholes, and pneumatic charging of stick cartridges and prilled ANFO have been tried in upholes. For reasons involving handling packaging waste, charging efficiency, and water resistance, all large blasting projects now use a repumpable emulsion. Currently a cavity pump attached to a pneumatic collar packer is used with bleeder tube arrangement. The system eliminates packaging waste and increases charging efficiency. To reduce preparation time, a mechanized hose-handling system that would allow upholes to be charged from toe to collar, thus eliminating the need for bleeder tube installation, is being considered.

Mucking of broken pillars has been handled in two ways. When pillars are blasted into the primary stopes where the original draw points can be maintained, mucking can be done through the primary development drifts. In cases where the primary draw points cannot be maintained, sublevels are developed 15 m (50 ft) below the designed blastline, and recovery crosscuts are excavated on 12-m (40-ft) intervals. Fan patterns are then drilled upward and angled toward the hanging wall from the face of the recovery crosscut. These fan holes are used to undercut the broken pillars, allowing them to cave. Draw control is maintained from west to east as mining retreats from

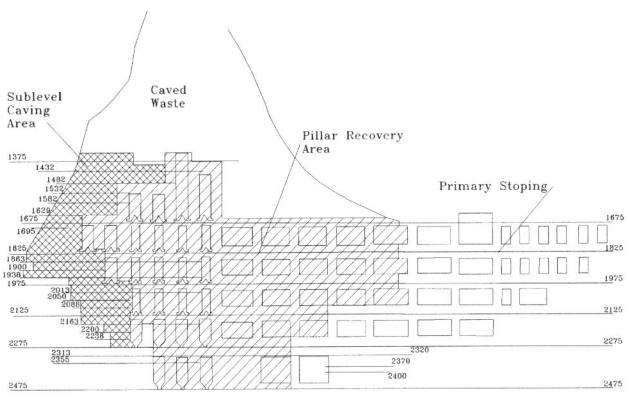

FIGURE 24.6   Longitudinal section through the Pea Ridge Mine

hanging wall to footwall. As many as 20 large pillar blasts have been executed since 1971, the largest of which broke nearly 2,721,500 tonnes (3,000,000 tons) of ore with 484 tonnes (533 tons) of explosives, making it one of the world's largest single blasts. (For additional information on pillar recovery by undercut-and-cave methods see Irvine 1982). The latest pillar shot (1995) required 334,800 kg (744,000 lb) of emulsion, and 4,700 man-hours to prep, charge, and blast. The shot fragmented 1,161,200 tonnes (1,280,000 tons) of ore.

This method of pillar recovery has proven to be technically successful. However, the current production rate of 453,600 tonnes (500,000 tons) per year does not support an attractive rate of return on the extensive drilling and blasting costs that must precede each shot. Although secondary blasting costs are expected to increase, pillar recovery will continue in the near future by simply undercutting to induce pillar caving.

As pillars between the stopes are recovered, all structural support is removed from those areas, allowing the ground to collapse. This has created a large caving zone that consists of a rubblized rock pile in excess of 366 m (1,200 ft) thick, an open cavity up to 90 m (300 ft) in vertical extent, and a solid arched beam 183 m (600 ft) long above the western workings of the mine (Figure 24.6). Although monthly borehole monitoring in the mid-1970s showed that the cavity was losing tens of meters each month, the cavity seems to have stabilized somewhat, forming a voussoir arch.

Panel Stoping. Panel stoping is a mining technique that evolved from a combination of sublevel caving and mechanized sublevel stoping with large-diameter blast holes. A typical panel stoping block is developed on two sublevels and a main haulage level (Figure 24.7). Panel stopes are oriented with their long axis parallel to the strike of the ore body. The upper sublevel provides panel drilling access, and the lower sublevel allows undercut development. Inclined fan drill holes of 64- to 76-mm (2.5- to 3-in) diameter are used to undercut the block in a fashion similar to that in stope undercutting. Blastholes of 165-mm (6½-in) diameter are then drilled in a direction parallel to the long axis of the undercut. The longholes are inclined from a maximum of 12° above horizontal to 45° below. The broken ore is then recovered from the main haulage level 10.6 m (35 ft) below the undercut. Again, fan drill undercutting is used to break the sill between the floor of the original undercut level and the back of the haulage level.

In an effort to maximize productivity, two panels are undercut, drilled, and blasted in each panel stoping cycle. A two-panel cycle will break between 90,700 and 122,500 tonnes (100,000 and 135,000 tons) of ore, depending on local conditions. At current production rates and mucking requirements from the panel recovery level, it takes between 2 and 2½ years to recover

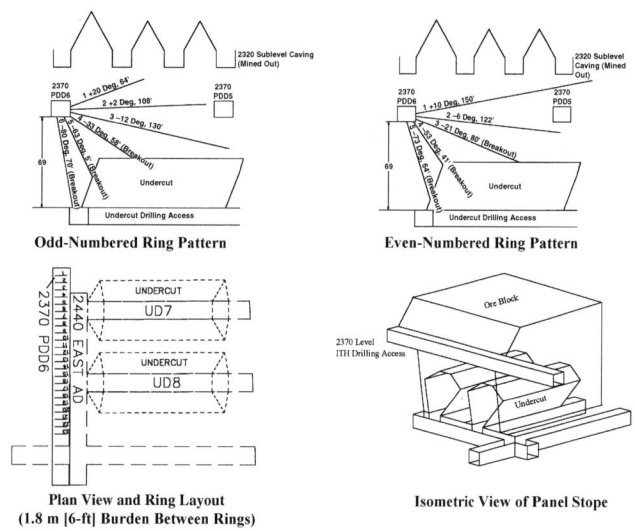

Odd-Numbered Ring Pattern   **Even-Numbered Ring Pattern**

Plan View and Ring Layout   **Isometric View of Panel Stope**
(1.8 m [6-ft] Burden Between Rings)

FIGURE 24.7 Panel stoping diagrams

the ore from two panels. At $0.60 a ton, a two-panel cycle costs roughly $80,000 to drill and blast, not including the costs of development and undercutting.

A new strategy that seems to be more cash-flow friendly is to undercut, drill, and blast one panel at a time. There is a loss in economies of scale, as well as a reduction in undercut blasting and mucking efficiency, but the reduction of operating cash tied up in broken reserves seems to justify this strategy at this time.

24.5.3 Haulage and Material Handling

Originally the mine was designed to hoist run-of-mine ore. Twenty-ton, bottom-dump Swedish cars pulled by electric locomotives on trolleys transported ore from stopes to a skip loading facility at the No. 2 shaft on the 2380 level. Swedish dumps were located on each level adjacent to the No. 2 shaft. These train dumps fed down to the No. 2 skip pocket through a series of conventionally constructed raises. There were also waste handling facilities adjacent to the No. 1 shaft on each level.

From 1968 to 1970, an underground crushing facility was operated on the 2475 level. During that time, a 1-m (42-in) gyrator crusher was installed, and train dumps were constructed on levels above the crushing station. These fed ore to the crusher via ore passes controlled by a 1.8-m (72-in) reciprocating plate feeder.

During the 1970s, a fleet of rubber-tired, diesel-powered, load-haul-dump (LHD) units have evolved to replace the trains. Original units (Wagner ST2-Bs, ST5-As, and ST5-Bs) were then utilized. Currently ST8s are the primary ore-handling unit. Ore is mucked from drawpoints, trammed to ore passes that feed into the train dump system, and fed to the crusher on the 2475 level. It is normally crushed to 15-cm- (6-in-) minus. From the crusher, ore is carried via a 549-m- (1,800-ft-) long, 1-m- (42-in-) wide conveyor to one of two available storage bins. From these bins,

ore is fed to the No. 2 skip pocket by another 1-m- (42-in-) wide conveyor 49 m (157 ft) long. For more information on the underground crushing station, refer to Irvine (1972).

24.6 FUTURE PLANS

Future plans focus on upgrading mine infrastructure and equipment. Recently (1998), the No. 1 shaft was refitted with a 30-cm- (12-in-) diameter, schedule 40XH, mine water discharge pipeline. Plans are currently being made to rehabilitate the central ore pass to the underground crusher and its feeding system. In 2000, the mine will acquire a mechanized ANFO charging vehicle, and financial ground is being laid for the acquisition of a new electric-hydraulic two-boom jumbo in 2001. Beyond these near-term plans, drawpoints are being moved farther away from the central ore pass, which requires that the LHD fleet tram distances beyond their most efficient capacities. Therefore, a system of secondary ore passes and truck chutes are being evaluated to reduce the cost per ton of ore transported.

24.7 REFERENCES

Emery, John A. 1968. Geology of the Pea Ridge iron ore body. In: *Ore Deposits of the United States, 1933–1967*, ed. John D. Ridge, Vol. 1, Chapter 18. New York: AIME.

Hussman, James R. 1989. Gold, rare earth element, and other potential by-products of the Pea Ridge iron ore mine, Washington County, Missouri. *Contributions to Precambrian Geology 21*. Missouri Department of Natural Resources, Division of Geology and Land Survey. Open File Report 89-78-MR.

Irvine, J.C. 1972. Development, installation, and effect of an underground crushing and conveying system at Pea Ridge. *SME Transactions* - Vol. 252. New York: SME/AIME.

Irvine, J.C. 1982. Recovery of pillars between blasthole shrinkage and sublevel stopes at the Pea Ridge mine. In: *Underground Mining Methods Handbook*, ed. W.A. Hustrulid. Chapter 10. New York: SME/AIME.

Kisvarsanyi, Eva, B. 1981. Geology of the Precambrian St. Francois terrane, southeastern Missouri. *Contributions to Precambrian Geology 8*. Missouri Department of Natural Resources, Division of Geology and Land Survey. Report of Investigations 64.

Marikos, Mark A., Laurence M. Nuelle, and Cheryl M. Seeger. 1989. Geology of the Pea Ridge Mine. In: *Olympic Dam-Type Deposits and Geology of Middle Proterozoic Rocks in the St. Francois Mountains Terrane, Missouri*, eds. V. Max Brown, Eva B. Kisvarsanyi, and Richard D. Hagni. Society of Economic Geologists. Guidebook Series Number 4, pp. 41–54.

Mason, J.M. 1973. Practical aspects of sublevel caving. *Mining Magazine*. January 1973, pp. 12–14.

Meramec Mining Company. 1977. *In situ stress field at the Pea Ridge Mine*. Final Report on USBM Contract J0265041. Bethlehem, PA: Bethlehem Steel Corporation.

Nuelle, L.M., C.M. Seeger, W.C. Day, and G.B. Sidder. 1991. Rare earth element- and gold-bearing breccia pipes of the Pea Ridge iron ore mine, Washington County, Missouri. Preprint Number 91-109, 1991 SME Annual Meeting, Denver, Colorado. Littleton, Colorado: SME.

Tucker, Larry J. 1981. Sublevel caving at Pea Ridge. In: *Design and Operation of Caving and Sublevel Stoping Mines*, ed. Daniel R. Stewart. Chapter 28. New York: SME/AIME., pp. 387–389.

Tucker, L.J., and D.P. Roberts. 1991. Development of open stoping for rare earths at Pea Ridge. Preprint Number 91-22, 1991 SME Annual Meeting, Denver, Colorado. Littleton, Colorado: SME.

CHAPTER 25

Sublevel Stoping at the Williams Mine

David Bronkhorst[*] and Greg Brouwer[*]

25.1 INTRODUCTION

25.1.1 Description of Property

The Williams Mine is one of three operating mines in the Hemlo gold deposit and is Canada's largest gold mine, with an annual production of 2.4 million tonnes and 424,000 ounces. The deposit is located 40 km east of Marathon, Ontario, along the north shore of Lake Superior (Figure 25.1). It consists of a highly mechanized underground operation and a small open pit.

25.1.2 Mining History

Production from the A-Zone open pit at the east end of the property began in mid-1985. The sinking of the A-Zone shaft augmented production from the A-Zone pit with underground ore from the same area to sustain an initial mining rate of 3,000 tonnes/day. The completion of the main shaft, the B-Zone infrastructure, and a mill expansion in 1988 facilitated an increase to 6,000 tonnes/day. In 1999, the mill was once again expanded and currently has a through-put of 8,500 tonnes/day. Mill feed includes ore from the Williams Mine's underground and open-pit operations as well as custom milling of ore from the neighboring David Bell Mine.

Current production is met from three mining blocks within the main zone (B-Zone). A fourth zone (C-Zone) is scheduled to come into production in 2001 (Figure 25.2). A small amount of ore is also supplied by the C-Zone pit at the west end of the property. The A-Zone shaft has been decommissioned and is now a ventilation exhaust raise.

25.1.3 Safety

In 1999, the mine achieved 1 million working hours without a lost-time accident, a significant milestone for the property.

25.2 GEOLOGY

The Williams Mine lies on the south side of the east-west-striking Heron belt of metamorphosed Archean-age rocks. The ore body lies along the contact between overlying metasedimentary rocks and underlying felsic volcanic rocks and dips north at 60° to 70° (Figure 25.3). The thickness of the ore body ranges from 3 to 45 m, depending on location. It is composed of a fine-grained, feldspar-rich ground mass with fracture-controlled pyrite and barite mineralization zones running parallel to foliation. The ore is intruded by late felsic dykes and may also contain muscovite schist bands.

Hanging wall rocks consist of banded, fine-grained, pelitic sedimentary rocks. A weak muscovite schist band that varies from 1 to 15 m thick often separates the ore body and the meta-sediments. The footwall rock is composed of quartzite-muscovite schist or felsic quartzite porphyry.

FIGURE 25.1 Location of Williams Mine

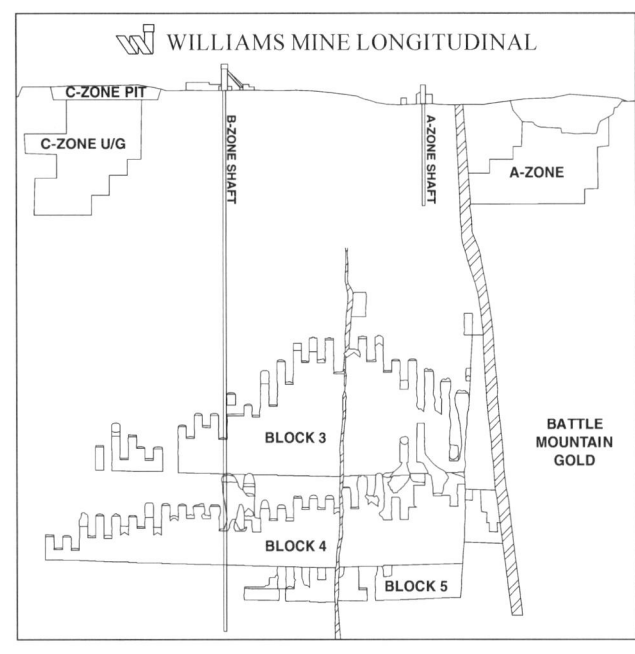

FIGURE 25.2 Longitudinal view of mine

---

* Williams Mine, Marathon, Ontario, Canada.

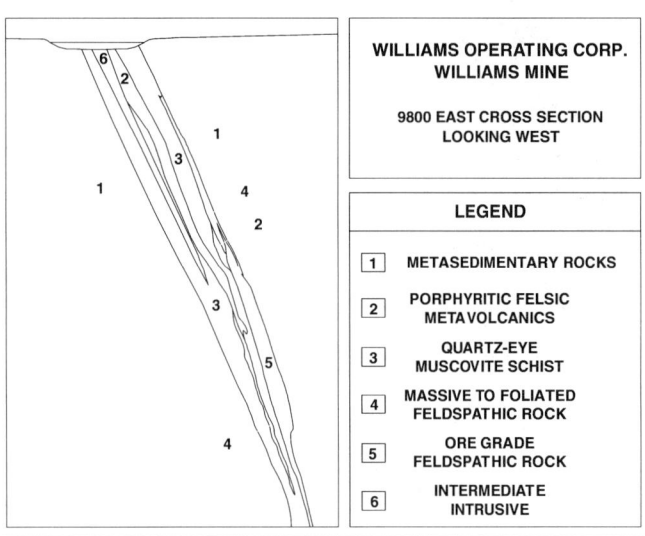

WILLIAMS OPERATING CORP.
WILLIAMS MINE

9800 EAST CROSS SECTION
LOOKING WEST

LEGEND

1	METASEDIMENTARY ROCKS
2	PORPHYRITIC FELSIC METAVOLCANICS
3	QUARTZ-EYE MUSCOVITE SCHIST
4	MASSIVE TO FOLIATED FELDSPATHIC ROCK
5	ORE GRADE FELDSPATHIC ROCK
6	INTERMEDIATE INTRUSIVE

FIGURE 25.3   Geologic section

25.3   GEOTECHNICAL

25.3.1   Rock Mass Classification

The initial rock mass classification program was undertaken in 1983. Discontinuities were mapped in outcrops, highway cuts, and surface sampling trenches. Three major discontinuity sets were recognized—east-west-striking foliation parallel to the ore body, a north-south-striking subvertical joint set, and a weaker, east-west-striking subhorizontal joint set. Mapping the underground openings as they were excavated confirmed the existence of these discontinuities throughout the ore body.

Laboratory tests to determine the strength of the intact rock was initially carried out on core selected from diamond-drill exploration holes. Subsequent tests have been carried out on core obtained at depth during mining.

The major sequences of hanging wall and footwall rocks show high strengths. The band of sericitic schists containing the ore body is only of medium strength, with a gradational increase toward the center of the ore body. Strength varies significantly within the same rock unit, depending on the degree of silicification.

Values for the Norweigan Geological Institute (NGI) Q value (Barton 1976) and the Council for Scientific and Industrial Research (CSIR) rock mass rating (RMR) (Bieniawski 1976) were determined from underground mapping and rock quality designation (RQD) values. Rock mass strength parameters "m" and "s" (Hoek and Brown 1980) were also determined. Representative rock mass properties are summarized in Table 25.1.

25.3.2   In Situ Stress

In situ stress measurements were made at depths of 900 and 1,000 m employing the overcoring technique and using diametral displacement gauges from the U.S. Bureau of Mines and hollow inclusion strain cells from the Commonweath Scientific and

Industrial Research Organization (CSIRO). The combined results from 16 successful tests showed that vertical stress is slightly greater than calculated overburden weight. Also, the ratio of the average horizontal stress to vertical stress ranges from 0.9–1.1 to 1, meaning that the in situ stress regime at depth in the Williams Mine is approximately isotropic.

25.3.3   Support

A stringent property-wide ground control standard has been established to ensure the safety of personnel, protect equipment, and assure continuity of operations. The support system varies according to the nature of the excavation, traffic flow, and local ground conditions.

All footwall access drives, footwall drives, cross cuts, and auxiliary workings are supported by 1.5-m-long, 19-mm-diameter grouted rebar. These supports are installed in a 1.1- by 1.1-m pattern in the back and upper walls. These areas are further secured by the installation of No. 6-gauge, 10-cm^2 mesh screen along with the rebar. All rebar and screens are installed during initial development.

Production stopes are also supported as development progresses. In these areas, in addition to the rebar and screen support, double-strand, 7- to 10-m-long, 1.6-cm-diameter cable bolts are installed on an engineered pattern that varies with ground conditions and stope size.

Rock support is installed using one of three types of equipment. A Tamrock H400/24 platform bolter and a Boart Stratamaster R712 basket bolter are used to install the resin rebar, and an Atlas Copco Simba H254 cable bolter is used for cable bolts.

25.4   MINING METHOD

25.4.1   Basic Mine Layout and Development

The Williams ore body is characterized by its uniformity, tabular shape, and steeply dipping orientation. As such, the ore body lends itself well to longhole open stoping.

Most mining activities are accessed by the main shaft, a 7-m-diameter, concrete-lined shaft collared at the 1,0340-m elevation and bottoming at the 903- m elevation. Total depth is 1,310 m. The shaft has five compartments that house two 22-tonne bottom dump skips, a 54-person main cage, an eight-person auxiliary cage, and a counterweight. The shaft accesses 10 main levels. The loading pocket is located on the 9065 level, and a ramp accesses the shaft bottom located at the 9030 elevation. The main level interval is typically 105 m in the production areas.

An internal ramp system interconnects all main levels and sublevels between the 9135 and 9870 levels. With an average grade of 15% and a typical profile of 4.8 by 3.5 m, this ramp facilitates efficient movement of workers, materials, and equipment between the main levels and sublevels. Materials are delivered to the main levels via the shaft, then distributed to the sublevel work areas through the internal ramp system.

The internal ramp also serves as an escapeway to the surface by way of the Golden Giant internal ramp to the A-Zone ramp and portal.

TABLE 25.1   Rock mass properties

| Location | Rock type | NGI (Q) | CSIR (RMR) | Hoek and Brown 1980 | | Unconfined compressive strength, MPa |
				m	s	
Footwall	Quartz-muscovite schist	3.9	56.3	2.52	0.0077	84
Ore	Baritic feldspathic	3.4	55.1	2.82	0.0068	100
Hanging wall	Muscovite schist	1.3	46.4	1.47	0.0026	85
Hanging wall	Banded sediments	3.5	55.4	3.05	0.007	115

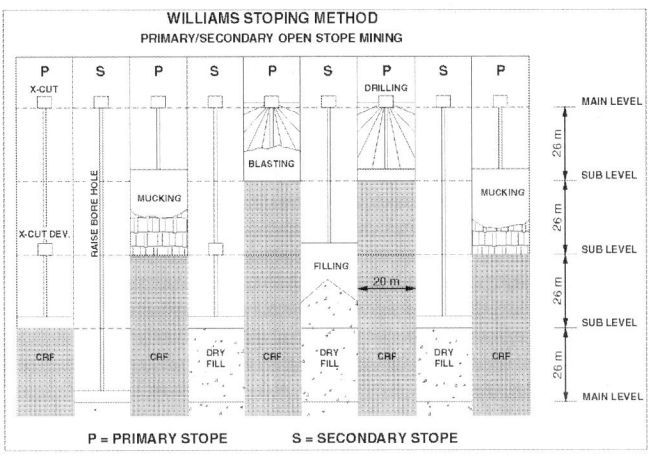

FIGURE 25.4 Stoping methods

25.4.2 Stoping Methodology

Early and ongoing geotechnical studies confirm optimum stope dimensions of 20 m wide and 25 m high for both primary and secondary stopes. Larger primary stopes are paneled, and smaller primary stopes and secondary stopes are mined as single panels. Cemented rock fill in the primary stopes and dry fill in the secondary stopes allow for 100% ore recovery (Figure 25.4).

Stopes are sequenced to maintain a triangular shape to the mined-out area by mining vertically with a lead stope, then outward along the rill of the triangle toward its base. By careful scheduling, adjacent primary stopes are mined and filled for two vertical lifts before mining of the secondary stope between them is started.

25.4.3 Drilling

Most of the production drilling requirements at the Williams Mine are met by a fleet of nine Cubex 5200- and 6200-series in-the-hole (ITH) drill rigs. Typically, six or seven of these rigs will be drilling in the mine at any given time. There are also two Atlas-Copco Simba electric hydraulic tophammer drills operating on-site. Both units (H4353 & H274) drill a 110-mm-diameter blasthole and utilize a rod-handling carousel. These units provide superior penetration rates and higher mobility than the ITH drills.

In most cases, the production drills operate on a fan pattern with a typical ring burden of 2.8 m and a toe spacing of 3 m. The average hole length is between 24 and 26 m. Based on the above design parameters, drill factors generally run at approximately 11.0 tonnes/m drilled in both primary and secondary stopes.

The ITH rigs are generally scheduled to drill 20,000 tonnes/month, whereas the tophammer drills are scheduled at 30,000 tonnes/month.

25.4.4 Blasting Description and Equipment

Production blasting is generally carried out in two separate lifts in both primary and secondary stopes. Blasts are initiated around a 1.1-m-diameter borehole located in the center of the stope. The hole initially serves as a fill raise for the stope below, then as a slot for blasting. Holes are charged with (ammonium nitrate and fuel oil) (ANFO) and tied-in immediately before the blast. Typical powder factors for production blasts range between 0.5 and 0.6 kg/tonne, depending on stope size and drill pattern.

Once a week, blasts over 1,000 kg are set off. No Saturday night shifts are scheduled in order to minimize lost time resulting from gas checks. A gas check is a safety procedure that follows all production blasts. Before the mine is re-entered, trained personnel use gas detection equipment to check air quality in all work areas to ensure that the mine is clear of blasting gases.

25.4.5 Mucking

Approximately 10 active stopes are required at any given time to meet production requirements. Six-yard JS600 and seven-yard Elphinstone R1500 scooptrams are used to muck the stope drawpoints and haul ore to one of five ore passes equally spaced along the strike of the ore body. To ensure the safety of the workers, scooptrams are operated by remote control in the stope once the brow is exposed at the drawpoint. JDT 426 trucks capable of hauling 16 tonnes are available for longer hauls that may occur from time to time.

25.4.6 Backfilling

The high-quality backfill placed in the primary stopes permits 100% ore recovery. All primary stopes are filled with cemented rockfill. When adjacent secondary stopes are being mined, fill walls in excess of 25 by 25 m are exposed and must stand until the secondary stope is completed. Similarly, when a primary stope footwall panel is being mined, the fill wall exposed at the hanging wall panel must remain competent throughout mining of the footwall panel.

Rockfill is generated at the west end of the property in the C-Zone open pit. Prior to delivery underground, an Allis Chalmers 42-65 gyratory crusher crushes this material. After crushing, the rockfill is delivered by conveyor to one of two surface backfill raises. The backfill raises deliver the rockfill to the top of each mining block. The material is then loaded into JDT426 trucks at the truck chute and dumped into 1.1-m-diameter boreholes positioned at the top of each stope to be filled.

The cement-flyash slurry is produced on the surface and delivered in batches by borehole and pipeline to the backfill truck chute areas underground. Slurry is sprayed onto the rockfill as each truck is loaded, and the mixture is delivered to the borehole for placement.

Optimum pulp density of the slurry is 56% by weight, of which 55% is flyash and 45% is portland cement. Overall binder content in the backfill is 4.2% by weight for primary stopes and 6.3% for any stope where mining may take place below it. Slurry is not generally required when filling secondary stopes, with the exception of a firm mucking floor. In such cases, a 5.0% binder content is used.

25.4.7 Crushing and Conveying Description

There are four Eagle jaw crushers on the 9175 sublevel that are responsible for sizing. Muck is reduced to 8-in nominal size while passing through the 36- by 48-in jaws. Once crushed, the material flows onto the conveyor system.

The conveyor system is located on the 9135 level. It consists of three sections of 1.1-m-wide conveyor and exceeds 750 m in overall length. Rated at 600 tonnes/hr, the system delivers ore from beneath the crushers to one of two 5,000-tonne-capacity storage bins near the shaft between the 9135 and 9065 levels.

Material from the bins is conveyed a short distance to the loading pocket along a 1.65-m-widebelt on the 9065 level. A fully automated hoisting system transports ore from the loading pocket to surface in two 22-tonne bottom dump skips.

25.4.8 Stope Development and Equipment

Mine development is an ongoing activity at the Williams Mine. About 50 development miners are involved in a continuing program to establish infrastructure and develop stope areas for production. Contractor services have been engaged for specialized projects and remote work areas as required.

Lateral development is composed primarily of main level and sublevel drifting. Main levels are driven from the shaft to within 30 m of the ore zone. At that point, a footwall drift is driven east and west along strike. The ore is accessed with cross cuts driven into the ore at 20-m intervals. Sublevels are identical to the main levels, but do not have a shaft access drift.

All lateral development drilling is done using Atlas Copco H227, two-boom electric hydraulic jumbos. Typical footwall drift dimensions are 4.8 m wide by 4.3 m high. Typical cross-cut dimensions are 4.7 m wide by 3.8 m high. Both drift types are drilled to a depth of 4 m. Development holes are charged using a Getman A-64 bulk ANFO truck and initiated using long-delay, nonelectric detonators.

All mucking is done using Jarvis Clark JS600 and Elphinstone R1500 scooptrams. Muck is hauled to either an empty stope, the shaft waste pass, or an ore pass, depending on the type of material. Jarvis Clark JDT426 trucks are available for longer hauls.

Lateral development and stope preparation crews attain a typical development efficiency of 0.55 m per manshift, with a total mine wide advance of 7,200 m per year.

Backfill raises and rock passes are excavated using Alimak raise climbers. A typical raise profile of 2.1 by 2.1 m is drilled to 2.4 m deep with a Secan stoper. Blasting is done using ANFO and nonelectric detonators.

When excavating ore passes, a typical raise length of 105 m will also involve excavating short 20-m finger raises at each sublevel to act as ore dumps. Finger raises are accessed by the Alimak as the main raise progresses, but are driven by conventional means.

A Redbore 40 production raise bore machine operates on site to bore 1.1-m-diameter raises in each stope. The unit operates from the top of each mining block and bores the full 105-m interval from the bottom of the mining block to the top. The upper part of each hole acts as a fill raise while the bottom acts as a slot and is blasted away with each successive production blast.

The raise bore productivity averages 10 m per shift for piloting and 3.5 m per shift for reaming.

25.5 PRODUCTION STATISTICS

The Williams Mine enjoys good productivity as a result of the longhole open stoping mining method. The average stope size mined in 1999 was 20,000 tonnes. To meet production targets of approximately 2,000,000 tonnes/yr from underground, 98 stopes were mined. An additional 400,000 tonnes of ore was supplied to the mill from the open pit. This productivity is achieved with a workforce of 605 people distributed as follows:

Administration	46
Engineering and geology	32
Mine (underground)	302
Maintenance	159
Open pit	29
Mill	37
Total	605

25.6 HEALTH AND SAFETY

The health and safety of employees is of paramount concern. Through the focused efforts of all employees, the mine incurred the lowest accident frequency of all mines in Ontario in 1999. As well, over 1,000,000 working hours without a lost-time accident was achieved early in 2000.

Department heads are responsible for safety in their area. To assist a department head in monitoring and controlling safety issues, two elected safety worker inspectors with full-time duties are assigned as part of a safety department. In addition, production and maintenance foremen have an elected health and safety representative on their crews. The purpose of these representatives is to deal with any safety issues that may arise

during the shift, conduct inspections with the foreman, and meet monthly with the department head to discuss concerns.

A five-point safety system is used at the mine in addition to a program called "100% Compliance." The 100% Compliance Program strives to have all employees—both management and workers—follow all regulations, standards, and procedures 100% of the time. Although the mine has yet to achieve the goal of 100% compliance, the program works well in reminding all employees to work safely.

To promote safety further, employees are paid a monthly incentive based on a number of factors, one of which is safety. Employees are assigned a base incentive related to their work environment. For example, an underground production driller has a base incentive of 40% of his hourly wage compared to a mill worker, who has 10%. The base incentive will increase or decrease each month as the mine performance factor changes. The mine performance factor is based on safety, costs, tonnage, and gold production relative to budget. Safety is an integral part of the mine performance factor and is compared to the average safety performance of Ontario's gold mining sector. Historically, safety has been a significant factor contributing to an employee's wages.

25.7 OPERATING COSTS

Operating costs for 1999 were approximately $300 Canadian per ounce (roughly $205 U.S. per ounce) of gold produced. Costs by area are as follows:

Mining (underground)	69%
Open pit	8%
Milling	14%
Administration	9%

Maintenance costs are allocated to the user department, and administration costs include engineering and geology costs. Total cost-per-tonne has remained relatively constant since 1990 at approximately $50 per tonne. This has been achieved through cost containment and continuous improvement programs.

25.8 FUTURE PLANS

As mining has progressed toward the boundaries of the ore body, the ore body itself is becoming narrower. The change in ore thickness has necessitated a re-evaluation of current mining practices. Cross-cut development can not be justified with smaller stopes and lower grades. Therefore, several areas of the mine are now under development for longitudinal retreat mining. One-hundred-percent ore recovery is still achievable with the use of cemented rockfill, but development costs drop drastically.

The primary disadvantage of longitudinal retreat is lower productivity because fewer stopes are available in a longitudinal sequence. However, because a relatively small percentage of ore produced from the underground operations will come from these longitudinal stopes, reduced productivity is not a concern. With this new method, the opportunity to mine out the remaining underground reserve profitably is now possible.

25.9 ACKNOWLEDGMENTS

The authors thank Williams Operating Corp., Teck Corp., and Homestake, Inc., for permission to prepare and publish this paper.

25.10 REFERENCES

Barton, N. 1976. Recent Experiences in the Q-System of Tunnel Support Design. Proc. Symp. on Exploration for Rock Engineering. Johannesburg, South Africa, pp. 107–117.

Bieniawski, Z.T. 1976. Rock Mass Classification in Rock Engineering. Proc. Symp. On Exploration for Rock Engineering. Johannesburg, South Africa, pp. 97–106.

Hoek, E., and E.T. Brown. 1980. Underground Excavations in Rock. London: Institution of Mining and Metallurgy.

CHAPTER 26

Sublevel Open Stoping—Design and Planning at the Olympic Dam Mine

Soma Uggalla*

26.1 INTRODUCTION

The Olympic Dam operation is located in the Stuart Shelf Geological Province approximately 560 km north-northwest of Adelaide in South Australia (Figure 26.1). The operation was named after a nearby livestock watering dam built by a local pastoralist to commemorate the 1956 Melbourne Olympic Games. The purpose-built town of Roxby Downs is situated 16 km to the south of the operation. The deposit was discovered by WMC in 1975. A joint venture was formed in 1979 between the BP Group (49%) and WMC (51%) to develop the discovery. In 1986, the joint ventures committed to Stage 1 of the development of the mine, that is, to produce approximately 45,000 tonnes of copper and 1,200 tonnes of uranium oxide (U_3O_8) a year. These facilities were commissioned in 1988, and underground production commenced using the sublevel open stoping mining method at a rate of 1.5 million tonnes a year.

Two small developments were undertaken subsequently to increase the underground production capacity to 3.4 Mtpa producing 85,000 tonnes of copper and 1,700 tonnes of U_3O_8 a year. In 1996, WMC committed to a further expansion to increase output to 200,000 tonnes of copper and 4,500 tonnes of U_3O_8 a year. The total capital investment of this expansion was A$1.94 billion. The underground expansion included 36 km of lateral development, a railway haulage system to deliver ore from a series of ore passes to a new crusher station, and a significant increase in ventilation capacity and mine services. In the latest underground expansion, the mining method essentially remained unchanged. The ore-handling system has incorporated some recent advances in technology, improving efficiency and safety. Planned underground production of 9.1 million tonnes of ore for the year 2000 has positioned Olympic Dam as the largest underground mine in Australia, with the ninth largest copper deposit and the largest single source of uranium in the world today (Table 26.1).

Access to the existing mine is through a surface decline and Whenan shaft. The Whenan shaft was originally sunk as an exploration access and later upgraded for hoisting. The 4-km-long service decline connecting to the Whenan shaft was later constructed to accommodate the increase in service demand. The Robinson shaft was constructed in 1995 to address the increase in development tonnage. As a part of the recent expansion, a new shaft named the Sir Lindsay Clark shaft was sunk and fitted with the largest mine winder in Australia in terms of power (6.5 MW) and hoisting capacity (1,375 tonne/hr).

26.2 GEOLOGY AND RESERVES

The ore body consists of a large number of semidiscrete mineralised zones scattered throughout an area 7 km long, 4 km wide, and over 1 km deep. The ore-bearing zone has a 350-m-

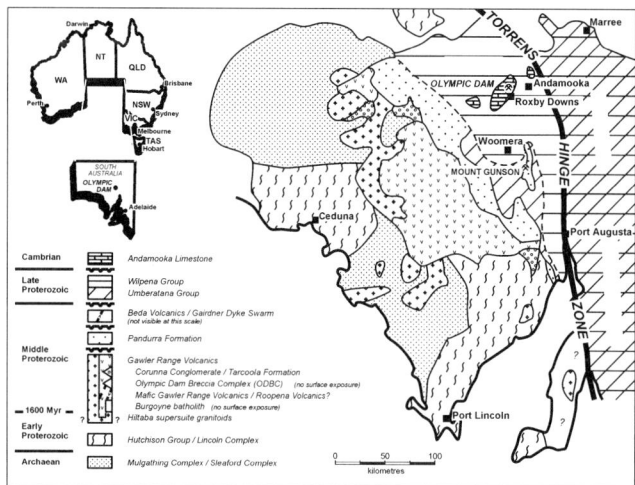

FIGURE 26.1 Geologic map of Olympic Dam Mine region

TABLE 26.1 Mining statistics (anticipated) for 2000

Description	Amount
Ore hoisted, tonnes/month	760,000
Underground development, m/month	1,100
No. of producing stopes per month	24
Stope filling rate, m³/month	158,000
Average stope size, tonnes	300,000
Average stope production rate, tonne/month	30,000
Average stope production time, months	10
Average stope filling time, months	1
Average curing time, months	3

thick overburden of barren sediments. As of December 1999, proven and probable ore reserves are 605 million tonnes averaging 1.8% copper. This means that a very long life can be expected from the Olympic Dam operation, even at the greatly expanded production rate.

Initial exploration was based on a conceptual model for the formation of sediment-hosted copper deposits and modelled gravity and magnetic anomalies. Based on drill hole data, point estimates of grades were generated for the centres of 5- by 5- by 10-m blocks throughout the ore body. Estimates were generated

---

* WMC Resources, Ltd., Olympic Dam Mine, Australia.

for copper, U_3O_8, gold, silver, and sulfur and for density. Once this process is completed, contours can be generated for individual metal concentrations and for total combined dollar values for the metal in situ. Based on long-term predictions of metal prices and exchange rates, individual metal concentrations in each block are factored and summed to give a combined dollar value. Currently the $70 in situ value contour is taken as the stope design cut-off value. The average dollar value of each stope has to be greater than $70 to be included in the proven ore reserves.

26.3 DESIGN AND PLANNING

26.3.1 Geotechnical Considerations

Rock Mass Conditions. Material properties of the intact rock of the Olympic Dam Mine area have been determined from more than 200 laboratory tests. Generally, increasing iron content in the granite-hematite breccia suite increases uniaxial compressive strength (UCS), stiffness, and P-wave velocity. The relationship between rock density (data collected from all drill core), hematite content, and UCS has been used to develop a three-dimensional model of estimated UCS for the resource area. Some localised variation in intact material properties may be related to rock matrix, sericite, or silica content.

Evaluation of drill core logs indicates that the mean structural spacing is greater than 6 m, which is consistent with underground mapping and observations. Because of the relatively low density of structures, the general rock mass condition for the resource area is "massive." Jointing is uncommon in most areas of the mine; however, discrete fault structures have been identified. The most significant set of these typically strike north, dip moderately west or east, and have sericite infilling at <1 cm. The most continuous of these structures has been traced more than 300 m. All structures logged in drill core have been assessed by their condition and categorised by their likely geotechnical significance. This information has been made available in a three-dimensional format for the resource area. The likely occurrence of structures for each 10 m of drill core has been established and used to develop a three-dimensional model of estimated relative structural density for the resource area.

Continuous natural structures that may reduce excavation predictability are increasingly being digitized for further analysis. A process to transfer data about geological structures from drafts to three-dimensional digital surfaces has been developed.

The major controlling structure at the mine is an unconformity lying approximately 345 m below the surface between weaker sediments above and stronger igneous host rock below. This has been modelled using information from more than 900 drillholes. The models have been used to gauge past performance under varying conditions to improve the predictability and efficiency of future stope and development design and also to refine reinforcement and support requirements.

Because of the massive nature of the rock mass, standard classification methods (Q, rock mass rating [RMR], rock quality designation [RQD], etc.) are limited when evaluating the performance of rock masses with varying excavation geometries and sizes. Therefore, the models are designed to characterise rather than classify ground conditions and include the two components of rock mass strength—material strength and the degree of natural breaks. However, the mean RMR classification would be greater than 80, which is "very good rock."

With few exceptions (related to greater structural density or weaker material properties), mine excavations are generally dry. In situ rock temperatures range from 30 to 45°C.

Stress Conditions. Virgin stress conditions in the rock of the Olympic Dam resource area have been assessed from nine measurements and over 30 observations of raisebore break-outs. They are comparable in magnitude with those of most Australian mines. The major principal stress is horizontal and approximately 2.5 times vertical stress (minor principal stress), which may be approximated from the weight of the overlying rock column. Indications are that the major principal stress is oriented north-northwest, subparallel to the ore body at depth. Analysis of mining-induced stress changes are undertaken as required using a boundary-element analysis code (map3d). A whole-mine indicative stress change model is being developed.

26.3.2 Computer-Aided Mine Design

The Olympic Dam operation currently has a combined mine planning system using Microstation and Datamine software. Both of these software packages run on Unix workstations. The Windows NT version of Datamine has been evaluated from both a user and a performance perspective, and as a result, all the Unix workstations are to be replaced with a new NT version. Considerable in-house programming has been undertaken to utilise the capabilities of both of these packages and to ensure the efficient exchange of design information between them. Datamine and Microstation designs can also be linked to the geology database and survey files to extract drilling and survey information.

Mine Works Planner and XPAC Autoscheduler, together with other built-in programs, are currently used for planning and scheduling purposes. Arrangements are being made to link the schedule to the Central Ore Reserves Database so that new additions or modifications made to the existing stopes will automatically be reflected in the production schedule.

All designs are currently done using the system as described above, thereby minimising the need for paper-based designs. Assessment of three-dimensional models, as well as the preparation and evaluation of alternate design options, are the major advantages of this system.

26.3.3 Stope Design Parameters

For the purposes of design and planning, the ore body has been divided into a number of mining areas. These areas have been defined by their relative geographic locations. In each mining area, stopes are identified by different colours, such as blue, green, red, purple, and brown.

Once the underground diamond-drilling and grade-analysis process is completed for each area, a Datamine model is generated. Based on this model, preliminary stopes, perimeter accesses, and other infrastructure, such as ore passes and ventilation raises, are designed.

Sublevel open stoping is the mining method used to extract ore. Intermediate-level intervals vary from 30 to 60 m. Stope length along strike is largely based on mineralisation, geologic structures, geotechnical issues (such as in situ stress distribution), stope geometry, and stope filling. The maximum transverse width (across strike) and length have been determined as 60 and 35 m, respectively. A typical stope indicating planned development is shown in Figure 26.2.

In general, the stope crowns are domed to maximise crown stability. A flat crown with a 10-m span is supported by cable bolts, if necessary. Perimeter drives are positioned a minimum of 15 m away from stopes.

Based on mine planning procedures, stopes are designed by mine design engineers in consultation with the area mine geologist and presented to the operating personnel. The presentation is required to gain formal approval from the underground production, development, and services departments and provides a forum for continuous improvement. A final document incorporating recommendations will then be issued. This document will include the following details:

- A three-dimensional view of the stope with planned and actual development.

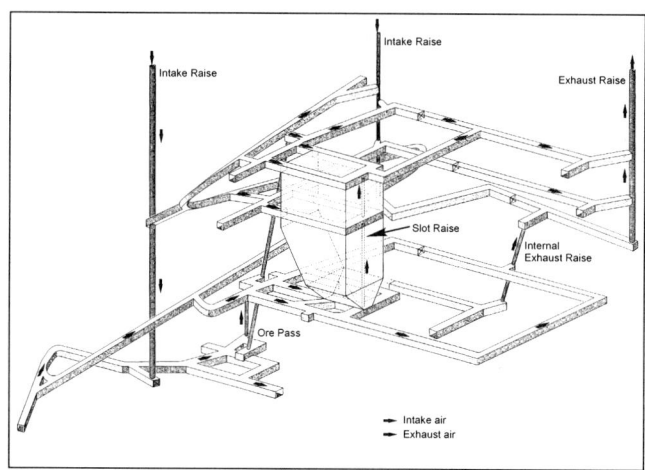

FIGURE 26.2 Stope with planned and actual development

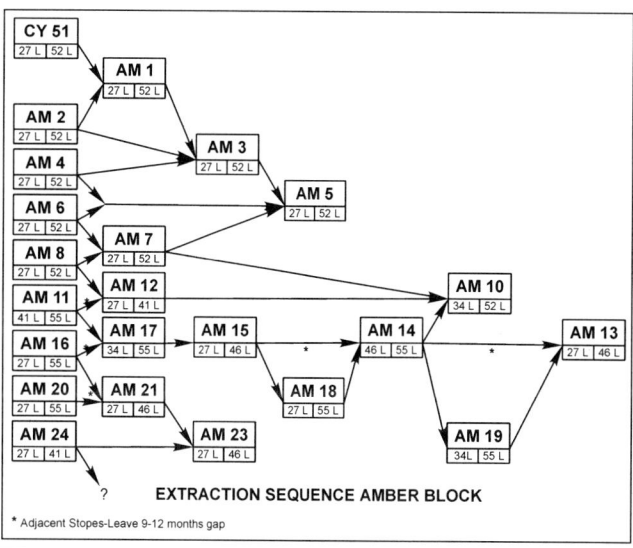

FIGURE 26.3 Typical stope sequence

- Development level plans indicating actual and planned development.
- A summary of lateral and vertical development.
- Stope sequencing in the area.
- Design considerations describing surrounding stopes and fill masses in detail.
- Geological information showing copper and U_3O_8 grades, copper:sulphur ratio, and structural controls (to be provided by the area mine geologist).
- Stope scheduling and concurrent activities.
- Preliminary ground support designs.
- Drilling and blasting design layouts and firing sequences.
- Ventilation during development and production (to be provided by the senior ventilation engineer).
- Preliminary backfill design.

26.4 MINING PRACTICE

26.4.1 Stope Sequencing

Upon the completion of preliminary stope designs, the sequence of stope extraction is determined. Mining, including backfilling, consideration of geologic factors such as mining-induced stresses and their regional behaviour, and ventilation controls are key factors to be evaluated at this stage. The XPAC Autoschedular is expected to improve the efficiency and effectiveness of the sequencing process.

Stope filling with cement aggregates (CAF) is a major component of mining cost. In many situations, primary stopes need to be filled with 100% CAF. For this reason, smaller primary stopes and larger secondary stopes are designed as standard practice. This is particularly important in areas where the ore body is relatively narrow.

Once primary stope bands are extracted, there are a number of alternate secondary stopes that are under pillar stress conditions. These stopes will be subjected to large and changing mining stresses. It is essential to understand the behavior of in situ stress in determining the successful extraction sequence of the ore body. Wherever possible, larger stopes along the main stress axis will be extracted first. This will effectively shadow the extraction of smaller stopes. The stoping sequence needs to be designed to manage such effects (Figure 26.3).

The selection of primary and secondary stopes and their sequence is essential. Consideration must be given to the

presence of structural features and infrastructure in the area. Currently, only a few major geological structures have been identified at Olympic Dam.

26.4.2 Mine Scheduling

Mine scheduling plays an important role in the mining process. The current production schedule at the Olympic Dam Mine is based mainly on a combination of copper and uranium grades, copper:sulphur ratio, ventilation, and ore pass use. The interaction of mining activities on each stope has to be carefully analysed to ensure efficiency in the production cycle.

In scheduling, each stope is taken as an individual project with the start and finish of production being the overriding driver of all other activities. A 5-year production schedule is produced in a spreadsheet format using area stoping sequence trees. This information is then used in the scheduling package to incorporate other activities, such as development, drilling, production, and filling.

The short-term scheduling is performed by the operations department on a rolling 3-month basis. There is also a 12-day forecast for production bogging (mucking) that describes ore sources and tonnages on a shift-by-shift basis.

26.4.3 Ventilation

Effective ventilation is a major consideration in a successful extraction strategy. Because of the presence of uranium, current underground mining practices at the Olympic Dam operation are primarily governed by the ventilation resources. The principal contaminants in the underground environment are heat, diesel fumes, dust, and radiation products. To create a satisfactory working environment, a number of ventilation design standards have been established by the Olympic Dam Mine. The overall design criteria are summarised in Table 26.2.

The mine operates under negative pressure with several return air raises located on the surface to exhaust air from underground. The overall ventilation requirement for the Olympic Dam Mine after the expansion is 2,915 m^3/s.

In terms of ventilation, the entire mine has been regionalised, allocating a specific intake (fresh air) and exhaust (return air) raise for each ventilation district. These intake and exhaust raises are linked via perimeter drives and stope drill drives. Based on current mining standards, each ventilation district has the capacity to operate two to four producing stopes at a time. Ventilation control in each district is achieved by the

TABLE 26.2   Mine ventilation design criteria

Requirement	Design criterion	Reason
Minimum air velocity in all transport and personnel access openings	0.5 m/s	Heat control, diesel exhaust dilution
Minimum air velocity in ore-producing areas (not including mined-out areas that will be sealed and kept under negative pressure)	1 m/s	Control of radon decay products
Minimum air velocity in development	0.5 m/s	Heat and control of radon decay products
Maximum air velocity in horizontal or decline intake airways where employees travel or work	Unlimited[1]	
Range of air velocity in horizontal or decline intake airways where employees travel or work	4 to 6 m/s	Economic design
Maximum air velocity in downcast service shaft	10 m/s	Movement of conveyances
Maximum air residence time after radon contact with uranium ore in production areas before reaching return airways	12 minutes	Control of radon decay products
Velocity in downcast air shaft	Up to 25 m/s	Drag constraints
Velocity in upcast air shaft	Below 7 m/s or above 12 m/s, but not more than 25 m/s	Rain drop entrainment
Re-entry time after blasting	>30 min	Fumes dilution/ removal
Ventilating air per kilowatt of rated diesel engine power	0.04 m³/s per kilowatt	Diesel exhaust dilution

[1]Except where >6 m/s and settled dust may become airborne, appropriate dust suppression measures must be taken.

use of regulators. The ventilation network has been designed so that stopes act under negative pressure, with air being drawn into drawpoints (when open) and drill levels, and exhaust through main ventilation returns.

26.4.4   Drilling and Blasting

Downholes 102 mm in diameter are drilled by Atlas Copco Simba 4356S electric hydraulic drilling rigs. Normal downhole parameters are 3-m burden and 4-m toe spacings. Uphole rings are dumped 20° forward and drilled to 89 mm in diameter by Solomatic jumbos with 2.5-m burden and 3.5-m toe spacings. A powder factor of 0.25 kg of explosives per tonne of ore is generally maintained. Stope slots are opened up by firing into a 1.4-m in diameter raisebore hole. Ring designs are generated by the planning engineers in consultation with the drill and blast engineer using the Datamine Rings package.

Stope charging is currently carried out by 2- by 2-men crews working 14 shifts a week. A forklift and an Elphinstone UC500 charging vehicle with twin kettles are used for stope charging purposes. Blasts range in size from about 500 tonnes when opening an undercut slot to 250,000 tonnes maximum in stope ring firing. Six to 10 blasts are performed a week.

Every blast is performed by a system called BlastPED (Figure 26.4), which was developed by Mine Site Technologies Pty. Ltd. in Australia in conjunction with mine operators. This system allows remote centralised blasting using a PED transmission system and eliminates both the high cost of maintaining underground blasting cables and lost production because of misfires. An IBM-compatible computer that runs the PEDCALL software controls the transmission system. The output of the transmitter powers a loop antenna. The signal passing through the antenna generates an electromagnetic field around the loop in a concentric pattern that, being ultralow frequency,

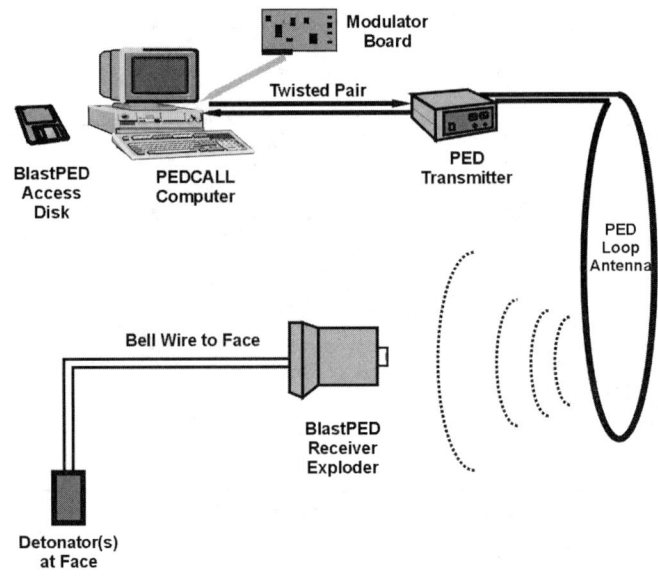

FIGURE 26.4   BlastPED system

travels directly through the strata. The BlastPED receiver is first "armed" by an initial signal that charges up internal capacitors. A "blast" signal is then sent that causes the charged capacitors to release their stored electric current into the firing circuit. The BlastPED system has improved the blast initiation process at Olympic Dam by allowing blasts to be fired reliably and safely.

26.4.5   Ore Handling

Design of Underground Ore Pass and Rail Haulage System. The ore pass system is a major component of the expanded operation. It is composed of 12 ore passes and 50 finger passes. Tramming distance, the amount of ore to be fed through each ore pass, and stoping configuration were the three main criteria considered in designing the ore pass system. Ore from each of the stopes is tipped through a grizzly that consists of four 1.2- by 1.2-m panels and into a 3-m in diameter ore pass. A portable rock breaker is used to break any oversized material. Each ore pass is connected to a 4.5-m in diameter surge bin below that has a capacity of around 1,000 tonnes.

An automated rail haulage system has been installed on 64L (−740 mRL) for ore transportation from surge bins to the crusher. This is based on the LKAB Kiruna Mine, which is the largest underground iron ore mine in the world. The rail haulage system is totally computer controlled with no personnel travelling on the train. Once a surge bin exceeds a set minimum level of ore, the train is directed to the respective ore chute.

Because of the nature of the ore body, ore passes have been designed at various angles. The minimum dip angle has been set at 65°. No ground support requirement for the ore passes has been anticipated at this stage. Because of radiation and ventilation issues, all ore passes have been designed in the fresh air side of the mine and operate under negative pressure. Normally an ore pass can have three to five tipping points. Only one tipping point can be active at any one time. A top exhaust ventilation system is used to control reverse flow resulting from tipping into an ore pass.

Initially, there are two trains in operation, each containing 14 cars, with a nominal capacity of 400 tonnes per train. Each mine car has a capacity of 14 m³. The rail system has a haulage capacity of 1,670 tonne/hr.

Primary Crushing. Based on current design parameters, a single trainload of ore can be unloaded into the dump station at a rate of 3,000 tonne/hr. This ore is fed from the gyratory crusher at the base of the dump station by a plate feeder and conveyor system into two crushed ore storage bins. The gyratory crusher has a design capacity of 2,000 tonne/hr, crushing to a maximum size of 300 mm.

Design of the crusher station was one of the major tasks undertaken by the underground expansion project. Prior to the commencement of the excavation, a detailed geotechnical evaluation was carried out in collaboration with Australian Mining Consultants and Barrett, Fuller and Partners, Australia. As a part of the programme, a series of underground holes were drilled and RMR was ascertained. Virgin stress tests were carried out to assess any significant influence of regional stress on the crusher complex.

Hoisting. Crushed ore from the two fine ore bins is discharged onto two conveyors via two vibrating feeders. Ore is then hoisted to the surface by the newly constructed Sir Lindsay Clark shaft, which is equipped with a Koepe double-compartment, four-hoist rope hoisting system. The skips are designed for bottom discharge with a payload of 36.5 tonnes (21.5 m^3).

26.4.6 Cavity Monitoring System

Once a stope is mucked empty, the stope void is surveyed using the cavity monitoring system (CMS). This is a system developed by the Canadian company Noranda Technology Centre. The CMS consists of a laser unit in a motorised head that is capable of being pushed 5 m into the stope void. Once the instrument is set up, it can locate up to 50,000 points on concentric 360° circles. The CMS has the capacity to pick up stope voids up to 100 m high when the visibility within the stope is clear. The major advantage of the system is that it gives the true volume of the void, allowing stope reconciliation and stope performance evaluation.

From the mine planning and design point of view, the CMS offers the following advantages:

- Based on the CMS survey, stope overbreak and under-break can be estimated accurately and adjacent stopes can be redesigned. This will be particularly helpful at the ring design stage to optimise stope recovery.

- Stope crown overbreak can be clearly identified. This will be helpful in assessing crown support requirements in nearby stopes.

- Possible sources of stope dilution can be identified.

- CAF and rock quantity requirements for stope filling can be estimated accurately.

26.4.7 Stope Backfilling

In the Olympic Dam operation, stope backfilling plays a critical role in stope sequencing and the design process. Following extraction and depending on the individual circumstances, stopes are filled with CAF, underground mullock, or a combination of these materials. A typical CAF mixture has 57% crushed rock by wet weight, 26.5% deslimed mill tailings and sands, 2.5% cement, 5% pulverised fly ash, and 9% neutralised tailings liquor, producing a nominal fill strength of 3 MPa. Specific CAF mix designs exist to allow strengths from 1 to 5 MPa to be delivered. The CAF plant has been upgraded and will mix CAF continuously in a pug mill at 6,000 m^3/d.

Filling cost is a significant portion of the mining cost. The possibility of reducing this cost while maintaining the minimum fill strength for the given stope geometry is currently being investigated. Some of the possibilities under investigation are a CAF mix, stope geometry CAF strength, and continuous CAF delivery systems. An overall review is being conducted to identify areas where the proportion of uncemented fill can be increased.

Regular communication between the engineers of the mine design and backfill departments occurs to ensure maximum utilisation of resources and to minimise the cost of fill placement.

The CAF plant has a maximum capacity of 350 m^3/hr and operates at 300 m^3/hr. This is achieved by continuous mixing in a pug mill that discharges into surge bins for loading into 15-m^3-capacity delivery trucks. CAF is delivered to a rock dropper borehole, which delivers from the surface to the top of the stope. Underground pipe work reticulation is currently being tested with a view to establishing continuous delivery from the CAF plant in the longer term.

26.4.8 Stope Review

Every effort is made to utilise information and experience acquired during the production life of a stope to improve the future design and planning process. Once the CMS is completed, the mine design engineer for the area conducts a stope review meeting. This meeting will investigate factors such as stope overbreak and underbreak, crown failure, stress-related problems, geologic structure, and variations of ring grades against actual tonnes produced. In addition, the stope review process will also collate information relating to drilling, blasting, ventilation, development, drainage, concurrent activities that took place, drawpoint condition, ground support strategy adopted, fragmentation, remote loader layout, trucking path, raise drilling, CMS survey location, and backfilling processes. Information gathered is presented to both planning and operating personnel. Following the meeting, the mine design engineer issues a recommendation-action statement. To ensure such a process is successful, each relevant officer is required to maintain a record of events that took place within his or her area of responsibility.

26.5 DEVELOPMENTS IN PROGRESS

26.5.1 Autotram Loader

Ore from the stope is currently mucked using conventional loader units. Remote mucking is used only to retrieve the broken ore from the open stope at the end of the production phase. WMC and Lateral Dynamics (Australia) have successfully tested a loader automation system at the WMC Leinster underground operation in western Australia. This system is composed of laser sensors and on-board computers that allow the machine to be guided from a designated point, as well as by tele-remote control. It has been successfully tested travelling with a minimum clearance of 50 cm against a wall. Direct comparisions of the Autotram technology with manual remote mucking during production trials at the Leinster operation have shown a 20% increase in productivity. Considering the reduced travel time and the ability to work through blasting times, the system is estimated to be able to achieve up to a 40% increase in productivity.

The project will now be tested in the Olympic Dam Mine using an Elphinstone R2900 loader mucking from stopes to ore passes. Commercial availability of this unit is expected in 2000. Automating and remote control of mobile equipment such as LHDs will be more challenging than operation of the track-bound train system. Such a system, however, will add flexibility to existing ventilation standards, improving safety and system efficiency.

26.5.2 Blast Design Optimization

Because of the large size and complexity of the deposit, it is essential to understand the geological and geotechnical characteristics of the ore body. Observations of blast performance have shown varying degrees of success with the current standard for blast design. In terms of drilling parameters, what seems adequate in one area does not necessarily perform satisfactorily in other areas. A detailed investigation has been carried out by the mining department in conjunction with the explosives supplier Dyno Nobel. As a part of a model of ground conditions, a geotechnical model incorporating rock mass strength and rock structures has been developed for the Olympic Dam resource area. This will now be used as a tool to regionalize areas with varying degree of rock blastability. Standard blast design can be modified accordingly.

26.5.3 Digital Mapping

Capturing data on the geology from underground and surface drill core is now computer based using a system designed and produced within WMC. The system has eliminated digitising of paper-based information and has speeded access to data by all system users. Underground mapping at the Olympic Dam operation is gradually changing from a traditional paper-based system to a computer-aided, pen-based digital mapping system. This is a joint project among Olympic Dam, WMC Group Technology, and the Earth Resource Center of the University of California. The mapping system consists of a pen-based computer, laser rangefinder, and GeoMapper and PenMapper software. The system allows recording and unloading of original mapping details in a form that is easily used by a variety of software types. The implementation of digital mapping will provide an opportunity to better visualize the Olympic Dam ore body and identify potential zones of mineralisation.

26.6 ACKNOWLEDGMENTS

The author wishes to thank the management of the Olympic Dam Mine for permission to publish this overview. The input to this paper by personnel from the Department of Technical Services is also acknowledged. The planning and design components of this paper were published in the Proceedings of the 10th Australian Tunneling Conference in March 1999.

26.7 REFERENCES

Adriana, P. 1998. Three Modern Metal Mines Using Rail System. *World Mining Equipment* 5:40–41.

Hardcastle and Richards. 1997. Engineering Specification for Automated Underground Rail Haulage System (internal document), 21 pp.

Jones, G. 1995. Optec Gives MIM a Head Start on Production. *Australian Mining* 9:60–61.

Kinhill Pty Ltd. 1997. Olympic Dam Operation Environmental Impact Statement (internal document), 10 pp.

Mine Site Technologies Pty Ltd. 1999. BlastPED- System Overview.

Olympic Dam Operation. 1993. Geology and Mining Technical Hand Book, 42 pp.

Olympic Dam Operation. 1998. Mine Planning Guide Lines (internal documents).

Olympic Dam Operation. (In press.) Geology and Mining Technical Hand Book.

Olympic Dam Operation-Radiation Management Plan for Mining Engineering and Ventilation Control.

Philpott, S. 1997. Geotechnical Programme Requirement (internal document).

Western Mining Corporation. 1999. Ore Reserve Statement, 19 pp.

William, A., and J. Bailey. 1999. Use of Blast Engineering Technology to Optimise Drilling and Blasting Outcomes at the ODM.

WMC Resources, Ltd. 1999. Technically Speaking (internal publication). April and December.

CHAPTER 27

Cannington Mine—30 Million Ounces of Silver per Year from a 300-m Cube

Paul Harvey, * **Martyn Bloss,** * **and Nicolas Bouliane** *

27.1 INTRODUCTION

The Cannington silver-lead-zinc deposit is located 135 km southeast of Cloncurry in Queensland, Australia (Figure 27.1). Similar in mineralisation style to Broken Hill, birthplace of the Broken Hill Proprietary Company Limited (BHP) in New South Wales in the 1880s, Cannington is located on semiarid grazing land at the headwaters of one of Australia's major watersheds feeding the Lake Eyre basin. Topographic relief in the area is of the order of only a few metres, with the mine being 250 m above mean sea level. The prospect area is occasionally subject to seasonal flooding.

The Cannington Mine is operated by BHP Minerals Pty. Ltd. (BHPM), a business group of Broken Hill Proprietary Limited (BHP) as a commute operation from locations within the Townsville-Mount Isa corridor.

Systematic exploration for Broken Hill-type silver-lead-zinc deposits in the Mount Isa region had been undertaken by BHPM beginning in the early 1980s on the basis of in-house models developed in the Broken Hill block (Skrzeczynski 1993). The deposit is concealed beneath 10 to 60 m of Cretaceous and Recent sediments. It was discovered through a percussion drilling program commenced in 1990 that targeted anomalous magnetic features identified by a detailed regional aeromagnetic survey completed in 1989. The third hole of the drilling program intersected a 20-m zone of mineralisation averaging 12.1% lead (Pb), 0.6% zinc (Zn), and 870 gm/tonne silver (Ag) in what is now referred to as the Southern Zone of the Cannington deposit.

Following prefeasibility studies in 1992, a feasibility study was commenced in July 1993 and included the development of an exploration decline. Approval to proceed with mine development and construction in March 1996 was followed by a 17-month construction period. The operation was commissioned in September 1997.

By the second year of operation (1999), mine production had stabilised at 1.5 million tonnes per annum (Mtpa) (the mine's nameplate capacity). Shaft hoisting performance averages 5,200 tonnes of ore per day on a 6-day-a-week basis. Waste hoisting and scheduled maintenance is conducted on the seventh day.

Processing plant facilities at Cannington consist of a 5.1-MW semiautogenous grinding mill feeding independent lead and zinc flotation circuits, and fluorine leaching and filtering of zinc and silver-rich lead concentrates for delivery into a fully enclosed storage facility. The processing plant at 1.5 Mtpa throughput produces 30 million ounces of silver contained in 290,000 dry metric tonnes of lead concentrate and 130,000 dry metric tonnes of zinc concentrate a year.

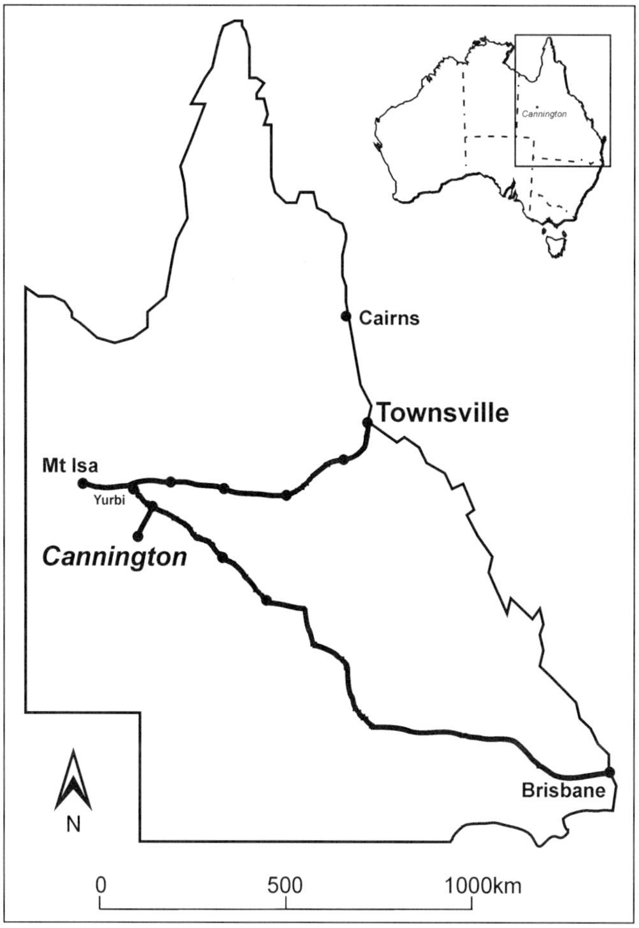

FIGURE 27.1 Location of Cannington Mine in Queensland

From the mine site, concentrates are hauled 200 km by road to the dedicated Yurbi covered storage and rail loading facility to the north. Concentrates are then transported by rail 700 km east to the port of Townsville for storage and blending prior to loading on ships through Cannington's specially built facility.

Setting and monitoring benchmark safety performance is one of the six guiding values of the Cannington business. The safety system framework utilised at Cannington is the National Occupational Safety Association (NOSA) system, which provides

---

* BHP Cannington Mine, BHP Ltd., Queensland, Australia.

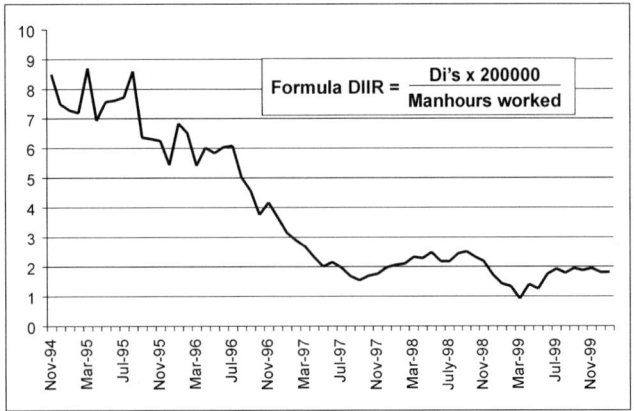

FIGURE 27.2   Disabling injury incidence rate at Cannington Mine since start-up

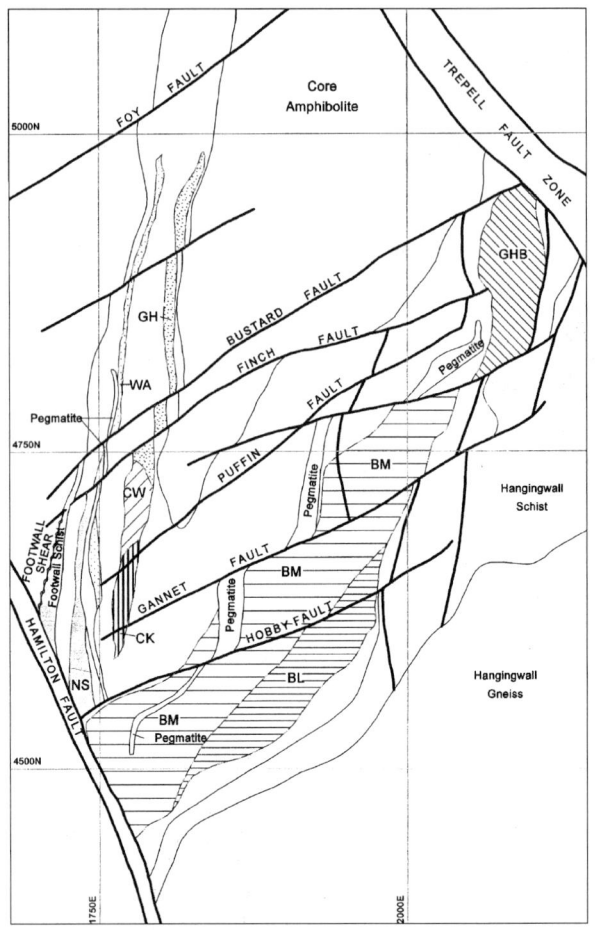

FIGURE 27.3   Interpreted geology and lode outlines at 400 m below the surface

a systematic and comprehensive safety management system. This system has been complemented by several other "safety tools," including the Positive Attitude Safety System™ (PASS) and Occupational Safety Performance Assessment Technology (OSPAT). Safety performance, expressed through the disabling injury incidence rate (DIIR), has progressively improved to around 2 in the 6 years since exploration decline development commenced, establishing Cannington as one of Australia's safest underground mining operations (Figure 27.2).

27.2   GEOLOGY

27.2.1   Regional Geology

The Cannington deposit is located within the Eastern Succession of the Proterozoic Mount Isa inlier in northwest Queensland (140° 55′ E, 21° 52′ S). Regionally, the Mount Isa inlier is divided into three broad stratigraphic and tectonic units: the Western Succession or fold belt, the Central belt (or Kalkadoon-Leichardt belt), and the Eastern fold belt. Within the Eastern Succession, several significant base and precious metals deposits have been recently discovered, including Cannington, Osborne, Ernest Henry, and Selwyn. The Cannington deposit is concealed beneath 10 to 60 m of Recent and Cretaceous sedimentary cover consisting of low-strength and high-slaking mudstones and siltstones. Beneath this cover, the Proterozoic sequence trends north-south and is offset by two major northwest-southeast-trending faults, the Hamilton and the Trepell (Roche 1994)

The Trepell fault separates the deposit into the deeper and higher grade Southern Zone where mining is in progress, and the lower grade, shallower Northern Zone currently under study. The deposit is hosted by a sequence of quartzites and garnetiferous quartzites with intercalated schistose horizons surrounding an amphibolite core. It lies within a quartz-feldspar-gneissic terrain. The deposit is characterised by intense deformation and metamorphism (Walters and Bailey 1998).

27.2.2   Deposit Geology

The geology of the Southern Zone is controlled structurally by a tight isoclinal synform that strikes north-south, dips from 40° to 70° to the east, and plunges to the south. The Cannington Mine has been developed in the larger, higher-grade Southern Zone of the deposit. The Southern Zone has a strike length of 500 m and is up to 400 m wide, with mineralised lodes between 8 and 70 m thick. The deposit is interpreted to be closed off 600 m below the surface by the keel of the isoclinal fold. A plan of the interpreted geology and lode horizon outlines at 400 m below the surface is

shown in Figure 27.3. Figure 27.4 shows a set of simplified east-west sections through the Southern Zone of the deposit and depicts the lodes, core amphibolite, and major faults.

Within the Southern Zone, mineralisation is classified into five main lode horizons containing nine defined mineralisation types. The lode horizons are the Footwall Lead, Footwall Zinc, Hangingwall Lead, and hangingwall zinc lodes, and the Brolga Fault Zone. Within these lode horizons, nine mineralisation types are defined based on zinc versus lead-silver sulphides combined with iron-rich versus siliceous gangue mineralogy (Walters and Bailey 1998) (Table 27.1). The predominant economic sulphides are galena and sphalerite.

Geological resources defined at Cannington as of May 1999 total 44 million tonnes (Mt) at 11.3% lead, 4.7% zinc and 511 gm/tonne silver. The split between Southern Zone and Northern Zone resources and established Southern Zone mining reserves is shown in Table 27.2. A large proportion of the high-grade Southern Zone resource is contained within a cube measuring 300 m on a side. Production mined since the commencement of operations from September 1997 to May 2000 totals 3.3 Mt.

27.2.3   Mine Geology

Following extensive surface delineation drilling during the project feasibility stage, ongoing underground diamond drilling has conducted to define resources further and to provide spatial information for stoping design. Mine design and stoping layouts

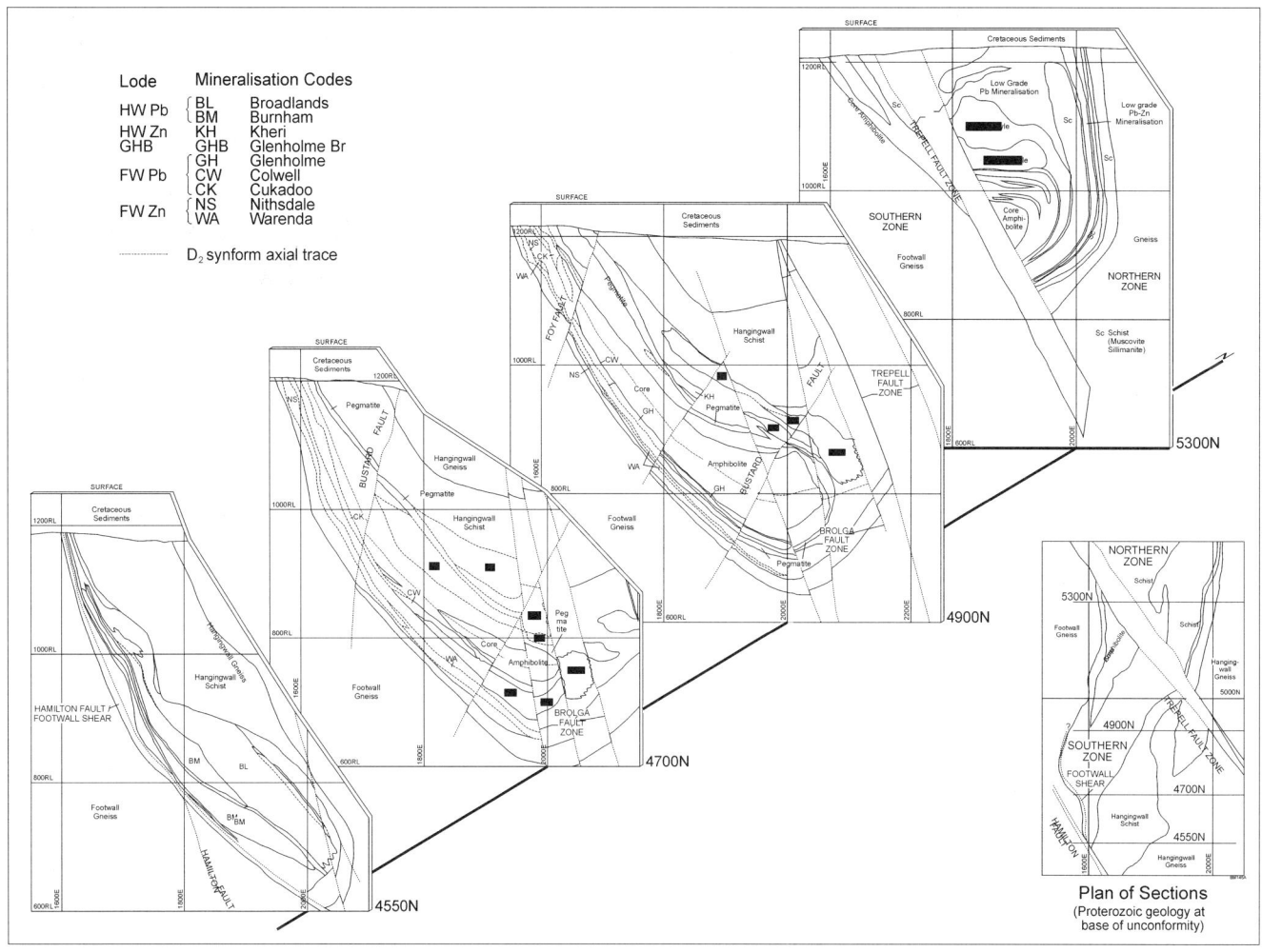

FIGURE 27.4   East-west section through Southern Zone of Cannington Mine

TABLE 27.1   Lode horizons and mineral types

Lode horizon	Mineralisation type	Gangue type	Ore type	Ore grade
Hangingwall Lead	Broadlands	Iron-rich with siliceous bands	Pb>>Zn	Medium/low
	Burnham	Iron-rich	Pb>>Zn	High
hangingwall Zinc	Kheri	Iron-rich	Zn>Pb	Low
Footwall Zinc	Colwell	Iron-rich	Zn>Pb	Medium/high
	Cukadoo	Siliceous	Zn>Pb	Low/medium
	Glenholme	Siliceous	Pb<>Zn	High
Footwall Lead	Nithsdale	Iron-rich	Pb>>Zn	Medium
	Warenda	Iron-rich with siliceous bands	Pb>Zn	Low
Brolga Fault Zone	Glenholme Breccia	Siliceous	Pb<>Zn	Very high

at Cannington are driven by complex lode structures and shapes. Narrower footwall lodes are mined by benching whilst the more massive Hangingwall and Glenholme Breccia lodes support larger sublevel open stopes.

Early mine production has focussed on the larger and higher-grade, lead-silver-rich Hangingwall Lead and the silver-lead-zinc-rich Glenholme Breccia lodes centred between 300 and 575 m below the surface. Careful blending of the lode types is required to balance lead-zinc ratios for steady-state processing plant feed and to manage fluorine concentrations that vary significantly both between adjacent lodes and between the adjacent stopes within lodes. Fluorine occurs as soluble salts such

as fluorite (calcium fluoride), apatite, and pyrosmalite, which are readily leachable, and as free-floating talc (magnesium silicate). Fluorine is managed by a combination of—

- Preproduction metallurgical tests for fluorine from stope delineation drill core.

- Daily control and blending of production sources.

- A prefloat for talc ahead of the lead and zinc flotation banks.

- High-temperature leaching of the soluble fluorine-bearing minerals.

TABLE 27.2 **Resources and reserves**

	Million tonnes	Lead, %	Zinc, %	Silver, gm/tonne
Geologic resources, total	44.0	11.3	4.7	511
Southern Zone	32.5	11.6	4.9	532
Northern Zone	11.5	9.8	4.1	450
Ore reserves in Southern Zone	8.1	11.1	5.7	477

Final fluorine content in concentrate products is kept beneath market specifications. Typical quality is 600 ppm fluorine in lead concentrate and 200 ppm fluorine in zinc concentrate.

The application of a modified sublevel open stoping mining method and subsequent individual stope designs at Cannington are driven primarily by geological and geotechnical constraints. For example,

- Stopes with very high ore grades (>600 gm/tonne silver) requiring maximum extraction have extremely poor schistose hangingwalls.

- The schistose hangingwall conditions require extensive cable bolt support.

- A sequence of late-stage, steeply dipping, cross-cutting faults trend northeast-southwest through all lode horizons.

- A suite of semiconformable, locally boudinaged pegmatite horizons through the Southern Zone varies from centimetres to tens of metres in thickness.

- Ground support in the widely varying geological domains ranges from pattern bolting to full profile, remotely applied fibrecrete reinforced with hollow groutable bolts and grouted cable bolts.

- High-density, cemented paste fill is applied immediately after mining to maximise resource extraction and provide regional support to the hangingwall.

Extraction of the first 35 sublevel open stopes indicates an average dilution of 1.6% of material from outside the designed stope outline. Approximately half of this is paste fill dilution when mining secondary and tertiary stopes with one to three fill walls. Geologic dilution is higher in stopes having muscovite sillimanite schist in the hangingwall or crowns and has reached up to 15% in two Glenholme Breccia breccia stopes mined thus far. The presence of faults and joint structures in stope walls within the ore body has resulted in overbreak of ore into adjacent stopes. The orientation of the stope design grid has been adjusted by 30° for future stopes to reduce the negative impact of this feature.

27.3 GEOTECHNICAL AND HYDROLOGICAL CONSIDERATIONS

27.3.1 Sedimentary Cover

The Cannington deposit is overlain with 10 to 60 m of Recent and Cretaceous sediments. This material consists of weak mudstones and claystones that display various states of weathering, from extremely weathered to fresh. The fresh mudstones have an average wet uniaxial compressive strength of 8 MPa. The mudstones are low density (1.72 tonne/m^3), have a high moisture content (23%), demonstrate high slaking characteristics, and break down readily when exposed to atmosphere or moisture. The sediments are moderately jointed and frequently lined with chloritic or kaolinitic clay infills. They demonstrate very low shear strengths when slickensided.

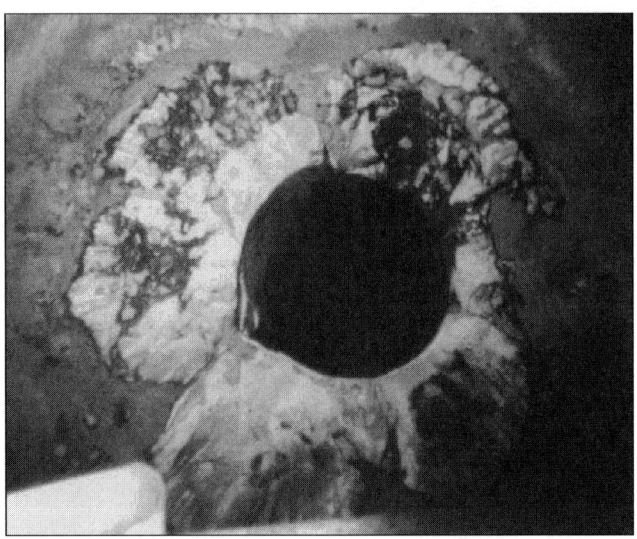

FIGURE 27.5 **Shaft piles during sinking**

A thin, unconsolidated fine sand horizon exists at the base of the Cretaceous in the southern and eastern parts of the prospect. This horizon required specific treatment during shaft sinking operations.

The jet grouting technique of preexcavation curtain piling was adopted to enable a safe transition from the surface sediments in the main hoisting shaft (645 m at 5.6-m finished diameter) and return air shaft (325 m at 5.6-m finished diameter). Jet grouting was conducted to depths in excess of 50 m on the hoist shaft site and represented the deepest application of this technique in the world when the shaft was constructed in 1996. The grout curtain enabled a 1.8-m in diameter raise bored pilot hole to be pulled through to the presink for conventional strip and line sinking of the remainder of each shaft. Figure 27.5 shows the jet-grouted pilot hole curtain at a depth of 45 m.

27.3.2 Mine Decline Access

The main decline access to the mine commenced with the mining of a 26-m-deep boxcut through the weathered sediments. The boxcut was subsequently equipped with a 6-m-wide, galvanised steel, multiplate arch prior to backfilling. This provided access to the unweathered mudstones for the commencement of roadheading using an AM75 Alpine miner through the remainder of the sediment cover. The New Austrian Tunneling Method (NATM) was adopted in mining this first 200-m section of the decline through the mudstones and incorporated the following features:

- Frequent convergence monitoring of tunnel spans.

- Quality-assured development and support techniques and procedures.

- Full-profile, remote fibrecreting immediately after each 1 to 2 m of excavation.

- Installation of a cement-stabilised invert roadway keyed into sidewall fibrecreting.

- Auger-drilled, patterned rock bolting with 3 m of cement-grouted rebar.

- Extensive cable bolting of fault structures.

- Use of grouted 3.6-m-long steel spiling bars on 400-mm crown spacings ahead of the face in extremely poor conditions.

TABLE 27.3 Summary of average ground conditions for rock and mineralisation types of the Southern Zone

Rock type	RQD	Foliations per metre	Fractures per metre	Uniaxial tensile strength, MPa	Uniaxial compressive strength, MPa
hangingwall gneiss	50	2	5	13.7	96
Footwall gneiss	31	5	5	13.7	96
Schists	48	1	6	11.2	97
Quartzite (foliated, schistose)	52	2	5	10.6	156
Quartzite (unfoliated, massive)	676	0	6	12.9	134
Quartzite (pyrobole, magnetite)	94	0	2.4	16.8	319
Lode mineralisation	91	0	2	11.3	183
Glenholme Breccia ore	89	0	3	9.2	101
Amphibolite	78	0	4.5	19.1	199
Pegmatite	69	0	5	9.2	101
Broken zone (faults, contacts)	15	0	20	—	—

27.3.3 Proterozoic Ground Conditions

The Cannington resource is hosted within and bounded by a faulted and highly jointed gneissic terrain of poor to very poor ground conditions (average rock quality designation [RQD] = 29%). As a result of the expected high cost and high risk, conventional placement of major infrastructure and accesses in the ore body footwall was not adopted. The main decline and level accesses were thus located within the amphibolite core between the hangingwall and the footwall lodes of the deposit.

Cannington uses the Norwegian Geological Institute's Tunnelling Quality Index 'Q' system (Barton et al. 1974) for rock quality classification. Along with geotechnical mapping at the face, the system is applied on a daily basis in the highly variable ground conditions to define the relevant ground support class for the prevailing conditions and for Cannington's modified Matthews stope stability analyses. Table 27.3 summarises average (unfaulted) ground conditions for the primary Southern Zone rock and mineralisation types.

The best ground conditions are found in the pyroboles and pyrobole quartzites associated with the mineralisation (Struthers et al. 1994). Because of spatial constraints, much of the access and infrastructure development was done through poor-quality schists and foliated quartzites that required extensive fibrecrete, cable bolt, and wire mesh support.

Geotechnical conditions have a significant impact on both mine design and ongoing mining operations.

- High silica- and garnet-content rock types are highly abrasive, which result in above-average drilling costs.

- Ore pass stability in poorer ground is a key operational issue. The best performance to date in sections of passes where rock mass quality is poor has been obtained from vertical passes fitted with high-abrasion-resistant steel liners.

- Sublevel intervals are limited to 25 m.

- Stope spans are typically limited to 20 m along strike and 20 to 30 m across strike to maintain stability whilst open.

- Extensive cable bolt support of stope hangingwalls and crowns is required.

- Backfilling with cemented, high-density paste fill is required to maintain rock mass integrity in the mine.

The in situ density of the ore varies from 3.35 to 3.85 tonne/m^3, depending upon base metal and iron content, with an average of 3.55 tonne/m^3. Gangue material densities range between 2.60 tonne/m^3 for pegmatite to 2.95 tonne/m^3 for amphibolite.

Development ground support consists primarily of point-anchored, full-column-grouted, 3-m-long hollow groutable bolts installed either by a development jumbo or a dedicated rock bolt rig. The standard pattern consists of eight bolts on a 1.5-m ring spacing. Friction anchors are used with wire mesh to support blocky zones or in ground rehabilitation, while steel-fibre-reinforced shotcrete is used extensively in highly jointed, foliated, and unravelling ground conditions. Trials of a single-pass resin-grouted bolt system are currently well advanced to improve development cycle times and efficiencies.

27.3.4 Hydrological Considerations

Underground hydrogeological conditions have minimal impact on mining operations. Total ground water flow from the current workings is 15 L/s. Water quality is of moderate salinity (total dissolved solids of between 700 and 10,000 mg/L) with a high sulphate content (1,200 to 4,000 mg/L). This requires full encapsulation of all installed ground support elements and ongoing management of corrosion on the mobile equipment fleet

Surface hydrological issues played a significant role in the original mine design. The mine is located about 30 km south of the western Queensland watershed with waters to the north of that line flowing to the Gulf of Carpentaria. Cannington lies at the headwaters of the huge Georgina-Diamentina catchment draining to Lake Eyre some 1,500 km to the south. The mine lies beneath the floodplain at the confluence of the Hamilton River and Trepell Creek. Both waterways are perennially dry and flow only on rare occasions every few years following strong monsoons. All mine openings are thus constructed on elevated pads some 3 to 4 m above the surrounding plain to provide protection from water inrushing. Mining method selection also considered the risk of subsidence, thus ruling out all caving methods in favour of a full-void-filling method.

27.4 MINE DESIGN AND MINING METHODS

27.4.1 Mine Access

The mine is accessed via a decline from the surface to a depth of 645 m. The decline is located in the amphibolite within the centre (or core) of the geological fold sequence and has a gradient of 1:8 from the surface to a depth of 450 m and 1:7 from 450 m to the bottom of the mine. The decline provides long-term access in the best available ground outside the mineralised host rocks. The decline was developed to its final depth before the full production rate was achieved to develop the ore handling and ventilation system fully. This arrangement also provided early production access to higher-grade zones lower in the mine. Construction of the decline was undertaken

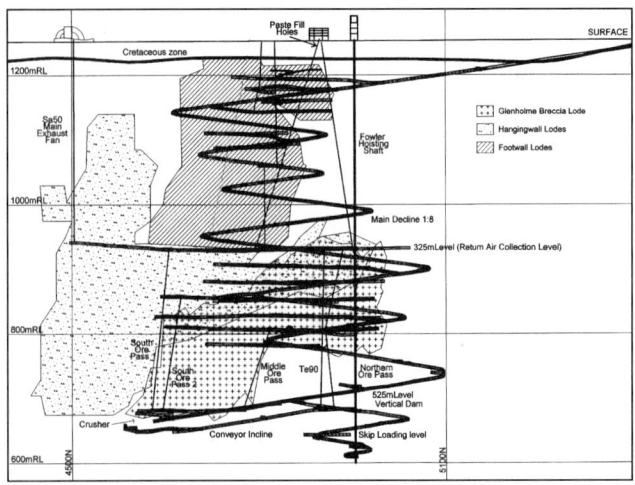

FIGURE 27.6   Schematic of mine longitudinal section looking west

FIGURE 27.7   Fowler headframe and paste fill plant

by mining contractors (Rossiter 1998; Gillespie 1998). Figure 27.6 is a longitudinal section through the mine looking west and shows the three lodes being mined as primary access and infrastructure development at year 3 of operations.

27.4.2   Mine Infrastructure

Cannington utilises conventional mobile diesel equipment to liberate and move production ore and development waste to an ore pass system for delivery to a collection and crushing level at the base of the mine (575-m level).

Ore handling is achieved using a fleet of four Toro 650 front-end loaders (6-m³ capacity) that tip ore into a series of 2.4- to 3-m in diameter vertical or steeply inclined ore passes that lead directly to the level where the underground jaw crusher is located 575 m below the surface. A Toro 2500E electric front-end loader (10-m³ capacity) hauls the ore from the bases of the ore passes to the crusher. The ore is then crushed to −400 mm and fed by conveyor to a fine ore bin adjacent to the hoisting shaft prior to being loaded into 9-tonne-capacity skips at the base of the hoisting shaft 645 m below the surface. The ore is then hoisted to the surface and discharged into trucks for hauling to the surface stockpiles. The hoisting shaft is fitted with an automatic, unattended tower-mounted friction winder running four head ropes and two tail ropes. The shaft is equipped for rock hoisting and emergency egress only, using rope guides. The shaft has a diameter of 5.6 m and a capacity of 2.2 million tonnes per annum. Figure 27.7 depicts the Fowler shaft headframe and the paste fill plant.

The key issues considered when designing the ore handling system were—

- The ore body is laterally confined (500 m along strike).

- Poor ground conditions outside the ore body sequence prevent optimal location of the ore handling system from a materials movement perspective.

- The crusher is located at the base of the centre of gravity of the ore body.

- Ore passes should be located so that the average stope-to-ore pass tramming distance did not exceed 150 m.

The primary ventilation system consists of fresh air intakes via the 5.5- by 5-m arched surface decline and the hoisting shaft. Venting of exhaust air is provided by a 5.6-m in diameter, 325-m-deep shaft to the 325-m level. Two 600-kW centrifugal fans connected in parallel provide 450 m³/s of air flow at around 1700 Pa. A number of 2.4- to 3.5-m in diameter internal fresh air and return (exhaust) air shafts completes the system. Fresh air is drawn off the decline to ventilate all working levels and exhausted up the return air shaft. Fresh air from the hoisting shaft is used to ventilate the crusher and hoisting infrastructure at the bottom of the mine. Air flow throughout the mine is controlled by a series of regulators at the connections between working levels and the return air shafts.

Mine dewatering consists of a network of steep, 100-mm diameter holes between the levels developed at a gradient of 1:50 to facilitate free drainage. All water is directed to the 525-m level along with decline inflows, where it is settled in a shallow sump. Sump overflow is directed through a sieve bend filter into a 35-m-high, agitated, vertical dirty water storage dam. The 2.2-m diameter dam is lined with steel.

Twin dirty-water GEHO positive-displacement piston pumps are capable of delivering water at 40 L/s to the surface from a depth of 600 m, although typical duty is 25 L/s. A linked emergency pumping arrangement has been established between the shaft bottom and the base of the main decline to provide emergency storage capacity.

A ladderway extends from the bottom of the mine to the surface to provide a secondary means of egress in one of the fresh air raises. Self-contained fresh air bases are located at regular intervals up the decline and throughout the production areas; these bases are fitted with personal rescue devices, communications equipment, first aid facilities, and other services.

27.4.3   Mine Layout

The ore bodies are accessed from the decline via working levels spaced 25 m apart and named according to their vertical depth below the surface. Ore bodies to the east of the decline and east of the geological synform fold axis are referred to as hangingwall ore bodies. These are generally thicker and of higher grade and were the primary source of production in the early years.

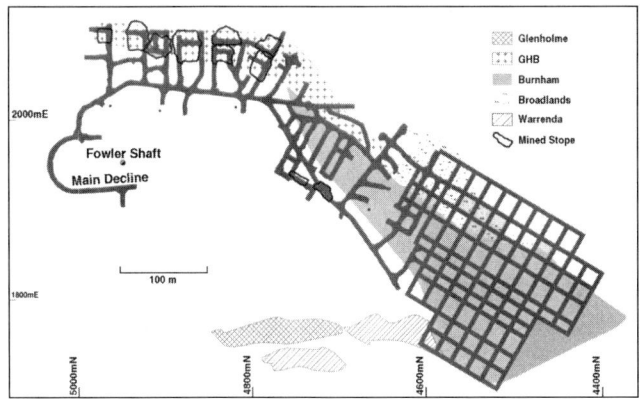

FIGURE 27.8 Plan view showing development and mineralisation types

The underlying ore bodies to the west are referred to as footwall ore bodies. They are generally narrow lenses with faulted structural offsets and generally of lower grade. The ore lodes are accessed on each level by a north-south-striking drive offset 15 to 30 m from the ore boundary; they are referred to as footwall drives or hangingwall drives.

East-west drives (referred to as crosscuts) branch off the main north-south drives and cut across the ore bodies to provide stope access. Figure 27.8 shows an example of an ore body access from the decline and the interpreted mineralisation type outlines for the 425-m level.

Drives are developed using a conventional drill-blast-load-haul cycle. Two Tamrock Minimatic electro-hydraulic development drill rigs are used to drill the development face. Ammonium nitrate-fuel oil (ANFO) explosives and perimeter blast techniques are used to charge up the face. Initiation of the blast is by nonelectric methods, with detonation achieved using a remote firing system based on very low frequency radio transmissions emitted from a buried surface cable loop. Blasting is conducted at the end of each 12-hr working shift. The blasted material is hauled to the ore passes either directly using loaders or with Toro 50D trucks if the haulage distance is great. Development ground support elements are a combination of tensioned bolts, grouted bolts, wire mesh, and fibrecrete (steel-fibre-reinforced shotcrete), depending on the stability of the rock surrounding the excavation.

27.4.4 Stoping Methods

Two primary stoping methods are used at Cannington. In the wider hangingwall ore bodies, sublevel open stoping is employed. Bench stoping is also used in the narrower sections of the footwall ore bodies.

Sublevel open stoping at Cannington is based on the classic form of an open stope where a cutoff raise is excavated, followed by opening of a cutoff slot, and finally firing of the main rings. In the case of exposing backfill in an adjacent stope, diaphragm rings are usually designed to protect the fill mass and are fired after the main rings. Figure 27.9 depicts a typical firing sequence for a 50-m-high stope in the Glenholme Breccia. The plan views and sections demonstrate the complex lode outlines and stope shapes required for controlled extraction against a very poor quality, schistose hangingwall contact.

Each stope is accessed from the sublevels and can be up to 150 m high, depending on the height of the ore body and local ground conditions. Typical widths of each stope range from 20 to 30 m in the east-west dimension (across the strike of the ore body) and 20 m in the north-south dimension (along strike).

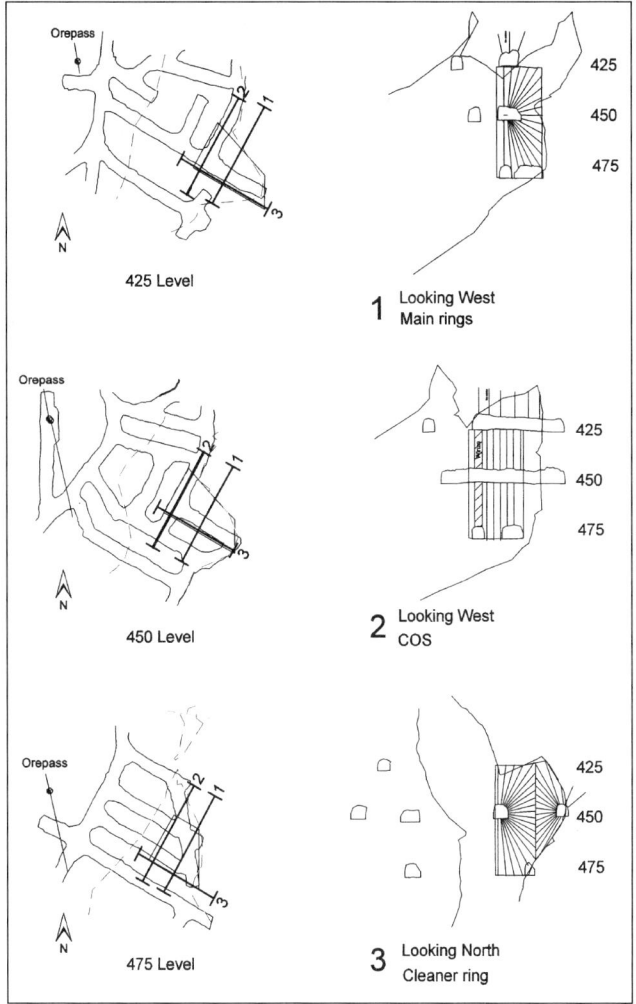

FIGURE 27.9 Firing sequence of a typical stope cross section

These dimensions are controlled by the stability of large geologic structures in the host rock and the nature of rock masses adjacent to the ore, such as poor-quality schists and gneisses.

Stope drilling in fans and rings is done using a Tamrock Solomatic 1020 tube-drilling system. Hole diameter is 89 mm and maximum length is 35 m. Ring spacing varies between 2.2 and 3 m with toe burdens between 2.0 and 3.5 m, depending upon local ground conditions and stope design. Stope ore is blasted in a controlled manner so that it is directed to drawpoints at the base of the stope. Toro 650 loaders equipped with tele-remote controls haul the ore to the nearest ore pass. Blasting the ore surrounding the drawpoints is done in sequence in order to form a trough through which the ore goes directly into the access drive. In this way, the ore can be loaded safely from the access drive without the loader having to advance into the stope to recover the ore. At the end of the stope, the trough is fired out and the remotely operated loader recovers the ore.

Mined-out stopes are filled with backfill so that adjacent stopes can be mined at a later date. High-density paste fill is used at Cannington and provides regional ground support for subsequent mining. Three to four percent cement is added to stabilize the fill when it is exposed in the adjacent stope.

Cut-off raises are typically 2 by 3 m in plan view and extend the full height of the stope. Raises are drilled by one of two methods. In the first method, a 600-mm diameter Terratec raise borer pilot hole is reamed downward and acts as the easer hole for the blast. A pattern of 89-mm holes are then drilled by the Solomatic 1020 and charged from the top access with explosives. The holes are fired with an appropriate delay sequence both vertically (using several firings from bottom to top) and horizontally (using inhole subsecond delays) to ensure successful excavation of the raise. In the second method, several 89-mm diameter holes (up to five) are drilled as easer holes in place of the Terratec hole.

Cut-off slots are extracted across the full width and height of the stope and form the void into which the main rings are fired. The slot is drilled with a pattern of 89-mm, parallel, and usually vertical, holes in a pattern similar to that used in bench stoping. The main rings are drilled as a sequence of fans that usually reach across the full width of the stope and are fired off one, or several, at a time.

Blasting of the raise, slot, and main rings is performed in a progressive sequence of firings (with a number of firings in each), and the broken ore produced after each firing is usually extracted from the stope prior to the next firing to prevent confined firing of the next ring.

Support of the stope walls or roof (crown) is sometimes required. Cable bolts up to 15 m in length are installed from the development access within (or sometimes adjacent to) the stope using a Tamrock Solo 620 drill rig. This machine also serves as a back-up production drill rig.

A variation on the open stoping method has recently been employed at Cannington with great success. In the Glenholme Breccia ore body, the hangingwall is particularly weak and the ore body is tall (up to 150 m high). Stopes are being designed that are diamond shaped in section, with the following advantages:

1. Stope heights can be made twice the height of the hangingwall.
2. Paste fill exposure on the hangingwall side will be more stable as it is inclined rather than vertical.
3. Flat-bottomed stopes are eliminated, which increases stability of the drawpoint area and significantly reduces remote loading.

In the wider sections of the ore bodies, the trend toward continuous extraction is progressing. Conventional primary-pillar extraction is proving difficult because of the relatively small stope widths in the spatially constrained geologic environment. Access to pillars is proving particularly difficult, especially when overbreak occurs in the adjacent primary stopes. Problems with access include—

1. Providing two drawpoints in each stope (one drawpoint generally does not provide sufficient security of production).
2. Creating unstable pillars between adjacent drives.
3. Limiting development of access well in advance of stoping because of uncertainties about the final stope design.

Continuous advance sequencing, where stopes are taken sequentially along a mining front, has several major advantages.

1. Good access to all stopes.
2. Regular design for all stopes, making the stoping process very repeatable.
3. The potential for development well in advance of stoping for all stopes.
4. Increased ore development and reduced waste development.

The third advantage is particularly important as it breaks the dependency between development and production, which has, to date, been a significant bottleneck in the stoping process. It is planned to provide access development on a regular grid pattern and then design each stope in the sequence to match the development. If significant overbreak occurs in any particular stope, then design of the next stope is adjusted to match the development accordingly. This is a change in the earlier stope design strategy where stopes were designed first and development was designed to match the stope geometry and fit the primary-secondary-tertiary extraction sequence to the complex ore body geometry at Cannington.

27.5 PRODUCTION STATISTICS

The mine operates on a continuous 12-hr-shift, 365-days-per-year basis. Because of the mine's remote location (800 km by road from the city of Townsville), a commute system with fully catered and self-contained, single-person accommodation is provided on site. Mining crews work a 2-weeks-on, 1-week-off schedule, which requires three full crews. Technical and support personnel work a 9-days-on, 5-days-off schedule.

The total Cannington work force is composed of 220 permanent employees and 180 contract partners. In the mine, each mining crew consists of one shift superintendent supervising a multiskilled team of 22 operator-miners. The mobile equipment fleet is maintained through a contract with the prime fleet supplier, Tamrock, with a crew of 24 with 8 per shift. Additional minor contractors are retained to provide fibrecrete supply and placement, underground diamond-drilling services, explosives supply, and construction services.

Mine operations management is provided by a team of seven mining engineers led by the mining manager. Mine planning, mine design, surveying, geologic studies, local exploration, and geotechnical services are provided by a technical services team composed of 15 technical personnel.

27.6 FILLING

The mine backfill system used at Cannington is based on production of paste fill material using process residues. It is the first of its type in Australia. The scarcity of alternative sources of backfill material and the world-class environmental standards at Cannington have combined to make paste fill the preferred choice. Tailings are delivered to a surge tank (1,500 m³ or 12 hr capacity). If paste fill is required, the tailings are then passed through two parallel disc filters at a capacity of 160 dry tonne/hr, where the material is discharged at 80% solids by weight. It is then mixed with cement (typically at an addition rate of 3.5%) in an 8-tonne-capacity mixer. Slurry is added (30 dry tonne/hr) to achieve 170- to 200-mm slump at approximately 77% solids. When capacity is achieved, the bottom of the mixer opens, and the paste fill is discharged to the underground reticulation system. If paste fill is not required and the surge tank is full, then the tailings are pumped directly to the tailings dam.

The underground reticulation system consists of two steeply inclined (75°) and cased 250-mm diameter boreholes and a network of near-horizontal 200-mm diameter pipes within the underground workings. An internal vertical cased borehole is currently under construction to feed the deeper sections of the mine. The surface boreholes enter the workings on the fill distribution level (325-m level), which is above the main two ore bodies. Raise-bored pilot holes up to 100 m long are drilled from the 325-m level to the top of the stope to provide access to the void for filling. Flanged pipes are used for reticulation on the levels as these are not prone to wear failure.

Cannington's license to operate requires that all waste rock be returned underground prior to the mine's closing, currently planned around 2020. To achieve this, a large proportion of the waste rock produced will need to be mixed with the paste fill to form a rock paste material capable of performing the same functions as the paste fill. A system has been designed to tip

TABLE 27.4   Equipment fleet information

Model	No. of units	Availability, %	Utilisation, %	Mean time between failures, operating hours	Mean repair time	Maintenance costs, A$
Toro 650D LHD	4	78	50	19.5	6.2	70/hr
Toro 50D ejector trucks	2	77	51	22.1	7.5	46/hr
Tamrock 2500E LHD	1	87	37	35	9	48/hr
Tamrock H205 Minimatic development jumbo	2	81	47	11.5	2.4	1.60/m
Tamrock Robolter development bolter	1	78	47	6.8	1.9	3.70/m
Tamrock Solo 1020, stope drilling, 89-mm holes	1	85	48	14.9	2.6	3.30/m
Tamrock Solo 620, stope/ cable drilling, 76-mm holes	1	85	34	11.7	2.6	5.10/m

underground development mullock into a significant proportion of the stope voids, together with the paste fill. Longhole winzes are fired from a level above the stope to the top of the stope. Trucks tip the rock into the stope only when paste is being placed through the same winze at a ratio of dry to wet material that provides adequate cementing throughout the fill mass. At this stage, the winzes are vertical and are located along the central axis of the stope to maximise mixing of the paste and rock and minimise any potential exposure of uncemented rock during adjacent stope extraction.

Ongoing and future research is required to improve the backfill system in the following areas:

- Rheology (flow characteristics) of the paste fill in the pipe network.

- Addition of waste material (crushed waste rock) into the pipe network.

- Quality of rock paste as a function of cement content and placement characteristics (winze location and angle, and operating procedures).

- Blast damage to fill resulting from adjacent stope extraction.

27.7   MINE COMMUNICATIONS AND SERVICES

Most established areas underground have two-way VHF radio coverage via leaky-feeder cable. Telephones are installed in all critical underground locations, including the mobile fleet workshop on the 450-m level, the crusher, the skip loading level, underground offices, etc.

Cannington also uses an ultralow-frequency transmitter for one-way-only communication from the surface to underground through short written messages. The transmitter uses a 3-km-long horizontal loop on the surface and a smaller loop on the 325-m level (vertically, in the centre of the mine) to achieve complete coverage. The receivers are mounted on every miner's cap lamp battery pack. Emergency messages can be sent quickly and accurately to any or all underground personnel. Thus, Cannington does not have a reticulated stench gas warning system. This transmitter is also used to stop and start secondary fans during firings, as well as for remote detonation of firings themselves, thus removing the need for a minewide firing line.

The Mine Information Centre, located in the mine processing plant production administration centre, is staffed 24 hr per day. All automated underground operations are monitored from this office. The PitRAM™ mine recording and reporting software is used to manage the data relating to the underground operations. Emergency communications for the entire Cannington business are maintained through this office, from where responses are coordinated.

27.8   MINE MAINTENANCE

Underground mobile fleet maintenance is provided on a contractor-partner basis by Tamrock as prime fleet supplier. All but major breakdowns and repair maintenance are conducted underground at the 450-m level workshop, which is equipped with refueling facilities, a welding bay, waste oil management, and a three-bay work area.

Table 27.4 depicts various statistics for the key production items of the equipment fleet for 1999. (Note: operating cost rates during 1999 are based on a total production year of 1.4 Mtpa, with rates reduced further since the attainment of a steady-state +1.5 Mtpa).

27.9   MINE DESIGN, PLANNING AND SCHEDULING

The framework for mine design and planning is based around regular updates of the mine's production plans. These plans describe the various capital and operating parameters required to sustain production at the required rate. These plans are—

- Life-of-Mine Plan (LOM). Updated annually. Describes in detail the long-term strategic direction and major capital expenditure required to provide for optimised and maximised exploitation of the resource.

- Five-Year Plan (5YP). Updated annually. Describes capital expenditure, future mining block access, and resource drilling demand required to sustain production rate over the medium term.

- Two-Year Plan (2YP). Updated half-yearly. Describes capital expenditure, future mining block access, and resource drilling and stope activity demand (development, rock support, production drilling, and backfilling) required to sustain production rate over the short to medium term. The first year of the 2YP forms the basis of the annual business budget.

- Three-month schedule (3MS). Updated fortnightly. Forms the basis of the daily and weekly operating schedule, describes all development and stoping activities, and details concentrate production forecasting for marketing purposes.

- Weekly schedule. Updated weekly. Describes specific development tasks on a shift basis in order to achieve necessary advance rates in priority development headings.

All plans contain production schedules at a level of detail appropriate for the timeframe considered in the plan. Information provided in each plan is rolled out into the following shorter-term plan in the sequence. The schedules within the LOM, 5YP, and 3MS are constructed on spreadsheet platforms, whereas the 2YP is developed with Mine Works Planner, a software package developed specifically for scheduling underground mining operations. The software provides for allocation of activities and resources, leveling of those resources, and activity tracking facilities.

Mine design follows the process of dividing the resource into manageable units of ore (stopes) at an increasing level of detail at each stage.

- Block design. Collection of stopes within a defined region of the resource.

- Conceptual stope design. Stope designs at a low level of engineering input.

- Final stope design. Approved final design for the stope.

- Ring design. Approved design for the blastholes used to blast the ore in the stope.

- Blast design. Approved explosives design and blasting sequence to blast the ore in the stope.

Mine design is structured so that it can be integrated with the above plans. Block designs are developed within the LOM and 5YP. Conceptual stope designs follow on from block designs and exist within the 5YP and 2YP. Final stope designs are included within the 2YP, and ring designs are included in the 2YP and 3MS. The information collected for each stope is stored in a stope atlas, and this information is used to reconcile stope performance. The conclusions are then fed back into future stope designs.

27.10 SAFETY AND HEALTH

One of the six key values that provide the foundation for operations at Cannington is "Simply Safe." Each individual accepts the responsibility for the well-being of themselves, their team members, others in the workplace, their families, and the equipment with which they work, thus maintaining a productive and rewarding work environment and lifestyle.

Cannington has adopted the National Occupational Safety Association's (NOSA) 5 Star Safety System as the framework for managing safety performance. The system has been modified to meet Cannington's needs as the "Can Work Safe" program. Cannington currently has a four-star rating.

A number of initiatives and tools have been developed with a view to creating a positive attitude toward safety awareness and continually improving safety performance. These include—

- Positive Attitude Safety System (PASS), which encourages daily employee participation in identifying and correcting hazards in the workplace.

- Safety committees composed of employee-elected representatives.

- HAZCAN, a customised hazard identification and control system.

- Job Safety Analysis (JSA), which is a set of procedures that enable the analysis of tasks so that safety measures can be established.

- Recognition awards for employees and teams that demonstrate a proactive approach to safety at work.

Current legislation describes the health and safety requirements that must be in place to protect people and the workplace. Reference is made where appropriate within the Can Work Safe program to mines acts, regulations, and any other relevant legislation or code of practice that ensures safety of the highest standard.

27.11 MINE OPERATING AND CAPITAL COSTS

Cannington Mine was accessed by a full-size exploration decline (5.5 m wide by 5.5 m high at a 1:8 gradient) to a depth of 325 m below the surface as part of the data acquisition and evaluation

TABLE 27.5 Mine capital construction costs

Item	Cost AU$ (1998)
Main decline development	$14.5 million
Lateral access and infrastructure development	$35 million
Ore handling system	$27 million
Hoisting shaft and return ventilation shaft	$18.9 million
Mobile equipment fleet	$16.6 million
Level and stope development	$4.8 million
Dewatering system	$2.9 million
Mining infrastructure (fans, workshops)	$4.5 million
Mine services (paste fill plant, electrical, communications,)	$15.2 million
TOTAL	$139.4 million

phase of a feasibility study conducted from 1993 to 1995. Following project approval early in 1996, construction of all mine development and infrastructure was completed for commencement of production in October 1997, 17 months after approval. The first production ore was trucked to the surface in May 1997, and trucking continued from an average depth below the surface of 425 m until May 1998, when the hoisting shaft was completed.

27.11.1 Mine Capital Cost

Diamond drilling and geological modeling were used to identify the deeper parts of the mine as having the better grade and thicker lode intersections suited to maximising investment return. These areas were targeted for initial production. The capital cost to establish the optimum mine start-up was AU$139M (1998) and included the following major items:

- Main access decline from the surface to the hoisting shaft bottom at 645 m.

- A concrete-lined, 5.6-m in diameter, 645-m vertical hoisting shaft and 325-m return ventilation shaft.

- Level and stope development access in the first 2–3 years of production.

- Purchase of a 17-piece prime mobile and drilling equipment fleet.

- Underground ore handling system that included a fully automated jaw crusher, inclined conveyor system, and shaft loading system.

- Underground workshop, electrical reticulation system, vertical dam, and mine dewatering system.

- Surface paste fill manufacturing plant and delivery boreholes.

- Other items, including two main centrifugal exhaust fans, mine communication system, escape ladderway, and electrical, compressed air, and service water reticulation throughout the mine.

All aspects of the planning, design, and operating systems have been integrated to provide a best-practice mining system targeting high efficiencies in a safe work environment. Proven technologies have been adopted and refined to provide secure production at the design rate (Lennox et al. 1998). Design and construction were based on a mine life in excess of 20 years. The breakdown of major capital construction cost items is shown in Table 27.5.

The capital cost was driven by the mine design which considered the following factors:

- Mine infrastructure sized to accommodate future development of the adjacent Northern Zone resource.

- Mine access completed to full depth preproduction to access higher-grade ore on the deeper levels.

- Highly variable and often very poor ground conditions, particularly in the shafts and ore handling system development at the base of the mine.

27.11.2 Operating Costs

Mine production commenced in September 1997 and achieved sustainable production at 1.5 Mtpa nameplate capacity in January 1999. Within 12 months, the decision was made to embark on a business improvement program, an integral part of which was a mine optimisation program. This study has indicated the viability of increasing mine throughput to 1.8 Mtpa with minimal additional investment. It is scheduled to be achieved by early 2001. The mine unit operating cost per tonne of ore mined before the impacts of the above improvement program was AU$27/tonne and is expected to be reduced to less than $25/tonne.

The largest component of operating costs is mine labour. Mine labour totals 140 personnel, including underground mining crews, services contractors, maintenance, engineering, management, and administration. Other key cost drivers are cement associated with paste fill, development and stope ground support, and maintenance.

At nameplate throughput of 1.5 Mtpa, the following typical monthly production performance applies.

- 130,000 tonnes of ore crushed and hoisted.

- 20,000 tonnes of development waste.

- Placement of 75,000 wet tonnes of paste fill in stope voids.

- 580 m of lateral and stope development (20% capital, 80% operating).

- 9500 m of production drilling.

- 3000 m of ore body delineation diamond drilling.

The split of mining costs by activity is shown below in Figure 27.10. The key cost drivers within each activity area are mine labour, cement consumption in paste fill, ground support, and mobile equipment maintenance.

27.12 REFERENCES

Barton et al. 1974.

Burns, J.S., 1996. PASS Process Guide. PASS Internal Document.

Farcich A., M.S. Kemp, and A.G.L. Pratt. 1998. Recruiting for an Underground Workforce at Cannington. *Australasian Institute of Mining and Metallurgy Underground Operator's Conference*, 1998. 25–28.

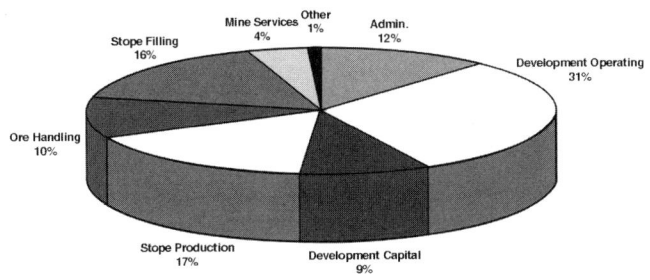

FIGURE 27.10   Mine costs by activity

Gillespie D.J., 1998. Contract Partnering–A Review of its Use in the Development of the Cannington Underground Mine. *Australasian Institute of Mining and Metallurgy Underground Operator's Conference*, 1998. 183–186.

Lennox A.W., P.H. McGuckin, and V. Seedwell. 1998. Learning about Safety. *Australasian Institute of Mining and Metallurgy Underground Operator's Conference*, 1998. 75–79.

Logan A.S. 1999. Integrated Ground Management–an Essential Component of our Licence to Operate. *Rock Support and Reinforcement Practice in Mining*. 259–265. Villaescusa, Windsor & Thompson (eds). 1999 Balkema, Rotterdam. ISBN 90 5809 045 0.

Roche M.T. 1994. The Cannington Silver Lead Zinc Deposit–at Feasibility. *The Australasian Institute of Mining and Metallurgy Annual Conference*. 193–196.

Rossiter A., 1998. Cannington Mine Development. *Australasian Institute of Mining and Metallurgy Underground Operator's Conference*, 1998. 277–285.

Skrzeczynski, R.H., 1993. From Concept to Cannington: A Decade of Exploration in the Eastern Succession. Symposium on Recent Advances in the Mt. Isa Block. *Australian Institute of Geoscientists Bulletin*, 13:35–38.

Stewart, S.B.V., and Forsyth, W.W., 1995. The Matthews Method for Open Stope Design, *CIM Bulletin July 1995*. 45–53.

Struthers M.A., M.F. Lee, and A. Bailey. 1994. Cannington Mine Feasibility Study Rock Mechanics, AMC Report No. AMC 192005. *BHP Minerals Cannington Feasibility Study documents*, Volume 6, Supplement 3.11.

Potvin, Y., 1988. Empirical Open Stope Design in Canada. PhD Thesis (unpublished), *The University of British Columbia*, 1988.

Potvin, Y., Hudyma, M., and Miller, H.D.S., 1988. The Stability Graph method for Open Stope Design, *90th CIM AGM*, Edmonton, May 1988.

Walters S., and A. Bailey. 1998. Geology and Mineralisation of the Cannington Ag-Pb-Zn Deposit: An Example of Broken Hill Type Mineralisation of the Eastern Succession, Mount Isa Inlier, Australia. *Economic Geology*. 93:1307–1329.

CHAPTER 28

Sublevel Stoping at Echo Bay's Lamefoot Mine, Republic, Wash.

Gordon L. Fellows

28.1 INTRODUCTION

The Lamefoot Mine is one of two gold mines that Echo Bay Minerals–Kettle River operates in northeastern Washington and is the fifth of six mines that have supplied ore to Echo Bay's Overlook Mill. The Lamefoot Mine is located in Ferry County, about 11 km northeast of the town of Republic and 153 km northwest of Spokane (Figure 28.1). At an elevation of 793 m, the mine occupies a northeasterly trending ridge in the western foothills of the Kettle Range. The area has been prospected and mined for gold since the turn of the century, and the nearby Republic District has produced over 62,200 kg of gold.

Echo Bay Minerals' exploration geologists discovered the Lamefoot Mine in 1989. The discovery was made using geophysical magnetic surveys and soil geochemistry, followed by rotary and core drilling. The first drill hole target, based on an overlap of the best magnetic and geochemical anomalies, proved to be ore. Exploration of the ore body continued through 1992, when mine development began.

Construction of the surface facilities began in May 1992 and the first blast for the portal was set off in early July 1992. By October 1992, the 671-m main haulage access, at an elevation of 808 m, was completed, and vertical expansion of the mine began. To date (the end of 1999), mining crews have excavated more than 16 km within the mine. Seven ore zones have been developed along a strike length of 500 m between elevations of 634 and 986 m (Figure 28.2). Mining levels have been established at vertical intervals of approximately 20 m. As of the end of 1999, the mine's ore reserves have been reduced to 360,890 tonnes at an average grade of 7.12 gm/t while 2,293,721 tonnes at an average grade of 7.45 gm/t gold have been produced since 1992 (Table 28.1).

Mining the ore body by trackless longhole stoping has allowed regular daily production to be as high as 1,800 t/d based on a schedule of two 10-hr shifts per day, 6 days a week. Mining is currently taking place at 1,100 t/d as the extremities and remnants of the ore body are mined. The ore is hauled 11 km to the Overlook Mill. Milling consists of a carbon-in-leach whole-ore treatment and a modified Merrill-Crowe recovery method.

28.2 GEOLOGY

The Lamefoot gold deposit is hosted by Permian siliciclastic rocks and limestone exposed within the Republic graben, a north-trending structural depression approximately 16 km wide. The graben is bounded on the east and west by the Kettle and Okanogan metamorphic core complexes, respectively. These Permian rocks accreted to the northwestern margin of North America in mid-Jurassic to early Cretaceous times. The Permian limestone is fine grained to micritic and usually massive. It strikes

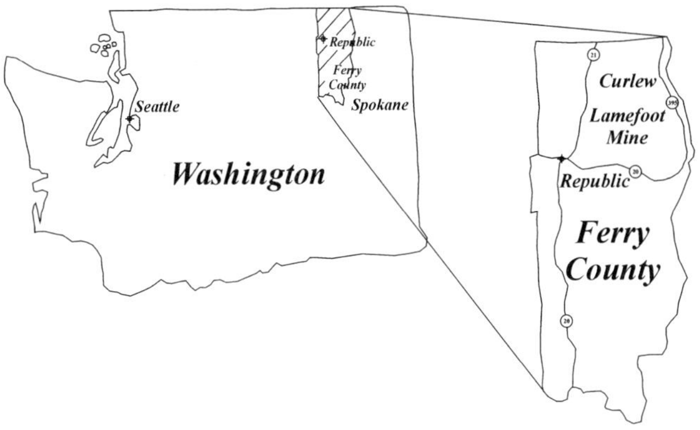

FIGURE 28.1 Map showing the location of the Lamefoot Mine

north-south and generally dips 50° to 70° west. Stratigraphically, beneath the limestone lies a package of Permian siliciclastic rocks ranging from siltstones to coarse-grained, heterolithic conglomerates. There are also interbeds of volcanic siltstones and greenstones. The clastic rocks have undergone widespread propylitic alteration and some bleaching in mineralized areas. The carbonates are relatively unaltered except where mineralized.

Gold mineralization occurs within a steeply dipping tabular body localized along the Anfo Fault, a north-trending, high-angle fault that juxtaposes massive limestone (footwall) against fine-grained, thin-bedded, siliciclastic rocks (hanging wall). The mineralized body extends 500 m along strike and 200 m along dip and ranges from 3 to 43 m wide. The Anfo Fault and the mineralized zone are cut and offset by a series of moderately northwest-dipping, northeast-trending cross faults, as well as several high-angle north-trending structures. Gold is generally associated with two types of host rock: a massive, pervasively silicified, magnetite-hematite-sulfide body in the immediate footwall of the Anfo Fault and quartz-pyrite-chalcopyrite veins within the siliciclastic rocks of the immediate hanging wall of the Anfo Fault. Mineralized rock close to the Anfo Fault can also contain silica veins and quartz vein breccia, which is usually associated with high-grade gold. The massive magnetite-hematite-sulfide body probably predates the gold mineralization, but later faulting and brecciation formed an excellent host for silica-gold deposition. The gold mineralization is probably the result of Tertiary epithermal activity within the graben and occurs as very fine, particulate free gold.

* Echo Bay Minerals, Republic, WA.

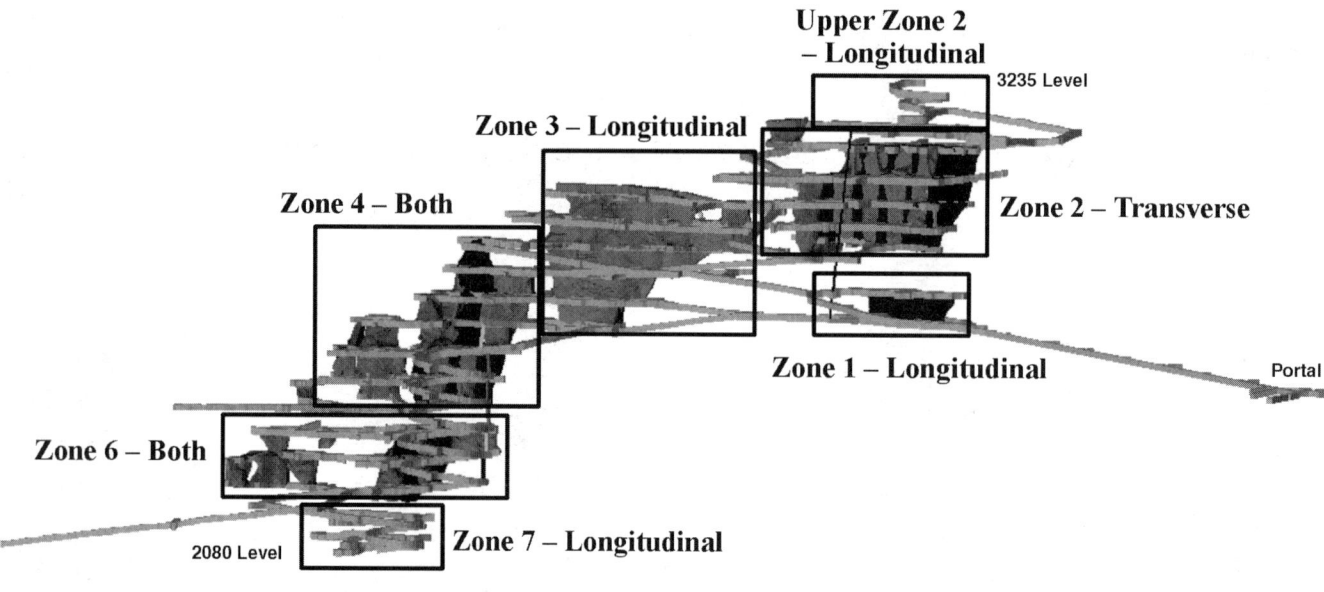

FIGURE 28.2   Seven ore zones along the strike length

TABLE 28.1   Historic production

Year	Tonnes	Kg, Au	Grade, gpT
1993	83,794	688	8.21
1994	32,153	231	7.19
1995	439,618	3,524	8.02
1996	496,652	4,220	8.50
1997	498,990	3,329	6.67
1998	395,603	2,786	7.04
1999	346,912	2,293	6.61
Total	2,293,721	17,071	7.44

TABLE 28.2   Basic data for transverse stoping

Stope Height	15.24 m
Crosscut—Waste	4.3 × 4.6 m
Topcut—Primary	7.3 × 4.6 m
Topcut—Secondary	6.1 × 4.6 m
Primary Width	7.3 m
Secondary Width	14.6 m
Primary—Number of Holes Per Fan	4
Primary—Drilled Meters Per Fan	61 m
Primary—Average Powder Factor	0.31 kg/tonne
Secondary—Number of Holes Per Fan	23
Secondary—Drilled Meters Per Fan	187 m
Secondary—Average Powder Factor	0.25 kg/tonne
Fan Burden	2.4 m
Hole Spacing	2.4 m
Hole Diameter	76 mm
Raise—Meters Drilled	351 m

28.3   MINING METHODS

The mine is entered via the 2650 level from the portal. A longitudinal, or strike, ramp leads to a spiral ramp up to the 3235 level at the top of the mine, while another spiral ramp goes down to the 2080 level at the bottom of the mine. The ramps are generally 5.5 m wide by 4.6 m high with a grade of 12% and a turning radius of the spirals of 20 m. Drifts are 5 m wide by 4.6 m high and are developed at 20-m levels, thereby defining a 15-m-high block of ore to be mined. Most level development takes place in the limestone footwall of the ore body. Delineation core-drilling of the ore body takes place from the levels. The ore body is drilled out on 8-m centers along strike and dip.

The primary method of mining at the Lamefoot Mine is overhand sublevel longhole stoping with delayed backfill. This is carried out in either transversely or longitudinally, depending on the width of the ore body in the area to be mined. The ore zones vary from 30 to almost 150 m in length and 3 to 43 m in width. Drift-and-fill mining is used as necessary to recover ore remnants.

28.3.1   Zone 2—Standard Transverse Stoping

In Zone 2 (between the 2830 level and the 3070 level), the width of the ore body exceeds 18 m, so transverse longhole stoping is used. Primary stopes are developed by crosscutting to and across the ore body from the levels in the limestone footwall. These primary crosscuts are 4.3 m wide by 4.6 m high in waste and are then expanded to be 7.3 m wide once in ore (Table 28.2).

Ground support consists of 2.1-m-long Split Set rock bolts and mats. Once the topcut is complete, then cable bolts are installed on a 2.4- by 2.4-m bolting pattern with 5.5-m-long holes being drilled by an Atlas Copco Simba H1354 cable bolter.

The crosscuts are on 22-m centers on each level and are aligned vertically with the crosscuts on the level below so that as the primary stopes are filled, vertical concrete pillars will be formed. The stope is then drilled out vertically between crosscuts using an Atlas Copco Simba H1354 longhole drill. A standard raise pattern is used (Figure 28.3). A 2.4-m spacing by 2.4-m burden pattern of 76-mm diameter holes is used with a five-spot or easer row running perpendicular from the raise to open up the slot against the hanging wall.

The 15-m-high raise is shot in one blast and cleaned up, after which the slot is blasted and opened up. Slices of ore 2.4 m thick can then be blasted off as space and production needs allow. All stope blasting is done with Dyno RU sensitized emulsion that is pumped from a Getman A-64 emulsion unit. All stope production holes are stemmed with 1.7 m of sand on the bottom

Lamefoot Mine
Standard Drop-Raise Pattern
Single Shot to 15.25 m Length

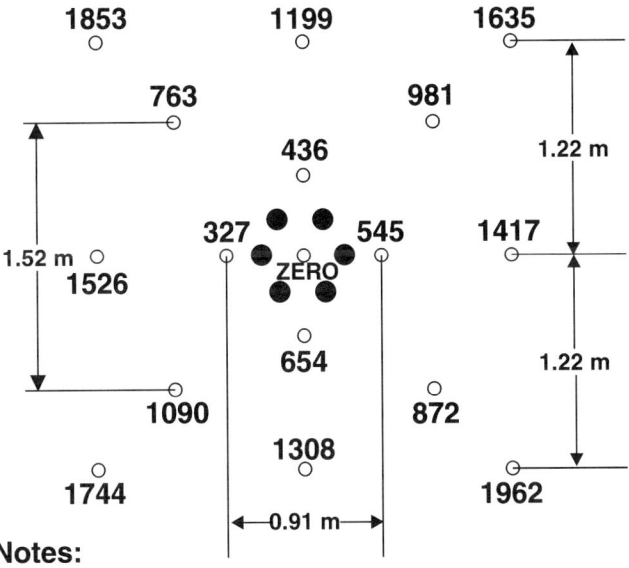

Notes:

*In-Hole Delay is #6LP or 2000 ms

*2 or 3 cap and boosters per hole

*23 holes total, 17 holes loaded

○ Loaded hole (76 mm diam.)

● Reliever hole (152.4 mm diam.)

FIGURE 28.3 Standard raise pattern

and 2.1 m of sand on the top and timed using MS Nonels. For stope production, an average powder factor is 0.3 kg/t, not including the raise.

The stopes are mucked out with a Wagner ST6-C Scooptrams fitted with ejector buckets. The stopes are mucked manually for as long as possible, but once the scoop must go beyond the brow, then remote mucking is done using Moog remote control units. In Zone 2, the stopes can be mucked directly into an 13.8-m² ore pass that runs from the 3005 level to the 2650 portal level.

Once a primary stope is mucked out and surveyed using Optech's Cavity Monitoring System (CMS), backfilling the stope with cemented rock fill must start. Approximately 11.8 tonnes of run-of-mine waste from the waste dump is placed into the rear portion of the box of a JCI 2604 C haul truck. This truck then goes to the batch plant just inside the portal, where 1,182 kg of cement slurry is delivered, via spraybars, over run-of-mine waste and into the front portion of the truck box. The result is a cemented rock fill with approximately 5% cement content by weight. The truck is then driven up the ramp to the top access into the stope and backs into the crosscut until it hits the bumper block, at which point it dumps its load. The slurry and run-of-mine waste have mixed on the trip up the ramp and mix further as they tumble into the stope. This process continues until the stope is filled, and the stope above is ready to be mined. Once the top of the zone is reached and the final stope in a column is filled, the top access is jammed tight to the back.

The automated and programmable batch plant is a Thiessen Tornado 1000 manufactured by Team Manufacturing, Ltd. The main components are a cement-weigh hopper, a water-weigh

hopper, cement discharge hopper, colloidal mixer, dust-collecting system, control system, and motor controls. The system has a cleanout and batch mixing time of about 4.5 min. Based on cylinder test samples, the compressive strengths of the cemented rock fill range from 1.4 to 12.1 MPa. An average strength of 6.3 MPa is possible. This mix will maintain a vertical wall for as long as needed and usually contributes less than 5% dilution to adjacent stopes.

Secondary stopes are developed between the now-cemented primaries. A crosscut 4.3 m wide by 4.6 m high is driven in waste midway between two primary crosscuts. Once this crosscut enters the ore zone, it is widened to 6 m, driven across the ore zone, and bolted with seven 5.5-m-long cable in every row. Three vertical bolts centered on the centerline of the stope are installed 2.4 m apart, and two other bolts, one rotated 20° from vertical and one rotated 40° from vertical, are installed in each upper corner.

The production longholes for the stope are then drilled. The raise is drilled with the same pattern as before (Figure 28.3). Drilling of the rings is next. The production mine engineer has used the CMS surveys of the two primaries on either side to design a fan pattern that will ensure almost 100% recovery and a minimum amount of dilution (Table 28.3). Holes approaching the cemented rock fill of the primaries are kept 1 m away, and all others are holed through to the level below. Once again the raise is dropped, the is slot opened up, and blasting of the rings or slices takes place as needed by production demands.

The secondary stopes are filled with uncemented rock fill that is either development waste from underground or material from a surface borrow. A running balance sheet is maintained to ensure that there is enough run-of-mine waste on the surface to produce a cemented rock fill, so any excess can be placed directly in the stopes. The borrow material consists of clay, gravel, and cobbles that come from a site about 0.4 km from mine and stockpiled at the portal. All the uncemented rock fill is dumped from a truck into the stope and leveled 1.5 m below the sill with a Caterpillar D-6. The remaining 1.5 m is filled with cemented rock fill, which then serves as a smooth mucking surface and minimizes dilution from below when the stope above is mined.

28.3.2 Zone 3—Longitudinal Stoping

Zone 3, between the 2680 and 2940 levels, rarely exceeds 18.3 m in width, so it is mined longitudinally. Stopes longer than 120 m are possible in this zone. As mentioned previously, level development takes place in the footwall for the length of the ore body. Three 4.3- by 4.6-m-high crosscuts, one at each end and one in the middle, are driven to access the ore body. These entry points allow for multiple mucking points, shorter mucking distances, the capability of panelizing the stope if ground conditions should warrant, and filling the stope while mining is on-going. The entire opening is supported with 5.5-m-long cable bolts installed on a 2.4- by 2.4-m pattern. Drifting takes place between the crosscuts so they are connected, and then the ore body is slashed to full width. The standard raise pattern (Figure 28.3) is drilled on the north end of the stope. The rings of the stope are drilled out with 2.4-m burdens between rings. Spacing of the hole toes averages 2.4 m, while the collars may be closer to fit in the width of the drift. Blind holes or holes that don't hole through into the undercut are subdrilled 1 m below the ore-waste contact. The production sequence is the same, that is, the raise is dropped, the slot opened up, and the rings blasted as needed.

Filling the stope is done with uncemented rock fill as described above. The sequence of mining and filling the stope depends on stope geometry and ground conditions. If there is a bend in the stope that would prevent the remote load-haul-dump (LHD) operator from having line-of-sight to the machine and muck pile, then the stope must be filled sufficiently to control the muck and keep it in sight. Filling also might be necessary to

TABLE 28.3 Zone 2—secondary stope analysis

Stope	CMS Predicted (tonnes)	Reported (tonnes)	Designed (tonnes)	CRF dilution (tonnes)	(%)	Other dilution (tonnes)	(%)	Overall dilution (tonnes)	(%)	Mining losses (tonnes)	(%)
2875/2810 C	35,430	38,417	33,955	689	1.9	0	0.0	689	1.9	859	2.5
2940/2875 C	12,967	11,305	12,364	141	1.1	685	5.3	825	6.4	521	4.2
2875/2830 E	15,705	17,420	18,109	269	1.7	0	0.0	269	1.7	1,582	8.7
2940/2875 E	18,031	14,961	16,174	358	2.0	785	4.4	1144	6.3	382	2.4
2875/2830 G	16,905	16,465	19,456	375	2.2	0	0.0	375	2.2	3,332	17.1
2940/2875 G	23,613	26,563	22,961	735	3.1	113	0.5	848	3.6	674	2.9
2875/2830 I	12,699	13,905	13,044	436	3.4	0	0.0	436	3.4	1,223	9.4
2940/2875 I	23,533	23,773	26,091	759	3.2	0	0.0	759	3.2	3,226	12.4
2940/2875 K	17,416	18,172	21,264	443	2.51	63	0.4	505	2.9	3,956	18.6
2940/2875 LMNO	18,890	20,510	17,727	0	0.0	1381	7.3	1381	7.3	918	5.2
3005/2940 G	26,138	26,909	24,868	962	3.7	589	2.3	1551	5.9	703	2.8
3005/2940 K	29,829	29,495	27,893	1,454	4.9	675	2.3	2128	7.1	607	2.2
3005/2940 LMNO	23,412	27,384	21,650	0	0.0	1350	5.8	1350	5.8	362	1.7
3070/3005 LMNO	16,823	18,725	16,030	0	0.0	988	5.9	988	5.9	548	3.4
3005/2940 E	18,158	20,115	16,583	744	4.1	442	2.4	1186	6.5	495	3.0
3005/2940 C	6,725	6,345	5,938	112	1.7	595	8.8	706	10.5	155	2.61
3005/2940 I	32,410	33,279	30,415	1,514	4.7	851	2.6	2365	7.3	2,041	6.7
3070/3005 E	19,853	19,869	20,132	209	1.1	573	2.9	782	3.9	1,143	5.7
3070/3005 K	23,531	24,283	22,014	744	3.2	508	2.2	1252	5.3	414	1.9
3070/3005 I	23,018	24,065	22,072	1,455	6.3	3474	15.1	4929	21.4	1,757	8.0
3070/3005 G	28,179	27,812	26,932	1,590	5.6	862	3.1	2452	8.7	709	2.6
Totals	443,265	459,772	435,669	12,989	2.9	13,932	3.1	26,922	6.1	25,606	5.9

control failures of the hanging walls or footwalls. If any failures are large enough or persistent enough, then a number of rings might be left in place to act as a mid-stope pillar, and a new raise would be drilled. Filling with cemented rock fill against the pillar might allow the pillar to be recovered at a later date. The uncemented rock fill material is capped with 1.5 m of cemented rock fill to provide a hard mucking surface and minimize dilution from below.

28.3.3 Lower Zone 4 and Lower Zone 6—Modified Transverse Stoping

During production from the last four stopes of Zone 2, there were failures that drastically reduced the productivity of the zone. The 3005 I raise was blasted and mucked clean, and then the slot was blasted open. At this time, almost the whole stope ore block dropped onto the undercut, a section about 30 m long. All the ore was recovered, but failures occurred in the remaining secondary stopes of the 3070 level during production blasting of the stopes. Two faults formed a wedge over the entrance to the undercut that was knocked loose each time. These faults existed during the mining of the primaries and two other secondaries, but showed no signs of failure.

A study of these failures concluded that there had been a decrease in the resistance to shear failure where the cemented rock fill pillars were in contact with the stope ore blocks. The multiple blasts during mining had caused the rock fill to break up along the ore contact, so that once the slot was mined in the 3005 I stope, it became a cantilevered beam with little, if any, side support. The beams broke off at the point where it could support itself. The fault-defined wedges in the 3070 stopes did not have any end support, so once blasting started, they vibrated loose. Mass blasting was also used to control these wedges as much as possible when they were anticipated.

Because of permitting complications and delays, Zones 2 and 3 were nearly mined out before mining extended to the bottom of Zones 4 and 6, which were the next areas where the width of the ore made it necessary to use transverse stoping methods. With the failures of Zone 2 in mind, it was necessary to come up with a new stope design that would eliminate future failures. A keystone or wedge method was designed in which the primaries were inverted wedges and the secondaries were shaped as keystones between the upside-down wedges. The primary and secondary crosscuts were kept the same size and on the same spacings (11 m). Just the drilling patterns were modified to induce the wedges and keystones. The outside holes of the primaries were rotated out 12.8°, which allowed more normal force to be applied to the cemented rock fill-ore contact and increased resistance to shear failure. A sample design can be seen in Figure 28.4.

At the same time, it was decided to use 5.2-m-long, 22.2-mm-diameter grouted rebar dowels instead of grouted cable bolts for support because negligible amounts of sag across the span were observed. The transverse stoping portions of Zones 4 and 6 were mined without failures. The remainder of the zones were mined with longitudinal stopes.

28.4 UNIT OPERATIONS

28.4.1 Development Drilling

Drilling of rounds is done primarily with an Atlas Copco H252 boomer capable of drilling holes 3.65 m long. A two-man development team composes each drilling crew. One miner operates the two-boom jumbo in one heading, while the other operates a rock bolter in another heading. If there are enough available faces and with proper support from the muck crews, more than two rounds of blasting per shift are possible. A standard round consists of 36 parallel, 44-mm-diameter holes with a burn consisting of five 44-mm-diameter holes and four 76-mm diameter holes. Waste rounds vary in size depending on the purpose of the heading.

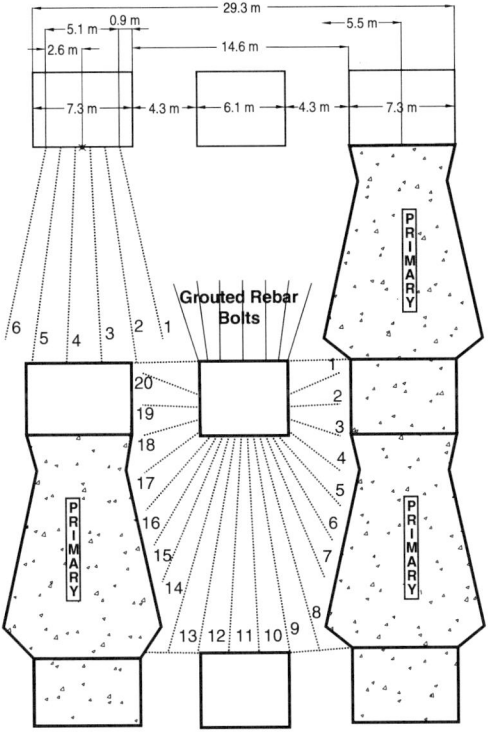

FIGURE 28.4 Keystone transverse stope design

28.4.2 Blasting

Development rounds are blasted with pumped emulsion from the Getman emulsion trucks and Dyno RU sensitized emulsion is supplied by a hose to the face. Each round is loaded by one of the two development crew members. Timing of the round is done with nonelectric LP caps with delays from 0 to 15. Use of trim powder is determined by ground conditions. Stope blasting is done by a two-man crew consisting of a longhole driller and a helper. The longhole driller from the previous shift will clean the holes to be shot, and the next shift will load and shoot the rings. All blasting takes place as the shift goes off.

28.4.3 Ground Support

Ground support in development headings is provided by four 30-cm-wide by 3-m-long mats held in place with 2.1-m-long Split Set rock bolts for every round of advance. A Tamrock Robolter H595 installs the rock bolts. Wire mesh is used only if ground conditions warrant; if so, it is placed with the bolter as well. Rebar resin bolts and double-strand cable bolts are also available in the mine warehouse if needed.

28.4.4 Mucking

Mucking the stopes and faces is done using 4.6-m³ LHDs. Most of the LHDs have EOD or ejector buckets, which allows the 26-ton haulage trucks from zones other than Zone 2 to be loaded easily in a 4.6-m heading. In Zone 2, an ore pass can be accessed from all levels, so mucking can be done directly from the stope to the ore pass with the LHDs. Ore from Zone 2 is hauled from the ore pass chute using 33-ton trucks. The smaller LHDs are used for clean up of faces and sumps and for construction.

28.4.5 Manpower and Shifts

Operations take place underground at the Lamefoot Mine 6 days a week. Two 10-hour shifts per day run from 07:00 to 17:00 and from 19:00 to 05:00. The 2 hours between shifts allow the mine to be aired out after blasting and permits daily maintenance and

TABLE 28.4 Mine crew #1 assignments

No.	Position	Shift schedule
1	Shift Supervisor	Rotating (D/N)*
1	Jumbo Driller	Rotating (D/N)
1	Jumbo Bolter	Rotating (D/N)
1	Longhole Driller	Mon–Thurs (D/N)
1	Remote Mucker Operator	Rotating (D/N)
1	Mucker Operator	Rotating (D/N)
4	Ore/Backfill Truck Drivers	Rotating (D/N)
1	Grader Operator/Dump Maintenance	Tues–Fri (D)†
1	Mine Repair/Pumps & Sump	Tues–Fri (D)
12		

*(D/N) Day and night shifts
†(D) Day shift only

TABLE 28.5 Mining equipment list

Load-haul-dumps (LHDs)		Utility vehicles	
1	Toro 400D 6 yd³	1	Caterpillar D4 H Dozer with Jamming Attachment
2	Wagner ST 6-Cs w/ Moog Remote Controls	6	Ford Tractor 1920
1	Wagner ST 6-C	1	Getman Scissor Lift A-64
1	JCI 600 M 6 yd³	2	Getman Emulsion Units A-64
1	JCI 922D 2.5 yd³	1	Getman Hiab Crane A-64
1	Wagner ST 3.5	1	Getman Trammer Personnel Carrier 799
		1	Toyota Land Cruiser BJ-745
Underground drills		1	Intermountain Man Carrier 1 MM 4x4-P
1	Tamrock Minimatic H 205 4.26 m	1	Getman Lube Truck A-64
1	Atlas Copco H252 Boomer 3.65 m	1	Getman Trammer Flatdeck with Man Bucket
2	Tamrock Robolter H595 with BH 24S	1	Caterpillar IT-18B
1	Tamrock Monomatic H105 3 m	1	Caterpillar D6D Dozer
1	Atlas Copco Simba H1354 Cable Bolter	1	Caterpillar 1406 Grader
1	Atlas Copco Simba H1354 Longhole Drill	1	Caterpillar 416B Backhoe with Rockbreaker
Trucks		1	Caterpillar 966 C Front End Loader
2	EJC 430 Underground Trucks		
5	JCI 2604 C Underground Truck		
2	Wagner MT426-30 Underground Trucks		
2	Wagner MT433 33 Ton Haul Trucks		

minor repairs on equipment. Three mine crews work on a rotating schedule (7-2 and 7-5) to provide the personnel for each shift. Table 28.4 lists positions. An underground supervisor or shifter provides supervision to each crew, and the mine superintendent is on site daily.

A listing of the equipment used at the Lamefoot Mine can be found as Table 28.5. These pieces of equipment were sufficient to provide 1,800 t/d of ore and were able to support a total of 3,650 t/d of material handling. Some of this equipment has been put to use at other mines as production has decreased at Lamefoot Mine because of decreased reserves.

TABLE 28.6   Forecasting parameters

Item	Units	2000 Average	2000 Max
Ore	Tonnes/Day	1036	1100
URF	Tonnes/Day	805	900
CRF	Tonnes/Day	95	450
Longhole Drilling	Meters/Day	200	305
Cable Bolt Drilling	Meters/Day	100	229
Development	Rounds/Day	1.3	3.0
Total Material Moved	Tonnes/Day	2021	2550

28.5 FORECASTING

The engineering department, with input from the production department, prepares all production forecasts for the Lamefoot Mine. Once a year, a life-of-mine plan is prepared, and from this plan, stope production, development footage, and other particulars are scheduled for each month of the following year. The parameters for planning and budgeting are given in Table 28.6. Quarterly plans for the following 2 years are also developed, as well as annual plans for the remainder of the mine's life.

Using the operating plan as the basis for expected tonnage and ounces, quarterly forecasts for the coming year are also prepared. These rolling quarterly forecasts show actual performance to date, as well as providing the basis for continual updating of monthly forecasts for the next 12 months. These monthly forecasts provide the mill with the newest data and expectations so that so it can make adjustments to meet the operating plan for ounces poured. Weekly forecasts are also prepared, but project only the sources of the week's tonnage and sources.

28.6 SUMMARY

Longhole stoping has been used at the Lamefoot Mine with great efficiency for all of the mine's life. Ground conditions and geology have changed with time and depth, so stope dimensions and processes have had to be adapted. Developing the mine as described above has allowed the effects of mining wider ore bodies to be minimized, because longitudinal stopes could be changed into transverse stopes. The success of the keystone transverse stoping method allowed us to eliminate failures that might have otherwise occurred in Zones 4 and 6 because of conditions similar to those in Zone 2. The Lamefoot Mine has been a success so far and will remain so for the remainder of its life.

28.7 ACKNOWLEDGMENTS

I would like to thank the staff and management of Kettle River Operations for the time and support given to me to write this chapter. Their aid was necessary for gathering data and editing that made this chapter possible. Junior Russell and Dave Riggleman contributed to the mining sections, while Max Lembeck and Rich Leep contributed to to the stoping and geology sections, respectively.

28.8 REFERENCES

Echo Bay Minerals–Kettle River Operations. *2000 Operating Plan.*
Echo Bay Minerals–Kettle River Operations. *2000 Ore Reserve Report.*
Fellows, G.L., and G. Stolp. December 1997. *The Integration of a Cavity Monitoring System and Computer Triangulations for Longhole Stope Analysis at Echo Bay's Lamefoot Mine.* Paper presented at the annual meeting of the Northwest Mining Association, Spokane, WA.
Leep, Rich. July 1997. Personal communication concerning geology at the Lamefoot Mine, Republic, WA.

CHAPTER 29

Sublevel Open Stoping at El Soldado Mine: A Geomechanic Challenge

Nolberto V. Contador* **and Marcelo F. Glavic†**

29.1   INTRODUCTION

El Soldado copper mine is located 132 km northwest of Santiago, Chile, on the western slopes of the Coastal range about 830 m above sea level (Figure 29.1). The mine is operated by Cia. Minera Disputada de las Condes Ltda. (CMD), an ExxonMobil Coal and Minerals Co. (EMCM) affiliate. Since the initiation of mining in 1842, production at El Soldado has been nearly continuous, although under a number of owners. EMCM assumed control of El Soldado in 1978 and since then has removed about 70 million tonnes of ore containing 1.8% copper using the sublevel open stoping mining method.

In 1989, an open pit began operating to support an increase in the production rate to the current production of 18,000 tonnes/d (Figure 29.2). Today, the underground mine provides 30% of the total concentrator feed.

Mineralization occurs as numerous isolated tabular ore bodies with variable dimensions from 100 to 200 m long, 30 to 150 m wide, and 80 to 350 m high. Ground conditions are classified according to Laubscher's rock mass classification method as competent and are characterized by a mining rock rating from 60 to 70. The stress regime is moderate and ranges from 15 to 30 MPa. These conditions facilitated the development of large open cavities with dimensions from 40 to 90 m wide, 50 to 290 m long, and up to 300 m high.

To maintain stability and maximize recovery, large ore bodies are divided into smaller production units or stopes, leaving temporary pillars to be recovered at the end of ore body life.

The mine has experienced a series of small-scale instabilities without affecting the mine plan significantly. The only large-scale collapse occurred under controlled conditions in one of the biggest underground stopes, coinciding with the start of open-pit mining.

A comprehensive instrumentation program is applied to monitor rock mass behavior. Stress changes and surface displacements are recorded continuously. Surface tension cracks are mapped and correlated with existing major faults. Instabilities are detected in advance, and the underground extraction rate is controlled in relation to rock mass response.

Mining at El Soldado is carried out under strict safety conditions. Corporate safety programs built upon involvement and risk mitigation have contributed to continuous improvements in safety performance. This has resulted in a safety record (as of 1999) of 3 years and over 4 million work hours without a lost-time injury among either employees or contractors.

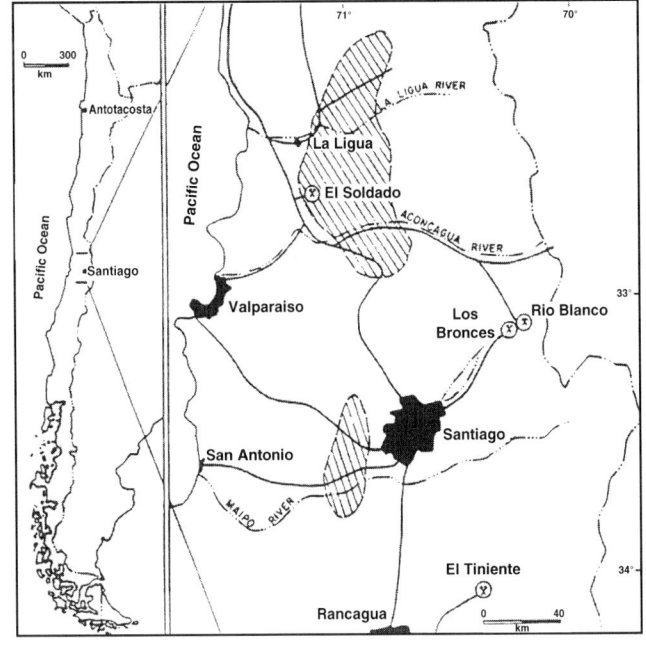

FIGURE 29.1   Location of El Soldado Mine, Chile

29.2   GEOLOGICAL AND GEOMECHANICAL ENVIRONMENT

29.2.1   Geologic Background

The deposit is thought to be of epigenetic origin and is placed in the Lower Cretaceous Lo Prado Formation. The Lo Prado Formation overlies the Upper Jurassic Horqueta Formation of continental volcaniclastic characteristics and underlies the Lower Cretaceous Veta Negra Formation composed mainly of flood basalts and intrusions of continental sediments. The Lo Prado Formation is composed of interlayered trachytic and andesitic flows, pyroclastics of acidic intermediate composition, and some fossiliferous sedimentary rocks associated with shallow marine deposition (Figure 29.3). Dikes of acidic and basic composition cut through the whole column. Some trachytic dikes are feeders for the corresponding trachytic flows.

---

* El Soldado Geomechanics Engineer, Chile.
† El Soldado General Manager, Chile.

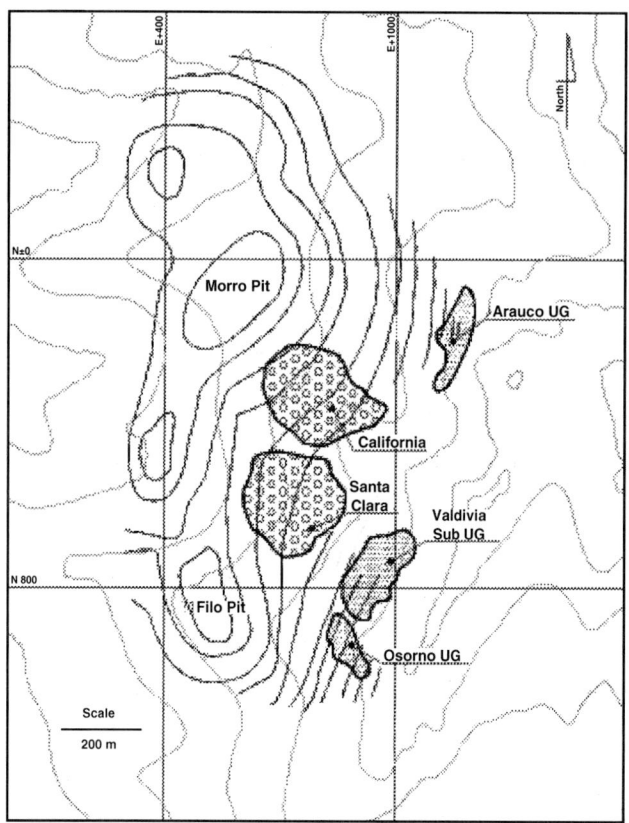

FIGURE 29.2   Overview of El Soldado Mine

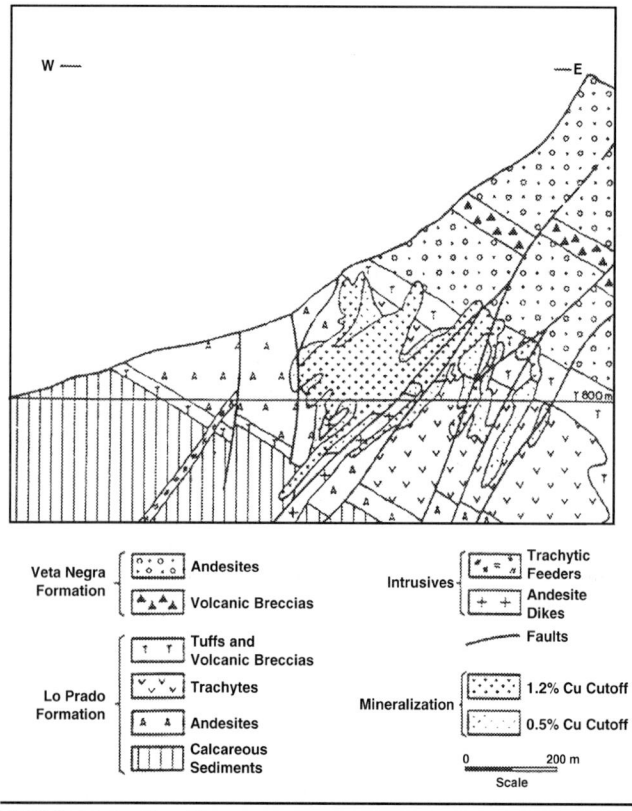

Veta Negra Formation	Andesites
	Volcanic Breccias

Lo Prado Formation	Tuffs and Volcanic Breccias
	Trachytes
	Andesites
	Calcareous Sediments

Intrusives	Trachytic Feeders
	Andesite Dikes

— Faults

Mineralization	1.2% Cu Cutoff
	0.5% Cu Cutoff

0   200 m
Scale

FIGURE 29.3   Simplified geologic cross section

The principal host rocks are trachytes, followed in importance by andesites. Tuffs of trachytic or andesitic composition are less common host rocks, as are the sedimentary intrusions. Copper mineralization of the El Soldado deposit has been recognized throughout an area 1,800 m long north-south and 800 m wide east-west in the upper member of the Lo Prado Formation. Mineralization is found as numerous isolated ore bodies with variable dimensions.

The ore bodies, with a strong structural control, are oriented parallel to north-south and east-west fault systems and discordant with stratification. Lateral limits of the ore bodies are characterized by abrupt variations in copper grade. The transition from high-grade mineralization (1.2% to 2% Cu) to low-grade (0.5% to 1.2%) areas outside the ore body takes place within a few meters.

Sulfide mineralization is of primary type. Ore bodies typically exhibit an outer pyrite-rich halo, followed inward by a core of abundant chalcopyrite and bornite with minor amounts of chalcocite and hematite. Main gangue minerals are calcite, quartz, chlorite, epidote, and albite.

29.2.2 Geologic Structure

Major structures (faulting and bed contacts) and rock fabric (jointing) within a rock mass play a critical role in determining the extent of any likely instability surrounding a proposed excavation at El Soldado Mine. The major structures have been mapped and identified on most levels of the mine. The geologic mapping campaign resulted in a complete three-dimensional picture of the localized network of major structures and provides a rich basis for understanding the mechanical behavior of the rock. Seven main fault systems and a system of bed contacts have been defined within the limits of the ore deposit (Figure 29.4).

Information on minor discontinuities, including joints, shear fractures, bedding planes, and fissures, has been collected in underground and open-pit exposures by using the cell mapping technique. These minor discontinuities were grouped into five sets of orientations similar to those of the main fault systems.

29.2.3 Rock and Rock Mass Properties

Rock mass property values are usually estimated using methods that relate rock mass properties to rock substance and discontinuity properties. The approach selected to obtain rock mass property values is that developed by Bieniawski (1978). The ratio of rock mass modulus to intact rock (substance) modulus (r) is related to the rock quality designation (RQD) (Deere 1968).

$$r = E_{mass}/E_i = 0.0225e^{0.013(RQD)}$$

where E_{mass} = Rock mass, Young's modulus

        E_i = Intact rock, Young's modulus

        r = Modulus ratio.

Through back-analysis of failures and fracture modeling studies, it was found that the strength of rock masses varies with the square of modulus ratio (r). The contributions of both intact rock and joints are combined to obtain the rock mass friction angle and cohesion. Thus, rock mass strength properties were calculated as follows:

$$\sigma c\ mass = r^2\sigma ci$$
$$\sigma t\ mass = r^2\sigma ti$$
$$\phi\ mass = r^2\phi i(1-r^2)\phi frac$$
$$C mass = r^2 Ci(1-r^2)Cfrac$$

where σc = compressive strength

        σt = tensile strength

        C = cohesion

        ϕ = friction angle

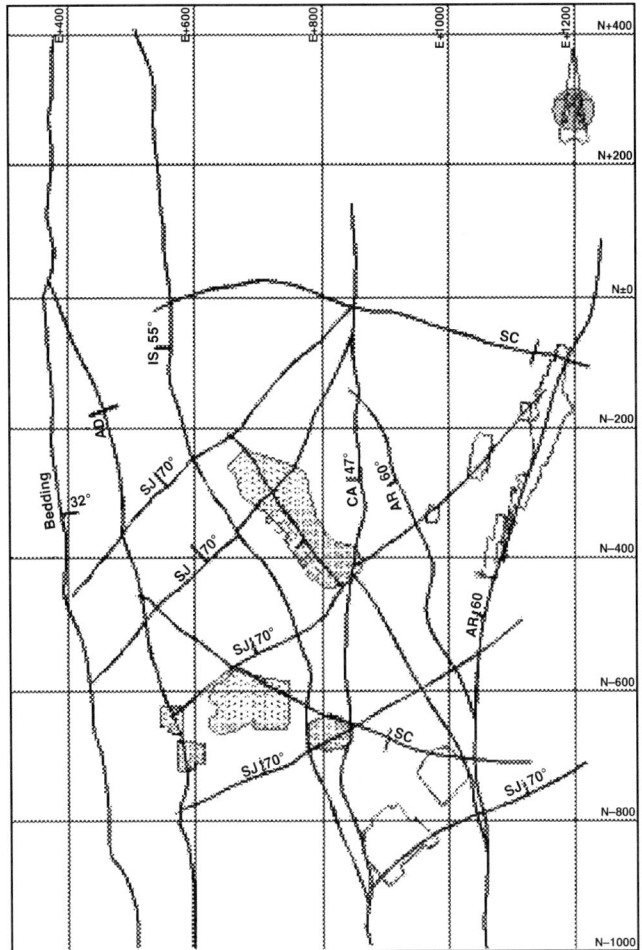

FIGURE 29.4   Major fault systems, El Soldado Mine

TABLE 29.1   Rock mass property values

| Rock type | Parameter rating | | | | Mining rock mass rating | Adjusted mining rock mass rating, 80% due to joint orientation |
	IRS	ROD	JS	JC		
Tuff	14	12	7.5	30.7	64.2	51.4
Trachyte	20	12	6.8	30.7	69.5	55.6
Andesite	16	12	8.1	30.7	66.8	53.4
Veta Negra	18	12	8.1	30.7	68.8	55.0

i = intact

mass = rock mass

and   frac = fracture.

No empirical relation is available to estimate the rock mass Poisson's ratio (v = nu). Based on the literature, at El Soldado, Poisson's ratio for the rock mass was estimated to be 1.1 times Poisson's ratio of the intact rock.

Table 29.1 shows rock mass properties for various ranges of RQD. Rock mass properties were validated by numerical analysis.

29.2.4   In Situ Stress

Currently at El Soldado, the induced state of stress (after excavation) is a major mine design criterion and monitoring objective. Mining has altered the rock mass, and it is difficult to find virgin stresses at the perimeter of the mine. In an attempt to

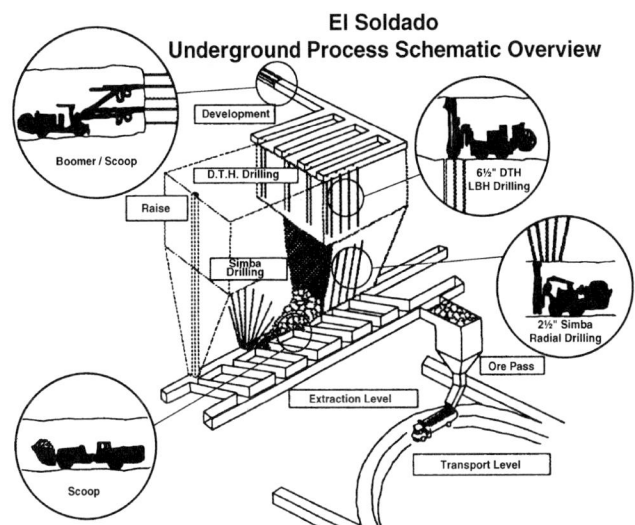

FIGURE 29.5   Isometric view of underground portion of the mine

obtain information on the in situ stress in critical areas of the mine, measurements have been collected using several different methods (CSIR, CSIRO, USBM). The results show that stress varies from one area to another both in orientation as well as in magnitude. In most of the mine, horizontal stresses are greater than vertical ones, and in general, stresses are lower in the hanging wall of the California fault than in the footwall. The stress difference is around 10 MPa.

29.3   UNDERGROUND MINE LAYOUT

Access routes to ore bodies are located on the slope of the Coastal range and several hundred meters above the valley floor. Today, the main entry is at the –100 level (730 m above sea level) and the haulage tunnel is at the –300 level. The mine is in a network of sublevels that represent the top and bottom of different extraction areas. Sublevels are interconnected by ramps with a maximum slope of 15%. Ore is loaded directly into ore passes with an overall capacity of 30,000 tonnes. Ore passes connect sublevels with the haulage level. The ore is transported to a crusher at the surface near the concentrator with 50-tonne trucks (Figure 29.5). Some ore bodies have been found below the haulage level. From here, the ore is transported directly by trucks from the mine to the crusher.

29.4   UNDERGROUND MINING METHOD

The massive but irregular ore bodies and the competent ground have made sublevel open stoping the historically preferred mining method. In 1983, a variation of the standard method was introduced, enabling an increase in production rates. A fully mechanized sublevel and large-diameter blasthole open stope (SBOS) resulted in the most productive mining method. Nominal stope dimensions are 30 to 60 m wide, 50 to 100 m long, and up to 100 m high. Large ore bodies are divided into several units, leaving rib and crown pillars as temporary support structures. Rib pillars are 30 to 50 m wide, and crown pillars are 25 to 40 m thick. The stopes are mined progressively downward by a traditional SBOS method and are not filled. Pillars are subsequently recovered by a mass blast technique (Figure 29.6). The last three mass blasts were designed to break more than 1 million tonnes of ore each.

Small isolated cavities left after the extraction of ore from the stopes have stable geometries with less than 5% dilution from back extension or wall failure. However, three large open stopes (Santa Clara, California, and Valdivia Sur) have experienced controlled structural caving, filling the existing void and breaking through to the surface.

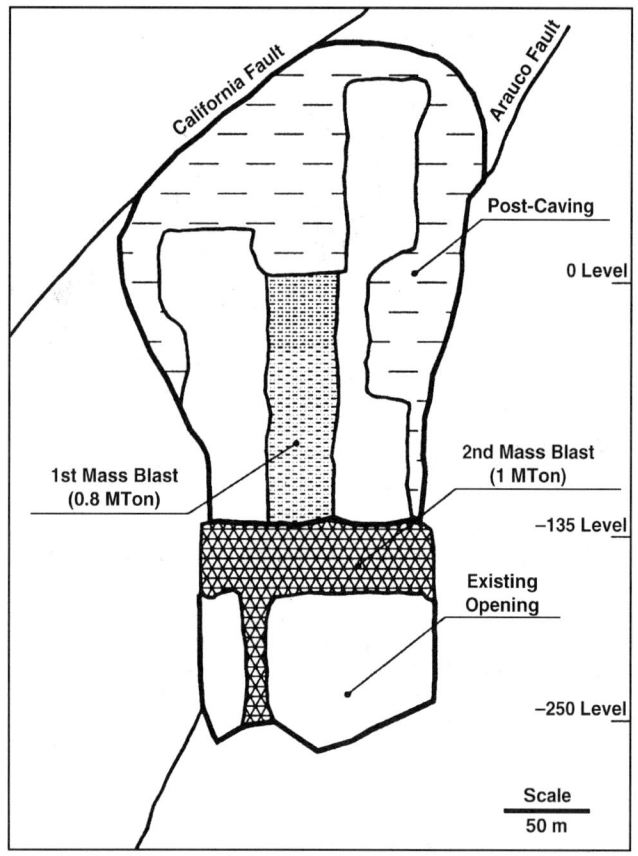

FIGURE 29.6   Mass blast stoping

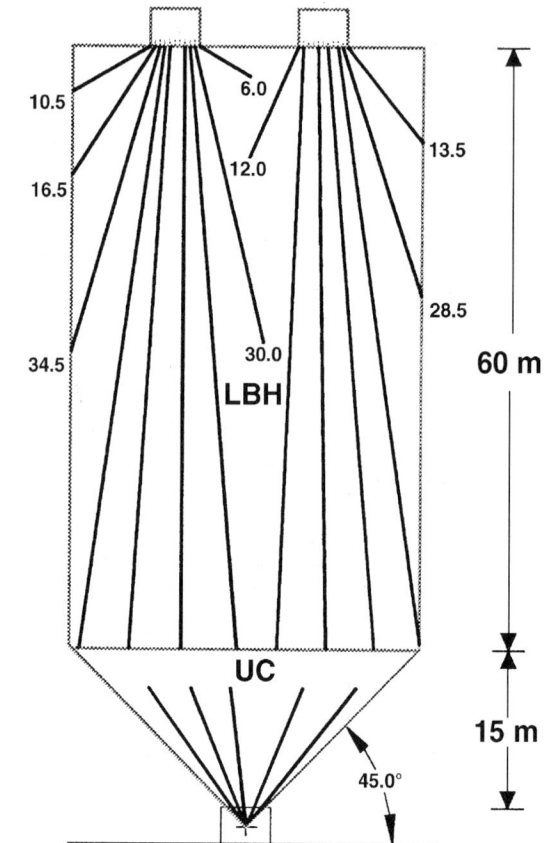

FIGURE 29.7   Stope drilling pattern

Stope block access is provided by drilling sublevels. A pattern of 5- by 3.7-m load-haul-dump (LHD) drawpoints are developed at the base of the stope. Block undercutting is accomplished with a fan pattern of 60- to 75-mm in diameter holes up to 25 m long loaded with ANFO and HE boosters. Slots are prepared by enlarging a 2.5- by 2.5-m blasthole slot raise situated either at the end or in the middle of the stope. Blastholes 165 mm in diameter and up to 80 m long are drilled with an underhand pattern (Figure 29.7).

Blast size and sequence are defined for each stope according to both major structural features and existing cavities. Dilution control is improved and blasthole losses are avoided by carefully considering the particular geometries created by the intersection of major discontinuities and free faces of planned excavations. Often faults present geometries that generate wedges that slide into the cavity, affecting fragmentation and resulting in oversized muck at drawpoints. The presence of cavities or simultaneous mining in nearby locations imposes restrictions in the mining sequence and the size of the blast.

Development headings average 18.5 m² in cross section. Drilling rounds are about 55 holes, 44 mm in diameter and 3.85 m in depth.

Production from stopes is mucked with 10-yd³ LHDs. One-way distances of 100 to 150 m are maintained to ore pass dumping points. Dump points are not equipped with grizzlies. Oversized drawpoint muck is drilled and blasted in place when necessary.

Ore passes end in hydraulic chutes at the haulage level (−300 level), where 50-tonne-capacity end-dump highway trucks are loaded with run-of-mine ore or development waste (Figure 29.8).

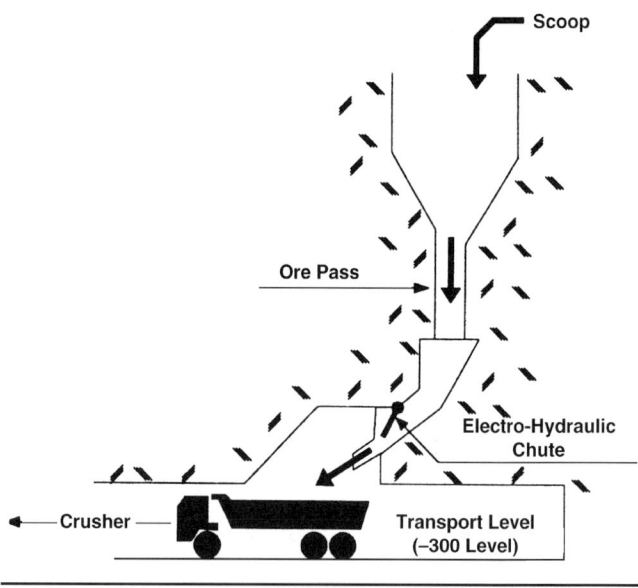

FIGURE 29.8   Ore pass and chute system

29.5   ROCK SUPPORT

A square pattern of 2-m-long split-set bolts in combination with wire mesh is used to maintain working areas free of rock. This approach to ground control is not intended for heavy rock loads or massive stress-induced instabilities. Cable bolts are used to

TABLE 29.2   List of machinery

Drilling	
Development	3 Boomer Atlas Copco H127, φ 45 mm
	1 Jumbo Tamrock Monomatic HS205D, φ 45 mm
Production	LBH, 3 DTH Atlas Copco H264, φ 165 mm
	UC, 2 Simba Atlas Copco H221-252, φ 65–75mm
Secondary	1 Jumbo Tamrock Monomatic, φ 38 mm
Blasting	4 PT-61 Atlas Copco
Loading & Transporting	
Production	5 Scoop Wagner ST813
Waste	2 Scoop Schopf L-272
Scaling	1 Scammer 1000 Normet
Support	1 Boltec 335-H Atlas Copco

TABLE 29.3   Technical data

Productivity	
Overall	50 tonne ore/man shift
Design Parameters	
Preparation	200–350 tonne/m
Drilling, φ 165 mm	25–35 tonne/m
Drilling, φ 75 mm	6–10 tonne/m
Explosive Consumptions	
LBH	220–280 gr/tonne
UC	250–300 gr/tonne
Secondary	4 gr/tonne
Raising	75 kg/m
Drifting	47 kg/m

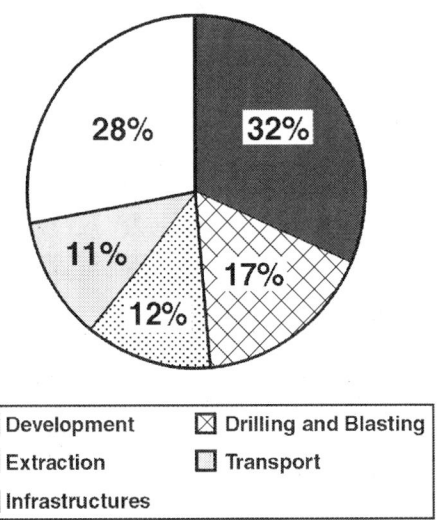

FIGURE 29.9   Cost distribution by process

support unfavorable geometries, such as large wedges or low dip bedding layers, and also to support drawpoints and ore passes where the rock conditions change dramatically. Occasionally cable bolts are used to minimize or prevent caving at the sublevel stopes.

29.6   TECHNICAL DATA ON SUBLEVEL BLASTHOLE OPEN STOPING

29.6.1   Machinery

The primary equipment used in development and production is listed in Table 29.2. Table 29.3 shows some key productivity data.

29.6.2   Cost Distribution

Cost distribution by process is shown in (Figure 29.9).

29.7   GLOBAL MINE STABILITY AND MINING SEQUENCE

Mine stability is a matter of prime importance in the planning process as the areas affected by mining activity have expanded over the years. The spatial superposition of the Morro open pit over the underground mine requires an integrated mine plan in which the sequence of extraction needs to satisfy safety and efficiency criteria. In addition, the collapse—under controlled conditions—of the California stope in 1989 affected the stability of the surrounding mine. Effective control measures, as described below, have been implemented, controlling vulnerability to instability.

The prime objective of mine engineering is the definition of a safe, technically feasible, and economically value-adding mine

plan in a continuously evolving geomechanical environment. The engineering strategy developed at El Soldado concentrated first on understanding the global failure mechanism by establishing a failure model, and second on defining design parameters and sequencing criteria conducive to development of a mine plan that provided safe and economic operation.

29.7.1   Large-Scale Failure Model

El Soldado's failure model is based on a broad geological database. Simplified and rationalized in situ information was elaborated into a conceptual model. Rock mass characteristics and geomechanical responses to mining were basic elements considered in this formulation. Past localized underground failures and a detailed revision of global stability phenomenon were included in the database used to formulate the conceptual failure model. In summary, the global failure mechanism at El Soldado can be characterized as the displacement of large blocks of rock bounded by major faults. Brady and Brown (1993) define this phenomenon as a "discontinuous subsidence" characterized by large surface displacements over limited surface areas and the formation of steps or discontinuities in the surface profile. At El Soldado, subsidence has been controlled by major faults with low shear-strength surfaces on which the undercut rock mass slides under the influence of gravity. The blocks undergo essentially rigid-body displacement without breaking up or dilating except in those areas where stress concentrations cannot be transferred out of the block efficiently (Figure 29.10).

29.7.2   Validation

A comprehensive verification of this conceptual failure model was performed with particular emphasis on modeling assumptions and rock mass behavior. The three elements considered for validation were monitoring results, field observations, and numerical model results. Surface displacement vectors were measured and compared with the orientation of expected block movement (Figure 29.11). The three-dimensional, distinct-element method 3DEC was used to model the whole mine area. The validation process concluded that the proposed failure model developed for El Soldado—that is, subsidence of large blocks bounded by major faults—closely reflected actual rock failure phenomenon. Furthermore, field observations matched the ground conditions anticipated by the 3DEC model very well, thus corroborating the conceptual model.

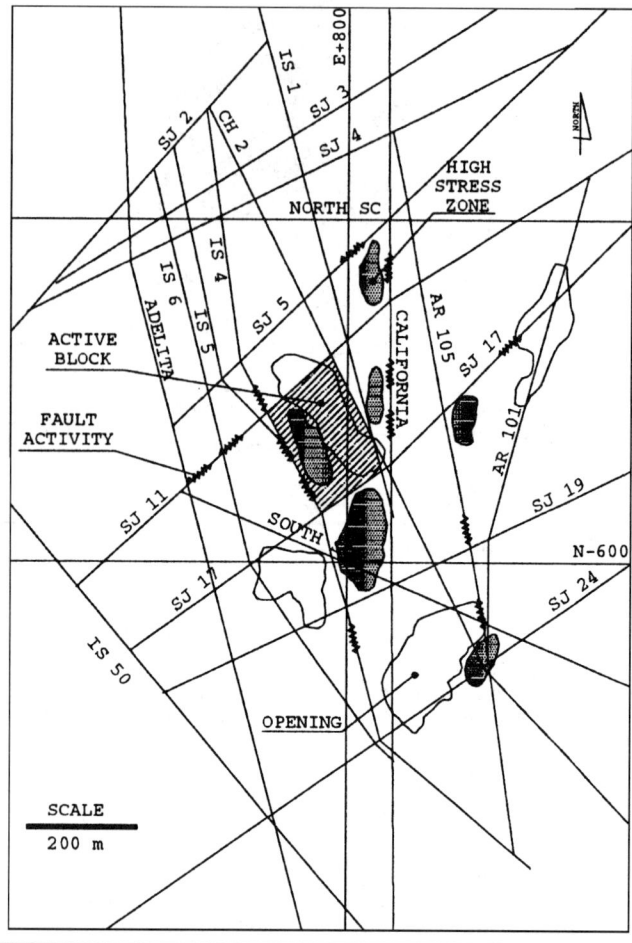

FIGURE 29.10   Block model for El Soldado Mine

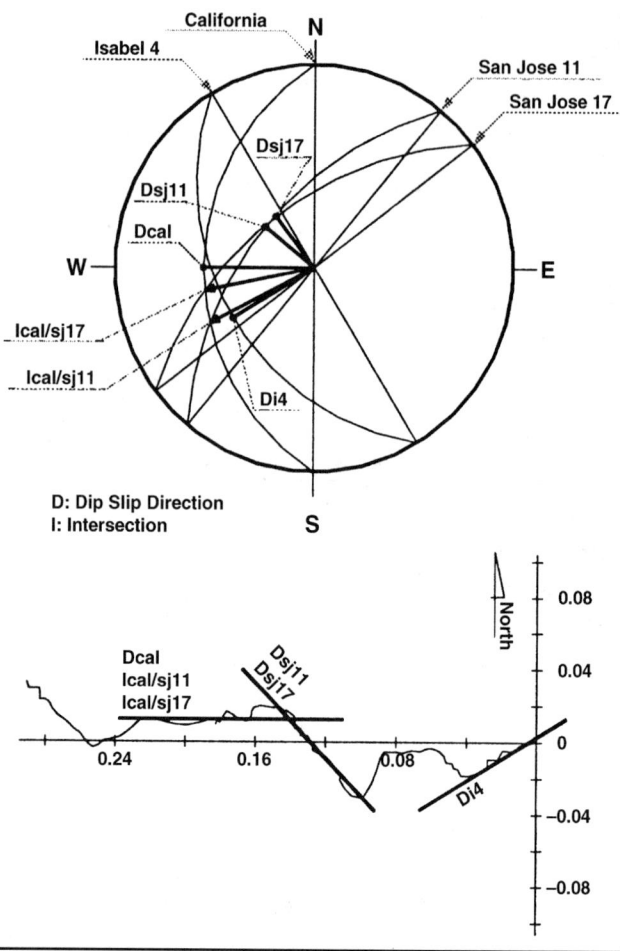

FIGURE 29.11   Measured displacement direction versus orientation of intersection vectors

29.7.3 Mining Sequence

E1 Soldado's underground mining sequence is based on conclusions provided by the validated conceptual failure model described above. Some of the key mine stability considerations are that—

- The extraction of the underground reserves must follow a sequence so that there are no impacts on the overlaying open-pit operations.

- The design and extraction sequence of underground stopes are to minimize disturbance of unmined areas, enabling maximum resource recovery.

- Sequencing of development and drawdown of individual stopes has to focus on zones where rock mass conditions remain almost unaltered by the global nature of the rigid body displacement mechanisms

29.8 MONITORING

Due to the geological nature of the rock mass, simultaneous underground and open-pit mining, and the complex layout of the underground mine, geomechanical monitoring is applied extensively. Monitoring is used to track rock conditions, detect and identify failures or instabilities, collect data for mine planning and stope design, and assess mine stability. Over the long term, the collected data provide ongoing control points to update the geomechanical database and to verify the assumptions made for design.

At El Soldado, there are three main monitoring systems deployed.

1. Surface displacement monitoring system. An automated total station (motorized theodolite with a distanciometer mounted on it) measures angles and distances to given field points simultaneously. The system is capable of searching, measuring, and recording the position of prisms installed at the mine surface continuously and automatically. Thus any movement of the prisms is measured. The information is recorded by a computer and transmitted via cable to the engineering planning office. Recorded surface movement is normally minimal but can accelerate because of rain or higher underground draw.

2. Stress changes monitoring system. Vibrating wire stress cells at different locations in the underground mine provide information on mining-induced geomechanical activity. The changes in stress data are collected by a datalogger. Periodic analysis of mining-induced stress enables early detection of any significant geomechanical event. Stress cells are installed mostly at key structures, such as pillars or rib pillars.

3. Progressive caving monitoring system. The time domain reflectometry (TDR) technique is used to detect caving around mining areas. Grouted coaxial cable is installed in drillholes around surface openings to estimate rate of caving and dilution.

29.9 FUTURE WORK

Continuing with El Soldado's tradition of excelling, safety at operations, cost competitiveness, and increased organizational capability are basic objectives to mine organization. Out of the broad scope of activities and within the context of this paper, a few items of future work are mentioned.

- Alternative, low-cost underground mining methods for the extraction of high-grade reserves contained in small ore bodies will increase resource recovery.

- Improved monitoring will lead to more precise geomechanical engineering. At present. a seismic monitoring system to detect sources of geomechanical activity remotely is being installed and validated. Also, improvements to the monitoring and data-collection systems are under investigation.

29.10 REFERENCES

Aliaga, A. 1998. Compañía Minera Disputada de Las Condes Mina El Soldado. *Revista Minerales*, Santiago, Chile. Vol. 53, No. 223, p. 29.

Bieniawski, Z.T. 1978. Determining rock mass deformability: experience from case histories. *Int. J. Rock Mech. Min. Sci.* Vol. 15, p. 237–247.

Brady, B.H.G., and E.T. Brown. 1993. *Rock Mechanics for Underground Mining*, 2nd ed. Chapman and Hall, 571 pp.

Contador, N.A. 1998. Rock Mechanics Study To Support the Long-Term Mining Plan at El Soldado Mine, Chile, South America. MSc. Thesis, Faculty of Engineering and Mines, University of Arizona, U.S.A.

Deere. 1968. Geological considerations. In *Rock Mechanics in Engineering Practice*, K.G. Stagg and O.C. Zienkiewicz (eds.), p. 1–20. London: Wiley.

Julia, E. 1984. Explotación con Tiros de Gran Diámetro en la Mina El Soldado. *Revista Minerales*, Santiago Chile. Vol. 39, No. 165–166, p. 5.

Julia, E., R.G. Hite, and J.G. von Loebenstein. 1988. Expanding and Mechanizing El Soldado. *Engineering and Mining Journal*, March, p. 26.

Klohn, E., C. Holmgren, and H. Ruge. 1990. El Soldado, a Stratabound Copper Deposit Associated with Alkaline Volcanism in the Central Chilean Coastal Range. In *Stratabound Ore Deposits in the Andes*, L. Fontbot G.C. Amstutz, M. Cardozo, E. Cedillo, J. Frutos (eds.), Springer-Verlag, pp. 435–448.

CHAPTER 30

Underground Mining in the Kola Peninsula, Russia

Anatoly A. Kozyrev,* Yuri V. Demidov,* Igor I. Bessonov,* Oleg Ye. Churkin,* Vladislav M. Busyrev,*
Vladimir N. Aminov,* and Victor A. Maltsev*

30.1 INTRODUCTION

The Kola Peninsula is situated in northwest Russia. It is washed by the Barents Sea to the north and the White Sea to the southeast and is bordered by Norway and Finland to the west (Figure 30.1). In the northern part of the Kola Peninsula, there are large deposits of copper-nickel ores (the Pechenga area); in the central part of the peninsula there are a number of iron ore deposits (the Olenegorsky basin); to the east of this group of deposits, there are the Lovozerskiye deposits of rare-metal ores; to the southwest of these, in the Khibiny region, there are the largest apatite-nepheline ore deposits in the world; and to the west of these deposits, there is the Kovdorsky massif with its iron ore and phlogopite-mica deposits. The deposits are mined by six large companies–GMK Pechenganikel, Olkon, Sevredmet, Apati, Kovdorsky GOK, and Kovdorslyuda.

The GMK Pechenganikel company mines the Pechenga copper-nickel deposits. The company is included in the Russian joint stock company Norilsky Nickel producing nonferrous and precious metals. The ore is mined in two open pits and two underground mines. The company is composed of a beneficiation plant, a roasting facility, a sulphuric acid plant, and other plants required to ensure production. Underground mining at the Kaula-Kotselvaara Mine was started in 1953. Due to the fact that the ore reserves will be depleted in the near future, the mine will be decommissioned. In 1972, the construction of the Severny Mine was started here. This mine was designed to apply the most effective mining techniques, including the use of trackless mining equipment.

The raw materials base for the mining company Sevredmet is the Lovozerskoye rare-metal deposit. The deposit is extracted from two underground mines, the Karnasurt (rather low level of mining mechanization) and the Umbozero (with the application of up-to-date trackless mining equipment). The company also has two beneficiation plants. The Karnasurt and the Umbozero mines were put into operation in 1951 and 1984, respectively. To date, these mines satisfy 70% of the rare metals and tantalum demand and 80% of the niobium demand.

The Apatit mining company, founded in 1930, is, at present, one of the largest suppliers of raw materials for mineral fertilizer production in the world. The company mines a number of thick apatite-nepheline deposits. Both open pit and underground mining are used. The company has two open pits, two underground mines, two beneficiation plants, railway and transport facilities, and a number of auxiliary services. Underground mining is mainly carried out at two highly efficient mines, the Kirovsky and the Rasvumchorrsky, with a capacity of 14 million tonnes of ore annually.

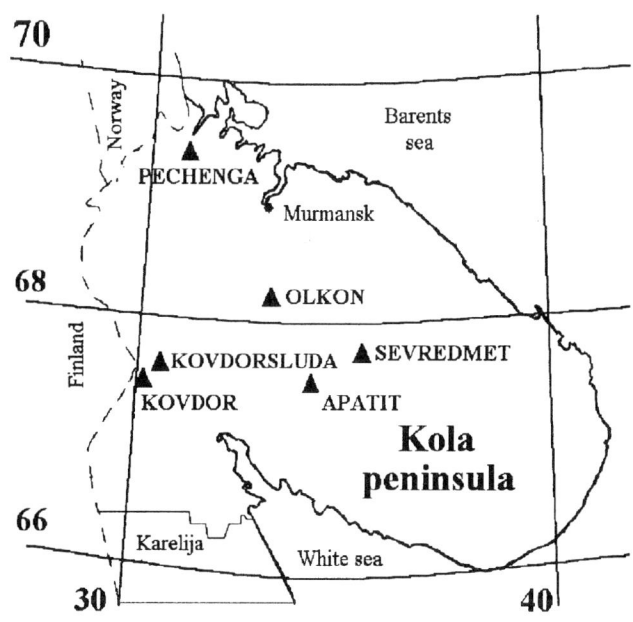

FIGURE 30.1 The mining companies in the Kola Peninsula

Underground mining at the Kovdor phlogopite-mica deposit started in 1933. At present, the Kovdorslyuda mining company has a phlogopite mine (open pit and underground mining), a vermiculite open pit, phlogopite and vermiculite beneficiation plants, and a mica comminution plant. The Kovdorslyuda mining company is the only supplier of mica products for the electronic, construction, rubber, and paint-and-varnish industries.

The Olkon and Kovdorsky GOK mining companies operate three large iron ore open pits. In the future, the Olkon Company plans to change to underground mining at the Kirovogorsky Mine.

30.2 GMK PECHENGANIKEL

30.2.1 Geology

The GMK Pechenganike mining company's raw-material base is eight copper-nickel sulphide ore deposits. Of these, four are being mined (Table 30.1) (Blatov et al. 1998).

The ore bodies are flat sheet or lenticular. The Zapolyarnoe and Sputnik deposits are from 400 to 1,200 m along strike and up to 2,000 m along dip. The thickness of the ore bodies varies from

---

* Mining Institute, Kola Science Centre, Russian Academy of Sciences, Apatity, Russia.

TABLE 30.1 Structure of the raw materials base, the GMK Pechenganikel company

Deposit	Average content, %	
	Nickel	**Copper**
Operated		
Zhdanovskoye	0.57	0.25
Zapolyarnoye	2.15	1.11
Kotselvaara-Kammikivi	1.18	1.18
Semiletka	0.77	0.77
Reserve		
Bystrinskoye	0.49	0.24
Tundrovoye	0.5	0.25
Sputnik	1.46	0.79
Verkhneye	0.5	0.24

1 to 40 m with the average being 6 to 8 m (Blatov and Sokolov 1996). The Zhdanovskoye, Bystrinskoye, and Verkhneye ore deposits are up to 1,300 m along strike, 1,500 to 2,500 m along dip, and up to 100 m thick. The dip of the ore bodies is 45°–65°. Two commercial types of copper-nickel ores are present: (1) rich, solid, brecciated, and multi-impregnated ores containing 1.5% to 2.5% nickel (the basic body of ores of the Zapolyarnoye and Sputnic deposit and the high-grade area of the Kotselvaara-Kammukivi deposit) and (2) run-of-mine ore with a 0.5% to 0.7% nickel content. Of the total, the high-grade ores account for 3.7% of the total ore, but the metal content accounts for 10.2% of the total. The average nickel, copper, and cobalt contents of the high-grade ore run 1.73%, 0.91%, 0.03%, respectively. The run-of-mine ore accounts for the remaining 96.3% of the reserves. The average nickel, copper, and cobalt grade of this material runs 0.61%, 0.26%, and 0.02%, respectively. The grades of the Kotselvaara-Kammikivi and Semiletka deposits are 2.8% nickel (high-grade ore) and 0.6% nickel run-of-mine (Blatov et al. 1995) in the reserve structure. The high-grade ore is selectively mined and is transferred to refining, whereas the run-of-mine is processed (94% of the total output). Ore density ranges from 2.9 to 3.14 tonne/m³, and the uniaxial compressive strength is 140 to 180 MPa. A limited quantity of platinoides, gold, silver, selenium, and tellurium is present in all the ores. However, the percentages of the mined value is not high, being equal to about 5% in high-grade ores and 1–2% in low-grade ores. The developed raw materials base allows the mine to be operated for the next 25 years.

30.2.2 Mining Method

The design output at the Kaula-Kotselvaara Mine is 1,700,000 tonnes annually. Because the ore base is nearly depleted, ore output has decreased from 1,481,000 (1990) to 789,000 tonnes (1998), and in the next few years, the mine will be closed. The ore mining technology is based on the use of mobile mining equipment and rail haulage. The mine extracts the dipping deposits of medium thickness in predominantly two stages. During the first stage, rooms are mined, and during the second stage, the interroom and interlevel pillars are mined and the roof is caved (Kamkin et al. 1996). The Severny Mine operates the Zapolyarnoye deposit. This is a steeply dipping sheet-type deposit 1 to 40 m thick. The initial design envisaged an annual production rate as high as 500,000 tonnes (Kuleshov et al. 1997). After a 7-year period from the start of the construction, the mine has reached its design output. The reserves are accessed by two ramps, one for loaded and the other for empty truck transport. The ramps are driven downward from the surface across the strike of the ore body at an angle of 6°. They cross the ore body at level –20 m, and then the ramps are driven along the ore body strike. After each level is opened, the ramps change their

direction by 180°. The ramp cross-sectional area of 22.1 m² is designed for the operation of 45-tonne dump trucks. The lower levels are accessed by vertical shafts sunk from the surface to a depth of 1,500 m. The production levels are accessed by transport crosscuts from the ramps. Connections are made to ventilation raises located at the deposit boundaries and to an air-rock fill raise in the middle. The sublevel stoping system is used in ore mining with the application of mobile equipment. The rooms are filled. The level is 60 m high; the sublevels are 15 m high.

The capacity of the fill complex is 150,000 m³/yr. Cemented fill made of screened rock is also used. The strength of the cemented fill is much higher than that of the sand fill. Primary and secondary stope mining is used. The secondary stopes are filled with dry backfill placed onto a cemented fill layer. If the ore body is less than 3 m thick, the rooms are mined by a shrinkage system with short-hole blasting. In the case where the ore body is over 3 m thick, a sublevel stoping system is used. The trackless equipment fleet is over 50 units. About 65% of the fleet is manufactured in other countries. These are drifting and production drilling machines, roof bolting machines, machines for scaling the roof, loading-hauling equipment, underground dump trucks, etc. The ventilation system consists of two major ventilation installations, each having a capacity of 223 m³/s, and two auxiliary installations. Forced-air ventilation is provided through the central and air-rock fill raises (cross-sectional area of 13 m³), with the air exhausted through boundary raises. Step-by-step water pumping through the ramps is carried out by the pumps from levels –500, –320, and +34 m to surface settling basins. The mine has underground facilities (700 m² in area) to carry out repair and maintenance operations. Due to ventilation restrictions, no more than eight dump trucks can operate simultaneously in the ramp. This is not sufficient to satisfy the 500,000 tonne/yr production requirements. Since 1992, the mine operates a transport system whereby 400,000 tonnes of ore is transported annually by ramps and about 100,000 tonnes of ore is transported annually by an inclined skip shaft.

The mine operation indices are ore loss = 5%, ore dilution = 20% to 25%, labour productivity of a face worker = 12 m³/manshift.

Underground Mining Potential. At present, the Severny Mine is operating according to an interim scheme of development with ore haulage via a 5-km-long ramp. Because of this, all technical capacities of the mine in terms of ventilation, transport, and mining fronts are exhausted, limiting the production rate of the mine to 400,000 to 500,000 tonne/yr. Some significant capital investments and time are necessary to complete the construction of new permanent hoisting facilities. Two of the three vertical shafts (skip and cage) have already been sunk. When completed, the mine's production rate will be 900,000 tonne/yr.

30.3 SEVREDMET

30.3.1 Geology

The Lovozersky alkali deposit consists of two main ore-bearing complexes: a differentiated one (82% of the total volume) and an eudialyte–lujaurite complex (15%). The Lovozerskoye deposit lies within the differentiated complex and the Chinglusuai deposit within the eudialyte complex. The Lovozerskoye deposit is composed of 12 ore areas, of which three are being mined (76% of the total reserves). The rest (24%) are reserved for the future. The main commercial mineral is loparite. The deposit is a series of layers 0.5 to 6 m thick, with the dip of the ore body being equal to 7° to 30° (combined into series I through IV). All in all, there are over 30 ore beds, of which four are being mined. The depth of mining is 50 to 500 m. The ores are complex and, in addition to loparite, contain aegirine, nepheline, feldspar, and apatite.

The Chinglusuai eudialyte ore deposit is a future reserve for the Sevredmet mining company. The Chinglusuai loparite juvite deposits (the Alluaiv area) are at a depth of 50 to 350 m below the surface. The ore body is a thin (1 to 5 m) interrupted bed dipping 3° to 17°. Underground mining can be used. The tantalum and niobium pentoxide ore grades are, on average, 1.5 times those of the analogous ones from the operating levels of the Lovozerskoye deposit. The raw materials can be used to produce loparite, aegirine, and nepheline concentrates. The deposit is not currently being mined. The Alluai eudialyte ores are near the surface and open-pit mining can be used. The ore body is 20 to 60 m thick and is of a nearly flat sheet formation. The ore is eudialyte-lujaurite grading 25% to 28% eudialyte, or 3.0% to 3.2% zirconium dioxide.

The potential of this raw materials base for the entire spectrum of products mentioned above is extremely high. For instance, the share of the A+B+C reserves is only 30% of the total reserves of eudialyte and loparite ores. The ore deposits mined by the Karnasurt Mine are confined to the upper horizons of series II. These are rather thin (0.6 to 0.9 m), gently dipping (8° to 18°), and persistent in strike. The main commercial horizon mined by the Umbozero Mine is confined to series III, which is located at a depth of 90 to 100 m lower than the ore horizons mined by the Karnasurt Mine. The horizon is characterized by a complicated mode of occurrence. Some areas pinch out near the surface, and in the contact zone, the dip of the ore body is much steeper, varying within the range of 25° to 60°. Away from the zone of contact, the ore bed gradually flattens out to 12° to 20°. The ore bed thickness varies within a wide range from 1.5 to 3.5 m.

30.3.2   Geomechanical Features

Based on measurement data, analysis of rock pressure occurrences, core drilling data, and other factors, the characteristics of the stress field in the rock mass gives the following: maximum horizontal stress is 40 to 60 MPa with an azimuth of 40° to 60°, the second (intermediate) horizontal stress is 20 to 30 MPa, and the vertical (minimum) stress is that due to gravity.

The measurements made in the deep levels of the Karnasurt Mine reveal a nonuniform state of stress within the different zones. Within some zones a significant nonuniformity of the horizontal stress component was observed. The orientation of maximum compressive stresses is not always clearly distinguished.

The orientation and magnitude of the highest stresses in the rock mass vary due to the surface relief, the presence of faults and other tectonic features that break the continuity, heterogeneity of rock properties, and some other factors. The stress deviations are ±30° in orientation and ±20 MPa in magnitude. Near the surface and near faults, this deviation may be greater.

All the basic varieties of rocks are considered to be strong (compressive strength σ_c = 140 to 230 MPa, tensile strength σ_t = 7 to 20 MPa), highly elastic (Young's modulus 50 to 60 GPa), brittle, and, if under high stress, potentially hazardous for rock bursts. Indices of potential hazard for rock bursts as characterised by the coefficient of brittleness, i.e., the σ_c/σ_t ratio, are 25 to 30. These significantly exceed their criterion values—the limiting coefficient of brittleness σ_c/σ_t = 10.

This is also confirmed by the forms of rock pressure effects at the mines—micro-rock bursts and intensive rock spalling, scaling, and slabbing, all testifying that, in some zones of the deposit, there is the hazard for rock bursts. In mining, it should be taken into account that the rock-burst hazard is higher near tectonic dislocations, faults, large fractures, and dikes, especially in cases where there are displacements along them. The potentially hazardous zone is within about 30 m from both sides of the geological discontinuity.

The tectonic dislocations of the Lovosero deposit are characterized mainly by small-scale block fracturing. Within the Karnasurt and Umbozero mining areas, seven systems of steeply and gently dipping fractures are identified in each area. The analysis of the fracture data obtained from the Karnasurt Mine area has shown that, of all of the systems distinguished, the most significant

are those with steeply dipping fractures. The System I–II fractures are a few tens to a few hundreds of meters long, and 0.5 to 3 mm wide. As a rule, the fractures are filled with secondary mineralization. The System I–II fractures are spaced at 0.5 to 1.5 m. Gently dipping fractures of various orientations are developed throughout the whole area.

Within the Umbozero Mine, there are two main systems of fractures: (1) steeply dipping, System I, and (2) gently dipping, System VI. The other fracture systems are locally distributed. The steeply dipping systems usually have flat, smooth walls and are filled with secondary minerals. The walls of the gently dipping fracture systems are rough and wavy, and mineralization is rare. Thus, the most probable structural surfaces for caving are the planes forming the steeply dipping fracture systems. The openings between the fractures of one and the same system are in the range of 0.05 to 1.7 mm.

The stability of mine workings is greatly dependent on the blocky nature, deformational properties, and the state of stress in the rock mass. The blocky nature is formed by three or more systems of fractures. The dimensions of a structural block, according to the classification of the International Society for Rock Mechanics, is usually characterized by the coefficient of volumetric fracturing (K_f), i.e., by a number of fractures crossing a unit volume of a rock mass. In accordance with this classification, structural blocks with K_f = 1–3, K_f = 3–10, K_f = 10–30 fr/m^3 are ranked as large, medium, and small blocks, respectively.

Structural blocks formed by the systems of fractures of the Lovozero deposit are mainly classified as large and medium since K_f lies within the range 1.5 to 3.8 fr/m^3. In some areas of the mine where closely spaced (0.2 to 1.0 m), steeply dipping fractures are found in combination with frequent gently dipping ones, the coefficient of fracturing is 12 to14 fr/m^3, which corresponds to small structural blocks. Within these zones, one should expect rock breaking and roof falls in the openings.

It is characteristic for the Lovozero deposit to be divided into a number of blocks of a few square kilometers to a few tens of square kilometres in area by tectonic features. These breaks in continuity are of the radial and ring type. Radial dislocations are, as a rule, represented by much deeper and more extensive rupture zones than those of the ring type. Within the Karnasurt Mine area, three large radial faults directed to the centre of the deposit occur. They have a fan pattern relative to each other, and the amplitude of fault displacement is 8 to 9 m.

30.3.3   Mining Method

The ore deposits extracted by the Karnasurt Mine are confined to the upper horizons of series II. They are rather thin (0.6 to 0.9 m), gently dipping (6° to 18°), and of persistent strike. The loparite content in the remaining reserves varies from 4.6% to 6.8%. The open-stope mining system with recoverable intermediate pillars is used. Excavation is accompanied by wall rock blasting, resulting in an increase of the stope height to 1.2 m. In the stoping operations, hand-held drilling equipment and scraper winches of up to 100-kW capacity are used. The ore is transported by electric transport. The dimensions of the unsupported spans in the mine workings as well as the pillar parameters are specifically designed for the different technological zones in the mine. The Karnasurt Mine is capable of producing for another 55 years, based on the explored reserves. The explored reserves of loparite ores within the Alluaiv area allow a doubling of the current mine life.

The main commercial horizon of the Umbozero Mine is confined to the series III beds and occurs at a depth of 100 m under the ore horizons. The horizon's mode of occurrence is complicated. In some areas near the surface, it is quite thin, while in the zone of contact, the dip of the ore body is steep, ranging from 25° to 50°. With increasing distance from the zone of contact, the ore bed gradually thins and flattens, reaching 12° to 20°. The thickness of ore beds of series II changes from 1.5 to

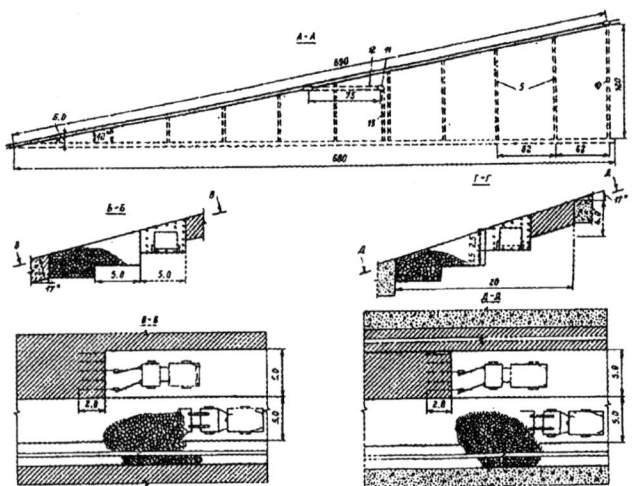

FIGURE 30.2 A slicing method at Umbozero mine

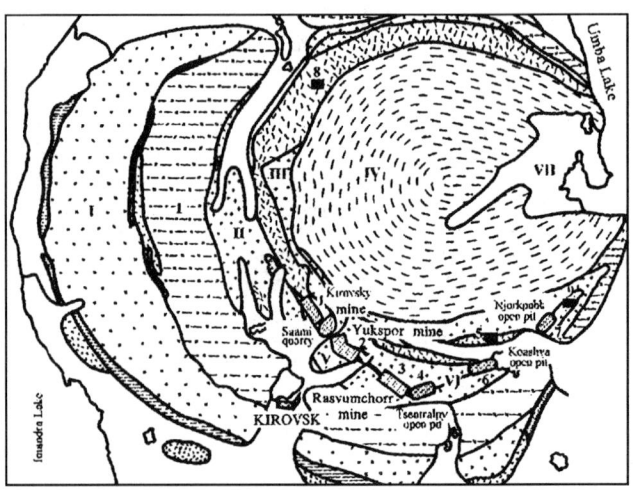

FIGURE 30.3 Scheme of the raw-materials base in the Khibiny mining industry area: 1–9 are apatite-nepheline ore deposits: 1—Kukisvumchorr, 2—Yukspor, 3—Apatite Cirque, 4—Rsvumchorr Plateau, 5—Eveslovchorr, 6—Koashva, 7—Njopakhk, 8—Partomchorr, 9—Olenij Ruchei, I–VII are rock types: I—khibinite, II—ristchorrite, III—lavchorrite, IV—foyaite V—ijolite-urtite, VI—apatite-nepheline ores, VII—carbonatite

8 m, with the loparite content equal to 2.5% to 3.5%. Variants of a two-stage, room-and-pillar filling method and slicing systems with open stopes and chain pillars are used at the Umbozero Mine (Figure 30.2).

The mobile equipment used in the stoping and mining operations are loading-hauling machines (Kavasaki, Toro) and drilling rigs (Paramatic, ВТФ-2Н). Ore is transported by МоА3-type dump trucks. Tailings and portland cement are used as components of the backfill. The explored reserves make it possible for the mine to operate during a 48-year period. If the reserves of a second ore field are brought into operation, this period will increase to more than 100 years.

Basic Directions of Mine Development. The future development of the company is associated with modernizing and re-equipping the present mining and processing facilities, leading to a decrease in the manufacturing cost of concentrate together with the creation of a loparite concentrate reprocessing operation. Questions connected with the construction at the Servedmet mining company of a plant for the chlorine-free chemical and metallurgical recovery of niobium, tantalum, rare earths, and titanium are under consideration. Of great potential will be bringing into commercial production the eulialyte ore reserves as a source of zirconium and yttrium, an unavailable element of the rare-earth group.

To promote the full use of the raw materials, to decrease costs, and to reduce the environmental impact, it is envisaged that—

- The operational conditions (excavation capacity) should be corrected for loparite ore in accordance with the ore bodies and the technological zones of the mine fields;

- Stoping parameters should be improved, including the introduction of large-hole drilling technology using the Rino-400 drilling rigs from Tamrock. It provides for concentrate output of 11,000 tonnes annually, with 400,000 tonnes of ore output, and decreases the solid waste output.

- The rational relationship between bulk and selective excavation technologies should be stipulated, taking into account the mine area technological zoning;

- The efficiency of using man-made cavities in mines for storing industrial waste should be assessed, thereby increasing both mining operations safety and the ecological purity of the environment;

- New production lines for aegirine and nepheline-feldspar concentrate should be introduced. At present, the Karnasurt concentrating plant operates one line, enabling it to produce 2,250 tonnes of aegirine concentrate annually. The Giredmet Institute, Russia, has developed the design documentation for producing nepheline-feldspar concentrate at the Umbozero beneficiation plant.

30.4 APATIT

30.4.1 Geology

The Khibiny massif is composed of nepheline-syenite and ijolite-urtite (and the associated apatite-nepheline ore deposits) and is, in terms of geology, the largest in the world. It is a complex multiphase alkali intrusion of the central type and is related to the regional tectonic fault responsible for the location of a number of other Finnish-Scandinavian deposits on a smaller scale. Oval in shape and 1,327 km² in area, its highest elevation is 1,191 m above sea level. It is composed of ancient Archean gneiss and Proterozoic schists.

The zonal inner structure of the deposit as determined by multiple intrusions of magmatic alkali melts of the nepheline-syenite composition through the system of ring faults is distinctly shown on the geological map (Figure 30.3). In the Khibiny massif formation scheme, several stages of magmatism are distinguished: nepheline syenite of different compositions, khibinite, ristchorrite, lavchorrite, foyaite as well as ijolite-urtite, apatite-nepheline ores, and carbonatite. The zonal position of intrusions as arching bodies replacing each other in the massif in the periphery-to-centre direction correspond, in general, to their age succession. In accordance with this rule, the earliest khibinites form the outer ring (zone) of the massif. From the inner side are located the much later-arriving ristchorites. Further to the centre are ijolite-urtites and, related to them, the apatite-nepheline ores. Finally, there is lavchorrite, which surrounds the core of the massif, which is composed of the youngest rocks–foyaite. The dimensions of the ring intrusive bodies are rather substantial; their length along strike (perimeter) is a few tens of kilometres, their width in plan (thickness) is 2 to 5 km. In the east-west direction, the northern and southern branches of the arches get narrower and eventually disappear near

TABLE 30.2 Mineral composition of the different textural types of apatite-nepheline ores

Textural types of apatite-nepheline ores	Abundance,%	Average content, %					
		Apatite	Nepheline	Aegirine	Sphene	Titanomagnetite	Feldspar
Mottled	6–18	69.2	19.9	5.3	1.6	0.5	1.7
Banded	45–56	39.9	46.1	7.2	2.1	2.5	0.4
Block	5–10	46.5	39.1	7.4	3.2	1.3	0.6
Massive	3–9	35.9	45.0	9.6	3.0	1.2	1.3
Net-like	9–14	17.5	54.2	19.8	4.7	2.9	9.9
Apatite urtite	4–10	15.0	58.7	15.9	4.1	0.8	2.9
Breccia	5–23	24.0	44.7	16.5	6.6	2.0	3.1
Sphene-apatite ores	2–6	20.4	31.4	17.5	18.6	6.2	0.9

the Umbozero coast. During the middle stage of the alkali magmatism development in the Khibiny, a complex three-stage ijolite-urtite intrusion took place. Genetically related to this event was the formation of the apatite-nepheline ores. The thickness of the ijolite-urtite intrusion is 500 to 1,000 m. Apatite nepheline ores in the Khibiny are presented either by 100- to 200-m-thick sheet bodies in the urtite hanging wall (the Kukisvumchorr-Yukspor-Rasvumchorr set of deposits) or by a system of 25 to 30-m-thick ore deposits inside the urtite formation (the Koashva, Partomchorr, and Njorkpakhk deposits). According to the generally accepted classification based on differences in mineral composition and structural peculiarities, the apatite-nepheline ores are subdivided into some natural (textural) types: mottled, banded, lenticular-banded, net-like, block, compact, and brecciated. These are related to the host urtite via a set of intermediate varieties. Each of the natural types of ores is found in the composition of all the Khibiny apatite deposits. However, these are not uniformly distributed throughout the volume of ore bodies. Within the Kukisvumchorr-Yukspor-Rasvumchorr deposits being mined, a regular distribution of ore types from top to bottom has been observed: sphene-apatite ores, mottled, banded, block, lenticular-banded, and net-like apatite urtite. In thick apatite deposits, mining is being carried out simultaneously in different horizons and over a wide area. The ores are blended to achieve a specified composition, and this blended ore mass is then transported to the beneficiation plant.

Thus, in terms of technology, the processed ores represent a technological type whose mineral composition (Table 30.2) is determined by actual conditions. All of the known apatite-nepheline ore deposits in the Khibiny are grouped within three ore fields: southwest, southeast, and northwest. Of most significance at the present time and in the near future is the southwest ore field, including the Kukisvumchorr, Yukspor, Apatite Cirque, and Rasvumchorr Plateau deposits. These are being operated at present.

30.4.2 Geomechanical Features

In situ investigations have shown that the state of stress in the Khibiny massif is characterised by a high level of horizontal stress. The maximum horizontal stresses are five to ten times as much as the stresses produced by the weight of the overlying rock. A relationship between the stresses measured and the strength properties of the rock has been observed. A higher level of stress corresponds to the more compact, highly elastic rocks. For instance, the stress values in the host rock are in the range of 40 to 60 MPa, whereas they are 20 to 40 MPa in the ores. Stresses oriented along geological distortions, such as near faults, increase two to three times. The measurement results show that the stresses increase with depth. For instance, on levels +320 m and +252 m of the Kirovsky Mine, the average stresses were equal to 30 and 40 MPa, respectively, and in near-shaft mine workings on levels +172 m and +92 m, the stresses were equal to 50 and 60 MPa. According to these data, at level

zero, the maximum horizontal stresses tend to values of 70 to 80 MPa in the host rocks and 40 to 60 MPa in ores. The other horizontal component of the stress tensor is about 0.5 to 0.7 that of the maximum stress for the Kirovsky and Yuksporsky mines, and 0.4 to 0.6 for the Rasvumchorr Mine in apatite ores. For the host rocks, the corresponding values are 0.4 to 0.6. For the Rasvumchorr Mine, the values are 0.3 to 0.5. The vertical component corresponds approximately to the values calculated by the weight of the overlying rock mass. There is a tendency for the horizontal stress to stabilize with depth. A decrease in the horizontal-to-vertical stress ratio from 5 to 10 to 2 to 3 is observed.

The rocks are characterized by high strength and elasticity indices. The compressive strength of ore is 80 to 150 MPa and that of rock is 120 to 250 MPa. The corresponding tensile strengths are 3 to 10 MPa and 5 to 20 MPa, respectively. The modulus of elasticity of ores is 30 to 60 GPa and that of rocks is 50 to 60 GPa. Both rocks and ores have high indices of the coefficient of brittleness as defined by the compressive strength-to-tensile strength ratio.

Within the deposits, rock fracturing is divided into categories of microfracturing, small-block, and large-block. Large-block fractures are on the order of a few tens or sometimes hundreds of meters in extent and are traced in some workings and on levels. The fractures vary in width within the range of 1 to 50 mm, and they are filled with hydrothermal minerals. Small-block fractures are, as a rule, 1 to 2 m in extent. Generally these fractures, from their start to their end, are traced within the limits of the mine workings and often branch into one another. Their width is as small as parts of a millimeter. Large-scale fracturing is represented by four systems of fractures: gently dipping and inclined fractures of System I (over 60%) and steeply dipping fractures of Systems 2–4 (total, less than 40%). The average distance between fractures is 2 to 30 m. Fracturing intensity decreases with depth, and the dip of System I fractures increases. Fracturing increases from the Kukisvumchorr to the Rasvumchorr deposit. Zones of ore and rock, oxidized and broken by fractures, do occur but their share is negligible—3% to 9%—and decreases with depth. Rocks oxidized and broken by fractures are relieved from tectonic stresses. Compacted and strong rocks prevail as a whole.

30.4.3 Mining Method

Apatite-nepheline ores are mainly mined at three deposits—the Kukisvumchorr, Yukspor, and Apatite Cirque. Taking into account the mountainous relief, the reserves occurring higher than the level of the valley bottom have been developed by tunnels and permanent adits as well as by permanent ore passes 250 m long driven from the tunnel levels to the upper operational horizons. Ore is transported by railway from the stoping blocks to ore passes. Electric locomotives with a gross weight of 14 to 25 tonnes and cars of 9-m³ capacity are used. On the tunnel level, using pneumatically

FIGURE 30.4   The ore-transportation inclined shaft equipped with two conveyors at the Kirovsky mine

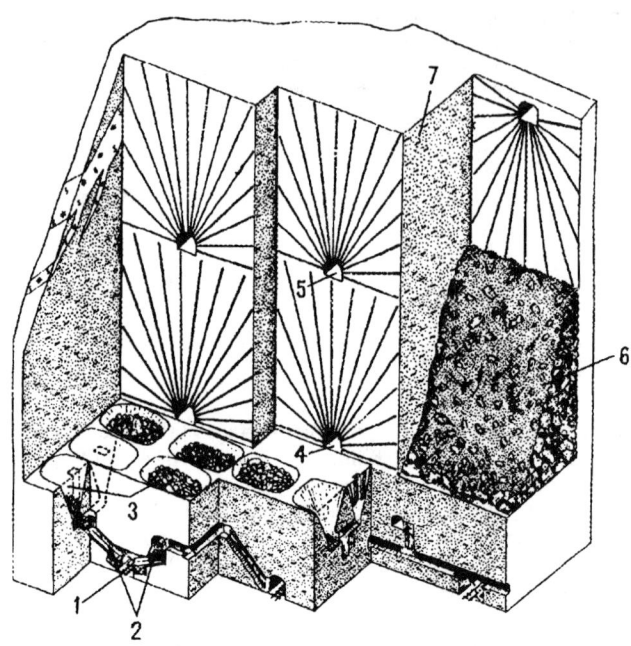

FIGURE 30.5   Induced block caving longhole method with ore drawing by vibrating conveyors: 1—haulage drift; 2—vibrating feeders; 3—ore drawing cone; 4—drilling-undercutting drift; 5—drilling drift; 6—blast-broken ore; 7—rock mass

operated chutes, the ore is loaded into dump cars with 105-tonnes load-carrying capacity and transported to the beneficiation plant. High efficiency and low material labour costs characterize this scheme.

The reserves of the Kirovsky Mine occur lower than the level of the valley bottom and are developed by the main conveyer shaft as well as by vertical auxiliary and ventilation shafts. An inclined conveyer shaft (28.5 m^2 in cross section and 1,030 m long) is driven at an angle of 15° from the load chamber on level +172 m to the loading hoppers on the surface. The shaft is equipped with two parallel belt conveyors (Figure 30.4) of 1,600 tonne/hr capacity. Taking into account the coefficient of nonuniform work being equal to 1.25, the complex capacity is 13 million tonnes/yr. Lower horizons of the Rasvumchorr Mine are developed by a ramp via a vertical shaft sunk from the level of the valley and equipped with a skip hoist and vertical and ventilation shafts. The capacity of the ore-transportation complex at the Rasvumchorr Mine is 6 million tonne/yr.

The induced block caving longhole method has been used predominantly at the underground mines since 1959. Two variants of the system are used: (1) slice blasting in stopes followed by pillar caving and (2) slice blasting in sections with shrinkage. The width of the extracted block is 64 to 72 m, and the length along ore body strike, depending on the horizontal thickness of the ore body, ranges from 100 to 300 m.

The block rock mass to be blasted is divided into sublevels 13 to 14 m high. On each sublevel, there are drilling drifts 3 by 2 m in cross section from which the rock mass is drilled by boreholes 105 mm in diameter. The blocks are mined from the centre to the flanks or from the footwall to the hanging wall. Two adjacent blocks with a 40-m distance between them are mined simultaneously. The boreholes were arranged in the section and room layers as vertical fans. The angle of the boreholes ranges from −10° to +35°. The boreholes are 40 to 45 m long, and the burden and the end spacing between the boreholes is 3.5 to 4.5 m. The specific explosive consumption is 220 to 240 gm/tonne. The volume of ore blasted at one bulk blast was 100,000 to 150,000 tonnes. Mine development rate per 1,000 tonnes of blasted reserves (specific development) is 5.6 m. In 1968, high-angle drilling was introduced. In this case, the height of the sublevel was increased to 30 to 35 m, allowing a reduction of the volume of mine-development operations by 30% and an improvement in the degree of fragmentation. In the application of these variants, an areal ore output by scraper winches of 100-kW capacity was used. The change to ore blasting using high-angle boreholes allowed increasing the bulk

blasting output to 250,000 to 300,000 tonnes. The successful introduction of vibrating feeders has created actual possibilities for an increase in intensive ore drawing from the block. The cyclic transport of ore within the mine was a problem to be solved.

The programme of underground technology improvement applied to the underground apatite mines was aimed at the solution of the following basic problems: to perfect the process of bulk ore caving to create great reserves of shrinkage ore in a stope; to perfect drilling and blasting operations in order to produce a blasted rock mass of a rather stable fragmentation with a minimum coarse fraction output; to provide for the intensive ore output by vibrating feeders; and to provide for highly effective inter-mine transport.

On this basis, a highly effective ore mining technology has been designed and introduced into production. This includes the bulk caving of sections containing 400,000 to 500,000 tonnes of reserves and the drawing of the broken ore by effective vibrating feeders with a technical loading capacity of 12 to 14 tonne/min to the cars. The general scheme of the mining system with vibrating feeders output is shown in Figure 30.5.

The block caving practice with ore drawing by vibrating conveyors has shown that vibrating feeder performance is significantly higher than that of ore transportation by electric train. To solve this problem, an attempt has been made to design a new scheme in which coarse-blasted ore is loaded directly onto the conveyor belt without additional breaking. To test this scheme under industrial conditions, the KLT-160 belt-car conveyor was used. The conveyor design is based on the principle of removing both static and dynamic loads from the conveyor belt at the expense of its being supported by special movable cars moving together with the belt.

The technical characteristics of the KLT-160 conveyor are—

Capacity, tonne/hr	2500
Width of belt, mm	1600
Travel speed, m/s	1

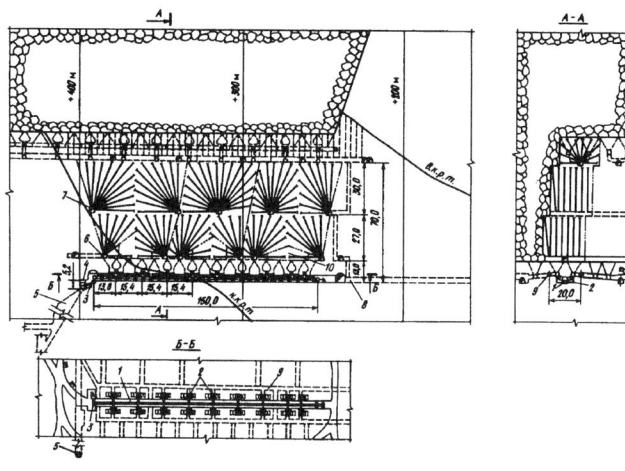

FIGURE 30.6   Schematic of a pilot conveyor panel: 1—haulage conveyor; 2—vibrating feeders; 3—reloading unit; 4—main conveyor; 5—orepass; 6 and 7—drilling-undercutting and drilling drifts, respectively; 8—service raise; 9—control-ventilation drift; 10—undercutting chamber

FIGURE 30.7   Simultaneous ore drawing to the conveyor belt

Support device for upper line	Carriage
Maximum size of ore lump transported, mm	1500
Drive drum diameter, mm	1250
Tension reel diameter, mm	1000
Electric drive power, kW	160
Drift conveyor length, m	150
Conveyor width, mm	2100
Conveyor height with edges, mm	2320
Conveyor mass, kg	121,000

In the test block's panel, 18 vibrating feeders and a conveyor 150 m long were mounted. The main drift was driven in the footwall of the ore deposit where a 61-m-long conveyor had been installed. To reload ore from the drift conveyor to the main one, a reloading unit composed of two vibrating feeders was mounted. From the main conveyor, the ore was transported into the ore chute equipped with a 5,600-tonne-capacity bunker (Figure 30.6).

A Noranda method was used in ore blasting with boreholes 105 mm in diameter. The percentage of blocks +1000 mm was 3.3%. For ore lumps 1,500 mm and over, the percentage was 0.8% of the total volume of ore broken. During the period of the experiments, which covered 412 shifts, ore output was 865,000 tonnes. An average technical output of the complex was equal to 1520 tonne/hr, with the maximum being equal to 1962 tonne/hr. Average shift output (with a 6-hr shift) was 4,200 tonnes per shift, while the maximum was 5,700 tonnes per shift. The coefficient of the intrashift use of the conveyor was on average 0.39, and the maximum was 0.564 (Figure 30.7).

As a whole, the conveyor complex was reliable with coarse ore (the maximum size of ore lump was 1200 mm) and during loading operations where there were a number of points along the conveyor.

At present, due to mine deepening and worsening mine-geological conditions because of a reduction in ore body thickness and an increase in stress in the massif, there is a trend away from the use of portable, stationary equipment and toward the use of mobile equipment in all technological processes of underground ore mining. The induced block caving method is gradually being changed to the sublevel caving method with face ore drawing.

Underground Operations Mechanization. Drift driving is mainly carried out using two-boom drilling rigs equipped with both hydraulic and pneumatic drills. In drilling blastholes, the drilling rig is powered either from a general-mine electric power network, 380 to 660 V, or from the main compressed air line.

Moving in the mine workings, the drilling rig is powered by a diesel engine. Blastholes are drilled 41 to 51 mm in diameter and 4 m long. In these operations, the UBSh, Minimatic GMS 205, and Paramatic (Tamrock, Finland) are used for drilling, and diesel-powered PD-8B load-haul-dump (LHD) machines are used for mucking. The ore is successfully transported over long distances (over 400 m) by a set of machines, including the loading machine PND-3B and MoAZ dump trucks or the USA underground dump trucks SDT-426 of 22-tonne load-carrying capacity. The loading-hauling machines are TORO-150E, TORO-350D, and TORO-350E.

The KPB-4 and the 2KB machines are used for raise boring. In stoping operations, blastholes for mass breaking are drilled using both the Russian-made NKR-100 m drilling rigs and the Atlas Copco-produced Solo G-808PA and Solo G-1008PA, Apa-Simba rigs.

In induced block caving, a blasting method involving parallel, 25-mm holes was tested. The holes were drilled with a BSh-200CA roller-bit drilling machine. The average shift output was 15.2 m in hard rock. In drawing and hauling operations, scraper winches with a power of 100 kW were used together with vibrating feeders of type VDPU-4TM. The scraper operator's labour productivity has been stabilised at 400 to 430 tonnes per shift with an explosive consumption for secondary breaking of about 100 gm/tonne. The vibrating feeder operator's labour productivity is 700 to 800 tonnes per shift with an explosive consumption for secondary breaking of 80 gm/tonne.

In sublevel caving, the ore is drawn and hauled by PD-8 LHDs with a scoop of 4-m³ capacity, as well as by Toro-400D and Toro-400E machines. In auxiliary operations, Robolt machines are used for installing support. Normet cassette rigs are used for scaling.

Geomechanical Aspects. The main principles for enhancing the stability of mine workings and for the prevention of rock bursts amounts to control over rock deformation and failure processes by (1) eliminating or minimizing stress concentrations within the zone of mining operations; (2) reducing the ability of the rock to deform and accumulate great amounts of elastic energy; and (3) controlling rock mass breaking and deformation. The differences in the principles for the prevention of local rock bursts or tectonic rock bursts is mainly in the scale of the rock masses to which the techniques are applied. In the case of local rock bursts, these rock masses are the zones adjacent to mine workings or to other construction elements of the mining system. In the case of tectonic rock bursts, the rock masses are the mined-out rock block, level, or mine field as a whole. There is, however, no distinct difference between prevention techniques used for the various types of rock bursts because, for example, the regional measures applied to destress the overall rock mass and prevent tectonic rock burst occurrence also removes the conditions for local rock burst occurrence. That's why local and regional rock burst prevention techniques are jointly considered.

To remove or decrease high concentrations of stress in rock masses adjacent to mine workings, rational drift orientation, special opening shapes, and smooth blasting techniques are applied at the apatite mines of the Kola Peninsula. If the horizontal stresses exceed the vertical stresses in a rock mass, the most stable mine workings are those in which the width exceeds the height and the roof curvature is minimum.

For vertical mine openings the most stable form is elliptical where the axis ratio is equal to the axis of the horizontal stress components. Under high stresses or under conditions when, for technical reasons, it is unreasonable to use such cross sections, a pointed or peaked (hip) roof is used in the mine workings to localize stress concentrations at the peak and to relieve stress in the adjacent zone.

Stress relief and the prevention of dangerous stress concentrations within the mined-out horizons are carried out by induced caving at the hanging wall, by a rational direction of mining fronts to keep them straight (parallel), and by a rational order of block-pillar mining.

To decrease stress concentrations in the zones adjacent to mine workings and in different pillars, destress blasting is used. The blasting-drilling directions and the technique to be applied under different geological conditions are specified. In the top-priority mine workings (tunnels, shafts, chambers for different purposes, and other permanent structures), the application of camouflet blasting is limited because it can cause significant damage to the zone adjacent to a working and exfoliation followed by significant rock falls. In these cases, it is preferable to drill a line of relieving boreholes in the most highly stressed zones adjacent to the mine workings, thereby excluding additional effects on the zone adjacent to the mine working.

The most radical technique enabling one to relieve a significant regional zone of the mined-out horizon from high horizontal stresses is making vertical or inclined relief zones in the hanging wall. This eliminates the necessity for the application of local measures in separate mine workings. Of practical significance is regional stress relief in thick deposits mined by the induced block-caving method combined with bottom ore drawing. It is difficult to apply local relief measures in complex mining workings that cross and influence each other. Under these conditions, regional relief measures are technically and economically preferable. For a horizon to be relieved regionally, it is necessary to blast the cut sections in the hanging wall along the length of the ore body and then to cut a relieving slot below the haulage horizon. The mining front should advance with some lag behind the operations described above. After that, the horizon is mined out from the cut section to the footwall. In the case of sublevel mining when mobile equipment is being used, the analogous relief zone is created during stoping operations by a certain order of the development of operations carried out on the sublevels.

FIGURE 30.8   The Saamsky open pit's flank caving in barrier pillar mining at the Yukspor deposit (Kirovsky mine)

Combined Method. The special feature of mining of apatite-nepheline ores is that underground mining is being carried out simultaneously with open-pit mining (Figure 30.8). Open-pit and underground operations are in direct contact or are separated from each other by temporary barrier pillars. An open-pit area is used to access the underground levels by adits (Kirovsky Mine) and to arrange the grounds for the main opening (Rasvumchorr Mine). As a result of this combination of open-pit and underground mining operations, the problem of extracting the barrier pillars produced between the open face and the underground workings becomes an actual one. Methods must be developed enabling one to predict the parameters involved with overburden rock caving and also the design parameters describing the risk zone in which the rock is ready to break out after caving.

Another important problem is dealing with the thermo-isolation of the areas of the ore body bordering the open pit. Thermo-isolation is necessary to (1) prevent ore blasted in the winter from freezing and (2) create a barrier preventing cold air from penetrating into the mine workings through the caving zone. To solve this problem, the Mining Institute KSC, RAS, has developed under-pit mining thermo-isolation technology using a snow-rock layer (Figure 30.9). During a long winter, as a result of subfreezing temperatures, the near-surface subquarry rock mass is frozen. After that, the bench surfaces are covered with a layer of snow. Then the snow layer is covered with overburden rock whose volume is sufficient to fill the underground mined-out areas.

Investigation results have shown that the snow reliably separates the ore from the overburden rock during ore drawing,

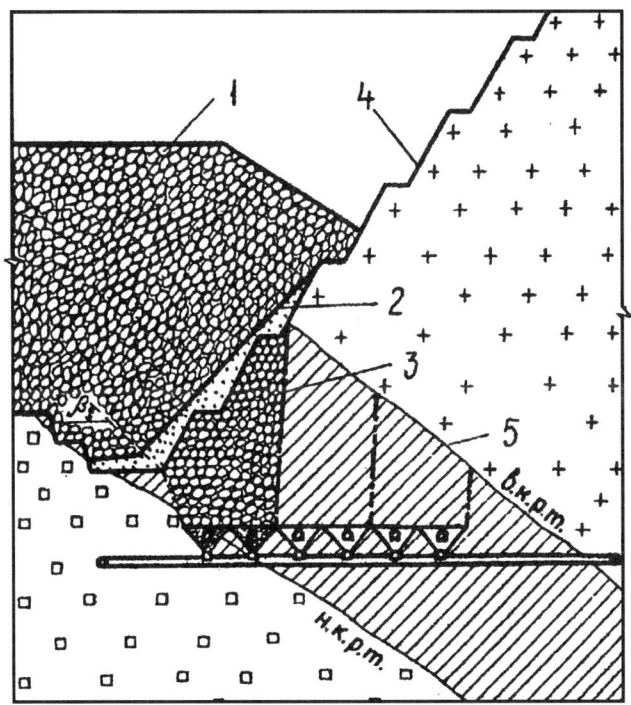

FIGURE 30.9 Technique for mining subquarry reserves using a snow-rock mass as a thermo-isolation layer: 1—blast-broken overburden layer; 2—compacted snow layer; 3—isolated underground mining excavation, 4—open-pit outline; 5—ore body

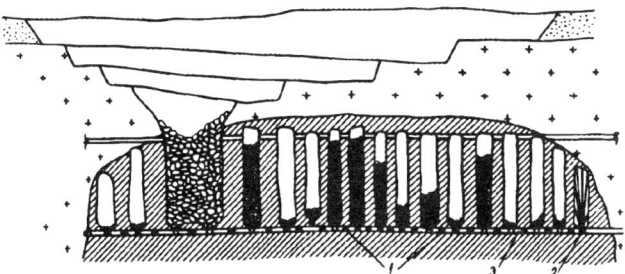

FIGURE 30.10 A shrinkage method: 1—haulage drift; 2—undercutting or drilling-undercutting drift; 3—loading cross-cuts

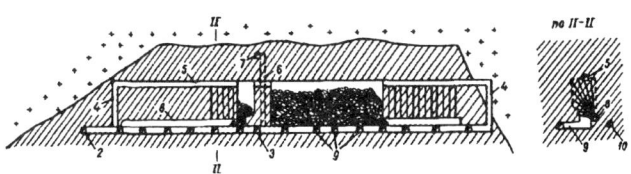

FIGURE 30.11 The Noranda shrinkage method

TABLE 30.3 Physical and mechanical properties of ores and rocks of the mica deposits

Density, kg/m³	Poisson's ratio	Modulus of elasticity, GPa	Tensile strength, MPa	Compressive strength, Mpa
3110–3150	0.16–0.27	63–116	6.8–12	82–257

improves the thermo-isolating properties of the layer, and preserves its stability during the whole period of mining the thermo-isolated area of the ore body.

30.5 KOVDORSLYUDA

30.5.1 Geology

The phlogopite-mica deposits are located mainly in the Kovdor massif of alkali-ultrabasic rocks and carbonatite (338–426 Ma). The massif has an area of 40.5 km² and is some kilometers in depth. Seven steeply dipping (40° to 80°) phlogopite deposits have been identified. They do not crop out. The Glavnaya and Zapadnaya deposits are being excavated. The former is 100 to 120 m thick and has a strike length of 400 m. The latter is up to 300 m thick and has a strike length up to 500 m. Maximum depth of mineralization is 800 m and the thickness of the overlying rocks is 100 to 200 m. From a structural-tectonic point of view, the phlogopite deposit is characterized by (1) the feldspar-ijolite and carbonatite dyke and vein intrusion, (2) filled and unfilled fractures, and (3) zones of fragmentation. As a whole, the ore massif is stable. Five rock types differing in physical and mechanical properties are distinguishable (Table 30.3).

30.5.2 Mining Method

The underground mine is accessed by two vertical shafts. The Kapitalnaya shaft is sunk to horizon +104 m and is equipped with two-skip hoisting (the design output is 105,000 m³/yr). The

Ventilyatsionnaya shaft is sunk to horizon +144 m and is equipped with one-skip hoisting. The reserves of the stopes in the upper part of the Glavnaya deposit (located above level +144 m) are mined with a shrinkage method. The ore is loaded by LHDs into electric cars (Figure 30.10). Stopes 7 m wide and 40 to 50 m high are arranged along the strike of the ore body. In some stopes, the Noranda shrinkage method from the drilling-undercutting drifts on the haulage horizon (Figure 30.11) is used.

The 14-m-wide interstope pillars were mined from the open pit after the mined-out space had been backfilled with sand through boreholes 250 mm in diameter drilled in the roof.

To mine the reserves in the stopes lying below the +104-m level, a chamber-and-block-caving method was applied in combination with the Noranda method. Holes 100 mm in diameter were drilled from the drilling-undercutting drifts on the haulage horizon (Figure 30.12). Blasting was carried out without preliminary undercutting. The drawing trench was made by blasting a fan pattern of holes.

Ore was drawn using loading machines. This technology made it possible to reduce the volume of labour-consuming operations (drift driving, block undercutting, trenching). Chambers were 20 m wide and in some cases up to 30 and 40 m. This system was also applied to mining the ore reserves in the Zapadnaya deposit. In some flank areas where the height of the ore layer overlying the haulage horizon did not exceed 10 m, the open stoping method with irregularly placed pillars was applied. The broken ore was drawn with scraper installations.

The mining of mica deposits is accompanied not only by ore loss because of incomplete recovery, but also by mica loss in the recovered ore as a result of damage to the crystals during the mining process.

On the basis of experiments carried out both under laboratory and industrial conditions (Busyrev 1987; Melnikov et al. 1998), a method has been designed to calculate the amount of loss depending on the conditions under which the ore mining technologies were implemented (Table 30.4)

30.6 CONCLUSIONS

In the Kola Peninsula, underground mining techniques are used to extract copper-nickel, rare metals, apatite-nepheline, and mica ore from a variety of deposits.

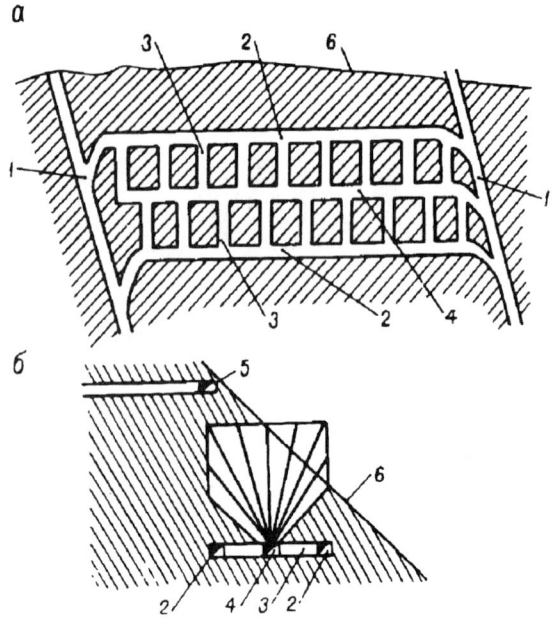

FIGURE 30.12 Chamber-and-block-caving method combined with the Noranda method: a—the haulage horizon plan; b—the chamber's cross section; 1—haulage drift; 2—haulage drift; 3—loading cross cut; 4—drilling undercutting drift; 5—haulage drift; 6—deposit outline

TABLE 30.4 Mica loss due to crystal deformation during mining operations, percentage of the initial mica mass

Operation	Industrial raw-mica	Face raw-mica
Small hole blasting to open mined-out area:		
One-row	$5.8q$	$0.6q$
Multi-row	$6.2q$	$0.8q$
Big hole blasting to open mined-out area		
One-row	$4.7q$	$0.4q$
Multi-row	$5.5q$	$0.5q$
Hole blasting in suppressed medium		
One-row	$3.5q$	$0.2q$
Multi-row	$4.8q$	$0.3q$
Magazine drawing	$0.34S^{0.17}H^{0.92}$	$0.05S^{0.17}H^{0.9}$
Loading by mine machines		
Phlogopite	$2.2S^{0.28}$	$0.15S^{0.28}$
Muscovite	$0.39S^{0.44}$	$0.03S^{0.44}$
Vibratory loading	$0.13S^{0.45}$	$0.026S^{0.45}$
Scraping	$0.6S^{0.75}$	$0.13S^{0.75}$
Loading by excavators	$9.75S^{0.22}$	$4S^{0.22}$
Moving open pit machines on the mined ground	$3–5$	$2–4$

Note: q—specific consumption of explosives, kg/m^3; S—crystal area, cm^2; H_{av}—average height of drawing equal to half the height of a stope, m.

The mode of occurrence of the ore bodies, their dimensions, and their geologic settings differ significantly. These determine the mining method selected and the difference in scale of underground mining. The annual productive capacity of the mines varies from some hundreds of thousands of tonnes per year for the mining of phlogopite-mica, copper-nickel, and the rare-earth ores to some millions of tonnes in the Khibiny apatite-nepheline deposits. For all the deposits, it is characteristic to apply effective ore body mining technologies and a high level of mining mechanization based on the powerful mobile equipment produced by leading world firms.

Problems arising in underground mining are primarily due to complicated geological settings induced by high in situ stresses and the complex structure of the ore bodies. As a result of their good strength properties and their hydrogeology, the ore bodies of all deposits are mined under rather favourable conditions, making it possible to drive openings without support or using only light support techniques. As a whole, the Kola region can be characterized as one of the most developed regions in the world within which intensive mining is carried out under various mining and geologic conditions.

30.7 REFERENCES

Blatov, I.A., V.S. Velim, A.A. Ivanov, et al. 1995. Basic problems of development of the raw-materials base of the GMK Pechenganikel mining company. *Ore Beneficiation (Rus)*. 4.

Blatov, I.A., and S.V. Sokolov. 1996. Mineral raw-materials base of the GMK Pechenganikel mining company. *Non-ferrous Metals (Rus)*. 5.

Blatov, I.A., I.R. Kakin, and A.A. Kuleshov. 1998. State of things and prospects of mining evolution at the GMK Pechenganikel mining company mines. *Mining Journal (Rus)*. 4.

Busyrev, V.M. 1987. *Rational Operation of Mica Deposits*. Leningrad: Nauka.

Kamkin, I.R., Z.L. Shvarts, and V.S. Velim. 1996. The basic directions in the development and improvement of mining at the mines of the GMK Pechenganikel mining company. *Non-ferrous Metals (Rus)*. 5.

Kuleshov, A.A., A.Ye. Popovich, and S.N. Trusov. 1997. Ways for improving vehicle equipment productivity at the Severny mine of the GMK Pechenganikel. *Mining Journal (Rus)*. 11.

Melnikov, N.N., V.M Busyrev, A.Sh Gershenkop, V.D. Pushka, and G.V. Cheremnykh. 1998. *Mica deposits of the Murmansk Region: Reality and Possibility of Mastering*. Apatity.

CHAPTER 31

Underground Mining Operations at the McArthur River Uranium Mine

Doug Beattie* and Chuck Edwards*

31.1   INTRODUCTION

The McArthur River Operation is jointly owned by Cameco Corporation (69.805%) and Cogema Resources, Inc. (30.195%). This underground uranium mine is currently in the initial stages of production and is projected to become the world's largest uranium mine in terms of output in 2000 with a very competitive cost structure.

Cameco, through its predecessor company, the Saskatchewan Mining Development Corporation, began exploring the McArthur River area of northern Saskatchewan in 1980 in search of uranium mineralization. In 1984, a 12-km-long basement conductor was identified at a depth exceeding 500 m. Systematic exploration drilling of this conductor resulted in the discovery of the P2 North deposit in 1988. Subsequent surface drilling programs from 1989 to 1992 delineated a narrow ore zone over a strike length of 1,700 m at a depth ranging from 500 to 600 m. At this time, geological reserves for the deposit were estimated to be 2,370,000 tonnes at an average grade of 5.0% U_3O_8 for a uranium content of 260 million pounds (Mlb) of U_3O_8.

To complete a feasibility study and environmental impact statement, an underground exploration program commenced in 1993. This program consisted of shaft sinking, lateral development, and diamond drilling. Shaft sinking reached the 530 level in 1994, and diamond drilling was conducted over a 300-m segment of the central portion of the ore body until August 1995.

Underground drilling identified two distinct ore zones (pods 1 and 2) composing a total proven reserve of 505,000 tonnes graded 22.15% U_3O_8, which equaled 246.5 Mlb of U_3O_8. Pod 1 (91,000 tonnes graded 17.46% U_3O_8 equaling 35.05 Mlb of U_3O_8) was not identified from surface drilling. One surface hole grazed the eastern edge of pod 2 (414,065 tonnes graded 23.17% equaling 211.5 Mlb of U_3O_8), but the significance of this intersection was not recognized at that time. The underground discoveries, therefore, largely represented additions to the existing resource base. Probable reserves and indicated resources total an additional 236 Mlb of U_3O_8.

Cameco's existing Key Lake Mine was scheduled for ore exhaustion in late 1999. To maintain continuity of production, the rapid development of the McArthur River deposit was necessary. An environmental impact statement was submitted in December 1995 and approval to proceed with construction was received in August 1997.

Construction was completed on time and on budget to allow initiation of mining operations in December 1999. The present paper has been written after 5 months of operation, during which over 2 Mlb of U_3O_8 has been extracted during the final stages of commissioning and the initial stages of production ramp-up.

Production rate is scheduled to increase to 18 Mlb/yr by 2002 at a nominal daily production rate of 150 tonnes of ore.

The high grade of the ore body has resulted in the development and implementation of unique mining and materials-handling methods to limit worker exposure to radiation. Furthermore, the presence of groundwater in some formations has resulted in the need to establish a freeze wall for the initial mining area of pod 2 to isolate flows from mine workings. This paper provides an overview of the key aspects of the mine design to accommodate the difficult challenges encountered.

31.2   GEOLOGY AND HYDROGEOLOGY

The McArthur River Operation is located in the southeastern portion of the Athabasca basin within the southwest portion of the Churchill Structural Province of the Canadian Shield. The crystalline basement rocks underlying the deposit are members of the Aphebian-age Wollaston Domain metasedimentary sequence. The Wollaston Domain consists of a lower assemblage of pelitic, semipelitic, and arkosic gneisses with minor amounts of interlayered calc-silicates and quartzites. Most of the Wollaston Domain rocks have been influenced by middle to upper amphibolite facies metamorphism.

The Wollaston Domain basement rocks are unconformably overlain by flat-lying, unmetamorphosed sandstones and conglomerates of the Helikian Athabasca Group. These sediments are over 500 m thick in the deposit area, as illustrated in Figure 31.1. Tectonically, the most significant feature in the region is the graphitic P2 fault.

Pods 1 and 2 are distinctly different ore bodies. Pod 1 mineralization was defined by underground drilling over a 70-m strike length and is considered to be typical of most of the deposit. Mineralization is preferentially found in the tectonically disrupted sandstone near the main zone of thrust faulting, as illustrated in Figure 31.1. In general, the mineralization in pod 1 follows the 45° dip of the main fault trend.

Hydrogeologically, the brittle, flat-lying sandstone has been well fractured by the tectonic forces of the thrust fault, and these fractures are water bearing. Drawdown tests have demonstrated that the fracture patterns, along with water-bearing joints and bedding planes, are directly connected to the surface groundwater table. The sandstone itself, however, is not porous. Groundwater within 20 m of the ore zones often contains high concentrations of radon gas, and this was a prime concern during the mine planning stage. One of the key design criteria with respect to mining method selection, therefore, was the elimination or minimization of the potential for groundwater entering mine workings.

---

* McArthur River Operation, Cameco Corp., Saskatoon, Saskatchewan, Canada.

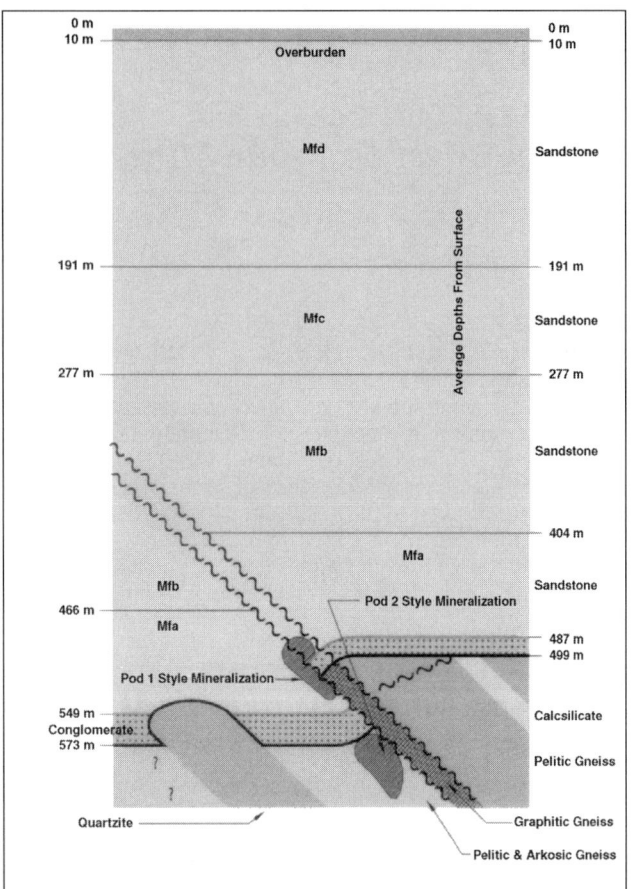

FIGURE 31.1 Schematic geological cross section

TABLE 31.1 Average rock strengths and rock classification

Rock type	Rock strength, MPa	Rock classification
Sandstone, silicified	220	Good to very good
Sandstone, unsilicified	90	Good
Quartzite	120	Fair to good
Perlite in proximity to ore zone	60	Fair to poor
Perlite not in proximity to ore zone	200	Good to very good
Pod 2 ore zone	30	Fair to poor
Pod 1 ore zone	120	Fair to poor

The pod 2 ore zone consists of an altered pelite and pitchblende matrix. The pod 1 ore zone consists of a silicified sandstone and pitchblende matrix. Alteration products in and near the ore zone, such as dravite and hematite, are present, which reduces overall rock strength. Sandstone below 400 m depth has been silicified. Sandstone above this depth is unsilicified and exhibits lesser rock strength.

The predominant geotechnical design consideration for pod 1 is the broken nature of the ground due to the P2 thrust fault. The predominant geotechnical design consideration for pod 2 is the VQ fault zone in the footwall.

In situ stress tests have not been conducted, but all evidence to date suggests a hydrostatic stress state with virgin stress levels around 15 MPa in the mining areas.

31.4 RADIATION DESIGN CRITERIA OVERVIEW

Annual worker exposure limits have been established by the Atomic Energy Control of Canada and take into account the cumulative exposure to alpha, gamma, and long-lived radioactive dust. Alpha radiation is generated by radon and radon decay progeny. Radon gas is typically derived from radon-bearing groundwater sources that enter mine workings. Radon emanating from ore is also a critical source of radiation, particularly where poor ventilation allows gas build-up. Gamma radiation is directly proportional to ore grade. Long-lived radioactive dusts are essentially airborne ore particulates that have been liberated during some aspect of mining or ore handling.

Wherever possible, discreet radon sources should be captured in suction ventilation ducts and delivered to nonentry return airways. Where this is not possible, air flow rates must be sufficient to both dilute the radon source and ensure that the time taken for this air to reach a nonentry return airway is minimized to limit the decay process.

Time, distance, and shielding are the key design factors used to limit gamma radiation exposure. Processes are designed to minimize the time personnel need to be in contact with ore sources. When interaction is necessary, the distance between personnel and ore source should be maximized. In addition, the surface area of the ore source should be minimized where possible. Finally, shielding, typically steel, lead, or concrete, is incorporated into processes as required. At McArthur River, much of the process piping is Schedule 160, and ore storage tanks typically have a suitable skin of +300-mm concrete.

Dust is controlled by using a wet process for ore handling as soon as practical and providing secondary process ventilation systems where necessary.

All personnel involved in ore extraction and handling carry direct reading dosimeters that provide a numeric readout of gamma radiation exposure. Personnel are also equipped with personal alpha dosimeters that measure alpha radiation exposure. Extensive dust sampling is also conducted.

A full appreciation of the implications of the potential for radiation exposure, provided by detailed modeling of each step of

Pod 2 mineralization is confined to the basement sequence. This ore body has a strike length of 100 m, is up to 100 m high, and has a thickness up to 25 m. A subvertical fault (VQ) is located adjacent to the ore body on the footwall side. This fault extends from the main overthrust block of the P2 fault through the underlying footwall quartzites and tapers out rapidly in a series of listric splays near the 640 level. The eastern side of this fault has been thrust downwards, resulting in folding of the quartzite, arkose, and pelite sequence. The intense structural disruption caused by this folding and faulting is believed to have provided a setting very conducive to ore emplacement.

Hydrogeologically, the vertical thrust fault in the quartzite contains water, clay, and unconsolidated rock up to 30 m below the overlying sandstone. Exploration drilling has proven to be difficult through this zone. Grouting has rarely been successful in providing enough ground stability to enable exploration drill holes to penetrate through the zone. Consolidation of the zone prior to production was identified as a key requirement for successful reserve extraction.

All diamond drilling conducted at McArthur River is performed through pressure-tested standpipes. Additionally, through the use of a Navi pump, groundwater static pressure in a hole can be neutralized to assist drilling performance. Cement grout is placed upon completion of all drill holes. Typically, a drill hole may encounter a groundwater flow on the order of 1.5 to 4 L/s during the course of drilling in the sandstone.

31.3 GEOTECHNICAL PARAMETERS

Average rock strengths and rock classifications are given in Table 31.1.

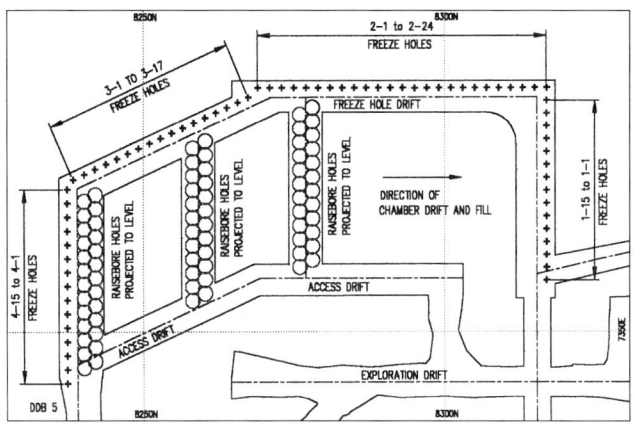

FIGURE 31.2 Plan view of initial mining area, 530 level

the process at McArthur River, led to the development of the nonentry mining method and ore-handling system described here.

31.5 GROUND FREEZING AND MINING METHOD DESCRIPTION

To meet the tight timeframe for commencement of production and to satisfy the constraints imposed by the need for nonentry mining methods and water control, the following decisions were made:

1. Concentrate on the initial extraction of pod 2.

2. Utilize ground freezing to cut off the groundwater flow path from the sandstone to the ore zone while providing consolidation of the VQ fault area.

3. Utilize raise boring as the mining method.

A 70-Mlb U_3O_8 portion of pod 2 was selected as the initial mining area for commencement of production. To isolate this ore from water-bearing sources in the sandstone, a total of seventy-six 105-m-long freeze holes were drilled from the 530 level to form a horseshoe surrounding the ore, as illustrated in Figure 31.2. Holes were drilled on 2-m centres. Dip varied from 65° to 75°.

It was necessary to develop a unique drilling system that could cope with the expected ground and hydrogeological conditions while maintaining adequate accuracy to ensure eventual freeze wall closure. The drilling system developed consisted of a Cubex Megamatic drill utilizing the Tubex overburden drilling system combined with the Sandvik G-Drill in-the-hole water-hammer system. The Tubex system allowed the final casing pipe to be advanced with the drill hole. This was a critical factor, since maintaining an open hole for casing insertion after drilling was not considered possible. Hole diameter was 143 mm.

The water-hammer system ensured that motive power to the in-the-hole hammer was maintained despite the potential for back pressure caused by static groundwater pressure. In addition, this water provided good hole flushing capability. In-the-hole hammer drilling was also recognized as being potentially more accurate than tophammer drilling.

All holes were drilled through pressure-tested standpipes inserted in the floor of mine development. Conventional blow-out-prevention equipment was attached to the standpipe. A unique down-the-hole packer was developed in-house to ensure that no groundwater could exit the drill string during rod addition and removal via the Tubex casing and drill string. An in-the-hole hammer check valve ensured that no groundwater could exit via the drill string.

Two Cubex drills were supplemented by two Morrisette LM150 diamond drills configured for high-pressure drilling conditions, and one pneumatic Cubex in-the-hole hammer drill

was used strictly for standpipe drilling. On numerous occasions, it was necessary to telescope the drill holes. Holes commenced by the Cubex drills that encountered excessive squeezing ground or unconsolidated material in the VQ fault area were left cased and grouted to that depth. The holes were then finished by the LM150 at PQ or HQ diameters at a later date by drilling inside and beyond the original casing.

Freeze hole drilling proved to be one of the most challenging aspects of the project, requiring over 1 year to complete. Drill deviation with the Cubex drills was minimal and exhibited a consistent bias. No redrilling was required for the holes commenced by the Cubex drills. Drill deviation by the LM150 drills was problematic at times, resulting in the need for five additional drill holes to ensure that wall closure would be obtained on time.

Concurrent with freeze hole drill efforts, a surface freeze plant and underground brine distribution system was constructed. The surface freeze plant is an 800-tonne-capacity ammonia refrigeration plant. Calcium chloride brine is delivered and returned from underground via two 250-mm insulated pipes installed in the Pollock shaft in a closed loop. Delivery temperature is typically −37° Celsius. On the 530 level, a series of heat exchangers interchanges with a low-pressure, closed-loop brine distribution system that delivers the brine laterally 400 m to the pod 2 area. Delivery temperature is typically −35° Celsius. Freeze pipes contain a small-diameter, inner polyvinyl chloride (PVC) feed pipe to allow the brine to travel to the bottom of the freeze pipe prior to flowing out of the feed pipe and contacting the outer casing, allowing heat transfer. Brine return temperature has typically varied from −20° to −28° Celsius, depending upon the relative growth of the freeze walls. A total of 150 m^3/hr of brine is delivered to the freeze holes. Two freeze holes are plumbed in series. Each pair of holes circulates approximately 4 m^3/hr.

Detailed in-house modeling estimated the time necessary to achieve freeze wall closure based upon initial ground temperature (10 to 25° Celsius), water content, and mineralogy. Quartzite and nonwater-bearing pelitic rock exhibited rapid freeze wall growth. However, the holes drilled through the ore zone and the water-bearing fault zone exhibited significantly slower growth rates.

Completed freeze holes were brought on line starting in January 1999. Upon commencement of production, the freeze wall in the quartzite was greater than 20 m thick at most locations, whereas the freeze wall in the ore zone was as little as 3 m thick. Freeze walls continued to grow during the mining phase at reduced freeze plant load. Over 200 temperature sensors were installed around the freeze wall to monitor wall growth. Ground temperature trends are carefully monitored to highlight any anomalies.

Raise boring was selected as the initial mining method for pod 2 because—

1. It is not possible to work in the ore zone because of the potential for radiation exposure,

2. Ground freezing, the need to control ventilation circuits extremely well, and low ore body rock strength eliminated the use of methods requiring explosives,

3. The lack of ventilation capacity during the construction phase eliminated methods involving multiple sublevels.

Raise boring at 2.4 m in diameter was deemed to be technically and commercial feasible. Raises extract, on the average, 200,000 lb of U_3O_8 each. Raises dip at 65° to 85° with pilot hole lengths varying from 115 to 125 m. Good ground stability to date has allowed the initiation of 3-m in diameter raises, resulting in an average production per raise of 300,000 lb of U_3O_8. The cycle time necessary to mine and fill a given raise varies from 2 to 3 weeks, depending upon raise length and diameter.

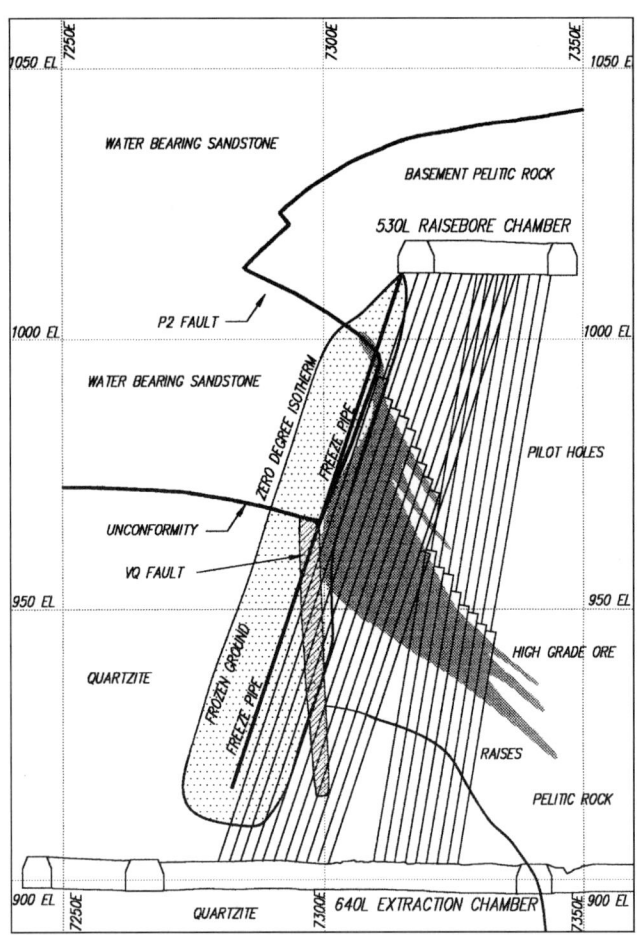

FIGURE 31.3   Cross section through initial mining area

FIGURE 31.4   Remote-controlled LHD below ore collection chute

Because of radon gas generation inside a given raise, precautions are taken to ensure the contents of the raise do not enter mine workings where personnel are situated. These precautions include—

1. Maintaining a good seal at the pilot hole collar and introducing compressed air down the pilot hole during reaming,

2. Utilizing a negative-pressure, 2.5-m^3/s, wet-bath dust scrubber at the bottom of the raise as part of the chute collection system, and

3. Ensuring the nonentry of personnel on the return air side of raise bores on the 640 level through the use of proper restrictive barriers.

Ore is collected at the bottom of the raises by line-of-sight LHDs. A simple chute arrangement is placed below the raise to direct material to a 1.2-m in diameter chute opening leading directly to the LHD bucket, as illustrated in Figure 31.4. Stationary cameras are used to alert the operator when the bucket appears to be 70% full. The operator then informs the raise bore operator to cease reaming. The operator, located 20 to 40 m away on the fresh air side, then backs the LHD away from the chute for a minimum of 15 m. The operator then boards the LHD and drives it to a scanning station where the ore grade is determined based upon gamma radiation response. The ore is then delivered to the location dictated by the scanner. Ore grade of less than 0.5% U_3O_8 is transferred via skip to up the Pollock shaft to the surface. Ore grade greater than 0.5% U_3O_8 is delivered to the ore-handling system.

If at any time it is necessary for a worker to approach within 10 m of the chute area, a radiation work permit is required. This is obtained by having a radiation technician conduct an alpha and gamma radiation survey. Based upon these results, the technician can establish the necessary guidelines and personal protective equipment required to allow work to proceed.

Upon completion of a raise, the reamer is lowered to the level, the area is cleaned up using line-of-sight LHDs and high-pressure fire hoses. The collection chute and dust scrubber are then removed. The reamer is lowered to the level, and a backfill gantry is slid into place below the raise. A reamer carrier has been developed for McArthur River that attaches to the LHD boom after the bucket is removed, allowing the reamer to be broken out from the drill string and removed remotely.

Concrete is used for backfilling. An initial 5-m-high plug is pumped via a feed pipe through the backfill gantry from below and allowed to cure for 48 hr. This is followed by a second pour

Three raise bore chambers were established on the 530 level inside the freeze wall horseshoe, as illustrated in Figure 31.2. Two rows of raises were bored from each chamber to the respective extraction chambers on the 640 level. Each row may contain from 10 to 16 raises, depending upon reaming diameter and local geology. All chambers are provided with a concrete floor for ease of cleanup and halogen lighting for better visibility. Figure 31.3 illustrates a typical cross section through the ore zone. A mineralization grade of 1% to 2% U_3O_8 extends 20 to 30 m below the higher-grade ore. This lower-grade ore, therefore, dictated the 640-level extraction horizon elevation.

Upon completion of reaming in a given chamber, the raise bore and extraction chambers will be backfilled with concrete, and adjacent chambers will be mined for reaming the next two rows of raises per chamber. A total of 26 rows are planned to recover the 70 Mlb of U_3O_8 contained in the first mining area.

Pilot hole deviation to date has averaged less than 0.5% because of careful alignment of the raise borers and standardization of pilot hole drilling parameters. Raises completed have yielded from 50,000 to 400,000 lb of U_3O_8 in raises averaging 65 m in length. Because of the generally soft nature of the ore zone, the production rate is constrained by the ore handling rate as opposed to the reaming rate. Reaming in excess of 50 tonne/hr is theoretically possible; however, 20 to 30 tonne/hr has proven more the norm.

The raise boring conducted can be considered to be essentially conventional. Special precautions are taken to collect pilot hole cuttings containing high uranium content.

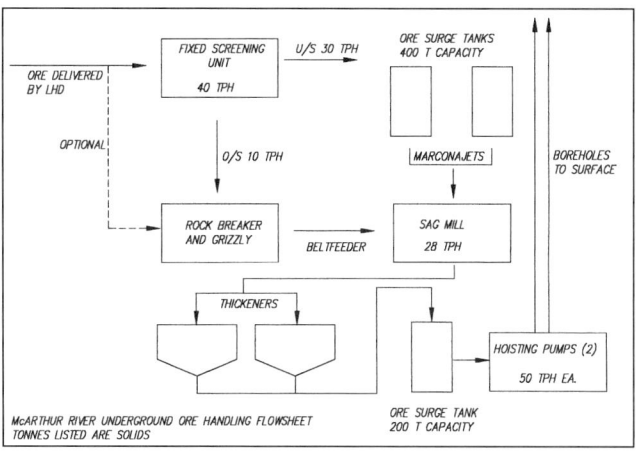

FIGURE 31.5   Flowsheet of underground ore-handling process

FIGURE 31.6   SAG mill station, 640 level

via the pilot hole and a final pour to complete raise filling 48 hr later. During this time, the raise borer is setting up to drill the next pilot hole.

Raises are designed to overlap slightly in order to achieve high ore recovery. Since backfill strength is similar to the in situ strength of the ore body, no material impacts from overlapping of raises have been noted during previous tests and current production. It is expected that 1 to 2 additional raises will be reamed per row to recover reserves missed during first-pass mining because of pilot hole deviation.

The mining equipment currently used consists of one 73RH and three 53RH Atlas-Copco-Robbins raise borers. Four MTI line-of-sight LHDs are utilized for ore handling. Four reaming heads, ore collection chutes, and backfill gantries are available. To reach full capacity of 18 Mlb U_3O_8 per year, an additional raise borer will be added to the fleet.

The freeze wall is being extended during 2000 to isolate a further 70 Mlb of U_3O_8 and initiate mining in an additional portion of pod 2 in 2001. Fifty-six 105-m-long freeze holes will be drilled. Also, a small 7-Mlb portion of pod 1 is being prepared for production utilizing standard cement grout for water control because of good cement grout acceptance in this area.

Underground exploration drilling is currently active in delineating ore identified from surface drilling. The bulk of this drilling will be completed by the end of 2001, allowing the development of detailed plans for extraction.

31.6   ORE-HANDLING METHOD OVERVIEW

Ore is not hoisted to the surface via the existing shafts. Instead, the ore is reduced to a slurry form and pumped to the surface. The ore-handling process design recognises the necessity of providing surge capacity to separate the inherently erratic mining production rate from the more smoothly operating ore grinding and subsequent processes. Ore is mined intermittently 24 hr/d. The mining rate varies, depending on the sequencing and boring rates of the mining machines. Two underground ore surge tanks provide a buffer between mining and milling. An underground grinding circuit is primarily fed from these surge tanks. Ground and thickened ore slurry is pumped to storage facilities on the surface. The ore loadout plant on the surface loads the stored ore slurry into specially designed containers for trucking to the Key Lake site for milling. The underground ore handling flow sheet is shown in Figure 31.5.

31.6.1   Ore Collection and Sizing

Ore collected by the LHDs at the bottom of the production raises is delivered to a fixed screening unit. This unit consists of

a hopper, belt feeder, vibratory screen, and pumpbox capable of sizing at 40 tonne/hr. Ore less than 25 mm is pumped to the ore surge tanks. Ore greater than this size is sent to rock boxes, which, when full, are delivered to a rock breaker station feeding the underground semiautogenous grinding (SAG) mill (Figure 31.6).

31.6.2   Cyclones and Ore Surge Tanks

Ore slurry pumped from the screening unit feeds dewatering hydrocyclones mounted over the ore surge tanks. Cyclone overflow carrying a few percent of fine solids is pumped to the pair of underground thickeners. Cyclone underflow at approximately 70% solids drops directly into the ore surge tanks, which together can hold approximately 400 tonnes of ore. This provides a surge capacity of approximately 12 hr at the average ore mining rate and 16 hr at the average ore grinding rate. Ore is reclaimed from each surge tank by reslurrying with a bottom-mounted Marconajet and is then pumped to the grinding mill.

31.6.3   Grinding

McArthur River ore is ground in the world's only underground SAG mill. The Marconajets are the primary source of mill feed. The secondary source is screened oversized ore from the rock boxes. The ore is dumped through a grizzly with 250-mm-square openings and fed to the SAG mill on a belt conveyor. The Svedala SAG mill is designed to grind 28 tonnes of ore an hour with a 200% circulating load. Ore slurry from the Marconajets feeds the scalping screen. Scalping screen oversize drops directly into the SAG mill, while the undersize material drops into the SAG mill discharge pumpbox, where it joins the ground material exiting the SAG mill. Slurry is pumped from the mill discharge pumpbox onto the twin classification screens. Classification screen oversize drops back into the SAG mill, while the undersize is pumped to the twin underground thickeners. The target ore grind is 100% passing 500 micrometers, 85% passing 300 micrometers, and 50% passing 75 micrometers.

31.6.4   Thickening

The twin underground 13-m in diameter Outokumpu thickeners have two process feeds: a cyclone overflow containing fine solids that originate from the raise borers and ground ore slurry from the grinding circuit. These feeds are divided evenly by distributors to the two thickeners. Thickener overflow collects in the thickener overflow tank, which is the source of recycling water for the underground ore-handling processes. Thickener underflow is pumped to a 200-m^3 thickener underflow tank that feeds the ore hoisting pumps.

31.6.5 Ore Slurry Hoisting

Ore slurry is pumped through one of two boreholes vertically 640 m to the surface by Wirth triplex piston diaphragms, positive-displacement pumps operating at a discharge pressure of 13 MPa. There are two installed hoist pumps, either of which can be used to hoist ore when required. The hoisting flow rate is such that the flow velocity in the pipeline is 3 m/s, well above slurry deposition velocity. Slurry is typically 45% to 50% solids by weight. More than 50 tonnes of solids are delivered to the surface per hour.

31.6.6 Surface Ore Storage and Loadout

Ore slurry pumped from underground is stored in four 650-m³, air-agitated pachuca tanks on the surface. The pachucas provide sufficient surge capacity (at average grade, about 12 days production) between the mining operations at McArthur River and the milling operations at Key Lake to smooth out ore mining rates and ensure the steady supply of ore to the Key Lake mill. They also allow for blending of ore to required grades (less than 30% U_3O_8) for shipment. From the storage pachucas, the slurry is pumped to an ore mixing tank for blending. As well as offering ore blending capabilities, the ore mixing tank provides surge capacity for feeding the loadout thickener. The Outokumpu loadout thickener supplies a blended ore slurry that has a consistently high solids content, thereby minimizing the shipment of water to Key Lake. Overflow from the thickener is sent to the primary water treatment feed tank. The 50% solids underflow from the thickener is pumped to the container feed tank in the slurry loadout area. The container filling pumps circulate ore slurry in a pipe header from the container feed tank to fill batch tanks in the container feed area. The slurry stored in the batch tanks is transferred to truck-mounted ore containers by introducing compressed air into the batch tanks. When the filling operation is successfully completed, the truck departs for the 80-km journey southward to the slurry-receiving facility at Key Lake. Each truck typically delivers 12 tonnes of ore to Key Lake. Therefore, 12 to 15 truck loads are required daily at full production rate.

31.7 MINE SERVICES

The three-compartment Pollock shaft is 5.5 m in diameter, concrete lined, and 684 m deep. The hoisting plant consists of a double-drum, Fullerton, Hogart, and Barclay dc-drive, 1,200-kW hoist. The main cage is used for personnel and material. Waste rock is hoisted by an 8-tonne skip loading from a 660-level loading pocket. The third compartment is used for personnel travel using a separate small auxiliary hoist.

The No. 2 shaft was sunk for mine exhaust purposes and is located 350 m south of the Pollock shaft. This 6-m in diameter, concrete-lined shaft extends to the 530 level. Ventilation capacity is currently 300 m³/s with future plans to increase capacity to 450 m³/s. Two Alphair, vane, axial-flow, 600-kW fans in parallel provide exhaust capacity. A third fan will be added in 2000. Sixty-million BTUs of mine air heating capacity is currently installed. A third shaft is currently being sunk 550 m south of the No. 2 shaft. This 60-m in diameter shaft will be used for fresh air intake and egress from the mine. Most of the resources identified from surface drilling are located between the No. 2 and the No. 3 shafts. All shafts have been sunk in 40- to 50-m lifts using extensive cement grouting to precondition the next stage of sinking below.

All ore production and handling areas underground are equipped with secondary exhaust duct systems to allow capture of radon prior to its entry into the general mine workings.

A conventional concrete batch plant rated at 50 m³/hr is utilised for backfilling and wet shotcrete production. The product is delivered to either the Pollock or the No. 2 shaft, where is it sent via slick lines to awaiting concrete trucks or LHDs below.

Mine dewatering is provided by three 40-L/s, seven-stage Mather and Platt centrifugal pumps on the 530 level. Four 30-L/s, single-stage Gould pumps on the 640 level lift water to the 530-level station. Current dewatering requirements are approximately 25 L/s.

An extensive data control system is incorporated to monitor and control the ore-handling and brine-distribution systems underground and the primary and secondary water treatment plants and the surface freeze plant on the surface. The system also provides important data with respect to air flow quantity and working level readings in key ventilation branches.

31.8 WORKFORCE

Currently, 50% of the total workforce at McArthur River originates from northern Saskatchewan communities. The site workforce numbers currently total 320. This includes over 160 short-term contract personnel conducting mine development, shaft sinking, underground exploration, and freeze hole drilling. Thirty long-term contractors are engaged for the ore truck haulage from McArthur River to Key Lake and site catering and janitorial requirements. Cameco personnel are largely engaged in ore production, ore handling, maintenance, technical, and administrative roles.

The McArthur River site operates on a fly in/fly out basis. Corporate staff work a 7-day-in/7-day-out schedule, whereas most contract staff work a 14-day-in/7-day-out schedule.

Cut-and-Fill Mining

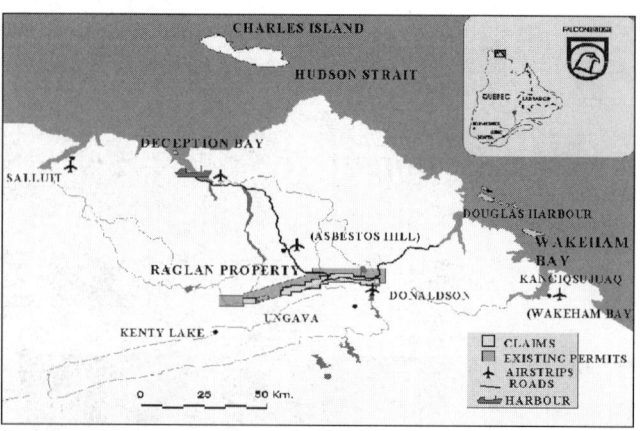

CHAPTER 32

Raglan, Underground Production in an Arctic Environment

F. Pronovost*

32.1 INTRODUCTION

The Raglan Mine is located at the northern tip of the New Quebec Peninsula, 1,930 km north of Montreal. The main administration office is located in Rouyn-Noranda, Quebec, which is the base for flight departure (Figure 32.1).

The site is accessible either by plane or by ship (Figure 32.2). The airstrip is located at Donaldson, the original (and still active) exploration camp at the eastern edge of the property, 22 km away from the main campsite at Katinniq. The surface of the airstrip is gravel and is equipped with navigational aids comparable to those at most southern airfields

All personnel and critical cargo are transported to the site utilizing a 737 Combi jet. The trip from Rouyn to the mine site takes 2 hours and 15 minutes. Most of the nonperishable cargo is shipped to site via the St. Lawrence Seaway from Montreal or Quebec City. The latter represents an 8- to 10-day trip to Deception Bay on the shores of the Hudson Strait. The ship returns with nickel concentrate (25,000 tonnes), which is delivered to Quebec City and shipped by rail to the Falconbridge smelter in Sudbury, Ontario.

The mine site is 100 km away from the docking facilities. A portion of the access road, as well as all docking installations, were built several years ago by the Asbestos Corporation. In 1995, an extension to the road was constructed by Société Minière Raglan du Québec. At the same time, the existing portion of the road was rehabilitated and the Asbestos Corporation mine site was dismantled.

The main property is 33 km in length, with additional claims 80 km west (Figure 32.3). The property hosts multiple ore bodies that total 20 million tonnes at 3.06% nickel and 0.87% copper that extend up to 22 km from Katinniq. The concentrator is also located here (Figure 32.4). Production began 3 months ahead of schedule in December 1997, following an 18-month preproduction period. The planned production rate for the year 2000 is 934,000 tonnes. Ore is now supplied from the Katinniq and Upper Southwest underground mines (420,000 and 48,000 tonnes) and the Zone 2 and Zone 3 open pits (93,000 and 241,000 tonnes). The Zone 2 pit is 3 km west of Katinniq, while the Zone 3 pit and the Upper Southwest mines are 2 and 2.5 km west of Zone 2, respectively. Pit work is contracted out and supervised by Falconbridge's staff, while underground work is carried out by Falconbridge employees.

The Raglan group of employees earned the 1998 O'Connell award from the Quebec Mining Association for having the best safety performance throughout all mines in the province of Quebe. The group should obtain the same award for 1999, as they also had the best performance in the province for this year.

Table 32.1 describes production to date from the Raglan property.

FIGURE 32.1 Location of Raglan Mine in northern Quebec

32.2 GEOLOGY

The Cape Smith/Wakeham Bay fold belt of northern Quebec hosts the Raglan ore bodies. It is early Proterozoic in age and is stratigraphically and geochronologically correlative to rocks of the Labrador Trough and the Thompson Nickel Belt. The Raglan property is located in the eastern part of the Cape Smith Belt along the contact between the Povungnituk and Chukotat groups and hosts important nickel-copper sulphide mineralization. The latter is associated within a peridotite complex that forms a slightly arcuate east-west trend of 55 km along the belt. The peridotite contains 1% fine disseminated iron-nickel-copper

* Société Minière Raglan du Québec ltée, Québec, Canada.

FIGURE 32.2   Local area of Raglan Mine

TABLE 32.1   Production from Raglan Mine

	1998	1999	2000 (plan)	Total
Underground development, m	1,990	2,167	4,775	8,932
Pit waste, tonnes	1,179,000	1,903,000	4,454,000	6,357,000
Underground ore, tonnes	337,000	448,000	600,000	1,385,000
Pit ore, tonnes	334,000	338,000	334,000	1,006,000
Total ore, tonnes	673,000	786,000	934,000	2,393,000
Nickel produced, tonnes	16,365	20,000	23,871	60,230
Copper produced, tonnes	4,365	5,280	5,872	15,517
Cobalt produced, tonnes	190	344	310	844

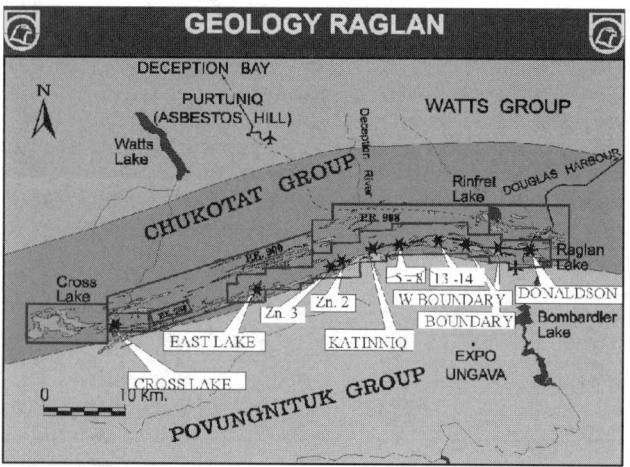

FIGURE 32.3   Property map

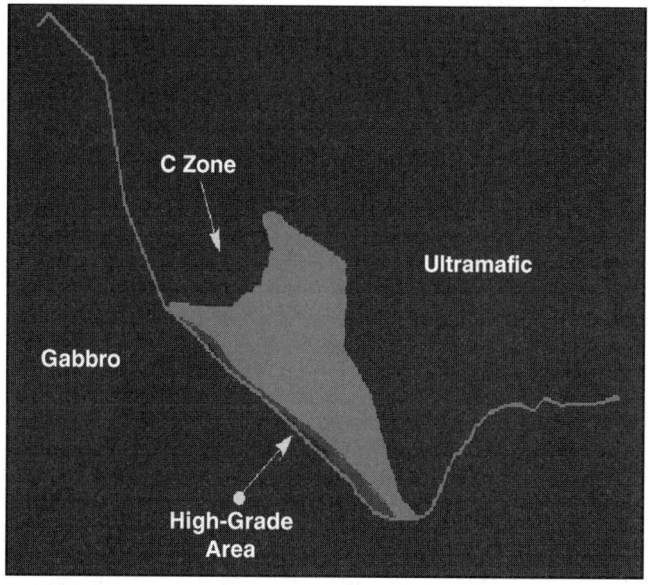

FIGURE 32.5   Typical Katinniq ore lens

FIGURE 32.4   Katinniq site

sulphides, predominantly pyrrhotite and pentlandite with minor amounts of chalcopyrite, accounting for the characteristic reddish-brown color of the weathered outcrop. The ore deposits themselves are composed of hexagonal pyrrhotite, pentlandite, chalcopyrite, magnetite, minor pyrite, trace sphalerite and local concentrations of ferrochromite in a peridotite gangue. Within

the ore zones are associated cobalt and platinum group elements. The Katinniq ore body consists of multiple ore deposits found in channels or podlike areas along footwall topographic lows (Figure 32.5).

At least 85% of the mineral inventory at Katinniq is located at or near the basal contact. The balance is located at or near the base of several overlying flow units. The footwall ore deposits are isolated from one another, with reserves ranging from 18,000 to 1,200,000 tonnes for an approximate original total of 7,000,000 tonnes at 3.09% nickel. These deposits extend from the surface to a depth of 350 m, are 1,400 m long on strike, and plunge northwest with an average dip of 45° to 50°. The ore body is open at depth (Figures 32.6, 32.7, and 32.8).

32.3   GEOTECHNICAL INFORMATION

32.3.1   Stope Dimensioning

The maximum span selected was established on the basis of rock mechanics data available prior to mining and will be subject to review once more data are compiled and analyzed. Rock quality designation (RQD) values from diamond-drill information were gathered and regrouped by families on sections. A three-dimensional model of these families was constructed and then sliced horizontally at each cut elevation (back height) to help locate sublevels. Data from the mapping completed in 1991 allowed the main families of joints to be identified. This information was, however, limited to the footwall material

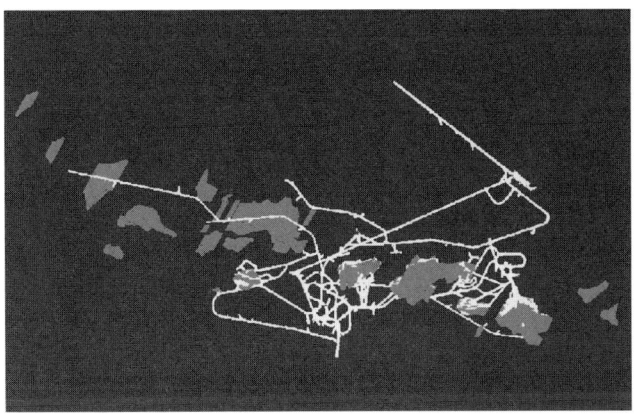

FIGURE 32.6   Plan view of Katinniq

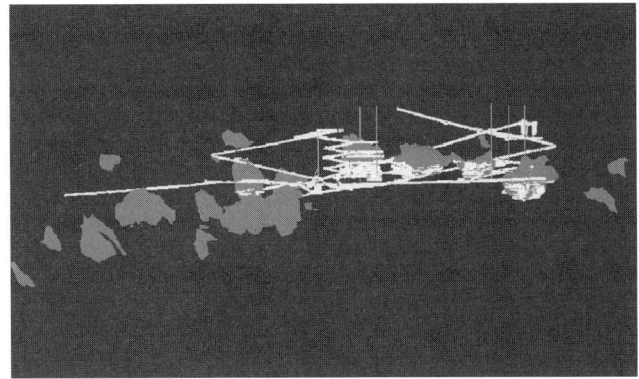

FIGURE 32.7   Longitudinal projection of Katinniq

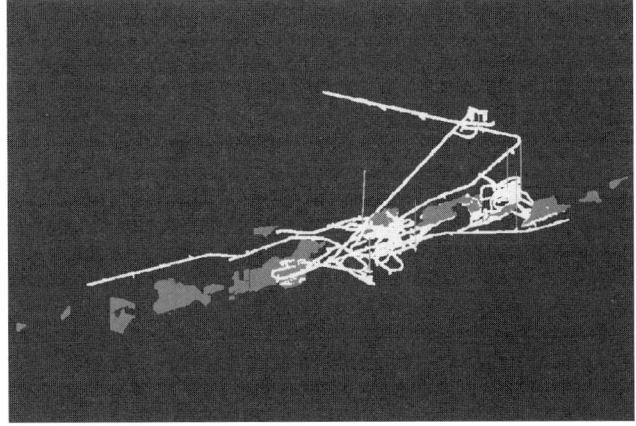

FIGURE 32.8   Isometric view of Katinniq

(gabbro). It was later confirmed by recent development in hanging wall material (peridotite) and in ore.

Core samples were analyzed for uniaxial compressive strength. Intact gabbro indicated 314 MPa while intact peridotite was 369 MPa. When failure occurred along jointing, these values dropped to 133 MPa and 144 MPa, respectively. At that time, there was no core available to do similar tests in ore. No in situ stress measurement campaign was done since the deposit is very

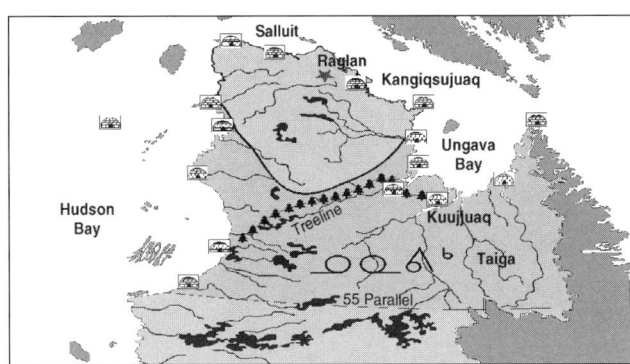

FIGURE 32.9   Limit of continuous permafrost in Quebec

close to the surface. This will be revaluated when mining progresses to lower elevations.

Simulation of the mining sequence in various cuts was done, and the maximum unsupported span (unsupported meaning no cable bolting) was established to be 20 m. This was determined using the Mathews/Potvin method as well as a cut-and-fill stability graph developed by Noranda based on observations and measurements from 172 different real cases throughout Canada and Australia. Both methods led to the same conclusion.

Once the maximum theoretical span was established at 20 m, a conservative design was utilized, i.e., 16 m to begin with. This design is to be confirmed later on as mining progresses.

32.3.2   Permafrost

The property lies within the limit of continuous permafrost (Figure 32.9). To preserve the integrity of the permafrost, fresh air delivered to the mine is not heated. This obviously results in very cold temperatures in main airways during the winter months. In 1991–1992, thermistors were installed in drill holes in the walls and backs at various depths (in relation to the surface) in the ramp and also at various depths in the holes. It was possible to determine that only 1.5 m of rock around the excavations would thaw and then freeze during a 1-year cycle. Even though the warmer summer weather influences the rock temperature around the excavations, this is offset by cooling in the winter months. These freeze-thaw cycles do not seem to have much effect on the stability of the excavations. Thermistors were also installed in drill holes to a depth of 275 m below the surface. Rock temperature underground is –5.5°C close to the surface and increases by about 1°C per 40 vertical meter for an expected limit of permafrost at 425 m below the surface (Figure 32.10).

32.4   MINING METHOD SELECTION

32.4.1   Original Design

Many factors had to be considered when proceeding with the design of the Katinniq underground mine. A primary concern was the availability of suitable material to use for backfill. It was obvious that it would be impossible to extract any zone in one large stope because of the span of the openings involved. This meant that panels had to be mined in sequence, each being filled prior to mining the adjacent one. The first mining method to be considered was blasthole with cemented rockfill (Figure 32.11). This was discarded because of the high cost of transporting cement to the property. It was then considered to try to take advantage of the ambient cold temperatures to replace cement by water, using frozen water as a binding agent. Other northern operations have done so successfully and, after a visit to Cominco's Polaris Mine, this was investigated even though our

Rock Temperature vs Depth

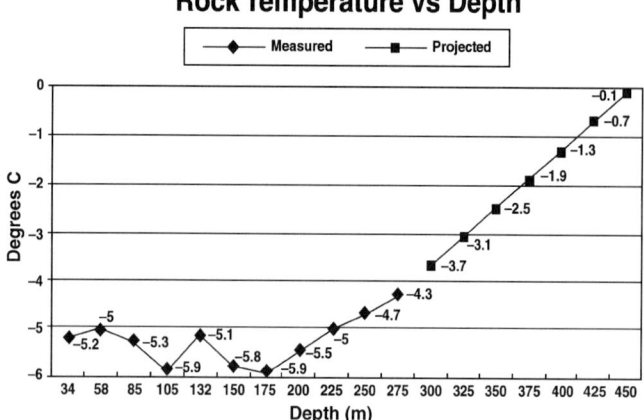

FIGURE 32.10 Limit of permafrost at depth

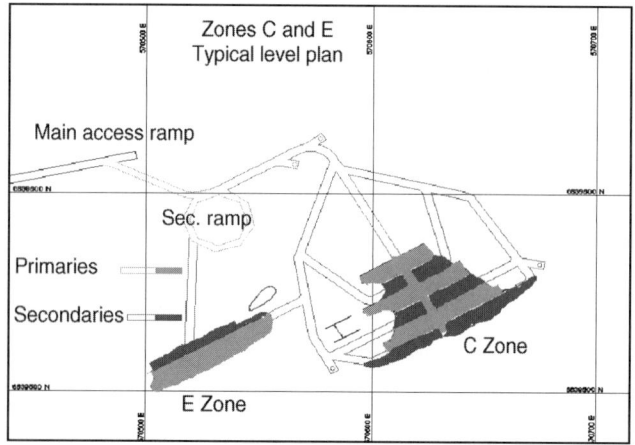

FIGURE 32.11 Blasthole with backfill

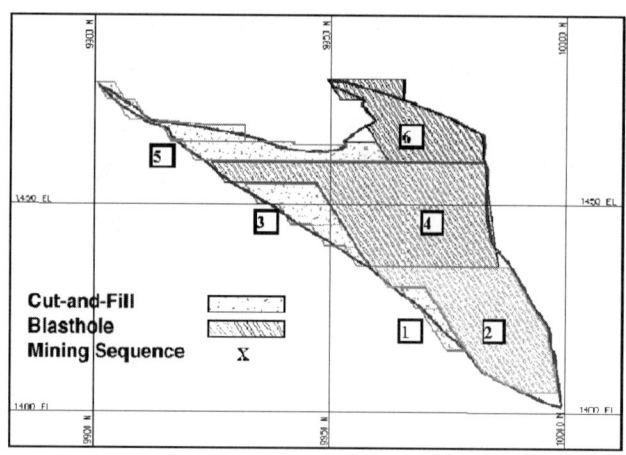

FIGURE 32.12 Combination method

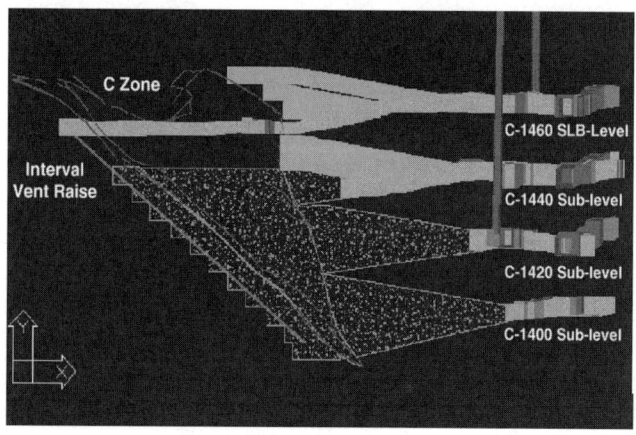

FIGURE 32.13 Sublevels and cut accesses

rock temperature is much warmer than at its site. A study was undertaken to determine the time required to allow the backfill to freeze in place and its optimum moisture content using stope dimensions of 10 m wide, 30 m high, and of variable lengths.

The results were discouraging since one could expect a maximum frozen band of 1.5 m around the backfill mass after 2 years. It would be difficult to develop a schedule allowing for 2 years wait before mining a secondary stope. In addition, there was no confidence that a 1.5-m frozen band would remain solid while mining the secondary. The next idea was to increase the amount of cold to be transferred to the backfill. The option considered was the installation of freeze pipes, which would allow a total freezeback in a short period of time. However, the high cost associated with this option made it uneconomic.

An attempt to develop a design that would account for the unconsolidated state of the backfill led to the development of what is termed a "combination method" (Figure 32.12). Between sublevels, a narrow area would be mined in a cut-and-fill fashion along the footwall contact, creating a fill wall inclined at 65°. That should allow the ore to flow to the drawpoints even though the footwall is 45° and also allow a more stable backfill.

This option was originally abandoned because it was development intensive and did not guarantee that the backfill would stay in place. It may, however, be considered again in the near future.

The next option considered was cut-and-fill mining with nonrecoverable pillars (5 by 5 m). This had the major advantage of eliminating 40% of the development required in the previous scenario, but the disadvantage of losing 7% of the ore. These pillars were designed on a regular grid in the stopes from one cut to the next, and some of them had to be located in the high-grade strip of ore that lies against the footwall. Although the ore loss was only 7%, the reality is that the loss of nickel was much higher.

The final design and one originally implemented at the site was single-pass, cut-and-fill mining. Each cut will be fully extracted by panels mined in sequence, allowing backfilling to be done in parallel with the mining on the active cut. Sublevels are now every 20 m, and each of these allows access to four mining horizons for a specific ore body. The access ramp is strategically located to serve up to three different ore bodies (Figure 32.13).

In the cuts, the primary opening (central panel) is driven directly across the stope until it reaches the far wall. At the same time, at the intersection with ore, development proceeds at 90° (fringe drift) on each side of this primary opening and follows the contact (geology control) until the far wall is intercepted again. These primary openings are 10 m wide by 5 m high.

As soon as the central panel is mined out, it is backfilled to support the back of the stope, thus reducing the final span by half. Once the fringe drifts are completed, a second panel (6 m wide) is opened on the inside of each of them (parallel to the

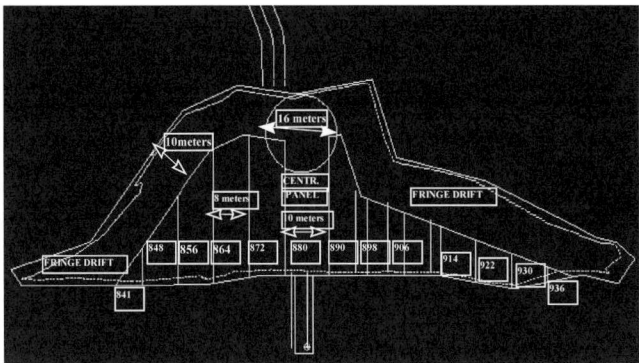

FIGURE 32.14A   Original in-stope panel arrangement

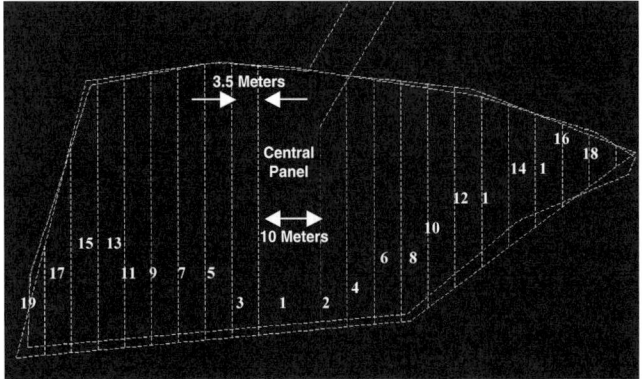

FIGURE 32.14B   Actual in-stope panel arrangement

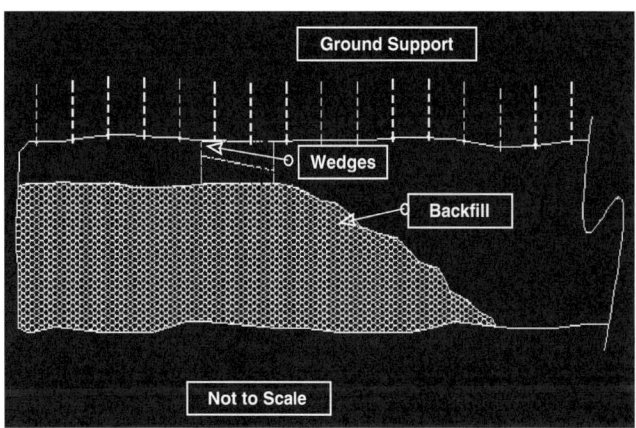

FIGURE 32.15   Concrete wedge arrangement

central panel). This results in a 16-m opening. From this point on, all subsequent panels will be 8 m wide, retreating toward the central panel. At any time, one panel will be under backfill, the adjacent one will be totally mined out, and the next one will be active (being extracted) (Figure 32.14A).

32.4.2   Actual Procedure

In reality, as mining progressed, some unexpected difficulties were encountered that forced us to revise and adapt the original design. Occasional wedges in the backs were encountered, and it was decided to reduce the width of our panels to a standard 5-m-wide excavation. A decision was also made to mine the panels in sequence, leaving every second one to be mined in a second pass. Any rock wall exposed over a previously mined area became an overhang that showed signs of instability. By narrowing the width of the panels, we had also multiplied these overhangs. The method also contributed to reducing the production rate that could be expected from the stoping areas.

It became a necessity to adapt the method again to ensure that work could be done in a safe fashion. We kept the central access concept as per design and decided to start mining outward from this central panel (Figure 32.14B). Occasional addition of support pillars is required (which will be discussed later), but the number of overhangs is reduced to a minimum. Drilling is done on either side of the central opening and full panels are taken in one blast. There are several advantages to this method. It generates an important quantity of ore at each blast. It reduces the travelling time of the equipment, as when the jumbo or the bolter starts drilling; it allows work for a longer period before moving somewhere else; it reduces the amount of bolting material required by minimizing wall area to be bolted; and most

important, it increases safety. By taking bigger blasts, it also gives us the ability to have certain reserves broken ahead.

Some smaller areas were also converted to longhole mining. Once we became confident that a certain span could be opened, we modified the design of the lenses with smaller dimensions and started drilling longholes to reduce production costs and improve the capability of feeding the concentrator. To date, three stopes have been mined successfully in this fashion. Whenever we reach the top cuts of cut-and-fill stopes, we also try to drill and blast uppers to reduce costs associated with development and mining of these areas.

32.5   BACKFILL

32.5.1   Original Design

The backfill is unconsolidated rockfill. It originates either from waste development or from the open pit (waste material crushed to 20 cm). As this rockfill cannot easily be packed against the backs, it does not give true support. As a consequence, a pillar strategy had to be developed to ensure local support of the backs. The original suggestion was to put circular shotcrete pillars in place prior to backfilling. These would have been sticking out of the backfill by approximately 1 m. However, when the superior cut was mucked, the pillars could have damaged the equipment. The pillar design was modified, and the possibility of using short (1 m) shotcrete pillars that would sit on the backfill was considered. These short pillars could be removed from the muck when mining the subsequent cut. It would, however, have been difficult to wedge them tightly against the back to provide efficient support without the use of more shotcrete, and shotcreting in such a confined space (1-m gap between the top of fill and the back) poses certain restraints. Because backfilling quickly and efficiently is of such vital importance, such a time-consuming job during the backfill cycle was not affordable. It was finally decided to use concrete wedges. These wedges (2.4 by 1.5 by 0.76 to 0.46 m) are prepoured on the surface, transported to the stope, and placed where required. The first wedge lies on the backfill, and once it is in place, the second one is wedged between the first and the back of the stope, providing support. These wedges can easily be removed when mining the subsequent cut and can be used elsewhere. Being prepoured on the surface, they minimize the time required to provide adequate support in the stopes (Figure 32.15).

32.5.2   Actual Procedure

The fill material and placement method stayed as per design in the cut-and-fill stopes. Longhole stopes are used for development waste disposal when this material is not required in the cut-and-

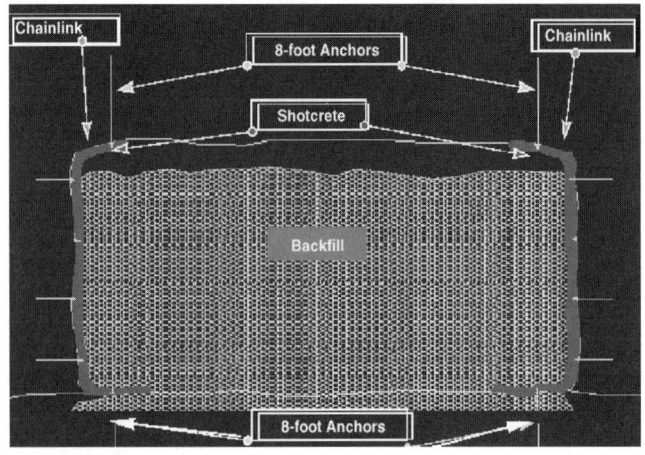

FIGURE 32.16   Shotcrete fences

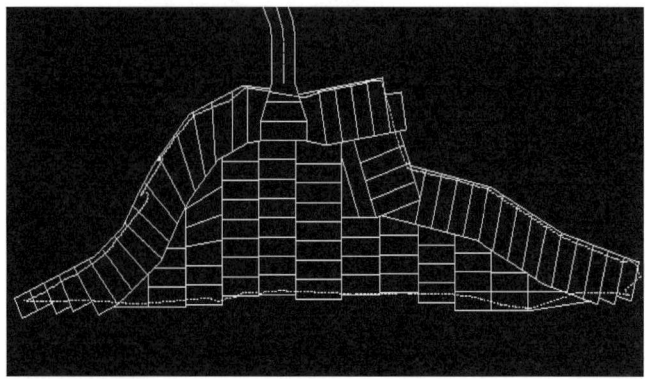

FIGURE 32.17   Standard planning

fill stopes, which means that backfill in the open stopes is a long-term process based on waste availability. As expected, it is impossible to place the rockfill tight against the backs, but the concrete wedges did not yield the expected results. Placement is not an easy task, and performance of the wedges was not as good as was expected. By positioning them strategically, we hoped to supply adequate support for the subsequent cut. Giving only point support, they do not contribute efficiently to eliminating unraveling in subsequent cuts and are not easily recoverable. It was decided to use full-height shotcrete pillars to reduce the spans. The unraveling problem was solved by modifying the in-stope mining sequence, as explained previously.

The next step in improving our ground support will be trying to use tailings to fill the gap between the backfill and the backs, which would eliminate the need for crushed rock (required to reduce loss of fines in the backfill). There would be many more advantages to doing this. The process would lead to savings in waste crushing and hauling, tailings filtering and hauling to the disposal area, and tailings pond reclamation (less volume). This would also contribute to fully supported backs in the stopes, i.e., improved safety.

32.6   DILUTION

32.6.1   Original Design

Another particularity was the use of shotcrete walls to minimize dilution from backfill. When the central panel is extracted prior to backfilling, screen mesh is installed on both walls. This screen is anchored in the back and in the backfill on the floor with 2.4-m rebar and also pinned to the walls with 0.6-m rock bolts. The mesh is then covered with a thin layer of shotcrete. This curtain will prevent the backfill from diluting ore from the adjacent panel (Figure 32.16). In further attempts to reduce dilution, large 2.5-cm-thick steel skids were fabricated. These skids are inserted under the breast being blasted (on top of the fill from the previous cut) and act as a steel mucking floor. The target is to achieve a maximum of 10% dilution in the stopes. Obviously, most of the dilution is picked up when excavating the fringe drift, which means that the bigger the cut is, the lower the dilution should be.

32.6.2   Actual Procedure

The limited use of the original mining concept led to the construction of very few of these shotcrete fences. Even though successfully implemented, they are not required in the existing mining plan and because they are expensive and time consuming to construct, they were abandoned early in the project. Our

dilution last year, even though slightly above expected value, averaged 12% for both cut-and-fill and longhole stopes together. In the open pits, dilution averaged 6.7% last year.

32.7   ORE SCHEDULING AND BLENDING

32.7.1   Grade Control—Original Plan

Scheduling and blending of the ore is the challenge of mining at Raglan. The objective is to get a mill feed grade of 3% nickel. Ore values can be as high as 14% nickel in some areas underground. In the pits, it can reach up to 6%. Sixty percent of the production comes from the underground stopes, while the remaining 40% is supplied by the pit. The grade underground not only varies from one ore body to another, but also within each ore body. There is always a rich strip of massive sulphides on the footwall that becomes disseminated toward the hanging wall. The actual production rate forces at least three ore bodies to be active at one time, with two workplaces in each cut-and-fill stope. Assuming that two of these six workplaces are on backfill, this leaves four different sources for the ore. It is essential that various grades of ore are available to ensure a proper blend. Within the pit, we have the same situation in which grade varies not only from one bench to another, but also within each bench. It is the engineering role to ensure that mill feed stays as close to 3% nickel as possible and that all workplaces are not on backfill at the same time. To achieve this, a 3-month planning schedule is made and is revised each month. A weekly planning is also issued on a round-per-round basis (Figure 32.17). This ensures availability of enough workplaces, a steady mill feed grade, and a mucking plan that is issued at the beginning of each shift. This plan states how many trucks are required from each workplace (including pits) to meet requirements. Two alternate mucking plans are made to ensure consistent mill feed. At the end of each shift, the actual mucking and blasting numbers are picked up by engineering. Geology applies grades to the fresh blasts, and a plan for the oncoming shift is prepared and presented to operations. This is normally done after a quick consultation with the mine captain to ensure that the coming plan will suit everyone.

32.7.2   Grade Control—Actual Procedure

Through production experience, it was learned that the issue of grade control at the mill was not as important as expected. The impact of mill head grade fluctuation on nickel recovery is not that severe. The same cannot be said, however, about the ratio between open pit and underground ore. The talc content in the pit ore is much higher than in the underground ore. Even though the mill can react quickly to variations in grade, it is much slower to respond to a fluctuation in the talc content of the feed, which leads to lower recoveries. This is the reason to keep the proportions of underground-to-pit ore as constant as possible. We were targeting an average grade of 2.95% nickel for 2000. Even though we were trying to achieve this average at year end, there were fluctuations on a monthly and weekly basis. However, one improvement was achieved early in 2000—we eliminated grade control on a daily basis. We are still producing the daily mucking plans, but do not force the mine to blend continuously. This translates into improvement in equipment efficiency. In the previous year, the mine was asked to try to mix, for example, two truckloads of ore from stope A at 2.5% with one truckload of ore from stope B at 4% to deliver a steady 3% to the mill. This lead to a lot of traveling from one place to the other for the underground equipment. Because the mine consists of small zones distributed over a large area, this meant an important amount of inefficient time spent traveling every day. We still try to manage very high-grade areas to avoid drastic fluctuations at the mill, but we are now more open to smaller variations.

32.8 EQUIPMENT FLEET

32.8.1 Production Fleet

- 4 Gardner Denver electric-hydraulic, two-boom jumbos equipped with Montabert HC-80 drifters
- 4 Wagner 436 haulage trucks
- 4 Wagner ST-7.5Z scoop tram. The 7.5-yd^3 buckets have been replaced by 6.5-yd^3 buckets on the units dedicated to ore production.
- 1 Toro 0010 scoop tram with 10-yd^3 bucket
- 3 MacLean bolters (scissor-lift type)
- 1 Secoma Bolter (boom-and-basket type)

32.8.2 Service Fleet

- 7 Toyota Landcruisers for personnel transportation (eight-passenger) for the mine
- 1 Toyota Landcruiser for personnel transportation (three-passenger) for geology
- 1 Toyota Landcruiser with boom and basket for surveyors
- 1 Kubota tractor equipped with small scissor platform
- 1 NCE tractor with forks/bucket/mancarrier
- 4 Toyota Landcruiser customized for maintenance crews

32.8.3 Support Fleet

- 2 Marcotte ANFO trucks
- 2 Marcotte scissor lifts
- 1 Marcotte fuel/lube truck
- 1 Marcotte flatbed/scissor truck
- 1 Marcotte boom truck
- 1 Marcotte mobile shotcreter
- 1 Caterpillar grader
- 1 Wagner ST-3.5 scoop tram
- 1 Wagner ST-2d forklift/dozer/forklift

32.9 OPERATIONAL RESTRAINTS ENCOUNTERED

Difficulties directly related to permafrost have been encountered. Most of them were dealt with and others are being solved systematically. Many difficulties have been overcome in learning how to work with Mother Nature.

32.9.1 Water Distribution

Water distribution underground could not be avoided because of the hard ground encountered. Dry drilling was attempted in 1991 during the exploration phase of the project. Even though regulations regarding dust counts were met, there were major

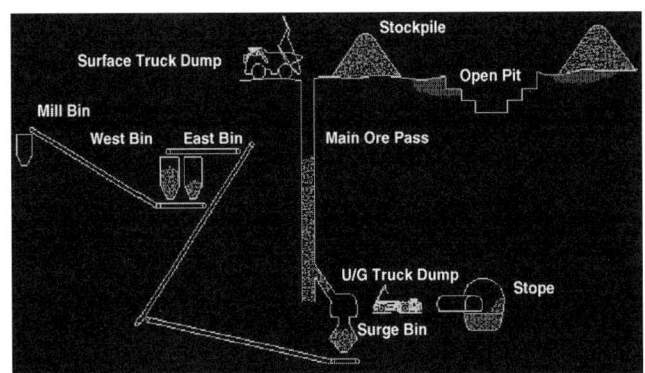

FIGURE 32.18 Ore-handling system

problems associated with overheating of bits and steel, and the resulting bit consumption was very high. It was quickly decided to abandon the concept and return to wet drilling. Fresh water cannot really be used because it freezes quickly; therefore, a mix of calcium chloride and fresh water is used at proportions of 1.3 lb calcium chloride per gallon of water (11%), which lowers the freezing point to around −7°C. During December, January, and February, this concentration is increased (tripled) to keep the water from forming slush and freezing in the pipes. The water distribution network is now fully insulated and heat traced, which guarantees a regular feed of water to workplaces, no matter what the outside temperature is. However, use of a brine mix at a minimum 11% is still necessary at any time to avoid freezing in the drillholes.

32.9.2 Resin Use

Installation of resin rebar was time consuming because of the use of cold resin. The setting time of the catalyst is directly related to the temperature of the product. Fast resin (0–30 s) would take up to 5 min before setting. Slow resin took up to 1 hr to set. A heated storage facility was built underground to keep both resin and rebar at temperatures within their normal recommended range. Even though it improved the situation, heated boxes (reusing the waste heat from diesel exhaust) had to be installed on each bolter to avoid the material becoming too cold while sitting on the machine. That allowed a normal installation procedure.

Mechanical rock bolts were used for a certain period of time, but major retorquing issues had to be dealt with. With time, it was concluded that the optimum bolting system in Raglan was Swellex bolts. Even though very expensive, they give a quality of support comparable to rebar, and the lower transportation costs (less heavy than rebar and resin) led to an almost complete conversion to this system. Installation is also faster, which ensures a better cycle time for the bolters, i.e., a more efficient use of these machines. Prior to the conversion, pull tests were done on both fresh and older Swellex bolts with conclusive results. Corrosion of bolts is not an issue, even though brine is used to inflate them.

32.9.3 Rehandling System

The ore-handling operation is complicated (Figure 32.18). The ore pass delivers ore from the pit to the underground crusher. It is a 3.7-m in diameter raise bore lined with a 3-m diameter steel liner embedded in concrete. At the bottom, a finger (not lined) directs ore to the underground dump located directly behind the crusher (Figure 32.19).

Obviously, the liner and the surrounding rock mass have reached equilibrium at approximately −5°C. The temperature of the ore delivered from the pit ranges between −30 and +15°C

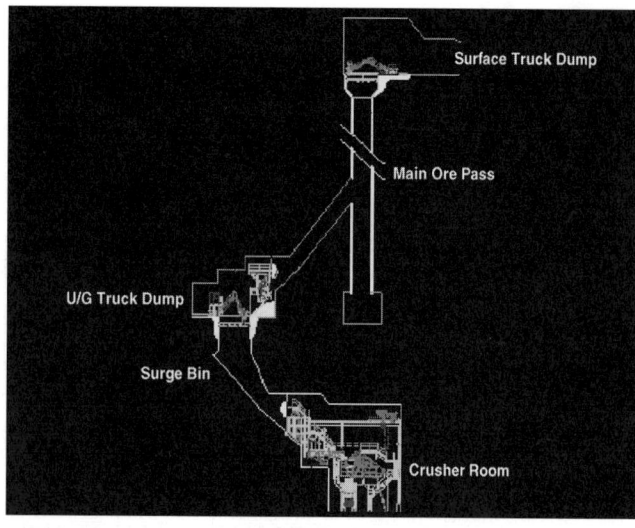

FIGURE 32.19   Ore pass operation

TABLE 32.2   Productivity and equipment availability

	Productivity (lb Ni/m-s)	Availability of jumbos and bolters	Availability, other
January	2,202	65.5%	86.8%
February	2,557	64.7%	89.5%
March	2,242	70.3%	90.3%
April	2,454	79.8%	85.6%
May	2,770	72.0%	87.2%
June	2,837	69.6%	85.0%
July	3,304	78.4%	87.5%
August	3,778	74.3%	85.8%
September	3,359	74.3%	88.0%
October	3,275	73.4%	79.0%
November	2,641	72.2%	81.6%
December	3,362	71.6%	85.0%
Average	2,898	72.2%	85.9%

depending on the time of year. In the wintertime, when dumped in the pass, the ore is subjected to condensation that quickly causes the muck to freeze back, thus creating hang-ups.

During summer, wet ore (from rain or melting snow) also freezes in place when dumped in a cold pass. Up to now, storing cold pit ore in the pass has led to hang-ups. We now spray brine on open pit trucks prior to dumping. It helps, but we still need to ensure that ore level remains below the steel liner (ore in the finger only) and that the finger is emptied at the end of each shift.

32.10   PRODUCTION AND MAINTENANCE STATISTICS

Table 32.2 shows productivities achieved during 1999. The Figures refer to the underground production crew only and take into consideration the whole mining group. This consists of 36 development miners, 7 construction miners, 3 shotcrete truck operators, 3 crusher operators, 8 equipment operators, 5 service men, 6 mine supervisors, as well as 12 and 7 people in engineering and geology. Most of these employees work on a rotation of 3 weeks on, 2 weeks off.

32.11   OPERATING COSTS

Tables 32.3, 32.4, and 32.5 show the distribution of on-site operating costs. One should note that these on-site costs include cost of shipping concentrate to the Quebec City facilities. Any cost incurred beyond this point (smelting, refining, selling, transportation, etc.) is considered an off-site cost. Table 32.3 shows the cost distribution for each department, while Table 32.4 concentrates on mining only and shows cost distribution between underground mining, open pit mining, engineering, geology, supervision, and mine maintenance.

Table 32.5 is a breakdown of costs within the underground mine itself. Development cost includes all lateral development. Stoping costs are a combination of cut-and-fill mining and longhole mining and include some narrow-vein mining. Backfill cost includes both preparation and distribution of backfill.

32.12   CONCLUSION

The Raglan mining operation is located within a harsh and yet highly sensitive environment. The area is undisturbed land. Our environmental policies and programs will ensure it remains as such. Our presence has benefited the local northern population in many ways and guarantees that wildlife and the environment will

TABLE 32.3   Distribution of 1999 on-site operating costs

	Mine	Mill	Services	Power plant	Personnel transportation	Supplies and concentrate freight	Accommodation catering	Other
January	23.4%	12.8%	15.6%	8.7%	9.2%	11.1%	2.2%	16.9%
February	22.1%	11.5%	15.0%	7.2%	3.9%	18.8%	2.8%	18.7%
March	31.3%	10.5%	13.6%	5.8%	7.0%	5.7%	2.7%	23.4%
April	26.7%	14.6%	11.1%	5.2%	5.9%	5.7%	2.9%	27.9%
May	25.2%	15.5%	11.7%	5.7%	8.2%	8.3%	2.9%	22.6%
June	25.6%	12.0%	13.9%	7.2%	6.2%	14.0%	1.6%	19.5%
July	25.1%	15.0%	11.2%	9.0%	7.6%	8.1%	3.9%	20.1%
August	26.9%	9.5%	10.7%	7.6%	4.7%	18.5%	1.9%	20.1%
September	31.4%	11.0%	11.6%	9.0%	6.2%	7.7%	3.9%	19.1%
October	31.6%	13.9%	10.4%	9.4%	6.0%	7.0%	2.7%	19.1%
November	24.0%	15.2%	11.2%	10.4%	6.2%	14.4%	2.8%	15.8%
December	22.2%	16.5%	12.7%	11.8%	7.5%	10.7%	3.2%	15.4%
Average 99	26.4%	13.2%	12.4%	8.0%	6.6%	10.7%	2.8%	20.0%
Average 98	28.9%	11.5%	12.6%	6.1%	6.8%	10.9%	3.2%	20.0%

TABLE 32.4   Distribution of 1999 mine operating costs

	Underground	Open pit	Engineering	Geology	Supervision	Mine maintenance
January	39.6%	8.6%	3.3%	2.9%	6.3%	39.2%
February	37.2%	13.8%	3.0%	6.7%	7.1%	32.2%
March	32.9%	34.4%	2.7%	2.0%	8.4%	19.7%
April	31.1%	32.0%	3.5%	0.7%	8.4%	24.3%
May	41.1%	33.3%	3.4%	2.6%	1.7%	17.9%
June	31.9%	13.2%	4.7%	9.7%	10.9%	29.7%
July	53.0%	5.3%	6.1%	1.0%	6.2%	28.4%
August	36.5%	31.6%	3.6%	2.7%	3.5%	22.2%
September	36.7%	28.9%	3.5%	2.4%	6.2%	22.2%
October	30.4%	37.2%	4.0%	2.4%	6.8%	19.2%
November	36.7%	25.7%	3.7%	2.4%	5.9%	25.6%
December	42.4%	20.4%	4.8%	3.4%	9.2%	19.8%
Average 99	37.2%	25.1%	3.8%	2.8%	6.7%	24.4%

TABLE 32.5   Distribution of 1999 underground mine operating costs

	Development	Stoping	Backfill	Ore haulage, crushing, and conveying	Construction services and training	Supervision and tech. services	Diamond drilling and fuel
January	16.5%	22.5%	5.1%	8.3%	22.7%	22.1%	2.8%
February	15.1%	26.2%	1.6%	7.0%	18.4%	21.3%	10.4%
March	10.1%	36.4%	3.7%	6.3%	15.9%	26.0%	1.6%
April	11.9%	28.6%	5.2%	6.7%	17.0%	26.1%	4.5%
May	28.4%	18.0%	6.7%	9.7%	19.3%	13.1%	4.8%
June	8.4%	21.8%	3.7%	7.0%	16.7%	30.6%	11.8%
July	15.1%	35.1%	1.5%	6.2%	11.5%	18.8%	11.8%
August	14.7%	33.6%	1.0%	7.3%	17.5%	20.5%	5.4%
September	16.5%	30.5%	5.2%	6.4%	16.3%	23.1%	2.0%
October	21.4%	30.0%	1.0%	6.2%	13.3%	25.8%	2.3%
November	22.0%	25.7%	2.9%	7.8%	17.0%	22.6%	2.0%
December	23.3%	20.3%	1.6%	8.9%	15.6%	28.1%	2.2%
Average 99	16.9%	28.8%	6.0%	7.8%	17.0%	21.1%	2.4%

not be subjected to any undue risk. It is now our role to monitor the results on a continuous basis. There is still a learning process on how to deal with the climate, but experience has proven that it is possible to operate smoothly. It has been necessary to adapt to poor weather conditions on the surface by developing a contingency plan to recoup on missed opportunities after being slowed down or totally stopped by blizzards or fog.

We firmly believe that teamwork is the key to the success of Raglan, and it is our goal to develop it to its maximum extent. We could not be as successful if decisions were made by single individuals and, with a work schedule where people have to live together 24 hr a day for 21 days in a row, it is necessary to develop a family spirit where everyone works together to meet objectives.

Our production rate is planned to increase beyond 1,000,000 tonnes a year, and it will be achieved through optimization of available human resources, teamwork, and motivation. All this will be done in harmony with the environment, the local population, and the local wildlife.

CHAPTER 33

Mining of PGMs at the Stillwater Mine

D. Bray,[*] **A.C. Alexander,**[*] **W. Strickland,**[†] **and D. Einarson**[‡]

33.1 INTRODUCTION

The Stillwater Mining Company (SMC) is engaged in the development, extraction, processing, and refining of palladium, platinum, and associated metals from the J-M Reef, a geological formation located in Stillwater, Sweet Grass, and Park counties, Montana (Figure 33.1). Associated by-product metals include rhodium, gold, silver, nickel, and copper. SMC is the only U.S. producer of palladium and platinum and the only significant primary producer of platinum group metals (PGMs) outside of South Africa. The Stillwater Complex is similar to South Africa's Bushveld Complex, which contains the well-known Merensky Reef, analogous to Stillwater's J-M Reef.

Exploration of the Stillwater Valley for copper and nickel began in 1893. This was then followed by exploration for other metals. The first chromium development was in 1905, and exploration for iron was carried out in the 1950s. In 1967, Johns-Manville began exploring for PGMs within the Stillwater Complex. From 1967 to 1972, low-grade PGM discoveries were made in several mineralized zones. In 1973, potentially economic platinum and palladium mineralization was discovered in the zone now known as the J-M Reef.

Initial development of the reef was in the Stillwater River valley near Nye, Montana. Test mining in the West Fork adit was conducted during 1975 and 1976. In 1979, a Johns-Manville subsidiary entered into a partnership arrangement with Chevron U.S.A., Inc., to develop the PGMs discovered in the J-M Reef. Test mining in the Frog Pond decline was conducted in 1980 and 1981 and in the Minneapolis adit (the 5150 level at the Stillwater Mine) in 1983 and 1984. Underground mining operations at the Stillwater property commenced near Nye in 1986. The mill was commissioned in March 1987 and the smelter in Columbus in August 1990. In 1992, Stillwater Mining Company was incorporated in Delaware, and in 1993, Chevron and Johns-Manville transferred substantially all assets, liabilities, and operations to SMC. In September 1994, SMC redeemed Chevron's entire 50% ownership, and Johns-Manville sold a portion of its shares, reducing its ownership of record to approximately 27%. In December 1994, Stillwater Mining Company became a stand-alone company with its initial public stock offering. In August 1995, Johns-Manville sold its remaining ownership interest in Stillwater Mining Company to institutional investors. The base metal refinery was commissioned in 1996.

The shaft at the Stillwater Mine was started in 1994 and was operational by April 1996. Future expansion is planned at the Nye operations and at a new facility (the East Boulder project) being developed at the East Boulder site at the western end of the J-M Reef.

33.2 STILLWATER GEOLOGY

The J-M Reef is located in the Beartooth Mountains in southern Montana along the northern edge of the Beartooth Plateau, which rises to elevations of over 3,000 m (10,000 ft). This plateau is deeply dissected by several rivers and their tributaries, including the Stillwater River, toward the eastern end of the deposit and by the Boulder River near the western end. Erosion by both rivers has cut deeply into the elevated plateau.

The J-M Reef is composed of an assemblage of basic and ultrabasic rocks derived from a single, large, buried magna body emplaced an estimated 2.7 billion years ago. The molten rock was sufficiently fluid at the time of emplacement to allow the heavier minerals of the magma chamber to cool to their temperatures of crystallization and form layers toward the bottom of the complex. The lighter, more siliceous, light-colored minerals crystallized out later to produce bands of norite, gabbro, anorthosite, troctolite, dunite, and pyroxenite that can be traced across most of the strike length of the complex.

The complex has been subdivided from bottom to top into three major series: (1) basal, with associated nickel-copper deposits, (2) ultramafic, with associated chrome deposits, and (3) banded or layered, with associated palladium and platinum deposits. Economic concentrations of PGMs, consisting of palladium, platinum, and rhodium, and small amounts of nickel, copper, silver, and gold, have been identified in one principal layer. Several noneconomic concentrations also exist in the area.

Over time the original gently dipping orientation of the reef was changed as the overlying strata were tilted at an angle of 50° to 90° to the north. The reef surface exposure has been identified for 40 km (25 miles) in an east-southeasterly direction.

The J-M Reef appears to form a continuous layer that is exposed from the highest ridges over 2,900 m (9,500 ft) above sea level to the deepest valleys almost 1.6 km (1 mile) below the surface of the plateau. Geological and geophysical evidence suggests that the J-M Reef extends downward beyond the limits of currently available mining practice. Geological mapping and gravity surveys also suggest that the dip of the J-M Reef flattens gently and may extend 48 km (30 miles) or more to the north. The Stillwater Mining Company controls the entire J-M Reef.

33.3 NARROW-WIDTH SUBLEVEL MINING AT THE STILLWATER MINE

33.3.1 Introduction

The platinum-palladium mineralized zone at the Stillwater Mine is a horizon enriched (0.25% to 5% by volume) in the yellow pathfinder sulfides pentlandite, pyrrhotite, and chalcopyrite. The

---

[*] Stillwater Mining Company, Columbus, MT.
[†] Consultant, Columbus, MT.
[‡] SNC-Lavalin America, Inc., Denver, CO.

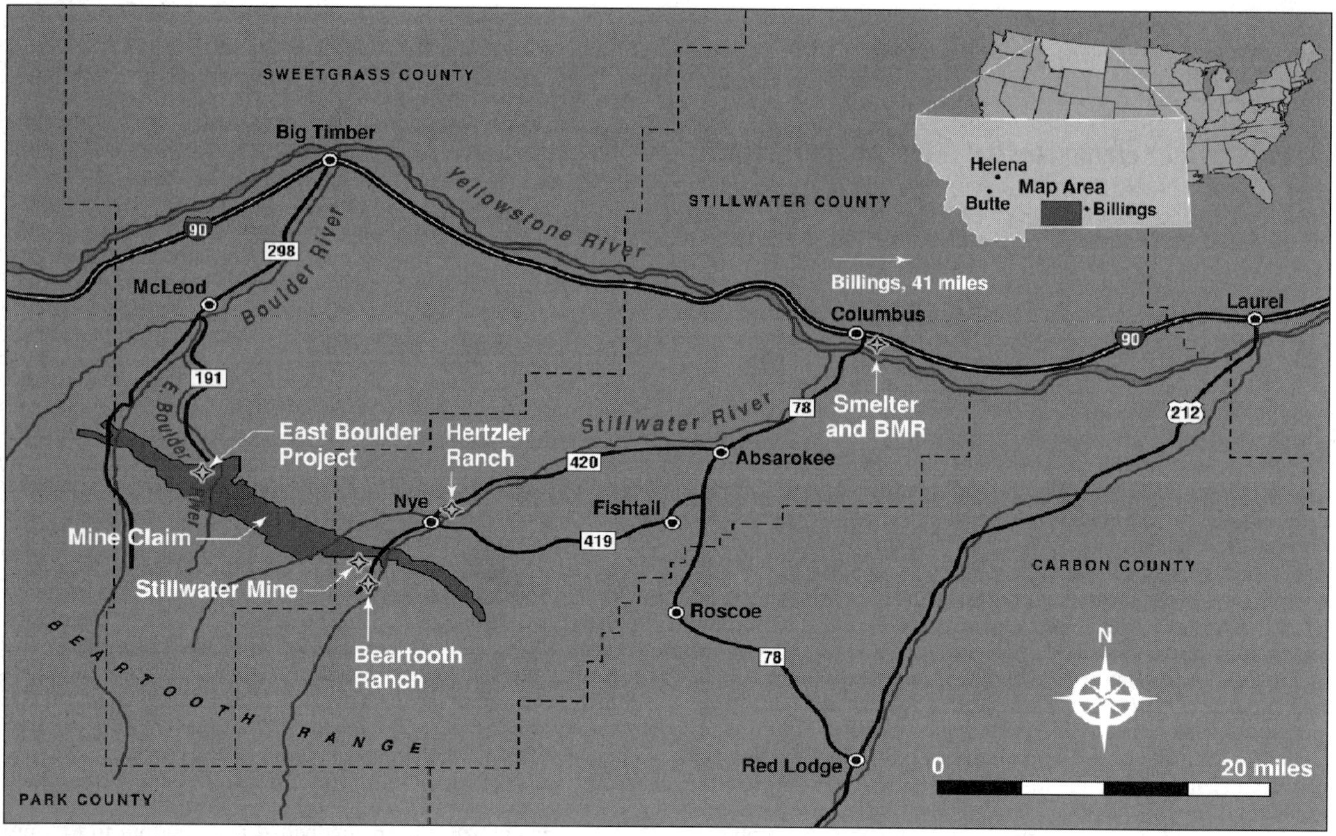

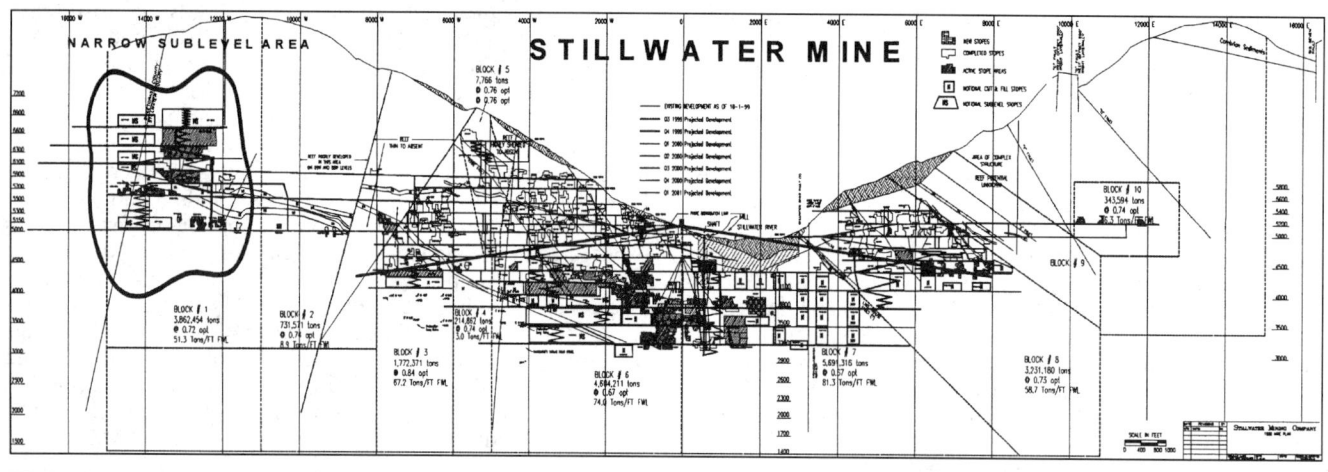

FIGURE 33.1   Location of Stillwater Mine

FIGURE 33.2   Overview of narrow sublevel mining

ratio of palladium to platinum within the J-M Reef is 3.4:1. The average width of the mined mineralized reef horizon is approximately 1.8 m (6 ft). Currently there is no closure at depth, with a maximum vertical extent of >2,520 m (8,300 ft).

The J-M Reef is typically narrow, with a long strike length. The dip varies from nearly vertical at the east end of the Stillwater Mine to approximately 55° in the East Boulder area. Access to the ore is from drifts, or laterals, driven along strike in the footwall of the ore body approximately 36 m (120 ft) off the ore zone. These footwall laterals are driven at 91-m (300-ft)

vertical intervals by conventional drill-and-blast methods. Stoping methods typically include conventional cut-and-fill, mechanized ramp-and-fill, and sublevel (longhole) mining. The mine is currently undergoing an expansion to increase the milling rate from 1,800 to 2,700 tonne/d (2,000 to 3,000 ton/d). The mine operates 7 days a week, and is currently utilizing two 10-hr production shifts a day and two 12-hr maintenance shifts a day on a 7-day-on, 7-day-off schedule.

Sublevel mining is used in the narrow, 0.6- to 2.0-m (2.0- to 6.5-ft), shallow-dipping (45°) portions of the reef (Figure 33.2).

Sublevel mining was first tested at Stillwater in 1993 in the lower west part of the mine. Sublevel mining in the shallow-dipping 45° ore was first utilized in 1995 on the 59W level. (Mine nomenclature addresses the mining level by its nominal elevation above sea level and its relative location by distance east or west of the shaft.) Since that time, ore has been extracted along this same strike length from the 57W level to the 63W level. Major increases in production from this mining area have occurred twice, first in July 1997, from 55 tonne/d (60 ton/d) to 205 tonne/d (225 ton/d), and the second in August 1999, from 205 tonne/d (225 ton/d) to 360 tonne/d (~400 ton/d).

The remainder of this section describes in some detail the various aspects of narrow-width sublevel mining. This mining method is expected to play a major role in long-term production expansion at the Stillwater Mine, as well as in the East Boulder project as it comes into production.

33.3.2 Footwall Development—Diamond Drilling

The main access drifts (footwall laterals) at Stillwater are driven approximately 36.5 m (120 ft) on the footwall side of the reef on 91-m (300-ft) vertical centers. The footwall lateral offset from the reef has historically encountered better ground conditions and provides an adequate distance from the reef for a proper penetration angle for diamond drilling. Pillars would have to be left above and below the access drifts if they were driven on-reef. The footwall laterals are driven 3.4 by 3.7 m (11 by 12 ft) with two-boom electric-hydraulic jumbos drilling an arched back and 4-m (13-ft) rounds. Load-haul-dump (LHD) machines with 3.1-m³ (4 yd³) buckets are used to muck out the heading and load into 15.5-tonne (14-ton) trucks for haulage to various borehole systems throughout the mine. Crews drill 44.5-mm (1.75-in) boreholes that are loaded with bulk emulsion under the pressure vessel system (PVS), which uses compressed air and water injection. Rounds are currently guarded and blasted at the end of shifts.

After approximately 152 m (500 ft) of the lateral is developed, the drift is turned over to the diamond core drillers. Drilling is done on 15.2-m (50-ft) centers horizontally and vertically using BQ core barrels. This core is logged with a bar coding system by the on-site geology staff. Following logging, plans and sections are developed, and calculations of the reserves are generated by a computer.

33.3.3 Sublevel Design

Sublevels at Stillwater are spaced 13 m (43 ft) vertically sill to sill. This vertical spacing sets the on-dip (45°) longhole lengths at 14.3 m (47 ft).

In 1997, sublevel access ramp designs were changed to spirals to position the reef access points near the centroid of the ore block. Figure 33.3 shows a spiral ramp layout with muck and vent passes and other features. The spiral ramps are driven at 15% grade and 15.2-m (50-ft) radius curves to the same dimensions as the footwall laterals. The spiral ramp design versus straight ramps with square corners reduces the amount of waste development required by 20%. The center access allows for ore, waste, and ventilation passes to be connected at each sublevel, which in turn allows truck haulage to be reduced except at horizontal transfer points. Horizontal transfers are required every 183 m (600 ft) to keep muck passes between 55° and 65°, as the ore dips at 45°. To ensure material flow, muck pass angles at Stillwater must be kept between 55° and 65°. If these between-sublevel sill passes are drilled as inverse longhole raises (or driven conventionally), they can be completed and ready to break into when development reaches the next sublevel. Center access also keeps the tramming distance equal for both development and extraction going east or west, as well as allowing for optimization of extraction sequencing.

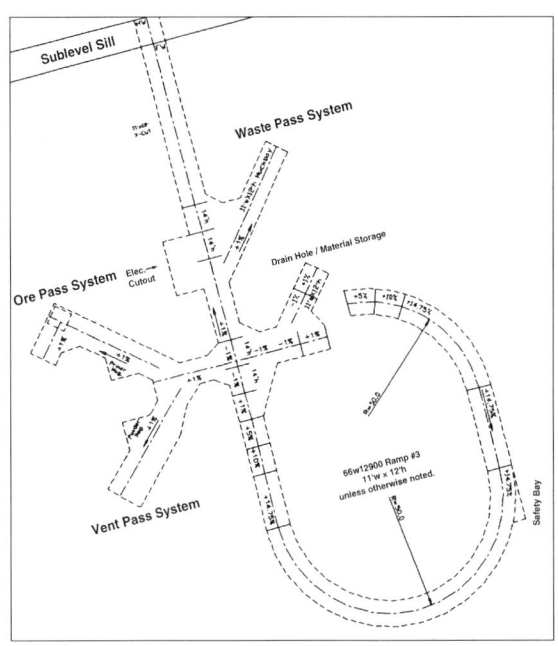

FIGURE 33.3 Spiral ramp layout at Stillwater Mine

33.3.4 Sublevel Sill Development

Sublevel sill drifts serve both as a top cut for downhole drilling and as a bottom cut for remote mucking. The sills are developed 3.1 m (10 ft) high with a shanty shape (hanging wall at a naturally occurring angle) and minimum width of 2.6 m (8.5 ft) when developing west and 2.9 m (9.5 ft) when developing east. Split shooting (double shooting) is done if the ore width is less than 1.85 m (6 ft) horizontally. The sills are driven using single-boom, electrohydraulic jumbos drilling 2.7-m (9-ft) rounds. LHDs (1.9 m³ [2.5 yd³]) are utilized to muck out the heading and haul to the ore and waste boreholes in the spiral ramp system. The sill crews drill 44.5-mm (1.75-in) boreholes that are later loaded with stick powder or an ammonium nitrate-fuel oil (ANFO) mixture. Rounds are currently guarded and blasted at the end of shift.

33.3.5 Panel Layout

The process begins when information on ore width, grade (from assayed channel samples), and faults is transferred onto a surveyed ground line of the sublevel sill drifts. Panel locations are determined from this.

33.3.6 Blasthole and Cable Bolt Layouts

Blasthole layouts are produced for each drill station using Pro-Mine, an AutoCAD-based mine design software package. Two-dimensional sections are cut, drill holes designed, and drill reports produced with a few simple steps (Figure 33.4). Blasthole layouts are based on ore width. In ore less than 0.9 m (3 ft) of horizontal width, a trenching (zipper) pattern is used. The trenching pattern consists of a single hole on the ore hanging wall at each drill station and a single hole on the ore footwall at the mid-point between each drill station. The zipper pattern is drilled as narrow as 0.75 m (2.5 ft). In ore greater than 1.1 m (3.5 ft) wide horizontally, drill holes are spaced at 0.9 m (3 ft) with station burden at 0.69 m (2.25 ft). One hole is dropped every other station (3-2-3 or 5-4-5, etc.)

Cable bolts are installed as needed. Rock mass parameters are measured to determine the need for additional support in the hanging wall. Cable bolt spacing is then determined by plotting the ratio of block size to span on a design curve. Using a

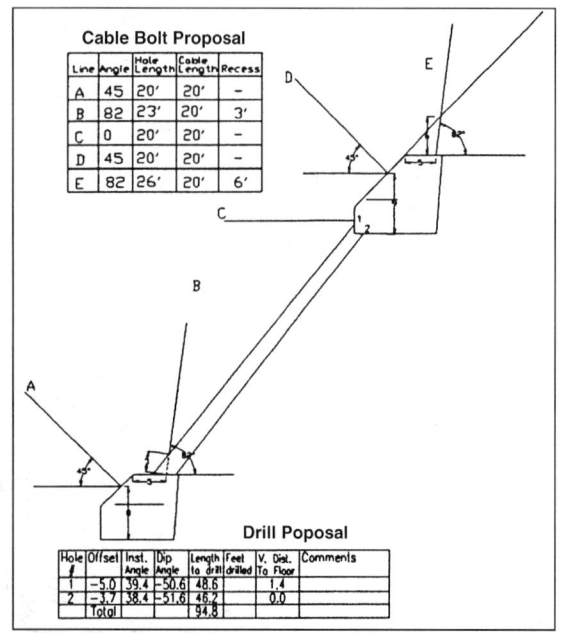

FIGURE 33.4   Pro-Mine computer program output for drill stations

being drilled, one is in the blasting-mucking cycle, and the third is being filled with gob.

Drilling is done with four longhole drills (currently three electrohydraulic and one pneumatic drill). Downholes are drilled 63.5 mm (2.5 in) in diameter, and upholes are drilled 50.8 mm (2 in) in diameter. Drillers check each breakthrough into the lower sublevel and redrill the hole if greater than 0.3 m (1 ft) from the geology department's painted outline of the ore. Remotely operated video cameras are being considered for assisting the drillers in checking the breakthroughs of drill holes.

Drop raises 1.2 by 1.2 m (4 by 4 ft) are drilled in the ore zone at the beginning of each panel. The drop raises are blasted in two or three shots, with the first blast being loaded under supported ground from the bottom and the others loaded from the top.

Blasting of the panels commences after the drop raise is completed or a muck window has been mucked (if the panel is a continuation of a previous panel and a new drop raise is not required). Dry blastholes are loaded with ANFO, and wet holes are loaded with stick powder, and both are blasted at the end of shift.

The ore is mucked out with 1.9- and 3.1-m³ (2.5- and 4-yd³) LHDs equipped with radio-remote control. The LHDs tram the ore to the ore passes on each sublevel for a maximum distance of 229 m (750 ft).

Cavity monitoring surveys (CMS) of open stopes have been done on occasion and work is underway toward making these surveys a routine function of sublevel extraction at Stillwater.

The open sublevels are backfilled with development waste rock using 1.9- and 3.1-m³ (2.5- and 4-yd³) LHDs. The LHDs tram the waste from the waste passes on each sublevel. Sand backfill from mill tailings is used to fill stopes throughout the mine and can be poured onto the waste backfill to close void spaces left by the waste rock. In the future, the feasibility of placing paste backfill into open panels in this area of the mine will be evaluated.

conservative design coupled with historical experience, a relatively constant 1.8-m (6-ft) spacing emerges. An actual spacing of 2 m (6.75 ft) allows greater drilling efficiency by staying on a multiple of the production hole spacing, thus minimizing drill moves (Figure 33.4). The vertical cables are recessed beyond the ore zone so they are not hanging in the ore after blasting.

33.3.7   Sublevel Extraction

Panel extraction is sequenced in a stair-step manner from bottom to top. Figure 33.5 shows a sublevel extraction arrangement. Three or more sublevel panels are required per sequence to optimize the extraction sequence. Of these three panels, one is

33.3.8   Ore Control

Ore grade control at the Stillwater Mine is conducted by on-shift grade-control geologists. Each production face round is examined by the geologists, who visually determinate mining azimuth and ore versus waste, as well as associated grade. These visual exams

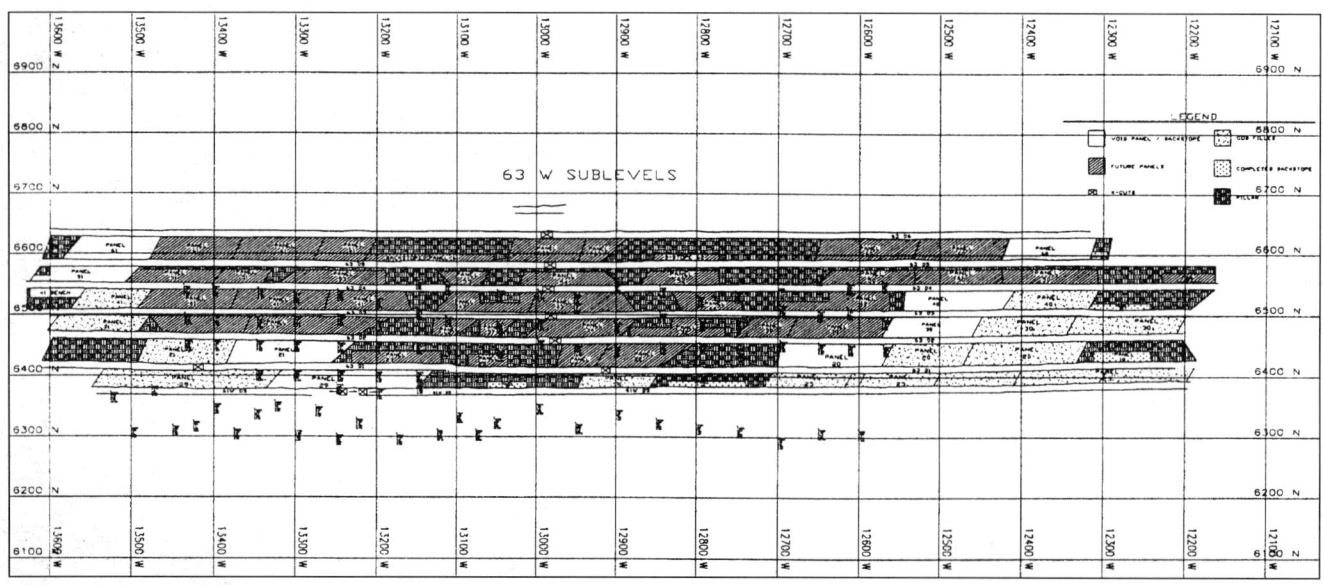

FIGURE 33.5   Panel extraction sequence at Stillwater Mine

are verified by channel samples that are assayed at SMC's lab. Channel samples are taken for each round in the sublevel development sills.

33.3.9 Development, Infrastructure, and Materials Handling

Development in the upper west (narrow, shallow-dipping sublevel area of the mine) has recently been accelerated. Development of the footwall lateral is proceeding on four different levels (50W, 60W, 63W, and 66W), as well as increased excavations for infrastructure. Major infrastructure improvements that have been completed or are under development include—

- Installation of 4,090 m (13,500 ft) of new tubular rail on the 50W haulage level.

- Implementation of Modular Mining's dispatch system.

- Raise boring muck and vent passes from 66W–61W and boring a third muck pass from the 61W–55W and 55W-50W levels.

- Installation of 3,903 m (13,000 ft) of new water and sand line, along with a new drill water pump and reservoir system.

- Development and construction of an expanded underground shop on the 61W level.

- Installation of a 559-kW (750-hp) fan on the 59W level to enhance ventilation in the upper west. This will switch the intake to the 59W level and exhaust to the 55W level, delivering two to 2 to $2\frac{1}{2}$ times the current volume of fresh air.

- Working toward a planned 549-m (1,800-ft) Alimak raise to the surface for exhaust ventilation.

33.4 EAST BOULDER PROJECT

33.4.1 Introduction

The East Boulder project provides a western access to the J-M Reef and is the second fully permitted access to the J-M Reef. The East Boulder Mine is located approximately 21 km (13 miles) northwest of the Stillwater Mine along the strike of the ore body. The nearest accessible point, and the location of the surface facilities, is in the East Boulder River valley, 5.6 km (3.5 miles) north of the ore zone and approximately 56 km (35 miles) south of Big Timber, Montana. The remoteness of the site, with 5.6 km (3.5 miles) between the surface support facilities and the underground production areas, and the very limited access from the high mountains above the ore zone, make the efficient design of this mine a challenge. Access into the East Boulder Mine was one of the most significant considerations to achieving an 1,800 tonne/d (2,000 ton/d) production rate.

Initial site work commenced in 1991. In 1996, SMC restarted work on the East Boulder project, including site preparation, construction of a power line, and procurement of a tunnel boring machine (TBM). During 1996, SMC invested $7.8 million in the East Boulder project, primarily for construction of the TBM and provision of an electrical power supply to the mine site. In October 1996, the project was placed on hold because of a downturn in palladium and platinum prices. However, permitting and environmental activities continued, with approximately $1.1 million expended during 1997. In November 1997, with the achievement of the Nye operation production goals and higher prices for palladium and platinum, SMC restarted the project. TBM 1 was delivered in the second quarter of 1998.

The decision to accelerate development of SMC's East Boulder Mine was made in March of 1998. It required the completion of conceptual designs by July of 1998. This was accomplished through teamwork and close cooperation among mine personnel, engineering firms, and vendors, all working toward the common goal of developing an efficient and workable design. This interaction, which included companies not normally associated with the mining industry, led to innovative approaches and designs.

33.4.2 Design Process

Initially, meetings were held with company personnel to take advantage of their expertise on existing logistic and mining inefficiencies at the Stillwater Mine and possible solutions. Further meetings were held in April 1998 with representatives of various engineering firms to discuss the options for development of the East Boulder project and address key project issues, such as access, materials and muck handling, personnel transportation, and infrastructure. Input from these sources led to designs that should provide efficient access and achieve accelerated development of the East Boulder Mine. SNC-Lavalin America (at the time, Kilborn SNC-Lavalin) was initially chosen to develop the preliminary designs and cost estimates. Later, final mine plans and detailed design of the underground mine infrastructure were developed by Stillwater Mining and SNC-Lavalin America through a collaborative process that included reviews and mine plan input by Mineral Resource Development, Inc. (MRDI).

33.4.3 Access

Originally, the 1998 mine plan called for a single, tunnel-bored, 4.6-m (15-ft) in diameter adit. A study of the logistics of moving all personnel, materials, and muck through the single adit showed the potential for creating a severe bottleneck. The only way to overcome this would be with a sophisticated and expensive control system. The limited cross-sectional area of the adit, along with the space requirements for pipes and cables, would present clearance problems for larger loads and equipment. A single adit to the mine would require, at a minimum, secondary excavation of the tunnel invert to change the shape of the tunnel from circular to horseshoe, or some similar profile. Options identified to modify the adit profile included excavation by drill-and-blast methods, use of road-headers, and use of hydraulic impact hammers. However, any enlargement methods would create schedule delays to the preproduction development and ultimately delays in meeting the production targets for the mine. It was decided instead to pursue the development of a second adit to prevent any transportation bottlenecks.

The first adit will penetrate the reef and enter the footwall at the 6450 level (measured in feet above sea level). The 6450 level will be the first production level and the location of the main shop and supplies handling and transfer station. This first adit will become a high-speed rail access for personnel and materials to the mine.

The second adit will also access the reef at the 6450 level. This level will initially serve as a mining level and will eventually serve as a dedicated rail haulage level as the mining operations move further out along the strike of the ore body. The second adit will be dedicated to muck haulage to the surface via conveyor.

33.4.4 Materials Handling

Early in the design process, it was recognized that the movement of materials would be a major issue in the economical and efficient operation of the East Boulder Mine. The long horizontal distances to be traversed required designs that duplicated the transportation efficiencies found in vertical shafts.

The cross section of both 4.6-m (15-ft) in diameter tunnels limited the size of any materials-handling equipment. The distances to be traveled required an efficient, relatively high-speed, transportation system. Plans initially focused on a system using 2.4- by 2.4- by 6-m (8- by 8- by 20-ft) Connex shipping containers to carry materials from the surface to an underground warehouse. These containers would handle all shop and normal warehouse materials that had to go into the mine and could be loaded at a central warehouse and shipped directly to the site and then underground.

Handling the various mining materials, including rock bolts, drill steel, explosives, and ventilation equipment, that would be going directly to the production and development areas in the mine was another problem. These materials not only had to go 5.6 km (3.5 miles) underground, they had to be distributed along the long strike length and vertical extent of the ore body.

Plans were later expanded to use a cassette-type (roll-on-off) system with dedicated rail flatcars and rubber-tired carriers to speed the delivery of materials from the surface to underground work areas. Materials could be bulk-loaded on pallets and loaded on the cassettes.

Cassettes could also be designed for bulk handling of explosives, fuel, trash, and sewage. The underground shop area was designed with a large cassette transfer area, complete with a 9-tonne (10-ton) overhead crane to lift the cassettes from the flatcars to cassette storage bays. Rubber-tired, diesel-powered carriers will be able to pick the cassettes up easily and transport them quickly to the work areas.

Two types of rail cars will be used to transport equipment. Standard flatcars will be provided for transportation of rubber-tired equipment into the mine, as well as supplies and supplies cassettes on pallets. The flatcars will be delivered to an underground warehouse-shop complex where overhead bridge cranes will be utilized to lift off equipment. A low-boy flatcar will also be available for over-height equipment and materials.

33.4.5 Muck Handling

Two problems confronted muck handling design at East Boulder: the long adits and the long strike length of the ore body. Originally, the plan was to use 9-tonne (10-ton) ore cars to move muck out of the mine through the primary adit. However, other options, including slurry, conveyor, and rubber-tired trucks, had some attractive features, especially when a second adit was planned.

Considerations for selecting the permanent muck-handling system included the long, straight, 5.6-km (3.5-mile) haul from the mine to the surface portal, the quantity of muck to be hauled (as much as 6,800 tonnes [7,500 tons] of ore and waste a day), the diameter of the adits, the amount of underground infrastructure required to support it, and the ability to increase capacity.

Slurry handling of muck was eliminated because of the large underground excavations required for a grinding circuit and potential adverse effects on the flotation circuits.

The size, circular profile, and length of the adits eliminated consideration of rubber-tired trucks for main haulage. The envelope of the access tunnels would limit the trucks to uneconomically small sizes.

Rail haulage of muck is a proven and efficient method and will be used initially to haul all development muck out of the mine. However, efficient long-term rail haulage to the surface would be difficult, due again to the envelope of the adits. Double-track or passing tracks would require extensive secondary excavation. A single-track system would cause muck, material, and personnel hauling bottlenecks and would require an elaborate and expensive signaling system.

A conveyor installed in the second adit, while requiring underground crushing, would reduce traffic and bottlenecks of material and personnel in the first rail adit. A conveyor could be designed to meet future production requirements with relatively minor modifications.

A factor in the selection of a conveyor was the advantage it provided during development. Large quantities of development muck could be moved from underground, and efficient muck haulage could be achieved during preproduction development. By including a temporary crusher, initially planned to go underground

but later moved to surface because of mine layout changes, the mine can reach its full production rate at the same time that a permanent underground crushing station and secondary conveyor ramps are excavated and permanent underground crushing and secondary conveying equipment is installed.

The final design for permanent muck haulage includes the use of rail haulage along strike at the 6450 level. The rail haulageway will initially be used to access the ore zones with both rail and rubber-tired equipment. This design was based on economics; it is not desirable from a long-term point of view. Plans are that later, as the ore zones are depleted, the level will become an isolated dedicated rail haulage. The trains will transport ore and waste from muck passes spaced at 1,000- to 1,500-m (3,280- to 4,920-ft) intervals to dump pockets at a permanent crusher station. The dump pockets will feed the muck into two 5.5- by 5.5- by 61-m- (18- by 18- by 200-ft-) deep bins above the crusher.

The underground crusher station will include a 1.5-m (60-in) apron feeder, 0.9- by 1.2- m (37- by 49-in) jaw crusher, a 1.3-m (54-in) crusher discharge conveyor, a belt magnet, and a 16-tonne (18-ton) overhead crane. Two transfer conveyors will carry muck from the crusher station via inclined ramps to the main conveyor.

The 0.9-m (36-in) main conveyor in the second adit tunnel will carry crushed minus 0.15-m (minus 6-in) rock from the underground crusher 5.6 m (3.5 miles) to the surface storage piles. Transportation of muck from the mining areas to the muck passes will be by rubber-tired LHDs or trucks.

33.4.6 Personnel Transportation

Efficient transportation of personnel over these distances will require speed. Speed will require specialized vehicles, high-quality track, and potentially high costs. Modified "transit-style" light rail vehicles and speeds up to 65 km/hr (40 mph) were considered in early studies carried out by Parsons-Brinkerhoff Transit and Rail Systems. The high cost of the vehicles, signaling equipment, and the need to convert the 0.9-m (36-in) track gauge to 1 m or the U.S. standard of 1.4 m (56.5 in) gauge convinced SMC that a more cost-effective system needed to be found.

Internal studies showed that speeds of 32 to 40 km/hr (20 to 25 mph) would provide the greatest benefit at the most reasonable cost. However, a rail vehicle that can safely carry large numbers of personnel at these speeds is still in the design stage.

Once the rail personnel carrier has reached the ore body, it will enter a station for rapid unloading. The station is situated within a short distance to a central ramp system for personnel transport to the upper levels of the mine.

33.4.7 Infrastructure

The mine's distance from the surface dictates a self-sufficient operation underground. A major piece of this operation will be the repair shops and supplies staging and storage areas. The need for underground shops and warehouse space was evident from the beginning. Equipment cannot be easily taken apart and moved 5.6 km (3.5 miles) by rail for repair in a surface shop. An underground shop and efficient underground staging will be essential for the efficient operation of this mine. The shop and materials staging and storage areas are designed so they can be accessed readily from all the production areas. The shops are designed to be capable of all repairs and will include limited space for full overhauls. They are also set up for fabrication, steam cleaning, fueling, and lubrication. The central crane bay includes a 23-tonne (25-ton) overhead crane.

The other major pieces of underground infrastructure include the main explosives magazine, a main substation, a compressor plant, a tailings backfill plant, and pumping stations.

33.4.8 Site Preparation

A 5,650-m (18,550-ft) long tunnel is required to gain access to the ore-bearing horizon (the J-M Reef). It was decided that access and primary mine development could be accelerated through the use of two 4.6-m (15-ft) in diameter TBMs. To prepare for tunnel boring and eventual mining, a number of site preparation actions were required.

Powerline. A 26-km (16-mile) 397 ACSR, 69-kV powerline with a fiber-optic ground wire (for communications) was constructed during the summer and fall of 1996. This was new construction that required acquiring right-of-way, removing timber, and rerouting existing powerlines. A 69-kV circuit breaker was required at the McLeod substation and a metering point at the Springdale substation. The project site substation consists a dual-voltage 50/69 MV 10/12.5/14 MVa-rated outdoor transformer. The project site substation was completed during the spring of 1998.

Roads and Bridges. Access to the East Boulder project site included upgrades to the Main Boulder, Elk Creek, and Mason Ditch bridges, which were completed during the summer of 1996. Minor road improvements were also included. Engineering for the widening of 10 km (6.5 miles) of U. S. Forest Service road 205 was completed during the spring of 2000. This construction will be completed in 2000. Several options are currently being evaluated for the upgrade or relocation of the county road.

Office and Change Trailers. Temporary offices and dry trailers were purchased and located on-site in 1996. Additional office and dry space was expanded during 1998 and 2000. A water well and distribution system was constructed in 1996. Holding tanks to contain sewage have also been installed. The sewage is transported and disposed off-site. A permanent site facility will be constructed in 2000. Communication is supplied through the fiber-optic ground wire located on top of the 69-kV powerline.

Shops and Warehousing. Temporary shops and warehouses were constructed by spanning 12-m (40-ft) Connexes with wooden trusses. The Connexes are utilized as warehouses and the spanned area as shop floor.

Rail and Trestle. A surface rail and trestle were constructed for muck handling. The trestle was constructed using reinforced earth retaining walls. A concrete surface pad was constructed to assemble the TBM and rail installed from the portal to the trestle.

Sediment and Percolation Ponds. Two sediment ponds and a percolation pond were constructed for handling adit water. The adit water is directed through an oil/water separator, then flows into one of the sediment ponds and then into the percolation pond. This system was designed to handle 4,500 L/min (1,200 gal/min).

33.4.9 Site Geology

The access tunnel is being driven approximately 5,650 m (18,550 ft) at an azimuth of 208° with a +0.5% vertical grade. Tunnel geology consists of approximately 1,006 m (3,300 ft) of sediments, 2,530 m (8,300 ft) of gabbro, 430 m (1,400 ft) of norite, 1,140 m (3,750 ft) of anorthosite, 335 m (1,100 ft) of olivine gabbro, and 210 m (700 ft) of troctolite. The overburden consists of talus that varies from 18 to 37 m (60 to 120 ft) thick.

- The *Mississippian Madison Limestone* is a massive cliff-forming gray limestone and dolomite with thin-bedded red and yellow argillaceous limestones in the upper part and gray limestone in the lower part (adit length 245 m [800 ft]).

- The *Upper Devonian Three Forks Dolomite and Shale and Jefferson Dolomite* consist of yellow calcareous shales, thin-bedded and massive limestones, and alternating massive gray and thin yellow limestones in the lower part (adit length 150 m [500 ft]).

- The partially dolomitized *Upper Ordovician Bighorn Dolomite* is a fine-grained, white limestone in the upper part, and a massive, porous, mottled gray limestone and yellow limestone in the lower part (adit length 80 m [270 ft]).

- *Upper and Middle Cambrian limestone and shale* consist of green and purple, thin-bedded shales, brown oolitic limestones, thin-bedded shales, gray limestones, and green fissile calcareous shales and mudstones that rest unconformably on the Stillwater Complex (adit length 520 m [1,700 ft]).

- The *Stillwater Complex* rocks consist of pigeonite gabbro zone VI with an estimated adit length of 885 m (2,900 ft), gabbro zone VI at 520 m (1,700 ft), troctolite-gabbro zone V at 90 m (300 ft), anorthosite zone V at 610 m (2,000 ft), troctolite-gabbro zone IV at 180 m (600 ft), anorthosite-troctolite zone IV at 210 m (700 ft), troctolite-gabbro zone III at 335 m (1,100 ft), anorthosite zone III at 400 m (1,300 ft), troctolite-anorthosite zone II at 60 m (200 ft), gabbro zone II at 885 m (2,900 ft), and norite zone II at 365 m (1,200 ft).

There are four known major faults along the adit trace—Rhino, Dew, Brownlee Creek, and Sam's. The faults may be grouped into two general types: transverse faults that cut across the trend of the layering and longitudinal faults that essentially parallel the trend of the layering. The transverse faults are generally steeply dipping (>70°), and the longitudinal faults are either parallel to or steeper than the layers. Numerous transverse faults of displacements less than 60 m (200 ft) occur. The proposed tunnel will be nearly parallel to the transverse faults. These faults have not been studied in particular detail, especially not along the length of the tunnel. Of all major structures, the longitudinal faults close to the mineralized zone have been studied most extensively. The Brownlee Creek fault is, for example, considered to be a broad fracture zone. The Dew and Rhino faults are expected to be equally major structures.

Major dikes up to 60 m (200 ft) wide and several thousand meters in length are prevalent in some areas. Numerous mafic dikes less than 0.3 m (1 ft) wide have been identified. Most of the dikes can be grouped into two major sets at 030° and 070° and two minor sets at 335° and 0°. The major sets tend to be more numerous, wider, and have longer strike lengths than the minor sets. The dikes appear to have been commonly emplaced along early faults and shears and are themselves sheared together with the adjacent Stillwater rock that indicates postemplacement movement along reactivated faults. The wider dikes have steep dips. The proposed tunnel will be approximately parallel to one of the major sets and close in orientation to one of the minor sets.

Rock strengths (unconfined compressive strengths) are as follows: limestone 60–40 MPa (8,900–20,000 psi), gabbro 70–190 MPa (10,000–27,000 psi), norite 100–150 MPa (15,000–22,000 psi), and anorthosite 60–190 MPa (8,900–27,000 psi). In general, the unconfined compressive strength test results show a wide range of values from 60 to over 190 MPa (9,000-27,000 psi). Most of the samples tested below 130 MPa (19,000 psi). Point load tests indicated that most of the material that tested in the upper range of compressive strength tended to fail at point load levels similar to samples in the moderate strength range. Punch test results confirmed the unconfined compressive strength and point load test results. In general, the rock is weak to moderate in strength. There are examples of moderate-strength rocks that gave high force to penetration relationships. This tends to indicate some areas of material may require higher thrust levels to generate rock fracturing than is indicated by rock strength alone. The composition of the rocks indicates low-to-moderate abrasiveness because of the absence of hardness 7 (Moh's hardness scale) and higher minerals.

33.4.10 Tunnel Boring

Introduction. The first TBM for the project was purchased from Construction and Tunneling Services (CTS) and was delivered to the site in June of 1998. Boring began in August of that year. A second 4.6-m (15- ft) TBM was purchased from BHP (Magma Copper-Lower Kalamazoo project) and was reconditioned by Stillwater Mining for this work.

Tunnel 1 Machine Specifications. Construction and Tunneling Services built a custom TBM for SMC. This TBM will be used for the initial East Boulder project access and for future use in mine development. Specific features included an assembled drive cartridge 3 m (10 ft) wide that allowed the cartridge to be lowered down the Nye shaft without disassembly and a short front shield configuration with a floating gripper that allowed a 60-m (200-ft) turning radius. All cylinder joints are ball sockets for easy reassembly. Face and crown support shields are included for encounters with faulted rock and contact zones that do not stand. The cutterhead has a full-diameter rotating rim bar with an extendable roof support behind the last gauge cutter. The cutterhead face is smooth and held close to the cutter tip profile. Ground support and probing are provided by a variety of methods: probing and grouting, radial drilling, ring beam installation, and crown strap installation.

The TBM is configured for a 61-m (200-ft) horizontal turn and a 305-m (1,000-ft) vertical turn. The ability to maneuver the machine through such a turn is a function of internal clearances and the ability of the front shield to adjust to a nonconforming profile. Horizontal clearances allow the conveyor and bridge to side-shift in the turn. The propel cylinder ball sockets are provided with sufficient angular rotation ability. The front shield accommodates the turn because it is not a fixed-diameter shield. The bottom of the front shield rides on wear plates that allow clearance between the skin and the bore and let the machine turn. The top shield is a three-piece construction that can pivot and slide in relation to the cutterhead drive. Thus it continues to support the ground and stabilize the cutterhead while it maintains alignment to the tunnel.

Specifications

Machine diameter	4.6 m (15 ft)
Cutter bearing	Tapered roller
Cutters	
Diameter	432 mm (17 in)
Number	32
Capacity	267 kN (60,000 lb)
Thrust System	
Cutterhead thrust	8.5 MN (1,920,000 lb)
Total thrust	15 MN (3,400,000 lb)
Cylinders, 4	406-mm (16-in) bore
Stroke	1,220 mm (48 in)
Cutterhead drive	
Type	Two-speed with clutches
Power	1,340 kW (1,800 hp)
Speed	11.6/3.8 rpm
Torque	1,100 kN-m (815,000 ft-lb)
Drive units, 4	335 kW (450 hp), water-cooled
Conveyor	
Drive	Hydraulic
Speed	120 m/min (400 ft/min)
Width	610 mm (24 in), trough belt
Hydraulic system	
Pump motors, 3	37 kW (50 hp)
Thrust/gripper	34.5 MPa (5000 psi) maximum
Other systems	20.7 MPa (3000 psi) maximum
Electrical system	
Motor circuits	460 V, three-phase, 60 Hz
Controls and lighting	120 V, 60 Hz
Primary	13,800 V
Transformers, 2	1,200 kVA
Weight	250 tonnes (275 tons)

Tunnel 2 Machine Specifications. The most critical element in the acceleration of the project schedule was the early establishment of a second means of egress and the ventilation required for underground development. Accordingly, a second TBM was purchased from Magma Copper. The Industrial Company of Wyoming (TIC) refurbished this machine in Billings, Montana.

Specifications

Machine diameter	4.6 m (15 ft, 2 in)
Cutter bearing	Three axis: two thrust, one roller
Cutters	
Diameter	432 mm (17 in)
Number	33
Capacity	222 kN (50,000 lb)
Thrust System	
Cutterhead thrust	7.3 MN (1,650,000 ft-lb)
Total thrust	8.6 MN (1,924,000 ft-lb)
Cylinders, 2	445-mm (17.5-in) bore
Stroke	1,550 mm (61 in)
Cutterhead drive	
Type	Two-speed with hydraulic clutches
Power	1,250 kW (1,680 hp)
Speed	12/4 rpm
Torque	983 kN-m (725,000 ft-lb)
Drive units, 4	315 kW (422 hp), water-cooled
Conveyor	
Drive	Hydraulic, direct drive
Speed	152 m/min (500 ft/min)
Width	762 mm (30 in)
Hydraulic system	
Pump motors, 2	75 kW (100 hp)
Thrust/gripper	29 MPa (4,200 psi)
Other systems	10 MPa (1,500 psi)
Electrical system	
Motor circuits	480/600 V, 60 Hz
Controls and lighting	120 V
Primary	13,800 V
Transformers, 2	1,300 kVA
Weight	225 tonnes (240 tons)

Results of Tunnel Boring. The use of the TBMs at East Boulder has allowed for accelerated access to the ore zone. The performance of the machines has been mixed due to a combination of factors, including mechanical breakdowns, ground and associated cutting conditions, machine configuration, impacts on the timing of ground support installation, and ease of operation. Table 33.1 summarizes machine performance.

In general, the number 2 tunnel boring machine has achieved significantly better performance. This can be attributed to experience gained in tunnel 1 with regard to encountering poor ground conditions first and development of support solutions, and identification of groundwater inflow areas requiring grouting.

33.5 CONCLUSIONS

In December 31, 1999, Stillwater had proven and probable reserves of approximately 33 million tonnes (36.3 million tons) of ore with an average grade of 24 gm/tonne (0.71 oz/ton) containing approximately 798 tonnes (25.7 million ounces) of PGMs. Based upon existing ore reserves and current production levels, the mine has an estimated life in excess of 40 yr. To utilize this reserve base, Stillwater Mining Company has been implementing a series of expansions since 1994 to increase its annual production. In 1998,

TABLE 33.1 Performance of tunnel boring machines

Month	Construction and Tunneling Services				Robbins			
	Footage	Hours	Use, %	Availability, %	Footage	Hours	Use, %	Availability, %
July 98	418	NA	NA	NA				
August	1,167	216	29	62				
Sept	1,358	244	34	70				
October	1,110	160	22	67				
November	373	67	9	50				
December	881	142	20	39				
Jan. 99	1326	180	25	51				
February	1352	176	26	61				
March	739	108	14	73	1,656	223	43	81
April	999	141	20	50	1,293	167	24	66
May	701	104	14	63	839	116	16	83
June	105	23	3	94	1,127	156	22	78
July	879	109	15	59	1,474	147	20	58
August	944	137	18	55	1,543	163	22	66
Sept.	624	100	14	38	950	117	16	78
October	679	130	17	58	760	123	17	72
November	1,334	181	25	59	1,481	181	25	72
December	1,108	151	21	49	1,993	237	33	61
Jan. 00	480	69	9	68	865	107	14	66
February	413	81	12	82	1,320	150	22	64
March	298	80	11	86	826	101	14	66
April	76	35	5	87	315	46	6	84
Average	17,364	2,629	17	63	16,442	2,030	20	71

Use = Time spent boring and regripping ÷ total time available (24 hr/d, 7 d/wk).
Availability = (Total time available–mechanical downtime) ÷ total time available.

PGM production increased to 13.7 tonnes (444,000 oz) from 11 tonnes (355,000 oz) in 1997 and from 7.9 tonnes (255,000 oz) in 1996. Stillwater is proceeding with the additional investment designed to increase the production rate at the Nye operations from 1,800 to 2,700 tonne/d (2,000 to 3,000 ton/d) and to develop the East Boulder project.

33.6 ACKNOWLEDGMENTS

The authors thank the management and personnel of Stillwater Mining Company for permission to publish this paper.

33.7 REFERENCES

Alexander, A.C. 1998. "East Boulder Project–#2 Portal Excavation," Stillwater Mining Company, Nye, Montana.

Alexander, A.C., 1998. "East Boulder Project–Portal Ground Support, Stillwater Mining Company, Nye, Montana.

Alexander, A.C. 2000. "Development of Stillwater Mining Company's East Boulder Project Using Tunnel Boring Technology," Stillwater Mining Company, McLeod, Montana.

BHP Copper, Inc. For Sale: Robbins Tunnel Boring Machine Model 156-275, pp 1.1–1.99.

Bray, D.L. 2000. "Narrow–Width Sublevel Mining at Stillwater Mine," Stillwater Mining Company, Nye, Montana.

Construction and Tunneling Services, Operation and Maintenance Manual, Volume 1, TBM Model 460-12, Machine Description, pp 2-2-2-8.

Czamanske, G.K. and M.L. Zientek. 1985. Montana Bureau of Mines and Geology, The Stillwater Complex, Montana: Geology and Guide, Special Publication 92.

Kilborn SNC-Lavalin, 1998. "Stillwater Mining Company, East Boulder Project, Preliminary Mine Design and Cost Estimate Report," Denver, Colorado.

Kirsten, H.A.D., D.A. St. Don, and R.B. Langston. 1996. "Rock Engineering Aspects Investigated During April Visit to Mine," Report #208592/2, Stillwater Mining Company, Nye, Montana.

Kirsten, H.A.D. and R.B. Langston. 1998. "Rock Engineering Aspects Investigated During April Visit to Mine," Report #208592/5, Stillwater Mining Company, Nye, Montana.

Mining Engineering, Tunnel Boring at Stillwater's East Boulder Project, September 1999.

Montana Bureau of Mines and Geology, Special Publication 92, The Stillwater Complex, Montana: Geology and Guide, 1985.

Nowak, D.E. 1996. "Proposal #5214 for Open Gripper Tunnel Boring Machine Model 460", Construction and Tunneling Services, Kent, Washington.

Nowak, D.E. 1988. "Jack Pine - C5740," Document #293, The Robbins Company, Kent, Washington.

Parsons–Brinkerhoff Transit and Rail Systems. 1998. "Concept Study for Stillwater Mining High Speed Transport System," Newark, NJ.

Stillwater Mining Co., Inc. "1995 Annual Report," Denver, Colorado.

Stillwater Mining Co., Inc. "1996 Annual Report," Denver, Colorado.

Stillwater Mining Company 1997 Annual Report.

Stillwater Mining Co., Inc. "1998 Annual Report," Denver, Colorado.

Strickland, W.A. and D.S. Einarson. 2000. "Design Considerations for Stillwater Mining Company's East Boulder Mine," Stillwater Mining Company, McLeod, Montana.

Todd, S.G. 1988. "Tabulation of Proposed Clear Cut Adit Rock Types," Stillwater PGM Resources, Billings, Montana.

CHAPTER 34

Mechanizing Sunshine Mine Operations

Michael E. McLean*

34.1 INTRODUCTION

The Sunshine Mine is located 3 km (1.9 miles) southeast of Kellogg, Idaho, in the center of the historic Coeur d'Alene Mining District. Discovered in 1884, the mine has produced over 10,886 tonnes (350 million oz) of silver, or approximately one-third of the district's total production. The current production rate of approximately 900 tonnes/day (1,000 st/day) is from narrow, silver-rich tetrahedrite-bearing veins. These steeply dipping (around 65°) veins occur in folded quartzites, argillites, and siltites of Precambrian age.

34.2 MINING

Until 1995, underground mining at the Sunshine Mine was generally done by conventional overhand cut-and-fill methods using timbered raise access from track levels. Pneumatic jackleg drills and slushers were used for drilling, moving ore, and creating access to and within stoping areas.

In recent years, economic factors forced re-evaluation of these methods. The high cost of labor and materials coupled with flat metal prices resulted in the adaptation of diesel-powered, rubber-tired equipment to narrow vein mining. This approach now prevails in most of the stopes in the mine.

Mobile underground equipment allows cost-effective development and mining of ore that would be uneconomic to recover using only conventional methods. The primary advantages are—

- Flexibility for accessing a wide horizontal and vertical range
- No fixed tie to track and level development
- Efficient deployment of labor and material resources among multiple workplaces
- Ability to respond to changing geologic and environmental conditions
- Ability to pursue below-track ore pillars and remnant reserves
- Freedom from restrictions on moving resources because of hoisting requirements

The flexibility of trackless mining allows it to be combined with conventional stoping in certain situations to mine lower grade ore and zones with limited heights or strike lengths.

While conventional stoping still has a place in some ore bodies, the knowledge and skills gained through trackless development and mining of the West Chance ore zone are, and will continue to be, vital to the success of the Sunshine Mine.

34.3 DEVELOPMENT

For exploitation of the West Chance ore zone, track drifts were driven west approximately 900 m (3,000 ft) from the existing Jewell and No. 12 shafts. This configuration allows for efficient track haulage of both ore and development waste to the Jewell shaft. Currently, all West Chance operations are served from the Jewell shaft only. Trackless development started from favorable locations directly off these track drifts. In other areas of the mine, trackless development from existing tracked drifts accessed isolated ore reserves and allowed ore pillars to be recovered below and above the track.

As quickly as development progress allowed, diamond-drill stations were excavated, and holes were drilled to define the geometry of the ore zone more closely. Based on drilling results, the center of the ore body was determined, and ramp design alternatives were considered. At the same time, a raise boring contractor was retained to excavate 1.5-m- (5-ft-) diameter bore-holes between the 3100 and 3700 levels for ventilation and rock transfer. This process was later repeated between the 2700 and 3100 levels.

General design criteria have been modified since inception of the West Chance ramp to accommodate changing ground conditions and further refine ore body geometry. Although development and stoping parameters can be considered dynamic overall, general criteria are used for design and scheduling purposes.

Primary ramps, either up or down, are driven at a nominal 17% grade, with flattening to 8% in corners with 12.2-m (40-ft) radii. Laterals are driven on designed elevations with openings to raise connections driven up at a nominal 5% to prevent water infiltration into the raises. Both ramps and laterals are driven with a nominal 2.4-m- (8-ft-) wide by 3-m- (10-ft-) high profile. Ramps are generally designed with a north-south orientation, while laterals are east-west. This orientation yields the best mining and ground support conditions while providing good access for stope development along strike. Bedding strikes are generally east-west with variable dips. Ramp system development is located up to 183 m (600 ft) into the hanging wall of the West Chance vein. Laterals are designed generally 61 m (200 ft) south of the expected vein system to allow for proper development of "attack ramps" into the ore. The vertical distance between laterals depends on the design height of stopes originating from the lateral. In general, this vertical distance is between 19 and 22 m (63 and 72 ft). Secondary excavations, such as substation cutouts, diamond-drill stations, sumps, storage areas, powder magazines, fuel storage, and shop areas, are excavated to support primary ramp and stope development.

34.4 STOPING

Stope access is driven from laterals down at a nominal −20% grade to within 8 m (25 ft) of the vein then flattened to cross at a designed I-drift elevation. These "attack ramps" are extended

---

* Sunshine Precious Metals, Kellogg, ID.

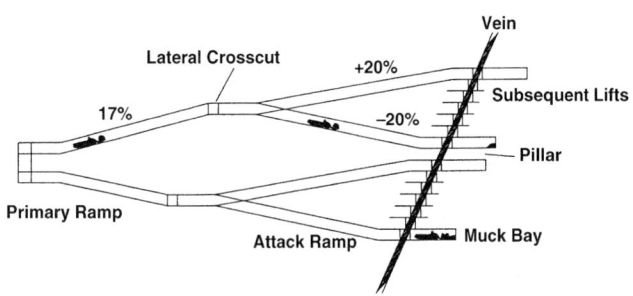

FIGURE 34.1   Typical attack ramp stope arrangement (cross section looking west)

into the footwall about 9 m (30 ft) to create a muck bay, which enhances rapid advance (Figure 34.1).

As subsequent cuts are taken, the operating grade of the ramp becomes steeper. At completion of generally six or seven prescribed 2.7-m (9-ft) lifts, the grade of the attack ramp is a nominal +20%. A minimum stope width of 2 m (7 ft) is maintained for the efficient operation of load-haul-dump (LHD) machines with minimum dilution. After the design or geologic limits of the ore are reached, the void is backfilled with broken gob (waste rock) using LHDs. After backfilling, or if no gob material is available, a wedge of waste is taken out of the back of the attack ramp to allow access to the next cut. Upon completing the wedge, the stope is filled with uncemented, hydraulically placed sand. Following sand placement and dewatering, the vein is crossed at the new elevation and the muck bay re-established. At this point, the vein is once again mined to its limits along strike, but with highly efficient breast down drilling and blasting.

A modification of the attack ramp stoping approach has been developed in response to grade variations within an individual stope. When zones of low-grade mineralization are encountered for distances of over 15 m (50 ft) along strike, in-stope wedging is often used to extract zones of higher grade. This is done by taking down a wedge of low-grade rock to create a ramp (+20%) into the ore zone, which is then mined along strike to the end of the zone or stope limit. This technique is particularly effective when reaching the updip extent of the stope or recovering high-grade zones while avoiding the cost of constructing main attack ramp wedges. The process can be continued upward after sandfilling, limited by the strike length of lower grade material available and distance to the main attack ramp.

Upon completion of an attack ramp stope, the remnant access ramp is used for storage, as a muck bay, or other beneficial purpose.

Conventional overhand cut-and-fill stoping is used for certain applications and is now being combined with ramp and lateral access to permit conventional mining in zones that are unsuitable for attack ramp development or are too remote from track drifts. Ramp and laterally accessed conventional stopes are serviced by LHD equipment to move ore to raises and bring supplies to stopes. These stopes are particularly useful in mining zones with limited strike length where the cost of repeated lateral access and attack ramp development cannot be supported. As with attack ramp stopes, in-stope wedging techniques, using electric slushers, have been successfully used to recover high-grade zones in conventional stopes where the cost of raising up with timber and mining low-grade ore or waste material is not justified. Typically this method is limited to three or four cuts, depending on the limitations of the slusher used and distance to the central timbered raise.

Trackless vein mining is done using pneumatic jackleg drills and 0.76- or 0.96-m³ (1- or 1.25-yd³) LHD units. Development and stope blasting are accomplished using long-period, nonelectric

blasting caps in conjunction with detcord initiated with fuse and cap. The primary explosive used is ammonium nitrate and fuel oil (ANFO). Cap-sensitive, nonnitroglycerine emulsion "stick" powder is used in certain situations, such as in wet ground. Because of ventilation and rock burst considerations, all blasting is scheduled at the end of mining shifts. This permits proper evacuation of gasses prior to arrival of the next shift. At the same time, blasting generally triggers rock burst activity and is the method demonstrating the highest degree of success in destressing rock-burst-prone areas.

34.5   EQUIPMENT

Trackless mining methods employ 0.8-, 1-, and 1.9-m³ (1-, 1.25-, and 2.5-yd³), diesel-powered, low-profile LHD front-end loaders. Future plans include using 4.5- to 6.3-tonne- (5- to 7-st-) capacity rear-end dump trucks as well. This equipment is particularly well suited to the Sunshine Mine. The small cross-sectional profile of equipment allows it to be lowered from the surface via the Jewell shaft, which has a 1.1- by 1.6-m (44- by 62-in) compartment profile. The larger 1.9-m³ (2.5-yd³) loaders are designed and built by Mining Technologies International (MTI) using bolt-together construction so they may be lowered in pieces and joints welded upon arrival at designated levels. The smaller loaders are lowered as integral units with tires, buckets, and cylinders removed. The current LHD fleet consists of fourteen 0.8- and 1-m³ (1- and 1.25-yd³) MTI loaders, three 0.8-m³ (1-yd³) Wagner HST1A loaders, and five 1.9-m³ (2.5-d³) MTI/JCI loaders.

Single-boom jumbo drill units equipped with hydraulic hammers are the primary ramp development drills. The highest productivity jumbo is an MTI electric-hydraulic CDJ-114N with a diesel-powered carrier. This jumbo is outfitted with an HC-40 third-generation Montabert hammer. This unit has demonstrated a routine advance rate of 150 to 180 m (500 to 600 ft) per month. Four Secoma Helios diesel-hydraulic units, one of which has been retrofitted with a Montabert boom and hammer, complete the drill fleet.

Kubota diesel-powered tractors are used for service vehicles in the ramp systems to transport supervisors, mechanics, tools, and supplies.

Diesel maintenance shops and fuel storage areas are located on each level adjacent to primary ramps and close to track access. Most shops are equipped to support engine, transmission, and tire replacement, as well as routine and preventative maintenance. A well-equipped surface shop provides additional support and provides for the routine needs of all other mine-related maintenance. A local fabrication and machine shop is regularly used for specialized needs. Another local firm is an authorized parts distributor and keeps a consignment of parts for jumbos and loaders, thus reducing on-site warehousing and inventory requirements.

34.6   GROUND CONTROL AND MONITORING

Ground control is accomplished using a variety of fixtures, including Split-Set rock bolts, point-anchor bolts, and Dywidag resin-rebar bolts. Cable bolting has become a standard method of ground control in the ramp systems to control hanging walls, stabilize pillars, and control wide spans, such as in diesel maintenance shops. Cyclone mesh fencing, steel mats (bolting straps), and 31- by 41-cm (12- by 16-in) "mini-mats" are used in conjunction with rock fixtures to laminate bedding, control spalling, and provide an increased bearing surface.

The Sunshine Mine, as other deep mines in the Coeur d'Alene Mining District, is subject to rock burst activity. An array of approximately 40 geophones is linked to a rock-burst monitoring system computer in the engineering office, which performs real-time monitoring of all seismic activity. In addition to data analysis, a sophisticated E.S.G. Canada software program, SeisVis v. 1.7e, also permits modeling seismic activity, including blasting, in three-dimensional space. A sensor and seismograph are also tied to the computer and are independent of the geophone array. Sunshine Mine works in collaboration with the National Institute for Occupational Safety and Health (NIOSH) and other mines in the district to collect and study data. The goal of this work is to move toward predicting seismic activity more reliably and consequently reduce the risks associated with rock bursts.

34.7 VENTILATION

Primary ventilation is accomplished with one 225-kW (300-hp) and two 150-kW (200-hp) fans located on the surface, the 3700 level, and the 4000 level, respectively. Fresh air is brought down the Jewell shaft at approximately 2,850 m^3/min (100,000 ft^3/min) and then split on the 2700 and 3100 levels. Intake (fresh) air is regulated on the 3100 level toward 10 shaft to ventilate the east side of the mine properly. On the 2700 and 3100 levels, fresh air enters the West Chance ramp system and is pulled down ventilation boreholes. The combined air reports to the 3700 level to ventilate ramps and stopes above and below that level. Most return air (exhaust) is pulled down the No. 12 shaft to the 4000 level by one of the 150-kW (200-hp) fans and up a system of raises, old workings, and boreholes to the surface by the 225-kW (300-hp) fan. A smaller amount of return air is pulled across the 3700 level by the second 150-kW (200-hp) fan and then into the same return system as the 4000-level exhaust system.

Air doors and regulators are used to channel air within the system to achieve desired results. In an emergency situation, the Jewell shaft intake can be sealed off, and fresh air can be pulled from the Con Sil Mine to the east by the 225-kW (300-hp) surface fan. Remotely operated fire doors are strategically located in the mine to control air movement in the case of a fire.

Secondary ventilation uses a variety of fans, generally in the 15- to 30-kW (20- to 40-hp) range. Air is picked up from the primary flow route and directed to individual development headings, stopes, shops, and other work areas as needed. Both hard fiberglass vent line and brattice cloth vent tubing are used to direct air.

34.8 INFRASTRUCTURE

Electrical power supplied by Avista Corp. is brought underground at 13,800 V to primary substations. It is then fed to secondary substations or disconnects as needed. Typical motor feeds are either 2,300 or 480 V. Additional reduction is made for lighting and other specialized needs. A redundant feed from the Con Sil Mine to the No. 10 shaft ensures the capability to operate this escapeway hoist during an emergency or a power system failure at the Sunshine Mine.

Compressed air is produced at approximately 690 kPa (100 psi) on the surface then piped down the Jewell shaft, where it is distributed to all working headings, shops, and other facilities as required. Three Fuller rotary compressors (6, 7, and 8) and four Ingersoll-Rand and Worthington reciprocating compressors (1, 3, 5, and 9) provide a total capacity of 2,831 m^3/min (100,000 ft^3/min). As a normal day proceeds, a base load of approximately 226 m^3/min (8,000 ft^3/min) is augmented by an additional 40 m^3/min (1,400 ft^3/min), for a typical operating load of 266 m^3/min (9,400 ft^3/min). Receiver tanks located on the surface provide ample capacity to ensure an uninterrupted supply during normal operations.

Water for mine operations is provided from Big Creek, which flows through the mine site. A treatment facility adjacent to Big Creek is used to chlorinate all water on the surface.

Noncemented hydraulic sandfill is produced in mill grinding operations and classified with cyclones. Two surface storage tanks with a combined capacity of 3,628 tonnes (4,000 st) feed an underground distribution system, which is extended as needed to individual stopes.

Two separate phone systems are used for primary underground communications. A dial system set up throughout

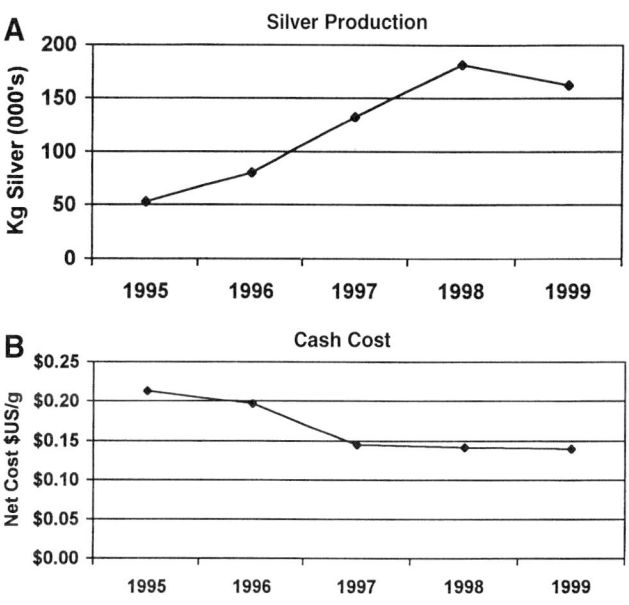

FIGURE 34.2   Sunshine Mine, 1995–1999. A. Silver production; B. Cash operating costs.

the mine can be accessed by the regular surface telephone system. A redundant pager-type voice system is also in place to provide backup communications in case of a dial system failure. A "leaky feeder" radio system has been installed to allow communication throughout the West Chance ramp system.

The Sunshine and Con Sil mines currently have five serviceable shafts outfitted with hoists. The Jewell shaft, which extends down to the 4000-level spill pocket, is the only shaft from the surface and is the primary entry into the mine. A Nordberg 520-kW (700-hp) double-drum hoist with a capacity of 3.6 tonnes (4 st) and a separate "chippy" hoist support the movement of ore, waste, personnel, equipment, supplies, and services in the Jewell shaft. The Con Sil shaft is connected to the surface by the Silver Summit tunnel. This is the Sunshine Mine's secondary escape route and has been maintained as such. Work is underway to upgrade the shaft and hoist to support a major exploration program and return the facility to rock hoisting capability. Three other internal shafts—the No. 10, No. 12, and No. 4—are used to varying degrees for access to the lower and intermediate levels, ventilation, and service distribution, and as secondary escape routes. As most operations were targeted for the West Chance ore zone above the 3700 level, the decision was made in early 1997 to allow both the No.10 and No.12 shafts to flood to slightly below the 4000 level. This action helps maintain the integrity of the shafts while reducing the effects of oxidation and deterioration of the lower levels. These shafts can be drained in the future if it becomes strategically prudent to do so.

34.9 PRODUCTION RESULTS

The overall effect of mechanization at the Sunshine Mine has been to increase productivity dramatically while reducing cash operating costs. The key to productivity is developing and maintaining an adequate number of individual stopes, thereby capitalizing on the flexibility of human, equipment, and material resources.

Figure 34.2 demonstrates the results of implementing mechanization to create this balance at the Sunshine Mine. Continued success in achieving the dual goals of cash cost reduction and increased production is dependent on successfully developing ores to which mechanization can be applied.

CHAPTER 35

Underhand Cut-and-Fill Mining at the Lucky Friday Mine

Clyde Peppin,[*] Tom Fudge,[*] Karl Hartman,[*] Doug Bayer,[*] and Terry DeVoe[*]

35.1 INTRODUCTION

The Lucky Friday Mine is located on the east end of the Coeur d'Alene Mining District in the northern Panhandle of Idaho. The mine site lies just east of the town of Mullan, 10 km (6 miles) west of the Montana border (Figure 35.1). The Coeur d'Alene Mining District has produced over 31,100 tonnes (1 billion ounces) of silver, 8 million tonnes (8.8 million tons) of lead and 4 million tonnes (4.4 million tons) of zinc from 1885 through 1999. The Lucky Friday Mine has produced 6,350,293 tonnes (7,371,181 tons) of ore to produce 3,570 tonnes (114,746,000 oz) of silver, 714,500 tonnes (787,600 tons) of lead, and 89,811 tonnes (99,081 tons) of zinc.

Production capacity of the Lucky Friday Unit in 2000 is over 1,000 tonne/d with a total work force of 210 employees. Mechanized underhand cut-and-fill and mechanized overhand cut-and-fill mining methods are used to extract silver, lead, and zinc ore from the Lucky Friday and the Gold Hunter veins. The ore-bearing minerals in both veins are tetrahedrite, galena, and sphalerite. In 1999, the Gold Hunter vein supplied 75% of the mill feed from two stopes mined using mechanized overhand cut-and-fill and two stopes mined using mechanized underhand cut-and-fill. Ore from the Lucky Friday vein was produced from two stopes using mechanized underhand mining methods. The average blended ore grade was 547 gm/tonne (16 oz/ton) silver, 10% lead, and 2% zinc. In 1999, 280,200 tonnes (309,000 tons) of ore was milled and produced 138 tonnes (4,441,000 oz) of silver, 25,000 tonnes (27,600 tons) of lead, and 1,720 tonnes (1,900 tons) of zinc in concentrates.

Underhand cut-and-fill mining is the method developed and used to mine the Lucky Friday vein. This method was extended to mining the Gold Hunter vein.

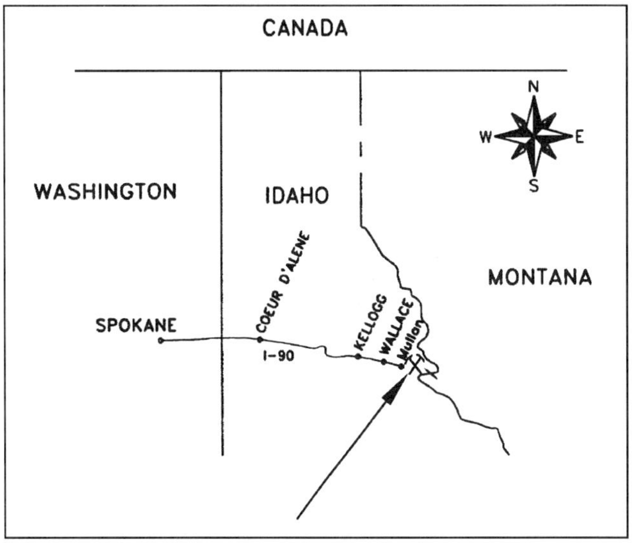

35.2 GEOLOGY

The Lucky Friday vein crops out as a thin, discontinuous vein in the St. Regis Formation. Between the 1800 and 2000 levels (approximately 600 m below the surface), the vein crosses into the Revett Formation and extends along strike for over 457 m (1,500 ft). Mineralization has been consistently strong at the Revett-St. Regis contact to the current mining horizon 1,800 m (5,930 ft) below the surface (Figure 35.2).

In the Revett Formation, the vein averages 1.2 to 1.8 m (4 to 6 ft) wide and dips nearly vertically to the southeast. From the surface to about the 2000 level (600 m below the surface), the Lucky Friday vein strikes east-northeast. Below the 2000 level to the 5840 level (1,770 m below the surface), the vein assumes a progressively northerly trend. The main Lucky Friday vein is constrained on each end by prominent west-northwest-trending strike-slip faults locally called the North and South Control faults (Figure 35.3). On the upper levels, mineralization was often very strong along these two controlling faults, but in the lower Revett

FIGURE 35.1 Location of Lucky Friday Mine

Formation, these faults contain no economic mineralization. A stratigraphic offset of 61 m (200 ft) or less is seen across the Lucky Friday vein. The vein dips at an angle of about 85°, which is about 10° steeper than the host rocks. In the lower levels, the vein locally reverses dip.

In plan view (Figure 35.3), the vein is S-shaped with a sinuous middle portion that coincides with a steeply plunging anticline. The vein takes this shape because of minor folds and major faults. Several persistent faults offset the vein laterally up to 6 m (20 ft). There are numerous narrow veins in the hanging wall and footwall of the main Lucky Friday vein; however, none has proven to be economic.

Lucky Friday ore mineralogy is argentiferous galena and sphalerite. The silver-bearing mineral is tetrahedrite that forms tiny discrete inclusions in the galena. Gangue minerals include quartz and siderite. The vein is easily identified, contrasting sharply with the quartzite. Silver and lead grades are higher in the quartzite of the Upper and Lower Revett Formation. Zinc values are increasing and silver and lead grades are decreasing slightly with depth. Average net smelter return (NSR) for 1999 from the Lucky Friday vein was $88.00 per ton.

35.3 MINE ACCESS AND SERVICES

The Lucky Friday and the Gold Hunter deposits are serviced from the surface through the Lucky Friday's Silver shaft. The Lucky Friday No. 2 shaft serves as a secondary access and exhaust route.

* Lucky Friday Mine, Hecla Mining Company, Mullen, ID.

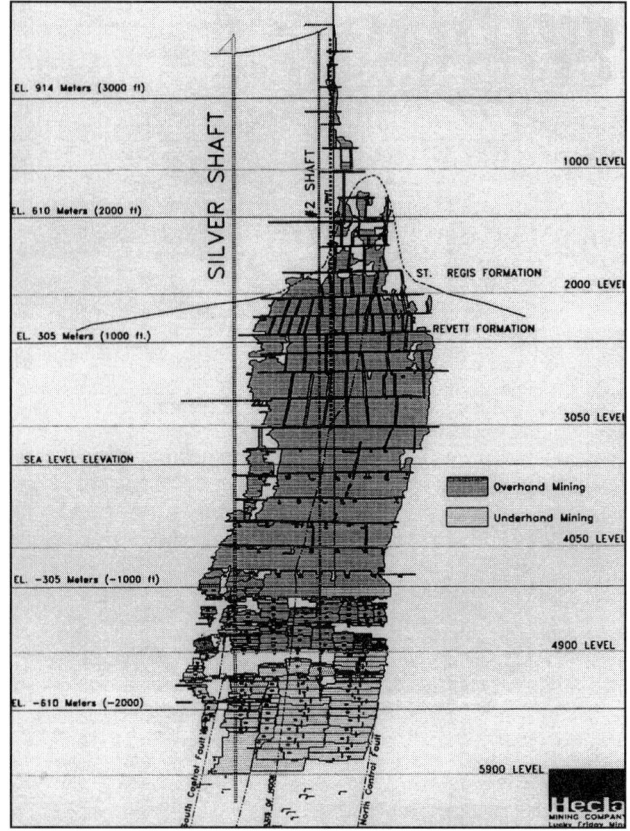

FIGURE 35.2   Longitudinal projection, Lucky Friday Mine

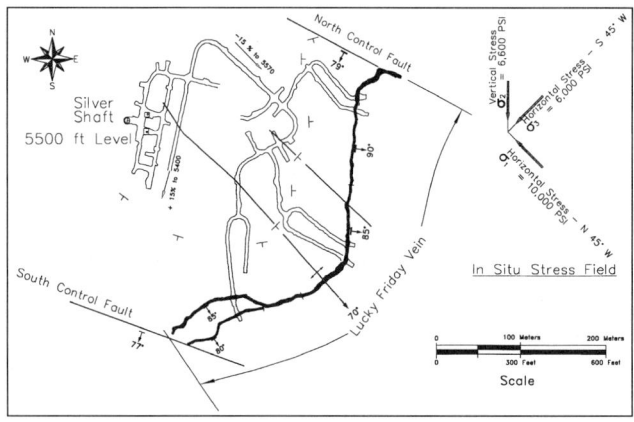

FIGURE 35.3   Lucky Friday Mine on the 5570 level

The Silver shaft is 5.5 m (18 ft) in diameter and concrete lined. Total depth is 1,890 m (6,200 ft) with stations at 2800, 4900, 5100, 5300, 5500, 5700, 5900, and 6100 levels. Skip loading facilities are at the 5370 and 5970 levels.

Hoisting is done with one Nordberg double-drum, double-clutch hoist 12 ft in diameter with an 8-ft face. There are two 1,118-kW (1,500-hp) dc motors driving the hoist. Muck hoisting speed is 11.6 m/s (2,300 ft/min). Personnel are hoisted and lowered at 7.6 m/s (1,500 ft/min). The hoist was designed to hoist a 9.1-tonne (10-ton) payload from 2,286 m (7,500 ft) without a cage attached. Currently, 4-m³ (140-ft³) bottom dump

skips are used with triple-deck cages beneath the skips. Each deck of the cage can hoist 20 people, and the decks can be removed to handle long loads. All supplies and equipment must fit through a 1.7- by 1.7-m (5.7- by 5.7-ft) opening.

Compressed air is supplied by three Ingersoll Rand XLE compressors. One I-R XHE is on standby. Total capacity is 142 m³/min at 7.6 bars (5,000 ft³/min at 110 psi). Normal usage is about 60% of that volume.

Pumping mine water and waste water from the refrigeration plants underground is done with Wilson Snyder plunger pumps at the 2,800- and 5,300-ft levels. Normal pumping rate is 38 to 76 L/s (600 to 1,200 gal/min). That water is made up of groundwater from the Lucky Friday vein at 9.5 L/s (150 gal/min) and water from chillers (depending on the time of the year) at 28.4 to 63 L/s (450 to 1,000 gal/min). Installed pumping capacity is 114 L/s (1,800 gal/min) with all three pumps operating at once on both the 2800 and 5300 levels.

35.4   VENTILATION

Rock temperatures have been measured at 40.5°C (105°F) at the 4900 level and 46°C (115°F) at the 5900 level. To cool the working areas to a maximum of 29°C (85°F) in the hottest part of the year, two 230 RT chillers are required to cool the Lucky Friday section of the mine. Other chillers are installed in the Gold Hunter section.

Total mine air flow is 6,650 m³/min (235,000 ft³/min) supplied by two 186-kW (250-hp) 130-mm (5-in) in diameter vane axial fans drawing from the No. 2 shaft and a ventilation raise and one 45-kW (60-hp) fan drawing from the No. 1 shaft. The 186-kW fans operate at approximately 5.4 in of water gauge. The 45-kW fan operates at about 4.5 in of water gauge.

There are also three 224-kW (300-hp), 1,500-mm (60-in), 1,770-rpm booster fans located underground. Two of the booster fans are in series on the 4050 level of the Gold Hunter vein to draw 3,700 m³/min (130,000 ft³/min) through the Gold Hunter, and the third is located on the 5660 sublevel of the Lucky Friday to draw 2,700 m³/min (95,000 ft³/min) through the Lucky Friday.

Auxiliary ventilation for each stope is supplied by two 56-kW (75-hp), 760-mm (30-in) in diameter fans in series supplying 850 m³/min (30,000 ft³/min) to each stope. Exhaust from the stopes flows to a ventilation raise system to the 5400 level then the ramp system to the bottom of the No. 2 shaft at the 5100 level.

Cooling is supplied by one 50 RT spray cooler for each stope. Chilled water at 9°C (48°F) from the chiller station at the 5300 level is piped to the sublevel being operated and into the spray coolers. Warm water at approximately 21°C (70°F) from the spray coolers is then pumped back to the chiller station. Waste heat from the chillers is carried by water that flows through the condensers on the chillers. This condenser water is piped to the main pump station on the 5300 level.

35.5   MINE PRODUCTION

The Lucky Friday vein has produced over 7 million tonnes (7.7 million tons) of ore at an average grade of 533 gm/tonne (15.49 oz/ton) silver, 10.68% lead, and 1.35% zinc. Production from the Lucky Friday vein from 1994 through 1997 averaged 544 tonne/d (600 ton/d). Production from the Lucky Friday vein in 1999 averaged 330 tonne/d (360 ton/d), which represented about 25% of total production. Most of the Lucky Friday Unit mill feed was produced from the Gold Hunter deposit located approximately 1,524 m (5,000 ft) to the northwest. Reduced production from the Lucky Friday vein after 1998 was a result of higher-value ore mined from the Gold Hunter deposit. Lucky Friday reserves are calculated to the 5930 level, with known resources extending down to the 6290 level.

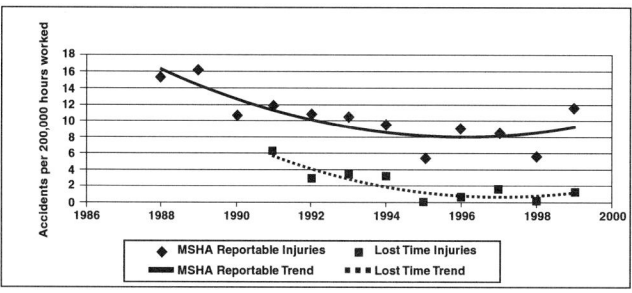

FIGURE 35.4   Lucky Friday incident rate

35.6   ROCK MECHANICS

The maximum principal stress in the Coeur d'Alene Mining District is horizontal and typically strikes N 45°W. In the Lucky Friday Mine at the 5100 level, 1,554 m (5,100 ft) below the surface, the magnitude of the major horizontal stress was measured to be some 69 MPa (10,000 psi) oriented almost perpendicular to the Lucky Friday vein. This high stress combined with the brittle nature of the host Revett Quartzite results in an environment that is prone to rock bursting.

To have a better understanding of the rock burst problem, Hecla installed a microseismic monitoring system in the early 1970s. The U.S. Bureau of Mines (USBM) had previously installed a seismograph to detect and study rock bursts and large seismic events. In 1989, the USBM installed a macroseismic system to allow waveform analysis of rock bursts and seismic events. A macroseismic event is generally considered to have Richter-scale magnitude larger than 1.0.

Prior to 1986, the mining method used was overhand cut-and-fill with raise access. Several levels were being mined at once using the overhand system. In the early 1980s, there were as many as 30 stopes in production to produce some 1,000 tonne/d (1,200 ton/d). As stoping progressed upward from level to level, the sill pillars below a level became smaller and smaller. These shrinking pillars had to support the high horizontal and mining-induced stresses. As a result of the very stiff and brittle Revett Quartzite host rock, sill pillars often failed catastrophically, causing major damage to levels and stopes.

To lessen the potential for these catastrophic failures, a new mining method was selected that eliminated the formation of highly stressed sill pillars. An underhand cut-and-fill mining method using cemented sandfill resulted from a joint research project by Hecla Mining Company, the Spokane Research Center of the USBM, and the University of Idaho. This new method shifted the high stresses to below the floor of the stoping horizon similar to a longwall in a coal mine. Hence, the acronym LFUL—Lucky Friday underhand longwall—was coined to describe it.

Because the new method no longer allowed mining to take place on multiple levels, only three or four stopes could be mined at any one time along the advancing longwall front. Therefore, each stope's productivity had to be increased from 15 to 20 tons per miner-shift to over 50 tons per miner-shift. This improvement in productivity was accomplished by the introduction of ramp access, load-haul-dump (LHD) machines, and trucks.

While the LUFL mining method results in working under an engineered and stable back, stope walls must be fully reinforced with friction bolts and chain link mesh in order to contain the effects of nearby rock bursts. Despite an increase in the incidence of seismic events with increasing depth, the LFUL mining method has greatly improved safety (Figure 35.4).

35.7   SAFETY

The underhand mining method and intense attention from management and the workforce improved safety performance.

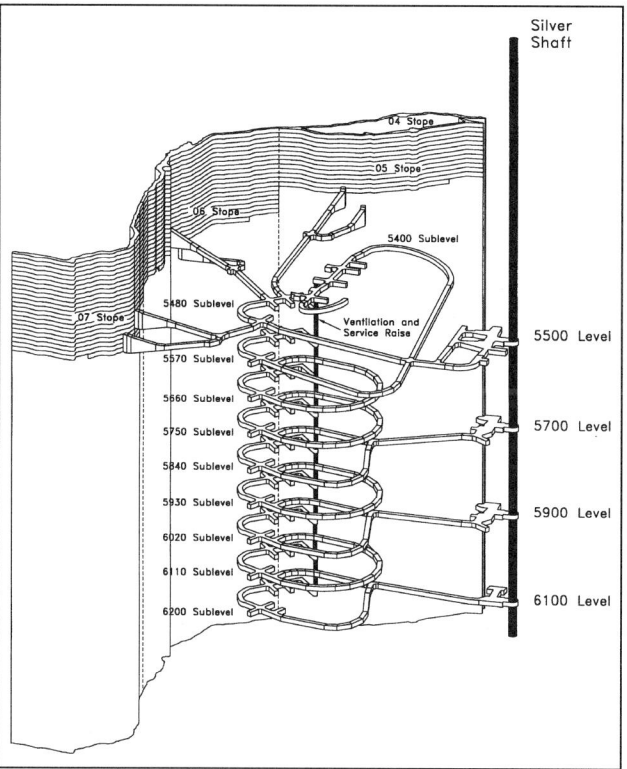

FIGURE 35.5   Sublevel plan, 5400 through 6200 levels, Lucky Friday Mine

During the period between 1995 and 2000, there have been two periods of more than 1 year with no lost-time accidents. Figure 35.4 shows the improvement in injury rates since 1988 and the inception of the LFUL mining method.

35.8   STOPE DEVELOPMENT

Access from the Silver shaft to the Lucky Friday vein is via ramps driven at 15% grade and a 3.2- by 3.3-m (10.5- by 11-ft) cross section (Figures 35.3 and 35.5). At 27-m (90-ft) vertical intervals, a sublevel is developed 100 m in the footwall of the vein. Each sublevel provides access to mine eight 3.3-m- (11-ft-) high cuts. Development is timed so that an entire sublevel is completed as the ore is mined from the sublevel above. Normal stoping progress is eight 3.3-m- (11-ft-) high cuts per year.

Eight-hundred-fifty meters (2,800 ft) of development are required each year to develop three stopes. An access ramp for each stope is driven flat from the sublevel to within 60 m (200 ft) of the vein. In this flat section is installed a spray cooler, a small storage area, and the controls for auxiliary ventilation of the stope. A muck bay and a back section 4.3 m (14 ft) high are cut in the sublevel so that 16-ton trucks can be side-loaded with 2- or 3.5-yd³ LHDs. A ramp is driven up at +20% grade to within 20 to 30 m (60 to 100 ft) of the vein. The grade of the access ramp is established so that there will be at least 3 m (10 ft) of flat ramp before the vein is intersected. A second ramp is driven down at −20% to within 35 m (120 ft) of the vein and flattened out for 5 m, or about two rounds. When all eight cuts accessed from the sublevel above are completed, the stope crew is moved to the predeveloped access and the +20% ramp is advanced until the elevation of the of the back of the advancing ramp is at the same elevation as the floor of the last cut mined from above. The ramp is then flattened out and driven across the vein and 7.5 m (25 ft) into the hanging wall to create a muck bay. The miners in the stopes will muck the rounds blasted on the previous shift into

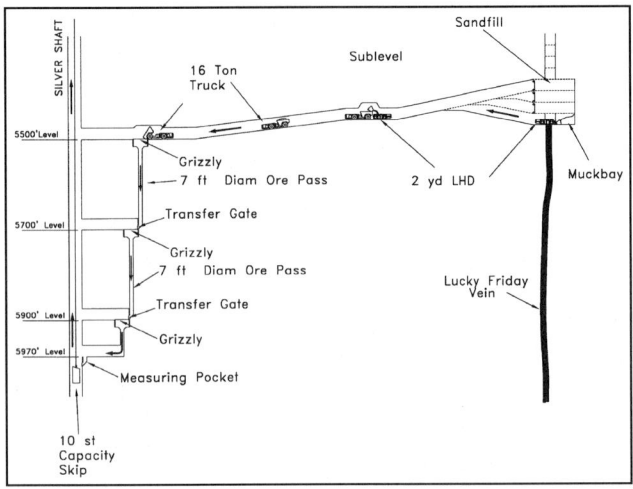

FIGURE 35.6   Muck flow and transfer system, Lucky Friday Mine

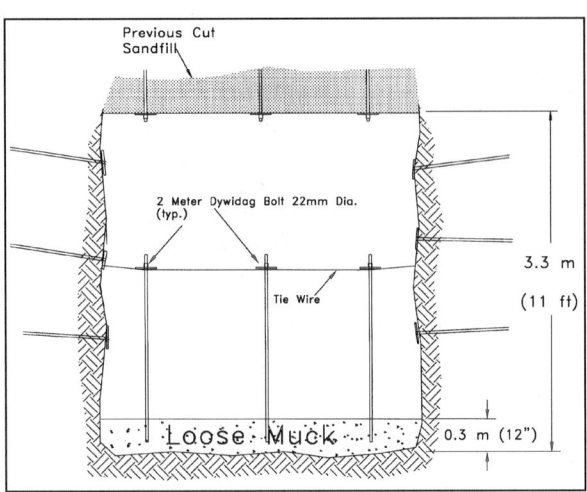

FIGURE 35.7   Backfill preparation, Lucky Friday Mine

this muck bay. Later, another crew will transfer the muck into trucks for haulage to the Silver shaft ore pass (Figure 35.6).

35.8.1   Mining a Cut

A stope crew consists of two miners per shift responsible for the complete cycle. Mining of the cut proceeds with each miner drilling and blasting one 2.4-m- (8-ft-) long round per shift each way from the access intersection. Before beginning to muck out the round blasted on the previous shift, the miners bar down any loose fill or rock above the muck pile. Then a strip of 10-gauge chain link wire mesh is hung from Dywidag rock bolts cast into the cemented sandfill on the previous cut. This mesh provides overhead protection for the miner as the round is being mucked out. Each miner mucks the round into the muck bay in the stope intersection. Once the round is mucked out, the miners then install additional 1-m- (3-ft-) long friction bolts in the cemented fill overhead as needed to tighten up the wire mesh. The walls of the stope are also supported with 1-m- (3-ft-) long friction bolts on approximately a 1- by 1-m (3- by 3-ft) pattern. When width allows and when ground conditions warrant, more aggressive support is installed, i.e., 1.2- and 1.8-m (4- and 6-ft) long friction bolts or 1.8-m (6-ft) grouted rebar bolts. All bolt installation is done with jackleg drills. When extremely poor ground conditions are encountered or when seismic activity is very high in a stoping block, shotcrete is used to support broken fill overhead or badly fractured walls. The shotcrete is applied immediately after mucking out, then bolts and wire are installed over the shotcrete.

Once the round is supported, the next 2.4-m- (8-ft-) deep round is drilled using a single-boom, diesel over hydraulic, narrow-vein jumbo. Eighteen 38- or 44-mm (1.5- or 1.75-in) holes are used to break a heading up to 3.1 m (10 ft) wide and 3.5 m (11.5 ft) high. An additional row of holes is added if the face is wider than 3.5 m (11.5 ft). The top two rows of holes intersect the loose muck laid on the bottom of the cut above. The top row is collared about 1 m (3 ft) below the bottom of the loose muck and angled up so that the holes intersect the loose muck about 1.5 m (4 ft) into the hole. Below this top row, 2.4-m (8-ft) long holes are drilled to intersect the loose muck at the end of the 2.4-m (8-ft) drill hole. Below this, four more rows of holes are drilled 2.4 m (8 ft) deep.

Paper-wrapped emulsion explosive primed with nonelectric blasting caps initiated by cap and fuse on detonating cord is used for explosive. Plastic-wrapped explosives have not proven to be efficient because the drill holes are extremely rough. ANFO is not used because of the drill holes are wet, and the mixture of ANFO with cement in the fill produces ammonia. The round is timed so that the middle hole in the top row is shot first.

Experience has shown that most rock bursts occur with the blast or within the first hour after blasting. Therefore, blasting is done only at the end of each shift. No blasting is done on shift in the Lucky Friday to minimize the chance of triggering a rock burst with a blast. Shifts are spaced at least 2½ hr apart to allow the seismic activity to settle down before the next shift comes on.

35.8.2   Backfill

The underhand cut-and-fill mining system depends on cemented backfill of consistent quality with uniaxial unconfined compressive strengths greater than 1.5 MPa (250 psi) and less than 4.5 MPa (650 psi), with the optimum being 3.4 to 3.8 MPa (500 to 550 psi). Fill will typically achieve a compressive strength of 1.4 to 1.9 MPa (200 to 300 psi) within 3 days. Average strengths at 7, 14, and 28 days are 1.9, 2.6, and 3.8 MPa (280, 380, and 550 psi), respectively. The fill strength is controlled by the amount of cement in the batch and the water-to-cement ratio. Pozzolith additives are being experimented with at the time of this writing to improve the flow characteristics of the paste in the delivery lines, reduce the amount of cement needed to achieve optimum strength, and improve the early strength of the fill.

Fill with compressive strength less than 1.5 MPa (250 psi) will not support itself for more than 1 or 2 days at widths over 2.5 m (8 ft). Fill with strengths over 4.5 MPa (650 psi) is susceptible to damage from seismic events and stope closure, causing it to break into large blocks. Fill with compressive strengths of 3.4 to 3.8 MPa (400 psi to 550 psi) will compress during stope wall closure and during a seismic event will fracture into smaller pieces and actually compress but not break into large blocks and fall out.

35.8.3   Preparation for Backfilling

Once mining of a cut is complete, the stope is cleaned out, and a 0.3-m (12-in) layer of broken waste rock, called prep rock, is placed on the bottom of the stope with a 1-yd^3 LHD. Dywidag bolts 2 m (6 ft) long and 22 mm (⅞ in) in diameter are driven into the loose prep rock on 1.2-m (4-ft) centers. A nut and plate are attached to the top of the bolt, and tie wire is used to stabilize the bolts vertically so they won't fall over during fill placement (Figure 35.7).

When the next lift below is mined and as each round is blasted, the loose muck cushions the fill from the blast and gives

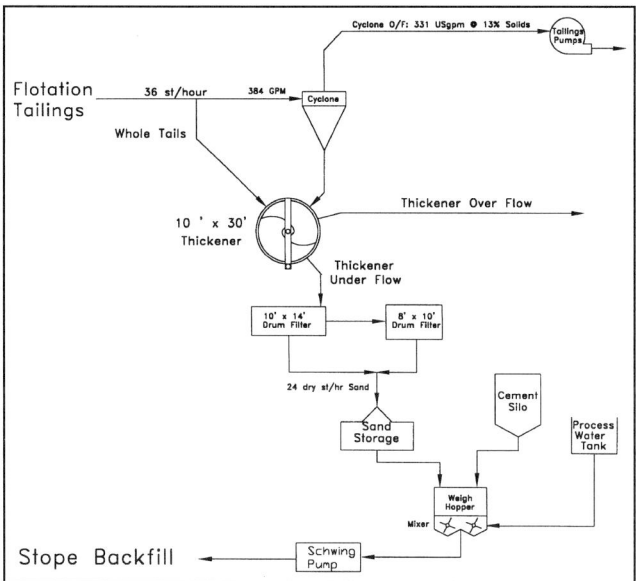

FIGURE 35.8 Backfill plant, Lucky Friday Mine

TABLE 35.1   Particle size distribution of backfill sand

Seive size	Micrometers	Whole tailings retained, %	Backfill sand retained, %
70	212	15	28
100	150	11	14
140	106	10	11
200	75	9	8
325	45	12	10
−325	−45	43	29

a little expansion room for the blast. The loose muck falls away from the fill and the bolts, leaving a short distance of bolt exposed. The miner can then put wire mesh over the end of the bolt sticking out of the fill, attach a nut and plate, and provide overhead protection immediately.

This preparation method is used for stopes up to 5 m (16 ft) wide. Any stope or area that exceeds 5 m (16 ft) in width will have timber supports added to the prep. The 10- by 10-in timbers are installed across the stope on 2.5-m (8-ft) centers. Holes are drilled in the timbers so the Dywidag bolts can be installed through the timber and into the loose muck every 1.2 m (4 ft) along the length of the timber. Cable is used to sling the end of the timber and secure it to the ribs. Care is taken to keep the ends of the timbers well away from the rock walls of the stope so that as the stope walls converge, the timbers will not buckle.

A wall of either muck or timber is built at the end of the section to be filled so that the filled section will be no longer than about 60 m (200 ft). Poor-quality fill is seen when sections longer than this have been excavated

When the sand fill preparation is complete and the wall is built, the stope is filled with cemented paste backfill and allowed to cure 3 days before mining of the next cut below is begun.

35.8.4   Paste Backfill System

The backfill plant has two independent functions. The sand plant makes the backfill sand from tailings from the flotation zinc circuit. The batch plant mixes the sand with water and cement to make a paste and pumps the paste to the Silver shaft collar for distribution to the stopes (Figure 35.8). Approximately 30% of the tailings from flotation go directly to a 9-m (30-ft) in diameter, 3-m (10-ft) deep thickener. The remaining 70% of the tailings are sent to a 25-cm (10-in) cyclone. Cyclone overflow is sent to the tailings impoundment and the underflow to the thickener. Thickener underflow at 65% solids by weight is pumped to an Eimco 3- by 4.3-m (10- by 14-ft) vacuum drum filter for dewatering. Overflow from the tub of the filter flows to an 2.4- by 3-m (8- by 10-ft) vacuum filter. The filter cake from the two filters averages 12% moisture and is transferred via 0.6-m (24-in) conveyor to a sand storage building. Backfill sand is made whenever the concentrator is operating. Table 35.1 lists the particle size distribution for the backfill sand.

The batch plant portion of the plant combines the backfill sand, cement, and water in batches that are pumped underground through a 15-cm (6-in) in diameter, schedule 80, pipe in the Silver shaft and a 10-cm (4-in) in diameter pipe on the level to the stope. The proportions of sand, cement, and water in each batch are computer controlled to ensure a consistent mix that meets requirements for the backfill needed. The dewatered sand is picked up with a bucket wheel reclaimer and dumped on a conveyor feeding a sand hopper. Cement is transferred from a silo to a cement weigh hopper. The weighed portions of sand and cement are dropped into a Nikko high-intensity concrete mixer where water is added under computer control. Each batch is mixed for approximately 2 min, then dropped into the pump hopper.

The most common batch used in the Lucky Friday is a mix with 10.5- to 11-in slump. The batch proportions are 2,558 kg (5,640 lb) of moist sand at 12% to 13 % moisture, 304 kg (670 lb) Type I/II portland cement, and 360 to 378 L (95 to 100 gal) of water. This mix yields a backfill with an unconfined uniaxial compressive strength of 3.4 to 4.1 MPa (500 to 600 psi) at 28 days.

A Schwing, positive-displacement concrete pump is used to pump the paste about 27 m (90 ft) vertically from the backfill plant to the shaft collar. The pump speed is controlled manually, and the process control computer starts and stops the reclaimer and conveyor to match the pump speed. The paste backfill can typically be pumped at 127 tonnes (140 tons) of paste per hour.

The 15-cm (6-in) paste fill line in the shaft is vented at the top to eliminate cavitation, which causes excessive pipe wear. The air that is pulled down with the fill has not proven to be a problem in the delivery lines and may help reduce the density of the paste and thus reduce line pressures. Lines in the stopes must be very well secured because of the high surging effect caused by the entrapped air. The point of highest pressure in the line occurs at the transition from the vertical section in the shaft to the horizontal section on the level. Line pressures of 41 to 62 bars (600 to 900 psi) are typical at this point. The pressure seen here is dependent on the length of the delivery line, the slump of the paste, and the height to which the fill is transported.

The operator in the batch plant monitors the line pressure and adjusts the mix and pump speed to maintain a safe line pressure. If the pressure at the transition rises above 83 bars (1,200 psi), the programmable logic controller (PLC) will stop the backfill pump until the pressure drops below 83 bars (1,200 psi). To reduce high line pressures, additional water may be added to the mix to reduce the slump, but additional cement must be added to maintain the proper water-to-cement ratio. The pump speed can also be reduced to lower the pressures in the delivery line.

Before backfill is mixed and pumped to a stope, flushing water is sent through the line to wet the line, ensure the line is not plugged, and provide a visual check for leaks. The stope miner notifies the backfill plant when the flushing water reaches the stope, then the batch plant is started. The flushing water before a pour is kept out of the stope by installing a valve in the delivery line just outside the stope. Telephone communication with the batch plant is very critical so the batch plant operator can judge the needs for cement and monitor progress.

When the depth of the fill in the stope is 0.6 m (2 ft) higher than the tops of the Dywidag bolts, the miner informs the batch plant operator that the stope is full enough. The batch plant then begins the shut-down sequence. The remaining fill in the pump hopper, the batch being mixed, and the batch being weighed must be completed and pumped out. Once these are pumped into the system, water is pumped into the line to flush the whole system clean. The batch plant operator informs the miner when the water is entering the system. From 15 to 30 min later, the water reaches the stope. When the miner sees flushing water at the stope valve, the valve is opened, the miner notifies the batch plant operator, and the batch plant operator turns off the water.

Experience has shown that excess water in the stope, either from groundwater or flushing water, creates very poor backfill quality. Before a fill is made, any water collected in the stope is pumped out and only enough flushing water is directed into the stope to be sure the pipe line from the flush valve to the end of the stope is clear at the beginning and end of the pour.

When a cut has been filled, the mining crew begins to drill and blast out the bottom of the access ramp beginning at a point 15 m (50 ft) from the sand wall. The gradient of the reentry is set at −20%. Within this 15 m (50 ft), the bottom has been taken up sufficiently to have a face of rock 3 m (10 ft) high when the face is just under the recently poured sandfill.

This reentry time period of 3 to 5 days combined with one or two shifts repairing ground support outside the stope allows enough time for the most recent pour to gain sufficient strength to stand up while mining under the fill.

35.9   COSTS

The highest single cost item for direct stoping cost is labor (Table 35.2). Miners' pay is based on a guaranteed hourly wage set by the labor-management agreement, and incentive is paid to miners on a total-cut incentive system. A value is set for the labor to mine a complete cut, and at the end of the cut, the value is divided by the miner shifts to complete the cut. Progress payments are made at the end of each 2-week pay period based on an estimate of the percentage completed. Hours spent during the filling of the stope and repair work resulting from rock bursts or poor ground conditions are not charged against the "contract."

Development costs are based on the cost of preparing the next block of ore to be mined as the current block is being mined and the cost of accessing a cut after filling a cut. Shared services are costs such as compressed air, hoisting, pumping, and refrigeration, main ventilation system, etc., that are shared among all stopes.

TABLE 35.2   Direct and indirect operating costs

Operating costs	Dollars per tons mined (US)
Labor	8.59
Explosives	1.62
Ground support	3.90
Bits and steel	0.45
Distributed equipment costs	6.68
Drill repair	0.79
Backfill	4.52
Other supplies	0.98
Total direct cost per ton	27.49
Development costs	9.96
Shared services	34.10
Total, direct and indirect costs	70.85

35.10   ACKNOWLEDGMENTS

Contributors to this section on the Lucky Friday have been Tom Fudge, Lucky Friday Unit manager; Randy Anderson, senior geologist; Steve Thomas, mine planner; Karl Hartman, ventilation engineer; Doug Bayer, senior engineer; Wilson Blake, consultant; and Clyde Peppin, engineering supervisor, all of the Lucky Friday Mine staff.

35.11   REFERENCES

Blake, W., and D.J. Cuvelier. Rock Support Requirements in a Rockburst Prone Environment: Hecla Mining Company's Lucky Friday Mine. Paper in Rock Support in Mining and Underground construction: Proceedings of the International Symposium on Rock Support, ed. By P.K. Kaiser and D.R. McCreath (Sudbury, ON, June 16–19, 1992). Balkema, 1992, pp. 665–674.

Bolstad, D.D., Rock Burst Control Research by the US Bureau of Mines, Proceedings of the 2nd International Symposium on Rockbursts and Seismicity in Mines, ed. by Charles Fairhurst (Minneapolis, June 8–10, 1988). Balkema 1990.

Lucky Friday Mine Staff.. Section 3, Chapter 6. In *Underground Mining Methods Handbook*, ed. by W.A. Hustrulid. AIME, 1982, pp. 549–557

Lucky Friday Unit, Hecla Mining Company. Policies and Guidelines Handbook. 1999.

Noyes, R.R., G.R. Johnson, and S.D. Lautenschlager. Underhand Stoping at the Lucky Friday Mine in Idaho. Paper presented at the 94th Annual Meeting, Northwest Mining Association, December 2, 1988, 14 pp.

Whyatt, J.K., T.J Williams., and W. Blake. In Situ Stress at the Lucky Friday Mine (In Four Parts): 4. Characterization of Mine In Situ Stress Field. USBM RI 9582, 1995.

. . . .
CHAPTER 36

Pasminco Broken Hill Mine

David R. Edwards* and Neil S. Rauert*

36.1 INTRODUCTION

Broken Hill is situated in far west New South Wales some 1,165 km west of Sydney and 508 km northeast of Adelaide. The Pasminco Broken Hill Mine is currently mining lead and zinc ore at a rate of 2.8 million tonnes per annum. The operation employs 650 personnel as of October 1999 and utilises a limited number of contract personnel. Presently operations extend to a depth of 1,200 m and are anticipated to reach 1,300 m. Longhole open stoping is now the predominant mining method. The Pasminco Broken Hill Mine operation is currently exploiting the southern end of the extensive Broken Hill ore body. The central and northern areas are now almost exhausted and largely abandoned.

Pasminco, Ltd., is the world's largest integrated producer of lead and zinc. As well as an international portfolio of smelters in Australia, the Netherlands, and the United States, Pasminco has the following mines:

- The Century Mine (northwest Queensland); said to be the largest single zinc mine producer in the world

- The Broken Hill operation

- The Rosebery Mine (Tasmania)

- The Elura Mine at Cobar, 500 km east of Broken Hill and now integrated into the same business unit as Broken Hill

- The Gordonsville and Clinch Valley mines in Tennessee, United States

Following the discovery of ore in the Broken Hill "Line of Lode" in 1983, numerous mining companies began mining in the north and central areas of the Line of Lode, including famous Australian companies like Broken Hill Proprietary and North Broken Hill, Ltd. The Pasminco Broken Hill operation has its origins back in 1905 when the Zinc Corporation, Ltd., was formed to recover zinc from tailings. In 1911, the company amalgamated with Broken Hill South Blocks, Ltd., Company, which had underground mining interests in the then-southern extremities of the Line of Lode. The Zinc Corporation commenced underground operations in 1912 when 141,000 tonnes of ore were mined (Smith and Spreadborough 1993).

In 1936, a major reconstruction and expansion phase began, including a new shaft known as Freeman's shaft, and a new concentrator plant. New Broken Hill Consolidated, Ltd. (NBHC), located on the southern extremity of the Line of Lode, was incorporated in 1936 and included the capital and leases of Barrier South, Ltd., acquired from the Zinc Corporation. NBHC did not commence until 1948 after World War II, but under management of the Zinc Corporation.

The NBHC service shaft was completed in 1953 and the NBHC haulage shaft in 1952.

Both the Zinc Corporation, Ltd., and NBHC, Ltd., became part of Conzinc Rio Tinto of Australia (CRA) in 1962, which continued until 1986 when the two operations combined as ZC Mines Pty. Ltd.

After the closure of Broken Hill South Limited's South Mine in 1972, only North Broken Hill's North Mine and the ZC and NBHC mines remained on the Line of Lode. On 22 June 1988, CRA, Ltd., and North, Ltd., merged their respective lead and zinc assets to form Pasminco, Ltd.

In February 1993 the Pasminco Northern Operation closed, leaving the Pasminco Broken Hill Mine operation in the southern area of the Broken Hill Line of Lode.

Initial ore extraction was carried out by cut-and-fill with backfill by dry fill placement by hand. The need for economic improvement has driven almost continuous evolution and changes in stoping methods. These methods have included—

- Flatback cut-and-fill

- Square-set timbered support cut-and-fill

- Mechanised cut-and-fill

- Shrinkage stoping

- Underhand timbered stope pillar extraction

- Overhand timbered stope pillar extraction

- Underhand cut-and-fill pillar extraction

- Mechanised underhand cut-and-fill pillar extraction

- Vertical crater retreat

- Longhole open stoping

- Uphole retreat stoping

- Mechanised pillar uphole retreat

A record production rate of 2,809,357 tonnes at mining grades of 4.3% lead, 44.8 gm/tonne of silver, and 7.9% zinc was achieved during the 1998–1999 financial year. This production came from the following sources:

- 2,260,000 tonnes from longhole open stoping

- 203,000 tonnes from the Potosi open-pit mine located 6 km northeast of Broken Hill

- 192,000 tonnes from development ore

- 155,000 tonnes from cut-and-fill pillar mining.

36.2 GEOLOGY

The Broken Hill ore body lies in a thick sequence of highly metamorphosed Proterozoic sedimentary and volcanic rocks known as the Willyama Supergroup. The sequence consists of

---

* Pasminco Broken Hill Mine, Broken Hill, New South Wales, Australia.

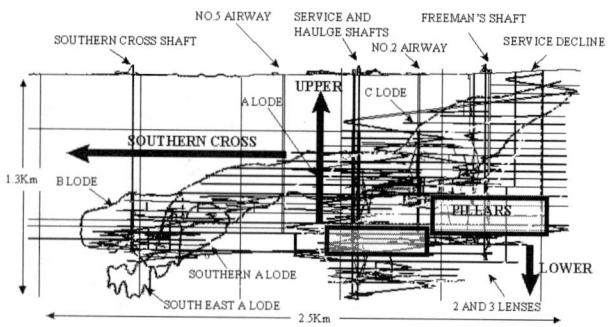

FIGURE 36.1 Longitudinal section of the Pasminco Broken Hill Mine showing production and geological areas

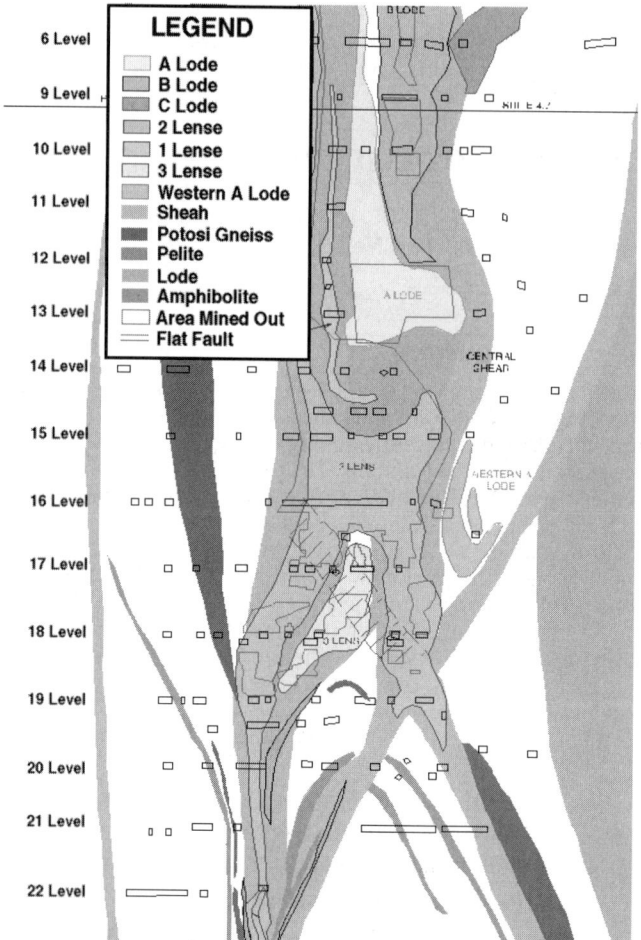

FIGURE 36.2 Geological cross section (Section 30) at the Pasminco Broken Hill Mine

10 separate, but closely related ore bodies that are stacked in a stratigraphic package: 3 Lens, 2 Lens, 1 Lens Lower, 1 Lens Upper, A Lode, A Lode Upper, Southern A Lode, Southern 1 Lens, B Lode, and C Lode.

The original unmined ore body was over 7 km in length and 250 m at its widest. It contained approximately 185 million tonnes of high-grade, economically mineable ore and another 100 million tonnes of mineralisation exceeding 3% combined lead and zinc (Lutherborrow and Edwards 1999). As of March 1999, reserves stood at 17.9 million tonnes at 4.0% lead, 41 gm/tonne silver, and 7.5% zinc. Pasminco Ltd. has a business plan for the operation until 2006, at which stage current economic reserves will be exhausted.

The mine is geographically divided into four production areas.

- **Upper,** composed of Lodes A and B (zinc) in the northern upper level areas.

- **Lower,** composed of 1 Lens, 2 Lens, and 3 Lens (lead) in the bottom area of the mine from the centre to the northern extremity.

- **Southern Cross,** composed of Lodes A and B (zinc) in the southern and lower area of the mine.

- **Pillars,** composed of the remnant and pillar areas left in ore bodies from cut-and-fill mining in the Upper Lode areas (lead).

Figure 36.1 shows a longitudinal section of the mine with the production areas shown. Figure 36.2 shows a typical geological cross section.

36.3 GEOTECHNICAL DESCRIPTION

Each of the four production areas has its own geotechnical description and challenges. Figure 36.3, a three-dimensional view of the operation, shows the large expanse of mined openings as well as a geotechnical summary for each production area.

36.3.1 Upper Area

This area consists largely of very competent rock (uniaxial compressive strength of 100 to 200 MPa) and some medium-level shearing on the east ore body contact. This area has a large unmined area surrounded by a large mined-out expanse. The result is a highly stressed area that is very microseismically active (Figure 36.3).

Stopes in this area have no significant wall or back overbreak problems and are medium-to-large in size (100 to 350 thousand tonnes). Stope size is limited to 100 m along strike.

36.3.2 Southern Cross

This area of the mine is highly stressed and exhibits microseismicity (Rauert and Tully 1998). In the lower areas of

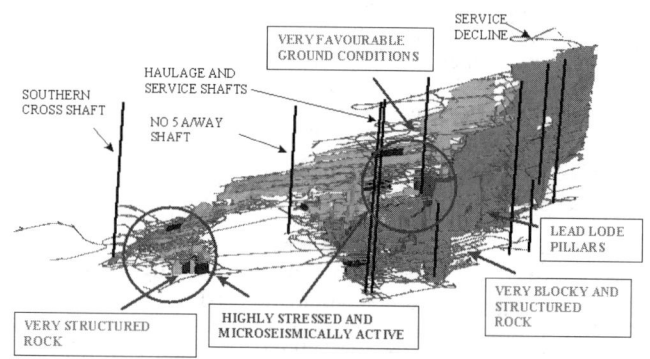

FIGURE 36.3 Isometric view of the Pasminco Broken Hill Mine showing major geotechnical issues in each area

the Southern A Lode and the South East A Lode, rock strength appears somewhat weaker, and the area is considerably influenced by the central shear, together with significant folding in the Southern A Lode area. This leads to the need for significant ground support and reinforcement in development and significant stope overbreak challenges in the Southern B Lode, Southern A Lode, and the newer South East A Lode areas.

36.3.3 Lower Area

This area has traditionally shown little sign of stress-related challenges, but a very weak contact and very blocky and heavily laminated structural conditions have led to the need for considerable ground support and reinforcement in development and significant stope overbreak challenges. Stope size is generally limited to 30 m along strike.

36.3.4 Pillar Area

The pillar area in the lead lode cut-and-fill area is usually of high-grade ore and adjacent to hydraulically placed sandfill. Access development is often through old sandfilled stopes and requires extensive use of shotcrete (Hewitt et al. 1998; Rauert 1992). The pillar stopes are mined in a top down sequence.

36.4 GROUND CONTROL

The full range of ground support and reinforcement controls are available at the operation, and over the years, considerable development in ground support and reinforcement technology (now used worldwide) has occurred at the operation. This includes the use of cable dowelling, which was largely developed at this operation in the early 1970s.

Ground support and reinforcement in use includes—

- Rock bolts—friction stabiliser and CT bolts (expansion-shell groutable rock bolt) in all development areas.

- Cable dowelling—reinforcement of poor ground and stope crown and wall reinforcement in development areas, including the Rib-Roc reinforcement technique (Rauert 1995).

- Shotcrete—for pillar mining and in heavily laminated ground, particularly in the Lower area.

The Rib-Roc stope reinforcement method (Rauert 1995), largely developed at the Broken Hill operation, utilises zones or ribs of superreinforced rock in the walls and back area of stoping.

36.5 GEOTECHNICAL MONITORING

In 1997, a ground support audit was carried out on the entire 90 km of accessible development (Butcher et al. 1999). The mine was divided into some 850 areas approximately 100 m long. Each area also had a risk rating applied using a well-defined risk matrix for ground fall probability and consequence.

As well as installed ground support and reinforcement, four types of ground condition were described.

1. Massive ground usually requiring spot rock bolting.
2. Jointed ground usually requiring patterned rock bolting.
3. Blocky ground usually requiring cable dowelling.
4. Very blocky and laminated ground usually requiring patterned rock bolting, cable dowelling, and shotcrete.

Installed ground support and reinforcement are continually monitored by area and compared to ground support and reinforcement requirements highlighted in the ground support audit. New areas are monitored as they are developed.

The operation also uses a microseismic monitoring system supplied by Integrated Seismic System International Pty. Ltd. for monitoring the risk of ground movement in microseismically active areas (Rauert and Tully 1998).

Every stope is surveyed after completion using an Optech cavity monitoring system. The results of these surveys are used for geotechnical monitoring and for blasting and stope grade reconciliation.

36.6 MINING METHOD

Currently two major mining methods are used, longhole open stoping and mechanised pillar recovery.

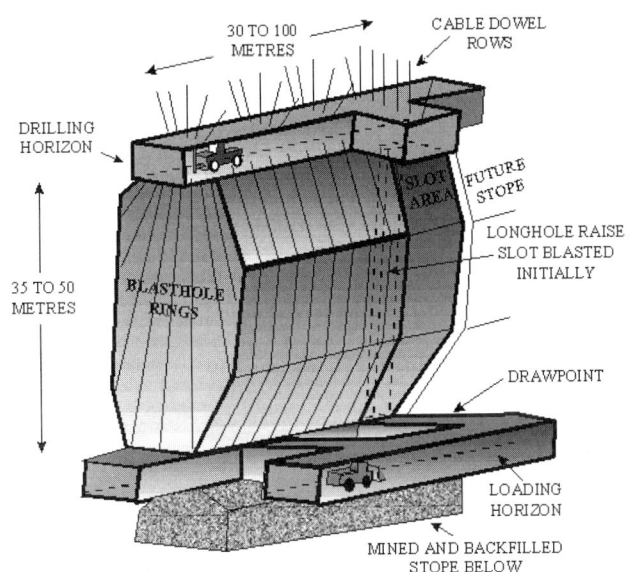

FIGURE 36.4   Schematic view of longhole stoping

36.6.1 Longhole Open Stoping

Over many years, longhole open stoping has evolved as the most suitable and economical method for primary ore recovery and has been adapted to the wide variety of ground conditions at the operation. A number of variations are used to suit a variety of ground conditions and stope geometries. Figure 36.4 shows the basic layout. To address the very different ground conditions, economics, and ore body shapes in each of the three primary stope production areas, differing stope dimensions and blasthole spacings are used.

Development. Development is carried out using twin-boom Atlas Copco 127 and 128 jumbos. Access development and most stope development is 5 m high by 5 m wide. Typically, 50 to 60 holes 45 mm in diameter are charged with a mixture of ammonium nitrate and fuel oil (ANFO). Low-density explosives are used in the wall holes.

Mucking is achieved using 7.1- and 6-m³-capacity load-haul-dump (LHD) machines to load 35-tonne-capacity diesel dump trucks.

Stope Reinforcement. Cable dowels are installed in nearly all stope backs in the Lower and Southern Cross areas to designs by geotechnical staff. These are either twin 15.2-mm in diameter cables installed in 9-m-long, 70-mm in diameter upholes spaced 2 to 3 m apart or rib-rock reinforcement for both backs and walls.

Production Drilling and Blasting. Production drilling is carried out by either Tamrock Data Solo or Atlas Copco production drill rigs. Blastholes typically range from 105 to 127 mm for downholes and 89 to 105 mm for upholes. In the Lower area and the lower parts of the Southern Cross, a burden of 2.5 m is used with a typical toe spacing of 3 m. In the Upper and Upper Southern Cross areas, the burden is 3 m.

A nine-hole pattern is used for the vertical longhole raise in the initial stope slot. A large primary slot is excavated across the strike of the stope, and the blasthole rings are then fired along strike into the excavated slot void. Longhole blasting is carried out with ANFO and emulsion explosives, which are supplied in bulk by the explosives manufacturer utilising a custom-built, bulk- explosives supply vehicle and taken directly to the stope. All production and development blasting is done via surface control between the end of shift and start of the oncoming shift when no personnel are in the underground workings. All blasts are monitored from various surface locations to evaluate the effects of blast vibrations.

Stope Production. Ore is mucked from the stope via stope drawpoints at the bottom sill of longhole stopes using 6.5- and 7.1-m³ LHDs loading into 35- and 40-tonne-capacity diesel dump trucks. Most production LHDs are fitted with radio-remote control, allowing operation of the unit in complete safety. An added benefit is that video images from the LHD can be viewed on the control unit.

Ore is trucked directly to the major ore pass system, where a grizzly and rockbreaker hammer are fitted. Ore is then fed by gravity to underground crushing stations and the skip loading area for transport to the surface and treatment at the concentrator plant.

Secondary breakage in the stope area is effected by—

- Drilling and blasting of stacked ore outside the stope on the mucking horizon, and

- Using a remotely controlled popping rig to drill water-filled holes into oversize material within the stope void. The water-filled holes are plugged, and a low-strength explosive is used to break the rock by dynamic hydraulic pressure.

Longhole Stope Backfill. The type of backfill is dependent upon sequencing, economics, and stope location. The types of backfill are—

- Mullock or waste development rock

- Hydraulic sandfill

- Hydraulic cemented sandfill

- A combination of waste rock and hydraulic cemented sandfill

- Tailings dam fines fed via a dedicated borehole to the stope void

Hydraulic sandfill is generated at a sandfill plant at the surface concentrator plant. Concentrator middlings in a water slurry are fed by gravity through four surface boreholes to the underground workings. When cement is required, it is added at the surface plant. The slurry is fed via a rubber-lined pipe and disposable polyurethane pipe to each stope location. Waste rock is trucked to the stope fill location via diesel dump trucks fitted with ejector trays. Prior to stope backfilling, brick drainage barricades are built in stope drawpoint areas. Geotextiles and slotted drainage pipes allow drainage during stope backfilling.

36.6.2 Mechanised Pillar Mining

Large tonnages of high-value remnant pillars remain between cut-and-fill stopes in the lead lodes. Previous pillar mining methods used have been underhand timbered stope pillar extraction, overhand timbered stope pillar extraction, underhand cut-and-fill pillar extraction, mechanised underhand cut-and-fill pillar extraction, and vertical crater retreat.

The current method involves the development of an uphole drill and mucking sill inside the vertical pillars of 3 to 6 m nominal width and uphole retreating for the bulk of the pillar resource in the vertical or crown pillar. In most cases, the pillars are surrounded by consolidated sandfill. Figure 36.5 shows typical sections.

Development. Development consists of drives 3.5 m wide and 4 m high inside the pillar and 5 by 5 m for access development. This is carried out using a single-boom Atlas Copco 126N jumbo. Ground support requirements are naturally high. All development inside the vertical pillars is shotcreted, with the primary concern being to contain sandfill in place and maintain the sand-to-rock friction contact in the vertical pillar block. Quite

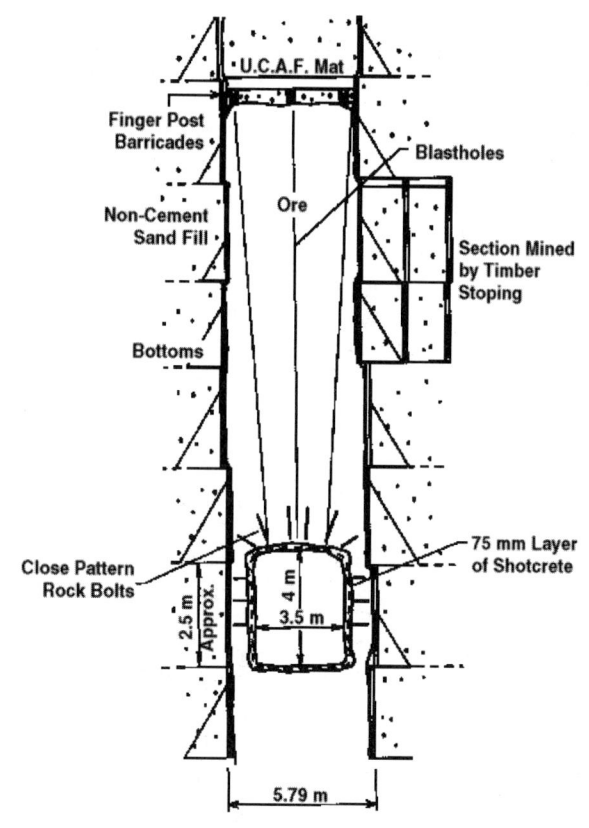

FIGURE 36.5 Sections showing mechanised pillar mining

often the pillars are only 3 to 6 m wide and are completely undercut by development. In areas where a skin of ore remains in the drive wall, 1.5- m-long friction stabiliser rock bolts are used to reinforce the walls prior to shotcreting.

Production Drill and Blast. Uphole blastholes are drilled by a Tamrock Solo 689 production drill. An uphole longhole raise slot is drilled at the furthermost extremity of the pillar stopes. Uphole rings are then blasted to this slot. Hole diameters of 70 mm are used with ring spacings of 1.5 m. Blasting is carried out using ANFO. Several rows are blasted at a time.

Stope Production. Stope mucking is carried out by remote-controlled, 4.8-m³-capacity LHDs. The operator can operate the LHD from well outside the pillar area. A combination of 6.5-m³ LHDs and 35- and 40-tonne-capacity diesel dump trucks are used to move the ore to the main ore pass system.

Backfill. Backfill is normally achieved by hydraulic sandfill placement. Often caving sandfill from the adjacent filled primary stopes necessitates refilling of these stopes.

Crown Pillar Mining. Most of the remaining lead lode cut-and-fill pillar reserve remains in crown pillars (Rauert and Tully 1992). These horizontal pillars are 5 to 8 m high and were formed from 10-m-wide cut-and-fill stoping panels. The final method for their extraction is still being developed, but various methods have been tried, including narrow-panel, cut-and-fill (Rauert 1995), uphole retreat using longhole drilling, and uphole retreat using a face jumbo back strip.

36.7 GENERAL MINE INFORMATION

36.7.1 Mine Design

A team of professional mining engineers, geotechnical engineers, and surveyors are responsible for designing 40 to 50 stopes annually. Geologists and mine design engineers use the Vulcan three-dimensional modelling software for resource definition and stope design. All mined-out areas are also modelled in Vulcan in a "mined openings" model.

Vulcan is also used for geotechnical analysis. Optech surveys are used for three-dimensional mined stope outlines in the mined openings model. Microseismic data are also analysed using Vulcan.

36.7.2 Ventilation and Mine Access

The operation is ventilated via three exhaust ventilation shafts and seven intake shafts. The mine is also serviced by a service decline from the surface that allows access by rubber-tired vehicles to all areas of the mine. The service decline also acts as an intake airway. The three surface fan installations are as follows:

1. No. 2 Airway. Three centrifugal fans with two operating and one spare capable of 600 m³/s
2. No. 3 Airway. Two centrifugal fans operating with a capacity up to 350 m³/s
3. No. 5 Airway. Two centrifugal fans operating with a capacity up to 530 m³/s

The major ventilation circuit is divided into four districts to suit each of the production areas. Air is directed to the working areas by ventilation regulators and booster fans with fabric ducts. Access to the mine is available as follows:

1. Service decline. Portal connection allowing personnel and material access into the mine (intake airway)
2. Service shaft. Caged shaft access for personnel and some materials (intake airway)
3. Southern Cross shaft. Periodic access for manpower and limited materials (intake and exhaust airway via divided shaft)

36.7.3 Manpower

Broken Hill employs approximately 650 personnel as of October 1999. The operation works 52 weeks a year. Mill and maintenance shifts work two 12-hr shifts a day, 7 days a week, while underground crews work two 11.5-hr shifts a day, 7 days a week.

36.7.4 Equipment

Major equipment as of October 1999 (Hawcroft 1999) is shown in Table 36.1.

36.7.5 Maintenance

Monitoring is carried out on all mobile equipment condition as follows:

- A prestart check list is gone through prior to machine start up.
- Oil in critical components on all LHDs and trucks is sampled and analysed once every 250 to 300 hr. Coolant is sampled and analysed once every 600 to 1200 hr.
- All LHDs and trucks fitted with fire detection and suppression systems have these systems checked once every 250 to 350 hr. These systems are released every 500 hr.
- All LHDs and trucks have a major overhaul once every 10,000 hr.
- All production drills and rock bolters are serviced once a week and rebuilt once every 26 weeks.
- Fire detection and suppression systems are checked once a week.
- All development drills are serviced once per 10 days (the fire detection and suppression systems are checked) and rebuilt once each 26 weeks.

TABLE 36.1   List of mine equipment

Location	Number	Type of equipment
Fixed plant		
No.2 Airway	3	Flat Richardson centrifugal fans
No.3 Airway	2	Flat Richardson centrifugal fans
No.5 Airway	2	Cemfec-Fantecnic centrifugal fans
21 level		Jaw crusher, 230-hp motor
24 level		Jaw crusher, 230-hp motor
Level loading station		Vibrating feeder, conveyors, measuring flasks
Level pumping station	4	Wirth TPM pumps
Haulage shaft		Skip winder
Service shaft		Cage winder
Southern Cross shaft		Cage winder
LHDs		
	1	Elphinstone R1500 (4.8 m³) with remote control
	1	Elphinstone R1600 (4.8 m³) with remote control
	2	Elphinstone R2800 (7.1 m³)
	5	Elphinstone R2900 (7.1 m³)
	3	Toro 501 (6.5 m³), two with remote control
Diesel dump trucks		
	3	Elphinstone AE40 (50 tonne), ejector-tray type
	3	Elphinstone AE40 (50 tonne), dump tray type
	2	Toro 40D (40 tonne), dump tray type
	1	German (13 tonne), dump tray type
Development drills		
	5	Atlas Copco H128, twin boom
	3	Atlas Copco 322, twin boom
	2	Atlas Copco M2C, twin boom
	1	Atlas Copco H126, single boom
Production drills		
	1	Atlas Copco 435
	2	Solo 60 Tamrock
	1	Solo 698 Tamrock
Misc. mobile equipment		
	2	Caterpillar 120G grader
	2	Shotcrete machines mounted on Atlas Copco and Tamrock carriers
	2	Elphinstone 925B scalers
	1	Ingersoll Rand roller
	1	Robbins mobile miner
	20	Caterpiller IT forklifts
	1	Popping rig (McClean)
	1	Atlas Copco Simba H45- twin-boom uphole drill (for drilling cable dowel holes)
Personnel vehicles		Nissan 4WD wagons and Traytop vehicles

For all fixed plant equipment, including crushers, winders, and surface fan installations, a condition monitoring program is carried out.

36.7.6 Mine Dewatering

All water in the mine gravitates to the bottom area of the mine, where it is directed to a prescreening plant on the 23 level. From there, it is fed via a 3.8-m in diameter raise-bored sump to the major pump installation on the 24 level. The pumping installation consists of four Wirth TPM positive-displacement (triplex piston) diaphragm pumps. Each pump has a capacity of

27.5 L/s (Duffield and Mitchell 1993). The pumps then deliver the water for 1080 m in a single lift via a 217-mm main raise to the surface discharge, where it is recycled in the concentrator or used for hydraulic sandfilling.

36.7.7 Communication

Throughout the mine, a leaky-feeder coaxial radio system has been installed with a base station at the Mine Communication Centre on the surface. The radio system is an integral facet of safe day-to-day mine operations.

The communications centre also uses a mine SCADA system for telemetry monitoring of fixed plant, including functioning of the ventilation fans and compressed air, fresh water, electrical, and sandfill reticulation systems within the mine.

All underground blasting is initiated from the communications centre at the end of shift. All personnel are registered via an electronic in-and-out tag system to ensure that all personnel are out of the underground operations during development and production blasting.

36.7.8 Environment

The operation utilises two full-time environmental scientists implementing a comprehensive corporate environmental policy and ensuring that the operation complies with detailed statutory environmental regulations.

36.8 OCCUPATIONAL HEALTH AND SAFETY

Occupational health and safety is the highest priority at all Pasminco sites. A team of occupational health and safety professionals ensure that the detailed corporate occupational health and safety policy systems and standards are implemented.

The current safety reporting system uses the Medical Referral Injury Frequency Rate. It was reported to be 94.8 for December 1999. The operation has systems in place for reporting and investigating all safety-related incidents and near misses. Safe work procedures have also been written for all major work-related tasks at the site.

In 1999, the mine was awarded the New South Wales Minerals Council's Gold Award for safety excellence, and in 1998 the mine was awarded the Bronze Award.

36.9 CONCLUSION

After some 92 years of continuous production, the Pasminco Broken Hill Mine will continue into the near future as a world-class mine still among the top 10 producers of lead and zinc metal.

36.10 REFERENCES

Butcher, P.W., Tully, K.P., Ormsby P.M., and Rauert, N.S. 1999. The Development of a Ground Support System at the Pasminco Broken Hill Mine. *Proceedings of the International Symposium on Rock Support and Reinforcement Practice in Mining.*

Duffield, S.B., and Mitchell D.S. 1993. Mine Dewatering Project— Pasminco Mining, Broken Hill. *Proceedings of the VIII Australian Tunnelling Conference.*

Hawcroft, D. 1999. Insurance Risk Survey Underground Operation. *Report to Pasminco Australia Ltd Broken Hill Mine by Damian Hawcroft and Associates.*

Hewitt, T., Ormsby P.M., and Price, A. 1998. Longhole Pillar Mining at Pasminco Broken Hill with Particular Reference to Driving Through Sandfill. *Proceedings of the Australian Institute of Mining and Metallurgy Underground Operators Conference.*

Lutherborrow, C.H., and Edwards, D.R. 1999. Pasminco Broken Hill Mine Resource and Reserve Estimate at 31 March *Pasminco Broken Hill Mine Report.*

Pasminco Limited 1999 Pasminco Annual Report. *Annual Report to Shareholders.*

Rauert, N.S. 1992. Evolution of Pillar Mining at the Pasminco Mining Broken Hill Operations Southern Operations. *Proceedings of the Australian Institute of Mining and Metallurgy Underground Operators Conference.*

Rauert, N.S. 1995. The Development of the Rib-Roc Cable Bolt Reinforcement System At the Pasminco Mining Broken Hill Operations. *Proceedings of the Australian Institute of Mining and Metallurgy Underground Operators Conference.*

Rauert, N.S., and Tully, K.P. 1998. Integration of a Microseismic Monitoring System in mining the Pasminco Broken Hill Southern Cross area. *Proceedings of the Australian Institute of Mining and Metallurgy Underground Operators Conference.*

Smith, I.K., and Spreadborough, M.J. 1993. Underground Mining at Pasminco Limiteds Broken Hill Southern Operations into the 21st Century. *Proceedings of the Australian Institute of Mining and Metallurgy Annual Conference.*

CHAPTER 37

Rock Support in Cut-and-Fill Mining at the Kristineberg Mine

Norbert Krauland,* Per-Ivar Marklund,† and Mark Board‡

37.1 BACKGROUND

Several cut-and-fill mines in Sweden experience severe ground conditions, even at shallow depths. In several operations, mining depth currently exceeds 1,000 m. Ground control problems, including roof collapse and wall slabbing, arise from a combination of variable rock quality and high in situ stresses. In Boliden AB's Kristineberg Mine, stoping conditions became so poor at a depth of 600 m that ore production declined from 450,000 to 370,000 tonnes annually in the middle of the 1980s.

In response to this situation, the following actions were taken: (1) immediate remedial measures to restore production and (2) investigations of the feasibility of cut-and-fill mining at greater depth under these conditions. These actions were carried out within the framework of G2000, a 5-yr research and development program of the Swedish mining industry. The project "Mining at Depth" was initiated with the objectives of (1) examining the basic question of the viability of cut-and-fill mining at depth, (2) predicting mining conditions at greater depth, (3) developing improved strategies for mining at depth, (4) estimating costs and production rates for these new strategies, and (5) estimating the ultimate extraction ratio and identifying methods for the recovery of sill pillars.

The Kristineberg Mine of Boliden AB was chosen as a test mine for this project because of the severe ground conditions that the mine was experiencing and because of the large amounts of observation and monitoring data that were available. The design considerations and practices for stope support that evolved during and after the project are the subject of this paper.

37.2 KRISTINEBERG MINE

The Kristineberg Mine is located in northern Sweden, approximately 130 km west of Skellefteå. The mine is owned and operated by Boliden AB and has an annual production (1999) of 550,000 tonnes.

37.2.1 Geology

The ore zone is a typical vein structure that strikes east-west with dip varying between 45° and 80°. The host rock is a schistose sericitic quartzite. The rocks immediately adjacent to the ore body are highly altered, often to very weak, talcy sericitic schists that vary in thickness between 0 and 3 m. Near the footwall plunge of the ore body, the footwall consists of chloritic schists, which often are heavily disturbed tectonically. The contacts between the ore body and the alteration zone are commonly planar, low-friction coatings. In the footwall contact, clay in-fillings up to 5 cm thick are often present. The hanging wall contact often consists of chlorite interlayered with pyrite with a thickness of up to 0.5 m.

TABLE 37.1 Rock mass classification

Rock unit	Low rating	High rating
Chlorite schist	0.1<Q<1 = very poor	1<Q<5 = poor
Sericitic quartzite	0.1<Q<1 = very poor	5<Q<10 = fair
Ore*	0.1<Q<1 = very poor	10<Q<50 = good

*Ore shows highly variable Q-rating because of frequent inclusions of chloritic schist within the ore body.

The ore bodies themselves consist of pyrite mixed with chalcopyrite and galena. The rock strength of both wall rocks and ore decreases from the hanging wall plunge toward the footwall plunge as a result of the greater geologic disturbance caused by folding and faulting. This disturbance has produced inclusions of chloritic schist and mylonite within the ore body that trend subparallel to dip.

A rock mass classification for the rocks at Kristineberg based on the Q-system is given in Table 37.1.

37.2.2 Mining Method

Cut-and-fill is the primary mining method. The ore body is developed with a series of crosscuts driven from a footwall ramp. From the crosscuts, stopes are mined in 5-m-high slices by breasting using drill jumbos and load-haul-dump (LHD) machines. After a stope has been mined out, development rock is placed in the stope to the extent it is available, followed by backfilling of the remaining volume with hydraulic tailings. When the ore body is more than 8 m wide, a drift-and-fill method is used. Mining is currently taking place from below 900 m to a depth of 1,150 m. In general, several stopes are mined simultaneously at different levels. The interval between levels (Figure 37.1) that was in use at the start of the research program varied between 70 and 100 m, but has been reduced to 50 m since then.

37.2.3 Ground Control Problems

The general failure patterns and ground stability problems at the Kristineberg Mine are closely related to the mining geometry and geological conditions. The backfilled stope constitutes a steeply dipping slot in a stress field with dominating horizontal compressive stresses. As a consequence, the ore in the roof of the stope is subjected to large horizontal stresses. This situation results in failure of both roof and sidewall in a pattern that was fairly common throughout the mine at the start of the project (Figure 37.2), that is, (1) slabs up to 0.5 m thick are formed parallel to the hanging wall, (2) intense stress-induced slabbing develops in the roof parallel to the free surface and, in combination with failure of the chloritic layer in the hanging

* Boliden Mining (retired), Boliden, Sweden.
† Boliden Mining, Boliden, Sweden.
‡ Itasca Consulting Group, Inc., Minneapolis, MN.

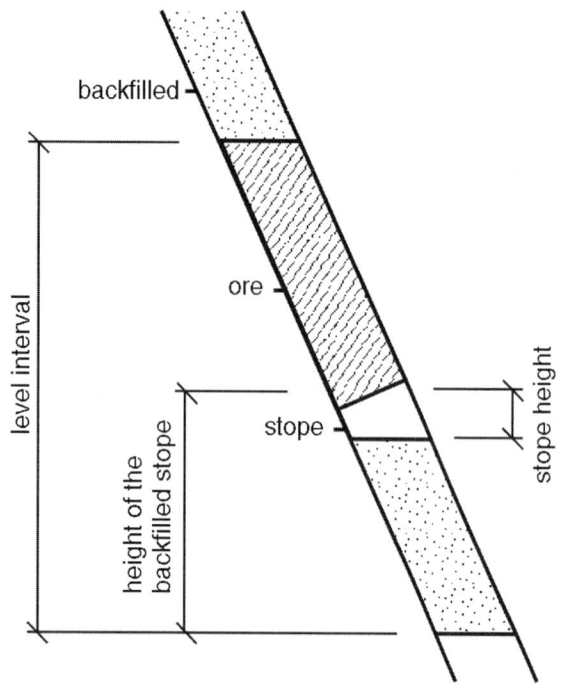

FIGURE 37.1   Cut-and-fill mining geometry and definition of terms

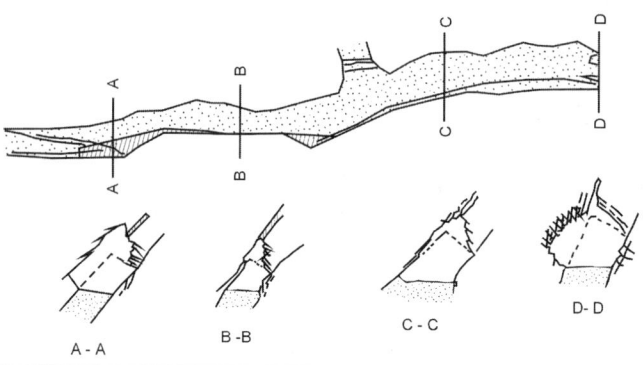

FIGURE 37.2   Failure observations typical for Kristineberg stopes (dashed lines indicate planned profile)

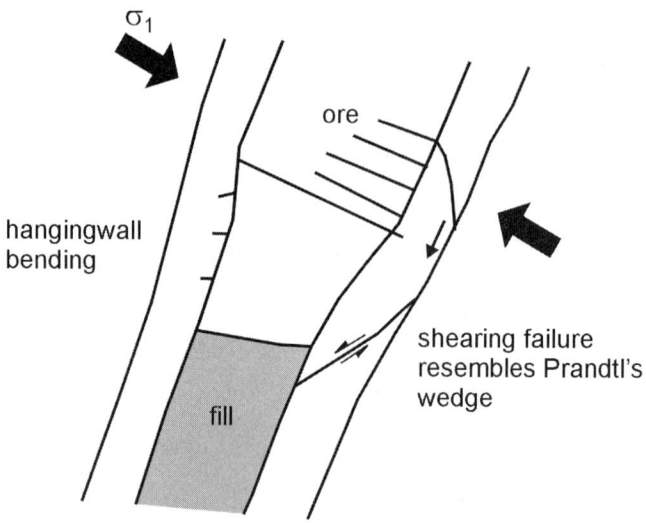

FIGURE 37.3   Schematic representation of failure mechanism showing stress fracturing of pillar and punching into the footwall

wall, sometimes creates a nearly vertical roof or high angle, and (3) slip is often observed along the footwall contact. Where the footwall consists of chlorite, deep-seated shear failure and disintegration of the footwall are common. The failure of the footwall induced by pillar punching results in shear on the contact and downward movements in the ore near the footwall. Eventually, the drag induced by the footwall pulls apart the roof along the horizontal stress fractures, resulting in roof falls that run up along the contact.

Where faults or other zones of weakness are present in the footwall, major roof falls may occur, as illustrated in Figure 37.2. These conditions led to severe production interruptions in the past, culminating in 1988.

37.3   IMMEDIATE REMEDIAL ACTION

In the past, the stopes were reinforced with fully grouted rebar bolts installed in a regular bolting pattern in the roof, but more widely spaced in the walls. Because of the dramatic increase in ground control problems, the mine embarked on a program of systematic bolting using fully grouted 20-mm in diameter rebar bolts on 1-m centers in the roof and walls, plus 5-cm, steel-fiber-

reinforced shotcrete where necessary. The term "systematic" refers to enforcement of a specified bolt pattern in which the separation distance at the ends of the bolts, as well as the collar spacings, are controlled. The shotcrete was primarily used at crosscut-stope intersections or in areas with particularly weak or raveling ground. At the same time, cable bolting of the roof with two cables (15 mm in diameter) per hole was introduced.

Specific support plans (on a stope-by-stope basis) were also introduced. These plans were based on experience gained during mining of the previous slice by mapping geology, roof and sidewall failure, and previously installed support. By applying these measures, control over ground conditions was regained; in general, stopes could be mined to the planned cross section.

37.4   SUPPORT CONSIDERATIONS

37.4.1   Technical Considerations

Loading Conditions Around Cut-and-Fill Stopes. The stress field to which a cut-and-fill stope is subjected not only depends on depth below the surface, but on the mining situation. Usually, cut-and-fill mining is carried out concurrently at several levels in a mine. As the height of the mined-out and backfilled stope increases, the stresses rise in the remaining ore pillar. When these reach the strength of the ore (or sidewall), failure occurs, and support is required to maintain stability. With continued mining, loading on the remaining ore increases, necessitating further support measures. When mining approaches the backfilled stope above, the remaining ore forms a sill pillar. At this stage, the stresses are high and increase rapidly with continued extraction. This usually results in either failure of the ore pillar, punching of the pillar into the sidewall, or both.

Geotechnical Considerations. *Failure Mechanism.* Field investigations in stope 804, carried out after control over ground conditions was regained, included rock failure mapping and deformation monitoring using extensometers and photogrammetry (Board et al. 1992). The proposed failure mechanism derived from these observations is shown in Figure 37.3. The high horizontal stresses cause the roof to "punch" into the footwall, the weak chlorite zone in the footwall is squeezed downward, and sliding occurs along the chlorite-country rock interface, which often consists of a clay seam. The high lateral stresses cause typical stress-induced, back-parallel fractures to form. Because of frictional drag along the ore-footwall contact, the roof tends to pull apart on the parallel fractures in the roof. This, in turn, results in propagation of roof falls that run updip. Chlorite zones within the ore have a similar effect on the adjacent ore. The failure in the roof progresses upward as the chlorite zone is squeezed out until equilibrium is achieved. The hanging wall material is somewhat stronger than the footwall material and tends to deform in a bending mode, with dilational separation occurring between the host rock and the schist or along the foliation of the schist itself.

Function of Support. The field investigations showed that the stope was surrounded by failed rock. Failure of the ore extends to more than 10 m above the roof. The alteration zones in both the footwall and hanging wall are in states of failure. Under these conditions, it is not feasible to support the failed ground by suspending it with bolts anchored in intact rock beyond the failure zone. Instead, the approach was to reinforce the fractured ground by means of rock bolts, thus creating a load-bearing structure within the failure zone itself. This structure provides support for the failed rock outside the bolted zone.

The load-bearing capability of bolted structures consisting of fractured rock was investigated by Lang (1961). Applying this approach to mining situations, the load-bearing capacity for different bolting patterns was evaluated to provide guidelines for the selection of bolt length and spacing (Krauland 1983).

Choice of Support Elements. *Rock Bolts.* The load-bearing capacity of bolted and fractured rock depends on the strength of the rock material and on the friction mobilized between rock blocks. Therefore, it is imperative that large displacements, shearing, or opening of joints not be allowed in order to maintain the frictional resistance that includes the interlocking of asperities.

Fully grouted, untensioned rebar bolts are the standard bolts in the Boliden mines. They are well suited for bolting fractured rock because of their high stiffnesses. The full nominal load is attained by only 0.2% elongation of the bolt (or 4 mm for a 2-m bolt), resulting in rapid development of confining load to the fractured mass around the excavation.

Steel-Fiber-Reinforced Shotcrete. The function of shotcrete in the cut-and-fill stopes in Kristineberg is to prevent raveling of the failed rock between the rock bolts, thus preventing a loss in the support function of the bolts. This support function is especially important in the weak sidewalls, which tend to break into small blocks. As there is practically no adhesion of the shotcrete to the weak talc and chlorites in the sidewalls, the shotcrete has to be anchored against the sidewalls by the rock bolts. This is achieved by plating the bolts on the outer surface of the shotcrete. In this application, the action of the shotcrete is similar to that of a membrane, which requires high tensile strength of the shotcrete. The shotcrete must also sustain a large amount of deformation; hence, the addition of steel fibers is necessary.

Cable Bolts. The function of long cable bolts installed in the roof in cut-and-fill mining is still a matter of discussion. The hypothesis that preinstalled support would decrease the amount of deformation and/or failure could not be verified, either in field observations or in modeling. However, evaluation of the effects of cable bolting at the Renström Mine showed that the incidence of damage to mining equipment, especially drilling jumbos, by falling rock decreased substantially. The savings in repair costs more than offset the cost for cable bolting. Thus, it is assumed that the main effect of the cable bolts is to provide support to the newly exposed roof skin immediately after blasting, but before the primary support is installed, thus allowing longer rounds than otherwise would be possible.

Originally, it was planned to use 22-m-long cable bolts. This would allow mining of four cuts with 6-m-long cables left in the roof of the last cut before new cable bolts would have to be installed. However, experience showed that, on average, only 15-m-long bolts could be installed because of drilling and installation difficulties.

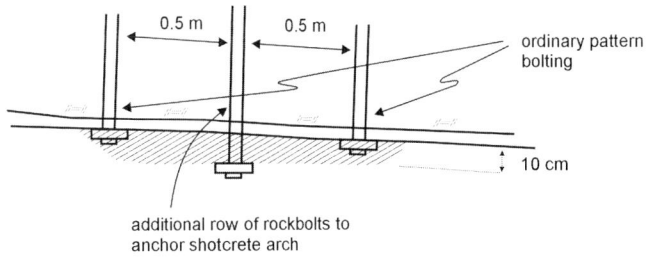

FIGURE 37.4   Cross section of shotcrete arch

Cable bolts are placed in the ore only. Attempts to fan out cable bolts into the sidewalls were unsuccessful, as the boreholes failed (to the extent they could be drilled) before cables could be installed.

Shotcrete Arches. Shotcrete arches are created by increasing the thickness of the steel-fiber-reinforced shotcrete by 10 cm over a length of 1 m along the stope, as shown in Figure 37.4. The arch is anchored by a row of rock bolts. The plates of these rock bolts are placed on the surface of the shotcrete.

The function of the shotcrete arches is similar to that of steel arches; that is, they provide additional local support in a cross section. Usually, they are placed at irregular intervals as needed. Shotcrete arches are always used as support for forepoling, where it is used.

Forepoling. As shown in Figure 37.5, forepoling is used as additional support for the roof in the intersection of a crosscut and stope to ensure the stability of the intersection during the lifetime of the stope. Rebar bolts 25 mm in diameter of required length (up to 10 m) are grouted into boreholes with spacings of approximately 0.5 m. The holes are drilled from the crosscut across the future stope into its hanging wall. In the crosscut, the forepoles are supported by a shotcrete arch. Forepoling is also used wherever exceptionally poor rock conditions are encountered.

37.4.2   Economic Considerations

The purpose of support is to maintain stability of the stopes (that is, to avoid disturbances that impede production). Thus, there is also a benefit side for support, namely, the avoided cost of a disturbance. It stands to reason that, by increasing the amount of primary support (support that is installed at the face), the amount of disturbance caused by stability problems will decrease. From an economic point of view, the optimum amount of rock support is achieved when the sum of support costs and disturbance costs is a minimum. This relation is shown schematically in Figure 37.6.

Both the costs for rock support installed at the face and the costs for operational disturbances depend on the amount of rock support and the operational and economic conditions in the mine. To obtain a realistic picture of costs, the total cost to the mine (not only the costs in the stope) needs to be considered. These include such items as relocation or creation of new working faces and the cost of production losses.

- Costs for support installed *at the face* include the costs for scaling, rock bolting, shotcreting, etc., and the cost for the time the stope is not available for production because of support installation.

- Costs for disturbances depend on the type and extent of the disturbance. Disturbances may range from scaling for safety, complementary bolting, or shotcreting, to the loss of equipment, extensive roof falls, loss of stopes, development of replacement stopes, and loss of production. In addition, the costs for nonavailability of stopes, equipment, and crew have to be considered.

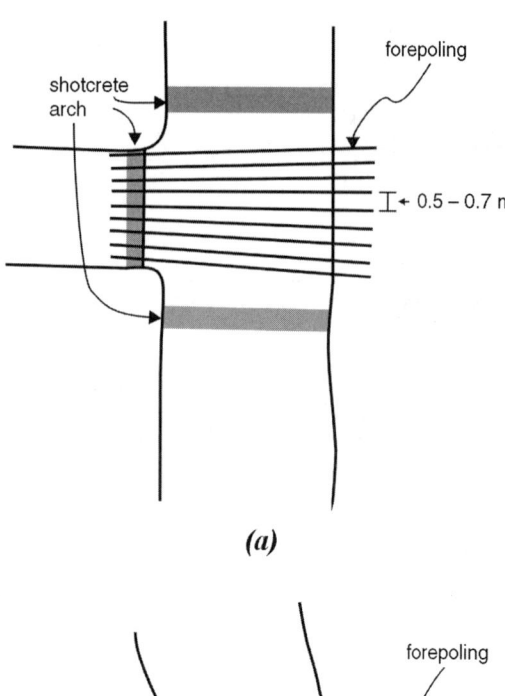

(a)

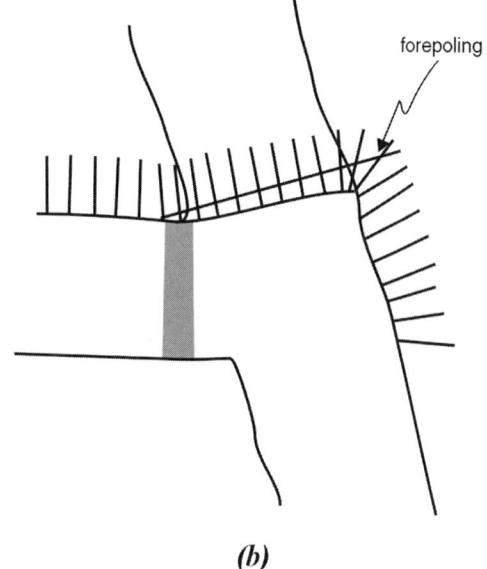

(b)

FIGURE 37.5   Forepoling and shotcrete arches for stope entrance; a, plan view; b, cross section

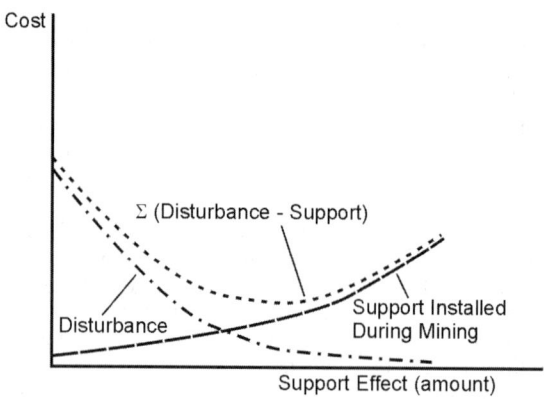

FIGURE 37.6   Schematic illustration of optimizing rock support economics. Plot shows optimum amount of ground support that will minimize the overall mining cost that includes the effect of failures (disturbance).

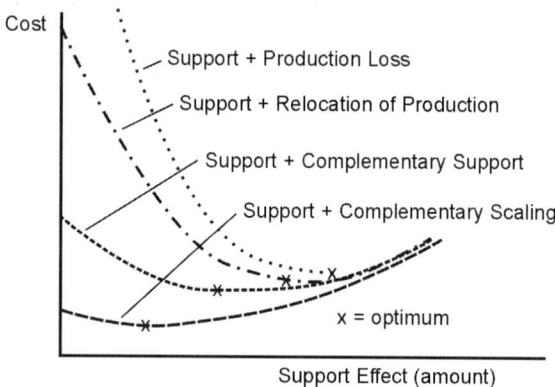

FIGURE 37.7   Dependence of economic optimum rock support on type of disturbance

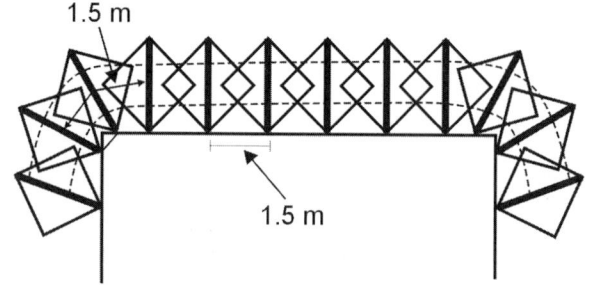

FIGURE 37.8   Geometry of rock bolt placement around corners to ensure interaction between rock bolts

Depending on the type of disturbance, costs will vary greatly. Figure 37.7 shows schematically the curves for the sum of costs for support at the face and the costs for different types of disturbances. If the cost for disturbances is high, increasing the intensity of primary support becomes economically reasonable. This implies that the most economic support is not governed by technical conditions only, but also by the economic risks associated with an operational disturbance.

One example for this consideration is the stope entrance; the stability of the entrance is critical for mining the whole slice. Experience has shown that the support required to maintain stable conditions and the first rounds is considerably less than the support required to maintain safe conditions at the stope entrance during the lifetime of the stope. This is because of the time-dependent deformation that occurs continuously. The installation of secondary support at a later stage is inefficient and implies disruption of stope production. It is far more efficient to install the total support required to maintain stability of the entrance for the lifetime of the stope when the entrance is mined.

37.5   SUPPORT DESIGN

37.5.1   Estimate of Required Support

Rock Bolting. *Geometry.* The purpose of bolting is to create a continuous load-bearing structure around the whole stope. This requires that interaction of hanging wall, roof, and footwall be ensured to achieve the required interaction between the bolts, including those in the corners between hanging wall and roof, and roof and footwall. The spacing of the bolts is measured at half-length of the bolts, as shown in Figure 37.8.

TABLE 37.2   Support required under different mining conditions determined from experience at Kristineberg Mine

| | | Support required to maintain stability | | |
Stage	Observed mining conditions	Rock bolts	Cable bolts in roof	Shotcrete
0	Surface spalling, support of single block	Spot bolting. Roof—1.2 to 1.4 m. Footwall >2.5 m. H. wall 1.5 to 2.5 m.		
1	Spalling in roof and sidewalls, time-dependent behavior	Systematic bolting. Roof and sidewalls— 1.5 by 1.5 m.	3 by 2 m	Thickness of roof and sidewalls >5 cm
2	Punching failure in sidewalls resulting in extensive roof failure by splitting	Systematic bolting. Roof and sidewalls— 1 by 1 m.	3 by 2 m	Thickness of roof and sidewalls >5 cm
3	Sill pillar stage			

Rock bolts = grouted rebar, 20 mm in diameter, 2.7 m long. Cable bolts = seven-wire strand, 15.2 mm in diameter, 20 m long, two cables per hole. Shotcrete = steel-fiber reinforced, 37 to 60 kg/m^3.

Bolt Length and Bolt Spacing. The bolting density previously applied corresponded to approximately 1.5- to 2-m spacings; this corresponded to a confining load to the rock surface of approximately 4 to 7 tonne/m^2. The potential confining load applied by the 1- by 1-m bolting pattern is about 15 tonne/m^2, a substantial improvement.

The load-bearing capacity of pattern-bolted rock structures is strongly dependent on the interaction between rock bolts. Guidelines for the selection of bolt length and spacing show that interaction requires $s < l$, where s is bolt spacing and l is bolt length. For the bolt length used in Kristineberg (l = 2.7 m), interaction is good for $s = 1$ m, but poor for $s = 1.5$ to 2 m.

37.5.2   Modeling of Support

Numerical modeling of the action of the ground support was undertaken in an attempt to understand the basic action of these supports under the specific loading conditions at Kristineberg and for the optimization of the support. Two types of numerical models were used to model support: (1) a large-scale elastic model to investigate, using a simple empirical approach, the required support as the height of the stope increased, as well as to optimize level spacings and (2) a detailed, geologically based model to study the interaction between support and the failed rock surrounding the stope.

Large-Scale Models. A large mine-scale model was used to predict the required support as the height of the stope increased and approached the sill pillar condition. A simple elastic model was used to predict stress concentrations at the stope as it is advanced updip, creating a sill pillar. These calculated stress levels were calibrated against past observations of failure intensity and the required support levels as derived from mining experience. Thus, a simple empirical rule relating stress level, stope height (pillar width), and expected failure and support level was constructed. This simplistic approach was found to work reasonably well to estimate past ground conditions (Borg 1983; Nyström et al. 1995). This approach implied that, for geologic conditions that are relatively constant, calibration against stress condition alone was reasonable for mine-scale planning purposes. This simulation was carried out for a mining depth down to 1,200 m.

The mining conditions and the corresponding support requirements were collected and classified into four stages as shown in Table 37.2. At depths below 1,000 m, most mining occurs under stage 1 and stage 2 conditions; stage 0 is not applicable.

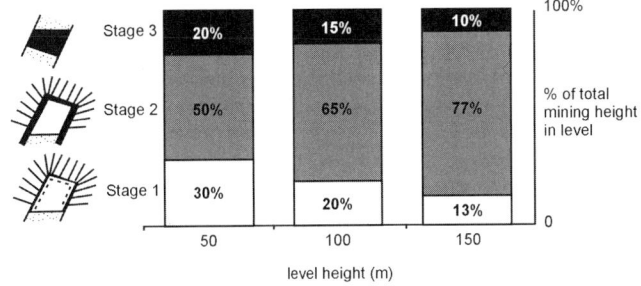

FIGURE 37.9   Predicted mining conditions as function of level interval. Each of the level intervals show the percentage of that level subject to a given support type. For example, for a 50-m interval, it is predicted that the initial 30% of the stope will require stage 1 support, 50% will require stage 3 support, and 20% sill-pillar conditions.

The effect of level height was examined by determining the percentage of a given level that would be subjected to stress conditions resulting in the various stages of ground support given in Table 37.2. The level height was optimized by minimizing the percentage of a level subject to higher support levels (thus minimizing costs). Level heights of 50, 100, and 150 m were examined for mining between depths of 930 and 1,200 m (Figure 37.9).

Figure 37.9 shows that an interval of 50 m gives the most favorable mining conditions and support costs. (The support costs for stage 2 are three times as high as those for stage 1.) The mining costs for the 50-m-interval are predicted to be 5% lower than for the 100-m-interval. The 50-m-interval allows a substantially higher production capacity as well, because of the increased potential for working faces. Thus, this simple calibrated modeling approach was used as a basis for decreasing level intervals with a resulting predicted decrease in support costs and increase in productivity. As discussed later, these predictions were, in fact, verified by mining results in the year following the change in level interval.

Detailed Model. A detailed, geologically based geo-mechanical model was developed to simulate the interaction between support and failing rock around a cut-and-fill stope. A test stope (804) was chosen for intensive instrumentation and observation for both model calibration and testing of various forms of rock support. The model of stope 804 was based on the two-dimensional distinct-element program UDEC (Itasca 1989). When the model was compared to observations and measurements of roof and wall deformation as the face was advanced, it was found to reproduce the failure mechanism observed in stope 804 (Figure 37.3) and the supporting action of fully grouted rock bolts.

The validity of the model was established by calibrating it against the deformation and failure phenomena observed underground. Figure 37.10 compares field observations, deformation, and failure modes, and the equivalent model output. As seen, the complex squeezing action of the footwall schist, bending of the hanging wall, and roof fracturing are reproduced reasonably well. A detailed description of the model and its calibration is given by Board et al. (1992) and Rosengren et al. (1992).

Once the validity of the model was established, it was used to investigate (1) various ground support options at the current mining level, (2) the effects of mining the remaining 40 m of ore above stope 804, including the sill pillar, slice by slice up to the backfilled stope above, and (3) the same mining steps repeated for future stopes at a depth of 1,000 m.

Figure 37.11 shows bolt axial loads developed in the reinforcement. Nearly all of the bolts are loaded to peak axial

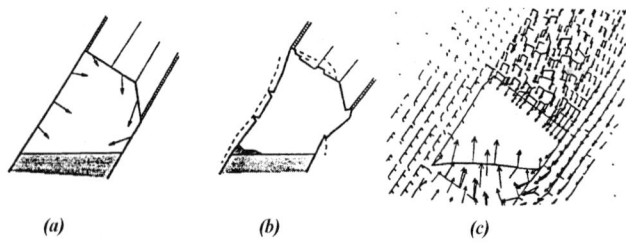

FIGURE 37.10   Comparison of field deformation measurements (a) and final stope appearance (b) to predicted model deformation mechanism (c). Note: Model predicts back-parallel fracturing, dragging-down of footwall-roof contact because of squeezing of weak footwall schist.

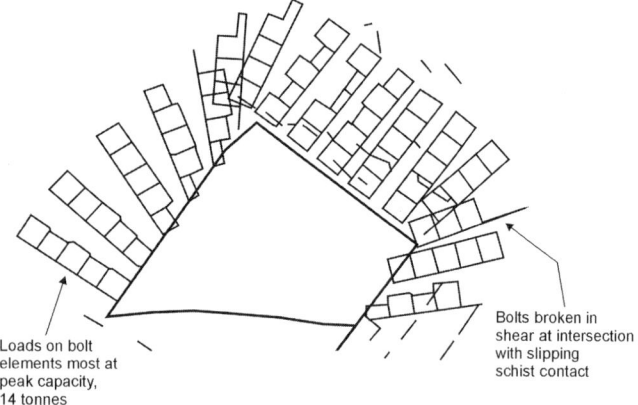

FIGURE 37.11   Axial forces in rock bolt elements for base case of reinforcement

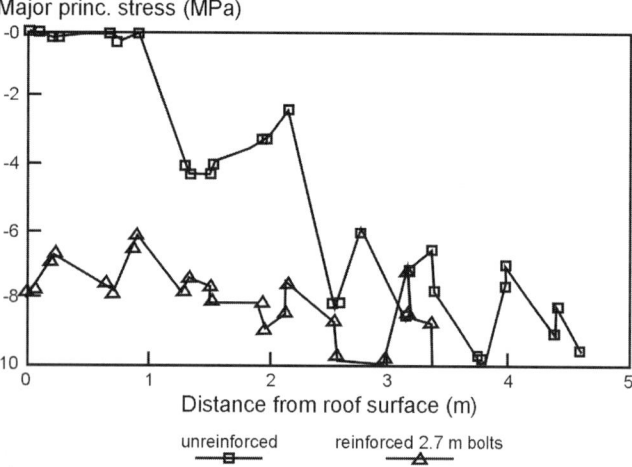

FIGURE 37.12   Predicted major principal stress with distance above roof for reinforced and unreinforced case. The model predicts a significant confining effect on the roof due to bolting in the first 2.5 m of the back. The confinement results in higher roof parallel (major) stress.

loads, with maximum axial strains of approximately 7%. (The extension limit for the bolts is approximately 10%.)

The stabilizing effect of the rock bolts developed in the model can be seen in the roof stresses plotted in Figure 37.12. The bolts provide a confining stress in the first 3 m of the roof, resulting in sustaining a roof-parallel stress of roughly 5 to 8 MPa. In contrast,

TABLE 37.3   Parametric variations of bolting patterns

Run	Description
1	Base case (current practice). Fully grouted rebar bolts, diameter = 20 mm, length = 2.7 m, 1-m spacing in roof and sidewalls
2	Unreinforced
3	Roof unreinforced; hanging wall and footwall reinforced as in 1
4	Footwall unreinforced; hanging wall and roof reinforced as in 1
5	Fully grouted rebar bolts; diameter = 20 mm, length = 2.7 m, 2-m spacings
6	Fully grouted rebar bolts; diameter = 20 mm, length = 3.5 m, 1-m spacings
7	Fully grouted rebar bolts; diameter = 20 mm, length = 2.3 m, 1-m spacings
8	Fully grouted rebar bolts; diameter = 25 mm, length = m, 1-m spacings
9	Diameter = 20 mm, length = 3.5 m, 1-m spacings; debonded in footwall chlorite schist, anchored in quartzite

an unreinforced model indicates a complete destressing of the immediate roof, followed by collapse of roof wedges. The amount of confinement to the footwall required to prevent squeezing out of the schist and dragging down the roof is quite small and can be provided by a dense pattern of bolting the walls.

Variation of Ground Support Options. The aim of the model parametric study was to predict rock mass and support performance for different support alternatives. The nine parameter variations shown in Table 37.3 were analyzed. These variations consist of two groups: (1) need for support of hanging wall, roof, and footwall and (2) the effect of bolt length and bolt diameter. The main findings are summarized as follows:

- The base-case (patterned bolting in roof, walls, and shotcrete) allows stability to be reached. As a check on the reality of this model, the same model with only spot bolting was unable to reach stability.

- Failure to reinforce any one of either hanging wall, roof, or footwall resulted in instability. Thus, the concept that the hanging wall, roof, and footwall together form a load-bearing structure around the stope was confirmed. To achieve stability, a dense pattern of bolts is necessary.

- If the support intensity is decreased below the base case (1 m), the stope becomes unstable. Increasing support intensity by increasing bolt length and/or bolt diameter does not result in significant benefits, such as decreased deformation or rock failure.

These findings indicate that the primary factor controlling stability is the density of bolt coverage of the walls and roof. Bolt lengths or diameters are of lesser importance. The model indicated that the primary action of shotcrete is merely to eliminate raveling of rock from between bolts. The shotcrete provides little structural resistance in itself in this type of schistose, talcy rock, although the function of eliminating ravelling is critical for the action of the rock bolts.

Maintaining the integrity of the footwall is critical for the stability of the stope. This is achieved by bolting on 1-m spacings, resulting in a confining load of 15 tonne/m^2. Two-meter spacings (4 tonne/m^2) are insufficient to stabilize the footwall.

In the footwall, anchorage of the bolts within the chloritic schists, as opposed to the stronger quartzites, appears to have little impact, as the bolts do not act as suspension members. The primary action here is to maintain the integrity of the footwall, not to inhibit shear along the chlorite-quartzite contact.

Sill Pillar Mining.   The reduction of the level interval from 100 to 50 m implies that the percentage of the ore reserve that falls under the category "sill pillar stage" increases from 15% to 20%. In the past, sill pillars were considered to be unmineable

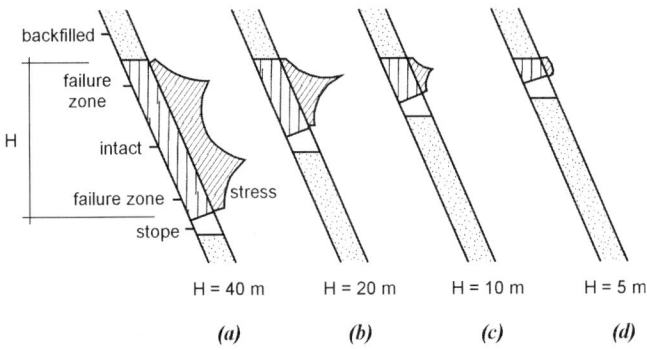

FIGURE 37.13 Extent of failure zone in sill pillar with decreasing height at 600-m depth: a, Failure zones resulting from current stope and backfilled stope above are separated by zone of intact ore; b, With continued extraction, lower failure zone approaches upper failure zone, stress levels peak, and mining conditions deteriorate, leading to c; d, Completely failed sill pillar with diminishing stress field and improved mining conditions.

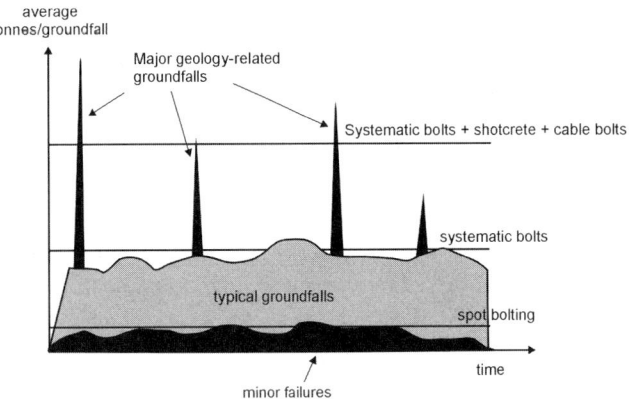

FIGURE 37.14 Schematic illustrating the frequency of large and small ground falls and the effectiveness of various support methods in stabilizing them. Note that systematic bolting will take care of "typical" ground falls. Standard support will stabilize most large potential ground falls that result in the greatest safety risks and costs

because of the extensive stability problems experienced. Because it is important to provide complete extraction of the available ore reserve, the detailed model was used to investigate the problems associated with sill pillar mining through a simulation of extraction of the remaining 40 m of ore of the standard 100-m-high level slice by slice up to the backfilled stope above.

The modeling results are summarized in Figure 37.13. Above the advancing stope, a failed zone approximately 10 m high is formed as a result of the squeezing action of the footwall. Similarly, the bottom cut of the stope above is surrounded by a failed zone. Between these failure zones, the ore body is confined and remains elastic and highly stressed. As the height of the pillar is reduced to 20 m, the two failure zones intersect; that is, the whole sill pillar moves into a failed condition. At this stage, mining conditions become very poor, with large amounts of deformation of the weak hanging wall and footwall schists as the pillar is punched into them. In the past, this stage was seldom reached because mining was stopped because of ground conditions resulting from inadequate support. However, the numerical analysis indicated that the stresses carried by the pillar are redistributed to the surrounding rock mass once the whole pillar is in a state of failure. As more slices are taken, pillar-punching is predicted to largely stop as a result of the lack of driving stresses. Thus, the predictions indicated that there should be a marked improvement in mining conditions once mining has passed this critical sill height.

These model results suggest that complete extraction of the sill pillars should be feasible provided that the critical mining stage at a pillar height of about 20 m can be handled efficiently. Special support measures, described later, are required during mining of the last slice under the uncemented fill to keep the fill in place.

Similar results were obtained for a mining depth of 1,000 m, but the extent of the failure zone in the roof increased to 15 m. Thus, the critical mining conditions occur when the height of the sill pillar is 30 m. Deformation increases somewhat, but stability can still be obtained with stage 2 support.

The development of mining conditions described above confirms that most of the ground conditions to be expected at depth and during the mining of sill pillars can be controlled by the use of stage 1 and stage 2 support.

37.5.3 Variability of Mining Conditions and Support

Figure 37.14 illustrates schematically the size of ground falls as a function of time and shows that there is a continuous level of small ground falls. The normal prevention method is spot bolting.

Next, a number of larger ground falls occurs due to the type of failure mechanism studied in the numerical models. These types of ground problems are handled efficiently by systematic support measures of types stage 1 and stage 2. However, there are still occasional large rock falls, even in systematically supported stopes. These ground falls are caused by conditions that deviate significantly from those represented by the models. The conditions are assumed to be caused by sudden changes in ore and/or sidewall geometry or rock strength, mostly resulting from tectonic activity such as faulting and folding.

By carefully mapping the mined-out stope (with attention to problem areas such as geologic and stress-related phenomena, mining conditions, and support), trouble spots for the next slice can be predicted. A support plan can then be developed that incorporates additional measures for the expected trouble spots. Tests carried out with such stope-support plans gave encouraging results, and it should be possible to reduce the frequency of major ground falls significantly.

37.6 PRESENT SUPPORT TECHNOLOGY AND MINING CONDITIONS

37.6.1 Support Technology and Installation

Since both modeling and field trials indicated that it is not possible to reduce the amount of support, it became necessary to lessen the unit cost of support and to increase the speed of installation. Therefore, rock support was mechanized as described below.

Rock Bolting. A bolt-setting rig has been developed for Kristineberg. After the holes for the rebar bolts are drilled by the jumbo in conjunction with round drilling, the "bolt setter" fills the holes with grout and installs the bolts.

Shotcrete. The cost of shotcrete has been reduced by 25%, and capacity has been doubled by means of an underground shotcrete plant and effective shotcreting equipment.

Cable Bolting. A cable bolting machine has been acquired. The "cable bolter" drills the holes, pumps grout into the holes, and installs cables.

Support Installation Procedure. Shotcrete is applied in two layers, each typically 3 cm thick. The first layer is applied at the face *before* the bolts are installed. At the same time, the second layer is sprayed on the *previous* round.

The combination of all these measures resulted in increased production capacity and reliability. The number of rounds per face and per week has increased from 1 to 2 to 4 rounds.

37.6.2 Present Mining and Support Practice

At the start of the research program, most of the stopes were at a depth of 600 m below the surface. Today, production comes from a depth of 900 to 1,150 m below the surface. Using the recommendations based on large-scale modeling, the interval between levels has been reduced to 50 m so that the amount of support required is reduced and production capacity is increased. On the basis of production capacity requirements, the height of the slices has been increased from 4 to 5 m.

With regard to stope support, 2.7-m-long rebar bolts are installed on a 1- by 1-m pattern as standard practice throughout the mine. Shotcrete is used in most stopes to a larger extent than predicted by modeling. Thus, stage 1 support is no longer used.

The mine has decreased the use of cable bolts and increased the use of shotcrete arches. The introduction of shotcrete arches and forepoling at stope entrances has increased the reliability of this critical area significantly. Prior to the introduction of these support elements, the lifetime of the stope entrances was limited, and it was not always possible to mine the whole length of the ore body.

Support is now being planned on a continuous basis, and support plans for each slice are being introduced for the whole mine. As a result of support planning, the frequency of rock falls was reduced significantly, which contributes to increased production.

Sill Pillar Mining. The results given by the detailed models suggested that the most difficult mining conditions would be experienced when the height of the sill pillar was 15 to 20 m, but that conditions would improve as extraction progressed beyond this point. These predictions have been verified by practical mining experience. Sill pillar mining was first tested successfully by mining out a sill pillar at a depth of 600 m. At a sill pillar height of about 15 m, deformation in both sidewalls and roof was substantial, requiring support in excess of stage 2. Deformation decreased during the following cuts as the pillar unloaded, and mining of the sill pillar was completed successfully, resulting in 100% extraction of the level. In this test, however, support measures were extensive and time consuming. Subsequently, several sill pillars were mined out. Again, mining conditions improved when sill pillar height was decreased below 20 to 25 m. In most cases, 3 to 5 m of ore was left, as the fill in the stopes above was not stabilized with cement. In recent years, the mine has introduced mining of sill pillars by uppers on retreat when the sill pillar is 12 to 17 m high. An attempt is made to leave 3 to 5 m of ore as a support to the uncemented fill back.

37.6.3 Further Development of Support Technology

The following are areas of current research at Kristineberg Mine.

Rock Bolts. Fully grouted rebar bolts have performed well in general; however, when large amounts of deformation occur, their yield is insufficient. Cone bolts, developed in South Africa, are quite stiff until the yield load of about 12 to 15 tonnes is reached. During yielding, the cone is pulled through the grout. Cone bolts were tested both in the lab and in the Kristineberg Mine, with promising results so far.

Shotcrete. To date, anchorage of the shotcrete against the rock surface by the plates of the rock bolts has not been satisfactory, as only the first layer is anchored. Alternatives are currently being tested.

Cable Bolts. The standard cables (seven-strand, 15.2 mm in diameter) have poor deformation properties (5% failure strain) compared to the rebar rock bolts (17% failure strain). The large amount of deformation in the roof of the stope during the lifetime of a cable indicates that the capability to undergo larger strains without breakage is desirable.

37.7 CONCLUSIONS

Despite the difficult conditions encountered at depths of 600 m, it has been possible to continue mining successfully to the present mining depth of 1,100 m as a result of improved mining and support methods. The key to these developments was the improved understanding of rock failure processes around the stopes and the stabilizing action of support. This understanding was obtained both from numerical models calibrated against field observations and from mining experience.

The main findings of the Kristineberg research are summarized below.

1. From a technical perspective, horizontal cut-and-fill mining was predicted to be feasible to depths beyond 1,000 m in the Kristineberg Mine. Present experience from mining below 1,000 m confirms this prediction.

2. Geology plays a major role in determining mining conditions. The weak alteration zones in the sidewalls govern failure behavior around the stopes by reducing maximum stresses, but giving rise to large amounts of deformation caused by failure. Knowledge of these processes is essential for efficient support design. Supports must be capable of sustaining large amounts of deformation while maintaining peak support pressures without failing.

3. Dense support is necessary to maintain stability. Efficient, mechanized support technology is necessary to reduce costs and maintain reasonable production capacity.

4. The combined effect of changes in support strategy, more efficient support capability, and reduced level intervals has resulted in increased production reliability and capacity, with favorable consequences for mining costs (Nyström et al. 1995).

5. Modeling demonstrated the feasibility of sill pillar mining. The most difficult mining conditions arise when failure extends over the whole height of the sill pillar, which occurs at a sill pillar height of 15 to 30 m, depending on depth. Once mining has passed this critical height, conditions improve as a result of decreasing stresses and deformation. This prediction was confirmed by practical mining experience.

37.8 REFERENCES

Board, M., N. Krauland, S. Sandström, and L. Rosengren. 1992. Analysis of ground support methods at the Kristineberg Mine in Sweden. *Proceedings of the International Symposium on Rock Support (Sudbury, Ontario, Canada, June 1992)*. 499–506. P.K. Kaiser and D.R. McCreath, eds. Rotterdam: A.A. Balkema.

Borg, T. 1983. *Prediction of rock failures in mines with application to the Näsliden Mine in northern Sweden*. Doctoral thesis, 1983:26D. Luleå University, Sweden.

Itasca Consulting Group, Inc. 1989. *UDEC (Universal Distinct Element Code)*, Version 1.5. Minneapolis: ICG.

Krauland, N. 1983. Rock bolting and economy. *Proceedings of the International Symposium on Rock Bolting (Abisko, Sweden, 1983)*. 499–507. O. Stephansson, ed. Rotterdam: A.A. Balkema.

Lang, T.A. 1961. Theory and practice of rock bolting. *AIME Transactions*. 220:333348.

Nyström, A., S. Sandström, N. Krauland, and M. Board. 1995. Prediction of mining conditions at depth in the Kristineberg Mine. Trans. IMM, Section A, 104:A169–A177.

Rosengren, L., M. Board, N. Krauland, and S. Sandström. 1992. Numerical analysis of the effectiveness of reinforcement methods at the Kristineberg Mine in Sweden. *Proceedings of the International Symposium on Rock Support (Sudbury, Ontario, Canada, June 1992)*. 507-514. P.K. Kaiser and D.R. McCreath, eds. Rotterdam: A.A. Balkema.

CHAPTER 38

Underhand Cut-and-Fill Mining at the Murray Mine, Jerritt Canyon Joint Venture

Carl E. Brechtel,* Greg R. Struble,† and Benjamin Guenther‡

38.1 INTRODUCTION

The Murray Mine is one of several underground mines at the Jerritt Canyon Joint Venture (JCJV) located approximately 80 km (50 miles) north of Elko, NV, in the heart of the Independence Mountains. The JCJV is operated by Anglogold (Jerritt Canyon) Corporation, which holds a 70% interest; the remaining 30% of JCJV is held by Meridian Gold.

The JCJV is a complex of open-pit and underground mines that began production in 1981. Mining of near-surface ore bodies began in 1981 and was carried through until November of 1999, when the DASH open pit was completed. Underground mining projects were identified as mineralization trends traced out of existing open pits, and underground production began in 1993 with construction of the West Generator and the Murray mines. Currently, three producing underground mines–Murray, SSX, and MCE–sustain JCJV production. Total gold production, including stockpiles, is nominally 350,000 oz/yr.

Mechanized cut-and-fill mining is used at the Murray Mine to assure recovery of a high percentage of the gold ore found in fractured, faulted rocks of low rock quality. Variations of the mining method are practiced in different areas of the mine, depending on rock quality. However, underhand cut-and-fill is the predominant approach. Total ore and waste production averages 1,400 tons per day with 1,200 tons per day ore. Backfill must be placed at a nominal rate of 900 tons per day to sustain the ore production rate.

Achievement of high degrees of safety during the operation of the mine has a very high priority throughout all levels of the organization. Proactive safety practice is achieved by using a five-point safety program in all areas and activities within the mine. The use of the five-point system requires each underground operator to perform key inspections at each workplace entered for the first time during the shift. This is especially significant in mechanized mining because operators move throughout the mine performing their functions, rather than being assigned a permanent workplace. Lack of "ownership" of any particular workplace can create neglect of unsafe conditions without the requirement for workplace inspection.

38.2 GEOLOGY AND HYDROLOGY

The geologic setting at Jerritt Canyon is similar to that of the Carlin Trend of northern Nevada, with gold ore found in deposits hosted by sedimentary rocks. The gold mineralization occurs within the carbonaceous limestone units of the Hansen Creek and Roberts Mountain formations. Faulting and deformation of the host rocks in the Basin-and-Range Province created extensive fracturing in the host sediments, thereby forming conduits for mineralizing fluids. Ore body geometries are controlled by faults and stratigraphic contacts and range from large tabular zones to narrow, vein-like zones along structural features.

Mineralization at the Murray Mine occurs primarily along the New Deep Fault, a large structural zone dipping at roughly 65° to the north. Mineralizing fluids moved along conduits formed by generally northeast-southwest-striking normal faults that cross cut the New Deep Fault and pooled at impermeable stratigraphic contacts in the Hansen Creek and Roberts Mountain formations. The extensive fracturing and deformation associated with the New Deep Fault enhanced porosity for formation of the ore.

38.2.1 Hydrology

Grabens to the east and west of the Independence Mountains form the lows for ground water elevations in the region. Ground water tables rise underneath the mountains, where snowmelt provides the chief source of recharge. The ore body ranges from 6,300 ft above mean sea level to 5,995 feet and intersects the water table at approximately 6,100 feet.

The major effects of ground water during excavation below the water table were on ground control; that is, the effects of low rock quality and sheared rock conditions on excavation stability were much more pronounced in the presence of water. The rock matrix is relatively impermeable, and water flow is predominantly through fractures. Inflow from excavation surfaces is generally small and can be handled with air-powered pumps. However, to improve ground conditions in the lower portion of the mine, a dewatering system has been installed.

38.2.2 Geotechnical Data

Fracturing and deformation of rock along the New Deep Fault and at intersections with cross-cutting normal faults have created variable rock quality throughout the ore body. Conditions range from blocky rock to intensely sheared and deformed fault gouge, and local variability in rock quality is high. The occurrence of higher-grade ore is generally associated with the lower range of rock quality. Mean rock quality designations (RQD) for 15-ft vertical levels from core drilling ranges from 15% to 35%. Mean rock mass quality as described by the NGI-Q system ranges from 0.1 to 0.8. Rock compressive strength is relatively low, ranging from 3,500 to 10,000 psi.

Jerritt Canyon is within the Basin-and-Range Province, an extensional structural province. It has been assumed that the minimum horizontal in situ stress is oriented normal to the mountain ranges and is generated by overburden pressure and elastic effects. Measurements of stress were attempted at JCJV

---

* Chief engineer, Anglogold (Jerritt Canyon) Cooperation, Elko, NV.
† Production manager, Anglogold (Jerritt Canyon) Cooperation, Elko, NV.
‡ Senior vice-president/general manager, Anglogold (Jerritt Canyon) Cooperation, Elko, NV.

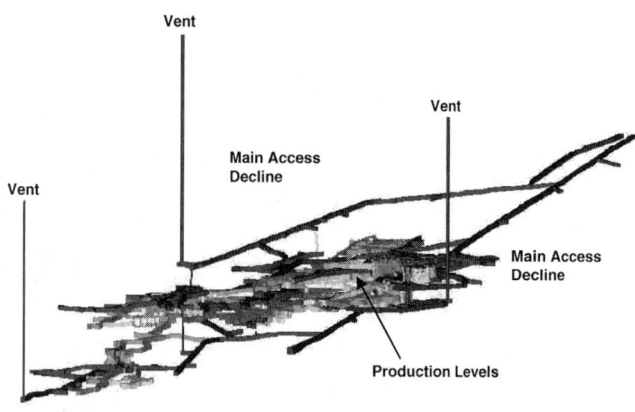

FIGURE 38.1 Isometric view of the Murray Mine showing main access drifts, ventilation boreholes, and production levels

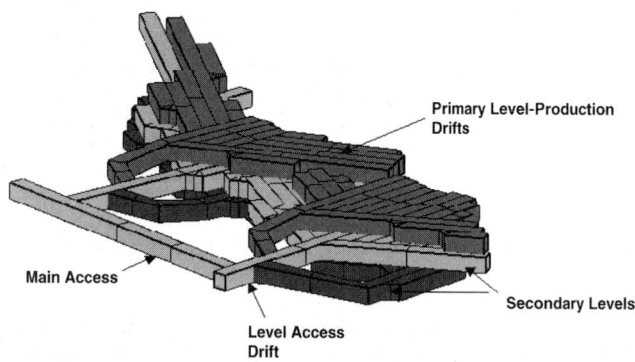

FIGURE 38.2 Isometric diagram of production areas showing primary level and underlying secondary levels to be mined under backfill capable of averaging 110 to 130 bolts per shift, with approximately 1.3 machines required to keep pace with production

but were unsuccessful because of the weak and fractured rock. This assumption is consistent with observations of gouge conditions observed in normal faults, which in many places appear to be undeformed, irregular debris.

38.3 MINING METHOD

All mining at the Murray Mine is performed using cut-and-fill techniques because of the relatively low rock mass quality. Initial production came from an area located well off the New Deep Fault, where conditions of higher rock quality allowed the application of an overhand sublevel stoping method. Zones located immediately along the fault are mined using underhand drift-and-fill mining.

Figure 38.1 shows an isometric view of the main access drifts and production levels throughout the mine. Higher rock quality in a zone off the fault allowed the formation of relatively large sublevel stopes 25 ft wide by 65 ft tall in overhand bench-and-fill mining. Underhand drift-and-fill mining is used in ore that is found immediately along the New Deep Fault.

Underhand drift-and-fill mining is illustrated in Figure 38.2, which shows an isometric view of three production levels. Ore production is accomplished in three different configurations.

- Primary level excavations where drifts are driven with rock forming the roof or back of the drift.

- Secondary level excavations where drifts are driven under backfill.

- Benches extracted on both primary and secondary levels where the roof may be either rock or backfill.

Ore is extracted in 15-ft levels. Primary extraction levels are developed every three to four levels, with the intervening levels then mined under high-strength, cemented rockfill.

38.3.1 Main Access

The mine is accessed by two portals excavated into the side of the mountain in the Jerritt Creek drainage. All main access drifts are excavated 15 ft wide by 15 ft high with an arched roof or back to improve excavation stability. The drifts are stabilized with a regular pattern of 7-ft-long Split-Set rock bolts installed through expanded metal mesh. Shotcrete is later applied to the access drifts out-of-cycle.

Production levels are accessed from declines that have been driven in the hanging wall of the ore. The declines have generally been developed with a 100-ft standoff distance from the ore. Ore access drifts are then driven off the declines at the primary production levels. Depending on the size of the production area, single- or multiple-access drifts may be driven into the ore. Secondary production levels may be accessed by separate drifts from the main declines or by taking up floor in existing primary access drifts.

38.3.2 Production Sequence

Production sequencing governs extraction rates and is a complicated part of the planning process. The rock in both the ore body and the surrounding country rock is black, carbonaceous limestone and silty limestones. Typically, there is no visible distinction between ore and waste, although the gold mineralization may be associated with the presence of realgar in some areas. Internal definition drilling is therefore required to establish the shape of the boundary of the ore body locally.

A central production drift is developed locally on each level to provide access for internal definition drilling, which is performed on 20-ft centers using a diesel tram, electrohydraulic percussion drill rig (Tamrock Solo). Holes are drilled with horizontal, inclined, and vertical orientations to define ore body shape for the current level and, to the extent possible, the next three underlying levels.

After definition drilling is completed, the final design of the level can be completed.

38.3.3 Primary Production Drifting

The height of the drift where the roof is formed by rock is generally limited to 15 ft, but has been as large as 20 ft in areas of higher rock quality. Drift orientations are established based on the shape of the ore body on each level. The most efficient orientation is parallel to the strike of the ore body, which allows the maximum heading length in each drift. Shorter drifts are locally developed off the long drives to maximize recovery of the ore and to minimize dilution at the boundary of the ore body.

Drift faces are drilled using diesel tram, electrohydraulic, two-boom jumbos (Atlas-Copco Boomers). Typically, three rounds must be drilled, loaded, and blasted during each shift to maintain production rates. Holes are drilled $1\frac{7}{8}$-in diameter using a burn cut. Round lengths are nominally 10 ft, but may be adjusted to 6 ft in rock of low quality.

Blasting is performed using emulsion explosives initiated by Nonel caps with boosters. An emulsion truck equipped with an articulated man-lift is used to transport and load the explosive. The emulsion is carried in a transport pot and placed into the hole by compressed air. The emulsion is metered into the hole using a timed cut-off.

Blasted muck is loaded and hauled using 6-yd^3 Wagner ST-6 load-haul-dump units (LHDs) teamed with EJC-430 haulage trucks with 25-ton capacities. Three 6-yd^3 LHDs are maintained to work with the fleet of four trucks. Trucks are loaded as close to the face as possible, with maximum LHD trams typically less than

100 ft. The truck and LHD crews work in two configurations, depending on the loading configuration of the specific production area. Typically, one loader operator will service two or three trucks working as a team. In areas that have been prepared for load-your-own operations, each truck operator will dismount from the truck and load it with the LHD.

Ground support, which consists of rock bolting, shotcreting, and backfilling, dominates the mining cycle in the primary levels. The primary rock support is 7-ft-long Split Set rock bolts installed on a minimum of 4-ft spacings and placed through expanded metal mesh. In areas of low rock mass quality, the installed bolt spacings can average 2 ft. All bolting is mechanized using two diesel tram, Altas-Copco Boltec machines. These machines are electro-hydraulic drill systems with six bolt carousels. The machines are capable of averaging 110 to 130 bolts per shift, with approximately 1.3 machines required to keep pace with production.

In many of the high-grade areas, the primary rock bolt support system must be augmented by the installation of up to 4 in of shotcrete. Shotcrete is typically installed off-cycle in drifts 30 to 60 ft long. The mine places approximately 500 yd^3 of wet shotcrete per month with a single crew of two or three people. The shotcrete is placed using a truck-mounted pump and spray unit with a boom-mounted spray nozzle. Two 4 yd^3 remix trucks transport the shotcrete mixture underground. The shotcrete is produced in the backfill batch plant using commercially available sand and pea gravel.

After a production drift has reached its designed length, it is backfilled with a high-strength cemented rockfill. Jamming the backfill tightly against the roof or back of the drift is a high priority to maintain excavation stability. Backfill strength averaged 720 psi during 1998. The backfill is produced from limestone aggregate mined and crushed on-site. Portland cement and Type F fly ash are added. A total of 6.5% cementious material is used in the mixture, with 20% of the cementious material being fly ash. A water-reducing admixture is used to improve strength characteristics.

Most of the fill is transported into the mine on the back haul during the production mucking operations. The fill is dumped at the fill face by haulage trucks and then jammed to the drift roof with a steel plate attached to an extension boom mounted on a 6-yd^3 LHD. Observations of the interface created at the roof suggest a zone of variable contact, with backfill touching the roof rock at some points and as much as 6 to 8 in away at other points.

When a production drift has been backfilled, an adjacent drift can be started within as little as 24 hours. Pillars are not generally formed because the relatively weak and fractured nature of the rock allows drift walls to slab when load is concentrated. This factor constrains mining by closely coupling the drifting and backfilling cycle in any local area.

38.3.4 Secondary Production Drifting

Mining of secondary levels (levels mined under backfill) can begin when a primary level is completed locally, and access can be established underneath the backfilled area. Productivity during mining under backfill is substantially higher because the rock bolting and shotcreting portions of the mining cycle are not generally required. The minimum delay between placement of backfill in a primary level and excavation of a production drift under the last backfill placed is 21 days. In practice, it is rare that a secondary level will undercut backfill that has been in place less than 3 months.

The backfilling operation is focused on maximizing the production advantages that occur underneath backfill by placing high-strength fill with good quality control and then taking advantage of the fill characteristics by increasing the span of production headings underneath the fill. Under-fill headings have nominal design spans of 21 ft but have been as large as 35 ft in

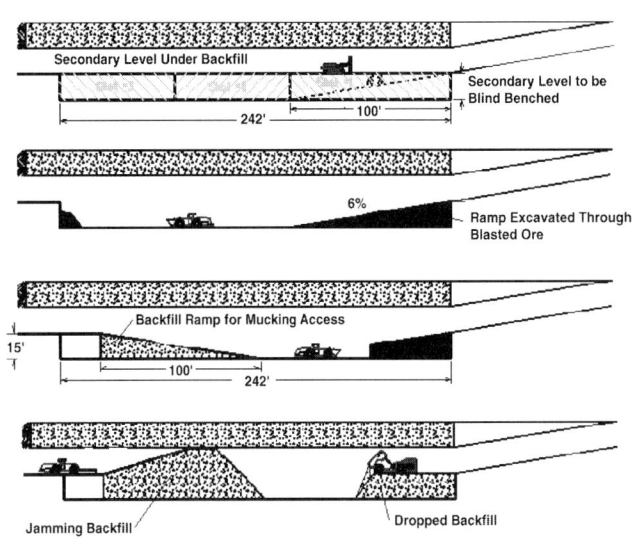

FIGURE 38.3   Cross-section views illustrating the sequence of blind benching under backfill with ramp access

specific headings. Engineering evaluations for increasing the allowable backfill span at the Murray Mine are described by Brechtel et al. (1999).

Production drifting operations underneath backfill are identical to the primary level operations with the exception that heading spans are larger and ground support installation is minimal. Rock bolting is performed only if fill quality is visibly below standards or in areas where the rock-fill interface appears to have the potential to cause instabilities. Shotcreting exposed backfill is performed if the exposed fill appears substandard or if a floor bench is to be extracted below the drift.

Backfill stability problems have been rare at the Murray Mine. In each instance, such problems have been related to the formation of cold joints because of delays in placement of fill. Operating procedures require that any backfill surface left exposed long enough to cure must be mucked to a vertical face to prevent the formation of inclined cold joints. Extensive excavation under the backfill has demonstrated that vertical interfaces between successive backfilled drifts are as strong as the backfill itself. Drifts under fill are driven parallel or normal to the overlying backfilled drifts with no stability problems due to cold joints.

38.3.5   Bench Production

Floor benching was initially performed in drifts under backfill, where the vertical loads on the bench walls had been removed by the previous mining of the primary level above. Low rock quality and weak rock conditions were expected to cause slabbing if wall heights were higher than the 15 ft produced in the primary production drifts. After successful bench excavations were created under backfill, the approach was extended into the primary levels where rock quality was suitable.

Productivity is further enhanced by increasing the total exposed wall heights to 30 ft by mining 15 ft of the floor of the production drifts. These blind benches are excavated in two configurations: access along the overlying drift by developing a ramp in the drift or access from the level at the floor of the bench elevation at the end opposite the overlying drift. Figure 38.3 shows a series of cross-sectional views to illustrate bench access from the overlying drift under backfill.

The floor benches ease production scheduling constraints by allowing access to additional tonnage without the need for further development drifting. Productivity is enhanced because

TABLE 38.1   Backfill dilution parameters estimated from survey data

Diluting surface	Equivalent thickness (ft)	Dilution (%)
Wall	0.43	—
Back	0.59	—
Floor	0.11	—
Total 1999	—	2.5

TABLE 38.2   Murray production and maintenance hourly manpower

Function	Activity	Total personnel
Operations	Production	55
	Shotcrete	3
	Surface support	2
	Roads and utilities	2
	Total operations	62
Maintenance	Breakdown	8
	Preventive maintenance	4
	Plant maintenance	6
	Electricians	4
	Total maintenance	22

TABLE 38.3   Productivity data for unit operations for 1999

Unit operation	1999 Average productivity	Coefficient of variation (%)
Drift drilling (tons/man-hour)	46.5	12.8
Sample and bench drilling (feet/man-hour)	20.2	20.4
Blasting (tons/man-hour)	48.3	14.7
Mucking (tons/man-hour)	22.2	11.8
Hauling (tons/man-hour)	24.9	6.1
Bolting (tons/man-hour)	24.2	17.4
Shotcreting (cu.yds./man-hour)	1.1	9.7
Backfill or shotcrete batching (tons/man-hour)	24.1	14.6
Backfill jamming (tons/man-hour)	49.4	7.7

TABLE 38.4   Maintenance and equipment average productivity data—1999

Unit operation	Equipment productivity (units/operating hr)	Maintenance requirements (operating hour/ maintenance hour)
Drift drilling (tons/operating hour)	52.4	2.3
Sample and bench drilling (feet/ operating-hour)	29.5	2.4
Mucking (tons/operating hour)	24.0	2.9
Hauling (tons/operating hour)	21.4	5.2
Bolting (tons/operating hour)	26.8	1.8
Shotcreting (cu.yds./operating hour)	1.1	15.7
Backfill or shotcrete batching (tons/operating hour)	53.6	2.8
Backfill jamming (tons/operating hour)	70.2	4.7

TABLE 38.5   Unit operations costs as a percentage of total mining ore maintenance costs for the Murray Mine—1999

Activity	Percentage of total operations cost	Percentage of total maintenance cost
Drift drilling	2.4	5.0
Sample and bench drilling	1.7	3.3
Blasting	8.1	—
Mucking	4.4	10.9
Hauling	5.8	10.3
Administration	33.8	17.1
Engineering/geology	6.6	—
Surface facilities	—	15.7
Bolting	6.0	10.3
Shotcreting	4.4	1.0
Batching	14.0	7.8
Backfill jamming	1.0	1.3
Drift repair	1.0	—
Underground utility	4.3	8.2
Surface utility	6.5	2.6
Utility vehicles	—	6.5
% of total mining cost	73.6	26.4

the rock bolting portion of the mining cycle is not required. Additional support is installed in the form of shotcreting the drift excavation above the floor to be excavated; however, shotcreting is much more productive per foot of drift than rock bolting.

Benches are drilled using a Solo production drill, and an open-pit-style drop-cut is used to create a free surface. ANFO or emulsion explosives are used in blasting, depending on the presence of water. The blasted muck is loaded using an LHD operated and controlled remotely so that personnel are not exposed to the resulting highwall.

38.3.6   Ventilation

Main ventilation is based on a system of booster fans positioned at the base of three 8-ft diameter ventilation boreholes to the surface. Air exhausts to the surface through three boreholes for a nominal total flowrate of 600,000 ft^3/min. Fresh air enters the mine through the two portals and two 8-ft-diameter boreholes.

Production areas are ventilated with auxiliary fans and ducting in a blowing mode. Production access drifts are equipped with 42-in-diameter fiberglass ducts, and several branches of collapsible 36-in-diameter ducts are used to ventilate the working faces.

38.3.7   Ore Grade and Dilution Control

Ore grade and dilution control are important aspects of the mining process because the ore at Murray is not visible. In addition, the ore body shape is very irregular locally and the ore-waste contact is very sharp at the boundary of the ore body. A regular pattern of sample drilling on each level is used to better define grade boundaries. Geology personnel construct a design boundary based on all data derived from diamond coring, sample drilling, and mining history. Three-dimensional modeling of grade data with excavation geometries allows construction of drift and bench designs that minimize edge dilution. It is not unusual for 30% of a month's production to be designed using sample drill data obtained during the same month.

Decisions to advance beyond design require evaluation by geology and ore control personnel and some evidence to indicate a high probability that material beyond the design boundary will meet the cut-off grade. The grade of the last round at the design boundary has been shown to be a poor decision criterion for advancing beyond the boundary because of the sharp gradation in ore grade at the boundary.

Backfill dilution has been estimated by superposition of excavation as-built survey data to develop prediction tools. Equivalent thicknesses of backfill that will become mixed into ore from walls, back, and floor have been established based on the as-built information. Table 38.1 lists the derived parameters and estimated total dilution for a historical reproduction of 1999 excavations.

38.4 PRODUCTION STATISTICS

The Murray Mine operates in a manpower-constrained mode to assure efficient labor costs. Production and maintenance personnel are itemized in Table 38.2. Managerial, technical, and supervisory staff increase total personnel by 23. Mine operations and maintenance are performed using two 11-hour shifts per day. Four production and maintenance crews are required for the 360 operating days per year. Shotcreting, surface support, and road and utility work are done on day shifts only.

Productivity data for unit operations for 1999 are listed in Table 38.3. Productivity data for the in-cycle face operations, drift drilling, sampling and bench drilling, blasting, mucking, hauling, bolting, and backfill jamming are used to develop production plans on a monthly basis and to develop annual plans and budgets for operations. Production planning for each month

is balanced to labor using a trailing 3-month average of productivity data. This process also forms the basis for production bonus criteria. Maintenance and equipment productivity data are listed in Table 38.4.

Cost data are presented as a percentage of mining costs in Table 38.5. These costs relate only to the Murray Mine operation and do not include site administration or milling costs.

38.5 CONCLUSIONS

Evolution of the underhand cut-and-fill mining method at the Murray Mine has led to substantial increases in productivity since the beginning of operations. Improvements have been made in many areas and have allowed reductions of face workers by 36%.

Key contributions have come from full mechanization of the rock bolting operation, an increase in spans under backfill, the addition of blind benching to mining practices, improved control over mining on ore boundaries, and critical evaluations of work practices in all functions.

38.6 REFERENCES

Brechtel, C.E., G.R. Struble, and B.W. Guenther. 1999. The Evaluation of Cemented Rockfill Spans at the Murray Mine. In: *Proceedings, 37th U.S. Rock Mechanics Symposium*, Vail, CO. Balkema.

CHAPTER 39

The Carlin Underground Mine

J. Gordon Sobering[*]

39.1 INTRODUCTION

The Carlin Mine is located 40 km (25 miles) northwest of the town of Carlin, Nevada, in the Lynn Mining District, Eureka County (Figure 39.1). Newmont Mining Corporation began exploration in northeastern Nevada in 1961, with gold production beginning in 1965. Through 1985, over 113,400 kg (4 million ounces) of gold were produced from the Carlin Pit. Following the cessation of open-pit mining, underground development began in the fall of 1993 to access, develop a trial stope, and delineate seven mineralized zones (Figure 39.2). The project was justified based on 1,690,000 tonnes (1,863,000 tons) and a grade of 0.41 oz/ton. Five years later, the total reserve (mined and reserves) had doubled. Ore production of 680 tonne/d (750 ton/d) was achieved early in 1995, and today the mine produces over 1,180 tonne/d (1,300 ton/d).

Mining methods include mechanized longhole benching with delayed fill and underhand drift-and-fill. Through the end of 1999, the Carlin underground mine has produced over 1,814,000 tonnes (2,000,000 tons) of ore grading 0.44 oz/ton and containing 870,000 oz. The ore is treated at Newmont's roasting facility.

39.2 GEOLOGY

The seven zones that make up the Carlin Mine have an approximate strike length of 915 m (3,000 ft). They are situated in an intense zone of fracturing that is bounded by the Leeville fault to the northeast, the Hardie fault to the northwest, the Castle Reef fault to the south, and the south-bounding Unk fault to the west. Mineralization is generally stratiform to podlike and is localized by northwest- and northeast-striking high-angle faults forming complex splays. The host rocks are dolomitic silty limestones in the upper Roberts Mountain Formation and are subdivided into two lithologies, Stls 1 and Stls2.

Stls 2, the dominant host rock, is a thin- to thick-bedded, light gray to black, variably carbonaceous, silty limestone with minor calcite layers. Overlying this is Stls 1, a thinly bedded, black, carbonaceous silty limestone, again with minor calcite. Stls 1 can also be planar laminated and may have a wavy or lenslike mottling of light and dark debris known in the district as "wispy lamination."

Bedding strikes northeast and dips 10° to 20° northwest. The Stls1 and 2 lithologies exhibit moderate to strong, pervasive decarbonatization and local weak to moderate silicification. The silicification occurs as selective replacement and is fault associated.

39.3 MINING

The Carlin underground mine is accessed via a portal at the bottom of the Carlin Pit. The cemented rock fill plant, shotcrete

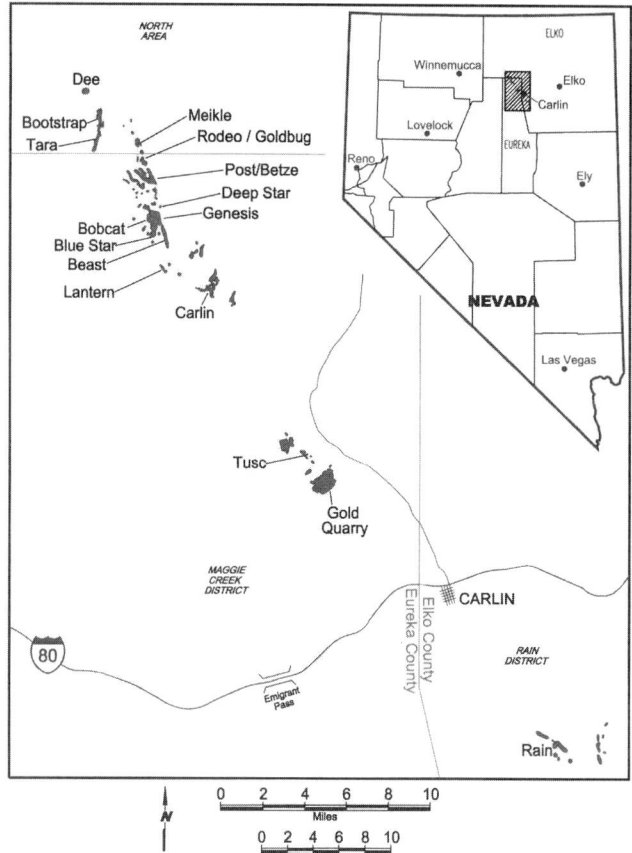

FIGURE 39.1 Location of Carlin underground mine

plant, maintenance shop, and laydown yard are also located here. Conventional drilling, blasting, and hauling are done with rubber-tired equipment. Table 39.1 lists the major equipment and its availability.

Three crews of 25 miners working 10½-hr rotating shifts, 6½ days a week, are employed at the Carlin Mine. In addition to the miners, personnel include 21 mechanics, 4 geologists, and 10 staff positions. Tuesday dayshift is reserved for road and equipment maintenance and utility upgrades. An incentive bonus is based on both productivity and safety goals. The bonus system is collective and proportions each task to the manpower required. Throughout the month, as production figures are shared with the

---

[*] Newmont Mining Corporation, Carlin, NV.

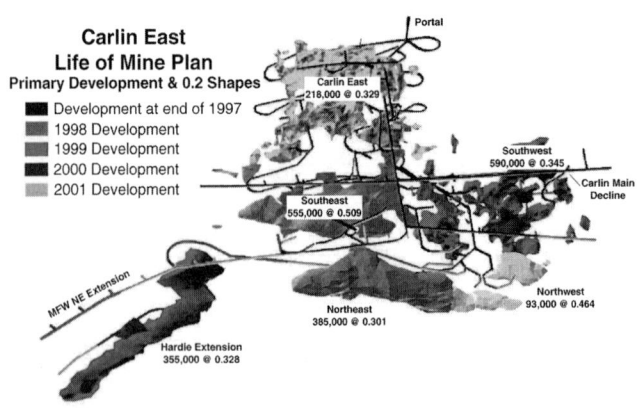

FIGURE 39.2 Map of current underground workings

TABLE 39.1 Equipment availability and use

Equipment	Number	Availability, %	Utilization, %
Single-boom jumbo drills	2	87	35
Twin-boom jumbo drills	2	93	9
Longhole drills	2	95	11
Bolter	1	86	50
Haulage trucks, 16 ton	7	93	47
Haulage trucks, 26 ton	5	90	54
Loaders, 4 yd	8	95	35
Loaders, 6 yd	3	85	41
Roadheader	1	96	23

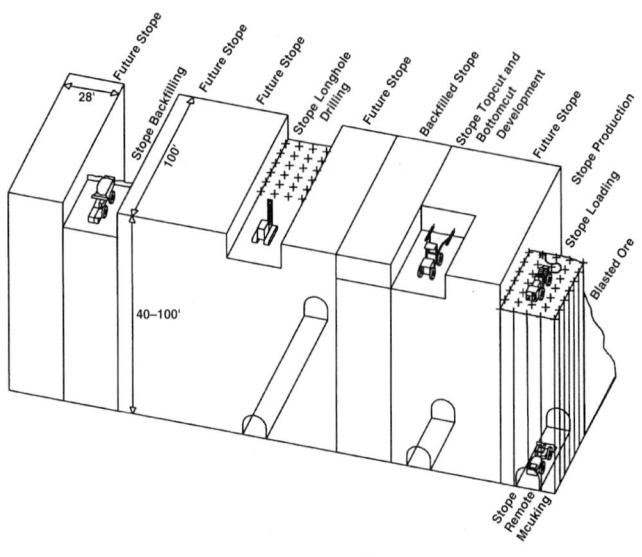

FIGURE 39.3 Longhole stope cycle

miners, crews can allocate resources effectively to ensure that all goals are achieved by month's end. The result is a more cooperative attitude among all hourly employees.

The mine produces 1,300 ton/d of mostly refractory ore using 50% longhole stoping and 50% underhand drift-and-fill methods. The average ore grade is 0.44 oz/ton. The mine has a horizontal extent of 3,000 by 3,000 ft and a vertical extent of 700 ft. Access is from a series of ramps driven in the footwall at a maximum 15% grade. Ore zones with a height in excess of 50 ft and steeply dipping (greater than 55°) are considered for longhole stoping. A typical stope measures 28 ft wide, 150 ft long, and 65 ft high when completed. Figure 39.3 shows the longhole mining cycle.

The Carlin underground mine is an assay-driven operation, as ore control cannot be done visually. All mined material is sampled once it is stockpiled on the surface with an assay turnaround time of 24 hr. Blast rounds grading greater than 0.15 oz/ton are shipped to Newmont's south area roaster while those with assays greater than 0.03 oz/ton are shipped to a refractory leach stockpile. Rotary and reverse circulation drilling is used extensively underground to supplement core drilling and to aid in the delineation and planning of ore mining horizons.

39.3.1 Longhole Stoping

Ore zones thicker than 7.6 m (25 ft) are mined using longhole stoping. The process begins with the mine planner contouring the geologic model on intervals of 0.10 oz/ton and vertical thicknesses of 4.5 m (15 ft). Access drifts and stopes designed using these contours have been successfully driven from either the hanging wall or the footwall, so they are not a design consideration. In cases where ore zones are between 7.6 to 9 m (25 to 30 ft) thick, the stope is blind benched, that is, one cut will be driven from the footwall and the other from the hanging wall.

This prevents the formation of a narrow pillar between levels and decreases the production footage needed for stope preparation.

Development of a longhole stope requires a top cut for drilling and blasting the slot raise and the longhole rings, and a bottom cut for extraction. Top cuts are typically 8.5 m (28 ft) wide by 3.7 m (12 ft) high with ground support installed after each blast round. Blast rounds of 2.4 m (8 ft) are standard but may be reduced as ground conditions dictate. Ground support in the top cuts includes 2.4-m (8-ft) friction sets in the back and 1.8-m (6-ft) sets in the rib (down to 1.8 m [6 ft] above the final sill elevation) in conjunction with wire mesh. Friction set patterns have spacings of no more than 0.9 m (3 ft) but may be reduced to 0.6 m (2 ft) in poorer ground. Point-anchored threaded rebar bolts with lengths of 3.7 m (12 ft) are installed on a 1.8- by 1.8-m (6- by 6-ft) pattern. Again, any negative changes in ground conditions require tightening the bolt pattern, increasing bolt length, and/or using shotcrete. Top cut development productivity is 0.5 m (1.7 ft) and 32 tonnes (35 tons) per man-shift.

Bottom cuts, typically 4.25 m (14 ft) high by 4.25 m (14 ft) wide, are driven below the top cuts. Ground support for bottom cuts is the same as for top cuts, but without the point-anchored rebar bolts. Bottom cut productivity is 0.8 m (2.7 ft) and 54 tonnes (49 tons) per man-shift.

Once top and bottom cut development is completed, an electric-hydraulic production drill is moved in to drill out the slot raise between the two levels. This raise is located nearest the hanging wall or footwall contact. Either one of two drills may be used for this task. One is equipped with a down-the-hole hammer and the other with a hydraulic top hammer. The slot raise is 2.4 by 2.4 m (8 by 8 ft) in plan and drilled with a hole diameter of 92 mm ($3\frac{5}{8}$ in). A burn cut consisting of nine holes is used with a burden and spacing of 0.6 to 0.9 m (2 to 3 ft), depending on rock hardness. Blastholes for the slot raise are loaded from the top cut level, with as many as four blasts required for breakthrough. A mixture of ammonium nitrate and fuel oil (ANFO) with cartridge emulsion primers is the blasting agent with 350 ms down-the-hole delays between holes and 25 ms surface delays. Powder factors are 0.83 to 1.24 kg/tonne (2 to 3 lb/ton).

Remote-controlled 6-yd³ load-haul-dump (LHD) machines remove the shot muck from the slot raise. Once this is completed, 92-mm ($3\frac{5}{8}$-in) production holes are drilled out on a 2.4-m (8-ft) burden on 2.4-m (8-ft) spacings. Blasting agents and

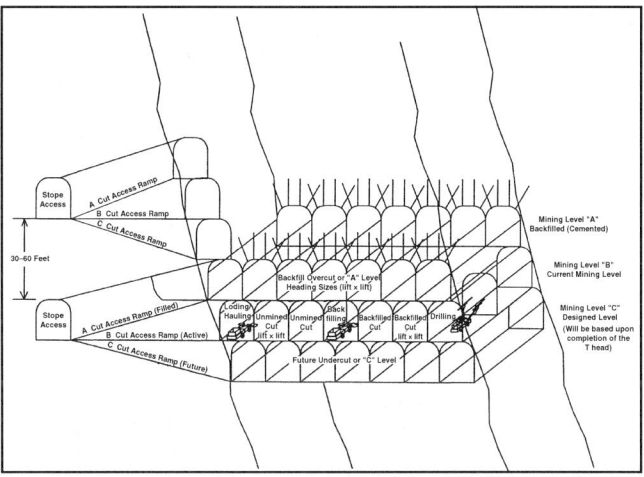

FIGURE 39.4   Underhand drift-and-fill cycle

FIGURE 39.5   AC-Eickhoff roadheader

delays sequences are as above; however, a powder factor of 0.28 kg/tonne (0.7 lb/ton) is desired. Normally, three longhole rings are blasted in one shot into the raise. Mucking into haul trucks is done remotely to ensure operator safety.

When a particular drift has been mined out, it is prepared for backfilling. This involves removing utilities and any muck or debris on the sills that may prevent a good bond from forming between the backfill and the rock. Backfill is dumped into the stope and then jammed along the entire length of the top cut. Typically, a ramp is made out of backfill so as to jam the backfill tightly to the stope back and avoid leaving horizontal cold joints that may negatively affect subsequent bottom cuts. Three days of curing are required before mining can begin adjacent to backfill, and 28 days are required before mining can commence underneath backfill. Cement content varies depending on whether subsequent mining is planned beside or underneath the stope. The top cut is jammed tight to the back with a LHD equipped with a boom and blade attachment.

39.3.2   Underhand Drift-and-Fill

Ore zones with thicknesses less than 7.6 m (25 ft) are mined using underhand drift-and-fill with either jumbo drills or a roadheader (Figure 39.4). These ore zones also have been successfully accessed from either a hanging wall or footwall decline with a cross-cut driven through the ore. Drifts are 4.2 m (14 ft) wide by 4.6 m (15 ft) high.

Mine planning is done in a way similar to that for longhole stoping. The ore zone is contoured based on the geologic block model and an initial sequencing plan. This is later updated from sample drill results. Cut-and-fill zones average 46 m (150 ft) long and 15 m (50 ft) wide, and for this reason, only one heading may be mined on a level at any given time. Normally, parallel cuts are taken in retreat from the hanging wall. These are later backfilled. In narrow ore zones, a central axis is driven with crosscuts mined in retreat. Crosscuts are driven at angles greater than 45° from the drift axis to minimize spans and reduce pillar stress. Backfilling is done similar to jamming a longhole stope's top cut, as discussed above.

The blast pattern used in production headings is a diamond cut with 45-mm (1.75-in) holes and 0.75-m (2.5-ft) burden and spacing dimensions. A powder factor of 0.45 kg/tonne (1.1 lb/ton) is achieved using gelatin stick powder. Blasts are initiated with detonation cord and fuse primers. The ground control plan for cut-and-fill drifts is the same as for the bottom cuts in longhole stopes.

Once the overcut level has been mined out, a drift is driven to access the next level. This may be from a new access or following blasting up the bottom. Subsequent underhand levels will have backfill above them, and so the drifts are designed to be offset or at an angle to the cut above. This minimizes the trace of cold joints, which are zones of potential weakness, from the backfill above. This process of underhand mining is repeated until the entire ore zone is extracted.

39.3.3   Roadheader

In 1997, Newmont Mining Company began the use of a 45-tonne- (50-ton-) class roadheader to supplement its cut-and-fill tonnage (Figure 39.5). Currently the roadheader contributes 20% of the mine's total ore production. The machine, an AC-Eickhoff ET 210, cuts rock using a transverse head with tungsten carbide picks. Once the rock is cut, it falls to an apron with a disk feeder and is transported through the machine on a chain conveyor to be loaded into a waiting haul truck.

Mining productivities are 41 tonnes (45 tons) and 0.6 m (2.11 ft) per man-shift. Although roadheader excavation costs are significantly less when compared to drill, blast, and load, the total costs are only 10% less than conventional drift-and-fill mining. This is due to the cost and maintenance of the components and the necessity for multiple cutting cycles to accommodate backfill and ground support installation. These factors limit the full potential of the machine.

To utilize the roadheader efficiently, a large drift-and-fill area is required to provide multiple headings. This allows the installation of ground support, backfilling, and mining to take place independently of each other. Since roadheader excavation is faster than the time required for haulage, ground support installation, and backfilling, a dedicated mechanical bolter is required to keep the machine productive. In addition, alternative headings must be close to minimize tramming time. Tramming distances vary from 75 to 110 m (250 to 350 ft) and are completed in less than an hour.

The mining sequence is generally on retreat with a two-cut pillar between active headings. Two independent access drifts are driven into the level whenever practical. Cuts off the access drifts are designed at angles greater than 45°, as in the case of conventional cut-and-fill. Drift height is kept at 4.6 m (15 ft) because of backfill jamming limitations and rib stability.

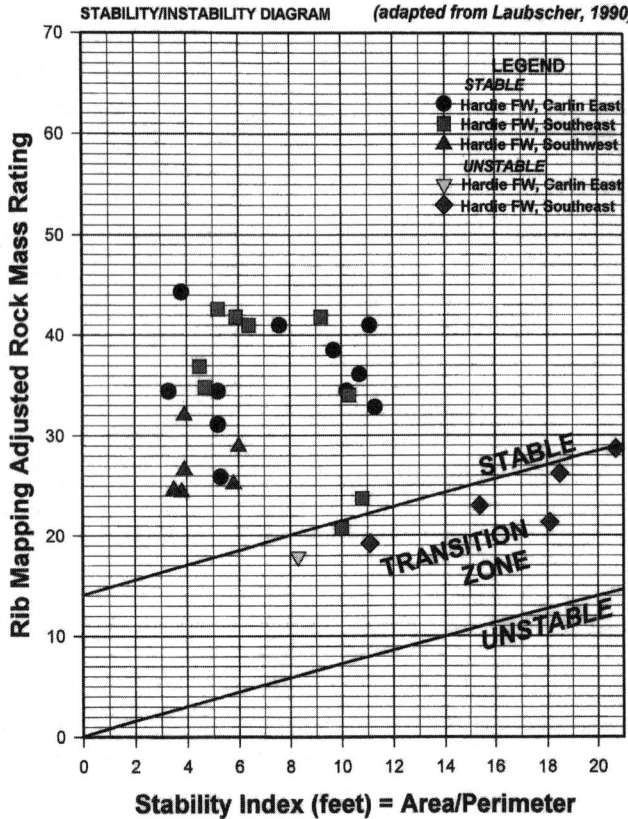

FIGURE 39.6   Carlin underground ARMR data

TABLE 39.2   Costs of mining method per ton, U.S. dollars

Activity	Longhole stoping	Drift-and-fill, 25%	Roadheader, 20%
Excavation and support	15.20	32.00	29.00
Backfilling	16.00	20.00	18.00
Access development	7.00	8.00	8.00
Other (repairs, utilities)	3.90	5.00	5.00

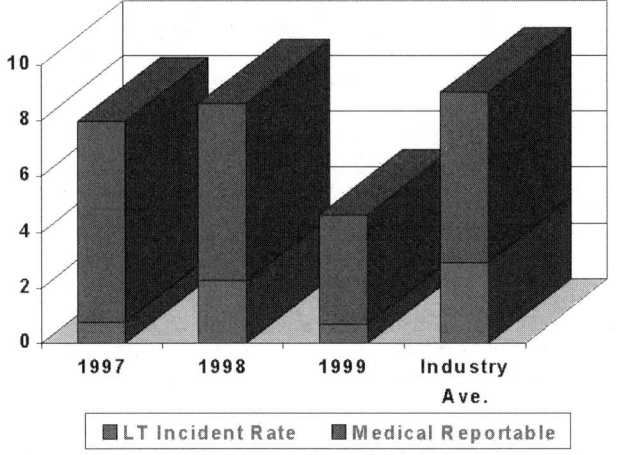

FIGURE 39.7   Newmont safety statistics

39.4   GEOTECHNICAL CONSIDERATIONS

Stope stability is a function of rock condition and geometry. At Carlin, the effect of a stope's dimensions are dealt with by using Laubscher's stability index (SI) method. This is defined as the area of the opening divided by its perimeter. The method provides three stability predictions: stable, unstable, and transition zone. Laubscher's original study was based on five working mines. The "stable" classification includes light, localized support, such as friction set bolts; the "transition zone" classification involves continuous ground support, such as either patterned rock bolts, cable bolts, bolts with straps and/or mesh, or bolts, mesh and shotcrete; and the "unstable" classification involves either heavy ground support or caving and surface subsidence. Figure 39.6 shows several ARMR versus stability index plots from rib mapping data at the Carlin Mine. Note the effective separation between the stable and unstable mine workings. Reading required ARMR values for a given stability index assists in stope design where a planning engineer and geotechnical engineer can determine if a stope needs to be taken in smaller stages and backfilled before subsequent cuts are made.

Laubscher's method of predicting ground stability has worked well at Carlin underground. Mining-related adjustments are essential to apply Laubscher's method and can reduce the rock mass rating (RMR) to less than half its initial value. Preliminary geotechnical evaluations should be periodically modified as additional information is produced by drilling and mine development.

39.5   BACKFILL

Since its beginning, cemented rockfill has been used at the Carlin Mine, both for longhole and underhand drift-and-fill stopes. The current mix uses 6.4% cement and a water-to-cement ratio

ranging from 0.8 to 1.2. Aggregate is supplied from open-pit sources. This rock, a diorite, is crushed to −2 in at Newmont's North Area leach crusher and hauled to the surface backfill plant, where it is fed by gravity into a hopper and conveyor system. A computerized batching system allows the operator to use a standard mix or to vary any of its components as required because of variations in moisture, size of the crushed aggregate, and other factors. A water-reducing admixture is added during the batching process to increase strength and improve workability.

Backfill is sampled every shift for each stope backfilled during that shift. A standard concrete cylinder is made and tested for uniaxial compressive strength after 28 days. Data are compiled in a database that tracks strength-versus-cement and water content to ensure the target strength of 5 MPa (735 psi) is met.

39.6   COSTS

Table 39.2 lists the costs per ton for each mining method for the Carlin Mine in 1999.

39.7   SAFETY

Accident rates at Carlin underground are below industry averages for similar mines (Figure 39.7). Newmont's safety program includes daily workplace inspections by the miners and checks by the foreman. Weekly crew safety meetings provide the opportunity to address topics from these daily inspections and bring forward other timely issues. The program also requires weekly safety inspections to be carried out by foremen and general foreman, and checked by the mine superintendent's monthly audit. Work policies and best-practice procedures are continually planned, reviewed, and updated with mining personnel. A safety incentive is designed in the bonus system to reward crews for working with no lost-time accidents.

39.8 CONCLUSIONS AND THE FUTURE

As the Carlin underground mine matures, both the operations and engineering departments are challenged to find ways to increase efficiencies and lower costs. Different technologies with respect to mining methods, backfilling, and ground support continue to be evaluated to keep the Carlin underground mine one of Newmont's lowest-cost producers.

39.9 ACKNOWLEDGMENTS

The author is indebted to Mark Odell, chief engineer, Doug Jones, mine superintendent, Joe Driscoll, general mine foreman, Matt Breitrick, senior engineer, Margie Lane, chief geologist, Mike Robinson, geologist, Bob Riley, chief surveyor, and Wes Leavitt, Ryan Mullin, and Tim Burns of the Underground Loss Control Department for their help and guidance.

39.10 REFERENCES

Abel, John A., Jr. Deep Star, Rain and Hardie Footwall Post Mortem Study. Report to Newmont Gold Company. November 15, 1997.

Breitrick, Matthew E. 1996. Breaking the Mold, Using Roadheaders for Production. In Hardrock Gold Mining.

Driscoll, Joseph M. 1997. From Surface to Underground: Newmont Gold's Carlin East Mine. *Mining Engineering*, August 1997.

Harris, Richard H. Tour Notes, Carlin Underground Gold Deposits. Newmont Exploration Limited. Carlin, Nevada. January, 1996.

Laubscher, D.H. A Geomechanics Classification System for the Rating of Rock Mass in Mine Design. *Journal of South African Institute of Mining and Metallurgy*. Vol. 90, No. 10. October 1990, pp. 257–273.

CHAPTER 40

Underhand Cut-and-Fill at the Barrick Bullfrog Mine

Dan Kump* and Tim Arnold*

40.1 INTRODUCTION

The Barrick Bullfrog Mine is located 6.4 km (4 miles) west of Beatty, Nevada, in the historic Bullfrog Mining District. The original Bullfrog Mine was discovered by Frank (Shorty) Harris and Ed Cross in the summer of 1904 and was located about 5.6 km (3.5 miles) west of the Barrick mine.

In August 1988, Bond Gold started construction of the Bullfrog open-pit mine and mill. The first bar of gold was poured on July 25, 1989. LAC Minerals acquired the Bullfrog Mine in November 1989, and in September 1994, Barrick Gold acquired LAC Minerals. From July 1989 through June 1991, all Bullfrog ore production came from the open-pit mine. The Bullfrog underground mine was put into production by LAC in July 1991. LAC's goal for its underground operation was to develop and operate a 1,000-tonne/d (1,102-ton/d) mine. After Barrick acquired the property, the goal was to further develop the underground mine and increase its production to 1,815 tonne/d (2,000 ton/d). Barrick mined out the underground mine in December 1998.

The purpose of this article is to relay the Bullfrog underground mine story. Several mining methods were used at the Bullfrog. Open stoping and overhand cut-andfill accounted for about 5% of the ore mined. Benching accounted for about 10%. The remaining 85% was taken using the underhand cut-and-fill mining method. The mining aspects of the underhand cut-and-fill method will be given. This will be followed with a discussion of the backfill. A table is presented with production, manpower, equipment availabilities, and costs.

40.2 BULLFROG MINE

40.2.1 Production History

In 1990, the North Extension was discovered by the Bullfrog exploration group. The ore body was delineated from the surface with a drilling program on 38-m (125-ft) spacings in 1990 and 1991. By January of 1991, enough drilling had been completed to evaluate the feasibility of mining the deposit based on an open-pit method and an underground method. The economics from feasibility studies were inconclusive as to proving a substantial benefit in mining with an open-pit method. The decision to go underground was based primarily on the fact that the mill would shut down in 1 yr because of the time to permit and strip the overburden, and because of potential difficulties with permitting open-pit waste disposal.

In February of 1991, test mining underground to define the mineralization, geologic structure, and geotechnical conditions of the vein breccia and footwall rock was proposed. This request was approved in April of 1991. Portal preparations began in May, and on June 18, 1991, the portal was established on the 1006 bench. Test mining proved that the proposed underhand open stoping method would not be economical, and an overhand cut-

and-fill method was adopted. The plan was to bring the underground production to 1,000 tonne/d (1,102 ton/d) by June of 1992. Following an evaluation of the ground support, costs incurred during the overhand extraction of the ore indicated a cost savings if an underhand technique were used. In 1992, this transition took place. Throughout the remainder of the mine's life, underhand cut-and-fill was the primarily mining method. From 1996 through 1998, where the geometry allowed, the more-productive benching and end-slicing methods were used.

40.2.2 Geology

The ore body dipped approximately 45° to the west, and the general strike of the ore body was north-south. Mineralization extended from an upper elevation of 1,010 m (3,314 ft) above sea level to a lower elevation of 775 m (2,543 ft). The strike length of the north ore area extended from the pit wall north for 650 m (2,133 ft); total strike length, including the Southwest Extension, which was primarily below the pit bottom, was 1,500 m (4,921 ft).

The ore zone was a vein breccia consisting of fissure veins and stockworks of quartz, calcite, and manganese oxides deposited through epithermal processes. The ore zone was very weak, with compressive strengths ranging from zero to 35 MPa (5,000 psi). The hanging wall was a silicified rhyolitic tuff and the footwall was predominantly latite. The latite footwall development was in a more competent rock ranging from 35 to 100 MPa (5,000 to 15,000 psi). Because of the extreme incompetence of the ore zone, backfill provided a better roof than the rock itself; therefore, the mining method of preference was underhand cut-and-fill.

During the feasibility work, the rock mass rating values were determined by Steffen, Robertson and Kirsten (1991) as follows: hanging wall = 39, ore = 23, and footwall = 44. Steffen, Robertson, and Kristen reported the average true thickness of the ore body equal to 5.6 m (18.4 ft) with less than 15% of the ore below 1.7 m (5.6 ft) and less than 15% thicker than 11 m (36 ft).

40.2.3 Mining Particulars

The original portal was on the 1,006-m (3,300-ft) elevation. The mine was accessed through a footwall decline driven from the north end of the open pit. The 5-m-wide by 4-m-high (16.4-ft by 13.1-ft) main ramp was driven at −15%. The ramp tended to run 40 m (131 ft) away from the ore body in the footwall in order to have the necessary room to access the ore body on several horizons from one ramp intersection. In August 1994, the 880-m (2,887-ft) north portal was established, and in April 1995, the 880-m (2,887-ft) south portal was established, both at the pit bottom on the 880-m (2,887-ft) elevation. Three raise systems exhausted a total of 14,000 m³/min (500,000 ft³/min) of air using two 500-hp fans and one 300-hp fan. This resulted in all portals being intake airways.

---

* Barrick Bullfrog Mine, Beatty, NV.

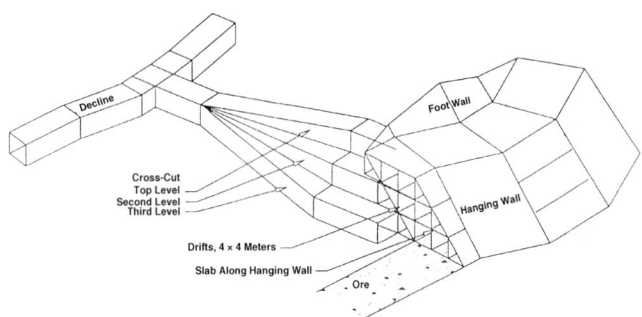

FIGURE 40.1   Typical mining method

Mining was conducted from the 1,006-m (3,300-ft) level to the 782-m (2,565-ft) level. The ore was drilled using single-boom jumbos and blasted with stick powder. The muck pile was ramped and bolted with jackleg drills. Full coverage of cyclone-type wire mesh to a 2.4-m (8-ft) spring line was attained using 1.8-m (6-ft) Split-Set bolts. The ore was supported with full coverage of mesh to springline (at least) and shotcrete. Mucking was carried out with 2.7-m³ (3.5-yd³) load-haul-dump (LHD) units. Ore was carried out of the mine in 16-ton mine trucks.

Development outside the ore was carried out using double-boom jumbos, 4.6-m³ (6-yd³) LHD units, and 24-tonne (26-ton) rear-dump haulage trucks. Ground support was primarily steel mats or wire mesh with 1.8-m (6-ft) Split-Set bolts and occasionally steel sets.

The underhand cut-and-fill mining technique included drifting along the footwall or hanging wall contact and then filling the drift with backfill. Another drift was then run beside the backfill in the ore. This was repeated until all the ore on that horizon was removed (Figure 40.1). Often the ore was slabbed on retreat if the ore thickness didn't warrant another full drift. Once the horizon was completed, another access beneath the last horizon was driven, and the process was repeated, only with a backfill cover this time.

40.2.4   Shotcrete, Geometry, and Other Productivity Improvements

Some of the most successful productivity improvements concerned heading geometry and shotcrete. The ore headings at the mine were in very incompetent rock. It could be held well, but the heading size had to be kept to a 4- by 4-m (13- by 13-ft) cross section and shotcreted quickly. The ore headings did not necessarily become problems because width of the drift was over 4 m (13 ft), but subsequent ore drifts next to or underneath the primary drifts were problematic because the backfill could not be placed tight to the back. The main focus in keeping the drift size down was to mine a uniform mining unit, thus assuring that all the other surrounding mining units would fit well together.

Shotcrete improved the ore headings immensely. The ground at the Bullfrog was not "heavy," nor was it blocky. It was simply very friable. Some areas mimicked full faces of wet sugar that had dried. The reason it became a problem was an "air slacking" type of effect that caused dribbling and small failures into the wire mesh. The shotcrete was applied immediately after bolting in problem areas and as much as 1 week later in better areas. This eliminated air slacking, the drifts stood for longer periods, and the need for rebolting was nearly eliminated.

The shotcrete was applied via a boom truck fed by a batch truck. The batch truck had separate compartments for the aggregate and cement and fed the shotcrete pot by an auger. It was a simple system and worked well. It was applied to a nominal depth of 5 cm (2 in). Very little fiber was applied to the shotcrete except in very problematic areas.

The trend in the industry is to shotcrete before installing wire mesh and bolts. This was considered a better system in most rock conditions. However, the Bullfrog Mine did not adopt this method out of concern about the friability of the ore. Even though the shotcrete could be applied to the rock, the rock may not have had the tensile strength to hold the shotcrete. The recourse was to apply several inches of shotcrete to create a beam of support. This was not feasible because of the availability and utilization of the shotcrete equipment.

40.2.5   Geometry

Staying on-grade and spending time to watch and enforce that we stayed on-grade improved productivity. Miners, supervisors, geologists, and surveyors were all responsible for keeping on-grade. The importance of staying on-grade was manifested when we reached a level with backfill above and below. If the operators dug too deeply, then the final drift was too thin to dump backfill with our trucks. If it undulated, the overhead backfill would be uneven.

The intersections accessing the ore to the north and south originally were excavated in a T configuration (Figure 40.2). After some problems required standing steel in these intersections, a Y access was designed that incorporated a section of the footwall rock in the pillar (Figure 40.3). The mining of the pillars was done on retreat and became a standard at the mine.

One of the greatest productivity gains made at the Bullfrog was the ore access philosophy. Originally, as the footwall decline crossed every 12-m (40-ft) horizon, an ore access was driven up at 15%. Since three ore drifts that were 4 m (13 ft) high were to be mined from this access, it split the mine into horizontal blocks of ore. Scheduling the blocks so that mining did not occur near other headings was relatively simple.

However, there was an intrinsic problem with this method. If the decline happened to access the ore body anywhere except near the middle, long ore drives to the north or south were created. This meant that some ore drives were open for as much as 2 yr. With the friability of the rock, even after the application of shotcrete this resulted in standing a lot of steel in ore headings.

The solution to mining in this manner was to redesign "subramps" or "mini-declines" in the footwall access. This split the ore body into as many as six ore headings, where once there would have been only one or two. Accesses were evaluated on their own merit by showing that the amount of ore retrieved from such an ore access and the increased productivity defended their cost.

Another system attempted was to mine the footwall drive first because the hanging wall generally ran at a lower grade than the footwall. Furthermore, the hanging wall was less competent. The theory was that if any problems with ground control were encountered, we could abandon the area, having taken most of the grade with the first pass. We could even elect to leave some of the ore near the hanging wall that was below cut-off grade.

While the theory was sound, the practice became a problem. As the drifts progressed laterally from footwall to hanging wall, arching across the drifts perpendicular to the strike took place. This also happened progressing from hanging wall to footwall. However, if the hanging wall (the less competent rock) was secured in the first pass instead of the last pass, fewer ground control problems occurred. The "footwall first" method was abandoned for the most part at the mine.

40.2.6   Dewatering

Groundwater was first encountered at the Bullfrog underground mine in early 1992 at an elevation of 934 m (3,064 ft). From April 1992 throughout the balance of the mine's life, dewatering was required. Initially, pumping was done with underground sumps. From 1993 through 1998, three dewatering wells were drilled and pumped. Except in 1996, sump pumping accompanied well

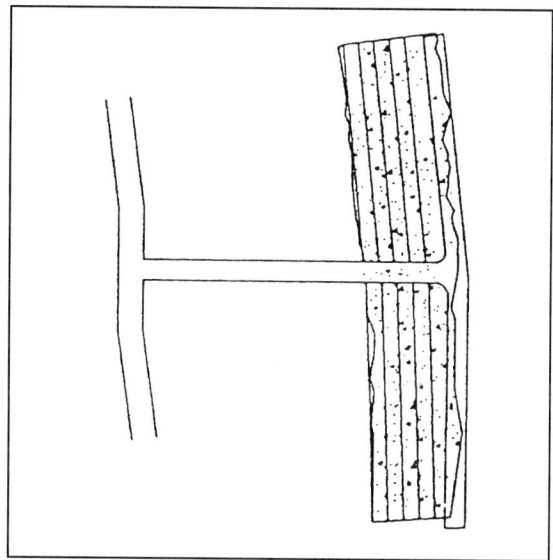

FIGURE 40.2 T intersection

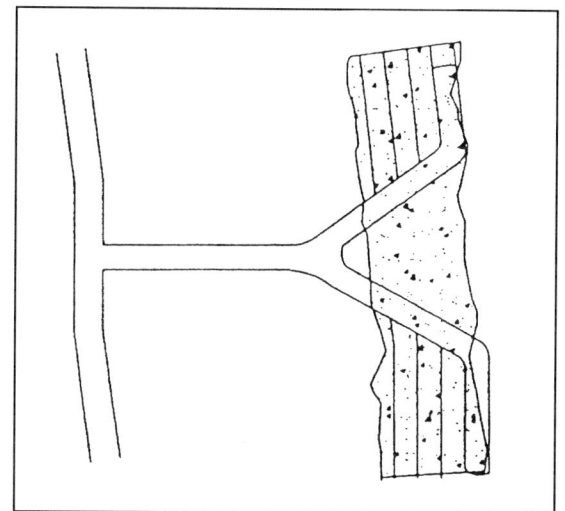

FIGURE 40.3 Y intersection

pumping. Usually, only one well was on-line at a given time. Two of the three wells were abandoned and subsequently mined through. The water was managed by using it for drill water in the underground mine, making backfill, spraying on surface and underground haulage roads for dust control, showers in the dries, processing water at the mill, and percolated in infiltration ponds. The final total drawdown of the water table was 152 m (499 ft). The average pumping rate was 2,426 L/min (641 gal/min).

40.2.7 Surface Haulage

The underground truck drivers hauled the ore out of the mine and stockpiled it on the surface near the portals. The ore was hauled to the mill by the open-pit contractor with surface haul trucks.

40.3 BACKFILLING AT THE BULLFROG MINE

40.3.1 Introduction

The Bullfrog backfill was a mixture of cement, aggregates, and water. Backfill was prepared at two surface batching plants. For

TABLE 40.1 Chronology of commissioning plant use at the Bullfrog Mine

Description	Time
Beginning of underground ore production	July 1991
1006 backfill plant	April 1992
1006 screening plant	July 1993
880 crusher	March 1995
880 backfill plant	May 1995
880 fines addition plant	November 1997

the most part, only one plant operated on any given day. The first plant, the 1006 plant, was located on the 1,006-m (3,300-ft) bench of the Bullfrog Pit and was commissioned in April 1992. The plant was used until its decommissioning in November 1997. The other plant, the 880 plant, was built in the bottom of the Bullfrog Pit at the 880-m (2,887-ft) elevation and was commissioned in May 1995. It was used throughout the life of mine. Twenty-five percent of the backfill was made at the 1006 plant, and 75% was made at the 880 plant. After the 880 plant was commissioned, the 1006 plant was rarely used.

Both plants batched high-strength (+7% cement) and low-strength (3% to 4% cement) backfill. High-strength backfill was placed in drifts where miners worked beneath the backfill and required a 30-day curing time before mining beneath it. Low-strength backfill was used in headings where miners did not work under the backfill, for example, in a footwall drift. Throughout the mine life, about two-thirds of the fill placed was high strength. Table 40.1 shows the chronology of the backfill plants.

40.3.2 1006 Plant

The 1006 plant was capable of producing about 900 tonne/d (1,000 ton/d). The plant was operated by the truck drivers. The plant consisted of a 45-tonne (50-ton) cement silo, a mixing tank to batch a cement-water slurry, and a pump to dispense the slurry into a haul truck containing aggregate. Partial mixing of the batch occurred during haulage to the fill site; final batching was effected by using a pushblade to mix and place the backfill at the fill location.

At first, the aggregate used at the 1006 plant was run-of-mine waste from footwall development. This was the main source until a screening plant was set up to prepare a minus 6-in aggregate from pit hanging wall waste in 1993.

40.3.3 880 Plant

To enable the planned increase in the ore production in 1995, a corresponding increase in backfill capacity was required, as well as a higher quality backfill. Capital funds were approved in late 1994 to build the 880 plant, which was designed to produce an improved product at a rate of 1,800 tonne/d (2,000 ton/d).

The 880 plant was located in the bottom of the south end of the Bullfrog Pit. Crushed aggregate was dumped from a catch bench that was about 54 m (177 ft) above the pit bottom onto a corrugated steel tunnel conveyor system that fed the mixer. The plant consists of a 10-yd³ tilt mixer, a 145-tonne (160-ton) cement silo, a tunnel conveyor system with two feeders, and a 360-tonne (400-ton) fines bin and its conveyor. The 880 plant had a batching drum that mixed the cement, water, and aggregates. The drum discharged into end-dump haul trucks that carried the mixed backfill to the headings. The backfill was dumped and rammed into the headings with a pushblade.

The 880 plant was run by an operator who generally followed recommended batch recipes. However, the operator could adjust the amount of water and fines based on the physical appearance of the backfill and input from the pushblade operators. Besides batching loads of backfill, the operator performed routine maintenance, clean-up, record-keeping, and sampling when required.

The 880 plant's approved capital budget included the funds to purchase a mobile crushing plant to prepare minus 3-in aggregate. Test work done from November 1995 through January 1996 led to a contract with a consulting firm to develop a model for fill strength and to evaluate the minus 3-in aggregate used at the 880 plant. An important issue resolved by the testing was the adoption of roller-compacted concrete aggregate as the basis for the Bullfrog backfill.

Run-of-mine waste from the hanging wall of the Bullfrog Pit was the source of feed for the backfill crusher. The mobile crusher system consisted of a 76.2- by 106.7-cm (30- by 42-in) jaw crusher that reduced run-of-mine waste to minus 6 in, a cone crusher that reduced the jaw product to minus 3 in, and a 91-cm by 3.4-m (36-in by 100-ft) radial stacker.

Screen analyses performed on the cone product showed a fairly well-graded material. A plot of the gradation curve paralleled the curve for roller-compacted cement aggregate. The Bullfrog crushed aggregate was about 15 percentage points below the gradation curve for the lower limit of the roller-compacted cement aggregate. Bullfrog used a modified gradation that was about 10 percentage points less than the lower limit based on consideration of both safety factors and costs.

The crusher product almost totally lacked 50-mesh (0.012-in) and 100-mesh (0.0059-in) particles. This problem was addressed by importing fine material from a local sand and gravel pit and feeding it into the mixer drum through the fines addition plant, which was commissioned in November 1997. With the addition of fines, the resulting gradation curve was about 10 percentage points below the lower limit of the roller-compacted cement aggregate.

40.3.4 Backfill Placement

Backfill material was brought to the workplace in the same trucks that were used for ore haulage. The backfill was hauled from the plants into the mine with end-dump haul trucks that carried 9-tonne (10-ton) loads. The fill was dumped in the headings and pushed into place with a 6-yd³-loader with a tamping plate in place of a bucket. There were two important aspects related to proper backfill placement: the drifts needed to be completely mucked out, and the fill needed to be shoved as close as possible to the back of the headings.

40.3.5 Backfill Strength

In 1996, before implementing a cement reduction cost savings initiative, the backfill was modeled to determine a minimum design strength for the 880 plant product. The model design strength for the Bullfrog Mine's backfill was 3.5 MPa (500 psi) in place. At various times, quality control cylinders were cast at the batch plant. Because of the 3-in top size aggregate particle, 22.8-cm (9-in) cylinders needed to be cast. (Cardboard tubes 25.5 cm (10 in) in diameter were readily available and were used to cast cylinders at the 880 plant.) Bullfrog did not have the capability to core samples from the placed, cured backfill to verify in-place strength. Consequently, a conservative safety factor of 1.3 was applied to the backfill cylinders cast at the batching plant. Based on this safety factor, the minimum breaking strength for the Bullfrog backfill cylinders was 4.5 MPa (650 psi). It was then assumed that if the backfill were properly placed, the in-place backfill would have a strength of 3.5 MPa (500 psi).

40.3.6 Batch Design

Tables 40.2 and 40.3 list batch design data for the 880 plant.

40.3.7 Quality Control and Testing

Between November 1995 and April 1998, three quality-control sampling campaigns were conducted at the Bullfrog. These tests

TABLE 40.2 Compressive strength data for the high-strength backfill from the 880 plant

Description	Cement, %	Strength, psi
Model sampling	9.2	870
May 1996 reduction	8.1	673
January 1998 reduction (with added fines)	7.8	1141

TABLE 40.3 Recipes for a cubic yard of backfill from 880 plant, Sept. 1998

	Low-strength		High-strength	
	Pounds	Percent	Pounds	Percent
Cement	150	4	272	7.2
Aggregate	3630	96	3508	92.8
Water	230		250	

required casting backfill cylinders to determine compressive strength. Quality was also verified during several other tests performed during evaluations of fines addition sources, fly ash sources, water-reducing agents, lime-kiln-dust-fly-ash mixtures, and a local crystalline silica product. Prior to any reductions in the addition of cement, compressive strength tests were performed to confirm their validity. The daily casting of compressive strength cylinders was never done at the Bullfrog. Instead, the major points of quality control at Bullfrog were to ensure that the headings were properly mucked out before backfill placement and that the backfill was placed as tightly as possible to the back.

40.4 MINING PRODUCTION AND COSTS

40.4.1 Miners and Bonus System

Much of the success of the Bullfrog Mine can be attributed to its miners, a good incentive bonus system, and a supervisory-management team that listened to and implemented improvements suggested by the miners. In 1998, the average bonus rate was $12.00/hr in addition to the hourly wage of $20.00 (an effective $32.00/hr). The bonus program coupled with the improvements to productivity caused by better mining techniques also had a positive impact on workplace safety. The workforce deserves credit for being flexible and working with management to achieve the common goal of a profitable and good place to work.

40.4.2 Mining Results

The initial reserve statement for the Bullfrog underground mine included 2,630,000 tonnes (2,899,000 tons) of ore at a grade of 7.92 gm/tonne (0.231 oz/ton), which would yield 18,909 (667,000 oz) contained kilograms of gold. The ore production rate increased from 590 tonne/d (651 ton/d) in 1992 to 1,714 tonne/d (1,889 ton/d) in 1998, and the ore productivity rate increased from 9.8 tonnes (10.8 tons) per shift in 1992 to 14 tonnes (15.5 tons) per shift in 1998. Table 40.4 summarizes the production by year throughout the mine life.

40.4.3 Production and Cost Data

Some of the items that led to an increase in mining and backfilling rates are—

- The addition of more operating days
- The addition of more miners
- The addition of more trucks and pushblades to the fleet
- The 880 plant
- The increased availability of the fleet

TABLE 40.4 Barrick Bullfrog underground mine—production and cost data

	Year and (Operating Days)								Total or Average
	1991 (126)	1992 (252)	1993 (306)	1994 (309)	1995 (345)	1996 (358)	1997 (357)	1998 (357)	(2,410)
Ore tons	30,953	163,973	248,469	219,429	329,433	542,326	638,879	674,517	2,847,979
Backfill tons	—	131,038	194,474	244,639	266,021	446,968	618,573	587,929	2,489,642
Total tons	30,953	295,011	442,943	464,068	595,454	989,294	1,257,452	1,262,446	5,337,621
Grade (oz/ton)	0.120	0.219	0.288	0.286	0.254	0.258	0.222	0.224	0.242
Contained Au ounces	3,650	35,887	71,478	62,734	83,764	139,266	142,022	151,139	689,940
Cost Au ounce	N/A	171.53	137.42	209.73	211.37	180.71	201.27	167.64	180.41
Ore tons/operating day	246	651	812	710	955	1,515	1,790	1,889	1,182
Backfill tons/operating day	—	520	636	792	771	1,249	1,733	1,647	1,033
Cost/ton (ore)	N/A	37.54	39.53	59.96	53.74	46.40	44.74	37.56	48.68
Cost/ton (backfill)	—	N/A	N/A	5.63	5.36	5.42	6.96	6.96	5.47
Workforce									
Hourly	40	64	82	96	178	182	170	132	129
Staff	8	8	8	12	21	20	19	17	15
Total	48	72	90	108	199	202	189	149	144
Ore tons/total 8 hr shift worked	N/A	10.8	12.5	12.4	9.4	10.3	11.9	15.5	12.1
Availabilities									
Scoops	N/A	N/A	N/A	N/A	N/A	73%	77%	86.1%	
Haul trucks	N/A	N/A	N/A	N/A	N/A	74%	76%	83.2%	
Jumbos	N/A	N/A	N/A	N/A	N/A	76%	84%	79.7%	
Pushblades	N/A	N/A	N/A	N/A	N/A	N/A	N/A	83.1%	
Development feet	2,490	2,753	2,572	3,878	9,548	6,495	2,224	—	29,960
Total costs	$5,063,570	$13,517,179	$10,052,014	$13,156,920	$17,705,308	$25,166,547	$ 8,585,375	$5,393,782	$138,640,695
Capital costs	$5,063,570	$ 7,361,581	$ 229,729	$ 1,283,660	$ 6,121,188	$ 2,774,500	$ 1,285,700	$ 56,172	$ 24,176,100
Labor and burden costs	$ 366,000	$ 2,678,000	$ 4,102,500	$ 4,087,000	$ 6,523,700	$10,031,075	$11,046,571	$9,748,300	$ 48,583,146
Production bonus costs	$ —	$ —	$ —	$ 86,218	$ 916,213	$ 1,651,913	$ 2,264,234	$3,016,200	$ 7,934,778

- A low miner turnover rate
- A good production bonus
- The addition of two portals in the Bullfrog Pit bottom
- Multiple accesses to ore horizons

40.5 CONCLUSIONS

40.5.1 Good Things

- The most important factor in expanding ore production at the Bullfrog Mine was achieving widespread acceptance of the criticality of maintaining a high backfill-to-ore replacement rate. To assure heading availability, it was critical to backfill mined-out areas as they became available. A good production bonus coupled with effective leadership fostered the miners' buy-in to the importance of increasing the backfill placement rate.

- Construction of the 880 plant and the 880 fines addition plant.

- Purchase of a spare mixer drum for the 880 plant. This greatly reduced downtime for mixer liner replacements. Having a spare drum allowed the maintenance department to replace worn liners at the shop and reduced downtime for liner change-outs at the plant from 2 days to 8 hr. (Because of the abrasiveness of the aggregate, the liners had to be replaced every 4 months.)

- Assigning full-time plant operators to the 880 plant. This allowed for greater control of the quality of the plant product.

- Installation of a rock-pick to clean out solidified remnants of backfill in haulage trucks. Periodic cleaning of the trucks was needed to keep the trucks capable of holding a 9-tonne (10-ton) load.

40.5.2 Not-So-Good Things

- The segregation of fines from the coarse aggregate that occurred during dumping of aggregate from the 934 bench to the 880 plant.

- Dumping fines on the top of loaded trucks of backfill. If the pushblade operator failed to mix the fines with the backfill, the fines acted as a joint and allowed backfill to fall out the back.

40.5.3 Other Things

In retrospect, what needed to be done, was done. The only negative aspect of the Bullfrog backfill program was how long it took to accept the importance of a well-graded aggregate. One thing that was not implemented and that might have been helpful was to reduce the maximum size of the coarse aggregate to minus 2 in. This would have made it easier to cast quality-control cylinders and it may have reduced the quantity of fines that had to be purchased.

40.5.4 Backfill Conclusions

- The Bullfrog experience suggests that there are two keys to a successful backfill program: A well-graded aggregate and proper placement techniques.

- The Bullfrog increased its backfill production rate from 470 tonne/d (520 ton/d) in 1992 to 1,494 tonne/d (1,647 ton/d) in 1998. (Table 40.4 presents backfill production data.)

40.6 SUMMARY

The Bullfrog underground mine produced 2,583,670 tonnes (2,847,979 tons) of ore at a grade of 8.30 gm/tonne (0.242 oz/ton), which yielded 19,559 contained kilograms (689,940 oz) of gold (the initial forecast was for 18,909 contained kilograms [667,000 oz] of gold).

The Bullfrog increased its ore production rate from 590 tonne/d (651 ton/d) in 1992 to 1,714 tonne/d (1,889 ton/d) in 1998, which was an increase in productivity rate from rate from 9.8 tonnes (10.8 tons) per shift in 1992 to 14 tonnes (15.5 tons) per shift in 1998.

The Bullfrog increased its backfill production rate from 472 tonne/d (520 ton/d) in 1992 to 1,494 tonne/d (1,647 ton/d) in 1998.

Management's priorities were to align goals with compensation, be honest on all issues, hold people accountable for their actions and performance, and treat people with respect. What resulted from focusing on these priorities was continuous improvement throughout the life of the mine. Eighteen months before shutting down, the miners were informed of the life-of-mine plan and were updated at least monthly. Barrick transferred a large portion of the workforce and went to great lengths to solicit other mining companies to find work for its employees. This mutual respect between management and the hourly workforce made the Bullfrog Mine one of the most successful operations of its time and a truly great place to work.

40.7 REFERENCES

Steffen, Robertson and Kirsten, "Bullfrog Mine Pre-feasibility Study Underground Resources," 1991.

CHAPTER 41

Case Study: Strathcona Deep Copper Mine

Hugh S. MacIsaac[*] and Graham Swan[†]

41.1 INTRODUCTION

Falconbridge Sudbury Operations, including Strathcona Mine, produce approximately 3 million tonnes a year from four mines operating in the Sudbury basin. Strathcona Mine[‡] is located 50 km northwest of Sudbury on the rim of the basin (Figure 41.1). The Deep Copper Zone began production in 1988 and currently produces approximately 450,000 tonnes of ore a year using a narrow-vein, mechanized cut-and-fill mining method. Copper constitutes 65% of the ore revenue and nickel 10%; the remaining 25% comes from precious metals.

41.1.1 History

The original Strathcona Mine began operation in 1968, producing 32 million tonnes of nickel ore before being mined out in 1992. The mine used two shafts: the original exploration shaft (a four-compartment timber shaft down to the 3,150-ft level) and a production shaft with five compartments to the 3,100-ft level.

In the early 1980s, the decision was made to deepen the original four-compartment shaft as a six-compartment shaft down to the 5,200-ft level to access the Deep Copper Zone. When this was completed, muck from both Deep Copper (450,000 tonne/yr) and Fraser (800,000 tonne/yr) could be hoisted to the 2700 level and be fed into the original ore-handling circuit (Figure 41.2). Currently, alternative loads of nickel ore from the Fraser Mine (approximately 2 km to the west) and copper ore from Strathcona are batched and hoisted up the Strathcona No. 1 shaft and into the original Strathcona production handling system, which feeds directly to the Strathcona Mill. The mine presently runs on a three-shift, four-crew schedule that covers 19 of 21 shifts a week.

41.2 GEOLOGY

Strathcona Deep Copper was discovered in 1979. It is a complex network of highly irregular, massive sulphide veins located in footwall rocks approximately 450 m from the norite contact of the Sudbury basin. These veins are predominantly composed of, in order of abundance, chalcopyrite, cubanite, and pentlandite, with lesser amounts of pyrrhotite, millerite, bornite, and magnetite. In width, the veins vary from fine veinlets up to a maximum of 6 m, and they are generally difficult to follow as they pinch, swell, and change direction from horizontal to vertical. The main zone as a whole strikes approximately east-west over a distance of approximately 350 m and plunges to the southwest at 30° to 40°. Total ore reserves (proven + probable) are currently estimated at 6.1×10^6 tonnes, grading at 6.7% copper.

Diamond drilling has had a greater focus in the last couple of years as reserves are depleted. Targets have been set and achieved utilizing two different drills. Stope drilling is done with a one-man "gopher" drill, which is a small portable diamond drill

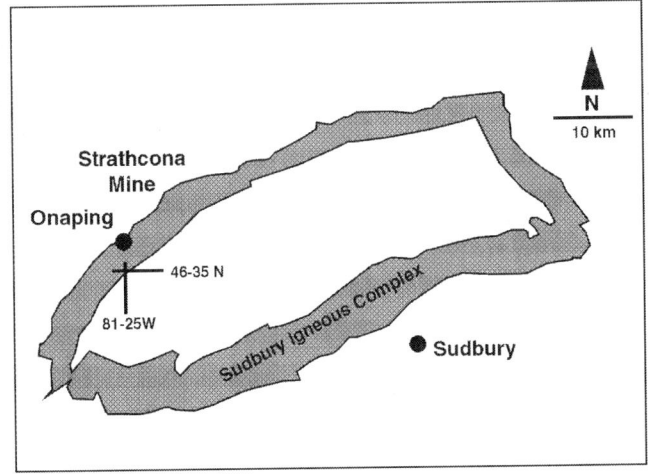

FIGURE 41.1 Sudbury basin showing Strathcona Mine

for holes up to 140 m in length. The drill is operated only during two of the four shifts. One man on each of the four shifts operates a larger Craelius 262. This machine is used outside the stope for both long and short holes. Both rigs are achieving over 17 m per manshift.

41.3 GEOTECHNICAL INFORMATION

While the complex geometry of the Deep Copper Zone is a strong factor in its geotechnical response to mining, an equal contribution is the extreme contrast in mechanical properties between the weak, friable ore and the strong, brittle host rock. A summary of the intact and rock mass strength properties of both is provided in Table 41.1. In modeling the sill pillar recovery stage of mining, it is apparent that both ore and waste have a brittle post-peak failure response such that strength drops to 25% of peak strength in gneiss and 50% of peak strength in ore over a 1% strain increment. All modeling (Board and Damjanac 1999) has been calibrated by field observations of depth of failure in stope backs.

Structural mapping showed the existence of a set of dominant joints in the waste surrounding the ore. These joints dip 45°, strike 235°, and are generally spaced at 20 to 50 cm. They persist over a distance of 10 to 100 m and are locally coated with chlorite-epidote. Recent microseismic and geotechnical analyses indicate that the source of several large-magnitude rock bursts (up to 3.0-M events) was fault-slip and related to these structures within the brittle felsic gneiss. The mine currently operates a

---

* Mine Captain, Craig/Onaping Mine, Falconbridge Limited, Sudbury Operations, Canada.
† Principal Rock Mechanics Engineer, Falconbridge Limited, Sudbury Operations, Canada.
‡ Recently combined with Fraser mine and now referred to as the Fraser Copper Zone.

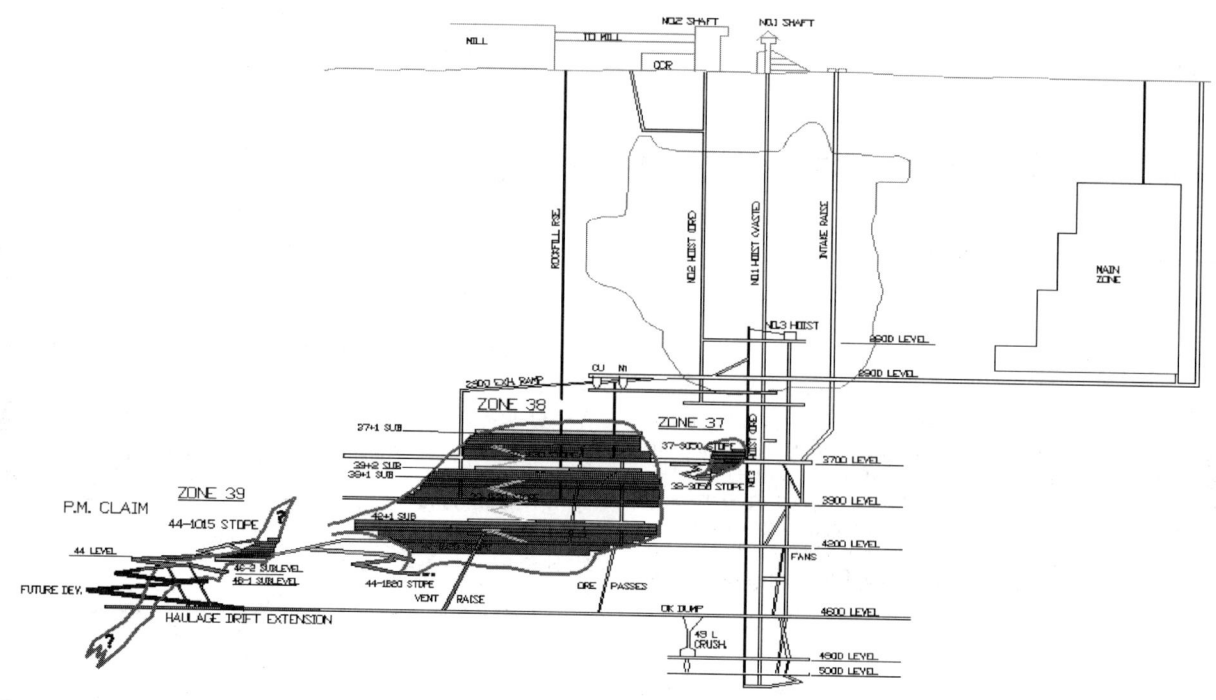

FIGURE 41.2   Section through zones and muck hoisting system

TABLE 41.1   Intact and rock mass properties for the Deep Copper Zone

	Intact rock properties				Rock mass properties					
Rock type	UCS, Mpa	E, GPa	ν	m_i	RMR	M_b	s	C, MPa	φ, deg	E, Gpa
Felsic gneiss (FW/HW)	290–360	60–78	0.22–0.25	27	60	6.5	0.012	12	50	18
Massive suphides	28–120	24–68	0.18–0.28	22	40	2.6	0.0013	4.5	30	5.5

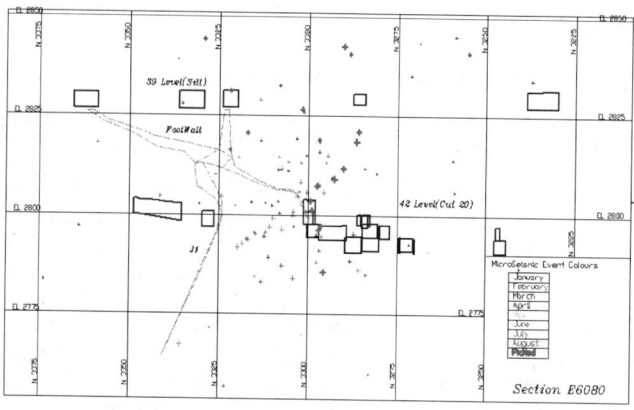

FIGURE 41.3   Vertical section through 3900 sill showing microseismic activity by month. Note trend of large events running diagonally and intersecting footwall.

FIGURE 41.4   Stoping area showing use of shotcrete posts to control span

32-channel, full-waveform ESG microseismic system, which, among other things, displays event location and time relative to mine excavations (Figure 41.3).

41.3.1   Ground Conditions

During the initial phase of mining and up to the formation of 30-m sill pillars, ground conditions remained quite good, with only occasional local instability. In these circumstances 12-ft Super Swellex

bolts were used, followed eventually, as the back began to show signs of more massive stress-induced failure, with the innovation of reinforced shotcrete posts (Swan et al. 1997) (Figure 41.4).

Numerical models have confirmed the existence of high stress concentrations in excess of 100 MPa on the footwall side of wide stopes. To counteract this problem, stope span has been restricted to 8 m before backfilling, using a drift-and-fill approach. Recently, moderate rock burst activity has been encountered in the 42 footwall stopes while mining the 39 sill pillar. The largest was a 3-M fault-slip-type event triggered by a blast (Figure 41.3). Rock-burst-damaged excavations are typically rehabilitated using a combination of mesh-reinforced shotcrete, resin-grouted rebar bolts, and 3.6-m plated Super Swellex, the later being particularly effective in areas where the immediate back comprises stress-failed ground.

41.4 MINING METHOD

The mining method in the Deep Copper Zone is ramp access, overhand, narrow-vein, cut-and-fill stoping on three levels (3700, 3900 and 4200) in the main ore zone. Details of the mining method have been well described elsewhere (Fuchs and McDonald 1992; MacIsaac 1997). Initially, while mining the sill cuts in the late 1980s, cemented rockfill was placed on the first three cuts above 3700 level and seven cuts above the 3900 level to form a stiff fill pillar (Swan et al. 1993). While the fill was placed with the intention of mitigating, at a late mining stage, closure-driven rock bursts (Swan et al. 1989; Hedley 1995), it also improved local ground conditions contemporaneously with mining. Today, while approaching the final stage of sill pillar recovery, all stopes are generally backfilled with uncemented classified tailings.

41.4.1 Development

All mine development consists of 4.6- by 4.6-m headings driven with an H227 Atlas Copco jumbo and an Emco Jarvis Clark 210 6-yd scooptram. Two-man crews on each shift cover the required advance of 1,300 m/yr. Re-mucks are driven every 150 m to improve face cycling time. A crew mucks waste from the re-muck to either the waste pass or a nearby stope that is ready to be filled. Ground support in development headings include 1.8-m mechanical bolts and 1.6- by 3-m sheets of No. 9 gauge screen. Headings are bolted and screened to the face, while walls are bolted and screened to within 2.3 m of the floor and within one round of the face. Intersections require 2.4-m supports with a 50/50 mix of rebar and mechanical bolts. Ten-centimeter air, 5-cm water, and 1-cm discharge pipes are installed, delivered in 3-m lengths (most suitable to fit in the cage). Currently, all permanent pipe installation is Schedule 40 steel pipe. At the time of writing, plans are being developed to reduce the width of stope access ramps to 4.3 m to help reduce the amount of waste produced.

41.4.2 Equipment

The development crew has its own set of gear that includes one 8B Wagner scoop with pneumatic tires. Drilling is done with an Atlas Copco H227 electric-hydraulic two-boom jumbo with longer feed beams for drilling 4-m rounds. A second jumbo is used by the stope crews for taking down backs for accessing new cuts from ramps. A Marcotte scissor truck is used for installing ground support, pipe, ventilation, and shotcrete. Production crews operate a fleet of seven one-boom Atlas Copco jumbos that are only 1.2 m wide. Three of these units (Model 126N) have been retrofitted with a B.U.T. 104 boom for increased flexibility and visibility, while the other four are Boomer H104s. The primary mucking unit in Deep Copper is the JCI 250M scoop-tram. Falconbridge has worked with the supplier to develop 1- and 2-yd scoops that meet the specific needs. These scoops use teeth on the buckets to muck through screen, bolts, and rebar. The tires on these scoops are core filled because of the harsh mucking conditions (tire life is short). A M-120 Caterpillar grader is operated on one shift by one operator and grades all the ramps.

41.4.3 Stoping Method

Stope access ramps are driven down at –15% to the footwall of the ore zones (Figure 41.5). Re-mucks are established in the ore zone when possible to reduce waste production. Mining height in

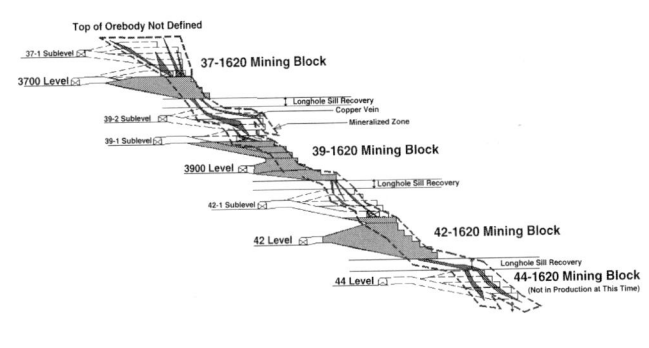

FIGURE 41.5 Schematic of vertical section showing mining method

the stopes is 3 m and follows the veins, breasting, slashing, or subdrifting as required. Following the ore is under the direction of the geology department, which visits the faces on a daily basis. Over time, the operation has changed to crews and supervisors directing themselves using their basic understanding of the cut-off between ore and waste; only headings that are borderline are left to the geologist.

Headings range from 1.8 m to a maximum of 11 m wide. Holes are drilled 38 mm in diameter and 2.9 m long using an AC one-boom jumbo with a "BUT 4" boom and 1032 drill. This short feed length is necessary to maintain flexibility and minimize dilution. Breasts, rounds, or slashes are loaded with ANFO using either Bex loaders or intermediate-sized loaders and Nonel caps. Blasting is initiated using an electric cap set off from blasting stations at the top of stope access ramps at lunch or quitting time. After blasting, muckpiles are washed, scaled, mucked down, and leveled to the proper height to accommodate the bolting and screening. The primary mucking unit in Deep Copper is the JCI 250M scoop in headings from 2.2 up to 11 m, while in the smaller headings, the JCI 125 scoop is used for leveling as well as mucking. With the muckpile leveled, all the gear and material required for bolting and screening are brought into the work area in a metal container whose size is matched to the scoop buckets.

Ground support is installed from the muckpile using jack-legs, stopers, and screen pushers. The ground support minimum in production is much the same as in development; however, much flexibility is given to the supervisor and crews as to additional support. Walls may require Split-Set bolts or shotcrete; backs may require Swellex bolts or shotcrete, and even shotcrete posts (Swan et al. 1997).

With ground support complete, headings are mucked out with either the 1- or 2-yd scoops to an established re-muck area at the entrance to the stope. To reduce the mucking distance of the smaller scoops, these re-mucks are sometimes moved into a more central location in the stope by means of double blasting (Figure 41.6). This can be done in areas where the vein is vertical and consistent. Crews drill off to the width required (4 m to accommodate the re-muck scoop), blast the ore, muck it out, and then blast the waste. Mucking from the re-mucks to the ore or waste pass is the responsibility of the re-muck crew using two Toro 400s, one EJC 210, one Wagner 8B scoop, and a JCI 26-ton truck.

41.4.4 Fill Preparation and Filling

When the mining cycle is complete, all floors are mucked out as thoroughly as possible. The remaining fines on the floors represent a significant value and must be recovered. Crews blow these fines into piles using a blowpipe. Other systems for picking up these fines have been tried (i.e., vacuum, water jet, small backhoe) without success. After the mined-out area is cleaned of

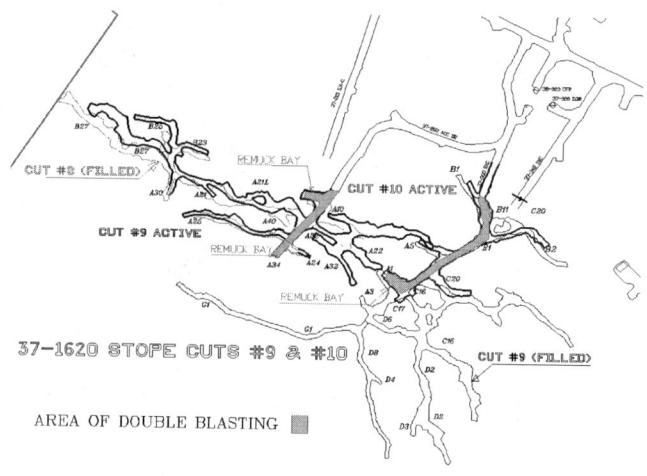

FIGURE 41.6   Level plan of 37-1620 stope

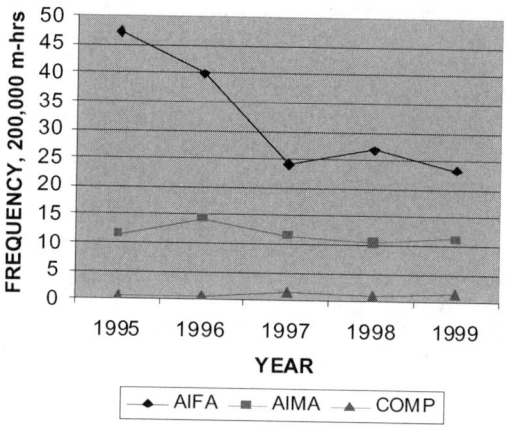

FIGURE 41.7   Mine safety statistics, 1995–1999

all fines, fill lines and fill fences are installed. Either timber or shotcrete fences are used depending on condition requirements and location.

As noted above, filling in Deep Copper has evolved from stiff, cemented rockfill to weakly cemented tailings to the current bulk fill with no cement. The reason for the use of cement in the hydraulic fill was to reduce fill dilution from floors, reduce slimes loss (which plugged up the drainage at the fill fences), and improve the flow of the fill. However, the recent addition of a flocculent called Percol 351* to the bulk fill has proven to be a cheaper and more effective alternative.

41.5   ORGANIZATION AND SAFETY

The front-line supervisor is responsible for all mining at Strathcona Mine as well as the mobile equipment maintenance shop. Much time and effort has been put into training operators and supervisors to achieve the required results. All equipment is serviced from one main central service shop on the 3900 level. To accommodate the ongoing requirement for training both mechanics and miners, an underground conference room has been set up. Moving from theory training in the conference room to hands-on training outside in the mine has proven to be a very effective training tool.

All mines work on safely while producing the best possible grade of ore at the lowest cost on a continuous basis. Narrow-vein mining is no different. Much time and effort is put into safety issues, dilution control, cost-cutting, and maintenance of a regular flow of muck to the mill. A formal morning meeting is held before day shift starts, with the two supervisors on night shift and the day shift supervisors to ensure continuity in mining and mechanical maintenance. This meeting may be extended to include current safety concerns with respect to injuries (Figure 41.7), related matters such as rock burst events and ground control, and even concerns over costs. Current net operating costs per tonne of ore hoisted run in the range of US$50-60.

---

* Supplied by Allied Colloids and dosed at rates between 35–50 g/tonne slurry.

41.6   CONCLUSIONS

Narrow-vein mining is, by its very nature, expensive. The challenge for the future is to continue improving safety and productivity while reducing the cost of copper per pound. As more individuals are exposed to the realities of narrow-vein mining, the more understanding they become. Cost per tonne or tonnes per manshift are relative terms, and those with first-hand knowledge of how and where these numbers are derived will have an appreciation for the ongoing challenge.

41.7   ACKNOWLEDGMENTS

The authors would like to thank Falconbridge Limited for permission to publish this paper. We would also like to acknowledge the many individuals at Strathcona Mine who directly or indirectly assisted with the engineering and operational content of this paper.

41.8   REFERENCES

Board, M., and B. Damjanac. 1999. Analysis of Deep Copper Rockbursting. *Itasca Report to Falconbridge, Strathcona Mine, December 1999.*

Fuchs, J., and J. McDonald. 1992. Mechanized Narrow Vein Cut-and-Fill Mining at Falconbridge's Deep Copper Project. *Proc. AIME-SME AGM, Phoenix, Arizona.*

Hedley, D.G.F. 1995. Stiff Backfill. *Canadian Rockburst Research Program 1990–1995, Chapter 4, p.77, CAMIRO Mining Division.*

MacIsaac, H.S. 1997. Strathcona Mine Deep Copper. *13th CIM Mine Operators Conference, Sudbury, February 1997.*

Swan, G., C. Steed, S.J. Espley, B. O'Hearn, and G.R. Allan. 1989. Strathcona Deep Copper Zone: Geomechanics Investigation. *Falconbridge Limited Internal Report.*

Swan, G., B. Arjang, and D.G.F. Hedley. 1993. On the Use of Rockfills in Overhand Cut-and-Fill Mining. *Proc. Int. Congress Mine Design, Kingston, Ontario, August 1993.* pp. 103–110.

Swan, G., G.R. Allan, M. Beaudry, and M. Board. 1997. Developments in Shotcrete Applications at Falconbridge Limited, Sudbury Operations. *13th CIM Mine Operators Conference, Sudbury, February 1997.*

CHAPTER 42

Evolution of Undercut-and-Fill at SMJ's Jouac Mine, France

Ronan Le Roy[*]

42.1 INTRODUCTION

The Bernardan uranium deposit is located in the eastern part of France, 50 km north of Limoges. It was discovered in 1968, and production began in 1973. The Bernardan is the main deposit being mined by the Societé des Mines de Jouac (SMJ). The company, which has been a subsidiary of COGEMA since 1993, produces 400 to 500 tonnes of yellowcake each year. Until recently, most of the 60,000 tonnes of ore mined annually were extracted using underhand cut-and-fill and complete filling with a concrete backfill. The search for cost-cutting measures identified a number of possibilities, including (1) optimizing the use of concrete backfill, (2) alternative mining methods, (3) partial backfilling, (4) drawing, and (5) selectivity. The adoption of these measures has allowed SMJ to continue mining its reserves despite the fall in the price of uranium.

42.2 GEOLOGY

The Bernardan deposit occurs as episienite pipes in a massive granite host rock. These ore masses appear like columns dipping 45° to 90° (Figure 42.1). They have an average horizontal cross section of 300 m^2 and a height which varies from 20 to 700 m. About 10 pipes have been discovered. The deposit will be exhausted in 2001. The granite host rock is competent with an unconfined compressive strength of 50 to 100 MPa. The episiyenite in which the mineralization occurs is much less competent with a compressive strength lying between 2 and 10 MPa. Because of the massive granitic host rock, water inflow is low.

42.3 INITIAL UNDERGROUND MINING METHOD

The deposit was first mined as an open pit. Since 1987, underground extraction has been taking over, and today the production is 300 tonne/d. Underhand cut-and-fill with concrete as the backfill material was the initial stoping method used at SMJ (Figure 42.2). A brief description of the method is as follows:

- A ramp is driven in the consistent barren rock, winding around the pipe.

- A first slice is mined in the ore using a drill-and-blast method. Roof support is required because of the poor ground conditions. Productivity is low (9.5 tonnes per manshift). Each drift making up the slice is backfilled with concrete as soon as it is completed.

- The second slice is extracted with the drift-driving direction perpendicular to that used on the level above. Extra reinforcement is not required, and a higher productivity (16.6 tonnes per manshift) is achieved. Each drift is backfilled with concrete.

- The third slice is then undercut, and the process continues.

The advantages of the method are that (1) worker safety is maximized since all mining is done under a concrete slab, (2) ore

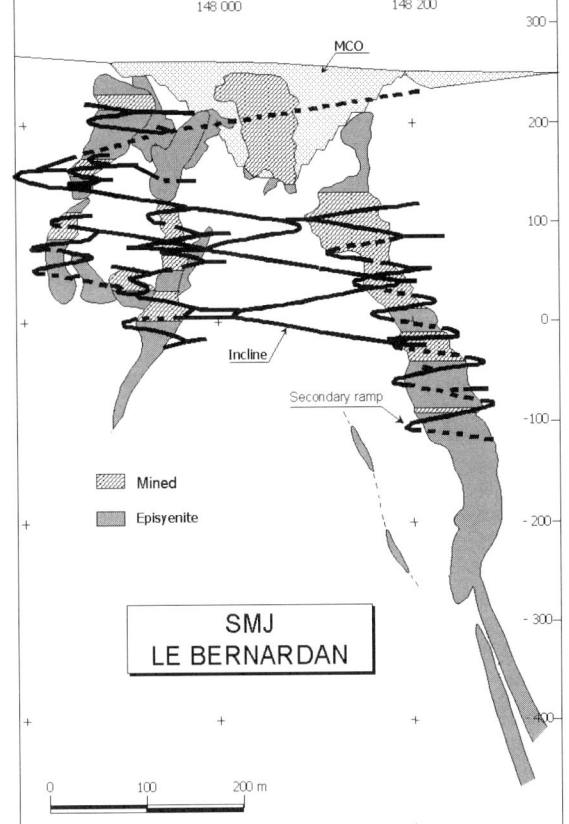

FIGURE 42.1 Bernardan deposit and general layout

recovery is nearly 100%, and (3) dilution is low. The main disadvantage is high cost, particularly because of the expense of the concrete and the backfilling operations.

The general underground mine layout consists of—

- A primary 15% ramp with a 20-m^2 cross section driven in the granitic host rock from the surface to the various ore bodies,

- Secondary ramps wound around the various vertical columns

- Ore haulage via 20- to 40-tonne trucks, and

- Ventilation at 250 m^3/s.

---

* Cogema Branche Uranium, Cedex, France.

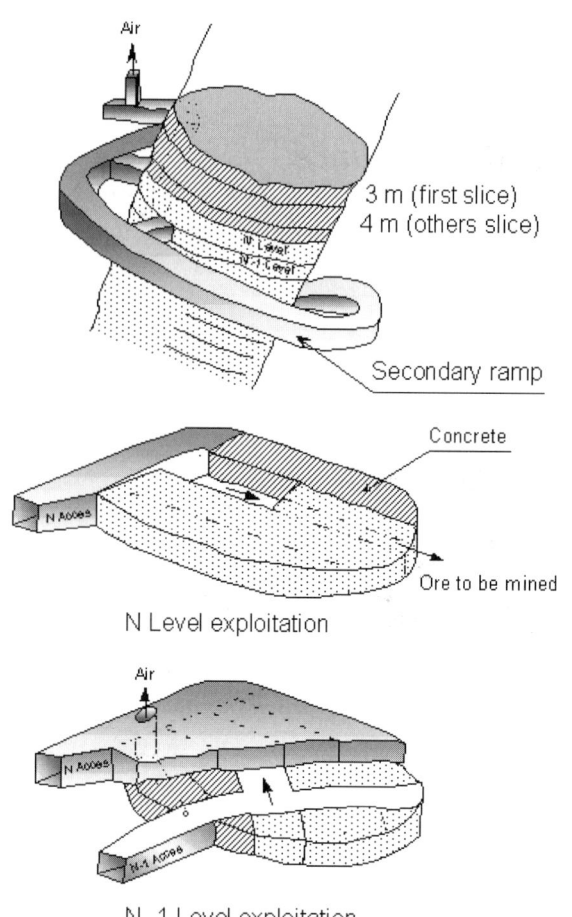

3 m (first slice)
4 m (others slice)

Secondary ramp

Concrete

N Level exploitation

Ore to be mined

N -1 Level exploitation

FIGURE 42.2   Initial top-slicing method

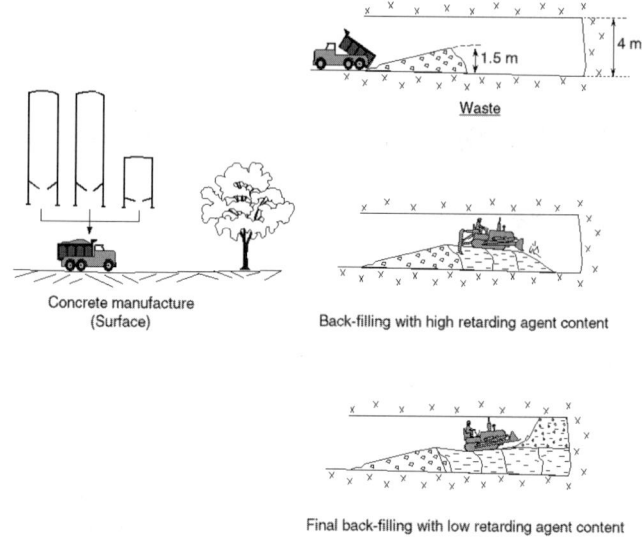

Concrete manufacture
(Surface)

Waste

Back-filling with high retarding agent content

Final back-filling with low retarding agent content

FIGURE 42.3   Dry method of backfilling small drives

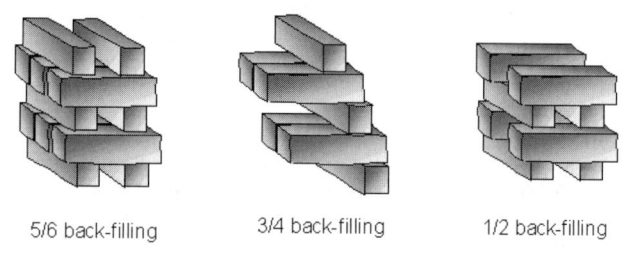

5/6 back-filling   3/4 back-filling   1/2 back-filling

FIGURE 42.4   Various backfilling ratios depending on vein geometry

42.4   EVOLUTION OF CURRENT MINING METHOD

The initial method was chosen on the basis of the geotechnical and geometric characteristics of the ore bodies. Considering the economic and geological changes, three major changes have been introduced. These are reductions in the cost of concrete and backfilling volume and introduction of a selective underhand cut-and-fill method.

Concrete is mixed on the surface using waste aggregate. It is then pumped to the drifts to be backfilled. To ensure concrete fluidity, the water content in the concrete is high, thereby reducing concrete strength. The reduction in concrete cost involves two main items.

1. **Concrete Formula.**   The results of geomechanical models of the mining system (concrete fill and the galleries) combined with in situ measurements indicated that the mine could reduce the required concrete strength from 15 to 12 MPa. By using additives, water content could be reduced. Based on these two factors, the cement ratio could then be reduced from 350 to 260 kg/m^3. The introduction of a quality-management system meant that variability in concrete quality was reduced, and therefore the required average strength could be reduced accordingly.

2. **Backfilling Operation.**   To backfill small drives, a dry method has been employed. To prevent the formation of an inconsistent bottom layer and because of the requirement to produce homogeneous concrete, the method

requires the use of slow-setting concretes, which is accomplished by adding decreasing amounts of concrete retardant as the slab is built up, as shown in Figure 42.3.

Reducing backfilling volume involves three geologic factors.

1. **Major Pipes.**   Depending on ore body geometry, either (1) the central gallery (that lying between the two previously backfilled adjacent drifts) is no longer backfilled, or (2) the footwall gallery is no longer backfilled. The backfilling ratio drops to three-quarters, five-sixths, etc., depending on the method. A one-half-ratio method, with stability being provided by pillars made up of backfilled drifts, was successfully tested. It could not be used routinely because of the geometric irregularities of the ore body (Figures 42.4 and 42.5).

2. **Veins.**   When width is less than 10 m, a single slab is preferred (backfilling ratio one-half)

3. **Small Masses.**   Using the original method, the ratio of barren rock to ore was too large to permit mining of small ore bodies. A drawing method was tested in 1992 to recover 3,400 tonnes of ore in such a mass and was then applied to other masses. A gallery with roof support is driven at the base of the ore. When drifting is complete, the roof support set is removed, and the ore is allowed to cave. Blasting using holes drilled on a 2- by 2-m pattern assists the caving process. A remotely operated load-haul-dump (LHD) machine removes the ore from the extraction gallery. Because of the poor ore

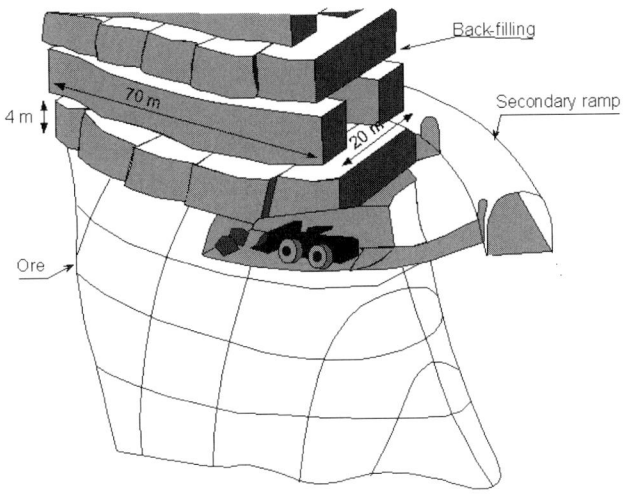

FIGURE 42.5   Five-sixths backfilling with top-slicing method

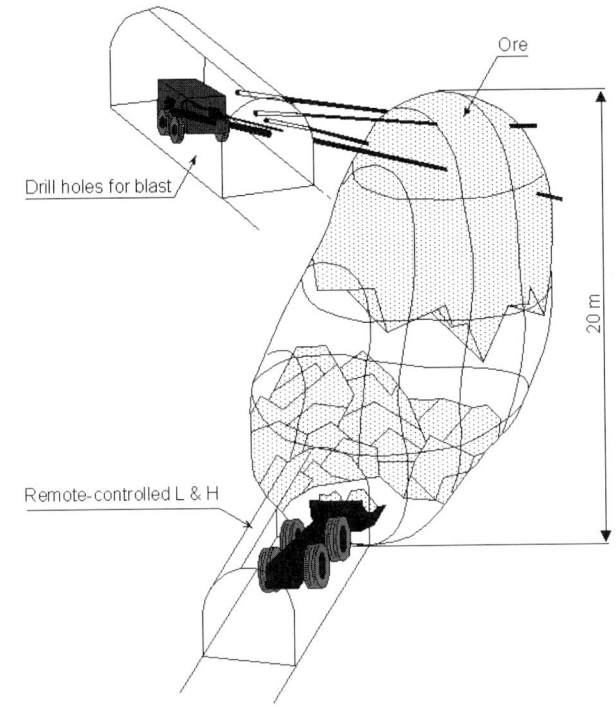

FIGURE 42.6   Mining small masses

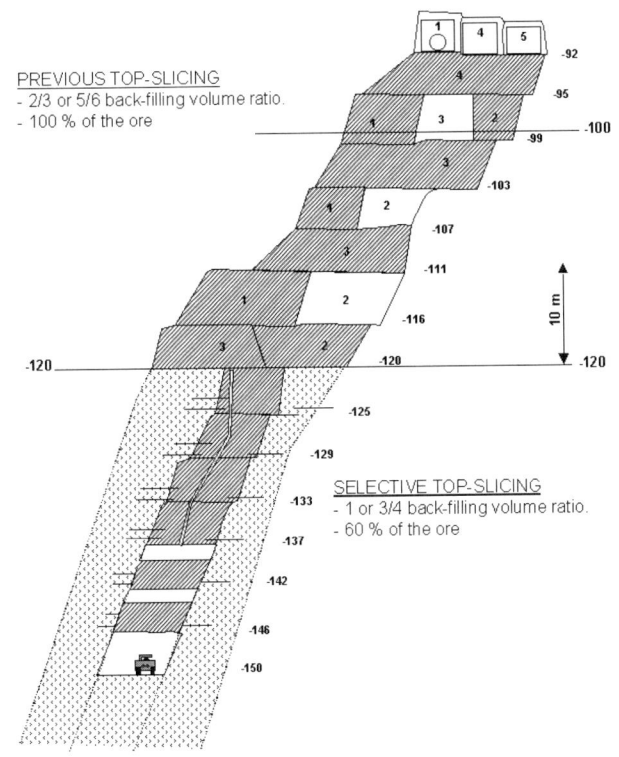

FIGURE 42.7   Selective top-slicing method

veins. A selective method has been introduced to prevent dilution from low-grade ore (Figure 42.7). This includes—

1. Limiting the undercut to the central high-grade body and
2. Driving the undercut under a concrete slab and reinforcing the hanging wall and footwall by bolting. Bolting the hanging wall is tricky and requires twice as many bolts as does bolting the footwall.

The selective top-slicing method allows deep, partially mineralized veins to be mined. Ninety percent of the total metal (which was contained in 60% of the ore) can be mined. This practice allows the operation to continue despite falling uranium prices.

42.5   CONCLUSION

Because of falling uranium prices, the reserves of SMJ were in jeopardy. Improvements made in the mining methods led to significant cost cuts.

1. Concrete operating costs were reduced by 15%,
2. The global backfilling volume ratio has dropped to 71%,
3. Selectivity has increased the average ore content of the deep pipes by 48% with only a small (11%) reduction in metal recovery.

These measures have increased the possibility for mining the reserves.

strength conditions and the competence of the granite, dilution was only 13% and ore recovery was acceptable (Figure 42.6).

Prior to 1996, all of the episienite pipes mined were fully mineralized with grades higher than the cutoff grade of 0.03%. In 1996, SMJ reached some deep and only partially mineralized

Sublevel Caving

CHAPTER 43

Underground Iron Ore Mining at LKAB, Sweden

C. Quinteiro,* M. Quinteiro,* and O. Hedström*

43.1 INTRODUCTION

Luossavaara Kirunavaara AB (LKAB) is a high-tech international mining company and one of the world's leading producers of upgraded iron ore products. LKAB's operations are conducted at five locations in northern Sweden and Norway and form a continuous flow from the mining and processing of ore to the transportation of finished products (by railway) from the mines to the harbors. The mines and ore processing plants are located in Malmberget and Kiruna, a pelletizing plant is located in Svappavaara, and shipping ports are located in Luleå, Sweden, and Narvik, Norway. The group head office and marketing unit is located in Luleå. Figure 43.1 is a map showing the location of the mines and ports. Sales offices are located in Luleå, Brussels, Essen, and Singapore. Iron ore mining is the main enterprise, but through subsidiaries, other operations related to ore extraction and treatment are carried out. These include, among others, the manufacturing of explosives (by KIMIT AB) and ore transportation (by MTAB).

The rich iron ore deposits in Kiruna and Malmberget have been mined by LKAB, founded in 1890, for more than 100 years. Industrial-scale extraction from the mines in Kiruna and Malmberget began in the early 1900s when the railway was completed from Luleå to Malmberget and Kiruna and further extended to the ice-free port of Narvik. At first, mining was done using open-pit methods. Underground operations started in Malmberget in the 1920s and in Kiruna in the 1950s. Sublevel caving has been the principal mining method at both mines since the 1960s.

Today, the Kiruna mine and the Malmberget mine produce some 23 and 13 million tonnes, respectively, of crude ore annually, which positions both of them among the largest underground mines in the world. All the crude ore is treated and upgraded. The main product is pellets (69%), while the rest is fines (31%). LKAB annually produces 21.2 million tonnes of finished products out of which 14.6 million tonnes is pellets. The ore reserves and annual production of LKAB are summarized in Table 43.1.

Maintaining a safe working environment is the top priority at LKAB. The safety and health work at LKAB has succeeded in reducing the number of accidents and cases of work-related illnesses by 70% in only 10 years. During the past 3 years, the number of accidents has been around 15 per million work-hours. The present goal is to have less than 10 accidents per million work-hours in the year 2000.

43.2 GEOLOGY

The ore body at Kirunavaara is approximately 4,500 m long. Width varies from a few meters up to 150 m, but averages about 80 m. The depth of the ore body is not known, but has been

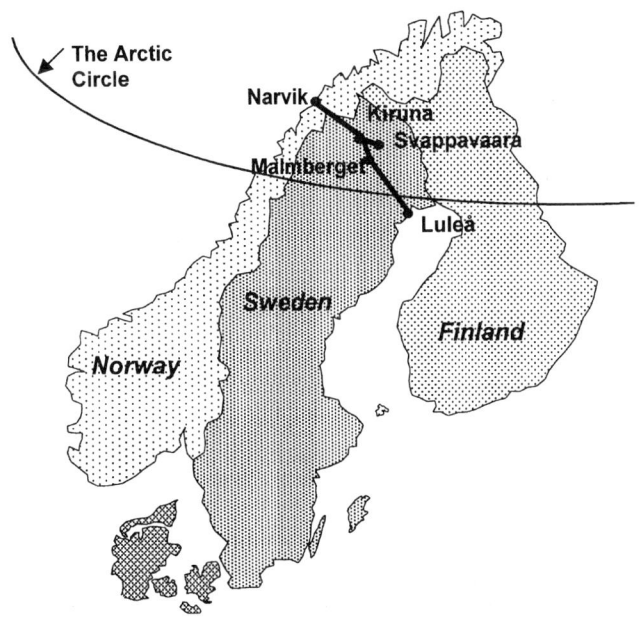

FIGURE 43.1   Location of LKAB activities

TABLE 43.1   Ore reserves and production at LKAB, million tonnes

	Kiruna	Malmberget
Planned crude ore production, 2000	23.3	13.8
Proven ore reserves	500 (to 1,045 m)	150 (to 800 m)
Probable ore reserves	350 (1,045–1,300 m)	145 (800–1,100 m)
Possible ore reserves	280 (1,300–1,500 m)	200 (1,100–1,500 m)
Total product capacity, 2000	13.6	7.6
Pellet capacity, 2000	10.4	4.2
Fines capacity, 2000	3.2	3.4

modeled to a depth of around 2,000 m. The ore is mainly a fine-grained magnetite with a varying content of fine-grained apatite. The ore contacts are distinct and sharp, but both the hanging wall and the footwall are brecciated and impregnated with magnetite. The ores are tectonically deformed with a lateral dip of 50° to 70°. There are two distinct ore grades at Kirunavaara: (1) a low-phosphorus ore with less than 0.1% phosphorus and (2) a high-phosphorus ore with more than 0.5% phosphorus. The iron

---

* LKAB, Sweden.

content is approximately 67% in the low-phosphorus ore and 60% in the high-phosphorus ore. The two different ore grades are made into different products, and major efforts are made to keep the different qualities separate during mining.

The footwall at Kirunavaara consists of syenite porphyry of good strength, while the hanging wall consists of quartz porphyry. There are a few veins of syenite porphyry that cut through the ore.

The ore reserve at Malmberget is distributed over some 20 large and small ore bodies, of which 13 are currently being exploited. The ore bodies are found in an ore field 5 km long by 2.5 km wide. In the western part of the field, the ores form a more-or-less continuous, undulating band of lens-shaped ore bodies. The ores in the eastern part of the field are intensively folded and tectonically deformed. The ore bodies have pronounced folds with a lateral dip of 45° to 70°. The width of the ore bodies varies from 20 to 100 m. About 90% of the ore reserve consists of magnetite, and the rest is hematite. The ores are mainly medium to coarse grained. The iron content varies between 54% and 63%. The hematite ore bodies are found in the western part of the ore field; these ores are also high in phosphorus. In the phosphorus-rich ores, the phosphorus usually occurs as apatite bands of varying width.

The country rocks at Malmberget consist of metamorphosed volcanic rocks such as gneisses and fine-grained feldspar-quartz rocks, called leptites. Granite veins often intrude the ore.

43.3 ROCK MECHANICS

The ore at Kiruna has a uniaxial compressive strength varying between 115 to 190 MPa, depending on the apatite content. The side rock has a uniaxial compressive strength between 90 and 430 MPa. This variation depends on the degree of weathering and on the distribution of included minerals. On a large scale, the ore body is situated within a tectonically relatively undisturbed triangle, which means that the ore body is not intersected by any regional thrusts or faults. Locally, some major structure systems are distinguishable. The "Captain's fault," which strikes northwest-southeast and dips steeply to the south, is the one that causes the most severe rock mechanics problems as it often contains a large amount of chlorite. The second group strikes and dips parallel to the ore body, and the third group strikes northeast-southwest and dips at a shallow angle. The rock quality designation (RQD) of the wall rock changes with depth. In the upper part of the mine (down to about 900 m), the average RQD is about 50%. In the lower part of the mine, RQD is about 75% to 80%.

At the Malmberget mine, the ore has a uniaxial compressive strength between 85 and 140 MPa, the red leptite a strength of 170 to 220 MPa, and the gray leptite a strength between 70 and 160 MPa. The gray leptite often carries a very high amount of biotite and kaolin concentrated in thin, spatially separated layers in this otherwise rather strong rock. The layering is very often parallel to the ore bodies.

Over the entire ore field, one structural group strikes and dips subparallel to the ore bodies. The RQD for the ore varies between 35% and 60% and for the side rocks between 55 and 75%.

43.4 MINING METHOD AND EQUIPMENT

43.4.1 Introduction

The main mining method used at LKAB mines is large-scale sublevel caving. Figure 43.2 shows a schematic view of this mining method. The choice between transverse and longitudinal sublevel caving depends on the thickness of the ore body. At Kiruna, the ore body is about 80 m thick, and transverse sublevel caving is used. In Malmberget, both transverse and longitudinal sublevel caving are used with sublevel intervals that can vary

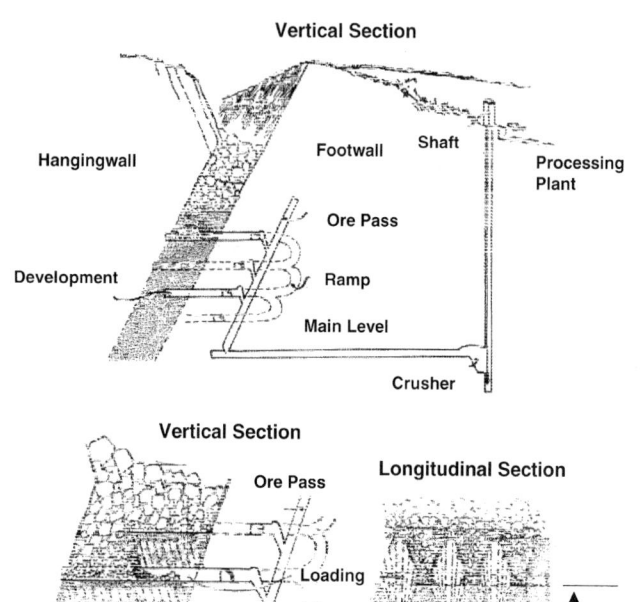

FIGURE 43.2   Sublevel caving layout at the Kiruna mine

from 20 to 30 m. Currently, there are 13 different ore bodies being mined at the Malmberget mine. In Kiruna, the sublevel intervals are 28.5 m and the production drifts are spaced at 25 m along ore body strike. The production drifts are 7 m wide and 5 m high. From these drifts, holes are drilled upward in fan-shaped patterns with a burden of 3 m. The fans are charged with explosive, and then blasted one by one. A typical fan contains about 10,000 tonnes of ore.

Load-haul-dump (LHD) machines transport the broken ore from the blasted fans (drawpoints) to the ore passes. A group of four ore passes is positioned about every 400 m along the ore body strike. The current production levels at Kiruna are 765 and 792 m. Level 820 is under development. Reference level 0 is located at the top of Kirunavaara Mountain. Thus, the mining depth at Kiruna is about 800 m today. The main haulage level is the 1045. Driverless trains transport the ore from the ore passes to the crusher stations. After crushing, the ore is hoisted in two stages to the surface. Figure 43.3 shows a schematic view of the ore body with ore passes, main haulage level, and hoisting system.

In Malmberget, truck haulage is used on the main level because of the scattered nature of the ore bodies. Figure 43.4 shows the layout of the new haulage level called M 1000 and the different ore bodies being mined.

43.4.2   Kiruna Mine

The drifting equipment used at the Kiruna mine consists of two Atlas Boomer H-353 S rigs with the capacity to drill 5-m-long holes and two Tamrock Para rigs with the capacity to drill 3.8-m-long holes. All these rigs have three booms and are equipped with instruments for controlling both horizontal and vertical angles (Bever or T-Cad). The Atlas rigs are used to drill in ore and the Tamrock rigs in waste. All drill holes are 48 mm in diameter. The planned amount of drifting for 2000 is 16,500 m.

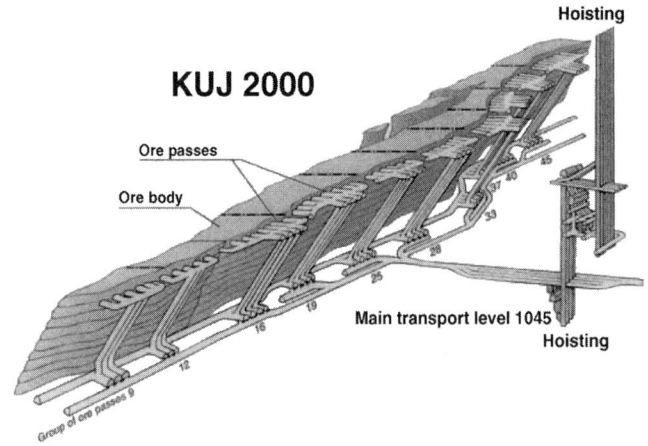

FIGURE 43.3   Mining layout at the Kiruna mine

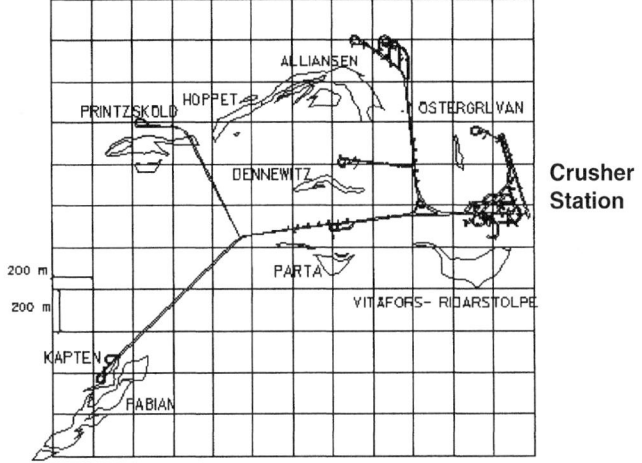

FIGURE 43.4   Layout of the 1,000-m haulage level at the Malmberget mine

KIMIT AB carries out the charging operation for both drifting and production in Kiruna. This company is owned by LKAB and manufactures and sells explosives and related technical services. Both the Kiruna and the Malmberget mines use the same explosive, a pumpable emulsion called Kimulux R. The amount of explosive used for drifting is about 3.7 kg/m^3. However, a current project at the mine aims to reduce this amount to about 2 kg/m^3.

The scaling and reinforcement equipment used in Kiruna is similar to that used in Malmberget Mine and is described in the section on the Malmberget mine.

For production drilling, there are five Atlas Copco Simba 469 W rigs, each equipped with one water-powered Wassara in-the-hole hammer supplied by G-Drill (a LKAB subsidiary). These machines are controlled remotely. After drilling a full ring in automatic mode, the operators must move the rig manually to a new set-up position and alignment. A Siemens programmable logic controller (PLC) S5 and Profibus are used for communication with the operator console.

These rigs were developed in a joint project between LKAB and Atlas Copco. The rigs have a maximum drill hole deviation of 1.5% at 30 m, a penetration rate of about 0.6 m/min, and an

average daily capacity of 317 m per machine. Every machine drills about 100,000 m of 115-mm in diameter hole annually. There are also two Tamrock Data Solo 1060 rigs. These rigs are also full-ring automated. Hole diameter is 115 mm, and each machine drills about 75,000 m of hole annually.

The loading operation at the Kiruna mine is carried out by electrically powered LHDs. Presently, there are nine Toro 2500Es having a capacity of 25 tonnes, one Toro 650D (diesel) with a capacity of 16 tonnes, and 11 Toro 500E/501Es with a capacity of 14 tonnes. The annual capacity of the LHD fleet is about 24 million tonnes of crude ore.

Automated loading at Kiruna has been a reality since 1999. Today four of the nine Toro 2500Es are fully automated and are operated from a control room on the 775-m level. The only operation that is carried out by remote control is filling the bucket at the drawpoint. Once this operation is completed, the automated system takes control and directs the LHD to the ore pass, where the ore is dumped, and then to the front for filling of a new bucket.

A wireless underground communications system (WUCS) has been developed for the transfer of images, audio, and data to make automation of autonomous vehicles at LKAB possible. This system consists of mobile terminals (vehicles), base stations, and control stations. A wireless spread-spectrum link at 900 MHz provides communications between the mobile terminals and the base stations. Spread-spectrum radio, digital image processing, image compression, and advanced high-speed data switching are the most significant technologies used in the WUCS. The system is designed to control up to 15 LHDs automatically and simultaneously from the six operator stations in the control room. The system reads data from the mine planning system and generates missions for the loaders. The production monitoring system receives information regarding every bucket dumped into the ore pass.

The current main haulage level is located at 1,045 m. This level is the third level with unmanned trains. LKAB has been using this technique since the 1970s. There are currently eight ore pass groups along the ore body, giving a total of 32 ore passes. With the development of the Lake ore, there will be 10 ore pass groups. The trains transport the ore from the ore passes to four gyratory crusher stations. The crushers have a discharge opening of 100 mm. At the skip level, located under the crusher station, the ore is distributed to a system of four shafts through which the ore is hoisted to crusher bins on the 775 level. From this level, the ore is distributed to a system of six shafts for further hoisting to the processing plant on the surface.

A total of seven driverless trains controlled by an Alfa Laval automation system are used on the main haulage level to transport the ore. The capacity of each train is 500 tonnes, and there are 24 wagons per train. Loading the trains in the ore passes is carried out by remote control from a control room. The operator has control of the train and the chute. Once the train is loaded, the system takes control and directs and empties the train into a crusher bin automatically. During unloading, a sample is taken automatically and analysed in a X-ray spectrometer for elements such as iron and phosphorus. Integrated into the system is a production-monitoring and material-tracking system. The ore for each train dumped into the crusher station is followed throughout the whole production flow up to the processing plants.

43.4.3   Malmberget Mine

Production capacity of the Malmberget Mine is about half that of the Kiruna Mine. Since both mines use the same mining method, they employ similar equipment with only minor differences.

The size of the drift is 6.5 m wide and 5 m high. Drifting operations are carried out by electrohydraulic Atlas Copco jumbos. There is one Boomer H-186 S and two Boomer H-353 S rigs; all three have Bever control (horizontal and vertical angles). One jumbo is equipped with automated rod-adding systems (RAS) for drilling 7.5-m-long holes using two rods. The other two rigs use one rod that is 4.3 m long. To direct the drill holes, portable lasers are used. These jumbos have an annual drifting capacity of about 11,500 m. Two types of firing patterns are used in drifting, one with a standard cut and the other with a 250-mm in diameter large-hole burn cut.

The explosive used in drifting is Kimulux R, a water-resistant emulsion that is pumped into the hole. The same explosive is used in all holes in the round. The perimeter holes are, however, "stringloaded," that is, the holes are decoupled by not filling the complete diameter of the holes. The charging vehicle is equipped with a platform and charging equipment.

Mechanized scaling is carried out with two electro-hydraulic trucks equipped with Jama SBU 800 scaling equipment and Tex 610 hammers. For manual scaling there are two Bask rigs with canopied working platforms.

Rock bolting is carried out with two Tamrock Robolt rigs and one Jama bolting rig. Rock bolts are 2.3 or 3.05 m long and are grouted in the holes with resin cartridges or cement grout. Bolting capacity is about 50 bolts per shift per rig.

Three rebuilt Volvo BM A 20 trucks equipped with Vertex spray robots carry out shotcreting. Two Scania trucks feed the spray robots with shotcrete. The shotcrete layer is about 3 to 5 cm thick, and shotcreting capacity is about 15 m³ per shift per robot. The planned volume of shotcrete to be sprayed for year 2000 is about 12,500 m³.

For production drilling, there are three two-boom AMW-24-SS rigs and three fully automatic one-boom Atlas Copco W 469 electrohydraulic rigs equipped with water-powered DTH W100 hammers. The Atlas Copco rigs are remote controlled. The annual capacity of these rigs is 420,000 m of 115-mm in diameter drill holes. The holes can be up to 40 m long. Every drill fan has about 9 to 10 drill holes inclined forward by 10° from vertical. The burden between fans is 3 to 3.5 m.

The production holes are charged with Kimulux R, a bulk emulsion explosive with high viscosity. For a hole 155 mm in diameter, this explosive has a linear charge concentration of about 12.5 kg/m. The specific charge of a fan is about 0.25 kg/tonne. The hose-feeding equipment is mounted on the tip of an articulated boom. The whole charging operation is automatic once the hose is in the hole. All operations are remote controlled from outside the RU-charging truck.

The loading operation is carried out by electric- and diesel-driven LHDs: five Bison electric loaders with a capacity of 22 tonnes, seven diesel-driven Toro 650 D loaders with a capacity of 16 tonnes, and one diesel-driven Toro 500 D with a capacity of 13 tonnes. The annual capacity of these LHDs is about 13 million tonnes of run-of-mine ore.

Boulders that are too large to be loaded into the ore passes are collected in special places for secondary blasting. The equipment used for drilling is a Bask TMH 400.

There are two main haulage levels in the Malmberget mine, one at the 600-m level and the other at the 815-m level. A new main level has just started production on the 1,000-m level, but it will not be fully operational until 2002.

All ore transport from the ore passes to the underground crushing stations is carried out by diesel-powered trucks. The annual haulage capacity is about 11 million tonnes of crude ore on the 815-m level and about 2.5 million tonnes on the 600-m level.

The current truck fleet is composed of the following: six Sisu SRH 450 Mammuts with a capacity of 120 tonnes, one Sisu SR with a capacity of 50 tonnes, two Sisu SR Bambinos with a capacity of 35 tonnes, one Scania with a capacity of 50 tonnes,

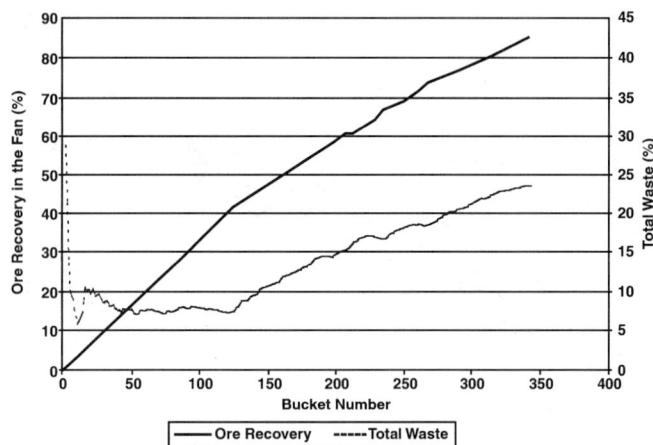

FIGURE 43.5 Dilution control during loading

three Scania Titans with a capacity of 70 tonnes, and one Volvo F12 with a capacity of 44 tonnes. Average haulage distance is about 1.7 km, and maximum speed is 50 km/hr.

There are 10 ore passes on the 815-m level and five ore passes on the 600-m level. Truck loading via a chute is controlled by the drivers remotely from inside their cabs.

On the 815-m level, two crushers with a capacity of 2,800 tonne/hr discharge material through a 130-mm in diameter opening. Two storage bins for crushed ore have a capacity of 4,000 tonnes each.

Two shafts with capacities of 950 tonne/hr each are available for transporting ore to the surface. They are equipped with a rope-guided, double-skip system with a maximum velocity of 16 m/s. Skip capacity is 23 tonnes.

43.5 DILUTION CONTROL

One of the main disadvantages of sublevel caving is the difficulty in controlling the inflow of waste during loading of the blasted fans. Since the fans are thin (3 m) and long (about 40 m) and blasted against caved rock, great care must be taken to assure that there is optimal ore recovery while keeping dilution under control. Planned ore recovery is 80% with a tolerated waste content of about 25%.

At LKAB mines, an on-line system measures, monitors, and evaluates the degree of dilution at the loading fronts. This system is based on measuring and analyzing the weight of every loaded bucket from the fans. The LHDs are equipped with a load cell device that can store information in a memory pack. At the end of the shift, the loader operator dumps this information into a database via computers linked to the mine network. The information is processed, and results are presented to the operator in terms of ore recovery and waste inflow per fan. Figure 43.5 is a diagram showing waste and ore recovery during loading of a fan.

43.6 VEIN MINING AT MALMBERGET

43.6.1 Introduction

In Malmberget, a number of smaller ore bodies having a relatively flat dip are situated some distance away from the other mining areas. In 1975, investigations began to evaluate ways to mine these ore bodies economically.

The final proposal was sublevel stoping, in which raises with roughly the same dip as the ore body are driven between an upper access drift and a lower loading drift. Drilling and charging the horizontal blastholes is carried out from platforms in the raises. The fragmented ore is loaded from the lower drift. This

TABLE 43.2   Vein mining at Malmberget mine

Ore plan area, m²	8,000
Level height, m	125
Stope area, m	50 by 12
Ore volume, m³	1,000,000
Ore tonnage, tonne	4,400,000
Iron content, %	58
Dip, °	50 (average)
Ore thickness, m	10–18
Compressive strength, ore, MPa	62.4
Compressive strength, hanging wall, MPa	115.6
Compressive strength, footwall, MPa	89.6

method, suitable for mining of thin ore veins, and hence called vein mining, was initially tested in the Indian ore body with satisfactory results.

The vein mining method was selected for mining the Baron ore body beginning in 1991. From a longitudinal drift on the 450-m access level, a crosscut was driven to each raise. Development on the loading level at 575 m consisted of three crosscuts for each room. A total of five inclined raises were excavated between the 575- and 450-m levels using the Alimak raise climbing method in which drilling and blasting were carried out from the raise climber. The raises are fully shotcreted and reinforced with grouted bolts where considered necessary. Current plans are for five rooms to be mined in the Baron area. No decision has been made as yet about how to mine out the 15-m-thick pillars between rooms. To date, about 850,000 tonnes have been extracted All mining operations are being performed by contractors, and LKAB is responsible for the planning. All ore is loaded with a Caterpillar 980F wheel loader into 35-tonne-capacity haulage trucks that transport the ore to the crusher on the 600-m main haulage level. Table 43.2 summarizes mining at Malmberget. Figures 43.6 and 43.7 show the mining layouts.

43.6.2   Geology

The footwall consists of red leptite with some granitic dykes and skarn. The occurrence of hematite veins and biotite-rich gray leptite increases towards the 600-m level. Toward room A, there is pure granite at the ore contacts. Beyond room A, hematite replaces the magnetite ore. Granitic dykes divide the ore body. The hanging wall consists of gray leptite, skarn, and dykes of granite and hematite. In the middle, close to room D, the hanging wall is pure granite.

43.6.3   Drilling

Drilling is carried out from a special raise climber equipped with a two-level platform. There is also a separate lift cage for personnel transport and a backup lift for emergencies. The drilling operation is fully remote controlled. The operator sits in the safety and comfort of a soundproof hut on the 450-m level. Three closed-circuit TV cameras on the drilling platform give the operator the ability to monitor and control drilling. The drilling system is custom built, with a pneumatically powered Atlas Copco ITH as its centerpiece. Hole diameter is 102 mm, and length varies from 10 to 25 m depending on the distance to the ore boundaries. The shortest holes are inclined slightly upward, and the longest slightly down (to ±10° from the horizontal plane). This results in a slightly domed surface.

Burden between each ring is usually 3 m, and the number of holes in each ring varies between 21 to 25, with an average drilled length per ring of 450 m. Drilling productivity averages

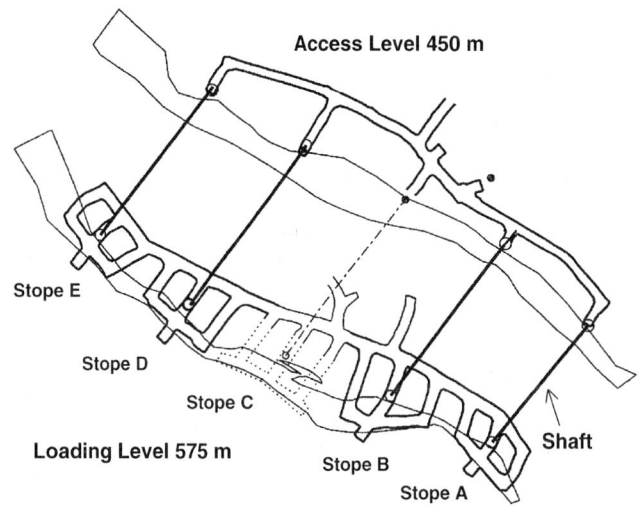

FIGURE 43.6   Horizontal vein mining layout at the Malmberget mine

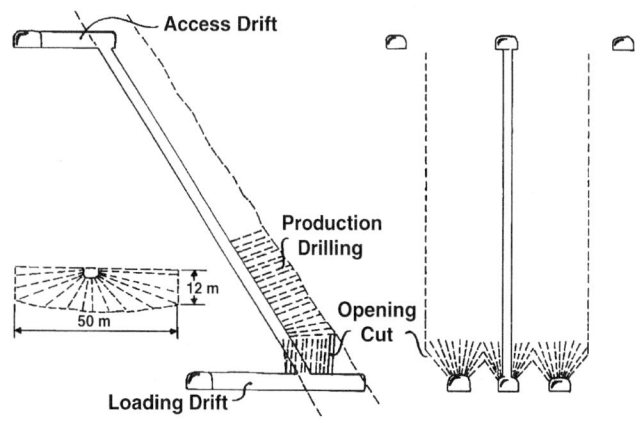

FIGURE 43.7   Vertical vein mining layout at the Malmberget mine

160 m/d over two shifts. The first room mined, room C, required 42 rings with a total drilled length of 19,000 m.

43.6.4   Blasting

When drilling has been completed in a raise, the equipment is changed, and the sequence of blasting and loading starts. Charging is performed pneumatically with prilled ANFO since slippery floors resulting from spillage of emulsion are considered a safety hazard. The long vertical distance from the access level to the lower rings makes handling of hoses impractical, and all explosives must be carried down in tanks on the charging platform.

Each hole is initiated with two Nonel blasting caps, one at the top and the other at the bottom of the hole, as a precaution against misfires. In each hole, the collar region is left uncharged. This length is 3, 6, and 9 m for three holes. The same sequence is repeated for the whole ring to maintain a uniform powder factor. The results of blasting are usually good, although it has happened that unfortunate fracture orientations have resulted in large back breaks between rings. The drilling pattern is designed to meet a theoretical specific charge of 1.2 kg/m³, or 0.27 kg/tonne of ore. The shortest holes are fired first, followed by the side holes. Several holes are fired on the same delay interval.

43.7 MAINTENANCE

After a recent reorganization, the company is now divided into divisions of production, service, marketing, etc., with no regard to geographical location. The difficult task of maintaining all production assets rests with the maintenance branch of the service division.

Mining equipment is being serviced and repaired in underground workshops (in Kiruna, this occurs on the 775-m level and in Malmberget on the 600- and 815-m levels) and by mobile service groups. All equipment has a schedule for maintenance, with service at regular intervals. The aim is to prevent breakdowns and emergency repairs, minimize interruptions to production, increase productivity, lower costs, and simplify planning.

LKAB staff in the underground workshops do the major part of maintaining mining equipment, although some contractors are using the same facilities. A few equipment suppliers with contracted maintenance of the delivered machines are running their own workshops underground. Since all mobile mining machinery moves on rubber tires, the infrastructure in the mines is based on road access from the surface to all active areas. This makes rapid field repairs possible and facilitates the relatively easy hauling of damaged machinery to well-equipped workshops. The most frequently needed spare parts are kept in depots co-located with the workshops or in the central storage facilities above ground.

All service and repair work is documented, and the statistics are utilized (1) to revise schedules continuously for maintenance and (2) to identify eventually needed modifications of machinery or service routines. The close cooperation (often on a daily basis) between maintenance staff, research institutions, other mining companies, and equipment manufacturers has resulted in a long list of new and/or improved machinery designs yielding better performance and a reduction in costs.

As machinery tends to get more complicated, the need for highly qualified maintenance staff increases. A slow but obvious trend at LKAB, as well as in other places in recent years, is that more and more of the maintenance duties are being performed by specialized contractors or by the equipment suppliers themselves. This trend is very likely to continue in the future.

Table 43.3 presents the mining equipment used at LKAB mines. Average availability as well as the size of the fleet are given for each piece of equipment at each mine.

43.8 SAFETY AND HEALTH

Risk management in the mines requires preventive measures in many different areas. Some are easily seen and understood while others are long-term and invisible. Every employee must be aware of the risks and take part in preventive risk management. The responsibility for maintaining a safe working environment is decentralized at LKAB and lies in the production organisation. This is because a safe working environment requires an overall view in which safety and health issues are a natural part of everyday work in the mines. This way of working with safety and health is based on Swedish legislation (AFS 1996:6) that prescribes a systematic and holistic approach. At LKAB the following risks in the mining environment have been identified:

- Falling rock
- Gas
- Radon
- Dust
- Remote-controlled vehicles, hoists, and trains
- Explosives

TABLE 43.3 Mining equipment at LKAB

Task	Number		Availability, %	
	Malmberget	Kiruna	Malmberget	Kiruna
Development				
Drifting				
Atlas Copco boomer H353	3	2	95	89
Tamrock		2		
Charging, KIMIT RT-charging vehicle	2	c	96	NA
Scaling				
Jama SBU 800	2	6	76	91
Bask	3		NA	
Bolting				
Tamrock Robolt	3	2	86	90
Jama	1		NA	
Shotcreting				
Jama	1	c	91	NA
Vertex	2			
Production				
Drilling				
Atlas Copco Simba W469	4	5	96	90
Solo 1000 sixty		2	92	NA
Loading				
Bison	5		65	
Toro 2500 E		9		88
Toro 500 E		11		NA
Toro 650 D	7	1	85	NA
Toro 500 D	1	1	NA	NA
Ore transport trucks				
Sisu SRH 450 Mammut	6		78	
Volvo 90-tonne	1		98	
Scania 70-tonne	3		98	

In addition, there are other risks that are common to all transport-related activities and other heavy industries. These are noise, working alone, stress, traffic-related risks, etc. All risks are dealt with in specific ways that are, above all, preventive.

Careful machine scaling of the drifts, which enables the scaler to work in safety, prevents rock fall accidents. Machine scaling was introduced in the early 1980s, and the number of scaling-related accidents thereby dropped from about 10 a year to none or 1 a year in the Kiruna Mine (Figure 43.8). In addition to this preventive work, every reported incident associated with falling rock is investigated so that scalers and other mine personnel can learn more about the causes of rock falls and how they can be prevented.

Nitrous oxide, carbon monoxide, carbon dioxide, and other gases that are mainly produced by blasting and diesel-driven vehicles are prevented from reaching dangerous levels by an effective ventilation system. Gas levels are carefully and continuously monitored to ensure that the levels do not exceed the limit values.

Radon is a long-term hazard. The alpha radiation emitted by the gas can, after many years of exposure, cause lung cancer. In some parts of the mines in Kiruna and Malmberget, there is radon-carrying rock. Radon exposure is avoided by effective ventilation and by continuous measuring of the radon levels.

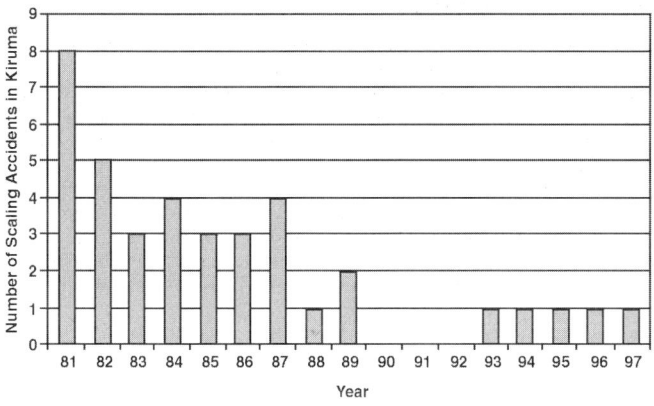

FIGURE 43.8   Accidents related to scaling at the Kiruna mine

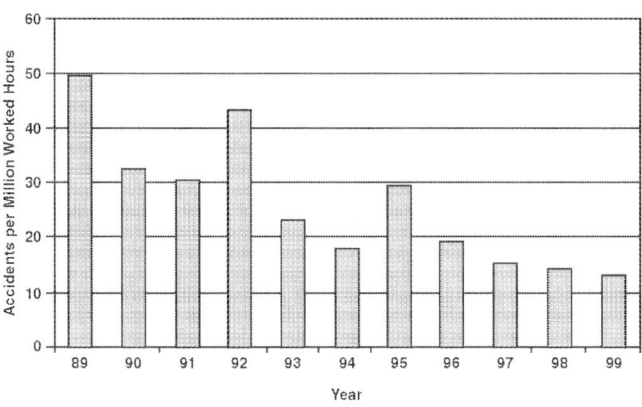

FIGURE 43.9   Accident rate at LKAB

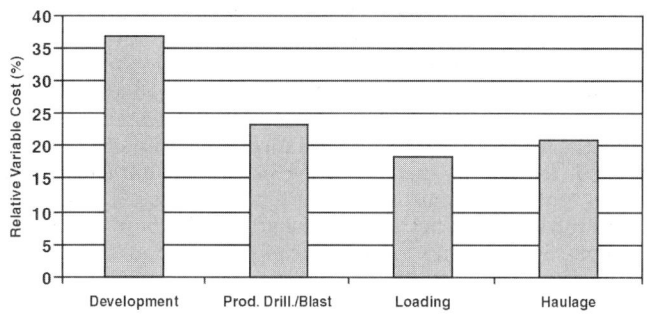

FIGURE 43.10   Relative variable mining cost at LKAB

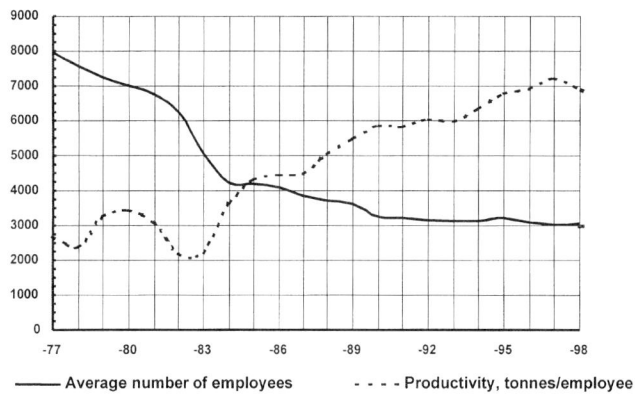

FIGURE 43.11   Employees and productivity at LKAB

Personnel working where there is a risk of radon exposure carry personal dosimeters to ensure that the annual maximum dosage of radon is not exceeded. If there is a risk that the annual dosage will exceed the permitted maximum, the person will be relocated to other work where there is no risk of radon exposure.

In the past, dust, especially from rock containing silica, was a major hazard for mine workers, causing the lung disease silicosis. Today, dry, dust-creating processes have been replaced by wet processes. This way, the dust is bound by water and cannot be a risk to the employees. However, measurements are conducted regularly to ensure that the levels of respirable dust are below the limit values. No cases of silicosis have occurred at LKAB since the early 1980s.

Remote-controlled vehicles, trains, and hoists are risks because an operator is a long way away and cannot see if someone is near the machine. These machines, therefore, require strict safety routines. The working areas around the machines are fenced off so that no one can accidentally be placed in danger. Maintenance personnel are required to have special training in safety routines and are given a certificate before they are permitted to do any work within these fenced-off areas.

Blasting and other explosives-related risks are also covered by strict safety routines. Blasting is permitted only at specified hours every night, and even then, any personnel in the mine are required to be at one of the predetermined meeting points.

The safety and health work at LKAB has succeeded in reducing the number of accidents and cases of work-related illness by about 70% in only 10 years (Figure 43.9). The present

goal is to have less than 10 accidents per million work-hours in the year 2000.

43.9   MINING COSTS

Relative variable mining costs for LKAB mines are presented in Figure 43.10. It is evident from this figure that development represents a major cost item in sublevel caving. It makes up about 37% of the variable mining costs at LKAB for 1999. This demonstrates the importance of scaling-up the sublevel caving layout, thus decreasing the amount of drifting, in order to reduce mining costs.

43.10   PRODUCTIVITY

Figure 43.11 shows the change in productivity and number of employees over the past 21 years at LKAB. Over this period, the productivity, calculated as the tonnage of products produced per employee, has increased from about 2,700 to about 7,000 tonnes per employee. This represents an increase by a factor of about 2.6, yielding an average annual productivity increase of about 4.8%. The scaling-up of sublevel caving through the years has had an important effect on the increased productivity at LKAB.

43.11   CONCLUSIONS

LKAB is striving to be the most attractive supplier of highly upgraded iron ore products to Europe and other prioritized markets.

LKAB's goal is to produce at the current production capacity of 26 to 28 million tonnes a year and to increase the percentage of pellets from 70% to 80%. Continued profitability is LKAB's overall economic objective. The long-term yield requirement is 15% on invested capital (before taxes) as measured over an economic cycle. LKAB will continue to improve the efficiency of the total mining system. The future competitiveness of our company depends upon a continuing program of rationalization leading to cost reductions along the entire pipeline, from the rock in situ to the shipping ports, while steadily improving an already high product quality. The primary mining cost reduction potential for LKAB does not lie in a further increase of either the scale of the mining method or the machinery. Rather it must occur through optimizing the current extraction design and utilizing the personnel and machinery to the best advantage possible. Sensible automation is one of the important ingredients in this overall process.

43.12 REFERENCES

AFS, Arbetarskyddsstyrelsens Författningssamling. 1996:6. Internal Control of the Working Environment (in Swedish).

Engineering Mining Journal, November 1996. LKAB Invests in the Future.

Marklund, I. 1982. Vein Mining at LKAB, Malmberget, Sweden. In *Underground Mining Methods Handbook*, W. A. Hustrulid, ed. New York: SME, AIME, pp 880–987.

Petersson, A., Halonen, T., and Hedström O. 1998. The Planning and Design of the New 1000 m Haulage Level at LKAB, Malmberget. Presentation at SME Meeting 9–11 March, Orlando, Florida,USA.

Samskog, P.O., Wigden I., and Tyni H. 1999. Process Automation in Ore Mining. International Congress Mining for Tomorrow's World, Dusseldorf, Germany, June 8–10.

CHAPTER 44

Sublevel Cave Mining Update at INCO's Stobie Mine

Harvey Buksa[*]

44.1 INTRODUCTION

Stobie Mine, a low-grade, nickel-copper underground bulk mining operation in the Sudbury basin in northern Ontario, continues its drive to reduce costs and improve productivity. After mining in the Sudbury basin for over 100 years, Stobie continues to enhance its operations to ensure that mining of the remaining ore continues to be practical and profitable. Advances in all areas of production–development, drilling, loading and mucking– combined with a vision to leverage teleoperation and information management will ensure the economic viability of Stobie Mine into the next decade.

The main focus of the improved process is to utilize effectively the highest degree of mechanization and teleoperation that is practical to improve recovery of the resources available to the mine. This, combined with planned quality improvements in all parts of the process, will allow for less rework, shorter cycle times, and higher running times for both the equipment and the process. Reengineering and redesigning the process to capitalize on the advances achieved in mechanization, automation, and teleoperation are an important part of the responsibilities of the Stobie management team.

44.2 STOBIE MINE HISTORY

The Stobie Mine ore body was discovered in 1885. Production from an open pit continued sporadically until 1901. In 1941, production was restarted from the open pit, and sinking the No. 7 shaft was initiated. The No. 7 shaft was completed to below the 2200 level by 1944, and preparations for underground production were started. Underground production commenced in 1948, and open-pit production was completed by 1950. The No. 8 shaft was started in 1951 and was completed the following year. Subsequently, in 1959, the No. 8 shaft was deepened to below the 2600 level. The No. 9 shaft development commenced in 1965, and by 1968 was completed to the 2600 level. First production through the No. 9 shaft was in 1971. The No. 7 shaft was deepened in 1975, and production from the 3950 loading pocket started in 1977.

Production at Stobie Mine to date has exceeded over 130 million tons at a grade of about 0.73% copper and 0.78% nickel. Several mining processes using both fill and open stoping have been used in the past to extract the ore.

44.3 STOBIE MINE ORE BODY

The Frood-Stobie ore body occurs in an offset in the footwall. It is a steeply plunging ore zone dipping at 70° to 75°. The offset is composed of a quartz diorite body in a zone of basic metavolcanics and metasediments, intruded by Murray granites, and brecciated by the Sudbury event to form an extensive zone of Sudbury breccia. Its surface dimensions are 9,600 ft along strike with a maximum surface width of 900 ft. The ore zone tapers out at various depths above the 4100 level.

Two main types of ore are found at Stobie: disseminated and massive sulphide. For mining purposes, the disseminated ore is largely mined by the sublevel caving method and the remainder is mined by vertical retreat methods.

The hanging wall and footwall range from hard, competent quartzite to a relatively soft and weak metagreywacke. Local shearing running parallel to the walls of the offset has strongly altered these rocks. Generally the ore body, the metabreccias, and the metavolcanics are hard, competent rocks with some areas displaying well-developed jointing.

44.4 STOBIE MINE AT NORMALIZED PRODUCTION

The ore body is divided into a low-grade, sublevel caving mining area and a high-grade vertical retreat mining zone. Total annual production has been 3.3 million tons, with about 70% of the production coming from sublevel caving.

In the next several years, vertical retreat mining will be phased out as the high-grade ore is depleted. Production needs to be replaced with additional ore production from sublevel caving to maintain a viable mining operation at current or higher production levels. This move to a bulk, high-volume mining method is necessary as it is the only way to improve productivity and lower costs sufficiently to continue mining this low-grade ore body.

The sublevel caving method at Stobie has been redesigned to take advantage of the improvements that have been made in equipment, teleoperation, and information management in the last few years. Drifts are supported by shotcrete so that bolts are used only in specified instances. This quality improvement in the drill drift support allows for teleoperation of the drills. It also largely eliminates rework and greatly enhances worker safety in the total production cycle.

A major change in processing time and the amount of labour required to support the roof during the development cycle is the result. For instance, instead of one miner to bolt and screen 12 ft of advance in a shift, it is possible to complete the support of a round in less than 30 min with shotcrete. This is over a four-fold increase in support productivity. In a multiheading environment, found in sublevel caving development, substantial direct development productivity improvements are realized when shotcrete is used as the primary support.

Electric-hydraulic drills equipped with straighter drilling tube rods allow for both longer and more accurate teleremote drilling. All drilling is now standardized at a nominal 4 in. in diameter. Emulsion explosives loaders allow these longer holes to be preloaded for subsequent priming and blasting as needed. This improvement in drilling and loading capability has, in turn, allowed mine design to change from 70-ft sublevel intervals to 100-ft sublevel intervals, with some apex holes up to 130 ft long.

---

* INCO, Ltd., Ontario Division.

Loading the production holes is carried out with a toe loading emulsion loader operated by one man capable of loading more than 10,000 lb of explosive emulsion per shift. The higher quality of the explosive allows preloading to take place, which in turn facilitates the production cycle by allowing blasting to take place as required, reducing delays. The 100-ft lifts not only mean fewer blasts with fewer related blasting delays, but fewer set ups (delays) for all elements of the drill, load, blast, and muck cycle. Two- to three-ring blasts now provide the total daily production requirement as compared to the previous requirement of four to six rings daily for 70-ft sublevels.

The standard crosscut is 16 by 16 ft, and the spacing between these crosscuts has been increased from 40 ft to between 50 and 60 ft. Ring spacing has increased to 8 ft, partially to encapsulate the enlargement in the pull envelope caused by the move to 100-ft sublevels. Toe spacing averages about 11 ft between holes in the same ring.

These design changes have reduced development requirements by over 30% and have increased available tonnage per ring blast by over 80%. Cycle time to develop and mine an ore zone is thus compressed, and profitability is therefore improved substantially in the 100-ft-sublevel area.

Design optimization on ring, crosscut, and hole spacing and relative position, explosives characteristics, hole size, ring angle, and crosscut size and configuration will be ongoing and will continue to evolve with experience and further on-site testing. Trade-offs among these variables are necessary to optimize the total process for muck size, recovery, dilution, and profitability. Variability in ore characteristics, water content, stress changes, etc., will continue to challenge the design team to balance the trade-offs and achieve the best results practical for the mine.

Based on model tests and actual experience to date, recovery and dilution rates are similar between the original 70-ft and the new 100-ft sublevels. Recovery ranges between 80% and 90%, and dilution varies between 10% and 20%. By simply applying different drawpoint cutoff parameters, both recovery and dilution can be varied at the discretion of the mine management team. An appropriate balance between recovery and dilution must be maintained for optimum results. A designed lower recovery on one level (deferred pull) allows a higher percentage of the ore to be recovered in subsequent levels, with the dilution being introduced from the deferred pull zone above having a higher grade.

Ore is handled with a fleet of 8-yd load-haul-dump (LHD) machines that haul to ore passes. These ore passes are covered with mantles with 36-in. in diameter holes to ensure that oversized material does not enter the pass, as this could cause hang-ups in the pass or other downstream blockage problems. The ore passes are distributed along the footwall access drift to limit the average haul to about 500 ft.

Ore is currently transported by a tipple-car track tram system on the 2200 level to either of two crushers for crushing to −6 in. before being hoisted to the surface and sent to the mill in standard surface rail cars.

The sublevel caving portion of the mine is approaching the main tram level on the 2200 level, so to continue production, a new ore handling system has been designed. This new system will be required to maintain production into the future. The main elements for the new system will include three new ore passes between the 2100 level and the 2600 level, a new truck transfer on the 2600 level, a single new crusher on the 2800 level, and a new conveyor to the 2400 level, No. 9 shaft.

44.5   PRODUCTION STATISTICS

Current overall mine-wide productivity in the sublevel caving zone at the mine averages better than 30 tons per manshift worked, which is two to three times more productive than other mining methods employed by INCO, Ltd., in the Ontario Division. This means that at 12,000 ton/d, operation labour requirements would be about 400 persons at work per day. To achieve this labour level for a continuous operation, about 590 people would have to be on the mine payroll.

Production scoops (8 yd) average about 100 tons per run-hour for the normal layout of ore passes and production stopes and drifts at Stobie Mine. The annual run time per scoop is 3,000 to 4,000 hr.

The electric-hydraulic drills average about 480 ft of 4-in. hole per day based on three shifts of operation on a 5-day week. The long-term average tons per foot drilled runs at 2.5 to 3.5 tons per foot. Therefore, for the same 12,000 ton/d operation, about 4,800 to 3,400 ft/d of drilling is required.

Production costs for sublevel caving run at about 1.5 to 3 times less than other typical mining methods.

44.6   SAFETY AND HEALTH

Sublevel caving is a very safe mining method. Shotcrete is used to provide ground support and basic protection for personnel and equipment in the mining area. Additional support, such as bolts, is applied only as necessary, based upon site-specific ground evaluations.

Production drilling is manned and operated from the surface so drillers are not exposed to the underground environment of vibration, noise, dust, dampness, etc., and the health of the worker is thus enhanced.

Personnel on-site carry out explosives loading with machines so exposure location and exposure time are both improved.

Overall, with fewer people exposed for much less time, the real safety afforded by sublevel cave mining results in dramatically improved health and safety for the workers.

44.7   CONCLUSION

This paper has discussed Stobie Mine as of today. The economic impact of new technology provides the opportunity for Stobie Mine to remain viable in the highly competitive world nickel market well into the future.

Sublevel cave mining with currently available modern technology is a highly productive, cost-effective mining method. Historically, sublevel caving has been viewed as a dirty (high dilution), low-recovery mining method that was considered more as a method of last resort than a preferred method. This view has limited the use of sublevel caving.

In reality, overall recovery and dilution for sublevel caving are not substantially different from those of other mining methods when both primary and secondary recovery are considered. In sublevel caving, all recovery is by primary mining, so secondary recovery is not required. It is a method that, due to the top-down mining required, makes eminent sense from both an investment and ground control point of view. Investment in development is required only as mining occurs.

Mining costs for modern sublevel cave mining operations, using technology available today, can approach open-pit mining costs. Low-cost, top-down mining combined with an early return on investment in development makes this a preferred method in appropriate ore bodies.

Mining Stobie low grade ore would not have been feasible with any other method. Advances in sublevel caving are continuing, and ongoing improvement will make this method even more economic in a wider range of ore bodies and circumstances.

CHAPTER 45

Longitudinal Sublevel Caving, Big Bell Mine

John Player[*]

45.1 INTRODUCTION

45.1.1 Property Information

The Big Bell gold deposit lies about 540 km north-northeast of the state capital of Perth at 27° 20′ S, 117° 40′ E (Figure 45.1) in the Murchison Province of the Yilgarn Craton, Western Australia. The mine is 30 km west-northwest of Cue and accessed via an unsealed road and a 1,500-m-long airstrip. Goods are transported by road 3 days per week to the mine from Perth.

45.1.2 Mining History

The mining history of the Big Bell deposit consists of four phases of mining and is summarised in Table 45.1. The four distinct phases identified are—

- Prospectors and early mining

- Underground mining operations managed by ASARCO

- Open-pit operations

- Recent underground mining operations[†]

The fourth phase of mining activity commenced in late 1994 with development accessing remnant mineralisation from the ASARCO phase of mining. The remnant material consisted of a low-grade halo about the previous mining operation, pillars between the stopes, and stope fill consisting of low-grade ore and mill tailings. For a detailed review of the remnant mining operations and problems at Big Bell Mine, see Player (1998).

The initial mine design called for primary stoping to be undertaken via the "core-and-shell" mining technique under a caving hanging wall. During 1997, difficulties associated with required production rates, blasthole closure, and pillar instability necessitated the conversion of the operation to an alternative mining technique, longitudinal sublevel caving.

45.1.3 Current Mine Production Rate

Development mining is undertaken by a mining contractor. Big Bell employees undertake all production activity except for loading explosives.

The production rate for the last 5 years is shown in Table 45.2. In 1999, 168,400 ounces were produced from the ore with an average mill grade of 3.1 gm/tonne. The daily production rate is 5,000 tonne/d using Elphinstone 73D trucks hauling up a 1-in-8 incline to the run-of-mine stockpile. Three production levels are required to meet the production schedules. At the time of writing, the deepest production level was 510 m below the surface.

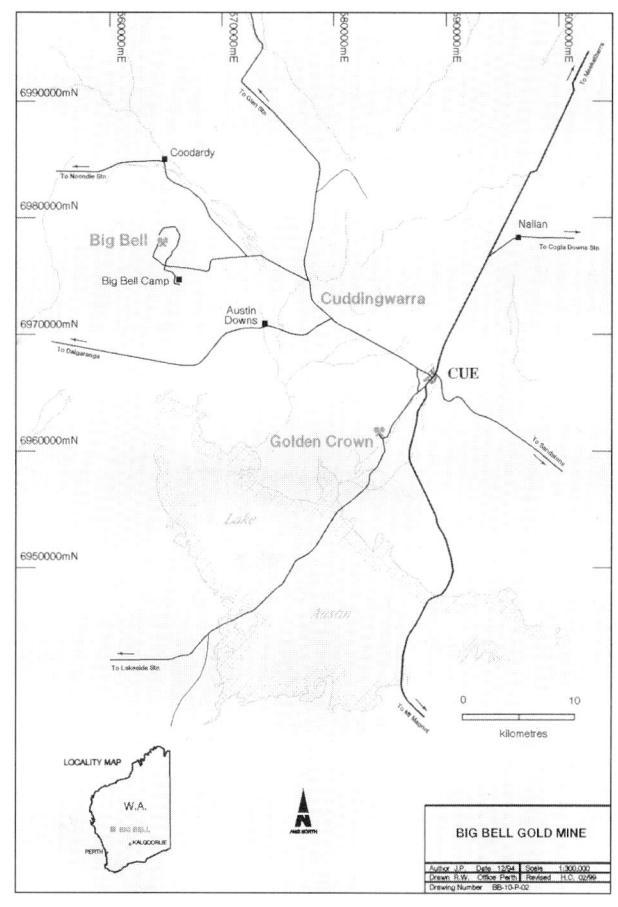

FIGURE 45.1 Location of Big Bell Mine

45.1.4 Safety Statistics

During 1999, 537,622 hr were worked with four lost-time incidents and 11 disabling incidents. For the year, the lost-time injury frequency rate was 9.3 per million man-hours. This figure is an improvement for the Big Bell operations from previous years and is an average performance for the Western Australian mining industry in general. For a small site such as Big Bell, a single lost-time injury results in a substantial change in the lost-time injury frequency rate.

---

[*] New Hampton Goldfields, Perth, Western Australia, Australia.

[†] Following two large seismic events on 17 June 2000 and 9 July 2000, it was decided to alter the mine layout and mining method significantly to mitigate the risk of another event. See the addendum following the conclusion.

TABLE 45.1   Mining and corporate history at Big Bell Mine

Period	Activity	Recovered tonnes	Grade, gm/tonne gold	Ounces
			Production	
Phase 1:				
1904	Gold discovered by prospector	330	29.0	308
1913–1923	Shaft and treatment plant operated by Chesson and Heydon.	64,482	5.21	11,000
Phase 2:		5,556,000	4.04	730,000
1937–1955	Underground mining by BBML. Operated through ASARCO.			
1969	ACM acquires title to area.			
1984	Joint venture with PPL..			
Phase 3:		11,710,000	2.20	847,000
1988–1991	Open-pit mining to 210 level, ACM-PPL..			
1991	NPL acquires ACM.			
Oct. 1992	NPL purchases PPL's share of joint venture.			
Feb. 1993	Open-pit mining ceases.			
Phase 4:				
Oct. 1994	Identification of total underground resources and submission of feasibility study. Reserves to 610 level. Underground mining commences.	12,000,000	3.50	1,350,000
Oct. 1997	Commencement of first level of longitudinal sublevel cave.			
Dec. 1997	Conclusion of remnant mining on all levels, but overdraw until mid-1998.	3,252,000	3.18	332,400
June 1999	Total underground recovery.			
	Total reserves	4,470,000	3.10	450,000
	Total resources	8,470,000	3.60	980,000
Jan. 2000	Sale of Big Bell to New Hampton Goldfields, Ltd. Normandy Mining takes a 40% stake of New Hampton as part of sale.			

BBLM = Big Bell Mines, Ltd. ASARCO = American Smelting and Refining Co. ACM = Australian Consolidated Minerals. PPL = Placer Pacific Pty. Ltd. NPL = Normandy Poseidon, Ltd.

TABLE 45.2   Production rate

Year	Tonnage	Grade	Ounces	Recovery, %
1994/1995	663,839	2.44	52,082	73.7
1995/1996	1,469,475	2.99	129,690	91.7
1996/1997	1,622,695	3.01	133,466	85.1
1997/1998	1,840,659	3.16	160,787	86.1
1998/1999	1,739,912	3.21	154,965	86.3

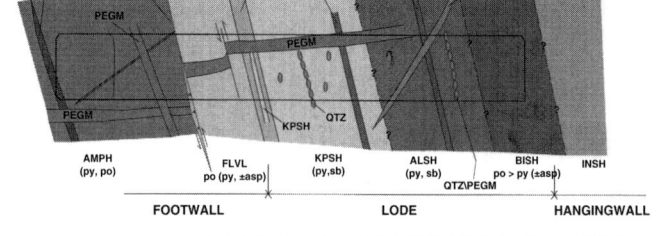

FIGURE 45.2   Geologic cross section through ore body

45.2   GEOLOGY

The Big Bell deposit is hosted by a greenstone sequence within the Murchison Province of the Yilgarn block. The greenstone sequence composes the western limb of a regional anticlinal structure that is about 1,500 m thick in the vicinity of the mine (Handley and Cary 1990). The deposit strikes at 030° from magnetic north and dips to the east at approximately 72°. Ore body dip varies from 55° to 80°. The deposit is pervasively foliated, and foliation parallels lithology. Cross-cutting features are few and are composed of intrusive pegmatite dykes of variable thickness. Little or no crosscut faulting is evident.

The geology of the deposit is illustrated diagrammatically in Figure 45.2. All access development is contained within the footwall amphibolite (AMPH), a basalt equivalent. The graphitic shear is a continuous graphitic breccia zone that parallels stratigraphy; a second graphitic shear structure is located 150 m west of the lode. The felsic volcanic (FLVL) is a banded quartz feldspar unit of uncertain origin. The corderite schist (CRSH) is

the footwall marker unit to the deposit and consists of biotite quartz and minor corderite. The Big Bell mineralisation is hosed within three units, K-feldspar schist (KPSH), altered schist (ALSH) and biotite schist (BISH). The three units consist of the following mineralogy:

KPSH—K-feldspar, quartz, white mica, aluminosilicates, pyrite.

ALSH—White mica, quartz, K-feldspar, aluminosilicates, pyrite.

BISH—Biotite, quartz, garnet, pyrite, pyrrhotite.

The Big Bell lode (KPSH, ALSH, BISH) system within the Big Bell Mine has been defined along strike for over 1,000 m and to a depth of 1,430 m. In plan view, the lode system is lenticular, varying from 5 to 8 m wide at the extremities and up to 50 m in the central area of the deposit. In longitudinal projection, a

TABLE 45.3   Big Bell rock properties

Rock type and statistical measure	Uniaxial compressive strength$_{50}$, MPa	Young's modulus, MPa	Poisson's ratio	Density, tonne/m^3
AMPH (footwall rock unit)				
Mean	178	72.0	0.28	2.85
Stand. deviation	62	6.9	0.07	0.05
Range	179	19.6	0.22	0.16
No. of samples	11	8	8	11
KPSH				
Mean	133	43.2	0.27	2.73
Stand. deviation	46	11.2	0.08	0.05
Range	140	34.6	0.27	0.18
No. of samples	17	14	14	17
ALSH				
Mean	112	44.2	0.21	2.79
Stand. deviation	48	13.9	0.11	0.06
Range	131	37.1	0.33	0.18
No. of samples	10	9	9	10
BISH				
Mean	103	51.4	0.23	2.90
Stand. deviation	30	13.2	0.06	0.07
Range	91	37.2	0.17	0.18
No. of samples	6	6	6	6

TABLE 45.4   Averaged in situ stress regime for footwall amphibolites on 365 level

Principal stress	Magnitude, MPa	Dip, degree	Azimuth, degree
Major	63.2	06	224
Intermediate	34.7	13	315
Minor	18.6	75	109

NOTE: Azimuth relative to mine grid north is 30.8°E east of Australian map grid. The surface is at zero level; therefore, the 365 level is 365 m below the surface.

TABLE 45.5   Load data for 485-level in situ stress regime

Principal stress	Magnitude, MPa	Dip, degree	Azimuth, degree
Major	69.1	27	274
Intermediate	34.3	6	007
Minor	29.9	63	109

NOTE: Azimuth relative to mine grid north is 30.8°E east of Australian map grid. The surface is at zero level; therefore, the 485 level is 485 m below the surface.

funnel shape is noted plunging near vertically. A 100-m-thick pegmatite zone cuts the ore body between 710 and 810 m below the surface.

Economic gold values are preferentially developed within the ALSH followed by the KPSH and BISH. Economic gold values are also associated with elevated pyrite content, but not exclusively so. The economic portion of the lode system is defined by grade; geological characteristics are inconclusive.

45.3   GEOTECHNICAL INFORMATION

45.3.1   Rock Properties

The pervasive foliation exhibits a significant influence on the response of the rock mass to stress redistribution and to test work on diamond-drill core. The schistose lode rocks are observed to exhibit nonlinear and anisotropic behaviour. Recording the relationship of foliation angle to loading direction, as well as failure mechanisms, is necessary when strength testing diamond-drill core. Rock properties at Big Bell are summarised in Table 45.3.

45.3.2   Rock Mass Characterisation

The rock mass at Big Bell can be subdivided into two broad domains: the footwall and the ore zone. The footwall is foliated but more massive, while the ore zone is schistose with mica well developed on foliation planes. From underground geotechnical mapping, seven joint sets have been identified, but at any particular locality two or three joint sets are generally present plus the ubiquitous foliation. Joints are usually planar, with rough but clean surfaces, and widely spaced. All rock units maintain a rock quality designation (RQD) between 90% and 100%.

The average Q-value (Barton et al. 1974) for the ore zone is 2.9, but can range from 0.4 to 12.5. The average Q-value for the footwall is 6.3, but can range from 2.1 to 15.0. The rock mass response to excavation and stress change for the two domains are

very different. Big Bell has notable differences in rock mass performance within the same domain for both footwall and ore body rocks, even where the Q-values are similar. The differences are considered to be due to the direction of the mining front and specific local rock properties.

45.3.3   In Situ Stress Regime

Three stress measurements have been undertaken at Big Bell (Tables 45.4 and 45.5). All have used the Commonwealth Scientific and Industrial Research Organization (CSIRO) hollow inclusion cell overcoring technique. At each site, a high degree of certainty was determined for orientation of the stresses. The results at 365 m (Table 45.4) are the average of two measurements from the 350- and 380-m levels within the footwall amphibolite. The measurement at 485 m (Table 45.5) was taken within the ore zone in proximity to the graphitic shear. The change in orientation is ascribed to the proximity of the graphitic shear with resultant local rotation of the stress field perpendicular to the structure.

45.3.4   Ground Deformation

Substantial stress redistribution is noted as the cave advances. Such redistribution directly influences the stability of the development openings. Stress redistribution, as well as local rock mass properties and structural features, affect the stability of development openings and the performance of the installed ground support regime. Once the cave is opened beyond a critical length (approximately 80 m), the footwall slowly starts to relax into the cave along tight, continuous faults that run between the lode and footwall graphitic shears. The relaxation is indicated by water coming from the backs of the previously dry footwall development behind the cave front. This movement can result in localised ground deterioration in unmeshed areas of the footwall development.

Stress loading through the lithological sequence may change the response of the footwall rock from being massive and stable with little dilation to seismic, as both the hanging wall and footwall rocks are relatively stronger and stiffer then ore body rocks. Shedding of load stress from ore zone rocks to the footwall is suggested as a role in the change of the footwall conditions from stable to seismic. This idea will be examined with the large area microseismic system.

Prior to the onset of seismically induced rock falls, the primary failure mechanism that reinforcement and support design were required to address was the intense fracturing above the backs of the ore drives caused by stress-induced failure of the rock. This was initially identified from exposures of the backs of old workings. The fracturing above the backs of the ore drives is

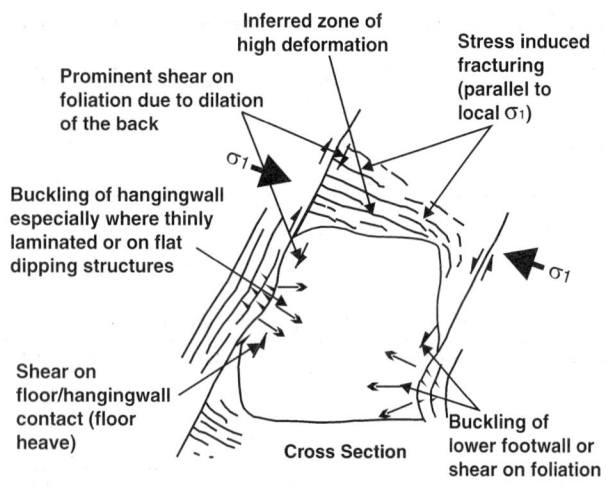

FIGURE 45.3   Nominal ground deformation behavior around ore drives

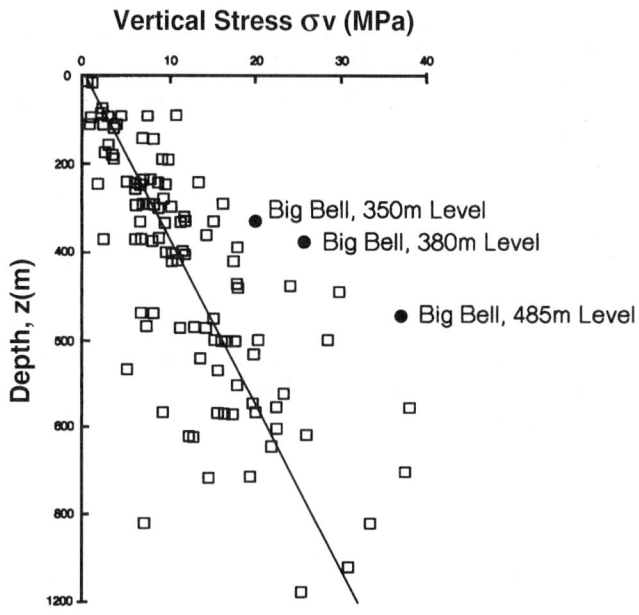

FIGURE 45.4   Measured vertical stress versus depth for Western Australia mines

accompanied by dilation, in turn promoting shearing of footwall foliation planes, particularly where these have been undercut by the drive profile. Dilation in the back also produces shear on foliation at the intersection of the back and hanging wall. An idealisation of these styles of ground behaviour is presented in Figure 45.3.

Near the cave front, a well-defined dilation zone can progress 25 to 30 m along the hanging wall corner of the ore drive with the advancing cave. Crack growth in the corner of the hanging wall is enhanced when either a weak, smooth foliation plane or lithological contact is in the hanging wall corner and acts as a release plane for the subhorizontal, stress-induced fractures in the drive back. Hanging wall shoulder dilation is not well defined until the first 30 m of caving has been advanced along the level.

At this time, hole closure problems will also become an issue. An increase in local seismicity in a drive during development has been correlated with large, shallow-dipping structures.

The main operational problems experienced following the introduction of sublevel caving were—

- Lost brows, preventing access to the collars of the next ring of blastholes, and

- Blasthole closure resulting from movement on structures near the collar of the hanging wall holes and stress-related damage of blastholes subparallel to foliation.

Loss of control at the brow resulted in back failures ranging from 0.2 to 8 m deep; approximately 1.5 m was typical. This adversely affected the production rate. The difficulties arising from brow failure are discussed in the section outlining the mine method. The problem of brow failure has been essentially alleviated by the use of strong brow support, which includes steel w-straps and grouted, mild steel threaded bars.

45.3.5   Seismicity

Stress measurements indicate a high and deviatoric stress regime at shallow operating depths (Figure 45.4). Seismicity that caused the first damage to the mine infrastructure began in February 1999. Table 45.6 lists the events with location and estimated magnitude. All but one event (the first) are located in the footwall amphibolite development. The mechanism is interpreted to be shear rupture along foliation, together with intact rock failure to the footwall of the shear plane (Figure 45.5).

TABLE 45.6   Seismic event history at Big Bell Mine

Date	Location	Level	Magnitude (M_L)*	Volume fallen/ ejected, m³
12 Feb. 99	Ore drive (3655 N)	460	NR	4
16 June 99	Footwall drive (3475 N)	435	NR	5
7 July 99	Footwall drive (2775 N)	485	NR	2
9 Aug. 99	Footwall drive (3790 N)	485	1.9	12
22 Aug. 99	Footwall drive (3805 N)	460	2.2	20
25 Nov. 99	Footwall drive (3820 N)	460	1.7	8
25 Nov. 99	Footwall drive (3845 N)	485	2.4	40
6 April 2000	Footwall drive (3775 N)	510	NR	3
11 April 2000	Footwall drive (3840 N)	485	NR	1

*Source: Australian Geological Survey Organisation. NR = Not recorded.

FIGURE 45.5   Shear failure in hanging wall plane

For seismicity to occur, it is considered that a number of criteria must be met (Mikula 2000).

- High stress conditions
- A stiff or burst-prone rock mass
- Relative geometry of the footwall development to the cave on the level above
- Release planes (either faults or foliation)
- Stress change resulting from a production blast
- Energy loading caused by production blasts
- Direction of mining retreat and stress concentration

Where a seismic risk is determined for the footwall development on the level underneath a production blast, then either a 12-hr or 24-hr exclusion zone is used. This depends on whether seismic-resistant support is installed. Necessary warning signs are noted underground and on the tag board before the firing of a shot. Exclusion zones are identified as the footwall development 25 m to the north and to the south of a production blast one level above. This was defined by correlating the location of production blasts and the occurrence of seismic damage, stress modeling of shear stress on foliation planes, and high principal stress, seismic monitoring, and analysis.

Stress modeling, via MAP3D (three-dimensional, indirect boundary-element modeling package using constant intensity fictitious force and displacement-discontinuity-type boundary elements) is used to refine knowledge of stress change that might indicate areas that might have an elevated seismic risk. Such areas are targeted for installation of seismic-resistant support. At Big Bell, seismic-resistant support is considered to be the installation of cone bolts in addition to the standard development drive support (Turner and Player 2000).

The seismic support regime follows from an examination of reinforcement and support performance at early seismic events. The reinforcement types in seismically induced rockfall areas consisted of plain, twin-strand, 15.2-mm cable bolts; plain, single-strand, 15.2-mm cable-reinforced and plated Split-Set rock bolts; plain Split-Set rock bolts; solid point-anchored bolts; and post-grouted point-anchor bolts. Mesh is installed as part of the development cycle; post-grouted rock bolts and mesh have been defined as shakedown-resistant reinforcement to a seismic event.

As a monitoring tool, Big Bell hired an AURA32 portable microseismic system to provide focused monitoring of areas of seismic risk. In early 2000, a wide-area real-time microseismic system was purchased and installed by ISS Pacific. It consists of 12 accelerometers (three triaxial and nine uniaxial) that monitor a larger area of the mine.

Big Bell has been working to develop an improved understanding of the mechanisms causing mining-induced seismicity to ensure that the workforce is minimally exposed to risk while maintaining cost-efficient production. A detailed analysis of the work undertaken may be found in Turner and Player (2000).

45.4 MINING METHOD

The mining method adopted at Big Bell is longitudinal sublevel caving under a caving hanging wall. The adoption of sublevel caving is discussed extensively in Player (1998), so only a brief description will be provided here. The initial mining method employed about the remnant stopes, from the ASARCO phase of mining, consisted of transverse sublevel caving. This was undertaken by retreating from the remnant stopes to the footwall or hanging wall of mineralisation. Concurrently with the transverse sublevel caving, preparations were undertaken for the core-and-shell technique under a caving hanging wall between the pit floor at a depth of 210 and 320 m below the surface. The core- and-shell mining technique was the initial method proposed

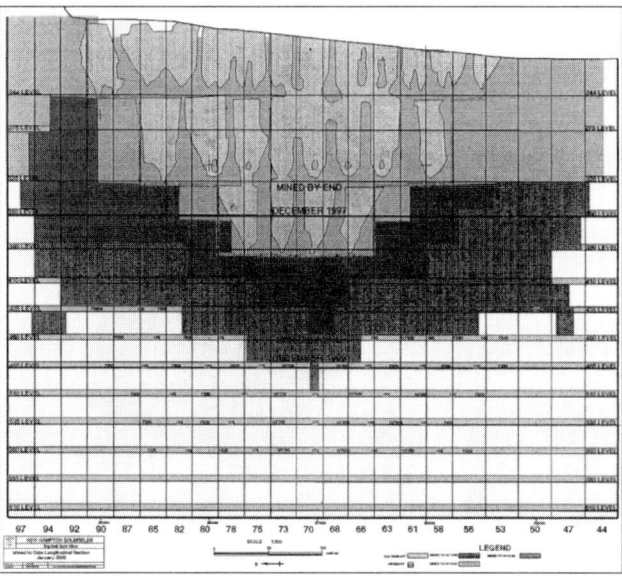

FIGURE 45.6   Longitudinal view of Big Bell Mine

for the primary ore outside the remnant area of the mine. However, because of production pressures, stress-induced ground control difficulties, and blasthole difficulties, the decision was made to convert to sublevel caving.

The sublevel caving method adopted at Big Bell is illustrated in Figure 45.6. The method is a top-down approach, working from the central area of the ore body out. The extremities of the economic mineralisation are developed and retreated back onto a rib pillar, termed a "limit retreat." Once both retreats close onto the pillar, a large blast is set off to retrieve ore within the pillar. During the northern and southern retreats to their final pillar positions, production from the limit retreat on the level below is minimised to reduce excessive concentration of stress in the pillars.

Material to fill the void created during mining originates from the caving hanging wall. No introduced fill is used.

Each level is commenced from a slot developed transverse to mineralisation and broken through to the level above from the hanging wall to the footwall. The slots are aligned from level to level. Once the slot is completed, the longitudinal retreats to the north and south can commence. It has been determined that the cave must be opened a minimum of 60 m prior to beginning the next slot. During stoping operations, a minimum of 30 m between brow positions from level to level must be maintained to minimise stress redistribution between levels.

Ore is blasted into the caved hanging wall waste and relies on high powder factors to produce good fragmentation to break out the toe of the ring. Good fragmentation enhances ore draw against the caved waste. As the ore is removed, waste rock is drawn down from the level above, replacing the extracted ore. Ring drilling is undertaken from the ore drives to break through to the level above. (The drilling and blasting parameters are discussed in the section on drilling and blasting.) The hanging wall progressively collapses once a critical strike length is exceeded and probably once stoping operations commence on a lower level.

Ore is extracted until the material reporting to the drawpoint is determined to be less then 1 gm/tonne, the drawpoint shut-off grade. The initial decision on drawpoint closure is made by visual inspection and confirmed by assaying. Throughout the sublevel caving experience at Big Bell, waste entry takes place early in the draw, and recovery between adjacent production blasts can be quite different, resulting in scheduling difficulties. The introduction of two-ring firing had an

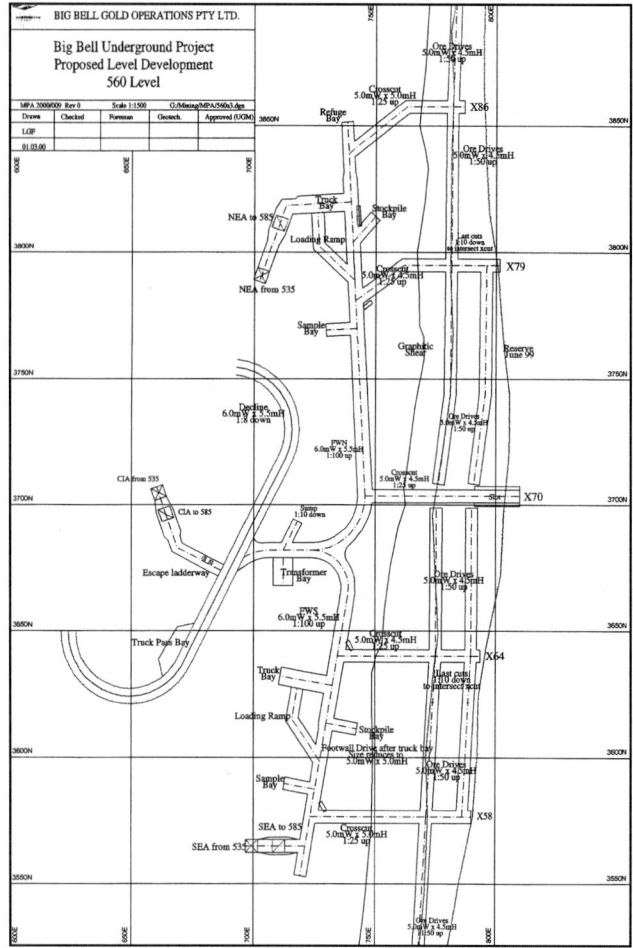

FIGURE 45.7   560-level development layout

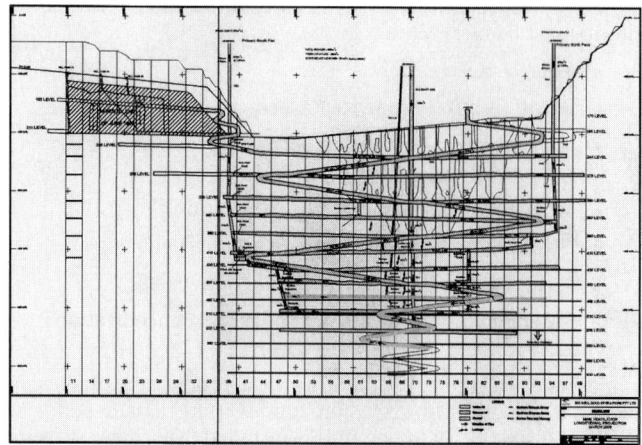

FIGURE 45.8   Longitudinal view of ventilation layout

initial negative impact on recovery, and ore was left behind, but it is likely that the ore was recovered from lower levels because it increased in later months. Two-ring firing enables simplification of the mining process.

The rock mass response to stress change is discussed in the geotechnical section. The development reinforcement installed is not designed for 100% survival capacity because of ore drive deformation, particularly the extreme stress load caused by blasting, hence the evolution of brow support.

- Gaining secure access to the blastholes for checking length prior to charging and cleaning out by the drill rig. On most occasions, the rill would prevent the drill rig from accessing the holes, thus requiring that the holes to be redrilled.

- A regular requirement for redrilling inaccessible stope rings involved dumping multiple rings forward, drilling from the footwall access drive, or drilling from one ore drive to the next. These options usually resulted in less favourable drilling patterns, increasing the risk of a blast malfunction.

45.4.1   Mine Layout

The typical mine level layout is shown in Figure 45.7. Levels have a vertical interval of 25 m floor to floor and are accessed near the centroid of the ore body to maintain an even level interval. The layout of the decline has changed from long straight passages parallel to the strike of the ore body. These resulted in access at

various northings along the lode and hence variable level intervals yielding a short elliptical configuration to maintain central access. This simplifies the layout of mine services, such as the central intake airway and drainage system. Figure 45.8 shows the longitudinal layout detailing the cave position, decline layout, ventilation intake, and exhaust.

To improve recovery, two ore drives are used where the ore body is wider then 22 m. The pillar between ore drives has been increased from 7 to 10 m to reduce stress influences and improve ground control. Ore drive location is determined by grade, dip, and strike of the ore body. Sections of the ore body steeper than 70° have the footwall shoulder of the ore drive on the footwall of the defined stope. Where the ore body is flatter then 70°, then the footwall ore drive undercuts the footwall of the defined stope to provide draw cones that increase recovery.

A change in ore drive size was implemented in 1999 with a reduction in size to 4.5 m high by 5 m wide by design. Changing to an arched profile for all development and tighter control on drilling the development round have significantly improved ground conditions. The use of arched profiles in all development is to minimise the influence of stress redistribution in the backs of the development. Table 45.7 describes the development profile, dimensions, and ground support used.

M-series welded mesh has smooth wires and a higher strength than conventional concrete reinforcing mesh. Both types use a 100-by 100-mm wire grid. Sheets are 2.4 m wide by 6 m long, but are cut in half to provide additional surface coverage in footwall and decline development. Mesh dimensions restrict bolt row spacings to 1.1 m between rows. For mesh and wall reinforcement, 1.8-m-long Split-Set bolts are used. Tubular groutable bolts are used for back reinforcement. An example of a design of ground support for footwall development is shown in Figure 45.9.

Because of the interpreted seismic risk, the footwall access development has been modified in conjunction with seismicity-resistant support to minimise seismicity-induced rockfalls. Features of the design are—

- Footwall development does not to run parallel to foliation over long distances, hence reducing the maximum failure potential. Footwall development is designed farther away from the ore body in areas of greater seismic risk.

- Truck loading bays and ramps are located in a position on the northern and southern sides of the ore body to minimise total tramming distances for the loaders. Staggering the intersections on footwall development ensures that four-way intersections are not created, thus minimising stress concentrations caused by development.

TABLE 45.7   Development opening and ground support

Development	Dimensions (width by height), m	Back profile	Welded mesh	Bolts
Decline	6 by 5.7	Elliptical	M61 galvanised, 8.8 m²/m advance	Nominal 7-bolt rows
Footwall access drive	6 by 5.5	Elliptical	M61 bright, 8.8 m²/m advance	Nominal 7-bolt rows
Crosscut to ore body	5 by 4.5	Arched shoulders	M71 bright, 6 m²/m advance	Nominal 5-bolt rows
Ore drive	5 by 4.5	Arched shoulders	M71 bright, 6 m²/m advance	Nominal 6-bolt rows

**Eight Bolt Pattern—Decline and Footwall Development
Semi-Elliptical Arch Profile**

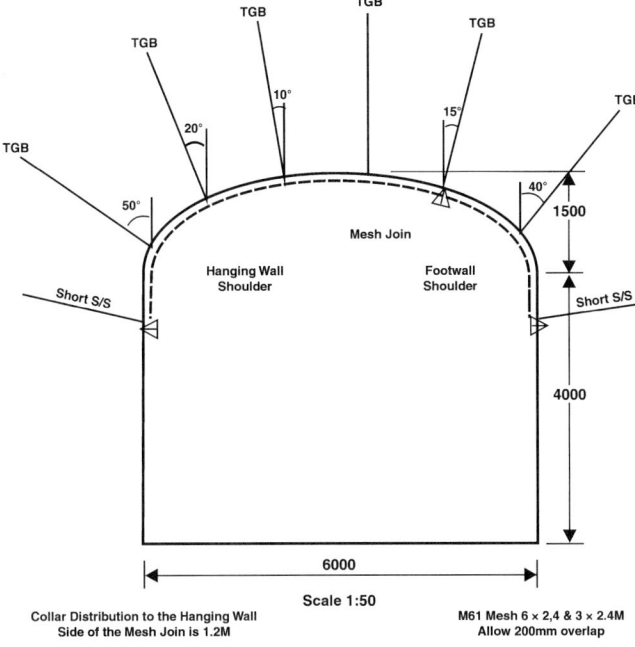

FIGURE 45.9   Arch profile for footwall access development

- The cross-sectional area of the footwall access development is reduced to 5 by 5 m past the truck loading bay.

- Cone bolts are installed on the footwall access development in the areas defined as having increased seismic risk. Such a zone is the footwall access development outside the middle 80 m around the center of the ore body. The cone bolt pattern is four 3-m-long bolt rows with 1.5-m spacings between rows. Bolts are located no closer then 1 m from the shoulders of the drive. Four-meter-long cone bolts are installed in the intersections on the footwall development.

Additional aspects of level design are—

- Truck loading ramps to allow higher tonnages (particularly with coarse fragmentation), less damage to trucks and loaders, and smaller excavations parallel to the strike of the ore body, which reduces seismic risks.

- Five crosscuts per level spaced 50 to 100 m apart. The central crosscut is used as the primary slot. Crosscut locations are a balance of ore body width, tramming distances for loaders, required number of working areas, and development schedule. Crosscuts are staggered from level to level to improve ore recovery.

- Loader stockpile bays for stockpiling while waiting for trucks and sampler bays for a safe work area.

- Sumps located on the access crosscut from the decline. The standard mine dewatering rate is 16 L/s, of which 50% is groundwater.

- Cable-bolted intersections in ore. The bolts are 4.5- and 6.5 m-long, twin-strand bulb type with one bulb per 2 m of strand. Cable bolt patterns are 2.7 by 2.5 m to suit production ring positions. For retreat pillars, cable bolts 10.5- 8.5- 6.5- and 4.5-m long on a 2.5- by 2.2-m pattern are used. Pillar and back stability have been acceptable using this reinforcement regime.

45.4.2   Stope Design

The estimated reserve is used as the basis for the stope design. Reserves are estimated at a 2.5 gm/tonne cut-off grade. Additional information obtained subsequent to estimating the reserve includes—

- Development face sampling

- Recent diamond core drilling

- Geological reinterpretation

- Comparisons with the level above to achieve a practical mining shape

- Evaluation of marginal material from potential draw cones and ore drives for inclusion in the stope design

Ultimately a consistent wire frame model in a practical mining shape must be achieved along both strike and dip.

45.4.3   Drilling and Blasting

The production blasting cycle is illustrated in Figure 45.10. The sequence is given below.

1. Install brow support 1 to 1.2 m behind the collar of the ring to be blasted.
2. Drill blastholes to a maximum of 30 m in advance of the cave face and a minimum of 5 m in front of the cave.
3. Prepare rings to be blasted.
4. Clean out blastholes with Atlas Copco Simba 357 if necessary. This process involves running a bit smaller than the original hole size up the hole to remove blockages.
5. Charge blastholes with bulk gassed emulsion.
6. Initiate blast at the end of shift. Clearing the mine is required.
7. Scale the drive and reinstall ventilation and drawpoint water sprays for dust suppression.
8. Once bogging cycle is complete, return to step 1, 2, or 3 if—
   - No additional brow support has been installed,
   - Brow support has been installed, but blasthole drilling is required, or
   - Brow support and blastholes are available.

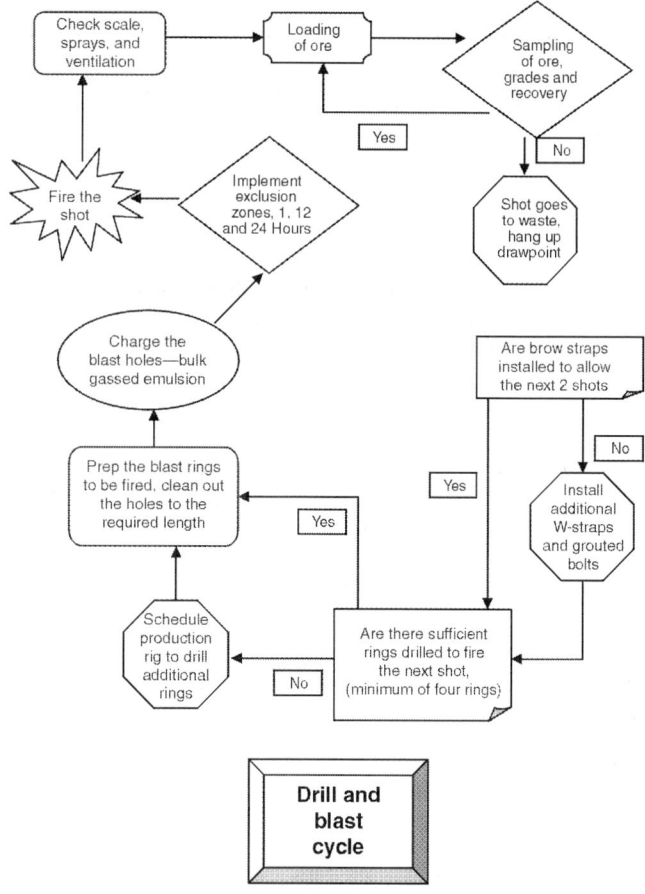

FIGURE 45.10 Production blasting cycle

TABLE 45.8 Drilling and blasting parameters and performance (Averages calculated on data from January 1999 to February 2000)

Hole diameter	102 mm nominal; range from 96 to 103 mm
Average tonnes per meter	12.7
Average shot powder factor	0.56 kg/tonne
Average tonnes fired per meter of cleanout	27.2
Average tonnes fired and mined per month	129,000 tonnes fired per 145,000 tonnes mined = 112% average recovery
Average meters required	10,200 m at 12.7 tonnes mined. Two drill rigs require 93 m per shift if there is an effective 110 drilling shifts per month.
Drilling	68-mm tube drilling
Ring dump from vertical	20°
Ring burden	2.7-m collars equate to 2.5 m of perpendicular distance between rings
Toe spacing	3.6 m
Maximum kilograms per delay	500
Product	Power bulk VE 1.2 density (a gassed bulk emulsion from Orica Explosives)
Leading holes	2
Timing	50 ms between first two holes and second holes, 25 ms between remaining holes, and 125 ms between rings
Rings fired	Two per shot in a 5-m slice of ore body
Minimum collar length	1 m on lead ring. Broken ground, structural features, brow strap position should be considered. 7 m on back ring.
Maximum uncharged collar spacing	2.8 m. Five-hole spread is used to determine charge length.
Explosive charging rate	Average of 67 tonne/month up to 8 tonnes in a single shift. Charge rate depends on frequency of drawpoints that go to waste.

Production blasting at Big Bell has evolved from blow loading of ANFO into 76-mm in diameter holes in the remnant mining blocks to gassed bulk emulsion with 102-mm in diameter blastholes. Production drilling is now standardised using 102-mm in diameter blastholes charged with gassed bulk emulsion, the benefits of which are outlined below (Yeung et al. 1999).

- 25% decrease in overall drilling and blasting costs
- Increased charging rates with only one operator required
- Improved charging quality control with simple and consistent documentation of exact charge length and weights
- Improved blasting performance from high-strength, waterproof explosive
- Reduced drilling requirements for higher tonnes per meter drilled
- Fewer survey markups needed
- Reduction in explosive stocks, as the emulsion is only an oxidiser until it is gassed up the hole, when it becomes an explosive
- Safer operation from reduced manual labour by using a mechanical hose pusher. No ANFO dust

All blasting practices are regularly reviewed and modified to accommodate changes in the mining environment. Modifications have been made to burden, spacings, ring dump, uncharged collars, single- or double-ring blasting, and timing of the shots. Most parameters are now settled (Table 45.8). The review process is ongoing. As drawdown and cave performance are better understood, drilling and blasting strategies can be further improved.

The mainstay of the production drilling fleet has been an Atlas Copco Simba 4356, which drills 102-mm in diameter upholes in tandem with contractor drill rigs. A cost analysis showed it was economic to hire and operate a Datasolo 1020 to replace the contractor's rig, which was done for 12 months. Standardising maintenance and further cost and drill rig performance analyses supported a decision to replace the Datasolo with a new Simba 4356S, which was acquired for a cost of $1.3 million, thereby reducing maintenance by specialists. All drill rigs use 1.5-m tubes to allow the rig to fit into the smaller sized ore drive.

Hang-ups in the cave require the use of a "cannon" to fire projectile charges at the ore hang-up to knock it down safely. The operator targets the hang-up and from a position of safety, fires the cannon. Bombing procedures are used for frozen ground in the cave with the construction of bund walls and the use of polyvinyl chloride pipe on guide wires to position the charge hose. A maximum of 250 kg of bulk emulsion is pumped into a hole or crack being targeted.

45.4.4 Grade Control

The grade control technique employed uses visual estimates and assays from two consecutive 5- to 6-kg bulk samples taken from consecutive buckets of a loader. The sampling frequency represents one sample per 80 to 100 tonnes trucked out of the mine. Each active production site is visited every hour. The sample is first described visually to assess the quality of the material from the drawpoint, which offers an on-the-spot opportunity to either close the drawpoint or keep it open. Samples are then submitted to an on-site laboratory that uses a 30-gm assay with an AAS finish analytical technique. Assay turnaround time varies from 12 to 24 hr. Drawpoint closure is finalised when all the laboratory assays are available, although in some instances, the next rings may be fired when the mining operation cannot wait the until the assays are returned.

Minimal surface stockpiling of underground ore is done because of mill overcapacity. However, blending of underground and open-pit ore is undertaken to achieve optimal mill performance.

45.4.5 Loading and Hauling

The original mining fleet consisted of five Elphinstone R2800 loaders with nominal 10-tonne load capacity and seven Elphinstone 73D trucks with a nominal capacity of 50 tonnes. However, the effective capacity of the haulage fleet varies from 42 to 45 tonnes. Capacity depends on fragmentation, loader operator experience, and the use of truck loading ramps. Of the seven-truck fleet, there is always one truck undergoing maintenance, either because of rebuilding, repairs or breakdowns, or servicing.

During 6 yr of production, all loading and haulage equipment has undergone one major rebuild. The bulk of this work was undertaken in 1997 on a rotation system. In 1999, because of under-utilisation and the high cost of fleet maintenance, one loader was removed, improving use and availability of the remaining fleet. During the minor overhaul period in 1999, an additional loader was hired to assist in maintaining production rates. Maintenance costs for the fleet, including drill rigs undergoing minor or major overhauls, totalled $3 million in 1999. A decision has recently been made to replace two of the existing loaders with two Elphinstone R2900 loaders in the middle of 2000 at a cost of $1.1 million each.

45.4.6 Ventilation

The source of fresh air is the central ventilation intake rises and the decline. Each development heading is ventilated from the central intake airway until the exhaust rises are established at the extremities of footwall development. All longhole vertical ventilation airways (intake and exhaust) have a minimum designed cross-sectional area of 36 m^2. Raise-bored exhaust shafts from the 380 level are 4.8 m in diameter. The surface fans are Howden Sirocco variable-speed, centrifugal fans with a capacity of 2.2 MW. The north fan has a higher capacity, running at 680 rpm, than the south fan running at 600 rpm. A total exhaust flow of approximately 600 m^3/s is required for production parameters, i.e., three production levels, one level development, and the decline development, without compromising air requirements when a diesel fleet is used. Sealing off completed levels to stop stope air leakage and regulating active levels is vital for good ventilation throughout the mine.

Production headings are all force ventilated with 45-kW, 1,067-mm in diameter fans. Water sprays have been installed at the brow of the cave for dust suppression. The brow is the principal source of dust in the mine. Timed fine-water sprays are also used for dust suppression along the decline as traffic movement causes the roadway to dry out, and dust is then easily transported by the air moving down the decline. This was an a more significant issue when the decline was the only primary source of air intake, but with the development of the central intake airway system, the air velocity in the decline has been reduced considerably.

45.5 PRODUCTION AND DEVELOPMENT STATISTICS

Big Bell Operations employs 364 people, of whom 279 are contract workers; 85 of these are associated with the open pit operations. The workforce recently increased from approximately 260 workers following the recommencement of open-pit mining.

TABLE 45.9 Fleet availability and use, 1999

Equipment	Availability, %	Use, %
Truck fleet	78	61
Loader fleet	74	45
Simba 4356	80	64
Datasolo	77	65

The mine operates 365 days a year, 24 hours a day with two 12-hr shifts. The majority of the employees work on a fly-in, fly-out basis from Perth and Geraldton. Two work rosters are applied: three production crews work a 2-week-on, 1-week-off schedule, while senior staff work 9-days-on, 5-days-off. The crews work 7 night shifts, receive a 24-hr break, and then work 7 day shifts.

The underground mine operates on an 18-man production crew basis. This staffing rate produces 33,300 tonnes of ore per man per year, equivalent to 135 tonnes per man-shift. However, when the productivity per man for the Big Bell underground operations (contractors included) is considered, then the rate is closer to 7,000 tonnes of ore per man per year.

The development advance rate for contractors using two jumbos to undertake all face boring and ground support averages 320 m/month, equating to 6 m per development contract employee on site.

The scheduled haulage rate of the truck fleet is 54 trucks per shift at 45 tonnes per truck to produce 5,000 tonne/d. The drilling and blasting parameters and productivity are described in Table 45.8, while maintenance reports for mobile fleet availability and use for 1999 are shown in Table 45.9.

45.6 SAFETY AND HEALTH ISSUES

Standard personal protective equipment applies in all areas of the mine except for the crib room. The main safety issues of concern to management are—

- Seismicity
- Hang-ups within the cave and brow integrity
- Dust levels and high silica content
- A generally inexperienced workforce

All issues are dealt with at weekly safety meetings where information is presented on hazards identified and incidents that have occurred. This is the main forum of communication between the underground workforce and management and where the entire crew will be presented with information. There is also the Site Safety Committee that meets to discuss issues that affect all personnel at the site, i.e., fitness-to-work policy, and to address items that may have been raised at the crew safety meetings.

Portable rescue chambers are located at each end of all production levels. The rescue chambers have the capacity to seat six people for 8 hr. A larger rescue chamber (converted sea container) is in place for development personnel and has the capacity for 12 people for 8 hr.

Underground communications consist primarily of a leaky-feeder radio system for general communications and a telephone backup system. Stench gas evacuation drills are undertaken yearly.

A key feature of seismic risk management is open communication and an education program with the workforce. This program includes presentations to operators about what seismicity is, the factors that influence seismicity, what mine management is doing to manage the risk, the introduction of exclusion zones, and how ground support works to withstand a seismic event.

TABLE 45.10   Half-yearly operating costs, Australian dollars

Period	Mining costs per tonne	Milling costs per tonne	Total cash cost per tonne	Cash cost per ounce
July 97 to Dec 97	16	12.5	35.3	390.6
Jan 98 to June 98	21.4	12.2	40.5	417
July 98 to Dec 98	18.6	12.7	42.9	377.4
Jan 99 to June 99	23	13.3	37.3	539.6
July 99 to Dec 99	17.6	10.7	34	394.1

45.7   MINE OPERATING COSTS

Mine operating costs during the 30 months from July 1997 to December 1999 are shown in Table 45.10. Development restarted in January 1999 after a period when no development was undertaken for a period of 16 months.

45.8   CONCLUSIONS

The Big Bell gold mine is a 1.7 million tonne/yr, low-grade, longitudinal sublevel caving operation that mines successfully in a high and deviatoric stress environment. The challenge of understanding and working with mine seismicity is being met and dealt with, and our knowledge continues to increase through the use of a real-time microseismic system.

The mining operation is a highly mechanised fleet of 50-tonne dump trucks matched to a fleet of 10-tonne tramming capacity loaders. A production rate of 135 tonnes per man-shift is standard. Production involves drilling 102-mm in diameter holes and contractor charging using bulk emulsion to achieve economies of scale and mechanisation.

Primary health and safety concerns for management are seismicity, cave hang-ups, dust levels, and an inexperienced workforce.

Mineralisation is open at depth; thus with favourable economics from recent cost-control exercises, the operation has the potential to continue for several levels below the 610 level.

45.9   ADDENDUM

Two large seismic events (greater than Richter magnitude 2.0) occurring on 17 June 2000 and 9 July 2000 damaged the footwall and ore drive development at the north ends of the 510 and 535 levels. Following these events, it was decided to alter the mine layout in an attempt to prevent the occurrence of similar large events in the future. These changes involved—

- Moving the footwall development 70 m to the west to be more then 100 m away from the ore body (it now comes off the northern and southern ends of the decline) with one crosscut ramping up and another down to access two levels

- Identifying development having an unacceptable probability of a significant event and supporting it with additional cone bolts and mesh

- Modifying the mining process and exclusion zones by firing all production headings at once to allow the mine to settle for 7 days before recommencing production

- Investigating the use of a hybrid longitudinal sublevel with barrier pillars to reduce the concentration of cave abutment stresses on lower-level development. Barrier pillars and accessways will be designed to make them recoverable at some stage in the future, but they will be sufficiently far apart to allow the hanging wall to cave and fill the void

Big Bell still has 5 yr of mineable resources left, and the evolution of a mining method is being discussed with international consultants to ensure the mine's future.

45.10   ACKNOWLEDGMENTS

The author would like to thank Big Bell gold mine for granting permission to publish this paper, and my colleges Don Barrett, Sig Slepecki, and Chris Woolford for their contributions, review of the paper, and answers to my questions. Thanks also to the publishers Australian Institute of Mining and Metallurgy (AusIMM) and Balkema that have given their permission for material to be reprinted.

The author can be contacted at Big Bell Gold Operations, GPO Box D170, Perth 6001, Western Australia, Australia, or John.player@newhampton.com.au.

45.11   REFERENCES

Barton, N., R. Lien, and J. Lunde. 1974. Engineering classification of rock masses for the design of tunnel support. *Rock Mechanics* 6:189–236.

Handley, G. A., and R. Cary. 1990. Big Bell Gold Deposit. *Geology Deposits of Australia and Papua New Guinea–AusIMM*, pp. 211–216.

Mikula, P. Feb. 25, 2000. Private communication, Kalgoorlie, Australia.

Player, J.R. 1998. Big Bell Mine Coming Back for Seconds, Thirds and Fourths. Proceedings of Underground Operators Conference–AusIMM, pp. 103–114.

Sandy, M.S., and J.R. Player. 1999. Reinforcement Design Investigations at Big Bell. Proceedings of the International Symposium of Ground Support–Balkema. Kalgoorlie, Australia

Turner, M., and J.R. Player. 2000. Seismic Reinforcement at Big Bell Mine. Proceedings of MassMine 2000–Brisbane, Australia

Yeung, C., J.R. Player, and K. Braddon. 1999. The Implementation of Bulk Emulsion Explosives at Big Bell Mine. Proceedings Explo 99–AusIMM, pp. 71–78.

BIBLIOGRAPHY

Barrett, D. 1999. Big Bell, Underground Again and Going Deeper. Mining in High Stress and Seismically Active Conditions. Australian Centre for Geomechanics–Workshop 9903.

Onederra, I., J.R. Player, P. Wade, and G. Chitombo. 1999. Mass Blast Design, Simulation, Optimisation and Monitoring at Big Bell Gold Mine. Fragblast 6.

Slepecki, S. 1995. Case Study : Smooth Wall Blasting at Big Bell Gold Mine Rock Slope Damage Control (Blasting). Australian Centre for Geomechanics–Workshop 9506.

Van Leuven, M.A. 1998. The Use of Risk Analysis in the Selection of a Mining Method. Proceedings of Underground Operators Conference–AusIMM, pp. 195–200.

REPRINTED MATERIAL

Sandy, M.S., and J.R. Player. 1999. Reinforcement Design Investigations at Big Bell.
Reprinted from: Rock support and reinforcement practice in mining– Proceedings of the international symposium on ground support, Kalgoorlie, Western Australia, 15–17 March 1999. Villaescusa, E., C.R.Windsor & A.G.Thompson (eds) 90 5809 045 0, 1999, 25 cm, 448 pp., EUR 105.00 / US$125.00 / GBP80.00. Please order from: D.A. Book (Aust.) Pty. Ltd., P.O. Box 163, Mitcham, Vic. 3132 (fax: (03)92107788; e-mail: service@dadirect.com.au).

Player J.R. 1998. Big Bell Coming Back for Seconds, Thirds and Fourths, Proceedings Seventh Underground Operators' Conference pp. 103–114 (The Australasian Institute of Mining and Metallurgy: Melbourne). E-mail: publications@ausimm.com.au.

Yeung, C., Player, J. and K. Braddon. 1999. The Implementation of Bulk Emulsion Explosives at Big Bell Mine Proceedings Explo 99. pp. 71–78 (The Australasian Institute of Mining and Metallurgy: Melbourne). E-mail: publications@ausimm.com.au.

CHAPTER 46

Theory and Practice of Very-Large-Scale Sublevel Caving

C.R. Quinteiro,* L. Larsson,* and W.A. Hustrulid†

46.1 INTRODUCTION

LKAB's two iron ore mines in northern Sweden have been using the sublevel caving mining method for nearly 40 years. In 1998, the ore production from the Malmberget Mine was about 12 million tonnes and from the Kiruna Mine about 21 million tonnes. In these mines, both longitudinal and transverse sublevel caving are used, depending upon local ore body geometry. In narrow ore bodies, the longitudinal method is employed. Because of tough international competition in the iron ore market, LKAB has been forced to pursue a relentless program of cost reduction in its operations. This has led to a continuous up-scaling of the sublevel caving method. In 1962, the sublevel height at Kiruna was 6 m. Today, the sublevel height is 27 m. This up-scaling has had a tremendous positive impact on the economics of the mining method, but a negative impact on ore recovery and waste dilution. In this paper, the current efforts being made by the Mining Research and Development team of LKAB toward improving the performance of its very large sublevel caving mining method at the Kiruna Mine will be discussed.

46.2 KIRUNA MINE

The ore body at the Kiruna Mine is about 4 km long and 80 m wide. It strikes N10°E and dips about 60° southeast. There are indications that the ore body extends to a depth of 2,000 m. The main transport level is located at 1,045 m, and currently, the deepest production sublevel is at 792 m. The ore is a rich, fine-grained magnetite with a waste content of about 5%. The presence of apatite in certain parts of the ore body causes problems in both mine planning and production. Low- and high-phosphorus ores must be kept separate during mining in order to meet production requirements. This, in turn, establishes an extra parameter to control during sublevel caving operations.

46.3 SUBLEVEL CAVING AT KIRUNA

Today, the main mining method employed at the Kiruna Mine is transverse sublevel caving. The transport of blasted ore from the drawpoint to the ore passes is done by electric load-haul-dump (LHD) units. Trains are used on the main haulage level for transport of the ore from the ore passes to underground crushers. A general explanation of the sublevel caving method, as well as its applicability, layouts, and technical requirements, has been given by Cokayne (1982). It is, however, important to mention here the advantages and disadvantages of this mining method. On the plus side are its suitability for mechanization, high ore recovery (no pillars are left behind), and high degree of safety. The disadvantages are high ore dilution (and the problems of controlling it) and high development costs (Cokayne 1982). Dilution can be controlled through brow support, good drilling and blasting practices, and strict draw control.

Over the years, LKAB has been using the advantages of this mining method to reduce mining costs. Mechanization has reached record levels. The main haulage level is fully automated with unmanned trains and remote loading from chutes into train cars. Drilling holes in an entire fan is done automatically. Drilling is supervised from a room located at the control center on level 775 m, which is kilometers away from the drill rig. Efforts are under way to automate loading of ore from drawpoints to the ore passes. Currently four semiautomatic LHDs are in use at the Kiruna Mine. LKAB has also been working to overcome the disadvantages of the sublevel caving method to reduce mining costs. Up-scaling the mine layout to reduce development has been continuous since the method's introduction in 1962 (Marklund and Hustrulid 1995). Today the sublevel height is 27 m, and the spacing of production drifts is 25 m. Burden is 3 m. Production drifts are 7 m wide and 5 m high. This layout means that only 5% of the total ore is extracted during development. This has resulted in a significant cost reduction for LKAB.

However, up-scaling has posed a challenge to engineers concerning how to keep ore recovery and dilution under control, as well as to keep the two types of ore with different phosphorus grades apart.

In 1995, the Mining Research and Development Department of LKAB started a project called Sublevel Caving 2000 aimed at tackling just the problems mentioned above by up-scaling the sublevel caving method. An area of the mine with a strike length of about 200 m and containing nine production drifts was selected to be the site for a series of experiments and collection of data. These activities were well planned and manned so that any disturbance to normal production was minimal.

46.4 SUBLEVEL CAVING 2000

This project, which had full support of the mining department, consisted of a series of experiments, some of which are mentioned below.

- Increase the burden from 3.0 to 3.5 m.
- Increase the width of the drifts from 7 to 11 m.
- Test different blasting delay intervals for the fans.
- Design and implement a new draw control procedure.
- Insert markers into the fans to define the ellipsoid of extraction.
- Film fans equipped with markers during drawing.
- Sample crude ore on-line and steer it to different ore passes based on ore quality.

---
* LKAB, Sweden.
† Dept. of Mining Engineering, University of Utah, Salt Lake City, Utah.

TABLE 46.1   Comparison of different ring burdens

	3 m	3.5 m
Extraction, percent	127	101
Ore recovery, percent	104	72
Dilution, percent	18	28

TABLE 46.2   Comparison of tests of different drift widths

	7 m	11 m
Extraction ratio	101	100
Ore recovery, percent	72	79
Dilution, percent	28	21

Mining this test area with its nine production drifts was completed in December 1997. The results achieved will be discussed in the following sections

46.5   KVAPIL'S DESIGN GUIDELINES

To see if the current layout at the Kiruna Mine satisfies the parameters of gravity flow, it was important to compare the sublevel caving layout used at Kiruna against the guidelines for designing sublevel caving proposed by Kvapil in 1982. Kvapil gives guidelines so that the parameters of gravity flow are satisfied. These guidelines suggest that for the given sublevel height of 27 m and an effective extraction width of 5 m, the spacing of production drifts should be less than 22 m and the burden less than 3.5 m. However, if the effective extraction width is considered to be the same as the width of the production drifts, i.e., 7 m, the dimensions change to a spacing of 25 m and a burden of 4 m. Thus, it seems that the dimensions used in sublevel caving at Kiruna (25 m for spacing of the drifts and 3 m of burden) are in general agreement with the theory of gravity flow. The experiments planned to increase the burden to 3.5 m and the width of the production drift to 11 m are also in the right direction to improve the efficiency of sublevel caving according to the theory of gravity flow.

46.6   RESULTS FROM EXPERIMENTS

The analysis of the results from all experiments is based on data collected by sampling during loading of the fans. The sampling technique consisted of taking a 1-kg sample from every tenth LHD bucket. Information on waste content and ore quality was recorded for the entire load-out of the fans. The number of boulders encountered was also recorded. A boulder is defined as a piece of rock that requires handling before it is dumped into an ore pass. In this way, the production from the fans could be classified according to ore recovery, dilution, ore quality, and number of boulders.

46.6.1   Increase Burden

Two production drifts were selected in which the burden was increased from 3 to 3.5 m. One drift had a typical width of 7 m. In the second drift, the width was increased to 11 m by slashing the walls. A total of 29 fans with a burden of 3.5 m were blasted and sampled. A comparison of the results obtained with burdens of 3 and 3.5 m for 7-m-wide drifts is presented in Table 46.1. The results show that for this experiment, the fans with a burden of 3.5 m yielded poorer performance than those with 3 m. Ore recovery was lower and dilution was higher.

46.6.2   Increase Width of Production Drift

Two production drifts were selected to investigate the effect of enlarging drift width. This enlargement was carried out by slashing both walls of the drift to a width of 11 m. For safety reasons, the length of the slashed region was kept to a maximum of 10 m. As the mining front retreated, the drift enlargement kept pace. (It should be mentioned that the fan drilling pattern was the same as that used for the 7-m-wide production drifts.) The objective was to increase the swelling volume available in the drift during blasting of the fans and also to improve the effective extraction width of the drifts. Table 46.2 compares results obtained with the drawing of ore from a drift with enlarged width and one with regular width. These two drifts had the same burden of 3.5 m.

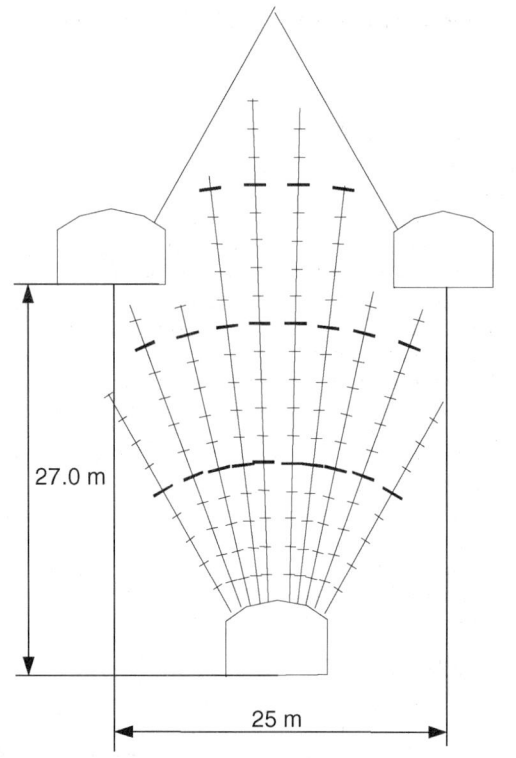

FIGURE 46.1   Fan drilling layout at Kiruna

The results shown in Table 46.2 are in accordance with the gravity flow theory. An increase in the effective extraction width will result in a wider ellipsoid of loosening and therefore a decreased passive zone. For the Kiruna layout, this signifies better ore recovery. The increase in the effective extraction width also has a positive effect on the delay of waste inflow because it creates a wider gravity flow motion of the broken ore.

46.6.3   Blasting Layout

The drilling pattern used for the sublevel fans in the test area is shown in Figure 46.1. The blasted ore in the fan was about 10,000 tonnes. The diameter of the holes was 114 mm, and the longest hole in the fan was about 40 m. The drill rigs used at LKAB have strict requirements for hole accuracy because this is vital to the success of the mining method.

The explosive used in the fans is Kimilux R, an emulsion that is pumped up the holes. One fan requires about 2,000 kg of explosives. The design takes into consideration the protection of the brow against damage. A damaged brow will encourage both ore loss and the early entry of dilution caused by a reduced extraction width.

The experiments carried out here were to change the sequence and timing for blasting the 10 holes of the fan. The results are presented in Figure 46.2. From this figure, it can be concluded that the layout with 300-ms delays yielded the least dilution when drawing. This layout consisted of first blasting the four middle holes using a short delay, and after a delay of 300 ms,

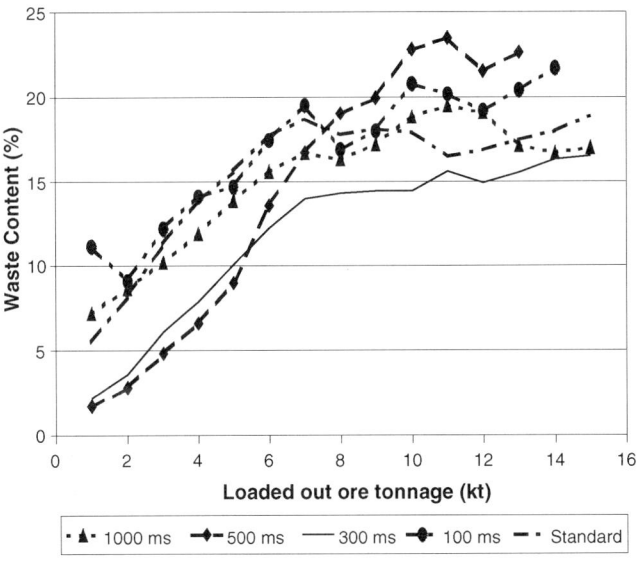

FIGURE 46.2   Influence of blasting delays on fan performance

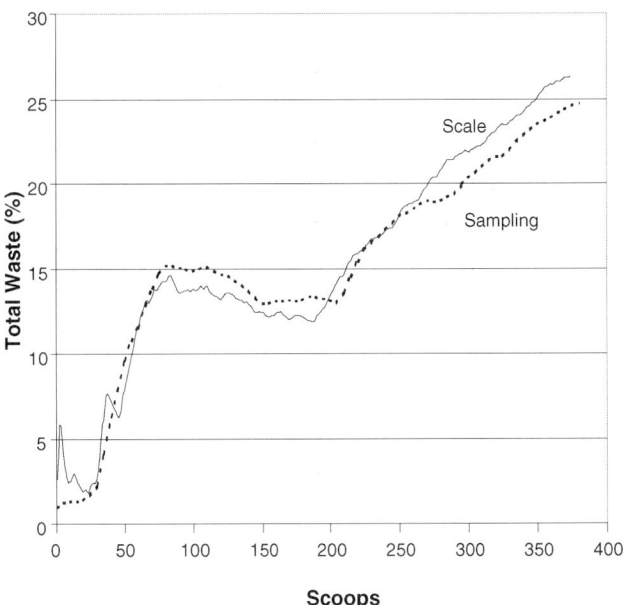

FIGURE 46.3   Comparison of calculations of waste content by sampling and scoop weighing

blasting the other holes in the ring. A delay of 50 ms was used between each hole.

46.6.4   Draw Control Procedures

The major disadvantage of the sublevel caving method is the potential for high dilution, i.e., waste rock in the cave is drawn together with the blasted ore during loading of the fans. The allowable percentage of dilution and the percentage of ore recovery from the fans are correlated. Permitting high dilution will result in high ore recovery. However, in reality, it is desirable to have a high ore recovery and to keep dilution under strict control. This, in turn, requires some kind of measurement of waste inflow during mucking of the blasted ore to quantify its magnitude. Once the dilution is quantified, then rules can be laid down to describe the loading strategy of the fans to meet the tolerated dilution. A systematic sampling of the loaded ore in the LHD bucket can be used to measure waste inflow.

As mentioned earlier, this sampling was carried out in the whole test area. This method works well, but unfortunately it demands manpower, which is a high cost factor. Thus, it was decided to test the possibility of quantifying the amount of contained waste by measuring the weight of the loaded bucket during mucking. The large difference in specific weight between ore and the waste rock facilitates this procedure. The in situ specific weight for ore is 4,600 kg/m^3 and for waste is 2,700 kg/m^3.

A load cell was installed on the bucket of a 25-tonne-capacity electric LHD, and algorithms were developed and tested to quantify the waste content loaded out of the fan. Figure 46.3 compares waste content as calculated using data from the LHD load cell and samples from the bucket during loading of a fan. The good agreement demonstrates the viability of the method. Rules for when to cease drawing and blast a new fan were also developed and tested. These rules are based on total dilution, dilution trend, and percentage of extraction.

46.6.5   Sublevel Caving Markers

The sublevel caving layout used at Kiruna has reached dimensions that are far beyond those that formed the basis for the development of the early design guidelines. Thus, there was a need to verify the gravity flow pattern for this very large sublevel caving area. It was decided to install markers in the fans so that one could estimate the ellipsoid of extraction. A total of 908 markers were

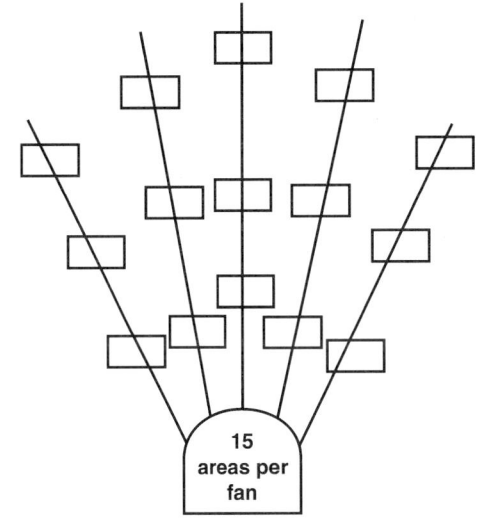

FIGURE 46.4   Layout for markers in fan

installed in 24 fans. Figure 46.4 depicts a layout for the location of the markers in the fan. The markers were installed in the middle of the fan in special holes drilled half way between the production rings. They were about 1 m long and had an identification number. During load-out of a blasted fan containing markers, special care was taken to collect the markers visible at drawpoint. A video camera was used to film loading of these fans and also to help locate the appearance of markers at the front. These films are an important source of data for future analysis of the draw.

A total of 272 markers was recovered from the fans. Figure 46.5 shows the results from a compilation of the information given by those 272 markers as a percentage. It can be seen that only a very small number of markers was recovered from the sides of the fan. This indicates that the ore flow coming from the sides of the fan was small. On the other hand, a large number of markers were recovered from the central part of the fan, indicating that the

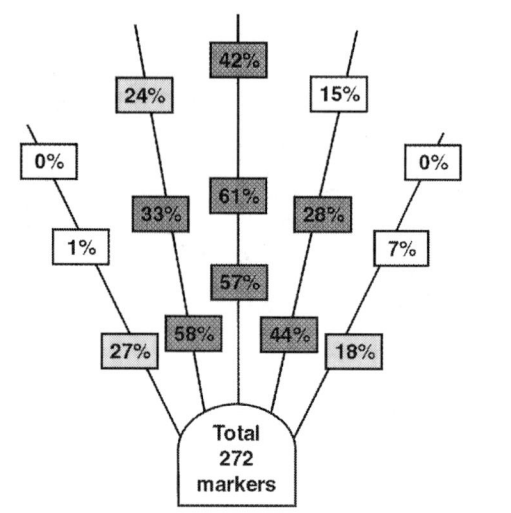

FIGURE 46.5 Recovery of markers from fans

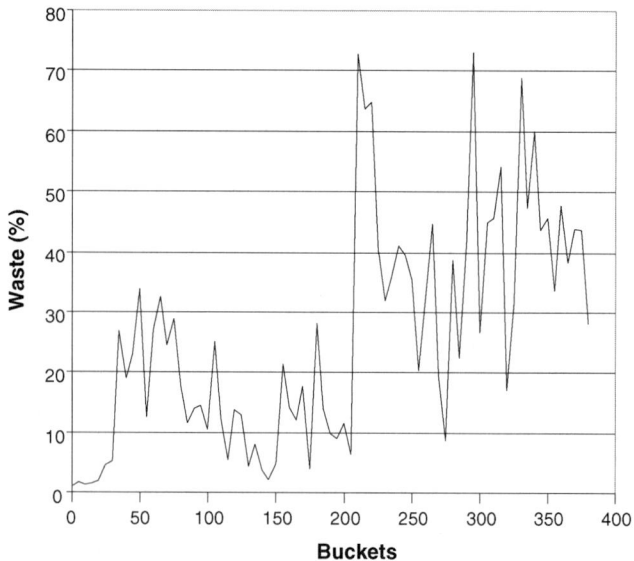

FIGURE 46.6 Waste inflow during drawdown

predominant ore flow pattern was in the center. This type of flow behavior will result in early dilution.

The results obtained from these marker tests indicate a need to improve the gravity flow pattern to achieve better ore recovery and less dilution.

46.6.6 Sampling the Sublevel Caving Drawing

The objectives with sampling during the loading out of the blasted fans were—

- To gather knowledge about the behavior of waste inflow and changes in ore quality

- To control drawing of the fans with the objective of optimizing ore recovery and separating the different ore qualities

- To test software developed for draw control procedures

The information collected by sampling was also used to evaluate the different experiments carried out in the test area. A total of 161 fans from nine different production drifts was sampled, with production from these fans about 1.7 million tonnes of crude ore.

This sampling procedure generated an invaluable database on gravity flow behavior for sublevel caving. The data show important differences in the behavior of waste inflow between laboratory tests and in situ measurements. The usual laboratory behavior of a smooth increase in waste percentage as the fan is drawn is more the exception than the rule in very large sublevel caving. The usual behavior observed is that of a pulsating waste inflow, i.e., high peaks of ore inflow are followed by high peaks of waste inflow. Figure 46.6 shows one example of waste inflow measured by sampling the drawing of a fan. The probable explanation for this behavior is that blasting produces a granular material with widely varying fragmentation sizes and mobility characteristics throughout the fan.

Normal laboratory gravity flow experiments do not account for this factor, i.e., the materials used, such as sand, have the same mobility across the fan. Thus, this pulsating behavior of waste inflow observed in situ makes the draw control more difficult to be optimized. Procedures have been developed, however, to follow the drawing of the fans to optimize ore recovery under pulsating waste inflow.

The results achieved in this experimental area of the mine by using these draw control procedures were an ore recovery of 93% and a dilution of 20%. This performance was very good, considering that the target for ore recovery was 80%. Also, because of systematic sampling of the draw, it was possible to optimize the tonnage and grades of the ore qualities produced.

46.7 CONCLUSIONS

The sublevel caving layout used at the LKAB iron ore mines has reached dimensions that are at the outer limits of earlier design guidelines. The use of very large scale sublevel caving by LKAB has resulted in very significant mining cost savings. However, this up-scaling has required continuous research to minimize the impact of the main disadvantages of the mining method in Kiruna, i.e., high dilution and/or ore loss and mixing of ore with differing amounts of phosphorus. In the full-scale experiments carried out by the Mining Research and Development Department of LKAB, it was shown that by exercising proper control, recovery can be high and dilution low, and that ore quality separation can be maintained. A pulsating waste inflow was observed as opposed to the continuously increasing inflow expected from laboratory tests, thus casting doubt on the applicability of the laboratory results on this scale. Very large scale sublevel caving is an excellent, low-cost, bulk mining method when applied under the proper conditions.

46.8 REFERENCES

Cokayne, E.W. 1982. Sublevel Caving: Introduction. In *Underground Mining Methods Handbook*, W.A. Hustrulid, ed. SME, AIME, pp. 872–879.

Kvapil, R. 1982. The Mechanics and Design of Sublevel Caving Systems. In *Underground Mining Methods Handbook*, W.A. Hustrulid, ed. SME, AIME, pp. 880–987.

Larsson, Lars. 1998. Slutrapport Projekt Skivras 2000–LKAB Internal Report.

Marklund, I., and W. Hustrulid. 1995. Large-Scale Underground Mining, New Equipment, and a Better Underground Environment: Result of Research and Development at LKAB, Sweden *Trans. Instn. Min. Metall. (Sect. A: Min. Industry)* 104, September-December.

CHAPTER 47

Sublevel Caving: A Fresh Look at this Bulk Mining Method

C.H. Page* and G. Bull†

47.1 INTRODUCTION

By some, sublevel caving is considered to be a development-intensive, high-dilution, and low-recovery mining method. Results from some recent operations, however, and changes the Scandinavians are making to very high level intervals would tend to contradict these assumptions. The clear advantage of sublevel caving is that it is a very predictable "factory"-type method with high production potential, reasonable costs, low up-front capital, and very little risk to the ore at any one time (a few thousand tonnes in an individual ring).

Sublevel caving seems to suffer from significant contradictions. The possibility of drawing a slice of broken material relatively cleanly while surrounded by broken waste (Figure 47.1) apparently defies logic. However, experience shows that it is possible. When a ring is blasted (typically on the order of 2,000 tonnes), the first part of the draw is reasonably clean. Waste from above and in front of the ring then starts to come into the drawpoint, and a mixture of ore and waste is drawn. The proportion of waste increases until a cut-off amount is reached. When the drawpoint is shut off, there will be ore left behind. This mixes with previous ore and waste in the cave, and the waste increases in grade as the cave matures. The objective is to keep the waste out for as long as possible, but to try to make the most out of the zones of higher grade dilution. For this reason, it is very important that a practical, common-sense model of how and why the method works should be agreed upon. This model will then be the basis for determining critical design and operating aspects.

The bases for any model are—

1. As ore is pulled from the blasted ring, it will be replaced by broken waste.

2. This waste will be mixed with some of the ore as the choked material is moving.

3. All design and operational efforts are directed toward extracting as much of the ore as clean as possible, delaying the appearance of dilution, and continuing the draw of an *economic* mixture of ore and waste for as long as possible.

The knowledge base for sublevel caving is comparatively small as very few mines use or have used the method. Most theory comes out of Scandinavia, is many years old, and is based on model studies and classical bin theory. Data on draw behavior are difficult to come by, and most of the data that are available comes from experience in iron mining. A major problem with applying such experience to other operations is that the dilution can carry very high iron grades, so it is difficult to assess the amount of true dilution. Here, by "true" dilution is meant the tonnage drawn from outside the current ring regardless of grade.

The problem of determining true dilution is also true for the current Australian "choke" method mines (Mt. Lyle, Mt.

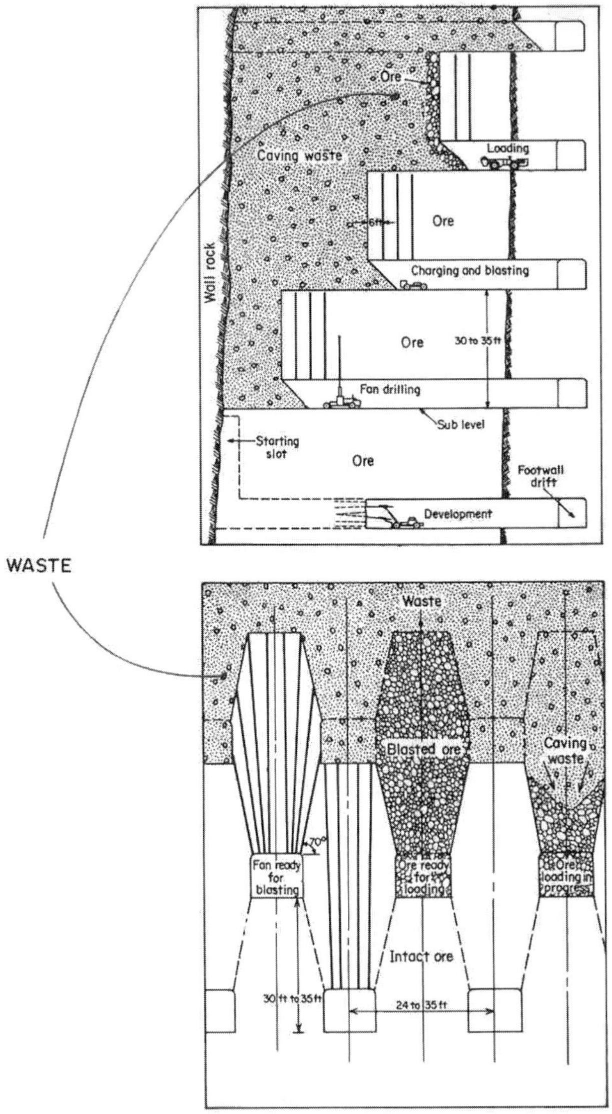

FIGURE 47.1 Overview of sublevel caving

Charlotte, Heller, Redeemer, and Perseverance), where very high recoveries and grade factors are quoted, but where the figures are not adjusted for the grade in the diluting material. A related problem in this regard is that many mines judge themselves

* SRK Consulting, Vancouver, BC, Canada.
† SRK Consulting, West Perth, Western Australia, Australia.

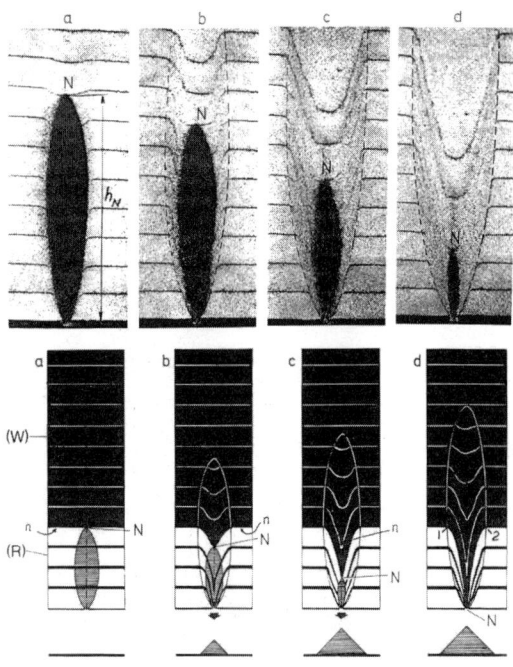

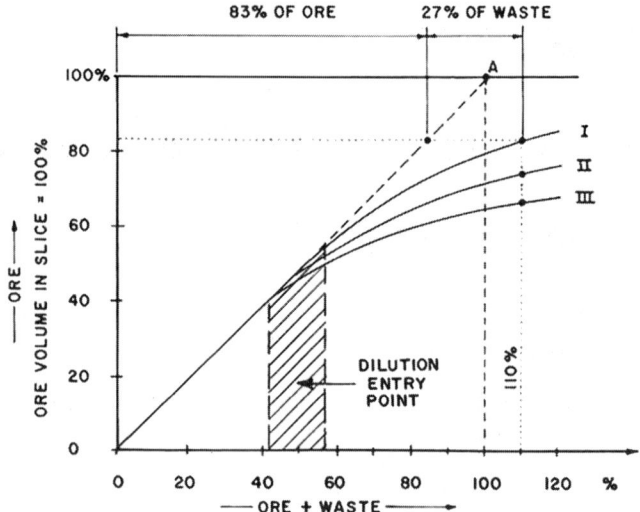

FIGURE 47.2 Sequence of drawing ore from a single drawpoint

against a reserve grade that already includes an estimate of mining dilution.

47.2 HOW DOES SUBLEVEL CAVING WORK?

The classic theory of sublevel caving (Kvapil 1982, 1992) is based on ellipsoids of motion and isolated draw from a single drawpoint (Figure 47.2). There are clearly some problems with this model as it ignores differences in fragmentation, material types, the significant weight of the cave (which compacts the caved material), and the action of the blast (which also consolidates the waste in front of the blasted ring). The conventional draw curves everyone has used in the past are shown in Figure 47.3. This suggests that in cave mining, dilution enters the drawpoint at low extractions and gives rise to the method's reputation as a high-dilution method. In practice, many operations have encountered dilution at 20% draw. At a major Chilean mine drawing beneath fine material, the waste has been observed to enter the drawpoint almost immediately. At more recent operations having different designs and procedures, it appears that dilution might enter the draw at much higher extraction levels (well above 50%). This results in much higher grade factors. The information used to justify this claim is very scarce and not well documented.

The updated model described below is based on the belief that delayed dilution entry is possible and could be the basis for some significant changes in the conventional model of sublevel caving.

47.3 CRITICAL ASPECTS OF THE NEW THEORY

The "new" theory of sublevel caving presented herein was initially based on the fact that the Redeemer Mine in Australia was breaking many of the classic rules, but achieving much better results than suggested by classic theory. The basic factors are illustrated in Figure 47.4.

1. *Interactive draw.* The drawpoints are retreated in a line and drawn in rotation so that they create a zone of low-density material that can flow at much lower angles than if a drawpoint were drawn in isolation from its neighbor.

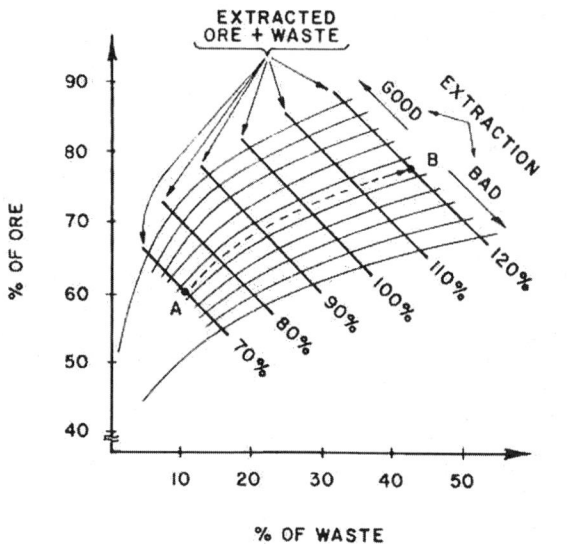

FIGURE 47.3 Conventional draw curves

2. *"Fluffed" ore and waste "consolidation."* These two factors express the concept that blasting thoroughly loosens the ore in the immediate ring (reducing density) while consolidating the waste behind the ring (making it more dense)

3. *Coarse material arching over a freshly blasted ring.* At the top of the ring, it is difficult to achieve fine fragmentation because of the extensive damage zone and significant loss of holes in the pillar area. Hence, this coarse material will arch more widely than the rest of the ring.

Although the model is unproven, it is a common-sense approach and is a reasonable basis for identifying the more critical aspects of sublevel caving. Several of the assumptions come from models of block caving and involve factors, such as interactive draw, that have been well documented.

Many of the critical aspects with the old theory remain important, but there are some new ones and some with a different emphasis in the new model. Questions are raised, however, about the importance of some of the assumptions, several of which have not been questioned until recently. What is clear is that the following are critical.

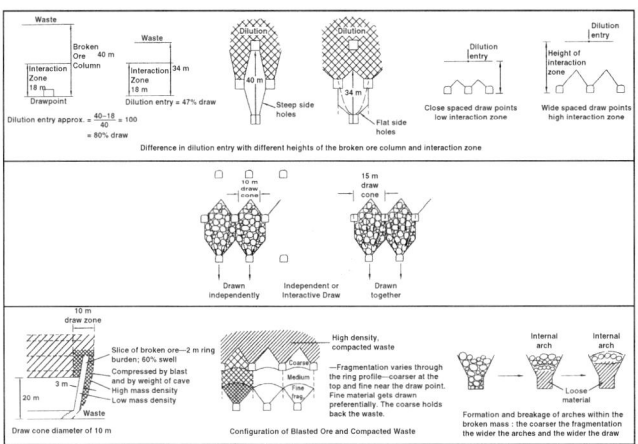

FIGURE 47.4   Schematic illustrating draw behavior in sublevel caving

- *Differential fragmentation.* Fine fragmented material flows more readily than coarse material. Dilution should be coarser than the ore since the fine material can flow through coarse material.

- *Compaction.* Compacted material does not flow as readily as freshly blasted material, and the blast must compact the "waste" in front of the fresh ore.

- *Temporary arching.* The coarser material at the top of the ring impedes the flow of waste, draws over a much wider arch than fine material, and can temporarily hang up while the finer material is drawn.

- *Draw coverage.* The more the ore is undercut by development, the more likely it is that the ore will flow into the drawpoint. Figure 47.5 shows model flow diagrams and dilution entry for different drawpoint spacings (D/Ps), isolated draw zone diameter (IDZ) ratios under interactive draw (A, B, and C), and isolated draw (D). The height of the interaction zone (HIZ) is also shown.

- *Interactive draw.* The simultaneous drawing of adjacent drawpoints will result in a much wider zone of moving material. Thus, one should try and draw from several drawpoints in a line. The Perseverance Mine has observed startling differences in extraction between isolated and interactive draw (Figure 47.6).

- *Ground support.* Ground support is installed primarily for the brow. The brow has to stay stable for the short period of draw and charging of the next ring. The support density will be much more than that required for normal tunnel stability since it (1) has to accept the blast damage and (2) has an extra degree of freedom.

- *Blasting.* Blasting has to be done to break the rock to the right size (uniform and not too fine) without causing too much damage. It should be sufficient to fluff the ore and compact the waste. The powder factors are generally more than 30% higher than for unchoked blasting.

- *Grade dilution.* To minimize dilution, extraction should be maintained within the mineralization envelope or mine where dips are steep. This affects the choice of vertical distance between the drawpoint and the waste (Figure 47.7).

The overall objective for a sublevel caving operation is delayed dilution entry. Classic theory and practice has dilution entering when 20% to 40% of the ring tonnage has been extracted. More recent operations have achieved 80% and better. By delaying the entry of dilution into the muckpile, both the proportion of clean ore and the grade factor for the same

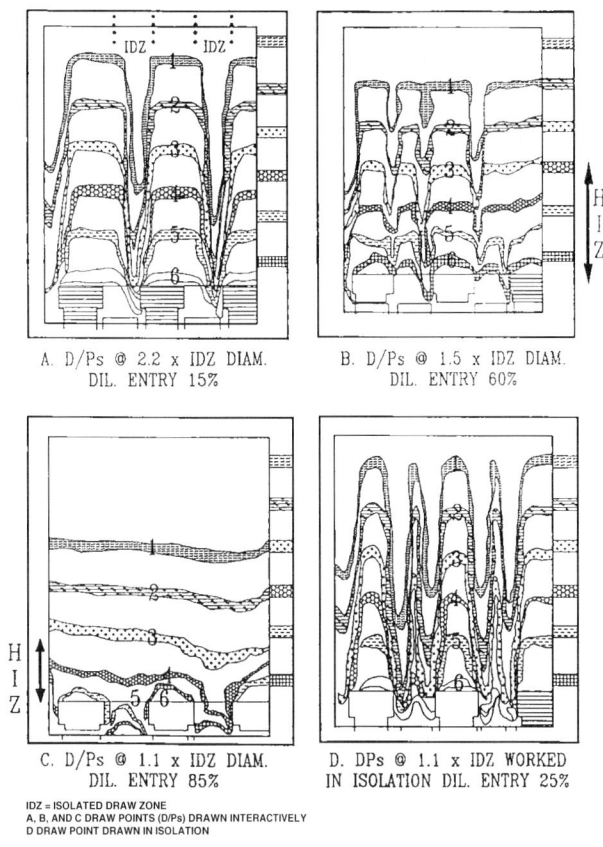

A. D/Ps @ 2.2 x IDZ DIAM.
DIL. ENTRY 15%

B. D/Ps @ 1.5 x IDZ DIAM.
DIL. ENTRY 60%

C. D/Ps @ 1.1 x IDZ DIAM.
DIL. ENTRY 85%

D. DPs @ 1.1 x IDZ WORKED
IN ISOLATION DIL. ENTRY 25%

IDZ = ISOLATED DRAW ZONE
A, B, AND C DRAW POINTS (D/Ps) DRAWN INTERACTIVELY
D DRAW POINT DRAWN IN ISOLATION

FIGURE 47.5   Model flow diagrams and dilution entry for different drawpoint spacings

extraction are increased. Dilution in this instance, as indicated earlier, is material outside the current ring. If it carries grade, this is clearly a plus.

47.4   IDEAL LAYOUT

The items listed in section 3 are very important when considering layout, conditions, equipment, and procedures that will achieve or promote ideal conditions. Everything in underground mining is a compromise between what looks good on paper and what an ore body will allow one to do. The compromise is one of reducing development and operating costs or having larger, more frequent development to achieve better grades. In the first case, there may be loss of tonnage and grade through earlier dilution entry and a lower grade factor. In the second case, large amounts may be spent on support and a reduction in production rate. One must be clear about what the compromises are and their consequences. Most mining engineers are happy to estimate costs from which they always conclude that less development is cheaper. However, they are very unhappy with estimates of head grade, which is mostly a function of planned and unplanned dilution, and loss of production. Dilution and production surely are, however, the two most important engineering considerations. There are a number of aspects that must be considered.

47.4.1   Longitudinal or Transverse Mining

A transverse layout (Figure 47.8) produces more drawpoints. The slot cuts off the major stress direction and the drifts are in the same direction as the major induced stress, so they are less affected. However, the long faces are more "relaxed" and can be less stable, and cutting the major principal stress can actually reduce the clamping that holds blocks in place. A longitudinal layout means a better development yield, i.e., fewer waste

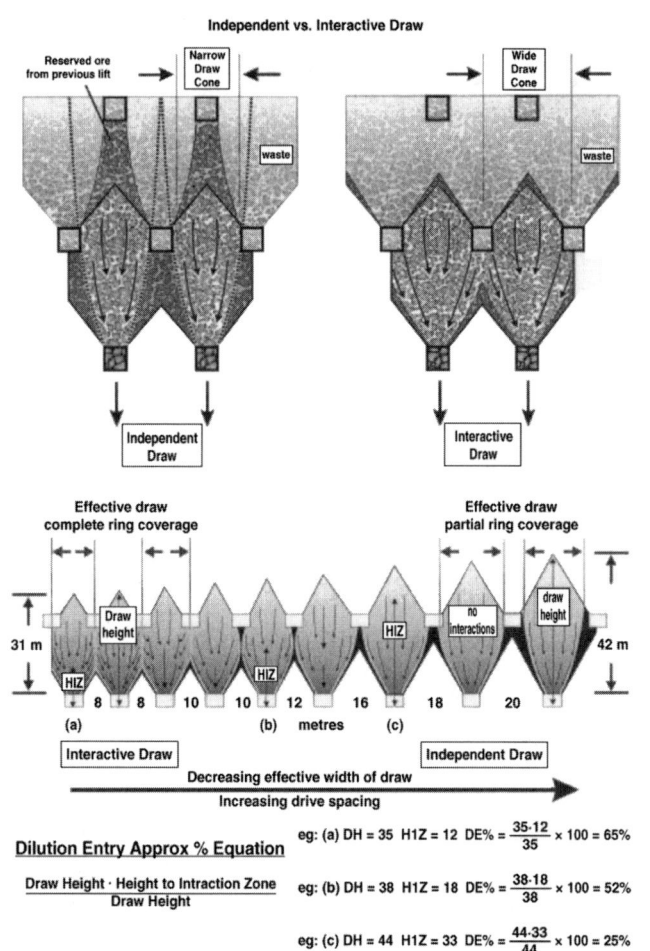

FIGURE 47.6 Comparison of independent and interactive draws. A, Effect of interactive draw; B, effects of drawpoint spacings

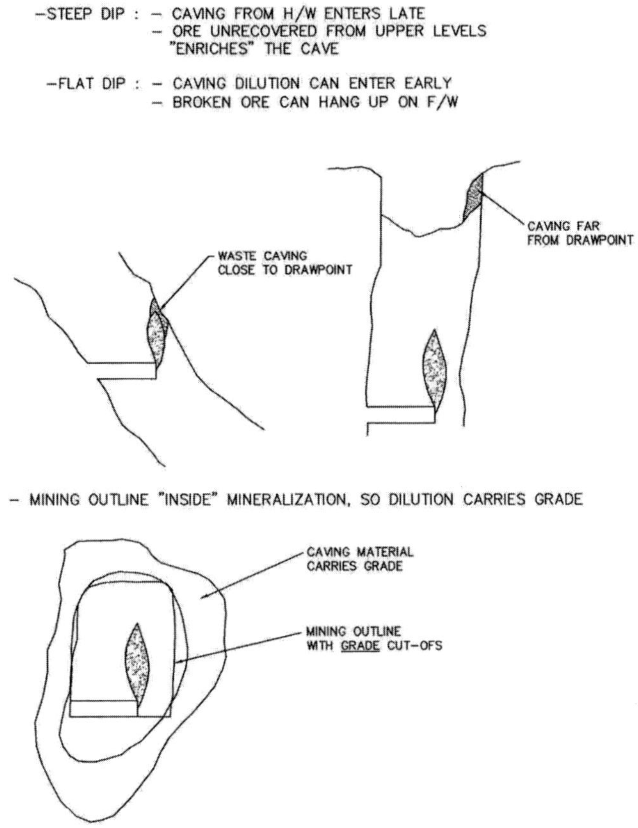

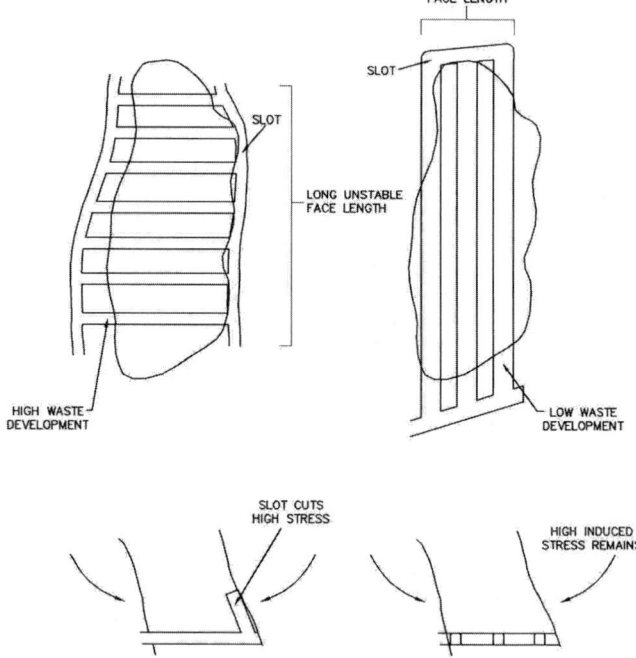

FIGURE 47.7 Minimizing dilution where dip is steep

connections. It also means potentially lower dilution from the hanging wall if the hanging wall is strong. The clamping stress from the hanging wall is retained, resulting in shorter, more stable extraction faces. On the other hand, these fronts stay in the "nutcracker" position. The longitudinal layout has less production potential and very long ventilation lengths.

47.4.2 Production Drift Size and Shape

The drifts should be as wide as possible to assure good draw coverage. They should also be as low as possible for stable pillar walls and shorter muckpiles for easier charging. In addition they should be as square as possible to widen the zone of moving material. This is an area of compromise, however, since the effectiveness of square brows is open to question. If an arched back results in much more stable development, then stability might take precedence (Figure 47.8).

47.4.3 Drift Spacing

The drifts should be as close together as possible. Spacing is dependent on pillar size and shape, which depends upon the loads being carried. In effect, the production level is a room-and-pillar layout and the pillars must be able to carry the vertical loads. These tend to be low because of the proximity of the cave (Figure 47.9).

47.4.4 Level Interval

Level interval is dependent on hole deviation. As common sense would suggest, the closer the levels are together, the less likely

FIGURE 47.8 Transverse or longitudinal layout

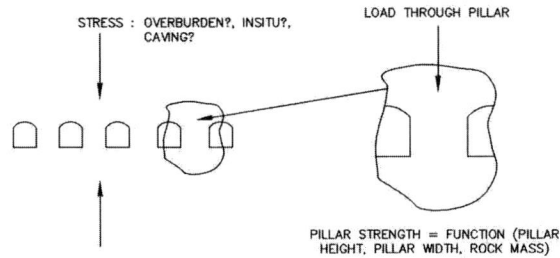

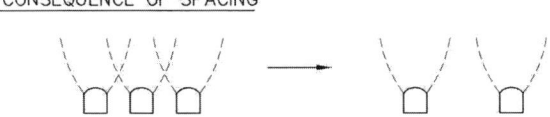

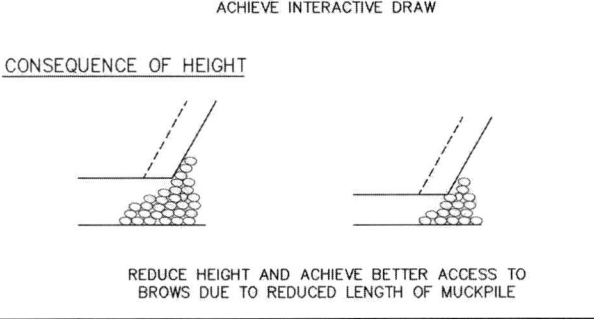

FIGURE 47.9   Considerations in drift spacing

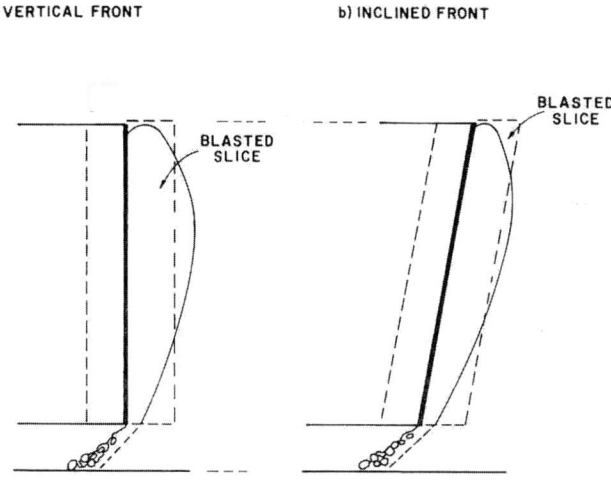

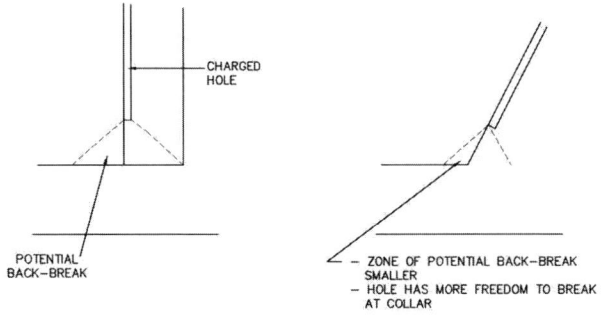

FIGURE 47.10   Ring inclination

that something will go wrong. Level interval has a significant effect on development cost per tonne. With improvements being made in drill technology, the industry is rapidly increasing level intervals.

47.4.5   Ring Burden

In the past, the ring burden has been related to the "dig depth"; however, this is doubtful since no bucket can dig anywhere close to the back of a muckpile. It is dependent on hole size and ring pattern. The burden must ensure both "fluffing" and "compaction." Furthermore, it must achieve the correct design powder factor of 0.9 kg/m^3 and minimize the incidence of freezing, which is an ever-present danger with choke blasting.

47.4.6   Hole Size

The largest possible holes are more economic and more accurate. Larger holes mean that level intervals can be increased. They are less likely to be closed and can have larger burdens, which makes it easier to get back under the brow and charge. However, they can do more damage to the next ring and can cause severe backbreak, affecting the collars of the next ring. Sticky ANFO can be used to charge holes with larger diameters.

47.4.7   Ring Pattern

Both fragmentation and damage are dependent on ring layout. Hole spacings should be much larger than the ring burden. The burden:spacing ratio at the toes of the holes should be at least 1:1.3.

47.4.8   Ring Inclination

The ring is inclined toward the cave to adjust the shape of the moving material, to shield the drawpoint from the waste above, and to protect the brow from backbreak (Figure 47.10). There is

very little information on ring inclination and the effect of different angles.

47.4.9   Column Height

As large a column height as possible is desired to keep pure waste well away from an active drawpoint. Height is a function of dip and size.

47.4.10   Short Hauls

Haulage distances should be kept as short as possible, and there should be as many tipping points as possible. Ore passes are cheap in terms of cost per tonne

47.4.11   Continuous Development

Crosscuts should not be mixed with drifts as it is very difficult to bring the rings through an intersection. In such cases, multi-ring blasts are often necessary, and there is a high frequency of freezing with multi-ring blasts in choke conditions.

47.4.12   Slots

A free face is needed before ring blasting can start. There are essentially four types of slot: (1) "individual" with a raise at the end of each drift; (2) "continuous" with a raise at one end, (3) "slashing" along the axis of the drift, or (4) "slashing" along a perpendicular drift (Figure 47.11).

47.5   WHAT CAN GO WRONG?

Before discussing the ideal conditions under which sublevel caving should be operated, it is useful to consider what can go wrong, remembering that the emphasis should be on achieving

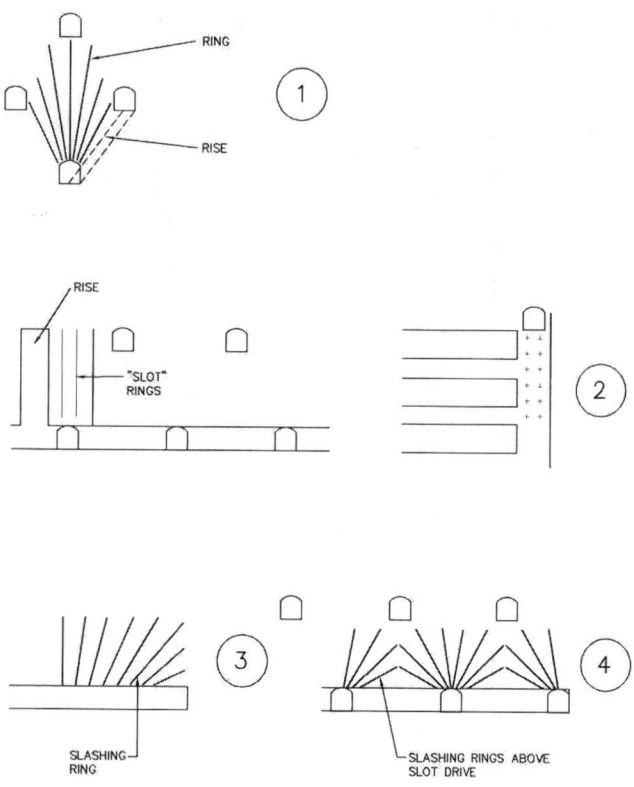

FIGURE 47.11 Slot types

(1) late dilution entry and (2) very high production rates per drawpoint. The items identified as contributing to the success of the new sublevel caving model will also detract from these results if they do not perform as planned. Some of the potential problems are indicated below.

- *Brows.* There are three potential problems with damaged brows: wedge failures narrow the draw, backbreak makes it difficult to charge the holes, and stress failure resulting from high horizontal stresses damages the hole collars.

- *Pillars.* The walls of the drifts can be lost because of high stresses and weak rock or the daylighting of structures. This can occur if development is too far ahead, the pillars are too small, and/or the walls are too high.

- *Bridges.* Bridge formation because of unbroken toes at the top of a ring or unbroken ore from above removes choke conditions. This allows waste to flow around. The bridges usually get thicker and thicker until they are down to the brow and the whole blast freezes (Figure 47.12a).

- *Walls.* Walls form when the front ring of a double-ring blast freezes, compacted waste freezes (over-compaction(?), or sticky material is present (Figure 47.12b).

- *Ring freezing.* Blasting problems or delays in operations allow the choke to overconsolidate.

- Ribs. Unbroken toes on the side of the ring leave unbroken ore between the drifts (Figure 47.12c).

- *Oversize.* Hole deviation and misfires can cause significant amounts of oversized rock. The oversize slows the draw process, affects production rate, and can result in premature dilution entry as it interrupts the draw process.

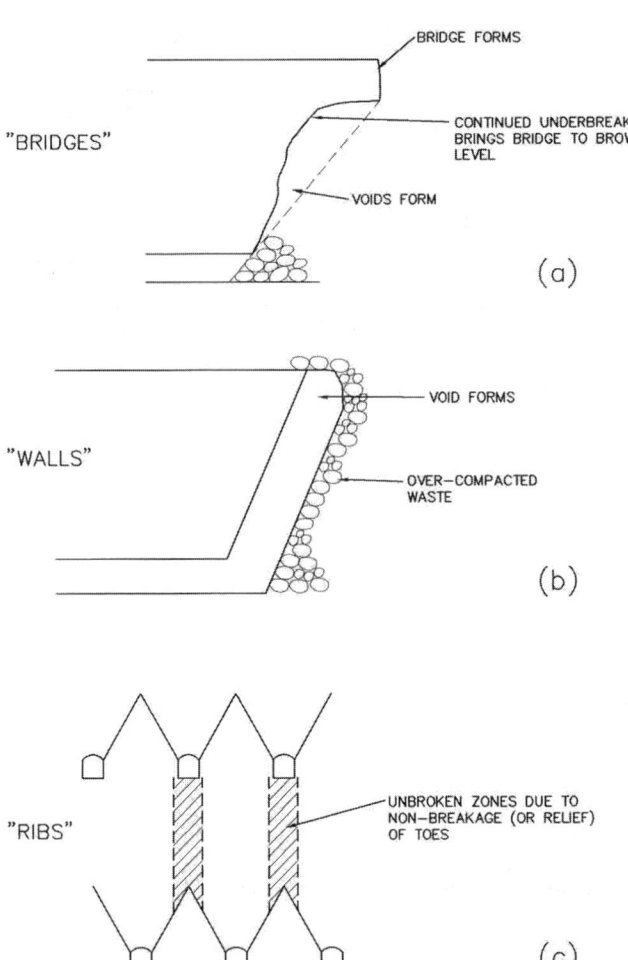

FIGURE 47.12 Examples of problems in sublevel caving

- *Wedge crushing.* Wedge crushing occurs with the day-lighting of large, continuous structures in the hanging wall and crushing of immediate development. The latter is rare.

- *Overhangs.* Dipping ore with delayed caving can result in significant stresses being induced because of an over-hang. This can result in crushing of development and/or loss of holes.

- *Incomplete slots.* If the sublevel caving production drive starts with an incomplete free face, the rapid formation of bridges and ribs can occur.

- *Loss of holes.* Drilling too far ahead and stress changes or relaxation can result in the deterioration of drill holes. Daylighting structures can cause block movement and cut-offs, loose material can rill into breakthrough holes, and blasting damage from the previous ring can cause hole loss.

47.6 IDEAL CONDITIONS

The ideal layouts for sublevel caving have been reviewed and what can go wrong during operations has been presented. As was discussed earlier, mining methods are a compromise between what looks good on paper and what the ore body will allow to be done. Determining what the ore body will permit is a technical consideration. Making sure that what should be *achievable* is actually *achieved* is management. The combination of ideal layouts and what can go wrong should indicate the ideal conditions for sublevel caving. What the ore body will allow one to do is a function of mining difficulty. This includes geometry, rock mass conditions, major structures, stresses, grade distribution, etc. Some of the factors to keep in mind are—

- *Strong rock.* Strong rock allows the use of small pillars (small drive centers).

- *Competent rock.* Rock competence is dependent on jointing and formation of wedges. Competent rock implies few joints and strong joint surfaces (with little infilling and irregular and rough surfaces). In competent rock, the brows are mostly disturbed by blasting and then by gravity. Such rock masses allow the creation of wide backs. Since the presence of joints may cause wedge failures in one direction but not in the other, a design decision might be to align development in the favorable direction.

- *Few major structures.* The lack of major structures minimizes massive wedge failures and/or hole cut-offs.

- *Steep dip.* A steep dip keeps the source of low-grade dilution farther away from current drawpoints. As a result, most dilution comes over the top of the mixture of ore and waste from caving at much higher levels than the current extraction level.

- *Massive ore body.* The ore body should be sufficiently massive to present a large footprint for high production rates. Most dilution comes from the boundary between ore and waste, and hence the more massive the deposit, the smaller the proportion of material from the boundary. The development yield (ore recovered ÷ waste development) is higher for massive deposits.

- *Fragmentation.* Fragmentation of both the blasted material and the caved material must be considered. Joint frequency, joint condition, and joint direction will all affect the degree of fragmentation. A competent rock mass is usually the most suitable because larger, widely spaced holes will still result in good fragmentation, but the caved material will be very coarse. This results in the blasted material being finer than that of the caved material and yields "even" pieces of fragmented material with a minimum of both oversize and fines.

- *Dry conditions.* A minimum of groundwater is positive both for the maintenance of roadways and the prevention of muck rushes. There should be a program to encourage drainage ahead of the cave face.

- *No muddy material.* There should be little, or preferably no, very weak or rapidly weathering material in order to avoid problems of muck rushes, over-compaction, and hang-ups.

- *Caving.* Caving is seldom an issue with sublevel caving as mining usually starts out of the bottom of an open pit. Lack of caving from the hanging wall is usually an advantage because it delays the beginnings of dilution. However, choke conditions must be maintained. At dips below 50°, for example, the occurrence of voids can be an issue. Cavability is a function of rock mass conditions and hydraulic radius (Figure 47.13).

47.7 PROCEDURES AND PROCESSES TO MAXIMIZE EFFECTIVENESS

"Ideal" is not found in mining, and compromises are inevitable. There are a number of processes and procedures that should be used to help overcome the possible consequences of making compromises and to enhance the effectiveness of sublevel caving.

The only methods in which most of the technical uncertainty is "designed out" of the system are those with cemented backfill.

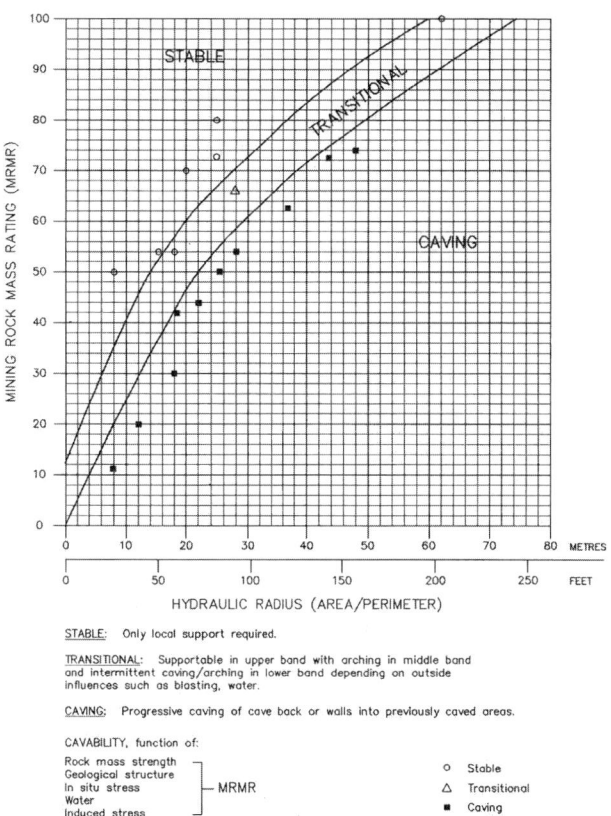

FIGURE 47.13 Graph showing cavability as a function of rock mass conditions and hydraulic radius

This, however, is done at great expense and with a significant reduction in production rate. There are a number of ways in which design (technical) and attention to detail (management) can be used to reduce uncertainty in sublevel caving.

47.7.1 Development

The excavation process can cause considerable damage. Irregularities in the size and shape of drifts can make them less stable and more difficult to support. Care must be taken in the design and management of the process. This means that any plan should include—

- Enough holes at the perimeter. Both the perimeter and the next row of holes in from the perimeter (if necessary) should be "cautiously" charged and timed.

- Good breaking faces for all holes (at least right-angle triangles).

- Timing to prevent out-of-sequence detonations.

- Standard patterns. Jumbo operators should not be allowed to "individualize" their patterns. The patterns should be marked in detail. Usually there are too few lines or marks on the face for a jumbo operator to follow.

- Autoparallelism. Too often the autoparallelism feature of modern drills is not used, and much too much look-out angle is put on the perimeter holes.

47.7.2 Ground Support

Roof support has to function when the brow has freedom to move and has to resist strong disturbances from blasting. It is not there to simply protect personnel in the drive. Whether precautionary or structural support is required should be established.

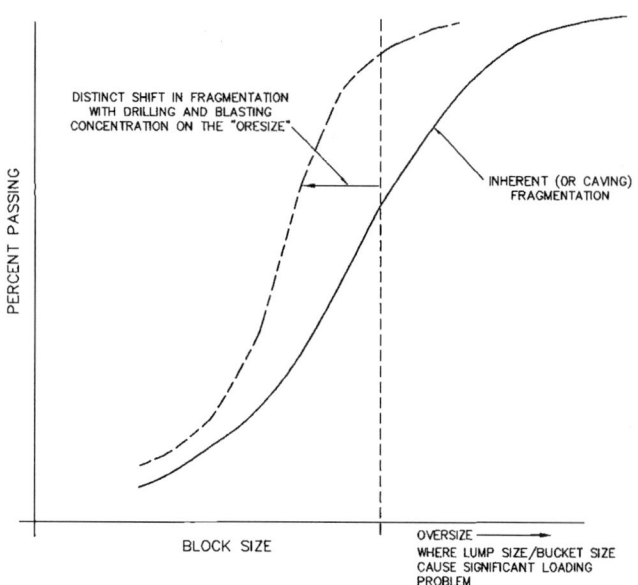

FIGURE 47.14   Distribution of material sizes

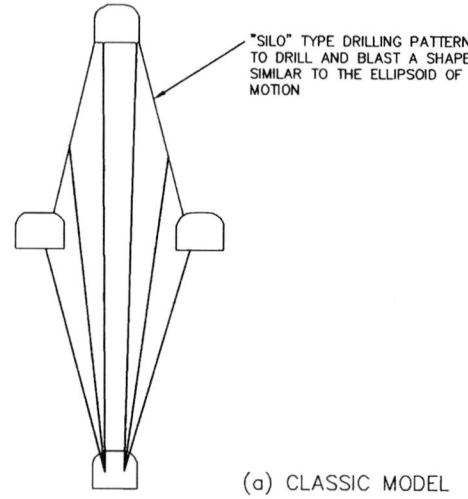

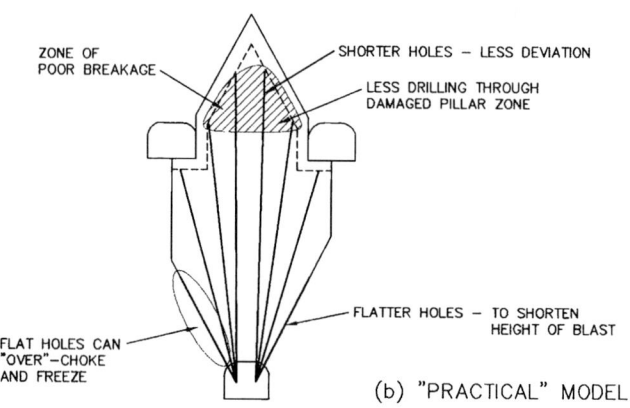

FIGURE 47.15   Effects of drilling patterns

Precautionary support can be Split-Set rock bolts and mesh, whereas structural support would normally include grouted bolts with strong plates. Bolt length is not usually important as all development is small. Increasing support intensity would include straps, etc., up through reinforced shotcrete. Support is concentrated at the brow position. Cables might be required to hold pillar walls. It should be remembered that the walls may often need more support than the backs. The objective is to retain brow shape, maintain stable collars, reduce backbreak, and produce safe charging conditions.

47.7.3   Ring Blasting

The objectives of the ring blasting program are "fluffing" ore, compacting waste, preventing misfires, creating good fragmentation that results in interaction and high productivity, and limiting damage.

- *Spacing and burden.* Spacings and the amount of burden are chosen to minimize oversized material that interrupts or slows draw. This is a function of drawpoint dimensions and bucket sizes of load-haul-dump (LHD) machines. A significant shift in the distribution of material sizes might need to be achieved, as illustrated in Figure 47.14. This is best done with small amounts of burden, large spacings, and longer delays. However, it is also necessary to be able to create swell by compacting the choke.

- *Pattern.* There is a need to flatten the lower holes as much as possible, but not so low that they choke and misfire. This is done to reduce hole length and to minimize the amount of hole that has to be drilled through the damaged pillar zone (Figure 47.15). The problem of relieving the flatter holes is reduced by interactive draw.

- *Charging pattern.* The charging pattern used must minimize sympathetic detonations, damage to the brow, and damage to the next ring. The charging density should be kept to below 1.5 times the design powder factor. Uncharged collar lengths should be at least 0.6 times the burden. If ledges should form at the brow, then these should be blasted cautiously with decoupled charges.

- *Timing pattern.* There should be no overlap in timing sequences used. Individual detonations should occur toward free faces. Sufficiently long delays should be used to achieve separation and movement.

- *Hole deviation.* Minimizing hole deviation is critical. The extra expense for stiffer drill assemblies, laser set-ups, and good roadbeds to assist with set up will always pay dividends. Hole deviation is so important that a system of hole checking is critical.

47.7.4   Draw Control

The objectives of draw control are to maximize the draw before dilution enters the muckpile, to maintain late dilution entry, and to correct quickly anything that is contributing to early dilution entry. Factors to consider in draw control are—

- *Interactivity of drawdown.* Drawpoints across a flat front should be drawn together. This is done on the basis of either so many buckets or so many trucks from each drawpoint. Too much or too little must not be drawn from any given drawpoint (for example, between several bucket loads to two truck loads from a drawpoint) before moving to the next position. In a dipping ore body, there may be a need to overdraw from the footwall side to prevent the ore from sticking on the footwall. This will be part of building up experience.

- *Even draw.* It is important to draw evenly from right across the muck pile and not to draw continually from one side or to draw around hang-ups or large blocks in the muckpile.

- *Secondary breakage.* Oversized material should be broken promptly when it cannot be quickly taken out of the muckpile. Leaving it in place disrupts the flow of material.

- *Hang-ups.* Very large oversize, bridges, ribs, etc., should be broken up immediately by hang-up drilling. The problem generally does not correct itself, and further ring blasting usually makes the problem worse.

- *Issuance of "bucket" sheets.* The available tonnage in a ring should be calculated and the LHD operator told how many buckets or truck loads can be drawn from a drawpoint before detailed visual sampling is done. This might be, for example, 90% of the likely extraction. The draw would be adjusted for potential extra tonnage left from the level above.

- *Visual checking.* The muckpiles should be regularly inspected to assess the proportions of fresh ore, old ore, and waste in the muckpile. In many instances, old ore, which is ore from above but not broken in current ring, cannot be distinguished from freshly broken ore.

- *Shut-off.* Careful checking at 90% draw must be done to ensure that the grade of the muckpile is above the shut-off value. If that is the case, extraction continues unless, of course, a dilution blanket is being preserved for final extraction. In such a case, in narrow, dipping ore bodies, there is a very real risk of losing this material as it hangs up on the footwall and dilution from the hanging wall cuts it off. Marginal costing should be used to adjust the shut-off grade continually. Care should be taken to assure that the predicted average grade is being achieved. Early dilution entry and high dilution grades can result in very high extraction amounts, but at a low average grade.

- *Monitoring.* It is most important that the draw be continually monitored to check the point of dilution entry and how the proportion of dilution changes with extraction. "Dilution" in this case is anything that was not in the freshly broken ring. This can only be done properly if fresh and old material can be distinguished. If this is not possible, then one simply distinguishes between ore and waste.

47.7.5 Floor Conditions

The floors should be dry and smooth. The roadbase should be engineered. There should be effective and rapid drainage. Grading is probably best done with a small dozer.

47.7.6 Repairs

Stability conditions should be carefully watched and repairs made before unstable conditions become a problem. The walls are of special importance. If repairs are becoming frequent, then support should be increased. Rehabilitating development is expensive, unsafe, and interrupts production.

47.7.7 Drilling Ahead

Production drilling should be done in advance, if possible. However, as soon as redrilling becomes a major interference with the progression of the blasting face, then coupled drilling and blasting should be examined.

47.8 EVALUATION (BUDGETING AND FORECASTING)

Assuming that the compromise between ideal layouts and conditions is a sensible one and that uncertainty will be well

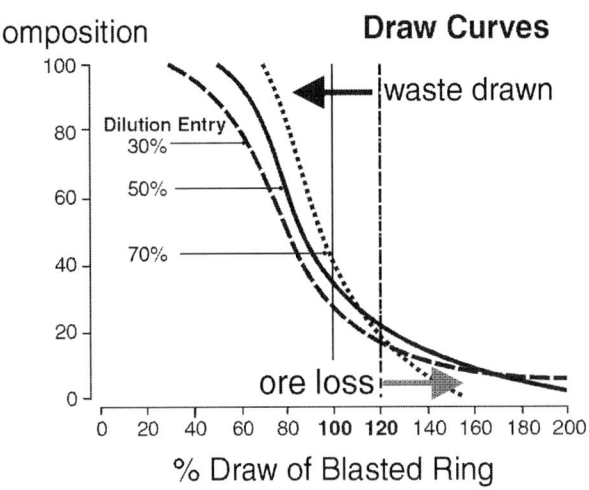

FIGURE 47.16   Model of mixture ore and waste in draw

managed by practical procedures, what will sublevel caving deliver? As with all mining methods, the costs per unit are easy to calculate. The difficulty is in predicting the productivity per unit (primarily the production rate per drawpoint) and the head grade. The latter is a function of planned and unplanned dilution, which in turn is dependent on the point of dilution entry, the degree of mixing, and the dilution grade. To add to this difficulty, there are very few sublevel caving operations worldwide, and the results from one of the Australian operations seem to be too good. This should not detract, however, from what is potentially a very productive and cost-effective method. It is a method that can deliver the lowest cost-to-metal under certain circumstances. In spite of the lack of good data, estimates and predictions must still be made.

With regard to grade prediction, a key element is to utilize the benefits of interactive draw fully. The draw model illustrated in Figure 47.16 has been based on models developed for block caving (Laubscher 1994), as well from previous experience with sublevel caving operations. The model shows the proportion of ore (below the line) and dilution (above the line) as extraction increases (along the x-axis). It can be seen that very high extraction can be achieved depending on the shape of the curve and the grade of the ore, as well as the grade of the waste.

Ideally we should also have a model that estimates the grade of dilution, which is a mixture of pure waste, mineralized waste, and ore left behind from each recovered ring. A method to estimate the dilution bin is illustrated in Figure 47.17. This is very simplistic, as we know that the dilution grade will decrease as the ring is drawn and the waste comes from further away. The draw nomogram shown in Figure 47.18 may be used at 60% dilution entry to predict tonnage and grade to shut off. It must be remembered that these curves are initially common sense, but with time will become real if the cave is correctly monitored. Monitoring is essential in this regard.

This method was used on a recent project and allowed reasonable sensitivities to be run. The results from the model agree reasonably well with the achievements of the operation. However, none of these operations is technically or operationally ideal. It is believed that a well-engineered and operated sublevel cave in ideal circumstances (geometry and rock mass) can deliver better results than predicted by the model.

Cut-off grades are determined by examining the cost-to-metal. The costs used are the *full* operational costs, including capital. The shut-off value should be continually re-estimated using marginal costs to ensure that the grade is sufficient to make

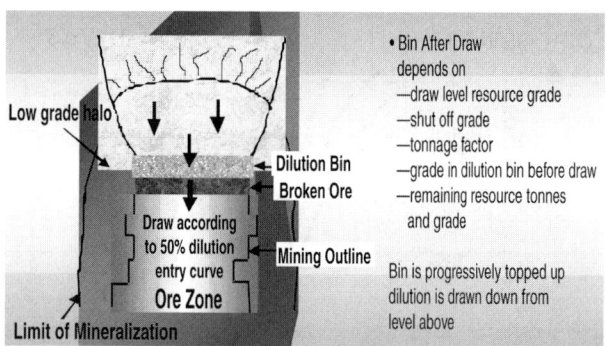

FIGURE 47.17   Mechanics of dilution grade estimation

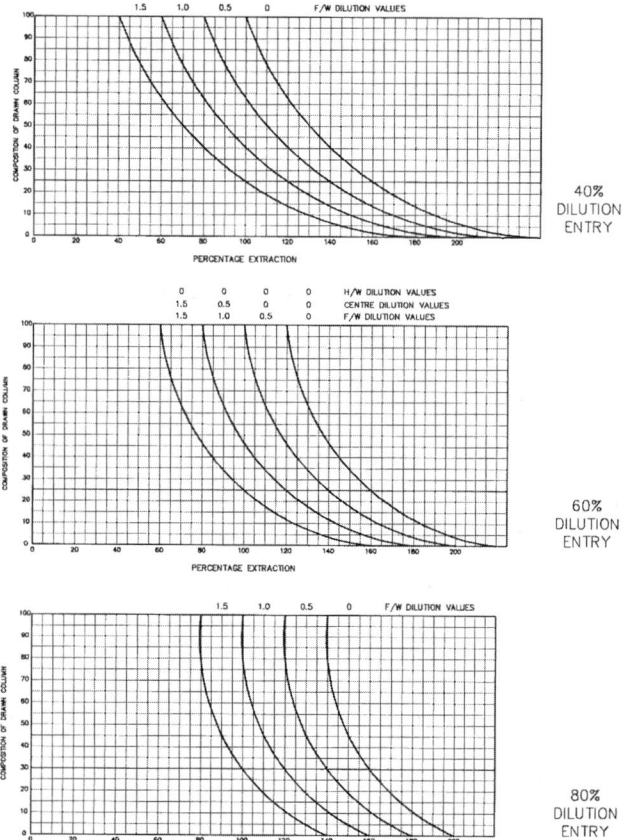

FIGURE 47.18   Draw curves at different dilution entries

a contribution to depreciation and amortization. It is not advisable to use marginal costing for development and stope boundary planning as the price is not normally known in advance. It is much better to work to a strategic cost-to-metal value. Hedging programs may change the cut-off and shut-off values, but should not change the common-sense approach of planning on the basis of cost and not on revenue.

To estimate production rate, the drawpoint is used as the key operational indicator. The number of available drawpoints will determine the overall production rate. To determine the required number of production drifts, the time for each dependent operational element at the drawpoint must be estimated. This includes drilling, charging, blasting (including any brow repairs and redrilling), mucking, secondary breakage,

and hang-up clearance. The influence of keeping the face flat and achieving interactive draw must also be considered.

With all of these conflicting activities, the long-term production rate from a set (or face) of interacting drawpoints must also be considered. The result is possibly 500 to 600 tonne/d from each drawpoint. Clearly, the haul distance should be kept as short as possible. In longitudinal retreat, this conflicts with minimizing the number of access crosscuts. Although fragmentation from sublevel caving is usually very good, there may be problems with ore passes and chutes.

The production LHDs should be as large as possible. Degree of fragmentation, road conditions, and haul distances should be such that an LHD is capable of at least 25 to 35 buckets per operating hour, assuming a straight run into the muckpile and filling the bucket in a single movement. A large LHD will require some four to five drawpoints to keep it supplied with broken muck. If interactive draw is also considered and the draw is across a group of four or five drawpoints, then at least 10 drawpoints may be required per LHD. If trucks are being loaded, then there might be only one LHD operating in a face. In such a case, the production rate will be a function of buckets per hour for that LHD.

The overall maximum production rate of any mining method is very difficult to estimate on paper, but typical rates of fall through the deposit under good conditions have been in the region of 65 m/yr.

47.9   CONCLUSION

Currently, the mining methods for the most massive deposits are either block caving or, if they will not cave readily, open stoping with cemented fill. Sublevel caving is generally not considered, or if considered, rejected based upon dilution and recovery concerns, that is, recovery will be higher and dilution lower with a filling method. This is partially correct, but recoveries are usually well below 100% and the dilution often runs at over 15%. Mature sublevel caving can achieve recoveries close to 100% with grade factors well over 80% (the proportion of pure waste is less than 20%). This assumes that there is mineralized waste around the ore (not unusual) and that the dilution grade steadily increases. The latter is the expected attribute of sublevel caving that must be maximized. In addition, sublevel caving can be very productive in terms of rate of fall through a deposit. It is generally double that which can be achieved with filling methods.

In a recent project, sublevel caving was shown to exhibit very superior economics over more conventional methods. Recoveries of over 100% recovery, grade factors over 80%, and operating costs less than $9/t were achieved.

Only a very few ore bodies are really suitable for sublevel caving. Such ore bodies should—

- Be strong and competent.

- Have a large footprint with a very steep dip.

- Preferably contain mineralized waste.

If these conditions apply, the only methods that can compete on a cost-to-finished-metal basis are possibly block caving and partial extraction.

For sublevel caving the major future advances are with (1) interactive draw, (2) understanding what contributes to the success of the method, and (3) ensuring that this success is achieved.

47.10   REFERENCES

Kvapil, R. 1982. The Mechanics and Design of Sublevel Caving Systems. *Underground Mining Methods Handbook*, W.A. Hustrulid, ed. SME, pp. 880–897.

Kvapil, R. 1992. Sublevel Caving. Chapter 20.2 in *SME Mining Engineering Handbook*, 2nd ed., H.L. Hartman, ed. SME, pp. 1789–1814.

Laubscher, D. H. 1994. Cave Mining B The State of the Art. *Journal of the SAIMM*, October, pp. 279–293.

Panel Caving

CHAPTER 48

Henderson Mine

William D. Rech[*]

48.1 INTRODUCTION

48.1.1 General Description

The mine site is located 80.5 km west of Denver, Colorado, and lies 3,170 m above sea level, on the eastern side of the Continental Divide. The ore body is over 1,000 m below the surface, and the lowest excavation is at a depth of 1,600 m, making Henderson one of the deepest caving operations in the world. The ore is transported by conveyor from a crusher complex at the 7065 level to the mill site 25 km away on the western side of the Continental Divide (Figure 48.1). The overall transfer of ore is achieved through the use of only three conveyor flights, one 1.2 km long in the mine, a second 16 km long through a former railroad tunnel, and a third 7 km long that travels overland to the mill. Until 1999, the ore transportation system consisted of a semiautomated train system that ran without operators for most of the 25-km trip between the mine and mill.

48.1.2 Mining History

The mine began operation in 1976, after a 10-year predevelopment program and $500 million investment. From 1976 through 1991, approximately 90 million tonnes of ore were produced from the 8100 level. In 1992, the 7700 level was brought into production, and over 45 million tonnes have been extracted so far from this level. The next production level will be the 7225 level, located 145 m below the 7700 level. It is important to note that the east region of the ore body has not been exploited through either the 8100 or 7700 level. As shown in Figure 48.1, very tall ore columns are available on the east side of the ore body at the 7225 level.

Throughout the past two decades, the molybdenum market has experienced varying states of over- and under-supply, and Henderson's production rates have varied accordingly. This includes a complete shutdown of the operation in 1983 for 15 months. As a result of these factors, the mining operation has had to be extremely flexible and allow for the movement of employees from mining to production functions as the business has cycled. In addition to the employee's need to be flexible, it has also been necessary for the caving activity to be flexible. Cave development has been stopped several times and restarted as markets have dictated. Many of the mining practices at Henderson have been developed to accommodate this contingency.

Improvement capital for the operation has been scarce for most of the past 20 years. Changes, therefore, have generally been in the form of continuous process improvement and not wholesale redesigns. A strong demand for molybdenum in 1995 provided the first opportunity since 1976 to make a significant change to the operation. The area needing the most attention was the ore transportation system. This modification to the operation

became a capital improvement project titled Henderson 2000. Henderson 2000 was engineered to address both improvement of ore transportation to the mill and development of the bottom and final production level for the mine. The conveying system associated with Henderson 2000 was completed in 1999. The development of the final production level will follow in 3 to 5 years, depending upon market conditions.

48.1.3 Mine Production

Mine production at Henderson, as previously described, has been extremely variable over the past 24 years. Figure 48.2 shows the history of both annual tonnages and employment levels. One of the opportunities of mass caving has been the ability to cut back operations in years of low demand and take advantage of "cave inventories." Conversely, one of the challenges has been that a large amount of capital is sunk prior to production, making it imperative that development rates be in accord with current market conditions.

48.1.4 Safety History

The mission to improve safety performance at Henderson continuously has paralleled the mining industry as shown in Figure 48.3. The caving operation has lent itself well to many of the technological improvements that have become available as Henderson has operated. The utilization of rock bolt jumbos in the late 1980s formed the basis for a significant improvement in safety performance. The use of hand-held jacklegs accounted for 10 reportable injuries the last year they were used. The most recent change—from rail haulage to truck and conveyor haulage—should provide a similar shift downward and allow the operation to keep pace with open-pit mining on an incident rate basis.

The most significant change through the years has been the quality of the employees at Henderson. The use of modern technology has made the job more demanding intellectually, while significantly reducing personal exposure to environmental hazards. The combination of these two features has made the mining operation more conducive to long-term employment, and the sophistication of the current employees rivals that of employees in any other industry.

48.2 GEOLOGY

The Henderson deposit is composed of two partially overlapping ore bodies that lie 1,080 m beneath the summit of Red Mountain. The ore bodies are entirely within a Tertiary rhyolite porphyry intrusive complex that has intruded Precambrian granite. The deposit is elliptical in plan, with overall dimensions of 670 by 910 m. In section, it is arcuate with an average thickness of 185 m. The top of the deposit is at an elevation of 2,610 m, while the lower limits range from 2,340 m on the west to 2,100 m on

---

[*] Climax Molybdenum Company, Empire, CO.

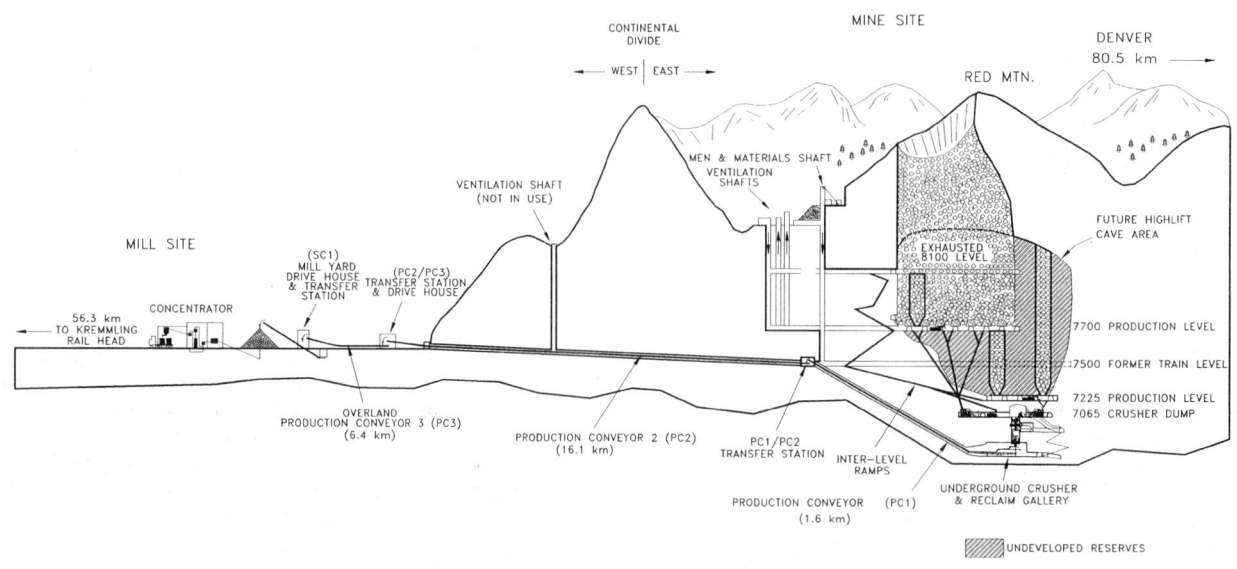

FIGURE 48.1   General cross section of Henderson Mine and Mill

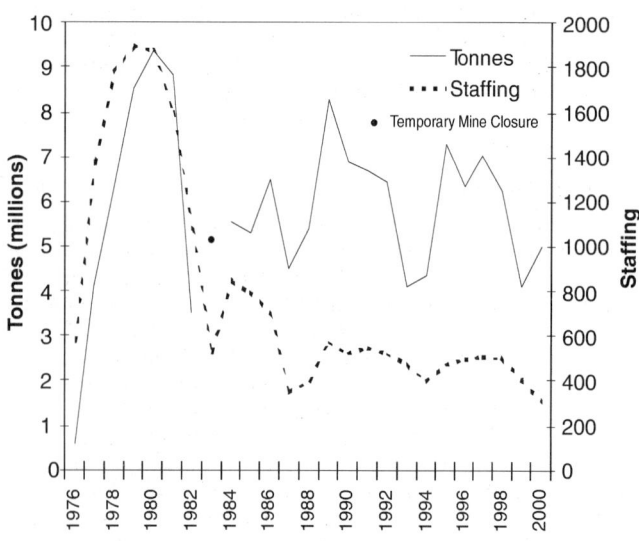

FIGURE 48.2   Employment and tonnage history of Henderson Mine

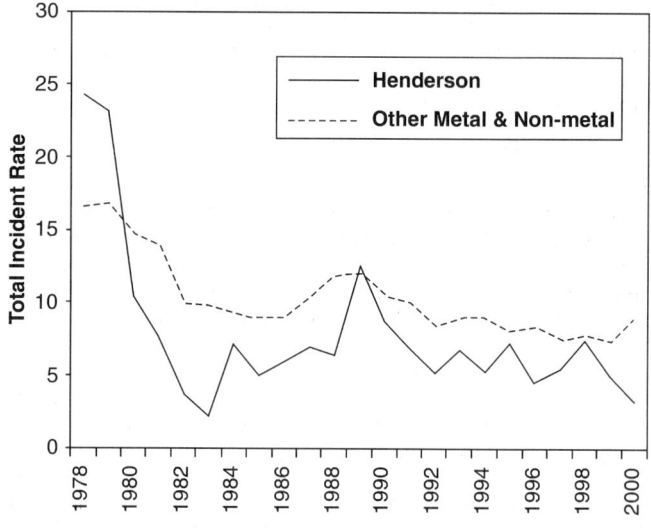

FIGURE 48.3   Reportable incident rate trend for Henderson Mine and all underground metal and nonmetal mines in the United States (incidents per 200,000 hr worked)

the east. The bottom of the eastern lobe of the ore body is 1,590 m. The mineralization is relatively continuous in the ore bodies and consists of molybdenite and quartz in random, intersecting, closely spaced veinlets.

48.3   GEOTECHNICAL INFORMATION

48.3.1   Rock and Ore Strength

The general nature of the ore body and the surrounding host rock is that of very competent granite with compressive strengths ranging from 100 to 275 MPa. Areas that have very little molybdenite do behave appropriately for medium-strength granite. Figure 48.4 illustrates a typical ore zone section showing the relationship of mineralization to mine workings and the need to have plans for excavations in both highly mineralized zones and barren zones.

However, the compressive strength designation may be misleading in the context of mass cavability. The very nature of molybdenite ore is that of a dry lubricant. In fact, 10% of the

mine's concentrates are sold as lubricant material. The low internal friction of the molybdenite leads to easy shear and tensile failure along mineralized veins. Ore grade at Henderson is, therefore, a good indication of both rock competency and cavability. Ore sections are a critical element to the mine planning effort when geomechanical issues are considered

Other rock strength designators of use are rock quality designation (RQD) and rock mass rating (RMR). Henderson's RQD ranges from 0 to 100, averaging 49, and its RMR ranges from 27 to approximately 60.

48.3.2   Geomechanical Modeling of the Ore Body

Geomechanical modeling of the gross excavations associated with the caving operation at Henderson has been a useful tool for forecasting problem areas and for strategic planning. By calibrating the model to historical caving events, in situ stress measurements, and measured rock strengths, these models

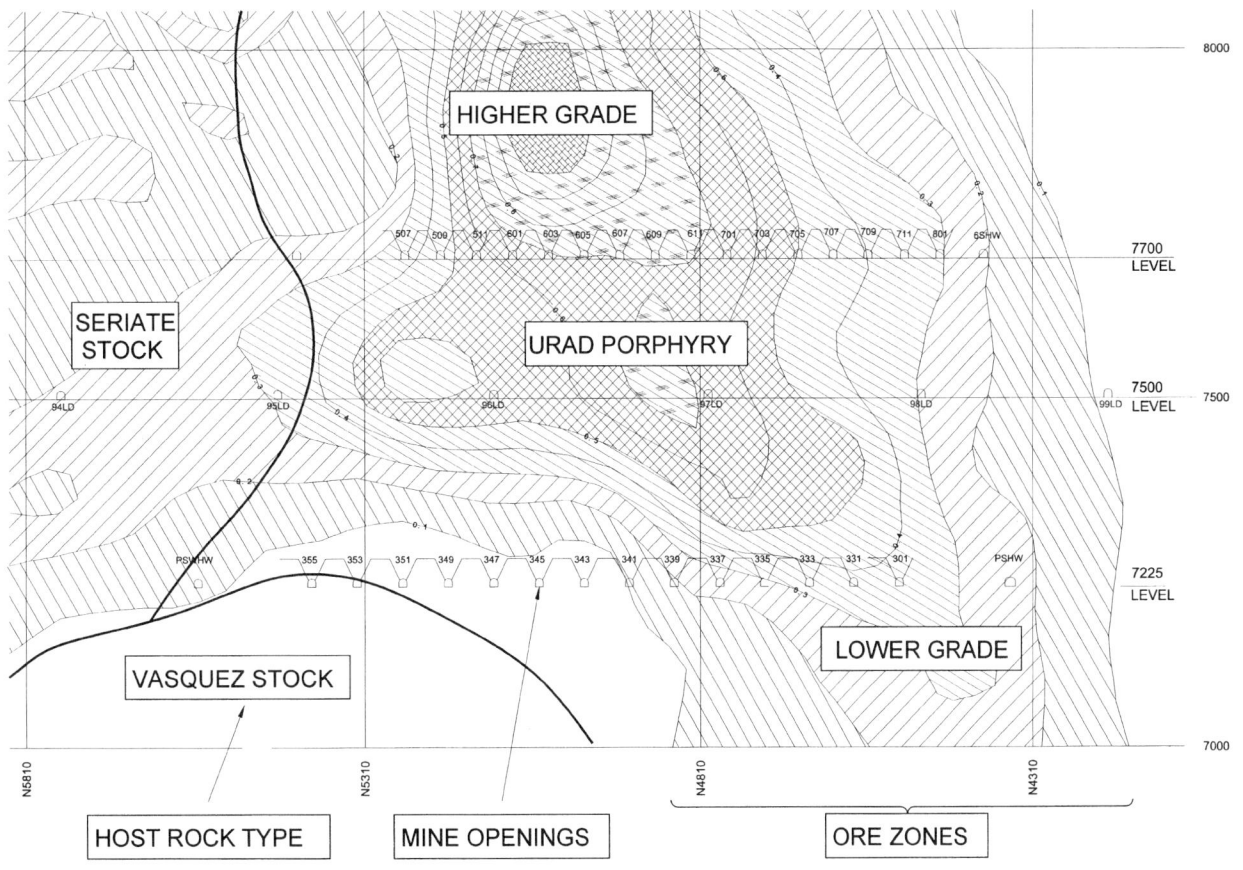

FIGURE 48.4   Typical section of rock types, ore zones, and mine openings

have had a high degree of reliability. Often the geometry associated with mass caving is the largest factor affecting rock stresses, and that feature of modeling is the least debatable, therefore leading to quick consensus as to the resolution of a problem. Both three-dimensional and two-dimensional models have been used effectively.

48.4   MINING METHOD

48.4.1   General Mine Layout

The layout of the mine today consists of five major levels. Figure 48.5 shows a typical isometric section incorporating all five levels. The top level is the "undercut" level located at an elevation of 2,364 m; this is where both the drawbells and the cave are developed. The next level is the "production" level, located at an elevation of 2,347 m. The drawpoints that provide access for the load-haul-dump (LHD) machines are located on this level. Immediately beneath the production level is the ventilation level at 2,333 m. Both intake and exhaust air are provided by this level. In addition, the ventilation level provides access for joining ore passes. The old train haulage level at 2,286 m provides a second access for joining the mechanically bored 2.4-m in diameter ore passes that feed the "truck haulage" level below. These ore passes are currently spaced about 100 m apart, with wider spacings being evaluated for future layouts. At an elevation of 2,153 m, the truck haulage level has center loading chutes that allow truck operators to self-load and dump at the gyratory crusher.

The undercut level presently consists of 3.7- by 3.7-m drifts on 24.4-m centers; future undercut spacings will be 30.5 m. Figure 48.6 shows the typical drill pattern used to develop both the

undercut and the drawbells below. These drill rings are on 2-m centers, utilizing 76-mm in diameter drill holes. This method of development has been in use for over 20 years. It should be noted that no other boundary cut-off levels are used, based upon caving experience on the upper level, where boundary weakening was unnecessary.

Also shown on Figure 48.6 is the "V-cut" drill pattern used to develop an open slot for the overhead ring pattern to be blasted into. This pattern is drilled from the production level 16.8 m beneath the undercut level. Figure 48.7 is a plan view of the drawpoint pattern currently in use. Henderson's drawpoints have evolved to the straight-through design shown here, with future spacings to be 30.6 by 20 m. Entry angles of 56° are the sharpest that can be effectively used with the current 7 m³ LHDs. The drawpoint itself is lined with concrete and utilizes a steel wear plate to protect the opening from erosion over the life of the opening (approximately 63,000 tonnes). The roadways and floors of the drawpoints are lined with 0.3-m-thick concrete. This provides a good surface for cleanup, which in turn reduces tire and articulation wear.

Ventilation to the production level is supplied by a multihorizon level 15 to 20 m below. Both intake and exhaust air are transported on two horizons to provide a general north-to-south fresh air- to-exhaust airflow. Each production ore pass is connected to exhaust air and has an associated intake raise from the intake drifts. This entire level is connected to a 8.5-m intake shaft and to 7- and 10-m exhaust shafts by way of several 5- by 5-m ventilation drifts. Approximately 3.7 million cubic meters per hour of air is moved through the mine.

The truck level consists of 6- by 6-m truck haulage drifts that provide access to center loading chutes, as shown in Figure 48.8.

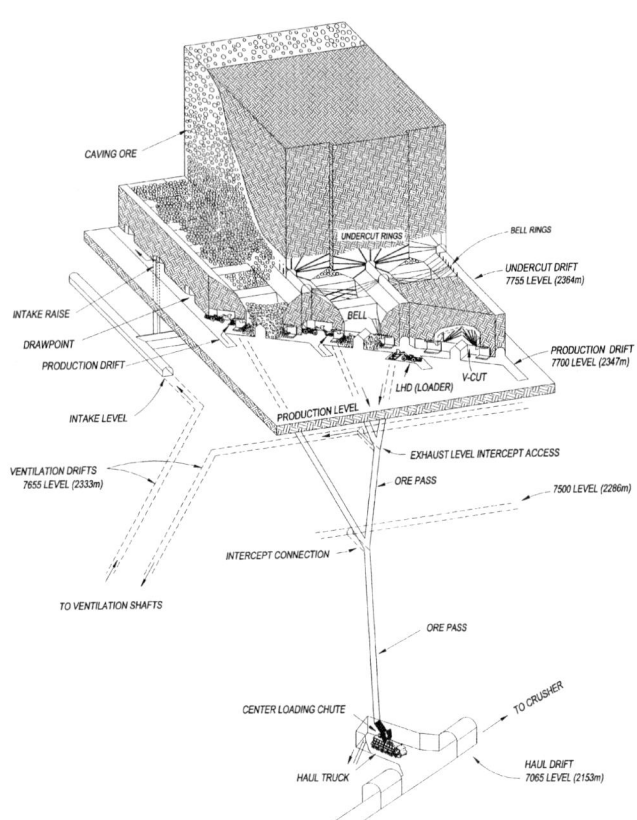

FIGURE 48.5   Isometric view of mining levels

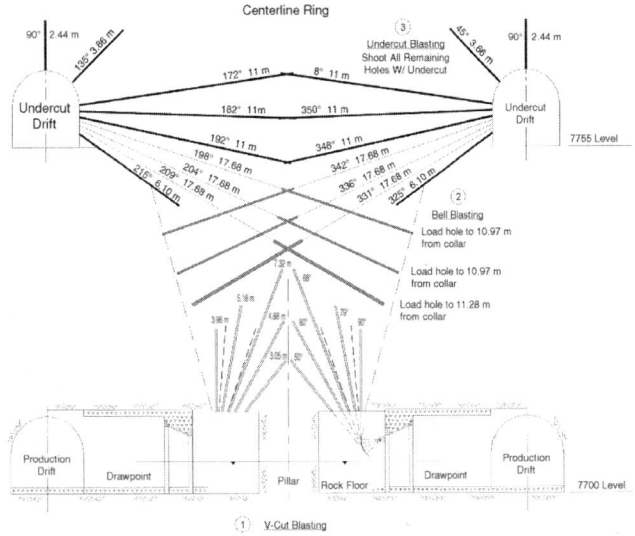

FIGURE 48.6   Section view of typical bell drill pattern and drawpoint

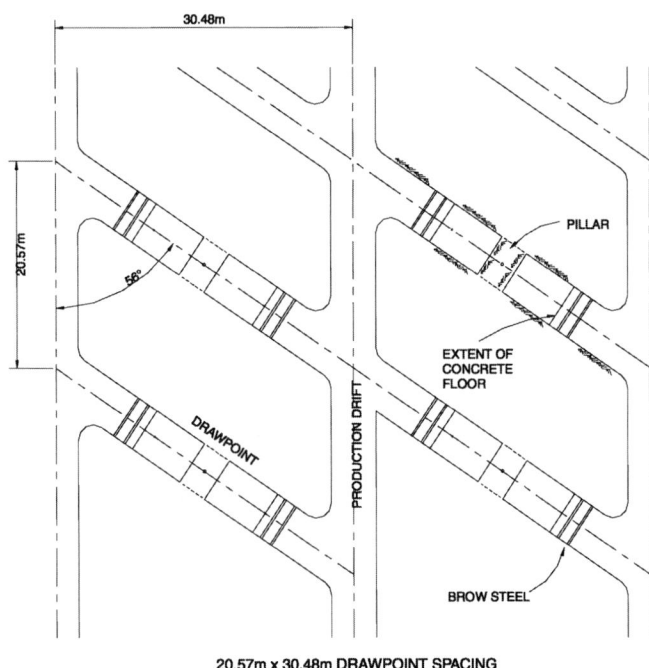

FIGURE 48.7   Plan view of latest drawpoint layout for Henderson Mine production level

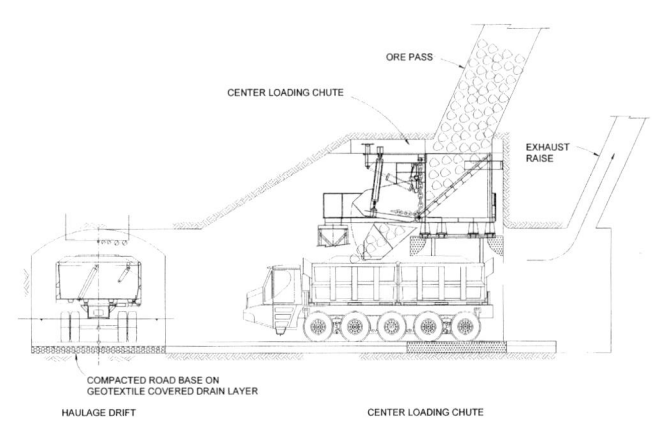

FIGURE 48.8   Typical truck loading chute and haulage drift

FIGURE 48.9   Photo of a 72.6-tonne side-dumping truck unloading into gyratory crusher

The roadways are constructed of a drainage layer and a compacted roadbase material, which is protected by a layer of geotextile on top of the drainage material. This provides a surface that is easily maintained by a grader.

Figure 48.9 is a photo of one of the 72.6-tonne side-dumping trucks unloading at the gyratory crusher.

48.4.2   Mine Development Sequence

The development of such a large series of interdependent mining levels requires a great deal of planning and coordination.

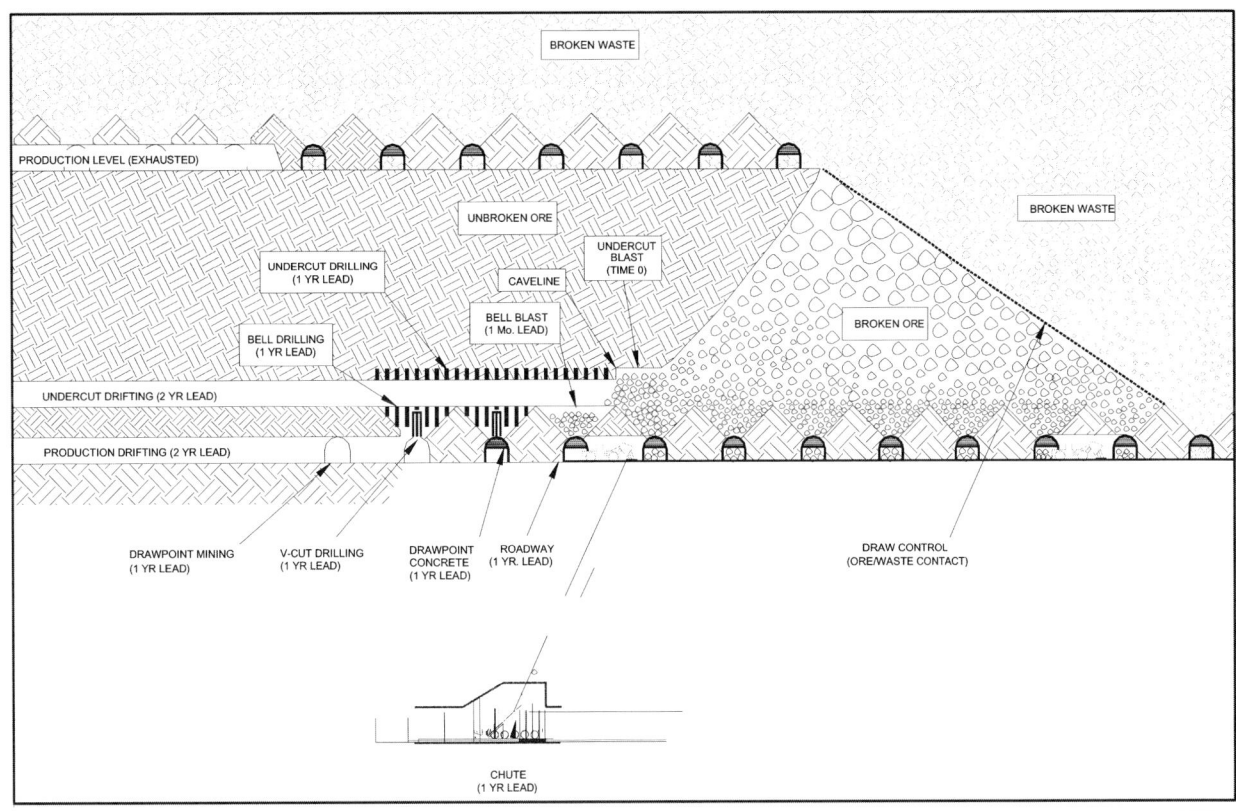

FIGURE 48.10 Typical mining sequence for a caving panel at Henderson Mine

Depicted in Figure 48.10 is the sequence of development for a generalized panel at Henderson. To date, panels have been eight to 12 production drifts in width. The leads shown are typical for an operation at full capacity and may vary as production rates are reduced. It is very important to have the entire 2-yr sequence on the same schedule, as many of the steps are of a series nature. For example, it is difficult to mine drawpoints adjacent to the drawbells that are being blasted because of the associated blast damage caused by the large shots. Finally, the multilevel nature of the ore pass system requires that mining on all levels be synchronized with one another. Otherwise there is no place to bore to, or from, if one level is behind the other.

48.4.3 Ore Grade Control

Ore grade control has been one of the highest priorities at Henderson throughout its history. A geologic reserve model, which is interfaced with the specific drawpoints and the theoretical column of in-place ore that lies overhead, provides the basic grade control. A computer model simulates ore being withdrawn from this column based upon projected operating parameters. Life-of-mine simulations are possible using various production rates.

In practice, the same model is used to assign numbers of buckets or tonnes to be withdrawn each operating shift. The operator tracks the actual tonnes totaled by a weightometer on the LHD and the location where the tonnage was extracted. If, for some reason, tonnage is higher or lower than assigned, this difference is tracked, and the operator is given an ahead or behind tally by drawpoint. The overall goal is to reconcile this difference each month. A weekly production meeting facilitates this process.

Ore samples are taken daily, and through the use of X-ray analysis in an underground assay lab, the grade is tracked. This information is used to track both dilution and the final

TABLE 48.1 Molybdenum ore recovery as compared to estimated diluted ore reserve, million tonnes (pounds)

Cave area	Diluted ore reserve estimate		Actual recovered ore		Percentage recovered	
	Tonnes	Pounds	Tonnes	Pounds	Tonnes	Pounds
8100	90	480	87	470	97	98
7700 to date	45	213	48	219	106	103

exhaustion of the column of ore. In addition to grade analysis, a visual estimate of dilution is made weekly by the mine geologist. These data are utilized in the weekly production meeting.

The earliest evidence of ore dilution has been with only 20% of the column removed. The primary response to the early dilution is to ensure slow and even withdrawal of material from the area. The overall ore inventory model assumes a loss of 7.5% of the ore and replaces it with a lower-grade dilutant material. To date, it has been possible to operate at or slightly above this performance level. Table 48.1 shows the overall performance for the mine to date.

48.4.4 Ore Transportation System

Because of the overall layout of the Henderson Mine and mill, ore transportation has played a large role in the performance of the mine. As noted in the introduction, the 25-km rail haulage system has recently been replaced with a conveyor system. As shown in Figure 48.1, this system consists of three flights of conveyors. The longest is over 15 km long and has a single drive station with four 2,000-kW drives. The belt is 1.2 m wide to accommodate a maximum lump size of 0.5 m. The surface belt incorporates nine vertical curves and seven horizontal curves, with the tightest horizontal curve having a radius of 1.5 km.

TABLE 48.2   Performance statistics for LHDs and haulage trucks

Type	Capacity, tonne	Fleet size	Productivity, tonne/hr	Availability, %	Cost per tonne, US$
LHD	9.5	7	318	80	$0.32
Truck	72	4	708	90	$0.16

It is important to note that the only sizing of ore ahead of the 1.4-m gyratory crusher is a single-beam grizzly with a spacing of 0.7 m. The optimum efficiencies for LHDs and trucks have been achieved with this grizzly spacing. No rock breakers are involved. Most of the secondary breakage required is done through the use of 1-kg, conically shaped explosive charges. This is accomplished with an emulsion-type explosive and a household funnel to provide the conical shape.

48.5   PRODUCTION STATISTICS

The Henderson Mine was originally designed around 4 m^3 LHDs and has evolved to 7 m^3 LHDs. This transition has been made possible by improved operator skill and more efficient equipment design. The net effect has been an improvement from 136 to 317 tonne/hr. The related fleet size has gone from 30 to only seven for the same production rate. Table 48.2 shows the general performance of this fleet along with total cost per tonne to operate, maintain, and own these LHDs.

The performance of the relatively new 72.6-tonne haul truck fleet is also shown on Table 48.2. With an average haul distance of 300 m each way, three of these trucks match the production capabilities of six LHDs. High-quality road maintenance is an essential component to the maintenance cost of this fleet. As previously mentioned, the road design allows for grading and continuous maintenance. The total cost of $0.18/tonne should be reduced as higher tonnages are handled. To date, the fleet has been constrained by start-up issues with the conveyor system and the ore passes.

48.6   SAFETY AND HEALTH

A large variety of improvements in safety and health has been achieved in recent years. Dedicated employees who have been involved in determining the optimum solution have made all of these improvements sustainable. The following are some of the highlights:

- Longhole drills that utilize rod handling and a drill tractor with a cab for remote operation and sound suppression
- Use of cabs on LHDs, concrete trucks, haulage trucks, and utility equipment
- Concrete form jumbos that minimize hand forming of concrete for the production drawpoints
- Rock bolt jumbos and screen-handling equipment for ground support
- Bulk material handling systems for ANFO, trash, and supplies
- Use of wet shotcrete and accelerators to improve ground support (which allows for the thicker application of shotcrete as well)
- Use of a leaky-feeder radio system throughout the mine, which has dramatically improved communication and response to emergencies
- Widespread use of multiplexing and solid-state controls for traffic control, ventilation, conveyor system, ore pass monitoring, and pumping systems

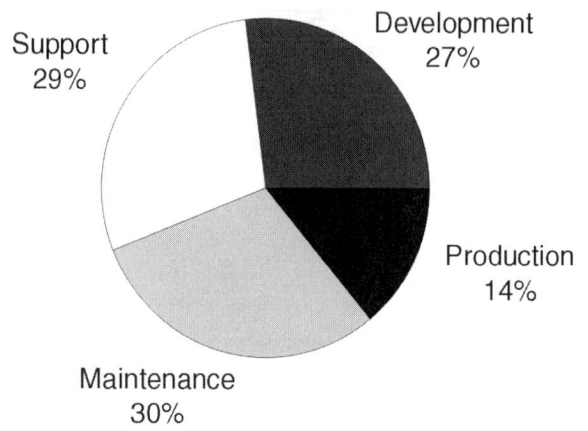

FIGURE 48.11   Breakdown of mining costs by function

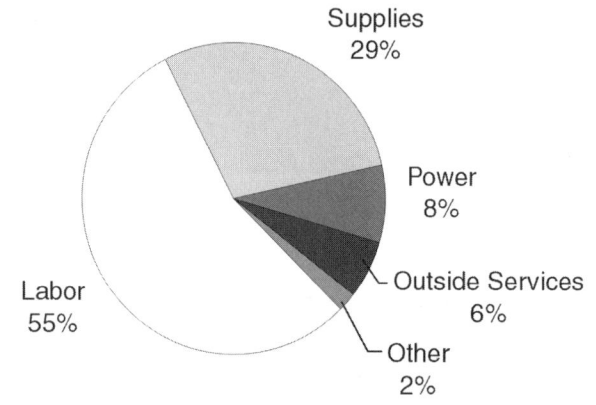

FIGURE 48.12   Breakdown of mining costs by category

The most recent safety statistics for the mine have shown an incident rate of less than 4.0 (incidents per 200,000 hr worked) and a lost-time accident rate of less than 1. All of the above improvements, along with a behavior-based safety program, have been essential to lowering the incident rate continuously. The current goal of the mine is to achieve zero reportable incidents.

48.7   OPERATING COSTS

Figures 48.11 and 48.12 are pie charts showing the general breakdown of mining costs for the operation. These costs do not include any milling, crushing, or conveying, or general and administrative costs. It must be noted that these costs are for a year with a large amount of mine development activity. As previously mentioned, development costs are a large driver to the overall mining costs. The support function is composed of general plant activities, not rock support. The categorical chart also reflects the high portion of labor costs associated with development mining. The proportions of these costs will vary from year to year as the mining activity changes and therefore should be used only as a general reference.

48.8   RECENT CAPITAL EXPENDITURES

Replacement of the rail haulage system with a crusher and conveyor system is the most significant capital expenditure in recent years. This investment totaled approximately $150 million with 20% of the cost relating to mine development and the remainder directly associated with the crush-and-convey system. As shown in Figure 48.1, the conveyor system is 25 km long, because of the unique location of the mill site.

Several technological advances have made this change practical today, whereas it may have been nearly impossible in the 1970s when the mine was built. The most significant of these advances are—

- Improved steel cord conveyor belting
- Dynamic simulation models for long conveyor systems
- Variable-frequency drive controls
- Increased experience with curved conveyor applications
- Fiber-optic data transmission and solid-state control

Ore transportation reliability and cost have been dramatically improved with this new system. Additionally, the skills required to maintain and operate a conveyor system are more readily available in today's labor pool. Overall, this system is more conducive to continuous safety and production improvement.

48.9 FUTURE PLANS AND CONCLUSION

The height of Henderson's caving panels, to date, has been dictated by ore zone dimensions. On the lowest level, there is an opportunity to increase this height dramatically and leverage the overall efficiency of the mine development. Figure 48.13 shows the potential for this increase in ore column height. The combination of height and the current drawpoint spacing will allow a threefold increase in the tonnage per drawbell.

Experience on the upper levels has shown that a quality drawpoint, with concrete floors and a steel wear brow, is essential to withstanding the abutment stresses associated with panel caving. In addition, a high-quality installation is the foundation to optimum equipment efficiencies and maintenance costs. All of these installations will be able to endure tripling the tonnage and duration of their use; this is a substantial opportunity for Henderson. Full utilization of every dollar invested in mine development is critical to future success.

In conclusion, the Henderson Mine has been able to keep pace with the changes technology has offered the mining industry. Employee involvement and skill have been key components to these improvements. The longevity of the mining levels has made it necessary to squeeze the largest equipment possible into the existing openings. Future pressures on minimization of investment capital will further the need to optimize existing openings. Redesigns are few and far between, and continuous improvement of efficiency and safety are the key to survival.

48.10 REFERENCES

Barfoot, G., and K.W. Keskimaki. 1998. The Henderson coarse ore conveying system. SME Annual Meeting. Orlando, FL.
Keskimaki, K.W. 1996. Productivity gains at the Henderson Mine. MINExpo 96. Las Vegas, NV.

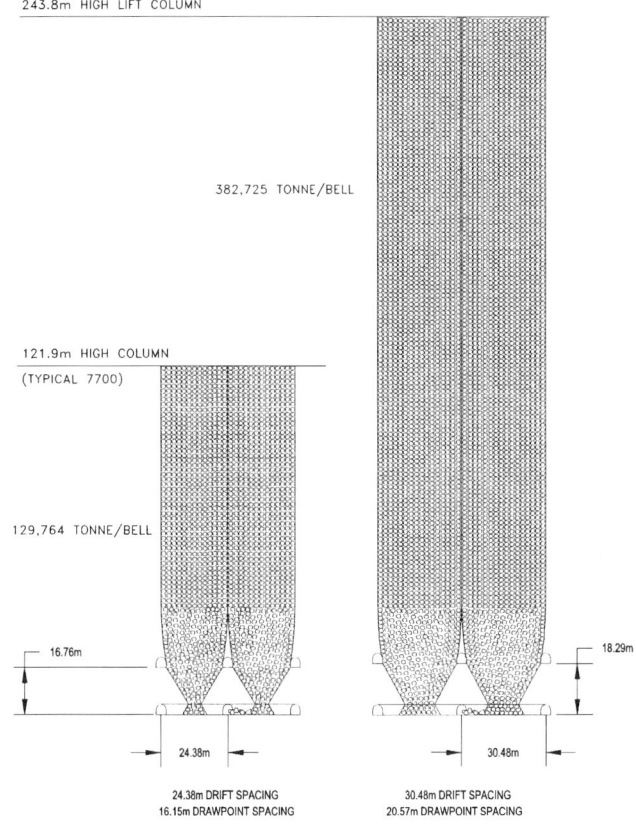

FIGURE 48.13 Comparison of current draw columns to draw columns planned for next production level

Keskimaki, K.W., and E.B. Jensen. 1992. Advances in equipment technology at the Henderson Mine. MASSMIN 92. Johannesburg, SAIMM: pp. 345–350.
Keskimaki, K.W., and R. Wagner. 1994. Secondary blasting at Henderson. SME Annual Meeting. Albuquerque, NM.
Nelson, B.V., and D.C. Harney. 1995. Remote control and monitoring at the Henderson Mine. SME Annual Meeting. Denver, CO.
Rech, W.D., E.B. Jensen, G. Hauk, and D.R. Stewart. 1992. The application of geostatistical software to the management of panel caving operations. MASSMIN 92. Johannesburg, SAIMM, pp. 275–281.
Rech, W.D., and L. Lorig. 1992. Predictive numerical stress analysis of panel caving at the Henderson Mine. MASSMIN 92. Johannesburg, SAIMM, pp. 55–62.
Rech, W.D., and D.K. Watson. 1994. Cave initiation and growth monitoring at the Henderson Mine. SME Annual Meeting, Albuquerque, NM.

CHAPTER 49

Palabora Underground Mine Project

Keith Calder,[*] Peter Townsend,[*] and Frank Russell[†]

49.1 INTRODUCTION

49.1.1 Location

Palabora Mining Company operates an integrated open-pit mine, concentrator, smelter, and refinery complex located in the Northern Province of South Africa about 560 km northeast of Johannesburg. Elevation of the pit rim is about 400 m above sea level. The climate is subtropical with an average annual precipitation of 480 mm.

49.1.2 History

Palabora Mining Company was established as a joint venture between Rio Tinto (formerly RTZ) and Newmont Mining Corporation in 1956 to exploit the copper resource identified around a hill known as Loolekop, where there was archaeological evidence of smelting from the 8th century. Construction of an open-pit mine started in 1963, and processing of ore began in 1966. Subsequent expansions increased milling capacity to the present 82,000 tonne/d ore and 135,000 tonne/yr of cathode copper. Capacity increases in the metallurgical plant allowed for economies of scale to be realized in the mining operation, and the cut-off grade has been lowered from the initial 0.30% copper to the present 0.10% copper. Life of the open pit was similarly expanded, ultimately to nearly double the originally planned 20 yr.

The present mining plan calls for open-pit operations to cease in the year 2003. While the ore body continues below the bottom of the final pit shell, the stripping ratio precludes further open-pit mining. Underground mining is the only viable alternative for the continuation of operations at Palabora.

Various studies into the viability of underground mining were carried out from the mid-1980s, culminating in a final feasibility study in 1996 with the project commencing in the same year.

49.2 GEOLOGY, HYDROLOGY, AND ORE RESERVES

The Palabora copper ore body is an elliptically shaped, vertically dipping volcanic pipe. The pipe measures 1,400 m and 800 m along the long and short axes, respectively. The ore body is open at depth with reserves proven to 1,800 m below the surface. Copper grades of approximately 1% are found in the central core of the ore body and decrease gradually toward the peripheries with no sharp ore-waste contact. Analyses during the feasibility study determined that the optimum grade boundary is 0.8% copper. This cut-off grade results in a mineable reserve of 245 million tonnes at 0.68% copper.

An additional resource of 467 million tonnes graded at 0.57% copper is situated adjacent to and below the 30,000 tonne/d reserve base. This ore is available for mining and constitutes a resource base for a life-of-mine extension or an expansion in production rate. The cut-off grade for the mine has been calculated at 0.31% copper.

Mineralization is hosted by three main rock types. Transgressive and banded carbonatites form the central core of the ore body and are made up of magnetite-rich sövite with minor amounts of apatite, dolomite, chondrodite, olivine, and phlogopite. Barren dolerite dykes with a steeply dipping northeast trend are present and account for approximately 8% of the 245-million-tonne resource.

49.3 GEOTECHNICAL INFORMATION

The average uniaxial strength of the carbonatites is about 120 MPa, with a variation in values between 90 and 160 MPa depending on mineralogy. Dolerite is a strong, brittle rock with a uniaxial strength of 320 MPa. Adjacent to the major faults, dolerite is locally weathered with a marked reduction in strength to around 80 MPa.

The in situ state of stress is assumed to be hydrostatic and approximately equal to the overburden load of 38 MPa. As part of the numerical and fragmentation analyses conducted, parametric studies were carried out between the limiting values of horizontal stress ratios of 0.75 and 1.5 to determine whether the solutions were sensitive to the inherent assumptions about state of stress.

The structure of the carbonatites is predominantly subvertical jointing. These joints are open or infilled with weak material, planar, and through-going. There are three steeply dipping sets striking approximately 010° (dip direction 290°), 310° (dip direction 040°), and 050° (dip direction 140°). The flat-lying joints have a different morphology from the vertical sets and are wavy, rough, and of limited continuity. There are two sets oriented approximately 20°/160° and 45°/350°. For the purposes of the fragmentation analysis, the rock fabric was idealised to three major sets.

Fractures in the dolerite are closely spaced and blocky. Within the cave area, fractures in dolerite are zoned by proximity to the major faults.

For the purpose of mine design, the ore body has been divided into less-jointed and well-jointed zones based on the formation of primary fragmentation at a size cutoff of 2 m^3. Rock mass rating (RMR) values were estimated from core samples (Table 49.1).

49.4 MINING

49.4.1 Summary

The underground mine will exploit ore below the final open-pit shell using mechanised block caving. The undercut level,

* Palabora Mining Co., South Africa.
† Rio Tinto Technical Services, England.

TABLE 49.1 Rock quality

Zone	Average RMR
Less jointed	70
Well-jointed plus dolerite	57
Ore body average	61

currently being mined at an elevation of 1,200 m below the surface and approximately 460 m below the ultimate pit bottom, is the uppermost level of the mine. The production level, which contains the drawpoints and other infrastructure, is being developed 18 m below the undercut.

During the early stages of cave propagation, it is anticipated that a significant percentage of the caved material will be too large to be handled by the ore extraction equipment, and specialised rigs will be used to drill and break the oversized material using both emulsion blasting and nonexplosive techniques. A fleet of 11 diesel-powered load-haul-dump (LHD) machines having a 14-tonne payload will muck from the drawpoints directly to four crushers along the northern periphery of the cave. The average one-way length of haul is 175 m.

The crushers will discharge onto sacrificial conveyors, which in turn will feed onto a horizontal section of a single, 2,000-tonne/hr inclined conveyor that delivers the ore to two 5,000-tonne-capacity production shaft silos. The ore will be hoisted out the mine using four 32-tonne payload skips. All major underground fixed equipment and the fully automated service and rock hoists will be monitored and controlled from a control room on the surface using tele-remote systems where applicable.

The underground project has four shaft systems, including the ventilation shaft. The first shaft to be sunk was the exploration shaft, which was initially sunk from bench 30 in the open pit to 889 m below the surface to develop a drive to facilitate exploration of the deeper-lying ore. After project approval, the shaft was deepened to the production level to allow an accelerated development program to be put in place concurrently with the sinking of the two main shafts. The exploration shaft is small (4.8-m in diameter) and equipped with two 6-tonne skips and a 27-person service cage. The shaft has a 4,000-tonne/d hoisting capacity and will remain in production until the end of 2000, after which it will be maintained as an emergency exit.

Sinking the new concrete-lined service and production shafts down to the final depth of 1,280 m below the surface was completed in the third quarter of 1999. The shafts are situated 72 m apart. The 10-m in diameter service shaft has a 86-m-high concrete headframe and is equipped with a large single-deck cage running on fixed guides and a 20-person-capacity auxiliary cage running on rope guides. The main cage has a 35-tonne payload and can accommodate 155 people. It is licensed to operate at 12 m/s. The auxiliary hoist is licensed to run at 8 m/s.

The production shaft has a diameter of 7.4 m and a 106-m-high concrete headframe. The shaft is being equipped with four 32-tonne payload skips running on rope guides. Maximum hoisting capacity will be on the order of 42,000 tonne/d. The production shaft, first crusher, and conveyor system will be commissioned early 2001. All the main hoisting systems use tower-mounted friction winders with integrated motors. The two rock hoist winders are each powered by 5,500 kW motors. Both shafts are sited outside the open-pit shell beyond the influence of the block cave. All hoists are fully automated and will be operated from the surface control room.

Intake air is downcast through the production and service shafts and is exhausted through a 5.76-m in diameter, 92-m-deep, raise-bored ventilation shaft. The two main 1,250-kW upcast ventilation fans are installed on bench 28 of the open pit. At present only one fan is being operated and is exhausting air at 340 m³/s. Both fans will exhaust a total of 500 m³/s of air via the ventilation shaft. Two 850-W booster fans have been installed underground that will enable 600 m³/s to be downcast through the production and service shafts and will put the cave area under positive ventilation pressure, minimizing the entry of dust and heat into the workings through the cave. An 18-MW refrigeration plant supplies chilled water to a bulk air cooler adjacent to the main shafts. The intake air to the underground workings is cooled to offset the high underground ambient rock temperatures of 50°C and heat generated by the diesel equipment.

Major underground infrastructure, including workshops and offices, are being constructed on the production level close to the mine workings, but outside the zone of abutment stresses.

Current production of 82,000 tonne/d from the open pit is processed at the concentrator, which consists of crushing, grinding, and flotation circuits. The concentrate is smelted in a conventional furnace and converter having a capacity of 135,000 tonne/yr of fine copper. The final process is the electrical refinement of the anode copper to produce cathode copper. When the underground mine is in production, surplus processing plant capacity will be used to increase overall recovery through changes in grind, increase retention time in the flotation circuit, and reduce concentrate grade

An environmental impact assessment (EIA) was completed as part of the feasibility study for the underground project. The conclusion was that, "No impacts were identified on the mine property, resulting from the proposed development of the (underground) mine that cannot be adequately mitigated." The key issues highlighted in the study were reduced water consumption, reduced waste disposal, and reduced emissions.

49.4.2 Undercut Design

The design and operation of the undercut is the key mechanism for initiating the cave. The features of the undercut design at Palabora are—

- An advanced undercut that is developed ahead of opening the drawbells to provide a "stress shadow" to protect the production level.

- A narrow undercut 4 m high. This is considered sufficient to initiate the cave while minimizing the amount of swell material that has to be removed during excavation using an advanced undercut.

- An inclined face over the major apex. This will create a chevron shape to the undercut to facilitate initiation of the undercut following construction of the drawbell by promoting the flow of blasted undercut material into the bell.

The resultant design is shown in Figures 49.1 and 49.2. Undercut development is supported with a combination of resin-grouted bolts, mesh, and shotcrete, taking into account the large stress changes that will occur as the undercut area is enlarged.

49.4.3 Production Level

The optimum drawpoint spacing is a balance between (1) obtaining good draw and recovery characteristics while (2) maintaining the strength of the drawpoint structure with a layout appropriate to equipment size and type. The offset herringbone layout was selected principally as a function of LHD maneuverability and the potential to use electric LHDs with trailing cables.

The drawpoint spacing was designed for 17-m centers as a function of the isolated draw zone characteristics of the coarse ore. The drawbells are rectangular with inclined walls and offset

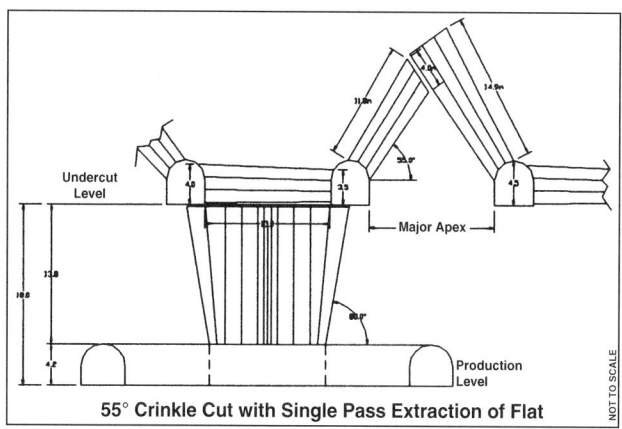

FIGURE 49.1 Undercut design

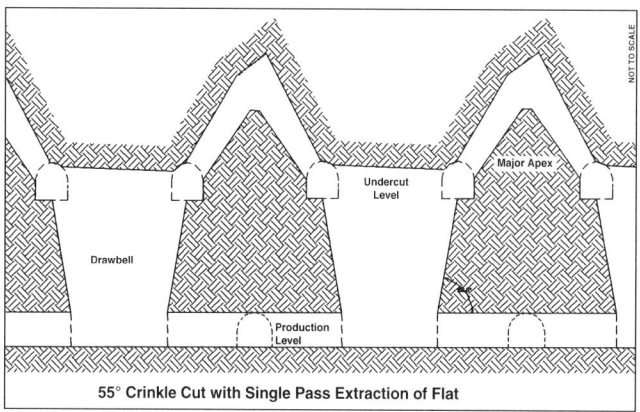

FIGURE 49.2 Undercut design

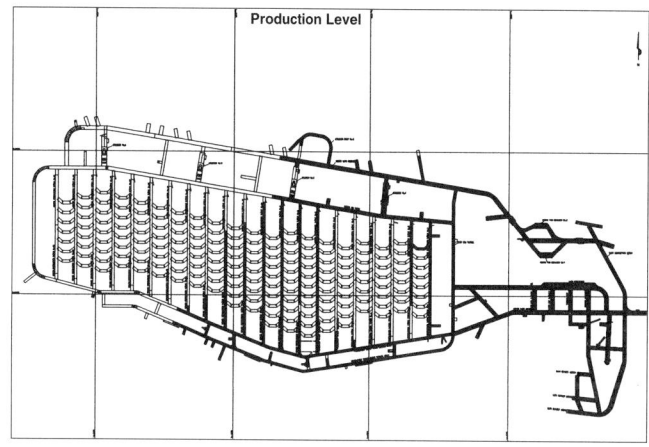

FIGURE 49.3 Plan of production level

between production drives spaced at 34 m. There is scope for minor modification of the drawbell shape if necessary to improve draw characteristics. The bell layout leaves substantial pillars for support and protection of the drawpoint structures. Numerical modeling has been carried out to check the integrity and the reinforcement for the drawpoints. Production tunnels will be 4.5 by 4.2 m with drawpoint crosscuts sufficiently long to allow the 6.5-m³ LHDs be to loaded without articulating. The design is shown in Figure 49.3.

The production level development is supported with a combination of resin-grouted roof bolts, cable bolts, and fibrecrete. Steel sets may be placed at the bow of drawpoints in weaker ground.

49.4.4 Loading

LHDs will be loaded from drawpoints and dump at crusher tips accessed from the main production drive on the northern periphery of the cave. Initially, LHDs will be manually operated, but work on the use of semiautonomous LHDs operated from the control room on the surface is well advanced. Secondary breaking equipment and services will access the production area via two service drives along the southern rim. The activities of the underground fleet will be monitored and directed from the central control room on the surface via an auto dispatch system. LHDs will be dispatched to drawpoints and crusher tips according to a production schedule that will be generated using information stored in a cave management database.

49.4.5 Crushers and Conveying Level

Each of the four crusher stations will have two tipping bays to enable two LHDs to tip simultaneously. Bays will be equipped with a scalping grizzly with a maximum aperture of 1.4 m. Oversized rock will be broken by a hydraulic rock breaker operated tele-remotely from the control room on the surface. Ore will be crushed to −200 mm by Krupp 1,700- by 2,300-mm jaw crushers. A total of 750 tonnes of live storage capacity is provided below each crusher. Ore will then be delivered via sacrificial conveyors onto the main conveyor inclined at 9° (reducing the production shaft depth of wind by 117 m) and discharging into two storage bins at the production shaft. The incline conveyor haulage will not be used as an intake airway to the mine workings because of the risk of dust pick-up and toxic fumes in the event of a fire. Ventilating air will be discharged directly into the return airway system.

49.4.6 Lateral Development

Mechanised trackless methods are employed for all mine development, and mining contractors are being used do the mining. The main infrastructure development will be completed by the second quarter of 2001 and the development of the 173 drawbells and 346 drawpoints by the end of 2003. Drilling is done with double-boom, electro-hydraulic jumbos, and mucking with 5.7-m³ diesel LHDs assisted by 40-tonne-capacity trucks on longer hauls. A fleet of ancillary units includes rock bolters, shotcreting machines, scissor lifts, utility vehicles, and a road grader. Steeply inclined excavations, such as ore passes, airways, and escapeways, are raise bored if over 25 m long; shorter passes are generally drop raised.

49.4.7 Dilution Control

The estimate of mineable reserve includes all material in an 85° draw angle from the cave perimeter and is low-grade ore rather than waste. To minimize any further dilution, it is planned to draw down all drawpoints evenly at a rate not to exceed 200 mm/d. A cave management system that includes a dispatch system linked to LHD tagging will be used to ensure that cave is drawn at the required rate.

49.4.8 Cave Height

The design column height for Palabora is approximately 460 m but varies according to the intersection with the pit walls. This is high in comparison to most caving operations; however, at these operations, the limitation of cave height is sometimes a function of ore body configuration and drawpoint life. In turn, drawpoint life is seen to be related to damage created by abutment stresses

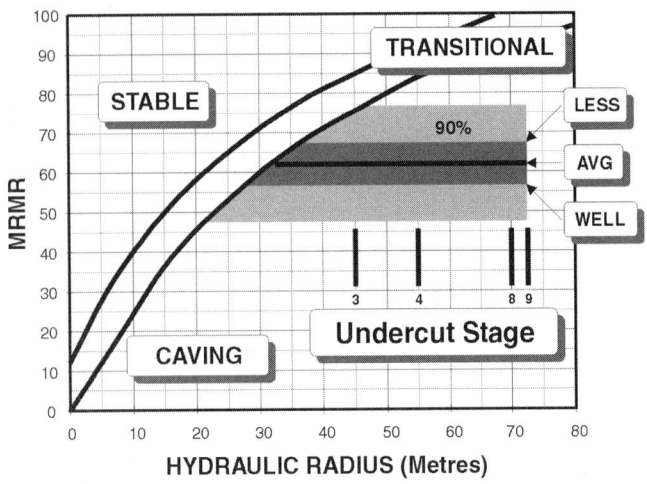

FIGURE 49.4   Laubscher stability diagram

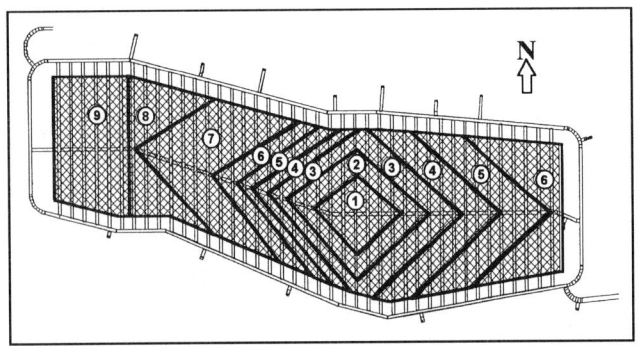

FIGURE 49.5   Plan of undercut level and undercut sequence

and subsequent damage by bad secondary blasting practice rather than to wear during operation. By using an advanced undercut and suitable secondary breaking equipment to minimize damage, it is not expected that the relatively high cave height presents a problem to the operation.

49.4.9   Caveability

With the rock quality being higher than that in any other block cave, extensive studies were carried out to gain confidence that the undercut area would cave. At the time of feasibility study, the model with greatest industry acceptance to predict caveability was the Laubscher stability diagram. This diagram (Figure 49.4) consists of a graph that plots rock strength, as measured by a mining rock mass rating (MRMR), against hydraulic radius, which is the size of the area available for undercutting.

The stability diagram is an empirical predictor that experience at other mines has proven to be conservative but reliable. Figure 49.4 shows the MRMR for the rock characteristics listed in Table 49.1. Assuming global average rock strength, the model indicates that caving will commence when a hydraulic radius of 35 m, equating to an area of 2 hectares, has been undercut. The total Palabora footprint comprises an area of 12.6 hectares. Figure 49.5 shows the plan of the undercut level and the sequence of opening the undercut.

49.4.10   Fragmentation

With the rock quality being higher than in any other block cave, studies were made to determine the size range of the rocks

TABLE 49.2   Underground mine staffing levels

Area	Number of personnel
Mining operations	210
Mining, technical	45
Engineering services	94
Total	349

transported to the drawpoint. The blockage of drawpoints resulting from large rocks is a potential constraint on the underground mine production rate. Fragmentation problems will be most severe when the cave is initiated. As the cave height is increased, the action of broken rock moving toward the drawbells will further reduce rock size. Estimates indicate that in the first year of production, over 70% of the rock will be greater than 2 m^3 and will therefore require secondary breaking before being loaded. Specific equipment will be used to handle the oversized material safely and efficiently. Fragmentation data, hang-up prediction, and clearance times were input into a dynamic simulation of the entire production process. Results of the simulation indicated that, in the year of worst fragmentation, the target production rate of 30,000 tonne/d could be achieved.

49.5   PRODUCTION STATISTICS

49.5.1   Staffing Levels

Labor levels forecast for the whole of the Palabora operation will be about 1,763 people, 39% less than the 2,859 employed in 1996. Achieving the targeted personnel levels is not expected to require involuntary retrenchments. The demographics of the workforce indicate that retirement, plus normal turnover, is expected to produce the necessary personnel reductions. Staffing levels for the underground mine are shown in Table 49.2.

49.6   SAFETY AND HEALTH-RELATED ISSUES

49.6.1   Drilling and Excavation

All main infrastructure development will be completed before the mine reaches full production. The only activities carried out during operation will be secondary breaking, loading, hauling, and dumping followed by crushing, conveying, and hoisting.

49.6.2   Secondary Breaking

Because there will be occurrences of drawpoint oversize and blockages beyond the reach of conventional equipment, a specialised high-reach drill rig has been developed to drill and charge these high hang-ups remotely without personnel entering the drawpoint. The high-reach rig has the capability of accessing blockages 21 m above the footwall elevation. It is equipped with a three-dimensional video system and is operated remotely from a mobile control module that detaches from the main unit. The rig is equipped with an emulsion charging system to load the drilled holes. Nonexplosive breaking techniques will be used to break drawpoint oversize, and these techniques may later be extended to the higher hang-ups.

49.6.3   Loading

LHD operations will be segregated from all other operations. The LHDs will operate between the drawpoints and the crushers located in the north main production drive. No personnel will be allowed access to areas where LHDs are operating. Secondary breaking and service units will enter the production crosscuts from the south access drift only when the LHD has completed production activities. A personnel detection warning system is already employed on the development fleet that warns the LHD operator that personnel are in proximity to the LHD. This will

TABLE 49.3   Estimated capital expenditures in US dollars (2000)

Item	Cost in millions
Exploration work	2.8
Mine surface facilities	15.5
Refrigeration and ventilation	13.6
Production shaft	48.6
Service shaft	36.9
Mining	142.9
Materials handling	20.4
Services	25.8
Control and instrumentation	14.4
Indirect costs	73.8
Subtotal	**394.7**
SA rand/US dollar exchange rate	15.5
Total	**410.2**

also be installed on the production fleet of LHDs. The potential use of a semiautonomous loading fleet greatly reduces the number of operators required and removes them from the underground environment to the comfort and safety of a surface control room.

49.6.4   Conveying and Hoisting

All crushing, conveying, and hoisting operations will be controlled from a central control room on the surface.

49.7   CAPITAL COSTS

The projected capital cost to construct and commission the underground mine is US$410.2 million in escalated terms (Table 49.3).

49.8   FUTURE DEVELOPMENTS

The mine design is based on existing technology, but the overriding philosophy is to use proven developments in technology and automation to the optimum benefit of the operation in terms of safety and productivity. Ongoing research and development in the automation of LHD operations could result in this technology being implemented within the next 18 months. Automation of many other routine tasks is being actively investigated on an ongoing basis.

It is believed that the mine being developed at Palabora will determine the viability of block caving in rock previously considered too difficult to cave. This has the potential to open many other deposits around the world using this low-cost, but safe, mining method.

. . . .

CHAPTER 50

Block Caving Lift 1 of the Northparkes E26 Mine

M. House, * **A. van As,**† **and J. Dudley**†

50.1 INTRODUCTION

The Northparkes Endeavour 26 (E26) porphyry copper-gold deposit was discovered in 1977 in central-western New South Wales, Australia, approximately 30 km northwest of the town of Parkes and 350 km west of Newcastle (Figure 50.1). The mine is a joint venture between North Limited (80%) and the Sumitomo Group (20%). North Limited manages Northparkes Mines.

This area has long been regarded as prospective because of the numerous historical small gold mining operations around Parkes and reports of copper oxide mineralisation in outcrops throughout the district. It was only in 1976 that a scout drilling program turned up distinctive green rock chips, which proved to be malachite. In the years that followed, a combination of drilling and geochemical and geophysical techniques outlined the four ore bodies–Endeavour (E) 22, E27, E26 and E48–on which the current operation is based.

Mining commenced with three copper-gold ore bodies, two of which are mined by conventional open cut methods. The third, E26, is mined by the underground block caving method. A fourth ore body, E48, has been identified in the centre of the lease. While further evaluation of this deposit is required, it appears to be amendable to block caving.

The E26 ore body is the largest and highest grade of the ore bodies discovered at Northparkes to date. The original mining reserve of 44 million tonnes at a grade of 1.5% copper and 0.5 gm/tonne gold will eventually yield 600,000 tonnes of copper metal and 20 tonnes of gold. The present delineated, pipelike ore body has dimensions of approximately 200 m in diameter and over 800 m in depth extending from just below the surface.

Block caving was selected as the most appropriate mining method because of the geometric and geophysical characteristics of the ore body. Northparkes E26 mine was thus heralded as the first Australian block cave mine. The E26 deposit has been divided into two blocks. The first block to be caved extends to 480 m below the surface and comprises 27 million tonnes of ore, whilst the second block (Lift 2) comprises an additional 24 million tonnes. Development for the first block (Lift 1) commenced in October 1993 and was fully commissioned by September 1997.

The underground mine has, up to December 1999, produced around 12,000 tonnes of ore per day, which equates to 4.1 million tonnes of ore per annum. The Northparkes copper concentrator produces around 440 dry tonnes of copper-gold concentrate per day, yielding 61,000 tonnes of copper and 40,000 oz of gold per year. In terms of annual ore tonnage, the underground mine is the largest underground metalliferous mine in New South Wales and the second largest underground mine in Australia. World-class productivity has been recorded at

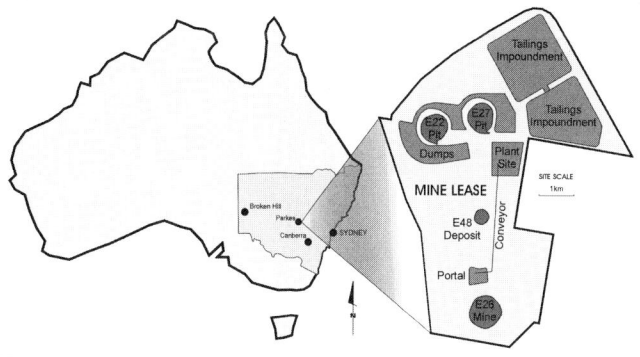

FIGURE 50.1 Location of Northparkes Mines

Northparkes' E26 underground mine with 42,686 tonnes of ore hoisted per underground employee per annum.

Northparkes Mines is committed to ensuring a safe and healthful work place and has been recognised by the Australian mining industry for achievements in safety.

50.2 GEOLOGY

50.2.1 Regional Geology

The Northparkes copper-gold deposits are hosted by the westernmost of three Early Ordovician to Early Silurian volcanic belts that form part of the Paleozoic Lachlan Orogen of New South Wales and Victoria. These belts may have formed as an intraoceanic island arc at an early Paleozoic convergent plate margin (Glen et al. 1998). Compositions in all three belts progress from calc-alkaline in their older parts to high-K calc-alkaline to shoshonitic in the youngest volcanics.

The Northparkes deposits are hosted within the high-K calc-alkaline to shoshonitic, Late Ordovician to earliest Silurian Goonumbla Volcanics. The Goonumbla Volcanics are predominantly composed of fine to extremely coarse deep-water volcaniclastic rocks that range in composition from basaltic andesite to trachyte. Minor ignimbrites, possibly emplaced as large slump blocks, occur high in the stratigraphy.

50.2.2 Local Geology

Copper-gold mineralisation at E26 occurs in stockwork quartz veins and disseminations associated with small, fingerlike monzonite porphyry stocks that intruded coeval Late Ordovician trachyandesites of the Wombin Volcanics. The mineralised body as defined by the 0.8% equivalent copper contour has maximum dimensions of 350 m in length, 180 m in width, and a known

---

* Deceased, formerly Northparkes Mines, Parkes, New South Wales, Australia.
† Northparkes Mines, Parkes, New South Wales, Australia.

411

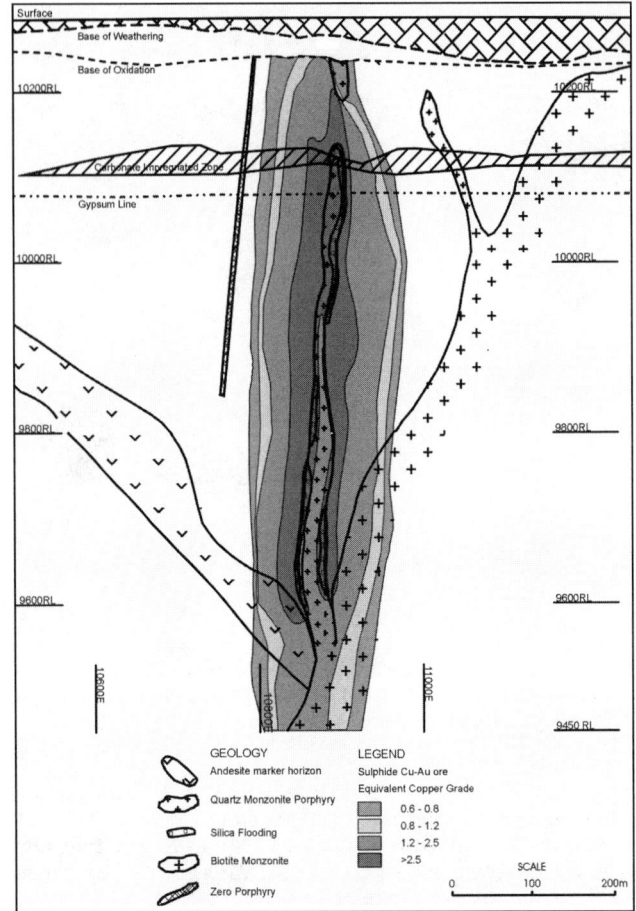

FIGURE 50.2   Longitudinal geologic section of E26

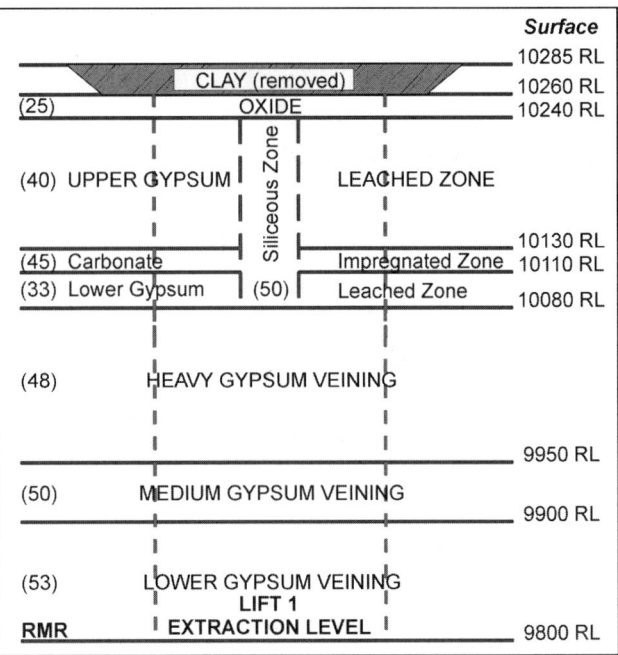

FIGURE 50.3   Geologic section of E26 Lift 1

depth of 850 m below the base of oxidation (Figure 50.2). The main/southern body has a strike length of about 200 m.

Mineralisation occurs both within the monzonite porphyry and in the surrounding volcanics. The volcanic rocks host the largest tonnage of ore. There is a strong correlation between the density of quartz veins and grade, particularly along the margin of the southern porphyry where thick quartz veining and pervasive silicification has produced a quartz-rich rock known locally as the "silica" zone.

The copper-gold mineralisation is associated with strong potassic alteration characterised by haematite, biotite, and potassium feldspar. This imparts a pervasive pink-red colouration to the volcanics and intrusive stocks.

Bornite is the dominant sulphide along with lesser amounts of chalcopyrite and chalcocite. Both occur principally as grains and clots in quartz veins and as disseminations in the ground mass. Gold occurs principally in bornite as small inclusions of free gold, electrum, and tellurides

Alteration, vein density, and copper mineralisation decrease gradually away from the intrusives. As a result, copper and gold grades are concentrically zoned around the intrusive stocks. At 300 to 400 m from the centre of mineralisation, pyrite becomes the dominant sulphide, often associated with magnetite. Hard grade boundaries do not occur naturally in this system, although the high-grade silica zone bounding the southern porphyry appears to be laterally restricted. A later monzonite porphyry intrusive occurs to the north (northern porphyry).

At E26, the ratio of gold to copper is concentrically zoned around the monzonite stocks. The gold:copper ratio is lowest on the

edges of the mineralised zone (i.e., low peripheral gold content) and increases inward. There is also a vertical zonation of gold content, and the gold:copper ratio is highest in the central portions of the main deposit and decrease dramatically between RLs 9,950 to 10,000 m and weakly with depth. The gold:copper ratio is about 0.1 or less in the lower-grade peripheries of the main body, increasing to 0.25 in the higher grade areas above RL 10,000 m, and to 0.5 to 1.0 in the high-grade core between RLs 10,000 and 9,700 m. The gold content of the Northern Porphyry mineralisation is low to insignificant throughout (<0.1 gm/tonne).

Mineralisation thins rapidly toward the surface, and as a result, only a small part of the system has been oxidised. Oxidation has occurred at depths of 25 to 60 m. The oxide zone consists of an upper layer of deeply weathered rock composed mainly of kaolin with lesser amounts of smectite clay minerals. This upper layer is termed the weathered zone and varies from 10 m thick to the southwest to 40 m to the northeast. Below the weathered zone is the hard oxide zone of hard, well-jointed rocks.

50.3   GEOPHYSICAL

The initial geophysical investigation at E26 was based on oriented diamond-drill core; however, geophysical information was continuously updated from mapping underground exposures as development progressed.

50.3.1   Geophysical Zonation

The post-mineralisation gypsum event (Heithersay and Welsh 1996) is unique to the E26 deposit and has pervaded to the upper regions of the fracture system, contributing strongly to the geophysical characteristics. The most distinct feature of the Lift 1 ore body is the gypsum line (located at RL 10,080 m), which is a planar surface above which the gypsum has been leached out by circulating groundwater, resulting in open fractures in the rock mass. Within this gypsum-leached zone, there lies a 10- to 35-m-thick subhorizontal layer that has been recemented by carbonate, commonly referred to as the carbonate impregnated zone. The intensity of gypsum veining decreases with depth from around 2% of the rock volume above RL 9950 m to less than 0.5% below RL 9800 m (Figure 50.3). Generally the rock mass strength, and

TABLE 50.1   Stress measurements at 450 m depth

Principal stresses	Magnitude, MPa	Dip, degree	Bearing, degree
σ_1	22.7	05	141
σ_2	15.0	15	049
σ_3	12.1	75	248

TABLE 50.2   Joint sets in the Northparkes ore body

Joint set	Dip direction, degree	Dip, degree
1	229	87
2	270	79
3	002	77
4	192	37

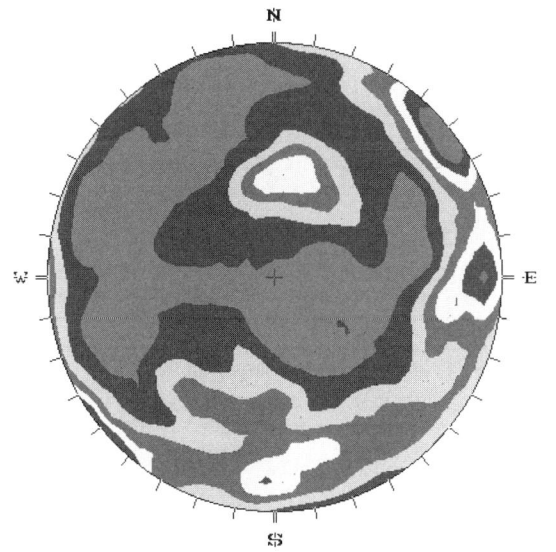

FIGURE 50.4   Stereographic projection of quartz sericite porphyry shears (poles plotted)

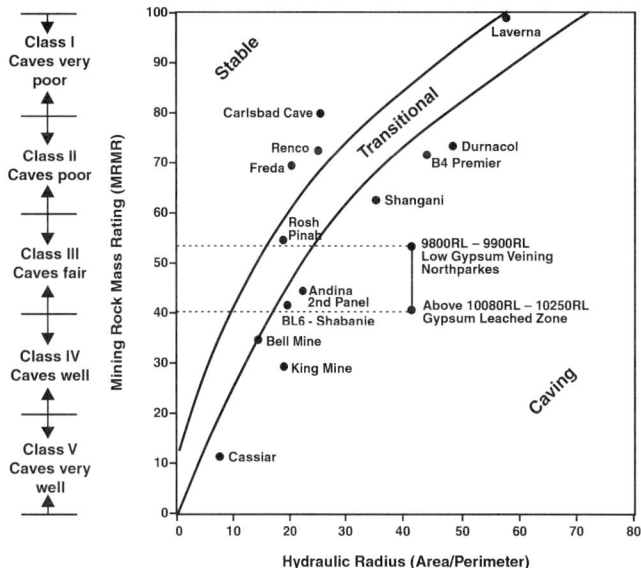

FIGURE 50.5   Empirical stability diagram (after Laubscher 1990)

hence the geophysical characterisation, is attributed to the density of gypsum veining within the ore body. Nine distinct zones have been defined within the rock mass, all of which are illustrated in Figure 50.3.

50.3.2   Field Stresses

Several in situ stress measurements were taken at various depths using the hollow inclusion cell (HICell) overcoring technique. All of the stress measurements indicated a low regional stress environment with a subhorizontal major principal stress component. The reliability of some of the stress measurements was questionable considering that none of the measurements provided comparable results. The reduced reliability of these stress measurements is largely attributed to the highly fractured nature of the rock mass, which made overcoring the HICells extremely difficult. The most reliable and accepted stress measurement was conducted 450 m below the surface, the results of which are summarised in Table 50.1.

50.3.3   Rock Strength

Rock strength tests were conducted on core samples taken from various lithologies within the deposit. The results show that the host rocks at E26 are generally strong with a uniaxial compressive strength ranging from 110 MPa in the volcanics to 143 MPa in the biotite quartz monzonite.

50.3.4   Structure

With the exception of one fault in the northeastern corner, there are no major faults or shear zones transecting the ore body. Occasional minor quartz sericite porphyry shears up to 300 mm wide have been mapped. These shears are generally steeply dipping with a northwestern trend (Figure 50.4). Four principal joint sets have been identified (Table 50.2) from geophysical

mapping. Most of the joints are well developed and rough walled, and their planarity varies from planar to stepped. The fracture frequencies are essentially related to the gypsum vein densities, which range from two per metre around RL 9,800 m to five per metre above the gypsum line.

50.3.5   Rock Mass and Cavability

Northparkes has adopted the Laubscher mining rock mass rating (MRMR) classification system (Laubscher 1990), which is widely used amongst caving operations. The system estimates the rock mass quality by assigning a rating to intact rock strength, fracture frequency per metre, and joint condition. The total sum of these ratings is referred to as the rock mass rating (RMR), which for the Lift 1 block ranged between 33 to 54 (Figure 50.5). The MRMR is the RMR adjusted for mining-induced effects, namely weathering, joint orientation, induced stresses, and blasting. The product of the adjustments made for Lift 1 was initially calculated as 1.0, thereby yielding an MRMR equal to the RMR.

The cavability of a rock mass is a function of the quality of rock mass and the hydraulic radius. The hydraulic radius is a ratio of the area over the perimeter of the undercut. The cavability of the E26 Lift 1 block was based on Laubscher's empirical stability diagram, and as can be seen from Figure 50.5, sustained caving was expected once the undercut development attained a hydraulic radius of between 20 and 25 (Laubscher 1990).

Cave propagation proved to be rapid as the undercut advanced, with initial caving commencing once the undercut attained a hydraulic radius around 23. Intermittent caving followed the advance of the undercut; however, once the entire undercut was developed (196 m long by 180 m wide), caving virtually ceased. On completion of the undercut, the cave had propagated to a maximum height of 95 m above the top of the undercut, yielding approximately 3 million tonnes of broken ore.

Increased production rates failed to induce further caving, but resulted in an increase in the air gap between the top of the broken ore and the cave.

It was obvious from the lack of caving that the MRMR values calculated for Lift 1 were incorrect and required reassessment. After considerable investigation and much debate, the RMR was adjusted by a factor of 1.2 to account for clamping stresses across the dominant subvertical joint sets, thus increasing the MRMR range to between 40 and 64 and placing Lift 1 near the transitional caving boundary (Figure 50.5).

50.4 MINING METHOD

50.4.1 Mine Layout

Access to the underground mine for all personnel and materials is via a 1-in-7 gradient decline with dimensions of 5 m wide by 5.5 m high. A hoisting shaft for ore production and development waste removal has been established to an initial depth of 520 m. The shaft is 5.3 m in diameter and concrete lined. It is equipped with a ground-mounted friction winder and rope guides. The headframe structure is 60 m high and contains a 100-tonne ore surge bin. A 5-m in diameter ventilation shaft has been established to a depth of 360 m. It was developed by a combination of blind boring from the surface and raise drilling and is not lined.

All shaft and decline development activities were carried out using mining contractors. Northparkes employees were used for long-term production and maintenance activities.

E26 Lift 1 operates as a mechanised block cave with the extraction level located 480 m below the surface. The extraction level comprises an offset herringbone layout of 130 drawpoints spread over seven extraction drives, each drive accessible via both inner and outer perimeter tunnels. The drawpoints are designed with a nominal 14- by 14-m spacing to cover the area of the ore body. Pairs of drawpoints are arranged in such a way as to connect into drawbells (or cones) that are mined upward into the undercut. The adoption of two perimeter tunnels surrounding the extraction level was to provide access for secondary breaking, maintenance, and nonproduction-related activities, thereby minimizing interference on tramming routes for load-haul-dump (LHD) equipment (Figure 50.6).

50.4.2 Mine Construction

Cave initiation was achieved by advancing two undercut levels with a combined height of 42 m. The lower undercut was located on RL 9,818 m and the upper undercut on RL 9,830 m (Figure 50.7). This unique double-undercut strategy was devised to provide approximately 4 million tonnes of well-fragmented ore, thereby eliminating the initial coarse fragmentation typical in caving operations and ensuring a speedy production ramp-up. An advanced undercut strategy was adopted whereby the upper undercut was excavated ahead of the extraction level development to provide protection from abutment stresses.

Ground support was integrated into the development cycle to provide early support to the rock mass using fully resin-grouted, 2.4-m-long, Gewi-type rock bolts. Campaigns of installation of Expamet expanded metal mesh straps followed rock bolting. The straps were installed longitudinally along the extraction drives and vertically on bullnoses and in drawpoints. Single-strand, 6-m-long, cement-grouted cable bolts were installed on a 2- by 2-m pattern throughout the extraction level. Along each extraction drive, there was effectively an 8-m-wide span between drawpoints. It was expected that the combination of shotcrete and mesh straps would provide a high resistance to shear failure. Steel-fibre-reinforced shotcrete (SRFS) was used as part of the mining cycle to supplement the rock bolt and mesh strap ground support. The shotcrete was specified at 40 MPa for

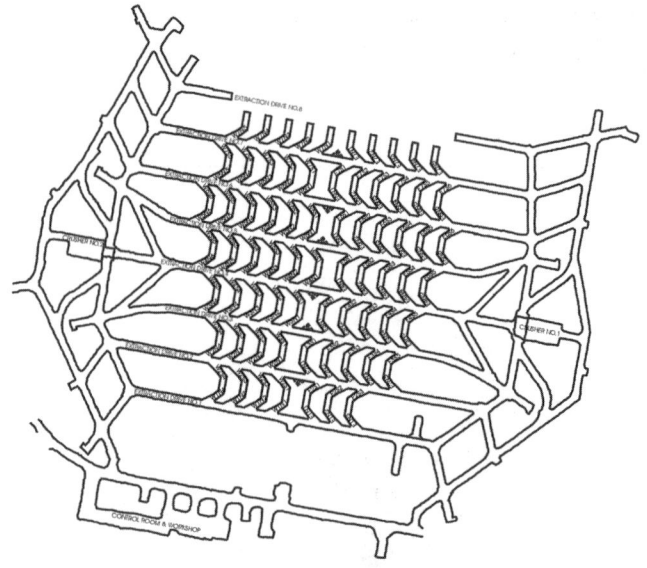

FIGURE 50.6   Extraction level layout, E26 Lift 1

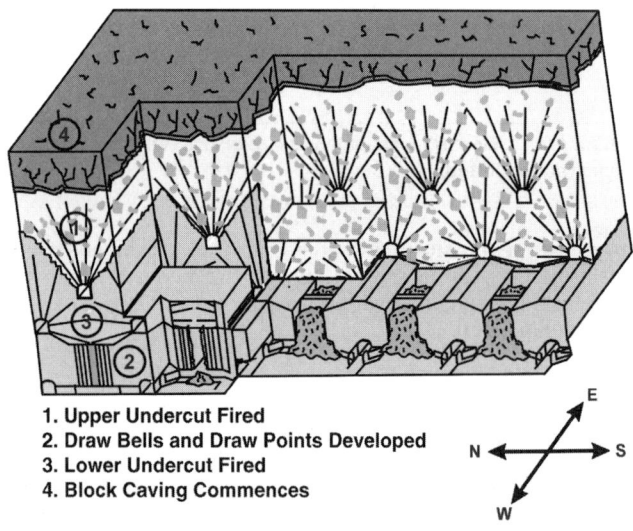

1. Upper Undercut Fired
2. Draw Bells and Draw Points Developed
3. Lower Undercut Fired
4. Block Caving Commences

FIGURE 50.7   Undercut sequence, E26 Lift 1

28-day compressive strength and 5 MPa for flexural strength. Steel fibres were specified as ranging in length from 25 to 40 mm. The amount of fibres was set at not less than 60 kg/m^3.

Drawpoint ground support specified for the drawpoint consisted of—

- Fully resin-grouted, 2.4-m-long rock bolts on 1- by 1-m spacings

- Expamet straps installed longitudinally into the drawpoint from the extraction drive and pinned vertically in brow areas

- Fully cement-grouted, 6-m-long cable bolts installed in brow areas

- Two steel sets fabricated from 250UC. The steel sets were encased in fibre-reinforced shotcrete to provide a tough, smooth lining system

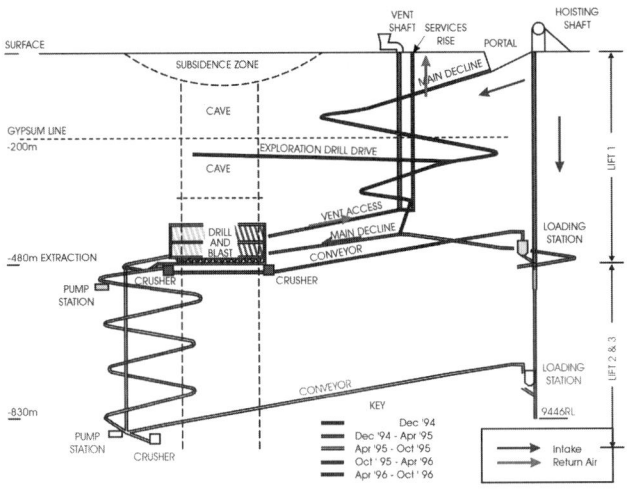

FIGURE 50.8   Schematic of E26 Lift 1 (proposed Lift 2 also shown)

The extraction level was surfaced with a high-abrasion-resistant concrete specified at a uniaxial compressive strength of 80 MPa and placed as a 200-mm-thick slab.

Lift 1 construction was judged as an example of engineering excellence in 1998 by the Institution of Engineers, Australia, and was one of only six projects awarded that year.

50.4.3   Mine Operations

Four LHDs, each of 6-m³ capacity, tram ore from the drawpoints to two crushers. The crushers are located either side of the ore body, resulting in an average tramming distance of 120 m. The rock is crushed to less than 180 mm in size and then conveyed by inclined belt conveyor to one of three 1,000-tonne ore bins adjacent to the loading station. The hoisting system, including the loading station, operates automatically with crushed ore delivered to the headframe surge bin (Figure 50.8). It is then transported by transfer and overland conveyors a distance of 3 km to the concentrator.

The use of advanced communication technologies allows remote monitoring and operation of all fixed plant and some mobile equipment through the extensive use of video and process logic control systems. Thus it is possible to operate the LHDs by remote control from the centralised control room on the extraction level. It is proposed that in the near future, the operator need only control the machine during the loading cycle followed by automatic control for hauling and dumping. LHD automation will have productivity, maintenance, and personnel benefits for the operation. The use of these and similar technologies has enabled Northparkes to achieve world-class productivity rates from its employees and to exploit relatively low-grade resources, such as Endeavour 26, and make them profitable mining operations.

Draw Control.   The draw control strategy at E26 Lift 1 was planned for a 3:2:1 ratio draw. This draw ratio takes advantage of the central higher grade area of the ore body to produce higher average grades over the greatest period.

Prior to cave propagation through to the surface, the draw strategy focussed on maximizing the grade in the caved ore, within geophysical constraints. Once the cave breached the surface in November 1999, ore extraction was simulated using Gemcom's PC Block Cave software (PCBC) to predict the best draw strategy to maximize grade and tonnage over the remaining mine life. Ore sampling and fragmentation analysis is used to calibrate the ore column model as drawdown of the ore column continues.

Monthly tonnages are extracted from PCBC and divided into weekly blocks, which are imported into the mine's SCADA control system. The SCADA control system, in turn, provides a 24- or 36-hr production schedule from which the tonnages and bogging order for this schedule (or call) are displayed graphically to the SCADA operator in the mine control room. The SCADA operator then communicates the draw requirements to each of the LHD operators on a drawpoint-by-drawpoint basis. The actual tonnage drawn from each drawpoint is recorded into SCADA, and at the end of each month, these daily tonnages are depleted from the ore column model in PCBC.

Secondary Breaking.   Northparkes E26 Lift 1 is designed to operate without centrally located, specialised secondary breaking equipment. The two crusher run-of-mine bins do not have a grizzly; therefore, LHD operators determine which oversize will be tipped into the run-of-mine bins. Initial forecasts of secondary breaking requirements for material in excess of 2 m³ indicated that 4,000 oversized rocks would need to be broken per month. The original goal for breaking oversize rocks centered on nonexplosive breaking methods and remote control of as many functions as possible to reduce risks to personnel.

A dedicated high-hang-up rig was designed and manufactured to break any extremely large rocks that bridged drawbells or jammed high in the drawbells. The benefits of this philosophy were to waste no LHD time moving rocks to central breaking bays, limit delays to production from firing times, reduce risk to personnel with no requirement to enter the drawbell, and keep costs low through limiting personnel numbers and using efficient breaking systems.

Largely inexperienced crews of secondary breaking personnel were introduced to the task by a group of specialised trainers. These trainers assisted the creation and documentation of tailored standard work practices aiming at best practice in secondary breaking. Workplace hazards are managed through workplace risk assessment and hazard identification systems. Standard operating procedures are in place with periodic reassessments to ensure personnel are up to date.

Some changes to the initial secondary breaking philosophy were necessary. Many trials with automated and semiautomated propellant-based secondary breaking systems were carried out. These systems were in their infancy—with prohibitive costs per rock broken.

Manual and mechanised boulder buster systems are used for most secondary breaking of oversized material at Northparkes. In addition, experience has shown that much of the gypsum-veined oversize will break when dropped from a LHD bucket onto the concrete floors. There are designated dropping points in each work area.

Although two Tamrock Monomatic-style secondary breaking rigs were purchased for specialised secondary breaking work, a hired Tamrock Commando 100 rig has been the most-used rig for secondary breaking. The specially designed high-hang-up rig has yet to be used for secondary breaking at Northparkes.

Hang-ups are usually dealt with by a combination of methods, commencing with bogging the drawpoint, then surrounding drawpoints if required, hosing, bombing, and then drilling. Explosive is loaded into hang-ups using a combination of extra-long wooden poles and low-static hose to deliver explosive to any holes. Start and end of shift are usually used for firing any hang-ups or oversize within the drawbell.

Experience has shown that LHD operators can effectively judge crusher oversize limitations. Any hang-ups in the crusher jaws are cleared using a manual boulder buster and a hand-held drill. In recent months, a hydraulic boom with a boulder buster and hydraulic drill have been obtained from Amquip and will be tested in one of the crushers.

Actual secondary breaking requirements were far less than anticipated with an average of 400 to 500 oversized rocks per month and 100–150 hang-ups per month. It is believed that Northparkes slow caving conditions caused caving material to fall some distance from the cave back to the muckpile, resulting in secondary fragmentation in the more competent sections of the ore body. Because of the low muckpile height (until caving through to the surface occurred in November 1999), limited communition within the muckpile occurred.

50.5   PRODUCTION STATISTICS

As of March 2000, Northparkes Mines employed 177 people and 85 major shutdown and permanent contractors. To the end of March 2000, 13.5 million tonnes had been mined from E26 Lift 1, inclusive of undercuts. With 98 Northparkes Mines' employees and contractors working underground, the annual production rate of 4.2 million tonnes equates to 42,686 tonnes per underground employee. This is inclusive of technical support staff and maintenance engineers. On a site-wide basis, 267 tonnes of copper are produced per person per annum.

Six Tamrock Toro 450Es are used in production at Northparkes. At any time three or four LHDs are operating. Productivity for the fleet has averaged 184 tonne/hr since June 1997. LHD availability has averaged 71% and use 51% in the same period.

A Siemens winder with a capacity of 815 tonne/hr is used for hoisting ore. Two 16-tonne skips are used in the shaft, passing mid-shaft at 15 m/s.

Two Krupp 84- by 66-in double-toggle jaw crushers are used for primary crushing of ore. They have crushed 5.8 and 6.5 million tonnes since being commissioned in late 1996 and early 1997, respectively. Combined crusher availability averages at 83% since commissioning with a use rate of 55%.

The Siemens winder has hoisted 12.3 million tonnes since June 1996, with an availability of 77% and use of 82%.

50.6   FINANCIAL DETAILS

It costs $AUD240,000 a day to operate the mine and plant. Of total site costs—

- 7% is spent on power (approximately $AUD7M per year), 10% on grinding media, 13% on labour, and 19% on other underground operating costs.

- Approximately $AUD5 million per annum is paid to the Government in royalties. In 1997, $AUD420,000 was paid to Parkes Council in rates.

- Export income from sale of concentrates averages $AUD215 million per annum.

- Underground unit mining cash costs are $AUD5.38/tonne, and mill unit cash costs are $AUD6.43.

50.7   MINE SAFETY

Northparkes has reporting system as the key control in managing site safety. Not only is the HIAES system used to report any issues concerning safety, the environment, and production, it is also used to identify and act upon hazards identified mine-wide. To ensure that a safe work ethic is maintained, the HIAES system is used to establish positive performance indicators on the basis of which both site and personal goals are set.

Northparkes Mines has been recognised by the mining industry for achievements in safety. These have included—

- The Highly Commended award for the HIAES reporting system, NSW Minerals Council Safety Innovation Awards 1998.

- Finalist, Minerals Council of Australia National Safety Innovation Awards 1999.

- Commendation for excellence in health and safety, Australian Minerals Council MINEX Awards 1999.

- One of the top three Australian sites identified by the Minerals Council of Australia Safety Culture Survey conducted in 1999.

50.8   FUTURE DEVELOPMENTS

A detailed feasibility study into the development of the lower lift (Lift 2) of the E26 ore body is currently under way and will be completed by the end of 2000. Lift 2 development is expected to commence by mid-2000 with initial production from the undercut and drawbell development during 2002.

The additional underground mining reserve at E48 (which is close to the existing hoist shaft) presents a number of potential options for the timing and sequencing of future underground developments. The E48 reserve is currently scheduled to follow on from the E26 Lift 2 reserve. Northparkes Mines is actively exploring in the mining lease and neighbouring exploration tenements. If the high prospectivity of the region yields further ore bodies, adequate room exists for project expansion. Project expansion could involve further processing of concentrate into various stages of metal production.

50.9   ACKNOWLEDGMENT

The authors gratefully acknowledge the contributions made by Messrs. C. Stewart, S. Smith, and S. Duffield to this article.

50.10   REFERENCE

Laubscher, D.H. 1990. A geomechanics classification system for the rating of rock mass in mine design. *J. S. Afr. Inst. Min. Metall.* 90:257–273.

CHAPTER 51

Preundercut Caving in El Teniente Mine, Chile

Eduardo Rojas, * **Rigoberto Molina,** * **and Patricio Cavieres** *

51.1 HISTORY

According to legend, El Teniente (Figure 51.1) was discovered by a fugitive Spaniard official in the 1800s. Exploitation first began in 1819, when the best minerals from what became the Fortuna sector were mined manually and transported on animals. In 1904, W. Braden, together with E. W. Nash, founded the first El Teniente company, Braden Copper Co., and built a road for carts and a concentration plant.

In April 1967, the Government of Chile bought 51% of the property from Braden Copper and founded the Sociedad Minera El Teniente. Under this agreement, major mine expansion was undertaken, and a new concentration plant was built in Colon, which increased total production to 63,000 tonne/d. As a result of constitutional reform, on July 11, 1971, El Teniente Mine became a state-owned company. In 1976, the Corporación Nacional del Cobre de Chile (Codelco) was formed, and El Teniente became part of it.

Currently, mine production is 100,000 tonne/d. By 1975, the mine had extracted 500 million tonnes of mineral, with another 500 million tonnes extracted between 1975 and 1995.

51.2 EVOLUTION OF EXPLOITATION METHODS

El Teniente Mine began operations in 1906. Since then, various exploitation methods have been used in productive sectors in secondary minerals. The methods range from "raised work over mineral" combined with shrinkage stoping and pillar recovery to block caving.

Later, as a consequence of deeper productive sectors and changes in the physical and mechanical properties of the rock, the exploitation of primary ore (of lower grade, stiffer, harder, and with coarser fragmentation than the secondary ore) has resulted in the mechanization of mining operations. This situation required a change from the standard block caving method used in secondary ore (primarily characterized by manual or semimechanized ore transfer) to the panel caving method, in which fully mechanized ore transfer is continuously incorporated into the production area, i.e., a dynamic caving face.

The panel caving method alternative most often used in the mine involves ore transfer via load-haul-dump (LHD) equipment. This method has been used since 1982 in Teniente-4 Sur, the first productive sector in primary ore. The exploitation sequence applied in this sector involves (1) development and construction of production level galleries, (2) undercutting at the undercut level, and (3) ore extraction. This method is called "conventional panel caving."

Knowledge gained over the years concerning primary ore exploitation with conventional panel caving (200 million tonnes extracted to date) has indicated that the advance of the caving face is the main cause of gallery damage in levels below the undercut level. Experience has also shown that a variation of

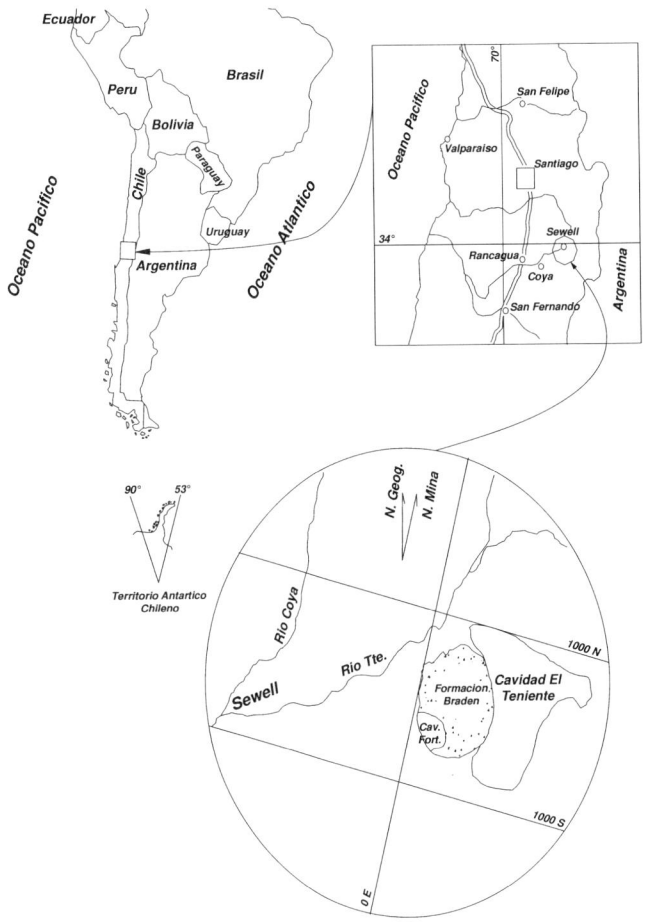

FIGURE 51.1 Location of El Teniente Mine

conventional panel caving (the "preundercut") reduces the degree of gallery damage in the levels below the undercut level, as well as the possibility of rock bursts associated with the advance of the undercut face. Preundercutting basically consists of advancing the undercut ahead of all development in the lower levels. All production level development is made behind the cave front and under the caved area.

51.3 CAVING METHODS

51.3.1 Undercutting

Exploitation methods currently used in El Teniente Mine are massive caving methods, such as block caving or panel caving.

---

* Codelco Chile, El Teniente Division.

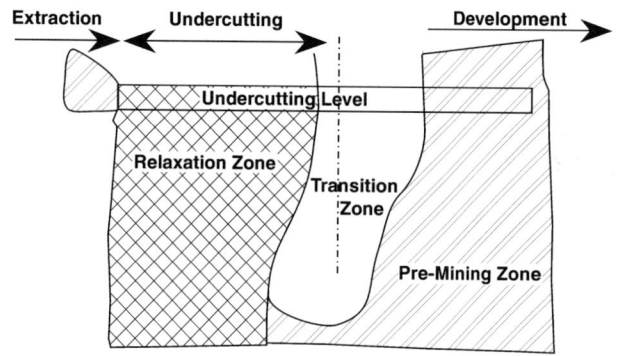

FIGURE 51.2 Distribution of zones according to stress state of rock mass

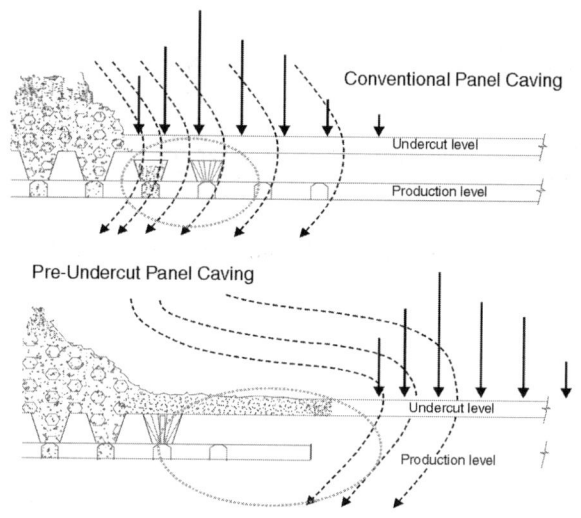

FIGURE 51.3 Conventional and preundercut panel caving

These methods involve undercutting the base of a block or panel of mineral, making sure that there are no supporting points. This is accomplished so that the lower part of the block behaves as a slab (simply supported or fixed), and the action of external forces, mainly gravitational, produces the progressive downfall of the block or panel as a result of an increase in the undercut area. Using such methods, the mineral fragments generated by caving can be extracted and transported according to the mine design of a particular sector.

51.3.2 State of the Rock Mass

Mining by panel caving is characterized by a dynamic caving face that modifies the condition of the rock mass. The method has the following rock condition and stress states (Figure 51.2).

- *Premining.* The stresses and the rock mass condition have not been affected by mining.
- *Transition.* Equilibrium is achieved. Continual stress changes (magnitude and orientation) occur as a result of mining activity, which affects the condition of the rock mass.
- *Relaxation.* The stresses have been modified, and there is a dramatic decrease in the magnitude of vertical stress.

Because of these conditions, it is necessary to establish a set of permissible distances as a integral part of the design. These distances must be set in such a way that safe working conditions will be provided or, conversely, the risks associated with each of the unitary operations are known in order to implement operational control measures.

Therefore, the exploitation sequence for pre-undercutting is to (1) undercut the level, (2) develop and construct the production level galleries, and (3) extract ore. This is the method currently used in the Esmeralda sector (Figure 51.3).

The engineering stage for the Esmeralda sector began in 1992. Planning included a change in the undercutting sequence from a conventional sequence to a preundercut sequence. Production started by the middle of 1997, with an investment of US $247 million and a total mineral reserve of 250 million tonnes, as detailed below.

51.4 GEOLOGY, ROCK MASS PROPERTIES, AND SEISMICITY

The predominant lithologies of Esmeralda sector are andesites, diorite breccias, hydrothermal breccias with anhydrite, hydrothermal breccias with tourmaline, diorites, and latite dikes (Figure 51.4). The Esmeralda sector has five predominant structural sets, and the rock mass is of fair-to-good geotechnical quality. Characteristics of the rock mass and typical stresses in the Esmeralda sector are provided in Tables 51.1 and 51.2.

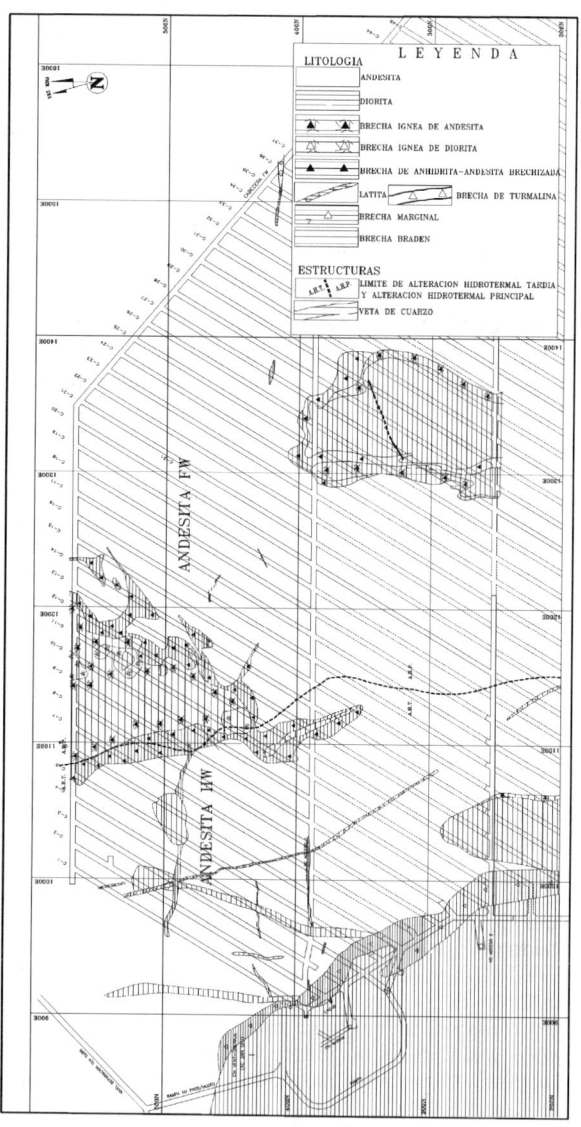

FIGURE 51.4 Plan view of geology in Esmeralda sector, undercutting level

TABLE 51.1   Rock mass characterization and properties

Property	Andesites	Breccias	Diorites
RQD, %	75–100	75–100	90–100
Fracture frequency per meter	0–4	1–3	0–2
[1]RML$_L$	53–68	58–68	65–74
Specific weight, tonne/m^3	2.70	2.70	2.70
Young's modulus, GPa	36–38	40–49	55
Poisson's ratio	0.18	0.15–0.18	0.15
Unconfined compressive strength, MPa	87–104	100–120	130
[2]Hoek-Brown parameter (m$_b$)	2.0	2.4–4.1	4.9
[2]Hoek-Brown parameter (s)	1.2E-3	2.3–4.2E-3	5.4E-3

[1]RML$_L$ = RMR from Laubscher.
[2]Hoek-Brown parameter values reflect disturbances due to abutment stresses and relaxation.

TABLE 51.2   Stress measurements, Esmeralda sector

Stress	Magnitude, MPa	Azimuth, °	Dip, °
S1	46	312	34
S2	35	50	11
S3	16	156	53

NOTE: Positive dip is below horizontal.

Conceptually, caving methods are based on the progressive rupture of a rock mass due to gravity. This rupture can be of sufficient size to release enough energy to damage underground workings. Therefore, rock mass response is directly related to mining activity. Slow extraction and undercutting rates may reduce the likelihood of major ruptures as the rock mass readjusts to mining.

In addition, empirical evidence indicates that there is greater risk of damage during the initial caving period (i.e., before the cavity generated by caving reaches the surface or previously caved areas). Experience also suggests that once the cavity reaches the surface or previously mined areas, expansion of the cave occurs with a more favorable seismic response.

For the Esmeralda sector, seismic activity is relevant in the initial stage, which is primarily associated with undercutting, and in the period before the cavity reaches the overlying, previously mined area.

51.5   CAVING, FRAGMENTATION, AND EXTRACTION RATE OF THE SECTOR

Production at the Esmeralda sector averaged 9,000 tonne/d at the end of 1999 (Figure 51.5), reaching a maximum of 12,000 tonne/d with 25,000 m^2 of available production area. Caving was achieved with 16,800 m^2 in production, once a problem of "support points" was solved. (Support points formed above the apex of the crown pillar and reduced the interaction between drawpoints, making the flow of ore from the undercut level difficult.) By December 1999, the amount of rock removed was equivalent to an extracted height of solid rock of about 50 m (Figures 51.6 and 51.7).

Observed fragmentation is slightly coarser than predicted, although it must be remembered that caving was initiated in an area that corresponds to igneous breccia with a fracture frequency of one to three fractures per meter.

The extraction rate defined for the Esmeralda sector was 0.14 to 0.44 tonne/d per square meter at the initial caving stage and reached 0.28 to 0.65 tonne/d per square meter at the steady-state caving stage (Table 51.3).

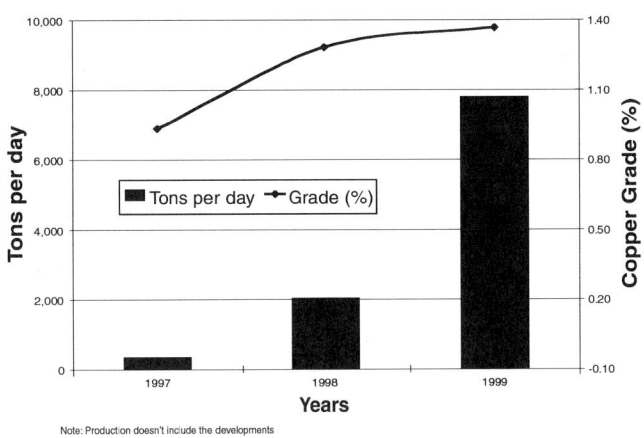

Note: Production doesn't include the developments

FIGURE 51.5   Mining production and grade from Esmeralda sector (1997–1999)

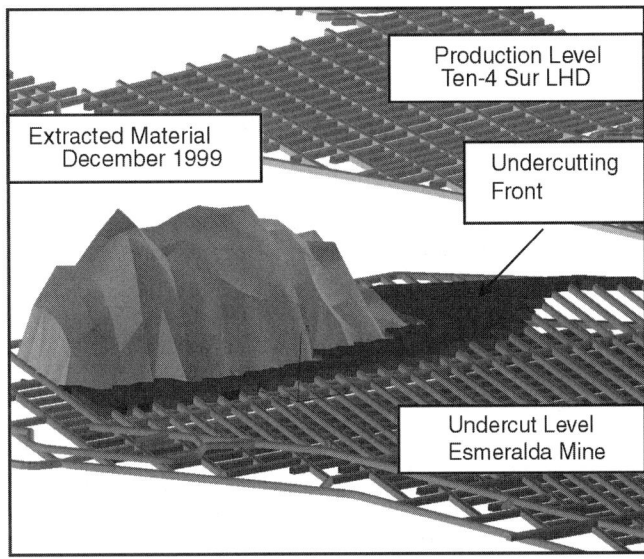

FIGURE 51.6   Height of solid extracted material as of December 1999

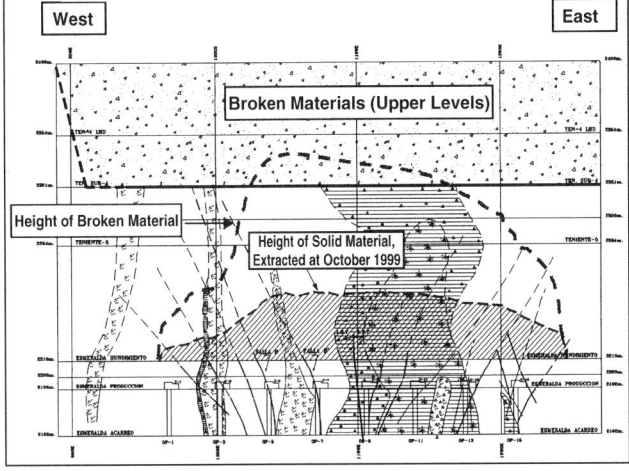

FIGURE 51.7   Height of extracted and broken material, east-west vertical section (looking north)

TABLE 51.3   Rates of average annual extraction 1997 to 1999

Year	Area, m²			Extraction rate, tonne/d per square meter	
	Open	Available	Utilized	Real	Effective
1997	3017	3017	2870	0.11	0.12
1998	13,326	10,635	12,556	0.14	0.15
1999	26,658	25,007	24,691	0.29	0.32
Average	14,334	12,886	13,372	0.18	0.19

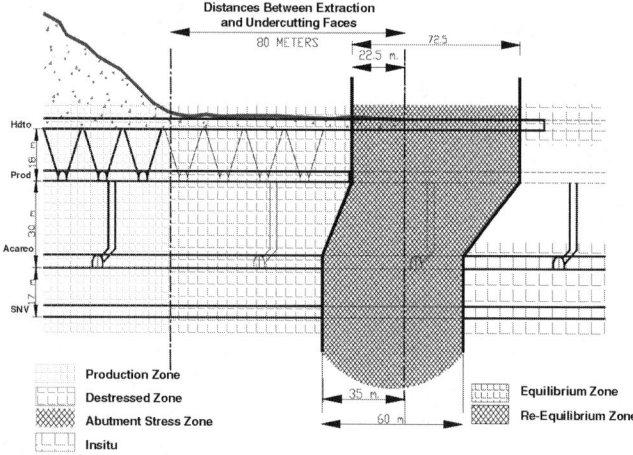

FIGURE 51.8   Distribution of zones according to the rock mass state

51.6   PERMISSIBLE DISTANCES BEHIND ADVANCING UNDERCUT FACE

The zone of abutment stress defines a spatial transition zone around the undercutting front and is determined by observed changes in magnitude and orientation of the stresses acting on the rock mass. These changes can increase the possibility of stability problems. The stability of the galleries is improved by developing and constructing within the relaxed zone and/or premining zone. The operational aspects of the preundercut sequence define relevant distances between the undercut, extraction, development, and construction faces (Figure 51.8).

> Undercutting face to development face = 22.5 m
>
> Development face to construction face = 30 m
>
> Construction face to constructed zone = 12.5 m
>
> Constructed zone to extraction face = 15 m
>
> Undercutting face to extraction face = 80 m, total

51.7   MINE DESIGN

The mine layout and design incorporate several rock mass strength aspects to reduce the likelihood of stability problems (e.g., orientation of galleries with regard to stresses and structures). The distance from floor to floor between the undercut level and the production level is 18 m, leaving an effective thickness of 14.4 m for the crown pillar. (The galleries have heights of 3.6 m.) The arrangements of the undercut, production, transport, and ventilation levels are shown in Figure 51.9.

51.7.1   Undercut Level

The drifts are separated by 15 m, with access crosscuts 200 m apart (Figure 51.10). The typical drift section is 3.6 by 3.6 m, which defines pillars with an effective width of 11.4 m. A typical section of an access crosscut is 4 by 4 m (Figure 51.11).

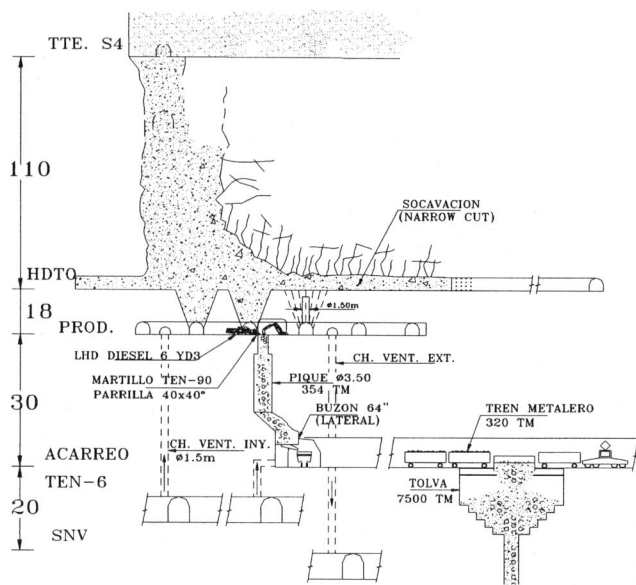

FIGURE 51.9   Typical preundercut panel caving, Esmeralda Mine

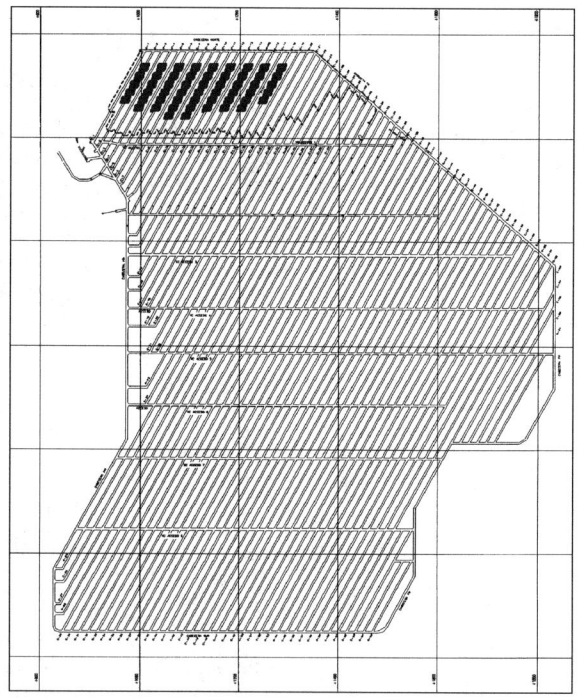

FIGURE 51.10   Undercutting level layout

Undercutting is typically done with half-pillar and complete-pillar blast designs (Figure 51.12), with the half-pillar blast design beinf the primary method.

51.7.2   Production Level

The production drifts are 30 m apart (Figure 51.13), with a typical section being 4.1 by 3.9 m (Figure 51.14). The crosscut drawbells (zanjas) are 15 to 17 m apart, with a typical section being 4.1 by 3.6 m. The production drifts make a 60° angle with the crosscuts (Figure 51.15).

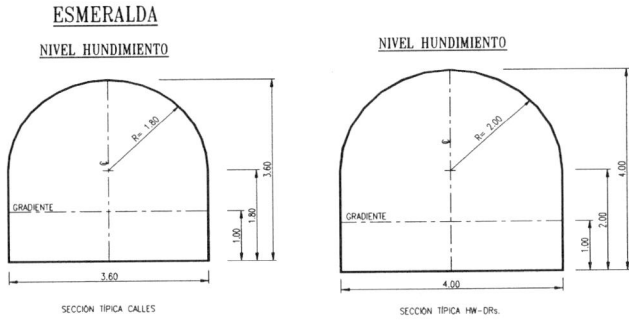

FIGURE 51.11   Geometries and sections of undercut level galleries

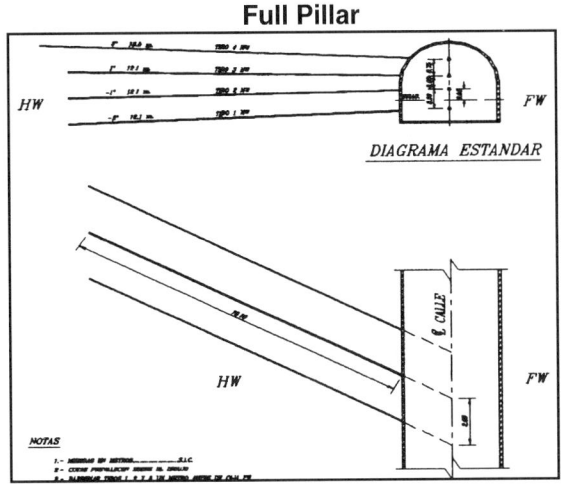

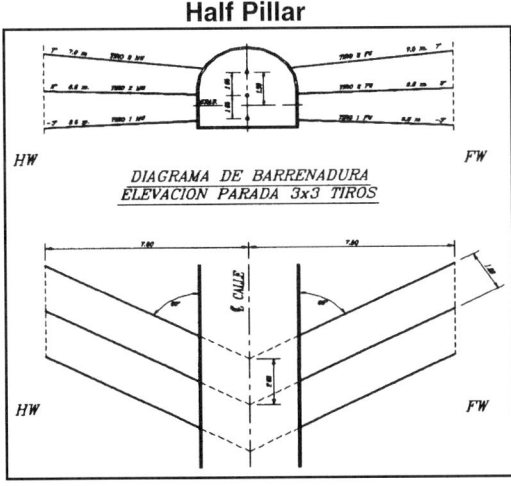

FIGURE 51.12   Undercutting diagrams

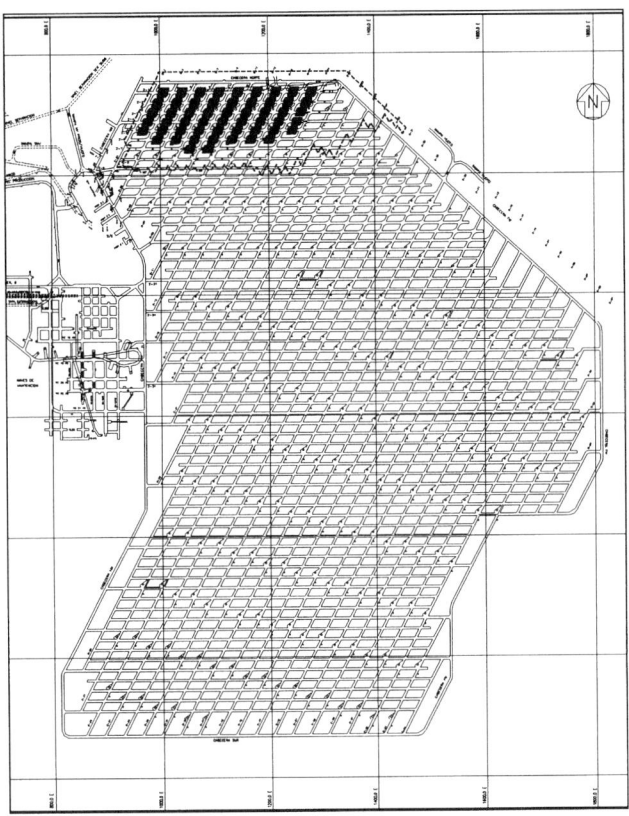

FIGURE 51.13   Production level layout

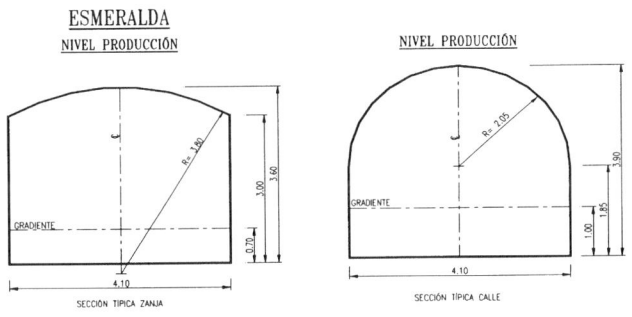

FIGURE 51.14   Geometries and sections of production level galleries

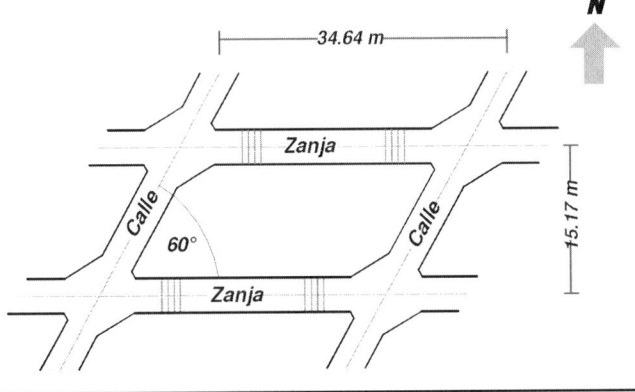

FIGURE 51.15   Typical production and drawpoint galleries

51.7.3   Drawbells

The preundercut method uses drawbell excavation totally accomplished from the production level in four blasting stages. The first two stages are blasted with free faces, and the last two stages are blasted into the broken material under confinement.

51.7.4   Transport Level

The transport crosscuts typically have sections of 4.8 by 4.8 m, while the connection drifts between crosscuts have sections of 4 by 4 m (Figure 51.16).

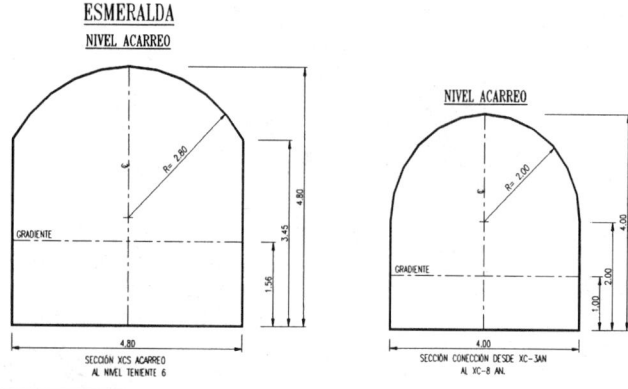

FIGURE 51.16 Geometries and sections of transporting level drifts

TABLE 51.4 Rock support

Undercutting	Fully grouted, 22-mm in diameter, 2.3-m long, rebar bolts; chain link mesh (10006); 1- by 1-m pattern.
Production	Development: Fully grouted, 22-mm in diameter, 2.3-m long, rebar bolts; chain link mesh (10006); 0.9- by 1-m pattern; 10 cm of shotcrete.
	Permanent: Fully grouted, 15.2-mm in diameter, long cable bolts in intersections of drifts and crosscuts.
	Steel sets and birdcage cables at drawpoints.
	Confining cables (zunchos) and straps on wall pillars.
Ventilation	Fully grouted, 22-mm in diameter, 2.9-m long, rebar bolts; chain link mesh (10006); 1- by 1-m pattern; 10 cm of shotcrete.
Transport	Fully grouted, 22-mm in diameter, 2.9-m long, rebar bolts; chain link mesh (10006); 1- by 1-m pattern; 10 cm of shotcrete.
	Fully grouted, 15.2-mm in diameter, long cable bolts.

51.7.5 Ventilation Crosscuts

Ventilation crosscuts have typical sections of 4 by 4 m, 5 by 4.5 m, and 5 by 5 m, with adits of 7 by 7 m and ventilation raises 1.5 m in diameter.

51.7.6 Rock Support

During the development of drifts and crosscuts, the support elements (fully grouted rebar and chain-link mesh) provide support. Later, the permanent support condition is achieved with shotcrete, steel sets, and cables. Also, the area in which drawpoint drifts intersect production drifts is reinforced with confining cables (zunchos) and straps (wall pillars). The typical support used in the different levels is summarized in Table 51.4.

51.7.7 Subsidence

Breakage angles have been estimated from experience, but recently available data on crater geometry and fracturing at galleries at different elevations have been used to calibrate two-dimensional, finite-difference numerical models. Model results define curves that predict these angles, as well as the extent of the zone of influence of the crater at different heights above the undercut level.

Figure 51.17 shows the curves used to estimate breakage angle as a function of height above the undercut level in a vertical section through the Esmeralda sector. The results of the numerical models, calibrated against observed behavior, were used to develop curves to predict the extent of the zone of influence of the crater.

Figure 51.18 shows a vertical southwest-northeast section that crosses the north side of the Esmeralda sector, where the

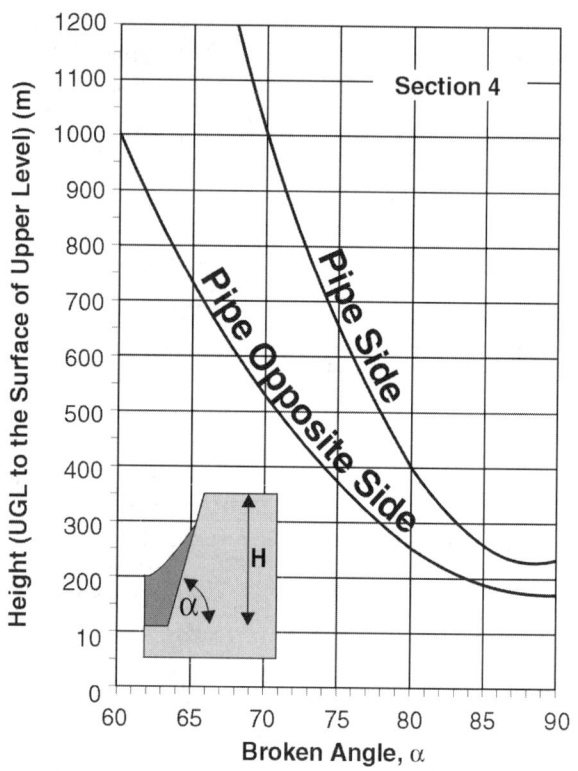

FIGURE 51.17 Variation of breakage angle with height over undercut level for section 4 (El Teniente Sur and Esmeralda sector)

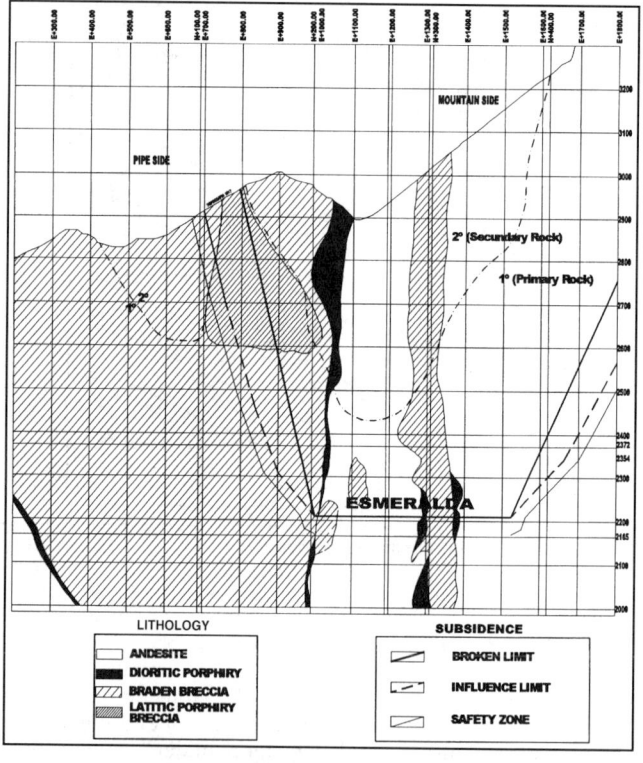

FIGURE 51.18 Southwest-northeast section through Esmeralda sector showing expected subsidence limits (defined by breakage angle and zone of influence)

subsidence limits were estimated using the curves previously mentioned. The figure shows that the breakage angles are larger on the pipe side (southwest) than they are on the mountain side (northeast). A safety zone of 40 m has been added to the zone of influence to account for unavoidable geotechnical uncertainties.

51.8 CONCLUSIONS

- The production area considers a planned production rate of 45,000 tonne/d. To date, the preundercut sequence has improved the stability of the galleries under the undercut level, increasing availability and use of the infrastructure.

- The distance between the extraction face and the undercutting face is the main variable controlling the stability of the undercutting level.

- One of the biggest difficulties when implementing this method has been the complex interaction among the development, construction, and extraction stages because of the limited area available between the extraction and undercutting faces.

- Mining-induced seismicity has been effectively controlled by regulation of undercutting and extraction rates.

51.9 ACKNOWLEDGMENTS

The authors thank the Superintendencia General Planificación Minco, Subgerencia Minco, Codelco Chile División El Teniente, for the authorization to publish this chapter.

51.10 REFERENCES

Betancourt A., M. Silva, and C. Valdivia. Resultados De Instrumentación Mina Esmeralda, Pl-I-021/99 Y Pl-I-066/99, Internal Reports, Area Ingeniería de Rocas, Codelco Chile División El Teniente, Chile, April and October 1999.

Cavieres, P. Evolución De Los Métodos De Explotación En Mina El Teniente, Internal Report, Area Ingeniería De Rocas, Codelco Chile División El Teniente, Chile, July 1999.

Cavieres, P., and E. Rojas. Experiencia En La Explotación De Roca Primaria En Mina El Teniente, Pl-I-050/99, Internal Report, Area Ingeniería De Rocas, Codelco Chile División El Teniente, Chile, July 1999.

Dunlop, R. Riesgo Sísmico en márgenes del Proyecto Esmeralda, PL-942/96, Internal Report, Area Ingeniería de Rocas, Codelco Chile División El Teniente, Chile, December 1996.

Jofre, J. Resumen Geomecánico Estudios Proyecto Esmeralda1994–1998, Pl-199/99, Internal Report, Area ingeniería de Rocas, Codelco Chile División El Teniente, Chile, April 1999.

Jofré, J., and J. Blondel. Geomecánica Conceptual Proyecto Esmeralda, EM-21/93, Internal Report, Departamento de Estudios y Métodos Operacionales, Subgerencia Mina, Codelco Chile Division El Teniente, Chile, June 1993.

Jofré, J., R. Dunlop, S. Gaete, and A. Karzulovic. Análisis Geomecánico Ingeniería Básica Proyecto Esmeralda, EM-17/94, Internal Report, Subgerencia Mina, Codelco Chile División El Teniente, Chile, August 1994.

Karzulovic, A. Caracterizacion Geomecanica Rocas Proyecto Esmeralda, Estudio DT-PE-97-001, A. Karzulovic & Asoc, Chile, January 1997.

Karzulovic, A. Distancias Permisibles Panel Caving Con Socavacion Avanzada Proyecto Reservas Norte, Estudio DT-CG-99-001, A. Karzulovic & Asoc, Chile, January 1997.

Karzulovic, A. Evaluacion Geotecnica Metodos De Socavacion Previa y Avanzada Mina El Teniente, Estudio DT-CC-98-003, A. Karzulovic & Asoc, Chile, December 1998.

Karzulovic, A. Evaluación Geotécnica Métodos de Socavación Previa y Avanzada Mina El Teniente, DT-CG-98-003, Karzulovic & Asoc, Chile, December 1998.

Karzulovic, A., P. Cavieres, and C. Pardo. Subsidencia Por Efecto Del Caving En Mina El Teniente, NI-Pl-410/99, Internal Report, Area Ingeniería de Rocas, Codelco Chile División El Teniente, Estudio DT-IG-99-003, A. Karzulovic & Asoc., Chile, August 1999.

Laubscher, D. Personal communication, Visit to El Teniente Division, July 1995.

Morales, A., and J. Seguel. Actualización De La Geología, Dominios Estructurales Y Geotécnia Al Norte De La Coordenada 300 N, Nivel De Socavacion Mina Esmeralda, Gl-216/98, Internal Report, Superintendencia Geología, Codelco Chile División El Teniente, Chile, November 1999.

Morales, A., and J. Seguel. Informe Caracterización De Estructuras Geológicas Mayores, Nivel De Produccion Mina Esmeralda, Gl-118/99, Internal Report, Superintendencia de Geología, Codelco Chile División El Teniente, Chile, November 1999.

Pereira, J, and A. Zamora. Estudio Granulométrico Mina Esmeralda, Gl-250/99, Internal Report, Superintendencia Geología, Codelco Chile División El Teniente, Chile, November 1999.

Rojas, E., and P. Cavieres. Hundimiento Avanzado: Una variante al método de explotación de Hundimiento por Paneles en Mina El Teniente, 44ª Convención Anual del Instituto de Ingenieros de Minas de Chile, Rancagua–Chile, 1993.

Rojas, E, M. Barraza, A. Bonani, A. Morales, R. Muñoz, O. Pastén, and M.A. Morales. Grupo Tarea Esmeralda Mediano y Largo Plazo Situación Sector Fw, Pl-I-005/2000, Internal Report, Codelco Chile División El Teniente, Chile, January 2000.

CHAPTER 52

Premier Diamond Mine

P.J. Bartlett*

52.1 INTRODUCTION

The Premier kimberlite pipe, situated 37 km east-northeast of Pretoria, is the largest known kimberlite in South Africa. It is one of 11 kimberlite diatremes found in the Cullinan-Raytonne area. Since mining operations started in 1902, a total of 326 million tonnes of ore have been mined yielding 113 million carats (22.6 tonnes) of diamonds for an average grade of 35 carats per hundred tonnes. Sixty-eight million tonnes of tailings material has been reprocessed since 1961, yielding an additional 15 million carats of diamonds. The tonnage mined on an annual basis has varied between 7 million and 2.4 million. The grades have varied between 29 and 60 carats per hundred tonnes.

Between 1902 and 1932, open-pit mining methods were used to extract ore down to a maximum depth of 189 m. The mine shut down in 1932, but in 1945, dewatering of the open pit started, and two shafts were sunk to allow the commencement of underground mining. Sublevel open-bench mining methods were used. Cave mining using scrapers was initiated in the early 1970s. Sublevel open stoping was attempted in the early 1980s, but was changed to cave mining using load-haul-dump (LHD) machines in the early 1990s.

Premier Mine currently produces 2.6 million tonnes per annum from two caves, one situated at a depth of 630 m below the surface and the other at 732 m below the surface. The mine employs 1,523 people.

The mine was graded by the National Occupational Safety Association (NOSA) during December 1999 and maintained its five-star rating for the 16th year in succession. The mine achieved 1,000,000 man-hours worked without a lost-time injury in February 1999.

52.2 GEOLOGY AND HYDROLOGY

The Premier kimberlite ore body is situated within the stable 3-billion-year-old Kaap-Vaal Craton. The pipe was intruded through 300 m of 1,700-million-year-old sediments of the Waterberg Supergroup (now eroded away) and an unknown thickness of underlying sediments of the Transvaal Supergroup (Raytonne Formation). The latter were originally laid down in a shallow marine environment 2,200 million years ago. The Transvaal Supergroup sediments were intruded 2,000 million years ago by a 350-m-thick felsite sill and 2- to 400-m-thick norite sills. The norite sills have been correlated with the lower part of the critical zone of the Bushveld Complex. The intrusion of the norite sills has resulted in varying degrees of metamorphism of the sediments into which they were intruded.

The Premier pipe had an elongated, kidney-shaped exposure on the surface with the east-west axis 880 m long and the north-south axis 450 m long at maximum. The pipe has a surface area of 32 hectares, decreasing progressively to 22 hectares 500 m below the surface and 12 hectares 1,000 m below the surface. The

geology is typical of the diatreme zone of a kimberlite pipe and remains consistent down to a depth of at least 1,000 m with little evidence of the geological complexities that characterise a root zone. Geologically, the region has been subjected to little tectonic activity since the intrusion of the pipe, with neither folding nor faulting affecting the ore body. It is estimated that 300 m of the original pipe have been eroded since intrusion 1,200 million years ago. The large volume and shape of the ore body allows low-cost, massive mining methods to be used in the exploitation of the mineral resource.

Three main kimberlite facies exist within the pipe.

1. The Brown kimberlite represents the first phase of intrusion and occurs on the eastern side of the pipe where it has a crescent-shaped outline in plan view. On the surface, the facies makes up 16% of the pipe area, increasing to 28% on the 1,000-m level. The Brown kimberlite has an average mining grade of 76 carats per hundred tonnes and disintegrates rapidly when exposed to water or moisture as a result of a high montmorrilonite clay content. Decomposition of the rock is accompanied by large volume and pressure changes that affect tunnel stability.

2. The Grey kimberlite represents the second phase of intrusion and occurs on the western side of the pipe. On the surface, the facies makes up 65% of the pipe area, decreasing to 57% on the 1,000-m level. The facies contains lesser amounts of montmorillonite clay and has a mining grade of 50 carats per hundred tonnes. It is further characterised by large rafts of country rock sediments that subsided into the diatreme at the time of intrusion. These rafts can be tens of metres in dimension and are found up to 1,000 m below their original stratigraphic horizon. The rafts dilute grade, but fortunately decrease in abundance with increasing depth. Both the Grey and Brown kimberlites are petrographically typical tuffisitic kimberlite breccias.

3. The Black kimberlite plug is made up of several minor facies of hypabyssal kimberlite intruded into and completely surrounded by the Grey kimberlite. The facies is intruded in turn by a number of low-grade, carbonate-rich dykes that extend into the surrounding kimberlites. The Black kimberlite makes up 15% of the ore body both on the surface and at a depth of 1000 m. It has a mining grade of 55 carats per hundred tonnes and is the most competent of the kimberlites, with a negligible clay content.

The pipe has an unusual diamond population with rare, very high-value, and occasionally very large, Type 2a and blue boron-bearing Type 2b diamonds. These are found only on the western side of the pipe. The mine has produced as many as 25% of the

---

* Priemier Diamond Mine, South Africa.

425

TABLE 52.1   Rock mass properties

Rock type	Tensile strength, MPa	Uniaxial compressive strength, MPa	Young's modulus, GPa	Poisson's ratio	Specific gravity	Rock mass rating
Brown kimberlite	7.9	50–80	16	0.27	2.67	45–55
Grey kimberlite	6.7	80–130	29	0.17	2.67	50–60
Black kimberlite	13.7	73–193	34–88	0.2–0.4	2.8	55–72
Gabbro sill	24.0	180–400	119	0.33	2.81	65
Felsite	16.8	240–300	62	0.29	2.6	72
Norite	16.8	140–220	74	0.25	2.8	45–60
Metasediments	5	60–240	30–167	0.15–0.33	2.5–2.8	40–65

world's large high-value stones of a size greater than 200 carats. These include the famous Cullinan diamond found in 1905 with a size of 3,106 carats. Type 2 diamonds make up less than 1% of the diamond population. Most of the diamonds are typical nitrogen-bearing Type 1 diamonds.

The Premier pipe is unique in that it has been intersected at a depth of between 350 and 525 m below the current surface by a dipping, 75-m-thick gabbro sill intruded 50 million years after the intrusion of the pipe. Within the confines of the pipe, the sill consists of 55 million tonnes of barren rock dipping from north to south at 30°. The sill has metamorphosed the kimberlite above and below it to a depth of as much as 15 m. Metamorphism has produced very hard, competent metakimberlite with an increased density and destroyed diamonds immediately adjacent to the sill. Mining below the sill creates complex mining challenges and has resulted in considerable waste dilution in the caves that operate below the now-collapsed sill.

Water has the potential to create major mining problems, as the ore decomposes when wet. Fortunately, the igneous rocks and sediments that surround the pipe have low pemeability and transmissivity. Such water as finds its way into the ore body moves toward the pipe only through joints and fractures. Nearly 100 years of mining have created a large drawdown cone around the pipe. Such water as does reach the pipe contact often moves along this shear zone, creating mining problems. Annual rainfall, mainly in the summer months, averages 700 mm. The 40-hectare catchment area of the open pit allows 28 million litres of water through to the underground workings each year. A negligible amount of water is pumped out of the mine in the dry months, rising to 4 ML/d in the wet season. The geology and layout of mining blocks are illustrated in Figure 52.1.

52.3   GEOPHYSICAL INFORMATION

The geophysical characteristics of the ore and surrounding rock are well correlated with the geology. The montmorillonite-rich Grey and Brown kimberlites decompose when wet and necessitate dry-drilling techniques and special support measures. The norite is well jointed, which results in key-block fallout and ore pass support problems. Table 52.1 gives the geophysical parameters of the rock mass.

There is a large variation in fracture frequency and jointing patterns within the various rock types and in the kimberlite. The norite, which forms the host rock to the pipe on the current mining levels, is well jointed with at least three near-vertical and one subhorizontal joint set. Fracture frequency varies between 0.5 and 2.5 joints per metre. Secondary mineralisation on joint planes, usually chlorite or serpentine, together with joints that extend over tens of metres in the vertical plane, create major problems for ore pass stability. In the gabbro sill, there are three well-defined joint sets—two subvertical and the other subhorizontal. The Grey and Brown kimberlites are poorly jointed commensurate with their gas-filled, low-temperature intrusive histories. Such joints as do exist are randomly oriented,

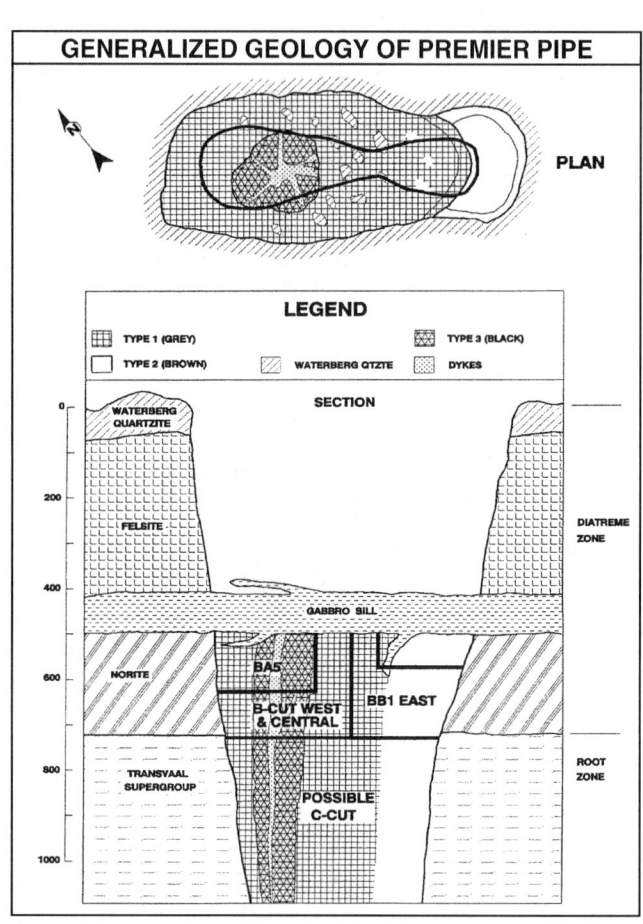

FIGURE 52.1   Diagrammatic plan and section of Premier pipe showing mining blocks

persistent over tens of metres, and often filled with calcite and other secondary minerals. The widely spaced jointing leads to coarse fragmentation in the caving process, but as the rock is soft, comminution is rapid. The Black kimberlite is well jointed and, although more competent, results in finer fragmentation in the caving process.

Tunnel support is often more expensive than tunnel development. Support is designed on the basis of experience, rock mass characterisation, and expected stress changes associated with cave mining as predicted by numercial stress modeling. Commonly used support elements include 1.8-m-long rock bolts and mesh-reinforced shotcrete. Final concrete and shotcrete linings are applied only after the abutment stresses associated with the undercut have abated.

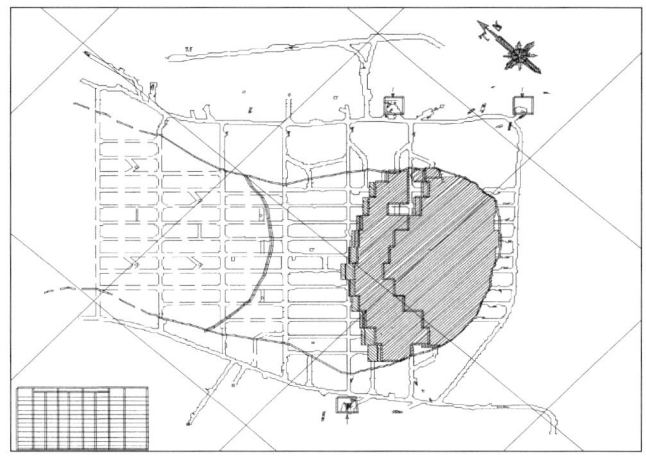

FIGURE 52.2   Plan of 717 BB1E undercut level

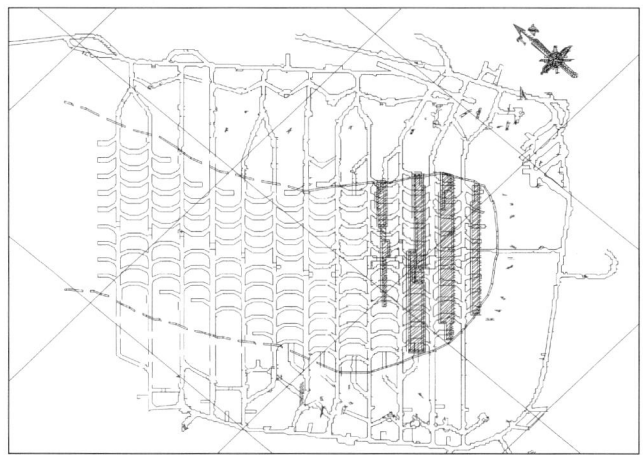

FIGURE 52.3   Plan of 732 BB1E extraction level

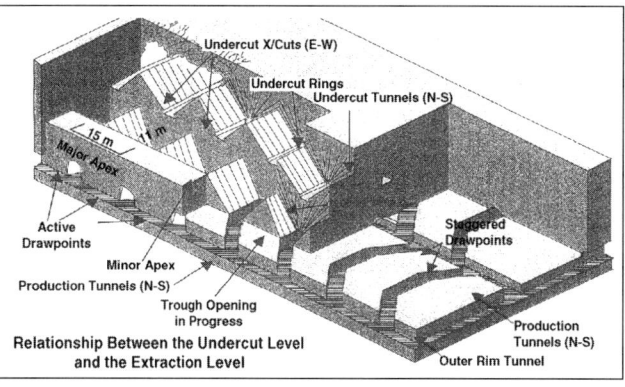

FIGURE 52.4   Isometric view of relationship between undercut and extraction level geometries

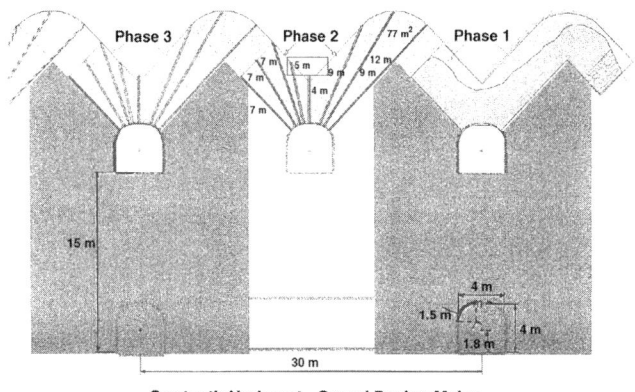

FIGURE 52.5   Diagrammatic section showing undercut drill ring design and relationship of undercut and extraction level

52.4   MINING METHOD

Panel retreat caving is used as the mining method in both mining blocks. The effects of high undercut abutment stresses with increasing depth of mining, substantial tunnel support, and coarse fragmentation resulting in a significant amount of secondary blasting are the major technical challenges associated with cave mining at Premier Mine.

An undercut level is developed to remove a slice of ore from the base of the block to be caved. The extraction level is developed 15 m below the undercut level to allow the caved ore to be extracted from the base of the cave through drawbells. LHDs are used to move the ore from the double-sided drawbells to ore passes sited in the country rock surrounding the pipe. Long tramming distances impact adversely on LHD productivity.

An advance undercut mining sequence is used. This means that as little development as possible is carried out on the extraction level prior to the undercut face passing overhead. Final development and support are completed only once destressing on the extraction level has occurred, and the undercut face with its associated high abutment stresses no longer affects excavations on the level below. Drawbells are developed into the base of the cave to access the broken, caved ore. The layout of the undercut and extraction levels in the BB1E mining block are illustrated in Figures 52.2 and 52.3, respectively. The undercut ring drill pattern and its relationship to the extraction level are shown in

Figures 52.4 and 52.5. Figure 52.5 shows that on the undercut level, the rings of adjacent tunnels are blasted successively. Cave mining parameters are summarized in Table 52.2.

The tunnels on the extraction level are spaced at 30-m intervals with drawpoints spaced at 15- and 18-m intervals. The ore column height above the drawpoints in the BA5 mining block varies between 80 and 140 m, depending on the height of the base of the overlying sill.

52.5   DRILLING AND BLASTING

Tunnels in both the ore and country rock range in size from 3.6 by 3.8 m to 4.2 by 4.2 m. Tunnel advance is effected with 1.8-m-long rounds using a electrohydraulic Bison or Tamrock single-boom rigs drilling 45-mm in diameter holes. Ring drilling on the undercut and extraction level is done with electrohydraulic Secoma-Mercury 14 and Tamrock LC22 drill rigs. Undercut rings comprise 64-mm in diameter holes drilled with a 1.5-m burden to a maximum toe spacing of 2 m and a maximum length of 15 m. In problem areas, bar-mounted S36 pneumatic drills are still used. The powder factor used for blasting in ring drilling is less than 350 gm/tonne.

Drawbell development is accomplished by creating a free face using a 1-m in diameter blind hole bored from the extraction level into the base of the caved ore above. The drawbell is then developed to its final dimensions of 15 by 13 m by ring drilling. Drill holes are usually predrilled, and electric delay detonators are used to accomplish blasting as timing of holes is important.

TABLE 52.2   Cave mining parameters

Parameter	BA5 mining block	BBIE mining block
Column height, m	80–140	148–164
Rock mass rating	45–65	45–55
Hydraulic radius	30	25
Mining sequence	Post and advance undercut	Advance undercut
Rate of undercutting, m²/month	900	1100
Tonnes in mining block	42 million	23 million
Tonnes per drawpoint	50,000 to 200,000	100,000 to 200,000
Drawpoint spacing, m by m	15 by 15	15 by 18
Distance across major apex, m	22.6	23
Average rate of draw, mm/d	270 (109 tonnes)	270 (120 tonnes)
Initial fragmentation, m³	30% > 2 m³	30% > 2 m³
Fragmentation after 20% drawn, m³	12% > 2 m³	7% > 2 m³
Drawpoint support	Cable anchors, rock bolts, mesh tendon straps, shotcrete	Cable anchors, rock bolts, mesh tendon straps, shotcrete
Brow wear, m	0 to 2 m wear after 50,000 tonnes drawn	1 to 3 m wear after 50,000 tonnes drawn
Tunnel size, m	4 by 4.2	4 by 4.2
LHD type: diesel and electric Toros	5 yd	5 and 3 yd
Tonnes per LHD per hour	118	131
LHD tramming distance, m	154	134
Hang-up frequency, percent per shift	30% of drawpoints	25% of drawpoints

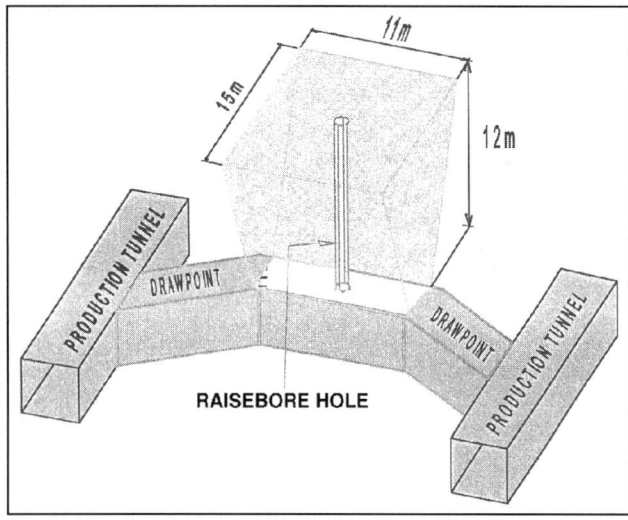

FIGURE 52.6   Isometric view of drawbell relative to production tunnel

Powder factors can be as high a 1,000 gm/tonne. Drawbell layout is illustrated in Figure 52.6.

Drill rings are usually pneumatically charged with Anfex and toe primed with Nonel detonator assemblies. Electric delay detonators are increasingly being used, especially for blasting complex drill patterns where timing delays are important. In wet conditions, emulsion cartridges are used instead of Anfex. All blasts are initiated electrically from a ring main system.

52.6   PRODUCTION AND GROUND HANDLING

Each production tunnel in the cave is worked by two LHDs operating from either end of the tunnel delivering ore to passes situated in the country rock outside the ore body. The LHD fleet includes both diesel and electrically powered units. At the ore passes, the LHDs tip ore onto grizzlies with openings measuring 660 by 660 mm. Each ore tip is equipped with an impact breaker. Passes vary in length from 30 to 130 m. All passes are concrete lined with 500 mm of concrete and are at least 3 m in diameter. Pass productivity averages 1,800 tonne/d, and passes have to be relined when they have handled between 500,000 and 1,000,000 tonnes of ore. Impact damage rather than abrasion is the primary cause of pass lining wear.

The bottom of the pass is equipped with a pneumatically controlled box front that allows the ore to be fed into the 6-tonne Granbys of an electric train that conveys the ore to centrally located jaw crushers situated at the north and south center of the pipe. Ore is crushed to a nominal ±150 mm before being conveyed via belt to the 522-m level, where it is fed into one of four 13-tonne skips that convey the ore to the surface.

The ore body has been sampled on two levels. The 50-tonne samples are taken from tunnels constructed in a grid system over the entire pipe area. Sampling involves the collection, transport, and treatment of thousands of tonnes of ore. Very large bulk samples are taken on an ad hoc basis for production and revenue control. Production-grade estimates are based on the sampling grades, production grades, and audits of production on a systematic basis. A draw control strategy is implemented that sets out the tonnage to be drawn from individual drawpoints on a daily basis to ensure that production targets are met and that individual drawpoints are not overdrawn, as this can lead to the entry of waste too early. Actual production from drawpoints is controlled by a vehicle monitoring system and the LHD drivers. Dense waste from the overlying gabbro sill creates problems in the recovery plant, and draw control officials inspect drawpoints regularly to determine their waste content.

52.7   MAINTENANCE

All underground vehicles are maintained in underground workshops situated on the 645- and 732-m levels. Maintenance is planned and controlled by a computer-based planning system. Key variables, such as safety and health, overtime worked, diesel and oil use, tire life, and average availability of vehicles maintained in the two workshops, are carefully monitored and controlled. The safety statistics of the maintenance section for 1999 and the availability and utilisation of vehicles maintained in the main 645-m-level workshop are plotted in Figures 52.7 and 52.8, respectively. The vehicles used in the mine are listed in Table 52.3. The productivity and efficiencies of the major vehicle-related mining activities are given in Table 52.4. The low level of use of vehicles is largely due to the fact that the mine works two 9-hr shifts during a 24-hr day.

52.8   SAFETY, HEALTH, AND ENVIRONMENT

In keeping with international trends, the Premier Mine is in the process of integrating its Health and Safety, Ventilation, and Occupational Hygiene and Environmental Departments to form a Department of Safety, Health, and Environment. This will allow for better utilisation of available resources and provide a more balanced service to employees.

With the promulgation of the South African Mine Health and Safety Act, the focus has expanded to include occupational diseases. This has highlighted the fact that there are a number of employees who suffer from occupational diseases, a legacy of

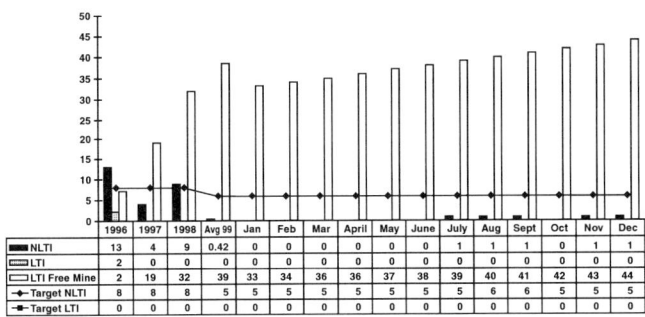

FIGURE 52.7   Trackless health and safety statistics

	1996	1997	1998	Avg 99	Jan	Feb	Mar	April	May	June	July	Aug	Sept	Oct	Nov	Dec
▬NLTI	13	4	9	0.42	0	0	0	0	0	0	1	1	1	0	1	1
▭LTI	2	0	0	0	0	0	0	0	0	0	0	0	0	0	0	0
▭LTI Free Mine	2	19	32	39	33	34	36	36	37	38	39	40	41	42	43	44
◆Target NLTI	8	8	8	5	5	5	5	5	5	5	5	6	6	5	5	5
▪Target LTI	0	0	0	0	0	0	0	0	0	0	0	0	0	0	0	0

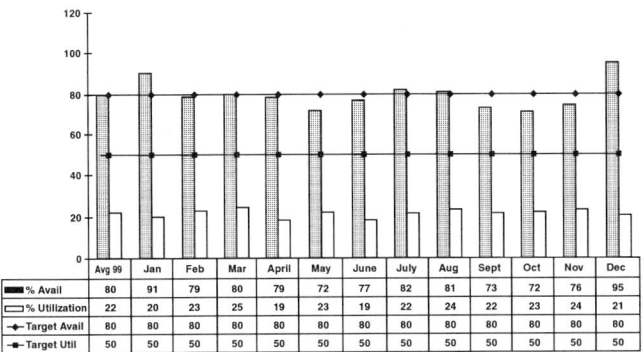

FIGURE 52.8   Vehicle availability

	Avg 99	Jan	Feb	Mar	April	May	June	July	Aug	Sept	Oct	Nov	Dec
▬% Avail	80	91	79	80	79	72	77	82	81	73	72	76	95
▭% Utilization	22	20	23	25	19	23	19	22	24	22	23	24	21
◆Target Avail	80	80	80	80	80	80	80	80	80	80	80	80	80
▪Target Util	50	50	50	50	50	50	50	50	50	50	50	50	50

TABLE 52.3   Vehicle fleet

Type of vehicle	1999	1998
LHD		
5 yd	11	14
3 yd	12	12
2 yd	1	1
Drill rigs		
Development	5	7
Longhole	4	4
Secondary blasting	6	6
Dump trucks	3	4
Light delivery vehicles	32	32
Utility vehicles	45	41
Surface vehicles	61	66
Concrete mixers	4	7
Total	184	194

earlier mining days at Premier and work histories on other mines. This phenomenon is aggravated by the fact that the average age of employees is relatively old, namely 47 years.

A brief review of health and safety achievements shows that—

- Premier has been instrumental in designing and implementing a fully electronic medical surveillance system. This system electronically links the risk exposure of every employee and this information is then stored for 40 years.

- The final audit will take place during the first half of this year for ISO14001 accreditation.

TABLE 52.4   Productivity and efficiencies

Parameter	Efficiency
5-yd LHDs, tonne/hr (average)	123
Tires, hr (average)	1,220
Availability, %	92
Use, %	52
Tamrock longhole rig, m per shift	114
Mercury-Secoma longhole rig, m per shift	54
S36 bar-mounted pneumatic longhole rig, m	59
Tonnes per manshift	18.11

TABLE 52.5   Safety statistics

	1999	1998
Lost-time injuries	9	11
Lost-time injury rate	0.47	0.62
Minor injuries	134	162

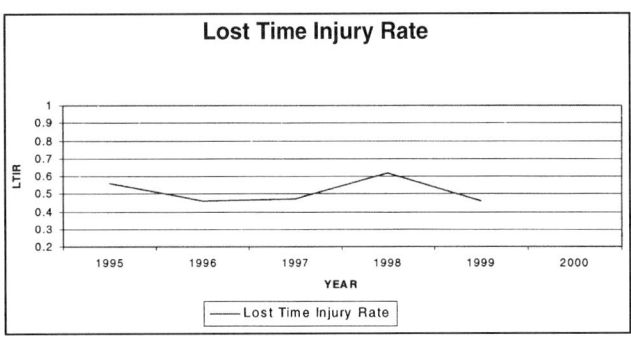

FIGURE 52.9   Summary of lost-time injuries, 1995–2000

TABLE 52.6   Fixed and variable costs

Cost element	Fixed costs, %	Variable costs, %
Labour	99	1
Stores	52	48
General expenses	36	62
Working cost expenses	49	51
Vehicles	34	66
Total	76	24

- For the second year running, the Premier Mine has been judged as the best NOSA five-star mine in the category of 500 or more employees. This is for the mines in the Northern Province of South Africa.

A summary of lost-time injuries is given below in Table 52.5 and Figure 52.9. Note that a lost-time injury is any injury that results in the injured person being unable to perform his or her normal duties on the day following the accident.

The lost-time injury rate is calculated by—

(Number of lost-time injuries × 200,000) ÷ Total number of man-hours worked.

52.9   COSTS

Variable and fixed costs for the major cost items associated with mining production are tabulated in Table 52.6. Capital costs vary widely on an annual basis as a function of projects undertaken.

TABLE 52.7 Mining production costs

Cost element	Cost, US$ (2000)	Percentage of total cost, %
Ground handling		
Labour	1.1	
Materials	0.1	
General expenses	0.1	
Working cost, miscellaneous	0.2	
Total	1.7	28
Mining production		
Labour	1.0	
Materials	0.2	
General expenses	0.2	
Working cost, miscellaneous	0.1	
Vehicle maintenance	0.4	
Total	2.0	33
Supervision and human resources	0.8	13
Control and instrumentation	0.2	3.3
Services	1.3	22
Total mining cost	6.0	100

Mining production costs (expressed in year 2000 US dollars) are set out in Table 52.7. Services include geology, survey, planning, salvage, shaft maintenance, and rock drill shop costs.

52.10 FUTURE PLANS

Premier Mine has shown that modern caving methods using relatively widely spaced drawpoints, an advance undercut mining sequence, and LHDs for ore extraction can be used to mine kimberlite ore in a cost-effective manner. The challenge posed by having to mine below the sill has been addressed and the mine has been able to meet tonnage and carat targets within budgeted cost in recent years. This has given the Premier Mine and the De Beers Group the confidence to embark on an expansion programme that could result in the establishment of a large cave at a mining depth in excess of 1,000 m, which could extend the life of operations by 17 or more years.

The mine is currently engaged in an evaluation of the resource inferred to exist in the pipe below current mining levels. Twelve thousand metres of large-diameter drilling (318 mm) is being used to sample a resource of at least 120 million tonnes of kimberlite ore. A further 17,000 m of core drilling and fine-diamond sampling have been undertaken to delineate the geological model. The evaluation programme is planned to define the mineral resource at the confidence level of at least an indicated resource.

The feasibility study being carried out in parallel with the evaluation programme suggests that it should be possible to mine the resource as a block cave at a rate of some 9 million tonnes per annum. The column height of the proposed cave would average 350 m. Total re-engineering of the mine, including development of new shafts and ground handling systems, could lower production costs considerably as a result of the economics of scale and by the judicious use of automation. A new treatment facility will be built.

The successful completion of the evaluation programme and feasibility study could lead to approval of the project by May 2001. Sinking a service shaft to provide rapid access to the proposed production level and ore body could start soon thereafter. Production from the 1082-m level could start in 2005, building up to 9 million tonnes per year by 2009. Re-engineering of the mine and treatment facility will aim at reducing working costs and exploiting the economies of scale as production increases from a current 3 million to 9 million tonnes per year. The skill levels of employees will be enhanced by training to meet the demands of a high-technology mine. The organisational design of the staffing structure will aim at a maximum of four levels of work. The total labour cost will be considerably reduced from the current high levels.

52.11 ACKNOWLEDGMENTS

The author would like to thank the general manager of Premier Mine and the De Beers Consolidated Mines geotechnical engineer for permission to publish this paper. The work done by colleagues in preparing diagrams and providing details of the operations as set out in the paper is gratefully acknowledged.

52.12 REFERENCES

Bartlett, P.J. 1992. The Design and Operation of a Mechanised Cave at Premier Diamond Mine. MASSMIN 92 SAIMM Publication Symposium Series S12.

Bartlett, P.J. 1994. Geology of the Premier Diamond Pipe. Twenty-Fifth CMMI Congress, Johannesburg, SA, SAIMM, H.W. Glen, ed. Vol. 3, pp. 201–213.

Bartlett, P.J. 1992. Support in a Mechanised Cave at Premier Mine. MASSMIN 92. SAIMM Publication Symposium Series S12.

Kirsten, H.A.D., and P.J. Bartlett. 1992. Rigorously Determined Support Characteristics and Support-Design Method for Tunnels Subject to Squeezing Conditions. SAIMM, Vol. 92, No. 7.

. . . .
CHAPTER 53

Block Caving the EESS Deposit at P.T. Freeport Indonesia

John Barber,* Suyono Dirdjosuwondo,* Tim Casten,* and Leon Thomas*

53.1 INTRODUCTION

PT Freeport Indonesia operates a copper and gold mining complex in the Ertsberg Mining District in the province of Irian Jaya, Indonesia (Figure 53.1). The Ertsberg district is located in the Sudirman Mountains at elevations from 3,000 to 4,500 m above sea level. The topography is extremely rugged. Rainfall in the mine area averages 3,000 mm per year.

Freeport began production in the district in 1972 when the mill began processing ore from the Ertsberg open pit. Underground mining began in 1980 when the GBT (Gunung Bijih Timur–Ertsberg East) was brought on line using block caving methods. The GBT reached a maximum production rate of 28,000 tonne/d in 1991 and was exhausted in 1994.

The Intermediate Ore Zone (IOZ) was brought into production in 1994, also using block caving methods, with a design rate of 10,000 tonne/d. The IOZ is currently producing at a rate of 18,500 tonne/d.

The Deep Ore Zone (DOZ) was discovered in the mid-1980s by deep drilling from the GBT. Portions of the DOZ were mined using open stoping methods from 1989 to 1992. In 1993, the first of several studies was completed indicating that portions of the DOZ could be successfully and economically mined using block caving methods. The DOZ is currently being developed and is scheduled to begin production in the second half of 2000. The DOZ is planned to produce 25,000 tonne/d of ore. A study is currently underway to determine if an ultimate production rate of 35,000 tonne/d is feasible.

The GBT, IOZ, and DOZ mines are stacked vertically on the Ertsberg East Skarn System (EESS). The EESS is open to depth and along strike (Figure 53.2). The DOZ ore body is situated in the lower portion of the EESS.

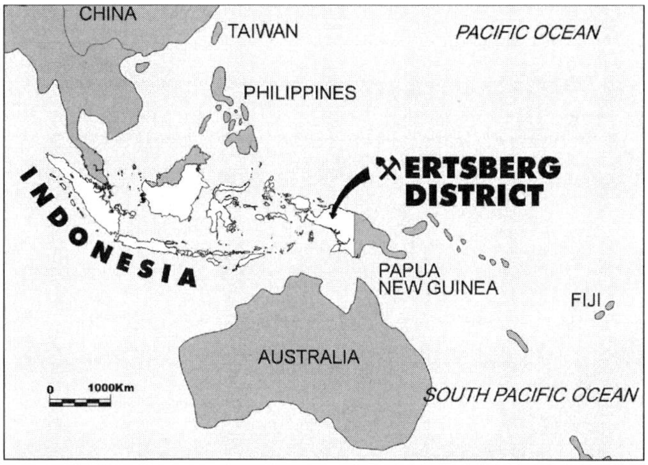

53.2 GEOLOGY AND ORE RESERVES

53.2.1 Geology

The EESS is hosted by Tertiary-age carbonates that have been altered to calcium-magnesium silicate skarn. The EESS is an essentially vertical tabular body with a vertical extent in excess of 1,200 m, a strike length of over 1,000 m, and an average width of 200 m. The northeast (hanging wall) contact of the EESS is a skarn reaction front in sudden contact with barren marble. This contact coincides with a zone of localized faulting and brecciation. The EESS is bounded to the southwest (footwall) by the Ertsberg Diorite intrusive.

The GBT ore bodies are dominated by calcium-magnesium skarn, such as monticellite and garnet. The GBT copper skarn ore bodies include magnetite and retrograde alteration, such as chlorite.

The IOZ ore bodies are dominated by magnesium-calcium skarn, such as forsterite and diopside. The IOZ ore bodies include

FIGURE 53.1 Ertsberg Mining District location map

magnetite and retrograde alteration, such as talc, serpentine, tremolite-actinolite, and chlorite.

Moving across the strike of the EESS from the footwall to the hanging wall, the specific rock units encountered in the EESS are—

- *Ertsberg diorite.* Generally a hard, competent rock unit with good ground conditions. Proximal to the skarn contact, the diorite has been locally altered and mineralized.

- *Forsterite skarn.* A massive unit adjacent to the Ertsberg Diorite contact, averaging 0.5% copper. Generally a hard, competent rock unit with good ground conditions.

- *Magnetite-forsterite skarn.* Grades vary between 0.5% to 2.0% copper. Often finely bedded. Generally a hard, competent rock unit with good ground conditions, but with localized zones exhibiting poor ground conditions

- *Massive magnetite.* Occurs mainly along the marble contact. Often strongly bedded. High-grade ore with grades ranging from 2% to 10% copper. Generally a hard, competent rock unit with good ground conditions, but with localized zones exhibiting poor ground conditions.

- *BAS breccia.* A contact breccia in the GBT and IOZ deposits that tends toward high copper grades (>3%) and very poor ground conditions.

- *DOZ breccia.* A lenticular zone that plunges westerly across the lower half of the DOZ, cross cutting all other

---

* PT Freeport Indonesia.

431

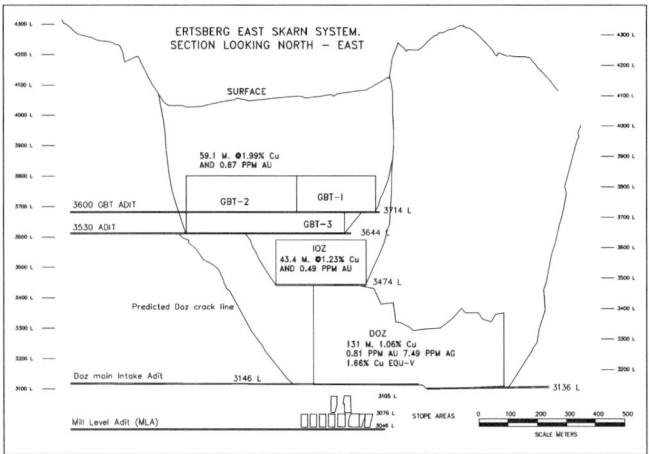

FIGURE 53.2   Longitudinal section of EESS

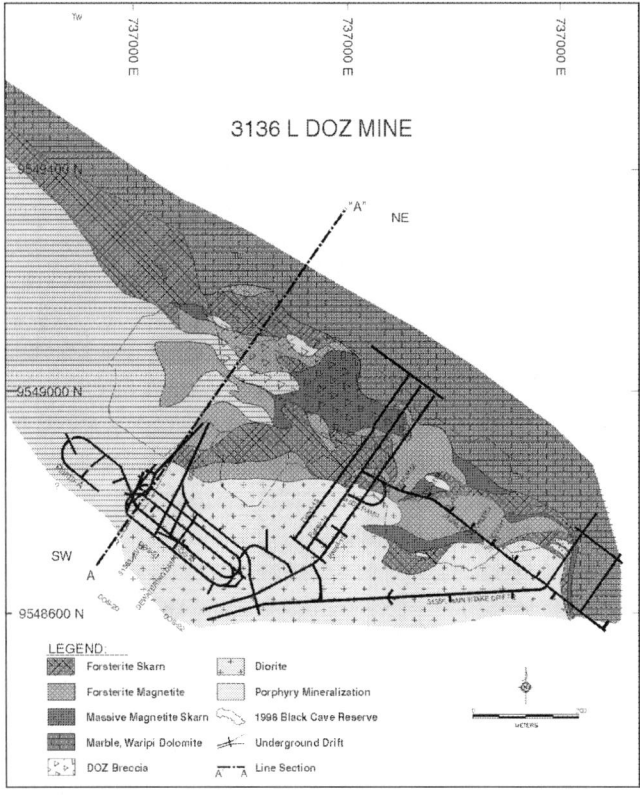

FIGURE 53.3   Geological plan of DOZ extraction level

units. Ore grades tend to be >2% copper and locally are greater than 4% copper. Almost without exception, ground conditions in this unit are very poor with a history of failure.

■ *Dolomite-marble.* Alteration extends 250 to 300 m from the skarn into the hanging wall and is generally barren of mineralization. Rock quality and ground conditions are highly variable and locally very poor proximal to the skarn-marble contact. Ground conditions in the marble contact zone deteriorate with depth.

Figure 53.3 shows a geological plan of the DOZ extraction level at 3,136 m.

53.2.2   Ore Reserves

Ore reserves on the EESS have increased since 1980. This is partially due to increased geologic information and partially to economies of scale made possible by the increased throughput of the mill. Total GBT production was 59.1 million tonnes at 1.99% copper and 0.67 ppm gold.

IOZ reserves (diluted) in 1994 were 21 million tonnes at 1.55% copper and 0.54 ppm gold. These were calculated at a cut-off grade of 0.98% copper. The IOZ production rate was calculated at 10,000 tonne/d to feed the concentrator operating at 60,000 tonne/d. During the fourth concentrator expansion project, economies of scale in the concentrator, open pit mines, and underground mining allowed P.T. Freeport Indonesia to reduce the cut-off grade at the IOZ to 0.90% copper equivalent. The copper equivalent calculations take into account copper grade, gold grade, silver grade, differences in metallurgical recovery, royalties, treatment charges, royalty charges, and overhead costs. The new expanded IOZ reserves were increased to 43.55 million tonnes at 1.23% copper, 0.49 ppm gold, and 7.7 ppm silver at a cut-off grade of 0.90% copper equivalent. Production was increased from 10,000 to 18,000 tonne/d. The majority of the increased IOZ reserves are located in the footwall of the deposit in the forsterite skarn and mineralized diorite.

Current reserves (diluted) for the DOZ block cave ore body are 131 million tonnes grading 1.06% copper, 0.81 gm/tonne gold, and 7.49 gm/tonne silver, or 1.66% copper equivalent, utilizing a cut-off grade of 0.90% copper equivalent.

53.2.3   Hydrogeology

The underground mines have a history of wet muck runs in the production areas. The problem first became serious in the IOZ production area. Wet muck runs as large as 2,000 m³ have occurred from individual drawpoints. These runs are a serious safety hazard and impediment to production.

Water enters the cave through rain, surface drainage, and groundwater. The water moves down through the cave and creates saturated conditions in the drawbells, which leads to the wet muck runs.

Two main sources of groundwater that must be intercepted have been identified.

1. Hanging wall water is impounded behind a sandstone aquaclude (Sirga Sandstone). As the cave cracks intercept the sandstone, the impounded water flows into the cave. The impoundment is continually recharged by rainwater from the surface (Figure 53.4).

2. Long-strike water flows into the cave along contacts and relict bedding in the skarn formations and along the skarn-diorite contacts. Some of the long strike water is perched water and some is recharge from the surface.

This situation is complicated by a series of high-angle cross-strike faults that provide some communication between the water sources and the cave. The model is further complicated by a number of karst features in the hanging wall.

Dewatering strategy is based on draining impounded water from behind the Sirga Sandstone so that the phreatic surface falls below the elevation at which the cave cracks intercept the sandstone. This will be accomplished by means of a drill gallery developed in the hanging wall outside the predicted ultimate crack line of the DOZ cave. Diamond drilling from the gallery will pierce the Sirga, allowing the impounded water to drain off through the drill holes. Continuous flows from the Sirga drainage drilling are predicted to be on the order of 315 L/s.

Long-strike water will be intercepted outside the predicted ultimate DOZ cave crack system by drilling from development in the hanging wall and footwall parallel to strike. The long-strike development will be on a number of different levels, utilizing existing workings as jumping off points. This method has been

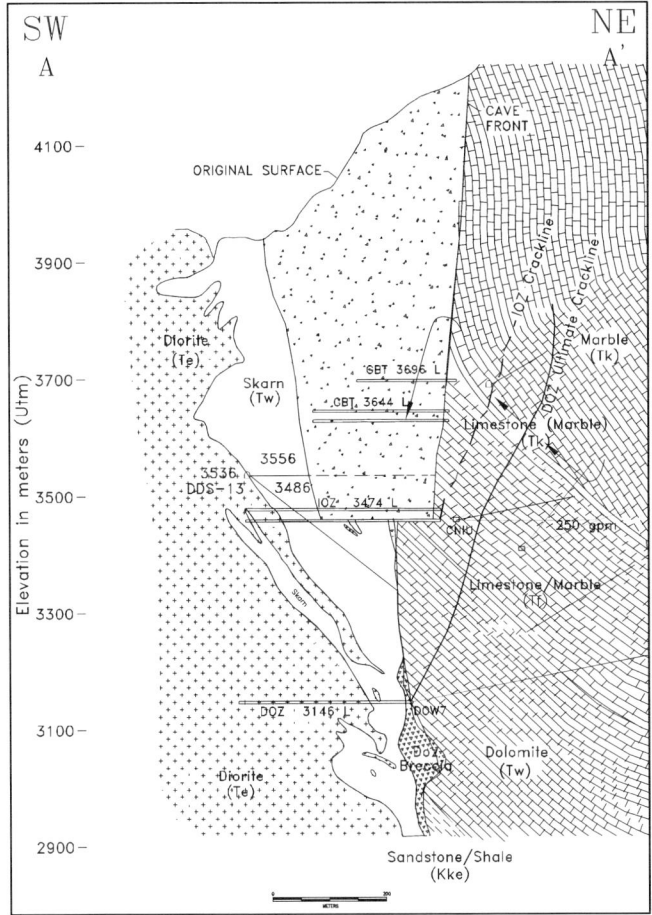

FIGURE 53.4   Geological section of EESS

successfully used in other areas of the EESS. Continuous flows from the long-strike drilling are also predicted to be on the order of 315 L/s.

53.2.4   Geophysical

Geophysical engineering and planning are based on diamond-drill holes, and on geological, scanline, and cell mapping of the workings. Until 1994, most drill holes were logged for rock quality designation (RQD) only; since then all drill holes have been logged for fracture frequency and rock mass rating (RMR) analysis.

Almost without exception, ground conditions in the EESS vary from very poor to very good as you proceed from hanging wall to footwall. Uniaxial compressive strength varies from a high of 219 MPa in some massive magnetite to less than 10 MPa in the DOZ breccia. RMR varies from a low of 25 in the poorest areas to a high of 65 in the most competent ground.

Cavability of these deposits was evaluated using a combination of RQD, RMR, and mining rock mass rating (MRMR) data. All cavability and fragmentation studies have been reviewed by outside geotechnical consultants. The hydraulic radius needed to sustain the cave ranges from a low of 10 m in the areas of poor ground to 30 m in the forsterite skarn. Footprints in the IOZ and DOZ have been in excess of the required hydraulic radius, so cave propagation has not been a problem in IOZ and will not be a problem in the DOZ.

Fragment size at the drawpoints has never been a problem in the IOZ until recently. Draw heights in the IOZ expanded reserve area are now reaching the elevation of the GBT mine. This is the footwall area of the GBT, where most of the

permanent facilities (shops, lunchroom, supply room, service facilities) were located.

As draw heights reached the GBT elevations, particle size at the IOZ drawpoints suddenly and unexpectedly increased. We think that as the cave front passes through the large pillars that protected the GBT facilities, the rock is not fragmenting on geological and geophysical controls, but rather the pillars are detaching from the intact ground and entering the cave as much larger masses. Comminution in the cave is not enough to reduce the large masses to manageable size before they reach the IOZ drawpoints.

For the DOZ mine, fragmentation studies predict that less than 30% of the ore body will exceed 1 m. Comminution within the draw column will further reduce the material size reporting to the drawpoints. We will apply the lessons learned in the IOZ to ensure that the IOZ pillars over the DOZ draw columns will be weakened or destroyed before abandoning the IOZ workings.

53.3   MINING METHOD

The mining method chosen for the IOZ, and later the DOZ mine, is mechanized block caving. The basic method is similar to the method successfully developed at the GBT mine.

The initial mining method used at the GBT, Area I, was block caving using slusher panels and rail haulage. Management soon realized that fragmentation was too coarse for this method to be used efficiently. An experimental panel using load-haul-dump (LHD) machines was opened. This proved so successful that all mucking since has been with LHDs.

Development of GBT Area II was started soon after Area I was put into production. Area II was designed using all LHD production and conveyor haulage. This was successful, and the GBT Area II layout and haulage system was used at the IOZ mine.

53.3.1   General Description, IOZ Mine

The IOZ mine footprint, expanded reserve, is 350 m long and 250 m across strike. Maximum draw height is 200 m. The draw columns have intercepted the production levels of GBT Area I and II.

Undercutting began in the northwest corner of the deposit, taking advantage of the weak, high-grade BAS. Cave propagation was sustained after a hydraulic radius of 10 m was achieved.

53.3.2   Extraction Level Design, IOZ

The extraction level of the IOZ mine is shown in Figure 53.5. Panels are transverse to the strike of the ore body on 30-m centers. Drawpoints are on 17.3-m centers. All mucking is done to the grizzly and rock breaker stations on the north and south fringe of the ore body (Figure 53.6).

Drawpoint Layout. Drawpoint layout has changed since the IOZ was first developed. The initial layout was the "El Teniente"-style layout. This was easy during the development of the mine, but created difficulties when the area entered production. The primary objection to the El Teniente layout is that mucker operators are forced to choose between a long (200-m) tram with the bucket behind the LHD and a shorter tram with loaded bucket in front of the operator. At times this has created difficulties with control of the draw.

A second objection to this layout is that a worker in one drawpoint is directly exposed to unanticipated muck slides from the drawpoint directly across the panel. This problem did not become apparent until the IOZ began experiencing severe wet muck problems.

We changed the drawpoint layout to an offset herringbone when the additional reserves were added. This eliminated the need to tram a loaded bucket ahead and provides a degree of protection to workers in a drawpoint from unanticipated slides in opposite drawpoints.

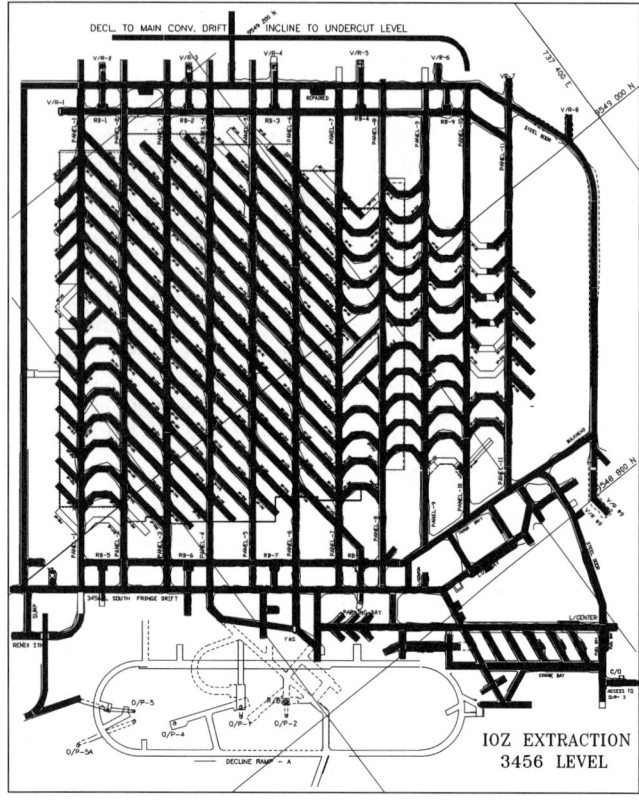

FIGURE 53.5   IOZ extraction level

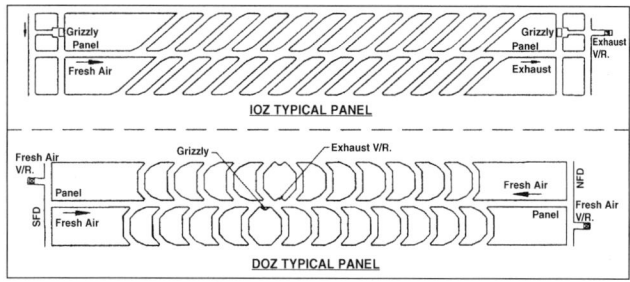

FIGURE 53.6   Typical panel layouts, IOZ and DOZ mines

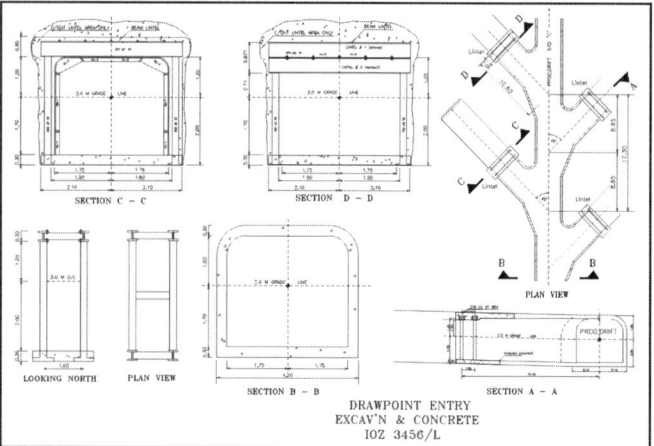

FIGURE 53.7   Concrete support in IOZ mine

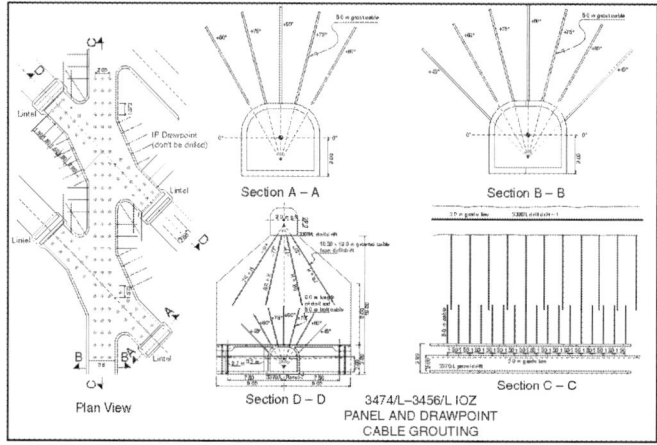

FIGURE 53.8   Cable bolt support in IOZ mine

Drawpoint Support. Drawpoint and panel support is a combination of mass concrete and cable bolts (Figures 53.7 and 53.8). The IOZ support system was designed using empirical methods based on support used at the GBT. This method has been moderately successful. The success of such a system is highly dependent on the quality of the concrete used, which, in turn, requires that full-time quality assurance and control people be assigned to the work.

53.3.3   Panel Ventilation, IOZ

At the GBT and IOZ mines, fresh air is delivered at the south fringe and exhausted from the north fringe. This is illustrated in Figure 53.6. The older method means that anyone working on the north side of the mine is always working in exhaust air, and that it is almost impossible to provide enough ventilation to operate two muckers in a single panel. This method was acceptable when using slushers and rail haulage, but was found to be inadequate when using diesel-powered LHDs.

53.3.4   Haulage System, IOZ

The haulage system used at GBT and IOZ consists of a pocket under each grizzly on the north and south fringes of the ore body (Figure 53.6). Grizzly size is usually set at 500 mm, but the grizzly panels can be changed if desired. Pocket capacity is 400 tonnes.

Each pocket discharges on to a 60-in coarse ore conveyor, which dumps into a 1,067- by 1,219-mm jaw crusher. There is a crusher on the north side and one on the south side of the ore body (Figure 53.9). Both crushers discharge on to a 48-in conveyor that delivers the ore to an ore pass in the footwall diorite. At the bottom of the ore pass, a hydraulic apron feeder directs ore onto a system of conveyor belts that deliver the ore to the surface and the mill stockpiles. Figure 53.10 is a cross section of the IOZ mine and clearly shows the relationship of the conveyor system to the extraction level.

53.3.5   Undercutting, IOZ

The undercut level is 18 m above the extraction level. Drill drifts are situated directly over the panel drifts.

After the drill drifts are developed, the major apex pillar is supported with cable bolts. Figure 53.8 shows the bolt pattern used. As with the panel concrete, this method requires a full-time quality control presence while the holes are drilled and the cables installed and grouted.

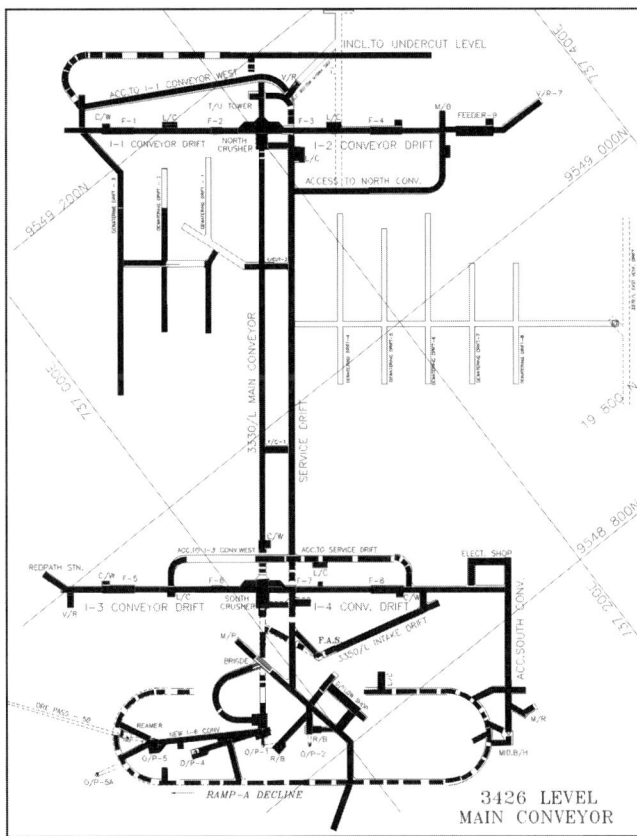

FIGURE 53.9 Conveyor level, IOZ mine

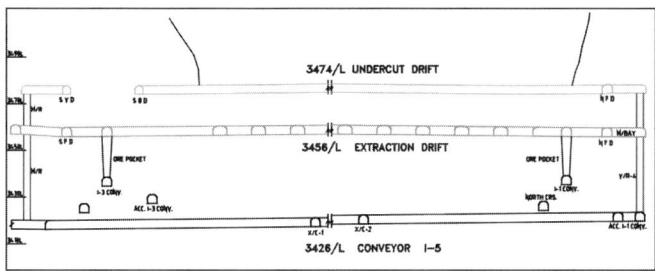

FIGURE 53.10 Cross section, IOZ mine

Drawbell blasting at the IOZ precedes undercut blasting. The drawbell drill pattern is shown in Figure 53.11. After the drawbell is drilled, it is blasted in three stages.

3. Load blast 1 and preload blast 2. Blast the center of the pyramid. Muck out.
4. Preload blast 3. Blast the pyramid to full height. Muck out.
5. Blast the remainder of the drawbell.

Undercut blasting follows the drawbell blast. The typical undercut drill pattern is shown in Figure 53.12. Drill patterns A and B are drilled over the drawbell and side walls of the drawbells. Pattern C is drilled over the minor apex between drawbells. Ring burden varies from 2 to 3 m and is adjusted to suit ground conditions.

Undercut rings are not blasted until swell muck has been pulled from the drawpoints, so that the ring is blasted into a loose muck pile, if not an open void. Rings are normally loaded with

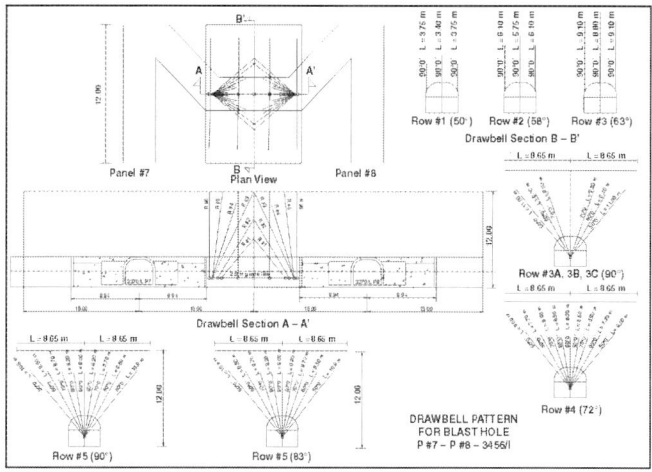

FIGURE 53.11 Typical drawbell drilling pattern

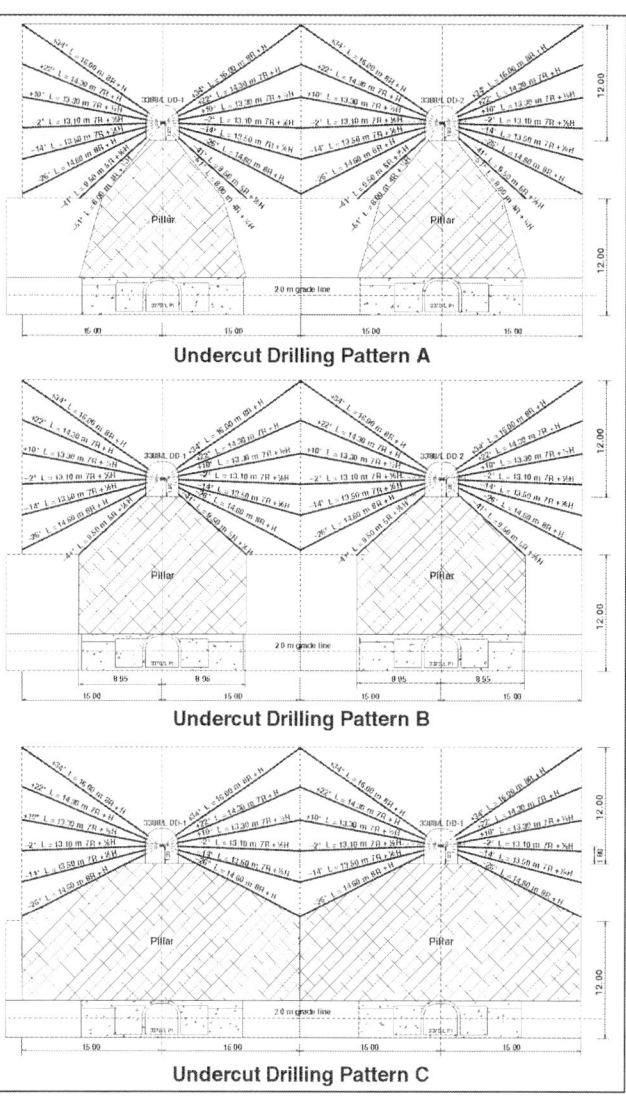

FIGURE 53.12 Typical undercut drill patterns

ANFO, and a collar is left to minimize brow damage on the undercut level. It is occasionally necessary to repair the brow before the next rings are loaded and blasted. This is usually done with timber sets.

53.3.6 Wet Muck Operations, IOZ

The wet muck situation in the GBT was less severe than in the IOZ for a number of reasons.

- The GBT draw columns were shorter than those in the IOZ, resulting in less comminution in the draw columns. This resulted in coarser muck reporting to the draw-points. Since coarse muck is more difficult to liquefy, there were not as many problems with wet muck runs.

- The GBT production levels are 150 to 200 m above the IOZ production level. The reduced depth results in reduced hydrostatic pressure, which results in lower water inflows into the cave.

- Production level geometry in the IOZ is different than in the GBT. The GBT used a "trench" undercut system. The trench was at the extraction level and connected all the drawpoints. This meant that water in the cave could not concentrate in a single drawpoint, but was distributed through all the drawpoints along the full length of the trench. The IOZ uses discrete drawbells that tend to concentrate water in a single drawbell. After the drawbell becomes saturated, the water spills over the minor apex into the adjacent drawbell.

- The IOZ is located directly below portions of the GBT. IOZ draw columns have broken into the GBT cave. The cut-off grade for the GBT was much higher than the IOZ (1.0% copper versus 0.90% copper equivalent), hence much of the "dilution" from the GBT cave is now classified as ore. As a result, production mucking continues long after most IOZ drawpoints are theoretically exhausted. In some instances, IOZ drawpoints have produced 175% of reserve tonnage. The GBT muck, saturated and drawn through an additional 150 to 200 m of draw column, is extremely fine grained when it reports to the IOZ drawpoints. The very fine muck is much more easily liquefied than the coarse muck from the GBT.

When the wet muck situation first appeared in the IOZ, operational controls were instituted that relied on rigid mucking schedules and close supervision. Conventional mucking equipment was used. This was not safe enough to meet the requirements of Freeport Indonesia or of government authorities.

As soon as it became apparent that the wet muck situation in the IOZ was more serious than in the GBT, outside hydrology and rock mechanics consultants were commissioned to determine the causes of the problem and recommend solutions. They have developed recommendations for the dewatering situations discussed above and for operating practices to help control the problem.

Mine management decided that the safest method to adapt was a remote-control system that would completely remove the operator from any danger. Because it would take some time to develop, install, and commission a complete remote-control system, we elected to use an interim method for wet muck operations. The interim method required the use of specially built closed (armored) cab muckers. These machines were built in our shops to provide heavy steel protection for the operators while pulling wet muck. The "closed-cab" muckers were used to pull coarse wet muck in certain areas of the production panels. Some areas were deemed too dangerous to pull even with the closed-cab machines and were shut down until remote-controlled equipment became available. The closed-cab

machines are still used on isolated, coarse wet muck drawpoints and for cleaning up after a wet muck run.

The ultimate solution to safely pulling wet muck has been tele-remote operations. This system has been developed in stages since 1997. The muckers are equipped with video cameras and remote-control electronics and controls. The operator runs the machine from a console located in a special room near the lunchroom on the south fringe of the mine area. The rock breakers on the north fringe (wet area) of the IOZ are also equipped with remote controls. All areas under remote-controlled operations are barricaded so that pedestrians cannot enter without stopping operations.

Remote-controlled operations give the IOZ mine the ability to safely pull wet muck, but at a price. Remote mucking is only about 75% as productive as manually operated mucking. Because of damage incurred by the remote muckers when they are involved in wet muck runs, availability of remote-controlled muckers is only 60% to 65%, compared to 80% for manually operated machines.

To minimize the problems of dealing with wet muck in the DOZ mine, an extensive dewatering program is underway. This will reduce the amount of wet muck reporting to the drawpoints in the DOZ.

53.3.7 Cave Management

Original System. The cave management-draw control system used at the IOZ was developed at the GBT. Paper draw orders were issued twice a week to the production superintendent. The draw order told the operations supervisor how many buckets of muck to pull from each individual drawpoint on each shift. The shift supervisors issued the paper draw sheets to the mucker operators. The operators were to pull the muck as ordered and record the number of buckets from each drawpoint. It was the responsibility of the supervisors to correct the shift draw orders to account for hung-up drawpoints, panel repair, or other operations problems.

The primary method of checking the accuracy of the reports was reconciliation against belt scale tonnage reports. This allowed a good check of the total bucket count, but did not help check against the drawpoint report. Spot checks and field bucket counts showed that draw report accuracy was not good.

The production reporting and recording system was primarily a manual system utilizing spreadsheets to manage the data. This was labor intensive and prone to errors during data manipulation and transfer.

The difficulties with field reporting and office data manipulation prompted a review of available technology. As a result, major modifications have been made to the cave management-draw control systems.

Current System. Improvements to the cave management system have been made in two areas.

- Automation of data management

- Dispatch system

Improvements to the data management system have been based on the development of a database, called Ubase, to manage all drawpoint data on a single platform. The database contains information on each drawpoint about daily draw order, actual reported draw, sampling data for copper and gold, drawpoint status, drawpoint condition, wet muck status, and initial drawpoint reserve. Ubase also records shift and daily production data and conveyor belt weightometer data.

Data are input only once, and the laboratory directly reports assay results over the Internet. These procedures minimize transposition errors. All daily, weekly, and monthly reports are generated from Ubase. Drawpoint history reports are much more readily run, which makes it much easier for the cave management engineers to analyze drawpoint grade trends, hang-up frequencies, etc.

A dispatch system from Modular Mining was commissioned in December 1999. The system provides real-time reporting of mucking activity to the dispatcher and allows the dispatcher to give draw orders to the mucker operators. The dispatcher can react to changing conditions in the production area (hang-ups, closed grizzlies, equipment, etc.) and issue changes to draw orders as needed.

We have already seen improvements in the accuracy of production reports from the field and in compliance with draw orders. Direct data transfer from the dispatch database to Ubase has reduced data processing time and transposition errors and given us the ability to update draw orders on a daily basis.

The dispatch system was initially installed to track production muckers only, but it is now being expanded to include jumbos for secondary blasting and other service equipment in production areas. The DOZ mine will have the system installed and operating when undercutting commences.

53.3.8 General Description, DOZ Mine

The production (extraction) level of the DOZ will ultimately be 900 m long and average more than 200 m wide, with the widest location being 350 m wide. The maximum height of draw will be 350 m. Production panels will be transverse to the ore body on 30-m centers. The undercut level will be 20 m above the extraction level. Drawpoints will be on 18-m centers along the panel drifts, yielding a draw column footprint of 15 by 18 m.

Undercutting will begin in the center of the ore body and progress eastward. This takes advantage of the extremely weak ground conditions and the higher-than-average grades in the DOZ breccia. After the IOZ has been depleted, undercutting in the DOZ will be advanced westward, underneath the IOZ.

53.3.9 Extraction Level Design

Extraction level design of the DOZ mine departs from previous Freeport Indonesia block caves (Figure 53.6).

Drawpoint Layout. Drawpoint layout is herringbone style. This was chosen because it allows us to use central muck raises, which is a major change from the IOZ and GBT where grizzlies were located between panels at the north and south fringe drift outside the ore body. This change will reduce average tramming distance by 40 m and maintain preferred mucker orientation for most drawpoints. Right and left drawpoints in each panel are staggered to minimize open spans and to reduce exposure of workers in a drawpoint to muck slides from opposite drawpoints.

Drawpoint Excavation and Support. Drawpoints are excavated with long-term ground support and stability in mind. Each round is line drilled and trim blasted to minimize pillar damage. After each round is advanced, the heading receives 125 mm of shotcrete as primary ground support before the heading is advanced another round. After the drawpoint has been excavated one round beyond the design lintel location, permanent ground support (3.5-m-long grouted rebar and monolithic concrete) is installed. This is the first time this procedure has been used at Freeport Indonesia; we are pleased with the results and have started using it in other areas that previously would have been supported with timber or steel sets.

53.3.10 Panel Ventilation

The panel ventilation system implemented at DOZ is designed so that all personnel working in a panel will be in fresh air. Ventilation in the panels is by means of fresh air delivered by both the north and south fringe drifts and exhausted through exhaust raises in the center of each panel. The exhaust raises discharge on a dedicated exhaust gallery that leads directly to the exhaust

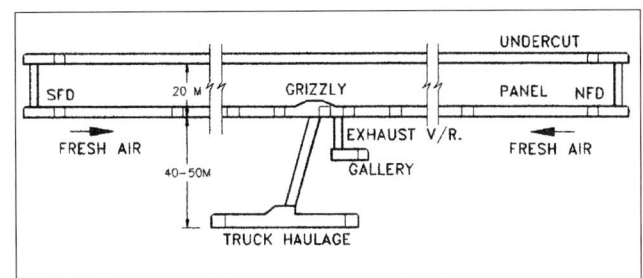

FIGURE 53.13 Typical cross section, DOZ mine

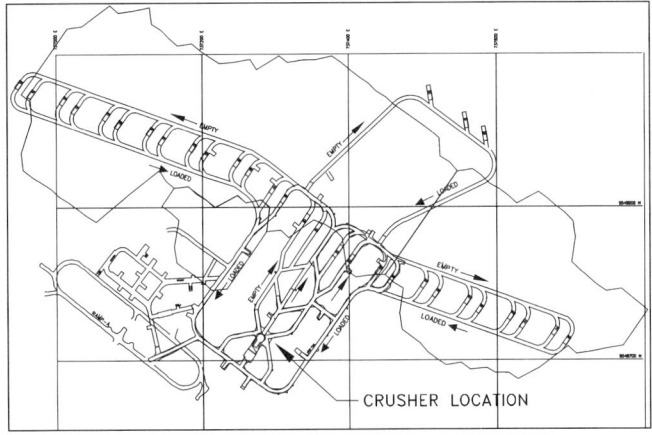

FIGURE 53.14 Truck haulage level

mains (Figure 53.13). This means that two muckers can operate at the same time in a single panel, each in its own split of air.

53.3.11 Haulage System

The haulage system built for the DOZ mine is a major departure from the haulage systems used at the GBT and IOZ mines. The system at DOZ uses a combination of trucks and chutes to deliver ore to the crusher.

Grizzly Muck Handling. Muck from the central grizzly in each panel is stored in a muck raise 4 m in diameter and 40 to 50 m long. Average raise capacity is 1000 tonnes. All panel muck raises bottom at chutes on the 3076 haulage level. This level is a limited access, one-way, racetrack-type truck loop with a chute for each panel muck raise (Figure 53.14). All roadways will be paved. Fifty-tonne-capacity trucks haul ore from the chutes to the direct dump crusher station.

Crushing and Conveying. The haul trucks will dump directly into the Fuller-Traylor 1372- by 1956-mm gyratory crusher installed just below the haulage level. Discharge from the crusher falls into a 1800-tonne, live-capacity ore bin. The bottom of the bin is equipped with a Jaques 18.29-cm by 11-m apron feeder that pulls the crushed ore from the bin and discharges it onto a 15.24-cm, 3,500-tonne/hr conveyor system that delivers the ore to the mill.

System Flexibility. The haulage system as installed at the DOZ is more flexible than the ones installed at IOZ and GBT. The new system can be readily expanded or changed to suit modifications in the mine plan or discovery of additional reserves. It can also easily support additional production from other areas. A further advantage of the DOZ haulage system is that it is sufficiently far below the extraction level that it will not be affected by mining-induced stresses and should be usable long after the DOZ is exhausted.

53.3.12 Ventilation Systems

The DOZ mine ventilation system includes a number of improvements over the ventilation systems built for the IOZ and GBT mines. All shops, storage facilities, conveyorways, crusher stations, ore dumps, compressor stations, powder magazines, etc., are ventilated from a fresh airway directly to a dedicated exhaust airway. All conveyor transfers and feeder stations are equipped with dust collection hoods and ducts to deliver the dust directly to dedicated exhaust airways. These two changes in design philosophy will result in a major improvement in the quality of the ventilation in the production areas.

The DOZ cave line will destroy the existing main fan installations. To ensure adequate mine ventilation throughout the life of the DOZ and beyond, two new ventilation shafts 800 m deep and 7 m in diameter are being excavated. Each shaft will be equipped with two centrifugal fans of 750 kW each. In addition, existing raises into the idle DOM mine have been converted into exhaust airways and equipped with a 450-kW centrifugal fan.

Overall ventilation of the DOZ is expected to be significantly better than that of the IOZ. Total ventilation throughput of the IOZ is currently 425 m³/s for a production rate of 18,000 tonne/d, giving a ventilation rate of 24 m³/ktonne. The DOZ will have throughput of 920 m³/s for a production rate of 25,000 tonne/d, giving a ventilation rate of 37 m³/ktonne.

53.4 SUMMARY

Since underground mining began at Freeport Indonesia in 1980, there has been a constant process of modifications and improvements to the mining methods and techniques used. This evolutionary process has resulted in improved safety, efficiency, and economics in the Underground Mines Division. We at Freeport Indonesia expect this process of continuous improvement to continue into the future as we meet the challenges of mining deeper, larger, and more complex ore bodies.

53.5 REFERENCES

Barber et al. 2000. Development of the DOZ Mine at PT Freeport Indonesia. SME 2000.

Coutts et al. 1999. Geology of the Deep Ore Zone, Ertsberg East Skarn System, Irian Jaya. Paper presented at PacRim '99, Bali, Indonesia, October 1999.

Casten et al. 2000 Excavation Design and Ground Support of the Gyratory Crusher Installation at the DOZ Mine, PT Freeport Indonesia. SME 2000.

Calizaya et al Commissioning of a 750-kW Centrifugal Fan at PT Freeport Indonesia's Deep Ore Zone Mine. SME 2000.

Calizaya et al New PT Freeport Mine Ventilation System–Basic Requirements (25 kt/day Plan). Mining Engineering, August, 1999, pp 54.

Call & Nicholas, Inc. Impact on Facilities Due To Mining the GBT, IOZ, and DOZ Deposits. 199904.

Call & Nicholas, Inc. 1997. Update to DOZ Cavability Study. 199804

PT Freeport Indonesia. 1998. Feasibility Study of the DOZ Block Cave. Internal Document, 19980727.

PTFI, U/G Mines Engineering Department & U/G Geology Department, IOZ Mine Expansion Study. Internal Document, 19970430.

Wallace et al. Ventilation System Review of PT Freeport Indonesia Company's DOZ Underground Mine. Mine Ventilation Services, Inc., 199711.

Wallace et al. June 1999. Ventilation System Review of PT Freeport Indonesia Company's DOZ Underground Mine. Mine Ventilation Services, Inc., 199907.

Wallace et al. Ventilation System Analyses of PT Freeport Indonesia Company's DOZ Underground Mine. Mine Ventilation Services, Inc. 199912.

CHAPTER 54

Finsch Mine: Open Pit to Open Stoping to Block Caving

Christopher Andrew Preece[*]

54.1 INTRODUCTION

Since 1966, De Beers Consolidated Mines, Ltd., has been exploiting a diamond-bearing kimberlite pipe. The pipe is known as the Finsch Mine after the prospectors who discovered it, Finscham and Schwabel. The mine is located in the Northern Cape in the Republic of South Africa (Figure 54.1). The pipe is elliptical and originally had an area of 17.9 ha on the surface, which is 1,590 m above mean sea level. The pipe is known to extend to more than 900 m below the surface.

Open-pit methods were first used to exploit the diamond pipe, but by 1976, it became apparent that the open-pit operation would reach its maximum economic depth toward the end of the 1980s. Planning and design of an underground mine were undertaken at that time to ensure continuity of operations, and sinking of the main shaft commenced in 1979. Two vertical shaft complexes, tunnels, and ground handling infrastructure were prepared for the continued exploitation of the diamond pipe with the use of highly mechanised underground methods.

The pipe has been divided into a series of blocks. Blocks 1 and 2 were mined by a combination of open-pit and blasthole open stoping methods. Block 3 is exclusively blasthole open stoping while block 4 will employ block caving. The reason behind the change in method is that significant failures are expected from the high, near-vertical faces of the country rock, and this would make the continuation of blasthole open stoping uneconomic. Block 5 has not as yet been fully delineated, but it is expected to be on the order of 200 m. The existence of further blocks has yet to be confirmed. Currently, each block consists of a drilling level and a loading level.

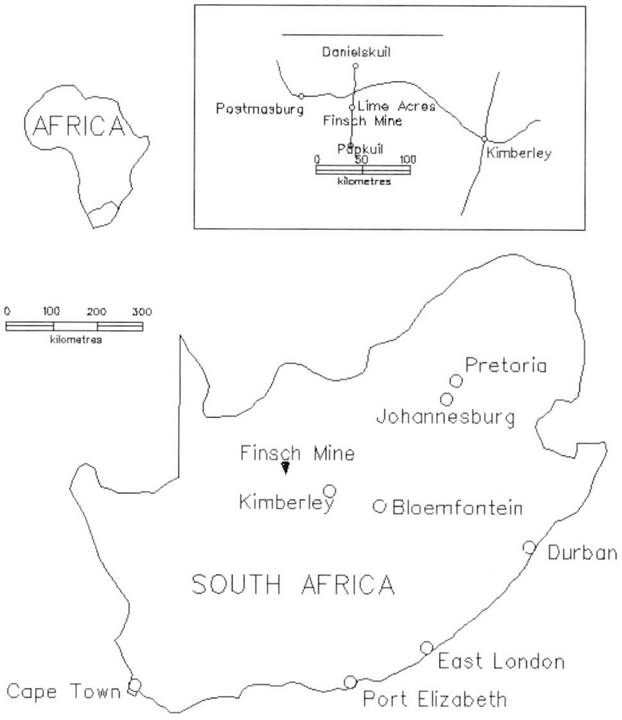

54.2 GEOLOGY

The Finsch kimberlite pipe is a near-vertical intrusion into the country rock, which consists of dolomite, dolomitic limestone with chert bands, and lenses of almost pure limestone. The pipe originally occupied 17.9 ha and was covered by rubble, infilling a topographic depression. It occurs on an external precursor dyke set striking approximately 50° east of north. Two minor pipes and two kimberlite dykes are known in the vicinity, making up the Lime Acres kimberlite cluster.

Eight different kimberlite types have been identified within the pipe. The most significant intrusion is designated F1 and is a diatreme-facies tuffisitic kimberlite breccia. It occupies 70% of the pipe area on the 350-m level (levels are designated by distance in metres below the surface), decreasing to 60% on the 630-m level. Large masses of Drakensburg Basalt, which makes up about 20% of the pipe volume, occur within the F1 kimberlite (Figure 54.2). A secretionary textured tuffistic kimberlite brecia, designated F8, is the second largest kimberlite type in the pipe. This kimberlite is similar to the F1 petrologically, but contains fewer inclusions of the country rock and may be part of the same

FIGURE 54.1   Location of Finsch Mine

major intrusive phase. In general, the F8 kimberlite is the high-grade area of the pipe.

An irregular, bulbous, satellite pipe is seen on the western side of the main pipe, giving rise to poor ground conditions in the area. Diamond content of this satellite pipe is moderate. Because its size is relatively small, mining has not been considered as a priority.

Proven ore reserves extend down to 630 m and will be exploited using the underground infrastructure. Below 680 m, there is an indicated resource over 200 m thick. Pre-1980 tailings, laid down before the modernisation of the treatment plant, are available for retreatment.

54.3 OPEN-PIT MINING

Production started in 1966 and had progressed to 364 m by the end of 1989. The final economic pit depth of 423 m was reached in September 1990. External waste stripping continued to approximately 244 m and was completed in 1986. After that time, benching was employed within the kimberlite, and only

[*] De Beers Finsch Mine, South Africa.

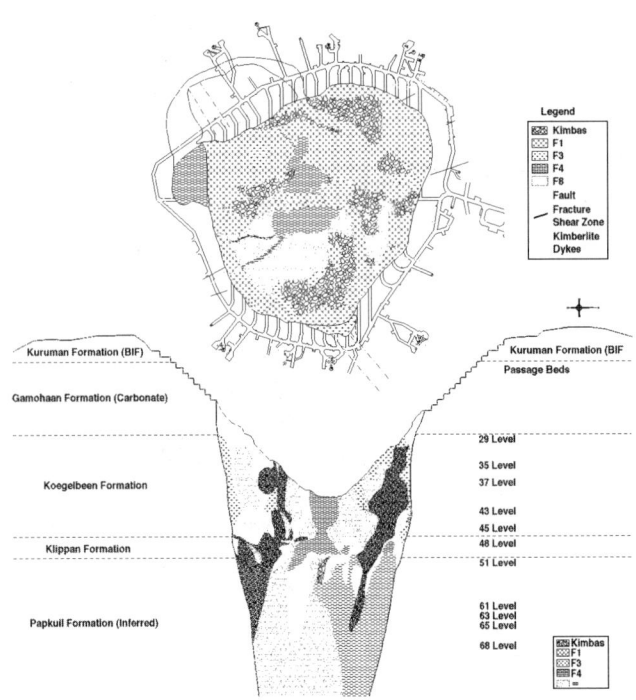

FIGURE 54.2   **Typical geological plan and section through Finsch pipe**

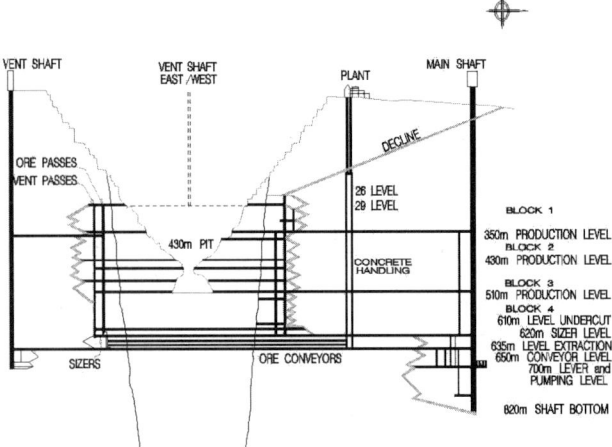

FIGURE 54.3   **Diagrammatic section through Finsch pipe showing infrastructure**

internal waste, consisting mostly of basalt, was separately removed at a rate of 1 million tonnes per year. The surface area at the perimeter of the pit excavation is 55 hectares.

In-pit crushing using a mineral sizer was used to reduce haulage costs. This gave the Finsch Mine an opportunity to test and make substantial modifications to the sizers before they were selected for use in underground production.

The inter-ramp angle in the country rock was 45°. This slope was designed to require no ground support. Calculations showed that if a 25-m-wide berm were left in the area where the haulage road was within 25 m of the country rock-kimberlite contact, an unsupported 45° slope could be maintained in the kimberlite below the berm. This would have resulted in the pit reaching the end of its life in 1988.

Economic considerations and the slump in the diamond market resulted in the need to find ways of delaying the acquisition of capital required for establishing the underground mine. Work undertaken by outside consultants and on-mine geotechnical staff showed that the wide berm could be eliminated and that the interramp angle could be steepened to 53° from the kimberlite contact down to four benches above final pit bottom and, if support were installed, to 57° thereafter.

The occurrence of joints, estimates of cohesive strengths (100 to 150 kPa), the angle of friction (32° to 35°), and predictions of likely failure modes were used to design the support requirements to achieve an acceptable factor of safety. Anchors tensioned to 600 kN were installed on 1-m spacings where the haulage road was closer than 25 m to the contact. After a number of small failures occurred, support density was increased, and in two areas where the failures resulted in a loss of access to the bench below, a system of 25-m-long spiles were installed vertically in the crest of the haulage road. These spiles were tied across the haulage road and into the country rock using tensioned rope anchors. A total of 370 anchors and 70 spiles have been installed. This work added 2 yr to the pit life and proved invaluable in designing slopes associated with the underground mining methods.

To achieve the above slopes and to facilitate the smooth transition to an underground operation, it was essential that good blasting results were achieved. After experimenting with a number of alternative designs, small-diameter holes for creating a presplit and for intermediate blasting achieved very good results with the use of very much reduced powder factors and carefully designed blast timing patterns.

During the last few years of the 1980s, 5 million tonnes of ore per annum were produced using conventional open-pit methods. Bench height was 12 m. Blastholes 250 mm in diameter were drilled on a 6.5- by 5.5-m pattern. Primary equipment consisted of 77-tonne diesel trucks, 10-m^3 front-end loaders, and 5-m^3 rope shovels. Access for haulage was by a single spiral haulage road starting at 8% and steepening to 11% for the final four benches.

54.4   UNDERGROUND MINING METHOD FOR BLOCKS 1, 2, AND 3

The selection of the underground mining method to be employed at the time of the changeover was governed by the following criteria:

- Advanced technology appropriate in the South African context commensurate with moderate capital investment and competitive working costs
- Consistent high tonnage output
- High productivity in terms of tonnes per man hour
- Effective waste control and minimum waste dilution
- High percentage of extraction
- Rapid tonnage build-up
- Flexibility

A block caving method was considered, but failed to meet the criterion of being able to separate out the bulk of the internal waste. Experience in block caving of kimberlite at other De Beers mines also indicated that the build-up to full output tonnage would be unacceptably slow. Blasthole open stoping, a variation of the sublevel open stoping method, was decided on as the most suitable method to meet the above criteria.

Underground mining has exploited remnants of the pipe left from the open pit from the 244-m elevation. A generalised section of the pipe, the mined-out open pit, the shaft systems, and the major development tunnels are shown in Figure 54.3.

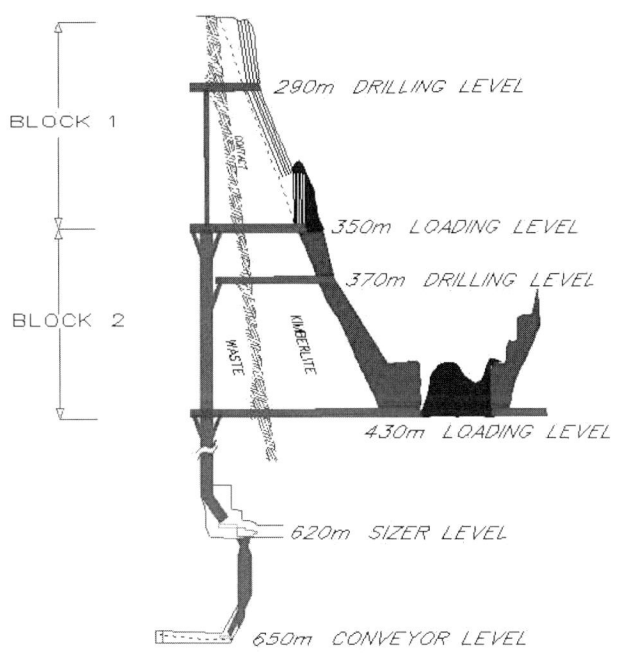

FIGURE 54.4   Schematic mining layout

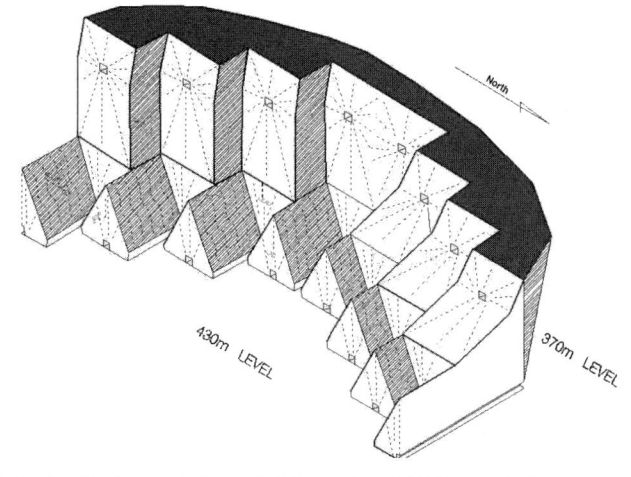

FIGURE 54.5   View of lead and lag loading tunnels

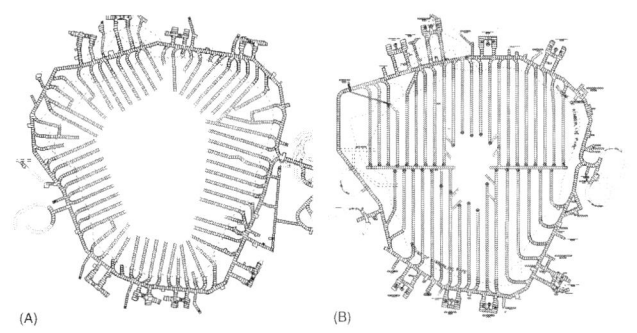

(A)   (B)

FIGURE 54.6   Plan of 37-m level (A) and 43-m level (B)

As mentioned previously, internal waste comprises about 20% of the pipe volume. It was estimated that 60% of this internal waste could be separated out and handled independently. The balance would be acceptable as waste dilution. However, experience has shown that it is not possible to readily distinguish kimberlite from waste rock in the underground drawpoints, and thus all ground is handled en mass.

The blasted ground reporting to the loading levels is loaded by 12-tonne-capacity load-haul-dump (LHD) machines at the drawpoint formed by the end of the loading tunnel. Figure 54.4 is a simplified layout showing concurrent operations on the 350- and 430-m levels. Some of the LHDs have remote capabilities, which allows them to be operated in unstable situations so the safety of the operator is not compromised.

The ground is drilled and blasted using a continuous cone (trough) pattern on the loading levels and a 360° composite ring pattern on the drilling levels. Alternate loading tunnels "lead" their neighbouring tunnels by 15 to 20 m, creating large troughs that are reduced to regular troughs as the "lag" tunnels progress (Figure 54.5). The solid ground between the large troughs provides a toe abutment for support of the face created by ring drilling above. The trough drilling patterns are drilled in the vertical plane to maximise the strengths of the drawpoint brows. Geophysical considerations of the face produced by ring drilling over a 50-m-high block led to the conclusion that a composite ring pattern would be required to maintain stability. The approved design calls for upholes 50° forward from the horizontal plane, as shown in Figure 54.4.

Open pit productivity during the last few months suffered from the restricted operation area, and it was necessary to supplement ore from the open pit with material from stockpiles that had been accumulated on the surface. These stockpiles were also used to supplement underground ore during the build-up phase of the first 6 months. The geometry of the final open pit, with its 18-m-wide spiral ramp to the pit bottom and benches up to 16-m-wide left for slope stability reasons, posed a problem of accessibility by underground drilling. Certain areas within the pipe could not be reached by blastholes from either the loading level or the drilling level.

To overcome this problem, the underground mining started by "scavenging" from the pit the remnants of ore left by the last open-pit blast. This gave sufficient time to drill off, using the open-pit surface drills, part of the access ramp leading to the floor of the pit. Once this had been blasted and made available to the drawpoints on the 430-m level, the ramps and wide benches were systematically wrecked on retreat to yield 1,065,000 tonnes of kimberlite ore. The remainder of the ore, which was then within reach of 30-m-long upholes from the loading levels, was drilled and blasted using carefully designed fan drill patterns.

54.4.1   Tunnel Layouts, Development, and Support

Block 1, the upper mining block, with its loading tunnels on the 350-m level, was laid out to take advantage of the extensive open-pit perimeter on this horizon and utilise it as a free face. A "radial retreat" concept was adopted, with the loading tunnels laid out like the spokes of a wheel (Figure 54.6A), resulting in 77% of the tunnels having the potential to produce as soon as the open-pit operation had stopped. The loading tunnels were spaced horizontally at 18 m, while the drilling tunnels were spaced at 36 m, so that two troughs collected the fragmented ground from a full ring blast.

For the 430-m loading level, an east-west orientation was considered appropriate (Figure 54.6B). The initial central tunnels had an immediate free face created by the bottom of the exhausted open pit. A north-south slot was then blasted to create a free face for the balance of the loading tunnels on the level. A similar layout of loading tunnels was used for block 3 on the 510-m level. Cutting of the required slot for this block had to be undertaken without the benefit of a preexisting open pit.

Development started in 1979 with the commencement of a decline from the surface. Until late in 1988, all development took place in the country rock. All rim tunnels, major access to ground handling infrastructure, and the ramp were developed 6 m wide and 5 m high. Nine electro-hydraulic face rigs were used for drilling in blocks 1 and 2 and five rigs in block 3. Emulsion explosives were initially used extensively, though this was changed to a mixture of ammonium nitrate and fuel oil (ANFO) and muck was transported using LHDs. Support was 2.6-m-long roof bolts spaced on 1-m intervals in the hanging wall only. Additional long anchors, mesh, and lacing, and in some cases shotcrete, were installed in the large excavations for tips, crushers, substations, conveyor transfer points, box fronts, and break-aways. A total of 52 km of development was in place when underground production started at the end of 1990, and of this, 6 km was within the kimberlite pipe. For block 1, a further 17 km of development was undertaken, of which 9 km was within the pipe and 4 km was small return airways.

Kimberlite's unfavourable weathering and relatively weak nature necessitated that in-pipe development be commenced as late as possible. The basic method was similar to that used on the country rock development except that a mist air system was used for drilling, thus minimising the amount of water used. Perimeter spacing was reduced, and small- diameter, low-strength explosives were used to achieve smooth blasting results. The size of the end was reduced to 4.6 by 4.8 m.

On-site geotechnical staff and external consultants carried out an extensive exercise to design the support requirements. The following factors were taken into account.

- Life of the tunnel
- Kimberlite types (of which there are eight)
- Weathering assessment
- Factor of safety

This resulted in the recommendation of a number of different support types in different areas. The requirements were as follows:

- Fully resin-grouted, 16-mm by 2.6-m-long roof bolts installed every 1 m down the footwall. This required 10 bolts every metre.

- Installation of 3- to 5-mm-thick, acrylic-based sealant. This material was sprayed onto the hanging wall and sidewall of the tunnels as soon as possible after exposure.

- The installation of 100 mm of mesh-reinforced shotcrete on the hanging wall and sidewall. This shotcrete was tied back to the bolting pattern.

The above systems were installed in 60% of the tunnels. In addition, when the country rock-kimberlite contact zone (the first 20 m) was developed, long, wooden dowel spiles were used in advance of development, and steel straps were installed over the shotcrete along the tunnel length. In the remainder of the tunnels, the requirement was less, but always included bolts and, in most cases, some combination of sealant, mesh, or straps. Quantitative geophysical measures such as rock mass classification were used to specify the support type and timing of its installation.

54.4.2 Ground Handling Infrastructure

Changing Finsch Mine from an open pit to an underground mine capable of producing at a similar tonnage rate required the use of many innovations and new technologies. The ground handling system was central to achieving this production requirement and thus required very careful designing. Major changes to this system would be extremely difficult after production start-up. Consideration had to be given to—

- Handling the material as delivered from the loading levels, thus reducing delays on these levels

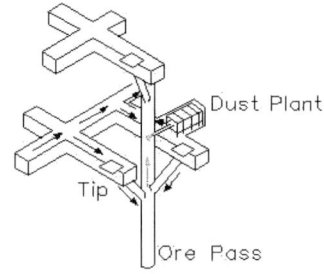

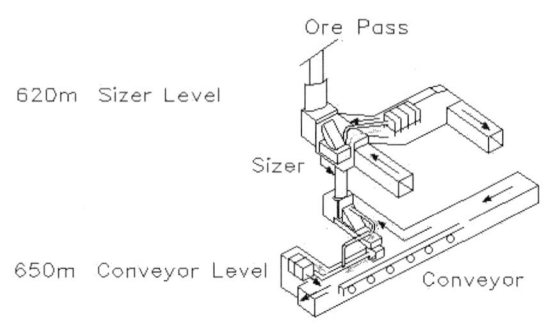

FIGURE 54.7   Ground pass arrangement

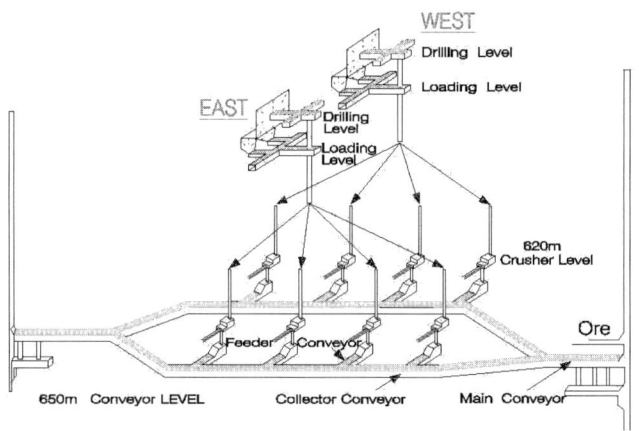

FIGURE 54.8   Schematic diagram of ground handling system

- Handling the tonnage at the required rate back to the shaft

- The separation of waste and ore as delivered to the passes

- Provisions for adequate surge capacity

- Consideration of all ventilation requirements when handling relatively dry kimberlite

The full ground handling system is shown in Figures 54.7 and 54.8. Central to the ground handling system are eight 6-m in diameter, 270-m-long ore passes. These have been developed in the country rock and are situated approximately 60 m from the kimberlite contact on the 35 level. Lump size in the passes is limited by the grizzly to 1.3 m. This gives a ratio of 1:4.6, which is well in excess of the 1:3 below which bridging should occur in the pass.

The passes were raise-bored to 2.9 m and then sliped to the 6-m size using a custom-built four-deck sliping rig. This method was chosen in preference to using a 6-m in diameter raise bore for the following reasons:

- Total cost was more favourable.

- A rough sidewall would be created, reducing the risk of hang-ups.

- Access would be available to undertake local cementation if water were encountered.

When kimberlite comes into contact with water, it weathers and can become very tacky, thus causing hang-ups. Mud rushes may also result. During subsequent operation of the passes, even small volumes of water were found to cause major problems, and thus all the passes had to be fully sealed. This was achieved by the use of chemical grouting from selected drill holes.

Tip legs 25 m long were drop-raised using diamond drilling equipment and conventional explosives. The tip construction consisted of a six-hole, 1.3- by 1.3-m grizzly with solid steel billets (370 kg/m) cast in 65 m³ of concrete and a fixed rock breaker capable of handling oversized material. Using a smaller number of mobile rock breakers would have caused interference with the flow of the LHDs. A subsequent improvement in this system was the development of a remote-control system for the rock breakers. This involved installing adequate lighting for the operation of closed-circuit television cameras and a remote-control system linked to a central control room on the 510-m level. From here, operators can watch all the operational tips and activate the appropriate rock breakers as and when required. This has reduced the number of operators required and allows those employed to work in a noise- and dust-free environment.

The automation process has not been without problems. Lighting was found to be critical, as was camera position. Initially, operators struggled to operate in a three-dimensional world while looking at a two-dimensional TV image. This was simplified by the use of only black-and-white images.

The 1.3-m grizzly installed on top of the pass is considered the smallest size that avoids excessive delays on the loading levels. Anything smaller would result in interference on these levels by LHDs waiting for the rock breaker to clear the grizzly. This size of rock and the high production requirement precluded a conventional track locomotive haulage system.

Three crushing systems were considered during the planning phase: (1) rock breakers that would reduce the size to minus-800 mm (intermediate breaking level) and a separate crusher level with eight jaw crushers to reduce the material to minus-300 mm, which is suitable for conveying; (2) the largest commercially available jaw crushers without an intermediate breaking level; and (3) a mineral sizer or roll tooth crusher at the bottom of each pass.

Method 1 was rejected because of the high cost of installing an additional level (3,500 m of development) and because this would have reduced the reserve available to this infrastructure horizon by 1.5 yr. Method 2 was rejected because acceptance size would have been reduced to 1.2 m, and this would have produced lumps up to 0.5 m, which would have been unsuitable for conveying. Consequently the third alternative was implemented. Mineral sizers were available that were capable of handling softer material such as coal, but that would be marginal when handling the harder kimberlite (40 to 60 kPa) and internal basalt (110 kPa).

Before fully committing to this system, extensive testing was undertaken, and a two-roll unit was commissioned as part of the in-pit crushing system. Each roll consisted of seven rings of four teeth. The rolls are positioned 1.15 m apart and are driven in

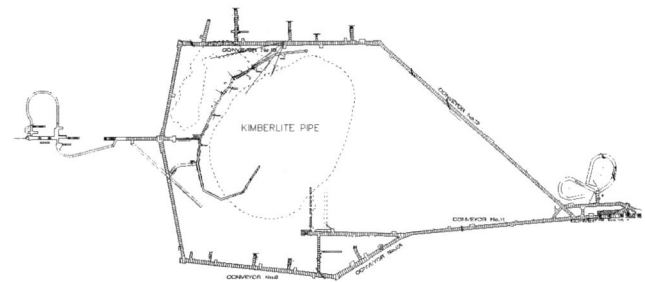

FIGURE 54.9 Plan of 65-m level conveyor level

opposite directions by two 275-kW motors through two gearboxes equipped with standard spur gears. Substantial modifications were made before a unit acceptable for underground was produced. These modifications included—

- Installation of a breaker bar below the crusher rolls

- Use of a number of longer "wild" teeth

- Modifications to the tooth scrolling arrangement

- Changing the feed control from large, hydraulically driven fingers to a 2.4-m-wide apron feeder

- Complete redesign of the gearboxes to incorporate epicyclic gears

These units are capable of producing 1,200 tonne/hr with a top acceptance size of 1.5 m and a nominal product size of 300 mm. The sizers are each installed in a 4,700-m³ excavation underground. The installation incorporates a large box front with a 4-m-wide chute to a 2.4-m-wide apron feeder. Eight 2.5-tonne, 9-m-long chains control feed onto the apron feeder. The apron feeder has been installed in such a way that the mineral sizer can easily be rolled out for maintenance.

The mineral sizer feeds a 30-m-deep, 600-tonne-capacity surge pass. Each installation consists of 1,250 m³ of concrete and 80 tonnes of structural steel. Under each surge pass is a box front beneath which is a 1.2-m-wide apron feeder capable of feeding material at a variable rate of up to 600 tonne/hr onto a short feeder conveyor. Each feeder conveyor has an overband magnet and metal detector to ensure a clean feed onto the longer transfer belts.

Situated on each side of the pipe and 30 m below the crusher level on the 65-m level are the two main conveyor tunnels (Figure 54.9). Two conveyor belts were installed on each, one running to the ore shaft and one to the waste shaft. However, once it was proven to be impracticable to separate waste and ore at the drawpoints and all rock was trammed as ore, the waste belts were decommissioned. The eight feeder conveyors had moving heads and could thus feed onto either the ore or the waste belt, depending on what was in the pass.

Each ore belt is thus fed by four feeder conveyors and is capable of handling 1,200 tonne/hr. The control system ensures that these belts are not overloaded. The two belts meet close to the shaft, and here a short 2,400-tonne/hr belt feeds a shuttle conveyor capable of feeding the three main 1,200-tonne surge passes situated 50 m above the main shaft loading level. On the shaft loading level, two conveyor belts feed a conventional shaft loading system and a single 5,500-kW Koepe hoist with two 28-tonne skips capable of hoisting the ore to the surface at a rate of 5 million tonnes per year. The main shaft feeds an 80,000-tonne surface stockpile that acts as a buffer between the mining operation and the treatment operations. The main shaft configuration is shown in Figure 54.10.

The waste belts were capable of feeding 600 tonne/hr to two surge passes situated above the waste shaft loading level. A 97-kW Koepe hoist with a 12-tonne skip hoisted the waste to the surface in

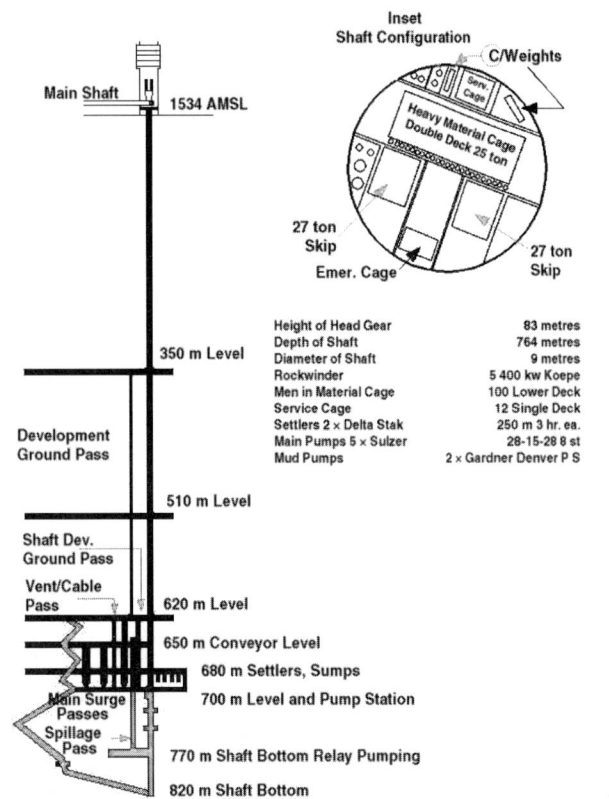

FIGURE 54.10   Main shaft configuration

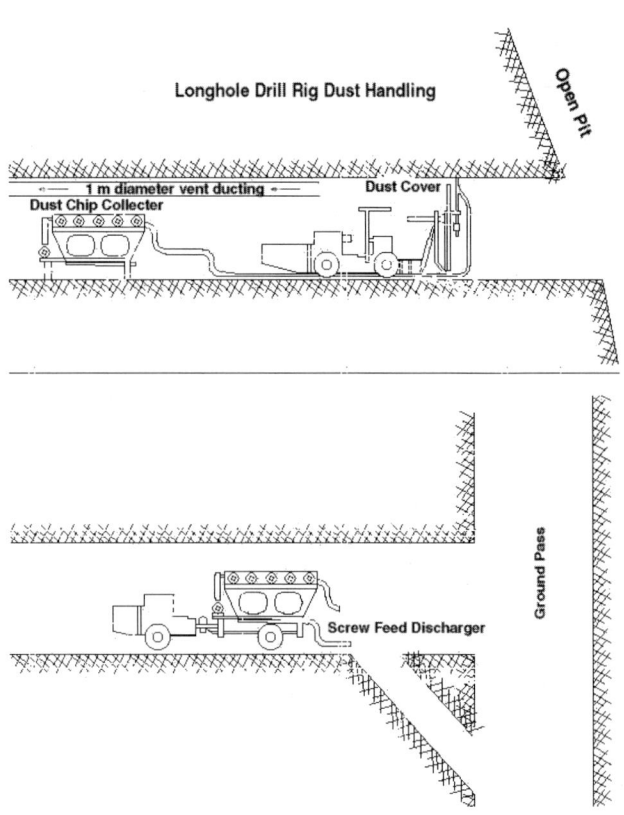

FIGURE 54.11   Drill rig and dust handling

rope guides at a rate of 1 million tonnes per year. This system was decommissioned in 1996.

The waste shaft was designed to feed a dump via a conventional belt spreading arrangement. As a contingency, a bypass system was installed so that ore could be trucked away from the shaft so that a reduced production capacity could be maintained in the event of a major delay on the ore side. This entire system was also decommissioned in 1996.

54.4.3   Equipment Selection

Being a fully trackless operation, all equipment used underground has to be fully mobile and, in order to fulfill the high production and labour productivity requirements, as large as practical. Access for this equipment is either by the decline development from the surface or on the large mancage in the main shaft, which has a payload of 25 tonnes and is 7.4 m long, 2.5 m wide, and 3.8 m high.

After extensive research, market surveys, and simulation exercises, it was decided to order sixteen 12-tonne-capacity LHDs. This was the smallest unit that could achieve the production requirement. Simulation exercises showed that additional smaller units would cause excessive interference. If substantially larger units were selected, supporting the tunnel excavations would have been a problem. In the current operation, only 13 units are used to achieve the required call.

The blasthole open stoping method required the following:

- Accurate downhole and uphole drilling

- Capability to drill up to 45 m

- Capability to drill dry (because of the weathering characteristics of kimberlite)

To maximise productivity, a hole diameter of 102 mm was required. Diameters any larger than this are liable to cause back

damage and may be problematic to charge. Five electro-hydraulic tube drilling units were selected for production drilling. The tube drilling system has proven to be very accurate for full 360° ring drilling requirements. The original accessories have now been replaced with tubes and bits to form 98-mm holes.

Four mobile ANFO units with on-board pumps and storage tanks ranging from 1.5- to 3.5-tonne capacity are used to charge development ends and ring holes. Initially emulsion was used. Twelve specially designed cassettes with a capacity of 2.5 tonnes transport ANFO from the surface to the various sites underground. Two oil cassettes transport oil from the surface to the underground workshops.

Dust cassettes (Figure 54.11) are used for dry longholes. These units consist of a dust extraction plant, a hopper capable of storing two shifts of drilling chips and dust (2.5 m^3), and a screw conveyor to empty the cassette. The dust plant is powered by a connection from the drill rig. When the cassette is full, a cassette carrier picks up the unit and transports it to a pass where it is emptied. A fourth type of cassette was used to transport and mix the dry products for underground shotcreting. A total of six 8-tonne cassette carriers are used to move these cassettes.

Other utility vehicles currently in use are six scissor-lift units, three forklifts, two mobile cranes, six container carriers, two flatbeds, and 36 light transport vehicles. Four concrete transporters and a tire handler were used in the past.

A fully equipped underground workshop (Figure 54.12) consisting of 17 bays and 1.5 km of development was established on the 35-m level for services, minor overhauls, and component changeouts on the 164-unit vehicle fleet. A number of small satellite service bays are also situated on the various levels to carry out minor services. Two vehicles are used for remote daily lubrication of the less-mobile drill rigs. A second workshop complex consisting of two large bays was established on the 51-m level for block 3.

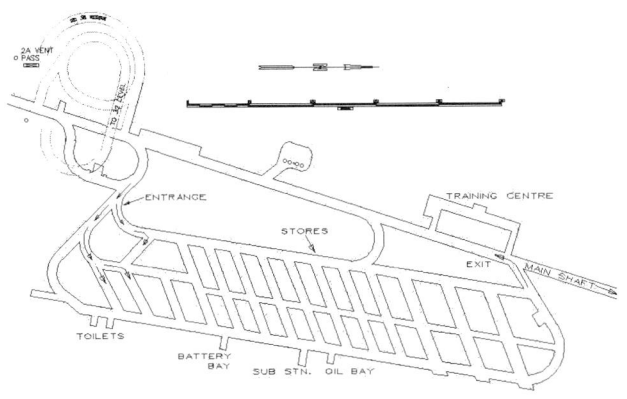

FIGURE 54.12   Workshop complex at 35-m level

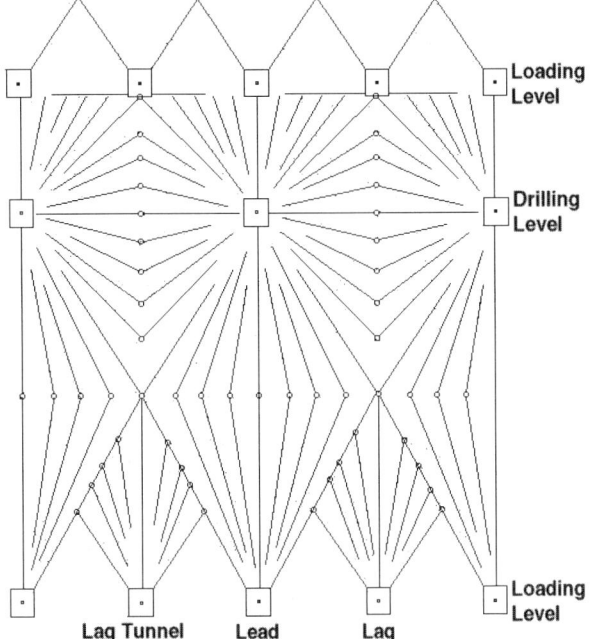

FIGURE 54.13   Schematic blasthole ring layout

54.4.4   Drilling and Blasting

Simulation exercises of the ground handling process indicated that the key to productivity was adequate fragmentation. Hence, great emphasis was placed on the design of drill patterns, blast initiation system, and choice of explosives. Little quantitative work had been done locally relating to the optimisation of blasting with regard to producing the desired degree of fragmentation while maintaining air-blast effects at acceptable levels, minimizing back damage, and controlling efficiency factors. Consequently, planning for drilling and blasting for blocks 1 and 2 had to rely heavily on experience within other De Beers mines, other mines, explosives manufacturers, and outside consultants. Some test work has been carried out in the open pit, but there is no doubt that methods will have to be modified as experience is gained in a practical situation. Since the inception of the underground operation, considerable attention has been paid to all aspects of drilling and blasting.

Ring patterns for the levels are shown in Figure 54.13. Burden and toe spacings are 3.2 and 4 m, respectively. This gives good fragmentation without major air-blast or back damage problems. The burden and spacings have to be varied for different ground conditions. Drilling accuracy is obviously a major concern. Taking into account the massive, nonstratiform nature of the ore body and the size of the hole and drill tubes, deviation of not more than 2% of the length of the hole is expected under normal conditions.

A locally based explosives manufacturer had developed a repumpable emulsion (now replaced with ANFO) with sufficient viscosity to be used with 102-mm in diameter upholes. It is transported as a nonexplosive material and gassed concurrently with the charging process. Charging lengths of each hole in the ring pattern vary to distribute the explosive across the area of the ring.

Initially, a high VOD booster and detonating fuse were used. Short-period delay detonators affected timing of the ring blast connected to a permanent blasting circuit. Initiation of the blast takes place from the surface control room. Blasting practices have been improved over the years, and in early 2000, the whole production blast was changed to electronic detonation.

54.4.5   Ventilation, Pumping, and Flood Control

At the commencement of the underground operation, four 970-kW fans situated on the 290-m level (top drilling level) exhausted 600 m³/s of air to the surface via 2- by 5.2-m in diameter ventilation shafts. These fans exhausted air from eight 2.9-m in diameter raise-bore return air passes situated around the kimberlite pipe (four on each side, as shown on Figure 54.6).

Nine 25-kW booster fans forced 800 m³/s of air into the mine. This ensured that when loading, if drilling tunnels were open to fresh air, the correct amount of air would still be entering the various working areas as required. These fans are situated at various intakes to the mine. Four are sited on the bottom of the four-corner ventilation passes. From that level up, the passes are return air passes. At each pass position on each of the two loading levels, a constant-pressure regulator was installed. A system of ducts fed this regulator and branched out and exhausted air from up to six loading tunnels. Where the ducts entered the tunnel, a regulator with three settings was installed. As a result of the constant pressure, these settings produced constant quantities. These were 20 m³/s for loading operations, 10 m³/s for any other work, and closed when no work was taking place. This system ensured that air was distributed only where needed, thus allowing optimisation of the total mine air requirements. For block 1, the system was modified, and the ducts were replaced by 1.8-m in diameter return airways. This is a more cost-effective option as it reduces the loading tunnel height, and the vent tunnels are not subjected to damage by vehicles, as is the case with ducts.

On the drilling levels, air is forced into the drilling tunnels. Air entering the ground handling levels is reused on the upper levels via dust plants and the booster fans. A system of smoke ducts allows blasting operations to take place at various points in the ground handling system. A system of dust plants is used at all ground handling transfer points. A total of 52 plants varying in size from 3 to 20 m³/s have been installed. Figure 54.7 shows the dust plants associated with one ore pass system. The overall system is shown in Figure 54.14.

The ventilation system was further modified in 1996 when tunnel instability necessitated moving the exhaust fans from the 29-m level to the 35-m level. The main exhaust ventilation shafts from the 29-m level to the surface were abandoned and replaced with exhaust fans on the 35-m level at the top of each of the return air passes. These vent directly into the open pit. Research has shown that by fitting variable-speed drives to these fans, considerable power savings can be achieved, and the necessary work has been scheduled for the second half of 2000.

Although the Finsch Mine is situated on the edge of the Kalahari Desert, with an average annual precipitation of only 350 mm, periods of torrential rain produce flooding. In February 1988,

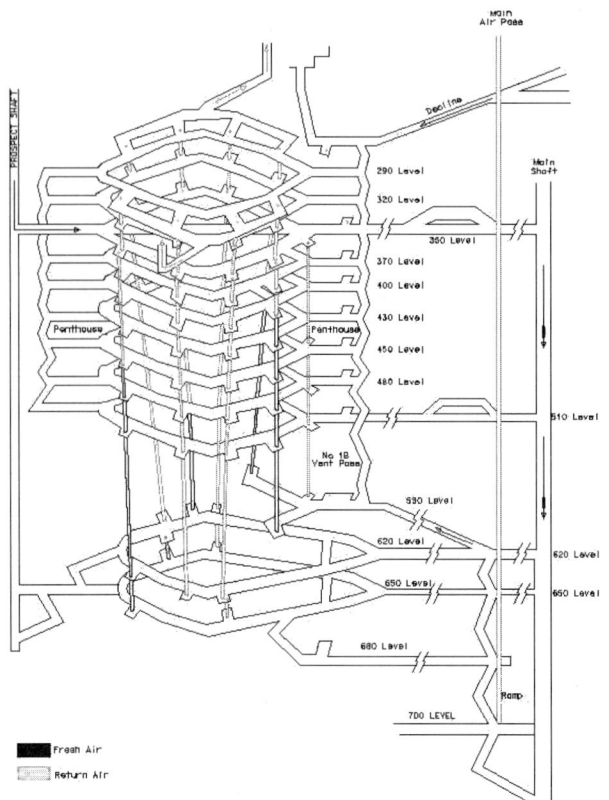

FIGURE 54.14   Schematic ventilation layout

over 300 mm of rain fell in a 4-day period, flooding the bottom bench of the open pit. Should such a rainfall event occur now that the underground mine is in full production, water collected by the catchment area of the defunct open pit could flow directly into the underground workings. Consequently, the design of the underground mine allows for all floodwater to be directed to the sump level and provides sufficient storage and pumping capacity to allow for one 100-yr flood event that could occur over the pit catchment area.

54.4.6   Computerised Monitoring and Control

Monitoring and control of the various complex facets of the underground operation have been achieved by a network of programmable logic controller (PLC) installations linked to a supervisory system housed in the control room on the surface. The supervisory system provides a "window" into the PLC system with colour graphic mimics and enables the operator to determine the status of the equipment being monitored and control its functioning

Hardware consists of minicomputers with colour operator stations, printers, and engineering terminals. The supervisory software is user-configurable to suit the application. Simulation software was used to validate and test the underground system control strategies.

The major areas in which the supervisory system is used to monitor and control the operation are—

- *Ground handling.* This involves the interlocking and control of the conveyors and apron feeders on the 65- and 70-m levels and includes belt rip detection.

- *Ventilation.* This involves monitoring of the main and booster fan dust plants.

- *Ground passes.* Monitoring the level of grounding each pass as well as low-level indicators will provide information on how much ore and waste are in storage. Interlocks between the low-level monitors and apron feeder delivery will also be installed.

Other monitoring and control functions of the system include—

- Pumping from the 70-m level pump station, the 68-m level settlers, and the shaft bottom

- Underground water and compressed air reticulation

- Fire detection

While computerised control of mobile equipment has been used in the open pit, very little experience existed with underground applications when Finsch moved underground in 1990. Following a feasibility study, the mine proceeded with the installation of a computer-based system to monitor and control the movements of LHDs underground. This was in addition to the leaky-feeder system installed for voice communication, which formed the backbone for data transmission. Hardware purchased consisted of microcomputers and infrared location beacons. Customised software seeks to optimise production by making LHD assignments to drawpoints and tips within the constraints imposed by the supervisor.

Additional benefits include tonnage, availability, and utilisation statistics; breakdown reports; and number of service and fuel calls.

54.5   UNDERGROUND MINING METHOD FOR BLOCK 4

The decision to change the mining method for block 4 from blasthole open stoping to block caving was as a result of the inherent risk for dolomite sidewall and internal kimberlite failures as well as the inability to separate waste underground. Subsequent to the completion of the block 4 prefeasibility study in 1994, it was decided to position the block cave undercut and extraction levels on the 61- and 63-m levels, respectively. The advantage was that much of the existing ground handling infrastructure would be incorporated into the design, thereby lowering capital expenditures.

Investigation of the undercut selection included three alternatives: a narrow, flat undercut; a narrow, inclined undercut; and a narrow, inclined undercut with observation tunnels at the apexes. Final selection between a narrow, flat, advanced undercut and a narrow, inclined, advanced undercut has been delayed pending the results of modelling presently being conducted by the geotechnical section. The tests on a narrow, inclined, advanced undercut presently being conducted at the Premier Mine will also be observed before a final decision is reached.

A low, narrow undercut is favoured because (1) it generates limited ground, thereby reducing loading requirements, (2) it reduces drilling and blasting requirements that lead to high undercut advance rates, and (3) it is simple to implement.

The undercut initiation position on the south side of the ore body (Figure 54.15) was selected on the basis of geotechnical data. The area consists of F8 kimberlite, which is a weak kimberlite having a rock mass rating (RMR) of 49, and hence should cave relatively easily. The hydraulic radius should be attained before intersecting any internal contacts. It has been decided to implement an advance undercut where only limited development will be completed on the production level prior to advancing the undercut overhead. The undercut operation will lead the extraction level drawbell development by at least 20 m so that the latter can take place in a destressed environment.

Drawbell design will be congruent with the present industry standard. It is crucial that the secondary breaking strategy is successful as the crusher has a limited acceptance size. The shear

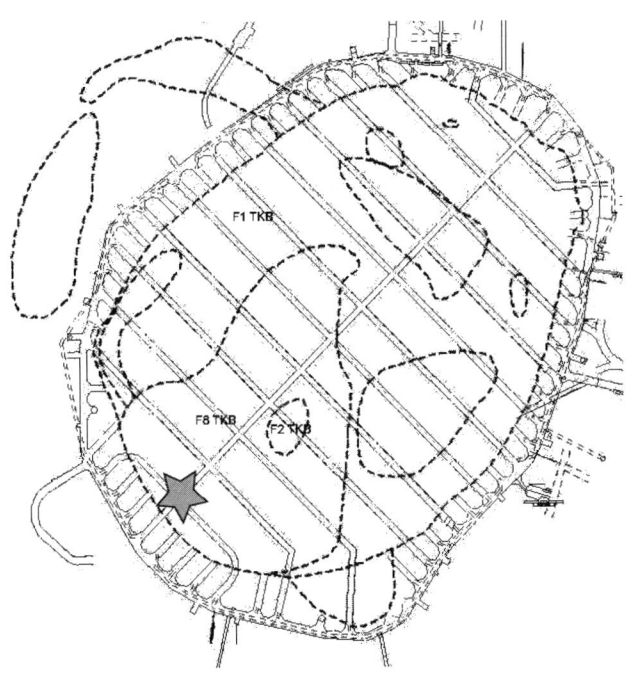

FIGURE 54.15   Geological plan of undercut level with undercut initiation position shown

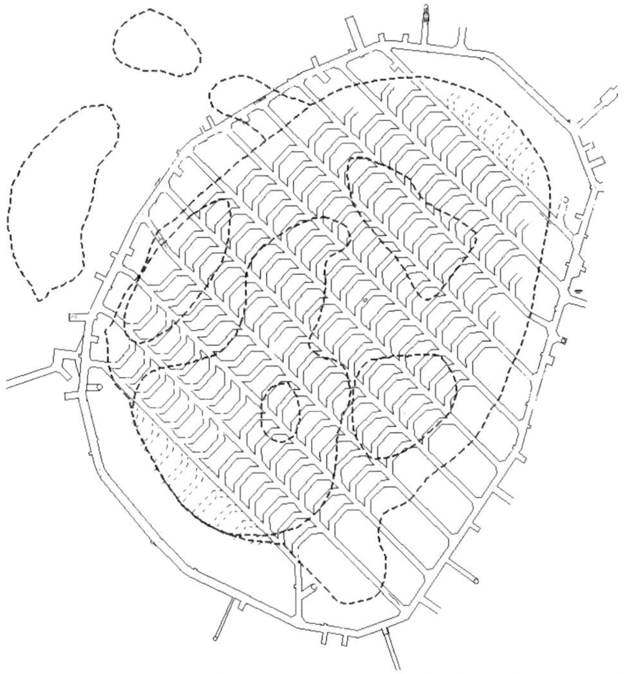

FIGURE 54.16   Undercut level layout and geology

weight and size of the available crushers in conjunction with shaft constraints have imposed these limitations.

For a draw control strategy to be successful, a number of requirements must be considered.

- Realistic production targets must be set.

- Geophysical constraints for the block cave must be known.

- An information system for gathering and distributing draw control data must be in place.

- Personnel need to be identified to manage the draw control system.

- Personnel must be trained in the theory of block caving.

Computer-based systems, which feed tonnage and location data to specialised draw control programmes, have become popular tools in draw control strategies. These include a system of identification beacons in drawpoints and radio storage units on LHDs that send data to the surface via a data highway. A system similar to this will be used in block 4, but with a more advanced vehicle monitoring and dispatch system than that used in block 3.

The "expected case" scenario for the block 4 resource has been calculated using the Laubscher dilution model and will amount to 36.8 million tonnes. It will include 28.4 million tonnes of kimberlite, 4.8 million tonnes of mixed material, and 1.7 million tonnes of dolomite dilution.

The interaction of the production and ground handling equipment and processes with each other in the block 4 operations will be complex. The Siman-Arena simulation package was therefore used to evaluate the various ground handling options and ensure that production requirements will be met.

According to the results and the high number of drawpoints contained in the ore body, it is possible to produce 5 million tonnes per year from block 4. This scenario was found to be unacceptable as it resulted in a peak in production in 3 yr, after which relatively new equipment would be mothballed as well as cause a retrenchment of 30% of the underground workforce. An annual production rate of 1.6 million tonnes has been selected as the optimum as it will not require changes to the present hoist arrangements.

With a hoisting speed of 12 m/s, the main shaft hoisting capacity is 4.3 million tonnes a year for a two-shift operation and 5.3 million tonnes a year for a three-shift operation. A maximum storage capacity of 2,000 tonnes exists between the production and hoisting processes and is sufficient to make up for lost production at shift changeover times. Shift changeover times will be used to ventilate the extraction level after blasting and should be about 15 to 30 min. It is, however, expected that the reentry time could be longer than this because of blasting the drawbells during the 5-yr construction phase, which will delay the production and hoisting processes.

In the expected case scenario, the life of the block will be 13 yr, commencing production in 2003 and ending in 2015.

54.5.1   Tunnel Layout, Development, and Support

The implementation of a complete "mirror-image" extraction pattern has been ruled out because of the presence of the southwest precursor and a set of dykes (Figure 54.16). Once the conceptual model for the undercut and extraction levels had been determined following geological and geotechnical analysis of the available data, primary development was commenced. This consisted of the rim tunnels on the 61- and 63-m levels and their accessways from the 62-m level.

Excavations of many different dimensions will be formed during the development for the block 4 operation. Laubscher's geomechanics classification system is considered adequate for support design in the country rock.

Layout of the required tunnels was not a simple process, as cognisance had to be taken of the existing 62- and 65-m level excavation and the possible detrimental interaction of these with the proposed 61- and 62-m level tunnels. A further complicating factor was the need for the permanent undercut and extraction level tunnels to be located at a sufficient distance from the pipe contact zone so as not to be affected by stress charges induced by the undercutting process. In many instances, the final tunnel positions

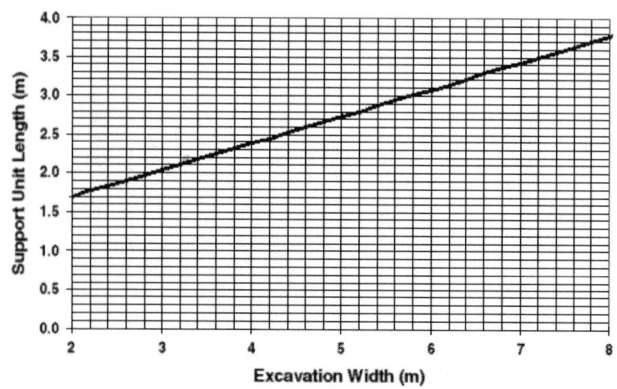

FIGURE 54.17   Support units lengths for excavations in country rock

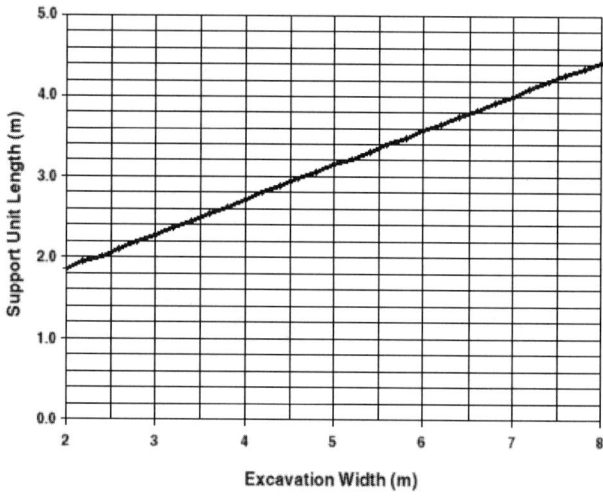

FIGURE 54.18   Support units lengths for excavations in kimberlite

were a compromise between the ideal and what was actually possible without negatively impacting the block 3 structures.

The nature of the carbonate rock mass below the 60-m level is well known and has a typical RMR of between 75 and 80. This becomes a mining rock mass rating (MRMR) of between 55 and 59 if good conventional blasting is employed. Laubscher recommends a unit spacing of 1 m, with the support unit length given by—

$$L = 1 + (0.33\,W \times F)$$

where   L = length of spacing,

               W = the span of the excavation in metres,

and   F = a factor dependent on the MRMR.

For an MRMR between 51 and 60, F = 1.05. These data have been used to compile the graph in Figure 54.17, which is applicable to all tunnels in the country rock.

In addition to normal rock bolt support, spiles will be required at the contact zone. Experience has shown that because of the limited cohesion on the interface between the country rock and kimberlite, movement and significant failure are common if support is inadequate. To date, a total of 4,884 m has been developed. A total of 17,367 m is required to install the block cave and infrastructure.

With respect to all the large excavations (e.g., crusher chamber and workshops), the support is specifically designed for each occurrence, taking into account rock mass conditions and the purpose of the excavation.

Kimberlite development will be driven by production requirements. Stress damage will be kept to a minimum by the postundercut installation of drawpoints and drawbells. The planned layout for the undercut features 12 tunnels spaced at 30-m intervals oriented in a northeast direction. The tunnel spacing allows flexibility in the final undercut method selection. The extraction level is based on a herringbone layout with 11 northeast-oriented tunnels spaced at 30-m intervals. This allows for the installation of 141 drawbells with the capacity to produce the required 3.6 million tonnes a year. During the initial stages of the undercutting process, few problems are likely to be encountered irrespective of the pillar dimensions, provided that complete extraction is achieved.

Because of the complex internal structure of the Finsch Mine pipe between the 51- and 63-m levels, support design must be based on the weakest rock type, i.e., F1 tuffisitic kimberlite breccia. Support lengths are calculated using the same principle as described above. The data have been used to produce the graph in Figure 54.18.

The mode of failure is expected to be brittle failure together with strain softening. Investigations have revealed that typically

in the kimberlite breccia, a fracture zone develops to a depth of between 200 to 800 mm and is accompanied by the frittering away of the ground between the rock bolts.

Only short-term activities will be undertaken in the undercut tunnels and so these tunnels can be considered as sacrificial, requiring only the minimal support necessary to ensure the safety of the workforce during undercutting.

Development Phase. Installation of 2.7-m-long, 20-mm in diameter, resin-grouted rock bolts on 1-m spacings continued down to the footwall to within 1 m of the development face.

Postdevelopment Phase. Installation of intrabolt support in the form of tendon straps from grade line to grade line for a maximum of 15 m behind the face. Spray with sealant.

The proposed postundercut development of the drawbells will result in development taking place in a destressed environment. Consequently, high stresses will not be encountered, although the effects of the undercut abutment stresses (i.e., extension fractures) will be evident to a greater or lesser degree depending on local rock mass strength. Thus the support system required will be one that can maintain a stable mass. In addition, provision must be made to reinforce the brow and thus minimise wear, which will otherwise be a major problem. In most cave mining operations, this situation is exacerbated by the secondary blasting necessary to break large rocks. Any design put forward now must allow for the effects of blasting in, or close to, the drawpoint.

Where the rock mass is found to have an exceptionally low RMR (<35), longer rock bolts and fully grouted cable anchors will be required. The spacing and number of such units will be individually specified as and when the need arises.

Investigations have concluded that the primary support in the production tunnels should consist of 2.9-m-long rock bolts. Previous experience at the Finsch Mine has indicated that support must be taken below grade elevation to the footwall, and there is no justification for changing this. A nominal rock bolt spacing of 1 m should be adopted. Shotcrete will be required, and it is recommended that a modification of the existing standard be employed, i.e., a multilayered construction consisting of sealant, 30 mm of shotcrete, mesh and tendon straps, and a final 70 mm of shotcrete. At the Finsch Mine, a complete lining is considered necessary because of the weak kimberlite and its propensity to weather rapidly. However, the application of shotcrete to the hanging wall should not take place until the undercut has passed over and the tunnel is effectively destressed.

In addition, two rows of 6-m-long anchors will be installed on a 1- by 1-m² pattern from footwall to footwall 3 m back from

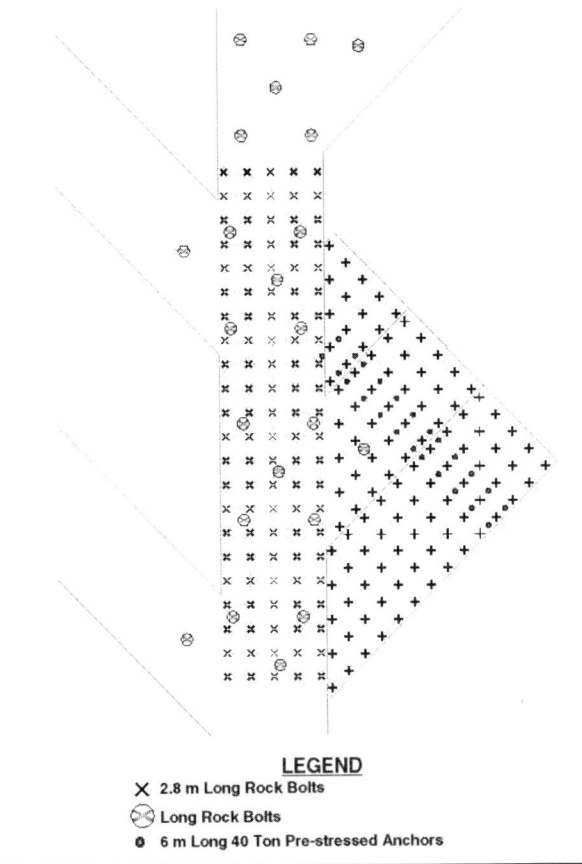

LEGEND

✗ 2.8 m Long Rock Bolts

⊗ Long Rock Bolts

⊙ 6 m Long 40 Ton Pre-stressed Anchors

FIGURE 54.19 Plan showing extraction level support

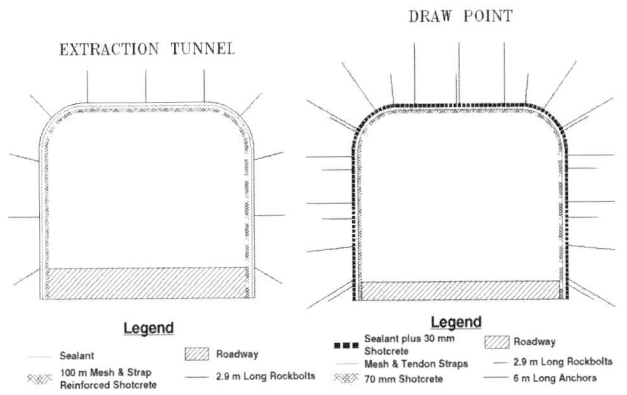

FIGURE 54.20 Sections showing extraction level support

the brow position. Figures 54.19 and 54.20 show the complete support system.

54.5.2 Ground Handling Infrastructure

Ground handling will be by means of LHDs tipping directly into dump trucks near the ore body. The dump trucks will tram the ore to a gyratory crusher situated at the shaft (Figure 54.21), after which the ore will be conveyed to the present storage passes on the 65-m level. The designed ground handling system is unique to the Finsch Mine. Seven 12-tonne LHDs will tram ore to nine 40-tonne trucks at four tipping points located on the extraction level in the country rock.

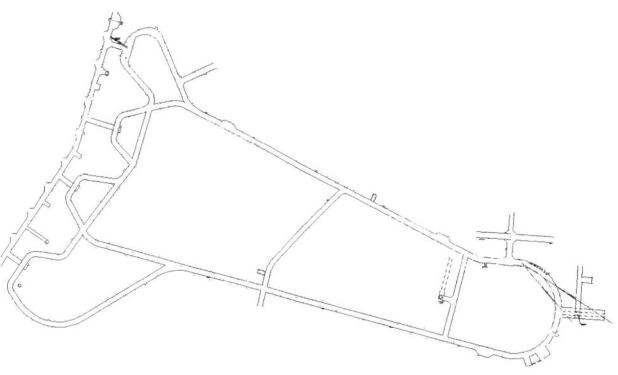

FIGURE 54.21 Layout for truck haulage

With respect to the undercut ground handling, a number of options have been considered, and the option of double-handling has finally been selected as it offers simplicity of operation, fewer processes, and few if any excavation stability problems. As 10% of the annual production of block 4 will be from the undercut level, the cost analysis underlines the financial viability of the double-handling option. At 20% of annual production, there will be little difference over the life of the block cave.

The brief for block 4 was to devise the simplest ore flow system possible. After fully investigating many options, we found that the best was to have LHDs tipping directly into trucks. Simulations have indicated that four loading points will be required to achieve the required production rates. Special bays will be constructed on the tramming loop with sufficient headroom for direct tipping. Dust collection systems will be installed so as to overcome the dust problem associated with working with dry kimberlite.

During the course of the simulation exercises, it became apparent that the performance of the various trucking options was highly dependent on the degree of interaction between the LHDs and the truck fleets. The direct tipping of an LHD into a truck is very dependent on having the correct number of trucks. Fairly inefficient use of the truck fleet was evident. Approximately 50% to 55% of the available truck time was spent in the loading bays rather than in the useful work of transporting ore. The only relief lies in the option of "over-trucking" (providing additional trucks) or the introduction of an intermediate bin or container to decouple operation of the two fleets.

The trucks will operate in a one-way circuit, discharging ore near the shaft into a single gyratory crusher, the detailed design of which is still in progress.

Of the methods compared for horizontal transportation, trucks were the preferred option because their use offered simplicity, flexibility, safety, and availability. Trucks are more flexible and more easily rerouted compared to conveyor systems. They have a low initial capital cost, and large rock sizes can be transported easily.

The Arena simulation exercise determined the trucking fleet requirements, and as a result, a financial comparison of the two systems was made. The fleet size required to meet block 4 production requirements will consist of eight 50-tonne-capacity trucks (two nonoperational) and nine 40-tonne-capacity trucks (two nonoperational).

In addition to the above factors, the following advantages for the trucking option have been identified.

- They are financially more feasible.

- They reduce the number of processes.

- A single crusher is required.

- Ore passes are eliminated.

- Supervision of production and ground handling sections will be concentrated on one level.

- Logistics for development and construction of the ground handling system are simplified.

It was decided to use fixed-bowl trucks for tramming ore from the ore body to a central crushing site near the main shaft area. Centralised crushing will reduce the number of crushers required and will maximise their use.

The fragmentation expected from the block 4 caving operations was based on a study completed for block 5. It is predicted that 30% of the rocks will be greater than 2 m³. The largest crusher that will satisfy shaft constraints and that can handle rocks 2 to 3 m³ in size with minimal presizing will be chosen.

The maximum production requirement from block 4 is 16,000 tonne/d (4 million tonnes a year) for two 9.2-hr shifts or 1,000 tonne/hr at 87% crusher utilisation. All crusher maintenance is scheduled for the third shift.

The following crushing options were considered.

1. *Existing mineral sizers*

   The existing mineral sizers at the Finsch Mine are prototypes and are no longer manufactured. An analysis showed that refurbishing these would cost more than buying new sizers.

   The use of new mineral sizers was not recommended, as they do not satisfy the acceptance criterion. Their working costs are considered high, the acceptance size is less than that required, and because of the varied rock types, the manufacturer does not recommend using them.

2. *Jaw crushers*

   Two crushers are required to meet the production requirements, which will result in an increase in support costs.

   A bin and apron feeder are required to provide continuous feed, resulting in more unit processes. The crusher produces slabby material, which is not favoured by the treatment process.

   A complicated layout is necessary.

3. *Gyratory crushers*

   Only those crushers that satisfied the following criteria were considered.

   - Weight of heaviest component = 28 tonnes
   - Acceptance size = 2.3 m³ (1.3 by 1.8 by 1 m)
   - Average capacity = 1,200 tonne/hr
   - Product size = < 250 mm (99%)

   The installation of this type of crusher requires a simple layout.

   Constant feed is not necessary; therefore, direct tipping into the crusher is feasible.

   Only five processes are required in the sequence from tipping the ore into the crusher to tipping the ore into the storage passes.

Subsequently, the manufacturers were requested to adapt their designs to satisfy the main shaft constraints using only one crusher. Based on the selection criteria, each crusher type was ranked. From the ranking, it was clear that a gyratory crusher would best meet the selection criteria even though it would contain the heaviest single component. The detailed design of such a crusher is currently in progress.

From the crusher discharge shoot, a short feeder conveyor will carry the crushed ore to the shuttle conveyor described in section 4.4.7. At this point, the process will be unchanged from that used in blocks 1, 2, and 3.

54.5.3 Equipment Selection

Drill Rigs. Primary drilling in block 4 will take place from the beginning of operations and will cease after the completion of the undercut and drawbells. Three longhole drilling rigs will be required for block 4. The final selection of drill rig make and model will be done later. The selection will, however, be based on the following requirements.

- Length of hole = up to 25 m
- Hole diameter = 56 to 76 mm
- Capability to drill parallel holes
- Mechanised rod change
- Facility to automate drilling process

LHD Selection. Various operational aspects were considered before reaching a final decision regarding the bucket size and type of LHD to be used for handling ground on the undercut and extraction levels. Aspects such as proven track record in the underground mining industry and advances in the field of diesel and electric LHD technology were considered. Consideration was also given to the cost-effective operation of the various technologies in terms of anticipated fleet size options. The major criteria for evaluation were identified, and the characteristics of diesel and electric LHDs were compared considering the Finsch Mine environment.

After all relevant information regarding the operation of block 4 was reviewed, the use of diesel-powered LHDs was recommended, mainly for the following reasons.

1. The anticipated maximum one-way tramming distances for LHDs in block 4 on the extraction level will be up to 275 m, but the maximum currently available lengths of trailing cables are approximately 240 m.

2. Although the operating cost of air-cooled diesel LHDs is higher than that of electric LHDs, recent developments in the field of water-cooled diesel engines have resulted in lower fuel consumption (28.7 L/hr versus 34 L/hr).

3. Less emission of gas and heat.

4. Lower noise levels (97 versus 103 dB).

5. Reduced operating and maintenance costs.

6. Although normally electric LHDs require less air volume and flow for acceptable working conditions, the high concentration of air-borne dust generated by LHD operations with kimberlite necessitates the use of even higher volumes of ventilating air in loading drives. Thus, this aspect of ventilation was not relevant to the selection process.

Specific criteria were considered regarding the size selection of LHDs selected for use in block 4 on the extraction level. Planned daily tonnage, the cost-effective operation of the smallest possible LHD fleet, as well as tunnel and drawpoint dimensions, were considered in detail. Bucket capacities of 10 and 12 tonnes were compared in terms of required fleet sizes and matched with selected truck bowl capacities to optimise the utilisation of trucks bowls fully.

Initial simulations using the Arena model have determined the fleet sizes of 10- and 12-tonne LHDs on the extraction level would require nine and seven operating machines, respectively.

The block 4 cave is of limited vertical extent (100 m) compared with contemporary block caves. The result is that relatively large fragments of ore are expected to report to drawbells because of lower stresses and the shorter time of communition among rock fragments. It is therefore important to avoid loading large rocks. Large rocks loaded at the drawbells by LHDs will also have to be handled by dump trucks. Momentum and impact generated by tipping these rocks will be costly to trucks.

After all relevant aspects were analysed and evaluated, 12-tonne LHDs were recommended for the extraction level for the following reasons:

- Larger LHDs will maximise the daily tonnages trammed per LHD loading cycle, thereby reducing truck loading times. The limited number of loading drives limit the size of the LHD fleet and will require that maximum tonnages be loaded per loading cycle.

- The bucket capacities of 12-tonne LHDs will more effectively match those of 40-tonne dump trucks.

- Fewer LHDs are required to achieve daily production requirements (coming from only 11 loading drives).

- A 12-tonne LHD fleet will be cheaper to operate per tonne trammed.

- 12-tonne LHDs presently being used can be used for block 4 as well (as part of the existing LHD replacement program).

It is currently envisaged that the same type of units currently used in block 3 will be used in block 4.

Truck Selection. Ore will be transported horizontally by trucks along a designated tramming loop to a crusher situated close to the shaft. An investigation into the size and type of dump truck was undertaken with the objective of achieving the optimum utilisation of truck tramming capacity with the lowest possible operating cost per tonne transported. A computer simulation model (Siman-Arena) was used to determine truck cycle times, as well as the fleet sizes required to achieve production targets. Two sizes of trucks were identified for use in block 4, namely 40 and 50 tonnes. These trucks were compared in terms of capital and operating cost per tonne and matched to selected LHD bucket capacities to ensure that the maximum use of trucks would be achieved.

Part of the truck selection process was to compare diesel-powered trucks to electrically powered trucks in terms of capital and operating costs. High emphasis was also placed on the operational efficiency of these trucks in the block 4 underground environment. Various truck models were compared in terms of the above criteria, but no final decision has been made regarding the preferred model. Tender specifications to truck manufacturers will include the ability of trucks to be operated fully autonomously or tele-remotely. The electronic monitoring of vital engine signs will also be included.

The following issues were identified as part of the investigation into the type of truck to be used.

- Electric trucks are, due to their special torque characteristics, ideally suited for steep ramps. Most haulageways in block 4 will be flat without any significant ramps.

- The capital cost of a 50-tonne electric truck is approximately R 7 million using a 1998 exchange rate of 6 SA rands to 1 US dollar, compared to R 4.1 million for a similar-sized diesel truck. The initial capital cost of electric trucks with all related infrastructures is approximately 40% higher than for a similar 50-tonne diesel fleet.

- The low operating costs of electric trucks compared to diesel trucks make them more suitable for long tramming distances. The minimum reported distances of operation in the industry are approximately 3,500 to 4,000 m, while the maximum anticipated travelling distance on block 4 will only be 1,500 m.

- A distinct advantage of electric motors is reduced airflow in terms of ventilation compared to diesel engines, which require high volumes of air to dilute and dispose of gasses and heat.

- Although the operating costs of electric trucks are approximately 30% lower than those for diesel trucks and ventilation requirements of electric trucks are reduced, the high initial cost of electric trucks makes them an expensive option for block 4. Electric trucks will, however, be easier to automate, which is a specific requirement for block 4 operation.

The use of diesel trucks for block 4 has been recommended based on the high capital cost of the electric truck option. The nature of the block 4 haulageway with its flat, short tramming distances negates one of the major advantages of electric trucks, namely the ability to negotiate steep ramps at high speed.

Bucket capacities of trucks need to be matched to those of production LHDs to ensure the optimum bowl use of trucks and the smallest possible fleet size. The effective bucket capacity of a 12-tonne LHD is 9.6 tonnes. It is anticipated that two LHDs will be loading trucks at each tipping bay and that one LHD will be assigned to a loading drive. By using a 12-tonne LHD for tipping into a 40-tonne truck, four LHD loads are required to fill the truck. In the case of 50-tonne trucks, an additional LHD load will be required to fill the truck to an acceptable level of bowl use.

It has been proven that in order to ensure a truck use of 95% or more using an LHD with a capacity of 9.6-tonnes (12-tonne LHD), the best match will be a truck with a capacity of either 40 or 50 tonnes.

With over 10 yr of trackless operation, Finsch has had many units modified to suit mine requirements. It is not anticipated that any new designs will be required for block 4. The range will remain as is currently in place for block 3, although with the operation centralised on two levels, the number of service vehicles will be significantly reduced.

The benefit of the on-site underground maintenance of vehicles has been proven over the Finsch Mine over the past 10 yr. Consequently, the same policy is to be adopted for block 4. The use of one central workshop located close to the ore body was proposed for the following reasons:

- All vehicles will be repaired and serviced centrally. This implies that other workshops and the rest of the mine can be sealed off, resulting in reduced ventilation requirements and eliminating required examinations of excavations (making safe).

- A central stores system can be operated, eliminating the unnecessary transport of spare parts between workshops.

- All resources will be located centrally, resulting in improved supervision.

- The once-off support of one large workshop is preferable to several smaller workshops. To be closer to the entire production fleet, the workshop will be located on the 63-m level west of the trucking loop and close to the shaft (Figure 54.22).

Extra development is required to ensure separate accesses for both LHDs and trucks on the 63-m level, as interference between LHD and truck fleets should to be eliminated, but no steep access ramps will need to be negotiated by vehicles entering the workshops.

The ventilation strategy for this option is to draw in air from the 62-m level station. The larger portion of the air will go to the production areas via the LHD access tunnel, while a smaller portion will be returned directly to a return airway on the 63-m level to ventilate a tire store, oil and battery bays, and toilets in the workshop. The proposed location of the workshops is therefore in a stable area on the 63-m level northwest of the ore body at a distance of approximately 160 m from the production area and approximately 180 m from the shaft. A width of 6 m is proposed for the LHD and truck access tunnels into the workshops with

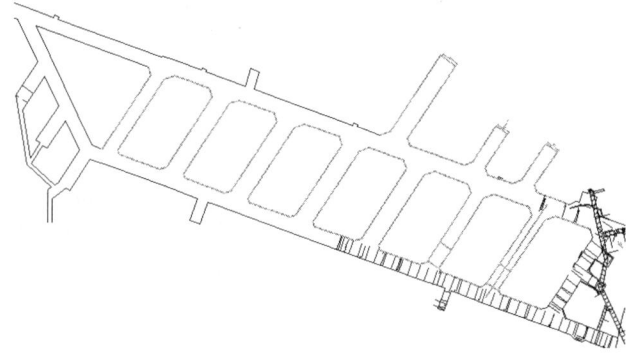

FIGURE 54.22   Layout for block 4 vehicle workshop

adequate laybys. LHDs and trucks can therefore not pass each other in access tunnels, but delays should be minimised by using an automatic vehicle monitoring and dispatch system.

54.5.4   Drilling and Blasting

The final details on the drilling and blasting system to be employed can only be decided once the undercut method has been selected. The general principle will be that the pillars created by the undercut development will be removed by retreat mining. If a flat undercut is selected, drilling horizontal holes through the entire width of the pillar half way from either side will be all that is required. For an inclined undercut, the system will be more complex, with inclined holes drilled to a position midway between a pair of drives. It is essential that complete extraction is achieved and that no crown pillar remains. To achieve this, it is envisioned that one set of holes will overlap the other with a very slight middling. Such a system would require the permission of the government mining engineer, as it would not be in accordance with current legislation.

It is currently planned to use ANFO. Charging will take place in a manner similar to that employed in block 3. Initiation will be with an electronic centralised system.

Strict procedures concerning concussion blasting must be introduced to ensure that miners do not use more explosives than necessary, thus minimising the effect of blast damage to drawpoint brows until such time as nonexplosive techniques are perfected for high hang-ups. Hang-up types must be properly identified, taking into account height and stability. In unstable conditions, a concussion blast will be performed to either stabilise conditions or bring a hang-up down. A trained and experienced miner will evaluate these parameters and conduct secondary breaking operations accordingly.

LHDs will be prevented from loading at drawpoints adjacent to a hang-up that has been drilled and charged or is in the process of being drilled or charged. This will be managed with the assistance of the dispatch system.

Concussion blasting is traditionally used to clear high hang-ups and rock jumbles, while lay-on charges are used to reduce oversized material to acceptable sizes. This may result in extended reentry periods and damage to equipment, brows, and tunnels, which will ultimately have to be reconstructed.

Companies such as Tamrock and Maclean Engineering are in the process of developing high-reach drilling and blasting rigs for the purpose of bringing down high hang-ups. These rigs are equipped with telescopic boom arrangements and should be able to reach hang-ups approximately 17 to 18 m above the footwall.

Oversize material and low hang-ups can, however, be broken by drilling and blasting or by means of the Ro-Bust nonexplosive technique. Both have been considered for the purpose of this exercise. If the Ro-Bust system is used, the life of a drawpoint

would be doubled if compared to conventional drilling and blasting. The Ro-Bust system is, however, R 0.39/tonne more expensive than drilling and blasting. The total cost difference between these two systems is approximately R 552,000/yr. The Ro-Bust system is safer than conventional techniques as it uses the minimum amount of energy to break rocks, and there is no resulting concussion, fly rock, dust, or toxic gas. Its operation is relatively simple, and there will be little interference with LHD operations because of the speed of the process. Fifty percent fewer personnel are required than when drilling and blasting. The need for activities such as blasting clearances and ventilation checks is also eliminated.

Most of the secondary breaking will be done by the Ro-Bust system. The success of this system is crucial to the success of block 4, and it is therefore important to continue to work in cooperation with the manufacturer (Maclean and Swartklip) on the development of the system. A method must be found to reduce the cost of the propellant cartridges, as this would significantly reduce the operating costs of the Ro-Bust system. When the remote cartridge system has been developed, it will be installed on the existing Ro-Bust unit. All subsequent units purchased will have an integrated remote cartridge system.

The full length of a Ro-Bust vehicle is 11.3 m, and the effective length of a drawpoint tunnel on the extraction level is 7 to 8 m. The Ro-Bust vehicle, while working in a drawpoint, will therefore protrude across the full width of an extraction tunnel. It has been planned that the Ro-Bust vehicle will not hamper LHD operations. In practice, there may be situations where a Ro-Bust rig will be required to work between an LHD and a truck loading area, and it is suggested that a shorter carrier be sought.

A rock breaker will be installed at each of the three tips on the undercut level. A static Ro-Bust and rock breaker, each with a 7.5-m-reach boom, are to be installed at the crusher.

54.5.5   Ventilation, Pumping and Flood Control

Return airway tunnels 1.8 m in diameter will be developed to link four separate ventilation districts to facilitate simultaneous loading and drilling operations on both sides of the pipe on the 61-m level. Dust extraction facilities at the tipping points on the extraction level will incorporate scrubbers and will eliminate the need for ventilation doors in the truck haulageway. The loading drives on the southwestern extent of the extraction level will be open for secondary breaking and construction activities by linking these tunnels to a dedicated return airway on the 65-m level. Open drawpoints will upcast from the 63- to the 61-m level by the creation of negative air pressure on the 61-m level.

Most of the existing ventilation appliances will be reused. These will include the 35-m level exhaust fans, the centrifugal booster fans, and the fire detection heads. The total air flow rate required for block 4 will be 614 m³/s.

The present water and mud handling systems will be used to handle the water and mud generated by block 4 operations. The capacities of the sumps and pumps are sufficient to meet block 4 requirements, and no changes are foreseen. The third settler may be needed to cope with additional mud generated by the scrubbers at the tipping points and at the crusher. Mud created by the scrubber installations will be routed to the sumps via a dedicated service tunnel graded to allow the unassisted flow of mud. The block cave rim tunnels and extraction tunnels have been graded toward the main water pass.

54.5.6   Computerised Monitoring and Control

The strategy adopted for the automation of block 4 operations is to integrate the operation of all major underground ground handling processes and mobile vehicles into a centrally controlled, mine-wide multimedia network. It is envisaged that all stationary processes (hoisting, conveying, ventilation, surface plant, crushing) will be fully automated and controlled from one centralised control centre on the surface. The wireless automatic dispatch system will be fully integrated with other equipment that will be operated from a supervisory control platform (SCADA).

Automation of ground handling has the potential to decrease operating costs of mining operations significantly. The use of fully autonomous production vehicles has been considered for block 4 as the final step. Currently, only a few mines in the world have adopted the use of tele-operated or fully automated production vehicles. Several new technologies incorporating wireless communication backbones and automatic vehicle guidance have been evaluated, but the operation of these technologies in the mining industry has not yet been proven. Provision has, however, been made in the design of automation networks for such a system.

The proposed control and instrument network for block 4 was designed around detailed user requirements identified by production and operations personnel. An extensive fibre-optic network management system will be installed to accommodate

the transfer of real-time video images, control and monitoring data, and production vehicle data over a single fibre-optic network at rates up to 155 megabyte/s. The use of asynchronous transfer mode switching was recommended for this purpose.

54.6 ACKNOWLEDGMENTS

The author acknowledges that the material contained in this paper is the work of many people and that he has merely had the opportunity to edit and revise the text. Sections 1 through 4 are based on the paper by Gould and Lea. The data for section 5 were taken from the block 4 feasibility report. The author would like to thank the director of Support Services, De Beers, and the general manager of Central Mines for permission to publish this paper.

54.7 REFERENCES

Finsch Mine. 1998. Block 4 Project Feasibility Report, Internal Report.
Gould, S.M., and R.F. Lea. 1990. Finsch Mine—The Change from a Conventional Open Pit to a Trackless Underground Operation.

CHAPTER 55

Cave Mining—The State of the Art

D.H. Laubscher[*]

55.1 INTRODUCTION

The expression "cave mining" will be used here to refer to all mining operations in which the ore body caves naturally after undercutting and the caved material is recovered through drawpoints. The term encompasses block caving, panel caving, inclined-drawpoint caving, and front caving. Caving is the lowest-cost method of underground mining, provided that drawpoint size and handling facilities are tailored to suit the caved material and that the extraction horizon can be maintained for the life of the draw.

The daily production from cave-mining operations throughout the world is approximately 370 kt/d. Table 55.1 shows production broken down by layout.

By way of comparison, the South African gold mines produce 350 kt/d.

Today, several open-pit mines currently producing in excess of 50 kt/d are examining the feasibility of implementing low-cost, large-scale underground mining methods. Several underground cave mines that produce high tonnages are planning to implement dropdowns of 200 m or more. This will result in a considerable change in their mining environments. These changes will necessitate more detailed mine planning, rather than the simple projection of current mining methods to greater depths.

As more attention is directed to mining large, competent ore bodies with low-cost underground methods, it is necessary to define the role of cave mining. In the past, caving has been considered for rock masses that cave and fragment readily. The ability to better assess the cavability and fragmentation of ore bodies, the availability of robust load-haul-dump (LHD) machines, an understanding of the draw control process, suitable equipment for secondary drilling and blasting, and reliable cost data have shown that competent ore bodies with coarse fragmentationl can be mined at a much lower cost using caving rather than with drill-and-blast methods. However, once a cave layout has been developed, there is little scope for change.

Aspects that have to be addressed are cavability, fragmentation, draw patterns for different types of ore, drawpoint or drawzone spacing, layout design, undercutting sequence, and support design.

Table 55.2 shows that there are significant anomalies in the quoted performance of different cave operations.

It is common to find that old established mines that have developed standards during the course of successful mining in the upper levels of an ore body are resistant to change and do not adjust to the ground control problems that occur as mining proceeds to greater depths or as rock types change. Mines that have experienced continuous problems are more amenable to adopting new techniques to cope with a changing mining situation. Detailed knowledge of local and regional structural geology, the use of an accepted rock mass classification to

TABLE 55.1 Production per day obtained using different mine layouts in kilotons

Grizzly	90
Slusher	35
Load-haul-dump machine	245
Total	370

TABLE 55.2 Anomalies in quoted performance of caving operations

Quote	Explanation
1.0 96% of ore recovered for 100% mineral extraction	**1.0** Underevaluation of ore body and dilution zone
2.0 Correct drawpoint spacing, but occurrences of 200% overdraw with 30% waste dilution entry	**2.0** Case of highly irregular draw and underevaluation of dilution
3.0 15% dilution entry in spite of correct drawpoint spacing and uniform fragmentation	**3.0** Drawpoints being drawn in isolation
4.0 Ore from lower 100 m of draw column still reporting in drawpoint, even though 260 m of ore has been drawn	**4.0** Large range in fragmentation and irregular, high values in dilution zone

characterize the rock mass, and knowledge of regional and induced stress environments are prerequisites for good mine planning. It is encouraging to note that these aspects are receiving more and more attention.

The Laubscher rock mass classification system provides both in situ rock mass ratings (IRMR) and rock mass strength. Such a classification is necessary for design purposes. The IRMR defines the geological environment, and the adjusted or mining rock mass ratings (MRMR) consider the effects of the mining operation on the rock mass. Figure 55.1 is a flow sheet of the MRMR procedure with recent modifications. The reader is encouraged to read the paper "The IRMR/MRMR Rock Mass Classification System for Jointed Rock Masses" by Laubscher and Jakubee, which is included in this volume. The ratings describe in detail cavability, subsidence angles, failure zones, fragmentation, undercut-face shape, cave-front orientation, undercutting sequence, overall mining sequence, and support design.

55.2 FACTORS AFFECTING CAVING OPERATIONS

The 25 parameters that should be considered before implementing any cave mining operation are set out in Table 55.3. The parameters in capital letters are a function of the parameters that follow in the same box. Many parameters are uniquely defined by the ore body and the mining system and are not discussed further. The parameters considered later are common to all cave-mining systems and need to be addressed if any form of cave mining is contemplated.

---

[*] Bushman's River Mouth, South Africa.

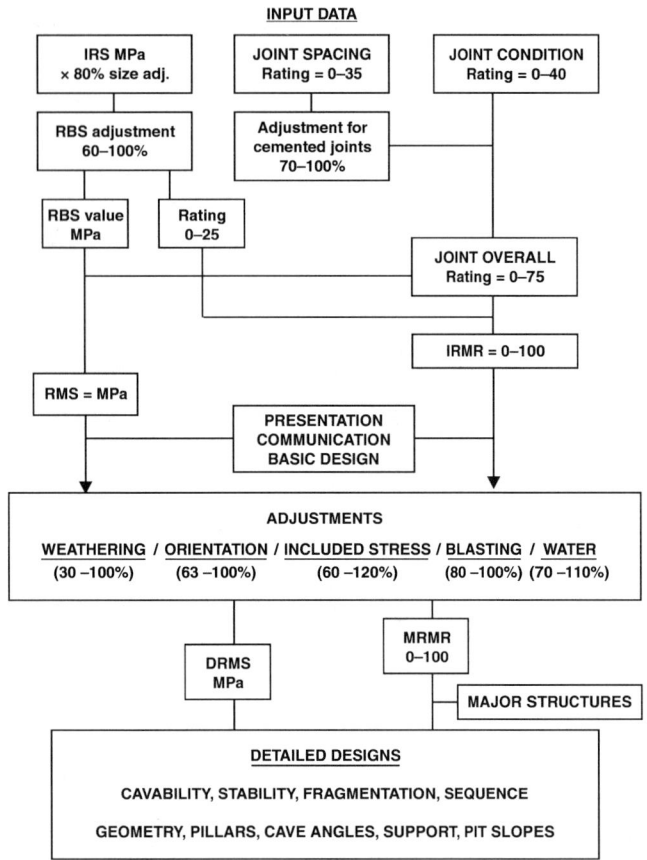

FIGURE 55.1   Flow sheet of MRMR procedure with recent modifications

55.2.1   Cavability

Monitoring a large number of caving operations has shown that two types of caving can occur—stress and subsidence caving. However, it is better to use the terms "vertical extension" to mean upward propagation of the cave and "lateral extension" to mean the propagation of the cave as a caved block is expanded.

Vertical extension caving occurs in virgin cave blocks when the stresses in the cave back exceed the strength of the rock mass. Caving can stop when a stable arch develops in the cave back. The undercut must be increased in size or the boundaries weakened to induce further caving. High horizontal stresses acting on steep dipping joints increase the MRMR. This was the situation in block 16 at the Shabanie Mine. A stable back was formed when the undercut had a hydraulic radius of 28 and the MRMR was 64. When block 7 (adjacent to block 16) caved, the horizontal confining stress was removed, which resulted in a reduction in the MRMR to 56 in block 16. At this point, caving occurred.

Lateral extension caving occurs when lateral restraints on the block being caved are removed by mining adjacent to the block. This often results in a large stress difference, leading to rapid propagation of the cave and limited bulking.

Figure 55.2, based on a worldwide experience base, illustrates caving and stable situations in terms of the hydraulic radius (area divided by perimeter) for a range of MRMR values. An additional curve has been added to account for the stability that occurs with equidimensional shapes.

All rock masses will cave. The manner of their caving and the fragmentation need to be predicted if cave mining is to be implemented successfully. The rate of caving can be slowed by control of the draw since the cave can propagate only if there is

space into which the rock can move. The rate of caving can be increased by a more rapid advance of the undercut, but problems can arise if an air gap forms over a large area. In this situation, the intersection of major structures, heavy blasting, and the influx of water can result in damaging airblasts. Rapid, uncontrolled caving can result in an early influx of waste.

In conventional layouts, the rate of undercutting (RU) should be controlled so that the rate of caving (RC) is faster than the rate of damage (RD) due to abutment stresses. Thus, RC > RU > RD.

However, in areas of high stress, the rate of caving must be controlled to maintain an acceptable level/amount of seismic activity in the cave back. Otherwise, rock bursts can occur in suitably stressed areas (pillars and rock contacts). As advance undercutting will be used in these situations, damage to the undercut and production levels will not be a problem.

The stresses in the cave back can be modified to some extent by the shape of the cave front. Numerical modeling can be a useful tool for helping an engineer determine the stress patterns associated with possible mining sequences. An undercut face that is concave towards the caved area provides better control of major structures. In ore bodies having a range of MRMRs, the onset of continuous caving will be in the lower-rated zones if they are continuous in plan and section. This effect is illustrated in Figure 55.3B, where the class 5 and 4B zones are shown to be continuous. In Figure 55.3A, the pods of class 2 rock are sufficiently large to influence caving, and cavability should be based on the rating of these pods. Good geotechnical information, as well as information from monitoring the rate of caving and rock mass damage, is needed to fine-tune this relationship.

55.2.2   Particle Size Distribution

In caving operations, the degree of fragmentation has a bearing on

- Drawpoint spacing
- Dilution entry into the draw column
- Draw control
- Draw productivity
- Secondary blasting and breaking costs
- Secondary blasting damage

The input data needed to calculate the primary fragmentation and the factors that determine the secondary fragmentation as a function of the caving operation are shown in Figure 55.4.

Primary fragmentation can be defined as the size distribution of the particles that separate from the cave back and enter the draw column. The primary fragmentation generated by subsidence caving is generally more coarse than that resulting from stress caving. The reasons for this are (1) the more rapid propagation of the cave, (2) the rock mass disintegrates primarily along favorably oriented joint sets, and (3) there is little shearing of intact rock. The orientation of the cave front or back with respect to the joint sets and the direction of principal stress can have a significant effect on the primary fragmentation.

Secondary fragmentation is the reduction in the size of the original particles as they move through the draw column. The processes to which particles are subjected determine the size distribution which reports to the drawpoint. A strong, well-jointed material can result in a stable particle shape at a low draw height. Figure 55.5 shows the decrease in particle size for different draw heights and coarse (less jointed) to fine (well jointed) rock masses. A range of RMRs will result in a wider range of particle sizes than that produced by rock with a single rating. This is due to the fact that the fine material tends to cushion larger blocks and prevents further attrition of these blocks. This difference is illustrated in Figure 55.3B, in which class 5 and class 4 material is shown to

TABLE 55.3 Parameters to be considered before implementation of cave mining

CAVABILITY	PRIMARY FRAGMENTATION	DRAWPOINT/DRAWZONE SPACING
Rock mass strength (RMR/MRMR)	Rock mass strength (RMR/MRMR)	Particle size of ore and overlying rock
Rock mass structure-condition geometry	Geological structures	Overburden load and direction
In situ stress	Joint and fracture spacing, and geometry	Friction angles of caved particles
Induced stress	Joint condition ratings	Practical excavation size
Hydraulic radius of ore body	Stress or subsidence caving	Stability of host rock mass (MRMR)
Water	Induced stress	Induced stress
DRAW HEIGHTS	**LAYOUT**	**ROCK BURST POTENTIAL**
Capital	Particle size	Regional and induced stresses
Ore body geometry	Drawpoint spacing and size	Variations in rock mass strength, modulus
Excavation stability	Method of draw—gravity or LHD	Structures
Effect on ore minerals	Orientation of structures and joints	Mining sequence
Method of draw	Ventilation, ore handling, drainage	
SEQUENCE	**UNDERCUTTING SEQUENCE (pre/advance/post)**	**INDUCED CAVE STRESSES**
Cavability: poor to good or vice versa	Regional stresses	Regional stresses
Ore body geometry	Rock mass strength	Area of undercut
Induced stresses	Rock burst potential	Shape of undercut
Geological environment	Rate of advance	Rate of undercutting
Rock burst potential	Ore requirements	Rate of draw
Production requirements	Completeness of undercut	
Influence on adjacent operations	Shape (lead, lag)	
Water inflow	Height of undercut	
DRILLING AND BLASTING	**DEVELOPMENT**	**EXCAVATION STABILITY**
Rock mass strength	Layout	Rock mass strength (RMR/MRMR)
Rock mass stability (drillhole closure)	Sequence	Orientation of structures and joints
Required particle size	Production	Regional and induced stresses
Hole diameter, lengths, rigs	Drilling and blasting	Rock burst potential
Patterns and directions		Excavation size (orientation and shape)
Powder factor		Draw point
Swell relief		Mining sequence
SUPPORT	**PRACTICAL EXCAVATION SIZE**	**METHOD OF DRAW**
Excavation stability	Excavation stability	Fragmentation
Rock burst potential	Induced stress	Practical drawpoint spacing
Brow stability	Caving stresses	Practical size of excavation
Timing of support: initial, secondary, and production	Secondary blasting	Gravity or mechanical loading
	Equipment size	
RATE OF DRAW	**DRAWPOINT INTERACTION**	**DRAW COLUMN STRESSES**
Fragmentation	Drawzone spacing	Draw-column height
Method of draw	Critical distance across major apex	Particle size
Percentage hangups	Particle size	Homogeneity of ore particle size
Secondary breaking/blasting	Time frame of working drawpoints	Draw control
Seismic events		Draw-height interaction
Air blasts-drawpoint cover		Height-to-short axis base ratio
		Direction of draw
SECONDARY FRAGMENTATION	**SECONDARY BLASTING/BREAKING**	**DILUTION**
Rock, block shape	Secondary fragmentation	Ore body geometry
Draw height	Draw method	Mining geometry
Draw rate, time-dependent failure	Drawpoint size	Particle size distribution
Rock block workability and strength	Gravity grizzly aperture	Range of particles, unpay ore and waste
Range in particle size, fines cushioning	Size of equipment and grizzly spacing	Grade distribution of pay and unpay ore
Draw control program	Ore handling system, size restrictions	Mineral distribution in ore
		Drawpoint interaction
		Secondary breaking
		Draw control (techniques, predictions)
		Draw markers
TONNAGE DRAWN	**SUPPORT REPAIR**	**ORE/GRADE EXTRACTION**
Level interval	Tonnage drawn	Mineral distribution
Shut-off grade	Point and column loading	Method of draw
Drawpoint spacing	Brow wear	Rate of draw
Dilution percentage	Floor repair	Dilution percentage
Controls	Secondary blasting	Cut-off grade to lant
Redistribution		Ore losses
	SUBSIDENCE	
RMR/MRMR	Minimum and maximum spans	Depth of mining
Height of caved column	Major geological structures	Topography

cushion the larger primary fragments from class 3 material. A slow rate of draw results in a higher probability of time-dependent failure as the caving stresses have more time to work on particles in the draw column.

Fragmentation is the major factor determining drawpoint productivity. Experience has shown that 2 m^3 is the largest block that can be moved by a 6-yd^3 LHD and still allow an acceptable rate of production to be maintained. In Figure 55.6, the productivity of a layout using 3.5-, 6-, and 8-yd^3 LHDs loading to a grizzly is related to the percentage of fragments larger than 2 m^3. The use of secondary explosives is based on the amount of oversized rock that cannot be handled by a 6-yd^3 LHD.

A computer simulation program has been developed for calculating primary and secondary fragmentation. The results

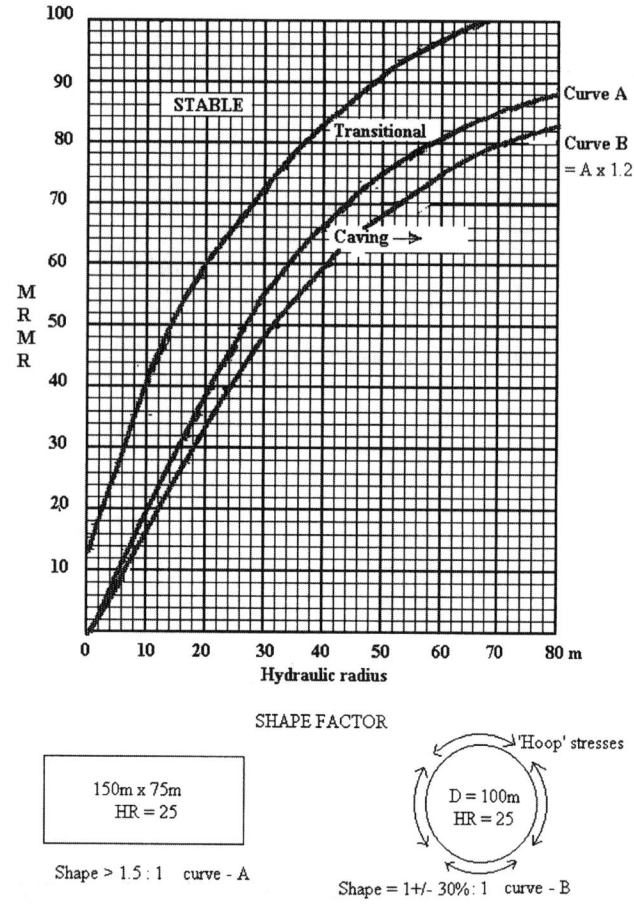

SHAPE FACTOR

150m x 75m
HR = 25

Shape > 1.5 : 1   curve - A

D = 100m
HR = 25

'Hoop' stresses

Shape = 1 +/- 30% : 1   curve - B

FIGURE 55.2   Stability diagram based on world-wide experience

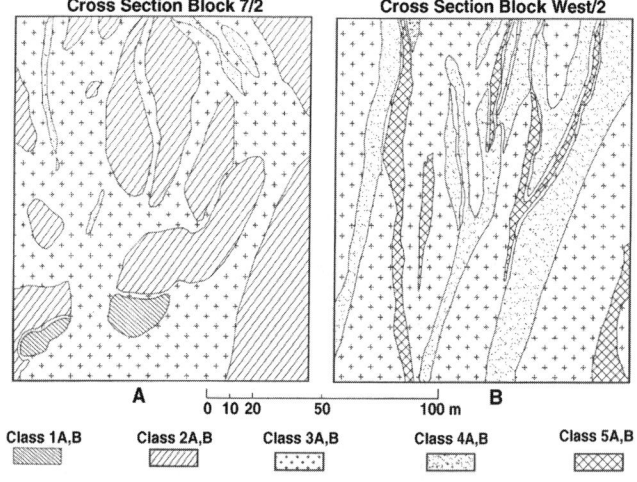

FIGURE 55.3   Geomechanics classification data

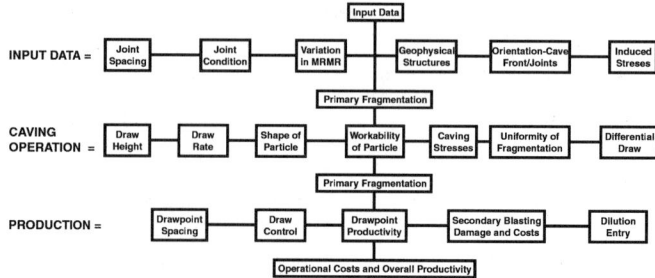

FIGURE 55.4   Input data for calculation of fragmentation

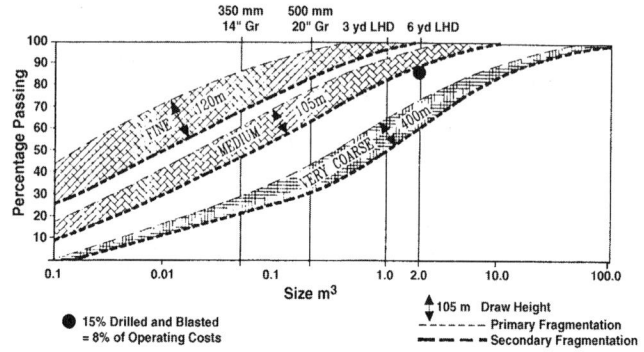

FIGURE 55.5   Size distribution of cave fragmentation

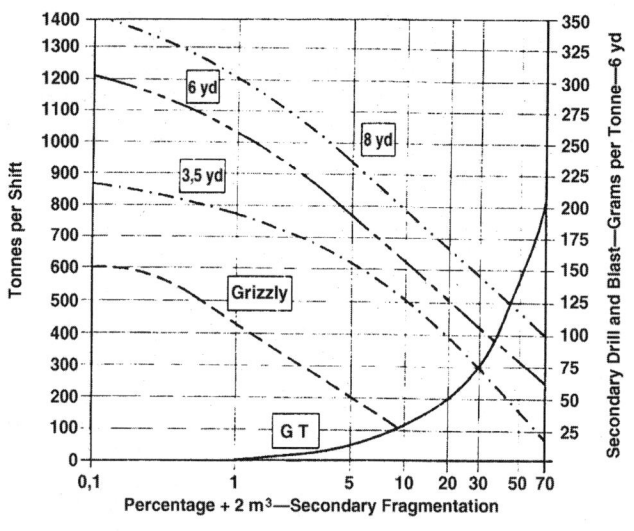

FIGURE 55.6   LHD productivity on a round trip of 100 m with 60% utilation

obtained from this program are confirmed by underground observations.

55.2.3   Drawzone Spacing

Drawpoint spacings for grizzly and slusher layouts reflect the spacing of the drawzones. However, in the case of LHD layouts with a nominal drawpoint spacing of 15 m, drawzone spacing can vary from 18 to 24 m across the major apex (pillar), depending

on the length of the drawbell. This situation occurs when the length of the drawpoint crosscut is increased to ensure that an LHD is straight before it loads. In this case, the major consideration (optimum ore recovery) is compromised by incorrect use of equipment or a desire to achieve ideal loading conditions. The drawzone spacings for 30 m center-to-center production drift spacings are shown in Figure 55.7, with the critical distance being across the major apex. It can be seen that the length of the drawbell can be increased from 8 to 10 m and the production drift spacing to 32 m without affecting the major apex spacing. There still will be interaction in the drawbell.

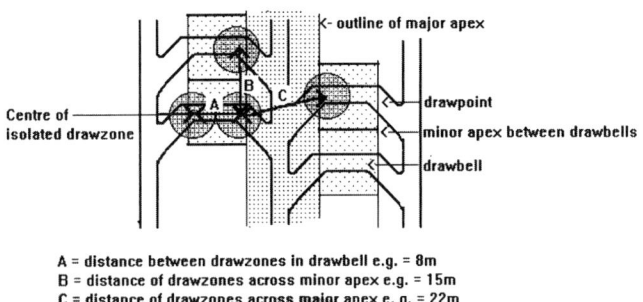

A = distance between drawzones in drawbell e.g. = 8m
B = distance of drawzones across minor apex e.g. = 15m
C = distance of drawzones across major apex e. g. = 22m

FIGURE 55.7 Maximum and minimum drawzone spacing (isolated drawzone = 10 m, area of influence = 225 m²)

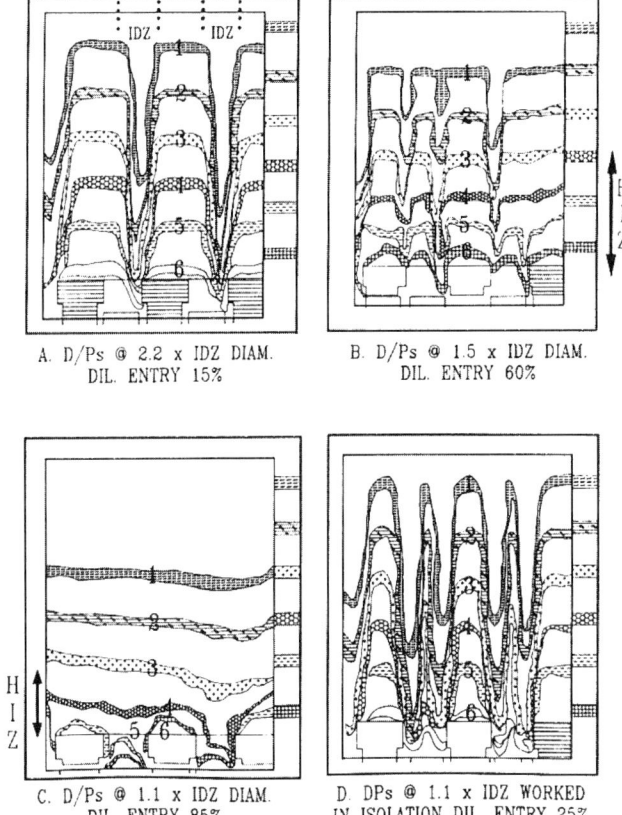

A. D/Ps @ 2.2 x IDZ DIAM. DIL. ENTRY 15%

B. D/Ps @ 1.5 x IDZ DIAM. DIL. ENTRY 60%

C. D/Ps @ 1.1 x IDZ DIAM. DIL. ENTRY 85%

D. DPs @ 1.1 x IDZ WORKED IN ISOLATION DIL. ENTRY 25%

FIGURE 55.8 Results of three-dimensional sand-model experiments

Sand model tests have shown that there is a relationship between the spacing of drawpoints and the interaction of drawzones. Widely spaced drawpoints develop isolated drawzones with diameters defined by the fragmentation. When drawpoints are spaced at 1.5 times the diameter of the isolated drawzone (IDZ), interaction occurs. Interaction also improves as drawpoint spacing is decreased, as shown in Figure 55.8. The flow lines and the stresses that develop around a drawzone are shown in Figure 55.9. The sand model results have been confirmed by observation of the fine material extracted during cave mining and by the behavior of material in bins. The question is whether this theory, which is based on isolated drawzones, can be wholly applied to coarse material, where arching (spans) of 20 m have been observed. The collapse of large arches will affect the large area overlying the drawpoint, as shown in Figure 55.10. The

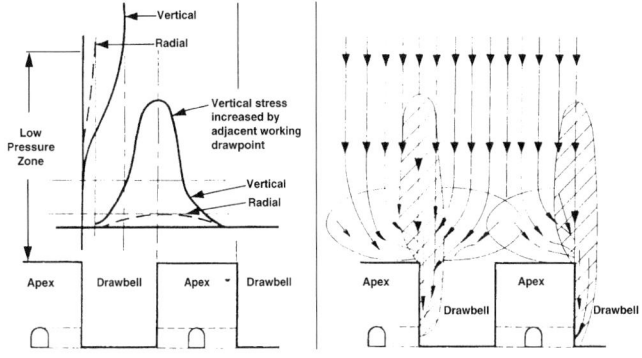

FIGURE 55.9 Flow lines and inferred stresses between adjacent working operations

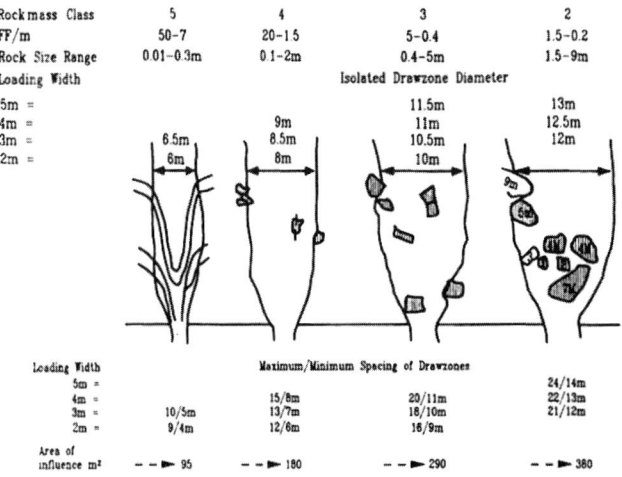

FIGURE 55.10 Maximum/minimum spacing of drawzones based on isolated drawzone diameter

formation and failure of arches lead to wide drawzones in coarse material, so that drawzone spacing can be increased to the spacings shown in Figure 55.10.

The frictional properties of the caved material must also be recognized. Low-friction material can flow greater distances when under high overburden load, and this can mean wider drawzone spacings. It is therefore logical to expect that when a line of drawbells is drawn this creates a large low-density zone. Similarly, when lines of drawbells are drawn on alternate shifts good interaction will result. The draw program at the Henderson Mine broadly follows this pattern.

There is a need to continue with three-dimensional model tests to establish some poorly defined principles, such as the interaction across major apexes when the spacing of groups of interactive drawzones is increased. Numerical modeling could possibly provide the solutions for the draw behavior of coarsely fragmented material.

55.2.4 Draw Control

Draw control requirements are shown in Figure 55.11 and Table 55.4. The grade and fragmentation in the dilution zone must be known if sound draw control is to be practiced. Figure 55.12 shows the value distribution for columns A and B. Both columns have the same average grade of 1.4%, but the value distribution is different. The high grade at the top of column B means that a larger tonnage of waste can be tolerated before the shutoff grade is reached.

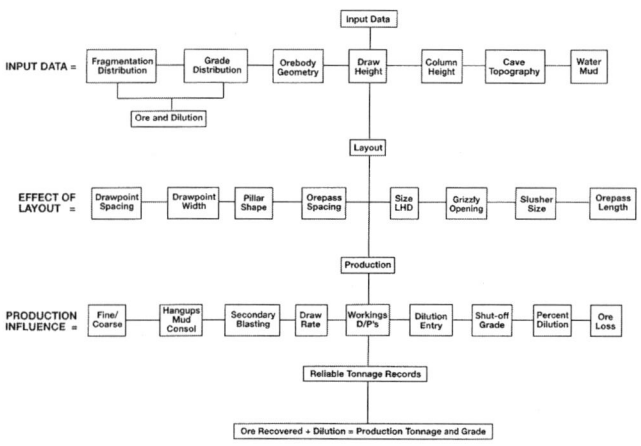

FIGURE 55.11 Draw control requirements

TABLE 55.4 Draw control

Objective	Application
Reduce dilution	Calculation of tonnage
Improve ore recovery	Recording of tonnages produced
Avoid damage to pillars	Controlling draw

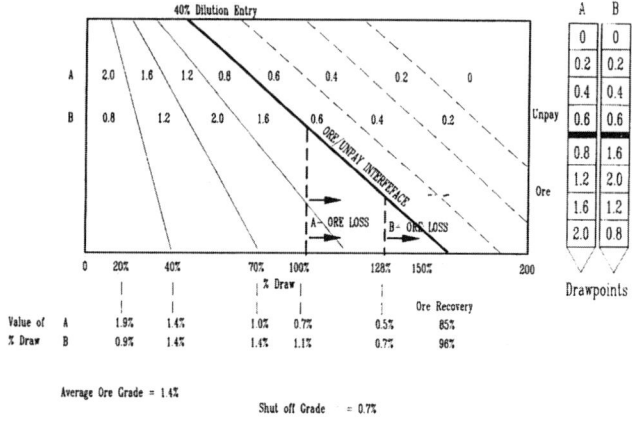

FIGURE 55.12 Grade analysis

In LHD layouts, a major factor in poor draw control is the drawing of fine material at the expense of coarse material. Strict draw control discipline is required so that the coarse ore is drilled and blasted at the end of the shift in which it reported in the drawpoint.

It has been established that the draw will angle towards less dense areas. This principle can be used to move the material overlying the major apex by creating zones of varying density through the differential draw of lines of drawbells.

55.2.5 Dilution

The percentage of dilution is defined as the percentage of the ore column that has been drawn before the waste material appears in the drawpoint. It is a function of the amount of mixing that occurs in the draw columns. Mixing is a function of—

- Ore draw height
- Range in fragmentation of both ore and waste

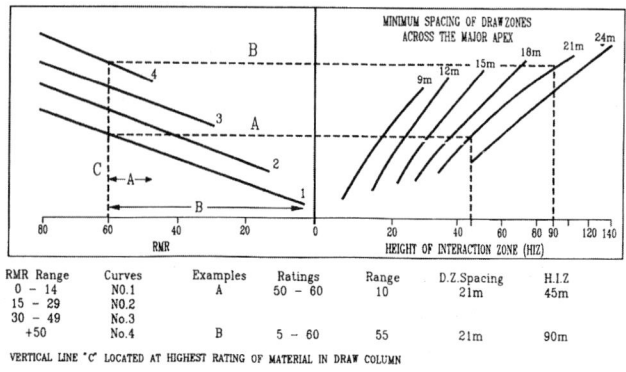

FIGURE 55.13 Height of interaction zone (HIZ)

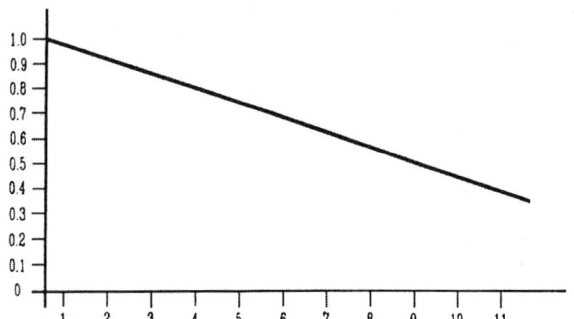

FIGURE 55.14 Draw control factor

- Drawzone spacing
- Range in tonnages drawn from drawpoints

The range in particle size distribution and the minimum drawzone spacing across the major apex will give the height of the interaction zone (HIZ). This is illustrated in Figure 55.13. There is a volume increase as the cave propagates so that a certain amount of material must be drawn before the cave reaches the dilution zone. The volume increase, or swell factors, is based on the fragmentation and is applied to column height. Typical swell factors for fine, medium, and coarse fragmentation are, respectively, 1.16, 1.12, and 1.08.

A draw control factor is based on the variation in tonnages from working drawpoints (Figure 55.14). If production data are not available, the draw control engineer must predict a likely draw pattern. A formula based on the above factors has been developed to determine the dilution entry percentage.

$$\text{Dilution entry} = (A - B)/A \times C \times 100,$$

where A = draw-column height H swell factor

B = height of interaction

and C = draw control factor.

The graph for dilution entry was originally drawn as a straight line, but underground observations show that, where the

$$\frac{A-B}{A} \times C \times 100 \qquad\qquad = \text{Dilution Entry}$$

A = Ore Draw Column Height x Swell Factor = 168m
B = Height of Interaction Zone = 90m
C = Draw-control Factor = 0.6

$$\frac{168 - 90}{168} \times 0.6 \times 100 \qquad\qquad = 28\%$$

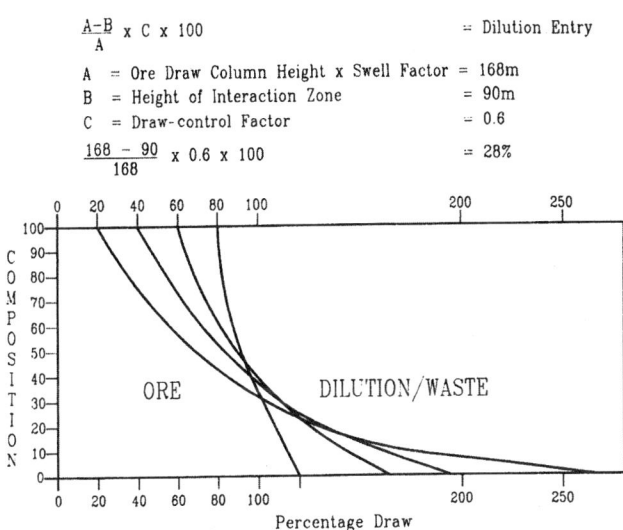

FIGURE 55.15   Calculation of dilution entry

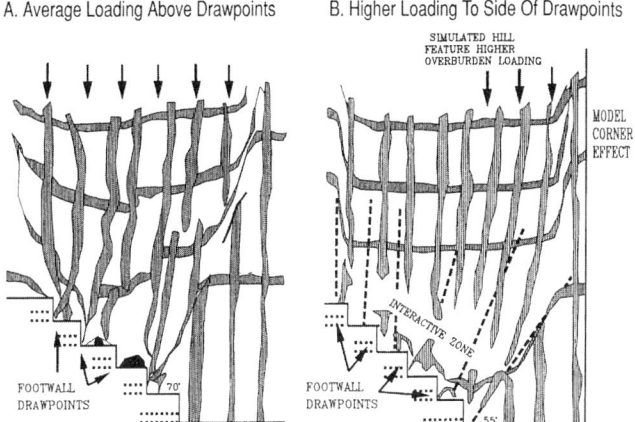

FIGURE 55.16   Inclined drawpoint layout showing effect of different overburden loading (three-dimensional sand-model experiments)

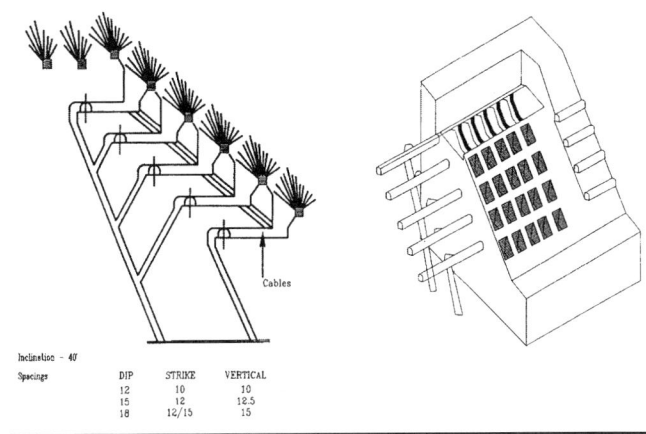

FIGURE 55.17   Example of layout of inclined drawpoint

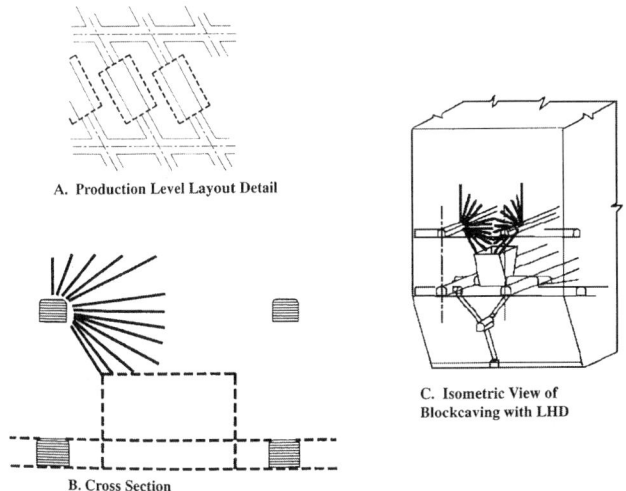

FIGURE 55.18   LHD layout at El Teniente

dilution occurs early, the rate of influx follows a curved line with a long ore "tail" (Figure 55.15). Figure 55.16 shows that dilution entry is also affected by the attitude of the drawzone, which can angle towards higher overburden loads.

55.2.6   Layouts

Eight different horizontal LHD layouts and two inclined drawpoint LHD layouts are used at various operating mines. An example of an inclined LHD layout is shown in Figure 55.17. The El Teniente layout is shown in Figure 55.18 and the Henderson Mine layout in Figure 55.19. Numerical modeling of different layouts showed that the Teniente layout was the strongest and had several practical advantages. For example, the drawpoint and drawbell are on a straight line, which results in better drawpoint support and flow of ore. If there is brow wear, the LHD can back into the opposite drawpoint. The only disadvantage is that the layout does not have the flexibility to accommodate the use of electric LHDs such as does the herringbone.

A factor that needs to be resolved is the correct shape of the major apex. It is thought that a shaped pillar will assist in the recovery of fine ore. Also, in ore bodies characterized by coarse fragmentation, there is less chance that stacking will occur

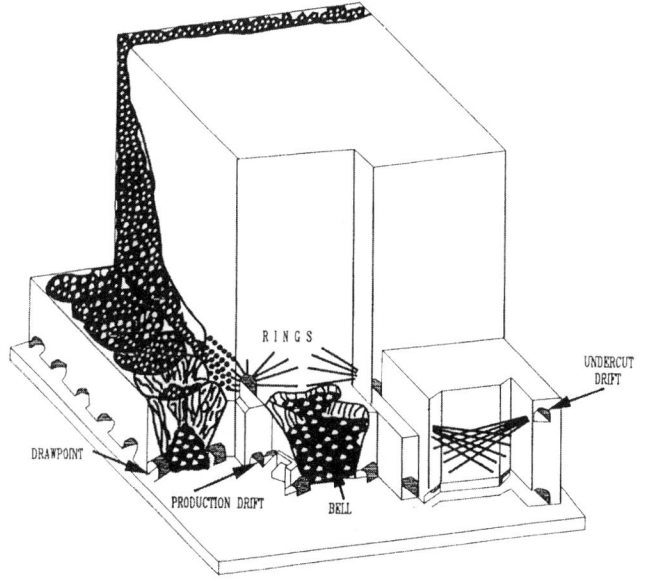

FIGURE 55.19   Isometric view of panel cave operation

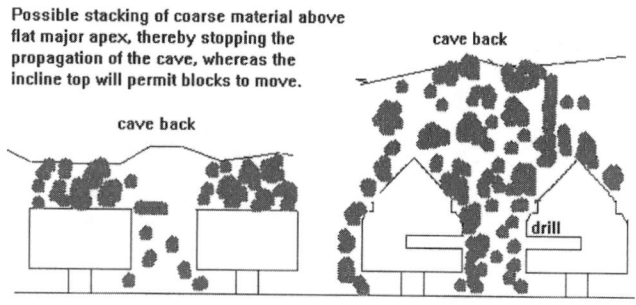

FIGURE 55.20 Effect of shape of major apex

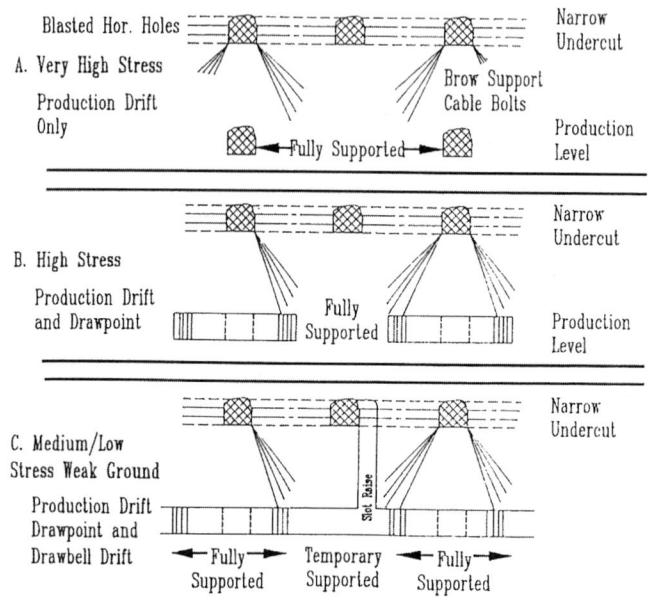

FIGURE 55.21 Different sequences of advanced undercutting

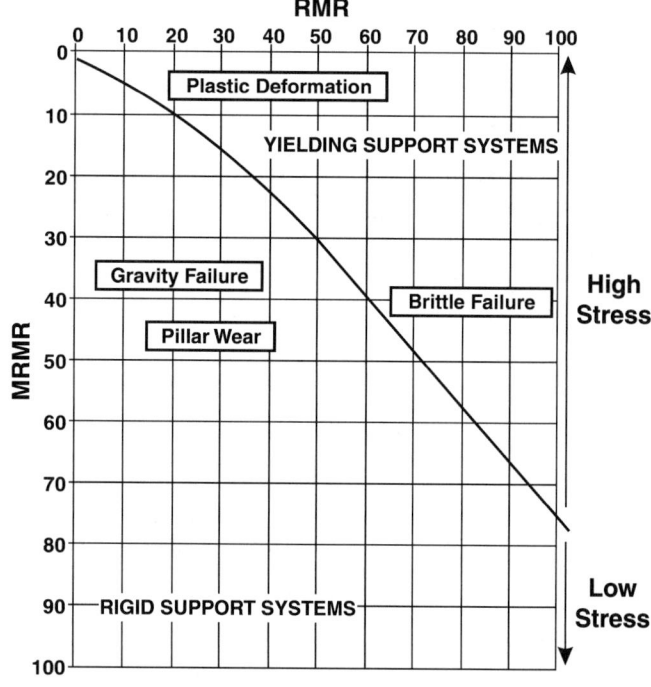

FIGURE 55.22 Support requirements for caving operations

(Figure 55.20). The main area of brow wear is immediately above the drawpoint. If the vertical height of the pillar above the brow is small, failure of the top section will reduce the strength of the lower section and result in aggravated brow wear.

More thought must be given to the design of LHD layouts to provide a maximum amount of maneuver room within a minimum size of drift opening. Thus larger machines can be used within the optimum drawzone spacings. Another aspect that needs attention is LHD design, that is, to reduce the length and increase their width. While the use of large machines might be an attraction, it is recommended that caution be exercised and that a decision on machine size be based on a correct assessment of the required drawzone spacing in terms of fragmentation. The loss of revenue that can result from dilution far exceeds the lower operating costs associated with larger machines.

55.2.7 Undercutting

Undercutting is one of the most important aspects of cave mining since not only is a complete undercut necessary to induce caving, but the undercut method can reduce the damaging effects of induced stresses.

The normal undercutting sequence is to develop the drawbell and then to break the undercut into the drawbell, as shown in Figure 55.18. In environments of high stress, the pillars and brows are damaged by the advancing abutment stresses. The Henderson Mine technique of developing the drawbell with long holes from the undercut level reduces the time interval and extent of damage associated with postundercutting. To preserve stability, the Henderson Mine has also found it necessary to delay the development of the drawbell drift until the bell must be blasted (Figure 55.19).

The damage caused to pillars around drifts and drawbells by abutment stresses is significant and is the major factor in brow wear and excavation collapse. Rock bursts also occur in these areas. The solution is to complete the undercut before development of the drawpoints and drawbells. The "advanced undercut" technique is shown in Figure 55.21.

In the past, it was considered that the height of the undercut had a significant influence on caving and, possibly, the flow of ore. The asbestos mines in Zimbabwe had undercuts of 30 m with no resulting improvement in caving or fragmentation. The long time involved in completing the undercut often led to ground control problems. Good results are obtained with undercuts of minimum height, provided that complete undercutting is achieved. Where gravity is needed for the flow of blasted undercut ore, the undercut height needs to be only half the width of the major apex. This results in an angle of repose of 45° and allows the ore to flow freely.

55.2.8 Support Requirements

In areas of high stress, weak rock will deform plastically and strong rock will exhibit brittle, often violent, failure. If there is a large difference between the RMR and MRMR values, yielding support systems are required. This is explained in Figure 55.22.

Prestressed cables have little application in underground situations unless it is to stabilize fractured rock in a low-stress environment. The need to constrain the rock laterally and for lining surfaces such as concrete cannot be too highly emphasized. Support techniques are illustrated in Table 55.5. The use of cable bolts in brows is common practice, but often these bolts are not installed in a pattern that takes into account joint spacing and orientation. Cable bolts should not be installed in highly jointed ground, as the bolts do not apply lateral restraint to the brow and serve only hold blocks in place.

TABLE 55.5 Support techniques

Support element	Low stress	High stress
Bolts:		
Length	1 m + (0.33 W × F)	1 m + (0.5 × F)
Spacing	1 m	1 m
Type	Rigid rebar	Yielding, e.g., cones
Mesh	0.5 mm × 100 mm square	0.5 mm × 75 mm or 50-mm aperture
Deep-seated support	Cables, 1 m + 1.5 W	Steel ropes, 1 m + 1.5 W Long cone bolts
Shotcrete linings	Mesh-reinforced shotcrete	Mesh-reinforced shotcrete
Arches	Rigid steel arches Massive concrete	Yielding steel arches Reinforced concrete
Surface restraint	Large washers (triangles) Tendon traps	Large plate washers Yielding tendon traps
Corners	25-mm rope, cable slings	25-mm rope, cable slings
Brows	Birdcage cables from undercut level Inclined pipes	Birdcage cables from undercut level Inclined pipes
Repair	Grouting Extra bolts, cables, plates, straps, and arches	Grouting Extra bolts, ropes, plates, straps, and yielding arches

W = span of tunnel.
F is based on MRMR = 0–20:F = 1.4; MRMR = 21–30:F = 1.3; MRMR = 31–40: F = 1.2; MRMR = 41–50:F = 1.1; MRMR = 51–60:F = 1.05; MRMR = >61:F = 1.0.

55.3 CONCLUSIONS

Cavability can be assessed provided accurate geotechnical data are available and geological variations are recognized. The MRMR system provides the necessary data for an empirical definition of the undercut dimension in terms of the hydraulic radius.

Numerical modeling can assist an engineer in understanding and defining the stress environment.

Fragmentation is a major factor in assessing the feasibility of cave mining in large, competent ore bodies. Programs are being developed for predicting fragmentation, and even the less-sophisticated programs provide good design data. The economic viability of caving in competent ore bodies is determined by LHD productivity and the cost of breaking large fragments.

Drawpoint and drawzone spacings for coarser material need to be examined in terms of recovery and improved mining environments. Spacings must not be increased to lower operating costs at the expense of ore recovery.

The interactive theory of draw and the diameter of an isolated drawzone can be used in the design of drawzone spacings.

Complications occur when drawzone spacings are designed on the basis of primary fragmentation and the secondary fragmentation is significantly different.

55.4 ACKNOWLEDGMENTS

This paper presents an update of the technology of cave mining. It is not possible to quote references since the bulk of the data supporting the contents of this paper have not been published, and the basic concepts are known to mining engineers. However, it is appropriate to acknowledge the contributions from discussions with the following people in Canada, Chile, South Africa, and Zimbabwe: R. Alvarez, P.J. Bartlett, N.J.W. Bell, T. Carew, A.R. Guest, C. Page, D. Stacey, and A. Susaeta. The simulation program for the calculation of primary and secondary fragmentation was written by G.S. Esterhuizen at Pretoria University.

Foundations for Design

......
CHAPTER 56

Rock Mass Properties for Underground Mines

Evert Hoek

56.1 INTRODUCTION

Any form of analysis used for designing underground excavations requires reliable estimates of the strength and deformation characteristics of rock masses. Hoek and Brown (1980a, 1980b) proposed a method for estimating the strength of jointed rock masses that is based upon an assessment of the interlocking of rock blocks and the condition of the surfaces between these blocks. This method was modified over the years to meet the needs of users who applied it to problems that were not considered when the original criterion was developed (Hoek 1983, Hoek and Brown 1988). Applying the method to very poor quality rock masses required further changes (Hoek, Wood, and Shah 1992) and, eventually, the development of a new classification called the Geological Strength Index (GSI) (Hoek 1994; Hoek, Kaiser and Bawden 1995; Hoek and Brown 1997; Hoek, Marinos and Benissi 1998). A review of the development of the criterion and of the equations proposed at various stages in this development is given in Hoek and Brown (1997).

This chapter presents the Hoek-Brown criterion in a form that has been found to be practical in the field and that appears to provide the most reliable set of results for use as input for methods of analysis currently used in rock engineering.

56.2 GENERALISED HOEK-BROWN CRITERION

The generalised Hoek-Brown failure criterion for jointed rock masses is defined by:

$$\sigma'_1 = \sigma'_3 + \sigma_{ci}\left(m_b\frac{\sigma'_3}{\sigma_{ci}} + s\right)^a \qquad (56.1)$$

where σ'_1 and σ'_3 are the maximum and minimum effective stresses at failure,

m_b is the value of the Hoek-Brown constant m for the rock mass,

s and a are constants that depend upon the rock mass characteristics, and

σ_{ci} is the uniaxial compressive strength of the intact rock pieces.

The Mohr envelope, relating normal and shear stresses, can be determined using the method proposed by Hoek and Brown (1980a). In this approach, Equation 56.1 is used to generate a series of triaxial test values, simulating full-scale field tests, and a statistical curve-fitting process is used to derive an equivalent Mohr envelope defined by the equation:

$$\tau = A\sigma_{ci}\left(\frac{\sigma'_n - \sigma_{tm}}{\sigma_{ci}}\right)^B \qquad (56.2)$$

where A and B are material constants

σ'_n is the normal effective stress, and

σ_{tm} is the tensile strength of the rock mass.

To use the Hoek-Brown criterion for estimating the strength and deformability of jointed rock masses, three properties of the rock mass have to be estimated. These are:

1. The uniaxial compressive strength σ_{ci} of the intact rock elements
2. The value of the Hoek-Brown constant m_i for these intact rock elements
3. The value of the *GSI* for the rock mass

56.3 INTACT ROCK PROPERTIES

For the intact rock pieces that make up the rock mass, Equation 56.1 simplifies to:

$$\sigma'_1 = \sigma'_3 + \sigma_{ci}\left(m_i\frac{\sigma'_3}{\sigma_{ci}} + 1\right)^{0.5} \qquad (56.3)$$

The relationship between the principal stresses at failure for a given rock is defined by two constants, the uniaxial compressive strength σ_{ci} and a constant m_i. Wherever possible, the values of these constants should be determined by statistical analysis of the results of a set of triaxial tests on carefully prepared core samples. When laboratory tests are not possible, Table 56.1 and Table 56.2 can be used to obtain estimates of σ_{ci} and m_i.

In the case of mineralised rocks, the effects of alteration can have a significant impact on the properties of the intact rock components, and this should be taken into account in estimating the values of σ_{ci} and m_i. For example, the influence of quartzseritic alteration of andesite and porphyry is illustrated in the Figure 56.1, based upon data provided by Karzulovic (2000). Similar trends have been observed for other forms of alteration and, where this type of effect is considered likely, the geotechnical engineer would be well advised to invest in a program of laboratory testing to establish the appropriate properties for the intact rock.

The Hoek-Brown failure criterion, which assumes isotropic rock and rock-mass behaviour, should be applied only to those rock masses in which there are a sufficient number of closely spaced discontinuities, with similar surface characteristics, where isotropic behaviour involving failure on multiple discontinuities can be assumed. When the structure being analysed is large and the block size small in comparison, the rock mass can be treated as a Hoek-Brown material.

Where the block size is of the same order as that of the structure being analysed, or when one of the discontinuity sets is significantly weaker than the others, the Hoek-Brown criterion

* Evert Hoek Consulting Engineer, Inc. North Vancouver, BC, Canada.

TABLE 56.1   Field estimates of uniaxial compressive strength

Grade*	Term	Uniaxial comp. strength (MPa)	Point load index (MPa)	Field estimate of strength	Examples
R6	Extremely strong	>250	>10	Specimen can only be chipped with a geological hammer	Fresh basalt, chert, diabase, gneiss, granite, quartzite
R5	Very strong	100–250	4–10	Specimen requires many blows of a geological hammer to fracture it	Amphibolite, sandstone, basalt, gabbro, gneiss, granodiorite, peridotite, rhyolite, tuff
R4	Strong	50–100	2–4	Specimen requires more than one blow of a geological hammer to fracture it	Limestone, marble, sandstone, schist
R3	Medium strong	25–50	1–2	Cannot be scraped or peeled with a pocket knife; specimen can be fractured with a single blow from a geological hammer	Concrete, phyllite, schist, siltstone
R2	Weak	5–25	†	Can be peeled with a pocket knife with difficulty; shallow indentation made by firm blow with point of a geological hammer	Chalk, claystone, potash, marl, siltstone, shale, rocksalt
R1	Very weak	1–5	†	Crumbles under firm blows with point of a geological hammer; can be peeled by a pocket knife	Highly weathered or altered rock, shale
R0	Extremely weak	0.25–1	†	Indented by thumbnail	Stiff fault gouge

*Grade according to Brown (1981).
†Point load tests on rocks with a uniaxial compressive strength below 25 MPa are likely to yield highly ambiguous results.

TABLE 56.2   Values of the constant m_i for intact rock, by rock group.*

Rock type	Class	Group	Texture			
			Coarse	Medium	Fine	Very fine
SEDIMENTARY	Clastic		Conglomerates (21 ± 3) Breccias (19 ± 5)	Sandstones 17 ± 4	Siltstones 7 ± 2 Greywackes (18 ± 3)	Claystones 4 ± 2 Shales (6 ± 2) Marls (7 ± 2)
	Non-clastic	Carbonates	Crystalline limestone (12 ± 3)	Sparitic limestones (10 ± 2)	Micritic limestones (9 ± 2)	Dolomites (9 ± 3)
		Evaporites		Gypsum 8 ± 2	Anhydrite 12 ± 2	
		Organic				Chalk 7 ± 2
METAMORPHIC	Non Foliated		Marble 9 ± 3	Hornfels (19 ± 4) Metasandstone (19 ± 3)	Quartzites 20 ± 3	
	Slightly foliated		Migmatite (29 ± 3)	Amphibolites 26 ± 6		
	Foliated†		Gneiss 28 ± 5	Schists 12 ± 3	Phyllites (7 ± 3)	Slates 7 ± 4
IGNEOUS	Plutonic	Light	Granite 32 ± 3	Diorite 25 ± 5 Granodiorite (29 ± 3)		
		Dark	Gabbro 27 ± 3 Norite 20 ± 5	Dolerite (16 ± 5)		
	Hypabyssal		Porphyries (20 ± 5)		Diabase (15 ± 5)	Peridotite (25 ± 5)
	Volcanic	Lava		Rhyolite (25 ± 5) Andesite 25 ± 5	Dacite (25 ± 3) Basalt (25 ± 5)	Obsidian (19 ± 3)
		Pyroclastic	Agglomerate (19 ± 3)	Breccia (19 ± 5)	Tuff (13 ± 5)	

*Note that values in parenthesis are estimates.
†These values are for intact rock specimens tested normal to bedding or foliation. The value of m_i will be significantly different if failure occurs along a weakness plane.

should not be used. In these cases, the stability of the structure should be analysed by considering failure mechanisms involving the sliding or rotation of blocks and wedges defined by intersecting structural features. Figure 56.2 summarises these statements in a graphical form.

56.4   GEOLOGICAL STRENGTH INDEX

The strength of a jointed rock mass depends on the properties of the intact rock pieces and also upon the freedom of these pieces to slide and rotate under different stress conditions. This freedom is controlled by the geometrical shape of the intact rock pieces as well as by the condition of the surfaces separating the pieces. Angular rock pieces with clean, rough discontinuity surfaces will result in a much stronger rock mass than one that contains rounded particles surrounded by weathered and altered material.

The GSI, introduced by Hoek (1994) and Hoek, Kaiser, and Bawden (1995), provides a system for estimating the reduction in

rock mass strength for different geological conditions. This system is presented in Table 56.3, for blocky rock masses, and Table 56.4 for schistose metamorphic rocks.

Once the GSI has been estimated, the parameters that describe the rock mass strength characteristics are calculated as follows:

$$m_b = m_i exp\left(\frac{GSI - 100}{28}\right) \tag{56.4}$$

For GSI >25, i.e., rock masses of good to reasonable quality:

$$s = exp\left(\frac{GSI - 100}{9}\right) \tag{56.5}$$

and

$$a = 0.5 \tag{56.6}$$

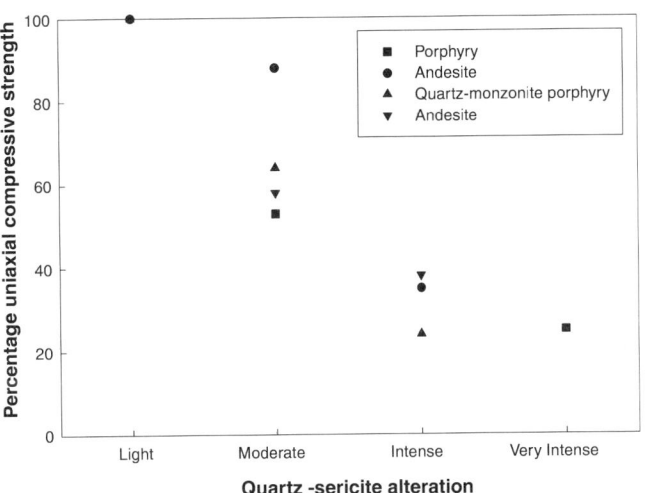

FIGURE 56.1 Influence of quartz-seritic alteration on the uniaxial compressive strength of "intact" specimens of andesite and porphyry (after Karzulovic 2000)

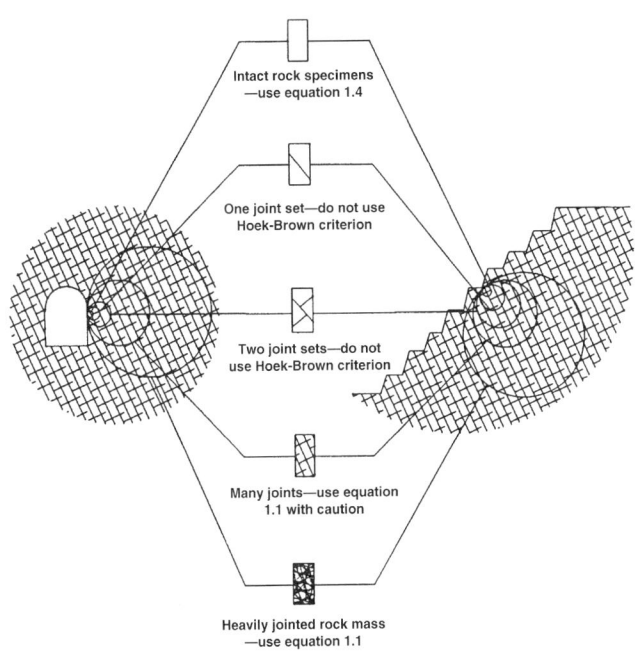

FIGURE 56.2 Idealized diagram showing the transition from intact to a heavily jointed rock mass with increasing sample size

For GSI <25, i.e., rock masses of very poor quality:

$$s = 0 \qquad (56.7)$$

and

$$a = 0.65 - \frac{GSI}{200} \qquad (56.8)$$

For better-quality rock masses (GSI >25), the value of GSI can be estimated directly from the 1976 version of Bieniawski's Rock Mass Rating (RMR), with the groundwater rating set to 10 (dry) and the Adjustment for Joint Orientation set to 0 (very favourable) (Bieniawski 1976). For very poor quality rock masses, the value of RMR is very difficult to estimate, and the balance between the ratings is no longer a reliable basis for estimating rock mass

TABLE 56.3 Characterisation of blocky rock masses on the basis of particle interlocking and discontinuity condition (after Hoek, Marinos and Benissi 1998)

GEOLOGICAL STRENGTH INDEX FOR BLOCKY JOINTED ROCKS — From a description of the structure and surface conditions of the rock mass, pick an appropriate box in this chart. Estimate the average value of GSI from the contours. Do not attempt to be too precise. Quoting a range from 36 to 42 is more realistic than stating that GSI = 38. It is also important to recognize that the Hoek-Brown criterion should only be applied to rock masses where the size of individual blocks or pieces is small compared with the size of the excavation under consideration. When the individual block size is more than about one quarter of the excavation size, the failure will be structurally controlled and the Hoek-Brown criterion should not be used.	SURFACE CONDITIONS	VERY GOOD Very rough, fresh unweathered surfaces	GOOD Rough, slightly weathered, iron stained surfaces	FAIR Smooth, moderately weathered and altered surfaces	POOR Slickensided, highly weathered surfaces with compact coatings or fillings or angular fragments	VERY POOR Slickensided, highly weathered surfaces with soft clay coatings or fillings
STRUCTURE		DECREASING SURFACE QUALITY ⇒				
INTACT OR MASSIVE - intact rock specimens or massive in situ rock with few widely spaced discontinuities	DECREASING INTERLOCKING OF ROCK PIECES	90 / 80	N/A	N/A	N/A
BLOCKY - well interlocked undisturbed rock mass consisting of cubical blocks formed by three intersecting discontinuity sets		70 / 60			
VERY BLOCKY- interlocked, partially disturbed mass with multi-faceted angular blocks formed by 4 or more joint sets			50		
BLOCKY/DISTURBED - folded and/or faulted with angular blocks formed by many intersecting discontinuity sets			40	30	
DISINTEGRATED - poorly interlocked, heavily broken rock mass with mixture of angular and rounded rock pieces					20
FOLIATED/LAMINATED - folded and tectonically sheared. Lack of blockiness due to schistosity prevailing over other discontinuities		N/A	N/A			10

strength. Consequently, Bieniawski's RMR classification should not be used for estimating the GSI values for poor-quality rock masses (RMR <25), and the GSI charts should be used directly.

If the 1989 version of Bieniawski's RMR classification (Bieniawski 1989) is used, then GSI = RMR$_{89}$' − 5 where RMR$_{89}$' has the groundwater rating set to 15 and the Adjustment for Joint Orientation set to zero.

56.5 MOHR-COULOMB PARAMETERS

Most geotechnical software is written in terms of the Mohr-Coulomb failure criterion, in which the rock mass strength is defined by the cohesive strength c' and the angle of friction ϕ'. The linear relationship between the major and minor principal stresses, σ'_1 and σ'_3 for the Mohr-Coulomb criterion is

$$\sigma'_1 = \sigma_{cm} + k\sigma'_3 \qquad (56.9)$$

where σ_{cm} is the uniaxial compressive strength of the rock mass, and k is the slope of the line relating σ'_1 and σ'_3. The values of ϕ' and c' can be calculated from:

$$sin\ \phi' = \frac{k-1}{k+1} \qquad (56.10)$$

$$c' = \frac{\sigma_{cm}(1 - sin\ \phi')}{2cos\phi'} \qquad (56.11)$$

TABLE 56.4   Characterisation of a schistose metamorphic rock masses on the basis of foliation and discontinuity condition (after M. Truzman 1999)

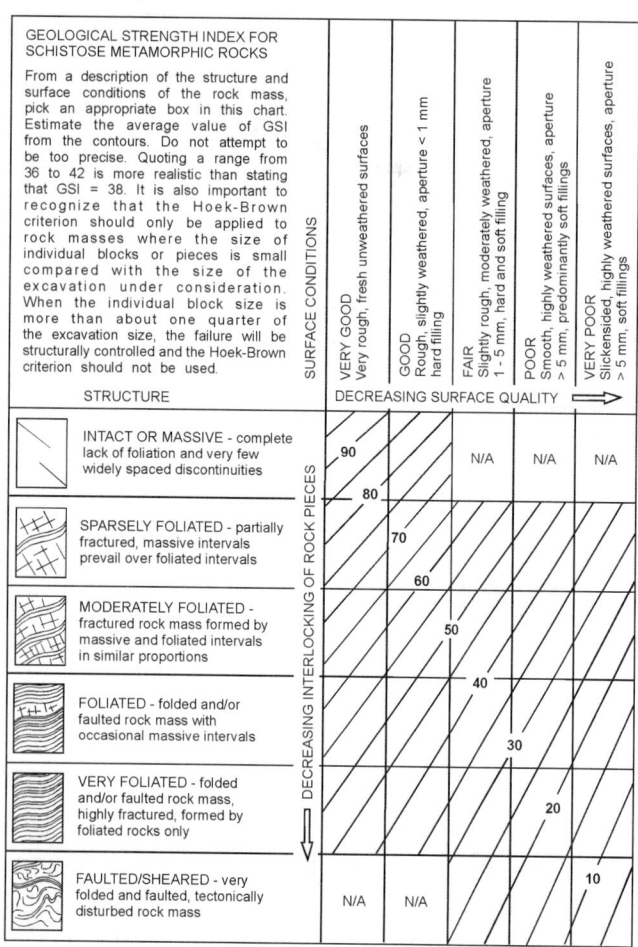

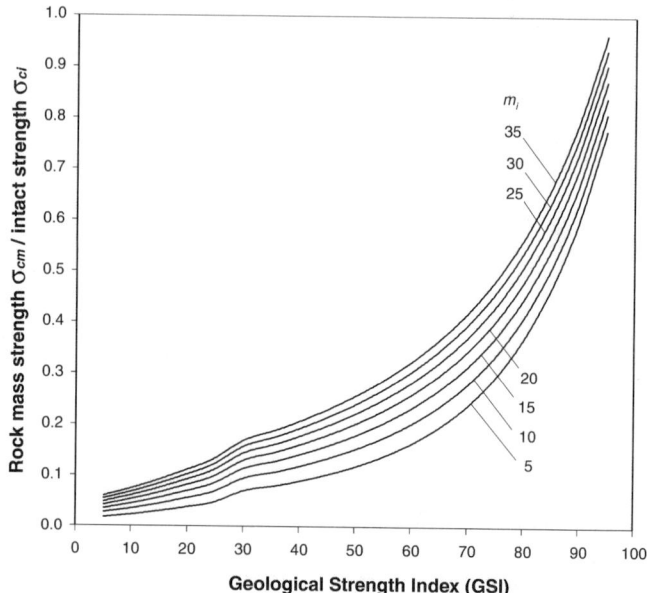

FIGURE 56.3   Ratio of uniaxial compressive strength of rock mass to intact rock versus Geological Strength Index (GSI)

tion. This relationship is based upon back analysis of dam foundation deformations, and it has been found to work well for better-quality rocks. However, for many of the poor-quality rocks it appears to predict deformation modulus values that are too high. Based upon practical observations and back analysis of excavation behaviour in poor-quality rock masses, the following modification to Serafim and Pereira's equation is proposed for $\sigma_{ci} < 100$:

$$E_m = \sqrt{\frac{\sigma_{ci}}{100}}\, 10^{\left(\frac{GSI-10}{40}\right)} \qquad (56.12)$$

Note that GSI has been substituted for RMR in this equation and that the modulus E_m is reduced progressively as the value of σ_{ci} falls below 100. This reduction is based upon the reasoning that the deformation of better-quality rock masses is controlled by the discontinuities while, for poorer quality rock masses, the deformation of the intact rock pieces contributes to the overall deformation process.

Based upon measured deformations, Equation 56.12 appears to work reasonably well in those cases where it has been applied. However, as more field evidence is gathered, it may be necessary to modify this relationship (Figure 56.5).

56.7   POST-FAILURE BEHAVIOUR

When numerical models are used to study the progressive failure of rock masses, it is necessary to estimate the post-peak or post-failure characteristics of the rock mass. In some of these models, the Hoek-Brown failure criterion is treated as a yield criterion, and the analysis is carried out using plasticity theory. No definite rules for dealing with this problem can be given but, based upon experience in numerical analysis of a variety of practical problems, the post-failure characteristics illustrated in Figures 56.6 to 56.8 are suggested as a starting point.

56.7.1   Very Good Quality Hard Rock Masses

For very good quality, hard rock masses, such as massive granites or quartzites, the analysis of spalling around highly stressed openings (Hoek, Kaiser, and Bawden 1995) suggests that the

There is no direct correlation between Equation 56.9 and the nonlinear Hoek-Brown criterion defined by Equation 56.1. Consequently, determination of the values of c' and ϕ' for a rock mass that has been evaluated as a Hoek-Brown material is a difficult problem.

Having considered several possible approaches, the most practical solution is to treat the problem as an analysis of a set of full-scale triaxial strength tests. Using the Hoek-Brown Equation 56.1 to generate a series of triaxial test values simulates the results of such tests. Equation 56.9 is then fitted to these test results by linear regression analysis, and the values of c' and ϕ' are determined from Equations 56.10 and 56.11. A full discussion on the steps required to carry out this analysis is presented in the Appendix, together with a spreadsheet for implementing this analysis.

The range of stresses used in the curve-fitting process described above is very important. For the confined conditions surrounding tunnels at depths of more than about 30 m, the most reliable estimates are given by using a confining stress range from zero to 0.25 σ_{ci}, where σ_{ci} is the uniaxial compressive strength of the intact rock elements. A series of plots showing the uniaxial compressive strength of the rock mass σ_{cm}, the cohesive strength c, and the friction angle ϕ are given in Figures 56.3 and 56.4.

56.6   DEFORMATION MODULUS

Serafim and Pereira (1983) proposed a relationship between the in-situ modulus of deformation and Bieniawski's RMR classifica-

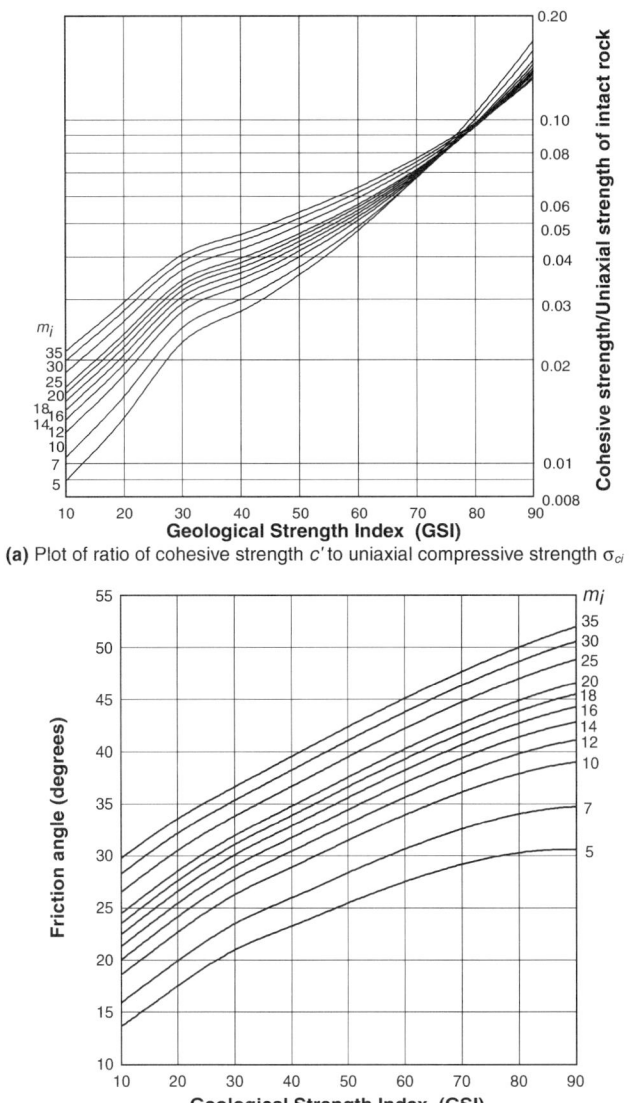

(a) Plot of ratio of cohesive strength c' to uniaxial compressive strength σ_{ci}

(b) Plot of friction angle ϕ'

FIGURE 56.4   Cohesive strengths and friction angles for different GSI and m_i values

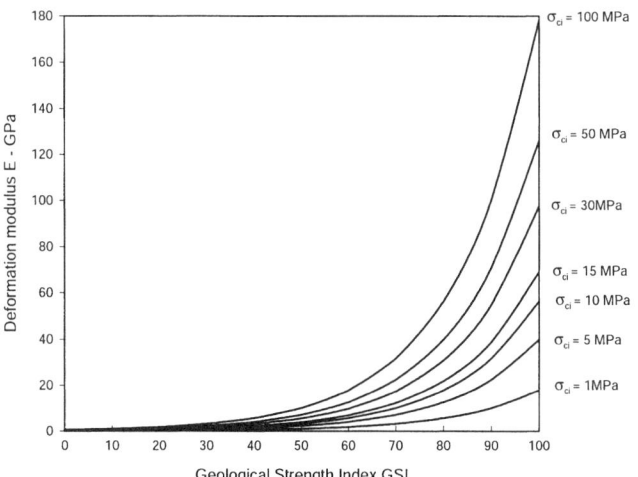

FIGURE 56.5   Deformation modulus versus Geological Strength Index (GSI)

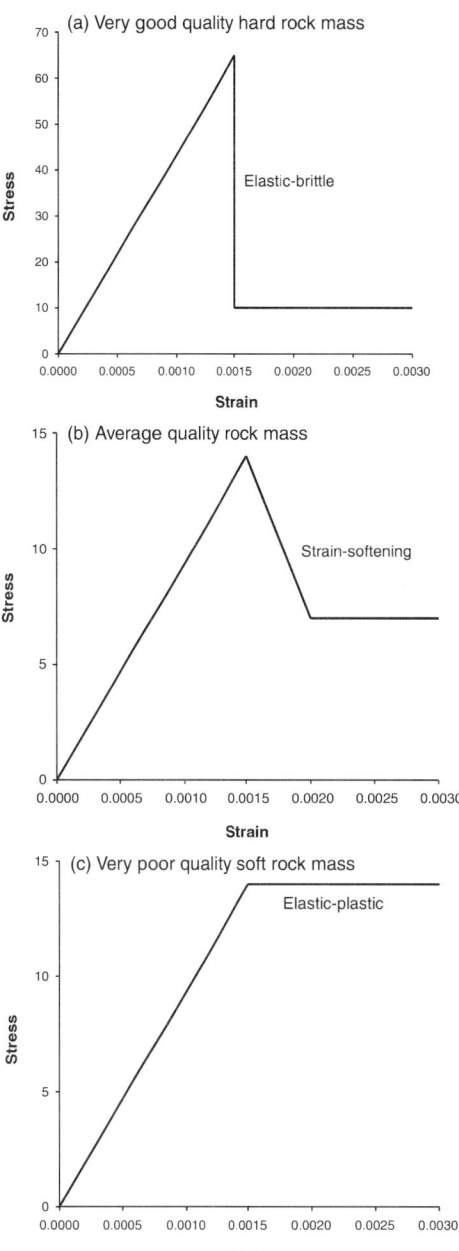

FIGURE 56.6   Suggested post failure characteristics for different quality rock masses

rock mass behaves in an elastic brittle manner as shown in Figure 56.6. When the strength of the rock mass is exceeded, a sudden strength drop occurs. This is associated with significant dilation of the broken rock pieces. If this broken rock is confined, for example by rock support, then it can be assumed to behave as a rock fill with a friction angle of approximately ϕ' = 38° and zero cohesive strength.

Typical properties for this very good, quality, hard rock mass may be as shown in Table 56.5. Note that, in some numerical analyses, it may be necessary to assign a very small cohesive strength to avoid numerical instability.

56.7.2   Average Quality Rock Mass

In the case of an average-quality rock mass, it is reasonable to assume that the post-failure characteristics can be estimated by reducing the GSI value from the in-situ value to a lower value that characterises the broken rock mass.

Hoek-Brown and equivalent Mohr-Coulomb failure criteria

Input:	sigci =	60	MPa		mi =	19			GSI =	50	
Output:	mb =	3.19			s =	0.0039			a =	0.5	
	sigtm =	-0.0728	MPa		A =	0.6731			B =	0.7140	
	k =	4.06			phi =	37.20	degrees		coh =	2.930	MPa
	sigcm =	11.80	MPa		E =	7746.0	MPa				

Calculation:

									Sums
sig3	1E-10	2.14	4.29	6.4	8.57	10.71	12.86	15.00	60.00
sig1	3.73	22.72	33.15	41.68	49.22	56.12	62.57	68.68	337.88
ds1ds3	26.62	5.64	4.31	3.71	3.35	3.10	2.92	2.78	52.45
sign	0.14	5.24	9.72	13.91	17.91	21.78	25.53	29.20	123.43
tau	0.70	7.36	11.28	14.42	17.10	19.49	21.67	23.68	115.69
x	-2.46	-1.05	-0.79	-0.63	-0.52	-0.44	-0.37	-0.31	-6.58
y	-1.93	-0.91	-0.73	-0.62	-0.55	-0.49	-0.44	-0.40	-6.07
xy	4.76	0.96	0.57	0.39	0.29	0.21	0.16	0.13	7.47
xsq	6.05	1.11	0.62	0.40	0.27	0.19	0.14	0.10	8.88
sig3sig1	0.00	48.69	142.07	267.95	421.89	601.32	804.50	1030.15	3317
sig3sq	0.00	4.59	18.37	41.33	73.47	114.80	165.31	225.00	643
taucalc	0.71	7.15	11.07	14.28	17.09	19.63	21.99	24.19	
sig1sig3fit	11.80	20.50	29.19	37.89	46.58	55.28	63.97	72.67	
signtaufit	3.03	6.91	10.31	13.49	16.53	19.46	22.31	25.09	

Cell formulae:

```
mb = mi*EXP((GSI-100)/28)
s = IF(GSI>25,EXP((GSI-100)/9),0)
a = IF(GSI>25,0.5,0.65-GSI/200)
sigtm = 0.5*sigci*(mb-SQRT(mb^2+4*s))
sig3 = Start at 1E-10 (to avoid zero errors) and increment in 7 steps of  sigci/28 to 0.25*sigci
sig1 = sig3+sigci*(((mb*sig3)/sigci)+s)^a
ds1ds3 = IF(GSI>25,(1+(mb*sigci)/(2*(sig1-sig3))),1+(a*mb^a)*(sig3/sigci)^(a-1))
sign = sig3+(sig1-sig3)/(1+ds1ds3)
tau = (sign-sig3)*SQRT(ds1ds3)
x = LOG((sign-sigtm)/sigci)
y = LOG(tau/sigci)
xy = x*y              x sq = x^2
A =    acalc =      10^(sumy/8 - bcalc*sumx/8)
B =    bcalc =      (sumxy - (sumx*sumy)/8)/(sumxsq - (sumx^2)/8)
k = (sumsig3sig1 - (sumsig3*sumsig1)/8)/(sumsig3sq-(sumsig3^2)/8)
phi = ASIN((k-1)/(k+1))*180/PI()
coh = sigcm/(2*SQRT(k))
sigcm = sumsig1/8 - k*sumsig3/8
E = IF(sigci>100,1000*10^((GSI-10)/40),SQRT(sigci/100)*1000*10^((GSI-10)/40))
phit = (ATAN(acalc*bcalc*((signt-sigtm)/sigci)^(bcalc-1)))*180/PI()
coht = acalc*sigci*((sign-sigtm)/sigci)^bcalc-signt*TAN(phit*PI()/180)
sig3sig1 = sig3*sig1         sig3sq = sig3^2
taucalc = acalc*sigci*((sign-sigtm)/sigci)^bcalc
s3sifit = sigcm+k*sig3
sntaufit = coh+sign*TAN(phi*PI()/180)
tangent = coht+sign*TAN(phit*PI()/180)
```

FIGURE 56.7   Spreadsheet for calculation of Hoek-Brown and equivalent Mohr-Coulomb parameters

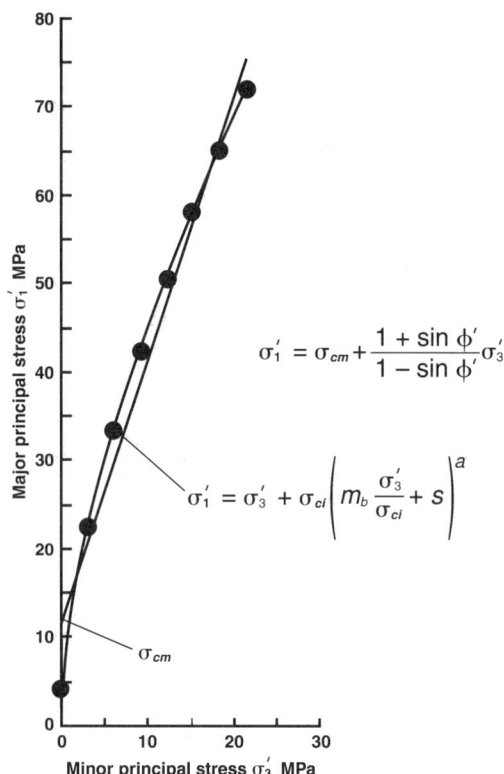

$$\sigma_1' = \sigma_{cm} + \frac{1 + \sin\phi'}{1 - \sin\phi'}\sigma_3'$$

$$\sigma_1' = \sigma_3' + \sigma_{ci}\left(m_b\frac{\sigma_3'}{\sigma_{ci}} + s\right)^a$$

σ_{cm}

Minor principal stress σ_3' MPa

Major principal stress σ_1' MPa

Reducing the rock mass strength from the in-situ to the broken state corresponds to the strain softening behaviour illustrated in Figure 56.6. In this figure, it is assumed that post-failure deformation occurs at a constant stress level, defined by the compressive strength of the broken rock mass. The validity of this assumption is uncertain.

Typical properties for this average quality rock mass may be as shown in Table 56.6.

56.7.3   Very Poor Quality Rock Mass

Analysis of the progressive failure of very poor quality rock masses surrounding tunnels suggests that the post-failure characteristics of the rock are adequately represented by assuming that it behaves perfectly plastically. This means that it continues to deform at a constant stress level and that no volume change is associated with this ongoing failure. This type of behaviour is illustrated in Figure 56.6.

Typical properties for this very poor quality rock mass may be as shown in Table 56.7.

56.8   REFERENCES

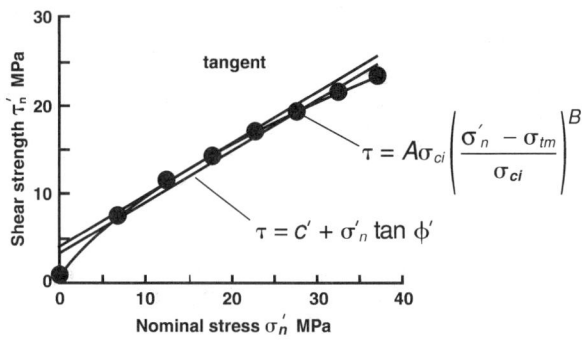

tangent

$$\tau = A\sigma_{ci}\left(\frac{\sigma_n' - \sigma_{tm}}{\sigma_{ci}}\right)^B$$

$$\tau = c' + \sigma_n' \tan\phi'$$

Nominal stress σ_n' MPa

Shear strength τ_n' MPa

FIGURE 56.8   Plot of results from simulated full scale triaxial tests on a rock mass defined by a uniaxial compressive strength σ_{ci} = 85 MPa, a Hoek-Brown constant m_i = 10 and a Geological Strength Index GSI = 45

Balmer, G. 1952. A general analytical solution for Mohr's envelope. *Am. Soc. Test. Mat.* 52, 1260-1271.

Bieniawski, Z.T. 1989. *Engineering Rock Mass Classifications*, New York, Wiley, p. 251.

Bieniawski, Z.T. 1976. Rock mass classification in rock engineering. *Exploration for Rock Engineering, Proc. of the Symp.*, 1, Z.T. Bieniawski, Ed. Cape Town, Balkema, 97-106.

Brown, E.T. (Ed). 1981. Rock characterisation, testing and monitoring - ISRM suggested methods. Oxford, Pergamon, p. 171-183.

Hoek, E. 1983. Strength of jointed rock masses. 23rd Rankine Lecture. *Géotechnique* 33(3), p. 187-223.

Hoek, E. 1994. Strength of rock and rock masses. *ISRM News Journal,* 2(2), p. 4-16.

Hoek, E. and E.T. Brown. 1980. *Underground excavations in rock*. London, Instn Min. Metall., p. 527.

Hoek, E. and E.T. Brown. 1980. Empirical strength criterion for rock masses. *J. Geotech. Engng. Div., ASCE*, 106 (GT 9), p. 1013-1035.

Hoek, E. and E.T. Brown. 1988. The Hoek-Brown failure criterion—a 1988 update. *Rock Engineering for Underground Excavations, Proc. 15th Canadian Rock Mech. Symp.* J.C. Curran, Ed. Toronto: Dept. Civil Engineering, University of Toronto, p. 31-38.

Hoek, E. and E.T. Brown. 1997. Practical estimates or rock mass strength. *Int. J. Rock Mech. & Mining Sci. & Geomechanics Abstracts.* 34(8), p. 1165-1186.

Hoek, E., P.K. Kaiser, and W.F. Bawden. 1995. *Support of underground excavations in hard rock*. Rotterdam, Balkema, p. 215.

Hoek, E., P. Marinos, and M. Benissi. 1998. Applicability of the Geological Strength Index (GSI) classification for very weak and sheared rock masses. The case of the Athens Schist Formation. *Bull. Engng. Geol. Env.* 57(2), p. 151-160.

TABLE 56.5 Typical properties for a very good quality hard rock mass

Intact rock strength	σ_{ci}	150 MPa
Hoek-Brown constant	m_i	25
Geological Strength Index	GSI	75
Friction angle	ϕ'	46°
Cohesive strength	c'	13 MPa
Rock mass compressive strength	σ_{cm}	64.8 MPa
Rock mass tensile strength	σ_{tm}	–0.9 MPa
Deformation modulus	E_m	42000 MPa
Poisson's ratio	ν	0.2
Dilation angle	α	$\phi'/4 = 11.5°$
Post-peak characteristics		
Friction angle	ϕ_f'	38°
Cohesive strength	c_f'	0
Deformation modulus	E_{fm}	10000 MPa

TABLE 56.6 Typical properties for an average rock mass

Intact rock strength	σ_{ci}	80 MPa
Hoek-Brown constant	m_i	12
Geological Strength Index	GSI	50
Friction angle	ϕ'	33°
Cohesive strength	c'	3.5 MPa
Rock mass compressive strength	σ_{cm}	13 MPa
Rock mass tensile strength	σ_{tm}	-0.15
Deformation modulus	E_m	9000 MPa
Poisson's ratio	ν	0.25
Dilation angle	α	$\phi'/8 = 4°$
Post-peak characteristics		
Broken rock mass strength	σ_{fcm}	8 MPa
Deformation modulus	E_{fm}	5000 MPa

TABLE 56.7 Typical properties for a very poor quality rock mass

Intact rock strength	σ_{ci}	20 MPa
Hoek-Brown constant	m_i	8
Geological Strength Index	GSI	30
Friction angle	ϕ'	24°
Cohesive strength	c'	0.55 MPa
Rock mass compressive strength	σ_{cm}	1.7 MPa
Rock mass tensile strength	σ_{tm}	-0.01 MPa
Deformation modulus	E_m	1400 MPa
Poisson's ratio	ν	0.3
Dilation angle	α	zero
Post-peak characteristics		
Broken rock mass strength	σ_{fcm}	1.7 MPa
Deformation modulus	E_{fm}	1400 MPa

Hoek, E., D. Wood and S. Shah. 1992. A modified Hoek-Brown criterion for jointed rock masses. *Proc. Rock Characterization, Symp. Int. Soc. Rock Mech.: Eurock '92*, J.A. Hudson, Ed. London, Brit. Geotech. Soc., p. 209–214.

Karzulovic, A. 2000. Personal communication.

Serafim, J.L. and J.P. Pereira. 1983. Consideration of the Geomechanics Classification of Bieniawski. *Proc. Intnl. Symp. Engng. Geol. And Underground Construction.* Lisbon, Portugal, Vol. 1, Part 11, p. 33–44.

Truzman, M, 1999. Personal communication.

56.9 APPENDIX—DETERMINATION OF MOHR-COULOMB CONSTANTS

The steps required to determine the parameters A, B, c' and ϕ' are given below. A spreadsheet for carrying out this analysis, with a listing of all the cell formulae, is given in Figure 56.7.

The relationship between the normal and shear stresses can be expressed in terms of the corresponding principal effective stresses as suggested by Balmer (1952):

$$\sigma_n' = \sigma_3' + \frac{\sigma_1' - \sigma_3'}{\partial\sigma_1'/\partial\sigma_3' + 1} \qquad (56.13)$$

$$\tau = (\sigma_1' - \sigma_3')\sqrt{\partial\sigma_1'/\partial\sigma_3'} \qquad (56.14)$$

For the *GSI* > 25, when a = 0.5:

$$\frac{\partial\sigma_1'}{\partial\sigma_3'} = 1 + \frac{m_b\sigma_{ci}}{2(\sigma_1' - \sigma_3')} \qquad (56.15)$$

For *GSI* < 25, when s = 0:

$$\frac{\partial\sigma_1'}{\partial\sigma_3'} = 1 + am_b^a\left(\frac{\sigma_3'}{\sigma_{ci}}\right)^{a-1} \qquad (56.16)$$

The tensile strength of the rock mass is calculated from:

$$\sigma_{tm} = \frac{\sigma_{ci}}{2}\left(m_b - \sqrt{m_b^2 + 4s}\right) \qquad (56.17)$$

The equivalent Mohr envelope, defined by Equation 56.2, may be written in the form

$$Y = \log A + BX \qquad (56.18)$$

where:

$$Y = \log\left(\frac{\tau}{\sigma_{ci}}\right), X = \log\left(\frac{\sigma_n' - \sigma_{tm}}{\sigma_{ci}}\right) \qquad (56.19)$$

Using the value of σ_{tm} calculated from Equation 56.17 and a range of values of τ and σ_n' calculated from Equations 56.13 and 56.14, the values of A and B are determined by linear regression where:

$$B = \frac{\Sigma XY - (\Sigma X\Sigma Y)/T}{\Sigma X^2 - (\Sigma X)^2/T} \qquad (56.20)$$

$$A = 10^{\wedge}(\Sigma Y/T) - (B\Sigma X/T) \qquad (56.21)$$

and *T* is the total number of data pairs included in the regression analysis.

The most critical step in this process is selecting the range of σ_3' values. As far as the author is aware, there are no theoretically correct methods for choosing this range and a trial-and-error method, based upon practical compromise, has been used for selecting the range included in the spreadsheet presented in Figure 56.7.

For a Mohr envelope defined by Equation 56.2, the friction angle ϕ_i' for a specified normal stress σ_{ni}' is given by:

$$\phi_i' = \arctan\left(AB\left(\frac{\sigma_{ni}' - \sigma_{tm}}{\sigma_{ci}}\right)^{B-1}\right) \qquad (56.22)$$

The corresponding cohesive strength c'_i is given by:

$$c'_i = \tau - \sigma'_{ni} \tan \phi'_i \qquad \textbf{(56.23)}$$

and the corresponding uniaxial compressive strength of the rock mass is :

$$\sigma_{cmi} = \frac{2c'_i \cos \phi'_i}{1 - \sin \phi'_i} \qquad \textbf{(56.24)}$$

The values of c' and ϕ' obtained from this analysis are very sensitive to the range of values of the minor principal stress σ'_3 used to generate the simulated full-scale triaxial test results. On the basis of trial and error, the most consistent results for deep excavations (depth >30 m below surface) are obtained when eight equally spaced values of σ'_3 are used in the range $0 < \sigma'_3 < 0.25\sigma_{ci}$.

An example of the results obtained from this analysis is given in Figure 56.4. Plots of the values of the ratio c'/σ_{ci} and the friction angle ϕ' for different combinations of GSI and m_i are given in Figure 56.5.

The spreadsheet includes a calculation for a tangent to the Mohr envelope defined by Equation 56.2. A normal stress has to be specified in order to calculate this tangent and, in Figure 56.7, this stress has been chosen so that the friction angle ϕ' is the same for both the tangent and the line defined by $c' = 3.3$ MPa and $\phi' = 30.1°$, determined by the linear regression analysis described earlier. The cohesion intercept for the tangent is $c' = 4.1$ MPa which is approximately 25% higher than that obtained by linear regression analysis of the simulated triaxial test data.

Fitting a tangent to the curved Mohr envelope gives an upper bound value for the cohesive intercept c'. It is recommended that this value be reduced by about 25% in order to avoid overestimation of the rock mass strength.

There is a particular class of problem for which extreme caution should be exercised when applying the approach outlined above. In some rock slope stability problems, the effective normal stress on some parts of the failure surface can be quite low, certainly less than 1 MPa. It will be noted that in the example given in Figure 56.8, for values of σ'_n of less about 5 MPa, the straight line, constant c' and ϕ' method overestimates the available sheer strength of the rock mass by increasingly significant amounts as σ'_n approaches zero. Under such circumstances, it would be prudent to use values of c' and ϕ' based on a tangent to the shear strength curve in the range of σ'_n values applying in practice.

CHAPTER 57

The MRMR Rock Mass Classification for Jointed Rock Masses

D.H. Laubscher* and J. Jakubec†

57.1 INTRODUCTION

Because the competency and engineering properties of jointed rock masses can vary greatly, there is a demonstrated need for a systematic numerical method of describing rock masses. Such a system can accomplish several things:

- Enhance communication between geologists, engineers, and operating personnel

- Provide the basis for comparing rock behaviour from project to project and over time

- Quantify experience for developing empirical relationships with rock mass properties and for guidelines for method selection, cavability, stability, and support design

This is a brief history of the Modified Rock Mass Rating (MRMR) system. In 1973, D.H. Laubscher met with Z.T. Bieniawski to discuss the rock mass classification system that Bieniawski was developing (the Rock Mass Rating [RMR]) for geotechnical investigations of civil engineering projects and to overcome communication problems (Bieniawski 1973). Bieniawski's approach was better than the system being developed in Zimbabwe by Heslop and Laubscher at that time (Heslop 1973). However, Laubscher decided that a lot more flexibility was required to adjust for different mining situations. The RMR concept was used for in-situ ratings, and adjustments were made for other situations. Thus, the MRMR system was developed (Laubscher 1975, Laubscher and Taylor 1976). Over the years, changes have been made to the ratings value as the relative importance of the items became apparent. For some time there has been concern that the role of fractures/veins and cemented joints were not properly included. The techniques developed to adjust for these items have been included here.

The overall objectives of this paper are (1) to show how the MRMR classification system can be applied to jointed rocks and (2) to indicate the changes made to the system over the years. Figure 57.1 is a flow sheet to help the reader follow the different parts of the system.

57.2 DEFINITIONS

The competency of jointed rock is heavily dependent on the nature, orientation, and continuity of the discontinuities in the rock mass. Figure 57.2 is a diagrammatic representation of the different structural features.

Faults and Shear Zones—Major features, large-scale continuity, and frequently very weak zones. Must be classified separately.

Open Joints—An easily identified structural discontinuity that defines a rock block.

Cemented Joints—A structural feature that has continuity with the walls cemented with minerals of different cementing strength. In high-stress environments, cemented joints can

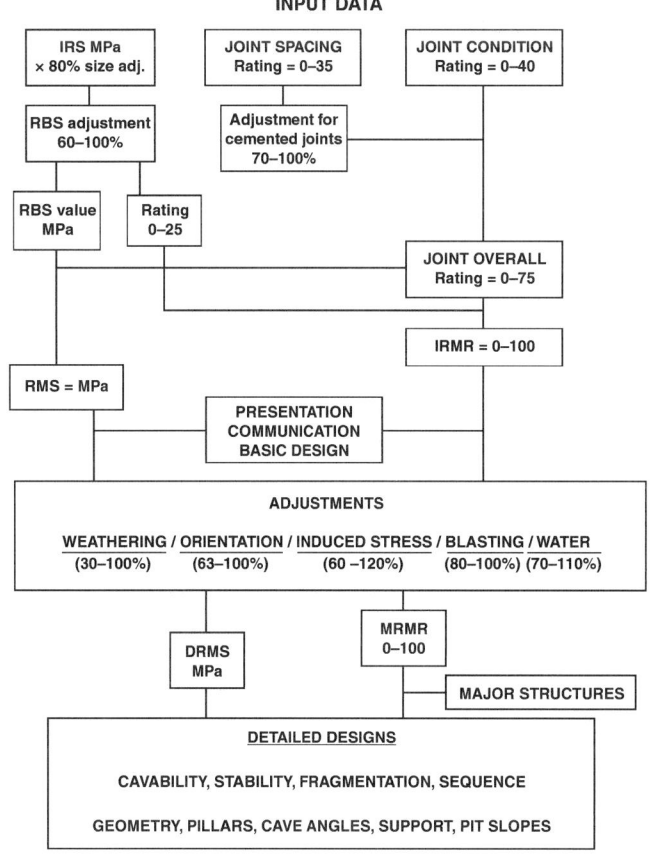

FIGURE 57.1   Flow sheet of the MRMR procedure with recent modifications

impact the strength of the rock mass. Therefore, the frequency and hardness of the cementing material must be recorded.

Fractures and Veins—Low continuities can occur within a rock block. The hardness number defines the fill material—open fractures have a hardness of 1.

Mapping and Core Logging—In scan-line mapping, it is essential to log the continuities of structures and to distinguish between fractures and joints and other structural defects. In drill-core logging, the geologist should attempt to classify the defects being logged. It should be noted that a joint might have several partings in close proximity; these will behave as a single joint and should be logged as such.

* Bushman's River Mouth, South Africa.
† SRK Consulting (Canada), Vancouver, BC, Canada.

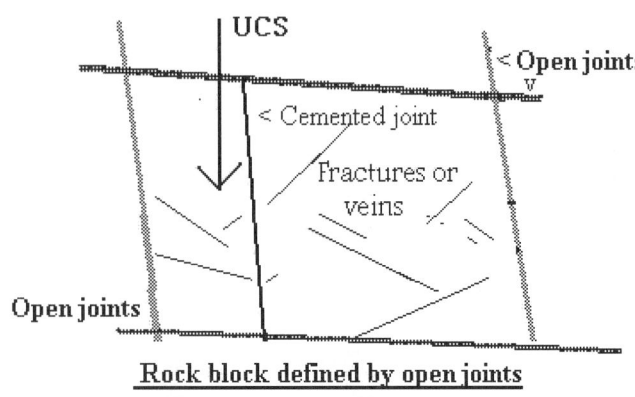

FIGURE 57.2 Definition of the structural terms

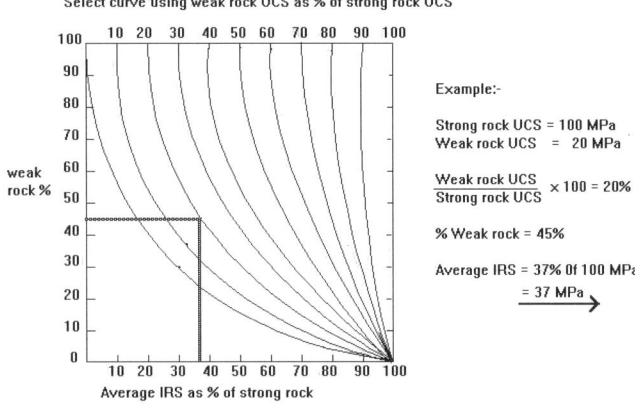

FIGURE 57.3 Nomogram for determining the "corrected" IRS value

57.3 INTACT ROCK STRENGTH (IRS) TO ROCK BLOCK STRENGTH (RBS)

57.3.1 Intact Rock Strength

The unconfined compressive strength (UCS) is the value derived from testing cores and is the value assigned to the IRS. The intact rock specimen may be homogeneous or have intercalations of weaker material. In the latter case, the procedure shown in Figure 57.3 should be adopted. Care must be taken in determining this value as the cores that are selected often represent the stronger material in the rock mass. To help the reader in this regard, an example is presented. As shown in Figure 57.3, the UCS values for the strong and weak rock are 100 megapascals (MPa) and 20 MPa, respectively. It is estimated that 45% is made up of weak rock. Using Figure 57.3, one locates this value on the Y-axis, moves horizontally to the curve representing the strength of the weak rock, and then drops down to the horizontal axis. In this case, the appropriate "corrected" IRS is 37 MPa.

57.3.2 Rock Block Strength

To obtain the RBS from the "corrected" IRS, various factors are applied depending upon whether the rock blocks are homogeneous or contain fractures and veins.

Homogeneous Rock Blocks—If the rock block does not contain fractures or veins, the RBS is the IRS value reduced to 80% to adjust for the small-to-large specimen effect.

Thus, RBS = 0.8 × "corrected" IRS.

Rock Blocks with Fractures and Veins—Fractures and veins reduce the strength of the rock block in terms of the number and

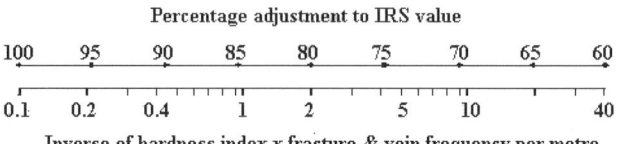

FIGURE 57.4 The nomogram relating the IRS adjustment factor to the hardness index and vein frequency

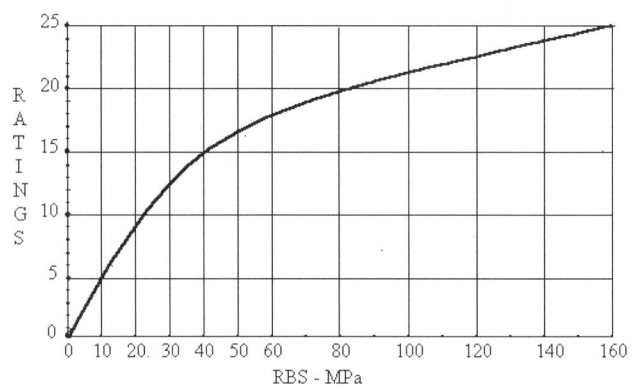

FIGURE 57.5 Rock block strength rating as a function of rock block strength

frictional properties of the features (see Figure 57.2). The Moh's hardness number is used to define the frictional properties of the vein and fracture filling. The standard hardness table is used only up to 5, because values greater than 5 are not likely to be significant. Open fractures or veins would be given a value of 1. The vein and fracture filling must be weaker than the *host* rock.

Index	1 = Talc, Molybd.	2 = Gypsum, Chlorite	3 = Calcite, Anhydrite	4 = Fluorite, Chalcopy.	5 = Apatite
Inverse	1.0	0.5	0.33	0.25	0.2

The procedure is to take the inverse of the hardness index and multiply that by fracture/vein frequency per meter, to arrive at a number that reflects the relative weakness between different rock masses. This number can then be used in Figure 57.4 to determine the percentage adjustment to the IRS value.

To obtain the RBS, the corrected IRS is adjusted by the size factor of 80% and then by the fracture/vein frequency and hardness adjustment.

RBS = IRS × 0.8 × fracture/vein adjustment (F/V) = MPa.

To illustrate this, consider the following example:

IRS = 100 MPa,

Gypsum veins: Moh's hardness = 2

ff/m = 8.0

The product of the inverse hardness and the fracture frequency is:

Inverse of hardness index × fracture frequency = 0.5 × 8 = 4.0

Using Figure 57.4, one finds that the adjustment is 0.75. Therefore:

RBS = 100 × 0.8 × 0.75 = 60 MPa.

The rating for the RBS can now be read from Figure 57.5. The slope of the curve is steeper for the lower RBS values, as

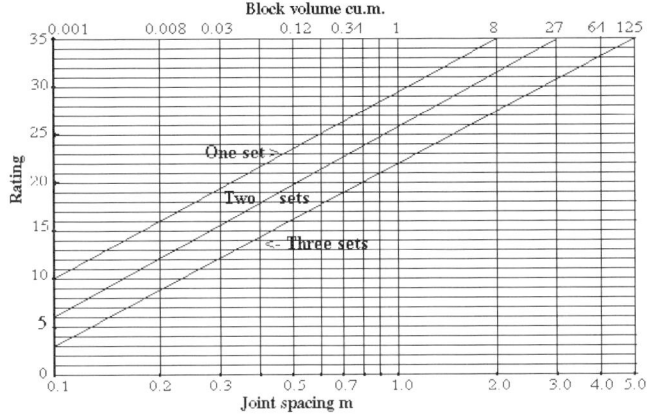

FIGURE 57.6   Joint spacing ratings

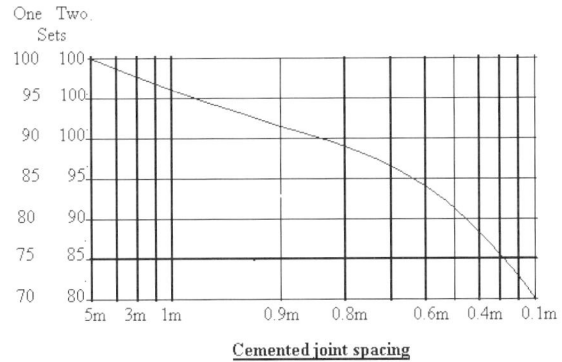

FIGURE 57.7   Adjustment factor for cemented joints

TABLE 57.1   Joint condition adjustments

A.	Large-scale joint expression	Adjustment % of 40
	Wavy—multidirectional	100
	Wavy—unidirectional	95
	Curved	90
	Straight, slight undulation	85
B.	Small-scale joint expression (200 mm × 200 mm)	
	Rough stepped/irregular	95
	Smooth stepped	90
	Slickensided stepped	85
	Rough undulating	80
	Smooth undulating	75
	Slickensided undulating	70
	Rough planar	65
	Smooth planar	60
	Polished	55
C.	Joint wall alteration weaker than sidewall and filling	75
D.	Gouge	
	Thickness < amplitudes	60
	Thickness > amplitudes	30

E. Cemented/filled joints—cement weaker than wall rock. The percentage in the column is the adjustment to obtain the cemented filled-joint condition rating

Hardness	Adjustment:
5	95%
4	90%
3	85%
2	80%
1	75%

small changes are significant. In this case, one sees that the RBS rating is

$$\text{RBS rating} = 17.5.$$

57.4   JOINTING

57.4.1   Open-Joint Spacing

In previous papers, one had the option of using the RQD and joint spacing or ff/m. However, the fracture/vein frequency and the condition are part of the rock-block strength calculation and, therefore, cannot be counted twice. Because of this, the joint spacing rating has been reduced to 35 and refers only to open joints. Although there are situations where there are more than three joint sets, for simplicity they should be reduced to three. The chart in Figure 57.6 is slightly different from the previous ratings chart in that the ratings for one and two sets are proportionately higher.

57.4.2   Cemented Joints

The cemented joints will influence the strength of the rock mass when the strength of the cement is less than the strength of the host rock. If the cemented joints form a distinct set, then the rating for the open joints is adjusted down according to Figure 57.7. For example, if the rating for two open joints at 0.5-m spacing were 23, an additional cemented joint with a spacing of 0.85 m would have an adjustment of 90%. The final rating would be 21, which is equivalent to a three-joint set with an average spacing of 0.65 m. The slope of the curve is increased to adjust for the significant influence of the closer joint spacing. Failure can often occur at the cemented joint contact under high-stress conditions or with poor blasting.

57.5   JOINT CONDITION

57.5.1   Single Joints

The RMR system has been revised to adjust for cemented joints and to have water as a mining adjustment. The maximum joint-condition (JC) rating remains at 40, but the JC adjustments have been changed to those given in Table 57.1.

57.5.2   Multiple Joints

Average JC ratings are required for RMR values; however, a significant variation in these ratings could be the result of trying to force dissimilar areas into one rating. *It is preferable to use the classification system to show variations in the rock mass as this zoning could influence planning decisions.* A weighted average of JC ratings can give the wrong result, particularly if the rating of one set is high. For example, the total rating for the case of a single joint set with three joints/m (a joint spacing [JS] rating of 22) and a JC rating of 20 would be 42.

If this set is combined with another set with a JC rating of 38 and seven joints/m, the weighted average JC rating is (3 × 20 + 7 × 38)/10 = 33. The JS rating for the 10 joints (two sets) is 13. Combining the JC and the JS ratings, one gets a total (combined) rating of 46. This is too high when compared with the 42 for one joint set. The addition of seven joints must weaken the rock mass.

Various procedures were tried to obtain a realistic average JC, and the chart in Figure 57.8 gives the best results by using the highest and lowest ratings. Therefore, if the diagram in Figure 57.8

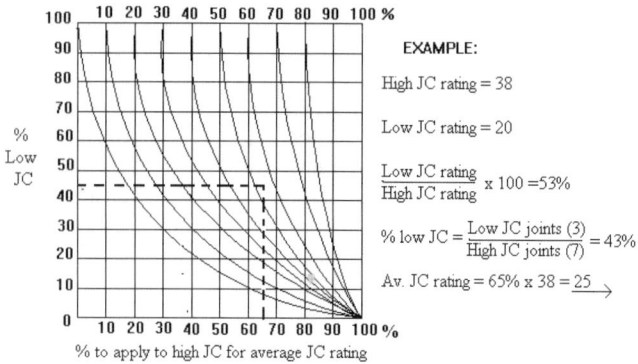

Select curve using lowest joint condition
rating as % of highest joint condition rating

EXAMPLE:

High JC rating = 38

Low JC rating = 20

$\dfrac{\text{Low JC rating}}{\text{High JC rating}} \times 100 = 53\%$

$\%\text{ low JC} = \dfrac{\text{Low JC joints (3)}}{\text{High JC joints (7)}} = 43\%$

Av. JC rating = 65% x 38 = 25 →

% to apply to high JC for average JC rating

FIGURE 57.8 Chart for considering multiple joint sets

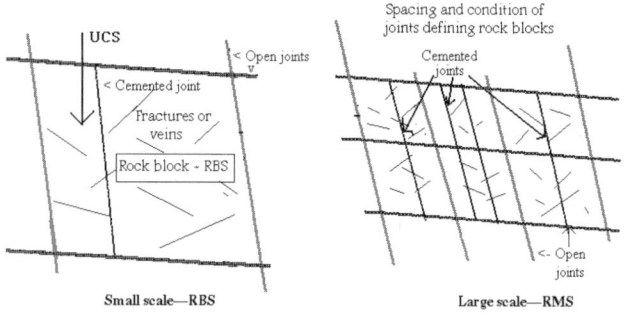

ROCK BLOCK STRENGTH - RBS AND ROCK MASS STRENGTH - RMS

Small scale—RBS

Large scale—RMS

FIGURE 57.9 Diagrammatic representation of small scale and large scale RBS

is used to average the JC ratings, this results in 25 (JC) plus 13 (JS) = 38, a more likely result when compared with the 42 for one joint set.

57.6 ROCK-MASS VALUES

57.6.1 Rock-Mass Rating

The RMR is defined as:

RMR = RBS rating + Overall joint rating

57.6.2 Rock Mass Strength

The rock mass strength (RMS) is derived from the RBS because the strength of the rock mass must recognise the role of the JS and the JC. This is shown diagrammatically in Figure 57.9. The principles of the original formula are:

RMS = (A – B)/70 (100 – RBS rating) × C

where:

A = RMR value

B = RBS rating

C = RBS value in MPa

still apply. For example, assume that:

RMR = 40

RBS rating = 6

RBS value = 60 MPa.

Then:

RMS = (40 – 6)/70 × 60 = 31 MPa

TABLE 57.2 Adjustments for weathering

Description	Potential weathering and % adjustments				
	6 months	1 year	2 years	3 years	4+ years
Fresh	100	100	100	100	100
Slightly	88	90	92	94	96
Moderately	82	84	86	88	90
Highly	70	72	74	76	78
Completely	54	56	58	60	62
Residual soil	30	32	34	36	38

57.7 MRMR ADJUSTMENT PROCEDURE

57.7.1 Introduction

The RMR rating is multiplied by an adjustment factor to give the MRMR rating.

The adjustment procedure has been described in previous papers, where it was stated that the adjustment should not exceed two classes. What was not made clear is that one adjustment can supersede another, and the total adjustment is not likely to be a multiplication of all the adjustments. For example, a bad blasting adjustment would apply in a low-stress area, but in a high-stress area, the damage from the stresses would exceed that of the blasting, and the only adjustment would be the mining-induced stress. The MRMR for a cavability assessment would not have blasting as an adjustment, nor would it have weathering unless the weathering effects were so rapid as to exceed the rate of cave propagation as a result of the structural and stress effects. The joint-orientation and mining-induced stress adjustments tend to complement each other. The object of the adjustments is for the geologist, rock mechanics engineer, and planning engineer to adjust the RMR so the MRMR is a realistic number, reflecting the RMS for that particular mining situation. Although expert systems are useful, the wide variety of features that have to be recognised in mine planning requires flexibility in assessing the situation. The complete dedication to computer-generated results has led to some major errors in the past, and one must not remove the human thought process. There is a better appreciation of the planning process and operation when personnel have to think in terms of adjustments.

57.7.2 Weathering

Certain rock types weather readily, and this must be taken into consideration in terms of life and size of opening and the support design. In the case of the fast-weathering kimberlites, for example, the rock surface needs to be sealed. The weathering adjustment refers to the anticipated change in RMS as the weathering process alters the exposed surfaces and joint fillings. It does not refer to the existing weathered state of the rock, as that would be adjusted for by the IRS and then by the RBS. The two items that are affected by weathering are the RBS and the JC. The RBS is affected by the weathering of fractures and veins and by the penetrative weathering of the intact rock. Borehole cores give a good indication of the weathering process, but the results are conservative, as the surface area of the core is high with respect to the volume of core. The weathering adjustment factors given in Table 57.2 cover known situations.

57.7.3 Joint-Orientation Adjustment

The shape, size, and orientation of the excavation will influence the behaviour of the rock mass in terms of rock-block stability. The attitude of the joints, with respect to the vertical axis of the block, the frictional properties of the joints, and whether the bases of the rock blocks are exposed, have a considerable influence on stability; and the RMR rating must be adjusted accordingly. The magnitude

TABLE 57.3   Joint-adjustment factors

Number of joints defining the block	Number of faces inclined from vertical	Orientation % adjustments for ranges in joint condition		
		0–15	16–30	31–40
3	3	70	80	95
	2	80	90	95
4	4	70	80	90
	3	75	80	95
	2	85	90	95
5	5	70	75	80
	4	75	80	85
	3	80	85	90
	2	85	90	95
	1	90	95	

TABLE 57.4   Blasting adjustment factors

Technique	Adjustment (%)
Boring	100
Smooth-wall blasting	97
Good conventional blasting	94
Poor blasting	80

TABLE 57.5   Water adjustment factors

Moist	Moderate pressure—1–5 MPa 25–125 1/m	High pressure — >5 MPa >125 1/m
95% to 90%	90% to 80%	80% to 70%

of the adjustment is a function of the number of joints, their dip, and their frictional properties. Obviously, a block with joints that dip at 60° is more likely to fail than one where the joints dip at 80°. Also, the joint adjustment cannot be looked at in isolation, as a low-dipping joint is liable to shear failure, whereas the steeply dipping joint could be clamped. A computer program could be developed to adjust for a variety of situations, but would only be valid if there were sufficient checks along the way. The joint-orientation adjustments in Table 57.3 have now been changed to reflect the influence of low-friction surfaces as defined by the JC rating.

The adjustment for the orientation of shear zones at an angle to the development is:

0°–15° = 76%

16°–45° = 84%

46°–75° = 92%

Advancing tunnel faces in the direction of the dip is preferable to advancing against the dip, as it is easier to support blocks with joints dipping in the direction of advance. An adjustment of 90% should be made when the advance is into the dip of a set (or sets) of joints.

57.7.4   Mining-Induced Stresses

Mining-induced stresses are the redistribution of field or regional stresses as a result of the geometry and orientation of the excavations. The orientation, magnitude, and ratio of the field stresses should be known either from stress measurements and/or stress analyses. If sufficiently high, the maximum principal stress can cause spalling, crushed pillars, the deformation and plastic flow of soft zones, and will ultimately result in cave propagation. The deformation of soft zones leads to the failure of hard zones at low stress levels. A compressive stress at a large angle to structures will increase the stability of the rock mass and inhibit caving. In this case, the adjustment factor is 120%. This was the situation in a caving operation where the back was stable and caving only occurred when adjacent mining removed the high horizontal stress. Stresses at a low angle will result in shear failure and have an adjustment of 70%.

The adjustment for high stresses that cause rock failure can be as low as 60%. A classic example of this was on a mine where the RMR was 60 in the low-stress area, but the same rock mass in a high-stress area was classified as having an RMR of 40. The 40 is not the RMR, but the MRMR, and the adjustment in this case is 40/60 = 67%.

When assessing mining-induced stresses, the following factors should be considered:

- Drift-induced stresses
- Interaction of closely spaced drifts
- Location of drifts or tunnels close to large stopes or excavations
- Abutment stresses, particularly with respect to the direction of advance and orientation of field stresses (an undercut advancing towards the maximum stress ensures good caving, but creates high abutment stresses and vice versa)
- Uplift as the undercut advances
- Column loading from caved ground caused by poor draw control
- Removal of restraint to sidewalls and apexes
- Increase in mining area and changes in geometry
- Massive wedge failures
- Influence of structures not exposed in the excavation but creating the probability of high toe stresses or failures in the back
- Presence of intrusives that might retain high stresses or shed stress into the surrounding, more competent, rock

The total adjustment varies from 60% to 120%.

57.7.5   Blasting

Blasting creates new fractures and opens up existing fractures and joints, thereby decreasing the rock-mass strength. Boring is considered to be the 100% standard in terms of the quality of the wall rock, but experience on several mines has shown that, while the rock mass might be stable at the face, deterioration occurs ±25 m back, and this is a stress-relief adjustment. Good blasting can allow for some stress relief, thereby improving the stability. The adjustments given in Table 57.4 are recommended.

57.7.6   Water/Ice Adjustment

Water will generally reduce the strength of the rock mass by reducing the RBS and friction across structures and by reducing effective stress. The adjustment factors for water are given in Table 57.5.

In the presence of ice in permafrost areas, the rock mass could be strengthened. This will depend on the amount of ice and on the temperature of the ice. Because of the creep behaviour of ice, the strength usually decreases with time. Adjustments will range from 100% to 120%

57.8   DESIGN RATINGS AND STRENGTHS

57.8.1   Design Rating MRMR

The MRMR is used for design. This is the RMR value as adjusted for weathering, orientation, induced stress, blasting, and water.

$$MRMR = RMR \times \text{adjustment factors}$$

57.8.2　Design Rock Mass Strength

The design rock mass strength (DRMS) is the RMS reduced by the same adjustment factor relating the RMR to the MRMR. In the case where:

RMS = 40 MPa

RMR = 50

MRMR = 40,

then the DRMS would be:

DRMS = RMS × MRMR/RMR = 40/50 × 40 = 32 MPa.

57.9　PRESENTATION

The RMR data should be plotted on plans and sections. The range of 0–100 covers all variations in jointed rock masses from very poor to very good. The classification is divided into five classes of 20 rating with A and B subdivisions. A colour scheme is used to denote the classes on the plan with full colour for the A subdivision and cross-hatched for the B subdivision. The colours are:

Class 1	Class 2	Class 3	Class 4	Class 5
(100–81)	(80–61)	(60–41)	(40–21)	(20–0)
Blue	Green	Yellow	Red	Brown

It is essential that classification data are made available at an early stage so the correct decisions can be made on the mining method, layout, and support design.

It must be stressed that every attempt should be made to zone the rock mass. Averaging a large range in ratings eventually leads to planning and production problems.

57.10　SCALE EFFECTS

Scale is a very important factor when one considers the behaviour of the rock mass, particularly when examining mass-mining methods. For example, the stability/cavability of a deposit cannot be based on the MRMR from a drift assessment alone, as widely spaced major structures play a significant role. That is, structures at a spacing of 10 m would have a marginal effect on the overall IRMR value obtained from drift mapping, but would have a large effect on the cavability of an orebody by providing planes along which displacements can occur. The MRMR value gives a hydraulic radius (HR) required for assuring cavability. This figure should be adjusted for the influence of major structures by using the following procedure to obtain an "influence" number. The various factors that contribute to the "weakness" of a major structure have been ranked as follows:

Rankings

A　**Dip:** 0°–20° = 6, 21°–40° = 4, 31°–40° = 2, 41°–60° = 1, >61° = 0

B　**Spacing:** 0–9 m = 6, 10–15 m = 4, 16–21 m = 3, 22–27 m = 1, >27 m = 0

C　**Joint Condition:** 0–10 = 6, 10–15 = 4, 15–20 = 2, 20–25 = 1, >25 = 0

D　**Stress/structure orientation:** 0°–20° = 7, 21°–30° = 9, 31°–40° = 6, 41°–50° = 3, 51°–60° = 2, 61°–70° = 1, >71° = 0

E　**Distance of major structures from undercut boundaries:** 0–9 m = 12, 10–20 m = 8, 21–30 m = 2, > 31 m = 0

F　**Stress values—Sigma 1 as % of RMS:** >100% = 14, 80%–99% = 12, 60%–79% = 8, 40%–59% = 4, 20%–39% = 2, <20% = 0

The rankings are plotted in Table 57.6. The highest likely ranking from the three sets is in the order of 100.

TABLE 57.6　Form for determining the major structure influence number

Major structures	A	B	C	D	E	F	Total
Set 1							
Set 2							
Set 3							
Total							

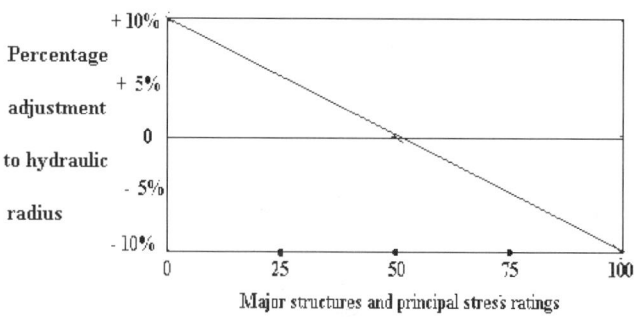

FIGURE 57.10　Adjustment factor for major structures

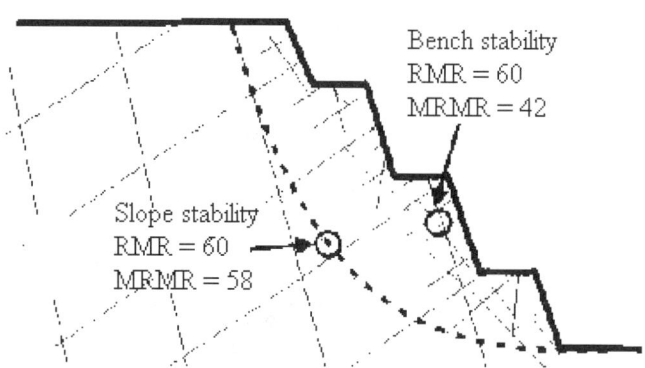

FIGURE 57.11　The difference in the MRMR values used for bench and slope design

This number, when plotted on Figure 57.10, will indicate whether the HR is acceptable, or whether it should be adjusted up or down. The adjustment to the HR is read from Figure 57.11.

If there are other features, such as internal silicified zones, that might contribute to stability, then a deduction should be made. The magnitude of the deduction should be 15% to 40% of the hydraulic radius.

In the case of pit slopes, the RMR of the bench and the overall slope will vary as shown in Figure 57.10.

57.11　PRACTICAL APPLICATIONS

The details of the practical applications can be found in the paper "Planning Mass Mining Operations" (Laubscher 1993). A summary of the applications is:

- Support design (Laubscher 1984)
- Cavability diagrams
- Stability of open stopes
- Pillar design
- Determination of cavability
- Extent of cave and failure zones

- Caving fragmentation
- Mining sequence
- Potential massive wedge failure

57.12 CONCLUSIONS

The object of this paper has been to show the changes that have been made to the original MRMR classification system. It must be stressed that when the system is properly applied, the results are good. Unfortunately, rock masses do not conform to an ideal pattern and, therefore, a certain amount of judgement or interpretation is required. A classification system can give the guidelines, but, as indicated, the geologist or engineer must interpret the finer details. It is important, for example, to divide the rock mass into zones in which there is not a great range in the ratings, and this can only be done if the geologist has an overall understanding. Because it is not possible to precisely define every mining situation, the engineer must use his judgement in arriving at the adjustment percentage. Where the system has been properly applied, it has proved to be successful as a planning and a communications tool. However, what has been found in several mines is that the geology departments dabble in different systems and, at the end of the day, they are masters of none. The numbers produced are incorrect because they do not have a feel for the rock mass. Unfortunately, the modern tendency is to have programs that do the thinking for the operator.

57.13 ACKNOWLEDGMENTS

A classification system can develop only if it is used and tuned to adjust for the many different situations that are encountered in designing mining operations. The writers wish to acknowledge the constructive suggestions from those people who are using the MRMR system and, in particular, the constructive suggestions from T.G. Heslop.

57.14 REFERENCES

Bieniawski, Z.T. 1973. Engineering classification of jointed rock masses *Trans. S. Afr. Inst. Civ. Eng.* 15.

Heslop, T.G. 1973. Internal company report.

Laubscher, D.H. August 1975. Class distinction in rock masses. *Coal, Gold, Base minerals S. Afr.* 23.

Laubscher, D.H. 1984. Design aspects and effectiveness of support in different mining conditions. *Trans. Inst. Min. Metall. Sect. A* 86.

Laubscher, D.H. 1993. Planning mass mining operations. *Comprehensive Rock Engineering,* Vol.2. Pergamon Press.

Laubscher, D.H. and H.W. Taylor. November 1976. The importance of geomechanics classification of jointed rock masses in mining operations. *Proceedings of the Symposium on Exploration for Rock Engineering.* Johannesburg, South Africa.

CHAPTER 58

Use of Numerical Modeling for Mine Design and Evaluation

Mark Board, * **Richard Brummer,** † **and Shawn Seldon** ‡

58.1 INTRODUCTION

Numerical models are routinely used to design mine and feasibility studies, to assess failure mechanisms, and to estimate geomechanical risk. This chapter discusses the techniques and methodology of applying numerical models to geomechanical mine design and mine viability studies, with particular emphasis on deep-mine failure mechanisms. In today's economic and safety environment, large capital expenditures for mine expansions or new mine development cannot be made without a technical consideration of geomechanical risk. In many cases, past experience does not provide adequate assurance that the risk has been investigated fully, and some form of analysis is necessary. In this case, geomechanical models may provide an excellent *tool* for supplementing the intuition and experience of the engineer. Models, however, should not be used in isolation from the interpretation of an experienced rock mechanics engineer.

A model used in this sense is a powerful tool that:

1. Can be used to predict the risk of potential ground-control problems for a given mining method and sequence

2. Provides a method to assess the nature and size of potential ground-control problems and their impacts on cost and production

3. Provides insight into the causes of observed failures

4. Assists in identifying critical geological or mining factors that control failure

5. Provides a simulation tool to assess measures for controlling or preventing ground problems

This chapter gives a brief review of the general methodology used to apply models to solve geomechanical mining problems. This review includes descriptions of modeling techniques and methods for estimating in-situ rock-mass properties, as well techniques for developing a database to assess site damage, production, and cost data. This database is used as a means of "calibrating" the model—i.e., providing a meaningful bridge between model stress and failure predictions and the reality of rock mass response and the production and cost implications underground. The deep extensions at the Kidd Mine are used as an example of the process of estimating geomechanical risk and its impact on mining cost.

58.1.1 Types of Models and Their Applicability

In general, numerical models can be subdivided into two basic classes: (1) those that assume the rock mass is elastic (i.e., there is no failure load limit, and stress concentrations are controlled by the extraction ratio and geometry of the excavations), and (2)

those that assume that the rock mass may fail and shed its load to surrounding regions. Typically, the elastic model is used for initial, nondetailed studies aimed at identifying approximate stress levels and at examining potential pillar sizing and stope sequencing by comparing average stress levels to empirical failure predictions. Elastic models can be good tools for predicting stresses in thin, tabular deposits where confinement is large and pillars remain elastic until the end of extraction.

However, when the orebody is thick or plug-like (e.g., several stope widths) and the depth is great, pillars (such as those created from a primary/secondary open-stoping sequence) are not expected to remain elastic; they are designed to yield in a nonviolent fashion. In this case, when the pillars yield, they shed a portion of their load to adjacent pillars or abutments. This actual response can vary dramatically from that predicted by an elastic approach, which can lead to erroneous and often highly conservative results. The type of model used depends, to a great extent, on the purpose of the analysis (e.g., a rapid study or a detailed attempt to simulate the expected rock-mass response), the extent of yield that is expected, and the impact this yielding has on mine stability as the extraction ratio increases. In any case, the model must be calibrated against field observations as a means of gaining confidence in its ability to adequately represent the mine behavior.

58.1.2 Approach to Using Geomechanical Models in Mining Evaluations

In general, the confidence in the application of a model is directly proportional to the detail used in calibrating it. The approach to applying numerical models is given in terms of the flow chart in Figure 58.1. Initially, it requires identifying the problem to be solved and determining the expectations from the analyses in as much detail as possible. For example, if only a rapid analysis of stope sequencing is required, only an elastic boundary element analysis is necessary. If a detailed assessment of the yield and seismic potential of pillars in a blasthole sequence is needed, a nonlinear analysis using a peak-and-residual-strength constitutive model would be useful. The requirements of the analysis will determine the type and detail necessary in the model.

To derive input properties for the model, a detailed geotechnical field assessment of the rock mass is required for all major rock types present. Experience has shown that characterization using the NGI "Q" (Barton et al. 1976) or rock mass rating (RMR) (Bieniawski 1976) systems (for a general estimate of rock-mass properties), along with detailed-line and gross geologic mapping of fracture characteristics (planarity, roughness, infilling, continuity, and orientation) and major continuous fault structures will provide an adequate basis for

* Itasca Consulting Group, Inc., Minneapolis, MN.
† Itasca Consulting Canada, Inc., Sudbury, Ontario, Canada.
‡ Falconbridge, Ltd., Kidd Creek, Ontario, Canada.

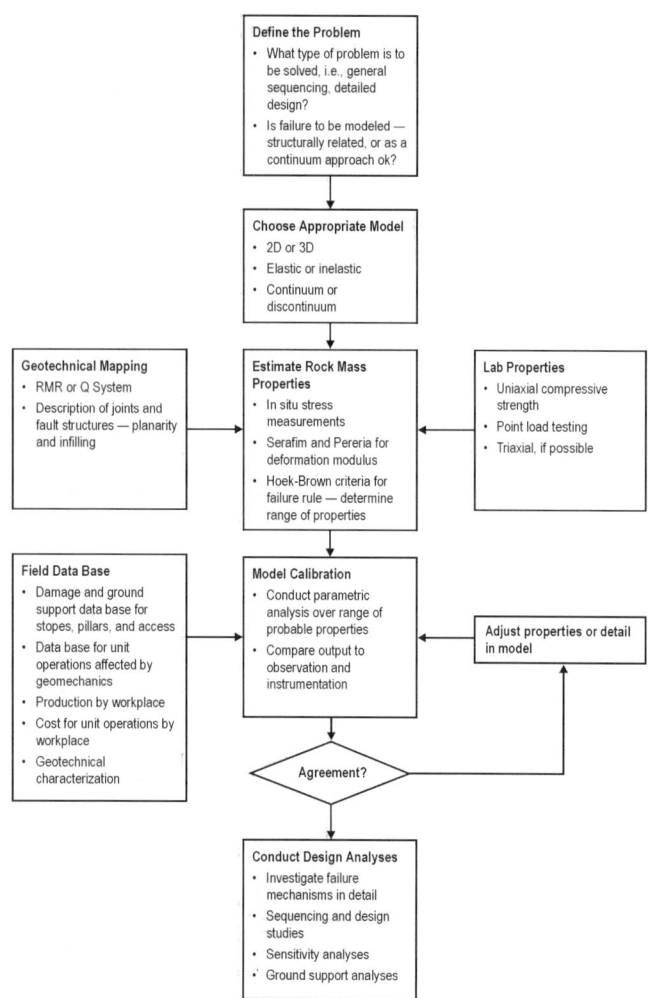

FIGURE 58.1   Flow chart showing the general methodology for applying numerical models to mining problems

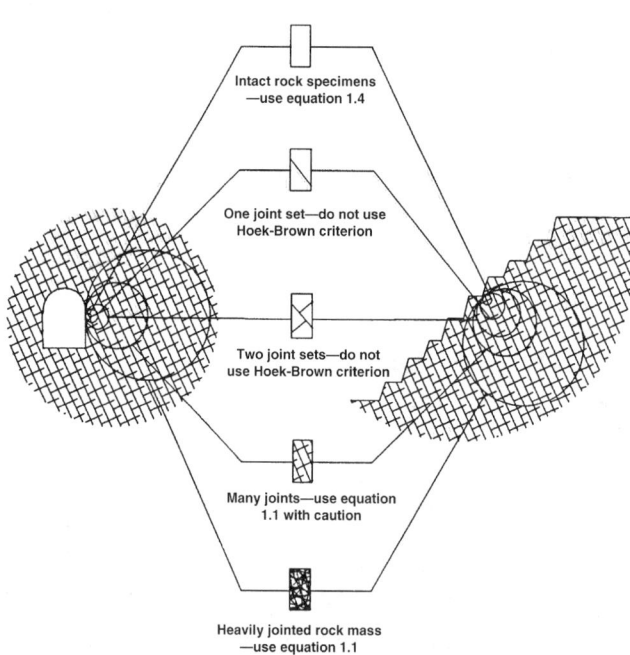

FIGURE 58.2   Schematic illustration of the scale of an analysis and the appropriate use of a nonelastic continuum constitutive model such as the Hoek-Brown model (Hoek 1998)

58.2   ESTIMATION OF IN-SITU PROPERTIES

58.2.1   Problem Scale and Choice of Material Model

In Figure 58.2, Hoek (1998) illustrates the problems of rock-mass scale and the impact of geologic structure on appropriate methods for representing its material behavior. This figure also illustrates that the type of constitutive model (e.g., elastic or some kind of yielding material) used needs to be commensurate with the scale of the problem to be examined, as well as the level of detail desired to represent potential failure mechanisms. When the scale of the problem is relatively large with respect to the spacing of structure in a well-jointed rock mass, it may be appropriate to use either an elastic model with an estimated deformation-modulus failure level determined empirically or a continuum-based constitutive model that represents the elastic and failure response of a jointed rock mass—i.e., the Mohr-Coulomb or Hoek-Brown model.

When the scale is small—perhaps on the scale of an individual drift in rock with one or two (or discontinuous) joint sets—using a plasticity model to represent the effects of the structure is not recommended. In these cases, wedge-analysis programs or the distinct-element method should be used to look at problems such as wedge stability. For most mining cases, we are dealing with the problem at a relatively large scale with heavily jointed rock, so using continuum elastic or inelastic constitutive models is appropriate, again depending on the purpose of the analysis. The important step in applying models is to make an estimate of the equivalent deformability and strength properties of the rock mass for all of the major rock units present.

58.2.2   Geotechnical Characterization for Initial Estimation of In-Situ Rock Mass and Joint Properties

It is impossible to perform controlled laboratory compression tests on large, representative, jointed rock samples, so estimates of the properties are typically made using rock-mass geotechnical classification combined with laboratory determination of intact rock strengths. Here, the technique developed by Hoek and

input to the numerical model. The average and range of the Q or RMR values is determined as a means of estimating the variability of the quality of each rock type. At the same time, a series of damage maps can be developed for numerous locations in the mine to document the mechanisms of drift and pillar failures. Our experience has shown that reasonable estimates of field properties can be obtained using the Hoek-Brown failure criteria in conjunction with laboratory values of the uniaxial compressive strength.

A database of case examples of stope and drift performance (both successful and unsuccessful) is developed as a means of model calibration. Preferably, these case examples should also include cavity surveys, detailed mapping of failure mechanisms, and instrumentation measurements.

The model is calibrated by simulating the mining sequence and comparing the model output to the database observations and instrumentation. Typically, this calibration is not highly detailed—i.e., exact comparisons of deformations are not necessary. However, the model should have the ability to reproduce the general failure mechanisms observed and be able to discriminate between regions that show intensive yield as well as those that do not. Greater confidence in the model is obtained when the damage extent can be approximately reproduced for a number of case studies using a narrow range of strength and deformation properties.

Brown (Hoek 1998) is used to estimate the in-situ strength properties. In cases where the strength of individual joints or faults is required, the work by Barton (1976) on the shear strength of discontinuities is used as a guide.

58.2.3 In-Situ Rock-Mass Strength Properties

The first task in applying these models is to make an initial estimate of the range of potential strength and stiffness properties for the various major rock units present. This is done by assuming a failure criterion for the rock and by estimating the strength properties using the geotechnical characterization and available laboratory-testing data.

Failure Criteria. The Hoek-Brown failure criterion is a commonly accepted method for estimating the relation of the principal stresses at failure for a rock mass. As discussed below, developing input properties for this criterion requires a disciplined approach of field geotechnical characterization as well as knowledge of intact rock properties. We have found that this method provides a good starting point for estimating in-situ properties, and, as an added benefit, it provides a known, common approach for determining properties that can be understood by all geotechnical engineers. As with any method, subsequent modeling needs to be compared to field observations and instrumentation to verify the rock mass properties.

The failure criterion relates the major principal stress (σ_1) to the minor principal stress (σ_3) at failure. The equation describing the criteria is given by:

$$\sigma_1 = \sigma_3 + \sigma_{ci}[m_b(\sigma_c/\sigma_{ci}) + s]^{1/2}$$

where m_b is the Hoek-Brown constant for the particular rock type, and s depends on the characteristics of the rock mass.

The value σ_{ci} is the uniaxial compressive strength of the intact rock.

The calculation of the m_b and s parameters is based on the degree of jointing and the alteration of joint surfaces reflected in the value of the RMR:

$$m_b = m_i \exp\left(\frac{RMR - 100}{28}\right)$$

and

$$s = \exp\left(\frac{RMR - 100}{9}\right)$$

m_i is based on the particular rock type and is shown in tables in Hoek (1998). Figure 58.3 presents the general classification range for rhyolites and andesites at one underground Canadian mine. As seen in this plot, there is a large range in rock quality, with the rock units ranging from very blocky to blocky-but-disturbed. The conditions vary from locally excellent, unaltered rock, to locally heavily altered, chloritized, and talcy rocks.

The resulting Hoek-Brown failure envelopes for the range of rock-mass conditions can be plotted in terms of the principal stresses at failure (Figure 58.4). The Hoek-Brown failure criteria can be used directly in many numerical models or can be expressed in terms of the average Mohr-Coulomb strength parameters: the cohesive shear strength, c, and the internal angle of friction, ϕ. Estimates for these strength parameters, given in graphical form, are presented in Figure 58.5. Here, an average set of properties from the previous field characterization is derived for both altered rhyolites and andesites, with RMR (GSI) values averaging around 35 to 45.

58.3 DEVELOPING A MINE-DAMAGE-ASSESSMENT DATABASE

The output of a numerical model is given in terms of stress and deformation, and—if a failure material model is used—an estimate of the type and extent of failure. This type of information is often of little practical use by itself and needs to be calibrated against a

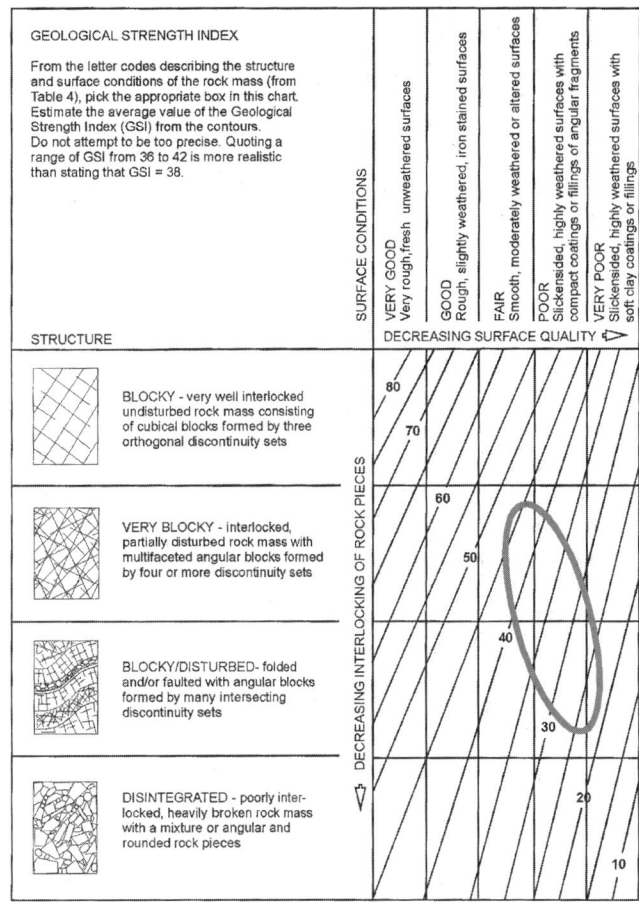

FIGURE 58.3 The relation of the Hoek classification to RMR (GSI) at a large Canadian mine. (Average values of GSI for altered andesites and rhyolites fall in the range of about 30 to 40, and may be as low as 25 locally, where gouge-filled faults are present, or may be locally of good quality [after Hoek 1998].)

database of information obtained from experience at the mine. To this end, it is recommended that a database be developed that includes information on the historical geomechanical response of stopes, pillars, and access drifts. This database may include:

- Rock-mass characterization for all major rock types and their variation throughout the mine

- In-situ stress measurements and observations of borehole breakouts in raises, orepasses, and boreholes

- Rock-mass-drift failure mapping and description, including the mechanism and intensity of failure

- Ground support (initial and rehab) practice in access and its relation to damage level

- Stope depth and extraction ratio at time of mining

- Stope and pillar failure mechanism and failure extent as determined from cavity surveys, planned excavation geometry, failed volume estimates, hydraulic radii of stope backs and walls at failure

- Data from rock instrumentation, including seismicity, stress change, and deformation measurements

- Production delays and cost impacts related to ground control problems, including drilling and redrilling, mucking of overbreak, fragmentation and secondary breakage, and ground support (initial and rehab)

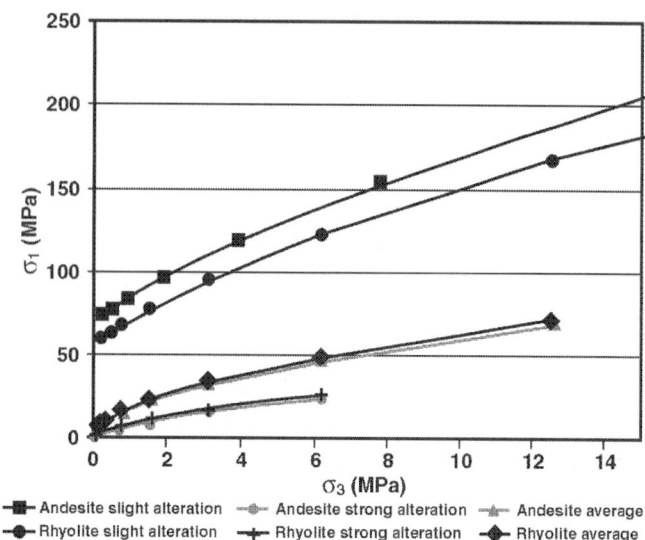

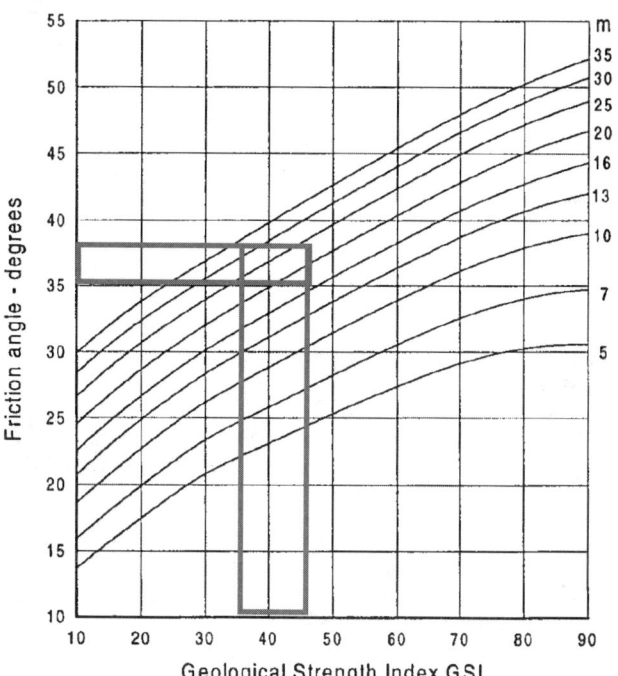

FIGURE 58.4 Estimated Hoek-Brown failure criteria for andesites and rhyolites of varying degrees of alteration. (The strongest is for highest quality (RMR >50) material with little alteration, the lowest for very poor (RMR~20), blocky material with strong alteration. The "average" values chosen are for blocky, altered rock of poor quality and RMR ~40.)

This database is then compiled into a series of case histories for individual stopes that documents the progressive mining history leading to extraction of that stope, and the resulting rock-mass failure conditions that accompany the mining. These case histories will provide a link between the reality of rock-mass behavior in the mine and the output generated by the numerical model. Ultimately, if cost and production figures are accounted for by individual working places, it is possible to directly *estimate* the cost, production, and grade impact of geomechanical damage for individual stopes. When this database is developed for several years of mining, it is possible to develop a history of the true geomechanical impact to the mine and how this impact has varied as the percentage of extraction increases.

Table 58.1 shows a general stope-damage-description method developed for this project. Developing the general geomechanics, production, and cost database was performed through a research project conducted in 1996–1997 by Kidd Mining Division. This project involved gathering the geological, damage, cost, and production data from the entire mine, followed by calibrating numerical modeling to this database. Here, a simple chart relating a damage level (seismic, moderate-to-high, moderate, low-to-moderate, nonviolent crushing failure with excessive deformation) for stopes and pillars is given.

For development headings, a drift-condition index is used as a basis for describing damage and ground-support requirements in terms of the peak applied stress and the rock-mass compressive strength.

Minimal Damage.   General slabbing along drifts, but not very deep. Affecting 20% to 30% of tunnel width or height. Standard support is adequate with only minor rehab required. This damage corresponds to a ratio of:

$$(\sigma_1 - \sigma_3)/\text{unconfined compressive strength (UCS)}$$
of approximately 0.2 or less.

Moderate Damage.   Heavier slabbing along drifts with walls broken into blocks affecting 30% to 50% of tunnel width or height. Will generally require some rehab, particularly at intersections or wide spans. Rehab may consist of scaling,

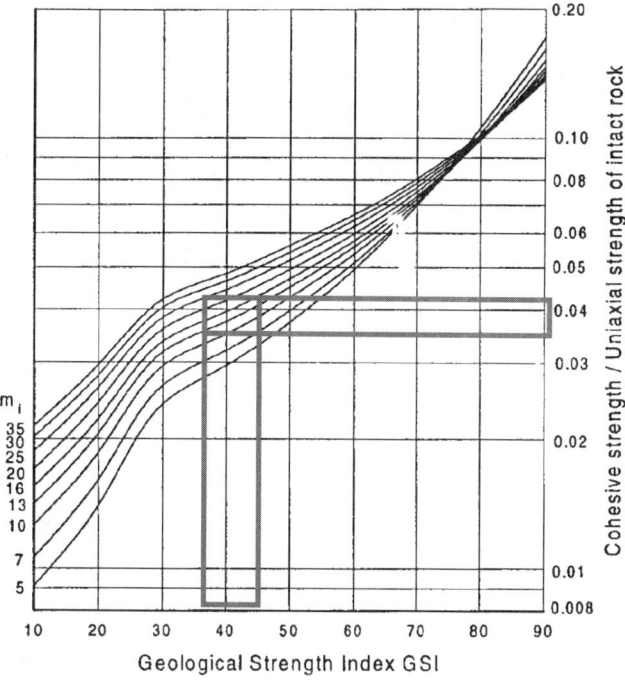

FIGURE 58.5 Relation of the internal friction angle and cohesive strength to the GSI (RMR) and the Hoek-Brown parameters. (Highlighted area gives average values used to represent poor-quality, altered rhyolite and andesite.)

additional bolts, and a thin layer of shotcrete. This damage corresponds to a ratio of:

$$(\sigma_1 - \sigma_3)/\text{UCS of approximately 0.2 to 0.4.}$$

High Damage Potential.   Here, the value of the major principal stress nears the uniaxial compressive strength of the rock mass. In rhyolites, this results in significant deterioration,

TABLE 58.1   Damage classification system for stopes and pillars based on historical data from Kidd Mine

Geomechanics risk category	Damage classification	Description of expected geomechanical response	General geomechanical risk
1	Seismic conditions expected	Seismicity expected, particularly in lead primary stopes of advancing wedge line during sill mining as well as stopes adjacent to faulted waste lens in narrow areas. Potential exists for extensive drill-hole shearing. Crosscut development will require heavy support for seismic conditions—cabled intersections and shotcrete. Extensive hourglassing of adjacent secondaries could occur; rapid backfilling is necessary. Mining rates will slow due to rehab and drilling difficulties.	Possible seismicity, primarily in sills and on poorly oriented faults
2	Moderate-to-high stress-related damage	This classification applies primarily to the lead primary stopes prior to sill breakthrough, as well as abutment stopes and squat, narrow primaries. Some moderate-level seismicity may occur. Stress-related problems will include hourglassing of secondaries and stress-spalling of crosscut development. Expect some hole re-drilling; rehab of crosscut development is likely. Lead primaries will require initially heavy support in drilling drifts.	Moderate to high
3	Moderate stress-related damage	Generally good conditions for primary stope mining. Minor seismicity, some pillar sloughing. Standard crosscut stope support, standard rehab probable.	Moderate
4	Low-to-moderate stress-related damage	Generally good conditions for secondary mining as long as rapidly confined with fill, and mined soon after primary mining. Mining of primaries causes stable yielding and destressing of secondary stopes. Dilution factors high where HW parallel faults or highly schistose ground present.	Low to moderate
5	Nonviolent failure, but excessive deformation of secondary pillars	Poor conditions for secondary mining. Excessive hourglassing and deformation of secondary pillars results in high risk of low recovery and high dilution factors. This classification applies primarily to secondaries allowed to become tall and narrow while waiting for primaries to be completed.	High risk of ore loss

large closures, and potential groundfall hazard. In massive sulphides or andesites, this condition means a high potential for seismicity. This damage corresponds to a ratio of:

$$(\sigma_1 - \sigma_3)/UCS \text{ of approximately } 0.4 \text{ or higher.}$$

58.4   RELATING STOPE DAMAGE CLASSIFICATIONS TO THE OUTPUT OF THE NUMERICAL MODEL—USING THE "STRESS PATH" APPROACH

Perhaps the most important phenomena to predict are the mode and violence of pillar failure in deep mining. That is, will the failure be accompanied by violent energy release, or will the failure be stable, dissipating the energy via crushing pillars and sliding on joint surfaces? An inelastic model will use a failure criterion to make estimates of the mode and extent of failure for a given stope and mining sequence. A technique termed the "stress-path" method can be used for predicting the violence of the failure of pillars, stope walls, or backs.

To use this approach, the principal stresses predicted by the model are sampled in pillars, stope backs, or walls at each mining step. The pairs of maximum (σ_1) and minimum, or confining (σ_3), stresses can be plotted on a standard stress-space plot upon which the rock-mass failure criterion has been overlaid (Figure 58.6). Prior to mining, the rock mass is in a highly confined, elastic (or unfailed) state. Mining will then proceed according to some stoping sequence (e.g., a series of primary and secondary stopes), creating pillars between backfill panels. As mining continues, the rock mass in close proximity to any given panel will show a decrease in the confining stress, and, generally, the applied maximum principal stress will increase due to stress concentrations. In other words, the shearing stress will increase as the principal components diverge. If the contours of the stress pairs for a given pillar or wall volume are plotted as the mining progresses, they will trace a path that will approach the failure criteria. Portions of the pillar or wall rock will fail when the stress state intersects the failure criteria. The path that the stress state takes in traversing the stress diagram reflects the general mining conditions that occur once a failure condition is reached.

The stress path denoted A in Figure 58.6 is indicative of a system that is storing strain energy due to confined stressing under a state of high confinement. This path is typical of abutments or wide pillars in which the confinement is unable to

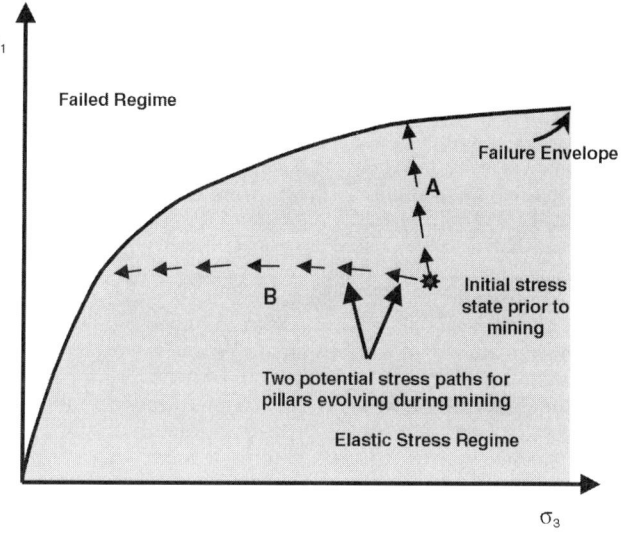

FIGURE 58.6   Concept of a "path" of the principal stress state within a pillar from initial (unmined) state to final state when a pillar is extracted. The path reflects the potential violence of failure.

relax. For brittle, high-strength rocks, such a loading system is more prone to violent energy release once failure does occur. Stress path B is indicative of the rapid loss of confinement accompanying mining–typical of the conditions for hanging walls or thinly cut secondary pillars. When this rapid drop in confinement occurs in pillars, nonviolent yielding is more probable.

At the Kidd Mine, after calibrating the model against approximately 80 case histories of stopes in the No. 1, 2, and 3 mines, covering a depth from ground surface to 6,000 feet, we found the following:

Hanging Wall Instabilities. Stopes exhibiting hanging wall failures and high dilution levels were in foliated rhyolites or had wall-parallel faults in combination with stress conditions typical of stress path B. The hanging walls are generally subjected to failure prior to extracting the stope, leading to degradation of the stope wall. After excavation, major portions of stope walls show a drop in the confining stresses to less than about 5 MPa, but with a majority in a tensile state of stress. In general, the large groundfalls that have occurred required the presence of a geologic structure such as a fault *and* conditions of low confinement. The propensity of the hanging wall to go into low confinement deep into the wall is a function of the mining sequence and the extent of stope span.

Pillar Stability. When pillars are created during primary stope extraction, their behavior is dependent on their width in relation to their hanging-wall-to-footwall thickness. In general, the typical 15-m-wide pillars tend to fail by following a stress path that reaches the failure criteria through loss of confinement rather than through a large stress buildup. This is a function of both the H/W ratio of the pillars and the oval shape of the orebody, which tends to shed load to the abutments as the pillars fail. This results in a nonviolent failure in shear throughout the pillar and creates few problems with mining. This type of behavior is illustrated by a path somewhere between paths A and B. Ground-control problems, such as crosscut instability, pillar-wall sloughing, and drillhole closure due to shear on joint surfaces have been encountered with varying degrees of severity during this failure mode. However, large increases in pillar width can result in a rapid change in stope behavior. For example, seismicity has occurred where using 30-m-wide secondaries has resulted in maintaining confinement and an elastic, highly stressed pillar core. The stress path for this type of pillar follows type A. This has also been the case where sill pillars are taken in a flat-back arrangement, thus maintaining internal confinement. Figure 58.7 shows typical, numerically generated, stress paths for narrow and wide pillars. The narrow pillar tends to yield in a nonviolent manner in a crushing mode due to loss of confinement, whereas the wide pillar tends to fail violently due to maintaining confinement.

In general, the results of calibration at the Kidd Mine showed that the behavior of stopes, given by the damage classifications in Table 58.1, correspond to the approximate stress-path predictions from the model, as shown in Figure 58.8. Using this general damage-classification method, a series of model simulations of various mining sequences, methods, or stope dimensions can be performed and predictions made about the stope-damage levels as a function of extraction ratio. It is, therefore, a relatively simple process to identify the extraction ratio possible prior to encountering major ground-control problems and to estimate the type of problems to be encountered—e.g., extensive pillar hourglassing to rockburst difficulties. If costs have been gathered showing the historical impact of the damage level on unit operations, a comparative estimate of geomechanical cost/ton can be developed for various mining options.

58.5 APPLYING THE MODELING METHODOLOGY TO RISK ASSESSMENT OF DEEP MINING AT KIDD MINE

58.5.1 Kidd Mine—Historical Overview

In 1963, the Kidd Mine orebody, a large-scale, copper-zinc deposit, was discovered near Timmins, Ontario, Canada. Estimates placed the ore reserves at 25 million tons, sparking a speculative frenzy on the floor of the Toronto Stock Exchange. (In its first hour of public trading, 11 million mine shares changed hands.)

Mining operations at the Kidd Mine went into full production in 1966. Initial open-pit operations began in the winter of 1964–1965. Open-pit operations continued for more than ten years. Shaft-sinking for No. 1 Mine began in 1969 and bottomed out at 930 m (3,050 ft) below the surface in March 1972. A simultaneous approach to mine development and production allowed initiation of mining on subsequent levels as

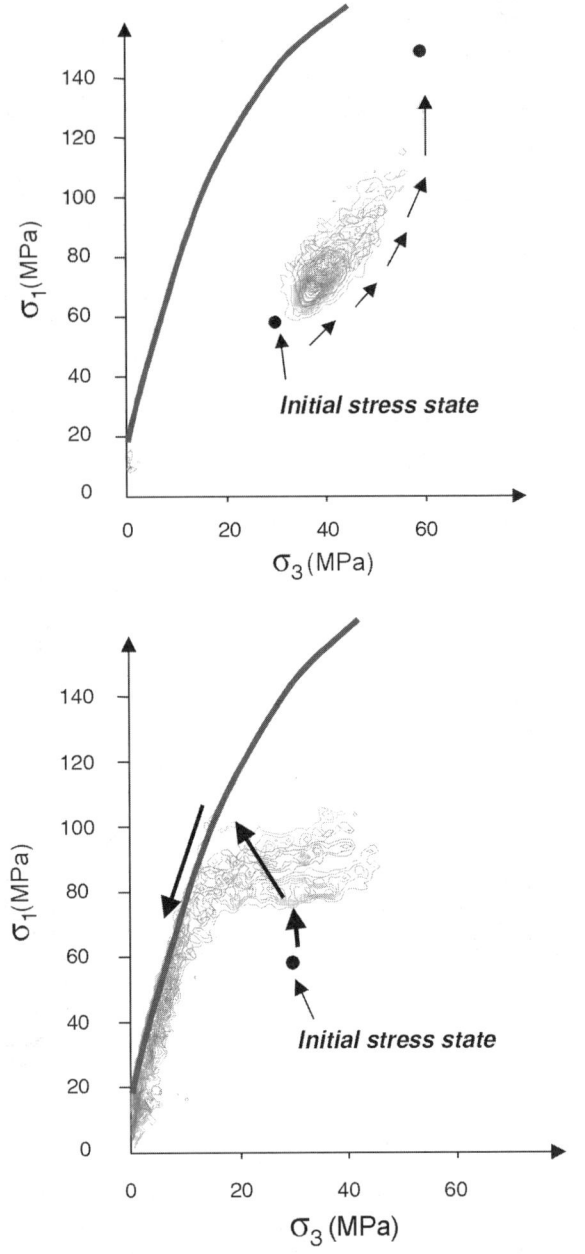

FIGURE 58.7 Example of two stress paths from a narrow (top) and wide (bottom) sill pillar subjected to the same in-situ conditions. The narrow sill shows a typical high seismic potential characterized by increasing confinement and driving stress. The wide portions fail early and continuously destress.

current levels progressed through primary and secondary stoping. By 1993, production from the sill pillar began at the 2500 and 2600 levels.

The head frame for the No. 2 shaft was begun and completed in 1974. Designed to be mined without pillars, stopes in the No. 2 Mine were mined directly adjacent to filled stopes, expanding panel by panel along the strike of the orebody. The mine's small stope size required rapid backfilling capability.

Development for No. 3 Mine began from the No. 2 Mine main service ramp, just above the 4600 level. Shaft sinking began below the 4700 level in October 1990. A 1-4-7 panel-mining system, new to operations at the Kidd Mine, was adopted to allow the reserves of No. 3 Mine to be exhausted as close as possible to

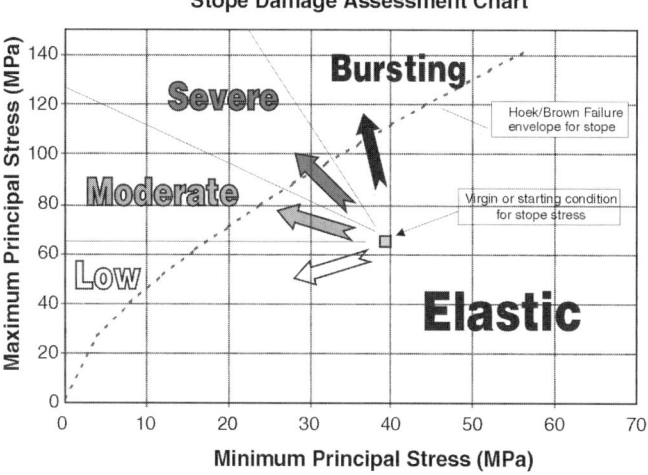

Stope Damage Assessment Chart

FIGURE 58.8 Generalized stress path plot for Kidd Mine showing the approximate correspondence of stress path to stope damage classification based on 80 stope-case histories

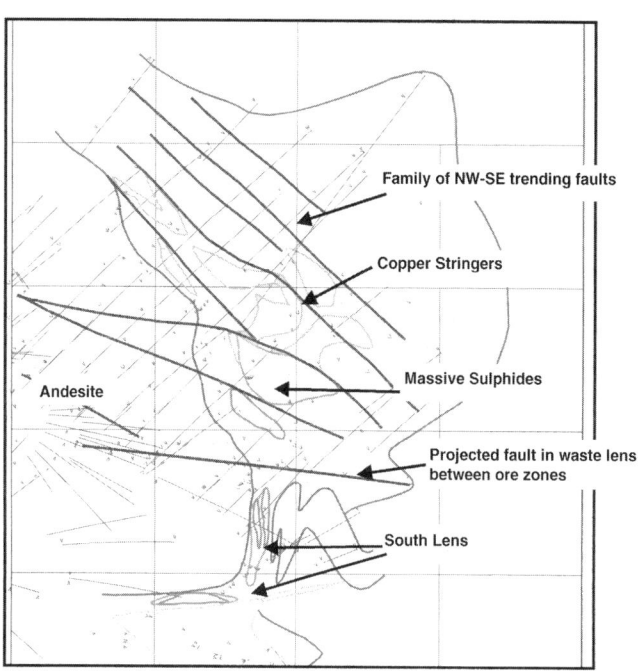

FIGURE 58.9 Plan view of the geology of the 7,000-ft level, Kidd Mine

the completion of No. 1 and No. 2 Mines. By 1994, 30 years after initial operations had begun at the Kidd Mine, mining in the No. 3 Mine had reached the 6000 level.

Located in a district with a history of mining that stretches back to the early 1900s, the Kidd Mine has been in continuous operation for 31 years. To date, 110 million tons have been extracted.

58.5.2 Mine Geology and Ore Reserves

Kidd Mine's main mineralized lenses are called the main (copper stringer and massive sulphides) and south lenses (Figure 58.9). These orebodies are located near the top of a locally thickened rhyolite, which is underlayed to the east by ultramafics and overlayed to the west by mafic flows and associated intrusions. The stratigraphy trends north-south, is overturned, and dips steeply to the east. All the lithologies in the Kidd Mines, including the ore, have been subjected to complex folding and faulting. Three phases of folding have made the Kidd orebody a tight, steeply northeast-plunging rod, or plug-shaped body, of astonishing size. The major faults that potentially affect mine-wide stability can be defined in two systems: the Gouge Fault and the south-dipping echelon faults. The south-dipping faults have been associated with the larger seismic events at Kidd, while the Gouge Fault and its splays primarily impact hanging wall dilution.

Primary ore-definition drilling of the deep extensions of the orebody has shown significant reserves of continuing good grades extending to depths in excess of 10,000 ft. The initial block of reserves identified for deep development, termed "Mine D," extends from the bottom of the current No. 3 Mine (6,800 ft) to the 8,200-ft depth. This block of ground is to be mined from two major silling horizons 2,286 m (7,500 ft) and 2,500 m (8,200 ft) using a series of levels developed from the main mine footwall ramp, which is located within the stronger andesites some distance from the orebodies.

58.5.3 Damage- and Cost-Based Risk Assessment

One of the major issues of any deep-mine planning is estimating the mining cost and the risk to the capital investment. In deep mining, the potential for geomechanical problems is a major consideration in the cost risk, and a means of assessing this risk in as quantitative a manner as possible is required. To accomplish this, a number of mining methods were examined using the

calibrated numerical model for predicting stresses and the stress-path approach for estimating the damage potential to stopes, pillars, and access. Because average unit costs were determined as a function of the geomechanical damage level, it is possible to make an estimate of the average geomechanical-related mining costs per ton for each candidate method. These can then be compared to current costs as a means of optimizing and assessing the risk of the various alternative methods.

58.5.4 Mining Methods Optimization for Mine D

An optimization study was performed to determine the mining method, layout, and sequencing that could potentially minimize both the geomechanical-related mining costs and risks. These risks include seismic potential, major production interruptions, and dilution potential. Several blasthole mining alternatives were examined for use in the 6800 to 8200 block of ground, including:

- Blasthole primary/secondary mining using a "flat-back" advance front, as currently employed at the Kidd Mine. (The primaries are mined through together, well in advance of the secondaries.)

- Blasthole primary/secondary mining using a chevron-shaped advance front in which primary stopes are filled with paste fill and secondary stopes are mined as quickly after primaries as possible. (This method is similar to that developed for the Williams Mine in the Hemlo area. Two approaches were examined in which the lead stope was centered in the main ore lens. Another approach was examined in which the lead stope was advanced along the southern boundary of the main lens in an attempt to fail a thin waste lens lying between the main lens and the south lens [see Figure 58.9.])

- Blasthole pillarless mining in both overhand and under-hand fashions, either as a single advancing pyramid-shaped front, or as multiple fronts.

The pillarless mining was examined from the geomechanical perspective and proved to have some advantages with respect to stress shading. However, it was eliminated from further

TABLE 58.2 Estimate of unit geomechanics-related cost as a function of damage classification compared to No. 3 Mine average base costs (Falconbridge 1996–1997)

	Low	Moderate	Severe	Bursting
Overcut and undercut development and support	100%*	128%	137%	165%
Stope support (cable bolts)	100%	100%	105%	110%
Drilling & redrilling	100%	105%	110%	120%
Primary blasting	100%	100%	102%	105%
Secondary blasting	100%	110%	120%	130%
Mucking	100%	100%	105%	110%
Rockfill	100%	100%	103%	105%
Dilution from fill (secondaries)	0%	3%	7%	10%
Dilution from back and walls (ore)	0%	7%	14%	20%
Dilution from hangingwall and abutments (waste)	0%	3%	7%	10%
Total	100%	124%	149%	174%

* 100% = cost of No. 3 Mine.

consideration because it lacked flexibility in production and fill scheduling. Also, the orebody width at depth varies significantly, making it difficult to develop this sequence while maintaining production.

58.5.5 Prediction of Stope Performance Using Calibrated Numerical Models

The optimization study was conducted using the numerical modeling that had been calibrated to past performance in the No. 1, 2, and 3 Mines. The stress-path method, described earlier, was used as a means of estimating the damage potential to stopes and pillars for each layout and sequence. Damage potential and support requirements for access were estimated from the predicted stresses and the damage classification presented earlier. The primary haulage lateral, orepasses, and ventilation-raise locations were adjusted until the mining influence was minimized.

Production costs for geomechanical-related items can be assigned to each of the damage classifications as given in Table 58.2. These costs were derived by relating stope damage levels to unit costs estimated for individual workplaces. Here, the "base cost" refers to the average geomechanical-related costs for No. 3 Mine, exclusive of the dilution costs.

Each alternative mining method was modeled, and a damage level was assigned to each stope. Figure 58.10 shows the three-dimensional model of the main ore zone as viewed from above and from the southeast. The orebody is divided into 20-m blasthole panels. The host rock has been hidden for viewing purposes. Figure 58.11 shows an example of a longitudinal section of the three-dimensional model of the main zone in which the major principal stresses and indicators of yielded regions have been superimposed on the stope and pillar geometry for one of the steps in the chevron advance sequence. The modeling showed that the sill begins failing when the lead primary stope has approximately two sublevels remaining. The lead stope will be subjected to high stresses in advance of the stope back as a result of the high level of confinement. This requires that full undercuts and overcuts not be used and that a seismic level of ground support be placed in the overcut drilling horizon. The lead primary panel creates a "bow-wave" effect that tends to destress adjacent stopes and shed stresses to the abutments. Secondary stopes are sized (20-m width) to yield when cut, rather than to burst. Because of the oval nature of the orebody,

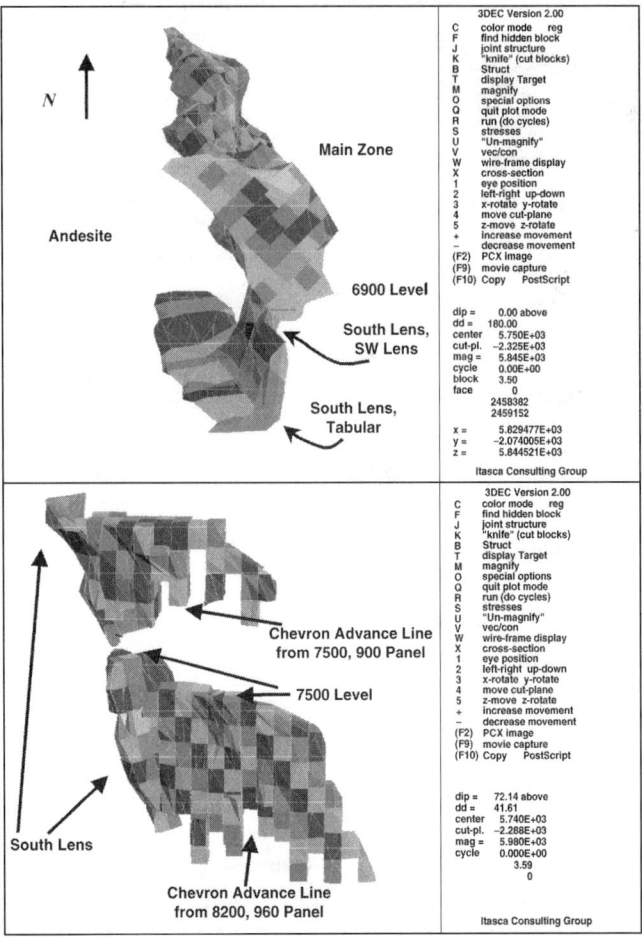

FIGURE 58.10 Plan and horizontal views of the orebody form a three-dimensional model of Kidd Mine. The orebody is subdivided into 20-m-wide panels, and the host rock has been hidden.

loads are shed to abutments, and the intensive pillar crushing typical of tabular deposits is not expected.

Unit costs for geomechanical-related items were assigned to each stope based on the estimated damage levels, and a comparative average life-of-mine total cost per ton was derived. Figures 58.12a and 58.12b show the result of the analyses of the three candidate approaches in terms of the predicted damage levels by total tonnage and a comparative cost estimate for geomechanical costs relative to the current No. 3 Mine. As seen, the chevron method with rapid secondary removal is superior, and the location of the lead stope along the strike of the orebody is not very important to the overall damage level of stopes.

58.5.6 Summary of Results

Based on the numerical analyses and damage assessment studies, a multiple-level blasthole mining method using a chevron-shaped front was chosen for Mine D. A major advantage of this method is its ability to rapidly fail sills and to extract yielded and de-stressed secondary pillars between paste fill primary panels. The V-shaped front allows incremental sill breakthrough without subjecting the entire sill to high stress, which is typical of a flat-back approach. Additionally, a "halo" of failed ground in the bow wave of the lead stope should allow improved ground conditions in subsequent panels. Rapid paste filling of primaries with removal of the yielded secondaries as soon as possible to minimize stand-up time is expected to provide high recoveries.

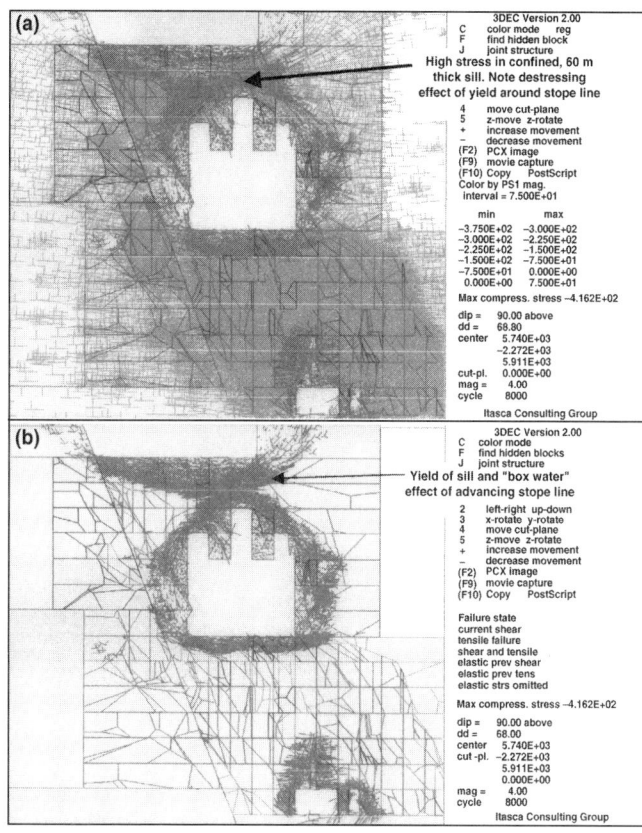

FIGURE 58.11 Longitudinal section of the main zone with super-imposed stress vectors (a) and regions of yield (b) for a step in the chevron sequence. Note that the lead primary stope creates a bow-wave effect that conditions the rock mass in advance of the following stopes. The orebody beneath the sill at 6800 is yielded for a considerable depth prior to mining.

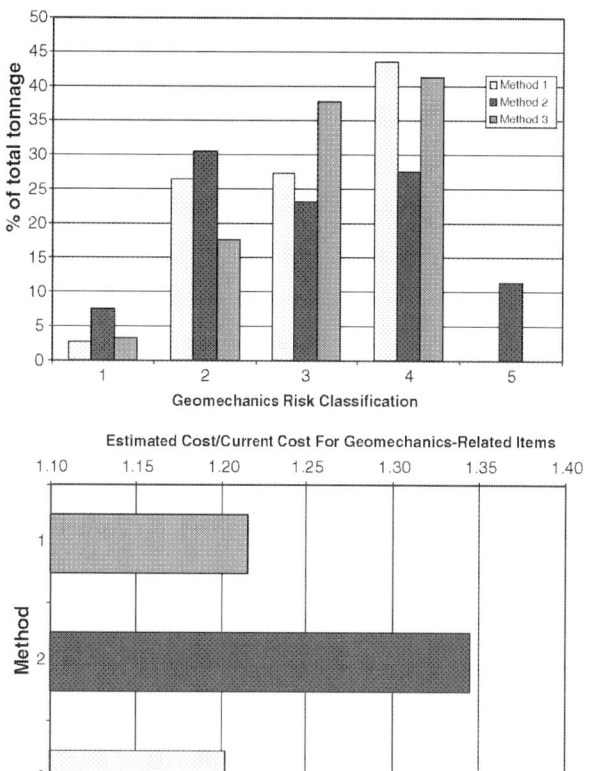

FIGURE 58.12 Results of the damage and cost assessments showing that the chevron methods are superior from both standpoints. Methods 1 and 3 use chevron-advance fronts with different lead stope portions; method 2 uses a flat-back advance. Damage classes (1–5) correspond to Table 58.1.

58.6 CONCLUSIONS

This chapter has presented a methodology for applying numerical models to assess the geomechanical risk of deep-mining operations. The important features of this methodology are:

- Initial detailed rock-mass characterization for developing a base set of rock-mass properties for all geologic units

- Development of a mine database that includes classification of observed damage levels in stopes and drifts as well as a means of relating production and cost impacts of the damage

- Calibration of the numerical model to provide a means of interpreting model output in terms that mine staff and management understand

- Parametric analysis of various mine designs and sequences, and comparison of damage- and cost-based risk assessment, as a basis for the structural design of the mine.

58.7 REFERENCES

Barton, N. 1976. The shear strength of rock and rock joints. *Int. J. Rock Mech. Min. Sci. & Geomech. Abstr.*, 13: pp. 255–279.

Barton, N., R. Lien, and J. Lunde. 1974. Engineering classification of rock masses for the design of tunnel support. *Rock Mech.*, 6: pp. 189–236.

Bieniawski, Z.T. 1976. Rock mass classifications in rock engineering. *Exploration for Rock Engineering (Proceedings of the Symposium on Exploration for Rock Engineering. Johannesburg, November 1976)*, 97–106. Rotterdam: A.A. Balkema.

Hoek, E. 1998. *Rock Engineering (Course Notes)*. University of Toronto: Department of Civil Engineering, November.

Falconbridge, Ltd., Kidd Mining Division. 1996–1997. *Risk Assessment Project, Phase 4: Preliminary Geomechanical Analysis of Alternative Mining Methods for the 6800-7800 Ore Block*, May.

CHAPTER 59

Pillar Design to Prevent Collapse of Room-and-Pillar Mines

R. Karl Zipf, Jr.*

59.1 INTRODUCTION

Room-and-pillar mining accounts for a significant portion of the total mineral production in the United States. As shown in Table 59.1, well in excess of $6 billion worth of mineral commodities is produced each year by this method. A substantial portion ($3.55 billion) of coal production still comes from room-and-pillar mining. Metallic minerals valued at about $1 billion, plus nonmetallic minerals valued well in excess of $1 billion, are also produced via room-and-pillar mining. A significant ($600 million) and growing portion of stone and aggregate production uses room-and-pillar mining. In addition, many other mineral commodities not noted in this table such as talc, iron, and copper are or have been produced in the United States using the room-and-pillar technique.

Comparing current data with 1973 data compiled by Bullock (1982) shows that production by room-and-pillar method has declined in some commodities and grown in others. For example, coal production by room-and-pillar method has declined from 90% of coal production then to 20% of the total today. However, since 1973, total coal production has almost tripled so that in terms of tons, room-and-pillar coal production has declined only from about 290 to 200 million tons today, which is still a very significant production amount. Prior to 1973, the room-and-pillar mining technique accounted for 60% of noncoal mineral production or about 80 million tons of material. Today, the method probably accounts for no more than about 20% of total noncoal production; however, as shown in Table 59.1, the tonnage and value is very significant. Large increases in noncoal mineral production using the room-and-pillar method have occurred in soda ash, potash, salt, and most recently in stone and aggregate.

One important parameter in the engineering of room-and-pillar mines is the pillar size, which coupled with the room width determines the achievable extraction percentage. Several opposing factors influence the choice of pillar size. Sizing pillars too large leaves valuable resources in the earth and risks poorer mining economics and waste of scarce mineral deposits. Sizing a pillar too small risks pillar failure and potential surface subsidence. Pillars that are too small can lead to a large-scale catastrophic collapse called a cascading pillar failure, which carries severe economic and health and safety risks (Zipf 1996 and Zipf and Swanson 1999). As shown in Table 59.1, with mineral production valued at over $6 billion coming from room-and-pillar mines, changing the extraction percentage by just 1% translates into over $60 million per year in resource conservation or mining revenue gains. Better understanding of pillar mechanics will therefore have significant impact on the economics of mines using the room-and-pillar method.

The objective of this paper is to present a design methodology aimed at eliminating the risk of large-scale pillar collapse in room-and-pillar mines. The design methodology involves evaluation of Salamon's local mine stiffness stability criterion. This section first summarizes traditional strength-based pillar design methods applicable to coal or hard-rock mines. Several examples of large-scale pillar collapse are reviewed, and the mechanics of these failures is presented. To decrease the risk of pillar collapse, three alternative design approaches are given—the containment approach, the prevention approach, and the full extraction approach. Finally, practical methods are presented to evaluate the local mine stiffness stability criterion and use results in practical room-and-pillar mine layout.

59.2 TRADITIONAL STRENGTH-BASED PILLAR DESIGN METHODS

Room-and-pillar mines may cover a small area several hundred feet square and contain just a few pillars as in a small zinc deposit. Alternatively, they may cover many square miles as is typical with coal, trona, or limestone mines. Large arrays of pillars are typically grouped into panels, which are in turn surrounded by barrier pillars. The small pillars within a panel are sometimes called panel pillars. Figure 59.1 shows typical layouts for coal, limestone, and lead room-and-pillar mines. In developing these layouts, the mining engineer must develop appropriate dimensions for the room spans, panel pillar widths, panel sizes and finally barrier pillar widths. In addition to strength considerations, developing these dimensions requires evaluation of the consequences of pillar failure, which could happen anywhere in the layout at any time.

Traditional strength-based pillar design first requires an estimate of pillar stress and then an estimate of pillar strength. The safety factor for the pillar is then evaluated as pillar strength over pillar stress. An acceptable safety factor depends on the tolerable risk of failure. A safety factor of 2 is typical for pillars in main development headings or panels during advance mining. Safety factors of 1.1 to 1.3 are typical for panel pillars after retreat mining. Safety factors much less than one are possible within panels where pillar failure is the intent.

59.2.1 Pillar Stress

As summarized by Farmer (1992), traditional pillar design for room-and-pillar mining begins by estimating the in situ vertical stress as

$$\sigma_z = \lambda z \qquad (59.1)$$

where λ is the unit weight of rock and z is depth to the mining horizon. The tributary area method then provides a first-order estimate of the average pillar stress. For the square room-and-pillar system shown in Figure 59.2, the average pillar stress is:

---

* NIOSH, Spokane Research Laboratory, Spokane, WA.

TABLE 59.1   Value and production by room-and-pillar mining in the United States

Mineral commodity	Totabl U.S. production (tons)	Approximate % by room-and-pillar	Room-and-pillar production (tons)	Value of room-and-pillar production	Typical extraction %
Coal (1)	1,014,000,000	20	202,000,000	$3,550 million	60
Lead (2)	493,000	90	444,000	$432 million	75
Zinc (2)	722,000	60	433,000	$491 million	75
Soda Ash (2)	10,100,000	80	8,000,000	$664 million	65
Potash (2)	1,300,000	100	1,300,000	$320 million	50
Salt (2)	40,800,000	60	32,000,000	$592 million	50
Gypsum (2)	19,000,000	50	9,000,000	$66 million	75
Stone and Agregate (3)	1,200,000,000	10	120,000,000	$600 million	75

(1) Coal Data 1999—National Mining Association.
(2) USGS Mineral Industry Surveys, 1998.
(3) The Aggregate Handbook—National Stone Association—1996.

$$\sigma_{pa} = \sigma_z \left(\frac{W_p + W_o}{W_p} \right)^2 \qquad \textbf{(59.2)}$$

where W_p is the pillar width and W_o is the opening width. For rectangular or irregular-shaped pillars, the average pillar stress is found using the extraction ratio as:

$$\sigma_{pa} = \sigma_z \left(\frac{1}{1-R} \right) \qquad \textbf{(59.3)}$$

where R is the extraction ratio.

R is found as:

$$R = \frac{A_M}{A_T} = \frac{A_T - A_P}{A_T} \qquad \textbf{(59.4)}$$

where A_T is the total area of the orebody, A_M is the area extracted and A_P is the pillar area.

The tributary area approach assumes that the mined area is extensive and that all the pillars have the same dimensions. It also ignores the deformation properties of the surrounding rock mass relative to the pillar rock. In general, pillars at the center of a panel have a higher stress than pillars at the edge of a panel. Coates (1981) solved part of this problem by developing the following relation for average pillar stress that accounts for the width and number of pillars across a panel and the relative mechanical properties of the pillar and rock mass:

$$\sigma_{pa} = \sigma_z \left\{ \frac{\left[2R - K_0 \dfrac{H}{L} \dfrac{(1-2\upsilon_{rm})}{(1-\upsilon_{rm})} - \dfrac{\upsilon_p}{(1-\upsilon_p)} K_0 \dfrac{H}{L} \dfrac{E_{rm}}{E_p} \right]}{\dfrac{H}{L} \dfrac{E_{rm}}{E_p} + 2(1-R)\left(1+\dfrac{1}{N}\right) + 2 \dfrac{RB}{L} \dfrac{(1-2\upsilon_{rm})}{(1-\upsilon_p)}} \right\} \qquad \textbf{(59.5)}$$

where H is the mining height, L is the lateral extent of the mined area, B is the individual opening width, N is the number of pillars across the panel, K_0 is the ratio of horizontal to vertical stress, E and υ are elastic constants, and subscripts rm and p indicate the rock mass and pillar, respectively. While this equation is based on two-dimensional elasticity theory and therefore only applies to long, narrow rib pillars, it illustrates the behavior of average pillar stress. As the E_{rm}/E_p ratio rises, the average pillar stress decreases due to the bridging effect of the stiff rock mass. Similarly as the panel width L decreases and the H/L ratio increases, average pillar stress decreases.

Similar to tributary area method, Coates' (1981) solution only gives average pillar stress for all panel pillars and does not give changes in pillar stress across the panel. Two-dimensional

boundary-element-method programs such as Examine[TAB] (2000) or quasi-three-dimensional BEM programs such as MULSIM/NL (Zipf, 1992a,b) and LAMODEL (Heasley 1997 and Heasley 1998) are needed to calculate changing pillar stresses across a panel or within an individual pillar. Figure 59.3 shows the changing stresses across a panel and within pillars using a BEM program.

59.2.2   Pillar Strength

Over the past several decades, a large amount of rock mechanics literature has addressed pillar strength, both in coal and metal/nonmetal mines (Obert and Duvall 1967; Hoek and Brown 1980; Bieniawski 1992; Brady and Brown 1993). Much of this work is empirical and has addressed two issues—the size effect whereby rock strength diminishes as specimen size increases and the shape effect whereby rock strength increases as width-to-height ratio increases. Using energy considerations, Farmer (1985) developed theoretical expressions relating strength to size. When failure occurs in a brittle manner, as it does in most rocks, strain energy within the specimen transforms to fracture surface energy, which is a constant for a particular rock. Based on energy conservation:

$$SED \times V = FE \times A \qquad \textbf{(59.6)}$$

where SED is the strain-energy-density of the rock, V is the rock volume, FE is the fracture-surface-energy, which is a material constant, and A is the fracture surface area.

Rearranging gives:

$$SED \times \frac{V}{A} = SED \times L = FE = \text{constant} \qquad \textbf{(59.7)}$$

where L is a characteristic dimension or length of the rock specimen.

Assuming that laboratory-scale failure is mechanistically similar to full-scale pillar failure, then

$$SED_S \times L_S = SED_P \times L_P \qquad \textbf{(59.8)}$$

where the subscripts S and P indicate laboratory-scale specimen and full-scale pillar, respectively. Since strain-energy-density at failure is proportional to the square of stress at failure,

$$\sigma_S^2 \times L_S = \sigma_P^2 \times L_P \qquad \textbf{(59.9)}$$

or

$$\frac{\sigma_P}{\sigma_S} = \left(\frac{L_S}{L_P} \right)^{1/2} = \left(\frac{V_S}{V_P} \right)^{1/6} = \left(\frac{V_S}{V_P} \right)^{0.17} \qquad \textbf{(59.10)}$$

where σ_S and σ_P are strength of the laboratory specimen and full-scale pillar, respectively and V is volume, which is proportional to L^3. This theoretical relationship accounts for the size effect in

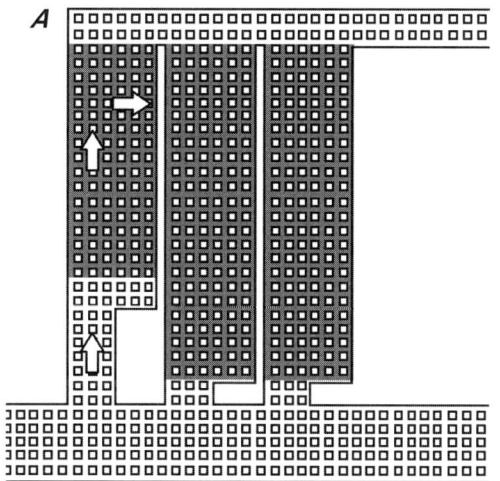

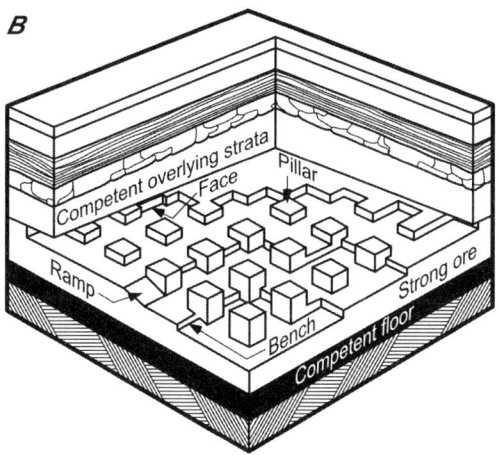

FIGURE 59.1   A: Typical room-and-pillar layout for coal mine with pillar extraction on retreat (Farmer 1992); B: Typical room-and-pillar layout for metal/nonmetal mine (Dravo 1974)

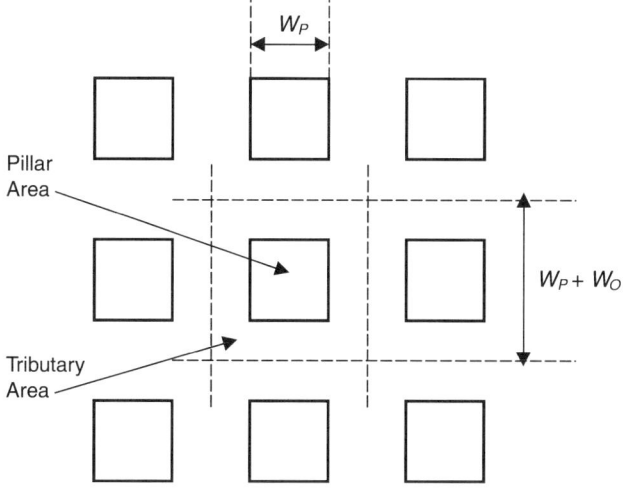

FIGURE 59.2   Plan view of room-and-pillar mine with dimensions for simple analysis

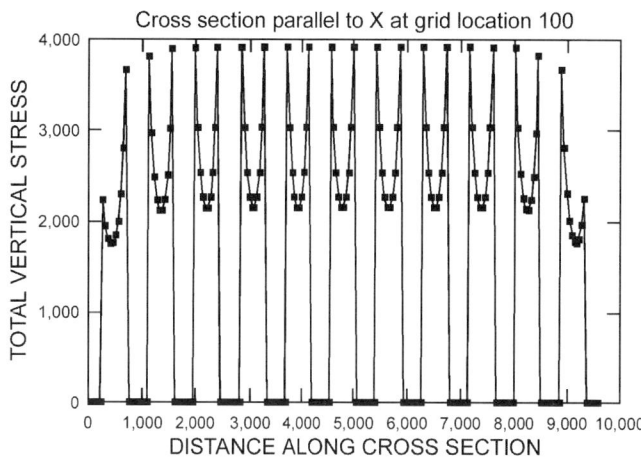

FIGURE 59.3   Pillar stresses across a panel and within pillars determined with the boundary element method probram MULSIM/NL

observed rock strength. The following empirical strength formula proposed by Hardy and Agapito (1977) for oil shale pillars follows this general theoretical form and provides some experimental confirmation.

$$\frac{\sigma_P}{\sigma_S} = \left(\frac{V_S}{V_P}\right)^{0.118} \left[\frac{W_P}{H_P}\frac{H_S}{W_S}\right]^{0.833} \quad \textbf{(59.11)}$$

where W and H are pillar and specimen width respectively. This empirical model also includes an additional term for pillar shape.

Classic empirical pillar strength formulas usually follow one of two general forms.

$$\sigma_P = \sigma_{S'}\left(a + b\frac{W}{H}\right) \quad \textbf{(59.12)}$$

$$\sigma_P = K\frac{W^a}{H^\beta} \quad \textbf{(59.13)}$$

Pillar strength formulas by Obert and Duvall (1967) and Bieniawski (1968a) follow the first form, whereas formulas by Salamon and Munro (1967) and Holland (1964) follow the second. In these forms, size effect is accounted for directly via the unit pillar strength $\sigma_{S'}$ or the rock constant K. $\sigma_{S'}$ is the strength of a cubical pillar ($W/H = 1$) at or above the critical size, and K is a constant characteristic of the pillar rock. The constants a, b, α and β in these equations account for the shape factor and show reasonable agreement as shown Table 59.2.

Several methods exist for estimating $\sigma_{S'}$ or K in equations 12 or 13, which is the strength of a cubical pillar ($W/H = 1$) at or above the critical size, where critical size is that size beyond which the rock mass strength remains relatively constant. For coal pillars, the critical size is widely recognized as about 0.9 meter or 36 inches. For U.S coal mines, the recommended strength value for a cube of coal this size is $\sigma_{S'} = 6.2$ MPa or 900 psi (Mark 1997b). This unit strength for a full-scale cube of coal can then be adjusted for shape effect using the recent Mark-Bieniawski relation (Mark 1999):

$$\sigma_P = \sigma_{S'}\left(0.64 + 0.54\frac{W}{H} - 0.18\frac{W^2}{HL}\right) \quad \textbf{(59.14)}$$

which reduces to the original Bieniawski relation when pillar width W equals pillar length L.

TABLE 59.2   Values for constants in empirical pillar strength formulas

Source	a	b	α	β	Comments
Bunting (1911)	0.7	0.3	—	—	Pennsylvania anthracite
Obert and Duvall (1967)	0.78	0.22	—	—	Laboratory rock and coal
Bieniawski (1968a)	0.64	0.36	—	—	South Africa coal
Skelly, Wolgamott and Wang (1977)	0.78	0.22	—	—	West Virginia coal
Greenwald, Howarth and Hartman (1939)	—	—	0.5	0.83	Pittsburgh seam mines
Holland (1964)	—	—	0.5	1	U.S. coal mines
Salamon and Munro (1967)	—	—	0.46	0.66	S.A. coal mines
Hardy and Agapito (1977)	—	—	0.60	0.95	U.S. oil shale mines

If laboratory-scale strength data (σ_S) is available, the full-scale strength of a cube of rock mass ($\sigma_{S'}$) can be found from equation 10 as:

$$\sigma_{S'} = \sigma_S \left(\frac{V_S}{V_{S'}}\right)^{0.17} \tag{59.15}$$

This unit strength for a full-scale cube of the rock mass can then be adjusted for shape effect using the Obert-Duvall relation:

$$\sigma_P = \sigma_{S'} \left(0.78 + 0.22\frac{W}{H}\right) \tag{59.16}$$

Finally, the Hoek-Brown failure criterion can also provide an estimate of the strength for a full-scale cube of the rock mass. For most rock masses with good to reasonable quality,

$$\sigma_1' = \sigma_3' + \sigma_c\left(m_b\frac{\sigma_3'}{\sigma_c} + s\right)^a \tag{59.17}$$

where m_b, s and a are constants which depend on the rock mass quality, σ_c is the uniaxial compressive strength of the intact rock pieces (equivalent to σ_S), and σ_1' and σ_3' are the axial and confining principal stresses.

The constants m_b and s are estimated using a rock mass classification index called GSI, which is equivalent to Bieniawski's Rock Mass Rating (RMR) assuming a dry rock mass. See Hoek, Kaiser and Bawden (1995) for a complete discussion of the Hoek-Brown failure criterion for rock masses.

59.2.3   Barrier Pillar Design

The above pillar stress and pillar strength formulas apply mainly to pillars within a large array or so-called panel pillars. The barrier pillars surrounding this array of panel pillars also require sizing. Tributary area method can provide a conservative estimate of barrier pillar stress, if we assume that the panel pillars have all failed or else are all mined and therefore carry no overburden stress. Alternatively, numerical methods such as MULSIM/NL or LAMODEL can provide an estimate of barrier pillar load without making the conservative assumptions inherent to tributary area method.

The empirical pillar strength formulas shown earlier apply to pillars with a width-to-height ratio less than 5. For barrier pillars and other pillars in coal mines with a W/H ratio greater than 5, Salamon's "squat" pillar formula applies:

$$\sigma_P = \sigma_{S'} \left(\frac{5^b}{V_P{}^a}\right)\left(\frac{b}{e}\left(\left(\frac{W/H}{5}\right)^e - 1\right) + 1\right) \tag{59.18}$$

where $e = 2.5$, $a = 0.0667$ and $b = 0.5933$.

For hard rock and other noncoal mines, an equivalent squat pillar formula does not exist. It is necessary to extrapolate the

Obert-Duvall relation to high W/H ratios or else use numerical methods and the Hoek-Brown failure criterion to estimate required barrier pillar size.

Koehler and Tadolini (1995) review no less than ten empirical and observational approaches to barrier pillar design in coal mines. Most methods calculate minimum barrier pillar width as a function of depth, and some include seam thickness. Coal strength is generally neglected. Equivalent empirical approaches to estimate barrier pillar size in noncoal mines do not appear to exist.

59.2.4   Summary of Traditional Strength-Based Pillar Design Methods

For room-and-pillar coal mining, the ARMPS method developed by Mark (1997a) applies. This method incorporates tributary area method and empirical strength formulas for sizing coal pillars during advance and retreat mining. The method also includes adjustments to the pillar stress for factors such as side loading from previously mined panels. The ARMPS method can also determine barrier pillar size. Again, an equivalent program does not appear to exist for noncoal room-and-pillar mining, but the ARMPS method would apply with suitable adjustments to the input parameters.

With the traditional strength-based pillar design methods, the user can determine panel pillar size and barrier pillar size, but these methods do not specify the panel dimension in any way. Operational considerations such as equipment and productivity set the panel width and usually set it as large as possible. Based on strength considerations alone, a narrow panel will require a narrow barrier pillar and a wide panel will require a wide barrier pillar. Rock mechanics factors do not enter the panel width determination with these methods. Maximum panel width may be determined by the size of air blast that an operation could withstand, but this is not a rock mechanics factor. Other rock mechanics considerations are needed to rationally determine maximum panel width along with panel pillar and barrier pillar sizes.

59.3   CASE HISTORIES OF PILLAR COLLAPSES

If the strength of a pillar in a room-and-pillar mine is exceeded, it will fail, and the load that it carried will transfer to neighboring pillars. The additional load on these pillars may lead to their failure and so forth. This mechanism of pillar failure, load transfer and more pillar failure can lead to the rapid collapse of very large mine areas. In mild cases, only a few tens of pillars might fail; however, in extreme cases, hundreds, even thousands of pillars can fail. This kind of failure has many names such as "progressive pillar failure," "massive pillar collapse," "domino-type failure," or "pillar run." Swanson and Boler (1995) coined the term "cascading pillar failure" or CPF to describe these rapid pillar collapses.

CPF can have catastrophic effects on a mine, and sometimes these effects pose a greater health and safety risk than the underlying ground control problem. Usually, the CPF induces a devastating air blast due to displacement of air from the collapse area. An air blast can totally disrupt the ventilation system at a mine by destroying ventilation stoppings, seals, and fan housings. Flying debris can seriously injure or kill mining personnel. The CPF might also fracture large volumes of rock in the pillars and immediate roof and floor, leading to the sudden release of large quantities of methane into the mine atmosphere. A methane explosion might result from the cascading pillar failure.

Figure 59.4 gives a simple illustration of cascading pillar failure. Figure 59.4A shows a series of 12 equal-sized pillars each subject to the same tributary area load. These pillars are assumed to be near their maximum load or strength. If two of the center pillars are weaker than average and fail, their load is transferred to the neighboring pillars causing their load to increase dramatically, as shown in Figure 59.4B. More pillars then fail as shown in Figure 59.4C, and even greater loads transfer to the surrounding

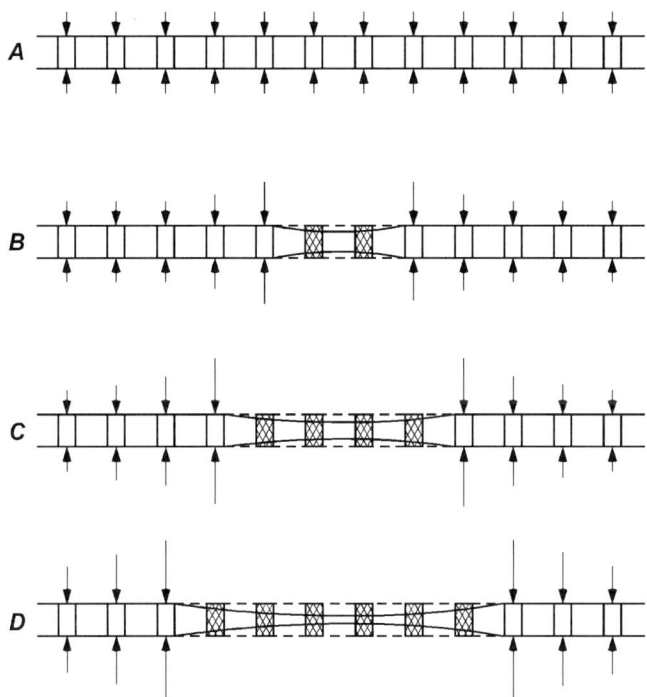

FIGURE 59.4 Simplified mechanics of cascading pillar failure or "CPF." Arrows indicate the relative magnitude of pillar load.

pillars as shown in Figure 59.4D. Thus, once the CPF has initiated, it becomes self-propagating. As shown in Figure 59.4, the loads on the intact pillars adjacent to the failure area increase as the extent of the failure increases, thereby driving the failure process even more. This process will continue until all pillars in the array have failed, or until a substantial barrier pillar or other solid abutment is reached.

The rapidity of these unstable pillar failures varies widely. At one end of the spectrum are slow "squeezes" that develop over days to weeks. Because of their slow progress, there is little immediate danger, and mining personnel and machinery have ample time to leave the vicinity safely. At the other end of the spectrum is cascading pillar failure, which can occur in a few seconds. With CPF, the failure occurs so rapidly that men and equipment do not have time to evacuate. Real danger of permanent entombment exists. Significant seismic energy is released as a result of the rapid failure and collapse. In addition, the rapid collapse is almost always associated with a potentially damaging air blast.

Many case histories exist of CPF in coal mines. The most infamous example is the Coalbrooke Colliery in South Africa where 437 miners perished when 2 square kilometers of the mine collapsed within a few minutes on January 21, 1960 (Bryan et al. 1966). Table 59.3 summarizes the mining dimensions of ten examples of rapid pillar collapse in coal mines. Most of these collapses occurred in the United States during the 1980s and 1990s except where noted. In all cases except Coalbrooke, these collapses happened suddenly or without any significant warning. All collapses resulted in substantial air blasts and severe damage to the ventilation system. Recent reports from India (Sheorey et al. 1995) and Australia (Galvin 1992) indicate that catastrophic pillar collapses and the associated air blasts have caused problems there as well.

Catastrophic pillar failures have also happened in many metal mines. Table 59.4 provides the mining dimensions of four room-and-pillar mines in the United States that most likely failed by the cascading pillar failure mechanism. The collapse areas can

be huge. Fortunately, some of these failures gave advance warning. At the Bautsch mine, slabbing from pillars and roof falls began four weeks prior to the main collapse. At the copper mine considerable rock noise and smaller failures preceded the main collapse by five days. Visible signs of pillar deterioration and rock noise also preceded the failure at the lead mine. All of the collapses induced an air blast; however, due to the large room dimensions, damage to the ventilation systems was never severe. The only known damage occurred at the copper mine where several air doors were bent and a few stoppings were blown down.

The largest and most devastating examples of CPF occurred in nonmetal mines. Table 59.5 documents the mining dimensions of seven nonmetal mines that probably failed by the CPF mechanism. Substantial air blasts resulted from most of these collapses. The occurrence of warnings was divided. Rock noise and other failure warnings preceded the silica mine collapse by three weeks. Evidently, both trona mine collapses occurred without warning.

Mines experiencing a cascading pillar failure generally exhibit the following characteristics.

1. Extraction ratios are usually more than 60%. A high extraction ratio will put pillar stress close to peak strength and provide ample expansion room for the failed pillar material.

2. Width-to-height ratio of pillars is always less than 3 for coal mine failures, usually much less than 1 in the metal mine failures and less than about 2 for the nonmetal mine failures. A low W/H ratio ensures that the failed pillar material can easily expand into the surrounding openings and that the failed pillar will have little residual load bearing capacity.

3. The number of pillars across the panel width is always at least 5 and usually more than 10, which typically ensures that pillars have reached their full tributary area load. Minimum panel widths for CPF are at least 80 m.

4. Substantial barrier pillars with width-to-height ratios more than 10 are absent from the mine layout.

5. Depth covers the full range of mining conditions. Although CPF seems more prevalent in shallow mines less than 100 m deep, this may be due to the prevalence of shallow room-and-pillar coal mines.

59.4 MECHANICS OF CASCADING PILLAR FAILURE

In the simple explanation for CPF given earlier, rapid load transfer away from failing pillars is important in the failure mechanism. However, the underlying mechanics of CPF are more complex. The nature of the pillar failure process depends on the relative magnitude of certain mechanical properties of the rock mass and the pillar. Salamon's local mine stiffness stability criterion accounts for these properties and predicts whether the failure process will occur in a stable, nonviolent manner or in an unstable, violent manner. This section will explain this criterion and the mechanics of stable and unstable failure of rock, first in the laboratory, and then in a mine.

59.4.1 Stable and Unstable Failure of Rocks in the Laboratory

Prior to the 1960s, compression testing of rock specimens produced a load convergence curve similar to that shown in Figure 59.5A. Typical tests ended suddenly and violently upon arrival at or shortly after reaching the "ultimate load." Beyond some critical applied displacement (i.e., convergence), the specimen had no load-bearing capacity, usually because it had disintegrated completely. Early tests were done with so-called "soft" testing

TABLE 59.3 CPF examples in coal mines

Reference and year of collapse	Collapse area (m)	Depth (m)	% Extr.	Mining height (m)	Room width (m)	Pillar size (m)	W/H ratio	Panel width (m)	Airblast damage
Bryan et al. (1964) Coalbrooke 1960	2 km²	140	56	4.3	6	12 × 12	2.8	210	437 miners killed
Ropchan (1991) Belina Mine 1991	300 × 450	170	56	5.8	6.1	12.2 × 12.2	2.1	170	Major, 1 injury
Chase et al. (1994) Case 1—1991	120 × 120	80	78	2.9	6.1	3 × 12.2	1.1	210	26 stoppings, 1 injury
Chase et al. (1994) Case 2—1990	90 × 120	75	78	3	6.1	3 × 12.2	1.0	150	40 stoppings
Chase et al. (1994) Case 3—1990	90 × 490	60	78	3	6.1	3 × 12.2	1.0	90–150	103 stoppings
Chase et al. (1994) Case 4—1992	120 × 150	70	64–75	3.3	6.1	6 × 6 9 × 9	1.8–2.7	180	37 stoppings
Mokgokong & Peng (1991) Emaswati Mine	60 × 140	60–140	75	3	6.1	12.2 × 12.2	2.0	80	Not known
Abel (1988) Roadside Mine 1983	200 × 400	210	84	2.1–2.4	6.1–12.2	3 × 24.4	1.3	430	Minor
Khair & Peng (1985)	120 × 120	90	75	2.1	6.1	6.1 × 6.1	2.8	120	Not known
Richmond (1998)	180 × 450	30–120	70	2.1	6.1	6.1 × 9.1	2.9	180	23 stoppings No injuries

TABLE 59.4 CPF examples in metal mines

Reference and year of collapse	Collapse area (m)	Depth (m)	% Extr.	Mining height (m)	Room width (m)	Pillar size (m)	W/H ratio	Panel width (m)	Airblast damage
Touseull & Rich (1980) Bautsch Mn. 1972	90 × 360	75	90	27	23	11 × 11	0.4	90	Minimal
Davidson (1987) Cu-Ag Mine 1987	90 × 120	300	60–65	20	14	9 × 90	0.5	150	Not known
Straskraba & Abel (1994) Copper Mine 1988	600 × 900	600	68	3.5–8.5	8.5	22 × 22 (A) 7 × 7 (R)	2.0–0.9	—	Minor
Dismuke et al. (1994) Lead Mine 1986	120 × 200	300	78	12	9.7	8.5 × 8.5	0.7	180	Minor

TABLE 59.5 CPF examples in non-metal mines

Reference and year of collapse	Collapse area (m)	Depth (m)	% Extr.	Mining height (m)	Room width (m)	Pillar size (m)	W/H ratio	Panel width (m)	Airblast damage
Swanson & Boler (1995) Trona Mine 1995	760 × 2,100	490	60–70	2.4–2.7	4.3	3.8 × 29	1.4	170	Major
Knoll (1990) Potash Mine 1989	6 km²	700–900	45	6.1	10.7	30 × 30	5	—	Not known
Spruell (1992) Silica Mine 1992	75 × 90	18	56	12.2	6.1	9 to 15 ave. 12	1	—	Not known
Denk et al. (1994) Salt Mine 1993	200 × 200	330	92	3.7	16.5	5.5 × 5.5	1.5	200	None
No Reference Limestone Mine 1975	300 × 600	<60	>80	>6	15–30	5 to 15 irregular	—	300	Not known
No Reference Trona Mine 2000	760 × 340	490	70	2.4–2.7	4.3	3.8 × 29	1.4	170	Not known
No Reference Limestone Mine 2000	75 × 75	40–60	> 90	4–5	10	5 × 5	1	—	None

machines, and they produced only part of the load-convergence relationship for the specimen.

Modern compression testing with so-called "stiff" testing machines produces a complete load-convergence curve similar to the one shown in Figure 59.4.5B (Cook and Hojem 1966; Bieniawski 1967; Wawersik and Fairhurst 1970). A typical test does not end suddenly and violently at the ultimate load. The load on the rock first increases to ultimate and then decreases gradually.

The rock specimen maintains its integrity and some load-bearing capacity even after the ultimate load is exceeded.

Jaeger and Cook (1979) discuss how confining pressure, temperature, loading rate, and other variables affect the shape of the stress-strain curve for a rock. For many practical mining engineering problems, the width-to-height ratio of a test specimen is of primary interest. Figure 59.6 from Das (1986) shows how the magnitude of peak strength, steepness of the post-failure portion

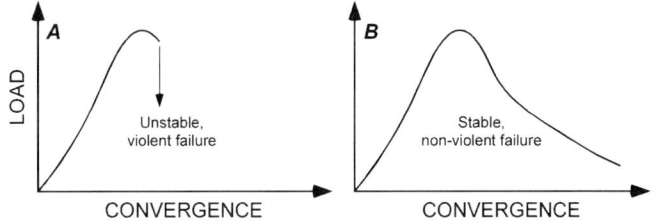

FIGURE 59.5 Typical load-convergence curves for rock from a "soft" testing machine (a) exhibiting unstable failure and from a "stiff" testing machine (b) exhibiting stable failure (Swanson and Boler 1995)

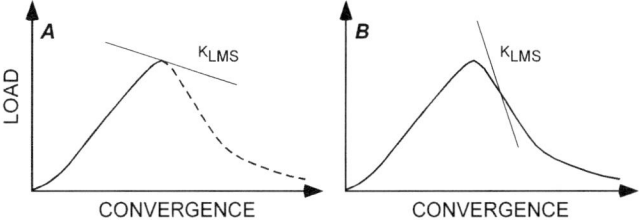

FIGURE 59.7 Unstable, violent failure versus stable, nonviolent failure. (A) loading machine stiffness is less than post-failure stiffnes sin a "soft" loading system. (B) loading machine stiffness is greater than post-failure stiffness in a "stiff" loading system (Swanson and Boler 1995).

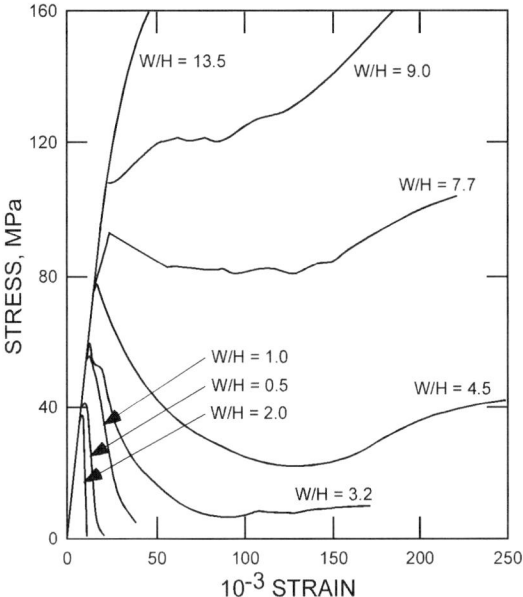

FIGURE 59.6 Complete stress-strain curves for Indian coal specimens showing increasing residual stength and post-failure modulus with increasing _W/H_ ratio (Das 1986)

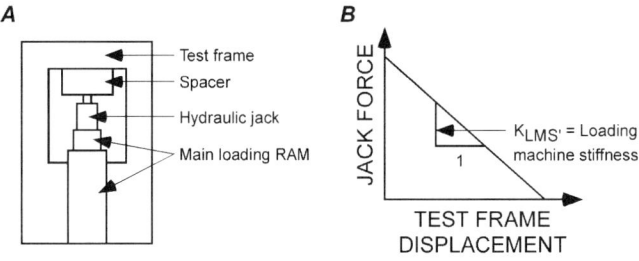

FIGURE 59.8 Illustration of scheme to measure loading machine stiffness. (A) pressurized jack replaces laboratory specimen in test frame; (B) loading machine stiffness is slope of jack force versus test frame displacement diagram.

of the stress-strain curve, and magnitude of the residual strength changes as the width-to-height ratio of a coal specimen increases. Specimen behavior is strain softening at first, then becomes elastic-plastic, and finally exhibits strain-hardening behavior as width-to-height ratio increases.

Salamon (1970) developed a criterion to determine whether the failure process observed in the laboratory occurs in a stable, nonviolent or in an unstable, violent manner. Figure 59.7 illustrates the criterion. Stable, nonviolent failure occurs when

$$|K_{lms}| > |K_p| \qquad \textbf{(59.19a)}$$

and unstable, violent failure occurs when

$$|K_{lms}| < |K_p| \qquad \textbf{(59.19b)}$$

where K_{lms} is the loading machine stiffness and K_p is the post-failure stiffness at any point along the load-convergence curve of the rock specimen. If the loading machine stiffness is less steep than the post-failure stiffness, as shown in Figure 59.7A, the failure is unstable and violent. However, if the loading machine stiffness is steeper than the post-failure stiffness, as

shown in Figure 59.7B, the failure is stable and nonviolent. If the above failure criterion is violated, the specimen disintegrates at or just after reaching the ultimate load and only part of the load-convergence relation is observed. Satisfying the above stability criterion can result in measurement of complete load-convergence behavior of the specimen.

"Soft" and "stiff" testing machines differ in much more than just outward appearance. A stiff loading machine has a much larger steel reaction frame as well as a much larger hydraulic ram. However, the fundamental difference between "soft" and "stiff" testing machines that leads to stable or unstable specimen failure is in the way the loading machine itself reacts to induced load and stores energy during a compression test. Both the test specimen and the loading machine deform as the applied load increases. A "soft" testing machine will deform more than a "stiff" testing machine and thereby store more energy in the testing machine itself. Energy is stored in the test frame (the steel posts, the crosshead and any steel platens), the hydraulic ram and the hydraulic fluid. Therefore, if the load on a specimen is the same, a "soft" testing machine will store more energy (load times deformation) than a "stiff" testing machine. When the specimen reaches its maximum load-bearing capacity, the "soft" testing machine will relax and displace inward and thereby transfer all its stored strain energy to the specimen, which will lead to an unstable, violent failure. With a "stiff" testing machine operating in displacement control, the testing machine itself will not displace inward sufficiently to destroy the test specimen. Failure will occur in a stable, nonviolent manner, and it is possible to measure the complete load-convergence relation for the specimen.

Figure 59.8 shows a simple analogy to measure loading machine stiffness. In 59.8A, the rock specimen in the test frame is replaced with an equivalent, pressurized hydraulic jack. Relaxing

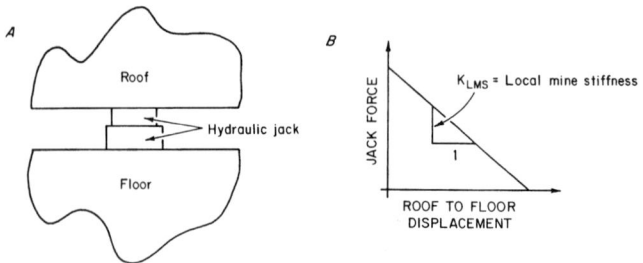

FIGURE 59.9 Illustration of scheme to measure K_{LMS}. (A) jack replces mine pillar; (B) K_{LMS} is slope of jack force versus mine roof-to-floor displacement diagram.

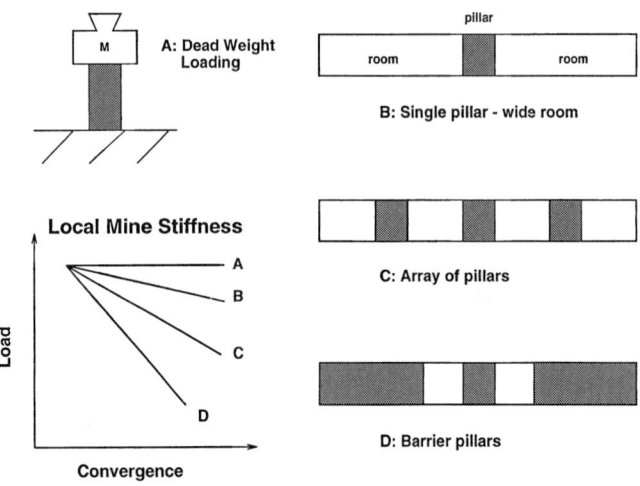

FIGURE 59.10 Relative local mine stiffness for various mine alyouts. Shaded pillar is analogous to hydraulic jack from Figure 59.9. Local mine stiffness magnitude as extraction increases (Swanson and Boler 1995).

the load on the jack causes the test frame to displace inward. The slope of the jack force versus test frame convergence relation, as shown in Figure 59.8B, is the loading machine stiffness.

59.4.2 Stable and Unstable Failure of Pillars in Mines

A rock specimen loaded in a laboratory test frame is analogous to a mine pillar loaded by the surrounding rock mass. Based on this analogy, Salamon's stability criterion (1970) also applies to mine pillars, and it will determine whether the failure process occurs in a stable, nonviolent or in an unstable, violent manner. Stable, nonviolent failure occurs when:

$$|K_{LMS}| > |K_P| \tag{59.20a}$$

and unstable, violent failure occurs when

$$|K_{LMS}| < |K_P| \tag{59.20b}$$

where K_{LMS} is the local mine stiffness and K_P is the postfailure stiffness at any point along the load-convergence curve of the pillar. The local mine stiffness, K_{LMS}, is an inverse load-convergence characteristic analogous to the loading machine stiffness, K_{lms}.

Similar to the means discussed earlier to measure loading machine stiffness, Figure 59.9 illustrates a means to measure K_{LMS} in a mine. The pillar is replaced by a large hydraulic jack (Figure 59.9A), which is then pressurized to simulate a loaded mine pillar. Relaxing the load on the jack causes the mine roof and floor to displace inward. The slope of this force-displacement relation, as shown in Figure 59.9B, is K_{LMS}.

The local mine stiffness depends on the modulus of the immediate roof, floor and pillar materials and the layout of pillars, mine openings and barrier pillars. Figure 59.10 shows how the local mine stiffness changes for different mine layouts, starting with a single pillar surrounded by wide openings, then an array of small pillars, and finally, a small pillar surrounded by barrier pillars. Consider the load-convergence response if a hydraulic jack replaces one small pillar and a unit load on the jack is released. For the case of a single pillar surrounded by wide openings, a unit-load decrease results in a large convergence increase. Therefore the local mine stiffness has a shallow slope as indicated by "B." In an array of small pillars, a unit-load decrease results in a smaller convergence increase, and the local mine stiffness has a steeper slope as shown by "C." Finally, if the small pillar is surrounded by large barrier pillars, a unit-load decrease results in a very small convergence increase, and the local mine stiffness is very steep as shown by "D." The worst case is that of "A" or dead weight loading, where the local mine stiffness is horizontal with zero slope. Such a local mine stiffness would always produce a violent failure if the pillars exhibit any strain softening behavior.

At this time, theory cannot predict the speed at which an unstable pillar failure will propagate. In other words, current rock mechanics theory does not predict whether an unstable failure will progress slowly, as a "squeeze," or rapidly as a CPF.

The speed of failure might depend in part on the degree in which the local mine stiffness stability criterion is violated.

59.5 STABILITY-CRITERION-BASED PILLAR DESIGN METHODS

The traditional strength-based pillar design procedures discussed earlier consider only the peak strength of the pillar. If the applied stress on the pillar reaches this level, its safety factor against strength failure is 1. There is great economic incentive to design pillars with a low strength-safety factor because as the strength-safety factor decreases the extraction ratio increases along with resource recovery, mining revenues, and potential profits. The most economic pillar designs are necessarily very close to the peak strength, but still on the prefailure side of the complete load-convergence curve for the pillar. Because of the inherent variability in pillar strength, it is also important to consider what might happen to a particular room-and-pillar mine plan if some of the pillars exceed the peak strength and enter the postfailure portion of their complete load-convergence curve. Such considerations are necessary especially when the number of pillars in an array is large. In that case even if the average strength safety factor of the pillars is large, say 1.5, and their probability of failure is small, the probability of pillar failure somewhere in the array is likely to be large. Failure of just a few pillars in an array may be sufficient to initiate a CPF even if their average strength safety factor is large.

Engineers have known for some time about the strain-softening, postfailure behavior of rock and the implications of this behavior on mine safety. However, it is another matter to translate that knowledge into efficient, economic mine designs for extraction of bedded deposits using room-and-pillar and related mining techniques. The proposed methodology goes beyond using only traditional strength-based pillar design criterion and incorporates Salamon's local mine stiffness stability criterion for both sizing pillars and determining mine layout. On the basis of this stability criterion, three different approaches are proposed to control CPF—containment, prevention, and full extraction approach.

59.5.1 Containment Approach

In the containment approach, shown in Figure 59.11, an array of panel pillars, that violate the local mine stiffness stability criterion and can therefore fail in an unstable, violent manner if their strength criterion is exceeded, are surrounded or "contained" by barrier pillars. The primary function of barrier pillars is to limit the potential failure to just one panel. Barrier pillars have a high W/H ratio, typically greater than about 10, and contain panel pillars with low W/H ratio, typically in the 0.5 to 2 range. It is a noncaving room-and-pillar method in that panel pillars are not meant to fail during retreat mining.

Two factors help decrease the risk of a instantaneous CPF. First, the barrier pillars tend to shield the panel pillars from full overburden stresses and thereby increase their strength safety factor and decrease their probability of failure. Second, if the panel pillars do fail, the failure will not propagate beyond the barrier pillars. A conservative approach is to size the barrier pillars on the assumption that all the panel pillars within have failed. Most important though, barrier pillars must have a sufficiently large W/H ratio to eliminate any strain-softening behavior.

If an engineer chooses to use panel pillars with a low W/H ratio that exhibit strain-softening behavior and can violate the local mine stiffness stability criterion, then it is imperative to limit panel sizes and surround all panels with adequate barrier pillars. The load transfer method presented by Abel (1988) provides an approach for estimating maximum panel width. Use of this method will help insure that panel pillars never experience full tributary area stress. Even though barrier pillars shield panel pillars from full tributary area stress, the small pillars should have the ability to withstand this maximum stress.

To summarize, the main design characteristics of the containment approach are:

1. Panel pillars have a low strength safety factor (usually between 1.1 to 1.5) and low W/H ratios (usually less than 3 or 4). The panel pillars violate the local mine stiffness stability criterion and can therefore fail violently.

2. Barrier pillars have high W/H ratios (usually greater than 10) and do not exhibit any strain softening behavior. Therefore, they cannot fail violently. Barrier pillars also have sufficient strength to remain intact even if all pillars within a panel should fail.

3. The minimum load transfer distance limits panel width to insure that panel pillar stresses are less than full tributary area stresses.

4. Panel sizes are also limited by the maximum air blast size that the mine layout can withstand should CPF occur within a panel.

59.5.2 Prevention Approach

In contrast to the containment approach, the prevention approach "prevents" CPF from ever occurring by using panel pillars that satisfy both the local mine stiffness stability criterion and a strength criterion. Therefore, the panel pillars cannot fail violently, and CPF is a physical impossibility. Strictly speaking, this approach may not need barrier pillars to insure overall stability against CPF; however, their use is still advisable. To satisfy the local mine stiffness stability criterion, the panel pillars will usually have high W/H ratios (greater than about 3 or 4) and high strength safety factors as well (greater than 2). Another approach to increase the local mine stiffness and satisfy the stability criterion is to limit the panel width with properly spaced and sized barrier pillars.

Figure 59.12 illustrates the prevention approach for another typical advance and retreat mining sequence. In this example, a modification to the retreat mining system leaves remnant pillars with higher W/H ratio and more desirable postfailure characteristics that satisfy the local mine stiffness stability criterion. As with the containment approach, it is a noncaving room-and-pillar system.

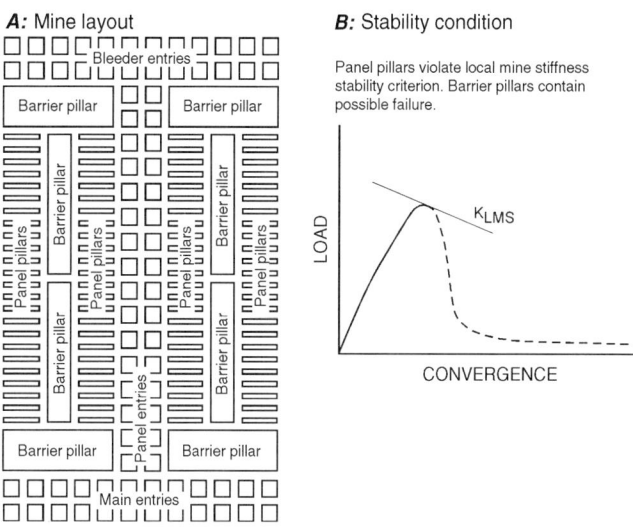

FIGURE 59.11   Containment approach to room-and-pillar mine layout: (A) pillar failure is 'compartmentalized'; (B) stability condition is such that pillar with low W/H ratio violate local mine stiffness stability criterion; panel pillars can, therefore, fail violently, but adequate barrier pillars that restrict spread of unstable failure surround them. Extraction for layout shown in 59%.

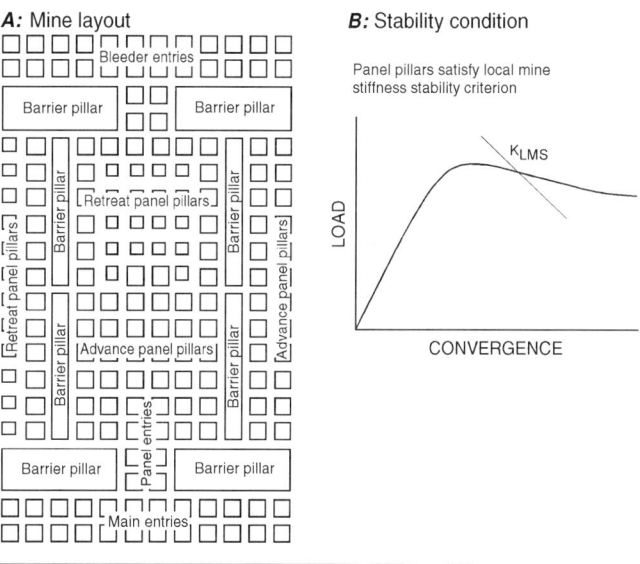

FIGURE 59.12   Full extraction approach: (A) failure of pillar remnants along with overburen occurs immediately after pillar extraction: (B) retreat mining must ensure development of sufficiently weak remnant pillars. Extraction for layout shown in 67%.

59.5.3 Full Extraction Approach

The full extraction approach shown in Figure 59.13 avoids the possibility of CPF altogether by ensuring total closure of the opening and full surface subsidence on completion of retreat mining. This approach does not require barrier pillars for overall panel stability; however, they are needed to isolate extraction areas and protect mains and bleeders. The main design characteristics of a full extraction approach are:

1. Panel pillars on advance must have adequate strength safety factors (greater than about 2).

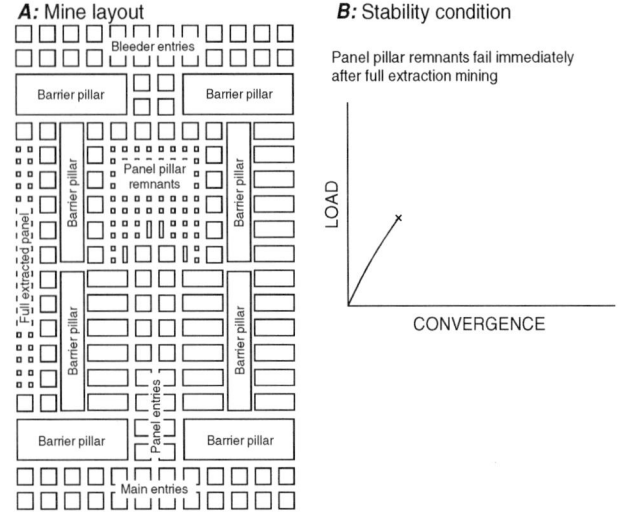

FIGURE 59.13   Full extraction approach: (A) failure of pillar remnants along with overburden occurs immediately after pillar extraction: (B) retreat mining must ensure development of sufficiently weak remnant pillars. Extraction for layout shown is 67%.

2. In addition, panel pillars should satisfy the local mine stiffness stability criterion during advance mining by having high W/H ratios (greater than about 4).

3. Panel pillars on retreat must have strength safety factors much less than 1 to ensure their complete collapse soon after retreat mining.

59.6   NUMERICAL SIMULATION OF CASCADING PILLAR FAILURE

59.6.1   Boundary-Element-Method Implementation

The NIOSH (former USBM) boundary element program called MULSIM/NL was modified to evaluate the local mine stiffness stability criterion and simulate the mechanics of CPF. MULSIM/ NL calculates stresses and displacements using a displacement-discontinuity approach, which applies especially well to thin, bedded-type deposits such as coal and certain metals. The in-seam materials can follow a strain-softening constitutive model. This strain-softening model enables the simulation of CPF.

MULSIM/NL calculates the local mine stiffness (K_{LMS}) around a pillar as

$$K_{LMS} = \frac{(S_u - S_p)A}{(D_u - D_p)} \qquad (59.21)$$

where

S_u = unperturbed stress at an element,

S_p = perturbed stress at an element,

D_u = unperturbed displacement at an element,

D_p = perturbed displacement at an element, and

A = element area.

The local mine stiffness for the pillar is the summation of all the stiffnesses composing the pillar. The unperturbed stresses and displacements are calculated in the usual way with MULSIM/NL. The perturbed stresses and displacements are then calculated by first removing the pillar where K_{LMS} is desired and then resolving for stresses and displacements. In this approach, S_p is identically 0.

TABLE 59.6   Unstable and stable model characteristics

	Unstable model	Stable model
Grid size	90 × 90	90 × 90
Element width (m)	3	3
Pillar height (m)	3	3
Depth (m)	80	160
Applied stress (MPa)	2	4
Rock nass modulus (MPa)	3,500	7,000
Coal modulus (MPa)	3,500	3,500
Pillar W/H ratio	1	3
Extraction on advance	59.1%	49.0%
Extraction on retreat	79.6%	81.6%

The minimum postfailure stiffness (K_P) for the pillar is calculated as

$$K_P = \frac{E_p A}{t} \qquad (59.22)$$

where

E_p = postfailure modulus for material model,

A = element area, and

t = element thickness.

The postfailure stiffness for the pillar is then the sum of all element stiffnesses making up the pillar.

Based on these calculations of local mine stiffness and post-failure stiffness, the modified MULSIM/NL program applies the local mine stiffness stability criterion and ascertains the nature of the likely failure process, either stable or unstable. Depending on whether the criterion is satisfied or violated, the stress and displacement calculations with MULSIM/NL behave in vastly different manners.

59.6.2   Example Unstable and Stable Numerical Simulations

Two simple models of coal mines, one stable and the other unstable, demonstrate the local mine stiffness criterion and its influence on numerical model behavior. These calculations demonstrate the feasibility of calculating pillar stresses for tradi-tional strength-based pillar design and evaluating the local mine stiffness for stability-criterion-based pillar design. Table 59.6 gives the basic model characteristics.

Figure 59.14 shows complete strain-softening, stress-strain curves for coal pillars with different W/H ratios as used in these simulations (Zipf 1999). The peak strengths in these curves are consistent with the coal pillar strength formulas summarized by Mark and Iannacchione (1992). The postfailure moduli approxi-mate field data presented by Bieniawski (1968b), Van Heerden (1975) and Wagner (1974). Finally, the residual strength values for these full-scale pillar stress-strain curves are an estimate based on limited field data and observed laboratory-scale data.

The mine geometry for the stable and unstable models is visible in the resulting stress and displacement analyses. In the unstable model (Figure 59.15), advance mining develops 6 m by 27 m pillars and 6-m-wide rooms. On retreat, 3 m by 27 m pillars are left for an overall extraction of almost 80%. These pillars have a W/H ratio of 1 and follow the corresponding stress-strain curve shown in Figure 59.14. In the stable model (Figure 59.16), advance mining creates 15-m-square pillars with 6-m-wide rooms. On retreat, 9-m-square pillars are left to achieve an overall extraction of about 82%. However, these pillars have a W/H ratio of 3, and as shown in Figure 59.14, have a higher strength and different postfailure characteristics.

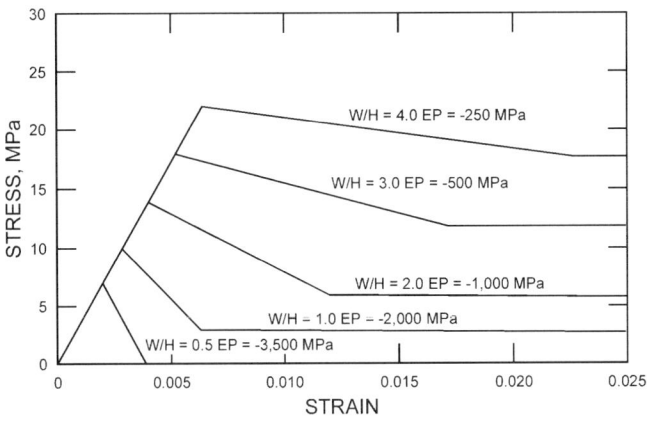

FIGURE 59.14 Strain-softening, stress-strain curves for coal pillars with increasing W/H ratio. E_P is minimum postfailure modulus.

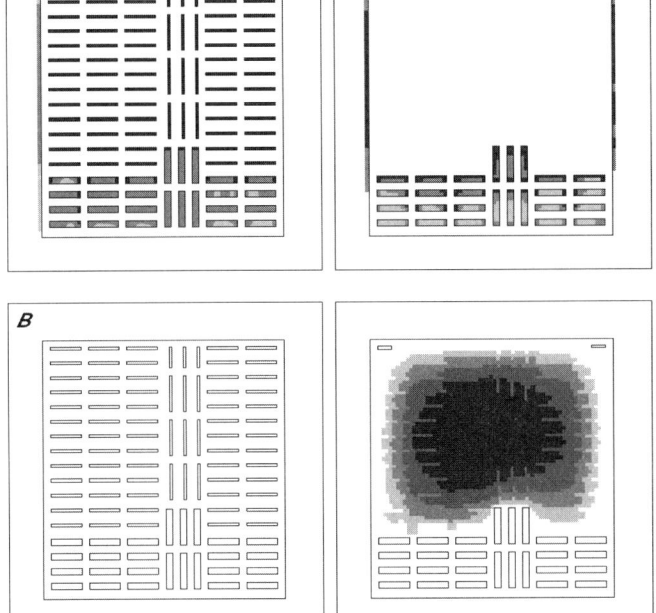

FIGURE 59.15 Unstable case, (A) stress and (B) convergence before and after pillar weakening. Light to dark gray indicates increasing magnitude of calculated vertical stress and convergence.

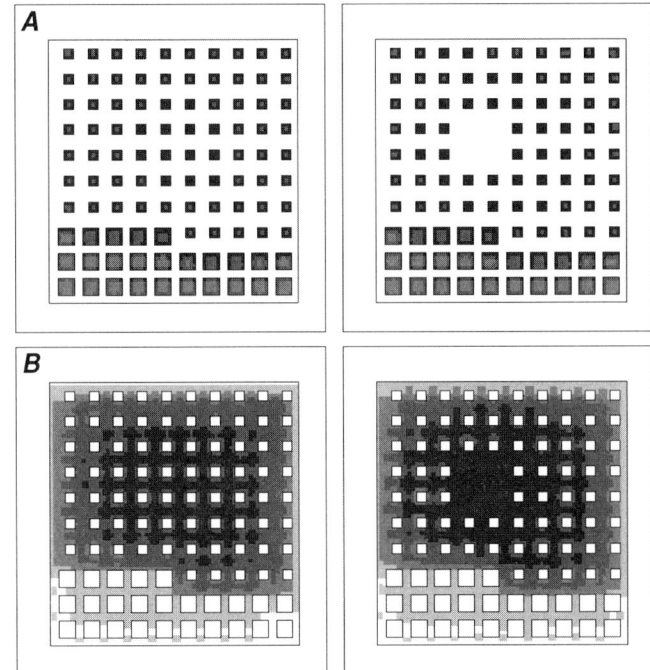

FIGURE 59.16 Stable case, (A) stress and (B) convergence before and after pillar removal. Light to dark gray indicates increasing magnitude of calculated vertical stress and convergence.

Each model has two mining steps. In the first step, the applied vertical stress brings the pillars close to failure. For the unstable case, a vertical stress of 2 MPa results in an average pillar stress of about 8.7 MPa, which implies a strength safety factor of about 1.15 for these pillars with a W/H ratio of 1. In the stable case, applying a vertical stress of 4 MPa results in an average pillar stress of 17.2 MPa and a strength safety factor of 1.05 for these pillars with a W/H ratio of 3.

In the second mining step, each model is perturbed artificially. For the unstable case, 3 pillars in the middle of the array of 77 pillars are weakened by decreasing their peak strength by 10% from 10 down to 9 MPa. In reality, this strength decrease could stem from creep or an undersized pillar. This small disturbance though is enough to trigger a CPF. Figure 59.15 shows the

computational results. The stresses in the small pillars shown in Figure 59.15A are near their peak stresses of 10 MPa prior to the CPF, and they decrease to their residual level of 3 MPa after the failure. Convergence shown in Figure 59.15B is small before pillar weakening and increases dramatically after the CPF. In an unstable-type failure, Figure 59.15 shows that a small disturbance or a small increment of mining can result in a much, much larger increment of failure.

The unstable case violates the local mine stiffness stability criterion given by Equation 19. In mining step 1 where all the small pillars are near failure, K_{LMS} calculated by removing one pillar is −7,700 MN/m while the post-failure stiffness for that pillar is −54,000 MN/m. Therefore, unstable failure is possible by the stability criterion. In mining step 2, failure is already complete. K_{LMS} for this pillar is computed as −21,500 MN/m. Again, the stability criterion is violated and unstable failure is possible; however, it has already occurred.

For the stable case, a more radical disturbance is perpetrated by removing 4 pillars from the array of 75 pillars. Even though this disturbance is far greater than the pillar weakening done in the unstable case, a CPF does not result. Rather, stable progressive failure of the surrounding pillars takes place. Figure 59.16 shows the numerical model calculations. Some of the stresses shown in Figure 59.16A are already post-peak before pillar removal. After pillar removal, the size of the area and the number of pillars in the postfailure regime increase; however, the increase is not radical. The convergence shown in Figure 59.16B increases after pillar removal; however, that increase is not catastrophic as in the unstable case shown in Figure 59.15B. In contrast to the previous case of unstable failure, Figures 59.16A and 59.16B show that in stable failure, a disturbance or some increment of additional mining results in a more or less equal increment of additional failure in the model.

The stable case satisfies the local mine stiffness stability criterion given by equation 19. In mining step 1, K_{LMS} calculated by removing one pillar is −34,400 MN/m, and the postfailure stiffness for that pillar is −13,500 MN/m, which implies stable failure. In mining step 2, K_{LMS} decreases in magnitude to −32,100 MN/m; however, stable failure still prevails.

59.6.3 Behavior of Local Mine Stiffness Calculations

Initial research with practical K_{LMS} calculations indicates that K_{LMS} tends to decrease in magnitude significantly as the peak strength of an array of pillars is approached, and then increase after failure is complete. The differences in K_{LMS} stem from whether the perturbation used for the K_{LMS} calculation is sufficient to initiate the CPF itself. At lower vertical stress and higher strength safety factor for an array of pillars, removal of one pillar is not sufficient to precipitate CPF, and K_{LMS} has a relatively high magnitude. When the pillar array is near failure and strength safety factor is near 1, removal of one pillar to calculate K_{LMS} initiates CPF in the model. Perturbed displacements calculated by the model increase dramatically; and therefore, K_{LMS} decreases in magnitude significantly. K_{LMS} calculations at a pillar within a failed array then seem to increase in magnitude back to the low stress, prefailure values. The magnitude of the decrease appears to be on the order of 3.

This decrease in K_{LMS} magnitude near and during failure has important implications in the mechanics of CPF and in design practices to eliminate it. As Swanson and Boler (1995) point out, once a CPF is initiated and K_{LMS} decreases in magnitude, violation of the local mine stiffness stability criterion becomes even more acute. Thus, once the CPF gets going, it becomes more self-propagating and more difficult to arrest the more it grows. It is possible that mine layouts that apparently satisfy the stability criterion, such as the stable case in the example above, could become unstable and CPF could result if the mine layout is disturbed by a sufficiently large "kick." In the unstable case, initial research shows that K_{LMS} could decrease in magnitude by a factor of 3 during the failure process. Accordingly, this decrease suggests that using a "safety factor" of at least 3 is necessary when applying the local mine stiffness stability criterion. Further research on K_{LMS} calculations is suggested in this area.

59.7 ANALYSES OF CASCADING PILLAR FAILURE CASE HISTORIES

Three recent case histories of CPF in mines are analyzed with numerical models, which calculate stress and convergence before and after catastrophic failure. In each model, best estimates and/or assumptions are made for the strain-softening properties of the pillars. Local mine stiffness is calculated, and the stability criterion is evaluated to show how each case most likely violated this criterion, and a CPF resulted.

59.7.1 Case History 1—Coal Mine Collapse

Chase et al. (1994) document this case history of massive pillar collapse or CPF in a coal mine. Table 59.7 summarizes the important characteristics of the boundary-element-method model of this mine. The stress-strain curves for coal shown in Figure 59.14 are applied in this model. Figure 59.17 shows the layout for the coal mine model. Using 6-m-wide rooms, a system of 12-m-square pillars with a W/H ratio of 4 were developed on advance. On retreat, these pillars were split down the middle with a 6-m-wide room to leave two 3-m-wide fenders with a W/H ratio of 1. The mine had split about nine rows of pillars with this method when CPF occurred causing seven rows of fenders to fail. The idealized numerical model contains just seven rows of fenders.

In the two-step model, the applied vertical stress of 2.3 MPa loads the fenders close to their peak strength of 10 MPa. Average pillar stress is about 9.3 MPa, which implies a safety strength factor of 1.08. In the second mining step, peak strength is decreased from 10 down to 9 MPa to weaken 8 fenders in the array of 138. This artificial weakening simulates time-dependent strength loss. This disturbance is sufficient to initiate the CPF.

TABLE 59.7   Case history 1—Coal mine collapse model characteristics

Grid size	90 × 100
Element width (m)	3
Pillar height (m)	3
Depth (m)	90
Applied stress (MPa)	2.3
Rock mass modulus (MPa)	6,000
Coal modulus (MPa)	3,500
Pillar W/H ratio—advance	4
Pillar W/H ratio—retreat	1
Extraction on advance	55.6%
Extraction on retreat	77.8%

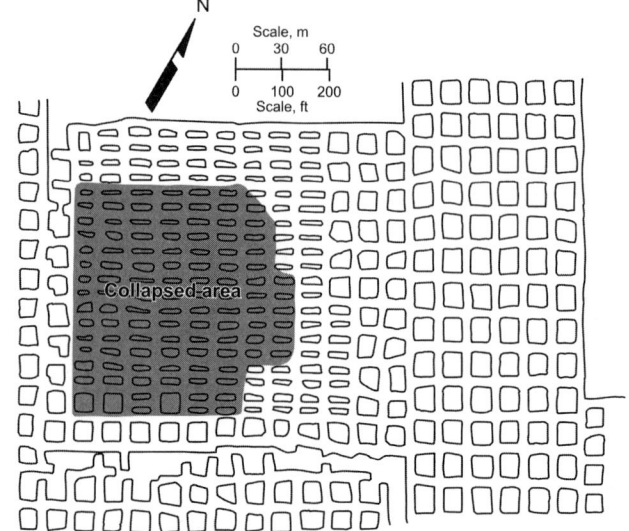

FIGURE 59.17   Mine layout for case history 1—coal mine collapse

Figures 59.18 and 59.19 show the stress and convergence before and after the failure. Stresses decrease from values near the peak stress to residual values while convergence increases dramatically. As in the previous unstable simulation, a small disturbance or increment of mining triggers a very large increment of failure in the model.

This case history apparently violates the local mine stiffness stability criterion. K_{LMS} is calculated as −6,600 MN/m during the first step and −19,700 MN/m during the second. Post-failure stiffness is −24,000 MN/m; therefore, unstable failure is possible by the criterion. Because the model is near failure under the 2.3 MPa vertical stress, computing K_{LMS} by removing a pillar precipitates the CPF. Again, the K_{LMS} value calculated during the failure process is about 1/3 less in magnitude than values calculated well before or well after failure.

59.7.2 Case History 2—Evaporite Mine Collapse

Swanson and Boler (1995) and Zipf and Swanson (1999) describe the mine geometry and aftermath of the collapse analyzed. Figure 59.20 shows the layout for this evaporite mine. During advance mining, a system of chain pillars is developed off a set of mains. At the same time, rooms are mined down the one side leaving long narrow panel pillars with a W/H ratio of 1.33 and a small interpanel pillar with a W/H ratio of about 2.66. On retreat, additional rooms are mined on the other side of the chain

Before Pillar Collapse

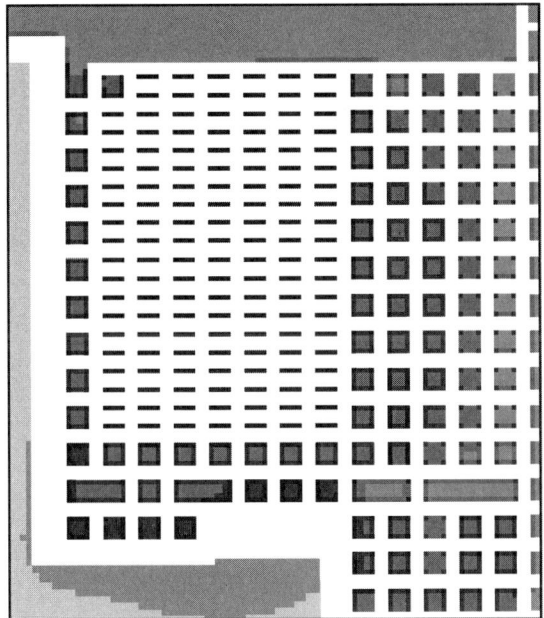

After Pillar Collapse

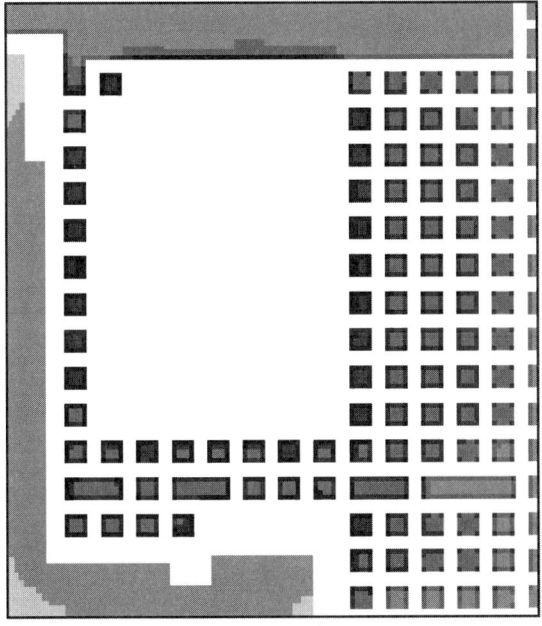

FIGURE 59.18 Coal mine stress before and after CPF. Light to dark gray indicates increasing magnitude of vertical stress.

Before Pillar Collapse

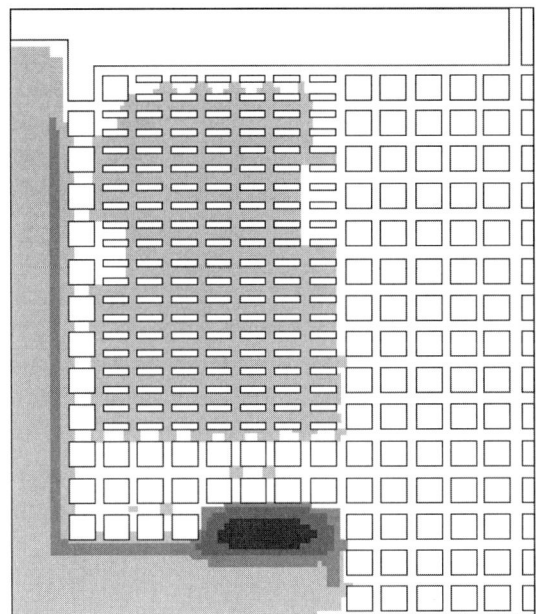

After Pillar Collapse

FIGURE 59.19 Coal mine convergence before and after CPF. Light to dark gray indicates increasing mgnitude of convergence.

pillars. In the actual mine layout, room width is 4.3 m, and the panel pillars are 3.8 m wide. The mine achieves an overall extraction of about 60% and extraction within a panel can be a little more than 70%.

Table 59.8 summarizes key input parameters for simple boundary-element-method models of this collapse. Due to the constant element size restriction, both room and panel pillar width are 4 m in the model. This approximation results in an extraction of 50% in the model.

In comparison to the available data for coal pillars, very little data exists for the complete stress-strain behavior of pillars in various metal and nonmetal mines. Direct measurements of the complete stress-strain behavior of pillars are difficult, very expensive, and often simply not practical. Laboratory tests on specimens with various W/H ratios can provide useful insights similar to the coal data shown earlier in Figure 59.6. MSHA (1996) used FLAC (Itasca 1995) to calculate the complete load-deformation behavior of the pillar-floor system in this evaporite mine. The objective of this modeling effort was to estimate post-failure stiffness of the pillar-floor system for a variety of pillar W/H ratios. Figure 59.21 shows the basic model geometry for these FLAC models. Each contained the same sequence of strong shale in the roof, strong

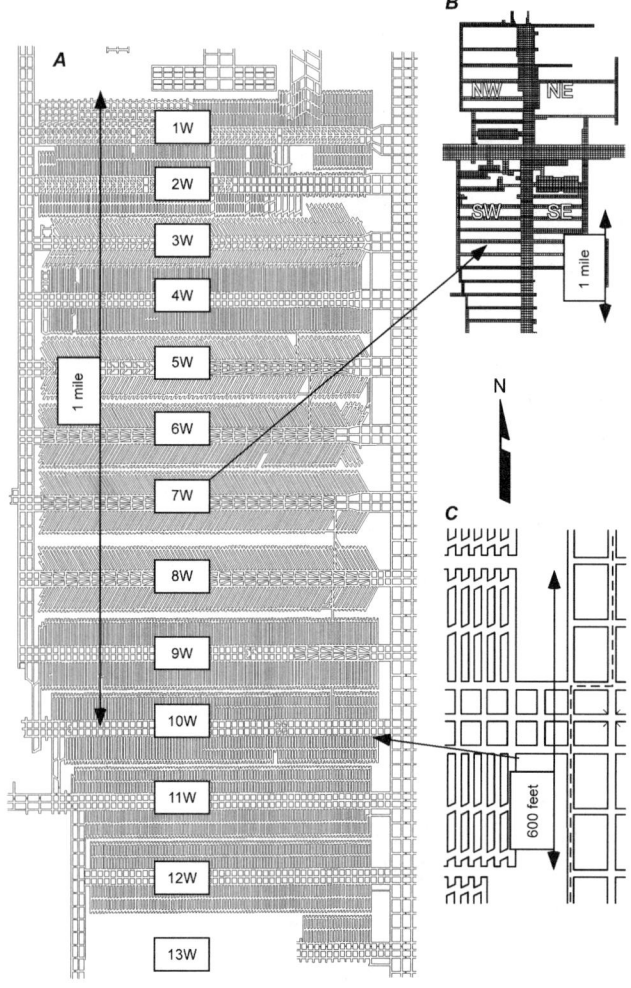

FIGURE 59.20   Mine layout for case history 2—evaporite mine collapse. (A) overall layout for part of the mine, (B) layout of southwest panels that collapsed and (C) details of a typical panel (MSHA, 1996).

TABLE 59.8   Case history 2—Evaporite mine collapse model characteristics

Grid size	120 × 120
Element width (m)	4
Pillar height (m)	3
Depth (m)	520
Applied stress (MPa)	13
Rock mass modulus (MPa)	3,500
Pillar modulus (MPa)	10,000
Pillar W/H ratio	1.33
Extraction (of model)	50%

and weak layers of the evaporite, a thin, strong oil shale layer in the immediate floor and a very weak mudstone deeper into the floor. Strain-softening material models were employed for each layer.

Figure 59.22 shows the computed rock movement after considerable deformation has occurred. The failure mechanism involves pillar punching through the thin, strong oil shale layer into the weak subfloor where a classic circular-arc-type failure

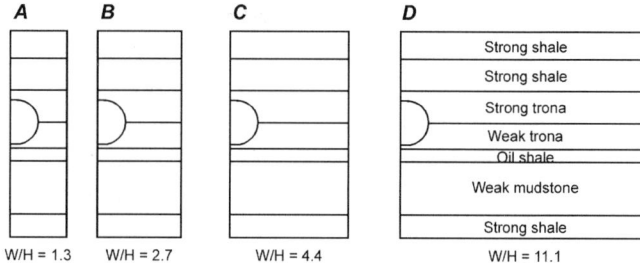

FIGURE 59.21   FLAC models of pillar-floor system for increasing pillar width and W/H ratio

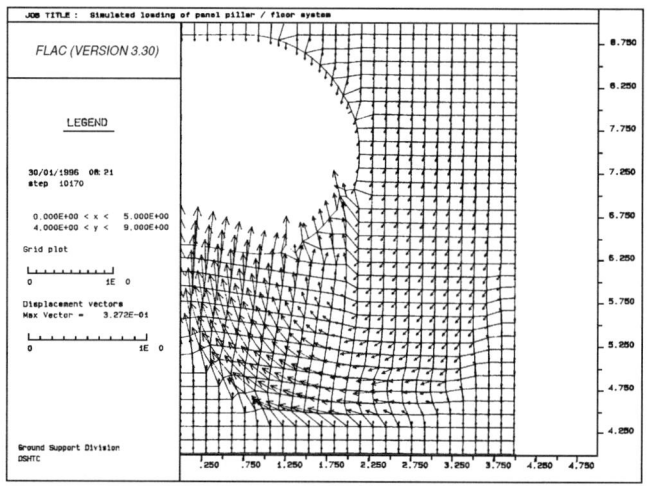

FIGURE 59.22   Computed rock movement after considerable deformation has occurred

develops. The computed failure mechanism matches field observations qualitatively; however, the computed deformations are much smaller than those observed in the field. This difference may arise because FLAC uses a continuum formulation to model a process that gradually becomes more and more discontinuous.

These computations provide an estimate of the complete stress-strain behavior of the overall pillar-floor system. Using the "history" function within FLAC, the model recorded average stress across the middle layer of the pillar and the relative displacement between the top and bottom of the pillar from which strain was computed. Figure 59.23 shows the effective stress-strain curves determined for the pillar-floor system from these four models. The initial post-failure portion of these curves provides an estimate of K_P for use in the local mine stiffness stability criterion. As stated earlier, computed displacements are smaller than observed, so the effective modulus and stiffness is also lower than indicated by these continuum-based computations. Figure 59.24 shows the strain softening, stress-strain relations as used in the boundary element models of the pillar collapse. These curves merely represent a best estimate of actual pillar-floor system behavior.

The BEM model in this analysis has two mining steps. The applied vertical stress of 13 MPa in the first mining step brings the long panel pillars close to failure. Average stress in these pillars is calculated as about 30 MPa. Since the peak strength of these pillars as shown in Figure 59.24 is 37 MPa, the safety factor of these pillars is about 1.2. In many circumstances, this safety factor may seem reasonable for many room-and-pillar mining applications. In the second mining step, 8 panel pillars in the array of 184 pillars are weakened by 15%. The artificial

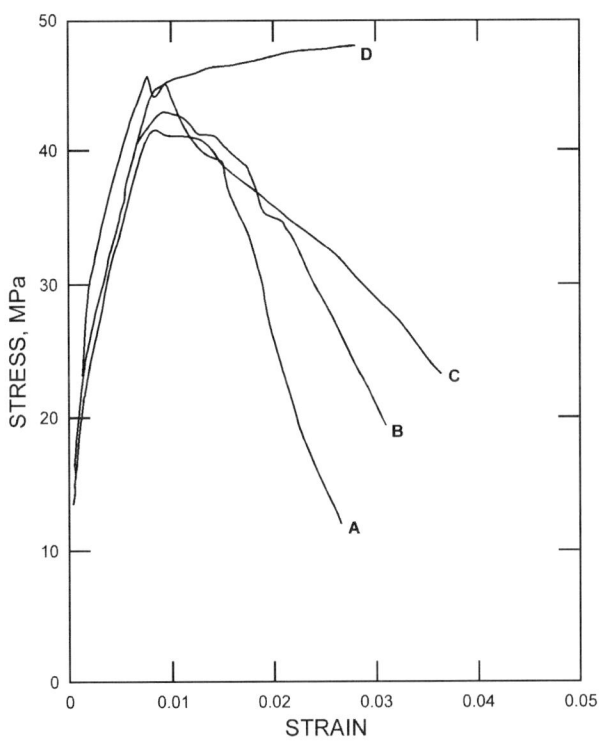

FIGURE 59.23 Stress-strain behavior of pillar-floor for increasing pillar width and *W/H* ratio

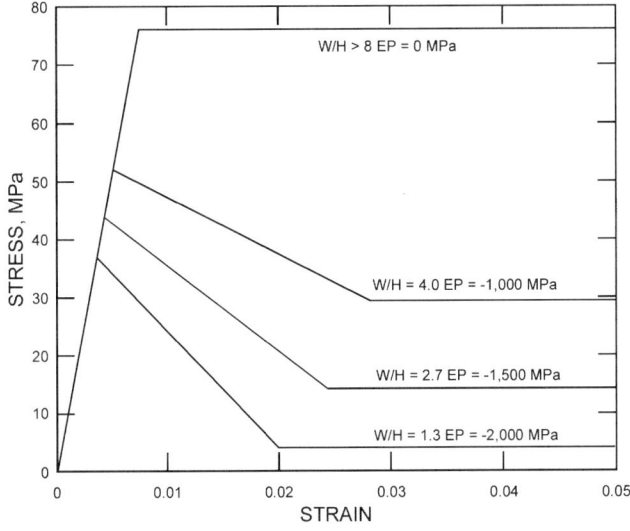

FIGURE 59.24 Strain-softening stress-strain curves for pillar/floor system with increasing pillar W/H ratio used in boundary-element-method analysis

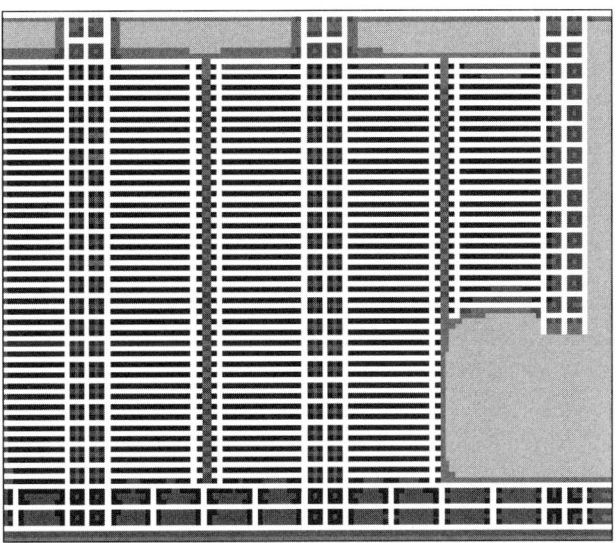

Before Pillar Collapse

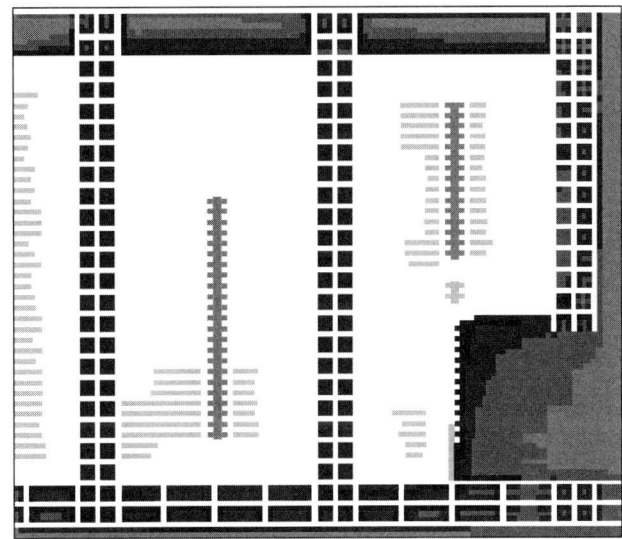

After Pillar Collapse

FIGURE 59.25 Evaporite mine stress before and after CPF. Light to dark gray indicates increasing magnitude of stress.

weakening approximates time-dependent strength loss. This small disturbance on about 4% of the panel pillars triggers the CPF.

Figures 59.25 and 59.26 present calculated stress and convergence before and after the CPF. The model shows that stresses shift to the perimeter of the collapse area while convergence increases dramatically in the middle. Once again, disturbance of a small area in the model leads to widespread failure.

K_{LMS} calculations suggest that this layout violates the local mine stiffness stability criterion. At one particular panel pillar, K_{LMS} is estimated as –60,420 MN/m during the first step before pillar weakening. After CPF is complete, K_{LMS} is –42,520 MN/m. Postfailure stiffness for the panel pillars is calculated as –170,700 MN/m. Therefore, unstable failure is a possibility by the local mine stiffness stability criterion. Again, good data on the postfailure behavior of the long panel pillars and the pillar-floor composite is limited. However, the simple assumptions and numerical analysis presented here support the hypothesis that CPF resulted because the local mine stiffness stability criterion was violated.

59.7.3 Case History 3—Metal Mine Collapse

Dismuke et al. (1994) describe this large pillar collapse in a major section of a room-and-pillar base metal mine. Figure 59.27 shows the collapse area. The failure began in four centrally located pillars and spread rapidly to include almost 100 pillars. Table 59.9

Before Pillar Collapse

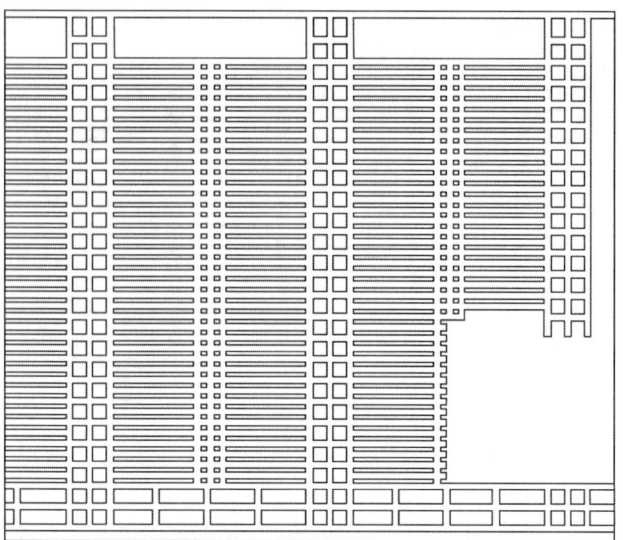

After Pillar Collapse

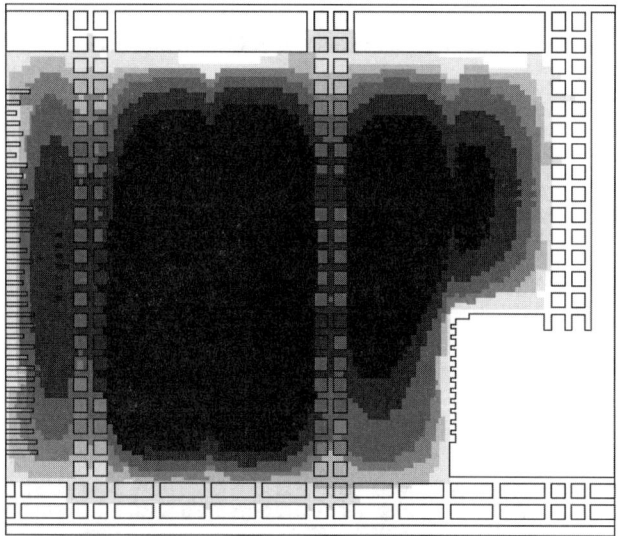

FIGURE 59.26 Evaporite mine convergence before and after CPF. Light to dark gray indicates increasing magnitude of convergence.

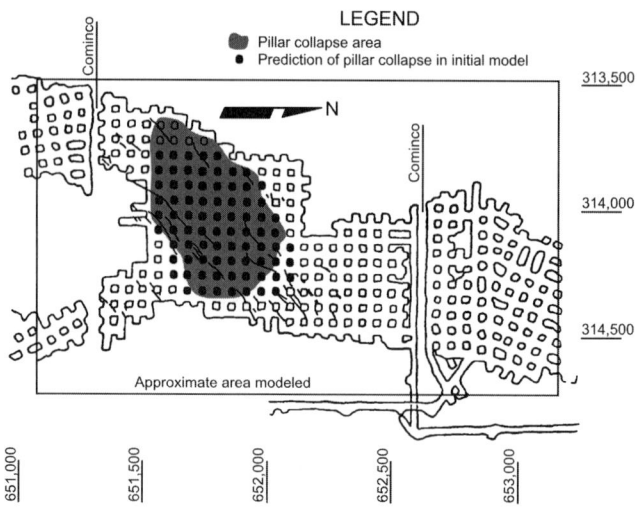

FIGURE 59.27 Mine layout for case history 3—metal mine collapse (Dismuke et al. 1994)

TABLE 59.9 Case history 3—Metal mine collapse model characteristics

Grid size	270 × 360
Element width (m)	1.21
Pillar height (m)	12.12
Depth (m)	300
Applied stress (MPa)	7.5
Rock mass modulus (MPa)	40,000
Pillar modulus (MPa)	40,000
Pillar W/H ratio	0.7
Extraction	80%

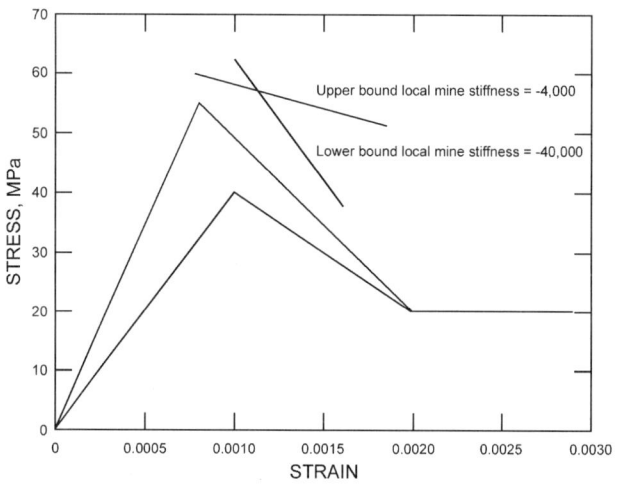

FIGURE 59.28 Stress-strain curves for 12.1- and 7.6-m-high pillars

summarizes the important characteristics of the boundary-element-method model of this mine. Pillar width is 8.5 m, and room width is 9.7 m. Two different pillar models are used, one for 12.1-m-high pillars in the central portion of the mine and the other for 7.6-m-high pillars around the perimeter. Figure 59.28 shows the stress-strain curves for these two pillars with W/H ratios of 1.17 and 0.70 respectively.

In the two-step model, the applied vertical stress of 7.5 MPa loads most of the pillars to about 37.5 MPa giving them a safety strength factor of about 1.06. In the second mining step, failure of a third pillar in the array leads to collapse throughout the entire area. As shown in Figure 59.27, the actual failure stopped far short of that predicted by these simplistic numerical model calculations, which may imply considerable conservatism in the models.

Figures 59.29 and 59.30 show the stress and convergence before and after the failure. Stresses decrease from values near the peak stress to residual values while convergence increases

dramatically. As in the previous unstable simulations, a small disturbance or increment of mining triggers a very large increment of failure in the model.

From the stress-strain curves input to the model, postfailure stiffness K_P is −240,000 MN/m. Local mine stiffness (K_{LMS}) calculated at central pillars in the array ranges from −200,000 MN/m

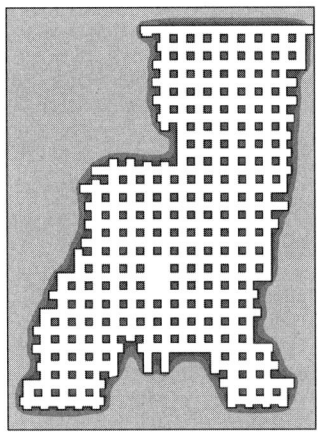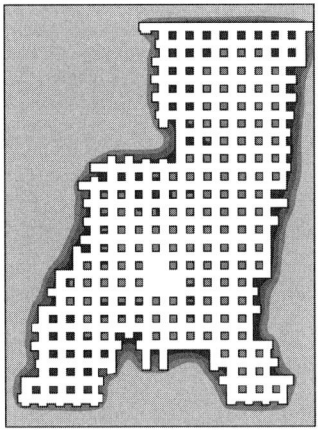

FIGURE 59.29 Metal mine stress before and after CPF. Light to dark gray indicates increasing vertical stress magnitude.

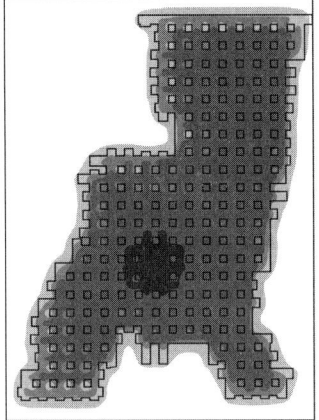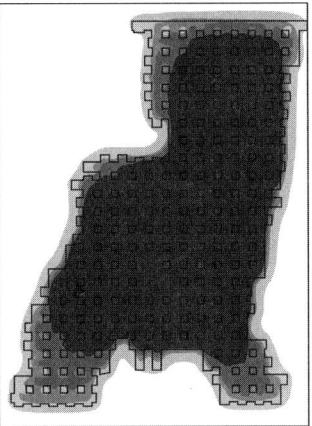

FIGURE 59.30 Metal mine convergence before and after CPF. Light to dark gray indicates increasing convergence stress magnitude.

prior to failure to as low as −27,800 MN/m at the moment of failure. Therefore, unstable failure is always possible by the criterion. As in previous examples, the K_{LMS} value calculated during the failure process is much less in magnitude than values calculated well before or well after failure. Unfortunately, very little is known of the postfailure behavior of mine pillars; however, calculations of K_{LMS} from collapse case histories can provide bounding estimates for K_P.

59.8 SUMMARY AND CONCLUSIONS

Catastrophic collapse or cascading pillar failure (CPF) is a potential problem faced by all room-and-pillar mining operations. CPF occurs when one pillar fails suddenly, which then overstresses the neighboring pillars causing them to fail, and so forth, in very rapid succession. Within seconds, very large mining areas can collapse via this mechanism while giving little or no warning. The collapse itself poses grave danger to miners. In addition, the collapse can induce a violent air blast that disrupts or destroys the ventilation system. Further grave danger to miners exists if the mine atmosphere becomes explosive as a result of CPF.

This paper has documented over 21 collapses that have occurred in the past 20 years mainly in U.S. room-and-pillar mines. Most of these collapses happened in coal mines since substantial production tonnage still comes from room-and-pillar mines; however, huge collapses have also occurred in various

metal mines (lead and copper) as well as nonmetal mines (trona, salt, and limestone). Many other similar collapses are known to have occurred around the world. CPF, also known as massive pillar collapse, domino-type failure, or progressive pillar collapse, is the likely mechanism underlying these mine failures. The three case histories given in the paper (Figures 59.17, 59.20 and 59.27) show that the risk of CPF is most acute where large arrays of developed pillars exist without interruption by substantial barrier pillars.

Traditional strength-based design methods are not sufficient to eliminate the possibility of CPF in room-and-pillar mines, and the number of documented collapses in the United States alone provides mute testimony to that statement. Pillar arrays with large average strength safety factors can fail in a domino-type failure (CPF) if just a few pillars in the array begin to fail. Pillars with large strength-based safety factors (for example 1.5) still have a finite probability of failure, and if the number of pillars in an array is large, failure somewhere in the array can become a near certainty, and that failure could in turn initiate CPF.

Traditional strength-based design begins by estimating pillar stress using tributary area method, boundary-element-methods or other numerical methods. Next, various empirical pillar strength formulas or rock mass failure criteria such as the Hoek-Brown criterion provide estimates of the peak pillar strength. Finally, a strength-based safety factor is computed as strength over stress. The traditional approach provides required panel pillar size and barrier pillar size for room and pillar layout; however, this approach does not provide panel pillar width nor does it give any consideration to what might happen if pillars somewhere in the array begin to fail. More advanced rock mechanics considerations such as the local mine stiffness stability criterion provide this design information and a rational basis to eliminate domino-type pillar failures or CPF.

The mechanics of CPF are well understood. Strain-softening behavior is the essential mechanical characteristic of pillars that fail rapidly via this mechanism. Pillars that exhibit strain-softening behavior undergo a rapid decrease in load-bearing capacity upon reaching their ultimate strength. The strain-softening behavior of pillars depends on both inherent material properties and geometry. Pillars with low W/H ratio exhibit a greater degree of strain-softening behavior than pillars with a higher W/H ratio, which typically have elastic-plastic or strain-hardening material behavior.

The local mine stiffness stability criterion developed by Salamon (1970) provides a means to distinguish between mine layouts that fail in a stable nonviolent manner and those that fail in an unstable violent manner via CPF. Simple quasi-three-dimensional boundary-element-method programs such as MULSIM/NL or LAMODEL with strain-softening material models can calculate local mine stiffness (K_{LMS}) and evaluate the stability criterion. These computer programs apply to a wide variety of thin, tabular, bedded-type deposits amenable to room-and-pillar mining methods.

Field data on the complete stress-strain behavior of full-scale mine pillars is limited. The best data is for coal (Zipf 1999). Nevertheless, enough is known to evaluate the stability criterion and assess various mine layouts for their potential to fail via CPF. Three case studies of CPF are examined, and all three probably violated the local mine stiffness stability criterion.

Three stability-criterion-based design approaches are suggested to minimize the risk of CPF, namely, containment, prevention, and full extraction. If an array of pillars violates the local mine stiffness stability criterion, the containment approach applies as shown in Figure 59.11. Low W/H ratio panel pillars that violate the stability criterion are surrounded by high W/H ratio barrier pillars that shield the panel pillars from full tributary area stresses and "contain" panel pillar failure should it initiate. However, if all the panel pillars in an array satisfy the stability criterion, then the prevention approach applies. The panel pillars do not exhibit much strain-softening behavior because their W/H ratio is sufficiently high (probably greater than 3 or 4). In the full extraction approach, the stability issue becomes a moot point, because complete and controlled opening closure occurs immediately after the completion of retreat mining.

Large mine collapses can pose enormous safety hazards to miners and room-and-pillar mining operations. Mining engineers can limit the danger of CPF through prudent application of the local mine stiffness stability criterion and the three stability-criterion-based design approaches suggested to decrease the risk of CPF.

59.9 REFERENCES

Abel, J.F. (1988). Soft Rock Pillars. *Int. J. Min. & Geol. Eng.,* Vol. 6, pp. 215–248.

Aggregate Handbook. (1991). National Stone Association, Washington, D.C.

Bieniawski, Z.T. (1967). The Mechanism of Brittle Fracture of Rock. *Int. J. of Rock Mech. and Min. Sci.* Vol. 4, pp. 396–435.

Bieniawski, Z.T. (1968a). The Effect of Specimen Size on Compressive Strength of Coal. *Int. J. of Rock Mech. and Min. Sci.* Vol. 5, pp. 325–335.

Bieniawski, Z.T. (1968b). In Situ Strength and Deformation Characteristics of Coal. *Engng. Geol.* Vol. 2. pp. 325–340.

Bieniawski, Z.T. (1992). Ground Control. *SME Mining Engineering Handbook,* 2nd Edition, H.L. Hartman, Ed., Society for Mining, Metallurgy, and Exploration, Inc., Littleton, CO, pp. 897–937.

Brady, B.H.G. and E.T. Brown. (1993). Rock Mechanics for Underground Mining. Kluwer Academic Publishers, Boston, 571 pp.

Bryan, A., J.G. Bryan, and J. Fouche. (1966). Some Problems of Strata Control and Support in Pillar Workings. *The Mining Engineer,* Vol. 123, pp. 238–254.

Bullock, R.L. (1982). General Mine Planning. *Underground Mining Methods Handbook,* W.A. Hustrulid, ed., SME of AIME, New York, pp. 113–137.

Bunting, D. (1911). Chamber Pillars in Deep Anthracite Mines. *Transactions,* AIME, Vol. 42, pp. 235–245.

Chase, F.E., R.K. Zipf, and C. Mark. (1994). The Massive Collapse of Coal Pillars—Case Histories from the United States. *Proc. 13th Int. Conf. on Ground Control in Mining,* West Virginia University, Morgantown, WV, pp. 69–80.

Chase, F.E. et al. (1996). Practical Aspects of Mobile Roof Support Usage. *Proc. 15th Int. Conf. Ground Control in Mining.* Golden, CO: Colorado School of Mines.

Coal Data 1999 Edition, National Mining Association, Washington, D.C.

Coates, D.F. (1981). Rock Mechanics Principles. CANMET Monograph 874, Ottawa, Canada.

Cook, N.G.W. and J.P.M. Hojem. (1966). A Rigid 50-Ton Compression and Tension Testing Machine. *J. S. Afr. Inst. Mech. Engr.* Vol. 1, pp. 89–92.

Das, M.N. (1986). Influence of Width/Height Ratio on Postfailure Behavior of Coal. *Int. J. of Mining and Geological Engineering.* Vol. 4, pp. 79–87.

Davidson, J. (1987). Ground Stability Evaluation—Troy Mine—ID No. 24-01467. Mine Safety and Health Administration, Denver, CO. 15 pp.

Denk, J.M., J.J. Jansky, M.T. Hoch, G.J. Karabin, and D.T. Kirkwood. (1994). Accident Investigation Report—Akzo Nobel Salt, Inc. ID. No. 30-00662. Mine Safety and Health Administration, Pittsburgh, PA. 141 pp.

Dismuke, S.R., W.W. Forsyth, and S.B.V Stewart. (1994). The Evolution of Mining Strategy Following the Collapse of the Window Area at the Magmont Mine, Missouri. *Proceedings of District 6 CIM Annual General Meeting.* Metal Mining, pp. 3–8.

Dravo Corp. (1974). Analysis of Large Scale Non-coal Underground Mines. U.S. Bureau of Mines Contract Report SO 122-059, 605 pp.

Examine[TAB]. 2000. Rocscience Inc., Toronto, Canada, (http://www.rockscience.com).

Farmer, I.W. (1985). Coal Mine Structures. Chapman and Hall, London, 310 pp.

Farmer, I.W. (1992). Room and Pillar Mining. *SME Mining Engineering Handbook,* 2nd Edition, H.L. Hartman, Ed., Society for Mining, Metallurgy, and Exploration, Inc., Littleton, CO, pp. 1681–1701.

Galvin, J. (1992). A Review of Coal Pillar Design in Australia. *Proc. of the Workshop on Coal Pillar Mechanics and Design,* U.S. Bureau of Mines IC 9315, pp. 196–213.

Greenwald, H.P., H.C. Howarth and I. Hartman. (1939). Experiments on the Strength of Small Pillars of Coal. U.S. Bureau of Mines, Technical Paper 605.

Hardy, M.P. and J.F.T Agapito (1977). Pillar Design in Underground Oil Shale Mines. *Proceedings 16th U.S. Rock Mechanics Symposium,* University of Minnesota, Minneapolis, MN, pp. 257–266.

Heasley, K.A. (1997). A New Laminated Overburden Model for Coal Mine Design. *Proceedings: New Technology for Ground Control in Retreat Mining,* NIOSH IC 9446, pp. 60–73.

Heasley, K.A. (1998). Numerical Modeling of Coal Mines with a Laminated Displacement-Discontinuity Code. Ph.D. Dissertation, Colorado School of Mines, 187 pp.

Hoek, E. and E.T. Brown. (1980). Underground Excavations in Rock. Institution of Mining and Metallurgy, London, 527 pp.

Hoek, E., P.K. Kaiser and W.F. Bawden (1995). Support of Underground Excavations in Hard Rock. A.A. Balkema, Rotterdam, 215 pp.

Holland, C.T. (1964). Strength of Coal in Mine Pillars. *Proceedings 6th U.S. Symposium on Rock Mechanics,* University of Missouri, Rolla, pp. 450–466.

Itasca. (1995). FLAC—Fast Lagrangian Analysis of Continua—Version 3.3—User's Manual. Itasca Consulting Group, Inc., Minneapolis, MN.

Jaeger, J.C. and N.G.W. Cook. (1979). Fundamentals of Rock Mechanics. Chapman and Hall, 3rd. ed. 593 pp.

Khair, A.W. and S.S. Peng (1985). Causes and Mechanisms of Massive Pillar Failure in a Southern West Virginia Coal Mine. *Mining Engineering,* Vol. 37, No. 4, pp. 323–328.

Knoll, P. (1990). The Fluid-Induced Tectonic Rockburst of March 13, 1989 in the "Werra" Potash Mining District of the GDR (First Results). *Gerl. Beitr. Geophysik–Leipzig,* Vol. 99, No. 3, pp. 239–245.

Koehler, J.R. and S.C. Tadolini. (1995). Practical Design Methods for Barrier Pillars. U.S. Bureau of Mines IC 9427, 19 pp.

Mark, C. and A.T. Iannacchione. (1992). Coal Pillar Mechanics: Theoretical Models and Field Measurements Compared. *Proc. of the Workshop on Coal Pillar Mechanics and Design,* U.S. Bureau of Mines IC 9315, pp. 78–93.

Mark C., F.E. Chase, and A.A. Campoli. (1995). Analysis of Retreat Mining Pillar Stability. *Proc. 14th Conf. Ground Control in Mining,* West Virginia University, pp. 63–71.

Mark, C. and F.E Chase. (1997). Analysis of Retreat Mining Pillar Stability (ARMPS). *Proceedings: New Technology for Ground Control in Retreat Mining,* NIOSH IC 9446, pp. 17–34.

Mark, C. and T.M. Barton (1997). Pillar Design and Coal Strength. *Proceedings: New Technology for Ground Control in Retreat Mining,* NIOSH IC 9446, pp. 17–34.

Mark, C. (1999). Empirical Methods for Coal Pillar Design. *Proceedings of the Second International Workshop on Coal Pillar Mechanics and Design,* NIOSH IC 9448, pp. 145–154.

Mine Safety and Health Administration, 1996. "Report of Technical Investigation, Underground Non-metal Mine, Mine Collapse Accident", Solvay Mine, Solvay Minerals Inc., Green River, Sweetwater County, Wyoming, February 3, 1995.

Mokgokong, P.S. and S.S. Peng (1991). Investigation of Pillar Failure in the Emaswati Coal Mine, Swaziland. *Mining Sci. and Tech.* Vol. 12, pp. 113–125.

Obert, L. and W.I. Duvall, (1967). Rock Mechanics and the Design of Structures in Rock. Wiley, New York, 650 pp.

Richmond, R. (1998). Report of Investigation (Underground Coal Mine), Nonfatal Roof-Fall Accident, Red Oak Mine, ID. No. 46-08135, Mine Safety and Health Administration, District 4, Mount Hope, West Virginia.

Ropchan, D. (1991). Ground Support Evaluation—Belinda No. 2 Mine—ID No. 42-01280. Mine Safety and Health Administration, Denver, CO. 12 pp.

Salamon, M.D.G. and A.H. Munro. (1967). A Study of the Strength of Coal Pillars. *Journal of the South African Institution of Mining and Metallurgy,* Vol. 68, pp. 55–67.

Salamon, M.D.G. (1970). Stability, Instability, and Design of Pillar Workings. *Int. J. of Rock Mech. and Min. Sci.* Vol. 7, pp. 613–631.

Sheorey, P.R., D. Barat, K.P. Mukherjee, R.K. Prasad, M.N. Das, G. Banerjee and K.K. Das (1995). Application of the Yield Pillar Technique for Successful Depillaring Under Stiff Strata. *Int. J. of Rock Mech. and Min. Sci. & Geomech. Abstr.* Vol. 32, No. 7, pp. 699–708.

Skelly, W. A., J. Wolgamott, and F. D. Wang. (1977). Coal Pillar Strength and Deformation Prediction Through Laboratory Sample Testing. *Proc. 18th U.S. Rock Mechanics Symposium*, Col. Sch. Mines, Golden, CO, pp. 2B5-1 to 2B5-5.

Spruell, J.L. (1992). Accident Investigation Report—Unimin Specialty Minerals, Inc. Birk 2A–ID No. 11-02598. Mine Safety and Health Administration, Duluth, MN. 12 pp.

Straskraba V. and J.F. Abel (1994). The Differences in Underground Mines Dewatering with the Application of Caving or Backfilling Mining Methods. *Mine Water and the Environment*, Vol. 13, No. 2, pp. 1–20.

Swanson, P.L. and F. Boler (1995). The Magnitude 5.3 Seismic Event and Collapse of the Solvay Trona Mine: Analysis of Pillar/Floor Failure Stability. U.S. Bureau of Mines OFR 86-95, 82 pp.

Touseull J. and C. Rich (1980). Documentation and Analysis of a Massive Rock Failure at the Bautsch Mine, Galena, Ill. U.S. Bureau of Mines RI 8453, 49 pp.

USGS Mineral Industry Surveys. (1998). U.S. Geological Survey, Reston, VA.

Van Heerden, W.L. (1975). In Situ Determination of Complete Stress-Strain Characteristics of Large Coal Specimens. *J. S. Afr. Inst. Min. and Metall.,* Vol. 75, No. 8, pp. 207–217.

Wagner, H. (1974). Determination of the Complete Load-Deformation Characteristics of Coal Pillars. *Proc. of 3rd Int. Congress on Rock Mechanics*, Natl. Acad. Sci. Vol. 2B, pp. 1076–1081.

Wawersik, W.R. and C. Fairhurst (1970). Study of Brittle Rock Fracture in Laboratory Compression Experiments. *Int. J. Rock Mech. and Min. Sci.* Vol. 7, pp. 561–575.

Zipf, R.K. (1992a). *MULSIM/NL Theoretical and Programmer's Manual.* U.S. Bureau of Mines IC 9321, 52 pp.

Zipf, R.K. (1992b). *MULSIM/NL Application and Practitioner's Manual.* U.S. Bureau of Mines IC 9321, 48 pp.

Zipf, R.K. (1996). Analysis and Design Methods to Control Cascading Pillar Failure in Room-and-Pillar Mines. *Milestones in Rock Engineering,* Z.T. Bieniawski ed. Rotterdam: Balkema, pp. 225–264.

Zipf, R.K. (1999). Catastrophic Collapse of Highwall Web Pillars and Preventative Design Methods. *Proc. 18th Int. Conf. on Ground Control in Mining*, West Virginia University, Morgantown, WV, pp. 18–28.

Zipf, R.K. and P.L. Swanson. (1999). Description of a Large Catastrophic Failure in a Southwestern Wyoming Trona Mine. *Proc. 37th U.S. Rock Mechanics Symposium*, Vail, CO, pp. 293–298.

CHAPTER 60

The Stability Graph Method for Open-Stope Design

Yves Potvin[*] and John Hadjigeorgiou[†]

60.1 INTRODUCTION

In the late 1970s to early 1980s, the underground metal-mining industry shifted its extraction strategy from highly selective "entry" methods, such as cut-and-fill, to "non-entry" methods such as open stoping. A review of Canadian practice has shown that 90% of the total production of underground metal mines, based on reported tonnage, rely on open-stope mining methods (Poulin et al. 1995). The popularity of open-stope operations can be attributed to the higher productions levels achieved by employing larger excavations and using mechanised equipment.

Considering the high cost of developing each stope, there is a significant incentive to produce a smaller number of large open stopes. The consequences, however, of exceeding the maximum critical stable dimensions of a stope can be disastrous. Instability around open stopes may require large remedial costs for ground rehabilitation, production delays, mining equipment loss, ore reserves loss, and, at the extreme, injuries or fatalities to mine workers.

Pakalnis (1986) reports that in a survey of 15 Canadian mines, almost half (47%) of the open-stope mines had more than 20% dilution with one fifth suffering excessive dilution of over 35%. Based on field data from 34 Canadian mines, Potvin (1988) demonstrated that open-stope design was based on past experience of mine operators in similar mining conditions and on trial and error. Consequently, it can be argued that the reported high dilution rates in the early 1980s could be attributed to the absence of comprehensive engineering design tools. It follows that there are significant economic gains to be made by improving open-stope stability.

60.2 EXCAVATION STABILITY

Evaluating the stability of a non-entry excavation such as a stope can be subjective. Unlike entry excavations in which mine workers have access, isolated rock falls in stopes are generally of no consequence, providing that they can be handled by mucking units. Therefore, a stope can be considered to be stable if it yields low dilution (less than 5%) and if there are no ground-fall-related operational problems. It has been argued (Pakalnis, Poulin, and Hadjigeorgiou 1995) that there is a unique acceptable dilution rate for every mine operation. This is defined as a function of ore grade, costs, grade of dilution material, and metal prices. Consequently, provided the operation remains safe and economical, it is possible to tolerate a level of dilution and a degree of instability for every stope. Open stopes that display excessive dilution and/or unmanageable stability problems are often referred to as caved. In this context, the term "caved" indicates major stability problems and should not be confused with the cave mining interpretation where it refers to orebody failure (cavings) after undercutting. This overlap of terminology has, from time to time, been the source of confusion amongst people not familiar with open-stope mining.

There are multiple and interrelated factors that potentially contribute to the instability of excavations. For convenience, they can be divided into two groups: the ones related to the in-situ conditions prevailing before mining, and the factors related to the disturbance of these conditions induced by mining.

The premining conditions can be characterised by rock-mass classification schemes and supplemented with structural geology data and an estimation of the in-situ stress field. The major factors related to mining are the size, shape, and orientation of excavations as well as the ground support used (including backfill). Blast damage and the effect of time in highly convergent rock may also affect the stability of excavations.

60.3 DESCRIPTION OF THE STABILITY-GRAPH METHOD

The stability-graph method is an empirical method for open-stope design (non-entry excavations). It aims to account for and quantify the major factors influencing the stability of open stopes. A stability index for each stope surface is subsequently traced against its dimensions. A series of empirically derived guidelines allow for predictions on the overall stability of a stope. Since its introduction (Mathews et al. 1981), it has gained wide acceptance and is used worldwide as a design tool. There are documented case studies of the method being used in Africa, Europe, and the United States, and extensive databases of case studies in Canada and Australia. In practice, the stability graph can be employed during three distinct mining stages. Its primary use is during the feasibility stage but it has also been found useful during individual stope planning. Finally, through the use of back analysis, it provides an index of stope performance and allows the mine operation to develop remedial strategies where warranted.

The method traces its origin to the recognition that traditional rock-mass classification and design tools were based on tunnelling case studies. A review of some case studies and engineering judgement resulted in the first version of the method, whereby a stability number (N) was traced against the hydraulic radius of a stope surface.

The stability-graph method uses the NGI tunnelling index Q (Barton, Lien, and Lunde 1974) as a basis for estimating rock-mass quality.

$$Q = \frac{RQD}{J_n} \times \frac{J_r}{J_a} \times \frac{J_w}{SRF}$$

where:

Q = NGI tunnelling index with

RQD = rock quality designation

---

* Australian Center for Geomechanics, Nedlands, WA, Australia.
† Laval University, Quebec City, Quebec, Canada.

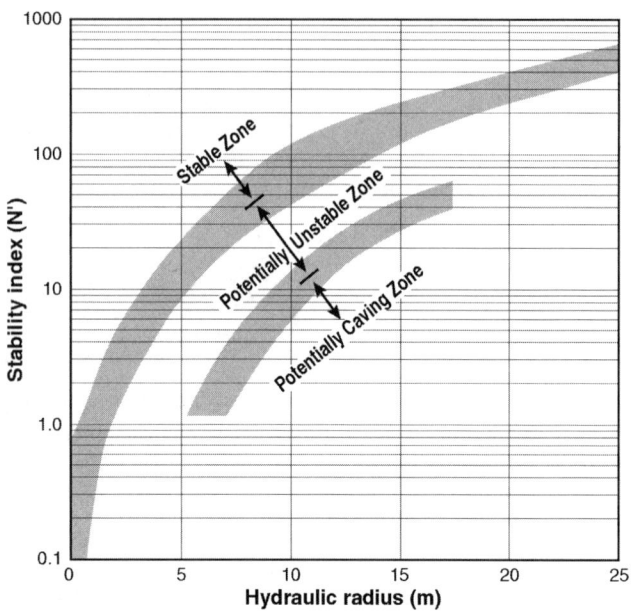

FIGURE 60.1   Stability graph, after Mathews et al. (1981)

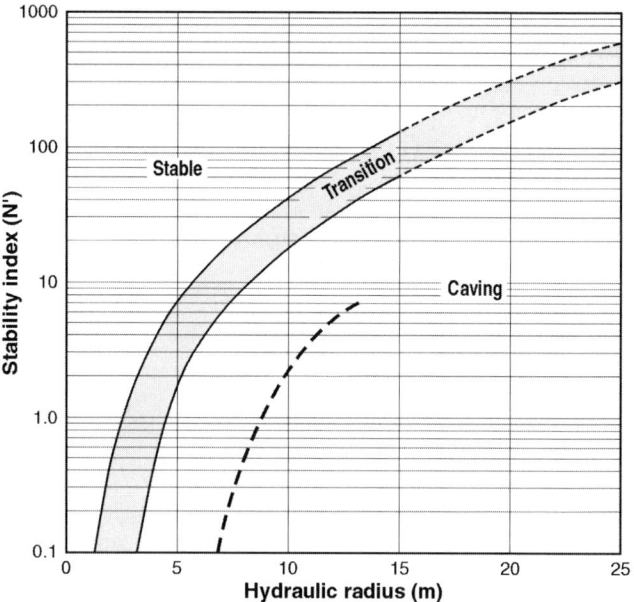

FIGURE 60.2   Stability graph after Potvin (1998)

J_r = joint roughness number

J_w = joint water reduction number

J_n = joint set number

J_a = joint alteration number

SRF = stress reduction factor

Using SRF equal to 1 is a departure from the original system (Barton, Lien, and Lunde 1974). This modified tunnelling index, Q, is further adjusted to account for stress, rock defect orientation, and design-surface orientation factors to arrive at a stability number N. The stability number was plotted against the hydraulic radius (surface area/perimeter) of the studied surface of an excavation (Mathews et al. 1981). Three zones of potentially stable, unstable, and caving were proposed with reference to the predicted stability of an excavation (see Figure 60.1).

In its early days, a major shortcoming of the method was that it was backed by limited field data—26 case studies from three mines. Once the database was expanded to 175 cases from 34 mines and the stability graph modified (Potvin 1988), the method rapidly gained wide acceptance in the Canadian mining industry. The transition zone from stable to unstable was reduced significantly, thus removing some of the subjectivity in using the design chart (see Figure 60.2). It should be noted that in the Potvin database, the adjustment factors were different than those proposed by Mathews et al. (1981). This resulted in what is commonly referred to as the modified stability-graph method using a stability index N'.

$$N' = Q' \times A \times B \times C$$

where:

N' = stability number

Q' = modified tunnelling quality index (NGI)

A = stress factor

B = joint orientation factor

C = gravity factor

A-Factor.   The A-factor is used to account for the resulting induced stress in the investigated stope surface. A series of charts that provided preliminary estimates of induced stresses for

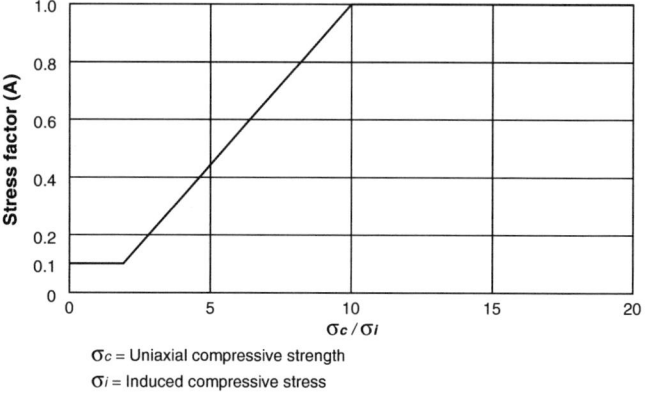

σc = Uniaxial compressive strength

σi = Induced compressive stress

FIGURE 60.3   Determination of the stress factor, after Potvin (1988)

different stope configurations and conditions were proposed (Mathews et al. 1981 and Potvin 1988) (Figure 60.3). These charts have now been superseded by ready access to 3-D numerical modelling stress packages.

B-Factor.   The B-factor considers the orientation of the most critical structure relative to the stope walls (Figure 60.4). The critical structure could be a joint set, bedding planes, or foliation. This is based on the observation that most structurally controlled failures occur when joints form a shallow angle with the stope surface.

C-Factor.   The C-factor accounts for the mode of failure. In open stopes, structurally controlled failures are recognised as gravity fall, slabbing, and sliding. Figure 60.5 demonstrates the influence of gravity for slabbing and gravity modes of failure.

As more case studies became available, it was noted that the adjustment factor for sliding could, under certain conditions, prove inadequate for footwall stability predictions. Following a mechanistic analysis and review of case-by-case field data, the following modifications, shown in Figure 60.6, were made (Hadjigeorgiou, Leclair, and Potvin 1995).

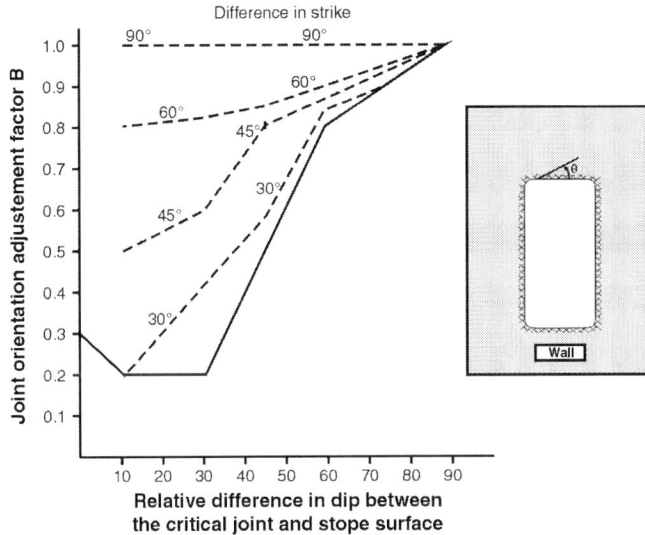

FIGURE 60.4 Determination of the Orientation Factor, after Potvin (1988)

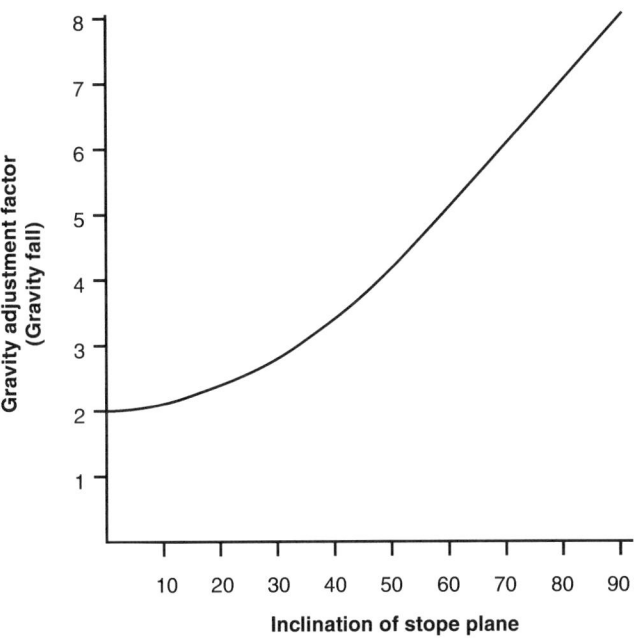

FIGURE 60.5 Influence of gravity for slabbing and gravity fall modes of failure

60.4 DISCUSSION OF THE INPUT FACTORS

As a result of wide dissemination in a number of textbooks (Hoek, Kaiser, and Bawden 1995, Hutchinson and Diederichs 1996) the input factors for the calculation of N' described above have now gained broad acceptance from practitioners and researchers. The applicability of the input methodology on a case-by-case analysis was reviewed and, with the exception of the minor modifications to the C factor shown in Figure 60.6, were found to be appropriate (Hadjigeorgiou, Leclair, and Potvin 1995). On the other hand, some authors (Stewart and Forsyth 1995; Trueman et al. 2000) have indicated their preference for the formulation of the input factors as originally proposed (Mathews et al. 1981).

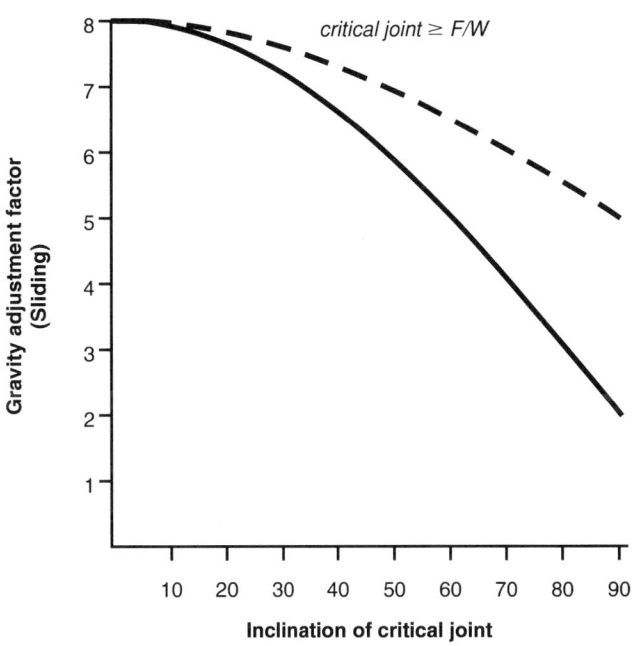

FIGURE 60.6 Influence of gravity for sliding mode of failure, after Hadjigeorgiou Leclair and Potvin (1995)

Several other modifications to the stability graph have been proposed during the last decade. The following offers a brief historical review. It should be noted that most of these proposals have not yet been extensively tested by case studies nor are they widely employed by practitioners.

Scoble and Moss (1994) suggested that there was merit in adding two further adjustment factors, D for blasting and E for sublevel interval rating with some tentative factors proposed. A fault factor was been developed that can be incorporated into the stability factor (Suorineni, Tannant, and Kaiser 1999). This fault factor accounts for the angles between fault and stope surface and the position where the fault intersects the stope surface. The fault factor was derived based on modelling and demonstrated that it could be critical for a series of documented case studies in Canada and Africa. At the Golden Giant Mine in Ontario, Canada, it was shown that under high-stress environments the introduction of a stress-damage factor merited attention (Sprott et al. 1999). Based on 3-D numerical modelling, they used the "extra stress deviator," the uniaxial resistance of the rock, and the hydraulic radius to define a stress-damage factor. It has been argued that the stability predictions of the stability-graph method may prove inaccurate due to the influence of rock-mass degradation and relaxation (Kaiser et al. 1997). It was recommended that stope sequencing be used as a tool to minimise stress-induced rock-mass degradation and to minimise stress relaxation. In their work, they defined rock-mass relaxation as stress reduction parallel to the excavation wall—not to stress reductions in the radial or a reduction in confinement. Rock-mass degradation was quantified as loss of rock-mass strength.

60.5 HYDRAULIC RADIUS

The term "hydraulic radius" has been used in the past to characterise the size and shape of stope surfaces (Laubscher 1977). This is the area over the perimeter of a given stope surface. It has also been demonstrated that, despite the advantages of hydraulic radius over span, it still has important limitations (Milne, Pakalnis, and Felderer 1996). In particular, when applied to irregularly shaped stope surfaces, it is possible to arrive at the

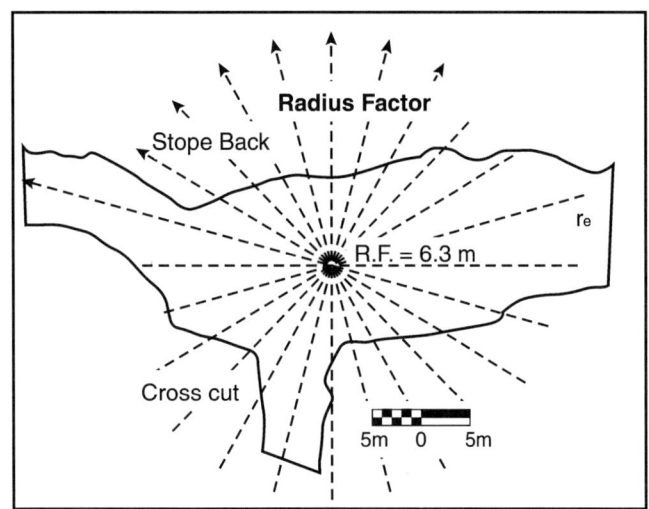

FIGURE 60.7 Determination of the radius factor, after Milne et al. (1996)

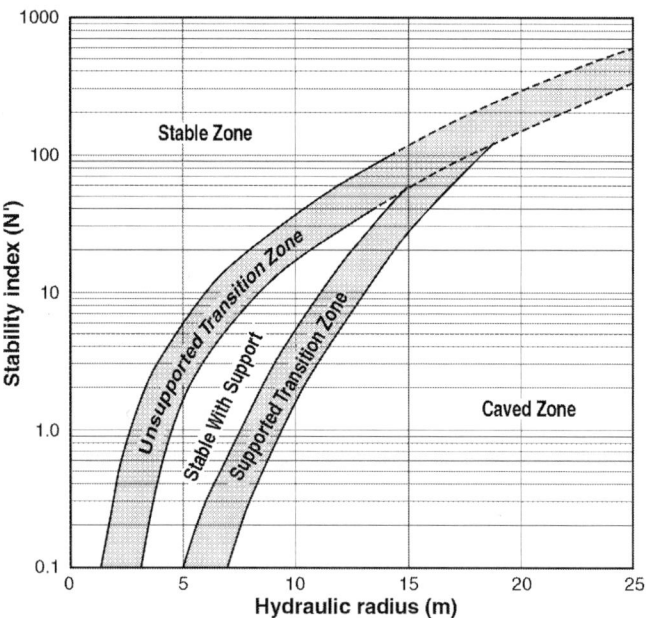

FIGURE 60.8 Stability graph, after Nickson (1992)

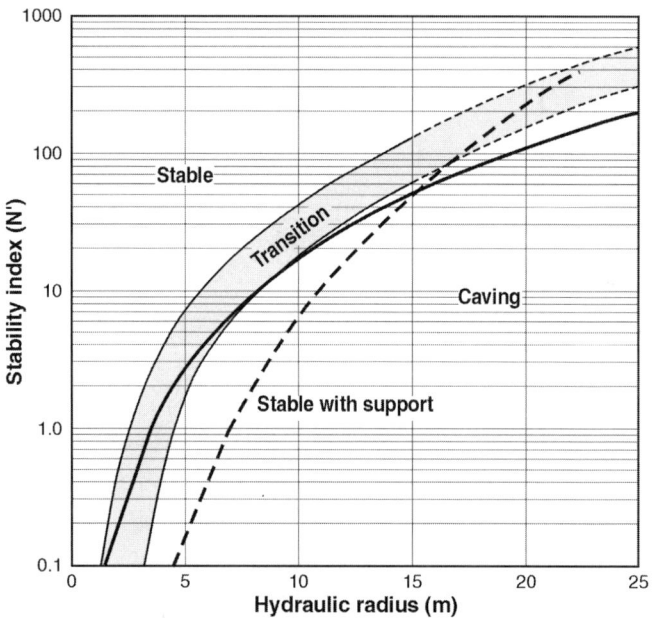

FIGURE 60.9 Stability graph design lines as developed by Hadjigeorgious et al. (1995)

same hydraulic radius. It has been put forward that a better way to describe the geometry of an irregularly shaped excavation is the radius factor (see Figure 60.7). This is determined by identifying the centre of any excavation and by taking distance measurements to abutments at small regular increments:

$$RF = \frac{0.5}{\frac{1}{n}\sum_{\theta=1}^{n} \frac{1}{r_{\theta}}}$$

where

r_{θ} = distance from the surface centre to the abutments at angle q

n = number of rays measured to the surface edge

In principle, the radius factor can be determined at any point on a surface. If the centre cannot be determined, a series of calculations are possible with the maximum value assumed to be the radius factor. Despite its somewhat cumbersome definition, the radius factor can easily be calculated by a routine integrated into a computerised design package.

60.6 DESIGN CHARTS

In reviewing the proposed chart, Figure 60.1 (Mathews et al. 1981), it can be seen that the developed guidelines were somewhat vague for design purposes. This was because there was insufficient data to provide more accurate zone definition. As more case studies became available, a narrower transition zone and a "support requirement zone" were defined (Potvin 1988). This has allowed for a calibrated and more versatile design tool (Figure 60.2). A more comprehensive statistical analysis further modified the support zones by introducing lines indicating where cable bolting could be used (see Figure 60.8) (Nickson 1992). A review of a larger database (Hadjigeorgiou, Leclair and Potvin 1995) confirmed the general validity of previous work (Potvin 1988, Nickson 1992) within statistical limits.

It should be noted that the work of Hadjigeorgiou et al. (1995) demonstrated that, for larger stopes with a hydraulic radius greater than 15, the design curve was in fact flatter (see Figure 60.9). More recent work in the United Kingdom by Pascoe et al. (1998) and in Australia by Trueman et al. (2000) has confirmed the same trends.

A series of design guidelines were proposed (Stewart and Forsyth 1995) that allowed for a finer definition of the types of stope failure, distinguishing between potentially unstable, potentially major, and caving failure separated by transition zones (see Figure 60.10). In their experience, the boundary between stable and unstable is clear cut, while the transition between unstable and major failure is not as well defined. It is of interest that the transition between a "potentially stable zone" and a "potentially unstable zone" is identical to Potvin's transition zone. In practice, it could be argued that, for open-stope design purposes, it is somewhat irrelevant to subdivide the area defining stope failure into three zones as the objective is to design stable stopes.

Cavity monitoring laser surveys have been employed to back-analyse the resulting volumetric measurements of overbreak/

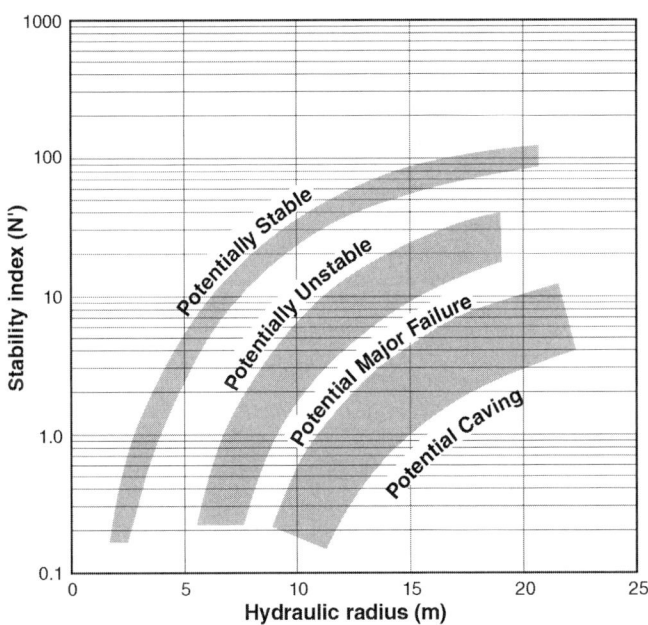

FIGURE 60.10   Stability graph, after Stewart & Forsyth (1993)

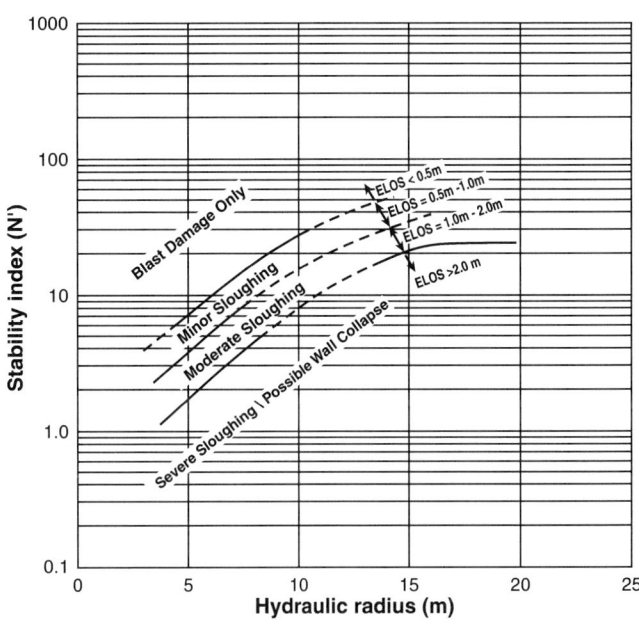

FIGURE 60.12   Estimation of overbreak/slough for nonsupported hangingwalls and footwalls, after Clark and Pakalnis 1997

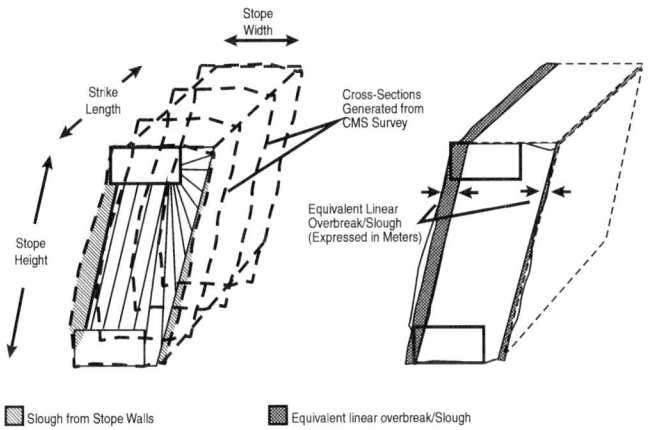

FIGURE 60.11   Schematic definition of the ELOS parameter, after Clark and Pakalnis 1997

slough and underbreak, and a new index has been proposed (Clark and Pakalnis 1997) (see Figure 60.11):

$$\text{ELOS} = \frac{\text{equivalent linear overbreak}}{\text{slough}}$$

$$= \frac{\text{volume of slough from stope surface}}{\text{stope height} \times \text{wall strike length}}$$

In Figure 60.12, ELOS has been integrated in the stability graph, providing a series of design zones (Clark and Pakalnis 1997). Although this data presentation does not account for the influence of support, it provides a useful back-analysis tool for hanging walls and footwalls in a low- or relaxed-stress state and with parallel geological structure being present.

All of the above graphs rely on arbitrarily drawn design curves. The first comprehensive statistical analysis of the then-available field data (Nickson 1992) clearly demonstrated the applicability of the modified stability graph (Potvin 1988) and laid the foundations for further statistical work (Hadjigeorgiou,

Leclair, and Potvin 1995, Pascoe et al. 1998, and Suorineni 1998).

Successful applications of the stability graph recognise that the method remains subjective. Despite using quantifiable values, the precise degree of inherent conservatism is not known. It reflects "current and past practice," which may have been influenced by legislation, local practices, or geological peculiarities, and does not necessarily constitute an optimum design methodology.

60.7   RISK ANALYSIS

It has been argued that the design of non-entry excavations lends itself to risk analysis much more than the design of access ways where worker safety is the major concern (Pine et al. 1996, Pascoe et al. 1998, Diederichs and Kaiser 1996). There are two basic elements in risk analysis. The first one deals with input variability and the second deals with calibration uncertainty. For practical purposes, the major challenge lies in defining how much risk is acceptable for design purposes. Using risk probability procedures for fine-tuning or calibrating site-specific design guidelines, while attractive, is hindered in that site-specific calibrated guidelines will be validated only towards the end of mine life. At that time, their impact will be limited to providing a better understanding of particular field conditions, but it may be too late to implement major design changes.

60.8   SUPPORT RECOMMENDATIONS

Potvin (1988) first addressed the influence of support on the stability of open stopes. The area of the graph that could successfully be supported by cable bolts was refined, and a series of design recommendations made on cable-bolting patterns (Nickson 1992). The basic concept is that there is a zone where support cannot be effectively used to stabilise the excavation. It has been shown (Hadjigeorgiou, Leclair, and Potvin 1995) that the actual supportable zone was smaller than predicted (Nickson 1992).

A design chart is available to select a suitable cable-bolt density as a function of relative block size (RQD/J_n) and the hydraulic radius (Potvin and Milne 1992). This graph, slightly modified in Figure 60.13, is most appropriate for stope backs. It has also been employed for hanging-wall reinforcement design, provided a systematic and regular cable-bolt pattern is used.

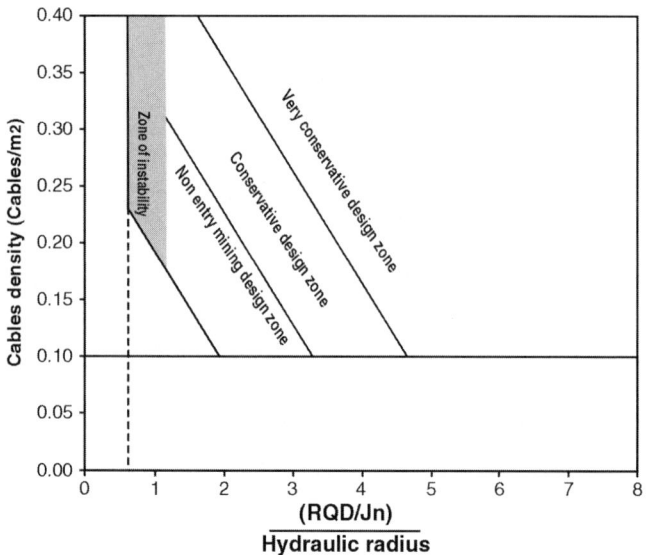

FIGURE 60.13   Determination of cable bolt density

At the time when the design curves for cable reinforcement were developed, support often consisted of single plain-strand cables. Recent years have seen a shift towards double-strand and modified-geometry cables as they provide higher support capacity. This is achieved in the presence of sufficiently high ground deformation whereby the strength of the steel is being mobilized. In other words, pattern reinforcement is designed to help the rock support itself and not necessarily to support the dead weight of the rock. A series of semi-empirical design charts to determine the design spacing for both single- and double-strand cable bolts has been proposed (Diederichs, Hutchinson, and Kaiser 1999). It should be noted, however, that there are no documented case studies confirming their application.

60.9   LIMITATIONS OF THE STABILITY GRAPH

All empirical methods are limited in their application to cases that are similar to the one used in the developmental database. Therefore, the stability graph is inappropriate in severe rock-bursting conditions, in highly deformable (creeping) rock mass, and for entry methods. Since its introduction, the stability graph method has been the subject of extensive efforts to expand its applicability to better account for the presence of faults, blast damage, and stress damage. Unfortunately, some of the proposed modifications are not supported by field data. Furthermore, when merging databases from diverse sources, it is necessary to verify the quality of collected data. In particular, the practice of using empirical correlations to convert from one rock-mass classification system to another should only be used as a last resort and even then with great caution.

For all practical purposes, the stability graph can be used during the feasibility stage, during individual stope planning, and for stope reconciliation or back analyses.

60.10   DESIGN CONSIDERATIONS DURING THE FEASIBILITY STAGE

The determination of adequate stope dimension is one of the most critical decisions to be made at the feasibility study stage of a mine. The profitability of an operation is directly linked to productivity, which in turn, is influenced by stope dimensions. Validation of stope-design methodology can begin once the first stope is extracted. By this time, however, the mine infrastructure is already in place, allowing for no or only minor modifications to

design stope dimensions. This emphasizes the importance of developing a reliable stope-design methodology at the earliest possible stage.

Many practitioners have reported on the reliability of the stability-graph method during the last 12 years (Reschke and Romanowski 1993, Bawden 1993, Pascoe et al. 1998, Dunne et al. 1996, and Goel and Wezenberg 1999). When properly used, the method provides a good "ball park" estimate of stable stope dimensions under different conditions. The major limitation at the feasibility study stage is the availability of quality geotechnical data. This is a concern for all design methods. Consequently, it is essential to optimise all available data while being fully aware of any limitations. The following guidelines can facilitate the estimation of realistic stability numbers during the feasibility stage.

An integral part of the stability method is the quantification of rock-mass quality based on the Q system. At the "green field" stage, the majority of geomechanical data are derived from boreholes. Consequently, it is possible to develop a comprehensive database of RQD readings, which can easily be integrated into geological models easily accessed by both planning and rock mechanics. It is strongly suggested that the number of joint sets be determined by using oriented diamond-drill cores in the orebody.

There are several case studies where core data were used to derive representative Q readings for underground mines (Milne, Germain and Potvin 1992, Germain, Hadjigeorgiou, and Lessard 1996). This has included a simplified approach to determine joint alteration, J_a. If the joint cannot be scratched with a knife, J_a is assumed to be equal to 0.75, and if it is possible to scratch, it varies from 1.0 to 1.5. When a joint feels slippery to touch and can scratched with a fingernail, J_a is equal to 2; and when it is possible to indent with a fingernail, J_a is equal to 4. The joint roughness parameter (J_r) is more difficult to assess on a small exposed surface of a core. However, in most cases, it is possible to estimate whether the surface is smooth or rough. In the absence of reliable data, joints are assumed to be planar. This allows for J_r values of 0.5 for slickenslide planar, 1.0 for smooth planar, and 1.5 for rough planar joints.

Factor A can generally be assumed to be equal to 1 for all stope walls, unless mining is to proceed very deep (say 1,000 m and deeper). As a first-pass estimation, the stress induced in stope backs could be assumed to be around 1.5 times the pre-mining horizontal stress for transversal mining (mining across the strike of the orebody). In longitudinal mining (mining along strike), a rough estimate of the induced stress in the back can be obtained by doubling the premining horizontal stress perpendicular to the orebody strike. The premining stress can be measured if underground access is available or otherwise based on regional data. The uniaxial compressive strength of rock is easily obtained by standard laboratory tests on cored rock. A larger database of UCS values can also be gathered at low cost using a standard point load test. When at least some oriented core is available, Factor B can be estimated. In the absence of joint orientation data, a minimum value of 0.2 is assumed. The estimation of factor C is independent of ground conditions and is, therefore, straightforward to determine, even at the feasibility stage.

A good methodology for the construction of a geomechanical model and the application of the stability graph method for mine feasibility assessment exists (Nickson et al. 1995). Stability numbers are calculated for back and walls and displayed on mine sections. For each stability number (N), a hydraulic radius (S) is determined from the stability graph in Figure 60.14, using the upper section of the transition zone. The option of increasing stope dimensions exists if a systematic pattern of cable-bolt support is to be used. This can be assessed using Figure 60.8.

Unless the ground conditions are consistent throughout the orebody, a number of stable hydraulic radii will be produced, which can be grouped into domains and displayed on a longitudinal section. Because most mines employ systematic layouts, a

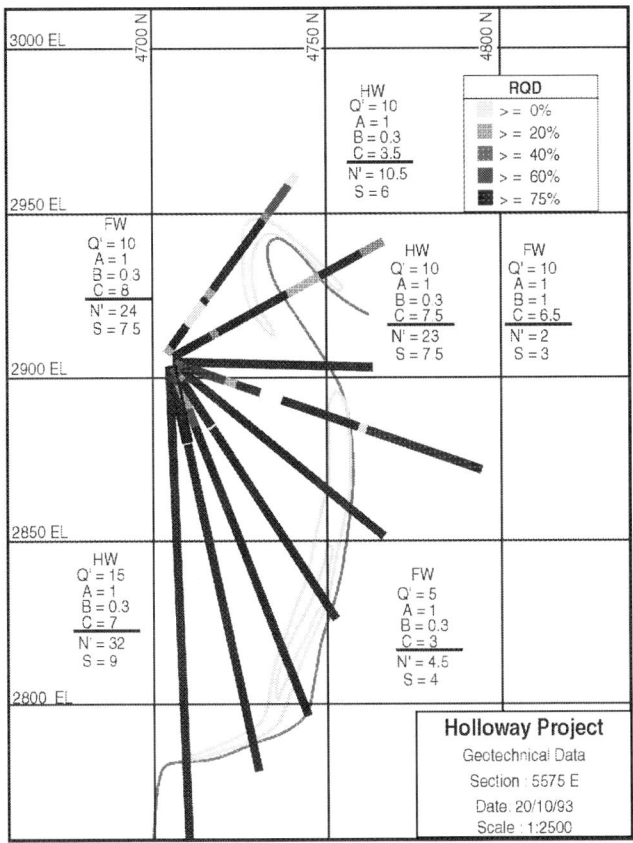

FW
Q' = 10
A = 1
B = 0.3
C = 8
N' = 24
S = 7.5

HW
Q' = 10
A = 1
B = 0.3
C = 3.5
N' = 10.5
S = 6

RQD	
	>= 0%
	>= 20%
	>= 40%
	>= 60%
	>= 75%

HW
Q' = 10
A = 1
B = 0.3
C = 7.5
N' = 23
S = 7.5

FW
Q' = 10
A = 1
B = 1
C = 6.5
N' = 2
S = 3

HW
Q' = 15
A = 1
B = 0.3
C = 7
N' = 32
S = 9

FW
Q = 5
A = 1
B = 0.3
C = 3
N' = 4.5
S = 4

Holloway Project
Geotechnical Data
Section : 5575 E
Date: 20/10/93
Scale : 1:2500

FIGURE 60.14   A case study during the feasibility stage, after Nickson et al. 1995

unique stope dimension is usually determined for each mine domain. Engineering judgement must be used in selecting the appropriate hydraulic radius. Selecting the smallest hydraulic radius would ensure that all stopes would be stable, but would not likely be the most economical option. Notwithstanding the value of ore, the impact of dilution, acceptable risks, and the operating philosophy, a good starting point for selecting a mean hydraulic radius for an entire domain would be to ensure that approximately 80% of the stopes are stable. The remaining 20% or so can then be dealt with individually using specific ground support or different extraction strategies.

The stope height, length, and width within each domain can be determined from the mean hydraulic radii (roof and walls). The orebody geometry obviously has an important influence on the determination of the stope geometry. In many cases, there will be an economic advantage to maximising the stope height as it has a major influence on the sublevel interval, and therefore, on the mine infrastructure cost.

As more and more operations integrate backfill in the extraction process, its impact on stope stability must be accounted for. The main function of mine backfill is to limit the exposure of stope surfaces during extraction by filling adjacent mined-out stopes. Provided a good quality-control program is followed, it can reasonably be assumed that backfill provides adequate support of adjacent mined out stopes. Consequently, the stability-graph method treats backfill as a rock material when calculating stope-wall dimensions. In practice, however, it is rare that a tight fill can be established against a stope back. As a result, in stope-back analysis, the influence of backfill is assumed to be minimal and ignored.

60.11   DESIGN CONSIDERATIONS FOR INDIVIDUAL STOPE PLANNING

It is good engineering practice to employ the stability graph at the planning stage to evaluate the stability of each stope. At this stage of development, there is usually underground access that allows for a revaluation of the rock-mass data collected during the feasibility study. Direct underground mapping can provide more reliable information than diamond-drill hole data. Another advantage of underground observations is that it can reveal early signs of stress. This can be complemented by stress measurements allowing for a better assessment of stress influence on the stopes. At this point, it is possible to integrate numerical modelling to investigate optimum sequencing.

Access to more quality data can allow for greater confidence in stope stability estimates than allowed during the feasibility study. Consequently, it is possible to consider modifications or fine-tuning to the ground support and extraction strategies. At this stage, it is also important to assess the influence on stope stability of some of the factors not well accounted-for in the stability-graph method. These can include faulting, shear zones, or areas susceptible to rock bursts.

One of the great benefits of using such a method at the planning stage is that it brings geomechanical considerations into stope design and increases the awareness of mine planners to ground-control issues. Modern stope designs require the close cooperation of geology, mine planning, and rock mechanics.

60.12   STOPE RECONCILIATION

Using the stability graph to assess and document stope performance is useful to build site-specific empirical knowledge that can be used in future design. Once a sufficient number of case histories have been collected, it may even be possible to refine the stability graph for a given site or extend its predictive capability to dilution (Pakalnis, Poulin, and Hadjigeorgiou 1995) or to quantify the probability of failure (Diederichs and Kaiser 1996). A major aid in stope reconciliation has been the introduction of cavity-monitoring systems.

These refinements are interesting and contribute to a better understanding of stope behaviour. However, the value added to operations from this effort remains limited in many cases because the initial stages of mining are completed, the mine infrastructure is in place, and opportunities for modifying stopes layout are restricted.

60.13   DISCUSSION AND CONCLUSION

By definition, empirical design is based on observation and experience. The stability-graph method owes its popularity in its ease of use, its application at early stages of mining, and the fact that it can provide a reference for stope performance. Invariably, it cannot provide a successful prediction for every stope at every operation because the complexity of ground conditions and operating practices can influence stope performance.

It has been argued in the past that the method only reaches its full potential when it is "site-calibrated." The basic assumption is that, as more data become available, the design recommendations can be modified through back analysis. Obviously, this is an important step in better understanding the site conditions. If the reconciliation exercise is done rigorously, it can reveal important information on the efficiency of mine practices such as blasting, prereinforcement, and sequencing.

However, this should not detract from the main objective of the method as a design tool at the feasibility stage when no such data is available, but when the critical decision must be made. The extensive calibration of the method worldwide makes it very robust and ideal for designing open-stope dimension at the feasibility-assessment stage. The added value of a very refined site-specific graph towards the end of a mine life can only be limited.

Over the years, there has been a proliferation of design charts aiming to refine the method or expand on its applicability. Design charts that are not backed by field data have limited use. Similarly, modifications that rely on limited sites should be viewed with caution.

Complex charts bring new and interesting ideas, but one should keep in mind that the method can be no better than the quality of input data available. This is particularly true at the feasibility stage where data are limited by access.

Introducing many zones to the graph has limited application at the design stage because the designer has to come up with a hydraulic radius number for each domain. The transition between stable and unstable and the potential for increasing dimensions by using pattern cable bolts remain the basis for stope design

60.14   REFERENCES

Barton, N., R. Lien, and J. Lunde. (1974). Engineering classification of rock masses for the design of tunnel support. *Rock Mechanics*, Vol. 6. No. 4, pp. 189–236.

Bawden, W.F., J. Nantel, and D. Sprott. (1989). Practical rock engineering in the optimisation of stope dimensions–Applications and cost effectiveness. *CIM Bulletin*, 82 (926), pp. 63–70.

Bawden, W.F. (1993). The use of rock mechanics principles in Canadian underground hard-rock mine design. *Comprehensive Rock Engineering*. Chapter 11, Vol. 5, pp. 247–290.

Clark, L.M., and R.C. Pakalnis. (1997). An empirical design approach for estimating unplanned dilution from open stope hangingwalls and footwalls. *99th CIM-AGM*, Vancouver, published on CD-ROM.

Diederichs, M., D.J. Hutchinson, and P.K. Kaiser. (1999). Cable-bolt layouts using the modified stability graph. *CIM Bulletin* November/December, pp. 81–85.

Diederichs, M., and P.K. Kaiser. (1996). Rock instability and risk in open stope design. *Can. Geotech. J.*, Vol. 33, pp. 431–439.

Dunne, K. and R.C. Pakalnis. (1996). Dilution aspects of a sub-level retreat stope at Detour Lake Mine. *Proc. 2nd NARMS Rock Mechanics Symp. On Rock Mechanics Tools and Techniques*, Montreal, Balkema, Vol. 1, pp. 305–313.

Dunne, K., R. Pakalnis, S. Mah, and S. Vongpaisal. (1996). Design analysis of an open stope at Detour Lake Mine. *98th CIM-AGM*, Edmonton.

Germain, P, J. Hadjigeorgiou, and J.F. Lessard. (1996). On the relationship between stability prediction and observed stope overbreak. *Proc. 2nd NARMS, Rock Mechanics Symp. on Rock Mechanics Tools and Techniques*, Vol. 1, pp. 227–283.

Goel, S.C., and U. Wezenberg. (1999). Stability of open stopes at Ashanti Goldfields-Obuasi Betriebe 1999. *Ninth International Congress on Rock Mechanics*. Vol 1, pp. 101–106.

Hadjigeorgiou, J., J.G. Leclair, and Y. Potvin. (1995). An update of the stability graph method for open stope design. *97th CIM-AGM, Rock Mechanics and Strata Control Session*, Halifax, Nova Scotia.

Hutchinson, D.J. and M. Diederichs. (1996). *Cable bolting in Underground Mines*. BiTech Publishers, Richmond, 406 pp.

Hoek, E., P.K. Kaiser, and W.F. Bawden. (1995). *Support of Underground Excavations in Hard Rock*. Rotterdam: A.A. Balkema, 1995, 215 pp.

Kaiser, P.K., V. Falmagne, F.T. Suorineni, M.S. Diederichs, and D.D. Tannant. (1997). Incorporation of rock-mass relaxation and degradation into empirical stope design. *CIM-AGM*, Vancouver.

Laubscher, D.H. (1977). Geomechanics classification of jointed rock masses-mining applications. *Transactions*, Institute of Mining and Metallurgy, Section A. Vol. 86, pp. A1–A7.

Mathews K.E., E. Hoek, D.C. Wyllie, and S.B.V. Stewart. (1981). Prediction of stable excavation spans for mining at depths below 1,000 m in hard rock mines. Canmet Report DSS Serial No. OSQ80–00081.

Milne, D., P. Germain, and Y. Potvin. (1992). Measurement of rock mass properties for mine design, *Proceedings of the ISRM-Eurock Symposium on Rock Characterization*, A.A. Balkema, Chester, England.

Milne, D., R.C. Pakalnis, and M. Felderer. (1996) Surface geometry assessment for open stope design. *Proc. North Americam Rock Mechanics Symposium, Montreal,* Balkema, pp. 315–322.

Nickson, S.D. (1992). Cable support guidelines for underground hard rock mine operations. M.A.Sc. Thesis, The University of British Columbia., 223 p.

Pakalnis, RC. (1986). Empirical stope design at Ruttan Mine. Ph.D. Thesis. The University of British Columbia.

Pakalnis R., R. Poulin, and J. Hadjigeorgiou. (1995). Quantifying the cost of dilution in underground mines. *Mining Engineering:* pp. 1136–1141.

Pascoe D.M., N. Powell, J.S. Coggan, K. Atkinson, and M. Owen. (1998). Methodology for underground risk analysis using the Canadian stability graph method. *Mine Planning and Equipment Selection.* pp. 199–205.

Pine R.J., K. J. Ross, P. Arnold, and M. Hodgson. (1995). A probabilistic risk assessment approach to hanging wall design. *IMM Conference on Health & Safety in Mining & Metallurgy,* London.

Potvin, Y. (1988). Empirical open stope design in Canada. Ph.D. Thesis. The University of British Columbia p. 350.

Potvin, Y. and D. Milne. (1992). Empirical cable bolt support design. *Proceedings of the International Symposium on Rock Support,* P. Kaiser and D. McCreath, Eds. Sudbury, Ontario, pp. 269–275.

Poulin R., R. Pakalnis, D.A. Peterson, and J.C. Kalynchuk. (1995). Evaluation of cable bolt applications for Canadian underground mines. *CIM Bulletin,* Vol. 88, No. 989 pp. 55–58.

Reschke A.E. and J. Romanowski. (1993). The success and limitations of Mathews analysis for open stope design at HBMS, Flin Flon operations. *95th CIM-AGM,* Calgary.

Scoble, M.J., and A. Moss. (1994). A. dilution in underground bulk mining: implications for production management, mineral resource evaluation II: methods and case histories. *Geological Society Special Publication* no. 79, 1994, p. 95–108.

Sprott, D.L., M.A. Toppi, X. Yi., and W.F. Bawden. (1999). The incorporation of stress damage factor into Mathew's stability graph. Canadian Institute of Mining Annual General Meeting, Calgary, published on CD-ROM.

Stewart, S.B.V., and W.W. Forsyth. (1993). The Mathews method for open stope design. *CIM Bull.*, Vol. 88 (992), 1995, pp. 45–53.

Suorineni, F.T. (1999). *Effects of Faults and Stress on Open Stope Design*. Ph.D. Thesis, Department of Earth Sciences, University of Waterloo, Waterloo, Canada, 345 pp.

Suorineni, F.T., D.D. Tannant, and P.K. Kaiser. (1999). Fault factor for the stability graph method of open stope design. *Trans. Instn. Min. Metall.* (Sect. A: Min. Industry) 108, pp. A92–A104.

Trueman R., P. Mikula, C. Mawdesley, and N. Harries. (2000). Experience in Australia with the application of the Mathews method for open stope design. *CIM Bulletin*, Vol. 93, No. 1036, pp. 162–167.

CHAPTER 61

Noranda's Approach to Evaluating a Competent Deposit for Caving*

Simon Nickson,[†] Adam Coulson,[‡] and Jeff Hussey[§]

61.1 INTRODUCTION

Mines Gaspé is located adjacent to Murdochville in Québec, Canada. The area is within the Shick Shock Mountains in the north central part of the Gaspé Peninsula (Figure 61.1), at an elevation of 575 metres (m) above sea level. Highway 198 connects Murdochville to the coastal village of Anse Pleureuse to the north (40 km) and the town of Gaspé to the east (95 km).

The porphyry/skarn copper (Cu) orebodies at Mines Gaspé were discovered in 1938. The mine began production in 1955 from the Needle Mountain open pit and produced more than 141 million tonnes of copper ore at an average grade of 0.9% Cu from two open pits and eight underground mining areas (A, B, C, D, E32, E29, E34, and E38). Mining operations ceased in late 1999. Open-pit mining included the Copper Mountain (porphyry Cu) and Needle Mountain (A-zone: skarn/stockwork) open-pit mines (Figures 61.2 and 61.3). Several underground skarn orebodies were originally mined by room and pillar and, more recently, by open-stope methods (Figure 61.3: Zones B, C, and E). A total of 1.1 million tonnes of copper metal were produced from Mines Gaspé and another 1.2 million tonnes from local and international concentrates (Hussey and Bernard 1998).

In 1994, a porphyry/skarn copper-molybdenum (Cu-Mo) deposit was discovered at a depth between 1,000 and 1,700 m below Mont Porphyre. By 1996, approximately 19,500 m of drilling in 12 diamond drill holes had outlined 200 million tonnes, grading 0.73% Cu and 0.08% Mo (Hussey and Bernard 1998).

Figure 61.3 illustrates a longitudinal view through the Mines Gaspé area and the various porphyry/skarn copper orebodies. The stratigraphy is composed of bedded sediments that are gently dipping between 20° and 30° to the north. The stratigraphic sequence, from the bottom up, consists of limestones, siltstones, and mudstones of the Forillon, Shiphead, and Indian Cove formations that are overlain by sandstones, siltstones, and mudstones of the York Lake and York River formations. The Forillon and Shiphead formations (Unit #3 in Figure 61.3) host the Mines Gaspé orebodies and consist mainly of mudstones with two limestone units (L1 and L2). Regional scale normal (70°) and tensional (340°) faults acted as conduits for the Copper Mountain and Porphyry Mountain intrusives. Initial mining operations began from the Needle Mountain open pit and subsequently advanced underground into the B, C, and E skarn orebodies using room-and-pillar and open-stoping methods. The Copper Mountain open pit was used to extract the top section of the Copper Mountain intrusive. The final underground mining was located in the deeper E-Zone area. The Mont Porphyre

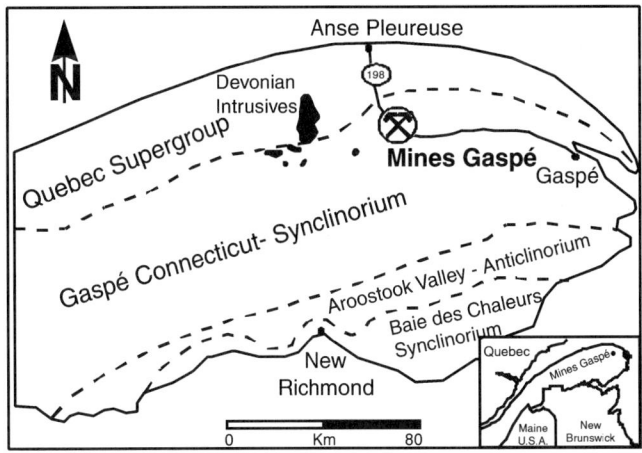

FIGURE 61.1 Location map that also outlines the four tectonostratigraphic assemblages of the Gaspé Peninsula

deposit is located north of the Mines Gaspé underground mining areas.

The alteration aureole that hosts the Mines Gaspé deposits surrounds the Copper Mountain and Porphyry Mountain quartz monzonite intrusives and is illustrated in Figure 61.2 at the L2 horizon. There is a skarn halo surrounding the intrusives that itself is surrounded by an outlying marble horizon and defines a total alteration halo of approximately 3 km × 3 km. The original limestones, mudstones, and siltstones were metasomatized to potassic porcellanite and diopsidic porcellanite. Diopsidic porcellanite hosts approximately 70% of the Mont Porphyre deposit tonnage, with the remaining 30% located within the intrusive unit itself. The reader is referenced to Hussey and Bernard (1998) for a more detailed discussion of the regional and mine geology.

The Mont Porphyre deposit is outlined with respect to the town of Murdochville and the smelting facilities in Figure 61.2. The deposit outlines an area of approximately 600 m by 450 m that is composed of mineralised intrusive and surrounding porcellanite. Noranda commissioned an internal scope study in late 1996 to look at various mining options for Mont Porphyre. The scope study looked at various derivatives of block caving and sublevel caving. Cavability was challenged by the apparent competency of the Mont

* This chapter was compiled and edited from previous presentations at the 100th and 101st Annual General Meeting of CIM (Montreal1998 and Calgary 1999). Reproduced with permission by the Canadian Institute of Mining, Metallurgy, and Petroleum.

† Noranda Inc., Noranda Technology Centre, Pointe-Claire, Quebec, Canada.
‡ Noranda, Inc., Toronto, Ontario, Canada.
§ Compañia Minera Antamina, Lima, Peru.

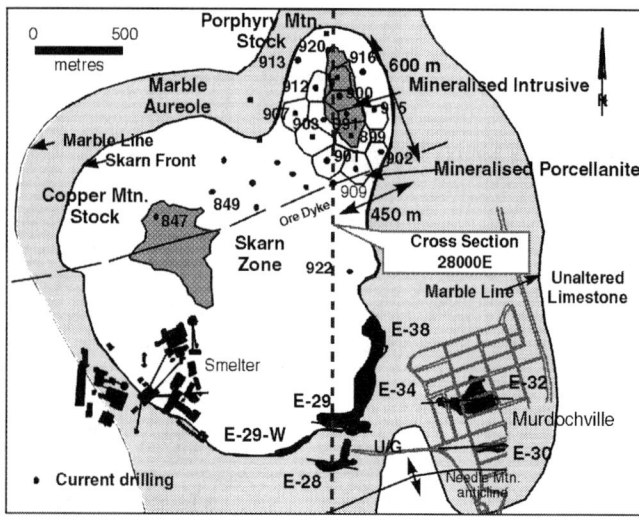

FIGURE 61.2   Plan view of a projection of the L2 skarn horizon, indicating location of Mont Porphyre, with respect to the E-zone orebodies. Location of the 280000E section line also indicated.

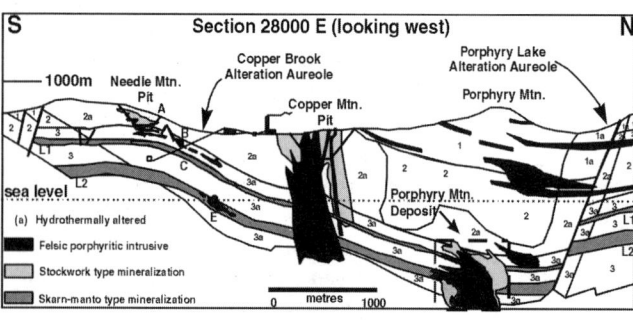

FIGURE 61.3   York Lake Fm. (1), Indian Cove Fm. (2), and Shiphead+Forillon Fm. (3). Mineable skarn/manto type mineralization (A,B,C, & E). The top of the deposit is approximately 1,000 m below any surficial exploration detection limit.

Porphyre orebody, but evidence suggested that current caving experience was moving to more competent rock-mass conditions. The scope study concluded that additional diamond drilling would be useful for looking at subvertical fracture frequency and the possibility of associated mineralization. It also suggested that diamond drilling would be useful to further define the extent of the lower grade intrusive.

Based on the recommendations of the scope study team, a drilling campaign was completed in the summer and fall of 1997. The main objectives were to improve the geotechnical characterisation of the subvertical jointing and to define the extent of mineralization. Previous surface diamond drilling was largely subvertical and, therefore, biased towards the definition of subhorizontal structure. One of the recommendations from the geotechnical portion of the Mont Porphyre Scope Study was to include some subhorizontal drilling and core orientation in any future diamond drilling.

Since the completion of the 1997 drilling campaign, work on the deposit has been limited, although there has been some work in the area of block-size characterisation. Noranda has also funded the International Caving Study, an internationally sponsored block-caving research project interested in pushing the limits of current experience. The underground operation at Mines Gaspé was closed near the end of 1999, and Mont Porphyre remains in Noranda's list

of resources. Each aspect of the history associated with Mont Porphyre will be reviewed in this paper. In 2000, activity is being refocused on the research developments that would be essential to making Mont Porphyre a mine.

61.2   SCOPE STUDY

In late 1996, Noranda assembled a multidisciplinary team to perform a Scope Study to evaluate mining the Mont Porphyre resource using derivatives of block-caving and sublevel-caving mining methods. The main challenge was the very competent rock mass that exhibited a joint spacing in excess of 1 m. It was recognised that experience with block caving under these competent rock conditions was very limited.

The study team was drawn from several areas of Noranda and, in some cases, combined with external expertise. A Project Manager with significant operating mining experience led the team. External resources were retained for the rapid development of project drawings as the study proceeded. On-site resources from Mines Gaspé were used for geological and mineral processing input. A project development office in Bathurst provided resources for geological reserve assessment and mine planning. The Noranda Technology Centre in Pointe-Claire, Québec, provided some resources on the geotechnical and mining automation side. Several other in-house experts were brought in from other Noranda operations to look at specialised topics such as hoisting and electrical power. A few external consultants were invited to review the project during the course of the study and to present their recommendations. Overall, the team assignment was to complete the study within three months and present the findings to Noranda with recommendations for future work. Elements of the study are explained in greater detail in this section.

61.2.1   Geology of the Deposit

Hussey and Bernard (1998) hypothesised that the Mont Porphyre deposit is not a classic porphyry deposit in that, during formation, most of the metalliferous fluids drained off up-dip into the receptive L1 and L2 skarn horizons, preventing a large fluid/temperature-convective cell. This resulted not only in low-grade mineralization of the porphyry (intrusive), but surrounding low-grade mineralization of the country rock—the L1 and L2 skarn horizons and porcellanite units of the Forillon formation (Figures 61.2 and 61.3). The intrusive porphyry accounts for approximately 40% of the ore deposit, with the remaining 60% being contained within the porcellanite P4, P5, and P6 units, the L1 and L2 skarn horizons (Figures 61.2 and 61.3). A base level of 1,700 m below surface is thought to define a reasonable grade cut-off, with the majority of mineralization above this level. The top of the deposit is located at a depth of 1,000 m below the surface (Figure 61.3) and is overlain by the York River and Indian Cove formations (1, 2, and 2a in Figure 61.3). The porphyry intrusive is capped by the T1, tuff unit, 25-m thick, which is very hard and void of mineralization, but highly fractured in comparison to the surrounding rock mass. The dominant upper limit of mineralization in the porcellanite is capped by the T2, tuff unit (50-m thick) that is similar to the T1 unit. This is separated from the T1 unit by the W1, Wollastonite unit (50 m thick) that exhibits a level of competence similar to the porcellanite, but contains very little mineralization. An important characteristic of the Forillon formation, which hosts the Mont Porphyre intrusive, is that its sediments are homogenous with very few bedding planes. This is quite different from the overlying Indian Cove and Shiphead formations that host the Copper Mountain ore deposit. The Forillon formation rocks were silicified by hydrothermal fluids, resulting in tight healing of the bedding planes. An additional T3 tuff is contained within the bulk of the orebody, just below the L2 horizon. However, this unit is only 8 m thick and has been incorporated as acceptable geological dilution in the orebody.

TABLE 61.1 In-situ stress gradient calculated based on stress measurements at Mines Gaspé

	Gradient (MPa/m)	Gradient (MPa/ft)	Trend (degrees)	Plunge (degrees)	Stress on the 1657-m level (MPa)
Sigma 1	0.0598	0.0182	103	18	99.1
Sigma 2	0.0300	0.0091	212	45	49.7
Sigma 3	0.0167	0.0051	357	39	27.7
Sigma Vertical	0.0275	0.0084	0	90	45.6

TABLE 61.2 Summary of average mechanical properties from testing performed at Mines Gaspé

Rock type and location	UCS (MPa)	Tensile strength (MPa)	Young's Modulus E (GPa)	Poisson's Ratio, ν	Density (g/cm^3)
E-32 Waste (Porcellanite)	213	17	52	0.27	2.70
E-38 Waste (Porcellanite P4, HW)	195	–	–	–	2.87
E-38 Waste (Porcellanite P5, FW)	340	–	–	–	2.97
Mont Porphyre (Indian Cove, IC)	262	–	71	0.18	2.75
Mont Porphyre (Tuffs, T1, T2)	454	–	91	0.25	2.80
Mont Porphyre (Porcellanite, P4, P5, P6)	211	–	59	0.25	2.80
Mont Porphyre (Intrusive)	195	–	43	0.18	2.70

61.2.2 Geomechanical Characteristics

The low-grade bulk tonnage of the deposit made it necessary to characterise the deposit in relation to parameters that are used to design block-caving and sublevel-caving mining methods. Block caving is the lowest-cost underground mining method currently used in the world. The cavability of a block-cave orebody is normally determined by characterising the quality of the rock mass using the Modified Rock Mass Rating (MRMR) developed by Laubscher (1990). This rating is a percentage scale (from 1% to 100%) that is similar to the CSIR Rock Mass Rating (RMR) classification (Bieniawski 1989), and incorporates factors for the intact rock strength (IRS), rock quality designation (RQD), joint spacing or fracture frequency, joint orientation, weathering, mining-induced stress, and blast damage. The last three parameters are adjustments not used in the CSIR RMR system. In the collection and characterisation of rock-mass quality information, the MRMR parameters were given strong consideration in conjunction with parameters normally collected within the Noranda Group for the modified (Q) system of rock-mass classification (Barton et al. 1974). The Q System is used regularly for open-stope design with the Modified Stability Graph Method (Potvin et al. 1989).

In-Situ Stress Regime. At Mines Gaspé, a number of in-situ stress measurement campaigns have been performed at varying depths within the existing mine: CSIR triaxial stress measurements by the University of Laval (1983); NTC slotter campaign E-29 (1993); and biaxial door-stopper tests performed by École Polytechnique for the E-38 (1996). From these measurements, a stress gradient was derived, which, when compared to stress measurements performed in the Canadian Shield (Arjang 1991), indicate an above-average gradient that corresponds to the upper limit of combined Canadian shield measurements. Although Mines Gaspé is not located within the Canadian shield, this was used as a reference to determine an outer limit, as no other information within the Gaspé peninsula was available. The simplified stress gradient (intercept through the origin), which was used for subsequent numerical modelling, is summarised in Table 61.1. The lowest mining level of the deposit was the 3100 L (1,657 m below surface). At this depth, in-situ principal stresses were expected to approach 100 Megapascals (MPa), with an approximate east-west orientation and a horizontal-to-vertical stress ratio of 2.2.

Mechanical Properties. A number of laboratory testing campaigns have been performed on Gaspé ore and waste over the years. When any geomechanical evaluation of a new deposit is performed, it is important to review all previous testing so the results can be put into perspective and potential similarities identified. Table 61.2 presents a summary of some of the average mechanical properties for Mines Gaspé. The intact uniaxial compressive strength (UCS) of most of the different ore types present at Mines Gaspé is exceptionally strong. The testing of the Mont Porphyre core indicates that the Porcellanite is marginally stronger than the intrusive, which compares well to previous Gaspé core testing. The maximum recorded UCS result was obtained from the tuffs at 619 MPa, which, to the author's knowledge, represents one of the strongest units in the world. Although this horizon was very strong, it was also the most fragmented of all the geological horizons investigated.

Rock-Mass Characterisation. The ore deposit in 1996 was outlined by 12 subvertical (>70°) diamond-drill holes from the surface. Rock-mass characterisation is used to determine the degree of natural fracturing and to develop an RMR for each rock type, using the Modified Q system (Barton et al. 1974), CSIR RMR classification (Bieniawski 1989), and MRMR Rating (Laubscher 1990). The original geotechnical data collected from these holes included RQD, fracture frequency, joint alteration (Ja), small-scale roughness (Jr), and joint orientation with respect to the core axis. The logging campaign also had provision for 1.5 m of core to be saved for every 30 m drilled to provide samples for mechanical testing before the core was split for assaying. To augment the drill-core data, in which it was difficult to clearly identify joint sets, some limited underground exposures, which exhibited similar geological characteristics and previous Copper Mountain open-pit mapping data, were reviewed. Each diamond-drill hole was analysed separately to determine a distribution of parameters across the ore deposit and along the holes. A global summary of averaged values for combined holes for each domain is presented in Table 61.3.

The majority of the rock types at Mont Porphyre fall into the upper class II (good rock) and verge on class I (very good rock) (Bieniawski 1989 and Laubscher 1990). In the area of the anticipated block caving undercut at the 3034 level (1,650-m depth), the average RMR (Laubscher 1990) was in the range of 75 to 91. Because of the high-stress environment and past experience of stress fracturing in deep-stope backs at Mines Gaspé, a 10% stress-reduction factor was applied as an adjustment. This resulted in a range of MRMR values between 67 and 82. The applied stress-reduction factor at this stage is based solely on experience and will require further appraisal in the future.

TABLE 61.3   **Summary of averaged rock-mass parameters**

Rock type	Length logged	No. of joints	FF/m	RQD %	Jn	Jr	Ja	Q'	RMR
IC	43	38	0.88	90	6	1.0	1.63	9	82
T1	45	54	1.24	67	9	1.0	1.21	6	70
W1	25.6	23	0.90	84	6	1.0	1.70	8	74
T2	125	267	2.13	61	9	1.1	1.56	4.5	70
L1	45	22	0.48	97	6	1.25	0.94	22	79
P4,P5,P6	1,558	1,232	0.81	98	6	1.34	0.94	21	79
IN	819	515	0.63	99	6	1.38	1.1	24	82

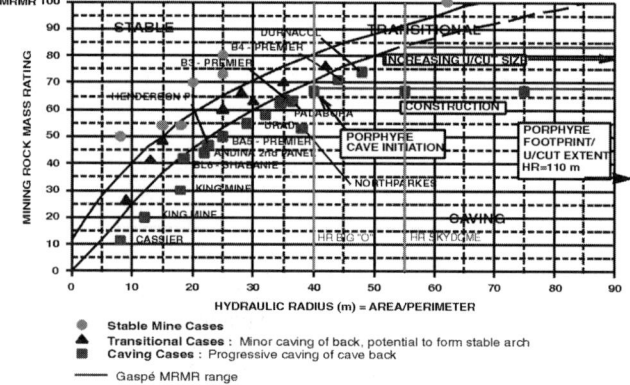

Stable Mine Cases
Transitional Cases : Minor caving of back, potential to form stable arch
Caving Cases : Progressive caving of cave back

— Gaspé MRMR range

FIGURE 61.4   Laubscher Caving Stability Graph (after Laubscher, 1990), indicating the relative location and points of potential cave initiation for the Mont Porphyre ore body, compared to other operations around the world

Cavability of the Ore Deposit.   Two major questions need to be answered to determine the cavability of an orebody (Laubscher 1995). First, is there sufficient area to undercut the orebody enough to cause sustained caving of the rock mass? Second, can the resulting fragmented rock mass be collected and handled with appropriate technology?

An answer to the first question can be determined using the empirically developed Laubscher Caving Stability Graph illustrated in Figure 61.4. This is a caving design graph similar in graphical representation to the Modified Stability Graph (Potvin et al. 1989) developed for open-stope design, but should not be confused as the design inputs and end results are different. The Laubscher Caving Stability Graph plots MRMR versus the hydraulic radius (HR). The caving zone represents a design region in which experience suggests that the complete and sustained caving of the ore column will occur, not just mass caving of stope walls, as is defined in the Modified Stability Graph. From the preliminary analysis, Mont Porphyre's MRMR rating ranges from 67 to 82 (Figure 61.4). Typical caving experience is related to mining large underground low-grade orebodies that mostly have an MRMR rating of less than 50. Mont Porphyre is far more competent than any existing block-caving operation. However, recent changes in mining methodology have highlighted the potential of these more competent orebodies, such as Palabora Mine (South Africa) and Northparkes Mine (Australia). Palabora has an MRMR ranging from 57 to 70 (Kear et al. 1996), and will be mined using a modified block-caving method, while Northparkes has an MRMR of 35 to 55 (Dawson 1995) and is Australia's first block-cave operation. The stability curve in Figure 61.4 illustrates the range of conditions for cavability at Mont Porphyre. Cave initiation could start somewhere between an HR of 40 and 55 m. This would represent a range of undercut areas from 160 m × 160 m to 220 m × 220 m. For illustrative purposes,

these dimensions were compared to the size of two well-known stadiums in Canada—the Olympic Stadium in Montréal and the Sky Dome in Toronto. This type of comparison is a good way to illustrate the required undercut excavation sizes for management personnel. The total available HR of the orebody footprint is 110 m (420 m × 420 m) and provides ample room for additional undercut expansion, if required, to initiate continuous caving of the ore column.

The answer to the second question requires determining the size of ore blocks reporting to the draw points. This is a function of the primary fragmentation (related to the in-situ discontinuity spacing) in the initial mining phase and the secondary fragmentation (related to mining-induced stresses and comminution of caved material in the broken ore column) during maturity of the cave. It is currently anticipated that, in the worst case, 70%–80% secondary blasting (ore blocks >2 m³, which could not be handled with an 8-yard scoop) will be required in the initial mining phase based on analysis of in situ FF distribution. The effects of stress fracturing have not been considered in much detail and are potentially an area of future research. Determination of secondary fragmentation was done by using an empirical fragmentation model and indicated that once the cave reaches maturity (cave height ±500 m), secondary blasting could reduce to 30%-40% due to self-comminution in the broken cave column (Gash 1997). This is an area that will need further investigation to more comfortably predict secondary fragmentation of competent orebodies. The ability to handle this large-sized material effectively on a continuous basis and maintain a smooth production cycle is one of the technical challenges for future consideration. Experience with oversize of this nature is currently being experienced intermittently by a number of mining operations worldwide.

61.2.3   Mine Design

Although the block-caving method is the primary method presented in this paper, the scope study reviewed six mining methods in all: block caving (modified block caving), sublevel caving, modified sublevel caving, large blasthole-induced caving (LBIC), horizontal hole-induced caving, and open-stoping. Because of the desire to maximise recovery and minimise cost, emphasis was mainly placed on caving methods, and a brief overview of the block-cave option is presented here. The other methods provided for induced fragmentation of the orebody, either by conventional sublevel-caving methods or more novel approaches. The LBIC method provided for the application of large-diameter blastholes from open-pit applications, not to fully blast the ore, but rather to induce additional fracturing.

Mine Layout.   An idealised representation of the block-caving mining option is presented in Figure 61.5. In this option, four levels would be required to mine the deposit.

1. The 1300 L (1,122-m deep) exploration level (Hussey and Bernard 1998) would be used for ventilation and instrumentation, incorporating microseismic monitoring, extensometers, TDR cables, and observation holes to monitor the cave advance.

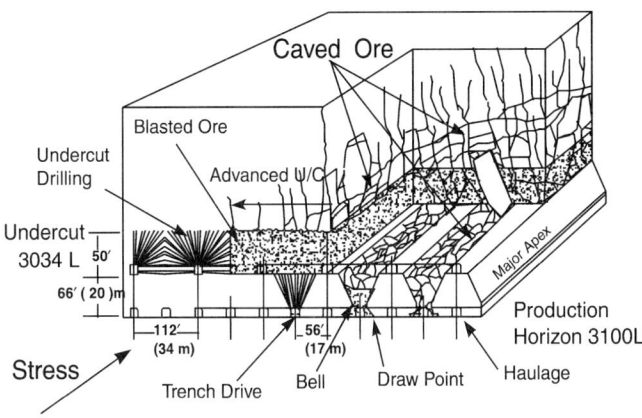

FIGURE 61.5   Idealized 3D schematic of the block caving option, showing the location of the undercut and the extraction level and illustrating the advanced undercut concept

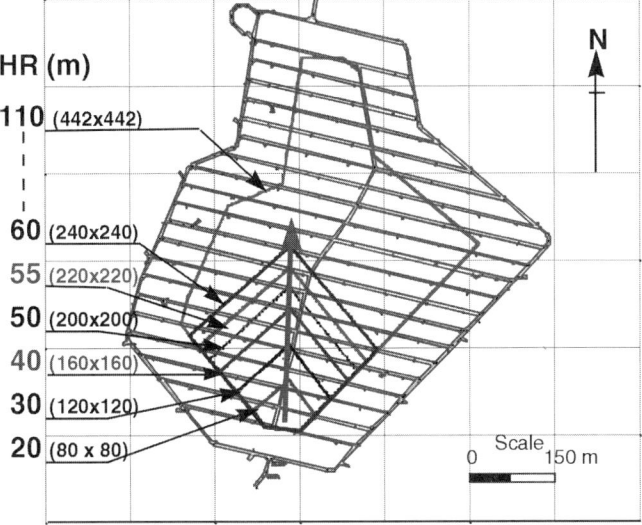

FIGURE 61.6   Plan view of the 3034L (undercut) showing the orientation of drill drifts and the expansion of the undercut. Also, indicated is the HR of the undercut at various stages.

2. The undercut level (3034 L–1,650 m deep) utilises an advanced undercut that would be developed before major development on the extraction level (Figure 61.6). The advanced undercut was proposed to limit stress degradation of draw bells and development, which would be affected by high induced stress at the advancing abutment.

3. The extraction level on the 3100 L (1,670 m deep) would be instrumented with convergence stations and stress monitoring in the major apex pillars and abutments (Figure 61.5).

4. The final level would be the 3200 L conveyor level at a depth of 1,700 m below surface.

The extraction level haulages and the undercut-level drill drives were designed to be oriented parallel to the principal stress direction to reduce stress degradation. The mining sequence on the undercut level that is illustrated in Figure 61.6 has been proposed to start from the weakest and most highly fractured region. By analyzing the drill-hole information, this region was

determined to be in the south end of the deposit. The undercut would advance perpendicular to the maximum principal-stress direction to take advantage of stress degradation in the back of the cave.

Subsidence.   One of the major concerns related to block caving is the effect of subsidence on the surface topography and infrastructure. Many case histories exist for near-surface cave operations, but there are few studies for cave mining at the depths associated with Mont Porphyre. It is anticipated that the cave will go to surface based on traditionally low swell factors in the range of 5% of the caved material. The low swell results from mass movement of large blocks, which tend to act as a piston. As such, it has been postulated that, in competent rock masses, cave angles will tend to be at steep angles between 75° to 90° (Nicholas 1997 and Laubscher 1990). A subsidence model was developed in which a cave angle of 85° was used from the undercut level and projected to surface. It was assumed that a glory-hole wall would form with a slope angle of 55° and an external relaxation angle of 45°. The depth of the glory hole was determined by calculating the removal of the orebody using a 5% swell factor. The projection of these expressions on the surface topography was used to identify potential areas of mass movement (tens of metres) and minor relaxation (tens of centimetres) within the glory hole and relaxation zone respectively. This type of subsidence model is supported by the subsidence failure at Henderson Mine (Stewart et al. 1981), in which, after only 15% draw of the column height, a steep-walled glory hole formed on the surface from an undercut at a depth of 1,280 m.

61.2.4   Other Aspects of the Scope Study

The scope study results were presented at the end of the three-month period during an all-day presentation and review with key Noranda personnel. The various team members made presentations on all aspects of the project. In general, it was concluded that a derivative of caving was applicable to Mont Porphyre, and it would likely have to include some way to induce fragmentation in advance. Recommendations for additional diamond drilling, as well as several options for accessing a potential underground exploration drive, were included in the presentations. Business cases for each caving method were presented with a base-case rate of return and several opportunities for increasing that return. Opportunities included a tax holiday, compressing the construction schedule, reducing operating costs, increasing the grade, increasing recovery, reducing capital expenditures, and reducing freight charges. Although the base-case return was below Noranda's requirement for a 12% return on equity, the opportunities provide a much brighter picture.

Various technologies that could contribute to the success of Mont Porphyre were identified during the presentations. These included:

- Hydraulic hoisting
- Automation technologies for mucking, drilling and blasting, high hang-ups, and communication
- Methods of inducing rock fragmentation
- Effects of stress fracturing and healed structure in porcellanite on primary fragmentation
- Block-size simulation
- Core orientation
- Handling large oversize in the drawpoint
- Rapid round development
- Cave monitoring technologies (microseismic monitoring and laser profiling)

Some additional consideration from the scope study will be presented and discussed in Section 61.3.

TABLE 61.4 Rock-mass domains and associated geotechnical core logging data extracted for the scope study

Rock unit	Description (original lithology prior to alteration)	Logged core (m)	RQD	Fracture frequency (per metre)
York River (YR)	Sandstone / mudstone	0	n/a	n/a
Indian Cove (IC)	Calcareous mudstone with limestone nodules	43	90	0.88
First Tuff Unit (TI)	Brown mudstone / tuff	44	67	1.24
First Porcellanite (P1)	Calcareous mudstone	12	87	1.35
Wollastonite (W1)	Mudstone and siliceous limestone concretions	26	84	0.90
Second Porcellanite (P2)	Calcareous mudstone	0	n/a	n/a
Second Tuff Unit (T2)	Brown mudstone / tuff + limestone nodules	125	61	2.13
Third Porcellanite (P3)	Calcareous mudstone	4	100	0.55
First Limestone (L1)	Argillaceous limestone	45	97	0.48
Fourth Porcellanite (P4)	Black calcareous mudstone and tuff	211	96	0.76
Second Limestone (L2)	Limestone	5	91	1.75
Fifth Porcellanite (P5)	Black calcareous mudstone	163	99	0.79
Third Tuff Unit (T3)	Tuffaceous mudstone	18	96	1.06
Sixth Porcellanite (P6)	Black calcareous mudstone	1,153	99	0.83
Intrusive Unit (In)	Quartz monzonite intrusive	819	99	0.63

61.3 MINE DESIGN CONSIDERATIONS FROM THE SCOPE STUDY

The success of caving mining methods depends highly on the primary and secondary fragmentation characteristics of the orebody. In caving terminology, primary fragmentation occurs at the cave front and is dependent on the nature of the rock fabric and associated stress-induced fracturing. Secondary fragmentation refers to the size distribution that reports to the draw bells as a result of comminution, breakage, and attrition during drawing of the caved material. The characterisation and evaluation of a rock mass for caving scenarios relates to the evaluation of primary and secondary fragmentation.

Production rates in a caving scenario rely heavily on the degree of secondary fragmentation, which, in the case of Mont Porphyre, was predicted to be 80% greater than 2 m³ in the initial phases of the caving process. Secondary fragmentation prediction relies on understanding the primary fragmentation that will result from the caving process in both the ore and the overlying waste rock. The T2 tuffaceous unit, located above the L1 horizon, was identified as being more fractured in the scope study and, therefore, less competent than the ore units. This increased fracturing implies a higher degree of primary fragmentation and an increased probability of the T2 unit percolating down into the ore column as the draw progresses. The issue of surface subsidence and glory-hole development requires a greater understanding of the Indian Cove rock unit overlying the deposit. The timing and location of glory-hole development are important considerations with any caving mining method.

The 1997 study was based on diamond-drill hole data obtained up to 1996 and included about 2,700 m of geotechnical core logging. Most of this information was extracted from within the ore horizon and very little applied to the overlying geology. The stratigraphy in the vicinity of the Mont Porphyre deposit can be determined from Figure 61.3; however, the rock-mass domains that were applied to rock-mass characterisation at Mont Porphyre are given in Table 61.4. The order in Table 61.4 follows the general stratigraphic column descending from the surface, with the exception of the intrusive unit that is listed last, but crosscuts many horizons. The basic conclusion from this work was that the intrusive unit exhibited a fracture frequency of 0.6 fractures/m and the porcellanite units were slightly higher at 0.8 fractures/m. Geotechnical information outside these units was limited; however, the tuff units exhibited a particularly low RQD and high fracture frequency. The tuff units, particularly the thicker T2 horizon, were identified

as exhibiting finer primary fragmentation and were therefore considered as a likely dilution source.

Several observations related to geotechnical trends observed during the scope study are listed as follows:

1. The study indicated that limited information existed in terms of the orientation of joint sets in any of the rock units, both within and overlaying the deposit. Assumptions for the scope study were drawn from unoriented diamond-drill core data, underground mapping of existing development exposures in similar rock units, and previous mapping within the Copper Mountain open pit. These assumptions needed to be verified with oriented core or underground exposures.

2. Analysing the fracture-frequency information suggested that the porphyry intrusive was slightly less jointed than the porcellanite. This has a potential impact on the primary fragmentation characteristics and the viability of a caving mining method.

3. Underground mapping at Mines Gaspé suggested that subvertical structure was more dominant than subhorizontal structure. All of the drill holes that have intersected the Mont Porphyre deposit were vertical to subvertical and potentially biased towards the definition of subhorizontal structure. For this reason, it was suggested that any further drilling include subhorizontal holes directed to intersect the heart of the deposit. This would assist in understanding the in-situ jointing and associated primary fragmentation distribution.

4. The fracture-frequency analysis from holes within the porcellanite units suggested that there might be more pervasive jointing in the south portion of the deposit. There also appeared to be some variability of depth in the amount and intensity of jointing that suggested some lack of homogeneity within the deposit.

5. A limited amount of geotechnical information was obtained from the drill-core-logging program that was completed for the scope study. Alternate assessment tools and more detailed information (such as logging in the overlying rock units) was suggested for any future drill-core-logging program to complete a more rigorous geotechnical logging exercise.

The 1997 study suggested that caving or some form of induced caving was potentially applicable to the Mont Porphyre deposit. The expected block size for normal caving methods was predicted to be initially very large, and it would be necessary to efficiently handle large oversize at the draw bell. The term "bell-blast caving" was coined to signify that the ability to efficiently handle large oversize at the draw bell would be of paramount importance. Induced-caving methods, such as sublevel caving and large-diameter-blasthole caving, were also included for consideration in the Mont Porphyre study. A stope-and-pillar-type mining method was eliminated due to low ore recovery and the potential problems of such mining at depth. The challenge for any method was initially to understand the fragmentation characteristics at Mont Porphyre.

61.4 1997 DRILLING CAMPAIGN

Based on the recommendations of the study team, a drilling campaign was completed during the summer and fall of 1997. The main objectives of the campaign were to improve on the geotechnical characterisation of subvertical jointing and to define the extent of mineralization. Previous surface diamond drilling was largely subvertical and, therefore, biased towards the definition of subhorizontal structure. One of the recommendations from the geotechnical portion of the Mont Porphyre study was to include some subhorizontal drilling and core orientation in any future diamond drilling.

The 1997 Mont Porphyre drilling campaign was planned to delineate ten targets at the caving undercut horizon and to obtain geotechnical information at subhorizontal inclinations. The campaign included the deepening of four existing holes, four new subvertical target holes, and five subhorizontal holes. Three diamond drills were engaged on site during the campaign. Subhorizontal drilling was oriented in two major directions to pick up geotechnical information perpendicular to the regional 070° and 340° sets and verify mineralization of fracture domains. Two of the subhorizontal holes were directionally drilled off an existing hole, and the other three were collared at a low angle on the surface. The drill program was not fixed in nature, but was "results" driven and modified as a better understanding of the directional drilling capabilities was attained. Detailed records of hole deviation and drilling conditions (such as stabilisation, bit changes, etc) were kept during the drilling campaign.

61.4.1 Geotechnical Logging Program

The 1997 drill campaign incorporated a revised geotechnical logging program that required the presence of one full-time person on site who was specifically dedicated to geotechnical logging and core orientation.

One point of interest in reference to geotechnical logging is the design implication of determining natural structure during logging for caving applications. In potential open-stoping situations, difficulties in determining what jointing is natural and what is induced by the drilling process can sometimes lead to incorporating induced drilling breaks within the logged data. In an open-stoping situation, this is absorbed into a slightly more conservative design approach as a result of additional structure. In caving applications, however, additional structure can be more favourable to the mining method and result in optimistic conclusions about cavability.

Geotechnical Logging Sheet. The geotechnical logging sheet that is illustrated in Figure 61.7 was developed during the drilling campaign. A more detailed description of the components of the geotechnical logging sheet is given in this section. Each column on the logging sheet is discussed with a single descriptive paragraph.

Box No. This column refers to the appropriate core box number, which can be useful if it is necessary to pull a particular drill interval at some point in the future. A note is also made if this box contains core that has been saved for rock mechanics testing with an "RX" inserted after the box number. The issue of

saving samples for rock mechanics testing will be discussed further in Section 61.4.2.

Core Interval. The logged core interval is noted (From − To) here in the same units that are used by the diamond-drill crews on their logging markers. A standard drill interval of 3 m (10 ft) is suggested as this normally coincides with the driller's markers.

Lithology. The lithology is noted from the geological drill logs. Geological logging was being done at the same time as the geotechnical logging in the case of Mont Porphyre, and lithological information was extracted at a later date.

NF—Number of Natural Fractures. The total number of natural fractures in the core interval is listed in this column. Natural fractures are generally identified as being relatively planar oblique fractures to the core axis and will generally have some minor alteration of the joint surface. Fractures that do not completely break the core are not usually included. It is sometimes difficult to identify the difference between natural fractures and driller-induced core breaks. For this reason, the breaks that are made at the end of the core run and at the end of each box row are not normally included in the count, as they are most likely not natural. If the core is handled well, then the number of artificial fractures can be kept to a minimum.

AF—Number of All Fractures (Natural and Artificial Breaks). This column represents the total number of artificial and natural fractures summed over the core interval, without artificial breaks at the end of the core run or at marker locations. The main purpose of this information is to monitor the total number of fractures, whether they are considered artificial or not. Artificial breaks can generally be identified as having very uneven fracture surfaces. When the uneven parts of the core are fitted together perfectly, a match can be made. Also, artificial breaks caused by removing the core from the core barrel or poor handling will not have any alteration or coating on the fracture surface.

RQD—Rock Quality Designation. RQD is a measure of the percentage of the lengths of all core pieces greater than 100 mm (4″) long and has been standardised for NQ core. If the core interval is standardised to 3 m (10 ft) and the recovery is 100%, the length of the interval measured will be 3 m. Artificial breaks are not included in the RQD measurement.

REC—Recovery of Core. Recovery is the percentage of the core that is recovered from the drill advance. It should be possible to take this directly from the driller's log; however, it is always useful to check. The recovery can often be reduced from 100% if a fault zone or vug is encountered.

JC and JO—Joint Coating and Joint Orientation. This column includes logging the orientation with respect to the core axis and joint coating or infilling. The information is logged for each natural joint. The joint orientation is the angle of the joint with respect to the axis of the core, such that a notation of '9' represents a joint cutting perpendicular to the core axis. The joint coating or joint infilling follows the same categories that were defined for geological logging purposes and a selected list for Mont Porphyre is given below.

C = chlorite	Cp = chalcopyrite
Q = quartz	Py = pyrite
F = Fluorite	Mo = molybdenite
Ca = carbonate	V = vide (void)
T = talc	G = fault gouge

The notation for each natural joint combines a numerical value that represents the joint orientation to the nearest tenth value combined with the appropriate joint-coating descriptor. Each natural joint is logged separately on a different row. The entry "9Mo" would represent a natural joint that is perpendicular to the core axis and exhibits a coating of molybdenite.

Box No.	Core Interval	Lithology	NF	AF	RQD	REC	JC + JO	T	Jr	Ja	Distances Between Natural Fractures / Comments
	10' or 3 m		#s	#s	(%)	(%)		(mm)	Range	Range	(N.B. core lengths should be measured even if they go into next interval)
	From: To:										

Geotechnical Core Logging Sheet

Hole Number ___   Date ___   Collar Easting : / Collar Northing :
Logged By ___   Hole Size ___   Collar Elevation : / Azimuth Dip :

NF = Number of Natural Fractures (Joints)
AF = Number of All Fractures (Joints + Core breaks)
RQD = % of core > 100 mm or 4"
REC = % recovery of core

JC = Joint Coating (C=Chlorite, Q=quartz, F=florite, Ca=carbonate, Cp=pyrite, Mo=molybdenite, V=vide, T=talc, G=gouge)
JO = Joint Orientation from Core Axis (1 = 10, 2 = 20, 3 = 10...etc.)
T = Thickness of fault/infill (mm)
Ja = Joint Alteration (0.75, 1.0, 1.5, 2.0, 3.0, 4.0)
Jr = Joint Roughness (P = polished, SR = slightly rought, R = rought, VR = very rough)

Page   of

FIGURE 61.7   Geotechnical core logging sheet

T—Thickness of a Fault or Major Infill (mm). If a fault or a joint with measurable infill (>2 mm) is encountered, this should be noted in the thickness column. The fault should also be identified with a "G" in the Joint Coating column.

JR—Joint Roughness. A range of joint roughness should be identified for the core interval. The definition of joint roughness can be determined from the examples given below. It is useful to identify a range of values and circle the most dominant range.

Definition	Feels to the touch
P = polished/smooth	Like a baby's bottom
SR = slightly rough	Like fine sandpaper
R = rough	Like coarse sandpaper
VR = very rough	Like coarse sandpaper with small-scale undulations (mm's)

JA—Joint Alteration. Joint alteration from the drill core is best determined from the physical characteristic of the joint surface. A simple guide is given below that is typically used with a knife and a fingernail to identify a range of JA over the core interval. The most dominant value is normally circled if a range of JA is given.

Physical characteristics	Observable characteristics	JA	ϕ_r Approx. residual friction angle
Very hard, cannot be scratched easily with a knife	Tightly healed, hard, non-softening, impermeable filling	0.75	
Can be scratched with a knife	Unaltered joint walls, surface staining, or hard infilling	1.0	25°–35°
Can be gouged (deep scratch) with a knife	Unaltered joint walls, surface staining, or hard infilling	1.5	25°–35°
Can be deeply gouged with a knife, scratched with a fingernail and feels slippery/soapy to the touch	Slightly altered joint walls, non-softening mineral coatings, sandy particles, clay-free disintegrated rock, etc.	2.0	25°–30°
	Silty- or sandy-clay coatings, small clay fraction (non-softening)	3.0	20°–25°
Can be deeply gouged with a knife and a fingernail, feels slippery/ soapy to the touch and has a decent thickness (i.e., 2–4 mm)	Softening or low-friction clay mineral coatings, i.e., kaolinite, mica. Also chlorite, talc, gypsum, and graphite, etc. (Discontinuous coatings, 1–2 mm or less in thickness)	4.0	8°–16°

Distances Between Natural Fractures. The distance between each natural fracture is noted in this column to determine a rough estimate of block size that can be related to primary fragmentation assessment. It is important that even if there is only one fracture in the core interval, the distance should be measured to the next natural fracture (measuring and summing the lengths of the core pieces) even if this is in the next core interval. Every natural fracture should, therefore, end up with a distance associated with that fracture. The first value that appears on the sheet should be the distance from the start of the core box to the first natural fracture, after which only the distances between natural fractures are noted. If the recoveries are always 100%, the sum of all of these distances should be equal to the length of core logged minus that distance of the last natural fracture to the end of the last core box.

General Comments. There is no "General Comments" column provided, but it is useful to add any comments or observations in the last column ("Distance between natural fractures") using as many rows as necessary. Particular items that are worth noting here are the presence of broken ground, faults, dykes, or shear zones, and any observations of core disking. The presence of "broken zones" or "rubble zones" would also be noted here. A "broken zone" would be classified as a core interval with pieces that are less than one core diameter, with 50% of the pieces greater than 50 mm (2″) in size. A "rubble zone" would be classified as a core interval with pieces that are less than one core diameter, with 50% of the pieces less than 50 mm (2″) in size.

Suggested Logging Procedure. The procedure for collecting the data identified on the *Geotechnical Core Logging Sheet* can be set up on site to meet the needs of the personnel involved. One such procedure is summarised in sequence as follows:

1) *Review geological logs to identify rock-mass domains to be logged.*

Geotechnical core logging does not have to be done for all the diamond-drill core obtained in a drilling program. However, from a geotechnical viewpoint this is the ideal situation, as the maximum amount of information is obtained for all potential rock-mass domains. In the caving options that were under consideration for Mont Porphyre, the primary source of geotechnical information was the orebody because the issue of cavability was of paramount importance. Information within the overlaying rock units was treated as a secondary source of information, but was also important because the initial core-logging campaign concentrated mainly on the ore horizon. The amount of core logged within each horizon can be modified to suit the time constraints involved as long as reasonable coverage is obtained within the different rock units.

2) *Lay out core boxes and note box numbers and hole depths involved.*

Core logging by its nature is best done with specific goals or lengths of core laid out in the core shack. This way, geotechnical logging is aimed at a specific number of core boxes and completed before additional core is reviewed.

3) *Identify and mark all natural fractures with a highly visible (yellow) wax marker pen.*

This involves reviewing all the core breaks and distinguishing between natural fractures and artificial ones that are induced by the drilling and handling process. This can be somewhat subjective, but basically core breaks induced by the drilling and handling process can be identified by very rough irregular surfaces that are often not following any specific structure or horizon. In addition, the break surface can appear fresh and should exhibit minimal alteration. Normally, an "X" with a circle, or some other appropriate symbol, is used to identify natural fractures. They should be marked with something highly visible and permanent so they will show up in photographs and can be referenced at a later date if the core is not split.

4) *Log the information identified on the Geotechnical Core Logging Sheet.*

The order of this work is up to the individual involved, but it is often best to get the Number of Natural Fractures (NF), Number of All Fractures (AF), Distance between Natural Fractures, RQD, and Recovery (REC) out of the way first as these tasks are relatively straightforward. RQD measurements are easily done by using a tape measure and summing the lengths of core that are greater than 50 mm (2″). In good-quality core, such as Mont Porphyre, it may be easier to sum the lengths of core that are less than 50 mm (2″).

The Joint Coating (JC), Thickness (T), Joint Alteration (JA), and Joint Roughness (JR) values are somewhat subjective and

may vary through the interval being logged. A range of values can be given with the preferred range for the interval circled for emphasis. A preliminary assessment of these values can be made during the initial identification of natural fractures, but a second review of each natural fracture is often required to assess these parameters.

Any observations of shear zones, faults, stress-induced disking, or abnormal alteration zones should also be noted separately on the logging sheet, using the last column for these comments.

5) *Photograph all core and reference core boxes to a particular film and photo number.*

After the logging process is complete, all core should be photographed as a permanent record before splitting. Photographs can include multiple core boxes, but an identifying tag should be included in the area to be photographed to note the hole number, core boxes, and hole depths. Also it is useful to place a scale, such as a brightly coloured ruler, to get a length reference within the photograph. Photographs are best taken from above and sometimes may require a series of overlapping photographs to get the area of interest. A log of the film number, picture number, and core identification parameters (hole number, hole lengths, box number) should be kept. The Mont Porphyre core was photographed with a digital camera and stored on CD-ROM. Try to ensure that the marks identifying the natural fracture locations are facing upwards and are visible in the photograph. When the films are developed, the photographs should be filed in an appropriate album.

6) *Identify samples to be saved for rock mechanics testing.*

In the rock units that will be split, it has been suggested to retain 1.5 m (5 ft) of every 30 m (100 ft) of core that will not be split and can be used for laboratory strength testing if necessary at a later date. These areas should be flagged in each core box so they will be easily identified during the splitting process. Laboratory strength testing requires good-quality core that does not have weak bands or structure that will affect the test results. Individual samples that are roughly 2.5 times the diameter will also be prepared from this core, so several samples can be obtained from core that is not broken up much. The location of the 1.5-m section may be moved around a bit to meet the above constraints.

Observations on Diamond Drilling. Whoever is doing the geotechnical logging and orientation should try to spend some time at the drill to observe conditions and procedures. It is also important to relate what is coming out of the core tube to what is seen in the core shack. During the Mont Porphyre drilling campaign, a great deal of time was initially spent at the drill to observe core-unloading procedures and to get an idea of natural fractures and artificial breaks that result from fitting the core into the core box. In the intrusive unit, artificial breaks were identified by a clean, rough, generally fresh-looking break with no alteration and were fairly easy to identify in the core shack. Certain varieties of porcellanite were found to have many healed structures that preferentially broke apart when fitting the core into the core box and often appeared natural. This was identified as a potential benefit to cavability.

In terms of core-removal procedures, two core springs rather than one were being used to break the core at the bottom of the hole. Core removal with two core springs was difficult and frequently accomplished by banging the end of the core against a metal trailer. This method had the potential of damaging the core, and a wooden block was substituted for the metal trailer to minimise core damage. Another method that also worked but was more time consuming was to manually loosen the core-spring tube segments and force the core back out into the core tube with a wooden stake. Occasionally, a metal ring around the core tube was used to absorb the impact of the hammer and ease the core removal. Using a rubber hammer was suggested if this procedure was deemed absolutely necessary. Productivity is usually the defining parameter and drill-core treatment is likely to vary significantly by crew and by day. Several visits were made to the drills just to observe and document the procedure during the drilling campaign. Optimal crews and procedures were identified for purposes of critical drilling and orientation trials.

Artificial breaks in the P4 rock unit, over approximately 30 m of logged core, were found to exceed 60% of all the fractures noted. Discussions on the importance of minimising artificial fractures were held with each drill crew, and a procedure established to have each drill helper mark natural breaks that come out of the core tube. This process was initially started with a marker pen but required special lumber crayons in wet conditions. The importance of minimising artificial breaks helped ensure the correct recording of natural fractures and also improved the ability to puzzle core together for orientation purposes.

61.4.2 Core Orientation Program

A core-orientation program was included in the 1997 drilling campaign for Mont Porphyre. The initial phase of this program was designed to try out different orientation techniques and adopt a preferred procedure for the remainder of the drill campaign. Different techniques that were identified for trial are listed below.

1. Cole directional-scribe method
2. Foster technologies acid-tube method
3. Noranda technology centre borehole camera
4. Clay imprint core orientor
5. Orienting to known bedding

No underground access was available for Mont Porphyre, and it was likely that the depth of the deposit would place heavy reliance on drill-core information. There was underground access available in the porcellanite unit close to the existing Mines Gaspé operations, but none was available into the Mont Porphyre intrusive unit. The Copper Mountain open pit provides some access to the Copper Mountain intrusive, but it is located near to the surface and far from the Mont Porphyre intrusive.

Cole Directional-Scribe Method. This orientation technique involves scribing three lines on the core as drilling progresses. One scribe line is referenced to an orientation lug that is attached to a Sperry-Sun Type B Multishot to reference the top of the hole. The technique was tested on site, but was not successful because of induced rotation of the core tube while coring and the resulting variable spiralling and slipping indicated by the scribe line. The hardness of the rock at Mont Porphyre and the difficulty in reducing the core-tube rotation were not favourable for this type of test. One drawback to this method is that the data-processing phase is fairly extensive and requires expertise with the Sperry Sun Multishot system.

Core Tech Canada Acid-Tube Method (Foster Test). A system was rented from Core Tech Canada (1988) that involved etching an acid tube that is keyed to a slotted core tube. The slotted tube will be referred to as the "Foster tube" in this text, and the overall procedure is noted as the "Foster test." A diamond-tip pen is used to scribe the core through the slots in the core tube after retrieval. The acid test gives the orientation of the high side of the hole (low point on the acid tube) and can be used to offset the scribed reference line. A simple goniometer is used in conjunction with the surveyed hole location to orient the structure in the hole.

Three trials were conducted to examine the applicability of this method. The first trial evaluated the logistics involved and familiarised site personnel with the procedure and equipment. The trial involved setting up a core barrel assembly with the Core Tech equipment and matching the length required by the drill crew, coring a 1.2-m (4 ft) section of rock, leaving the acid to etch, retrieving the core barrel, and then scribing a line on the

retrieved core. The Foster tube will allow for approximately a 1.4-m (4.5 ft) cored section before the core would impact on a pin that holds the acid tube in place. The test was completed in 3 hours and 15 minutes at a depth of 1,159 m (3803 ft), but this included extra time to allow the acid to etch and organise various equipment requirements. The results of the trial suggested that a completed Foster Test would likely take about 3 hours in the 1,200-m (4000 ft) range.

For the Foster test to be successful, the core must not rotate in the core tube before the line is scribed on the core after retrieval. The possibility of rotation was looked at in a second trial that involved completing two separate Foster tests that were separated by a cored length of 34 m (110 ft). The top-of-hole orientation was determined from the results of both Foster tests. A top-of-hole line was scribed on the core from the first Foster test to the second by puzzling together each successive piece of core and using a special scriber prepared at the Noranda Technology Centre (NTC). The top-of-core scribe line matched for both Foster tests and suggested that core rotation was not occurring prior to breaking. A Sperry Sun Multishot survey by Cole Directional was incorporated into both Foster tests to monitor potential rotation. The Multishot results for the first test, using a film speed of 1 frame per minute, did not indicate any rotation. The second Foster test did not show any rotation until the very last frame before breaking occurred, where a 14° change was noted. The actual timing of the last frame before the break has a degree of uncertainty attached and likely occurred after the core break had occurred.

The last-minute uncertainty provided by the Sperry Sun results in the second trial led to a third trial with the Foster test to increase the level of confidence with this orientation method. The third trial consisted of three Foster tests with a 4.9 m (16 ft) cored interval between each test. The third Foster test was necessary because it was impossible to piece together a section of broken core in the first 4.9 m coring interval. The top-of-hole scribe line was again successfully matched between the second and third Foster tests.

Assuming the core can be successfully pieced together, which was very likely in all rock units except the tuff horizons in the case of Mont Porphyre, two separate Foster tests that line up would suggest a good test. Core can then be pieced together and scribed as far as possible on either side. A procedure for core orientation based on two successive tests that line up was developed and adopted for Mont Porphyre. The procedure relied on the ability to successfully piece together core between successive tests.

The Noranda Technology Centre Borehole Camera. The NTC borehole camera can be used as a tool to provide information on joint orientation and rock-mass condition. The camera is designed to fit in a 76-mm-diameter borehole, but unfortunately can only be used to a maximum depth of 600 m (2,000 ft). It incorporates a head assembly that is designed to rotate 360° and provide the ability to locate and orient structure within a borehole. Open structure, stress-induced spalling, fracturing, and joint orientation, location, and condition can all be visually assessed with the camera. Borehole camera logging of structure within the Indian Cove unit at shallower depths was originally proposed as part of the core orientation program. This work was not completed during the drill campaign due to time constraints; however, this work can still be completed at a later date if required.

Clay-Imprint Core Orientor. The clay-imprint core orientor (Call, Savely, and Pakalnis 1982) is weighted by a steel tube half filled with lead and is able to find gravity as it travels down the hole. It requires a drill hole that is inclined between 40° and 70° (Call 1993), so it is not good for subvertical drilling. The Foster test was found to be readable up to 80°, at which point reading the acid tube became difficult. The method involves taking a clay imprint of the core left at the bottom of the hole and then matching the imprint with the top of the next core run. Several trials were conducted to evaluate the clay orientor procedure, and one test was successful at obtaining an imprint at a depth of 1,768 m. The clay imprint method has a lot of potential as it reduces the time of an orientation test by approximately 60% at a depth of 1,200 m. Obtaining a good imprint however was found to be difficult in the deeper holes and the "feel" of the driller was difficult to utilise. With deeper holes, the pressure built up quickly, and the imprint had to be taken quickly in order to avoid blowing a hose. Taking an imprint with new drill bits, when the diamond-impregnated portion was at its thickest, seemed to be more difficult. Adding more clay was tried, but it became unstable and was knocked off. Although the clay-imprint core orientor seems to have had reasonable success at shallower depths (Call, Savely, and Pakalnis 1982), the deep holes at Mont Porphyre presented some problems. Orientation imprints also provided an imprint of the inside of the drill bit, which was useful for evaluating bit wear without pulling the rods.

Orientation to Known Bedding. The stratigraphy at Mines Gaspé can be related to over 100 tuffaceous marker horizons that have been identified. Diamond drilling over the years has identified the regional variation in dip and dip direction of these marker horizons. It was thought possible to orient core from the identification of a known marker horizon, given a rough idea of its dip and dip direction in a particular area. The key to the success of this method is identifying the correct marker horizon and extracting a dip and dip direction from the regional-scale information.

Pin-Imprint Orientor. The use of a pin-imprint orientor was considered, and several contacts made to discuss the applicability of the method. It is similar to the clay-imprint technique except the core imprint is taken by a number of pins. It was felt that it would be difficult to interpret the imprint with the small number of pins usually incorporated into pin-imprint orientation devices on the market. One group, however, had used a 12-pin orientor, which sounded a bit more promising.

Core-Orientation Strategy. The fracture frequencies that were logged during the scope study were used to provide an indication of the required amount of oriented core. It was assumed that at least 100 fractures per rock-mass domain would have to be collected. Initial efforts were to concentrate on the porcellanite, intrusive, and T2 (tuff) horizons.

An orientation procedure was developed based on the Foster test trials and completed with all three crews on one drill to familiarise everyone with the procedure. The procedure incorporated two separate Foster tests that were completed one after the other. A successful test was indicated by the top-of-hole line for each test being within 10° of each other. Once a good top-of-hole scribe line was obtained, core was pieced together as far as possible, and individual structures were oriented to the scribe line. The success of this orientation procedure depended on the ability to piece core together between individual tests and required particular care to be placed on core handling. Piecing together core in the T2 unit was difficult and limited the amount of data collected from this horizon. Core-orientation data was recorded on a separate logging sheet created for the purpose. Oriented core was initially saved from splitting and assaying until the orientation and geotechnical logging ran smoothly. Some lengths of oriented core were kept on a longer-term basis in case additional information was required later.

One Foster test, as outlined in the orientation procedure, took approximately 6 hours at a depth of around 1,200 m. There was a significant amount of "wait time" as the Foster tube was pumped down or retrieved by the wire line. Various operational problems were encountered with broken acid tubes, apparent slipped core on retrieval, and piecing together core between tests. Complete orientation tests frequently required several orientation runs to get

subsequent tests that lined up correctly. However, the orientation program did eventually provide more than 400 oriented joints and much of the credit for this goes to the diligence of several Mines Gaspé personnel involved over the course of the drill campaign.

The orientation of natural fractures was the initial goal of the orientation program. Some of the porcellanite units were noted to have healed structures, which were considered as a potential source of weak break points during the caving process. Some orientation of these structures was completed for reference. Simple drop tests, or "belt-buckle" tests, were also used to evaluate the effect of healed structure. These tests involved dropping samples from belt-buckle height and observing the core breakage after dropping. Drop tests in the P6 unit showed that breaks preferentially followed healed structure first, while tests in the P4 unit showed breaks crosscutting both intact rock and healed structure.

61.4.3 Geotechnical Data Processing

Geotechnical core-logging and orientation data was collected and entered into a spreadsheet for future manipulation. The data display and geostatistical tools available through the Gemcom software that was used at Mines Gaspé was considered for producing the geotechnical sections' associated geostatistical analysis. A compilation of the geotechnical logging data from the 1997 drill campaign is given in Table 61.5.

The geotechnical-logging program from the 1997 drill campaign did not improve the primary and secondary fragmentation predictions from the scope study. It did, however, provide a much better database of structural data for future reference and identified opportunities in the areas of stress fracturing and healed structure. Block-size assessment was one area that could be explored in more detail with the updated structural database and oriented-core information.

Block-Size Assessment from Core. In the geotechnical-logging sheet discussed in Section 61.4.1, there was a reference to recording the distances between natural fractures. This information was collected to create distributions of block lengths, examples of which are illustrated in Figures 61.8 and 61.9. This information is collated from the "distances between natural fractures" and manipulated to provide a distribution of potential block sizes assuming equal dimensions in all directions (Nicholas 1997). It can also be used to extract fracture frequency and RQD information from core without having someone particularly log this information. Both fracture frequency and RQD could be determined from this information since it is related to core depth, recovery and the notation of broken or rubble zones.

The examples in Figures 61.8 and 61.9 show the difference in size distribution using this method for the intrusive and porcellanite rock units. The block-length estimation for porcellanite suggested a primary fragmentation block-size distribution that was 30% finer than 1.2 m. The intrusive unit, which was understood to be more competent from the rock-mass characterisation data, exhibited a block-size distribution that was 12% finer than 1.2 m. Both of these results were in close agreement with an empirical prediction using a software called Block Cave Fragmentation (BCF). BCF was developed by E. Esterhuizen at the University of Pretoria and is commonly used for caving fragmentation prediction. The 2-m³ cutoff was used to relate to the maximum size that could be loaded in a 8-yd³ scooptram.

Oriented Core Data. The oriented core data included 403 joints with the distribution illustrated in Table 61.6. The joints were logged as either natural breaks (n), artificial breaks that looked natural (dn), or natural breaks that did not look natural (nn). The main reason for differentiating between natural and artificial breaks was to provide information on the susceptibility of a particular rock mass to induced breakage along healed or internal structure. It also provided some information on the frequency of artificial and natural breaks. Natural breaks that

TABLE 61.5   Rock-mass domains and associated geotechnical core logging data from the 1997 drill campaign

Rock unit	Description (original lithology prior to alteration)	Logged core (m)	RQD	Fracture frequency (per metre)
York River (YR)	Sandstone/ mudstone	1,924	96	1.30
Indian Cove (IC)	Calcareous mudstone with limestone nodules	4,391	97	1.30
First Tuff Unit (T1)	Brown mudstone/tuff	131	89	1.80
First Porcellanite (P1)	Calcareous mudstone	85	96	1.20
Wollastinite (W1)	Mudstone and siliceous limestone concretions	356	98	0.80
Second Porcellanite (P2)	Calcareous mudstone	71	98	1.10
Second Tuff Unit (T2)	Brown mudstone/tuff + limestone nodules	456	87	2.20
Third Porcellanite (P3)	Calcareous mudstone	70	83	2.50
First Limestone (L1)	Argillaceous limestone	276	97	0.70
Fourth Porcellanite (P4)	Black calcareous mudstone and tuff	1,680	97	0.90
Second Limestone (L2)	Limestone	98	96	0.60
Fifth Porcellanite (P5)	Black calcareous mudstone	314	98	0.70
Third Tuff Unit (T3)	Tuffaceous mudstone	30	94	1.40
Sixth Porcellanite (P6)	Black calcareous mudstone	2,274	98	0.90
Intrusive Unit (In)	Quartz monzonite intrusive	1,559	99	0.50

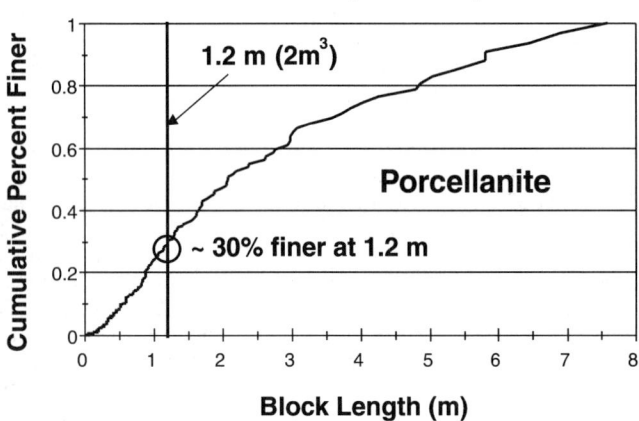

FIGURE 61.8   Block length estimation from porcellanite core

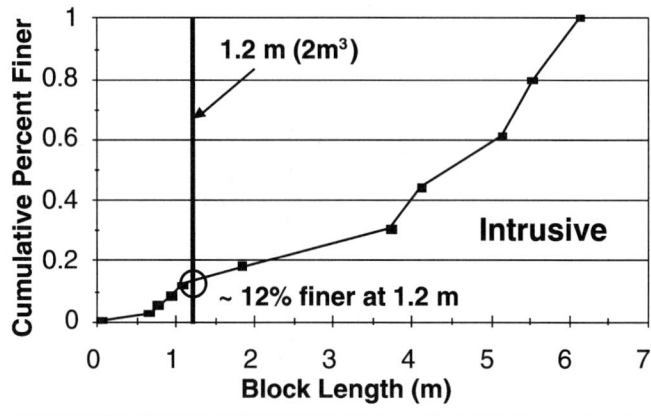

FIGURE 61.9   Block length estimation from intrusive core

were observed during core unloading, but did not look natural, were logged separately and noted with a "nn." They may be related to breaks that occur during the drilling or core retrieval process. Some of this data was collected from observations at the drill by geotechnical logging personnel, but the drill crew eventually marked natural breaks themselves as the core was unloaded. The latter procedure was not as accurate but was considered worthwhile.

An exhaustive analysis of the oriented core data has not been fully completed at this stage and is an opportunity to explore further. Additional work on block-size assessment is proceeding, but at the moment is not first priority. The importance, however, was placed on collecting the information for future use as opportunities develop. Basic stereonets of the oriented core have been created by rock type to identify the potential number of joint sets in each rock unit. A basic premise of cavability in the scope study was that basic block formation would require three joint sets. Where less than three joint sets exist, cavability would rely heavily on stress-induced structure or a mining method that incorporates some form of drilling and blasting to induce additional structure.

TABLE 61.6   Oriented core data

Rock unit	Natural break (n)	Natural break (nn)	Artificial break (dn)	Total
IC	34	16	4	54
T1	10	1	4	15
L1	12	1	13	26
P4	65	14	44	123
P5	19	0	0	19
P6	82	2	8	92
In	31	18	25	74
Total	253 (63%)	52 (13%)	98 (24%)	403

Figure 61.10 illustrates the stereonet analysis performed for the most dominant oriented-core data collected during the orientation campaign. The major plane data extracted from these stereonets is summarised in Table 61.7. These data suggest that there may be three sets in the porcellanite and intrusive units, both of which make up the orebody. The Indian Cove unit and all the data treated together exhibit only two joint sets.

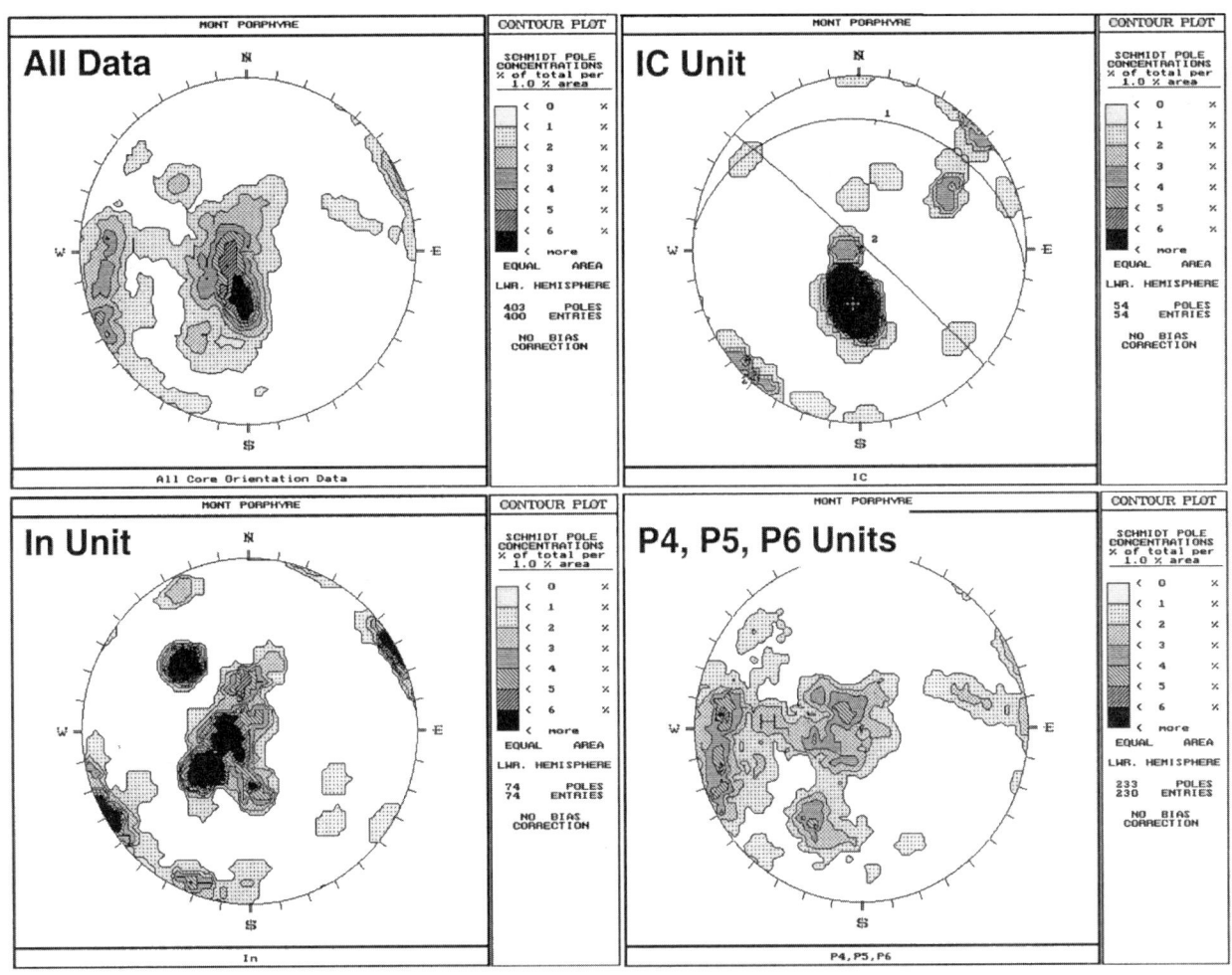

FIGURE 61.10   Stereonets for all oriented data, Indian Cove (IC) unit, intrusive (In) unit and the porcellanite (P4, P5, P6)

TABLE 61.7   Summary of major planes extracted from oriented core data

Data set	No. of poles	Set #1 (dip, dip dir., strike*)	Set #2 (dip, dip dir., strike*)	Set #3 (dip, dip dir., strike*)	No. of sets
All oriented data	403	22/007/277	81/071/341	n/a	2
Natural breaks only	249	22/007/277	85/070/340	n/a	2
Indian cove unit	54	26/007/277	88/042/312	n/a	2
Intrusive unit	74	16/063/333	87/061/331	46/136/046	2–3
P4, P5, P6 units	233	10/132/42	75/089/359	52/028/298	2–3

*Strike is based on the right hand rule.

61.5   CONCLUSIONS

This paper has presented a discussion on the geotechnical investigation—including geotechnical logging and orientation data—of potential cave mining of the Mont Porphyre resource. To date, the investigation has been completed using conventional methods, but using geostatistical analysis to compile geotechical data is also being reviewed. The advantages of using geostatistical methods are that the same tools that permit the visualisation of diamond-drill-hole geological and assay data could be used for geotechnical purposes. Sections can be produced that illustrate parameters such as rock-mass rating or RQD and provide the ability to compile a true 3D representation.

The geotechnical investigation of Mont Porphyre revealed a very competent orebody that would normally eliminate any

further consideration of a caving mining method. Current trends in practice and theory, however, suggest that there may indeed be a possible application of caving techniques to Mont Porphyre. The recent trend in caving experience is towards the exploitation of more competent ore deposits. Northparkes, in Australia, and Palabora in South Africa are two particular operations that are pushing current experience and have or will shortly initiate caving applications in competent rock. Noranda has joined the three-year International Caving Study that has been jointly developed by the Julius Kruttschnitt Mineral Research Centre and Itasca Consulting Group. The main motivation for the project came from the growing worldwide interest in applying caving mining methods to stronger massive low-grade deposits that lie outside current caving experience.

The basic premise of worldwide caving expertise is that anything will cave if sufficient area is available for undercutting. The important issues then become the size of the broken ore and the ability to handle large-size material at the drawpoint. The benefits of reduced mining costs should provide the incentive to further study the application of caving to competent orebodies. Identifying new technologies and methods that can be applied to traditional caving techniques has been a part of the work associated with Mont Porphyre. Methods of inducing and maintaining the cave using open-pit blasting approaches and new technologies associated with the handling of large block sizes are two examples that have been considered.

Subsurface diamond-drill investigations at Mont Porphyre has included detailed geotechnical logging and a successful core-orientation program. The associated data analysis discussed in this paper is far from exhaustive, but suggests that Mont Porphyre will be a challenging application of cave-mining methods. Opportunities for technology development were identified as part of the scope study. Efforts in core orientation and block size characterisation have since complemented the case for Mont Porphyre.

The geotechnical-logging program was designed to obtain as much data as possible and was particularly directed at caving-type applications. Not all of the collected data has been fully explored, but the information remains available for future use. Core orientation can provide valuable information, especially where underground access is not available. The experience with core orientation at Mont Porphyre demonstrated that several orientation techniques can be incorporated into a geotechnical logging program. The geotechnical logging and orientation programs developed for Mont Porphyre can also be applied to the delineation of future deposits.

This paper illustrates that there are many different core-orientation techniques. One technique that could be used at Mont Porphyre for relative core orientation is simply piecing together core in the core shack and orienting structure to a random scribe line. Although this technique does not provide true oriented-core information, it can provide a reasonable representation of joint families if sufficient lengths of core can be puzzled together. In the case of Mont Porphyre, the general competence of the rock units allowed for long lengths of core to be assembled.

Processing tools for dealing with large amounts of geotechnical information, such as the Mont Porphyre data, are not well developed. In the early 1990s, the Noranda Technology Centre had developed a psion-based geotechnical logging software that was incorporated into underground geotechnical mapping work. This technology has since become dated, and recent efforts are associated with utilising pen-based systems and the Microsoft Access database product. Gemcom software has become widely used within Noranda for drill-hole database information, reserve calculation, and geological modelling. The incorporation of geotechnical data processing is within the capabilities of a product like Gemcom, but is only now being aggressively pursued. An alternative is Gocad, which is a 3-D geological modelling software that allows for sophisticated 3-D visualisation of complex data sets and quantitative data integration. This tool is being explored further for handling rock-mass characterisation data.

One effort that grew out of the Mont Porphyre geotechnical logging and orientation program is a project on fracture simulation. This effort is working on the development of a geostatistical method to assess the block-size distribution of a deposit. Geotechnical logging of fracture frequency and core orientation data will serve as input to this technique. The work is currently in progress and has utilised the Mont Porphyre data for model formulation.

61.6 ACKNOWLEDGMENTS

The authors greatly appreciated contributions during the scope study of the project team members and associates involved. The group was led by Bill Rogers and included Gaston Morin, Jacques Gagné, Pierre Bernard, Phil Gaultier, Claude Jacob, Robert Vaillancourt and J.C. Bélanger. We would especially like to acknowledge the efforts of all the personnel who contributed to the geomechanical study, including Paul Germain, J.P. Basque, David Gaudreau, Lionel Catalan, Luc St. Arnaud, and Ken Liu from the Noranda Technology Centre; the Mine and Geology Departments of Mines Gaspé, especially Peter Marenghi and Harold Vachon; external consultants Peter Gash, Dave Nicholas, Maged Rizkalla; and Laval University. During the 1997 drill campaign, special thanks must be given to Mike Cole, and particularly the field assistance from C. Tremblay, E. Stephenson, K. Janssen, and V. Vezeau.

Visits to several caving operations—De Beers Premier Mine, Palabora Mining Company, and the Northparkes Mine—were conducted as part of the work associated with Mont Porphyre. The assistance of various personnel at these operations and the welcome extended by the associated companies is greatly appreciated.

61.7 REFERENCES

Arjang, B. (1991). Pre-mining stresses at some hard rock mines in the Canadian Shield. *CIM Bulletin*, pp. 80–85.

Barton, N., R. Lien, and J. Lunde. (1974). Engineering classification of rock masses for the design of tunnel support. *Rock Mechanics,* 6 May, pp. 186–236.

Bieniawski, Z.T. (1989). Engineering rock mass classifications. New York: Wiley.

Call, R. (1993). *Clay Imprint Core Orientor Manual.* Call and Nicholas. Revised April 1993.

Call, R.D., J.P. Savely, and R. Pakalnis. (1982). A simple core orientation technique. *Proceedings of the third international conference on stability in surface mining,* June 1–3, 1981, in Vancouver. C.O. Brawner, Ed. Society of Mining Engineers of AIME, New York. pp. 465–481.

Core Tech Canada. (1988). Manual of instruction for Core Tech Canada Diamond Drill Core Orientation System.

Coulson, A., S. Nickson, and J. Hussey. (1998). "A geomechanical investigation of the application of caving mining methods to competent rock at the Gaspé Mont Porphyre deposit, Québec, Canada." Presented at the 100th annual general meeting of CIM in Montréal, Québec from May 3–7, 1998.

Dawson, L.R. (1995). Developing Australia's first block caving operation at Northparkes Mines—Endeavour 26 Deposit. *Proceedings of the 6th Underground Operators Conference,* Kalgoorlie, November 13–14, 1995, pp. 155–164.

Gash, P. 1997. (1997). Review of block caving potential. Internal report to Noranda Mines & Exploration Inc., Mines Gaspé Division, Mont Porphyre Project. February 1997.

Hussey, J. and P. Bernard. (1998). Exploration of the Porphyry Mountain Cu-Mo deposit. *Mining Engineering,* Vol. 50, No. 8 (August 1998), pp. 36–44.

Kear, R.M., F. Fenwick, and R.L. Kirk. 1996. The sizing of Palabora Underground Mine. SAIMM Colloquium: Massive Mining Methods, Johannesburg, 1996.

Laubscher, D.H. (1990). A geomechanics classification system for the rating of rock mass in mine design. *Journal of the South African Institute of Mining and Metallurgy,* Vol. 90, No. 10, Oct., pp. 257–273.

Laubscher, D.H. (1995). Cave mining - the state of the art. *Proceedings of the 6th Underground Operators Conference,* Kalgoorlie, November 13–14, 1995, pp. 165–175.

Nicholas, D. (1997a). Review of block caving and induced blast hole caving methods. Internal report to Noranda Mines & Exploration Inc., Mines Gaspé Division, Mont Porphyre Project.

Nicholas, D. (1997b). Personal communication.

Nickson, S., J. Hussey, and A. Coulson. (1999). "Rock mass characterisation for block caving fragmentation assessment at Mont Porphyre." Presented at the 101st annual general meeting of CIM in Calgary, Alberta from May 2–5, 1999.

Potvin, Y., M. Hudyma, and H. Miller. (1989). Design guidelines for open stope support. *CIM Bulletin,* 82, No. 926, pp. 53–62.

Stewart, D., R. Rein, and D. Firewick. (1981). Surface subsidence at the Henderson Mine. *Design and operation of caving and sublevel stoping mines.* D. Stewart, Ed. New York: Society of Mining Engineers of AIME, pp. 203–212.

CHAPTER 62

Rock Support in Theory and Practice

Håkan Stille[*]

62.1 INTRODUCTION

This chapter discusses some aspects on the art of supporting rock to assure safe tunnels and tunnelling. The intention is not to cover the subject fully, but to emphasise the more modern rock-support methods such as untensioned grouted rock bolts (dowels) and steel-fibre-reinforced shotcrete.

The main text is taken from a keynote lecture the author gave at the International Conference on Rock Support in Mining and Underground Construction (Stille 1992). The text has been completed with some new research work that has been judged to be of interest.

The art of supporting rock to achieve the necessary stability and safety in caverns and tunnels has been the subject of innumerable articles, conference papers, textbooks, and theses around the world. Together, these publications describe the complexity of the problem (Gerard 1983).

There are major theoretical problems in describing both the complex *interactions* between the rock mass and the rock support and the *properties* of the rock mass and the rock support.

There are also many practical difficulties involved in identifying the stability problems and describing the rock mass in the correct way. In many cases, the support measures are installed under difficult conditions that affect the quality of the work performed.

In this chapter, three common stability problems (see for example Hoek and Brown (1980), Brady and Brown (1985), Feder (1986), or Maury (1987)) are dealt with. The three problematic areas are:

1. A fallout of blocks generated by existing joints and weaknesses in the rock. The driving force is gravity, and the deformations are restricted in certain directions due to the local stress field.

2. General shear failure in the rock mass caused by overloading from the existing stress field. Shear failure of the type bearing capacity failure in weak sidewalls, or roof failure due to incomplete arch formation come under this category.

3. Problems of instability due to high stresses. The intensity may vary from surface splitting, spalling, bending and buckling of slabs, up through explosive failure rockbursts, the effects of which extend far into the rock.

The failure mechanisms that are the underlying cause of these problems are discussed as well as the possibility of stabilising the rock with bolts and surface support.

62.2 POSSIBILITIES FOR INFLUENCING THE DEVELOPMENT OF FAILURE MECHANISMS

62.2.1 Introduction

In this section, we analyse the available options for influencing the failure mechanisms and failure development. Each failure situation or emergence of a failure mechanism is due to a load effect, S, being greater than the bearing capacity (resistance), R. Therefore, the emergence of the failure mechanism or its development can, in principle, be influenced by changing the load effect or the resistance capacity. The term "rock support" usually implies increasing the capacity of the rock to resist loads. However, both factors (i.e., reducing the load effect as well as increasing the resistance capacity) are discussed in this section. In many cases, the load effect, S, and the resistance, R, are dependent on the deformations that have arisen. The classical illustration of this is the ground-reaction curve. This describes how the required load decreases with the increased deformation of a cavern or a tunnel for example, while at the same time the resistance provided by the support increases. See for example Brown et al. (1983).

This basic concept is important to properly understand the rock-support problem, and it is, in principle, applicable to all three of the failure mechanisms described here. The concept has gained its widest acceptance in connection with the analysis of the general shear failure around circular tunnels in an isotropic stress field. The concept is usually illustrated as in Figure 62.1.

62.2.2 Possibilities of Influencing Block Falls

Block falls are caused by weakness zones and the existing joint system. The driving forces are dead weight and the local stress field around the block. A block that is located near the contour of a future cavern may be subjected to a shear movement when the cavern is excavated. This occurs when the local stress conditions, after excavating the cavern, are such that the shear strength is exceeded in one or more of the joint systems adjacent to the block (see Figure 62.2).

One method of reducing the load on the future support is to dislodge loose blocks of rock by scaling. This is a suitable method when there are occasional loose blocks. If the block serves a so-called "locking effect" (i.e., the bearing capacity of an arch is at risk if the block falls out) scaling may cause catastrophic consequences and initiate the development of progressive failure. In such cases, the block must be held in place by support.

The load that is relatively well defined is the weight of the block. The support must at least be capable of bearing this load. If shearing is present along any of the joint systems surrounding the block, the support will be subjected to additional loads that are

---

[*] Division of Soil and Rock Mechanics, Royal Institute of Technology, Stockholm, Sweden.

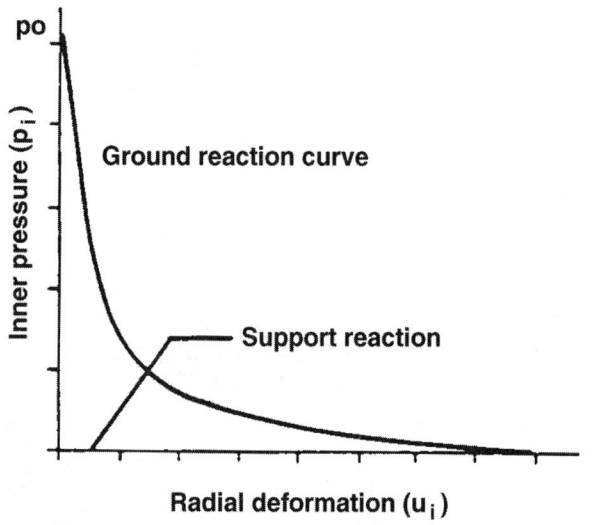

FIGURE 62.1 Example of a ground reaction curve

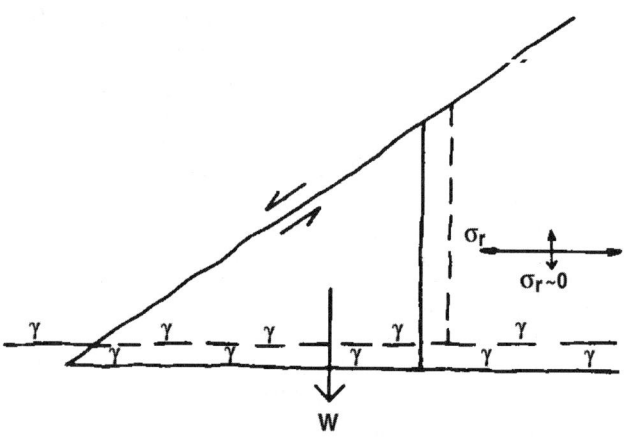

FIGURE 62.2 Block movement during excavation of a rock cavern

difficult to define. The size of the additional load depends on the size of the shear deformation and the stiffness of the support in relation to the rigidity of the rock. In most cases, applying support will not prevent the type of deformation that is restricted in certain directions. The support must, therefore, be capable of withstanding this displacement and retaining sufficient bearing capacity to enable it to carry the load of the block. There are many examples of rock sections that have fallen down in spite of the fact that the support was fully capable of carrying the dead weight of the block in the rock section. This applies to the rock support provided by bolts as well as by shotcrete.

62.2.3 Possibilities of Influencing General Shear Failure

Shear failure occurs in the rock mass when the shear stress exceeds the strength of the material (i.e., when the difference between the maximum and minimum principal stresses becomes too large). If the strength of the rock mass can be described by Mohr-Coulomb's failure criterion, then the following relationship between the maximum and minimum principal stresses at failure is valid

$$\sigma_1 = \sigma_{c,m} + ((1 + \sin \phi)/(1 - \sin \phi))\sigma_3 \qquad (62.1)$$

where $\sigma_{c,m}$ is the uniaxial compressive strength of the rock mass. The factor $((1 + \sin \phi)/(1 - \sin \phi))$ is called the pressure coefficient,

K_p, for passive pressure and is in the region of 3–9. Thus, an increase of σ_3 gives an advantage in the form of increased bearing capacity.

Failure of Bearing Capacity. The ground-reaction-curve concept can also be used to describe bearing capacity failure. At increasing deformations, the need for resistance decreases until the load-bearing capacity of the rock is reached (i.e., bearing-capacity failure has occurred in the rock).

According to the classical theory, three types of bearing failure under pillars are identifiable. These are general shear failure, local shear failure, and punching shear failure; see Terzaghi (1943) or Lambe and Whitman (1969).

General shear failure is characterised by a well-defined ultimate load. The ultimate load is only slightly higher than the load that causes local shear failure. Penetration of the pillar corresponds to the heave of the surface.

Punching shear failure is characterised by compression of the material under the loaded area. The heave of the surface is minor.

Local shear failure is characterised by a failure sequence that lies somewhere between general shear failure and punching shear failure. Shear failure starts below the pillar and comes to a halt out in the rock mass.

The development of a bearing failure can be influenced by measures taken on the load side. These include cutting slots for relieving pressure, altering the geometry of the cavern, and providing extra supporting pillars.

Resistance can be influenced by altering the geometry so the shape of the slip surface is affected. By bolting it together, the slip surface can be forced to go deeper, thus increasing the bearing capacity. It is also possible to increase the bearing capacity by installing lateral support in the form of fill, a strut, or bolts anchored outside the slip surface. The strength along the slip surface can also be increased by absorbing part of the shear force in the tensile force reaction of the bolts. The strength is also increased by the dowel effect. In all of these techniques for influencing the shape of the slip surface, it is essential to ensure that the supported surface is stable locally. A local failure in the surface may initiate a progressive development ending in a major bearing-capacity failure.

Bearing capacity that corresponds to general shear failure in a material having both cohesion, c, and friction, ϕ, can be described by the following equation

$$q_u = c N_c + q_0 N_q + 0.5 \, \gamma \, g \, B \, N_\gamma \qquad (62.2)$$

The factors N_c, N_q and N_γ are functions of the angle of internal friction, ϕ, of the material according to Figure 62.3.

A moderate support effort q_0 can mean a great increase in bearing capacity. For example, backpressure can be achieved by backfilling or support by bolts (see Figure 62.4).

The presence of the bolts in the rock mass will also affect its properties by impeding the occurrence of sliding along the joint system and rotation of blocks. In this way, the rock-mass cohesion is improved, resulting in increased bearing capacity.

Reduction of arching capacity due to shear failure. Arching implies that a pressure-head curve develops so the load is transferred to the surroundings. A usual cause of failure is that shear failure occurs due to (a) an unfavourable angle between the pressure-head curve and the weakness zone joints in the rock, or (b) the rock blocks are broken by high pressure. The bearing capacity of the arch can be described, in principle, as follows:

$$q_{max} = min \begin{cases} 8H f_{max}/L^2 - q_0 \\ 2H (\tan\phi)/L - q_0 \end{cases} \qquad (62.3)$$

where H is the horizontal component of the resultant pressure, f_{max} is the height of the arch, L is the span width of the arch, and

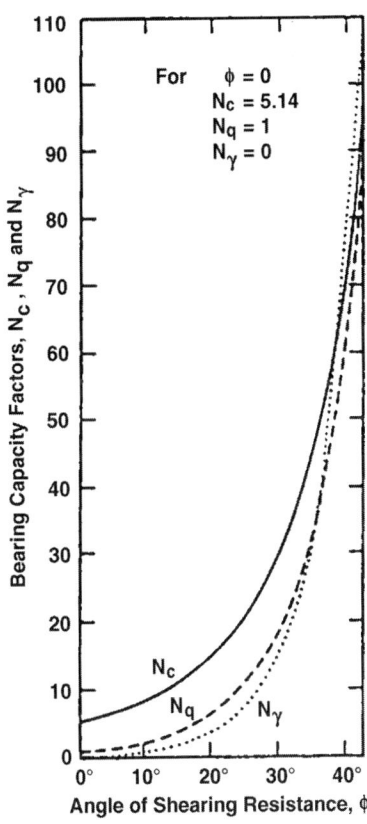

FIGURE 62.3   Bearing capacity

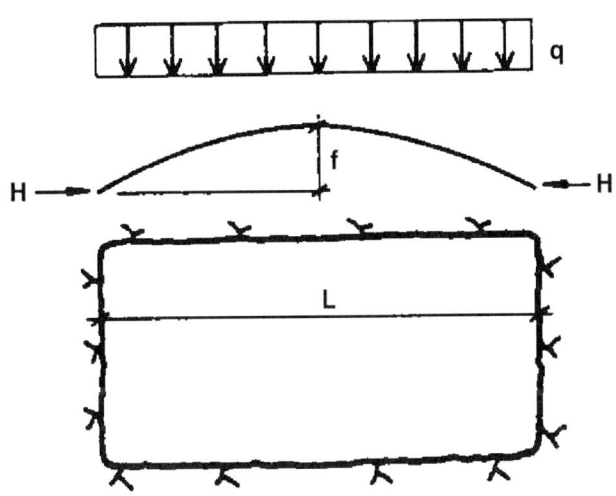

FIGURE 62.5   Example of arching

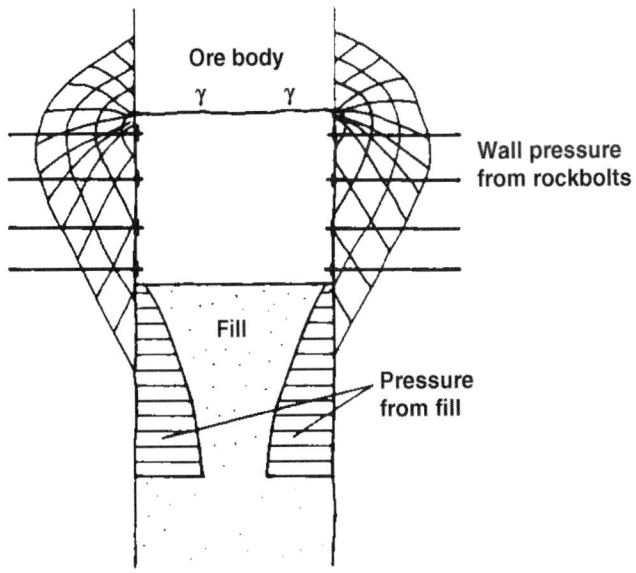

FIGURE 62.4   Different types of back pressure to increase bearing capacity

ϕ is the effective angle of friction. The capacity of an arch to carry loads in excess of its own dead weight also depends on the weight of the arch q_0.

The stress resulting from the compressive stresses passing through the arch must be less than the compressive strength of the rock mass. An example of arching is given in Figure 62.5.

Altering the geometry of the cavern has an influence on the load effect, but also, and to a greater extent, on the dead weight of the arch, the span of the arch, and the horizontal force.

It is important to consider the abutment's capability to absorb the necessary horizontal forces from the arch. To a certain extent, it is possible to use rock support to assure the arch's capacity to absorb load. Firstly, parts without bearing capacity can be suspended in stable arches located higher up. Secondly, by creating a radial pressure (σ_ρ) it is possible to increase the capacity of the rock mass to take up the compressive stresses (σ_{arch}) that are found along the pressure-head curve, so that the following is valid:

$$\sigma_{max} = \sigma_{c,m} + k_p\,\sigma_r > \sigma_{arch} \qquad (62.4)$$

where $\sigma_{c,m}$ is the uniaxial compressive strength of the rock mass. Radial pressure can be created by means of struts, bolts, or a shotcrete arch.

To change the shape of the pressure-head curve by means of support—e.g., by significantly altering the friction capacity of joints and weakness zones—usually requires so much support that it is unrealistic.

Because confinement increases with distance from the opening at the same time as the load decreases, it is usually feasible to get stable arches at a distance of about one time the height of the opening. This means that the rock just above the opening must carry a corresponding load.

When supporting the arch, account must be taken of the fact that the rock must not break up and gradually loosen round the installed support.

62.2.4   Possibilities of Influencing Instability Problems

To avoid problems with instability, there are several measures that can be taken to decrease the load on the unstable rock section. These include, for example, altering the geometry so the load is reduced or cutting slots that relieve the pressure on the critical area of the rock.

Spalling.   Primary fracturing with micro cracks parallel to the direction of load starts when the tensile strain in the rock exceeds a critical value, ε_c, that is typical for the type of rock (70–170 micro strain) (see Stacey 1981). Once micro cracks have formed, the failure may develop into an instability failure such as spalling and buckling of slabs or shear failure.

To determine whether there is a risk of the rock mass around a tunnel or a rock cavern splitting, the magnitude of the tensile strain occurring in the rock mass with excavation of the opening can be calculated.

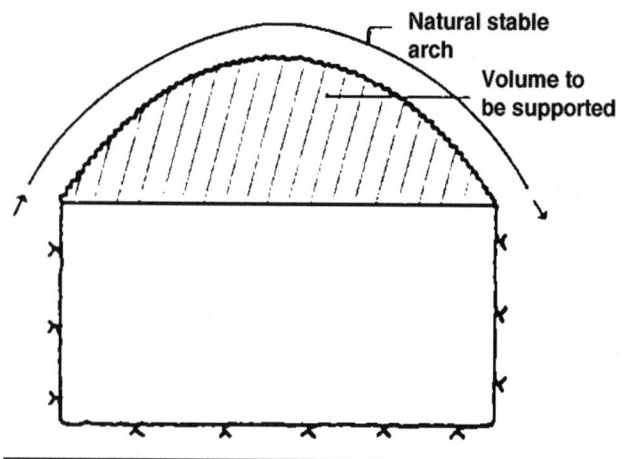

FIGURE 62.6 Size of the zone to be supported

In plane strain conditions, the following relation for calculating the tensile strain, ε_3, is valid. To avoid splitting, ε_3 must be less than the critical strain ε_c:

$$\varepsilon_3 = \frac{1}{E}[(1 - v^2)\sigma_3 - v(1 + v)\sigma_1] < \varepsilon_c \qquad \textbf{(62.5)}$$

The tensile strain around an opening can be reduced by making the contour more curved or by applying a surface pressure. The surface pressure also helps the rock mass absorb a higher pressure parallel to the rock surface. If the resistance generated by the surface pressure is sufficiently large, there is no splitting, but other types of failure (such as shear failure) may occur.

Applying sufficient radial pressure with just the aid of bolts to prevent splitting is not feasible given a realistic distance between the bolts. The rock must be allowed to split. The task of the support is to keep the rock mass in place once it does split.

The volume of the rock between the natural, stable arch and the actual contour of the opening must then be carried by arching or by suspension with the aid of support (see Figure 62.6).

The bolts and the surface support must also withstand the deformations that occur when the rock mass splits and breaks up.

In the case of brittle types of rock, splitting and spalling occur explosively in connection with release of energy (rock burst). This should be especially noted.

Slab Buckling. If the rock mass originally contains weakness zones that are parallel to the contours of the cavern, this may result in the formation of thick slabs extending over the entire span of the opening.

If the stress parallel to a slab is high, the slab will buckle. A differentiation is made between elastic buckling, which is an instability problem, and buckling due to bending, which occurs much more slowly. Elastic buckling occurs when the load reaches a certain critical level. The slab is subjected to instantaneous major bending. Bending or buckling occurs when the slab is initially bent to a certain extent, Y_0, for example from a load perpendicular to the plane of the slab. The load may be caused by the dead weight of an overlying, nonload-bearing layer. Bending will increase as the load on the plane of the slab increases. Bending in the centre of the plate increases according to the following equation:

$$Y_{middle} = Y_0 / (1 - P/P_k) \qquad \textbf{(62.6)}$$

where P_k is the elastic buckling load and P the actual load. If the material in the slab has unlimited strength, the maximum bearing capacity will be equal to the elastic buckling load (see Figure 62.7).

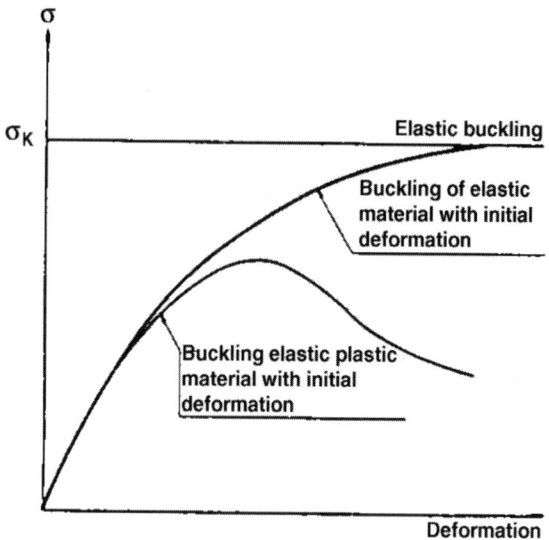

FIGURE 62.7 Relation between load and bending at elastic buckling and bending/buckling

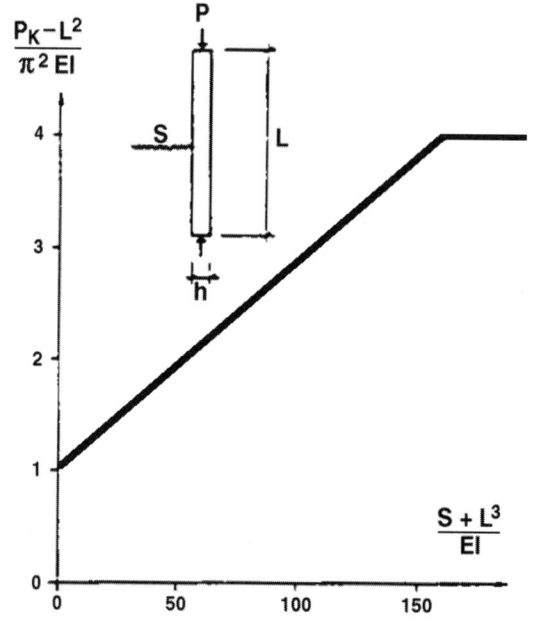

FIGURE 62.8 Variation of buckling load with stiffness of support

In connection with bending the slab, the moment in the slab will increase according to:

$$M_{max} = M_0 / (1 - P/P_k) \qquad \textbf{(62.7)}$$

where M_0 is the moment corresponding to the bending Y_0. The maximum stress in the slab will be:

$$\sigma_{max} = P/A + M_0 / (W(1 - P/P_k)) \qquad \textbf{(62.8)}$$

where A is the cross-sectional area of the slab, and W is its section modulus. As the strength of the rock is limited, failure occurs in the slab at a certain degree of bending. Failure can occur as tensile failure, compressive failure, or shear failure depending on which strength is exceeded first. These failures can limit the bearing capacity of the slab parallel to its plane.

Placing support at the centre of the slab can increase the buckling load (see Figure 62.8). This applies to elastic buckling as well as to bending/buckling. The extent of the increase in the buckling load depends on the elastic properties of the support. If the stiffness of the support is above a certain level, buckling will take place between the new support and the end of the slab. The buckling load cannot be increased further without introducing more support at different points.

The effective stiffness of the bolts may be low if the layer in which the bolts are anchored has low rigidity and low bearing capacity.

The bolt load can be calculated according to

$$P_{bolt} = SY_{middle} = SY_0 / (1 - P/P_k) \qquad \text{(62.9)}$$

where P_k is calculated according to Figure 62.8, and Y_0 represents the slab's bending due to the load applied perpendicular to the plane of the slab.

62.3 BOLTS—BEARING CAPACITY AND ABILITY TO WITHSTAND DEFORMATION

62.3.1 Introduction

This section reviews the performance of bolts used as support for the failure mechanisms discussed in Section 62.2.

Bolts that are intended to control the rock mass in its post-failure stage cannot be regarded as isolated bolts. The rock mass, bolts, and surface support interact as a unit. The bolt plates and surface support must be capable of withstanding deformation and be of equal strength, otherwise the bolt plates punch into the rock mass and it flows and falls out around the bolt plates.

The active part in this system is the rock mass. The deformation and failure behaviour of the rock mass determine the strain on the bolts (see Stille 1987). Splitting, shear failure, and joint movements impose different loads on the bolts.

In Section 62.2, it was shown that a block can be subjected to deformations in the form of shear movements along the surrounding joint system. These movements may be restricted in certain directions. The bolts that are to hold the block in place will, therefore, be subjected to both a shear movement and to an axial load, T_a. The shear movement gives rise to a lateral force, T_s, in the bolt.

If the failure development results in shearing of the rock mass along existing or new joint systems, as is the case with bearing capacity and shear failure in an arch, then the bolts must (a) be capable of withstanding these shear movements and (b) allow the rock mass to shear to reduce the stresses. At the same time, the bolt must retain its load capacity in the axial direction.

If the dominant failure mechanism is the splitting of the rock, the bolts must be capable of withstanding the major axial strains that arise when the rock is split and the rock slabs are pushed or buckle outwards. The increase in volume that this involves must be accommodated, or the forces acting on the support will be too great, and the bolts will either be pulled out or the bolt plates will punch into the rock mass.

When the rock splits, the deformations may be rapidly transferred to the bolt when the joint is formed. In the case of a fully grouted bolt, this deformation must be taken up in a short stretch above the joint. The high load velocity and the limited area of strain may cause brittle failure.

Stjern (1995) studied the effect of blasting and loading on untensioned cement-grouted bolts. He found that, after three days of curing, there is no difference in the pull-out strength of bolts installed close to the blast, and those more distant. Furthermore, there is no difference in the long-term pull-out strength for bolts with young grout compared to bolts with cured grout, when subjected to blast loading. Repeated pulling of previously pulled bolts show a healing effect of the grout. If the grout was subjected to large deformations after the true hardening of the grout had started (after 12–24 hours) a decrease in the long-term pull-out strength was recorded. Because the anchor length is nearly always designed with a very high safety factor, this reduction will

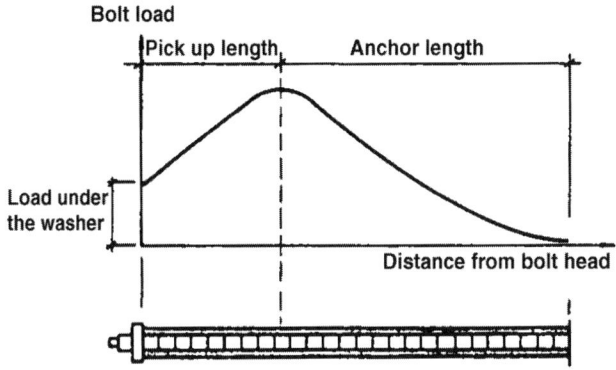

FIGURE 62.9 Load distribution along the bolt according to Farmer (1975)

not have any significant influence on the behaviour of grouted bolts in practical tunnelling. These results confirm the observations of other authors such as Krauland (1993), Stillborg (1984), and Littlejohn et al. (1987) who have researched the subject.

The following general demands can be made on a rock bolt.

- It must be capable of accepting a substantial tensile load.

- It must withstand the deformations restricted in certain directions to which it is subjected, both in an axial direction and by shearing along the joint system.

Tensile Loads on Bolts. The primary function of the bolt—to take up tensile loads—means that the load distribution in the bolt can be divided into two parts. The first part is where the load develops in the bolt (i.e., the bolt supports the rock). In the second part, the load is transferred from the bolt to the rock, which is the actual anchoring zone, as shown in Figure 62.9 (Farmer 1975). The ability to transfer load between the bolt and the rock depends on the type of bolt, the quality of the rock and, where relevant, the quality of the grouting. The presence of the nut and plate and the degree of contact affect the way in which the load develops in the bolt. The importance of good contact has been shown by Stille, Holmberg, and Nord (1989). The transference of load between the bolt and the rock has been studied both theoretically and by experiment. A variety of pullout tests have been performed and reported on in the literature. Stillborg (1991) has done a comparative study on different types of bolts. The interesting point about Stillborg's pullout tests is that they have been done without resistance acting as a load on the joint surface. From this point of view, the load situation in the tests is similar to that which pulled-out bolts are subjected to in reality. Some of the test results are presented in Figure 62.10. These test results are typical of what can be found in the literature. Grouted bolts normally show great rigidity and break when the bolt is pulled out. Anchor bolts show major deformations caused by sliding in the expander. Yielding Swellex and Splitset are of the friction bolt type and normally have a marked slip load.

A theoretical study of the load development along the bolt was conducted by Farmer (1975) and Aydan et al. (1985) among others. The elastic behaviour of grouted bolts can be described by the following equation:

$$T_t = \frac{E \, \pi r_b^2 \cdot \alpha}{\lambda_{el}} \cdot U_t \qquad \text{(62.10)}$$

$$\alpha = \sqrt{\frac{2K_g}{E r_b}}$$

where E is the modulus of elasticity of the steel, r_b is the radius of the bolt, K_g is a stiffness factor, and λ_{el} is the empirical correction factor.

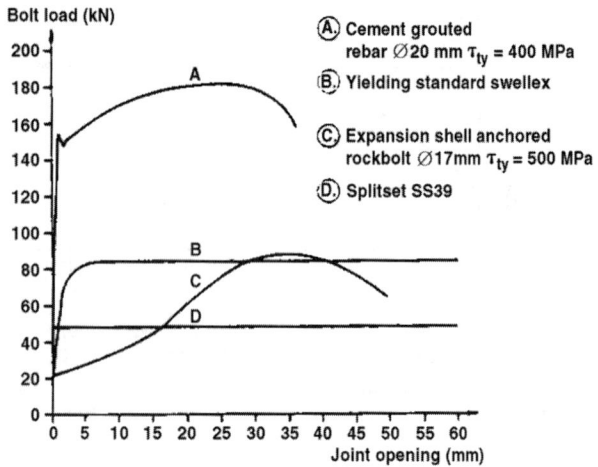

FIGURE 62.10 Tensile loading tests on different rock bolts according to Stillborg (1991)

TABLE 62.1 Capacity of bolts to take up loads

Type of bolt	Strength	Pullout load	Load capacity before greater deformation (0–5 mm)	Deformation failure
Cement-grouted rebar	Steel bar	200–600 kN/m*	Steel yielding strength	30–40 mm
Resin-grouted rebar	Steel bar	500–1000 kN/m	Steel yielding strength	10–20 mm
Swellex: Standard Yielding	Steel tube Friction Capacity	100–200 kN/m 100–200 kN/m	Steel yielding Strength	10–25 mm sliding
Splitset	Friction Capacity	20–50 kN/m	Friction capacity	Sliding
Expansion shell	Shell capacity	60–180 kN	Prestress load	30–50 mm

*W/C ratio 0.30–0.32

The corresponding equation for anchor bolts is:

$$T_+ = \frac{U_t}{\dfrac{\ell}{E\,\pi\,r_b^2} + \dfrac{\ell}{K}} \qquad (62.11)$$

where K is the spring stiffness of the expander and $\ell\,1$ is the free length of the bolt (see Hoek and Brown 1980). Analyses of the pullout tests point to the fact that the stiffness of the bolt is usually 3–10 times lower than is implied by the actual properties of the bolt steel.

Two reasons for this could be (1) slipping at the point of fixture or in the grout, and (2) the lack of appropriate theories.

A summary of the most important properties of some different types of bolts is given in Table 62.1.

When grouted bolts are used, the quality of the grouting is important as adhesion decreases at higher water-to-cement ratios. The quality of the rock is also important as lower rock strengths usually mean lower pullout loads. Most bolts show large load capacities even at relatively small deformations. To enable expansion-shell-anchored bolts to be utilised to their maximum capacity, large deformations are required, which may mean that the rock has started to break up. It should be noted that resin-grouted rebars and standard Swellex bolts have the lowest resistance to deformation.

Bolts Subjected to Shear Loads. Besides the intended tensile load, many bolts are subjected to shear deformations in

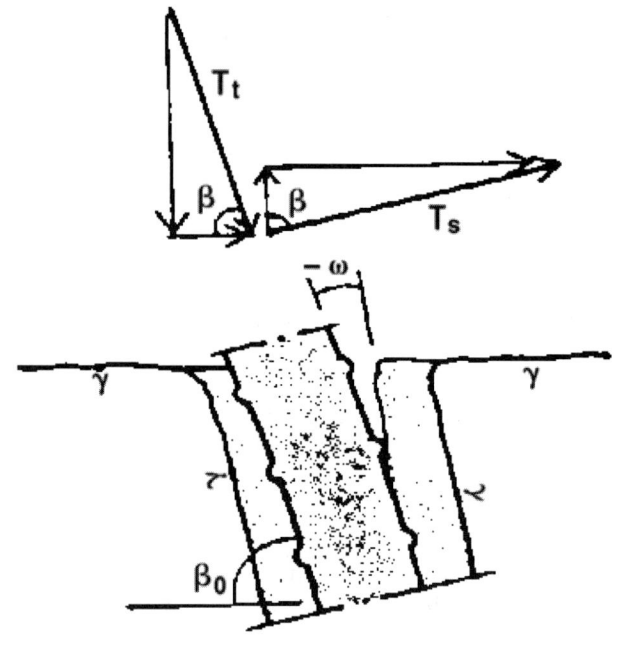

FIGURE 62.11 Load situation in a bolt at shear

certain directions. Sometimes the bolts are cut off when deformations are relatively small, and sometimes a crank shape develops and the bolt can withstand very large deformations. Several researchers have carried out experiments relating to this problem. Since the early studies of Bjurström (1973), many shear tests have been performed. Special mention can be made of Dight (1982), Ludwig (1983), Spang and Egger (1990), and Schubert (1984). Theoretical studies have also been performed—for example, Holmberg (1991), Schubert (1984), and Azuar and Panet (1980).

These tests and theoretical analyses have made it possible to describe and explain the very complex interplay of forces that control the bolts' capacity to take up shear.

In connection with shearing, both the tensile loads T_t and the shear forces (true dowel effect) T_s occur in the bolt. Depending on the strength of the rock and the grouting, the quality and geometry of the bolt, and the inclination of the bolt and the tensile load, the load and deformations in the bolt will develop differently.

The interplay of forces in the bolt is shown in Figure 62.11. With the designations according to the figure, the contribution T_0 to the joint's ability to take up shear is obtained according to the following:

$$T_0 = T_s(sin\beta - cos\beta \cdot tan\gamma) + T_t(cos\beta + sin\beta \cdot tan\gamma) \qquad (62.12)$$

where T_s = true dowel effect

T_t = tensile load in the bolt

β = angle between the bolt and a normal to the joint

γ = friction angle of the joint

β = angle between shear plane and bolt

In weak rock, the bolt will gradually crush the rock. The tensile load component will develop while the dowel component will decrease due to the deformation of the bolt and the formation of a crank shape. The bolt will break due to the fact that the tensile strength is exceeded (see Figure 62.12).

In hard rock, the shear force becomes so large due to the high shear resistance that the bolt will cut off before a crank

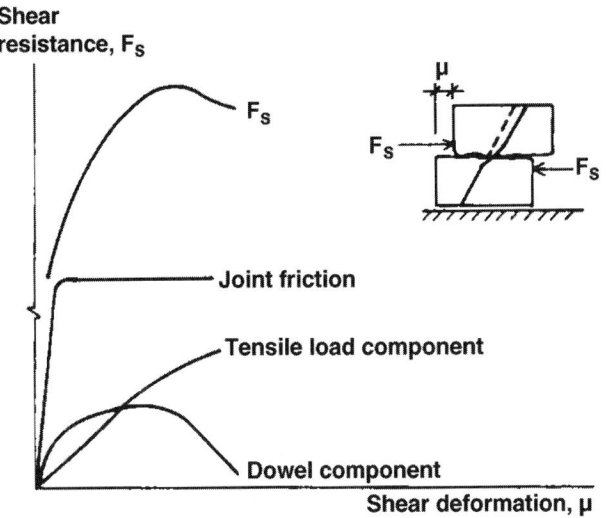

FIGURE 62.12 Diagram showing principle of relationship between different components

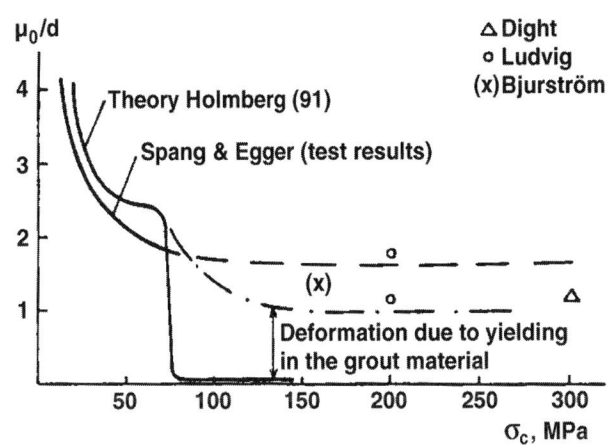

FIGURE 62.13 Relative ability of fully grouted rebars to take shear deformation for different rock strengths

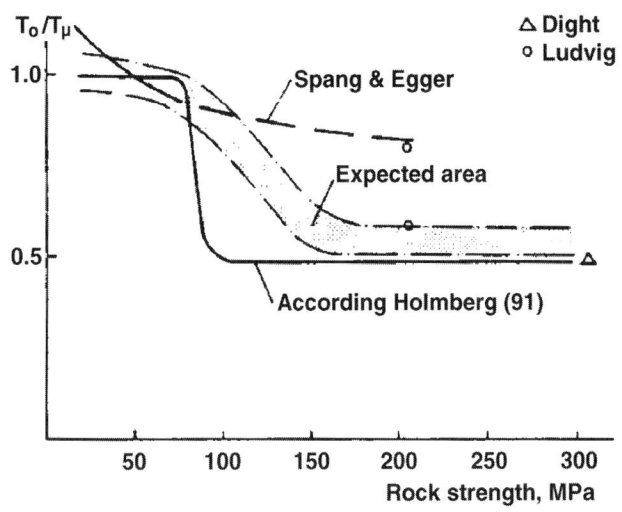

FIGURE 62.14 Additional shear resistance capacity from the bolt in relation to rock strength for perpendicular bolts

shape has had time to develop. The dowel effect is great here, whereas the tensile force has not had time to develop. The ability of the bolt to take up shear deformations is greatest in bolts installed perpendicular to the joint.

Spang and Egger's comprehensive tests enabled them to establish an empirical connection between relative shear deformation at failure (U_o/d) and the strength of the rock, σ_c'.

$$U_o/d = (15.2 - 55.2\sigma_c^{-0.14} + 56.2\sigma_c^{-0.28}) \cdot$$

$$\left(1 - \left(\frac{70}{\sigma_c}\right)^{0.125} \cdot \frac{\tan\beta}{\sqrt{\cos\beta}}\right)$$

(62.13)

where β is the angle between the bolt and a normal to the joint. The relationship is valid for strengths of less than 70 Mega Pascals (MPa).

Holmberg (1991) studied the behaviour of untensioned grouted bolts subjected to shearing along a joint. His theories take into account the fact that shear deformation at failure depends on the failure strain of the steel, which is not considered in the empirical equation. However, his equations do not take into consideration the deformations that occur as a result of crushing the grouting material. The strength of the cement mortar is normally 50–70 MPa. The usual diameter of the borehole is about double that of the bolt. This means that additional deformation up to the size of the bolt diameter can be expected in hard rock. Shear deformations in the actual bolt are usually negligible (a few millimetres).

Holmberg's analyses, therefore, show too rigid a behaviour at higher rock strengths. Figure 62.13 presents Spang and Egger's empirical relation together with calculations based on Holmberg's theories for ϕ20 mm rebar and some other test results.

The figure can be used to assess the extent of the shear deformations that a bolt is capable of withstanding. This applies to the case of perpendicular bolts not subjected to tensile loads. With regards to other types of bolts, experience is limited. The lower stiffnesses of both Swellex and Splitset bolts indicates that their ability to crush rock is less, which means that lower values of relative shear deformations are to be expected. According to Ludwig, the tests show that the bolts usually break at a deformation corresponding to the diameter of the drill hole.

The influence of the bolt on the shear resistance of the joint also depends on the strength of the rock. At low strengths, a

crank shape is formed, which means that the shear resistance of the joint increases with the tensile strength of the bolt T_u. According both to Holmberg and to Spang and Egger, this has been verified.

According to Holmberg's theories, there is a well-defined limit between the formation of a crank shape and the point at which shearing occurs. In the case of pure shearing, a load capacity of about half the tensile strength can be expected. However, additional deformations from the cement grouting have a compensatory effect.

According to Spang and Egger, the maximum bolt contribution will be as follows:

$$T_o = T_u[1.55 + 0.11\sigma_c^{+1.07}\sin^2(\beta + i)] \cdot$$

$$\sigma^{-0.14}(0.85 + 0.45\,tan\,\gamma)$$

(62.14)

Figure 62.14 shows the way in which the contributory load varies with the strength of the rock, as determined by different researchers. The anticipated contribution is also shown in the figure. More studies are needed, however, especially in the sensitive area of 100–150 MPa.

With increasing angle (β) of the bolt with the normal to the joint) the shear contribution increases. However, at the same time,

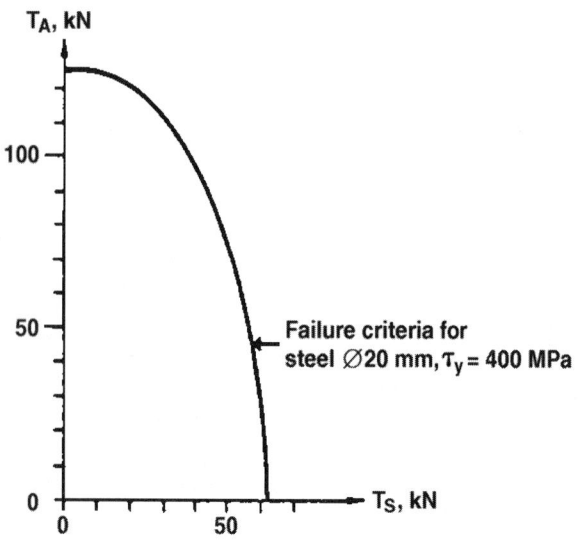

FIGURE 62.15   Limit curve for yield load in a bolt subjected to both axial and laterial load

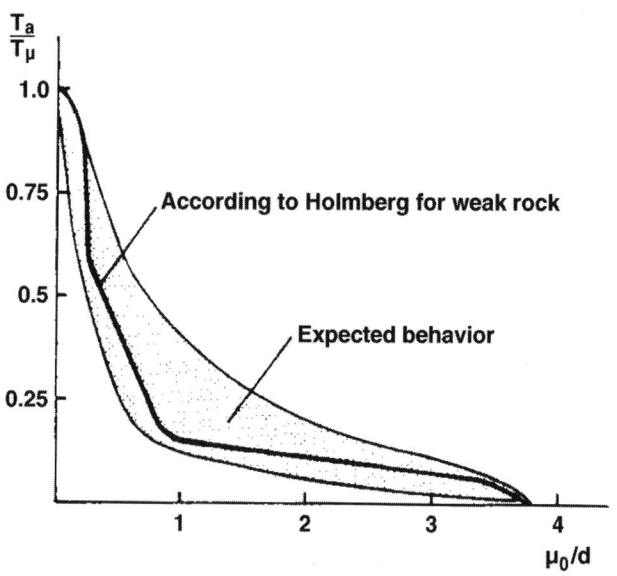

FIGURE 62.16   Possibility of taking different shear deformations for different outer load T_a in the bolt, for weak rock

the bolt can withstand lower shear deformations. This is true provided that the bolt is not required to take up other tensile loads.

Bolts with Combined Tensile and Shearing Loads.   Bolts that are already subjected to tensile loads when shearing starts can withstand minor deformations. The background to this is that the failure conditions are linked. In a bolt subjected to both a tensile force T_a and a lateral force, T_s, shear failure occurs when the following conditions are fulfilled:

$$(T_a / T_{ty})^2 + (2T_s / T_{ty})^2 = 1 \qquad (62.15)$$

T_{ty} is the load corresponding to the tensile strength.

The failure condition is represented in Figure 62.15. Because the initial stiffness is normally greater in connection with shearing than with pulling, relatively minor deformations give rise to large shear loads. If the bolt is already subjected to a considerable tensile load, shear failure will occur. In the case of minor tensile loads, however, a crank shape may develop. In that case, larger shear deformations can be taken up. Using Holmberg's theories, it is possible to calculate a relationship between shear deformation and outer tensile load in the bolt. In Figures 62.16 and 62.17, a theoretical account of the expected behaviour is given for both soft rock ($\sigma_c \approx 50$ MPa) and hard rock ($\sigma_c = 150$ MPa). In both cases, the bolt becomes much more sensitive to shear deformations if it is axially loaded right from the beginning.

Due to local failure in the grouting and the rock, a somewhat more flexible behaviour is expected. However, the problem needs to be studied more closely.

Requirements and Choice of Bolt Support for the Different Failure Mechanisms.   Despite intensive research in this field, there are several important factors still to be investigated. However, some general conclusions can be drawn.

- The ability to withstand deformation decreases with the strength of the surrounding rock.

- In hard rock, the bolts are cut off at a load that may be considerably lower than that which a bolt in soft rock is capable of taking up as an outer shear load.

- If the bolt is utilised for an outer load, its ability to withstand shear deformations is drastically reduced. The bolt's capacity to bear shear loads does not decrease as much.

- Inclined bolts that are loaded both axially and transversely in connection with joint deformation have greater

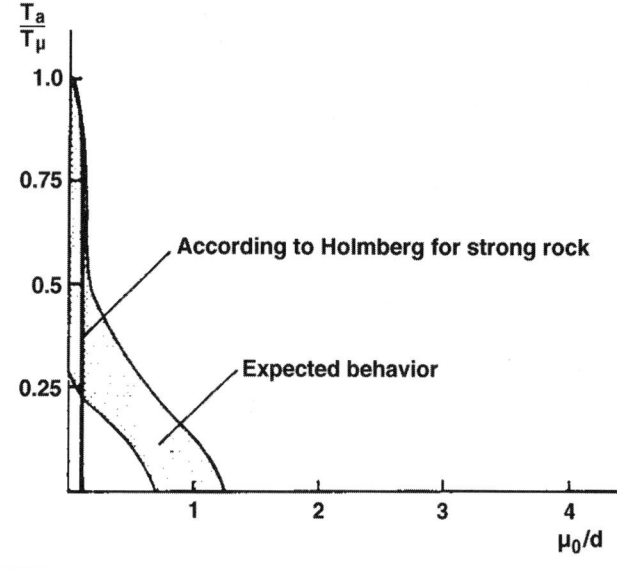

FIGURE 62.17   Possibility of taking different shear deformations for different out load T_a, in the bolt for strong rock

stiffness and are, therefore, less able to withstand deformations.

- Anchor bolts normally do not have sufficient stiffness to prevent local loosening of the rock around the bolt.

Friction bolts of the standard Swellex and Splitset type may be less capable of withstanding deformation than grouted bolts in medium-hard rock.

62.4   BEARING CAPACITY AND DEFORMATION RESISTANCE OF SURFACE SUPPORT

62.4.1   Introduction

The bearing capacity of surface support and its deformation resistance (toughness) have a decisive effect on the ability of the surface support to interact with the rock and stabilise the surface.

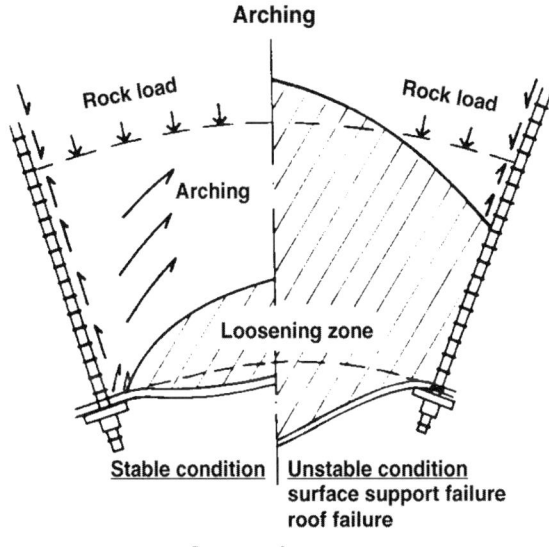

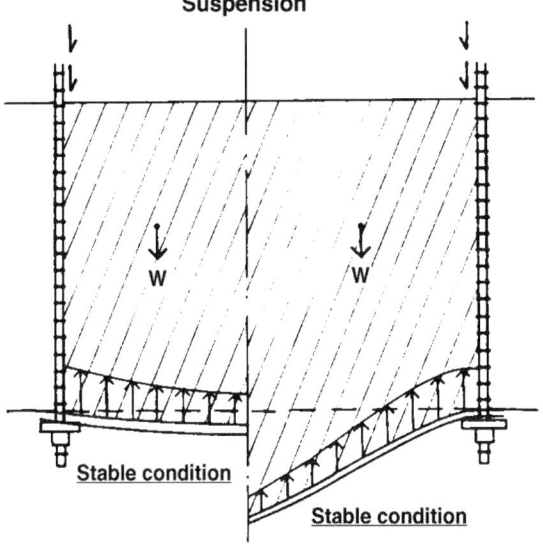

FIGURE 62.18 Toughness requirements of surface support for arching and suspension

Surface support (strong arches of concrete lining or steel ribs) may be sufficient in itself to stabilise the tunnel. The bearing capacity of this type of surface support is mainly connected to compression stresses in the arch. It is a clear tendency in modern tunnelling to replace the arduous work associated with steel ribs with the more flexible and mechanised shotcrete lining. This will be studied in this section.

To avoid excessive loads in a too-stiff arch or to tolerate larger deformations, open slots can temporarily be used in a shotcrete lining (e.g., Rabcewicz 1975 and McWilliam 1992). In other cases, it is possible to use surface support of lower strength, but it must then interact with the bolt support to achieve the necessary stability of the entire tunnel. The most common surface support methods of this type are shotcrete and mesh.

The role of the surface support in this connection is to give stability to the area between the bolts so the bolts function in the intended way (see Figure 62.18). This implies a very complex interplay of factors. However, the bearing capacity of this type of surface support is connected to its ability to take up bending moments or tensile loads due to membrane action.

If large deformations arise in the surface support, it could lead to a situation where the bolts cannot take up the load from the rock as intended. The load will then be transferred to the surface support, which is not designed to carry this load, and the tunnel will collapse. In other load situations and for other failure mechanisms, the bolt support will function despite the fact that large deformations occur and the rock breaks up. Large deformations in the surface support are not a disadvantage here.

Research has also been carried out to study the sensitivity of young shotcrete to vibration from blasting. Ansell (1999) reported that a thin layer of young shotcrete is more sensitive to vibrations than a thicker layer. The most sensitive age is between 2 and 12 hours. In a field test, young shotcrete lining was able to resist vibrations of at least 500 mm/s. Hulshizer (1996) reported that a young concrete may survive vibrations as high as 250 mm/s without seriously being damaged during its first 24 hours after casting. Old shotcrete linings (cured) can tolerate much higher velocities without being cracked. Values between 700 mm/s to 1,450 mm/s have been reported in the literature (e.g., McCreath et al. 1994).

Tests carried out in connection with a large underground construction project Sturk (2000) verify that no significant reduction in bearing capacity was observed for shotcrete applied close to the face and exposed to high vibrations at an early age.

62.4.2 Shotcrete Lining as Arch

In poorer quality rock, a shotcrete lining is normally used as an arch. The shotcrete is applied close to the front and the load will come on the young shotcrete quite early. Moussa (1993) reported a laboratory test series on the effects of early loading. His conclusion was that reduction in long-term compressive strength is negligible for shotcrete if the preload in an early stage is less than about 70% of the compressive strength at the loading stage. Pöttler (1990 a and b) studied the load development close to the face. Chang (1994) has developed a theory based on the ground reaction curve concept to calculate the load in a young shotcrete lining.

The conclusion is that attention must be given to the early loading conditions on a young shotcrete lining, but that the very complex interaction is not yet fully understood. Therefore, an observational approach is recommended. The design of such shotcrete linings aims to give the lining a shape that provides mainly compression stresses in the lining.

An interesting aspect is how irregularities in the lining affect the bearing capacity. A waved and uneven tunnel surface will normally be the result when the tunnel is excavated by the drill and blast method. Very little research work has been carried out to study this problem.

To study the effects of irregularities, Chang (1994) performed a series of model tests. He compared the bearing capacity between a circular, a simply curved, and a doubly curved steel fibre shotcrete lining. Chang's test conclusion was that a simply curved lining, with the irregularities parallel to the tunnel alignment, had little supporting effect compared to a smooth circular lining. The doubly curved lining had the same or even a better overall load-bearing capacity when compared to the circular lining. The tests indicated that a doubly curved lining could have a more ductile behaviour than a corresponding smooth circular lining. However, this effect has to be studied further.

62.4.3 Shotcrete as a Beam or Membrane

Shotcrete can be applied both in reinforced and unreinforced form. The unreinforced type of shotcrete has only very slight resistance to deformations. Failure normally develops as an adhesion failure and at deformations of about a few millimetres.

This type of support is useful for stabilising isolated loose blocks in an otherwise stable rock mass. However, it is important to obtain satisfactory adhesion. The reinforcement component in shotcrete consists either of individual bars, usually rebar with

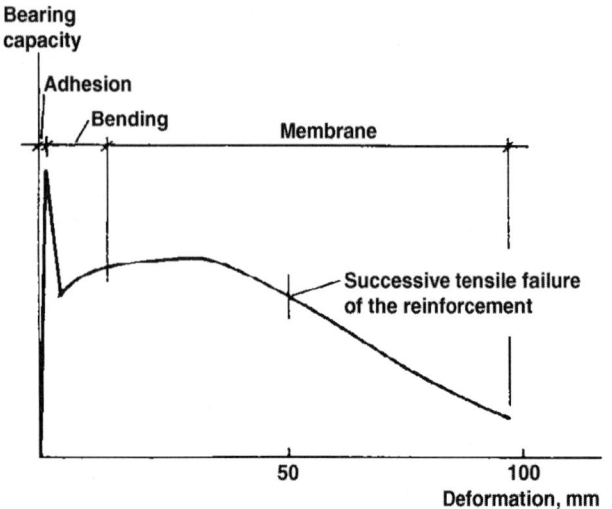

FIGURE 62.19 Different stages of a shotcrete layer resistance curve

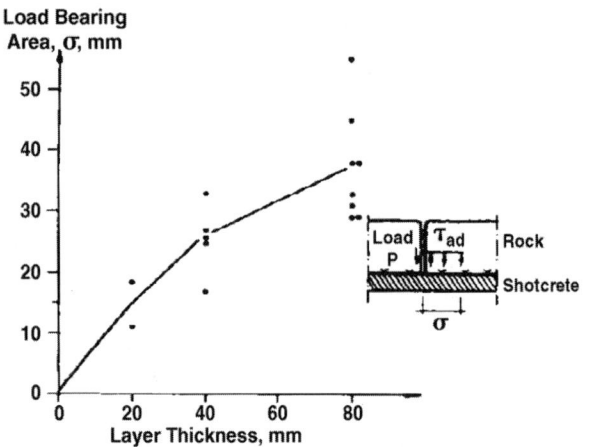

FIGURE 62.20 Bearing capacity due to adhesion for different thicknesses of the shotcrete layer

TABLE 62.2 Values of the factor k for different fibre contents

μ	k
0.5%	1.8
0.75%	2.2
1.0%	2.6
1.25%	2.9
1.5%	3.3

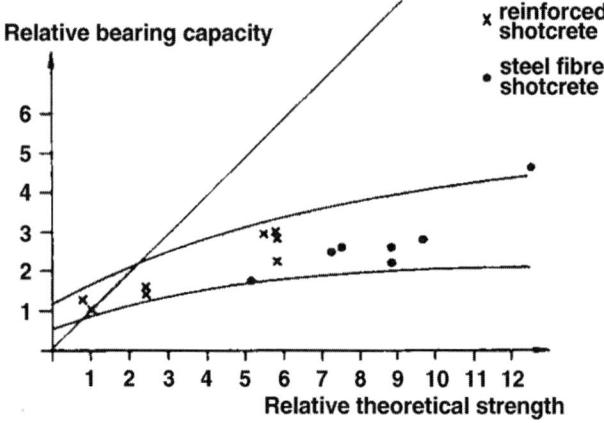

FIGURE 62.21 Relative bearing capacity of a reinforced shotcrete layer as a function of relative bending capacity

diameters of 6–8 mm, or mesh with a diameter of 5–6 mm and a distance between centres of 100–200 mm, or steel fibres.

The comprehensive testing performed to examine the function of shotcrete (e.g., Holmgren 1979, 1985 and Opsahl 1982) provides a good account of the load-bearing capacity of shotcrete and its ability to withstand deformations. Three different stages have been identified: first adhesion, then the beam effect, and finally the membrane effect. These different stages are illustrated in Figure 62.19.

Adhesion depends first and foremost on the rock quality and the cleanliness of the rock surface. The total bearing capacity increases somewhat with increasing thickness of the shotcrete. This is illustrated in Figure 62.20. The figures are based on test results taken from Holmgren (1979).

The beam effect has its maximum bearing capacity at angle changes of about 1/100, and bending deformations of 5–10 mm.

The bearing capacity can then be calculated by the following equations:

$$\text{Bar reinforcement } M_f = \frac{Aa}{c} \cdot 0.4h \cdot \sigma_{ty} \qquad (62.16)$$

$$\text{Steel fibre reinforcement } M_f = \frac{h^2}{6} \cdot \mu_s \cdot \sigma_{ty} \cdot \frac{1}{k} \qquad (62.17)$$

where Aa is the cross section of the bar, c is the centre distance, h is the thickness of the shotcrete, μ_s is the fibre content (volume %), k is the empirical factor that describes the degree of utilisation of the fibres, and T_{ty} is the tensile strength.

Based on the results reported in Bekaert's handbook, Vandewalle (1992) for steel-fibre-reinforced shotcrete, values of the factor k have been calculated for different fibre contents (see Table 62.2).

The table shows that the fibres are more fully utilised at lower-fibre contents, and this value decreases with increased fibre content.

The comprehensive testing carried out by Holmgren (1985) and Opsahl (1982) enables an evaluation to be made of how measured load-bearing capacity in bending varies with tensile strengths. Figure 62.21 shows the relative load-bearing capacity as a function of the relative flexure strength.

A conventional reinforced shotcrete (6-mm diameter at 150-mm centres = 220 MPa) was chosen as a reference. The lever arm of the reinforcement is 2.0 cm, which corresponds to a theoretical concrete thickness of 4.0 cm (Test No. F22, according to Holmgren). The figures show that, with increasing amounts of reinforcement, an increased load capacity is obtained, but less than is suggested by the flexure strength. The problem is to utilise the flexure strength, as other types of failure may occur, such as punching failure around the bolts.

Holmgren (1992) suggested using yield-line theory in designing steel-fibre-reinforced shotcrete slabs. He suggests the following relationships between the residual stress factors:

$$\frac{R_{5,10}}{100} \cdot f_s \text{ and } \frac{R_{10,30}}{100} \cdot f_s \qquad (62.18)$$

determined by testing and the yield-point bending moment, M_y:

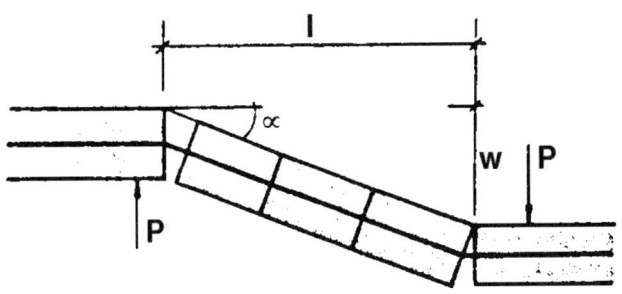

FIGURE 62.22   Shotcrete layer as a membrane

TABLE 62.3   Deformation resistance in different types of shotcrete

	α	∈ b
Fibre shotcrete	5–10°	1%
Mesh-reinforced shotcrete	5–0°	1%
Conv. reinforced shotcrete	15–20°	5%

$$M_y = 0.9\frac{R_{5,10} + R_{10,30}}{200} \cdot f_s \frac{h^2}{6} \qquad (62.19)$$

where f_s is the first crack strength (flexural strength).

For this type of shotcrete support, some testing and analysis has been carried out to evaluate the effect of irregularities on the load-bending capacity. The analysis carried out by Chang (1994) indicates that the bearing capacity in bending increases about 30% for a doubly curved shotcrete lining as compared to a flat lining.

The ultimate failure strength and deformation resistance of the reinforced shotcrete is determined by its ability to form a membrane to take up the loads.

The load-bearing capacity, P, and the corresponding deformation resistance, w, may be written as:

$$P = \frac{Aa}{c} \cdot \sigma_{ty} \cdot sin\alpha$$

$$sin\alpha \approx \sqrt{\frac{2\varepsilon_b}{1 + \varepsilon_b}} \qquad (62.20)$$

$$w = \ell \cdot \tan\alpha \qquad (62.21)$$

where ε_b is the failure strain of the grouted reinforcement, and the other designations are as in Figure 62.22. Based on investigations (Holmgren 1979 and other tests) of the failure strain of grouted reinforcement bars, the approximate values of ε_b and α are given in Table 62.3.

62.4.4   Surface Support with Mesh

Mesh is applied to roofs and walls to prevent rock-falls and stabilise the rock. The broken pieces of rock must, however, be larger than the holes in the mesh. The behaviour of mesh is completely different from that of shotcrete. There must be a relatively large deformation before the mesh starts to take up the load. Ortlepp (1983) tested several different types of mesh. All the tests showed a relatively flexible behaviour that gradually becomes more rigid until the mesh breaks due to the tensile forces (see Figure 62.23). The failure strength and the failure deformations can be relatively well described by the membrane theory given above.

The bearing capacity is normally around 10 to 40 kN/m, depending on the type of mesh. The values are somewhat lower

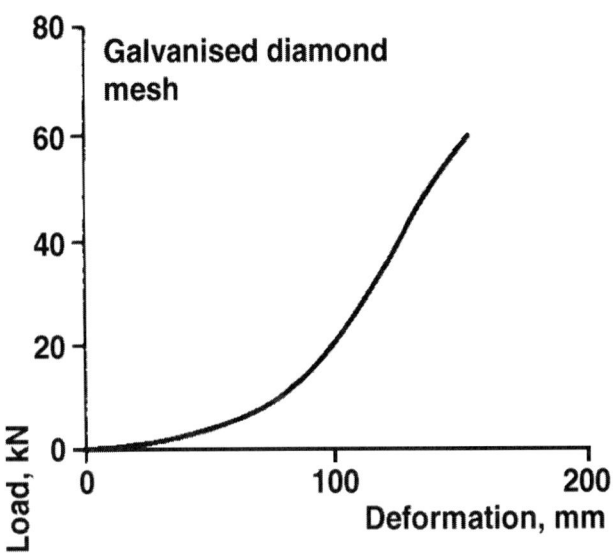

FIGURE 62.23   Bearing capacity and toughness of a diamond mesh according to Ortlepp (1983)

TABLE 62.4   Mesh-testing evaluation (see Ortlepp 1983)

Type of mesh	P	α	∈ b
Diamond mesh: φ3.2 c50, σ_{ty} = 450 MPa	48 kN/m	25–30°	15–20%
φ4.0 c105, σ_{ty} 450 MPa	37 kN/m	25–30°	15–20%
Square woven: φ2.3 c60, σ_{ty} = 1000 MPa	26 kN/m	25–30°	15–20%
Weld mesh: φ3.2 c100, σ_{ty} = 485 MPa	10 kN/M	25–30°	5%

than those shown for reinforced shotcrete, which, according to the tests, has a load-bearing capacity of 20 to 100 kN/m. The failure values of some different types of mesh are shown in Table 62.4.

62.4.5   Need for Surface Support in Connection with the Different Failure Mechanisms

Both mesh and shotcrete can support single blocks of rock. Shotcrete normally has better long-term properties, as the risk of corrosion is less. The shotcrete can be applied without reinforcement, but on the condition that the adhesion is satisfactory. However, unreinforced shotcrete cannot withstand extra shear deformations. In such cases, reinforced shotcrete is preferred.

Bolting to achieve an arching effect and prevent load-bearing failure is based on an interaction with the rock. To prevent loosening and impaired interaction, the surface support between the bolts must be given the necessary resistance. In these cases, reinforced shotcrete is usually required.

If, on the other hand, a stable and natural arch is present higher up, and the bolts are designed to carry the entire loose core, then mesh reinforcement may be suitable to support the surface and prevent falls of rock.

In the case of instability problems, not even shotcrete has the necessary stiffness to prevent splitting, spalling, or buckling. Both mesh and reinforced shotcrete have been used to reduce outward bending and to prevent fallout of rock. The choice of which to use may depend on the degree of splitting. In cases of deep splitting, mesh is perhaps preferable in that it can withstand larger deformations.

62.5   REFERENCES

Ansell, A. (1999). Dynamically Loaded Rock Reinforcement. PhD thesis, Inst. för Byggkonstruktion, KTH, Stockholm.

Aydan, T., Y. Ichikawa, and T. Kawamoto. (1985). Load-bearing capacity and stress distribution in/along rockbolts with inelastic behaviour of interfaces. *Proc. of 5th Int. Conf. on Numerical Methods in Geomechanics,* Nagoya, pp.1281–1292.

Azuar, J.J. and M. Panet. (1980). Shear behaviour of passive steel in rock masses. Rock bolting revue Industrie Minerale, Saint Etienne, pp. 85–90.

Bjurström, S. (1974). *Bolted hardjointed rock* (in Swedish). Fortifikations-förvaltningen, rapport nr. 121:3, Stockholm.

Brady, B.H.G. and E.T. Brown. (1985). Rock mechanics for underground mining.

Brekke, T.L., H.H. Einstein, and R.E. Mason. (1976). State-of-the-Art Review on Shotcrete. California Univ., Berkeley, Army Engineer Waterways Experiment Station, Vicksburg, Massachusetts Inst. of Tech., Cambridge, A.A. Mathews, Inc., Rockville. Contract Report S 76-4.

Brown, E.T., J.W. Bray, B. Ladanyi, and E. Hoek. (1983). Ground response curves for rock tunnels. *Journal of Geotech. Eng.* ASCE, Log, pp. 15–39.

Chang, Y. (1994). Tunnel support with shotcrete in weak rock—a rock mechanics study. Doctoral thesis, Division of Soil and Rock Mechanics, Royal Institute of Technology, Stockholm, Sweden.

Dight, P.M. (1982). Improvement to the stability of rock walls in open pit mines. Ph.D. Thesis, Monash University, Australia.

Farmer, T.W. (1975). Stress distribution along a resin grouted anchor. *Int. J. Rock Mech. Min. Sci. & Geomech.* Abstr. 12, pp. 347–352.

Feder, G. (1986). 10 Jahre Gebirgsmechanik aus dem Institut Fur Konstruktiven Tiefbau der Montana-Universität Leoben. Berg-und Huttenmännische Monatshefte.

Gerard, C. (1983). Rock bolting in theory: A keynote lecture. *Proc. of the Int. Symp. on Rock Bolting,* pp. 353–365. Abisko.

Hoek, E. and E.T. Brown. (1980). Underground Excavations in Rock. Inst. of Mining and Metallurgy, London.

Holmberg, M. (1991). The mechanical behaviour of untensioned grouted rock bolts. Ph.D. Thesis. Dept. of Soil and Rock Mechanics. KTH, Stockholm.

Holmqren, J. (1979). Shotcrete, punchloaded shotcrete linings on Hard Rock. BeFo, Nr 7:2/79, Ph.D. Thesis.

Holmgren, J. (1985). Bolt Anchored Steel Fibre *Reinforced* Shotcrete Linings. BeFo, 73:1/85.

Holmgren, J. (1992). Rock support with shotcrete. Swedish State of Power Board, Stockholm (Swedish).

Hulshizer, A.J. (1996). Acceptable shock and vibration limits for freshly placed and maturing concrete. *ACI Materials Journal,* 93, pp. 524–533.

Krauland, N. (1993). Personal communication, Boliden Mineral AB, Sweden.

Lambe, T.W. and R.V. Whitman. (1969). Soil Mechanics. John Wiley and Sons Inc., New York.

Littlejohn, G.S, A.A. Rodger, D.K.V. Mothersville, and D.C. Holland. (1987). Monitoring the influence of blasting on the performance of rock bolts at Penmaenbach tunnel. Int. Conf. on Foundations & Tunnels. Vol. 2. pp. 99–106.

Ludwig, B. (1983). Shear tests on rock bolts. Proc. Int. Symp. on Rock Bolting, Abisko.

Maury, V. (1987). Observations, researches and recent results about failure mechanisms around single galleries. *Proc. of the 6th ISRM Int. Congress,* Montreal, Canada, Vol. 2, pp 1119–1128.

McCreath, D.R., D.D. Tannant, C.C. Langille. (1994). Survivability of shotcrete near blasts. Rock Mechanics, Nelson and Laubach, Eds., Balkema, Rotterdam, pp. 277–284.

McWilliam, F. (1992). Innsbruck Tunnel Slots into Place. *Tunnels & Tunnelling,* Jan. Vol. 24, No. 1.

Moussa, A. (1993). Finite Element Modelling for Shotcrete in Tunnelling. PhD Thesis, Innsbruck University, Austria.

Opsahl, O.A. 1982. Stålfiberarmert sprutebetong til fjellsikring. Rapport: NTNF-Prosjekt 1053.09511.

Ortlepp, W.D. (1983). Considerations in the Design of Support for Deep Hard-Rock Tunnels. *ISRM,* 5th International Congress on Rock Mechanics, Melbourne, Australia.

Pöttler, R. (1990a). Time-Dependent Rock – Shotcrete Interaction – Numerical Shotcrete. *Computer and Geotechnics,* 9, pp. 149–169.

Pöttler, R. (1990b). Green Shotcrete in Tunnelling: Stress-Strength deformation. Shotcrete technology, Papers presented at the 3rd Conf. Shotcrete Technology, Innsbruck, Austria.

Rabcewicz, L. (1975). The New Austrian Tunnelling Method, Parts I, II, III, Water Power, London.

Rosengren, L. (1990). Surface support (in Swedish). Litteraturstudie. Gruv 2000 90:39, Luleå.

Schubert, P. (1984). Das Tragvermögen des mörtelversetzten Ankers unter aufgezwungener Kluftverschiebung. Ph.D. Thesis, Montan-Universität, Leoben.

Spang, K. and P. Egger. (1990). Action of fully grouted bolts in jointed rock and factors of influence. *Rock Mech and Rock Eng.* 23, pp. 201–229.

Stacey, T.R. (1981). A simple extension strain criterion for fracture of brittle rock. *Int. Jn. Rock Mech. Min. Sci.,* Vol. 18, pp. 469–474.

Stillborg, B. (1984). Experimental investigation of steel cables for rock reinforcement in hard rock. Doctoral thesis. Luleå University, Luleå, Sweden.

Stillborg, B. (1991). Rock and cable bolt tensile loading across a joint. (Bergmekanikdagen 1991). Swedish Foundation of Rock Engineering, *BeFo, Stockholm.*

Stille, H. (1987). Rock Bolting—Analysis Based on the Theory of Ground Reaction Curve. Swedish Research Engineering Foundation 151:1, Stockholm.

Stille, H., M. Holmberg, and G. Nord. (1989). Support of weak rock with grouted bolts and shotcrete. *Int. J. Rock Mech. Min. Sci. & Geomech.* Abstr. 26, pp. 99–113.

Stille, H., R. Johansson, and G. Nord. (1988). Rock Support and Excavation under Various Conditions. Int. Symposium on Tunnelling for Water Resources and Power Projects, New Delhi, India.

Stille, H. (1992). Keynote lecture: Rock support in theory and practice. *Proc. of the Int. Symposium on Rock Support,* Sudbury, Ontario, Canada, 16–19 June, 1992, pp 421–438.

Stjern, G. (1995). Practical performance of rock bolts. PhD thesis. Inst. for Geologi og Bergteknikk, NTH; Trondheim, Norway.

Sturk, R. (2000). Personal communication.

Terzaghi, K. (1943). Theoretical Soil Mechanics, John Wiley and Sons Inc., New York.

Vandewalle, M. (1992). Tunnelling in the world. N.V. Baekert S.A.

CHAPTER 63

Rock Bolting for Underground Excavations

John Hadjigeorgiou* and François Charette†

63.1 INTRODUCTION

In most underground mines, rock bolts are the primary means of rock reinforcement. Rock bolts reinforce the rock mass by one or more of the following methods: beam building, suspension of weak fractured ground to more competent layers, pressure arch, and support of discrete blocks (see Figure 63.1). Over the years, new bolts and techniques have been developed to cover the possible range of applications for different ground conditions.

A successful rock-bolting system is safe, economical, and meets production requirements. The economics of rock bolting have been addressed by classifying costs due to support (ground preparation and bolting) as well as disturbance costs that account for rehabilitation and production loss (Krauland 1983).

63.2 ROCK-BOLTING SYSTEMS

It is customary to classify rock bolts as mechanical, resin- or cement-bonded, and friction. However, bolts are now available that use a combination of these mechanisms. Figure 63.2 summarises some of the major rock-bolt types.

63.2.1 Mechanical-Point Anchors

A mechanical-point anchor bolt uses an expansion shell to secure an anchor point at the end of a drilled hole in competent ground. Installation is relatively simple, with the bolt pushed up the drilled hole and subsequently tensioned. Bolts are installed normal to the rock surface. The bolts are installed manually with a stopers or impact wrench; pneumatic bolters are used for mechanical installation. Bolts are tensioned to 135 to 230 N.m for 16-mm diameter. Several adapters are available for use with portative drills.

Correct placement and adequate tensioning assure optimum performance of mechanical rock bolts. It is important to provide adequate torque during installation (170 to 270 N.m) and to select appropriate expansion shells for the ground conditions. Pull tests are recommended for these purposes. Hole diameter and length are important and should be specified. In a short hole, it is not possible to torque the bolt. If the hole diameter is too small, the plug cannot seat inside the shell; and if it is too big, the holding strength is reduced. The type of shell used also influences hole tolerance. Using suitable installation adapters avoids damage to the plates and nut.

The most common cause of tension loss is bolts installed at an angle less than 80° to the rock face (see Figure 63.3). Spherical washers, however, can tolerate some variations. Bolts should also be arranged to intersect any joints at an angle of more than 45°. A series of examples meeting these requirements are shown in Figure 63.4 (Choquet 1991).

It is important to develop a monitoring program for the bolts after installation. Blasting vibrations can lower bolt tension and require that bolts close to the blasting face be retensioned to the

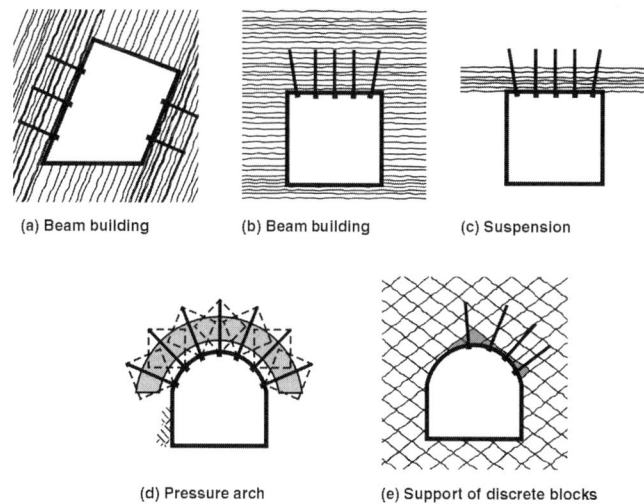

FIGURE 63.1 Modes of support for rock bolts

(a) Beam building   (b) Beam building   (c) Suspension

(d) Pressure arch   (e) Support of discrete blocks

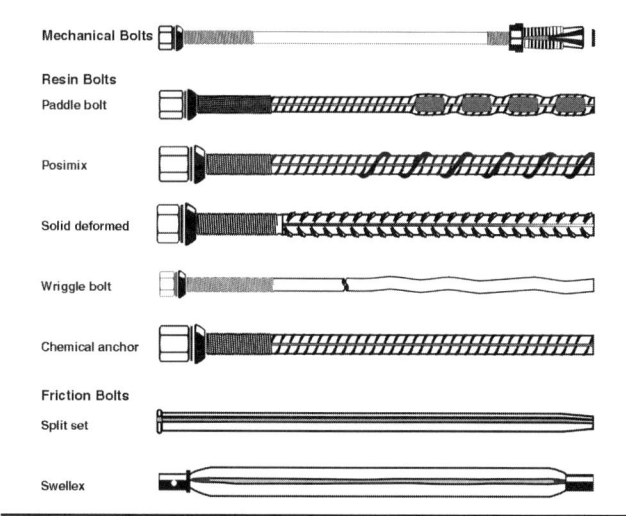

FIGURE 63.2 Types of rock bolts

required levels. Visual signs of poor installation are loose bolt heads and plates. Crumbling rock under the faceplate can also lead to load loss. Using larger plates and wire-mesh support can alleviate this. Ductile or brittle failure indicates that the bolt capacity was exceeded. Corrosion should also be monitored.

* Laval University, Quebec City, Quebec, Canada.
† Atlas Copco, Montreal, Quebec, Canada.

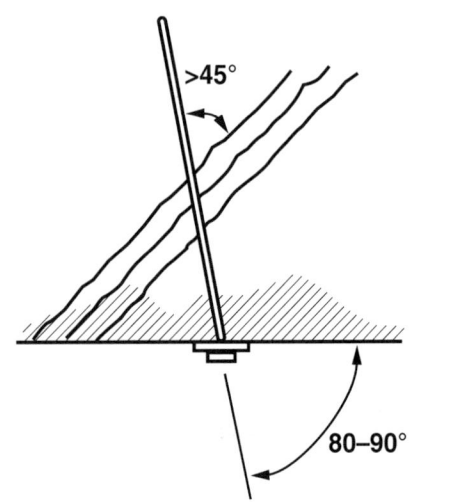

FIGURE 63.3 Optimal bolt inclination

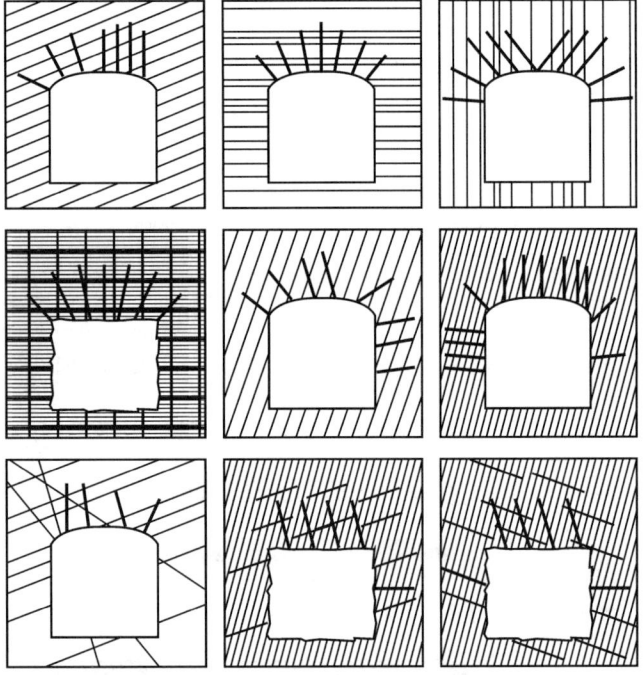

FIGURE 63.4 Systematic bolting patterns in jointed ground, after Choquet (1991)

Mechanical bolts are the least expensive and provide immediate support. They are, however, severely limited for weak supporting rock and are susceptible to corrosion. A systematic quality-control program should monitor frequent loosening due to blasting or ground relaxation.

63.2.2 Grouted Bolts

Resin- and cement-grouted bolts rely on the bonding strength of the cement or resin to transfer rock loads.

Resin Bolts. Resin-grouted bolts use several sealed polyester-resin and catalyst cartridges that are mixed within the hole by spinning the bolt. Different types of rods can be used. A nut and friction-reducing washer holds a plate and provides tension to the system. Recent developments, such as the Posimix

and Paddle bolts, allow for improved resin mixing in the hole and can arguably result in better quality control. The Posimix resin bolt employs a spring-wire welded to the toe of the bolt, while the Paddle bolt uses hot forged paddles on the uphole end of the bolt. Both bolts are receiving attention in Australia but have yet to be introduced in North America.

After drilling a hole of suitable length and diameter, resin cartridges are inserted in the correct sequence into the hole. Both fast- and slow-set resins are available. The bolt is inserted on a spinning dolly and pushed through the cartridges. Spinning the bolt mixes the resin. After a specified time, spinning is stopped, and the resin is allowed to set. The set resin forms an intimate bond between the bolt and the drill-hole walls. Tensioning the bolt may take place within a few seconds of the resin setting, depending on the setting rate of the admixture. Resin suppliers provide charts to assess anchorage capacity, setting time, and strength variation for their respective products. The recommendations providing for the grouting length for a given set of hole, bar, and cartridge diameters should be followed. Grouted length should be at least 180 mm, although a safety factor is often employed. The theoretical minimum grouted length should always be verified by pull tests, especially in the case of very broken or loose rock mass. It should be noted that a short length of resin can sustain a relatively high load.

Maximum performance is obtained when resin rebars are grouted along the entire length. This is still not possible to verify after installation; consequently, installation procedures must be regularly verified. Attention should also be given to the age, type, and temperature of employed resin cartridges. The equipment used must provide adequate mixing of cartridges. In very fractured holes, resin can migrate into the joints, resulting in poor support. If it becomes necessary to tension a bolt, adequate equipment and accessories must be used. On threaded bolt heads, the resulting tension can be checked with a torque wrench, although this test is not always reliable for fully grouted rebars. The most significant quality-control problems can be traced to incomplete bolt grouting.

Resin bolts provide high load-bearing capacity and are suitable for the majority of ground conditions. In general, they are highly resistant to corrosion. A rigid quality-control program is necessary during installation. Other limitations include limited shelf life and setting-time variation with temperature.

Cement-Grouted Bolts. The reinforcement action of cement-grouted bolts is similar to that of resin-grouted bolts. The hole should be completely filled with grout. Cement-grouted bolts require a rigid quality control of the grout. A high water:cement ratio, while easier to pump, results in a weakened grout. A water:cement ratio around 0.4 should be maintained at all times. Spot testing the cement quality is recommended. Cement cartridges are also available but are less popular.

Cement-grouted bolts can be used under most ground conditions. They provide good protection against bolt corrosion, superior to that of resin bolts (Hoek, Kaiser, and Bawden 1995). Furthermore, they are relatively inexpensive. Limitations include longer installation and slower setting time. Consequently, they do not provide immediate support. Grout mix should meet the specifications and additives used as required. It should be noted that grout has a limited shelf life and, in very broken and loose ground, grout loss can be a problem.

63.2.3 Split Set

Split Set bolts consist of a split hollow tube rolled from a steel strip with a tapered end. The bolt collar has a steel ring welded across the rolled strip. Split Set bolts are inserted in a drill hole of slightly smaller diameter using a percussive rock drill for manual installation and a hydraulic impact hammer when an automated bolter is used. Split Sets are frictional bolts that provide continuous anchorage over the whole bolt length. The most important parameter to monitor during installation is hole diameter. A hole that is too large results in low holding force, while one that is too small can damage the bolt as it is pushed in. Bit wear also influences hole diameter. The hole length should be 150 mm to 200 mm longer than the bolt. Insertion time is indicative of holding capacity and hole diameter. If pull tests are performed, it is necessary to use a special holding device that has to be installed during bolt installation.

If loose rock is observed under the plate, it is usually a sign of poor installation or rock fracture. Using larger plates, wire mesh, or shotcrete can control systematic scaling. If the bolt head is observed failing, it could be due to excess deformation at the collar. The width of the slot provides an indication of the confinement of the bolt.

Split Set bolts are popular with miners because they provide immediate support and are easy to install. Shear mobilisation will rapidly increase the mechanical interlock and within a week bolts will develop their maximum holding capacity. Relative softness and low capacity, relative to other bolts, can be an important limitation. Split Sets are susceptible to corrosion, although minor surface corrosion can provide more resistance in the short term.

Grouted Split Sets. A relatively recent practice in some operations is to grout the inside of the split sets. This modifies the anchoring mechanism and results in a significant increase in bolt stiffness (Villaescusa and Wright 1997). It also permits the mobilisation of the full strength of the steel over shorter bond lengths, resulting in a higher load-bearing capacity. On the other hand, the ability of split sets to sustain large displacements may be significantly reduced. Furthermore, it has not yet been demonstrated that the grouting process increases bolt resistance to corrosion considerably.

63.2.4 Swellex

A Swellex bolt is made of a steel tube, folded in, and expanded in a smaller hole through high water pressure. Bushings are welded at both ends; one end is sealed and the other receives the inflation bushing. An integral part of the Swellex system is a water pump that provides 30 megapascals (MPa) of pressure through a pressure intensifier. The source air pressure must be at least 515 kPa, and a minimum water flow must be available. Swellex bolts provide a range of nominal capacity and employ a friction and mechanical interlock anchoring mechanism.

The bolt is inserted into a drilled hole, either manually or using mechanised or semimechanized systems. The head is kept at the rock face, and the bolt is inflated using adequate high-pressure water and a dedicated inflation chuck. Faceplates can be used but must be put onto the bolt prior to installation. For rehabilitation or postmeshing application, friction plates can be applied on the head of already installed bolts.

During installation, it is vital to keep the inflation pressure constant. Variations in hole-diameter size, within the recommended limits, are not critical. Hole length must be sufficient to completely insert the bolt; if the end sticks out of the hole, the bolt can fracture the rock at the collar of the hole. Water and air are filtered as dirt reduces the performance of pumps and increases maintenance. It is advisable to implement a water-monitoring program to identify the potential for corrosion.

Failure of the rock at the hole collar is usually caused by bolts not being fully inserted in the hole. Bolt-head failure of a standard bolt can be caused by high deformation at the rock face, and a yielding bolt should be considered as a better option. In burst-prone ground—conditions encountered in some hard, massive orebodies—the Swellex bolts may induce high horizontal stresses that could be detrimental to roof stability. However, after fracturing has occurred, Swellex bolts can create a supporting arch in the roof. Swellex bolts are recommended for rehabilitation work, very fractured ground, or under low to medium stresses. High water flow and the presence of open joints increase the risk of corrosion. A borehole camera can detect evidence of corrosion in the bolt.

The Swellex bolt benefits from a rapid and simple installation process that provides immediate support. It can be applied to different ground conditions, and quality control is limited to insertion and inflation control. Limitations include the relatively high purchasing cost and the need for a special pump. Swellex bolts are susceptible to corrosion.

63.2.5 Tubular Bolts

Over the years, different variations of tubular bolts have been employed at several operations. Such bolts differ from mechanical bolts in that the steel rod is hollow and the system can be grouted with cement using a grouting tube. A rigid quality-control program is necessary to ensure that tubular bolts are grouted as designed and use a 0.4 water:cement ratio.

One advantage is the combined immediate support provided by the anchor and the long-term efficiency of the bolt by grouting. Failure can result in corrosion-related problems. Even then grouted tubular bolts are still susceptible to corrosion. This is the case when the rock mass relaxes, resulting in cracked grout that exposes the thin wall bolts. Tubular bolts are a two-pass system and are perceived to be weak in shear.

63.2.6 CT-Bolt

The CT-Bolt consists of a 20-mm-diameter steel bar installed with point-anchor expansion shells in conjunction with faceplates (Villaescusa and Wright 1999). The tension to the bolt is provided by tightening a nut-hemispherical washer and a plate against the rock on the exposed end of the bolt. The bolt is installed in a specially designed corrugated polyethylene tube to facilitate grouting following installation. The hemispherical washer is hollow, and the grout is injected through a 16-mm hole in the washer. This system is relatively expensive. It provides an immediate support, but the need for post grouting makes it a two-pass system. In field trials, it has been reported that the complex installation can result in a high proportion of spoiled bolts (Potvin et al. 1999).

63.2.7 Screen

Screens are used in conjunction with rock bolts to support loose material. Chain-link screens tolerate more deflection, while welded screens provide stiffer support (see Figure 63.5). Wire size is dictated by projected service life, with #9 gauge often used for temporary headings and #6 or even #4 for permanent excavations. The screen should be placed as close as possible to the rock surface. It is good practice to ensure that there is an overlap between sheets.

As screen failure appears to be progressive, regular inspections identify loose accumulation, snapped wires, and the onset of corrosion. In access excavations once excessive loose accumulates (bagging), the screen should be cut, the loose scaled, and the screen replaced.

A major production inconvenience is the time it takes to apply a screen. Furthermore, screens are prone to corrosion. Although galvanised screens are available, their long-term resistance to corrosion is not established.

63.2.8 Straps

Metal straps are generally 102-mm wide and are available in various lengths and thicknesses. They are used in areas with dominant planes of weaknesses or schistocity and for pillar reinforcement. For the straps to be effective, they must be oriented in such a way to intersect the planes of weakness. Straps are also useful in aiding screen support bagging load. Some mines employ straps to increase the support surface of bolts.

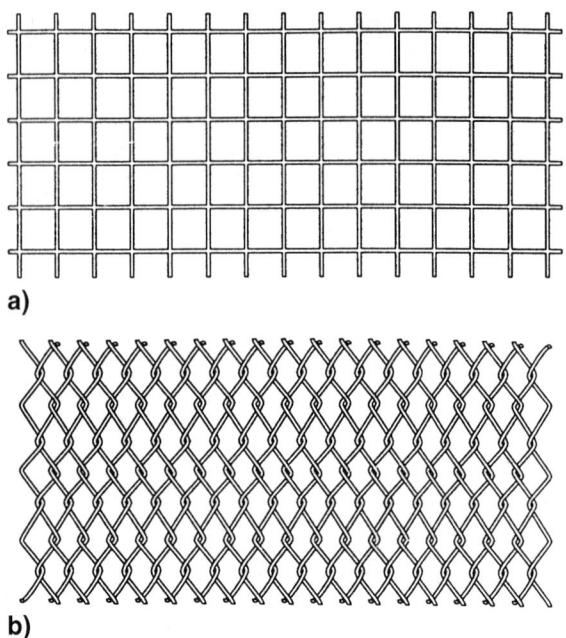

a)

b)

FIGURE 63.5 a) Welded screen type WWF, b) interlaced screen type WF

63.2.9 Shotcrete

Shotcrete is mortar or concrete pneumatically projected at high velocities. Recent years have seen a considerable increase in the use of shotcrete in underground mines, particularly in poor ground conditions. Both wet-mix and dry-mix applications are popular. The choice is usually based on local conditions and preferences. When shotcrete is used to provide initial support, it can be applied to the rock surface and the bolts are subsequently installed through it. A second shotcrete layer then seals the bolts. Shotcrete requires a good quality-control program that should address aggregate quality, admixtures, and strength, as well as application procedures and resulting rebound. Compliance to the design thickness is a priority.

In poor or squeezing ground, it may be desirable to reinforce concrete. Until recently this was achieved by using screen, as it tolerates high displacements. Because of the difficulties associated with spraying though screen, time to install, and production costs, fibre-reinforced shotcrete provides an alternative (Vandewalle 1998). Using remote robot arms increases personnel safety. A successful application of shotcrete may justify reducing the bolt requirements. When shotcrete is to be applied at depth, attention must be given to implementing the necessary infrastructure.

63.2.10 Membranes

More recently, the use of membranes has received attention in mines as a means of screen and shotcrete replacement. A comprehensive review of several membrane systems, as well as work undertaken at INCO mines in Canada is available (Espley 1999). The implementation of a water-based liner system at Falconbridge provides an interesting case study (Swan and Henderson 2000) as well as work in South Africa (Wojnol and Toper 1999) and Australia (Finn, Teasdale, and Windsor 1999). The main drive behind the introduction of membranes is the potential benefits to be derived from using less time-consuming and labour-intensive support systems, such as screen. At the present time, several systems are being tested at several hard-rock mines. One limitation in assessing and comparing the perform-

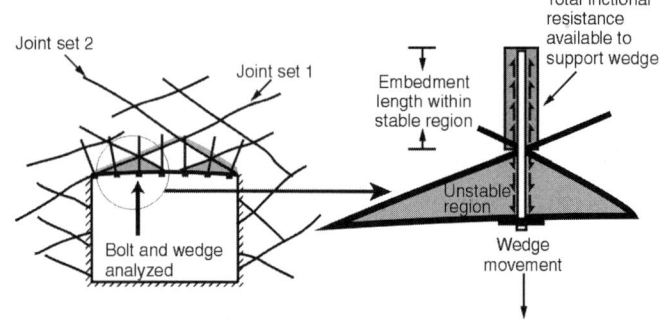

FIGURE 63.6 Load transfer and embedment length, after Windsor and Thompson (1993)

ance of membranes is that, at the present time, there are no universally accepted testing procedures. It is expected that membrane systems will become widely used in the next few years.

63.3 ROCK-BOLTING DESIGN PROCEDURES

There are several reinforcement design methodologies available (Hoek, Kaiser, and Bawden 1995). Essentially most processes aim to provide reinforcement under structurally or stress-controlled conditions. An integral part of any design is rock-mass characterisation, identification of possible failure modes, selection of a reinforcement system, design, implementation, and monitoring of performance.

The reinforcement action of a bolt involves three basic steps (Windsor and Thompson 1993). In the first step, ground movement at the boundary results in load transfer from the unstable rock to the reinforcement element. This load is transferred from the reinforcing element in the unstable area to a stable zone; the final step involves transferring the reinforcement element load to the rock in the stable interior zone. This implies the need for a minimum embedment length of the bolt in the stable zone (see Figure 63.6). A minimum embedment length of 1.0 m is recommended (Schach, Garshol, and Heltzen 1979).

63.3.1 Wedge Support

Rock bolting can be used to stabilise a sliding block or a block that is susceptible to falling under gravity (see Figure 63.7). A basic assumption is that the blocks are discretely defined. A three-dimensional limit equilibrium analysis is available to determine the required bolting support (Hoek and Brown 1980). Several commercially available software programs facilitate this type of analysis.

63.3.2 Reinforcement for Bedded Rock with Overlying Competent Strata

Stillborg (1994) proposed a methodology to determine the bolting requirements to support an unstable layer of rock into a solid rock (see Figure 63.8). Depending on the type of application, a safety factor in the order of 1.5 to 3.0 is used.

63.3.3 Rock Reinforcement Unit

It has been suggested that every bolt installed in a back could be treated as a reinforced rock unit defined by its zone of influence (see Figure 63.9) (Lang and Bischoff 1982). A series of reinforced rock units constitute a reinforced rock structure, and stability is ensured if the minimum rock-bolt tension is greater than a minimum rock-bolt tension.

If the rock reinforcement is installed prior to the occurrence of significant deformation, it is considered to have an active

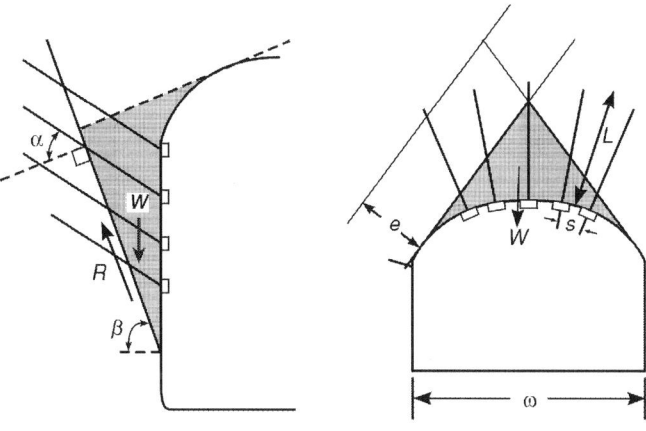

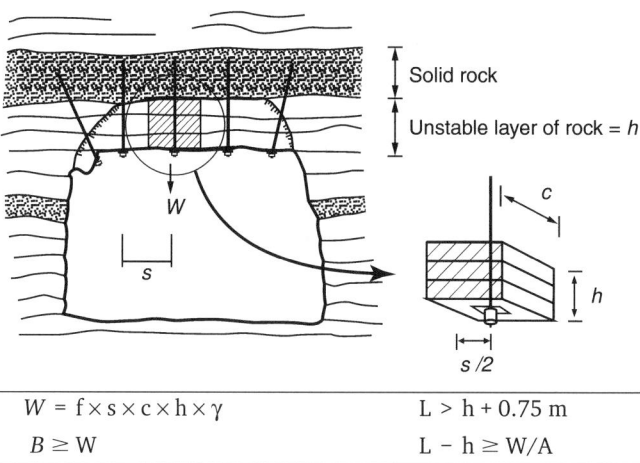

Wedge susceptible to slide along a discontinuity or along the intersection of two discontinuities	Wedge susceptible to fall under the effect of gravity
$N = \dfrac{W(f \sin \beta - \cos \beta \tan \phi) - cA}{B(\cos \alpha \tan \phi + f \sin \alpha)}$	$N = \dfrac{Wf}{T}$
$R = cA + W \cos \beta \tan \phi$	$s \le 3e$
$f = 1.5$ for grouted bolts	$p \ge L + 1.0\ m$
$f = 2.0$ for mechanical bolts	$2 \le f \le 5$

Legend:

N = number of bolts

W = weight of the wedge

f = safety factor

A = surface area along the sliding plan

β = orientation of the sliding surface

e = joint spacing

s = spacing between bolts

L = bolt length

α = angle between the plunge of the bolt and the normal to the sliding surface

c = cohesive strength of the sliding surface

ϕ = angle of friction of the sliding surface

R = resistance to sliding

ω = excavation span

B = bearing capacity of bolt (tonnes)

FIGURE 63.7 Reinforcement of a distinct wedge, after Choquet and Hadjigeorgious (1993)

contribution to the excavation stability (a = 0.5). If passive reinforcement is used, a = 0. For initial design purposes, cohesion should be taken as equal to 0.

63.3.4 Reinforcement Based on the Rock-Arch Concept

The introduction of rock bolts close to each other results in a load-carrying arch within the rock mass stabilizing the excavation back. Figure 63.7 provides design guidelines for moderately and heavily jointed rock masses.

63.3.5 Rock-Mass Classification

There are two ways to employ rock-mass classification systems. In the first case, such systems are used to select a support system,

$W = f \times s \times c \times h \times \gamma$	$L > h + 0.75$ m
$B \ge W$	$L - h \ge W/A$

Legend:

W = weight of tock to be supported by a single bolt (metric ton)

f = safety factor

s = bolt spacing perpendicular to the axis of excavation (m)

B = load bearing capacity of bolt (metric ton)

c = bolt spacing along the axis of excavation (m)

h = thickness of unstable layer of rock (m)

γ = rock density (tonne/m^3)

L = bolt length (m)

FIGURE 63.8 Reinforcement of an unstable layer of horizontal bedding planes overlaid by solid rock, after Stillborg (1994)

and, in the second, to determine the support pressure. In practice, the majority of operations rely on the Q system (Barton, Lien, and Lunde 1974) and the Rock-Mass Rating (RMR) system (Bieniawski 1973). An extensive review of rock-mass classification systems is available as well as a discussion on the advantages and inconveniences of these systems with reference to mining (Milne, Hadjigeorgiou, and Pakalnis 1998).

The Q system, arguably one of the more popular rock-mass classification systems, employs six parameters: RQD (Deere et al. 1967), Jn (the joint set number), Jr (a joint roughness index), Ja (an alteration index), Jw (a water index), and SRF (a stress reduction number). These parameters are combined to determine Q (see Figure 63.10). The Q index is linked to the excavation span to provide guidelines on the stability of an excavation as well as to select an appropriate support system. The more recent guidelines (Barton and Grimstad 1994) are presented in Figure 63.11. Using the Equivalent Support Ratio (Table 63.1) allows the user to select the level of safety required based on the type of excavation.

The RMR (Bieniawski 1973) has evolved over the years by modifying the weight assigned to its basic constitutive parameters. Due to its relative ease of application, it is widely used to characterise the rock mass (Bieniawski 1989). It has been demonstrated that, depending on the version of the system employed, there can be important differences in the predicted stability of an excavation (Milne, Hadjigeorgiou, and Pakalnis 1998). Both RMR and Q allow for a zoning of equivalent ground conditions at a mine. As such, they contribute to defining appropriate local support standards.

63.3.6 Empirical Rules

While classification systems incorporate design-support recommendations of an empirical nature, several rules of thumb also exist that are applicable to selecting the reinforcement type and

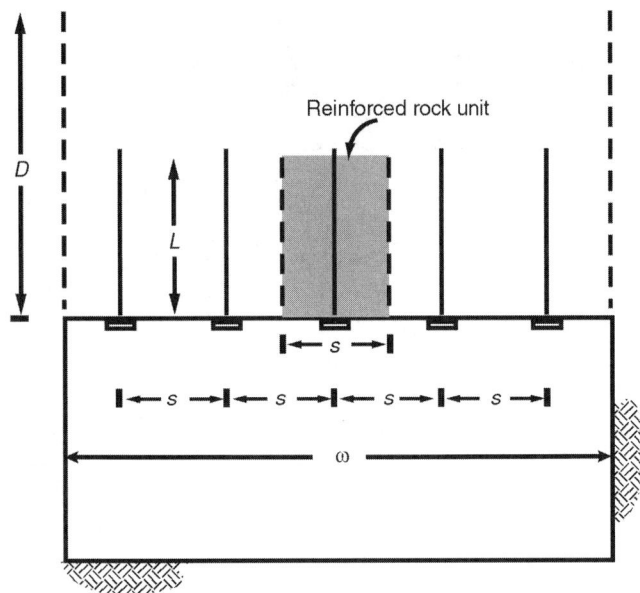

Minimum rock bolt tension to ensure stability

$$T = \frac{a\gamma AR}{(tan\phi)k}\left(1 - \frac{c}{\gamma R}\right)\left[\frac{1 - exp\left[-(tan\phi)(kD)/R\right]}{1 - exp\left[-(tan\phi)(kD)/R\right]}\right]$$

Determination of the length of the rock bolts

$$L = \omega^{2/3}$$

Legend:

T = minimum rock bolt tension

A = area of roof carrying one bolt ($s \times s$)

P = shear perimeter of reinforced rock unit ($4s$)

R = shear radius of the reinforced rock unit ($A/P = s/4$)

ϕ = angle of internal friction of the rock mass

k = ration of horizontal to vertial stresses

ω = width of excavation

L = bolt length

s = bolt spacing

c = apparent cohesion of the rock mass

a = factor depending on time of installation

γ = unit weight of the rock

D = height of destressed zone

FIGURE 63.9   Reinforced rock unit, after Lang and Bischoff (1982)

TABLE 63.1   Equivalent support ratio for different support systems

Excavation type	ESR
Temporary mine openings	2–5
Permanent mine openings	1.6–2.0
Storage caverns, access tunnel	1.2–1.3
Power stations, portals, intersections	0.9–1.1
Underground nuclear power stations	0.5–0.8

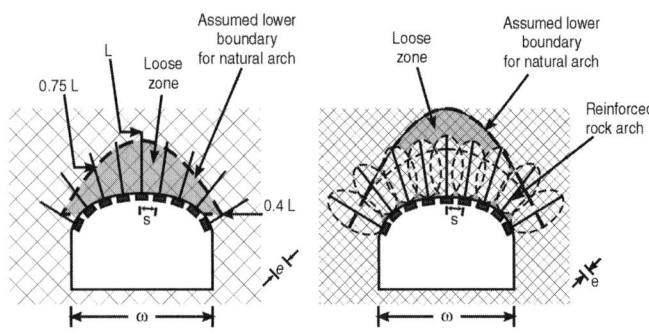

Moderately jointed rock	Heavily jointed rock mass
Use nontensioned bolts	Use tensioned rock bolts
L = 1.40 + 0.184ω	$L = 1.60 + \sqrt{1.0 + 0.0012\omega^2}$
Bolt lengths can be reduced based on the above diagram	In order for a compression zone to be developed: L/s ≥ 2 s ≤ 3e 0.5B < T < 0.8B Shotcrete and wire mesh reinforcement

FIGURE 63.10   Reinforcement based on rock arch theory, after Stillborg (1994), Schach et al. (1979)

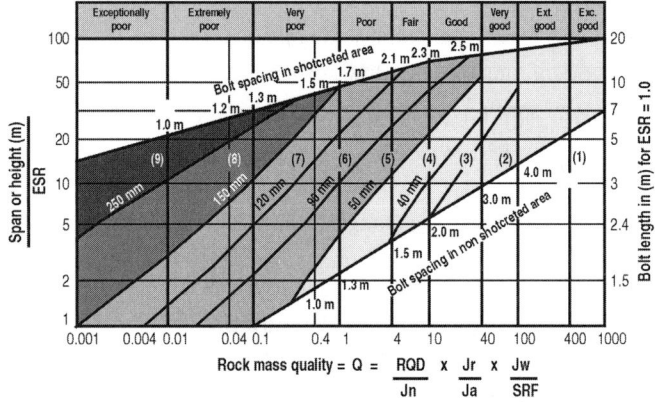

Legend:

1) Unsupported

2) Spot bolting

3) Systematic bolting

4) Systematic bolting (and unreinforced shotcrete, 4–10 cm)

5) Fibre reinforced shotcrete and bolting, 5–9 cm

6) Fibre reinforced shotcrete and bolting, 9–12 cm

7) Fibre reinforced shotcrete and bolting, 12–15 cm

8) Fibre reinforced shotcrete and bolting, >15 cm, reinforced ribs of shotcrete and bolting

9) Cast concrete lining

FIGURE 63.11   Design chart based on the Q system, after Barton and Grimstad (1994)

TABLE 63.2   Minimum length and maximum spacing for rock reinforcement

Minimum length:
Greatest of:
  A. Two times the bolt spacing
  B. Three times the width of critical and potentially unstable rock blocks
  C. For elements above the spring line:
    1. Spans less than 6 m – ½ span
    2. Spans from 18 m to 30 m – ¼ span
    3. Spans 6 m to 18 m – interpolate between 3 m and 4.5 m lengths, respectively
  D. For elements below the spring line:
    1. For openings less than 18 m high – use lengths as determined in C above
    2. For openings greater than 18 m high – 1/5 the height

Maximum spacing:
Least of:
  A. ½ the bolt length
  B. 1½ the width of critical and potentially unstable rock blocks
  C. 1.8 m
    Minimum spacing 0.9 to 1.2 m

a. Where the joint spacing is close and the span is relatively large, the superposition of two bolting patterns may be appropriate, e.g., long heavy bolts on wide centres to support the span and shorter and thinner bolts on closer centres to stabilize the surface against raveling due to close jointing.
b. Greater spacing than 1.8 m would make attachment of surface treatment such as chain link fabric difficult.

TABLE 63.3   Rock support systems suitable for rock-burst-prone conditions. After Kaiser, McCreath, and Tannant 1996

Mechanism	Damage severity	Load (kN/m²)	Displ. (mm)	Energy (kJ/m²)	Examples of suggested support systems
Bulking without ejection	Minor	50	30	Not critical	Mesh with rock bolts or grouted rebars (and shotcrete)
	Moderate	50	75	Not critical	Mesh with rock bolts and grouted rebars (and shotcrete)
	Major	100	150	Not critical	Mesh and shotcrete panels with yielding bolts and grouted rebars
Bulking causing ejection	Minor	50	100	Not critical	Mesh with rock bolts and Split Set bolts (and shotcrete)
	Moderate	100	200	20	Mesh and shotcrete panels with rebars and yielding bolts
	Major	150	>300	50	Mesh and shotcrete panels with strong yielding bolts and rebars (and lacing)
Ejection by remote seismic event	Minor	100	150	10	Reinforced shotcrete with rock bolts or Split Set
	Moderate	150	300	30	Reinforced shotcrete panels with rock bolts and yielding bolts (and lacing)
	Major	150	>300	>50	Reinforced shotcrete panels with strong yielding bolts and rebars and lacing
Rockfall	Minor	100	N/A	N/A	Grouted rebars and shotcrete
	Moderate	150	N/A	N/A	Grouted rebars and plated cable bolts with mesh and straps or mesh-reinforced shotcrete
	Major	200	N/A	N/A	As above plus higher density cable bolting

dimensions, (Farmer and Shelton 1983; Laubscher 1984). These can be used independently of classification systems and analytical design, but are best used as comparative tools to allow the choice of spacing and bolt length. An example of such rules allowing for the estimation of length, spacing, and support pressure is provided in Table 63.2 (U.S. Corps of Engineers 1981). Any empirical rules, however, give only a preliminary configuration for rock reinforcement, which must be checked, analyzed and, as necessary, modified to meet the requirements of a specific rock-reinforcement design. It should also be noted that most empirical rules are based on past practice, which is often influenced by a variety of other nontechnical factors such as legislation and preferences.

63.3.7   Design of Support for Rock-Burst Conditions

Given that most empirical systems have been developed for low-stress conditions, it is questionable how well they apply to rock-burst conditions. Recently, the NGI system was modified, allowing for very high Stress-Reduction Factors for rock-burst conditions (Barton and Grimstad 1994). It has been argued that for rock-burst conditions, RMR can be reduced by 20 points (Lang et al. 1991). In practice, more and more mines that operate in rock-burst environments have developed their own guidelines.

The choice of support for rock-burst-prone conditions has been linked to the mechanism and severity of expected damage (see Table 63.3) (Kaiser, McCreath, and Tannant 1996).

The results of a rock-bolt testing programme under dynamic conditions are available (Stacey and Ortlepp 1999; Kaiser, McCreath, and Tannant 1996).

63.4   STANDARDS

It is now expected and, in certain jurisdictions, explicitly legislated that all operations have a set of design standards. These are developed to ensure a certain level of quality control both for the design process, installation, and follow up. Ideally these standards are arrived at after consulting both engineering and operations to ensure their proper implementation. Standards allow for ground conditions, required life of the excavation, stress environment, personnel exposure, proximity to other openings, and local experience.

Rock-reinforcement standards should be reviewed because ground conditions change during the life of a mine, and new reinforcement equipment and technology is constantly being developed. The development of ground-control standards should provide the background for selecting a support system, designing an installation procedure that includes training, the timing of the

Foundations for Design

installation and scaling, and a quality-control programme (Potvin et al. 1999).

63.5 GROUND-SUPPORT EVALUATION

The ultimate objective of performing quality control on the rock reinforcement is to make sure that the design specifications are promoting the expected performance of excavation and support. Quality control applied only to the specifications may fall short of giving the feedback needed for a successful rock-support system. As rock-mass conditions are growing different from the basis used for the support design, so is the validity of quality control. Even perfectly installed, an inadequately designed rock-bolt system will still be inadequate.

To be fully effective, quality control must apply to site specifications, hardware, specifications, procedures, and expected performance. Performance follow up is a form of monitoring, but testing for long-term compliance to specifications is also a form of quality control. Design parameters must also be checked during quality control operations. Bolt types, length, and accessories, as well as bolt spacing, placement, and orientations should comply with specifications. Any changes must be submitted to engineering for approval or rejection. In addition, changes in rock-mass conditions and loading conditions should be part of the quality control.

An important element of a support-evaluation system is the implementation of a report system. This should clearly establish a communication system where limitations are noted as well as all remedial action taken and feedback.

63.6 GROUND-CONTROL AUDITS

Ground-control audits provide an independent review of ground-control practice at a particular site. A successful audit has a very specific mandate and is undertaken based on established current technical terms. By its nature, it relies on evidence in documentation and implementation of the process in the field. Examples include the availability of standards, the background to develop these standards, the implementation of these standards, quality-control issues, reporting ground falls, and dissemination of information.

63.7 REFERENCES

Barton N.R., R. Lien, and J. Lunde. (1974). Engineering classification of rock masses for the design of tunnel support. *Rock Mech.* 6, pp. 189–239.

Barton N. and E. Grimstad. (1994). The Q-System Following Twenty years of Application in NMT Support Selection. *Felsbau* 12 Nr. 6, pp. 428–436.

Bieniawski, Z.T. (1973). Engineering Classification of Jointed Rock Masses. *Transaction of the South African Institution of Civil Engineers*, 15, pp. 335–344.

Bieniawski, Z.T. (1989). *Engineering Rock-Mass Classifications.* John Wiley & Sons.

Charette, F. and J. Hadjigeorgiou. (1999). Guide Pratique du Soutènement Minier. *Association Minière du Québec Inc.*, pp. 141.

Choquet, P. (1991). *Rock Bolting Practical Guide.* Minister of Supply and Services Canada, Ottawa, Canada. pp. 160.

Deere, D.U., A.J. Hendron, Jr., F.D. Patton, and E.J. Cording. (1967). Design of Surface and Near Surface Construction in Rock. *Failure and Breakage of Rock.* C. Fairhurst, Ed. Society of Mining Engineers, AIME, New York, pp. 237–302.

Espley, S. (1999). Thin Spray-On Liner Support and Implementation in the Hardrock Mining Industry. M.Sc. Thesis, Laurentian University.

Finn, D.J., P. Teasdale, and C.R. Winsor. (1999). In situ trials and field-testing of two polymer restraint membranes. *Proceedings Rock Support and Reinforcement in Mining*, Kalgoorlie WA. Balkema, pp. 139–153.

Hoek, E., P.K. Kaiser, and W.F. Bawden. (1995). *Support of Underground Excavations in Hard Rock.* Balkema, pp. 215.

Hoek, E., and E.T. Brown. (1980). *Underground Excavations in Rock.* The Institution of Mining and Metallurgy, London. pp. 527.

ISRM. (1981). *Rock Characterization Testing and Monitoring.* ISRM Suggested Methods. Brown E.T., Ed. Pergamon Press, pp. 211.

Kaiser, P.K., D.R. McCreath, and D.D. Tannant. (1996). *Canadian Rockburst Support Handbook.* Geomechanics Research Centre, Sudbury, Ontario.

Krauland, N. (1984). Rock Bolting and Economy. *Proceedings of the International Symposium on Rock Bolting.* Balkema, pp. 499–507.

Lang, B., R. Pakalnis, and S. Vongpaisal. (1991). Span design in wide cut and fill stopes at Detour Lake Mine. 93rd Annual Meeting. Canadian Institute of Mining AGM, Vancouver. Paper # 142.

Laubscher, D.H. (1984). Design aspects and effectiveness of support systems in different mining conditions. *Trans. Inst. Min. Metall.* 93, A70-A81.

Milne, D., J. Hadjigeorgiou, and R. Pakalnis. (1998). Rock mass characterization for underground hard rock mines. *Tunneling and Underground Space Technology*, Vol. 13, Issue 4, pp. 383–391.

Potvin, Y., D.B. Tyler, G. MacSporran, J. Robinson, I. Thin, D. Beck, and M. Hudyma. (1999). Development and implementation of new ground support standards at Mount Isa Mines Limited. *Proceedings: Rock Support and Reinforcement in Mining*, Kalgoorlie WA. Balkema, pp. 367–371.

Schach, R., K. Garshol, and A.M. Heltzen. (1979). *Rock Bolting: a practical handbook.* Pergamon Press, p. 84.

Stillborg, B. (1994). *Professional Users Handbook for Rock Bolting.* Second edition, Trans Tech Publications, Series on Rock and Soil Mechanics, Vol. 18, p. 164.

Swan, G. and A. Henderson. (2000). Water-based sprayon liner implementation at Falconbridge Limited. CIM Annual General Meeting, Toronto, CD-ROM.

Vandewalle, M. (1998). The use of steel fibre reinforced shotcrete for the support of mine openings. *The Journal of the South African Institute of Mining and Metallurgy*, pp. 113–120.

Villaescusa, E. and J. Wright. (1997). Permanent excavation reinforcement using cement grouted split set bolts. *Aus IMM Proceedings*, 1, pp. 65–69.

Villaescusa, E. and J. Wright. (1999). Reinforcement of underground excavations using the CT Bolt. *Proceedings: Rock Support and Reinforcement in Mining*, Kalgoorlie WA. Balkema, pp. 109–115.

Wojnol, L. and Z. Toper. (1999). In-situ trials for structural membrane support. *Proceedings Rock Support and Reinforcement in Mining*, Kalgoorlie, WA. Balkema, pp. 131–138.

CHAPTER 64

Cable Bolting

Christopher Windsor[*]

64.1 INTRODUCTION

Cable bolts are long, grouted, high-tensile-strength elements used to reinforce rock masses around surface and underground mining and civil engineering excavations. Cable-bolting practice contributes significantly to the reinforcement of surface and underground excavations and has done so since its initial applications some 35 years ago. In underground mining, the early successes of excavating through arrays of cable bolts previously placed in the backs of cut-and-fill stopes was followed by using cable bolts to prereinforce the larger spans associated with open stoping. This practice enabled bulk extraction mining methods to be developed resulting in increased productivity without excessive dilution. In surface excavations, using systematic cable bolting has resulted in more stable slopes in civil engineering and enabled steeper pit slopes, which leads to reduced stripping ratios in mining.

This chapter is a brief discussion of three areas thought to be of most benefit to the reader who has little or no experience with cable bolting:

Terminology for Cable-Bolting Practice. This section will clarify a number of mechanical concepts and aid communications during design and installation.

The Development of Cable Bolt Devices. This section will describe the many types of cable bolts that have evolved out of the necessity for solving different rock excavation stability problems. The referenced papers are particularly important in that they indicate the original circumstances for which the particular type of cable bolt was designed, and they contain many hard-learned lessons on the appropriate and inappropriate selection of cable-bolt type.

Installation Issues. There are many installation areas that are critical. Two are chosen here—grouting and stressing. In recent years, both areas have seen a move away from the original practices copied and modified from civil engineering ground-anchor technology. The new procedures bring logistical and economic advantages to mine operations but, unless fully understood and properly implemented, can have disastrous effects on the in situ performance of cable bolts.

Although references to important works are given throughout this report, there are three valuable texts that are recommended immediately:

1. Kaiser and McCreath (1992)
2. Hutchinson and Diederichs (1996)
3. Villaescusa, Windsor, and Thompson (1999)

These texts cover the topics selected here in much greater detail as well as issues not discussed here that are associated with the mechanical behaviour, design, testing, and performance monitoring of cable bolts.

64.2 TERMINOLOGY FOR CABLE-BOLTING PRACTICE

Cable-bolting practice is a subset of rock-reinforcement practice. The application of rock reinforcement began in the early 1900s and became a systematic practice in the late 1950s, approximately 100 years after the invention of reinforced concrete. Similarly, the practice of applying ground anchors for civil excavations began in the early 1950s, approximately 60 years after the patent for prestressed concrete was awarded. However, it was not until the mid-1960s that high-tensile-strength, steel elements used in the prestressed concrete industry were installed as long, fully grouted cable bolts. Thus, it is fitting that any terminology for cable-bolting practice be consistent with the disciplines from which the technology was derived.

64.2.1 Rock Support and Rock Reinforcement

The terms *support* and *reinforcement* are often used interchangeably. However, it is useful to consider the two terms as being explicitly different due to the method by which they stabilise the rock adjacent to an excavation. Essentially, support is the application of a reactive force at the face of the excavation and includes techniques and devices such as fill, timber, steel or concrete sets, and shotcrete. Reinforcement is considered to be improvement of the overall rock-mass properties from within the rock mass and will, therefore, include all techniques and devices installed within the rock mass such as rock bolts, cable bolts, and ground anchors.

64.2.2 Pre-Reinforcement and Post-Reinforcement

Pre-reinforcement is the application of reinforcement prior to creating the excavation. Post-reinforcement is the application of reinforcement at an appropriate time after creating the excavation.

64.2.3 Pre-Tensioned and Post-Tensioned Reinforcement

Pre-tensioning is the application of an initial tension to the reinforcing system during installation. Post-tensioning is the tensioning, or re-tensioning, of reinforcement systems subsequent to installation.

64.2.4 Cable-Bolt Devices and Components

Cable bolts are usually manufactured from the high-tensile-strength, steel elements used in the prestressed concrete, the steel structures, and the materials-hoisting-and-conveying industries. The evolution of cable bolting has led to a variety of terms that are used to describe similar and sometimes different devices. To avoid misinterpretation, it is useful to define a limited terminology consistent with the original technology:

- Wire—single, solid section element
- Strand—set of helically spun wires

---

[*] Rock Technology, Perth, Western Australia, Australia.

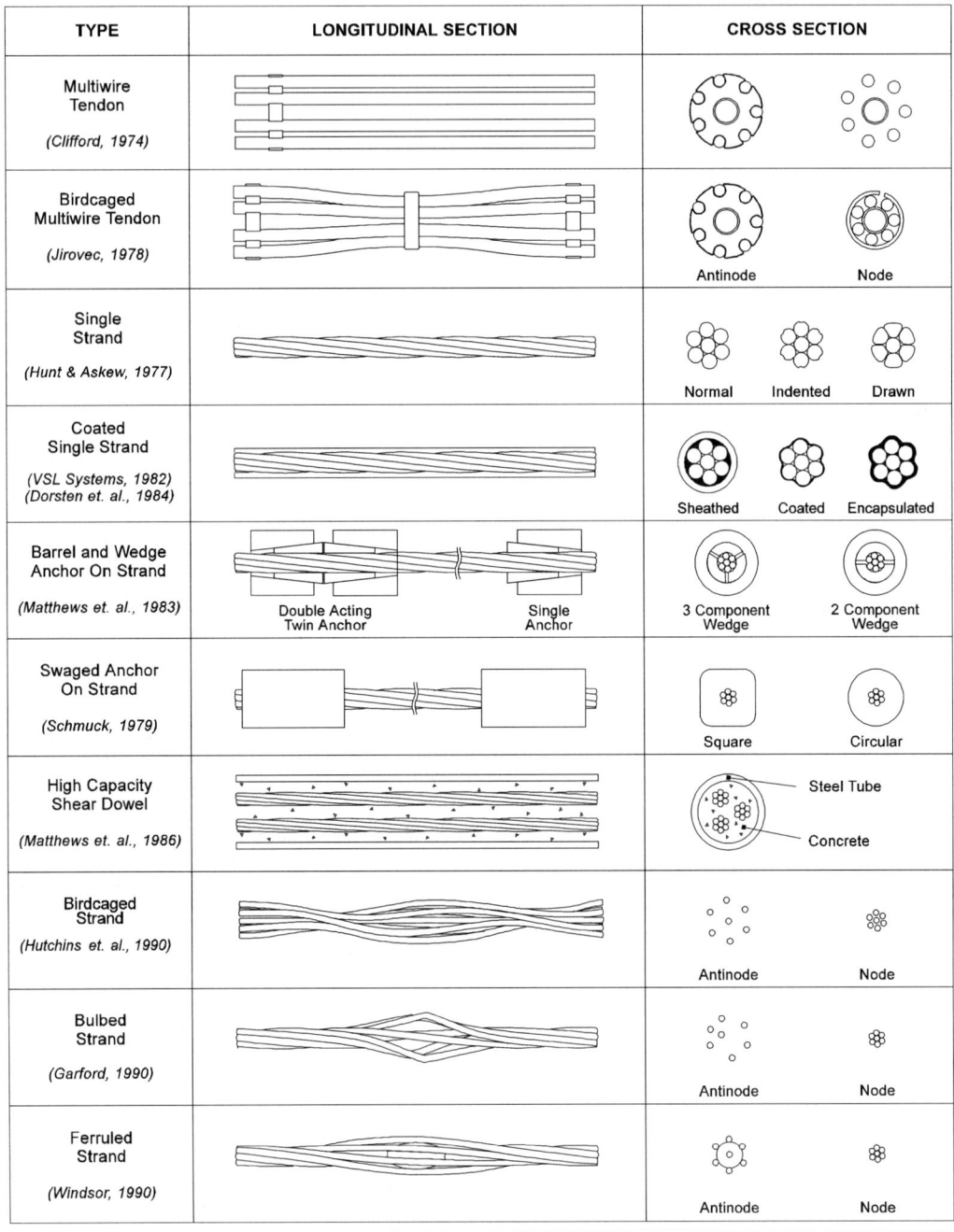

TYPE	LONGITUDINAL SECTION	CROSS SECTION
Multiwire Tendon *(Clifford, 1974)*		
Birdcaged Multiwire Tendon *(Jirovec, 1978)*		Antinode Node
Single Strand *(Hunt & Askew, 1977)*		Normal Indented Drawn
Coated Single Strand *(VSL Systems, 1982)* *(Dorsten et. al., 1984)*		Sheathed Coated Encapsulated
Barrel and Wedge Anchor On Strand *(Matthews et. al., 1983)*	Double Acting Twin Anchor Single Anchor	3 Component Wedge 2 Component Wedge
Swaged Anchor On Strand *(Schmuck, 1979)*		Square Circular
High Capacity Shear Dowel *(Matthews et. al., 1986)*		Steel Tube Concrete
Birdcaged Strand *(Hutchins et. al., 1990)*		Antinode Node
Bulbed Strand *(Garford, 1990)*		Antinode Node
Ferruled Strand *(Windsor, 1990)*		Antinode Node

FIGURE 64.1   A summary of the development of cable bolt configurations

- Cable—arrangement of wires or strand
- Tendon—pre-tensioned wires or strand
- Dowel—un-tensioned wires or strand

Consequently, cables that are pre-tensioned are termed "cable tendons," and cables that are un-tensioned are termed "cable dowels." Clearly, if cable bolts are pre-placed to reinforce the rock mass prior to excavation, they are termed pre-reinforcement cables (tendons or dowels), and when placed after excavation, are termed post-reinforcement cables (tendons or dowels). The term "cable bolting" is now generally used to describe the complete practice of using cable bolts.

64.3   THE DEVELOPMENT OF CABLE-BOLT DEVICES

The earliest uses of cable-bolt devices are believed to have been at the Willroy mine in Canada (Marshall 1963) and at the Free

State Geduld Mines, Ltd., in South Africa (Thorn and Muller 1964). The main reason for the early applications of cable bolts was the realization that pre-stressing wires could be supplied in long, flexible lengths. This would enable bolts to be anchored deep within the rock mass without the need to couple threaded bars together. A summary of cable-bolt configurations and how they have developed is given in Figure 64.1.

Early cable bolts consisted of discarded winder rope and smooth pre-stressing wires. Although unused winder rope is relatively cheap, the labour involved in unwinding and degreasing it, plus its relatively short supply in terms of mine-wide cable bolting, reduced its attractiveness. The pre-stressing wire was readily available and was made into cable bolts consisting of seven, straight, 7-mm diameter, high-tensile, steel wires arranged with plastic spacers. The practical applications have been described by Clifford (1974) and Davis (1977), and a field investigation of their use as pre-reinforcement for cut-and-fill stopes is described in detail by Fuller (1981). The load-transfer characteristics of plain wires is quite poor due to their smooth, straight profile and the Poisson effect, which causes radial contraction. The load transfer could have been improved markedly using the spacer-and-tie arrangement described by Jirovec (1978). However, the installation and tensioning process for the plain-wire system was also quite complicated, and they were replaced by a seven-wire pre-stressing strand.

The first use of strand cable bolts is thought to have occurred in the early 1970s in Broken Hill, Australia, and its first systematic application in cut-and-fill stopes has been described by Hunt and Askew (1977). The conversion from plain wire to strand was revolutionary and gave marked improvements in productivity, adaptability, and mechanical performance. Its relatively rough profile compared to the smooth wires gave it a much higher load transfer, and its axial rigidity allowed it to be pushed into 30-m upholes. Although, some operations have experimented with steel rope (Stheeman 1982) and fibreglass systems (Fabjanczyk 1982), seven-wire, 15.2-mm nominal diameter strand is still the most common material used for cable-bolting throughout the world.

Strand consists of one straight, central "kingwire" and six, slightly smaller peripheral wires that are helically wound longitudinally about the kingwire. Different manufacturing processes can be used to produce strands with a variety of enhanced properties. These include normal- and low-relaxation strand, indented strand, and compacted, or drawn, strand. Strand is also available in stainless steel or may be galvanised, sheathed, coated, or encapsulated for corrosion protection. "Monostrand" (VSL Systems 1982) is sleeved in a polypropylene sheath filled with grease that provides debonding and corrosion protection. Epoxy-coated and encapsulated strands are available with and without a bond-enhanced surface. Tests have shown that the epoxy coating does not inhibit the mechanical performance, and it is impervious to most commonly encountered corrosive media (Dorsten et al. 1984).

Throughout the 1970s, plain strand performed well in relatively narrow cut-and-fill stopes (Hunt and Askew 1977 and Fuller 1981). In the late 1970s, the load-transfer characteristics of plain strand were improved by adding steel ferrules at intervals on the strand (Schmuck 1979 and Cassidy 1980). These modified systems proved useful in wider cut-and-fill stopes, but the material and fitting cost of internal anchors and the need for larger-diameter holes made them relatively expensive. They were phased out as mines economised by buying bulk strand that could be delivered underground in coils and cut to length on site.

Cable-bolt technology, developed initially for cut-and-fill mining, was applied to the support and reinforcement of larger open-stope spans in the late 1970s and 1980s. A review by Fabjanczyk (1982) and a second by Fuller (1983a, 1983b) indicated that, although collapse of many of the wider spans was common, there were very few cases of ruptured cable bolts associated with these failures. The rock was simply stripping off the strands, leaving them bare and without their full load-carrying capacity being utilised. This generally poor performance of plain strand in larger spans has been repeatedly confirmed by later reviews (Wyllie 1986).

By 1981, two modified designs comprising strand with supplementary anchors were attempted for the post rock-mass-yield reinforcement of highly stressed crown pillars at Broken Hill in Australia. These trials were very successful, whereas three previous crown pillars reinforced with conventional cable dowels had failed (Matthews et al. 1983). The anchors consisted of double-acting, barrel, and wedge anchors or rectangular, steel swages. The intervals between anchors were debonded using either polyethylene tubes or simply painted. At about the same time, surface-restraint systems comprising anchors, plates, mesh, and straps were also starting to be used with increasing success in wide open-stope spans (Bywater and Fuller 1983). Some of these systems were arranged with short debonding sleeves near the collar, which enabled the restraint system to be held tight against the face.

In 1983, the birdcaged strand was developed for use in cut-and-fill mining at Mt. Isa in Australia. It consists of an unravelled and rewound strand that results in an open-weave cross section with greatly enhanced load-transfer characteristics (Hutchins et al. 1990). This meant the strand could be used in multiple cuts without the need to fit expensive face plates and anchors to the strand after each lift. Plating can still be achieved because the open weave can be terminated back into a normal profile.

In the last few years, a number of other "modified" weave strands have appeared—"bulbed" strand (Garford 1990) and "ferruled" strand (Windsor 1990). Bulbed strand is formed by gripping the strand and compressing it axially to separate and deform the wires over a small interval. Ferruled strand may be formed by spinning the peripheral wires during manufacture over a ferrule placed on the kingwire.

A formidable range of advantages are often highlighted for cable bolts made from materials other than steel (e.g. fibreglass, Mah et al. 1991). Cable bolts may be designed using polymeric materials with a similar strength to steel, but the penalty is usually a brittle failure at relatively low ultimate strains. Viable alternatives certainly do exist (e.g. carbon-fibre composites), but costs are currently prohibitive.

The development of hardware for cable bolting has been matched by an improved design philosophy and technique. It is important to note that design includes choosing a suitable type of cable bolt, an appropriate installation procedure, and deciding on whether to use pre- or post-reinforcement in conjunction with pre- or post-tensioning. The decisions are generally based on logistics, equipment, and experience. The choice and arrangement of suitable cable-bolt elements, the orientations, lengths, and density of elements in the cable-bolt array often evolve on each particular site into a scheme that works (Lappalainen et al. 1983, Cullum and Nag 1984, and Hutchison 1989). Similarly, the installation procedure also evolves to suit the chosen cable-bolt type, the equipment available, and the training of the work force (Thompson et al. 1986 and Zarichney 1990).

64.4 CABLE-BOLT INSTALLATION ISSUES

The installation procedure for cable bolts is the most important component in cable-bolting practice and has the potential to completely dictate the in-situ mechanical performance. The length and transverse flexibility of cable bolts (in comparison with rock bolts) create a number of difficulties in ensuring a quality installation. Figure 64.2 sets out the aspects of installation that can adversely affect the performance of a cable bolt. The important point to note is that detecting negative installation events after installation is not guaranteed. Similar problems are encountered in ground-anchor practice. However, adopting strict guidelines for hole-cleaning, installation, grouting, and stressing operations (British Standards Institution 1989) will take care of most of these problems. Furthermore, there are stringent requirements for conducting proof-loading tests and for testing the ability of ground anchors to sustain elevated service loads for a number of cycles, at certain durations, and within prescribed relaxation and load-loss limits. However, simply transferring these requirements to cable-bolting practice in the more competitive mining industry environment is rarely possible. Two aspects of installation where cable-bolting practice has demonstrably evolved away from ground-anchor technology—grouting and stressing—will be discussed here.

Aspect of installation	May be identified	Detection guaranteed
HOLE:		
Correct position/orientation	Yes	Yes
Correct length diameter	Yes	No
Flushed clean	Yes	No
CABLE:		
Correct type/capacity	Yes	Yes
Correct length	Yes	Yes
Anchors attached correctly	Yes	No
Spacers attached correctly	Yes	No
Clean before insertion	Yes	No
Clean after insertion	Yes	No
Cable to end of hole	Yes	No
Cable central in hole	No	No
GROUT:		
Correct cement type	Yes	Yes
Correct water content	No	No
Correct additive	Yes	Yes
Correct mixing	Yes	No
Breather tube to end of hole	Yes	No
Breather/grout tubes filled	Yes	No
Hole filled with grout	No	No
Segregation or loss of grout	No	No
TENSIONING:		
Correct anchor fittings	Yes	Yes
Correct surface hardware	Yes	Yes
Correct angle with face	Yes	Yes
Correct procedure	Yes	No
Correct installed tension	No	No

FIGURE 64.2   Aspects of poor cable bolt installation

64.4.1   Cable-Bolt Grouting

The original procedures for grouting cable bolts essentially comprised drilling and flushing the hole, pushing the cable into the hole, and then grouting. The cable bolt was fitted with a system of plastic spacers and separators that separated multiple strands or wires and held them in the center of the hole and provided alignment for an air-bleed tube that ran to the end of the hole. The cable was also provided with a hook arrangement to suspend the cable under its own weight, which also tended to keep it straight. The collar was sealed with a plug that included a short grout-delivery tube. The borehole was filled with a grout with a water:cement ratio of about 0.40 to 0.45, together with an anti-bleed agent. Grouting from the collar up using grout of the correct consistency ensured a complete filling of the hole and pneumatic air bleed from the bleed tube positioned just above the end of the cable. Using a water:cement ratio of about 0.45 meant that mixing and pumping was reasonably straightforward using fairly robust, primitive equipment; full hydration of the grout occurred; and the additive minimised the loss of water from the mix due to bleed. It is worth noting that a theoretical minimum water:cement ratio of about 0.4 is required to ensure complete hydration and strength development. However, one must consider here how much water will be scavenged from the mix during "wetting" of the borehole walls (critical in hot, dry rock masses) and how much is lost by "bleed" into cracks (critical in highly fractured, open rock masses). Note that, in civil engineering prestressed concrete practice, grouting strand tendons is usually conducted with the hole or ducts initially full of water rather than air.

The advent of cable-insertion equipment and cable-bolting machines means that completely mechanised hole drilling, cable insertion, and grouting are now possible. With these machines,

more viscous grouts are used, and the cable bolts are usually "plunged" into a pre-grouted hole. The productivity gains are high, but there are some disadvantages. One disadvantage is a loss of flexibility in choosing the cable bolt because most machines will handle only plain strand (although later modifications to pusher assemblies now allow bulbed strand to be accommodated). It is also difficult to install multiple cables in long holes, and spacers and separators cannot be used. The lack of choice of cable-bolt type is important in some circumstances, especially where a stiffer cable-bolt response is required (i.e., birdcage or multiple-bulbed strand). Furthermore, losing the ensured separation of multiple strands and their centrality in the borehole means that individual strands may actually be in contact with one another and the borehole wall over a considerable length of the borehole. This is especially true in non-vertical holes and can drastically reduce cable-bolt capacity.

Other procedures have been developed in an attempt to capitalise on the benefits of using stronger and thicker grouts. The cable is installed without a bleed tube and collar plug, and the grout is pumped into the hole from the top downwards. Some of the disadvantages with this system are that, again, spacers and separators are not used because of the hole-diameter requirement to accommodate the larger grout delivery tube; and the thicker grout tends to push the strands together and against the borehole wall. The greatest disadvantage is the possibility of not completely filling the borehole, leaving voids that will ultimately interfere with load transfer between the rock and the cable bolt.

There are a number of structural and logistical advantages in using the more viscous, stronger, lower water:cement grouts; and it is possible, with the correct mixing and pumping equipment, to use the modern methods of grouting and achieve a good cable-bolt installation. However, the civil engineering industry learned that pursuing stronger concrete mixes introduced workability and placement problems that had the overall effect of reducing the concrete strength. Thompson and Windsor (1998) attempted to review the consequences of moving away from the original grouting procedures and using the lower water:cement ratio grouts. In that work, and using the results of other researchers, the authors were able to show both theoretically, and with experimental results, that the average strength and stiffness of the grout certainly increased with lowering the water:cement ratio. Unfortunately, they also found that the variability of strength and stiffness increased and was dependent on the gel:space ratio or the percentage of air voids introduced into the grout during mixing and pumping. It is much more difficult to wet all the cement particles and pump the thicker grouts. The results of this study are shown in Figure 64.3 and imply that, unless very good mixing and pumping equipment are used, the same lesson learned in civil engineering applies—the variability is such that overall only a marginal increase of in-situ strength and stiffness can be assured. The results of these investigations highlighted the conflicting requirements on grout for optimum reinforcement performance (high strength and stiffness) and the requirements for placement (mixability and flowability).

A new concept, which describes grouting methods in terms of the direction of grout flow relative to gravity was introduced. This enables a uniform terminology that is independent of up or down holes and is independent of whether the application is related to surface or underground excavations. This concept allows grout flow to be termed either "gravity retarded" or "gravity assisted." These two different types of flow are shown in Figure 64.4a for the situation where the grout tube is stationary in the borehole.

When the flow of the grout in the borehole is retarded by gravity, the grout emerges from the tube into a column in which there is positive resistance due to the height of the grout above the outlet and friction at the borehole wall. Depending on the water:cement ratio and viscosity of the grout, there will be a

(a)

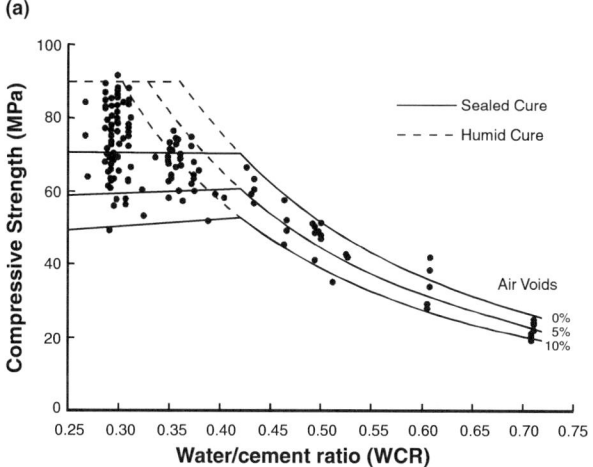

(b)

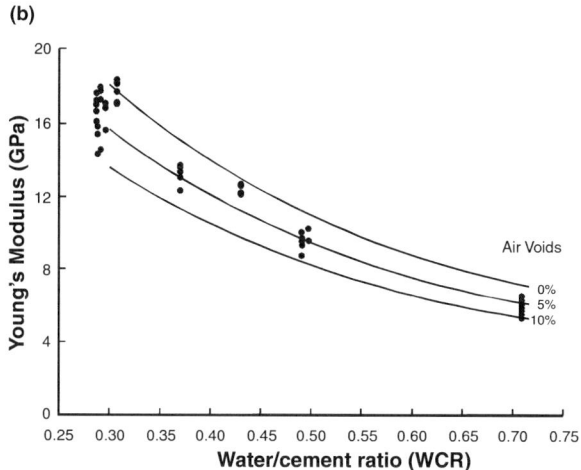

FIGURE 64.3 Variation of (a) compressive strength and (b) stiffness for different water/cement ratios

(a)

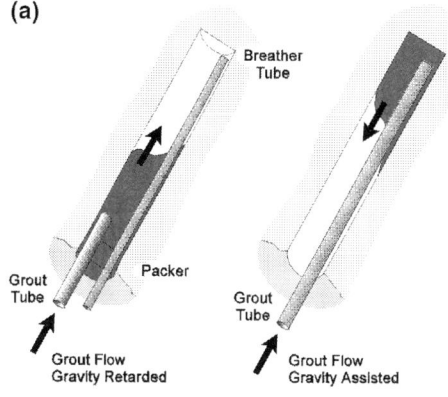

(b)

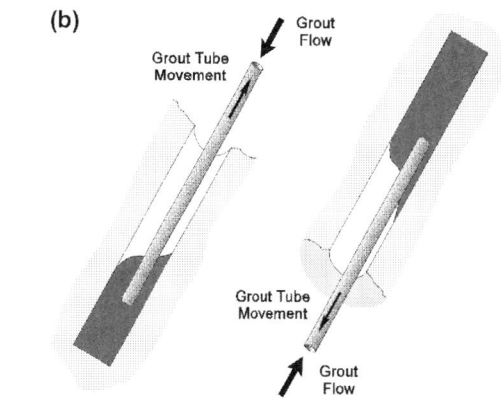

FIGURE 64.4 Schematic representation of (a) "gravity retarded" and "gravity assisted" flow, and, (b) grout placement using withdrawal of the grout tube

definite tendency for grout mixing to occur; and the hole will initially be completely filled with grout, irrespective of the water:cement ratio.

When the flow of the grout is assisted by gravity, the grout emerges from the tube into the borehole with friction only at the borehole wall. In some cases, friction at the reinforcement element surface and obstructions such as spacers will also resist the flow. In this case, filling the borehole may be problematic if the grout viscosity and cohesion are too low, causing the flow in the borehole to be faster than the flow from the grout-delivery tube. There will be a definite tendency for gaps to be created in the grout flow. In addition, the grout may separate and flow around obstructions and not fill the void in the "dead spot."

When the grout delivery tube is withdrawn during pumping, the grout is not required to flow the complete length of the borehole. However, the end of the grout tube must remain within the grout column as shown in Figure 64.4b. In both instances, extreme care is required to prevent air voids being introduced into the grout column due to withdrawal of the grout tube at a greater rate than the filling rate of the borehole, particularly in a very low water:cement ratio grout with high viscosity and cohesion.

In conclusion, the situation often arises where a low water:cement ratio grout may be specified in ignorance of the equipment required to enable proper mixing and the procedure for proper pumping and placement. In this situation, the

operators responsible for grout placement will be required to improve the mixability and pumpability. In ignorance of the adverse consequences on reinforcement performance caused by reducing grout strength and stiffness, pumpability is achieved simply by arbitrarily increasing the water:cement ratio. This situation must be avoided with the modern grout-placement techniques and the grout mix design should specify appropriate water-reducing admixtures to ensure that the grout may be placed at the required water:cement ratio using the available equipment. The results from studies on the physical and mechanical properties of cement grout in both fluid and hardened conditions lead to the following conclusions.

Reducing the water:cement ratio of cement grout used in conjunction with rock reinforcement aims to:

- Improve productivity of grout placement
- Increase grout strength and stiffness
- Improve reinforcement system performance

but requires:

- Using larger, more efficient mixing equipment.
- Using more powerful continuous-feed pumps

or leads to:

- Difficulty placing the grout in the borehole
- Requiring admixtures to reduce viscosity and improve pumpability
- Introducing air voids and incomplete hydration of the grout

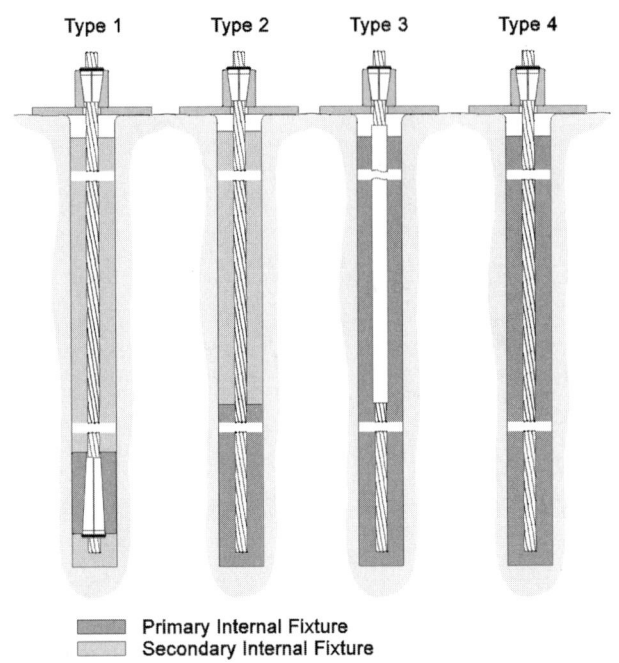

FIGURE 64.5   Possible configurations for tensioned cable installations

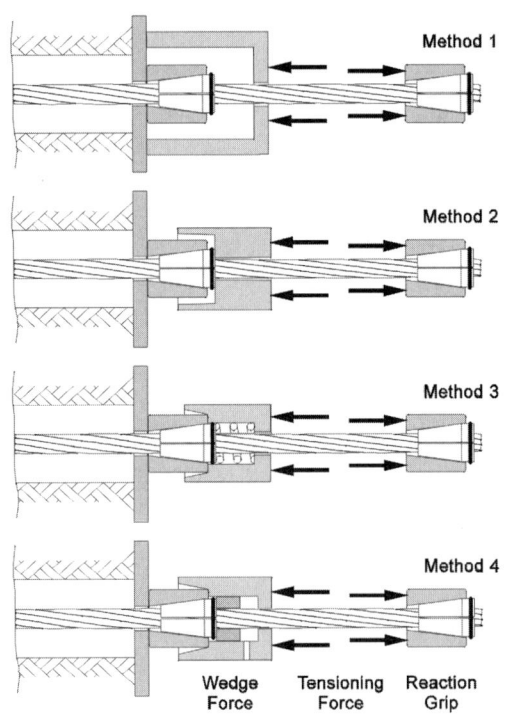

FIGURE 64.6   Methods of tensioning cable bolts

It is recommended that interested readers refer to the texts by Littlejohn (1982), Nonveiller (1989), and Houlsby (1990) for more information on grout-mix design, mixing, and pumping.

64.4.2   Cable-Bolt Stressing

The original practice of pre-tensioning cable bolts to produce a cable tendon was copied and modified from ground-anchor technology. This practice ensured that the resulting tension in the cable tendon was predictable and consistent. However, the formal procedure of pre-tensioning was abandoned about 25 years ago when it was realised that tension quickly develops in un-tensioned pre-reinforcement in response to rock movement induced by blasting or stress redistribution effects during or after excavation. Later, the recognition of the need for surface-restraint fittings (plates and anchors), particularly for plain strand cable bolts, saw the reintroduction of a so-called "post-tensioning" process that might be better described as stressing. Stressing involves less exacting procedures than pre-tensioning and have evolved to address the basic requirements of productivity and safety associated with installing large numbers of cable bolts in sometimes hazardous conditions.

Variations on stressing now include placing the cable bolt in the hole, grouting an anchorage at the far end, fitting a face restraint, pre-stressing the cable bolt, and then grouting the balance of the hole. Some operations replace this two-stage grouting process by pre-stressing and then grouting a cable bolt that has been fitted with a special expansion shell anchor (Rock Engineering 1985). Most operations simply install a short debonding tube near the collar, grout the full length of the hole, attach the restraint system, and then stress the short decoupled length which, pulls the restraint system tight against the face. The various arrangements of cable-bolt systems currently in popular use are shown in Figure 64.5.

In general, two stages are involved in the stressing process:

- **Tensioning.**   This involves applying a tensioning force to the cable to establish an initial tension. During tensioning, the surface hardware is pushed against the rock face and held tightly.

- **Relaxation.**   This involves securing the external fixture to the cable and removing the tensioning force. This second stage is usually includes reducing the cable tension to leave a residual tension in the cable bolt.

When pre-tension is applied to a fully grouted strand to hold the face restraint, only a small residual pre-tension remains in the system. The initial tension that can be produced in the element and the residual tension left after stressing depend on the cable-bolt system configuration, the installation procedures, the stressing equipment, and the values of several critical parameters. In almost all cases, the external fixture is a barrel-and-wedge fitting. The properties of the external fixture and its placement have the most influence on the final results of the stressing process.

The stressing equipment for cable bolts consists essentially of a hollow hydraulic cylinder, an anchor to grip the cable, and an auxiliary assembly to push on the bearing plate. Four alternative assemblies currently in use are shown in Figure 64.6. These assemblies result in the tensioning force being distributed differently between the barrel and the wedge of the external fixture. The assembly design influences the initial tension produced during tensioning and the residual tension after relaxation. Relaxation is due mainly to the displacements between the cable and the external fixture and to "pull-in" of the wedges once the active pressure is released. The effectiveness of the assembly increases in the order of the method given in Figure 64.6. Consequently, for situations requiring a consistent, positive pre-stress to hold the external fixtures (the author recommends 50 kN minimum residual pre-stress), it is critical to use either Method 3 with the correct stiffness spring, or Method 4 with the correct secondary pressure in the nose cone, because both methods push the wedges into the barrel independent of loading the face plate assembly and stressing the cable. Methods 1 and 2 give inconsistent and unreliable results.

The technology of stressing cable bolts and the design-calculation procedures to calculate the residual tension and proof of their validity have been researched and described by

Thompson (1992) and revisited by Thompson and Windsor (1995). It is also worthwhile to consider the formal procedures of prestressing required in ground-anchor practice, and with this in mind, the interested reader is referred to the texts by the British Standards Institution (1989), Habib (1989), and Xanthakos (1991).

64.5 CONCLUSIONS

The key issues in cable-bolting practice today are not design or analysis, but selecting the appropriate type of cable bolt for the particular mining engineering problem and implementation of the correct installation procedure. It is extremely important that design engineers understand both the reasons for and the circumstances under which the different varieties of cable bolts were developed. It is equally important that installation staff receive the proper training courses to provide them with some reasonably stringent but sensible installation guidelines and an understanding of the consequences of the common installation problems.

64.6 ACKNOWLEDGMENTS

The author wishes to thank his colleagues, Alan Thompson and Glynn Cadby, at Rock Technology Pty, Ltd., for their constant support and advice.

64.7 REFERENCES

British Standards Institution. (1989). British standard code of practice for ground anchorages (BS 8081:1989). London: BSI.
Bywater, S. and P.G. Fuller. (1983). Cable support of lead open stope hangingwalls at Mount Isa Mines Limited. Proc. Int. Symp. on Rock Bolting. Rotterdam: Balkema, pp. 539–555.
Cassidy, K. (1980). The implementation of a cable bolting program at the Con Mine. Proc. 13th Can. Rock Mech Symp. (Underground Rock Engineering). Montreal: CIM., pp. 67–72.
Clifford, R.L. (1974). Long rockbolt support at New Broken Hill Consolidated Limited. Proc. AusIMM Conf., No. 251, pp. 21–26.
Cullum, A.J. and D.K. Nag. (1984). King Island Scheelite. Engineering and Mining Journal, July: pp. 34–43.
Davis, W.L. (1977). Initiation of cablebolting at West Coast Mines, Rosebery. Proc. The AusIMM Conf. Tasmania: pp. 215–225.
Dorsten, V., F.H. Frederick, and H.K. Preston. (1984). Epoxy coated seven-wire strand for prestressed concrete. Prestressed Concrete Inst. J. 29(4): pp. 1–11.
Fabjanczyk, M.W. (1982). Review of ground support practice in Australian underground metalliferous mines. Proc. AusIMM Conf. Melbourne: pp. 337–349.
Fuller, P.G. (1981). Pre-reinforcement of cut and fill slopes. Proc. Application of Rock Mechanics to Cut and Fill Mining. London: IMM., pp. 55–63.
Fuller, P.G. (1983a). The potential for cable support of open stopes. Proc. 5th ISRM Congress on Rock Mechanics. Melbourne: ISRM and AGS. D39–D44.
Fuller, P.G. (1983b). Cable support in mining: A keynote lecture. Proc. Int. Symp. on Rock Bolting. Rotterdam: Balkema, pp. 511–522.
Garford Pty Ltd. (1990). An improved, economical method for rock stabilisation. Perth: Australia.
Habib, P. (Editor) (1989). Recommendations for the design, calculation and construction of ground anchorages. 3rd Edition. Rotterdam: A.A. Balkema.
Houlsby, A.C. (1990). Construction and Design of Cement Grouting. New York: John Wiley & Sons.
Hunt, R.E.B. and J.E. Askew. (1977). Installation and design guide-lines for cable dowel ground support at ZC/NBHC. The AusIMM (Broken Hill Branch) Underground Operators' Conference: pp. 113–122.

Hutchins, W.R., S. Bywater., A.G. Thompson, and C.R. Windsor. (1990). A versatile grouted cable dowel reinforcing system for rock. The AusIMM Proceedings 1: pp. 25–29.
Hutchison, B.J. (1989). Recent developments in rock reinforcement at BHP-Utah's Yampi Sound Operation. MMIS/IMM Joint Symposium. Kyoto: MMIS.
Hutchinson, D.J. and M.S. Dierderichs. (1996). Cablebolting in Underground Mines. BiTech Publishers Ltd. Richmond, B.C., Canada.
Jirovec, P. (1978). Wechselwirkung zwischen anker und gebirge. Rock Mechanics, Suppl. 7: pp. 139–155.
Kaiser and McCreath (Editors). (1992). Rock Support. Proc. Int. Symp. on Rock Support, Sudbury, Canada. Rotterdam: A.A. Balkema.
Lappalainen, P., J. Pulkkinen, and J. Kuparinen. (1983). Use of steel strands in cable bolting and rock bolting. Proc. Int. Symp. on Rock Bolting. Rotterdam: Balkema, pp. 557–562.
Littlejohn, G.S. (1982). Design of cement based grouts. Proc. Conf. On Grouting in Geotechnical Engineering, W.H. Baker, Ed. New Orleans: ASCE. pp. 35–48.
Mah, P., R. Pakalnis, and D. Milne. (1991). Development of a fibreglass cable bolt. Proc. 93rd AGM CIM, Vancouver, Paper 43.
Marshall, D. (1963). Hangingwall control at Willroy. CIM Bulletin 56: pp. 327–331.
Matthews, S.M., V.H. Tillmann, and G. Worotnicki. (1983). A modified cable bolt system for the support of underground openings. Proc. AusIMM Annual Conf., Broken Hill.
Matthews, S.M., A.G. Thompson., C.R. Windsor, and P.R. O'Bryan. (1986). A novel reinforcing system for large rock caverns in blocky rock masses. Proc. Int. Symp. on Large Rock Caverns. Oxford: Pergamon. pp. 1541–1552.
Nonveiller, E. (1989). Grouting Theory and Practice. Amsterdam: Elsevier. pp. 243–255.
Rock Engineering. (1985). Cablebolting Brochure. Melbourne.
Schmuck, C.H. (1979). Cable bolting at the Homestake gold mine. Mining Engineering, December: pp. 1677–1681.
Stheeman, W.H. (1982). A practical solution to cable bolting problems at the Tsumeb Mine. CIM Bulletin 75(838): pp. 65–77.
Thompson, A.G. (1992). Tensioning reinforcing cables. Rock Support. Proc. Int. Symp. on Rock Support., Sudbury, Canada. Rotterdam: A.A. Balkema, pp. 285–291.
Thompson, A.G., S.M. Matthews., C.R. Windsor., S. Bywater, and V.H. Tillmann. (1986). Innovations in Rock Reinforcement Technology in the Australian Mining Industry. Proc. 6th ISRM Congress on Rock Mechanics, Montreal. Rotterdam: A.A. Balkema, pp. 1275–1278.
Thompson, A.G. and C.R. Windsor. (1995). Tensioned cable bolt reinforcement—an integrated case study. Proc 8th ISRM Congress on Rock Mechanics. Tokyo, Japan. Rotterdam: A.A. Balkema. V2, pp. 679–682.
Thompson, A.G. and C.R. Windsor. (1998). Cement grouts in theory and reinforcement practice. Proceedings of the 3rd North American Rock Mechanics Symposium, NARMS'98, Cancun, Mexico. Compact Disk, Paper No. AUS-330-2.
Thorne, L.J. and D.S Muller. (1964). Prestressed roof support in underground engine chambers at Free Geduld Mines Ltd. Trans. Assoc.Min.Mngr. Sth.Afr.: pp. 411–428.
Villaescusa, E., C.R. Windsor, and A.G. Thompson, Editors. (1999). Rock Support and Reinforcement Practice in Mining. Proc. Int. Symp. on Ground Support., Kalgoorlie, Australia. Rotterdam: A.A. Balkema.
VSL Systems, Ltd. (1982). Slab post tensioning. Switzerland.
Windsor, C.R. (1990). Ferruled Strand—Unpublished memorandum, Perth: CSIRO.
Wyllie, R.J.M. (1986). Cable bolting. Engineering and Mining J. 187(2): pp. 38–40.
Xanthokos, P.P. (1991). Ground anchors and anchored structures. New York: John Wiley and Sons.
Zarichney, G. (1990). Ground control at the Golden Giant mine. Proc. 92nd AGM CIM, Ottawa: p. 13.

CHAPTER 65

Shotcrete as an Underground Support Material

Sam Spearing[*]

65.1 INTRODUCTION

According to the American Concrete Institute, shotcrete (sprayed concrete) is defined as pneumatically applied mortar or concrete, projected at high velocity. The principles applied to standard concrete technology are no different in shotcrete. This means that shotcrete should not be "randomly designed," and consideration should be given to achieving a technically acceptable and a cost effective mix. This implies that the following technical aspects must be considered.[*]

- Material components and overall composition
- Application conditions (including access to and the availability of services)
- Application mode (dry or wet process)
- Logistic constraints (mainly as they influence material handling)
- Health and safety requirements

The true final (placed) costs should also always be considered, and this seldom occurs, at present, in the mining industry. The key considerations are—

- Material costs (an easy cost to establish)
- Equipment, capital, and operating costs
- Labour costs (including transportation to site)
- Application efficiencies (wastage, especially rebound)
- Time-related costs (the influence on the overall mining cycle time)

The first machine to spray concrete was developed in Pennsylvania, USA, in 1907 by Carl Akeley for use in construction. This concept was later improved in 1915 by the Cement Gun Company, which later became the Allentown Gun Company.

For shotcrete to be the productive and efficient support that it can and should be, all aspects must be considered. Too often in the mining industry, technically inferior or inappropriate support systems are used because of convenience, ignorance, resistance to change, and/or expediency.

65.2 SHOTCRETE MATERIALS

The material constituents of the shotcrete are important in order to achieve a desired target-placed performance at the lowest cost. Such constitutents are described below.

65.2.1 Binders

Cement Binders. Cement is the bonding material (glue) that holds a cementitious material together. For most shotcreting applications, portland cement is used. This cement was invented by an English bricklayer, Joseph Aspdin, in 1824. The name was derived from the set material's color and texture, which resembled a local limestone from the Isle of Portland off the British coast called portland stone. Cement is produced from mainly a mixture of klinker and gypsum. Klinker is typically produced in a rotary kiln from lime, silica, alumina, and ferric oxide.

Other types of cement that are commonly used in shotcrete include—

- Sulfate-resisting cement, which typically has a lower tricalcium aluminate content than portland cement, and
- High-alumina cement (HAC) produced by fusing (melting) a mix of bauxite and limestone. HAC is therefore not a portland cement derivative and is often used in refractory applications where it can be troweled in place or shotcreted.

Cement Extenders. Cement extenders are commonly used in shotcrete to reduce costs. Fly ash is the material most commonly used and is obtained as a waste product from thermal power plants. It is supplied primarily as a blend with portland cement in bags or bulk.

Ground, granulated, blast furnace slag (GGBS) is another cement extender. GGBS is a glassy, granular material usually produced in a blast furnace as a by-product of iron production. The molten slag is rapidly quenched and then finely ground.

Fumed silica, silica fume, or microsilica, as it is often called, is also a cement extender. This material is described under the section on "Admixtures" because of its unique and desirable properties.

Cement extenders are commonly used to replace about 30% (but sometimes up to 50%) of the cement used, but generally the early strength gain of the shotcrete is reduced, which is frequently undesirable in mining because of safety considerations.

65.2.2 Aggregates

As in concrete, aggregates are used to provide dimensional stability and reduce the cost. The main characteristics to consider are—

- Grade
- Particle shape
- Presence of reactive chemicals and minerals (for possible alkali-acid reactions)
- Compressive strength

When applying shotcrete with the dry process, aggregates can be selected so that voids are minimized, which is the common approach when concrete is used. As a general guide, the use of

---

+25-mm material is to be avoided; the modern trend is to use material with a maximum size of 8 mm.

A typical dry process base mix would therefore consist of—

- 20% to 25% cementitious binder
- 15% to 20% coarse aggregate
- 55% to 65% sand (natural, washed sand is preferred)

This approach needs to be modified when the wet process is used because pumpability is a major issue, and generally more fines are needed to reduce the chance of line blockages and make pumping easier.

65.2.3 Water

Water quality can be important. The water used should be free from oils and not heavily acidic.

65.2.4 Admixtures

The American Concrete Institute in its *Manual of Concrete Practice* (ACI 1999) defines an admixture as "A material other than water, aggregates, hydraulic cement, and fiber reinforcement, used as an ingredient of concrete or mortar, and added to the batch immediately before or during its mixing."

Many different types of admixtures can be beneficial in shotcrete, depending on the specific application and requirements. Most admixtures are used in the wet process only (discussed later), but generally a set accelerator (usually in powdered form) can be used when dry shotcreting. To achieve the quantifiable benefits from admixture usage, it is critical that the dosing equipment is reliable and appropriate.

Accelerators. Shotcrete accelerators generally fall into the following categories:

- Silicates (e.g., water-glass or sodium silicate)
- Sodium or potassium aluminates
- Alkali-free accelerators

Silicates are not really true accelerators because they only create a gelling effect rather than a rapid early strength gain. Accelerators also tend to reduce the final strength of the shotcrete. This is partly because a slower rate of strength gain results in a finer and more dense crystalline growth that creates a stronger final product.

In a typical dry mix, a powdered accelerator addition of between 2% and 5% based on the weight of the total cementitious addition would be reasonable. Accelerator dose rates are normally expressed as a percentage of the total cementitious content.

In wet shotcrete, the dose range would be around 2% to 6%. The latest trend is toward the (noncaustic) alkali-free accelerators because they are more environmentally safe and tend to cause significantly less long-term strength loss. With such an accelerator, a dose rate of up to 8% is not uncommon.

Water Reducers (Plasticizers and Superplasticizers). There are various types of water reducers available, and they tend to fall into three broad categories.

- Low-range water reducers (e.g., lignosulphonates) give about a 15% water reduction, but some can retard strength gain.
- Medium-range water reducers (e.g., melamines) give about 25% water reduction.
- High-range water reducers (e.g., polycarboxylates) give about a 45% water reduction.

These admixtures work by charging each cement particle ionically and causing them to separate, thereby effectively lubricating the mix and enabling the amount of water to be reduced while obtaining the same consistency (i.e., slump).

Microsilica (Fumed Silica). Microsilica is a very fine, spherical material with a high pozzolanic reactivity. A typical

dosage would be 5% to 10% by weight of the cementitious binder. The use of microsilica in shotcreting has the following benefits:

- Improved durability (more resistance to freeze/thaw cycles and improved resistance to sulphate attack)
- Improved bonding to substrates
- Higher strengths
- Reduced rebound
- Improved flow in the delivery hose (wet process)
- Reduced wear in the pump and nozzle (wet process)

Curing Agents and Concrete Improvers. There is an incorrect perception that an underground environment provides good curing conditions for shotcrete. This is incorrect because the ventilation tends to cause premature drying of the shotcrete surface, which can result in—

- Poor hydration, which results in a weaker final product
- Reduced substrate bonding if the shotcrete layer is less than 75 mm
- Significant shrinkage cracking

Solutions to these problems include—

- Regular wetting of the placed shotcrete
- Application of an external curing agent (such as a spray-applied wax)
- Inclusion in the mix (wet process only) of a concrete improver that has the potential to improve curing and increase the bond with the substrate

Regular wetting of the placed product is often impractical and too time consuming. The application of an external curer involves a second (albeit simple) operation and makes it difficult to apply another layer of shotcrete at a later time (for whatever reason) unless the coating is removed.

Consistency Controllers. Under certain conditions, the use of a consistency control system can help in the application of wet shotcrete. The first component is added into the mix before pumping to keep the open time and improve pumpability. The second component is added at the nozzle and stiffens the mix and aids in strength gain.

Hydration Controllers. The useful life of a wet shotcrete mix can be a limiting factor in underground applications because of logistics considerations. A typical batch of untreated shotcrete should be discarded as waste after a period between 1 and 2 hours, depending on the ambient temperature. Conventional retarders can extend this to about 4 hours, but hydration controllers can effectively put the mix "to sleep" for up to 72 hours. This technology is of great advantage in many underground wet shotcreting applications because it helps resolve logistics problems.

Hydration controllers function in two ways: By acting as an effective dispersant, thus keeping hydrating particles apart, and by forming a barrier around all the cementitious particles, thus stopping the hydration process altogether (unlike conventional retarders).

This effect is overridden when shotcreting by adding an effective accelerator, and the hydration controller has no adverse effect on the rate of strength gain and ultimate strength provided that an adequate amount of accelerator is added.

65.2.5 Fibers

Concrete is by nature a brittle product and is weak in tension. In shotcreting, reinforcement can be provided by the use of screen mesh or fibers. Fibers have obvious advantages over screens (Figure 65.1).

- Fibers are more evenly distributed throughout the shotcrete.
- Mesh is difficult and labour intensive to apply.

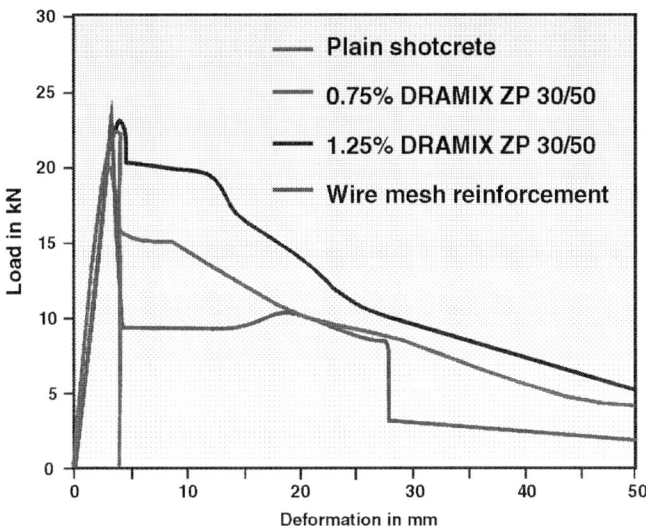

FIGURE 65.1 The load-bearing capacity of plain, mesh, and steel fibre reinforced shotcrete (after Vanderwalle)

- Fiber-reinforced shotcrete effectively lines the tunnel periphery, whereas additional shotcrete is needed to fill depressions when a screen is used, as the mesh is usually fixed at the high points on the tunnel periphery.

- Mesh can increase shotcrete rebound significantly due mainly to screen vibration during spraying.

Fibers are mainly used to improve durability, increase impact resistance, and reduce surface cracking in concrete. The application of fibers in construction dates back centuries to the use of horse hair, jute, sisal, and cotton.

Natural fibers have been replaced by engineered materials, such as polypropylene (monofilament or fibrilated), carbon fiber, glass fibers, drawn steel wire, slit sheet, milled steel pieces, and melt extract pieces. The most commonly used are the drawn wire and polypropylene fibers.

The most important characteristics that fibers should have are aspect ratio (overall ratio of fiber length to diameter), tensile strength, and shape.

An ideal fiber should have the following characteristics:

- A length such that it can overlap and bridge at least two of the largest aggregate particles used in the mix (typically a length between 25 and 40 mm)

- High aspect ratio (i.e., thin)

- High tensile strength

- A shape that results in a good anchor, particularly at the fiber ends in the shotcrete

The use of steel fibers in dry shotcrete is not generally recommended because of the high fiber loss found in the rebound (significantly more fiber loss than the overall measured rebound). Where steel fibers are used, a dosage rate of between 30 and 50 kg/m³ (0.4% to 0.6 % by volume) is generally used.

The use of polypropylene fibers in shotcrete for mining applications has increased dramatically with the development of high-performance polymer fibers. Their high tensile strength and crimped shape have resulted in performances very similar to those obtained with steel. Dosage rates of between 9 and 18.0 kg/m³ (1% to 2% by volume) are typical. In the future, more use will be made of polymer fibers, possibly blended with steel fibers, in high-performance shotcrete applications.

65.3 SHOTCRETE DESIGN

The final strength of shotcrete, like concrete, depends mainly on the water-to-cement ratio and air content after placement. Much information is available concerning the design of shotcrete, but simple factors are frequently overlooked initially, and these can create major losses, costly delays, and final sprayed linings that do not meet the desired performance requirements. Shotcrete design must include more than creating a laboratory mix that meets strength gain requirements with locally available raw materials (such as cement, sand, stone, and water) in adequate supply. While these factors are important, three other equally vital aspects must not be ignored.

- The fact that strength must be achieved on the rock and not in a laboratory

- The time available for spraying a given volume

- The placed cost of the mix

65.3.1 Placed Strength

Strength gain (and general performance) of shotcrete needs to be reliably achieved as sprayed. This means that the mix must be pumpable, bond well to the substrate (with a minimum of rebound), build up desirable thickness in a few passes, and usually gain strength rapidly. These requirements generally imply that the need is for a cohesive mix with an initially high slump and a final low slump on placement. Such needs are unlikely to be evident during lab trials. Achieving such requirements is governed by the use of admixtures and additives.

- Microsilica for cohesion, rebound reduction, and durability

- Superplasticizers to control slump and the water-to-cement ratio (in the wet mix process)

- Accelerators for early strength development and high single-pass application thickness (e.g., with alkali-free accelerators, overhead single-pass thicknesses of between 30 and 50 cm are possible)

- Concrete improvers (internal curing admixtures) to achieve long-term strength, better bonding, fewer cracks, and improved durability (in the wet mix process)

65.3.2 Time

Time is often a scarce commodity in mining, and when shotcreting, lack of time presents two problems: Supplying adequate volumes of material to the shotcrete site for spraying and actually spraying the area in the time available to fit with the mining cycle.

With dry mix machines, up to 8 m³ of shotcrete can be placed in an hour, and with wet shotcrete machines, up to 25 m³ of shotcrete can be sprayed in an hour. Therefore, the time needed to actually spray the area is not a problem, *if* the shotcrete machine can be adequately supplied with material.

Adequate amounts of material can be obtained by many means, but the most exciting is the transport of the material as a slurry down a pipe, as most successfully undertaken by shaft sinkers at the Moab No.11 shaft at the Vaal River Operations (Buckley 1998). Such a transportation method permits an almost limitless quantity of quality wet shotcrete mix.

Untreated shotcrete has an open (useful) time of only 1 to 2 hours, and hence it is often essential to use a hydration control admixture to increase the open time to up to 72 hours.

65.3.3 Cost

Material costs can be judiciously reduced by careful selection of raw material, blending, and use of admixtures (e.g., water reducers). It must, however, be noted that other very significant

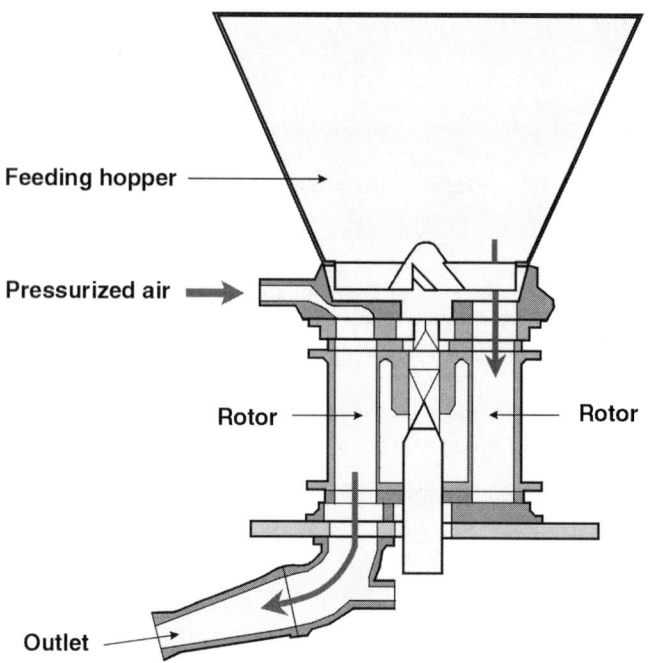

FIGURE 65.2   The rotor principle of a typical dry-spraying machine (after MEYCO)

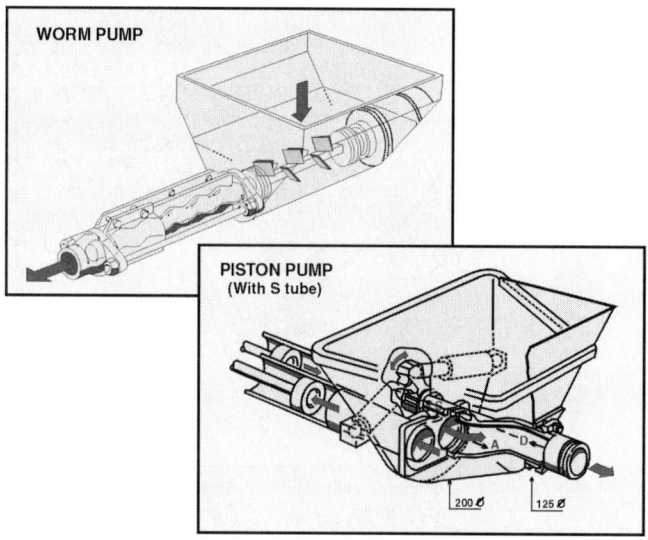

FIGURE 65.3   Wet shotcrete pump types (after Vanderwalle)

TABLE 65.1   Comparison of wet and dry processes

Wet process	Dry process
Little dust	Considerable dust
Low maintenance cost	High maintenance cost
High capital cost	Low capital cost
Low rebound, typically about 5% to 10%	High rebound, usually more than 25%
Moderate to high placement rate, between 4 and 25 m3/hr	Low to moderate placement rate, up to 6 m3/hr
Low transport distance, up to 200 m	High transport distance
Moderate to high placed quality	Moderate placed quality

cost elements are frequently overlooked or ignored. These include—

- Rebound that is seldom if ever measured, but generally lies between 5% and 10% (for wet shotcrete) and 25% to 35% (for dry shotcrete), depending on the application process

- Material transport, not just for shotcrete, but for any support system and ancillaries

- Equipment capital and operating costs

65.4   SHOTCRETE APPLICATION

65.4.1   Application Processes

There are two application processes in shotcreting, and selecting which one to use depends on the specific application and site conditions.

Dry Process.   In this application, the particulate material is conveyed pneumatically in a basically dry state from the pump (Figure 65.2) to the nozzle, where the water is added. The nozzleman is the key to successful application because he controls addition of the water. Too much water causes the shotcrete to sag away from the rock and reduces strength, and too little water causes higher rebound and possibly strength reduction. The moisture content of the mix prior to adding water at the nozzle should be between 2% and 5% to minimize dust at the pump. More than 5% water can cause blockages in the line.

In a typical dry application, the water-to-cement ratio should be in the 0.40 to 0.45 range.

Wet Process.   Until the last decade, the dry process was the most common method of application, but the wet process is rapidly gaining popularity in mining, in line with the general move in mines to mechanize underground operations for safety and productivity reasons. In the wet process, the entire mix, including the total amount of the water, is fed into a hopper and then pumped to the nozzle (Figure 65.3). The mix needs to be fluid enough to be pumped, and therefore the introduction of water reducers into the mix is generally essential to maximize strength gain and reduce overall costs.

The nozzle design in wet shotcrete is important because compressed air is added to improve spray velocity; generally, accelerators are added to improve the early strength gain.

Comparison of Wet and Dry Processes.   Table 65.1 compares the main differences between dry and wet processes.

There can be no doubt that the overwhelming trend is toward wet shotcreting, due mainly to performance and overall cost considerations. This trend has occurred only because of improvements in equipment and development of high-range water reducers and hydration control admixtures.

There is no truth in the commonly held view that dry shotcreting is the better option where overall worker skill levels are low (i.e., in developing and Third World countries). Wet shotcrete has proven to be a cost-effective method all over the world if placement volumes are high and performance requirements high.

The main factors to consider when selecting the more appropriate shotcrete application process for a specific situation are—

- Overall volume needed for the situation and the time available to spray it

- Logistics (for example, can bulk bags be handled, or can wet shotcrete be transported via pipeline?)

- Performance required (for example, use of fiber would tend to favor selection of the wet process)

- Overall costs

65.4.2 Substrate Preparation

Substrate preparation is a key element in the successful application of shotcrete. The substrate should be free from loose materials, dust, and films, such as oils. This can generally be achieved by using a water jet. Adhesion to weak materials, such as shales and mudstones, is frequently poor, and this factor should be considered when designing an appropriate support system. Spraying onto a surface that can vibrate (such as a screen or mesh) can cause problems, such as poor placed density (and even voids), as well as increased rebound.

65.4.3 Equipment and Services

For a successful shotcrete application, the following are needed:

- Adequate mix design for spraying and performance
- Adequate material supply at the machine
- Equipment and infrastructure matched to the application
- Trained crews
- Correct preparation prior to spraying
- Adequate services (e.g., power, air, and water)
- Correct application technique
- Appropriate quality control and remedial actions

65.4.4 Training

An aspect that is frequently overlooked is that if spraying with a hand-held nozzle is to be successful, the volume sprayed should be limited to between 6 and 8 m^3/hr. Shotcrete nozzlemen should also be rotated regularly during a shift, as the work is arduous. Should higher volumes be required, the use of a spray manipulator, or robot, is essential and offers other benefits, such as higher quality after placement.

65.5 QUALITY CONTROL

Quality control is an important and generally overlooked parameter needed to ensure safe, consistent, and cost-effective support. Quality control tests should be—

- Reliable
- Meaningful
- Timely
- Simple
- Relatively inexpensive

The main targets to be checked in any shotcrete quality program should be associated with design compliance (bond and strength) and sprayed design thickness. It is, however, totally unacceptable to have a quality control system in place but fail to take adequate actions if noncompliance is identified.

The absolute minimum parameters that should be checked regularly during a spraying shift are—

- Mix design (including water content)
- Services (e.g., air volume and pressure)
- Strength
- Thickness

Typical air requirements, which are the most critical service, are as follows:

- In dry shotcreting at 5 m^3/hr, air consumption will be about 15 m^3/hr at a pressure between 3 and 6 bars.
- In wet shotcreting at 15 m^3/hr, air consumption will be about 12 m^3/hr at a pressure of about 6 bars

Good application techniques (Figure 65.4) are the key to ensuring compliance with target specifications. Caution must be taken not to incorporate rebound lying on the tunnel floor against the sidewall into the shotcrete applied to the sidewall as

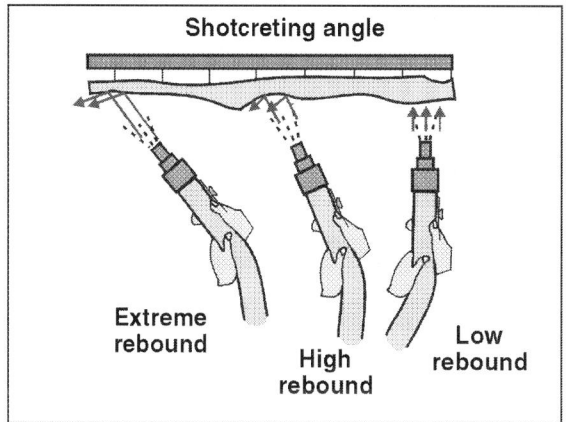

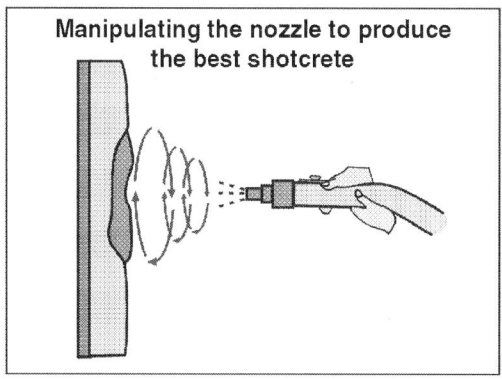

FIGURE 65.4 Correct shotcrete application

this affects in situ shotcrete strength very negatively. To avoid this, it is common practice to start shotcreting on the sidewalls and move upward to the tunnel roof. Rebound should always be discarded and never considered for reuse.

A machine (Tschumi 1998) that should have a most significant impact on shotcreting in the future, particularly on quality control, is the current development of a semiautomatic (or automatic) spray manipulator that can—

- Measure a tunnel profile using a laser scanner,
- Apply shotcrete to a desired thickness automatically and keep spray angle and distance at ideal settings, thus minimizing rebound, and
- Check that design thickness has been achieved and respray any areas that have been undersprayed.

This significant innovation should help resolve the most difficult quality control issue, that of achieving the design thickness of the placed shotcrete. A prototype is currently being evaluated by INCO at its test mine in Canada.

65.6 SHOTCRETE COSTS

The costs of placing shotcrete are seldom, if ever, accurately estimated. The same is evident for most underground support systems. Some of the more significant cost elements are described below.

65.6.1 Material Costs

Material costs are generally relatively easy to establish and are often higher for wet mixes.

65.6.2 Equipment Costs

Capital costs for the shotcrete and ancillary equipment are also easy to determine and vary significantly depending on the capacity of the equipment and the application process. Wet machines are always significantly more costly than dry machines. Maintenance costs are generally overlooked, but can be substantial. Costs of maintaining dry-process equipment per cubic meter sprayed are between two and four times more than costs for maintaining wet-process equipment. In Canada and the United States, maintenance (and replacement) costs for the dry-process machine, hoses, and nozzle are typically around $14/m³ sprayed.

65.6.3 Material and Equipment Transportation

This important cost is generally ignored or avoided, but can be very significant and should be considered. Mines that bother to investigate this aspect are generally able to justify the installation of a shaft pipeline with pumps or "agitator-cars" for horizontal transportation of shotcrete. This assumes that the monthly volumes needed are regular and relatively large.

65.6.4 Labour Costs

The true labour cost involved with the entire process needs to be considered. When comparing the costs of shotcrete against the costs of other types of supports, any rehabilitation costs should also be estimated.

65.6.5 Application Efficiencies

Rebound must be considered when shotcreting and is not too difficult to estimate, i.e., in quality wet shotcreting, rebound should be 10% or less, and in quality dry shotcreting, rebound should be 25% or more. These costs must be accounted for in any shotcrete costing exercise.

65.6.6 Time-Related Costs

Time-related costs need to be considered if the mining cycle is critical for development of a specific site. The effects of blasts lost because support installation took too long are very significant in many cases.

65.7 EXAMPLES OF MIX DESIGN

Examples of typical wet mixes per cubic meter for different applications are shown below.

1. Where high early strength and thick single-pass layers of shotcrete are required, the following proportions could serve as a guideline.

Portland cement	520 kg
Microsilica	25 kg
Aggregate (0 to 8 mm)	1,700 kg
Water reducers	6.5 to 7.5 kg
Internal curer	5 kg
Hydration controller	2 kg
Accelerator	8%
Steel fiber (25 mm)	50 kg
Water:cement ratio	0.45
Average thickness	20 cm

The above mix was used at North Cape Tunnel in Norway (Melbye 1997) and produced a compressive strength of over 2 MPa at 1 hour with a total rebound of less than 5%.

2. For shaft sinking operations, a typical shotcrete should have high cohesiveness (especially if transported in a pipeline), a long open time at elevated temperatures, high durability, and rapid strength gain.

Portland cement	450 kg
Microsilica	40 kg
Stone (8 mm)	1,400 kg

Sand	320 kg
Water reducer	10.7 kg
Internal curer	5.0 kg
Hydration controller	2.8 kg
Accelerator	5% (on total binder content)
Steel fiber	50 kg

The above mix was used for sinking to 2,300 m below the surface at the No. 11 shaft of the Vaal River Operations in South Africa (Buckley 1998). The mix was batched on the surface and sent underground in a nominal 150-mm diameter pipe.

3. A mix that has been successfully used to replace bolting and screening in drifts at INCO'S Stobie Mine in Sudbury, Ontario (O'Hearn and Buksa 1997) is as follows:

Portland cement	400 kg
Microsilica	40 kg
Coarse aggregate	350 kg
Fine aggregate	1,275 kg
Water reducer	4.0 to 5.0 kg
Steel fiber (30 mm)	50 kg
Water:cement ratio	0.40 to 0.45
Average thickness	65 mm

4. Telfer Gold Mine, part of Newcrest Mining, Ltd., in Australia, compared the economics of both wet and dry shotcreting (Cepuritus 1996). The wet shotcrete process was about 10% cheaper than the dry process, mainly because of transportation costs as the mine is very remote.

Portland cement	425 kg
Microsilica	40 kg
Aggregate (−7 mm)	500 kg
Coarse sand	1,000 kg
Dune sand	250 kg
Water reducer	4 L
Stabilizer	3 L
Water:cement ratio	0.49
Average thickness	+50 mm

65.8 REFERENCES

American Concrete Institute. 1999. Manual of Concrete Practice.

Buckley, J.A. 1998. The Application of Wetcrete as a Support Medium. *In* Shotcrete and Its Application. SAIMM, Johannesburg, South Africa

Cepuritus, P.M. 1996. Economic Advantages of On-site Wet-mix Shotcreting. *In* Shotcrete: Techniques, Procedures and Mining Applications. Kalgoorlie, Australia.

Galinat, M.A. 1998. High Performance Polymer Fiber Reinforced Shotcrete. Australian Shotcrete conference, Sydney, Australia.

Melbye, T.A. 1997. Sprayed Concrete for Rock Support. Master Builders Technologies, Zurich, Switzerland.

O'Hearn, B., and H. Buksa. 1997. Boltless Shotcrete. 1st South African Rock Engineering symposium, Johannesburg, South Africa.

Spearing, A.J.S. 1998. Practical Guidelines on Shotcrete Application. *In* Shotcrete and Its Application. SAIMM, Johannesburg, South Africa.

Spearing, A.J.S., and N. Chittenden. 1998. Design, Application and Quality Control To Ensure Safe and Cost Effective Shotcreting. *In* Rock Mechanics and Productivity. SANGORM, Johannesburg, South Africa.

Spearing, A.J.S., and P.J.L. Nel. 1999. The Design, Transportation and Application of Wet Shotcrete for the Support of Vertical Shafts and Related Development. 3rd International Conference on Sprayed Concrete, Gol, Norway.

Tschumi, O. 1998. State of the Art of the Latest Generation Concrete Spraying Robot. 100th CIM conference, Montreal, Quebec.

Vandewalle, M. 1997. Tunnelling the World. N.V. Bekaert S.A.

CHAPTER 66

Shotcrete Linings in Hard Rock

B. Jonas Holmgren[*]

66.1 INTRODUCTION

In shotcrete design systems, it is common to specify an increased thickness of the lining when rock conditions become worse. A closer analysis of the prerequisites for these specifications is worthwhile. Many questions can be raised in connection with the design of rock support using shotcrete: What influence has an increased thickness of the shotcrete in flat or nearly flat roofs? In shotcrete arches without end supports? With end supports? What is the use of reinforcement under the above-mentioned conditions? How is a proper interaction between shotcrete and rock bolts obtained?

In Sweden, research on shotcrete as a means of support for hard rock started in 1973. This research is still going on. This paper will give an overview of the most important results of this research. Where references are not given, the results emanate from research by the author.

66.2 FUNDAMENTAL STRUCTURAL PROPERTIES OF SHOTCRETE LININGS

66.2.1 Basic Load-Carrying Mechanisms for Shotcrete in Hard-Rock Tunnelling

Three basic mechanisms can be distinguished if rock burst problems are excluded.

1. Gravity load from loosening blocks is transferred to the surrounding rock by means of shotcrete that adheres to the rock surface (Figure 66.1A and 66.1B).

2. Gravity load from loosening blocks is transferred to the rock above the loosening part by means of shotcrete interacting with rock anchors. The shotcrete must be reinforced in order to carry the load between the anchors (Figure 66.1C).

3. The gravity load from loosening blocks is transferred to the tunnel floor by means of shotcrete acting as an arch (Figure 66.1D and 66.1E).

In research within Sweden, large-scale laboratory tests have been performed in a 5-tonne test rig (Figures 66.2 and 66.6). Some fundamental findings are described below.

- The primary failure of a good-quality shotcrete lining on hard rock is adhesion failure (Figure 66.3).

- Primary failure load is not dependent on the thickness of the layer when dealing with ordinary thicknesses.

- Adhesion stresses are concentrated in a narrow band along cracks where transverse load is transferred.

- After adhesion failure, the lining might not collapse if it is reinforced and anchored by rock bolts or if it acts as an arch, which can transfer the load properly all the way down to a fixed support (Figures 66.4 and 66.5).

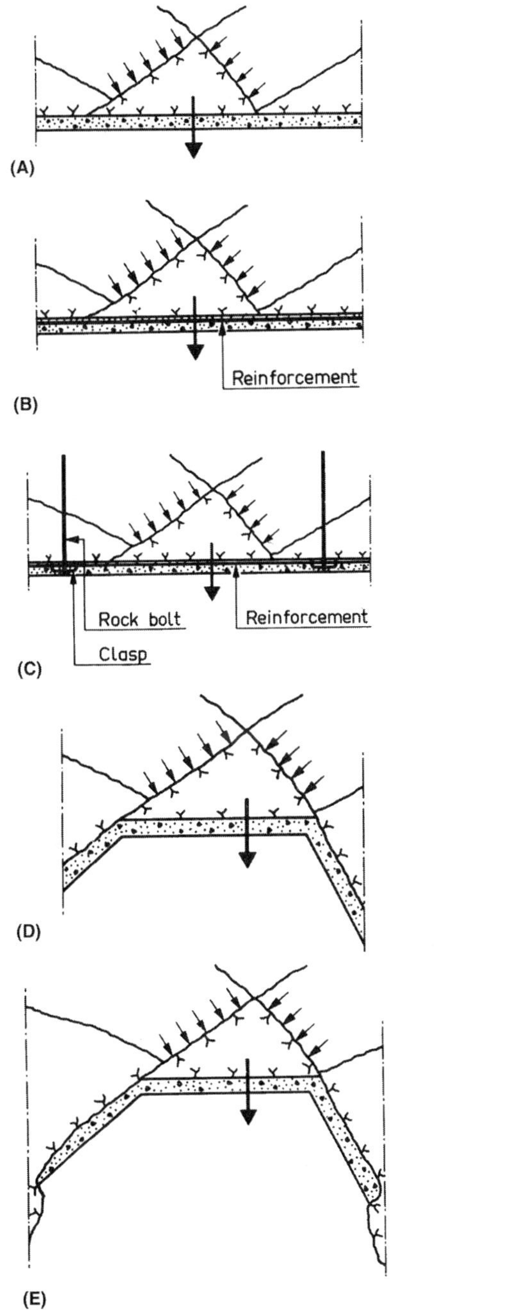

FIGURE 66.1 Principles of punch loading of shotcrete in hard rock

---

* Concrete Structures, Royal Institute of Technology, Stockholm, Sweden.

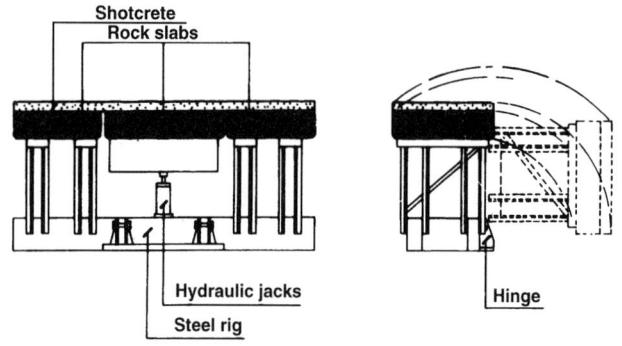

FIGURE 66.2   Test rig for flat shotcrete layers

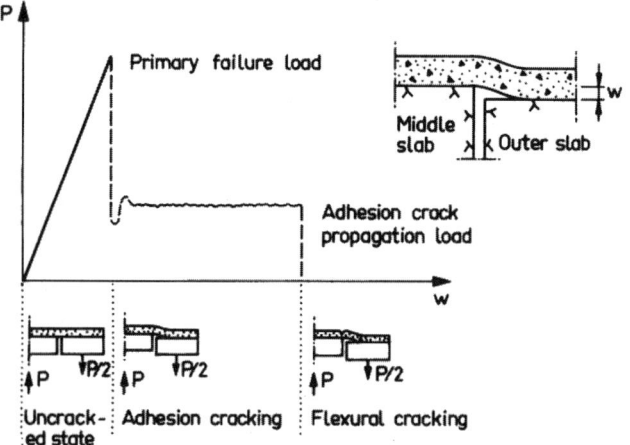

FIGURE 66.3   Principal load-displacement curve for flat shotcrete layers

- The load transfer from shotcrete to rock anchor is of utmost importance for the load-carrying capacity of the lining and thus the total economy of the design (Figure 66.5).

- Steel-fibre reinforcement gives better performance than bar reinforcement.

- If bar reinforcement is used, it should be a mild steel to avoid brittle failure in a collapse situation.

66.2.2   Adhesion Strength of Shotcrete on Hard Rock

A thorough investigation of the adhesion strength of shotcrete applied to different types of rock was performed by Hahn [1]. In his investigation, two factors were studied.

1. Influence of mineralogical composition of the rock
2. Influence of the roughness of the rock surface

All other important parameters were kept constant in the test program. Shotcrete was applied to rock slabs, and adhesion was tested through pull tests on cores drilled through the shotcrete into the rock slab.

The most important findings (Figure 66.7) were—

- High adhesion strengths were found for sandstone, limestone, marble, quartz-rich gneiss (cut perpendicular to the 001 direction), and gabbro.

- Slate, mica shist, granite with large feldspar-porphyroblasts, and gneiss (cut parallel to the 001 direction) showed low values of adhesion strength.

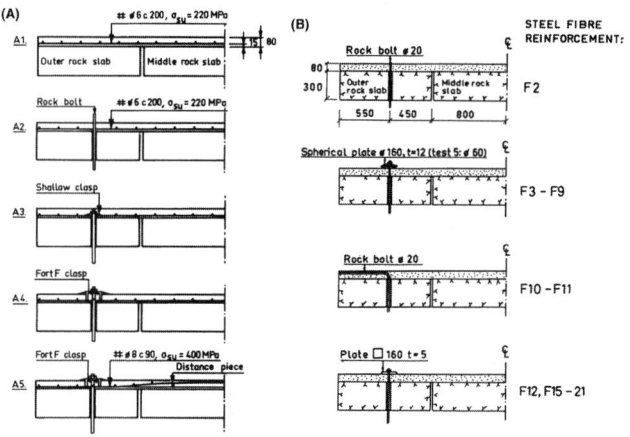

FIGURE 66.4   Bolt-anchored, reinforced test specimens. A, bar reinforcement; B, fibre reinforcement.

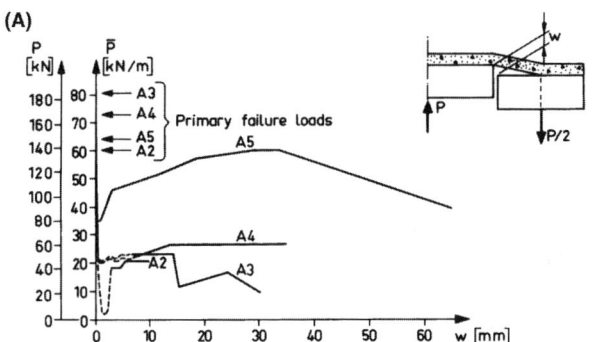

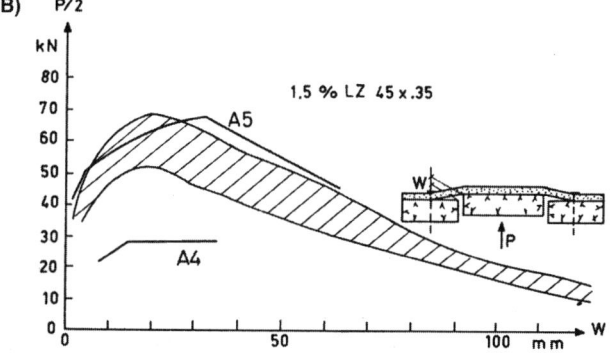

FIGURE 66.5   Load-deflection curves for test specimens in Fig. 66.4A and example from 66.4B. N.b. Different definitions of vertical scales.

- Adhesion strength was somewhat increased along rough surfaces.

66.2.3   Reinforced, Bolt-Anchored Shotcrete Linings Subjected to Dynamic Loading

Falling weight tests were performed on unreinforced, adhering shotcrete and reinforced, bolt-anchored, reinforced slabs. Reinforcement material was ordinary bars in some tests and steel fibres in others. The most important findings were–

- Primary failure was adhesion failure, after which dynamic energy had to be absorbed by the reinforced shotcrete and/or the rock anchors.

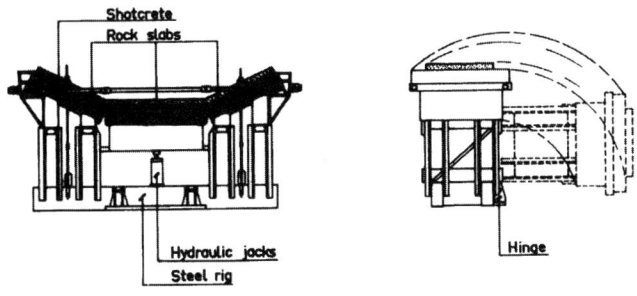

FIGURE 66.6 Test rig for punch-loaded shotcrete arch

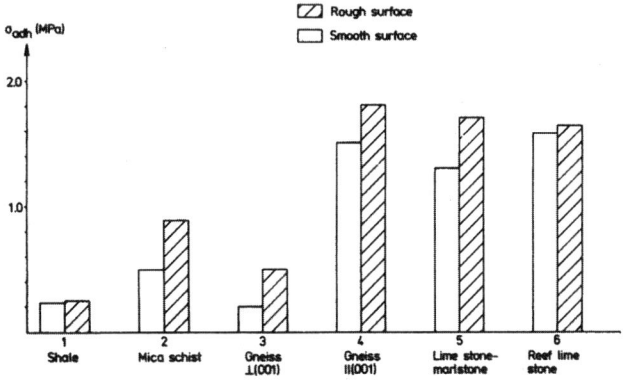

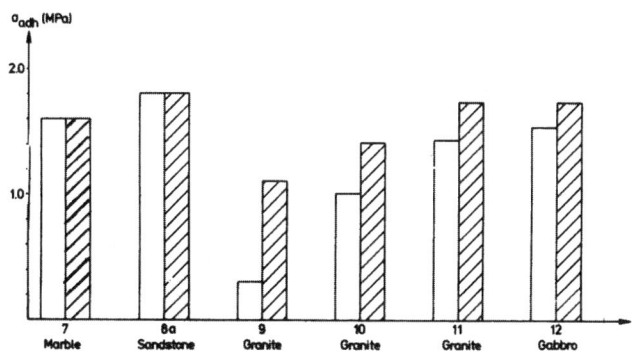

FIGURE 66.7 Adhesion strength for different rock materials according to Hahn and Holmgren (1979)

- Energy absorption capacity of the reinforced shotcrete was small compared to the capacity of a rock anchor allowed to yield over a large part of its length.

- Steel-fibre-reinforced shotcrete performed better than bar-reinforced shotcrete.

66.3 SIMPLE ANALYTICAL MODELS FOR THE DESIGN OF SHOTCRETE LININGS ON HARD ROCK

This section is intended to give some guidance to those who work with the design of shotcrete linings in hard rock. In conventional buildings, the dimensions of the structural system are determined from thorough calculations that follow detailed regulations for loading, material properties, and even for method of analysis. The design of rock reinforcement is generally a less-regulated process. Loading is difficult to determine. Gross simplifications have to be made in modeling rock behaviour. Often linings are designed on the basis of experience. The author prefers to design shotcrete linings as ordinary concrete structures, taking earlier experience into consideration.

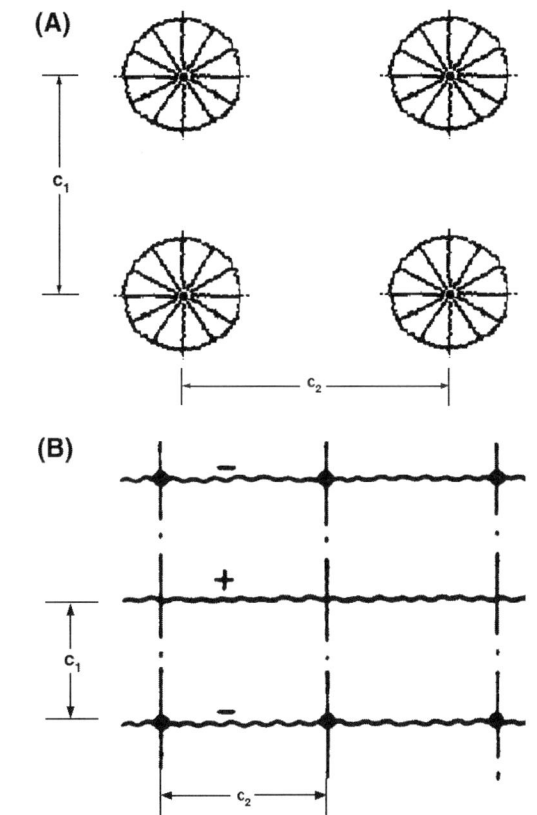

FIGURE 66.8 A, Local yield line pattern; B, global yield line pattern

Which lining should be designed on the basis of calculations? Dividing linings into the categories of simple and complex seems rational. Simple linings need not be designed using calculations. Their dimensions can be determined on the basis of judgment only. One layer of unreinforced shotcrete is an example of a simple lining. Complex linings normally consist of thick layers of reinforced shotcrete interacting with rock bolts. They should be designed using a structural model and calculations. It is not possible to determine the loads acting on the lining, but the method forces the designer to establish a consequent model and thus to fulfil the equilibrium conditions of the model. This in turn makes it natural to balance the strength and the economy of all parts of the structure.

Below, a few simplified structural models are shown. Formulas are developed for statically and dynamically loaded, rock-anchored, reinforced shotcrete linings. Design diagrams for economically optimized structures are presented.

66.3.1 Statically Loaded, Rock-Anchored Shotcrete Lining

Anchoring a loosening rock mass above a tunnel roof is an example of a static loading situation. If it is believed that the rock mass cannot carry itself even if an adhering shotcrete lining is applied, a rock-anchored, reinforced shotcrete lining has to be adopted. Methods for determining the volume of the loosened rock mass will not be treated here, but should always be determined in cooperation with an engineering geologist. Below, gravity load from the loosened rock mass is considered as a uniform load, q N/m^2.

Design of Rock Anchors and Shotcrete Layer Subjected to Static Loading. Bolt force is given by $F_b = q\, c_1\, c_2$, where c_1 and c_2 are the distances between bolts in two perpendicular directions. Reinforcement and thickness of the shotcrete layer are determined using yield line theory. For $0.8 < c_1/c_2 < 1.25$, the sum of positive and negative bending moments is obtained from $m' + m = 0.16\, q\, c_1\, c_2$ (local yield line pattern). For other values of this quotient, a global yield line pattern is more likely, and $m' + m = 0.125\, q\, c^2$, where c is the maximum value of c_1 and c_2. For fibre-reinforced shotcrete, $m' = m$, and so $m' + m$ can be replaced by $2m$ in the above formulas (see Figure 66.8). Mesh-reinforced shotcrete is not treated in this paper. The formulas can easily be adapted for that case, however.

The value of m obtained by the formulas above must not exceed the yield line moment capacity obtained by the expression

$$m = \frac{R_{5.10} + R_{10.50}}{200} f_{cr} \frac{d^2}{6} \qquad (66.1)$$

where $R_{5.10}$ and $R_{10.50}$ are residual strength factors according to ASTM C1018, f_{cr} is the bending stress at the first crack according to the same standard, and d is the thickness of the steel-fibre-reinforced shotcrete layer.

Outside the shotcrete layer, a nut and a washer are placed on the rock bolt. The radius of the washer is b. The value of the nominal local shear stress according to the expression

$$\tau_{nom} = \frac{q c_1 c_2}{\pi d (2b + d)}$$

must not exceed the value of τ_b as given in the chart below. Normally a 10-mm-thick washer with a diameter of 150 mm will be sufficient. It must be covered with at least 20 mm of shotcrete for rust protection.

Allowable shear stress around rock anchor

Compressive strength for 150-mm cubes, MPa	30	40	50
τ_b, Mpa	1.6	2.0	2.4

Economic Optimization of Statically Loaded Rock-Anchored Shotcrete Lining. Rock-anchored steel-fibre-reinforced shotcrete linings are well suited for economic optimization since there is no minimum thickness of the layer as is needed for mesh-reinforced shotcrete. The following assumptions are made: (1) Cost of rock bolt in place $= C_B$ \$/unit and (2) cost of shotcrete in place $= C_S$ \$/m^3. Bolt distance is assumed to be c in both directions.

This gives a cost of rock bolting where $K_B = C_B/c^2$ \$/m^2. The cost of shotcrete is $K_S = C_S\, d$ \$/m^2. The expressions for the bending moment, $2m = 0.16\, q\, c^2$, and bending tensile stress (in this situation equal to the strength, $f_{cbt} = 6m/d^2$) give

$$d = c \sqrt{\frac{0.48q}{f_{cbt}}} ,$$

which is inserted into the expression for K_S. The function $K_B + K_S$ has a minimum for $c = c_{opt}$, where

$$c_{opt} = \sqrt[3]{2C_B \frac{\sqrt{\frac{f_{cbt}}{0.48q}}}{C_S}} \qquad (66.2)$$

Figure 66.9 shows optimum bolt distance, c_{opt}, and optimum thickness of the shotcrete layer, d_{opt}, for different combinations of load, cost, and flexural strength of the shotcrete.

66.3.2 Dynamically Loaded, Rock-Anchored Shotcrete Lining

Bomb detonations or blasting generate shock waves in a rock mass. This means that the rock mass particles start vibrating. This vibration propagates through the rock mass. When reaching a free surface, the wave turns and goes back. This generates tensile stresses in the rock mass, and tensile failure may occur. This only happens when particle velocity is very great. If the rock mass, however, is fractured from the beginning, even small velocities theoretically may cause ejection of rock pieces. The kinetic

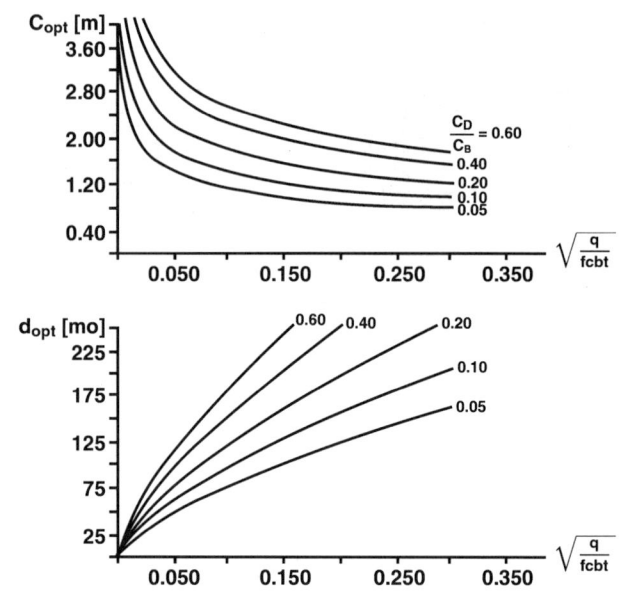

FIGURE 66.9 Economic bolt distance, c_{opt}, and thickness of the shotcrete layer, d_{opt}, at different relationships between the cost of bolting (C_B \$/unit) and the cost of steel-fibre-reinforced shotcrete (C_S \$/m^3). q is the static load from the rock mass in N/m^2, and f_{cbt} is the design value of the bending tensile strength of the shotcrete in N/m^2.

energy of the ejected rock mass has to be absorbed by a plastic structure. Such a structure may consist of a rock-anchored, steel-fibre-reinforced shotcrete lining. The velocity of the ejected rock mass can be judged from existing formulas for the vibration velocity caused by blasting. The thickness, t_u, of the ejected volume has to be judged from the rock structure and the geometry of the tunnel. Some vibration energy is absorbed when the rock mass bursts. This loss of energy is very difficult to calculate and is therefore often neglected.

The design principle is to take care of the kinematic energy in the rock bolts and to design the shotcrete layer for a load that corresponds to the yield load of the rock bolts. By choosing a utilized elongation of the rock bolt at each loading occasion, the structure can be made to survive a certain number of loadings.

Systematic bolting with equal bolt distances, c, in both directions gives the most unfavourable yield line pattern, i.e., $m = B/4\pi$, where $B = f_{sy}\pi\phi^2/4$ is the yield load of a rock bolt with diameter ϕ. The thickness of the shotcrete layer is obtained from $f_{cbt} = 6m/d^2$. Combining the equations above gives—

$$d = \frac{\phi \sqrt{\frac{6 f_{sy}}{f_{cbt}}}}{4} \qquad (66.3)$$

The kinetic energy of the ejected mass will be absorbed by the rock bolts. The energy balance equation becomes—

$$\frac{\gamma_m t_u c^2 v^2}{2} = \frac{f_{sy}\pi\phi^2 \varepsilon L}{4} \qquad (66.4)$$

where γ_m = density of the rock mass,

t_u = thickness of the ejected layer,

g = acceleration due to gravity,

v = velocity of the ejected layer,

ε = relative elongation of the rock bolt,

and L = bolt length.

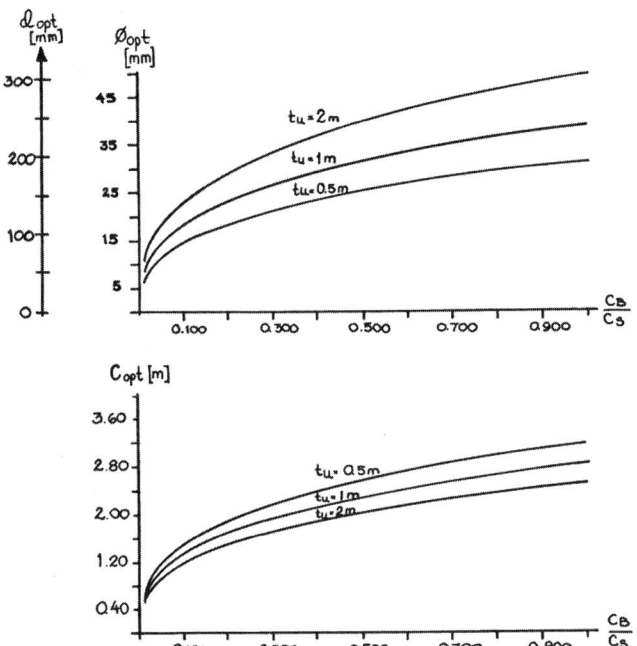

FIGURE 66.10 Economic dimensions of a steel-fibre-reinforced, rock-anchored shotcrete lining subjected to dynamic load from an ejected rock mass. C_B is the cost of the rock bolt ($/unit); C_S is the cost of the shotcrete ($/m³); t_u is the thickness of the ejected rock layer; c_{opt} is the optimum bolt distance. d_{opt} is the optimum thickness of the shotcrete layer; and ϕ_{opt} is the optimum bolt diameter. Assumed values are f_{sy} = 400 MPa, f_{cbt} = 4 MPa, L = 2.5 m, v = 750 mm/s, and ε = 0.5 %. The diagram can be used for cases with relatively small charges at moderate distance.

γ_m = 2700 kg/m³ gives–

$$c = \frac{0.024\phi \sqrt{\dfrac{f_{sy}\,\varepsilon\,L}{t_u}}}{v} \qquad (66.5)$$

Using these expressions, the required thickness of the shotcrete layer and the bolt distance can be obtained when the bolt diameter is known. ε should be chosen so that the required number of loadings can be absorbed without excessive deformation. The diameter of the washer should give $\tau_{nom} < \tau_b$ in accordance with earlier expressions.

There must not be any cut threads or other weaknesses that can cause fracture initiation in the rock bolt. Otherwise the bolt will not reach the yield stage and energy absorption capacity will be substantially decreased. A good solution to that problem is to fix the washer to the bolt by welding a knob to the rock bolt.

Economic Optimization of Dynamically Loaded, Rock-Anchored Shotcrete Lining. The expressions for d and c above are inserted into expressions for the cost of the bolt and the shotcrete. The minimum cost for the lining is obtained for

$$\phi_{opt} = 17.7 \sqrt{\frac{C_B v^2 t_u \sqrt{\dfrac{f_{cbt}}{f_{sy}}}}{C_S f_{sy}\,\varepsilon\,L}} \qquad (66.6)$$

Figure 66.10 gives the economic dimensions of the lining for a "civil case," e.g., a blasting situation in a mine.

The calculations (or the diagram) might give a smaller bolt dimension than what is practical. If bolt diameter is increased, layer thickness must also be increased according to the formula

$$d = \frac{\phi \sqrt{\dfrac{6 f_{sy}}{f_{cbt}}}}{4} \qquad (66.7)$$

If the shotcrete layer does break before the rock bolt yields, the energy absorption capacity will be considerably decreased.

66.4 SPECIFICATION AND QUALITY CONTROL OF SHOTCRETE

66.4.1 Introduction

Each country has its own standards for concrete specifications and quality control. Typically, shotcrete standards are adapted from standards for ordinary concrete. The reason is simply that shotcrete is a concrete material. However, some differences exist because some production techniques affect the properties of the concrete. In this paper, only adhesion and fibre reinforcement are treated.

66.4.2 Specification and Control of Adhesion Strength of Shotcrete to Hard Rock

It has been pointed out in previous sections that the adhesion of shotcrete to rock is an important property of the lining. When conditions are good, shotcrete adheres without reinforcement, and rock anchors provide high load-carrying capacities. Such capacity is not only dependent on the mineralogical composition of the rock, but also the workmanship of the contractor. This may result in conflicts between the owner and the contractor since the owner "delivers" the rock material, and the contractor delivers the work. There are no evident solutions to this problem. The author recommends that a required adhesion strength, e.g., 0.5 MPa, be stated in the contract for sections where adhering shotcrete is judged to be sufficient, but also that the contractor be obliged to report if he judges that this level cannot be reached. In such a case, a discussion between the two parties has to take place.

For linings that are acting as bolt-anchored reinforced structures, it is recommended that no requirements for adhesion strength be made. The responsibility for ensuring that the shotcrete does not fall down before it has been anchored lies with the contractor since labour protection is his exclusive domain.

For linings acting as arch structures, adhesion is not required if the arch goes down to the tunnel floor, which it should.

If there are requirements for adhesion strength, then adhesion must be checked. In Sweden, it is normally specified that three cores shall be pulled in a certain area or a certain tunnel length. The average adhesion strength for the accepted cores (see below) is calculated. It is typically stated that this average shall exceed the required value. No single value must be lower than half the required value. The reason why this type of tensile test should not be treated like an ordinary tension test on concrete is that rock properties vary widely even within a limited area.

Experience has shown that there is a great risk in using only a few cores to determine adhesion strength in a way that is acceptable for a contractor, who stands to risk being forced to take costly measures if adhesion strength is unacceptably low.

Certain rules should be followed in order to avoid dubious test results.

- Only shotcrete on sound rock should be tested for the purpose of checking a contractor's workmanship.

- The person who drills the cores must not be one who normally drills holes, but the person must be aware of the purpose of the core drilling.

- Pulling equipment must be newly checked and adjusted.

- Equipment must ensure that the pulling force coincides with the core axis.

- Shotcrete must be checked to ensure that it is not loose from the rock. (A core that is loose from the rock from the beginning cannot be treated statistically together with other cores.)

- If failure is noted, it must be checked to determine if it really has taken place at the interface between the shotcrete and the rock. Mixed failures do not give true adhesion strength.

- The inclination angle between the core axis and the rock surface must be checked to ascertain that it is larger than 75°. Smaller inclinations might result in mixed failure.

66.4.3 Beam Test For Quality Control of Steel-Fibre-Reinforced Shotcrete

In this section, the flexural properties of steel-fibre-reinforced shotcrete are treated. Other properties do not differ much from those of ordinary concrete.

In Sweden, testing is undertaken according to a modification of ASTM C1018. Much can be said about flexural testing of steel-fibre-reinforced concrete, but the topic will not be discussed here. A few important principles will be formulated, however.

- The test method should reveal something about the strength of a cracked and deformed lining, since imposed deformation is a realistic loading case for a lining in rock.

- The test method should give the flexural strength of the concrete itself.

- The test method should give strength values suitable for design.

- The test method should be suitable even in cases when strain hardening is to be specified.

According to ASTM C1018, the beam test is not ideal in all the above cases; on the other hand, no other established method is ideal either.

It is recommended that a certain value for the residual strength of the shotcrete be specified. For plastic design it has been found relevant to specify that

$$f_{flres} = \frac{f_{5.10} + f_{10.50}}{2} \geq f_{req}$$

where $f_{a,b} = \dfrac{R_{a,b}}{100} f_{cr}$ according to the ASTM C1018 standard.

f_{req} can be used as a "yield" strength in calculations according to yield line theory for slabs. Since beam testing is rather expensive, the contractor is often allowed to replace beam tests with periodic measurements of fibre contents if he can prove that the production of fibre-reinforced shotcrete has stabilized at the correct quality level.

The dimensions of the test beam for the fibre-reinforced shotcrete used in Sweden is shown in Figure 66.11.

Criticism has been leveled against the ASTM method because of difficulties in determining load and deflection at the first crack. The author is convinced that the accuracy of the method can be increased, but the disadvantage of testing a beam instead a slab remains.

One reason for large variations in test results is that a failure crack falls randomly between point loads, but deflection is always measured at mid-span.

Load-deflection curves may vary considerably depending on where the dominant crack is situated. In one study, the ratio between the highest value of I_{60} and the lowest was as much as 1.4!

To obtain a deflection value that can be connected with the mid-span deflection, measured deflections should be corrected after cracking according to the formula—

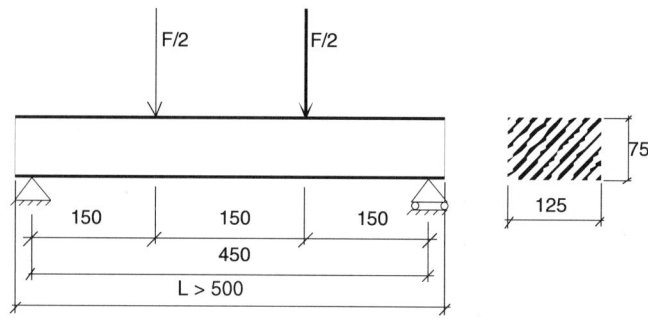

FIGURE 66.11   Test beam used in Sweden for flexural testing of steel-fibre-reinforced shotcrete

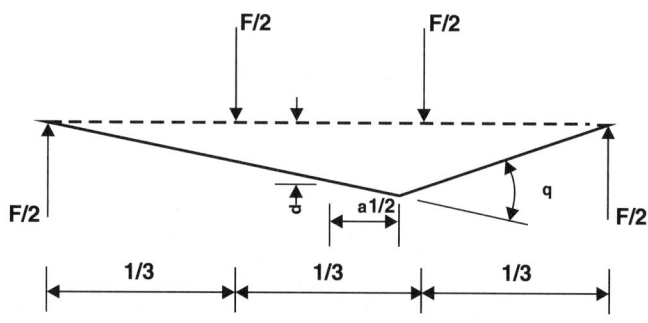

FIGURE 66.12   Drawing of ASTM test beam after development of dominant bending crack

$$\delta_{corr} = \frac{1}{1 - \alpha_{obs}} \cdot (\delta_{obs} - \delta_{el}) + \delta_{el} = \frac{\delta_{obs} - \alpha_{obs} \cdot \delta_{el}}{1 - \alpha_{obs}} \quad \textbf{(66.8)}$$

where α_{obs} is obtained from observation of the tested specimen in accordance with Figure 66.12. The elastic part of the deflection can be obtained from

$$\delta_{el} = \frac{F}{F_{cr}} \cdot \delta_{cr} \quad \textbf{(66.9)}$$

At this stage, it is sufficient to use an estimate of crack load (F_{cr}) and crack deflection (δ_{cr}). A method for evaluating crack load and deflection automatically has been developed, but is not shown here.

66.5   ACTIONS FOR INADEQUATE SHOTCRETE QUALITY

66.5.1   Introduction

A railway tunnel contract in Sweden consisted of six single-track tunnels and 10 double-track tunnels. The total length of the single-track tunnels was about 2.5 km, and the total length of the double-track tunnels was about 5.6 km. The span of the single-track tunnels is 7 m, as is the distance from the top of the rails to the crown. The corresponding dimensions for the double-track tunnels are 11.5 and 7.45 m.

To avoid regular maintenance scaling, the decision was made to provide the tunnels with minimum strengthening by shotcreting the roofs and the upper parts of the walls, even if selective rock bolting would have been sufficient from a load-bearing point of view. On the basis of experience, unreinforced shotcrete was considered insufficient because of the risk of pieces bursting loose as a result of frost action in the relatively short tunnels. Simplicity and speed of execution favoured steel-fibre-reinforced shotcrete.

The minimum reinforcement consisted of 40 mm of steel-fibre-reinforced shotcrete with a 20-mm-thick covering layer of unreinforced shotcrete and selective rock bolting. Where greater strength was required and where adhesion could not be counted on, layer thickness was increased to 70 mm plus 20 mm, or even 90 mm plus 20 mm in some cases. Where these thicker layers of shotcrete were used, rock bolts were installed systematically with a washer to provide interaction between the shotcrete and the bolts.

66.5.2 Contract Specifications for Shotcrete
The most important parameters specified for the shotcrete were—
1. Tests at intervals corresponding to 500 m^2 of shotcreted surface
2. Thickness of shotcrete layer
3. Compressive cube strength ($\geq$ 35 MPa)
4. Maximum bending tensile strength ($\geq$ 5.0 MPa)
5. Residual bending tensile strength
$$\left(\frac{R_{10.30}}{100}f_{cr} \geq 2.5 \text{ MPa}\right)$$
6. Adhesion strength between shotcrete and rock ($\geq$ 0.5 MPa)
7. Adhesion strength between shotcrete layers ($\geq$ 1.0 MPa)
8. Density of shotcrete ($\geq$ 2200 kg/m^3)

66.5.3 Geological Mapping
During excavation, rock conditions, such as rock type, joints, tectonic zones, clay alteration, and water leakages, were continually mapped by a geologist. The purpose of the mapping was to–
1. Serve as the basis for evaluating necessary reinforcement, grouting, etc.
2. Serve as documentation of rock conditions.

Rock conditions and information on the performance of reinforcement, grouting, etc., were put together on drawings as documentation for future use. Results from tests of the reinforcement were also marked on the drawings.

66.5.4 Actions to Address Inadequate Shotcrete Quality
Most of the first 250 m of the tunnel did not meet the required specifications and was not approved by the owner. The actions required to meet specifications were adapted to the actual rock situation in each section and the quality of the shotcrete. This section describes the analysis that led to the demands placed on the contractor.

Philosophy. When starting this work, the philosophy was that rock conditions, shotcrete quality, and existing rock reinforcement should be considered when decisions were made concerning what actions the contractor would be enjoined to take. The contractor's explanations as to why shotcrete quality was less than the quality specified in the contract were also to be included.

The principal aim was to secure adequate rock reinforcement rather than to punish the contractor. This did not mean, however, that the owner would pay for the actions taken.

66.5.5 Analysis of Different Tunnel Sections
Section 1. *Geological Conditions.* The length of this section was 36 m. The predominant rock type is gneiss with sheets and lenses of pegmatite. The foliated bedrock is oriented nearly parallel with the tunnel and dips steeply to the south. The brittle rock is dominated by fractures along the foliation and horizontal-subhorizontal structures. Moreover, steeply dipping fractures occasionally cross the tunnel at an obtuse angle.

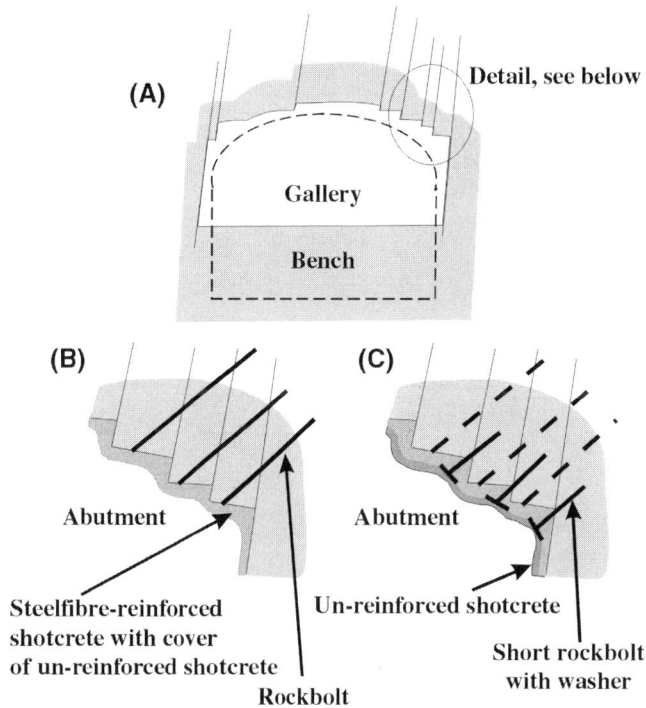

FIGURE 66.13 Section 1. A, Tunnel shape after blasting; B, intended reinforcement; C, additional reinforcement in abutments

Due to rock falls controlled by fractures parallel to foliation and subhorizontal structures, the abutments and tunnel walls display a blocky, staircaselike form. The roof is generally of a higher quality with a fair arch shape. However, in connection with the horizontal-subhorizontal fracture planes, instable sections are present. There is partial water leakage, particularly adjacent to the tunnel portal.

Achieved Shotcrete Quality. The intended shotcrete lining was 80 mm of steel-fibre-reinforced shotcrete plus 20 mm of unreinforced shotcrete in the abutment and 40 mm of steel-fibre-reinforced shotcrete plus 20 mm of unreinforced shotcrete in the crown (Figure 66.13).

Two adhesion tests showed zero adhesion. One test showed 0.65 MPa. Furthermore, the residual bending tensile strength varied between 1.04 and 2.16 MPa.

Other Rock Reinforcement. The tunnel was bolted selectively with approximately 3.8 bolts per metre of tunnel.

Claimed Actions Adapted to Achieved Shotcrete Quality. Because of the unsatisfactory adhesion strength the shotcrete layer in the abutment should be anchored to the rock with 1-m-long, 25 mm in diameter rock bolts and washers. The bolts should be placed at a distance of 1.5 m between centres, but the position of the bolt should be adapted to local conditions. The washers must be placed tightly to the shotcrete layer. Bolt ends and washers should be covered with 20 mm of unreinforced shotcrete.

Section 2. *Geological Conditions.* The length of this section is 36 m. In the first 12 m, bedrock conditions are similar to those described for section 1. After 17 m (left side wall), a somewhat blocky greenstone dike crosses the tunnel at a moderate angle and intersects the right tunnel wall approximately 10 m away from its crossing on the left side. In the last 24 m of this section, the stability of the abutments is better than in the previous sections. The staircase-like formation is less accentuated than previously.

Achieved Shotcrete Quality. The right abutment was reinforced with 80 mm of steel-fibre-reinforced shotcrete plus 20 mm of unreinforced shotcrete. The blocky part in the vicinity of the

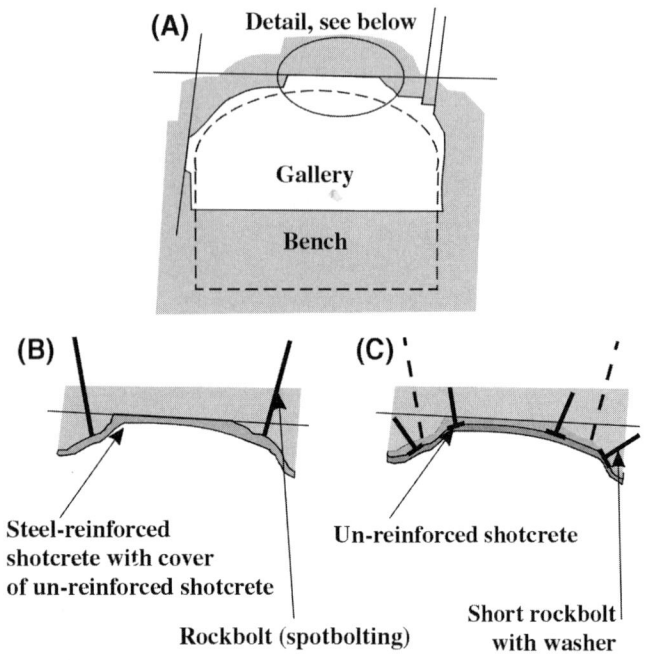

FIGURE 66.14   Section 2. A, Tunnel shape after blasting; B, intended reinforcement; C, additional reinforcement in whole gallery

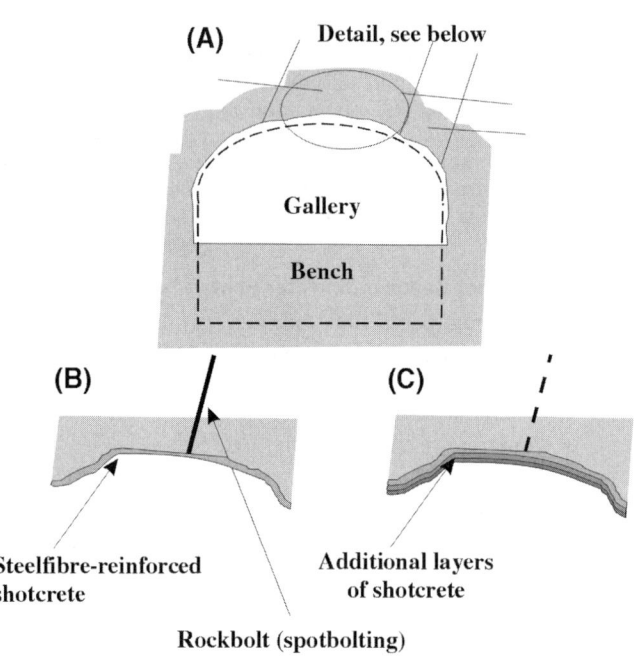

FIGURE 66.15   Section 3. A, Tunnel shape after blasting; B, intended reinforcement; C, additional reinforcement in the whole gallery

greenstone was reinforced with 60 mm of steel-fibre-reinforced shotcrete plus 20 mm of unreinforced shotcrete. The remaining part of section 2 was reinforced with 40 mm of steel-fibre-reinforced shotcrete plus 20 mm of unreinforced shotcrete (Figure 66.14).

No acceptable results have been obtained from the adhesion tests. All test cores showed failure in the rock at low strength levels. All other tests showed acceptable results. It was judged that the rock surface had been improperly scaled before shotcreting operations had taken place.

Other Rock Reinforcement. The tunnel was bolted selectively with approximately 1.6 bolts per metre of tunnel.

Claimed Actions Adapted to the Achieved Shotcrete Quality. The same as for section 1.

Section 3. *Geological Conditions.* The length of this section is 36 m. The predominate rock type in the first 16 m is gneiss with sheets and lenses of pegmatite. Beyond 16 m, the bedrock shows a higher degree of complexity and contains gneiss, granite, amphibolite, and pegmatite.

The bedrock is foliated. In some places, the foliation crosses the tunnel at a somewhat oblique angle, in contrast to sections 1 and 2, where the foliation is nearly parallel with the tunnel. The brittle rock is dominated by fractures along the foliation and horizontal-subhorizontal structures. Furthermore, steeply dipping fractures occasionally cross the tunnel at an obtuse angle. Elsewhere, the bedrock shows a somewhat sheety or blocky structure. Water-conducting fractures occur fairly often. Stability in the abutments is generally better than in sections 1 and 2, especially where foliation crosses the tunnel at an oblique angle. The tunnel roof is generally good with a relatively well-developed arch shape (Figure 66.15).

Achieved Shotcrete Quality. The gallery was reinforced with 40 mm of steel-fibre-reinforced shotcrete.

One of the adhesion test cores showed an extremely low value (0.15 MPa). The residual bending tensile strength was too low (0.81–1.57 MPa). The thickness of the shotcrete layer averaged 29 mm.

Other Rock Reinforcement. The tunnel was bolted selectively with approximately 2 bolts per metre.

Claimed Actions Adapted to the Achieved Shotcrete Quality. Fortunately, the cover of 20 mm of unreinforced shotcrete had not yet been applied, so the thickness of the reinforced layer could easily be increased to 50 mm. Where the tunnel contour curved inward, the shotcrete should be anchored with short bolts as in section 1.

Section 4. *Geological Conditions.* In principle, the bedrock displays the same complex conditions as described for the interval after the first 16 m of section 3.

Achieved Shotcrete Quality. The gallery was reinforced with 40 mm of steel-fibre-reinforced shotcrete plus 20 mm of unreinforced shotcrete.

Compressive strength was 25 to 38 MPa, and maximum bending tensile strength of the shotcrete was 4.0 to 4.3 MPa. The thickness of the shotcrete layer averaged 45 mm.

Other Rock Reinforcement. The tunnel was bolted selectively with approximately 1.7 bolts per metre.

Claimed Actions Adapted to the Achieved Shotcrete Quality. At least three cores shall be drilled from the shotcrete lining for compression tests. If the compressive strength of these cores is too low, a special investigation regarding the durability of the shotcrete shall be performed. Further actions will be decided upon later.

66.6   CONCLUSIONS

A contract contains many conditions. Very often, required qualities are not achieved in one respect or another. It does not seem reasonable to force the contractor to take away large parts of a shotcrete lining because of inadequate adhesion. Such actions will definitely delay the whole project, but they will not result in better quality in the end. On the other hand, it does not seem reasonable to allow the contractor to give the owner economic compensation for diverging from the original requirements. That would decrease respect for the contract in a devastating way.

The method described above takes into consideration the positive results that have been achieved already. The contractor is enjoined to supplement what already has been done with actions that will ensure that the rock reinforcement works as it was intended from the beginning.

A brief summary of actions follows.

Quality inadequacy	Recommended action	Remarks
Adhesion strength too low	Anchor with short rock bolts and washers	If shotcrete is unreinforced, a reinforced layer must be placed on old shotcrete.
Bending tensile strength of steel-fibre-reinforced shotcrete too low.	Increase thickness of shotcrete layer to get proper moment capacity.	Might not be usable if chemicals have been put on surface. Adhesion testing might be necessary.
Compressive strength too low.	Check more cores and density of shotcrete.	Durability of shotcrete may be at stake.
Residual bending tensile strength of steel-fibre-reinforced shotcrete too low.	Increase thickness of shotcrete layer to get proper moment capacity.	Might not be usable if chemicals have been put on surface. Adhesion testing might be necessary. If residual strength is far too low, see below.
Residual bending tensile strength of steel-fibre-reinforced shotcrete far too low	New layer of steel-fibre-reinforced shotcrete with better properties must be put on old one. Depending on adhesion conditions, etc., old layer might be removed.	Very low residual strength indicates shotcrete far more brittle than intended.
Layer too thin.	Increase thickness.	Might not be usable if chemicals have been put on surface. Adhesion tests might be necessary.

66.7 REFERENCE

Hahn, T., and J. Holmgren. 1979. "Adhesion of Shotcrete to Various Types of Rock Surfaces." International Society of Rock Mechanics, Proc. of Int. Symp. on Rock Mechanics, Montreux, pp. 431–439.

CHAPTER 67

The Role of Shotcrete in Hard-Rock Mines

P.K. Kaiser* and D.D. Tannant†

67.1 INTRODUCTION

Shotcrete is sprayed concrete consisting of aggregates (generally with <10% of size >8 mm) and portland cement, as well as various admixtures: microsilica, accelerators or retarders, plasticizers, and reinforcing components (fibres) specifically formulated to adhere to rock surfaces and to allow build-up with minimal rebound. Recent trends in civil engineering applications favour wet shotcrete over dry. Shotcrete has many advantages—such as complete aerial coverage—over other support systems. Shotcrete is quick to apply, and the application equipment can be fully mechanised. Shotcrete completely coats the rock surface, locks in rock fragments, reduces rock movement, prevents loosening around the excavation, and promotes overall stability. When applied as an arch or support ring, shotcrete can also provide a support reaction pressure. In argillaceous rock, shotcrete reduces slaking and retains the original rock-mass moisture conditions. While our knowledge of how shotcrete works is still incomplete, if shotcrete is applied correctly, it almost always performs better than anticipated.

For mining, however, material-handling costs often provide a disincentive for using shotcrete over other support systems. Nonetheless, the accelerated usage of dry shotcrete since 1992 (Figure 67.1) in one mining camp in the Sudbury basin demonstrates an increasing reliance on shotcrete for ground control and other applications (e.g., fill fences). Recently, innovative applications such as shotcrete posts have been introduced (Swan et al. 1997). These posts have found wide application because they provide a cost-effective and reliable method to reduce hanging-wall convergence and deterioration. In many cases, shotcrete posts have replaced cable bolts.

Shotcrete support design is well established for civil engineering excavations where shotcrete is used as a support arch or structural liner. In these applications, the shotcrete typically functions as a stiff shell and designs (shotcrete thickness and strength) are based on keeping bending and hoop stresses in the liner below critical values. For shotcrete to be treated like a liner, the shape of the excavation needs to be smooth; and the shotcrete acts either as a continuous (closed) ring or as a support arch because of the way it is tied into the excavation wall. This chapter provides guidance on how to design and use shotcrete in hard-rock mining where conditions often differ from those encountered in civil construction.

In mining, shotcrete is often applied to flat surfaces (rectangular drifts or pillar walls) or the drift profile is so irregular, relative to the applied shotcrete thickness, that thrust forces cannot be continuously transmitted without creating excessive shear or bending stresses. Furthermore, due to mining-induced stress changes, rock often fails long after the support has been installed, imposing much larger deformations on the

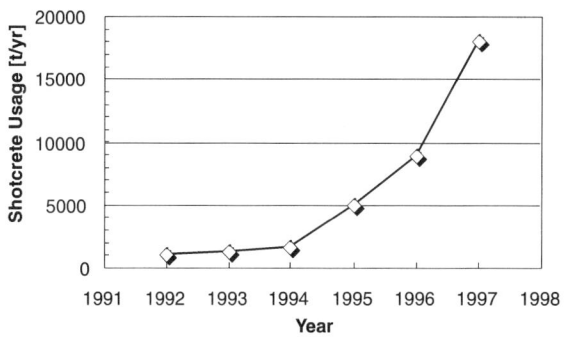

FIGURE 67.1 Innovative use of shotcrete as shotcrete posts, and annual consumption of dry shotcrete at Falconbridge Ltd. Sudbury Operations (after Swan et al. 1997)

support than conventionally tolerated in civil engineering structures. For these and other reasons, civil engineering design methods cannot be readily transposed to mining. Hence, the challenge in mining is to utilise shotcrete in such a manner that it can perform well at the mining stage when stress changes occur, or when rock fractures and large deformations accumulate.

* Geomechanics Research Center, MIRARCO-Mining Innovation, Laurentian University, Sudbury, Ontario, Canada.
† School of Mining and Petroleum Engineering, University of Alberta, Edmonton, Alberta, Canada.

Shotcrete design in mines is complex because (1) rock-mass characteristics are often difficult to quantify, (2) the stresses around the excavation change over time and often negatively affect the rock mass, and (3) little or no measurements for design verification are available. Because of these inherent complexities, shotcrete design is still based on empirical and observational methods, often combined with simplified analytical analyses.

The most fundamental goal of shotcrete design is to create a self-supporting arch of shotcrete, bolts, and rock, a reinforced structure that can survive imposed loads and deformations. Consequently, shotcrete design must consider other support components such as rock bolts or grouted rebar and mesh. Shotcrete must be designed as an integrated support system that includes shotcrete as only one component. For this reason, most analytical shotcrete design methods—and most laboratory and field shotcrete loading tests—are also limited in their applicability.

No matter how well shotcrete is designed, successful application depends on a sound understanding of the anticipated rock-mass behaviour and the role of shotcrete in a particular situation. Most important, the shotcrete user must understand how it functions, what its role is in a support system, and how it can be applied to survive large deformations.

Costly mistakes are often made in practice—not because the quality of the shotcrete is inadequate or the installation is poor—but because the role of shotcrete is not understood or because the shotcrete is not properly integrated into the support system (not tied in properly or installed at the wrong time). So, first we need to understand what we are trying to achieve, then we have to evaluate whether shotcrete is suitable (should be used as part of a support system), and finally, we have to install it so it can survive mining-induced loads and deformations.

This chapter focuses on using shotcrete in hard-rock mines where two primary instability modes must be dealt with: (1) gravity-induced wedge-type failures, and (2) stress-induced rock fracturing in a gradual (yielding) or violent (rock burst) manner. In other words, "how to use shotcrete to control potentially unstable rock blocks and bad ground in a high-stress environment." This chapter does not deal with mix design or application techniques (the reader is referred to handbooks such as Melbey and Garshol 1999). We assume that good quality shotcrete is obtained and installed with consistent quality.

67.2 ROCK BEHAVIOUR NEAR EXCAVATIONS

67.2.1 Instability Modes

The rock-mass behaviour or the possible instability modes for an underground excavation depend upon the magnitude of stress relative to the rock-mass strength and structure (i.e., degree of jointing and joint persistence). Six distinctly different behaviour modes are illustrated in Figure 67.2, ranging from elastic behaviour of essentially intact rock in a low-stress environment to plastic behaviour of a highly jointed rock mass under high stresses.

What is not shown in Figure 67.2—but is implied—is that the mining engineer must also anticipate how changes in mining-induced stress affect the rock mass. In other words, the "in-situ stress" near a drift or tunnel may change over time; and a tunnel that was excavated and supported, for example, in a low-stress environment, may gradually move to an intermediate-stress environment, or vice versa. Furthermore, from a support-design perspective, conditions are not the same when a tunnel is advanced in an intermediate-stress field or when it is advanced in a low-stress field and then followed by mining-induced stress changes. In the former case, much of the deformation due to rock fracturing may have dissipated between the tunnel face and the location of the support installation, whereas in the latter case, all deformation due to rock fracturing is imposed on the support.

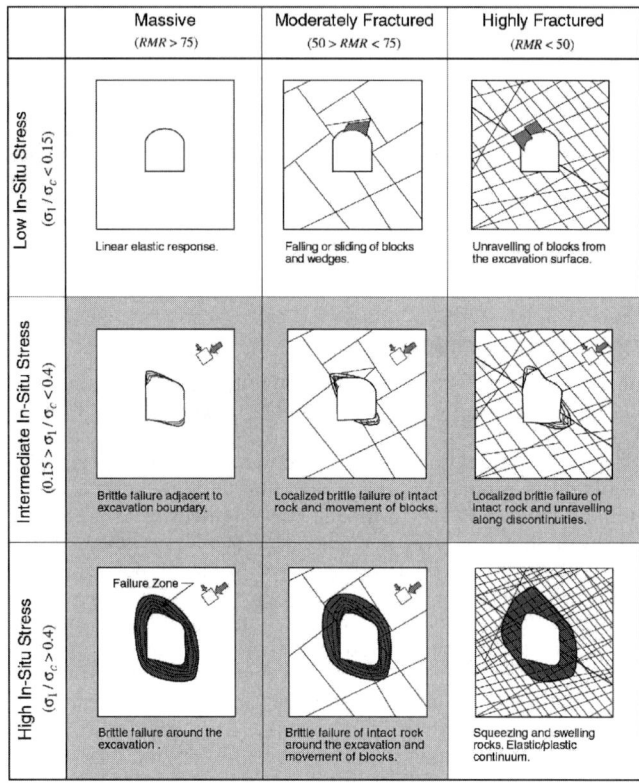

FIGURE 67.2 Rock mass behaviour matrix (Martin et al. 1999)

67.2.2 Low to Moderately Stressed Rock (Wedges)

When mining near the surface in low to moderately stressed rock, underground excavations are either stable or the instability modes are largely controlled by the rock-mass structure consisting of joints, laminations, bedding, and other weakness planes. Common instability modes in this type of environment include wedge-type failures, lamination peeling, or ravelling, which can be caused by the gradual removal of key blocks. Controlling these processes involves stabilising the ground at small displacements, and limit-equilibrium analyses are commonly used to design support for this class of ground behaviour. Various types of shotcrete, including fibre-reinforced shotcrete, have found wide application in these situations and often provide a very effective means for ground control. Because deformations are not an issue for this class, shotcrete can be applied as a continuous arch, which, due to its stiffness and strength, can provide substantial support capacity against dead-weight loading (see Section 67.4.2).

It is important to remember that only rock blocks that can kinematically fall need to be analysed. These blocks may either fall vertically or slide on one or more discontinuities under the influence of gravity. The orientation, spacing, persistence, and shear strength of the discontinuities are important input parameters. The volume or mass of rock that must be supported depends on the orientation and spacing of rock discontinuities combined with the size, shape, and orientation of the excavation. However, the actual block sizes are generally unknown. The maximum size that can fall is limited by the size of the excavation. A common assumption when using shotcrete in conjunction with rock bolts to support potentially unstable blocks is that the shotcrete need support only the weight of blocks that can fall between the bolts. If large, unstable blocks are expected, the rock bolts or cable bolts support them. In terms of shotcrete,

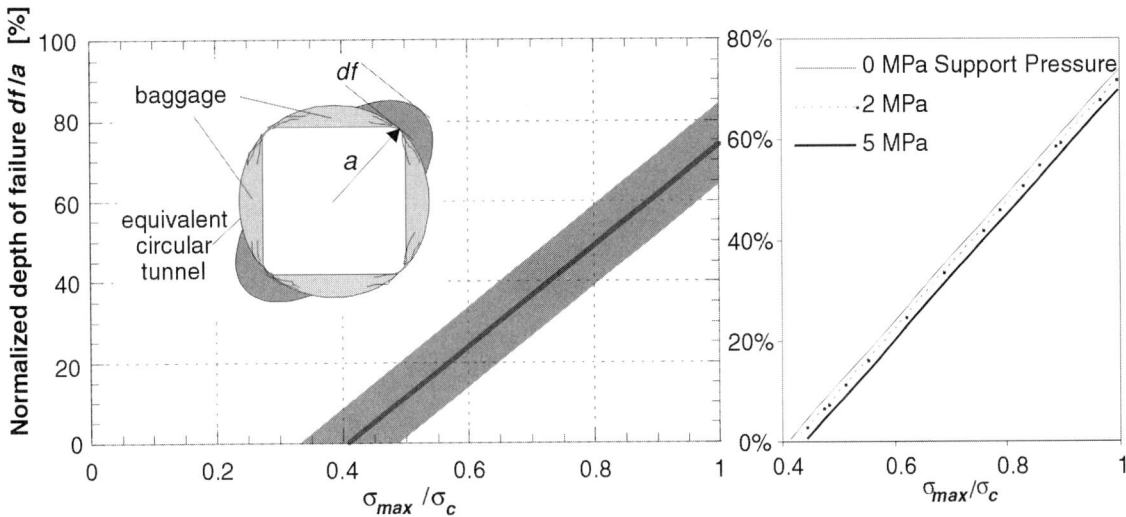

FIGURE 67.3   Depth of failure of stress level (σ_{max}/σ_c) based on case studies (Kaiser et al. 1996; Martin et al. 1999) and (in-)sensitivity to support pressure for range of 0 to 5 MPa

this limits the block size to one with a basal area that lies between a pattern of bolts. A software program (UNWEDGE) produced by the Rock Engineering Group at the University of Toronto, and our companion program (AWEDGE, used for mapping and critical block identification) provide an effective method to design for this class of ground behaviour.

Barrett and McCreath (1995) suggested that a reasonably conservative assumption for weight calculation is a pyramid-shaped wedge of loose rock, with side angles of 60° and a basal area defined by the bolt spacing. This approach ignores frictional resistance along the sides of the block, which may significantly reduce the load that could eventually be applied to the shotcrete liner. The load development (the average pressure over a footprint area) for a 60° wedge spanning the full tunnel width (infinitely long) or pyramid (equal sided) is shown in Figure 67.16 and for a wedge/pyramid limited by bolt spacing is shown in Figure 67.17. These figures show that the pressure on bolted shotcrete will typically be less than 0.2 megapascals (MPa), whereas practically feasible shotcrete ring capacities of 2 MPa may be exceeded when a tunnel span exceeds 6 m.

Mining-induced stress changes are of particular concern because reducing the confinement (e.g., by mining a nearby shrinkage stope) will lower the stabilising interface friction at the block boundaries. Similarly, at intersections, removing the constraints in four directions leads to conditions favouring wedge-type failures. These effects often increase the demand on shotcrete over time and have been extensively investigated by Diederichs (1999).

67.2.3   Intermediately Stressed Rock (Loose, Fractured Rock)

When mining at depth, or when the extraction ratio increases beyond a critical value, various forms of rock-mass instability may be encountered. Often, they cannot be prevented and must be controlled by effective support systems that retain the broken rock, control the dilation process of the fractured rock mass, and tie the broken rock back to stable ground. Practical experience shows that shotcrete can greatly assist in controlling the ground if it is used as an integral part of a support system and as long as deformation compatibility between the various support components and the ground is maintained.

Depth of Failure.   In hard, brittle rock, when the stresses near an excavation exceed the rock-mass strength, the failure

process creates and propagates new fractures, eventually leading to the disintegration of the rock mass and a gradual transition from continuum to discontinuum behaviour. Rock fractures usually start from a stress raiser (corner of the excavation) or from the point of maximum tangential stress concentration. Rock-mass failure may either be gradual or violent (bursting) presenting as surface parallel-slabbing or localised notch formation. The depth of failure can be predicted from the relationship shown in Figure 67.3, where σ_c is the intact rock strength (UCS) and σ_{max} is the maximum tangential stress calculated for a circular excavation at the same location in the virgin, in-situ, or mining-induced stress field (Kaiser et al. 1996 and Martin et al. 1999). Again, it is important to consider the ultimate state of stress when designing support. From Figure 67.3, the anticipated deepening of the failure zone can be estimated by comparing in-situ and mining-induced stress levels. Furthermore, in hard, brittle rock, the depth of failure is essentially independent of the support pressure applied at the excavation wall (Figure 67.3, right figure).

For comparison, the load development (average pressure over the extent of the breakout) for a depth of failure of 50% of the tunnel radius, which corresponds to a stress level of about 0.8 (see Figure 67.3), is shown in Figure 67.16. At this stress level, the gravitational loading pressure from the broken rock is less than that for a 60° full-tunnel-width wedge.

If a violent (bursting) fracture occurs, not only must the gravitational loads be carried, but the energy released during the rock-fracture process must also be dissipated (Kaiser et al. 1996).

Rock-Mass Bulking.   The process of stress-induced fracturing near excavations is associated with substantial dilation (volume increase) resulting from two sources: dilation due to shear at fractures or joints, and, more important, dilation due to geometric incompatibilities that occur when blocks of broken rock move relative to each other as they are forced into the excavation. This dilation process, called rock-mass bulking, produces large radial deformations in the fracture zone and consequently at the excavation wall. Figure 67.4 illustrates the bulking process near an excavation damaged by a rock burst. The rock mass was supported by light bolts and mesh. The rock was loaded tangentially, creating surface-parallel fractures that opened during the fracture process.

Contrary to the depth of failure, the amount of bulking inside the failure zone strongly depends on the support pressure

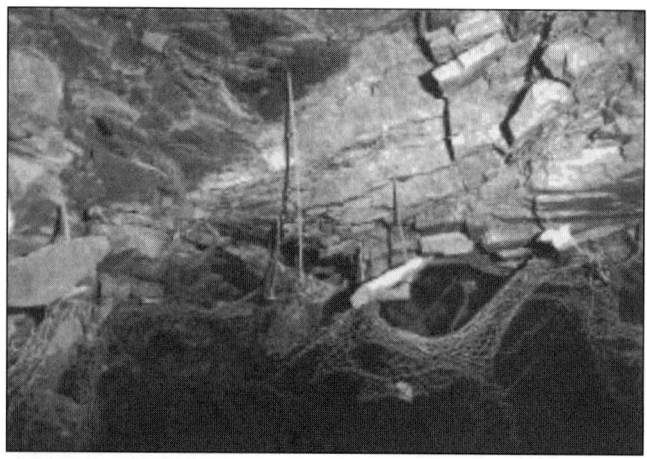

FIGURE 67.4 A localized volume of moderately jointed rockmass that bulked due to damage by a small strainburst behind a light support system of welded-wire mesh and mechanical bolts

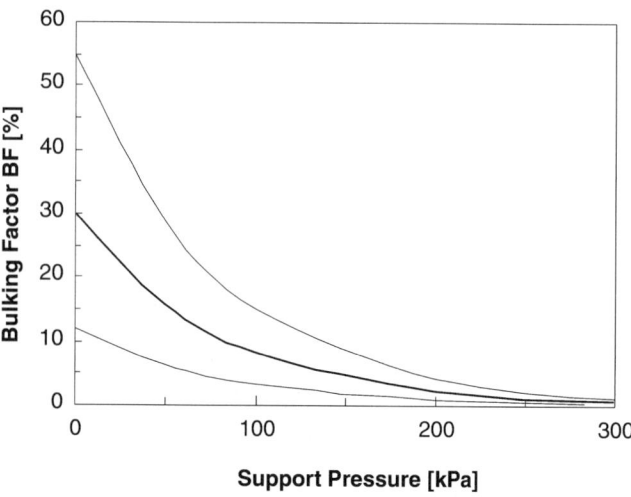

FIGURE 67.5 Bulking factor as a function of average support pressure (trend line and range of ±3 standard deviations)

as illustrated by Figure 67.5, which is based on a limited number of quantitative field observations (Kaiser et al. 1996). Unconfined rock can increase in volume by as much as 30% to 60% (e.g., in the floor of a tunnel), but confinement and rock reinforcement drastically reduces bulking. An effective reinforcement system that provides an equivalent support pressure of >200 kPa can reduce bulking to less than 3%.

It is important to realize that it is not the support pressure alone that controls the bulking process. As illustrated in Figure 67.6, rock reinforcement (e.g., by grouted rebar) prevents a fracture from opening, which more effectively controls bulking. Consequently, in intermediate to highly stressed rock, shotcrete should always be combined with rock reinforcements. By reducing rock-mass bulking, the shotcrete will be deformed less and thus be stressed to a lesser extent. Even if the rock bolts fail from excessive straining inside the failed rock zone, they are still very beneficial as they restrain bulking everywhere except at the fracture point.

From a practical perspective, it is most important to recognise that the depth of failure is limited, even if the tunnel is unsupported, and that the depth of failure cannot be significantly reduced by the support. However, the bulking process and thus

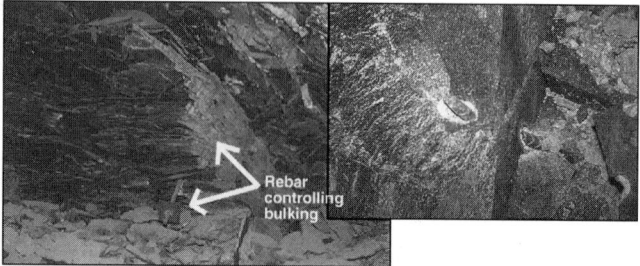

FIGURE 67.6 Grouted rockbolt (rebar) limiting rock mass dilation (bulking) during failure of brittle, laminated rock. Such rebars are effective in controlling bulking, even if they fail to fulfill their holding function.

the wall deformations can be reduced significantly by a combination of rock reinforcement and shotcrete support. In other words, in moderately stressed ground, an excavation can be stabilised if the support has sufficient capacity to hold the weight of the broken rock and has sufficient flexibility to deform in response to bulking. The rock-fracture process cannot be prevented, but it can be controlled.

67.2.4 Highly Stressed Rock (Deep Rock Failure)

At great depth or in highly stressed ground (high extraction ratios or slender pillars), the rock-mass strength may be exceeded around the entire excavation, and the failure zone involves the entire excavation. In ductile rocks, such as clay-shale or friable sandstone, this failure process can be described as yielding and can be simulated reasonably well by applying conventional failure criteria (e.g., commonly adopted yield criteria: Mohr-Coulomb, Hoek-Brown, or others). In brittle rock, however, we recommend the approach proposed by Martin et al. (1999). In either case, it is very important to obtain a reliable estimate of the anticipated wall convergence and its circumferential distribution (relative displacements). The best way to obtain this is by numerical modelling, although flow-rule assumptions will dictate the outcome. As indicated above, rock-mass bulking is very sensitive to support pressure and rock-mass reinforcement, but very few models properly simulate the bulking process. Furthermore, there is little information available to use in selecting appropriate dilation parameters. Also, the depth of failure is rarely uniform around an excavation and, when combined with variability in bulking, significant non-uniform wall movement (closure) must be anticipated. It is this differential straining that may damage shotcrete. Consequently, when large deformations are anticipated in highly stressed rock, the engineer is well advised to find a practical means to deal with large deformations (Kaiser and Tannant 1999). As will be illustrated later, the deformation limits of shotcrete arches or rings are very limited, and the designer must revert to bolted shotcrete panels (slotted ring support) to prevent premature shotcrete failure.

67.3 SHOTCRETE AS A COMPONENT OF A SUPPORT SYSTEM

Support systems are commonly composed of various components: bolts, mesh, shotcrete, and spray-on linings. Bolts are used to hold rock in place or to reinforce the rock mass, mesh is used to retain loose or broken rock, and shotcrete performs a combination of these functions. It holds by adhesion, strengthens the rock by preventing relative movements at the shotcrete/rock interface, and acts as a "supermesh" by providing a stiff retaining component with substantial bending or flexural capacity.

Before designing a shotcrete support system, its function, role, and mode of operation must be clearly defined. When

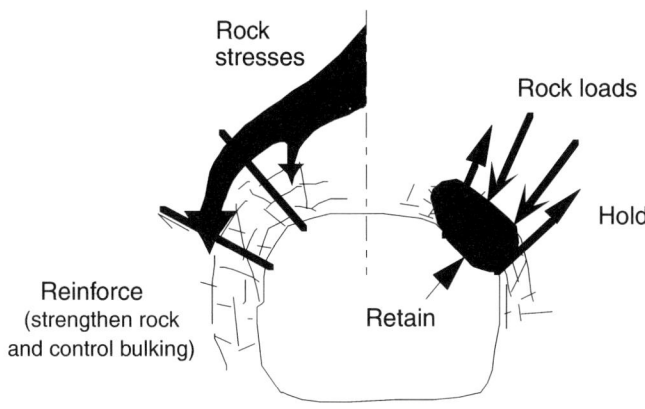

FIGURE 67.7 Three primary functions of support elements (Kaiser et al. 1996)

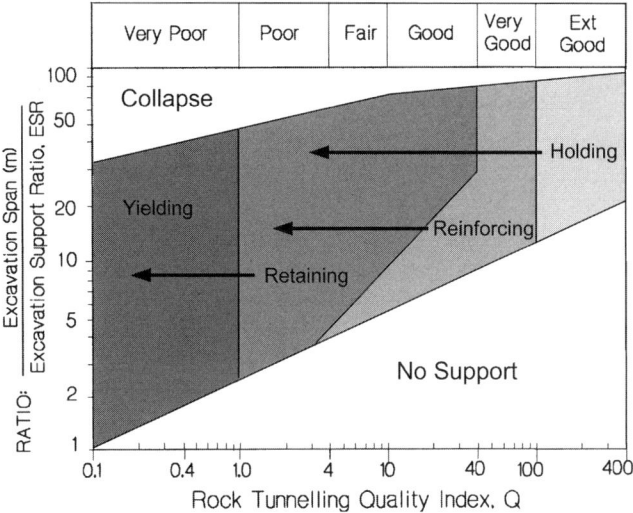

FIGURE 67.8 Support functions as a function or rock mass quality: hold → hold and reinforce → hold, reinforce, and retain → fulfill these functions while yielding

shotcrete performs unsatisfactorily, it can often be attributed to either inappropriate application or a lack of understanding of its role in a support system.

67.3.1 Support Functions and Role of Shotcrete

Each support component is intended to perform one of three functions as illustrated in Figure 67.7: (1) hold loose rock, key blocks, and other support in place; (2) reinforce the rock mass and control bulking; and (3) retain broken or unstable rock between the holding and reinforcing element to form a stabilised arch.

In good rock, the holding and reinforcement functions are the most important, whereas in fair to poor rock, all three functions must be integrated (Figure 67.8). In addition, in very poor ground (Q<1; Barton et al. 1974), large deformations must be anticipated; and support components must retain their functionality over a large displacement range (i.e., must yield). In this domain, it is necessary to design the support system so it can yield without building up excessive stresses; most important, however, the support system must contain reinforcement elements that can control the bulking process.

When designing support for mines, it must be recognised that mining-induced stresses alone can reduce the rock-mass quality from its virgin state by more than one order of magnitude, thus changing the support function requirements in otherwise identical rock conditions. To define an appropriate support function for mining applications, it is therefore paramount to select an appropriate stress-reduction factor (SRF; Barton 1994) that takes mining-induced stresses into account (see later Figure 67.22).

Directly or indirectly, shotcrete can contribute to all three of the support functions noted above but only if it is properly integrated into the support system.

Holding. Shotcrete resists load by adhesion, thrust, and bending resistance. If shotcrete forms a continuous-support arch or ring, it can hold rock in place by building thrust forces in the arch; and if it adheres well to the rock mass, its holding capacity is substantially increased by the combined arching action of the rock and shotcrete.

Reinforcement. Shotcrete as a surface-support component does not directly reinforce rock. However, if we define the effect of reinforcement as a strengthening the broken rock, then shotcrete has an equivalent effect. It increases the rock-mass strength by increasing the confining stresses and, more important, by preventing the rock mass or broken rock from loosening at the surface.

Retention. When combined and connected to proper holding elements (rock bolts or cables), shotcrete provides a stiffness and high-capacity retaining function. In well-jointed or blocky rock masses, shotcrete prevents ravelling between bolt heads, locks key blocks in-place, helps transfer the load onto bolts, and helps maintain the integrity of rock blocks. In this case, shotcrete works as a high-capacity, very tight mesh ("supermesh"). However, unlike mesh, it does prevent relative shear of rock blocks, creating a major stabilising effect.

In burst-prone mines, the shotcrete has to assume one additional function: energy dissipation. Because energy dissipation is, in part, proportional to the mass of the support, shotcrete dissipates large amounts of energy (see later) and can be utilized as an effective component of a burst-resistant support system.

Shotcrete may also serve other functions that are not covered here, e.g., protecting rocks susceptible to rock-mass degradation (weathering or slaking) and reducing wall roughness (for ventilation or sealing).

Seldom does shotcrete fulfill all functions and for an economic design, the designer must decide which function the shotcrete is expected to perform. For example, a heavy shotcrete arch may serve to hold the loose rock in place, whereas a thin, bolted shotcrete layer may serve only to retain rock (the bolts are expected to carry the load). A poor understanding of which function the shotcrete is intended to serve causes many failures.

67.3.2 Operation and Failure Modes

Shotcrete can fulfil these different roles by:

Adhesion. Resisting load by adhering to the excavation surface (as long as there is not surface-parallel slabbing or spalling).

Shear Resistance. Resisting the punching through of rock blocks (as long as there is sufficient adhesion outside the perimeter of the block).

Thrust Capacity. Resisting ground movements and loads by transforming radial pressures into thrust forces or hoop stresses. Closed-ring support or arches are intended to utilise the primary quality of shotcrete, which is its compressive strength. Transferring loads to shotcrete footings or abutments is possible only as long as the flow of stresses is ensured without creating excessive bending or shear stresses.

Flexural Strength. Resisting loads by bending moments. This mode of operation is primarily activated in flat-wall support systems and to bridge between bolts ("supermesh" function).

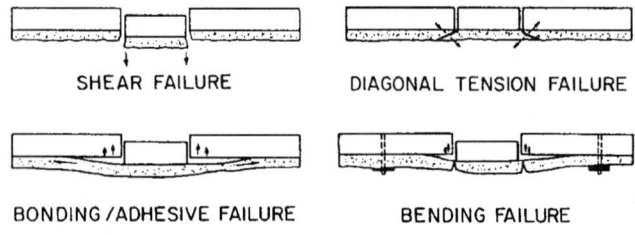

FIGURE 67.9 Shotcrete failure modes (after Rose 1985)

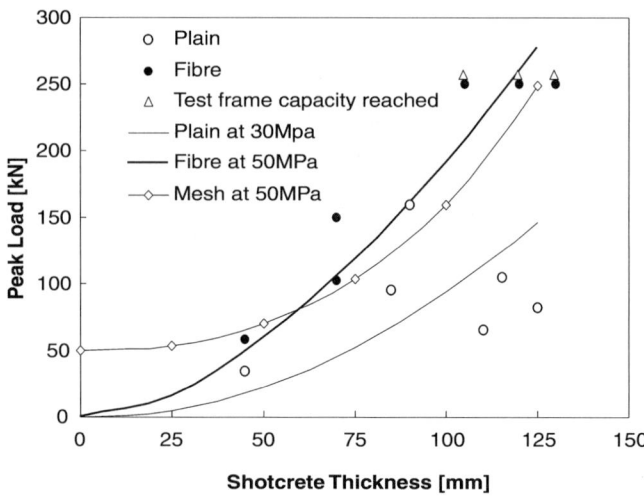

FIGURE 67.10 Plate (diameter = 250 mm) pull-test results on plain and fibre-reinforced shotcrete compared to capacity predictions for various types of shotcrete assuming 30 or 50 MPa compressive strength

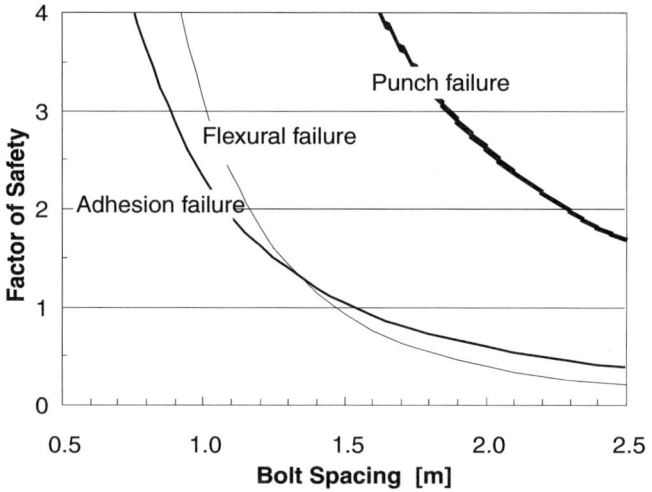

FIGURE 67.11 Factor of safety for various shotcrete failure modes (thickness = 75 mm, adhesive strength = 0.5 MPa over a 25 mm bond width, and shotcrete tensile strength = 1.5 MPa)

Interlocking. Creating an interlocked system of broken rock and shotcrete by preventing relative movement between the shotcrete and the rock or between fractured rock blocks. As long as the interlocked system maintains its integrity, shotcrete will not act alone as a support. It will largely ensure the self-supporting capacity of the rock mass.

The manner in which shotcrete operates as a support varies from case to case, and this must be considered to arrive at a proper design.

Shotcrete may fail to perform its desired role for a number of reasons. Figure 67.9 illustrates some of the commonly observed failure mechanisms in shotcrete.

Direct Shear Failure (or Punch-Through Failure). This type of failure is seldom seen, except in thin shotcrete applications (≤50 mm), because other failure mechanisms are usually more critical or occur first (Figure 67.11).

Adhesive Failure. Loss of the adhesive bond between the shotcrete and the rock is a common problem if the rock surface is not well prepared or dirty (mud, dust, and oil) or because the rock itself is weak in tension (highly foliated, closely bedded, or spalling). This latter failure mode is typically encountered at depth, where stress-induced slabbing occurs, or in pillar applications (hourglassing leading to surface-parallel slabbing).

Diagonal Tensile Failure. This type of failure occurs when a block of rock is punched through the shotcrete. High displacement gradient generates tensile stresses in the shotcrete around the perimeter of the displaced block and leads to tensile fractures that radiate outwards around the perimeter of the displaced rock block. O'Donnell and Tannant (1998) tested the diagonal tensile capacity of shotcrete by field pull tests with 250-mm-diameter plates (Figure 67.10).

Flexural Failure. Flexural or bending failure occurs when the rock fractures and bulks behind the shotcrete layer. Generally, adhesive failure precedes flexural failure, and both mechanisms are commonly seen together (Figure 67.11). When shotcrete bends, it can fail in tension. In these cases, using shotcrete reinforcements (fibre for small strains and wire mesh for larger strains) is essential.

Compressive Failure. Crushing of shotcrete is encountered when the hoop stresses exceed the shotcrete strength. This failure mode is common when large deformations occur, when geometric incompatibilities between the shotcrete and the deformed excavation shape exist, or when shotcrete is dynamically loaded (rock bursting) (Figure 67.12).

Figure 67.10 demonstrates generally superior diagonal failure resistance and less variability for fibre-reinforced shotcrete. Also shown are model predictions for plain, fibre- and mesh-reinforced shotcrete. These curves illustrate a very important aspect for thin shotcrete applications in mining. Because of the residual capacity of mesh at zero shotcrete thickness, mesh-reinforced shotcrete may be advisable, unless a minimum thickness of 50 mm can be guaranteed, to resist punch-type failures when less than 50–75 mm of shotcrete are applied.

However, punch-type failure may not be critical as illustrated by Figure 67.11 where, for the selected parameters shown in the figure caption, both adhesive and flexural failure are much more likely to occur.

Compressive failure indicates that the shotcrete is attracting too much load as illustrated by Figure 67.12. In such situations, increasing the thickness or reinforcement of the shotcrete will not alleviate the problem because the deformation capacity of shotcrete arches is rather limited and is essentially independent of its thickness (Figure 67.13). In such situations, longitudinal slots or intentional (mesh-reinforced) shotcrete weakness zones must be integrated into the design. Alternatively, on a flat wall, bolted shotcrete panels with mesh-reinforced shotcrete in excavation corners may be used.

67.4 DESIGN APPROACHES FOR SHOTCRETE SUPPORT

The design of shotcrete as a construction material is well established today (e.g., Melbey and Garshol 1999). As indicated

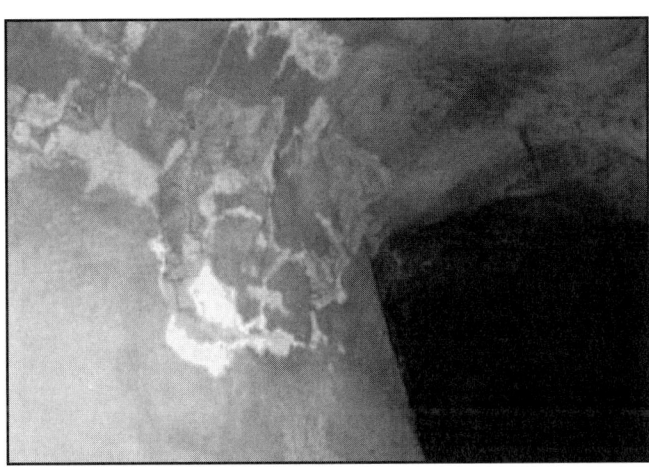

FIGURE 67.12 Damage to a shotcrete arch due to deformation incompatibility during minor rockburst leading to rock mass fracturing and bulking behind the shotcrete, which created high compressive hoop stresses within the shotcrete

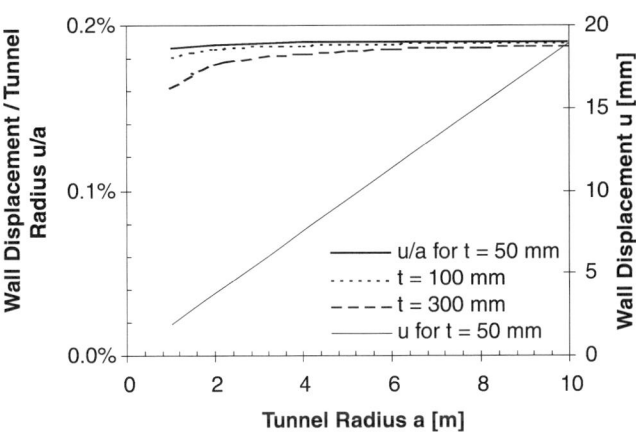

FIGURE 67.13 Deformation capacity at peak load (without shrinkage allowance)

earlier, designing shotcrete mix is not the focus of this chapter. For the following charts, unless indicated otherwise, a minimum 28-day strength of 40 MPa with a modulus of 21 GPa was assumed. Designing shotcrete as an underground support or a component of a support system is still rather imprecise and is often based on a combination of empirical and analytical or numerical design considerations. Fundamentally, shotcrete can be designed to carry loads or to resist and control deformation.

To carry loads, the shotcrete must act as a support ring or arch, or must transfer loads to other holding elements such as rock bolts, anchors, or cables. The design approach then consists of determining a realistic design load and comparing this design load with the support capacity of the shotcrete or the support system including shotcrete. At shallow depths, the load can be determined from the geometry of potentially unstable rock blocks (wedges) or volumes of broken rock bound by an arch of self-stabilised rock (Section 67.4.2). In highly stressed rock, the weight of fractured material can be estimated from the anticipated depth of failure (Section 67.5.4).

When the anticipated loads exceed 1 or 2 MPa, or when large deformations—particularly differential deformations—are anticipated, shotcrete must be designed to move with the ground, or it may fail prematurely because of excessive stresses induced by geometric incompatibility.

What are large deformations? First, large deformations are those that lead to excessive stresses in the shotcrete. For the stiffest system—a closed shotcrete ring—the deformations at peak capacity are shown in Figure 67.13. Accordingly, closure after applying shotcrete in excess of about 0.2%, (i.e., 10 mm) for a tunnel with a radius of 5 m, will fail a closed shotcrete ring.

An open support ring (or shotcrete arch) will be able to tolerate significantly higher deformations because thrust forces do not build up as rapidly due to slip at the shotcrete/rock interface. In this case, the deformation capacity of the shotcrete will be significantly higher and will increase rapidly with tunnel radius. The deformation capacity for flat walls is essentially infinite because no thrust or bending is induced by uniform wall deformation. However, in this situation differential displacements govern (i.e., relative displacements between bolts and midspan). Laboratory and field-test results indicate that relatively large differential deformation (on the order of 5% to 20% of the span over which the differential movement occurs) can be absorbed by steel-fibre- or mesh-reinforced shotcrete as long as there is not simultaneous tangential (thrust) loading and

as long as the deformation gradient is uniform (flexure failure). If there is strong adhesion outside the loading area, localised differential displacements of 1% to 3% may be sufficient to cause punch-type failure.

It is often not practical to adopt rigorous support design procedures, and empirical design guidelines based on experience are often adopted. However, their limitations and range of applicability must be well understood.

67.4.1 Empirical Design of Rock Support with Shotcrete

Traditionally, empirical design methods for rock support using shotcrete are based on simple rules of thumb that stated the required shotcrete thickness (and often bolt spacing) for a given size of excavation and ground condition. The emphasis on design related to shotcrete has often been the material itself, (i.e., a mix of design and application techniques and equipment). Barrett and McCreath (1995) present a historical summary of empirical support designs and stated that there was rarely a clear rationale behind many of the "rules."

While many stability diagrams based on rock-mass conditions can be found in the literature, relatively few go so far as to recommend support for a specific situation. In terms of shotcrete, two complementary and practically useful empirical approaches are those given by Hoek et al. (1995) and Grimstad and Barton (1993). They suggest that the support requirements be evaluated with both systems and then checked analytically for both load capacity and deformation compatibility.

Hoek et al. (1995) provide a table of recommended shotcrete applications in underground mining for different rock-mass conditions (Appendix A). This table provides a simple link between rock-mass description and behaviour and recommendations for shotcrete. It can also serve to check designs obtained by other means. The support recommendations cover the whole spectrum of anticipated excavation behaviour including wedge-type instability in low-stressed rock to moderately violent rock fracturing during rock bursts. Support recommendations for burst-prone ground are covered in more detail later, and the reader is also referred to Kaiser et al. (1996) for details.

Based largely on data from tunnelling projects, Barton et al. (1974) developed a now widely utilised empirical stability chart based on the rock-mass quality index (Q). The excavation support ratio (ESR) was introduced to extend the applicability of this empirical chart to temporary excavations and mining conditions. The ESR is a factor used by Barton to account for different degrees of allowable instability (risk) based on the excavation service life and usage. ESR ranges from about 1–3 for permanent

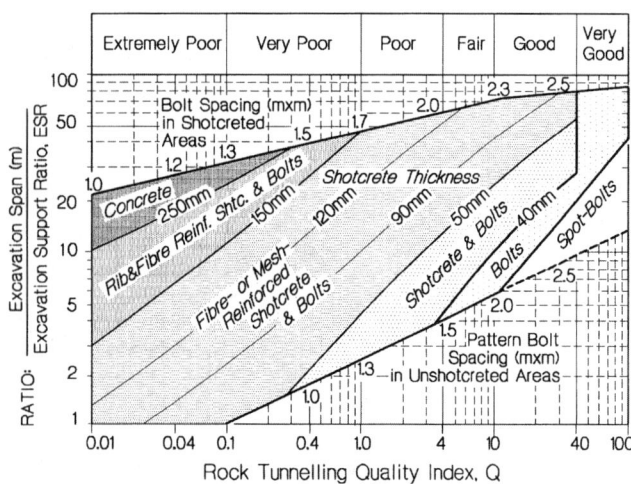

FIGURE 67.14 Shotcrete support recommendations based on Q, span, and ESR (after Grimstad and Barton 1993 by Hutchinson and Diederichs 1996)

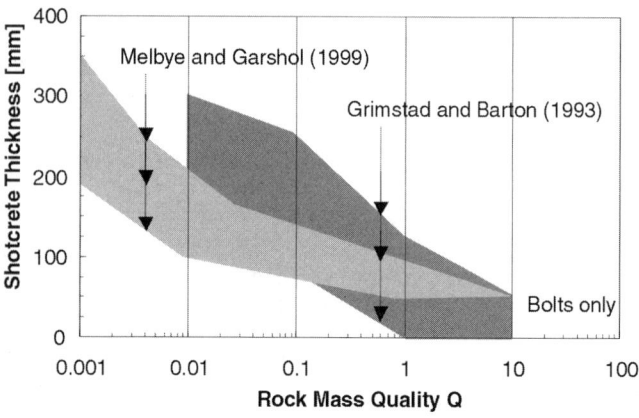

FIGURE 67.15 Recent trends in high performance wet-mix sprayed concrete thickness (Melbye and Garshol 1999) compared with recommendations based on Grimstad and Baton (1993)

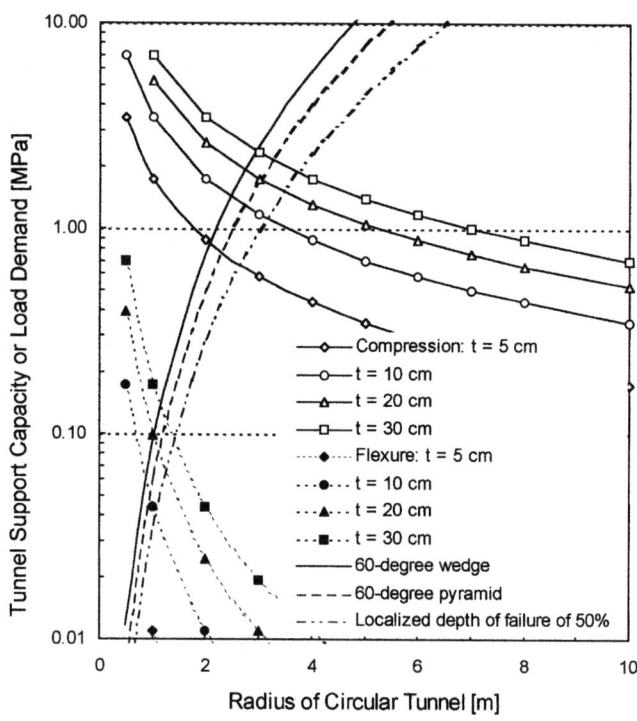

FIGURE 67.16 Tunnel support capacity for a closed ring support (labeled: Compressive t = 5 to 40 cm), and a flexible arch (labeled: Flexure t = 5 to 40 mm). Also, shown are loads generated by two wedge geometries and a localized failure zone penetrating to a depth of 0.5 times the tunnel radius.

or temporary mine openings and up to 300 for severe rock-bursting conditions. A version of the stability chart based on Grimstad and Barton (1993) is reproduced in Figure 67.14. These support recommendations are largely applicable for low- to moderately high-stress conditions, from which the chart itself was developed.

Since the publication of this design chart, much progress has been made in developing high performance shotcrete. Melbey and Garshol (1999) summarised their recent experience with high-performance, wet-mix sprayed concrete with steel fibre and presented a chart relating rock-mass quality (Q) and shotcrete thickness. Their experience is compared in Figure 67.15 with data from Grimstad and Barton (1993) (Figure 67.14). The central line of each range from small (lower limit) to large (upper limit) spans represents tunnels with a span of about 10 m. From this figure, it can be seen that today's shotcrete technology allows stable excavations with significantly less shotcrete than proposed by Grimstad and Barton (1993). For example, for a 10-m tunnel at Q = 0.1—although 140 mm is 40% more than 100 mm, 100 mm is about 29% less than 140 mm—would be required if top-quality shotcrete can be applied and if this shotcrete is properly integrated into a bolted support system.

67.4.2 Analytical Design of Wedge Support with Shotcrete

Various analytical approaches exist for designing shotcrete (see Barrett and McCreath (1995) for summary). In each case, the design is based upon a postulated failure mechanism for the shotcrete and the rock mass (wedge-type failure). For example, force-equilibrium methods may be used to design shotcrete to support potential wedge-type instabilities. The stresses or loads that may apply to a specific shotcrete design can be determined from the failure modes described earlier. The objective is to determine the force required to hold the wedge in place with a satisfactory safety factor. The loads carried by the shotcrete can be estimated by knowing the position and orientation of the discontinuities that define blocks; or they can simply be assumed to be no larger than the weight from a wedge- or pyramid-shaped block with a basal area defined by the bolt spacing as discussed earlier. The frictional resistance provided along the sides of the blocks or wedges contributes to the stability of a wedge but is difficult to predict accurately. More important, in mining, nearby stopes often cause a loss of confinement (clamping stress reduced to zero) that creates an elevated hazard. Hence, it is often meaningful to design for the worst-case scenario of full gravity loading as adopted here.

The support capacity of a perfectly circular shotcrete ring is very high compared to the capacity of an arch or panels that are failing by flexure. Figure 67.16 presents the approximate support capacities for a circular tunnel (Hoek and Brown 1980, Kaiser 1989). As shown, even under almost perfect conditions (circular, closed ring), an economically viable shotcrete support system can, at best, provide about 1 to 2 MPa support pressure. If bending stresses are induced (i.e., the shotcrete resists by flexure), corresponding capacities are almost two orders of magnitude lower (0.01 to 0.02 MPa).

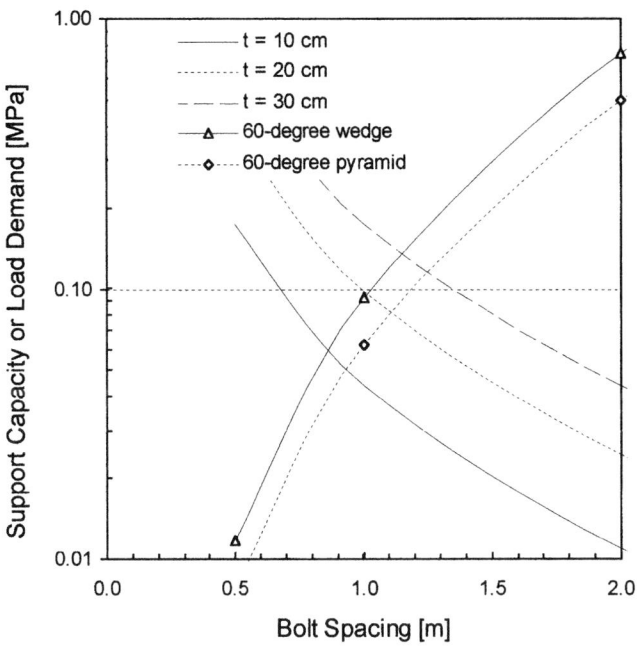

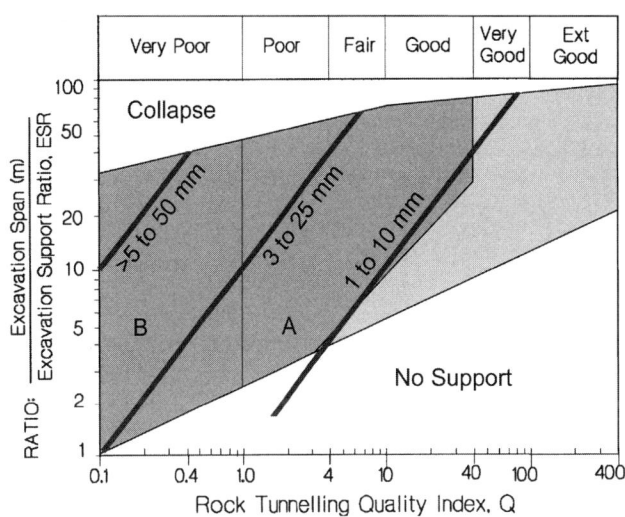

FIGURE 67.18 Deformation limits observed from well supported excavations superimposed on stability chart (after Kaiser 1986)

FIGURE 67.17 Support capacity (flexure mode) and anticipated loads for bolted shotcrete panel (same parameters as for Figure 67.16)

In the most extreme situation, full-width wedges may have to be supported or a body of broken rock extending to the depth of failure may have to be stabilised. The resulting loads for specific situations are shown for comparison. It can be seen that a closed shotcrete ring can support full-span wedges or fractured, loose rock (at a depth of failure of 0.5 times the radius) for a tunnel width of about 6–8 m. However, one open ring would only be stable to a width of about 2–4 m. Since it is seldom economically viable to install closed rings in mining (except in shafts), it follows that bolts and cables must be utilised to carry such loads. The corresponding support capacities and load demands are presented in Figure 67.17 as a function of bolt spacing. It should be noted that the underlying assumptions are rather conservative and that much more favourable conditions may be encountered in many situations. Nevertheless, this figure, for an assumed failure mode of flexure, illustrates the interdependence of bolt spacing and shotcrete thickness. For example, for the assumed pyramidal wedge geometry, a bolt spacing of 1 m and 5 cm shotcrete would be equivalent to a 1.5-m bolt spacing and 40 cm of shotcrete. As indicated earlier, a number of possible shotcrete failure mechanisms must be considered to determine which one is most critical in a given situation. Adhesion failure or failure by bending are most likely dominant.

For more detailed analyses, the reader is referred to Barrett and McCreath (1995), who present a summary of analytic designs for shotcrete support of loose rock. Other scenarios can be easily assessed for criticality (see for example Figure 67.11).

67.5 SHOTCRETE SUBJECT TO LARGE DEFORMATIONS AND ROCK BURSTS

67.5.1 Deformation Compatibility

Kaiser (1986) presented deformation limits from field observations on well-supported caverns and large tunnels. These limits are reproduced in Figure 67.18 illustrating zones in the stability chart where deformations are expected to exceed 1–10 mm, 3–25 mm, and 5–50 mm, respectively. Consequently, when Q falls below 1, the wall displacements must be expected to exceed 0.2% of the

excavation radius, and a closed ring would not be able to withstand the anticipated deformations. Hence, excavations falling to the left of Q = 1 must be designed such that the support system is compatible with large deformations (see also Figure 67.8).

When large deformations occur, the deformed shape of an excavation no longer matches the undeformed shape of the installed support/shotcrete, and debonding or flexural failure is induced. In other words, for Q < 1, shotcrete should be applied in such a manner as to prevent deformation incompatibility. Appropriate installation techniques preventing "self-destruction" of the shotcrete and deformation compatibility are (1) longitudinal slots or gaps in shotcrete arches to break the flow of hoop stresses, or (2) flat-wall shotcrete panels that are free to move without causing constraints at panel boundaries (Kaiser and Tannant 1997). When shotcrete is applied as panels or slotted arches, it must be held in place by rock bolts or cables that are capable of surviving the deformations imposed on them by the bulking rock mass and the deforming shotcrete.

In mining, matters are complicated by the fact that mining-induced stress changes and related rock-mass failure causes additional deformations, ones that are not encountered in civil engineering excavations (caverns and single- or twin-tunnel situations). Unless a drift or tunnel is remote from a mining zone or the mining-induced stress changes do not cause rock-mass failure, additional deformations must be anticipated. Failure to account for these post-support deformations is often the reason that shotcrete performs poorly, particularly in burst-prone ground.

While it is difficult to accurately predict these deformations, their impact can be estimated by selecting two stress-reduction factors, one for premining and one for postmining. If a tunnel is supported with a stiff shotcrete ring—e.g., at Q = 2 for a 4-m-wide tunnel (A in Figure 67.18)—and the SFR suddenly increases from 2 to >20, the Q drops to 0.2 or less (B in Figure 67.18), and the shotcrete cannot deform unless it was designed for deformation compatibility. This method is particularly useful when assessing deformation compatibility in burst-prone ground as described later.

67.5.2 Deformation Control

As introduced earlier, the primary source of deformation in over-stressed rock at depth, in pillars, or at high extraction ratios, is rock bulking near the excavation wall. Unfortunately, it is difficult, if not impossible, to predict the deformations and the anticipated loads because the relevant material model parameters are not known. However, two important observations, described earlier, provide sufficient insight to give practical guidelines for deformation control. First, the bulking factor is very sensitive to confinement (support pressure) and, second, rock reinforcement by grouted rebar provides an effective means to prevent bulking. Consequently, when large deformations due to rock-mass bulking are expected, a support system consisting of rebar, rock bolts or cables, and shotcrete should be applied. The roles of the various components are:

- Minimise bulking by using frictional bolts or grouted rebar to prevent joints and fractures from opening. Even if the rebars break at some depth and cannot be relied upon to hold the broken rock, they still provide an effective means for dilation control and protect the shotcrete from excessive deformations.

- Minimise bulking by applying continuous surface pressure. Even relatively minor support pressures (see Figure 67.5) are sufficient to significantly reduce bulking. However, this pressure must be applied uniformly and before the rock mass is allowed to loosen. Even a relatively thin layer of reinforced shotcrete (~50 mm), providing little support capacity, has a very beneficial effect on wall deformations.

Because these support components are intended to control bulking and may be over-stressed (fail) in the process, they may not hold with an adequate safety factor. Additional support components are then needed to:

- Hold the reinforced broken rock in place using long, untensioned mechanical bolts or yielding bolts (Super-Swellex™, Conebolts™, and plain cables). Unless the bulking is effectively controlled by the above-mentioned means, these holding components could be excessively strained and fail prematurely.

- Hold the reinforced broken rock in place using a heavy shotcrete ring or arch. If the bulking can be effectively controlled by rock reinforcement, so the remaining deformations will not damage shotcrete, a thicker layer of shotcrete can be applied for the holding function. Ideally, shotcrete should be applied so that it remains unstressed until nearby mining affects the drift. In many mining situations, a second shotcrete layer that is intended to fulfil the holding function should be applied only shortly before mining-induced stresses affect an area.

When deformations are under control, most shotcrete types perform well.

67.5.3 Shotcrete Reinforcement

When should fibre- or mesh-reinforced shotcrete be used? Much has been published about the pros and cons of various types of shotcrete reinforcement (fibres or mesh). The apparent controversy of which type is more appropriate can be resolved if it is accepted that short fibres have a limited extension capacity and mesh performs poorly when sheared or stressed locally. The selection of an appropriate reinforcement depends largely on the anticipated type and amount of deformation. If shotcrete is applied so later straining is limited and uniform, fibres will perform well. On the other hand, when adhesion failure occurs and shotcrete acts as a strong retaining system, mesh-reinforced shotcrete will perform better. Furthermore, when shotcrete is expected to sustain large deformations and fail in compression (e.g., at stress raisers), mesh reinforcement works better.

Punch Failure or Pull-Out Test Results. Under laboratory testing conditions, steel-fibre-reinforced shotcrete is stronger

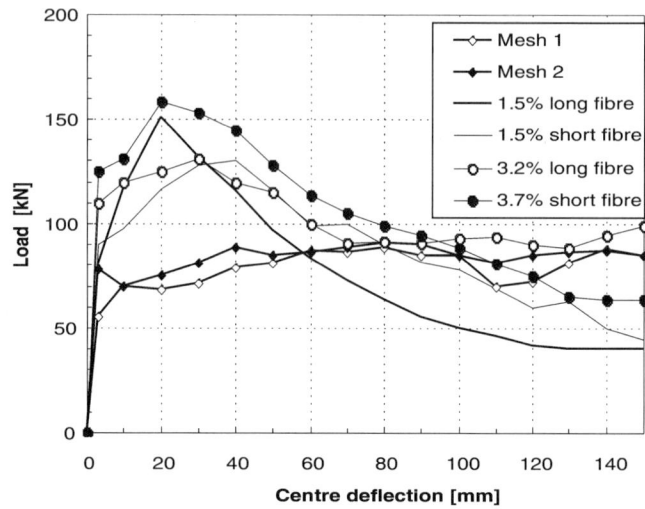

FIGURE 67.19 Comparison of mesh and fibre-reinforced shotcrete in distributed load tests (unpublished data; pers. com.)

than plain shotcrete. More important, adding the fibres greatly increases the shotcrete's toughness and post-peak strain capacity. One can expect steel-fibre shotcrete to have superior performance characteristics in the field as well. However, the steel fibres that reinforce the shotcrete are typically quite short (about 25–40-mm long), and this places a displacement or strain limitation on the shotcrete that can be easily exceeded in certain mining environments. For example, in-situ pull tests on 250-mm-diameter sections of plain and steel-fibre-reinforced shotcrete have shown that both types failed after about 10–15 mm of displacement (O'Donnell and Tannant 1998). These observations make sense given that "pull-out" of the steel fibres is common along the boundaries of shotcrete that is failing in diagonal tension. Given that roughly one half of the fibre length remains embedded in the shotcrete, the maximum allowable displacement across a fracture is one half the length of the fibre or 10–20 mm.

During these tests, the plain shotcrete was able to tolerate almost as much displacement as the steel-fibre shotcrete because more favourable conditions were created by slightly different failure mechanisms. The pull tests also showed that adhesion loss followed by flexural bending was a dominant failure mode when the thickness of plain shotcrete exceeded about 50–60 mm (see Figure 67.10).

Bending Failure or Flexure Test Results. The failure process in shotcrete is often progressive in nature, especially when adhesion loss and flexural bending occur. If this process is gradual, leading to a uniform loading of the shotcrete, fibre-reinforced shotcrete will demonstrate much superior behaviour (as illustrated by Figure 67.19) where the fibre-reinforced shotcrete exhibits almost double the capacity at centre deflections (1-m span) of 20–40 mm. However, at very large deflections exceeding 60–100 mm, the mesh-reinforced shotcrete continues to perform better, while the fibre-reinforced shotcrete loses its load-bearing capacity.

It is important to realise that tests involving distributed loads favour fibre-reinforced shotcrete because distributed-load tests cause uniform loading conditions. When loaded locally (by point loads), strains are localised, and the capacity is significantly compromised. This is illustrated by Figure 67.20, which presents results from a simulation of two loading geometries. For point loads, the peak capacity is roughly 30% lower and the post-peak strength loss is continuous, leading to about 50% capacity at centre deflections exceeding 100–150 mm.

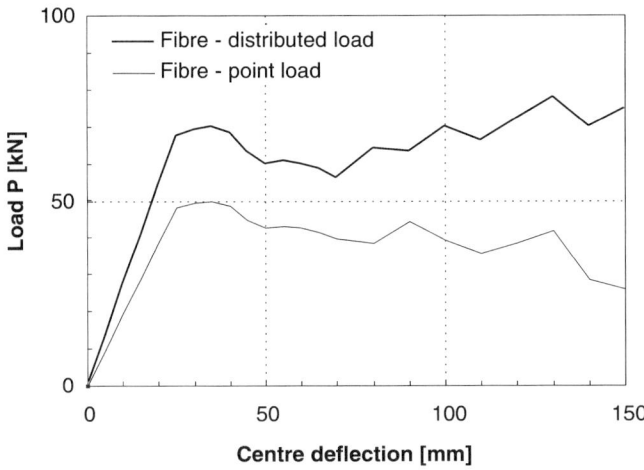

FIGURE 67.20   Point load versus distributed load performance (simulation at low fibre content)

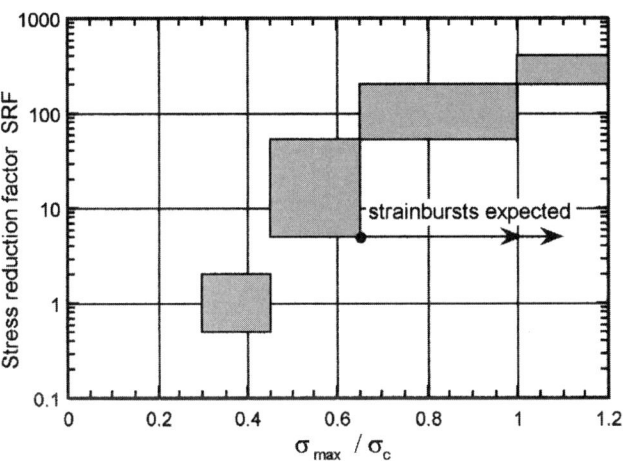

FIGURE 67.22   Recommended stress reduction factors according to Barton (1994) (Kaiser et al. (1996); use 0.75 σ_c for strongly anisotropic in-situ stresses at $5 \leq \sigma_1/\sigma_3 \leq 1$)

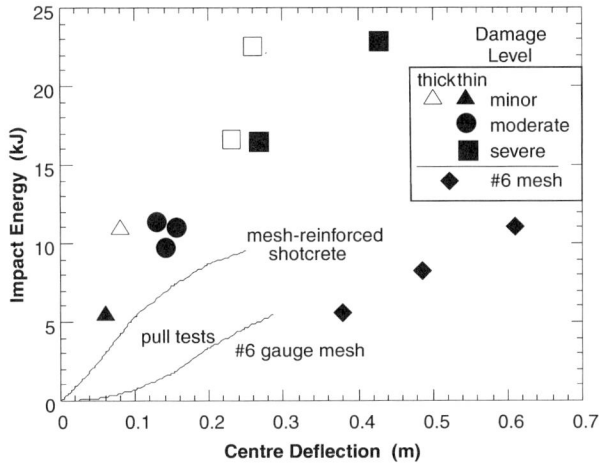

FIGURE 67.21   Kinetic energy at impact versus final deflection from impact tests compared to the energy versus displacement curves from pull tests (Kaiser et al. 1996)

These observations are not just important for test data interpretation but are also of practical relevance. If shotcrete is allowed to debond and sag, such that central, point loading becomes dominant, the performance will move from distributed toward point loading, and the shotcrete capacity is significantly reduced. It is, therefore, paramount that shotcrete, particularly thin layers, be well bolted and, ideally, applied in a slight arch shape such that sag and localised loading is prevented.

67.5.4   Shotcrete Subject to Rock Bursts

When dynamically loaded during a rock burst, incremental stresses may cause further rock fracturing and related bulking, imposing further deformations on the support system. It is, therefore, necessary to ensure that the support system can survive these additional, dynamic deformations. Furthermore, stored strain energy may be released during the fracturing and momentum transfer may impose additional forces.

Kaiser et al. (1996) published the results of an extensive investigation on the energy dissipation capacity of shotcrete, and the reader is referred to this publication for details. Nevertheless, Figure 67.21 summarises the essence of these tests (Tannant and Kaiser 1997). Mesh-reinforced shotcrete can dissipate substantial

amounts of energy long before it is severely damaged. When compared to static-loading tests, it was able to dissipate roughly twice as much energy. Most important, it retained a significant self-supporting capacity even after dissipating more than 10 kJ/m^2 (Tannant et al. 1996). While only a limited amount of testing was conducted on fibre-reinforced shotcrete, its ability to dissipate energy was significantly lower. For this reason, we favour mesh-reinforced shotcrete for support in burst-prone ground of moderate to severe burst potential.

The impact of a rock burst largely depends on the stress level before the event. A stable excavation will be less damaged than one that is already experiencing instability under static loading. To assess the potential impact of a rock burst on the support requirements, a relationship between stress level and stress-reduction factor (SRF) is presented in Figure 67.22 (relate to Figure 67.3 for definition of stress level). It can be seen that strain bursting is expected when the stress level exceeds about 0.6, and the SRF should be increased by about one order of magnitude at this point (more in highly stressed situations when stress levels exceed 1.0).

The requirement for yielding support will increase rapidly as the rock-burst severity increases. This is illustrated in Figure 67.23 by example for a 4-m-wide excavation in very good rock (Q (static) = 40). Under normal, static conditions (at SRF = 1), this excavation would not need to be supported because it falls well under the no-support limit defined by the stability chart. In areas of minor rock-burst severity (SRF = 25), this excavation would demand standard, light support (<50 mm of shotcrete and bolts), mostly to control minor slabbing and spalling. No excessive deformations would be anticipated, and a continuous shotcrete layer would be expected to perform well. Fibre-reinforced shotcrete would be most appropriate. At a stress level of about 0.7 or SFR = 50 and beyond (moderate to major rock-burst severity), yielding support will be required to survive relatively large deformations. At SRF = 100, the deformation demand will be relatively limited and fibre-reinforced shotcrete should still perform well, as long as constructive means are implemented to prevent build-up of hoop stresses. At SRF = 300, however, the deformations will likely be very large, and mesh-reinforced shotcrete would be most appropriate. It is interesting to note that even under most severe burst conditions, a shotcrete thickness of only 100 mm is recommended. This is appropriate because the bolting system will have to control bulking, contribute to the dissipation of the burst energy, and hold the shotcrete in place.

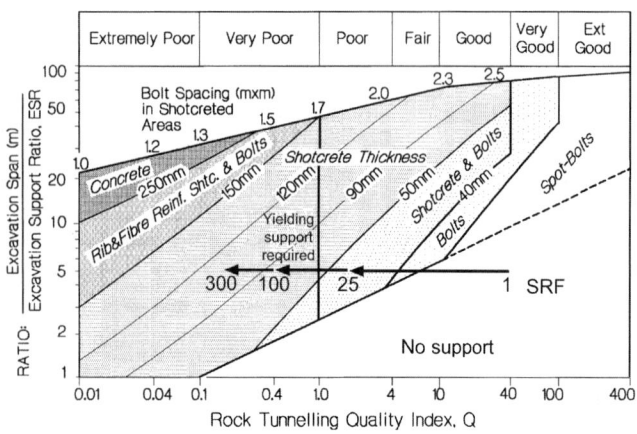

FIGURE 67.23 Example illustrating increasing support requirements for an excavation experiencing progressive higher stress levels and associated rockburst potential. Starting at SRF = 1, a 5 m wide excavation in Q = 40 does not require support. At SRF = 25 for minor burst severity, it can be controlled with standard support. For SRF > 50, or for moderate to major rockburst severity, this excavation needs to be supported with a yielding support system (Kaiser et al. 1996).

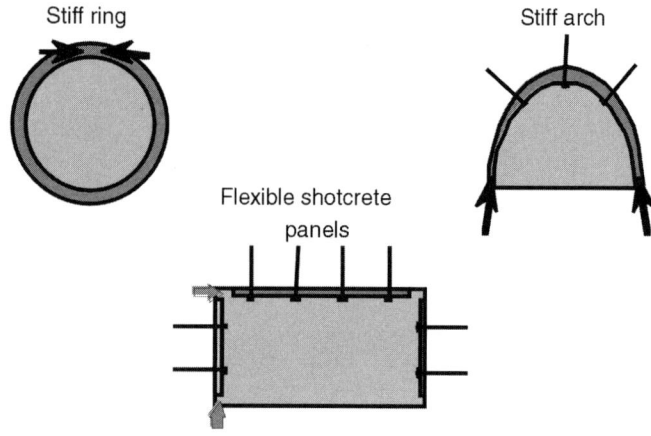

FIGURE 67.24 Shotcrete as a support ring, support arch, or flexible retaining panels

67.6 CONCLUSION

Design approaches for shotcrete used in hard-rock mines differ from those used for civil engineering applications. Underground excavations created during mining typically have flat excavation walls and are often exposed to mining-induced stress changes after the shotcrete is applied. These conditions may cause the rock to fracture (spall) and bulk behind the shotcrete, imposing large deformations or strain. Appropriate shotcrete design for mining requires an understanding of (1) the expected behaviour of the excavation over its design life, (2) likely rock-mass instability modes and the anticipated depth of failure, and (3) how shotcrete works as part of a support system. Shotcrete can perform different support roles depending on how it is reinforced, whether it is applied as flat panels or thick arches, and how it interacts with other support components.

Shotcrete thicknesses exceeding 60–75 mm provide a very effective means for dead-weight loading by loose wedges or volumes of broken rock. However, adhesion loss (often followed by flexural rupture) is a common failure mode for shotcrete. Therefore, it is important to ensure that the rock surface is clean before spraying the shotcrete or that short bolts are used to prevent debonding.

Empirical designs demonstrate the need to increase the shotcrete thickness and to ensure that the rock mass is well reinforced for poorer quality rock masses or as the rock-burst hazard increases. In very poor ground conditions, below Q = 1, it is also important that the shotcrete support system has the ability to yield or deform with the ground. To control brittle rock-mass failure, shotcrete must be applied in such a manner that deformations do not cause detrimental (tangential) stresses that exceed the strength of the shotcrete lining. Shotcrete must be able to deform freely. Hence, shotcrete must not be applied as stiff support arches but as a supermesh, or must retain element deformation capacity to control rock-mass bulking. The three diagrams in Figure 67.24 illustrate this. For example, to enhance deformation compatibility, high stress concentrations in shotcrete—caused by convergence in rectangular-shaped drifts—can be minimised by not spraying shotcrete in corners and by allowing independent movement of large, bolted shotcrete panels on the flat sections of the excavation surface.

Steel-fibre-reinforced shotcrete has good physical properties and works well as long as the deformation gradient is uniform. When shotcrete experiences large deformations (>100 mm) and sudden localised, differential deformations are anticipated, using mesh-reinforced shotcrete is recommended. Steel-fibre-reinforced shotcrete is expected to perform well in conditions of minor to moderate burst severity, but mesh-reinforcement is recommended for areas of major rock-burst potential.

Experience shows that support in highly stressed rock has little effect on the depth of rock fracturing around the excavation. However, using appropriate support components (reinforcements in particular) minimises bulking and wall deformations. In highly stressed rock, it is therefore important to use good reinforcement, such as grouted rebar, to suppress excessive wall displacements. In this manner, shotcrete performance can be greatly enhanced.

67.7 ACKNOWLEDGMENTS

The content of this chapter results from many years of research at the Geomechanics Research Center at Laurentian University, and the authors acknowledge financial support from the Natural Sciences and Engineering Research Council (NSERC), a grant to the Chair for Rock Mechanics and Ground Control at Laurentian University, and funding by the Canadian Mining Research Organisation (CAMIRO) and its many supporting mining companies. Many individuals have contributed at various stages of this work. We particularly wish to acknowledge the contributions of Drs. C.M. Martin, D.R. McCreath, M.S. Diederichs, S. Maloney, and S. Yacici, as well as numerous students and research staff. They all have made valuable contributions to the advancement of our understanding of the role of shotcrete in ground control.

67.8 REFERENCES

Barton, N., R. Lien, and J. Lunde. (1974). Engineering classification of rock masses for the design of tunnel projects. *Journal of Rock Mechanics*, 6(4): pp. 189–236.

Barton, N. (1994). A Q-system case record of cavern design in faulted rock. *Proc. Symp. On Tunnelling in Difficult Conditions*, Turino, 16: pp. 1–14.

Barrett, S.V.L. and D.R. McCreath. (1995). Shotcrete Support Design in Blocky Ground: Towards a Deterministic Approach. *Tunnelling and Underground Space Technology*, 10 (1): pp. 79–89.

Diederichs, M.S. (1999). *Instability of Hard Rock Masses - The Role of Tensile Damage and Relaxation*. Ph.D. Thesis submitted to Department of Civil Engineering, University of Waterloo, Waterloo, Ontario, 566 p.

Grimstad, E. and N. Barton. (1993). Updating of the Q-System for NMT. *Proc. Int. Symp. On Sprayed Concrete–Modern Use of Wet Mix Sprayed Concrete for Underground Support,* Kompen, Opsahl, and Berg, Eds. Norwegian Concrete Association.

Hoek, E., P.K. Kaiser, and W.F. Bawden. (1995). *Support of Underground Excavations in Hard Rock.* Balkema, 215 pages.

Hoek, E. and E.T. Brown. (1980). *Underground Excavations in Rock.* Institute of Mining and Metallurgy, London, UK.

Hutchinson, D.J. and M.S. Diederichs. (1996). *Cable Bolting in Underground Mines.* BiTec Publishers, Ltd.: Richmond, BC, Canada, 406 p.

Kaiser, P.K. (April 1989). *Use of Shotcrete in Highly Deforming Ground.* Seminar on Shotcrete Technology for the Mining Industry, Laurentian University, Sudbury, Ontario, 15 p.

Kaiser, P.K., D.D. Tannant, and D.R. McCreath. (1996). Drift support in burst-prone ground. *Canadian Mining and Metallurgical Bulletin,* 89 (998): pp. 131–138.

Kaiser, P.K., D.D. Tannant, and D.R. McCreath. (1996). *Canadian Rock Burst Support Handbook.* Geomechanics Research Center, Sudbury, Ontario, 385 p.

Kaiser, P.K. and D.D. Tannant. (1997). Use of shotcrete to control rock mass failure. *Proc. Symp. on Rock Support,* Lillehammer, Norway, pp. 580–595.

Kaiser, P.K. and D.D. Tannant. (1999). Lessons learned for deep tunnelling from rock burst experiences in mining. *Proc. Vorerkundung und Prognose der Basistunnels am Gotthard und am Lötschberg,* Balkema, pp. 325–337.

Martin, C.D., P.K. Kaiser, and D.R. McCreath. (1999). Hoek-Brown parameters for predicting the depth of brittle failure around tunnels. *Canadian Geotechnical Journal,* 36 (1): pp. 136–151.

Melbey, T. and K.F. Garshol. (1999). *Sprayed Concrete for Rock Support.* MBT International Construction Group, 7th Edition, Zurich, Switzerland, 229 p.

O'Donnell, J.D.P. and D.D. Tannant. (1998). Field pull tests to measure in-situ capacity of shotcrete. *CIM Annual General Meeting,* Montreal, 8 p.

Rose, D. (1985). Steel-fibre-reinforced shotcrete for tunnel linings: the state of the art. *Proc. 1985 Rapid Excavation and Tunnelling Conference.* Society of Mining Engineers of the American Institute of Mining, Metallurgical, & Petroleum Engineers, Vol. 1, pp. 392–412.

Swan, G., G. Allan, M. Beaudry, and M. Board. (1997). Developments in shotcrete application at Falconbridge Ltd., Sudbury Operations. *Proc. Mine Operators Conference,* Sudbury, Ontario, Canada, 18 p.

Tannant, D.D., D.R. McCreath, and P.K. Kaiser. (1996). Impact tests on shotcrete and implications for design for dynamic loads. *2nd North American Rock Mechanics Symp.,* Montreal, Vol.1: pp. 367–373.

Tannant D.D. and P.K. Kaiser. (1997). Evaluation of shotcrete and mesh behaviour under large imposed deformations. *Proc. Symp. on Rock Support,* Lillehammer, Norway, pp. 782–792.

Bibliography—Papers on Related Topics

Kaiser, P.K, V. Falmagne, F.T. Suorineni, M. Diederichs, and D.D. Tannant. (1997). Incorporation of rock mass relaxation and degradation into an empirical stope design. CIM Annual General Meeting, Rock Mechanics and Strata Control Session, Vancouver, published on CD-ROM, 18 p.

Martin, C.D., D.D. Tannant, S. Yazici, and P.K. Kaiser. (1999). Stress path and instability around mine openings. *9th ISRM Congress on Rock Mechanics,* Paris, Balkema, pp. 311–315.

McCreath, D.R., D.D. Tannant, and C. Langille. (1994). Survivability of shotcrete near blasts. *1st North American Rock Mechanics Symp.,* Austin, pp. 277–284.

Pritchard, C., G. Swan, A. Henderson, D. Tannant, and D. Degville. (1999). TekFlex as a spray-on screen replacement in an underground hard rock mine. CIM Annual General Meeting, Calgary, published on CD-ROM, 6 p.

Tannant, D.D., G. Swan, S. Espley, and C. Graham. (1999). Laboratory test procedures for validating the use of thin sprayed-on liners for mesh replacement. CIM Annual General Meeting, Calgary, published on CD-ROM, 8 p.

Tannant, D.D., S. Espley, and R. Barclay. (1999). Two field tests of Mine-Guard. CIM Annual General Meeting, Calgary, published on CD-ROM, 8 p.

Tannant, D.D., S. Espley, R. Barclay, and M. Diederichs. (1999). Field trials of a thin sprayed-on membrane for drift support. *9th ISRM Congress on Rock Mechanics,* Paris, Balkema, pp. 1471–1474.

Tannant, D.D., P.K. Kaiser, and S. Maloney. (1997). Load-displacement properties of welded-wire, chain-link, and expanded metal mesh. *Proc. Symp. on Rock Support,* Lillehammer, Norway, pp. 651–659.

Tannant, D.D., G.M. McDowell, R.K. Brummer, and D.R. McCreath. (1994). Shotcrete performance during simulated rock bursts. Workshop on Applied Rock Burst Research, IV South American Congress on Rock Mechanics, Santiago, pp. 221–230.

Wood, D.F. and D.D. Tannant. (1993). Optimisation of shotcrete support systems for use in a Canadian gold mine. Shotcrete for Underground Support VI, Niagara-on-the-lake, Engineering Foundation, 12 p.

Wood, D.F. and D.D. Tannant. (1995). Blast damage to steel fibre-reinforced shotcrete. *Fibre Reinforced Concrete—Modern Developments,* UBC Press, Vancouver, pp. 241–250.

APPENDIX A

Shotcrete recommendations in hard-rock underground mining (after Hoek et al. 1995)

TABLE A1.   Summary of recommended shotcrete applications in underground mining, for different rock mass conditions

Rock mass description	Rock mass behaviour	Support requirements	Shotcrete application
Massive metamorphic or igneous rock. Low stress conditions.	No spalling, slabbing or failure.	None.	None.
Massive sedimentary rock. Low stress conditions	Surfaces of some shales, siltstones, or claystones may slake as a result of moisture content change.	Sealing surface to prevent slaking.	Apply 25 mm thickness of plain shotcrete to permanent surfaces as soon as possible after excavation. Repair shotcrete damage due to blasting.
Massive rock with single wide fault or shear zone.	Fault gouge may be weak and erodible and may cause stability problems in adjacent jointed rock.	Provision of support and surface sealing in vicinity of weak fault of shear zone.	Remove weak material to a depth equal to width of fault or shear zone and grout rebar into adjacent sound rock. Weldmesh can be used if required to provide temporary rockfall support. Fill void with plain shotcrete. Extend steel fibre reinforced shotcrete laterally for at least width of gouge zone.
Massive metamorphic or igneous rock. Moderate to high stress conditions.	Surface slabbing, spalling and possible rockburst damage (see also last two classes).	Retention of broken rock and control of rock mass bulking.	Apply 50 mm shotcrete over weldmesh anchored behind bolt faceplates, or apply 50 mm of steel fibre reinforced shotcrete on rock and install rockbolts with faceplates: then apply second 25 mm shotcrete layer. Extend shotcrete and bolt application down sidewalls where required.
Massive sedimentary rock. High stress conditions.	Surface slabbing, spalling and possible squeezing in shales and soft rocks.	Retention of broken rock and control of squeezing.	Apply 75 mm layer of fibre-reinforced shotcrete directly on clean rock. Rockbolts or dowels are also needed for additional support.
Metamorphic or igneous rock with a few widely spaced joints. Low stress conditions.	Potential for wedges or blocks to fall or slide due to gravity loading.	Provision of support in addition to that available from rockbolts or cables.	Apply 50 mm of steel-fibre-reinforced shotcrete to clean rock surfaces on which joint traces are exposed.
Sedimentary rock with a few widely spaced bedding planes and joints. Low stress conditions.	Potential for wedges or blocks to fall or slide due to gravity loading. Bedding plane exposures may deteriorate in time.	Provision of support in addition to that available from rockbolts or cables. Sealing of weak bedding plane exposures.	Apply 50 mm of steel-fibre-reinforced shotcrete on clean rock surface on which discontinuity traces are exposed, with particular attention to bedding plane traces.
Jointed metamorphic or igneous rock. High stress conditions.	Combined structural and stress controlled failures around opening boundary.	Retention of broken rock and control of rock mass dilation.	Apply 75 mm plain shotcrete over weldmesh anchored behind bolt faceplates or apply 75 mm of steel fibre reinforced shotcrete on rock, install rockbolts with faceplates and then apply second 25 mm shotcrete layer. Thicker shotcrete layers may be required at high stress concentrations or, alternatively, intentionally weak zones in shotcrete may be created (slots) and sprayed later after deformations have stabilized.
Bedded and jointed weak sedimentary rock. High stress conditions.	Slabbing, spalling and possibly squeezing.	Control of rock mass failure and squeezing.	Apply 75 mm of steel fibre reinforced shotcrete to clean rock surfaces as soon as possible, install rockbolts, with faceplates, through shotcrete, apply second 75 mm shotcrete layer.
Highly jointed metamorphic or igneous rock. Low stress conditions.	Ravelling of small wedges and blocks defined by intersecting joints.	Prevention of progressive ravelling.	Apply 50 mm of steel fibre reinforced shotcrete on clean rock surface in roof of excavation. Rockbolts or dowels may be needed for additional support for large blocks.
Highly jointed and bedded sedimentary rock. Low stress conditions.	Bed separation in wide span excavations and ravelling of bedding traces in inclined faces.	Control of bed separation and ravelling.	Rockbolts or dowels required to control bed separation. Apply 75 mm of fibre-reinforced shotcrete to bedding plane traces before bolting.
Mild rockburst conditions in massive to moderately jointed rock subjected to high stress conditions.	Localized spalling and slabbing to depth of about 0.25 m.	Retention of broken rock and control of rock mass degradation. Little energy dissipation capacity.	Apply 50 to 75 mm of steel-fibre-reinforced shotcrete or shotcrete over mesh or cable lacing that is firmly attached to the rock surface by means of (mechanical) rockbolts or cablebolts.
Moderate rockburst conditions in massive to moderately jointed rock subjected to high stress conditions.	Localized spalling and rock fracturing to depth of about 0.75 m.	Retention of broken rock and control of rock mass bulking. Moderate energy dissipation capacity many be needed.	Apply 50 to 100 mm of shotcrete over mesh or cable lacing which is firmly attached to the rock surface by means of rockbolts or cablebolts. Use short grouted rebar in combination with mechanical or cable bolts is recommended. Bolting through at least part of shotcrete is preferred.
Severe rockburst conditions in massive to moderately jointed rock subjected to high stress conditions.	Wide spread spalling and rock fracturing to depth of less than 1.5 m.	Retention of broken rock, control of rock mass bulking, and ability to yield. Moderate to high energy dissipation capacity.	Apply 75 to >100 mm of shotcrete over mesh or cable lacing which is firmly attached to the rock surface by means of yielding/ frictional rockbolts or plain cablebolts. Use short grouted rebar in combination to control bulking. Prevent stress raisers by introducing longitudinal slots or thin mesh-protected zones of weak shotcrete. (If shotcrete spalling is encountered, mesh over shotcrete may have to be added.)

CHAPTER 68

Support and Structural Use of Shotcrete in Mines

Richard K. Brummer* and Graham R. Swan†

68.1 INTRODUCTION

Shotcrete usage in mines has increased significantly during the last ten years. In most of these applications, the shotcrete is not used for cosmetic purposes, but is required to have structural strength. Shotcrete walls or bulkheads are used to carry loads from hydraulic fill, and pillars are used to carry loads from stope backs. Shotcrete is also used as a support element for backs and walls in poor ground, and has been evaluated as a replacement for screen in stopes and drifts. Shotcrete has also recently been used as an orepass lining. While design assumptions can be made that allow one to calculate the strength of these structures, actual underground measurements often show that the behaviour is much more complicated than has been assumed. Thus, it is essential to verify the design assumptions by measuring the real behaviour of shotcrete underground.

68.2 HISTORICAL USE OF SHOTCRETE

68.2.1 Canadian Mines

Before the 1990s, Canadian mines had made relatively minor use of shotcrete underground.

INCO's Creighton mine in Sudbury used shotcrete (applied over mesh) to protect the 15-ft failed pillars between the 15-ft by 15-ft drifts of the top and bottom sills of VRM stopes on the 7200 level. Due to destress blasting, the pillars were usually heavily destressed, and their condition was generally poor and extensively fractured. The life expectancy of these excavations is usually six months to one year (three years maximum). The shotcrete was found to minimise corrosion of the mesh and bolts. Although no significant local rock bursts occurred during the trials, VRM blasting—and even secondary blasting at the brow—took place within the shotcreted drifts. It was noted that when similar blasting is done without using shotcrete, the brow often needs rehabilitation, whereas this was not necessary with the shotcreted brows. (The brows were cable-bolted prior to shotcreting.) While there were some doubts about the blast- and burst-resistance of the shotcrete, this experience indicated that shotcrete is able to successfully withstand some degree of local rock bursting or blasting. INCO found it impossible to successfully shotcrete a fresh heading, because a large amount of deformation took place soon after excavation. INCO permitted most of the convergence (and minor dynamic deformation) to occur before shotcreting. The shotcrete also performed well under rock-bursting conditions.

INCO also used shotcrete for passing through sand-fill areas at the 7,000-foot level. The shotcrete was applied directly to the sand fill as soon as it was excavated. Twenty-four hours later, mesh was installed using short split sets, and then a finishing layer of shotcrete was applied.

During the early 1990s, one of INCO's main concerns with shotcrete was that it was difficult to schedule the application and still maintain an acceptable advance rate. Mining bonuses are based on the rate of advance. When shotcrete was required, a maximum advance of 30 feet ahead of the last shotcrete application was allowed. Crews had to wait 24 hours for the shotcrete to set up before being able to re-enter the heading. This caused considerable delays in a single-heading operation. In a multiple-heading operation, this was not a problem because the multiple headings can be sequenced. Accelerators have now considerably shortened this waiting period.

Minnova's Lac Shortt Mine in Quebec (now closed) used blast-hole stoping. Shotcrete was used as support for draw-point brows and in access drifts. The mine used a 50-mm by 50-mm-square welded mesh with the shotcrete. This is smaller than conventional mesh and usually results in rebound problems, although it did not cause any problems at Lac Shortt, where a rebound figure of 20% was reported. Mechanical bolts, split sets and Swellex were used as support units, and, in some cases, steel straps were used between the various bolts.

Some of the shotcreted areas were severely damaged as the breaking front advanced along the length of the access drifts, but, in most cases, the shotcrete performed well in maintaining the integrity of the back. The shotcrete seemed to stiffen the mesh so it could span between support units and even act as a cantilever, supporting severely loosened and fractured ground in the back. This action took place even though the integrity of the shotcrete had been damaged in several places, including the spring line at the sidewall abutment. In this case, the shotcrete clearly did not benefit from any arching action, and yet still performed a useful support function. Extensive seismic activity occurred during the mining at Lac Shortt, and although the recorded events were generally small, the damage was usually confined to the immediate area of the burst site. Using shotcrete in bursting ground at Lac Shortt indicated that the shotcrete appeared to work well in stiffening the mesh and controlling the deformation of the fractured ore.

68.2.2 Shotcrete in Burst-Prone South African Mines

The deep gold mines of the Witwatersrand in South Africa have employed shotcrete in a limited way as a last-resort support element in burst-prone mines. The most common type of support system currently used in burst-prone South African mines consists of a primary support system of rock bolts or grouted shepherd's crooks, together with a permanent support system consisting of grouted shepherd's crooks, mesh (usually "diamond mesh" or chainlink type), and lacing (destranded or unravelled hoist rope or scraper rope, usually 12 mm to 16 mm in diameter).

Shotcrete is used infrequently because it is expensive and is considered by most practitioners to be too brittle to be effective in

---

* Itasca Consulting Canada Inc., Sudbury, Ontario, Canada.
† Falconbridge Limited, Sudbury, Ontario, Canada.

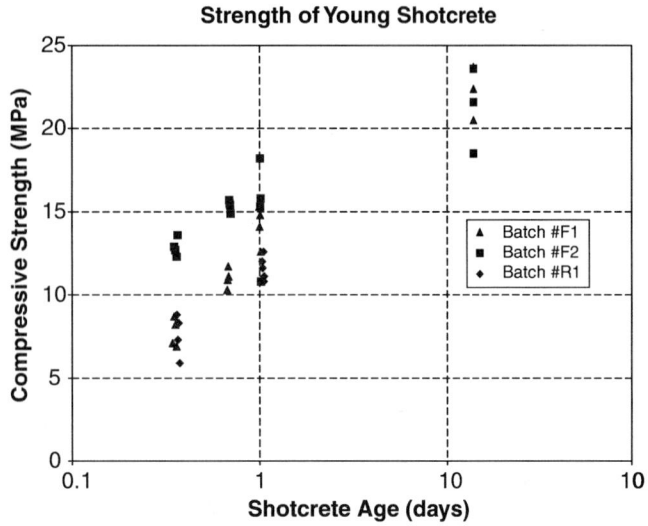

FIGURE 68.1 Strength gain of different blends of shotcrete (Tannant 1994)

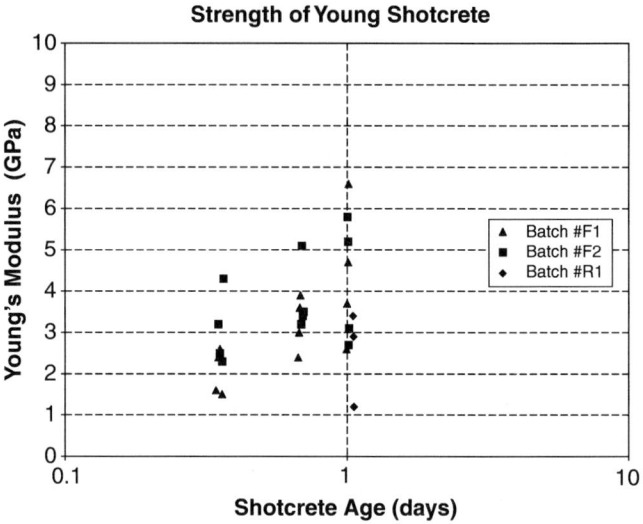

FIGURE 68.2 Modulus gain of different blends of shotcrete (Tannant 1994)

rock-burst conditions. Shotcrete has been found to break up badly in tunnels where large deformations occur (these deformations could amount to as much as 500 mm of radial inward movement). For this reason, shotcrete is virtually never used without some form of mesh reinforcing. Because the mesh reinforces the shotcrete, steel fibres are almost never used under these circumstances.

In spite of this, shotcrete is used as a "last-resort" form of support in circumstances where immediate support is necessary before drilling can be done, or for certain high-stress tunnels, where the shotcrete is applied over weld mesh and cable lacing. This support is rehabilitated if its condition deteriorates. The weld mesh and shotcrete combination appears to work well: the shotcrete stiffens the mesh, and the mesh reinforces and provides ductility to the shotcrete.

68.3 PROPERTIES OF YOUNG SHOTCRETE

One of the difficulties of using shotcrete for drift support is that, in some cases, it is necessary to wait for several hours for the shotcrete to gain strength. Tests were done at Laurentian University in the mid-1990s for local mining companies and shotcrete suppliers to establish the early strength and modulus gain of shotcrete. Some of these results are shown in Figures 68.1 and 68.2.

These figures show that shotcrete gains strength considerably faster than ordinary concrete, and can have most of its long-term strength within a day or so of being shot. During the testing described here, it was also found that the fastest strength gain occurs with the lowest water:cement ratios (i.e., the driest shotcrete).

68.4 SHOTCRETE FOR DRIFT SUPPORT

68.4.1 Structural Behaviour of Shotcrete for Drift Support

Shotcrete forms a "skin" and provides areal coverage to the inner surface of a drift. Reinforced shotcrete (especially mesh-reinforced shotcrete) has very good load-bearing capacity, even at small deformations. It is obviously stronger and tougher and absorbs more energy than mesh, as can be seen in Table 68.1. It may also perform a reinforcing function by preventing the displacement of surface blocks. This prevents the loss of friction between these blocks and stops the loss of cohesion of the rock mass at the surface of the excavation. The shotcrete also protects the mesh from the effects of corrosion. It spreads the load imposed by individual blocks and improves the connection to the

TABLE 68.1 Load-displacement parameters for shotcrete panels (60 mm thick; 1.2-m diamond bolting pattern) (from Canadian Rock Burst Research Programme, CAMIRO, 1990–1995)

Type	Lp (kN)	Dp (mm)	Du (mm)	Ep (kJ/m²)	Eu (kJ/m²)	E* (kJ/m²)
#6 mesh-reinforced	45–55	70–90	100–150	3–5	6–9	6
Steel-fibre-reinforced	20–30	30–40	na	< 1	2–3	1

*Suggested energy absorption values for design purposes.

holding elements. Fibre-reinforced shotcrete can have properties similar to mesh-reinforced shotcrete at small deformations, but it is not as ductile as mesh-reinforced shotcrete. At small deformations, fibre-reinforced shotcrete resists cracking very well, but once cracks have opened more than a few mm, its resistance decreases rapidly.

Mesh- and fibre-reinforced shotcrete seems to be able to survive ground motions of around 1.5–2.0 m/s in lab test rigs and in underground burst simulations, depending on the deformations experienced and the area of the disturbance.

68.4.2 Shotcrete as Replacement for Screen (Falconbridge's Sudbury Mines)

Almost all Ontario mines use bolts and screen for back support. To improve productivity, it is very desirable to have a support system that can be mechanised. Unfortunately, screen is very difficult to mechanise, and there is some interest in developing alternative support systems using shotcrete or other spray-on liners. During 1995, a trial of the effectiveness of shotcrete for stope and drift support was carried out (Swan, Ng, and Brummer 1996). Tests were conducted with different support systems in such a way that differences between "screen and bolt" systems and "shotcrete and bolt" systems could be identified. Expanded bolt patterns (with the shotcrete) were also tried. The deformation of the back was measured with GMMs, which were installed right at the advancing face (see Figure 68.3) and were protected using rubber mats, armoured cable, urethane foam, and polyester roof-bolt resin.

Test results are shown in Figures 68.4 through 68.7, where it can be seen that the back deformations are considerably reduced when shotcrete is used, even where the bolt pattern is less than half the standard bolt density.

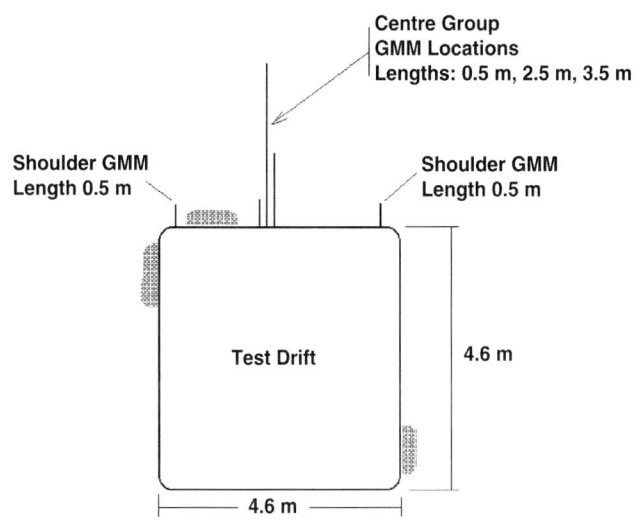

FIGURE 68.3   Layout of GMMs for drift trial

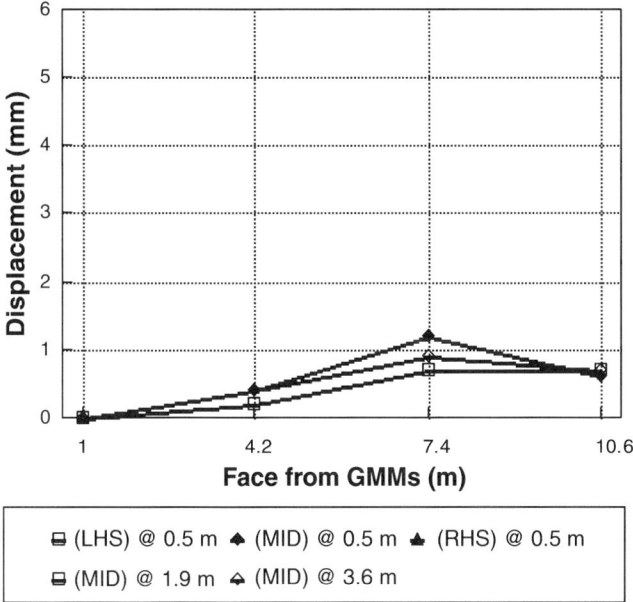

FIGURE 68.4   Drift back deformation, 6′ mechanical bolts and screen, base case

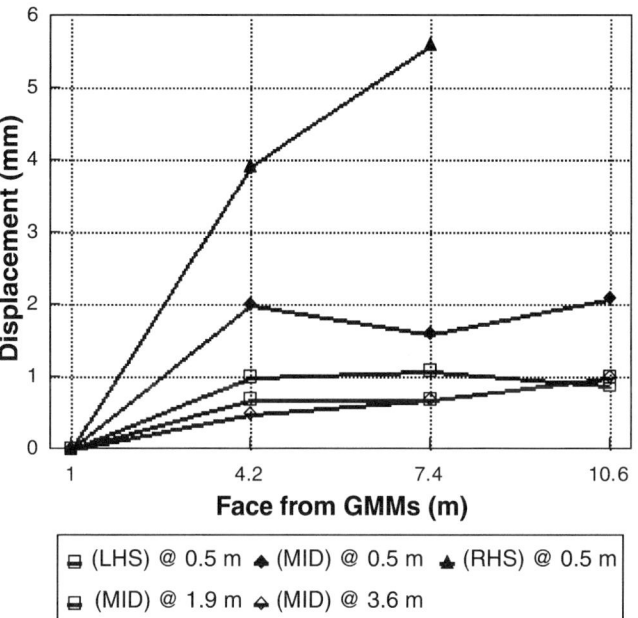

FIGURE 68.5   Drift back deformation, 6′ rebar and screen, 1.1 bolts/m²

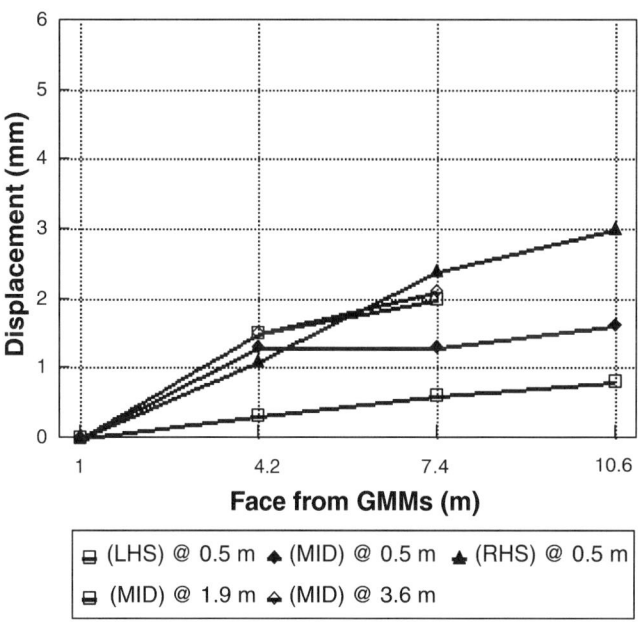

FIGURE 68.6   Drift back of deformation, 6′ mechanical bolts, 1.1 bolts/m² and plain shotcrete, 5 cm average thickness

This action of the shotcrete lining cannot be explained or analysed as an arch, because the back at this location (and in the cut-and-fill stope of the stope trial) was flat. The only mechanisms that would cause this reduced deformation to occur must rely on the shotcrete *adhering* to the intact rock, and using its *tensile* and *shear* strength to limit the deformation. This is in contrast to conventional design assumptions for concrete materials, which ignore any tensile strength and rely only on the compressive strength of the shotcrete. Personnel involved in this trial were also of the opinion that the shotcrete effectively *sealed* the rock surface and prevented blast gases from penetrating and opening up fractures or previously existing joints.

68.4.3   Shotcrete-Only Support (Boltless Shotcrete)

INCO operates the Stobie Mine in Sudbury. The mine uses a sublevel caving method, but has also begun to support drifts using a boltless steel-fibre-reinforced shotcrete system. The mine is also unique in Sudbury in that it uses wet-mix shotcrete that is delivered from the surface through a slick line. The change to a wet-mix system was introduced mainly as a method to reliably incorporate the fibres as well as to suppress the dust. Stobie Mine currently produces about 8,000 tons per day (tpd) of low-grade ore from the sublevel cave operations.

The slick line is a full-column, grouted 6-inch pipe installed in a 9-inch hole drilled from the surface to the 1,800-ft level. According to INCO's figures, this system enables the mine to

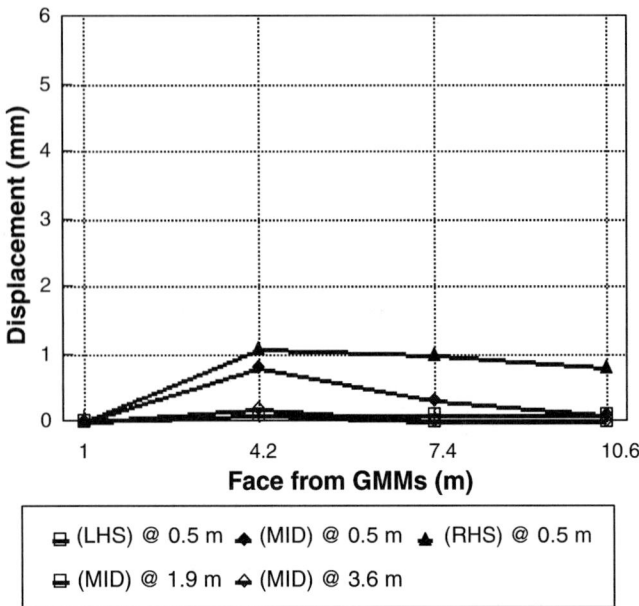

FIGURE 68.7   Drift back deformation, 6′ rebar, 0.44 bolts/m², and plain shotcrete, 5 cm average thickness

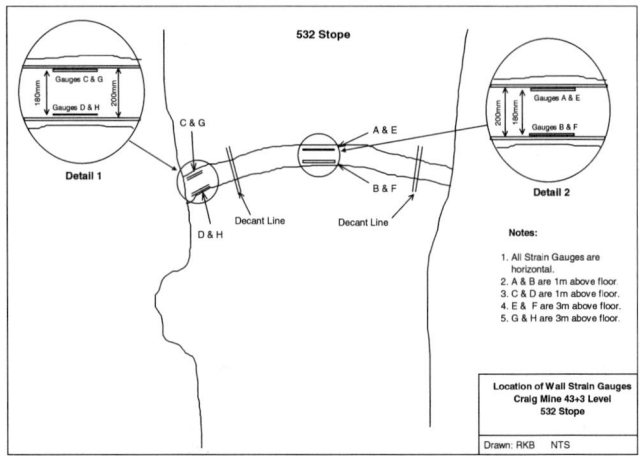

FIGURE 68.8   Typical fill fence instrumentation

provide about 3 m³ of wet-mix shotcrete per minute, with a slick-line velocity of about 3 m/s.

The drifts are 14 ft by 14 ft and are entirely boltless, except for intersections, where 8-ft rebars are installed through the shotcrete as second-pass support. The SFRS is nominally sprayed at a thickness of 4 inches, but tests have shown that in many places only 3 inches is achieved. A liquid accelerator is introduced at the spraying nozzle.

INCO has been very careful to maintain the quality of the shotcrete at the Stobie site. A contractor delivers the actual wet mix to the slick line on the surface in response to a telephone call to the contractor's batching plant a few kilometres (km) from the headframe. Tests are done to maintain the correct consistency (slump) and temperature of each batch, as well as the age of the mix. Batches that fall outside the agreed limits, or are too old, are rejected. This tight quality control was necessary because of some initial problems with adhesion.

The sublevel cave blast holes (rings spaced at 9 ft) are drilled through the shotcrete, so blasting takes place very close to the shotcrete support, resulting in significant dynamic and vibration effects. These dynamic effects, as well as the more gradual mining-induced deformations, were investigated during the proving stage of the project (Epps 1997). However, the SFRS has been very successful and has provided productivity and safety improvements.

68.5   SHOTCRETE FILL FENCES

Falconbridge first used shotcrete fill fences in Sudbury. These are superior to plain or cabled timber-fill barriers mainly because they result in far better sealing of the fence/rock interface and, therefore, minimise leakage and the inevitable cleanup required. Swan et al. (1996) also reported substantial overall cost savings for these fences. However, as part of their development, it was necessary to demonstrate that the fences are capable of with-standing the pressures exerted on them during a fill pour.

A design based on arch theory would place the fence in compression and would result in a relatively economic design. However, this ignores the out-of-plane bending caused by uneven loading during filling, and assumes the walls and floor are stiff. Therefore, a series of fences was instrumented during 1995 and 1996 to measure the pressures exerted during a fill pour, as well

as the behaviour of the shotcrete fence. A typical instrumentation layout is shown in Figure 68.8.

Because of their good long-term performance and robust characteristics, vibrating-wire strain gauges were selected for the strain monitoring in the fill fence. These devices also provide a temperature measurement, which was felt to be desirable for diagnostic purposes during the trial. A Campbell Scientific CR-10 datalogger was selected because it is capable of reading vibrating-wire devices. Because the monitoring trial was short term, communication options were not installed, and storage of the measurements was all done within the memory of the CR-10.

The monitoring trial indicated that the initial hydration of the shotcrete generates significant temperature gains of up to 20°C above ambient. The temperature profiles were also extremely useful diagnostic indicators of activities that took place in the stope. Ambient temperature at the time of datalogger installation was about 25°C. During the spraying of the shotcrete fill fence between December 11 and 12, the sensor temperatures in the bottom of the fence rose dramatically, due to the heat caused by the hydration of the cement in the shotcrete. The gauges in the top of the fence also rose in a similar way, but were delayed by about one day, since the top of the fence was evidently sprayed about 24 hours later than the bottom. Fill placement began on December 13, with the fill reaching the bottom gauges late on December 13, and reaching the top gauges in the after-noon of December 15. The temperatures fell significantly as the fill reaches the gauges. Hydrating the fill produced a gradual increase in temperature of about 3°C–from 20°C on December 14 to about 23°C two weeks later (see Figure 68.9).

The field trials showed that the pressures on the fill fence are low compared to the static head of water or fill which (for design purposes) could be assumed to act on the fence. Each episode of fill introduction results in increasing pressure on the fence, but this pressure usually decreases after about four days, as can be seen in Figure 68.10. In the long term, the pressure exerted on the fence by the fill appears to relax back to zero. Drainage and consolidation appear to play a role in reducing the pressures on the fence, but it appears that some mechanism is at work that allows the fence itself to relax (creep) and shed load. This reduces the pressure in the fill immediately behind the fence and allows the fill to stand in a self-supporting manner. It would appear that in the long term (after a few weeks), relatively little pressure is exerted on the fill fence by the fill, and any designs using static heads are likely to be fairly conservative.

The results of the first trial revealed some interesting aspects in the fence's behaviour (the fence appeared to be acting in a

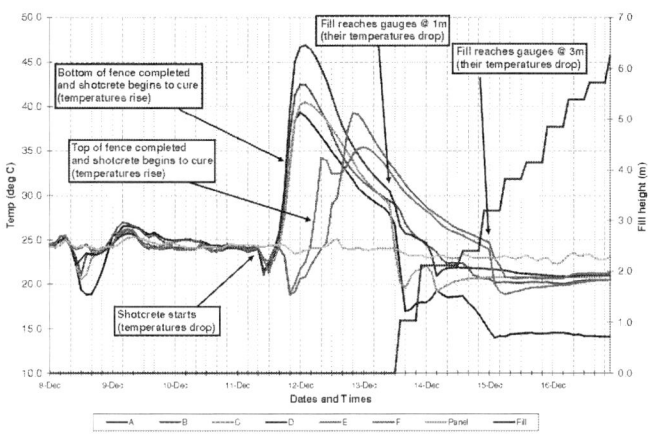

FIGURE 68.9   Temperatures measured in fill fence

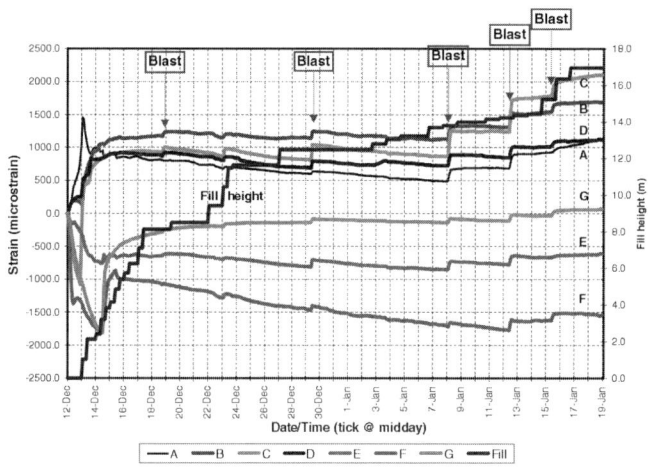

FIGURE 68.11   Strains measured in fill fence

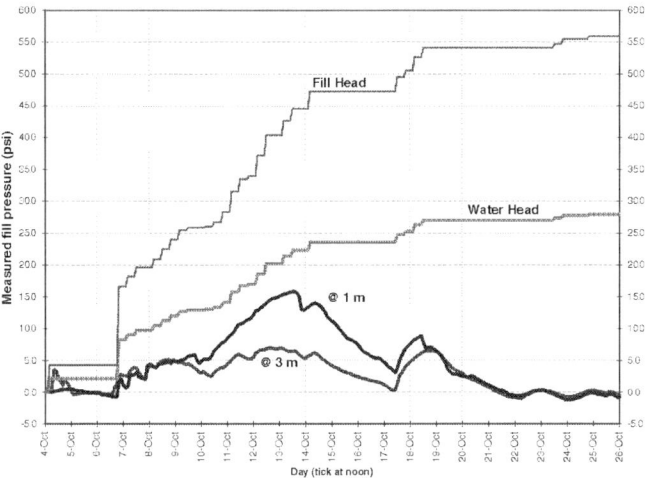

FIGURE 68.10   Pressures measured on fill fence compared to head in stope

both trials, the arch was subject to uneven loading because the fill height increases gradually and allows the lower portion of the fence to be loaded before the upper region. Evidently, this results in significant out-of-plane bending. The trial in 532 Stope also showed that the fence was significantly affected by blasting in the adjacent stope, and all locations experience sudden increases in strain due to these blasts (see Figure 68.11). This effect should be kept in mind when blasts are set near a fence that has just been filled with hydraulic fill.

The conclusion is that the fill fences behave in a complicated manner and are not amenable to simple analyses as are pinned or fixed arches. Out-of-plane bending, plate action, partial yield, flexible abutments, dynamic loading, changing loads as the fill drains, temperature changes in both the fences and the fill, and perhaps even other factors, all contribute to the complex behaviour. It is also noted that, although some yielding in the fences can take place (under bending action), the only mechanism by which complete failure of a fence can occur is an arching action, where the fence fails in compression. Evidently this arching action takes place only after significant bending has occurred.

68.6   SHOTCRETE PILLARS (ALSO CALLED SHOTCRETE POSTS)

Falconbridge exploits massive sulphide orebodies in Sudbury using "post" pillar cut and fill. These "post" pillars are part of the support for cut-and-fill stope backs for most ground conditions (see Figure 68.12). Other types of support are used for difficult ground conditions (e.g., cable bolts), but it is difficult to provide equivalent support to a post pillar. One unit that seems to be a viable alternative is the shotcrete pillar or shotcrete post. These are constructed of shotcrete and founded on solid rock or on a consolidated fill-mucking floor. These pillars can be constructed relatively quickly and easily in areas of poor ground and have proved useful in situations where emergency support must be provided to an unstable back.

68.6.1   Support Mechanism

The load capacity of a shotcrete post depends on:

- The post itself (i.e., the material from which the post is constructed)
- The roof deflection
- The foundation, which provides a bearing support for the post

Where a post is placed directly on a rock footwall, relatively little roof deflection is needed to develop load. However, in

bending manner instead of as an arch) and it was, therefore, decided to repeat the trial by instrumenting a second fill fence in an adjacent stope. This second trial confirmed that the fences are initially acting in a bending manner before they act as arches.

As shown in Figure 68.11, the strains in the fence induced by the first introduction of fill (to a height of about 1.5 m) are such that the lower gauges—A, B, and D—are placed in tension. Gauge C is initially placed in compression, then reverses and behaves similarly to Gauges A, B, and D. All of these locations undergo tensile strains, eventually reaching strains between 1,000 and 2,000 me (microstrains) in mid-January. Gauges F and G, at a height of 3 m at the abutment, reach compressive strains of up to 1,800 me on December 14, then suddenly reverse, as if the behaviour of the fence suddenly changes at this time. This could be caused by yielding, cracking, or some other form of nonlinear behaviour. These gauges subsequently strain incrementally in tension within a time period of a few hours. In the longer term, both of these gauges remain in compression (about 1,500 me and 500 me, respectively).

In most cases, strains in the fence at the measured points were tensile and can be explained only by the action of the fence, not as a compressive arch, but with a significant degree of bending at midspan. The mesh acts like reinforcing steel in an "underreinforced" concrete beam or slab. It is also noted that, in

FIGURE 68.12   Shotcrete pillar in stope

Geokon vibrating wire strain gauges

Example calculation: Converting from gauge readings to strain

Model 4200

Gauge factor is   3.304

Equation: microstrain = ($f^2 \times 10^{-3}$) x Gauge Factor

Note that the GK-401 box gives a reading in ($f^2 \times 10^{-3}$) directly, so this "digits" reading is converted directly to microstrain by multiplying the reading by the Gauge Factor and adjusting for the initial reading.
Note also that no adjustment is made here for temperature or creep effects since elastic behavior is assumed.

Example: Post No 2

Initial readings ($f^2 \times 10^{-3}$)	Date	N	E	S	W
	14-Nov	6969	5529	6361	6418
Reading	5-Dec	6709	5295	6178	5772
Difference (positive = tensile)		-260	-234	-183	-646
microstrain (x GF)		-859.0	-773.1	-604.6	-2134.4

Average strain on post   -1092.8 microstrain

Young's modulus of shotcrete E (estimate)   20 Gigapascals

Average stress in post (strain x E / 1000)   -21.86 Megapascals

Area of post (in square metres)   0.785

Total load carried by post (area x stress)   -17.17 Meganewtons (negative = compression)

Load in "tonnes" (neg = compression)   -1750 metric tons

Load in "tons" (i.e., 2,000 lb tons)   -1587 Imperial (short) tons

FIGURE 68.13   Sample calculation sheet for determining load in shotcrete pillar

places where it is necessary to place a post on a fill floor, the stiffness of the fill plays a key role in determining the capacity of the post, as well as how soon it develops load. To be effective, a post must be installed early, before extensive loosening of the rock mass occurs.

A commonly used design for a 1.0-m-diameter post provides vertical rebars and screen reinforcement, together with accelerated dry shotcrete, to achieve about 27 MPa after 28 days, 14 MPa after 24 hours. This design also uses a reinforced 2.4-m base and should sustain a theoretical peak load of about 2,200 tonnes at 28 days.

68.6.2   Support Modes

Ground support in stopes is usually either for local support (to control problems such as brows or wedges), or for areal support (general roof control of slabby, loosened ground). Interior supports such as cribbing or steel sets are generally installed to provide local support. They work by applying "point" forces to local loosened material, but must be closely spaced to provide effective areal support. Areal support, on the other hand, is placed following a predetermined pattern without any particular foreknowledge of geological structure. Rock bolts, cables, and shotcrete are examples of areal support. Areal support works by densely covering an area with support placed close to the working face prior to extensive loosening of the ground.

An obvious application of shotcrete posts is as local support for identified unstable wedges or blast-damaged brows. These posts can also be installed to support such structures and even shaped to fit openings by locating tight to walls. This allows the open areas to be kept open wide enough to allow vehicle access. Pillar loads as high as 600 tonnes have been recorded in these types of applications.

Another possible use for shotcrete pillars is as a replacement for post pillars in large cut-and-fill stopes, but this has not yet been accepted. Post pillars may be considered active support because they pre-exist the excavation and therefore act to reduce span and prevent a deep loosening zone through a true stress arching action. Because shotcrete pillars are installed after stope excavation, they are obviously not "active" or preloaded. Mine operators are gradually increasing their use of shotcrete posts by trying them in selected areas and monitoring the loads generated. Falconbridge has been using them in cut-and-fill stopes where spans are limited to relatively narrow, single-post situations such as occur occasionally at Strathcona's Deep Copper mine (MacIsaac 1997). Brunswick Mine has also adopted these posts in situations where extra support is needed to hold up backs that could become unstable due to ground falls.

For design purposes, it is therefore necessary to know the load-bearing capabilities of these shotcrete pillars. Simple calculations based on the strength of shotcrete material and the cross-sectional area (adjusted for the amount of reinforcing steel and the slenderness) will indicate the ultimate capacity of the pillar. However, allowance must also be made for the foundation or footing of the pillar, and it must also be noted that such a pillar is *passive* and will not carry load unless the back deforms sufficiently to load it. The load actually carried by such a post is therefore determined by the strength of the shotcrete material, the reinforcing, the age of the pillar, the nearby mining sequence, and the strength of the floor below the pillar.

The load capacity of these pillars, allowing for the internal screen and rebar reinforcement, together with accelerated dry shotcrete, which achieves about 27 MPa after 28 days, should be approximately 2,200 tonnes (Swan et al. 1997). Underground trials have been carried out on Falconbridge mines to measure the load in these pillars and verify that the pillars can, in fact, carry the assumed load. The underground trials used vibrating-wire strain gauges arranged at mid-height of the pillars. Figure 68.13 shows a sample calculation sequence for determining the load in the pillar using the vibrating-wire strain gauges, while Figures 68.14 and 68.15 show sequences of load measurements in several posts installed in stopes in underground situations.

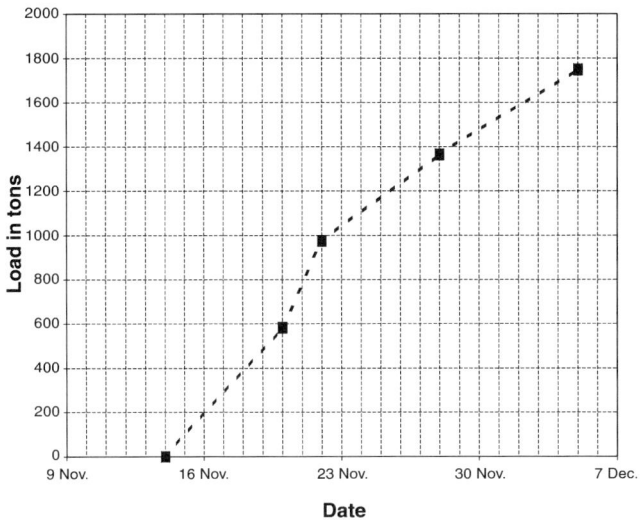

FIGURE 68.14   Load development in shotcrete post over a 21-day period

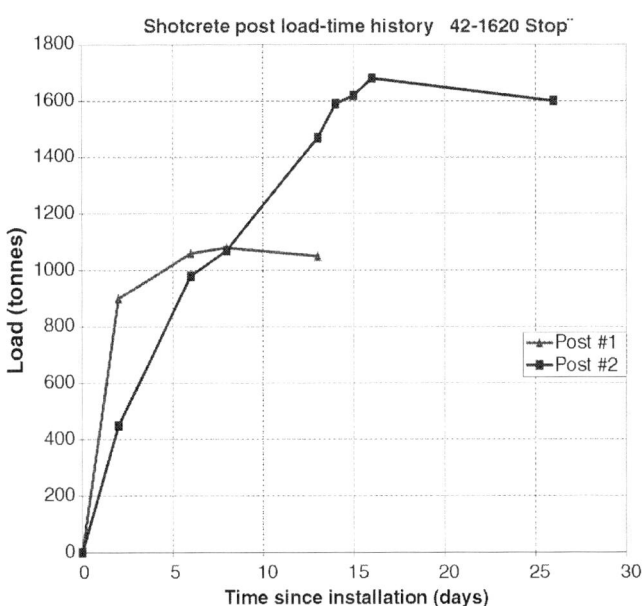

FIGURE 68.15   Load-time history for two posts installed in a stope

These measurements demonstrate that the pillars are capable of carrying substantial loads, and that the design assumptions are realistic. Work is currently in progress to assess the stiffness and bearing capacity of fill floors because, in some cases, these may be the factors that determine the load-bearing capacity of shotcrete pillars in cut-and-fill mining.

The stiffness of various fill floors has been determined by full-scale testing (Beaudry et al. 1998). This shows that it is possible to achieve loads of more than 1,000 tons with relatively small deflections (see Figure 68.16).

68.6.3   General Guidelines for Shotcrete Posts

Based on local experience, as well as the measurements described, general guidelines have been developed for using shotcrete posts on mines. Among other things, special attention should be paid to the following aspects:

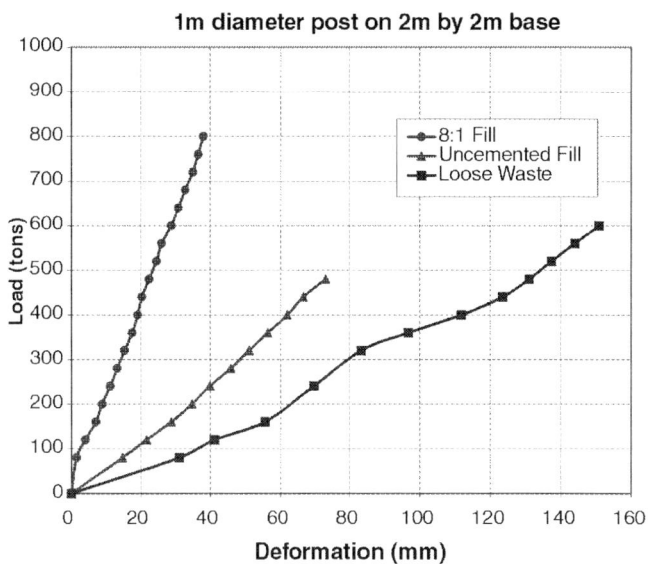

FIGURE 68.16   Shotcrete foundation stiffness (after Beaudry et al. 1998)

- Shotcrete posts placed on broken or loose rock may be ineffective because of the unpredictable nature of some waste rock. If this is unavoidable, the loose can be compacted by driving a scoop over it several times and/or flooding it locally with cement grout.

- Shotcrete posts will perform best if they are placed on the best possible foundation. However, this will depend on the expected roof deformation, and it may be possible to "design" a floor stiffness depending on the roof sag expected.

- Rebound material should not be incorporated into a post because this will weaken it locally and may cause it to deflect prematurely to one side or even buckle.

- The slenderness ratio of shotcrete posts should be limited to about 5:1 or 6:1 (i.e., a 1-m-diameter post should not be made longer than about 6 m). However, the buckling behaviour of posts is particularly sensitive to the way they are constructed. If the post is well made, vertical, founded on a flat floor, and built to a flat roof, it may be possible to make more slender posts.

- Vibrating-wire strain gauges with at least a 150-mm gauge length should be used to measure loads. At least two VW gauges (diametrically opposed) or three gauges at 120 degrees should be used. The gauges must be installed at mid-height.

- It has been found that pre-cast stub columns are efficient at saving transport costs. This means that only the top half of the post is shot in place, together with the expanded foundation. One stub column will typically save four or five bags (tonnes) of dry-mix shotcrete.

- Shotcrete will, in some cases, develop significant temperatures (up to 40° above ambient), and this must be allowed for. This effect is worse in larger posts.

- Shotcrete will also shrink by up to 6% in some cases as it cools, hydrates, or loses water. This effect must also be accounted for when measuring post loads. Fortunately, most vibrating-wire strain gauges also incorporate temperature compensation to account for this effect.

68.7 CONCLUSIONS

Shotcrete is being increasingly used in mines as more experience is gained with the material and equipment needed. Some mines use shotcrete as part of the overall support system, but there are places where shotcrete is the complete support system.

Results from field trials of shotcrete, where used as a support element and as a substitute for screen, show that the back deformations are considerably reduced when shotcrete is used, even where the bolt pattern is less than half the standard bolt density. This action of the shotcrete lining cannot be explained or analysed as an arch, because the backs at the trial sites (drift and stope) were flat. The only mechanisms that would allow this reduction in deformation to occur must rely on the shotcrete *adhering* to the intact rock, and using its *tensile* and *shear* strength to limit the deformation. This is in contrast to conventional design assumptions for concrete materials, which ignore any tensile strength and rely only on the compressive strength of the shotcrete. Personnel involved in this trial were also of the opinion that the shotcrete effectively *sealed* the rock surface, and prevented blast gases from penetrating and opening up fractures or pre-existing joints.

Strains measured in shotcrete fill fences are in many cases tensile, and can only be explained by the action of the fence, not as a compressive arch, but with a significant degree of bending at mid-span. The mesh acts like reinforcing steel in an "under-reinforced" concrete beam or slab. The arch is subject to uneven loading, because the fill height increases gradually and allows the lower portion of the fence to be loaded before the upper region. Evidently, this results in significant out-of-plane bending. Fences may also be significantly affected by blasting in nearby stopes. This effect should be kept in mind when one blasts near a fence that has just been filled with hydraulic fill. It has also been found that the highest pressures are exerted on fences for relatively short periods of time, and that these pressures dissipate within days of stopping a fill pour. In most cases, pressures are well below static hydraulic or slurry heads. The conclusion is that the fill fences behave in a complicated manner and are not amenable to simple analyses as pinned or fixed arches. Out-of-plane bending, plate action, partial yield, flexible abutments, dynamic loading, changing loads as the fill drains, temperature changes in both the fences and the fill, and perhaps other factors, all contribute to this complex behaviour. It is also noted that, although some yielding in the fences can take place (under bending action), the only mechanism by which complete failure of a fence can occur is by means of an arching action, where the fences fail in compression. This arching action can evidently take place only after significant bending has occurred in a fence. To date, hundreds of these fences have been constructed without mishap.

Shotcrete posts appear to be capable of carrying substantial compressive loads, and much work has been carried out to assess the stiffness and bearing capacity of fill floors. Measured pillar loads come close to theoretical pillar strengths, thus demonstrating that the pillars are strong enough to withstand these loads, and also that sufficient back deformation can occur to load the pillars in cut-and-fill situations. Shotcrete posts will perform best if they are placed on the best possible foundation, but to some degree the floor can be compacted by driving a scoop over it several times and/or flooding it locally with cement grout. It may be possible to design a shotcrete post that accounts for the expected floor stiffness and roof sag. The slenderness ratio of shotcrete posts should be limited to about 5:1 or 6:1, but the buckling behaviour of posts is particularly sensitive to the way they are constructed. Vibrating-wire strain gauges can be used to measure loads.

Pre-cast stub columns are efficient at saving transport costs, and one stub column will typically save four or five bags (tonnes) of dry-mix shotcrete.

In some cases, shotcrete will develop significant temperatures, and this must be kept in mind. It is even possible that the temperature gain can be used as a diagnostic measure of how successfully the cement was hydrated. Shrinkage of up to 6% can also occur in extreme cases, and this effect must also be allowed for in the design.

68.8 REFERENCES

Beaudry, M., G. Swan, and M. Mooney. (1998). Properties of shotcrete post foundations. Ground Control Study August–October 1998.

Epps, A. (1997). Boltless development at Stobie Mine. Sudbury Mine Operators Conference–"Back To Basics–Practical Uses Of Technology," February 16–19, 1997.

MacIsaac, H.S. (1997). Strathcona Mine–Deep Copper. Sudbury Mine Operators Conference–"Back To Basics–Practical Uses Of Technology," February 16–19, 1997.

Swan, G.R., G. Allan, M. Beaudry, and M. Board. (1997). Developments in Shotcrete Applications at Falconbridge Limited, Sudbury Operations. Sudbury Mine Operators Conference–"Back To Basics–Practical Uses Of Technology," February 16–19, 1997.

Swan, G.R., L. Ng, and R.K. Brummer. (1996). Shotcrete as a primary support system for cut-and-fill mining. *NARMS '96—Second North American Rock Mechanics Symposium,* Montreal, June 19–21 1996.

Tannant, D.D. (1994). Personal communication.

CHAPTER 69

Backfill in Underground Mining

Dave Landriault*

69.1 INTRODUCTION

Backfill can fulfill several roles at an underground mine site. It can be used as:

- A construction material
- A major ground-support tool
- A primary mine-waste-disposal method

The materials used for backfill at most underground mines generally consist of mill tailings or waste rock from underground or open-pit mining. At sites where these materials are not available in sufficient quantities to meet the mining-method requirements, alternative sources must be found. Alternative sources such as alluvial sands, quarried rock, or air-cooled smelter slag are used to replace or supplement the mine waste as underground backfill.

As an "underground construction material," mine backfill is used as:

- A floor to mine from
- A wall to mine next to
- A roof to mine under

In most mines, a binder (cement, slag cement, and/or fly ash) must be mixed with the backfill material to give it the strength required for the above construction purposes.

Backfill is the "major ground-support tool" in most underground mines. When placed in an excavation or open stope, backfill provides confinement for the walls and back (when tight filled). This confinement significantly improves the overall stability of the rock mass by limiting closure and preventing unraveling. In most mines, long-term ore removal would be impossible without using backfill to support the rock mass.

Mine backfill, placed as an underground construction material, most often plays a dual role and, consequently, contains a binder. However, if a backfill is being placed only for ground support (i.e., secondary or bulk-mining slopes), a binder is not added. Under certain circumstances, a small amount of binder is occasionally added to paste backfill to guard against liquefaction.

Today, the mining industry uses mine-waste materials for backfill as its primary mine-waste-disposal method. Using this method has become a priority because of the increasing perception among the general public that disposing of mine wastes (tailings, slurry containment, cyanide/arsenic ground water contamination, and acid generating from tailings/waste rock) on the surface is physically and environmentally unstable. The small percentage of surface mine-waste-storage facilities that have failed in recent years have had a substantial negative economic impact on the mining companies involved and on the mining industry in general. As a result of public and political pressure, regulations for surface mine-waste storage have

become more onerous, and the potential liability associated with storage-system failure has increased substantially. These factors have caused corporate management and project financiers to push for reducing the storage of mine waste on the surface and to insist on more conservative surface-disposal designs that are more cost effective in the long term

This has led to a greater emphasis in the mining industry on maximizing the return of the waste to the underground work place in the form of backfill. This emphasis has added to the increasing acceptance of the paste-backfill method as an economic alternative to hydraulic slurry and rock backfill methods. The paste-backfill method, unlike hydraulic slurry placement, can use mill tailings without fine particle removal, maximizing the amount of tailings that can be used regardless of the size distribution.

Economic and environmental issues have also increased the acceptance of alternative surface storage methods for mine waste. Dry (filter cake) and semi-dry (thickened) tailings storage and co-disposal with mine-waste rock are gaining prominence as safe and economical waste disposal methods. The increasing economic importance attached to waste disposal is slowly transforming the focus of the mining industry as it evaluates future ventures. Historically, this focus has been towards ore production, but economic and environmental pressures are changing the focus towards long-term waste management.

69.2 TERMINOLOGY DEFINITIONS

Before proceeding with a detailed discussion about mine backfill, it is important to first define the terminology typically used by the mining industry.

Porosity	Volume voids/(volume voids + volume solids)
Wt% solids	Wt. solids/(wt. solids + wt. water) = pulp density
Wt% moisture	Wt. water/(wt. water + wt. solids)
Wt% binder	Wt. binder/(wt. binder + dry wt. backfill)
Wt% passing	Wt. passing given size/total wt. of dry sample
Slump	Slump is a measure of paste consistency. It is the number of inches a paste will slump under its own weight using the standard North American 12-inch-high concrete industry cone slump test (according to CSA standard A23.2-5C or ASTM standard C143-90A).

69.3 HISTORY OF UNDERGROUND MINE BACKFILL

This history of backfill describes the developments as they occurred in North America. While many of the innovations are described from a North American perspective, there is no doubt that similar developments occurred in other hard-rock mining communities throughout the world (Australia, Europe, etc).

In early mining years, typical mining methods required no backfill. The most common methods used in subvertical

---

* Golder Associates, Sudbury, Ontario, Canada.

orebodies were open stoping and shrinkage stoping. In sub-horizontal hard-rock mines, the room-and-pillar method was the most common, followed by square-set stoping. In coal mines, seams were essentially self-filling, often resulting in surface subsidence. In the case of open stoping of subvertical orebodies, stopes were also often self-filling, with the wall rocks collapsing into the stope while the ore was drawn out below the barren material. When the waste rock began to appear in the material drawn out of the stope, this section would be closed down and mining would progress along strike and downwards to the next horizon. This mining method often resulted in noticeable subsidence on the surface.

In the case of shrinkage, stoping the wall rocks was generally stronger, and the stopes would remain open after the ore was removed. Once the stope was empty, mining would progress along strike and down to the next horizon, leaving a small pillar between the new stope and the empty one. As the mines became deeper and more extensive along strike, unexpected collapses became increasingly common, and the only solution was to abandon the area surrounding the caving stope and begin mining elsewhere. As the caving areas became larger and more frequent, it became obvious that it would be more cost-effective to fill the shrinkage stopes with waste material of some kind rather than abandon valuable ore resources.

A square-set stoping method was commonly used in shallow-dipping orebodies with wider, more extensive areas of roof or back requiring support. In this method, the roof or back of the stope was supported by timber-sets or timber cribs. Unconsolidated material was introduced into the stopes to confine the timbering and provide a floor for the men to work from. While the timbering provided some support to the roof of the stope, it did not always prevent it from collapsing. However, because most of the stope was filled, the collapse was limited; and it was possible to continue mining, either right on top of the collapsed area or very close to it.

Initially, unconsolidated material such as development-waste rock or surface sand and gravel was supplied through raises and mine cars to the stopes. It was fed into the stopes through raises and then was moved into place manually. This backfill method greatly increased labor requirements, backfill cycle time, and ore recovery

Cost savings and increased revenues resulted in advances in backfill methods. Because cycle time and labor are the main cost factors for backfill, reducing these factors resulted in more advanced systems. Similarly, recovery of ore resulted in backfill advances through increased revenues.

69.3.1 Development of Modern Backfill Systems

One of the first innovations in consolidated backfill technology took place in Canada in 1933. At Noranda's Horne Mine, granulated furnace slag was mixed with pyrrhotite tailings to form a backfill material very similar to today's cemented tailings. The oxidation of the pyrrhotite consolidated the tailings, and it was found that drifts could be driven through the material without support, and stopes could be mined up against it with very little dilution. However, this kind of mixture is very sensitive to the material used, and many other mines were unable to repeat Noranda's success.

The next major step forward in underground backfill took place in the late 1940s when a hydraulic slurry system was introduced into North American hard-rock mines. In this case, the system supplied classified tailings (fine size fraction removed) to square-set stoping areas. Because the volume of fill height was small, the risk of the fill becoming mobilized through liquefaction was minimal, although the rate of drainage was slow by today's standards. However, even allowing for the slow rate of drainage through material with a relatively high fines content, the speed at which stopes could be filled was much greater than the rate of supplying rock fill. Another advantage was that the fill could be directed to different areas of the stope relatively easily by setting up a series of pipes to direct the flow of material.

Mining up against this unconsolidated material remained a problem and became particularly difficult when mining the pillars between old filled stopes. With the introduction of rock bolts, cut-and-fill mining methods largely replaced the square-set stoping method. Stopes became much more open and allowed larger equipment (such as slushers) to be introduced. However, timber was still used to form gob fences or slats on the walls of the stope to restrain the fill when mining the adjacent pillar stopes. In the early 1960s, mines began using an undercut-and-fill method to recover longitudinal stopes, although working under the unconsolidated backfill material required a great deal of timbering.

Around the same time, several mines began tests to demonstrate that a small amount of cement could improve the performance of backfill, and that tailings should be deslimed to achieve a minimum percolation rate of 10 cm/hr. Thereafter, cemented tailings with about 3%–4% cement were used to consolidate fill. In cut-and-fill stopes, a 10% cement layer was in the mucking floor to improve the recovery of the fine (often high-grade) ore material, which became the standard in the cut-and-fill stoping areas. A similar high-cement-content layer was used in underhand cut-and-fill stopes to provide a more stable roof and reduce the amount of overhead timber support. Even very low cement content eliminated the need for gob fences to control fill walls and made it possible for classified tailings fill walls to be self-supporting up to heights of 160–200 ft (50–60 m).

The next major innovation was mechanizing cut-and-fill with jumbo drills and scooptrams, which began to replace jackleg and slusher techniques from the late 1960s. This did not affect backfill, however, as it was essential to have a good mucking floor to support the heavy equipment.

In the 1970s, bulk mining techniques were being introduced. These techniques took advantage of the ability of layered fill to be self-supporting to a height of 200 ft (60 m). This allowed the mining of secondary pillars between primary stopes with little dilution. Since then, primary and secondary mining with bulk techniques has largely replaced selective methods such as cut-and-fill and undercut-and-fill in sub-vertical base-metal mining operations.

The only significant changes to the backfill system in these bulk operations was in the design of stable freestanding backfill heights and in the operational control over the filling process itself. Operational controls include conservative bulkhead designs, drainage system designs, and stope sequencing to allow for percolation and backfill set times. High-grade, shallow-dipping base-metal mines continue to use selective methods profitably, using highly mechanized operations, which take advantage of the greater flexibility and selectivity that these methods offer.

The bulk-mining operations also allowed for supplying various forms of rock fill in combination with cemented tailings and cement slurries without the need to physically rehandle the rock material in the stope. In the 1960s, this technique was introduced at several mines where rock fill was supplied into the stope on top of the ore. As the ore was extracted, the stope filled up with waste. Once the stope was completely filled with rock fill, cement slurry was added to produce a cemented rock fill that was self-supporting and did not result in significant dilution. Experience indicated that the final product performed better if the slurry was added at the same time as the rock fill. These findings led to the development of the cemented rock-fill methods commonly used in many mines today (Kidd Creek in Canada and Mt. Isa in Australia).

The next innovation was replacing the cement binder in cemented tailings fill with cheaper alternatives. The initial trials involved replacing some of the cement with finely ground iron furnace slag and fly ash, as well as a few others. The slag was cementitious while the fly ash had a pozzolanic effect, but both resulted in low early strength with a higher strength over the long term. Both types of material became common components of cemented tailings slurry and rock-fill systems in moderate-depth mines.

In the early 1980s, the development of paste backfill began with the first active system being commissioned in Germany at Preiessage's Bad Grund Mine. Paste backfill allows full plant tailings with a high solids content to be transported at a paste consistency (7 in. [180 mm] to 10 in. [250 mm] slump) to underground workings. Paste backfill is transported through a borehole/pipeline underground distribution system similar to what is used to transport hydraulic classified tailings backfill. The higher viscosity paste material, as compared to hydraulic backfill, produces a greater resistance to flow and consequently higher pipeline pressures. However, paste backfill has lower cement consumption because it eliminates cement segregation upon placement and because it has a lower water:cement ratio.

While paste backfill requires less cement to produce a given backfill strength, it cannot be placed underground in an uncemented form like classified tailings or rock fill. The high fine-material content (15 wt% passing 800 mesh or 20 microns) required to produce a colloidal water retention necessary for paste transport also makes uncemented paste backfill susceptible to liquefaction. Even with the need for low cement contents in secondary and tertiary stopes, paste backfill generally produced a 30% to 60% reduction in cement consumption over alternative backfill methods.

Developing paste backfill also allowed mines with high fines-content tailings to use their tailings as the primary or sole backfill material. This often resulted in a significant reduction in the use of costly alternative materials (alluvial sand or rock) as an underground fill.

The paste backfill method gained increasing acceptance in the 1990s as a cost-effective alternative backfill method to hydraulic slurry and rock fill and as a method of maximizing mine-waste placement underground. At the present time, there are 23 operating paste backfill plants throughout the world, and several others are under construction. Paste backfill has allowed many operations with the appropriate conditions to reduce fill cost and maximize tailings disposal underground.

The most important innovations in underground mine backfill are listed below:

Prior to 1930s	– Unconsolidated rock fill
1930s	– First consolidated backfill
1940s	– Hydraulic slurry backfill systems
1950s	– Addition of cement to hydraulic slurry backfill
1960s	– Addition of cement to rock fill
1970s	– Bulk stoping – free standing height
	– bulkhead design
	– saturation control
	– alternative binders – iron blast furnace slag
	– fly ash
1980s	Research and development of paste backfill
1990s	Implementation of paste backfill

69.4 UNDERGROUND MINE BACKFILL METHODS

As previously discussed, not all mines or mining methods require backfill. Selecting a mining method for a given mine is influenced by the geometry and grade of the orebody and the stability and properties of the rock mass. The potential requirement of backfill

TABLE 69.1 Rock fill versus slurry fill versus paste fill

Properties	Rock fill	Slurry fill	Paste fill
Placement State	Dry	60–73 wt% solids	65–85 wt% solids
Transport System	Raise, mobile equip., separate cement system	Borehole/pipeline via gravity	Borehole/pipeline via gravity
Cemented vs. Uncemented	Cemented or uncemented	Cemented or uncemented	Cemented only
Water:Cement (w:c) Ratio	Low w:c ratio, high binder strength	High w:c ratio, very low binder strength	Low to high w:c ratio, low to high binder strength
Placement Rate	100 to 400 tons/hour	100 to 200 tons/hour	50 to 200 tons/hour
Segregation	Stockpile segregation, reduced strength and stiffness	Slurry settlement segregation, low strengths	No segregation
Stiffness	High stiffness if correctly placed	Low stiffness	Low or high stiffness
Tight Filling	Hard to tight fill	Cannot tight fill	Easy to tight fill

is inherent in any mining method. The type of backfill used by an operation will be dependent on several factors:

- The configuration of the mining process
- The stope sequences and excavation sizes determined by the mining method
- The depth and the orientation of the orebody
- The materials available to use as a backfill, focusing on the mine waste management requirements over the life of the orebody.

The three most important types of modern backfill used in hard-rock mining are rock fill, slurry fill, and paste fill. The choice between these three types is site specific and will depend on the particular requirements of each mining operation. The three types of backfill have different characteristics, which are presented in Table 69.1.

The different backfill systems have different capital and operating costs attached to them. However, these costs must be considered in the context of the cost structure of the particular mining operation. The different characteristics of the three types of fill can be described as advantages and disadvantages in terms of the configuration of the particular mining operation. It is these advantages and disadvantages, along with the mine's overall waste management requirements, that determine the most suitable type of backfill for that mine, rather than the generic operating cost of the system.

69.4.1 Hydraulic Slurry Backfill

The hydraulic-slurry-backfill placement method was developed in the 1940s and has evolved over time to be the most widely used backfill method in the mining industry. It consists of mixing an appropriate-sized granular material with water on the surface to produce a slurry that can be transported underground through a borehole/pipeline distribution system.

The size distribution of hydraulic-slurry placed backfill is governed by two parameters. First is the maximum size and volume of coarse particles that can be transported above the critical flow velocity in a given distribution system; and second, the maximum size and volume of fine particles that will allow an adequate water percolation or drainage rate to maintain the scheduled mining cycle (Thomas, Nantel, and Notley 1978).

Figure 69.1 shows the wt% finer size distribution range of granular materials that are suggested for use with a hydraulic-slurry backfill placement method. Particles above the mesh size

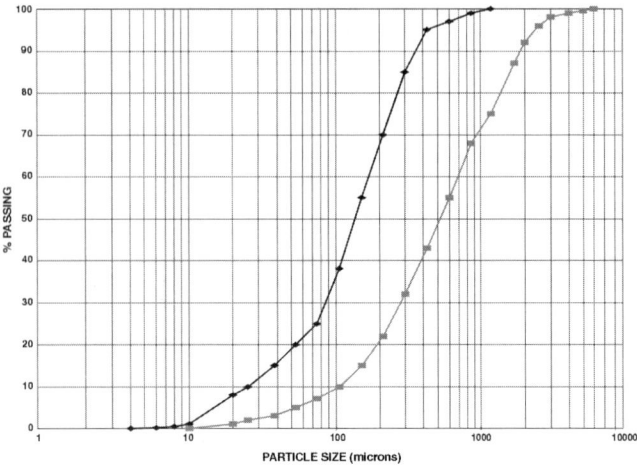

FIGURE 69.1 Size distribution range for hydraulic-slurry backfill materials

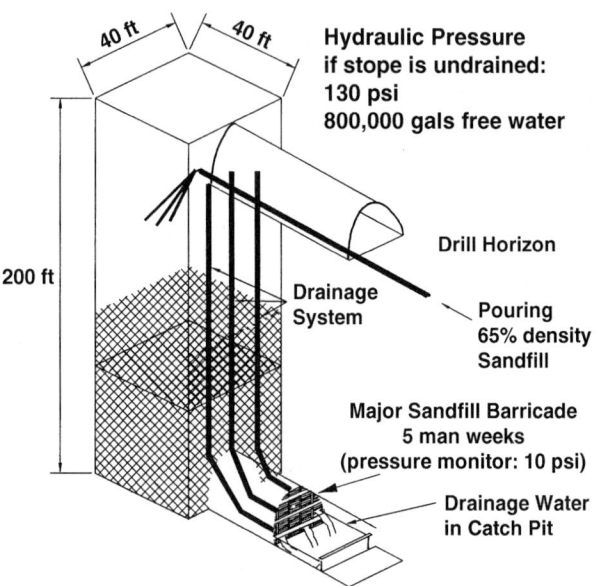

FIGURE 69.2 Hydraulic slurry backfill, auxiliary drainage

are difficult to keep in suspension in slurry, even when transported above the normal critical flow velocity. The coarse particles also increase the dynamic pipe wear exponentially. Decades of operating experience in mines worldwide have led to this recommended restriction on coarse particle size.

The fine particle fraction, as indicated earlier, dictates the rate at which water will percolate through the backfill mass allowing the removal of the excess slurry water that was used to carry the backfill solids to the stope. A binder-free backfill percolation rate of no less than 10 in./hr (2.5 cm/hr) is recommended to allow the water to drain through the fill mass at a sufficient rate to maintain production in most mines. Lower percolation rates will result in water pooling on top of the fill for several shifts after placement has been suspended.

It is common practice to suspend slurry backfill placement in a stope if the water level on top of the fill is greater than 4 ft (1.3 m).

Greater water levels will segregate the granular backfill as gravity settles through the pooled water to the solids/liquid contact in the stope. This is extremely detrimental if a binder has been added to the slurry backfill because bands of concentrated fine binder particles will be formed throughout the fill mass, with other sections being left with little binder content and little strength.

A hydraulic head of pooled water can cause large sections of the fill mass to be totally saturated as the wave of water slowly flows through the granular mass to a drainage location below. This condition could cause an unconsolidated (no binder) fill mass, especially when the fine content is high, to liquefy under dynamic loading from blasting or rock-burst vibrations. A consolidated fill mass, because of the increased cohesion that results from adding the binder, will not liquefy even if it is saturated with water.

Most hydraulic-slurry placed backfill materials (no binder) are sized to have no greater than 8 wt% passing the 20-micron (800-mesh) size range. Higher fine-material content generally reduces the water percolation rate below the accepted 4 in./hr (10 cm/hr). If the fine-material content is kept below this level—allowing a percolation rate or higher to be maintained—placing the backfill into underground stopes without adding binder (unconsolidated) is acceptable. Laboratory testing to confirm this percolation prior to production use of the backfill material is essential.

Unconsolidated hydraulic-slurry backfill is placed in stopes that will not be mined next to, or under. Some mines do use unconsolidated slurry backfill as a mucking floor in low-grade

stopes where broken ore is left on the backfill floor to prevent fill dilution during ore removal.

In many mines, especially when a binder is added to the slurry backfill, auxiliary drainage facilities (perforated piping) are installed in the stopes to accelerate the removal of free water (Figure 69.2). While these systems greatly increase water removal rates from the stopes, care must be taken to ensure that excessive amounts of fine solids from the backfill are not removed with the water. This is extremely important when a binder has been added to the backfill, which generally consists primarily of fine particles (>20 microns). Most mines cover the drainage pipes with a fine mesh sock to prevent the fines from entering the drainage system, allowing only water to be removed from the stope.

The underground borehole/pipeline distribution system of a hydraulic-slurry backfill is generally designed to allow a transport flow velocity above 5.0 ft/s (1.5 m/s) or higher to be maintained. This is typically the critical flow velocity between laminar and turbulent flow transport for most granular material suitable for hydraulic-slurry-backfill placement. Below this flow velocity, solids settle out of slurry suspension and eventually plug the borehole/pipeline transport system.

Transporting hydraulic-slurry backfills typically uses the hydraulic head generated by gravity to provide the energy required in the given diameter distribution system to move the slurry above the critical flow velocity. In some shallow layering or laterally extensive orebodies, the vertical head is insufficient to provide adequate energy. Centrifugal pumps (in the form of booster pumping stations) are generally used to add energy to the system on the surface or underground.

Designing all underground hydraulic-slurry transport systems must be conducted with a sound understanding of the specific slurry-transport properties. The friction-loss flow properties, potential transport-system wear problems, and energy versus borehole/pipeline diameter versus flow velocity balance must be defined to ensure trouble-free transport to all the work places throughout the mine. A design error can result in a hydraulic-slurry backfill system that is plagued with transport problems and can result in major production losses and eventual destabilization of the entire rock mass.

Mill Tailings as a Hydraulic Slurry Backfill. The majority of hydraulic-slurry backfill placed in underground mines throughout the world utilizes classified mill tailings as the fill material. All mill tailings have too high a fine-material content to satisfy the recommended percolation rate for hydraulic backfill. Consequently, the tailings are classified through cyclones to remove the fine fraction. In many cases, however, the tailings have such a high fine-material content that, after removing the fine fraction to produce a fill material with only 8 wt% of its particles finer than 20 microns, there is an insufficient volume of tailings to meet the underground demand for backfill. In these situations, a mine will implement an alternative backfill method or supplement the lack of available coarse mill tailings with an alluvial sand source or turn totally to alluvial sand as the hydraulic-slurry backfill material.

When mill tailings are used as the backfill, the material must be carefully analyzed for chemical and mineral composition. There are several minerals (zinc, lead, and some pyrites) that can affect the binder reaction, resulting in strength retardation, reduction, and long-term deterioration. Laboratory short- and long-term binder-strength testing is required before the tailings can be used underground as a production backfill.

Consideration must also be given to the health and safety of underground workers when a potential tailing material is analyzed for use as a backfill. Many tailings contain health-hazardous contaminants (i.e., cyanide, arsenic, etc.) from the milling process. There are health and safety standards in most countries for underground backfill, and for underground contaminants in general, that must be reviewed to determine if they will be exceeded by using a given tailings material as an underground backfill.

Another concern with pyritic materials is the exothermic potential of certain tailings. Some forms of pyrrhotite and pyrite can chemically react under the proper underground moisture and oxygen conditions, internally heating to temperatures that can ignite their sulfur content and produce a self-sustaining underground fire. The sulfide gas produced from such fires can be very toxic and hazardous to health in a confined underground environment.

Alluvial Sands as a Hydraulic Backfill. Many mines are remote from the milling operations. Others are associated with a milling process that produces high fines-content tailings. After cyclone classification, these tailings cannot produce a sufficient quantity of coarse tailings suitable for hydraulic-slurry backfill placement (only 8 wt% finer than 20 microns). Such mines supplement or replace the tailings with alluvial sand from a natural deposit accessible to the mine site.

Used for this purpose, alluvial sands (whether blended with mill tailings or used as a stand-alone hydraulic-slurry backfill) must still have a size distribution (Figure 69.1) suitable to allow trouble-free borehole/pipeline slurry transport and adequate in-place water percolation or drainage rates. When using alluvial sand for a slurry backfill deposit, those containing clays and/or micas should be avoided. Both of these minerals are flat and platy in particle shape, which tends to reduce percolation rates. Furthermore, because of their long surface area, such particles generally require higher binder contents to produce a given backfill strength.

There are other minerals that are found naturally in alluvial deposits that can also affect hydraulic-slurry backfill drainage and binder strength. Any materials being considered for use as a slurry backfill (whether it be tailings or an alluvial deposit) must undergo vigorous laboratory testing. These tests must define the material's transport, drainage, and binder-strength properties before it is used underground as a mine backfill.

69.4.2 Rock Fill

Rock fill has been used to provide underground support in the mining industry for more than 80 years. The main attraction of this material has always been its availability at mine sites. In the 1960s, cemented rock fill was introduced to the industry and has since had a great impact on many mining operations. Immediate benefits were realized through improvements to bulk-mining practices such as pillar recovery and void-filling operations. Cemented rock fill is now being used successfully in many mines worldwide.

Rock fill generally refers to waste rock used for underground filling. Rock fill may be used in the form of cemented (CRF) or uncemented (URF) material. To give the rock sufficient cohesion to be self-supporting, CRF is generally placed in an underground stope when mining will be conducted either next to or under the stope. In some high-grade orebodies, CRF will also be used in a stope floor to insure maximum ore recovery with minimum fill dilution during ore removal.

CRF strength is controlled by a number of factors including grading of fill particles, cement content, rock type, angularity of fill particles, time in place before exposure, placement techniques, segregation, and moisture content (i.e., excess water). The required fill strength is a function of the mining methods in use, the dimensions of the stopes, and the stope cycle times. CRF strength is mainly controlled by the amount of cement added to the rock fill. The strength, however, can be influenced by the self weight of the fill, the degree of arching between solid rock walls, blast damage during adjacent stope mining, abrasion or attrition of the rock during raise transport, and rock-mass ground movements.

The quantity of cement in a rock fill is the main strength modifier that can be easily controlled in the fill system. Typically, the cement content for rock fills lies in the range of 4% to 8% by weight. However, because cement is expensive, there are economic incentives to find acceptable substitute materials, such as slag and fly ash, which are used at a number of operations.

Swan (1985) has examined the effects of cement content and aggregate size using test results available from several mines (see Figure 69.3). He determined and proposed that, for any given backfill material, the 28-day laboratory unconfined compressive strength (σ_c) is related to cement content (C_v) by volume through the expression:

$$\sigma_c = \alpha\, C_v^{2.36}$$

Typical strength and deformation characteristics determined from both small- and large-scale tests on cast samples of cemented rock fill are summarized below:

- Unconfined compressive strength: 1 to 11 MPa

- Unconfined deformation modulus: 300 to 1,000 MPa

Experience at a number of mining operations has shown that the fill strength obtained in sample tests can be significantly greater than that of the placed CRF. This difference is due to a number of factors including scale effects, segregation, and quality control. A variety of studies (e.g., Barrett and Cowling 1983) suggest that the in-situ strength of a CRF may be about 50% of that measured in laboratory tests.

A number of other factors also influence the strength of CRF. Fill grading is an important factor in determining fill strength, as illustrated in Figure 69.4. An excessive fraction of coarse particles will result in a "loose" material that is susceptible to blast damage because it relies largely on point-to-point contact. It may also allow excess water to percolate through the fill mass, washing out cement. Excessive fines tends to consume cement due to the large surface area:weight ratio. Because of the high relative consumption rate, fills with high fine contents often have poor particle-to-particle bonding.

The grain-size distributions for cemented rock fills used at a number of mines throughout the world are shown in Figure 69.5. Generally, the upper boundary for particle size, as delivered to the top of the fill raise, is set within the 150-mm to 200-mm range.

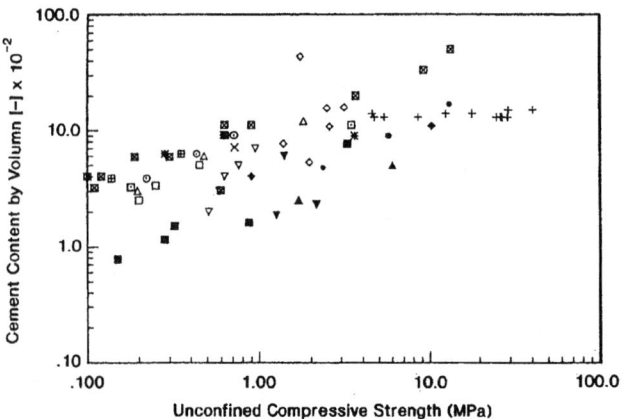

Unconfined compressive strength of selected mine backfills and concretes versus cement content by total weight per cent. Filled symbols designate cemented rockfill (see Table 69.2). Curing conditions: 28 days, 100% humidity.

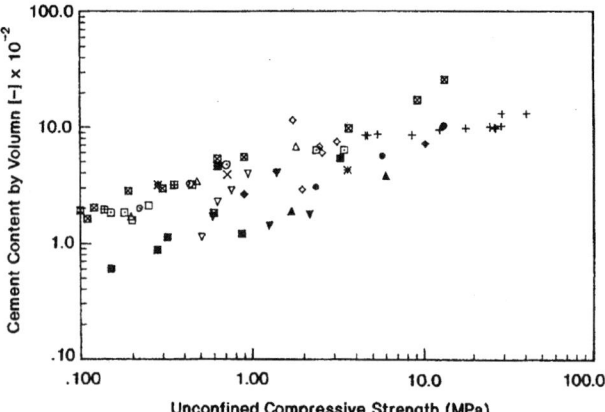

Unconfined compressive strength of selected mine backfills and concretes versus total cement content by volume per cent (see Table 69.2 for symbol code).

FIGURE 69.3   Strength of cemented backfills

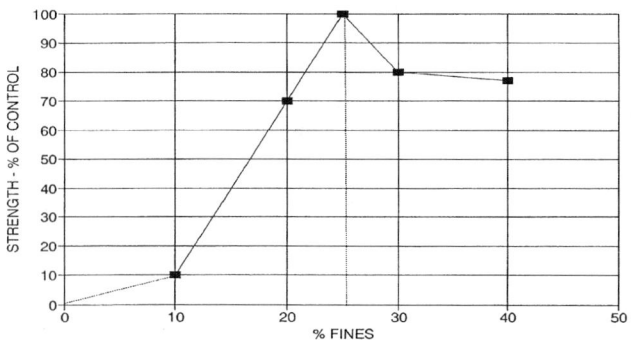

C.R.F strength as a function of the fines content. Control samples had a gradation of 75% coarse/25% fines. Minus 150 mm rockfill aggregate utilizing 450 mm diameter cylinders with 5% cement content by weight of aggregate.

NOTE: Fines defined as particles less than 10 mm size (from Quesnel and de Ruiter, 1989)

FIGURE 69.4   Effect of gradation on rockfill strength

A number of mines have observed rock-fill attrition when transporting the fill from the surface to the stope. This reduces the size of the fill particles and increases the fines content. In general, the maximum particle size is reduced by approximately 50% per 300 m of vertical distance traveled. The primary factors

TABLE 69.2   Key to mine backfills

Symbol	Mine, material*	Reference
+	Concrete	(1), (8)
X	Black Mount, CTSF	(5)
⊕	Warrego, CGF	(19)
⊠	Strathcona, CHF & CTSF	(12), (14)
⊞	Garson, CSF	(21)
⬢	Levack, CSF	(21)
⊙	INCO, CHF	(14)
◇	Lockerby, CHF	(14)
△	Falconbridge, CHF	(13)
▽	Mount Isa, CHF	(16), (18)
□	Selbaie, CHF	(7)
⊡	Kiena, CTSF	(22)
★	Coal Washery, CRF	(20)
▲	Kidd Creek, CRF	(11)
▼	Mount Isa, CRF	(16), (17)
■	Selbaie, CRF	(7)
●	Uludag, CRF	(6)
◆	Cavorrano, CRF	(15)

*NOTE: CTSF = cemented tailings–sand fill
CGF = cemented gravel fill
CSF = cemented sand fill
CHF = cemented hydraulic (tailings) fill
CRF = cemented rock (tailings secondary component) fill

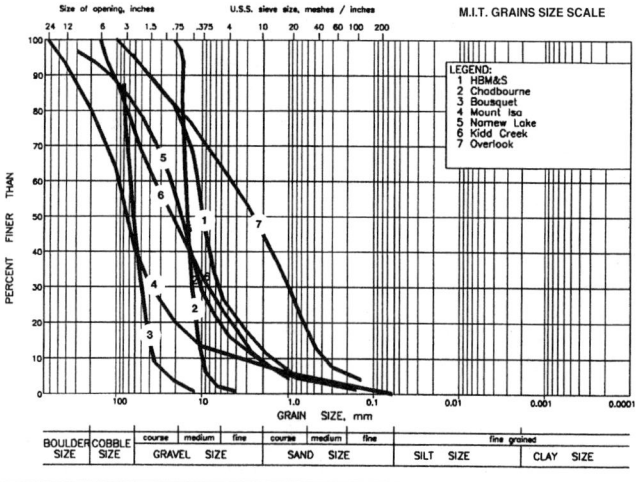

FIGURE 69.5   Grain size distribution of rockfills used at a number of mines

influencing the degree of attrition are raise inclination, distance of free fall, and rock type (Gignac 1978). As shown in Figure 69.6, similar results were found at the Kidd Creek Mine (Yu 1989).

The water content or amount of free water in the fill mass has a strong influence on strength. If excess water exists, it will tend to percolate through the fill mass upon placement, removing cement in suspension. Excess water may be present in the fill for a variety of reasons. Common reasons include the groundwater in the fill raise wetting the rock fill, water seeping into the stope, and poor quality control during mixing. Any of these reasons can cause excess water and adversely affect the fill properties.

Water quality of both the mix water and any potential inflow water must be checked for potential chemical reactions that may affect the cementing process.

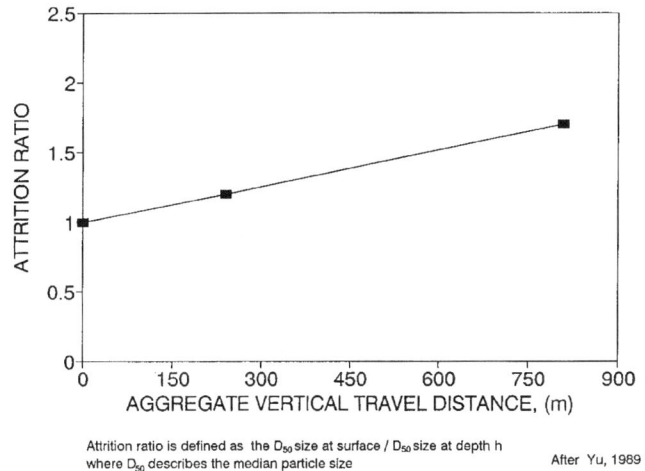

Attrition ratio is defined as the D_{50} size at surface / D_{50} size at depth h
where D_{50} describes the median particle size

After Yu, 1989

FIGURE 69.6 Attrition rate of aggregate in a rockfill raise

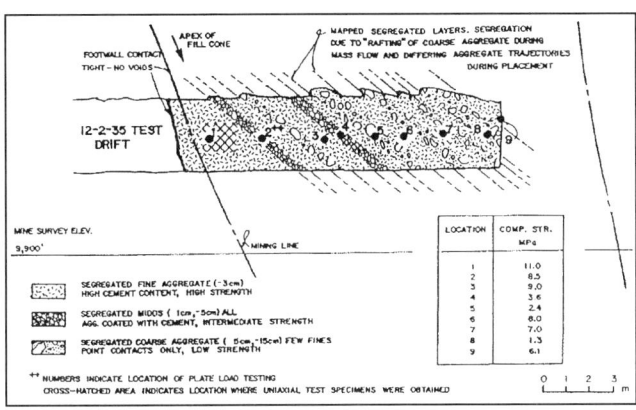

FIGURE 69.7 Rockfill segregation

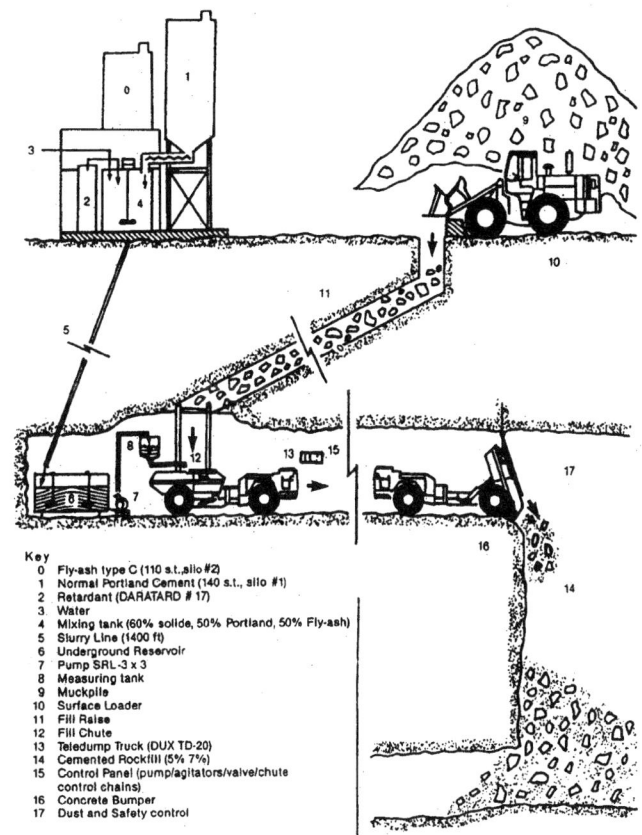

FIGURE 69.8 Gravity placement of rockfill

The strength of the rock aggregate influences the basic fill strength. If it is too weak, the rock will break down during the mixing, transportation, or placement process and result in a poor-quality fill. Blasted rock tends to give stronger fills than those derived from natural gravel because of the effect of particle angularity.

Segregation cannot be avoided when fill is dumped into a stope. Larger particles generally travel farther, having greater momentum. The result is a gradual coarsening away from the dump point. As shown in Figure 69.7, studies at the Kidd Creek Mine (Yu and Counter 1983) found "that a zone of fine aggregate tends to occur near the impact area (i.e., below the dump point). Most of the cement is consumed in this zone leaving a low cement content rock fill at the perimeter of the fill cone." Adopting good placement techniques (multiple dump points) and controlling the water content can minimize segregation.

Rock aggregate is usually transferred underground dry. When truck transportation to the stope is used, mixing is generally carried out as the rock fill and cement slurry are discharged into the truck at the underground batch plant. When conveyors are used, mixing is generally carried out at the edge of the stope as the aggregate discharges from the conveyor.

Most of the mines using CRF send the cement binder underground in slurry form, via a borehole. The pulp density of this slurry typically ranges from 55% to 60%. The correct proportion of cement is obtained in the fill when the cement

slurry is added to the rock fill. It is critical that no free water drainage from the fill occurs once it is placed in the stope.

Some surface cement batch plants use PLCs (programmable logic controllers) to allow the entire batching process to be controlled from underground by the operator at the slurry/fill mix point. Experience at a number of these mines indicates that there is a considerable learning curve during start-up. It is important that rigorous quality control be maintained during start-up to identify and resolve "bugs" within the system.

Placement methods can be divided into three groups:

- Spread placement
- Gravity placement
- Consolidation placement.

Spread placement refers to directly dumping the fill in the stope and spreading it using a bulldozer. This method usually results in an evenly compacted fill. This approach tends to be used in cut-and-fill, drift-and-fill, and low-lift bench mining.

Gravity placement involves dumping the fill (either from a truck or conveyor) down a raise or from an access drift and allowing the fill to fall over some vertical distance. This approach is the most common placement method, particularly for high-volume filling of large open stopes (Figure 69.8). Unfortunately, segregation invariably occurs. This segregation is controlled by the location of the dump point with respect to the stope geometry and the angle that the fill enters the stope. Ideally, these stopes would be leveled and topped-up by spread placement.

Consolidation placement is used in a limited number of cases and refers to either blowing the fill into position (pneumatic placement) or throwing it into position using belt-slinger methods. At the Meggen Mine in West Germany, for example, cemented rock

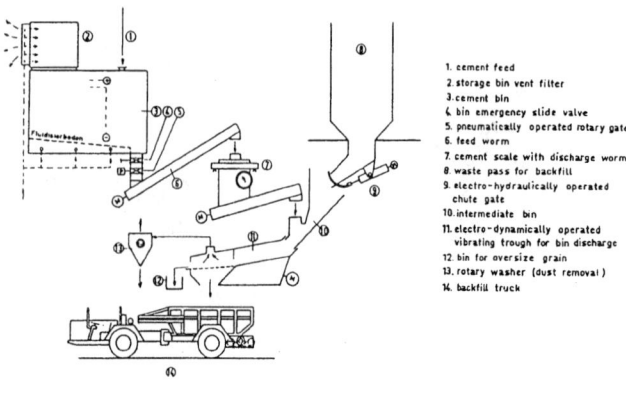

1. cement feed
2. storage bin vent filter
3. cement bin
4. bin emergency slide valve
5. pneumatically operated rotary gate
6. feed worm
7. cement scale with discharge worm
8. waste pass for backfill
9. electro-hydraulically operated chute gate
10. intermediate bin
11. electro-dynamically operated vibrating trough for bin discharge
12. bin for oversize grain
13. rotary washer (dust removal)
14. backfill truck

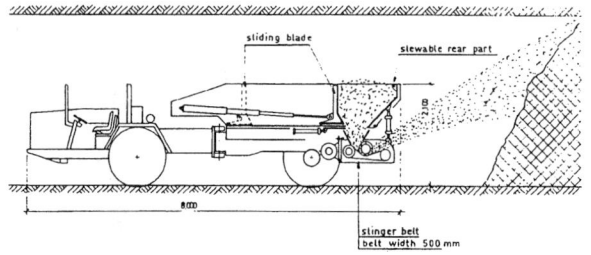

FIGURE 69.9 Rockfill placement using a slinger truck

TABLE 69.3 Design criteria for fill raises

Ratio of orepass dimension to particle	Relative frequency of dimension interlocking
D/d > 5	Very low; almost certain flow
5 > D/d > 3	Often, flow uncertain
D/d <3	Very high, almost certain no-flow

where: D = minimum span/diameter fill raise/borehole
        d = particle dimension

fill is placed in room-and-pillar stopes by 6 and 10 m³ trucks, using belt-slinger technology (see Figure 69.9). This produces a reasonably uniform, compact fill material.

The fill raise must be sized so that transferring the material from the surface to underground is not restricted. The size of the opening required is a function of the size of fill to be transferred and the volume of fill required. Rock-fill raises from surface to underground are typically 2 m and 3 m in size. Two types of blockage can occur:

- Hangups due to interlocking arches

- Hangups due to cohesive arches.

Interlocking arches form as a result of large-sized boulders becoming wedged together to form an obstruction. Such blockages will occur at changes of direction or reductions in cross-sectional area of raises or boreholes. Screening oversize boulders on the surface is the main method of minimizing this type of blockage. Hambley (1987) presented empirical design criteria for avoiding interlocking arches due to oversize as outlined in Table 69.3.

Cohesive arches form as a result of sticky fine particles adhering to each other. Whether a cohesive arch is formed depends upon the span of the raise. If the opening is sufficiently large, gravity forces will exceed the cohesive and frictional forces, and an arch will not form.

Hambley (1987) proposed that, to prevent cohesive arching, the minimum dimension of a raise be determined from:

$$D > 2(G/\delta)(1 + 1/r)(1 + \sin \phi)$$

where   D = ore pass dimension

       G = cohesion of fines (fines defined as particle size less than 0.25 mm)

       δ = density of fines

       r = length/width ratio of opening

       φ = internal angle of friction of lines

An important aspect of the raise is water inflow. Above all else, the potential for blockages to occur in a raise increases as both the percentage of fines and the water content of the fill material increase. Furthermore, if the raise makes any water, this will increase the moisture content of the rock fill and result in a change to the water:cement ratio if not otherwise adjusted for. Excessive water in the raise can cause a poor quality CRF.

Water flow into the raise can also lead to an uncontrolled out-rush of rock if a hydraulic head of water is allowed to build above the rock and/or the rock becomes saturated and liquefies due to dynamic vibration from blasting or rock bursts. Such conditions are very dangerous and have resulted in fatalities in several mines.

Raises can be operated in a choked condition, an empty condition, or somewhere between these two limits. Choked raises reduce the wear and attrition of the fill materials, but the risk of blockages increases. Because of the need to maintain surge capacity and the difficulty of monitoring the fill level within a raise, some compromise is usually reached, with the fill level fluctuating between one-third and two-thirds full.

Quality control testing and monitoring is required to assess the performance of a rock fill system. Routine laboratory and in-situ material testing should become an integral part of any rock fill system design. Such testing is the only means of quantifying the actual "in-place" material properties and the impact of placement processes (drainage, segregation, curing, and dynamic loading).

69.4.3 Paste Backfill

A paste is a granular material mixed with sufficient water to fill the interstices between the particles so that the material behaves as a fluid. The granular material retains all the water between the particles because of its colloidal electrical particle charge that bonds the solid particles to the water molecules. In this state, paste can be transported through a pipeline but has no critical flow velocity (i.e., the velocity at which the solid and liquid components separate into two distinct phases). If more water than can be held between the particles is added to the paste, it becomes slurry, and the material does have a critical flow velocity. In this state, the material will flow through the pipeline but will separate out into two distinct phases if the pipeline velocity drops below the critical value.

In general, a granular material must have at least 15 wt% of its particles finer than 20 microns for the colloidal properties of the material to retain sufficient water to form a paste. Granular materials with less than 15 wt% of fine material will not possess the colloidal properties to form a paste and cannot be transported as such. The colloidal properties of a material are governed not only by the size of the particles, but also by their chemical content and mineralogical composition. This means that different materials will form a paste with different size distributions. Each granular material must be tested independently to determine its properties and its behavior as a paste. Both the production plant and the pipeline distribution system have to be designed according to these specific properties.

Paste backfills should never be placed underground without adding a binder. Uncemented paste backfill is very prone to liquefaction and will remain in a fluid state for days, weeks, and even years after being placed underground. If the ore being

Colloidal properties   Strong colloidal properties

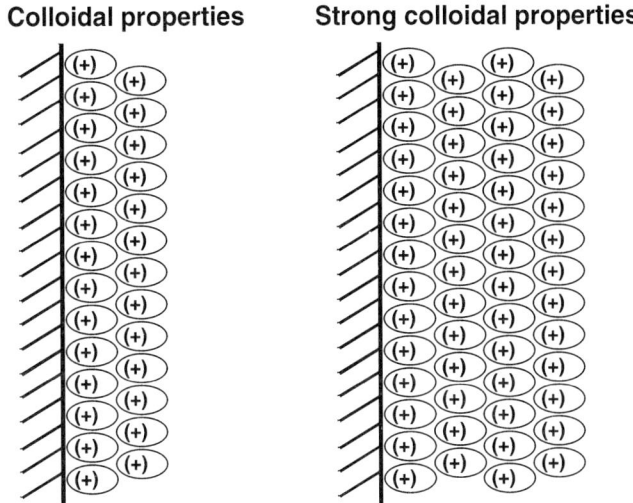

FIGURE 69.10   **Paste backfill colloidal chemistry**

mined will not economically support binder addition to the backfill, then paste backfill should not be used in that mine.

Full Plant Tailings Mix Design. There are three size-distribution categories for paste-backfill-mix design for most hard rock mine tailings throughout the world. These are coarse, medium, and fine tailings (Landriault 1995).

Coarse tailings contain between 15 wt% and 35 wt% minus 20 micron content. At a 7-inch slump, paste backfill will have a pulp density ranging from 78 wt% to 85 wt% solids depending on the specific gravity of the tailings material. For a given cement content, the high solids content provides a good water:cement ratio and results in backfill strengths at least double that of comparable hydraulic-slurry backfills.

Medium tailings contain between 35 wt% and 60 wt% minus 20 micron content. At a 7-inch slump, paste backfill will have 70 wt% to 78 wt% solids, again depending on the solids' specific gravity. These tailings generally produce a good paste fill, but typically have lower strength than the coarse tailings because of a higher water:cement ratio.

Fine tailings, which come from a mill process that grinds for recovery, contain 60 wt% to 90 wt% finer that 20 microns. High water retention is expected with these tailings, which usually produces a paste fill that is good for flow transport but poor for strength because of a very high water:cement ratio. Fine tailings at a 7-inch slump will have pulp densities between 55 wt% and 70 wt% solids.

The amount of water required to produce a given slump consistency in a paste backfill is dependent on the colloidal properties of the particles. These properties are a measure of the Zeta potential charge of the fine particle in the paste (Figure 69.10). The Zeta potential charge causes the water molecules to bond to the fine particles, giving the paste the ability to retain water within its particle matrix. The stronger the Zeta potential charge, the greater the water retention of a paste material at a given slump. As the fine material content increases, the amount of water retained by the paste at a given slump increases.

Soft-rock tailings generally contain much a higher water content (a 7- to 10-inch slump consistency) regardless of their size distribution. This is primarily because of their mineralogy, which often consists of high calcite, clay, or similar highly colloidal, high water-retention minerals. These tailings can range from 50 wt% to 30 wt% solids at a 7-inch slump consistency depending on the fines material content and mineral composition.

As with hydraulic-slurry backfill, the mill tailings used in paste backfill must be analyzed for mineral content (zinc, lead, and some pyrites) that can affect the binder reaction. Certain minerals can produce strength retardation, reduction, and long-term deterioration. Laboratory testing of the short- and long-term binder strength is required before a backfill mix consisting of such tailings is used underground as a paste backfill.

The health and safety of the underground workers must be considered when designing a paste backfill for a given mine. Tailings can contain health-hazardous contaminates (i.e., cyanide, arsenic, etc.) from the milling process. There are health and safety standards in most countries for underground backfill, and underground contaminates in general, that must be reviewed.

When pyritic tailings are used as a paste backfill, the exothermic properties must be investigated. As stated earlier, some forms of pyrrhotite and pyrite can chemically react under the proper underground moisture and oxygen conditions. Such reactions can cause internal heating to temperatures that will ignite the sulfur content of the material and result in a self-sustaining underground fire that produces toxic sulfide gas.

Blended Paste-Fill-Mix Design. Some mines have found that it was to their advantage to blend a coarse material with their mill tailings to produce a blended paste backfill (Landriault 1995). This advantage comes from the strength gains produced by the lower water:cement ratio created with blended paste backfills. By adding coarse material to the tailings, the particle size distribution of a paste fill is widened, which produces a lower material porosity. This results in less water being required to fill the lower amount of voids between the particles. This also allows a 7-inch slump to be obtained at a much higher solids content, producing a lower water:cement ratio and greater strength for a given cement content. Blended paste backfills are most attractive when the mine produces only fine tailings, which generally yield low cemented-paste backfill strengths.

Blended paste backfills are used in some mines because of their improved ground-support properties. The lower porosity created in blended paste fill results in a higher-modulus material. This improves the load response of the backfill as the result of rock-mass closure. Thus, less closure can occur before the backfill begins to carry some of the stress. This stress would normally be transferred through pillars and mining abutments.

In small mines with low production rates, blended paste fill allows them to produce backfill at a rate higher than their mill tailing production rate. This allows these mines to backfill stopes at a faster rate than if they were using a full-plant-tailings paste backfill. In mines with ground support problems, backfilling open stopes as quickly as possible is very important to keeping the rock mass stable.

If tailings are used as part of the blended paste backfill, the strength and health and safety concerns associated with the chemical/mineralogical composition of the underground backfill material (lead, zinc, cyanide, arsenic, exothermic properties) should still be determined.

69.4.4   Paste Backfill Rheology

Paste backfills produce a plug flow when transported through a borehole or pipeline (Figure 69.11). The annulus of fine particles that naturally forms in the plug flow condition acts as a lubrication layer along the pipe wall, reducing flow resistance and pipe wear. The coarse particles are naturally forced to the center of the pipeline and are transported in the fine-material carrier. This natural phenomenon allows very coarse material to be transported through a borehole/pipeline by a fluid material with paste flow properties.

Most paste materials can be categorized as Bingham fluids, but several have also demonstrated pseudoplastic flow properties (Landriault 1992). They all possess a yield stress that varies greatly in magnitude for different tailings paste materials.

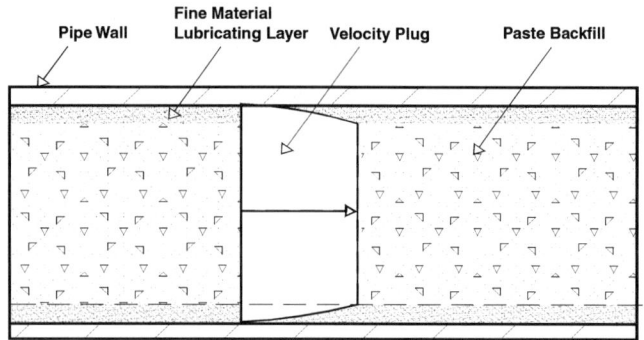

FIGURE 69.11   Velocity profile for plug flow

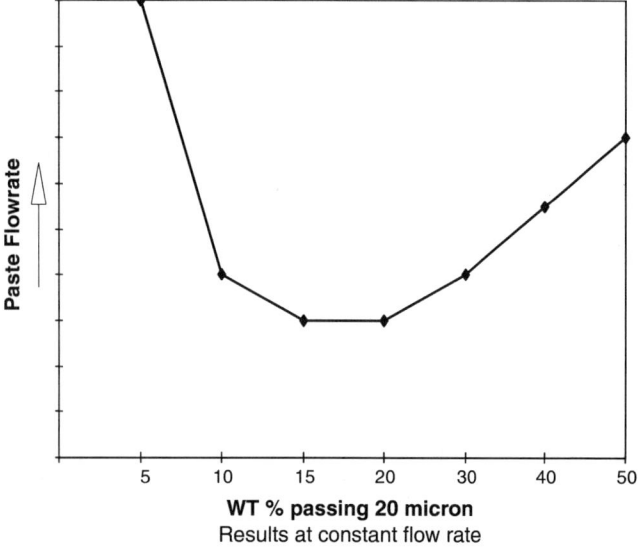

WT % passing 20 micron
Results at constant flow rate

FIGURE 69.12   Paste backfill resistance versus 20

In general, a granular material must have at least 15 wt% finer than 20 microns to produce sufficient colloidal water retention to create paste-flow properties. Many paste materials demonstrate an increase in flow resistance with increasing 20-micron content (Figure 69.12). The resistance to flow, as with all fluids, increases with decreasing transport borehole/pipeline diameter.

Paste material properties, such as water retention and flow resistance, are not controlled by particle-size distribution alone. The chemical content and mineralogical composition of the paste material are as important to paste-material behavior as fine-material content and particle-size distribution.

69.5   MINING METHOD BACKFILL STRENGTH REQUIREMENTS

As previously described, backfill is used with several different mining methods. The backfill strength required underground is dependent on the construction and geotechnical needs for the specific mining method and rock mass in that given section of the mine. The following section outlines the basic approaches used to determine the backfill properties required to safely mine the orebody using different mining methods.

69.5.1   Bulk Mining (Long Hole, Shrinkage, Vertical Retreat)

Backfill in bulk-mining stopes is placed primarily as a ground support tool to minimize hanging-wall dilution, to reduce foot

wall/hanging wall closure, and to stabilize the overall rock mass. Backfill also plays a major role as a construction material fulfilling three functions in most bulk stopes.

The first function is to act as a consolidated (cemented) backfill bulkhead in the mucking level draw point to contain the fill that will be placed above the draw-point horizon. With hydraulic-slurry and paste backfills, the backfill bulkhead can also eliminate the potential for liquefaction of freshly placed cemented fill.

If rock fill is used to backfill the bulk stope rather than hydraulic slurry or paste backfill, the need for a backfill bulkhead is not necessary because the rock fill is placed unsaturated and will not liquefy. Consolidated hydraulic slurry and paste backfills, even at low cement concentrations, typically produce sufficient strength after 24 hours of curing to contain the weight of fill above the draw point and to prevent liquefaction.

A typical backfill bulkhead is constructed by building a timber and/or waste-rock barricade in the draw point drifts accessing the bulk-mining stope. These barricades must be water retaining, and they usually incorporate an internal stope-drainage system to remove excess slurry water through the barricade and bulkhead out into the draw-point drift. If paste backfill is used, the barricades do not have to be water retaining, and auxiliary stope drainage is not required because paste backfill does not produce any noticeable water runoff. The high viscosity of the paste backfill will allow development rock to be used as the barricade without the potential of leakage from the stope.

The barricade is normally located back from the draw-point brow a distance that is twice the largest drift dimension. The purpose of the barricade is to act as a low-strength (40 kPa) construction form designed to contain the consolidated backfill until it sufficiently hardens to the strength required to withstand the potential head created by freshly placed backfill above it.

Backfill is placed into the stope from the drill drift on the level above or, if slurry or paste fill is used, through a borehole accessing the stope from some upper level. Slurry and paste fills are allowed to build up behind the barricade to a height of approximately 3 meters above the draw-point brow. The backfill bulkhead can normally be placed in one to three days, depending on stope size, and is typically allowed to cure to a uniaxial strength of 25 psi (170 kPa) before any subsequent backfill is placed in the stope. The time period required to reach this backfill strength depends on the type of backfill and its cement content.

The second function of backfill in bulk-mining stopes is as a freestanding wall that will be self supporting while the adjacent stope is mined. Prior to secondary stope removal, the primary stope is filled with cemented-paste backfill. After a period of time (at least 28 days), the secondary pillar will be taken. To ensure minimal dilution and to improve the regional ground support after the mining of that area is complete, the backfill must remain self-supporting when exposed.

The design requirements of backfill serving this function, regardless of type, are dependent on the dimensions (length, height, and depth) of the freestanding backfill wall that will be exposed during subsequent mining. The method used for calculating the necessary backfill properties for a competent freestanding backfill wall for the various bulk-mining stopes dimensions are based on work conducted by R. J. Mitchell of Queens University in Ontario, Canada.

Figure 69.13 shows the proposed stope size and the extent of the exposed face that must remain intact.

The equation used to determination the strength required is based on a wedge-type failure of the backfill. The parameters required are the stope dimensions, density of the fill, and the internal angle of friction. The internal angle of friction is generally determined through triaxial tests of consolidated backfill, but does not have a great influence on the final strength

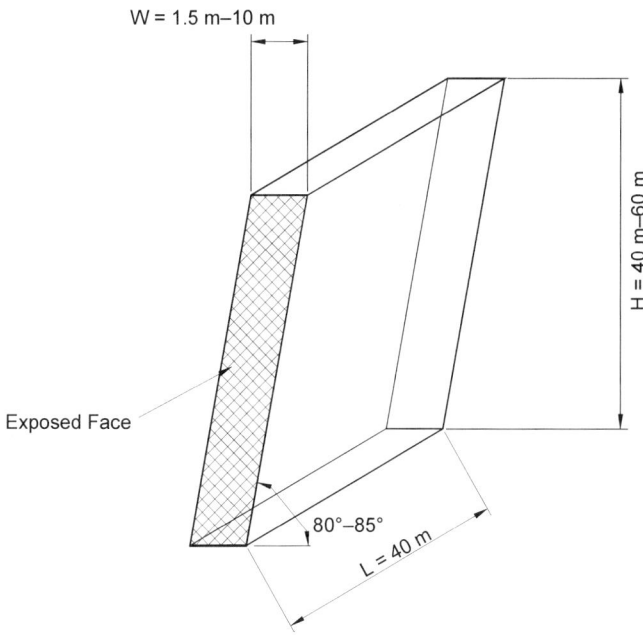

FIGURE 69.13 Proposed stope size and the area that must remain intact

given the above geometry. Most consolidated backfill has an angle of internal friction that ranges from 20–35 degrees.

Mitchell's method of determining the backfill properties required for a given bulk-mining situation have been used in the North American mining industry for more than 20 years and have proved to be very reliable in more than 100 mines.

The basic formula requires the height, length, and depth of the freestanding wall to be defined along with the in-place bulk density or weight of the backfill to be determined. The angle of internal friction of the backfill is also required for input into the equation. Based on hundreds of triaxial tests conducted on mine backfills throughout the world, a generic internal friction value of 30° is used since the values of the tests varied only between 28° and 30°.

The third function of backfill is as a cohesive soil in the secondary or tertiary stopes of a bulk-mining operation. The backfill placed in these peripheral stopes is primarily a ground-support tool. These stopes will not have the fill exposed, but the fill must still have sufficient cohesion to prevent liquefaction. Paste backfill is the only fill type that cannot be placed in the secondary or tertiary stopes in an uncemented form. The natural water retention properties of the high fine-material-content paste backfills (15 wt%, 20 micron content or greater) make them susceptible to liquefaction if placed under a dynamic load from blasting, rock burst, or a major ground fall. Hence, paste backfill placed in the peripheral stopes must have sufficient binder added to give the fill the necessary cohesion to eliminate any potential of liquefaction. The traditional paste backfill design strength to eliminate liquefaction potential in underground North American mines is 25 psi (170 kPa) after 28 days of curing.

When it is necessary for the efficient disposal of mine waste, uncemented rock will be dumped into the primary bulk-mining stopes as consolidated hydraulic slurry or paste fill is being placed. If conducted properly, this is a common and successful method of waste-rock disposal at most mines.

Waste rock should be dumped into the stope only while slurry or paste backfill is being poured. The rate of waste-rock placement should not exceed the rate of slurry/paste fill placement into the bulk stope. The rock should be dumped into a

pool of fluid fill and should not be allowed to build up above the level of the paste in the stope against a stope wall that will later be exposed by subsequent mining. Slurry and paste backfills will not penetrate the interior of a dry pile of waste rock. Exposing the unconsolidated waste rock will result in wall failure and dilution of the ore in the adjacent stope.

69.5.2 Drift-and-Fill, Cut-and-Fill, and Uppers-Retreat Mining Methods

Mining using the drift-and-fill, cut-and-fill, or uppers-retreat method requires the backfill to act as the mucking floor for each sequential cut. With the bottom-up mining approach using these methods, stopes are often filled to within 0.3 to 0.5 meters from the back. This opening will be left to provide a free face for the next cut of ore to be blasted into.

Rock fill can generally be placed to within 1.5 meters from the back with a loader and then requires dozing and/or ramming to place it closer to the back of the drift. Producing a level mucking floor with rock fill would be difficult, if not impossible, using this placement method. Adding a moderate amount of cement would be required to produce a hard-rock-fill mucking floor of 150 psi (1 MPa). The uneven/non-smooth nature of the floor, however, would not be conducive to a high/undiluted recovery of high-grade ore fines that will be produced from blasting in higher-grade stopes. No backfill barricades are required for rock fill placement.

Hydraulic-slurry and paste backfills can be easily placed to within 0.5 meters from the drift back and, with the addition of a moderate amount of cement, can produce a hard mucking floor of 150 psi (1 MPa). The normal method of placing these types of backfills tight to the drift back utilizes pipelines that are secured approximately 0.5 meters from the back of the drift. A small raise can be created near the end of the drift to allow the pipeline to discharge at an elevation slightly higher than the normal drift back.

As paste backfill is discharged into the drift, it builds up with an angle of repose that can range from 1–10 degrees, which will be controlled by the slump of the paste fill being placed at any given time. Floors are poured at a high slump, resulting in floors from 1° to 3°. The paste fill will flow like lava from the discharge point back to the backfill barricade, which is normally placed about 60 meters apart to accommodate the placement angle of the paste-fill floor.

Hydraulic-slurry backfill typically produces a floor angle of 1–2 degrees, but because of the excess slurry-water removal from the stopes, substantial segregation of the particle sizes occurs. The result is a high fine-material and binder content near the drain locations and a variation in the backfill floor throughout the stope.

These mining methods are also used where the backfill is to be placed 2.4 meters from the back. The distance from the backfill to the stope back must be sufficient to allow the up-holes for the subsequent round to be drill, loaded, and blasted. As with the drift-and-fill methods, the same basic backfill mucking floor requirements (150 psi) are necessary.

69.5.3 Mining Under Backfill (Undercut-and-Fill, Sill-Pillar Recovery)

To facilitate production requirements, mining generally progresses on several different horizons in the mine. This results in mining that progresses from lower stopes up into areas that have already been mined. This will create a situation where sill-pillar ore will have to be removed from under the previously backfilled stopes, creating the need for backfill sill pillars.

This will create a similar situation that exists with undercut-and-fill mining, where men and/or equipment are exposed under a backfill head cover. Undercut-and-fill mining traditionally uses timber post and lagging support to contain the backfill but, with the introduction of paste backfill, some mines are using unsupported backfill with this mining method.

To safely recover the ore and to minimize dilution, the backfill above the ore being mined must be strong enough to be self-supporting over the span where it will be exposed. To determine the required strength, a safety factor of 1.2 is generally used if only equipment (remote mucking) is exposed under the backfill; a safety factor of 2.0 is used if both men and equipment will be working below the fill.

With these mining methods, the possible failure modes of the exposed fill are associated with arching, flexural or bending failure, block sliding due to side-shear failure, rotational failure at the hanging wall contact, and sloughing of wall rock. The stability of fill backs has been researched by Mitchell and Roettger (1989) and calibrated with laboratory testing. The following back-stability calculations are based on their work. The analytical design methods are based on typical sections. If there are significant variations in the geometry of the orebody, the calculations become very complex and should be implemented with extreme care and confirmed by numerical modeling.

69.5.4 Mining Through Backfill

Access into secondary mining areas often requires some development through in-situ backfill. Where this access has been anticipated, high-strength backfill can be placed to accommodate drift development through the fill and safe access to the ore with minimal mechanical support being required (typically friction bolts with welded wire mesh screen for a span of 3.0 meters or less). In areas where access has not been anticipated, and where the fill strengths are lower or drift spans of greater than 3.0 meters are excavated, a combination of shotcrete reinforced with friction bolts and welded wire-mesh screen will be used to support the backfill.

Typically, a 50-mm layer of shotcrete is applied to the backfill over a 2-meter advance. Wire-mesh screen is placed over the shotcrete and held in place with 0.3-meter friction bolts. A second 50-mm layer of shotcrete is then applied over the screen to provide added support. This approach to backfill support has been successfully used in North American mines for more than a decade.

69.6 BACKFILL AS GROUND SUPPORT

One of the most important functions of backfill is to provide support to the surrounding rock mass. It is important to establish at the outset what kind of support the backfill can and cannot be expected to provide.

Backfill can provide confinement to pillars and to stope walls and help prevent the progressive deterioration of exposed and unsupported rock surfaces. Backfill has a limited direct effect on the mining-induced stresses.

69.6.1 Effect on Stresses

Backfill cannot significantly alter the mining-induced stress conditions because of three factors. These factors include the contrast between the deformation modulus of the backfill and the excavated rock, the time lapse between the excavation of the rock and the placement of the backfill, and the tightness of the placement of the backfill.

Except in a few particular circumstances, the kinds of backfill that are placed in typical underground hard-rock mines are too weak to carry the kinds of load previously borne by the excavated material. Typical mine backfill, even cemented backfill, provides only passive support, so its capacity to resist wall convergence increases as it is compressed. In most cases, a great deal of the total convergence occurs before the backfill can be placed and consolidated. However, the compressibility of the fill also has an effect: cemented backfill provides more confinement than uncemented slurry fill which, in turn, provides more support than loose sand and gravel fill.

One case where backfill can carry significant stresses is in a narrow, steeply dipping orebody. The amount of convergence is dependent on the stress normal to the orebody, the properties of the wall rocks, and the area of exposed hanging wall and footwall. The resistance of the fill is determined by the total amount of void space in the fill, which determines how quickly the fill will begin to resist the convergence. In orebodies wider than 15 or 20 ft., the fill never gets to the point where it will offer significant resistance to convergence; therefore, the adjacent rib pillars and sill pillars become highly stressed. Only in the narrow orebodies can the fill offer resistance and reduce the stress concentrations in the adjacent pillars. The support pressure offered by backfill in such cases can be as high as 700–800 psi (5 MPa).

The deformation modulus of fill increases with the cement content of hydraulic-slurry backfill and is the most important factor in resisting wall convergence. Cemented hydraulic-slurry fill has a modulus of around 7,500 psi (50 MPa). Cemented rock fill has a higher stiffness or deformation modulus, ranging from 15,000–150,000 psi (100 to 1,000 MPa), depending on the density and cement content. Paste fill can achieve a modulus equal to the low end of cemented rock fill. A modulus of 150,000 psi (approximately 1 GPa) is sufficient to alter the energy balance and reduce stress concentrations in adjacent areas so the strain energy stored in the surrounding rock is reduced.

69.6.2 Effect of Confinement

Backfill enhances the behavior of the surrounding rock mass by providing a confining pressure to the exposed rock surfaces. The most obvious example of improving the capacity of rock by applying confinement is the case of post-pillar cut-and-fill mining. In this case, an array of pillars with a width:height ratio of about 1 are left in place to support the roof or back of the stope. The extraction ratio of post-pillar cut-and-fill stopes is about 85%–90%. Once the first cut of ore has been removed, the floor is filled with cemented tailings and the second cut extracted, leaving an identical array of pillars on top of those of the previous cut. With each successive cut, the width:height ratio decreases so that the pillars inevitably fail, usually on the second or third cut (width:height ratio = 2 or 3). However, the fill maintains the integrity of these failed pillars and, as the pillars dilate during the failure process, the surrounding fill begins to resist the dilation. The confining stress applied by the fill maintains the residual strength of the pillars as the width:height ratio becomes precariously small. Left unconfined, these pillars would undoubtedly collapse on the third or fourth cut, but in actual stoping situations, they continue to provide support for cuts as high as 14 or 15.

If the pillars are made of weak rock, a fill with little or no cement will increase both the peak and residual strength of the pillars. If the pillars are made of strong rock, such as a quartzite, a high-cement-content fill is required to increase the residual strength. It is possible to generate a residual strength of around 70% of peak with confinement provided by a 10% cement backfill. The same pillar surrounded by uncemented-tailings fill will have a residual strength of only 25% of peak, while an unconfined pillar will have almost no residual strength.

The strength of the confining material also has an effect on how violently the pillars fail. In highly stressed conditions, pillars confined with weak backfill will fail violently, while those surrounded by much stronger, high-cement-content backfill fail passively. In general, increasing the confining pressure on any failing rock mass allows it to behave in a less brittle, more ductile way, releasing less seismic energy during the failure process. The effect of high-strength backfill has important implications for mining in highly stressed conditions.

This confining effect is less obvious in bulk-stoping operations, but is equally effective. First, the backfill prevents the progressive collapse of the hanging wall. In plan, hanging wall

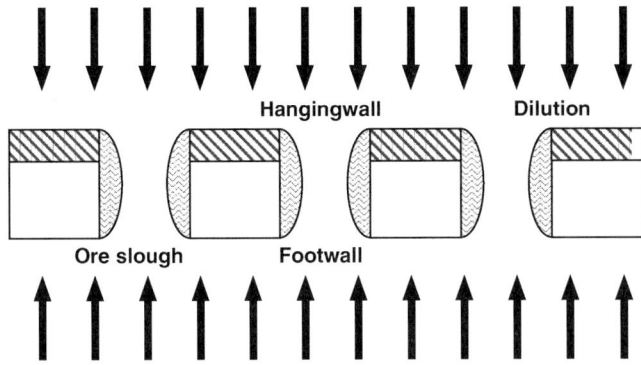

FIGURE 69.14   Plan of bulk stopes

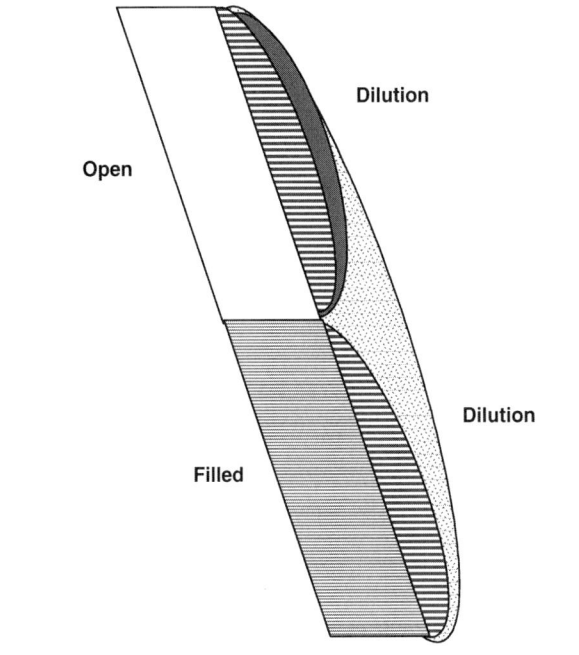

FIGURE 69.15   Cross section of bulk stopes

collapse changes the slenderness ratio of the ore pillars in a horizontal stress field and will contribute to slough ore off the pillar walls (Figure 69.14). In many cases, this is not considered a problem because the additional material is ore. However, this over-break causes the stope walls to become concave, and when these expanded stopes are filled and later mined, the fill walls are convex in shape, which results in fill dilution during mining of the pillars.

Second, this slough extends the failed zone into the hanging wall, and when this occurs at each of the primary stopes, the hanging wall of the secondary stopes become less stable, even before mining begins. The same is true for the hanging wall of the primary stopes on the next upper horizon. Hanging wall dilution in the lower stope effectively undercuts the hanging wall of the stope on the next horizon (Figure 69.15). Excessive dilution off the hanging wall of the primary stopes creates a patchwork of failed zones extending into the hanging wall and causes even greater dilution when these areas are mined.

There is little effective difference in the amount of confinement provided by the different types of backfill as long as they are cemented. Uncemented fill has a much lower stiffness

and is relatively easily compressed and provides little support. Cemented rock fill can provide the greatest support, but it is also the most variable of the three cemented-backfill types because of the degree of percolation of cemented slurry through the rock fill and the potential for segregation during placement.

69.6.3   Backfill Rate of Delivery

A major difference between the types of backfill is the rate at which the backfill begins to provide support. There is virtually no difference between the rate of delivery of slurry fill and paste fill. The material handling setup required for cemented rock fill placement (haulage or conveying) is much more extensive than slurry or paste pipeline placement. If the material handling system is in place on the mining level, the placement rates can be twice as high as pipeline transport. However, if the placement system is not in place, a considerable delay could occur, resulting in more stope dilation or convergence occurring before the rock fills can provide any support at all.

In addition to the delivery rate, there is the time required for the backfill to become capable of carrying load (set-up time). Both slurry and paste fill can be supplied at about the same rate (100–200 ton/hour). However, the actual rate of delivery is often not the most critical factor. In bulk-mining operations, using slurry fill requires a major effort to construct water-retaining barricades, which contain the saturated material while it drains. The construction time and the relatively slow pouring rate in the early stages of backfilling are the most important factors. In the case of cemented rock fill, unless the entire backfill system is set up with conveyors or a large vehicle haulage system, the delayed and/or low rate of rock fill placement to the stopes will produce the same result. Paste fill requires minimal barricade construction to contain the material, and there is no water entrained in the system, except the amount required to hydrate the cement.

While the pouring rate of slurry and a paste fill is almost the same, the overall rate of delivery is much slower for the slurry fill. This is because most slurry-fill operations have to stop pouring until the saturation level of the slurry falls below the critical level as the material drains. Because there is virtually no free water in a paste-fill system, there is no need to delay the fill pours. Finally, the consolidation rate of a slurry fill is much slower than for paste fill and rock fill. Both fill types reach the required strengths days or weeks before that of slurry fill because of their higher water:cement ratios.

The rate of completely filling a stope has an effect on the stability of the rock mass surrounding it. During the backfill preparation period and the early stages of placement, the stope walls will be deteriorating; and the cumulative effect of this deterioration will result in even more dilution in later stopes. It is possible to design the stopes slightly more conservatively to ensure that the rock mass remains stable during filling, but more conservative designs are less cost effective than optimally designed stopes. In the case of slurry fill, the excess water lubricates the joint systems and further weakens the rock mass. In intensely jointed conditions, minor collapses during the fill cycle are common in cut-and-fill stopes. Even when the stope is filled with slurry fill, it takes some time before the fill gains sufficient strength to resist the dilation of stope walls or pillars.

69.6.4   Mining in High-Stress Conditions

One of the main reasons behind the development of paste fill was the expectation that because of its overall lower porosity it would reduce the release of seismic activity in deep, highly stressed mines. Unfortunately, it has now been established that, apart from narrow orebodies, no backfill material can provide sufficient resistance to wall convergence to prevent rock bursts. However, implementing high backfill-placement rates (i.e., paste or rock) in deep, highly stressed mining operations can

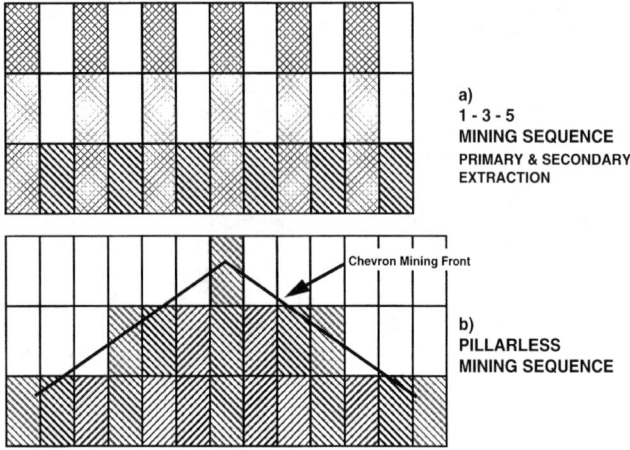

FIGURE 69.16 Alternative stope sequences

significantly improve the management of rock burst problems by allowing for a more appropriate mining sequence.

The most economic mining sequence for bulk stopes is the primary-secondary stope-pillar sequence, with equal-sized stopes and pillars (Figure 69.16). However, in highly stressed situations, it becomes very difficult to mine the secondary pillars because they have become so highly stressed that violent failure is inevitable as soon as production blasting begins. The situation is worse if the stopes and pillars are unequal in size. The best solution to this problem has been the introduction of the center-out mining sequence, where the next stope to be mined is immediately adjacent to the most recently filled stope. The disadvantage of this sequence is the small number of stopes that can be brought into production early in the schedule and the time delay in mining the next stope in the sequence. In a slurry-fill operation, this sequence is viable only if there are several blocks or areas of the mine available for production at any one time. A paste- or rock-fill system can allow for a much shorter cycle time and makes it easier to adopt a center-out mining strategy in highly stressed conditions.

Reducing the cycle time in a mining operation by placing backfill at a high rate has the potential to radically change the approach to all mining operations, not just those in highly stressed conditions. In the past, mining operations have attempted to achieve economies of scale by mining a few very large stopes at one time. As mines become deeper and extraction ratios increase, the stability of large stoping blocks and the surrounding rock mass decreases. Unfortunately, reliance on these stopes is relatively high, compared to the selective small-scale mining operations that were prevalent 20 years ago (drift-and-fill, cut-and-fill). A large stoping area can represent as much as 20% of the mine's production and any interruption to the flow of ore from these stopes has serious implications for the mine as a whole.

At the same time, the very large size of each individual stoping area generally allows less control over the production process rather than more. The cost of conservative designs for ground support and blasting tends to erode the economies of scale but leads to problems with stope stability and fragmentation. The variations in the mining process (i.e., dilution and secondary blasting) necessitate costly remedial activity and elaborate contingency plans to cover for production short falls. The economies of scale that were essential because of the long cycle time of an individual stope are now being eroded by the increasing cost of contingency planning and remedial activities.

As the physical conditions in deeper, older mines deteriorate, these problems will worsen and reduce the cost effectiveness of this approach to bulk mining.

69.7 ACKNOWLEDGMENTS

The author would like to acknowledge the contribution of other Golder Associate employees in the production of this document. Several sections were borrowed from work previously completed by various colleagues. This document not should be viewed as a contribution to the handbook exclusively from the author, but more as a joint effort contribution from Golder Associates in general.

69.8 REFERENCES

Barrett and Cowling. (1983). Investigation of Cemented Fill Stability in 1100 Orebody, Mount Isa, Ud, Queensland, Australia, Trans. Aust. Ins. Min. Metall, 89.

Bodi, Laszlo, G. Hunt, and T. Lahnalampi. (1996). Development and Shear Strength Parameters of Paste Backfill. Proceedings of the Third International Conference on Tailings and Mine Waste '96. Fort Collins, Colorado, January 16–19. Rotterdam: A.A. Balkema, pp. 169–178.

Cincilla. W.A., D.A. Landriault, and R. Verburg. (1997). Application of Paste Technology to Surface Disposal of Mineral Wastes. Proceedings of the Fourth International Conference on Tailings and Mine Waste '97. Fort Collins, Colorado, January 13–16. Rotterdam: A.A. Balkema, pp. 343–356.

Cincilla, W.A. and D.R. East. (1989). The Design and Construction of Fully Drained Systems for Storage of Mineral Wastes. Proceedings of the 21st Mid-Atlantic Industrial Waste Conference. Harrisburg, Pennsylvania, June, pp. 247–258.

Gignac, L. (1978). Filling Practice at Brunswick Mining No. 12 Mine. Mining with Backfill. Proc. 12th Can. Rock Mech. Symp. Sudbury, Can. Inst. Min. Metal.

Grice, A.G. (1989). Fill Research at Mount Isa Mines Limited. Innovation in Mining Backfill Technology, Hassani et al. Ed.

Hambley, D.F. (1987). Design of Ore Pass Systems for Underground Mines. Bull. Can. Inst. Metall.

Landriault, D.A. and R. Tenbergen. (1995). The Present State of Paste Fill in Canadian Underground Mining. Proceedings of the 97th Annual Meeting of the C.I.M. Rock Mechanics and Strata Control Session. Halifax, Nova Scotia, May 14–18, 1995.

Landriault, D.A. (1995). Paste Backfill Mix Design for Canadian Underground Hard Rock Mining. Proceedings of the 97th Annual General Meeting of the C.I.M. Rock Mechanics and Strata Control Session. Halifax, Nova Scotia, May 14–18, 1995.

Landriault, D.A. and W. Lidkea. (August 1993). Paste Fill and High Density Slurry Fill. International Congress on Mine Design. Queens University, Kingston, Ontario, Canada.

Landriault, D.A. (September 1992). Paste Fill at Inco. 5th International Symposium on Mining with Backfill. Johannesburg, South Africa.

Mitchell, R.J., R.S. Olsen, and J.D. Smith. (1981). Prediction of Stable Excavation Spans to Mine Backfill, Report for CANMET, DSS File 18SQ-23440-9-9077.

Quesnel W.J.F., and H. deRuiter. (1989). The Assessment of Cemented Rockfill for Regional and Local Support in a Rockburst Environment. Lac Minerals Ltd., Macassa Division. Innovation in Mining Backfill Technology, Hassani et al., Eds.

Robinsky, E. (1979). Tailings Disposal by the Thickened Discharge Method for Improved Economy and Environmental Control. Tailing Disposal Today, G. Argall, Ed.. San Francisco: Miller Freeman.

Swan, G. (December 1985). A New Approach to Cemented Backfill Design. Bull. Can. Inst. Min. Metall.

Yu, T.R. (1983). Ground Support Consolidated Rockfill. Proc. Symp. Underground Support Systems, Sudbury, Can. Inst. Min. Metal., pp. 85–91.

Yu, T.R. and D. Counter. (1983). Backfill Practice and Technology at Kidd Creek Mines. Can. Inst. Min. Metal.

Yu, T.R. (1989). Some Factors Relating to the stability of Consolidated Rockfill at Kidd Creek. Innovation in Mining Backfill Technology, Hassani et al., Eds.

CHAPTER 70

Mining Dilution in Moderate- to Narrow-Width Deposits

Charles R. Tatman*

70.1 INTRODUCTION

Mine dilution has historically been defined as "the contamination of ore with barren waste wall rock." The term "below-cutoff-grade material" is an appropriate replacement for the term "barren waste wall rock" to accommodate instances where ore-waste boundaries are not well defined. Some authors have referred to external and internal dilution. External dilution has the same meaning as presented above, while internal dilution refers to below-cutoff-grade material within reserve blocks. Only external dilution is considered here. Internal dilution is a factor to be considered when estimating ore reserve blocks.

A survey of 22 mines throughout Canada identified nine variations on a definition of dilution (Pakalnis 1986). These are presented in Table 70.1.

Dilution is most commonly defined by the descriptions given in Equations 1 and 2 according to a review of Canadian mining practices (Scoble and Moss 1994). However, Equation 2 does not fully define the magnitude of dilution. For example, a 2:1 slough-to-ore ratio would produce a dilution factor of 50% as compared to a dilution factor of 200% as expressed by Equation 1. Thus, Equation 1 will be used throughout this paper.

By definition, dilution is calculated as a percentage:

$$\frac{\text{units of dilution} \times 100}{\text{units of ore}} = \text{percent dilution}$$

Olsson and Thorshag (1986) recommend that dilution be described as primary and secondary dilution. Annels (1996) refers to intentional and unintentional dilution. Pakalnis et al. (1995a) categorize dilution as internal and external. Olsson and Thorshag's terminology has been adapted in this paper and is inclusive of Annels' intentional and Pakalnis et al.'s external dilution. Thus,

primary dilution + secondary dilution = total dilution

Primary dilution is dilution inherent in the mining method selected. For example, a vein 0.8 m (2.6 ft) wide being extracted in a shrinkage stope designed to be mined at a width of 1.2 m (4.0 ft) has a primary dilution factor of 50%. *Secondary dilution* is dilution incurred beyond the planned stope dimensions. Using the previous example, hanging-wall rock of poor quality yields 0.6 m (2.0 ft) of hanging-wall slough, which results in a secondary dilution of 75% for a total dilution of 125%, that is [(0.4 m + 0.6 m)/0.8 m] × 100.

Nearly all underground mines experience some form of dilution resulting from mining beyond the ore boundaries, caving roofs, heaving floors, and caving and sloughing hanging walls and footwalls. The amount of dilution can range from 5% to 15% in well-engineered and managed stratabound deposits and up to

TABLE 70.1 Definitions of dilution*

Equation (1)	Dilution = (Tons waste mined)/(Tons ore mined)
Equation (2)	Dilution = (Tons waste mined)/(Tons ore mined + tons waste mined)
Equation (3)	Dilution = (Undiluted in-situ grade as derived from drill holes)/(Sample assay grade at draw points).
Equation (4)	Dilution = (Undiluted in-situ grade reserves)/(Mill head grades obtained from same tonnage)
Equation (5)	Dilution = (Tonnage mucked – tonnage blasted)/(tonnage blasted)
Equation (6)	Dilution = Difference between backfill tonnage actually placed and theoretically required to fill void.
Equation (7)	Dilution = Dilution visually observed and assessed
Equation (8)	Dilution = ("X" amount of meters of foot wall slough + "Y" amount of hanging wall slough)/(ore width)
Equation (9)	Dilution = (Tons drawn from stopes)/(Calculated reserve tonnage) over the last 10 years

*Note: Dilution is generally expressed as a percentage.

50% to 60% in poorly engineered and managed narrow vein deposits.

The application of an appropriate dilution factor to convert a resource block to a reserve block is an important component of the ore reserve estimate. For an undeveloped underground deposit, the dilution factor is the best estimate made by an experienced engineer and is nearly always open to discussion. For an operating mine, the dilution factor should be carefully measured and adjusted through actual experience.

By industry definition, the control of dilution in operating mines is most often referred to as the grade control program. As will be demonstrated in the following discussion, the grade control program is often the most important controllable component of mine profitability.

Errors in reserve grade estimates are often concealed in the dilution factor of an operating mine. In developing mines, negative errors in the reserve grade can be a major contributor to an apparent excessive dilution rate during the early stage of mining. A second major contributor to apparent high dilution rates is inaccurate determination of the head grade and tonnage at the processing plant. Any evaluation of mine dilution rates should include studies of the statistical accuracy of the processing plant's sampling system and metallurgical balance.

The net effect of dilution is to reduce cash flow from operations. Mining, transporting, and processing waste are costly, and in fact are often the most costly factors in an underground mine. Each unit of dilution (waste) that is mined, transported, and processed replaces a unit of profit-generating ore in the production capacity of the mine. This is demonstrated in Table 70.2.

---

* Behre Dolbear & Company, Walkerville, Montana.

TABLE 70.2 Effect of dilution

	15% dilution	30% dilution
Tonnes processed per year	360,000	360,000
Grade (0.35 ozAu/t reserve)	0.304	0.269
Process plant recovery	95.0%	94.4%
Gold sales—ounces	103,968	91,417
Revenue (US$350/ozAu)	US$36,388,800	US$31,995,950
Operating costs (US$83.68/ton)	US$30,124,800	US$30,124,800
Cash flow from operations	US$6,264,000	US$1,871,150

The information presented in Table 70.2 shows that a 100% increase in dilution over and above the 15% dilution allowed when converting resources to reserves has a cost of over US$4 million per year in reduced cash flow. Note that processing plant ore recovery is reduced from 95% to 94.4%, which is a seldom-considered effect of increased dilution.

Table 70.3 shows that a 5% increase in dilution from 15% to 20% will reduce cash flow from operations by 15.2%. Table 70.4 shows that a 5% decrease in mining costs increases cash flow by 14.7%. Reducing dilution is a shorter path to profit than decreasing mining costs or increasing through-put. However, optimization of all factors is important to mine profitability.

Estimates of dilution factors during the feasibility phase of mine development are a significant variable in any analysis of cash flow. The project is subject to high risk if the dilution factors selected are not the result of underground test stoping.

70.2 MYTHS AND CERTAINTIES

The period between the 1970s and late 1980s saw a rapid decline in underground mining in North America, followed by a rapid increase in the number of open-pit mines as low-grade gold deposits became economic. As underground mining became increasingly important in the late 1980s, many of the management and technical personnel employed to develop these new underground mines were hired from open-pit operations. Unfortunately, many of the management and technical skills developed over many years of underground mining were no longer available, and practices that had proven effective in open-pit mines were applied directly to the new underground mines. While some open-pit practices work underground, others do not, and these are referred to below as myths. Likewise, experience dictates that there are certain fundamental practices associated with minimizing dilution in underground mining.

70.2.1 Myths

1. Mechanizing a mining operation and/or increasing production capacity results in lower costs and increased profits.
   Only if the dilution factor can be held constant.

2. Sufficient development ahead of production requirements is a luxury that business types consider an unwarranted expenditure of reportable profits.
   Insufficient development ahead of production often leads to mining below-cutoff-grade stopes and less-than-prudent monitoring of dilution by mine supervisors.

TABLE 70.3 Effect of dilution on cash flow

Percent dilution	5%	10%	15%	20%	25%	30%	35%
Dry tons processed	360,000	360,000	360,000	360,000	360,000	360,000	360,000
Concentrator head grade	0.333	0.318	0.304	0.292	0.280	0.269	0.259
Concentrator recovery	95.5%	95.3%	95.0%	94.8%	94.6%	94.4%	94.2%
Gold sales—ounces	114,485	109,099	103,968	99,654	95,357	91,417	87,822
Revenue-$350/oz (000)	$40,069	$38,185	$36,389	$34,879	$33,375	$31,996	$30,741
Operating Costs—(000)	$30,125	$30,125	$30,125	$30,125	$30,125	$30,125	$30,125
Cash Flow—(000)	$9,944	$8,060	$6,264	$4,754	$3,250	$1,871	$616
Percent decrease in cash flow	0%	19.0%	37.0%	52.2%	67.3%	81.2%	93.3%

TABLE 70.4 Effect of mining costs reduction on cash flow

	Dilution remains constant at 15%; Concentrator recovery remains constant at 95.0%						
Percent of mining costs	100%	95%	90%	85%	80%	75%	70%
Dry tons processed	360,000	360,000	360,000	360,000	360,000	360,000	360,000
Concentrator head grade	0.304	0.304	0.304	0.304	0.304	0.304	0.304
Gold sales—ounces	103,968	103,968	103,968	103,968	103,968	103,968	103,968
Revenue—$350/oz (000)	$36,389	$36,389	$36,389	$36,389	$36,389	$36,389	$36,389
Operating costs*—(000)							
Mining @ $51.08/ton	$18,389	$17,469	$16,550	$15,630	$14,711	$13,792	$12,872
Concentrator @ $19.10/ton	$6,876	$6,876	$6,876	$6,876	$6,876	$6,876	$6,876
General & Admin @ $5.85/ton	$2,106	$2,106	$2,106	$2,106	$2,106	$2,106	$2,106
Indirect @ $7.65/ton	$2,754	$2,754	$2,754	$2,754	$2,754	$2,754	$2,754
Total operating costs—(000)	$30,125	$29,205	$28,286	$27,366	$26,447	$25,528	$24,608
Cash flow from operations*—(000)	$6,264	$7,184	$8,103	$9,023	$9,942	$10,861	$11,781
Percent increase in cash flow	0%	14.7%	29.4%	44.1%	58.7%	73.4%	88.1%

*Note: Total may not be exact because of rounding.

3. Mechanized operations require a less intense level of supervision and grade control resulting in substantial savings in salaries and sampling costs.
   True only if effective grade control can be maintained.

4. A mechanized operation is always more cost effective than a labor-intensive manual operation.
   Only if the cost savings in labor are not offset by increased costs of dilution.

70.2.2 Certainties

1. Every underground mining operation has a natural production rate and a natural mining method. These natural rates are the optimum profitability rate per unit. If attempts are made to exceed the natural production rate or an incorrect mining method is chosen, the inevitable result will be intolerable levels of dilution.

2. The lack of adequate supervision and management diligence will result in increased dilution. Incentive plans do not replace adequate supervision.

3. The failure to reconcile extracted ore reserves with production practices in which dilution is measured through systematic sampling at each step of the process will result in a loss of management control and the inability to forecast financial projections accurately .

The magnitude of the dilution problem varies greatly with the type of deposit. The engineers and geologists at Behre Dolbear have arbitrarily selected three general descriptions of deposits to show the difficulties inherent in estimating dilution factors.

Bulk Deposits. These are large replacement- and massive-sulfide-type deposits where the source of dilution is usually confined to the margins of the deposit and fill contacts. Examples of the mining methods employed in extracting these types of deposits include vertical crater retreat (VCR), block caving, sublevel caving, and sublevel blasthole stopes.

Stratiform Deposits. These are usually bedded deposits where the source of dilution is most likely to be from falling roof and heaving floor material and/or use of oversized equipment. Examples of the mining methods employed include room-and-pillar and longwall.

Moderate- to Narrow-Width Deposits. These are veins and replacement-type deposits where the sources of dilution include footwall and hanging wall overbreak, structural failures, and grade boundaries. Examples of mining methods include cut-and-fill (either overhand or underhand), drift-and-fill, shrinkage stoping, and sublevel stoping.

It is the moderate-to-narrow-width-deposit category that presents the greatest challenge to an engineer who must estimate dilution prior to placing the deposit into production. The other two categories of deposits, bulk and stratiform, will not be given further attention in this discussion.

70.3 EXAMPLES

Several examples are presented to demonstrate the types of problems encountered in estimating and controlling dilution in underground mines.

70.3.1 Example 1

Preproduction planning for a new gold mine was based on a combination of open blasthole and shrinkage stoping. Diamond-drill holes were spaced on 12.5-m (40.75-ft) centers. The deposit dipped 50° to 60°. The resource blocks were diluted to a minimum of 1.53 m (6.0 ft) wide for the proposed shrinkage stopes (primary dilution). A secondary dilution rate was calculated by adding 0.33 m (1.07 ft) to the perimeter of the shrinkage stopes and 0.5 m (1.63 ft) to the blasthole stopes, which resulted

in an average secondary dilution rate of 17.5%. The test mining program consisted of accessing one stope block on two levels and driving a raise through the block. The ore from the access drifts and the raise was processed as a bulk sample. This secondary dilution rate appeared reasonable after completion of the underground test stoping; however, the actual amount of dilution in the underground test may have been masked by the adjustment of sample grades because of the large nugget factor present.

The blasthole stopes were changed to shrinkage stopes after work in the development drifts showed that the geological environment made blasthole stoping unfeasible. The stopes were advanced with breasting rounds using 2.4-m (8-ft) drill steel because of planar discontinuities and undulations (en echelon deposition) in the deposit. The mine had an excellent grade control program in which geologists on each shift mapped, sampled, and marked each face for the miners. Grade control records were excellent. Muck samples were taken as the stopes were pulled. The test stope proved to be the most consistent grade and had the best ground conditions of the eight ore shoots in the deposit. When the ore shoots were developed, it was found that they undulated in all three directions and had many minor fault offsets. Attempts to follow the ore closely resulted in bulges and bellies in the hanging wall that caved when the stope was drawn.

The preproduction reserve grade of 0.31 oz of gold per ton (Au/t) was never achieved, and in fact, the processing plant head grade averaged 0.21 oz Au/t during the 6 months of production before the mine was closed as being uneconomic. The minimum shrinkage stope width (primary dilution) plus the secondary dilution rate of 17.5% used in the reserve calculation plus the shortfall in head grade of 32% resulted in a total actual dilution rate of 47.5%.

70.3.2 Example 2

A second example is provided by a base metal vein deposit exploited by a combination of shrinkage stoping in the narrow sections and cut-and-fill stoping in the wider sections. This was an old mine with a classic grade control program managed by the geology department and maintained at a reasonable level until the mine was nationalized. However, producing concentrator feed became the primary production criterion. The grade control geologist complained that the miners, with the concurrence of the production supervisors, were not following the face ore markup. In addition, the length of the shrinkage stope had remained constant as the mine deepened, resulting in excessive hanging wall dilution when the stopes were drawn down. Problems in the grade control program were exacerbated by the fact that development had fallen behind production requirements, and wide stopes were required to provide concentrator feed. Reconciliation of the stope reserve estimates and production was no longer maintained. A sample of grade control records showed that overbreak in the shrinkage stopes ranged up to 62%. The comparison of average reserve grade to average production grade on an annualized basis showed a dilution rate of 56%.

70.3.3 Example 3

A third example shows that a good grade control system and excellent operations management can turn a dilution disaster around to generate a profitable mining operation. In this mine, historic mining records showed that pre-1942 operations had a dilution factor of 24.6% when working veins averaging 1.4 m (4.7 ft) wide in stopes averaging 1.8 m (6.0 ft) wide. The mine was reopened in 1996. The operator elected to access the mine with a new decline and to mechanize mining with trackless haulage and sublevel and longhole stoping. The initial three stopes showed an extraction of 128% of the reserve tonnage containing 78% of the reserve metal.

Management recognized the magnitude of the dilution problem and elected to return to manual shrinkage stoping. The reserve calculation methodology was changed from a minimum stope width (primary dilution) of 1.2 m (4.0 ft) to a minimum of 1.5 m (5.0 ft) and adding 0.30 m (1.0 ft) of dilution to all reserve blocks over 1.5 m (5.0 in) thick (secondary dilution). The mine now takes a 2.45-m (8.0-ft) back round in the stopes and has an excellent grade control program on all shifts. The next four shrinkage stopes had an average extraction of 102% of the reserve tonnage containing 116% of the reserve metal. However, the mine is again behind on development and short of active stopes. This has required that the manager maximize production from each stope in the short term; as a result, in the manager's opinion, the mine currently suffers from excessive dilution. He plans to increase grade control supervision as development catches up. He also has the option of reducing the back rounds to a 1.8-m (6-ft) steel, which should decrease overbreak.

70.3.4 Example 4

The fourth example is an innovative small Latin American mine. The veins dip 42° and average 0.5 m (1.6 ft) in true width. Grade averages 1.2 oz Au/t. The mining method is manual random room-and-pillar stoping with manual and mechanized (slusher) stope cleaning. The prior operator realized a 0.36-oz Au/t mill head utilizing a stope width of 1.2 m (4.0 ft). Based on stope width, the calculated dilution would be 150% of the tonnage mined and a decrease in the mill feed grade from 1.2 to 0.34 oz Au/t. Actual mining practice showed a dilution rate of 230%. The mine was sold to a new operator who reduced the stope width to 0.9 m (3.0 ft) and the dilution rate to 88%, which should have yielded a processing plant feed grade of 0.42 oz Au/t. Because of the size difference between broken ore and waste, the mine has found that wet screening at 50 mm (2 in) increases grade to 0.57 oz Au/t. A further hand-sorting step increases the mill feed grade to 0.65 oz Au/t. The grade of the screen reject is not known; however, visual inspection of the dark-colored waste shows little or no gold-bearing white quartz.

70.4 CAUSE AND EFFECT

Dilution of the reserve grade and tonnage during extraction is a function of three primary factors: geological environment, mining method, and mine supervision.

70.4.1 Geological Environment

The thickness of the ore, geomechanical characteristics of the ore and enclosing waste rock, boundaries of the mineralization, and dip or inclination of the deposit are important geological conditions that control the amount of dilution to be expected during extraction. The geological environment often leaves few options in the selection of the mining method.

Logging diamond-drill core from underground deposits requires greater attention to detail than what is usually afforded drill core from open-pit deposits. Descriptions of the hanging wall and footwall become as important as descriptions of the mineralization. Structural details often become key elements in estimating dilution. Diamond-drill core recovery also is a key criterion; low core recovery rates usually equate to high dilution rates.

To an engineer attempting to estimate the dilution rate for a yet-to-be-developed deposit, a scientific measurement of the geological environment is a critical criterion. The rock quality designation (RQD) data obtained from diamond-drill core are minimal requirements.

The Norweigan Geological Institute's (NGI) Quality Index (Mathews et al. 1981) classifies rock quality in terms of rock quality as presented in Table 70.5.

A more inclusive description of the geological environment that can be obtained from diamond-drill core is the rock mass strength (RMS) index. The RMS consists of four measurements:

TABLE 70.5 NGI rock quality index

Rock quality	RQD
Very Poor	0–25
Poor	25–50
Fair	50–75
Good	75–90
Excellent	90–100

intact rock strength (IRS), rock quality designation (RQD), joint spacing (Js), and joint condition-water (Jc) (Laubscher 1990). These measurements can then be converted by a geotechnical engineer to the design rock mass strength (DRMS) to form the basis for an engineered opening design and an estimate of dilution (Laubscher 1990).

Well-developed cross sections are a necessity for mine planning and dilution control, and physical and/or computer-simulated models of deposits can be very helpful. Three-dimensional, finite-element modeling is often a useful tool for large bulk deposits.

Each mineral deposit is unique and, as a result, the amount of dilution resulting from the geological environment is likewise unique. The geologic factors important to estimating future dilution when developing a deposit are—

1. Characteristics of the ore zone,
2. Nature of the boundary between ore and waste, and
3. Structural integrity of the surrounding rock.

The characteristics of the ore zone are of critical importance in estimating dilution. Of primary consideration is the nature of the mineralization. Can the grade be estimated by visible ore minerals or by secondary accessory minerals? If the grade cannot be visually estimated, increased levels of dilution are common, often resulting from delays in receiving sample assays. Do the values include hard minerals surrounded by soft gangue, which makes accurate sampling difficult? Ore zones are commonly found in fault structures associated with structurally incompetent rock and alteration zones of no value. Many times the entire structurally weak zone must be taken during stoping, so the reserve block estimate should include the diluting waste material.

The nature of ore-waste boundaries for both walls is critical to estimating dilution. An abrupt, free-breaking, structural contact implies less dilution. A gradational grade boundary implies a greater dilution rate, as does an ore zone that undulates between footwall and hanging wall contacts or that has planar discontinuities paralleling the contact.

The structural integrity of the ore and surrounding rock determine the stope and pillar dimensions in balance with an acceptable dilution factor. Several engineered systems are available for predicting dilution factors controlled by the structural integrity of the ore and surrounding rock. These design systems are discussed later.

70.4.2 Mining Method Selection

The selection of a mining method is determined by the structural integrity of the ore and enclosing waste. The basic mining method(s) can usually be selected adequately from the parameters defined by the diamond-drill core, RQD logs, and test stopes.

Careful attention should be given to stoping methods and stope design parameters in older mines. The mining methods and stope design parameters used in the upper levels cannot always be carried downdip without modification if acceptable dilution rates are to be maintained. Increasing ground pressure with depth will usually require reducing stope strike length if hanging wall sloughing is to be avoided or converting the stope to longwall undercut-and-fill, such as was done at the Lucky Friday Mine (but for different reasons).

Most mines today use cut-and-fill, shrinkage stoping, or sublevel longhole stoping to extract moderate-to-narrow-width deposits. The significant sources of dilution with these mining methods are overbreak to permit equipment clearance, planar continuities, grade boundaries, sloughing walls, drilling and blasting overbreak, and contamination of the ore with fill material.

Each mining method potentially has some unique dilution problems.

Cut-and-Fill Stoping. Diesel or electrically powered microstopers with capacities of 680 to 1,360 kg (1,500 to 3,000 lb) permit mechanized cut-and-fill mining down to an effective stope width of 1.1 m (3.9 ft) using hand-held pneumatic rock drills (Cardenas 1983). Fully mechanized cut-and-fill stopes can be operated in widths as narrow as 2.0 m (6.6 ft). Cut-and-fill stopes, when advanced with short drill rounds, can provide moderate selectivity in deposits with planar discontinuities. Following grade boundaries can prove difficult in mines that operate 21 shifts per week and where assay results are often not available on a timely basis.

Many older cut-and-fill mines are operated without cemented backfill. This often results in mixing of high-grade fines with the fill and results in either a loss of ore or dilution of ore with backfill when the upper layer of fill is mined to recover the high-grade fines. In some mines with high-grade fines, large vacuum machines are being used to clean the cemented fill floor.

Shrinkage Stoping. As with cut-and-fill stoping, shrinkage stoping can be moderately selective, and dilution control is usually a function of miner discipline. Mechanically leveling the working floor before drilling can improve drill hole alignment. Since a major portion of the broken ore is kept in the stope as a working floor, sloughing of the hanging wall and footwall over time can be a major contributor to dilution. Such dilution is difficult to measure without careful material accounting. Many mines with high-grade fines are washing down the stopes before stope closure.

Several operators have semimechanized their shrinkage stopes to improve productivity. Mucking with LHDs from draw-points or loading trucks from air gate chutes are becoming common practices in shallow mines accessed by declines. To improve productivity, one operator has converted from conventional raises to Alimak raises in shrinkage stopes so that one raise provides access and services to two adjacent stopes.

Sublevel Longhole Stoping. Sublevel stoping has become popular in recent years as a replacement for labor-intensive shrinkage stoping. One Scandinavian country prohibits the use of hand-held drills because they exceed noise restrictions, and many companies are concerned about carpal tunnel syndrome problems related to hand-held drills. Mechanized drill units with enclosed operator cabs are used in sublevel stoping to mitigate these concerns.

There have been a number of well-documented test programs in recent years in which shrinkage stoping has been replaced with sublevel stoping in narrow deposits. Many of these tests have been less than successful, primarily because of excessive levels of dilution caused by planar discontinuities. A number of mines are successfully using sublevel stoping where the deposit is as narrow as 1.00 m (3.3 ft). The controlling factors are the number and magnitude of planar discontinuities. Where the structure of the deposit can be well defined before stoping and the sublevel interval matched to the discontinuities, the method has been successfully applied with acceptable dilution levels.

A frequent cause of excessive dilution in sublevel stoping has been overbreaking, which results from high powder factors associated with using blasthole drills to drill larger-diameter holes. Several unique drill patterns have been developed to reduce the powder factor. The use of low-strength explosives has been successfully applied at several mines.

Horizontal blastholes drilled from an Alamik Raise Climber have been tested at several mines with some success in narrow

deposits. The governing factors are the length of the holes that can be drilled and the number of raises required to match the discontinuities.

Drift-and-Fill Stoping. Since its introduction in Japan, drift-and-fill stoping is rapidly gaining acceptance in North America and was implemented recently in a new mine in Peru. It is difficult to estimate secondary dilution for this mining method, and each use should be considered as unique. Secondary dilution is a factor of entry angle, size of heading, and length of rounds. It is common for 0.15 m (0.5 ft) of the concrete from adjacent rounds to slough into the new round during blasting.

70.4.3 Mine Supervision

The control of dilution is a top-down management prerogative. Some managers place maximum importance on meeting daily tonnage goals. The good manager places maximum importance on meeting daily saleable product goals. As was shown in Tables 2, 3, and 4, minimizing dilution is the fast track to profitability. A quality dilution control program is a team effort: the geologist marks the ore limits; the miner drills the rounds; the front-line supervisor monitors the miner's performance; the mine manager demands goal achievement. In addition, an incentive program rewards saleable-value production. A high degree of confidence in the metallurgical balance by the entire staff is a prerequisite.

Supervisor and equipment operator training is a key ingredient in minimizing dilution. Supervisors must be trained to recognize dilution-generating work practices and have the leadership skills to take corrective action. Equipment operators must be trained to recognize grade boundaries and proper drilling and blasting practices to prevent overbreak. For a new mine, there is a learning curve, and the expectation of an additional 5% of dilution in the first operating year and $2^{1}/_{2}\%$ in the second year is not unreasonable.

The control of dilution is normally assigned to the grade control group, which is typically under the geology department because sampling is involved. It is important that every working face be inspected each shift by a supervisor, as well as by the grade control geologist or a technician, to ensure that drilling is within the marked ore limits and that the drill holes do not wander outside the ore limits. The grade control geologist or technician should also note and record the effectiveness of the previous blasted round and pass the information along to that shift's supervisor. The time to prevent dilution is before the round is fired.

Grade control programs should be evaluated to determine if there is an adequate number of front-line supervisors to provide supervision *at the face* after travel time and paperwork constraints are taken into account. There should be sufficient producing stopes to meet production requirements without resorting to hot muck practices. Stope preparation should provide an adequate number of developed stopes to allow for early stope closure because of poor grade or ground conditions.

Some mines do not have adequate waste-handling facilities. Some facilities are designed to backfill waste into worked-out stopes, a system that isn't always possible or convenient. Mines with inadequate waste-handling facilities often mix waste with ore, thus contributing to dilution.

Underground mines have historically used contractors for major development projects. On a few occasions, underground mines have used contractors for production mining. While this practice has been highly successful in surface mines, the divergent goals and unique supervisory requirements of the grade control program in underground mines often leave both the contractor and owner less than satisfied with the results.

A recent discussion with the general manager of a 1,800 mt/d (2,000 st/d) cut-and-fill mine emphasizes the importance of mine supervision. A program of weekly stope visits by the general manager to inspect the working face and discuss the effects of dilution on incentive program payments with miners has, in his

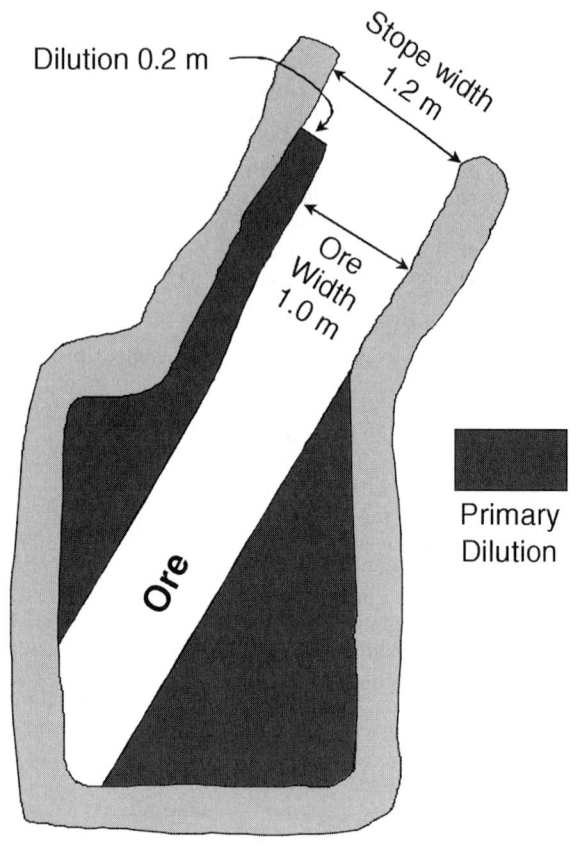

FIGURE 70.1   Primary dilution

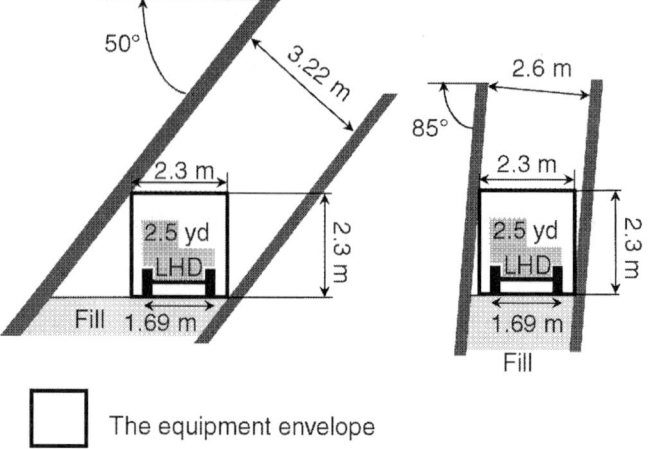

The equipment envelope

FIGURE 70.2   Equipment envelope. The box illustrating the equipment envelope is bigger than the dimensions of the equipment to provide adequate operating and maneuvering room.

TABLE 70.6   Manual stope width primary dilution

Ore width	Dilution width	Minimum stope width	Dilution
0.15 m (0.5 ft)	1.05 m (3.5 ft)	1.20 m (4.0 ft)	700%
0.30 m (1.0 ft)	0.90 m (3.0 ft)	1.20 m (4.0 ft)	300%
0.45 m (1.5 ft)	0.75 m (2.5 ft)	1.20 m (4.0 ft)	167%
0.60 m (2.0 ft)	0.60 m (2.0 ft)	1.20 m (4.0 ft)	100%
0.75 m (2.5 ft)	0.45 m (1.5 ft)	1.20 m (4.0 ft)	60%
0.90 m (3.0 ft)	0.30 m (1.0 ft)	1.20 m (4.0 ft)	33%
1.05 m (3.5 ft)	0.15 m (0.5 ft)	1.20 m (4.0 ft)	14%
1.20 m (4.0 ft)	0	1.20 m (4.0 ft)	0%

opinion, reduced dilution from 30% to 15%. In fact, after the first several months, the processing plant head grade was increased by an average of 14.2%.

Carefully constructed incentive programs can be a major tool in reducing dilution. The recent trend toward rewarding team performance based on meeting cost, metal production, and safety targets provides both supervisors and stope miners with the necessary incentives to minimize dilution. The old-style incentive programs that were based on tonnes broken or cubic units mined often had the opposite effect.

70.5   ESTIMATING DILUTION

Estimating dilution prior to actual mining is a difficult task, and the mining engineer preparing the estimate must use his best judgment. Information from adjacent mines or mines with similar conditions is often invaluable. The most reliable preproduction estimate of dilution is that gained from a well-engineered underground test program. The engineered dilution estimate has two components that should be calculated separately.

Total dilution = primary dilution + secondary dilution

70.5.1   Primary Dilution

Primary dilution is usually confined to narrow deposits. The thickness of the ore zone becomes the governing criterion for dilution as the ore narrows to less than the minimum stope width. Sources of primary dilution are shown in Figure 70.1, in which the shaded areas are waste rock that must be mined to maintain a minimum mining width.

Manual Stoping. Where stopes are mined by a miner with a jackleg or stoper drill, the stope must be wide enough for the miner and the machine. How wide is wide enough depends upon the miner and the inclination (dip) of the deposit. It requires

approximately 2.5 m (8 ft) of vertical standing room for a miner to work a stope taking a back round. Somewhat less head room is required for taking horizontal rounds

Near-vertical shrinkage stopes have been worked when they were as narrow as 0.5 m (1.6 ft). However, these are the exception and not the rule. To work successfully at these widths, miners must be carefully trained and not have previously experienced the luxury of working in wider stopes. Manual sublevel stopes have been worked down to 0.6 m (2 ft); again, careful training, close supervision, and often an incentive plan that rewards performance are required. A 1.2-m- (4.0-ft-) wide stope is an industry-accepted standard in steeply dipping deposits; wider stope widths are required as the dip flattens. At 50°, the nominal 1.25-m- (4.0-ft-) wide stope is increased to 1.50 m (5.0 ft) to provide adequate working room. Figure 70.2 illustrates the rapid increase in width with decreasing inclination of the stope. The additional width required should be carefully considered during equipment selection.

As easily discovered surface deposits are depleted, many companies are exploring for and developing underground mines. Many of these mines are narrow deposits where high unit values appear extremely attractive. A prudent predevelopment assumption is that manually extracted stopes should be a minimum of 1.2 m (4.0 ft) wide. This is applicable to manual shrinkage and cut-and-fill mining methods.

Table 70.6 shows the applicable primary dilution factor for ore zones less than 1.2 m (4.0 ft) wide.

A track drift 1.8 m (6.0 ft) wide by 2.5 m (8.0 ft) high is commonly used in narrow manual stopes. When the drift development muck is trammed as ore, it can have a higher dilution

TABLE 70.7   Mechanized stoping: minimum working widths

Equipment type	Equipment width	Operating space	Minimum stope width
0.5 yd³ LHD	1.0 m (3.2 ft)	0.60 m (2.0 ft)	1.6 m (5.2 ft)
1.0 yd³ LHD	1.20 m (4.0 ft)	0.60 m (2.0 ft)	1.80 m (6.0 ft)
1.5 yd³ LHD	1.30 m (4.5 ft)	0.60 m (2.0 ft)	1.90 m (6.5 ft)
2.5 yd³ LHD	1.70 m (5.4 ft)	0.60 m (2.0 ft)	2.30 m (7.4 ft)
Brand "A" jumbo	1.40 m (4.6 ft)	0.60 m (2.0 ft)	2.0 m (6.6 ft)
Brand "A" blast hole drill	1.40 m (4.6 ft)	0.60 m (2.0 ft)	2.0 m (6.6 ft)
Brand "B" jumbo	1.20 m (4.0 ft)	0.60 m (2.0 ft)	1.80 m (6.0 ft)
Brand "B" blast hole drill	1.20 m (4.0 ft)	0.60 m (2.0 ft)	1.80 m (6.0 ft)

factor than stope ore. For example, in a mine with an level interval of 46 m (150.0 ft), 5.3% of the ore is extracted by the level drift. If the drift material is classified as ore, then a separate, and different, drift dilution factor is often appropriate.

Narrow Mechanized Stoping. Primary dilution in mechanized stopes is principally a function of the width of the headings for the safe operation of the selected equipment. In practical terms, the minimum equipment working space is the width of the unit plus 0.3 m (1.0 ft) on each side and above the unit. Some countries require greater equipment clearance widths; 0.4 m (1.30 ft) is required in some Canadian provinces.

There is a limited selection of equipment available for narrow mechanized stopes. Table 70.7 presents a list of equipment and operating widths for selected narrow mechanized stoping equipment. Table 70.7 assumes that 0.3 m (1.0 ft) of operating space is required on each side of the equipment.

Assuming an ideal geological environment, sharp ore waste boundaries, and no planar undulations in the ore between levels, a 1.8-m- (6.0-ft-) wide mechanized stope is marginally feasible, and primary dilution can be calculated using the above table. In less than ideal conditions, a minimum 2.0-m (6.6-ft) stope width should be assumed for calculating primary dilution.

Much attention has been given in recent years to mechanization. Little attention has been given to the potential effects of mechanization on dilution. In sublevel mining systems, the sublevels are often driven on 7.7-m (25-ft) centers, resulting in 30% to 40% of the production coming from the sublevel development. Haulage level development headings are often driven 3.0 to 4.0 m (10 to 13 ft) wide to accommodate haulage trucks. Some mines have successfully utilized taking split rounds to prevent excessive dilution of the development ore from haulage and sublevel drifts. Close attention should be given as to whether the development material is above cutoff grade and should be treated as ore or whether it should be classified as waste. If the development material is treated as ore, the amount of primary dilution for the stope should be adjusted.

70.5.2 Secondary Dilution

Secondary dilution is dilution incurred beyond the planned stope dimensions (Figure 70.3). The most common causes of secondary dilution are sloughing and drilling and blasting practices. However, secondary dilution can be caused by a number of factors, including ground conditions, planar discontinuities and undulations, mining method, equipment, and work practices.

Guidelines for estimating secondary dilution prior to test stoping are discussed in the following paragraphs. These guidelines are generic and should be adjusted for site-specific conditions.

Sloughing is the collapse of the hanging wall and/or footwall because of rock failure. These failures can occur at high-stress points generated by planar discontinuities or stope spans exceeding the support capacity of the rock.

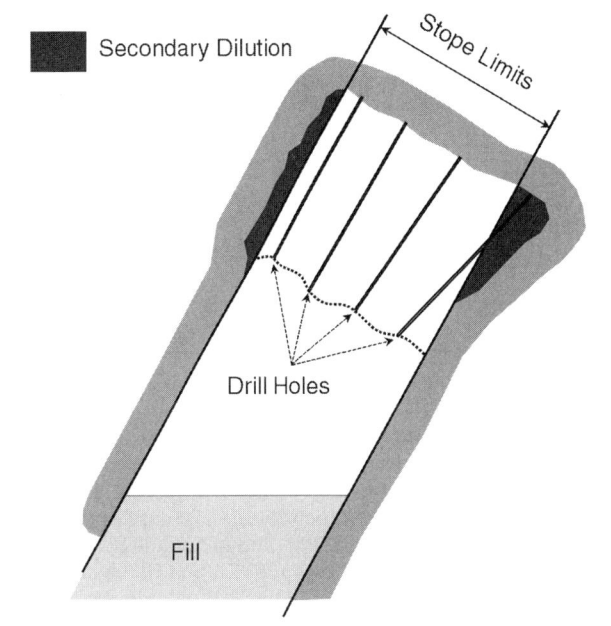

FIGURE 70.3   Secondary dilution

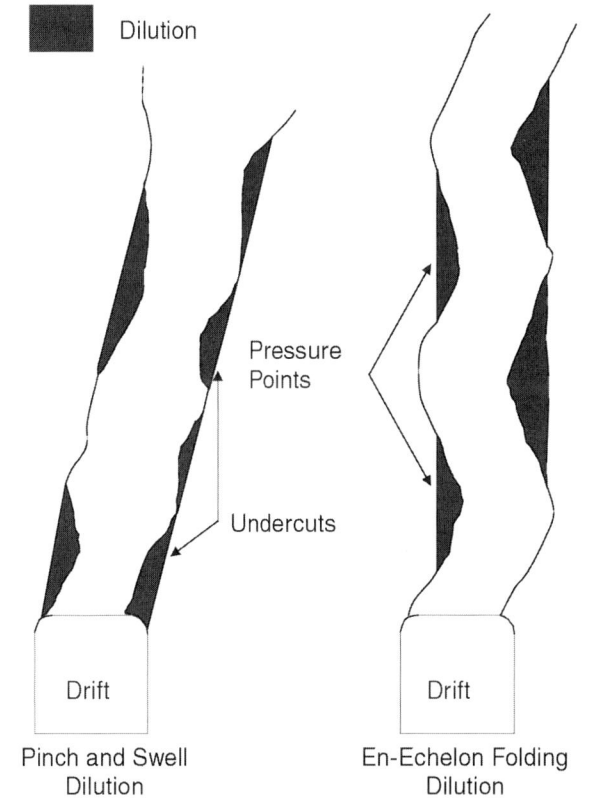

FIGURE 70.4   Planar discontinuities. The shaded areas are likely to slough, creating dilution even when drilling and blasting have avoided breaking these areas.

Planar discontinuities are commonly found in pinching and swelling deposits; they are less common in deposits that are tightly folded or contain en echelon mineralization. This type of mineralization is shown in Figure 70.4 and can occur in both the horizontal and vertical directions. The effect of en echelon mineralization on secondary dilution factors is described in example 1.

TABLE 70.8   Secondary dilution: the effect of overbreak

Designed stope width	0.15 m (0.5 ft) Overbreak on each wall	0.30 m (1.0 ft) Overbreak on each wall
1.20 m (4.0 ft)	25%	50%
1.50 m (5.0 ft)	20%	40%
1.80 m (6.0 ft)	17%	34%
2.15 m (7.0 ft)	14%	28%
2.45 m (8.0 ft)	13%	26%
2.75 m (9.0 ft)	11%	22%
3.00 m (10 ft)	10%	20%
Over 3.00 m (10 ft)	5%	10%

TABLE 70.9   Svedefo study: frequency of mining methods

Mining method	Distribution %
Cut and fill, manual	40
Cut and fill, mechanized	5
Sublevel stoping	20
Shrinkage stoping	15
Others	20

TABLE 70.10   Svedefo study: Zinkgruvan Mine test area parameters

Length of the test area	150 meters
Vertical height between sub-levels	50 meters
Average width	1.3 meters (varied from 0.5 m to 3.5 m)
Ore content	29,250 tonnes
Average dip	74 degrees

TABLE 70.11   Svedefo secondary dilution estimates

Cut and fill stoping	10% to 20% dilution
Shrinkage stoping	20% to 40% dilution
Sublevel stoping	30% to 40% dilution
Raise stoping	30% to 40% dilution

Short drill rounds can facilitate following the ore. However, the resulting bulges of the hanging wall become high-stress points that often fail over time. Pinching and swelling deposits have a high rate of undercut failures.

Span failures have been the subject of intense investigations in South Africa by Laubscher(1990) and in Canada by Mathews et al. (1981), Potvin (1988), Nickson (1992), and Hadjigeorouou and Leclair (1994). Pakalnis et al. (1995b) provide an excellent summary of the state of the art in span design and failure analysis. While these works are primally concerned with large open stopes, they are equally applicable to shrinkage, narrow sublevel, and cut-and-fill stoping methods.

Collection of the baseline geological data required for span design and failure analysis is best done with fresh diamond-drill core. The field logging geologist can usually be adequately trained by a specialist early in the drilling program to avoid costly relogging of the core by a geotechnical expert at a later date, often after the cores have degraded during storage.

Drilling and Blasting Practices.   A primary cause of secondary dilution is stope overbreak from drill and blasting. This can result from improperly aligned drill holes, high powder factors, exceeding equipment limitations, and less-than-ideal footwall and hanging wall conditions. A 0.15-m (0.5-ft) overbreak on the footwall and hanging wall is a commonly accepted normal occurrence. Table 70.8 shows the effect of overbreak for various stope widths.

The appropriate dilution for ore zones with a free-breaking identifiable hanging wall and footwall averages 0.15 m (0.5 ft) of overbreak on each side of the ore zone. Where there is a grade boundary contact, 0.30 m (1.0 ft) of overbreak on each side of the ore zone is appropriate. Deposits with one free-breaking, identifiable contact and one gradation contact should use an average of the two dilution rates as appropriate for the designed stope width.

Where the ore zone undulates and/or is discontinuous, overbreak rapidly increases. These features are very difficult to detect from diamond-drill core. Careful mapping of underground openings and test stopes is required to estimate dilution with any accuracy. Historic dilution factors for mines with similar types of mineralization and/or adjacent mines in the district can provide a realistic estimate of the dilution factor in a new mine.

70.5.3   Test Stoping

While secondary dilution factors can be estimated in the prefeasibility phase of development without accessing the deposit and conducting tests, the risk of underestimating the effects of dilution on cash flow during the final feasibility phase is very high. The degree of risk increases with the complexity of the geological environment. Drifts along the mineralized zone provide knowledge about strike axis, and a raise between levels provides knowledge about dip axis, but only a test stope can provide an engineered estimate of the dilution factor. In large, massive deposits, test stopes are often not practical, and a reasonable estimate of the secondary

dilution factor is sufficient, since the relative area of the ore boundary is small.

There is no substitute for underground exploration of the deposit prior to preparation of the final feasibility study and production decisions. Underground exploration should include exposing the deposit along both strike and dip axes, as well as the surrounding waste rock. The preferred underground exploration program would uncover the deposit on two levels and then connect them by several raises to expose the dip axis. Underground diamond drilling is often a necessary component of the underground exploration program. A well-engineered program will provide data to enable the deposit to be upgraded to the reserve category. A well-engineered test stope provides the most accurate preproduction estimate of the degree of secondary dilution expected.

Test stopes should be carefully mapped and sampled round by round, and the volume carefully calculated. Laser mapping of the completed stope provides excellent data. A comparison of the mined volume to the extracted volume often provides a measure of dilution from a sloughing hanging wall and/or footwall during drawdown. Careful sampling of each ore rail car or ore truck provides a estimate of the extracted grade. Bulk sampling of the extracted ore is a much more accurate measurement of grade and tonnage.

70.5.4   Experience with Dilution

Olsson and Thorshag (1986) studied 30 narrow mines in their Svedefo study. They classified narrow ore bodies as those less than 4 m (13.2 ft) wide, having a dip of more than 45°, and extending no deeper than 1,000 m (3,260 ft). The distribution of mining methods reviewed is presented in Table 70.9. The Svedefo study also included a comparative analysis of four mining methods used in the Zinkgruvan underground mine in Sweden. The basic parameters of the test areas are shown in Table 70.10, and the results of an analysis of secondary dilution is presented in Table 70.11.

A review of recent literature on four Canadian mines using shrinkage stoping reported secondary dilution rates of 10% to 15% with averages of 12.5% when mining widths varied from

TABLE 70.12   Primary dilution calculation

Sample location draft mark	Raw sample data		RQD >75 Corrected to 1.2 m minimum mining width		RQD <75 Corrected to 2.0 m minimum mining width	
	Width m	Grade g/t	Width m	Grade g/t	Width m	Grade g/t
DM 120	1.6	7.64	1.6	7.64	2.0	6.11
DM 122	1.4	10.19	1.4	10.19	2.0	7.13
DM 124	1.1	12.32	1.2	11.29	2.0	6.78
DM 126	0.8	16.14	1.2	10.76	2.0	6.46
DM 128	1.6	5.95	1.6	5.95	2.0	4.76
DM 130	2.0	4.67	2.0	4.67	2.0	4.67
DM 132	1.8	9.77	1.8	9.77	2.0	8.79
DM 134	1.4	10.19	1.4	10.19	2.0	7.13
DM 136	0.8	26.75	1.2	17.83	2.0	10.7
DM 138	0.9	14.01	1.2	10.51	2.0	6.30
DM 140	1.2	14.44	1.2	14.44	2.0	8.66
DM 142	1.7	7.22	1.7	7.22	2.0	6.14
DM 144	1.3	4.67	1.3	4.67	2.0	3.04
DM 146	0.7	22.10	1.2	12.89	2.0	7.74
DM 148	0.9	17.41	1.2	13.06	2.0	7.83
DM 150	1.2	15.29	1.2	15.29	2.0	9.17
Over 30 meters	1.28	10.92	1.40	9.95	2.0	6.96

1.5 to 2.0 m (4.9 to 6.5 ft). The same data sources showed that five mines using either conventional or mechanized cut-and-fill experienced secondary dilution rates ranging from 3% to 220% when mining widths varied from 1.8 to 3.5 m (6.0 to 11.4 ft). The highest dilution rate, 220%, was associated with the 1.8-m (6.0-ft) mining width. The nominal secondary dilution rate for the four mines was 12%.

70.6   CALCULATIONS OF DILUTION FACTORS

In this age of mean and lean staffing levels and reliance on complex computer programs to estimate resources and reserves, basic errors in applying dilution factors are not uncommon and can have a significant impact on estimates of reserves and processing plant feed grade. The result is distortions of mine economics. The correct application of primary and secondary dilution factors is illustrated in a model of a narrow vein gold deposit. The model represents one dimension of the stope block in a single-level development heading. The model uses shrinkage stoping with a minimum stope width of 1.2 m (4.0 ft). Nongrade boundaries are assumed; therefore, the dilution has no grade.

It is important that any grade capping be performed on individual samples before the addition of primary and secondary dilution.* Applying grade capping to the grade of the total stope width will result in an inflated estimate of resource-reserve grade.

70.6.1   Primary Dilution

Primary dilution should be applied to each sampled width, not to the average width of the stope block. The correct application of primary dilution factors is illustrated in Table 70.12. The table is based on two considerations: whether RQD is equal to or greater than 75, which is good to excellent rock quality, or whether RQD is less than 75.

The "raw sample" column shows actual grades and widths of a series of samples taken at 2-m (1.8-ft) intervals over a total interval of 30 m (98 ft). In this example, if the raw sample data are averaged without adjusting to the minimum mining width required by the RQD, the average stope width of 1.3 m (4.2 ft)

exceeds the 1.2-m- (4-ft-) minimum stope width and will not require the addition of primary dilution. However, averages are not mined. Diluting each raw sample to the required 1.2-m- (4-ft-) minimum stope width (for an RQD > 75) produces an entirely different average width and grade (1.4 m of 9.95 gm/mt [4.6 ft of 0.32 oz/st], 9.4% additional tonnes at 8.9% less grade). For the case where RQD < 75, all widths must be increased to a minimum of 2 m (6.6 ft), so the average grade is even lower (6.96 gm/mt [0.22 oz/st]). The correct primary dilution factor must be applied to each sample interval to arrive at a correct primary diluted tonnage and grade.

70.6.2   Secondary Dilution

Secondary dilution should be added to the stope width after primary dilution. If the ore has grade boundaries, then the sample grade dilution for each interval should be added individually to stope width and grade. Accuracy is sacrificed if an average grade for an increment of hanging wall and footwall is added as secondary dilution.

Computer programs facilitate calculating and applying average dilution grades; however, real increment values are required. The results can be that marginal-grade stopes can become ore-grade stopes, while in reality they are below cutoff grade. Contrary to the claims of several software providers, most computer programs, especially modified open-pit programs, can handle only internal dilution. The results of any computer-generated estimate of underground reserve, therefore, should be checked by calculating randomly selected stope blocks independently to ensure that primary and secondary dilution factors have been correctly applied.

70.7   GRADE CONTROL PROGRAMS

Most underground mines have some type of a grade control program. The importance of such a program to the profitability of a mine generally increases with the complexity of the geological environment. Canadian mines are known worldwide for their excellent grade control programs. These programs have enabled many low-grade underground gold mines to continue production during periods of depressed gold prices.

---

* Whether or not grade capping should be used is a highly debated issue not addressed in this paper. The comment applies where grade capping is used.

Grade control programs generally utilize a group of geologists assisted by samplers under the direction of a senior technical supervisor. The function is commonly assigned to the geology department.

The purpose of a grade control program is to provide–

- Technical assistance to the operating staff to control dilution,
- Data for reserve reconciliation,
- Data for estimating mine production grade, and
- A comparison of mine grade to processing plant head grade.

70.7.1 Technical Assistance

Grade control personnel generally provide assistance to the operators by defining the limits of the ore zone and guiding operators through planar discontinuities in stopes and development faces for each face advance. This service normally consists of marking ore zone limits before a round is drilled by mapping, measuring, and sampling each stope and development face. These activities should follow a tightly documented procedure to ensure continuity between the personnel involved and to provide long-term records that can be used to forecast future production.

The importance of the grade control program's technical assistance is greatly increased in deposits with planar discontinuities and undulations. Here the grade control geologist and/or engineer is called upon to provide operations with directions as to where the face should be advanced.

Where ore boundaries are not well defined or consist of assay walls, sampling plays a key role in maintaining the stope within the cutoff-grade ore zone. In these instances, it becomes increasingly important to coordinate assay turnaround time closely with stope operations to ensure that the stope is not advanced before the grade control department has the requisite assay data to mark the ore zone adequately before the next round is drilled.

In blasthole stopes, the grade control personnel are often called upon to sample the blasthole cuttings to determine the location of ore boundaries. Cuttings are sometimes used to train new drill operators so they will learn how to identify the ore boundary through color or texture changes.

Many well-designed grade control programs are used to define subore-grade zones within stope blocks so that they may be left, if the mining method permits, as pillars. A well-designed program can also provide operations with timely information on stopes that are not yielding the desired ore grade so that the stope can be discontinued.

Mine operations often use grade control programs to monitor and control stope dilution. Daily reports on overbreak, as shown by face samples and measurements, provide mine management with a current measure of the effectiveness of the grade control program.

70.7.2 Reserve Reconciliation

A comparison of the stope production grade with the reserve block grade provides a gross check of the accuracy of the reserve calculation; however, it does not provide detailed knowledge about the source of any discrepancies. Careful sampling and measuring can be used to estimate dilution more accurately and provide the data required to determine the validity of grade assignments for the reserve block. Data from the high number of samples collected during the grade control program should be used to identify trends in mineralization and values distribution, thus providing a valuable check on the average grade of the reserve block as calculated from samples from drill holes and/or development headings.

70.7.3 Production Grade Estimates and Comparison with Processing Plant Head Grade

Many mines require a daily, weekly, and/or monthly estimate of mine production grade. These estimates can be prepared from samples from the stope face, broken ore, and/or production drill hole cuttings as appropriate for the mining method used.

The grade control department often assists the mine planning department in preparing mine production forecasts. Accurate, current stope assay data and dilution measurements compiled by grade control personnel increase the credibility of the production forecasts.

The estimated mine grade also serves as a benchmark for comparing processing plant head grade. Particularly in operations where the head grade of the mill is back-calculated through a metallurgical balance, estimating mined grade serves as a check. Although mine grade and processing plant head grade may differ, the difference should be consistent within acceptable limits. If it is not, then further inquiry into why is warranted.

70.7.4 Measuring Dilution

Measuring dilution can be either quite inexpensive or very expensive, depending upon the degree of control and financial forecasting requirements. Some companies require detailed and accurate financial forecasting on a daily basis. This in turn requires detailed blending and monitoring of the processing plant head feed. Measuring and controlling dilution is the key element in any forecasting system, since it is the largest variable affecting mine output grade.

Unacceptable Dilution Measurements. The least costly but least accurate method of measuring dilution is to compare average reserve grade against processing plant head grade on an annual basis. This method involves a minimum of engineering effort and does not require any underground sampling, but neither does it provide an accurate measurement of dilution. This method tends to remove too little metal from the reserve, resulting in a gradual inflation of the remaining reserve grade. This is not an industry-accepted practice, nor does it meet the requirements of a reserve audit.

Minimal Dilution Measurements. A common practice is to calculate the tonnage and grade of ore removed from the reserve blocks on a monthly or annual basis and to compare the tenor of the material removed with the tenor of the processing plant feed. Any difference is classified as an adjustment often referred to as the "mine call factor." This is a relatively low-cost option since it requires mainly that the surveying department do the appropriate underground measurements and then calculate that portion of the reserve blocks that have been removed during the year. However, the method assumes that reserve estimates are accurate, which is often unwarranted in narrow vein deposits. Furthermore, a comparison of estimated reserve tonnage and grade with processing plant feed does not allow for ready identification of inappropriately designed or sloppy mining practices, again a problem more common in narrow vein deposits.

70.7.5 Preferred Dilution Measurements

Bulk Mining. For bulk mining methods such as block caving, vertical crater retreat, and large-block blasthole or sublevel stoping, each stope should be considered as a reserve block and the tonnage and grade to be extracted should be carefully calculated before mining commences. The tonnage of blasted ore is carefully estimated as the broken ore is extracted from the stope. A system of grade sampling of the car(s) and truck(s) is designed that is statistically valid. These calculations provide a running average grade to determine if secondary dilution is becoming a problem. It also provides a means for a limited amount of blending of underground production from several stopes to obtain a desired processing plant head grade. If sampling is performed correctly, an operator will be able to stop production from a subcutoff-grade stope. It is not uncommon for a stope reserve block of marginal grade to produce subcutoff-grade material because of sampling errors or unforeseen dilution.

Some mines drill sampling patterns upon completion of sublevel development to provide a higher sampling density prior to production drilling. The additional sampling data are combined with the original data and averaged or kriged to arrive at a more accurate estimate of the grade and volume of the reserve block. Other mines collect cuttings from blasthole drilling to adjust the estimated grade of the reserve block.

As a check, the volume of these types of stopes can be accurately measured upon completion by using a laser cavity monitoring system. The more usual technique is to estimate the volume removed from the stope through car or truck counts in conjunction with appropriate load factors.

Narrow Vein Mining. Many mines sample each lift in a shrinkage stope by collecting samples from the back or muck pile at statistically determined intervals. Observed sampling intervals range from 1.5 to 5.5 m (5.0 to 18 ft). If a muck pile sample is taken, it represents the grade of the broken ore. If a back sample is taken, and ore and overbreak are sampled separately, a measure of both ore and overbreak (dilution) is obtained. The width of ore and overbreak should be carefully measured and not estimated. Some mines measure widths in 0.5-m (1.5-ft) increments where measurements in centimeters would provide a more accurate measure of dilution. Each car or truck load drawn from the stope should have its weight estimated and its grade sampled. From the calculation of the tonnage and grade of material broken and the tonnage and grade drawn down, the staff can estimate the amount of sloughing from the hanging wall and the footwall during drawdown. A drawdown model is often prepared to predict the grade of material extracted during the drawdown.

In cut-and-fill stoping, each lift should be mapped, measured, and sampled at statistically determined intervals. The sampling and measuring procedures appropriate for shrinkage stoping are also applicable to cut-and-fill stopes. In addition, if uncemented sandfill is used, the sand floor should be periodically sampled to determine the amount of ore left in the sandfill (particularly heavy fines) and/or dilution from taking a portion of the sandfill as ore during the mucking cycle. As with shrinkage stopes, each (or a statistically representative number of) car or truck load of production should be sampled and recorded.

In sublevel stopes, each round in the sublevel development should be careful mapped, measured, and sampled for both ore and overbreak. Stope production grades can be adjusted during the development cycle for overbreak in the subdrifts. Measuring and sampling the subdrifts is very important because they can represent up to 30% of the reserve block. These data also furnish the basis for adjusting the reserve block grade if there is significant variance before stoping is started. On occasion, sublevel sampling may result in reclassifying the block as uneconomic. As in shrinkage and cut-and-fill stopes, car or truck loads of production ore should be recorded and sampled.

70.7.6 Sampling Practices

No discussion of dilution would be complete without comments on sampling practices. At many mines, practices have degenerated from taking samples with a 50- by 100-mm- (2- by 4-in-) size hammer and maul, to collecting shallow, diamond-sawn channel samples with full or partial removal on the channel between the saw cuts, to collecting random chips. Spaces between samples are often increased along with the degeneration of sample integrity. This evolution is often justified as a cost savings. In many cases, these shortcuts result in inaccurate measurements of the true grade of the face and may result in increased costs because of questionable mine planning decisions and uncontrolled dilution.

An example of the deterioration of a grade control program was observed over several years. Analysis of data from an underground gold mine that initially used shrinkage and cut-and-fill stoping and that is now being converted to mechanized sublevel stoping demonstrates the costs of modernization and the influence of an open-pit operating philosophy.

This mine was opened in the mid-1980s. The reserve and grade control programs were designed on the basis of statistical studies performed by a South African engineering group. From the mine's inception through the early 1990s, the reconciliation of mine production grade to processing plant feed grade ranged from 98% to 103%. Stope and drift samples were taken on 3-m (10-ft) centers. The stope development samples were combined with diamond-drill samples to estimate stope grade.

In 1996, sample spacing was increased to 6 m (20 ft) to reduce sampling labor and assaying costs. At this time, the mine also changed over to mechanized sublevel stoping. Mine production-to-processing plant feed grade reconciliation dropped to as low as 78% and rarely exceeded 90%. The reconciliation problem was attributed to increased dilution from the mechanized sublevel stoping method. A review of the sampling data strongly suggested that the reconciliation differences were directly attributable to the increase in sampling interval.

In mid-1997, a geostatistical study of the sampling interval was conducted by a consultant who made recommendations that the sampling interval be increased to 12 m (39 ft). A review of the geostatistical study by a recognized geostatistical expert raised serious questions as to the methodology used to arrive at the 12-m (39-ft) sampling interval. A review of the sampling data using kriging techniques developed by Francis Pitard suggests that the 3-m (10-ft) sampling interval was correct, and that the new 12-m (39-ft) sampling interval may be little better than no sampling at all.

No discussion of face sampling would be complete without a brief mention of unique sampling tools. These would include radiometric sampling in radioactive deposits such as uranium, color-enhancing chemicals used to detect chalcocite and molybdenite, black lights to detect scheelite, and ion absorption to detect and measure some precious metals. The Russians use a plutonium-based ion absorption device to measure lead and zinc grades in boreholes and mine cars. With proper calibration, these devices can provide a cost-effective grade-control tool.

Statistical tools are now available to design the required sampling frequency and size required to obtain a representative face sample. Modern spreadsheet programs contain many routine statistical functions and can calculate to a precision not matched by mainframe computers of just a few years ago. Modestly priced statistical packages are readily available. The method of face, muck pile, and car/truck sampling, combined with sample preparation procedures and analytical procedures, is referred to as the "sampling protocol." There are several eminent consultants available to supplement in-house capabilities. These tools should be used to their fullest. A regular system of check sampling should also be employed to verify that sampling procedures are being followed.

Each sample bag should be numbered and should contain location data and a code or name of the sampler. To be accurate, sampling must be a structured program with a written procedure. Samples should be weighed before assay preparation as a measure of adherence to the established procedures. Periodic checks should be made by screen-size analysis to verify that samples are representative and that procedures are being followed.

A systematic program of resampling (check samples) should be instituted that periodically verifies the techniques used by each sampler. This is a basic quality control-quality assurance program issue. Exploration programs are expected to duplicate 1 in 20 samples (5%) at random. This is a high percentage for a producing mine, but 1 in 50 or 1 in 100 is not unreasonable; a testing program can be run to determine the percentage required. Things change, and these regularly taken duplicate samples provide the data to either confirm no change or identify change when it occurs.

Extra caution should be used in developing sampling practices where there is a significant difference in friability between the valuable mineral(s) and gangue in the ore zone. Random chip samples often selectively oversample the friable material, resulting in a situation where the sample is adversely weighted in favor of the grade of the friable materials.

The foregoing sampling and estimating methodology is rapidly being neglected in many new mines or eliminated from older mines as a cost-savings measure. Admittedly, it is a costly function; however, it provides the engineering staff with an increased level of confidence in the reserve estimate and management with the tools to forecast production and measure performance. One can also expect that the routine quality assurance-quality control procedures being increasingly demanded of exploration programs by the financial community will result in the expectation that similar procedures will be required of producing mines. It will not longer be sufficient to say that "We tested that 8–10 years ago." Auditors will demand that the results be confirmed on an ongoing basis.

70.8 DILUTION FACTOR ADJUSTMENTS

An adjustment to the dilution factor that is used to convert resources to reserves is always controversial and is a red flag to reserve auditors. Adjustments should be fully documented and receive senior management approval before they are implemented. Adjustment of less than 10% of the total dilution should be substantiated over several years before a new factor is applied to ensure that it is not a one-time event. Changed conditions should be carefully documented if the increase is 10% or more.

The preproduction estimate of dilution is seldom correct and usually understates actual dilution experienced in production. The first year of production seldom yields a representative dilution rate as both miners and supervisors are climbing the learning curve. Many preproduction feasibility studies provide for a modest 5% to 10% of additional dilution in the first production year, a number that has proven to be reasonable. For new mines with complex geological environments, the application of an additional 10% of dilution in the first production year and 5% in the second production year is a reasonable consideration.

A modification of the mining method normally results in a change in the dilution factor. These periods should be carefully monitored and data collection closely supervised to ensure that any change in dilution rate is fully documented and understood.

70.9 ACKNOWLEDGMENTS

These comments were prepared with the assistance of Mr. Michael Easdon, who reviewed many drafts commenting on content, Mr. Rachal H. Lewis, Jr., who contributed to the content and accuracy of the paper, and Mr. David M. Abbott, Jr., who pulled it all together. This paper is a compilation of information from the sources listed in the references and the experiences of the author as a member of the Behre Dolbear's technical due diligence audit team.

70.10 REFERENCES

Annels, A.E. 1996. Ore Reserves: Errors and Classification. *Institution of Mining and Metallurgy, Section A,* Vol. 105 (A137–83, Sept.–Dec.), pp. A150–A162.
Cardenas, R.C. 1983. Huaron Mines Reserves in Narrow and Winding Veins. *World Mining,* June.
Hadjigeorgiou, J., and Leclair. 1994. Stability Methods: Internal Report. University of Laval.
Laubscher, D.H. 1990. A Geomechanics Classification System for the Rating of Rock Mass in Mine Design. *Journal of the South African Institution of Mining and Metallurgy,* Oct., pp. 257–273.
Mathews, K.E., E. Bose, D.C. Wyllie, and S.B.V. Stewart (Golder Associates). 1981. Prediction of Stable Excavation Spans for Mining at Depths below 1,000 Meters in Hard Rock. Canada Center for Mineral and Energy Technology, Serial No. OSQ80-00081, April.
Nickson, S. 1992. Cable Support Guidelines for Underground Hard Rock Mine Operations. M.S. thesis, University of British Columbia, Vancouver, BC.
Olsson, M., and H. Thorshag. 1986. Mining Technique for Narrow Orebodies. Swedish Detonic Research Foundation.
Pakalnis, R. 1986. Empirical Stope Design at Ruthan Mine. Ph.D. thesis, University of British Columbia, Vancouver, BC, 276 pp.
Pakalnis, R.C., K. Dunne, and K. Cook. 1995a. Design Guidelines for Sub-Level Retreat Mining Method. Canada Centre for Mineral and Energy Technology, Energy, Mines and Resources, DSS file No. 15 SQ.23440-2-9142.
Pakalnis, R.C., R. Poulin, and J. Hadligeorgiou. 1995b. Quantifying the Cost of Dilution in Underground Mines. *Mining Engineering,* Dec.
Potvin, Y. 1988. Empirical Open Stope Design in Canada. Ph.D. thesis, University of British Columbia, Vancouver, BC.
Scoble, M.J., and A. Moss. 1994. Dilution Underground Bulk Mining: Implications for Production Management. Mineral Resource Evaluation II: Methods and Case Histories. Geological Society Spec. Publi. 79, pp. 95–108.

CHAPTER 71

Design of Ore Passes

M.J. Beus,* W.G. Pariseau,† B.M. Stewart,* and S.R. Iverson*

71.1 INTRODUCTION

Ore passes are underground conduits for the gravity transport of broken ore, waste rock, and fill from one level of a mine to a lower level. Inclination of ore passes typically ranges from vertical to 30° (Pariseau 1966), and cross sections may be square, rectangular, or circular. Historically, cross-sectional areas were generally less than 5.6 m² (60 ft²) (Peele 1947). In recent years, much larger ore passes have come into use, especially in conjunction with open-pit mines where run-of-mine rock may contain boulders that are more than 1.5 m (5 ft) in size. Ore pass lengths range from 18 to over 180 m (60 to 600 ft), and undoubtedly there are even shorter and longer passes in existence. Fill passes from the surface may be exceptionally long. A level interval of 46 m (150 ft) is perhaps a representative ore pass length.

Ore passes serve two important purposes: transport and storage. The latter is essential to efficient mine operation. Without adequate storage, any slowdown or stoppage in one part of a mine's transport system (for example, a train derailment or conveyor belt tear) could bring the entire system to a costly halt. If ore passes intended for transport and collection of ore do not provide adequate underground storage, then additional storage facilities must be installed. Surface storage bins also help to buffer mine and mill. Total storage needed depends mainly on the production schedules of mine and mill and are thus site specific.

Components of the ore pass system include (1) the ore pass itself connecting (2) two or more levels in a mine; (3) top-end facilities incorporating material size and volume control mechanisms, such as grizzlies, crushers, and surge chambers; and (4) bottom-end structures to control material flow and enable loadout. Figure 71.1 shows a typical ore pass for load-haul-dump (LHD) machines.

71.2 ORE PASS DESIGN FACTORS

71.2.1 Introduction

Design of ore passes for transport requires that broken ore, waste rock, or fill will flow when the ore pass outlet is activated. Flow is essentially a process of continuous shear failure of the muck. The process is driven by gravity and resisted by friction and cohesion. Ore pass muck is thus an engineering material with physical properties that need to be considered in the design of ore passes for flow and stability. Two malfunctions of ore pass operations need to be prevented: (1) failure to flow, which results in hang-ups, and (2) failure to flow over the entire cross section of the ore pass, often referred to as "piping" or "rat-holing." A third design consideration is the stability of ore pass walls.

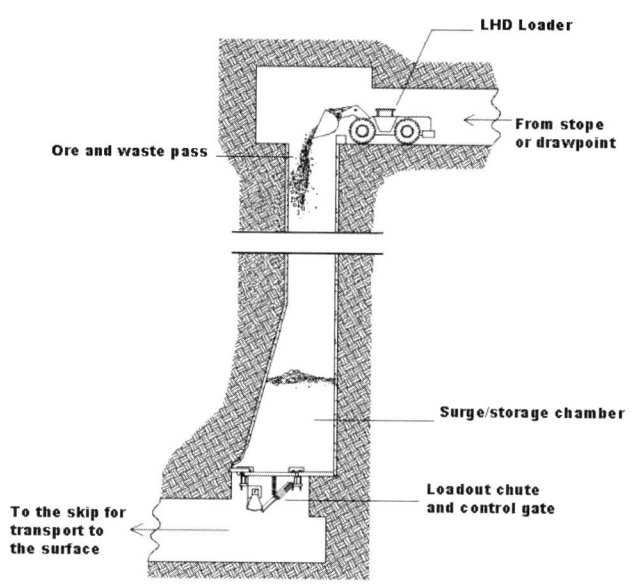

FIGURE 71.1 Typical ore pass for LHDs

71.2.2 Hang-Ups

Two major types of hang-ups in ore passes are boulder arches and cohesive arches. The same types of hang-ups occur in chutes and other transition sections of an ore pass, usually with much greater frequency.

Boulder Arches. A review of the literature, including references from industrial operations handling bulk materials in surface bins and hoppers and laboratory model experiments with sand, indicates that ratios of ore pass diameter to maximum particle diameter of 1:3 (Peele 1947), 1:4.2 (Aytaman 1960), 1:4 to 1:6 (Zenz and Othmer 1960), 1:5 (Jenike 1961), 1:3.6 to 1:45 (Kvapi 1965), and 1:4 to 1:6 (Li et al. 1980) will likely ensure against the formation of boulder arches. A ratio of ore pass diameter to maximum particle size of 5 is likely to result in flow, whereas a ratio of less than 3 is likely to result in a hang-up. A ratio between 3 and 5 is somewhat likely to result in flow. In this regard, ore pass and particle diameters are characteristic linear dimensions when the ore pass is not circular and particles are not spherical.

The distribution of particle shapes and sizes significantly affects the likelihood of hang-ups as well. Particle-size distribution

---

* National Institute for Occupational Safety and Health, Spokane, Wash.
† University of Utah, Salt Lake City, Utah.

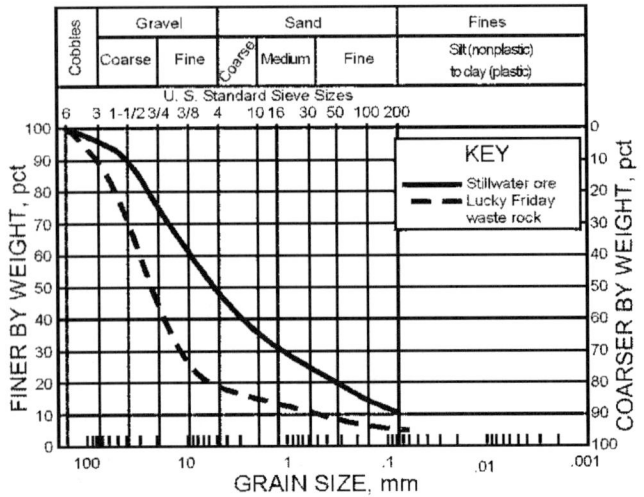

FIGURE 71.2   A typical particle distribution curve from two mines

should be determined from a representative sample of ore pass material after blasting and before the material is dumped into the ore pass. A typical particle distribution curve from two mines is shown in Figure 71.2. If many large boulders are present, then the likelihood of hang-ups is certainly greater than if only a few boulders are present. Hence, these simple, empirical rules-of-thumb for avoiding boulder arches need to be applied with some caution.

The use of a very coarse screen or grizzly that limits the size of the particle entering an ore pass at a dump point is a practical method of ensuring that the maximum particle size is less than one-fifth the ore pass diameter. If oversize material is not prevented from entering the ore pass, then the occurrence of hang-ups will certainly increase, as will operating costs associated with the dangerous task of removing hang-ups. Another factor to consider is that extraneous materials, such as timbers, wire mesh, or rock bolts, often get dumped into ore passes, and these are just as likely to cause hang-ups as oversized rock. Considerable care should be exercised to avoid such a situation.

Cohesive Arches.   Cohesive arches are more easily analyzed than boulder arches, but designing for their prevention is also more difficult. The cause of a cohesive arch is usually the stickiness or cohesion of the fine fraction of the broken ore, waste rock, or fill. Fines are subsand-sized particles, that is, those that pass a 200-mesh screen (about 0.072 mm [0.003 in]). The coarse fraction of gravel-sized and larger particles would be retained on an 8-mesh screen (2.4 mm [0.093 in]). Sand-sized particles fall between these limits. Different soil classification schemes for engineering mechanics have slightly different boundaries between gravel, sand, and fines (silt and clay sizes), but the differences are of little practical consequence. Electrostatic surface forces dominate the behavior of fines, while gravity and frictional contact forces govern the behavior of sand-sized and larger particles.

Cohesion of the fine fraction holds the larger particles together to form a continuous arch across the ore pass. The greater the fine fraction, the greater the potential for cohesive arch formation. A rule of thumb in soil mechanics is that when the fine fraction reaches about 20% by weight, the fines may form a continuous matrix in which the coarser particles are embedded. Thus, measurement of soil cohesion is readily done in a direct shear test of the fine fraction, and a representative sample of mine muck containing gravel- and boulder-sized particles is not needed to determine cohesion. Both cohesion c and angle of internal friction ϕ can be determined in the laboratory by a simple standard direct shear test.

Production of fines during ore flow adds to the potential for hang-ups, as does segregation and accumulation of fines, especially at changes in ore pass direction or shape. The cohesion of the fine fraction is seldom constant; extended time between draws often allows for additional consolidation, changes in water content, and increases in cohesion. An increase in moisture introduced by water sprays for dust control and moisture losses during gravity drainage in the ore pass also change cohesion. *Excessive water or uncontrolled water flowing in an ore pass can result in catastrophic inundation and unexpected muck flow at the load-out level and should be avoided at all times.*

Impact loading by ore falling from a dump point may compact ore already in an ore pass and further increase the likelihood of cohesive arch formation before the next draw. Keeping the ore pass active, that is, frequent drawdown, assists in the prevention of cohesive arch formation in cases where cohesion increases with time. What the draw frequency should be necessarily depends on site-specific conditions and experience.

A cohesive arch will fail if the weight of arch W exceeds shearing resistance T where the arch abuts against the ore pass walls. If the arch is of a unit thickness in the vertical direction, then flow occurs provided $\gamma A > \tau P$, where A is the cross-sectional area, P is the perimeter, and τ is the resisting shear stress. Any additional load on the arch from superincumbent material will add to the tendency to flow. Because the arch is unconfined from below, τ is no greater than one-half the unconfined compressive strength of the arch material. Hence, flow occurs if $A/P > (c/\gamma)\cos(\phi)/[1 - \sin(\phi)]$. If D is the diameter of a circular ore pass, the least dimension of a rectangular ore pass, or a long slot, then flow occurs when $D > (1+1/r)(c/\gamma)\cos(\phi)/[1 - \sin(\phi)]$, where $1/r = 1$ for a circular ore pass, D/L for a rectangular ore pass, and 0 for a long slot (Pariseau 1983).

71.2.3   Piping

Piping is not often a problem in an ore pass unless the ore pass is unusually wide. "Wide" in this context means relative to maximum fragment size. A 9-m (30-ft) in diameter ore pass handling dry sandfill would be wide. Some care needs to be given to ensure flow over the entire cross section, that is, flow on a first-in, first-out basis (also called mass flow). If flow does not occur over the entire cross section of the ore pass, then much of the capital cost of excavating the ore pass is wasted. In the event that piping is a concern, then prevention lies in making the transition from the ore pass itself to the chute gate relatively steep and smooth, thus ensuring mass flow action.

Piping or rat-holing occurs when a cohesive material forms a stable annulus between the pipe and the ore pass walls after the material in the flowing portion of the pipe is withdrawn. Such a rathole is stable provided the weight of material above does not overcome the compressive strength at the bottom, thus preventing the more desirable condition of mass flow. The material at the pipe wall is unconfined, and the stress at the pipe bottom is directly proportional to pipe height. *Thus, if the height of the pipe is sufficient, the annulus will fail under its own weight.* The critical pipe height h in this case is given by the ratio of unconfined compressive strength to specific weight of material, that is, by C_o/γ. The unconfined compressive strength may be estimated from cohesion c and angle of internal friction ϕ assuming a Mohr-Coulomb failure criterion. In that case, $C_o = [2c\cos(\phi)/(1 - \sin(\phi)]$. Hence, if $h > [2c\cos(\phi)/(1 - \sin(\phi)]/\gamma$, then flow of the material in the annulus may be initiated. Initiation of flow may destroy or greatly reduce cohesion, so that flow may well continue with further draw of the material even though the height of the material drops below the original critical height. Indeed, with complete loss of cohesion, the critical height for a stable pipe is reduced to zero. However, increased pipe height and therefore weight may increase cohesion by compaction, so there are trade-offs to consider when running an ore pass full to reduce rathole formation.

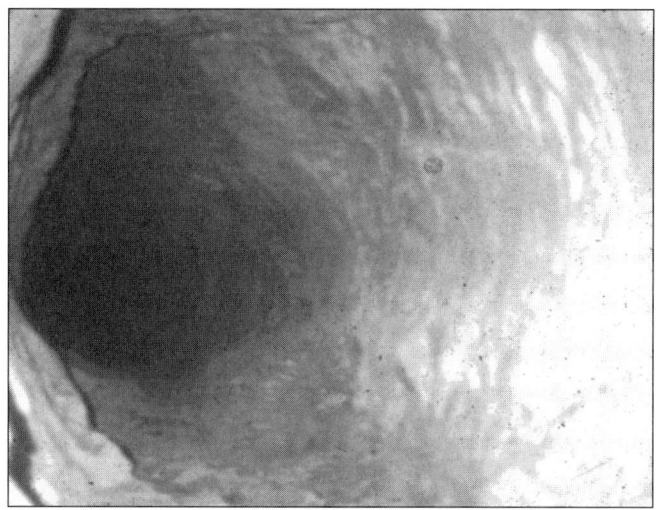

FIGURE 71.3   Extreme case of ore pass borehole breakout

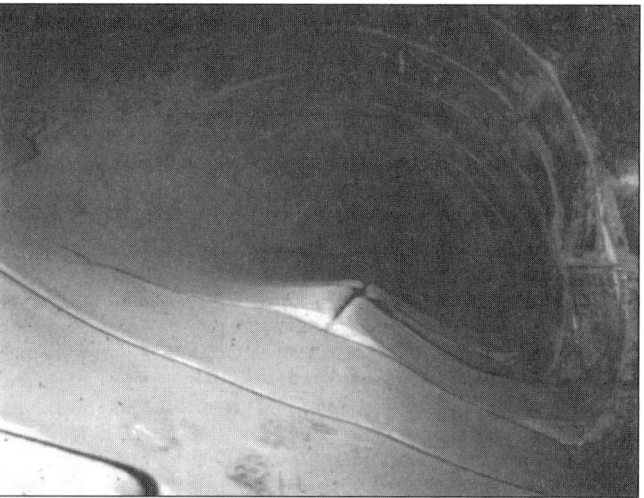

FIGURE 71.4   Steel-lined ore pass that has been subjected to squeezing ground and will require repair or abandonment

71.2.4   Ore Pass Wall Stability

Stability of rathole walls brings into question stability of the ore pass walls proper. Wall stability is a rock mechanics problem that may be addressed by consideration of rock mass strength and stress. Both depend on the properties of joints, fractures, faults, and intact rock between. A convenient index to ore pass wall safety or stability is the traditional engineering factor of safety, a ratio of strength to stress. Thus, factor of safety = C_o/σ_c, where σ_c is peak compressive strength of the ore pass wall. If an appropriate stress concentration factor K is known, then $\sigma_c = K\sigma_o$, where σ_o is a reference stress, usually the premining major principal compression σ_1.

This approach to wall stability assumes that tension at the ore pass walls is absent, a reasonable assumption in most cases. If tension should be present, then a similar analysis may be done to establish a factor of safety with respect to rock mass tensile strength and peak tensile stress concentrated at the ore pass wall. Shape, size of the ore pass cross section, and orientation with respect to in situ stresses are important determinants of stress concentration. Thus, there is an opportunity for engineering design to maximize ore pass wall stability by selecting an optimum shape, orientation, and location. However, what is optimum from a purely rock mechanics view may not be a total optimum. The overall design must consider cost of excavation, lining (if any), and maintenance, especially if the shape selected is unusual—an ellipse, for example.

Ore passes in regions of high stress may be subject to breakouts and excessive wear. Figure 71.3 shows an extreme case of ore pass borehole breakout. It may then be necessary to install a lining for wall stability and proper functioning of the ore pass. Wood linings were often used in the past, but shotcrete and steel liner plates are more often used today where wear is high and erosion of the ore pass cannot be tolerated; for example, in the immediate vicinity of a dump point or chute. Linings generally require repair, periodic maintenance, and sometimes complete replacement. Figure 71.4 shows a steel-lined ore pass that has been subjected to squeezing ground and will require repair or abandonment.

71.3   CHUTE DESIGN

71.3.1   Introduction

The ore pass chute is of critical importance. The design of chutes involves geometric considerations to accommodate transitions in shape and direction, prevention of hang-ups, and structural elements. Chutes are usually sized to fit ore cars or conveyor belts for transport of the ore to shaft loading pockets or the surface. Because chute width is generally less than ore pass width, hang-ups are much more likely to occur in the vicinity of a chute rather than in the ore pass itself. Hydraulic or pneumatic control of the chute gate is a common way of regulating the flow of the muck from the chute. Arc gates and undershot guillotine gates are examples of chutes gates used widely in underground vein mining. Anchor chains are often used to regulate the flow of muck during the loading cycle. Although the type of chute used varies from mine to mine, the common design element is chute width relative to particle size.

The same guidelines for preventing hang-ups in ore passes apply to chutes. A chute width-to-maximum-particle-size ratio of 5 or greater is desirable. Rational design of an associated grizzly then requires that grizzly spacing be no more than five times chute width rather than five times ore pass width. A lesser ratio increases the risk of hang-ups at the chute. In caving operations, where control of maximum particle size is difficult, a relatively close spacing of grizzly bars would lead to the need for secondary breakage of oversized material at dump points, but this is preferable to an increased number of hang-ups at chutes. Well-designed blast patterns produce much less oversized material and the need for secondary breakage at grizzlies.

71.3.2   Static Loads

The main task for structural design is estimating what loads the chute support will take during operation. Both static and dynamic loads need to be considered. A simple equilibrium analysis provides some guidance for estimating static load. An equilibrium analysis of a horizontal layer of material in an ore pass shows that the vertical stress averaged over the cross-sectional area of the ore pass is given by—

$$\sigma_V = \left(\frac{\gamma(A/P) - cM}{M\ tan(\phi)} \right)\{1 - exp[-zM\ tan(\phi)/(A/P)]\}$$

where
  σ_V = average vertical stress,
  z = depth of muck in the ore pass,
  A = cross-sectional area,
  P = perimeter,

γ = specific weight of muck,

c = cohesion,

and $M = 1/[1+2(\tan\phi)^2]$.

This formula is an extension of the Janssen (1895) formula and takes into account any tendency for fines to adhere to ore pass walls, as well as friction of muck on walls (Pariseau 1983). The Jannsen formula was originally derived for cohesionless grains in surface silos and forms the basis for silo design in many countries. The derivation is straightforward and has been repeated by many investigators and tested repeatedly in model studies and by measurements on full-scale silos. Most laboratory models use cohesionless model materials, mainly sand. An example within a mining context is the work of Blight and Haak (1994), who report on model studies of inclined ore passes and chute loading. Adhesion to and friction on ore pass walls are likely to be less than cohesion and friction angle of muck on muck. Because flow requires continuous shear failure of the muck regardless of whether slip at the walls occurs, the muck-on-muck cohesion and friction angle are used in this formula rather than the corresponding muck-on-wall properties.

The equilibrium condition can be expressed more compactly as

$$\sigma_V = (C_2/C_1)\{1 - \exp[-zC_1]\}$$

where $C_1 = M\tan(\phi)/(A/P)$

and $C_2 = \gamma - cM/(A/P)$.

In this compact form, one clearly sees that there is an asymptotic limit to the vertical stress exerted by the muck column on the material below. A chute positioned directly below a vertical muck column at depth h sees only a fraction of the total column weight. The bottom load exerted by the muck column is simply $\sigma_V A$. The difference is transferred via shear to the ore pass walls. If W is muck column weight, then the total load delivered in shear to the ore pass walls is $T = W - \sigma_V A$. At a muck depth of about $z = 1/C_1$, the bottom stress σ_V is two-thirds the stress that would be possible in an infinitely deep ore pass. If the ore pass is circular with diameter D and the friction angle is $35°$, then $M\tan(\phi)$ is about one-third, and σ_V approaches two-thirds the limit stress at a depth of about three times ore pass diameter.

The greatest limit stress at the column bottom occurs when cohesion is nil. In this case, the limit stress is $3\gamma D$. Because ore pass depth is usually much greater than its diameter, the bottom stress does not increase much with depth beyond $3D$. Hence, for static load chute design, one is justified in using a conservative value, that is,

$$\sigma_V = \gamma(A/P)/M\tan(\phi).$$

A simple adjustment for inclined ore passes is to use a component of specific weight normal to the ore pass cross section, that is, $\gamma\sin(\delta)$, where δ is ore pass inclination. This adjustment should be acceptable within a practical range of gravity flow ore pass inclinations, generally less than $45°$.

71.3.3   Dynamic Loads

A dynamic load is delivered to muck at the top of an ore pass during dumping, that is, transmitted through the muck to the chute and also to the ore pass walls. Dynamic loads that arise during dumping are much more difficult to estimate and are often taken to be some small multiple of the static load. Dynamic chute loads may be several times the static load (Blight and Haak 1994; Beus et al. 1998). An approximate approach to quantifying the dynamic effect of dumping is to consider the impact at the top of the muck in the ore pass as an additional stress σ_o. Dynamic equilibrium in a vertical ore pass may then be expressed as

$$\sigma_d = (C_2/C_1)\{1 - \exp[-z_o C_1]\} = \sigma_o \exp(-z_o C_1)$$

where z_o is the muck depth at the time of impact. Thus, the effect of impact loading treated in this quasi-static manner diminishes exponentially with depth, while the limit stress remains the same as in the static case. This allows one to quantify the concept of leaving a muck cushion in an ore pass to protect a chute directly below. Of course, the muck forces are concentrated in the structural supports of the chute and gate, and the actual distribution of axial and bending stresses in the supports must take into account how the muck forces are tranmitted to these supports. Chute design is thus a challenging combination of muck mechanics and traditional structural analysis.

Another source of dynamic chute loading is the sudden release of muck from a hang-up above a chute. In this case, there is likely to be very little muck left in the chute to offer protection against the impact of the muck falling from above. The potential for severe structural damage may be quite high under direct impact of a released hang-up. The dynamic stress, σ_d, produced in a beam loaded at midspan from the impact of a body of weight W falling from a height h is greater than the static stress σ_{st} according to—

$$\frac{\sigma_d}{\sigma_{st}} = 1 + [1 + 2h/\delta_{st}]^{1/2}$$

where δ_{st} is the deflection at midspan under gradual (static) application of the same load (Gere and Timoshenko 1997). If h is zero, the dynamic stress produced by a sudden release of the weight that is just in contact with the beam is twice the stress produced by gradual release of the same weight. Therefore, if the energy losses of material falling down an ore pass are very high, then the dynamic stress on the chute gate in an empty ore pass may be approximated by dynamic stress greater by a factor of 2 over a static stress produced by the same weight. However, if h is large compared to deflection and energy loss is negligible, then the dynamic stress is approximately—

$$\sigma_d = \sigma_{st}\sqrt{2h/\delta_{st}}$$

By off-setting the chute to one side of an ore pass, a common practice, the damaging effects of impact loading may be reduced. Bends and knuckles serve the same purpose. Inclined ore passes may also experience less impact chute loading as muck ricochets from ore pass walls during its fall. Particle trajectories initiated during dumping, even in a vertical ore pass, will almost certainly strike the ore pass wall unless muck is present near the dump point.

An elementary analysis of impact shows that the average force of particle-wall impact is directly proportional to particle velocity and particle weight and inversely proportional to duration of impact (Pariseau 1983). If ore pass wear is caused mainly by impact during dumping rather than by abrasion during draw, then large boulders should be made to move more slowly. As a practical matter, this objective may be approached by leaving an adequate muck cushion in the ore pass and by using a properly sized grizzly. Unfortunately, the potential benefits of a protective muck cushion and of running an ore pass full are not always available. For example, in case of ores liable to sulphide oxidation, with attendant cementation and the creation of fire hazards, residence time in the ore pass is limited and may require drawdown before the cycle of mining and dumping is completed.

71.3.4   Detection and Removal of Hang-Ups

Methods of locating hang-ups include visual inspection, lowering a camera down the ore pass from the top, running a long pole through the ore pass, or sending a helium balloon up the ore pass. Various sensors that use laser, radar, microwave reflection, or ultrasonic technologies to delineate material level and location from the top are also on the market. Investigators at the Spokane Research Laboratory of the National Institute for Occupational

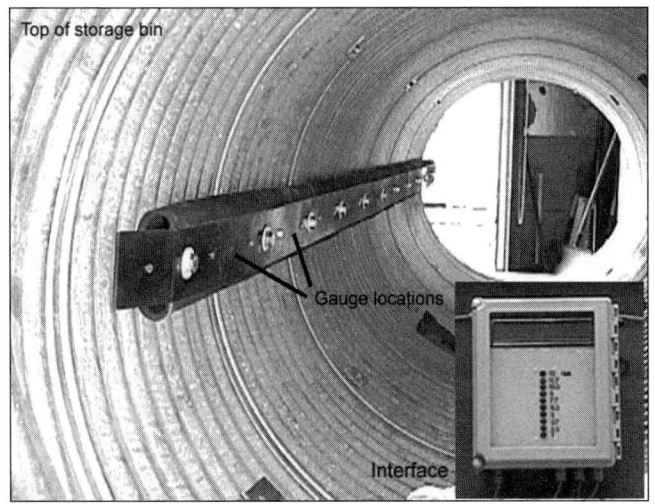

FIGURE 71.5 Ore pass level and blockage indicator

Safety and Health (SRL/NIOSH) have developed an "ore pass level and blockage indicator" (patent pending) based on strain gauge technology. This device can detect bin or ore pass levels as well as voids within the muck column (Figure 71.5). Currently, hang-ups are most commonly detected by a trammer or chute-puller when a lack of muck is noticed or or an operator attempts to dump muck into a filled ore pass.

A number of methods are used for hang-up removal. The most common and effective method to remove hang-ups is blasting. The method of getting explosives to a hang-up depends on location in the ore pass. For hang-ups high in the ore pass, various devices may be used to "float" an explosive package or "bomb" to the hang-up using air pressure. Another method is to fire a projectile at the hang-up from the bottom of the ore pass. Many hang-ups occur just above the chute gate where flow is constricted and the direction of the ore pass changes. These hang-ups can be removed by mechanical devices, hand tools, or explosives placed with plastic pipe or powder-tamping poles taped together.

Because of the hazards to personnel and the potential for chute and ore pass damage, other safer, nondestructive methods are desirable. Blasting to remove cohesive hang-ups may prove ineffective and only compact the material more. High-pressure air injection directed at or just above the point of arching or controlled water injection directed to undermine the arch may be more effective. Water flushing from a dump point is also used, but care needs to be exercised to avoid impounding water in the voids of the muck that may suddenly break free in a mudflow that can cause considerable damage and possible injury to operators.

71.4 SAFETY CONSIDERATIONS

The hazards related to the operation of ore and waste rock passes have been identified as a significant safety problem in underground metal mines in the United States. Applicable ore pass safety and design criteria, as defined in the Code of Federal Regulations 30 CFR, parts 57.9310 and 9309, state—

a. Prior to chute-pulling, persons who could be affected by the draw or otherwise exposed to danger shall be warned and given time to clear the hazardous area.

b. Persons attempting to free chute hangups shall be experienced and familiar with the task, know the hazards involved, and use the proper tools to free material.

c. When broken rock or material is dumped into an empty chute, the chute shall be equipped with a guard or all

persons shall be isolated from the hazard of flying rock or material.

d. Chute-loading installations shall be designed to provide a safe location for persons pulling chutes.

Evaluation of statistics and other information from the Mine Safety and Health Administration (MSHA), particularly narratives from investigative reports, are useful for identifying the underlying cause of accidents. Several accident narratives that illustrate the nature of ore pass accidents are given below.

Fatality: [T]he miners began banging the chute gate against the chute lip...material broke loose and impacted the chute with enough force to separate the chute assembly from the steel beam supports. The material fatally engulfed one of the miners as they tried to escape.

Fatality: Victim was loading ore train using pneumatically operated chute door when very wet ore...broke the chute structure burying the victim.

Fatality: While the skip tender was loading a muck skip, the heavy muck came loose [in the ore pass] causing an overflow of water...overflow washed the miner over and through the shaft guards.

Fatality: Employee was fatally injured when a chute filled with ore blew out covering him with approximately 5 feet of ore.

71.5 RESEARCH AND DEVELOPMENT AT SRL/NIOSH

71.5.1 Introduction

Ore pass research and development activities have been ongoing at SRL/NIOSH for several years and involve laboratory tests, field tests, and computer modeling to validate theoretical formulations for static and dynamic loads and to test new methods of prevention, detection, and removal of hang-ups. Laboratory and field tests have been conducted to validate these theoretical formulations.

71.5.2 Laboratory Tests

A laboratory ore pass facility has recently been completed to facilitate tests in a controlled environment. (Beus and Ruff 1996). Figure 71.6 shows this fully automated facility utilizing an 18.3-m hoist tower to simulate the headframe and shaft. A 5.5-m- (18-ft) deep underground "shaft" lined with concrete sections houses a loading pocket and a measuring cartridge. The ore-hoisting skip has the capacity of about 450 kg (1,000 lb). The ore pass is a 1.03-m (3.5-ft) in diameter corrugated culvert that can be inclined 65° to 90° vertically in 5° increments. Design of the ore pass, chute support frame, I-beams, hanger bolts, and saddle beams are one-third the size of an actual ore pass and chute control gate currently used at an underground mine.

Ore or waste material is loaded through a grizzly and into the skip for hoisting. The skip hoists the ore to the top of a headframe, where it is discharged into a hopper and chute assembly, which routes it to the top of the ore pass. Two types of gravel were tested using the ore pass in a vertical configuration. The impact period at the chute was about 5 s due to the stream effect of discharge into the ore pass. Strain data were collected 60 times per second from each of the eight supporting bolts. Pea gravel of uniform distribution indicated an impact load factor ranging from 1.24 to 1.39 when it was dumped into the empty chute. Coarse gravel of uniform distribution indicated an impact load factor from 1.22 to 1.39. Figure 71.7 illustrates a typical impulse curve from one skip dump of gravel. Several large plugs of material were also dropped to simulate a large boulder or released hang-up directly impacting the control gate. These tests resulted in dynamic loads of up to nearly 30 times the static load.

A full-scale mockup of the reduced-scale chute and gate assembly was also constructed. Three impact tests were conducted in which loads of mine waste rock weighing 808, 1,000, and 563 kg (1,778, 2,200, and 1,238 lb) were dropped

FIGURE 71.6 Automated ore pass test facility with an 18.3-m hoist tower to simulate a headframe and shaft

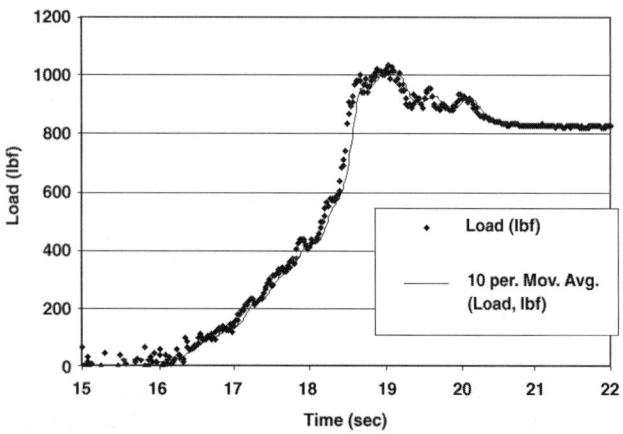

FIGURE 71.7 Typical impulse curve from one skip dump of gravel

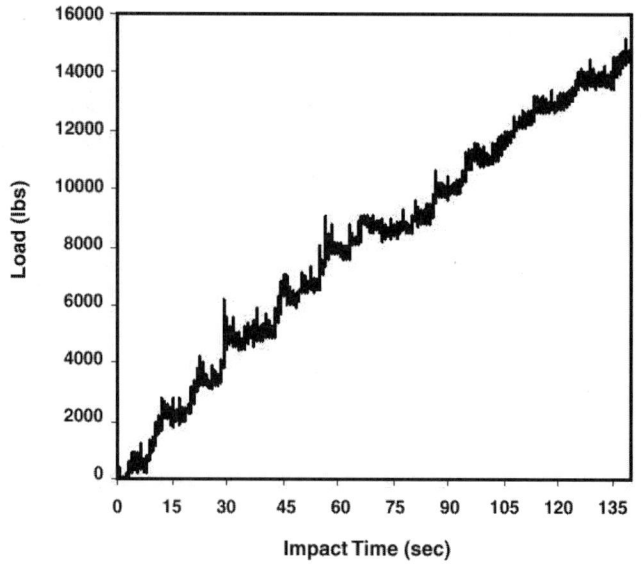

FIGURE 71.8 Typical dump pattern as shown by the total force on all support bolts

from a height of 1.8 m (82 in) into a 2.4-m- (8-ft-) wide, steel-reinforced container sitting on the mockup chute assembly. A front-end loader with a clam shell bucket was used to drop the material into the container. After each drop, the material was removed and weighed. The impact period was about 1 s, and the maximum impact load factor was 2.2.

71.5.3 Field Tests

Field tests were also conducted in mines to determine actual static and impact loads. The ore pass chute and gate systems provided by the cooperators for testing contain offset boreholes and doglegs that direct falling material to the floor of the ore pass to absorb the initial impacts from the falling column of ore or waste. The initial field test was conducted in a deep silver mine in northern Idaho, USA. The approach was based on the system developed and tested on the full-scale mockup. The centerline of the gate assembly was offset 2.4 m (8 ft) from the longitudinal axis of the ore pass so that falling ore did not directly hit it.

Tensile strains produced from eight strain-gauged 3.8-cm (1.5-in) in diameter bolts that support the ore pass chute and gate provided a measurement of the total vertical force acting on the structure as material was dumped into the ore pass. Fourteen loads of damp waste rock from LHD units at 1.53 m³ (54 ft³) per load were dumped into an empty ore pass. Figure 71.8 shows a typical dump pattern as shown by the total force on all support bolts. Load measurements were collected for 10 s for each dump. Twelve of the dumps averaged from 2,270 to 2,730 kg (5,000 to 6,000 lb) of material; two of the dumps (8 and 9) were material from cleaning up the drift and weighed about 270 kg (600 lb) each.

The weight of material dumped in the ore pass was in excess of 27,300 kg (60,000 lb); however, a maximum static load of only 6,800 kg (15,000 lb) was measured. This load was approximately the weight of waste material required to fill the chute. The rest of the static load was carried by timber adjacent to the chute assembly and the ore pass walls. Impact load factors ranged from 1.06 to 1.33 on the chute and gate assembly and were reduced significantly because the chute was offset from the ore pass.

A second field test at another underground mine involved a chute and gate support structure. Figure 71.9 shows four of eight instrumented chute support bolts, after tensioning, using load

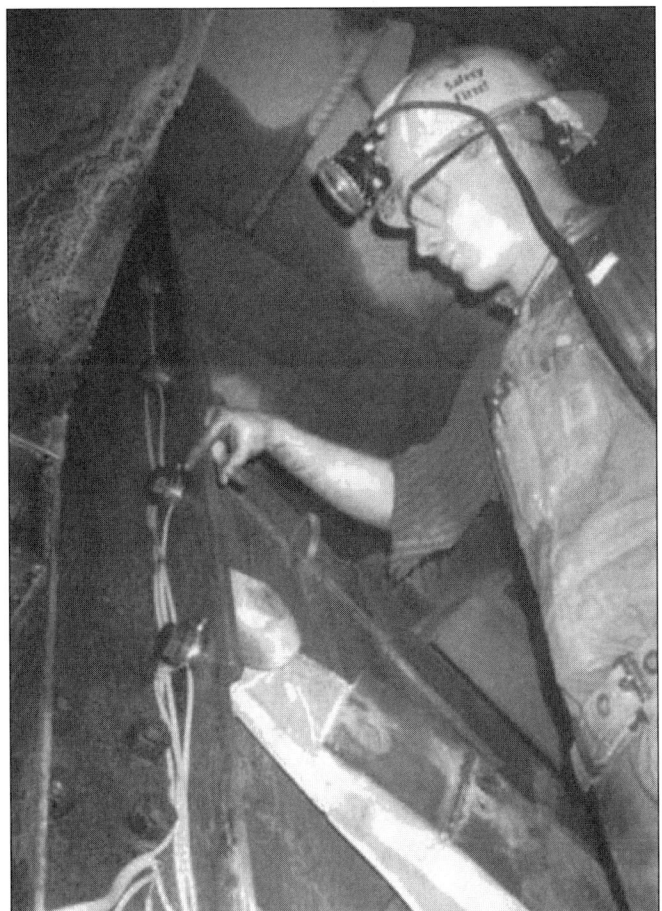

FIGURE 71.9 Four of eight instrumented chute support bolts, after tensioning, using load washers to determine total horizontal force on the gate

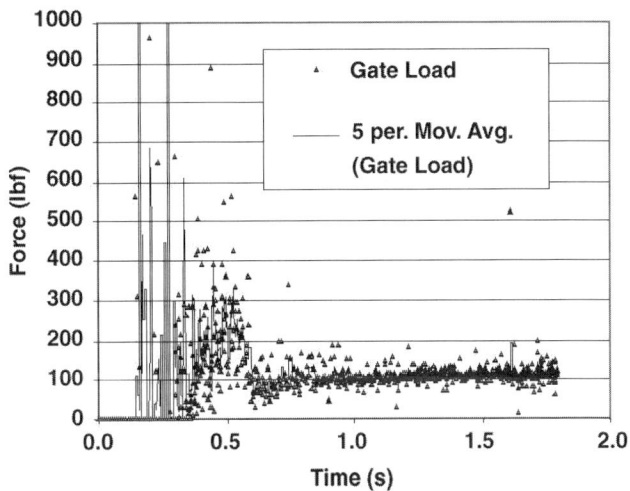

FIGURE 71.10 Dynamic load response on chute gate over impact period

washers to determine total horizontal force on the gate. During the experiment, the ore pass was consistently blocked and required direct blasts in the chute at the bottom of the ore pass to free the material. Comparing initial bolt tensions with tension data after 24 hr of chute operation, nearly 90% of the initial gate support bolt tension load was lost. It is possible that large dynamic components from blasting may have overloaded the bolts. Dynamic loads on structures from secondary blasting in ore passes is unknown. The shockwave from blasting may have had more effect on chute structural integrity than rock impacts.

71.5.4 Computer Modeling

Computer modeling aims to simulate dynamic and static forces measured in the field on the chute structure, as well as overall particle flow phenomena and the potential for hang-ups. Three-dimensional particle flow PFC^{3d} models (Itasca Consulting Group 1995a, 1995b) were used to simulate flow through the ore pass and truck chute instrumented in the field tests. The static load effect of 40 LHD dumps of about 7 tonnes (7.7 tons) of material each for a total weight of about 280 tonnes (308 tons) were simulated. The rate of increase in gate loads dropped dramatically after about one-third (10 to 15) of the LHD loads had been delivered. Analysis indicates that dynamic loads were a factor on the control gate only during the first three to five dumps.

Comparison of measurements and computer results using particle flow codes indicate that several difficulties remain before realistic determination and modeling of the dynamic effects of particle flow in ore passes and impact loads on the gates can be achieved. Computer analyses can overestimate the dynamic impact load from the rock compared to impact loads measured in field tests if energy loss during impact is not accounted for in the simulation.

Obviously uniformly graded, smoothly rounded particles used in the PFC^{3d} code are not found in a freshly blasted muck pile prior to transport to the ore pass. The distribution of particle shapes and sizes affects all the behavioral characteristics of the falling muck column. Incorporating more realistic rock particle shapes and distributions of particle sizes in numerical models improves rocklike characteristics during fall and impact. Other factors difficult to model are rock durability, angularity, and rebound characteristics following impact (Larson et al. 1998).

Single dumps in a PFC^{3d} model of the one-third-scale ore pass were compared with load data collected at the ore pass test facility. Shaped particles and appropriate stiffness and friction properties were used from Larson et al. (1998). A damping constant was included to account for energy loss during particle impacts. The particles were formed by clumping four balls together in a tetrahedral geometry. Realistic impact rebound trajectories and particle rotations were achieved. An improved impulse curve compared well with actual dynamic load data from the ore pass test facility. Figure 71.10 illustrates the dynamic load response on the chute gate over the impact period. Note that the first particle impacts have very high dynamic loads, while the later particle arrivals are cushioned by the initial material.

71.6 ACKNOWLEDGMENTS

The authors gratefully acknowledge the assistance of Hecla Mining Company and Stillwater Mining Company in providing field test sites and technical assistance for ore pass research.

71.7 REFERENCES

Aytaman, V. 1960. Causes of Hanging in Ore Chutes and Its Solution. Canadian Mining Journal. Vol. 81, pp. 77–81.

Beus, M., and T.L. Ruff. 1996. Development of a Mine Shaft and Ore Pass Research Facility. U.S. Bureau of Mines RI 9637, 12 pp.

Beus, M.J., S.R. Iverson, and B.M. Stewart. 1998. Design Analysis of Underground Mine Ore Passes: Current Research Approaches. Presentation at 100th Can. Instit. Min., Metall., and Petro., Montreal, PQ, May 2–8, 1998, 8 pp. Available on CD-ROM from CIM, Montreal, PQ.

Blight, G.E., and B.G. Haak. 1994. A Test on Model Underground Ore Passes. Bulk Solids Handling, Vol. 14, No. 1, pp. 77–81.

Brauer, R. 1990. Safety and Health for Engineers. Van Nostrand Reinhold, pp. 551–555.

Gere, J.M., and S.P. Timoshenko. (1997) Mechanics of Materials. PWS Publishing, pp. 658–660.

Itasca Consulting Group, Inc. 1995a. PFC2D (Particle Flow Code in 2 Dimensions), Version 1.1, Minneapolis, MN.

Itasca Consulting Group, Inc. 1995b. PFC3D (Particle Flow Code in 3 Dimensions), Version 1.1, Minneapolis, MN.

Janssen, H.A. 1895. On the Pressure of Grain in Silos. Inst. Civ. Engr., Vol 124, pp. 553–555.

Jenike, A.W. 1961. Gravity Flow of Bulk Solids. Utah Engr. Exp. Sta. Bull. 108, 309 pp.

Larson, M.K., S.R.Iverson, B.M. Stewart, and K. Walker. 1998. Preliminary Assessment of Particle Flow Code as a Tool to Assess Ore Pass Safety. Presentation 98th NARMS Congress, Cancun, Mexico, June, 1998.

Kvapil, R. 1965. Gravity Flow of Granular Materials in Hoppers and Bins in Mines—II Coarse Material. Intl. J. Rock Mech. and Min. Sci., Vol. 2, pp. 277–304.

Li, D., Z. Zhang and B. Guo 1980 The Design and Practice of Tunnel-Ore Pass System in Open Pit. Reprint First International Mine Planning and Development Symposium, Deidaihe, China, Sept. 18–27, 20 pp.

Pariseau, W.G. 1966 The Gravity Induced Movement of Materials in Ore Passes Analyzed as a Problem in Coulomb Plasticity. Ph. D. Thesis, University of Minnesota, 218 pp.

Pariseau, W.G. 1983 Rock Flow in Ore Passes. In: Guidelines for Open-pit Ore Pass Design. Vol. I: Final Report U.S. Bureau of Mines. Contract J0205041. Engineers International, 259 pp.

Peele, R. 1947. Mining Engineers Handbook, 3rd ed. Wiley and Sons, Vol. 1., 403 pp.

Pfleider, E.P. and C.A. Dufresne 1961 Transporting Open Pit Production by Surface-Underground Haulage. Mining Engineering, Vol., 13, No. 6, pp. 592–598.

Zenz, P.A. and D.F. Othmer. 1960. Fluidization and Fluid-Particle Systems. Rheinhold Publishing, pp. 75–168.

CHAPTER 72

Blast Design for Underground Mining Applications

Roger Holmberg,[*] William Hustrulid,[†] and Claude Cunningham[‡]

72.1 INTRODUCTION

Good blast design and execution are essential ingredients for successful underground mining. Poor blasting practices can have a severely negative impact on the economics of mining. Military blasting, rather than precision blasting, can result in overbreak and dilution of high-grade ores. Military blasting can damage sensitive or tender rock structures that make up the hangingwall or footwall so unwanted caving occurs with the possibility/probability of ore loss and/or dilution. Poor design and execution when ring drilling can mean that succeeding rings are damaged and unchargeable. Failure to complete undercuts can mean the transmission of very high loads to the underlying structures and their subsequent failure. The results are lost revenues and added costs.

Good drilling and good blasting go hand in hand. If the drilling is poor, there is little one can do to correct the rest of the job. It is similar to building a house. If the foundation is poorly done, there are major problems along the rest of the way. A discussion of drilling practices is beyond the scope of this chapter, but the blast design must begin there. Prior to designing the blasting, one must be sure that the miners have the machines and the capabilities to build the design. If the design cannot be built with the tools at hand, then it is no design. Hence, one starts the design process by carefully examining the drilling capabilities. Fortunately, great improvements have been made in machine-based drilling precision capabilities over the past few years. These include laser alignment, using tubular steel, in-hole guidance, and boom alignment instruments and techniques. However, the miners must then use the capabilities of these improved machines to the fullest.

Although a good blast design base does exist for underground mining applications, it has often been poorly explained and documented in the literature. This chapter tries to correct this situation, at least to some degree, by offering some initial design guidance for:

- Bench blasting
- Ring drilling
- Crater (VCR) blasting
- Drifting

In this short chapter it is impossible to provide full coverage of such a broad topic as underground blast design. Several different approaches are described, and examples are given as appropriate. A number of simplifications have been made to facilitate the presentation. This is to be regarded as providing an introduction to the topic and not as a cookbook. The reader is advised to contact an explosive supplier and/or other specialists for help in solving complex blasting problems.

72.2 EXPLOSIVE PROPERTIES

There are a large number of properties that one considers when choosing an explosive. In terms of doing a blast design, however, the properties are relatively few. They can be simplified to:

- Density
- Weight strength
- Specific gas volume

For the rock, the properties are also relatively few:

- Density
- Fracture conditions
- Rock strength, modulus, and toughness
- Water conditions

Figure 72.1 shows the process involved in developing a blast design. One starts with the results that one wants to achieve and then works to the final design. These results may be expressed in the form of a certain desired fragmentation, minimum blast damage to the surrounding rock, vibration limits of sensitive structures, and/or minimum overbreak. Here, for simplicity, it will be assumed that the explosive-rock interaction can be defined in terms of a "powder factor." Although the term "powder factor" is widely used in mining, it must be used with care. It may be expressed in terms of the amount of explosive required to fragment either a certain volume or mass of rock. Thus, it is sometimes expressed in the units of kg/m^3 or kg/ton. The difference between these is the density of the rock. For the same generic rock type, the required powder factor to yield the same fragmentation may be quite different, depending on the initial fracturing condition. When discussing powder factor, one must also indicate the explosive that has been used, because the breaking characteristics depend upon the characteristics of the particular explosive in that rock. For example, one kilogram of ANFO has a different energy content than does one kilogram of an emulsion explosive. Even for explosives having the same energy content as expressed in kcal/kg or MJ/kg, the distribution of the energy in terms of the shock energy and the gas expansion energy can be rather different. This partitioning depends upon the characteristics of the rock mass as well as the explosive, so the process is complicated. In performing an actual design, one might consider several different explosives. Then it is important to have a way to examine their relative performances and associated costs on

* Dyno Nobel, Gyttorp, Nora, Sweden.

† Dept. of Mining Engineering, University of Utah, Salt Lake City, Utah.

‡ AECI Explosives Ltd., Modderfontein, South Africa.

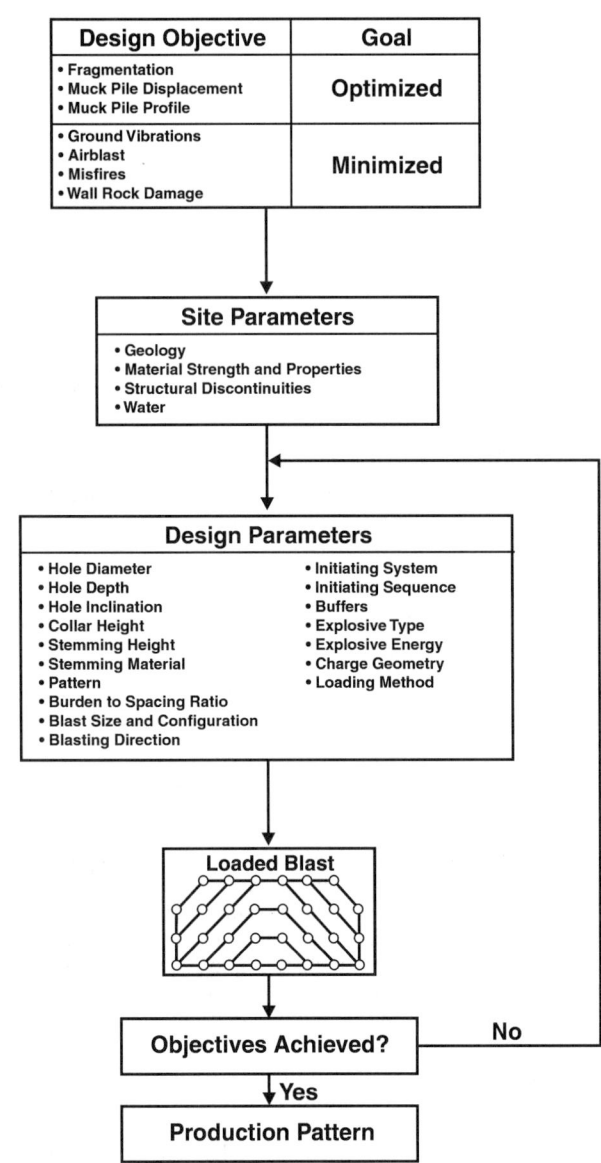

FIGURE 72.1   The blast design process, Atlas Powder Company (1987), Hustrulid (1999)

paper. Langefors and Kihlström (1963) suggested that, when comparing the relative strengths of explosives, one should consider both the weight strength and the gas volumes produced. The relative strength of explosive A with respect to ANFO is expressed by:

$$S_{ANFO} = 5\ Q_A/(6Q) + V_{GA}/(6V_G) \qquad (72.1)$$

where S_{ANFO} = relative strength of explosive A with respect to ANFO

Q_A = weight strength of explosive A (kcal/kg)

Q = weight strength of ANFO (kcal/kg)

V_{GA} = gas volume of 1 kg of explosive A at STP (l)

V_G = gas volume of 1 kg of ANFO at STP (l)

The values of weight strength are often given in the specification sheets for different explosives. The values for the gas volumes are seldom given; hence, one is often forced to use the

simplification that the relative weight strength of explosive A compared to ANFO is

$$S_{ANFO} = \frac{Q_A 645}{Q} \qquad (72.2)$$

Unfortunately, in the literature there is no standard value for Q. Here, to provide an index, Q will be assumed to be 912 kcal/kg or 3.82 MJ/kg. The bulk strength (BS) is the weight strength times the density and has the units of kcal/cm³, MJ/cm³ or equivalent. The relative bulk strength of explosive A having density ρ_A compared to ANFO having density ρ is:

$$BS_{ANFO} = \frac{\rho_A \times Q_A}{\rho \times Q} \qquad (72.3)$$

where BS_{ANFO} = relative bulk strength of explosive A compared to ANFO

ρ_A = density of explosive A

ρ = density of ANFO

As an example of the application, consider

ρ = 800 kg/m³

ρ_A = 1200 kg/m³

Q = 912 kcal/kg

Q_A = 850 kcal/kg

The weight strength of explosive A relative to ANFO is:

$$S_{ANFO} = \frac{850}{912} = 0.93$$

The bulk strength of explosive A relative to ANFO is:

$$BS_{ANFO} = \frac{1200 \times 850}{800 \times 912} = 1.4$$

Because it is a particular hole volume that is to be filled with explosive, the bulk strength is of major importance when considering the blast design.

From a drilling point of view one must look at the ability to drill-out the design. The reasons for the problems are:

- Collaring inaccuracies

- Set-up errors

- In-hole deviation

As will be seen, these problems affect all of the different operations from blasting the cut in a drifting round to stoping patterns.

72.3   BENCH-BLAST DESIGNS

72.3.1   Introduction

Bench-type blast designs are used in a number of different underground mining applications. They include:

- Stoping with parallel holes (Figures 72.2 and 72.3)

- Drifting in overhand cut-and-fill (Figure 72.4)

- Up-holes in shrinkage stoping

- Benching in room-and-pillar mining (horizontal or vertical) (Figure 72.5)

- The stoping holes in drifting

Normally a square pattern of holes (rows and columns aligned) rather than a staggered pattern is used (Figure 72.6) because the final opening shape is normally square. The hole pattern itself can be rectangular or square. There are two approaches to arriving at a design. Both start with knowing the hole diameter (D) to be drilled.

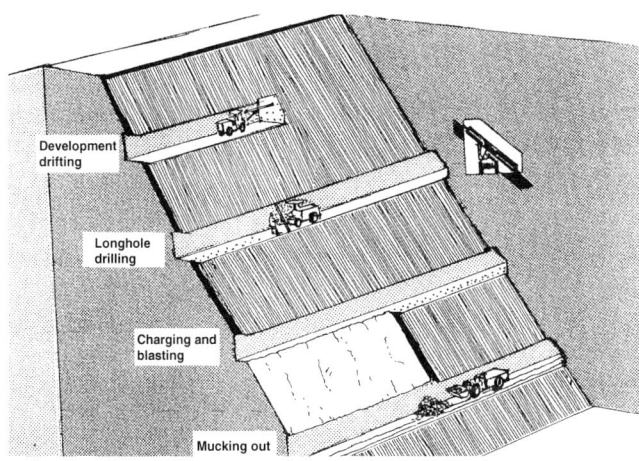

FIGURE 72.2 Narrow vein stoping using longholes, Atlas Copco (1997)

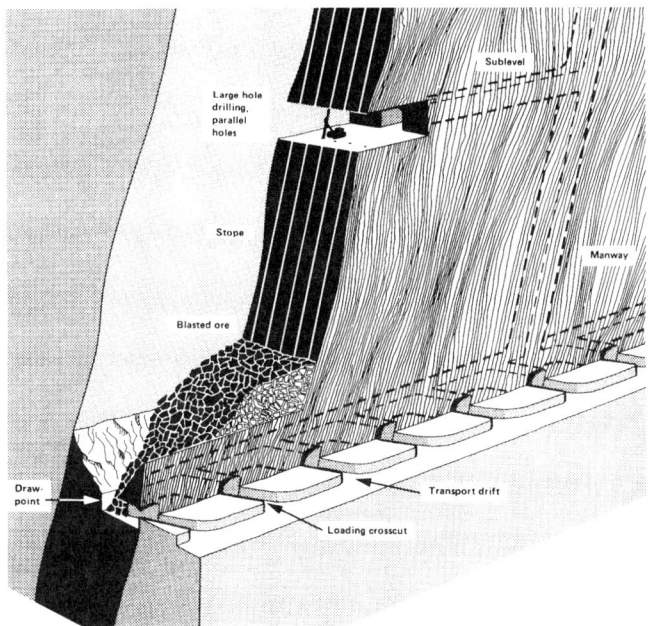

FIGURE 72.3 Blasthole stoping with parallel holes, Atlas Copco (1986)

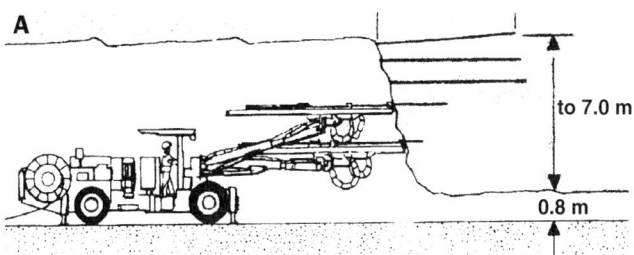

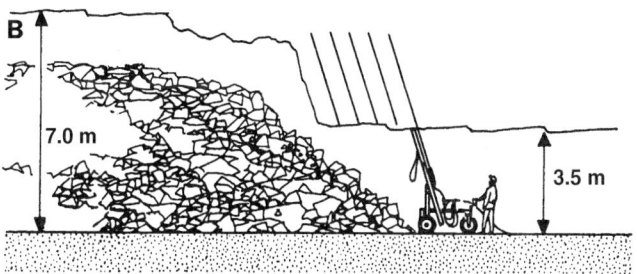

FIGURE 72.4 Cut-and-fill mining using (A) horizontal drilling and (B) uppers, Atlas Copco (1986)

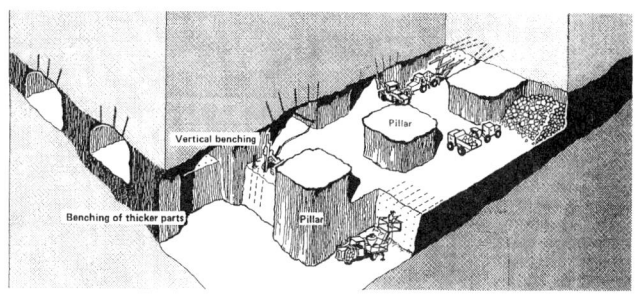

FIGURE 72.5 Benching in room-and-pillar mining, Atlas Copco (1986)

72.3.2 The Ash Approach

A standard approach to designing open-pit blasts is based upon the design guidelines of Ash (1963). The same approach, with some minor modifications, can be applied to bench-type blasting in underground mining. It can be shown that the burden B, the distance from the hole to the closest free surface, is directly related to the diameter of the blasthole (D). This is expressed as:

$$B = K_B D \qquad (72.4)$$

The hole spacing (S) is related to the burden by:

$$S = K_S B \qquad (72.5)$$

and the stemming (T) can also be related to the burden through:

$$T = K_T B \qquad (72.6)$$

When examining actual operations, Ash found that for ANFO used in rocks having a density of about 2.5 g/cm³:

$$K_B = 25$$

The design-spacing factor K_S was in the range of (1 to 2) B. The practical range for open-pit mines is from 1 to 1.3, based on energy distribution considerations. The average stemming factor is 0.7 B. Hustrulid (1999) has shown that the factor K_B can be related to other explosives using the relative bulk strength relationship:

$$K_B (\text{explosive A}) = K_B (\text{ANFO}) (BS_A)^{1/2} \qquad (72.7)$$

For the example given above:

$$K_B(\text{explosive A}) = 25(1.4)^{1/2} = 29.6$$

The value used for K_B also depends upon the density of the rock mass. For rock densities significantly above or below 2.5 g/cm³, this factor must be considered.

For underground applications, the spacing ratio is about the same as used for open pit mining:

$$K_S = 1 \text{ to } 1.3$$

Assume that in an open-pit mine, that holes with a diameter of 4 in (100 mm) are used with ANFO (density of 0.8 g/cm³) in a rock with a density 2.5 g/cm³. For simplicity, it will be assumed that:

$$K_B = 25$$
$$K_S = 1.3$$

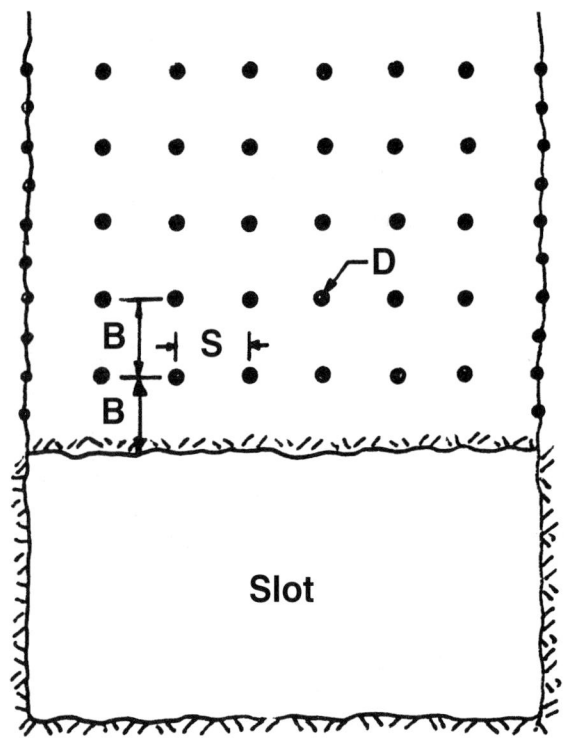

FIGURE 72.6 Plan view of a stope showing blasthole placement with regard to the opening slot

Using these together with the hole diameter, the burden and spacing become:

$$B = 2.5 \text{ m}$$

$$S = 3.25 \text{ m}$$

The explosive charge/meter of hole (Mc) is given by:

$$Mc = \pi/4 \, D^2 \, \rho_{ANFO} = \pi/4 \, (0.100)^2 \, 800 = 6.28 \text{ kg/m} \quad \textbf{(72.8)}$$

The volume of rock broken per meter of hole is:

$$V = B \times S \times 1 = (2.5)(3.25)(1) = 8.125 \text{ m}^3/\text{m} \quad \textbf{(72.9)}$$

The amount of rock broken per meter of hole is given by:

$$W = SG_{rock} \times V = 2.5 \times 8.125 = 20.3 \text{ tonnes/m} \quad \textbf{(72.10)}$$

where W = mass broken/m of hole

   V = volume of rock broken/m of hole

   SG_{rock} = specific gravity of rock.

The powder factor (ANFO) expressed as required charge per mass then becomes:

$$PF(ANFO) = W/L = 6.28/20.3 = 0.31 \text{ kg/tonne} \quad \textbf{(72.11)}$$

The use of (ANFO) in the expression simply means that it applies for ANFO. In terms of volume, the powder factor (ANFO) becomes:

$$K(ANFO) = L/V = 6.28/8.125 = 0.77 \text{ kg/m}^3 \quad \textbf{(72.12)}$$

This is very similar to the typical ANFO powder factors used in open-pit mining. However, in open-pit mining the loading equipment is often large and there is lots of room to handle oversize. The same is not true underground, where the presence of oversize can cause significant handling problems. To overcome this, the powder factor is increased to produce a finer average fragmentation. The authors have found that this can be taken

into account by simply reducing the K_B values in the Ash formulas when applying them underground. The following value is used as a first approximation:

$$K_B = 20 \text{ for ANFO}$$

In rock with a density of 2.5g/cm^3. If the above example is repeated with this value one finds that:

$$B = 20 \, (0.1) = 2 \text{ m}$$

$$S = 1.3 \, B = 2.6 \text{ m}$$

$$V = 5.2 \text{ m}^3/\text{m}$$

$$W = 13 \text{ tonnes/m}$$

and the powder factor becomes:

$$PF \, (ANFO) = 6.28/13 = 0.48 \text{ kg/tonne}$$

As will be seen later, this is very similar to that experienced in practice.

72.3.3 The Powder Factor Approach

In the powder factor approach, one begins with a given hole diameter and assumes (or knows) the required powder factor. One can then write the volume of rock that is broken for a fully charged hole of length L with a burden B and spacing S. This is:

$$V = B \times S \times L \quad \textbf{(72.13)}$$

The mass of rock broken is:

$$W = B \times S \times L \times SG_{rock} \quad \textbf{(72.14)}$$

where L = charged length

   SG_{rock} = the specific gravity of the rock

The amount of explosive in the hole is given by:

$$V_{expl} = \pi/4 \, D^2 \, \rho_{expl} \quad \textbf{(72.15)}$$

Using the known powder factor PF, one can combine equations (72.14) and (72.15) to yield:

$$B \times S = \frac{\pi D^2 \rho_{expl}}{4(PF)(SG_{rock})} \quad \textbf{(72.16)}$$

Assuming that the hole diameter is 4 in (100 mm), the explosive is ANFO with a density of 800 kg/m^3, and the required powder factor is 0.48 kg/tonne. Substituting these values into Equation (72.16), one finds that:

$$B \times S = \frac{\pi D^2 \rho_{expl}}{4(PF)(SG_{rock})} = \frac{\pi(0.10)^2(800)}{4 \times (0.48)(2.5)} = 5.236 \text{ m}^2$$

Now one must choose either B or S to solve for the other. Assume that:

$$S = 1.3 \, B$$

Equation (16) then becomes:

$$1.3 \, B^2 = 5.236 \text{ m}^2$$

Solving for B, one finds that:

$$B = 2.0 \text{ m}$$

Thus, the pattern would be:

$$B = 2.0 \text{ m}$$

$$S = 2.6 \text{ m}$$

and the powder factor is that required.

72.3.4 Fragmentation

The fragmentation depends somewhat on the burden-spacing ratio, and the same powder factor will give different fragmentation

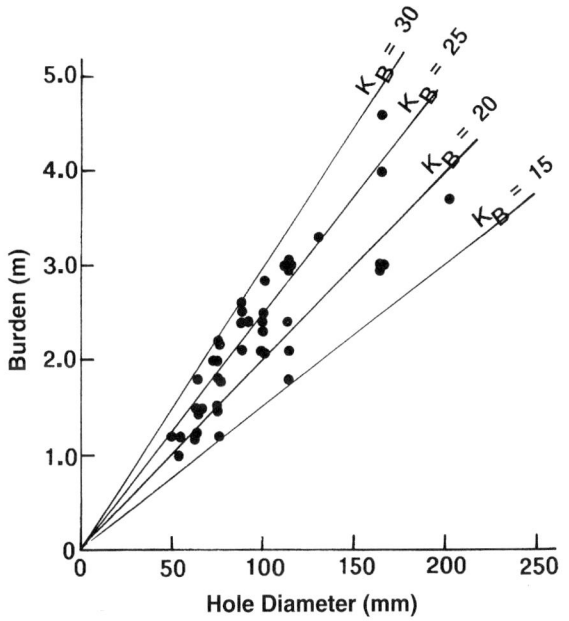

FIGURE 72.7 The relationship between burden and hole diameter from the Canadian data

depending upon the rock-mass condition and the explosive used. The timing is extremely important to the final results achieved. However, these two approaches are a way to begin the design. For further information, the interested reader is referred to the papers by Cunningham (1992) and the books by Persson, Holmberg and Lee (1994) and by Hustrulid (1999).

72.3.5 Wall Damage

Along the walls, special designs are used to minimize unwanted damage. This procedure will be discussed under the drifting section.

72.3.6 Data from Canadian Mines

In the Canadian Mining Journal's "2000 Mining Sourcebook," a large amount of data collected from Canadian underground mining operations is presented. The blast-pattern data for the open-stoping and blasthole-stoping operations are summarized in Table 72.1. One can see that a wide range of hole diameters and explosive types are included. The spacing:burden ratio varies from 1 to 3.5, with the most common values in the range of 1 to 1.3. Figure 72.7 is a plot of burden versus the hole diameter. Due to a lack of complete information, the different explosive types were not taken into account. Superimposed on the data set are lines corresponding to different K_B ratios. As can be seen, the data fall between the lines of $K_B = 15$ and $K_B = 30$. The spread is natural considering the different explosives and rock densities involved. Assuming that $K_B = 20$ as a first approximation is probably good enough for starting the design. Based on the results, the pattern can then be spread or drawn in. Table 72.2 is a summary of the powder factors and the host rock types involved with the different operations. The values range from 0.27 to 1.05 kg/tonne.

72.3.7 Summary

As a first approximation, the design approach used in surface mining can be applied to bench-blasting geometries underground. The principal change involves increasing the powder factor to provide the required finer fragmentation.

TABLE 72.1 Summary of blasting patterns (Canadian Mining Journal)

Mine	Explosive	Diameter (mm)	Pattern B(m) × S(m)	S/B
Agnico-Eagle	ANFO	100	2.1 × 2.1	1.0
Barrick, Bousquet	AMEX	100	2.4 × 2.5	1.0
Barrick, Holt-McDermott	ANFO and	54	1 × 2.5	2.5
	watergel	76	2.2 × 2.5	1.2
Boliden, Westmin	80% ANFO,	76	2.2 × 2.4	1.1
	20% emulsion	89	2.5 × 3.2	1.3
		64	1.8 × 2.3	1.3
Cambior, Bouchard-Hebert	AMEX	165	3 × 3.4	1.1
Cambior, Langois	ANFO	115	1.8 × 2.2	1.2
		63	1.2 × 1.2	1.0
Cominco, Polaris	ANFO	76	2.2 × 2.2	1.0
		89	2.6 × 2.6	1.0
		102	2.85 × 2.85	1.0
Falconbridge, Fraser	ANFO	100	2.4 × 2.4	1.0
Falconbridge, Lockerby	Watergel,	76	1.2 × 4.2	3.5
	ANFO	115	3 × 3	1.0
Hudson Bay, Callinan		64	1.5 × 1.5	1.0
		76	1.5 × 2.5	1.7
Hudson Bay, Ruttan	Nilite, Lorite	114 & 92	2.4 × 3	1.3
	Tovan	76	2 × 2	1.0
Hudson Bay, Trout Lake	ANFO & Iremite	76	2 × 2.5	1.3
Inco, Coleman		100	2.1 × 3.4	1.7
Inco, Copper Cliff North	ANFO	89	2.4 × 2.7	1.1
	Canamex 550	165	3 × 3	1.0
		203	3.7 × 3.7	1.0
Inco, Copper Cliff south	Numex &	165	3 × 3	1.0
	Canamex	165	4.6 × 4.6	1.0
	Numex, others	89	2.1 × 2.1	1.0
Inco, Stobie	Emulsion	165	3 × 3	1.0
Inco, T-1	ANFO and	165	3 × 3	1.0
	slurry	115	2.1 × 2.1	1.0
Niobec	ANFO & Tovan	165	4 × 4.9	1.2
Noranda, Brunswick	Slurry &	115	3 × 4	1.3
	emulsion	130	3.3 × 4.3	1.3
Noranda, Gaspe	ANFO & Tovan	115	3 × 3	1.0
Noranda, Heath Steele	Emulsion &	76	1.8 × 2.4	1.3
	Tovex	64	1.5 × 2.1	1.4
		115	3 × 3.7	1.2
Placer Dome, Campbell	Superfrac &	54	1.2 × 1.5	1.3
	ANFO	64	1.5 × 2.1	1.4
		76	1.8 × 2.4	1.3
Placer Dome, Musselwhite		100	2.3 × 2.4	1.0
Richmont, Francouer	ANFO	50	1.2 × 1.2	1.0
Teck-Corona, David Bell	ANFO & emulsion	76	2 × 2	1.0
TVX Gold, New Brittania		76	1.5 × 1.5	1.0
		64	1.2 × 1.2	1.0
Williams		115		

TABLE 72.2   **Summary of powder factors for open stoping and blast-hole stoping operations (Canadian Mining Journal)**

Mine	Explosive	Host rock type	PF (kg/t)
Agnico-Eagle	ANFO	Feslic volcanics	0.65
Barrick, Bousquet	AMEX	Dacite and rhyolite	0.50
Barrick, Holt-McDermott	ANFO and watergel	Basalt	0.4
Cambior, Bouchard-Hebert	AMEX	Pryoclastic rocks/ rhyolite	0.35
Cominco, Polaris	ANFO	Dolomite	0.45
Falconbridge, Fraser	ANFO	Granite, gneiss, and norite	0.9
Hudson Bay, Callinan		Quartzphyric rhyolite	0.5
Hudson Bay, Trout Lake	ANFO and Iremit	Quartz porphyry	0.3
Inco, Copper Cliff North	ANFO Canamex 550	Quartz diorite	0.65–0.8
Inco, Copper Cliff south	Numex and Canamex	Quartz diorite	0.75
Inco, Stobie	Emulsion		0.75
Inco, T-1	ANFO and slurry	Felsic schist	0.35–0.5
Niobec	ANFO and Tovan	Carbonatite	0.4
Noranda, Brunswick	Slurry and emulsion	Metasedimentary and volcanic	0.27
Noranda, Gaspe	ANFO and Tovan	Skarn porcellanite	0.4
Noranda, Heath Steele	Emulsion and Tovex	Felsic metavolcanics and metasediments	0.45
Placer Dome, Campbell	Superfrac and ANFO	Mafic and ultramafic volcanics	0.5
Richmont, Francouer	ANFO	Volcanic	0.95
Teck-Corona, David Bell	ANFO and emulsion	Metasediments and metavolcanics	0.95
TVX Gold, New Brittania		Mafic volcanics and quartz carbonate	0.75–1.05
Williams		Felsic porphyry	0.4

72.4   RING BLASTING DESIGN

72.4.1   Introduction

Ring blasting is fundamentally different from all other forms of blasting because the holes are drilled from a central point in and radiate outwards to the limits of the ore block being mined. As shown in Figure 72.8, the technique involves three distinct operations:

- Driving a tunnel (or tunnels) along the axis of the proposed excavation. This is the "ring drive."
- Creating a vertical slot at the end of the ring drive(s) to the full width of the excavation.
- Drilling sets of radial holes called "rings" parallel to the slot. These rings are then blasted progressively into the slot.

Figures 72.9–72.13 show some typical designs and demonstrate the basic benefits of the technology.

- The method is very flexible in coping with variations in block size and shape.
- It enables large blocks of ground to be blasted from a few access points, thereby improving both development costs and ground stability.

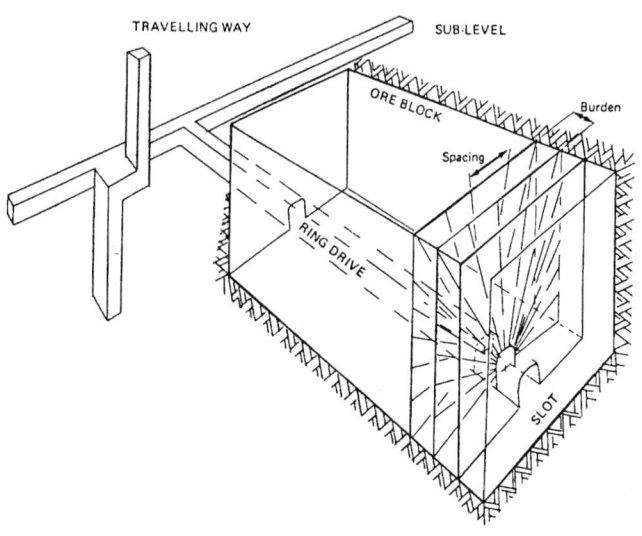

FIGURE 72.8   **The three operations involved in ring blasting, AECL (1980 a,b)**

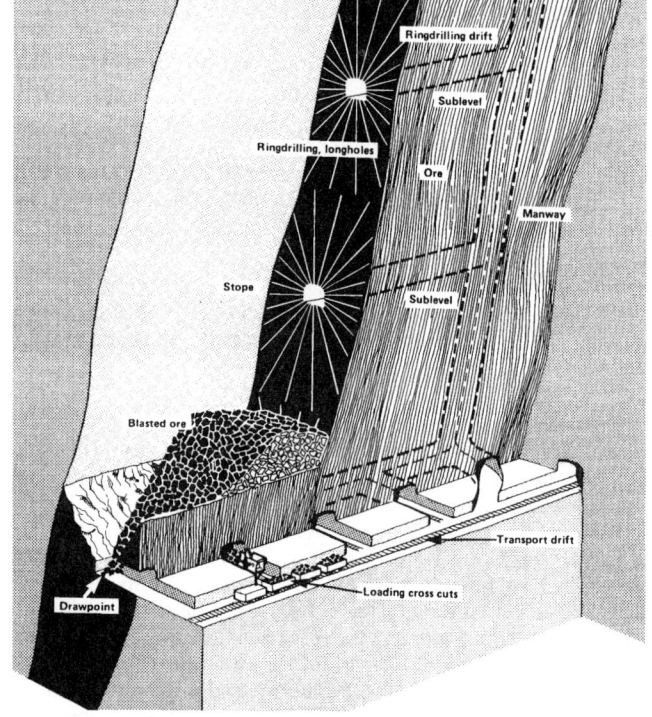

FIGURE 72.9   **Sublevel stoping using fan patterns, Atlas Copco (1986)**

- It offers high productivity and insurance against breakdowns because the holes can be drilled and charged well in advance of blasting operations.
- It is safe, as men do not enter the stope.

In fact, as underground mines seek to meet the challenges posed by the new century, many are seeking to maximize the benefits of ring drilling by harnessing modern drilling technology to extend its application and achieve previously undreamed-of efficiencies. Unfortunately, there is also a downside to ring blasting

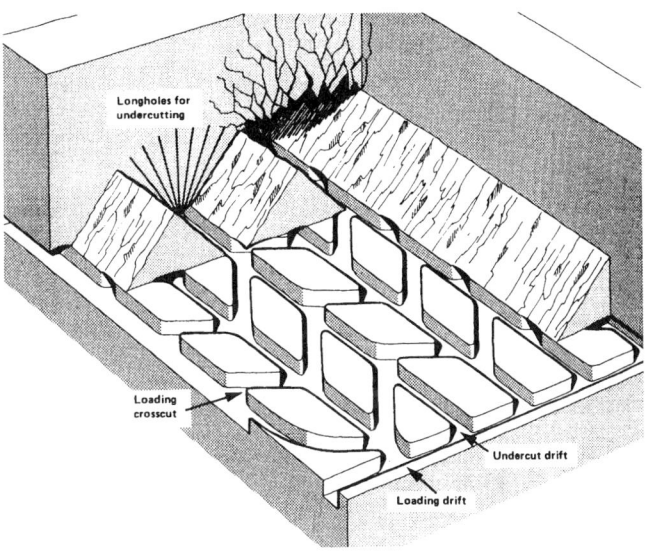

FIGURE 72.10 Ring blasting used in creating extraction troughs in panel caving, Atlas Copco (1986)

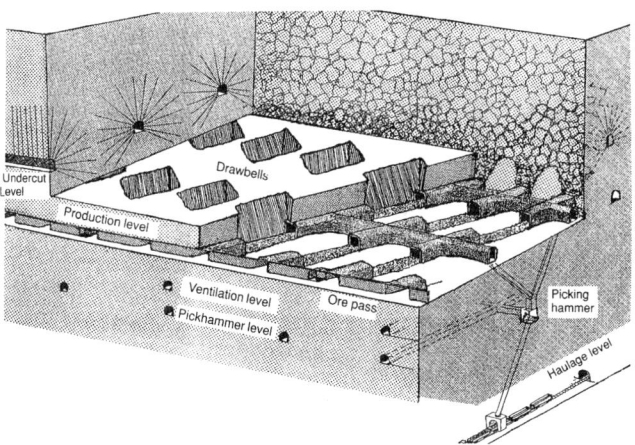

FIGURE 72.11 Drawbell creation using fan drilling, Atlas Copco (1997)

■ Because the blastholes radiate from a central drilling position, the coverage of the block of ground varies from minimum at the toe position to maximum at the collar position. This leads to continuously varying powder factors, greater explosive consumption than is theoretically necessary, and complications in charging the holes so as to minimize overcharging.

■ The holes are drilled at different angles, leading to varying deflection forces for the holes and a potential for gross discrepancies between the designed and achieved drilling patterns. This can lead to poor blasting results and failure to reap the potential production from each ring.

■ The holes meet the block outline at different angles, making it difficult to ensure that an appropriate and equitable concentration of explosives is achieved in the critical breaking area at the toe.

■ The holes are drilled at different angles and to different lengths, being collared at close intervals. This imposes a

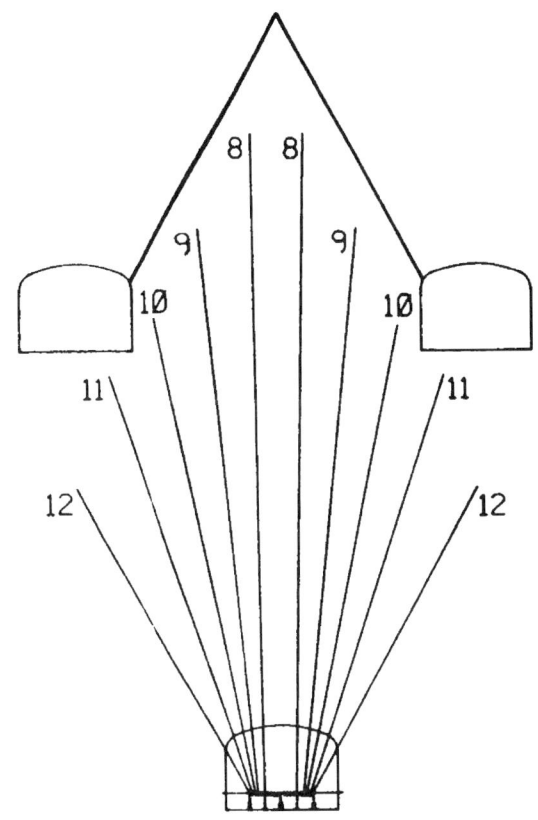

FIGURE 72.12 Typical ring blast design for large-scale sublevel caving, LKAB (1996)

strain on the production team in terms of achieving the planned layout. In addition, if in-hole initiators are used, stock controls over lengths and delays can be difficult.

■ Up-holes are normally used; these present added costs and problems in terms of retaining explosive, especially where large-diameter holes are concerned.

■ Ring holes are very prone to misfires caused by cutoffs in the collar area, where the holes are closely bunched and excess energy tends to be available. On the other hand, inter-hole delays are necessary, especially with large blastholes, to reduce the vibration levels. Therefore, selecting and implementing the initiating system is critical.

The extent to which the above challenges can be met depends both on an understanding of the extent of the problem posed, and on ways in which good design can reduce or eliminate each problem. The major recurring problems related to blast design is as follows:

■ Failure of up-holes to break out at the toe, leaving 'crown' pillars

■ Poor drawpoint availability caused by the enormous boulders created by back-damage, allowing blocks behind the ring to fall into the drawpoint

■ Excessive secondary breaking due to a poor distribution of explosives related to both design and implementation problems.

In this section, an approach to ring design is presented that will implement intrinsically productive designs, or at least give advance warning of problem areas.

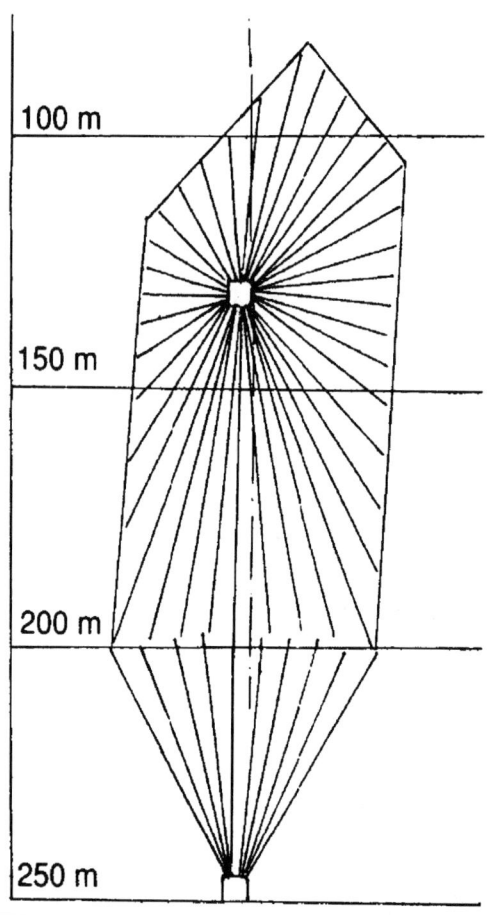

FIGURE 72.13   Ring design for stoping and undercutting

TABLE 72.3   Drilling angles and maximum deflection for 3.5-m spacing and 25% deviation at the toe

Hole length (m)	Maximum deflection (degrees)	Inter-hole angle (degrees)
15	3.34	13.49
25	2.00	8.05
40	1.25	5.01
60	0.84	3.34

72.4.2   Basic Mine Layout

There is a general desire to increase the rate of bringing a mine into production at minimum cost by having very great vertical separation between the mining levels. This means that rings need to be drilled so as to reach between levels, and the longest holes will be those that just reach to the furthermost corner of a block. First, serious consideration needs to be given to the inter-level spacing because the longer the holes, the greater will be their deviation, and the greater the problem if the drill string should get stuck during drilling, or the hole should become blocked prior to blasting. In addition, very large ore blocks will call for more holes to be drilled from the same position, multiplying the bunching effect around the ring drive. It is difficult to believe that angular increments of fractions of a degree are meaningful in this context, yet this is what is called for if reasonably uniform drilling patterns are required. Ideally, the holes should not be more than 25% from their planned positions, let alone intersecting each other. Table 72.3 gives different lengths for holes designed to have a spacing of 3.5 m and an angular deviation of 25%, owing to both setup and wander.

Table 72.3 shows that, above 40 m, exceptional drilling equipment, setup procedures, and operator commitment is required. A further consideration is that very large rings reduce the ability of a mine to source production from different areas in the interests of grade, or of keeping production up in the event of problems with the production ring.

The decision as to the relative areas to be assigned to up-rings as opposed to down-rings needs to be seriously considered. Up-holes should be drilled as short as possible, because their cost and the work involved in charging them is greater than for down-holes, and the differential increases with length. In addition, the ability to load to a controlled collar distance is much greater with down holes than with up-holes. Apart from operational factors, spillage of explosive from up-holes is inevitable, the extent depending on the borehole conditions, the skill in the handling of loading equipment, and the variability of the explosive being use. Waste reduces efficiencies, but the safe and hygienic disposal of spilled explosive also impacts on the overall productivity. Thus, while it is clear that the effective loading of long up-holes is technically possible, it is highly desirable to facilitate charging operations by locating the ring drive high in the ore block. If the ring drive is no more than 15 m below the top of the block, loading operations are much enhanced. The shape of the ore blocks—especially their placement opposite one another and other profiles—is extremely important. It is easy to create "shadow" areas inaccessible by drilling from the available ring drives, or areas that result in an uneconomic drilling pattern owing to indentations or "nodules" at critical points. These may be unavoidable, but can sometimes be prevented during the design stage.

72.4.3   General Design Principles

Good fragmentation can result only if sufficient explosive is used in correctly drilled holes, with initiation arranged to ensure that all the explosive detonates in the proper sequence. Owing to the complexity of ideal layouts for ring blasting, however, compromises have to be reached between what is theoretically desirable and what can reasonably be implemented.

Blasthole Length.   When laying out rings, one of the main considerations is to limit the maximum hole length. To minimize development, holes are often laid out over excessive lengths simply to span the distance between sublevels. This creates the following problems:

1. Inaccurate drilling—the small-diameter holes used in this technique can seldom be drilled with reasonable accuracy over more than about 20 meters with conventional equipment. Beyond this length, and especially with angled holes, severe departures from the planned burden and spacing are likely, resulting in poor fragmentation, toe formation, and overbreak.

2. Gauge reduction—used as the hole results in reduced critical toe area. Smaller drill bits are often used as the length increases, with explosive loading in the critical toe region.

3. Drilling inefficiency—the efficiency of blow transmission in the drill string decreases by about 10% across each extension coupling and this results in reduced penetration rates and increased equipment wear over long holes. In ring blasting, the situation is even worse because of the short extension steels necessarily used in constricted ring drives.

Ideally, both up-and down-holes should be drilled to double the distance between sublevels, but in sublevel caving, for example, only up-holes can be drilled. This factor should be taken

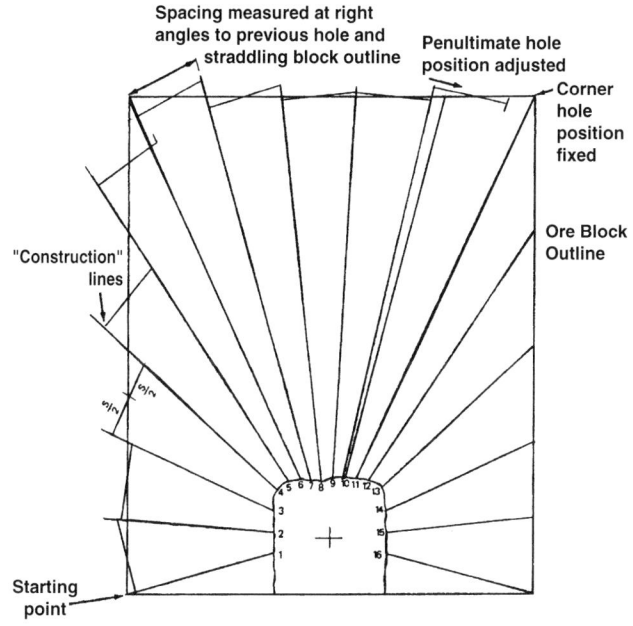

FIGURE 72.14   The layout process for a typical ring design. AECL (1980 a,b).

into account when laying out sublevels. As a rule, the majority of holes should not exceed 30 meters unless specialized equipment (operated by properly trained drilling crews) is used. With longer holes, poor fragmentation is likely to be inherent to the system and can only be compensated for by adopting a higher explosive-loading ratio than would normally be considered.

Determining Drilling Pattern. The terms "spacing" and "burden" require definition because of a lack of general agreement among ring-blasting operators. "Burden" means the distance between two consecutive rings. "Spacing" means the distance between the ends (or "toes") of neighboring holes in one ring, measured at right angles and straddling the outline of the ore block, using construction lines (see Figure 72.14). The correct drilling pattern is one that delivers the appropriate energy at the toes of the blastholes. If it can be shown that a particular drilling pattern in a parallel-hole benching operation gives the required blasting results, this pattern should be adopted for ring blasting. An example of this is tests that were performed in kimberlite. The ring-drilling pattern was derived from preliminary blasting trials in the last months of the open-pit operation, with fragmentation as the criterion. In the absence of such guides to hole spacing, a powder factor approach can be taken. Care must be exercised, however, because it does not allow for special geological conditions, specific breaking results, the blasting geometry, or the effect of explosive type.

The following method of calculating burden and spacing ensures good breaking in the critical toe area around the ends of the blastholes. For simplicity, the calculation assumes that all holes are parallel, with the explosive column reaching to a collar length of twenty charge diameters. The hole length used in the calculation is the average length in the ring. For ANFO or slurry, the explosive diameter equals the hole diameter. The fact that the powder factor increases towards the ring drive through convergence of the holes is compensated for in practice by using different uncharged lengths in each hole. Alternatively, spaced charges or charges of lower strength can be used, but these solutions are not recommended because of the complications they introduce.

The formula relating the burden and spacing to the powder factor is:

$$B \times S = \frac{L\,Mc}{H\,K} \qquad \textbf{(72.17)}$$

where   B = nominal ring burden (m)

   S = nominal toe spacing (m)

   L = length of explosive column (m)

   H = average hole length for ring drilling (m)

   Mc = explosive mass per unit length (kg/m)

   K = powder factor (kg/m^3)

For simplicity, we will assume that all holes are parallel, and the uncharged portion is equal to 20 charge diameters. The charged hole length becomes:

$$L = H - 20D \qquad \textbf{(72.18)}$$

where   D = explosive diameter (m)

In general, the toe spacing of holes should exceed the burden on the ring, but the exact ratio of spacing to burden is not critical. The spacing:burden ratio is normally assumed to be 1.3, but can be as high as 1.5.

In normal ring blasting, good fragmentation is required and fairly high powder factors are necessary, whatever the ground hardness. A starting figure of 0.8 kg/m^3 would suit most types of weak rock, while 1.2 kg/m^3 would suit a dense, strong rock. The design powder factor depends on the toughness and blockiness of the ground and whether or not tight breaking conditions prevail. It is good practice to allow for at least 30% expansion from the solid when blasting into a restricted slot.

A major error to be avoided is using a planned powder factor across the whole ring. This approach is likely to result in excessive hole spacing, since convergence of the holes always causes overcharging inside the pattern. It is patently poor blasting practice to tolerate inadequate energy at the toe portion just so an average powder-factor can be obtained.

To indicate how the calculation process works, consider using an emulsion explosive with a density of 1,180 kg/m^3 in a 102-mm-diameter blasthole. The powder factor is 0.8 kg/m^3. For simplicity, we will assume that the explosive extends to the collar and that the hole length does not have to be considered. In Equation 72.17:

$$Mc = 9.64 \text{ kg/m}$$

$$K = 0.8 \text{ kg/m}^3$$

and thus

$$B \times S = \frac{L\,Mc}{H\,K} = \frac{9.64}{0.8} = 12.05 \text{ m}^2$$

Assuming that:

$$S = 1.3\,B$$

then

$$1.3\,B^2 = 12.05$$

and

$$B = 3.05 \text{ m}$$

Therefore, the "nominal" drilling pattern is 3.0 m × 4.0 m. The rings should, therefore, be drilled 3 m apart along the ring drive and the holes angled to terminate with their toes 4 m apart.

Laying Out the Ring. Once the "nominal" drilling pattern has been obtained, it must be applied to the block outlines. The spacing is measured perpendicular to the blasthole. Because most of the holes will not meet the block outline at right angles, the holes have to be fitted so that no energy starvation takes place between them along the oblique parts of the outline. This is done

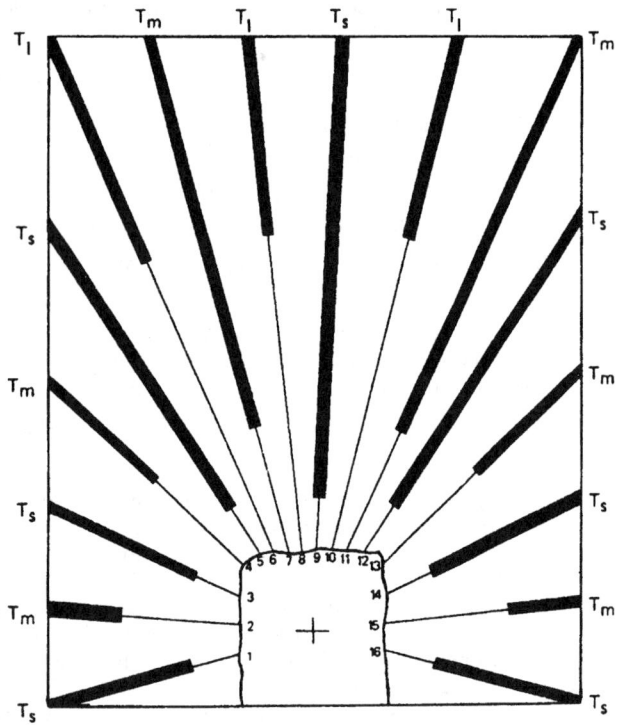

FIGURE 72.15   The charged lengths for the completed design, AECL (1980 a,b)

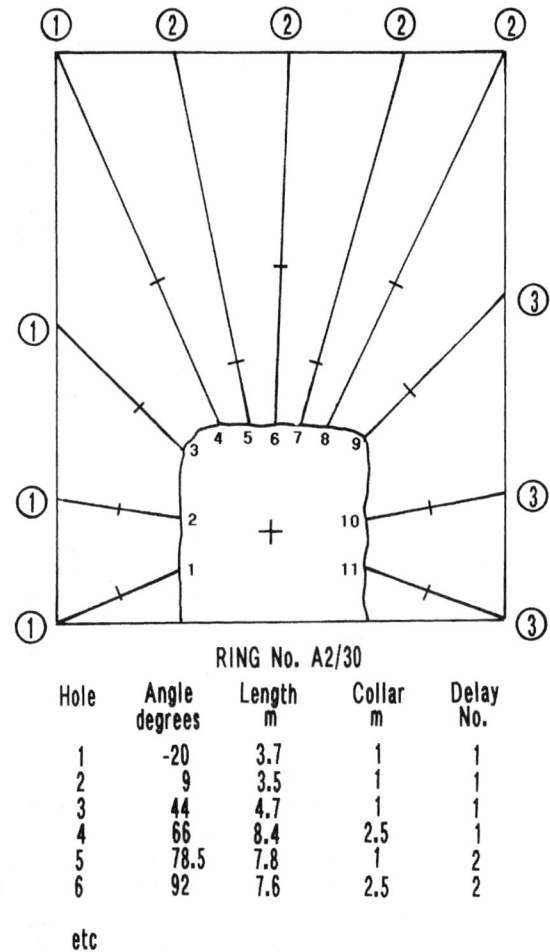

RING No. A2/30

Hole	Angle degrees	Length m	Collar m	Delay No.
1	-20	3.7	1	1
2	9	3.5	1	1
3	44	4.7	1	1
4	66	8.4	2.5	1
5	78.5	7.8	1	2
6	92	7.6	2.5	2

etc

FIGURE 72.16   The ring design as transmitted for drilling and blasting, AECL (1980 a,b)

by plotting the hole positions beyond the outline and imposing the designated spacing between the holes so as to straddle the outline equally (Figure 72.14). Each variation in the ore block requires a new layout. The exercise is easily and quickly performed by hand as follows:

1. Holes are drawn in from the drilling position to the corners of the ore block.

2. Using a scale and 90° offset, the toes of intervening holes are marked off consecutively, using the calculated spacing, as shown in Figure 72.14.

3. As the intermediate holes approach fixed corner holes, small adjustments are made. The hole before the corner hole is (a) either centralized between the corner hole and the next one back or (b) omitted. This is done at the discretion of the person doing the layout. When it is omitted, the next hole back should be centralized, if necessary. This practice rarely results in more than 10% deviation from planned spacing. When this process is complete, the last hole will very seldom fit exactly on the last part of the outline, and the spacing should be readjusted, reducing or increasing the spacing slightly until all the holes are equally spaced (Figure 72.15).

4. On completion of the ring layout, the actual angles and lengths of the holes are measured and entered on the instruction sheet (Figure 72.16).

If the ring is not vertical, the exercise must be done in the plane of the ring.

Charging Pattern.   The blast holes converge towards the ring drive and, to avoid serious overcharging, require an alternating pattern of uncharged collar lengths. This pattern has to be simple and repetitive if charging crews are to adhere to it meaningfully, and the following compromise arrangement is suggested.

Assume three uncharged collar lengths:

T_s = 20 explosive diameters

T_m = 50 explosive diameters

T_l = 125 explosive diameters.

The holes are numbered beginning at a readily recognizable hole, e.g., the lower left-hand side. The uncharged lengths are then specified in that order:

$T_s, T_m, T_s, T_m, T_s, T_l, T_m, T_l, T_s, T_l, T_m, T_s, T_m, T_s, T_m, T_s$

Prior to charging up, a responsible person attaches tags indicating whether T_s, T_m, or T_l apply to the collar length for each hole. It should be noted that no uncharged collar should be more than two-thirds of the hole. If T_m exceeds two-thirds of the hole length, use T_s. If only T_l exceeds two-thirds of the hole length, alternate T_s and T_m until holes longer than 1.5 T_l are encountered. The end holes (which only occur where full rings are not drilled) are always given T_s. When the planned powder factor and the actual projected explosives used per cubic meter broken are finally compared, there may be a significant discrepancy, caused by using approximations to specify the uncharged lengths. This is unavoidable without either sacrificing fragmentation or imposing a more complex charging system. It can be justified on the grounds that a higher powder factor will result in finer overall fragmentation, which is usually welcome in ring-blasting operations. In trying to reduce the eventual overall powder factor, it is important not to decrease the powder factor in the original calculation. This would result in poor breaking in the critical region of the ring perimeter.

If a large number of ring designs must be made, the process is very tedious and time-consuming if done by hand. Software has been developed to perform this task. The software "Ring" developed by AECI's Blast Consult Group is one such example. It provides optimized ring-drilling designs in a minimum of time. This is extremely useful in comparing the effects of different hole diameters and checking the achieved hole spacing against the design. The package normally uses the nominal spacing as a base to find the number of holes in the ring and the actual spacing. It can also be given the required number of holes and can derive the spacing from this. Once the correct outline pattern is achieved, the spacing can be offset against the designed powder factor to correct for excessive or inadequate energy in the toe area. For example, if the exercise in the above example yielded a final average toe spacing of 3.7 m, instead of 4.0 m, Equation 72.17 could be used, which would derive a ring burden of 3.25 m instead of 3.0 m.

72.4.4 Optimizing Results

The basic approach to designing the layout of ring-blast holes and distributing explosives within each ring was discussed in the previous section. This section shows how blasting results can be optimized by suitable choice of hole diameter, explosive type, and initiation system. Priming and initiation of holes is also discussed.

Choice of Hole Diameter. Hole diameter and explosive density basically determine the hole spacing. An increase in diameter increases the spacing and enables fewer holes to be drilled. This means that large-diameter holes are a good choice for large ore blocks. The largest feasible hole diameter should be used for ring drilling. This has the following benefits:

- Less drilling—Large holes take more explosive, which can break to wider patterns. This means the holes are less crowded around the ring drive and charging operations are faster, and greater production is obtained from the available equipment.

- Easier loading—Larger holes tend to suffer less from obstructions that prevent cartridges (or loading hoses) from being pushed to the back of the hole. Using hole diameters below 45 mm may necessitate charging with 32×200-mm cartridges, which have the added disadvantage of being shorter and taking longer to load.

The largest diameter that should be considered for up-holes is about 100 mm. Above this, retention of the explosive is a major problem.

It often happens, however, that the same drilling equipment used for large blocks is also used for small blocks, e.g., production cones. It is patently inappropriate to use a hole diameter that calls for a hole spacing of 3.5 m if the available space is 4.5 m, or even 9.0 m. Each available block accommodates a particular number of equally spaced holes and the smaller the block, the less the ability of large holes to deliver the ideal hole spacing. This can be particularly so when the outline includes "notches" because of the intrusion of other openings or geological features.

A further consideration is the extent to which up-holes will be required. Holes in excess of 64 mm are difficult to load with cartridged explosives, while those in excess of 89 mm require special skills with ANFO. For holes in excess of 115 mm, the effort needed to retain any explosive is so great that only short, relatively infrequent holes should be considered.

If rock deterioration is a serious concern, it should be remembered that large-diameter holes deliver greater vibration energy to the rock. The explosive energy can be reduced to counter this by using smaller-diameter holes.

Selecting Explosive. From the foregoing, it will be apparent that selecting the explosive is likely to follow naturally from the choice of hole diameter, which itself is largely determined by considerations of ore-block size and the extent to which up-holes are to be drilled. Almost any explosive is suitable for near-vertical down-holes, but an in-hole transport system is necessary as the inclination to the horizontal decreases below 50°. The same restrictions apply for up-holes. In larger holes, ANFO and pumped emulsion products are the only option; the actual choice will depend on parameters beyond the scope of this paper, not least of which is the availability of suitable loading equipment.

High-density, high-energy, well-coupled explosives have the best potential for ensuring good fragmentation in ring blasting. ANFO can only be used in dry blastholes. Its low price and convenience in pneumatic loading, together with its effective breaking, make it a good choice in most conditions, but it is not always used to best advantage. In particular, the ANFO column should be boostered at 5-m intervals to ensure the maintenance of a stable detonation velocity over the entire charge length. Any cap-sensitive cartridged explosive can be used for boosters. Failure to observe proper boostering is bound to result in substandard performance.

Charging Holes. Because the blastholes converge, grossly inefficient blasting will occur unless the uncharged collar is varied from hole to hole. What length to leave uncharged is not a trivial matter, because it is related not only to the planned hole spacing, but also to the size and shape of the ore block. The primary criterion is that holes should be charged to the point at which the tangential distance from the end of the charge to the next hole is half the designed hole spacing. In practice, even skilled charging crews are unlikely to be able to control charging lengths to accuracies better than whole meter lengths, and design considerations should take this into account. When deciding on the uncharged lengths, the software programs are extremely useful. They can quickly determine which regions are under- or over-charged.

Initiation. Initiation is a subject of its own. Technically, ideal initiation is basically implemented when each hole is initiated from a point near its toe with very rapid firing of holes spreading from a central point downward on each side. This results in optimized fragmentation while limiting the vibration levels and avoiding cutoffs near the ring drive. Unfortunately, the practical problems with ideal initiation sometimes lead to pragmatic solutions, the chief aim of which is merely to ensure that all the holes fire while the vibration is limited as far as possible. In planning a blast, delay units and primers are allocated to each hole, and judgment must be used as to whether backup units are needed. Ground conditions and loading methods will determine the likelihood of initiation failures, and any sign that a hole may misfire calls for a backup system. Collar priming imposes real limits on timing options because cutoffs automatically result if some holes fire earlier than their neighbors. Delay detonators inside the blastholes are the most commonly used means of initiating ring blasts because they are less likely to be cut off by rock movement and concussion. Continuity of detonation is assured by lining each blasthole with detonating cord.

Selection of Delays. Theoretically, a different delay should be used in each hole to improve fragmentation and limit vibration effects, but this has some disadvantages:

- Limitation of blast size—The range of delays is limited and assigning different delays to each ring would severely restrict the number of rings that could be blasted at one time. Large blasts are usually desirable as they result in fewer oversize slabs of rock

- Misfiring of later delays—The collars of the holes converge on the ring drive and with the common practice of collar priming, the detonators are located close together in the rock mass.

As a result, the first hole to detonate is likely to break out the collars of adjacent blastholes, complete with detonators. This can happen even when detonators are quite deep inside the blast-holes, but the likelihood is much reduced if all the delays are the same.

In view of these problems, and because it is simpler to implement, it is preferable to specify only one delay per ring. The detonator lead wires should be not less than 4.5-m long, as most cutoffs take place within 3–4 m of the collar, and the detonator should therefore be located at greater depth.

Notwithstanding the above, it is sometimes essential to use two or even three different delays per ring to reduce concussion. To minimize the danger of cutoffs, the delays are not alternated between holes, but are apportioned to whole sections of the ring as shown in Figure 72.16.

Delay Range. Using alternate rather than consecutive delay numbers is recommended to eliminate the possibility of both out-of-sequence shots and "crowding" between consecutive delays, due to inherent scatter about the nominal firing times of delay detonators.

In some cases, large numbers of rings have to be fired in sequence, requiring in excess of the entire range of delay numbers. No attempt should be made to increase the delay coverage by firing pairs of rings on the same delay: choking, over-break, and poor fragmentation are likely to result.

The maximum range of delays is:

SPD numbers 1 – 20

LPD numbers 3 – 14

Thus, there is a total of 32 delays. If only some SPD numbers have to be used consecutively in a medium-scale blast, delays 1–6 should be the first ones run consecutively, as there is minimal chance of overlapping with these units.

Instantaneous detonators are not recommended for series-in-parallel circuits, as their detonation may cause premature dislocation of the blasting circuit, resulting in misfires. Naturally, if only single rings are blasted, these can be primed simply with detonating cord and initiated using an electric detonator or capped fuse. The advent of the electronic delays means that there are many more possibilities for initiating the holes and in the number of rings that can be fired.

72.4.5 Priming Ring Holes

ANFO. Pneumatic ANFO loading, generates high-voltage static electricity, which can prematurely fire normal electric detonators. Three solutions to this hazard exist as follows:

- "NONEL" detonators can be used with confidence in pneumatically charged ANFO. The detonator is initiated by a shock wave transmitted through a long, non-destructing plastic tube, which may, however, constitute a contaminant in certain types of ore.

- Static-safe electric detonators. These detonators require a 60 MJ/ohm firing impulse (as against 4 MJ/ohm for normal detonators) and can be safely used in pneumatically charged ANFO. They require more powerful exploders and heavier blasting cables.

- The simplest, but least satisfactory solution because of the danger of cutoffs is to collar-prime all ring holes with normal electric detonators. The detonators should not be attached to the detonating cord down-line for at least an hour following the completion of loading operations, to allow for the dissipation of any static accumulation.

Cartridged Explosives. The detonator is normally fixed inside a primer cartridge, which is then pressed into place using a

TABLE 72.4 Idealized approach to ring-blasting design

Design parameter	Recommendations
Block shape	Avoid "shadow" and "nodule" areas if possible
Block dimensions	Restricted by up-holes to 15 m vertically up, by down-holes to realistic limit of drilling accuracy under production conditions, possibly 40 m.
Up-hole length	Loading considerations in large holes suggest no more than 15 m vertically upwards, 25 m inclined. Small diameter holes similarly restricted by drilling accuracy.
Hole size	Should be related to block size and the need to control breakage limits, but in up-holes, no greater than 115 mm owing to charging problems. Tight breaking and ground stability problems call for smaller blastholes. Hole size needs to be iteratively compared with drilling pattern so that it matches the block size and requirements.
Drilling pattern	Ideal drilling pattern should be related to required fragmentation and acceptable powder factor. Iteration with different hole diameters will show hole size that permits best matching between ideal and achieved drilling pattern.
Explosive	(a) Retainable in up-holes, (b) clean delivery system with controllable waste, (c) high-speed loading. Few restrictions with down-holes.
Charging	(a) No area between charges to exceed toe spacing, (b) no area to be covered by more than two fracture envelopes.
Initiation	(a) All holes primed within 3 m of toe position, (b) one hole per delay, (c) delay interval close enough to avert cutoffs.

buffer cartridge or plastic "spider." Where the uncharged collar length exceeds that of the leading wires, the detonator is attached to the detonating cord down line.

Toe Priming. Because of problems associated with long lead wires, it is not advisable to toe-prime ring holes with electric detonators. This is easily achieved, however, if "NONEL" delay detonator assemblies are used.

72.4.6 Blasting Circuit

Current leakage from the blasting circuit is a common cause of misfires in ring blasting, especially under the following conditions:

- Where it is wet and conductive salts are leached out of the explosive

- In conductive orebodies

- Where high resistance iron leading wires are used

Therefore, precautions should be taken to maintain effective insulation between the blasting circuit and the country rock. When charging up, the PVC insulation on detonator lead wires is prone to damage from sharp edges or tight kinking. Care must be taken to avoid damage while unraveling the lead wires, making up primers, pushing primers into blastholes, and loading explosive over lead wires. Insulating putty should be used to enclose the bare wires at each connection. Neatly suspended blasting cables are easily inspected and are less vulnerable to accidental damage. An earth-leakage tester is invaluable for checking circuit insulation. Furthermore, all modern detonators are designed to minimize current leakage by operating at a relatively low voltage.

72.4.7 Conclusion

A review of the foregoing indicates that a technically optimized ring-blasting layout can be proposed. The idealized design, then, would be as given in Table 72.4.

While ring blasting techniques offer substantial benefits, there are intrinsic problems that are exacerbated when large blocks of ore are mined in attempts to improve the production economics. These problems relate to both the convergence of the holes around the ring drive and the charging of the up-holes. The size and shape of the ore block concerned should be carefully

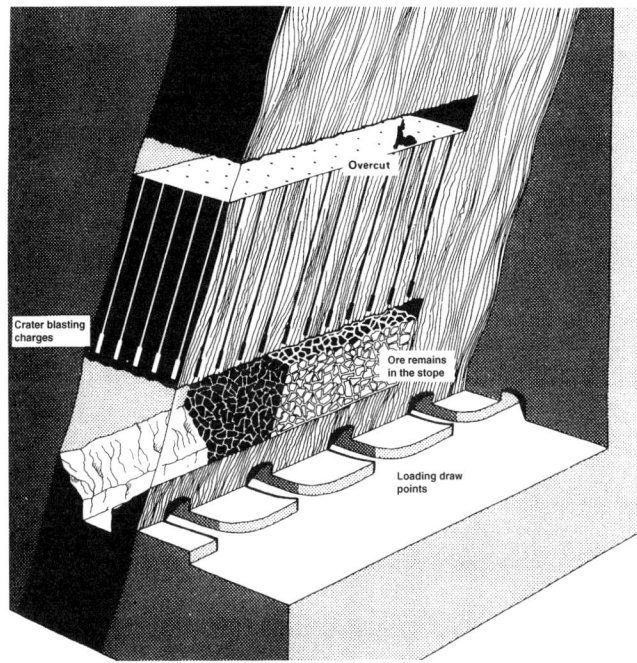

FIGURE 72.17   The vertical crater retreat mining method, Atlas Copco (1986)

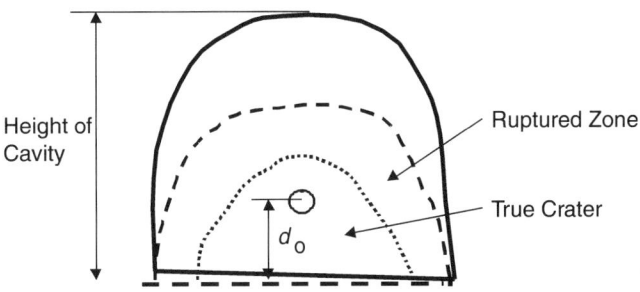

FIGURE 72.18   Principle for the VCR blasting method

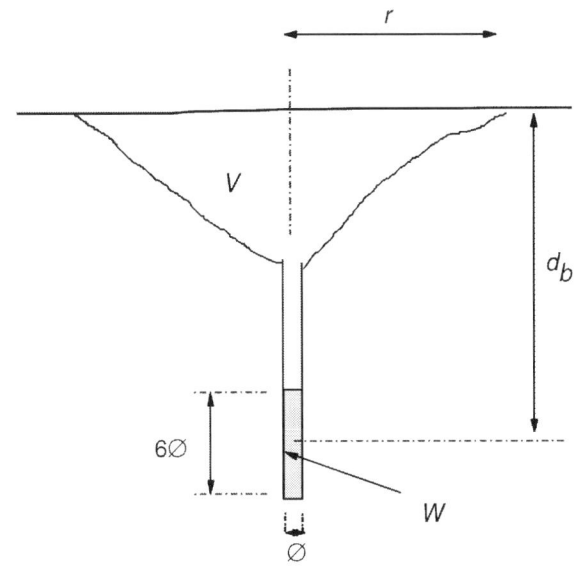

FIGURE 72.19   Cratering dimensions

considered so that good drilling patterns can be achieved. The hole diameters chosen both affect, and are affected by, the size of the ore block. The explosive selected depends on the hole diameter and, together with the diameter, determines the nominal drilling pattern. Ring-drilling patterns are best derived through the use of computer software together with engineering judgment. Explosive loading is necessarily high, and charging patterns require careful thought if overcharging is to be minimized and adequate energy is to be available at all parts of the ore block. Initiation systems are critical to the blasting results, but sub-optimal systems must be adopted if collar priming is practiced. In view of all these factors, idealized guidelines can be given for ring design, but it should be recognized that developments in technology might in time change these parameters.

A blasting development program that focuses on determining the real effectiveness and productivity of ongoing blasting operations is a worthy investment at any mine that uses ring blasting. The design process for ring blasting is exceptionally demanding, and good software is invaluable for exploring the various options in a timely fashion. Such software is particularly useful if it also provides drilling, charging, and initiation instructions and if it can show the efficiencies and costs for any particular layout.

72.5   CRATER-BLASTING THEORY AND APPLICATION

72.5.1   Introduction

In Canada, a new underground mining method, the Vertical Crater Retreat (VCR) or Vertical Retreat Mining (VRM) method (see Figure 72.17) was developed in 1975 for primary stoping, pillar recovery, and drop raising. This was made possible by the introduction of 165-mm-diameter holes to underground mining.

When vertical (or inclined) large-diameter holes are drilled on a designed pattern from a cut over a stope or pillar to bottom in the back of the undercut, and spherical charges of explosives are placed within these holes at a calculated optimum distance from the back of the undercut and detonated (see Figure 72.18), a vertical thickness of ore will be blasted downwards into the previously mined area. Repeating this loading and blasting procedure, mining of the stope or pillar retreats in the form of

horizontal slices in a vertical upwards direction until the top sill is blasted and the mining of a stope or pillar is completed.

The VCR method has been and currently is being practiced in various mines in Canada, the United States, Europe, Central America, and Australia.

In this section, the theory of the VCR method will be discussed and then applied to the Luossavaara mine in northern Sweden.

72.5.2   Cratering Theory

Introduction.   The concept of cratering and its development may be attributed to C.W. Livingston. It is a versatile tool for studying the blasting phenomenon, and its application has resulted in the development of a new underground mining method, the VCR method of primary stoping, pillar recovery, and drop raising. A crater-blast is a blast where a spherical or near spherical charge (1:6 diameter-to-length ratio) is detonated beneath a surface that extends laterally in all directions beyond the point where the surrounding material would be affected by the blast.

Figure 72.19 shows the nomenclature used in VCR, and they are described below.

$\varnothing$ = Hole diameter

$6\varnothing$ = Charge length

d_b = Depth of burial. Distance from surface to center of charge

d_o = optimum depth of burial. The depth of burial at which the greatest volume of rock is broken

N = Critical distance. The depth of burial at which the effects of a cratering charge are just noticeable on the surface

r = Radius of crater

r_o = Radius of crater formed at optimum depth of burial

V = Crater volume

W = Charge weight

There is a definite relationship between the energy of the explosive and the volume of the material that is affected by the blast. This relationship is significantly affected by the placement of the charge. Livingston determined that a strain-energy relationship exists, as expressed by an empirical equation:

$$N = EW^{1/3} \qquad \textbf{(72.19)}$$

where   N = the critical distance at which breakage of the surface above the spherical charge does not exceed a specified limit.

E = the strain energy factor, a constant for a given explosive-rock combination,

W = the weight of the explosive charge

The same equation may be written in the form of:

$$d_b = \Delta \, EW^{1/3} \qquad \textbf{(72.20)}$$

where d_b = the distance from the surface to the center of gravity of the charge, i.e. depth of burial

$\Delta = d_b/N$ which is a dimensionless number expressing the ratio of any depth of burial compared to the critical distance

When d_b is such that the maximum volume of rock is broken to an excellent fragment size, this burial is called the optimum distance: d_o. For further study of the cratering theory, see Lang (1983).

Choosing the Best Explosive for VCR Mining. When the material that is to be blasted remains constant, but several different explosives are considered, the cratering theory may be used to determine the most suitable explosive through the application of Livingston's Breakage Process Equation:

$$V = ABCWE^3 \qquad \textbf{(72.21)}$$

where: W = Charge weight

V = Crater volume

E = Strain energy factor

A = Energy utilization number

B = Material's behavior index

C = Stress distribution number

Inasmuch as V, W, and E can be measured with certainty, it remains for the observer to isolate the variables A, B, and C.

The energy utilization number, A, is the ratio of the volume of the crater within limits of complete rupture at any depth, to the volume at optimum depth, where the maximum proportion of the energy of the explosion is utilized in the failure process:

$$A = V/V_o \qquad \textbf{(72.22)}$$

The maximum value of A is equal to 1.0 at optimum depth (where fracturing reaches the most efficient development). Accordingly, the numerical values of A are less than 1.0 at other charge depths.

The material behavior index, B, is a constant for a given type of explosive and weight of charge in a given material. B is measured at optimum depth and:

$$B = V_o /N^3 \qquad \textbf{(72.23)}$$

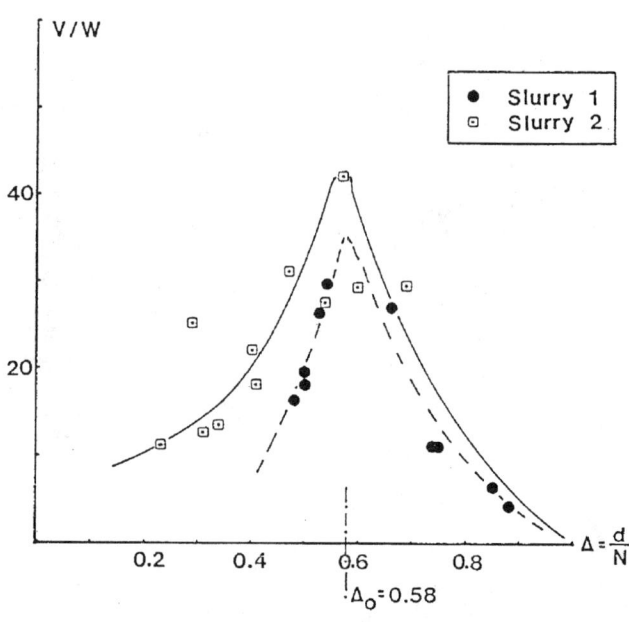

FIGURE 72.20   *V/W (crater volume/charge weight) versus* Δ *(depth of burial d/critical distance N)*

It has been derived from:

$$V_o = B(WE^3AC) \qquad \textbf{(72.24)}$$

where   A = 1 at optimum depth d_o

C = 1 if the charge is spherical

One can conclude that both A and B describe the effect of the explosive upon the failure process in blasting. The value 'A' best describes the effects of the variation in energy density with distance, and B best describes the effects of the variation in energy density accompanying changes in the stress-strain relations as measured at a given reference energy level. The following example will demonstrate the application of the breakage process equation for the comparison of the performance of the explosives in the same rock.

Basic cratering research was conducted in a hard, cherty magnetic-iron formation with two types of slurry: Slurry 1 (Selleck 1962) and Slurry 2 (Lang 1962). The curves of V/W versus Δ for the two experiments are plotted in Figure 72.20. The optimum depth ratio was found to be the same for both explosives: Δ_o = 0.58, but E and N were different.

The values of A were calculated for each crater, and the results were plotted against depth ratio (see Figure 72.21). The two curves are similar to those of V/W versus Δ. This diagram clearly indicates that in the case of Slurry 2, more energy is being utilized in the secondary fragmentation range and in the flyrock range than in the case of Slurry 1. This is responsible for better fragmentation and more gas energy. Production-scale blasts confirmed the results of these cratering experiments.

Material behavior index values for both explosives were calculated at optimum depth and found to be:

Slurry 1. B_o = 0.42

Slurry 2. B_o = 0.33

Higher values of B are characteristic of brittle-type failure. Experiments show that B decreases as the material becomes more plastic-acting, which was true in this experiment as well. Slurry 1 had a high detonation velocity; thus the material was acting in the brittle manner. Slurry 2, due to the 10% Al content, had a lower detonation velocity and the load was a slower and more

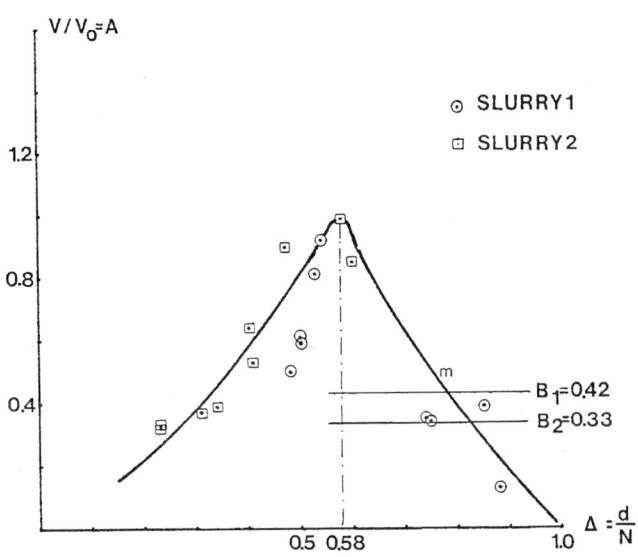

FIGURE 72.21 "A" plotted against depth ratio from Slurry 1 and Slurry 2

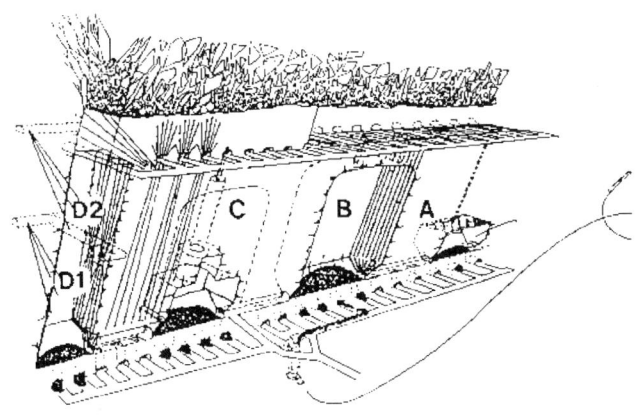

FIGURE 72.22 Schematic layout of planned mining activities in the Luossavaara Research Mine

sustained type. Hence, the same rock behaved in a rather plastic manner.

The stress distribution number, C, was 1 because both experiments employed spherical charges. One may conclude that when comparing different explosives in the same material, the comparison must be made keeping the geometry of blast constant; otherwise, the results will be misleading.

1. Separate cratering experiments should be conducted with the different explosives in the same material.

2. Determine N, Δ_0 for each experiment.

3. If this information is for designing VCR-type blasting, then d_0 and optimum spacing should be calculated for each explosive and ore combination. These criteria should be used in each respective stope.

Small-Scale Cratering Tests. The purpose of performing small-scale crater tests is to obtain the data required to make qualified predictions of the blasting results in the full production stope. It is necessary to conduct the crater tests as close as possible to the stope where the VCR method will be used. Different rock properties and structural geology may cause an over- or under-estimation of the depth of burial for the production blasts.

If the depth of burial is less than optimum, it will result in a satisfactory breakage, but the cost for drilling and explosives will be too high. If the depth of burial is larger than the optimum one, bells or unsatisfactory fragmentation may occur.

Due to development work in the stope, it is sometimes possible to carry out the test in the stope (in the undercut).

72.5.3 Application of Crater Blasting to the Luossavaara Mine

Introduction. When the activities in the Luossavaara Research Mine were planned and outlined, it was decided that new mining methods with large-diameter holes should be tested. Discussions during the planning of the Luossavaara Mine resulted in a small test stope D1 (see Figure 72.22) where the VCR method could be evaluated under Swedish conditions. Mr. Leslie Lang, of L.C. Lang & Associates, Inc., in Canada, was engaged as a consultant to the Research Mine and SveDeFo when the test shots for the future design were planned. This section describes

the results of the test shots and the proposal that was made for the full-scale production of stope D1.

Small-Scale Crater Tests. In Luossavaara, a field mapping procedure was carried out to investigate whether the abandoned mine area above production stope D would be suitable for crater tests (Mäki 1982). Through comparison to available data, it was found (Mäki 1982 and Röshoff 1981) that "On the basis of the analysis of structure densities, structure lengths and RQD values it may be concluded that no major variation of rock properties exists within the orebody." The data from the test level at 250 m was compared with data from an access drift between the 270- and the 290-m level. However, as the major part of this drift was located in the footwall, it was possible that the conclusion could change somewhat when data concerning the stope itself became available. Cores from profiles in the test area indicated good-to-excellent rock. The RQD values were in the range of 80%–100%. The primary rock types of the side walls were breccia, quartz porphyry, and syenite porphyry. A total of 23 holes were drilled with hole diameters of 38 and 102 mm.

The tests were located at the 250-m level above stope D where the rock properties and geological structure were similar to what was expected in the production stope. Eleven 102-mm-diameter holes were drilled and blasted with a non-cap-sensitive TNT-slurry, Reolit. Two 102-mm-diameter holes were blasted with ANFO, and six holes with a diameter of 38 mm were drilled and blasted with ANFO. The reason for blasting the 38-mm-diameter holes was to investigate the scale effect. The test holes were horizontally drilled in the rib, perpendicular to the drift. The face was relatively smooth without large hills or valleys. The collars of the 102-mm-diameter holes were drilled 1.6–2.0 m above the floor, and the distances between the holes were not less than 4.5 m. The 38-mm-diameter holes had a spacing of 2.5 m. The length of the holes used for the 102-m-diameter TNT-slurry tests were 3.0, 2.8, 2.6, 2.0, 1.8, 1.6, 1.6, 1.2, 1.1, 1.0, and 0.85 meters. Only two 102-mm-diameter ANFO test holes were drilled and blasted. One hole was 1.1 and the other hole was 0.9 m deep. In the ANFO test, holes 38 mm in diameter and 1.3, 1.0, 0.8, 0.7, 0.6, and 0. 5 meters long were used. All holes were drilled in the footwall. The holes were flushed clear and measured after drilling.

The explosives used in the small crater tests were Reolit, manufactured by Nitro Nobel AB, and ANFO K2Z, manufactured by Kimit AB. To initiate the explosives, a plastic PETN explosive NSP-71, developed by Bofors, was used. Reolit is a non-cap-sensitive TNT-slurry explosive containing 22 % TNT and 3% Al. The density is 1,450 kg/m³. The weight strength relative to ANFO is 1.20. ANFO K2Z consists of 47.3 % prill, 47.3 % crystalline ammonium nitrate,

TABLE 72.5   Data for crater tests with 102-mm hole diameter and TNT-slurry

Hole #	Hole depth m	Stemming m	Charge length m	Charge weight kg	Depth of burial m	Crater depth m	Crater radius m	Crater volume m³	VOD km/s	d_b/N	V/w	Comment
1	3.0	2.4	0.6	6.8	2.7	0	0	0	5.6/4.0	1.08	—	no crater
2	1.6	1.0	0.6	6.8	1.3	0.8	0.5	0.5	1.8/1.0	0.52	0.07	low VOD
3	2.8	2.2	0.6	6.8	2.5	0.75	1.7	1.7	4.5/4.5	1.00	0.25	N = 25
4	1.8	1.2	0.6	6.8	1.5	1.1	2.25	2.25	5.1/4.8	0.60	0.33	
5	2.6	2.0	0.6	6.7	2.3	0	0	0	4.5/5.0	1.00	—	no crater
6	2.0	1.4	0.6	6.8	1.7	0.6	0.65	0.65	4.7/4.7	0.68	0.10	
7	1.6	1.0	0.6	6.7	1.3	0.8	1.8	1.8	3.7/3.6	0.52	0.27	
8	1.2	0.6	0.6	6.7	0.9	1.0	1.0	1.0	5.0/5.2	0.36	0.15	
9	1.1	0.5	0.6	6.75	0.8	1.0	1.1	1.1	4.5/5.8	0.32	0.16	full crater
10	1.0	0.4	0.6	6.8	0.7	1.1	2.65	2.65	4.5/5.8	0.28	0.37	full crater
13	0.85	0.25	0.6	6.8	0.55	0.6	0.68	0.68	5.2/5.0	0.22	0.10	full crater

and 5.4 % fuel oil. The density is 1,000 kg/m³. The charge weight in the 102-mm-diameter holes was 6.7 kg for the TNT slurry and 4.75 kg for ANFO. Using 38-mm-diameter holes, the charge weight was 0.25 kg/hole. The 102-mm-diameter charges were primed with 250 g of NSP 71 Bofors and the 38-mm-diameter holes with 25 g of the same explosive. To be able to load the horizontal holes the explosives were packed in 100-mm-diameter plastic bags.

To check the performance of the explosive, the velocity of detonation was measured in the 102-mm-diameter holes. The VOD measurement setup consists of probes, a pulse former unit, and a transient recorder, a Nicolet Explorer 1090 A. When the detonation front reaches a probe, the circuit is shorted, and a pulse in the pulse former unit is generated. This pulse has a defined RC constant. A transient recorder registers the positive or negative pulse. By positioning a number of probes along the travel path of the detonation front, a number of pulses, each with its own RC characteristic, can be registered by the transient recorder. By measuring the time lapse between the pulses and by knowing the distances between the probes, the velocity of the detonation wave can be calculated. Three probes for VOD measurements were taped on a PVC rod with a diameter of 5 mm. By pushing the PVC rod into the very bottom of the hole, the distance between the hole bottom and the first probe was fixed. Half the charge was then placed and tamped at the bottom of the hole. The primer was pushed in with a loading stick and with great care tamped close to the inner part of the charge. The remaining part of the charge was finally placed and tamped. The three probes for VOD measurement were now placed in the outer half of the charge. The length from the collar to the charge was measured and a three-part wooden plug kept the charge in position. Stemming consisted of a plug of bentonite and gravel (0–30 mm). A Nonel-cap initiated the primer.

After firing the shot, scaling of the crater walls was kept to a minimum. All structural weakness planes, which may have influenced the size or shape of the crater, were noted. Photographs of the craters were taken after each shot. Crater depth as a function of position was determined using a sliding ruler attached perpendicular to a 2.5 × 2.5-m vertically mounted aluminum frame. From measurements made on a 25-cm grid, the maximum depth and radii could be determined and the volume calculated. A total of 11 craters were formed at different depths of burial while keeping the charge weight of Reolit slurry constant: W = 6.8 kg. The results are given in Table 72.5.

The critical distance (N) was determined to be N = 2.5 m, and the calculated strain energy factor (E) was:

$$E = N/W^{1/3} \approx 1.32$$

The plot in Figure 72.23 indicates some scatter. This is due to minor geological discontinuities present in the rock mass. The structural geology will generally have a more overwhelming

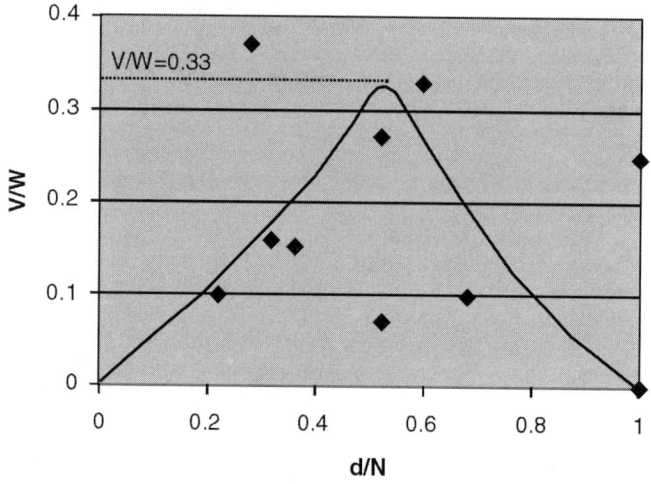

FIGURE 72.23   Crater curve of V/W versus Δ = d/N for the TNT-slurry

influence on the cratering results when using relatively small charges than it will with larger charges. The optimum depth ratio (Δ_o) appears to be in the range 0.52–0.6. Additional tests would be needed to reduce the interval. To be on the safe side Δ_o is estimated to be Δ_o = 0.52 in this test. This suggests a predominantly shock type failure of the ore when using Reolit. For Δ_o = 0.52 and N = 2.5 m, the calculated depth of burial for a 6.8 kg charge is:

$$d_o = \Delta_o N = 0.52 \times 2.5 = 1.3 \text{ m}$$

From Figure 72.23, the value of V/W corresponding to the optimum depth ratio is V/W = 0.33 and hence V = 0.33 × 6.8 = 2.24 m³. Representing the crater by a cone with apex at the bottom of the charge, one may calculate the radius r_o.

$$r_o \times \pi(d_o + 6\pi/2)/3 = 2.24$$

or

$$r_o = 1.2 \text{ m}$$

The following basic data have been obtained:

W = 6.8 kg
N = 2.5 m
E = 1.32
Δ_o = 0.52
d_o = 1.3 m
r_o = 1.2 m

TABLE 72.6 Data for crater tests with 38- and 102-mm hole diameters and ANFO K2Z

Hole #	Diameter mm	Hole depth m	Stemming m	Charge length m	Charge weight kg	Depth of burial m	Crater depth m	Crater radius m	Crater volume m³	DOC km/s	Comment
14	38	1.3	0.5	0.24	0.25	1.18	0.6	0.85	0.45		fractured rock
18	38	1.0	0.5	0.24	0.25	0.88	0.35	0.85	0.26		
20	38	0.8	0.5	0.24	0.25	0.68	0.5	0.75	0.29		two boulders
21	38	0.7	0.45	0.24	0.25	0.58	0.2	0.6	0.08		
22	38	0.6	0.35	0.24	0.25	0.48	0.1				
23	38	0.5	0.25	0.24	0.25	0.38	0.5	0.7	0.26		full crater
12	102	0.9	0.3	0.6	4.75	0.6	0.55	1.2	0.67	2.7/3.0	almost full
11	102	1.1	0.5	0.6	4.75	0.8					major joints

The Production Scale Design. The next step is to scale up these cratering results for a production-scale blast in VCR stopes when the charge weight of the same explosive is increased to W_2 = 31 kg and the hole diameter is 6.5 in. Following the Livingston theory, it is assumed that E = 1.32 remains constant. The critical distance for the 31-kg charge weight is:

$$N = EW_2^{1/3} \approx 4.15$$

The center of this charge should be at the optimum distance D_0 from the back of the stope:

$$D_0 = \Delta_0 N = 0.52 \times 4.15 \approx 2.2 \text{ m}$$

The U.S. Army Corps of Engineers, Nuclear Cratering Group, uses another form of the calculation. They calculate a scaling factor (F) of:

$$F = (W_2/W)^{1/3} = (31/6.8)^{1/3} \approx 1.66$$

The optimum distance for this larger charge is:

$$D_0 = d_0 F = 1.3 \times 1.66 \approx 2.2 \text{ m}$$

The crater radius becomes:

$$R_0 = r_0 F = 1.3 \times 1.66 \approx 2.0 \text{ m}$$

It is important to ensure complete breakage of the rock between two adjacent holes in the stope by designing an optimum spacing between holes. The recommended hole spacing (S_0) should be in the range of 1.2 R_0 to 1.6 R_0. In the case of Reolit it is:

$$S_{min} = 1.2 \times 2.0 \approx 2.4 \text{ m and}$$

$$S_{max} = 1.6 \times 2.0 \approx 3.2 \text{ m}$$

It is more prudent to design the first stope using S_{min} and then increase it gradually to S_{max}. In further stopes, if the results are satisfactory, the pattern can be expanded. However S = 3 m will probably not cause any problem in the Luossavaara type of ore.

The advance A will be:

$$A = D_0 + 6 \oslash/2 = 2.2 + 0.5 = 2.7 \text{ m}$$

The specific charge (q) for S_0 = 3 m is:

$$q = 31/(A \times 3^2) = 1.3 \text{ kg/m}^3$$

Comments on the Reolit Tests. As can be seen from Figure 72.23 where the smooth curve has been fitted by eye to the experimental points, it is obvious that the curve could have been drawn in various ways. For example, if the curve had been fitted using the least-square method, it would definitely not have looked the same. The interpretation of the results appears to require a considerable amount of subjective assessment on the part of the person conducting the small-scale tests. This means that cratering should preferably be carried out only by persons with previous experience in crater testing and, if possible, production blasting. For instance,

the values for shot No.10 were omitted when plotting the curve because the geological mapping indicated that the rock was fractured to a greater extent around this hole than around the rest of the holes. Shot No. 2 had a low VOD, which might explain its small crater volume. If the value for shot No. 10 had been taken as the peak of the curve (yielding the optimum ratio A_0 = 0.28), a strong shock-type failure would have been indicated with all its consequences (d_0, V_0). This would not be consistent with the majority of the information.

The ANFO Cratering Results. Tests were carried out with ANFO charges to estimate the breaking capacity with an explosive having a smaller energy density than a TNT-slurry. The plan was to carry out experiments with both 102-mm and 38-mm hole diameters. The 38-mm hole diameter was to be used to estimate the effect of the rock structure on the breakage. Unfortunately, all of the $\oslash$102-mm holes that had been drilled could not be used because of potential interference with the production demands in the mine. Therefore, only two shots could be made in 102-mm holes.

Of the 38-mm diameter shots, six holes yielded questionable results. The results reveal that the rock structure plays a major role in the breakage process. Because of the small number of test shots carried out, no evaluation of the proper scaling for production blasts could be made. The tests, however, did show that the hole diameter used for the tests should be chosen as close as possible to the hole diameter used for production blasting. The rock structure obviously influences the breakage process in a dominant way. However, for guidance, the test shots are reported in Table 72.6.

72.5.4 The Production Blasts

The Proposal. This part outlines a proposal based on the performed crater tests. As the reader may have noticed, the experiments with ANFO charges were not carried out to such extent that proper scaling could be done with this type of explosive. However, because of its low price, ANFO is and will be a very competitive explosive whenever dry conditions can be achieved in blasting. Therefore, ANFO should be tested in the Luossavaara mine to make a proper evaluation of the blasting performance.

Although there were not enough ANFO tests for proper scaling, the results achieved, combined with the experience of Leslie Lang, allowed an initial alternative hole pattern relative to the TNT-slurry to be considered. ANFO has lower energy density than a TNT-slurry but produces a larger gas volume per weight of explosive. Because the charge weight only will be around 20 kg, both the energy content and the gas volume release will be much higher with the TNT-slurry. This indicates that the specific drilling will be much higher with the ANFO explosive. Experience from earlier experiments carried out by Lang indicates that the ANFO explosive will need a spacing of 2 m with a loading depth of 1 m. The crater curve for this type of explosive also indicates a more plastic type of breakage behavior (Figure 72.24).

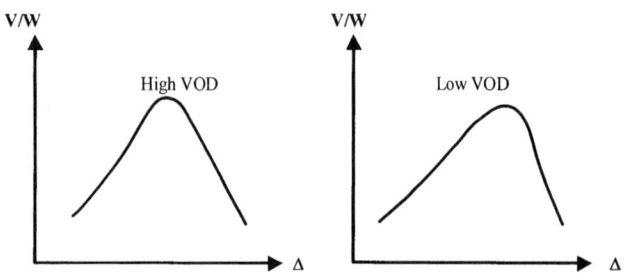

FIGURE 72.24   Schematic sketch of breakage performance for a high velocity/high energy explosive and a low energy explosive

Today, in addition to the high energy TNT-slurries and the ANFO type of explosives, a new generation of explosives exists—the emulsion slurries. Emulsion slurries have been introduced throughout the world and are priced somewhere between the TNT-slurry and the ANFO explosives. The energy density and the gas release energy for this type of explosive are also somewhere between the explosives mentioned. They can be easily loaded up-hole, and they are water-resistant. Today there are three types of explosives, each with a different breaking performance, to study for production blasting—the TNT-slurry, the emulsion slurry and ANFO.

If the optimum explosive-geometry-ore combination is to be found, then the following procedure should be followed:

1. Explosive Type 1 (TNT-slurry), Type 2 (Emulsion), and Type 3 (ANFO) are considered for the VCR stope. Explosive Type 1 has a higher density and energy ratio than explosive Type 2, which has higher ratings than explosive Type 3. Consequently d_0 and spacing will be greater for Type 1 than for Type 2 and Type 3.

2. Three different hole patterns are drilled in the stope Dl. Type 1 should be used for the test stope with a hole pattern equal to 3×3 m. Type 2 should be used for a pattern 2.5×2.5 m, and Type 3 should be used for a pattern 2×2 m.

3. The Type 1 explosive should also be tried in the stopes drilled for Type 2 and Type 3 to examine whether the higher specific charge will result in such a good fragmentation that the lower cost for loading, hauling, and crushing will pay for the extra costs caused by the higher specific drilling and specific charge resulting in a lower total mining cost.

4. If the Type 1 explosive pulls satisfactorily in the lower part of the stope with hole pattern 3×3 m, it would be wise to see whether a larger d_0 can be used (a smaller specific charge). The breakage process must be somewhat different when several charges are detonated at the same level and at the same time compared to the crater experiments where just a single shot is made.

5. To evaluate the results, it is important to follow up the fragmentation distribution achieved and the costs for drilling, blasting, loading, hauling, and crushing.

Table 72.7 and Figure 72.25 give the details of the proposal. The suggested values for the Type 2 explosive (emulsion) are uncertain and should probably be checked by a test in 4-in hole diameter with loading depth of 0.6, 0.7, 0.8, and 0.9 m.

Results from the VCR Production Blasts. The main objective of this section was to show a way to measure, perform, and evaluate crater tests. The results from the production stope are summarized in Tables 72.8 and 72.9 to give the reader a feeling of what the outcome became when the test data were applied as design parameters for scaled-up blast.

TABLE 72.7   Proposed tests for Type 1, Type 2 and Type 3 explosives in 165-mm hole diameter

	Type 1 TNT-slurry	Type 2 Emulsion	Type 3 ANFO
Test stope (see Figure 72.22)	A,B,C	B	C
Density, kg/m³	1450	1250	900
Charge weight, kg	31	28	20
Weight strength (rel. ANFO)	1.06	0.86	1.0
d_0	2.2, 2.5	1.8	1.0
Spacing	3.0×3.0 2.5×2.5 2.0×2.0	2.5×2.5	2.0×2.0
Advance, m	2.7, 3.0	2.3	1.5
Specific charge, kg/m³	1.1, 1.3, 1.8, 2.9	1.9	3.3

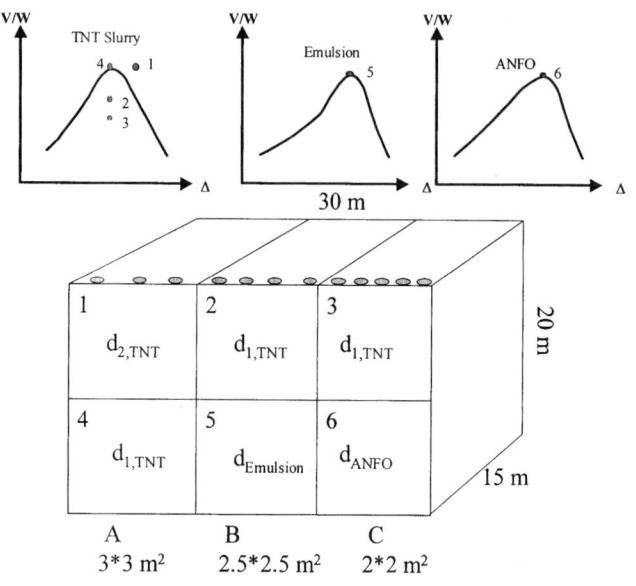

FIGURE 72.25   Final proposal for stope

TABLE 72.8   Explosive data

Explosive	Density (kg/m³)	Strength per vol. (rel to ANFO)	Energy content (MJ/kg)	Gas volume (STP) (m³/kg)	Estimated detonation pressure (Gpa)
ANFO	825	100%	3.9	0.97	3
Emulsion	1280	157%	4.2	0.83	11
Reolit	1450	161%	3.7	0.76	9

The explosive used in the production stope had the data found in Table 72.8.

72.6   BLAST DESIGN FOR DRIFTING

72.6.1   Introduction

Fully automated drifting is not yet a reality, and it will take some time before it is developed to its full potential. Manual work is still needed to charge the rounds. In drilling, scaling, rock support, and mucking, mechanization has improved. In modern mines, hydraulic drilling equipment has taken over after the

TABLE 72.9   Summarized results from the production stope

	ANFO	Reolit	Emulsion	Emulsion
Geometry	2 × 2 m	3 × 3 m	2.5 × 2.5 m	3 × 3 m
Tons of ore blasted	4090	5302	7441	3950
Boulder %	5.6	5.0	2.2	0.5
No. of hang-ups	0	5	0	0
Specific charge, kg/ton	0.58	0.36	0.30	0.43
Specific drilling, m/ton	0.065	0.030	0.035	0.040
Average advance, m	2.4	3.1	3.0	2.6
Average width of stope, m	12	11	12	8
K_{50}, m	0.18	0.15	0.10	0.15
Relative cost per ton	1.60	1.07	1.00	1.30
Overbreak, % of theoretical advance	32	25	21	4

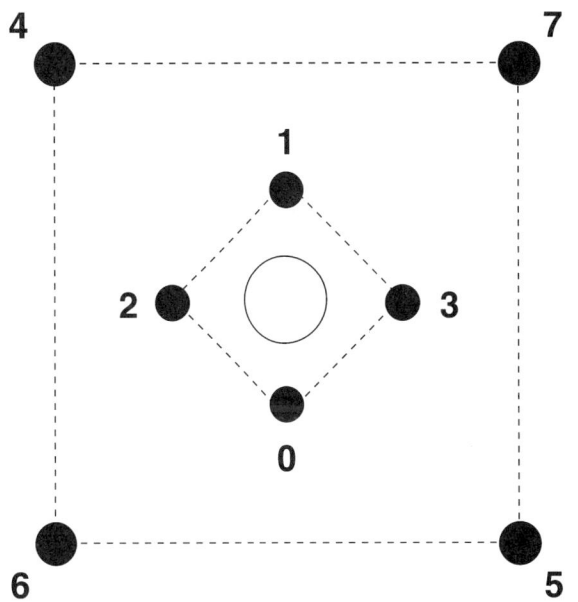

FIGURE 72.26   Parallel hole cut with two quadrangles

pneumatic. This has meant enormous capacity increases as the penetration rates have gone up by a factor of 3–4.

Most new drill rigs are equipped with air-conditioned and noise-insulated cabs where the operator may comfortably sit and listen to his choice of music, if so inclined. Computers improve the accuracy and help drill the holes in the face right at the spot where they should be. The earlier, time-consuming survey work and mark-up of the holes to be drilled is, or will soon be, gone.

Today, the charging work is not as highly mechanized as the drilling, but there has been considerable development work to increase safety and provide a better work environment. Many mines still use fuse-and-cap or electric detonators, but in modern mines shock-tube detonator systems like Nonel™ are a must due to safety aspects. With electric detonator systems, there are, unfortunately, still too many accidents due to electrical interference. Many mines perform the drift-charging with scissor-lifts placed on a truck. The charging is done with sticks of NG explosives such as dynamites and cartridges of watergel or emulsion. In dry conditions ANFO is usually charged by means of pneumatic equipment. Formerly, at larger tunnel excavations, one could see drill rigs with one or two booms plus a telescopic working basket that made it possible to charge the holes at the same time as the drilling. In many countries this is considered unsafe and has been abandoned. The risk of falling rock was large and mistakes drilling into a charged hole caused unintentional detonations.

Modern mines have invested in dedicated charging equipment, which can visit the face after the drilling is performed, and from where the crew can safely charge the round. The trend in recent years has been away from NG-explosives and toward emulsion explosives. ANFO, however, is still overwhelmingly used. Many emulsion-charging trucks use a repumpable, ready-made emulsion that can be pumped into the boreholes. The most modern of these can bring with them an unsensitized emulsion matrix that will not be sensitized until the emulsion leaves the nozzle of the charging hose. Through chemical gassing, small gas bubbles (hot spots) are introduced that transform the unsensitized matrix to an explosive with excellent characteristics. The advantage of the emulsion is that it is a very good water-resistant explosive. When the detonation is ideal, it produces far less toxic fumes than other explosives, the breakage performance is excellent, and, through smart equipment, the linear charge strength can be made flexible for performing cautious blasting of the contours.

72.6.2   The Design Process

Dividing the Tunnel Face Area in Design Sections. The basic principles for charge calculations are still based upon the work by Langefors and Kihlström (1963). When the charge calculation of the drill and blast pattern is performed, it is normal to divide the face into five separate sections:

- Cut section
- Stoping holes breaking horizontally and upwards
- Stoping holes breaking downwards
- Contour holes
- Lifter holes

The most important operation in the blasting procedure is to create an opening in the face to develop another free surface in the rock. This is the function of the cut holes. If this stage fails, the round can definitely not be considered a success. In the worst case, the rest of the round freezes and cannot be mucked or reshot safely.

Advance. Figure 72.26 shows a cut design in which one large-diameter empty hole has been used. The advance is restricted by the diameter of the empty hole and by the hole deviations for the smaller diameter holes. For economy, the entire hole depth must be utilized. Drifting becomes very expensive if the advance (I) is much less than 95% of the drilled hole depth (H).

$$H = 0.15 + 34.1\varnothing - 39.4\varnothing^2 \qquad \textbf{(72.25)}$$

where $\varnothing$ is the large hole diameter expressed in meters ($0.05 < \varnothing < 0.25$ m) and H is the drilled depth (m).

The advance I for 95% advance then is

$$I = 0.95H \qquad \textbf{(72.26)}$$

Equations 72.25 and 72.26 are valid for a drilling deviation not exceeding 2%.

Hole depths needed for a $\varnothing$102-mm-large hole in a parallel cut would be about 3.2 m. A $\varnothing$120mm hole would need a hole length of 3.7 m and a $\varnothing$150 mm will need 4.4 m. For the large hole diameters of $\varnothing$250 and $\varnothing$300 mm used in the LKAB tests presented in sections 6.3 and 6.4, the advances were 6.1–6.9 m before scaling for the $\varnothing$250 mm hole and 7.1–7.5 m for the 300 mm.

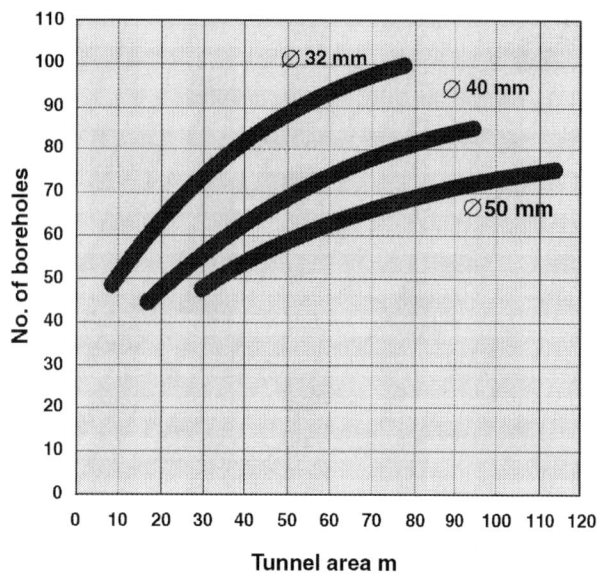

FIGURE 72.27   Number of boreholes as a function of the tunnel area for hole diameters ⌀32 to ⌀50 mm

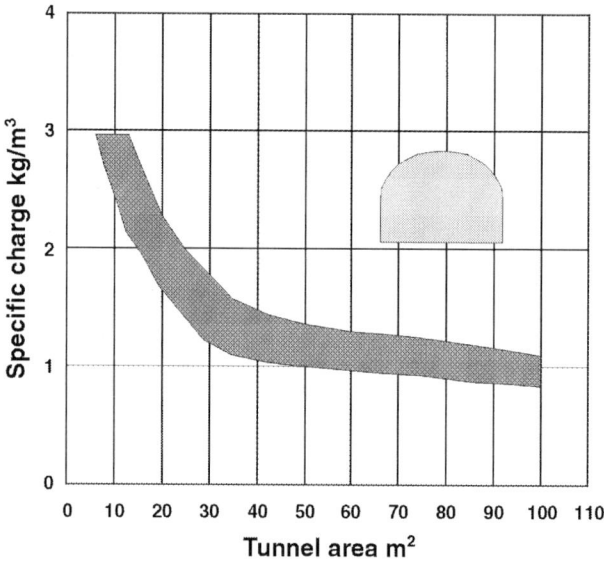

FIGURE 72.28   Specific charge as a function of the tunnel area for the hole diameter of 45 mm

Equations 72.25 and 72.26 would predict 5.9 m for the ⌀250mm and 6.4m for the ⌀300 mm hole, indicating that the equations underestimate the advances when very large hole diameters are used. The calculations of each individual burden in the quadrangles, the stoping holes, the lifters and the contour holes are described in detail by Persson, Holmberg, and Lee (1994).

Number of Blastholes.   The number of blastholes necessary to provide a balanced distribution of explosive energy is dependent upon the rock type, the geology, the stress field, the explosive characteristics, blasthole diameter, and the contour blasting requirements.

In tunneling and drifting, the number of blastholes and the specific charge used is a function of the drift area. Figure 72.27 gives an indication of the number of boreholes needed for various tunneling areas. Figure 72.28 indicates the required specific energy.

Cut.   In the cut, the holes are arranged geometrically in such a way that firing the charges in sequence creates an opening, which becomes wider and wider until the stoping holes can take over. The cut holes can be drilled to form a series of wedges (V-cut), to form a fan (fan cut), or in a parallel-hole geometry they may be drilled in a pattern close to and parallel with an empty, large central hole (parallel-hole-cut or parallel cut).

The choice of cut must be made with an eye toward what drilling equipment is available, how narrow the tunnel is, and the desired advance. With V-cuts and fan cuts (where angled holes are drilled), the advance is strictly dependent upon the width of the tunnel. The parallel hole cut with one or two centered large-diameter empty holes is being used extensively with large, mechanized drilling rigs.

The advantages are obvious: in narrow tunnels, the large booms cannot be angled sufficiently to create the necessary V-cut angles; it is easier to maintain good directional accuracy in the drilling when all holes are parallel so there is no need to change the angle of the booms.

In the parallel cut, standard-diameter holes are drilled with high precision around a larger hole usually with diameters of 65 to 175 mm. The large, empty hole provides a free surface for the smaller holes to work toward, and the opening is enlarged gradually until the stoping holes can take over the breakage.

Stoping Holes.   After the cut has been shot, the stoping holes will successively enlarge the excavation opening,. The stoping holes have a much easier job to do than the cut holes and the burdens can be increased considerably as the free face to shoot towards is wider.

Lifters.   When the lifters, the wall holes, and the back holes are drilled, the lookout angle should be considered. For an advance of 4 m, a lookout angle of 3° should be enough for providing space to drill the next round.

The floor is seldom dry and as the holes are angled downwards, they are often filled with water. Therefore, a water resistant explosive should be used. It is important to achieve good heave and fragmentation in this part of the round to provide for an acceptable mucking operation.

Methods for Minimizing the Damage to the Walls. Optimum results (with respect to cautious blasting through which unwanted damage and smooth perimeter walls are produced) will be achieved when drilling holes are placed at the intended place and when the perimeter holes are shot simultaneously. As indicated, experiments have shown that if adjacent holes are separated in time more than 1 ms, the result deteriorates. Such precise timing will require the use of electronic detonators.

Many techniques are used to reduce the linear charge concentration in the contour row and in the buffer row. For example:

- Decoupled plastic pipe charges
- Detonating cord
- String-loaded bulk emulsion
- Low density/strength bulk explosives (e.g., ANFO or emulsion with Polystyrene)
- Notched holes together with a very light charge

Successful smooth-blasting requires extremely good precision drilling and a fairly good rock quality. However, it is worth noting that even if the rock mass contains such structural features as bedding planes, joints, and fractures, or if the rock mass contains some poorly consolidated material, a cautious blasting

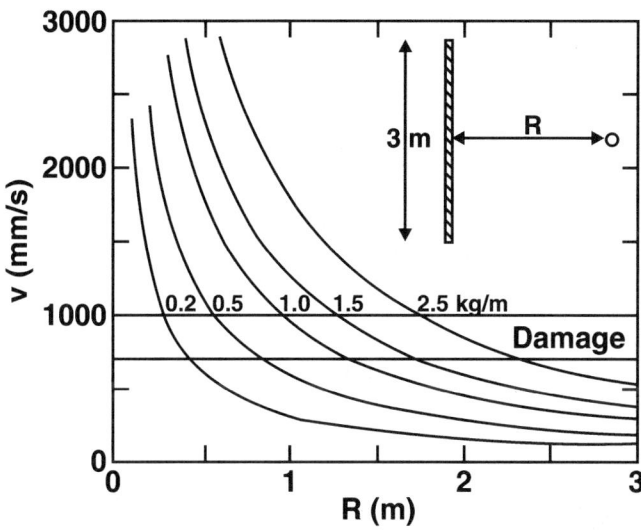

FIGURE 72.29 Estimated peak particle velocities as a function of distance for various linear charge concentrations

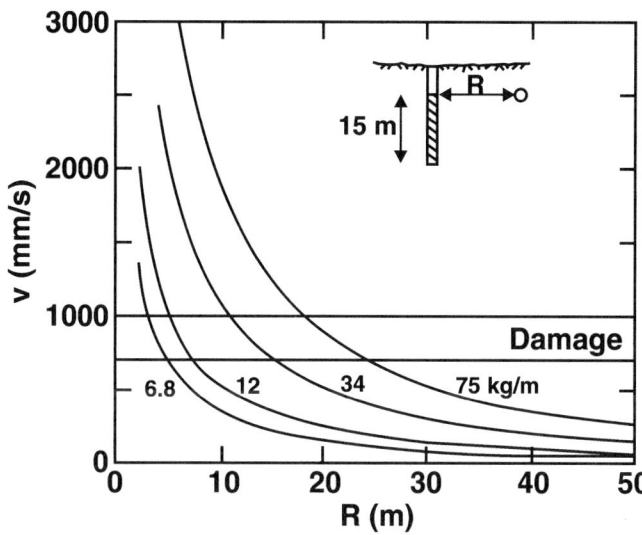

FIGURE 72.30 Estimated peak particle velocities as a function of distance for various hinear charge concentrations in stoping

method will always result in less overbreak and less disturbance of the rock mass. Whether the damage affects the stand-up time of the rock contour or not depends upon the character of the damage, the rock structure. the groundwater flow and, last but not least, the orientation of the damaged weakness planes in relation to the contour and the gravity load.

The rock damage can be described by the induced peak particle velocity. This is proportional to the induced rock strain, and it becomes a measure of the damage potential of the wave motion. The surrounding rock mass, of course, contains a number of potential weak planes, each of which is able to withstand a different level of peak particle velocity.

It is not unusual for blasters to fail to consider the effects of the charges in the rows adjacent to the often well-planned smooth-blasted contour row. Charging the adjacent rows with a heavy charge results in cracks spreading further into the remaining rock than from the smooth-blasted row. It is better to optimize the charge calculations such that the damage zone from the contour holes is limited. This can easily be done by use of Figure 72.29, where the damage zone is given for different linear charge concentrations. Persson, Holmberg, and Lee (1994) provide the equations for calculation of the damage curves.

A burden of 0.8 m is normal for a hole diameter of 48 mm with ∅17 mm Gurit pipe charges. From Figure 72.29 it can be seen that this charge results in a damage zone of about 0.3 m. Choosing a fully charged hole of ANFO (charge concentration of l = 1.6 kg/m) in the next row with a damage zone of 1.5 m is of no value because this results in a damage zone that extends 0.4 m further into the rock (1.5 – 0.8 – 0.3 = 0.4 m) than to use a charge concentration that results in a damage zone equal to that caused by the Gurit plus the burden, i.e., 1.2 m.

It is apparent from this example that a reduction of the damage zone can be obtained by reducing the charge concentration per meter of drill hole (such a charge should have l = 0.8 – 1.2 kg/m). This obviously results in increased costs for drill and blast operations, but these are balanced, for example, in tunneling by the advantage of a safer roof and decreased costs for grouting and maintenance.

The same exact approach can be used for designing the blasts near stope perimeters. In this case, the curves shown in Figure 72.30 can be used.

SveBeFo has performed extensive tests (Olsson and Bergqvist 1993) where they have directly studied the crack

lengths from blastholes. Today, several hundreds of boreholes have been blasted in the Vånga granite dimensional stone quarry in southern Sweden. In a 5-m-high bench, three or four identically charged and simultaneously initiated holes were shot. Each hole was primed, the charge length was 4.5 m and the top was unstemmed. Electronic delay detonators (EDS) from Dyno Nobel were used throughout.

After the blast, large blocks were carefully removed and cut horizontally using a large diamond circular cutoff saw. The crack pattern was highlighted using a conventional dye penetrant at one or several places along the hole axis.

By this direct method SveBeFo has tested the influence of the burden and the spacing, the hole size, the charge concentration, the decoupling ratio, the VOD, and the initiation delay time between holes. The hole diameters used have been mainly ∅38, ∅51 and ∅64 mm.

The explosives used were primarily the Swedish contour blasting explosives like Gurit, Kimulux 42, Detonex 80 (80 g/m PETN cord) and Emulet 20 (an emulsion styropore mix with 20% of the volume strength relative to ANFO). The explosives tested have large differences in their VOD values, ranging from about 2,000 m/s for Gurit, to about 4,800 m/s for Kimulux 42, to about 6,500 m/s for Detonex 80.

Figure 72.31 gives a comparison of single hole blasting and multi-hole blasting with simultaneous initiation of the holes. Here the basic blasting pattern with a burden (B) and spacing (S) of B × S = 0.5 × 0.5 m was used.

Some of the observed results were:

- Simultaneous initiation with electronic detonators gives much shorter cracks. Thus the cooperation between the different charges has a positive effect on the crack lengths. This is in contrast to the far field vibrations.

- The crack length decreases with decreasing coupling ratio. A bulk explosive that completely fills the hole gives the longest cracks.

- The crack lengths increase with increasing charge concentration.

- When delay times used were as low as 1 ms, the crack lengths still looked more like the cracks from single hole blasts.

- Traditional smooth-blasting procedures with conventional LPs give unnecessarily long cracks.

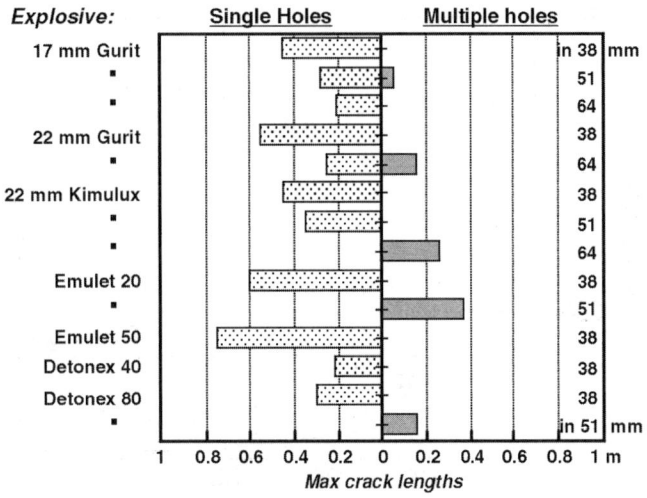

FIGURE 72.31   Results from the Vånga tests

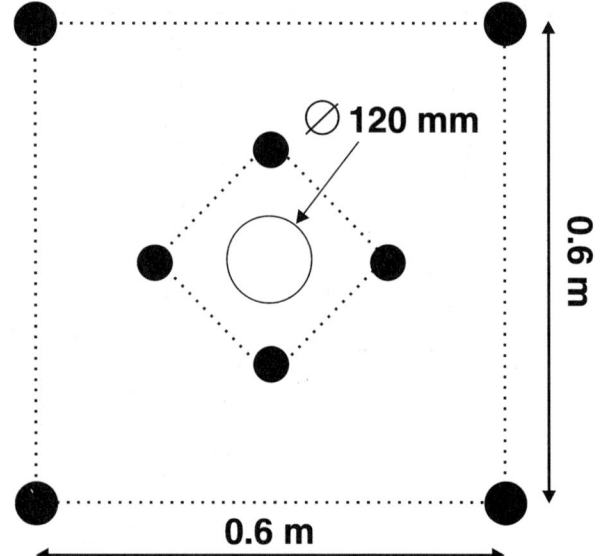

FIGURE 72.32   The standard Kiruna parallel hole cut

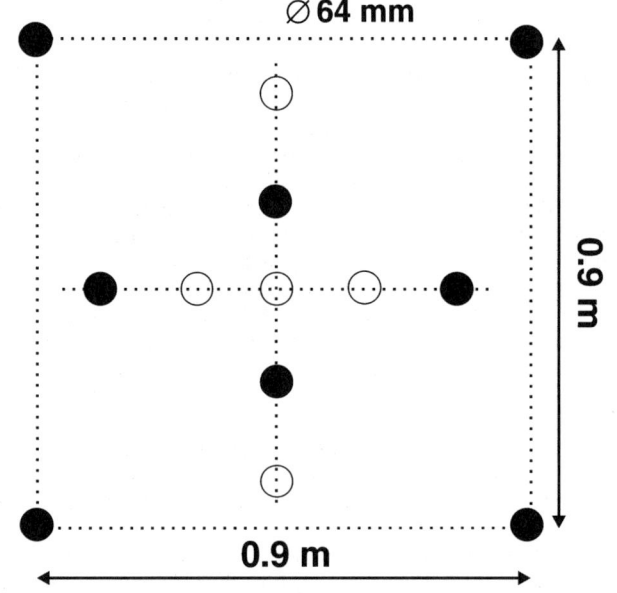

FIGURE 72.33   Cut with 64 mm hole diameters

Blasthole Sequencing. The charges in a tunnel blast must be initiated in such a sequence that the opening produced by a previous hole can be utilized by the following holes. The initiation sequence is normally as follows:

1. Cut (in the following order: first quadrangle, second quadrangle, third quadrangle, fourth quadrangle).
2. Stoping (stoping towards the cut and stoping downwards).
3. Contour holes, wall. These holes are shot with the same interval number.
4. Contour holes, roof. These holes are shot with the same interval number.
5. Lifters except corner holes. These holes are shot with the same interval number.
6. Lifter corner holes.

Because the rock removed between each hole in the cut and the central empty hole must be blown out to provide expansion room for the rock removed by the next charge, a long enough time interval between these holes is needed for this ejection to occur. The delay times in the cut are usually 75 to 100 ms. For the rest of the round, where the holes have a larger burden, the delay intervals are of the order of 500 ms.

72.6.3 Review of Long Rounds Drifting at the LKAB Malmberget Mine

Background. Tests with long rounds were conducted in the late 1980s at LKAB in Malmberget (Niklasson, et al. 1988). These tests indicated that it was possible to drill and blast normal drift rounds with a length up to 7.4 m and, therefore, it was recommended that long rounds be introduced, if it could be done with reasonable operating costs.

From these tests, the specification for a suitable drill rig was established. This rig would then be tested under operating conditions in the Sofia project during 1990 and 1991 at LKAB Kiruna. The Sofia project has been reported by Niklasson and Keisu (1991). A total length of 1,260 m of drifts were blasted. The project looked at improving the cuts, the rounds themselves, and improving the contour blasting. The Sofia project was divided into two main parts. The first part dealt with short rounds (4.4 m hole depth) and the second part dealt with long rounds (7.8 m hole depth).

The standard rounds in Kiruna utilize ∅48 mm blastholes and a ∅120 mm cut hole (Figure 72.32). Nearly thirty rounds were blasted in part 1 of the project, to serve as a reference for the continuing tests with ∅64 mm blastholes.

The second part of the project had the goal of developing a cut so that entire rounds could drilled and blasted with only ∅64 mm and no large center hole in the cut. Many types of cuts were tested. As the research work continued, a cut was developed that functioned well for long rounds. Figure 72.33 shows a developed cut providing good advance.

In the Sofia project, Dyno Nobel electronic detonators were used to refine the contour blasting. It was shown that a higher quality of contour was achieved when the contour holes were fired instantaneously. In the contour, the delay time was 5,500 ms with a maximum scatter of 1 ms. Notched holes were also tested in the contour with good results. The notching was performed with a water-jet nozzle.

After the Sofia program was completed, the Atlas Copco rig with mechanized rod adding system was introduced into production in the Malmberget Mine for drilling 7.8 m long holes.

The tests by Niklasson and Keisu (1993) in Malmberget were carried out in parallel with production. Drifting was made both in ore and waste rock with considerable variation in rock quality. Of the 220 long rounds drilled and blasted, 115 were monitored in detail.

Fjellborg and Olsson (1996) reported additional tests of the long round concept with a large center hole in the cut. The tests conducted at the LKAB Malmberget mine were very encouraging. Parts of this project are described in Section 72.7.

Drilling. The drilling pattern in Malmberget was projected on the face with a standard slide projector and manually marked. A portable laser used by the drilling operator gave the reference direction. The drilling pattern was the same as for conventional rounds (short round, ∅48 mm). Consequently there was no reduction in the number of holes. The short rounds with ∅48 mm holes used a parallel hole cut with a ∅102 mm center hole. See Figure 72.32.

Malmberget used the cut that was developed at the Sofia project, a ∅64 mm opening without a large center hole. See Figure 72.33.

Cross bits were used in the ore and button bits in waste rock. Tube steel was used on the middle boom during some weeks of the tests. This gave a stiffer drill string and the accuracy in drilling improved.

Charging. As mechanized equipment using a hose feeder had not yet been developed, the conventional charging method was used. Normally in the ∅48 mm holes, ∅22 mm and ∅32 mm pipe charges are used. Pipe charges adjusted to ∅64 mm diameter boreholes were not available because this diameter is very unusual in drifting. This resulted in very decoupled charges with a tendency to be easily blown out. A number of undetonated pipe charges were found on the muck pile after blasting.

The best result was achieved with the ANFO back blowing technique. This method makes it possible to fill just part of the borehole and thereby reduce the linear charge concentration. Unfortunately this could not be used at all times as ANFO can only be used in dry holes.

Kimulux with a diameter of 29 mm was tested in the contour in some of the ∅64 mm rounds. The function of the charge was better than when Kimulux 22 mm was used. Less undetonated pipe was found on the muck pile but the blasting seemed to be too powerful.

For lifters, when not using ANFO, ∅39 mm Dynamex was used.

The timing sequence of the cut holes differed from the traditional short rounds. Delays between the holes were much longer to avoid "line up" problems.

Scaling. Scaling costs were calculated to be reduced by half when pulling the long rounds. However, the scaling of the face increased considerably, even though the round pulled to full length.

Even though the specific charge was much higher for a long round than for a conventional one, no sign of increased damage in the roof or walls was present. These observations were based only on visual observations and on the amount of scaling work that was required.

Advance per Round. The authors' report that the performance varied between good to excellent. In summary, 40% of the long rounds functioned very well showing an advance of more than 93%; 40% were fairly good with 90%–93% advance; and remaining rounds were acceptable. The mean advance per round was 7.0 m. Poor advance mainly depended on some factors such as defects in the rock or an abnormal amount of water. Hole deviations were reported to be of no problem when ∅64 mm Retrac bits were used. Figure 72.34 shows the advances for long rounds in Malmberget.

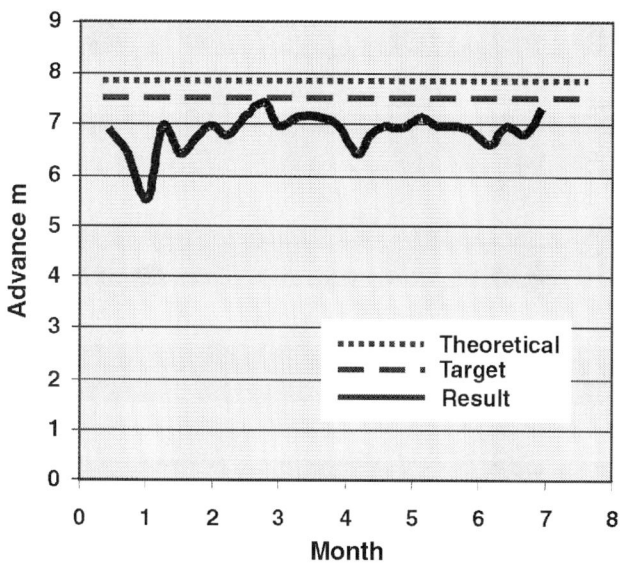

FIGURE 72.34 Advances for long rounds

Conclusions from the Results of the Malmberget Test.

- Opening cuts with only ∅64 mm holes work well, just as well as standard cuts with large-diameter empty center holes. These tests apply for short rounds with 4.0–4.5 m drilled depths, both in ore and in bedrock.

- Long rounds, 7.8-m long, were found to be economically feasible for introduction at LKAB in Malmberget.

- By precisely delayed intervals or radial notching, the quality of the contour could be improved considerably.

- The contour test showed that standard explosive products suitable for the ∅64 mm holes do not yet exist. This is to be expected because this diameter is not normally used, and this might be the reason why the contours of the 48 mm rounds generally were of better quality than those with ∅64 mm holes.

- A precise laser reference for the alignment instruments and an accurate marking of the drilling pattern on the face are very important factors, not only to keep the overbreak low but also to keep the drift heading in the right direction at the right level.

- A clean floor in front of the face is required to obtain a correct lookout angle for the lifter holes.

72.6.4 Substitution of the ∅64 mm Cut by a ∅300 mm Center Hole

Background. Introducing very large-scale sublevel caving has led to higher productivity and lower costs for LKAB's mines in Kiruna and Malmberget.

Development is still the most expensive unit operation. Large scale means that the number of available development faces on each level is very limited, which means that the requirement for effective drift driving is pronounced. In the work cycle for drifting, up to ten different activities are needed. The benefit of using long rounds is obvious as the total set-up time for pulling twice the conventional advances is reduced by 50%.

To increase the effectiveness, the length of the round has been increased as mentioned in section 72.6.3. In 1993 trials were

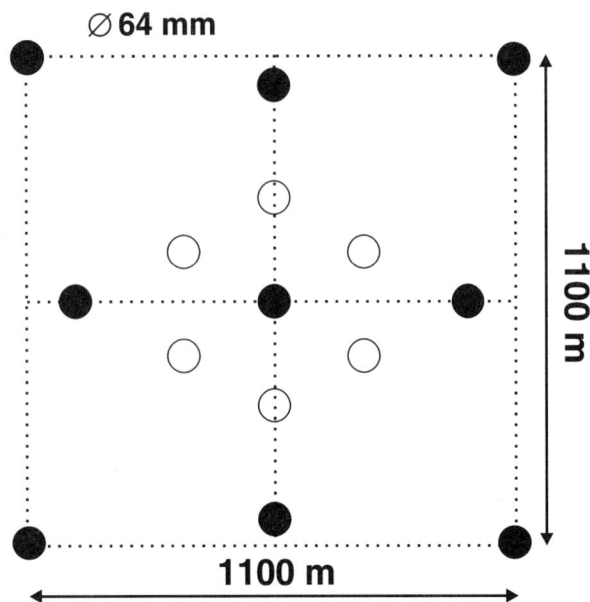

FIGURE 72.35   The cut based on the use of ⌀64 mm diameter holes

FIGURE 72.36   The AMV drilling rig for the center holes

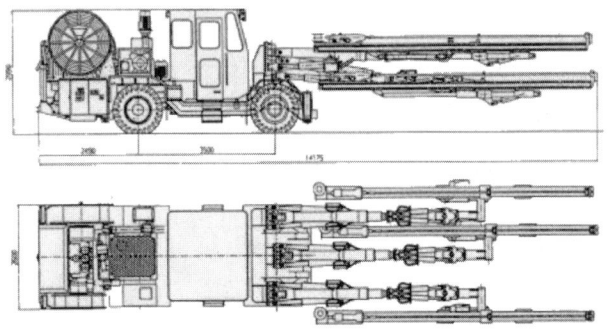

FIGURE 72.37   Drawing of the Atlas Copco Rocket Boomer 353S drift drilling rig for ⌀64 mm

conducted with various cut geometries based upon the use of ⌀64 mm holes. One of the cuts (Figure 72.35) provided advances up to 95% of the drilled length. This cut had six uncharged ⌀64 mm holes to provide for a better swell volume. The geometry of the cut also meant that the distance between charged holes increased and thereby the risk for dynamic dead-pressing of the explosive was reduced.

Cut performance tests clearly showed that when a ⌀300 mm empty hole was introduced in the cut the advance per round increased to 100%. The cost for the large-diameter central hole could be balanced by the cost savings associated with less drilling, explosive use, face scaling, and a better quality of the contour.

In 1994, LKAB decided to purchase a special drill rig for drilling this large-diameter cut hole. A project was established around the use of this drilling machine to:

- Optimize the diameter of the large diameter hole
- Find the best blasting plan
- Determine the blast damage zone
- Reduce scaling and reinforcement

Project Goal.   The main goal of the project was to use a predrilled large-diameter cut hole to optimize the length of the rounds, refine the work-cycle, maximize the advance, and minimize the blast damage zone. The target was to achieve 99% advance in 90% of the rounds. Drilling error for the contour holes should be limited to 20 cm outside the planned contour, and the scaling and reinforcement costs should be reduced by 30%.

Test Area.   The main testing area was in the Norra Alliansen orebody on the 790-m level. Four 150-m-long drifts mainly in the ore were allocated for the tests. The magnetite orebody dipped at 45°. The footwall waste rock was a leptite and a low-strength biotite schist with layers having a thickness of a few centimeters up to 1.5 meters.

Drilling Equipment.   The drilling rig purchased for the large-diameter cut hole was an AMV equipped with a 6-in Wassara ITH water-powered machine (Figure 72.36). The large-diameter hole was drilled in two steps. A ⌀165 mm pilot hole was drilled first and then reamed to a diameter of 250 or 300 mm. The maximum hole length was determined to be 32 m based upon an estimated maximum hole deviation of 1%. The tube

magazine contained 25 pieces of 2-m-long drill tubes. The drill penetration rate for the ⌀165-mm-diameter pilot hole in magnetite was 0.3–0.4 meter/min. The drill penetration rate when reaming the pilot hole to full size was 0.17 meter/min for the ⌀250 mm and 0.11 meter/min for the ⌀300 mm large hole.

An Atlas Copco Rocket Boomer 353S (Figure 72.37) equipped with an automatic rod adding system (RAS) and Bever Control tunneling position system were used to drill the long drift rounds. The average depth of drilled boreholes was 7.5 m.

Part 1—Optimal Diameter of the Large Hole.   The standard drilling pattern for the ⌀64 mm holes was used (see Figure 72.38). The standard drift round has a size of 6.5 × 5.0 m and contains 57 holes. The predicted advance is 7.5 m. All holes except the back holes are charged with the non-cap-sensitive pumpable water-resistant emulsion explosive Kimulux R and a KP primer (VOD 7500 m/s). The back holes are charged with a small 0.5 m long bottom charge of emulsion plus a 40 g/m detonating cord KSP40.

The objective of this part of the project was to compare the advance between the standard long drift rounds and rounds containing a large center hole of ⌀250 mm or ⌀300 mm. When testing the large-diameter hole, the center of the standard cut was replaced by the large hole (Figure 72.39).

These tests included 14 standard rounds, 7 rounds with the ⌀250-mm-diameter hole and 5 rounds with the ⌀300 mm hole.

Part 2—Optimal Blasting Plan.   This part of the project included tests with a drill pattern where every borehole is located based upon the expected rock removal produced by that hole. Each hole should have its own optimal burden. The hole positions and the delays around the cut are set so the blasting sequence becomes "corkscrew-shaped" (see Figure 72.40).

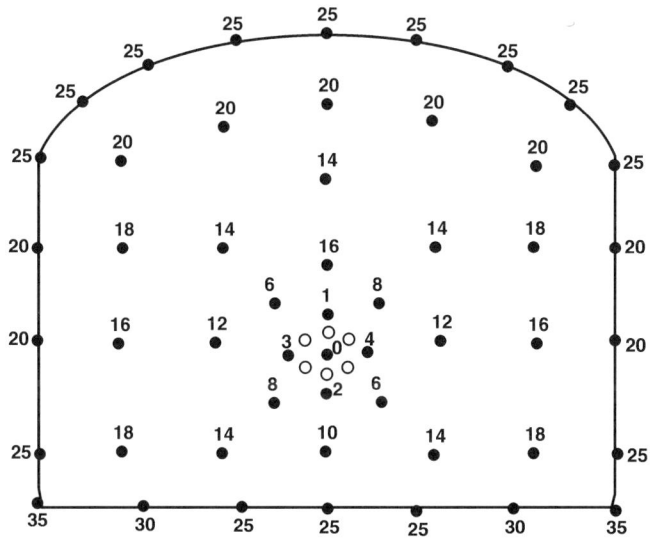

FIGURE 72.38 Standard long drift drilling plan

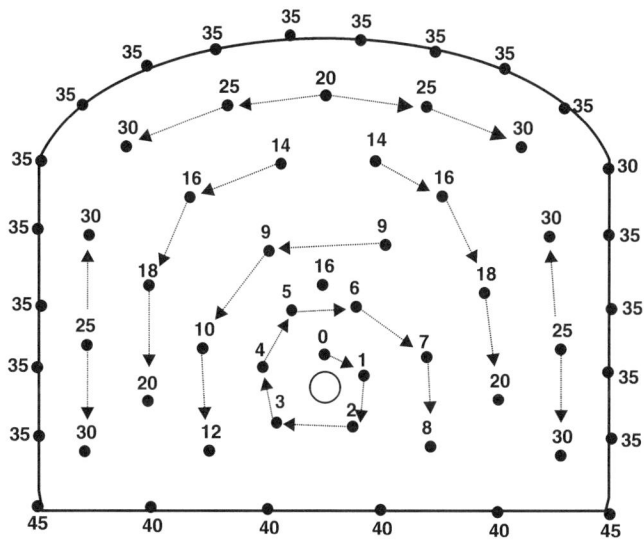

FIGURE 72.40 Cork screw drilling/ignition pattern

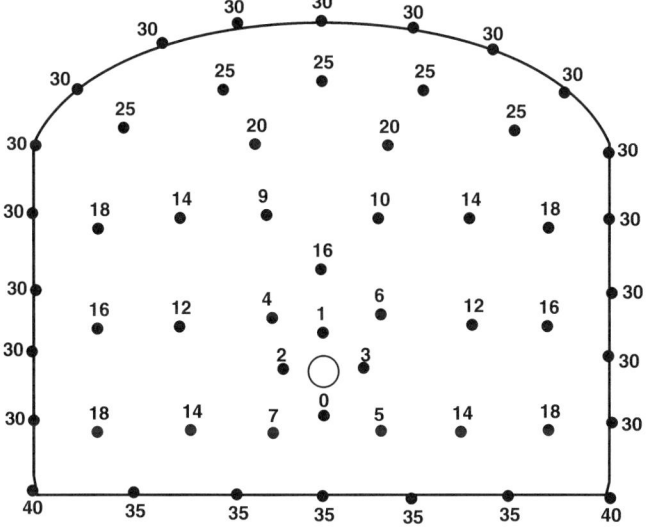

FIGURE 72.39 Drilling plan for rounds with a large center hole

TABLE 72.10 Tested rounds for optimal blasting plan

Contour charging method	Rounds with ∅250 mm large hole	Rounds with ∅300 mm large hole
Cord 40g/m or 80 g/m	6	4
String loaded emulsion	12	3
Cord 40g/m or 80 g/m + EDS	3	
String loaded emulsion +EDS		4
Total	21	11

TABLE 72.11 Data for used contour blasting explosives

Explosive	Density (kg/l)	VOD (m/s)	Gas volume STP (l/kg)	Energy (MJ/kg)	Linear charge conc. (kg/m)
Cord 40 g/m	1.05	6500	780	5.95	0.04
Cord 80 g/m	1.05	6500	780	5.95	0.08
Kimulux R	1.21	5500	906	2.94	3.86
String loaded Kimulux R	1.21	5500	906	2.94	0.55

Tests were made using electronic detonators (EDS), 40 or 80g/m detonating cord and string-loaded Kimulux R in the contour holes (Table 72.10). The tests with electronic detonators included tests of the two systems developed by DNAG and Dyno Nobel.

The Associated Blast Damage Zone. Both borehole logging and the slot technique were performed to determine the amount of damage induced in the contour rock (Table 72.11). Borehole logging was applied in four diamond-drill holes and cracks were observed before and after drifting by a borehole logging TV camera. The tests using the slotting technique were performed by SveBeFo who have applied this technique (Olsson and Bergqvist 1996) over many years to study blast damage.

After blasting, a special diamond saw was used to make a series of vertical cuts 2 m long and 0.5 m deep.The rock between the cuts was then removed and the rock surface perpendicular to the bore hole axis could be examined for radial blast-induced cracks. A dye penetrant was sprayed on the surface and photos were taken of the crack patterns.

Results. The advances were measured before and after scaling and compared with the standard reference 65-mm long drift rounds. Scaling was performed with a Montabert BRP 30 hydraulic hammer and with water at 100 bars pressure. The results clearly showed that using a large-diameter cut hole increased the advance when using the standard drill pattern (Figure 72.41). The advance for the ∅300 mm holes, even before scaling, had an average advance of 97%. It should also be noted that the rounds with large-hole-diameter central holes have a better advance before scaling compared to the standard reference round after scaling.

In all of the blasted rounds, hole deviation was measured in about 20 bore holes per round (Table 72.12). The deviation varied from zero to a maximum deviation of 0.4 m. The deviation is equally spread over the face but with the largest deviation observed in the contour holes. The average is 2.8% for the standard rounds while that for the large center cut hole rounds is 2.2%.

The appearance of half casts varies greatly with the type of explosive, the ignition system, and their different combinations

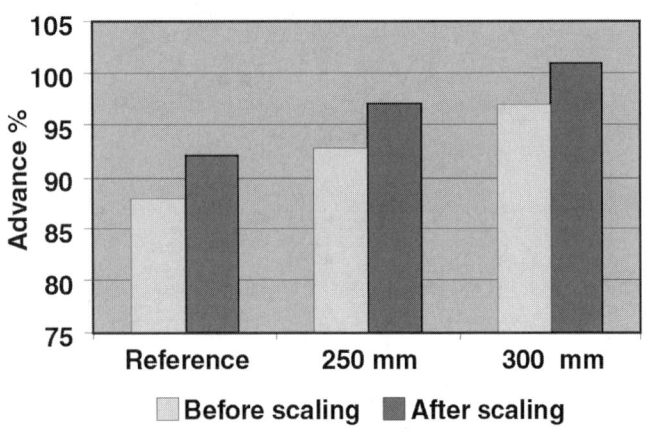

FIGURE 72.41   Advances achieved with a large diameter central hole

TABLE 72.12   Results of the hole deviation measurements

	Part 1		Part 2	
Type of round	All holes (%)	Contour holes (%)	All holes (%)	Contour holes (%)
Reference rounds	2.8	4.8	—	—
Large center hole	2.2	5.7	3.1	5.2

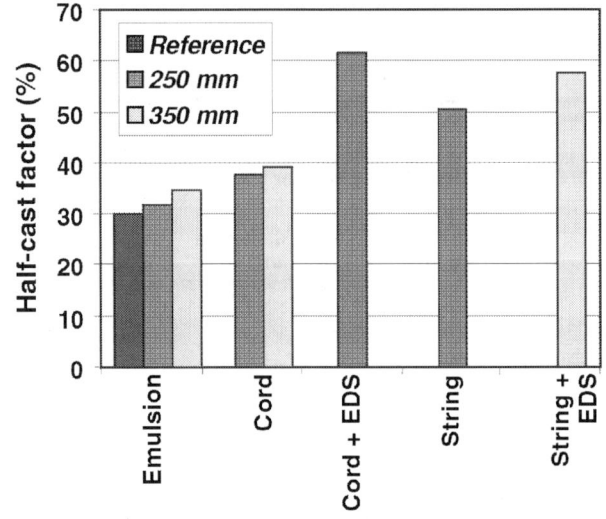

FIGURE 72.42   Half cast factor for the various tests

(Figure 72.42). Contour holes fully charged with emulsion and contour holes charged with 80g/m detonating cord when initiated by long period delay (LP) caps showed almost the same result of half casts. Standard LP caps with a delay of 3,500 ms were used in the whole contour. Detonating cord used in combination with electronic caps (EDS) gave results that were significantly better due to the instantaneous ignition. The best overall results based upon half cast observations were obtained when using electronic caps (EDS).

The method with the slot combined with dye penetrants was successful in examining blast damage. The results could very clearly distinguish crack patterns with the different explosives

and initiation combinations. Contour blasting with a decoupled string of emulsion initiated instantaneously with electronic detonators resulted in no blast initiated cracks. Contour holes fully charged with emulsion ended up with radial crack lengths of at least 0.5 m. The results of the cautious contour blasting tests indicate that, when electronic detonators are used, the type of explosive used has a minor influence on the results. The excellent results depend mainly upon the instantaneous ignition with scatter below one millisecond.

The amount of overbreak was measured for every round. The amount of overbreak depends on a number of factors such as the alignment of the drilling rig, the drilling accuracy, the method of scaling, and the geology. Therefore, it cannot be used as a true measure of the blasting quality, but it does gives a good indication of how the overall operation is developing. The average overbreak at the beginning of the project was greater than 15%, sometimes reaching 30%. At the end of the project, overbreak had decreased to an average of 12%.

This project, during the period of September 1994 through November 1995, included a total of 53 drift rounds involving the use of a large-diameter hole. By achieving an average advance of 99.5% and by cutting the need for scaling by 50% when using a large-diameter cut hole in combination with the modified drilling pattern, the main goals of the project were fulfilled.

Based upon the project results, LKAB has introduced the predrilled large-diameter cut hole method in the Malmberget mine.

72.6.5   Some Environmental Issues

Good Blasting Practice Is Important. To minimize any spillage during the charging and blasting, it is essential to be aware of some factors that may affect the aquatic environment. Operators and management should pay attention to the following:

- A good drilling accuracy is important. Too short distances between the holes can result in dynamic dead pressing of the adjacent hole causing undetonated explosives in the muck pile.

- Do not charge nonwaterproof explosives such as ANFO into wet or water-filled boreholes.

- Bulk explosive may be discharged when one carelessly moves loading hoses from hole to hole.

- When ANFO is being loaded pneumatically, avoid any blowback during the charging or any spillage when the hose is transferred to the next hole.

- Be aware of excess ANFO at the collar that will get into the muck pile.

- Spillage may occur when charging or recharging loading equipment on site or cleaning the equipment after performing the operation.

- Do not empty loading hoses by flushing the explosive onto the ground.

- Proper timing is essential as cutoff boreholes can result in undetected explosives in the muck pile presenting a potential contamination hazard.

Long exposure to water and warm temperatures increases the rate of the reaction from ammonium nitrate to ammonia and nitrite. Therefore, it is of utmost importance to test the quality of water on blasting sites. Water of doubtful quality should never be released into natural water systems.

All civil explosives contain ammonium nitrate. Products originating from chemical reaction of ammonium nitrate may present a risk to the aquatic environment if nondetonated explosive comes into contact with water and the ammonium nitrate dissolves. The dissolved ammonium nitrate may be transformed into nitrite or ammonia. At mines, quarries, and tunneling work usually there are limits to what the owner/contractor can release to the environment and it is always a good practice to pay attention to the drilling and blasting practice and the explosives used in order to minimize any contamination.

Depending on their nitroglycerine content, gelatinous explosives have good-to-excellent water resistance. The water resistance of emulsion and watergel explosives is excellent. In addition, extra protection can be obtained from the cartridge wrapping. The higher the degree of water resistance of an explosive, the lower is the risk of contamination. Spillage is not a problem when using cartridged explosive.

Powder explosives containing nitroglycerine or TNT form the majority of nonwaterproof cartridge explosives. The explosives' compositions themselves are not inherently waterproof although the cartridge wrapping affords some protection but, in general, cartridging only affords a small measure of protection, since the cartridge may be damaged during the loading operation

Waterproof bulk explosives include watergels and emulsions, and these may resist water for several weeks or even months. Spilled emulsion or watergel explosives will dissolve slowly. Dissolving will be faster after the emulsion has been exposed to mechanical stresses. In the case of emulsion explosive, such stresses may break down the emulsion and separate salts such as ammonium nitrate from the oil and water. Subsequently the salts may dissolve in water.

The main nonwaterproof bulk explosive in this category is ANFO, which will dissolve easily in water and should not be used in wet or water-filled holes.

Toxic Fumes. Fumes are the gases resulting from detonation. Typically, one kilogram of explosives will produce between 700 to 1,000 liters of such gases. The stable products of detonation are nitrogen, carbon dioxide, and water, but in addition small quantities of carbon monoxide and nitrous gases are produced. Toxic fumes amounts to around 4% of the after-detonation gases. CO typically ~3% and NO_x ~1%.

The amount of nonideal detonation products formed depends on a number of factors: the type of explosive, the water resistance, type of cartridge wrapping, the VOD, the charge diameter, loading density, type of initiation, and especially the confinement of the explosives.

Some recommendations to minimize the toxic fumes:

- Be careful to drill the holes at the right position according to the drilling plan. This gives less toxic fumes and best blasting results.

- Use alignment devices when drilling so most holes are slanted slightly upwards. This prevents water from accumulating in the drill hole, contaminating the explosives and affecting the detonation properties.

- Use an oxygen-balanced explosive with good fume characteristics.

- Leave an unloaded hole length or stem the holes!

- Explosives in the collar increase the amount of toxic fumes—but not the breakage!

- Avoid cord in ANFO as it might not initiate ANFO to complete reaction, resulting in toxic fumes. The cord itself is strongly oxygen deficient, and by itself generates about 3-l CO per meter cord.

- Explosives in "air" increase the fumes.

- Avoid spacers between cartridges!

- Considerable quantities of after-detonation fumes can become trapped in the muckpile. A good practice is to flush the muckpile with water to remove the dust and the trapped gases before mucking and hauling start.

- Good shot-firing practice contributes towards balanced fumes from blasting operations but does not remove the need for proper and adequate ventilation.

- Measure the toxic fume concentration before entering the mine after a blast

72.7 ACKNOWLEDGEMENTS

The authors wish to thank the management of AECI Explosives & Chemicals Limited for permission to include the section on ring blasting.

72.8 REFERENCES

AECI. (1980a). Ring Blasting:Design of Ring Patterns. Explosives Today, Series 2, No. 21, September.

AECI. (1980b). Ring Blasting:Optimizing Results. Explosives Today, Series 2, No. 22, December.

Almgren, G. and R. Benedik. (1968). How Boliden's Crater Cut Slashes. Raise Cost, World Mining, Feb. 1968, pp. 38–42.

Ash, R.L. (1963). The Mechanics of Rock Breakage. Parts I–IV. Pit and Quarry, Volume 56, Nos. 2–5, Aug.–Nov., pp. 98–100; 112; 118–123; 126–131; 109–111; 114–118.

Ashbridge, M.P. Optimization of ring blast designs at Finsch Mine. AECI

Atlas Copco. (1986). Guide to Underground Mining Methods and Applications.

Atlas Copco. (1997). Guide to Underground Mining Methods and Applications

Atlas Powder Company. (1987). Explosives and Rock Blasting. 622 pages.

Canadian Mining Journal. (1999). 2000 Mining Sourcebook. Southam Publications.

Chung, S.H., Lee, N.H., and Hunter, C.J. (1991). A blast design analysis for optimizing productivity at INCO Limited's Thompson Open Pit. Proceedings 17th Conference on Explosives and Blasting Techniques. Las Vegas. Society of Explosives Engineers, pp. 119–127.

Crocker, C.S. (1979). Vertical Crater Retreat Mining at the Centennial Mine of Hudson Bay Mining and Smelting Co., Limited, CIM Bulletin, Jan. 1979, pp. 90–94.

Cunningham, C.V.B., 1992. The design of ring blasting in the 1990s. Proceedings of MASSMIN '92, Johannesburg, SAIMM. pp. 263–267.

Explosives & Chemicals Limited, Blast Consult report FIN/2/89, 1989.

Fjellborg, S. and M. Olsson. (1996). Successful long drift rounds by blasting to a large diameter uncharged hole. Proceedings, Fragblast 5, Rock fragmentation by blasting. B. Mohanty, Ed. Balkema, Rotterdam, pp. 397–405.

Fjellborg, S. and M. Olsson. (1996). Long drift rounds with large cut holes at LKAB. SveBeFo Report 27, Stockholm, Sweden, (In Swedish).

Goodier, A. (1982). Mining Narrow Veins by Vertical Crater Retreat at the Radiore No. 2 Mine, CIM Bulletin, June 1982.

Holmberg, R. (1997) Environmental aspects on the use of initiation systems and explosives. Discussion meeting of the Swedish Rock Construction Committee, Stockholm, Sweden, (In Swedish) pp. 163–169.

Holmberg, R., A. Rustan, T. Naarttijärvi, and K. Mäki. (1980). Driving a raise with VCR in the LKAB mine in Malmberget, SveDeFo Report DS1980:12, Stockholm, Sweden. (In Swedish).

Hustrulid, William. 1999. Blasting Principles for Open Pit Mining (Volumes 1 and 2). A.A. Balkema, Rotterdam.

Johnson, S.M. (1971). NCG Technical Report No. 21,Explosive Excavation Technology. U.S. Army Engineer Nuclear Cratering Group, National Technical Information Service, U.S. Department of Commerce, Springfield, Virginia 22151, United States.

Lang, L.C. (1983). A Brief Review of Livingston's Cratering Theory, SveDeFo Report DS 1983:1, Stockholm, Sweden.

Lang, L.C. (1981). Driving Underground Raises with VCR, SEE News, Volume 6, No. 3, Sept. 1981, United States.

Lang, L.C. (1981). VCR Used Successfully from Surface in Underground Stoping in Australia. SEE News, Volume 6, No. 4, Dec. 1981, United States.

Lang, L.C. (1978). Cratering Theory Evolves into New Underground Mining Technique, Rock Breaking-Equipment and Techniques. The Australasian Institute of Mining and Metallurgy, Australia, pp 115–124.

Lang, L.C. (1976). The Application of Spherical Charge Technology in Stope and Pillar Mining. E/MJ, May 1976.

Lang, L.C. (1962). A Blasting Theory and its Application, PR(R) 10/62, Iron Ore Company of Canada, Internal Report.

Lang, L.C., W. Comeau, and M. Sampara. (1981). New Underground Drilling Blasting and Mining Methods at Manic-5 Additional-Power Hydro-Electric Project. Proceedings of the Seventh Conference on Explosives and Blasting Technique, Calvin J Konya, Ed. Society of Explosives Engineers, Ohio, pp. 113–131.

Lang, L., R. Holmberg, and B. Niklasson. (1982). A Proposal for the Design of a VCR Stope at the Luossavaara Research Mine, SveDeFo Report 1982:20, Stockholm, Sweden.

Lang, L.C., R.J. Roach, and M.N. Osoko. (1977). Vertical Crater Retreat, an Important Mining Method, Canadian Mining Journal, Sept. 1977.

Langefors, U. and B. Kihlstrom. (1963). The Modern Technique of Rock Blasting, John Wiley & Sons, Inc., New York, USA, and Almqvist & Wiksell, Stockholm, Sweden

Lindqvist, P-A. (1978). Proposal for the Research Mine, Teknisk rapport 1978:63 T, Högskolan i Luleå, Luleå, Sweden. (In Swedish).

LKAB. (1996). Personal communication with Ulf Enback, LKAB, Kiruna, Sweden.

Mäki, K. (1982). Characterization of rock structures at crater blasting experiments in the Luossavaara Mine, SveDeFo Report DS 1982:17, Stockholm, Sweden.

Miller, R.E. (1979). Vertical crater retreat mining method as applied to L 519 slot 14B and 15D Mount Isa Mine, Australian Mineral Foundation Workshop 120/79, June 25–29, Adelaide, Australia.

Monahan, C.J. (1979). The Crater Blasting Method. Applied to Pillar Recovery at Falconbridge Nickel Mines Limited, CIM Underground Operators Conference, Feb. 19–21, Timmins, Canada.

Niklasson, B. (1982). Report from some Australian mines using the VCR-method, SveDeFo Report 1982:8, Stockholm, Sweden (In Swedish).

Niklasson, B. (1979). Large Hole Diameter Mining at Sullivan Mine, Cominco Ltd., Kimberley, BC, Canada. Bergsskolan i Filipstad, Sweden. (In Swedish).

Niklasson, B., R. Holmberg, K. Olsson, and S. Schorling. (1988). Longer rounds to improve tunneling and development work. Tunnelling-88, London, UK, pp. 213–221.

Niklasson, B. and M. Keisu. (1993). New techniques for tunnelling and drifting. Proceedings, Fragblast 4, Rock fragmentation by blasting. H-P Rossmanith, Ed. Balkema, Rotterdam, pp. 167–174.

Niklasson B. and M. Keisu. (1991). New technology for drifting and tunneling—The Sofia-project. 1991. SveDeFo report DS 1991:10, Stockholm, Sweden. (In Swedish).

Olsson, M. and I. Bergqvist. (1996). Crack lengths from explosives in multiple hole blasting. Proceedings, Fragblast 5, Rock fragmentation by blasting. B. Mohanty Ed. Balkema, Rotterdam, pp. 87–91,

Persson G. (1983). Methods to reduce the emission of air pollutants at underground blasting. SveDeFo Report DS 1983: 18, Stockholm, Sweden, (in Swedish.).

Persson, P.A., R. Holmberg, and J. Lee (1994). Rock Blasting and Explosives Engineering. CRC Press, Inc., Boca Raton, Florida.

Röshoff, K. (1981). Structure geological examinations at the Luossavaara Research Mine, Swedish Mining Research Foundation, Report FB 8109, Kiruna, Sweden. (In Swedish).

Rowlandson, P. (1979). Applications of DT Drilling at Pamour Porcupine Mines, CIM Underground Operators Conference, Feb. 19–21, Timmins, Canada.

Selleck, D.J. (1962). Basic Research applied to the Blasting of Cherty Metallic Iron Formation, International Symposium on Mining Research, Volume 1. George B. Clark, Ed., Pergamon Press, pp. 227–248.

Underground Mining Looks to the Future

CHAPTER 73

Small Resource and Mining Companies—Present and Future

P.C. Jones* and H.B. Miller†

73.1 INTRODUCTION

The fascination surrounding small mines and mineral exploration is deeply rooted in North American history. For hundreds of years, men have labored in the pursuit of that elusive strike that would bring fame and economic fortune. Such aspirations were built upon the fundamental belief that through hard work, skill, and a little luck, an individual could determine his own destiny. This basic philosophy underwrites much of the U.S. mining law and is a driving force behind the entrepreneurial nature of many small resource and mining companies. While this inherent appeal is tempered by the economic and regulatory realities of a modern industrialized world, there are few exploration geologists and mining engineers who do not harbor the occasional thought of operating their own small mining company.

Despite traditional misconceptions, the "small miner" of today plays an important and well-defined role in the mining industry. By their very nature, these small opportunistic companies are highly adept at operating in environments that are usually too risk-intensive or uneconomic for large corporations. While the term "small miners" is inherently ambiguous, it effectively describes a broad spectrum of small- to medium-sized companies engaged in the discovery, delineation, exploitation, and marketing of mineral resources and deposits. These companies are normally global in character and range in structure from simple proprietorships to publicly traded corporations. They are representative of highly dynamic and productive enterprises that often, because of limited economic utility, become specialized in specific niche activities, mineral commodities, and geographic locations. Regardless of whether a company's expertise lies in grassroots exploration or in operations, today's small miner faces a myriad of challenges virtually unheard of just two decades ago. With these challenges come fundamental changes in both the strategic objectives and management of these companies as well as in the types of economic opportunities that are likely to emerge.

The primary intent of this chapter is to present a brief description of the role that small miners play in today's economic and political environment, and to address some of the daunting challenges these companies now face. The chapter will conclude with a discussion of how these factors are contributing to the recent trends in domestic and international investment by small resource and mining companies.

73.2 INDUSTRIAL ROLE OF THE SMALL RESOURCE AND MINING COMPANIES

Independent of their limited financial capacity, small miners possess some advantages over major mining companies by virtue of their size, structure, and flexibility. The diminished need for successive layers of management and decision-making bureaucracy, the hands-on involvement required by company executives, and the economic resourcefulness of these small companies allow them to be extremely flexible and capable of reacting quickly to new and changing investment opportunities. Given these attributes, small resource and mining companies have found economic niches that have helped define their role in today's mineral industry. These areas include:

- Exploration network for major mining companies
- Exploitation of small- to medium-sized deposits
- "High risk" ventures

Each area has its own inherent characteristics, economic appeal, and specific challenges.

73.2.1 Exploration Network for Major Mining Companies

Historically, a primary role of small resource companies, often referred to as "juniors," has been in the exploration and early-stage development of deposits that may ultimately be ventured to, or acquired by, a major mining company. The economic and socio-political requisites required to permit, construct, and operate a profitable mine dictate that a great number of prospects be evaluated before an economically viable mineral target can be identified. It has been published that, on average, more than 5,000 exploration prospects must be evaluated to locate a single economic property (USDA 1977). As internal production and revenue requirements increase, large companies are forced to scrutinize even more exploration prospects to find deposits that meet their required standards. It would be economically prohibitive, if not physically impossible, for the exploration staffs of these major companies to independently evaluate all the properties required to find those few deposits that warrant their investment. Hence, many of the major companies rely extensively upon junior companies to discover and advance, to some degree, resources that may be identified as worthy of large-scale development. In essence, junior companies provide potential investment opportunities to these large companies in a cost-efficient manner and mitigate the need for the major company to conduct the massive grassroots exploration programs and the systematic evaluation of numerous prospects that would otherwise have to be performed.

The effectiveness of small resource companies to perform exploration and early-stage development stems from their flexible and highly efficient organizational structure. Most major companies follow some type of administrative protocol that passes important decisions up through successive levels of management. In contrast, the absence of an extensive corporate structure allows junior companies to react much faster to potential opportunities and to the changing economic and technical conditions that are endemic to all exploration

* Sovereign Gold Company, Ltd., Golden, CO.
† University of Arizona, Tucson, AZ.

programs. The ability to make rapid decisions regarding financial obligations and strategic planning permits junior companies to be much more effective at coping with the inherent risks of exploration, resource assessment, and property acquisition.

73.2.2 Exploitation of Small- to Medium-Sized Deposits

There are numerous high-grade mineral deposits that are not attractive to major mining companies simply because they are not big enough to meet corporate investment objectives. This can be largely attributed to the cost of the administrative infrastructure and overhead inherent to any large corporation. While difficult to ascertain, these indirect costs compel companies to establish standards for the minimum acceptable earnings required from any investment. As a company increases in size and sophistication, the indirect cost attributed to administrative overhead and management generally increases accordingly. The inherent magnitude of these expenditures for a large multinational mining company necessitates a significant production capacity to financially offset these expenditures. Such production requirements have a direct consequence on the size and type of exploration targets that can be economically justified by a company. These requisites generally limit major mining companies to pursuing larger deposits amenable to lower cost surface- and bulk-mining methods. Consequently, a large number of known deposits exist that, because of limited tonnage and/or constraints limiting production, are inappropriate for many large mining companies. These are the deposits that have traditionally been exploited by small miners.

Successful small resource and mining companies are extremely efficient and incur a far smaller percentage of overhead and administrative costs compared to their larger counterparts. In most cases, company personnel serve a multitude of roles: the responsibility of corporate officers may extend from company management, finance, and stock promotion to that of exploration geologist, mining engineer, or metallurgist. In addition, compensation is usually heavily tied to company performance, either through stock options or some form of deferred remuneration. This is particularly the case in small public companies without operating mines, where direct compensation to officers, in the form of salaries, is often capped by exchange regulations. Another effective strategy routinely used by juniors to maximize economic utility is the sharing of resources. Often driven by necessity, companies optimize economic synergy through cooperative arrangements for such things as office space, equipment, and professional and secretarial services. The consequence of these characteristics is the creation of small, cost-efficient, and performance-driven companies capable of profitably exploiting mineral deposits that are well below the economic thresholds of larger mining companies.

73.2.3 "High Risk" Ventures

An important—and often overlooked—role of small resource and mining companies is their willingness to assume risk. Unlike large corporations, that are generally averse to risk, many junior companies will readily undertake ventures in which the risks are poorly quantified or ambiguous if the potential exists for an economic windfall. While most companies routinely engage in some sort of cost-benefit analysis during project evaluation and capital budgeting, the organizational structure of juniors allows them to tolerate a far greater amount of early-stage risk than most major mining companies are willing to accept. In fact, it is a common strategy among many junior companies to aggressively pursue exploration targets that are deemed too high-risk for major corporations. Through development and due diligence, juniors will often attempt to mitigate the risks until these targets become attractive for investment.

Traditionally, one of the focal areas of project risk that is characteristic of small resource and mining companies is the

exploration and development of properties in foreign countries exhibiting political, social, or economic instability. There are numerous examples in which junior companies have established the precedence and framework within a country that has made entry by major mining companies more attractive. Other forms of risk commonly exploited by junior companies include ores with complex metallurgy, properties in remote or hostile environments, mineral products with highly volatile or poorly defined markets, and prospects in areas with a high potential for large antecedent liabilities.

73.3 CURRENT CHALLENGES FACING SMALL MINES AND OPERATORS

There are three major challenges facing junior companies in today's social, economic, and business environment. They are:

- Permitting and regulatory compliance
- Risk capital and financing
- Labor productivity and technology

Each of these issues can critically influence the success or failure of a small resource or mining company. For that reason, these issues are discussed in detail below.

73.3.1 Permitting and Regulatory Compliance

The laws and regulations governing mine development and operation in the United States and Canada make no statutory distinction between large and small companies. While "small miner" provisions are often contained in the language of many state regulatory programs pursuant to exploration and operating permits, these provisions are principally formulated to accommodate recreational mining interests and have little bearing on commercial ventures by small private or public corporations. This lack of distinction creates economic inequities for junior companies, in which they must shoulder a financial burden to achieve regulatory compliance that is disproportionate to their overall assets and project expectations. This trend is a relatively recent development beginning in the late 1970s. In the United States, the financial commitments required by small companies, in order to comply with the ever-increasing number and sophistication of regulatory demands, are growing at an exponential rate. It is estimated that 5% of the gross revenues generated by a mining company go toward servicing the costs of environmental permit compliance (pers. commun.). It is important to note that this estimated cost does not include the direct and administrative expenses incurred to fulfill reclamation commitments and comply with the myriad of other regulations overseeing labor, safety, and operating issues.

Because much of the cost of environmental compliance is fixed and not proportionate to resource size or production capacity, most junior companies are at an inherent economic disadvantage, as they must amortize permitting costs over the relatively small deposits they normally target. The small miner also faces economic inequities in bonding requirements, mid-stream regulatory changes, and unexpected environmental or regulatory compliance problems. Because these issues are of current relevance, each will be individually addressed.

A major economic issue that disproportionately impacts small resource and mining companies is the requirement to post bonds for exploration and mining operations. Most juniors, whether public or private, do not have the underlying financial means required to commercially purchase reclamation bonds. Surety companies require an adequate financial guarantee and/or collateral in the event the bond is revoked. For these junior companies, securing bonds can be quite difficult and extremely expensive. If a bond cannot be purchased, a company must either post cash security or participate in a bonding pool such as the one organized by the State of Alaska. While a company may be able to raise sufficient capital to post the initial bond as well as finance permitting and mine development, it is generally difficult to find additional investment to fund subsequent adjustments in the bond once the mine is in operation. Such amendments to reclamation bonds are becoming increasingly common, if not expected, and can severely cripple the cash flow of highly leveraged operations or of those operations that have not adequately planned for the financial aspects of postmining closure.

The concept of perpetual financial responsibility for site maintenance after mine closure has recently come under regulatory scrutiny. Principally directed towards the future treatment of possible long-term surface and subsurface water contamination, companies may be required to purchase bonds and/or allocate monies in escrow to cover the potential remediation of these disturbances after reclamation and closure. Unlike other types of bonding provisions, the uncertain size of this financial liability could place the cost of acquiring these bonds beyond the means of most junior companies.

One of the most serious economic challenges facing small resource and mining companies operating in the United States and Canada has been the significant growth in the promulgation of new permitting policies and regulations. This is particularly the case on U.S. public lands where the permitting process is becoming ever more influenced by external politics, special interest groups, the discretionary interdiction of individual regulators and land managers, and litigation. While the technical aspects of these processes can be quite formidable, the primary concern is the time and expense customarily being spent in securing permits to conduct exploration and resource development.

Mining projects that required permitting periods of 24 to 36 months in the late 1980s are now taking 6 to 8 years to complete. Commensurate with this trend is the tremendous proliferation in cost that must be absorbed by companies in order to secure these permits. This condition developed not because of increasing project complexities or onerous changes in environmental permitting requirements, but is largely due to bureaucratic delays and litigation spawned by special interest groups. Regulatory agencies on both state and federal levels frequently discount the economic constraints under which companies must operate and expect applicants to possess the financial resources to incur a lengthy process fraught with ambiguous costs and time restraints. This situation is compounded by the universal use of litigation by project opponents and the absence of regulator accountability.

It is difficult for a major mining company, let alone a small mine operator, to economically justify the initial capital expenditure associated with project planning and permitting that can easily top $20 to $30 million for a 2,000-ton-per-day open-pit gold project. For most of the small deposits commonly targeted by junior companies, there is simply insufficient return to warrant the investment, particularly given a lengthy permitting period. When the risk-discounted, time-value of the capital required to plan and permit an operation is considered, even at modest discount rates, these long preproduction periods necessitate that junior companies seek exploration targets with potential revenues large enough to generate an acceptable net present value. This effectively eliminates the small- to medium-sized orebodies traditionally sought by juniors. Furthermore, in situations where the investment in such orebodies is justified, it is inevitable that market and financial conditions will change over the course of these long permitting periods. As has happened, these market changes can ultimately result in economic conditions that prohibit the exploitation of the resource once it has finally been permitted. These delays, coupled with the cost ambiguity associated with the permitting process, constitute a significant risk that did not exist a mere decade ago.

The financial commitment attributed to permitting deters most junior companies from advancing more than one or two prospects much beyond prefeasibility. The limited capital resources of these companies also dictate how long a preproduction period they can endure. Few small companies can survive an 8-year permitting period and the related costs of such a lengthy regulatory approval process. This situation is further exacerbated if the property possesses a high holding cost. As such, a junior assumes a tremendous financial risk once a decision has been made to proceed to permitting. For these companies, a single poor capital budgeting decision or incorrect economic forecast can spell economic ruin.

73.3.2 Risk Capital and Financing

As mentioned above, one of the niche roles of small resource and mining companies has historically been the discovery and delineation of economic deposits that will eventually be advanced to the "majors" for development. This process only works when there is a pool of risk capital from which these juniors can draw in order to conduct exploration. As with most economic factors, the availability and cost of risk capital are market driven and hence, dynamic.

Since debt-financing through conventional sources (i.e., banks and lending institutions) does not exist for exploration, smaller companies are by necessity reliant upon equity investment. In periods where venture capital is readily available, exploration activities by junior companies increase and ultimately provide a greater number of potential investment opportunities for major companies. Inversely, when the supply of risk capital is limited, junior companies downsize their exploration programs and the number of potentially viable properties that reach an advanced stage of feasibility decline. During these lean periods, the situation possesses negative consequences not only for the small mining and resource companies who are fighting for their survival, but also for numerous other businesses that are dependent upon these small companies for finding and developing exploration targets or the purchasing of goods and services.

Unlike major mining companies who may be capable of funding exploration programs internally, most juniors only possess sufficient working capital to sustain a limited number of projects for any length of time. These small companies, therefore, rely extensively on infusions of capital through investment or earned revenues in order to conduct their exploration activities. When risk capital becomes tight, they are forced to consider creative alternatives to conventional equity financing. One innovative arrangement is to obtain major mining company support for the exploration efforts of a "related" junior company. This type of agreement is mutually beneficial to both companies and has been successfully employed by several of the major Canadian and European metal mining companies.

The lack of properties reaching a bankable stage during periods of insufficient equity funding has created a relatively recent trend by venture capitalists and the banking community to create "equity funds" that assist companies in advancing highly attractive properties through feasibility. While investment from these funds provides opportunities for fast-tracking a property to project financing, they are very difficult to obtain, quite expensive, and seldom, if ever, used for grassroots exploration or for projects in prefeasibility. Consequently, in periods where the availability of risk capital is limited, a great deal of time and effort is spent by juniors soliciting funds rather than exploring for deposits.

Once an economically viable orebody has been located and the decision to proceed into production has been made, many junior companies, by virtue of their size, face fundamental obstacles in financing mine development. Because most lending institutions and banks view small mining companies as high-risk

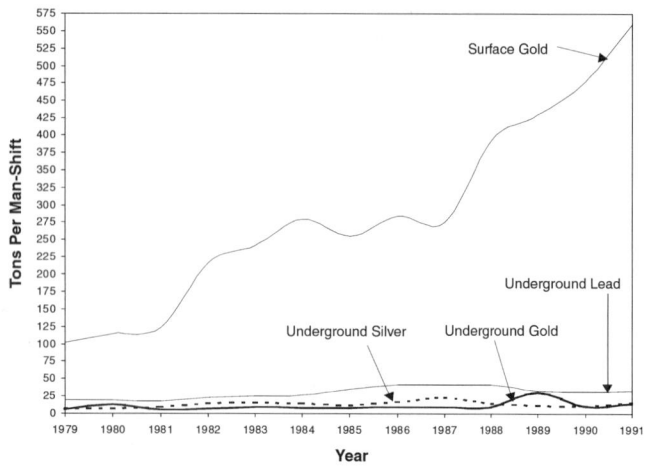

FIGURE 73.1 Trends in mine labor productivity (after Miller and Hrebar, 1995)

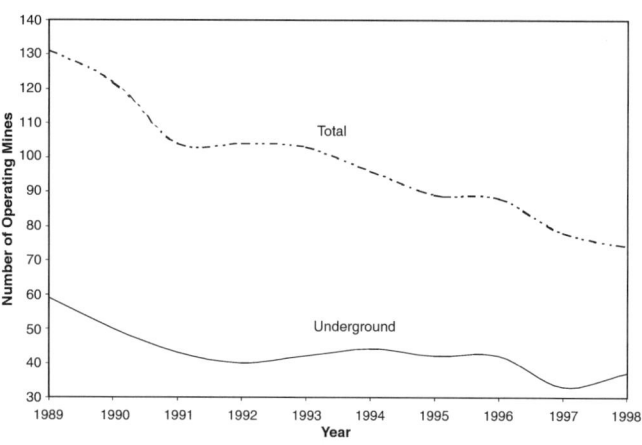

FIGURE 73.2 U.S. operating metal mines with between 20–150 employees (after MSHA)

ventures, they will often attach additional conditions to the debt that may not be compulsory for large companies. These provisions generally include higher ratios of cash flow coverage for debt servicing, increased loan margins, and low debt/equity ratios. Such provisions usually afford the lending institution greater control over the management and operation of the venture, including issues related to scheduling, construction, production, and metal marketing. Because of the perceived risk, these loan requirements have a disproportional impact on the junior through higher financing costs as well as contractual limitations on corporate decision-making authority.

73.3.3 Labor Productivity and Technology

One area of traditional interest to the small mining company has been the exploitation of tabular hardrock deposits by underground mining methods. The challenges posed by geometric and logistical constraints, equipment availability, high labor dependence, and small-scale project economics make this type of underground mining unique and unlike any other sector of the industry. Conventional underground mining methods are extremely dependent upon the productivity of skilled miners to perform labor-intensive tasks in fairly adverse environments. This is particularly true when mining narrow veins, where there have been few substantive advances in equipment and operating systems over the last 30 years. Unlike bulk-mining and surface operations, which have experienced phenomenal gains in labor productivity through mechanization and equipment economies of scale, productivity in underground metal mines has remained essentially flat. These trends are illustrated in Figure 73.1.

There have been periodic attempts by many companies to introduce new technology and increase equipment utilization in labor-intensive underground activities with varying degrees of success. Geometric constraints, dilution, increased development requirements, and high equipment capital and operating costs are major factors that erode the economic utility of large-capacity equipment in these applications. The costs related to these innovations usually offset the financial benefits attributed to these efforts.

While economic benefits can be realized by optimizing equipment selection, stope layouts, and operating procedures, substantial advances in productivity are restricted by the limitations attributed to conventional underground equipment (e.g., LHDs, jumbos, and hand-operated drills) and the cyclical nature of unit operations. To achieve the sustained growth in productivity, which has been experienced in surface mining,

fundamental changes in technology and mining systems are required. However, the relatively modest economic stature of these operations presents a financial disincentive for equipment manufacturers to make a serious commitment to the research and development of new technologies applicable to small underground deposits. As such, technologic advancements are largely adapted from surface mining and underground construction and tunneling. With few exceptions, the large capital investment normally associated with these technologies is economically prohibitive for use in small underground mines given their inherent production and cash flow constraints. Good examples include mechanical excavators, such as roadheaders, whose initial capital cost can exceed $1 million per unit. For North American operators, the ominous consequence of these factors is that advances in productivity and cost-efficiency have failed to outpace increases in the cost of labor, equipment, and consumable supplies by a substantial margin (Miller and Hrebar 1995). When coupled with the continued erosion of metal prices and the high dependency of skilled labor, many junior companies feel they possess no economic alternative except to invest in foreign countries with relatively low labor rates.

73.4 TRENDS IN DOMESTIC AND INTERNATIONAL RESOURCE DEVELOPMENT

Independent of their size, mining operations in the United States, and, to a lessor extent, Canada, continue to be impacted by a litany of stringent environmental and land-use regulations and related litigation. The implementation of these regulations often appears ambiguous and counterproductive. The federal government of the United States has not had a coherent, unified mineral and energy policy since the early 1970s. These issues, coupled with large increases in the cost of permitting, bonding, and litigation, have created an operating environment that has contributed to a continuing erosion of the economic attractiveness of mineral investment within the United States.

While small resource and mining companies share the same legal, regulatory, and financial framework that large companies face, they possess far less financial resources to deal with these challenges. The disproportionate nature of these impacts, the extraordinarily long preproduction periods attributed to permitting, and the lack of substantive improvements in technology and/or productivity have further compounded the economic situation assailing these junior companies. The consequence of these detrimental factors, shown in Figure 73.2, has contributed to the dramatic decline in the number of small- to medium-sized mines operating in the United States and constitutes a real economic threat to the viability of existing and proposed mining projects in North America (USDL).

In light of the politically inspired disincentives to exploration and mining that currently exist within the United States, most junior companies have elected to invest in foreign exploration and resource development opportunities. Numerous countries with high resource potential are also promoting this relatively recent trend by adopting policies and legislative reforms to encourage foreign investment in resource exploration and mine development. In 1994, roughly 20% of the worldwide exploration expenditures by North American companies were within the United States. In 1998, this value dropped to less than 7% (Metals Economic Group). Given the proclivity of recent administrative policies restricting exploration and resource development on public lands, it is expected that such trends will continue until remedied by some sort of legislative relief.

Canadian, Australian, and more recently, U.S. junior companies have been quite active in the exploration of mineral targets throughout Latin America, Africa, and Asia. Much of this activity reflects the recent "opening of borders" and privatization policies in many formerly socialist countries that possess mineral wealth. Beginning with Latin America in the late 1980s and moving swiftly to Africa and Asia in 1992–1994, many countries started actively soliciting, and in effect competing for, investment capital associated with resource exploration and development. This movement has been, at least partially, fostered by international development banks and similar multinational aid organizations to improve the economies of underdeveloped countries. Good examples of these efforts include Chile, Tanzania, and more recently, Mongolia. International agencies encourage select countries to adopt democratic reforms through financial incentives and foreign aid to promote an attractive environment for international investment. Coupled with these economic incentives are environmental protection requirements designed to achieve similar goals as those in use in North America, but generally without the restrictive "command and control" infrastructure used by the regulatory agencies in the United States and Canada. The success of these programs is apparent, as several major mining companies and banking entities have publicly acknowledged that they now view the political risk of project financing in many third-world countries less than that of the United States.

These types of foreign ventures are extremely amenable to the inherent strengths and expertise normally associated with junior companies. As such, these small companies have established a definitive role for themselves as the "prospectors" of the world, particularly specializing in under-explored locations and countries without an appreciable infrastructure or an existing mining industry. In most cases, juniors perform the grassroots exploration required to delineate potential economic deposits. Once found, these small companies normally advance attractive properties to some stage of feasibility before ultimately venturing them to a larger company. In effect, this provides major mining companies with a cost-efficient approach to finding foreign investment opportunities without incurring the expense and risk attributed to running simultaneous exploration programs in multiple countries.

73.5 SUMMARY

The role of the small resource or mining company in today's regulatory, business, and financial climate is considerably different than it was 20 years ago. Viewed from a global prospective, small companies possess a very definite role as "project generators" for the major mining companies. It is through their efforts that many, if not most, "new" deposits are found and advanced toward a bankable stage. These junior companies have become, in effect, international prospectors in the pursuit of resources as well as the creation of potential investment opportunities. Similarly, the role of these small companies as mine operators is also rapidly becoming international. While the repercussions of the current permitting and regulatory climate are quickly diminishing the economic appeal of junior companies operating in the United States and Canada, many other countries readily possess the resources and operating requirements necessary to attract mine investment. The trends towards international resource exploration and development are likely to continue due to the shear quantity of attractive foreign investment opportunities and the continued promulgation of detrimental regulatory and land management policies domestically.

73.6 REFERENCES

Metals Economic Group, Mining Engineering, SME Annual Feature on Exploration, December 1994–1999.

Miller, H.B. and M.J. Hrebar. March 1995. "The State of Underground Metal Mining," SME Annual Meeting Proceedings. Preprint Number 95-232, Denver, CO.

USDA Forest Service. 1977. Anatomy of a Mine from Prospect to Production, Intermountain Forest and Range Experiment Station, General Technical Report, INT-35, pg. 2.

USDL Mine Safety and Health Administration. Injury Experience in Metallic Mineral Mining, Informational Reports, 1972–1998.

CHAPTER 74

Telemining™ Systems Applied to Hard Rock Metal Mining at Inco Limited

G.R. Baiden[*]

74.1 INTRODUCTION

Inco Limited is a metal-processing company that produces 17 minerals and chemicals from sulphide and laterite deposits. Sulphide Ni-Cu deposits have enjoyed a competitive advantage over Ni laterites primarily because of the lower cost of ore processing. Recent advances in processing laterite deposits have put considerable pressure on the sulphide producers to lower their costs to remain competitive. The main difference between sulphide producers and laterite producers is the cost of mining. Laterite producers are low-cost, open-pit mining operations, while the sulphide producers are burdened with higher cost underground extraction methods. This paper describes tele-remote techniques that have the potential to close the gap in mining costs between the laterite and sulphide producers.

The Mining Automation Program (MAP) began as an idea several years ago with the fundamental question "Can an orebody be profitably mined without operator presence in the workings?" The concept was to design mining equipment and systems that would allow the teleremote operation of a mine. Automation was only one component of this much larger teleremote mining issue. To start with, proving or disproving the feasibility of this concept was essential. The fundamental requirements are:

- Telecommunications
- Positioning
- Navigation
- Equipment
- Software
- Electronics
- Mining Engineering
- Organization.

Each of these requirements had to be considered. Technically, the feasibility of the project hinged on the successful development of a high-capacity communication network, an accurate underground positioning system, a navigation system and, finally, the electronics and software that could integrate these. Mining equipment that could perform all the mining functions was the next requirement, along with explosive supplies that could be delivered to the holes teleremotely. Whatever systems were developed, they must be able to work in both bulk and selective mining situations and in varying conditions within a mine. Finally, we had to understand the organizational issues of changing from our current method of operation to teleremote mining.

74.2 DEFINITIONS

Teleremote operation is the operation of mining equipment over a network from any location out of the line-of-sight of the machine.

FIGURE 74.1 Tamrock DataSolo™

Equipment automation is the addition of technology to a machine that allows the unmanned operation of a machine and enhances the productivity of the operator. For example, a guidance system that allows a load-haul-dump (LHD) to drive itself, freeing the operator for other tasks.

Telemining™ describes a mining process that combines the use of teleremote operation (potentially with some automation), positioning and process engineering, monitoring and control.

Tamrock DataSolo™ is a computerized drilling rig as shown in Figure 74.1.

Tamrock DataMini™ is a computerized development drill rig used for drifting and construction tunneling as shown in Figure 74.2.

Tamrock Toro 450™ is an LHD (Figure 74.3).

DYNO NOBEL ROCMEC™ is an explosive-loading machine that allows an operator to perform the entire loading function from the cab (Figure 74.4).

[*] Inco Limited—Mines Research, Copper Cliff, Onario, Canada.

FIGURE 74.2   Tamrock DataMini™

FIGURE 74.3   Tamrock Toro 450™

FIGURE 74.4   Dyno Nobel Rocmec™

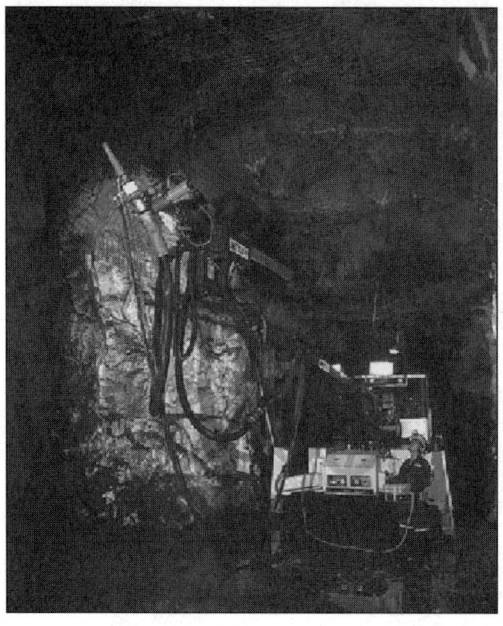

FIGURE 74.5   Meyco Spraymobile

FIGURE 74.6   Automated Diamond Drill

Meyco Spraymobile is a shotcrete sprayer with a laser measurement device that ensures quality control (Figure 74.5).

Automated Diamond Drill is a drill capable of computer-controlled drilling and rod changing (Figure 74.6).

MOS is short for the Mine Operating System that is composed of software that facilitates the operation of mining processes.

74.3   TELEMINING™

Telemining™ is the application of remote sensing, remote control, and the limited automation of mining equipment and systems to mine mineral ores at a profit. The main technical elements are:

- Advanced underground mobile computer networks
- Underground positioning and navigation systems
- Mining process monitoring and control software systems
- Mining methods designed specifically for Telemining
- Advanced mining equipment.

Telemining™ reduces cycle times, improves quality, and increases the efficiency of equipment and personnel, which results in increased revenue and lower costs. Figure 74.7 conceptually represents the key technological ingredients.

Advanced underground mobile computer networks form the foundation of teleremote mining (Figure 74.8). The underground mine may be connected via this telecommunication system so mines can be run from surface operations centers. Inco, in conjunction with IBM and Ainsworth Electric, developed an advanced mobile computer network in the early 1990s. It consists of a high-capacity CATV network backbone linked to 2.4-GHz-capacity radio cells that are located in central areas on each level of the mine. The high capacity allows the operation of mobile telephones, handheld computers, mobile computers on board machines, and multiple video channels to run multiple pieces of mining equipment from surface operation centers.

To apply mobile robotics to mining, accurate positioning systems are absolutely necessary. Inco has developed underground positioning systems that have sufficient accuracy to locate the mobile equipment in real time at the tolerances necessary for mining. Practical uses of such systems include machine setup, hole location, and remote topographic mapping. These systems function similar to Global Positioning Systems (GPS). The positioning equipment consists of a Ring-Laser-Gyro (RLG) and accelerometers. Units are mounted on all types of drilling machines so surface operators can position the equipment without going underground and without conventional surveying. The RLG systems track the location of mobile

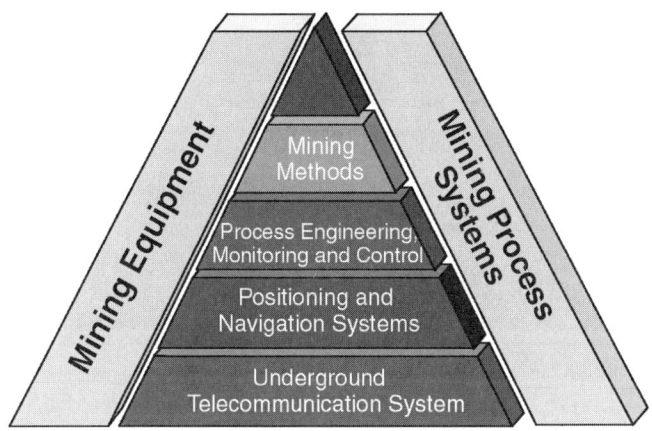

FIGURE 74.7   Telemining™ concepts

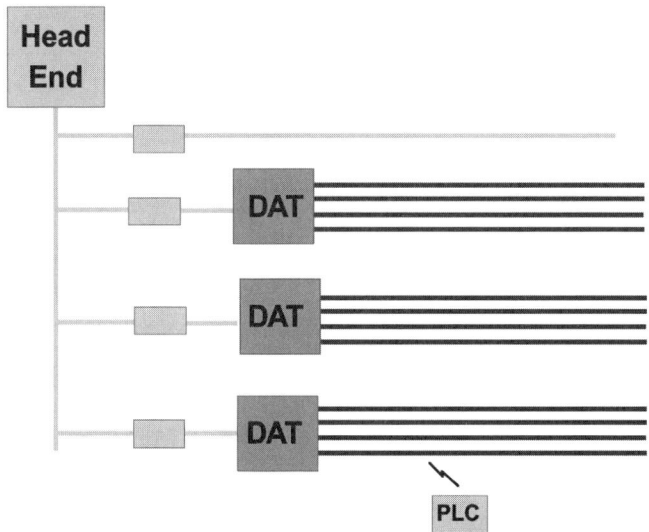

FIGURE 74.8   Underground cellular telecommunications concept

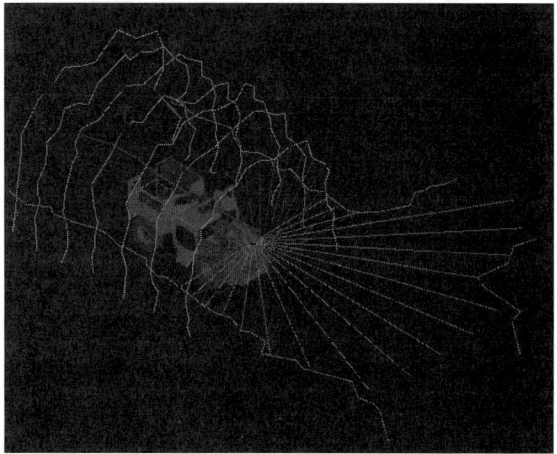

FIGURE 74.9   RLG and test-bed surveying unit

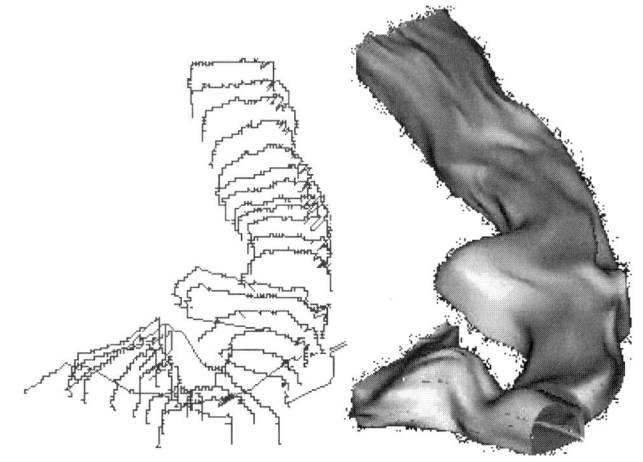

FIGURE 74.10   Software-generated drift from test-bed machine data

machinery in the mine. Accurate positioning systems mounted on mobile equipment enable the application of advanced manufacturing robotics to mining. In advanced manufacturing, robotic equipment is usually fixed to the floor according to very accurate surveying. The positioning systems developed for Telemining™ allow mobile underground equipment to locate itself for mining as if it were fixed robotics for manufacturing.

As a test of these systems, a prototype machine was developed as shown in Figure 74.9. This unit consists of an RLG and a laser scanner mounted on a mobile machine. It is driven along the drift collecting data. The data is then modeled using software to create a "virtual drift." The output from this unit can be incorporated directly into our existing computer-based mine plans as shown in Figure 74.10. At present this unit is capable of surveying a 1-km drift (tunnel) in a few hours as opposed to several days using current work practices.

Mine planning, simulation, and process control systems using the foundations of telecommunications, positioning, and navigation are the next logical step in applying advanced manufacturing systems to mining. Linking engineering directly to operations is essential to the successful application of teleremote mining. To complete the advance to teleremote mining, mine planning systems need to be capable of supplying data directly to the machines doing the work and vice versa. The mine-planning system must be coordinated with the overall strategic mine plan. The output from the system would be integrated into the equivalent of a Manufacturing Resource Planning (MRP) II system. Current practice requires that information be transferred in several steps. For example, a ring layout is currently generated in the mine engineering office, handed to the foreman and then to the driller; the driller assesses the environment for the drilling and attempts to drill the ring as laid out. The on-line information

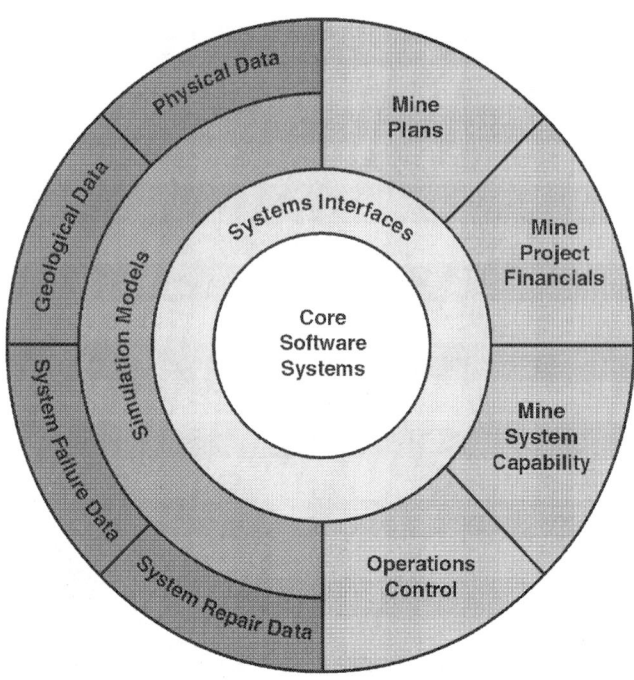

FIGURE 74.11   Software systems for Telemining™

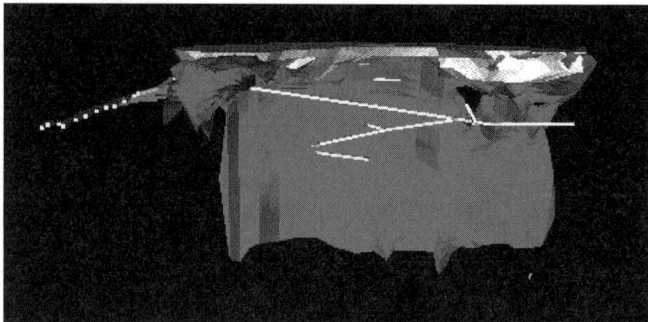

FIGURE 74.12   Research telemine

FIGURE 74.13   Teleremote operation chair

and the actual location of the drill holes are usually not captured. Under the MOS™ currently being developed at Inco, the transfer of information is instantaneous.

The core of the MOS™ consists of spatial databases, network management, and machine control systems as shown in Figure 74.11. The core systems work in real time, linking all the pieces of mining equipment to the engineering systems shown in the outer portion of the figure. Because these systems are directly connected, coordinates for drill layouts may be communicated rapidly to the machines. This system can provide a greater quantity and precision of data (location, rock strength, geometry) that can be integrated into the mine models to improve the efficiency of the mine operation and to allow for more accurate simulation.

74.4   THE RESEARCH MINE

Increased mining speed and lower mining costs significantly enhance the value of an orebody. The major driving factors in increasing the mine value is, therefore, the reduction of cycle time.

The Inco Limited Research Mine is presented as a case study to describe how these systems work. In 1997, the 175 orebody was opened as a research mine to test the concepts of Telemining™. The mine provided an opportunity to test a complete suite of software and mining tools in an operating environment. Our objective was to test the performance all of the systems described above to operationally run the research mine's development and production processes. Planning and simulation are discussed in the Benefits section.

The 175 orebody has a total reserve of approximately 7 million tonnes (Figure 74.12) grading 0.5% nickel and 0.5% copper. The challenge is to use Telemining™ techniques to raise the efficiency of ore production and increase the reserves by mining lower-grade material. Controlling the process with computers improved the quality through more precise control, which reduces costs and enhances the mine reserves because lower-grade material can be mined profitably.

The mining process consists of four components: ore delineation, development, production, and materials handling.

The technologies discussed here are applicable to all four components, each of which will be operated teleremotely. The necessary equipment consists of diamond drills, drifting drills, LHD machines, ground support units, production drilling rigs, explosives-loading machines and trucks, and materials-handling systems, all of which have already been developed to be run remotely.

Operating teleremote equipment is similar for all processes and equipment. An operator station, as shown in Figure 74.13, is connected to the machine via the telecommunications system. This, together with positioning and navigation systems, will allow the operator to run several machines simultaneously and to instantaneously move from machine to machine across multiple mine environments. All the processes described in the following sections will operate in this fashion.

Delineation is a process to establish the grade location and shape of mineralized bodies. The first step in this process is drill sampling, followed by collecting, logging, and assaying samples. Immediate access to information allows the cycle time to be reduced, which in turn improves the efficiency of the ongoing definition of the orebody. Currently, acquiring the information used in delineation takes about 3 months. Monitoring drill performance from the surface may reduce the cycle time to days. In the future, introducing Spatial Information Systems (3-Dimensionsional Geographical Information S), together with near-infrared (NIR) and visible infrared (VIR), may provide assaying as the hole is drilled.

Development cycle times, using conventional equipment such as the two-boom electric/hydraulic jumbo, loading explosives by hand, mucking, and ground support (using bolting and screening) are typically 24 hours for one round (16 ft × 16 ft) at Inco's Ontario Division. Teledevelopment will provide reduced

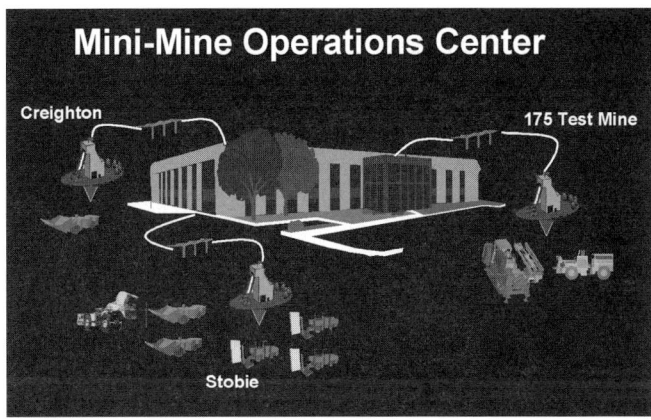

FIGURE 74.14   Mines operation center

FIGURE 74.15   Mine geometry

cycle times and greater flexibility; the "available time" for personnel will be increased by at least 30% due to surface teleoperation, and the ability to run more than one machine simultaneously will increase productivity. We project that teleoperating the jumbo, the explosives loader, and the LHD; using fast-setting ground support coatings; and better synchronizing the development cycle will increase development by 30%. Similarly, the utilization of people and equipment will be enhanced through the surface teleoperation of the machinery and automation. An increase of 30% in the development-rate has the potential to significantly change the profitability of the mine as illustrated in the following section.

Production cycle time has less impact on the mine cash flow than mine development. Production cycle time may be improved significantly with a consistent fragmentation-size profile, which avoids the many problems caused by oversized material. Accurately locating drill holes and consistently loading explosives produce a more consistent profile. Currently, longhole drills, explosive-loaders (typically loading ANFO), and LHDs are used, followed by filling if required. Teleproduction at the research mine will be done with Tamrock Data Solo drills specifically designed for teleoperation, by Dyno Nobel's latest emulsion explosive using microprocessor-based detonators, and LHDs and trucks for material handling.

Software tools, such as MOS™, assist in the simultaneous operation of the mining process and in assimilating, analyzing, and distributing relevant data in real time. The consequent acceleration of the mining rate together with just-in-time techniques that balance mine development and stope inventory, provide further performance enhancement. At Toyota, the KANBAN production-order-processing system is analogous to the MOS™ and has reduced Toyota's cycle time from a week to 46 hours.

74.5   MINE OPERATION CENTER

Several mines may be operated from one mine operating center (MOC). There are several advantages to this: the connection to several mines can be maintained from one location, it is possible to operate several pieces of mining equipment simultaneously, and the equipment can be used to the full potential of the mining method. Each one of these has the potential to significantly reduce costs. In terms of people, "hot changes" are instantaneous because of the telecommunication connection, and work time is maximized as change times and travel times are reduced.

A prototype MOC that connects the Stobie Mine, the Creighton Mine, and the research mine is shown in the picture below (Figure 74.14). As seen in this picture, all the mines are connected to the MOC. At this time, three Tamrock Data Solo drills and five LHDs of various types are working or have worked

from the MOC since its inception. Several other pieces of mining equipment are being readied for this style of operation. An MOC can be expanded to centralize mine engineering and maintenance functions, providing a means of dispatching personnel to the work as needed.

74.6   BENEFITS

Some significant benefits of this style of operation are safety, productivity, and value-added time. Operators spend less time underground, reducing their exposure to underground hazards, and productivity is improved from the current one person per machine to one person per three machines. Initial tests indicate that an LHD can operate for 23 continuous hours during a 24-hour period, which is significantly better than the current 15 hours. Clearly, capital requirements in the latter situation are reduced.

The effectiveness of teleremote mining may be analyzed in the short term using computer-based simulation systems, which are powerful quantification and visualization tools for technology and operations. They have been used effectively in open-pit mining and small-scale transportation surface material handling (Sturgul 1997). The following examples show the impact of the teleoperated mining technology to throughput, mine life, resource utilization, and value generation for the organization.

74.7   SIMULATION MODEL LOGIC

74.7.1   Capturing the Geometry

Ore and waste in the deposit are represented as multiple blocks, each of which is identified and characterized by a number of properties, such as grade and tonnes, which are included in the model. Figure 74.15 is a section showing stope and drift representations.

The representation of the mine plan as three-dimensional model geometry is a lengthy process that involves the following steps:

- Develop a three-dimensional representation, i.e., center lines in AutoCad™ and/or DataMine™ for the drifts and stopes on each level.

- If AutoCad™ is used, import them to DataMine™, then import the level geometry into AutoMod™ as a vehicle guide-path system (at this point we are using AutoMod™ only as a geometry data manager).

- Name each segment in AutoMod™ using a naming convention to incorporate certain drift properties (direction, orebody, level, workplace, and type).

- Repeat for each level.
- Combine the geometry data into one file.

A separate preprocessor program was developed to read the geometry data for drifts and to separately maintain a spreadsheet containing a list of all stopes and their geometry and other properties. The preprocessor program generates data on all the mine openings (connectivity and adjacency data) and writes this information to the textual data files used to run the model. The preprocessor program also generates geometric descriptions that are grafted into the actual AutoMod™ simulation model to allow the mine operation to be animated. As long as the underlying geometry does not change, the complete sequence of steps described above is required only once.

74.7.2 Implementing the Geometry to Represent Different Mining Methods

The ranking and sequencing of the drifts and stopes are modeled to effectively represent the mining methods and resource requirements. Drift scheduling is based on the stope sequence. At the beginning of the model run, every stope and some drifts are initially assigned a ranking number (not necessarily unique) that is specified in the input data set as a sequence number. Rankings are propagated from stopes to drifts using what are called "predecessor relationships" as described below.

The development sequence of the mine is represented as a "tree structure" with each branch representing a stope or a drift. Every working has a predecessor that must be "complete" or "already accessible." All drifts on a given level that are tagged as "infrastructure drifts" in the input data set are collected and made logical predecessors to every stope on the level. Finally, additional logical predecessor relationships defined by the model user, if any, are read in from a data file. All of these relationships together form a single logical predecessor tree.

Before simulated mining starts in the model, the ranking values are copied from branch-to-root throughout the logical predecessor tree, as follows. Each material block calls a function that looks at its immediate predecessors one by one. If it has a higher (more urgent) ranking than the predecessor being examined, it copies its ranking value to the predecessor and calls the function recursively for that predecessor. (Actually, in the model, a lower-valued sequence or ranking number means "more urgent," but we use the term "higher ranking" here to avoid confusion.)

The result of this ranking scheme is that the development resource assignment rule will always choose faces that allow development to progress toward the highest-ranking remaining stopes (or along or toward the highest ranking drifts, if rankings have been attached to drifts in the input data set). For the purpose of this discussion of predecessors and ranking, a raise is treated as a drift segment.

For the test mine undertaken in this study, the simulation of sublevel retreat and blasthole methods requires the following:

- A bottom sill is required for access to the stope.
- Sequencing begins from north and south extremities, mining stopes by retreating to the center. No stope can start until the stope above is completed, therefore ensuring top-down mining.
- Ventilation raises are always developed first.

Figure 74.16 shows the sequencing adopted for a type of blasthole mining

Drift segments and stopes are brought into the simulation model as "Queues." The status of the material blocks is defined as "Waiting" (waiting for development), "Accessible" (development has accessed), "Complete" (all processes are complete), "Blocked" (blocked from work because of other work), "Not

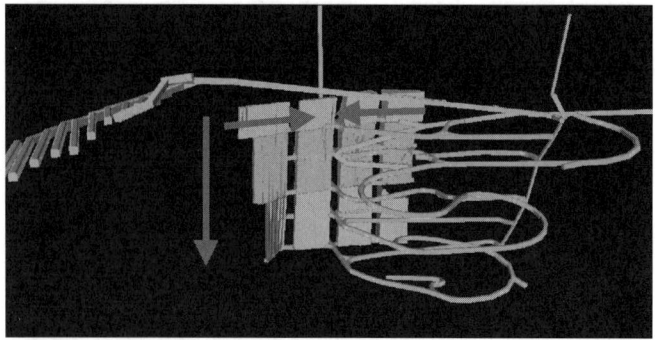

FIGURE 74.16   Sequencing of blasthole mining

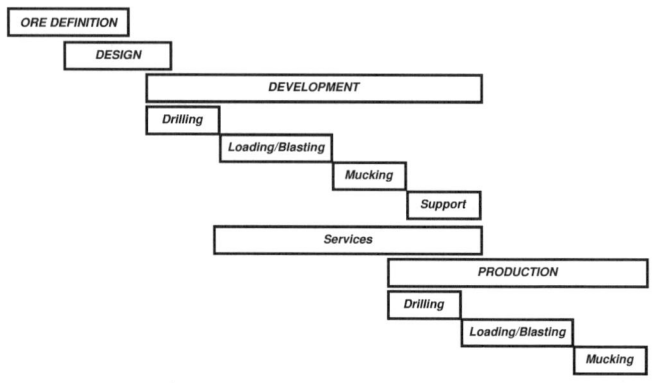

FIGURE 74.17   Process-based modeling of simulation

Economical" (cannot be mined profitably), and/or "Eligible" (eligible but not resources).

74.7.3   Simulation Model Logic

Simulating a mine requires an understanding of how mining processes operate. Ore definition, design, mine development, and mine production are the steps required for mining. Ore definition and design are done and converted into the model as described in previous sections. The development and production processes illustrated in Figure 74.17 are discussed in detail in the next sections

74.7.4   Development Process (DP) Logic

In the earliest simulation model, the development process logic (DP) was modeled in a very simple way. A fixed number of "development resources" were deployed. (For a definition of a simulation "resource" see Schriber and Brunner 1998.). A development resource is a unit containing all of the crew and equipment needed to advance a single face. Any face that is determined to be physically and logically accessible is assigned one of these resources or, if none is available, is added to a waiting list. Whenever a crew resource becomes available, the program scans the list for the highest-ranking drift to reassign the "crew." Once a drift segment has a development resource, the development proceeds. The rate can be adjusted to take into account shift-scheduling inefficiencies but the long-term effect of short-duration downtimes, both scheduled and unscheduled, are not accommodated as specific random failures in the earlier model.

Moreover, in the original DP model, there is a provision for each crew to have a different development rate but there is no way to assign the crews to other parts of the mine or otherwise influence which crew will work in which drift. In the original DP model, there is also no provision for the increased efficiency that results from having a development resource (equipment and people) simultaneously work on two or more faces that are close to one another in the mine.

In later versions of the DP model, development resources are allowed to have up to three faces to work on, and the development rate depends on the number of faces in a nonlinear way (user-specified). In addition, a modification of the model, incorporated in 1996, added a feature such that crews are not restricted to specific areas of the mine.

74.7.5 Process/Resource Input Logic

In 1997, the detailed process/resource was added as follows:

- Detail was added to the original process flow: development was broken into the individual subprocesses of drilling, loading/blasting, mucking, support, and services installation.

- Equipment was defined and declared by type.

- Equipment was certified to run a subprocess of development and a corresponding process time was defined as an input.

- Downtime detail was added so that failed equipment could either be repaired at the face or returned to the shop for repair or service depending on user-defined thresholds.

- Stochastic downtime of equipment was represented by "Time Between Failures," "Time to Repair," and "Waiting for Repair" probability distributions.

- Equipment used for support can also be used for services, or a separate piece of equipment for services is allocated.

- Individual crewmembers are separated out as resources that must be jointly allocated with appropriate specific equipment to perform a task. A development crewmember was allowed to perform different development tasks, such as drilling, loading, mucking, or services.

74.7.6 General Process Logic

For the simulation process to function, both the equipment and crew need to be available to run the process. The user inputs the number of crewmembers per shift. Once a crew and equipment are allocated, they stay within the development area until all the jobs in that area are completed. For a given ore or waste block to be mined, five requirements must be met:

- Equipment must be available (for the next process step).

- Manpower must be available (for the next process step).

- The block must be accessible both physically (no material in the way) and logically (e.g. for a stope, any required level infrastructure must be complete).

- Restrictions on nearby simultaneous activity must be satisfied.

- There must not be higher-priority blocks that meet the other constraints and are competing for the same resources.

74.7.7 Controllers of the Model

There are two levels of management decision in the model: one at the superintendent level, and the other at the foreman level. These management levels are called "controllers" (Inco 1998). The team created several flow charts of the decision process, which were used to determine the controller logic in the model.

As shown in Figure 74.18, a "management controller" was implemented to direct the initial allocation and long-term movement of equipment and crew among development

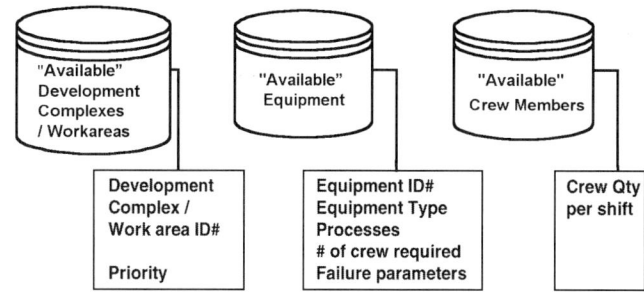

FIGURE 74.18 Elements of management controller

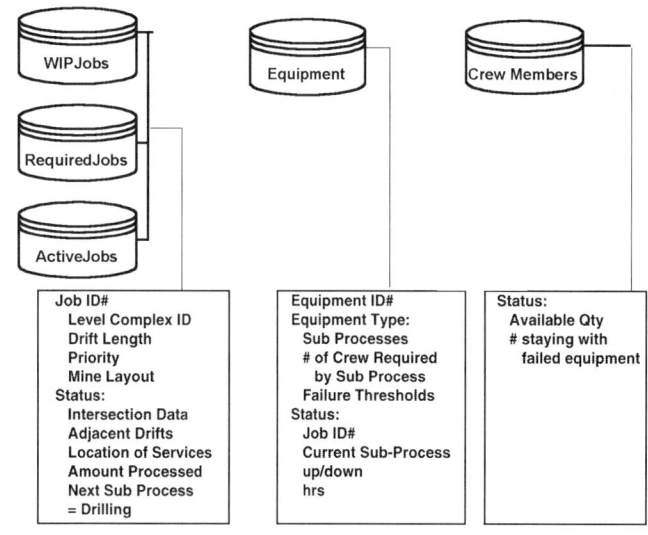

FIGURE 74.19 Elements of the foreman controller

complexes. Currently, the user can define different complexes based on orebody boundaries, and the management controller allocates a set of equipment (as defined in equipment list files) and the number of crewmembers to the appropriate development complex.

Once the equipment and crew is allocated to a development complex, the "foreman controller" takes over (see Figure 74.19). A "foreman controller" makes crew and equipment allocation decisions at the beginning of each shift and also when other state changes occur such as the freeing of a resource, return of equipment from repair, sharing of equipment or completion of a job. Each development complex has its own foreman controller. The foreman controller is active at the beginning of each shift.

The foreman controller checks the list of jobs before allocating resources, and keeps track of what subprocess is performed and what the next process is. The jobs are defined as:

- Work-in-process jobs, started but not completed

- Required jobs, services that need to be advanced before any development job

- Active jobs

In addition to development, the foreman controller is also active in the production process (PP) logic.

74.7.8 Production Process Logic

The PP is similar to the DP described above in that each equipment type is certified to perform a specific operation

TABLE 74.1   Process input and output variables of the model

Input	Output
Operating shift time for development and production	Sub-process cycle time /drift/ stope
Number of crew members per shift	Ore tons throughput
Required times to start a new development process before the end of the shift	Rock tons throughput
Blast schedule for development	Production drill footage/month
Sub-process times/equipment type	Development advance footage/month
Advance and requirement service thresholds	Production start/end, development end days
Equipment downtime thresholds	Equipment utilization
Tons per blast	Crew utilization
Number of equipment per DC/work area	Face utilization

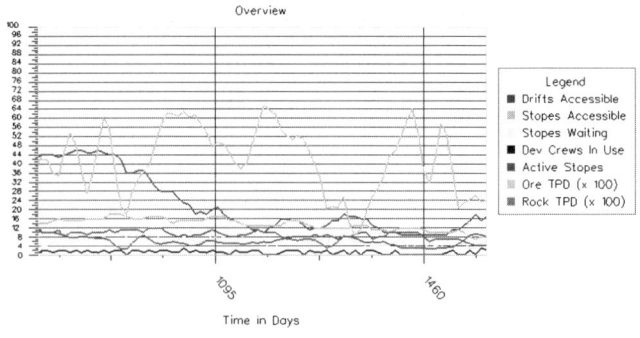

FIGURE 74.20   Sample time plot from the DP model

(drilling, loading, mucking, barricading, and sandfill). Unlike development, production crewmembers are certified to do a specific process; production crewmembers are not modeled separately—they are assumed to be attached to the equipment.

One significant resource allocation logic is to share LHDs between development and production processes using the appropriate mucking rates. This also frees up a development crewmember, increasing crew utilization, as well as LHD utilization over the mine life.

The management controller allocates equipment and crews of a fixed size to work areas. In cases of multiple work areas, each one is allocated separate resources, but if equipment is idle for a certain time specified by the user, the management controller reallocates the equipment to another work area. The foreman controller in PP acts in the same manner as the development foreman controller, assigning already allocated equipment to the production complexes, in the context of work in process and active jobs. The foreman controller is started at the beginning and end of the shift, when the job is completed or equipment needs to be repaired.

74.8   MODEL OUTPUTS AND PLANNING DECISIONS

Data files are controllable by the user, who may modify the three-dimensional geometry and process data. A significant variable is "number of operating hours in a shift." Currently, the model runs with 8-hour shifts. In a conventional mining setting, only 5 hours out of 8 are considered productive due to crew travel time and equipment as well as process delays and level-up keep work. A summary of input and output variables is given in Table 74.1.

74.8.1   Model Outputs

Given the crew and equipment resources and allowing for stochastic breakdown of equipment, one of the significant outputs is to observe mine life and throughput. Throughput, utilization, time/number in queue, and time/number in system were considered important outputs in evaluating manufacturing processes in other industry examples (Chen et al 1998). By evaluating tons produced and footage advanced per day, per week, and per month, decisions on resource allocations, such as varying quantities of equipment and crew per shift, shift schedules, and speed of equipment can be changed to improve throughput and mine life.

Mine life is significant for mine planning decisions. As an output of the model, the number of accessible faces and the number of accessible stopes, both for the whole mine and for each orebody can be generated. In Figure 74.20, the graph provides a quick visual confirmation of the number of accessible stopes over time, the end of development process shown as the

rock tons, and the start and end of the production process, specified with ore tons. The costing of the project as well as financing decisions are based on the duration of the development and production. These constitute input for the cost model.

Other significant outputs of the models are:

- Ore tons per day and per month produced from sills and stopes
- Rock tons produced from drift and raises
- Utilization of equipment and crew resources per shift; equipment utilization is calculated for each piece of equipment.
- Face utilization during the drifting process
- Drift and stope cycle times, cycle time of each subprocess per drift segment and per stope and waiting times associated with each process. This allows evaluation of the impact of equipment breakdowns as well as impact of geometry on process productivity. The model records elapsed times for each process.

The simulation model allows the team to run quick experiments, such as increasing the number of development resources deployed. Some of the typical questions asked are: "Could a predicted dip in production be reduced or eliminated by adding more resources?" or "What would be the impact on the mine life of changing the number or efficiency of development crews?"

Evaluation of Teleoperated Mining. The simulation model is used to evaluate the impact of teleremote operations on mine life and provide the outputs required to make planning decisions. In an operating mine, teleremote operations have been shown to be capable of 7 to 7.5 hours of operation per 8-hour shift as compared to 5 hours in a conventional mine. Other significant differences between conventional and Telemining™ mining are increased flexibility and safety. Blasting and other subprocesses can be done at any time as opposed to the start and end of the shift. Underground personnel are needed to move and/or service the equipment. Once the equipment is in place, it is controlled from the surface, which reduces the need for underground staff and also reduces the risk.

Although, teleremote crew settings are not currently simulated in the model, the ability to use more operating hours and flexibility in process schedules showed a great advantage in reducing mine life, increasing throughput, and maximizing resource utilization. Some comparative graphs are shown in Figure 74.21.

Mine life is reduced by 38% using teleremote versus conventional mining because of the higher mining rate from improved throughput and face utilization. Moreover, utilization of LHD equipment is increased by 80% in teleremote mining compared to conventional settings. With a total of two LHDs,

Conventional Blasthole Mining

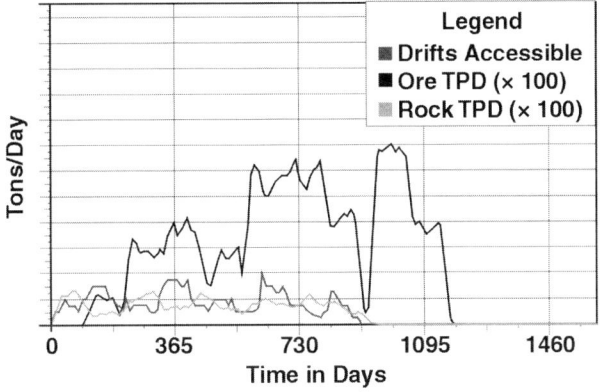

Teleoperated Blasthole Mining

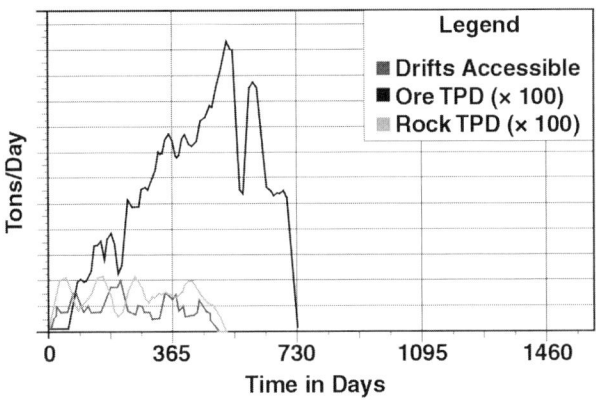

FIGURE 74.21 Conventional and teleremote mine life comparisons

high rates of production were achieved. Simulation also allows optimization to smooth out production peaks.

74.9 MINE MANAGEMENT WITH TELEMINING™

Telemining has the potential to radically change the efficiency of underground mining operations. The MOC will be the center for most operational decisions in engineering, equipment maintenance, and planning. However, it will remain under the control of local management. New management structures will be required that are compatible with the new technology and with a 24 hour-a-day, 7-days-a-week operation. Training will be required at all levels to operate Telemining effectively.

Engineers will need to work directly with on-line engineering and simulation systems linked directly to underground equipment possibly in several separate mines. They will supply on-line information to the machines via the MOS. Teleoperators will work from the MOC in a safe and comfortable environment on the surface and will operate several machines simultaneously. These operators will require a high level of skill to be able to operate this advanced technology and to adapt to frequent changes in operating conditions from mine to mine. It is likely that operators will occupy a more senior role than is customary in current mining practice.

Maintenance will be will be managed by the MOC and will include maintenance of system software; expansion and servicing of telecommunication, positioning, and navigation systems; and

computer hardware and software. The maintenance planning function will be centralized. A work-order system will ensure that the mine is kept in good working order and has sufficient headings and stopes to run effectively.

Selecting personnel to work in this environment will be an important issue. The people selected need to have the technical capabilities to work in a computerized "virtual world," and the agility to function in a high-volume information environment that will require a high degree of concentration and accuracy in the decision-making process. This type of work is similar to that of an air-traffic controller. As operator experience is gained at the MOC, the shift scheduling required in this intense environment remains to be determined to ensure reliable systems operation.

Compensation schemes will be important to an operation using Telemining™. Several new career opportunities will exist, including functions such as Mining Systems Engineers, Mechatronics Specialists, Mining Process Engineers, Teleoperators, and Underground Logistics Specialists. Compensation schemes for these highly skilled workers need to be responsive to the skill and education level, the ability to work in a highly electronic environment, and the ability to work collectively in small, focused teams that could span several mines.

74.10 SUMMARY

Telemining™ offers a huge potential opportunity to lower mining costs through increasing quality and reducing cycle times. Telecommunication core technologies have been developed, tested, and are working in four of Inco's operating mines. Positioning and navigation systems have been developed and implemented on a test vehicle and are showing promising results. MOS™ concepts have been developed, and initial prototypes have been demonstrated. Mining equipment has either been developed, piloted, is in production, or is under development within Mines Research or as part of the MAP consortium.

Simulation tools have been developed to estimate the benefits of telemining systems as compared to conventional mining techniques. The Inco team has used the model to compare conventional, futuristic mine design and technology alternatives with various combinations of scenarios (plans, methods, equipment types, and staffing strategies). The outputs of the simulation show substantial benefits to implementing Telemining™ mining techniques to mining operations.

74.11 ACKNOWLEDGMENTS

The author would like to thank the members of the project teams from all companies and organizations for their hard work in establishing the Mining Automation Program. Particular thanks go the Senior Management of Inco Limited, the Mines Research Staff, especially Dr. Hulya Yazici and the personnel at Stobie Mine and Creighton Mine for their belief and vision in the development of this project. The members of the management committee of MAP, Seppo Seppala, and the MAP Project Managers deserve particular credit for their efforts.

74.12 REFERENCES

Chen, S., L. Chen, and L. Lin. May 1998. "A Simulation and Knowledge-Based System for the Improvement of Manufacturing Cell Performance," in *IERC-98*, Alberta, Canada.

Inco Limited. 1998. DP Model Documentation (unpublished).

Schriber, T.J., and D.T. Brunner. "Inside Simulation Software: How It Works and Why It Matters," *Proceedings of the 1998 Winter Simulation Conference*, p. 77–85.

Sturgul, J. 1997. "History of Discrete Mine Simulation," *Proceedings of the First International Symposium on Mine Simulation Via the Internet.* G.N. Panagiotou and J.R. Sturgul, Eds., p. 27.

. . . .
CHAPTER 75

Mine Automation at LKAB, Kiruna, Sweden

Jan-Olov Nilsson,* Irving Wigdèn,* and Håkan Tyni*

75.1 INTRODUCTION

LKAB celebrated its 100th birthday on December 18, 1990. In spite of its years, LKAB must remain fresh, vigorous, and energetic in its outlook, ideas, and actions to profitably extract iron ore from ever-increasing depths in an extremely competitive iron ore market when using underground mining methods. LKAB must, quite simply, be on the cutting edge of mining technology. To accomplish this, the company invests between 1% and 2% of its annual turnover on research and development (R&D). A great deal has been written about LKAB over the past several years and in this short paper we do not intend to repeat what can easily be found elsewhere (see the references). The primary concentration here will be on the direction the company is taking with respect to automation.

75.2 SOME IDEAS REGARDING TECHNOLOGY LEVEL

75.2.1 Productivity Improvements and Technological Changes

Prior to discussing the details of automation at LKAB, a few words will be spent describing the process of technological development. Figure 75.1 shows an idealized set of technology development versus time curves that can be applied to mining operations.

A mine may spend its entire life in one stage or it can progress through a series of stages. Within each stage there is a region in which improvements are gained quite rapidly followed by a slower growth phase. Eventually a plateau is reached. Depending upon where an individual mine is in the development picture, different opportunities for improvements are presented. In some cases, the discontinuous jumps between surfaces may be accomplished relatively easily; for example, through the purchase of available technology. This may, however, be accompanied by some difficult side effects such as:

- New mining system development
- Rationalization of work procedures
- Reduction in work force
- Integration of new technology with older technology

On the other hand, some entirely new technology (machines, etc) may have to be developed to accomplish the jump. In general, the height of these jumps becomes smaller with each increasing step, the time required for the next step becomes successively longer, and the effort required (measured in time and/or money) to achieve the step increases. When viewed from a distance, the development curve may appear as the overall curve in Figure 75.1, where now the individual steps and plateaus are not seen. When the curves (both for the individual

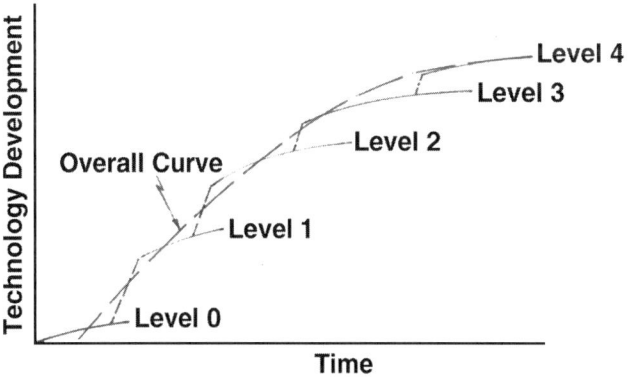

FIGURE 75.1   An idealized set of technology development versus time curves applied to mining operations

technological steps and the overall) begin to plane out, the process is said to be "mature." This does not mean that new advances cannot occur, but greater efforts are required.

In summary, when studying the evolution of productivity one finds that improvements are generated by two kinds of actions:

- Introduction of new methods or equipment allowing important progress; a breakthrough of productivity is thereby achieved.

- Continuous R&D to improve specific points; the resulting progress is steady but instantaneously weak.

The continual improvements, while extremely necessary, are only enough to account for things such as increased haulage distances, higher pumping costs, greater ventilation demands, and additional reinforcement associated with increasing mining depth. The real productivity changes happen only through the introduction of new technology.

75.2.2 The Time to Introduce a New Technology Level

There are two different connotations implied by the word *time* as used in the title of this section. The first is determining the "right time" to introduce the new technology. The second is the "amount" or "length" of time required to successfully introduce and implement the technology. A strategy covering both the development of such technology and how and when it should be implemented is critical. Much can be learned in this respect from examining new product development curves such as shown in Figure 75.2.

* LKAB, Kiruna, Sweden.

681

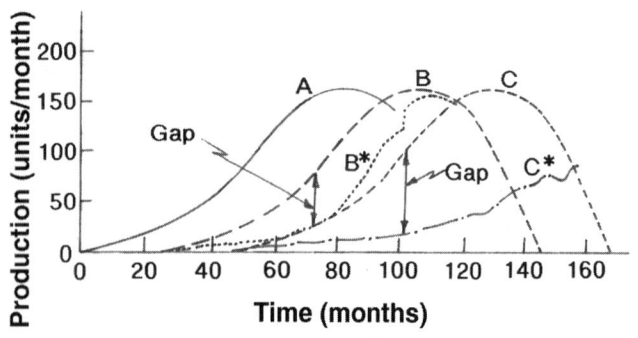

FIGURE 75.2   New product development curves

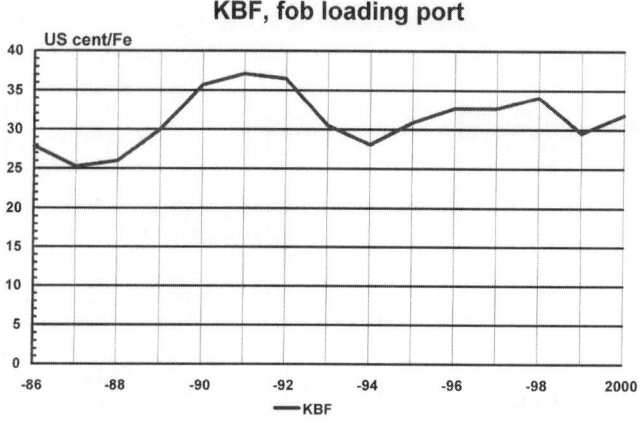

FIGURE 75.3   Price development for low phosphorus fines

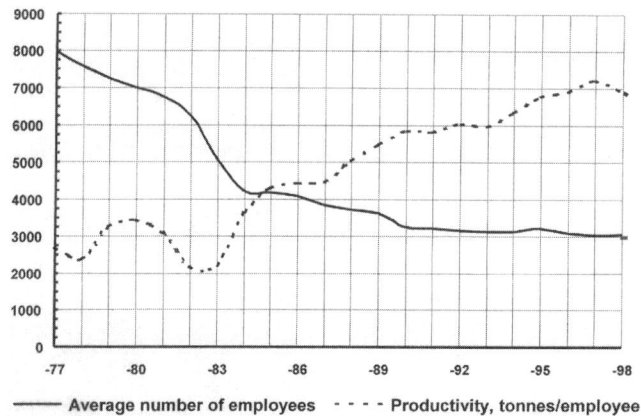

FIGURE 75.4   Employees and productivity at LKAB

Curves A, B, and C represent the ideal development chain for new products of a company so that the sales remain at a given level. Product C is to replace Product B, which replaces Product A. Curve B* represents the actual development path for Product B. Poor planning and execution has resulted in late development. This has affected sales performance in this stage and has affected even more severely the development of Product C. For a mine, the units per month shown in Figure 75.2 can be likened to tons per year or product per year produced. Failure to incorporate the appropriate improvements in the proper time frame can have severe competitive consequences. On the other hand, introducing an immature technology too early or introducing a mature technology into an immature organization can be disastrous to a mining company.

For those attempting to push the cutting edge of mining technology, there are few, if any, off-the-shelf products to accomplish the task at hand available from some other industry or application. There is no product because there has been no demand and vice versa. Something must occur to break this chain. For the suppliers, the potential market demand for these cutting-edge products must be carefully assessed. If the demand is judged small, the development costs high, and the possibility of failure or at least partial failure high, then they are probably not willing to take the risk—at least not alone. In addition, given the often-severe conditions under which mining systems are expected to function, high demands are placed on the ruggedness and reliability of the machines. Having as broad a cutting edge as possible is important for all concerned so the development risks and expenses can be shared and so the manufacturers can hope for a larger potential market in the end. Therefore, there is a pressing need for those working at the highest technological levels to seek forums for cooperation and standardization. Experience to date has shown that this is not all that easy. Some of the ideas for future mining development already discussed within LKAB (Sandberg, 1987) were:

- Increase mining dimensions with a resulting reduction in specific development
- Time utilization will be better with more shifts/week for better capital usage
- Automation will be more frequent
- Mobile electric machines will replace many diesel machines
- Long drifting rounds
- Computer-controlled drill jumbos
- Production drilling will probably be completely computerized
- Straight-hole drilling

- Communication technology facilitating production planning, control, and follow-up
- Maintenance systems with advanced diagnostics for greater productivity, machine availability, and economy

Although most of these have now been, or are on the way to being, implemented, some are still waiting in the wings.

75.2.3   The Application of These Concepts to LKAB

Figure 75.3 shows the price development for low phosphorus fines over the past 15 years. Although there has been some cyclic behavior that is typical of metals prices, the overall trend has been flat. Thus, the income per Fe unit has remained essentially constant. Over this same period, the labor costs associated with any given employee, for example, have steadily increased.

The discrepancy between constant price and increasing costs must be overcome by increasing labor productivity. Figure 75.4 shows the total employment and the derived productivity versus time curves for LKAB over the same time period. First, it can be seen that the total number of employees has decreased significantly over this period. Most recently, the number has remained relatively constant at about 3,200. The productivity curve follows the theoretical curve presented in Figure 75.1. It consists of a series of steps followed by periods of slow but steady growth. Looking back over the time period, it is difficult to couple a specific step in the curve with a specific technological advance because there are activities happening on many fronts (in the mine, the plant, service activities) at the same time. From the mining side, a major contributor to productivity improvement has been the increase in mining scale from the 12-m sublevel intervals predominant in 1985 to today's 27-m interval. How do we envision the future mine design and extraction process at LKAB? Does the next technological level involve a further increase in scale? The answer is probably not. A modest increase in sublevel spacing from 27 m to 28.5 m, with a corresponding change in other dimensions, is being considered. Therefore, if one cannot expect a productivity increase through increasing scale, and we already have a highly mechanized operation, then where will the new step come from? Is our near-term, or even long-term, vision to turn Kiirunavaara into a completely automated mine? The answer is no. We believe that it lies in a prudent introduction of automation, or more probably, semi-automation in appropriate unit operations.

75.3 MINE AUTOMATION REALITIES AND PLANS

75.3.1 Introduction

LKAB has a long history of mine automation. The present 1,045-m haulage level is the third level with automatic driverless trains and remote tapping of the orepasses from a central control room. In this application of automation, LKAB has more than 20 years of experience. LKAB has been at the forefront in other areas of mine automation as well. More than 10 years ago, at the end of the 1980s, LKAB, in cooperation with Tamrock, conducted full-scale tests with an LHD equipped with a guidance system, four video cameras, and a communications system. During that test period, nearly one million tons of iron ore were carried by the LHD to the orepasses. The production method used during the test was sublevel stoping and not the present sublevel caving method. Using sublevel stoping, an LHD is loading in the same production area for a period of 6 months after every blast. With sublevel caving, the life of a production block is only about 18 months. One main conclusion from this test was that it would not be cost-effective to implement a technology heavily based on infrastructure. The communications and navigation methods used during the test did not fulfil this requirement.

The test provided input for economic simulations. The simulations that LKAB made were very conservative from the cost (high) point of view as well. There is certainly a reduction in the workforce when one operator can control up to three or four LHDs and, of course, there is an improvement in the working conditions. The main conclusion was that there are profits to be made with LHD automation, especially profits from better machine utilisation. The automated LHD doesn't stop for lunch and keeps on working the entire night. An important parameter is also the possibility of reducing the operating costs. With regard to communications, a conclusion from the project was that a digital video solution is preferable to an analogue solution because of specific conditions in the mining environment, such as high propagation attenuation and strong multi-path propagation. Although the test results were encouraging, they did not fulfil the goals set up for production and availability. LKAB decided to wait for better solutions to appear on the market.

In 1995, LKAB began a new LHD-automation project (SALT 3), with the goal being to develop a system for all of the production loading in Kiruna. The general characteristics of that system, broken down by major subareas can be described as follows:

1. *Traffic Control System*. This is an overall system that performs all entry and parameter determinations. The system can be used for traffic management, and also permits the LHDs to be remotely controlled.

2. *Communications System*. This is used for transferring information between the LHDs and the Traffic Control System.

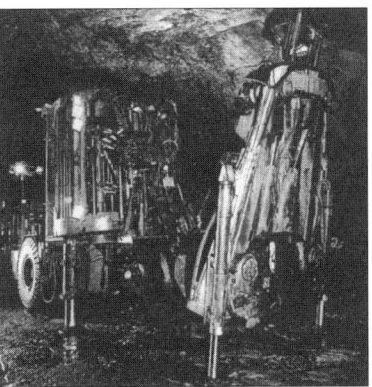

FIGURE 75.5   A Simba 469W rig is remote controlled from the Control Center on the 775-m level and is full-ring automated

3. *LHD System*. This system handles the autonomous and remotely controlled state.

On the basis of the experience from the different SALT projects, LKAB went forward with LHD automation. The results are discussed in Section 3.4 and Section 5.

75.3.2 Development Drifting

The need for development drifting per year to satisfy a crude-ore production of 26 Mt in the new KUJ 2000 system in Kiruna totals 20,300 m. Approximately half of the total meters are drifts made in the footwall rock. This includes access to orepasses and footwall drifts. The drifts are generally 7.0 m × 5.0 m (width × height). For all drifting, conventional drill and blast methods are used. Although it is possible to drill out a round in automatic mode, because of the short times required and the assurance provided should something go wrong, the machines will be manned.

75.3.3 Long-Hole Production Drilling

A decision was taken in the early 1990s to scale up the sublevel caving system to the 28–30-m sublevel interval range with corresponding maximum hole lengths of up to 50 m. Experience showed very clearly that the performance of the existing drill rigs could not meet the hole-deviation specification. Even though considerable efforts had been made over the years to improve the drilling quality, it wasn't enough for the longer holes.

The first of the new generation of drilling rigs, the Simba 469W, equipped with one water-powered Wassara in-the-hole hammer, was put into operation in March 1995. Today there are five Atlas Copco Simba 469W rigs and two Tamrock DataSolo 1069 rigs. The Simba 469W rigs (Figure 75.5) are remote controlled from the Control Center on the 775-m level and are full-ring automated. After drilling a full ring of 8–10 holes in automatic mode, the operators must manually move the rig to a new setup and alignment.

Operating drilling rigs requires different solutions than operating LHDs. One obvious difference is that, while an LHD is constantly moving, a drilling rig only moves every 10th hour when a drilling ring is completed. The drilling rig's lack of mobility caused disturbances that hadn't been predicted. In some parts of a drift, for example, the picture can be worse than it is just three meters further on. When the drill is moved, the picture gets better. This phenomenon is probably caused by parts of the tunnel having more multipath propagation and the fact that the rig is standing still. With an LHD, this part of the drift just causes some bit errors when the LHD drives ahead. The multipath problem that disturbs high-speed links was solved with diversity receivers in both the Base Station (BS) and the Mobile Terminal (MT).

FIGURE 75.6   An electrically powered 25-ton-scoop-capacity Toro 2500E with conventional drive

FIGURE 75.7   Full production requires six trains operating at the same time carrying 500 tonnes of ore per train

Illuminating the drilling rig is another area that needs tuning. Early tests indicate that illumination is an important parameter for good picture quality. However, the mining environment—with vibrations, water, and falling rocks—is a hard one for ordinary lamps. Another important area to study is the dynamics of the camera, which can improve the picture quality. Other important issues are the operation and maintenance of the network itself. This experience and knowledge is, in some cases, new to mining operations.

75.3.4   Loading

The concentration of development drifting, using longer rounds, and the good machine mobility achieved by using diesel generators, makes it possible to do almost 100% of the loading by electric machines. The remaining part—for example, loading in declines—is done by a contractor. When the 7-m-wide drifts came into use, the need for bigger machines to increase loading productivity became clear. In 1994, Tamrock Loaders started to develop a bigger machine to meet LKAB's needs and specifications. This is an electrically powered 25-ton-scoop-capacity machine with conventional drive. The demands for availability and MTTF (Mean Time To Failure) for the Toro 2500E (Figure 75.6) were set high with the expectation of using this machine in the fully automatic mode.

Today, the production loading fleet is made up of 9 Toro 2500Es, each with a capacity of 25 tonnes; 1 Toro 650D (diesel), with a capacity of 16 tonnes; and 11 Toro 500Es with a capacity of 14 tonnes.

Since March 1999, loader automation is a reality at Kiruna. Today, four of the nine Toro 2500Es are fully automated and are being operated from the Control Center on the 775-m level. The only operation not carried out by remote control is filling the bucket at the drawpoint. Once this operation is completed, the automated system takes control, guides the LHD to the orepass, dumps the ore, and returns it to the front to fill a new bucket. The details of this system are described in Section 75.3.5.

75.3.5   Automation in Transportation and Hoisting

The haulage and hoisting system is designed for 26 Mt of hoisted crude ore per annum, which corresponds to a daily production of 75,000 tonnes. This system, which provides a continuous supply of ore to the processing plants located aboveground, operates 24 hours a day.

The haulage level at 1,045 m is a "shuttle train system." The trains transport crude ore from eight groups (ten groups in the

FIGURE 75.8   The consoles for controlling the hoist system are located in the Central Control Room

near future) of orepasses, with four shafts in each group, distributed along the length of the approximately 4-km-long orebody. Full production requires six trains (Figure 75.7) operating at the same time (each with 24 cars) carrying 500 tonnes of ore per train.

Four dumping stations with associated crushers and skip-loading stations have been built at the central hoisting area on the 1,045-m level. Hoisting to the ore-processing plants, situated some 800 m above the main level, is done in two stages. Using the new hoisting plant consisting of four internal blind shafts, the ore is hoisted to the existing crusher bins located below the 775-m level. From there the ore is skipped to the surface through six shafts. Haulage on the new 1,045-m main level is computer-controlled and automated, just as on the 775-m level. Loading the trains is done as described earlier by remote control from consoles located in the existing control room. The consoles for controlling the hoist system have also been moved to the Central Control Room (Figure 75.8) on the 775-m level. The hoisting from both levels is fully automated.

75.3.6 Other Unit Operations

LKAB's total mining operation is very highly mechanized. As indicated, the development drilling is done with personnel on board the rigs. The same is true for the rigs used for scaling operations, installing rock reinforcement, shotcreting, and charging blastholes, both for drifting and production. Currently, there are no plans to try to automate these operations.

75.4 COMMUNICATIONS 2000

75.4.1 Introduction

Information technology (IT) is being used more and more in the mining environment. Today, electrical-power, water-pumping, and ventilation systems are all being controlled with different IT implementations. The Communications 2000 (COM2000) project was originally formed around requirements derived from automation and technology-driven trends. The basic idea behind the project was to create a multipurpose infrastructure to carry different types of digital information: data, audio, and video. The project was also divided in two different parts: (1) a radio frequency (RF) part and (2) a cable-based network infrastructure. An early decision was made to find a solution to the LHD RF communications. We thought that if a RF solution would fit the LHD requirements, it would also fit the requirements for the drilling rigs. Another early decision was to implement an ATM Network. In short, the main purpose of the COM2000 project was to introduce automatic operation into two key parts of the mining process, namely the LHD and the drilling rig operations. The COM2000 design parameters came from:

- SALT II
- The mining method
- The mine architecture
- The mine operation

The basic requirements passed on from the SALT II project were:

- Small latencies 100–150 ms are needed.
- No (or little) infrastructure to keep running costs low
- High availability (99.8%)

Earlier experience indicated that the 900-MHz frequency band was one of the best propagating frequencies for mine RF communications. This was also confirmed via channel measurements that where carried out.

The other design parameters considered by the COM2000 project have come from (1) the mine layout itself and (2) the optimum number of vehicles needed to obtain the planned production results. The length of the iron orebody is approximately 4 km. The orebody will be divided into ten production blocks (today there are eight). LKAB intends to have three levels "in working mode" in each production block at any given time. One level is for drilling and two levels are for loading. This gives a total of 30 production blocks in use at the same time. The sinking rate with today's production is about 24.75 m/year, which means that a new level is taken into production each year. The distance from the control room on the 775-m level to the production blocks is approximately 1.5–5 km. A maximum of three units will be used in a production block: two LHD's and one drilling rig or vice versa. LKAB will have a total of 11 LHDs (Tamrock 2500) in production and is planning on using eight drilling rigs, six Atlas Copco Simba 469W rigs, and two Tamrock DataSolo 1060 rigs.

The overall objectives of the project were:

- To transfer video, voice, and data from mobile remote machines in the mine to operators located in a control room

- To ensure interconnectivity and interoperability through the use of a widely accepted standard
- To support different kinds of traffic such as data, video, and voice within the same network
- To create an infrastructure that is easy to expand and configure, and easy to operate, maintain, and support

Associated with the introduction of certain automation elements are the side benefits associated with the presence of the necessary communication links throughout the mine. These lead to improvements in other nonautomated actions through better information collection, analysis and processing, and presentation so that better decisions can be made in real time. In the short term, these spin-offs may be at least as valuable, possibly even more so, than those elements directly associated with the changes to automated or semiautomated processes.

To fully understand the final decisions made regarding the mine communications system, it is necessary to discuss the different basic technical requirements regarding latencies, picture quality, RF method, and the communications architecture that (1) were formulated in the early stages of the project and (2) resulted from the experience that the project acquired during the different stages.

75.4.2 Latencies

Latency is defined as the time from which a command is performed until it appears on the operator's monitor. Instead of transmitting the analogue video signal directly on the air, digital image processing is used for mobile video communications to reduce bandwidth. The analogue video signal bandwidth uses approximately 5 MHz of bandwidth on a link. Digitizing, filtering, and compressing the image can reduce the information bandwidth to 10% of the original signal bandwidth, although it is questionable if that amount of compression provides a satisfactory image quality. The backside of digital image processing is that it introduces latencies into the system. Teleoperation sets special requirements for the latency over the communications link. From various discussions, a latency of between 100 and 150 ms was defined as tolerable. The main reason for the latency is the time to digitize the video picture. These values were also set with consideration to additional latencies caused by, for example, the hydraulic system of the vehicle. One problem was that there was not that much experience as to how a latency of 200 or 300 ms would affect the operation of a teleoperated vehicle. Driving loaders with a control stick and a video monitor may, however, require some training even with a short latency. Later it turned out that the calculated latency for the Simba 469W drilling rig with a cabled analogue video solution and separate Profibus communication link from the operator to the rig was nearly 230 ms. This latency is still tolerable for the operator. In a test performed with an ordinary wireless Ethernet LAN (Wavelan) and a Bitfield Video Card, the latency was around 500 ms. This is not acceptable. The solution is to consider the whole operation cycle in order to minimize latencies. The design of control panels, etc., should be a part of the design process. This means that different suppliers have to cooperate to create the optimal solution. An example of such cooperation could be to design the operator interface in an automated environment.

75.4.3 Picture Quality

The quality of the picture is defined by a set of parameters:

- Frame speed
- Available bandwidth
- Picture coding algorithm
- Size of monitor

There are a several image compression algorithms used in image transmission. In an early stage the following algorithms were considered:

- H.261/H.263
- MPEG-1/MPEG-2
- Wavelet

Some of the algorithms (H.261/H.263 and MPEG-1) were developed for applications such as video conferencing and are based on block DCT encoding. An image frame is divided into small blocks, which are then encoded. After encoding, the results are compared to the previous frame. Most of the time, only the changes in blocks between successive frames are sent and the required bandwidth is reduced drastically. There are also other, even more effective, compression methods that are not based on block coding. However, in case of bit errors the block-encoded images are only partially disturbed while the other algorithms may lose entire frames. Wavelet coding is one example of this coding method. On the other hand, broadcasting image quality could be obtained by extremely complicated algorithms such as MPEG-2, but the hardware required to run this kind of compression algorithm is too expensive and complicated. MPEG-2 also requires a lot of bandwidth (8 Mbit/s). Experiments conducted during 1996 indicated that the use of the H.261 algorithm with a resolution of 352 by 288 pixels was good enough for the teleoperation of mining machines.

75.4.4   RF Method

Another early decision in the COM2000 project was to use spread-spectrum technology for RF links. The spread-spectrum technology offers a powerful way of communicating over the radio channels present in a harsh mining environment. The technology is a widely used broadband digital communications technology (radio LANs, satellite communications) and was originally developed for the military in the 1960s. The spread-spectrum radio offers several advantages over a conventional narrow-band radio system including:

- Excellent performance in multipath radio channels, such as in mines
- Robustness and immunity to interfering signals
- Security due to pseudonoise codes
- Lower signal power density generating less interference to other radio systems

75.4.5   The Communications Architecture

Automation and teleoperation set new requirements for a mining communications infrastructure. A conventional leaky feeder-based communications system is not capable of providing the data transfer capacity required for teleoperation. The basic requirement for a communications system is the ability to simultaneously carry video, audio, and data (condition monitoring, production information, statistical information, teleoperation data) between production areas in the mine. The communications system should also, at some point, interface with the company's local area network (LAN). As we stated earlier, the early objectives of the communications project were to create a multipurpose communications network based on ATM. However, ATM is a complicated communications technology. Another disadvantage with ATM was the lack of standards in certain areas such as traffic control. Nevertheless, the decision was made to go with ATM. Other requirements defined from various needs in the mine were:

- The ability to access maintenance systems and the company LAN from the mine

- The need to connect other types of systems and devices to the network. For example, PLCs and ore pass level monitoring

The next consideration was to design the architecture of the network. There are three possible ways of designing the ATM infrastructure.

1. A dedicated automation network
2. A dedicated network with tunnelling functions (e.g., Ethernet)
3. A general purpose network

Based on the installed links in the Kiruna mine, the following conclusions can be made in comparison between analogue and digital video:

- Clear, disturbance-free image over the entire communications range
- No problems with reflections (no ghosting)
- Noise-free image (no "snowfall")
- No synchronisation problems caused by strong multipath propagation

The final decision was based on the assumption that it is easier to keep high availability with a dedicated automation network than with a general-purpose network. From the original idea of creating a multipurpose network for audio, data, and video, in the end, the project implemented two separate networks for the Kiruna mine: an automation network based on ATM, and a switched Ethernet network with an ATM backbone. Another one was already implemented on the haulage level.

75.4.6   The Communications System Selected

LKAB chose the Finnish company Elektrobit Oy to develop and deliver the RF communications solution. The name of the system is the Wireless Underground Communication System (WUCS).

The WUCS transfers data, video, and audio in digital format between the systems in the Control Center on Level 775 m and the systems on the LHDs. It is even used for communication between the Control Center and the remotely controlled production drilling machines. The WUCS consists of mobile terminals (MT) installed on the vehicles, base stations (BS) installed at fixed locations in the production areas, and control stations (CS) located in the Control Center. The BS and the CS communicate via an ATM network basically using private virtual circuits (PVC's) and multicasting. The connection between the MT and the BS is provided by a wireless spread-spectrum link at 900 MHz. One BS can communicate with three MTs simultaneously without interference. The most significant technologies utilized in the WUCS are:

- Spread-spectrum radio—a digital image processing and image compression (H.261)
- Advanced high-speed data switching (ATM)

The WUCS provides multiple transparent data interfaces such as RS-232, RS-485, and Profibus between the operators and the underground machinery. In addition to data, the WUCS provides interfaces for video and audio transfer.

An overview of the WUCS system (Figure 75.9) is the following:

- The CS communicates with the BS via a fiber-optic cable. This is an internal WUCS network.
- The BS communicates via a radio connection (wireless) with the MT. The video signal is compressed in the MT before the video data is transferred to the BS, which, in turn, decompresses the video data before it is transferred further to the CS.

Optofiber

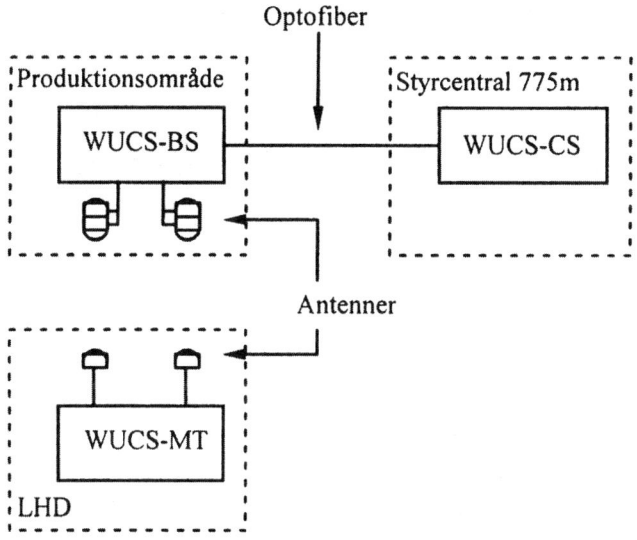

FIGURE 75.9 An overview of the WUCS system

- Antennas connected to the BS are positioned along the drift roof. Antennae on the LHD are connected to the MT.

The CS has three ports and the MT has one. A port consists of an audio channel, a video channel, and three series channels for data communication. The WUCS can transfer 1 Mb/second per port (audio, video, three data channels) from MT to CS. The transfer rate in the opposite direction (from CS to MT) is significantly lower because video data are not sent this way. The largest part of the transfer capacity from MT to CS is reserved or taken up by the digital video information. Some of the technical data for the system are included below:

Radio Link

Transfer technique:	spread spectrum
Modulation, data:	DQPSK (BQPSK)
Modulation, spreading:	BPSK
Radio path symbol rate:	512 ksym/s
Spreading method:	Direct sequence
Chip rate:	5.632 Mchips/s
Frequency band:	800-1,000 Mhz
Transmission power, attn:	5W

Video/Audio

Video system:	PAL/NTSC
Compression algorithm:	H.261/MPEG-1
Video data speed:	Max 935 kb/s
Bandwidth, audio:	300–4,000 Hz
Frame rate:	25/30 Frames/s
Picture resolution:	352 by 288 (FCIF)

Delays

Video:	150 ms max
Data:	10 ms average

User Interface

Windows-NT PC

To date, the operation of the remote mining equipment using RF and digital video has been proven feasible, although some important issues have to be further defined:

- Standards for communication equipment interfaces
- Cooperation between mining automation suppliers and communication technology suppliers to optimize solutions and operation

- Introducing new technology requires new skills to operate and maintain the technical solutions
- New communications technology offers new opportunities for mining suppliers. Using the Internet and automation networks, suppliers can provide real-time maintenance functions to rolling stock

75.5 SEMI-AUTOMATED LOADING AND HAULING (SALT4)

Currently, SALT4 is the program that controls the electrically powered loading machines (LHDs) in the production areas of the mine. When completed, it will consist of six operator places and provide control for up to 15 LHDs and 20 production areas. A production area is about 400 m long and consists of a number of crosscuts on a certain level. The crosscuts are connected to a longitudinal footwall drift. From each production area, there are one or two connections to the ramp system. Within the production area, the ore is transported from the loading places to one of the orepasses, belonging to an orepass group, by an LHD. An orepass group can consist of up to four orepasses, which lead down to the transport level on level 1045 m. There, driverless trains transport the ore to crushers for further transport to the surface via skips.

The loading cycle consists of the following:

- Filling the bucket
- Transporting the loaded bucket to the orepass
- Emptying the bucket into the orepass
- Returning the empty bucket back to the loading place

Via the operator's panel in the Control Center, the operator can control all of the LHD's movements to fill the buckets by remote control. To assist the operator, there are video pictures from the cameras onboard the LHD. The operator's panel is equipped with controls corresponding to those in the cab of the LHD.

Transporting the bucket from the loading place to the orepass and back again, together with emptying the bucket into the orepass, is done automatically. The maneuvering is carried out by the computers and the navigation system onboard the LHD.

A central computer in the Control Center gives the commands to the LHD (choice of route from the loading place to the orepass and back) and also handles traffic control in the case of two LHDs in the same production area.

Whereas the focus is often upon the hardware and software changes directly associated with the item to be automated—in this case, the LHD—there many other system changes that occur as well. Today, general access to a production area is available almost anytime by simply making contact with the loader operator. For the automated loading operation, no one is allowed in the area when the loading machines are operating in either the remote or the automatic mode. Support operations (production hole cleaning; road maintenance; boulder drilling, blasting, and breaking; repairing and installing ventilation tubes and other media; hole charging and preparation for firing; installing reinforcement; and scaling backs and brows associated with production) must all be scheduled and carried out in a completely different way than is being done today. This new way of operating does not occur by chance, and the total system is being thoroughly studied and planned.

In the automated mode, the production in an area occurs in cycles. Currently, a production cycle means working in an area for eight days, followed by a two-day stop for preventive maintenance. Maintenance includes charging the rings, road maintenance, etc. The maintenance on the LHD (oil and filter changes) will require less time, and the machine can be moved to another production block.

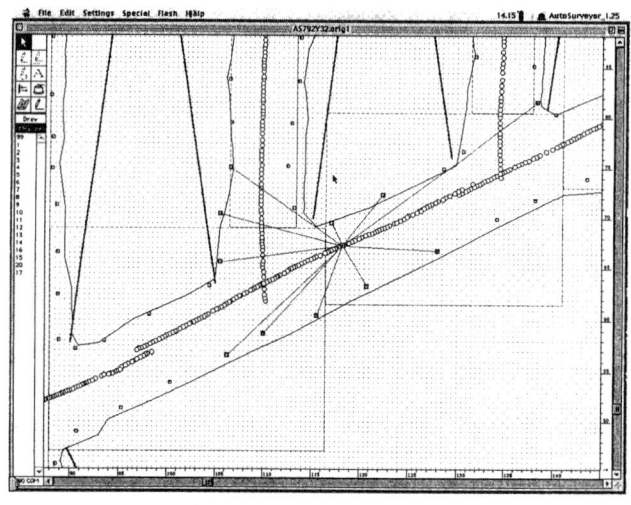

FIGURE 75.10   The LDS is used mainly for the design and definition of the production areas

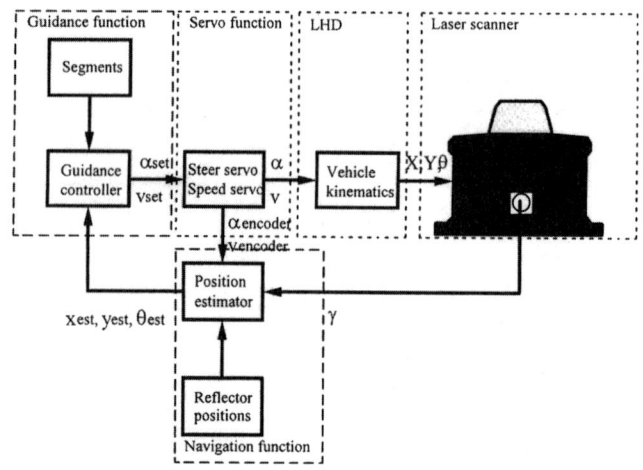

FIGURE 75.11   The HUNS has replaced the driver in the cab of the LHD

FIGURE 75.12   The HUNS laser scanner

Shorter production stops (1–4 hours) are planned during the production day. During these stops, other maintenance, such as cleaning the production holes, is carried out.

Personnel safety is very important when operating the LHDs in automatic mode. There are a number of techniques and procedures in place to assure the safety of personnel; only a few will be mentioned here. One requirement is that the LHD be prevented from moving either under automatic or remote control when someone is in the production area. Gates isolate the production area where an unmanned loading machine is operating. When an unmanned machine is operating, all access to the area is forbidden. If a gate is opened incorrectly, the LHDs stop because the power to the area is shut off. The access to each orepass can be closed to prevent the LHD from entering and dumping its load, which would be done in situations where work on the orepass was being done on underlying levels. A wire connected to the personnel safety system will, if broken, shut off all power to the area.

The central control system (designated LUCS for LHD Underground Control System) is located in the Control Center on level 775 m. It handles the following tasks:

- All communication with the personnel-safety system
- All order-giving to the LHDs in automatic mode
- All traffic control in an area where two LHDs are working
- All connections between the operator and the LHD for the remote-controlled bucket filling
- All communication with loading-related systems

The LHD Definition System (LDS) is used mainly for the design and definition of the production areas (Figure 75.10), but is also used for changes (e.g., driving routes) during a production period. Drift contours and ring positions for the actual area are imported into the LDS. The drift contours are used as a visual reference for drawing the driving routes. The ring positions are coupled to the route network. The LDS contains a number of different "tools."

The High-Speed Underground Navigation System (HUNS) is the system that replaces the driver in the cab of the LHD (Figure 75.11). It drives the LHD according to the orders sent by the LUCS. The control values calculated by HUNS to carry out the orders are sent to another system on the machine that is directly coupled to the hydraulic system, the transmission, the motor, etc. In addition to controlling the machine, HUNS also

has a number of machine-safety functions that prevent damage to the LHD (e.g., monitoring oil pressure). HUNS consists of two units: the navigation computer and the laser scanner (Figure 75.12). The laser scanner measures the angles via a rotating laser beam to reflectors consisting of reflective strips of tape mounted on the sides of the tunnel walls (Figure 75.13). The low power laser scanner is approved for use in a factory environment by Swedish authorities.

The angles that the laser scanner measures are used in the navigation computer for determining the LHD's position. This position determination function is called the "navigation function." The HUNS navigation computer performs the navigation calculation 40 times per second. The LHD's midpoint (the pivot point on the LHD), angle, and speed, together with the angle measurements to the reflectors, are used to calculate the navigation. The navigation function in HUNS functions, in

FIGURE 75.13   A scoop using a laser scanner to navigate the tunnels of the mine

principle, the same way as a GPS receiver with the only difference being that the satellites are replaced by reflectors. The angles to the reflectors are measured with a laser scanner that rotates an infrared laser beam in the plane parallel to the road. When the laser contacts a reflector, the beam is reflected back to the laser scanner. When the reflected beam is detected in the laser scanner, the angle is read with high precision on the rotating head. The route-following function in the HUNS navigation computer does the following:

- Calculates the control signal for controlling the pivot-point angle so the LHD will follow the pre-defined route

- Calculates the control signal for the gear position and the brakes so the speed agrees with that given and so the LHD stops at the correct point

The scoop function in the HUNS navigation computer does the following:

- Calculates the desired scoop height and scoop angle as a function of LHD position

- Regulates the scoop and scoop angle so they correspond to the desired values

There is also a data collection and processing unit (DCPU), which checks the different measuring devices on the machine, accumulates maintenance information, controls the video exchange, and performs other communication functions.

75.6   STEPWISE INTRODUCTION OF AUTOMATION

Implementing new technology into normal production is often the most difficult part of development. One way to overcome this is to plan the implementation step by step. Another problem is that investments made in new technology may not always turn out to be as cost effective as planned, and wrong decisions might be made. It is always easier to take a small step backward than a big step backward. For both technical and economic reasons, introducing new development often goes slowly. One example that can be used to illustrate this is with respect to production drilling. One-hole automation was first established several years ago, and only now, some years later, has the upgrade to full-ring automation and remote-controlled operation become real. Developing LHD automation and using the 25-ton loaders will be a step-by-step process over a period of 3–4 years. In Kiruna, there are a total of about 95 loader-operators to cover all of the shifts. When the new technology is truly introduced, only about half of

these operators will be needed. These development phases provide us with the chance to handle the human capital, our employees, in a responsible way by reeducating them for other jobs and/or by reducing the number of employees through normal retirements. Training and educating people in new technology cannot be done overnight. All of our employees know that new development will mean a reduction in jobs; however, they are also aware of the necessity of this because of the competition in the iron ore market. Trying to coach people and develop goal-oriented personnel when the cost pressure increases and the number of jobs decrease is most difficult. We must handle these changes concerning our employees with care and in a responsible and acceptable way. Through this step-by-step strategy in introducing all this new technology, we might have a chance to manage these human factor situations and make full use of the developments.

75.7   THE FUTURE

The future competitiveness of our company depends upon a continuing program of rationalization, leading to cost reductions along the entire pipeline—from rock in-situ to the shipping port—while steadily improving an already high product quality. The primary cost-reduction potential for LKAB does not lie in further increasing either the scale of the mining method or the machinery. Rather, it must occur through optimizing the current extraction design and utilizing the personnel and machinery to the best advantage possible. Sensible automation is one of the important ingredients in this overall process.

75.8   REFERENCES

Björnström, G. 1994. Mobil och fast multimedia-kommunikation i gruvmiljö - En förstudie. Luleå Technical University.

Eriksson, G. and A. Kitok. 1991. Automatic loading and dumping using vehicle guidance in a Swedish mine. *Proceedings, International Symposium on Mine Mechanization and Automation,* Golden Colorado. Vol. 2, pp. 15–33 to 15–40.

Hulkkonen, A. and T. Poutanen. 1997. *Proceedings, International Symposium on Mine Mechanization an Automation,* Brisbane, Queensland. Vol. 2, pp. B3-7–15.

Hustrulid, W. 1987. General Report for Theme 3: Capital Requirements, Organization and Productivity in Mechanized Mining (Industrial Minerals and Metal Mining). 13th World Mining Congress, Stockholm, Sweden.

Hustrulid, W., and J-O. Nilsson. 1998. Automation and productivity increases at LKAB, Kiruna, Sweden. *Proceedings, CIM '98,* Montreal Canada.

Marklund, I., and W. Hustrulid. 1995. Large-scale underground mining, new equipment and a better underground environment—result of research and development at LKAB, Sweden. *Transactions of the Institution of Mining and Metallurgy.* Vol. 104, A164–168. September-December.

Nilsson, J-O. 1996. Vision or reality? The concept behind the future production system: A case study from the LKAB Kiruna mine, Kiruna, Sweden. *Proceedings, Conference on Large Scale Mining,* Sydney, Australia.

Rönnback, K. 1997. LKAB's role in the world pellet market. *Skillings Mining Review.* January 18. pp. 4–11.

Samskog, P-O, I. Wigdén, and Håkan Tyni. 1999. Process Automation in Ore Mining. *Proceedings, MineTime 99.* June 8–10, 1999.

Samskog, P-O, I. Wigdén, H. Tyni, and J. Björkman. 2000. Company-wide Automation Strategy at LKAB. SME Annual Meeting, Salt Lake City, Utah. February 28–March 1, 2000.

Sandberg, N., 1987. Highly mechanized underground mining for high productivities in LKAB's mines in Sweden. Proceedings, 13th World Mining Congress, Stockholm, Sweden. pp. 563–566.

Wigdén, I. and R. Bergström. 1997. LKAB Communication. *Proceedings, CIM.'98,* Montreal, Canada.

Wyllie, R.J.M. 1996. LKAB invests in the future. *Engineering and Mining Journal.* Vol. 197, No. 11, November 1996, pp. 35–67.

CHAPTER 76

Methods to Mine the Ultra-Deep Tabular Gold-Bearing Reefs of the Witwatersrand Basin, South Africa

Fernando M.C.C. Vieira,* David H. Diering,† and Raymond J. Durrheim‡

76.1 INTRODUCTION

In 1886, gold was discovered in the Archean conglomerates that crop out near Johannesburg, South Africa. Since then, the Witwatersrand Basin has produced half the gold mined worldwide. The conglomerates, known locally as reefs, vary in thickness from a few centimetres to several metres, and extend over wide areas. Consequently, the mining excavations (stopes) formed while extracting the ore have a tabular geometry. The reefs were traced around the perimeter of the basin, where they are mostly concealed by younger strata, leading to the discovery of several new gold fields. The reefs were also followed down-dip, the mines becoming the deepest in the world, with stoping currently taking place at depths exceeding 3,500 m. Each year about 30 million square metres of reef is mined, and some 800 km of new tunnels are developed in addition to the 10,000 km already in use.

76.1.1 Scattered Mining Methods

Scattered mining methods were initially used to mine the gold-bearing reefs. A typical layout is shown in Figure 76.1. As the mining depth increased, so did the stress-related problems of fracturing, mobilisation, and overall instability, especially in cases where relatively small areas of unmined reef, known as remnants, were formed. In some cases, tremors occurred as a result of the violent failure of the stressed rock, causing damage to the excavations and injury to workers. In the mid-1920s, a statutory Rockburst Committee issued a number of recommendations aimed at reducing the incidence of rockbursts in South African gold mines. The most significant recommendations, which remain valid today, are:

- The formation of isolated areas of reef (i.e., remnants) should be avoided.

- Stoping operations should be concentrated as far as practicable.

- Working faces should be advanced rapidly and continuously.

76.1.2 Longwall Mining

The longwall method was devised in an attempt to improve mining conditions. The first systematic application of this method was at the East Rand Proprietary Mines (ERPM) during 1941 (Ortlepp and Steele 1975). Early ERPM longwalls were designed to have overall face lengths between 900 and 1,200 m (Figure 76.1). However, extensive ground falls occurred, and it soon became apparent that these very long longwalls did not solve the rock burst problems at depth. Because stress concentrations at stope faces

* CSIR:MiningTek, Auckland Park, South Africa.
† Anglo American Technical Services, Marshalltown, South Africa.
‡ DEEPMINE, Auckland Park, South Africa.

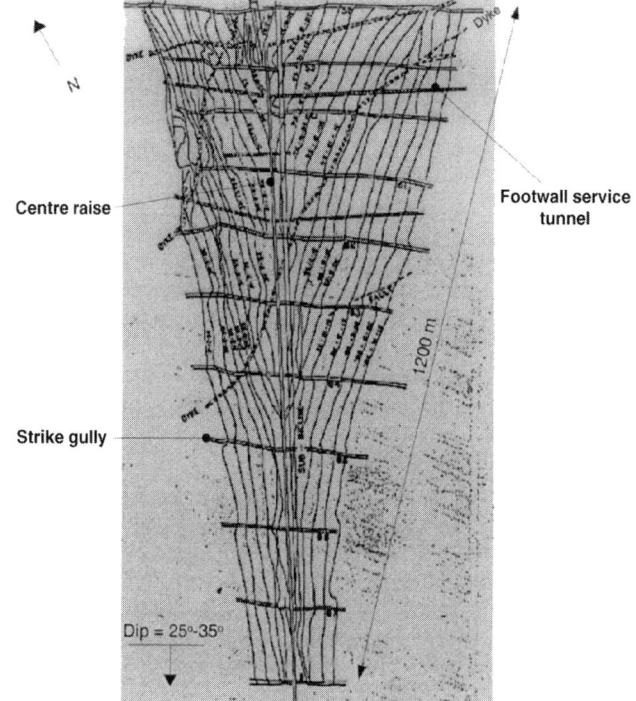

FIGURE 76.1 Typical longwall section at ERPM during late 1950s at a depth of about 2,100 m below surface. Note the extremely long overall face length (1,200 m).

increased as the longwall stope spans increased, there was a greater probability of hanging wall failure and rock bursts.

The ideal orebody for the longwall mining method is one that has a consistent grade distribution, with few disruptions of the reef by faulting and intrusions, conditions that do not always occur. Some of the benefits of longwalling are fewer remnants; fewer lead-lags between faces, which were shown to be associated with rock bursts; better utilisation of ventilation; and the concentration of mining activities leading to improvements in supervision and the provision of services (Beck et al. 1961).

76.1.3 Strike Stabilising Pillars

The next significant advance in mine design came through the realisation that the rock-burst hazard could be reduced either by

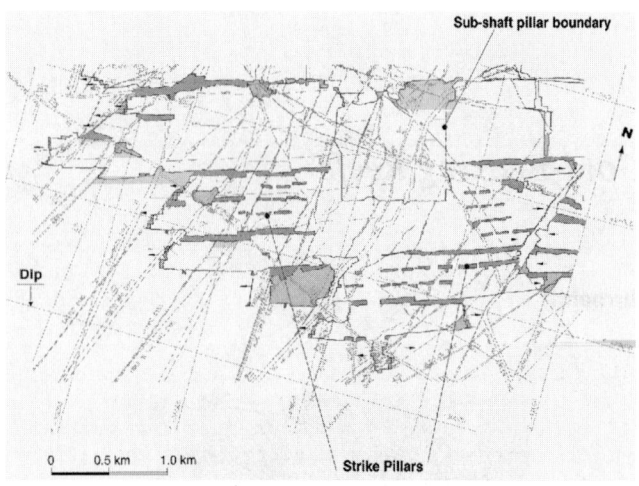

FIGURE 76.2   Actual longwall layout with strike-stabilising pillar at Western Deep Levels, Carbon Leader Reef. Note the complexity of geology on which the method is implemented.

increasing the rate at which energy is dissipated in a nonviolent manner, or by reducing the rate at which the energy is released (Cook et al. 1966). Three practical methods were proposed that would limit convergence and, consequently, also limit the rate of energy release:

1. **Reducing the stope width.** Operational constraints, however, dictated that the stoping width could not be made significantly less than 1 m.

2. **Filling the stope.** Early attempts at waste filling had not been effective in limiting closure, but it was believed that hydraulic backfilling could lead to appreciable benefits.

3. **Partially extracting the reef.** Cutting the pillars offered, by far, the best chance for reducing the amount of energy released (Cook and Salamon 1966).

Strike-stabilising pillars were introduced at ERPM in the 1960s, where stoping was already approaching depths of 3,000 m. These pillars were about 80 m wide, with an accompanying longwall dip span of 270 m. In the late 1970s, this design was reviewed, and the pillar width reduced to 60 m. This design afforded a high extraction ratio while limiting the average pillar stress to 400 MPa. Strike-stabilising pillars were subsequently introduced into several other deep Witwatersrand gold mines (e.g., Blyvooruitzicht, Western Deep Levels, Kloof). A typical longwall layout incorporating strike-stabilising pillars in a Carbon Leader Reef horizon (CLR) is shown in Figure 76.2.

Empirical rock engineering criteria are applied to the design of regional stability pillars. Parameters commonly considered are the average pillar stress (APS), the width:height ratio, and the energy release rate (ERR). Pillar stability is improved by keeping foundation stresses and stress changes at a minimum. The understanding is that the APS should not exceed 2.5 times the uniaxial compressive strength of the country rock. The probability of pillar failure increases sharply for ratios less than 15:1 (Salamon and Wagner 1979). To avoid failure by punching, the minimum principle stress, σ_3, in the zones above and below the pillar must remain compressive, and

$$\sigma_1 - k\sigma_3 < \sigma_c \qquad (76.1)$$

where σ_1 is the maximum principle stress, σ_c is the uniaxial compressive strength (UCS), and k is a factor that describes the effect of confining stress on rock strength (typically $6 < k < 10$).

Case History: Western Deep Levels. A comprehensive review of the development and implementation of longwalling with strike-stabilising pillars at Western Deep Level mine is given by Hagan (1987) and Diering (1987). In 1979, fatalities in the mine were running at double the industry average, and rock bursts accounted for a large proportion of fatalities and accidents. The energy release rate (ERR) for the mine was estimated at 40 MJ/m². In an attempt to halve the ERR, stabilising pillars were introduced to improve mining conditions. The pillar width:height rule of 15:1 was applied. On the CLR, where the stoping height is of the order of 1 m, 20-m-wide pillars were spaced at 133 m centre-to-centre, while on the Ventersdorp Contact Reef (VCR), where the stoping height may be as great as 1.5 m, 30-m pillars were implemented. Pillar widths of 20 m or more also met the criterion that the average pillar stress should be less than 2.5 times the uniaxial compressive strength of the surrounding strata to avoid the risk of pillar punching (Cook et al. 1973). These layouts yielded an extraction ratio of about 85%.

Cutting the 20-m-wide pillars on the CLR longwalls began early in 1980. Initially, the pillars had 8-m-wide ventilation slots cut through them at 120-m intervals, but these slots soon proved to be too narrow and were increased to 16 m. The position of the stabilising pillars also had to be taken into account when placing service tunnels, travelling ways, and other auxiliary developments. In the six years following the introduction of the new design, 26 new pillars were cut on the CLR. Although the number of rock bursts was reduced, several serious problems were associated with the narrow (20-m) pillars (Diering 1987; Hagan 1987; Lenhardt and Hagan 1990):

1. The area of the stope immediately up-dip of the pillar suffered severe and regular damage from rock falls and rock bursts.

2. From within the ventilation slots, it was observed that the pillars were pervasively fractured, leaving no solid core, and that appreciable deformation was taking place.

3. It was very difficult to cut ventilation slots, and rock bursts often occurred when doing so.

4. It was difficult to maintain pillar widths, and pillars often ended up being less than 20 m wide.

5. Some of the pillars failed violently, with the seismic source parameters indicating a crush-type failure mechanism.

As a result of these problems, the pillar width was increased to 40 m, with a 280-m centre-to-centre spacing. This layout design still enabled an 85% extraction ratio to be achieved. It was observed that the 40-m pillars had an unfailed core, indicative of greater stability. Furthermore, the layout using 40-m pillars is less development intensive and eliminates the need for ventilation slots to be cut into the pillars.

76.1.4   Backfill

The need to apply good quality backfill in deep-level mines was highlighted by Diering (1987). The idea was that backfill could partially replace pillars and increase the extraction ratio. Filling also had the added advantages of improving ventilation and reducing the fire risk. A system of 40-m-wide strike pillars at 280-m centres can maintain ERRs below 30 MJ/m² at the working faces of a longwall up to 3,200 m below the surface (Hagan, 1987). At 4,000 m below surface, 90-m-wide strike pillars centred at 280 m are required to maintain ERRs below 30 MJ/m², but the extraction ratio would be only 68%. Hagan (1987) concluded that a system that integrates pillars and backfill would give the maximum benefit.

Hydraulic backfill became widely used in the early 1980s, and it was observed that backfilled stopes had significantly lower

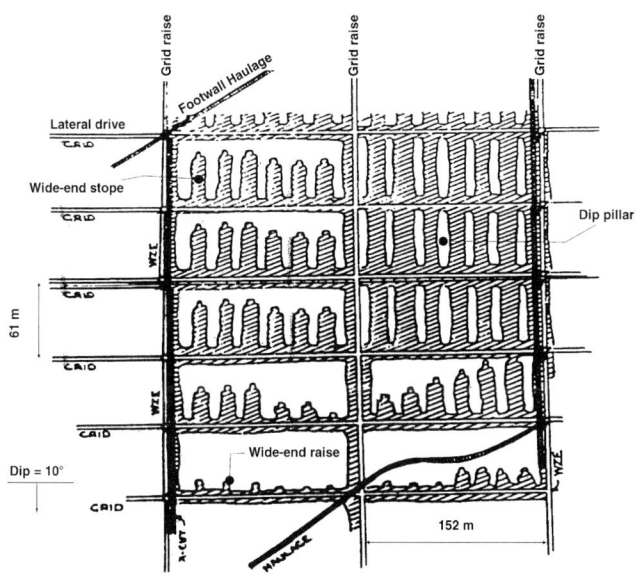

FIGURE 76.3   Grid of primary and reef development with stoping in progress in a *wide-end stoping* layout with dip stability pillars applied by the Government Mining Areas (after Hindle 1957). Note that mining proceeded up dip.

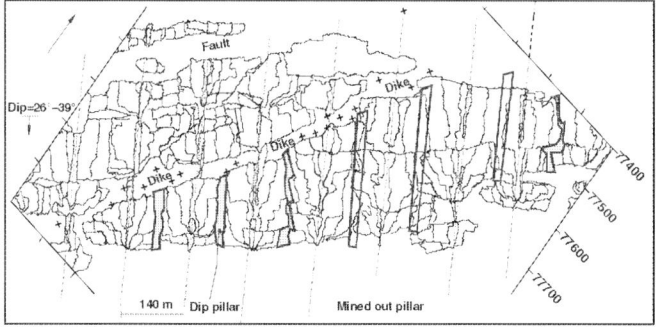

FIGURE 76.4   Dip pillars in the Strathmore-South area (SMS), Buffelsfontein Gold Mine. Face outlines as of February 1998.

closure rates than unfilled stopes (Gürtunca et al. 1989). Back-filling is now well established in a number of South African deep mines. Piper and Ryder (1988) state that using backfill in combination with stabilising pillars can provide a reduction of up to 30% in both the rate of energy released and the average pillar stresses of regional layouts.

76.1.5   Dip Pillar Systems

Dip-orientated pillar systems were first used in the South African gold mining industry during the 1950s. For example, a method called *wide-end stoping* was successfully used to limit subsidence at shallow depth (Figure 76.3). The problem of subsidence became less relevant as the mining depth increased, and these early dip-pillar type layouts were discontinued. In 1987 another dip-orientated pillar system was introduced in a section of the Buffelsfontein Gold Mine that was experiencing acute seismic activity related to a major dyke. The scattered mining practised at Buffelsfontein was considered to be unsuitable for stoping the ground adjacent to this active structure, and a dip-orientated pillar system was designed which provided sufficient regional support and also bracketed the dyke (Figure 76.4).

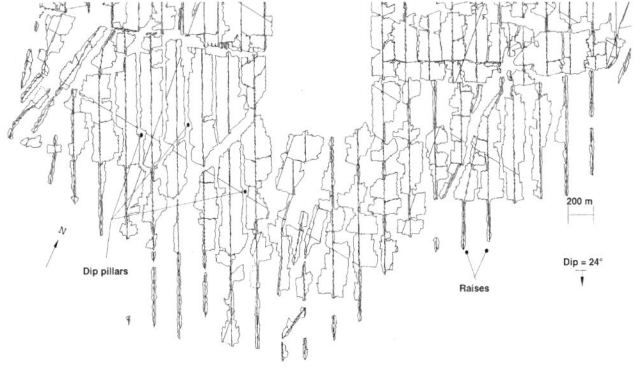

FIGURE 76.5   Macro plan of the sequential grid-mining layout with dip stabilising pillars at Elandsrand. Face outlines as of March 1998 are shown.

A new type of dip-pillar system was introduced at Elandsrand Gold Mine in the 1990s. The large number of dykes and faults and the erratic grade distribution led to the conclusion that longwall layouts were not optimal for mining the orebody. A mining method was devised that utilised a regular grid of dip pillars for both regional support and to bracket most faults and dykes. A grid of footwall tunnels is developed prior to stoping. Haulages are typically placed between 70 and 80 m vertically below the reef, and crosscuts to the reef are developed at roughly 200-m intervals, followed by raises off each crosscut. These raises are ledged and equipped prior to beginning full stoping operations. This method, termed *sequential grid mining* (Applegate 1997) was first implemented 2,200 m below surface and has since been extended to a depth of 2,800 m (Figure 76.5).

In the late 1990s, two new methods were implemented that employ closely spaced narrow dip pillars to keep stope closure and ERRs low. One method, employing down-dip mining (termed the *sequential down-dip* mining method), was implemented at Kloof Gold Mine to extract a VCR orebody. Strike spans are small enough (about 75 m) to limit closure to the point where it is possible to dispense with backfill. The second method, employing breast mining (termed the *closely spaced dip pillar* mining method), was implemented at Driefontein Consolidated Gold Mine to extract a CLR orebody. This method allows the orebody to be accessed rapidly, because it is possible to begin stoping immediately after the reef is intersected by a crosscut driven from a hanging-wall access haulage.

76.2   MINING AT ULTRA-DEPTH

76.2.1   DEEPMINE Research Programme

The Witwatersrand Basin of South Africa has produced more than 50,000 tons of gold since the discovery of the auriferous reefs in 1886. Annual gold production reached a peak of 1,000 tons in 1970, but has since fallen to less than 500 tons. The main reasons for declining production are a decrease in gold prices coupled with an increase in production costs, owing to, among other factors, the depletion of the shallower and high-grade reserves.

Reflection seismic imaging, deep drilling, and the extrapolation of pay shoots from areas of current mining indicate that the gold-bearing horizons persist to depths in excess of 5 km. If a favourable gold price is maintained, and challenges such as high virgin rock stress and temperature are overcome, the depths at which mining occurs will steadily increase. While only about 5% of production currently occurs below 3 km, this could increase to more than 40% by 2015 (Willis 1997). The gold resource that resides at these depths is thought to be similar to the amount of gold recovered from the reefs of the Witwatersrand basin during the last century. The great challenges of mining at ultra-depth has led to unprecedented cooperation between mining companies, research institutions, government, labour, and universities and to the establishment of the DEEPMINE Collaborative Research Programme in 1998 (Diering 1997; Gürtunca 1998).

The DEEPMINE Programme aims to create a technological and human resources platform that will make it possible to mine gold safely and profitably at depths of 3–5 km (the term "ultra-deep" is used for this depth range). For several decades, the deep tabular reefs of the Central, West, and Far West Rand were predominantly mined using layouts characterised by longwalls, with strike pillars providing regional support. During the past few years, however, a new generation of layouts has been introduced, based on the concept of scattered mining with dip pillars for regional support. One of the most important issues to be addressed is the choice of mining method to be adopted at ultra-depth, and a study was undertaken to analyse and compare the performance of these mining methods.

76.2.2 Assessment of Alternative Layouts

For this study, the model used was of a typical ultradeep VCR orebody, derived from the interpretation of a 3-D reflection seismic survey constrained by information from shallower mining and boreholes (Gibson et al. 1999). Several reflectors, representing major stratigraphic breaks, were mapped. Minor internal reflectors were also interpreted and together helped constrain fault geometry. Faults with throws greater than 18 m clearly offset the reflectors and can be easily mapped, while faults with throws less than 6 m cannot be detected. Faults with intermediate throws are identified using seismic attribute analysis. The orebody considered is 24 km^2 in extent, ranging from 3,800 m to 5,000 m below surface, with an average dip of 23° (Figure 76.6a). As the seismic survey could only detect faults with throws exceeding 6 m, several smaller faults were added to the model. To give more realism to the geological model, and to test the ability of the different methods to mine difficult geologic conditions, a graben-type structure was created by manipulating the displacement directions of two adjacent major faults. This model orebody, with a total mineral resource area of 19.7×10^6 m^2, was named "Iponeleng," a Setswana word meaning "discover me" or "explore me." This model (Figure 76.6a) was used to evaluate the four mining methods currently used in an ultra-deep environment. A reef thickness of 1.5 m was assumed, typical of the VCR. The four methods are longwalling with strike stability pillars (LSP), sequential grid mining (SGM), the sequential down-dip (SDD), and the closely spaced dip pillar (CSDP) methods.

Some operational assumptions were made and applied consistently to all four layout assessments (Table 76.1). First, access to the ultradeep reserves is through an existing main shaft and a newly sunk subshaft system, the latter being the infrastructure requiring new capital investment. The position for the subshaft infrastructure, assumed to be common for all mining methods, was selected on the following considerations (Figure 76.7b):

- The shaft pillar is circular with a radius of 500 m and considered to be mined out and with all station infrastructures completed.

- The subshaft site should have as few geological disturbances as possible.

- The main shaft intersects the reef at a depth of –3,500 m. The full subshaft column would thus exist from a depth between –3,500 m to –5,020 m.

The mining rules, scheduling rates, performance criteria, and design specifications for each layout on both micro- and macro-scale were defined by teams of experts (Table 76.2). A *micro layout*

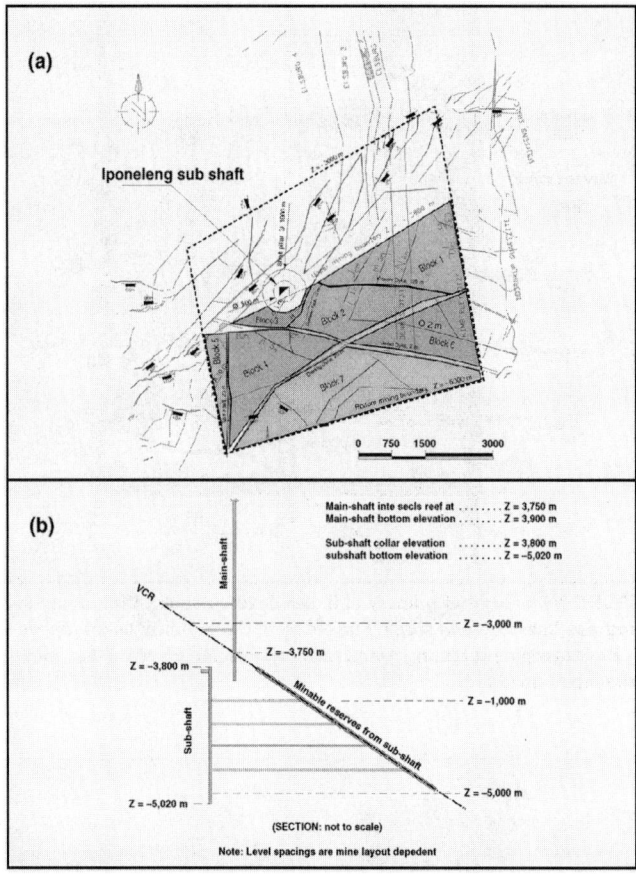

FIGURE 76.6 Iponeleng mine geological model with (a) identified mining blocks and (b) vertical limits of the minable reserves

TABLE 76.1 Macro design guidelines and assumptions

Orebody structure	based on 3D vibroseismics
Depth range	3,000 m–5,000 m
Size of orebody to evaluate (plan distances)	6 km dip × 6 km strike
Assume constant grade of:	15 g/ton
Mine (Shaft) production output	start from 0 building up to 45,000 m^2/month
Average face advance rate	15 m/month
Inter-level spacing	mining method dependent
Minimum number of levels	5
Development rates (max. values)	twin-ends: 100 m/mth/twin
	raise: 40 m/mth (from 0–50 m long)
	raise: 30 m/mth (from 50–100 m long)
	raise: 25 m/mth (for >100 m long)
Mine out shaft pillars first?	yes
Average stope-width	1.5 m
Average stope face length	to suit the method
Working days per month	25
Areal extraction ratio	method dependent
Length of stope backs	method dependent
Tramming width	1.3/1.8 m

TABLE 76.2 Expert design specifications for the various layouts

		Mining methods			
	(units)	LSP	SGM	SDD	CSDP
Generic macro design parameters					
Raise-to-raise spacing	m	n/a	200	100	180
Maximum length of back	m	240	240	240	175
Dip pillar's maximum deviation from true dip	degrees	n/a	45		
Dyke and fault bracket pillar width (on either side)	m	20	10	20	20
Preliminary vertical distance between levels	m	<120	92	70	68
Width of dip-stabilising pillar for depth range < 4 km	m		30		
Width of dip-stabilising pillar for depth range between 4 km and 5 km	m		40		
Width of dip-stabilising pillar up to a max. depth of 4.6 km	m			25	40
Width of strike-stabilising pillar for depth range between 3 km and 4 km	m	40			
Width of strike stabilising pillar for depth range between 4 km and 5 km	m	50			
Design of flat development infrastructure					
Minimum vertical distance of haulage / RAW below reef	m	90	120	50	90
Minimum distance to reef of replacement haulages	m	90			
RAW-HLGE connecting crosscut spacing	m	250		200	180
Footwall drives/Haulage and RAW tunnel dimensions (height × width)	m²	3.0 × 3.5	4.5 × 4.0	4.5 × 4.0	4.0 × 3.0
Replacement haulage and RAW tunnel dimensions (height × width)	m²	4.0 × 3.5			
Minimum distance from a HLGE - RAW connecting to reef X/cut breakaway	m		25	25	25
Minimum distance between opposite breakaways (longwall)	m	15			
Haulage and RAW spacing (centre to centre)	m	30	30	30	30
Minimum middling of follow-behinds from reef plane	m	20			
Maximum middling of follow-behinds from reef plane	m	40			
Minimum middling of incline man & material way from reef plane	m	60			
Follow-behind tunnel dimensions (height × width)	m²	3.0 × 3.5			
Crosscut-to-crosscut spacing (grid definition)	m	63	200	100	180
Crosscut to reef dimensions (height × width)	m²	3.0 × 3.5	3.5 × 3.5	3.5 × 3.5	3.0 × 3.0
Minimum crosscut distance below reef (perpendicular) at reef intersection	m	10	12		15
Design of stoping-related infrastructure					
Panel face length between 3 km and 4 km	m	40	30-40	46	25
Panel face length between 4 km and 5 km	m	35			
Maximum number of panels per longwall section	unit	6			
Maximum number of panels per raise	unit	n/a	7	2	7
Average stope width (VCR considered)	m	1.5	1.5	1.5	1.5
Raise dimensions (height × width)	m²	2.0 × 1.8	3.0 × 1.5	2	1.8 × 2.5
Strike gullies dimensions (height × width)	m²	1.6 × 2.0	1.8 × 2.8		1.8 × 2.0
Ledging distance from centre line (on either side of raise)	m		10	6	15
Maximum strike scraper distance	m	100	80		70
Maximum on-dip distance for continuous scrapping	m				150
Maximum length of winze	m		75		
Depth of wide raise (on either side from centre of raise)	m			6	

(continues)

consists of a plan and a section of a raise connection and its footwall infrastructure showing all the components of an operational raise connection and its service elements (e.g., timber bays, travelling ways, boxholes, cooling loops and cubbies, ventholes, etc.). This was to provide an understanding of how the layout actually works. The design parameters include level spacing, back length, crosscut spacing, pillar sizes, panel lengths, panel shape, mining direction, face orientation (overhand, underhand), and leads lags. A *macro layout* expands the micro-mine design over four or five mining levels. This enables the mine designers to determine whether or not a mine-wide layout is indeed practicable (e.g.,

whether or not the desired airflow is actually possible) or whether the overall mining sequence is achievable for a realistic orebody such as the Iponeleng model.

Once the micro and macro concepts of the layouts and scheduling rates were defined, the mining of the ultradeep entire Iponeleng orebody was simulated using the CADSmine package. The program set an initial production target of 30,000 m² per month (equivalent to approximately 150,000 tons), which is similar to the output of some current deep-level mines. The target was subsequently raised to 45,000 m² per month in view of preliminary economic considerations. The four conceptual mine

TABLE 76.2   Expert design specifications for the various layouts (continued)

		Mining methods			
	(units)	LSP	SGM	SDD	CSDP
Design of vent control infrastructure					
Bulk-cooling loop dimensions (height × width)	m²	3.0 × 3.5			
Cooling car cubby (one side extra width, in excess of x/cut width)	m	4 + 4		5	
Cooling car cubby length	m	6	15	20	
No. of dedicated vent holes per x/cut	unit	1	1		
Design of ore-handling infrastructure					
Conventional stope box hole dimensions (height × width)	m²	2.0 × 1.8	1.8 × 1.4	1.4 × 1.5	1.8 × 2.5
Raise bore stope orepass diameter	m		1.4 m		2.1
Box hole stub length (extra width in excess of x/cut width)	m				5
Box hole stub length	m		12		20
Maximum raise bore length	m		120		
No. of conventional box holes/box fronts per X/cut	unit		2	2	2
No. of Y-leg box holes/box fronts per X/cut	unit		3		
Inclination of conventional stope box hole	degrees		55	55	65
Maximum inclination of raise bore orepass	degrees		45	15	65
Design of access-ways infrastructure					
Incline man and material way (longwall) (height × width)	m²	4.0 × 3.0			
Length of the station for the incline man and material way (longwall)	m	25			
Travelling way dimensions (height × width)	m²	2.8 × 2.0	2.4 × 2.4	2.4 × 2.4	2.0 × 2.5
Travelling way length	m		15		15
Travelling way inclination	degrees		34	34	34
Design of materials-handling infrastructure					
Timber bay dimensions (height × width)	m²		3.0 × 4.0	3.0 × 3.0	3.0 × 2.0
Timber bay length	m		30	20	30
Timber bay distance from reef	m		20		10
Generic scheduling rates and mining constraints					
Normal haulage and RAW development	m/mth/end	35	30	35	35
High speed haulage and RAW development	m/mth/end	70	70	70	70
X/cut development	m/mth/end	35	30	35	35
Raise development	m/mth/end	30	25	12	30
Winze development	m/mth/end		15		
Box hole development (conventional)	m/mth/end	20	10	35	20
Drop-box hole development (conventional)	m/mth/end				30
Raise bore development	m/mth/end	35			35
Delay in stripping of services at a raise	months		1		
Delay attributed to stope equipping	months		6		
Lodging face advance	m/mth/face		25		
Stoping face advance	m/mth/face	12	15	15	15

designs were applied to arrive at an extraction path of the Iponeleng orebody and then subjected to exhaustive evaluation with respect to:

- Efficiency and cost of providing ventilation and refrigeration
- Efficiency and cost of transporting men, material, and rock
- The degree to which rock engineering criteria, such the energy release rates, average pillar stress, excess shear stress, and stresses in footwall excavations were met

The program assumed that the ultradeep virgin stress regime is determined solely by the weight of overburden rock, and that the rock mass is an isotropic continuum. The virgin stress conditions at a point below surface are then defined by the vertical σ_v and the horizontal σ_h components, given by Equation 76.2 and Equation 76.3, respectively:

$$\sigma_v = \rho h g \qquad (76.2)$$

$$\sigma_h = k \sigma_v \qquad (76.3)$$

where ρ is the rock density, h is the distance below surface, g is the acceleration due to gravity, and k the ratio of horizontal to vertical stress. A k-ratio of 0.5 was used to model the rock of the Witwatersrand Basin. The virgin stress at the depths considered in the current analysis are indicated in Table 76.3. The ERR distribution ahead of mining faces was calculated using the MINSIM 3-D elastic boundary element method (Napier and Stephenson 1987). ERRs ahead of mining faces were calculated assuming that no backfill had been placed.

TABLE 76.3 Theoretical virgin stress values in the far field at various ultra-deep elevations (assuming a rock density of 2,780 kgm^{-3} and g = 9.81 ms^{-2})

Depth [m]	σ_v [MPa]	σ_h [MPa]
3,000	82	41
4,000	109	55
5,000	136	68

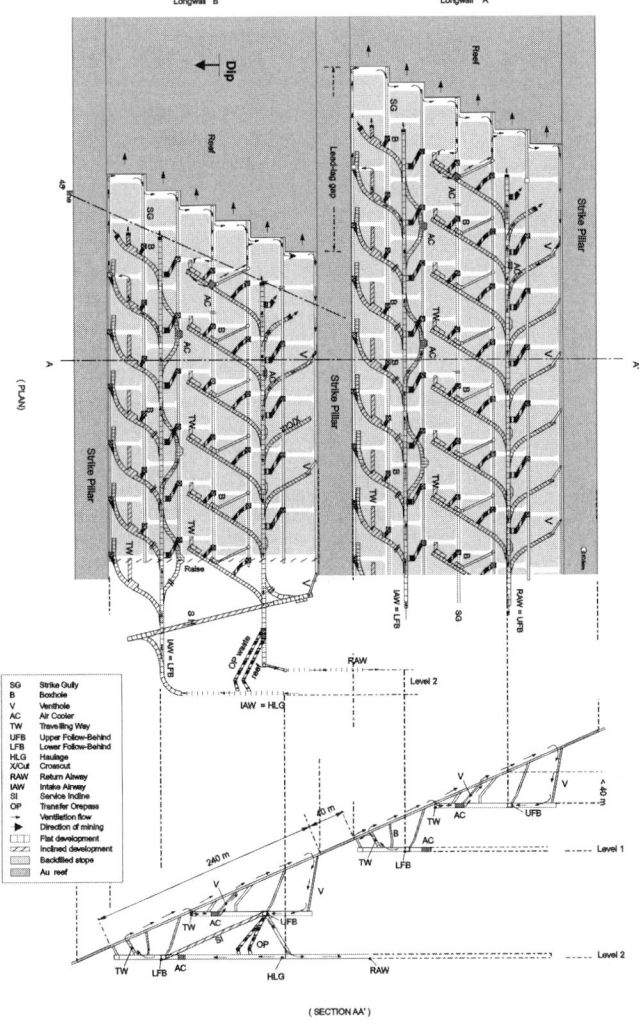

FIGURE 76.7 Micro concept of a longwall layout with strike-stabilising pillars. Longwall A is depicted as a mature longwall production unit. Longwall B is depicted with all footwall infrastructure required for longwall establishment.

76.2.3 Longwall with Strike Stabilising Pillars (LSP)

Micro Layout. Twin haulages, one for intake air (IAW) and the other for return air (RAW), are driven from the subshaft to the block of ground to be mined (Figure 76.7). To establish a longwall, a single on-reef raise is developed between strike pillars, providing the platform for ledging and stoping. The panels are mined in the strike direction, with the lower panels leading, yielding an overhand configuration. A service incline (SI) is established from the level of the intake haulage to an intermediate elevation (see section of longwall B in Figure 76.8).

Two flat and parallel tunnels are established from the flat connections of the SI and are driven along strike beneath and to within 40 m of the mined-out reef plane. These are known as *follow-behind* (FB) tunnels because they lag sufficiently far behind the advancing faces to avoid the extreme abutment stresses. The lower follow-behind (LFB) is on the same level as the main access haulage (IAW or HLG). Crosscuts are established at 63-m intervals on either side of the FB tunnels. Travelling-ways are established at the end of the reef crosscut to provide access to the stope. The broken rock is moved from the face area to the strike gully developed at the foot of each panel either by scraping or water jetting and then scraped along the gully to the nearest box hole. From the box holes, the rock is transported by train to the main transfer orepass.

The lower and upper follow-behinds are used as intake and return airways to the stopes. Loops for bulk air-cooling are developed along the LFB after every second crosscut. Recooling the air in the lower and upper crosscuts is necessary because return air from the LFB headings is mixed with the intake air to the stopes, and the return air from the stopes is used to ventilate the upper FB headings. Careful attention must be given to the airflow at the top of a longwall to ensure that no methane accumulates. The top of the longwall is connected to the upper crosscut by a box hole or vent hole. The high virgin rock temperatures at ultra-depth will probably make in-stope cooling necessary.

Macro Layout. The upper longwalls advance ahead of the lower longwalls, yielding an overall underhand mining configuration (Figure 76.8). Twin replacement tunnels are typically placed 145 m below the reef, largely beyond the influence of mining-induced stresses. Slots are cut into the strike pillars to reduce the stress on the replacement tunnels passing beneath them. Each longwall unit is an independent ventilation district with its own intake and return airways, giving a fair amount of flexibility to the ventilation system as borehole connections can be used to adapt the system as and when required.

Simulated Mining of the Iponeleng Model Orebody. The orebody model in Figure 76.6 is first used to define viable longwall blocks. Note that it is assumed that the shaft and all service excavations within the shaft pillar are already established. The major decisions made by the design team are the span between pillars, the number of panels (and the length of each panel), and the width of the strike-stabilising pillars. The geometry of the longwall unit will also define the shaft station vertical interval. The design team specified a longwall layout composed of six 40-m-long panels between 3 and 4 km deep, with 35-m-long panels between 4 and 5 km deep; and 40-m-wide strike-stabilising pillars between 3 and 4 km deep, with 50-m-wide pillars between 4 and 5 km deep (Table 76.2 and Figure 76.9). This design gives a back length of about 240 m. The ERR ahead of the mining faces is a function of the local geometry of the longwall and depth. At 3,000 m, the probability for the ERR to exceed 30 MJ/m^2 on any face is 0.2, while at 5,000 m the probability is 0.6, indicating an increase in the potential for rock bursting. The placement of backfill is considered essential to reduce the ERR values. Furthermore, backfill assists in ventilation control and also reduces the potential heat load originating from worked-out areas. The APS on the reef plane ranges from 450 MPa at 3,000 m to 695 MPa at 5,000 m. At great depth, the APS exceeds 2.5 times the UCS of the host rock, and the stability of the pillar may be compromised. Tunnels passing near or under pillars or abutments experience stresses between 140 MPa at 3,000 m and 260 MPa at 5,000 m.

Eleven levels are required to extract the model orebody. The LOM is expected to be 45 years, during which time a total area of 13.1 × 10^6 m^2 out of a 19.7 × 10^6 m^2 resource is mined, yielding an overall extraction ratio of 67%, with 4.1 × 10^6 m^2 being left locked up in stabilising pillars and ground adjacent to geological features. The total length of off-reef development is 552 km.

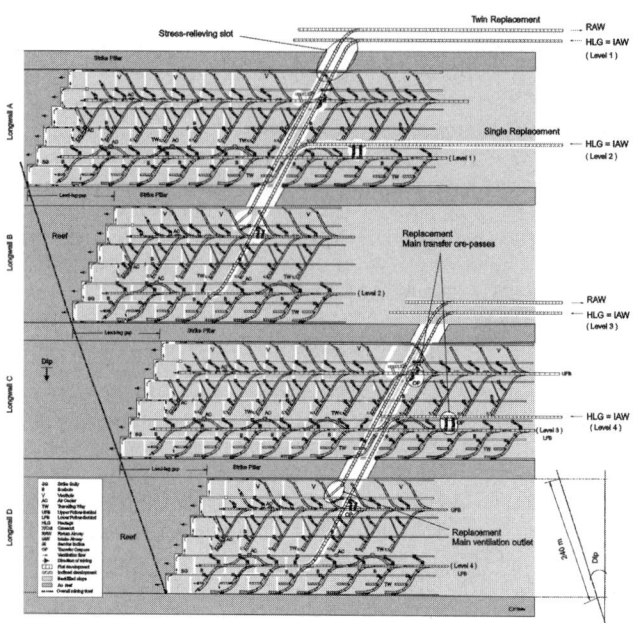

FIGURE 76.8 Macro concept of a longwall layout with strike-stabilising pillars. Three longwall units are depicted. Replacement haulages are shown, these being developed roughly every 750 m.

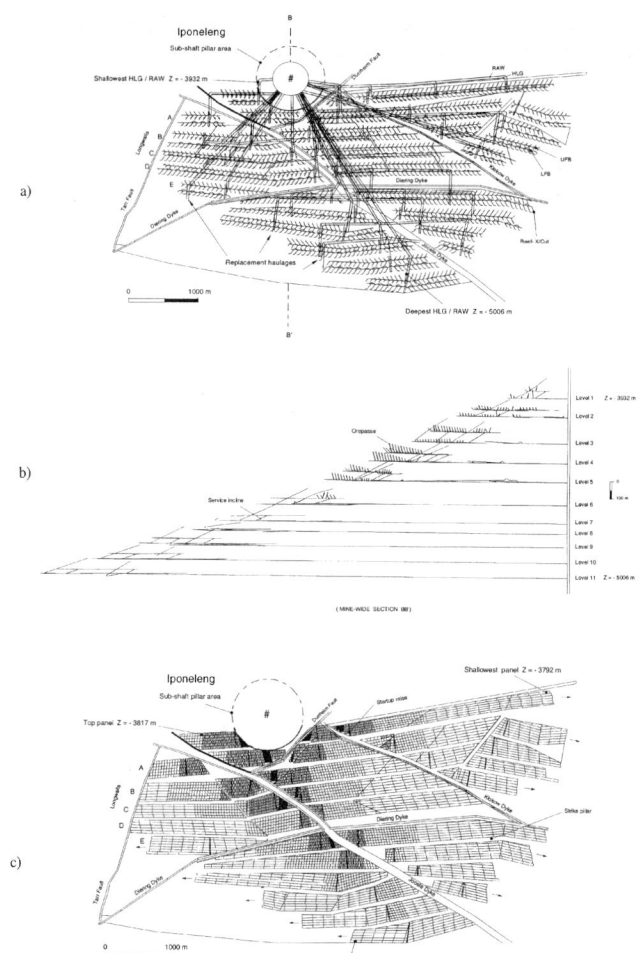

FIGURE 76.9 Application of the LSP micro- and macro-layout design concepts to the extraction of the Iponeleng orebody. Outlines represent Year 20 of 45 years of life of mine (LOM). (a) depicts the extent of the development infrastructure, (b) shows the number and relative position of service levels, and (c) the shows the regional stoping pattern.

Evaluation. The longwall method works well where the grade is consistent and where dykes and faults are sparse. Ventilation is also easy to manage, very little reconditioning of the ventilation air is necessary, and only an average amount of ventilation air is required.

However, there are several disadvantages to the longwall method. First, the servicing of an advancing longwall demands the development of a substantial infrastructure (FB tunnels, crosscuts, loops and cubbies for air coolers, timber bays, box holes) that takes significant time and money to establish. Furthermore, FB tunnels must lag behind the advancing face to ensure that they are situated in stress-relieved ground, and the maximum practical pulling distance for a scraper from face to box hole is considered to be 100 m. These considerations, if current technology is applied, limit the stoping face advance rate in a longwall unit to 12 m/month/face. Second, as the longwall method cannot mine selectively, much low-grade ore may be mined in areas of erratic grade distribution. Lastly, it is a major effort to establish a new longwall, and it is often necessary to mine through potentially hazardous structures such as dykes and faults. If there is a significant throw on the fault, substantial time and effort must be spent to reestablish the stope on the reef horizon.

76.2.4 Sequential Grid Method (SGM)

Micro Layout. As for the longwall method, twin haulages are driven from the shaft to access the block of ground to be mined (Figure 76.10). These haulages, as well as the reef-access crosscuts, are developed prior to stoping, allowing faults and dykes to be located and pillars to be designed to bracket any hazardous structures. Haulages are situated deep in the footwall (about 120 m) to reduce their vulnerability to the strain and stress changes associated with stoping.

Crosscuts are developed at regular intervals to give access to the orebody, and the raises are developed, ledged, and equipped. Mining is carried out according to a strict sequence to ensure low ERRs and pillar stresses (Applegate 1997). Panels are first mined, in an underhand configuration, from the raise line towards the

planned pillar position closest to the shaft. Only after this is complete is the other half of the stope area mined. This is termed "single side mining," to distinguish it from similar methods where mining may take place concurrently on both sides of the raise mine. The broken rock is moved from the face area to the strike gully at the foot of the panel either by scraping or water jetting, and then scraped along the gully to the nearest box hole. One of the most demanding aspects of the SGM is the development of the orepasses servicing the uppermost part of the stope, as the vertical distance from crosscut to reef may be as great as 100 m, depending on the dip of the reef and the back length.

The microventilation system is quite simple, with a dedicated intake at the bottom of the stope, a maximum of seven panels, and a return airflow through a ventilation box hole at the top of the stope to the lower-level return airway. Ventilation control is relatively easy due to the required use of backfill and a maximum airflow strike distance of only 80 m (i.e., single-side distance). Although backfill will reduce the heat load from worked-out areas, *in-stope cooling* will still be required at ultra-depth, which might complicate the control of ventilation air in stopes. The SGM layout requires ventilation air to be returned from the top of a stope to the lower-level return airway via a

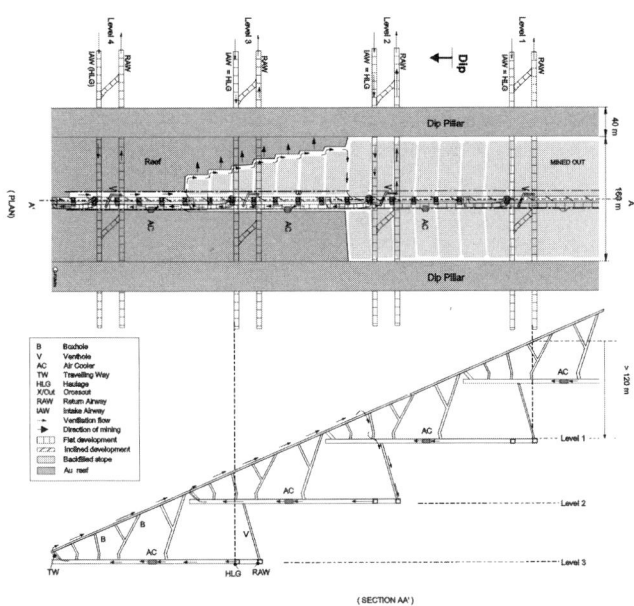

FIGURE 76.10   Microconcept of a sequential-grid mining layout with dip-oriented pillars

disused box hole, which may be a restriction in the ventilation system. Also, care must be taken to avoid the accumulation of methane in the top section of the stope.

Macro Layout.   Stoping may take place concurrently on two adjacent raise lines, while the next raise is being ledged, and a fourth raise is being developed or equipped (Figure 76.11). If stoping is occurring in two adjacent raise lines, it is important that the two faces are never closer than 70 m to avoid placing an abnormal load on the pillars.

The primary ventilation system consists of twin airways (intake and return) on each level connected to each stope through crosscuts, box holes, or vent holes. Bulk air-cooling is required at the top of the subshaft to ensure acceptable conditions at the stations. Sealed-off HLG-RAW connecting crosscuts are used as cubbies for closed-circuit coolers in the intake airways, as and when required (Figure 76.12).

Simulated   Mining   of   the   Iponeleng   Model Orebody.   The orebody model in Figure 76.6 is first used to define viable mining blocks for this method. The major decisions made by the design team are the span between pillars (which dictates crosscut spacing), the width of the dip-stabilising pillars, the length of back (which dictates the position of haulages), and the depth of the main haulages beneath the reef. The design team specified a 200-m crosscut spacing; seven panels per raise, each with a face length of 30 to 40 m, yielding a back length of about 240 m; stabilising pillars 40 m wide between 3 and 4 km deep, and 50 m wide between 4 and 5 km deep; and main haulages at least 120 m below the reef plane (Table 76.2 and Figure 76.10). There was a calculated probability of 0.1 for the ERR ahead of any stope face to exceed 30 MJ/m^2 at a depth of 3,000 m, increasing to 0.35 at 5,000 m. The placement of backfill is considered essential to reduce the ERR values ahead of the working faces. The footwall tunnels were placed 146 m below the reef, and they all pass beneath dip pillars where the near-field stress ranges from 93 MPa at 3,000 m to 170 MPa at 5,000 m. The surface stress concentrations on the tunnel surface are higher (Figure 76.13). The average pillar stress on the reef plane ranges from 320 MPa at 3,000 m to 520 MPa at 5,000 m.

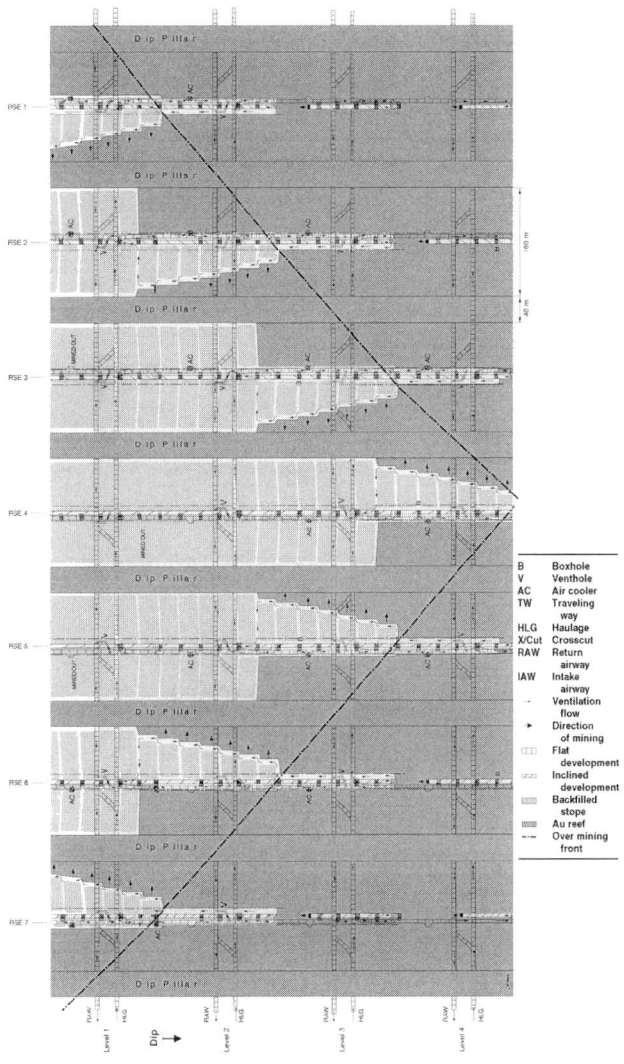

FIGURE 76.11   Macro concept of a sequential grid mining layout with dip-stabilising pillars

Twelve levels were required to extract the Iponeleng orebody, and the LOM is expected to be 31 years. A total area of 11.7×10^6 m^2 of the 19.7×10^6 m^2 resource is mined during this period, yielding an overall extraction ratio 59%, with 7.8×10^6 m^2 being left locked up in pillars and ground adjacent to geological features. The total length of off-reef development required is 290 km.

Evaluation.   The main advantage of the SGM is that the development of tunnels ahead of the stoping provides foreknowledge of potentially hazardous structures and enables them to be incorporated into stabilising pillars. It is also possible to practice selective mining, leaving behind areas of low-grade ore. One disadvantage of the layout is that the service excavations connecting the crosscut and reef are subject to large stress changes, which may compromise their stability. Changes as great as 190 MPa are possible at a depth of 4,000 m.

The sequential grid method requires a relatively high volume of ventilation air. This system requires fairly long ventilation holes, usually boxholes or boreholes, from the top of each stope or from the crosscut on the upper level to the return airway at the intake level (Figure 76.10). The ventilation system is, however, flexible and relatively easy to manage.

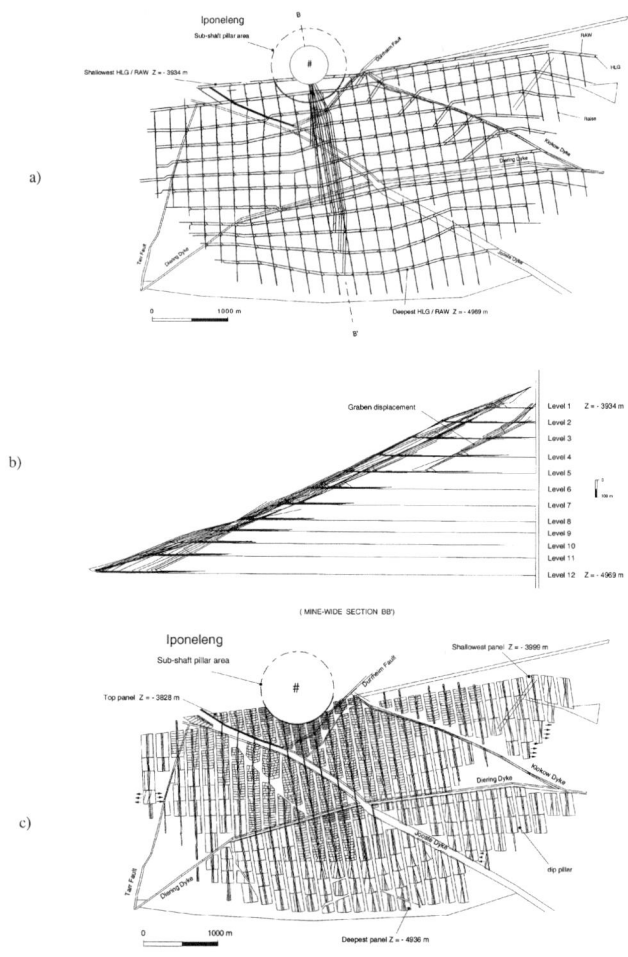

FIGURE 76.12 Application of the SGM micro and macro layout design concepts to the extraction of the Iponeleng orebody at Year 20 of 31 years of LOM. (a) depicts the extent of the development infrastructure, (b) shows the number and relative position of service levels, and (c) shows the regional stoping pattern.

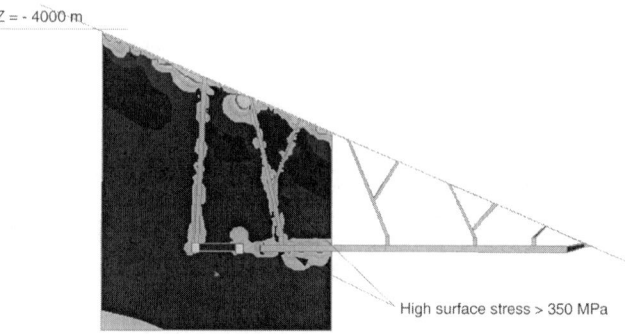

FIGURE 76.13 Stress environment in footwall excavation surfaces when placed underneath an SGM raise at a depth of 4,000 m

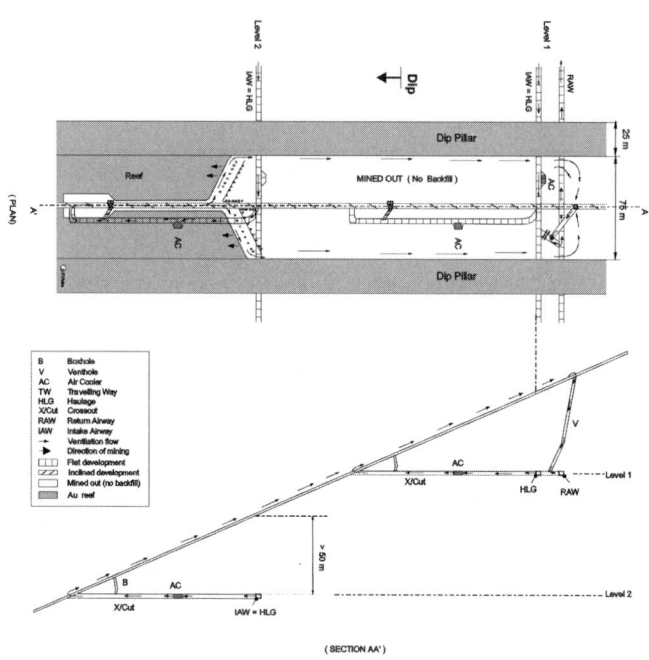

FIGURE 76.14 Micro concept of a sequential down-dip mining layout with dip-oriented pillars

76.2.5 Sequential Down-Dip Method (SDD)

Micro Layout. Haulages are driven on each level from the shaft to the block of ground to be mined (Figure 76.14). Alternate levels have both intake and return airway tunnels, while other levels have only intake airways. These haulages are developed prior to stoping, which allows faults and dykes to be located and pillars to be designed to bracket any hazardous structures. Crosscuts are driven until the reef is intersected at the foot of the planned stope. Reef drives are developed for 10 m along strike on either side of the raise line. A 20-m-wide raise is mined up-dip, overstoping the area above the timber and cooling car bays. Once the raise is 6 m beyond the box hole, the span of the wide raise is reduced to 12 m. The 240-m-long wide raise is mined at a rate of 20 m/month, until the crosscut on the level above is reached. A further 2 months is allowed for services such as winches to be installed before down-dip stoping begins. Two panels, one on each side of the raise and each with a face length of 46 m, are mined down dip at a face advance rate of 15 m/month/face. The broken rock is moved from the face area to the wide raise either by scraping or water jetting. It is then scraped to the box hole at the foot of the raise. Men and materials enter the stope from the upper crosscut.

The microventilation system is fairly simple. It consists of a dedicated intake at the bottom of the stope through a 12-m-wide raise, a 46-m-long down-dip panel to each side of the raise, and a return through the worked-out top section of the stope to the upper-level return airway. Ventilation control is somewhat difficult as faces are underhand and no backfill is used, which makes it difficult to channel the airflow close to the face. The build-up of pollutants in SDD stopes is not a major problem.

Macro Layout. A macro layout is shown in Figure 76.15. Mining rules do not permit stoping to take place concurrently on the same level on adjacent raise lines. While one raise line is being stoped, the next crosscut is developed. Because it cannot be assumed that the orebody is planar, development of the haulage beyond the crosscut breakaway is halted until a prospecting borehole has been drilled and the reef located, so that the precise position of the next crosscut breakaway is established. This process limits the rate at which haulages are developed and new raise lines are opened. The optimum global mining configuration is underhand, with stoping starting first on the upper levels, and expanding to either side of the centre of the mining block.

The macroventilation system is largely dictated by the production requirements. The airway system is fairly rigid and is relatively difficult to manage (Figure 76.16).

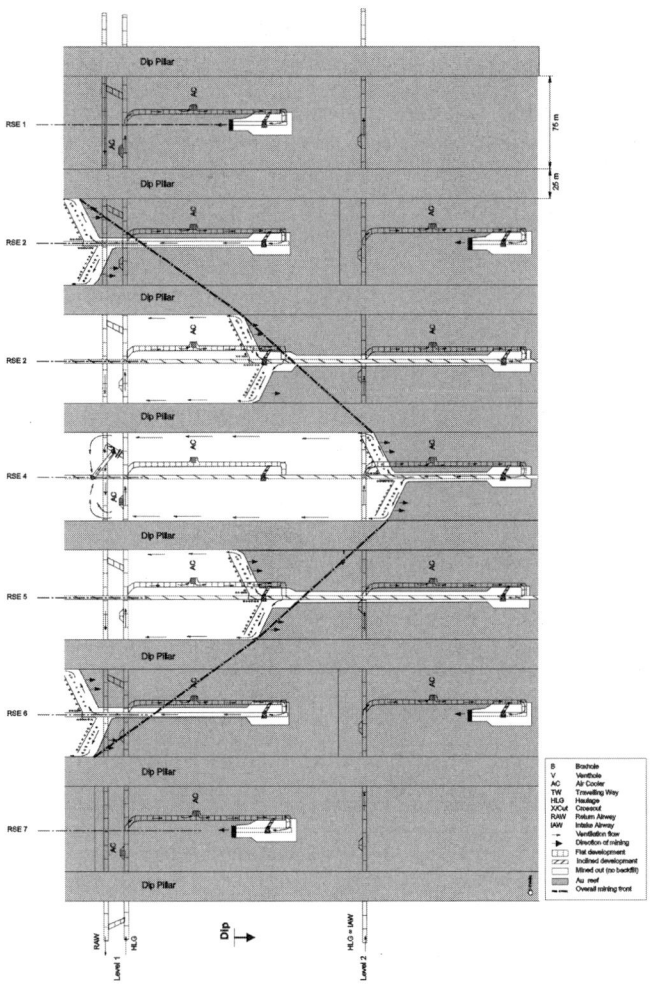

FIGURE 76.15 Macro concept of a sequential down-dip mining layout with dip-stabilising pillars

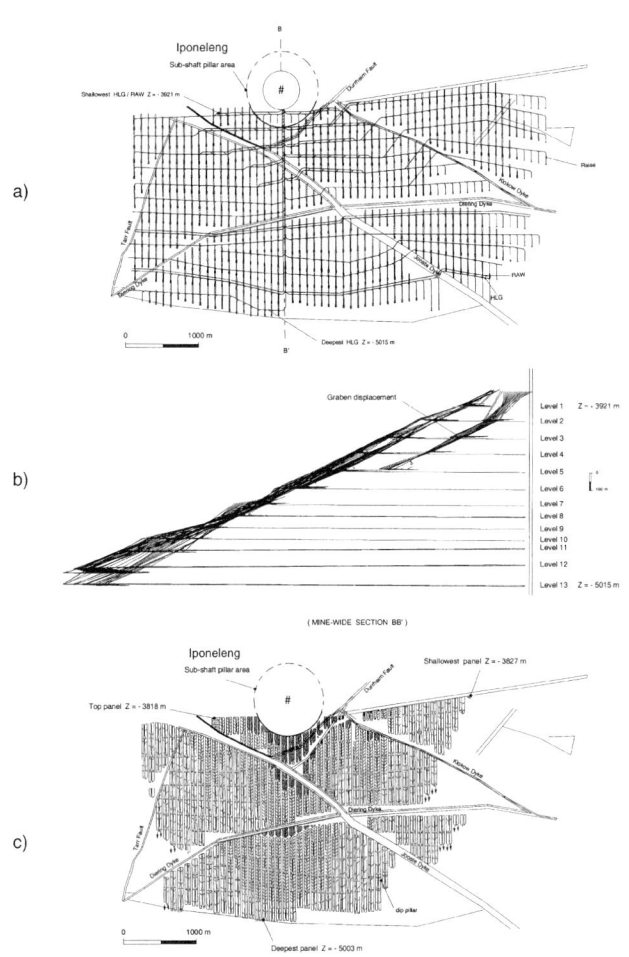

FIGURE 76.16 Application of the SDD micro and macro layout design concepts to the extraction of the Iponeleng orebody at Year 20 of 37 years of LOM. (a) depicts the extent of the development infrastructure, (b) shows the number and relative position of service levels, and (c) shows the regional stoping pattern.

Simulated Mining of the Iponeleng Model Orebody. The orebody model is first used to define viable mining blocks. The major decisions made by the design team are similar to those made for a sequential grid layout: the haulage and crosscut spacing, the width of the dip-stabilising pillars, and the depth of main haulages beneath the reef. The design team specified a 100-m crosscut spacing; haulages about 220 m apart yielding a back length of about 240 m; 25-m-wide dip-stabilising pillars; and main haulages at least 50 m below the reef plane (Table 76.2 and Figure 76.14). This layout ensures that ERR will not exceed 30 MJ/m². The maximum ERR value at 3,000 m depth is 15 MJ/m², and only 25 MJ/m² at 5,000 m. Because of the narrowness of the mined-out spans, very low closure rates occur and these closure rates increase only marginally with depth. It is unlikely that total closure will occur during the phase of mining, unless abnormal strata conditions exist. As a result, backfill is considered to be unnecessary. Haulage tunnels have been placed about 88 m below reef, and may experience some stress changes due to mining. The tunnel stresses range from 50 MPa at 3,000 m to 91 MPa at 5,000 m. While the layout gives rise to significant stress reversals in the footwall rock mass, little infrastructure is affected (Figure 76.13). The average pillar stress on the reef plane ranges from 310 MPa at 3,000 m to 510 MPa at 5,000 m.

Thirteen levels are required to extract Iponeleng orebody when applying the SDD method. (Figure 76.16). The LOM planned using this method is expected to be 31 years. During this period, a total area of 11.4×10^6 m² of the 19.7×10^6 m² resource is mined, with 4.7×10^6 m² being left locked up in pillars and ground adjacent to geological features, yielding an overall extraction ratio of 58%. The total length of off-reef development required is 344 km.

Evaluation. The main advantages of the SDD method are the very low ERRs, which make backfilling unnecessary, and the physical separation of the rock transporting function from the men and materials transporting function. However, this has serious implications for ventilation. First, very large volumes of air are required because it is difficult to keep the airflow close to the face in the absence of backfill. Second, the return airways have to be cooled because they are used as travelling ways, and the air exiting the stope would have gained heat while passing through the worked-out areas. One of the major disadvantages of this method is the dependency on one single short orepass with limited storage, which may constrain production output.

76.2.6 Closely Spaced Dip Pillar Method

Micro Layout. As in the preceding methods, haulages are driven from the shaft to the block of ground to be mined. Haulages used as intake airways are developed on every level, with a twin return airway on every second level (Figure 76.17). All footwall tunnels are developed prior to stoping, enabling

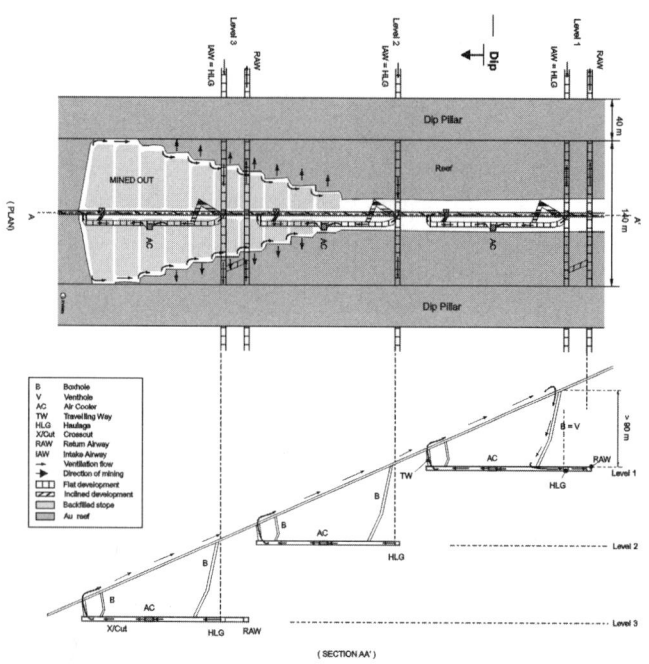

FIGURE 76.17   Micro concept of a closely spaced dip pillar mining layout

FIGURE 76.18   Up-dip continuous scraping facility currently on trial in a deep CSDP stope

faults and dykes to be located and pillars to be designed to bracket any hazardous structures. Crosscuts are driven until 12 m below reef, with a bay for timber and cooling cars on the side of the crosscut. A travelling way to the reef horizon is excavated, and a 20-m-wide raise is developed and then ledged in a down-dip direction. Only two orepasses service the entire back: the box hole near the foot is only used during the excavation of the raise, while the long orepass at the top is the passage for all the ore once production begins. Stoping only begins once the entire raise line has been ledged, supported, and equipped. Panels that are 25 m long are mined on breast towards the planned position of the dip-stabilising pillar at a face advance rate of 15 m/month/face. Panels on either side of the raise are mined simultaneously (i.e., double-side mining), beginning at the bottom of the raise. The start of the adjacent up-dip panel is delayed until a 5-m lag has been established, yielding an overall overhand configuration. The broken rock is moved from the face area to the raise either by scraping or water jetting, and then moved to the box hole at the top of the raise by means of a continuous scraper pulling up-dip (Figure 76.18). This method is the only one that considers some degree of continuous mechanisation in the cleaning operations.

The ventilation system has a dedicated intake at the bottom of the stope, a maximum of 24 overhand panels, and the ventilation air return is through a 10-m-wide ledged raise to the next return airway. Ventilation control is not difficult because of the use of backfill, a maximum strike distance of 70 m, and the overhand face orientation. Although backfill will reduce the heat load from worked-out areas, in-stope cooling will still be required at ultra-depth, which might complicate the control of ventilation air within the stopes. The build-up of pollutants in CSDP stopes could occur due to long residence time of the ventilation air, but this is not seen as a major problem.

Macro Layout.   A macro layout is shown in Figure 76.19. The mining rules of this method permit only one level along the same raise line to be stoped at any time. Preparing the excavations and installing services is a time-consuming process that must be taken into account during scheduling. One month is

allowed from the completion of the crosscut to the start of the wide raise; two months is allowed for the raise to be cleaned and services installed so ledging may begin; and six months is allowed between completion of the ledging and the beginning of stoping for the equipment to be installed.

Simulated Mining of the Iponeleng Model Orebody.   As in all the other methods, the orebody model (Figure 76.6) is first used to define viable mining blocks. The major decisions made by the design team are similar to those made for the sequential grid and sequential down-dip methods: the haulage and crosscut spacing, the width of the dip-stabilising pillars, and the depth of the main haulages beneath the reef. The design team specified a 180-m crosscut spacing; haulages about 220 m apart on plan yielding a back length of about 240 m; 40-m-wide dip-stabilising pillars; and main haulages at least 90 m below the reef plane (Table 76.2 and Figure 76.17). The ERRs ahead of the faces were determined. Modelling indicates that the ERR ahead of all working faces will be less than 25 MJ/m^2 at a depth of 3,000 m, and at a depth of 5,000 m the probability is high (0.96) that all ERRs ahead of working faces will be less than 30 MJ/m^2. The placement of backfill is considered essential to reduce the ERR values ahead of working faces. Haulage tunnels have been placed about 94 m below reef and experience only a small degree of mining-induced stress. The tunnels pass beneath dip pillars, where near-field stress regimes between 50 MPa at 3,000 m and 91 MPa at 5,000 m are possible. The average pillar stress on the reef plane ranges from 300 MPa at 3,000 m to 500 MPa at 5,000 m.

As many as 17 levels are required to extract the Iponeleng ore body (Figure 76.20). This method required substantially more levels than the other methods because the back length of a CSDP raise is substantially less. The expected LOM is expected to

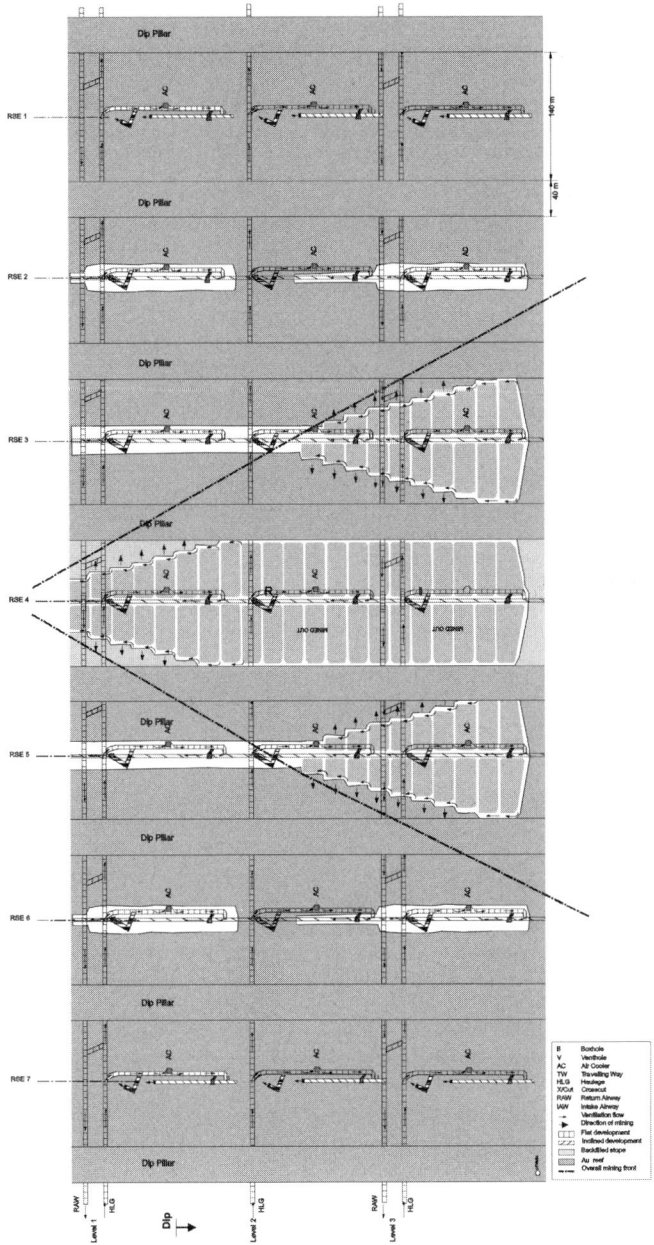

FIGURE 76.19   Macro concept of a closely spaced dip pillar mining layout

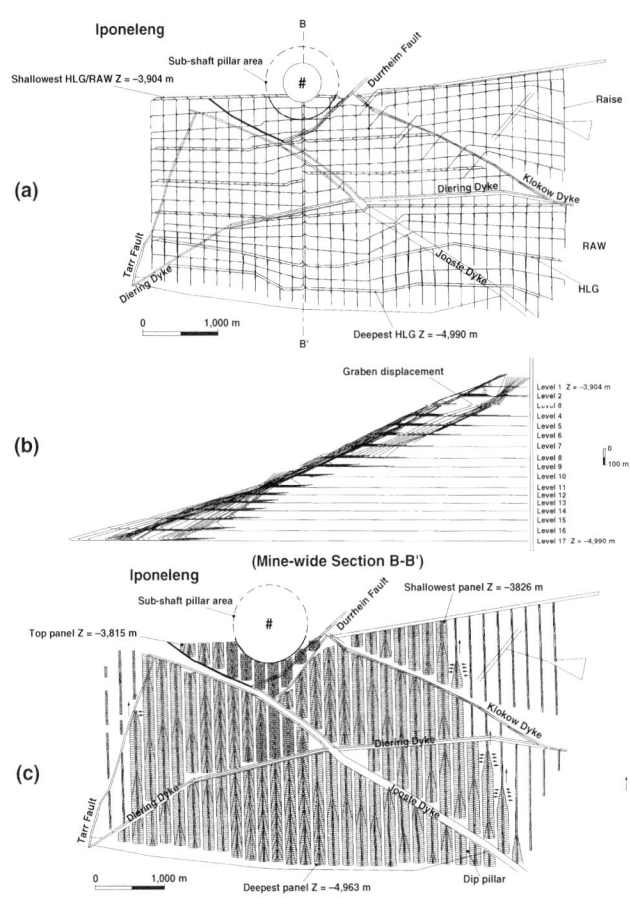

FIGURE 76.20   Application of the CSDP micro and macro layout design concepts to the extraction of the Iponeleng orebody at Year 20 of 26 years of LOM. (a) depicts the extent of the development infrastructure, (b) shows the number and relative position of service levels and (c) shows the regional stoping pattern.

be the least of all four methods, as only 26 years are required to extract the orebody. During this period, a total area of 12.2×10^6 m^2 of the 19.7×10^6 m^2 resource is mined, yielding an extraction ratio 62%, with a total of 5.2×10^6 m^2 being left locked up in pillars and ground adjacent to geological features. The total length of off-reef development required is 381 km.

Evaluation.   The closely spaced dip pillar method is characterised by highly concentrated stoping, with mining taking place on both sides of the raise and up to 24 panels per stope. Relatively high production rates are achieved because of double-sided mining on each raise line, and adjacent raise lines being stoped concurrently. Development is also fairly limited due to the limited number of stoping areas required for this method. These factors result in a reasonably low ventilation-air requirement, and the management of the primary ventilation system is simple.

76.3   CONCLUSIONS

This study of the mining of the ultra-deep Iponeleng model orebody has identified some strengths and weakness of each method. The study has demonstrated the importance of simultaneously evaluating the impact of a particular layout on the efficiency of cooling; the efficiency of transporting men, material, and rock; and the meeting of rock engineering criteria. Trade-offs will inevitably have to be made. This study has also highlighted the importance of carrying out multidisciplinary LOM simulations to assist in determining the economic viability of a mining venture.

The longwall method is ideal for a large mining block with a regular grade distribution, while the other methods offer greater flexibility where faults and dykes are prevalent or where the grade is erratic. The choice of mining method ultimately depends on the nature of the orebody, the critical mine design factors being the size of viable mining blocks and the regularity of the grade. It is conceivable that more than one method could be used on a single mine depending on the characteristics of the different mining blocks.

76.4   ACKNOWLEDGMENTS

The authors thank the DEEPMINE Collaborative Research Programme for permission to publish this paper, and also thank the many experts from industry and researchers who contributed to the definition and evaluation of the mining methods.

76.5   REFERENCES

Applegate, J.D. 1997. The successful introduction of sequential grid mining at Elandsrand G.M. Co.Ltd. Ass. of Mine Managers of South Africa, Papers and Discussions. pp.132–145.

Beck, A.I., G.H. Henderson, R.N. Lambert, and R.A. Mudd. 1961. Stoping practice on the Transvaal and Orange Free State goldfields. *Trans. 7th Commonwealth Min. Metall. Congr. Papers and Discussions.* SAIMM , Vol. II, p 655–697.

Cook, N.G.W. and M.D.G. Salamon. 1966. Report on the use of stabilising pillars for stope support. C.O.M. Ref.: Project No. 107/66. Res. Rept. 43/66.

Cook, N.G.W., E. Hoek, J.P.G. Pretorius, W.D. Ortlepp, and M.D.G. Salamon. 1966. Rock mechanics applied to the study of rock bursts. *J. S. Afr. Inst. Min. and Metall.*, pp. 436–528.

Cook, N.G.W., J.W. Klokow, and A.J.A. White. 1973. Practical Rock Mechanics for Gold Mining. P. R .D. Series No. 167, Chamber of Mines of South Africa, Johannesburg.

Diering, D.H. 1987. Regional support at Western Deep Levels, Limited. Ass. of Mine Managers of South Africa, Papers and Discussions. pp. 1–74.

Diering, D.H. 1997. How the ultra-deep levels will be mined. *Mining Weekly*, as reported by Martin Creamer (ed), Nov. 28–Dec 4. Johannesburg. South Africa.

Gibson, M.A.S., S.J. Jolley, and A.C. Barnicoat. 1999. Interpretation of the Western Ultra Deep Levels 3-D Seismic Survey. South African Geophysical Association 6th Biennial Conference and Exhibition (28 September–1 October 1999).

Gürtunca, R.G., A.J. Jager, D.J. Adams, and M. Gonlag. 1989. The *in situ* behaviour of backfill material and the surrounding rock mass in South African gold mines. *Innovation in Mining Backfill Technology.* Balkema 1989. Rotterdam, pp. 187-197.

Gürtunca, G. 1998. Mining below 3000 m and challenges for the South African gold mining industry. In *Mechanics of Jointed and Faulted Rock,* (ed. Rossmanith, H-P) (Balkema, Rotterdam), pp. 3–10.

Hagan, T.O. 1987. An Evaluation of Systematic Stabilising Pillars as a Method of Reducing the Seismic Hazard in Deep and Ultra-deep Mines. Ph.D. Thesis, University of the Witwatersrand, Johannesburg.

Hindle, E.V. 1957. Description of the mining and sand-filling of the Black Reef deposit of Government Gold Mining Areas (M) Consolidated, Limited, with a note on the method of sluicing sand out of tanks. Association of Mine Managers of South Africa, Papers and Discussions 1956–1957.

Lenhardt, W.A., and T.O. Hagan. 1990. Observations and Possible Mechanisms of Pillar-associated Seismicity at Great Depth. International Deep Mining Conference: Technical Challenges in Deep-Level Mining, Johannesburg, S. Afr. Inst. Min. Metall., pp. 1183–1194.

Napier, J.A.L., and S.J. Stephansen. 1987. Analysis of Deep-level Mine Design Problems Using the MINSIM-D Boundary Element Program. *Proc. 20th Int. Symposium on Application of Computers and Mathematics in Mineral Industries.* Vol. 1: Mining. Johannesburg, SAIMM. pp. 3–19.

Ortlepp, W.D. and K.E. Steele. 1975. The nature of the problem and the countermeasures on East Rand Proprietary Mines, Ltd. Ass. of Mine Managers of South Africa, Papers and Discussions. pp. 225–278.

Piper, P.S. and J.A. Ryder. 1988. An assessment for regional support in deep mines. Backfill in South African Mines. *J. S. Afr. Ins. Min. Metall.*, Johannesburg, pp. 111–136.

Wagner, H. 1975. The application of rock mechanics principles to strata control in South African gold mines. *Proc. 10th Canadian rock mechanics symposium.* Dept. Mining Engineering, Queens University. pp. 247–280.

Willis, R.P.H. 1997. Towards an integrated system for deep level mining using new technology. *Proc. 4th Int. Symp. on Mine Mechanisation and Automation,* Brisbane, Vol. 2, pp. A9–A14.

CHAPTER 77

Simulation of Underground Mining Operations

Daniel T. Brunner*

77.1 INTRODUCTION

The type of simulation discussed in this chapter is more formally known as "discrete-event system simulation." What is discrete-event simulation, and how does it work?

77.1.1 Characteristics

Comparison with Other Types of Simulation. One way to characterize discrete-event simulation is to compare it with other types of simulation. The goal of any simulation is to mimic something. Typically, there is a stimulus/response interaction. In a flight simulator, the goal is to mimic the visual and tactile interaction between a person and a physical system (the aircraft). In a circuit simulator, the stimulus is the input signal, and the goal is to mimic the response of physical circuit components. Manufacturers of complex parts or assemblies often want to simulate just the visual appearance of the item. But for some parts (e.g., a gear), it is also necessary to simulate the physical stresses (using finite element analysis); and for assemblies (e.g., an automobile), engineers might simulate the response of the suspension to road bumps, the way the vehicle moves through the air, and many other physical factors. All of these examples are primarily *physical* models.

The Time Element. Some simulation types are static—meaning the effect (appearance, load-bearing capacity) is observed only at a single instant of time. But for discrete-event simulations, the goal is almost always to study the behavior of the system over a period of *time*. For this reason, discrete-event system simulations are sometimes called "dynamic simulations."

Focus on Logic. Discrete-event system simulations focus on the *logical* behavior of a system more than on its physical behavior or visual appearance. This is illustrated in Sections 77.2 and 77.3 by the lists of typical inputs and outputs. Usually discrete-event models are built to run with built-in logic (i.e., without human intervention).

Discrete Events. As the name implies, discrete-event simulations typically model systems whose states change at discrete points in time. The beginning of a drilling cycle, a skip reaching the surface, and the contents of a pass reaching a critically low level are all examples of discrete events. This is in contrast to *continuous* system simulations, where nonlinear behaviors (chemical reactions, heat transfer) must be modeled using equations. Interestingly, material-flow systems, where liquids or bulk solids move at constant rates, *are* candidates for discrete-event simulation. The events are the beginning and ending times of a particular flow.

Randomness. Discrete-event simulations also typically include representations of randomness. Simulations where multiple trials are made with different outcomes depending on inputs that randomly take on different values from trial to trial are called *stochastic* simulations. (Not all stochastic simulations are dynamic. The *Monte Carlo simulation*, not otherwise discussed in this chapter, is a commonly used name for static stochastic simulation. If you know the probabilities that various components of an LHD will independently fail today, and if you want to know the probability that the whole LHD will fail today, you could run a static experiment a thousand times (or a million times) to see how likely it is that the LHD will fail. This particular problem might also be solved mathematically, but a stochastic Monte Carlo simulation might be easier to set up for some types of situations.

A comparison of various types of simulations appears in Table 77.1. This is presented to help clarify the role of discrete-event system simulation.

77.1.2 Simulation Basics

Simulation can be thought of as a *framework* for describing a system's operation. Describing a system using this framework can sometimes be very beneficial even if the model is never used for experiments—and could even be very beneficial if the model is never implemented! It is worthwhile to explore the framework.

Entities and Resources. In a simulation-oriented system description, *entities* flow through a series of *resources*. Entities are units of work or units of traffic. They queue up for and use resources according to logical rules. Resources are generally constrained.

In a simplistic model, each entity might represent some quantity of material (ore or waste), and the resources might be people, LHDs, or storage areas. In a more complex model, there might be entities that represent the logical controllers of the system (human managers as well as automated controls). Equipment and operators might also be represented as entities.

For more information on how entities and resources are handled in the inner workings of simulation software, see Schriber and Brunner (1999).

Interactions. Another distinguishing element of discrete-event simulation, as mentioned previously, is the passage of *time*. This type of simulation captures interactions in a way no static technique can. There are many, many dynamic cascading events in an underground mine. A given task is subject to waiting for blasts, waiting for repairs, waiting for adjacent work to be completed, and so forth.

Random Variables. Finally, most discrete-event simulations include representations of random variations. Using point estimates (averages) as inputs to any kind of model can cause interactions to be overlooked. This is typically most obvious in the case of failure modeling, but many other aspects of mine behavior can be represented by sampling from a probability distribution.

---

* Systemflow Simulations, Inc., Indianapolis, Indiana.

TABLE 77.1   Types of Simulation

Description	Is time-based?	Has a human in the loop?	Focuses on effects of laws of physics?	Includes logical detail (rules)?	Produces system visualization?	Includes randomness?
Flight, Driver, Operator	Yes	Yes	Yes	Yes	Intrinsic	Usually
Training game	Yes	Yes	Usually not	Yes	Sometimes	Sometimes
Circuit	Sometimes	No	Yes	No	Usually not	Usually not
Finite element analysis	Sometimes	No	Yes	No	Sometimes	Sometimes
Continuous systems (e.g. ballistics, chemical reaction, heat transfer)	Yes	No	Yes	Sometimes	Sometimes	Sometimes
Monte Carlo	No	No	Sometimes	No	Usually not	Yes
Discrete event	Yes	*Usually not*	*Usually not*	Yes	*Usually*	*Usually*

As a result of random inputs, model outputs are also typically random variables that need to be considered from a statistics point of view for proper interpretation.

77.1.3   Simulation Inputs

Model inputs fall into these broad categories: logical data, system description data, process data, and demand data. Examples from mining are presented below.

Logical Data.   Logical data includes all rules for operating the system. What activities must halt when a blast is scheduled? Who decides which scoop to assign to a pending mucking task, and how is that decision made? What shift schedule do the jumbo drill operators follow? Is there a stope sequence that is to be followed in the model, or will the model make sequencing decisions dynamically; and if so, how?

System Description Data.   System description data represents the physical system. This includes mine geometry, material properties, and equipment lists. It is often useful to break the geometry into individual *material blocks* that may have unique properties. (This is not the same as modeling small bits of material as individual entities that move through the system—a computationally expensive approach that can usually be avoided.)

Process Data.   Process data is the rates and speeds that constrain system performance. Operator and equipment performance, equipment failure, hoisting rates, conveyor speeds, and similar data fall into this category.

Demand Data.   Demand data drives the model. In other contexts, the demand data is how many cars are we trying to make or how many boxes need to be shipped today. Demand data is not typically a major factor in a mine model, because the model is generally set to go all out given the other constraints in place. However, there may be cases when the model is set to start only certain material blocks at certain times, or to stop before the end of any shift once a certain tonnage is hoisted; and the goal is to see how the equipment and operators are utilized.

Data for a mining model is often difficult to gather and reduce to a usable form. The logical data may be incompletely understood by a single individual, and a team of people may have trouble agreeing on what practices are or will be followed underground. Different miners have different ways of doing things. System description data may exist but may be difficult to translate into a form usable by the model—some automated way of doing this must often be found or developed. To be used in the model, process data—particularly failure data—must not only be gathered in raw form if it does not already exist, but must also be analyzed statistically so that it is validated and properly prepared.

77.1.4   Simulation Outputs

Are simulation outputs always statistical? No, for two reasons. First, an important benefit of a well-done simulation project may be the insights gained by describing the system in the framework

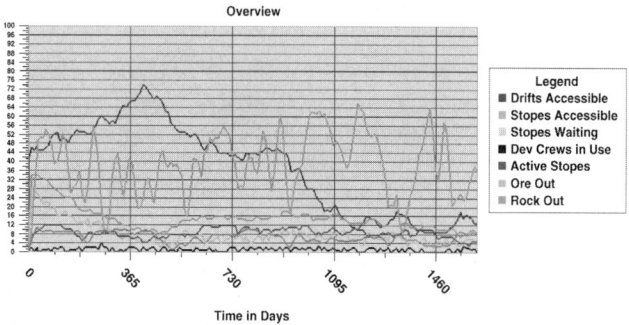

FIGURE 77.1   Sample time plot from a long-range planning simulation, showing an underdeveloped condition and slightly deminised ore output about one year into the simulation. The process detail in this geometry-focused model was minimal.

of simulation. This includes gathering and analyzing the raw process data as well as developing the logical rules for system operation. Second, *animation* is used to learn even more, and demonstrates to analysts, operators, management, and others that the model is valid.

Statistics.   Statistical outputs generated by a model can be derived from a standard output report or customized to suit the model user. Typical outputs include resource statistics, queuing statistics, and other summary information. The model can also produce any other measurable statistic that is consistent with the model's level of detail (e.g., the duration of various activities, time series for plotting, trace files for understanding the model logic, and resource tables [schedules]).

For resource statistics, *utilization* statistics must be well defined. As a simple example: Would the utilization of a drill be the operating time divided by (a) total time, (b) total time minus maintenance shifts and shutdowns, (c) time when crew members are available to run the drill, (d) operating time divided by the sum of operating and repair time, or (e) something else? If an organization has a standard way of presenting this information in current reports on actual mine activity, the simulation can easily be tailored to match this system. One should not assume that everyone agrees on what "utilization" means if this subject has not been addressed.

Time plots, whether generated by the simulation software or plotted from data produced by the model, can provide many insights. A sample time plot is in Figure 77.1.

Animation.   A simulation model can be animated in many ways. Animation can be 2-D or 3-D; it can show people and equipment moving or only show material state changes; it can be pictorial or schematic; it can be to scale or not to scale; and it can be delivered with or separately from the model.

All these forms of animation are valuable for verifying and validating the model. They are also extremely useful for achieving buy-in from other operating personnel as well as top management. A model isn't going to be considered useful unless everyone understands that it is valid; and it is not going to be considered at all if people do not understand what it is.

77.1.5 Steps in a Simulation Project

A typical simulation project consists of the following steps:

1. Define the simulation project objectives.

2. Define the project scope. What will be the deliverables? What level of detail will be required to achieve the project objectives?

3. Plan the project. Figure out who is going to do what, and when.

4. Formulate the model. Plan how the various system components will be represented. Develop the input data requirements. Begin planning the experiments, which may drive the output requirements.

5. Collect the data. Data is both physical (speeds, failure rates, quantities to be processed) and logical (the entire logical specification of the system). This phase may overlap with model construction, as not all data (particularly physical data) is immediately required to construct most models.

6. Construct the model. If the logic is complex (typical in a mine simulation), a parallel design specification and/or flowchart development effort can be done at this time and will later serve as the basis for the project documentation.

7. Verify the model. Verification is the process of making sure the completed model is logically correct—that it does what the modeler intended (i.e., correctly implements the specified logic). Verification can be performed during model construction (i.e., testing the logic piece by piece) as well as after the model has been built (testing the overall model logic in a planned way, with varying inputs). (For more on simulation verification as well as validation, which is discussed in the following paragraph, see Balci 1998.)

8. Validate the model. Validation is the process of making sure the completed and verified model is an accurate representation of the system being modeled—that the logic specification is correct or, more precisely, correct enough for the problem at hand. It is normally difficult to estimate the time required for this step because "go back to step 6" or even "go back to step 4" is a possible outcome. Note: If whole-system verification was not performed—and often it is not cost-effective or reasonable from a schedule standpoint to perform exhaustive verification—then the process of validating the model may also uncover some logic problems, particularly in the case of heavily data-driven models, where virtually unlimited combinations of input data can be presented.

9. Perform experiments. Sometimes the early experiments are ad hoc, in which case this step could be considered part of the validation process. The experimentation phase may also trigger new rounds of model changes or additional experiments, so this phase can also be difficult to estimate at the beginning.

10. Prepare and deliver the final documentation and/or presentation.

11. Implement the project's recommendations. Sometimes implementation can begin before the simulation project is complete. The insights gained from Steps 1 through 6 can be substantial even before the model is completed. That is the value of the simulation framework in studying any problem.

Alternative lists of the steps in a simulation project appear in general simulation textbooks such as Banks, Carson and Nelson (1995) and Law and Kelton (1999).

77.2 SIMULATION OF UNDERGROUND MINING OPERATIONS

This section is a general discussion of the various aspects of underground mining simulation. The first four sections discuss ways in which one model can differ from the next, and the other four sections provide other useful information.

77.2.1 Uses and Benefits

Simulation has many applications in mining and mineral processing. In underground mining, simulation has the following uses:

- Analysis of proposed capital expenditures
- Analysis of operating procedures
- Analysis of plans and schedules
- Understanding and communication of system behavior
- Day-to-day decision support

Benefits include increased production, capital cost savings, operational cost savings, and improved forecasting. Simulation achieves these benefits by:

- Identifying and helping eliminate bottlenecks
- Enhancing equipment and staff utilization
- Generating plans, schedules, and forecasts that can be implemented with confidence
- Helping to develop more effective operating procedures

There are many ways simulation can be used in planning and operating underground mines. Variables include the model time frame, the model focus, and the model objectives.

77.2.2 Model Time Frame

Another way one model differs from the next is in the scope of its time frame. The time frame of a model depends in part on its intended use.

- Short-term models (which might run for up to a few simulated months) are used to evaluate operating policies, to schedule the operations directly, or to assess the impact of exception conditions such as failures or absenteeism.
- Medium-term models (which might run for up to a few simulated years) are used to evaluate equipment plans and scheduling and operating policies.
- Long-term models (which might run for a simulated decade or more) are used to evaluate long-range mine plans, look for development bottlenecks, and so forth.

Of course, simulated time is not the same as real time. A properly designed model can usually execute one replication in a matter of minutes on a personal computer, even when the focus is long term. (Longer-term models usually require proportionally less detail than shorter-term models.)

77.2.3 Model Focus

Along with the many different objectives possible in underground mine modeling, the areas of focus can also vary widely. Some general areas of focus include:

- The development process
- The production process
- Supporting processes such as services, materials (including fill material if applicable), and repairs
- Material handling including muck movement by vehicle; bin flow with passes, crushers, hoppers, and other intermediate storage; and hoisting and removal
- Trucking operations
- Operator training

It is important to limit the focus of the model, particularly if the analysts and end users involved (see Section 77.3 for definitions of these terms) are new to simulation. A long, drawn-out first project can be difficult to complete and nearly impossible to verify and validate.

The challenge is to be able to limit the focus without sacrificing the desired level of fidelity. For example, in building a 20-year mine model, if the objective is to confirm qualitatively a perceived underdeveloped condition, the focus on things like equipment failure or even equipment allocations can be very light. But if that model is to be adapted for a detailed comparison of two mining methods or two models of LHD, then the required level of detail increases. If the model is to be used for *absolute* answers such as "what is the rate of return on this project," then even more attention to detail is required—not only to the detail put into the model, but also to the technical correctness of the experiment design, output analysis techniques, and results presentation.

77.2.4 Model Objectives

The objectives for modeling an underground mining operation can vary widely. Typical objectives may include some or all of the following:

- Equipment type comparisons
- Specific capital purchase decisions
- Mine plan analysis (discussion of integrated planning and modeling)
- Mining method comparisons
- Operating policy evaluation and improvement (equipment deployment decisions, equipment and crew assignment decisions, sequencing)

A screen snapshot from a mine plan analysis model is shown in Figure 77.2. This particular model has been used and adapted in a variety of ways to study short- and long-term planning decisions.

There are animation screen shots and clips available from other models, but they are not included in this paper.

77.2.5 Geometry and Database Issues

Much of mine modeling is data driven, although the logic can also be complex. The steps for modeling an underground mining operation include:

1. Define the mine geometry.
2. Create a database based on the geometry (location, dimensions, mass, volume) of each defined block of material.

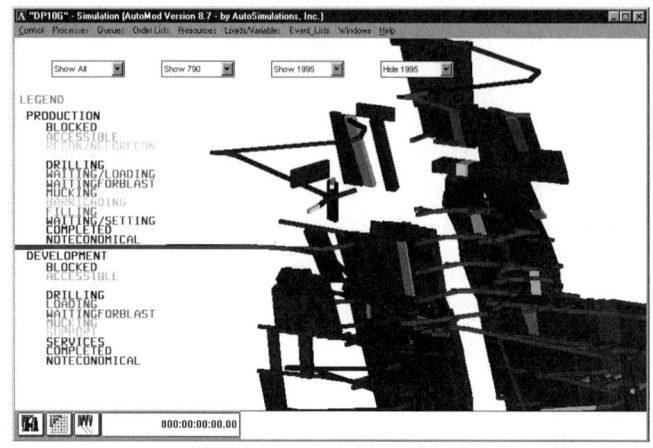

FIGURE 77.2 Mine plan analysis model

3. Populate the database with attributes (material properties, priorities, cost codes).
4. Populate a resource database (equipment, manpower, shifts) with related tables (e.g., equipment failure data by type or by age).
5. Define and code the rules of operation.

Managing the data is one of the two major efforts required, and it is ongoing throughout the course of a project. It is important to separate the data from the logic and to develop a method for naming and tracking various data and logic versions.

77.2.6 User Education Issues

While discrete-event simulation is widely accepted in some areas, such as automotive and semiconductor manufacturing, as well as distribution and warehousing, and is well known in others areas such as telecommunications and transportation, it is just now becoming more widespread in mining. The author knows of six individuals in North America who are currently outside consultants specializing in simulating mining operations—compared to two or three just a few years ago. (There are others who specialize in mineral processing simulation or who do both. And, of course, an indeterminate, but surely growing, amount of simulation is being performed internally by staff personnel.)

This early stage of acceptance means that organizational commitment—one of the four keys to simulation success—may take a little more work in the typical mining organization than in some other fields. The key to gaining organizational commitment is education. A person who intends to champion simulation in an organization where it is new should be prepared to educate the various constituencies described above to bring success to the organization. Fortunately, new tools such as web-based interactive mine simulation examples are becoming available to help with this. See Brunner (2000) for a web-based example.

77.2.7 Miscellaneous Lessons Learned

- Take the time to know and to document the rules of operation.
- Avoid detail inappropriate to the time frame and objectives.
- Be sure organizational commitment is present to support simulation data requirements.
- Be sure operating personnel and other decision or policy implementers have bought into the process.
- Expect to spend some money.
- Use animated graphics where possible and feasible.

77.2.8 Key Fact

Allocating and assigning equipment (and units of work–i.e., "which face will this crew work on during this shift") in an underground mine can present major challenges in formulating the model framework. Experience shows that this is more difficult in mine simulation than in many other discrete-event simulation applications. Often the rules by which decisions are made underground are not well known or understood and can be inconsistently applied in practice. Clearly, the process of trying to capture this information is valuable in itself.

77.3 UNDERGROUND MINE SIMULATION EXAMPLES

This section contains some examples from the author's personal experience in simulating underground mining operations. There are other examples described in other sources, such as Sturgul (1995).

Each example is presented with the project goals, model structure overview, description of animated graphics, notable challenges, results summary, and, if applicable, lessons and other benefits gained from the project.

77.3.1 Bulk Material Handling

Goals. The objective was to assess the tramming and hoisting capacity of a particular mine where the primary method was VRM. The time frame was originally long term, and the focus was on developing a highly detailed model.

Structure. System elements included underground rail vehicles and their track configurations, trucks, passes, conveyors, and a two-skip hoisting operation.

Graphics. The material being processed was not displayed graphically *in situ*, but did appear on bulk conveyors, in passes, and in skips. Individual vehicles were shown as appropriate. The animation was three dimensional and to scale, which made watching it more difficult in some ways because of the large area covered by the mine and the relatively small size of the animated elements.

Challenges. The big challenge in this model was how to model the detailed management rules used to operate the mine. This includes decisions such as when to break balance and when to haul ore versus when to haul waste in the various handling subsystems. The solution was unique–let the person operating the model make these decisions.

Results. This model was eventually abandoned, because it proved too difficult to operate due to the requirement for a person in the loop. Also, experimental replication is impossible for such a model.

Lessons and Other Benefits. The development of this model did generate experience with bulk-material-handling simulation (conveyors and vessels). That experience became the basis for several future models and also for one commercial simulation software vendor's implementation of bulk-handling vessels in their software.

77.3.2 Long-Range Planning

Goals. The objective was to assess the 20-year mine plan for a particular mine having VRM and URM stopes. The time frame was long term, and the major output was yearly tonnage statistics and long-term time plots of mine activity.

Structure. System elements included "material blocks," each representing a drift segment or major stope. The primary model input was a database of these material blocks and their desired sequence. In what was initially a two-step process, the model would preprocess the block database to develop physical predecessor and adjacency relationships. There was no equipment or crew. Development progressed in "feet per shift" and various production subprocesses were handled in a similarly coarse way.

Graphics. This animation was three-dimensional, but equipment and other resources were not shown. Instead, the animation showed a to-scale representation of each material block, and used color changes of these blocks to depict the progress of the simulation over time. This is the animation depicted in Figure 77.2.

Challenges. The big challenges in this model were to develop the database of stopes and drift segments in the model and to represent those stopes and drift segments graphically on the screen. The (2-D) mine plans existed in AutoCAD, but a database of lines does not contain a lot of other needed information. The solution was somewhat labor-intensive. An operator placed centerlines on the floor of each drift segment, and the geometry was then converted from AutoCAD to a simulation-compatible format. From within the simulation software, an operator tagged each line with various attributes—ore vs. waste, ramp vs. cross-cut—and adjusted the geometry to a 3-D representation. Meanwhile, an engineer developed the stope database without a graphical tool. Grade data, mining method, desired sequence, and geometry were among the elements of this database. The preprocessor (mentioned above) took the drift and stope databases and created the geometric shapes displayed in the model.

Results. With its deliberately simplified process detail, this model quickly executed 20 years of operation for the mine for a given data set and produced the desired results. Figure 77.1 (in Section 77.4) shows the output from this model.

Lessons and Other Benefits. Although the data mechanism effectively tied this version of the model to the mine being studied, the project provided many insights for future work in integrating other data sources with the same model so other mines could be modeled. However, further use of the model for the original mine has been limited to date, perhaps in part because mine-operating personnel were not initially involved with the project.

77.3.3 Narrow-Vein Mining

Goals. The objective in this case was to study the productive capacity of a small orebody. The time frame was short term. The focus was on rapid model development.

Structure. The system consisted of a fixed set of six production faces. In the model, the faces stepped through nine different production states (a mixture of "wait-for" and "working" states) in a continuous loop (periodically pausing for cleanup, filling, and development activities). The model contained individual miners, each capable of doing any task, as entities. Each miner, following a fixed shift schedule, would look for work according to a list prioritized first by task type and second by which face was waiting (each face had a preassigned priority). Equipment was constrained for each of the four production operations (drill, load, muck, bolt) and the sandfill source was also constrained.

Graphics. The animation was two-dimensional and schematic (showing each face hopping through its sequential states and each miner either at a face or in an idle queue). This very simple approach proved quite adequate for the task at hand.

Challenges. This was a rapidly developed model with many simplifying assumptions. The only challenge was for a model with no representation of physical space constraints, operator or equipment movement times, equipment failures, material movement constraints, or mine configuration evolution, to be sufficiently useful for experimentation.

Results. This model helped the mine firm up rates for long-range planning purposes and to find balanced combinations of face count and head count. It showed daily variations in tonnage in a time plot inside the model.

Lessons and Other Benefits. This model got a surprising amount of use considering its short model development cycle. In a general sense, it demonstrated the ability of a rapidly developed model to provide answers given a sufficiently narrow scope and agreed set of assumptions.

77.3.4 Trucking and Material Flow

Goals. The mine operators wanted to know how much benefit could be gained from adding another truck to an existing system under various real-life operating conditions.

Structure. As with many models, the physical scope is difficult to restrict to one subsystem. We decided early on that the supply side would be driven by a production schedule in a data file (generating quantities of ore and waste according to a defined plan), and that the dumping constraints would be realistically modeled by modeling the network of passes and bins all the way to the surface using vessels and flow rates.

Graphics. Truck movement was animated as well as the current level of each pass and storage bin. The trucking system was three-dimensional (the paths had to be to-scale anyway) while the bin network was a separate two-dimensional schematic representation. Animation was an important verification and validation tool in this model.

Challenges. This model was really two models in one: a trucking system and a material-flow system. Both presented many challenges, and the project had limited funding considering the detail involved. With the trucking system, the path geometry was imported from AutoCAD already in 3-D and translated into the simulation software without much difficulty. But the trucking logic—which resembles some types of rail simulation because many sections are one-way—took some time to define and implement. Under exactly what circumstances, for example, does a downward-bound truck decide to pull into a passing bay to wait for a full upward-bound truck to go by? The rules also required some definition for the pass/bin/hoist network. When and how should ore/waste and waste/ore boundaries migrate through the network?

Results. Validation of the material movement network was difficult because of the ways small changes in the flows could break the defined logic (e.g., could spotlight loopholes or special cases previously unimplemented in the logic). Eventually these difficulties were overcome, but the model's tendency to run into problems with ore/waste boundaries persisted. It turned out that these problems were also being experienced in the actual mine, which was a valuable insight into the mine's problems. Using the model to test proposed solutions (new rules) and extensive truck experimentation both had to wait, however, for additional funding.

Lessons and Other Benefits. One of the difficulties with this model was that some of the logic must change as trucks are added. This can be time consuming. Then, when the model is complete, it is difficult to reapply it to a different mine because the operating constraints as well as the physical geometry are all different, and the trucking system logic depends, at least to some extent, on both.

77.3.5 Comparative Analysis

Goals. The model developed in Section 77.3.2 has been reapplied to a number of different mines with different goals each time. Sometimes different mining methods were compared. Other times the logical rules, equipment mix, and shift schedules were varied.

Structure. Various new features were added to the basic structure of the Section 77.2 model during repeated enhancement cycles. Detailed crew and equipment modeling, including failures, is a significant enhancement as the model time frame has shifted from 20 years to closer to 1–5 years. Other mining methods were added in varying degrees of detail including sublevel retreat, blast hole, and mechanized cut-and-fill. More logical control was added for the model users.

Automated stope sequencing was implemented. The data interface was adapted to use a commercial mine design package as the initial data source (although this is not fully automated yet).

Graphics. As with previous versions of this model, the animation graphics consist of three-dimensional material block representations that change color, although the number of states depicted via color changes has increased.

Challenges. It is challenging to adapt and enhance a single model for use in many different mines and contexts over a five-year life (at this writing). Methods for managing the growing code base evolved as different people worked on it, but it inevitably became more difficult to make major changes without requiring major rework elsewhere in the model. Unscheduled downtime modeling presented some major challenges both because the available data needed to be carefully cleaned up and analyzed for proper use, and because the repair operation (as of this writing) was not included in a detailed way.

Results. This model has been successfully applied in several different mine designs to prove the advantages of one mining method over another or one set of equipment specifications over another.

77.4 GUIDELINES FOR SIMULATION SUCCESS

What elements must be in place for an organization to use simulation successfully? This section presents four essential components. The application must be appropriate; sufficient resources must be in place; the organization must be committed to the technique; and simulation projects must be managed effectively. A different but useful treatment of this subject is in Cesarone (2000).

77.4.1 An Appropriate Application

First, the application itself must be appropriate. Do discrete units of traffic compete for scarce resources over time? (Probably.) Is everything that can be known about the system is already known? (Of course not.)

Is the level of detail in the model in line with the project goals? This is an area than can trip up first-time simulation users. The author built a 20-year planning model with *no* moving vehicles, *no* randomness in the process times, *no* discrete pieces of equipment, and a host of other simplifying assumptions. But the goal was to demonstrate the degree of underdevelopment in the mine, and the output was highly useful. Had the detail in this first model been too much, everyone might have gotten discouraged. The modeling costs would have skyrocketed and useful results might never have seen the light of day. (This is the model described in Section 77.2.)

Finally, the simulation must be timely. One of the problems in some nonmining simulation applications, such as distribution center design, is that even though the cost of a simulation can be easily justified, there simply is not time to do one. A good simulation study can require anywhere from two weeks to two years (or more) to complete. (Why is there such a wide range? See the above comment about using the appropriate level of detail.)

77.4.2 The Right Resources

Second, the right resources must be in place. Simulation resources include software (and computer hardware, which is not discussed here) and qualified personnel.

Simulation Software. Discrete-event simulation software generally runs on standard PCs, but they should be on the fast side and, depending on the software, may benefit from the latest graphics hardware. Simulation software tools that are known by the author to have been used in mining applications include—from among off-the-shelf simulation products—AutoMod,

AweSim (formerly SLAMSYSTEM), GPSS/H,[*] Proof Animation, SLX, ProModel, WITNESS, and Arena. Other general- or special-purpose simulation tools could certainly be used. A complete list of standalone discrete-event simulation software tools is in Swain (1999). The Swain survey is updated every year or two and is available online.

Some modelers may use general purpose programming languages (such as C, C++, or BASIC) or other tools (such as a spreadsheet) for developing discrete-event simulations. However, most dedicated simulation software (such as the tools listed in the previous paragraph) has many important and difficult-to-recreate features built in, such as clock management, event management, list management, model debugging, random variate generation, statistics gathering and support, and animation support. As a result, simulation software will generally provide much faster model development times and frequently faster simulation execution times than general-purpose tools.

Others may use multipurpose tools such as mine design and planning software that have some simulation functionality built in (although the incorporation of these features was not common as of this writing).

Important features of the software include:

- Strong set of simulation constructs
- User interface for model building
- Programmability
- Model delivery features (for allowing nondevelopers to manipulate data and run the model)
- Interactive and debugging tools
- Statistical features and graphics
- Animation
- Execution speed
- Geometry interface with the simulation (this will be discussed later)

Model Designers and Programmers. The qualified personnel includes many people (see Section 77.3) but here the focus is on the person who is *designing and programming* the model. Simulation modeling requires the logical thought process of a programmer (even if the modeler does not happen to be an experienced programmer). Experience with the software being used counts, as does simulation expertise in general. Finally, knowledge of the system being modeled is also important. The person building the simulation will become a system expert if not one already, so it is important to think up front about how to transfer this expertise to others. The model designer/builder/end-user team should include:

- Programmer's thought process
- Engineer's problem-solving approach
- Statistical grounding to avoid errors in input or output data analysis or presentation
- Organizational skills to keep all interested parties involved and believing in the face of possible conflicts of opinion
- Sales skills to present the end results

77.4.3   Organizational Commitment

The third component required for successful simulation is an organizational commitment. If the organization (or relevant part of the organization) is not committed to the simulation, it will probably fail. Constructing and documenting the model may be impeded, and useful insights may never be disseminated, understood, or used.

There are many people who need to be involved in a successful modeling project. These include:

- System designers (those who plan and/or engineer the system)
- System operators (those who work in the system)
- Other experts in the system logic
- Information technologists (may be needed for gathering data)
- Outside entities that provide key data (e.g., equipment suppliers)
- Person or persons constructing the model
- Person or persons performing experiments and analysis
- End users of the information being produced by the simulation project

The term "analysts" refers here to the people building the model and running the experiments. The term "end users" refers to the people benefiting from the information produced by the simulation project. Sometimes one person or team might serve in both roles.

The key, if one is faced with a skeptical organization, is to start with small pilot projects whose validity can easily be demonstrated at all levels of the organization.

77.4.4   Project Management

The fourth and final component is effective management of the simulation project. Project management itself is outside the scope of this chapter, but this is a good place to mention the importance of (1) integrating the simulation effort with the planning or process redesign effort that may also be underway and (2) agreeing on deliverables (which also means agreeing on the scope of the model and the extent of any experimentation). Possible deliverables include documentation, animations (delivered live or in computer-executable form), a working model that can be used on an ongoing basis, and training in use of the model.

77.5   CONCLUSION

Simulation is a *proven* tool widely used in the design, analysis, explanation, and operation of complex underground mining systems. It is an *understandable* technique that provides a valuable framework (there's that word again) for describing a complex system's actual behavior. Finally, simulation is *available and accessible* to any organization if the costs and benefits are understood and the guidelines are followed.

77.6   ACKNOWLEDGMENTS

The author is grateful to John Sturgul of the University of Idaho, whose long-term advocacy of applying simulation in mining has borne much fruit and who has been a personal source of encouragement and inspiration in this area. He and Edgardo Chilviet provided direct feedback during the preparation of this chapter.

People who participated deeply in some of the projects described here and thereby contributed to this work include Greg Baiden, Tom Corkal, Samantha Espley, Terry Villeneuve, and John Galbraith of Inco Limited; Hulya Yazici of the University of

---

[*] A textbook (Sturgul 2000) exists that focuses on mining simulation applications developed using GPSS/H and Proof Animation.

Wisconsin – La Crosse; Peter Rutherford of Falconbridge Limited; and Edgardo Chilviet and Joseph Brill of Systemflow Simulations.

77.7   REFERENCES

Balci, O. 1998. Verification, Validation, and Accreditation. *Proceedings of the 1998 Winter Simulation Conference*, Piscataway, NJ: IEEE, 41–48.

Banks, J., J.S. Carson II, and B.L. Nelson. 1995. *Discrete-Event System Simulation*. 2nd Ed. Upper Saddle River, NJ: Prentice-Hall.

Brunner, D.T. 2000. An Interactive Mine Simulation Example. Published at www.systemflow.com/minesim.

Brunner, D.T., H.J. Yazici, and G.R. Baiden. 1999. Simulating Development in an Underground Hardrock Mine. *SME 1999 Annual Meeting Preprints*. Littleton, Colorado: SME.

Cesarone, J. 2000. Don't just simulate, solve!. *IIE Solutions* 32(5): 44–48.

Law, A.M., and W.D. Kelton. 1999. *Simulation Modeling and Analysis*. 3rd Ed. New York: McGraw-Hill.

Schriber, T.J., and D.T. Brunner. 1999. Inside Simulation Software: How It Works and Why It Matters. *Proceedings of the 1999 Winter Simulation Conference*, Piscataway, NJ: IEEE, 72–80.

Sturgul, J.R. 1995. Simulation and Animation Come of Age in Mining. *Engineering and Mining Journal* 196(10): 38–42.

Sturgul, J.R. 1997. History of Discrete Mine Simulation. *Proceedings of the First International Symposium on Mine Simulation Via the Internet*, G.N. Panagiotou and J.R. Sturgul, Eds.

Sturgul, J.R. 2000. *Mine Design: Examples Using Simulation*. Littleton, Colorado: SME.

Swain, J.J. 1999. Imagine New Worlds. *OR/MS Today* 26(1): 38–41.

Swain, J.J. 1999. 1999 Simulation Software Survey. *OR/MS Today* 26(1): 42–51.

Index